S0-AKY-175

The chemistry of the **quinonoid compounds** Volume 2

Part 1

Edited by

SAUL PATAI

ZVI RAPPOPORT

The Hebrew University, Jerusalem

1988

JOHN WILEY & SONS

CHICHESTER – NEW YORK – BRISBANE – TORONTO – SINGAPORE

An Interscience ® Publication

Library of Congress Cataloging-in-Publication Data:

The Chemistry of the quinonoid compounds.

(Chemistry of functional groups)
'An Interscience publication.'
1. Quinone I. Patai, Saul
II. Rappoport, Zvi
III. Series
QD341.Q4C47 1987 547'.633 86-32494

ISBN 0 471 91285 9 (Part 1)
ISBN 0 471 91914 4 (Part 2)
ISBN 0 471 91916 0 (Set)

British Library Cataloguing in Publication Data:

The chemistry of the quinonoid compounds.—
(The chemistry of functional groups)
Vol. 2
I. Patai, Saul
II. Rappoport, Zvi
547'.636 QD341.Q4

ISBN 0 471 91285 9 (Part 1)
ISBN 0 471 91914 4 (Part 2)
ISBN 0 471 91916 0 (Set)

Typeset by Macmillan India Ltd, Bangalore 25
Printed and bound in Great Britain by Bath Press Ltd, Bath, Avon.

Contributing Authors

Hans-Dieter Becker	Department of Organic Chemistry, Chalmers University of Technology and University of Gothenburg, S-412 96 Gothenburg, Sweden
S. Berger	Fachbereich Chemie der Universität Marburg, D-355 Marburg, Germany
Jerome A. Berson	Department of Chemistry, Yale University, P.O. Box 6666, New Haven, Connecticut 06511-8118, USA
Peter Boldt	Institut für Organische Chemie der Technischen Universität Braunschweig, D-3300 Braunschweig, FRG
Eric R. Brown	Color Negative Technology Division, Photographic Research Laboratories, Eastman Kodak Company, 1669 Lake Avenue, Rochester, New York 14650, USA
James Q. Chambers	Department of Chemistry, University of Tennessee, Knoxville, TN 37996, USA
Tze-Lock Chan	Department of Chemistry, The Chinese University of Hong Kong, Shatin, New Territories, Hong Kong
M. Catherine Depew	Department of Chemistry, Queen's University, Kingston, Ontario, Canada K7L 3N6
K. Thomas Finley	Department of Chemistry, State University College, Brockport, New York 14420, USA
Colin W. W. Fishwick	Organic Chemistry Department, The University, Leeds LS2 9JT, UK
Karl-Dietrick Gundermann	Institut für Organische Chemie, Technische Universität Clausthal, 3392 Clausthal-Zellerfeld, Leibnizstrasse 6, FRG
P. Hertl	Institut für Organische Chemie der Universität Tübingen, D-7400 Tübingen, Germany
Hiroyuki Inouye	Faculty of Pharmaceutical Sciences, Kyoto University, Sakyo-ku, Kyoto, Japan
Shouji Iwatsuki	Department of Chemical Research for Resources, Faculty of Engineering, Mie University, Tsu, Japan
David W. Jones	Organic Chemistry Department, The University, Leeds LS2 9JT, UK
Marianna Kańska	Department of Chemistry, University of Warsaw, Warsaw, Poland
L. Klasinc	Department of Chemistry, Louisiana State University, Baton Rouge, Louisiana 70803, USA
Eckhard Leistner	Institut für Pharmazeutische Biologie, Rheinische Friedrich-Wilhelms-Universität Bonn, Bonn, FRG

Dieter Lieske	Institut für Organische Chemie, Technische Universität Clausthal, 3392 Clausthal-Zellerfeld, Leibnizstrasse 6, FRG
Tien-Yau Luh	Department of Chemistry, The Chinese University of Hong Kong, Shatin, New Territories, Hong Kong
Kazuhiro Maruyama	Department of Chemistry, Faculty of Science, Kyoto University, Kyoto 606, Japan
S. P. McGlynn	Department of Chemistry, Louisiana State University, Baton Rouge, Louisiana 70803, USA
Richard W. Middleton	The Cancer Research Campaign's Gray Laboratory, Northwood, Middlesex, UK and Brunel University, Uxbridge, Middlesex UB8 3PH, UK
Roland Muller	Institut für Organische Chemie, Universität Tübingen, Auf der Morgenstelle, D-7400 Tübingen, FRG
K. A. Muszkat	Department of Structural Chemistry, The Weizmann Institute of Science, Rehovot, Israel
Yoshinori Naruta	Department of Chemistry, Faculty of Science, Kyoto University, Kyoto 606, Japan
P. Neta	Center for Chemical Physics, National Bureau of Standards, Gaithersburg, Maryland 20899, USA
Atsuhiro Osuka	Department of Chemistry, Faculty of Science, Kyoto University, Kitashirakawa Oiwakecho, Kyoto 606, Japan
John Parrick	Brunel University, Uxbridge, Middlesex UB8 3PH, UK
A. Rieker	Institut für Organische Chemie der Universität Tübingen, D-7400 Tübingen, Germany
John R. Scheffer	University of British Columbia, Vancouver, Canada, V6T 1Y6
Lawrence T. Scott	Department of Chemistry and Center for Advanced Study, College of Arts and Science, University of Nevada-Reno, Reno, Nevada 89557, USA
Anne Skancke	Department of Chemistry, University of Tromsø, P.O.B. 953, N-9001 Tromsø, Norway
Per N. Skancke	Department of Chemistry, University of Tromsø, P.O.B. 953, N-9001 Tromsø, Norway
Howard E. Smith	Department of Chemistry, Vanderbilt University, Nashville, Tennessee 37235, USA
John S. Swenton	Department of Chemistry, The Ohio State University, Columbus, OH 43210, USA
James Trotter	University of British Columbia, Vancouver, Canada, V6T 1Y6
Alan B. Turner	Department of Organic Chemistry, Chalmers University of Technology and University of Gothenburg, S-412 96 Gothenburg, Sweden
Jeffrey K. S. Wan	Department of Chemistry, Queen's University, Kingston, Ontario, Canada K7L 3N6
Henry N. C. Wong	Department of Chemistry, The Chinese University of Hong Kong, Shatin, New Territories, Hong Kong
Klaus-Peter Zeller	Institut für Organische Chemie, Universität Tübingen, Auf der Morgenstelle, D-7400 Tübingen, FRG
Mieczysław Zieliński	Isotope Laboratory, Faculty of Chemistry, Jagiellonian University, Cracow, Poland

Foreword

The first volume on quinones in 'The Chemistry of Functional Groups' appeared (in two parts) in 1974. In Supplement A (1977) there was no new material on quinones. However, in the decade which has passed since, much new information has accumulated, on quite new subjects as well as regarding rapid and significant developments on subjects which were already included in the main quinone volume.

Hence we decided that it would be timely to publish a second volume on quinones and indeed this has turned out to be a weighty tome, even though we attempted to avoid duplication as far as possible between the two volumes.

Several subjects were intended to be covered but the invited chapters did not materialize. These were updates on quinone methides, on complexes and on rearrangements of quinones as well as a chapter on quinonoid semiconductors and organic metals.

Literature coverage in most chapters is up to 1986.

SAUL PATAI
ZVI RAPPOPORT

Jerusalem
August 1987

The Chemistry of Functional Groups
Preface to the Series

The series 'The Chemistry of Functional Groups' is planned to cover in each volume all aspects of the chemistry of one of the important functional groups in organic chemistry. The emphasis is laid on the functional group trated and on the effects which it exerts on the chemical and physical properties, primarily in the immediate vicinity of the group in question, and secondarily on the behaviour of the whole molecule. For instance, the volume *The Chemistry of the Ether Linkage* deals with reactions in which the C—O—C group is involved, as well as with the effects of the C—O—C group on the reactions of alkyl or aryl groups connected to the ether oxygen. It is the purpose of the volume to give a complete coverage of all properties and reactions of ethers in as far as these depend on the presence of the ether group but the primary subject matter is not the whole molecule, but the C—O—C functional group.

A further restriction in the treatment of the various functional groups in these volumes is that material included in easily and generally available secondary or tertiary sources, such as Chemical Reviews, Quarterly Reviews, Organic Reactions, various 'Advances' and 'Progress' series as well as textbooks (i.e. in books which are usually found in the chemical libraries of universities and research institutes) should not, as a rule, be repeated in detail, unless it is necessary for the balanced treatment of the subject. Therefore each of the authors is asked *not* to give an encyclopaedic coverage of his subject, but to concentrate on the most important recent developments and mainly on material that has not been adequately covered by reviews or other secondary sources by the time of writing of the chapter, and to address himself to a reader who is assumed to be at a fairly advanced post-graduate level.

With these restrictions, it is realized that no plan can be devised for a volume that would give a *complete* coverage of the subject with *no* overlap between chapters, while at the same time preserving the readability of the text. The Editor set himself the goal of attaining *reasonable* coverage with *moderate* overlap, with a minimum of cross-references between the chapters of each volume. In this manner, sufficient freedom is given to each author to produce readable quasi-monographic chapters.

The general plan of each volume includes the following main sections:

(a) An introductory chapter dealing with the general and theoretical aspects of the group.

(b) One or more chapters dealing with the formation of the functional group in question, either from groups present in the molecule, or by introducing the new group directly or indirectly.

(c) Chapters describing the characterization and characteristics of the functional groups, i.e. a chapter dealing with qualitative and quantitative methods of determination including chemical and physical methods, ultraviolet, infrared, nuclear magnetic resonance and mass spectra: a chapter dealing with activating and directive effects exerted by the group and/or a chapter on the basicity, acidity or complex-forming ability of the group if applicable).

(d) Chapters on the reactions, transformations and rearrangements which the functional group can undergo, either alone or in conjunction with other reagents.

(e) Special topics which do not fit any of the above sections, such as photochemistry, radiation chemistry, biochemical formations and reactions. Depending on the nature of each functional group treated, these special topics may include short monographs on related functional groups on which no separate volume is planned (e.g. a chapter on 'Thioketones' is included in the volume *The Chemistry of the Carbonyl Group*, and a chapter on 'Ketenes' is included in the volume *The Chemistry of Alkenes*). In other cases certain compounds, though containing only the functional group of the title, may have special features so as to be best treated in a separate chapter, as e.g. 'Polyethers' in *The Chemistry of the Ether Linkage*, or 'Tetraaminoethylenes' in *The Chemistry of the Amino Group*.

This plan entails that the breadth, depth and thought-provoking nature of each chapter will differ with the views and inclinations of the author and the presentation will necessarily be somewhat uneven. Moreover, a serious problem is caused by authors who deliver their manuscript late or not at all. In order to overcome this problem at least to some extent, it was decided to publish certain volumes in several parts, without giving consideration to the originally planned logical order of the chapters. If after the appearance of the originally planned parts of a volume it is found that either owing to non-delivery of chapters, or to new developments in the subject, sufficient material has accumulated for publication of a supplementary volume, containing material on related functional groups, this will be done as soon as possible.

The overall plan of the volumes in the series 'The Chemistry of Functional Groups' includes the titles listed below:

The Chemistry of Alkenes (*two volumes*)
The Chemistry of the Carbonyl Group (*two volumes*)
The Chemistry of the Ether Linkage
The Chemistry of the Amino Group
The Chemistry of the Nitro and Nitroso Groups (*two parts*)
The Chemistry of Carboxylic Acids and Esters
The Chemistry of the Carbon–Nitrogen Double Bond
The Chemistry of the Cyano Group
The Chemistry of Amides
The Chemistry of the Hydroxyl Group (*two parts*)
The Chemistry of the Azido Group
The Chemistry of Acyl Halides
The Chemistry of the Carbon–Halogen Bond (*two parts*)
The Chemistry of the Quinonoid Compounds (*two parts*)
The Chemistry of the Thiol Group (*two parts*)
The Chemistry of Amidines and Imidates
The Chemistry of the Hydrazo, Azo and Azoxy Groups (*two parts*)
The Chemistry of Cyanates and their Thio Derivatives (*two parts*)
The Chemistry of Diazonium and Diazo Groups (*two parts*)

The Chemistry of the Carbon–Carbon Triple Bond (two parts)
Supplement A: The Chemistry of Double-bonded Functional Groups (two parts)
The Chemistry of Ketenes, Allenes and Related Compounds (two parts)
Supplement B: The Chemistry of Acid Derivatives (two parts)
Supplement C: The Chemistry of Triple-Bonded Functional Groups (two parts)
Supplement D: The Chemistry of Halides, Pseudo-halides and Azides (two parts)
Supplement E: The Chemistry of Ethers, Crown Ethers, Hydroxyl Groups and their Sulphur Analogues (two parts)
The Chemistry of the Sulphonium Group (two parts)
Supplement F: The Chemistry of Amino, Nitroso and Nitro Groups and their Derivatives (two parts)
The Chemistry of the Metal–Carbon Bond (four volumes)
The Chemistry of Peroxides
The Chemistry of Organic Se and Te Compounds Vol. 1
The Chemistry of the Cyclopropyl Group (two parts)
The Chemistry of Organic Se and Te Compounds Vol. 2

Titles in press

The Chemistry of Sulphones and Sulphoxides
The Chemistry of Organosilicon Compounds
The Chemistry of Enones
Supplement A2: The Chemistry of the Double-Bonded Functional Groups, Volume 2.

Titles in Preparation

The Chemistry of Enols
The Chemistry of Sulphinic Acids, Esters and Derivatives
The Chemistry of Sulphenic Acids and Esters.

Advice or criticism regarding the plan and execution of this series will we welcomed by the Editor.

The publication of this series would never have started, let alone continued, without the support of many persons. First and foremost among these is Dr Arnold Weissberger, whose reassurance and trust encouraged me to tackle this task. The efficient and patient cooperation of several staff-members of the Publisher also rendered me invaluable aid (but unfortunately their code of ethics does not allow me to thank them by name). Many of my friends and colleagues in Israel and overseas helped me in the solution of various major and minor matters, and my thanks are due to all of them, especially to Professor Z. Rappoport. Carrying out such a long-range project would be quite impossible without the non-professional but none the less essential participation and partnership of my wife.

The Hebrew University
Jerusalem, ISRAEL

SAUL PATAI

Contents

The Chemistry of Quinonoid Compounds, Vol. II
Edited by S. Patai and Z. Rappoport

CHAPTER 1

General and theoretical aspects of quinones

ANNE SKANCKE and PER N. SKANCKE
Department of Chemistry, University of Tromsø, P.O.B. 953, N-9001 Tromsø, Norway

LIST OF ABBREVIATIONS

A. Names of Chemical Compounds

DDQ	2,3-dicyano-5,6-dichlorobenzoquinone
MBQDM	*meta*-benzoquinodimethane
MBQM	*meta*-benzoquinomethane
NMP	*N*-methylphenazine
OBQ	*ortho*-benzoquinone
OQDI	*ortho*-quinone diimine
OQDM	*ortho*-quinodimethane
OQI	*ortho*-quinone imine
OQM	*ortho*-quinomethane
PBQ	*para*-benzoquinone
PQDI	*para*-quinone diimine
PQDM	*para*-quinodimethane
PQI	*para*-quinone imine
PQM	*para*-quinomethane
TCNQ	tetracyanoquinodimethane
TTF	tetrathiofulvalene

B. Names of Experimental and Theoretical Methods

AO	atomic orbital
MO	molecular orbital
SCF	self-consistent field
ΔSCF	difference in SCF energy between different species or different states of a given species
CBL	constant bond length
VBL	variable bond length
REPA	resonance energy per atom
REPE	resonance energy per electron
HOMO	highest occupied molecular orbital
LUMO	lowest unoccupied molecular orbital
LCAO	linear combination of atomic orbitals
FMO	frontier molecular orbital
CI	configuration interaction
STO-3G	minimum Gaussian basis set used in *ab initio* calculations
4-31G	medium-sized Gaussian basis set used in *ab initio* calculations
CNDO MNDO INDO MINDO HAM HMO EHMO PPP	semi-empirical molecular orbital methods
IR	infrared
UV	ultraviolet
EPR	electron paramagnetic resonance
NMR	nuclear magnetic resonance
PES	photoelectron spectroscopy

VTPES	variable temperature electron spectroscopy
ETS	electron transmission spectroscopy
ENDOR	electron nuclear double resonance

I. INTRODUCTION

Molecules exhibiting a quinonoid structure constitute a large and important class in the chemistry of organic compounds. Extensive treatments of chemical and physical properties of quinonoid compounds are available in the literature. The two volumes on these compounds edited by Patai[1] give a broad survey of the state-of-the-art in the field up to 1974, and the bibliography in the series *Houben-Weyl* also has two volumes dedicated to quinones[2].

We will focus on contributions to the field that have appeared after the publication of the first relevant volume in this series[1]. Furthermore our presentation will put emphasis on general and theoretical aspects of the chemistry of quinonoid compounds. For detailed discussions of particular aspects of specific groups within the family of these compounds, we refer to subsequent chapters of the present volume.

The outline of our presentation will be the following. In section II we give a brief survey of the available experimental data on the molecular structures of quinonoid systems. Section III contains a short discussion of symmetries and orbital topologies for the appropriate symmetry groups (D_{2h} and C_{2v}). In Section IV we focus on physical properties of molecular ground states, and discuss explicitly resonance energies and aromaticity, thermodynamic and kinetic stabilities. We also include a brief discussion of electron affinities, and a comparative study of open and bicyclic forms in the case of *ortho*-compounds. In this section we also touch briefly on the problem related to the description of the *meta*-form, referring to Chapter 10 for a full treatment of this class of compounds. A few comments on molecular polarizabilities are included. In Section V we give a short presentation of certain aspects of the spectroscopy of some of the quinonoid compounds focusing on the interesting problem of Koopmans and shake-up states which are of vital importance in the interpretation of photoelectron spectra (PES). For a full account of this topic we refer to Chapter 5 of this volume. However, UV and IR spectra are also discussed to some extent, the main theme being current interpretations of the properties of the near-degenerate $n\pi^*$ states of *para*-benzoquinone (PBQ). In Section VI we give some brief comments on current models used in the interpretation of cycloaddition reactions, referring to separate chapters for comprehensive presentations of chemical reactions in general. Finally in Section VII we describe some of the current contributions to polymerization reactions.

We emphasize that our list of references does not pretend to be exhaustive. In the selection of papers to be quoted we have adopted the following guidelines: (1) as a rule attention is given mainly to recent papers; (2) the papers should allude to theoretical aspects of the molecular properties and not be of a purely technical nature; (3) extensive lists of papers referring to any particular special issue should be looked for in the pertinent chapters of this volume.

The types of compounds included in our treatment has been limited. We have put emphasis on molecules containing one benzene ring, and have included naphtho- and anthraquinones only when demanded in comparative discussions. The same applies to other derivatives of compounds containing a single benzene ring.

The main theme of our presentation will be a discussion of chemical and physical properties of the *para*-compounds (**1a**–**1g**) and the corresponding *ortho*-compounds (**2a**–**2g**):

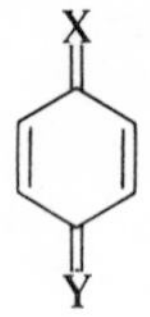

(1) 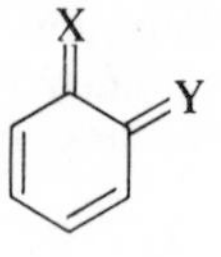 (2)

where

Y \ X	O	CH_2	NH	S
O	a	b	d	f
CH_2		c		
NH			e	
S				g

The entries in the table above indicate compounds that have been synthesized and/or detected as highly reactive intermediates. We have adopted the following names in full and abbreviated forms for the *para*-compounds.

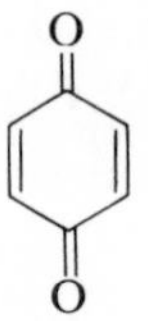

p-benzoquinone PBQ (**1a**) *p*-quinomethane PQM (**1b**) *p*-quinodimethane PQDM (**1c**)

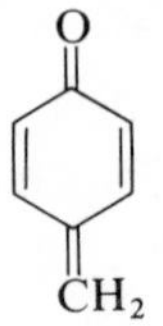

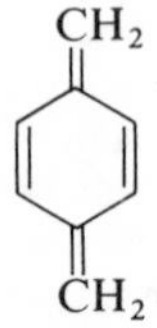

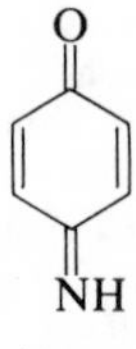

p-quinone imine PQI (**1d**) *p*-quinone diimine PQDI (**1e**) monothio-PBQ (**1f**) dithio-PBQ (**1g**)

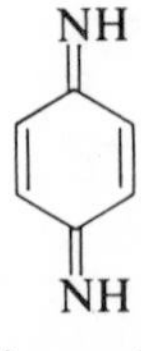

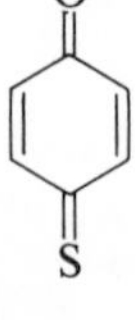

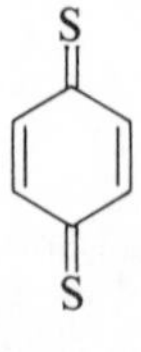

For the *ortho*-compounds (**2a–2g**) and the *meta*-compounds (**3a–3g**) a completely analogous naming and labelling system has been adopted (the letter P substituted by O and M respectively).

Of the compounds listed above PBQ (**1a**) and OBQ (**2a**) are comparatively stable solids at room temperature[3–5]. The remaining species are very unstable and highly reactive. PQM (**1b**) has been produced by flash thermolysis and isolated as a solid film at 77 K[6], whereas the isomer OQM (**2b**) has been trapped as a solid at liquid nitrogen temperature[7]. More recently this molecule has also been detected in the vapour phase by the VTPES technique[8]. The hydrocarbon PQDM (**1c**) was known only as a reactive intermediate[9,10] until it was isolated in the solid state at 77 K[11]. The other isomer OQDM (**2c**) has not been isolated but observed in rigid glass at 77 K as a result of photolysis[12–14] and trapped in an argon matrix in the temperature range 8–30 K[15]. It has also been postulated to occur as an intermediate in a sonochemical reaction in solution at room temperature[16]. The unsubstituted species PQI (**1d**) and PQDI (**1e**) are rather unstable at room temperature[2], and their *ortho* counterparts OQI (**2d**) and OQDI (**2e**) are even more reactive. However, due to the basic character of the imine nitrogen, stable salts, e.g. chlorides, have been prepared and studied[17]. The sulphur-containing analogues monothio-PBQ (**1f**) and dithio-PBQ (**1g**) have recently been detected as products by pyrolysis[18].

The monothio OBQ (**2f**) has been postulated as a possible intermediate in the pyrolysis of 1,2,3-benzoxadithiole-2-oxide and 1,3-benzoxathiole-2-one[19]. MNDO calculations[19] indicate that the molecule has a strong preference for a closed shell polyene structure as opposed to a bicyclic isomer.

The *meta* analogues of (**1**) and (**2**) are expected to have diradical nature since no classical

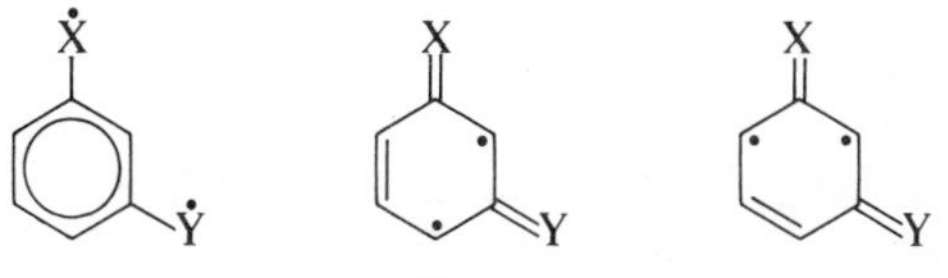

X Y X Y X Y

(**3**)

Kekulé structure may be written for these cases. Some of their properties will be commented on briefly in Section IV. For a full account we refer to Chapter 10 in this volume.

II. GEOMETRICAL STRUCTURES

A limited number of experimental investigations on the molecular structures of quinones and quinonoidal systems are available.

PBQ has been studied by electron diffraction in the vapour phase[20–22], and by X-ray diffraction in the crystal[23]. The geometries obtained from the crystal data[23] and from the most recent electron diffraction data[22] are in very good agreement except for the C=C bond which is found to be somewhat longer in the vapour experiment (see Table 1). The

TABLE 1. Geometrical parameters for PBQ (**1a**) obtained by electron diffraction[22] and X-ray diffraction[23]

Parameter[a]	Electron diffr.[b]	X-ray diffr.[c]
C=O	1.225 ± 0.002	1.222 ± 0.008
C=C	1.344 ± 0.003	1.322 ± 0.008
C–C	1.481 ± 0.002	1.477 ± 0.006
C–H	1.089 ± 0.011	—
∠C(2)C(1)C(6)	118.1 ± 0.3	117.8 ± 0.6

[a] Distances in Ångstrom, angles in degrees.
[b] Error estimates are 2σ.
[c] Error estimates are σ.

length of this bond is a good indicator to the possible importance of benzenoid contributions to the structure of this molecule. As revealed by the data in Table 1, the molecule displays a pronounced single-bond double-bond alternation both in the vapour phase and in the solid state, the C=C bond in the solid state being even shorter than the one in ethylene which is 1.336 Å[24]. These data indicate clearly that in using structural information as an indicator, this molecule has a typical quinonoid structure. In both investigations[22,23] the molecule was found to have a planar structure displaying D_{2h} symmetry.

Also for derivatives of PBQ, structural investigations have confirmed an alternation between typical single and double carbon–carbon bonds in the ring. This is the case for reported studies of *p*-chloranil[25], *p*-fluoranil[26,27] and duroquinone[27,28].

p-chloranil *p*-fluoranil duroquinone

OBQ has been reported to exist in two different forms[4]. The red crystalline form that is stable at room temperature has been determined structurally by X-ray crystallographic methods[5]. This is the quinonoid form **2a** with pronounced single-bond double-bond alternation in the ring. However, as revealed by Table 2, two of the distinct single bonds in the ring are significantly longer than in PBQ. This is an indication of a slightly more benzenoid-like structure of this molecule. The single bond C(1)–C(2) is found to be abnormally long considering trigonally hybridized carbon atoms. Moreover, small, but significant deformations from planarity have been observed in this molecule, the ring having adopted a slightly twisted boat conformation with the oxygen atoms displaced on either side of the mean ring plane[5].

TABLE 2. Geometrical parameters for OBQ obtained by X-ray diffraction[5]

Parameter[a]	Value[b]
C=O	1.220 ± 0.002
C=C	1.341 ± 0.002
C(1)–C(6)	1.469 ± 0.002
C(4)–C(5)	1.454 ± 0.004
C(1)–C(2)	1.552 ± 0.003

[a] Distances in Ångstrom.
[b] Error estimates are σ.

A microwave study of OBQ in the vapour phase did not result in a set of structural parameters[29]. However rotational constants derived from the spectrum were in agreement with the corresponding constants derived from the X-ray analysis[5]. This excludes the presence of significant amounts of the valence tautomer

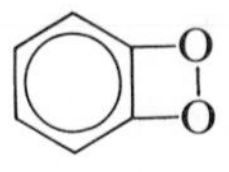

in the vapour phase (*vide infra*).

Even for the very unstable PQDM, which has been detected as a reactive intermediate in the polymerization of the pyrolysis product of *p*-xylene[9] and of [2.2]paracyclophane[10], and which has been isolated in the solid state only at 77 K[11], electron diffraction data from the vapour phase are available[30]. The structural parameters obtained by this study do support a model having an alternation between long and short carbon–carbon bonds although not as pronounced as in the case of PBQ. The value reported for the long C–C bond is 1.451 Å as opposed to an average value of the C–C short bonds of 1.381 Å[30]. The latter value, which is larger than one corresponding to a normal double bond, demonstrates that PQDM, although having an alternating structure, is not composed of pure single-bond and double-bond fragments.

With the exception of the slight distortion of OBQ observed in the crystalline state[5], we may conclude that the structural information quoted above support the idea of planar quinonoid molecules. This has an important bearing on any discussion related to the distribution of the outer valence electrons, making possible a clear-cut distinction between σ and π electrons.

III. SYMMETRIES AND ORBITAL TOPOLOGIES

Referring to the experimental results discussed in the previous section we assume that the *para*-compounds (**1**) all belong to the symmetry group D_{2h} whereas the *ortho*- (**2**) and *meta*-compounds (**3**) are classified according to the group C_{2v}.

The atomic orbitals (AOs) constituting the basis for the π-electron system span the following representations in the group D_{2h}.

$$a_u(\pi) \ : a - b + c - d$$
$$b_{2g}(\pi) \ : a + b - c - d$$
$$b_{1g}(\pi) \ : a - b - c + d$$
$$b_{3u}(\pi) \ : a + b + c + d$$

where orientation of coordinate system and labelling of AOs are as follows.

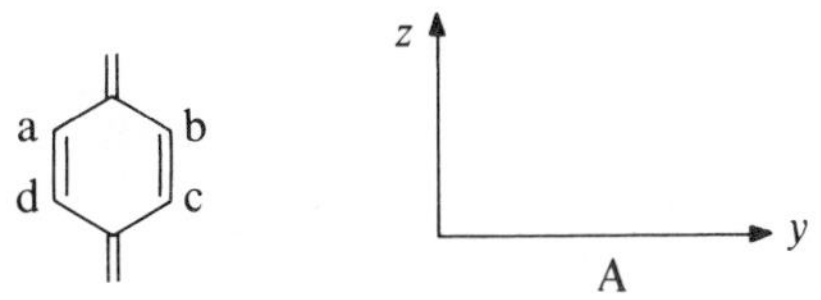

a b d c z y A

If we employ the alternative coordinate systems also used in the literature

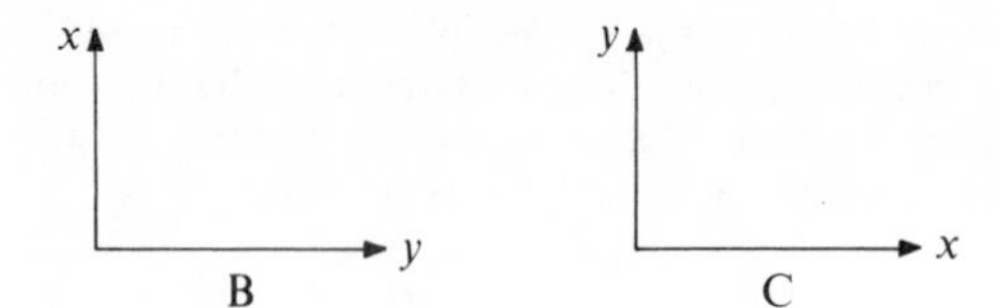

we obtain the following relations between the irreducible representations.

A	B	C
$a_u(\pi)$	$a_u(\pi)$	$a_u(\pi)$
$b_{2g}(\pi)$	$b_{2g}(\pi)$	$b_{3g}(\pi)$
$b_{1g}(\pi)$	$b_{3g}(\pi)$	$b_{2g}(\pi)$
$b_{3u}(\pi)$	$b_{1u}(\pi)$	$b_{1u}(\pi)$

In our presentation we will label our symmetry orbitals according to reference system A.

Qualitative lobe diagrams giving the phases of the AOs in the different representations are shown below. It is worth mentioning that the exocyclic AOs of π symmetry do not

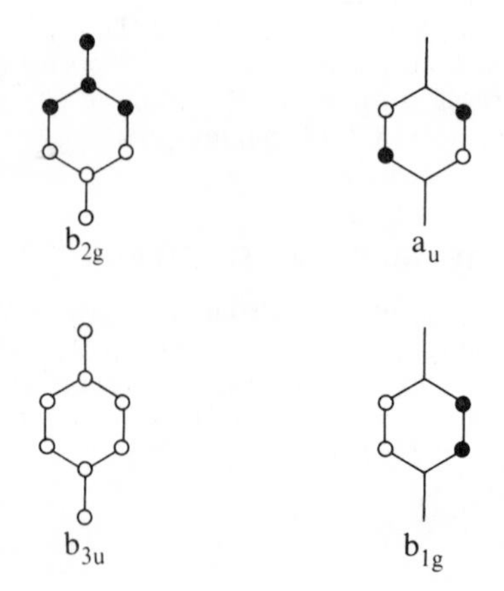

participate in the molecular orbitals (MOs) belonging to the representations b_{1g} and a_u.

In PBQ and analogues having the same symmetry (D_{2h}), AOs describing the exocyclic lone pairs constitute symmetry orbitals belonging to the representations $b_{2u}(n_1+n_2)$ and $b_{3g}(n_1-n_2)$:

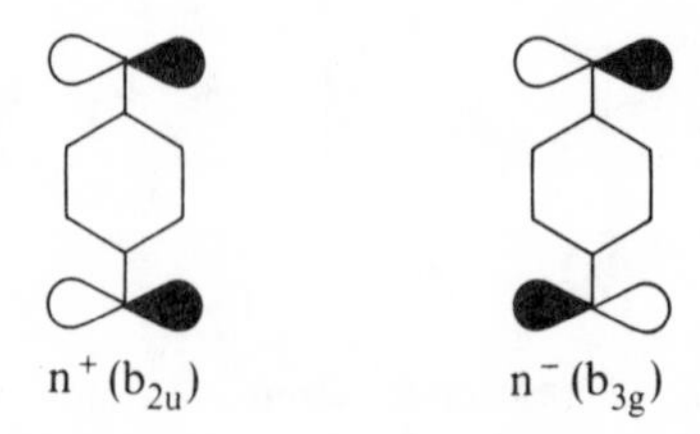

This implies that for symmetry reasons the lone-pair orbitals are prevented from mixing with the π system. The lone-pair orbitals are usually labelled as:

$$n^{+} = \frac{1}{\sqrt{2}}(n_1 + n_2)\ b_{2u}$$

$$n^{-} = \frac{1}{\sqrt{2}}(n_1 - n_2)\ b_{3g}$$

By employing the rotated coordinate systems described above we obtain the following relations between the representations to which the lone-pair orbitals belong.

A	B	C
$b_{2u}(n)$	$b_{2u}(n)$	$b_{3u}(n)$
$b_{3g}(n)$	$b_{1g}(n)$	$b_{1g}(n)$

For the *ortho* and *meta* isomers the appropriate outer valence orbitals of π type span the following representations of the symmetry group C_{2v}.

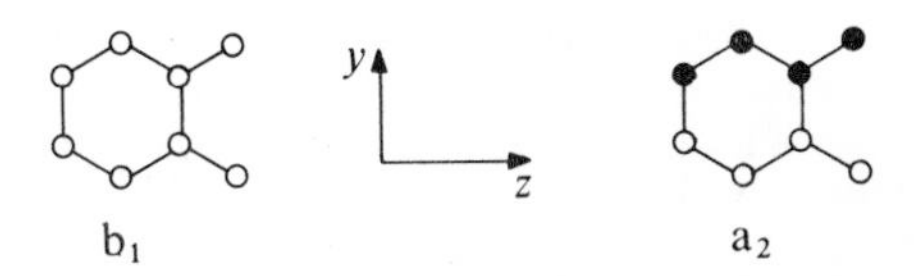

where the *ortho* isomer has been chosen to illustrate the phases of the AOs. In the case of quinones and analogues having exocyclic lone pairs we obtain the following lobe diagrams:

$n^{+}(a_1)$ $n^{-}(b_2)$

where

$$n^{+} = \frac{1}{\sqrt{2}}(n_1 + n_2)\ a_1$$

$$n^{-} = \frac{1}{\sqrt{2}}(n_1 - n_2)\ b_2$$

IV. PHYSICAL PROPERTIES OF GROUND STATES

Aspects of ground state properties relevant for molecular spectra are presented in Section V (*vide infra*).

A. Relative Stabilities

1. Aromaticity and resonance energies

A measure of the relative stabilities of the quinones and quinodimethanes would be their relative resonance energy or aromatic character. The former may be taken relative to hypothetical structures defined on the basis of experimental data. A study by Herndon[31] aiming at determining resonance energies on the basis of photoelectron spectra and graph theoretical methods gives us a starting point for this discussion, since both series of compounds have been investigated by the same procedure. In that work the arbitrariness in the choice of reference structure is avoided by using the same experimental data for parametrization of both the reference and parent system. For a series of hydrocarbon systems there is a reasonable agreement with experimental counterparts. A resonance energy of +0.10 eV for PQDM puts this species in the category of non-aromatic compounds. In a similar way a value of − 0.27 eV is assigned to PBQ, placing it among the anti-aromatic compounds. It must be remarked, however, that the presence of a heteroatom does make an extra uncertainty here. This point was taken up in some detail in a previous chapter in this series[32].

Gleicher[33] has calculated stabilities of various hydrocarbon quinonoid compounds following the method of Hess and Schaad[34, 35]. This method is based on the Hückel wave function, and a measure of the resonance energy, REPE (resonance energy per π electron) is obtained by introducing a localized structure characterized by standard bond energies for various carbon–carbon single and double bonds.

Whereas earlier calculations on the quinodimethane series tended to predict a substantial thermodynamic stabilization for quinodimethanes[36, 37] the calculation by Gleicher[33] moderated this view, since REPE values of only 0.005 β and 0.006 β were found for the 1,2 and 1,4 isomer respectively. These results are in accordance with the prediction by Herndon for the 1,4 isomer. Another interesting result emerged from the calculations by Gleicher: annelation of benzenoid units to the basic structures seemed to have a strongly stabilizing effect. For instance, 9,10-anthracenequinodimethane was found to have a REPE value of 0.040 β, and similar values were reported for other systems where both ring double bonds from the prototype were incorporated in the aromatic unit. Because of an apparent inconsistency for derivatives of the 1,2 isomer, the same authors also carried out SCF CBL calculations in a slightly modified version of the method of Dewar and Gleicher[38, 39]. The results predicted an anti-aromatic character to all quinodimethanes for which only a single classical structure may be written. However, a third set of calculations by these authors based upon the SCF VBL technique of Dewar (where the σ energies of the carbon–carbon bonds are calculated explicitly, avoiding implementation of empirical parameters) finds the 1,4 isomer to be slightly anti-aromatic (REPE = − 0.004 β) while the 1,2 isomer is predicted to be somewhat aromatic (REPE = 0.012 β).

In the same line, Gutman and Bosanac[40] have calculated the 'cycle energy' as a measure of thermodynamic stability. Their approach is based upon a graph theoretical treatment of the Hückel π-electron energy. In that work, only the *meta* and *para* quinodimethanes were considered, both showing some aromatic stabilization, but less than other systems of the $(4n + 2)$ ring size. The cycle energies reported were +0.0958 β and +0.0612 β respectively. Note the more stabilized *meta* form.

In an investigation of the aromatic properties of polycyclic benzenoid compounds, Kruszewski[41] has assigned an index of aromaticity to each ring. The index is based upon

Julg's definition of aromaticity[42] although instead of a distance criterion, a bond order criterion is used. The bond orders were calculated by HMO technique. In this scheme the reference molecule benzene is assigned an index value of zero, and an increase in the index value implies a decrease in aromaticity. We present below some results of interest. Note the large difference between *d* and *e*. (The single bonds emanating from the rings indicate

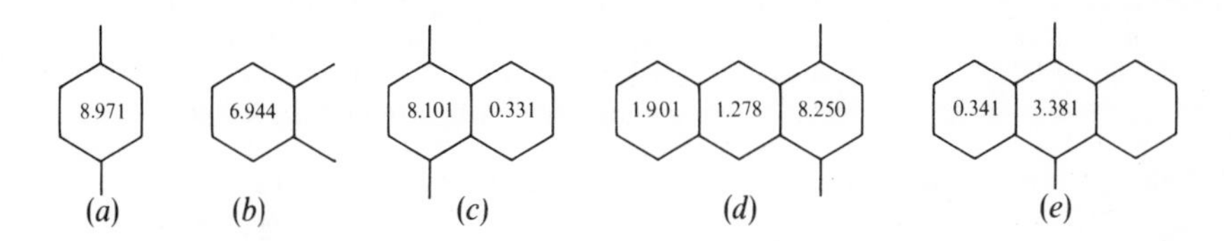

double bonds to CH_2). In *e*, the substituted ring has a lower degree of aromaticity than the side ring, but the index value of 3.381 is a moderate value. In this respect this series of compounds differs from many polycyclic aromatic hydrocarbons, where the central rings are so reactive that they are prone to undergo reactions that will remove them from conjugation[43]. The findings pertaining to the quinodimethanes seem to agree qualitatively with the previously mentioned findings by Gleicher[33] where aromatic stabilization was predicted when two bonds of the basic structures were shared with benzene rings. One might conclude from the works by Herndon[31], Gleicher[33] and Gutman[40] that both OQDM and PQDM should be nearly non-aromatic. As for the *meta* isomer, having no classical structures, we have treated this species separately (see Section IV.D).

The predicted thermodynamic stability of the quinodimethanes relative to the quinones seems to be in contrast with the ease of formation of these compounds. Whereas the *ortho* and *para* quinones are stable compounds, the corresponding quinodimethanes are highly unstable having resisted isolation until recently trapped at 77 K[11].

2. *Kinetic vs. thermodynamic stability*

As pointed out by Hess and Schaad[44], the ease of isolation of a system depends not only on its thermodynamic stability, but also on its kinetic stability. Thus, a high energy compound may be isolable if it has no favourable conversion routes, and, vice versa, a stable compound may not be isolable if it has favourable routes to still more stable compounds. The kinetic instability of a series of non-benzenoid hydrocarbons has been discussed in detail by Aihara[45], who pointed out several examples of kinetic instability in thermodynamically stable compounds.

As a measure of kinetic stability towards electrophiles and nucleophiles, Aihara has calculated reactivity indices based upon Wheland's localization energy[46,47] and Fukui's superdelocalizability[48,49].

The line of reasoning involves a characterization of the transition state. For instance, an electrophilic substitution results in a new conjugated system of one less atom and two less electrons (a σ complex or Wheland intermediate). The localization energy is then a measure of the energy in going from a given hydrocarbon to the σ complex. The smaller the localization energy, the more susceptible the carbon, and an index is assigned to each non-equivalent carbon in the system.

When applied as a measure of the reactivity towards nucleophilic substitution and addition reactions, the localization energy is calculated for a σ complex having a -1 formal charge. For alternant hydrocarbons (like the *o*- and *p*-quinodimethanes) predictions based upon localization and superdelocalizability leads to an equal amount of nucleophilic and electrophilic reactivity because of the symmetry of the π-orbital energies.

The work of Aihara predicts both *o*- and *p*-quinodimethanes to fall in the category of non-aromatic compounds having a small per cent resonance energy. All carbon atoms were predicted to be susceptible, the ring carbons to the same extent and slightly more so than the exocyclic carbons.

From a very different approach, Jug has reached the conclusion that PBQ should be non-aromatic[50]. His method is based upon the ring current concept. A bond order is defined as the weighted sum of eigenvalues of the two-centre parts of the density matrix of the pair of atoms considered. The minimal bond order in a ring system in its equilibrium is then defined as an index of aromaticity.

Table 3 summarizes some of the predictions regarding aromaticity.

TABLE 3. Aromatic properties of PQDM, OQDM and PBQ predicted in literature

Method[a]	PQDM	OQDM	PBQ	Ref.
PES + graph theory	Non-arom.		Anti-arom.	31
Huckel ('REPE')	Non-arom.	Non-arom.		33
SCF-CBL	Anti-arom.	Anti-arom.		33
SCF-VBL	Anti-arom. (slightly)	Aromatic		33
Huckel ('cycle energy')	Aromatic (slightly)			40
Localization energy	Non-arom.	Non-arom.		45
Ring current concept			Non-arom.	50

[a] See text.

None of the cited theoretical works has compared the stability of OBQ to the other systems in the table, but evidence from experimental electrochemistry gives us some information on this point. Indeed, experimental redox potentials offer a straightforward way of comparing stabilities of quinone derivatives. Since the *ortho* species has a higher electrode potential (0.792 V) than the *para* form (0.715 V), the former species should be the less stable of the two[51]. The redox potentials for a series of quinones have been measured[51] and calculated values have been obtained by Schmand and coworkers[52] who made use of the postulated (by Dewar and Trinajstic[53]) linear relationship between the redox potentials and the difference in the heat of atomization between the quinone and the hydroquinone. There is a good overall agreement between the experimental and calculated values. A particularly low redox potential was found for 9,10-anthraquinone (calculated value 0.25 V, experimental value 0.154 V). This is in line with the previously mentioned high REPE value for 9,10-anthracenequinodimethane and the general trend extracted from the work of Gleicher[33] that annelation of benzene rings has a stabilizing effect.

Of interest in this context is also the stability considerations carried out in a work by Banks and coworkers[54]. On the basis of an *ab initio* STO-3G calculation PQDM is found to be more stable than the most stable conformer of the open-chain cross-conjugated isomer 3-methylene-1,4-pentadiene by 4.35 kcal mol^{-1}. From arguments involving homodesmotic criteria (comparisons between compounds having an equal number of carbon atoms with a particular hybridization and an equal number of carbon atoms with a given number of hydrogens attached), PQDM was found to be slightly stabilized as compared to other open-chain systems, leaving no doubt that the strain energy of PQDM must be negligible.

B. Electron Affinities and Properties of Anionic States

Much interest has been focused on the electron affinities of this class of compounds. This is due to the ability of quinonoid systems to form charge transfer complexes by acting as electron acceptors. Several of these complexes are good conductors.

The electron affinity of PBQ has been calculated by *ab initio* ΔSCF calculations augmented by a semiempirical estimation of correlation energy by Millefiori and Millefiori[55]. The latter contribution was calculated according to a method proposed by Sinanoglu[56] and is based upon CNDO/2 wave functions. The molecular electron correlation energy is given in terms of matrix elements that may be partitioned into one- and two-centre terms. The electron affinities for PBQ as calculated at different levels of sophistication were: on the basis of Koopmans' theorem, −0.07 eV; on the basis of ΔSCF calculations (taken as the difference in total energy between PBQ and the corresponding anion), 0.81 eV; and ΔSCF augmented by electron correlation contribution as explained above, 2.20 eV. All these calculations were carried out at the 4-31G level. An experimental value of the electron affinity of PBQ is found by Cooper and coworkers[57] by caesium ion electron impact to be 1.89 eV. These authors also provide evidence for the existence of a long-lived metastable state of the PBQ^- ion at 1.4 eV. This ion is claimed to be generated by direct attachment of the incident electron into the b_{3u} orbital followed by internal conversion to the $^2B_{2g}$ ground state of the anion[57]. Electron affinity values of 1.98 eV and 1.83 eV have furthermore been inferred from half-wave reduction potential and charge transfer complex data[58]. Millefiori and Millefiori also were able to deduce a value for the disproportionation energies of the reaction

$$2PBQ^{1-} \rightarrow PBQ + PBQ^{2-}$$

including the correlation term to be 6.43 eV at the 4-31G level, in reasonable agreement with the value of 5.97 eV that may be inferred from electrochemical data[59–61].

MNDO calculations by Dewar and Rzepa[62] yielded an adiabatic electron affinity value of 1.88 eV and a vertical value of 1.51 eV. This large difference is in itself indicative of a substantial geometry change in going from the neutral molecule to the anion. Indeed, the calculated geometry of the anion reported in the work by Dewar and Rzepa shows a more benzenoid character for this species than for the neutral molecule.

The high electron affinity of PBQ is enhanced by suitable substitution. As pointed out by Cooper and coworkers[63], the perhalo derivatives, i.e. *p*-fluoranil, *p*-chloranil and *p*-bromanil, should have stronger affinity for electrons than PBQ, and indeed values of 2.92 ± 0.2 eV, 2.76 ± 0.2 eV and 2.44 ± 0.2 eV were found for these systems by means of collision technique. Electron affinities for a number of substituted quinones have been found also on the basis of gas phase equilibrium constants for reactions of the type $A^- + B = A + B^-$ using an electron beam high ion source pressure mass spectrometer[64]. The electron affinity for PBQ determined in this way was found to be 1.81 eV. Annelation of one benzene ring to PBQ was found to reduce the electron affinity slightly, while chlorine substitution was found to increase the electron affinity. For chloranil, the value was found to be 2.68 eV. These values are comparable to the values for electron affinity which have been reported for TCNQ. Much attention has been focused on this latter species because of its potential as a superconductor and its tendency to form charge transfer complexes. Two *ab initio* calculations have been carried out on this species[65, 66]. These works predict electron affinities for TCNQ of 2.83 eV and 2.63 eV respectively, the latter value being determined with a slightly larger basis set and a structure closer to the X-ray structure of the species[67].

In conjunction with electron transmission spectroscopy (ETS) measurements of PBQ, PBQ derivatives, and related molecules Modelli and Burrow[68] have used MO theory in order to pin down the sequence of the virtual orbitals of PBQ. They find the LUMO to be of b_{2g} symmetry, and the subsequent π orbitals to follow in the order b_{3u}, a_u, b_{2g} with energies of 0.69 eV, 1.41 eV and 4.37 eV, respectively. The sequence b_{3u}, a_u is apparently not unequivocally established, as the metastable state at 1.4 eV reported by Cooper and coworkers[57] was assigned to be a $^2B_{3u}$ state. Semiempirical calculations of PPP-type[69], of CNDO-type[70] and $X\alpha$ calculations[72] indicate a larger stability of the b_{3u} orbital.

A curious point in this connection is mentioned in a review article on crystalline π compounds by Herbstein[73]. According to data given in that article, it looks as if mixing substituents produces the most powerful acceptors, thus 2,3-dicyano-5,6-dichlorobenzoquinone (DDQ) is reported to be a better acceptor than either chloranil or cyananil (all of which are more powerful than PBQ). Although there are large uncertainties and small differences in electron affinity reported in that article (0.25 eV between cyananil and DDQ), it may be that unsymmetrical substitution favours the formation of molecular complexes due to dispersion forces and dipole moments in addition to electrostatic forces. However, the usefulness of this in forming charge transfer complexes may be limited due to an increased tendency for DDQ to form complexes with itself.

Although the high electron affinity is involved in any rationalization of the conductivity properties of these systems, single molecule calculations alone or calculations on corresponding anions are apparently not sufficient to explain the high conductivities, and at least two units would have to be considered to predict the collective behaviour in the crystalline phase[66]. The work by Johansen[66] gives an interestingly low value of 4.50 eV for the disproportionation reaction

$$2\mathrm{TCNQ}^- \rightarrow \mathrm{TCNQ} + \mathrm{TCNQ}^{2-}$$

and suggests that these data and the change from quinonoid to benzenoid structure in going from the neutral species to the anion are among factors responsible for the high conductivity.

A review by Andre and coworkers[74] summarizes the existing data up to 1976 on physical properties of highly anisotropic systems like the charge transfer complexes formed with TCNQ (e.g. TTF–TCNQ and NMP–TCNQ).

Within the framework of EHMO/LCAO-MO-CI Zhang and Yan[75] have calculated the energy band structure of the triclinic crystal form of NMP–TCNQ. This form is reported to have an electric conductivity comparable to a metal[76] whereas the monoclinic form is reported to be a semiconductor[77].

A study by Mirek and Buda[78] gives MNDO values for adiabatic electron affinities, LUMO energies and ionization potentials for a number of polycyano compounds, among them the di- and tetra-substituted cyano derivatives of PQDM. An increase in electron affinities and ionization potentials and a decrease in LUMO energies was found as the number of cyano groups was increased.

The relative electron affinities of a number of molecules including substituted quinones have been reported by Fukuda and McIver[79]. The results were derived from pulsed ion cyclotron resonance spectrometry. For a whole series of molecules, there is a marked additivity in substituent effects. Thus, there is a decrease in electron affinity of PBQ upon successive methylation, and an increase upon fluorination. The effect of successive chlorination upon the electron affinity is somewhat more complex, since steric factors are also important.

C. *Ortho* Form vs. Bicyclic Form

The *ortho* forms of the quinonoid systems pose a special structural problem, because of the possible conversion to benzocyclobutene:

X ⟶ X

The intramolecular $(2\pi + 2\sigma)$ cycloaddition of **2** may be useful in synthetic chemistry. Kolshorn and Meier[80] have calculated relative stabilities of **1** and **2** for the cases X = CH_2, NH, O and S using EHMO technique. For all these systems, the quinonoid form was found to be the more stable one, and increasing thermodynamic stabilities were found when going from X = CH_2 to S, NH and O. Furthermore, the activation energy for the ring-closure process was particularly high for the oxa compound. In other words, the more electronegative X favours the open, quinonoid structure because of resonance stabilization of the dipole form

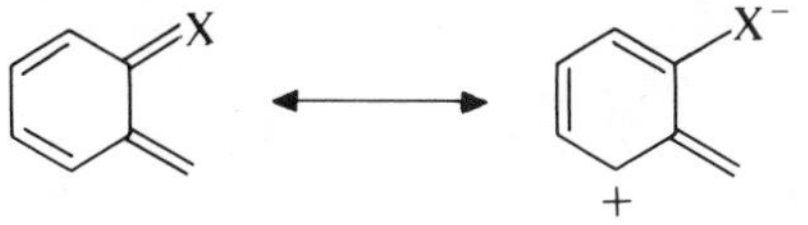

Ring-closure from the first excited singlet was predicted to occur for X = CH_2 and S but not for NH or O. Although these results are inferred from calculations neglecting configuration interaction, it is worth noticing that several derivatives of the closed form for X = CH_2, S and NH are known, but derivatives of the oxa analogue are hitherto unknown.

Kinetic work by Roth and coworkers[81] unambiguously favours the closed form of the hydrocarbon, and the energy profile for the conversion is deduced from thermochemical data. The enthalpy of formation of the closed form was found to be 47.7 kcal mol^{-1}, the corresponding values for OQDM and the transition state were 58.2 and 86.6 kcal mol^{-1} respectively, all values referring to the gas phase. The data were only slightly modified in a later shock tube technique work by the same group[82]. MINDO/3 calculations by Bingham and coworkers[83] yielded a value of enthalpy of formation of benzocyclobutene of 59 kcal mol^{-1}, but complete geometry optimization using the same procedure[78] yielded a value of 48.5 kcal mol^{-1} in excellent agreement with existing experimental values.

A novel application of the open form–closed form equilibrium is found in a flash vacuum pyrolysis experiment of benzocyclobutene by Trahanovsky[84]. The reaction was found to give anthracene among other high molecular weight products, the reaction is proposed to go via the (4 + 4) dimer form of OQDM followed by loss of two carbon atoms and six hydrogens. The reaction is highly regiospecific.

D. The *Meta* Isomers

As mentioned in Section I we will allude only briefly to the properties of the *meta* isomers (**3**), as their characteristic features are discussed extensively in a separate chapter.

The *meta* species most thoroughly discussed in the literature appears to be the hydrocarbon *meta*-benzoquinodimethane (MBQDM), on which a series of both experimental and theoretical investigations have been carried out. However, *meta*-benzoquinomethane (MBQM) has also been claimed to occur as a result of irradiation of 6-methylenebicyclo[3.1.0]hex-3-ene-2-one in a solution of 2-methyltetrahydrofuran at 11 K[85].

The hydrocarbon MBQDM has been prepared from dehydro-*m*-quinodimethane in a host matrix at 77 K[86]. Very recently Goodman and Berson were able to isolate the hydrocarbon at low temperature, and to study its reactivity in solution[87].

The dominant feature of the *meta* isomers as compared to the *ortho* and *para* counterparts is their diradical nature. This has been confirmed experimentally by electron paramagnetic resonance (EPR) spectra[85,86] demonstrating triplet ground states. Furthermore, theoretical calculations at different levels of sophistication have predicted a triplet ground state of MBQDM[14,88–90]. The only *ab initio* study[90] among these calculations led to a $^3B_2$ ground state of the molecule, the $^1A_1$ state being 10 kcal mol^{-1} higher in energy. The predicted energy difference between the $^3B_2$ ground state of MBQDM and the ^1A_g ground state of PQDM was found to depend strongly on the method used for recovering the molecular correlation energy. A lower bound of 24 kcal mol^{-1} to the energy difference $^3B_2 - {}^1A_g$ was suggested[90]. This value is in accordance with a value of 26 kcal mol^{-1} estimated indirectly by Hehre and coworkers[91].

The electronic and fluorescence spectra of MBQDM have been recorded at 77 K[92], and electronic transitions have been estimated by accompanying MO calculations[92].

E. Polarizabilities

In a series of papers Hameka and coworkers have calculated both linear and third-order non-linear electric susceptibilities of aromatic and conjugated hydrocarbons[93-97], among which both PBQDM and OBQDM were included. The numerical values obtained for these species were dramatically different between the Hückel and the PPP methods[97]. The discrepancies were interpreted in terms of different estimates of the lowest excitation energies within these two approximations.

Electric polarizabilities and diamagnetic susceptibilities have been related to the quinonoid character of some conjugated systems in papers by Luzanov's group[98,99]. In the studies, which were performed within the π-electron approximation, it was found that molecules having quinonoid features as measured by other properties, also displayed comparatively large π-electron polarizabilities[99]. Thus for PBQDM a value of 11.35 Å^3, and for PBQ a value of 6.94 Å^3 were obtained. The corresponding value for benzene is 4.0 Å^3. Furthermore, a pronounced anisotropy of the π-electron polarizability was predicted for the quinonoid compounds. This has been interpreted in terms of double bonds located along the long axis of the molecules. The slightly lower value for PBQ as compared with PBQDM has been attributed to the high electronegativity of oxygen[99]. It should, however, be mentioned that a polarizability study based on the δ-function model[100] does not lead to significant anisotropy of the π-electron polarizability of PBQ[101].

V. SPECTROSCOPIC PROPERTIES AND EXCITED STATES

A. General Considerations

Experimental results available for the molecules belonging to the *ortho* and *para* groups, (**1**) and (**2**), respectively indicate pronounced polyene structures, and closed shell electronic ground states. Thus for PQDM electron diffraction data[30], photoelectron spectra[102], low temperature IR and UV spectra[11], NMR spectrum at low temperature of pyrolysis product[103], and other spectroscopic studies[104,105], all support the assumption of a closed shell ground state as opposed to the alternative biradical form. An estimate made by homodesmotic reactions indicates that there is no strain in the molecule[54]. Molecular orbital calculations within the CNDO/S and INDO approximations lead to the conclusion that the energetically favourable configuration is a spin-paired singlet[106]. PPP calculations with full CI in the π space do, however, indicate a certain amount of biradical character[89]. A similar conclusion is obtained by valence-bond calculations[107].

The species PQM is very reactive and unstable. It has been characterized by low temperature NMR, IR, and UV spectra[6]. Its UV spectrum in dilute solutions has also been recorded[108]. The spectra confirm the quinonoid structure of the molecule. So do semiempirical calculations both of Hückel type and at the SCF level[109,110]. Its UV spectrum has been rationalized by semiempirical SCF calculations[111].

The species OQDM has not been isolated. It has been studied in organic matrices by fluorescence, by fluorescence excitation and by UV absorption[12-14]. Spectra with higher resolution have been obtained in Shpolskii matrices[92], and in argon matrices IR, Raman, UV, fluorescence and fluorescence excitation spectra have been obtained[15]. The spectroscopic data are consistent with a closed shell molecular ground state. This conclusion has furthermore been confirmed both by Hückel-type calculations[80] and by SCF studies within the PPP approximation[88, 112].

For the isomer OQM an IR spectrum has been observed of the matrix isolated molecule[113]. A UV photoelectron spectrum of the molecule has also been recorded in the temperature range 100–600°C, and its electronic structure has been discussed[8]. The data available support the assumption of a closed shell structure with alternating single and double bonds.

For the very unstable species PQI a UV spectrum has been recorded in solution[114]. Similar spectra are available also for PQDI[114, 115], for which also IR information[115] and NMR data are available[115]. For the *ortho* counterparts OQI and OQDI, UV spectra in solution have been recorded[116].

Very recently monothio-PBQ and dithio-PBQ have been produced[18], and their photoelectron spectra have been recorded[18]. Both IR and UV spectra of these compounds have been obtained in argon matrices[18]. The results of MNDO calculations indicate an electron configuration analogous to PBQ[18].

B. Photoelectron Spectra, Koopmans' Theorem and Shake-up States

An important source of information regarding the occupancy and sequence of the outer valence orbitals is provided by photoelectron spectra (PES) in conjunction with theoretical calculations. Of central importance in the interpretation of PES is the assumed validity of Koopmans' theorem[117] which implies a frozen orbital approximation in the description of the ionization process. As pointed out by Richards[118] the success of Koopmans' theorem in quantitative discussions of vertical ionization energies may be traced back to a fortuitous cancellation of errors due to reorganization of electrons and neglect of electronic correlation effects. In a series of recent investigations, strong evidences for a breakdown of Koopmans' theorem have been accumulated[119–122]. An example of particular importance in this context is the case of PQDM. The PES of this molecule and its 3,7-dimethyl derivative were recorded and interpreted by Koenig and coworkers[102, 123]. The samples of the compounds were generated by flash vacuum pyrolysis of [2.2]paracyclophane and its dimethyl derivative, respectively, and the spectra were analysed in terms of the structure representation (SR) method[124–126]. The first band peaking at around 7.9 eV was assigned to a pure Koopmans' configuration ($^2B_{3u}$) of the ion, and the second observed peak at around 9.7 eV was interpreted in terms of an ionic Koopmans' configuration of symmetry $^2B_{1g}$[102]. Arguments in support of this assignment were found by comparison with observed spectra for related molecules[127]. The third band at about 10 eV having an apparently low intensity was rationalized by an ionization process of a non-Koopmans nature. It was suggested that this band could be described by a $^2B_{2g}$ ionic state having a large coupling between $1b_{2g}(\pi)^{-1}$ Koopmans hole state and the non-Koopmans valence orbital excitation $2b_{3u}(\pi)^{-1} \rightarrow 2b_{2g}(\pi^*)$[102]. Support for this interpretation was obtained by a photoelectron study of the 2,5-dimethyl derivative of PQDM[123]. This highly unconventional assignment invoking a HOMO–LUMO shake-up state was challenged by Dewar[128]. On the basis of MNDO calculations it was claimed that the energy difference between the hole state and the non-Koopmans excitation was too large for an effective coupling[128]. Results of additional recent calculations[129–132], however, have given support to the original assignment of Koenig and coworkers[102, 123].

The sequence of the outer valence orbitals in PBQ is still somewhat uncertain in spite of a series of experimental and theoretical investigations. The PES of PBQ and many of its derivatives have been recorded by different groups[71, 133–138]. A UV spectrum of PBQ in the vapour phase has also been reported[70]. A multitude of semiempirical methods have been applied in efforts made to pin down a unique assignment of the highest filled orbitals[71, 136–145]. Several *ab initio* studies have been made on the electronic ground state of PBQ[146–148], but they have only partially settled questions related to the assignment of the low energy bands in the PES. However, the rather extensive calculations by Ha[148] have shown that a striking change in the ordering of state energies takes place after inclusion of configurational mixing. This is a demonstration of the insufficiency of the orbital model.

Interestingly, the supposedly best quality *ab initio* calculation[145] on PBQ agrees with CNDO/2[130, 134, 141] and HAM/3 calculations[71] in placing the n states at higher energies than the π states. One of the crucial requirements in the assignment of the orbital sequence, imposed by the UV spectrum[70], is that the energy splitting between $n^+ (b_{2u})$ and $n^- (b_{3g})$ should not exceed 0.3 eV[137].

TABLE 4. Suggested sequences for the four highest filled orbitals of PBQ*

Method	I(1)	I(2)	I(3)	I(4)	Ref.
CNDO/2	$n^- (b_{3g})$	$n^+ (b_{2u})$	$\pi_1 (b_{3u})$	$\pi_2 (b_{1g})$	133, 137, 144
INDO	$n^- (b_{3g})$	$\pi_1 (b_{3u})$	$n^+ (b_{2u})$	$\pi_2 (b_{1g})$	136, 142, 144
CNDO + CI[a]	$^2B_{3g}(n^-)$	$^2B_{2u}(n^+)$	$^2B_{3u}(\pi_1)$	$^2B_{1g}(\pi_2)$	138
HAM/3	$n^- (b_{3g})$	$n^+ (b_{2u})$	$\pi_2 (b_{1g})$	$\pi_1 (b_{3u})$	71
CNDO/S	$n^- (b_{3g})$	$\pi_1 (b_{3u})$	$\pi_2 (b_{1g})$	$n^+ (b_{2u})$	134, 141
	$\pi_1 (b_{3u})$	$\pi_2 (b_{1g})$	$n^- (b_{3g})$	$n^+ (b_{2u})$	145
	$n^- (b_{3g})$	$\pi_1 (b_{3u})$	$\pi_2 (b_{1g})$	$n^+ (b_{2u})$	143
Ab initio	$\pi_1 (b_{3u})$	$n^- (b_{3g})$	$\pi_2 (b_{1g})$	$n^+ (b_{2u})$	146
Ab initio	$\pi_1 (b_{3u})$	$\pi_2 (b_{1g})$	$n^- (b_{3g})$	$n^+ (b_{2u})$	147
Ab initio[a]	$^2B_{3g}(n^-)$	$^2B_{2u}(n^+)$	$^2B_{3u}(\pi_1)$	$^2B_{1g}(\pi_2)$	148

* Ascending ionization energies from left to right.
[a] Due to strong mixing of configurations we use state symmetries rather than orbital symmetries.

In Table 4 we present a survey of some of the suggested orbital sequences in PBQ based on different semiempirical and theoretical calculations. As revealed by the table the different approaches used are still in mutual conflict regarding the interpretation of the PES. It is worth mentioning that also for this molecule a consistent prediction of the orbital sequence is dependent on the use of configurational mixing, i.e. the allowance for non-Koopmans states. This was emphasized by Schweig and coworkers[138] in 1975 using their own CI approach[149] based on CNDO wave functions. Very recently their conclusion was confirmed by Ha in *ab initio* calculations[148].

Turning to the *ortho* isomers, the species OQDM has so far resisted isolation, but its PES has recently been recorded by Schweig and coworkers[150] using their variable temperature photoelectron spectroscopy (VTPES) technique[151, 152]. They succeeded in generating and detecting OQDM by choosing a precursor compound different from one previously used[153]. Semiempirical MO calculations including configurational mixing both for the molecular ground state and the singly ionized states have given an ionization spectrum of the species[130]. The predicted assignments of the ionized states are $^2A_2 (\pi)$, $^2B_1 (\pi)$, $^2A_2 (\pi)$, $^2B_1 (\pi)$, $^2A_1 (\sigma)$ in order of ascending energy.

In these calculations it was found that the first excited ion state, $^2B_1 (\pi)$, was strongly mixed with the HOMO–LUMO shake-up configuration[130]. The energy predicted for the first non-Koopmans state was found to depend rather strongly on the particular calculational approach used (8.6 eV by PPP–CI, 10 eV by PERT–CI)[130]. The PES recently recorded[150] has given strong evidence for an ionization process involving low energy shake-up states.

The PES of OBQ has been recorded by Schweig and coworkers[154]. They have also made assignment of the spectrum by applying their perturbation method (PERT–CI) on MNDO and CNDO/S calculations[154]. The ionic state sequence arrived at is $^2A_1\,(n^+)$, $^2A_2\,(\pi_1)$, $^2B_2\,(n^-)$, $^2B_1\,(\pi_2)$ in order of increasing energy. As expected the energy splitting between the lone-pair states A_1 and B_2 (1.2–1.8 eV)[154] is substantially larger in this molecule than in PBQ (0.3 eV)[137] due to the proximity of the oxygens.

C. UV Spectra, IR Spectra and Structure of Excited States

Derivatives of quinomethanes and quinone imines show strong absorption in the visible part of the spectrum, and constitute an important group of dyes such as fuchsone, Homolka-base, crystal violet, phenolphthalein and others[3,189]. In our survey we will not pursue this part of quinone chemistry, however important it might be. We will primarily focus on publications concerned with problems associated with the interpretation of the electronic spectrum of PBQ, and with some properties of its lowest electronically excited states. The problems arise mainly from the fact that excitation from the symmetrized lone-pair orbitals in D_{2h} symmetry leads to two nearly degenerate states $B_{1g}(n^- \rightarrow \pi^*)$ and $A_u(n^+ \rightarrow \pi^*)$, the π^* being the LUMO, having b_{2g} symmetry[68,155]. The energy splitting between these states is a crucial measure of the lone-pair lone-pair interaction. Transitions from the ^1A_g ground states to these states are symmetry forbidden in the electric dipole approximation.

Since the first detailed analysis of the vapour phase absorption spectrum of PBQ in the UV region was made by Hollas and coworkers[156–158], a wealth of theoretical and experimental studies related to the low energy transitions in this molecule have appeared.

In a series of papers spectroscopic measurements of PBQ single crystals and of PBQ in host lattices have been carried through at very low temperatures[159–169]. Spectral studies of PBQ in the vapour phase have also been reported[170–173]. Furthermore several theoretical studies, both semiempirical ones and *ab initio* calculations, describing spectral properties of PBQ and structural features of its lowest electronically excited state have appeared[174–186].

The detailed analysis of the vapour phase absorption spectrum performed by Hollas and coworkers[156–158] led to the conclusion that the majority of observed bands in the singlet system of PBQ could be interpreted in terms of only the $^1A_g \rightarrow {}^1B_{1g}$ transition. The origin of this, electric dipole forbidden, transition was located by means of a vibrational mode of b_{3u} symmetry which is a 'boat' vibration.

An alternative band assignment locating the origin as the magnetic dipole allowed transition $^1A_g \rightarrow {}^1B_{1g}$ was suggested later in a supersonic jet experiment[171]. The origin of this transition was found to be at 20 045 cm^{-1}, and the energy splitting between the states $^1B_{1g}$ and ^1A_u was reported to be only 54 cm^{-1}, the ^1A_u state being the lower one. The same ordering with a splitting of 64 cm^{-1} in the singlet system has been observed[167] in a Ne host at 4.2 K. This value for the energy of the origin requires a lowest frequency vibration in the vapour of only about 87 cm^{-1}, a value which is extremely low, and which deviates appreciably from values obtained by IR measurements in solution[187,188], viz. 108 cm^{-1}. Furthermore this value appears to be independent on the solvent. A far-IR study of the vapour phase has indeed confirmed the presence of the very low vibrational frequency. The actual values reported were 88.9 cm^{-1} and 87.5 cm^{-1} for PBQ-h_4 and PBQ-d_4 respectively. The nature of this vibration was found to be a b_{3u} mode boat-type distortion[173]. The large frequency shifts observed in going from vapour to solution were interpreted in terms of intermolecular forces[173]. The origin of this $^1A_g \rightarrow {}^1B_{1g}$ transition was located at 20 047 cm^{-1} for both PBQ-h_4 and PBQ-d_4, in agreement with the value obtained in the supersonic jet experiment referred to above[171].

Problems encountered in MO descriptions of the excited states of PBQ have been thoroughly discussed by Martin and coworkers[182,183]. Their main conclusion is that a description in terms of symmetrized (in D_{2h} symmetry), delocalized molecular orbitals leads to a qualitatively incorrect picture. A proper interpretation of these states would be a symmetry-broken structure resulting from a localized excitation. This description is in accordance with the dimer model of the excited states. The dimer model is also invoked in the interpretation of low-temperature measurements of proton ENDOR spectra of substituted PBQ[166]. It has been shown that the C_{2v} symmetry of the unsubstituted half of the molecule is virtually intact[166]. Stark measurements on crystals of toluquinone[169] have confirmed the description of the nπ* excited states in terms of local excitations of the C=O groups. Furthermore it has been claimed that in this molecule the excitation to the lowest state resides on the C=O group not adjacent to the methyl group for both the singlet and the triplet system[169].

Regarding the structure of the $^3(n\pi^*)$ state, spectral interpretations have led to conflicting evidence. One model emerges from the assumption that the splitting between ^3A_u and $^3B_{1g}$ is around 300 cm^{-1}, and that the lower state ($^3B_{1g}$) is distorted from D_{2h} symmetry along a mode of b_{1u} symmetry[70]. The barrier between the two equivalent C_{2v} minima generates a pair of inversion doublets split by 10 cm^{-1}. The alternative model is based on the assumption that the triplet–triplet splitting is 10 cm^{-1} and that both states (^3A_u and $^3B_{1g}$) have roughly D_{2h} symmetry[167]. Valence-bond calculations were not conclusive regarding the presence of a global C_{2v} minimum in the $^3B_{1g}$ state[183].

VI. CYCLOADDITION REACTIONS

Chemical reactions of quinones and quinonoid systems constitute a vast and complex field in chemistry. For a broad and thorough treatment of addition and substitution reactions of quinones we refer to the Chapter by Finley in this series[1]. Furthermore, a good introduction is found in the basic textbook by Tedder and Nechvatal[3].

By including a brief section on cycloaddition reactions on quinonoid systems we intend to put emphasis on some of the underlying theoretical models that have been developed and utilized in the interpretation of such processes. Also we wish to focus on cycloaddition reactions, since these have been applied in a number of recent, elegant syntheses of a number of different products. In particular, there has been a major effort in the field of preparing various anthracycline antibiotics which are effective in cancer chemotherapy[190–195].

Although the efforts in these investigations have been directed towards the potential usefulness of the products, the vast amount of information obtained has also been important for understanding the regioselectivities of these reactions.

Theoretical and mechanistic aspects of cycloadditions to PBQ and substituted PBQs have been widely discussed. One point of central importance in this context has been the analysis of the competition between two reactive sites in PBQ, the carbon–carbon double bond and the carbon–oxygen double bond.

A systematic study of this problem has recently been published by Shiraishi and coworkers[196] who carried through both experimental and theoretical studies of the reactions between substituted nitrile oxides and substituted PBQs. This study and a related theoretical one by Houk and coworkers[197] have demonstrated the versatility of some of the simple interaction models for qualitative predictions. But at the same time they have revealed the shortcomings of these models when small energy differences are decisive.

Kelly and coworkers have applied resonance theory[198] to account for the site of attack by nucleophiles on quinones[190–195]. The resonance picture accounts satisfactorily for the impact of donor substituents on remote unsubstituted bonds, and also for the impact of acceptor substituents on neighbouring bonds. For other important cases like the influence of conjugating substituents, and the acceptor substituent effect on remote bonds the results are less convincing[197].

The Frontier Molecular Orbital (FMO) theory developed by Fukui[199] is a powerful tool for explaining and predicting reactivities and selectivities in organic reactions[200]. In cases where quinones are involved we have commonly addition of nucleophiles to the quinonoid system. Substituent effects on reaction rates and selectivities are in this framework discussed in terms of the perturbation of the substituent on the LUMO of the quinonoid system.

The works by Shiraishi[196] and Houk[197] are based upon STO-3G calculations and HOMO–LUMO energy differences are computed. The type of information to be inferred from these calculations may be exemplified by the work of Shiraishi[196]. In that work, treating reactions of nitrile oxides with substituted PBQs, distinction is made between two main reaction types: C=O addition (leading to spiro(1,4,2)dioxazole) and C=C addition (leading to an isoxazoline). The former route is determined by the HOMO (nitrile oxide) and the LUMO (quinone) energy differences while the latter is determined by the LUMO (nitrile oxide) and HOMO (quinone) energy differences. The borderline energy difference between two interactions was found to be around 4 eV, although exceptions were found. In view of the simplicity of the method and the inherent shortcomings of the basis set, lack of configuration interaction, consideration of lower-lying orbitals and lack of geometry optimizations, the overall results must be regarded as very good.

VII. POLYMERIZATION REACTIONS

Whereas the quinone family of molecules consists of relatively stable members, the isoelectronic QDMs are highly reactive. For instance, PQDM although made and characterized from pyrolysis of *p*-xylene, exists only in the vapour phase and in low temperature solutions[11]. Upon heating, polymerization takes place[123], and Figure 1 shows the multitude of reaction paths that may occur when different reagents are added. Since diradical structures are essential in the polymerization process, the ease of polymerization should parallel the ability to form diradicals. A narrow gap between the lowest singlet and the lowest triplet state is one criterion for biradicaloid structure, and indeed the energy difference between the singlet ground state and the first triplet excited state for PQDM has been calculated by Coulson and coworkers[37] to be 8–9 kcal mol^{-1}. (The similar energy difference for ethylene is reported to be 82 kcal mol^{-1} [201]). This small energy difference is in accordance with the high reactivity of PQDM. Koutecky and Döhnert[89] have pointed out the lack of precision in this criterion, and suggest the use of occupation numbers in natural orbitals as a measure of biradical character. Whenever two non-bonding occupation numbers are different from zero or two, the system exhibits biradical character, and also by using this criterion PQDM belongs to the category of biradicaloids. Similar results have been found by Flynn and Michl[14]. We wish to point out, however, that the most direct experimental technique for investigating quinonoid vs. biradical character is magnetic studies, and already 50 years ago the dominating quinonoid character of PQDM was established by Müller and Müller-Rodloff[202]. Only sterically hindered and twisted Chichibabin hydrocarbons (see below) were found to be paramagnetic[203]. For a further discussion of electronic properties we refer to Section IV.

As early as in 1904 Thiele obtained poly-PQDM in an attempt to obtain the parent monomer[204]. Although the polymer was initially an undesired product, it has received growing attention in recent years because of its use as plastic coating and other industrial uses. We refer to several review papers[205–208] that give account of the development in the field. Here we will focus our attention on some key points rather than give an extensive account of the area.

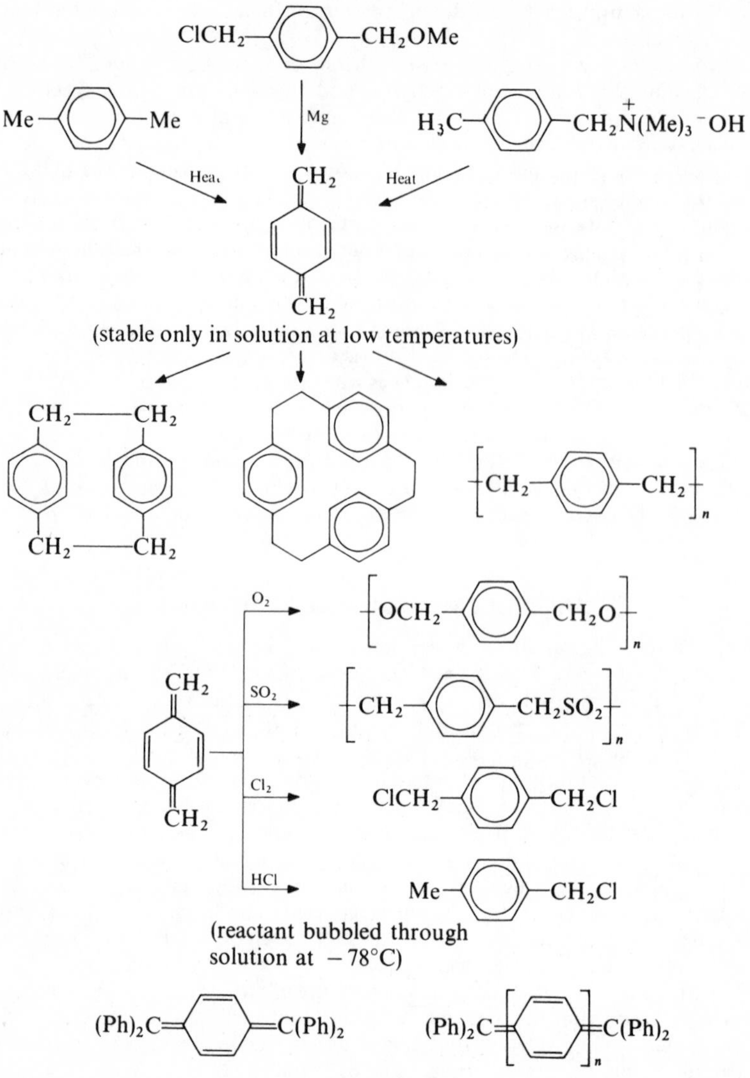

(stable only in solution at low temperatures)

(reactant bubbled through solution at $-78°C$)

FIGURE 1. Some of the possible reaction paths of PQDM. *Adapted from* Basic Organic Chemistry *by J. M. Tedder and A. Nechvatal, part* 2, *p.* 248 (*Ref.* 3), *published by John Wiley and Sons*

The work by Thiele[204] initiated a series of interesting studies that provided new insight into a variety of fields. For instance, the existence of a diradical was demonstrated in the tetraphenyl derivative of PQDM (usually referred to as Thiele's hydrocarbon, **4**).

Although the equilibrium is strongly shifted to the left (less than 0.2% of diradical form was believed to exist), the possibility of diradicals in reaction mechanisms was established. Furthermore, a number of analogue compounds were found to have a large fraction of

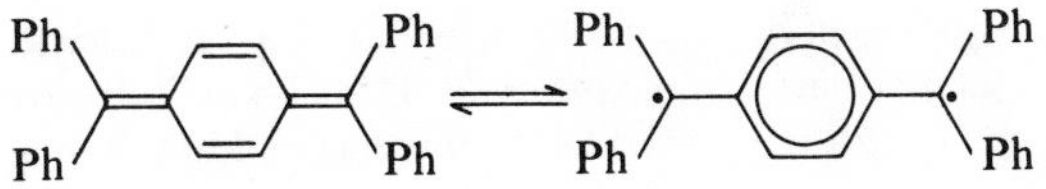

(4)

diradical character in the ground state. Notable examples are the Chichibabin (5) and Schlenk (6) hydrocarbons, the former of which has been found (on the basis of ESR measurements) to be a paramagnetic species in the dimer form[209]. With an increasing number of rings attached, the fraction of diradical form increases, paralleling the increase in electrode potentials with an increasing number of rings in *p*-quinones (see Section IV). As for the Schlenk hydrocarbons, the stability of free radicals is in accordance with the impossibility of writing classical resonance forms. A modern discussion of the diradical vs. diradicaloid nature of these compounds is given in a review by Platz[210].

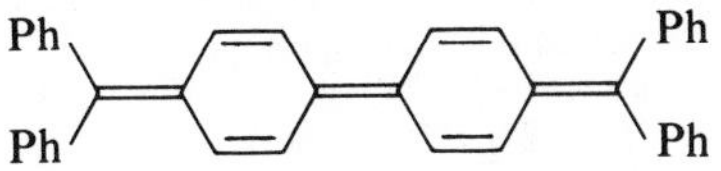

(5)

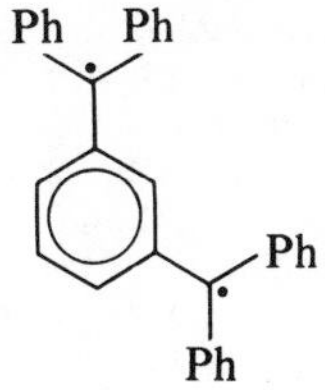

(6)

Several methods for the preparation of poly-PQDM are now available, and different pyrolytic methods are being used. The review by Iwatsuki[208] gives reaction mechanisms for both isothermal low temperature reactions and a non-isothermal process. Also, the same review mentions a number of different reaction routes giving various yields and qualities of polymers.

An especially pure linear polymer with high molecular weight has been obtained in high yields by Gorham[205] who used a vacuum pyrolysis technique at 600°C. The key intermediate was identified as (2,2)*p*-cyclophane, which was quantitatively cleaved to PQDM which spontaneously polymerized. The method was adapted to a number of substituted poly-PQDMs, many of which have not been prepared by other methods. The products were particularly pure and free of cross-linked and low molecular weight products. A kinetic study has been carried out on the process[211]. The polymers have low permeability to moisture and other corrosive gases, and are useful for the coating of electrical assemblies. The physical and electrical properties of three of these polymers (which are now commercially available) are given in the review by Iwatsuki[208].

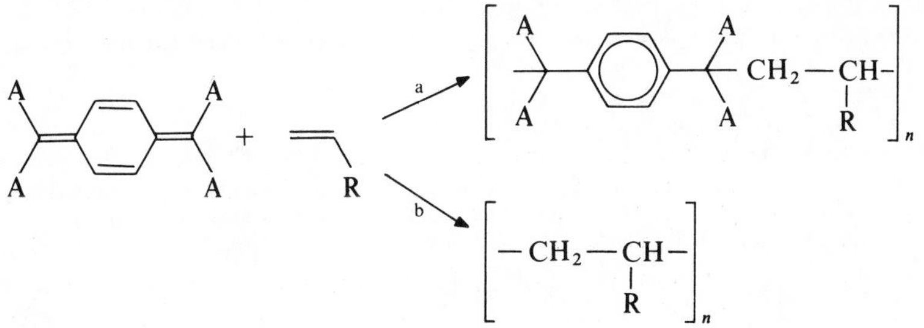

Because of the high reactivity of PQDM, copolymerization with vinyl monomers requires a large excess of the latter. If, however, PQDM is substituted with highly electron-withdrawing groups, alternating copolymers (route a) or homo vinyl polymers may be formed (route b)[206].

The propagating species is found to be a diradical formed from the *p*-cyclophane in all cases where the product contains a PQDM unit. Since *p*-cyclophane is a rather strained species, this parallels findings for reactions involving a number of other strained species[212].

For the case of homo vinyl polymerization, a cationic species is believed to be the propagating species[206].

VIII. REFERENCES

1. S. Patai (Ed.), *The Chemistry of the Quinonoid Compounds*, parts 1 and 2, John Wiley and Sons, 1974.
2. C. Grundmann, Chinone I and II, in *Houben-Weyl*, vol. 7, parts 3a and 3b, Thieme Verlag, 1979.
3. J. M. Tedder and A. Nechvatal, *Basic Organic Chemistry*, part 2, John Wiley, 1967.
4. W. M. Horspool, *Quart. Rev.*, **23**, 204 (1969).
5. A. L. Macdonald and J. Trotter, *J. Chem. Soc. Perkin Trans. 2*, 476 (1973).
6. M. C. Lasne, J. L. Ripoll and J. M. Denis, *Tetrahedron*, **37**, 503 (1981).
7. P. D. Gardner, H. Sarrafizadeh and R. L. Brandon, *J. Am. Chem. Soc.*, **81**, 5515 (1959).
8. V. Eck, A. Schweig and H. Vermeer, *Tetrahedron Lett.*, 2433 (1978).
9. L. A. Errede and J. M. Hoyt, *J. Am. Chem. Soc.*, **82**, 436 (1960).
10. W. F. Gorham, *J. Polym. Sci., Polym. Chem. Ed.*, 3027 (1966).
11. J. M. Pearson, H. A. Six, D. J. Williams and M. Levy, *J. Am. Chem. Soc.*, **93**, 5034 (1971).
12. E. Migirdicyan, *C. R. Hebd. Séances Acad. Sci., Ser. C*, **266**, 756 (1968).
13. C. R. Flynn and J. Michl, *J. Am. Chem. Soc.*, **95**, 5802 (1973).
14. C. R. Flynn and J. Michl, *J. Am. Chem. Soc.*, **96**, 3280 (1974).
15. K. L. Tseng and J. Michl, *J. Am. Chem. Soc.*, **99**, 4840 (1977).
16. B. H. Han and P. Boudjouk, *J. Org. Chem.*, **47**, 751 (1982).
17. P. Venuvalalingam, U. C. Singh, N. R. Subbaratnam and V. K. Kelkar, *Spectrochim. Acta*, **36A**, 103 (1980).
18. H. Bock, S. Mohmand, T. Hirabayashi, G. Maier and H. P. Reisenauer, *Chem. Ber.*, **116**, 273 (1983).
19. R. Schulz and A. Schweig, *Angew. Chem.*, **93**, 603 (1981).
20. S. M. Swingle, *J. Am. Chem. Soc.*, **76**, 1409 (1954).
21. M. Kimura and S. Shibata, *Bull. Chem. Soc. Japan*, **27**, 163 (1954).
22. K. Hagen and K. Hedberg, *J. Chem. Phys.*, **59**, 158 (1973).
23. J. Trotter, *Acta Crystallogr.*, **13**, 86 (1960).
24. K. Kuchitsu, *J. Chem. Phys.*, **44**, 906 (1966).
25. K. Hagen and K. Hedberg, *J. Mol. Struct.*, **49**, 351 (1978).
26. A. Meresse, C. Courseille and N. B. Chanh, *Acta Crystallogr.*, **30B**, 524 (1974).
27. H. Schei, K. Hagen, M. Traetteberg and R. Seip, *J. Mol. Struct.*, **62**, 121 (1980).
28. D. Rabinovich, G. M. J. Schmidt and E. Ubell, *J. Chem. Soc. B*, 131 (1967).
29. G. L. Blackman, R. D. Brown and A. P. Porter, *J. Chem. Soc. Chem. Commun.*, 499 (1975).
30. P. G. Mahaffy, J. D. Wieser and L. K. Montgomery, *J. Am. Chem. Soc.*, **99**, 4514 (1977).
31. W. C. Herndon, *Pure appl. Chem.*, **52**, 1459 (1980).
32. G. J. Gleicher, in *The Chemistry of the Quinonoid Compounds* (Ed. S. Patai), John Wiley and Sons, 1974, Chapter 1.
33. G. J. Gleicher, *J. Am. Chem. Soc.*, **95**, 2526 (1973).
34. B. A. Hess Jr and L. J. Schaad, *J. Am. Chem. Soc.*, **93**, 305 (1971).
35. B. A. Hess Jr. and L. J. Schaad, *J. Am. Chem. Soc.*, **93**, 2413 (1971).
36. A. J. Namiot, M. E. Dyatkina and Ya. K. Syrkin, *Dokl. Akad. Nauk*, **48**, 267 (1945).
37. C. A. Coulson, D. P. Craig, A. Maccoll and A. Pullman, *Disc. Farad. Soc.*, **36** (1947).
38. M. J. S. Dewar and G. J. Gleicher, *J. Am. Chem. Soc.*, **87**, 692 (1965).
39. M. J. S. Dewar and G. J. Gleicher, *J. Am. Chem. Soc.*, **87**, 685 (1965).
40. I. Gutman and S. Bosanac, *Tetrahedron*, **33**, 189 (1977).

41. J. Kruszewski, *Societatis Scientarum Lodziensis, Acta Chim. Lodz.*, **16**, 77 (1971).
42. A. Julg and Ph. Francois, *Theor. Chim. Acta*, **7**, 249 (1967).
43. M. J. S. Dewar and C. de Llano, *J. Am. Chem. Soc.*, **91**, 789 (1969).
44. B. A. Hess Jr and L. J. Schaad, *Austr. J. Chem.*, **25**, 2331 (1972).
45. J. Aihara, *Bull. Chem. Soc. Japan*, **56**, 1935 (1983).
46. G. W. Wheland, *J. Am. Chem. Soc.*, **64**, 900 (1942).
47. A. Streitwieser Jr, *Molecular Orbital Theory for Organic Chemists*, Wiley, New York, 1961, Chapter 11.
48. K. Fukui, T. Yonezawa and C. Nagata, *Bull. Chem. Soc. Japan*, **27**, 423 (1954).
49. K. Fukui, T. Yonezawa and C. Nagata, *J. Chem. Phys.*, **27**, 1247 (1957).
50. K. Jug, *J. Org. Chem.*, **48**, 1344 (1983).
51. E. J. Moriconi, B. Rakoczy and W. F. Connor, *J. Org. Chem.*, **27**, 2772 (1962).
52. H. L. K. Schmand, H. Kratzin and P. Boldt, *Liebigs Ann. Chem.*, 1560 (1976).
53. M. J. S. Dewar and N. Trinajstic, *Tetrahedron*, **25**, 4529 (1969).
54. A. Banks, G. J. Mains, C. W. Bock, M. Trachtman and P. George, *J. Mol. Struct.*, **56**, 267 (1979).
55. S. Millefiori and A. Millefiori, *J. Mol. Struct.*, **104**, 131 (1983).
56. O. Sinanoglu and H. O. Pamuk, *J. Am. Chem. Soc.*, **95**, 5435 (1973).
57. C. D. Cooper, W. T. Naff and R. N. Compton, *J. Chem. Phys.*, **63**, 2752 (1975).
58. E. C. M. Chen and W. E. Wentworth, *J. Chem. Phys.*, **63**, 3183 (1975).
59. M. E. Peover, *J. Chem. Soc.*, 4540 (1962).
60. M. E. Peover, *Trans. Farad. Soc.*, **58**, 2370 (1962).
61. F. A. Matsen, *J. Chem. Phys.*, **24**, 602 (1956).
62. M. J. S. Dewar and H. S. Rzepa, *J. Am. Chem. Soc.*, **100**, 784 (1978).
63. C. D. Cooper, W. F. Frey and R. N. Compton, *J. Chem. Phys.*, **69**, 2367 (1978).
64. E. P. Grimsrud, G. Caldwell, S. Chowdbury and P. Kebarle, *J. Am. Chem. Soc.*, **107**, 4627 (1985).
65. H. T. Jonkman, G. A. van der Velde and W. C. Nieuwpoort, *Chem. Phys. Lett.*, **25**, 62 (1974).
66. H. Johansen, *Int. J. Quant. Chem.*, **9**, 459 (1975).
67. R. E. Long, R. A. Sparks and K. N. Trueblood, *Acta Crystallogr.*, **18**, 932 (1965).
68. A. Modelli and P. D. Burrow, *J. Phys. Chem.*, **88**, 3550 (1984).
69. T. L. Kunii and H. Kuroda, *Theor. Chim. Acta*, **11**, 97 (1968).
70. H. P. Trommsdorff, *J. Chem. Phys.*, **56**, 5358 (1972).
71. L. Åsbrink, G. Bieri, C. Fridh, E. Lindholm and D. P. Chong, *Chem. Phys.*, **43**, 189 (1979).
72. J. E. Bloor, R. A. Paysen and R. E. Sherrod, *Chem. Phys. Lett.*, **60**, 476 (1979).
73. F. H. Herbstein, in *Perspectives in Structural Chemistry*, Vol. 4 (Eds J. D. Dunitz and J. A. Ibers), Wiley, New York, 1971.
74. J.-J. Andre, A. Bieber and F. Gautier, *Ann. Phys.*, **t.I.** 145 (1976).
75. Q. Y. Zhang and J. M. Yan, *Int. J. Quant. Chem.*, **27**, 407 (1985).
76. A. J. Epstein, E. M. Conwell, D. J. Sandman and J. S. Mill, *Solid State Commun.*, **23**, 355 (1977).
77. B. Morosin, *Acta Crystallogr.*, **B32**, 1176 (1976).
78. J. Mirek and A. Buda, *Z. Naturforsch.*, **39a**, 386 (1984).
79. E. K. Fukuda and R. T. McIver Jr, *J. Am. Chem. Soc.*, **107**, 2291 (1985).
80. H. Kolshorn and H. Meier, *Z. Naturforsch.*, **32a**, 780 (1977).
81. W. R. Roth, M. Biermann, H. Dekker, R. Jochems, C. Mosselman and H. Hermann, *Chem. Ber.*, **111**, 3892 (1978).
82. W. R. Roth and B. P. Scholz, *Chem. Ber.*, **114**, 3741 (1981).
83. R. C. Bingham, M. J. S. Dewar and D. H. Lo, *J. Am. Chem. Soc.*, **97**, 1294 (1975).
84. W. S. Trahanovsky and B. W. Surber, *J. Am. Chem. Soc.*, **107**, 4995 (1985).
85. M. Rule, A. R. Matlin, E. F. Hilinski, D. A. Dougherty and J. A. Berson, *J. Am. Chem. Soc.*, **101**, 5098 (1979).
86. B. B. Wright and M. S. Platz, *J. Am. Chem. Soc.*, **105**, 628 (1983).
87. J. L. Goodman and J. A. Berson, *J. Am. Chem. Soc.*, **107**, 5409 (1985).
88. J. Baudet, *J. Chim. Phys. Phys.-Chim. Biol.*, **68**, 191 (1971).
89. D. Döhnert and J. Koutecky, *J. Am. Chem. Soc.*, **102**, 1789 (1980).
90. S. Kato, K. Morokuma, D. Feller, E. R. Davidson and W. T. Borden, *J. Am. Chem. Soc.*, **105**, 1791 (1983).
91. S. K. Pollack, B. C. Raine and W. J. Hehre, *J. Am. Chem. Soc.*, **103**, 6308 (1981).
92. E. Migirdicyan and J. Baudet, *J. Am. Chem. Soc.*, **97**, 7400 (1975).

93. H. F. Hameka, *J. Chem. Phys.*, **67**, 2935 (1977).
94. E. F. McIntyre and H. F. Hameka, *J. Chem. Phys.*, **68**, 5534 (1978).
95. E. F. McIntyre and H. F. Hameka, *J. Chem. Phys.*, **69**, 4814 (1978).
96. O. Zamani-Khamiri and H. F. Hameka, *J. Chem. Phys.*, **71**, 1607 (1979).
97. O. Zamani-Khamiri and H. F. Hameka, *J. Chem. Phys.*, **73**, 5693 (1980).
98. A. V. Luzanov and Yu. B. Vysotskii, *Zhurn. Strukt. Khim.*, **17**, 1111 (1976).
99. V. E. Umanskii, A. V. Luzanov and I. V. Krivoshei, *Zhurn. Strukt. Khim.*, **15**, 1024 (1974).
100. E. R. Lippincott and J. M. Stutman, *J. Phys. Chem.*, **68**, 2926 (1964).
101. L. Dixit, S. K. Kapoor, I. D. Singh and P. L. Gupta, *Indian J. Pure Appl. Phys.*, **14**, 648 (1976).
102. T. Koenig, R. Wisselek, W. Snell and T. Balle, *J. Am. Chem. Soc.*, **97**, 3225 (1975).
103. D. J. Williams, J. M. Pearson and M. Levy, *J. Am. Chem. Soc.*, **92**, 1436 (1970).
104. I. Tanaka, *Nippon Kagaku Zasshi*, **75**, 218 (1954).
105. J. R. Schaefgen, *J. Polym. Sci.*, **15**, 203 (1955).
106. R. W. Bigelow, *J. Mol. Struct. THEOCHEM.*, **94**, 391 (1983).
107. P. C. Hiberty and P. Karafiloglu, *Theor. Chim. Acta*, **61**, 171 (1982).
108. J. Pospisek, M. Pisova and M. Soucek, *Coll. Czech. Chem. Commun.*, **40**, 142 (1975).
109. M. Pisova, L. Musil, B. Koutek, J. Pospisek and M. Soucek, *Coll. Czech. Chem. Commun.*, **41**, 2919 (1976).
110. S. G. Semenov and S. M. Shevchenko, *Zhurn. Struct. Khim.*, **24**, 25 (1983).
111. L. Musil, M. Pisova, J. Pospisek, and M. Soucek, *Coll. Czech. Chem. Commun.*, **41**, 2238 (1976).
112. G. J. Gleicher, D. D. Newkirk and J. C. Arnold, *J. Am. Chem. Soc.*, **95**, 2526 (1973).
113. C. L. McIntosh and O. L. Chapman, *J. Chem. Soc. Chem. Commun.*, 771 (1971).
114. J. F. Corbett, *J. Chem. Soc. B*, 207 (1968).
115. D. Sellmann and J. Müller, *J. Organomet. Chem.*, **277**, 379 (1984).
116. T. Nogami, T. Hishida, M. Yamada, H. Mikawa and Y. Shirota, *Bull. Chem. Soc. Japan*, **48**, 3709 (1975).
117. T. Koopmans, *Physica*, **1**, 104 (1934).
118. G. Richards, *Int. J. Mass Spec. Ion Phys.*, **2**, 149 (1969).
119. E. Haselbach, T. Bally, R. Gschwind, U. Klemm and Z. Lanyiova, *Chimica*, **33**, 405 (1979).
120. T. Bally, S. Nitsche, K. Roth and E. Haselbach, *J. Am. Chem. Soc.*, **106**, 3927 (1984).
121. T. Shida, E. Haselbach and T. Bally, *Acc. Chem. Res.*, **17**, 180 (1984).
122. J. Spanget-Larsen, *Croat. Chem. Acta*, **57**, 991 (1984).
123. T. Koenig and S. Southworth, *J. Am. Chem. Soc.*, **99**, 2807 (1977).
124. W. Simpson, *J. Am. Chem. Soc.*, **75**, 597 (1953).
125. W. Simpson and C. Looney, *J. Am. Chem. Soc.*, **76**, 6285, 6793 (1954).
126. R. Wielesek, J. Huntington and T. Koenig, *Tetrahedron Lett.*, 2429 (1974).
127. T. Koenig, T. Balle and W. Snell, *J. Am. Chem. Soc.*, **97**, 662 (1975).
128. M. J. S. Dewar, *J. Am. Chem. Soc.*, **104**, 1447 (1982).
129. T. Koenig, C. E. Klopfenstein, S. Southworth, J. A. Hoobler, R. A. Wielesek, T. Balle, W. Snell, and D. Imre, *J. Am. Chem. Soc.*, **105**, 2256 (1983).
130. R. Schulz, A. Schweig and W. Zittlau, *J. Am. Chem. Soc.*, **105**, 2980 (1983).
131. R. W. Bigelow, *Chem. Phys.*, **80**, 45 (1983).
132. R. W. Bigelow, *Chem. Phys. Lett.*, **100**, 445 (1983).
133. D. W. Turner, C. Baker, A. D. Baker and C. R. Brundle, in *Molecular Photoelectron Spectroscopy*, Wiley, New York, 1970.
134. C. R. Brundle, M. B. Robin and N. A. Kuebler, *J. Am. Chem. Soc.*, **94**, 1466 (1972).
135. D. O. Cowan, R. Gleiter, J. A. Hashmall, E. Heilbronner and V. Hornung, *Angew. Chem. Int. Ed. Engl.*, **10**, 401 (1971).
136. T. Kobayashi, *J. Electron Spectrosc. Relat. Phenom.*, **7**, 349 (1975).
137. D. Daugherty and S. P. McGlynn, *J. Am. Chem. Soc.*, **99**, 3234 (1977).
138. G. Lauer, W. Schafer and A. Schweig, *Chem. Phys. Lett.*, **33**, 312 (1975).
139. P. E. Stevenson, *J. Phys. Chem.*, **76**, 2424 (1972).
140. M. F. Merienne-LaFore and H. P. Trommsdorff, *J. Chem. Phys.*, **64**, 3791 (1976).
141. R. W. Bigelow, *J. Chem. Phys.*, **68**, 5086 (1978).
142. N. J. Bince, J. E. Ridley and M. C. Zerner, *Theor. Chim. Acta*, **45**, 283 (1977).
143. J. T. Gleghorn and F. W. Conkey, *J. Mol. Struct.*, **18**, 219 (1973).
144. J. E. Bloor, R. A. Paysen and R. E. Sherrod, *Chem. Phys. Lett.*, **60**, 476 (1979).
145. P. Jacques, J. Faure, O. Chalvet and H. H. Jaffe, *J. Phys. Chem.*, **85**, 473 (1981).

146. M. H. Wood, *Theor. Chim. Acta*, **36**, 345 (1975).
147. H. F. Jonkman, G. van der Velde and W. C. Niewport, *SRC Atlas Symp.*, **4**, 243 (1974).
148. T.-K. Ha, *Mol. Phys.*, **49**, 1471 (1983).
149. G. Lauer, K. W. Schulte and A. Schweig, *Chem. Phys. Lett.*, **32**, 163 (1975).
150. J. Kreile, N. Münzel, R. Schulz and A. Schweig, *Chem. Phys. Lett.*, **108**, 609 (1984).
151. A. Schweig, H. Vermeer and U. Weidner, *Chem. Phys. Lett.*, **26**, 229 (1974).
152. W. Schäfer and A. Schweig, *Z. Naturforsch.*, **30A**, 1785 (1975).
153. T. Koenig, D. Imre and J. A. Hoobler, *J. Am. Chem. Soc.*, **101**, 6446 (1979).
154. V. Eck, G. Lauer, A. Schweig, W. Thiel and H. Vermeer, *Z. Naturforsch.*, **33A**, 383 (1978).
155. M. Guerra, D. Jones, G. Distefano and A. Modelli, *Chem. Phys.*, **85**, 389 (1984).
156. J. M. Hollas, *Spectrochim. Acta*, **20**, 1563 (1964).
157. I. G. Ross, J. M. Hollas and K. K. Innes, *J. Mol. Spectrosc.*, **20**, 312 (1966).
158. J. Christoffersen and J. M. Hollas, *Mol. Phys.*, **17**, 665 (1969).
159. H. P. Trommsdorff, *Chem. Phys. Lett.*, **10**, 176 (1971).
160. H. Veenvliet and D. A. Wiersma, *Chem. Phys.*, **2**, 69 (1973).
161. H. Veenvliet and D. A. Wiersma, *Chem. Phys. Lett.*, **22**, 87 (1973).
162. R. M. Hochstrasser, L. W. Johnson and H. P. Trommsdorff, *Chem. Phys. Lett.*, **21**, 251 (1973).
163. T. M. Dunn and A. H. Francis, *J. Mol. Spectrosc.*, **50**, 1 (1974).
164. H. Veenvliet and D. A. Wiersma, *Chem. Phys.*, **8**, 432 (1975).
165. J. H. Lichtenbelt, J. G. F. M. Fremeijer, H. Veenvliet and D. A. Wiersma, *Chem. Phys.*, **10**, 107 (1975).
166. J. H. Lichtenbelt and D. A. Wiersma, *Chem. Phys.*, **34**, 47 (1978).
167. J. Goodman and L. E. Brus, *J. Chem. Phys.*, **69**, 1604 (1978).
168. J. P. Galaup and H. P. Trommsdorff, *J. Mol. Struct.*, **61**, 325 (1980).
169. J. P. Galaup and H. P. Trommsdorff, *Chem. Phys.*, **85**, 461 (1984).
170. T. M. Dunn and A. H. Francis, *J. Mol. Spectrosc.*, **50**, 14 (1974).
171. G. Ter Horst and J. Kommandeur, *Chem. Phys.*, **44**, 287 (1979).
172. T. Itoh, *Mol. Phys.*, **55**, 799 (1985).
173. H. P. Trommsdorff, D. A. Wiersma and H. R. Zelsmann, *J. Chem. Phys.*, **82**, 48 (1985).
174. J. R. Swenson and R. Hoffmann, *Helv. Chim. Acta*, **53**, 2331 (1970).
175. J. H. Lichtenbelt, J. G. F. M. Fremeijer and D. A. Wiersma, *Chem. Phys.*, **18**, 93 (1976).
176. V. Galasso, *Gazz. Chim. Ital.*, **106**, 571 (1976).
177. I. Fischer-Hjalmars, A. Henriksson-Enflo and C. Herrmann, *Chem. Phys.*, **24**, 167 (1977).
178. J. H. Lichtenbelt, D. A. Wiersma, H. T. Jonkman and G. A. van der Velde, *Chem. Phys.*, **22**, 297 (1977).
179. J. Fabian and A. Mehlhorn, *Z. Chem.*, **18**, 338 (1978).
180. R. Hilal, *Int. J. Quant. Chem.*, **15**, 37 (1979).
181. M. Nepras and M. Titz, *Int. J. Quant. Chem.*, **16**, 543 (1979).
182. R. L. Martin, *J. Chem. Phys.*, **74**, 1852 (1981).
183. R. L. Martin and W. R. Wadt, *J. Phys. Chem.*, **86**, 2382 (1982).
184. A. Kuboyama, Y. Kozima and J. Maeda, *Bull. Chem. Soc. Japan*, **55**, 3635 (1982).
185. B. Dick and B. Nickel, *Chem. Phys.*, **78**, 1 (1983).
186. R. D. Tewari and P. C. Mishra, *Indian J. Pure Appl. Phys.*, **22**, 491 (1984).
187. E. Charney and E. D. Becker, *J. Chem. Phys.*, **42**, 910 (1965).
188. B. Wyncke, F. Brehat and A. Hadni, *J. Phys.*, **36**, 159 (1975).
189. K. Høyland, *Ottar* (Ed. Tromsø Museum), **152**, 5 (1985).
190. T. R. Kelly, J. W. Gillard, R. N. Goerner Jr and J. M. Lyding, *J. Am. Chem. Soc.*, **99**, 5513 (1977).
191. T. R. Kelly, R. N. Goerner Jr, J. W. Gillard and B. K. Prazak, *Tetrahedron Lett.*, 3869 (1976).
192. T. R. Kelly, J. W. Gillard and R. N. Goerner Jr, *Tetrahedron Lett.*, 3873 (1976).
193. T. R. Kelly, *Tetrahedron Lett.*, 1387 (1978).
194. T. R. Kelly and J. Montury, *Tetrahedron Lett.*, 4309 (1978).
195. T. R. Kelly and J. Montury, *Tetrahedron Lett.*, 4311 (1978).
196. T. Hayakawa, K. Araki and S. Shiraishi, *Bull. Chem. Soc. Japan*, **57**, 1643 (1984).
197. M. D. Rozeboom, I.-M. Tegmo Larson and K. N. Houk, *J. Org. Chem.*, **46**, 2338 (1981).
198. R. H. Thomson, *J. Org. Chem.*, **16**, 1082 (1951).
199. K. Fukui, *Acc. Chem. Res.*, **4**, 57 (1971).
200. K. N. Houk, *Acc. Chem. Res.*, **8**, 361 (1975).
201. D. F. Evans, *J. Chem. Soc.*, 2753 (1959).

202. E. Müller and I. Müller-Rodloff, *Ann. Chem.*, **517**, 135 (1935).
203. E. Müller and E. Tietz, *Chem. Ber.*, **74**, 807 (1941).
204. J. Thiele and H. Balhorn, *Chem. Ber.*, **37**, 1463 (1904).
205. W. F. Gorham, *J. Polym. Sci.*, Part A-1, **4**, 3027 (1966).
206. J. E. Mulvaney, in *Polymer Science and Technology* (Plenum), **25**, 311 (1984).
207. L. A. Errede and M. Schwarc, *Q. Rev.* (*Lond.*), **12**, 301 (1958).
208. S. Iwatsuki, *Adv. Polym. Sci.*, **58**, 93 (1984).
209. S. A. Hutchinson Jr, A. Kowalsky, R. C. Pastor and G. W. Wheland, *J. Chem. Phys.*, **20**, 1485 (1952).
210. M. S. Platz, in *Diradicals* (Ed. W. T. Borden), John Wiley & Sons, New York, 1982, Chapter 5.
211. A. Kumar, A. N. Rao and S. K. Gupta, *Indian Chem. Engineer.*, **24**, 38 (1982).
212. H. J. Reich and D. J. Cram, *J. Am. Chem. Soc.*, **91**, 3517 (1969).

The Chemistry of Quinonoid Compounds, Vol. II
Edited by S. Patai and Z. Rappoport

CHAPTER **2**

Physical and chemical analysis of quinones

S. BERGER
Fachbereich Chemie der Universität Marburg, D-355 Marburg, Germany
P. HERTL and A. RIEKER
Institut für Organische Chemie der Universität Tübingen, D-7400 Tübingen, Germany

I. INTRODUCTION

About 22000 papers dealing with one or more aspects of spectroscopy and/or chemical determination of quinones have appeared since 1973. Most of these, however, just collect singular data on special compounds without presenting any evaluation or processing of these data in analytical or structural respects. Although of high value in a special context, they may be of minor interest in a survey like this. Hence, single compounds are discussed here only

(1) if their analytical features violate the rules derived in our original chapter on the subject[1];
(2) if they are examples of new trends within known methods or even demonstrate the application of methods not yet mentioned in Ref. 1; and
(3) if, in the authors' opinion, they are of high importance for mechanistic, preparative or applied chemistry and/or biochemistry.

We will restrict the matter to be treated in this chapter mainly to the discussion of the monocyclic systems **A** and **B**, showing the typical 1,4-(*para*) or 1,2-(*ortho*)-quinonoid π systems. For an interesting 'chemical definition' of the quinone conception see Ref. 2. Although homo- and hetero-condensed quinones (**C**, **D**) may be of high interest as natural products, they can be discussed only in context with those analytical methods which allow a detection of the typical quinonoid substructure. Thus, for example, ^{1}H NMR data of **C** and **D** mainly characterize the periphery, and are not included in this report.

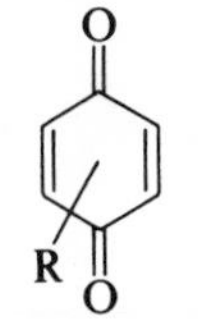

(A) 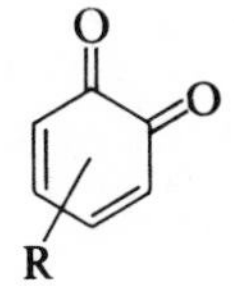 (B)

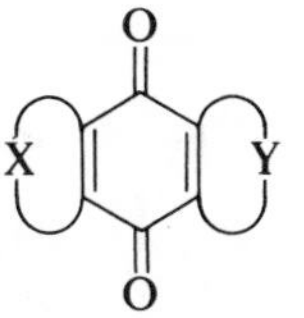

(C) 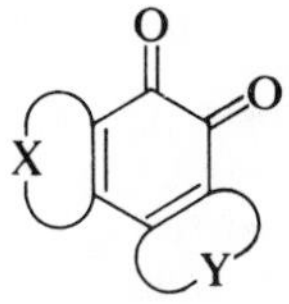 (D)

X, Y = carbon or heteroatoms or fragments containing them

Moreover, emphasis is laid on methods which allow a qualitative or quantitative analysis of quinones and, eventually, a distinction between the *para*- or *ortho*-type without sophistication.

Thus, each section will take a certain method and describe the parameters underlying the qualitative and quantitative analysis in some detail (especially in dependence of structural and methodological facts), illustrated by some applications of practical interest.

II. NMR SPECTRA OF QUINONES

A. General Remarks

As already pointed out in the original chapter in this series[1], NMR of quinones is in most cases not very specific for the quinonoid state and only in simple monocyclic benzoquinones are typical 'quinonoid' NMR absorptions shown both for ^{1}H- and ^{13}C-NMR. After the first systematic ^{13}C-NMR study on quinones[3] and our compilation[1], there appeared numerous papers dealing with quinonoid systems. Today, one might find assigned ^{13}C-NMR spectra for nearly all reported quinones in the literature.

B. Proton Magnetic Resonance

There have been numerous papers, which report on routine ^{1}H-NMR chemical shifts of the aromatic or side-chain protons of benzo-[4,5], naphtho-[4d,u,5b,6], anthraquinones[6a,g,p,z,7] or anthracyclines[7h,i,m,p,8], and other condensed quinones (including homo[6v,7n,9] or hetero[4a,l,n,5c,k,6e,v,y,z,7h,10] rings), as well as heterocyclic quinones[11]. However, there have been only few fundamentally new additions to the literature dealing with proton NMR of quinones. The H,H couplings in 1,4-benzoquinone (**1**) and 2-bromo-1,4-benzoquinone (**2**) have been remeasured and correlated with results from IR spectroscopy[12]. The stability of benzoquinone in superacid solutions was investigated by proton NMR[13]. Only the diprotonated species **3** was detected with δ-values of 8 ppm for the quinonoid protons and 14.2 ppm for the protons at the oxygen atoms. The solutions were stable at -40°C. A lanthanide-induced shift study revealed that quinones such as 2,6-dimethylbenzoquinone (**4**) bind the lanthanide ion with both carbonyl oxygens[14a]. Lanthanide-induced shift data of 38 naphthoquinones allow the differentiation of the

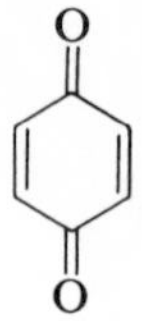

(1) 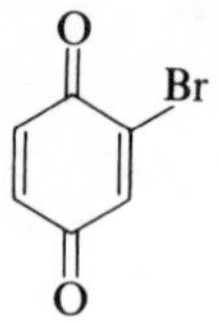 (2) 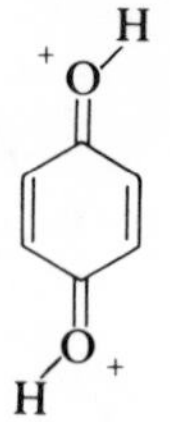 (3) 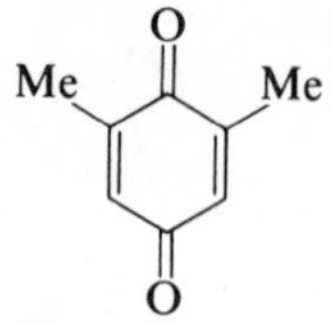 (4)

quinone protons H(2)/H(3) in most cases of unsymmetric substitution of the benzenoid ring[14b].

Porphyrins can be capped with a quinone unit. Up to 4 ppm the quinonoid protons are shielded by the ring current of the porphyrin ring system[15], and the influence of the porphyrin metal ions on the quinone ring was investigated[16] (see also Ref. 17). There have been reports on CIDNP of quinones during photolysis[18–20]. For ^{1}H-NMR spectra of quinone compounds of interest in special fields see the following references: cyclophanes[21], crown ethers[71, 22], triptycenes[23].

C. ^{13}C Magnetic Resonance

With the development of better assignment techniques in ^{13}C-NMR spectroscopy, it is now possible to assign all carbon atoms even of higher condensed quinones fairly safely. Several groups point out that the use of long range C, H spin–spin coupling constants from gated decoupled spectra is most essential for the correct analysis[24–26]. Some assignment errors in our first compilation[1] had to be corrected on this basis[25, 27]. A full analysis of the C, H spin–spin coupling constants of **1** itself has been published[28].

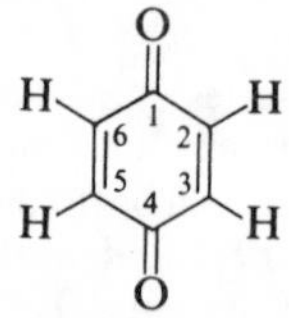

$^{1}J_{C(2), H(2)} = 168.4$ Hz
$^{3}J_{C(1), H(3)} = 10.3$ Hz
$^{3}J_{C(2), H(6)} = 4.5$ Hz

(1)

Very recently a method was published to measure carbon, carbon spin–spin coupling constants in symmetrical molecules[29]. With this method the ^{13}C, ^{13}C spin–spin coupling constants of benzoquinone (**1**) and hydroquinone (**5**) were measured[30]. It is interesting though not unexpected that the spin–spin coupling constant of the carbon atoms of the 2–3 bond does not change much from benzoquinone to hydroquinone. Instead, the ^{13}C, ^{13}C spin–spin coupling constant between C(1) and C(2) shows a considerable change. This

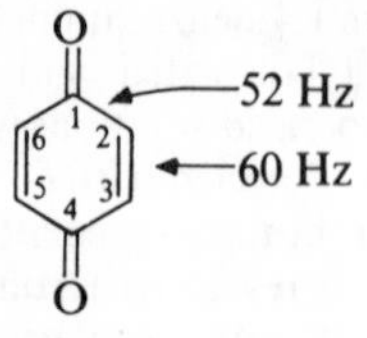

52 Hz
60 Hz

66 Hz
58 Hz

(1) 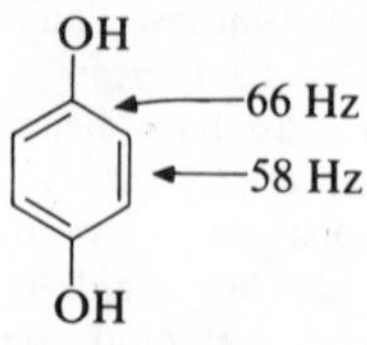 (5)

points to the fact that ^{13}C, ^{13}C spin–spin coupling constants are affected more by local hybridization effects than by delocalization[31].

Whereas the main body of the ^{13}C-NMR literature reports the routine data of various quinonoid systems[4d, j, k, o, s, w, 5f, i, j, n–p, t, 6b, c, z, 7h, i, r, 8c, e, 9d, g, 10a, d, j, m, q, 17c, 32–36], some groups have tried to develop increment systems, which allow a fast calculation of expected ^{13}C-NMR chemical shifts for a given quinone[37–40]. However, it has been shown that the quinonoid ring does not resemble a unique entity for ^{13}C-NMR spectroscopy. The substituent-induced chemical shift changes of the olefinic carbon atoms of a quinone correlate well with those in simple olefins and, not unexpectedly, there is virtually no difference between the substituent chemical shifts in the aromatic rings of a condensed quinone and simple aromatic molecules[41]. Furthermore, there is very little 'crosstalk' with respect to the ^{13}C-NMR chemical shifts between the aromatic and quinonoid ring systems in condensed quinones.

Temperature-dependent ^{13}C-NMR spectroscopy was carried out on perezone (**6**) and its derivatives in order to demonstrate the tautomeric equilibrium[39], and to study the hindered rotation of coumarin residues in homologues of juglone[42].

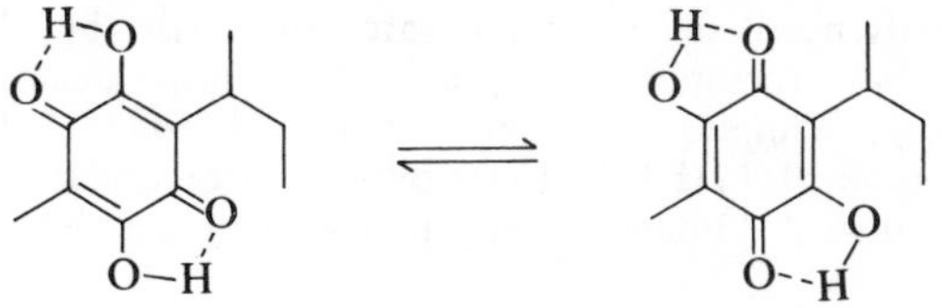

(6)

The complexation of quinones with organoplatinum compounds (**7**) was studied[43] and the solution structure of the complexes investigated with low temperature ^{13}C-NMR spectroscopy; however, a clear distinction between η^2 and η^4 structures was not possible.

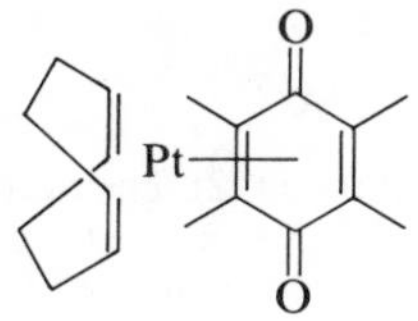

Pt

(7)

The effect of strain imposed by mono- or bis-annulation of the cyclobutene ring on 1,4-benzoquinones (**8**, **9**) and on naphthazarin (**10**) was observed by ^{13}C-NMR[44, 45]. Clamping

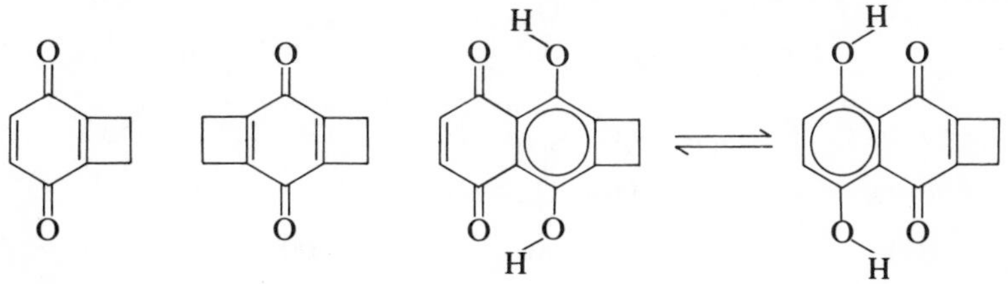

(8) **(9)** **(10)**

in anthraquinonophane causes distinct shifts in the ^{13}C-NMR spectra with respect to the unclamped anthraquinones ($\Delta\delta_{CO}$ up to 3 ppm)[46].

The solid state ^{13}C-NMR spectra of quinones have been measured with the cross-polarization/magic angle spinning (CP/MAS) technique, and the solid state reaction between quinones was followed by this method[47]. The solution and solid state spectra of simple quinones are rather similar.

D. ^{17}O Magnetic Resonance

^{17}O-NMR spectroscopy was performed on several simple quinones. The linewidths are acceptable and the chemical shift dispersion is sufficient to be of some diagnostic value[48].

E. Analytical Applications

Quinones may be easily recognized by the carbonyl resonance near 180 ppm if no other carbonyl groups are present in the molecule. However, the distinction between 1,2- and 1,4-quinones is not always straightforward[1] (for a successful case see Ref. 5i).

Although, presently, nearly all new quinones are characterized by ^{1}H- and/or ^{13}C-NMR spectroscopy, essential structural assignments or quantitative determinations are rarely based on NMR only. Examples are the application of long-range ^{1}H/^{13}C heteronuclear correlation spectroscopy (LR HET COSY) to the structure elucidation of murayaquinone (**11**)[49a] and cervinomycin[49b] or the investigation of the regiochemistry of nucleophilic

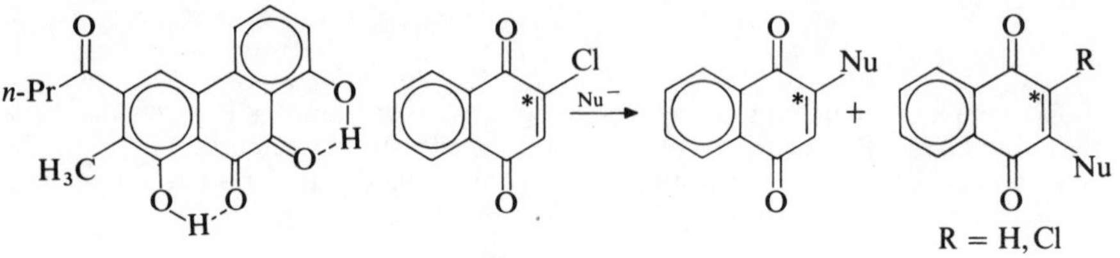

n-Pr H$_3$C Cl Nu$^-$ Nu + R Nu

R = H, Cl

(**11**) (**12**) (**13**) (**14**)

displacements in chloronaphthoquinone (**12**). The ratios of the products, **13** and **14**, were determined by comparison of the ^{13}C-NMR spectra of labelled and unlabelled material, starting with 2-[^{13}C]-2-chloro-1,4-naphthoquinone (**12**)[50].

The biosynthesis of mollisin (**15a**) has been studied, using sodium [1-^{13}C]-, [2-^{13}C]-, and 1,2-[di-^{13}C] acetate[51a]. High levels of acetate incorporation and conversion of mollisin into its acetate (**15b**) allowed the assignment of all ^{13}C-chemical shifts and ^{13}C, ^{13}C coupling constants of the latter compound. Likewise, the biosynthesis of the

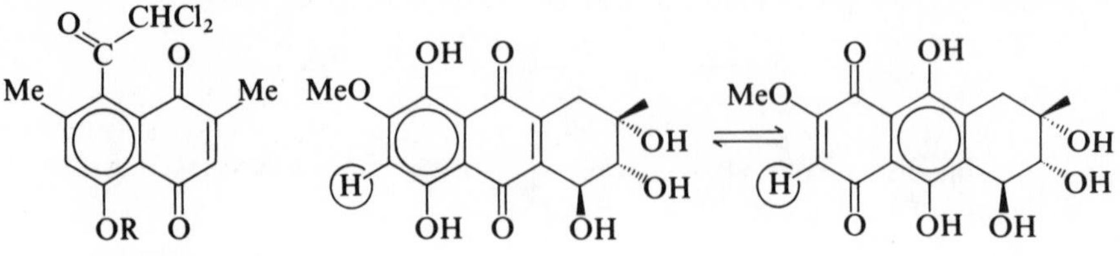

CHCl$_2$ Me Me OR MeO OH OH OH OH MeO OH OH OH OH OH

(**15**) (**a**) (**16**) (**b**)

a: R=H
b: R=COMe

kinamycin antibiotics (heterocondensed naphthoquinones) has been investigated by ^{13}C-NMR analysis of the enrichment of carbon atoms after feeding of [1,2-$^{13}C_2$]-acetate to cultures of *Streptomyces murayamaensis*[51b].

^{1}H-NMR has been used to settle the question of 'naphthazarin tautomerism' in bostrycin (**16**) in favour of structure **16b** by comparing the shift of the circled proton (δ = 6.45, DMSO-d_6) with that of the corresponding proton in model compounds[52a]. For the investigation of a similar tautomerism in 1,4-dihydroxy-9,10-anthraquinon-9-imines by ^{1}H-NMR see Ref. 52b.

The structures of rubellins A (**17a**) and B (**17b**), two novel anthraquinone metabolites isolated from *Mycosphaerella rubella*, have been assigned, mainly by detailed analysis of

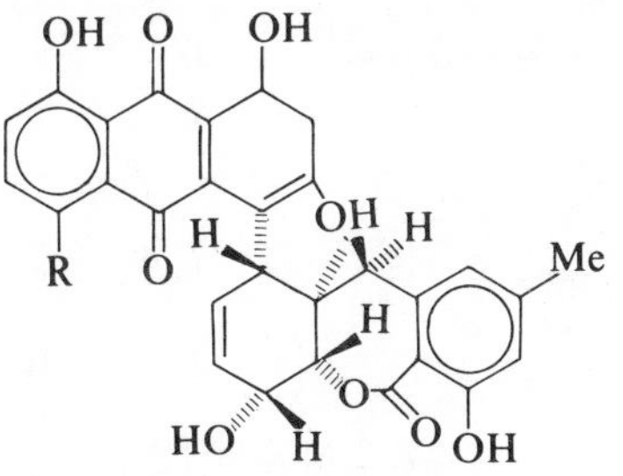

a: R=H (**17**)
b: R=OH

their ^{1}H- and ^{13}C-NMR spectra[53a]. Thermochromism of 1,4-benzoquinones substituted with a R–N–CH_2CH_2OH group is caused by equilibration with the corresponding quinone ketals, as shown by ^{13}C-NMR investigation[53b]. Finally, analysis of high-boiling fractions of shale oil, containing quinones, by combined ^{1}H-NMR and IR spectroscopic methods is also possible[54].

III. IR SPECTRA OF QUINONES

A. General Remarks

Only few new compilations and discussions of IR data of quinones are available[55–57], although complete calculations and analyses of fundamentals of several quinones have been reported since 1973: 1,4-benzoquinone[58], 1,4-naphthoquinone[58a, 59], 9,10-anthraquinone[58a, 59b, c, 60], chloro-1,4-benzoquinones[61] and (in part) their fully deuterated analogues, with some differences in the assignments given by different authors. Vibrational analysis was also suggested for vitamins K_1, K_3, K_4, K_5 involving fundamental and combination tones showing a preponderance of quinonoid structures in all of them[62].

Besides IR, Raman spectroscopy could equally well be used. However, only few investigations are known to us: chlorinated 1,4-benzoquinones[61], naphthoquinone[57a] and anthraquinone[57a].

Before reading the following text or using Tables 1 and 2, the general remarks on pages 186, 187 of Ref. 1 should be considered. Again, the discussion will be restricted to the position of the double-bond vibrations near 6 μm. All new data should be treated with caution, since there is no complete set of $\tilde{\nu}_{C=O}$ measured in the same solvent under identical conditions.

B. Benzoquinones

A wealth of data has appeared[4, 5, 6n, 10f, i, p, 21, 22, 44, 56, 63], single data or series of them, with simple or very complicated, even exotic, substitution patterns. Some examples for 1,4-

TABLE 1. Carbonyl absorption of 1,4-benzoquinones

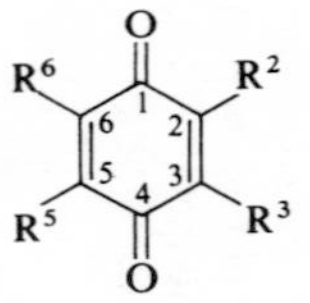

Compound no.	R^6	R^5	R^3	R^2	$\tilde{\nu}_{C=O}$ (cm^{-1})	$\bar{\tilde{\nu}}_{C=O}$[a] (cm^{-1})	Solvent	References
(9)	$-CH_2-CH_2-$		$-CH_2-CH_2-$		1670	1670	CCl_4	44
(18a)	c-C_3H_5	H	H	c-C_3H_5	1654	1654	[b]	63l
(18b)	H	c-C_3H_5	H	c-C_3H_5	1640	1640	[b]	63l
(19)	$C_{12}H_{25}$	H	H	$C_{12}H_{25}$	1655	1655	CCl_4	4t
(20)	$C_{12}H_{25}$	OH	OH	$C_{12}H_{25}$	1645	1645	CCl_4	4t
(21)	$C_{12}H_{25}$	OMe	OMe	$C_{12}H_{25}$	1670	1670	CCl_4	4t
(22)	Allyl	OH	Me	OH	1610	1610	KBr	10f
(23)	Allyl	OMe	Me	OMe	1650	1650	Neat	10f
(24)	H	H	OMe	OMe	1680, 1670, 1635	1662	KBr	63m
(25)	H	OPh	H	OMe	1680, 1613	1646	[b]	4b
(26)	H	OH	H	OH	1604	1604	Paraffin	4c
(27)	–O–(CO)–O–		–O–(CO)–O–		1685, 1662	1673	THF	63i
(28)	H	H	SMe	SMe	1655	1655	$CHCl_3$	5d
(29)	H	H	H	$N(Me)CH_2R$	1675, 1640	1658	Film	4h
(30)	H	$N(Me)CH_2R$	H	$N(Me)CH_2R$	1630	1630	KBr	4h
(31)	H	NHPh	H	NHPh	1640	1640	Paraffin	4c
(32)	NHAc	H	H	Cl	1684, 1628	1666	Nujol	5a
(33)	$CON(Et)_2$	NH_2	$CON(Et)_2$	NH_2	1605	1605	KBr	5g
(34)	H	H	H	$COCH(Me)_2$	1670	1670	KBr	5h
(35)	H	H	H	COOH	1630	1630	$CHCl_3$	4e
(36)	H	H	OH	COOH	1630	1630	Nujol	4e
(37)	H	COOMe	H	COOMe	1666	1666	KBr	63j
(38)	CN	Cl	Cl	CN	1710, 1680	1695	KBr	5f

[a] The arithmetical average of the two $\tilde{\nu}_{C=O}$ values given; see Ref. 1.

[b] Solvent not stated.

benzoquinones (to supplement Table 8 of Ref. 1) are given in Table 1, whereas 1,2-benzoquinones (to supplement Table 10 of Ref. 1) may be found in Table 2. By and large, the rules given in Ref. 1 concerning the appearance of the spectra as well as the site and type of substitution are still valid. However, the details presented by different authors do not always harmonize.

The examination of asymmetrically substituted quinones revealed that the carbonyl frequencies are split for all classes of compounds, representing the vibrations of the two distinct carbonyl groups (LFER equations of $\tilde{\nu}_{C=O}$ being determined)[64]; see, however, doubling by Fermi resonance[1].

If we compare the 1,4-benzoquinones of Table 1 with **1** ($\bar{\tilde{\nu}}_{C=O}$ = 1671 cm^{-1} in solution)[1], we can again conclude that electron-donating groups generally lower $\bar{\tilde{\nu}}_{C=O}$ (consider e.g., the pairs **1/18**, **1/20**, **1/28**, **1/29**, **1/31**), whereas electron-withdrawing groups raise it (e.g. **1/38**). However, it seems that MeO and COOH/COOMe were exceptions. The effect of the methoxy group[1] is still not clear-cut. Comparing **19** with **21**, we would expect a $\Delta\bar{\tilde{\nu}}_{C=O} = +7.5$ cm^{-1} for OMe, comparable to that of Cl. The pair **1/24** would also suggest a shift to higher wave numbers by OMe, as long as the two higher absorptions (1680, 1670 cm^{-1}) are used for calculating $\bar{\tilde{\nu}}_{C=O}$. If the third (1635 cm^{-1}) is also considered, a small shift in the opposite direction can be traced. Contrary to this situation, $\Delta\bar{\tilde{\nu}}_{C=O}$ for SMe (-8 cm^{-1} from **1/28**) and NR_2 (> -13 cm^{-1}; **1/29**; **29/30**, see also Ref. 41) are straightforward. Comparison of **9** and **19** shows that internal strain caused by annelation of a cyclobutene ring increases $\tilde{\nu}_{C=O}$.

Again, the frequency-lowering effect of hydrogen bridging between a hydroxy group in the *ortho* position and the quinone carbonyl group on $\bar{\tilde{\nu}}_{C=O}$ is obvious, e.g. **20/21**, **22/23**, **1/26**, with $\Delta\bar{\tilde{\nu}}_{C=O}$ ranging from 25 to 65 cm^{-1} (part of the shift may be due to differing of solvents). Hydrogen bonding also seems to occur between amino groups and the quinone carbonyl (e.g. in **33**), whereas in **36** the hydroxy group is more likely to prefer bridging with the carboxyl group (**35/36**). The difference in $\bar{\tilde{\nu}}_{C=O}$ of the aminoquinones **29** and **30** can be explained by a suggestion of Dähne and coworkers[4m]. Accordingly, **30** has the structure of

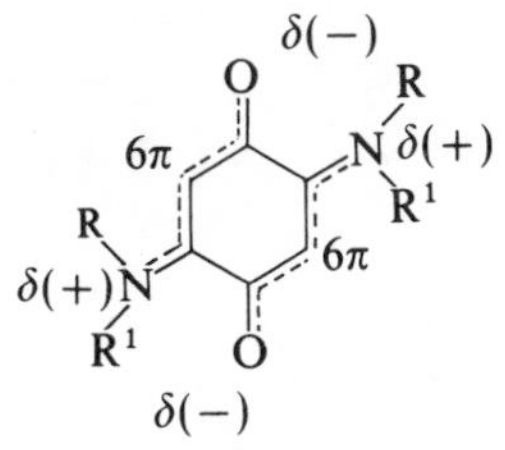

δ(−) O R N δ(+) R¹ 6π 6π R δ(+)N R¹ O δ(−)

(E)

O R N δ(+) R¹ 6π O δ(−)

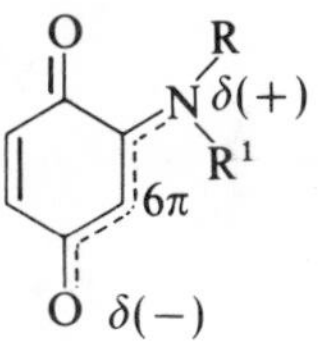

(F)

a coupled 'merocyanine' (**E**), whereas **29** shows one 'merocyanine' and one 'polyene' structural element (**F**). This leads to the polymethine carbonyl absorption at 1630–1640 cm^{-1} for both compounds and to an additional polyene carbonyl absorption at 1675 cm^{-1} for **29**.

An 'intermolecular' hydrogen bridge between a quinone and a hydroquinone being fixed in cyclophanes is nicely demonstrated comparing compounds **39/40** ('parallel' arrangement), whereas in the case of compounds **41/42** ('vertical' arrangement) the effect is much smaller[21b, d, g]. Mono- and polyquinones fixed by crown ether or alkyl bridges seem to show no unusual $\tilde{\nu}_{C=O}$ values[21a, 22].

In the case of 1,2-benzoquinones (Table 2) again only few new examples have been reported. Whereas Table 10 of Ref. 1 mainly contains 4,6- and 3,4,6-substituted quinones,

TABLE 2. Carbonyl absorption of 1,2-benzoquinones

Compound no.	R^6	R^5	R^4	R^3	$\tilde{\nu}_{C=O}$ (cm^{-1})	$\bar{\tilde{\nu}}_{C=O}$[a] (cm^{-1})	Solvent	References
(43)	Me	H	H	Me	1681, 1659[b]	1670	CH_2Cl_2	63 h
(44)	$-O-CH_2-O-$		H	Me	1620	1620	KBr	4 y
(45)	$-O-CH_2-O-$		Me	Me	1625	1625	KBr	4 y
(46)	$-O-CH_2-O-$		*n*-Pr	*n*-Pr	1655	1655	$CHCl_3$	4 y
(47)	$-O-CH_2-O-$		H	$C(Me)_3$	1668	1668	$CHCl_3$	4 y
(48)	H	–O–C(Me)Et–O–		H	1660	1660	Paraffin	4 x
(49)	H	NHPh	NHPh	H	1615	1615	Paraffin	4 c
(50)	H	H	N(Me)Ph	H	1680[b], 1631 (1616)	1658	KBr	4 m
(51)	Br	Br	OMe	Me	1689	1689	Nujol	4 k

[a] The arithmetical average of the two $\tilde{\nu}_{C=O}$ values given.
[b] Higher intensity.

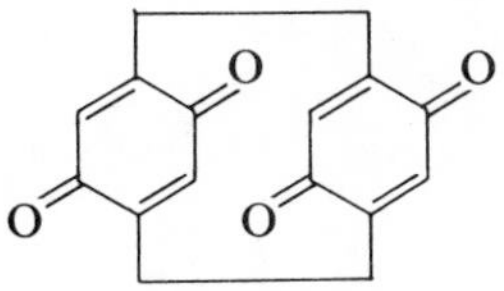

1660 cm^{-1}(KBr)

(39)

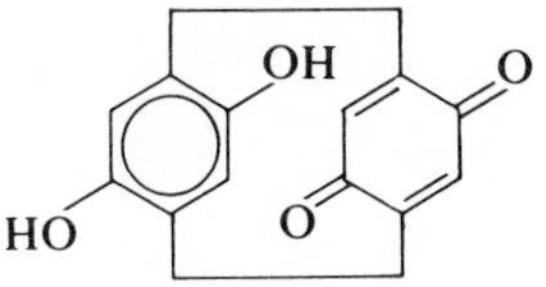

1620 cm^{-1}(KBr)

(40)

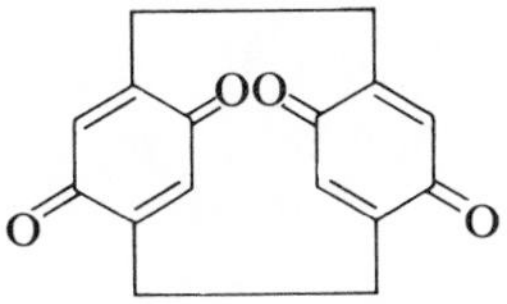

1660 cm^{-1}(KBr)

(41)

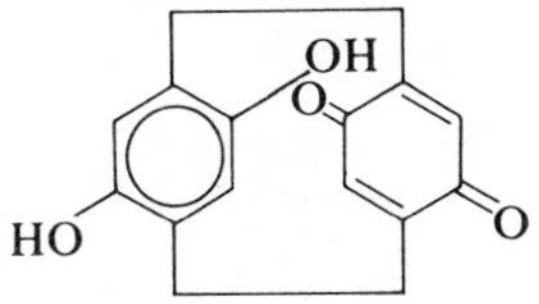

1645 cm^{-1}(KBr)

(42)

Table 2 of the present report also includes other substitution patterns. However, still, too few compounds have been measured to derive substituent shift rules. Whereas **43** corresponds well to 1,2-benzoquinone itself ($\tilde{\nu}_{C=O}$ = 1680, 1658 cm^{-1})[1], it is surprising that the methylenedioxy-substituted derivatives **44**–**48** show only one carbonyl absorption at relatively low frequencies (1620–1668 cm^{-1}). In the case of **49**/**50**, Dähne's explanation[4m] again holds: **50** contains one polyene and one merocyanine structure (**G**), **49**

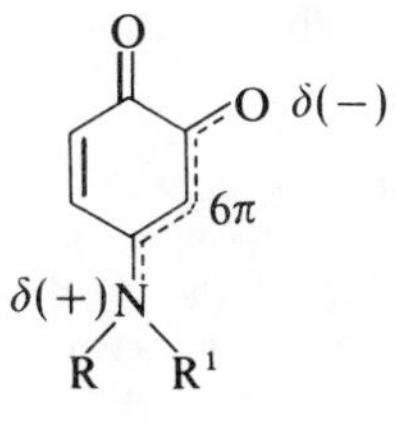

(G)

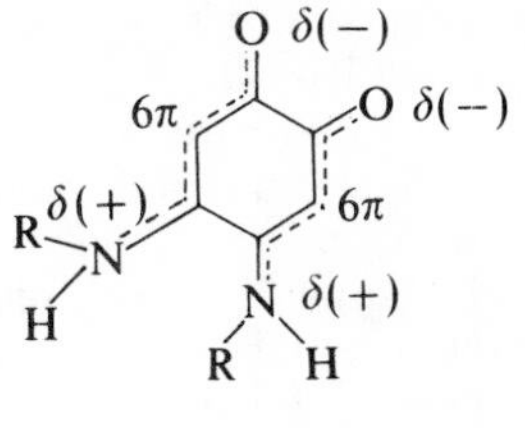

(H)

reveals two coupled merocyanine structural elements (**H**). Hence, two $\tilde{\nu}_{C=O}$ values are observed for **50** (polyene: 1680; polymethine: 1631 and/or 1616? cm^{-1}), and one $\tilde{\nu}_{C=O}$ for **49** (coupled polymethine: 1615 cm^{-1}).

C. Condensed Quinones

A wealth of IR data of homo- and heterocondensed quinones is available, which is far beyond the scope of this article: naphthoquinones[4d, 5b, 6, 7q, 8c, f, g, 9h, 10l, o, 14b, 20e, 52a, 65–72], anthraquinones[5t, 6a, p, 7, 8f, 52a, 65, 72, 73], anthracyclines[7h, i, m, p, r, 8, 74, 75], other homocondensed quinones[6v, 7n, 9, 65, 71, 76, 77] and heterocyclic or heterocondensed quinones[4l, 5c, 6l, v, 10, 11, 32b, 49b, 63g, 66, 70, 78–80]. A collection of the carbonyl absorptions of some parent homocondensed quinones is given in Table 11 of Ref. 1. Homocondensed 1,2-quinones seem to absorb at higher frequencies than 1,4-quinones, however, the

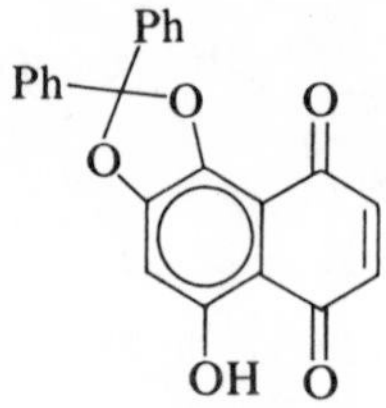

(52)

1670, 1645, 1610 cm^{-1}
(KBr)[52a]

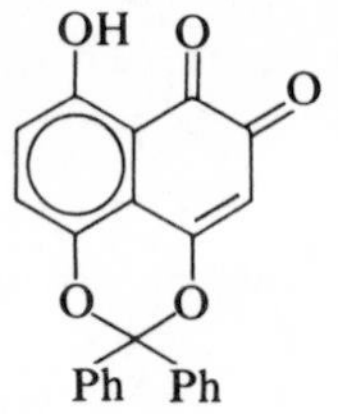

(53)

1670, 1650, 1628 cm^{-1}
(KBr)[52a]

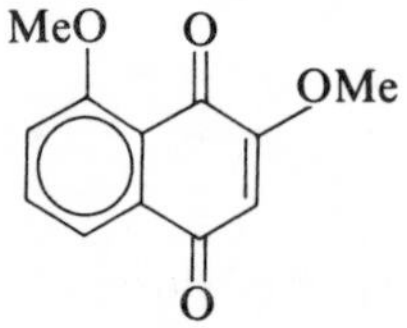

(54)

1670, 1645 cm^{-1}
(KBr)[6h]

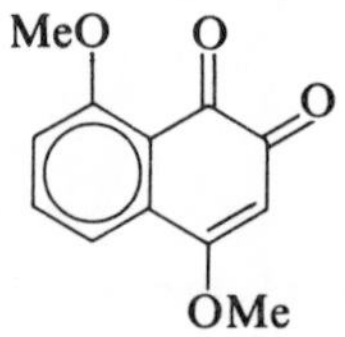

(55)

1680, 1640 cm^{-1}
(KBr)[6h]

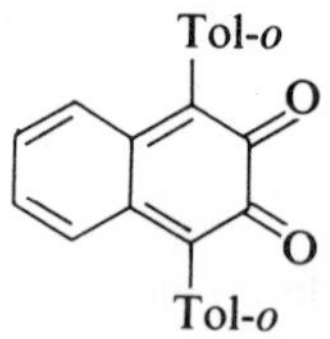

(56)

1655(m), 1629(s) cm^{-1}
(CH_2Cl_2)[71]

difference may be small, e.g. **52/53** and **54/55**. For comparison, the values for a 2,3-naphthoquinone (**56**) are also given.

A correlation of $\tilde{\nu}_{C=O}$ with the bond order $P_{C=O}$, calculated by the SCF method for 36 parent quinones (mainly condensed quinones), proved to be poor[65].

As already demonstrated[1], the influence of substituents (even in the non-quinonoid rings), in general, follows the rules given for benzoquinones, the effect of the methoxy group again being not straightforward. No systematic investigations of the substituent effects are available, and, indeed, they would involve tremendous synthetic efforts. Even in the simplest cases of 1,4- and 1,2-naphthoquinones, the number of positional isomers is very large (Table 3). Some comparable measurements on di- and trisubstituted naphthoquinones (Me, OH, OMe, OAc, F) are available[6o, x].

TABLE 3. Positional isomers of naphthoquinones

Substitution type		1,4-Naphthoquinone	1,2-Naphthoquinone
Mono	R	3	6
Di	R, R	9	15
Di	R^1, R^2	15	30
Tri	R, R, R	10	20
Tri	R^1, R^2, R^3	60	120

As regards the other homo- and heterocondensed quinones the reader is referred to the literature quoted above.

For the lowering of $\tilde{\nu}_{C=O}$ due to the 'naphthazarin tautomerism', (see page 155 of Ref. 1 and compounds **10** or **16** in this report), many new examples in the naphthoquinone[6y, 8f, g, 101], anthraquinone[6a, 7d, h–j, m, o, 8f, g], anthracycline[7h, i, m, p, r, 8b, d, f, g, j–m, 74] and naphthacenequinone[9c, d] series have been reported. Anthraquinones/anthracyclines may

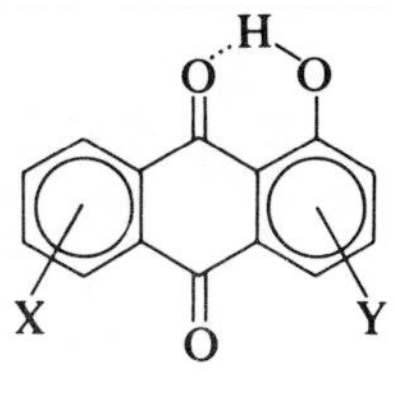

(I) 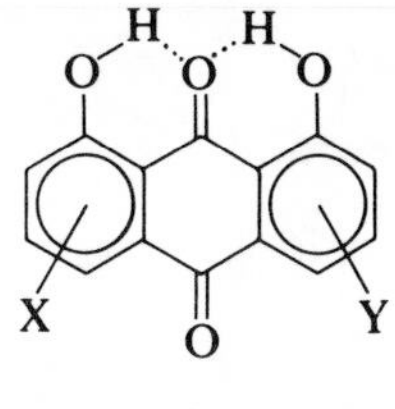 **(K)**

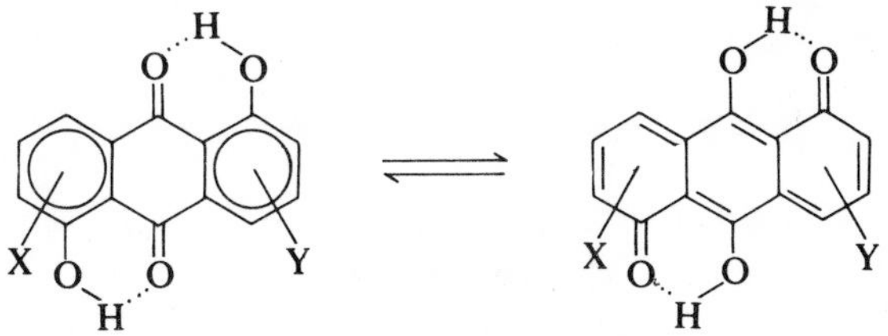

a **(L)** **b**

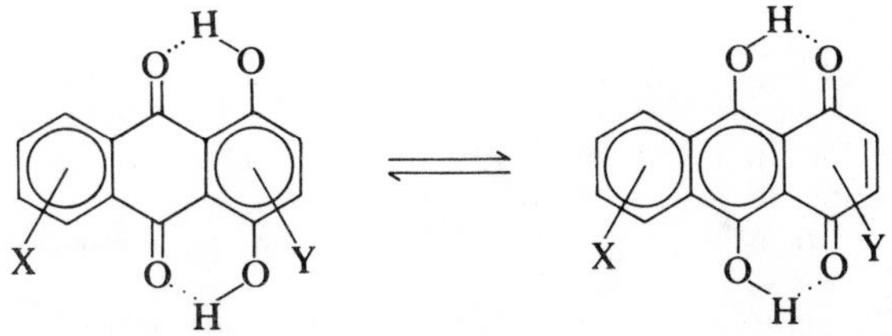

a **(M)** **b**

serve as examples. The most interesting combinations of one or two hydroxy groups are (**I**)–(**M**), X and Y being any substituent.

Compounds **I** show a $\bar{\nu}_{C=O}$ which is ca. 20–30 cm^{-1} lower than in the case of the corresponding methoxy compound, which may be attributed to an intramolecular hydrogen bridge.

A bifurcated hydrogen bridge as in 1,8-dihydroxyquinones (**K**) does not lower $\bar{\nu}_{C=O}$ noticeably with respect to **I**. On the other hand, 1,5- and 1,4-dihydroxyquinones **L** and **M**, with a possible bridging of both carbonyl groups, show very low $\bar{\nu}_{C=O}$ values (1605–1630 cm^{-1} for **L**, 1609–1630 cm^{-1} for **M**, with an accumulation each between 1610 and 1620 cm^{-1}). Therefore, as in naphthazarin[1], a tautomerism involving the structures **a** and **b** might be envisaged in these cases, although the structures **b** are energetically less favourable than the structures **a**, both for **L** and **M**. Whereas **L** and **M** generally show only one carbonyl absorption, there are two in the case of **I** and **K**, one lying around 1655–1680 and the other around 1620–1635 cm^{-1}. Therefore, it has been pointed out[81] that the absorption at lower frequency should be ascribed to the hydrogen-bridged carbonyl group. Due to the possible occurrence of Fermi resonance[1], this conclusion has to be further confirmed. For the analogous effect in 1,8-dihydroxyphenanthraquinones, such as **11**, and in heterocondensed quinones, such as **57**, see Ref. 49a and Ref. 10h, respectively.

Most heterocondensed or heterocyclic quinones investigated are related to natural products and are derived from indole (e.g. derivatives of mitomycine C)[5c, 10b–g, o, q, 78],

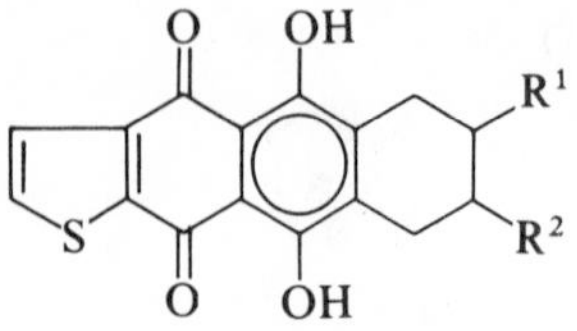

(57) 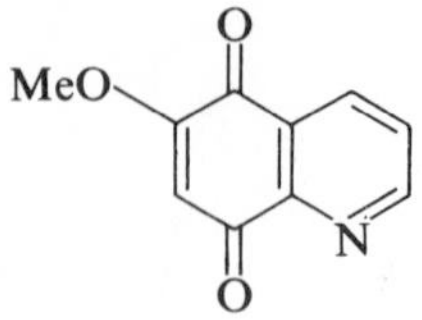 **(58)** 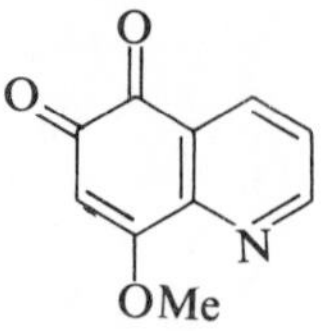 **(59)**

a: $R^1 = Ac$, $R^2 = H$
b: $R^1 = H$, $R^2 = Ac$

1590 cm^{-1}(KBr)[10h] 1685, 1670 cm^{-1}(KBr)[63g] 1700, 1642 cm^{-1}(KBr)[63g]

isoindole (38 compounds in $CHCl_3$, $\bar{\nu}_{C=O} = 1649$–$1670\ cm^{-1}$, correlations of $\tilde{\nu}_{C=O}$ with Hammett's σ[57c]), indazole[10d, 1], quinoline/isoquinoline[5c, 10f, h, k, n, 63g, 69] (e.g. **58**, **59**[63g]), thiophene[10h, 32b], and carbazole[79, 80].

D. Analytical Applications

The distinction of 1,2- and 1,4-benzoquinones using $\bar{\nu}_{C=O}$ is less straightforward than has been supposed in Ref. 1 since the new data of Table 2 extend the region of $\bar{\nu}_{C=O}$ for 1,2-benzoquinones to lower frequencies (ca. 1620 cm^{-1}). Although IR spectroscopy is an excellent method for the investigation of hydrogen bridging in quinones, it has not acquired any essential importance in their general structural determination in the last decade. A certain lack of clarity of the IR spectra and the complexity of $\tilde{\nu}$/structure relations may be some of the reasons.

IR spectroscopy combined with other methods, on the other hand, has been used for analytical purposes, e.g. to study high-boiling fractions of shale oil (IR, NMR)[54] and to determine the rate of dehydrogenation of phenol and quinone in soil (IR, voltammetry)[81]. Quantitative determination of quinones applying Lambert–Beer's law to the C=O absorption band are possible, although not in use.

Alternatively, quinones may be characterized indirectly by IR or Raman spectroscopy of their reduction products (semiquinones, dianions)[56, 82].

IV. UV/VIS SPECTRA OF QUINONES

A. General Remarks

UV /vis is the classic spectroscopy in the realm of quinones. It is therefore not surprising that most of the publications in the field since 1973 deal with UV/vis in one way or the other. Reviews[55, 83, 84] as well as an atlas[85], have appeared. In the present authors' opinion, there have been two crucial developments in UV/vis spectroscopy of quinones since 1973: The first of these concerns theoretical calculations of electronic energy levels and of the spectra derived from them, as well as complete analyses of the absorptions in many individual compounds, especially by Nepraš, Fabian and coworkers[86], Kuboyama

TABLE 4. UV spectra of benzoquinones and

Compound no.	R^6	R^5	R^4	R^3	R^2	Absorptions λ_{max}(nm) (log ε)	Solvent	Ref.
(61)	H	H	—	–CH=CH–$(CH_2)_4$–		249(4.20), 368(3.25), 439[a](2.34)[b]	Cyclohexane	4g
(62)	H	H	—	–O–$(CH_2CH_2$–O$)_4$–		253(4.11), 400(3.15) [c]	CH_2Cl_2[d]	22a
(63a)	H	$C(Me)_3$	—	H	$C(Me)_3$	253(4.25), 305(2.47), 458[e]	*n*-Heptane	87a, 87c
						261(4.18)		
(63b)	H	$C(Me)_3$	—	NH_2	$C(Me)_3$	273(4.13), 465(3.21) [c]	$CHCl_3$	4v
(64)	$NHC(Me)_3$	H	—	H	$C(Me)_3$	278(4.24), 486(3.47) [c]	Ethanol	4v
(65)	H	Cl	—	H	Cl	270(4.42), 325(2.53) [c, f]	*n*-Heptane	87a
(35)	H	H	—	H	COOH	255(3.84), 338(3.38) [c]	MeCOOH	4e
(34)	H	H	—	H	$COCH(Me)_2$	246(4.28), 300(2.66)[a], 450(1.48)	MeCN	5h
						248(4.32), 291(2.75)[a] [c]	CH_2Cl_2	4j
(66)	H	H	—	H	COMe	250(4.26) [c] [c]	CH_2Cl_2	4j
(67)	Me	Me	—	–$(CH_2)_3$–C(O)–		264(4.19), 364(2.95) [c]	CH_2Cl_2	4j
(68)	H	H	—	H	CHO	288(3.59), 355(3.05) [c]	CH_2Cl_2	4j
(69)	CN	CN	—	Cl	Cl	270(4.06), 372(2.94) [c, g]	MeCN	103a
						280(4.01)		
(48)	H	–O–C(Me)Et–O–		H	—	293(4.23), 395(3.23) [c]	Ethanol	4x
(70)	H	OMe	H	OMe	—	300(3.51), 350(3.77), 490(2.70)	Water	103b

[a] Shoulder.
[b] Shoulders at 460 (2.09) and 492 nm (1.64).
[c] Not stated.
[d] A. Merz, private communication; see also Ref. 22c (MeCN).
[e] Further absorption at 175.5 nm (4.37).
[f] Further absorption at 186.5 nm (4.27).
[g] Further absorptions at 209 (4.20), 216 (4.18), and 226 nm (4.16).

and coworkers[87] and by others[88]. The second area involves investigations of molecular and charge transfer complexes, often in connection with the design of organic conductors, or for biochemical reasons (benzoquinones[63j, 89–94], condensed quinones[94, 95], triptycene[23, 96], cyclophane[21a, d–f, i–k, 63h, 97, 98] and porphyrin[5q, r, 99, 100] quinones).

Neither field can be treated in great detail in this chapter, which is devoted to the analysis of quinones as such. Besides UV/vis, photoelectron spectroscopy[88d, 101] has become a tool for the investigation of quinones, and will be treated below in Section IV. E. Fluorescence, luminescence, and phosphorescence studies can be no more than mentioned[86c, i–m, 88a, 102].

B. Benzoquinones

In spite of the existence of many single UV/vis data of quinones, only very few are suitable to supplement Tables 12 and 13 of Ref. 1. Often, the substituents are rather complicated, only one absorption is given, the intensities are lacking, or the data are hidden in a figure. Data on some new compounds are collected in Table 4. From the three tables mentioned above, the absorption characteristics of 1,2- and 1,4-benzoquinones were computed and arranged in Table 5. In addition to the three absorptions given for each

(1) **(60)**

benzoquinone type, there is a strong $\pi \rightarrow \pi^*$ absorption at shorter wavelengths [(**1**: 171.5 nm, $\log \varepsilon = 4.19$, 1,1,1,3,3,3-hexafluoroisopropanol (HFP))[87a]; **60**: 200 nm[86d], 'strong'[88c], ethanol], which have not been recorded in Tables 4 and 5 since most authors did not take measurements below 220 nm. The $n \rightarrow \pi^*$ band seems to be a superposition of two nearly degenerate transitions[88e].

Several calculations using different quantum mechanical methods, which cannot be discussed here, prove the assignments of Table 5. They are, however, at variance with each other as to the symmetry species[86d, 87a, 88a–f, h, l].

TABLE 5. Absorption characteristics of 1,4- and 1,2-benzoquinones

Type	λ_{max}(nm)	$\log \varepsilon$	Assignment
1,4-	① 240–300	3.8–4.5	$\pi \rightarrow \pi^*$
(*para*)	② 285–485	2.4–3.5	$\pi \rightarrow \pi^*$
	③ 420–460	1.2–2.3	$n \rightarrow \pi^*$
1,2-	① 250–300	2.6–4.2	$\pi \rightarrow \pi^*$
(*ortho*)	② 370–470	2.8–3.5	$\pi \rightarrow \pi^*$
	③ 500–580	1.4–1.8	$n \rightarrow \pi^*$

Unfortunately, there is no series of substituted benzoquinones measured in the same solvent. Nevertheless, some conclusions may be drawn from Table 4 (in agreement with Ref. 1).

(1) Substitution causes bathochromic shifts of all absorptions (a small hypsochromic shift of the $n \rightarrow \pi^*$ transition by Me and Cl has been reported, see Ref. 1).
(2) In 1,4-benzoquinones, substitution mainly influences the second $\pi \rightarrow \pi^*$ transition. The difference in λ_{max} (Δ nm) between substituent and parent compound is small for electron-withdrawing substituents [COOH (**35**): $\Delta = 50$ nm; R–C=O (**34**, **67**, corrected for the two methyl groups): $\Delta = 5$–10 nm] and large for electron-donating (Δ nm see Ref. 1) substituents. As to additivity rules or influence of substitution position, see also Ref. 1, page 197.

The difference in magnitude of Δ nm for H–C=O and R–C=O (**68**, and **34**, **67**) is surprising. It may be due to conformational factors. Thus, the COR group is perpendicular

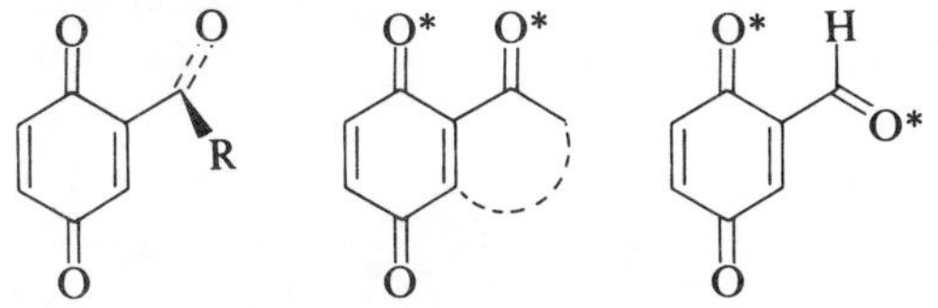

(**N**) (**O**) (**P**)

to the ring in **34** (**N**)[4j] (see also COOR[88m]), planar with a '*syn*' conformation of the two marked carbonyl groups in **67** (**O**)[4j] (see also Ref. 63e), and with an '*anti*' conformation of those carbonyl groups in **68** (**P**)[4j].

The absorption areas of Table 5 are supported by many additional compounds not mentioned here (see Refs[4d, g, v–x, 5f, h, k, t, v, 21g, 22, 63b], for solvent effects see Ref. 104).

There are also quinones which do not fit into the absorption domains of Table 5. These are mainly amino and polyalkoxy (polyhydroxy) quinones; for onio-substituted quinones see Ref. 5o. Since the substituents involved cause large Δ nm values (apparently for all three bands), a strong bathochromic shift combined with a change in extinction is not unexpected[4h, m, p, s, 5j, n, u, 10f, 49, 105, 106]. Furthermore, it should be mentioned that the spectra of polyhydroxy quinones are subject to solvent and pH variation[107].

Two types of compounds will be discussed in more detail. Dallacker and coworkers[4y] reported the UV/vis spectra of eight 1,2-benzoquinones **71**, all measured in MeCN. Three absorptions, I–III, are found, which are puzzling, when correlated to the absorption areas of Table 5 or to those of 3- and 4-methoxy-1,2-benzoquinone (Table 13, Ref. 1). However, it would be reasonable to assume that band I corresponds to the high frequency absorption of 1,2-benzoquinone (**60**), being red-shifted by about 10 nm in **71**. Accordingly, II would correspond to ①, and III to ② of Table 5, a correlation which is also suggested by band intensities and by the strong substitution dependence of band III, characteristic of the second $\pi \rightarrow \pi^*$ transition (see above). However, a clear-cut substituent/shift relation for **71** cannot be derived at present, and the $n \rightarrow \pi^*$ absorption may also contribute to band III. The strong bathochromic effect of the methylenedioxy group $-OCH_2O-$ on bands II and III is also evident in 1,4-benzoquinones. Here the following gradation, reflecting a certain ring-strain effect, has been observed[108]: 2,3-dimethoxy < $O-(CH_2)_3-O$ < $O-(CH_2)_2-O$ < $O-CH_2-O$.

A second interesting compound is azidanil (tetraazido-1,4-benzoquinone, **72**)[88k]. Its UV/vis spectrum shows four absorption bands I–IV, the variation widths being due to the

(71)

λ_{max}(nm)	log ε
I : 208–214	3.47–4.21
II : 232–288 (347)	1.83–4.05
III: 435–580	1.26–3.81

R^1 = H, Me, n-Pr, t-Bu, Br,
R^2 = H, Me, Et, n-Pr

(72)

λ_{max}(nm)	log ε
I : 205–215	4.18–4.47
II : 268–281	4.40–4.55
III: 355–382	4.11–4.26
IV: 540–570	2.39–2.50

solvents (*n*-hexane, cyclohexane, MeCN, MeOH) used. Again, three $\pi \rightarrow \pi^*$ transitions I, II (corresponding to ① of Table 5), III (corresponding to ②), and one $n \rightarrow \pi^*$ transition IV (corresponding to ③) are observed. Compared to **1**, the azido groups exert their strongest effect on band III, which is expected, as long as only $\pi \rightarrow \pi^*$ transitions are considered. Most surprising here is the strong red shift of the $n \rightarrow \pi^*$ transition of about 110–120 nm. The assignment was achieved by calculations using the PPP–CI method[88k] indicating also that the charge density distribution in **72** is not much different from that of chloranil (tetrachloro-1,4-benzoquinone). Indeed, the bands II and III of **72** correspond nicely to those (i.e. ① and ②) of the perchlorinated quinone (see Ref. 1, Table 12).

C. Condensed Quinones

1. The parent compounds

As already pointed out[1], the spectra of condensed quinones are rather complex. Only the most important homocondensed parent compounds will be discussed here. Many quantum chemical calculations (PPP, PPP/CA, INDO/S–CI) have been made on these systems[86b, c, e–h, m, 87a, b, 88b, c]. Although, in principle, the π system is delocalized, PPP calculations with configuration analysis (CA) have revealed that (especially in higher annelated quinones) the electronic transitions may be treated as local transitions in 'fragments'[86h]. One possible way of dividing a quinone into fragments is shown in **Q**, which allows local transitions in the enedione fragment, in the acene fragment, and charge transfer (CT) transitions *between* these two fragments[86h]. According to this procedure, it is generally possible to identify 'benzenoid' transitions with the dominant character of L_b or L_a transitions of acenes (plus some CT character), 'quinonoid' transitions with a localization in the benzoquinone (endione) part, and CT transitions[86g, h].

In the case of 1,4-naphthoquinone (**73**), the assignment given in Table 14 of Ref. 1 has been completely confirmed as regards the $\pi \rightarrow \pi^*$-, $n \rightarrow \pi^*$- and the benzenoid/quinonoid

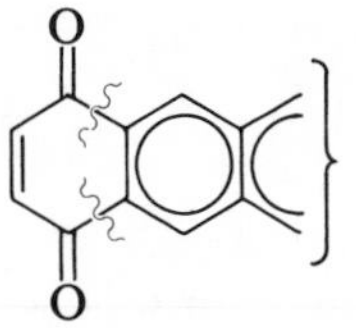

(Q) 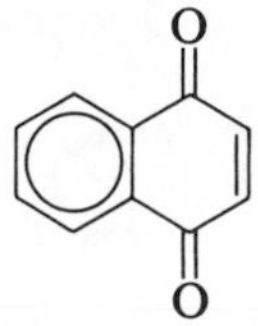 (73) 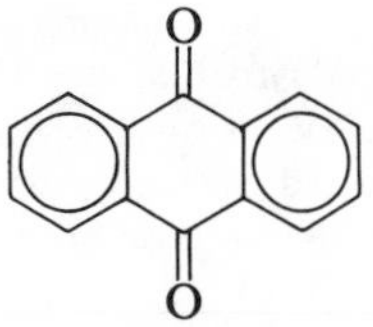 (74)

band notation[86b, g, 87a, 88b, c]. In addition, a fifth (high-frequency) $\pi \rightarrow \pi^*$ transition, which had been predicted on the basis of calculations, was also observed experimentally at 193.5 nm ($\log \varepsilon = 4.58$) in *n*-heptane[87a].

9,10-Anthraquinone (**74**), the parent compound of many synthetic and natural dyes, has been extensively studied since 1973[86b, f, 87a, 88c, 109]. A comprehensive literature review is presented in Ref. 86f; some features will be summarized in the following.

The assignments of the five absorptions given in Table 14 of Ref. 1 in terms of $\pi \rightarrow \pi^*$-, $n \rightarrow \pi^*$- and benzenoid/quinonoid band classification have been confirmed[86b, f, 87a, 88c], although the benzenoid band at 250 nm is of more complex origin, involving also benzene-to-carbonyl charge transfer[86f]. Interestingly, the second quinonoid $\pi \rightarrow \pi^*$ transition at about 340 nm, which can usually be found only in substituted anthraquinones, has now been detected in *n*-heptane solution ($\lambda_{max} = 319.5$ nm, $\log \varepsilon = 3.68$)[87a]. In addition, two high-frequency $\pi \rightarrow \pi^*$ transitions were found at 204 nm ($\log \varepsilon = 4.57$) and 181 nm ($\log \varepsilon = 4.61$)[87a].

In contrast to **74**, the lowest frequency $\pi \rightarrow \pi^*$ transition of 1,4-anthraquinone (**75**) is located at much longer wavelengths (ca. 400 nm), therefore, the $n \rightarrow \pi^*$ transition is only observed as a shoulder on the long-wavelength tail of the $\pi \rightarrow \pi^*$ band. For the other bands see Refs[86b, 88c].

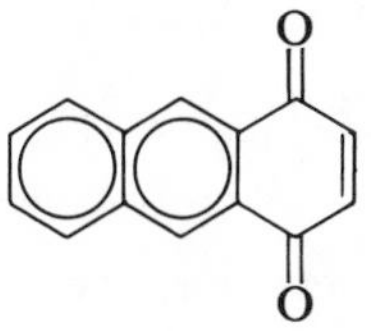

(75) 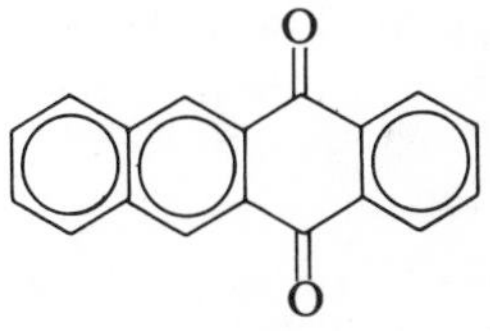 (76)

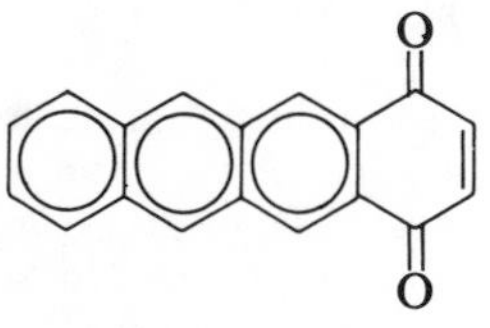

(77) 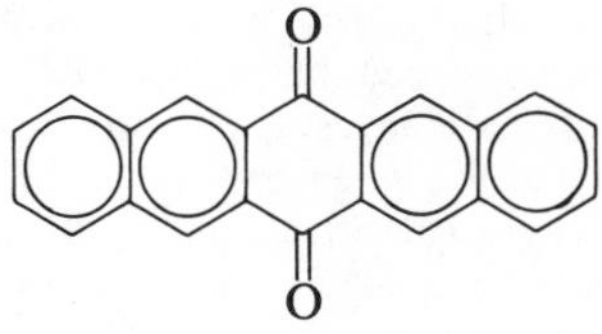 (78)

Higher-condensed 1,4-quinones have also been investigated, experimentally and theoretically: 5,12-tetracenequinone (**76**)[86b, 109b], 1,4-tetracenequinone (**77**)[9f, 86m], 6,13-pentacenequinone (**78**)[86b, 109b], other linear condensed *p*-quinones[9f, g] and some angular condensed *p*-quinones[9h, 86c, 88c]. CNDO/S–CI calculations[109b] show that $n \rightarrow \pi^*$ transitions are practically independent of the extension of the ring system.

$\pi \rightarrow \pi^*$ Transitions, on the other hand, experience a bathochromic shift, which may reverse the usual order of $n \rightarrow \pi^*/\pi \rightarrow \pi^*$ transitions.

The UV/vis spectra of singly and doubly protonated species derived from **73**–**78** can be interpreted on the basis of the PPP method[86a, n].

Investigations on homocondensed 1,2-quinones are much more scarce. The assignments given in Table 15 of Ref. 1 have been confirmed for 1,2-naphthoquinone (**79**)[87b, 88b, c], as

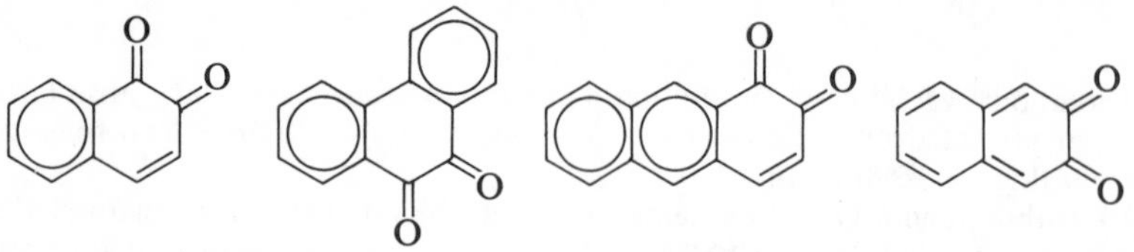

(**79**) (**80**) (**81**) (**82**)

regards the $\pi \rightarrow \pi^*$ and $n \rightarrow \pi^*$ categories. Two further high frequency $\pi \rightarrow \pi^*$ absorptions were calculated and observed in 1,1,1,3,3,3-hexafluoro-2-propanol (HFP) (196 nm, $\log \varepsilon = 4.39$; ~ 170 nm, $\log \varepsilon = 4.35$)[87b]. Since no configurational analysis[86] has been performed, the assignment of 'benzenoid' and 'quinonoid' bands given in Table 15 of Ref. 1 should still be regarded as tentative.

For 9,10-phenanthraquinone (**80**) the former and the new[87d, e] data are collected in Table 6. The remarkable changes of λ_{max} in the low-frequency $\pi \rightarrow \pi^*$ bands in moving from heptane to HFP solutions are attributed to a strong hydrogen-bond formation between **80** and HFP, a strong proton donor[87d]. Calculations (PPP) of the absorptions of **80**[87d, 88c] do not always satisfactorily fit with the experimental data[87d]. For diazaphenanthrenequinone (phenanthrolinequinone) see Ref. 110.

PPP calculations are also available for the $\pi \rightarrow \pi^*$ transitions of 1,2-anthraquinone (**81**)[88c] and INDO/S–CI calculations for the $\pi \rightarrow \pi^*$ transitions in 2,3-naphthoquinone (**82**)[88b].

2. *Substituent effects*

Substituents render the spectra of condensed quinones even more complex, whereby type, number and position play an important role. Polarity of the solvent used, hydrogen bridging, special steric interactions or charge transfer from substituents to the carbonyl groups are further causes. Absorptions may overlap, accidentally coincide or overtake each other as a result of hypso- or bathochromic substituent shifts. Moreover, the bands may not only be perturbed, but new bands may arise.

Since systematic investigations on mono- and/or disubstitution are lacking (and also would involve too great expense, cf. Table 3), no general rules can be derived. Even semiempirical calculations often do not allow unequivocal assignments of *all* absorptions of a compound in question: for special information, the reader is referred to pages 201, 204 of Ref. 1 and to the literature compiled in the following: 1,4-naphthoquinones[6a, c, i, m, o, t, w, y, z, 8f, g, 51, 52, 87a, 88i, 111, 113, 114]; 1,2-naphthoquinones[6b, o, t, y, 52, 68c, 87b, 112, 116]; 9,10-anthraquinones[6a, 7h, m–o, q, 8f, g, 46, 52, 86f, i–l, 113, 115]; anthracyclines[8f, g, i, k, l, 74]; phenanthraquinones[87d]; other homocondensed quinones[6v, 9]; heterocondensed quinones[4p, 6c, v, 10b–d, f, h, j, p, 49b, 79, 102a, 111, 117–122].

Alkoxy, hydroxy, alkylamino and amino groups, in the case of polysubstitution, cause bathochromic shifts of all, or, at least, of the lowest frequency $\pi \rightarrow \pi^*$ transition (hereafter abbreviated as LFPT)[6a, c, i, o, w, 8f, g, 10p, 53, 74, 86f, i–l, 113, 114, 117].

TABLE 6. UV/vis spectra of 9,10-phenanthraquinone (**80**)

Absorptions λ_{max}(nm)/(log ε)									Solvent	References
178	209.5	256	263		314	~ 332	398	~ 500[a]	*n*-Heptane	87d, e
	210	256	264.5		321		414.5		Ethanol	87d
		256	265		322		410		Methanol	1
		(4.46)	(4.49)		(3.62)		(3.13)			
182	210	253.5	261	~ 275	333.5		442	sh 490[a]	HFP	87d
(4.50)	(4.57)	(4.50)	(4.50)	(4.13)	(3.78)		(3.14)			

[a] $n \rightarrow \pi^*$, all others $\pi \rightarrow \pi^*$ transitions.

TABLE 7. LFPT[a] of 2,3-disubstituted 1,4-naphthoquinones

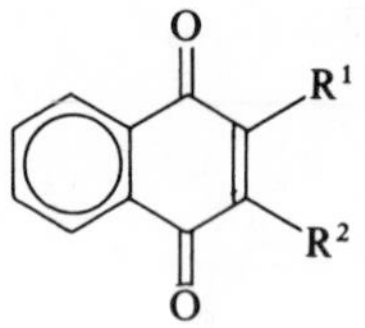

Compound no.	R^1	R^2	Ethanol λ_{max}(nm)	Cyclohexane λ_{max}(nm)/(log ε)
(73)	H	H	ca. 330	
(83a)	OMe	H	333	328 (3.53)
(83b)	OMe	OMe	335	329 (3.46)
(83c)	OMe	Cl	338	334 (3.49)
(83d)	NHMe	H	452	423 (3.45)
(83e)	$N(Me)_2$	H	471	444 (3.53)
(83f)	NHMe	Cl	477	454 (3.04)
(83g)	NHMe	OMe	492	484 (3.45)
(83h)	NHMe	NHMe	544	525 (3.34)

[a] Lowest-frequency $\pi \rightarrow \pi^*$ transition.

In 1,4-naphthoquinones, methoxy groups in position 2 and/or 3 have practically no influence on the 'quinonoid' LFPT (**73** → **83a** → **83b** → **83c**, Table 7), which is hidden in **73** itself below the 'benzenoid' band at 335 nm[114]. This is further demonstrated by tri- and tetrasubstituted 1,4-naphthoquinones of type **84**[6o], if we assume the (337–408 nm) absorption to be the LFPT. Moreover, the absorption pattern as a whole is relatively constant in methanol.

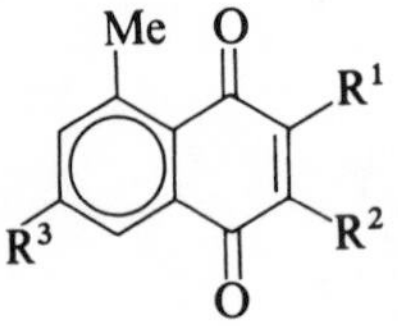

(84)

	R^1	R^2	R^3
a	OMe	H	OMe
b	H	OMe	OMe
c	OMe	OMe	OMe
d	H	OMe	OH
e	H	OH	OMe
f	H	OH	OH

262–266 nm (log ε = 4.25–4.39)
290–298 nm (log ε = 4.09–4.17)
337–348 nm (log ε = 3.36–3.62)
400–408 nm (log ε = 3.10–3.25)

Alkylamino groups in position 2 (**83d, e**) seem to shift LFPT to 420–470 nm[6c, 114], which appears to be reasonable in the light of PPP SCF-MO calculations[114]. Moreover, the extinction would be too high for the $n \rightarrow \pi^*$ transition, usually found in this range.

Remarkable is the further shift of LFPT up to 550 nm in the case of 2,3-disubstitution with strong donors (e.g. **83g, h**), in agreement with a high CT character of this absorption[6i, 114] (*vide supra*).

It may be of interest that naphthazarin-type (see Section III.C) quinones (**85**) also reveal strong absorptions in the visible area (520–550 nm), even if only one donor substituent X is present[6w].

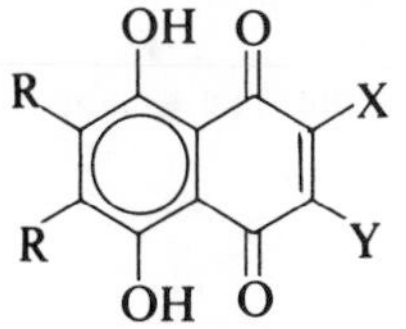

(85) 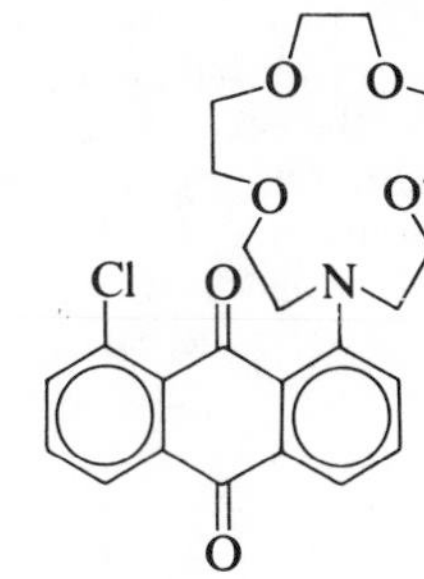 **(86)** 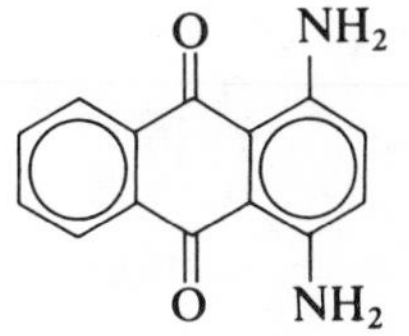 **(87)**

R = H, Cl
X = *O*-Alkyl, *N*-Alkyl
Y = H, Cl

Even more work has been carried out with donor-substituted 1,4-anthraquinones, which are important as dyes. Spectra of 1- and 2-aminoanthraquinones and the whole series of diaminoanthraquinones have been measured and calculated (PPP, PPP-CA)[86i–l, 123, 124] (for a short summary see Ref. 86i). Again, the LFPT band apparently involves strong CT character[86i, 123, 124] and therefore is shifted bathochromically (410–470 nm for 1-, 2-aminoanthraquinones[86i], 493 nm for 1-dialkylamino-anthraquinone **86** of the crown ether type[113]). Compound **86** shows hypso- or bathochromic shifts in the light absorption upon addition of salts, which has been discussed in view of the crown ether complexation in the donor or acceptor part of the molecule[113].

Strong red shifts are calculated (PPP-CA) and observed for diamino-substitution[86j–l]. **It may be of interest that 1,4-diaminoanthraquinone (87) with a naphthazarin-type system exerts the strongest red shift of the LFPT band of all diaminoanthraquinones.**

In principle, the same gradation is observed with dihydroxyanthraquinones: 1,4-dihydroxyanthraquinones (including the 'naphthazarin system') of the general formula **Ma** (Section III.C) show LFPT bands at 520 nm and beyond[8g, 53], whereas 1,5-dihydroxyanthraquinones (**La**) and 1,8-dihydroxyanthraquinones (**K**) reveal these bands at 440–460 nm[6a, 8f, 53, 74]. This is in contrast to the effect on the carbonyl absorption $\bar{\nu}_{C=O}$ in the IR region (Section III.C), where **M** and **L**, but not **K**, give rise to a remarkable frequency lowering.

This unique effect of the 'naphthazarin unit' is further demonstrated for heterocondensed 1,4-quinones[10h, 117]. Systems like **88–90** (X = S or Se; Y = H or F_4), so called IR dyes

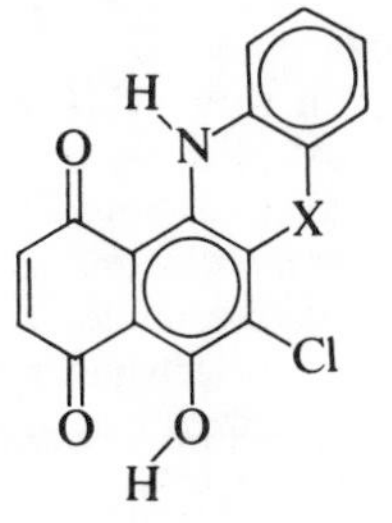

(88) 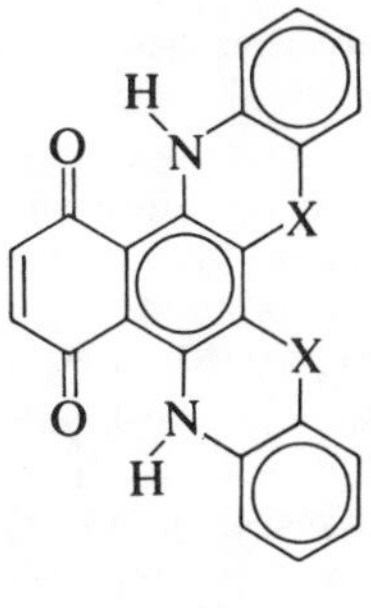 **(89)** 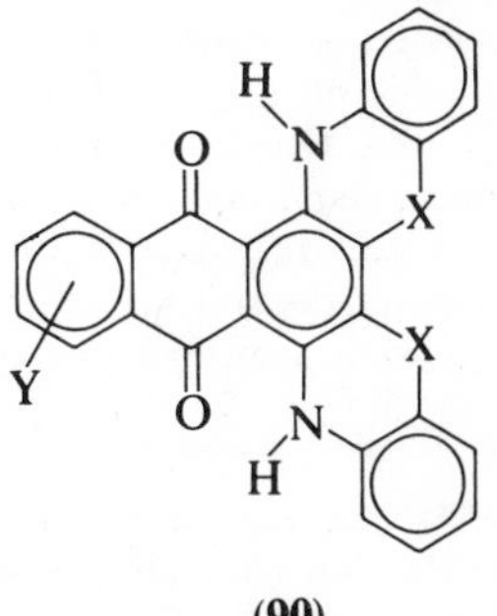 **(90)**

for optical recording media, reveal two strong absorptions in the visible region: 616–725 nm ($\log \varepsilon = 3.83$–4.25) and 642–780 nm ($\log \varepsilon = 3.66$–4.19) in $CHCl_3$[117a]. We suppose that at least one or both of these bands are due to charge transfer from the S,Se rings to the carbonyl groups (see also Ref. 117b). For related compounds see Ref. 117b–e.

In the series of condensed 1,2-quinones, 1,8-dihydroxyphenanthraquinone would correspond to 1,4-dihydroxyanthraquinone, and, indeed, the derivative **11** shows an absorption of $\lambda_{max} = 524$ nm ($\log \varepsilon = 3.65$) in methanol[49a], which might be ascribed to the red-shifted LFPT.

D. Charge Transfer Spectra

In sections IV.B, C we have already mentioned that the LFPT band of a quinone may show charge transfer (CT) character, if an intramolecular CT transition is defined as 'an electronic transition from an occupied molecular orbital localized in one part of a conjugated system to an unoccupied molecular orbital localized in the other part of the same conjugated system'[89c]. Charge transfer electron donor–acceptor complexes in a more concise sense, however, possess *insulated* donor and acceptor groups or areas in two different molecules (intermolecular CT) or within the same molecule (intramolecular CT). They give rise to a new band in the visible spectrum, which is usually broad and featureless, and not always well separated from the quinone spectrum.

The following compounds were mainly used as donors for *intermolecular* quinone CT complexes: aromatic hydrocarbons[4j, 63j, 90b, e, 91, 92, 95], amines[63j, 90c, i–k, q, 93, 96a], amino acids and proteins[90d, p], nitrones[90f, 92d], *N*-heterocycles[63e, j, 90h, i, m, q–s], phenols (quinhydrone)[90l], sulphides[90a], tetrathiofulvalene[63j, 90n], hexamethylenetetratellurafulvalene[90o] and ferrocene[21k, 90g]. The acceptor quinones were preferentially halogenated and cyano-substituted 1,4- and 1,2-benzoquinones, as well as the parent quinone **1**, but such complicated systems as 'poly' triptycene quinones[96a] have also been used.

Among others, the frequency ν_{CT} of the CT band is determined by the ionization potential I^D of the donor and the electron affinity E^A of the acceptor. Series of relative electron affinities of quinones and other acceptors are available[125, 126]. These are given detailed treatment in Ref. 127; for special information see the references collected in Section IV.A.

In the case of intramolecular CT transition, there is continuous change from planar conjugated π systems (like condensed quinones, see section IV. C) via conformational non-rigid or twisted π systems, like **91**[89c], to insulated π systems, as the cyclophanes **40**, **42**[21d] and **92**[21i] or the triptycene **93**[23] and finally to spacer-separated π systems, as in porphyrin-linked quinones[99, 100].

Considerable effort has been involved in synthesizing cyclophane quinones[21, 97, 98]. The compounds **40** and **42**, as examples, reveal drastic differences in their CT bands: the absorption intensity of **40** is about ten times higher than that of **42**, with the absorption maximum being blue-shifted by only 20 nm. Since **40** and **42** have the same E^A of the acceptor and the same I^P of the donor, this observation clearly demonstrates the orientational dependence of the CT absorption (**40**: pseudogeminal; **42**: pseudo-*ortho*)[21d]. This was satisfactorily explained by simple HMO theory and using π-electronic methods explicitly allowing for electron interaction[98b]. The differences in ν_{CT} and extinction are due to the lower transannular interaction between the frontier orbitals (relevant for the CT band) of donor and acceptor in **42** with respect to **40**. On the other hand, a through-bond homoconjugative interaction between the donor and acceptor rings has been proposed for CT in the triptycene quinone **93**[23].

(91)
$\lambda_{CT} = 431$ nm (CH_2Cl_2)
505 nm (CF_3COOH)

(42)
$\lambda_{CT} = 515$ nm (MeOH)

(40)
$\lambda_{CT} = 495$ nm (MeOH)

(92)
$\lambda_{CT} = 517$ nm (EtOH)

(93)
$\lambda_{CT} = 430$ nm (Me_2SO)

E. Photoelectron Spectra

The He(I) photoelectron spectra (UPS) of **1**[88d, 101a–c] and **60**[101d, e] are easily distinguishable in the region of lowest ionization potentials (9–13 eV), as may be seen from the stick spectrum in Figure 1[101f]. The molecular orbitals from which ionization occurs (given in Figure 1) have been assigned on the basis of MO calculations (MINDO/3) and

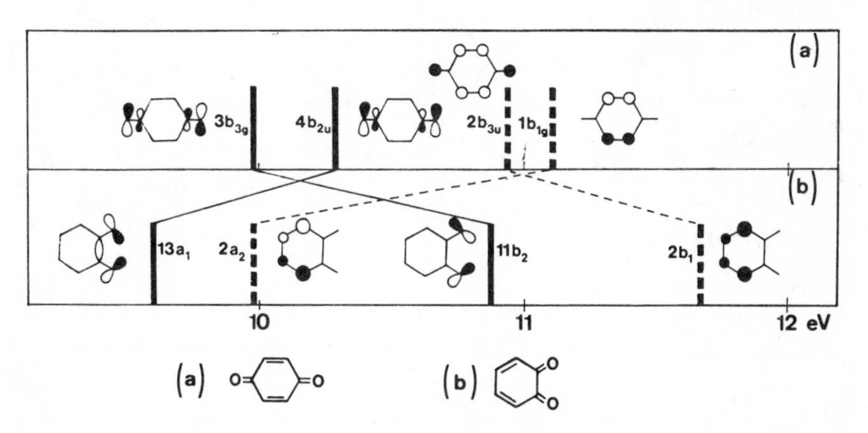

FIGURE 1. He(I) photoelectron stick spectra (UPS) according to Ref. 101f (a) of **1**, (b) of **60**

qualitative bandshape criteria (for **60**)[101d] and empirically (vibronic structure of UPS bands and substitution effects, for **1**)[101c]. The latter assignment is still a matter of controversy[88d,e,101a,b,g–k]. The influence of substituents (alkyl[101c,d,e,l,q], halogen[101c,g,m] and of condensation (naphthoquinones[101h,n,p]) has also been investigated. For X-ray emission spectra (XPS) of **1** and condensed quinones see Refs 101o, p; for electron transmission spectroscopy (ETS) see Ref. 101q.

F. Analytical Applications

UV/vis spectroscopy is easily performed with μg samples. In particular, the rather unique absorptions in the near UV/vis region may be of diagnostic value. Even a distinction between 1,2- and 1,4-benzoquinones is possible, since the *o*-quinones usually show their LFPT and the $n \rightarrow \pi^*$ transition at lower frequencies. In every case, the effect of substituents, especially of OR and NHR/NR_2 groups, should be carefully analysed. The quantitative determination of quinones via application of the Lambert–Beer law on their characteristic absorption maxima is a standard method.

Examples of the analytical use of UV/vis spectroscopy have already been mentioned throughout this section. Further examples are the application of isoprenoid quinone analysis for bacterial classification and identification[128], the spectrophotometric determination of **1** and **60** in mixtures with their dihydroxy compounds[129], the characterization of the quinonoid oxidation products of hydrocarbons in the presence of antioxidants[130] and the identification of natural quinones[131]. The long-wavelength absorption (and CD bands) or rifampicin (**94**) and rifampicin quinone (**95**) were used to monitor the effects of *Escherichia coli* RNA polymerase binding of DNA, dinucleotides and nucleoside triphosphates[132].

New developments include the combination of UV/vis spectroscopy with electrochemical methods (spectroelectrochemistry) for the investigation of electrode processes, e.g. occurring in the diffusion layer in a DMSO solution of **1**[133] and for monitoring

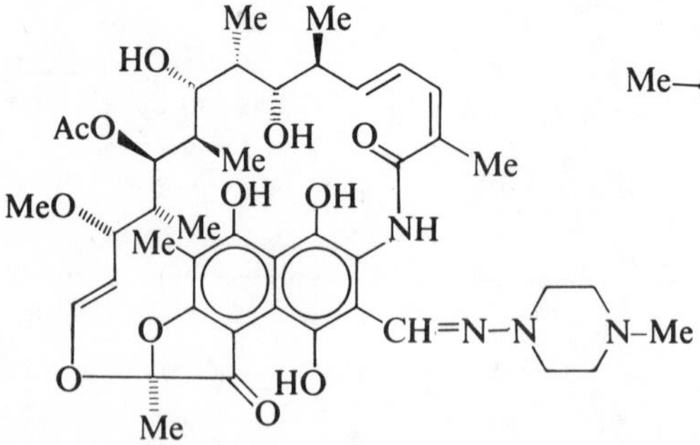

(**94**)

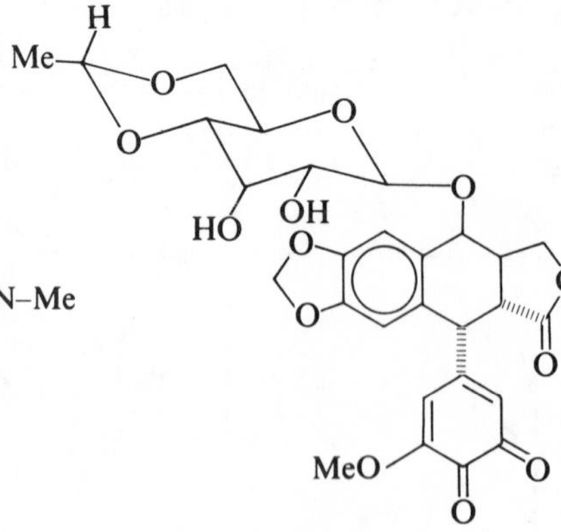

(**96**)

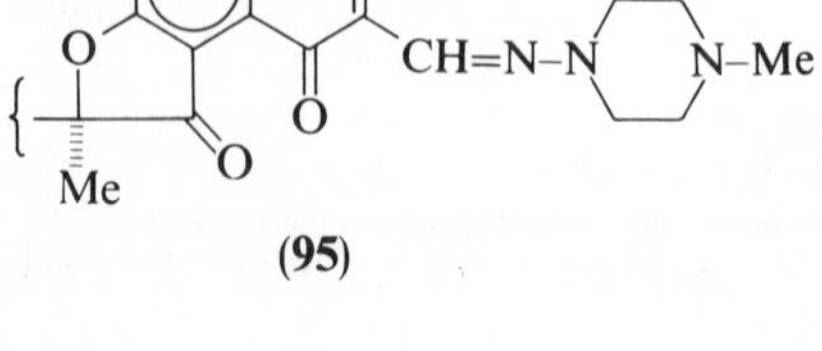

(**95**)

the formation of electrochemically generated products, e.g. in the oxidation of the antineoplastic agent etoposide to its 1,2-quinone (**96**)[134]. The application of HPLC with UV/vis detection has become a powerful analytical tool for quinones and will be discussed in Section VII.A.

Furthermore, CT absorption may be used for the determination of quinones[135], as well as of the donors, e.g. amines[136]. The structure of polymer film coatings obtained by electropolymerizing of naphthol derivatives (containing quinonoid units) has been elucidated by XPS[137]. Finally, quinones may be analysed by the UV/vis spectra of their corresponding semiquinones obtained by reduction[138].

V. ESR/ENDOR SPECTRA OF QUINONE ION RADICALS

A. General Remarks

As pointed out earlier[1], ESR investigation of quinones is only feasible when the diamagnetic parent compounds are converted to paramagnetic species. This transformation can be done by oxidation of the quinone to the corresponding radical cation[139]; by reduction of the quinone to radical anions (semiquinones)[1]; by photoexcitation of the quinone and quenching of the resulting triplet state[140b]; or by combination of the parent compound with a paramagnetic molecule[138d] (**R**). We will restrict the discussion to the first

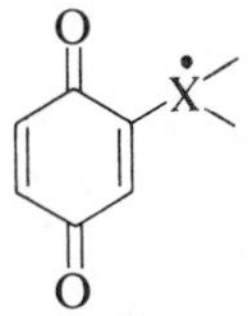

(**R**)

two methods. The main spectroscopic methods pertinent to the spin density distribution in free radicals are ESR, ENDOR and CIDEP.

There have been a few reports on CIDEP of quinones during photolysis[140, 141]; the method, however, is too complicated to serve as a standard analytical technique. ESR and ENDOR, on the other hand, are now standardized, and numerous reports on hyperfine splitting (HFS) constants a and g factors of radicals derived from quinones have appeared since 1973[142]. These spectra may therefore be used to identify the corresponding quinones.

B. Cation Radicals

Only two papers deal with ESR investigations of quinone cation radicals[139, 141a]. Photooxidation of **1** at room temperature as well as electrochemical oxidation of tetrakis-(dimethylamino)-1,4-benzoquinone (**97**) produced the corresponding cations (**1a**, **97a**).

The g values and coupling constants of these one-electron oxidation products are collected in Table 8 together with the data of the semiquinones (**1b**, **97b**) obtained by one-electron reduction of the same quinones. The ESR spectrum of **97b** was poorly resolved since the small proton HFS and the line width were of comparable magnitude. The a_H value was determined from an ENDOR experiment; coupling with N could not be observed at all. These findings are in accordance with McLachlan calculations[139], which predict a very small spin density at the N atoms in the anion (1/14 of that of the corresponding cation) due to the different localizations and symmetries of the singly

TABLE 8. ESR spectra of 1,4-quinone cation (**a**) and anion (**b**) radicals

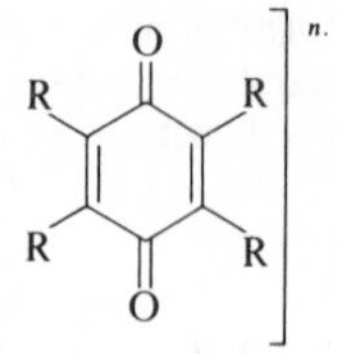

Compound no.	R	a (Gauss)	Charge n	g Value	Solvent	References
(**1a**)	H	2.99 (4H)	+1	2.0038	CF_3COOH	141a
(**1b**)		2.37 (4H)	−1	2.0047	EtOH	1
(**97a**)	$N(Me)_2$	3.32 (24H) 3.32 (4N)	+1	2.0032	DMF	139
(**97b**)		0.112 (24H) 0.0 (N)	−1	2.0046	DMF	139

occupied molecular orbitals in **97a** and **97b**[139] (for the oxidation of a quinone closely related to **97**, leading to a diamagnetic dication, see Ref. 143).

Obviously, the spin density distribution as well as the g value depend heavily on the oxidation state of the paramagnetic species. However, not enough data are available to derive rules for the variation of these properties.

Oxidation of 1,4-dihydroxybenzenes in H_2SO_4 also yielded cation radicals[149]. These species are, however, not oxidation products of quinones and may rather be classified as diprotonated semiquinone anions, i.e. phenol cation radicals. Nevertheless, it is of interest that the g values are smaller than 2.004 and therefore comparable to those observed in the cations obtained by quinone oxidation. Moreover, they can be linearly related to the π-electron spin density at the O atoms.

C. Anion Radicals

Various reducing agents (Na, K, amines, H_2/Pt, $Na_2S_2O_3$ or amalgams)[1,150] and electrolytic reduction were used to prepare semiquinones from the corresponding quinones. The most universal method, however, proved to be the autoxidation of alkaline solutions of hydroquinones[151].

1. g Values

In semiquinone radicals, the spin density on the oxygen atoms particularly contributes to the g values. On the other hand, the oxygen atoms are the basic sites of the radicals and are therefore prone to interact with solvent and/or counter-ions. For this reason, solvent polarity as well as temperature and counter-ion concentration influence the magnitude of the g values[152,153]. The addition of crown ethers with strong affinity to metal ions leads to complexed ion pairs which show alkali metal HFS even at room temperature[146]. These observations point to fast temperature-dependent equilibria between free radicals and those forming hydrogen bonds to the solvent or building ion pairs with the counter-ions. Therefore, the data derived are time-averaged values of these different states. Consequently, the identification of semiquinones using g values only is unreliable.

An exception was observed for tetrahalogenated semiquinone anions where a substantial contribution to g originates in the spin-orbit interaction of the halogens[154]. The values increase significantly with an increasing atomic number of the halogen and may be used to identify these species.

2. *Hyperfine structure constants*

In Tables 9 and 11 the HFS coupling constants *a* are collected for some of the new semiquinones in order to supplement Tables 17 and 18 of Ref. 1. Data derived from semiquinones with larger substituents as in L-adrenalinesemiquinones[155] or heterocyclic compounds such as triazolo-[144], quinoline-[156], isoquinoline-[156], quinoxalinesemiquinones[156] and diaziquone[157], were not included.

TABLE 9. ESR spectra of 1,4-semibenzoquinones

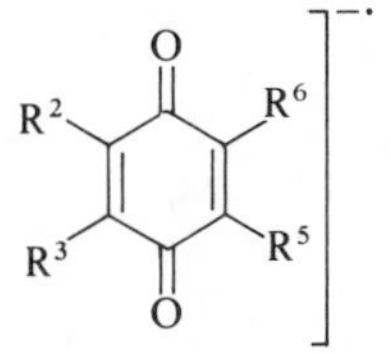

No.	R^2	R^3 *a* (Gauss)	R^5	R^6	Solvent	References
(**98b**)	NH_2 1.85(N) 0.75(2H)	H 0.18	Me 4.65	H 0.75	DMF/H_2O	144
(**99b**)	NH_2 2.5(N) 0.90(2H)	H 0.7	NH_2 2.5(N) 0.90(2H)	H 0.7	DMF/H_2O	144
(**24b**)	OMe 0.0	OMe 0.0	H 2.65	H 2.65	H_2O	145
(**100b**)	NH_2 0.6(N) 0.0(H)	NH_2 0.6(N) 0.0(H)	H 3.0	H 3.0	EtOH	146
(**101b**)	–O–CH_2–O– 0.18		H 2.63	H 2.63	H_2O	145
(**102b**)	OMe 0.0	OMe 0.0	Me 2.45	H 1.98	H_2O	145
(**103b**)	OMe 0.0	OMe 0.0	Me 2.15	Me 2.15	H_2O	145
(**104b**)	NH_2 1.75(N) 0.90(2H)	H 0.2	H 4.9	H 1.52	EtOH	146
(**105b**)	NH_2 1.0(N) 0.25(2H)	H 1.3	H 1.3	NH_2 1.0(N) 0.25(2H)	EtOH	146
(**106b**)	CN 0.61(N)	CN 0.61(N)	H 1.05	H 1.05	DME*	147
(**66b**)	COMe 0.25(3H)	H 3.89	H 1.25	H 2.49	EtOH	142

* DME, dimethoxyethane.

A serious problem arising in the characterization of quinones from their corresponding anion radicals is the dependence of *a* on the polarity of the solvent, the temperature and the counter-ion. A detailed investigation on a series of various substituted 1,2- and 1,4-semibenzoquinones in solvents with decreasing solvation capability (H_2O, EtOH, DMSO, DMF and HMPA) revealed that the changes of *a* are at a maximum for all radicals when the data determined in H_2O and HMPA were compared[148] (Table 10). This is due to the fact that aprotic solvents (like HMPA) solvate the ions only weakly, whereas protic

TABLE 10. Solvent effect on a of 1,4-semibenzoquinones[148]

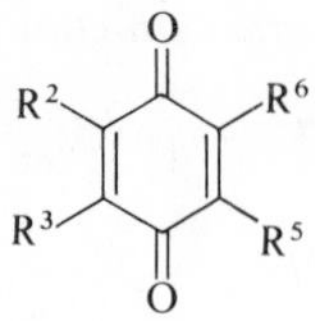

Compound no.	R^2	R^3 a (Gauss)	R^5	R^6	Solvent
(107b)	C(Me)$_3$	H	H	C(Me)$_3$	
	0.0	1.31	1.31	0.0	H_2O
	0.0	2.25	2.25	0.0	HMPA
(4b)	Me	H	H	Me	
	2.1	1.87	1.87	2.1	H_2O
	1.85	2.22	2.22	1.85	HMPA
(108b)	OMe	H	H	OMe	
	0.80	1.47	1.47	0.80	H_2O
	0.50	1.92	1.92	0.50	HMPA

solvents form hydrogen bonds to the basic sites (O atoms). In the latter case, the electron distribution is perturbed, which is manifested as changes in a with the solvent. As a general trend of the solvent effect we may notice that protic solvents decrease the spin density at positions 2, 3, 5 and 6 in 1,4-semibenzoquinones and at positions 3 and 6 in 1,2-semibenzoquinones; the opposite effect was observed at positions 4 and 5 of the latter.

The applicability of ESR spectroscopy to the analysis of quinones requires well resolved spectra, which can be interpreted with limited expense. This feature is often met with symmetrically substituted species, where only one or two different nuclei contribute to the splitting pattern. However, the effect of solvent variation on the spin density distribution is of the same magnitude as the substituent effect. Therefore, data determined in different solvents must be examined critically.

From a comparison of the compounds of Tables 9 and 10 we may derive general trends for the variation of the proton HFS caused by different substituents.

(1) In 2,3-disubstituted 1,4-semibenzoquinones, electron donating substituents such as NH_2 (**100b**), or OMe (**24b**), increase a_H at positions 5/6 compared to **1b**; CN (**106b**) has the opposite effect.
(2) In 2,6-disubstituted 1,4-semibenzoquinones (**105b, 4b, 107b** and **108b**), these effects are reversed; the more electron releasing the substituent the smaller a_H at positions 3/5 becomes.
(3) In 2-substituted 1,4-semibenzoquinones (**66b, 104b**), substituents of opposite electronic properties (COMe and NH_2) also cause opposite shifts of the three non-equivalent a_Hs at positions 3, 5 and 6: NH_2 increases a_{H5} and decreases a_{H3} and a_{H6} compared to a_H in **1b**, for COMe the reverse behaviour is noticed.

Steric effects may also play an important role in perturbing the spin density distribution. In the case of vicinal OMe groups, this has been observed by Gascoyne and Szent-Györgyi[158]. Whereas the HFS of the two OMe groups in **108b** is well resolved, semiquinones with two *vicinal* methoxy groups in no case revealed HFS of the OMe protons. The authors attributed this surprising effect to the steric hindrance, breaking the hyperconjugation between the quinone ring and the methyl moiety of the methoxy groups. Vicinal amino groups (**100b**) also suffer a significant decrease in spin density compared to

2,6-di- or 2-monosubstituted species (**105b**, **104b**). These effects may also be observed in 1,2-semibenzoquinones (Table 11); however, the spin density of the unsubstituted species is five times larger at positions 4/5 than at 3/6. HFS of OMe groups at positions 3 or 6 therefore could not be observed (HFS of NH_2 decreased significantly), when other substituents were vicinal. Substituents at positions 4/5 also suffer a significant decrease in *a* when the vicinal positions 3/6 were additionally substituted.

TABLE 11. ESR spectra of 1,2-semibenzoquinones

Compound no.	R^3	R^4 *a* (Gauss)	R^5	R^6	Solvent	References
(**109b**)	NH_2	H	H	H	EtOH	150
	1.65(N)	0.0	5.1	0.85		
	0.70(2H)					
(**110b**)	H	NH_2	H	H	EtOH	150
	1.90	2.10(N)	3.10	0.85		
		3.95 (2H)				
(**111b**)	NH_2	NH_2	H	H	EtOH	150
	1.15(N)	1.60(N)	3.20	0.89		
	0.0 (H)	0.95(2H)				
(**112b**)	H	NH_2	NH_2	H	EtOH	150
	0.2	2.25(N)	2.25(N)	0.2		
		0.90(2H)	0.90(2H)			
(**113b**)	OMe	OMe	H	H	H_2O	145
	0.0	0.44	3.92	1.70		
(**114b**)	OMe	OMe	H	Me	H_2O	145
	0.0	0.53	2.95	1.85		
(**115b**)	OMe	Me	H	OMe	H_2O	145
	0.0	5.61	1.70	0.61		
(**116b**)	H	OMe	OMe	H	H_2O	145
	0.32	1.10	1.10	0.32		
(**117b**)	H	–O–CH_2–O–		H	H_2O	145
	0.33	4.22		0.33		
(**118b**)	OMe	OMe	OMe	H	H_2O	145
	0.0	0.10	1.20	0.60		
(**119b**)	OMe	OMe	H	OMe	H_2O	145
	0.0	0.90	1.10	0.68		
(**70b**)	OMe	H	OMe	H	H_2O	145
	0.75	1.70	1.00	0.95		
(**120b**)	$C(Me)_3$	H	H	$C(Me)_3$	EtOH	142
	—	3.3	3.3	—		
(**121b**)	COOEt	H	H	H	EtOH	142
	0.092	1.445	2.537	3.727		

Several authors[159–162] designed additivity rules and increment systems to compute the proton coupling constants of various substituted 1,4-[160,161] and 1,2-semibenzoquinones[159], or semianthraquinones[162]. The correlation of experimental data and theoretical results obtained from the derived models is good. The range of substituents under consideration is not complete, however, and therefore the applicability of these systems is limited.

3. Secondary products

A problem arises in the preparation of semiquinones in alkaline media because of the formation of secondary products. Quinones having free positions on the quinonoid ring react with hydroxide ions[163] or alkali hydroxides in alcohols[164] to give the hydroxy- or alkoxy-substituted semiquinones. The formation rate of these products increases with increasing alkali concentration of the solvent. When OH^- is the attacking species, the products of 1,2- and 1,4-quinones are identical and reveal the 2-hydroxy-1,4-semiquinone structure[163]. The spectra obtained from these product mixtures are unsymmetrical and dependent on time due to the different *g* values and the reaction kinetics of the (at least) two species.

4. Metal complexes

A large group of researchers used various metals salts, organometallic reagents or transition metal complexes to stabilize semiquinones by complexation with the metal[165, 166].

The addition of diorganothallium hydroxide to solutions of catechols yielded persistent 1,2-semiquinones; in ion pairs consisting of radical anions and organothallium counter-ions, the metal HFS could be observed[159, 167]. When group IIIa elements (Al, In, Ga) or Sn were reacted with 3,6-di-*t*-butyl-1,2-benzoquinone, hexacoordinated triradical complexes were formed[168]. The variety of semiquinone complexes was much larger when transition metal salts or complexes were used. Hexacoordinated structures in different oxidation states with three 1,2-semiquinone ligands as well as complexes with mixed ligands were observed[169]. The *g* values of these complexed species differed significantly from those of the free semiquinone radicals; often a strong coupling with the metal ions was found[169, 170].

1,4-Semiquinones react with metals and metal complexes in a less defined manner. Decacarbonyldimanganese and 2,6-di-*t*-butyl-1,4-benzoquinone (**107**) thus give an 'α-keto'-type spin adduct (**S**)[171a]. The spin density distribution is comparable to that observed

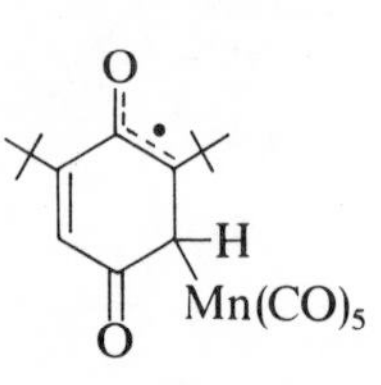

H $Mn(CO)_5$ OMRn

(**S**) 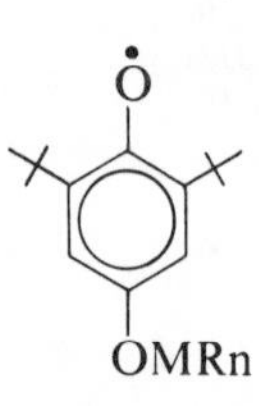 (**T**)

M = Si R = Me[172]

M = Ge, Sn, Pb R = Me[171a]

M = Ge R = Et[171b]

in the 2-cyclohexanoyl radical and therefore not of the 1,4-semiquinone type. The quinone **107** was also used as spin trap for various organometallic radicals[172] to give the 4-substituted phenoxy radicals (**T**).

D. Analytical Applications

ESR spectroscopy is a useful tool in the analysis of many naturally occurring semiquinones. Pedersen was able to identify quinones and quinols in plant extracts by their

HFS and g values, and quantitative analysis was performed by comparing the ESR signal intensity with standard curves[142, 173]. A review, collecting the various applications of ESR in medicine, reports (among others) on semiquinone studies[174].

Chemical reaction pathways may also be investigated by ESR spectroscopy, as shown by Bubnov and coworkers[175] for the photoreduction of the sterically hindered 3,6-di-*t*-butyl-1,2- and 2,6-di-*t*-butyl-1,4-benzoquinone (**120**, **107**).

Determination of biochemical pathways and the verification of reaction models may be facilitated using the sensitive ESR technique. Semiquinones were detected in fulvic and humic acids; from the change of the spin concentration by electrolysing the samples, it was concluded that they are at least partially responsible for the reducing capability of these materials[176].

The primary step of photosynthesis consists of a photochemical one-electron transfer to an electron acceptor (often a quinone). Many authors have investigated the photochemically induced generation of semiquinone radicals in biological materials and model systems by ESR[177].

VI. ELECTROCHEMICAL METHODS

A. General Remarks

Electrochemistry is another discipline usually thought to be predestined for the analysis of quinones, since a typical feature of the latter is their cathodic reduction proceeding as a single two-electron or as two separated one-electron transfers, depending on the conditions, and often being 'quasi-reversible'. For the reasons given below, and since the electrochemistry of quinones is treated in detail in another chapter of this book, neither a detailed discussion of the principles underlying the applied methods, nor a collection of data is presented in this section.

B. Voltammetric Methods

Mainly, linear sweep voltammetry (LSV)[178], especially in the form of polarography (i.e. LSV at the dropping mercury electrode[178-180]) or cyclic voltammetry (CV)[178, 181, 182] were used for the investigation of quinones. The applied potential is altered linearly (LSV, polarography) or triangularly (CV) with time, and the resulting current is recorded. A more expensive, newer polarographic method also used in quinone analysis is pulse polarography (PP), above all in the form of differential pulse polarography (DPP)[178-180]. In principle, a rectangular voltage pulse is applied to the mercury-drop electrode within a short time interval at the end of the lifetime of the drop (PP). In the case of DPP, the current is measured a few ms before applying the pulse to the mercury drop, stored, and then subtracted from the value measured in the second half of the pulse period. In this way, concentrations of 10^{-8} M may be detected in comparison with 10^{-7} M in PP.

The resulting experimental data are half-wave potentials $E_{1/2}$ (polarography) or peak potentials E_p (LSV, CV and DPP), and the current values at these potentials. In the case of reversible redox processes, $E_{1/2}$ of LSV and E_p of CV are simply related to each other and to the formal potential E° (equation 1).

$$E^\circ = E_{1/2} = 1/2 \cdot (E_p^{ox} + E_p^{red}) \qquad (1)$$

The value of the potentials depends on the reference electrode and may be influenced by electrode materials, supporting electrolytes, and diffusion potentials (connection to the reference electrode).

$E_{1/2}$ and E_p are characteristically different for 1,2- and 1,4-quinones. Moreover, the peak and half-wave (or wave) heights as current quantities are proportional to the concentration of the electroactive quinones, and allow their quantitative determination. It should be clearly stated, however, that many other compounds are also electroactive in the potential area in question, a fact which may hamper quinone analysis involving $E_{1/2}$ and E_p values only.

Investigation of electrode kinetics, the genuine domain of LSV and CV and highly interesting for reaction mechanisms, is of no practical importance for diagnostic purposes, since the relevant data (e.g. k values) are not easily accessible.

For compilations of potential data of quinones and for discussions see the following references: LSV[1, 6b, 57c, 77, 110, 158, 183, 184], CV[7a, 1, 9a, 22b, 46, 184h, v, 185].

Besides the usual metal electrodes, thin-layer electrodes[186a] have been used. Several papers deal with chemically modified electrodes, either using quinones as substrates to modify electrodes[187] or investigating the behaviour of quinones on modified electrodes[188]. These are interesting new developments, but not yet suitable for a general, easy analytical approach.

Thin-layer voltammetry of quinones[186b] subjects a very thin layer ($\delta = 10$–$50\ \mu m$) of solution to electrolysis in order to eliminate diffusion effects (restricted diffusion).

C. Coulometry

Coulometric techniques determine the quantity of material electrolysed from the amount of charge passed through the electrochemical cell, on the basis of Faraday's law (equation 2):

$$N = Q/n \cdot F = (1/n \, . \, F) \cdot \int_0^t i \cdot \mathrm{d}t \qquad (2)$$

where N = number of moles of the substance being electrolysed; n = number of electrons transferred per molecule; Q = total charge passed (Coulombs); i = current (Amperes); t = time (s); F = Faraday's constant (96.485 Cb mol^{-1}).

In controlled-potential coulometry (CPC), a constant potential is applied to the working electrode in order to electrolyse the substrate completely. The current is integrated to give the charge needed for an exhaustive electrolysis. In constant-current coulometry (CCC)[189], the experimental parameters are now current i (set constant) and time t, i.e. $Q = i \cdot t$. For quinone analysis, mainly CCC was used, especially in the form of secondary coulometric titration[189], i.e. the quinone is not reduced directly at the electrode but rather by an electrogenerated reagent.

D. Other Methods

Besides the hitherto mentioned 'dynamic' methods ($i \neq 0$), potentiometry as a 'static' method ($i = 0$) has also been applied to quinones in the form of the well-known potentiometric titration.

Electrochemical detection (ELD)[190] has been used in the chromatography of quinones (see Section VII). The underlying electrochemical principle is direct-current amperometry (the measurement of electrochemical current in response to a fixed electrode potential) in a special form of chrono-amperometry (measuring the current as a function of time). Since the mobile phase must be able to carry an ionic current, only reversed phase (RP) chromatography with aqueous solutions (with or without methanol, acetonitrile, or tetrahydrofuran) can be used. Of course, ELD *per se* cannot discriminate quinones from other substances (see Section IV.B).

Finally, it should be stated that electrochemical methods can be combined with other spectroscopic methods in a static or hydrodynamic way to determine quinones. ESR spectroscopy is mostly used for this purpose (see Section V for examples, Ref. 191a for a general discussion), but UV/vis is also a candidate (see Section IV.F) and ^{1}H-NMR has been chosen to determine the electron exchange rate between **1** and **1b** or ubiquinone and its semiquinone[191b].

E. Analytical Applications

Electrochemical methods have been mainly applied to the analysis of natural quinones or the investigation of quinonoid compounds as models for biochemical reactions. Besides the literature already mentioned in Section VI. A–D, a few more examples are given below, starting with voltammetry.

1,2-, 1,4- and 2,3-Quinones and hydroquinones may be discriminated using CV in the presence of sodium borate in an alkaline medium[192]. The borate interacts with 1,2-dihydroxybenzenes and the resulting complex can be oxidized at more positive potentials than the free form. An example of the application of CV to mechanistic problems is the redox system etoposide/etoposide quinone (**96**), with an extensive investigation of the pH influence[134].

Quinones formed during oxidation of phenols in waste water may be determined polarographically in the presence of excess phenols making use of a change of $E_{1/2}$ values with the quinone concentration in relation to the pH[193].

The cathodic reduction of anthracyclines (especially of daunomycin), potential antitumour antibiotics, has attracted much interest[194]. It has turned out that glycoside elimination occurs only after a two-electron reduction[194d]. Possible relations between redox potentials and cytotoxic[195] as well as antineoplastic[196] activity of simpler quinones have been investigated. In the case of 1,4-benzo- and naphthoquinones the compounds of the lower (i.e. more negative) redox potentials generally possess the most potent antitumour properties towards a Sarcoma 180 test system. Benzoquinones with a less negative redox potential than naphthoquinones were in general inactive against Sarcoma 180[196]. The reaction of phenanthraquinone (**80**) with Ehrlich ascites tumour cells of mice[197a] and the electrochemical reductive activation of mitomycin C[197b] have been further targets of voltammetry.

Another item investigated was the electron transport in model compounds for biological systems: hydroxyquinones[198], ubiquinones[199], or linked porphyrin quinones (PQ)[200]. Thus, the energies of the charge separated $P^{+\cdot}Q^{-\cdot}$ states of the amide-linked ('spacered') PQ molecules **122** and **123** in CH_2Cl_2 amount to 1.37 ± 0.02 eV, taken as the difference in redox potentials for the formation of $P^{+\cdot}Q$ and $PQ^{-\cdot}$ [200].

Besides CV, differential pulse voltammetry (DPV) with alternating current was used for these investigations. Examples for the application of differential pulse polarography (DPP) are the determination of quinonoid and other functional groups of enomelanin from *Vitis vinifera*[201], the analysis of non-volatile water pollutants, as quinones, without

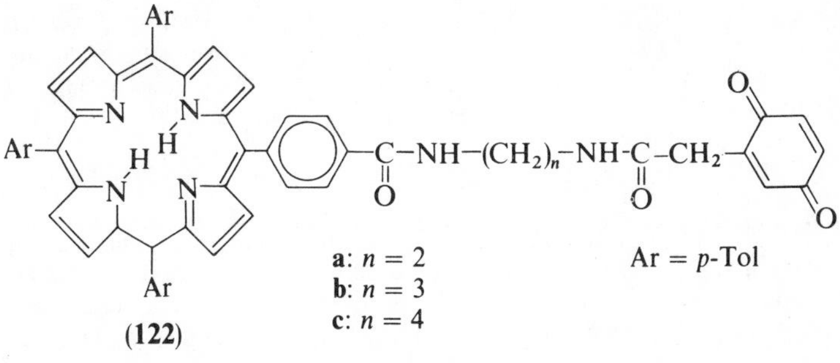

a: $n = 2$
b: $n = 3$
c: $n = 4$

Ar = *p*-Tol

(**122**)

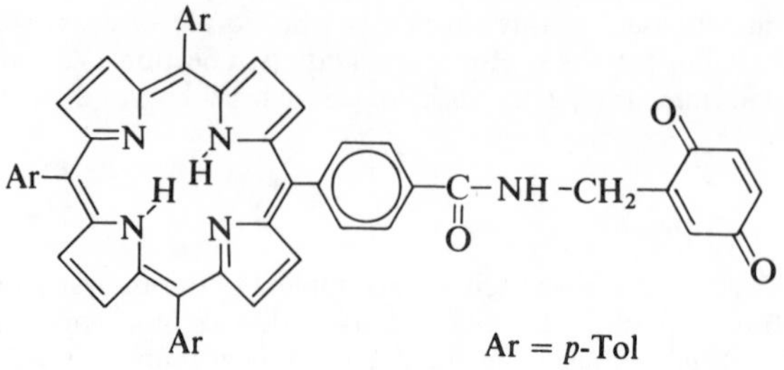

Ar = *p*-Tol

(123)

separation[202], or the indirect microdetermination of quinone **1**[203]. For this purpose, **1** was reduced by I^- in the presence of acid. The liberated I_2 was then extracted into $CHCl_3$ and reduced to I^- with $NaHSO_3$. The iodide was afterwards oxidized with Br_2/H_2O to IO_3^-, which was determined by DPP, the detection limit being 1.2×10^{-6} M quinone.

Finally, reports on voltammetric analysis of pharmacologically active[204] and clinically relevant[205] quinones are also available.

Direct coulometric investigations (reductions) have been performed on simply substituted 1,4-benzoquinones (CPC)[184e, h, 206], tetrahydroxy-1,4-benzoquinone[198], ubiquinones (CPC)[184r], and condensed quinones (CPC)[184e]. Indirect coulometric titrations used electrogenerated Cu^+ from a copper electrode[207a], Fe^{2+} (CCC)[207b], Ti^{3+} [208], and Ce^{4+} [209] for the quantitative determination of quinones and for drug control[210].

Examples of potentiometric titrations (end point detection) of quinones are the direct reaction of quinones with Ti^{3+} [211], Fe^{2+} [212], or V^{2+} [213], or the indirect titration of a mediator which reacts with the quinone[214].

VII. CHROMATOGRAPHY OF QUINONES

A. General Remarks

As other organic molecules, quinones may be purified, detected and determined by chromatographic methods. Gas chromatography (GC), liquid chromatography (LC), high-performance liquid chromatography (HPLC) with normal or reversed phase (NP or RP), thin-layer chromatography (TLC), or paper chromatography (PC) have been used. Since quinones are coloured, they may be easily detected in LC, TLC, or PC even by the naked eye; and for GC the usual detectors are suitable.

As a rule, however, special detection techniques were used together with LC, HPLC, or TLC, the three chromatographic methods chiefly applied to quinones. In the case of HPLC and LC, all physical methods discussed in this chapter, i.e. ^{1}H-NMR, ^{13}C-NMR, IR, Raman, UV/vis spectroscopy (and luminescence[215]), ESR spectroscopy, as well as electrochemistry (ELD), are suitable, although UV detection was the method of choice.

Supercritical fluid chromatography/Fourier-transform-IR spectrometry with an automatic diffusive reflectance interphase ensures high sensitivity; it was applied to acenaphthenequinone[216a].

HPLC/ESR needs paramagnetic species and has been used by Creber and Wan[216b] to isolate and characterize rhenium carbonyl quinone complexes.

Few HPLC/ELD investigations on quinones are known, since substituted quinones will not always dissolve in the polar mobile phase used for reversed phase (RP) chromatography and necessary for ELD (see Section VI.D). Hg-film[217] and glassy carbon[218] electrodes have been tested. The observed oxygen overpotential on glassy carbon proved to be more negative (relative to Hg/Au), thus allowing the determination of quinones (2.5 pmol) via reduction in the presence of dissolved O_2.

In TLC the suppression of fluorescence of added special indicators is familiar, but spray reagents (see Table 12) or vapour development (e.g. with OsO_4[219]) have also been used.

Retention times (GC, LC, HPLC) or R_F values (TLC) are characteristic and identify the quinone. Intuition and empiricism are still requested in this field, and a collection of chromatographic data is not provided here. Peak areas of recorded or computer-processed chromatograms (GC, LC, HPLC) allow an easy quantitative determination, whereas the densitometric[220] evaluation of TLC spots provides some problems.

The present authors recommend TLC as an inexpensive, simple, fast and effective method for a qualitative characterization of quinones. It should be stated, however, that some quinones are highly unstable in the adsorbed state, especially when exposed to light, hence there may be 'tailing' due to the formation of hydroquinones which, in turn, are prone to be partially air-oxidized, when subjected to TLC.

Reviews on chromatography treating also quinones (e.g. LC[221]) or specialized on certain classes of quinones (e.g. anthraquinones[222]) are available.

B. Coupling with Spectroscopic Methods

After a chromatographic purification, the products are usually further characterized or structurally investigated by spectroscopic methods. An advantage over this procedure would be attained by an immediate spectroscopic investigation of the compounds as they are eluted, either in a direct or by-pass flow system or by a type of stop-flow procedure.

In contrast to the detection methods mentioned in Section VII.A which just monitor the elution profile at a singular point of the spectral domain (e.g. at a singular wavelength in UV), complete spectra of the eluted (and separated) compound would be available before isolating the sample. The most effective coupled system, gas chromatography–mass spectroscopy (GC–MS) has, indeed, proved to be of high value also for quinones[223], and even TLC–MS may be of advantage in special cases[224]. Other combinations (such as HPLC coupled to UV or NMR spectroscopy or to CV, installed in an electrochemical detector) seem not to have been met with interest in quinone chromatography, so far. Only a type of LC–voltammetry has been investigated for the quinone-like 6-methyldihydropterins[225].

C. Analytical Applications

A selection of applications of chromatographic methods to quinones is presented in Table 12, and a few special aspects of natural quinones are enumerated below. Further literature may be found in the references quoted in Table 12.

Naphtho- and anthraquinones of the bark of *Rhamnus fallax*[241] (HPLC, TLC), α-tocopheryl quinone in blood[242] (HPLC–UV) and prenyl quinones (0.1 nmol) from spinach leaves[243] (RP–HPLC–ELD, MeOH/EtOH/0.05 M $NaClO_4$) have been determined. The investigation of metabolization of aromatics was another item: 1,4-naphthoquinone from 1-naphthol by liver microsomes (RP–HPLC–ELD, MeOH/EtOH/0.05 M $NaClO_4$)[244]; benzo[*a*]pyrene quinones from the corresponding hydrocarbons also by liver chromosomes (HPLC–TLC) or human lymphocytes (HPLC)[245].

Isoprenoid quinones in bacteria[128a, 246] or in marine diatoms[247] may also be chromatographically isolated and identified (NP and RP TLC, HPLC, UV, MS).

TLC is also a versatile method for checking the purity of drugs. Thus, rifampicin quinone (**95**) can be detected in the antibiotic rifampicin **94** and quantitatively determined with UV/vis after extraction of the TLC zones[248].

TABLE 12. Chromatographic analysis of quinones

Method (adsorbent)	Detection[a]	Type of quinone	Remarks	References
HPLC (silica gel, LiChrosorb ® -Si60, -diol, -RP8 ® ; Permaphase ODS ®)	UV	**1**, **73**, **74**, toluquinone, prenylquinones		226, 227
(HP) LC (silica gel/octadecyl)	UV	**73**, **79**, naphthoquinones	Polarity effect of mobile phase	228
HPLC (silica gel, μBondapak CN ®)	UV	Hydroxy-, methylnaphthoquinones		229
HPLC (silica gel/RHg^+)	UV	Miscellaneous quinones	Selectivity for S, O, N donor ligands	230
LC (cyclodextrin)	UV	**1**, **73**, **74**, **79**, 2,5-dimethyl-**1**		231
HPLC (RP/ion-pair chromatography)	UV	Pyrroloquinoline quinones	(From enzymes)	232
TLC (silica gel)	Spray reagents: RB, 4-A, DBDC, P, I_2, 1 N NaOH	**1**	(0.2–10 μg)	233

TLC (silica gel, cellulose)	FQ, (spray) reagents: RB, 4-A, P, DBDC 1 N NaOH, H_2SO_4, UV light	**1**, **73**, **74**, naphthoquinones		234, 235
TLC (silica gel)	FQ	3,4-Benzpyrene quinones		236
TLC (silica gel/$AgNO_3$)	Spray reagents: RB, anilino-naphthalene sulphonic acid, long-wavelength UV	Prenylquinones		237
GC (silicone oil or grease, OV 17, DC 550)	FID	**1**, methyl-, methoxy-1,4-benzoquinones	(< 0.1 μg/μl)	223a
GC (Chromosorb WAW DMCS ® coated with apiezon L ® or diethylene glycol succinate polyester/H_3PO_4	ECD, FID	**65**, chlorinated benzoquinones		238
Paper chromatography	Spray reagents: Na nitroprusside/NH_4OH, Na vanadate, NH_3-vapour, 1 % NaOH/methanol	Miscellaneous hydroxyquinones	(10–25 μg) ($\geqslant 5$ μg)	239 240

[a] Abbreviations: RB = rhodamine B; 4-A = 4-aminoantipyrine; DBDC = 2,6-dibromoquinonechloroimide; P = phosphomolybdic acid; FQ = fluorescence quenching; FID = flame ionization detector; ECD = electron-capture detector.

Finally, it has been demonstrated by HPLC and GC–MS that **74** and 1-chloroanthraquinone (among about 200 other organic compounds) are formed by incineration of municipal waste[249] and by means of gel chromatography that **73** and **80** are present in weathered Ecofisk crude oil[250]. Application of HPLC (silica gel) and TLC [polyamide or $Mg(OH)_2$] combined with UV/vis fluorescence spectroscopy and MS to air pollution analysis[251] revealed the presence of **74**, 3 benzo[*a*]pyrene quinones, and of dibenzo[*b*,*def*]chrysene-7,14-quinone.

VIII. CHEMICAL METHODS

A. General Remarks

In principle, all chemical reactions of quinones described in this book may be used for a qualitative and, in some cases, quantitative analysis of quinones, as well as of the reactants. Originally, colour reactions and spot tests[1] were of main importance in qualitative analysis. They are still in use, although mainly in combination with chromatographic separation techniques (especially TLC) and spectroscopic methods (preferentially UV/vis). Some colour reactions will be discussed in more detail in Section VIII.B.

UV/vis spectroscopy of coloured and uncoloured reaction products also allows a quantitative determination (photometry). Certain reactions of quinones give rise to luminescence[102a,215], or fluorescence[252,253] which allow a quantitative analysis of quinones. Quinones are easily reduced to semiquinones or dihydroxybenzenes (or the corresponding anions), which may be determined by ESR (see Section V) or UV spectroscopy[254]. Finally, redox titration[212,213,255] is also widely used, with a variety of end-point detectors: colour indicators, coulometry (see above), and potentiometry (see above).

B. Colour Reactions

1. Introduction

The term colour reaction is used in this section to mean reactions in which the colour changes (to another colour or to colourless). The different types of colour reactions of quinones and the various tests based on them have been already classified in Table 20 of Ref. 1. Meanwhile, a lot of new colour reactions have been reported, which may be divided into reactions of quinones with bases (anion formation[71]), with free or incipient carbanions[256,257], with amines and their derivatives[4c,5u,258], *N*-heterocycles[259] and sulphinic acids[260]. Some are described below in greater detail.

2. Anion formation

Anion formation from hydroxyquinones as a colouring principle has been demonstrated earlier[1]. The analogous reaction occurs also with aminoanthraquinones[7e]. Thus, for example, 1-amino-9,10-anthraquinones (**124**) produce green amides (**125**) (λ_{max} = 695 nm) with KOH/DMSO in air. The formation of amide anions is demonstrated: (1) by recovering **124** after quenching with water, (2) by the formation of *N*-alkylated products at room temperature after the addition of alkyl halides, (3) by the absence of a typical colour change, when 1-dimethylamino-9,10-anthraquinone (**127**, see below) is treated with KOH/DMSO. Under an atmosphere of nitrogen, the anions **125** are further reduced via a one-electron transfer from OH^- to give the corresponding dianion radicals (λ_{max} = 560–570 nm; ESR signal).

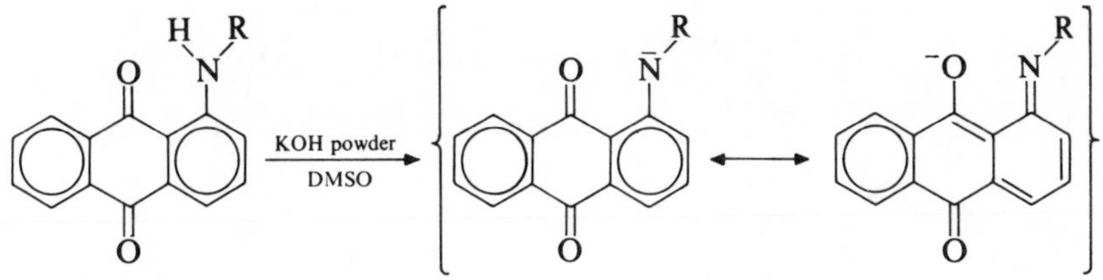

KOH powder / DMSO

(124) (125)

a: R = H
b: R = Me

3. Reactions with carbanions

Quinones are able to add carbanions formed from CH-acids (like acetoacetic ester, malononitrile or nitromethane) and a base (like ammonia in ethanol) via their α,β-unsaturated carbonyl systems. The substitution products, resulting after subsequent reoxidation, e.g. **U**, are even stronger CH-acids and dissociate into their blue-green or

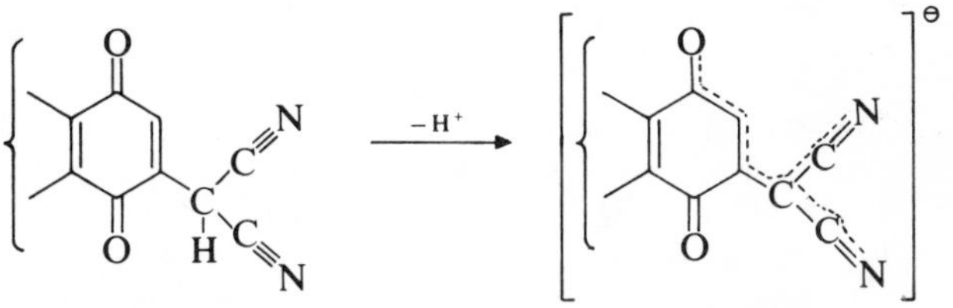

$-H^+$

(U) (V)

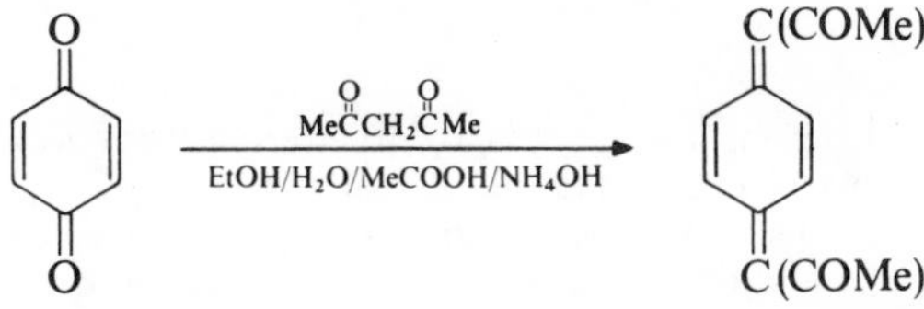

$MeCOCH_2COMe$ / $EtOH/H_2O/MeCOOH/NH_4OH$

$C(COMe)_2$

(1) (126)

violet-blue anions, e.g. **V**. This is the typical feature of the Kesting–Craven test[1]. For a new application to 2-methyl-1,4-naphthoquinone see Ref. 256. Recently it has been shown that addition to the carbonyl group of quinones is also possible under the reaction conditions[257a]. The resulting quinodimethane **126** can be used for a spectrophotometric determination of **1** in water[257a]. If there is no quinonoid position free, the Kesting–Craven test does not work, since adducts like **U** cannot be formed, and addition to the carbonyl group seems to occur[7e]. The resulting quinol derivative **128** gives rise to a hypsochromic shift (510 → 420 nm).

Another interesting reaction involves the addition of pyridinium ylides to quinones[68b], e.g. to **129**. The Kesting–Craven adduct **130** may be transformed via bromination and dehydrobromination to **131**, containing the basic skeleton of fungal and aphid pigments.

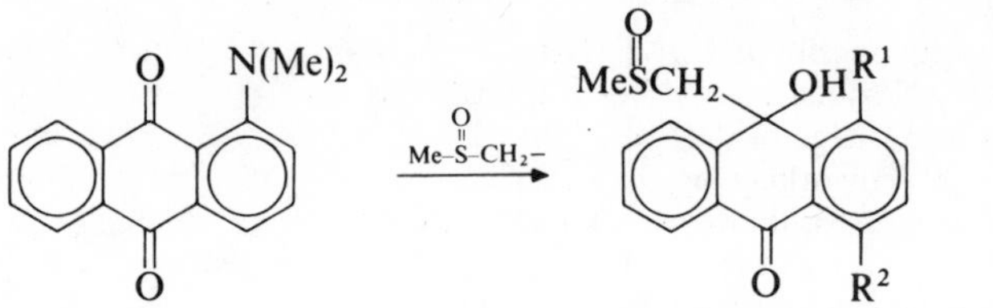

(127) (128)

$R^1 = H, R^2 = N(Me)_2$
or $R^1 = N(Me)_2, R^2 = H$

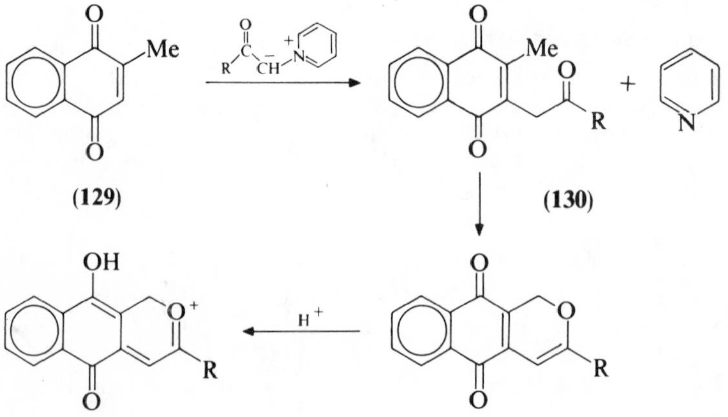

(129) (130)

(132) (131)

R = Ph

In trifluoroacetic acid, **131** is instantly protonated to give a brilliant purple solution slowly changing to inky blue-black. The purple cation, assumed to be **132**, has $\lambda_{max} = 310$ and 529 nm ($\log \varepsilon = 4.31, 3.29$) and may be considered as the cationic counterpart of **125**.

Closely related to the addition of carbanions is the reaction of **1** and **73** with the nucleophilic carbon of the isocyano group in organic isonitriles; however, the primary adduct adds a second isonitrile molecule to give dark blue 4,7-isoindolediones. In the case of **1**, the same reaction sequence may occur twice, ending up in the formation of 1,5- and 1,7-bis(arylamino)benzodipyrrolediones. For more details see Ref. 257b.

4. Reactions with amines

The vast majority of colour reactions of quinones involves amines and their derivatives. In principle, the primary reactions may be: complex and charge transfer complex formations; electron transfers; and addition/elimination or addition/oxidation reactions involving C=C or C=O bonds. Some readers may have missed categories like 'oxidation' or 'substitution' in the above enumeration. However, the oxidation of amines by quinones is no simple and uniform process, it may start with electron transfer or with addition. A detailed discussion of this problem, relying on redox potentials, is provided for 1,2-

phenanthrolinequinones by Eckert and Bruice[110]. Finally, substitution at the quinonoid system is mostly a sequence of addition and elimination reactions.

A profound knowledge of quinone–amine colour reactions has been accumulated by Kallmayer and his coworkers and by other authors[4c, 5u, 258]. Their investigations often originate in pharmaceutical interests and necessities. However, a complete elucidation of the mechanisms or even of the structure of the coloured reaction products is not always essential for a quantitative (photometric) determination.

We will now concentrate on additions; charge transfer complexes have already been mentioned in Section IV. Consider the reaction of an amine with the α,β-unsaturated quinonoid system (**W**). 1,4-Addition will first lead to the adduct (**X**) or (after proton

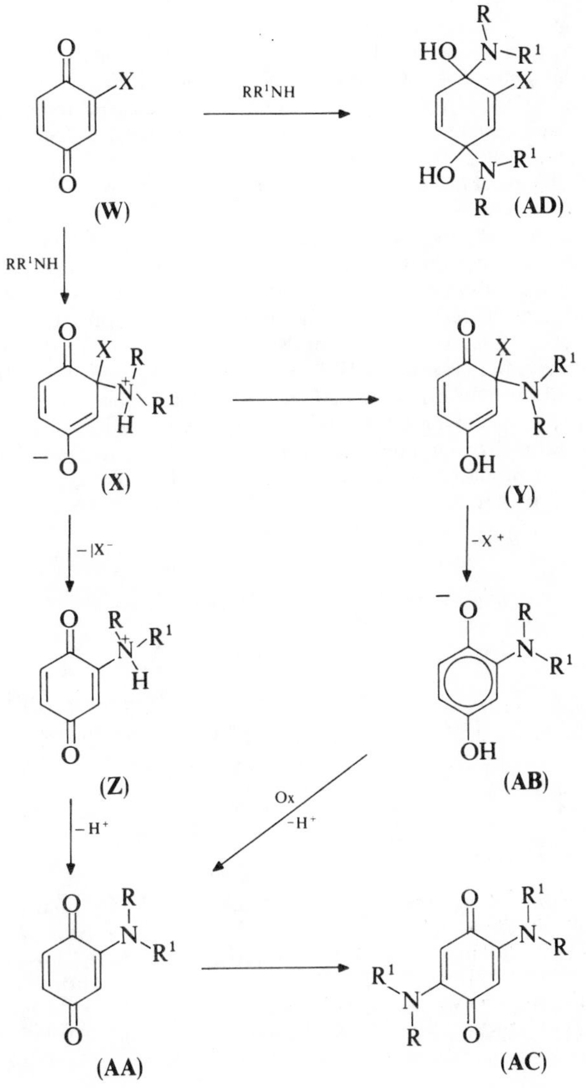

(**W**) —RR¹NH→ (**AD**)

(**W**) —RR¹NH→ (**X**) → (**Y**)

(**X**) —(−|X⁻)→ (**Z**) —(−H⁺)→ (**AA**) → (**AC**)

(**Y**) —(−X⁺)→ (**AB**) —(Ox, −H⁺)→ (**AA**)

transfer) to the equivalent **Y**. If X is prone to leave as an anion, we arrive directly at the quinone **Z** or (after deprotonation) the quinone **AA**. The result is a formal *substitution* of X via an addition–elimination mechanism. The aminoquinone **AA** absorbs at longer wavelengths than the quinone **W** (see Section IV) and may be used for an indirect photometric determination of **W** and/or the amine, provided the reaction proceeds quantitatively.

Chlorine and SO_3H (or the salts $SO_3^- M^+$) seem to be suitable leaving groups X in the sense of the outlined scheme. This is demonstrated by the reaction of 2,3,5,6-tetrachloro-1,4-benzoquinone (chloranil) with primary, secondary and tertiary aliphatic and aromatic amines[258a] or of sodium rifamycin-3-sulphonate with aromatic and aliphatic secondary amines (with primary aliphatic amines, the $SO_3^- Na^+$ group is not substituted, but the *ansa* ring of rifamycin is cleaved instead)[6i].

The reaction **W** → **AA** may alternatively proceed via **Y** and leaving of X as X^+ to give the anion **AB** of a hydroquinone, which would be deprotonated (by the amine) and oxidized (e.g. by **W**) to the quinone **AA**. Quinol halides as **Y** are indeed known to act as 'positive halogen' donors[261]. Moreover, a primary single electron transfer from the amine to the quinone and the subsequent recombination of the semiquinone and aminium radical ions is a reasonable third alternative.

For X = H, the reaction presumably takes the course **W** → **Y** → **AB** → **AA**, since hydrogen usually leaves as H^+, especially in the presence of a base. Examples are the reactions of **1** with a large series of primary and secondary amines[4c,h,258j,262], or of alkylated 1,4-benzo- and naphthoquinones with secondary amines[5u,258i]. In the latter case, additional substitution of a hydrogen atom of a quinone ring methyl group by the amine, or photochemical dealkylation of the amino function to give the monoalkylamino quinone have been observed[5u,258i] (see also Kallmayer and Tappe[6y]). It should be mentioned that the reaction sequence **W** → **AA** may occur a second time, starting with **AA** as substrate, to produce the bis-adduct **AC**[4c]. However, a 1,2-bisaddition product of type **AD** (addition to the carbonyl groups) has also been obtained with an excess of the amine, starting with the dihydroxyquinone **26**, which dehydrates to **AC** in boiling aniline[4c].

In the case of a bisamine like ethylene diamine, the primarily formed substitution product **AE** (equivalent to **AA**) reacts further via 1,2 addition of the second amino function

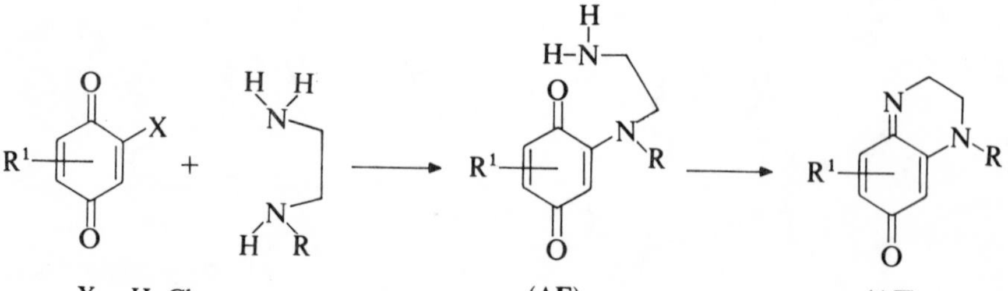

X = H, Cl
R = Alkyl, not Aryl

(AE)

(AF)

to the adjacent carbonyl group and subsequent dehydration. The final reaction product has the quinoxalinone structure **AF**[6s,258b–f,k]. Due to a retro-Mannich reaction, 2,3-dimethyl-1,4-naphthoquinone undergoes dealkylation to 2-methyl-1,4-naphthoquinone under the reaction conditions[258k]. For the corresponding (naphtho)thiazinones from naphthoquinones and cysteamine see Ref. 258g.

Quinoxalines themselves are suitable to characterize 1,2-quinones[1,6b,258h]. Most of the other quinone–amine reactions discussed here for 1,4-quinones are also observed with 1,2-

quinones. For the sake of brevity, the reader is referred to the relevant literature: 1,2-benzoquinone (**60**)[4c]; 1,2-naphthoquinone-4-sulphonates (like **133**)[6b, 68a]. Reactions of 1,2-quinones (**60**, **133**) with amino acids seem to be of the same category[258l, m].

It is interesting to note that **133** reacts with secondary aliphatic as well as with primary and secondary aromatic amines to yield directly the yellow substitution product **AG**, which might exist in a tautomeric form **AH** in the case of primary amines. With primary aliphatic amines possessing at least one hydrogen at the α-C atom, however, complex violet mixtures are formed, containing **AI** and other products besides **AG**[68a, 258n]. In spite of this, the reaction could be standardized to an effective procedure for determining primary aliphatic amines in the presence of secondary aliphatic amines, which is of importance for pharmaceutical preparations (e.g. noradrenaline in adrenaline)[258o].

$SO_3^-Na^+$ (**133**) $\xrightarrow{NHR^1R^2}$ (**AG**) $\underset{\text{for } R^1 = H}{\rightleftharpoons}$ (**AH**)

(**133**) $\xrightarrow{NH_2CHR^1R^2}$ (**AI**) [NH–CHR^1R^2; OH; H; N^+; SO_3^-]

For a quantitative photometric determination by means of colour reactions involving so many different pathways and reaction products, it is essential to make effective use of calibration.

5. *Other reactions*

Reactions of quinones with pyrrol[6b], 1,2,4-triazole[6b], methimazole[259a], thiamine[259b], 8-hydroxyquinoline[259c], 3-phenylthiazolidine-2,4-dione[259d] and benzenesulphinic acid[260] have been investigated. The reactions of *N*-heterocycles with quinones resemble those of amines discussed above; however, some heterocycles (pyrrol) may attack the quinone via a carbon instead of a nitrogen atom.

C. Analytical Applications

Chemical reactions of quinones are easily performed, proceed quickly and do not need special equipment. As already mentioned, they may be used for the determination of the quinones as well as of the substrates. Most investigations have been done to determine the reaction partners and not the quinones. Examples, especially in the field of pharmaceutical preparations and in biological materials, have already been mentioned in Section B and may be found additionally in the references presented there. For a review on the detection and determination of psychopharmaceuticals of the desipramine type by quinone–amine reactions see Ref. 263. Some of the applications to quinones are reported below.

Menadione (2-methyl-1,4-naphthoquinone, **129**) was detected by TLC in the presence of other 1,4-naphthoquinones (even of 2-ethyl-1,4-naphthoquinone) after reaction with 4-aminophenol[264]. The colour reaction with thiosemicarbazide[265], 8-hydroxyquinoline[259c], 3-phenylthiazoline-2,4-dione[259d] or benzenesulphinic acid[260] allows a detection of quinones (5–80 μg). The thiosemicarbazide reaction is specific in the presence of aliphatic and aromatic aldehydes, aliphatic ketones, phenols, 1,3-dihydroxy- and 1,3,5-trihydroxybenzenes. However, 1,2-dihydroxy- and 1,2,3-trihydroxybenzenes also produce colours[265]. 8-Hydroxyquinoline can also be applied to a detection of quinones on paper and TLC chromatograms[259c]. A characteristic fluorescence occurs in the reaction of 3-aminothiocarbostyril with the 1,4-quinones **1** and **73**, which allows a distinction from the corresponding 1,2-quinones **60** and **79**, not producing such fluorescence[253]. Reviews are available for the analysis of isoprenoid quinones[266] and anthraquinone drugs[267].

IX. CONCLUSIONS

This survey has reviewed several new results relevant to the physical and chemical analysis of quinones. Nevertheless, the conclusions and suggestions for a qualitative and quantitative determination given in Ref. 1, especially with respect to the distinction between 1,4- and 1,2-quinones, are still valid in the main. The emphasis has clearly shifted in favour of the physical methods. This is further demonstrated in some recent reviews on the analysis of isoprenoid quinones[266] and anthraquinones[268, 269]. However, the full potential of these methods for quinone analysis is still far from having been exhausted. Many authors were satisfied to present the results of one or two methods, although combinations were also used, e.g. (UV, IR, MS, ^{1}H-NMR)[128b], (UV, MS, ^{13}C-NMR)[33b], (LSV, IR, UV, ^{1}H-NMR, ^{13}C-NMR)[4j], to give just a few examples. UV/vis spectroscopy as a versatile method is still very popular, although its disadvantages become especially evident with condensed quinones. It is astonishing to observe that hardly any systematic investigations into the effects of substituents or other structural parameters on spectroscopic data have been undertaken in recent years. This may reflect the general tendency in chemistry to investigate applications in technology or biology rather than to do basic research.

Due to the pitfalls of every method we suggest that all methods available should be applied, if the structure of a quinone is to be determined; even a crystal structure determination may be necessary. To our knowledge, about 100 crystal structures of quinones have been solved since 1973, comprising simple systems like **60**[270], as well as complicated natural quinones. If, on the other hand, a known quinone is simply to be detected, e.g. as impurity in a preparation, a simple chemical colour test may still be the method of choice.

X. ACKNOWLEDGEMENTS

The authors are greatly indebted to Professors R. Gleiter, (Heidelberg) and F. Dallacker (Aachen) for supplying Figure 1 and unpublished material, respectively. We wish also to thank Prof. H. Pauschmann, Dr B. Speiser, and Dr G. Koppenhöfer (Tübingen) for many discussions and for reading parts of the manuscript.

XI. REFERENCES

1. S. Berger and A. Rieker, in *The Chemistry of the Quinonoid Compounds* (Ed. S. Patai), John Wiley and Sons, London, New York, Sydney, Toronto, 1974, pp. 163–229.
2. H.-J. Kallmayer, *Pharm. Unserer Zeit*, **12**, 145 (1983).
3. S. Berger and A. Rieker, *Tetrahedron*, **28**, 3123 (1972).
4. (a) K. J. Shaw, J. R. Luly and H. Rapoport, *J. Org. Chem.*, **50**, 4515 (1985); (b) K. F. West and H. W. Moore, *J. Org. Chem.*, **49**, 2809 (1984); (c) C. R. Tindale, *Aust. J. Chem.*, **37**, 611 (1984); (d) Y. Takizawa, T. Munakata, Y. Iwasa, T. Suzuki and T. Mitsuhashi, *J. Org. Chem.*, **50**, 4383 (1985); (e) T. J. Holmes, Jr, V. John, J. Vennerstrom and K. E. Choi, *J. Org. Chem.*, **49**, 4736 (1984); (f) F. Farina and T. Torres, *J. Chem. Res. (S)*, 202 (1985); (g) Y. Kanao and M. Oda, *Bull. Chem. Soc. Japan*, **57**, 615 (1984); (h) H.-J. Kallmayer and C. Tappe, *Arch. Pharm. (Weinheim, Ger.)*, **318**, 569 (1985); (i) A. Fischer and G. N. Henderson, *Synthesis*, 641 (1985); (j) J. M. Bruce, F. Heatley, R. G. Ryles and J. H. Scrivens, *J. Chem. Soc., Perkin Trans. 2*, 860 (1980); (k) J. R. Luly and H. Rapoport, *J. Org. Chem.*, **46**, 2745 (1981); (l) G. Cajipe, D. Rutolo and H. W. Moore, *Tetrahedron Lett.*, 4695 (1973); (m) S. Dähne, D. Leupold and R. Radeglia, *J. pr. Chem.*, **314**, 525 (1972); (n) L. Jurd and R. Y. Wong, *Aust. J. Chem.*, **33**, 137 (1980); (o) A. Nishinaga, T. Itahara, T. Matsuura, S. Berger, G. Henes and A. Rieker, *Chem. Ber.*, **109**, 1530 (1976); (p) W. Schäfer and A. Aguado, *Tetrahedron*, **29**, 2881 (1973); (q) J. M. Singh and A. B. Turner, *J. Chem. Soc., Perkin Trans. 1*, 2556 (1974); (r) L. Syper, J. Młochowski and K. Kloc, *Tetrahedron*, **39**, 781 (1983); (s) G. Schill, E. Logemann, B. Dietrich and H. Löwer, *Synthesis*, 695 (1979); (t) G. Schill, C. Zürcher and E. Logemann, *Chem. Ber.*, **108**, 1570 (1975); (u) T. Torres, S. V. Eswaran and W. Schäfer, *J. Heterocycl. Chem.*, **22**, 697 (1985); (v) I. Baxter and W. R. Phillips, *J. Chem. Soc., Perkin Trans. 1*, 268 (1973); (w) B. Carté, C. B. Rose and D. J. Faulkner, *J. Org. Chem.*, **50**, 2785 (1985); (x) E. R. Cole, G. Crank and H. T. Hai Minh, *Aust. J. Chem.*, **33**, 527 (1980); (y) F. Dallacker, P. Kramp and W. Coerver, *Z. Naturforsch.*, **B38**, 752 (1983); (z) K. Maruyama, H. Iwamoto, O. Soga, A. Takuwa and A. Osuka, *Chem. Lett.*, 595 (1985).
5. (a) T. R. Kelly, A. Echavarren and M. Behforouz, *J. Org. Chem.*, **48**, 3849 (1983); (b) T. Itahara, *J. Org. Chem.*, **50**, 5546 (1985); (c) L. S. Hegedus, T. A. Mulhern and A. Mori, *J. Org. Chem.*, **50**, 4282 (1985); (d) B. Wladislaw, L. Marzorati and C. Di Vitta, *Synthesis*, 464 (1983); (e) O. Reinaud, P. Capdevielle and M. Maumy, *Tetrahedron Lett.*, **26**, 3993 (1985); (f) R. Neidlein and R. Leidholdt, *Chem. Ber.*, **119**, 844 (1986); (g) R. Neidlein and U. J. Klotz, *Z. Naturforsch.*, **B40**, 429 (1985); (h) Y. Miyagi, K. Maruyama, N. Tanaka, M. Sato, T. Tomizu, Y. Isogawa and H. Kashiwano, *Bull. Chem. Soc. Japan*, **57**, 791 (1984); (i) R. H. Burnell, M. Jean, D. Poirier and S. Savard, *Can. J. Chem.*, **62**, 2822 (1984); (j) M. Chang, D. H. Netzly, L. G. Butler and D. G. Lynn, *J. Am. Chem. Soc.*, **108**, 7858 (1986); (k) F. Dallacker and G. Sanders, *Chem.-Ztg.*, **110**, 372, 405 (1986); (l) D. A. Dorsey, S. M. King and H. W. Moore, *J. Org. Chem.*, **51**, 2814 (1986); (m) K. Kloc, J. Młochowski, L. Syper and M. Mordarski, *J. Prakt. Chem.*, **328**, 419 (1986); (n) P. Rüedi, *Helv. Chim. Acta*, **69**, 972 (1986); (o) R. Weiss, N. J. Salomon, G. E. Miess and R. Roth, *Angew. Chem.*, **98**, 925 (1986); (p) P. F. Wiley, S. A. Mizsak, L. Baczynskyi, A. D. Argoudelis, D. J. Duchamp and W. Watt, *J. Org. Chem.*, **51**, 2493 (1986); (q) A. D. Joran, B. A. Leland, G. G. Geller, J. J. Hopfield and P. B. Dervan, *J. Am. Chem. Soc.*, **106**, 6090 (1984); (r) Y. Sakata, S. Nishitani, N. Nishimizu, S. Misumi, A. R. McIntosh, J. R. Bolton, Y. Kanda, A. Karen, T. Okada and N. Mataga, *Tetrahedron Lett.*, **26**, 5207 (1985); (s) J. A. Schmidt, A. Siemiarczuk, A. C. Weedon and J. R. Bolton, *J. Am. Chem. Soc.*, **107**, 6112 (1985); (t) A. C. Alder, P. Rüedi, R. Prewo, J. H. Bieri and C. H. Eugster, *Helv. Chim. Acta*, **69**, 1395 (1986); (u) H.-J. Kallmayer and C. Tappe, *Arch. Pharm. (Weinheim, Ger.)*, **319**, 421 (1986); (v) K. Hayakawa, M. Aso and K. Kanematsu, *J. Org. Chem.*, **50**, 2036 (1985).
6. (a) V. Guay and P. Brassard, *Tetrahedron*, **40**, 5039 (1984); (b) Y. Asahi, M. Tanaka and K. Shinozaki, *Chem. Pharm. Bull.*, **32**, 3093 (1984); (c) A. V. Pinto, V. F. Ferreira, M. C. F. R. Pinto and L. U. Mayer, *Synth. Commun.*, **15**, 1181 (1985); (d) K. A. Parker and M. E. Sworin, *J. Org. Chem.*, **46**, 3218 (1981); (e) B. Morgan and D. Dolphin, *Angew. Chem.*, **97**, 1000 (1985); (f) J. Weiser and H. A. Staab, *Angew. Chem.*, **96**, 602 (1984); (g) N. Benfaremo and M. P. Cava, *J. Org. Chem.*, **50**, 139 (1985); (h) G. Wurm and U. Geres, *Arch. Pharm. (Weinheim, Ger.)*, **318**, 664, 931 (1985); (i) M. Taguchi, T. Kawashima, N. Aikawa and G. Tsukamoto, *Chem. Pharm. Bull.*, **33**, 1982 (1985); (j) I. R. Green, R. G. F. Giles, M. H. Pay and D. W. Cameron, *S. Afr. J. Chem.*, **31**, 37 (1978); (k) K. Maruyama and S. Tai, *Chem. Lett.*, 681 (1985); (l) A. J. Hamdan

and H. W. Moore, *J. Org. Chem.*, **50**, 3427 (1985); (m) R. P. Shishkina, V. I. Mamatyuk and E. P. Fokin, *Izv. Akad. Nauk SSSR, Ser. Khim.*, 855 (1985); *Chem. Abstr.*, **103**, 160189 (1985); (n) R. Cassis, M. Scholz, R. Tapia and J. A. Valderrama, *Tetrahedron Lett.*, **26**, 6281 (1985); (o) V. Guay and P. Brassard, *J. Org. Chem.*, **49**, 1853 (1984); (p) S. J. Danishefsky, B. J. Uang and G. Quallich, *J. Am. Chem. Soc.*, **107**, 1285 (1985); (q) M. F. Semmelhack, L. Keller, T. Sato, E. J. Spiess and W. Wulff, *J. Org. Chem.*, **50**, 5566 (1985); (r) H. Uno, *J. Org. Chem.*, **51**, 350 (1986); (s) H.-J. Kallmayer and K. Seyfang. *Arch. Pharm. (Weinheim, Ger.)*, **317**, 743 (1984); (t) E. Ghera and Y. Ben-David, *J. Org. Chem.*, **50**, 3355 (1985); (u) R. W. Franck and R. B. Gupta, *J. Org. Chem.*, **50**, 4632 (1985); (v) K. Maruyama, T. Otsuki and S. Tai, *J. Org. Chem.*, **50**, 52 (1985); (w) M. Matsuoka, K. Hamano, T. Kitao and K. Takagi, *Synthesis*, 953 (1984); (x) D. W. Cameron, G. I. Feutrill, P. G. Griffiths, and K. R. Richards, *Aust. J. Chem.*, **35**, 1509 (1982); (y) M. F. Aldersley, F. M. Dean and B. E. Mann, *J. Chem. Soc., Perkin Trans. 1*, 2217 (1986); A. Bekaert, J. Andrieux and M. Plat, *Bull. Soc. Chim. France*, 314 (1986); W. M. Best and D. Wege, *Aust. J. Chem.*, **39**, 647 (1986); F. Dallacker, H.-J. Schleuter and P. Schneider, *Z. Naturforsch.*, **B41**, 1273 (1986); F. Farina, M. C. Paredes and V. Stefani, *Tetrahedron*, **42**, 4309 (1986); F. Farina, R. Martinez-Urilla, M. C. Paredes and V. Stefani, *Synthesis*, 781 (1985); A. T. Hudson, M. J. Pether, A. W. Randall, M. Fry, V. S. Latter and N. McHardy, *Eur. J. Med. Chem.-Chim. Ther.*, **21**, 271 (1986); H.-J. Kallmayer and C. Tappe, *Arch. Pharm. (Weinheim, Ger.)*, **319**, 791 (1986); T. Kamikawa and I. Kubo, *Synthesis*, 431 (1986); H. Laatsch, *Ann. Chem.*, 1847, 2420 (1985); 1669 (1986); *Z. Naturforsch.*, **B41**, 377 (1986); N. E. Mackenzie, S. Surendrakumar, R. H. Thomson, H. J. Cowe and P. J. Cox, *J. Chem. Soc., Perkin Trans. 1*, 2233 (1986); B. C. Maiti and O. C. Musgrave, *J. Chem. Soc., Perkin Trans. 1*, 675 (1986); K. Maruyama, S. Tai and H. Imahori, *Bull. Chem. Soc. Japan*, **59**, 1777 (1986); A. Takuwa, O. Soga, H. Iwamoto and K. Maruyama, *Bull. Chem. Soc. Japan*, **59**, 2959 (1986); A. Takuwa, H. Iwamoto, O. Soga and K. Maruyama, *J. Chem. Soc., Perkin Trans. 1*, 1627 (1986); A. Takuwa, Y. Naruta, O. Soga and K. Maruyama, *J. Org. Chem.*, **49**, 1857 (1984); M. Watanabe, S. Hisamatsu, H. Hotokezaka and S. Furukawa, *Chem. Pharm. Bull.*, **34**, 2810 (1986); G. Wurm and U. Geres, *Arch. Pharm. (Weinheim, Ger.)*, **319**, 97, (1986); G. Wurm, H.-J. Gurka and U. Geres, *Arch. Pharm. (Weinheim, Ger.)*, **319**, 1106 (1986); (z) K. Bock, N. Jacobsen and B. Terem, *J. Chem. Soc., Perkin Trans. 1*, 659 (1986); H. Laatsch, *Ann. Chem.*, 1655 (1986); T. Naito, Y. Makita, S. Yazaki and C. Kaneko, *Chem. Pharm. Bull.*, **34**, 1505 (1986).

7. (a) R. L. Blankespoor, A. N. K. Lau and L. L. Miller, *J. Org. Chem.*, **49**, 4441 (1984); (b) J. Thiem and H.-P. Wessel, *Z. Naturforsch.*, **B40**, 422 (1985); (c) A. P. Krapcho, K. J. Shaw, J. J. Landi, Jr and D. G. Phinney, *J. Org. Chem.*, **49**, 5253 (1984); (d) F. Setiabudi and P. Boldt, *Ann. Chem.*, 1272 (1985); (e) S. Arai, S. Kato and M. Hida, *Bull. Chem. Soc. Japan*, **58**, 1458 (1985); (f) J. O. Morley, *J. Appl. Chem. Biotechnol.*, **27**, 143 (1977); (g) A. Andreani, M. Rambaldi, D. Bonazzi and G. Lelli, *Arch. Pharm. (Weinheim, Ger.)*, **318**, 842 (1985); (h) I. K. Boddy, R. C. Cambie, N. F. Marsh, P. S. Rutledge and P. D. Woodgate, *Aust. J. Chem.*, **39**, 821 (1986); (i) R. C. Cambie, D. S. Larsen, C. E. F. Rickard, P. S. Rutledge and P. D. Woodgate, *Aust. J. Chem.*, **39**, 487 (1986); (j) A. Echavarren, F. Farina and P. Prados, *J. Chem. Res. (M)*, 3137, 3162 (1986); (k) F. Farina, M. T. Molina and M. C. Paredes, *Synth. Commun.*, **16**, 1015 (1986); (l) D. A. Gustowski, M. Delgado, V. J. Gatto, L. Echegoyen and G. W. Gokel, *J. Am. Chem. Soc.*, **108**, 7553 (1986); (m) K. Krohn and W. Baltus, *Synthesis*, 942 (1986); (n) H. Laatsch, *Ann. Chem.*, 839 (1986); (o) G. A. Morris, K. B. Mullah and J. K. Sutherland, *Tetrahedron*, **42**, 3303 (1986); (p) K. Ravichandran, F. A. J. Kerdeskey and M. P. Cava, *J. Org. Chem.*, **51**, 2044 (1986); (q) B. Simoneau and P. Brassard, *Tetrahedron*, **42**, 3767 (1986); (r) H. Wild and W. Steglich, *Ann. Chem.*, 1900 (1986).
8. (a) A. Vigevani, M. Ballabio, E. Gandini and S. Penco, *Magn. Reson. Chem.*, **23**, 344 (1985); (b) I. K. Khanna and L. A. Mitscher, *Tetrahedron Lett.*, **26**, 691 (1985); (c) H. Uno, Y. Naruta and K. Maruyama, *Tetrahedron* **40**, 4725 (1984); (d) J. S. Swenton, J. N. Freskos, G. W. Morrow and A. D. Sercel, *Tetrahedron*, **40**, 4625 (1984); (e) J. S. Swenton, D. K. Anderson, C. E. Coburn and A. P. Haag, *Tetrahedron*, **40**, 4633 (1984); (f) K. Krohn and E. Broser, *J. Org. Chem.*, **49**, 3766 (1984); (g) K. Krohn, K. Tolkiehn, V. Lehne, H. W. Schmalle and H.-F. Grützmacher, *Ann. Chem.*, 1311 (1985); (h) A. S. Kende and S. Johnson, *J. Org. Chem.*, **50**, 727 (1985); (i) P. N. Preston, T. Winwick, and J. O. Morley, *J. Chem. Soc., Perkin Trans 1*, 39 (1985); (j) Y. Tamura, M. Sasho, H. Ohe, S. Akai and Y. Kita, *Tetrahedron Lett.*, **26**, 1549 (1985); (k) J. G. Bauman, R. C. Hawley and H. Rapoport, *J. Org. Chem.*, **50**, 1569 (1985); (l) D. J. Brand and J. Fisher, *J. Am. Chem. Soc.*, **108**, 3088 (1986); (m) R. A. Russell, R. W. Irvine and R. W. Warrener, *J. Org. Chem.*, **51**, 1595 (1986); (n) A. A. Abdullah, J.-P. Gesson, J.-C. Jacquesy and M. Mondon, *Bull. Soc. Chim. France*, 93 (1986).

9. (a) J. Bindl, T. Burgemeister and J. Daub, *Ann. Chem.*, 1346 (1985); (b) H. Laatsch, *Ann. Chem.*, 605 (1985); (c) Y. Tamura, F. Fukata, M. Sasho, T. Tsugoshi and Y. Kita, *J. Org. Chem.*, **50**, 2273 (1985); (d) P. Cano, F. Farina, M. D. Parellada, C. Pascual and P. Prados, *J. Chem. Soc., Perkin Trans. 1*, 1923 (1986); (e) W. C. Christopfel and L. L. Miller, *J. Org. Chem.*, **51**, 4169 (1986); (f) J. G. Smith, P. W. Dibble and R. E. Sandborn, *J. Org. Chem.*, **51**, 3762 (1986); (g) A. D. Thomas and L. L. Miller, *J. Org. Chem.*, **51**, 4160 (1986); (h) R. Bunte, K.-D. Gundermann, J. Leitich, O. E. Polansky and M. Zander, *Chem.-Ztg.*, **110**, 157 (1986).
10. (a) P. R. Sleath, J. B. Noar, G. A. Eberlein and T. C. Bruice, *J. Am. Chem. Soc.*, **107**, 3328 (1985); (b) J. Rebek, Jr, S. H. Shaber, Y.-K. Shue, J.-C. Gehret and S. Zimmerman, *J. Org. Chem.*, **49**, 5164 (1984); (c) J. B. Hendrickson and J. G. de Vries, *J. Org. Chem.*, **50**, 1688 (1985); (d) N. L. Agarwal, H. Bohnstengel and W. Schäfer, *J. Heterocycl. Chem.*, **21**, 825 (1984); (e) Y. Kitahara, S. Nakahara, R. Numata and A. Kubo, *Chem. Pharm. Bull.*, **33**, 2122 (1985); (f) P. R. Weider, L. S. Hegedus, H. Asada and S. V. D'Andreq, *J. Org. Chem.*, **50**, 4276 (1985); (g) M. Bean and H. Kohn, *J. Org. Chem.*, **50**, 293 (1985); (h) M. F. Harper, J. O. Morley and P. N. Preston, *J. Chem. Res. (M)*, 3533 (1985); (i) P. Germeraad, W. Weyler, Jr and H. W. Moore, *J. Org. Chem.*, **39**, 781 (1974); (j) W. L. Lütolf, R. Prewo, J. H. Bieri, P. Rüedi and C. H. Eugster, *Helv. Chim. Acta*, **68**, 860 (1985); (k) D. L. Boger, S. R. Duff, J. S. Panek and M. Yasuda, *J. Org. Chem.*, **50**, 5782, 5790 (1985); (l) F. Farina, M. C. Paredes and V. Stefani, *Tetrahedron*, **42**, 4309 (1986); (m) C.-H. Lee, J. H. Gilchrist and E. B. Skibo, *J. Org. Chem.*, **51**, 4784 (1986); (n) K. T. Potts, D. Bhattacharjee and E. B. Walsh, *J. Org. Chem.*, **51**, 2011 (1986); (o) V. S. Romanov, A. A. Moroz and M. S. Shvartsberg, *Izv. Akad. Nauk SSSR, Ser. Khim.*, 1090 (1985); *Chem. Abstr.*, **104**, 109409 (1986); (p) T. Torres, S. V. Eswaran and W. Schäfer, *J. Heterocycl. Chem.*, **22**, 705 (1985); (q) W. Verboom, E. O. M. Orlemans, H. J. Berga, M. W. Scheltinga and D. N. Reinhoudt, *Tetrahedron*, **42**, 5053 (1986).
11. M. A. Ansari, *Indian J. Chem.*, **B24**, 972 (1985).
12. L. Czuchajowski and W. Pietrzycki, *Pr. Nauk. Uniw. Slask. Katowicach*, **205**, 111 (1978); *Chem. Abstr.*, **90**, 71305 (1979).
13. A. Ben Hadid, P. Rimmelin, J. Sommer and J. Devynck, *J. Chem. Soc., Perkin Trans. 2*, 269 (1982).
14. (a) M. C. Aversa, P. Giannetto and A. Saija, *Magn. Reson. Chem.*, **23**, 322 (1985); (b) D. W. Cameron, G. I. Feutrill and A. F. Patti, *Aust. J. Chem.*, **32**, 575 (1979).
15. (a) K. N. Ganesh and J. K. M. Sanders, *J. Chem. Soc., Chem. Commun.*, 1129 (1980); (b) K. N. Ganesh, J. K. M. Sanders and J. C. Waterton, *J. Chem. Soc., Perkin Trans. 1*, 1617 (1982).
16. (a) P. Leighton and J. K. M. Sanders, *J. Chem. Soc., Chem. Commun.*, 24 (1985); (b) K. N. Ganesh, *Proc. Indian Acad. Sci., Chem. Sci.*, **93**, 647 (1984); *Chem. Abstr.*, **101**, 87595 (1984).
17. (a) A. Osuka, H. Furuta and K. Maruyama, *Chem. Lett.*, 479 (1986); (b) J. Weiser and H. A. Staab, *Tetrahedron Lett.*, **26**, 6059 (1985); (c) W. J. Albery, P. N. Bartlett, C. C. Jones and L. R. Milgrom, *J. Chem. Res. (M)*, 3801 (1985).
18. E. G. Bagryanskaya, Yu. A. Grishin, R. Z. Sagdeev, T. V. Leshina, N. E. Polyakov and Yu. N. Molin, *Chem. Phys. Lett.*, **117**, 220 (1985).
19. H. D. Roth, M. L. M. Schilling, T. Mukai and T. Miyashi, *Tetrahedron Lett.*, **24**, 5815 (1983).
20. (a) K. Maruyama and H. Kato, *Mem. Fac. Sci., Kyoto Univ., Ser. Phys., Astrophys., Geophys. Chem.*, **36**, 463 (1985); *Chem. Abstr., 104*, 42950 (1986); (b) J. Bargon and K. G. Seifert, *Ber. Bunsenges. Phys. Chem.*, **78**, 1180 (1974); (c) P. P. Levin, A. P. Darmanyan, V. A. Kuz'min, A. Z. Yankelevich and V. M. Kuznets, *Izv. Akad. Nauk SSSR, Ser. Khim.*, 2744 (1980); *Chem. Abstr.*, **94**, 138892 (1981); (d) K. Maruyama, H. Furuta and A. Osuka, *Chem. Lett.*, 475 (1986); *Tetrahedron*, **42**, 6149 (1986); (e) K. Maruyama, H. Imahori, A. Osuka, A. Takuwa and H. Tagawa, *Chem. Lett.*, 1719 (1986).
21. (a) T. Shinmyozu, T. Sakai, E. Uno and T. Inazu, *J. Org. Chem.*, **50**, 1959 (1985); (b) W. Rebafka and H. A. Staab, *Angew. Chem.*, **85**, 831 (1973); (c) H. A. Staab, C. P. Herz and H.-E. Henke, *Tetrahedron Lett.*, 4393 (1974); (d) H. A. Staab and W. Rebafka, *Chem. Ber.*, **110**, 3333 (1977); (e) H. A. Staab, W. R. K. Reibel and C. Krieger, *Chem. Ber.*, **118**, 1230 (1985); (f) H. A. Staab, M. Jörns, C. Krieger and M. Rentzea, *Chem. Ber.*, **118**, 796 (1985); (g) M. Tashiro, K. Koya and T. Yamato, *J. Am. Chem. Soc.*, **104**, 3707 (1982); (h) H. Tatemitsu, T. Otsubo, Y. Sakata and S. Misumi, *Tetrahedron Lett.*, 3059 (1975); (i) M. Stöbbe, S. Kirchmeyer, G. Adiwidjaja and A. de Meijere, *Angew. Chem.*, **98**, 162 (1986); (j) H. Machida, H. Tatemitsu, T. Otsubo, Y. Sakata and S. Misumi, *Bull. Chem. Soc. Japan*, **53**, 2943 (1980); (k) A. Kasahara, I. Shimizu and K. Ookura, *Chem. and Ind.*, 285 (1986).
22. (a) F. Dietl, G. Gierer and A. Merz, *Synthesis*, 626 (1985); (b) K. Maruyama, H. Sohmiya and H. Tsukube, *Tetrahedron Lett.*, **26**, 3583 (1985) *J. Chem. Soc., Perkin Trans. 1*, 2069 (1986);

(c) K. Hayakawa, K. Kido and K. Kanematsu, *J. Chem. Soc., Chem. Commun.*, 268 (1986).
23. H. Iwamura and K. Makino, *J. Chem. Soc., Chem. Commun.*, 720 (1978).
24. G. Castillo, G. J. Ellames, A. G. Osborne and P. G. Sammes, *J. Chem. Res. (S)*, 45 (1978).
25. G. Höfle, *Tetrahedron*, **32**, 1431 (1976); **33**, 1963 (1977).
26. I. A. McDonald, T. J. Simpson and A. F. Sierakowski, *Aust. J. Chem.*, **30**, 1727 (1977).
27. S. Berger and A. Rieker, *Chem. Ber.*, **109**, 3252 (1976).
28. J. Dabrowski, D. Griebel and H. Zimmermann, *Spectrosc. Lett.*, **12**, 661 (1979).
29. S. Berger, *J. Magn. Reson.*, **66**, 555 (1986).
30. S. Berger, Abstract of Chemiedozententagung Würzburg, 1986.
31. H. O. Kalinowski, S. Berger and S. Braun, *^{13}C-NMR Spectroscopy*, Thieme Verlag, Stuttgart, 1984, Chapter 4.
32. (a) Y. Kitahara, S. Nakahara, R. Numata, K. Inaba and A. Kubo, *Chem. Pharm. Bull.*, **33**, 833 (1985); (b) T. Gillmann and K. Hartke, *Chem. Ber.*, **119**, 2859 (1986).
33. (a) V. P. Papageorgiou, *Planta Media*, **40**, 305 (1980); (b) F. Scherrer, H. A. Anderson and R. Azerad, *J. Chem. Soc., Chem. Commun.*, 127 (1976).
34. M. Kobayashi, Y. Terui, K. Tori and N. Tsuji, *Tetrahedron Lett.*, 619 (1976).
35. R. J. Wikholm, *J. Org. Chem.*, **50**, 382 (1985).
36. R. Radeglia and S. Dähne, *Z. Chem.*, **13**, 474 (1973).
37. Y. Berger, A. Castonguay and P. Brassard, *Org. Magn. Reson.*, **14**, 103 (1980).
38. Y. Berger and A. Castonguay, *Org. Magn. Reson.*, **11**, 375 (1978).
39. P. Joseph-Nathan, D. Abramo-Bruno and D. A. Ortega, *Org. Magn. Reson.*, **15**, 311 (1981).
40. R. Neidlein, W. Kramer and R. Leidholdt, *Helv. Chim. Acta*, **66**, 2285 (1983).
41. Y. Berger, M. Berger-Deguée and A. Castonguay, *Org. Magn. Reson.*, **15**, 303 (1981).
42. P. G. Waterman, S. Zhong, J. A. D. Jeffreys and M. Bin Zakaria, *J. Chem. Res. (S)*, 2 (1985); *(M)*, 0101 (1985).
43. M. J. Chetcuti, J. A. K. Howard, M. Pfeffer, J. L. Spencer and F. G. A. Stone, *J. Chem. Soc., Dalton Trans.*, 276 (1981).
44. Y. Kanao, M. Iyoda and M. Oda, *Tetrahedron Lett.*, **24**, 1727 (1983).
45. T. Watabe and M. Oda, *Chem. Lett.*, 1791 (1984).
46. R. Wingen and F. Vögtle, *Chem. Ber.*, **113**, 676 (1980).
47. J. R. Scheffer, Y.-F. Wong, A. O. Patil, D. Y. Curtin and I. C. Paul, *J. Am. Chem. Soc.*, **107**, 4898 (1985).
48. S. Chandrasekaran, W. D. Wilson and D. W. Boykin, *Org. Magn. Reson.*, **22**, 757 (1984).
49. (a) Y. Sato, R. Kohnert and S. J. Gould, *Tetrahedron Lett.*, **27**, 143 (1986); (b) S. Omura, A. Nakagawa, K. Kushida and G. Lukacs, *J. Am. Chem. Soc.*, **108**, 6088 (1986).
50. D. W. Cameron, P. J. Chalmers and G. I. Feutrill, *Tetrahedron Lett.*, **25**, 6031 (1984).
51. (a) M. L. Casey, R. C. Paulick and H. W. Whitlock, Jr, *J. Am. Chem. Soc.*, **98**, 2636 (1976); (b) Y. Sato and S. J. Gould, *J. Am. Chem. Soc.*, **108**, 4625 (1986).
52. (a) T. R. Kelly, J. K. Saha and R. R. Whittle, *J. Org. Chem.*, **50**, 3679 (1985); (b) F. Farina, M. T. Molina and M. C. Paredes, *Tetrahedron Lett.*, **26**, 111 (1985).
53. (a) A. Arnone, L. Camarda, G. Nasini and G. Assante, *J. Chem. Soc., Perkin Trans. 1*, 255 (1986); (b) H. Jancke, R. Radeglia, D. Tresselt and H. Berg, *Z. Chem.*, **17**, 105 (1977).
54. K. B. Chernysheva, N. G. Radchenko, E. E. Nikitin and I. A. Polovnikova, *Khim. Tverd. Topl. (Moscow)*, 55 (1983); *Chem. Abstr., 98*, 110139 (1983).
55. R. H. Thomson, in *Chem. Biochem. Plant Pigm.* (Ed. T. W. Goodwin), 2nd edn, Vol. 2, Academic Press, London, 1976, pp. 207–232.
56. B. R. Clark and D. H. Evans, *J. Electroanal. Chem.*, **69**, 181 (1976).
57. (a) M. Kirszenbaum, *Report*, CEA-BIB-219, 23pp. (1976); *Chem. Abstr.*, **86**, 42560 (1977); (b) I. M. Roushdi, A. A. Mikhail and I. C. Ahmed, *Acta Pharm. Jugosl.*, **26**, 287 (1976); *Chem. Abstr.*, **87**, 22826 (1977); (c) E. Müller and W. Dilger, *Chem. Ber.*, **107**, 3946 (1974).
58. (a) A. Girlando, D. Ragazzon and C. Pecile, *Spectrochim. Acta*, **A36**, 1053 (1980); (b) B. Lunelli and C. Pecile, *Spectrochim. Acta*, **A29**, 1989 (1973); (c) T. M. Dunn and A. H. Francis, *J. Mol. Spectrosc.*, **50**, 1 (1974); (d) D. M. Chipman and M. F. Prebenda, *J. Phys. Chem.*, **90**, 5557 (1986).
59. (a) N. S. Strokach and D. N. Shigorin, *Zh. Fiz. Khim.*, **51**, 2722, 3036 (1977); (b) T. Murao and T. Azumi, *J. Chem. Phys.*, **70**, 4460 (1979); (c) K. K. Lehmann, J. Smolarek, O. S. Khalil and L. Goodman, *J. Phys. Chem.*, **83**, 1200 (1979).
60. N. S. Strokach, E. A. Gastilovich and D. N. Shigorin, *Opt. Spektrosk.*, **30**, 43, 433 (1971).
61. (a) A. Girlando and C. Pecile, *Spectrochim. Acta*, **A34**, 453 (1978); (b) A. Girlando and C. Pecile, *Spectrochim. Acta*, **A31**, 1187 (1975).

62. B. P. Singh and R. S. Singh, *Indian J. Phys.*, **B51**, 99 (1977); *Chem. Abstr.*, **88**, 165755 (1978).
63. (a) J.-C. Blazejewski, R. Dorme and C. Wakselman, *Synthesis*, 1120 (1985); (b) J. Einhorn, S. Halut-Desportes, P. Demerseman and R. Royer, *J. Chem. Res. (S)*, 98 (1983); (c) B. Errazuriz, R. Tapia and J. A. Valderrama, *Tetrahedron Lett.*, **26**, 819 (1985); (d) A. Gaudemer, K. Nguyen-Van-Duong, N. Shahkarami, S. S. Achi, M. Frostin-Rio and D. Pujol, *Tetrahedron*, **41**, 4095 (1985); (e) K. Hayakawa, M. Aso and K. Kanematsu, *J. Org. Chem.*, **50**, 2036 (1985); (f) N. Jacobsen and K. Torssell, *Acta. Chem. Scand.*, **27**, 3211 (1973); (g) A. Kubo, Y. Kitahara, S. Nakahara and R. Numata, *Chem. Pharm. Bull.*, **31**, 341 (1983); (h) Y. Miyahara, T. Inazu and T. Yoshino, *Tetrahedron Lett.*, **23**, 2189 (1982); (i) C. Nallaiah, *Tetrahedron*, **40**, 4897 (1984); (j) A. B. Padias and H. K. Hall, Jr, *J. Org. Chem.*, **50**, 5417 (1985); (k) Y. Tsuji, T. Ohta, T. Ido, H. Minbu and Y. Watanabe, *J. Organomet. Chem.*, **270**, 333 (1984); (l) R. Victor, R. Ben-Shoshan and S. Sarel, *Tetrahedron Lett.*, 4211 (1973); (m) M. Matsumoto, H. Kobayashi and Y. Hotta, *J. Org. Chem.*, **50**, 1766 (1985); (n) J. O. Karlsson, N. V. Nguyen, L. D. Foland and H. W. Moore, *J. Am. Chem. Soc.*, **107**, 3392 (1985).
64. M. L. Meyerson, *Spectrochim. Acta*, **A41**, 1263 (1985).
65. G. J. Gleicher, D. F. Church and J. C. Arnold, *J. Am. Chem. Soc.*, **96**, 2403 (1974).
66. H. Laatsch, *Ann. Chem.*, 1367 (1984).
67. B. Achari, S. Bandyopadhyay, K. Basu and S. C. Pakrashi, *Tetrahedron*, **41**, 107 (1985).
68. (a) K. Hartke and U. Lohmann, *Chem. Lett.*, 693 (1983); (b) M. F. Aldersley, T. F. Dean and A. S. Hamzah, *Tetrahedron Lett.*, **27**, 255 (1986); (c) A. N. Grinev and I. K. Sorokina, *Zhur. Org. Khim.*, **18**, 2363 (1982).
69. A. Kubo, S. Nakahara, R. Iwata, K. Takahashi and T. Arai, *Tetrahedron Lett.*, **21**, 3207 (1980).
70. A. T. Hudson and M. J. Pether, *J. Chem. Soc., Perkin Trans. 1*, 35 (1983).
71. D. W. Jones and A. Pomfret, *J. Chem. Soc., Chem. Commun.*, 703 (1983).
72. D. A. Koshman, O. M. Bukachuk and M. I. Shevchuk, *Zh. Obshch. Khim.*, **55(177)**, 1738 (1985).
73. H. Inoue, T. Togano, K. Ikeda, H. Mihara and M. Hida, *Nippon Kagaku Kaishi*, 2023 (1985).
74. K. Krohn and W. Priyono, *Angew. Chem.*, **98**, 338 (1986).
75. C. Lenzen and L. Delcambe, *Inf. Bull., Int. Cent. Inf. Antibiot.*, **12**, 178pp (1975); *Chem. Abstr.*, **84**, 95583 (1976).
76. Y. M. Sheikh, N. Ekwuribe, B. Dhawan and D. T. Witiak, *J. Org. Chem.*, **47**, 4341 (1982).
77. J. Daub, L. Jakob, J. Salbeck and Y. Okamoto, *Chimia*, **13**, 393 (1985).
78. M. L. Casner, W. A. Remers and W. T. Bradner, *J. Med. Chem.*, **28**, 921 (1985).
79. R. M. Issa, A. A. El-Samahy, I. M. Issa and S. H. Etaiw, *Acta Chim. Acad. Sci. Hung.*, **92**, 211 (1977); *Chem. Abstr.*, **87**, 183621 (1977).
80. I. M. Issa, A. A. El-Samahy and S. H. Etaiw, *Egypt. J. Chem.*, **19**, 441 (1976), publ. 1978; *Chem. Abstr.*, **91**, 131471 (1979).
81. V. A. Medvedev and V. D. Davydov, *Pochvovedenie*, 133 (1974); *Chem. Abstr.*, **80**, 94580 (1974).
82. S. M. Beck and L. E. Brus, *J. Am. Chem. Soc.*, **104**, 4789 (1982).
83. J. E. Bell and C. Hall, in *Spectrosc. Biochem.* (Ed. J. E. Bell), *UV and Visible Absorbance Spectroscopy*, Vol 1, CRC Press, Boca Raton, Florida, 1981.
84. R. Baker, W. F. Beech, E. Hemingway and T. J. Smith, *Rep. Progr. Appl. Chem.*, **53**, 141 (1968).
85. (a) V. A. Koptyug (Ed.), *Atlas of Spectra of Aromatic and Heterocyclic Compounds, No. 10: Absorption Spectra of 1,4-Naphthoquinone Derivatives in the Infrared, Ultraviolet, and Visible Regions*, Novosibirskii Institut Organicheskoi Khimii, Novosibirsk (USSR), 1976; (b) V. A. Koptyug (Ed.), *Atlas of Spectra of Aromatic and Heterocyclic Compounds, No. 13: Absorption Spectra of 9,10-Anthraquinone Derivatives in the Infrared, Ultraviolet, and Visible Regions*, Novosibirskii Institut Organicheskoi Khimii, Novosibirsk (USSR), 1977; (c) V. A. Koptyug (Ed.), *Atlas of Spectra of Aromatic and Heterocyclic Compounds, No. 19: Absorption Spectra of Heterocyclic Derivatives of 9,10-Anthraquinone and 1,4-Naphthoquinone in Infrared, Ultraviolet, and Visible Regions*, Novosibirskii Institut Organicheskoi Khimii, Novosibirsk (USSR), 1980.
86. (a) M. Nepraš, M. Titz, D. Šnobl and V. Kratochvíl, *Collect. Czech. Chem. Commun.*, **38**, 2397 (1973); (b) A. Novák, M. Titz and M. Nepraš, *Collect. Czech. Chem. Commun.*, **39**, 1532 (1974); (c) A. Novák, M. Nepraš and M. Titz, *Collect. Czech. Chem. Commun.*, **41**, 2669 (1976); (d) J. Fabian and A. Mehlhorn, *Z. Chem.*, **18**, 338 (1978); (e) M. Nepraš and M. Titz, *Int. J. Quant. Chem.*, **16**, 543 (1979); (f) J. Fabian and M. Nepraš, *Collect. Czech. Chem. Commun.*, **45**, 2605 (1980); (g) J. Fabian, *Z. Chem.*, **20**, 299 (1980); (h) M. Nepraš, J. Fabian and M. Titz, *Collect. Czech. Chem. Commun.*, **46**, 20 (1981); (i) M. Nepraš, J. Fabian, M. Titz and B. Gaš, *Collect.*

Czech. Chem. Commun., **47**, 2569 (1982); (j) M. Nepraš, B. Gaš, J. Fabian and M. Titz, *Collect. Czech. Chem. Commun.*, **47**, 2583 (1982); (k) M. Nepraš, M. Titz, B. Gaš and J. Fabian, *Collect. Czech. Chem. Commun.*, **47**, 2594 (1982); (l) M. Nepraš, M. Titz, J. Fabian and B. Gaš, *Collect. Czech. Chem. Commun.*, **47**, 2604 (1982); (m) B. Gaš, V. Štěpán, M. Titz and V. Kratochvíl, *Collect. Czech. Chem. Commun.*, **48**, 538 (1983); (n) M. Nepraš, V. Kratochvíl, V. Štěpán, M. Titz and B. Gaš, *Collect. Czech. Chem. Commun.*, **48**, 976 (1983).

87. (a) A. Kuboyama, S. Matsuzaki, H. Takagi and H. Arano, *Bull. Chem. Soc. Japan*, **47**, 1604 (1974); (b) A. Kuboyama and H. Arano, *Bull. Chem. Soc. Japan*, **49**, 1401 (1976); (c) Y. Jinnouchi, M. Kohno and A. Kuboyama, *Bull. Chem. Soc. Japan*, **57**, 1147 (1984); (d) A. Kuboyama, F. Kobayashi and S. Morokuma, *Bull. Chem. Soc. Japan*, **48**, 2145 (1975); (e) A. Kuboyama and M. Anze, *Nippon Kagaku Kaishi*, 229 (1972); *Chem. Abstr.*, **76**, 105994 (1972).
88. (a) T. M. Dunn and A. H. Francis, *J. Mol. Spectrosc.*, **50**, 14 (1974); (b) V. Galasso, *Gazz. Chim. Ital.*, **106**, 571 (1976); (c) H. Hoffmann and G. Rasch, *Z. phys. Chem. (Leipzig)*, **257**, 689 (1976); (d) P. Jacques, J. Faure, O. Chalvet and H. H. Jaffé, *J. Phys. Chem.*, **85**, 473 (1981); (e) M. F. Merienne-Lafore and H. P. Trommsdorff, *J. Chem. Phys.*, **64**, 3791 (1976); (f) P. C. Mishra and A. K. Pandey, *Spectrosc. Lett.*, **8**, 953 (1975); (g) P. C. Mishra, *Indian J. Pure Appl. Phys.*, **16**, 787 (1978); (h) M. Remko and J. Polčin, *Monatsh. Chem.*, **108**, 1313 (1977); (i) S. G. Semenov, V. V. Redchenko, J. Freimanis and V. S. Bannikov, *Zh. Obsch. Khim.*, **54**, 2330 (1984); *Chem. Abstr.*, **102**, 61541 (1985); (j) D. N. Shigorin and N. A. Shcheglova, *Zh. Fiz. Khim.*, **52**, 1126 (1978); (k) U. C. Singh, M. S. Ramachandran, N. R. Subbaratnam and V. K. Kelkar, *Spectrochim. Acta*, **A35**, 663 (1979); (l) M. H. Wood, *Theoret. Chim. Acta (Berl.)*, **36**, 345 (1975); (m) D. Pitea and G. Moro, *J. Mol. Struct., THEOCHEM.*, **105**, 55 (1983); (n) R. D. Tewari and P. C. Mishra, *Indian J. Pure Appl. Phys.*, **22**, 491 (1984); *Chem. Abstr.*, **101**, 210130 (1984); (o) P. P. Levin, V. B. Luzhkov and A. P. Darmanyan, *Izv. Akad. Nauk SSSR, Ser. Khim.*, 1542 (1980); *Chem. Abstr.*, **93**, 220023 (1980).
89. (a) J. Aihara, G. Kushibiki and Y. Matsunaga, *Bull. Chem. Soc. Japan*, **46**, 3584 (1973); (b) J. Aihara, *Bull. Chem. Soc. Japan*, **47**, 2063 (1974); (c) K. Chiba, J. Aihara, K. Araya and Y. Matsunaga, *Bull. Chem. Soc. Japan*, **53**, 1703 (1980).
90. (a) G. G. Aloisi, S. Santini and S. Sorriso, *J. Chem. Soc., Faraday Trans. 1*, **70**, 1908 (1974); (b) A. V. Bratchikov, D. M. Kizhner, G. L. Ryzhova and B. F. Minaev, *Zh. Fiz. Khim.*, **50**, 563 (1976); (c) M. J. M. Campbell and B. Demetriou, *J. Chem. Soc., Perkin Trans. 2*, 917 (1983); (d) A. K. Chattopadhyay and S. C. Lahiri, *J. Indian Chem. Soc.*, **57**, 705 (1980); *Chem. Abstr.*, **93**, 174679 (1980); (e) A. K. Das and R. Basu, *Indian J. Chem.*, **A22**, 845 (1983); (f) A. M. N. El-Din, *Indian J. Chem.*, **A24**, 511 (1985); (g) M. F. Froix (Xerox Corp.), US-Patent 3975289, 17.8.1976; *Chem. Abstr.*, **85**, 114570 (1976); (h) Z. Gasyna, W. R. Browett and M. J. Stillman, *Inorg. Chem.*, **24**, 2440 (1985); (i) A. S. Ghosh and S. B. Saha, *Indian J. Chem.*, **A15**, 378 (1977); (j) Y. M. Issa, A. E. El-Kholy and A. L. El-Ansary, *Acta Chim. Hung.*, **118**, 43 (1985); *Chem. Abstr.*, **103**, 141369 (1985); (k) D. M. Kizhner, B. F. Minaev, S. A. Matasova and G. L. Ryzhova, *Zh. Fiz. Khim.*, **49**, 1369 (1975); (l) G. G. Lazarev, Ya. S. Lebedev and M. V. Serdobov, *Izv. Akad. Nauk SSSR, Ser. Khim.*, 2520 (1978); *Chem. Abstr.*, **90**, 102904 (1979); (m) N. Mataga, A. Karen, T. Okada, S. Nishitani, Y. Sakata and S. Misumi, *J. Phys. Chem.*, **88**, 4650 (1984); (n) S. Matsuzaki, T. Moriyama, M. Onomichi and K. Toyoda, *Bull. Chem. Soc. Japan*, **56**, 369 (1983); (o) G. Saito, H. Kumagai, J. Tanaka, T. Enoki and H. Inokuchi, *Mol. Cryst. Liq. Cryst.*, **120**, 337 (1985); (p) R. S. Snart, *Bioelectrochem. Bioenerg.*, **2**, 154 (1975); (q) G. Tosi, P. Bruni and L. Cardellini, *Gazz. Chim. Ital.*, **114**, 125 (1984); (r) T. Nyokong, Z. Gasyna and M. J. Stillman, *Inorg. Chim. Acta*, **112**, 11 (1986); (s) S. Fukuzumi, N. Nishizawa and T. Tanaka, *J. Org. Chem.*, **49**, 3571 (1984).
91. (a) P. C. Dwivedi and A. K. Banga, *Indian J. Chem.*, **A19**, 908 (1980); (b) P. C. Dwivedi, A. Gupta and A. K. Banga, *Curr. Sci.*, **51**, 651 (1982); (c) P. C. Dwivedi and R. Agarwal, *Indian J. Chem.*, **A24**, 1015 (1985).
92. (a) A. F. E. Mourad and A. M. Nour-el-Din, *Gazz. Chim. Ital.*, **113**, 213 (1983); (b) A. F. E. Mourad, *Spectrochim. Acta*, **A39**, 933 (1983); (c) A. F. E. Mourad and E. S. H. Eltamany, *Gazz. Chim. Ital.*, **115**, 97 (1985); (d) A. M. Nour-el-Din and A. F. E. Mourad, *Monatsh. Chem.*, **114**, 211 (1983).
93. (a) U. Muralikrishna and M. Krishnamurthy, *Indian J. Chem.*, **A21**, 1018 (1982); (b) U. Muralikrishna and M. Krishnamurthy, *Spectrochim. Acta*, **A40**, 65 (1984).
94. A. Malmanis, J. Eiduss, J. Freimanis and J. Dregeris, *Spectrochim. Acta*, **A35**, 1229 (1979).
95. V. Kampars and O. Neilands, *Zh. Obshch. Khim.*, **49**, 2558 (1979).
96. (a) E. Lipczynska-Kochany and H. Iwamura, *Chem. Lett.*, 1075 (1982); (b) K. Yamamura, K. Nakasuji, I. Murata and S. Inagaki, *J. Chem. Soc., Chem. Commun.*, 396 (1982).

97. (a) T. Shinmyozu, T. Inazu and T. Yoshino, *Chem. Lett.*, 1347 (1977); (b) T. Shinmyozu, T. Inazu and T. Yoshino, *Chem. Lett.*, 1319 (1978).
98. (a) R. Reimann and H. A. Staab, *Angew. Chem.*, **90**, 385 (1978); (b) H. Vogler, *Tetrahedron Lett.*, **24**, 2159 (1983); (c) H. Vogler, L. Schanne and H. A. Staab, *Chem. Ber.*, **118**, 1254 (1985).
99. (a) A. C. Chan, J. Dalton and L. R. Milgrom, *J. Chem. Soc., Perkin Trans. 2*, 707 (1982); (b) J. S. Connolly, K. L. Marsh, D. R. Cook, J. R. Bolton, A. R. McIntosh, A. Siemiarczuk, A. C. Weedon and T.-F. Ho, *Sci. Inst. Pap. Phys. Chem. Res. (Jpn)*, **78**, 118 (1984); *Chem. Abstr.*, **103**, 186735 (1985); (c) T.-F. Ho, A. R. McIntosh and A. C. Weedon, *Can. J. Chem.*, **62**, 967 (1984).
100. N. Mataga, A. Karen, T. Okada, Y. Sakata and S. Misumi, *Springer Ser. Chem. Phys.*, **38** (Ultrafast Phenom. IV), 365 (1984).
101. (a) D. O. Cowan, R. Gleiter, J. A. Hashmall, E. Heilbronner and V. Hornung, *Angew. Chem., Int. Ed. Engl.*, **10**, 401 (1971); (b) T. Kobayashi, *J. Electron Spectrosc. Relat. Phenom.*, **7**, 349 (1975); (c) D. Dougherty and S. P. McGlynn, *J. Am. Chem. Soc.*, **99**, 3234 (1977); (d) P. Schang, R. Gleiter and A. Rieker, *Ber. Bunsenges. Phys. Chem.*, **82**, 629 (1978); (e) V. Eck, G. Lauer, A. Schweig, W. Thiel and H. Vermeer, *Z. Naturforsch.*, **A33**, 383 (1978); (f) R. Gleiter, W. Dobler and M. Eckert-Maksić, *Angew. Chem., Int. Ed. Engl.*, **21**, 76 (1982); *Nouv. J. Chim.*, **6**, 123 (1982); (g) L. Åsbrink, G. Bieri, C. Fridh, E. Lindholm and D. P. Chong, *Chem. Phys.*, **43**, 189 (1979); (h) G. Lauer, W. Schäfer and A. Schweig, *Chem. Phys. Lett.*, **33**, 312 (1975); (i) C. R. Brundle, M. B. Robin and N. A. Kuebler, *J. Am. Chem. Soc.*, **94**, 1466 (1972); (j) J. E. Bloor, R. A. Paysen and R. E. Sherrod, *Chem. Phys. Lett.*, **60**, 476 (1979); (k) R. W. Bigelow, *J. Chem. Phys.*, **68**, 5086 (1978); (l) R. Gleiter, G. Jähne, M. Oda and M. Iyoda, *J. Org. Chem.*, **50**, 678 (1985); (m) N. Sato, K. Seki and H. Inokuchi, *J. Chem. Soc., Faraday Trans. 2*, **77**, 47 (1981); (n) V. V. Redchenko, J. Freimanis and J. Dregeris, *Zh. Obshch. Khim.*, **50**, 1847 (1980); (o) M. Guerra, D. Jones, G. Distefano and A. Modelli, *Chem. Phys.*, **85**, 389 (1984); (p) Yu. A. Zhdanov, L. N. Mazalov, A. T. Shuvaev, B. Yu. Khel'mer, P. I. Vadash and O. E. Shelepin, *Dokl. Akad. Nauk SSSR*, **232**, 85[Chem.], (1977); *Chem. Abstr.*, **86**, 154759 (1977); (q) A. Modelli and P. D. Burrow, *J. Phys. Chem.*, **88**, 3550 (1984).
102. (a) J. L. Burguera and M. Burguera, *Talanta*, **31**, 1027 (1984); (b) R. H. Dekker, J. A. Duine, J. Frank, Jr, P. E. J. Verwiel and J. Westerling, *Eur. J. Biochem.*, **125**, 69 (1982); (c) S. R. Flom and P. F. Barbara, *J. Phys. Chem.*, **89**, 4489 (1985); (d) J. C. Mialocq, D. Doizi and M. P. Gingold, *Chem. Phys. Lett.*, **103**, 225 (1983); (e) R. N. Nurmukhametov, P. D. Mil'khiker, G. T. Khachaturova and Z. M. Baskakova, *Zh. Prikl. Spektrosk.*, **43**, 54 (1985); *Chem. Abstr.*, **103**, 112614 (1985); (f) V. V. Nekrasov, R. N. Nurmukhametov and N. S. Strokach, *Dokl. Akad. Nauk SSSR*, **276**, 412 (1984); *Chem. Abstr.*, **101**, 170535 (1984); (g) C. Krieger, J. Weiser and H. A. Staab, *Tetrahedron Lett.*, **26**, 6055 (1985); (h) S. Nishitani, N. Kurata, Y. Sakata and S. Misumi, *Tetrahedron Lett.*, **22**, 2099 (1981).
103. (a) J. S. Miller, P. J. Krusic, D. A. Dixon, W. M. Reiff, J. H. Zhang, E. C. Anderson and A. J. Epstein, *J. Am. Chem. Soc.*, **108**, 4459 (1986); (b) E. Adler and G. Andersson, *Ann. Chem.*, 1435 (1976).
104. F. M. Abd El-Halim, R. M. Issa and A. A. Harfoush, *Egypt. J. Chem.*, **19**, 203 (1976), publ. 1978; *Chem. Abstr.*, **91**, 201449 (1979).
105. (a) L. V. Volod'ko, A. I. Komyak, A. A. Min'ko, B. A. Tatarinov and P. A. Matusevich, *Zh. Prikl. Spektrosk.*, **24**, 1009 (1976); *Chem. Abstr.*, **85**, 142171 (1976); (b) L. V. Volod'ko, A. A. Min'ko, A. I. Komyak, P. A. Matusevich and B. A. Tatarinov, *Zh. Prikl. Spektrosk.*, **26**, 691 (1977); *Chem. Abstr.*, **87**, 52543 (1977).
106. (a) M. S. Ramachandran, U. C. Singh, N. R. Subbaratnam and V. K. Kelkar, *Proc. Indian Acad. Sci.*, **A88** (Pt. 1, No. 3), 155 (1979); *Chem. Abstr.*, **91**, 81213 (1979); (b) P. D. Mize, P. W. Jeffs and K. Boekelheide, *J. Org. Chem.*, **45**, 3543 (1980).
107. (a) A. M. Hammam and S. A. Ibrahim, *J. Indian Chem. Soc.*, **58**, 178 (1981); (b) J. Kalamar, E. Steiner, E. Charollais and T. Posternak, *Helv. Chim. Acta*, **57**, 2368 (1974).
108. F. Dallacker and E.-M. Both-Pollmann, unpublished results; E.-M. Both-Pollmann, Thesis, Aachen 1985.
109. (a) T. G. Edwards, *Theoret. Chim. Acta (Berl.)*, **30**, 267 (1973); (b) G. Olbrich, O. E. Polansky and M. Zander, *Ber. Bunsenges. Phys. Chem.*, **81**, 692 (1977).
110. T. S. Eckert and T. C. Bruice, *J. Am. Chem. Soc.*, **105**, 4431 (1983).
111. M. L. Jain and R. P. Soni, *J. Prakt. Chem.*, **325**, 353 (1983).
112. A. N. Grinev and I. K. Arsenichev, *Zh. Org. Khim.*, **21**, 1315 (1985).
113. J. P. Dix and F. Vögtle, *Chem. Ber.*, **114**, 638 (1981).
114. K.-Y. Chu and J. Griffiths, *J. Chem. Soc., Perkin Trans. 1*, 1083 (1978).

115. L. M. Gornostaev and V. A. Levdanskii, *Zh. Org. Khim.*, **20**, 2452 (1984).
116. S. Knapp and S. Sharma, *J. Org. Chem.*, **50**, 4996 (1985).
117. (a) M. Matsuoka, S. H. Kim and T. Kitao, *J. Chem. Soc., Chem. Commun.*, 1195 (1985); (b) M. Matsuoka, S. H. Kim, Y. Kubo and T. Kitao, *J. Soc. Dyers Colour.*, **102**, 232 (1986); (c) K. Takagi, M. Kawabe, M. Matsuoka and T. Kitao, *Dyes and Pigments*, **6**, 177 (1985); (d) S. H. Kim, M. Matsuoka and T. Kitao, *Chem. Lett.*, 1351 (1985); (e) S. H. Kim, M. Matsuoka, Y. Kubo, T. Yodoshi and T. Kitao, *Dyes and Pigments*, **7**, 93 (1986).
118. (a) A. S. Hammam, *Egypt. J. Chem.*, **15**, 391 (1972); (b) A. S. Hammam, *J. Appl. Chem. Biotechnol.*, **26**, 667 (1976).
119. (a) R. M. Issa, A. S. Hammam, I. I. Ezzat and R. Abd El-Hamid, *U. A. R. J. Chem.*, **14**, 425 (1971); *Chem. Abstr.*, **79**, 136051 (1973); (b) R. M. Issa, F. M. Abdel-Halim and A. A. Harfoush, *Egypt. J. Chem.*, **19**, 185 (1976), publ. 1978; *Chem. Abstr.*, **91**, 148734 (1979).
120. L. I. Kosheleva, Yu. S. Tsizin and N. B. Karpova, *Khim. Geterotsikl. Soedin*, 1559 (1974); *Chem. Abstr.*, **82**, 72297 (1975).
121. H. Nakazumi, K. Kondo and T. Kitao, *Bull. Chem. Soc. Japan*, **54**, 937 (1981).
122. R. P. Soni and J. P. Saxena, *J. Indian Chem. Soc.*, **58**, 885 (1981).
123. I. M. Issa, R. M. Issa, K. A. Idriss and A. M. Hammam, *Indian J. Chem.*, **B14**, 117 (1976).
124. Y. Kogo, H. Kikuchi, M. Matsuoka and T. Kitao, *J. Soc. Dyers Colour.*, **96**, 475 (1980).
125. E. K. Fukuda and R. T. McIver, Jr, *J. Am. Chem. Soc.*, **107**, 2291 (1985).
126. E. P. Grimsrud, G. Caldwell, S. Chowdhury and P. Kebarle, *J. Am. Chem. Soc.*, **107**, 4627 (1985).
127. R. Foster and M. I. Foreman, in *The Chemistry of Quinonoid Compounds* (Ed. S. Patai), John Wiley and Sons, London, New York, Sydney, Toronto, 1974, pp. 257–333.
128. (a) M. D. Collins, *Soc. Appl. Bacteriol. Tech. Ser.*, **20** (Chem. Methods Bact. Syst.), 267 (1985); *Chem. Abstr.*, **103**, 84504 (1985); (b) K. Aoki, Y. Yamada and Y. Tahara, *Agric. Biol. Chem.*, **44**, 1693 (1980).
129. G. V. Ratovskii, L. I. Belykh, S. S. Timofeeva, A. M. Panov and D. I. Stom, *Zh. Anal. Khim.*, **32**, 133 (1977); *Chem. Abstr.*, **86**, 199429 (1977).
130. K. Miglierini and V. Kello, *Ropa Uhlie*, **23**, 359 (1981); *Chem. Abstr.*, **95**, 149598 (1981).
131. M. Vincenzini, R. Materassi, A. Ena and G. Florenzano, *Microbiologica (Bologna)*, **7**, 41 (1984).
132. R. R. Reisbig, A. Y. M. Woody and R. W. Woody, *Biochemistry*, **21**, 196 (1982).
133. J. Pawliszyn, M. F. Weber, M. J. Dignam, A. Mandelis, R. D. Venter and S. M. Park, *Anal. Chem.*, **58**, 239 (1986).
134. J. J. M. Holthuis, W. J. van Oort, F. M. G. M. Römkens, J. Renema and P. Zuman, *J. Electroanal. Chem.*, **184**, 317 (1985).
135. C. S. P. Sastry and B. G. Rao, *Natl. Acad. Sci. Lett. (India)*, **6**, 127 (1983); *Chem. Abstr.*, **100**, 79225 (1984).
136. (a) U. Muralikrishna, M. Krishnamurthy and Y. V. S. K. Seshasayi, *Indian J. Chem.*, **A22**, 904 (1983); (b) U. Muralikrishna, M. Krishnamurthy and N. S. Rao, *Analyst (London)*, **109**, 1277 (1984).
137. Minh-Chau-Pham, A. Hachemi and M. Delamar, *J. Electroanal. Chem.*, **184**, 197 (1985).
138. (a) P. Hobza, P. Čársky and R. Zahradník, *Collect. Czech. Chem. Commun.*, **38**, 641 (1973); (b) R. Mitzner, H. Dorst and D. Frosch, *Z. Chem.*, **15**, 400 (1975); (c) E. McAlpine, R. S. Sinclair, T. G. Truscott and E. J. Land, *J. Chem. Soc., Faraday Trans. 1*, **74**, 597 (1978); (d) G. A. Abakumov, V. K. Cherkasov, V. A. Muraev and S. A. Chesnokov, *Izv. Akad. Nauk SSSR, Ser. Khim.*, 2785 (1980); *Chem. Abstr.*, **94**, 138931 (1981).
139. H. Bock, P. Haenel, W. Kaim and U. Lechner-Knoblauch, *Tetrahedron Lett.*, **26**, 5115 (1985).
140. (a) H. Murai, M. Minami, T. Hayashi and Y. J. I'Haya, *Chem. Phys.*, **93**, 333 (1985); (b) J. E. Pedersen, C. E. M. Hansen, H. Parbo and L. T. Muus, *J. Chem. Phys.*, **63**, 2398 (1975).
141. (a) M. C. Depew, L. Zhongli and J. K. S. Wan, *J. Am. Chem. Soc.*, **105**, 2480 (1983); (b) K. S. Chen, J. K. S. Wan and J. K. Kochi, *J. Phys. Chem.*, **85**, 1726 (1981); (c) B. B. Adeleke and J. K. S. Wan, *J. Chem. Soc., Faraday Trans. 1*, **72**, 1799 (1976).
142. J. A. Pedersen, *Handbook of EPR spectra from Quinones and Quinols*, CRC Press, Boca Raton, Fla., 1985, 382 pp.
143. R. Gompper, R. Binder and H.-U. Wagner, *Tetrahedron Lett.*, **27**, 691 (1986).
144. W. T. Dixon, P. M. Hoyle and D. Murphy, *J. Chem. Soc., Faraday Trans. 2*, **74**, 2027 (1978).
145. D. M. Holton and D. Murphy, *J. Chem. Soc., Perkin Trans 2*, 1757 (1980).
146. S. Konishi, S. Niizuma and H. Kokubun, *Chem. Phys. Lett.*, **71**, 164 (1980).
147. C. Corvaja, L. Pasimeni and M. Brustolon, *Chem. Phys.*, **14**, 177 (1976).

148. D. M. Holton and D. Murphy, *J. Chem. Soc., Faraday Trans. 1*, **78**, 1223 (1982).
149. C. C. Felix and B. S. Prabhananda, *J. Chem. Phys.*, **80**, 3078 (1984).
150. J. D. R. Clay and D. Murphy, *J. Chem. Soc., Faraday Trans. 2*, **77**, 1589 (1981).
151. P. Ashworth and W. T. Dixon, *J. Chem. Soc., Perkin Trans. 2*, 739 (1974).
152. G. Krishnamoorthy and B. S. Prabhananda, *J. Magn. Reson.*, **24**, 215 (1976).
153. G. R. Stevenson, A. E. Algeria and A. McB. Block, *J. Am. Chem. Soc.*, **97**, 4859 (1975).
154. B. S. Prabhananda, C. C. Felix and J. S. Hyde, *J. Chem. Phys.*, **83**, 6121 (1985).
155. H. B. Stegmann, K. Stolze, H. U. Bergler and K. Scheffler, *Tetrahedron Lett.*, **22**, 4057 (1981).
156. J. D. R. Clay and D. Murphy, *J. Chem. Soc., Perkin Trans. 2*, 1781 (1984).
157. P. L. Gutierrez, B. M. Fox, M. M. Mossoba and N. R. Bachur, *Mol. Pharm.*, **26**, 582 (1984).
158. P. R. C. Gascoyne and A. Szent-Györgyi, *Int. J. Quant. Chem., Ser. Quant. Biol. Symp.*, **11**, 217 (1984).
159. H. B. Stegmann, K. Stolze and K. Scheffler, *Z. Naturforsch.*, **B38**, 243 (1983).
160. J. A. Pedersen, *Mol. Phys.*, **28**, 1031 (1974).
161. P. Ashworth, *Mol. Phys.*, **30**, 313 (1975).
162. J. A. Pedersen and R. H. Thomson, *J. Magn. Reson.*, **43**, 373 (1981).
163. J. A. Pedersen, *J. Chem. Soc., Perkin Trans. 2*, 424 (1973).
164. N. M. Atherton and P. A. Henshaw, *J. Chem. Soc., Perkin Trans. 2*, 258 (1975).
165. M. I. Kabachnik, N. N. Bubnov, S. P. Solodovnikov and A. I. Prokof'ev, *Usp. Khim.*, **53**, 487 (1984).
166. F. J. Smentowski, in *Characterization of Organometallic Compounds*, Vol. 26, *Pt. 2*, (Ed. M. Tsutsui), Interscience, New York, 1971, pp. 481–651.
167. M. G. Peter, H. B. Stegmann, H. Dao-Ba and K. Scheffler, *Z. Naturforsch.*, **C40**, 535 (1985).
168. A. I. Prokof'ev, N. N. Bubnov, S. P. Solodovnikov and M. I. Kabachnik, *Dokl. Akad. Nauk SSSR*, **245**, 1123 (1979); *Chem. Abstr.*, **91**, 131650 (1979).
169. R. M. Buchanan, J. Claflin and C. G. Pierpont, *Inorg. Chem.*, **22**, 2552 (1983).
170. M. E. Cass, D. L. Greene, R. M. Buchanan and C. G. Pierpont, *J. Am. Chem. Soc.*, **105**, 2680 (1983).
171. (a) T. Foster, K. S. Chen and J. K. S. Wan, *J. Organomet. Chem.*, **184**, 113 (1980); (b) K. S. Chen, T. Foster and J. K. S. Wan, *J. Chem. Soc., Perkin Trans. 2*, 1288 (1979).
172. H. Chandra, I. M. T. Davidson and M. C. R. Symons, *J. Chem. Soc., Perkin Trans. 2*, 1353 (1982).
173. J. A. Pedersen, *Phytochemistry*, **17**, 775 (1978).
174. N. J. F. Dodd, *Electron Spin Resonance*, **8**, 445 (1984).
175. N. N. Bubnov, A. I. Prokof'ev, A. A. Volod'kin, I. S. Belostotskaya and V. V. Ershov, *Dokl. Akad. Nauk. SSSR*, **210**, 100 (1973).
176. S. A. Wilson and J. H. Weber, *Anal. Lett.*, **10**, 75 (1977).
177. (a) M. Brok, F. C. R. Ebskamp and A. J. Hoff, *Biochim. Biophys. Acta*, **809**, 421 (1985); (b) A. W. Rutherford, *Biochim. Biophys. Acta*, **807**, 189 (1985); (c) W. F. J. Vermaas and A. W. Rutherford, *FEBS Lett.*, **175**, 243 (1984); (d) G. C. Dismukes, H. A. Frank, R. Friesner and K. Sauer, *Biochim. Biophys. Acta*, **764**, 253 (1984); (e) A. R. McIntosh, A. Siemiarczuk, J. R. Bolton, M. J. Stillman, T.-F. Ho and A. C. Weedon, *J. Am. Chem. Soc.*, **105**, 7215 (1983).
178. W. R. Heineman and P. T. Kissinger, in *Laboratory Techniques in Electroanalytical Chemistry* (Ed. P. T. Kissinger and W. R. Heineman), Marcel Dekker, New York, Basel, 1984, pp. 51–127.
179. M. Geissler, *Polarographische Analyse*, Verlag Chemie, Weinheim (Ger.), Deerfield Beach (Fla.), Basel, 1981.
180. A. M. Bond, *Modern Polarographic Methods in Analytical Chemistry*, Marcel Dekker, New York, Basel, 1980.
181. B. Speiser, *Chem. Unserer Zeit*, **15**, 21, 62 (1981).
182. J. Heinze, *Angew. Chem.*, **96**, 823 (1984).
183. J. Q. Chambers, in *The Chemistry of Quinonoid Compounds* (Ed. S. Patai), John Wiley and Sons, London, New York, Sydney, Toronto, 1974, pp. 737–792.
184. (a) G. Klopman and N. Doddapaneni, *J. Phys. Chem.*, **78**, 1820 (1974); (b) F. Dallacker, G. Löhnert and I. Kim, *Chem. Ber.*, **107**, 2415 (1974); (c) M. K. Kalinowski and B. Tenderende-Guminska, *J. Electroanal. Chem.*, **55**, 277 (1974); (d) J. Brisset, *J. Electroanal. Chem.*, **60**, 217 (1975); (e) J. P. Masson, J. Devynck and B. Tremillon, *J. Electroanal. Chem.*, **64**, 175 (1975); (f) H. Heberer, H. Schubert, H. Matschiner and B. Lukowczyk, *J. Prakt. Chem.*, **318**, 635 (1976); (g) F. E. Dinkevich, A. S. Vovk and O. S. Ksenzhek, *Vopr. Khim. Tekhnol.*, **57**, 42 (1979); *Chem. Abstr.*, **93**, 122373 (1980); (h) J. Bessard, G. Cauquis and D. Serve, *Electrochim. Acta*, **25**, 1187 (1980); (i) C.-Y. Li, L. Caspar and D. W. Dixon, Jr, *Electrochim. Acta*, **25**, 1135 (1980); (j) J. S.

Jaworski, E. Lesniewska and M. K. Kalinowski, *Pol. J. Chem.*, **54**, 1313 (1980); (k) R. Vallot, A. N'Diaye, A. Bermont, C. Jakubowicz and L. T. Yu, *Electrochim. Acta*, **25**, 1501 (1980); (l) T. Nagaoka, S. Okazaki and T. Fujinaga, *J. Electroanal. Chem.*, **133**, 89 (1982); (m) D. H. Evans and D. A. Griffith, *J. Electroanal. Chem.*, **136**, 149 (1982); (n) T. S. Eckert, T. C. Bruice, J. A. Gainor and S. M. Weinreb, *Proc. Natl. Acad. Sci. U.S.A.*, **79**, 2533 (1982); (o) A. Ben Hadid, J. Devynck and B. Trémillon, *Bull. Soc. Chim. Fr.*, I-177 (1983); (p) O. S. Ksenzhek and S. A. Petrova, *Bioelectrochem. Bioenerg.*, **11**, 105 (1983); (q) R. Narayan and K. L. N. Phani, *Trans. SAEST*, **19**, 177 (1984); *Chem. Abstr.*, **101**, 159998 (1984); (r) M. Shanshal and K. H. Hassan, *Stud. Biophys.*, **105**, 59 (1985); (s) B. D. Sviridov, T. D. Nikolaeva, V. F. Venderina and S. I. Zhdanov, *Zh. Obshch. Khim.*, **55**, 821 (1985), *Chem. Abstr.*, **103**, 141211 (1985); (t) K. L. N. Phani and R. Narayan, *J. Electroanal. Chem.*, **187**, 187 (1985); (u) T. Konse, H. Kano and T. Kubota, *Bull. Chem. Soc. Japan*, **59**, 265 (1986); (v) A. Aumüller and S. Hünig, *Ann. Chem.*, 165 (1986).

185. (a) T. W. Rosanske and D. H. Evans, *J. Electroanal. Chem.*, **72**, 277 (1976); (b) M. Iyoda, Y. Onishi and M. Nakagawa, *Tetrahedron Lett.*, **22**, 3645 (1981); (c) D. H. Evans and D. A. Griffith, *J. Electroanal. Chem.*, **134**, 301 (1982); (d) S. I. Bailey and I. M. Ritchie, *Electrochim. Acta*, **30**, 3 (1985); (e) M. P. Soriaga, J. H. White, D. Song, V. K. F. Chia, P. O. Arrhenius and A. T. Hubbard, *Inorg. Chem.*, **24**, 73 (1985); (f) J. H. White, M. P. Soriaga and A. T. Hubbard, *J. Phys. Chem.*, **89**, 3227 (1985).
186. (a) M. P. Soriaga, E. Binamira-Soriaga, A. T. Hubbard, J. B. Benziger and K.-W. P. Pang, *Inorg. Chem.*, **24**, 65 (1985); (b) S. A. Petrova and O. A. Ksenzhek, *Elektrokhimiya*, **22**, 137 (1986); *Chem. Abstr.*, **104**, 97902 (1986).
187. (a) K. Ravichandran and R. P. Baldwin, *J. Electroanal. Chem.*, **126**, 293 (1981); (b) M. Fukui, A. Kitani, C. Degrand and L. L. Miller, *J. Am. Chem. Soc.*, **104**, 28 (1982); (c) M. Fujihira, S. Tasaki, T. Osa and T. Kuwana, *J. Electroanal. Chem.*, **150**, 665 (1983); (d) G. Arai and M. Furui, *Nippon Kagaku Kaishi*, 673 (1984); *Chem. Abstr.*, **101**, 109968 (1984); (e) C. Degrand, L. Roullier, L. L. Miller and B. Zinger, *J. Electroanal. Chem.*, **178**, 101 (1984); (f) L. L. Miller, B. Zinger and C. Degrand, *J. Electroanal. Chem.*, **178**, 87 (1984); (g) C. Degrand, *Ann. Chim. (Rome)*, **75**, 1 (1985); (h) P. M. Hoang, S. Holdcroft and B. L. Funt, *J. Electrochem. Soc.*, **132**, 2129 (1985); (i) G. Arai, M. Matsushita and I. Yasumori, *Nippon Kagaku Kaishi*, 894 (1985); *Chem. Abstr.*, **103**, 61333 (1985); (j) G. Arai, K. Akiba, K. Yanagisawa and I. Yasumori, *Nippon Kagaku Kaishi,* 1867 (1984); *Chem. Abstr.*, **102**, 45324 (1985); (k) D. K. Smith, G. A. Lane and M. S. Wrighton, *J. Am. Chem. Soc.*, **108**, 3522 (1986); (l) P. Audebert, G. Bidan and M. Lapkowski, *J. Chem. Soc., Chem. Commun.*, 887 (1986).
188. (a) P. C. Lacaze, Minh Chau Pham, M. Delamar and J. E. Dubois, *Electroanal. Chem.*, **108**, 9 (1980); (b) R. C. M. Jakobs, L. J. J. Janssen and E. Barendrecht, *Electrochim Acta*, **30**, 1313 (1985).
189. D. J. Curran, in Ref. 178, pp. 539–568.
190. P. T. Kissinger, in Ref. 178, pp. 611–635.
191. (a) I. B. Goldberg and T. M. McKinney, in Ref. 178, pp. 675–728; (b) V. G. Mairanovsky, L. Yu. Yusefovich and T. M. Filippova, *J. Magn. Reson.*, **54**, 19 (1983).
192. M. O. F. Goulart, A. E. G. Sant'ana and V. Horak, *An. Simp. Bras. Electroquim. Electroanal.*, *4th*, 395 (1984); *Chem. Abstr.*, **101**, 162983 (1984).
193. (a) S. N. Suslov and D. I. Stom, *Zh. Anal. Khim.*, **33**, 1423 (1978); *Chem. Abstr.*, **90**, 109440 (1979); (b) D. I. Stom, S. S. Timofeeva, N. F. Kashina, L. I. Belykh, S. N. Suslov, V. V. Butorov and M. S. Apartsin, *Acta Hydrochim. Hydrobiol.*, **8**, 213 (1980).
194. (a) B. A. Svingen and G. Powis, *Arch. Biochem. Biophys.*, **209**, 119 (1981); (b) D. Tresselt, W. Ihn, G. Horn and H. Berg, *Pharmazie*, **39**, 417 (1984); (c) S. Sakura and H. Imai, *Anal. Sci.*, **1**, 413 (1985); (d) A. Anne and J. Moiroux, *Nouv. J. Chim.*, **9**, 83 (1985).
195. R. Pethig, P. R. C. Gascoyne, J. A. McLaughlin and A. Szent-Györgyi, *Proc. Natl. Acad. Sci. U.S.A.*, **80**, 129 (1983).
196. A. J. Lin and A. C. Sartorelli, *Biochem. Pharmacol.*, **25**, 206 (1976).
197. (a) R. Naumann and D. Kayser, *Bioelectrochem. Bioenerg.*, **4**, 171 (1977); (b) P. A. Andrews, S. Pan and N. R. Bachur, *J. Am. Chem. Soc.*, **108**, 4158 (1986).
198. M. B. Fleury and G. Molle, *Electrochim. Acta*, **20**, 951 (1975).
199. P. R. Rich, *Biomed. Clin. Aspects Coenzyme Q*, **4**, 25 (1984).
200. J. H. Wilford, M. D. Archer, J. R. Bolton, T.-F. Ho, J. A. Schmidt and A. C. Weedon, *J. Phys. Chem.*, **89**, 5395 (1985).
201. Yu. L. Zherebin, V. L. Kuev and A. V. Bogatskii, *Zh. Obshch. Khim.*, **51**, 2767 (1981).
202. S. H. Eberle, C. Hösle, O. Hoyer and C. Krückeberg, *Vom Wasser*, **43**, 359 (1974).

203. D. Amin and F. M. El-Samman, *J. Indian Chem. Soc.*, **60**, 502 (1983).
204. J. L. Vandenbalck, J. C. Vire, C. A. Mairesse-Ducarmois and G. J. Patriarche, *Analusis*, **7**, 88 (1979).
205. J. P. Hart and A. Catterall, *Anal. Chem. Symp. Ser.*, **2** (Electroanal. Hyg., Environ., Clin. Pharm. Chem.), 145 (1980).
206. N. A. Butakova and L. B. Oganesyan, *Elektrokhimiya*, **21**, 1535 (1985); *Chem. Abstr.*, **104**, 41848 (1986).
207. (a) A. I. Kostromin and R. M. Badakshanov, *Zh. Anal. Khim.*, **27**, 2046 (1972); *Chem. Abstr.*, **78**, 52060 (1973); (b) A. I. Kostromin, L. L. Makarova and L. I. Il'ina, *Zh. Anal. Khim.*, **31**, 240 (1976); *Chem. Abstr.*, **85**, 116088 (1976).
208. (a) S. Mitev and P. K. Agasyan, *Zavod Lab.*, **41**, 396 (1975); *Chem. Abstr.*, **83**, 125868 (1975); (b) V. Stuzka, I. Lukas and J. A. Jilek, *Acad. Univ. Palacki. Olomuc., Fac. Rerum. Nat.*, **49** (Chem. 15), 93 (1976); *Chem. Abstr.*, **87**, 77974 (1977); (c) J. C. Vire and G. J. Patriarche, *Anal. Lett.*, **A11**, 307 (1978).
209. M. Ignaczak and J. Dziegiec, *Chem. Anal. (Warsaw)*, **20**, 229 (1975); *Chem. Abstr.*, **83**, 22043 (1975).
210. J. C. Vire, M. Chateau-Gosselin and G. J. Patriarche, *Mikrochim. Acta*, **1**, 227 (1981).
211. E. Ruzicka and Z. Cermakova, *Chem. Zvesti.*, **33**, 612 (1979); *Chem. Abstr.*, **92**, 173985 (1980).
212. K. N. Murty and P. M. D. Murthy, *Talanta*, **29**, 234 (1982).
213. N. K. Murthy and P. M. D. Murthy, *Proc. Natl. Acad. Sci., India*, **A54**, 172 (1984).
214. (a) Y. Matsumura and H. Takahashi, *Carbon*, **17**, 109 (1979); (b) S. S. M. Hassan and M. B. Elsayes, *Mikrochim. Acta*, **2**, 333 (1978); (c) A. M. Talati, N. D. Godhwani, A. D. Sheth, K. B. Shah and Y. K. Joshi, *Indian J. Technol.*, **22**, 468 (1984).
215. R. W. Frei, N. H. Velthorst and C. Gooijer, *Pure Appl. Chem.*, **57**, 483 (1985).
216. (a) K. H. Shafer, S. L. Pentoney, Jr and P. R. Griffiths, *Anal. Chem.*, **58**, 58 (1986); (b) K. A. M. Creber and J. K. S. Wan, *J. Am. Chem. Soc.*, **103**, 2101 (1981).
217. H. Gunasingham, B. T. Tay and K. P. Ang, *Anal. Chem.*, **58**, 1578 (1986).
218. K. Bratin and P. T. Kissinger, *Talanta*, **29**, 365 (1982).
219. K. Krohn, *J. Chromatogr.*, **130**, 327 (1977).
220. E. Berndorfer-Kraszner and L. Telegdy Kovats, *Prikl. Biokhim. Mikrobiol.*, **12**, 392 (1976); *Chem. Abstr.*, **85**, 43196 (1976).
221. J. Churacek, *J. Chromatogr. Libr.*, **3** (Liq. Column Chromatogr.), 455 (1975); *Chem. Abstr.*, **84**, 12065 (1975).
222. (a) P. Kusz, *Chemik*, **37**, 91 (1984); *Chem. Abstr.*, **101**, 122273 (1984); (b) M. Wang, *Yaoxue Xuebao*, **21**, 230 (1986); *Chem. Abstr.*, **104**, 213335 (1986).
223. (a) H. Röper and K. Heyns, *Z. Naturforsch.*, **C32**, 61 (1977); (b) D. R. Choudhury, *Environ. Sci. Technol.*, **16**, 102 (1982); (c) J. König, E. Balfanz, W. Funcke and T. Romanowski, *Anal. Chem.*, **55**, 599 (1983).
224. L. Ramaley, M.-A. Vaughan and W. D. Jamieson, *Anal. Chem.*, **57**, 353 (1985).
225. C. E. Lunte and P. T. Kissinger, *Anal. Chem.*, **56**, 658 (1984).
226. U. Prenzel and H. K. Lichtenthaler, *J. Chromatogr.*, **242**, 9 (1982).
227. Y. Yamano, S. Ikenoya, M. Anze, M. Ohmae and K. Kawabe, *Yakugaku Zasshi*, **98**, 774 (1978); *Chem. Abstr.*, **89**, 117665 (1978).
228. R. Vespalec and J. Neča, *J. Chromatogr.*, **281**, 35 (1983).
229. A. Marston and K. Hostettmann, *J. Chromatogr.*, **295**, 526 (1984).
230. J. Chmielowiec, *J. Chromatogr. Sci.*, **19**, 296 (1981).
231. D. W. Armstrong, A. Alak, W. DeMond, W. L. Hinze and T. E. Riehl, *J. Liq. Chromatogr.*, **8**, 261 (1985).
232. J. A. Duine, J. Frank, Jr and J. A. Jongejan, *Anal. Biochem.*, **133**, 239 (1983).
233. H. Thielemann, *Z. Chem.*, **14**, 28 (1974).
234. H. Thielemann, *Sci. Pharm.*, **40**, 291 (1972).
235. H. Thielemann, *Sci. Pharm.*, **47**, 246 (1979).
236. H. Thielemann, *Z. Wasser Abwasser Forsch.*, **5**, 176 (1972).
237. H. K. Lichtenthaler, K. Börner and C. Liljenberg, *J. Chromatogr.*, **242**, 196 (1982).
238. S. Onodera, M. Tabata, S. Suzuki and S. Ishikura, *J. Chromatogr.*, **200**, 137 (1980).
239. M. Sugumaran and C. S. Vaidyanathan, *J. Indian Inst. Sci.*, **60**, 51 (1978); *Chem. Abstr.*, **89**, 142745 (1978).
240. K. P. Tiwari and P. K. Minocha, *J. Indian Chem. Soc.*, **57**, 110 (1980).
241. H. W. Rauwald and H. Miething, *Dtsch. Apoth.-Ztg.*, **125**, 101 (1985).
242. D. D. Stump, E. F. Roth, Jr and H. S. Gilbert, *J. Chromatogr.*, **306**, 371 (1984).

243. S. Okayama, *Plant Cell Physiol.*, **25**, 1445 (1984).
244. M. T. Smith, D. S. Fluck, D. A. Eastmond and S. M. Rappaport, *Life Chem. Rep.*, **3**, 250 (1985).
245. (a) S. K. Yang, J. K. Selkirk, E. V. Plotkin and H. V. Gelboin, *Cancer Res.*, **35**, 3642 (1975); (b) C. Raha, R. Hines and E. Bresnick, *J. Chromatogr.*, **291**, 231 (1984); H. L. Gurtoo, J. B. Vaught, A. J. Marinello, B. Paigen, T. Gessner and W. Bolanowska, *Cancer Res.*, **40**, 1305 (1980).
246. (a) P. Gast, T. J. Michalski, J. E. Hunt and J. R. Norris, *FEBS Lett.*, **179**, 325 (1985); (b) D. E. Minnikin, A. G. O'Donnell, M. Goodfellow, G. Alderson, M. Athalye, A. Schaal and J. H. Parlett, *J. Microbiol. Methods*, **2**, 233 (1984); (c) C. W. Moss, W. F. Bibb, D. E. Karr and G. O. Guerrant, *Inst. Natl. Sante Rech. Med.* [*Colloq.*], **114** (Baccilles Gram Negat. Interet Med. Sante Publique), 375 (1983); *Chem. Abstr.*, **101**, 106990 (1984).
247. K. Shimazaki, K. Takamiya, and M. Nishimura, *J. Biochem. (Tokyo)*, **83**, 1639 (1978).
248. (a) A. F. Zak, L. P. Snezhnova, V. M. Shatrova, R. D. Polisar, N. A. Il'inskaya and E. G. Pozdnyakov, *Antibiotiki (Moscow)*, **27**, 569 (1982); *Chem. Abstr.*, **97**, 133646 (1982); (b) G. Ovcharova, *Probl. Farm.*, **13**, 30 (1985); *Chem. Abstr.*, **104**, 116142 (1986).
249. H. Y. Tong, D. L. Shore, F. W. Karasek, P. Helland and E. Jellum, *J. Chromatogr.*, **285**, 423 (1984).
250. T. Barth, K. Tjessem and A. Aaberg, *J. Chromatogr.*, **214**, 83 (1981).
251. R. C. Pierce and M. Katz, *Environ. Sci. Technol.*, **10**, 45 (1976).
252. D. E. Ryan, T. P. Meyerhof and M. T. Fairhurst, *Anal. Chim. Acta*, **86**, 195 (1976).
253. T. Yoshida and T. Uno, *Bunseki Kagaku*, **28**, 351 (1979); *Chem. Abstr.*, **91**, 116920 (1979).
254. S. I. Obtemperanskaya and V. K. Zlobin, *Zh. Anal. Khim.*, **31**, 1205 (1976).
255. D. Amin and F. M. El-Samman, *Mikrochim. Acta*, **1**, 467 (1983).
256. H.-J. Kallmayer, *Arch. Pharm. (Weinheim, Ger.)*, **306**, 257 (1973).
257. (a) J. Jenik, *Chem. Prum.*, **35**, 154 (1985); *Chem. Abstr.*, **103**, 58959 (1985); (b) W. Ott, V. Formaček and H.-M. Seidenspinner, *Ann. Chem.*, 1003 (1984).
258. (a) R. E. Smith and W. R. Davis, *Anal. Chem.*, **56**, 2345 (1984); (b) H.-J. Kallmayer and K. Seyfang, *Arch. Pharm. (Weinheim, Ger.)*, **319**, 52 (1986); (c) H.-J. Kallmayer and K. Seyfang, *Dtsch. Apoth.-Ztg.*, **123**, 2147 (1983); (d) H.-J. Kallmayer and K. Seyfang, *Arch. Pharm. (Weinheim, Ger.)*, **317**, 329 (1984); (e) H.-J. Kallmayer and K. Seyfang, *Arch. Pharm. (Weinheim, Ger.)*, **318**, 607 (1985); (f) H.-J. Kallmayer and K. Seyfang, *Arch. Pharm. (Weinheim, Ger.)*, **317**, 855 (1984); (g) H.-J. Kallmayer and K. Seyfang, *Arch. Pharm. (Weinheim, Ger.)*, **318**, 360 (1985); (h) H.-J. Kallmayer and K. Seyfang, *Arch. Pharm. (Weinheim, Ger.)*, **316**, 283 (1983); (i) H.-J. Kallmayer and C. Tappe, *Arch. Pharm. (Weinheim, Ger.)*, **319**, 607 (1986); *Pharmazie*, **41**, 29 (1986); (j) H.-J. Kallmayer and C. Tappe, *Arch. Pharm. (Weinheim, Ger.)*, **314**, 884 (1981); (k) H.-J. Kallmayer and K. Seyfang, *Arch. Pharm. (Weinheim, Ger.)*, **318**, 865, 1100 (1985); (l) M. N. Gupta, R. Gopinath and K. Raju, *Anal. Biochem.*, **118**, 227 (1981); (m) E. V. Butrov, *Byull. Vses. Nauch.-Issled. Inst. Fiziol., Biokhim. Pitan. Sel'skokhoz. Zhivotn.*, **6**, 77 (1972); *Chem. Abstr.*, **81**, 22757 (1974); (n) U. Lohmann and K. Hartke, *Arch. Pharm., (Weinheim, Ger.)*, **317**, 313 (1984); (o) K. Hartke and U. Lohmann, *Dtsch. Apoth.-Ztg.*, **123**, 1013 (1983).
259. (a) N. Gallo, P. Bianco, R. Tapino and G. Luisi, *Minerva Med.*, **74**, 875 (1983); *Chem. Abstr.*, **99**, 81922 (1983); (b) N. Gallo and V. DeMola, *Experientia*, **33**, 411 (1977); (c) V. G. Rani, M. Sugumaran, N. A. Rao and C. S. Vaidyanathan, *J. Indian Inst. Sci.*, **60**, 43 (1978); *Chem. Abstr.*, **89**, 142795 (1978); (d) M. T. M. Zaki and A. G. Abdel-Rehiem, *Microchem. J.*, **29**, 44 (1984).
260. (a) D. I. Stom, *Acta Hydrochim. Hydrobiol.*, **3**, 39 (1975); (b) A. S. Stadnik, Yu. Yu. Lur'e and Yu. M. Dedkov, *Zh. Anal. Khim*, **32**, 1801 (1977); *Chem. Abstr.*, **88**, 109990 (1978); (c) D. I. Stom, S. S. Timofeeva, N. F. Kashina, L. I. Belykh, S. N. Suslov, V. V. Butorov and M. S. Apartsin, *Acta Hydrochim. Hydrobiol*, **8**, 203 (1980).
261. A. Rieker, J. Bracht, E.-L. Dreher and H. P. Schneider, in *Houben-Weyl-Müller: Methoden der organischen Chemie* (Ed. C. Grundmann), Thieme Verlag, Stuttgart, 1979, pp. 763.
262. M. A. Korany, A. M. Wahbi and M. H. Abdel-Hay, *J. Pharm. Biomed. Anal.*, **2**, 537 (1984).
263. H.-J. Kallmayer and C. Tappe, *Dtsch. Apoth.-Ztg.*, **126**, 11 (1986).
264. H.-J. Kallmayer and A. Hund, *Sci. Pharm.*, **47**, 240 (1979).
265. K. L. Bajaj, *Ann. Chim. (Paris)*, **1**, 5 (1976).
266. M. D. Collins, *Methods Microbiol.*, **18**, 329 (1985).
267. J. A. J. M. Lemli, *Pharmacology*, **14**, Suppl. 1 (Anthraquinone Laxatives, Proc. Symp. 1975), 62 (1976); *Chem. Abstr.*, **91**, 216848 (1979).
268. (a) I. Chvatal, *Chem. Prum.*, **35**, 638 (1985); *Chem. Abstr.*, **104**, 161250 (1986); (b) S. Ebel and M. Kaal, *Dtsch. Apoth.-Ztg.*, **120**, 1412 (1980).
269. J. H. Zwaving, *Pharmacology*, **20** (Suppl. 1), 65 (1980).
270. A. L. Macdonald and J. Trotter, *J. Chem. Soc., Perkin Trans. 2*, 476 (1973).

The Chemistry of Quinonoid Compounds, Vol. II
Edited by S. Patai and Z. Rappoport

CHAPTER **3**

Mass spectra of quinones

KLAUS-PETER ZELLER and ROLAND MÜLLER
Institut für Organische Chemie, Universität Tübingen, Auf der Morgenstelle, D-7400 Tübingen, FRG

I. INTRODUCTION

A detailed description of the positive-ion electron impact mass spectra of quinones was given in this series several years ago[1]. In the meantime mass spectral data of quinones have been reported in a large number of papers for the sake of characterization of synthetic or naturally occurring quinones. Concerning the basic behaviour of quinones under electron impact, relatively little new information can be extracted from these publications.

The nature of the $C_4(Ph)_4{}^+$ ions formed by consecutive loss of two CO molecules from tetraphenyl-1,4-benzoquinone is still a matter of interest and has been reinvestigated by more advanced mass spectrometric techniques.

The importance of mass spectrometry for structure elucidation and identification of naturally occurring quinones is illustrated in this chapter by recent examples. Bowie and coworkers demonstrated that negative-ion mass spectra of substituted quinones, not dealt with in the earlier review, yield valuable structural information. This technique is treated in some detail in Section III.

Throughout this chapter, fragmentations substantiated by a metastable transition in the second field-free region or by the metastable defocusing technique in the first field-free region are indicated by an asterisk.

II. POSITIVE-ION MASS SPECTRA

The positive-ion electron impact spectra of quinones exhibit two characteristic features[1], namely (1) the stepwise loss of two molecules of carbon monoxide, and (2) the formation of peaks two mass units higher than the molecular mass due to partial reduction before the ionization step. The latter is particularly pronounced in the case of *ortho*-quinones but also in *para*-quinones having high redox potentials.

A. Electron Impact Induced Decarbonylation

For anthraquinone it could be shown that the second elimination of CO (m/z 180 → m/z 152) in the second field-free region is not a truly unimolecular process but caused by collisional activation due to sample leakage or residual background gas[2]. This follows from the pressure dependence of the second process. Probably, this finding is of general implication for consecutive metastable reactions.

The stepwise loss of two molecules of carbon monoxide is also found in the non-benzenoid quinones **1**[3], **2**[4] and **3**[4] related to azulene. In the mass spectra of these molecules

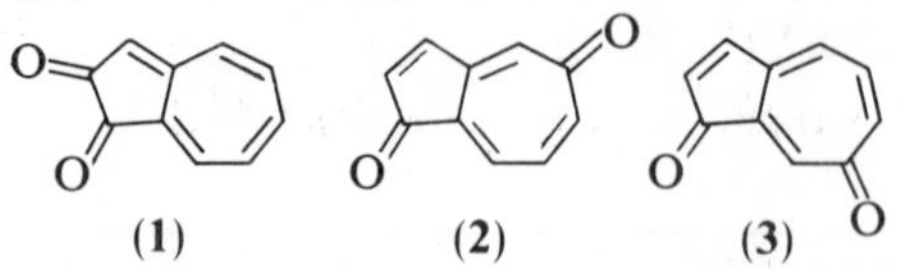

(**1**) (**2**) (**3**)

the $[M-2CO]^{+\cdot}$ ions give rise to the most intensive peaks. In accordance with the high tendency for loss of carbon monoxide the mass spectra of pentacene-5,7,12,14-diquinone (**4**) and its derivatives are dominated by four decarbonylation steps[5]. Some years ago the

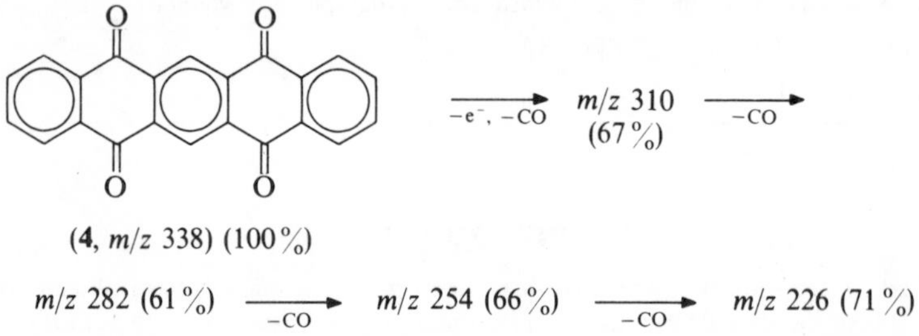

(**4**, m/z 338) (100%) $\xrightarrow{-e^-,\ -CO}$ m/z 310 (67%) $\xrightarrow{-CO}$

m/z 282 (61%) $\xrightarrow{-CO}$ m/z 254 (66%) $\xrightarrow{-CO}$ m/z 226 (71%)

structure of the $[M-2CO]^{+\cdot}$ ion of tetraphenyl-1,4-benzoquinone was investigated by labelling of the aromatic rings with the *para*-fluoro substituent[6]. The key compound in this study was the unsymmetrically substituted 2,5-diphenyl-3,6-bis(4-fluorophenyl)-1,4-benzoquinone (**5**). It was reported that the $[M-2CO]^{+\cdot}$ ion of **5** yields on further decomposition $C_{14}H_{10}^{+\cdot}$, $C_{14}H_{9}F^{+\cdot}$ and $C_{14}H_{8}F_{2}^{+\cdot}$ ions.

A cyclobutadiene structure **a** would only predict monofluoro-labelled product ions. Therefore, Bursey and coworkers[6] concluded that the $[M-2CO]^{+\cdot}$ ions either decompose as a tetrahedrane-like structure **b** or the decomposing species consist of isomeric substituted tetraphenylcyclobutadiene cation radicals undergoing interconversions via **b**. If the tetrahedrane intermediates (or transition states) are undistorted, the intensity ratio of unlabelled, singly labelled, and doubly labelled product ions should be 1:4:1. The experimental ratio obtained by measurement of conventional metastable ion intensities in the spectrum of **5** generated by a single focusing magnetic deflection instrument was

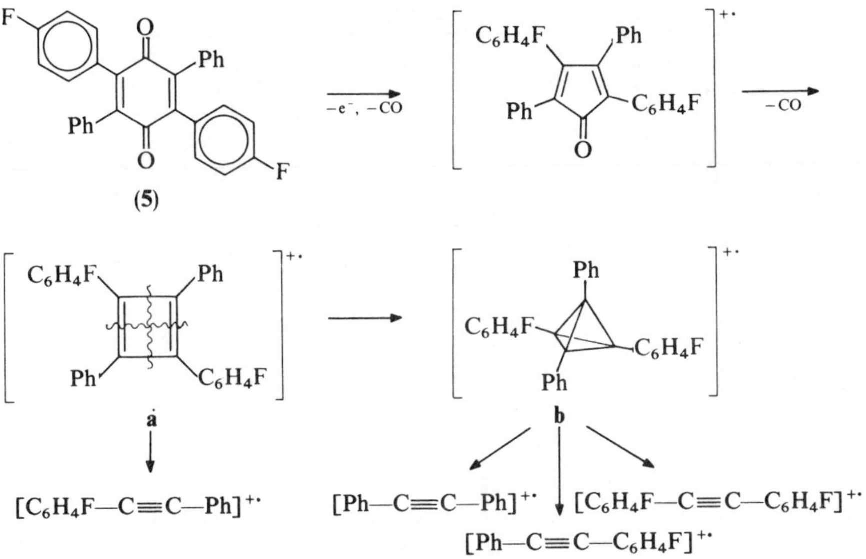

$[C_6H_4F—C\equiv C—Ph]^{+\cdot}$ $[Ph—C\equiv C—Ph]^{+\cdot}$ $[C_6H_4F—C\equiv C—C_6H_4F]^{+\cdot}$ $[Ph—C\equiv C—C_6H_4F]^{+\cdot}$

1:4.8:0.84. The divergence between the experimental and the theoretically predicted ratio was rationalized by a distorted tetrahedrane structure. Very similar results were obtained from fluoro-substituted $C_4Ar_4^{+\cdot}$ ions produced from tetracyclone derivatives and other suitably fluoro-labelled precursors.

The tetrahedrane problem was again studied by Schwarz and coworkers[7] using $[2,4\text{-}^{13}C_2]$tetraphenylcyclopentadienone (**6**). In contrast to the earlier results, the collision-induced fragmentation of the $[M\text{–}CO]^{+\cdot}$ ion of **6**, mass-selected in a triple quadrupole instrument operated under MS/MS conditions, indicated that only one product, the singly labelled ion $^{12}C_{13}{}^{13}CH_{10}{}^{+\cdot}$ was formed. This finding excludes an intermediate tetrahedrane ion.

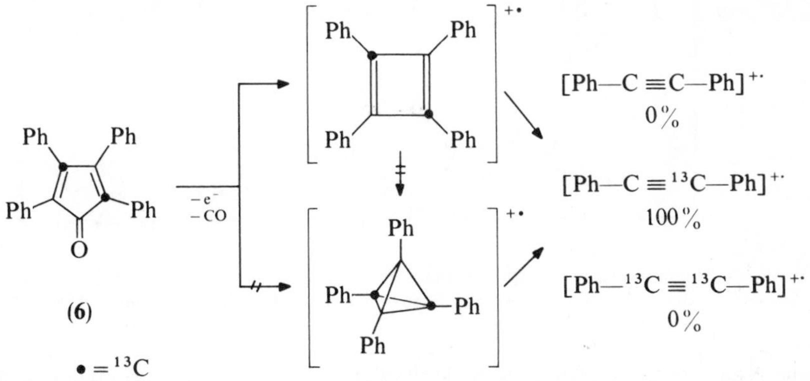

$[Ph—C\equiv C—Ph]^{+\cdot}$ 0%

$[Ph—C\equiv {}^{13}C—Ph]^{+\cdot}$ 100%

$[Ph—{}^{13}C\equiv {}^{13}C—Ph]^{+\cdot}$ 0%

(**6**)

• = ^{13}C

These contradictory results stimulated a reinvestigation of the F-labelled compounds on a reversed Nier-Johnson geometry mass spectrometer[8]. In Figure 1 the unimolecular MIKE spectrum and the collisionally activated decomposition of the m/z 392 ion

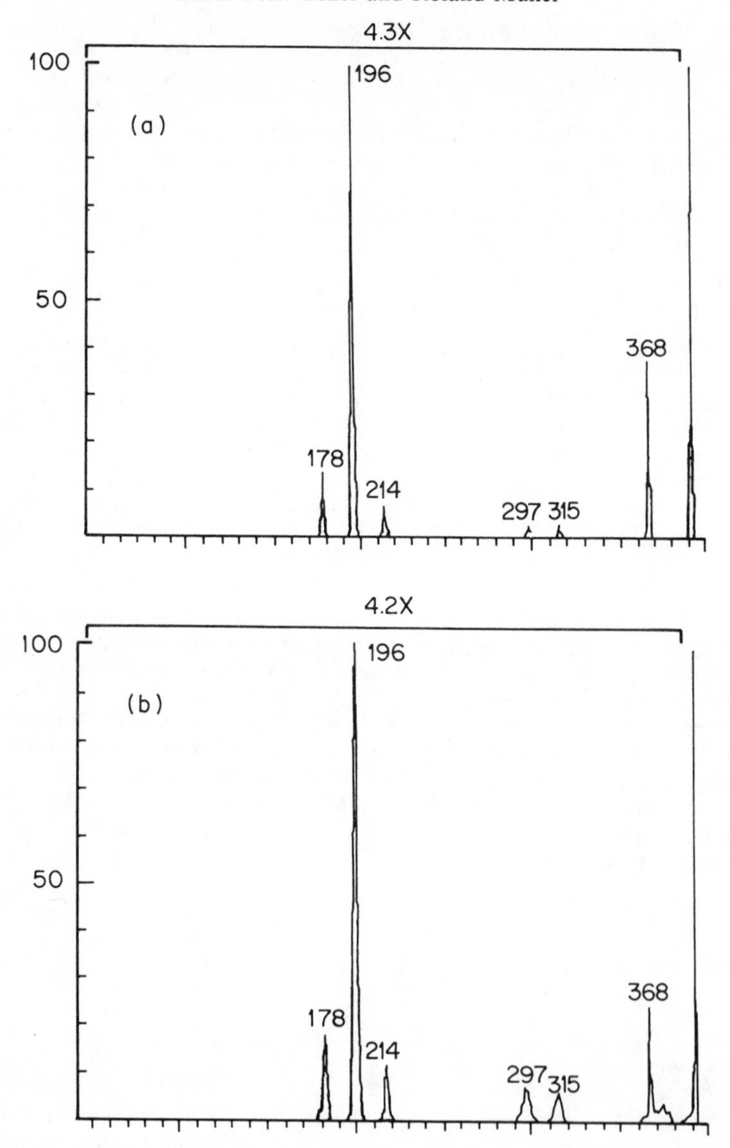

FIGURE 1. Unimolecular MIKE spectrum (a) and collisionally activated decomposition (b) of the *m/z* 392 ion from 2,5-diphenyl-3,6-bis(4-fluorophenyl)-1,4-benzoquinone (**5**). Reproduced from Ref. 8

produced by loss of 2 CO from the molecular ion of **5** are given. Both contain peaks at *m/z* 178, 196 and 214. Thus, they confirm the previous results based on assignments of metastable peaks. However, the intensities of peaks diagnostic for the tetrahedral form **b** are lower than found originally. It should be remembered that the previous measurements on metastable peaks correspond to a shorter time-scale. This points to the possibility that the part of fragmenting ions passing through a tetrahedral form **b** depends on the lifetime (and therefore internal energy) of the observed ions[8]. In this context it is of interest that the conflicting ^{13}C-labelling results were obtained on a quadrupole instrument with a longer time-scale.

Alternatively, an influence of the *para*-fluoro substituent on the population of excited levels has been discussed[8] to account for the presence of peaks diagnostic for the tetrahedral intermediate in the fragmentation of the $[M-2CO]^{+\cdot}$ ion of **5**, whereas corresponding peaks are absent in the decomposition pattern of the $[M-CO]^{+\cdot}$ ion of the ^{13}C-labelled compound **6**.

B. Chemical Changes prior to Ionization

The formation of $[M+2]^{+\cdot}$ peaks often obscures mass spectrometric investigation of quinones[1]. An interesting example has been reported for derivatives of polyporic acid **7a**[9]. The diacetate **7b** and the dimethoxy compound **7c** give m/z values corresponding to the molecular masses but ions two mass units larger than these, often of greater intensities, are also observed. The abundance of such ions depends on the sample temperature, the ion source pressure, and the partial pressure of water in the instrument. In addition, ions at m/z 420 and 462 are found in the spectrum of **7b** which are parents of the $[M+2]^{+\cdot}$ ions at m/z 378. Similarly, $[M+17]^{+}$ and $[M+32]^{+\cdot}$ ions are observed in the spectrum of the dimethoxy derivative **7c**. In view of the low pressure applied it seems unlikely that the reactions observed are bimolecular. The participation of water in the reaction suggests that it occurs on the metallic surfaces of the source, where water is known to be adsorbed. The transfer of acetyl and methyl groups, respectively, may occur at the same site but may also be the result of a solid state reaction or even reaction in the gas phase after vaporization of quinhydrone dimers. Although the exact origin of these peaks is not known, any explanation of the phenomena requires the reduction of the benzoquinone moiety for the formation of tri- and tetraacetates. These findings are of relevance for the establishment of the presence of 2,5-dihydroxy-1,4-benzoquinones in plant extracts.

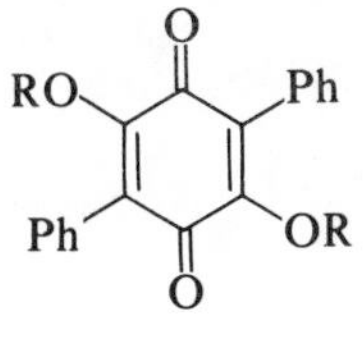

O
RO Ph
Ph OR
O

(**7a**: R = H)

(**7b**: R = COMe)

(**7c**: R = Me)

(**7**)

A phenomenon reminiscent of the appearance of $[M+2]^{+\cdot}$ peaks in the mass spectra of certain quinones[1] has been reported for 5,14-dihydroxypentacene-7,12-quinone (**8**)[5]. The 40 eV-EI spectrum of **8** shows no molecular ion peak. However, typical peaks for the dehydroproduct **4** ($M^{+\cdot}$, m/z 338) and 5-hydroxypentacene-7,12-quinone (**9**, $M^{+\cdot}$, m/z 324) are found. Whereas the formation of $[M+2]^{+\cdot}$ peaks of other quinones results from reduction prior to ionization by residual water, a disproportionation process must be operative to produce **4** and **9**.

C. Structure Determination of Naturally Occurring Quinones

Mass spectrometry proved to be a powerful tool in the structural elucidation of quinones originating from natural sources[1,10,11]. In the following, some selected examples are given.

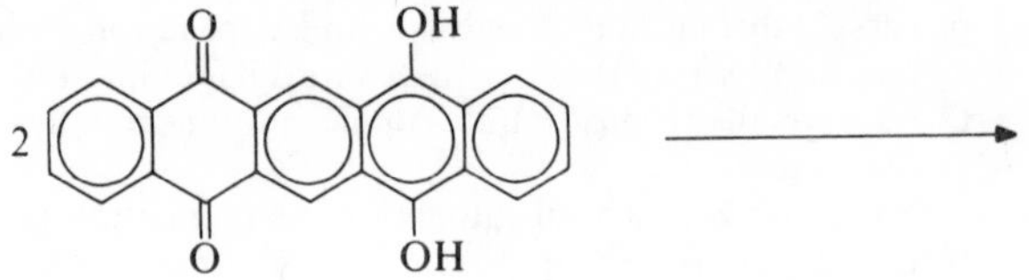

(**8**)

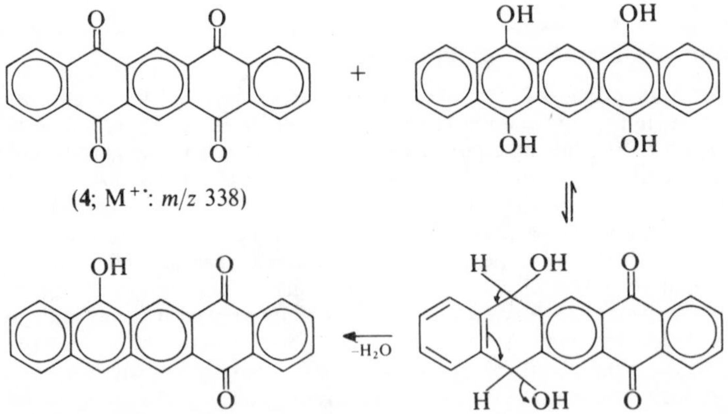

(**4**; $M^{+\cdot}$: m/z 338)

(**9**; $M^{+\cdot}$: m/z 324)

In contrast to the biogenetically related boviquinones (**11**) the mass spectra of tridentoquinone (**10**) and its derivatives isolated as pigments from *Suillus tridentinus* yield intensive molecular ion peaks[12]. This has been rationalized by a pronounced electron impact induced degradation of the isoprenoid side chain in **11**, whereas similar fragmentations are less likely in the molecular ion of the *ansa*-1,4-benzoquinone (**10**).

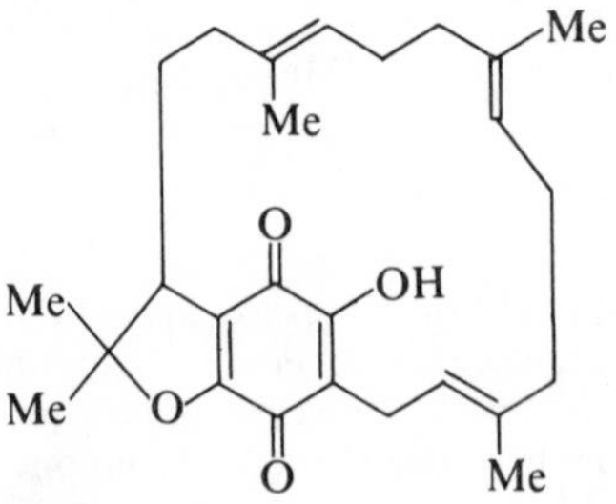

(**10**)

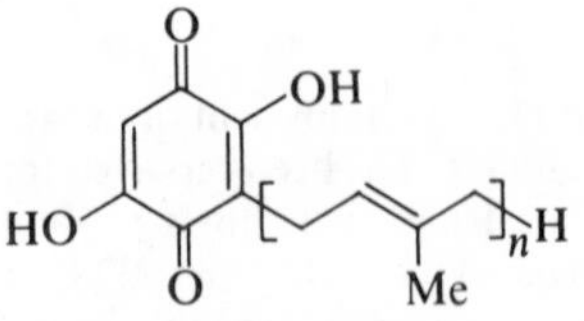

(**11**; n = 3, 4)

The mass spectrum of isolapachol (**12**) is dominated by fragmentations involving the side chain[13]. The same fragments are formed from the molecular ion of lapachol (**13**), indicating isomerization to a common structure prior to fragmentation.

Mass spectrometry has been involved in the structure elucidation of 5,8-dihydroxy-1,4-naphthoquinones of the general type **14**[14,15] and **15**[15] isolated from *Macrotomia euchroma* and *Arnebia nobilis*.

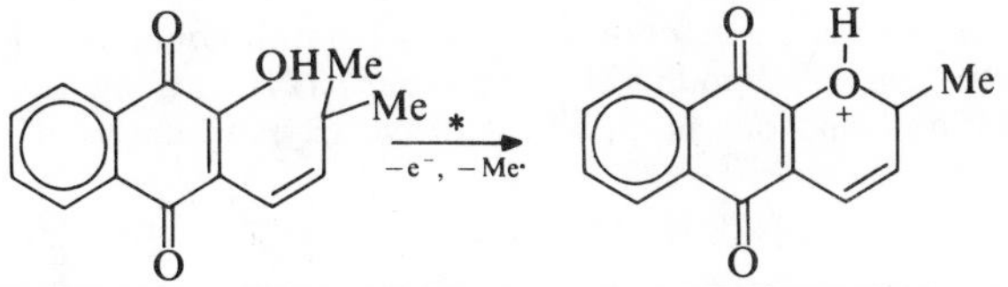

(**12**; $M^{+\cdot}$: m/z 242) → ($-e^-$, $-Me^\cdot$) m/z 227

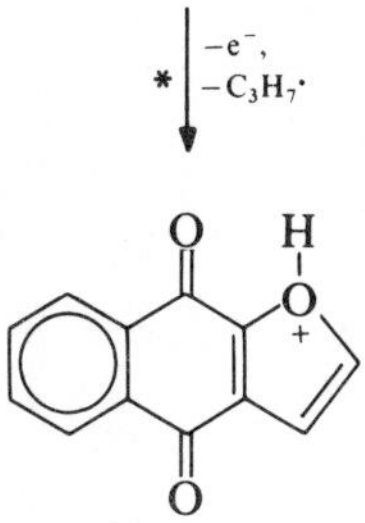

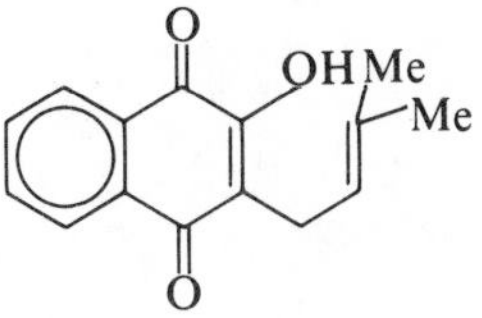

(**12**) → ($-e^-$, $-C_3H_7^\cdot$) m/z 199

(**13**)

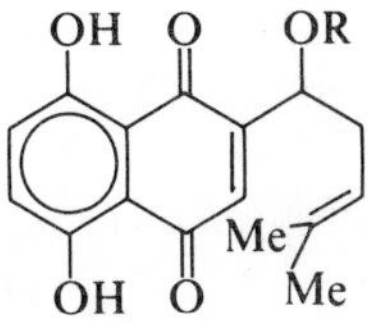

(**14**) 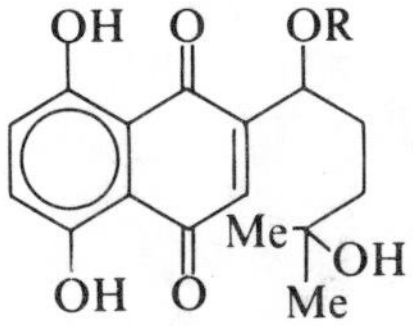 (**15**) 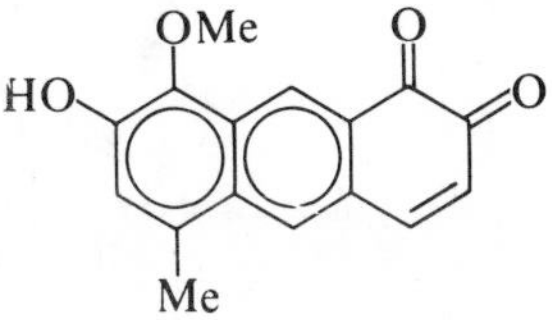 (**16**)

Similarly, the mass spectra of 5-hydroxy-1,4-naphthoquinones found in sea urchins have been investigated[16]. The mass spectrum of the red pigment hallachrome (**16**) obtained by extraction of the seaworm *Halla parthenopeia* with chloroform played an important role in recognizing the *ortho*-quinonoid structure[17]. As expected for an *ortho*-quinone, the molecular ion and the $[M-Me]^+$ ion suffer from pronounced decarbonylations and a prominent $[M+2]^{+\cdot}$ peak is formed.

Naturally occurring anthraquinones are the active principles of many plant drugs. Amongst other recent examples, mass spectrometry assisted the structure determination of the antitumour agent morindaparvin B (**17**)[18] from *Morinda parvifolia*.

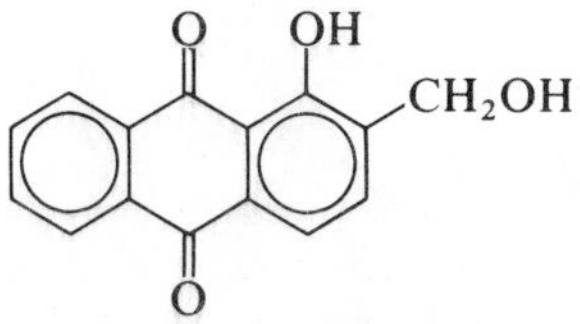

(**17**)

The electron impact mass spectra of the natural 1,8-anthraquinones chysophanol (**18**), aloe-emodin (**19**), rhein (**20**), emodin (**21**) and catenarin (**22**) show intensive molecular ions and a fragmentation behaviour as expected for hydroxylated anthraquinones[19].

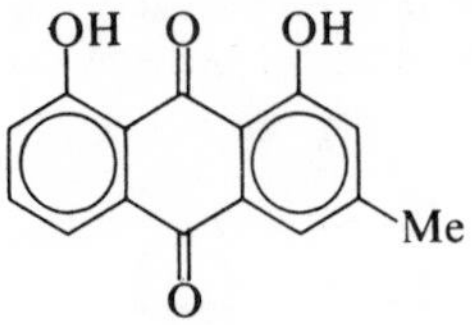

(**18**) 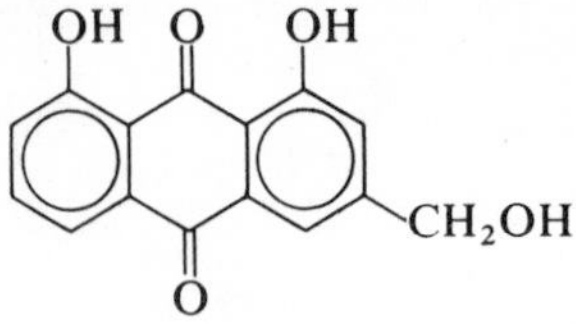 (**19**) 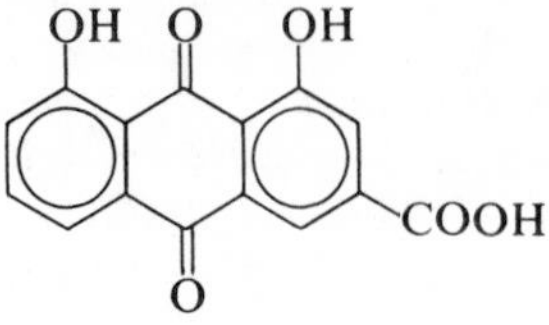 (**20**)

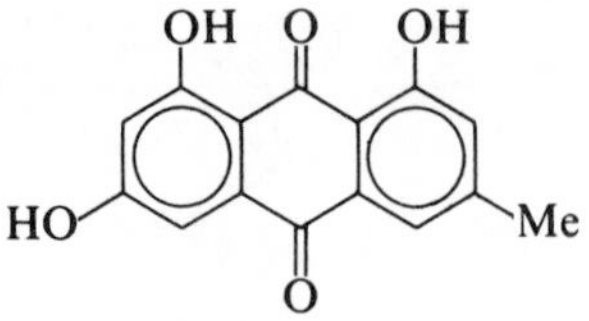

(**21**) 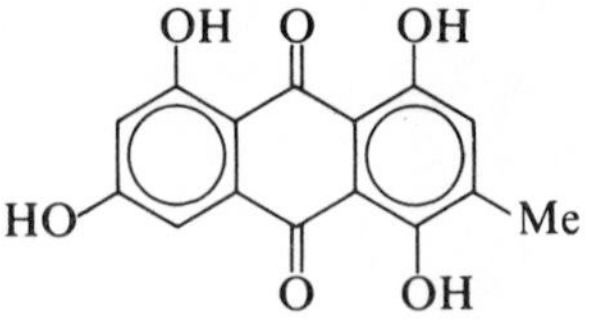 (**22**)

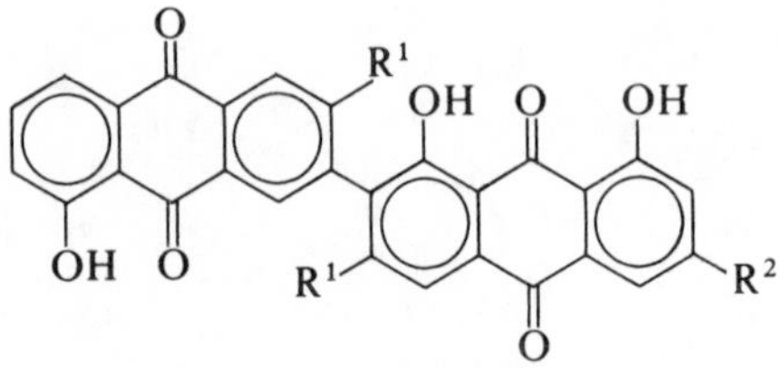

(**23a**; R^1 = Me, R^2 = OH)
(**23b**; R^1 = Me, R^2 = H)

Certain *Cassia* species produce dianthraquinones in addition to the monomeric compounds. As shown for cassiamin A (**23a**) and cassiamin C (**23b**) the EI spectra of these dimers are characterized by intensive molecular ion peaks enabling ready distinction from their monomeric counterparts[19]. This is of considerable utility since ^{1}H-NMR spectra often fail in this context. Furthermore fragments are found corresponding to cleavage into the monomers. This feature is particularly useful when the dimer is of the mixed type as cassiamin A.

Anthraquinones bearing phenolic hydroxy groups are often acylated prior to spectroscopic and analytical characterizations. The positive-ion EI mass spectra of *O*-acylated

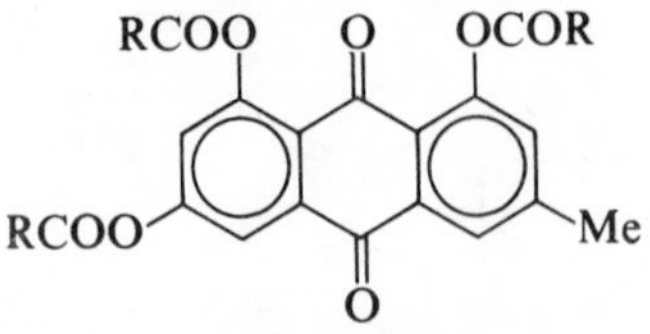

(**24a**: R = Me)
(**24b**: R = Ph)

hydroxyanthraquinones exhibit marked differences depending on the nature of the acyl groups. This has been exemplified with the emodin derivatives **24**[20]. Whereas the triacetyl compound **24a** gives three consecutive electron impact induced ketene eliminations, the tribenzoyl derivative **24b** produces an intensive peak at m/z 105 for the benzoyl ion and no further fragment ions are detected in the region between m/z 105 and the molecular ion.

In contrast, in the mass spectrum of the structurally related hexabenzoyl derivative of skyrin (**25**) (from *Hypericum perforatum* L.) an additional peak at $[M–226]^{+\cdot}$ appears, corresponding to a formal loss of benzoic acid anhydride[20]. Studies with model compounds confirmed that a proximity effect leading to the resonance stabilized fragment ion **c** is responsible for this behaviour.

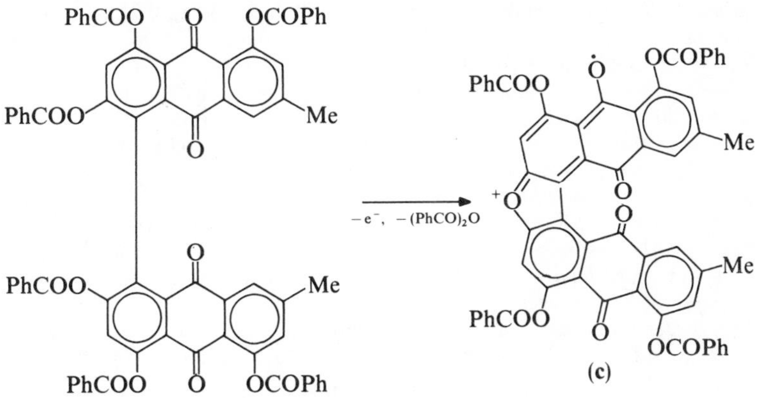

PhCOO O OCOPh
PhCOO Me
O
O
PhCOO Me
PhCOO O OCOPh
(**25**)

$\xrightarrow{-e^-,\ -(PhCO)_2O}$

PhCOO O OCOPh Me O O Me PhCOO O OCOPh
(**c**)

Glycosides of polyhydroxyanthraquinones are considered as the pharmacologically active constituents of many crude plant drugs. Their EI mass spectra do not exhibit molecular ion peaks; the base peaks usually correspond to the respective aglycones. Ammonia chemical ionization mass spectrometry has been used as a complementary technique for the identification of these glycosides[19]. However, in the case of aloe-emodin-8-glucoside (**26**) no molecular or quasi-molecular ion could be detected. The chemical ionization spectrum is dominated by an ion which may arise by attachment of ammonia to the glycosidic portion with transfer of a proton onto the aglycone. In addition, an abundant ion for the protonated aglycone is present.

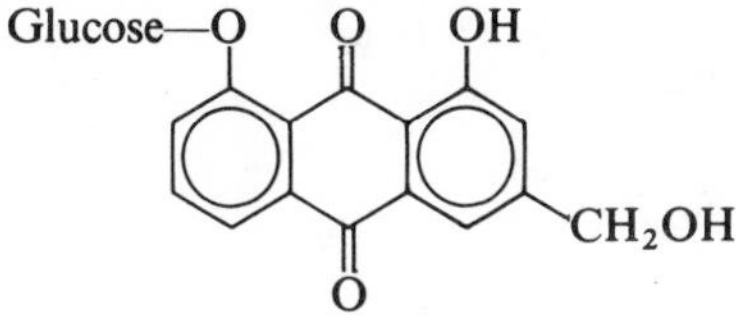

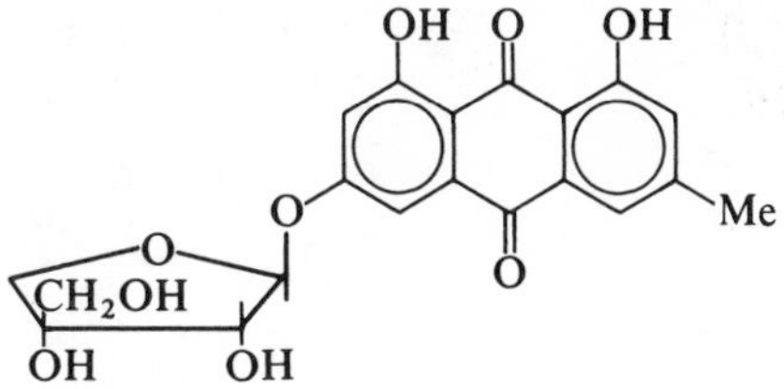

Glucose—O O OH
CH_2OH
O
(**26**)

OH O OH
O Me
O
CH_2OH
OH OH O
(**27**)

In contrast, the field desorption mass spectrum of **26** yields a molecular ion and a peak of the composition $[M + Na]^+$ as quasi-molecular ion originating from the attachment of sodium present as traces in the sample[19]. Similar results have been obtained with rhein-8-glucoside.

The structure of 6-*O*-(D-apiofuranosyl)-emodin (frangulin B, **27**) isolated from *Rhamnus frangula* L. has been elucidated by combined electron impact and field desorption mass spectrometry[21]. The field desorption mass spectrum of the pentaacetate shows a quasi-molecular ion peak at *m/z* 614 ($[M+2]^{+\cdot}$) and the sugar part obtained by hydrolysis has been recognized by field desorption and electron impact mass spectrometry of the trimethyl derivative as apiose.

D. Analysis of Quinones by Combined Chromatography/MS Methods

Combined chromatography/mass spectrometry techniques have found wide applications in biochemistry, natural product, environmental, agricultural and technical chemistry to identify and analyse quinones. The well documented fragmentation pattern of quinones[1] facilitates the assignment of chromatographic peaks when the mass spectrometer is used as a specific detector.

Combined gas-liquid chromatography/mass spectrometry has been used to identify the defence secretion components from members of the diplopod order Julia (Anthropoda)[22]. *Julus nitidus* produces a defensive secretion that consists exclusively of 2-methyl-3-methoxy-1,4-benzoquinone. The expelled defensive secretions of *Unciger foetidus, Cyclindroiulus coeruleocinctus, C. punctatus, C. luridus* and *Phyiulus psilosus* consist of mixtures of methyl-1,4-benzoquinone and 2-methyl-3-methoxy-1,4-benzoquinone. Similarly, it has been shown that the secrets of the pygidial glands of staphylinids of the genus *Bledius mandibularis* and *B. spectabilis* contain methyl-1,4-benzoquinone in addition to terpenes, undecene and γ-dodecalactone[23].

An analytical method for identification of minute amounts of hexahydroubiquinones in lipid extracts with the aid of gas-liquid chromatography/mass spectrometry has been developed[24].

The separation and identification of 37 naturally occurring hydroxyanthraquinones as the corresponding trimethylsilyl derivatives by capillary gas chromatography/mass spectrometry has been described[25].

An interesting application has been reported for the diagnosis of gangrene in potato tubers. The causal organism of this plant disease, *Phoma exigua* var. *foeata*, contains pachybasin (**28**) as anthraquinone pigment. The detection of specific ions from **28** by pyrolysis mass spectrometry, direct probe mass spectrometry and gas chromatography/mass spectrometry as diagnostic approaches offer advantages over other methods[26]. In particular, the GC/MS method has been suggested as a promising technique because of its high specificity, sensitivity, suitability for automation and reduction of total expertise time.

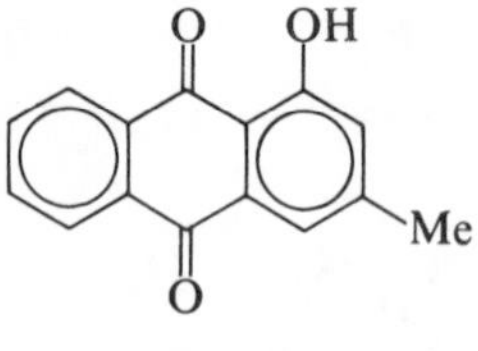

O OH Me O O NHR NHR

(28) 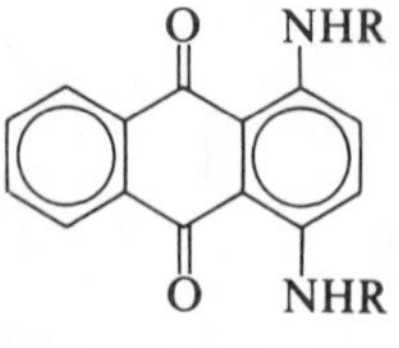 **(29**; R = H—$C_{18}H_{37}$)

Quinones derived from polycyclic aromatic hydrocarbons have been identified by glass capillary gas chromatography/mass spectrometry in diesel exhaust particulate matter[27, 28].

Amongst other dyes, 1,4-dialkylaminoanthraquinones (**29**) are added to gasoline to identify manufacture and to trace petroleum contamination in soil, groundwater, wastewater, etc. Several mass spectrometric and chromatographic techniques have been suggested for this purpose[29]. Due to the low volatility of the anthraquinone dyes (**29**), mass spectrometry as a specific and sensitive detection technique should be combined with high-performance liquid chromatography as the separation method. This has been successfully demonstrated[30] using the thermospray HPLC/MS interface. This device is able to handle high quantities of aqueous solvents and the ionization process is 'soft', resulting in abundant $[M + H]^+$ ion formation. By this method it has been shown that the commercial blue dye is composed of the various alkyl-substituted aminoanthraquinones (**29**) with R ranging from H to $C_{18}H_{37}$. Furthermore, the thermospray HPLC/MS method has been demonstrated to detect dyes in water, soil and gasoline at the low parts-per-million levels.

Recently, a new thin-layer chromatogram scanner/mass spectrometer system has been introduced using 1,4-benzoquinone, 1,4-naphthoquinone and anthraquinone as testing materials[31]. The scanner moves the TLC plates past a pulsed CO_2 laser as a source of desorption energy. The chemical ionization reagent gas sweeps the material desorbed from the TLC spots into the ion source of the mass spectrometer. Reasonable chromatographic peaks have been produced by monitoring mass chromatograms ($[M + H]^+$) from TLC plates spotted with 10 μg of samples of benzoquinone, naphthoquinone and anthraquinone.

III. NEGATIVE-ION MASS SPECTRA

In an electron impact source negative ions can be produced by three main processes[32, 33].

(1) resonant electron capture of secondary (thermal) electrons:

$$AB + e^- \rightarrow AB^{-\cdot}$$

(2) dissociative attachment:

$$AB + e^- \rightarrow [AB^{-\cdot}]^* \rightarrow A + B^{-\cdot}$$

(3) ion-pair production:

$$AB + e^- \rightarrow A^+ + B^- + e^-$$

In general, in an electron impact source operated at 70 eV little or no molecular anions are generated, and fragment ions may be produced by different processes. The problems inherent in negative-ion mass spectrometry may be overcome by using a structural moiety which is able to produce stabilized molecular anions without utilization of non- or antibonding molecular orbitals. The quinone and in particular the anthraquinone system

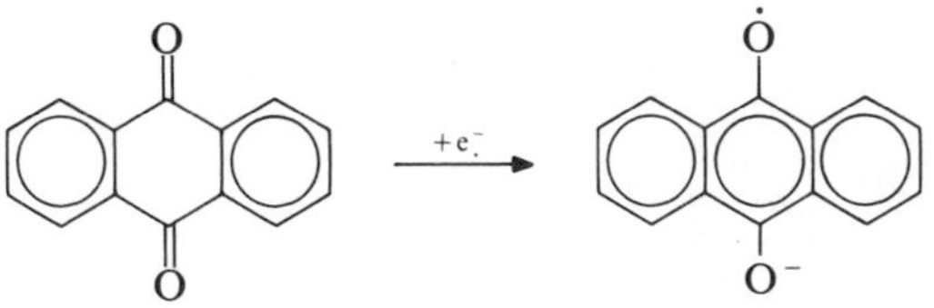

(**d**)

yields such stabilized molecular anions of the semiquinone type **d**. The stable unsubstituted semiquinone system **d** shows no decomposition; however, attached substituents may cause unimolecular fragmentations[34].

The molecular anions of 2-alkoxycarbonylanthraquinones (**30**) eliminate alkyl radicals from the ester group[34]. This process increases with the size of R and when a secondary

TABLE 1. Relative intensities of the molecular anion and of the fragment ions at m/z 251 and 207 in the negative-ion spectra of 2-alkoxycarbonylanthraquinones (**30**)

30	R	$M^{-\cdot}$	$[M-R]^-$ m/z 251	$[M-R-CO_2]^-$ m/z 207
a	Me	100	5	7
b	Et	100	7	12
c	*n*-Pr	100	10	16
d	*i*-Pr	100	22	25
e	*n*-Bu	100	12	18
f	*n*-Hex	100	16	20

Reproduced from Ref. 34 *by permission of the Royal Society of Chemistry.*

radical is eliminated. The elimination of an alkyl radical is followed by loss of carbon dioxide (Table 1). Figure 2(a) illustrates this sequence for 2-ethoxycarbonylanthraquinone (**30b**). This fragmentation pattern, which has no counterpart in the positive-ion spectra, is substantiated by appropriate metastable peaks and the metastable defocusing technique. As indicated in the scheme by an asterisk a metastable transition is also observed for the loss of the alkoxycarbonyl group.

(**30**$^{-\cdot}$) —(*, $-R^\cdot$)→ m/z 251

(**30**$^{-\cdot}$) —(*, $-\dot{C}O_2R$)→ m/z 207

m/z 251 —(*, $-CO_2$)→ m/z 207

In addition to similar cleavages, the spectrum of 1-ethoxycarbonylanthraquinone (**31**) contains a $[M-EtO]^-$ ion (Figure 2(b)). This is probably the result of a proximity effect which cannot be operative in the 2-substituted isomer.

(**31**$^{-\cdot}$; m/z 280) —(*, $-EtO^\cdot$)→ m/z 235

The negative-ion spectra of 1- and 2-acetoxy-substituted anthraquinones are also quite different (Table 2)[34]. The spectrum of the 2-isomer (**32**) is dominated by the loss of an acetyl radical from the molecular anion. On the other hand, the molecular anion of the

1-isomer **33** eliminates ketene, which demonstrates the occurrence of a proximity effect. The hydrogen rearrangement can either proceed to a radical or to an anionic centre.

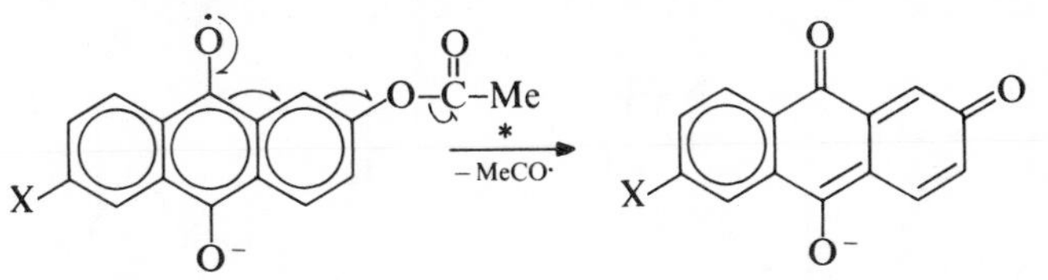

O—C—Me
$\xrightarrow[-\text{MeCO}\cdot]{*}$

(**32**$^{-\cdot}$; X = H)
(**37**$^{-\cdot}$; X = OCOMe)

X = H; m/z 223
X = OCOMe; m/z 281

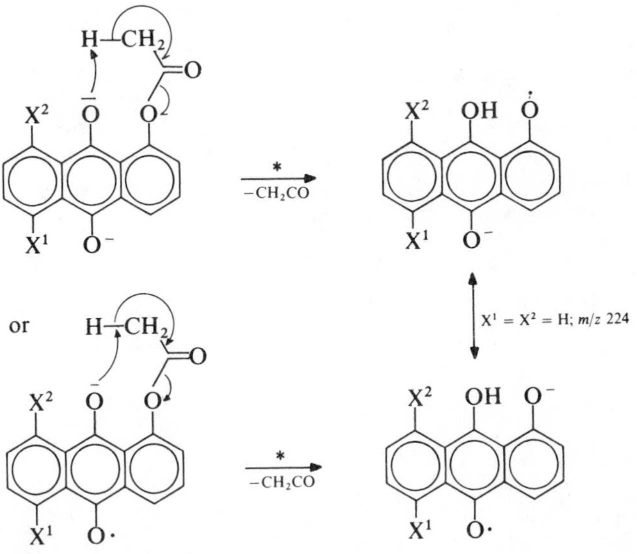

$\xrightarrow[-\text{CH}_2\text{CO}]{*}$

$X^1 = X^2$ = H; m/z 224

or

$\xrightarrow[-\text{CH}_2\text{CO}]{*}$

(**33**$^{-\cdot}$; $X^1 = X^2$ = H)
(**34**$^{-\cdot}$; X^1 = OCOMe, X^2 = H)
(**35**$^{-\cdot}$; X^1 = H, X^2 = OCOMe)

In the negative-ion spectra of the disubstituted compounds **34** and **35** the stepwise elimination of two molecules of ketene is observed. In the case of compound **34** the second loss of ketene requires a prior hydrogen transfer if the above mechanism is applicable.

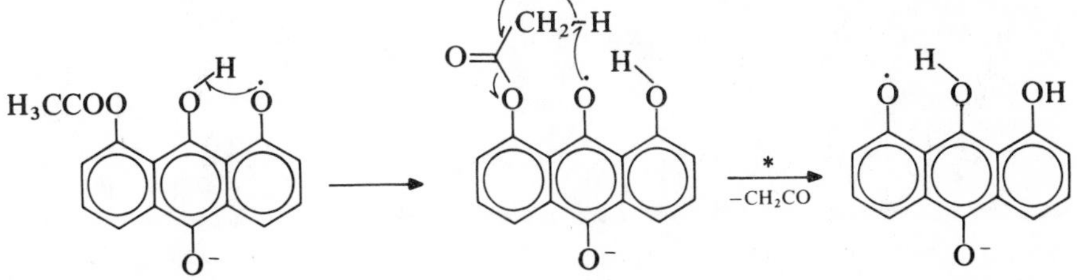

H_3CCOO

$\xrightarrow[-\text{CH}_2\text{CO}]{*}$

m/z 282 m/z 282 m/z 240

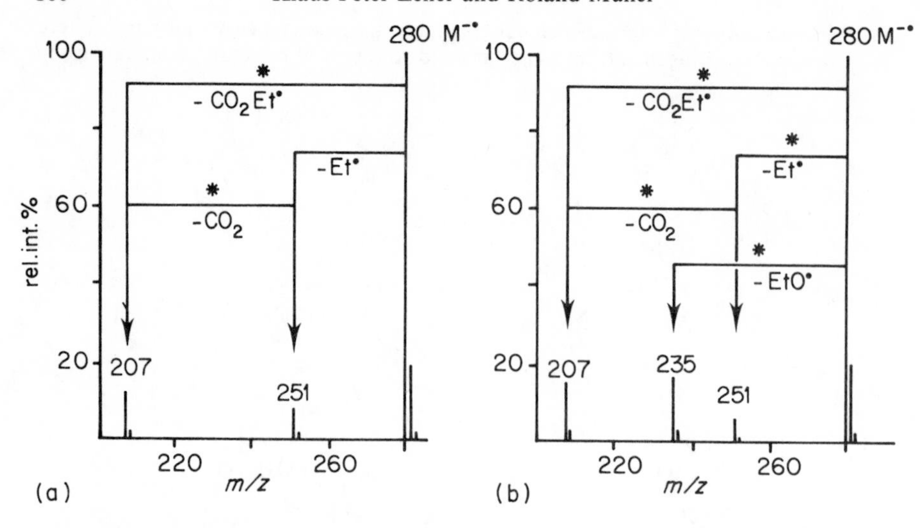

FIGURE 2. Negative-ion electron impact mass spectra of 2-alkoxycarbonylanthraquinone (**30b**) (a) and 1-ethoxycarbonylanthraquinone (**31**) (b). *Reproduced from Ref. 34, by permission of the Royal Society of Chemistry*

TABLE 2. Relative intensities of molecular and fragment ions in the negative-ion spectra of mono- and diacetoxyanthraquinones

Compound	$M^{-\cdot}$	$[M–CH_2CO]^{-\cdot}$	$[M–MeCO]^-$	$[M–2CH_2CO]^{-\cdot}$	$M–CH_2CO–MeCO]^-$
32	40	0	100	–	–
33	18	100	9	–	–
34	20	58	15	100	6
35	39	100	4	43	14
36	30	100	10	30	14
37	60	0	100	0	15

Reproduced from Ref. 34 by permission of the Royal Society of Chemistry.

The negative-ion spectrum of 1,2-diacetoxyanthraquinone (**36**) discloses two fragmentation pathways. The major process involves two proximity effects which result in the elimination of two ketene molecules. As demonstrated by appropriate deuterium labelling the first ketene elimination takes place from the 1-position. Additionally a sequence $M^{-\cdot}–CH_2CO–MeCO^{\cdot}$ is observed.

$OCOCH_3$ $\xrightarrow[-CH_2CO]{*}$ $\xrightarrow[-CH_2CO]{*}$

(**36**$^{-\cdot}$; m/z 324) m/z 282 m/z 240

The semiquinone radical anion formed from 2,6-diacetoxyanthraquinone (**37**) contains no phenoxide radical (or anion) centre adjacent to an acetoxy group. Consequently, no elimination involving a hydrogen transfer is possible and the only process observed is loss of an acetyl radical. The fragment formed is an anion with an even number of electrons. Therefore, the loss of a second acetyl radical is impossible if the fragmentation is triggered by the radical site. However, ketene as a neutral molecule can be eliminated leading to a further even-electron ion at m/z 239.

The different behaviour of α- and β-acetoxy groups bound to anthraquinones can be applied in the analysis of positional isomers. This information is not forthcoming from positive-ion spectra, as all arylacetate molecular cations eliminate ketene.

Similar rearrangement processes including elimination of ketene are found in the negative-ion spectra of the naphthoquinones **38** and **39**[34].

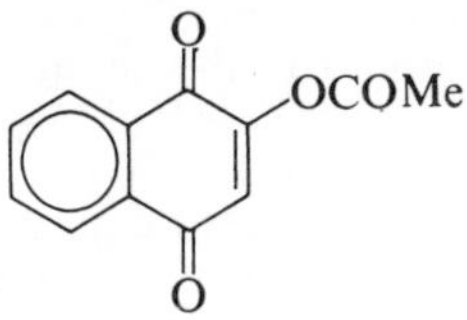

OCOMe

(**38**)

MeCOO

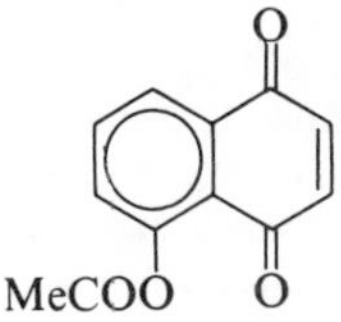

(**39**)

The basic fragmentation of the molecular anions of alkoxy-substituted anthraquinones **40** and **41** involves loss of an alkyl radical to produce resonance-stabilized fragments (Table 3)[35]. Proximity effects which should produce $[M\text{–alkene}]^{-\cdot}$ ions from α-substituted compounds **41** (R $\geqslant$ Et) are not observed.

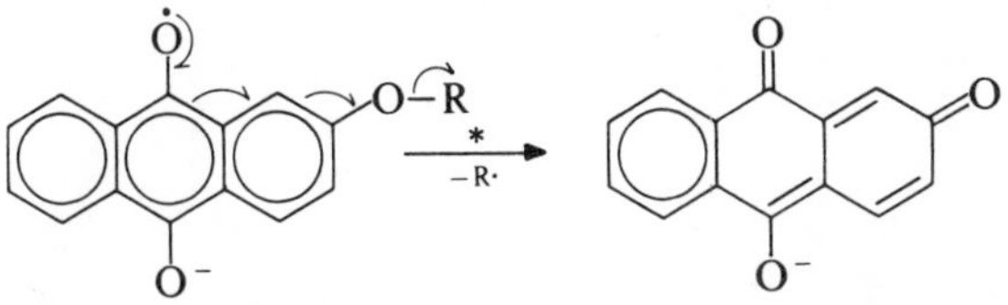

$\xrightarrow[-R\cdot]{*}$

(**40a**$^{-\cdot}$; R = Me)
(**40b**$^{-\cdot}$; R = n-Bu)

(**e**)

$\xrightarrow[-R\cdot]{*}$

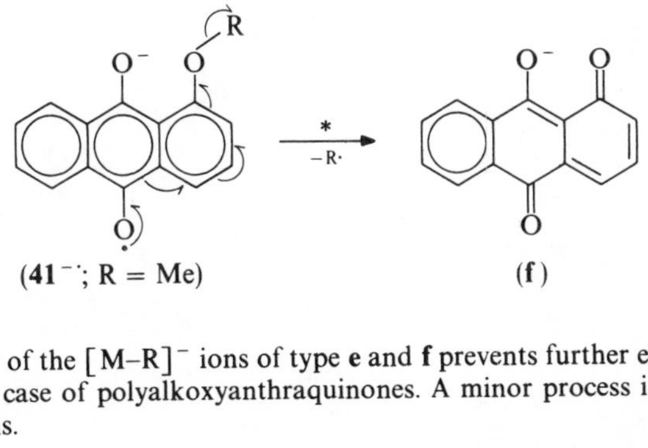

(**41**$^{-\cdot}$; R = Me)

(**f**)

The stability of the $[M\text{–R}]^-$ ions of type **e** and **f** prevents further elimination of alkyl radicals in the case of polyalkoxyanthraquinones. A minor process is the formation of $[M\text{–OR}]^-$ ions.

TABLE 3. Relative intensities of molecular anions and fragment ions in the negative-ion spectra of alkoxy-substituted anthraquinones

Compound	$M^{-\cdot}$	$[M-R]^-$	$[M-OR]^-$
40a	100	20	1
40b	38	100	3
41	100	21	1

Reproduced from Ref. 35, by permission of the Commonwealth Scientific and Industrial Research Organization, Australia.

Many naturally occurring anthraquinones contain both methoxy and hydroxy groups. By converting the latter into acetate functions a differentiation between positional isomers is possible. The spectra of **42** and **43** illustrate this point (Figure 3(a), (b))[35]. From the known behaviour of alkoxy and acetoxy groups in the negative-ion spectra of anthraquinones the following sequences are expected; (**42**): $M^{-\cdot}$–$Me^{\cdot}$–CH_2CO and/or $M^{-\cdot}$–$MeCO^{\cdot}$, (**43**): $M^{-\cdot}$–$Me^{\cdot}$–CH_2CO and/or $M^{-\cdot}$–CH_2CO–$Me^{\cdot}$.

Both fragmentations are expected to occur for **42**. Competitive studies with the derivative monodeuteriated at the acetoxy group yields a primary deuterium isotope effect

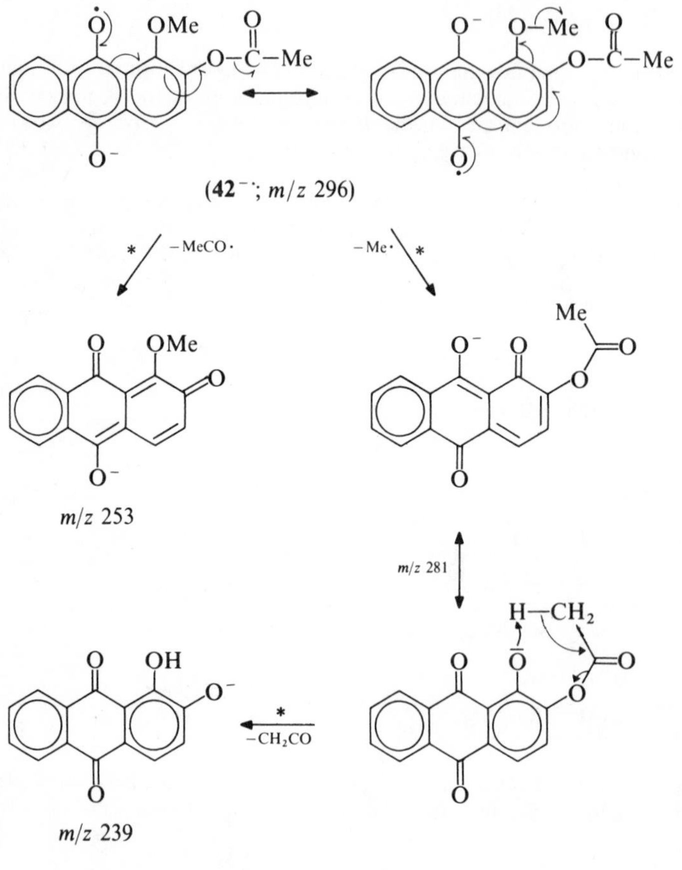

(**42**$^{-\cdot}$; m/z 296)

$-MeCO\cdot$ $-Me\cdot$

m/z 253

m/z 281

$-CH_2CO$

m/z 239

k_H/k_D of 2.1 in the ion source and 2.4 in the first field-free region[36]. The close similarity of both values is in accordance with the fact that molecular anions in contrast to molecular cations have uniformly low internal energies since they are formed by secondary (thermal) electron capture.

In the case of **43** the fragmentation almost exclusively occurs by initial elimination of ketene (which is not possible with the molecular anion of **42**) and subsequent loss of a methyl radical.

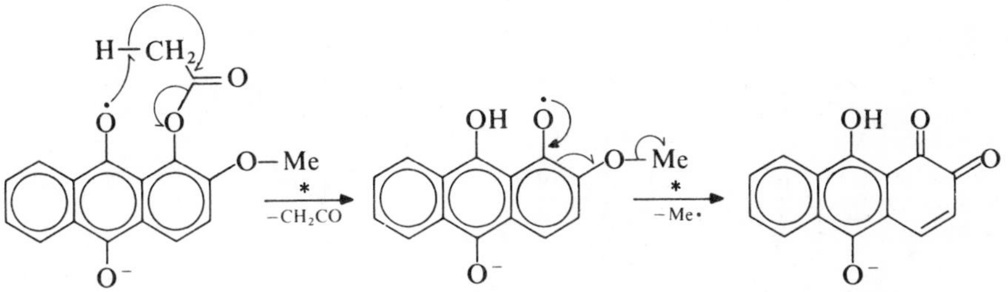

(**43**$^{-\cdot}$; m/z 296) $\xrightarrow[-CH_2CO]{*}$ m/z 254 $\xrightarrow[-Me\cdot]{*}$ m/z 239

The positive-ion spectra cannot afford such ready differentiation between two anthraquinone isomers. The spectrum of 1,8-diacetoxy-3-methyl-6-methoxyanthraquinone (**44**) (Figure 3(c)) is more complex. However, it clearly indicates the presence of one methoxy group together with either two *peri*-acetoxy (1,4; 1,5; 1,8) or two vicinal acetoxy (1,2) functions.

The molecular anions of 1- (**45**) and 2-nitroanthraquinone (**46**) represent the base peaks in the negative-ion mass spectra[37]. To some extent these species decompose by elimination of NO$^\cdot$ to produce fragments with an even number of electrons. Their relative intensities are 5 and 3 %, respectively. These skeletal rearrangements are accompanied by flat-topped metastable peaks.

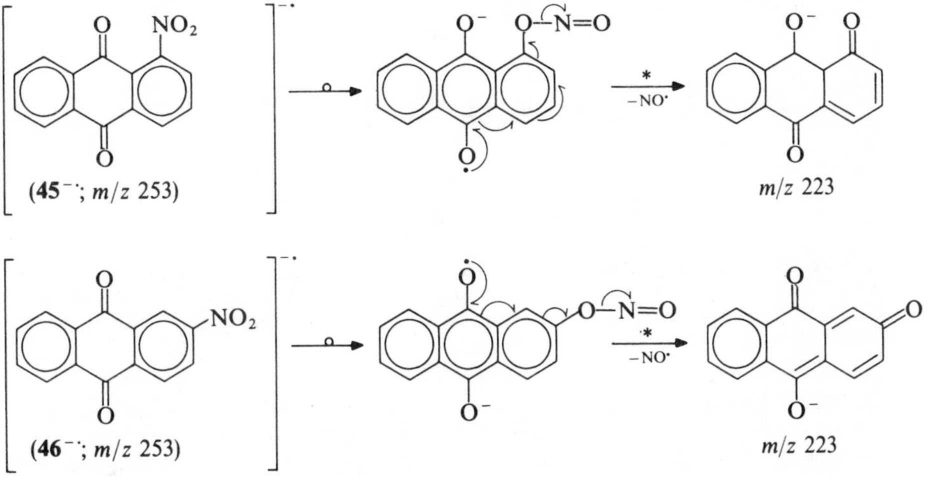

(**45**$^{-\cdot}$; m/z 253) $\rightarrow$ $\xrightarrow[-NO^\cdot]{*}$ m/z 223

(**46**$^{-\cdot}$; m/z 253) $\rightarrow$ $\xrightarrow[-NO^\cdot]{*}$ m/z 223

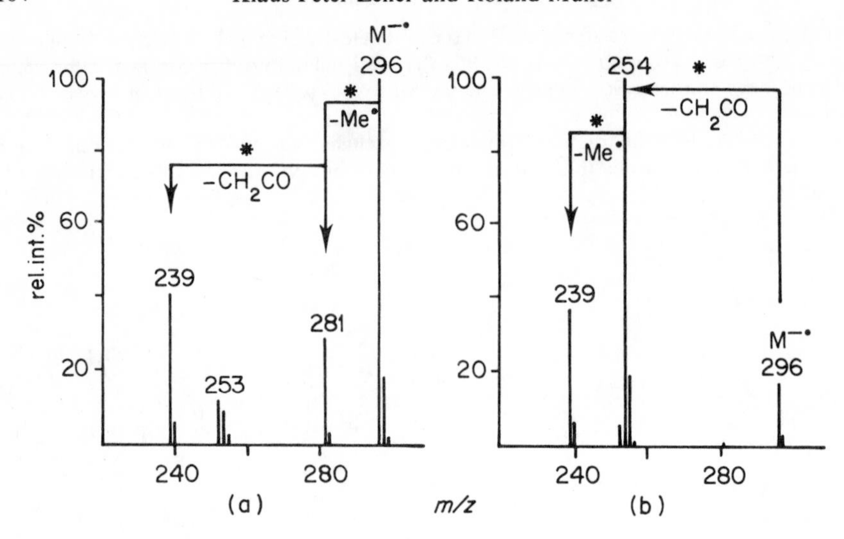

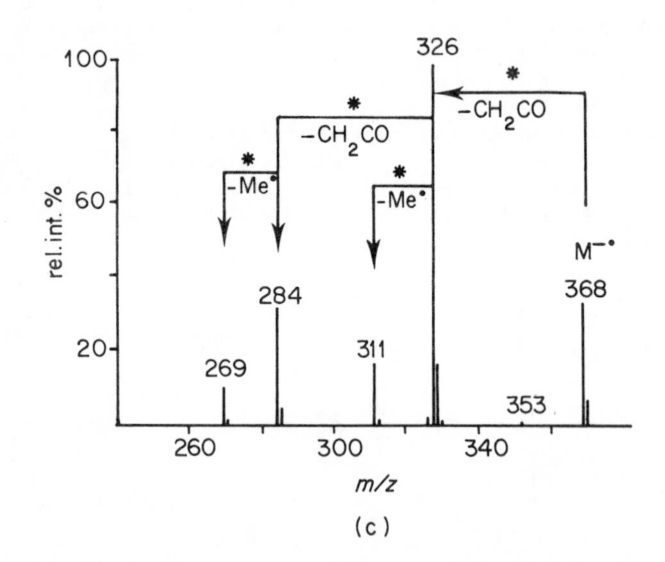

FIGURE 3. Negative-ion mass spectra of 2-acetoxy-1-methoxyanthraquinone (**42**) (a), 1-acetoxy-2-methoxyanthraquinone (**43**) (b) and 1,8-diacetoxy-3-methyl-6-methoxyanthraquinone (**44**) (c). *Reproduced from Ref.* 35, *by permission of the Commonwealth Scientific and Industrial Research Organization, Australia*

Unsubstituted quinones exhibit no fragmentation in the negative mode due to the high stability of the semiquinone radical anions. However, collisional activation with an inert gas in the field-free regions of the mass spectrometer imparts further internal energy to the molecular anion. These ions then undergo collision-induced decompositions which may be detected in either of the field-free regions, applying techniques originally developed to investigate the collision-induced dissociation of positively charged ions. The collision gas should not yield negative ions and should not react with negative ions to produce ion–molecule product ions. Krypton, nitrogen, benzene and toluene have successfully been applied at pressures of ca. 3×10^{-5} torr. Under these conditions the naphthoquinone molecular anions give pronounced collision-induced peaks in both field-free regions originating from the unusual decompositions $M^{-\cdot}$–$CHO^{\cdot}$ and $M^{-\cdot}$–($CHO^{\cdot}$ + CO)[38].

The negative-ion spectra of anthraquinone 1- (**47**) and 2-carboxylic acid (**48**) show only molecular anions when measured under normal conditions (70 eV, source pressure ca. 10^{-6} torr)[39]. The collision-induced fragmentation with toluene as collision gas (3×10^{-5} torr) has been detected in the second field-free region by standard magnetic scan and in the first field-free region by the metastable defocusing technique. The molecular anions of both isomers yield identical collisional activation spectra with no proximity effect allowing differentiation between **47** and **48**. The main fragmentations are $M^{-\cdot}$–$^{\cdot}COOH$ and $M^{-\cdot}$–CO_2. The former represents α-cleavage to the charged carboxy group. Driving force for the elimination of CO_2 is the generation of the very stable semiquinone radical

anion. A third peak observed corresponds to the process $M^{-\cdot}$–$C_2HO_3^{\cdot}$. This process must consist of the consecutive loss of CO_2 and $CHO^{\cdot}$ involving one of the carbonyl centres of the quinone moiety.

Plots of the ionization efficiencies of the molecular anion and the collision-induced $[M–CO_2]^{-\cdot}$ and $[M–CO_2H]^-$ ions of anthraquinone 2-carboxylic acid (**48**) against the nominal electron beam energy are shown in Figure 4[39]. From the ionization efficiency curve of the molecular anion it follows that this ion may be produced in two ways: (1) capture of low energy primary electrons at nominal 5–10 eV and (2) capture of

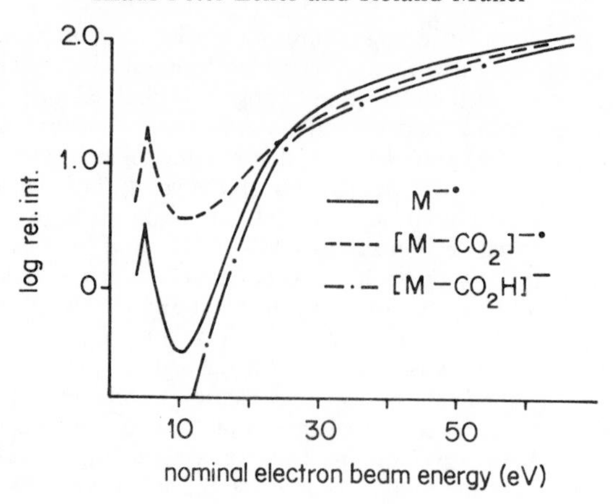

FIGURE 4. Ionization efficiency curves of $M^{-\cdot}$, $[M\text{–}CO_2]^{-\cdot}$ and $[M\text{–}CO_2H]^-$ from anthraquinone 2-carboxylic acid (**48**). The relative intensity of each ion is arbitrarily taken as 100 % at 70 eV. *Reproduced from Ref.* 39

secondary (thermal) electrons when the nominal beam energy is greater than 15 eV. Obviously, the collision-induced loss of CO_2 is possible from molecular anions produced by either process. In contrast, elimination of $^\cdot CO_2H$ occurs only by collisional activation of molecular anions formed by secondary electron capture.

It has been demonstrated by Bowie and Blumenthal[40] that non-decomposing molecular anions can undergo charge inversion by high-energy ion molecule reactions in the analyser regions of the mass spectrometer. This process is accompanied by efficient conversion of translational energy to internal energy leading to subsequent decomposition. Thus, spectra of positive ions are produced from molecular anions.

$$M^{-\cdot} + N \rightarrow [M^{+\cdot}]^* + N + 2e^-$$
$$\downarrow$$
$$\text{fragments}$$

The spectra obtained are named $+E$ spectra because the electric sector in a double focusing instrument is operated at the reverse potential to that $(-E)$ used to measure conventional negative-ion spectra. $+E$ spectra may be determined for dissociations in either field-free region of the mass spectrometer. Decompositions in the first field-free region are detected by the ion kinetic energy technique and in the second field-free region by carrying out a magnetic scan with the electric sector at $-E$ and the magnet set to transmit positive ions. The target gas may be some species which does not produce negative ions in the source (e.g. He, N_2, C_6H_6), or the sample itself may be used as the target gas (pressure in the collision region ca. $3\text{–}5 \times 10^{-5}$ torr).

A number of benzoquinones, naphthoquinones, and higher quinones have been investigated by this technique[40,41]. The $+E$ mass spectrum from the naphthoquinone molecular anion is given in Figure 5(a)[40]. It yields very similar information as the conventional positive-ion spectrum (Figure 5(b))[1,42]. In the latter spectrum the peaks are produced by a series of consecutive and competitive unimolecular decompositions; in the $+E$ spectrum the same peaks are obtained via collision-induced dissociations. Differences are seen concerning the abundances of fragments; e.g. in the normal positive-ion spectrum the rearrangement ion $[M\text{–}CO_2]^{+\cdot}$ at m/z 114 is of low relative intensity[42], whereas in the

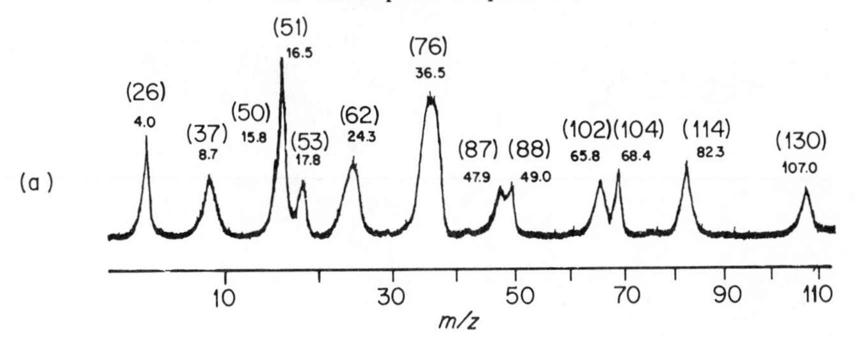

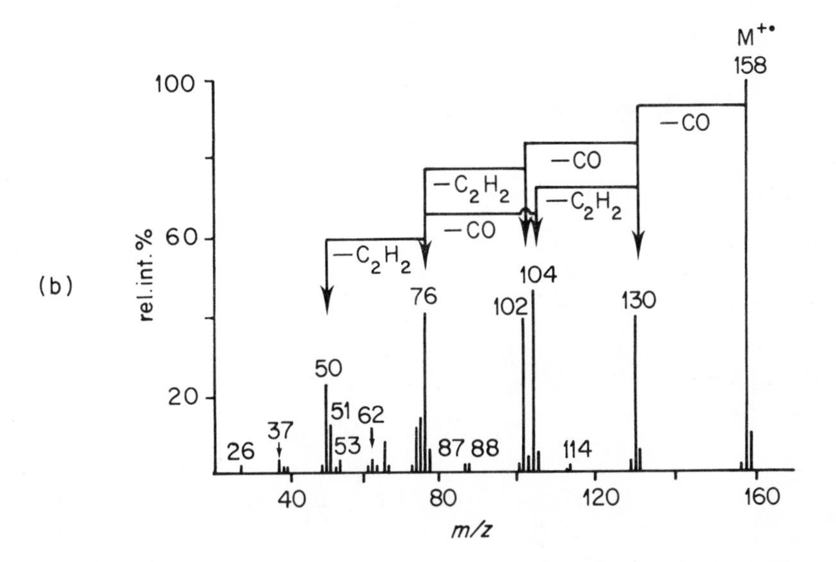

FIGURE 5. (a) $+E$ mass spectrum derived from the 1,4-naphthoquinone molecular anion in the second field-free region (sample pressure ca. 3×10^{-5} torr). The values given are readings on the mass scale corresponding to m_2^2/m_1; values in parenthesis refer to the fragment ion mass m_2. (b) Conventional positive-ion mass spectrum of 1,4-naphthoquinone. *Reproduced from J. H. Bowie and T. Blumenthal,* J. Am. Chem. Soc., **97**, 2959 (1975), *by permission of the American Chemical Society. Copyright* (1975) *American Chemical Society*

$+E$ spectrum this process contributes more substantially. A particular example is afforded by 2-hydroxy-1,4-naphthoquinone (**49**) and 2-amino-1,4-naphthoquinone (**50**). The positive-ion spectra of **49**[1] and **50**[41] are dominated by the process $(M^{+\cdot}–CO) \rightarrow (M^{+\cdot}–CO)–C_2HO^{\cdot}$ and $(M^{+\cdot}–HCN) \rightarrow (M^{+\cdot}–HCN)–C_2HO^{\cdot}$, respectively. In the $+E$ counterparts peaks due to the final product ion of these sequences at m/z 105 are present, however, of low abundances.

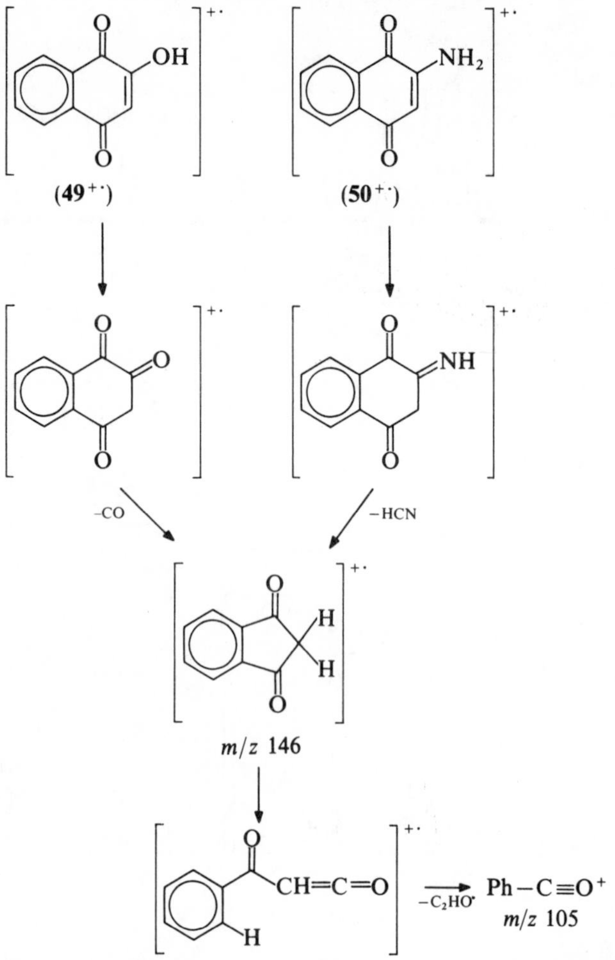

(**49**$^{+\cdot}$) (**50**$^{+\cdot}$)

−CO −HCN

m/z 146

CH=C=O $\xrightarrow[-C_2HO^{\cdot}]{}$ Ph−C≡O$^+$ *m/z* 105

Benzoquinone when ionized by fast atom-beam bombardment (FAB) in a glycerol matrix is able to capture an electron to produce the semiquinone radical anion (rel. intensity ca. 60 %)[43]. This process is followed by hydrogen abstraction from glycerol to yield the monoanion of hydroquinone representing the base peak. The formation of these ions is of general importance for the theory of ion formation under fast atom-beam bombardment, since it indicates that in addition to protonation and proton abstraction electron capturing and hydrogen abstraction may also contribute to the formation of ions in a FAB source.

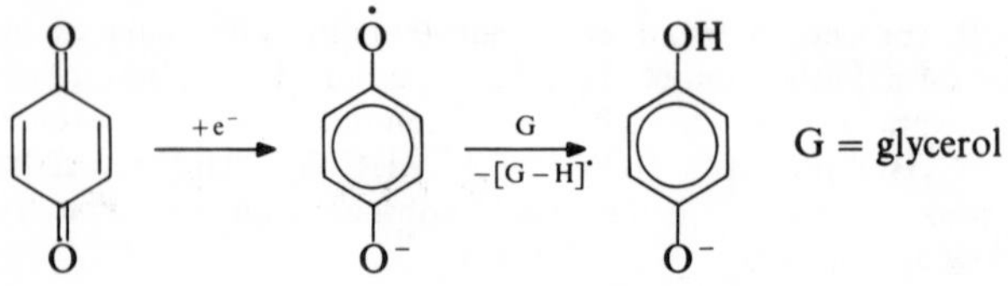

$\xrightarrow{+e^-}$ $\xrightarrow[-[G-H]^{\cdot}]{G}$ G = glycerol

IV. REFERENCES

1. K.-P. Zeller, *The Chemistry of Quinonoid Compounds*, Part 1, p. 231 (Ed. S. Patai), Wiley-Interscience, 1974.
2. J. E. Szulejko and M. M. Bursey, *Org. Mass Spectrom.*, **20**, 374 (1985).
3. T. Morita, M. Karasawa and K. Takase, *Chem. Lett.*, 197 (1980).
4. L. T. Scott and C. M. Adams, *J. Am. Chem. Soc.*, **106**, 4857 (1984).
5. A. D. Vasileva, I. A. Rotermel, N. N. Artamonova, L. A. Gaeva, A. A. Dubrovin and B. N. Kolokolov, *Zh. Org. Khim.*, **8**, 2109 (1972); Engl. translation 2156 (1972).
6. For a discussion of these investigations see Ref. 1, p. 234.
7. W. Blum, H. Kurreck, W. J. Richter, H. Schwarz and H. Thies, *Angew. Chem.*, **95**, 59 (1983); *Angew. Chem. Int. Ed. Engl.*, **22**, 51 (1983).
8. M. M. Bursey, D. J. Harvan and J. R. Hass, *Org. Mass Spectrom.*, **20**, 197 (1985).
9. R. D. Grigsby, W. D. Jamieson, A. G. McInnes, W. S. G. Maass and A. Taylor, *Can. J. Chem.*, **52**, 4117 (1974).
10. R. H. Thompson, *Naturally Occurring Quinones*, Academic Press, London, New York, 1971, p. 78.
11. G. R. Waller and O. C. Dermer (Eds), *Biomedical Applications of Mass Spectrometry*, First Supplementary Volume, Wiley Interscience, New York, 1980.
12. H. Schwarz, V. Pasupathy and W. Steglich, *Org. Mass Spectrom.*, **11**, 472 (1976).
13. T. A. Elwood, K. H. Dudley, J. M. Tesarek, P. F. Rogerson and M. M. Bursey, *Org. Mass Spectrom.*, **3**, 841 (1970).
14. P. Cong, *Yaoxue Xuebao*, **19**, 450 (1984); *Chem. Abstr.*, **103**, 104350 (1985).
15. Y. N. Shukla, J. S. Tandon, D. S. Bhakuni and M. M. Dhar, *Phytochem.*, **10**, 1909 (1971).
16. T. L. Folk, H. Singh and P. J. Scheuer, *J. Chem. Soc. Perkin Trans. 2*, 1781 (1973).
17. G. Prota, M. D'Agostino and G. Misuraca, *Experientia*, **27**, 15 (1971).
18. P. Chang and K. H. Lee, *Phytochem.*, **23**, 1733 (1984).
19. F. J. Evans, M. G. Lee and D. E. Games, *Biomed. Mass Spectrom.*, **6**, 374 (1979).
20. G. Schwinger and D. Spitzner, *Org. Mass Spectrom.*, **13**, 177 (1978).
21. H. Wagner and G. Demuth, *Tetrahedron Lett.* 5013 (1972).
22. H. Röper and K. Heyns, *Z. Naturforsch.*, **32c**, 61 (1977).
23. J. W. Wheeler, G. M. Happ, J. Aranjo and J. M. Pasteels, *Tetrahedron Lett.*, 4635 (1972).
24. J. Gürtler and R. Blomstrand, *Int. J. Vitamin Nutr. Res.*, **41**, 204 (1971).
25. G. W. van Eijk and H. J. Roeijmans, *J. Chromatogr.*, **295**, 497 (1984).
26. A. C. M. Weijman, G. W. van Eijk, H. J. Roeijmans, W. Windig, J. Haverkamp and L. J. Turkensteen, *Neth. J. Plant Pathol.*, **90**, 107 (1984); *Chem. Abstr.*, **102**, 44470 (1985).
27. J. Schulze, A. Hartung, H. Kiess, J. Kraft and K. H. Lies, *Chromatographia*, **19**, 391 (1984).
28. H. Y. Tong, J. A. Sweetman, F. W. Karasek, E. Jellum and A. K. Thorsrud, *J. Chromatogr.*, **312**, 183 (1984).
29. T. L. Youngless, J. T. Swansiger, D. A. Danner and M. Greco, *Anal. Chem.*, **57**, 1894 (1985).
30. R. D. Voyksner, *Anal. Chem.*, **57**, 2600 (1985).
31. L. Ramaley, M.-A. Vaughan and W. D. Jamieson, *Anal. Chem.*, **57**, 353 (1985).
32. C. E. Melton, in *Mass Spectrometry of Organic Ions* (Ed. F. W. McLafferty), Academic Press, New York, London, 1963, Chap. 4.
33. M. von Ardenne, K. Steinfelder and R. Tümmler, *Elektronenanlagerungsmassenspektrographie Organischer Substanzen*, Springer-Verlag, Berlin, Heidelberg, New York, 1971.
34. A. C. Ho, J. H. Bowie and A. Fry, *J. Chem. Soc. (B)*, 530 (1971).
35. J. H. Bowie and A. C. Ho, *Aust. J. Chem.*, **24**, 1093 (1971).
36. J. A. Benbow, J. C. Wilson and J. H. Bowie, *Int. J. Mass Spectrom. Ion Phys.*, **26**, 173 (1978).
37. J. H. Bowie, *Org. Mass Spectrom.*, **5**, 945 (1971).
38. J. H. Bowie, *J. Am. Chem. Soc.*, **95**, 5795 (1973).
39. J. H. Bowie, *Org. Mass Spectrom.*, **9**, 304 (1974).
40. J. H. Bowie and T. Blumenthal, *J. Am. Chem. Soc.*, **97**, 2959 (1975).
41. J. H. Bowie and T. Blumenthal, *Aust. J. Chem.*, **29**, 115 (1976).
42. J. H. Bowie, D. W. Cameron and D. H. Williams, *J. Am. Chem. Soc.*, **87**, 5094 (1965).
43. E. Clayton and A. J. C. Wakefield, *Chem. Commun.*, 969 (1984).

The Chemistry of Quinonoid Compounds, Vol. II
Edited by S. Patai and Z. Rappoport

CHAPTER 4

Chiroptical properties and absolute configurations of chiral quinones

HOWARD E. SMITH
Department of Chemistry, Vanderbilt University, Nashville, Tennessee 37235, USA

I. INTRODUCTION

In his classical work on the chiroptical properties of chemical substances, Lowry[1] reviewed the early history of rotatory power measurements. Thus from the very beginning Biot and Louis Pasteur made such measurements using light of various wavelengths in the visible part of the spectrum, the rotatory power of a substance over a spectral range being known as its optical rotatory dispersion (ORD)[1]. After Biot's death in 1862 and the introduction about 1866 of the Bunsen burner as a light source, it became much easier to work with the nearly monochromatic light of the sodium flame (589 nm), and the more laborious study of ORD was largely abandoned. As a matter of convenience then, the chiroptical property most frequently reported for chiral substances is their rotatory power for sodium D light[2].

Useful compilations of this chiroptical property have appeared, and Thomson[3] in his very important review of naturally occurring quinones has presented their physical properties, including the specific rotations of naturally occurring chiral quinones. More recently the absolute configurations of a host of chiral substances, including chiral quinones, have been given in two collections[4,5]. In one of these[4], the absolute configurations of approximately 6000 compounds, the method by which each configurational assignment was made, the sign of the rotatory power for a particular enantiomer, and appropriate literature references are given. In the other[5], the absolute configurations, the sign of the rotatory powers for given states (liquid or as solutions in various solvents) and literature references are tabulated for nearly 6000 compounds. This latter collection has limited value for chiral quinones since it is restricted to compounds with one chiral center (asymmetric carbon atom), and many chiral quinones have more than one such center or are chiral as the result of restricted rotation about a single bond (atropisomerism[6]) and thus do not have any chiral center.

The rotatory powers (with sodium D light) of chiral organic compounds, including chiral quinones, are useful as the means for their characterization, both as a physical property which distinguishes one enantiomer from the other but also as the means by which the optical purity (percent enantiomeric excess, % ee) of particular samples can be easily determined.

Prior to 1950 rotatory powers were occasionally used for the determination of absolute configurations (rotatory power comparisons)[7], but most of these assignments were made by chemical correlations[4,5]. Certainly rotatory power measurements are not as reliable as ORD and circular dichroism (CD) methods[8] for absolute configurational assignments by comparison, but before 1950, ORD curves were measured only with ease in the visible spectral region (380–780 nm) and with great difficulty in the near ultraviolet region (200–380 nm)[1]. In 1953, a commercially manufactured spectropolarimeter capable of routine ORD measurements from 700 to 280 nm became available[9] and now measurements are easily made to 185 nm[8]. Circular dichroism (CD) curves were measured in the visible spectral region during the nineteenth century[1], but they were not common before 1960[9]. The description of the first recording circular dichrograph[10] led to a rapid development in this field and to commercial availability of instruments capable of routine CD measurements in the visible and the near ultraviolet spectral regions (180–600 nm)[8,11].

The main focus of this chapter then is a brief outline of the use of ORD and CD measurements in the visible and near ultraviolet spectral region for the establishment of the absolute configurations of chiral quinones, occurring for the most part as natural products[3]. Other chiroptical measurements such as far ultraviolet circular dichroism[12] (FUVCD), vibrational (infrared) circular dichroism[13] (VCD) and Raman optical activity[14] (ROA) are just beginning to have an impact on stereochemical problems, but they have not been used with chiral quinones and will not be discussed here. Magnetic circular dichroism

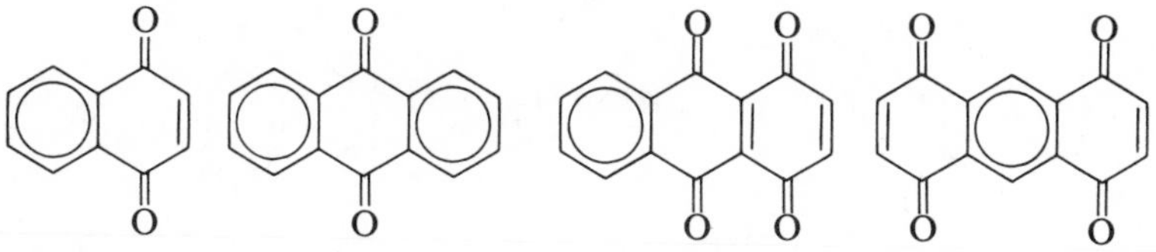

(1) (2) (3) (4)

(MCD) measurements using 1,4-naphthoquinone[15] (**1**), anthraquinone[16] (**2**) and 1,4,9,10- (**3**) and 1,4,5,8-anthracenetetrone[15] (**4**), linear dichroism studies of 1,4-naphthoquinone and several substituted 1,4-napthoquinones[17] and induced circular dichroism observations on 1,4-naphthoquinone in β-cyclodextrin[18] have been reported but only in connection with studies of the electronic absorption (EA) spectra of quinones[15, 17, 18] and of the MCD of aromatic carbonyl compounds[16].

It is to be recognized, however, that the absolute configuration of a chiral quinone can be assigned with ORD and CD measurements only on the basis of empirical comparison with model compounds of similar structures of known absolute configurations. In what follows, the chiral quinones are grouped for convenience into the usual types of quinones and for each type as classes of natural products, and part of the discussion for each class concerns the establishment of the absolute configurations of model compounds by rotatory power comparison, chemical correlation, or X-ray techniques. For completeness, the absolute configurations of a few types of important naturally occurring quinones are discussed although these absolute configurations were not determined by chiroptical methods or were not used for other chiroptical studies.

II. BENZOQUINONES

A. Synthetic *p*-Benzoquinones

Usually chiral benzoquinones occur as natural products[3], but a few chiral *p*-benzoquinones have been prepared and studied as their enantiomers. Thus oxidation of 1-(*m*-hydroxyphenyl)ethanol (**5**) and 1-(*m*-hydroxyphenyl)propanol (**6**) with Fremy's

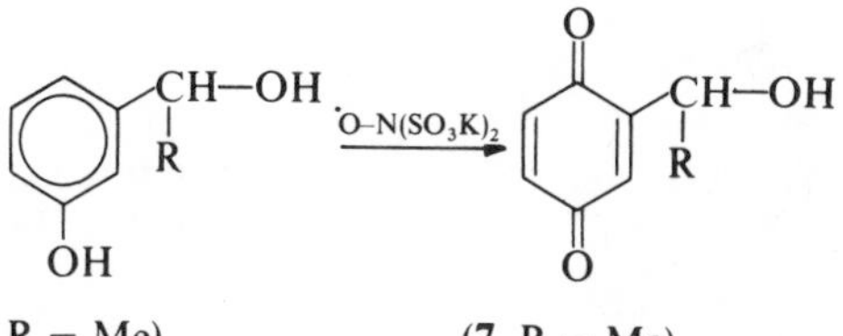

(**5**, R = Me) (**7**, R = Me)
(**6**, R = Et) (**8**, R = Et)

salt[19, 20] gave the respective racemic 1′-hydroxyethyl- (**7**) and 1′-hydroxypropyl-*p*-benzoquinone (**8**)[21] and using the β-cyclodextrin inclusion complexes, one enantiomer of each was obtained in low enantiomeric excess, 27 % for **7**[21]. Using Fremy's salt and a pure enantiomer of a chiral phenol such as **9**, the optically pure *p*-benzoquinone was obtained in very high chemical yield, 98 % for **10**. The reaction proceeded in almost as high a chemical yield (91 %) with the phenol diol corresponding to **9**[22]. Hydrogen peroxide trifluoroacetic

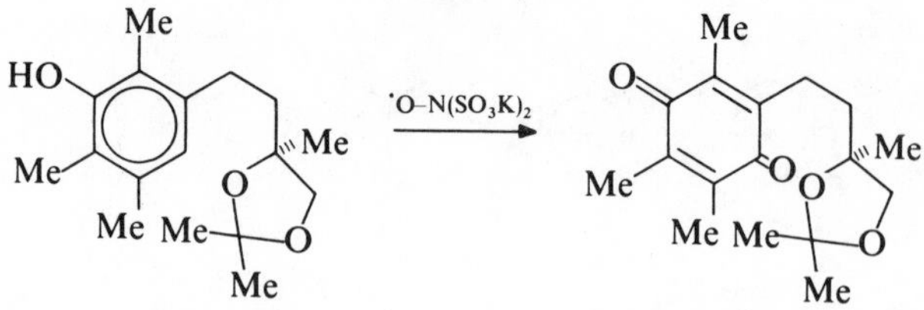

(9) (10)

acid oxidation of the methyl ester of the dehydroabietic acid **11** and of 17β-acetoxy-3-methoxyestrane (**12**) in chloroform gave the corresponding *p*-benzoquinones, **13**, **14** and

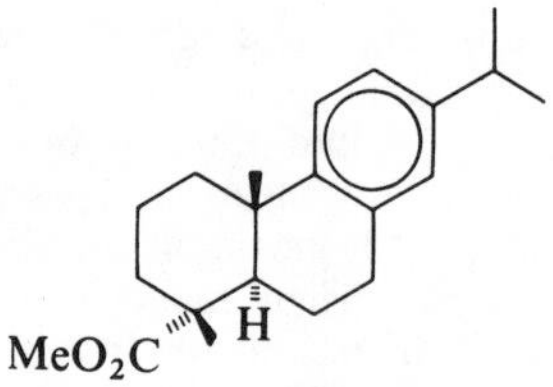

(11) 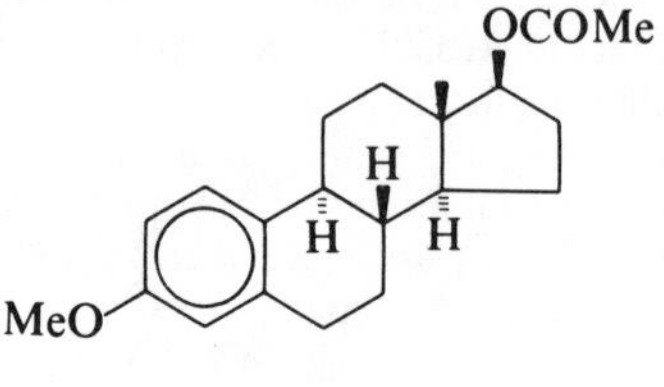 (12)

15 in low chemical yield, 10%, 10% and 1% respectively[23]. Compound **12** gave both **14** and **15**[23]. Benzoquinone **13** was prepared in connection with the possibility that the acid **16** is an intermediate in the degradation of dehydroabietatic acid[23].

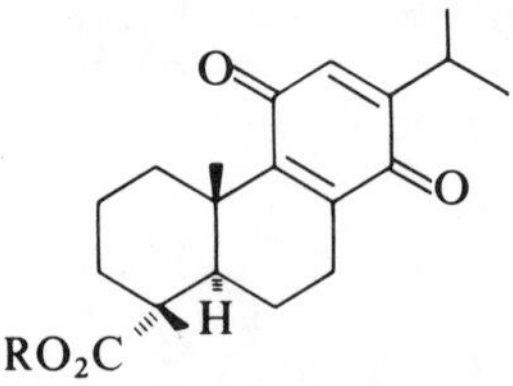

(**13**, R = Me)
(**16**, R = H)

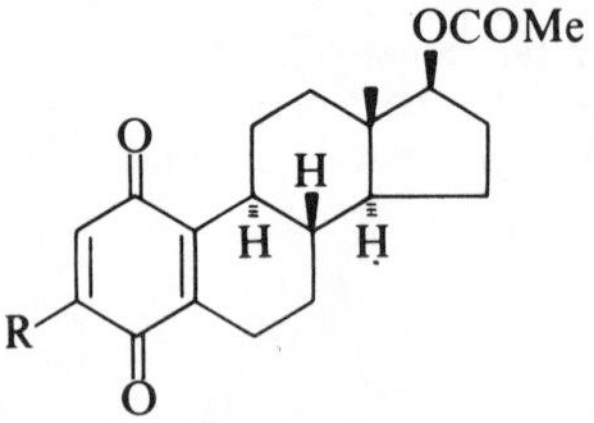

(**14**, R = MeO)
(**15**, R = H)

The EA spectra of **13**–**15** (Figure 1) show maxima near 450, 340 and 260 nm which are assigned, respectively, to an n $\rightarrow \pi^*$ of the carbonyl groups, an electron transfer transition and an $\pi \rightarrow \pi^*$ transition[3]. The CD spectra of **13**–**15** (Figure 2) also show similar Cotton effects (CEs), a negative maximum near 460 nm followed by a positive maximum in the 340–380-nm region[23]. That the CD spectra of the three *p*-benzoquinones are very similar can be seen by comparison of the non-traditional representation of **13** as stereoformula **17** with those of **14** and **15**. In all of these, the chiral centers alpha and beta to the *p*-benzoquinone chromophore have the same absolute configuration. Thus these CD spectra constitute a set of model spectra for the assignment of the absolute configurations of *p*-benzoquinones similar in structure to **13**–**15**[23].

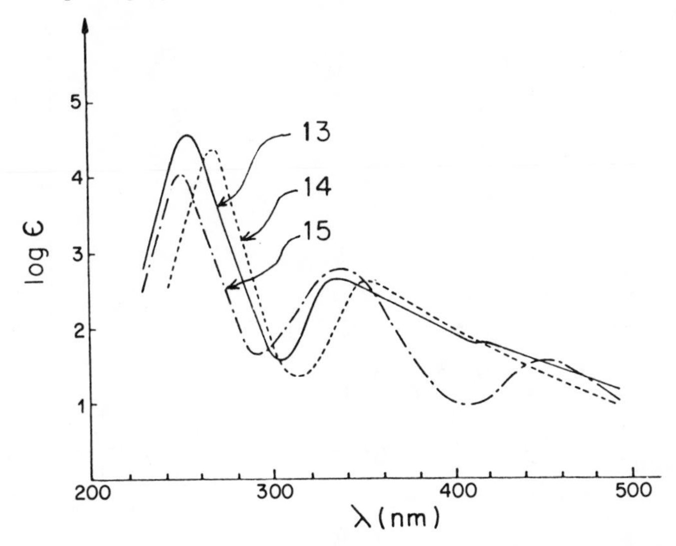

FIGURE 1. Electronic absorption (EA) spectra for the *p*-benzoquinones **13**, **14** and **15** in ethanol. *Adapted from Ref. 23 by permission of the French Chemical Society*

Since the signs of the observed CEs in the CD spectra of **13**–**15** reflect the chirality of the half-chair cyclohexene ring attached to the *p*-benzoquinone chromophore (**18**) as imposed

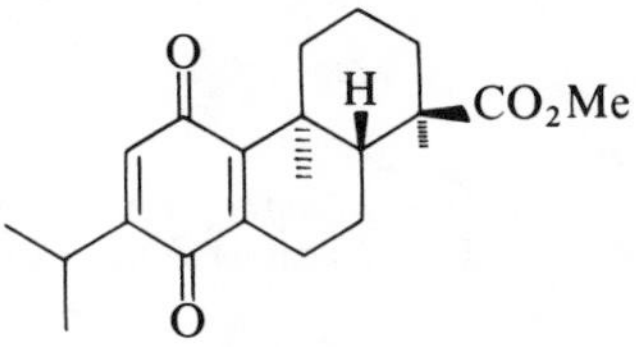

(**17**)

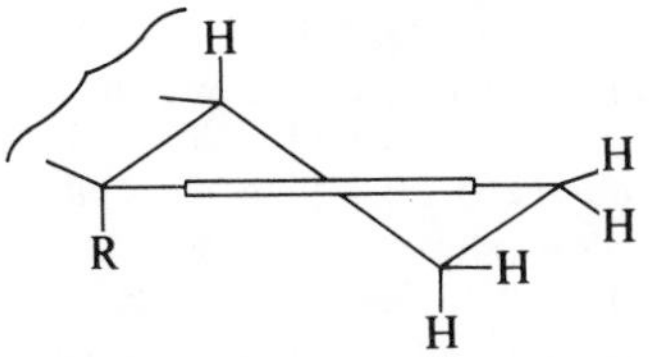

(**18**)

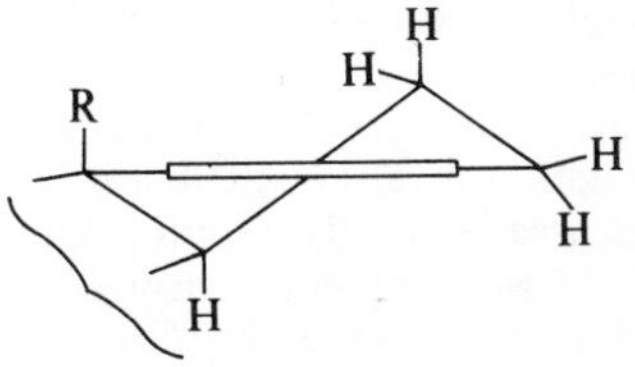

(**19**)

by the alpha and beta chiral centers, the benzoquinone **16** should show a CD spectrum similar to those of **13**–**15**. Reversal of the chirality of the half-chair cyclohexene ring with respect to the benzoquinone chromophore (**19**) will result in a change in sign of the observed Cotton effects near 460 and 360 nm.

B. Neoflavanoids. Dalbergiones

A number of chiral *p*-benzoquinones was isolated from the heartwood of trees (African Blackwood, Senegal Ebony, Mozambique Ebony) of various species of the genus

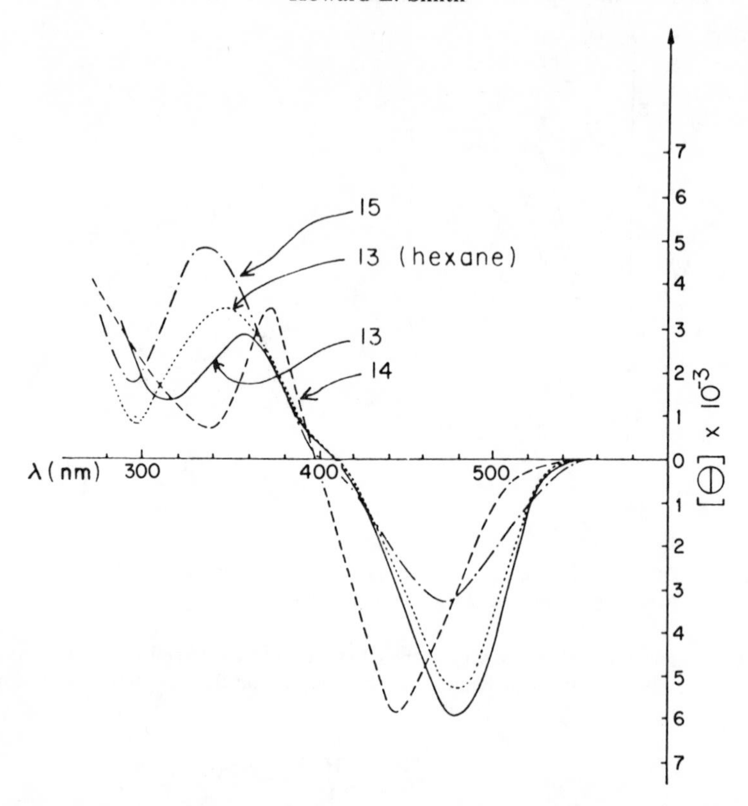

FIGURE 2. Circular dichroism (CD) spectra of the *p*-benzoquinones **13**, **14** and **15** in ethanol and of **13** in hexane. *Adapted from Ref. 23 by permission of the French Chemical Society*

Dalbergia[24]. These compounds were given the general name dalbergiones[24] and the simplest member of the group, initially called dalbergione[24], was shown to have the structure and absolute configuration shown in stereoformula **20**[24]. These *p*-benzoquinones, although not heterocyclic compounds, belong to a class of natural products called neoflavanoids to emphasize their relationship to the heterocyclic natural products with the 4-arylchroman structure (**21**) to which has been given the name neoflavanoid[25]. Since the first report of the structures and configurations of the

(20) **(21)**

dalbergiones[24], the nomenclature has been changed, and **20** is now named (*R*)-4-methoxydalbergione[25]. Other members of the group, isolated from *Dalbergia*[25, 26] and *Machaerium*[27] species, are (*S*)-4-methoxy-[25], (*S*)-4,4′-dimethoxy-[25], (*S*)-4′-hydroxy-4-methoxydalbergione[25] (**22**–**24**, respectively) and (*R*)-3,4-dimethoxydalbergione[27] (**25**). Catalytic reduction of **20** and **23** gave the corresponding dihydro compounds, **26** and

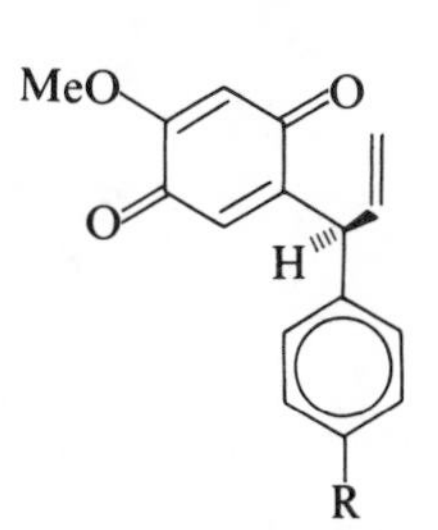

(**22**, R = H)
(**23**, R = OMe)
(**24**, R = OH)

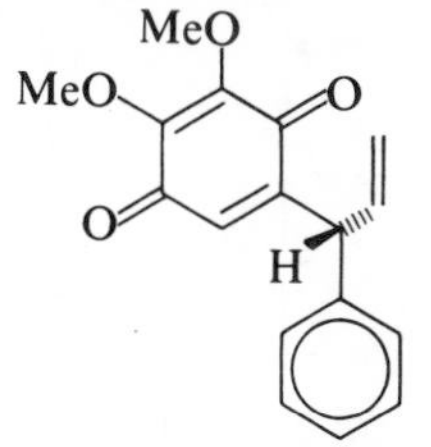

(**25**)

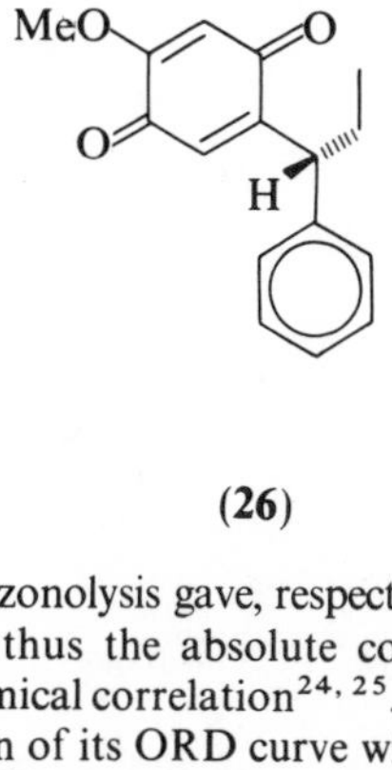

(**26**)

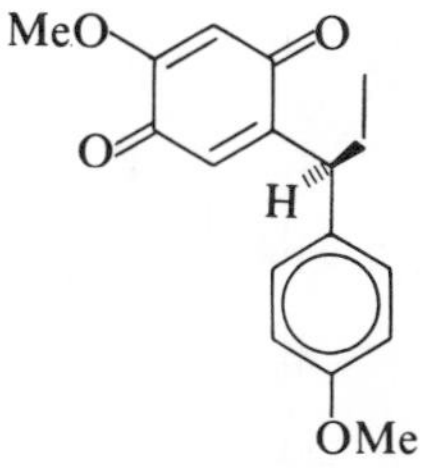

(**27**)

27[24, 25] which on ozonolysis gave, respectively, (*R*)-α-phenyl and (*S*)-α-(*p*-methoxyphenyl)-*n*-butyric acid, and thus the absolute configurations of **20**, **23**, **26** and **27** were firmly established by chemical correlation[24, 25]. The absolute configuration of **24** was established as *S* by comparison of its ORD curve with those of **22** and **23**[25]. As seen in Figure 3, the dalbergiones (**20**, **23**) and dihydrodalbergiones (**26**, **27**) with the *R* and *S* configurations give essentially enantiomeric ORD curves, the presence of a methoxyl (or an hydroxyl group[25]) at C(4′) having little effect on the ORD curves[25]. The same is true for the EA and CD spectra[25]. The absolute configuration of **25** was also established by comparison of its ORD curve and CD spectrum with the ORD curve of **20**[27].

Another isoflavanoid isolated from a *Dalbergia* species is latifolin (**28**), and although not

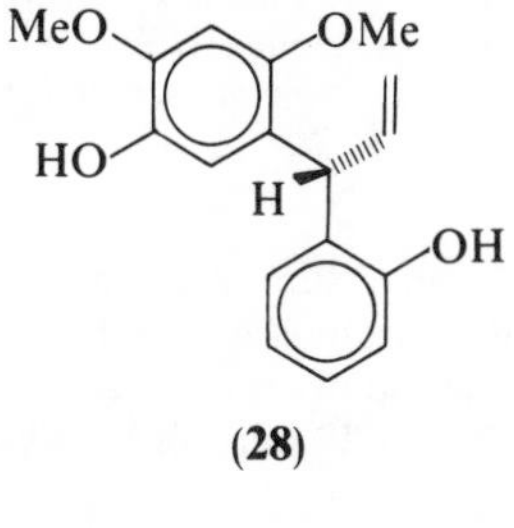

(**28**)

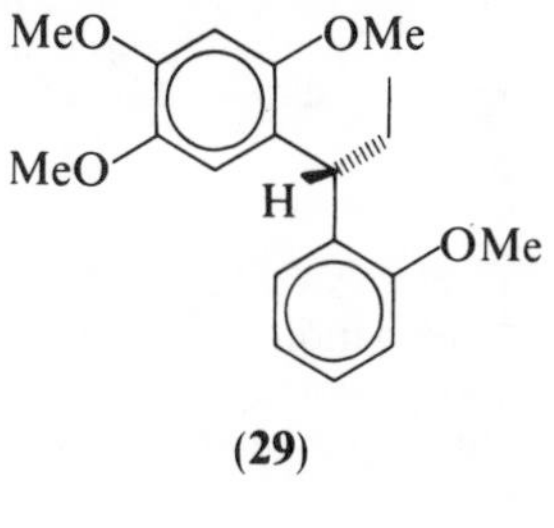

(**29**)

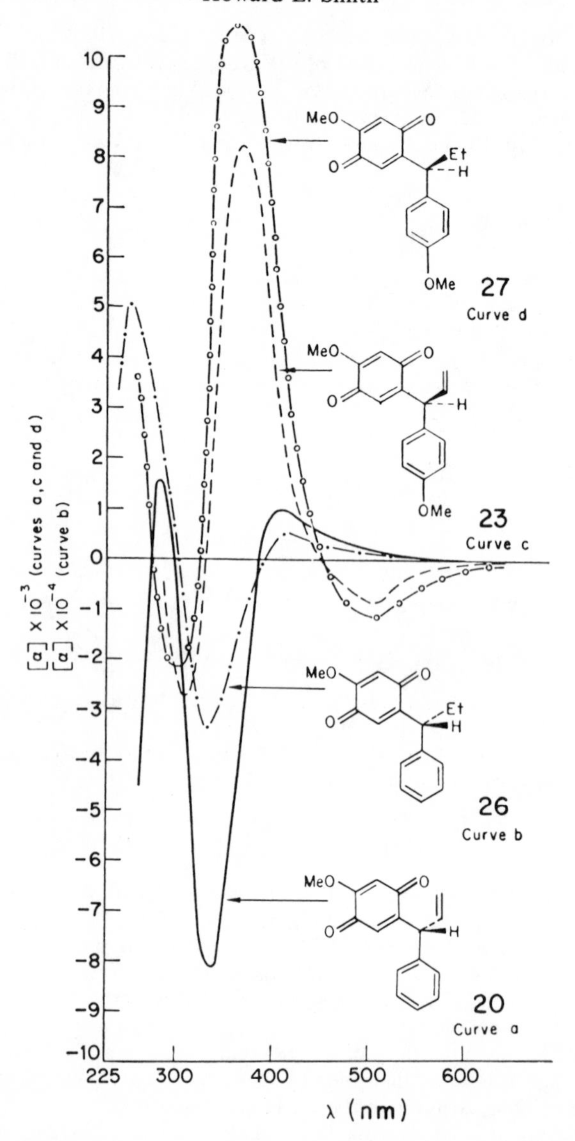

FIGURE 3. Optical rotatory dispersion (ORD) curves of (*R*)-4-methoxydalbergione (**20**), (*S*)-4,4′-dimethoxydalbergione (**23**), (*R*)-dihydro-4-methoxydalbergione (**26**) and (*S*)-dihydro-4,4′-dimethoxydalbergione (**27**) in dioxan. *Adapted from Ref. 25 by permission of Pergamon Press*

a dalbergione, it is a close structural relation[28]. Conversion of **28** to its dihydrodimethyl derivative (**29**) and oxidation of the latter with chromic acid in acetic acid gave (*R*)-dihydro-2′,4-dimethoxydalbergione (**30**) and the isomeric *o*-benzoquinone **31**[28]. The absolute configurations of **30** and **31** and thus **28** and **29** were all established as *R* on the

(30) **(31)**

basis of a comparison of the ORD curves of **30** and **31** with that of (*R*)-4-methoxydalbergione (**20**). Finally the absolute configuration of latifolin (**28**) was independently established by comparison of its ORD curve and those of some of its derivatives with those of related quinol diacetates derived from various dalbergiones[29]. As part of this work, the ORD curves of the four dalbergiones **23**–**25** and **30** were examined in some detail. In these curves, at least three apparent CEs occur which overlap one another and thus the curves can only be compared on an empirical basis. It is not possible to interpret the results without analysis of the curves into their single transition components. The CE centered in the ORD curve at about 336 nm (Figure 3), positive for those with the *S* configuration (**23**, **24**) and negative for those of the *R* configuration (**25**, **30**), however, can be used with confidence to assign the absolute configuration of chiral *p*-benzoquinones of similar structure[29].

C. Isoflavanoids. Mucroquinone and Claussequinone

Also isolated from a *Machaerium* species, *M. mucronulatum*, is a levorotatory, orange compound to which was assigned a methoxy-*p*-benzoquinone structure and the name mucroquinone[30] (**32**). Since **32** has the 3-arylchroman structure, it is a member of the class of natural products to which has been given the name isoflavanoid.

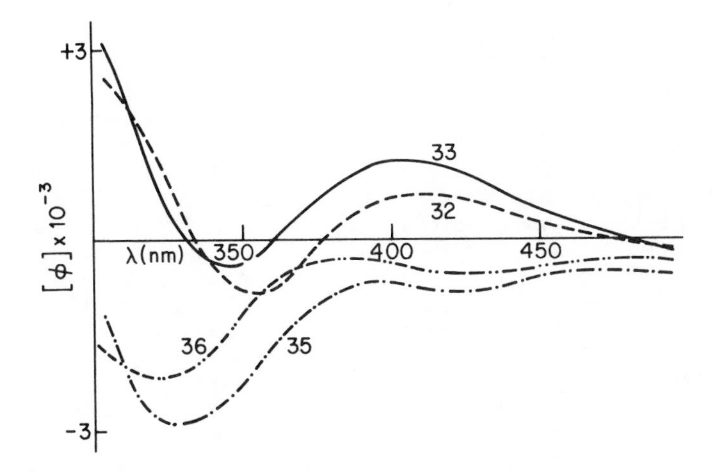

FIGURE 4. Optical rotatory dispersion (ORD) curves of mucroquinone (**32**), (*S*)-7,4′-dimethoxyisoflavan-2′,5′-quinone (**33**), (*S*)-7-hydroxy-8,4′,6′-trimethoxyisoflavan-2′,5′-quinone (**35**) and (*S*)-7,8,4′,6′-tetramethoxyisoflavan-2′,5′-quinone (**36**). *Adapted from Ref. 32 by permission of the Royal Society of Chemistry*

The absolute configuration of levorotatory mucroquinone (**32**) isolated from *M. mucronulatum*[30] and *M. villosum*[31] was established as *S* by comparison of its ORD characteristics with those of the isoflavanquinone **33**, the ORD curves for both **32** and **33** being very similar from 330 to 450 nm (Figure 4)[32]. The absolute configuration of **33** follows from its preparation by oxidation of the corresponding isoflavan **34** which was shown to have the *S* configuration[32]. Although the ORD curves for the configurationally

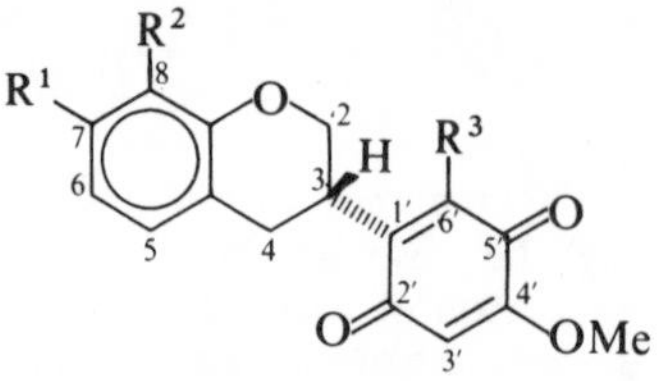

(**32**, R^1 = OH, R^2 = OMe; R^3 = H)
(**33**, R^1 = OMe, R^2 = R^3 = H)
(**35**, R^1 = OH, R^2 = R^3 = OMe)
(**36**, R^1 = R^2 = R^3 = OMe)

related isoflavans such as **34**, **37** and **38** display a consistent pattern, caution must be exercised in the interpretation of the ORD curves for the isoflavanquinones and

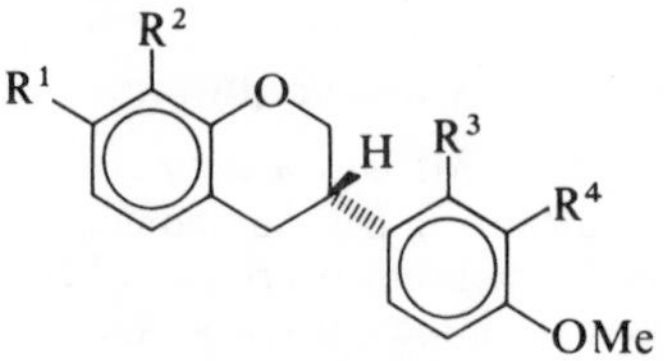

(**34**, R^1 = OMe; R^2 = H; R^3 = OH; R^4 = H)
(**37**, R^1 = OH, R^2 = R^3 = OMe; R^4 = OH)
(**38**, R^1 = OMe; R^2 = R^3 = R^4 = H)

comparison of curves is clearly limited to closely related structural types[32,33]. The isoflavanquinones **35** and **36**, although known to have the *S* configuration from their preparation from the corresponding isoflavan **37**, have ORD curves (Figure 4) in the 400–500-nm region which are almost enantiomeric to those of **32** and **33**[32,33]. This at first sight could be just a consequence of the different chromophores of the two pairs of quinones, but the proton nuclear magnetic resonance (^{1}H-NMR) spectra of the dihydropyran ring protons show that there are also conformational differences between the pairs of compounds[32,33]. The dihydropyran ring in **32**, **33**, **35** and **36** is expected to assume a preferred half-chair conformation. It is not possible on the basis of conformational arguments to state which of the two conformers, **39a** or **39b**, is of lower energy, since the axial 3-quinonyl substituent in **39b** does not result in the destabilizing 1,3-diaxial interaction of axial-substituted cyclohexane and cyclohexene systems. The vicinal coupling constants for the protons of the dihydropyran ring for the isoflavanquinones **35** and **36** are consistent with those expected for the half-chair conformation **39a** in which the 3-quinonyl substituent occupies an equatorial position[33]. The observed vicinal coupling constants for the corresponding protons in the isoflavanquinone **32** are, however, different and are consistent with the values expected from a conformational equilibrium in which **39b** with an axial 3-quinonyl substituent is the major contributor[33]. This conformational

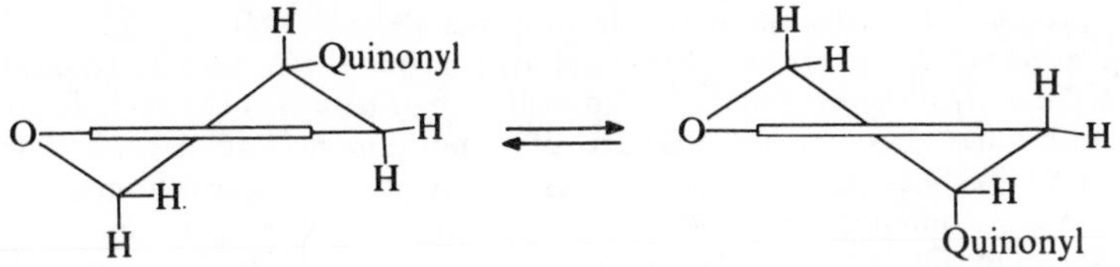

(**39a**) (**39b**)

difference between quinones **32** and **33**, which lack the 6′-methoxy group, and quinones **35** and **36**, which have a 6′-methoxy group, is clearly related to the difference in the ORD curves shown by **32** and **33** on the one hand and those shown by **35** and **36** on the other (Figure 4)[32].

(*R*)-Mucroquinone (**40**), the enantiomer of **32**, and claussequinone (**41**) were both isolated from species of the genus *Cyclolobium*, **41** so named because of its occurrence in the heartwood of *C. clausseni*[34]. The absolute configurations of both of these isoflavanquinones were established by comparison of their ORD curves with that of (*S*)-mucroquinone (**32**)[34], the curves of **40** and **41** being essentially enantiomeric to that of **32**[32].

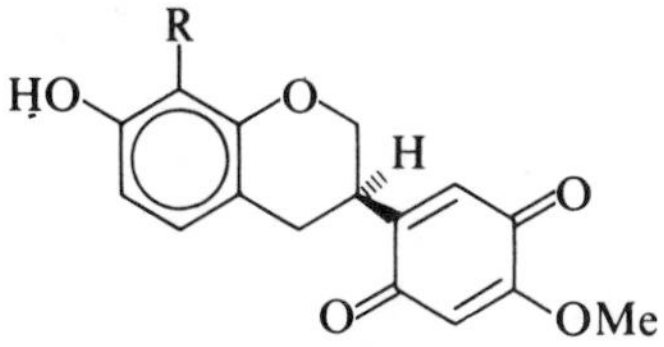

(**40**, R = OMe)
(**41**, R = H)

D. Sesquiterpenes. Perezone

Although monoterpene *p*-benzoquinones occur as natural products, none so far known is chiral, but chiral sesquiterpene *p*-benzoquinones are known[3, 35]. An important one with the bisabolene structure is perezone (**42**) which was found as a plant constituent and

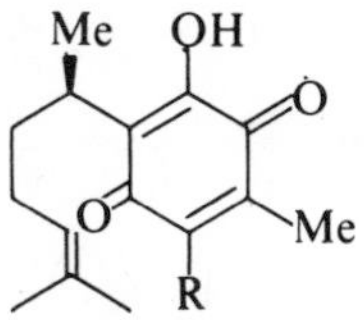

(**42**, R = H)
(**43**, R = OH)

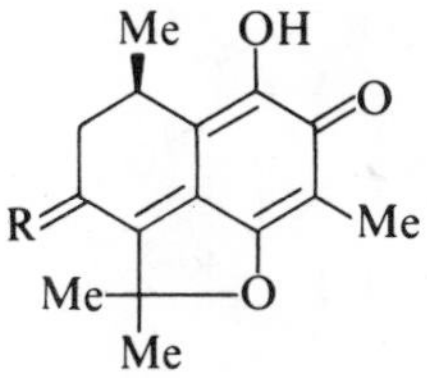

(**44**, R = H_2)
(**45**, R = O)

characterized as a *p*-benzoquinone many years ago[36, 37]. The structure for perezone as originally proposed[37] was subsequently revised[38], but the earlier work in connection with its absolute configuration[37, 39] still establishes its absolute configuration as *R*[39].

Optical rotatory dispersion measurements were reported for perezone as well as a number of its derivatives[40, 41]. Among the latter was hydroxyperezone (**43**), isolated as a natural product[3] but also prepared from perezone[37,42] (**42**). Others were the *p*-benzoquinone methides, perezinone (**44**) and oxoperezinone (**45**), formed, respectively, by dehydration of hydroxyperezone[37,42] and oxidation of perezinone[43]. Included in one of these chiroptical studies was the ORD curve of the norquinone[41] **46**, an oxidation product of cacalol[44] (**47**), the latter a sesquiterpene with a structure[44] different from that assigned earlier[45]. The *R* configuration for **46** and for **47** was originally assigned on the basis of a comparison of the ORD curve of **47** with those of a number of chiral 1-substituted indans[41]. It is to be noted, however, that although **47** and the indans all incorporate benzene chromophores, the substantial difference in the structure of **47** from those of the indans suggests that comparison of the ORD curve of **47** with those of the chiral 1-substituted indans is not justified. Hence the configurational assignment for **47** and also for **46** must be recognized as not firmly established.

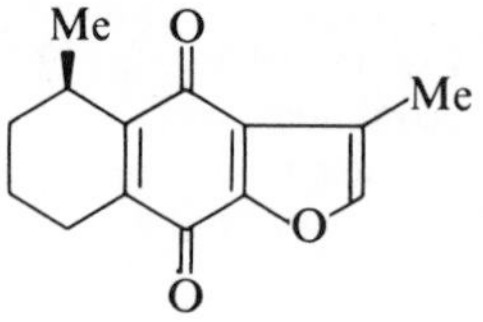

(**46**)

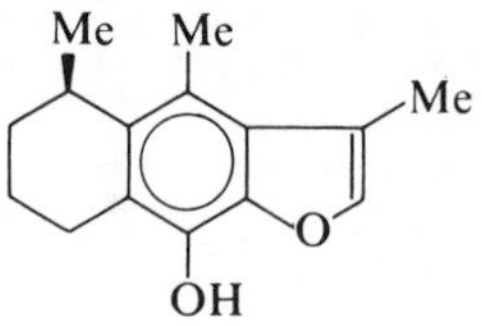

(**47**)

E. Diterpenes

1. Royleanones

As reviewed by Eugster[35], a vast number of diterpenoid pigments occurs in nature, some of which incorporate the *p*-benzoquinone chromophore. Important among these is royleanone (**48**), a benzoquinone with the abietane carbon skeleton and isolated from the root of *Inula royleana*, a perennial shrub that grows in the western Himalayas[46]. Royleanone has been found in other plants, and it has given its name to a group of related diterpenes, the royleanones[35]. The approximately sixteen compounds which comprise this group have the basic royleanone carbon skeleton and chromophore but have hydroxyl

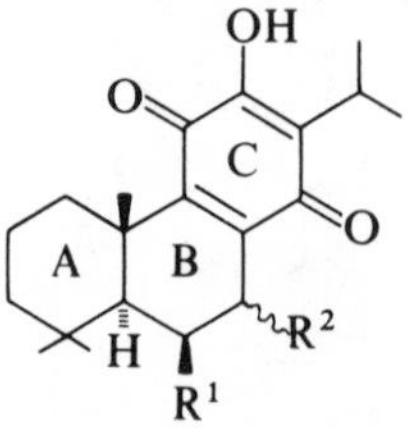

(**48**, $R^1 = R^2 = H$)
(**49**, $R^1 = H$; $R^2 = \beta$-OH)
(**50**, $R^1 = H$; $R^2 = \alpha$-OH)
(**51**, $R^1 = OH$; $R^2 = \alpha$-OH)

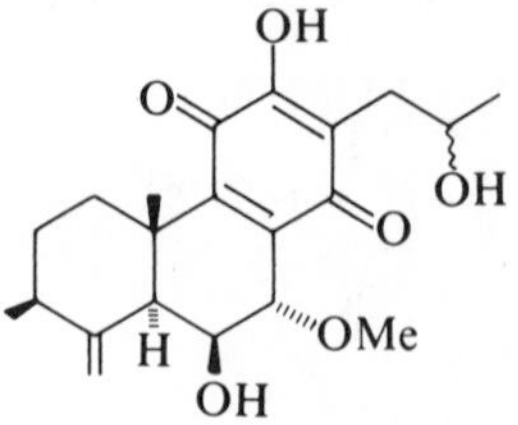

(**52**)

substituents on various positions of rings A and B[35]. Important ones are taxoquinone (**49**), horminone (**50**) and 6*β*-hydroxyhorminone (**51**)[35]. Occasionally a rearranged abietane skeleton (**52**) is encountered[35].

The structure and configuration of royleanone (**48**) follows its formation by oxidation of ferruginol (**53**) with hydrogen peroxide in acetic acid-sulfuric acid[46]. The CD spectrum

(**53**)

(Figure 5) of **48** was reported in connection with stereochemical studies of related compounds[47] and may be compared with those of **13**–**15** (Figure 2) (Section II.A). The spectra of **48** and of **13**–**15** show substantial differences in both the wavelength and intensity of their respective CD maxima. This again emphasizes the requirement that, for comparison of ORD curves and CD spectra in configurational studies, comparison must be made between compounds of very similar structures unless differences in the structures are known to have no substantial effect on the ORD curves and CD spectra as among **22**–**24**[25].

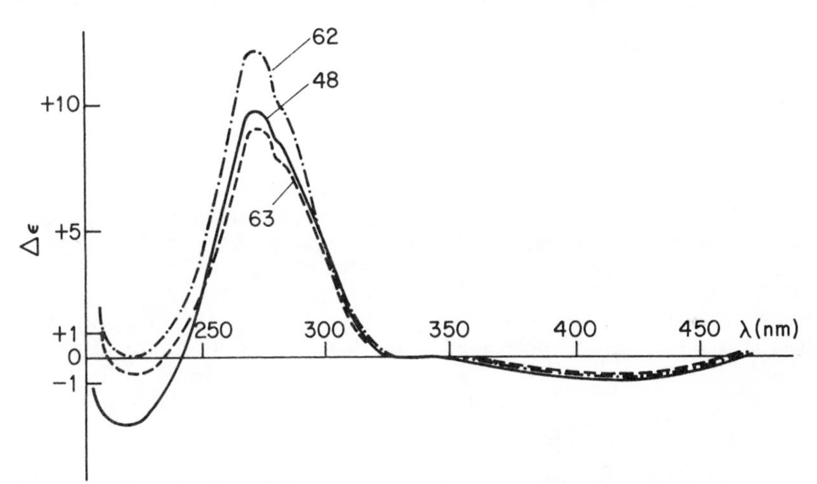

FIGURE 5. Circular dichroism (CD) spectra of royleanone (**48**) and the *p*-benzoquinones **62** and **63** from coleone E (**61**) in dioxan. *Adapted from Ref. 47 by permission of the Swiss Chemical Society*

The CD spectra for royleanones other than **48** have not been reported, but it is anticipated that an hydroxyl, methoxyl, or acetoxyl group on rings A or B will have only a small effect on their CD spectra.

2. Coleones

Another group of diterpene pigments is the coleones, so named because they were first found in the glands of *Coleus* species[48]. These substance, by 1977 over 70 compounds[35], have a great structural diversity, but almost all can be derived from abietane. Among these is the hydroquinone coleone B[49] (**54**), so named because it was the second coleone found[35]. Coleone Z was reported in 1979[50].

Coleone B (**54**) on oxidation yields coleone-B-quinone (**55**)[49]. The absolute configuration of **54** and thus that of **55** were assigned on the basis of the similarity of the ORD curves of the trimethyl ethers **56** and **57**, derived, respectively, from royleanone (**48**) and coleone B (**54**)[49].

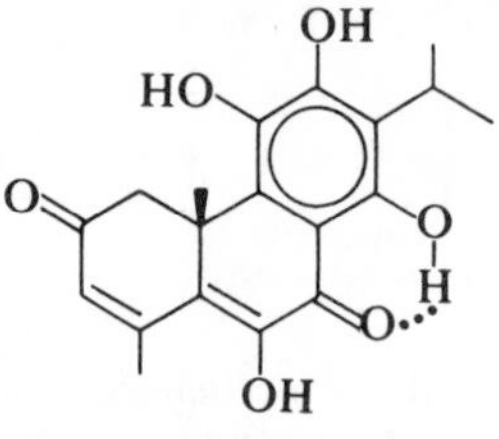

(**54**)

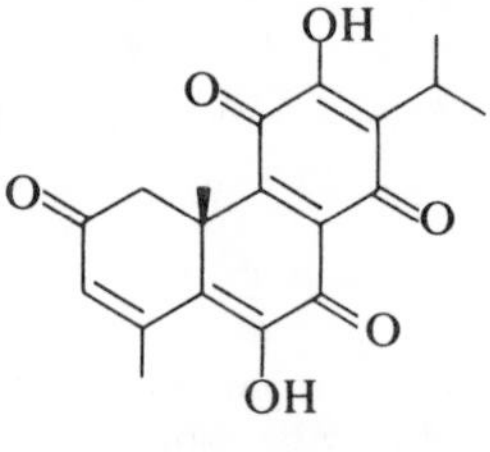

(**55**)

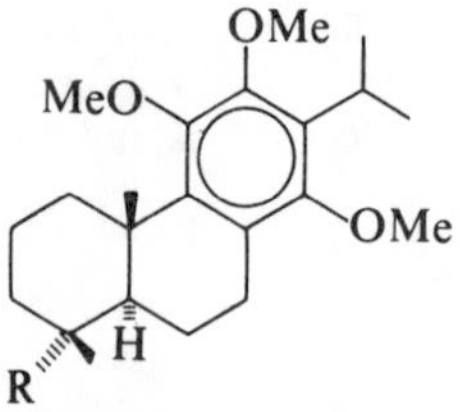

(**56**, R = Me)
(**57**, R = H)

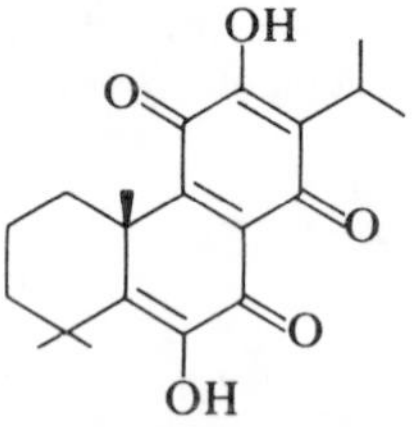

(**58**)

The CD spectrum of coleone-B-quinone (**55**) has not been reported, but a quinone similar to **55**, coleone-U-quinone (**58**), has recently been isolated from the red leaf glands of *Plectranthus agentatus*[51]. The absolute configuration and CD spectrum of **58** were reported[51].

A substance related to coleone B (**54**) is coleone C (**59**) which on isolation is a bright yellow, sharp melting, crystalline solid[52]. On the basis of its electronic absorption

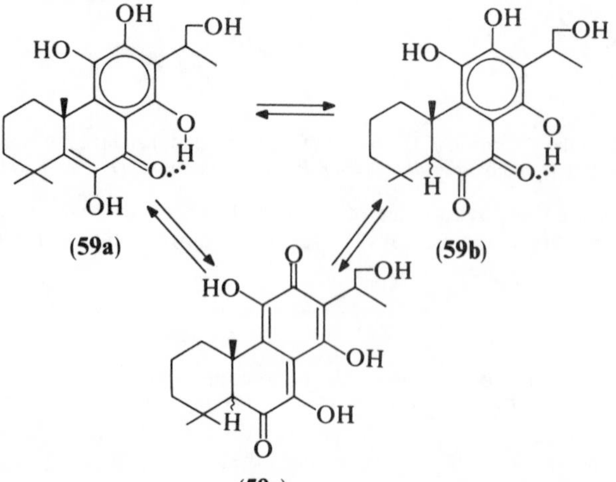

(**59a**)

(**59b**)

(**59c**)

(60)

spectrum in ethanol with a strong absorption maximum near 400 nm and other spectral and chemical evidence[52], **59** in solution is assigned tautomeric structures such as **59a–c**, including the quinone methide **59c**. The absolute configuration of **59** follows from a comparison of its CD spectrum in dioxan (Figure 6) with that of dihydrocoleon-B-alcohol (**60**) which on the basis of its spectral properties must also have a tautomeric structures in dioxan similar to those of **59**[52].

Some of the members of the coleone group are in fact fully conjugated quinone methides[47, 53]. Coleone E[47] (**61**) is a deeply red substance, and in its EA spectrum has a

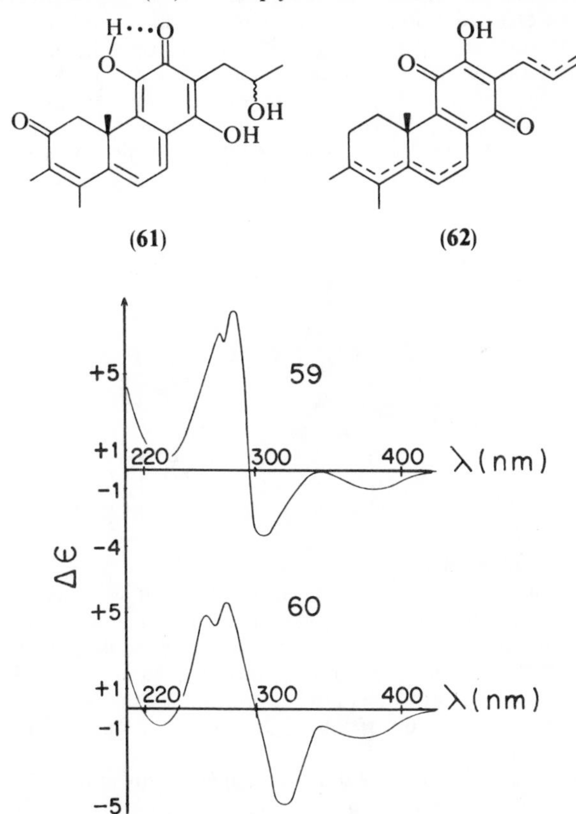

FIGURE 6. Circular dichroism (CD) spectra of coleone C (**59**) and dihydrocoleone-B-alcohol (**60**) in dioxan. *Adapted from Ref. 52 by permission of the Swiss Chemical Society*

strong absorption maximum near 440 nm[47]. Conversion of **61** to the *p*-benzoquinone **62** and then reduction to **63**, and comparison of the CD spectra of **62** and **63** with that of

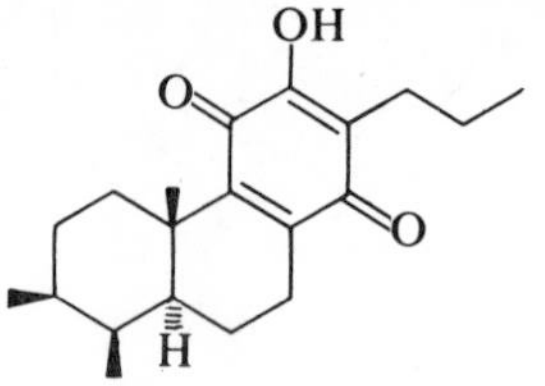

(**63**)

royleanone (**48**) (Figure 5) fixes the configuration of the angular methyl group at C (10) in both **62** and **63** and thus also in coleone E (**61**)[47].

F. Mitomycin Antibiotics

In the search for new antibiotics produced by actinomycetes, a group of closely related substances was isolated from *Streptomyces caespitosus*[54]. This mixture had substantial antitumor activity and was called mitomycin[54]. Chromatography and recrystallization gave pure mitomycins A and B[54]. These substances are reddish violet and violet, respectively, and have electronic spectra with an absorption maximum near 550 nm indicating the presence of a *p*-benzoquinone chromophore[54]. Mixtures of four mitomycins, mitomycins A, B, C and porfiromycin (**64**–**67**, respectively) are also produced by strains of *S. verticillatus*[55] and their structures and relative configurations were established

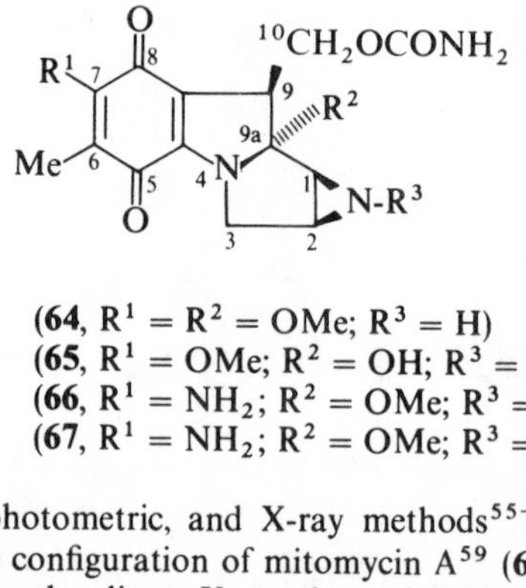

(**64**, $R^1 = R^2 = OMe$; $R^3 = H$)
(**65**, $R^1 = OMe$; $R^2 = OH$; $R^3 = Me$)
(**66**, $R^1 = NH_2$; $R^2 = OMe$; $R^3 = H$)
(**67**, $R^1 = NH_2$; $R^2 = OMe$; $R^3 = Me$)

by chemical, spectrophotometric, and X-ray methods[55–58]. An early X-ray determination of the absolute configuration of mitomycin A[59] (**64**) was later corrected[60]. This revision was based on the direct X-ray determination of absolute configuration of mitomycin C[60] (**66**) which has been stereochemically related to **64** and porfiromycin (**67**). Since mitomycin B (**65**) has not been derived directly from or converted to any of the others, its absolute configuration was also determined by the X-ray method using 7-demethoxy-7-*p*-bromoanilinomitomycin B[61]. Thus, all of the mitomycin antibiotics have the same absolute configuration.

The importance of the mitomycin antibiotics and especially mitomycin C (**66**) is their antitumor activity, with **66** currently in clinical use. It interacts with DNA *in vivo* and *in vitro*, resulting in covalent linkage of the drug to DNA as well as in the formation of covalent cross-links between the two complementary DNA strands[62]. These modifications of DNA are generally thought to be primary events in the antibiotic and antitumor activity of **66**. The drug requires acidic or reductive activation, and it is known that high guanine and cytosine content of DNA promotes cross-link formation[62]. In model studies, the reaction of **66** with 2′-deoxyguanylyl-(3′ → 5′)-2′-deoxycytidine [d(GpC)] (**68**) was studied in some detail[63]. Enzymatic degradation of the mitomycin-d(GpC) adducts yielded one major mitomycin–deoxyguanosine adduct. A combination of ^{1}H-NMR, Fourier transform infrared, and circular dichroism spectroscopic techniques established its structure

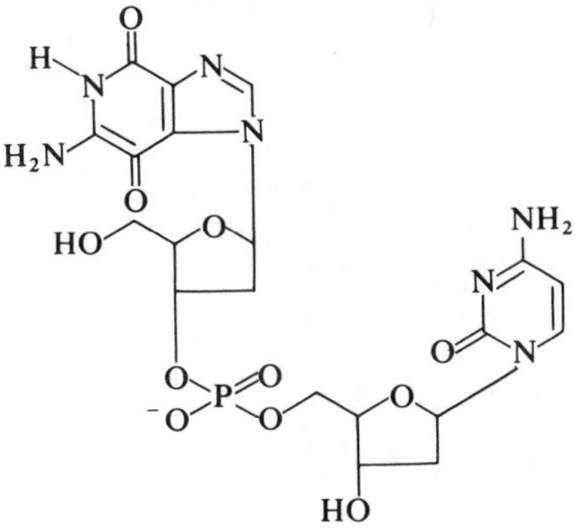

(68)

and configuration as 1,2-*trans*-1-[O^6-(2′-deoxyguanosyl)]-2,7-diaminomitosene (**69**)[63], compound **70** having the trivial name 2,7-diaminomitosene[56]. The stereochemistry of the adduct **69** was determined to be *trans* by comparison of its CD spectra with those of the diastereomeric 2,7-diamino-1-hydroxymitosenes (**71** and **72**)[63] (Figure 7). The relative stereochemistry at C(1) and C(2) in **71** and **72** had been previously established[64], but the

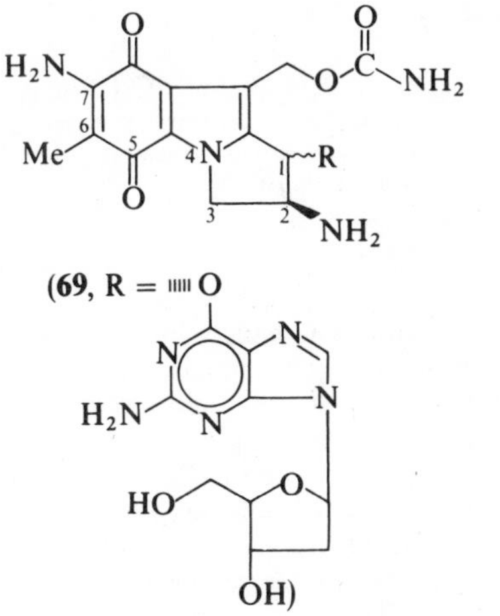

(**69**, R = ⦙⦙⦙O-deoxyguanosyl … OH)

(**70**, R = H)
(**71**, R = ◄OH)
(**72**, R = ⦙⦙⦙OH)

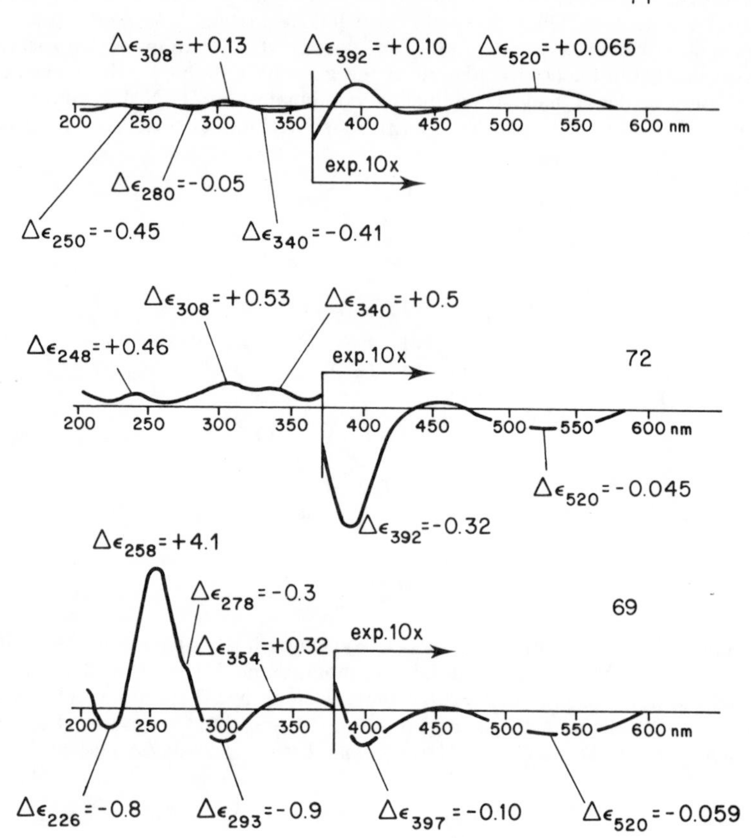

FIGURE 7. Circular dichroism (CD) spectra of 1,2-*trans*-1-[O^6-(2′-deoxyguanosyl)]-2,7-diaminomitosene (**69**), *cis*-2,7-diamino-1-hydroxymitosene (**71**), and *trans*-2,7-diamino-1-hydroxymitosene (**72**) in methanol. *Adapted with permission from M. Tomasz, R. Lipman, J. K. Snyder and K. Nakanishi,* J. Am. Chem. Soc., **105**, *2059 (1983). Copyright (1983) American Chemical Society*

absolute configuration of the mitosene moiety in **69**–**72** is different from that reported earlier[63]. In **69**–**72**, the absolute configurations conform to those required by the recent X-ray studies on the absolute configuration of mitomycin C (**66**)[60].

The mitosene and guanine chromophores in adduct **69** both have EA in the 200–300-nm region[63] and hence the intense CD Cotton effects in **69** in the region below 400 nm (Figure 7) can certainly be ascribed to coupled oscillator interactions[65]. However, since the directions of the electric transition moments of the two interacting chromophores are unknown, and the conformation of the guanosine moiety with respect to the mitosene chromophore is not clear, the CD maxima can not be used for absolute configurational assignments[63]. A more extensive study of the weak CD Cotton effects displayed by 1,2-*cis*/*trans* isomers of 1-substituted 2-aminomitosene in the 530-nm region (Figure 7)[66] shows that a positive CE is associated with a 1β configuration (**71**) while a negative one is associated with a 1α configuration (**72**). When C(1) is unsubstituted, as in **70**, the 530-nm CE is absent, indicating that the substituent at C(2) exerts no influence. These results establish that observation of the sign of the 530-nm CE constitutes a non-ambiguous method for deducing the C(1) configuration of mitosenes. This CD method was then used to establish the stereochemistry of mitosenes derived from the acid-catalyzed solvolysis of porfiromycin[66] (**67**) and applied[66] to a minor mitosene–deoxyguanosine adduct formed on reductive alkylation of d(GpC) by mitomycin C (**66**)[63]. The CD method was thus shown to be applicable to mitosene products from metabolic reactions of **66** as well as to nucleoside–mitosene adducts[66].

G. Fungal Quinone Methides

Although the mitomycins are the most important benzoquinone fungal metabolites, those with the benzoquinone methide structure have been of interest for many years[67]. The structure of citrinin[68] (**73**) as a *p*-benzoquinone methide has recently been definitively established by an X-ray diffraction study[69] and its absolute configuration was established earlier by application of the Prelog atrolactic acid method using citrinin degradation

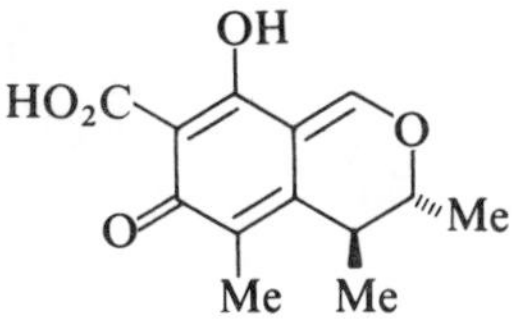

(**73**)

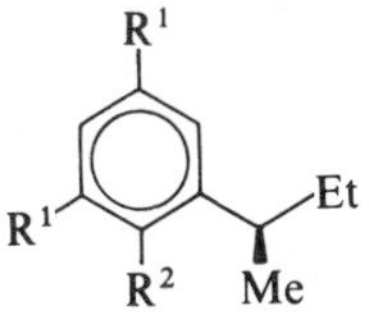

(**74**, $R^1 = R^2 = H$)
(**75**, $R^1 = OMe$; $R^2 = Me$)

products[70]. This assignment has some, but not unequivocal confirmation by comparison of the ORD curve of (*R*)-2-phenylbutane (**74**) with that of 2-(3,5-dimethoxy-2-methylphenyl)butane (**75**)[71], also obtained as a degradation product of **73**[71]. Since both **74** and **75** show plain negative dispersion curves from 600 to 350 nm, **75** was assigned the *R* configuration and thus the corresponding chiral center in **73** has the *S* configuration. Correlations of this type must be used with extreme caution since the ORD and CD associated with the longest wavelength benzene transition (^1L_b) at about 260 nm depend on the spectroscopic moments and the positions of the benzene ring substituents[72].

The configuration of the chiral center in **73** was, however, firmly established using the

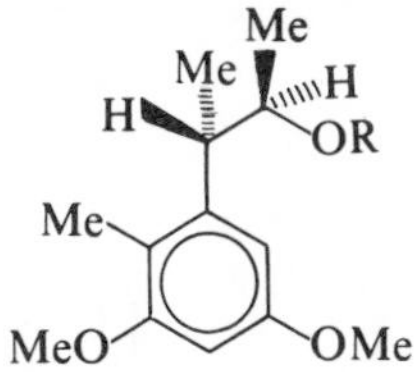

(**76**, $R = CS_2Me$)
(**77**, $R = H$)

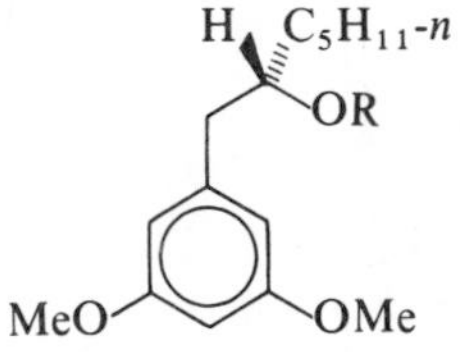

(**78**, $R = CS_2Me$)
(**79**, $R = H$)

CD of the methyl xanthate derivative **76** of the alcohol **77**, a degradation product of **73**[73]. The CD of the methyl xanthate derivative **78** of the alcohol **79** derived from pulvilloric acid (**80**) was also used to establish the absolute configuration of this *p*-benzoquinone methide.

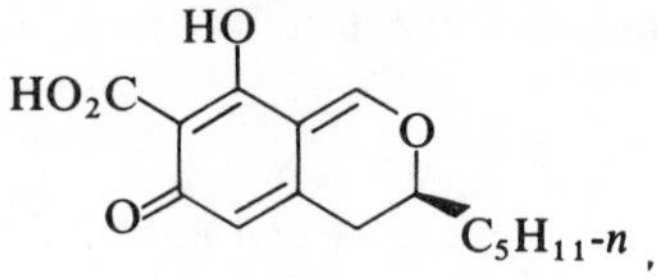

(**80**)

Thus each methyl xanthate exhibits a strong Cotton effect at 355 nm, positive and negative for **76** and **78**, respectively.

III. NAPHTHOQUINONES

A. Tetrahydroanthraquinones

A few chiral tetrahydroanthraquinones of the basic emodin structure (see Section IV.A) are known. Notable are altersolanols A[74,75] (**81**) and B[75] (**82**) and bostrycin[76] (**83**), but no

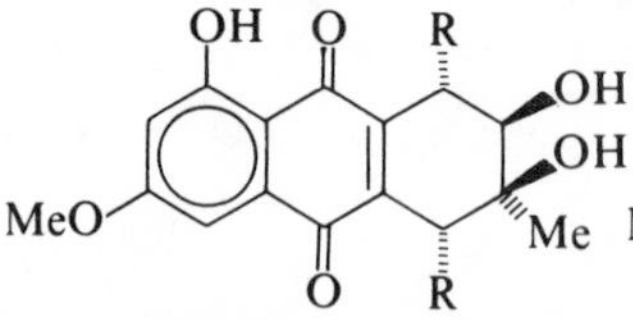

(**81**, R = OH)
(**82**, R = H)

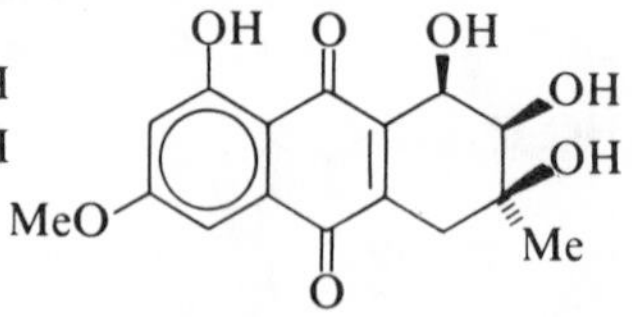

(**83**)

CD spectrum or ORD curve has been reported for these substances. For the altersolanols (**81, 82**), both red pigments from *Alternaria solani*, a pathogen of solanaceous plants[75,77], the absolute configurations have not been reported, but their relative configurations were assigned on the basis of chemical and proton nuclear magnetic resonance evidence[77]. The absolute configuration of bostrycin (**83**) rests on an X-ray study of a *p*-bromobenzoate derivative[78].

The antiviral antibiotic julimycin B-II[79] (**84**) also incorporates reduced tetrahydroanthraquinone moieties. An X-ray study using a bromine derivative[80] gives the absolute configurations at its six chiral centers. There is no indication, however, if there is substantial restricted rotation (atropisomerism[6]) about the single bond connecting the two naphthoquinone groups so that **84** has a particular configuration about this bond much the same as is encountered with the bianthraquinones (see Section IV. B).

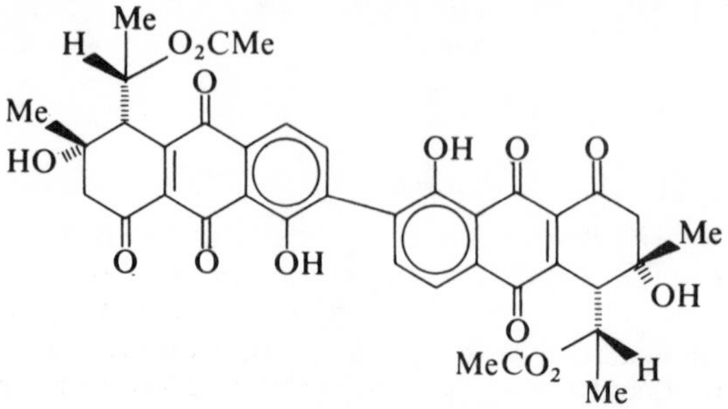

(**84**)

B. Pyranonaphthoquinones

1. Eleutherins

The first pyranonaphthoquinones (benzoisochromanquinones) to be reported were eleutherin (**85**) and isoeleutherin (**86**)[81, 82]. Compounds **85** and **86** are diastereomers since

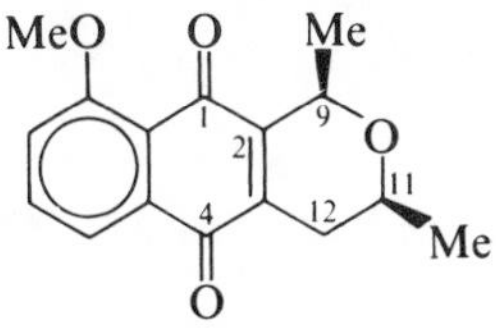

(**85**) 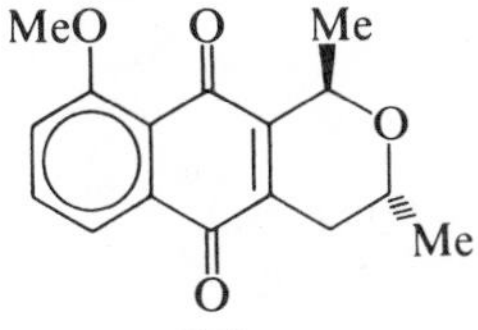 (**86**)

in phosphoric acid **85** gives the enantiomer of **86** (alloeleutherin)[82]. Under similar conditions, the enantiomer of **85** (alloisoeleutherin) can be obtained from **86**[82]. That **85** and **86** epimerized at C(9) was established by the observation that on Clemmensen reduction, both **85** and alloisoeleutherin (enantiomer of **86**) gave the same enantiomer of

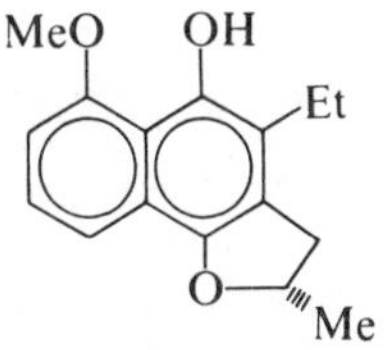

(**87**) 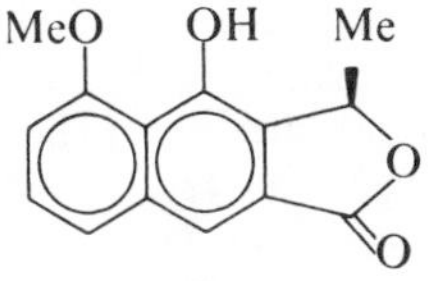 (**88**)

the dihydrofuran **87**[83]. The absolute configuration at C(11) was established by degradation of **87** to (*S*)-*β*-hydroxybutyric acid. The configuration at C(9) was deduced by reference to (*R*)-eleutherol (**88**) which was degraded to D-lactic acid[83]. Correlation of the molecular rotations of various *O*-methyl and *O*-acetyl derivatives of leucoeleutherin (**89**) and leucoisoeleutherin (**90**) with similar derivatives of (*R*)-eleutherol (**88**) indicate that **89** and **90** have the same absolute configuration at C(9) as does **88**[83]. Subsequent ^{1}H-NMR spectral measurements support the relative configurational assignments for **85** and **86**[84].

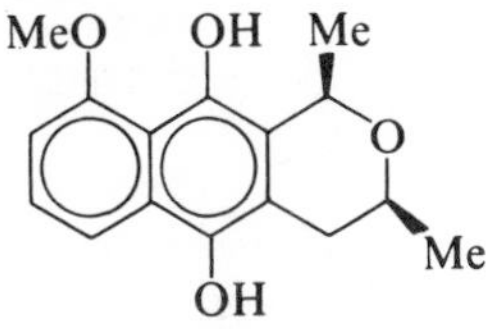

(**89**) 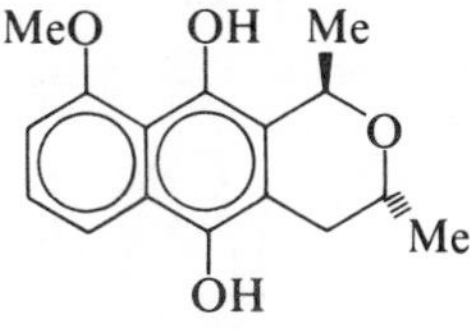 (**90**)

2. Nanaomycin antibiotics

A number of antibiotics also have the pyranonaphthoquinone structure. Kalafungin[85] (**91**) and nanaomycins A[86, 87] (**92**), B[86, 87] (**93**), C[88] (**94**) and D[89] (enantiomer of **91**) are closely related to the eleutherin pyranonaphthoquinones. That kalafungin and nanaomycin D are enantiomers was shown by the coincidence of their physical properties except that the ORD curve of nanaomycin D shows a trough at 355 nm and a peak at

292 nm while the curve of kalafungin is enantiomeric with a peak at 355 nm and a trough at 292 nm[89]. Thus the genus *Streptomyces* produces the two enantiomers of the same antibiotic.

(**91**, R = H)
(**95**, R = Me)

(**92**, R = OH)
(**94**, R = NH_2)

(**93**)

The structure and relative configurations of the three chiral centers in kalafungin were established by X-ray studies[90] and chemical and other physical properties support these assignments[90]. The absolute configuration of kalafungin (**91**) was assigned on the basis of the comparison of the ORD curves of 9-*O*-methylkalafungin (**95**) with that of isoeleutherin (**86**) (Figure 8), both curves showing significant similarities[90]. Since the C(1) methyl group in **95** and the corresponding one in isoeleutherin (**86**) at C(9) are, on the basis of conformational arguments[90], pseudoaxial, the two ORD curves then should be similar if

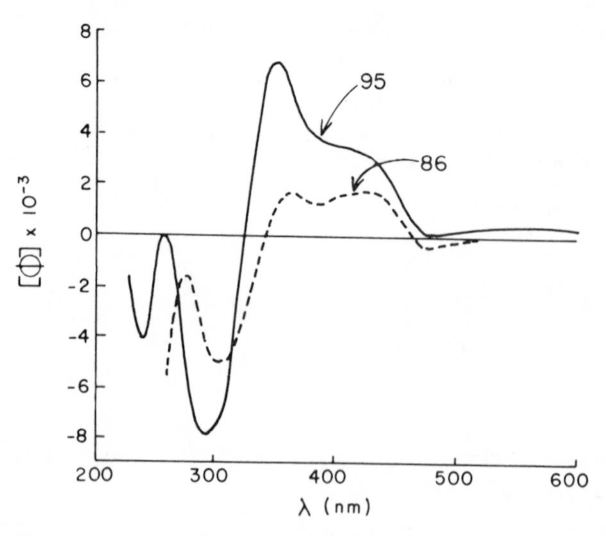

FIGURE 8. Optical rotatory dispersion (ORD) curves of isoeleutherin (**86**) and 9-*O*-methylkalafungin (**95**). *Adapted from Ref. 90 by permission of the Japan Antibiotics Research Association*

95 has the 1*R*,3*R*,4*R* configuration[90]. If 9-*O*-methylkalafungin (**95**) has the 1*S*,3*S*,4*S* configuration, its ORD curve would be enantiomeric to that observed in Figure 8.

The structure of nanaomycins A (**92**) and B[91] (**93**) and nanaomycin C[88] (**94**) were established on chemical and spectroscopic grounds. Their absolute configurations follow from the air oxidation of nanaomycin A (**92**) to nanaomycin D (enantiomer of **91**)[89]. Since **93** and **94** have been configurationally related to **92**[88,91], the absolute configurations of **93** and **94** were also established.

It is to be noted, however, that the absolute configurations established for the nanaomycin antibiotics (**91–94**) all depend on the configurational correlation of 9-*O*-methylkalafungin (**95**) with that originally established for isoeleutherin (**86**). This latter assignment was made only on the basis of a comparison of the rotatory powers (sodium D line) of *O*-methyl and *O*-acetyl derivatives of leucoeleutherin (**89**) and leucoisoeleutherin (**90**) with the rotatory powers of similar derivatives of (*R*)-eleutherol (**88**)[83]. Verification of these configurational assignments for **91–94** by another method would thus be of value.

3. Griseusin antibiotics

Two other pyranonaphthoquinone antibiotics, also isolated from a genus of *Streptomyces*, are griseusins A[92] (**96**) and B[92] (**97**). The structure and relative configurations of **96** and **97** were established on chemical and spectroscopic evidence[93]. The absolute configurations were originally assigned incorrectly, however, on the basis of a

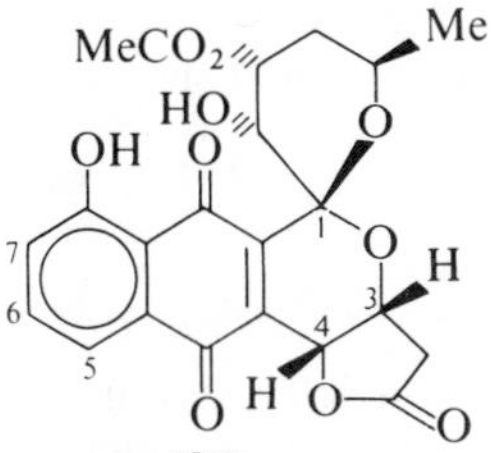

(**96**)

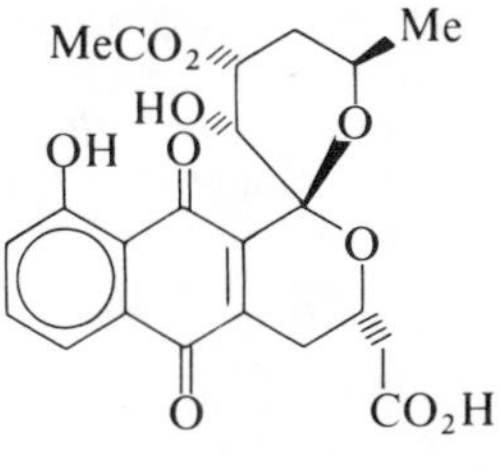

(**97**)

comparison of their respective CD spectra with that of actinorhodinindazolquinone[93] (**98**) (Figure 9). The absolute configuration of this latter compound was assigned by a chemical correlation with (*R*)-3-(1-carboxyethoxy)glutaric acid (**99**)[94]. The configuration of **99** follows from its ORD spectrum which is enantiomeric to that of (*S*)-lactic acid[94]. The absolute configurations of **96** and **97** have subsequently been established correctly by an X-ray study utilizing 5,7-dibromogriseusin A[95].

That actinorhodinindazolquinone (**98**) can not serve as a model compound for the assignment of the absolute configurations of **96** and **97** by a CD comparison may be due to a substantial difference in the chromophore in **96** and **97** as compared to that in **98**. An

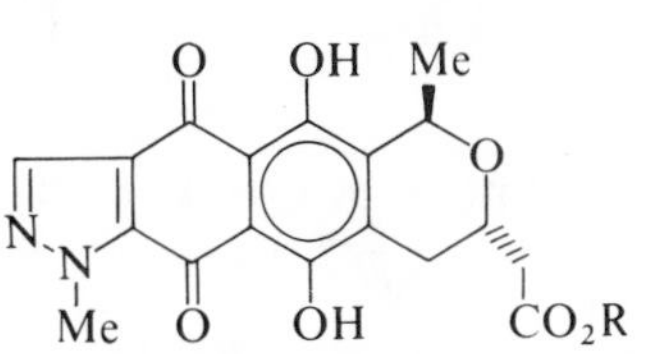

(**98**, R = H)
(**102**, R = Me)

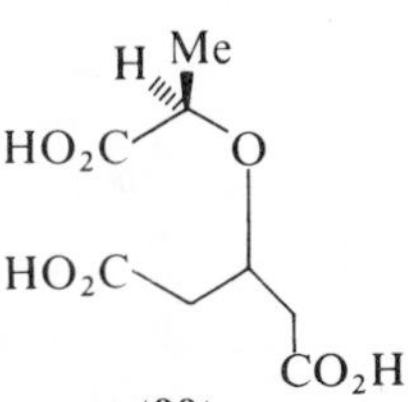

(**99**)

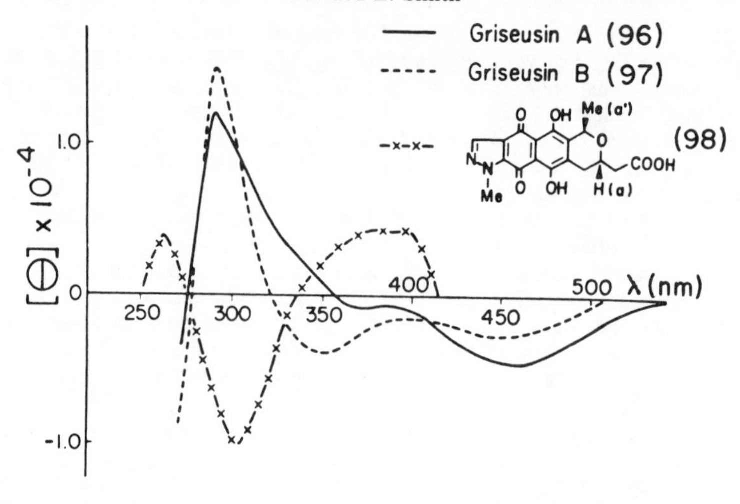

FIGURE 9. The circular dichroism spectra of griseusin A (**96**), griseusin B (**97**) and actinorhodinindazolquinone (**98**). *Adapted from Ref. 93 by permission of Pergamon Press*

alternative difficulty may be that for corresponding absolute configurations at C(1) and C(3) in **96**–**98**, a preferred chirality of the half-chair pyrano ring with respect to the naphthoquinone chromophore in **96** and **97** may be opposite to that in **98** because the substituents at C(1) in **96** and **97** are different from those at C(1) in **98**.

4. *Actinorhodin antibiotic*

The structure and configuration of actinorhodinindazolquinone (**98**) is of some interest in that it is a degradation product of another pyranonaphthoquinone antibiotic, actinorhodin[94] (**100**). For this binaphthoquinone, each naphthoquinone moiety has a structure similar to nanaomycin A (**92**), but the absolute configurations at C(1) and C(3) in **100** are opposite those in **92**. It is not known, however, if **100** is the 6,6′- or 7,7′-dimer. If the former, the diazomethane adduct with the dimethyl ester of **100** has structure **101**.

(**100**) (**101**)

Oxidation of **101** and then *N*-methylation gives the methyl ester of **98** (**102**). If **100** is the 7,7′-dimer, the structure of the diazomethane adduct of the dimethyl ester of **100** is **103**. Oxidation of **103** and then *N*-methylation gives **104**.

In connection with the configurational assignments for other isochromanquinone antibiotics[96], the respective indazolquinones degradation products similar to **102**/**104** were assigned their respective absolute configurations at C(3) on the basis of their CD spectra,

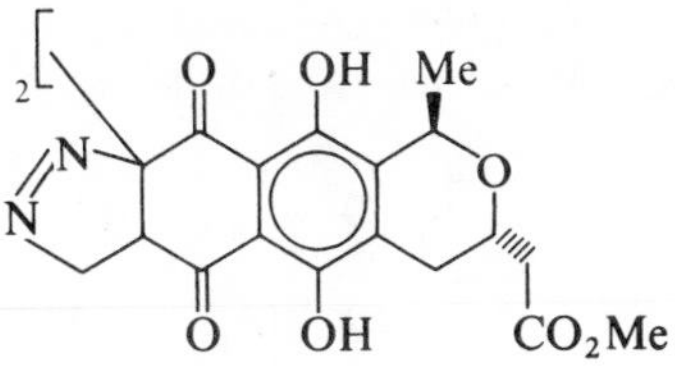

(103)

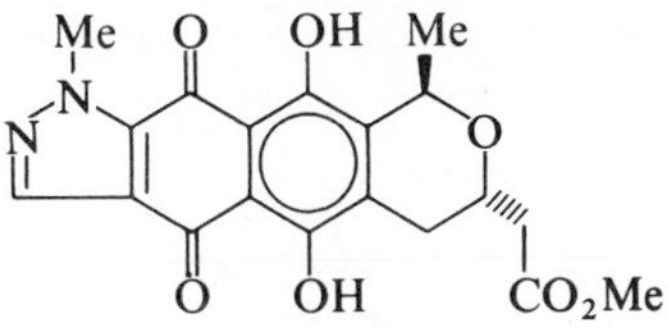

(104)

each showing a strong positive or negative maximum at 300 nm depending on the configuration at C(3)[96].

5. *Protoaphins and aphins*

Some of the most highly condensed aromatic compounds found as natural products are the extended quinones[3]. One group of these is the aphins, formed on intramolecular condensation of the protoaphins[97]. The latter are yellow water soluble substances which occur in aphids and incorporate a pyranonaphthoquinone moiety (**105**–**107**). The chirality of some of their chiral centers are preserved in the aphin group of extended quinones.

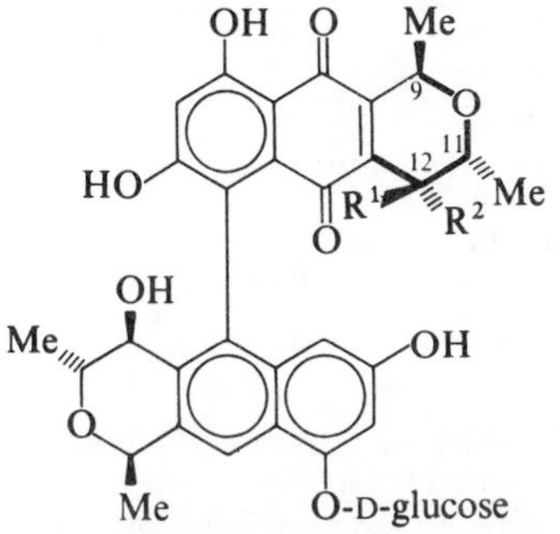

(**105**, R^1 = OH, R^2 = H)
(**106**, R^1 = H, R^2 = OH)
(**107**, R^1 = R^2 = H)

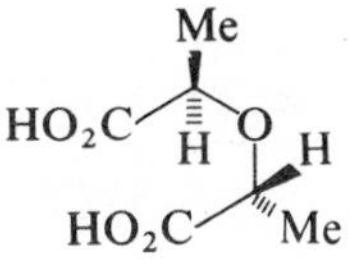

(**108**)

The three protoaphins so far reported are protoaphin-*fb*[98] (**105**), protoaphin-*sl*[98] (**106**) and deoxyprotoaphin[99] (**107**), the suffix in **105** and **106** indicating the species from which the protoaphin was first isolated, *Aphis fabae and Tuberolachnus salignus*, respectively[3]. Deoxyprotoaphin (**107**) was isolated from various aphid species of the genus *Dactynotes*[99]. The structure and absolute configuration of each of the tricyclic systems in **105** and **106** were established by degradation, the first step of which was mild reductive cleavage of the binaphthyl system[97,98]. The same glucoside was obtained from each protoaphin, but a different quinone, each of the latter having the same carbon skeleton and the same absolute configurations at C(9) and C(11) as was found for isoeleutherin (**86**). Hydrolysis of the glucoside obtained on reductive cleavage of the binaphthyl system and oxidation of the aglycone gave the same quinone as was obtained directly from **105**. The absolute configurations of the chiral centers at C(9) and C(11) in the three quinones were established by their oxidation to (*R*, *R*)-dilactic acid (**108**)[98]. Nuclear magnetic resonance studies established other structural and configurational features[84].

Under the influence of an enzyme system, each protoaphin is converted to a particular series of aphins, successively a xanthoaphin, a chrysoaphin, and an erythroaphin[99]. In the

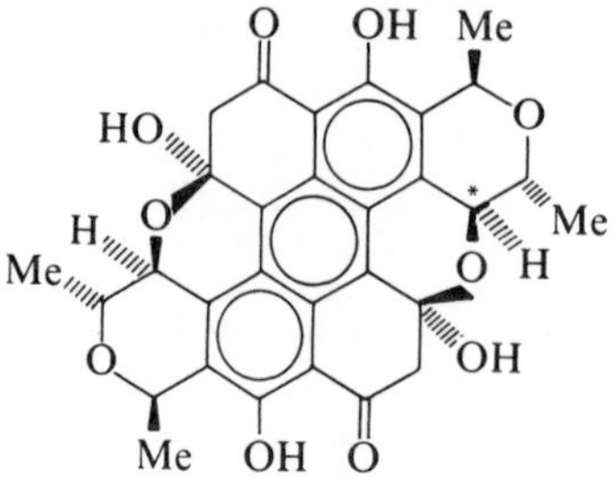

(109)

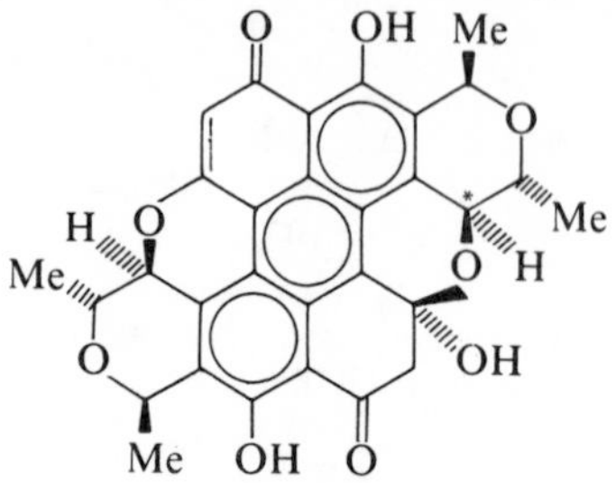

(110)

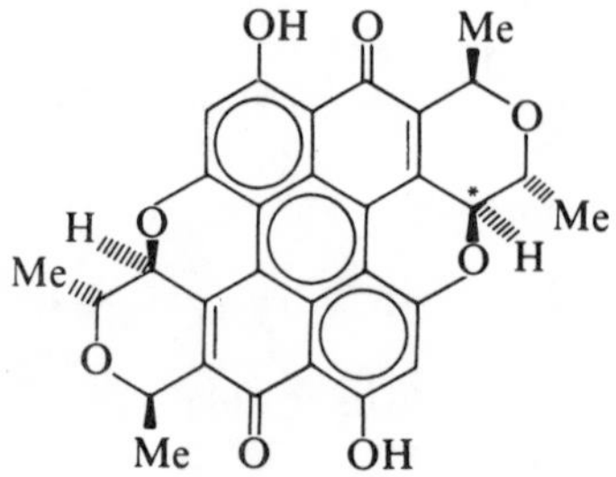

(111)

case of protoaphin-*fb* these are xanthoaphin-*fb* (**109**), chrysoaphin-*fb* (**110**) and finally erythroaphin-*fb* (**111**)[100]. The corresponding aphins from protoaphin-*sl* have a different configuration at a single chiral center shown with an asterisk in **109**–**111** because of the difference in the chirality of the corresponding center in the protoaphin[101].

It is to be noted that in the protoaphins **105**–**107**, the two tricyclic units cannot attain coplanarity and this introduces into these molecules an additional element of chirality (atropisomerism[6, 100]). On the basis of an analysis of the steric demand for the cyclization of protoaphin-*fb* (**105**) and protoaphin-*sl* (**106**) to xanthoaphin-*fb* (**109**) and xanthoaphin-*sl*, respectively, the chirality of the restricted rotation in both **105** and **106** was assigned as *S*. This chirality is shown in **105** and **106** with the non-aromatic ring of the naphthoquinone moiety out of the plane of the paper toward the observer[100]. Because of the condensation of deoxyprotoaphin (**107**) by an enzymic extract of *A. fabae* to a compound analogous to a xanthoaphin, it is presumed that the absolute configuration of **107** about the binaphthyl linkage is the same as that in protoaphin-*fb* (**105**)[99].

C. Ansamycin Antibiotics

A somewhat large group of antibiotics some of which incorporate a naphthoquinone or naphthoquinone derived group are the ansamycins[102], so called because each contains an aliphatic bridge connecting two non-adjacent positions of an aromatic nucleus. The term *ansa* is used to designate *meta*- and *para*-bridged aromatic, usually benzene, compounds.

Various types of this class of antibiotics have been identified: the rifamycins, the streptovaricins, the tolypomycins and the naphthomycins[102, 103]. The rifamycins and naphthomycins incorporate a naphthoquinone group[103] while the streptovaricins have a naphthoquinone methide moiety[102]. An important one of the latter type is streptovaricin C (**112**) for which the stereochemistry and circular dichroism have been studied in some

(**112**, $R^1 = R^2 = R^3 = H$)
(**113**, $R^1 = MeCO$; $R^2R^3 = {>}BC_6H_4Br$-*p*)

detail[104]. Stereoformula **112** shows the absolute configurations for the various chiral centers in streptovaricin C as well as that of the helix with *P* configuration for which the acetoxyl group at C(24) is below the *ansa* bridge[104]. When streptovaricin C (**112**) was boiled in toluene overnight, an equilibrated mixture of **112** and atropisostreptovaricin C was obtained[104]. Atropisostreptovaricin C is a diastereoisomer (atropisomer[6]) of **112** in which the chiral centers have the same absolute configurations but the acetoxyl group at

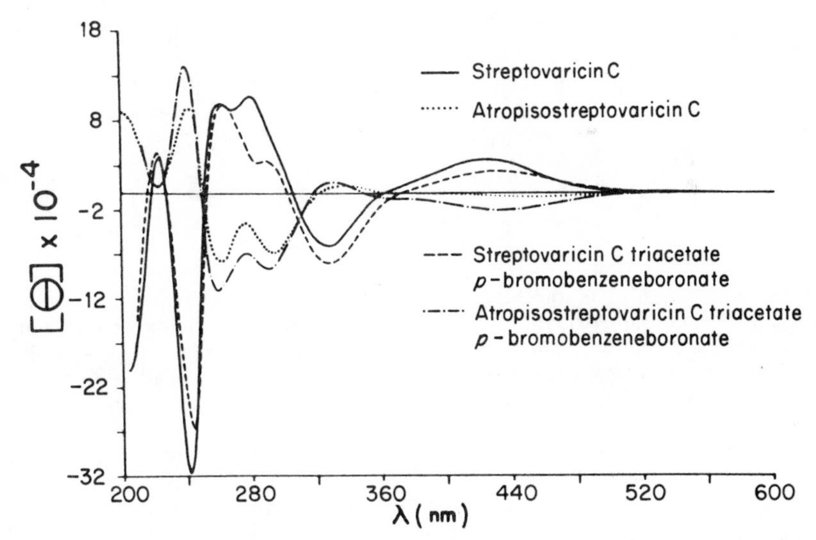

FIGURE 10. Circular dichroism (CD) spectra of streptovaricin (**112**), atropisostreptovaricin C, streptovaricin C triacetate *p*-bromobenzeneboronate (**113**) and atropisostreptovaricin C triacetate *p*-bromobenzeneboronate. *Adapted with permission from K. L. Rinehart, Jr., W. M. J. Knöll, K. Kakinuma, F. J. Antosz, I. C. Paul, A. H.-J. Wang, F. Reusser, L. H. Li and W. C. Krueger,* J. Am. Chem. Soc., **97**, *196 (1975). Copyright (1975) American Chemical Society*

C(24) now is above the *ansa* ring. The helix so formed has the *M* absolute configuration. The nature of the isomerism of **112** and atropisostreptovaricin C was established by the nearly mirror image relationship of their CD spectra (Figure 10)[104]. A similar equilibration of streptovaricin C triacetate *p*-bromobenzeneboronate (**113**) with its atropisomer was also obtained and the atropisomers also show enantiomeric CD curves (Figure 10). An X-ray investigation was used to establish the absolute configuration of **113**. On the basis then of the respective CD curves, the *P* helicity of streptovaricin C (**112**) and the *M* helicity of its atropisomer were also assigned.

The apparent CD couplets in the spectra of **112** and **113** and their atropisomers in Figure 10 indicate that the source of the high rotational strength in the 200–300-nm region is the chirality of the *ansa* bridge unsaturated polarizabilities with the naphthoquinone methide polarizabilities. A striking confirmation of this was given in a recent CD study of rifamycin *S* and its tetrahydro and hexahydro derivatives[105]. Thus in the 200–300-nm spectral range, the CD intensity of the hydrogenated derivatives is much lower in comparison to the intensity shown by rifamycin *S*.

IV. ANTHRAQUINONES

A. Simple Anthraquinones

Simple anthraquinones form the largest group of naturally occurring quinones, and their occurrence and structures have been reviewed by Thomson[3]. The majority of these substances, which are assumed to be elaborated by the acetate–malonate pathway, conforms to the emodin (**114**) pattern formed by folding and condensation of a polyketide chain derived from eight acetate units. In endocrocin (**115**), the terminal carboxyl group is retained.

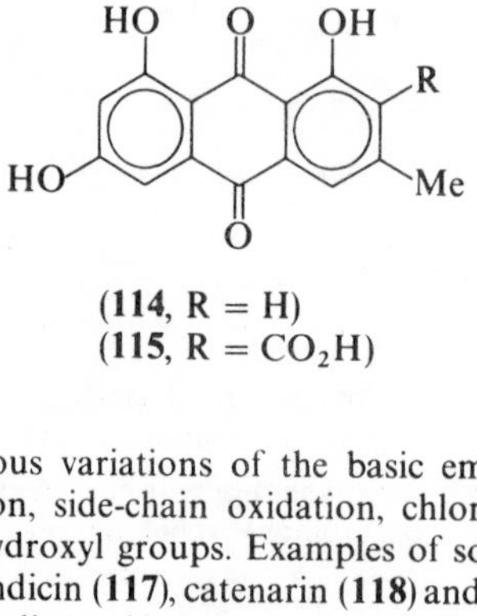

(**114**, R = H)
(**115**, R = CO_2H)

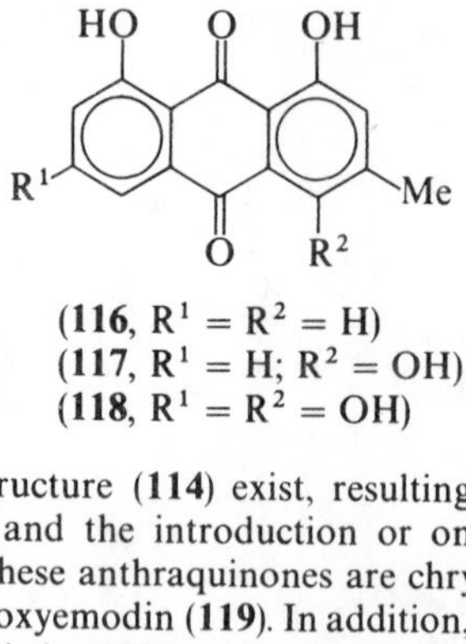

(**116**, $R^1 = R^2$ = H)
(**117**, R^1 = H; R^2 = OH)
(**118**, $R^1 = R^2$ = OH)

Numerous variations of the basic emodin structure (**114**) exist, resulting from *O*-methylation, side-chain oxidation, chlorination and the introduction or omission of nuclear hydroxyl groups. Examples of some of these anthraquinones are chrysophanol (**116**), islandicin (**117**), catenarin (**118**) and ω-hydroxyemodin (**119**). In addition, three 1,4-anthracenedione pigments, two of which are viocristin (**120**) and isoviocristin (**121**), have

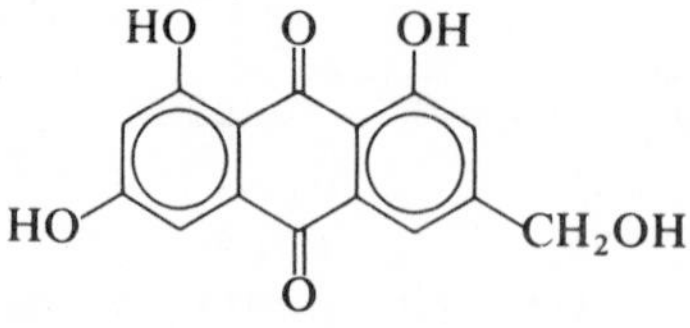

(**119**)

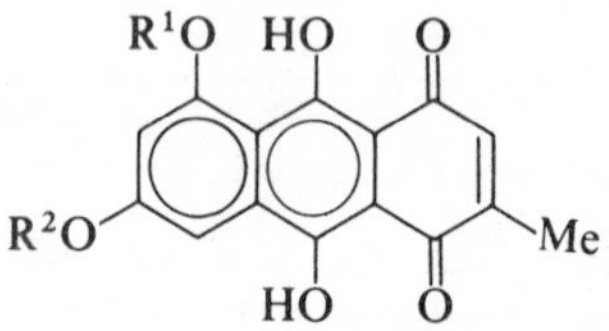

(**120**, R^1 = Me; R^2 = H)
(**121**, R^1 = H; R^2 = Me)

recently been found as natural products[106]. These anthraquinones and 1,4-anthracenediones (**114**–**121**), however, are symmetrical and show no chiroptical properties.

In a few cases the usual β-methyl group is replaced by a three- or a five-carbon substituent[107,108] and in rhodoptilometrin[107,108] (**122**) and isorhodoptilometrin[107] (**123**), this side chain is oxidized such that both of these anthraquinones contain a chiral center. In rhodoptilometrin (**122**), isolated from a crinoid[107], the configuration was

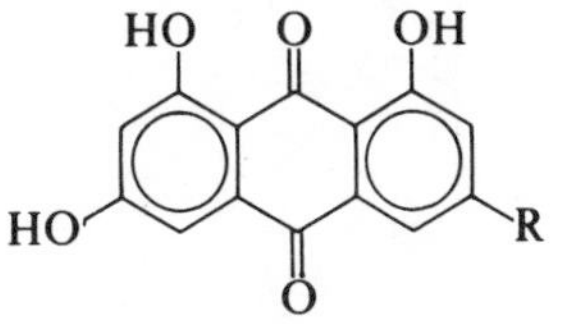

(**122**, R = CH(OH)Et)
(**123**, R = $CH_2CH(OH)Me$)

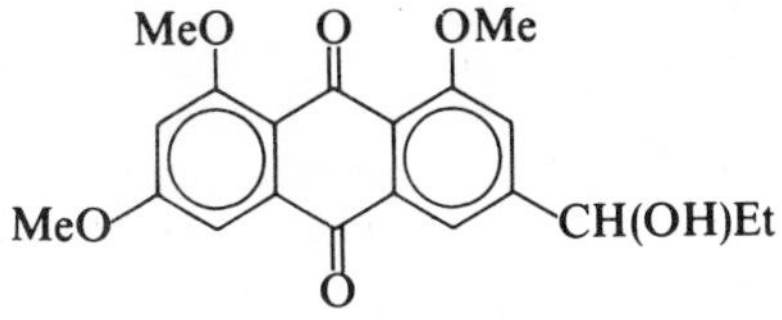

(**124**)

assigned as *S* on the basis of an empirical rule relating the configuration of the side chain to the rotatory powers of the trimethyl ether (**124**) and the so-called trimethyl ether leucoacetate (**125**)[107]. Lacking confirmation of the configurational assignment for **124** or **125**, the configurational assignment for **122** is not conclusively established.

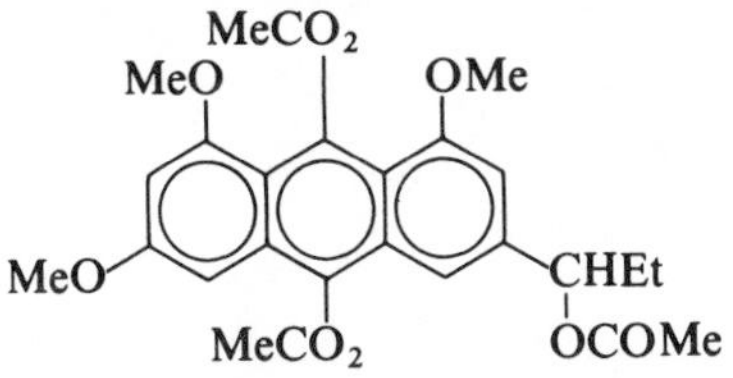

(**125**)

Other anthraquinones with chiral centers in attached side chains have been reported[3,109] but aside from their rotatory powers at the sodium D line their chiroptical properties have not been studied. As noted above (Section I), magnetic circular dichroism (MCD) of anthraquinone (**2**), however, has been reported as part of a general study of the electronic absorption of aromatic compounds[16].

B. Bianthraquinones

One interesting group of pigments derived from simple anthraquinones are the bianthraquinones isolated from several strains of *Penicillium islandicum*[110,111]. The structures of these natural products were established as 1,1′-dimers of simple anthraquinones by the reductive cleavage with alkaline sodium dithionate and ^{1}H-NMR spectral analysis of the respective acetates. Thus (+)-dianhydrorugulosin [(+)-**126**)] yields two equivalents of chrysophanol (**116**) on reductive cleavage[110]. Four other bianthraquinones are 1,1′-dimers of one particular simple anthraquinone with structures **114**, **117**, **118** and **119**[111]. Seven others, such as (+)-roseoskyrin [(+)-**127**] are 1,1′-dimers of two different simple anthraquinones. Thus on reductive cleavage, (+)-roseoskyrin [(+)-**127**] gives one equivalent each of chrysophanol (**116**) and islandicin (**117**)[111].

All of the dimeric anthraquinones from *P. islandicum*[110] and from other fungi and lichens[111] are optically active as the result of the restricted rotation about the C(1)–C(1′) bond (atropisomerism[6]). Chirality due to a similar steric effect is well known[6] and both

(**126**) (**127**)

(*R*)- and (*S*)-2,2′-dimethyl-1,1′-bianthraquinonyl [(*R*)- and (*S*)-**128**] have been prepared[112, 113].

The ORD curves and CD spectra of the bianthraquinones isolated from *P. islandicum* show the same features with Cotton effects associated with the EA spectrum from 350 to 500 nm (Figure 11). In each ORD curve, the extremum at the longest wavelength is positive[110, 111]. That this extremum is positive is indicated by the attachment of a

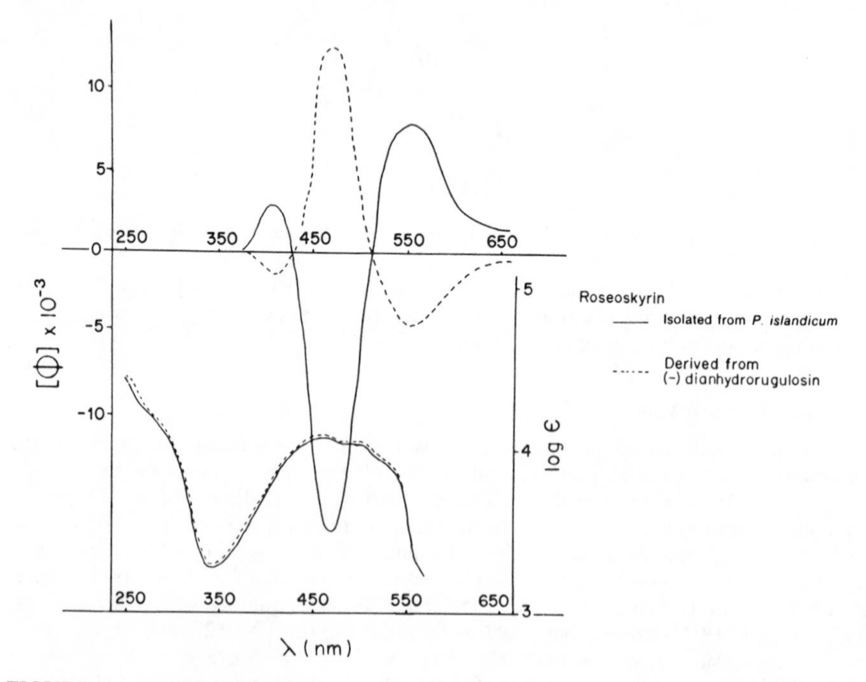

FIGURE 11. Electronic absorption (EA) spectra and optical rotatory dispersion (ORD) curves in dioxan of (+)-roseoskyrin [(+)-**127**] from *Penicillium islandicum* and (−)-roseoskyrin [(−)-**127**] derived from (+)-rugulosin [(+)-**129**] by way of (−)-dianhydrorugulosin [(−)-**126**]. *Adapted from Ref. 111 by permission of Pergamon Press*

(+) prefix to the name. For the bianthraquinones isolated from all *Penicillium* species, the similarity of their ORD curves suggests that they all have the same absolute configuration

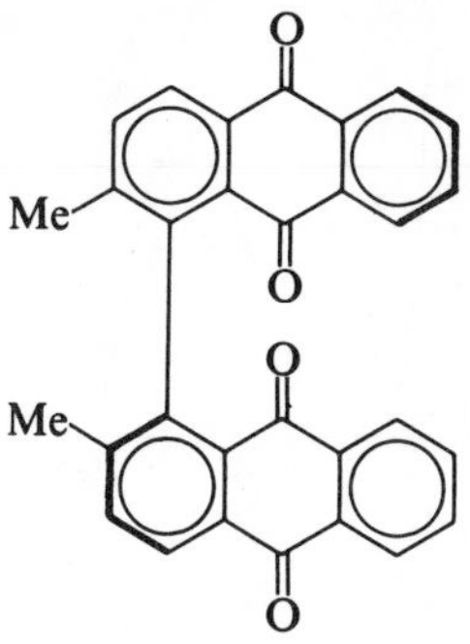

[(*R*)-**128**]

and arise by a stereospecific oxidative coupling of the monomeric anthraquinones, the latter also occurring in these fungi.

The ORD curves of these pigments[110, 111] (Figure 11) show two Cotton effects in the 350–550-nm region which are due to exciton interaction[65] between the two anthraquinonyl chromophores. The nature and direction of the relevant transition moments giving rise to these Cotton effects, however, are not established[114] and hence the absolute configuration of a bianthraquinone can not be determined on the basis of the chiral exciton coupling mechanism[65]. Further, neither the ORD curve nor CD spectrum of (*R*)- or (*S*)-**128** has been reported, and thus model spectra are not available for the possible assignment of absolute configuration of the natural products by a comparison of their ORD curves or spectra CD with that of a model.

Also occurring in *P. islandicum* and other fungi is a group of modified bianthraquinones[111,115], which in fact are not anthraquinones at all but are included here because of their structural and biogenetic relationship to the bianthraquinones. An important modified bianthraquinone is (+)-rugulosin [(+)-**129**] for which the absolute configuration was established by an X-ray study of its dibromodehydrotetrahydro-derivative[116] and which could be formed biogenetically by the base-catalyzed Michael-type intramolecular condensation of a hypothetical, partially reduced bianthraquinone (**130**).

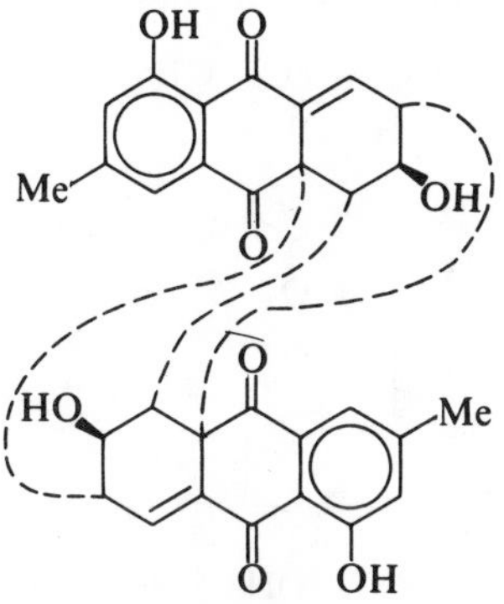

[(+)-**129**]

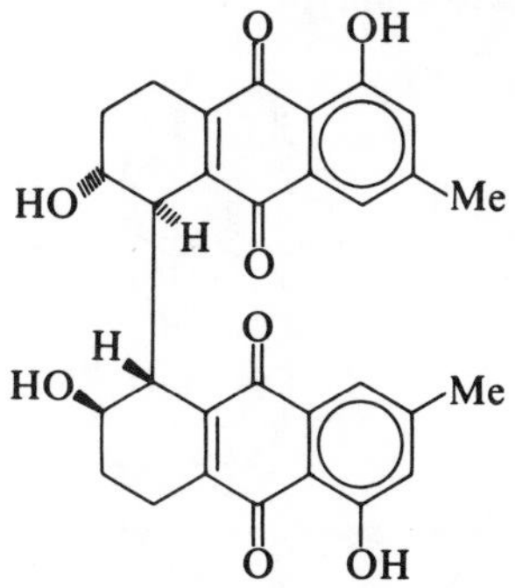

(**130**)

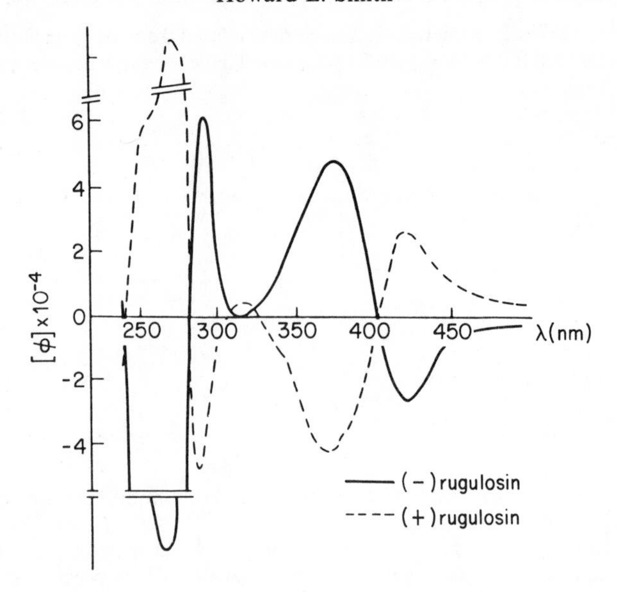

FIGURE 12. Optical rotatory dispersion (ORD) curves in dioxan of (+)-rugulosin [(+)-**129**] isolated from *Penicillium brunneum* and (−)-rugulosin [(−)-**129**] derived from (−)-flavoskyrin [(−)-**132**]. *Adapted from Ref. 117 by permission of Pergamon Press*

The stereochemistry of (+)-**129**, which shows a strong positive Cotton effect centered at 400 nm and a strong negative one at about 280 nm in its ORD curve (Figure 12)[117], is of interest. Thus (+)-**129** on dehydration with thionyl chloride and pyridine at 0°C affords (−)-dianhydrorugulosin [(−)-**126**] which on oxidation with pertrifluoroacetic acid in methylene chloride at 0°C gives (−)-iridoskyrin [(−)-**131**] and (−)-roseoskyrin [(−)-(**127**)][111]. The ORD curve of (−)-**127** is enantiomeric to that of the naturally occurring (+)-**127** (Figure 11)[117] and thus the chiralities about the respective C(1)–C(1′) bonds are enantiomeric. The respective chiralities of the related bianthraquinones, however, can not be inferred from this transformation in combination with the known absolute configuration of (+)-**129**.

(−)-Flavoskyrin [(−)-**132**] is another type of modified bianthraquinone also occurring

(**131**)

[(−)-**132**]

in *P. islandicum*[117]. This compound, with a positive Cotton effect centered at 415 nm and a negative one at 276 nm[117], almost enantiomeric to that of (+)-rugulosin, forms, on treatment with pyridine, (−)-rugulosin [(−)-**129**] and (+)-dianhydrorugulosin [(+)-**126**]. It is to be noted that these chemical transformations relate the absolute configuration of (−)-**129**][111] to that of (−)-**132** but not to that of (+)-**126**. Thus the absolute configurations of (−)-flavoskyrin is established as in stereoformula (−)-**132**[117], but the absolute configurations of (+)-**126** and the other chiral bianthraquinones from *P. islandicum*[110, 111] remain unknown.

C. Anthracyclinones

In the search for new antibiotics, the screening of many hundreds of *Streptomyces* bacteria revealed numerous pigments. One group of about 30 anthraquinones, studied extensively by Brockmann and his associates[118], is the anthracyclinones, occurring either free or as the glycosides (anthracyclines) of various sugars including amino sugars such as rhodosamine[3, 118](**133**). While these substances are found mostly in *Streptomyces*, one has been found in another genus[119]. Only a few anthracyclinones, such as η-pyrromycinone (**134**), are fully aromatic and in most cases ring A of the naphthacenequinone system is such that the anthracyclinones are in effect 2,3-dialkylpolyhydroxyanthraquinones with chiral centers in ring A.

Brockmann[118] introduced a trivial nomenclature which is used in association with the number and position of the hydroxyl substituents of the anthraquinone moiety (cf. **135**–**140**). Individual members of the various groups are distinguished according to the

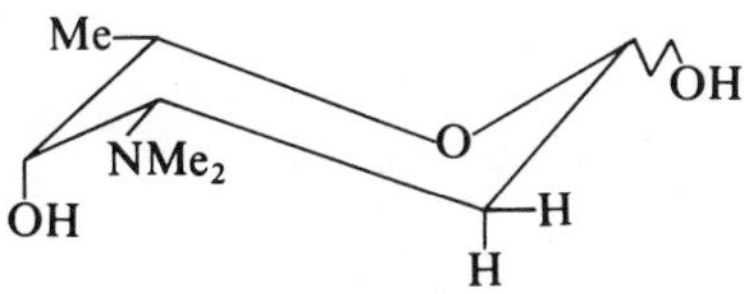

(**133**)

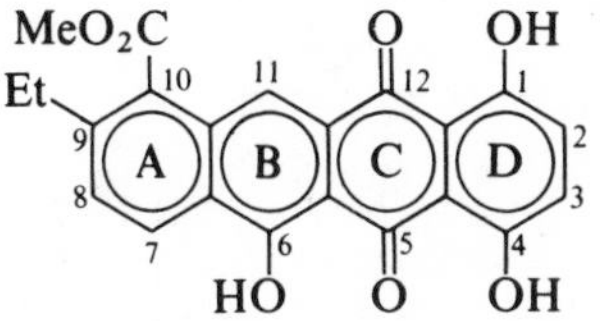

(**134**)

order of increasing R_f values. Originally α-rhodomycinone was the rodomycinone (**137**) with the lowest R_f value, but other still lower values were found later and were designated α_1-, α_2- and α_3-rhodomycinones.

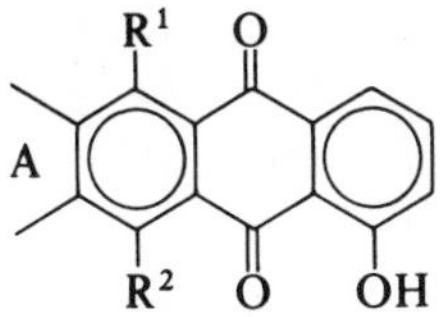

(**135**, R^1 = H; R^2 = OH)
Aklavinones

(**136**, R^1 = OH; R^2 = H)
Citromycinones

(**137**, R^1 = R^2 = OH)
Rhodomycinones

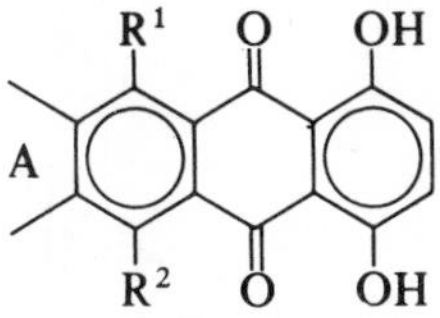

(**138**, R^1 = OH; R^2 = H)
Rhodomycinones

(**139**, R^1 = H; R^2 = OH)
Pyrromycinones

(**140**, R^1 = R^2 = OH)
Isorhodomycinones

The structure and 7S, 9R, 10R configuration of ring A in β-rhodomycinones (**141**) is the same as that of other anthracyclinones with a few variations: the hydroxyl group at C(10) may be replaced by a methoxycarbonyl group as in ε-rhodomycinone (**142**) or be absent; the hydroxyl group at C(7) may be absent as in γ- (**143**) and ζ-rhodomycinone (**144**) or may have the 7R configuration as in α-rhodomycinone; and the ethyl group at C(9) may be replaced by an acetyl group as in daunomycinone (**145**) or by an hydroxyacetyl group as in adriamycinone (**146**)[118].

(**141**, R^1 = OH; R^2 = OH)
(**142**, R^1 = CO_2Me; R^2 = OH)
(**143**, R^1 = OH; R^2 = H)
(**144**, R^1 = CO_2Me; R^2 = H)

(**145**, R = MeCO)
(**146**, R = $HOCH_2CO$)

On the basis of chemical evidence and the CD spectra of various pyrromycinones (**139**), rhodomycinones (**137**) and isorhodomycinones (**140**), Brockmann Jr and Legrand[120, 121] suggested that all of the anthracyclinones had the same absolute configuration for the chiral centers in ring A. Thus for the compounds examined, the CD spectra all showed the same characteristic S-shape curve in the 270–390-nm region (Figure 13), a spectral region where

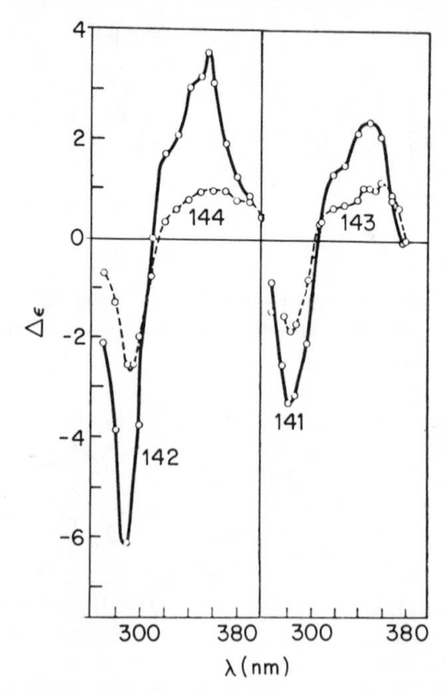

FIGURE 13. Circular dichroism (CD) spectra of β-rhodomycinone (**141**), ε-rhodomycinone (**142**), γ-rhodomycinone (**143**) and ζ-rhodomycinone (**144**) in dioxan. *Adapted from Ref. 121 by permission of Pergamon Press*

all of these pigments show similar absorption characteristics[121]. Initially it was recognized that the CD spectra for a number of anthracyclinones were similar because of the same absolute configuration and the same preferred half-chair conformation of ring A[120]. Thus in **141** and **142**, the ethyl group at C(9) occupies an equatorial position (**147**, **148**) while the

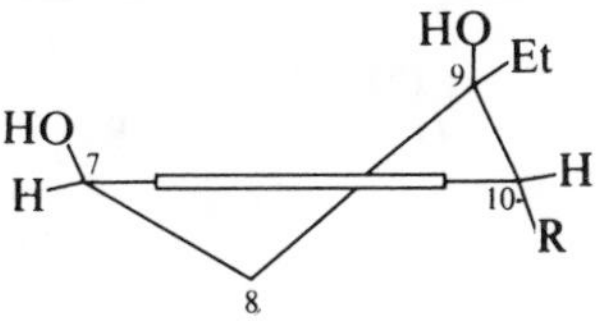

(**147**, R = OH)
(**148**, R = CO_2Me)

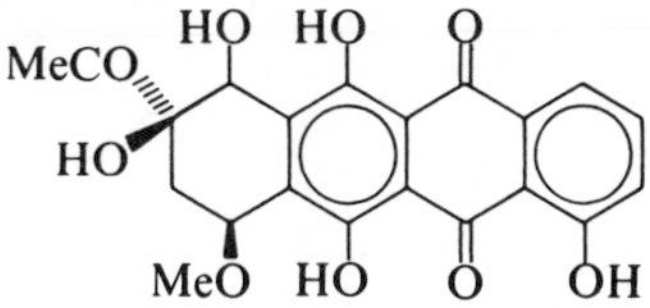

(**149**)

conformation of the hydroxyl groups at C(7) and C(10) in **141** are both pseudoaxial (**147**) and the hydroxyl group at C(7) and the methoxycarbonyl group at C(10) in **142** are also both pseudoaxial (**148**)[121]. The same preferred conformation of ring A exists for those anthracyclinones in which the ethyl group at C(9) is replaced by an acetyl or an hydroxyacetyl group (**145**, **146**) and results from the preference of an ethyl or acetyl group at C(9) to be equatorial. A preferred pseudoaxial conformation for a substituent at C(7) and C(10) is suggested by ^{1}H-NMR studies with 3,5-dichlorobenzylamine[122] and 3,5-dibromoethylbenzene[123] and calculations using empirical potential functions[124] for the conformational distribution in 1-phenylethanol. These studies show that for these compounds the conformation of the substituent group about its attachment bond in which one hydrogen atom eclipses the plane of the benzene ring is lower in energy than those in which an amino, methyl or hydroxyl group eclipses the ring plane.

The CD spectra of an anthracyclinone then depends only on the substitution pattern of the saturated ring A and not on the number and positions of the phenolic hydroxyl groups in the anthraquinone part of the molecule[125]. The relative configurations of the chiral centers at C(7), C(9) and C(10) are supported by extensive chemical and ^{1}H-NMR spectral evidence[125]. It is to be noted, however, that the absolute configurations of the anthracyclinones was not and in fact could not be established on the basis of their CD spectra. Rather the absolute configuration of this group of natural products was based on the oxidative degradation of 7-*O*-methyldesmethyldaunomycinone (**149**) to (*S*)-methoxysuccinic acid[126]. The configurational assignment for the anthracyclinones was later confirmed by X-ray anomalous scattering using γ-rhodomycinone[127] (**143**).

The CD spectrum of an anthracyclinone can be used as an aid for the assignment of the relative configurations of the substituents on ring A[125]. Thus, on the assumption that the conformation of ring A (**147**, **148**) does not depend on whether a C(7) hydroxyl group is present or not, the relative configuration of an hydroxyl group at C(7) can be deduced by comparison of the CD spectrum with that of the corresponding 7-desoxy compound[125], an hydroxyl group at C(7) with the *S* configuration causing the S-shaped CD curve in the 270–390-nm region to be increased in intensity[125] (Figure 13).

As outlined above, the early efforts for structural and stereochemical characterization of the anthracyclines, the glycosides of the anthracyclinones, began in the 1950s. The discovery of adriamycin (**150**) and recognition of its unusual potential efficacy in treatment of human cancers in 1969[128, 129] have spawned an enormous interest in anthracycline synthesis. This was stimulated by a quest for synthetic anthracyclines which retain the efficacy of adriamycin but lack its severe toxicity in cancer chemotherapy[130]. The synthesis of anthracyclines can be divided into three parts: formation of the anthracyclinone,

(150)

(151, R^1 = H; R^2 = OMe)
(152, R^1 = Me; R^2 = H)

preparation of the sugar, and coupling of the two. It is the synthesis of the anthracyclinones, however, which has particularly engaged the interest of synthetic organic chemists, and has been the subject of a recent symposium-in-print[131] and another review[132]. In many of these syntheses one enantiomer of the anthracyclinone is prepared[133] and the rotatory power at the sodium D line is used for evaluating the optical purity of these preparations[134]. The specific rotations (sodium D line) of anthracyclinones, however, exhibit dramatic variation induced by protic solvents and there are reports of purportedly optically pure anthracyclinones having dramatically different reported specific rotations when measured under seemingly identical conditions[134]. The origin of this variation was traced to the purity of the solvent used in the determination of specific rotations[134]. Figure 14 shows that the presence of polar organic oxygen solvents, in particular ethanol and methanol, can dramatically alter the magnitude of specific rotations determined for

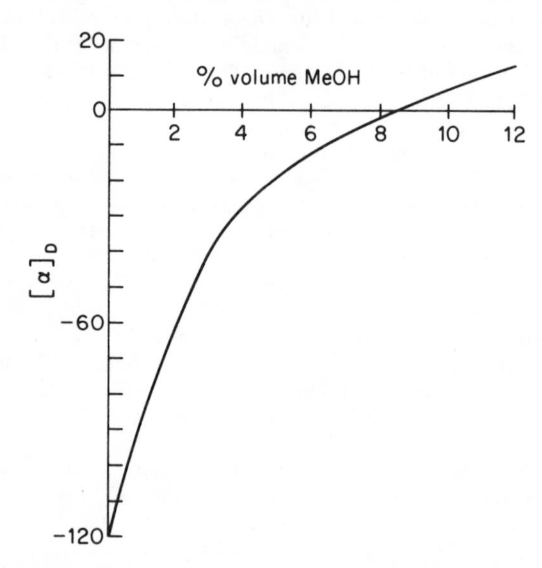

FIGURE 14. Variation of $[\alpha]_D$ for 7-desoxy-11-*O*-methyldaunomycinone (**151**) in chloroform with increasing concentration of methanol. *Reproduced from Ref. 134 by permission of Pergamon Press*

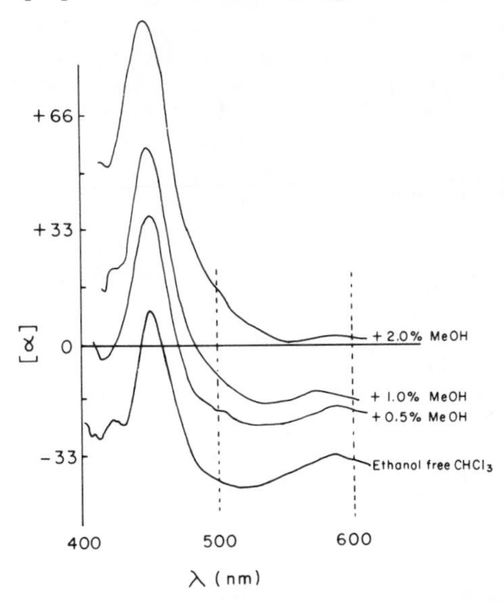

FIGURE 15. Optical rotatory dispersion (ORD) curve of 4,7-didesoxy-6-*O*,11-*O*-dimethyldaunomycinone (**152**) in chloroform and chloroform–methanol. *Revised from Ref. 134 by permission of Pergamon Press*

chloroform solutions of 7-desoxy-11-*O*-methyldaunomycinone (**151**). This problem can be overcome by determining specific rotations in a 1 : 1 methanol–chloroform mixture[134]. The origin of the solvent dependency is reflected dramatically throughout the visible ORD curve of 4,7-didesoxy-6-*O*,11-*O*-dimethyldaunomycinone (**152**) (Figure 15), this curve being a revision[135] of that reported earlier[134]. It is not clear, however, whether the effect originates from conformational changes or molecular aggregation[134].

D. Anthracyclines

Although much earlier work focused on the structure and stereochemistry of the anthracyclinones, the structure and stereochemistry of the anthracyclines were also of interest[118]. The structure[136, 137] and the absolute configuration[126] of the antitumor antibiotic daunomycin (**153**) were of prime importance. Since for the establishment of the

(**153**)

(**154**)

absolute configuration of **153**, the latter was converted by hydrolysis to 7-*O*-methyldesmethyldaunomycinone (**149**) and in **149** it is possible that the configuration at C(7) was altered in the conversion of **153** to **149**. That this was not the case is indicated by comparison of the CD curve of daunomycin and daunomycinone which are very similar in the 280-nm region[137]. Removal of the hydroxyl group at C(7) gives 7-desoxydaunomycinone, which exhibits a very weak dichroic effect, and implies that the substantial effect in the CD curves of daunomycinone and daunomycin is largely contributed by the substituent at C(7)[137].

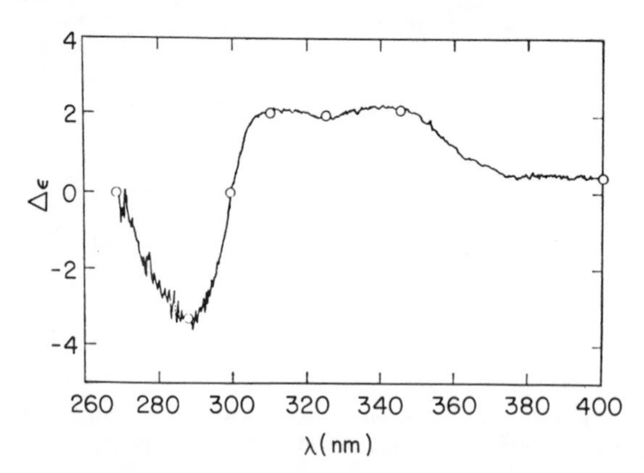

FIGURE 16. Circular dichroism (CD) curve of daunomycin (**153**) in dioxan. *Adapted from Ref. 138 by permission of the Royal Society of Chemistry*

The CD spectra of daunomycin (**153**) was also reported[138] (Figure 16) and was used to establish the configuration at C(7) of two D-glucosides (**154**) of the anthraquinone analog of daunomycinone. These isomers were prepared to explore the structure–activity relationship in the anthracycline cytotoxic antibiotics. Reaction of the racemic aglycone of **154** with acetobromoglucose in the presence of mercuric cyanide gave the two diasteromeric tetraacetates of **154**. Removal of the acetyl groups with methanolic sodium methoxide and then preparative thin layer chromatography gave the two stereoisomers with structure **154**[138]. The CD spectra (Figure 17) of these isomers are essentially enantiomeric and reflect the enantiomeric nature of two half-chair conformations **155** and **156**. In both isomers, the substituent at C(7) is preferentially either pseudoequatorial (R^1 in **155** and **156**) or pseudoaxial (R^2 in **155** and **156**) and the ring conformation depends only on the configuration at C(7). That isomer with its CD spectrum showing a negative

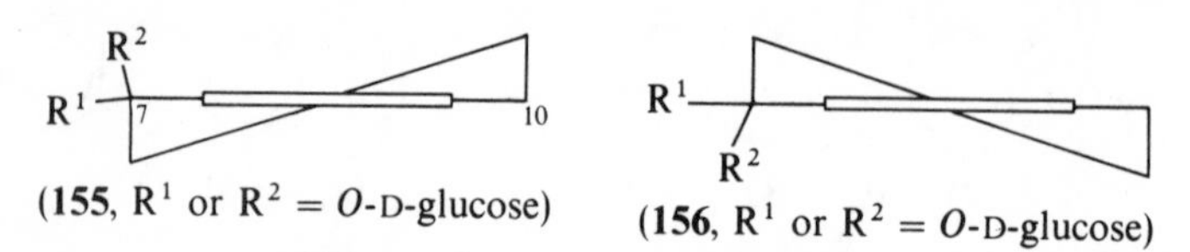

maximum near 290 nm, similar to that for daunomycin (Figure 16), was assigned the 7*S* configuration[138]. Thus the glucoside substituent, represented by R^2, is pseudoaxial in both **155** and **156**, the conformation at C(7) also being pseudoaxial in daunomycin[121]. If, however, the glucosidic substituent in the isomers of **154** is pseudoequatorial (R^1 in **155** and **156**), that isomer with a positive maximum near 290 nm would have the 7*S* configuration.

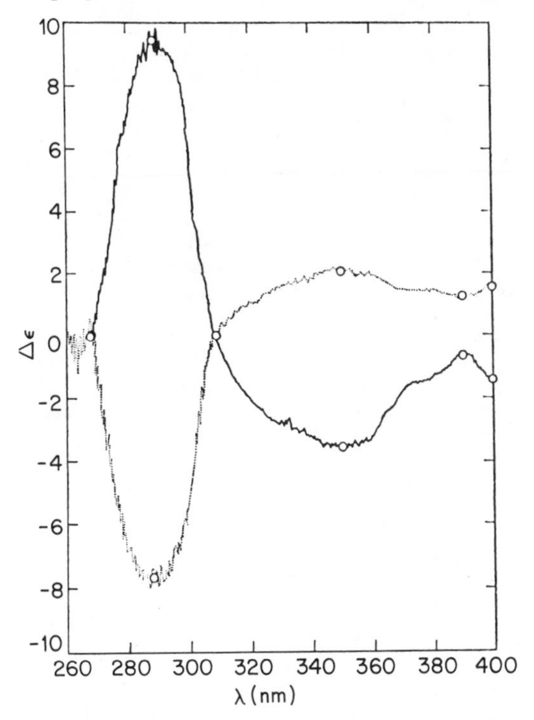

FIGURE 17. Circular dichroism (CD) spectra of the glucoside of diastereomeric anthraquinone analogs of daunomycinone (**154**) in dioxan. *Adapted from Ref. 138 by permission of the Royal Society of Chemistry*

As discussed above in connection with preferred half-chair conformation for anthracyclinones (Section IV.C), ^{1}H-NMR measurements[122,123] and calculations using empirical potential functions[124] indicate that the pseudoaxial conformation (R^2 in **155** and **156**) for a substituent at C(7) and C(10) is slightly lower in energy than is the pseudoequatorial conformation (R^1 in **155** and **156**). Thus, the assignment of the 7*R* and 7*S* configurations to the D-glucosidic isomers which show positive and negative CEs near 290 nm, respectively, has some experimental and theoretical justification.

V. DEDICATION AND ACKNOWLEDGEMENT

This chapter is dedicated to my late brother, Charles A. Smith, Jr. He was a good friend to me, and I will miss him. Also, I thank the many authors and copyright holders for permission to reproduce the respective figures.

VI. REFERENCES

1. T. M. Lowry, *Optical Rotatory Power*, Longmans Green, London, 1935; republished by Dover, New York, 1964.
2. W. Heller and D. D. Fitts, in *Techniques of Organic Chemistry*, 3rd edn, Vol. 1 (Ed. A. Weissberger), Interscience, New York, 1959, Chap. 33, pp. 2147–2333.
3. R. H. Thomson, *Naturally Occurring Quinones*, 2nd edn, Academic Press, New York, 1971.

4. W. Klyne and J. Buckingham, *Atlas of Stereochemistry*, 2nd edn, Vol. 1, Oxford University, New York, 1978; Vol. 2, Chapman and Hall, London, 1978.
5. J. Jacques, C. Gros and S. Bourcier, in *Stereochemistry*, Vol. 4 (Ed. H. B. Kagan), Georg Thieme, Stuttgart, 1977.
6. E. L. Eliel, *Stereochemistry of Carbon Compounds*, McGraw-Hill, New York, 1962, Chap. 6, pp. 156–179.
7. J. A. Mills and W. Klyne, *Prog. Stereochem.*, **1**, 177 (1954).
8. M. Legrand and M. J. Rougier, in *Stereochemistry*, Vol. 2 (Ed. H. B. Kagan), Georg Thieme, Stuttgart, 1977, pp. 33–183.
9. C. Djerassi, *Optical Rotatory Dispersion*, McGraw-Hill, New York, 1960, Chap. 1, pp. 1–10.
10. M. Grosjean and M. Legrand, *C. R. Hebd. Séances Acad. Sci.*, **251**, 2150 (1960).
11. L. Velluz, M. Legrand and M. Grosjean, *Optical Circular Dichroism*, Academic Press, New York, 1965.
12. C. Bertucci, R. Lazzaroni and W. C. Johnson Jr, *Carbohydr. Res.*, **133**, 152 (1984).
13. P. L. Polavarapu, L. P. Fontana and H. E. Smith, *J. Am. Chem. Soc.*, **108**, 94 (1986).
14. L. D. Barron, J. F. Torrance and J. Vrbancich, *J. Raman Spectrosc.*, **13**, 171 (1982).
15. M. Fukuda, A. Tajiri, M. Oda and M. Hatano, *Bull. Chem. Soc. Japan*, **56**, 592 (1983).
16. H. H. Dearman, *J. Chem. Phys.*, **58**, 2135 (1973).
17. G. Gottarelli and G. P. Spada, *J. Chem. Soc. Perkin Trans. 2*, 1501 (1984).
18. M. Ata and H. Yamaguchi, *Nippon Kagaku Kaishi*, 297 (1983); *Chem. Abstr.*, **98**, 160127y (1983).
19. L. F. Fieser and M. Fieser, *Reagents for Organic Synthesis*, Wiley, New York, 1967, pp. 940–942.
20. J. M. Bruce, D. Creed and K. Dawes, *J. Chem. Soc. C*, 2244 (1971).
21. J. Jurczak and B. Gankowski, *Pol. J. Chem.*, **56**, 411 (1982).
22. N. Cohen, R. J. Lopresti and G. Saucy, *J. Am. Chem. Soc.*, **101**, 6710 (1979).
23. J.-F. Biellmann and G. Branlant, *Bull. Soc. Chim. Fr.*, 2086 (1973).
24. W. B. Eyton, W. D. Ollis, I. O. Sutherland, L. M. Jackman, O. R. Gottlieb and M. T. Magalhães, *Proc. Chem. Soc.*, *Lond.*, 301 (1962).
25. W. B. Eyton, W. D. Ollis, I. O. Sutherland, O. R. Gottlieb, M. T. Magalhães and L. M. Jackman, *Tetrahedron*, **21**, 2683 (1965).
26. B. J. Donnelly, D. M. X. Donnelly, A. M. O'Sullivan and J. P. Prendergast, *Tetrahedron*, **25**, 4409 (1969).
27. W. B. Eyton, W. D. Ollis, M. Fineberg, O. R. Gottlieb, I. Salignac de Souza Gumâres and M. T. Magalhâses, *Tetrahedron*, **21**, 2697 (1965).
28. D. Kumari, S. K. Mukerjee and T. R. Seshadri, *Tetrahedron Lett.*, 3767 (1966).
29. D. M. X. Donnelly, B. J. Nangle, P. B. Hulbert, W. Klyne and R. J. Swan, *J. Chem. Soc. C*, 2450 (1967).
30. K. Kurosawa, W. D. Ollis, B. T. Redman, I. O. Sutherland, A. B. de Oliveira, O. R. Gottlieb and H. M. Alves, *Chem. Commun.*, 1263 (1968).
31. K. Kurosawa, W. D. Ollis, I. O. Sutherland, O. R. Gottlieb and A. B. de Oliveira, *Phytochemistry*, **17**, 1405 (1978).
32. K. Kurosawa, W. D. Ollis, B. T. Redman, I. O. Sutherland, O. R. Gottlieb and H. M. Alves, *Chem. Commun.*, 1265 (1968).
33. K. Kurosawa, W. D. Ollis, B. T. Redman, I. O. Sutherland, H. M. Alves and O. R. Gottlieb, *Phytochemistry*, **17**, 1423 (1978).
34. O. R. Gottlieb, A. B. de Oliveira, T. M. M. Goncalves, G. G. de Oliveira and S. O. Pereira, *Phytochemistry*, **14**, 2495 (1975).
35. C. H. Eugster, in *Pigments in Plants*, 2nd edn. (Ed. F.-C. Czygan), Gustav Fischer, New York, 1980, pp. 149–186.
36. F. G. P. Remfry, *J. Chem. Soc.*, **103**, 1076 (1913).
37. F. Kögl and A. G. Boer, *Recl. Trav. Chim. Pays-Bas*, **54**, 779 (1935).
38. E. R. Wagner, R. D. Moss, R. M. Brooker, J. P. Heeschen, W. J. Potts and M. L. Dilling, *Tetrahedron Lett.*, 4233 (1965).
39. D. Arigoni and O. Jeger, *Helv. Chim. Acta*, **37**, 881 (1954).
40. J. Padilla, J. Romo, F. Walls and P. Crabbé, *Rev. Soc. Quím. Méx.*, **11**, 7 (1967).
41. P. Joseph-Nathan and Ma. P. González, *Can J. Chem.*, **47**, 2465 (1969).
42. F. Walls, J. Padilla, P. Joseph-Nathan, F. Giral, M. Escobar and J. Romo, *Tetrahedron*, **22**, 2387 (1966).
43. P. Joseph-Nathan, J. Reyes and Ma. P. Gonzáles, *Tetrahedron*, **24**, 4007 (1968).

44. H. Kakisawa, Y. Inouye and J. Romo, *Tetrahedron Lett.*, 1929 (1969).
45. P. Joseph-Nathan, J. J. Morales and J. Romo, *Tetrahedron*, **22**, 301 (1966).
46. O. E. Edwards, G. Feniak and M. Los, *Can. J. Chem.*, **40**, 1540 (1962).
47. P. Rüedi and C. H. Eugster, *Helv. Chim. Acta*, **55**, 1994 (1972).
48. C. H. Eugster, H.-P. Küng, H. Kühnis and P. Karrer, *Helv. Chim. Acta*, **46**, 530 (1963).
49. M. Ribi, A. Chang Sin-Ren, H.-P. Küng and C. H. Eugster, *Helv. Chim. Acta*, **52**, 1685 (1969).
50. T. Miyase, F. Yoshizaki, N. Kabengele, P. Rüedi and C. H. Eugster, *Helv. Chim. Acta*, **62**, 2374 (1979).
51. A. C. Alder, P. Rüedi and C. H. Eugster, *Helv. Chim. Acta*, **67**, 1523 (1984).
52. P. Rüedi and C. H. Eugster, *Helv. Chim. Acta*, **54**, 1606 (1971).
53. P. Rüedi and C. H. Eugster, *Helv. Chem. Acta*, **56**, 1129 (1973).
54. T. Hata, Y. Sano, R. Sugawara, A. Matsumae, K. Kanamori, T. Shima and T. Hoshi, *J. Antibiot. Ser. A*, **9**, 141 (1956).
55. D. V. Lefemine, M. Dann, F. Barbatschi, W. K. Hausmann, V. Zbinovsky, P. Monnikendam, J. Adam and N. Bohonos, *J. Am. Chem. Soc.*, **84**, 3184 (1962).
56. J. S. Webb, D. B. Cosulich, J. H. Mowat, J. B. Patrick, R. W. Broschard, W. E. Meyer, R. P. Williams, C. F. Wolf, W. Fulmor, C. Pidacks and J. E. Lancaster, *J. Am. Chem. Soc.*, **84**, 3185 (1962).
57. J. S. Webb, D. B. Cosulich, J. H. Mowat, J. B. Patrick, R. W. Broschard, W. E. Meyer, R. P. Williams, C. F. Wolf, W. Fulmor, C. Pidacks and J. E. Lancaster, *J. Am. Chem. Soc.*, **84**, 3187 (1962).
58. T. Tulinsky, *J. Am. Chem. Soc.*, **84**, 3188 (1962).
59. A. Tulinsky and J. H. van den Hende, *J. Am. Chem. Soc.*, **89**, 2905 (1967).
60. K. Shirahata and N. Hirayama, *J. Am. Chem. Soc.*, **105**, 7199 (1983).
61. R. Yahashi and I. Matsubara, *J. Antibiot.*, **29**, 104 (1976).
62. W. Szybalski and V. N. Iyer, in *Antibiotics*, Vol. 1, *Mechanism of Action*, (Eds D. Gottlieb and P. D. Shaw), Springer, New York, 1967, pp. 211–245.
63. M. Tomasz, R. Lipman, J. K. Snyder and K. Nakanishi, *J. Am. Chem. Soc.*, **105**, 2059 (1983).
64. W. G. Taylor and W. A. Remers, *J. Med. Chem.*, **18**, 307 (1975).
65. N. Harada and K. Nakanishi, *Circular Dichroic Spectroscopy—Exciton Coupling in Organic Stereochemistry*, University Science Books, Mill Valley, Calif., 1983.
66. M. Tomasz, M. Jung, G. Verdine and K. Nakanishi, *J. Am. Chem. Soc.*, **106**, 7367 (1984).
67. W. B. Whalley, *Prog. Org. Chem.*, **4**, 72 (1958).
68. A. C. Hetherington and H. Raistrick, *Phil. Trans. Roy. Soc. Lond.*, **220B**, 269 (1931).
69. O. R. Rodig, *Chem. Commun.*, 1553 (1971).
70. P. P. Mehta and W. B. Whalley, *J. Chem. Soc.*, 3777 (1963).
71. R. K. Hill and L. A. Gardella, *J. Org. Chem.*, **29**, 766 (1964).
72. H. E. Smith, *Proc. of Fed. Eur. Chem. Soc. Int. Conf. Circular Dichroism, Sofia, Bulgaria*, September 1985, pp. 250–269.
73. G. C. Barrett, J. F. W. McOmie, S. Nakajima and S. W. Tanenbaum, *J. Chem. Soc. C*, 1068 (1969).
74. A. Stoessl, *Chem. Commun.*, 307 (1967).
75. A. Stoessl, *Can. J. Chem.*, **47**, 767 (1969).
76. T. Noda, T. Take, M. Otani, K. Miyauchi, T. Watanabé and J. Abe, *Tetrahedron Lett.*, 6087 (1968).
77. A. Stoessl, *Can. J. Chem.*, **47**, 777 (1969).
78. A. Takenaka, A. Furusaki and T. Watanabé, *Tetrahedron Lett.*, 6091 (1968).
79. N. Tsuji, *Tetrahedron*, **24**, 1765 (1968).
80. H. Nakai, M. Shiro and H. Koyama, *J. Chem. Soc. B*, 498 (1969).
81. H. Schmid, A. Ebnöther and Th. M. Meijer, *Helv. Chim. Acta*, **33**, 1751 (1950).
82. H. Schmid and A. Ebnöther, *Helv. Chim. Acta*, **34**, 561 (1951).
83. H. Schmid and A. Ebnöther, *Helv. Chim. Acta*, **34**, 1041 (1951).
84. D. W. Cameron, D. G. I. Kingston, N. Sheppard and Lord Todd, *J. Chem. Soc.*, 98 (1964).
85. M. E. Bergy, *J. Antibiot.*, **21**, 454 (1968).
86. S. Ōmura, H. Tanaka, Y. Koyama, R. Ōiwa, M. Katagiri, J. Awaya, T. Nagai and T. Hata, *J. Antibiot.*, **27**, 363 (1974).
87. H. Tanaka, Y. Koyama, J. Awaya, H. Marumo, R. Ōiwa, M. Katagiri, T. Nagai and S. Ōmura, *J. Antibiot.*, **28**, 860 (1975).

88. H. Tanaka, H. Marumo, T. Nagai, M. Okada, K. Taniguchi and S. Ōmura, *J. Antibiot.*, **28**, 925 (1975).
89. S. Ōmura, H. Tanaka, Y. Okada and H. Marumo, *Chem. Commun.*, 320 (1976).
90. H. Hoeksema and W. C. Krueger, *J. Antibiot.*, **29**, 704 (1976).
91. H. Tanaka, Y. Koyama, T. Nagai, H. Marumo and S. Ōmura, *J. Antibiot.*, **28**, 868 (1975).
92. N. Tsuji, M. Kobayashi, Y. Wakisaka, Y. Kawamura, M. Mayama and K. Matsumoto, *J. Antibiot.*, **29**, 7 (1976).
93. N. Tsuji, M. Kobayashi, Y. Terui and K. Tori, *Tetrahedron*, **32**, 2207 (1976).
94. A. Zeeck and P. Christiansen, *Justus Liebigs Ann. Chem.*, **724**, 172 (1969).
95. N. Tsuji, T. Kamigauchi, H. Nakai and M. Shiro, *Tetrahedron Lett.*, **24**, 389 (1983).
96. A. Zeeck, H. Zähner and M. Mardin, *Justus Liebigs Ann. Chem.*, 1100 (1974).
97. K. S. Brown Jr, *Chem. Soc. Rev.*, **4**, 263 (1975).
98. D. W. Cameron, R. I. T. Cromartie, D. G. I. Kingston and Lord Todd, *J. Chem. Soc.*, 51 (1964).
99. H. J. Banks and D. W. Cameron, *Aust. J. Chem.*, **25**, 2199 (1972).
100. A. Calderbank, D. W. Cameron, R. I. T. Cromartie, Y. K. Hamied, E. Haslam, D. G. I. Kingston, Lord Todd and J. C. Watkins, *J. Chem. Soc.*, 80 (1964).
101. D. W. Cameron, R. I. T. Cromartie, Y. K. Hamied, P. M. Scott and Lord Todd, *J. Chem. Soc.*, 62 (1964).
102. K. L. Rinehart Jr, *Acc. Chem. Res.*, **5**, 57 (1972).
103. W. Keller-Schierlein, M. Meyer, A. Zeeck, M. Damberg, R. Machinek, H. Zähner and G. Lazar, *J. Antibiot.*, **36**, 484 (1983).
104. K. L. Rinehart Jr, W. M. J. Knöll, K. Kakinuma, F. J. Antosz, I. C. Paul, A. H.-J. Wang, F. Reusser, L. H. Li and W. C. Krueger, *J. Am. Chem. Soc.*, **97**, 196 (1975).
105. P. Salvadori, C. Bertucci, C. Rosini, M. Zandomeneghi, G. G. Gallo, E. Martinelli and P. Ferrari, *J. Am. Chem. Soc.*, **103**, 5553 (1981).
106. H. Laatsch and H. Anke, *Justus Liebigs Ann. Chem.*, 2189 (1982).
107. V. H. Powell and M. D. Sutherland, *Aust. J. Chem.*, **20**, 541 (1967).
108. G. L. Bartolini, T. R. Erdman and P. J. Scheuer, *Tetrahedron*, **29**, 3699 (1973).
109. C. Bassett, M. Buchanan, R. T. Gallagher and R. Hodges, *Chem. Ind. (Lond.)*, 1659 (1970).
110. Y. Ogihara, N. Kobayashi and S. Shibata, *Tetrahedron Lett.*, 1881 (1968).
111. N. Takeda, S. Seo, Y. Ogihara, U. Sankawa, I. Iitaka, I. Kitagawa and S. Shibata, *Tetrahedron*, **29**, 3703 (1973).
112. S. Yamada and H. Akimoto, *Tetrahedron Lett.*, 3967 (1968).
113. H. Akimoto and S. Yamada, *Tetrahedron*, **27**, 5999 (1971).
114. A. I. Scott, *Interpretation of the Ultraviolet Spectra of Natural Products*, Pergamon Press, Oxford, 1964, Chap. 3, 123–126.
115. U. Sankawa, S. Seo, N. Kobayashi, Y. Ogihara and S. Shibata, *Tetrahedron Lett.*, 5557 (1968).
116. N. Kobayashi, Y. Iitaka and S. Shibata, *Acta Crystallogr.*, **B26**, 188 (1970).
117. S. Seo, U. Sankawa, Y. Ogihara, Y. Iitaka and S. Shibata, *Tetrahedron*, **29**, 3721 (1973).
118. H. Brockmann, *Prog. Chem. Nat. Prod.*, **21**, 121 (1963).
119. M. C. Wani, H. L. Taylor, M. E. Wall, A. T. McPhail and K. D. Onan, *J. Am. Chem. Soc.*, **97**, 5955 (1975).
120. H. Brockmann Jr and M. Legrand, *Naturwissenschaften*, **49**, 374 (1962).
121. H. Brockmann Jr and M. Legrand, *Tetrahedron*, **19**, 395 (1963).
122. T. Schaefer, W. Danchura and W. Niemczura, *Can. J. Chem.*, **56**, 2229 (1978).
123. T. Schaefer, W. Niemczura and W. Danchura, *Can. J. Chem.*, **57**, 355 (1979).
124. O. Červinka, W. Král and P. Maloň, *Coll. Czech. Chem. Commun.*, **41**, 2406 (1976).
125. H. Brockmann, H. Brockmann Jr and J. Niemeyer, *Tetrahedron Lett.*, 4719 (1968).
126. F. Arcamone, G. Cassinelli, G. Franceschi, P. Orezzi and R. Mondelli, *Tetrahedron Lett.*, 3353 (1968).
127. D. W. Engel, K. Zechmeister and W. Hoppe, *Tetrahedron Lett.*, 1323 (1972).
128. A. Di Marco, M. Gaetani and B. Scarpinato, *Cancer Chemotherapy Reports* (Part 1), **53**, 33 (1969).
129. F. Arcamone, G. Franceschi, S. Penco and A. Selva, *Tetrahedron Lett.*, 1007 (1969).
130. I. K. Khanna and L. A. Mitscher, *Tetrahedron Lett.*, **26**, 691 (1985).
131. T. R. Kelly (Ed.), Recent aspects of anthracyclinone chemistry, *Tetrahedron*, **40**, 4537–4793 (1984).
132. M. J. Broadhurst, C. H. Hassall and G. J. Thomas, *Chem. Ind. (Lond.)*, 106 (1985).

133. M. Suzuki, T. Matsumoto, R. Abe, Y. Kimura and S. Terashima, *Chem. Lett.*, 57 (1985).
134. R. A. Russell, R. W. Irvine and A. S. Krauss, *Tetrahedron Lett.*, **25**, 5817 (1984).
135. R. A. Russell, University College, The University of New South Wales, personal communication, 1986.
136. F. Aroamone, G. Franceschi, P. Orezzi, S. Penco and R. Mondelli, *Tetrahedron Lett.*, 3349 (1968).
137. R. H. Iwamato, P. Lim and N. S. Bhacca, *Tetrahedron Lett.*, 3891 (1968).
138. J. P. Marsh Jr, R. H. Iwamoto and L. Goodman, *Chem. Commun.*, 589 (1968).

The Chemistry of Quinonoid Compounds, Vol. II
Edited by S. Patai and Z. Rappoport

CHAPTER 5

Photoelectron spectra of quinonoid compounds*

L. KLASINC† and S. P. McGLYNN
Department of Chemistry, Louisiana State University, Baton Rouge, Louisiana 70803, USA

* This work was supported by the U.S. Department of Energy.
† On leave of absence from the Rudjer Bośković Institute, Zagreb, Croatia, Yugoslavia.

I. INTRODUCTION

In the original Quinone volume, the leading chapter ('Theoretical and general aspects' by G. J. Gleicher) cited just two experimental ionization energies for quinonoids: 9.68 eV for *p*-benzoquinone (PBQ) and 9.34 eV for 9,10-anthraquinone (AQ), both taken from Vilesov and coworkers[1]. Furthermore, only two theoretical works, one by Newton and coworkers[2] and the other by Aussems and coworkers[3a], both of which used SCF-MO theory to predict the ionization potentials of PBQ[3], 1,4-naphthoquinone (NQ)[3b] and AQ[3], were quoted†. Today, merely a dozen years later, we find it necessary to write an entire chapter on the photoelectron spectroscopy of quinonoid compounds! Nothing less would do justice to the enormous growth of the field or to the pertinence of the electronic structure of these molecules to modern organic chemistry. Hence, this essay.

The interpretation and assignment of the spectra of quinonoid compounds has caused much headache for both photoelectron spectroscopists and quantum chemists. For example, it is now known that the two lowest-energy ionization events in PBQ are connected with the removal of oxygen lone-pair electrons and are best denoted $I(n_{\pm})$. However, the path leading to this conclusion is exceedingly tortuous[1–27], so much so that it provides a chastening example for all spectroscopists of theoretical bent. For that reason, we have illustrated the twists and turns, along a time-line beginning in 1966, of the spectroscopic assignments for the four lowest-energy ionization events in *p*-benzoquinone. The total number of possible assignments is 4! The reader will note that most of these 4! permutations are represented in Figure 1.

The four relevant ionization events of Figure 1 correspond to the removal of electrons from two oxygen lone-pair orbitals, $I(n_+)$ and $I(n_-)$, and two benzenoid (or ring) orbitals, $I(\pi_+)$ and $I(\pi_-)$. All assignments of Figure 1 embrace these four orbital characterizations, virtually all of them being heavily influenced by the results of quantum chemical calculations. Such calculations have served very well in organic chemistry. Yet, in *p*-benzoquinone, they fail dramatically.

The reasons for all these difficulties seem to be the following:

(1) Quinones are oxidants; that is, they tend to accept (or extract) electrons from their environment. Photoelectron spectroscopy, on the other hand, measures the exact opposite process, namely the release (or injection) of electrons into the continuum. This disparity is the source of some problems.

(2) As with most electron acceptors, quinonoid compounds usually possess tightly-bound virtual orbitals, a point that is amply substantiated by the absorption spectroscopy[28–34] of these compounds. These low-lying virtual orbitals generate low-energy excited states and play an important role in the photoionization act, in the sense that they and the continuum may be simultaneously populated by a single X-ray excitation photon.

† Incidentally, both of these calculations predicted, quite wrongly it should be emphasized, that the lowest-energy ionization events were of π nature. Despite the mis-assignments, 'good' agreement was obtained between the measured[1] and calculated values[3b] for PBQ and AQ. Thus, the prediction of energy and assignment for the unknown, lowest-energy ionization event in NQ was taken, again quite wrongly, to be reliable. Finally, as if to add insult to injury, the experimental value $I = 9.68$ eV for *p*-benzoquinone is now known to be erroneous.

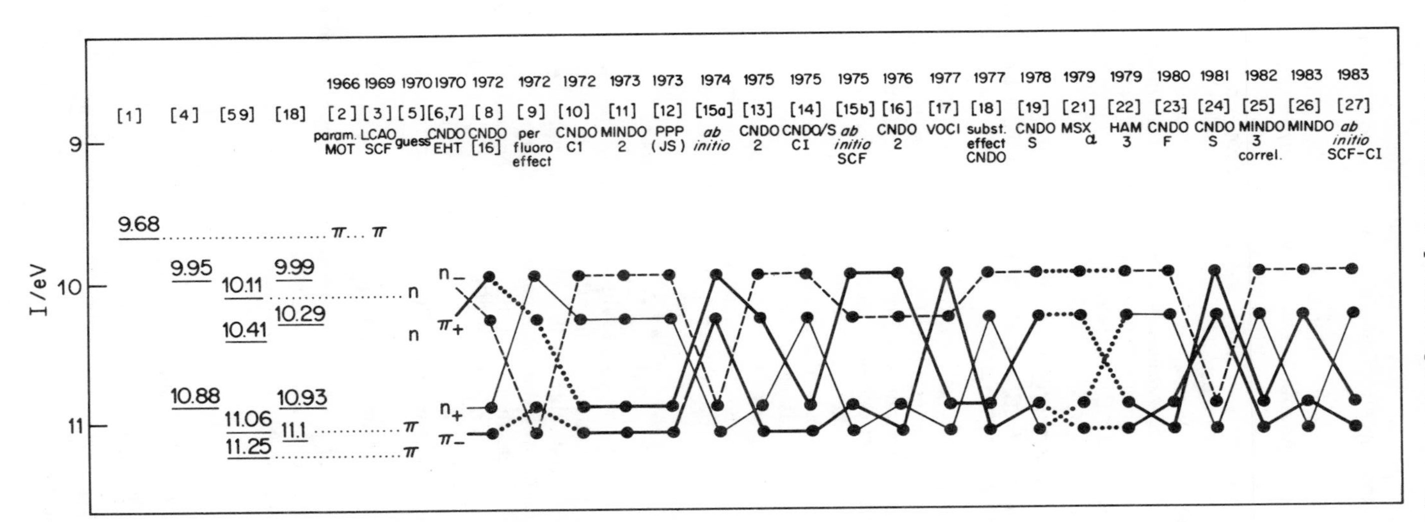

FIGURE 1. Assignments of ionization energies in *p*-benzoquinone by year from 1966 to 1983. References are bracketed. The comments under the references refer to the quantum chemical method used in the cited work except for that under [9], which refers to the empirical use of the perfluoro effect. The variation of the experimental measurements is represented in the first 4 columns, the data under [18] being now generally accepted. Differentiated assignments begin in 1970. The correct assignments are those of [10], [11], [12], [13], [18], [22], [25] and [27]

That is, the photoelectron spectrum may contain 'shake-up' events. The presence of such events in the photoelectron spectrum imposes serious difficulties on the empirical assignment process.

(3) The presence of low-lying virtual orbitals, as discussed in (2), implies a high density of low-energy electron configurations. This, in turn, implies a high probability of Koopmans' breakdown. That is, the extensive configuration interactions that may occur as a result of the energetic proximity of singly and doubly excited configurations to the ground configuration invalidates almost all the approximations inherent in the Koopmans derivation. Consequently, ionization energy no longer relates to an orbital energy. In specific, Koopmans' theorem[35] is certainly inapplicable to *p*-benzoquinone and certain other quinonoid compounds. Since this point is so important for the interpretation of the valence photoelectron spectra of quinonoids, we will discuss it briefly in the next section.

II. KOOPMANS' THEOREM: ITS CONSEQUENCES FOR PHOTOIONIZATION PROCESSES, PHOTOIONIZATION ENERGIES AND ELECTRONIC STRUCTURE[35–37]

The release of an electron from any chemical system constitutes ionization of that system. We confine ourselves here to the ionization of free atoms and molecules in the gas phase caused by electromagnetic radiation (photons). Such events are known as 'photoionization' processes, the released electrons are termed 'photoelectrons' and the technique used to determine their excess kinetic energy is referred to as 'photoelectron spectroscopy', or PES. If the photon energy exceeds the ionization energy, the surplus excitation is transferred to the ejected electron as a kinetic energy, E_k. The techniques that measure E_k are referred to as 'X-ray or UV photoelectron spectroscopy' or, for short, XPS and UPS, respectively. When E_k is plotted against the number of ejected photoelectrons, the result is a photoelectron (PE) spectrum, from which ionization energies I_k may be obtained using the equality

$$I_k = h\nu - E_k;\ k = 1, 2, \ldots w;\ I_w \leqslant h\nu$$

In the MO picture, the photoelectrons are supposed to originate in individual electronic orbitals of the molecular ground state configuration. If spin-orbit coupling is small, each non-degenerate orbital is occupiable by two electrons of opposite spin (Figure 2). However, the photoionization spectrum need not consist solely of the set of single-event processes that supposedly describes the UPS and XPS processes. Indeed, electron excitation can accompany electron ejection ('shake-up'); two electrons can be ejected simultaneously ('shake off'); or a second electron can be subsequently ejected from the original highly-excited ion produced in the XPS process (Auger event).

A. The First Part of Koopmans' Theorem

The connection of theory and experiment is given by Koopmans' theorem, which states that the electronic wave function of a singly-ionized state is adequately described by Slater determinants based on the set of $N-1$ ground state, self-consistent field (SCF), molecular spin orbitals (MSOs). That is, if

$$\Psi^N = |\phi_a(1)\phi_b(2) \ldots . \phi_n(N)|$$

is a good descriptor for the ground state, then

$$\Psi^{N-1} = |\phi_a(1)\phi_b(2) \ldots . \phi_m(N-1)|$$

is a good descriptor for the singly ionized state. The ionization energy, then is

$$I_n = E(\Psi^{N-1}) - E(\Psi^N) = \langle \Psi^{N-1}|H^{N-1}|\Psi^{N-1}\rangle - \langle \Psi^N|H^N|\Psi^N\rangle = -\varepsilon_{nn}$$

where H^{N-1} and H^N are the SCF hamiltonians for the $(N-1)$ and N electron systems, respectively. This statement, namely that the ionization energy equals the negative of the orbital energy of the ejected electron, $-\varepsilon_{nn}$, is the first part of Koopmans' theorem.

B. The Second Part of Koopmans' Theorem

Now, the function Ψ^{N-1} is by no means optimal. The optimal function may be written as the CI (configurational interaction) expansion

$$\Psi^{N-1} = \sum_{k \in \text{occ}} \Psi_{-k} C_k + \sum_{\substack{k,l \in \text{occ} \\ u \in \text{unocc}}} \Psi^{u}_{-kl} C^{u}_{kl}$$

where, for example, Ψ_{-k} denotes a determinant Ψ_0 in which spin-orbital ϕ_k has been deleted; Ψ^u_l is one in which ϕ_l has been replaced by ϕ_u; and where we have dropped the $N-1$ superscripting to avoid crowding. The function Ψ^u_{-kl}, as is obvious, is a shake-up configuration (Figure 2). Simplification of Ψ^{N-1} might consist of truncation to

$$\Psi^{N-1} = \sum_{k \in \text{occ}} \Psi_{-k} C_k$$

However, what we really desire is

$$\Psi^{N-1} = \Psi_{-k}$$

This gross simplification is equivalent to the demand that we find an orthogonal transformation of the set of Hartree–Fock MSOs so that the cationic state can be represented by one single determinant constituted from this set, namely Ψ_{-k}, and the neutral ground state can be represented by one single determinant constituted from the same set, namely Ψ_0. Koopmans' theorem asserts this possibility and, furthermore, it identifies the appropriate MSO set as the canonical Hartree–Fock set.

This latter assertion is the second, and more important, part of Koopmans' theorem. It may be rephrased alternatively: the only allowed ionizations are those which remove an

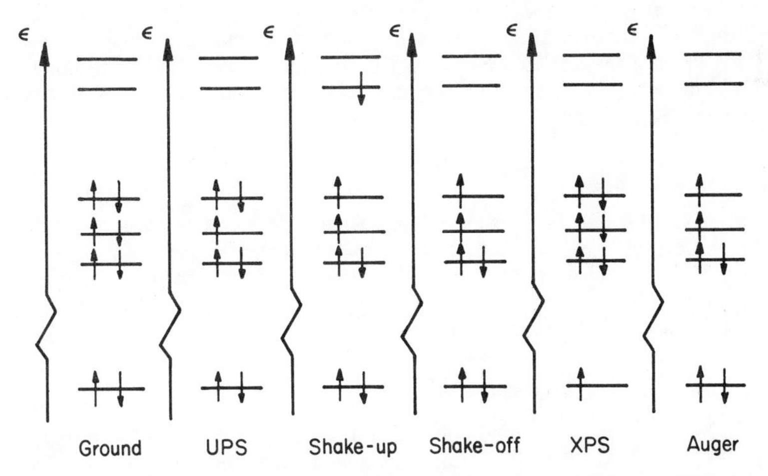

FIGURE 2. A depiction of various types of ionization processes in an MO format. Orbital energy is denoted by ε. The jogs in the vertical arrows connote a large energetic separation of core and valence orbitals. The two topmost orbitals are the virtual orbitals

electron from an MSO (or shake-up and shake-off transitions are forbidden). If spin-orbit coupling is small (< 20 meV), a further restatement becomes possible: the only allowed ionizations are those which remove one electron from an MO.

C. The Deficiencies of Koopmans' Theorem

Koopmans' theorem provides a salient experiment/theory interface. Consequently, it is well to specify the approximations inherent in its derivation. These are as follows.

1. *Fixed-nuclei approximation*

The Born–Oppenheimer approximation permits the notion of 'molecular geometry'. It is understood that the cationic $N-1$ electron system, which is the immediately terminal state of the process

$$N \text{ electron system} + h\nu = (N-1) \text{ electron system} + e^-$$

is identical in all geometric detail to the initial state of the N electron system. This, of course, is the Franck–Condon approximation. Consequently, Koopmans' theorem applies only to vertical ionization events.

2. *The correlation energy*

The neglect of correlation energy is intrinsic to the Hartree–Fock approximation. The correlation energy is caused by the fact that electrons adjust their motions to the instantaneous charge distribution, and not to an average charge distribution (as is assumed in the Hartree–Fock equations). In fact, the correlation energy is the difference between the correct energy and the Hartree–Fock energy associated with any given Hamilton operator. If relativistic effects are small, the latter is well known; and the 'correct energy' is the same as the experimental energy. Thus, correlation energies are often readily determinable.

Electrons of opposite spin usually tend to stay considerably further apart (i.e. correlate their motions better) than a single determinantal wave function will allow and, as a result, correlation energies can be quite substantial. Nonetheless, while large for any one state, it is only the difference between two states, namely between the initial N and terminal $(N-1)$ electron states, which is of significance to photoelectron spectroscopy. This difference may well be small. Koopmans' theorem implies that it is zero.

3. *The relaxation energy*

The same set of spin-orbitals is used to construct the Slater determinants for the N and $(N-1)$ electron systems. This supposition implies that the electrons of the cation do not adjust or relax in order to accommodate the changes of interelectronic repulsions which characterize the $(N-1)$ electron system. This supposition is known as the 'frozen-core (fc)' or 'frozen orbital (fo)' approximation.

4. *The non-relativistic approximation*

This approximation is not a consequence of deficiencies in the wave functions; it is, rather, a defect caused by the omission of relativistic terms from the Hamilton operator. We have omitted these terms solely for convenience. The various relativistic terms—for example, spin-orbit or spin–spin interactions—might have been included in the Fock operator in a way which would not have altered any of our conclusions. In fact, in his original paper, Koopmans included relativistic effects explicitly—and to no ill effects whatsoever.

5. Restriction to closed-shell systems

Koopmans' theorem is restricted to closed-shell N electron systems. Thus, at least in the form expressed here, it is specifically inapplicable to non-closed-shell systems (e.g. most transition metal complexes).

The relationships between the experimental ionization energy, the MO energy and the Hartree–Fock ionization energy are schematized in Figure 3. The correlation energy is always negative and is shown to be slightly larger for the system with the larger number of electrons. The reorganization energy for the cation (i.e. ΔE^+ (fc) $\equiv$ [E^+ (Koopmans) $- E^+$ (HF)]) will almost always be positive. Hence, there is a tendency for $|\Delta E^+ \text{(fc)}| + |\Delta E^+ \text{(corr)}|$ to equal (approximately, of course) $|\Delta E_0 \text{(corr)}|$. It is this tendency which is responsible for the moderate successes of Koopmans' theorem.

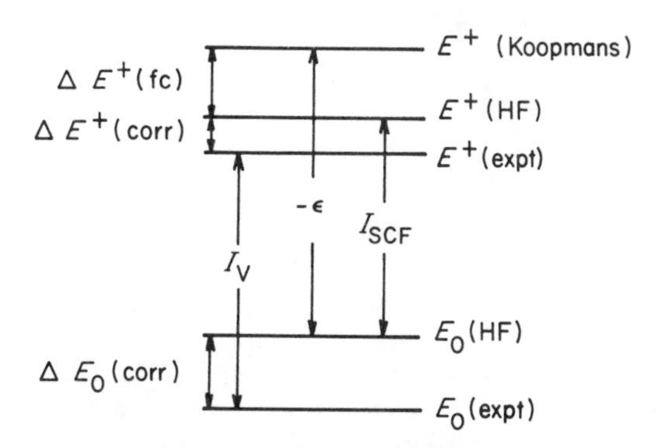

FIGURE 3. The relation between experimental vertical ionization energy, I_v, the Koopmans MO energy, ε, and the Hartree–Fock self-consistent field ionization energy, I_{SCF}. The symbol E^+ denotes cation energy; E_0 denotes the neutral molecule energy; ΔE(corr) denotes the correlation energy; and ΔE(fc) denotes the frozen-core energy correction

D. The Interface of Experiment and Koopmans' Theorem

The second part of Koopmans' theorem implies an isomorphism of the set of ionization potentials of the neutral molecule and the set of canonical MO energies. The first part of Koopmans' theorem specifies the connectors for this isomorphism. This, of course, is a very desirable state of affairs: it dictates the mapping of the UPS data set onto the HF-SCF MO set as an order-isomorphism in the energies. Not only that, it is well established that this particular isomorphism is usually correct; that is, the assignment of a given ionization event as the removal of an electron from an MO of given charge distribution (i.e. symmetry) is usually correctly rendered.

Such a rendering, however, is not always correct. First, the deficiencies of Koopmans' approximation can readily induce an uncertainty of ± 1 eV in the energy match and, second, the MO computations may not (in fact, are usually not) of HF-SCF quality. Nonetheless, a very large category of publication exists in which a set of UPS energies is mapped onto a set of MO energies obtained by some algorithmic quantum chemical scheme. We think this to be unfortunate. Such a mapping is valid only when vested in other empirical informations. Among these, we list:

(1) cross-sections as a function of the ionizing photon energy.
(2) vibronic structure of the UPS band (i.e. normal modes active in the ionization event).
(3) energies and intensities of Rydberg series which terminate on specific cationic states.
(4) perturbation techniques based on relatively innocuous chemical substitution, such as substitution of a methyl group, an SiH_3 group, or even a fluorine atom (i.e. the perfluoro effect). Such substituents usually exert the largest effect on the ionization event associated with a specified MO when substituted at that locus at which the MO in question possesses the largest electron density. Indeed, the nature of the substituent can be used to gauge relative σ/π contributions to the electron density at the locus in question. Thus, an adroit use of substituent effects can be used to define shape and constitution (i.e. σ, π or relative σ/π admixture) of an MO. In a sense, then, this technique can be considered to be a chemical MO mapping procedure. It is also expected that even an algorithmic quantum chemical calculation, even if only moderately 'good', should mimic the observed perturbative effects. Hence, a new projective element is added to the Koopmans' isomorphism, namely MO shape and constitution.
(5) correlative studies based on the supposition that the spectrum of certain molecules should be a composite of the spectra of their constituent parts. For example, it is reasonable to suppose that the set of ionization events for nitroaniline should be related in some way to those of NH_3, HNO_2 and benzene.

The higher the ionization energy (that is, the deeper the MO from which electron ejection occurs), the greater is the chance for breakdown of this simple one-electron picture. Therefore, different levels of sophistication for the assignment of PE spectra using quantum chemical calculations do exist. These are:

(1) At the first stage, the canonical MO energies are correlated directly to ionization energies. Both parts of Koopmans' theorem are applicable at this stage.
(2) At the second stage, SCF calculations are performed for both the ground and the radical cation states and the energy differences of these states, which we denote ΔE_{SCF}, are correlated with the ionization energies. Only the second part of Koopmans' theorem is applicable at this stage.
(3) At the third stage, extensive configurational mixing is imposed. One usually includes configurations that are singly and doubly excited relative to all electron configurations of primary interest. No part of Koopmans' theorem retains validity at this level of sophistication. Indeed, finite intensity predictions for non-Koopmans' transitions (shake-up and shake-off) are intrinsic to this level and these predictions are of considerable help in tracking these low-intensity PES bands in an otherwise very intense spectrum.
(4) At the fourth stage, the calculations focus on the dynamic nature of the PES transition, even to the extent of including the effects of nuclear motion (i.e. vibronic and Coriolis coupling). This approach, which was elaborated by Cederbaum and coworkers[38], is well documented and there is no doubt that it is especially important for the determination of PES band shapes and intensities. It is also pertinent in those situations in which studies at any of the prior three stages suggest serious non-Koopmans behavior of the ionization energies. Fortunately, these latter instances are not very plentiful and they appear to be largely confined to small molecules—fortunately, that is, because the magnitude of the computational effect makes this approach impractical for large molecules.

In a somewhat naive picture, the canonical MOs and their energies may be supposed to represent the 'molecular electronic structure' of the molecule of interest. If this be so, then one of the consequences of being forced into a stage four discussion is that the PES technique will have lost much of its ability to probe this structure, becoming no more than a means of measuring ionization energies. Luckily, however, the applicability of Koopmans' theorem seems to be reasonably general, with the result that the PES technique retains its virtues as a good probe of 'molecular electronic structure'.

III. QUINONES, QUINONOIDS AND QUINOID STRUCTURES

The contemporary reference literature (e.g. *Chemical Abstracts*) does not contain much information under the subject listing 'quinones'. The reason lies in the fact that the term 'quinone', as well as a number of other related terms, is not uniquely defined. In a chemical sense, a 'quinone' is readily defined as a cyclic dicarbonyl compound, Q, which, (*a*), by addition of two electrons (and two protons) is converted into a hydroquinone, HQ, as follows

$$Q + 2e^- + 2H^+ \rightarrow HQ$$

and (*b*) which possesses a positive electron affinity, EA[39]. It is not generally agreed that the resulting hydroquinone should be conjugated or even that it should possess more than one resonant (Kekulé) structure. We advocate the definition: a *quinone is either a cyclic or polycyclic conjugated diketone*. This definition introduces certain complications. For example, such a well-known compound as camphorquinone is not a quinone and is misnamed. On the other hand, this definition does ensure that every cyclic diketone is not, wrongly, considered to be a quinone.

Those quinones and their analogs in which one or both of the oxygens in the carbonyl group have been replaced by a valence-equivalent atom or group (e.g. $=CH_2$, $=NH$, $=S$ etc.) will be termed 'quinonoids'. *A quinonoid, then, is a cyclic or polycyclic conjugated compound with two exocyclic double bonds.*

We wish to reserve the term 'quinoid' for structures as opposed to molecules. Thus, we define a *quinoid as a cyclic or polycyclic conjugated structure that must contain two exocyclic double bonds.*

It must be emphasized that the ability to write a quinoid structure for a given compound does not imply that the compound in question is quinonoid. That is a matter that must be decided by appropriate computation and/or experiment. A good example of the computational approach happens to be available for the xylylenes, for which one may write at least 4900 structural formulas. It has recently been shown by sophisticated quantum chemical calculations that the quinoid structure constitutes 55% of the ground state of *o*-xylylene and 64% of the ground state of *p*-xylylene[40]. In other words, if the ground state is represented as a linear combination of valence bond structures, there being at least 4900 such structures, the weighting factor of the quinoid structures in this state is 55%/64% for *o*-xylylene/*p*-xylylene. Therefore, it would not be erroneous to refer to *p*-xylylene as quinonoid. In order to make the discussion more concrete, the most important valence structures for *o*- and *p*-xylylene are shown in Figure 4.

Some quinones have been found to possess considerable anti-cancer activity[41]. Therefore, the question of the number and relative stability of the quinones that might be generated from a polycyclic aromatic is not simply an academic exercise. Graph theoretical and computer enumeration techniques (e.g. those used above to determine the number of Kekulé structures pertinent to a given molecule; those used to specify the number and structure of alkane and polycyclic aromatic hydrocarbon isomers; those used to enumerate the total set of alkyl-substituted polyaromatic hydrocarbons[42]; etc.) could contribute much to the solution of this problem[43].

IV. X-RAY PHOTOELECTRON SPECTROSCOPY OF QUINONOIDS

We have found only two reports on the XPS of quinonoids[44,45]. The few available spectra contain no surprises and they also suggest that none is expected.

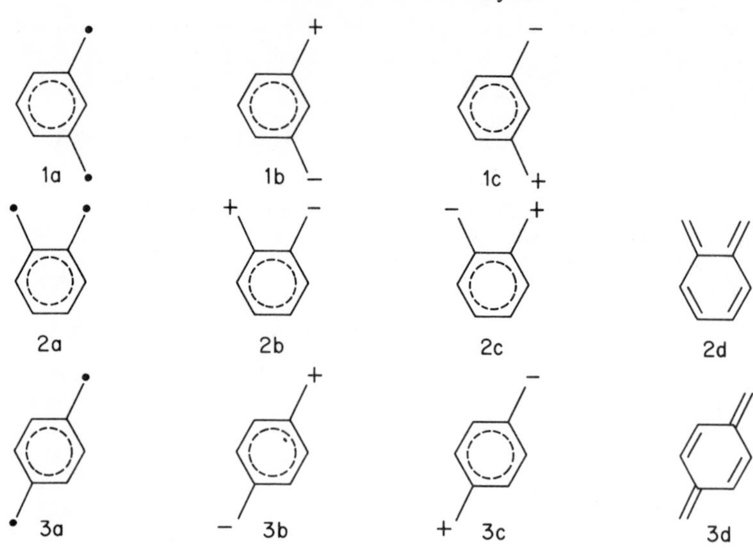

FIGURE 4. Structural formulas important to the ground state of *m*-xylylene (top), *o*-xylylene (middle) and *p*-xylylene (bottom). Quinoid structures are important in *o*-xylylene (see structure 2d) and *p*-xylylene (see structure 3d) but not in *m*-xylylene (see Ref. 40)

The XPS technique is used to determine ionization energies of inner, core electrons associated with specific atomic nuclei of the molecule in question. The experimental quantity of primary interest is known as the 'chemical shift', namely the change of ionization energy relative to that for the free unbound atom. A knowledge of this shift leads to considerable information about the composition and structure of the molecule. In the case of *p*-benzoquinone and its tetrahalogenated derivative, two C_{1s} peaks of intensity ratio 2:1 (Figure 5), one O_{1s} peak and several halogen ns, np, and nd peaks were observed. The energies for these peaks are given in Table 1. Application of the Gelius method[46] permits a ready conversion of the chemical shifts into a theoretical molecular charge distribution.

The correlation of the calculated shifts, ones based on charge densities calculated by the CNDO/2 algorithm, with binding energies yields the linear relation of Figure 6. This linearity also encompasses the C_{1s} chemical shift for triquinoyl (TQ).

```
       O       O
       ‖       ‖
       C———————C
      /         \
  O=C            C=O
      \         /
       C———————C
       ‖       ‖
       O       O
```

Because of the low resolution that still limits XPS accuracies, even simple computational schemes usually provide good correlation with experiment[47] and good predictive power.

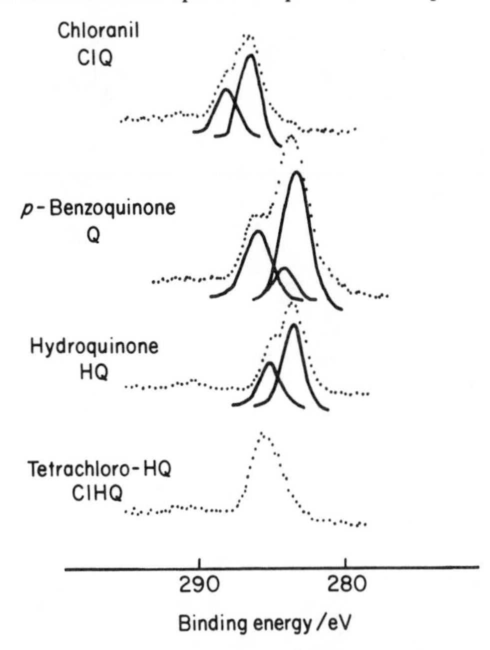

FIGURE 5. Carbon 1s X-ray photoelectron spectra taken from reference 45. The two main peaks, ratio 1/2, refer, in PBQ, to carbons with oxygen substituents and carbons with hydrogen substituents, respectively (i.e. a 2/4 ratio). Chlorine substitution for the hydrogens, as in chloranil, ClQ, moves the more intense peak to higher energies. Hydrogenation of the oxygens, as in HQ, moves the less intense peak to lower energies. Thus, it is not unexpected that both peaks should become coincident in tetrachlorohydroquinone, ClHQ

TABLE 1. XPS energies for some quinones, hydroquinones and triquinoyl*

	C_{1s}					
Molecule	C_O	C_X	O_{1s}	ns	np	nd
Chloranil	287.7	286.1	531.5	270.9	200.2	
Bromanil	287.5	285.6	531.4		183.4	70.1
Iodanil	286.7	284.4	531.0	187.2	123.0	50.1
p-Benzoquinone	286.0	283.4	530.8			
Hydroquinone	285.1	283.5	532.2			
Tetrachlorohydroquinone	285.5		532.5	270.7	200.0	
Triquinoyl	288.5					

* Energies are in eV. The C_O peak is the C_{1s} energy for the carbons to which an oxygen atom or hydroxyl group is attached. The C_X peak is the C_{1s} energy for the carbons to which an atom X = H, Cl, Br or I is attached. The principal quantum number is n = 2, 3 & 4 for chlorine, bromine and iodine, respectively. The C_O and C_{Cl} peaks are accidentally degenerate in tetrachlorohydroquinone and truly degenerate in triquinoyl.

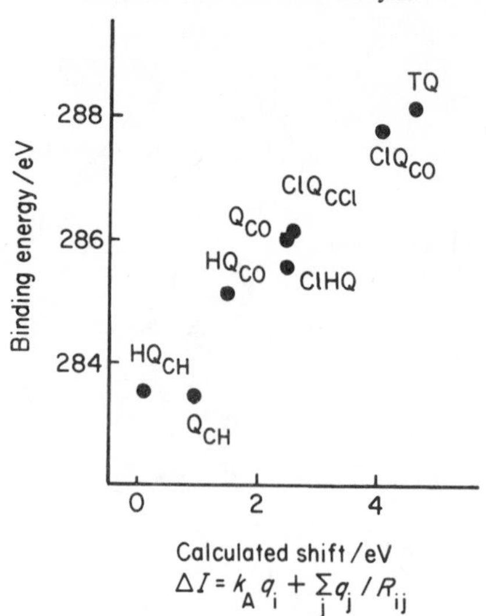

FIGURE 6. Binding energy (eV) for C_{1s} core electrons versus calculated chemical shifts (eV) for *p*-benzoquinone (PBQ), hydroquinone (HQ), tetrachlorohydroquinone (ClHQ), chloranil (ClQ) and triquinoyl (TQ). The circles refer to C_X and to C_O values of Table 1. This material is taken from Refs 44 and 45 of the text. The chemical shift at center i is $\Delta E_i = k_A q_i + \sum_j q_j / R_{ij}$ where q_i is the charge on center i, q_j that on center j and R_{ij} the distance between centers i and j

V. THE INTERACTIONS OF VALENCE ELECTRONS IN QUINONES

Quinones, as stated, are cyclic, conjugated, dicarbonyl compounds. The composite molecule method provides an apt means to discuss the electronic structure of such molecules: a knowledge of the low-energy levels and their electronic structure for the constituent parts (i.e. the carbonyl group and an unsaturated cyclic hydrocarbon) can be used to reconstitute the quinone and, in the process, to induce considerable insight into its electronic structure.

An isolated carbonyl group is characterized by two lone-pair orbitals, one of p-type and the other of s-type, of widely different energy. Only the orbital of low binding energy, the p orbital (which also is the highest-energy occupied σ orbital in simple ketones and aldehydes), is of interest for the present considerations. This p orbital is referred to also as an n orbital, where n denotes 'non-bonding'.

If a molecule contains more than one equivalent carbonyl group, the group orbitals must be combined into molecular orbitals in order to fit the requirements imposed by the point symmetry of the molecule. Unfortunately, these symmetry impositions, while mandatory, provide little or no information about either the extent of the group orbital interactions, the degree of their delocalization over the entirety of the composite molecule or even the magnitude by which the initial zero-order degeneracies are split in the actual molecule.

Quinones also possess group orbitals of a cyclic hydrocarbon π-orbital nature, and these orbitals are energetically very similar to the lone-pair n orbitals. It is our aim, in this section, to schematize the qualitative factors that influence the relative energetic disposition of the various orbitals. As examples, we will choose *p*-benzoquinone (PBQ) and *o*-benzoquinone (OBQ).

A. *p*-Benzoquinone

In PBQ, the two oxygens are situated far apart. Thus, although we can anticipate the need for n_+ and n_- linear combinations of the σ lone pairs, their direct through-space (TS) interaction should be quite negligible. To the extent that any splitting occurs, it is expected that n_- should be of lower binding energy than n_+. To this point, the interactions of the $n_\pm$ orbitals with the σ orbitals of the six-membered ring have been neglected. Since some of these σ orbitals are of appropriate symmetry to interact with the n_+/n_- orbitals, such neglect is unjustified. The σ-ring orbitals are well known in benzene or 1,4-cyclohexadiene[48]. They and the $n_\pm$ orbitals are schematized below.

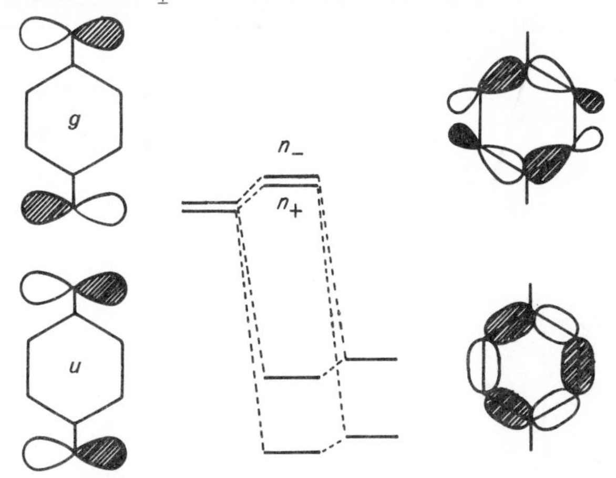

The orbital of g symmetry, which corresponds to the ~ 11.5 eV ionization event in benzene, is of lower binding energy than the orbital of u symmetry which appears at $I > 12$ eV. Since the g orbital can interact only with n_-, which already lies above n_+, the final result of this interaction is expected to maintain the original energetic order and to contribute only moderately, perhaps a few tenths of an eV, to the final splitting. This ring-orbital/n-orbital interaction is often referred to as through-bond (TB) interaction.

B. *o*-Benzoquinone

In the case of OBQ, the pure TS interaction should cause the antisymmetric (A) combination to be of lesser binding energy than the symmetric (S) combination. However,

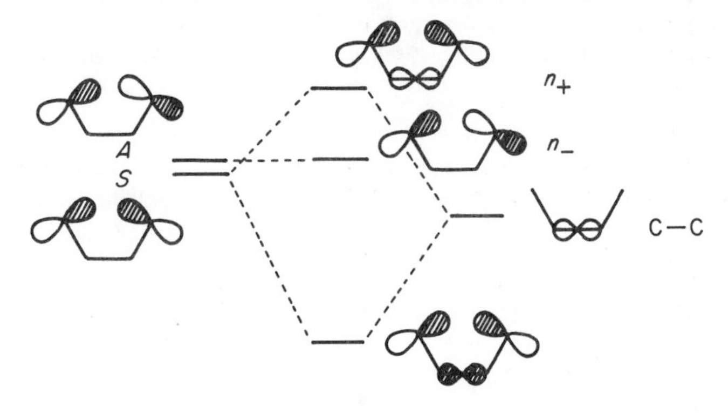

since this interaction should be quite weak, the two levels should be almost degenerate. The additional TB interaction which operates through the joint C–C bond and, for symmetry reasons, affects only the (S) combination of n orbitals causes a reversal of the zero-order situation. This situation is schematized above.

The TB interaction in OBQ is expected to be much stronger than in PBQ. Thus, there is little doubt that the n_+ orbital is the highest-energy occupied MO in OBQ and that the n_+/n_- split is quite large. As a result, the $I(\pi)$ events, which are largely unperturbed and associated with the conjugated hydrocarbon part of the molecule, usually lie between the $I(n_+)$ and $I(n_-)$ lone-pair ionization events.

The effects of variation of the angle between the two coplanar carbonyl groups on the zero-order I(n) events is of some interest and is shown below[6]. From the initial ideal angle

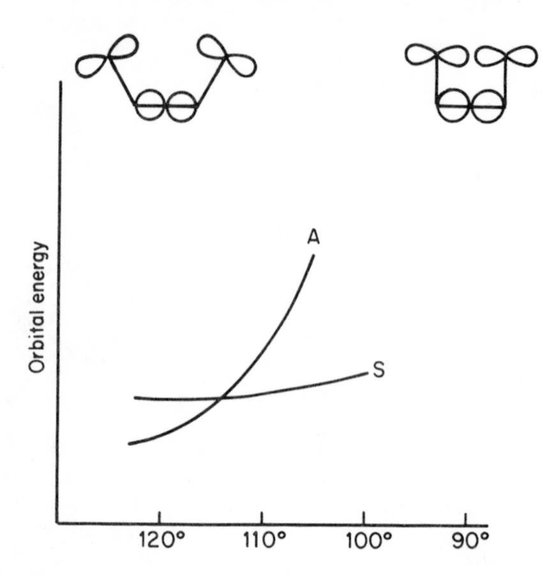

of 120° to the limiting 90° case, in which the two p orbitals point directly at each other, the binding energy of the A combination is expected to decrease consistently. The energy of the S (antibonding) combination, on the other hand, experiences compensation between increased bonding of the oxygen orbitals and increased antibonding carbon–oxygen interactions, the result being that the energy of this level remains largely invariant to angle. Calculations also predict that, at some angle close to 90°, the energy of A will cross over that of the π^* virtual antibonding MO, providing another example of the inability to form a four-membered ring by thermal reaction methods (Woodward–Hoffmann rules).

C. The π-MOs

The nature of the π-MOs of PBQ are best visualized as follows. Consider two p orbitals, one on each of the two oxygen centers. These are shown below, the positive lobes lying

below the plane of the paper. Consider also two cyclohexadiene group orbitals, one symmetric and one antisymmetric, as shown below. The interactions of these three group

orbitals is quite straightforward, the cyclohexadiene group orbital denoted A (or π_-) converting directly into a molecular orbital without any significant change of energy. The interaction of the S group orbitals, however, is extensive and it is this mixing which yields $I(\pi_+) < I(\pi_-)$. These mixings are shown below.

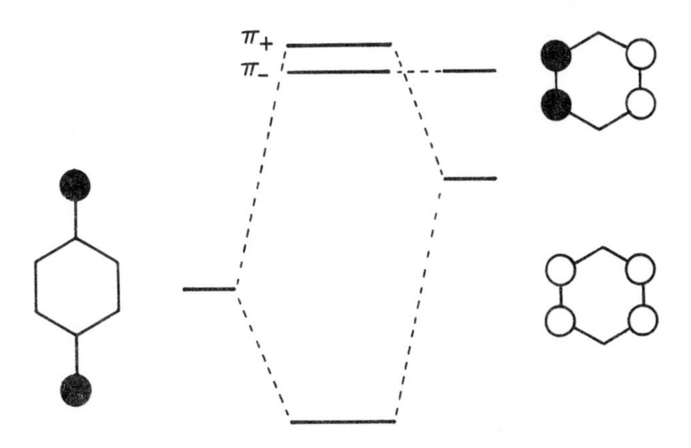

D. Summary

In sum, then, one predicts for PBQ that $I(\mathrm{n}_-) < I(\mathrm{n}_+)$ and $I(\pi_+) < I(\pi_-)$. The shapes of these orbitals, as obtained from CNDO/2 calculations, are shown in Figure 7. However, qualitative considerations cannot specify the $\mathrm{n}_+/\mathrm{n}_-$ splitting, the π_+/π_- splitting or even the relative order of the n and π subsets. Indeed, as will be discussed *in extenso* in the next section, we will show that simple computations tend to predict that $I(\pi) < I(\mathrm{n})$ whereas experimental arguments usually bespeak a reverse ordering.

VI. THE *p*-BENZOQUINONE STORY

The history of PES assignments for *p*-benzoquinone has been long and tortuous. Specifically, while the assignment of the four lowest-energy PES events to the ionization processes $I(\mathrm{n}_+)$, $I(\mathrm{n}_-)$, $I(\pi_+)$ and $I(\pi_-)$ has been generally accepted, virtually every one of the 4! permutations of these processes has been advocated at one time or another. This situation has produced much confusion and some discord. In a sense, the turns and twists of the PBQ PES assignments provide an object lesson in the need for scientific sobriety, particularly the desirability of leavening semiempirical quantum chemical results with large doses of empiricism. For that reason, we will devote some space to history.

The lowest-energy ionization of PBQ was first reported as $I_1 = 9.67$ eV in 1957 by Vilesov and coworkers[1]. Dewar and Worley[4] reported a set of ionization energies, $\{I_j\} = \{9.95, 10.88, 13.26, 14.05, 16.44, 18.65, 19.19\ \mathrm{eV}\}$, in which I_1 differed substantially from that of Vilesov. Turner and coworkers[5], using higher resolution, showed that each of

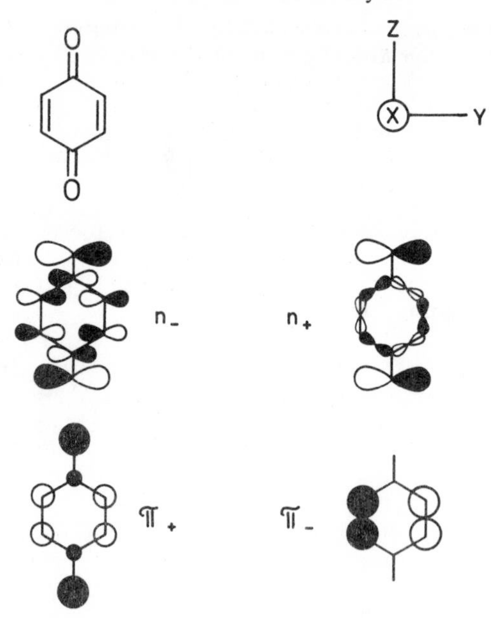

FIGURE 7. CNDO/2 computed molecular orbitals[18] for PBQ. Only the n_+, n_-, π_+ and π_- molecular orbitals are shown

the two lower Dewar–Worley values actually consisted of two close-lying bands, one pair at 10.11 and 10.41 eV and the other at 11.06 and 11.25 eV. Turner and coworkers also provided a correct assignment for the lowest set of ionization energies, namely $I(\text{n}), I(\text{n}) < I(\pi), I(\pi)$, but they gave no information concerning the method used to make these attributions. These assignments were all the more remarkable because prior calculations had unanimously predicted the topmost-filled MO (HOMO) of PBQ to be a π-MO[2,3]. As will be seen, this predictive tendency is a common but wrong characteristic of most quantum chemical calculations for PBQ.

Swenson and Hoffmann[6] investigated the TS and TB lone-pair interactions of the oxygens in 1970 and correctly predicted that $I(\text{n}_-) < I(\text{n}_+)$ for PBQ. However, their CNDO/2 and EHT calculations overestimated the degree of interaction of the lone pairs and, as a result, they predicted an $I(\text{n}_+)–I(\text{n}_-)$ gap much greater than the 0.3 eV posited by Turner. This, in turn, led Cowan and coworkers[7] to an assignment $I(\text{n}_-) < I(\pi_+) < I(\text{n}_+) < I(\pi_-)$, all four transitions fitting into the $I_4–I_1$ gap of ~1 eV observed by Turner.

Trommsdorf[8], Brundle and coworkers[9] and Stevenson[10] independently proposed very different assignments in 1972. Trommsdorf's assignment, based mainly on a CNDO calculation by Merienne-Lafore and Trommsdorf[16], yielded the order $I(\pi_+) < I(\text{n}_-) < I(\text{n}_+) < I(\pi_-)$. Brundle and coworkers[9] obtained improved spectra for PBQ and its tetrafluoro derivative but, unfortunately, they seem not to have calibrated their instrument. Furthermore, the PES of tetrafluoro-PBQ was so complex that the use of the perfluoro effect as an assignments device was largely vitiated. Nevertheless, on the assumption that the perfluoro effect stabilized the σ-MOs by as much as 5 eV and the π-MOs but slightly, they made the following assignments: $I(\text{n}_+) = 10.11$, $I(\text{n}_-) = 11.5$, $I(\pi) = 10.41$, $I(\pi) = 11.06$ eV and $I(\sigma) = 13.43$ eV. The assignments $I(\text{n}_+)$ and $I(\text{n}_-)$, tentative to begin with, were reversed in all later quotations of this work[9] so that the ordered set became $I(n_-) < I(\pi) < I(\pi) < I(n_+)$. However, the availability of the Brundle spectrum rapidly led to three correct assignments: (1) the Stevenson[10] assignment, $I(n_-) < I(n_+) < I(\pi_+) < I(\pi_-)$, based on CNDO-CI calculations; (2) the Gleghorn–McConkey[11] assignment based on a MINDO/2 calculation, which even gave numerical results {9.90, 10.58, 11.16 and 11.47 eV} in remarkable agreement with experiment; and the Hojer and coworkers[12] assignment based on a Pariser–Parr–Pople calculation (with Jensen–Skancke parameters for oxygen[49]) which yielded remarkable numerical agreement with experiment, namely 10.21, 10.51, 11.03 and 11.12 eV.

Kobayashi[13] initiated a rash of new work with the publication of improved, calibrated spectra for PBQ, *p*-toluquinone and 2,5-dimethyl-PBQ. Kobayashi paid particular attention to the band shifts induced by methylation and he concluded that the order was $I(n_-) < I(\pi_+) < I(n_+) < I(\pi_-)$ for both PBQ and methyl-PBQ and $I(n_-) < I(n_+, \pi_+) < I(\pi_-)$ for 2,5-dimethyl-PBQ. These assignments produced rather large $I(n_+)$–$I(n_-)$ splittings of 0.95, 0.59, and 0.48 eV, respectively. These splittings were immediately questioned by Lauer and coworkers[14] who had just recorded new spectra for PBQ, NQ and AQ, from which they concluded that the lone-pair splitting was less than 0.4 eV—a conclusion that also agreed with deductions from the electronic spectroscopy of neutral PBQ, namely that $I(n_+)$–$I(n_-) < 0.25$ eV[8, 50].

Lauer and coworkers[14] also pointed out the invalidity of Koopmans' theorem for the $n_-/^2B_{3g}$ and $n_+/^2B_{2u}$ ionic states of PBQ. They showed that configuration interaction of

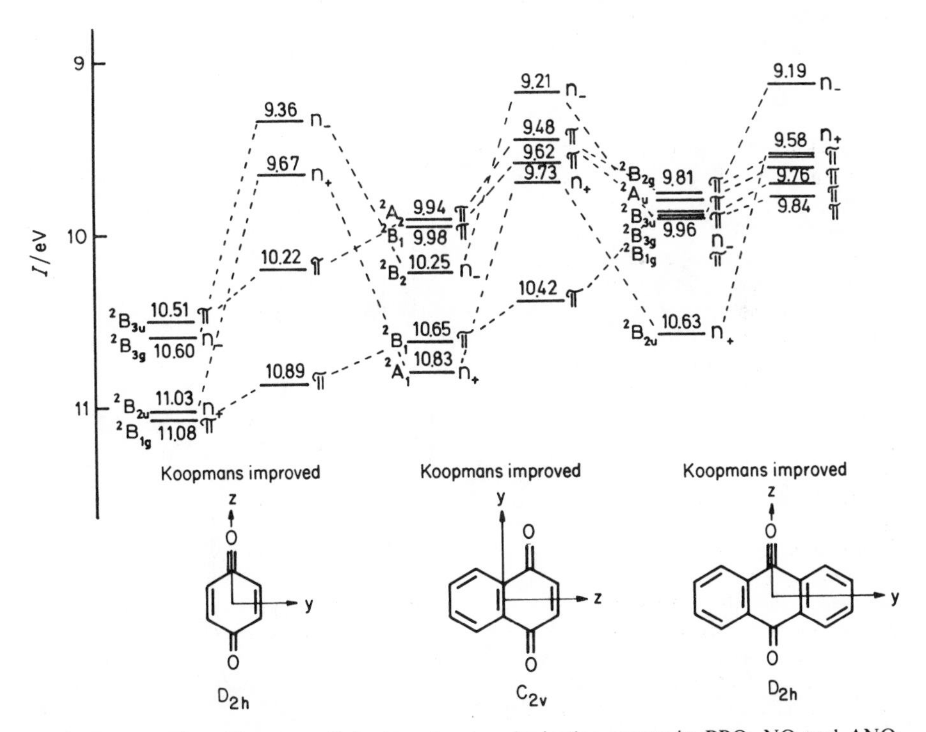

FIGURE 8. The assignment of the lowest-energy ionization events in PBQ, NQ and ANQ, according to Lauer and coworkers[14]. The 'improved' set refers to a situation in which configuration interaction was introduced

the ionic states was prerequisite to generation of the correct order of ionzation energies $I(n_-) < I(n_+)$ (Figure 8). Finally, a series of extensive computations pointed up the sensitivity of the computed results to the assumptions of the computational method.

Two *ab initio* SCF studies of PBQ[15a, b] produced the quite different MO sets $\{-10.91$ (π_+), -11.37 (π_-), -11.75 (n_-), -12.35 $(n_+)\}$ and $\{-8.30$ (π_+), -9.23 (n_-), -9.67 (π_-), -9.95 $(n_+)\}$, respectively. Since a small gaussian basis and a minimal Slater orbital basis were used in these calculations, it was not surprising that the agreement with experiment should be unsatisfactory. However, the discord of the two sets was surprising. This surprise was emphasized when it was found that contraction of the already small (6, 3/3) gaussian basis to a (4, 2/2) basis and the application of a restricted SCF method in the full D_{2h} symmetry yielded $\{-10.42$ (π_+), -10.53 (π_-), -11.04 (n_-), -11.23 $(n_+)\}$ whereas a local description in C_{2v} symmetry yielded $\{-8.60$ (n_- and n_+), -10.06 (π_- and π_+)$\}$.

Bunce and coworkers[17] investigated the excited states of several *p*-quinones (PBQ, duroquinone, AQ, chloranil and NQ). They used the virtual orbital configuration method (VOCI), which gives a good description of excited states and is an INDO variant[51]. However, it yields an ionization set for PBQ, namely $\{I(\pi_-) < I(n_-) < I(\pi_+) < I(n_+)\}$, which is clearly discordant with those discussed above.

Dougherty and McGlynn[18], in 1977, recorded the He(I) PE spectrum of PBQ in both low and high resolution; they reported spectra for several derivatives (i.e. tetrafluoro-PBQ, 2,5-dimethyl-PBQ, tetramethyl-PBQ (duroquinone), tetrachloro-PBQ (chloranil) and 2,3-dichloro-5,6-dicyano-PBQ); they provided a generally accepted recalibration of the low-energy PBQ bands; and they analysed a number of vibrational structures. Using a composite molecule approach and empirical data for band shapes, vibronic structures and

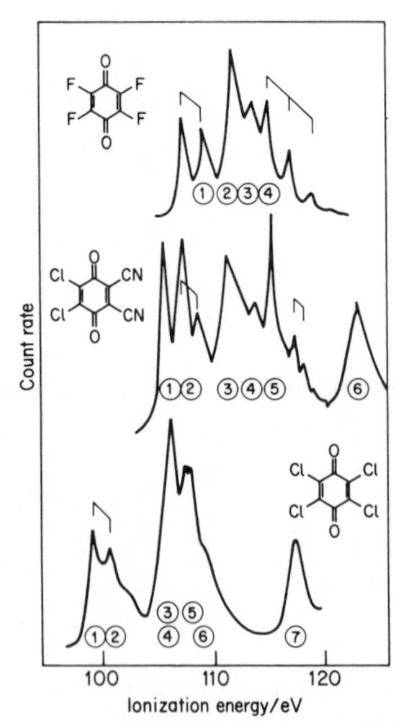

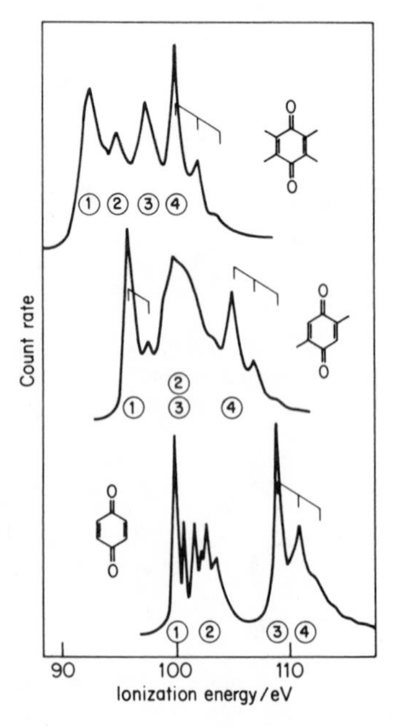

(a)

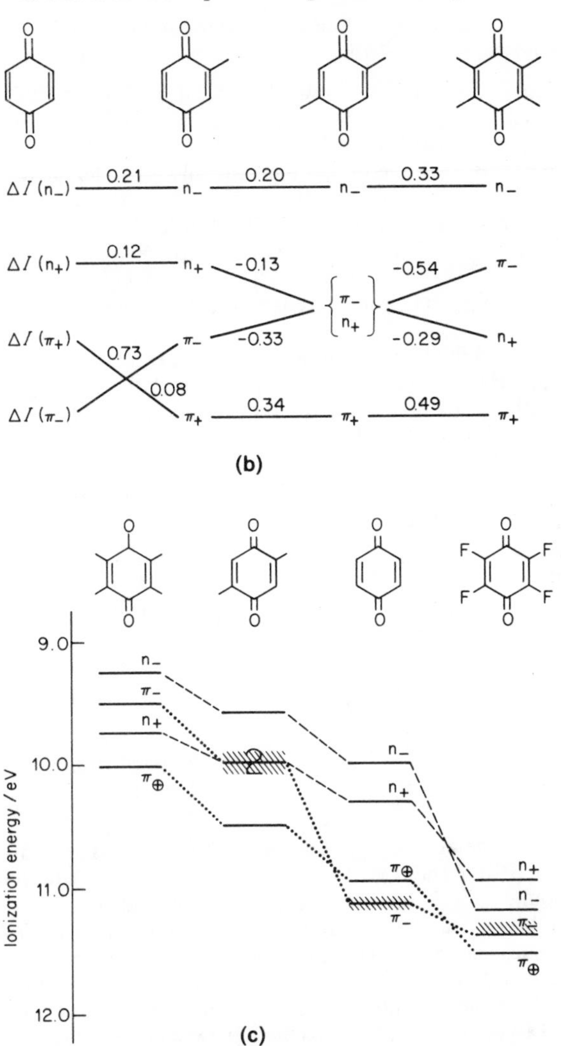

FIGURE 9. (a) The expanded scale He(I) photoelectron spectra of 1,4-benzoquinone, 2,5-dimethyl-1,4-benzoquinone, and tetramethyl-1,4-benzoquinone are given in the right side rectangle. Those of tetrafluoro-1,4-benzoquinone, 2,3-dichloro-5,6-dicyano-1,4-benzoquinone, and tetrachloro-1,4-benzoquinone are given in the left side rectangle. Vibrational spacings are indicated on the spectra[18]. (b) Schematic of the ionization energy differences between corresponding orbitals for 1,4-benzoquinone, toluquinone, 2,5-dimethyl-1,4-benzoquinone, and tetramethyl-1,4-benzoquinone[18]. (c) UPS correlation diagram for tetramethyl-1,4-benzoquinone, 2,5-dimethyl-1,4-benzoquinone, 1,4-benzoquinone and tetrafluoro-1,4-benzoquinone[18]

substituent shifts (Figure 9), these authors established the assignments $I(n_-) = 9.99 < I(n_+) = 10.29 < I(\pi_+) = 10.93 < I(\pi_-) = 11.1$ eV. In doing so, these authors aborted the details of their own CNDO/S calculations and placed primary reliance on empirical data correlations. This happenstance, however, seems not to have had much influence on succeeding computationists. As witness:

(1) Bigelow[19], in 1978, performed a CNDO/S calculation for PBQ and tetrafluoro-PBQ. He concluded, wrongly, that the assignment for PBQ was $I\,(n_-) < I(\pi_+) < I(\pi_-) < I\,(n_+)$ and for tetrafluoro-PBQ was $I\,(\pi_-) < I\,(\pi_+) < I\,(n_+) < I\,(n_-)$.

(2) Goodman and Brus[20] questioned the assignment of Dougherty and McGlynn[18] since they had found $|\Delta E\,(n_+ - n_-)| < 10^2\,\mathrm{cm}^{-1}$ in the UV absorption spectrum, which they thought to be inconsistent with the 0.3 eV splitting observed (and assigned) in PES.

(3) Bloor and coworkers[21], using the MSX$_\alpha$ method, found the order $I\,(n) < I\,(\pi) < I\,(n) < I\,(\pi)$ for PBQ. They also noted that 'for the CNDO/S method, slight changes in geometry and/or the oxygen parameters enable one to change at will the order of the top four MOs'.

The PES spectra of PBQ and tetrafluoro-PBQ were recorded in 1979 by Åsbrink and coworkers[22] using He(II) excitation. They interpreted the spectrum using the semi-empirical MO method, HAM/3. For PBQ they found $I\,(n_-) < I\,(n_+) < I\,(\pi_+) \leqslant I\,(\pi_-)$ and for tetrafluoro-PBQ they found $I\,(n_+) < I\,(\pi_-) < I\,(n_-) < I\,(\pi_+)$, in agreement with Dougherty and McGlynn[18]. Bock and coworkers[23] reported spectra for the isoelectronic molecules 1,4-difluoro-2,3,5,6-tetramethyl-1,4-dibora-2,5-cyclohexadiene (B) and duroquinone (i.e. tetramethyl-PBQ):

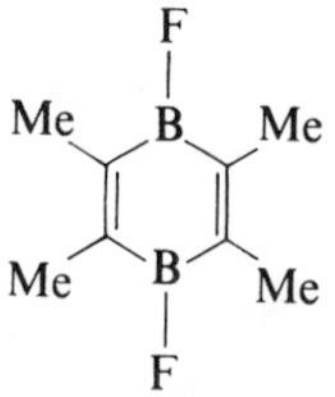

B

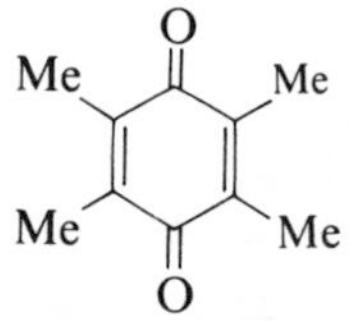

tetramethyl-PBQ
(duroquinone)

Based on CNDO/F calculations, they obtained the following assignments $I\,(n_-) < I\,(n_+) < I\,(\pi_-) < I(\pi_+)$ (PBQ) and $I(n_-) < I\,(\pi_-) < I\,(\pi_+) < I\,(n_+)$ (duroquinone, B and its parent compound 1,4-difluoro-1,4-dibora-2,5-cyclohexadiene).

The ability of MO calculations to reproduce the UV spectra of PBQ was critically evaluated by Jacques and coworkers[24], their aim being to reparametrize the CNDO/S method and to resolve disagreements in the assignments of the low-energy ionizations. The parameters, which seemed to provide an adequate representation of the whole UV spectrum of PBQ and to improve the agreement between the calculated and experimental spectra of other carbonyl compounds as well, yielded the sequence $I\,(\pi_+) < I(\pi_-) < I\,(n_-) < I\,(n_+)$. The assignments of the PBQ spectrum, as given in Refs 19, 21 and 24, are compared in Figure 10.

Gleiter and coworkers[25a, b] discussed the relationship between tropoquinones and benzoquinones and quote the order $I\,(n_-) < I\,(n_+) < I(\pi_+) < I(\pi_-)$ for PBQ.

Bock and coworkers[26] produced the first evidence of PES nature for the formation of *p*-thiobenzoquinones. The thermal decomposition of 1,4-di(heteroallyl)-substituted benzenes ($H_2C{=}CH{-}CH_2X{-}C_6H_4X{-}CH_2{-}CH{=}CH_2$, with $X = O$ or S) produces PBQ, monothio- or dithio-PBQ and 1,5-hexadiene in the gaseous phase. Because of the overlapping 1,5-hexadiene spectrum, only the two lowest-energy ionization events in dithio-PBQ, namely 8.5 and 9.1 eV, could be determined. Based on MNDO calculations, which for PBQ gave the order 10.9(n_-), 11.0(π_+), 11.2(π_-), 11.6(n_+), the assignment for the dithio-PBQ was given as $I\,(\pi_+) < I\,(n_-) < I(n_+) < I\,(\pi_-)$.

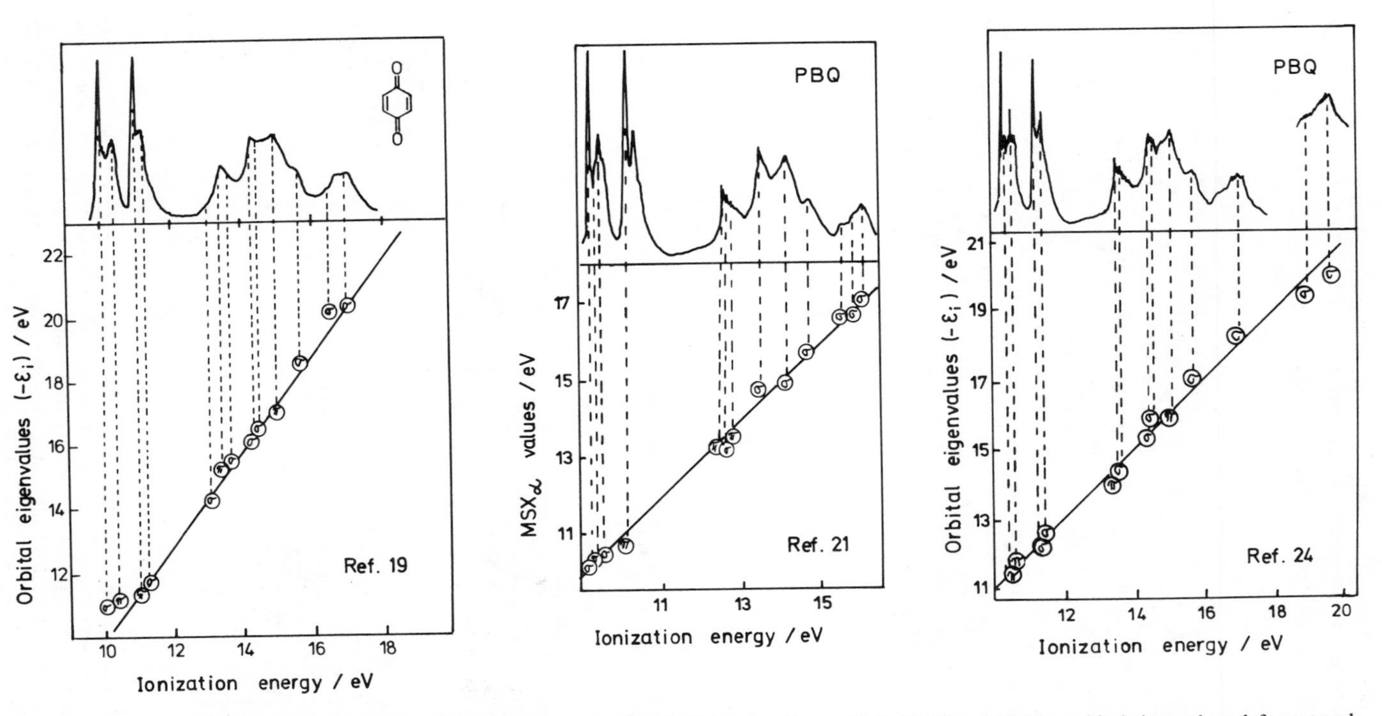

FIGURE 10. Alternative assignment of the photoelectron spectrum of PBQ, according to Refs. 19, 21 and 24. The orbital eigenvalues, leftmost and rightmost, are CNDO/S in nature. Those in the middle are MSX_α 'state' differences

Ha[27] reported extended CI calculations employing a gaussian set of double-zeta (DZ) quality for the ground, excited and ionized states of PBQ. The SCF calculations for the ground state (^1A_g) yielded the configuration

$$[\ldots (1b_{3u})^2 (8a_g)^2 (3b_{3g})^2 (1b_{2g})^2 (5b_{2u})^2 (4b_{3g})^2 (1b_{1g})^2 (2b_{3u})^2]$$

and, therefore, $I(\pi_+) < I(\pi_-) < I(n_-) < I(n_+)$. A CI calculation was carried out for each state of interest, all single and double excitations of the valence shell electrons being included. The results are shown in Figure 11, in which the experimental ionization energies (EXPT) are compared with the calculated orbital energies, ε_i; with the SCF energy differences between the cations and ground state, ΔSCF; and with the CI energy differences between the cations and the ground state, ΔCI. The CI stage of calculation indicates that the effect of electron correlation is least for the two top π levels and that it effects a change of ionization order to one which remains unchanged in all multi-reference CI schemes which incorporate higher-order correlation (Davidson correction). This ordering, which is supported by experimental evidence, may not be obtained at the SCF level employing either Koopmans' theorem or the direct SCF differences between the ground state and the various cationic states. Thus, Ha's calculation indicates that the PBQ spectrum may be assigned $I(n_-) < I(n_+) < I(\pi_+) < I(\pi_-)$ when CI is imposed but that,

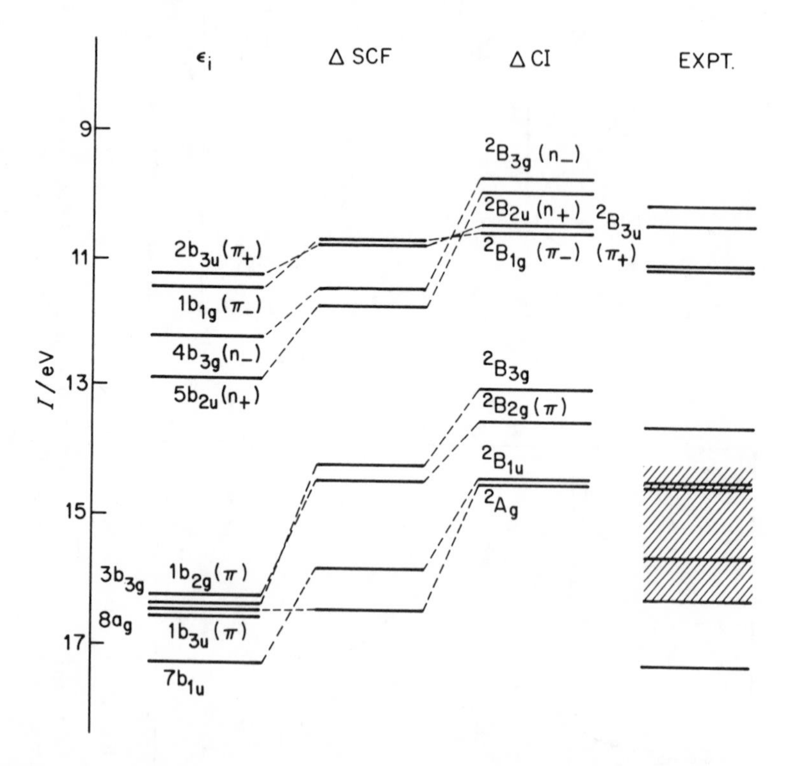

FIGURE 11. A comparison of the calculated ionization potentials in the SCF and CI levels with the experimental ionization energies, EXPT. ε_i is the orbital energy; ΔSCF is the SCF energy difference between the cations and the ground state; ΔSCI is the CI energy difference between the cations and the ground state. The hatched area stands for the PES region between 14.0 and 16.0 eV, which has an irregular fine structure

in the absence of CI, the MO order is probably $\varepsilon(\pi_+) > \varepsilon(\pi_-) > \varepsilon(n_-) > \varepsilon(n_+)$. Similar large changes of level ordering also take place in *p*-benzoquinodimethane as a result of CI (*vide infra*). Thus, to some extent at least, the root of the assignment problem has been uncovered.

VII. ALL ABOUT *p*-QUINODIMETHANE

The *p*-quinodimethanes can be defined[52] as even-membered, conjugated hydrocarbons. They are identical to the quinones except that $=CH_2$ groups replace the oxygens. *p*-Quinodimethane, for example, is understood to have the structure

CH_2

CH_2

(1)

It should be emphasized that *p*-quinodimethane is a colloquialism in the same way that 'quinone' is vernacular fo '*p*-benzoquinone'; other names are *p*-xylylene and *p*-benzoquinodimethane.

The He(I) photoelectron spectrum of *p*-(benzo)quinodimethane (**1**), as obtained by flash vacuum pyrolysis of [2,2]paracyclophane, was first analysed by Koenig and coworkers[53]. The PES of [2,2]paracyclophane had been reported by Pignataro and coworkers[54] and by Boschi and coworkers[55]. Contrary to [2,2]paracyclophane, which has a broad low-energy composite band (maximum at 8.00, shoulder at 8.5, sharp peak at 9.52 eV), the spectrum of **1** exhibits a sharp band at (7.87 ± 0.05) eV and a broad band at (9.7 ± 0.1) eV, the intensity ratio of these bands being 1:1.2. These latter bands have been assigned to the $^2B_{3u}$ and ($^2B_{1g} + {^2B_{2g}}$) states of the radical cation of **1**. However, an unperturbed $^2B_{2g}$ state, such as that assigned above, should lie at higher energy than a $^2B_{3g}$ state. Consequently, Koenig and coworkers[53] were forced to postulate that the energy of the $^2B_{2g}$ state was lowered by an interaction between ionic forms derived from both the quinoid and diradical structures that are present in the ground state. In effect, part of this broad band was assigned to a non-Koopmans, shake-up transition. This important and novel assignment was supported by Allan and coworkers[56], who showed that the valence bond (VB) scheme used by Koenig[53] was equivalent to a linear combination of bond orbitals (LCBO) model. In fact, the LCBO model yielded an excellent set of assignments for **1**.

State of $\mathbf{1}^+$	I_v (calc)/eV	I_v (obs)/eV
$^2B_{3u}$	7.94	7.87 ± 0.05
$^2B_{1g}$	9.80	9.7 ± 0.1
$^2B_{2g}$	10.00	
$^2B_{3u}$	12.86	(12.5?)

The LCBO model also performed equally well for anthraquinodimethane[56].

Tetracyanoquinodimethane (TCNQ) was the first derivative of **1** investigated by PES techniques: Ikemoto and coworkers[58] reported the gas phase spectrum; Grobman and coworkers[59] reported the solid-state He(I) spectrum; and numerous authors[60,65,72] performed computations in an attempt to assign the spectrum. The available data for TCNQ are summarized in the Appendix. The agreement between experiment and theory seems to be satisfactory. The spectrum of TCNQ appears much like that of **1** but is displaced from it by ~ 1.4 eV because of the inductive effects of the CN groups[57].

The three lowest-energy ionizations of TCNQ appear to be of $I(\pi)$ nature. In addition, recent open-shell CNDO/S-CI calculations[72] indicate that, as in **1**, two non-Koopmans transitions, both of which arise from a mixing of the one-electron $b_{2g}(\pi)^{-1}$ Koopmans state and the non-Koopmans $b_{3u}(\pi)^{-1} \rightarrow 2b_{2g}(\pi^*)^{-1}$ valence excitation, should be observed at 11.44 and 12.84 eV. Being aware that such an assignment needs additional support, Koenig and Southworth[66] investigated the PES of 2,5-dimethyl-**1**, in which the accidental degeneracy of the second band might be expected to be removed. According to MO and structure representation (i.e. VB) arguments, dimethyl substitution should lower the energy of the first two ionization events (i.e. $^2B_{1u}$ and $^2B_{3g}$) but not that of the 'non-Koopmans' $^2B_{2g}$ state; thus, a weak band near 9.8 eV should evolve as a result of dimethylation. The PES spectrum of 2,5-dimethyl-**1** is shown in Figure 12. It actually exhibits the expected low-intensity band. However, these deductions did not go uncriticized: Dewar[67], for example, preferred the supposition that the weak bands were impurity related. However, in the interim, the evidence for non-Koopmans transitions has grown, even in simple conjugated hydrocarbons[68] such as were particularly questioned by Dewar.

In a subsequent work, Koenig and coworkers[69] showed that their assumption of non-Koopmans behavior for **1** and 2,5-dimethyl-**1** was supported by HAM/3-CI calculations. Bigelow[70–72] also studied the breakdown of Koopmans behavior in the PES of selected organic molecules such as 2,5-dimethyl-**1**, **1** and TCNQ as well as benzene, xylene, *s-trans*-1,3-butadiene, *p*-nitroaniline, *trans*-stilbene and acrolein, all within a semiempirical, open-shell RHF-CNDO/S(CI) approximation. Schulz and coworkers[73] investigated the low-energy non-Koopmans ion states of unsaturated hydrocarbons using a semiempirical PERTCI approach for **1**, *o*-xylylene, *iso*benzofulvene, *iso*benzofulvalene, 2,2-dimethyl-*iso*indene, benzocyclobutadiene, butadiene and 1,1,4,4-tetrafluorobutadiene. Their results[73] support the existence of non-Koopmans' shake-up ionizations in **1** and tetrafluorobutadiene and suggest their occurrence in certain new cases (*iso*benzofulvene, *iso*-benzofulvalene and benzocyclobutadiene, for example) in which, unfortunately, the predicted intensities are so low as to inhibit observation. Finally, it has been shown for 2,2-dimethyl*iso*indene[74] and for 4,4-dimethyl-1-methylidene-2,5-cyclohexadiene[75] that the lowest-energy excited ion state is almost surely of non-Koopmans nature.

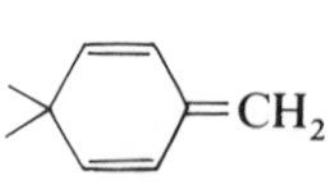

$=CH_2$

An *ab initio* SCF CI study of the ground and excited states of **1** was recently performed by Ha[76]. The results demonstrate that configuration mixing is as important to proper interpretation of the low-energy PES region as it is to the inner-valence, high-energy XPS region. A comparison of experiment with the calculated ionization energies at the SCF and CI stages is shown in Figure 13. Thus, the first ionization potential corresponds to

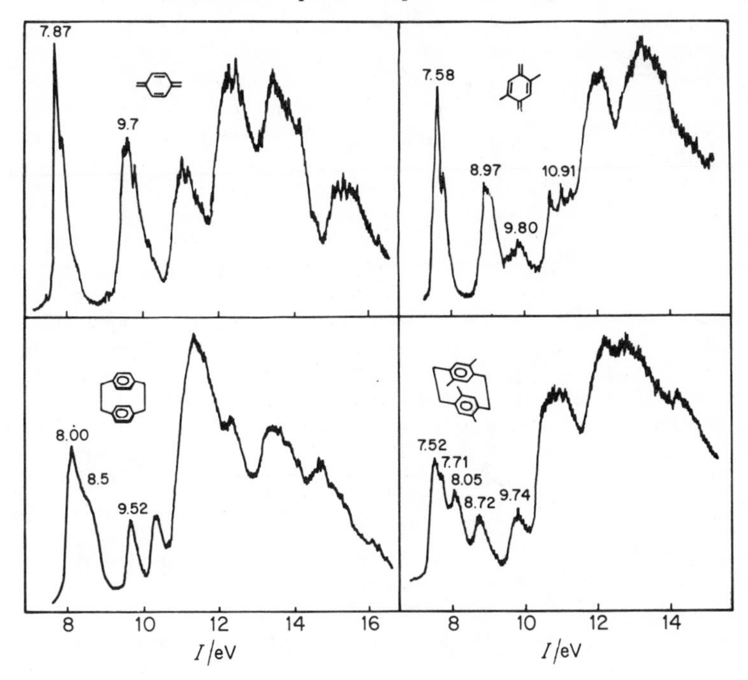

600°C

D_{2h}

z

y

1

FIGURE 12. He(I) photoelectron spectra of *p*-benzoquinodimethane (**1**, upper left), 2,5-dimethyl-**1** (upper right), and of their corresponding paracyclophane precursors (below), according to Refs 53 and 66

ionization from a π-type b_{3u} orbital (ΔCI = 7.12 eV; EXPT = 7.78 eV). Of the next two bands, the low-energy $\sim$ 10 eV component corresponds to a π ionization from a $1b_{1g}$ orbital (ΔCI = 9.27 eV) and the higher-energy 11 eV component to a σ ionization from the $4b_{3g}$ orbital (ΔCI = 11.68 eV). The remaining two components represent the first[53] and second non-Koopmans states of PBQ. The *ab initio* CI results indicate clearly that the high-energy, low-intensity flanks of the second and third PES bands are best ascribed to shake-up ionizations. Thus, the PES problem for **1** is resolved, much as for PBQ.

VIII. THE *o*-QUINONOIDS

The PES of the *o*-quinonoids are simple and readily interpretable. The reasons are:

(1) The spectra of the *o*-quinones are closely related to those of the unsaturated α-diketones: the dicarbonyl and conjugated hydrocarbon moieties behave almost additively and interact little if at all.

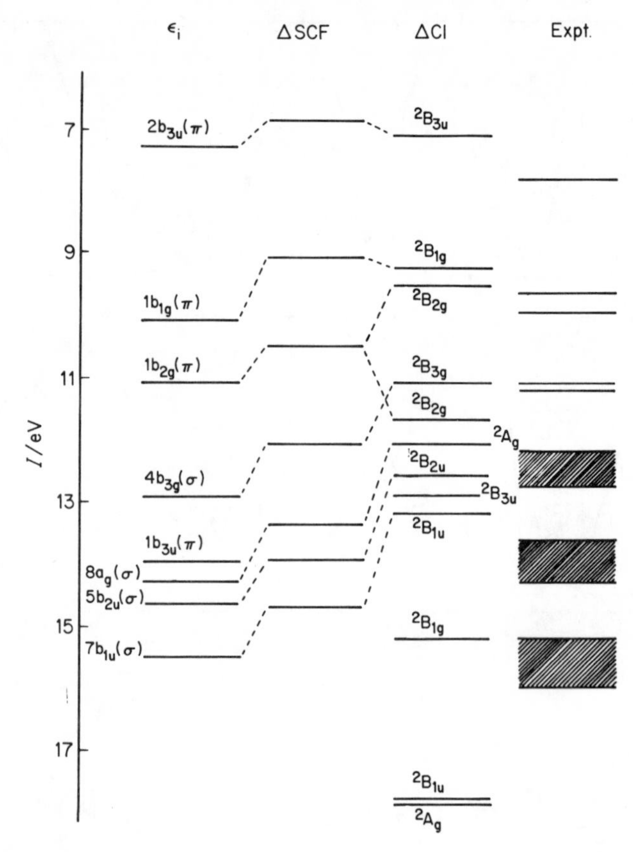

FIGURE 13. Comparison of experiment (EXPT) with the calculated ionization energies, where the ε_i are orbital energies, the ΔSCF are the energy differences between the SCF ground and cationic states, and the ΔCI are energy differences between the configurationally interacted ground and cationic states[76]

(2) A vast PES literature for α-dicarbonyls, saturated and unsaturated, has been amassed during the last decade[77] and the spectra are well understood.

(3) Most PES for *o*-quinonoids were recorded latterly, at a time when the occurrence of non-Koopmans transitions was already widely known. That is, the *o*-quinonoids—being unstable, highly reactive or transient species in thermal and photochemical reactions—required special synthetic, handling and recording techniques prior to PES measurements.

Three research groups have been heavily involved in the study of *o*-quinonoids: those of Professors Koenig (Eugene), Gleiter (Heidelberg) and Schweig (Marburg). Each of these groups have developed sophisticated methods for the preparation and measurement of such compounds, among them vacuum flash pyrolysis[78] and variable temperature spectroscopy (VTPES)[79]. However, the first PES report for an *o*-quinonoid was the study of the orbital sequence in fulvene and 3,4-dimethylcyclobutene by Heilbronner and coworkers[80]. Clark and coworkers[81] also reported the PES of benzo(c) thiophene, 2,1,3-benzothiadiazole and benzofurazan.

Flynn and Michl[82a] were the first to prepare *o*-xylylene. Although they did not make PES measurements, their synthesis, calculations and electronic spectroscopic data were important to all the later work on *o*-quinonoids. The PES spectrum of *o*-xylylene was not reported until 1984 by Schweig's group[82b] since earlier attempts to generate and detect it by VTPES produced benzocyclobutene; the problem of the fast unimolecular consecutive reaction to this end product was solved by using 5,6-dimethylene bicyclo[2.2.1]hept-2-en-7-one as the precursor compound.

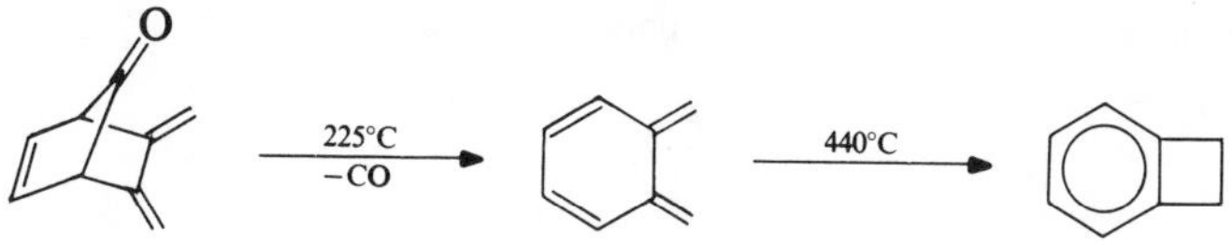

$$\xrightarrow[-CO]{225°C} \qquad \xrightarrow{440°C}$$

Tri*iso*propylidenecyclopropane (**2**) may be supposed to be a multiple *o*-quinonoid. Its PES was recorded, discussed and compared with the MO data for the parent unsubstituted 3-radialene by Bally and Haselbach[83].

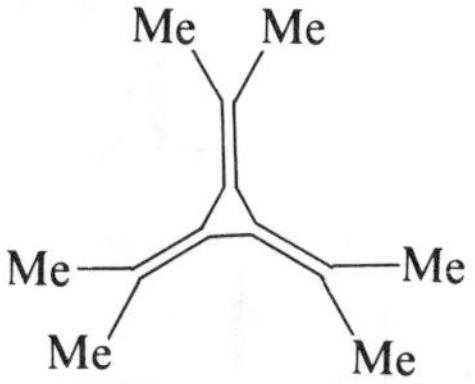

Me Me Me Me Me Me

(**2**)

Palmer and Kennedy[84] reported PES data for benzo(c)furan (**4**), benzo(c)thiophene (**5**), and *N*-methyl*iso*indole (**6**), which are the quinonoid isomers of the benzo derivatives of five-membered ring heterocycles related to naphthalene (**3**):

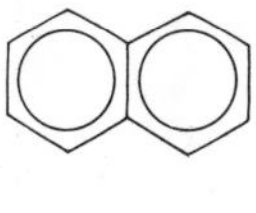

O S NMe

(**3**) 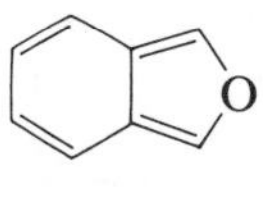 (**4**) 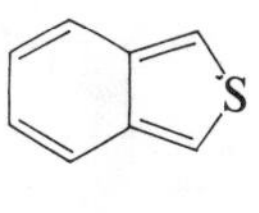 (**5**) 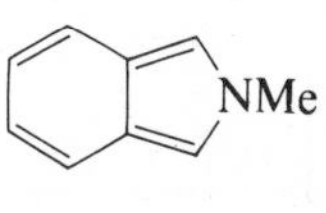 (**6**)

Correlation with the PES of naphthalene led to assignment of the first three π ionizations in **3**–**6** as

$$I_1(\mathbf{6})<I_1(\mathbf{5})<I_1(\mathbf{4})<I_1(\mathbf{3})<I_2(\mathbf{6})<I_2(\mathbf{3})=I_2(\mathbf{5})<I_3(\mathbf{6})=I_2(\mathbf{4})<I_3(\mathbf{5})=I_3(\mathbf{3})<I_3(\mathbf{4})$$

Rettig and Wirz[85] investigated the PES of **6** and its higher analogs **7** and **8**:

NMe NMe

(7) (8)

They concluded that any description based on a single *o*-quinoid structure was inadequate and that the inclusion of zwitterionic structures such as

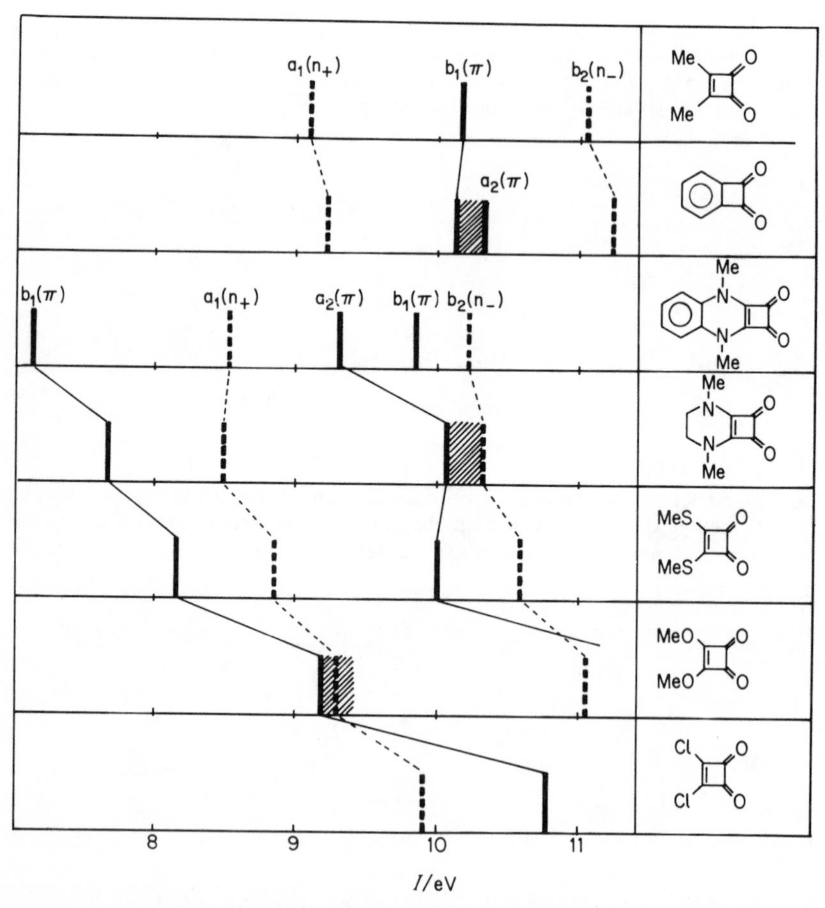

FIGURE 14. The assignment for and correlation diagram of some cyclobutene-1,2-dione derivatives[91]

was mandatory.

The PES of *o*-benzoquinone (OBQ), first reported by Koenig and coworkers[86] was later studied by Eck and coworkers[87] and by Schang and coworkers[88]. The assignment, using MNDO(PERTCI) and MINDO/3 calculations, yields the order[87,88]: $I(n_+) < I(\pi) < I(n_-) < I(\pi)$.

Schang and coworkers[88] also reported the PES of cyclobutene-1,2-dione (CBD), some of its derivatives and certain OBQ derivatives:

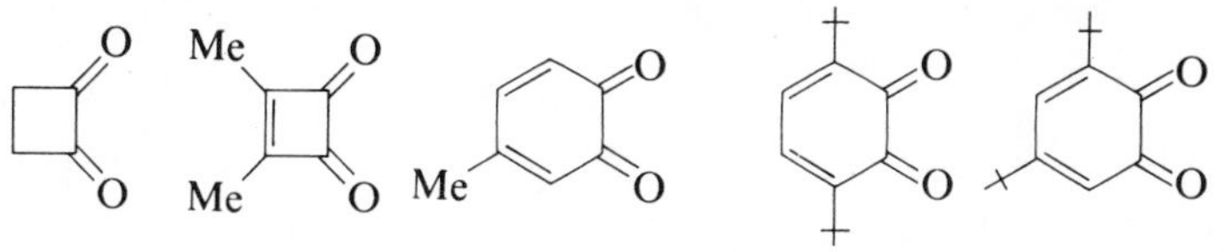

The assignments[89,90] of some of these PES data are shown in Figure 14. Gleiter and coworkers[91] also reported PES data for several squaric acid derivatives:

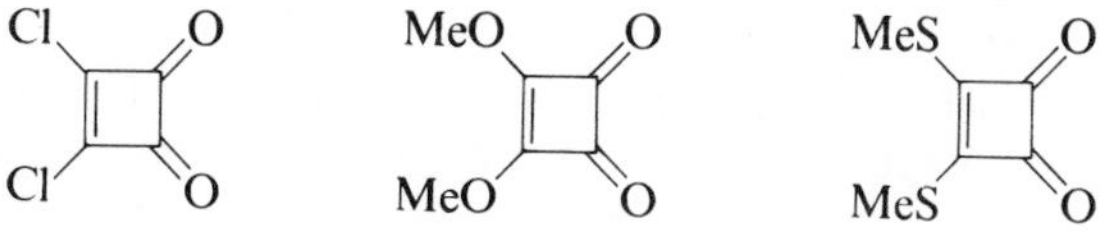

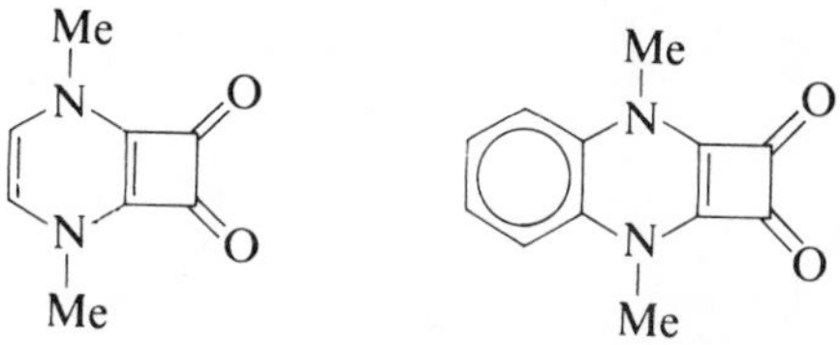

as well as that of benzocyclobutenedione and CBD. The interpretation of these[89,91] is also shown in Figure 14.

Eck and coworkers[92] reported the PES of *o*-benzoquinone methide generated by pyrolysis of 2-hydroxybenzyl methyl ether in a heated and temperature-controlled target chamber:

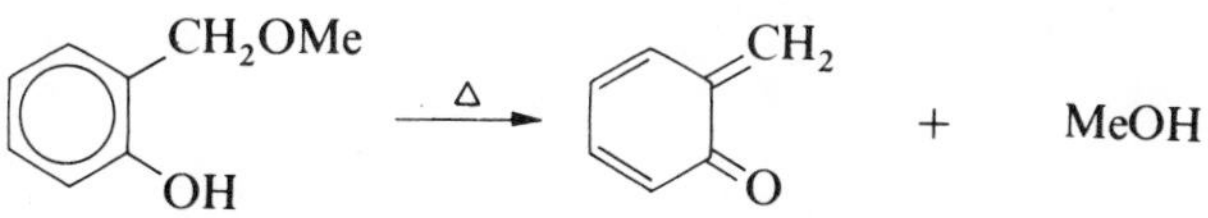

The spectra of the pyrolysis products, recorded over the range 100–600 °C, led, at 500 °C, to a new spectrum with bands at 8.80(π), 9.37(π), 10.63(π) and 12.02(π) eV.

Schulz and Schweig[93] carried out the following reactions in the PES mode:

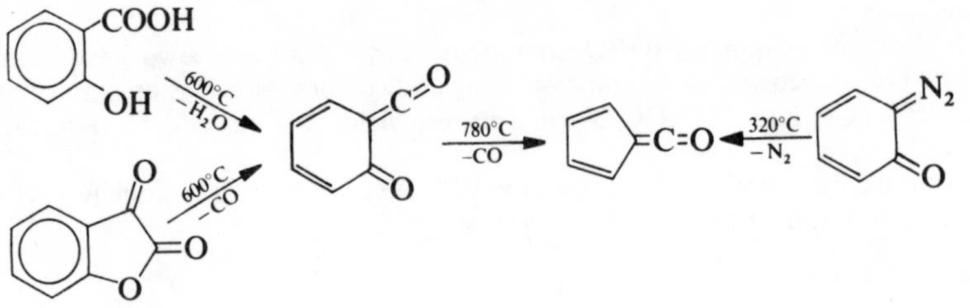

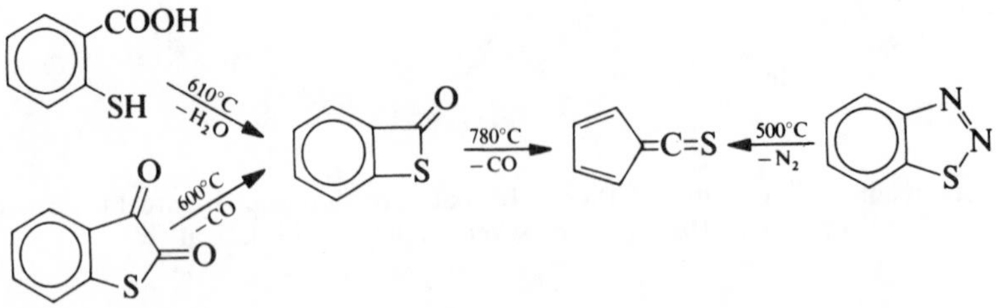

The spectra of the reaction products were recorded and their structure confirmed using computed[94] MNDO-optimized structures and large configuration interaction (PERTCI) calculations based on CNDO/S wave functions. The possible reaction products

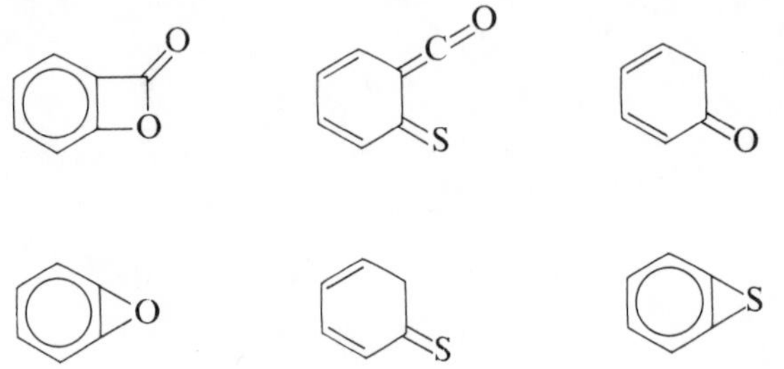

were excluded on the basis of disagreements between the calculated and observed spectral patterns.

Schweig and coworkers investigated the reactions:

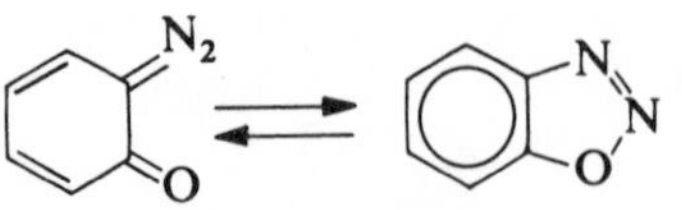

Ref. 95

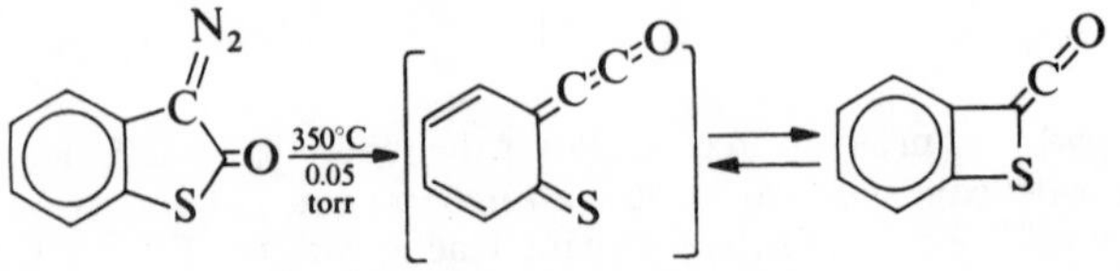

350°C 0.05 torr

Ref. 96

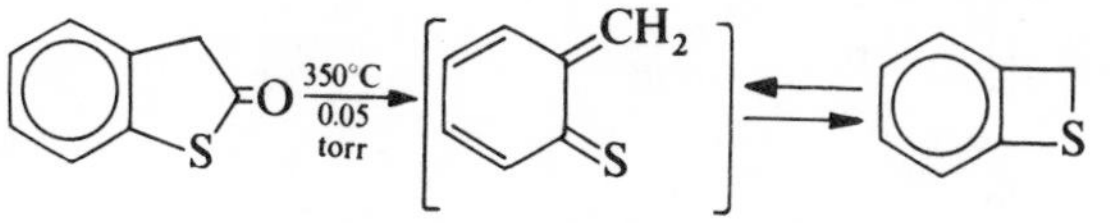

O S 350°C 0.05 torr [CH_2 S] S Ref. 96

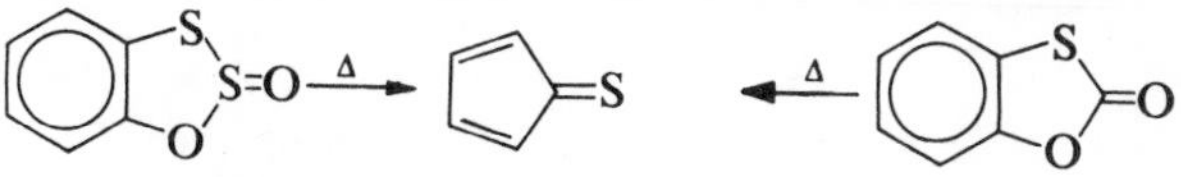

S S=O O Δ S Δ S O O Ref. 97

with

S O

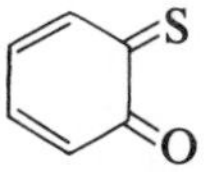

as a postulated reactive intermediate[97]. They also investigated the processes

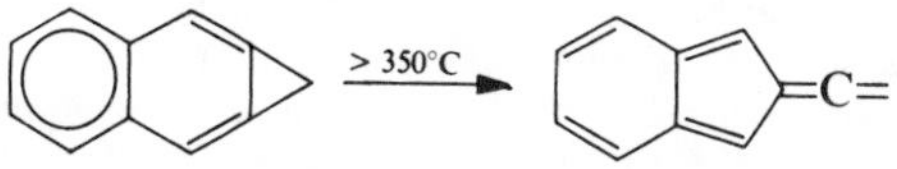

> 350°C =C= Ref. 98

S O S

≥ 690°C

S S=O S ≥ 540°C S S and not S S

≥ 720°C

S S

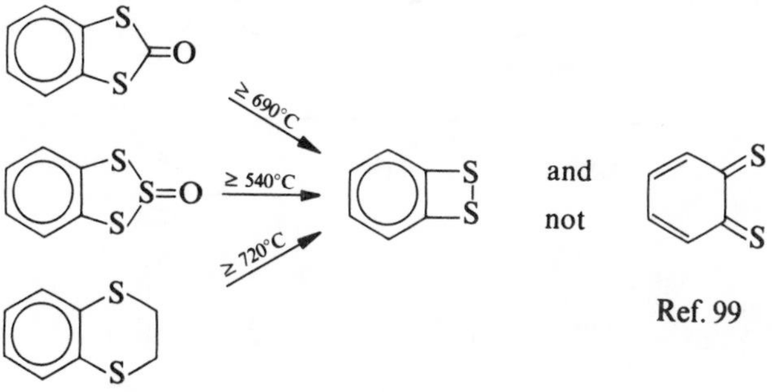

Ref. 99

S S ⇌ S C C S Ref. 100

The formation of an *o*-quinonoid compound was generally observed for quinones and quinone methides whereas formation of a four-membered ring was preferred when a sulfur atom was present.

Schweig and coworkers investigated the PES of the *o*-quinonoids: bicyclo[3.2.0]hepta-1,4,6-triene (BHT), its analog furocyclobutadiene (FCB)[101] and *iso*benzofulvene (IBF)[102]. The PES of these and a large number (> 100) of other molecules have recently[103] been calculated using the semiempirical methods: MNDO, MNDO PERTCI, CNDO/S, CNDO/S PERTCI, LNDO/S and LNDO/S PERTCI; and the results have been correlated with the experimental data.

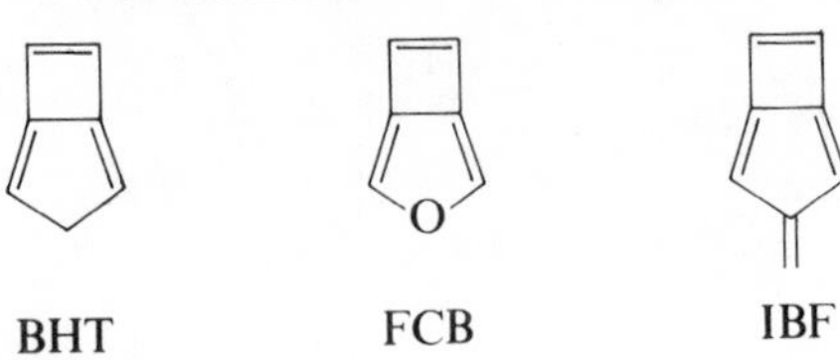

Gleiter and coworkers[25a,b] measured the PES of cyclopentene-1,2,3-trione, *o*-tropoquinone and *p*-tropoquinone. By comparison with related molecules and MINDO/S calculations, the low-energy ionization events in these compounds were assigned. These assignments are given in Figure 15. They provide an interesting insight into the distinct relatedness of the electronic structures of the benzoquinones and tropoquinones. Gleiter and coworkers[104-107] have also discussed the PES of several other quinonoid molecules.

Very little research has been invested in higher 'classical' quinones, ones such as those that would be generated by the oxidation of polycyclic aromatic hydrocarbons (PAHs). This, perhaps, is not too surprising. Most of these compounds are thought to be potent carcinogens and, in addition, their melting points may be too high to permit easy PES measurements. One such rare investigation is the work of Bigotto and coworkers[108] on acenaphthenequinone (ANQ) and naphthalic acid (i.e. naphthalene-1,8-dicarboxylic acid) derivatives, namely naphthalic anhydride, naphthalimide and *N*-methylnaphthalimide. The lowest-energy π-ionization events in these compounds were assigned using a Pariser–Parr–Pople SCF approach. The assignment of the two n ionizations then followed from band intensity and band shape data and by reference to results for phthalic anhydride, phthalimide and various cyclic 1,2- and 1,3-dicarbonyl compounds. For ANQ, the following results were obtained.

ANQ

Expt/eV	Calc/eV	Assignment
8.77	8.69	$\pi(a_2)$
8.77		n_+
9.53	9.48	$\pi(b_1)$
10.60	10.44	$\pi(b_1)$
10.60		n_-
11.48	11.40	$\pi(a_2)$

IX. CONCLUSION

Although much work has been invested in the photoelectron spectroscopy of quinonoids, one retains the impression that the task is grossly incomplete and heavily concentrated on but a few topics (e.g. *p*-benzoquinone and *p*-(benzo)quinodimethane and their derivatives; small, mostly unstable compounds or transients of quinoid structure). However, despite the paucity of XPS data, one may conclude that quinonoids behave normally. That is, their XPS spectra, for the most part, are predictable using standard computational schemas.

The more extensive parts of the UPS work has been put to some stringent tests: (1) to determine the electronic structure of a given compound; (2) to test calculation methods; and (3) to identify new (unknown) molecules and their structure from PES spectra. While good success has been achieved with (3) and moderate success with (2), topic (1) has proven to be highly unsatisfactory. There is no doubt that *o*-quinonoids, being essentially α-diketones (or analogs of them) are easy to assign because the interactions are short-range

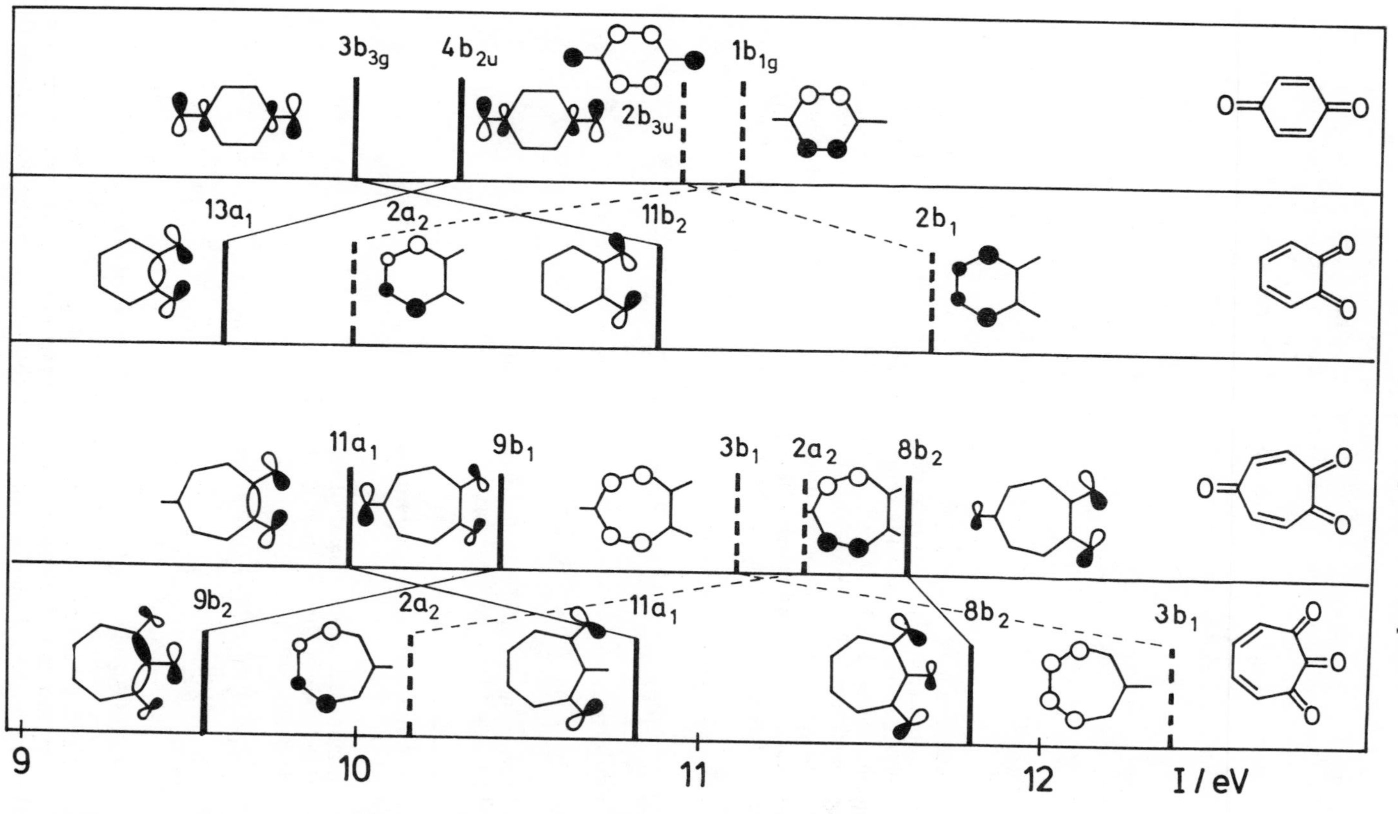

FIGURE 15. Relatedness of electronic structures of the benzoquinones and tropoquinones

and quantum chemistry knows how to handle them. However, severe problems surface in the *p*-quinonoids. So much so, that after 30 years of work on PBQ, a leading researcher feels moved to write: 'The assignment of the PE spectrum of PBQ is, however, still a matter of controversy and therefore our assignments must remain tentative'[109]. Finally, it must be stressed that the *o*- and *p*-quinonoids are but a small subdivision of this whole class of molecules and that other very interesting questions can be asked: what, for example, is the nature of the interaction of lone pairs in

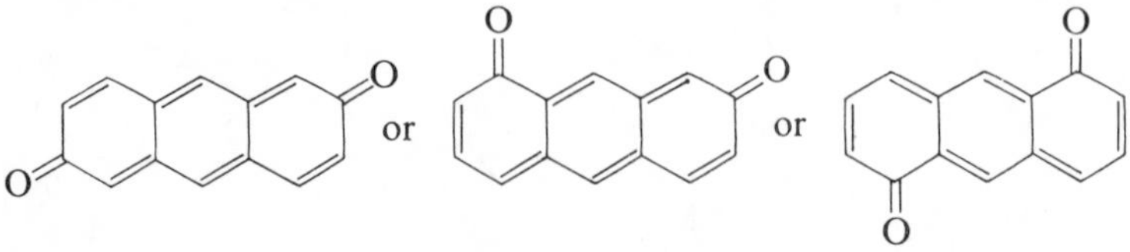

or or

and is it large or small?

PES data for steroids[109, 110] suggest that long-range interaction is strongly promoted by the localized double bonds that lie between the carbonyl groups. Indeed, there is some experimental evidence that such interactions can be promoted over numerous saturated (single) bonds[111–121]. No such work, interesting though it might be, exists for quinonoids.

A compilation of data are given in the Appendix. This compilation includes the experimental ionization data, the assignments made on the basis of various computational schemas and experimental data relating to vibrational structure (where available).

X. APPENDIX

A. Compilation of Data

(in cooperation with Dr. B. Kovač, Rudjer Bošković Institute, Zagreb)

This part in the form of a table covers the data published on the subject UPS of quinonoids using the following format:

Structural formula of compound	**Sum formula**	**Symmetry**	
I/eV: ionization energy in electronvolts (*comment on experimental details*)	Reported values and assignment if given empirically	Reference number	
$-\varepsilon_i$/eV: Negative calculated MO energy in electronvolts (*basis for assignment*)	Reported values and assignment if given in the publication	Type of calculation	Reference number
ν/cm^{-1}: Observed vibrational structure in wavenumbers	Reported values of vibrational spacings		

If several entries are from same reference number this is indicated only on the first entry.

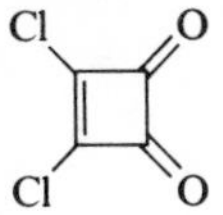

$C_4Cl_2O_2$ C_{2v}

I/eV:	9.89	10.78	12.04			91
$-\varepsilon_i$/eV:	9.65 $a_1(n_+)$	10.21 $b_1(\pi)$	10.81 $b_2(n_-)$		MINDO/3	

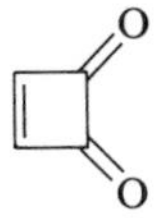

$C_4H_2O_2$ C_{2v}

I/eV:	9.79	11.55	11.87	13.61		88
$-\varepsilon_i$/eV:	9.35 $a_1(n_+)$	11.05 $b_1(\pi)$	10.71 $b_2(n_-)$	13.74 $a_2(\pi)$	MINDO/3	
	8.22 $a_1(n_+)$	8.92 $b_1(\pi)$	11.08 $b_2(n_-)$	11.88 $a_2(\pi)$	STO-3G	

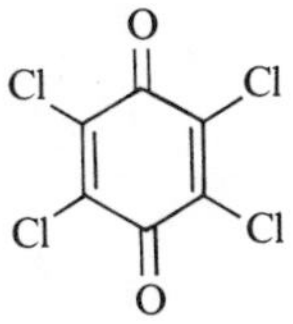

$C_6Cl_4O_2$ D_{2h}

I/eV:	9.90 $b_{3g}(n_-)$	10.1 $b_{2u}(n_+)$	10.65 b_{1g} & $b_{3u}(\pi)$	10.8	11.0	18
v/cm^{-1}:	1300					

(correlative study; substituent effects)

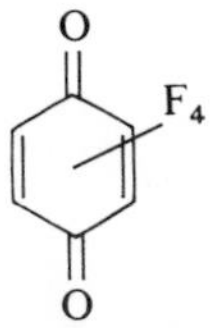

$C_6F_4O_2$ D_{2h}

I/eV:	10.72 π	11.18 π	14.82	15.0 n_u	15.96 n_g		5, 9

(assignment based on perfluoro effect)

I/eV:	10.96 $b_{2u}(n_+)$	11.21 $b_{3g}(n_-)$	11.31 $b_{1g}(\pi)$	11.53 $b_{3u}(\pi)$		18
v/cm^{-1}:	1650	1350		1700		

(correlative study; substituent effects)

$-\varepsilon_i$/eV:	11.47 $b_{1g}(\pi)$	11.58 $b_{3u}(\pi)$	12.17 $b_{2u}(n_+)$	12.63 $b_{3g}(n_-)$	CNDO/S	19
$-\varepsilon_i$/eV:	11.1 $b_{1g}(\pi)$	11.3 $b_{3g}(n_-)$	11.3 $b_{2u}(n_+)$	11.4 $b_{3u}(\pi)$	HAM/3	22

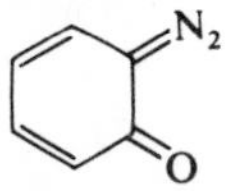

$C_6H_4N_2O$ C_s

I/eV:	8.20					95
$-\varepsilon_i$/eV:	8.10 a (π)	9.18 a′ (n_-)	10.15 a (π)	11.82 a (π)	CNDO/S	
	8.50 a (π)	9.79 a′ (n_-)	9.95 a (π)	11.30 a (π)	MNDO–PERT–CI	

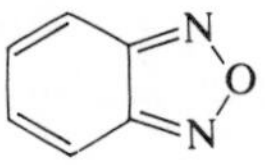

$C_6H_4N_2O$ C_{2v}

I/eV:	9.37 $a_2(\pi)$	10.44 $b_2(\pi)$	11.18 $a_1(n_+)$	11.58 $a_2(\pi)$	12.0 $b_1(n_-)$	81

(assignment based on band shapes and semiempirical calculations)

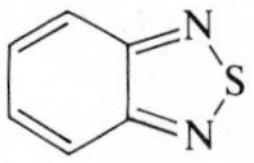

$C_6H_4N_2S$ **C_{2v}**

I/eV:	8.98 $a_2(\pi)$	9.52 $b_2(\pi)$	10.64 $a_1(n_+)$	11.31 $a_2(\pi)$	11.6 $b_1(n_-)$		81

(assignment based on band shapes and semiempirical calculations)

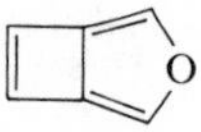

C_6H_4O **C_{2v}**

I/eV:	8.05 $b_1(\pi)$	8.95 $a_2(\pi_2)$	11.40 $a_1(\sigma)$	11.85			101

(based on MNDO PERTCI)

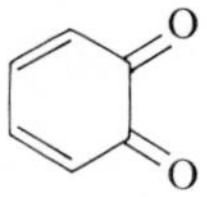

$C_6H_4O_2$ **C_{2v}**

I/eV:	9.6	9.99	11.0	12.3			86
I/eV:	9.55	9.89	10.78	12.12			87
$-\varepsilon_i$/eV:	9.72 $a_1(n)$	10.07 $a_2(\pi)$	11.67 $b_1(n)$	11.95 $b_1(\pi)$		MNDO	
	9.20 $a_1(n)$	9.63 $a_2(\pi)$	10.98 $b_2(n)$	12.27 $b_1(\pi)$		CNDO/S	
I/eV:	9.60	9.98	10.88	12.16			88
$-\varepsilon_i$/eV:	9.18 $a_1(n_+)$	9.84 $a_2(\pi)$	10.84 $b_1(n_-)$	12.25 $b_1(\pi)$		MINDO/3	

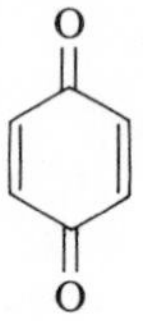

$C_6H_4O_2$ **D_{2h}**

I/eV:	9.67						1a
I/eV:	9.68						1c
I/eV:	10.11 n_g	10.41 π_4	11.06 n_u	11.5 π_3			9
	(assignment based on the perfluoro effect and CNDO/2)						
$-\varepsilon_i$/eV:	9.92 $b_{1g}(n)$	10.61 $b_{2u}(n)$	10.91 $b_{1u}(\pi)$	11.03 $b_{3g}(\pi)$	14.31 $a_g(\sigma)$	CNDO–CI	10
$-\varepsilon_i$/eV:	9.89 $b_{1g}(n)$	10.58 $b_{2u}(n)$	11.16 $b_{1u}(\pi)$	11.48 $b_{3g}(\pi)$	11.83 $a_g(\sigma)$	MINDO/2	11
$-\varepsilon_i$/eV:	10.21 $b_{3g}(n)$	10.51 $b_{2u}(n)$	11.03 $b_{3u}(\pi)$	11.12 $b_{1g}(\pi)$	13.72 $b_{2g}(\pi)$	PPP	12
I/eV:	10.01 $b_{3g}(n)$	10.29 $b_{3u}(\pi)$	10.96 $b_{2u}(n)$	11.16 $b_{1g}(\pi)$		CNDO/2	13
	(assignment based on CNDO/2)						
I/eV:	10.11	10.41	11.06	11.25			14
$-\varepsilon_i$/eV:	10.51 $b_{3u}(\pi_1)$	10.60 $b_{3g}(n_-)$	11.03 $b_{2u}(n_+)$	11.08 $b_{1g}(\pi_2)$		CNDO/S	
$-\varepsilon_i$/eV:	9.36 $b_{3g}(n_-)$	9.67 $b_{2u}(n_+)$	10.22 $b_{3u}(\pi_1)$	10.89 $b_{1g}(\pi_2)$		CNDO/S-CI	
$-\varepsilon_i$/eV:	10.91 b_{1u}	11.37 b_{2g}	11.75 b_{1g}	12.35 b_{3u}	14.91 b_{3g}	*ab initio*	15a
	10.42 b_{1u}	10.53 b_{2g}	11.04 b_{1g}	11.23 b_{3u}	14.20 b_{3g}	ΔSCF–D_{2h}	
	8.60 b_1	8.60 a_2	10.06 a_1	10.06 b_1	13.40 a_1	ΔSCF–C_{2v}	
$-\varepsilon_i$/eV:	8.30 b_{1u}	9.23 b_{1g}	9.67 b_{2g}	9.95 b_{3u}		*ab initio*	15b
$-\varepsilon_i$/eV:	11.62 $b_{3g}(\pi)$	11.90 $b_{3g}(n_-)$	12.17 $b_{2u}(n_+)$	12.50 $b_{1g}(\pi)$	15.58 $b_{3g}(\sigma)$	CNDO/2	16
I/eV:	9.99 $b_{3g}(n_-)$	10.29 $b_{2u}(n_+)$	10.93 $b_{3u}(\pi)$	11.1 $b_{1g}(\pi)$			18
ν/cm^{-1}:	725	700	1600				
	1500						

(correlative study; substituent effects)

$-\varepsilon_i$/eV:	11.17 $b_{3g}(n_-)$	11.17 $b_{3u}(\pi)$	11.20 $b_{1g}(\pi)$	11.76 $b_{2u}(n_+)$	14.35 $a_g(\sigma)$	CNDO/S	19
I/eV:	9.99	10.29	10.93	11.0	13.5		22
	14.3	14.8	14.9	15.0	15.5		
	16.2	16.7	17.0	19.5	20.1		
	(He(I) and He(II) values)						
$-\varepsilon_i$/eV:	10.04 $b_{3g}(n_-)$	10.50 $b_{2u}(n_+)$	10.95 $b_{1g}(\pi)$	11.01 $b_{3u}(\pi)$	13.53 $b_{3g}(\sigma)$	HAM/3	
$-\varepsilon_i$/eV:	11.14 $b_{3u}(\pi)$	11.40 $b_{1g}(\pi)$	11.90 $b_{3g}(n_-)$	12.58 $b_{2u}(n_+)$	14.06 $b_{2g}(\pi)$	opt. CNDO/S	24
$-\varepsilon_i$/eV:	10.9 n_-	11.0 π_1	11.2 π_2	11.6 n_+	14.5 π_3	MNDO	26
$-\varepsilon_i$/eV:	10.64 $b_{1g}(\pi)$	10.67 $b_{3u}(\pi)$	11.44 $b_{3g}(n_g)$	11.67 $b_{2u}(n_u)$		ΔE(SCF)	27
	9.59 $b_{3g}(n_g)$	9.85 $b_{2u}(n_u)$	10.35 $b_{3u}(\pi)$	10.55 $b_{1g}(\pi)$		ΔE(CI)	
	9.84 $b_{3g}(n_g)$	10.10 $b_{2u}(n_u)$	10.32 $b_{3u}(\pi)$	10.50 $b_{1g}(\pi)$		ΔE(CI′)	
	9.06 $b_{3g}(n_g)$	9.26 $b_{2u}(n_u)$	10.35 $b_{3u}(\pi)$	10.50 $b_{1g}(\pi)$		ΔE(CI″)	

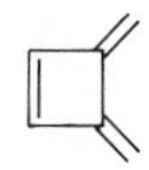

C_6H_6 C_{2v}

I/eV:	8.8	9.44	11.5	12.3	13.3	14.1	80
$-\varepsilon_i$/eV:	11.02 $b_1(\pi)$	11.66 $a_2(\pi)$	14.86 $a_1(\sigma)$	15.56 $b_2(\sigma)$	16.41 $b_1(\pi)$	SCF-LCAO-MO	

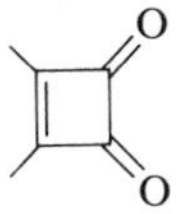

O O

$C_6H_6O_2$ C_{2v}

I/eV:	9.10	10.18	11.05	13.08		88.91
$-\varepsilon_i$/eV:	9.11 $a_1(n_+)$	10.11 $b_1(\pi)$	10.24 $b_2(n_-)$	13.49 $a_2(\pi)$	MINDO/3	
	7.78 $a_1(n_+)$	7.87 $b_1(\pi)$	10.59 $b_2(n_-)$	11.56 $a_2(\pi)$	STO-3G	

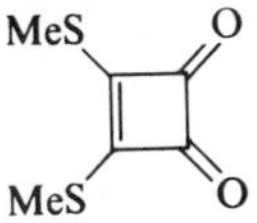

MeS O MeS O

$C_6H_6O_2S_2$ C_{2v}

I/eV:	8.18	8.87	10.02	10.60		91
$-\varepsilon_i$/eV:	8.91 $b_1(\pi)$	8.91 $a_1(n_+)$	10.04 $a_2(\pi)$	9.48 $b_2(n_-)$	MINDO/3	

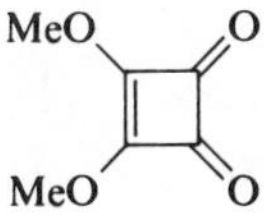

MeO O MeO O

$C_6H_6O_4$ C_{2v}

I/eV:	9.20	9.30	11.05		91
$-\varepsilon_i$/eV:	8.80 $b_1(\pi)$	9.02 $a_1(n_+)$	10.48 $b_2(n_-)$	MINDO/3	

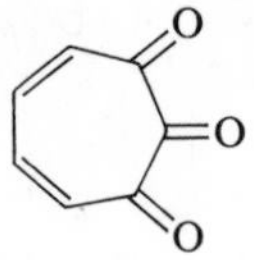

$C_7H_4O_3$ C_{2v}

I/eV:	9.55	10.16	10.82	11.8	12.4		25a, b
$-\varepsilon_i$/eV:	9.21 b_2(n)	10.07 $a_2(\pi)$	10.67 a_1(n)	11.59 b_2(n)	12.43 $b_1(\pi)$	MINDO/3	

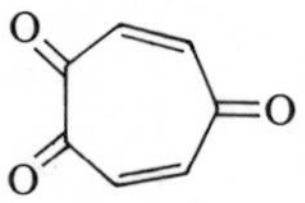

$C_7H_4O_3$ C_{2v}

I/eV:	9.98	10.42	11.1	11.3	11.6		25a, b
$-\varepsilon_i$/eV:	9.67 a_1(n)	10.24 b_2(n)	11.12 $b_1(\pi)$	11.16 $a_2(\pi)$	11.31 b_2(n)	MINDO/3	

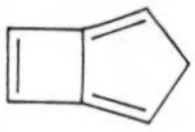

C_7H_6 C_{2v}

I/eV:	8.41 $b_1(\pi_1)$	8.93 $a_2(\pi_2)$	10.75 $a_1(\sigma)$	>12.0	101

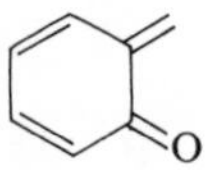

C_7H_6O C_s

I/eV:	8.80	9.37	10.63	12.02		92
$-\varepsilon_i$/eV:	8.56 π	9.40 n	10.78 π	11.85 π	PERTCI(CNDO/S)	

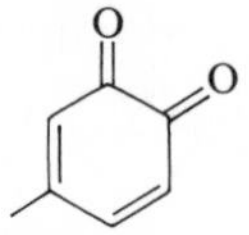

$C_7H_6O_2$ C_s

I/eV:	9.78 n	10.17 n	10.37 π	10.85 π	13.1		13
	(based on correlation with PBQ and 2,5-dimethyl-PBQ)						
$-\varepsilon_i$/eV:	9.8 $b_{3g}(n_-)$	10.3 $b_{2u}(n_+)$	10.4 $b_{1g}(\pi)$	10.8 $b_{3u}(\pi)$		HAM/3	22
	(symmetry of PBQ)						
I/eV:	9.40	9.72	10.70	11.65			88
$-\varepsilon_i$/eV:	9.12 a′ (n+)	9.72 a″ (π)	10.76 a′ (n−)	11.60 a″ (π)		MINDO/3	

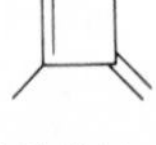

C_7H_8 C_s

I/eV:	8.34 a″ (π)	9.12 a″ (π)	11.1	104
	(supported by the results of MINDO/3)			

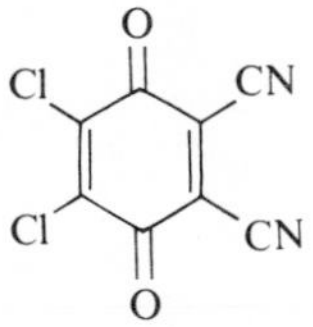

$C_8Cl_2N_2O_2$ C_{2v}

I/eV:	10.58 $b_2(n_-)$	10.76 $a_1(n_+)$	11.20 $b_1(\pi)$	11.4 $b_1(\pi)$	11.58	12.4	18
v/cm^{-1}:	1200/1450	1200			1350, 1750		

(correlative study; substituent effects)

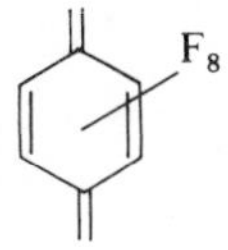

C_8F_8 D_{2h}

$-\varepsilon_i$/eV:	9.0 $b_{3u}(\pi)$	10.3 $b_{1g}(\pi)$	12.0 $b_{2g}(\pi)$	12.4 (π)	13.6 (σ)	ΔE_{CI}	69

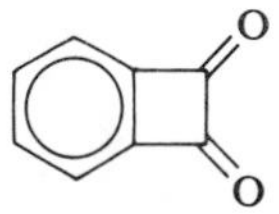

$C_8H_4O_2$ C_{2v}

I/eV:	9.23	10.14	10.43	11.23		91
$-\varepsilon_i$/eV:	9.03 $a_1(n_+)$	9.77 $b_1(\pi)$	9.93 $a_2(\pi)$	10.25 $b_2(n_-)$	MINDO/3	

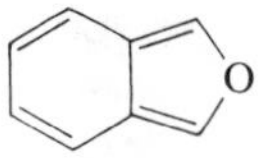

C_8H_6O C_{2v}

$-\varepsilon_i$/eV:	8.06 $a_2(\pi)$	10.89 $b_1(\pi)$	11.53 $a_2(\pi)$	14.58 $b_1(\pi)$	14.71 $b_2(\sigma)$	*ab initio*	84

(also given PE spectrum without values)

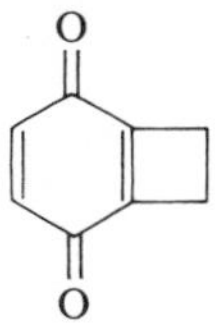

$C_8H_6O_2$ D_{2h}

I/eV:	9.64 n_-	9.9 n_+	10.1 π	10.78 π		107

(assignment based on correlation with PBQ and duroquinone)

$-\varepsilon_i$/eV:	9.30 $b_2(n_-)$	9.81 $a_1(n_+)$	10.05 $b_1(\pi)$	10.86 $b_1(\pi)$	MINDO/3	

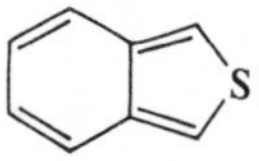

C_8H_6S C_{2v}

I/eV:	7.75 $a_2(\pi)$	8.9 $b_2(\pi)$	9.9 $a_2(\pi)$	11.33 $a_1(\sigma)$		81

(assignment based on band shapes and semiempirical calculations)

$-\varepsilon_i$/eV:	7.84 $a_2(\pi)$	9.69 $b_1(\pi)$	11.29 $a_2(\pi)$	13.16 $b_1(\pi)$	*ab initio*	84

(also given PE spectrum without values)

$C_8H_6S_2$ **C_{2v}**

						Method	Ref.
I/eV:	7.42	8.52	9.75	10.7			105
$-\varepsilon_i$/eV:	7.48 $b_1(\pi)$	8.43 $a_2(\pi)$	10.16 $b_1(\pi)$	11.07 $a_2(\pi)$		HMO	
	8.58 $b_1(\pi)$	9.30 $a_2(\pi)$	10.69 $b_1(\pi)$	11.39 $a_2(\pi)$		MNDO	
	7.84 $b_1(\pi)$	8.51 $a_2(\pi)$	9.64 $b_1(\pi)$	11.13 $a_2(\pi)$		MINDO/3	

C_8H_8 **C_{2v}**

						Method	Ref.
I/eV:	7.70 $a_2(\pi)$	9.6 $b_1(\pi)$	10.05 $a_2(\pi)$	10.49 $b_1(\pi)$	11.44 $a_1(\sigma)$		82b
	(*assignment according to the PERTCI calculation*)						82c

C_8H_8 **D_{2h}**

						Method	Ref.
I/eV:	7.87	9.7, 9.8	(12.5 ?)				53, 56, 66
$-\varepsilon_i$/eV:	7.98 $^2B_{3u}(\pi)$	-9.7 $^2B_{1g}$ & $^2B_{2g}(\pi)$		10.2–10.5 $^2B_{2g}(\pi)$		SR	53
$-\varepsilon_i$/eV:	7.94 $b_{1u}(\pi)$	9.8 $b_{3g}(\pi)$	10.0 $b_{2g}(\pi)$	12.86 b_{1u}		LCBO	56
I/eV:	7.87	9.7	11.1	12.2–12.8	13.6–14.3		67
$-\varepsilon_i$/eV:	8.18 $b_{1u}(\pi)$	9.94 $b_{3g}(\pi)$	10.73 $b_{2g}(\pi)$	11.76 $b_{1g}(\pi)$	12.70 $b_{1u}(\pi)$	MNDO	
$-\varepsilon_i$/eV:	7.85 $b_{3u}(\pi)$	9.51 $b_{1g}(\pi)$	10.30 $b_{2g}(\pi)$	10.41 $b_{2g}(\pi)$	11.58 (σ)	HAM/3	69
	7.92 $b_{3u}(\pi)$	9.41 $b_{1g}(\pi)$	9.70 $b_{2g}(\pi)$	11.02 $b_{2g}(\pi)$	11.37	ΔE_{CI}	
$-\varepsilon_i$/eV:	9.02 $b_{3u}(\pi)$	11.58 $b_{1g}(\pi)$	12.74 $b_{2g}(\pi)$	13.10 $b_{3g}(\sigma)$	13.28 $a_g(\sigma)$	CNDO/S	70
	7.90 $b_{3u}(\pi)^{-1}$	9.92 $b_{1g}(\pi)^{-1}$	9.96 $b_{2g}(\pi)^{-1}$	11.19 $b_{3g}(\sigma)^{-1}$	11.33 $a_g(\sigma)^{-1}$	CNDO/S–CI	
$-\varepsilon_i$/eV:	8.18 $b_{3u}(\pi)$	10.54 $b_{1g}(\pi)$	11.74 $b_{2g}(\pi)$	12.11 $b_{3g}(\pi)$	12.28 $a_{1g}(\pi)$	CNDO/S	71
I/eV:	7.87	9.70	-9.8	11.10	-12.5		57, 72
$-\varepsilon_i$/eV:	7.87 $b_{3u}(\pi)$	9.89 $b_{1g}(\pi)$	9.93 $b_{2g}(\pi)$	11.16 $b_{3g}(\sigma)$		CNDO/S	72
$-\varepsilon_i$/eV:	7.12 $b_{3u}(\pi)$	9.27 $b_{1g}(\pi)$	9.50 $b_{2g}(\pi)$	11.00 $b_{3g}(\sigma)$	12.0 a_g	ΔE-CI	76
	7.35 $b_{3u}(\pi)$	10.10 $b_{1g}(\pi)$	11.12 $b_{1g}(\pi)$	12.93 $b_{3g}(\sigma)$	14.28 a_g	*ab initio*	
	6.78 $b_{3u}(\pi)$	9.10 $b_{1g}(\pi)$	10.51 $b_{2g}(\pi)$	12.08 $b_{3g}(\sigma)$	13.4 a_g	ΔE-SCF	

$C_8H_8O_2$ **C_{2h}**

						Method	Ref.
I/eV:	9.60 $a_g(n_-)$	10.05 $b_g(\pi)$, $b_u(n_+)$		10.51 $a_u(\pi)$			18
ν/cm^{-1}:	1500			1600			
	(*correlative study; substituent effects*)						
I/eV:	9.58 n	10.06 n	10.58 π	13.0 π			22
$-\varepsilon_i$/eV:	9.7 $b_{3g}(n_-)$	10.2 $b_{2u}(n_+)$	10.1 $b_{1g}(\pi)$	10.4 $b_{3u}(\pi)$		HAM/3	
	(*symmetry of PBQ*)						

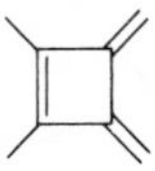

C_8H_{10} C_{2v}

I/eV:	8.05 $b_1(\pi)$	8.87 $a_2(\pi)$	10.75				104

(supported by the results of MINDO/3)

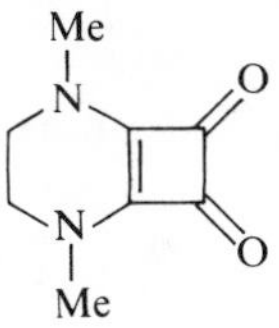

$C_8H_{10}N_2O_2$ C_{2v}

I/eV:	7.68	8.48	10.33	11.08			91
$-\varepsilon_i$/eV:	7.64 $b_1(\pi)$	9.07 $a_1(n_+)$	10.49 $b_2(n_-)$	10.21 $a_2(\pi)$		MINDO/3	

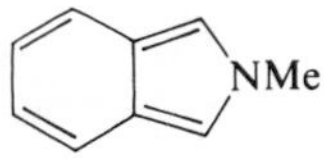

C_9H_9N C_s

$-\varepsilon_i$/eV:	7.14 a″ (π)	9.26 a″ (π)	10.55 a″ (π)	12.71 a′ (σ)	13.16 a″ (π)	*ab initio*	84

(also given PE spectrum without values)

I/eV:	7.12 $a_2(\pi)$	8.35 $b_1(\pi)$	9.42 $a_2(\pi)$	11.0 (σ)			85

(assignment based on HMO and PPP-SCF-CI)

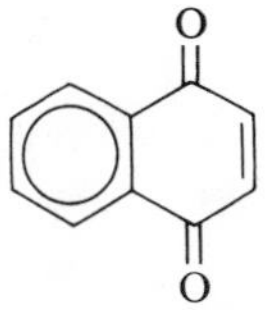

$C_{10}H_6O_2$ C_{2v}

I/eV:	9.49	9.62	9.82	10.53			14
$-\varepsilon_i$/eV:	9.94 $a_2(\pi_1)$	9.98 $b_1(\pi_2)$	10.25 $b_2(n_-)$	10.65 $b_1(\pi_3)$	10.83 $a_1(n_+)$	CNDO/S	
	9.21 $b_2(n_-)$	9.48 $a_2(\pi_1)$	9.62 $b_1(\pi_2)$	9.73 $a_1(n_+)$	10.42 $b_1(\pi_3)$	CNDO/S–CI	

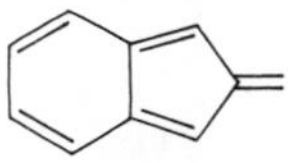

$C_{10}H_8$ C_{2v}

I/eV:	7.32 $a_2(\pi)$	9.10 $b_1(\pi)$	9.90 $a_2(\pi)$				102

(based on LNDO/S–PERTCI)

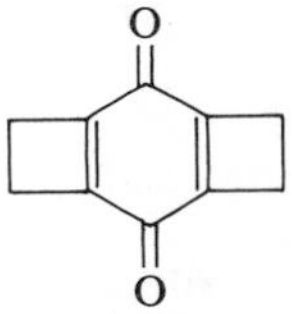

$C_{10}H_8O_2$ D_{2h}

I/eV:	9.4 n_-	9.5 π	9.7 n_+	10.3 π	10.5		107

(assignment based on correlation with PBQ and duroquinone)

$-\varepsilon_i$/eV:	9.01 $b_{3g}(n_-)$	9.54 $b_{2u}(n_+)$	9.84 $b_{1g}(\pi)$	10.15 $b_{3u}(\pi)$		MINDO/3	

$C_{10}H_{12}$ C_{2h}

I/eV:	7.58 $a_u(\pi)$	8.97 $b_g(\pi)$	9.80 $b_g(\pi)$	10.91 $a_u(\pi)$	−12		66
	(assignment based on SR method)						
I/eV:	7.58	8.97 (9.80)	10.70	11.7–12.5	13–14		67
$-\varepsilon_i$/eV:	8.18 $a_2(\pi)$	9.76 $b_2(\pi)$	10.69 $b_2(\pi)$	11.71 $a_1(\sigma)$	12.27 $a_2(\pi)$	MNDO	
$-\varepsilon_i$/eV:	7.77	8.91	9.62	10.78	11.09	ΔE_{CI}	69
I/eV:	7.58	8.97	9.80	10.91			72
$-\varepsilon_i$/eV:	7.58 $a_2(\pi)$	9.17 $b_2(\pi)$	9.57 $b_2(\pi)$	10.6 $a_1(\sigma)$	11.15	CNDO/S–CI	

$C_{10}H_{12}O_2$ D_{2h}

I/eV:	9.25 $b_{3g}(n_-)$	9.50 $b_{1g}(\pi)$	9.75 $b_{2u}(n_+)$	10.02 $b_{3u}(\pi)$			18
ν/cm^{-1}:				1600			
	(correlative study; substituent effects)						
$-\varepsilon_i$/eV:	9.5 $b_{3g}(n_-)$	9.7 $b_{1g}(\pi)$	10.0 $b_{2u}(n_+)$	10.0 $b_{3u}(\pi)$		HAM/3	22
$-\varepsilon_i$/eV:	9.35 (n_-)	9.53 (π)	9.83 (π)	10.10 (n_+)		CNDO/F	23

$C_{11}H_8$ C_{2v}

I/eV:	7.25 $a_2(\pi)$	8.83 $b_1(\pi)$	9.63–9.89 $b_2(\pi)$, $a_2(\pi)$		11.02 $a_1(\sigma)$		98
	(based on CNDO/S PERTCI)						

$C_{12}H_4N_4$ D_{2h}

I/eV:	9.61	10–12	12.41	12.68	12.90		57, 58
$-\varepsilon_i$/eV:	9.01 $b_{1u}(\pi)$	10.44 $b_{3g}(\pi)$	11.0 $b_{2g}(\pi)$	11.85 $b_{1g}(\sigma)$	11.95 $b_{2u}(\sigma)$	HAM/3	57
$-\varepsilon_i$/eV:	9.95 $b_{1u}(\pi)$	15.31 $b_{3g}(\pi)$	15.86 $b_{3u}(\sigma)$	15.88 $b_{2u}(\pi)$	16.05 $b_{2g}(\pi)$	CNDO/2	58
$-\varepsilon_i$/eV:	8.94 $b_{1u}(\pi)$	11.69 $b_{3g}(\pi)$	12.14 $b_{2g}(\pi)$	13.66 $b_{2u}(\pi)$	13.65 $b_{1g}(\pi)$	*ab initio*	60
$-\varepsilon_i$/eV:	9.30 $b_{1u}(\pi)$	9.7 $b_{3g}(\pi)$	11.2 $b_{2g}(\pi)$	12.4 $b_{1u}(\pi)$	12.8 $b_{1g}(\sigma)$	X_α	61
$-\varepsilon_i$/eV:	10.06 $b_{1u}(\pi)$	13.90 $b_{1g}(\sigma)$	14.82 $a_g(\sigma)$	15.37 $b_{2g}(\pi)$	15.89 $b_{2u}(\sigma)$	CNDO/2	62
	9.26 $b_{1u}(\pi)$	10.44 $b_{1g}(\sigma)$	10.75 $b_{2u}(\sigma)$	10.81 $a_g(\sigma)$	11.22 $b_{3u}(\sigma)$	MINDO/2	62
$-\varepsilon_i$/eV:	9.52 $b_{1u}(\pi)$	12.06 $b_{3g}(\pi)$	12.53 $b_{2g}(\pi)$	14.11 $b_{1g}(\sigma)$	14.14 $b_{2u}(\sigma)$	*ab initio*	63
	8.90 b_{1u}	10.81 b_{3g}	12.04 b_{2g}	13.73 b_{1g}	13.74 b_{2u}	ΔSCF	
$-\varepsilon_i$/eV:	7.85 $b_{1u}(\pi)$	10.32 $b_{3g}(\pi)$	10.51 $b_{2g}(\pi)$	11.12 $b_{1g}(\sigma)$	11.33 $b_{2u}(\sigma)$	INDO/S	65
	9.54 $b_{1u}(\pi)$	11.80 $b_{3g}(\pi)$	11.88 $b_{2g}(\pi)$	13.03 $b_{1g}(\sigma)$	13.40 $a_g(\sigma)$	CNDO/S2	64
	9.65 $b_{1u}(\pi)$	13.28 $b_{1g}(\pi)$	14.14 $a_g(\sigma)$	14.49 $b_{2u}(\sigma)$	14.95 $b_{2g}(\pi)$	INDO	65
I/eV:	9.75	11.4	11.4	$>$12.50			72
	9.75 $b_{3u}(\pi)$	11.44 $b_{2g}(\pi)$	11.77 $b_{1g}(\pi)$	12.66 $b_{3g}(\sigma)$	12.84 $b_{2g}(\pi)$	CNDO/S	

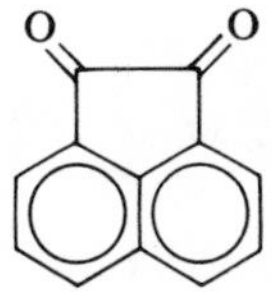

$C_{12}H_6O_2$ C_{2v}

I/eV: 8.77 9.53 10.60 11.48 108
$-\varepsilon_i$/eV: 8.69 $a_2(\pi)$ & n_+ 9.48 $b_1(\pi)$ 10.44 $b_1(\pi)$&n_- PPP
(assignment of n ionizations based on correlation with α-dicarbonyls)

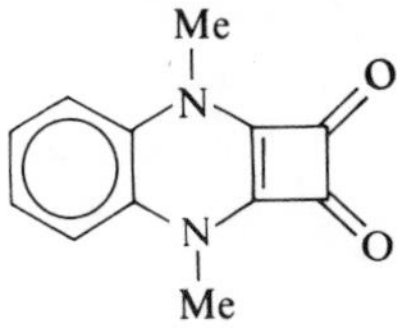

$C_{12}H_{10}N_2O_2$ C_{2v}

I/eV:	7.13	8.53	9.31	9.86	10.23		91
$-\varepsilon_i$/eV:	7.02 $b_1(\pi)$	9.06 $a_1(n_+)$	9.22 $a_2(\pi)$	9.76 $b_1(\pi)$	10.16 $b_2(n_-)$	MINDO/3	

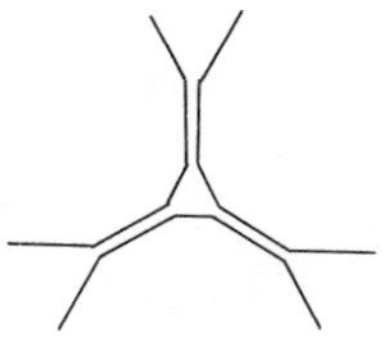

$C_{12}H_{18}$ D_{3h}

I/eV: 7.49 $e''(\pi)$ 9.71, 10.45 e' (Walsh) 11.50 $a_2''(\pi)$ 83
(assignment based on MINDO/3)

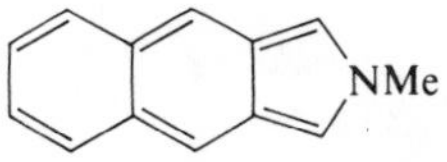

NMe $C_{13}H_{11}N$ C_{2v}

I/eV: 6.57 $a_2(\pi)$ 8.13 $b_1(\pi)$ 8.55 $a_2(\pi)$ 9.74 $a_2(\pi)$ 10.2 $b_1(\pi)$
(assignment based on HMO and PPP SCF CI) 85

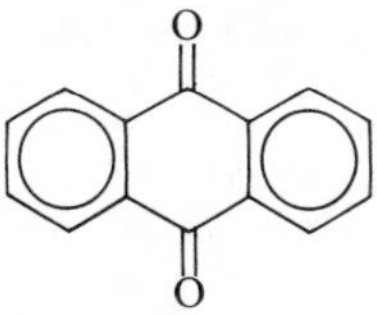

$C_{14}H_8O_2$ D_{2h}

I/eV: 9.34 1c
I/eV: 9.25 9.63 14
$-\varepsilon_i$/eV: 9.81 ($b_{2g}(\pi_1)$: $a_u(\pi_2)$: $b_{3u}(\pi_3)$: $b_{3g}(n_-)$: $b_{1g}(\pi_4)$) 10.63 $b_{2u}(n_+)$ CNDO/S
9.19 ($b_{3g}(n_-)$: $b_{2u}(n_+)$: $b_{2g}(\pi_1)$:$a_u(\pi_2)$) 9.76 ($b_{1g}(\pi_4)$: $b_{3u}(\pi_3)$) CNDO/S–CI

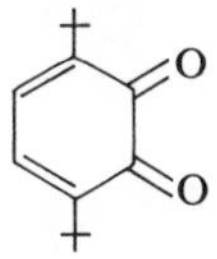

$C_{14}H_{20}O_2$ C_{2v}

I/eV:	8.71	8.93	10.03	11.34		88
$-\varepsilon_i$/eV:	8.93 $a_1(n_+)$	9.33 $a_2(\pi)$	10.69 $b_2(n_-)$	11.66 $b_1(\pi)$	MINDO/3	

			$C_{14}H_{20}O_2$		C_s		
I/eV:	8.81	9.02	10.01	11.03			88
$-\varepsilon_i$/eV:	8.97 a′ (n_+)	9.47 a″ (π)	10.61 a′ (n_-)	11.50 a″ (π)		MINDO/3	
			$C_{16}H_{12}$		D_{2h}		
I/eV:	7.95	8.66	9.1		10.7		56
	7.87 $b_{1u}(\pi)$	8.62 $b_{2g}(\pi)$	8.99 $b_{3g}(\pi)$	9.17 $a_u(\pi)$	10.69 $b_{1u}(\pi)$		
I/eV:	8.0	8.7	9.1	10.7			67
	planar						
$-\varepsilon_i$/eV:	8.27 $b_{1u}(\pi)$	9.0 $b_{2g}(\pi)$	9.28 $b_{3g}(\pi)$	9.46 $a_{1u}(\pi)$	10.91 $b_{1u}(\pi)$	MNDO	
	non-planar						
	8.44 $b_{1u}(\pi)$	9.05 $b_{2g}(\pi)$	9.29 $b_{3g}(\pi)$	9.38 $a_{1u}(\pi)$	10.81 $b_{1u}(\pi)$	MNDO	
			$C_{17}H_{13}N$		C_{2v}		
I/eV:	7.15 $a_2(\pi)$	7.61 $b_1(\pi)$	8.14 $a_2(\pi)$	9.20 $a_2(\pi)$	9.35 $b_1(\pi)$		85
	(assignment based on HMO and PPP-SCF-CI)						

XI. REFERENCES

1. S. Patai (Ed.), The Chemistry of Quinonoid Compounds, Vol. 1, Wiley, Chichester, 1974.
2. (a) F. I. Vilesov and A. N. Terenin, *Dokl. Akad. Nauk SSSR*, **115**, 744 (1957); (b) F. I. Vilesov, *Dokl. Akad. Nauk SSSR*, **132**, 632 (1960); (c) F. I. Vilesov, *Soviet Physics*, **6**, 888 (1964).
3. (a) M. D. Newton, F. P. DeBoer and W. N. Lipscomb, *J. Am. Chem. Soc.*, **88**, 2367 (1966); (b) C. Aussems, S. Jaspers, G. Leroy and F. Van Remoortere, *Bull. Soc. Chim. Belges*, **78**, 487 (1969).
4. M. J. S. Dewar and S. D. Worley, *J. Chem. Phys.*, **50**, 654 (1969).
5. D. W. Turner, C. Baker, A. D. Baker and C. R. Brundle, *Molecular Photoelectron Spectroscopy*, Wiley, London, 1970, Fig. 9.15 and p. 252.
6. J. R. Swenson and R. Hoffmann, *Helv. Chim. Acta*, **53**, 2331 (1970).
7. D. O. Cowan, R. Gleiter, J. A. Hashmall, E. Heilbronner and V. Hornung, *Angew. Chem.*, **83**, 405 (1971); *Angew. Chem. Int. Ed.*, **10**, 401 (1971).
8. H. P. Trommsdorf, *J. Chem. Phys.*, **56**, 5358 (1972).
9. C. R. Brundle, M. B. Robin and N. A. Kuebler, *J. Am. Chem. Soc.*, **94**, 1466 (1972).
10. P. E. Stevenson, *J. Phys. Chem.*, **76**, 2424 (1972).
11. J. T. Gleghorn and F. W. McConkey, *J. Mol. Struct.*, **18**, 219 (1973).
12. G. Hojer, S. Meza and M. E. Ruiz, *Acta Chem. Scand.*, **27**, 1860 (1973).
13. T. Kobayashi, *J. Electron Spectrosc.*, **7**, 349 (1975).
14. G. Lauer, W. Schäfer and A. Schweig, *Chem. Phys. Lett.*, **33**, 312 (1975).
15. (a) H. T. Jonkman, G. A. van der Velde, W. C. Nieuwport, *Proceedings of S. R. C. Atlas Symp. No.* 4, 1974; (b) M. H. Wood, *Theor. Chim. Acta*, **36**, 345 (1975).
16. M. F. Merienne-Lafore and H. P. Trommsdorf, *J. Chem. Phys.*, **64**, 3791 (1976).

17. N. J. Bunce, J. E. Ridley and M. C. Zerner, *Theor. Chim. Acta*, **45**, 283 (1977).
18. D. Dougherty and S. P. McGlynn, *J. Am. Chem. Soc.*, **99**, 3234 (1977).
19. R. W. Bigelow, *J. Chem. Phys.*, **68**, 5086 (1978).
20. J. Goodman and L. E. Brus, *J. Chem. Phys.*, **69**, 1604 (1978).
21. J. E. Bloor, R. A. Paysen and R. E. Sherrod, *Chem. Phys. Lett.*, **60**, 476 (1979).
22. L. Åsbrink, G. Bieri, C. Fridh, E. Lindholm and D. P. Chong, *Chem. Phys.*, **43**, 189 (1979).
23. H. Bock, W. Kaim, P. L. Timms and P. Hawker, *Chem. Ber.*, **113**, 3196 (1980).
24. P. Jacques, J. Faure, O. Chalvet and H. H. Jaffe, *J. Phys. Chem.*, **85**, 473 (1981).
25. (a) R. Gleiter, W. Dobler and M. Eckert-Maksić, *Nouv. J. Chim.*, **6**, 123 (1982); (b) R. Gleiter, W. Dobler and M. Eckert-Maksić, *Angew. Chem.*, **94**, 62 (1982).
26. H. Bock, S. Mohamand, T. Hirabayashi, G. Maier and H. P. Reisenauer, *Chem. Ber.*, **116**, 273 (1983).
27. T.-K. Ha, *Mol. Phys.*, **49**, 1471 (1983).
28. W. Hug, J. Kuhn, K. J. Seibold, H. Labhart and G. Wagniere, *Helv. Chim. Acta*, **54**, 1451 (1971).
29. R. M. Hochstrasser, L. W. Johnson and H. P. Trommsdorf, *Chem. Phys. Lett.*, **21**, 251 (1973).
30. H. Veenvliet and D. A. Wiersma, *Chem. Phys.*, **2**, 69 (1973).
31. H. Veenvliet and D. A. Wiersma, *Chem. Phys.*, **8**, 432 (1975).
32. D. Cooper, W. T. Naff and R. N. Compton, *J. Chem. Phys.*, **65**, 2752 (1975).
33. G. Ter Horst and J. Kommandeur, *Chem. Phys.*, **44**, 287 (1979).
34. R. L. Martin and W. R. Wadt, *J. Phys. Chem.*, **86**, 2382 (1982).
35. T. Koopmans, *Physica*, **1**, 104 (1933).
36. K. Wittel and S. P. McGlynn, *Chem. Rev.*, **77**, 745 (1977).
37. L. Klasinc, H. Güsten and S. P. McGlynn, *Mathematics and Computational Concepts in Chemistry* (Ed. N. Trinajstić), Ellis Horwood Ltd, Chichester, 1985, Chapter 15, pp. 155–170.
38. L. S. Cederbaum, W. Domcke, J. Schirmer and W. von Niessen, *Phys. Scripta*, **21**, 481 (1980); W. Domcke, H. Köppel and L. S. Cederbaum, *Mol. Phys.*, **43**, 851 (1981); L. S. Cederbaum, H. Köppel and W. Domcke, *Int. J. Quant. Chem. Symp.*, **15**, 251 (1981); H. Köppel, L. S. Cederbaum and W. Domcke, *J. Chem. Phys.*, **77**, 2014 (1982); L. S. Cederbaum, W. Domcke, J. Schirmer and W. von Niessen, *Adv. Chem. Phys.*, **65**, 115 (1986).
39. For determination of EA of quinones see e.g. A. L. Farragher and F. M. Page, *Trans. Faraday Soc.*, **62**, 3072 (1966); E. P. Grimsrud, G. Caldwell, S. Chowdhury and P. Kebarle, *J. Am. Chem. Soc.*, **107**, 4627 (1985).
40. P. C. Hiberty and P. Karafiloglou, *Theor. Chim. Acta*, **61**, 171 (1982).
41. M. T. Smith, private communication; B. Pullman, *Int. J. Quant. Chem.*, **QBS13**, 95 (1986).
42. See e.g. N. Trinajstić, *Chemical Graph Theory*, CRC, Boca Raton 1983; J. V. Knop, K. Szymanski, N. Trinajstić and P. Krivka, *Computers and Mathematics*, **10**, 369 (1984); J. V. Knop, K. Szymanski, L. Klasinc and N. Trinajstić, *Computers and Chemistry*, **8**, 107 (1984); J. V. Knop, K. Szymanski, Z. Jeričević and N. Trinajstić, *J. Comp. Chem.*, **4**, 23 (1983); J. V. Knop, V. R. Muller, K. Szymanski and N. Trinajstić, *Computer Generation of Certain Classes of Molecules*, SKTH, Zagreb, 1985.
43. L. Klasinc, S. P. McGlynn and N. Trinajstić, paper in preparation.
44. U. Gelius, P. F. Heden, J. Hedman, B. J. Lindberg, R. Manne, R. Nordberg, C. Nordling and K. Siegbahn, *Phys. Scripta*, **2**, 70 (1970).
45. T. Ohta, M. Yamada and H. Kuroda, *Bull. Chem. Soc. Japan*, **47**, 1158 (1974).
46. U. Gelius, *Phys. Scripta*, **9**, 133 (1974).
47. Z. B. Maksić and K. Rupnik, *Z. Naturforsch.*, **35a**, 988 (1980); Z. B. Maksić, K. Rupnik and M. Eckert-Maksić, *J. Electron Spectrosc.*, **16**, 371 (1979); Z. B. Maksić and K. Rupnik, *Nouv. J. Chim.*, **5**, 515 (1981); Z. B. Maksić, K. Rupnik and N. Mileusnić, *J. Organomet. Chem.*, **219**, 21 (1981); Z. B. Maksić, K. Rupnik and A. Veseli, *J. Electron Spectrosc.*, **32**, 163 (1983).
48. See, e.g. K. Kimura, S. Katsumata, Y. Achiba, T. Yamazaki and S. Iwata, *Handbook of HeI Photoelectron Spectra of Fundamental Organic Molecules*, Halsted Press, New York, 1981.
49. H. Jensen and P. N. Skancke, *Acta Chem. Scand.*, **22**, 2899 (1968).
50. T. M. Dunn and A. H. Francis, *J. Mol. Spectrosc.*, **50**, 14 (1974).
51. J. E. Ridley and M. C. Zerner, *Theor. Chim. Acta*, **32**, 111 (1973).
52. A. Pullman, G. Berthier and B. Pullman, *Bull. Soc. Chim. France*, **15**, 450 (1948). A. Pullman, B. Pullman, E. D. Bergmann, G. Berthier, E. Fischer, D. Ginsburg and Y. Hirshberg, *Bull. Soc. Chim. France*, **18**, 707 (1951).
53. T. Koenig, R. Wielesek, W. Snell and T. Balle, *J. Am. Chem. Soc.*, **97**, 3225 (1975).

54. S. Pignataro, V. Mancini, J. N. Ridyard and H. J. Lempka, *Chem. Commun.*, 142 (1971).
55. R. Boschi, E. Clar and W. Schmidt, *Angew. Chem. Int. Ed.*, **12**, 402 (1973).
56. M. Allan, E. Heilbronner and G. Kaupp, *Helv. Chim. Acta*, **59**, 1949 (1976).
57. L. Åsbrink, C. Fridh and E. Lindholm, *Int. J. Quant. Chem.*, **13**, 331 (1978).
58. I. Ikemoto, K. Samizo, T. Fujikawa, K. Ishii, T. Ohta and H. Kuroda, *Chem. Lett.*, 785 (1974).
59. W. D. Grobman, R. A. Pollak, D. E. Eastman, E. T. Maas Jr and B. A. Scott, *Phys. Rev. Lett.*, **32**, 534 (1974).
60. H. T. Jonkman, G. A. van der Velde and W. C. Nieuwpoort, *Chem. Phys. Lett.*, **25**, 62 (1974).
61. F. Herman, A. R. Williams and K. H. Johnson, *Phys. Rev. Lett.*, **33**, 94 (1974); F. Herman and I. P. Batra, *Nuovo Cimento*, **23**, 182 (1974).
62. J. Ladik, A. Karpfen, G. Stollhoff and P. Fulde, *Chem. Phys.*, **7**, 267 (1975).
63. H. Johansen, *Int. J. Quant. Chem.*, **9**, 459 (1975).
64. N. O. Lipari, P. Nielsen, J. J. Ritsko, A. J. Epstein and D. J. Sandman, *Phys. Rev.*, **B14**, 2229 (1976).
65. K. Krogh-Jespersen and M. A. Ratner, *Theor. Chim. Acta*, **47**, 283 (1978).
66. T. Koenig and S. Southworth, *J. Am. Chem. Soc.*, **99**, 2807 (1977).
67. M. J. S. Dewar, *J. Am. Chem. Soc.*, **104**, 1447 (1982).
68. L. S. Cederbaum, W. Domcke, J. Schirmer, W. von Niessen, G. H. F. Diercksen and W. P. Kraemer, *J. Chem. Phys.*, **69**, 1591 (1978); W. von Niessen, G. Bieri, J. Schirmer and L. S. Cederbaum, *Chem. Phys.*, **65**, 157 (1982); L. S. Cederbaum, W. Domcke, J. Schirmer and W. von Niessen, *Phys. Scripta*, **21**, 481 (1980); E. Haselbach, U. Klemm, R. Gschwind, T. Bally, L. Chassot and S. Nitsche, *Helv. Chim. Acta*, **65**, 2464 (1982); W. von Niessen, G. Bieri, J. Schirmer and L. S. Cederbaum, *Chem. Phys.*, **65**, 157 (1982); G. Kluge and M. Scholz, *Int. J. Quant. Chem.*, **20**, 669 (1981).
69. T. Koenig, C. E. Klopfenstein, S. Southworth, J. A. Hoobler, R. A. Wielesek, T. Balle, W. Snell and D. Imre, *J. Am. Chem. Soc.*, **105**, 2256 (1983).
70. R. W. Bigelow, *Chem. Phys.*, **80**, 45 (1983).
71. R. W. Bigelow, *J. Mol. Struct.* (*Theochem.*), **94**, 391 (1983).
72. R. W. Bigelow, *Chem. Phys. Lett.*, **100**, 445 (1983).
73. R. Schulz, A. Schweig and W. Zittlau, *J. Am. Chem. Soc.*, **105**, 2980 (1983).
74. P. Forster, R. Gschwind, E. Haselbach, U. Klemm and J. Wirz, *Nouv. J. Chim.*, **4**, 365 (1980).
75. T. Bally, L. Neuhaus, S. Nitsche, E. Hasselbach, J. Janssen and W. Lüttke, *Helv. Chim. Acta*, **66**, 1288 (1983).
76. T.-K. Ha, *Theor. Chim. Acta*, **66**, 111 (1984).
77. For some relevant publications on this subject see: Refs 6, 7 and D. Chadwick, D. C. Frost and L. Weiler, *Tetrahedron Lett.*, 4543 (1971); W. R. Harshbarger, N. A. Kuebler and M. B. Robin, *J. Chem. Phys.*, **60**, 345 (1974); W.-C. Tam and C. E. Brion, *J. Electron. Spectrosc.*, **3**, 467 (1974); W. Schäfer, A. Schweig, G. Maier and T. Sayrac, *J. Am. Chem. Soc.*, **96**, 279 (1974); G. Hentrich, E. Gunkel and M. Klessinger, *J. Mol. Struct.*, **21**, 231 (1974); J. Kelder, H. Cerfontain, B. R. Higginson and D. R. Lloyd, *Tetrahedron Lett.*, 739 (1974); J. C. Bünzli, D. C. Frost and L. Weiler, *J. Am. Chem. Soc.*, **96**, 1952 (1974); J. L. Meeks and S. P. McGlynn, *Spectrosc. Lett.*, **8**, 439 (1975); S. P. McGlynn and J. L. Meeks, *J. Electron Spectrosc.*, **6**, 269 (1975); P. Bischof, R. Gleiter and P. Hofmann, *Helv. Chim. Acta*, **58**, 2130 (1975); J. L. Meeks, Y. F. Arnett, D. B. Larson and S. P. McGlynn, *J. Am. Chem. Soc.*, **97**, 3905 (1975); J. L. Meeks and S. P. McGlynn, *J. Am. Chem. Soc.*, **97**, 5079 (1975); J. L. Meeks, H. J. Maria, P. Brint and S. P. McGlynn, *Chem. Rev.*, **75**, 603 (1975); N. Bodor, M. J. S. Dewar and Z. B. Maksić, *Croat. Chem. Acta*, **48**, 9 (1976); S. P. McGlynn and J. L. Meeks, *J. Electron Spectrosc.*, **8**, 85 (1976); R. Hernandez, P. Masclet and G. Mouvier, *J. Electron Spectrosc.*, **10**, 333 (1977); D. C. Frost, C. A. McDowell, G. Pouzard and N. P. C. Westwood, *J. Electron Spectrosc.*, **10**, 273 (1977); D. Dougherty and S. P. McGlynn, *J. Chem. Phys.*, **67**, 1289 (1977); R. Gleiter, R. Bartetzko, P. Hofmann and H. D. Scharf, *Angew. Chem.*, **89**, 414 (1977); B. Kovač and L. Klasinc, *Croat. Chem. Acta*, **51**, 55 (1978); H.-D. Martin, H.-J. Schiwek, J. Spanget-Larsen and R. Gleiter, *Chem. Ber.*, **111**, 2557 (1978); P. Masclet and G. Mouvier, *J. Electron Spectrosc.*, **14**, 77 (1978); R. Bartetzko, R. Gleiter, J. L. Muthard and L. A. Paquette, *J. Am. Chem. Soc.*, **100**, 5589 (1978); D. Dougherty, P. Brint and S. P. McGlynn, *J. Am. Chem. Soc.*, **100**, 5597 (1978); D. Dougherty, R. L. Blankespoor and S. P. McGlynn, *J. Electron Spectrosc.*, **16**, 245 (1979); R. Gleiter, P. Hofmann, P. Schang and A. Sieber, *Tetrahedron*, **36**, 655 (1979); L. Klasinc, N. Trinajstić and J. V. Knop, *Int. J. Quantum Chem.*, **QBS7**, 403 (1980); S. Chattopadhyay, J. L. Meeks, G. L. Findley and S. P. McGlynn, *J. Phys. Chem.*, **85**, 968 (1981); G. Jähne and R. Gleiter, *Angew. Chem.*, **95**, 500 (1983); R. Gleiter and W. Dobler, *Chem. Ber.*, **118**,

1917 (1985); R. Gleiter, P. Schang, M. Bloch, E. Heilbronner, J.-C. Bünzli, D. C. Frost and L. Weiler, *Chem. Ber.*, **118**, 2127 (1985); V. Balaji, K. D. Jordan, R. Gleiter, G. Jähne and G. Müller, *J. Am. Chem. Soc.*, **107**, 7321 (1985).
78. E. Hedaya, *Acc. Chem. Res.* **2**, 367 (1969); F. P. Lossing and G. Zemeluk, *Can. J. Chem.*, **48**, 955 (1970); T. Koenig, T. Balle and W. Snell, *J. Am. Chem. Soc.*, **97**, 662 (1975); T. Koenig, R. Wielesek, W. Snell and T. Balle, *J. Am. Chem. Soc.*, **97**, 3225 (1975).
79. A. Schweig, H. Vermeer and U. Weidner, *Chem. Phys. Lett.*, **26**, 229 (1974); W. Schäfer and A. Schweig, *Z. Naturforsch.*, **30a**, 1785 (1975); C. Muller, W. Schäfer, A. Schweig, N. Thon and H. Vermeer, *J. Am. Chem. Soc.*, **98**, 5440 (1976).
80. E. Heilbronner, R. Gleiter, H. Hopf, V. Hornung and A. de Meijere, *Helv. Chim. Acta*, **54**, 783 (1971); L. Praud, P. Millie and G. Berthier, *Theor. Chim. Acta*, **11**, 169 (1968).
81. P. A. Clark, R. Gleiter and E. Heilbronner, *Tetrahedron*, **29**, 3087 (1973).
82. (a) C. R. Flynn and J. Michl, *J. Am. Chem. Soc.*, **96**, 3280 (1974); (b) J. Kreile, N. Münzel, R. Schulz and A. Schweig, *Chem. Phys. Lett.*, **108**, 609 (1984); (c) R. Schulz, A. Schweig and W. Zittlau, *J. Am. Chem. Soc.*, **105**, 2980 (1983).
83. T. Bally and E. Haselbach, *Helv. Chim. Acta*, **58**, 321 (1975).
84. M. H. Palmer and S. M. F. Kennedy, *J. Chem. Soc. Perkin Trans. 2*, 81 (1976).
85. W. Rettig and J. Wirz, *Helv. Chim. Acta*, **59**, 1054 (1976).
86. T. Koenig, M. Smith and W. Snell, *J. Am. Chem. Soc.*, **99**, 6663 (1977).
87. V. Eck, G. Lauer, A. Schweig, W. Thiel and H. Vermeer, *Z. Naturforsch.*, **33a**, 383 (1978).
88. P. Schang, R. Gleiter and A. Rieker, *Ber. Bunsenges. Phys. Chem.*, **82**, 629 (1978).
89. R. C. Bingham, M. J. S. Dewar and D. H. Lo, *J. Am. Chem. Soc.*, **97**, 1285 (1975).
90. W. J. Hehre, R. F. Stewart and J. A. Pople, *J. Chem. Phys.*, **51**, 2657 (1969).
91. R. Gleiter, P. Schang and G. Seitz, *Chem. Phys. Lett.*, **55**, 144 (1978).
92. V. Eck, A. Schweig and H. Vermeer, *Tetrahedron Lett.*, 2433 (1978).
93. R. Schulz and A. Schweig, *Tetrahedron Lett.*, 59 (1979).
94. M. J. S. Dewar and W. Thiel, *J. Am. Chem. Soc.*, **99**, 4899 (1977); M. J. S. Dewar, M. L. McKee and H. S. Rzepa, *J. Am. Chem. Soc.*, **100**, 3607 (1978).
95. R. Schulz and A. Schweig, *Angew. Chem.*, **91**, 737 (1979); *Angew. Chem. Int. Ed.*, **18**, 692 (1979).
96. R. Schulz and A. Schweig, *Tetrahedron Lett.*, 343 (1980).
97. R. Schulz and A. Schweig, *Angew. Chem.*, **93**, 603 (1981); *Angew. Chem. Int. Ed.*, **20**, 570 (1981).
98. R. Schulz, A. Schweig, C. Wentrup and H.-W. Winter, *Angew. Chem.*, **92**, 846 (1980); *Angew. Chem. Int. Ed.*, **19**, 821 (1980).
99. M. Breitenstein, R. Schulz and A. Schweig, *J. Org. Chem.*, **47**, 1979 (1982).
100. R. Schulz, A. Schweig, K. Hartke and J. Köster, *J. Am. Chem. Soc.*, **105**, 4519 (1983).
101. C. Müller, A. Schweig, W. Thiel, W. Grahn, R. G. Bergman and K. P. C. Vollhardt, *J. Am. Chem. Soc.*, **101**, 5579 (1979).
102. G. Gross, R. Schulz, A. Schweig and C. Wentrup, *Angew. Chem. Int. Ed.*, **20**, 1021 (1981).
103. R. Schulz, A. Schweig and W. Zittlau, *J. Mol. Struct.* (*Theochem.*), **121**, 115 (1985).
104. K. Gubernator, J. Spanget-Larsen, R. Gleiter and H. Hopf, *J. Org. Chem.*, **48**, 2097 (1983).
105. R. Gleiter, G. Krennrich, D. Cremer, K. Yamamoto and I. Murata, *J. Am. Chem. Soc.*, **107**, 6874 (1985).
106. R. Gleiter, W. Schäfer and H. Wamhoff, *J. Org. Chem.*, **50**, 4375 (1985).
107. R. Gleiter, G. Jähne, M. Oda and M. Iyoda, *J. Org. Chem.*, **50**, 678 (1985).
108. A. Bigotto, V. Galasso, G. Distefano and A. Modelli, *J. Chem. Soc. Perkin Trans. 2*, 1502 (1979).
109. N. Bhacca and L. Klasinc, *Z. Naturforsch.*, **40a**, 706 (1985).
110. L. Klasinc, B.Ruščić, N. S. Bhacca and S. P. McGlynn, *Int. J. Quant. Chem.*, **QBS 12**, 161 (1986).
111. L. T. Calcaterra, G. L. Closs and J. R. Miller, *J. Am. Chem. Soc.*, **105**, 670 (1983).
112. P. Passman, F. Rob and J. W. Verhoeven, *J. Am. Chem. Soc.*, **104**, 5127 (1982).
113. J. R. Miller, K. W. Hartman and S. Abrash, *J. Am. Chem. Soc.*, **104**, 4296 (1982).
114. P. Pasman, N. W. Koper and J. W. Verhoeven, *Recl. Trav. Chim. Pays Bas*, **101**, 363 (1982).
115. R. K. Huddleston and J. R. Miller, *J. Chem. Phys.*, **79**, 5337 (1983).
116. C. A. Stein, N. A. Lewis and G. Seitz, *J. Am. Chem. Soc.*, **104**, 2596 (1982).
117. A. D. Baker, R. Scharfman and C. A. Stein, *Tetrahedron Lett.*, **24**, 2957 (1983).
118. C. A. Stein, N. A. Lewis, G. Seitz and A. D. Baker, *Inorg. Chem.*, **22**, 1124 (1983).
119. M. N. Paddon-Row, *Acc. Chem. Res.*, **15**, 245 (1982).
120. M. N. Paddon-Row, H. K. Patney, R. S. Brown and K. N. Houk, *J. Am. Chem. Soc.*, **103**, 5575 (1981).
121. F. S. Jorgenson, M. N. Paddon-Row and H. K. Patney, *J. Chem. Soc. Chem. Commun.*, 573 (1983).

The Chemistry of Quinonoid Compounds, Vol. II
Edited by S. Patai and Z. Rappoport

CHAPTER 6

Photochromism and thermochromism in bianthrones and bianthrylidenes*

K. A MUSZKAT
Department of Structural Chemistry,
The Weizmann Institute of Science, Rehovot, Israel

*Dedicated to Professor Ernst Fischer, on the occasion of his 65th birthday, as a token of personal friendship and of admiration for his contributions to our understanding of reversible photochromic molecular transformations.

I. INTRODUCTION

Several striking colour phenomena are associated with bianthrone (**1**), the tetra benzo-annelated diphenoquinone (**1a**), and with several bianthrylidene (**2**, X = CH) related systems (e.g., X = CH_2, CHOH, O, S, and *N*-Me). These colour phenomena are largely

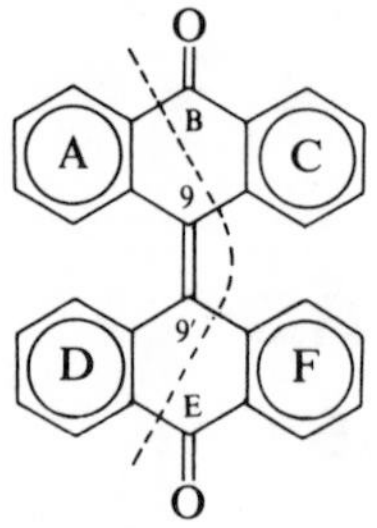

(**1**) 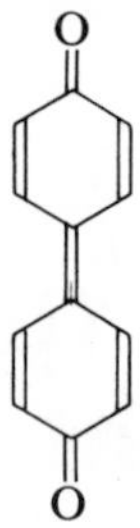 (**1a**) 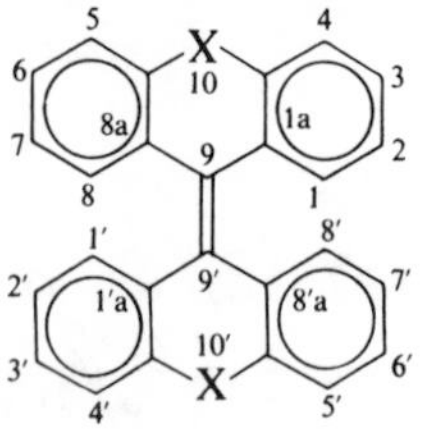 (**2**) (*E*, D_2) 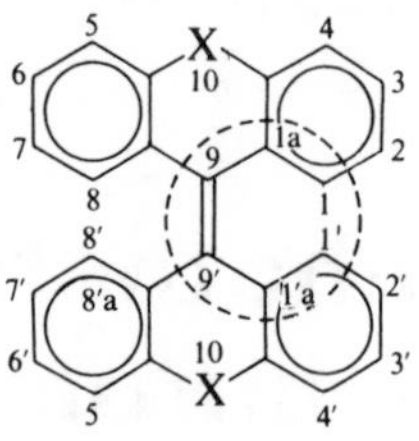 (**2**) (*Z*, C_2)

reversible, and can be induced by a variety of physical agents providing for different means of molecular excitation: light[1], exposure to higher temperature[2], application of high pressure[2] and ionizing radiation[3]. The first three of these effects are widely known as photochromism[1], thermochromism[2] and piezochromism[2], respectively, and form the subject of this chapter.

The explanation of these colour changes in terms of well accepted classical molecular structures has been a matter of considerable scientific challenge and much debate over a long period, since the original description of thermochromism and of piezochromism of bianthrone by Meyer in 1909[2]. Only in the last decade, after the advent of new and powerful methods of structure determination could significant progress be made in the definitive elucidation of the molecular structures responsible for these colour changes[4,5]. Photochromism and thermochromism, the two principal reversible colour changes, have been presently associated with definite but nevertheless labile molecular modifications formed in chemically reversible processes[5]. All these labile modifications have singlet electronic ground states[6].

As we shall have ample opportunity to see later in detail, many of the unusual physical, chemical and photochemical characteristics of bianthrones and of all the other members of the bianthrylidene series can be traced to their fundamental structural unit, the central 9,9′ double bond. Obviously this bond is very much different from the isolated double bond of ethylene, as it is significantly perturbed by the surrounding atomic network. In addition, the central double bond is part of a smaller *cis*-1,2-diphenyl ethylene unit (i.e. ring A, 9–9′ double bond, ring D) as shown in **1**, itself containing as subunit the appropriately substituted hexatriene (e.g., carbon atoms 1, 1a, 9, 9′, 1′a and 1′, as shown in **2**). In addition several of the physically realizable conformations of the 9,9′ double bond of these molecules can be and also are effectively constrained and stabilized by the molecular skeleton and by its internal non-bonded interactions, as we shall see below.

There are three main molecular transformations (see Scheme 1) giving rise to labile modifications in the bianthrylidene series. (The designation of the stable modifications as **A**, of the thermochromic as **B**, and of the photolabile as **C** is due to Kortüm[7,8].)

(1) Concerted photocyclization similar to that of *cis*-stilbene (**3**), to give 4a,4b-dihydrophenanthrene (**4**)[9,10], yielding the **C** photochromic modification (between centres 1–1′ or 1–8′ in **2** or **2′**)[7,8]. The absorption spectrum of one such photointermediate, the **C** modification of 1,1′,3,3′-tetramethylbianthrone is shown in Figure 1.

hv / hv,Δ

(3) (4)

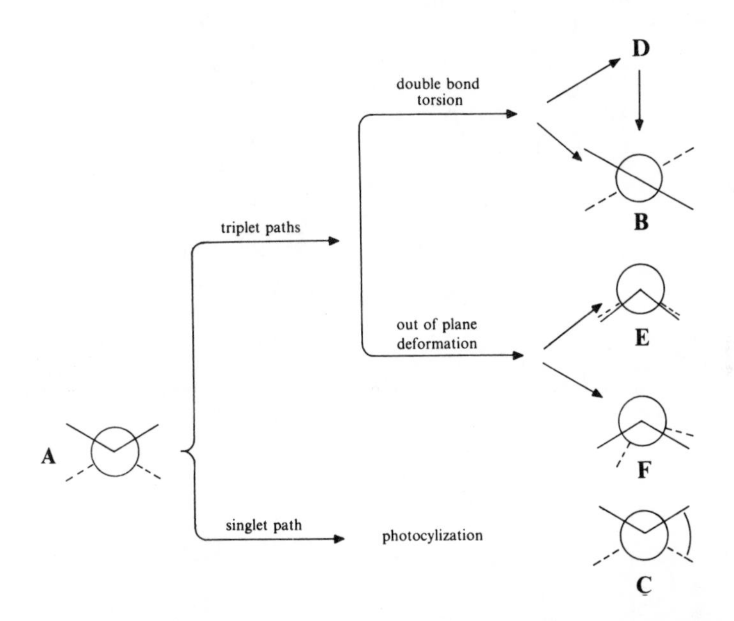

SCHEME 1: Reversible photoproducts of bianthrones and of related bianthrylidene-derived systems. Conformations (Newman projections along 9,9′ double bond) and photochemical paths. **A**, fundamental conformation; **B**, double bond twisted photo and thermo conformer; **D**, labile precursor of B; **E**, out of plane deformed photoconformer; **F**, same as **E**, but with double bond twist; **C**, 1,1′ or 1,8′-photocyclization product, photolabile

(2) Torsion about the central 9,9′ double bond, to give double bond conformers, as in the photochromic and thermochromic **B** modification, and its labile precursor **D**[5,11,12]. The absorption spectra of two such intermediates, the **B** and **D** modifications of 2,2′,4,4′-tetramethylbixanthylene[11] are given in Figure 2.

(3) Out of plane deformations about the four central single bonds between atom 9 and atoms 1a and 8a and between atom 9′ and atoms 1′a and 8′a (see e.g., **2** and **2′**) to give single bond conformers, as in the photochromic **E** and **F** modifications[12,13]. Absorption spectra of such forms in 10,10′-dimethylbiacridan and in 1,1′,10,10′-tetramethylbiacridan are shown in Figure 3.

The bianthrylidene atomic network (see, e.g., Figures 4 and 5) possesses a highly successful built-in molecular mechanism ensuring the freezing-in of labile (metastable) conformers reached thermally or photochemically. This is the strong 1–8′ and 1′–8 (in D_2 symmetry species; or 1–1′ and 8–8′ in C_2 species, cf. **2′** (*E*) and **2** (*Z*); here we shall limit ourselves to the D_2 case) non-bonded interaction coming into effect in all molecular

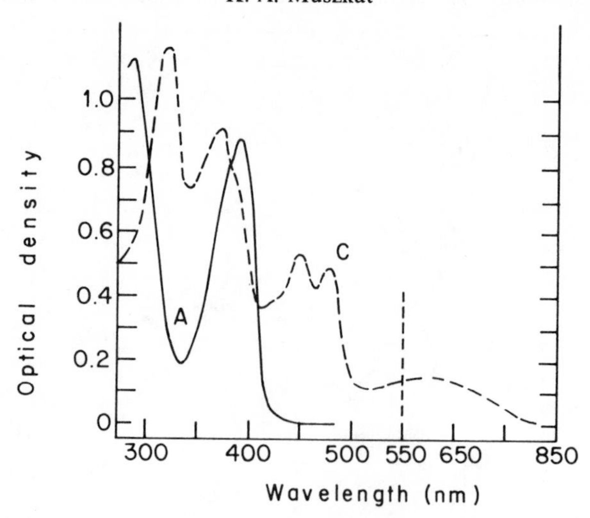

FIGURE 1. Absorption spectra of 1,1′,3,3′-tetramethylbianthrone in 2-methyltetrahydrofuran solution at 183 K[6]. Full curve A form; broken line, C form, obtained by 405 nm irradiation. This is a computed spectrum, as only partial conversion of **A** into **C** is realizable in practice. Reproduced by permission of the Weizmann Science Press of Israel from Ref. 6

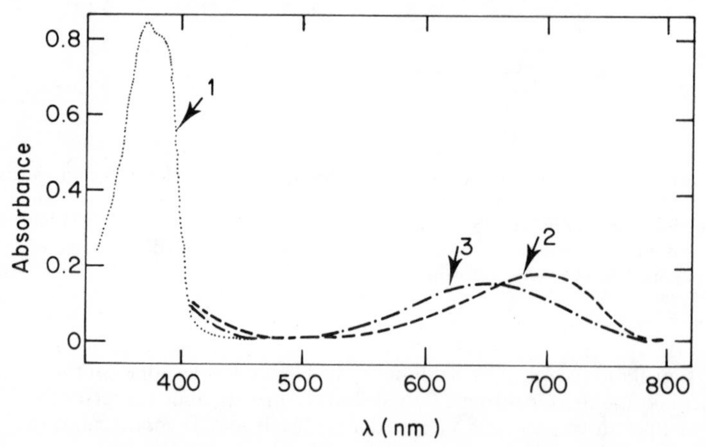

FIGURE 2. 2,2′,4,4′-Tetramethylbixanthylene in methylcyclohexane/2-methylpentane 1 : 1 solution, at 83 K[11]. Curve 1, A form; curve 2, D form, obtained by 365 nm irradiation. Curve 3, B modification, formed by thermal decay of D form. Reproduced by permission of Elsevier Sequoia SA from Ref. 11

structures in which coplanarity of the two anthryl units is attempted. This interaction forms the steep walls of the potential well which oppose the various vibrations of the 9–9′ bond along the reaction coordinates of the reversal processes leading to the fundamental modification (see Scheme 1). In addition to this kinetic factor, the 1–8′ and 1′–8 non-bonded interactions determine also the thermodynamics of the various conformers. Thus the exact structure of each form (for definition see caption to Scheme 1) is due to a delicate

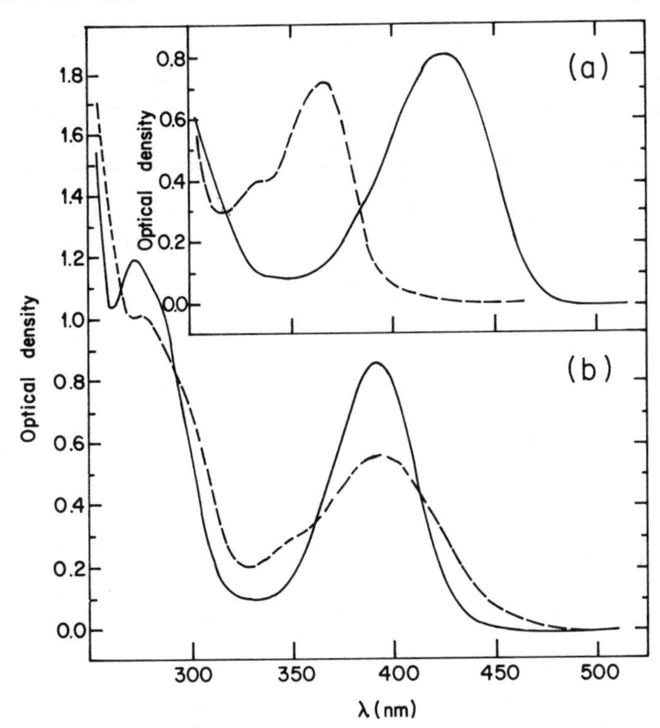

FIGURE 3. Absorption spectra of 10,10′-dimethylbiacridan (a) and of 1,1′,10,10′-tetramethylbiacridan (b), in methylcyclohexane-3-methylpentane 1:1 solution at 93 K. (a) Full curve, A form; broken line, E form. (b) Full curve, A form; broken line, F form[13]. Reproduced by permission of the Royal Society of Chemistry from Ref. 13

interplay of opposing energy contributions operating within the constraints imposed by the rigid molecular skeleton.

These considerations (kinetic and thermodynamic) determine not only the existence of conformations of isocovalent modifications but also play a decisive role in explaining kinetics and thermodynamics of covalent photoisomers such as the **C** forms. In these the potential barriers preventing thermal decay to the fundamental modification contain additional contributions, from the antibonding interactions of a forbidden conrotatory ground state electrocyclic process in a hexaene-derived system[9,66–68].

To illustrate the operation of the thermodynamic factors let us consider the case of the **B** forms. Here the non-bonded interaction between the atoms at the 1 and 8′ (and 8–1′) positions is minimized by the 9–9′ double bond torsion. This deformation acts however to destabilize the ground state, decreasing all stabilizing interactions such as the bonding π-system interaction[5].

The double bond twisted **B** modifications of bianthrones and bianthrylidene-related systems provide spectroscopists and chemists alike with the unique opportunity of studying a twisted essential double bond. The MO theoretical analysis of the spectral consequences of this unusual deformation have been predicted in the earlier work of Coulson and Longuet-Higgins[14].

The diphenoquinone moiety (**1a**) does not play a cardinal role in determining the existence of photo- or thermochromism, for these phenomena are observable in other

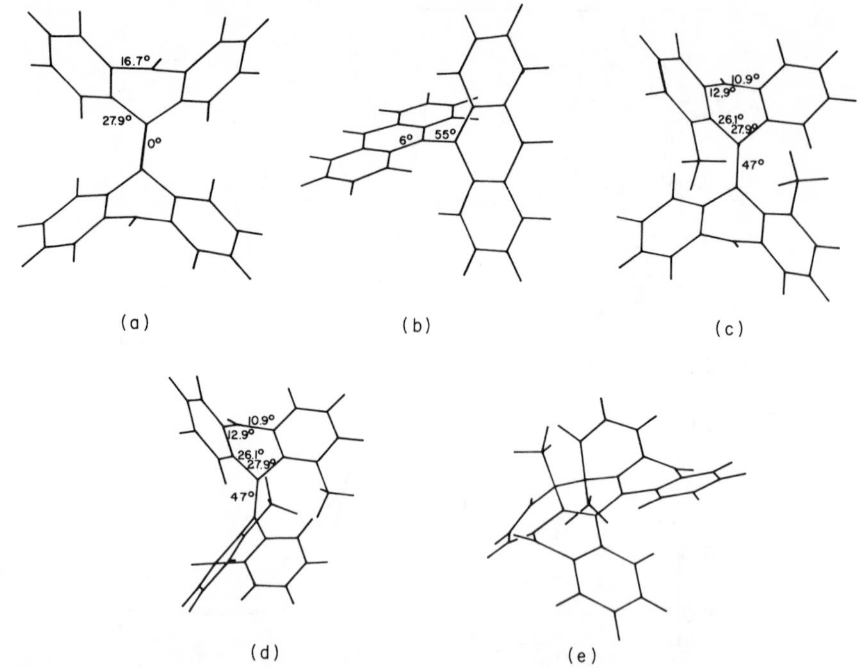

(a) (b) (c)

(d) (e)

FIGURE 4. Space models built from computed minimum energy geometries of the modifications of bianthrylidene-type systems. (a) Fundamental form, A; (b) 9–9′ bond twisted thermochromic and photochromic **B** form, in systems unsubstituted at 1, 1′, 8 and 8′ positions; (c) and (d) **B** form in 1- and 8′-disubstituted bianthrylidenes; (e) **C** form (4a,4b-dihydrophenanthrene-like) in 1,1′-disubstituted bianthrylidenes. Reproduced with permission from Korenstein, Muszkat and Sharaly-Ozeri, *J. Am. Chem. Soc.*, **95**, 6177–81. © (1973) American Chemical Society

dianthrylidene-derived systems (**2**, X = CH_2, CHOH, O, S, and N–Me[11,13,15–22]). Nevertheless the irreversible processes accompanying photocyclization[23] and probably thermal cyclization in bianthrone proper and in bianthrones not substituted at the 1 and 8′ positions depend on this structural element, as also does the efficient singlet to triplet intersystem crossing process of all bianthrones[13,21].

As far as the interesting history of the effects is concerned, thermochromism in bianthrone was first described by Meyer in 1902[2]. The range of known systems was much enlarged by Schönberg[24,26] and by Bergmann[27,28]. It was he who introduced these topics to Rehovot[29] and later, to Jerusalem. The modern phase of experimental research on thermochromism starts undoubtedly with the important works of Grubb and Kistiakowsky[30] and of Theilacker and coworkers[31]. Photochromism was first described by Hirshberg in 1950[1]. In the fifties, significant progress was made, in the parallel investigations conducted by Hirshberg in Rehovot[32–34], and by Kortüm and collaborators in Tübingen[35]. This situation led to a bitter and very regrettable personal controversy[35–37]. These studies resulted, however, in the clear distinction, between the photostable **B** modification, and the photolabile **C** modification, on both spectrophotometric and photochemical grounds[7,8,33,34].

Our own research on the reversible photochromism and thermochromism of bianthrones and of related systems was started in 1968[6,15]. At that time our interest in this field was motivated by the emergence of clear possibilities for achieving significant

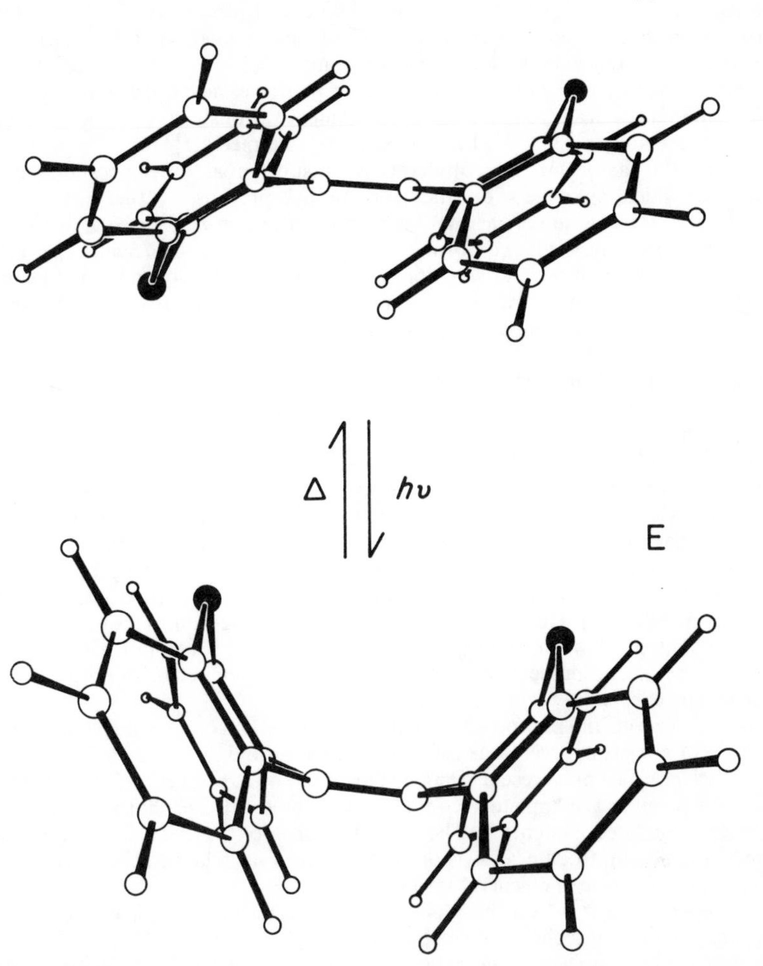

FIGURE 5. Molecular conformations of A and E forms (e.g. in bixanthylidenes, bithioxanthylidenes, biacridans and 10,10′-di-H-bianthrylidenes). Reproduced by permission of Elsevier Sequoia SA from Ref. 21

progress as indicated by the just previously completed work on the 4a,4b-dihydrophenanthrenes[9, 10]. The approach adopted in that work consisted of:

(1) development of suitable molecular architecture to achieve the highest stability and chemical inertness[10, 38].
(2) application of NMR measurements to determine molecular structure.
(3) application of MO theory and related theoretical procedures to the study and correlation of the electronic spectra of the labile intermediates. In particular, the Exciton Theory of molecular electronic spectra of conjugated polyenes proved to be very useful for the analysis of the spectra of the 4a,4b-dihydrophenanthrenes[39]. This method was later supplemented by the π-electron MO computational analyses.

Several new factors proved to be of particular importance in our investigations of bianthrones and of related systems. First and foremost proved to be the possibility of securing separate existence of each photochromic intermediate by taking advantage of effects such as differences in thermal stability or photolability[5] and differences in solvent or temperature dependences of photoformation or photocleavage quantum yields.

Such differences lead to what can be termed variously as kinetic, thermal, or photochemical isolation[5,19,20] of a desired labile intermediate[11].

Another new factor was the observation of substantial differences in the NMR diamagnetic shielding effects on the ring methyl protons of the various modifications[5,13,20,21]. The numerical analysis of such effects proved to be a rich source of information on the molecular conformation of photointermediates. Such experimentally derived geometry could be, and was, compared with theoretically computed geometries generated first[5] using our strain energy minimalization program CONFI[40], and then also[11] with geometries computed by the Consistent-Force-Field-π-electron model of Warshel and Karplus[41].

Finally, we should mention that the theoretical analysis of the energy levels of the twisted double bond formed the essential basis for our understanding of the electronic spectra of the **B** modifications[5].

While the study of labile photochromic modifications and of their electronic spectra might seem at first glance to be a frankly esoteric subject, it has nevertheless considerable importance as a model for one molecular geometry change which could result in spectral changes similar to those observed in the vision cycle of rhodopsin[19]. Much support was obtained in recent years for a process of reversible proton transfer in a protonated retinal Schiff base as a mechanism responsible for the geometry change of rhodopsin in the vision cycle (see, e.g. Ref. 42). However, the possibility of double bond torsion in the retinal moiety of rhodopsin, much similar to the transition to the **B** form of bianthrone, is an interesting alternative source of large scale molecular geometry changes producing strong bathochromic effects[19].

Thermochromism and photochromism of bianthrones and of related systems have been the subject of numerous reviews, as befits this intriguing and rather mysterious chapter of chemical science. The most recent (1984) is the review of E. Fischer[43]. The PhD thesis of R. Korenstein gives a complete account of our studies of the photochromism in the bianthrylidene-related systems[44]. The introductions to the papers by Agranat and Tapuhi[45] provide up to date reviews on thermochromism. The earlier literature on this subject is referred to in the general review of Day[46].

Our companion review[9] on the 4a,4b-dihydrophenanthrenes deals exhaustively with that aspect of the photochemistry of bianthrone-related systems. Our 1970 report on bianthrone photochromism[15], the paper by Becker and Earhart[47], as well as Kortüm's definitive account of his work[48] provide descriptions of the development of this field just at the inception of the NMR period. A survey of Schönberg's contributions is to be found in his 1946 paper[49].

Many different hypotheses have been advanced over the years to explain the molecular structures of the thermochromic and of the photochromic modifications of bianthrones and of the related bianthrylidene systems. Up to the late 1960s, the experimental means available to prove or disprove these hypotheses were as a rule far too short for such ambitious attempts.

The first hypothesis concerning thermochromism was put forward in 1910 by Padova[50] who suggested a depolymerization process. Wizinger, in 1927[51] suggested a zwitterion (**5a**) as the thermochromic form. Schönberg[25] and later Bergmann and Engel[27] and Bergmann and Corte[28] all suggested a biradical structure,**5b**. Grubb and Kistiakowsky[30] proposed a double bond twisted structure, with one anthrone moiety perpendicular to the other. Its ground state would be a triplet. Matlow carried out π-electron–LCAO–MO computations

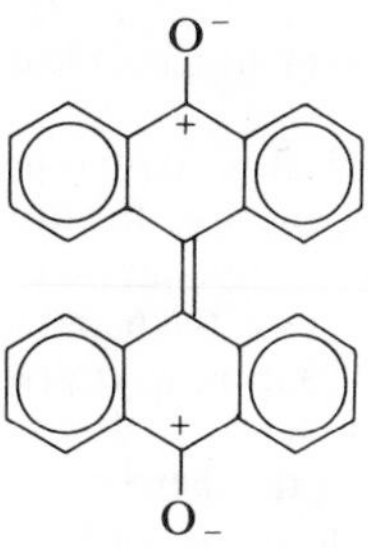

(5a) 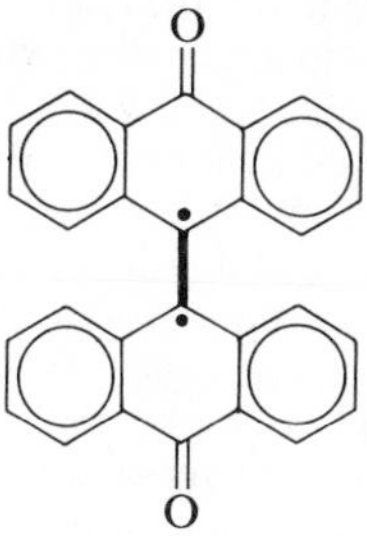 (5b) 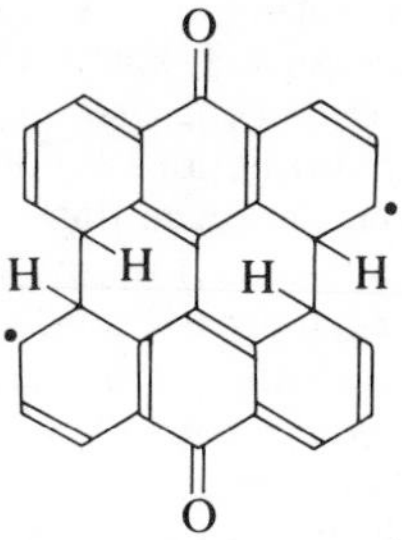 (5c) 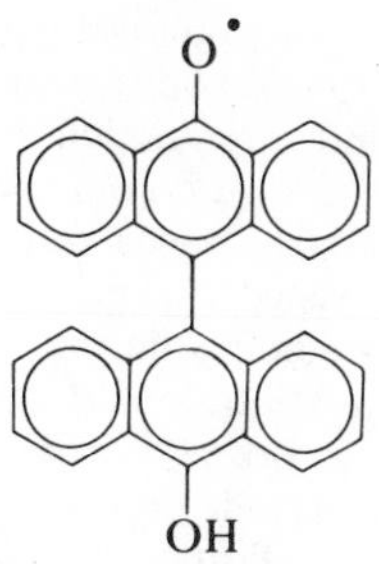 (5d)

for the planar ground state, the planar triplet and the perpendicular triplet and concluded in support of Grubb and Kistiakowsky that the thermochromic form had the perpendicular triplet structure[52]. In fact, several ESR studies of thermochromism from that period (Nielsen and Fraenkel[53], Wasserman and Woodward[54]) attempted to attribute it to paramagnetic species, e.g. such as **5c**.

However, later studies by Hirshberg and Weissman, by Harrah and Becker and by Kortüm and Koch have clearly shown that the **B** form is diamagnetic[55–57]. We could fully confirm such observations for **B** forms obtained either photochemically or thermally[4]. Nevertheless the formation of paramagnetic species upon heating solutions of bianthrone or of bixanthylidene (**2**, X = O) is a genuine observation. However, this free radical is formed in an irreversible reaction, and is not involved in the reversible thermochromism. Agranat's group have identified this free radical as the phenoxyl species **5d**[58].

In juxtaposition to the hypotheses on paramagnetic structures, several cyclic diamagnetic structures were variously assigned to the thermochromic modifications. The cyclic structure **6a** (now undoubtedly known to be that of the photochromic **C** isomer) has

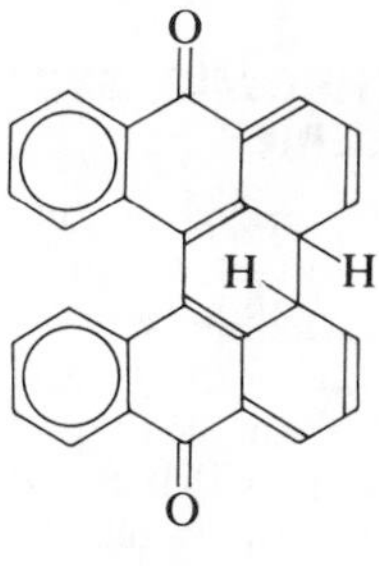

(6a) 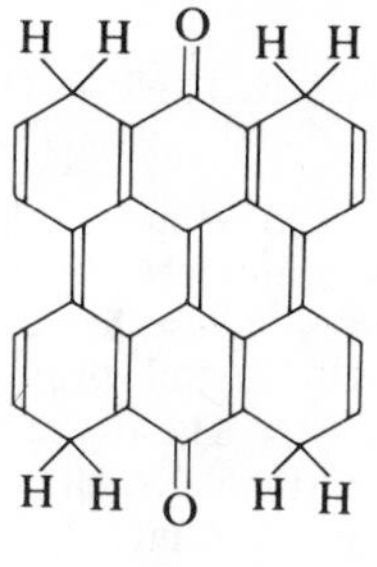 (6b)

been tentatively assigned by Lorentz and coworkers[59] to the **B** form. Similarly, Philips and Schug[60] suggested a double cyclization to the quinonoid structure, **6b**. The presently available experimental evidence concerning **B** (NMR measurements as well as chemical and photochemical inertness) speaks clearly against these two possibilities.

Torsion about the 9–9′ double bond, but in the reversed sense to what we accept as correct today, has been invoked by Heller[61]. On the basis of his computations he suggested that the fundamental form (**A**) is twisted by 76 degrees, while the thermochromic form (**B**) is twisted only by 14 degrees.

Very interestingly, Bergmann in 1948 (quoted in the 1951 note)[29] has arrived at the 9,9′ double bond twisted structure. It seems he meant a partially twisted structure, his ideas much resembling our actual concepts.

However, our present day notions about the structure of the **B** modification should be clearly traced to the specific suggestion of Harnik[62a] that the thermochromic form is twisted by some 60 degrees about the central double bond. This conclusion was reached by assuming the same closest approach non-bonded distances of the 1–8′ atoms in the A and B modifications. Important support for Harnik's idea was provided by Grabowski and coworkers[63] from the conclusions of their polarographic and voltammetric studies of electrode processes of bianthrone. The type of isomerism existing among the then known modifications of bianthrone has been very appropriately termed by Harnik as 'chromic isomerism'[62a].

Before concluding this introduction we would like to mention some points about the nomenclature of the molecules and of the effects. In addition to the common ring atom numbering system (see e.g., **2** and **2**′) there is another numbering system used, e.g. by Becker and coworkers[47, 56] and by Strong and coworkers[64], in which the numbering of positions 1 and 4 is interchanged. Great care is needed to avoid confusion, as the numbering convention used is never self-evident.

In the typical photochromic or thermochromic systems the labile isomers absorb at longer wavelengths than the parent, usually a colourless molecule. These are the bathochromic effects observed in the **B**, **C** and **D** modifications. However, the reverse change, i.e. the disappearance of a strong absorption band because of photochromic or thermochromic effects, is not uncommon. In photochromic bianthrylidene-related systems this situation is observed with the **E** and **F** modifications[11, 13, 22, 23]. In some thermochromic systems a reversed effect is also possible, as in the rhodamine lactone–zwitterion system where the low temperature form is deeply coloured[65]. Obviously the decisive factors defining the spectral effects of both thermochromism and photochromism are the exact molecular structures which determine the effect on the energy and intensity of the electronic transition.

II. THERMOCHROMISM AND THERMOCHROMIC SYSTEMS IN THE BIANTHRYLIDENE SERIES

Bianthrylidene-derived systems (**2**, X=CO and O) are as a rule thermochromic. Thermochromism is prevented by bulky substituents at the 1, 1′, 8 and 8′ positions[1b, 62b]. For reasons of molecular mechanics it is not observed in dithioxanthylidene or in 10,10′-dimethylbiacridan[31, 69]. Substitution by fluorine at position 1 does not prevent thermochromism[1b, 62b]. The absence of thermochromism in the 1 (and 1′, 8 and 8′) as well as otherwise substituted derivatives is not an indication that the **B** form should be incapable of existence. Quite to the contrary, this form is readily obtained by the photochemical route. However, the thermal path leading to it requires surmounting too high potential barriers due to the strong non-bonding 1–8′ interactions and is therefore unobservable.

In addition to the symmetric thermochromic systems in which the groups at the 10 and 10′ positions are identical, several mixed thermochromic bianthrylidenes have been reported, e.g. 9,9′-fluorenylidene anthrone, **7a**[70], 9,9′-diphenylmethylene anthrone, **7b**, and its substituted derivatives; xanthylidene anthrone, **7c**, and diphenyl methylene xanthene, **7d**[49].

Solutions of difluorenylidene, **8a**, and of its benzo-annelated derivatives, e.g. [*a,a*′]-dibenzo, [*b,b*′]-dibenzo and [*a,f*]-dibenzo, as well as [*a,a*′,*f,f*′]-tetrabenzo difluorenylidene, are deeply coloured at room temperatures[71]. Here the fundamental modifi-

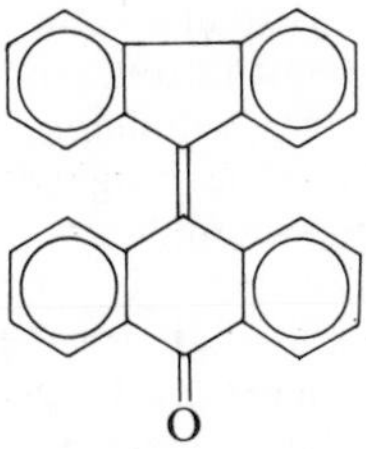

(7a) 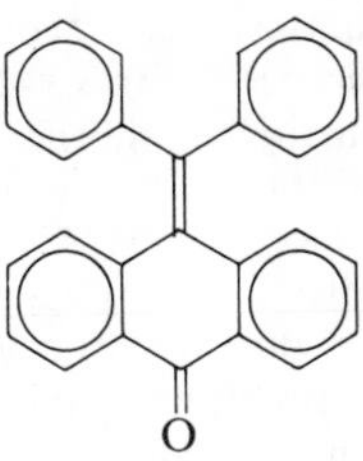 (7b) 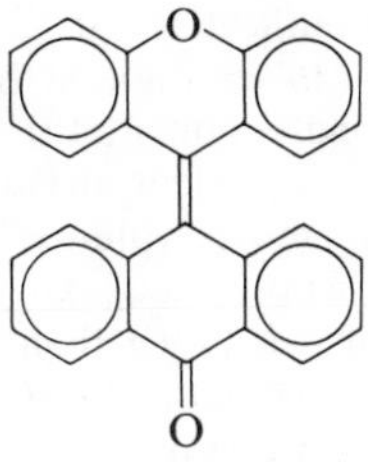 (7c) 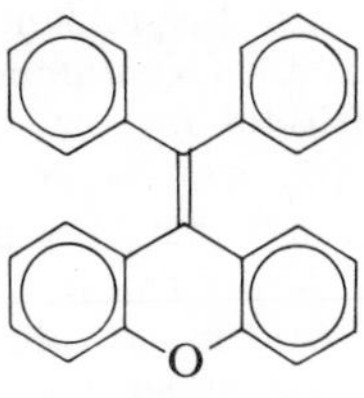 (7d)

cation seems to be already twisted significantly about the 9,9′-double bond, resembling in this sense the **B** conformers of bianthrylidenes.

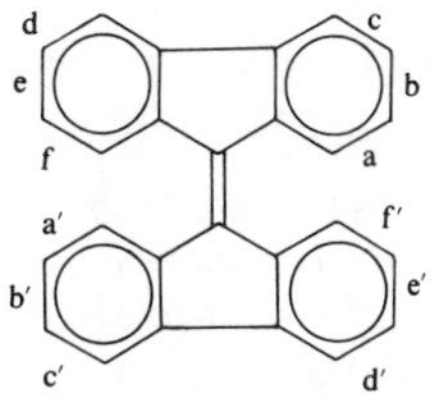

(8a)

From the change (increase) of the equilibrium constant $K = [\mathbf{B}]/[\mathbf{A}]$ with temperature it is seen that the thermal conversion $\mathbf{A} \rightarrow \mathbf{B}$ is slightly endothermic. Values of this enthalpy change falling within the range of 2–4 kcal mol^{-1} can be variously estimated[15,30,31,45f,48]. This narrow range, little affected by substitution pattern (for the less sterically hindered molecules) certainly merits some special attention. Conversion extents of up to 0.1 are obtained at atmospheric pressure. The higher conversion extents (0.05–0.1) are observed at temperatures around 500 K in the not excessively sterically hindered bianthrones, e.g. the parent molecule[31] and 3,3′-dicarboxybianthrone[48]. The conversion extents obtained in the 2,2′-dimethylbianthrone are much lower, 0.0013 at 330 K and 0.008 at 490 K[45f], because of additional non-bonded interactions of intermediate magnitude. Such values are estimated by assuming either that the ε_A and ε_B extinction coefficients at the respective maxima of **A** at 380 nm and of **B** at around 700 nm are approximately equal, or from the ε_B values deduced under conditions of complete photoconversion to **B**, $\varepsilon_B = 15\,500$[45f]. The thermochromic **B** form can be also obtained by sublimation, e.g. by condensing bianthrone vapour (at ca. 0.001 mmHg pressure) on a surface cooled to 90 K[72].

III. PIEZOCHROMISM

Pressure and strong shearing such as grinding result in the formation of the coloured form (now known to be the **B** form) identical with the photo- and thermochromic forms[2,49]. Piezochromism is temperature dependent, being stopped by cooling to 90 K[73]. Schönberg's conclusion[73] that such temperature dependence is an indication that piezochromism is to be attributed to thermochromism seems, however, unwarranted. In fact much insight into the nature of the piezochromic transition of bianthrones has been obtained in high pressure studies[74]. Such studies on bianthrone and its substituted derivatives, embedded in polymethyl methacrylate, show very significant extents of conversion into the **B** modification. At 120 kbar, the observed conversion extent amounts to 0.3 in bianthrone, 0.13 in 3,3′-dimethylbianthrone, 0.16 in 3,3′-dibromobianthrone, and 0.08 in the more strongly hindered 1,1′-dimethylbianthrone. The energy difference ΔH_A–ΔH_B is pressure dependent decreasing three- to fourfold over the range of 0–120 kbar. The partial molar volume difference for the **A** → **B** process amounts to ca. −0.90 ml for bianthrone at 60 kbar and drops to −0.50 ml at 120 kbar. That the partial molar volume of **A** is larger than of **B** is the obvious macroscopic origin for the pressure-dependence effect. The molecular interpretation of this difference seems to rest on the number and magnitude of the various non-bonded interactions in the region of closest non-bonded approach. Apparently similar differences have been previously invoked in other photoconformers, such as in the *cis* and *trans* isomers of stilbenes. There steric hindrance is the origin of the 1–α single bond torsion[75–77] as it is of the 9–9′ double bond torsion in the **B** conformer.

IV. CHEMICAL AND ELECTROCHEMICAL PATHS LEADING TO THE *B* MODIFICATION

We have already mentioned the paths to the **B** form involving the thermal, high pressure, ionizing radiation, photochemical and sublimation processes. The crystallization (precipitation) step in the chemical synthesis of bianthrone has already been reported by Meyer[2] to lead to crystals of the metastable green coloured modifications (presumably the **B** form, cf. also Ref. 7). Two other paths leading to the **B** form, one involving a chemical process[7], the other an electrochemical process[63], have been carefully studied. The chemical process[7] consists of the low temperature (183 K) hydrolysis of the (diprotonated) adduct of 1,3,1′3′-tetramethylbianthrone (= 1,3,6′,8′-tetramethylbianthrone) with sulphuric acid in an ethanol–water mixture. The absorption spectrum clearly shows[7] efficient (but partial) conversion into the **B** form. Oxidation of the corresponding 1,1′-dimethyl dianthranols leads also to the **B** forms[7], observable as metastable products[48]. Good evidence connecting the green form **B** to the 9,9′-double bond twisted conformer is provided by the electrochemical results of Grabowski and coworkers[63a, b] and of Peover[78]. The principle governing the electrochemical reactivity of two species K and L converted one into the other by an electrode process is the least structural change rule[63]. Their rule states that the overall rate of an electrode process is high provided the structure difference between K and L is minimal. Thus bianthranol (**9**) is reversibly converted into the **B** form by anodic oxidation on a frozen mercury drop electrode at 188 K. The rate of the thermal process **B** → **A** is sufficiently slowed down at this temperature so that in the oxidation stage of the cyclic voltammetry experiment the **B** form will accumulate in the vicinity of the electrode.

The reduction of the A form (reduction potential $E_{1/2} = -1.25$ V against saturated Hg sulphate electrode) to bianthranol is a diffusion-limited irreversible process. The reduction of the **B** form to bianthranol takes place at a much less negative potential, $E_{1/2} = -0.55$ V. This reduction is kinetically controlled. The reduction current for this process is temperature dependent, increasing strongly upon going up from 288 to 333 K. This current is kinetically limited by the **A** → **B** equilibrium (which is heavily in favour of **A** at these temperatures). To conclude, the electrode reduction of **B** and the oxidation of bianthranol form a reversible redox system. Thus according to the least structural change rule we can deduce that the **B** form and the necessarily twisted (about 9,9′ bond) bianthranol have similar nuclear conformations.

These findings of Grabowski and coworkers and of Peover have been amply confirmed in the recent comprehensive studies of Evans and coworkers on electrode processes of bianthrones in non-aqueous solvents[79, 80].

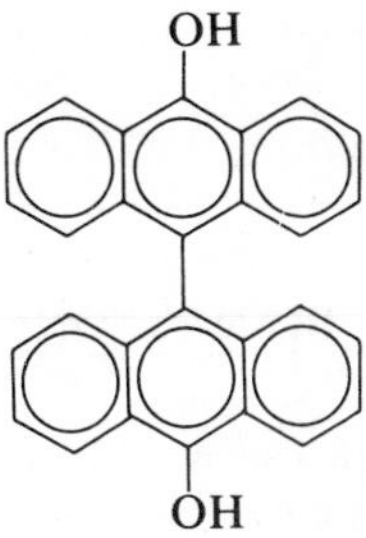

(9) 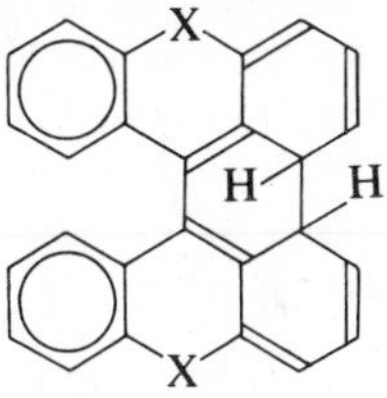 (10) 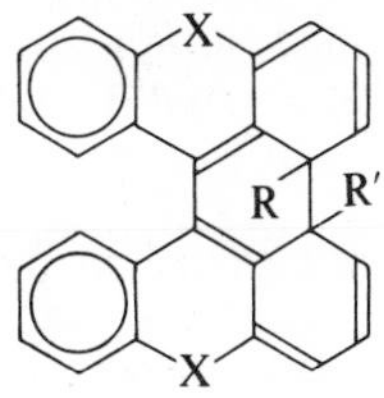 (10a)

V. INTERCONVERSIONS AND EQUILIBRIA BETWEEN *A* AND *B* MODIFICATIONS IN IONS OR RADICALS

We have already mentioned that the bond order (and electronic density) of the unique 9–9′ double bond and the electronic densities at the 9 and 9′ atoms are of prime importance in determining both the interconversion kinetics of the **A** and **B** forms and their relative thermodynamics as well. For this reason it is to be expected that the relative energetics and the interconversion rates and barriers would be deeply affected by electronic processes and reactions which change the electronic density at the 9–9′ bond: ionization (or oxidation) to form the molecular cation $\mathbf{A}^{+\cdot}$; reduction to give the anion radical $\mathbf{A}^{-\cdot}$ or its protonated counterpart $\mathbf{AH}^{\cdot}$ and twofold reduction to give the dianion $\mathbf{A}^{2-}$; protonation at the O atoms to give successively $\mathbf{AH}^{+}$ and $\mathbf{AH}_2^{2+}$; and undoubtedly excitation to give the excited singlet and the excited triplet, either by intersystem crossing or by triplet sensitization. These processes are likely to weaken this double bond either by removing an electron from the bonding HOMO; by inserting one or two electrons into the mirror symmetry antibonding LUMO; by promoting an electron from a HOMO into a LUMO; and by perturbing the nuclear core topology (by positive charges as in the diprotonated dication $\mathbf{AH}_2^{2+}$). We have already seen how many of these processes (e.g. excitation, diprotonation and oxidation) worked in practice, and in subsequent sections we shall consider the effects of some of the other processes. In general, as the interconversion $\mathbf{A} \rightarrow \mathbf{B}$ for the neutral ground state configuration is only weakly exothermic, $\Delta H = 2$–5 kcal, we expect any weakening of the 9,9′ C=C bond to result in further molecular deformation to allow better reduction of the 1–1′ and 8–8′ (or 1–8′ and 1′–8) non-bonded repulsions. This situation should give rise to further preference for the **B** form. In addition to the above-mentioned results, information on some processes has become available from pulse radiolysis and cyclic voltametry experiments [79–81]. The pulse radiolysis experiments in 2-propanol solution involve the radiolytic preparation of the acetone ketyl radical, $Me_2C^{\cdot}OH$, and the study of its subsequent reaction with the **A** modification to give within 4–5 μs the anion radical $\mathbf{A}^{-\cdot}$ which is converted into $\mathbf{B}^{-\cdot}$ within the next 25–30 μs[81]. The interpretation of these experiments rests on the assumption that the absorption maximum of $\mathbf{A}^{-\cdot}$ is at 560 nm and that of $\mathbf{B}^{-\cdot}$ is at 460 nm and on other assumptions as well. For the $\mathbf{A}^{-\cdot} \rightarrow \mathbf{B}^{-\cdot}$ step in bianthrone they obtain a rate constant $k_{AB^{-\cdot}} = 7 \times 10^4\ s^{-1}$. In 1,1′-dimethylbianthrone they get $k_{AB^{-\cdot}} = 1.1 \times 10^3\ s^{-1}$. For the 3,3′-disubstituted bianthrones they estimate $k_{AB^{-\cdot}}$ roughly similar to those for bianthrone (for the dimethyl compound, $k_{AB^{-\cdot}} = 1.7 \times 10^4\ s^{-1}$, for the dimethoxy compound, $k_{AB^{-\cdot}} = 7 \times 10^4\ s^{-1}$, as in bianthrone)[81]. The $k_{AB^{-\cdot}}$ values obtained from the cyclic voltametry experiments are $k_{AB^{-\cdot}} = 2 \times 10^6\ s^{-1}$ for bianthrone and $k_{AB^{-\cdot}} = 5 \times 10^4\ s^{-1}$ for 1,1′-dimethylbianthrone[79,80]. While the trend of these values is entirely reasonable, there is still the discrepancy between the two sets of experiments to be explained. The above values for $k_{AB^{-\cdot}}$ in the radical anion of bianthrone should be compared to the corresponding value of $k_{AB} = 8 \times 10^{-3}\ s^{-1}$ for ground state singlet bianthrone. This value can be deduced from the values of $K_{AB} = [\mathbf{B}]/[\mathbf{A}] = 2.2 \times 10^{-3}$ and the accepted value of $k_{BA} = 3.7\ s^{-1}$ [79b].

VI. IRREVERSIBLE PHOTOCHEMICAL PROCESSES OF BIANTHRONE AND OF 1,1′-UNSUBSTITUTED DERIVATIVES

A very significant aspect of the molecular transformations of the bianthrylidene series is that of high chemical reversibility, implying that transformation of one conformation or structure into the other and back can be carried out repeatedly without occurrence of side reactions which would yield stable side products, and consequently lead to depletion of the starting parent molecules. The crucial factor governing reversible behaviour in the bianthrone series is the presence or absence of at least two substituents (such as alkyl, alkoxy or benzo annelation) at the 1 and 1′ (or 8 and 8′) positions[4,6,23,48]. Such substituents determine the resistance or reactivity of the 4a,4b-dihydrophenanthrene-like C photoisomers (**10** and **10a**) in irreversible processes such as oxidation or hydrogen shift of the angular H atoms at the C centres forming the new bond[4,6,9–11,15,16,18,22].

The processes leading to and from the **D**, **B**, **E** and **F** photoisomers are highly reversible in all bianthrones and in the other bianthrylidene series (**2**, X = O, CH_2, CHOH, NMe and S). As all irreversible behaviour is to be traced to the **C** modification, high extent of chemical reversibility will be observed in all members of these series under conditions under which no **C** modification is formed, such as at very low temperatures[23]. High extent of reversibility is observed also in compounds such as **2**, X = NMe in which no **C** modification is formed at all. In the **C** modifications of bixanthylidenes and of 10,10′-di-*H* bianthrylidenes in the absence of 1 and 1′ substituents, the source of irreversibility is the oxidation of these hydrogens by molecular oxygen as in other 4a,4b-dihydrophenanthrenes[9]. However, in the **C** modifications of bianthrones (again, in the absence of 1 and 1′ substituents) other avenues of irreversible chemical reactivity are open as well: intramolecular H atom shift followed by intermolecular H shift or oxidation by molecular oxygen, and further photochemical and dark processes leading to a very complex photochemical system[4,6,23]. The products of these processes were correctly figured out in 1949 by Brockmann and Mühlmann[82]. Their conclusions about the identity of the products have withstood the test of time, though much additional information was obtained in later studies[23] of bianthrone. Here, the **B** modification is formed exclusively at temperatures below 140 K, or at higher temperatures, ca. 200 K and above, by biacetyl triplet sensitization. The **B** modification of bianthrone is chemically inert and goes reversibly into **A**. The **C** modification (**10**) can be observed at low temperatures in polar (protic or aprotic) solvents, as a short-lived transient (half lifetime of 80 ms at 200 K in 2-propanol)[23]. This intermediate undergoes an irreversible 1,5 H atom shift to give dihydrohelianthrone, **11**. Dihydrohelianthrone is highly reactive chemically. It reacts with oxygen at low temperatures to give helianthrone, **12**, and hydrogen peroxide. This reaction is not inhibited by 2,6-di-*t*-butyl-4-methylphenol[9,83–85].

The reasons for this difference could be either very fast reaction rate of **11** with oxygen or simultaneous reaction of the two 1 and 1′ hydrogens. In the absence of oxygen **11** will react by intermolecular hydrogen atom transfer with the parent molecule **2** to give one molecule each of **12** and of dianthranol **9**. These are also the products of the photolysis of **2** in non-polar solvents such as hydrocarbons or CS_2. Here no dihydrohelianthrone can be observed, probably because its reaction with bianthrone is too fast. No details are known about the molecular paths of the rearrangement of **10** to give **11**, and of the dismutation of **11** with **1** to give **11** and **12**. The *o*-quinonoid double bond topology of **11** (rings B, C, E, and F) seems to be at the origin of its unusual properties. First, the considerable reactivity

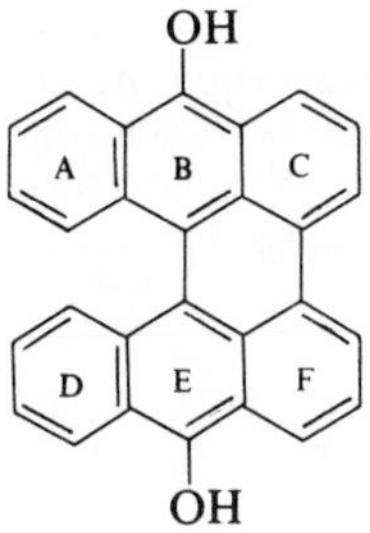

(11) 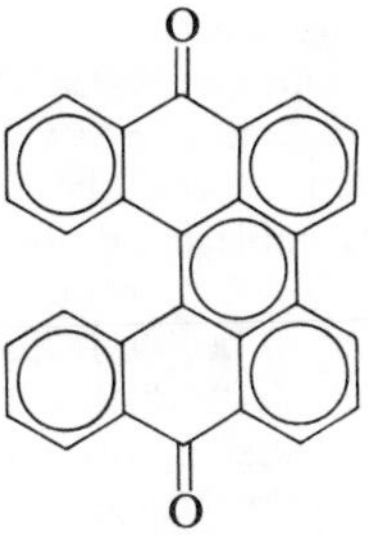 (12) 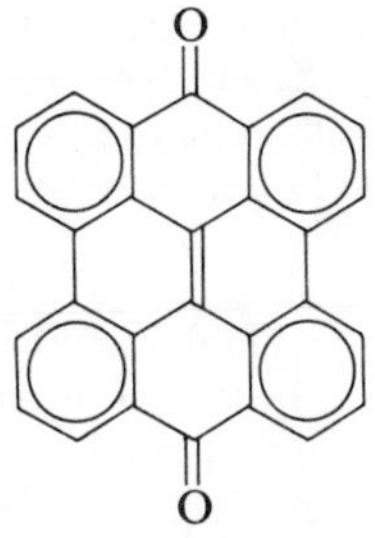 (13)

we mentioned already, and then also the blue colour of this compound, due to strong absorption maxima in the visible at 610 and 575 nm. **11** is thermally stable at room temperature in the absence of oxygen. It is also stable photochemically. Further irradiation of helianthrone **12** results in the formation of mesonaphthobianthrone, **13**. In the absence of oxygen, the two angular H atoms in the primary cyclization product preceding **13** apparently can be transferred to **12** to give **11**[82].

VII. THE PHOTOCHROMIC *C* MODIFICATION

Our initial proton NMR studies on the **C** forms of 1,8′-dimethylbianthrone and of 1,3,6′,8′-tetramethylbianthrone[4, 15] and our subsequent study of the C form of 1,4,5′,8′-tetramethylbixanthylene[11] clearly indicates that these **C** forms are photocyclization products of the 4a,4b-dihydrophenanthrene type[9, 10, 38]. In such photoisomers[9] the two groups attached at the atoms forming the new bond (i.e. R and R′ in **10a**) have been shown to be in the *trans* conformation. We assume that this is also the case for the **C** photoisomers. This is in fact the steric outcome predicted by the orbital symmetry conservation rules (see, e.g. Refs 9, 10 and 66) for the case of the concerted conrotatory first excited state photocyclizations of a hexatriene or a derived system to give a cyclohexadiene-related product[66]. Our studies[4, 11, 15] show also that in the case of 1,8′-dimethyl substitution in the educt, the new bond of the product is always formed between two methyl-bearing ring atoms (Me–C–C–Me mode, i.e. in **10a**, R = R′ = Me). None of the other possible cyclization products (H–C–C–H mode, $R = R^1 = H$ or H–C–C–Me mode, R = H, R′ = Me) is formed. The calculated energy difference[11] for 1,8′-dimethylbianthrone between the Me–C–C–Me mode and the H–C–C–H mode amounts to 27 kcal in favour of the former. This difference is to be attributed to the much higher molecular strain imposed by the large non-bonded methyl–methyl repulsion in the H–C–C–H mode product.

All **C** modifications undergo the reverse 1–1′ bond cleavage, both thermally (dark reaction) and photochemically (excited singlet process). The photochemical conrotatory process (ring-opening) is allowed by orbital symmetry rules and takes place without requiring any activation energy[4, 9, 10, 15]. The thermal process is a forbidden conrotatory ground state ring-opening of a D_2 symmetry system related to cyclohexadiene[66, 86]. That the thermal ring-opening in 4a,4b-dihydrophenanthrenes[9, 10, 66] and in **C** photoisomers of the bianthrylidene series[4, 15] takes place quite readily is chiefly due to the significant destabilization of the ground state, a fact which decreases the potential barrier for ring-opening[66, 86]. The measured values of the ring-opening activation energies of the **C** photoisomers all fall in the very narrow range of 12–15 kcal[4, 10, 11, 15, 22]. The **C** photoisomers are less stable (towards thermal ring-opening to **A** form, in the absence of oxygen) than typical 4a,4b-dihydrophenanthrenes[9, 10]. The least stable (apart from the **C** form of bianthrone itself) are the **C** forms of 10,10′-di*H*-bianthrylidenes (**2**, X = CHOH, or X = CH_2) where the 1 and 1′ positions are unsubstituted. In such photoisomers thermal half lifetimes τ are of the order of 0.5–2 s at 253 K. **C** forms of bianthrones with dimethyl substitution at 1 and 1′ positions have τ values of the order of 1–2 h at 215 K[22]. The **C** form of bixanthylene is however unusually stable, resembling 4a,4b-dihydrophenanthrenes. It is stable at 273 K, while at 328 K τ is 14 s[18]. Obviously not all factors governing thermal stability of the **C** photoisomers seem to be completely understood.

The **P** photoisomers of bixanthylidene[18] and of bithioxanthylidene[21] seem to be closely related to the **C** forms. Preliminary results indicate that the **P** isomers originate from the excited singlet. It is yet unclear whether the **P** forms are precursors of **C** forms. There is also a possibility that **P** and **C** photoisomers are formed concurrently. Unlike the **C** modifications, **P** forms are light stable. They were observed at temperatures in the range of 250–360 K. The **P** photoisomer of bixanthylidene has been detected in flash photolysis experiments at 293 K. Its absorption spectrum resembles that of **C**, but its first absorption band maximum at 520 nm is much more intense, and thus it is quite probable that this is an unstable conformeric isomer (and precursor) of **C**, the two differing in some conformational details. The thermal decay activation energy of **P** amounts to 14 kcal. Its thermal half lifetime value τ is ca. 11 s at 250 K.

The **A** and **C** forms constitute a photoreversible photochromic system as light converts one form into the other. The composition of such a system (concentration c_i of each isomer) at photostationary equilibrium is determined by the extinction coefficient ε_i at the irradiation wavelength, and by the quantum yield ϕ_i so that the product $\varepsilon_i \cdot \phi_i \cdot c_i$ for each isomer is constant (cf. Refs 9 and 10). Thus visible light absorbed only by **C** results in complete reversal to **A**. On the other hand short wavelength light absorbed by **A** (and so also by **C**) leads to a photostationary equilibrium. We should note however that system **A** + **C** does not constitute a perfect photoreversible system as excitation of **A** leads to the parallel formation of **C** and of the triplet product **B** which does not undergo the photoreversal reaction to give back **A**. Formation of the other triplet products **E** and **F** leads to the same consequences. Thus the time evolution of a prolonged irradiation of **A** at temperatures where both **B** and **C** are stable is characterized by the initial simultaneous formation of **C** and **B**. The concentration of **C** then passes through a maximum and decreases gradually till complete conversion into the photostable isomer is obtained[4,11,15,18,22].

However, when the photostable intermediate (**B**, **E** or **F**) is less stable thermally than the **C** form we should expect high conversion into this form and complete reversibility. This situation is achieved, e.g. with 1,1′-dimethylbianthrone in methylene chloride[4,15]. It seems that the **C** form is the only photoisomer observed in 10,10′-dihydroxy-10,10′-di*H*-1,1′-dimethylbianthrylidene[22]. While all wavelengths absorbed by **A** are effective in producing some conversion into **C**, the highest extent of conversion (cf. Ref. 10) at photostationary equilibrium is achieved at irradiation wavelengths for which the extinction coefficient of **C** is a minimum while that for **A** is a maximum. With the excitation frequencies choice provided by the mercury arc, the highest extent of conversion is usually obtained using the 405 nm line. As in most 4a,4b-dihydrophenanthrenes[9,10], the *photocyclization* quantum yields $\phi(\mathbf{A} \rightarrow \mathbf{C})$ of all **A** forms to give the **C** forms show a strong temperature dependence[4,15,18,22]. Thus in the 10,10′-di*H*-bianthrylidenes, quite typically, these quantum yields are high, ca. 0.5, and constant, down to 150 K. They drop to zero below 110 K[22]. Obviously the excited state cyclization process requires surmounting a small potential barrier (3–4 kcal) (cf. Refs 9, 66). In 1,4,1′,4′-tetramethylbianthrone[16], $\phi(\mathbf{A} \rightarrow \mathbf{C})$ is even higher, ca. 0.9. The temperature dependence of $\phi(\mathbf{A} \rightarrow \mathbf{C})$ is a genuine temperature effect and is not due to a transmitted viscosity dependence[75,76]. The photocyclization process shows marked solvent preferences[7,8,15]. For 1,3,6′,8′-tetramethylbianthrone and

1,8′-dimethylbianthrone highest initial conversions **A** → **C** (and thus probably highest quantum yields) are obtained using 2-methyltetrahydrofuran and methylene chloride as solvents[6, 15].

The pronounced resemblance of the photochemical behaviour here and in the 4a,4b-dihydrophenanthrenes extends also to the *photochemical* ring-opening process **C** → **A**. Its quantum yield is high, 0.2–0.3 in the 10,10′-di*H*-bianthrylidenes[22]. It is temperature independent, to be expected of an allowed excited state conrotatory cyclohexadiene-type ring-opening[66, 68]. In the **C** forms of bianthrones ϕ(**C** → **A**) is lower (0.08–0.05) and shows a weak temperature dependence at low temperatures[6, 15].

The absorption spectrum of the **C** forms[4, 9, 15, 16, 18, 22] resembles closely that of 4a,4b-dihydrophenanthrene and of its derivatives[9, 10], both in terms of transition energies and intensities. This similarity is in fact a powerful piece of independent evidence for assigning the 4a,4b-dihydrophenanthrene structure to the **C** forms. The first long wavelength band (I) of the **C** forms has its maximum at 500–600 nm, depending on the exact system[4, 15, 22]. It is a low intensity band (e.g. $\varepsilon = 3360$ in 10,10′-di-*H*-bianthrylidene[22]). According to the exciton theory of the polyene spectra[9, 10] this band is the fundamental, in-phase transition. Its low intensity is due to the partial cancellation of the unitary ethylene transition moments (M) in the vectorial summation of the unitary moment to give the molecular transition moment of a s-*cis*,s-*trans*,s-*trans*,s-*trans*,s-*trans*,s-*cis*-hexaene. In this case the molecular transition moment $\sqrt{I}$ is $\sqrt{I_1} = 0.465$ M. For the sake of comparison, in the all-*trans* linear polyene this transition is much stronger, $\sqrt{I_1} = 2.03$ M. The large width of the first transition, $\Delta\nu_{1/2} \sim 4500\ \mathrm{cm}^{-1}$, indicates a significant geometry change going from the ground to the first excited state. In some systems[9] it was possible to identify a progression in the first excited state totally symmetric stretching mode of the C=C double bonds at ca. 1200–1300 cm^{-1}[9]. The second transition (II, Ref. 9) of the **C** modifications, in the 380–480 nm range has also a number of typical features. It is much stronger than the first transition (e.g. $\varepsilon = 17\,700$ in 10,10′-di*H*-bianthrylidene[22]. It shows a very distinct 2–3 vibrational component progression in the 1200–1300 cm^{-1} mode. The first component is the strongest, implying a small geometry change. According to the exciton model the transition moment of this transition (first harmonic band) is $\sqrt{I_2} = 1.30$ M, i.e. much larger than for the first transition. Two shorter wavelength transitions, at 370 and 320 nm, have been observed in the **C** forms of 1,3,6′,8′-tetramethylbianthrone (in 2-methyltetrahydrofuran or methylene chloride solutions). It seems uncertain whether one of these should be identified with the second harmonic band of the hexaene system (at 237 nm in 4a,4b-dihydrophenanthrene[9, 10].

VIII. THE TRIPLET STATE OF BIANTHRONES AND OF BIANTHRYLIDENE-DERIVED SYSTEMS

Several detailed studies of the bianthrylidene series have shown that photoproducts **D**, **B**, **E** and **F** are formed through the intermediacy of a distinctly observable triplet state T[13, 15–22, 87, 88]. In fact, the bianthrylidene series is one of the rare photochemical systems in which exact information on the course of molecular events leading to photoproducts could be obtained. The common triplet state, observable by both optical and magnetic resonance methods is in all probability the first triplet state T_1 and is formed directly by activated singlet-to-second triplet intersystem crossing ($S_1 \rightarrow T_2$). In several systems (e.g. bixanthylidenes, dithioxanthylidenes and 10,10′-dimethylbiacridans) the $S_1 \rightarrow T_2$ process is incomplete, as the formation of the triplet product (**B**, **D**, **E** or **F**) can be strongly enhanced by external spin-orbit coupling enhancing agents such as molecular oxygen, alkyl halides, carbon disulphide and xenon[17–19, 21, 22]. The exact mechanism of this enhancement process is unclear at this moment. One possible description of this enhancement is by the acceleration of an unactivated path to first (lower energy) triplet T_1. However, this possibility is not unique, as several theories of radiationless transitions (see, e.g. Ref. 22) predict a transition to a single triplet state showing non-activated behaviour at low temperatures but requiring activation at higher temperatures. The lifetimes and chemical evolution of bianthrylidene triplets T are all very strongly controlled by the viscosity of the medium[15–22]. Very high viscosities, as achieved in hydrocarbon glasses around 90 K, in alcohol glasses around 120 K or in triacetin glasses around 200 K will stabilize these triplets to an extent allowing time-resolved optical or ESR studies on a time scale extending from tens of microseconds to tens of milliseconds. In fluid solutions at 300 K these triplets can be observed by nanosecond flash photolysis[13, 15–22, 87–89]. The optical absorption of bianthrylidene series triplets shows a maximum at around 500 nm (e.g. 490 nm in 1,3,6′,8′-tetramethylbianthrone at 100 K, 450 nm at ~ 300 K; 540 nm in dibenzo[*a,g*] bixanthylidene, and 400–450 nm in 10,10′-dihydroxybianthrylidene). The triplet–triplet absorption spectra extend to much longer wavelengths beyond the ~ 500 nm maximum. In the 1,3,6′,8′-tetramethylbianthrone triplet a long wave absorption maximum at 890 nm was reported[90].

The ESR spectra of the bianthrylidene series triplets T_1 provide a definitive identification of the optical transients. These spectra have established positively that the optical transients are genuine triplet spin species and not free radicals (doublets) or diamagnetic transients. The crucial experimental evidence consists of the strong two quantum ESR transition $\Delta m = 2$ observable at half field, e.g. at 1525 *g* in 1,3,6′,8′-tetramethylbianthrone[15] and at 1528 *g* in dibenzo[*a,g*]bixanthylidene[16]. The rates of disappearance of the optical absorption of the triplets, the rates of formation of the triplet manifold products and finally the rates of disappearance of the ESR signal at half field are equal within experimental error.

The formation of **B**, **E** and **F** can be sensitized by triplet energy transfer from biacetyl triplet (see, e.g. Refs 13, 21 and 91). Triplet sensitization provides access to triplet products under conditions where the singlet manifold product **C** is formed preferentially by direct excitation. This is the situation observed with 1,8-dimethylbianthrone in methylene chloride at 203 K. Direct excitation leads to **C** while biacetyl triplet energy transfer results in the formation of **B**[87].

We mentioned previously that all molecular transformations and photophysical processes of the triplets of the bianthrylidene series are strongly controlled by viscosity (in the range of very high viscosities). Genuine temperature effects seem to be factors of lesser importance in this context (cf. Ref. 75 for the related phenomena in the stilbene series). This point is made clear by noting the critical temperature (t_1) for the complete reversal of the triplet ^{3}**A** to the ground state **A**, (^{3}**A** → **A**) and the critical temperature (t_2) for the complete conversion of the triplet ^{3}**A** into the triplet product, ^{3}**A** → Product (e.g. **D**, **E** or **F**). For 1,3,6′,8′-tetramethylbianthrone t_1 is 87 K in hydrocarbon (methylcyclohexane/2-methylpentane) glass; 104 K in alcohol glass (1-propanol/2-propanol) and 195 K in a triacetin glass. Below t_1 complete reversal to **A** takes place and above t_2 some product is obtained. In the same derivative, t_2 is 100 K for the hydrocarbon glass, 119 K in an alcohol glass and 213 K in the triacetin glass. Above t_2 the extent (e.g. quantum yield) of photoproduct formation from ^{3}**A** is maximal.

The stability of the triplet manifold metastable products, such as **D** forms, which convert into the **B** forms[15, 16] is also completely controlled by the viscosity of the medium. In fact the apparent activation energies E_c for the process ^{3}**A** → **D** and for the process **D** → **B** seem to be sensibly equal and very similar to the apparent value of the potential barrier for the viscosity determining diffusion process, E_v. Thus in the same bianthrone derivative, in a triacetin glass medium at 214 K, E_c for the ^{3}**A** → **D** process amounts to 36 kcal, E_c for the **D** → **B** process amounts to 46 kcal, and E_v amounts to 46 kcal (at 214 K). The viscosity of this medium at this temperature is 6×10^8 cp. For comparison, the approximately viscosity-independent ^{3}**A** → **A** process has under the same conditions an activation energy E_c of 5 kcal[15]. These results are of great importance as they provide as yet the only approach of arriving at an educated guess about the conformations of the metastable ^{3}**A** and **D** intermediates. To explain the differences in the viscosity effects on ^{3}**A** → **A**, ^{3}**A** → **D** and on **D** → **B** in terms of molecular geometry changes we should note first that at such high viscosities any large scale internal motions of our molecules are severely restricted, even more than are the internal motions of the solvent molecules. First we can conclude that the ^{3}**A** → **A** process involves only minor geometry change. Thus the triplet and ground state of these molecules should have roughly similar conformations. This is indeed a surprising but none the less rigorous conclusion as the perpendicular conformation is usually assumed for the bianthrone triplet. The opposite applies to **D** and **B**. These two when taken together, as well as **A**, have all different conformations. We know already that this conclusion applies to **A** and **B** when taken together. However, the present results imply also that **D** and **B** have significantly different conformations. While that of **B** seems to be quite well established[5], that of its metastable form **D** is as yet known only in general terms. Like **B**, it should involve some large scale twisting about the 9–9′ double bond. However, considering its origin from the **A** form we anticipate a lesser extent of folding about the 9–10 and 9′–10′ axes of the anthrone subunits than in the **B** form.

IX. THE PHOTOCHEMICAL INERTNESS OF THE TRIPLET MANIFOLD PHOTOINTERMEDIATES *D, B, E* AND *F*

The triplet manifold photointermediates of the bianthrylidene series (**D**, **B**, **E** and **F**) enjoy the unique distinction of absolute photochemical inertness towards the reverse photochemical process (**D**, **B**, **E**, **F** $\xrightarrow{h\nu}$ **A**). This photostability (which to the best of our knowledge has not been explicitly considered before) is in direct contrast with the high photochemical reactivity of the fundamental modifications **A**. Compared with other photochromic systems this stability is no less remarkable. Thus it sharply contrasts the ring-opening photoreactivity of the **C** photoisomers of bianthrylidene-type molecules and of the whole large group of the closely related 4a,4b-dihydrophenanthrenes. To understand this special situation we need to consider first the conformation of the core region in the four triplet products **D**, **B**, **E** and **F**. This core region (consisting of the 9,9′ double bond and of the four adjoining single bonds) is very strongly deformed, having been subjected to strong torsion about the 9,9′ bond in **D**, **B** and **F** and to significant folding along the 9,9′ axis in **E** and in **F** as well. All the torsional deformations and most of the folding deformations are absent in the corresponding **A** forms. In the absence of 1 and 1′ substituents the core region in the **A** forms is planar, while in the 1,1′-substituted molecules it is folded to some extent about the 9,9′ axis, though less than in the photoisomers **D**, **B**, **E** and **F**. In **D**, **B**, **E** and **F**, upon excitation, the four essential single bonds contract while the central 9,9′ double bond expands. Large changes in geometry occur going from ground to excited state (or falling back) as in these molecules changes in bond length, torsional angles and out of plane deformations are very strongly coupled to strong non-bonded repulsions on one hand and to electronic density changes on the other. As a result of the large geometry changes accompanying the excitation (or de-excitation) a large number of low frequency totally symmetric normal modes acquire large Franck–Condon overlap factors (cf. Ref. 77). This phenomenon was previously treated in the related case of sterically hindered stilbenes[77]. As a result, the **B**, **D**, **E** and **F** forms as other sterically hindered non-planar aromatics undergo fast and efficient radiationless transitions from their excited states to their ground states. To recapitulate, we consider radiationless transitions (this description applies certainly to aromatics and to polyenes) as taking place by crossing from the upper electronic state (relaxed at low viscosities, constrained at very high viscosities) to a virtual isoenergetic highly excited vibrational substate of the lower electronic state. The rate constant (k) of a radiationless transition (in a common formulation of the theory) is given by the product of the density of states (ρ), the square of the electronic coupling factor (J) between the two states, and the total mean Franck–Condon factor $\bar{F}$ for an energy gap E[77].

$$k = (2\pi/h)\, J^2 \cdot \rho \cdot \bar{F}(E)$$

Thus, unlike the situation in planar molecules, severely deformed non-planar systems such as **D**, **B**, **E** and **F** command a considerable number of large (one mode) Franck–Condon factors for the excited singlet or triplet to ground state gap[77]. In addition, the density of vibronic states in the lower electronic state (isoenergetic with lowest vibrational level of upper state) is also increased by the presence of low frequency out of plane deformations. All these factors combine to increase strongly the rates of radiationless internal conversion processes, deactivating efficiently the excited singlet and triplet states of **D**, **B**, **E** and **F**.

X. ACKNOWLEDGEMENTS

Much of this review is based on the research on photochromism and thermochromism in bianthrones and in related systems carried out at the Weizmann Institute of Science, between 1968 and 1977. It is a great pleasure to acknowledge the major contributions to this phase of our research of my former associates R. Korenstein, G. Seger and S. Sharafi-Ozeri, of my old-time friend E. Fischer, and of my colleague T. Bercovici. What we could achieve owes certainly much to the fine synthetic work of Mr M. Kaganovitch and to the excellent experimental skills of Mrs N. Castel and Mrs M. Kazes.

XI. REFERENCES

1. (a) Y. Hirshberg, *Compt. Rend.*, **231**, 903 (1950); (b) Y. Hirshberg and E. Fischer, *J. Chem. Soc.*, 629 (1953).
2. H. Meyer, *Monatsh. Chem.*, **30**, 165 (1909); *Chem. Ber.*, **42**, 143 (1909).
3. (a) Y. Hirshberg, *J. Chem. Phys.*, **27**, 758 (1957); (b) Y. Hirshberg, *Proc. 24th Meeting Israel Chem. Society, Bull. Res. Counc. Israel*, **7A(4)**, Sept. (1958).
4. R. Korenstein, K. A. Muszkat and E. Fischer, *Helv. Chim. Acta*, **53**, 2102 (1970).
5. R. Korenstein, K. A. Muszkat and S. Sharafi-Ozeri, *J. Am. Chem. Soc.*, **95**, 6177 (1973).
6. R. Korenstein, K. A. Muszkat and E. Fischer, *Israel J. Chem.*, **8**, 273 (1970).
7. G. Kortüm and G. Bayer, *Ber. Bunsenges. Phys. Chem.*, **67**, 24 (1963).
8. G. Kortüm and H. Bach, *Zeit. Phys. Chem. N. F.*, **46**, 20 (1965).
9. K. A. Muszkat, *Topics in Current Chemistry*, **88**, 89 (1980).
10. K. A. Muszkat and E. Fischer, *J. Chem. Soc.*, 662 (1967).
11. R. Korenstein, K. A. Muszkat and E. Fischer, *J. Photochem.*, **5**, 447 (1976).
12. (a) T. Bercovici, and E. Fischer, *Israel J. Chem.*, **7**, 127 (1969); (b) T. Bercovici, G. Fischer and E. Fischer, *Israel J. Chem.*, **7**, 227 (1969).
13. R. Korenstein, G. Seger, K. A. Muszkat and E. Fischer, *J. Chem. Soc. Perkin Trans. 2*, 550 (1977).
14. (a) C. A. Coulson and H. C. Longuet-Higgins, *Proc. R. Soc. London*, **A191**, 39 (1947); (b) H. C. Longuet-Higgins, *J. Chem. Phys.*, **18**, 265 (1950).
15. T. Bercovici, R. Korenstein, K. A. Muszkat and E. Fischer, *Pure Appl. Chem.*, **24**, 531 (1970).
16. R. Korenstein, K. A. Muszkat and E. Fischer, *Mol. Photochem.*, **3**, 379 (1972).
17. R. Korenstein, K. A. Muszkat and E. Fischer, *Chem. Phys. Lett.*, **36**, 509 (1975).
18. R. Korenstein, K. A. Muszkat, M. A. Slifkin and E. Fischer, *J. Chem. Soc. Perkin Trans. 2*, 438 (1976).
19. R. Korenstein and K. A. Muszkat, in *Environmental Effects on Molecular Structure and Properties* (Ed. B. Pullman), D. Reidel, Dordrecht, 1976, pp. 561–571.
20. R. Korenstein, K. A. Muszkat and G. Seger, *J. Chem. Soc. Perkin Trans. 2*, 1536 (1976).
21. R. Korenstein, K. A. Muszkat and E. Fischer, *J. Photochem.*, **5**, 345 (1976).
22. R. Korenstein, K. A. Muszkat and E. Fischer, *J. Chem. Soc. Perkin Trans. 2*, 564 (1977).
23. R. Korenstein, K. A. Muszkat and E. Fischer, *Helv. Chim. Acta*, **59**, 1826 (1976).

24. A. Schönberg, *Chem. Ber.*, **58**, 580 (1925).
25. A. Schönberg and O. Schutz, *Chem. Ber.*, **61**, 478 (1928).
26. A. Schönberg and S. Nickel, *Chem. Ber.*, **64**, 2323 (1931).
27. E. D. Bergmann and L. Engel, *Zeit. Phys. Chem.*, **8b**, 137 (1930).
28. E. D. Bergmann and H. Corte, *Chem. Ber.*, **66**, 39 (1933).
29. Y. Hirshberg, E. Löwenthal and E. D. Bergmann, *Bull. Res. Counc. Israel*, **1**, 139 (1951).
30. W. T. Grubb and G. B. Kistiakowsky, *J. Am. Chem. Soc.*, **72**, 419 (1950).
31. W. Theilacker, G. Kortüm and G. Friedheim, *Chem. Ber.*, **83**, 508 (1950).
32. Y. Hirshberg, *Bull. Chem. Res. Counc. Israel*, **1**, 123 (1951).
33. Y. Hirshberg, *Bull. Chem. Res. Counc. Israel*, **2** (3) (1952).
34. Y. Hirshberg, *J. Am. Chem. Soc.*, **78**, 2304 (1956).
35. G. Kortüm, W. Theilacker and V. Braun, *Zeit. Phys. Chem. N. F.*, **2**, 179 (1954).
36. (a) Y. Hirshberg and E. Fischer, *J. Chem. Phys.*, **23**, 1723 (1955); (b) G. Kortüm, W. Theilacker and V. Braun, *J. Chem. Phys.*, **23**, 1723 (1955).
37. (a) Y. Hirshberg and E. Fischer, *Angew. Chem.*, **70**, 573 (1958); (b) G. Kortüm, *Angew. Chem.*, **70**, 573 (1958).
38. K. A. Muszkat, D. Gegiou and E. Fischer, *Chem. Commun.*, 447 (1965).
39. J. N. Murrell, *The Theory of the Electronic Spectra of Organic Molecules*, Methuen, London, 1963, Chapt. 7.
40. S. Sharafi-Ozeri and K. A. Muszkat, Program CONFI, Quantum Chemistry Program Exchange, Program QCPE 334, 1977.
41. A. Warshel and M. Karplus, *J. Am. Chem. Soc.*, **94**, 5612 (1972); **96**, 5677 (1974).
42. R. S. Liu, H. Matsumoto, A. E. Asato and D. Mead. *J. Am. Chem. Soc.*, **108**, 3796 (1986).
43. E. Fischer, *Rev. Chem. Intermed.*, **5**, 393 (1984).
44. R. Korenstein, Ph.D. Thesis, The Weizmann Institute of Science, Rehovot, 1976.
45. I. Agranat and Y. Tapuhi, *J. Am. Chem. Soc.*, (a) **98**, 615 (1976); (b) **100**, 5604 (1978); (c) **101**, 665 (1979); (d) *J. Org. Chem.*, **44**, 1941 (1979); (e) *Nouv. J. Chim.*, **1**, 361 (1977); (f) Y. Tapuhi, O. Kalisky and I. Agranat, *J. Org. Chem.*, **44**, 1949 (1979).
46. J. H. Day, *Chem. Rev.*, **63**, 65 (1963).
47. R. S. Becker and C. E. Earhart, *J. Am. Chem. Soc.*, **92**, 5049 (1970).
48. G. Kortüm, *Ber. Bunsenges. Phys. Chem.*, **78**, 391 (1974).
49. A. Schönberg, A. F. A. Ismail and W. Asker, *J. Chem. Soc.*, 442 (1946).
50. R. Padova, *Ann. Chim. Phys.*, **19**, 386 (1910).
51. R. Wizinger, *Zeit. Angew. Chem.*, **40**, 939 (1927).
52. S. L. Matlow, *J. Chem. Phys.*, **23**, 152 (1955).
53. G. Nielsen and G. K. Fraenkel, *J. Chem. Phys.*, **21**, 1619 (1953).
54. R. B. Woodward and E. Wasserman, *J. Am. Chem. Soc.*, **81**, 5007 (1959); E. Wasserman, *J. Am. Chem. Soc.*, **81**, 5006 (1959).
55. Y. Hirshberg and S. I. Weisman, *J. Chem. Phys.*, **28**, 739 (1958).
56. L. A. Harrah and R. S. Becker, *J. Phys. Chem.*, **69**, 2487 (1965).
57. G. Kortüm and K. W. Koch, *Chem. Ber.*, **100**, 1515 (1967).
58. I. Agranat, M. Rabinovitz, H. R. Falle, G. R. Lockhurst and J. N. Ockwell, *J. Chem. Soc. B*, 294 (1970).
59. R. Lorenz, U. Wild and J. R. Huber, *Photochem. Photobiol.*, **10**, 233 (1969).
60. D. H. Phillips and J. C. Schug, *J. Quantum Chem.*, 221 (1971).
61. A. Heller, Ph.D. Thesis, The Hebrew University, Jerusalem, 1960.
62. (a) E. Harnik, *J. Chem. Phys.*, **24**, 297 (1956); (b) E. Löwenthal, quoted in Ref. 62a, p. 299.
63. (a) Z. R. Grabowski and M. S. Balasiewicz, *Trans. Faraday Soc.*, **64**, 1 (1968); (b) Z. R. Grabowski, B. Czochralska, A. Vincenz-Chodkowska and M. S. Balasiewicz, *Disc. Faraday Soc.*, **45**, 145 (1968).
64. L. J. Dombrowski, C. L. Groncki, R. L. Strong and H. H. Richtol, *J. Phys. Chem.*, **73**, 3481 (1969).
65. I. Rosenthal, P. Peretz and K. A. Muszkat, *J. Phys. Chem.*, **83**, 350 (1979).
66. K. A. Muszkat and W. Schmidt, *Helv. Chim. Acta*, **54**, 1195 (1971).
67. K. A. Muszkat and S. Sharafi-Ozeri, *Chem. Phys. Lett.*, **20**, 397 (1973).
68. K. A. Muszkat, G. Seger and S. Sharafi-Ozeri, *J. Chem. Soc. Faraday Trans. 2*, **71**, 1529 (1975).
69. G. Kortüm, W. Theilacker and G. Schreyer, *Zeit. Phys. Chem. N. F.*, **11**, 182 (1957).
70. A. F. A. Ismail and Z. M. El-Shafei, *J. Chem. Soc.*, 3393 (1957).

71. (a) E. D. Bergmann, E. Fischer, Y. Hirshberg, D. Lavie, Y. Sprinzak and J. Szmuszkovicz, *Bull. Soc. Chim. Fr.*, **20**, 798 (1953); (b) E. D. Bergmann and Z. Pelchowicz, *Bull. Soc. Chim. Fr.*, **20**, 809 (1953).
72. E. W. Wassermann and R. E. Davis, *J. Chem. Phys.*, **30**, 1367 (1959).
73. A. Schönberg and E. Singer, *Tetrahedron Lett.*, 1925 (1975).
74. D. L. Fanselow and H. G. Drickamer, *J. Chem. Phys.*, **61**, 4567 (1974).
75. D. Gegiou, K. A. Muszkat and E. Fischer, *J. Am. Chem. Soc.*, **90**, 12 (1968).
76. S. Sharafi and K. A. Muszkat, *J. Am. Chem. Soc.*, **93**, 4119 (1971).
77. G. Fischer, G. Seger, K. A. Muszkat and E. Fischer, *J. Chem. Soc. Perkin Trans. 2*, 1569 (1975).
78. M. E. Peover, *Disc. Faraday Soc.*, 177 (1968).
79. (a) D. H. Evans and R. W. Busch, *J. Am. Chem. Soc.*, **104**, 5057 (1982); (b) B. A. Olsen and D. H. Evans, *J. Am. Chem. Soc.*, **103**, 839 (1981).
80. D. H. Evans and N. Xie, *J. Am. Chem. Soc.*, **105**, 315 (1983).
81. P. Neta and D. H. Evans, *J. Am. Chem. Soc.*, **103**, 7041 (1981).
82. H. Brockmann and R. Mühlmann, *Chem. Ber.*, **82**, 348 (1949).
83. A. Bromberg, K. A. Muszkat and E. Fischer, *Chem. Commun.*, 1352 (1968).
84. A. Bromberg and K. A. Muszkat, *J. Am. Chem. Soc.*, **91**, 2860 (1969).
85. A. Bromberg, K. A. Muszkat and A. Warshel, *J. Chem. Phys.*, **52**, 5952 (1970).
86. K. A. Muszkat, S. Sharafi-Ozeri, G. Seger and T. A. Pakkanen, *J. Chem. Soc. Perkin Trans. 2*, 1975 (1975).
87. T. Bercovici, R. Korenstein and E. Fischer, *J. Phys. Chem.*, **80**, 108 (1976).
88. E. W. Förster and E. Fischer, *J. Chem. Soc. Chem. Commun.*, 1315 (1972).
89. J. R. Huber, U. Wild and H. H. Günthard, *Helv. Chim. Acta*, **50**, 589, 841 (1967).
90. K. H. Gschwind and U. P. Wild, *Helv. Chim. Acta*, **56**, 809 (1973).
91. H. H. Richtol, R. L. Strong and L. J. Dombrowski, *Israel J. Chem.*, **12**, 791 (1974).

The Chemistry of Quinonoid Compounds, Vol. II
Edited by S. Patai and Z. Rappoport

CHAPTER 7

Chemiluminescence of quinonoid compounds

KARL-DIETRICH GUNDERMANN and DIETER LIESKE
Institut für Organische Chemie, Technische Universität Clausthal, 3392 Clausthal-Zellerfeld, Leibnizstrasse 6, FRG

I. INTRODUCTION

Chemiluminescence is the emission of light from electronically excited molecules whose excitation is produced from chemical reactions by direct energy transformation, and not, or nearly exclusively not, from heat. The most general mechanistic theory of the formation of such excited states is that of appropriate electron transfers[1-3]: a reaction product is formed in a way that at least part of the electrons are transferred to the lowest unoccupied molecular orbital (LUMO). It depends on the energetics of the chemical reactions involved, whether a first excited singlet state (leading to fluorescence emission) or a first excited triplet state (yielding phosphorescence) is formed as the primary excited species. This primary excited species can be the light emitter only when it luminesces in the visible range of the spectrum ('direct chemiluminescence')—otherwise a fluorescer must be present to which the excitation energy can be transferred ('indirect chemiluminescence'). For further details see Ref. 4. This chapter is restricted to chemiluminescence in the visible range of the spectrum and does not deal with IR or UV chemiluminescence.

Quinone/hydroquinone systems are characterized by electron transfer processes, since they are redox systems. One should therefore expect in these systems chemiluminescence reactions, and they have actually been observed although some of them exhibit very low chemiluminescence quantum yields. Moreover, if oxygen is present in such chemiluminescent reactions of quinone/hydroquinone systems there appears to be the possibility of formation of excited singlet oxygen species.

Some quinone derivatives such as the anthraquinone dicarboxylic acid hydrazide (1)[25, 26] are chemiluminescent on oxidation—evidently not as quinones but as hydrazides of the well-known luminol type.

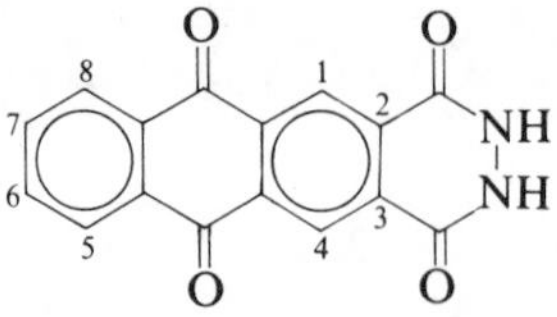

(1)

The oxidative elimination of the hydrazide nitrogen in **1** follows a reaction path involving a quinodimethane peroxide (see Section VI). This chapter will discuss the chemiluminescence of the following types of compounds: quinones; semiquinones and hydroquinones; quinone derivatives with cyclic hydrazide groups of the luminol type; diazaquinones; quinone imines and related compounds, e.g. flavins; and quinodimethanes.

II. QUINONES, SEMIQUINONES AND HYDROQUINONES

The earliest observations of the chemiluminescence of a quinone was reported in 1883[5], when it was observed that a colourless light emission occurs when ethanolic potassium hydroxide solutions of phenanthrene quinone (**2**) are shaken with air. It was proposed that **2** is transformed to diphenic acid:

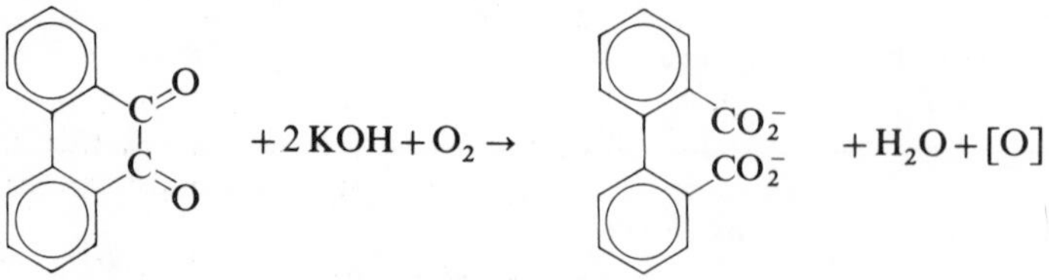

$$(2) + 2\,KOH + O_2 \rightarrow \text{diphenate } (CO_2^-)_2 + H_2O + [O]$$

(2)

The dichloro derivative (**3**) also yielded light emission under the same reaction conditions, probably via **2** formed by hydrolysis in the alkaline medium. The quantum yield as well as the excitation mechanism have been unknown.

A. Violanthrone

Reacting violanthrone (**4**) with dibenzal diperoxide (**5**) in paraffin oil at 250 °C produces a light-red chemiluminescence[6]. A brilliant red chemiluminescence is also given by **4** on

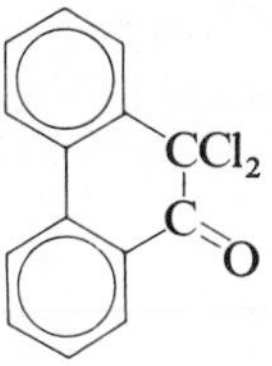

(3)

treatment of its chloroform solution with aqueous alkaline hydrogen peroxide and gaseous chlorine, i.e. on oxidation with singlet oxygen[7a]. This rather complicated example

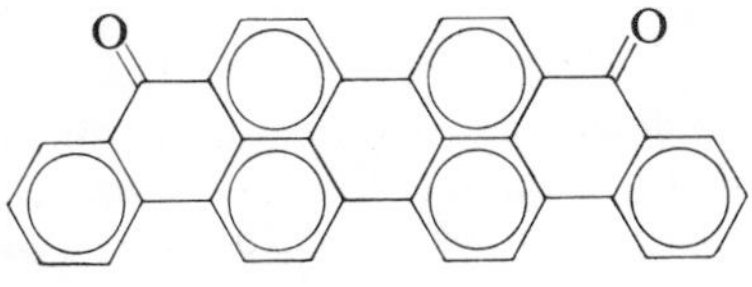

(4) 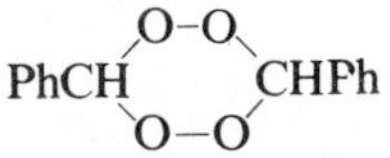 (5)

of a quinone chemiluminescence is reported before the discussion of simpler quinones because it has been investigated especially thoroughly using modern methods and mechanistic concepts, and also taking into account the role of excited singlet oxygen in the chemiluminescence.

The first investigator of the chemiluminescence of **4** suggested the following mechanism[6]:

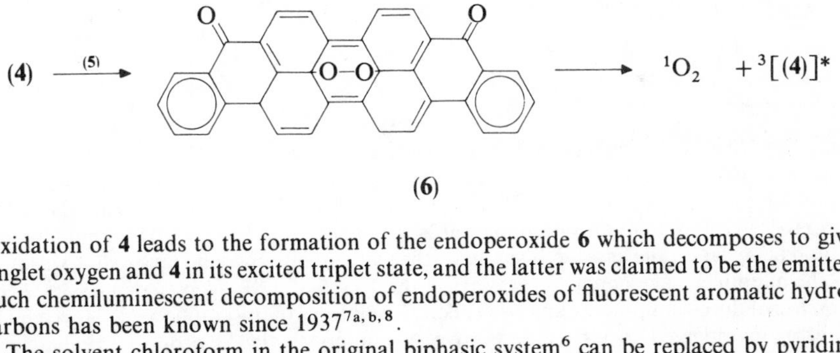

$$(4) \xrightarrow{(5)} (6) \longrightarrow {}^{1}O_2 + {}^{3}[(4)]^*$$

(6)

Oxidation of **4** leads to the formation of the endoperoxide **6** which decomposes to give singlet oxygen and **4** in its excited triplet state, and the latter was claimed to be the emitter. Such chemiluminescent decomposition of endoperoxides of fluorescent aromatic hydrocarbons has been known since 1937[7a, b, 8].

The solvent chloroform in the original biphasic system[6] can be replaced by pyridine which simultaneously serves as base so that the chemiluminescence of **4** can be performed in a homogeneous phase.

In subsequent investigations[9, 10] the role of singlet oxygen was firmly established in the violanthrone chemiluminescence. It was stated that the same red emission can be obtained by treatment of **4** in dibutylphthalate solution with singlet oxygen produced by microwave discharge, and that the emitting species is not an excited triplet but the first excited singlet state of **4**. This species is also the emitter in the above-mentioned chemiluminescent reaction of **4** with dibenzal diperoxide (**5**)[11]. The excitation of **4** occurs by energy transfer on a transient excited triplet state of **4** by triplet benzaldehyde and/or singlet oxygen monomers and collisional pairs.

B. Other Quinones, Hydroquinones and Semiquinones

A large number of simple quinones and hydroquinones have been described to yield chemiluminescence on oxidation or reduction. The quantum yields were always low or very low ('ultraweak' chemiluminescence, i.e. chemiluminescence with quantum yields of 10^{-8} or less).

In most of these cases the formation of singlet oxygen and the chemiluminescence emission of the latter appears to be the decisive mechanistic pathway. It appears reasonable to think that intermediate reduction products of quinones such as semiquinones or even hydroquinones are the electron donors. Stauff and coworkers[12] and other authors[13] previously proposed the following electron transfer steps:

$$O_2 + D \rightarrow O_2^- + D^+ \qquad (D = \text{electron donor})$$

$$O_2^- + O_2 \rightarrow (O_4)^-$$

$$(O_4)^- + O_2^- \rightarrow (O_2O_2)^* + O_2^{2-}$$

The following quinones exhibited chemiluminescence on treatment of their aqueous alkaline solutions with oxygen or by electrochemical reduction: duroquinone[12], benzoquinone[14], methylbenzoquinone[14], 1,4-naphthoquinone[14], 9,10-anthraquinone[14], 9,10-anthraquinone-2-sulfonate[15].

Whereas in most cases of reduction of quinones the above-mentioned mechanism of the formation of singlet oxygen appears rather clear since the quinones are transformed first into the corresponding semiquinones or hydroquinones which in turn could act as electron donors, the situation is more difficult in the case of duroquinone where it was postulated[12]

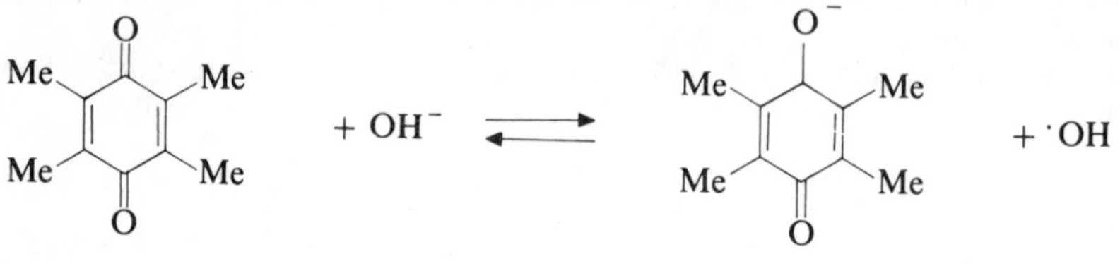

Me Me Me Me (7) + OH⁻ ⇌ Me Me Me Me O⁻ O + ·OH

(7)

that a redox reaction should occur between **7** and hydroxyl ion to give the semiquinone and hydroxyl radicals. However, since the reaction takes place in an alcoholic medium, the actual mechanism may involve $^-$OR ions as the electron donors*. Comparing the redox potentials of the systems duroquinone/durosemiquinone and $HO^-/\cdot OH$ the equilibrium should be by far on the left side. That, nevertheless, semiquinones are involved in this type of chemiluminescence could be proved by the fact that ESR spectra demonstrate the presence of durosemiquinone anions in alkaline solutions of duroquinone[15]. Moreover in all cases investigated so far the respective semiquinones always produced higher light yields than the parent quinones.

However, there are examples where chemiluminescence of quinones does not require oxygen at all, or where at least the light emission is caused both by excited singlet oxygen and by other compounds formed in certain oxidation reactions. Thus, the semiquinone of 9,10-anthraquinone-2-sulfonate (**9**) produced by reduction of 9,10-anthraquinone-2-sulfonate (**8**) at pH 12.5 using a platinum cathode yields chemiluminescence not only by its reaction with oxygen but also with ferricyanide[16]. When the hydroquinone (**10**) is oxidized by ferricyanide the light yield is twice that of the corresponding reaction of the semiquinone (**9**). It was suggested that excited forms of the quinone (**8**) and of the

* We are indebted to Prof. Joachim STAUFF for this remark.

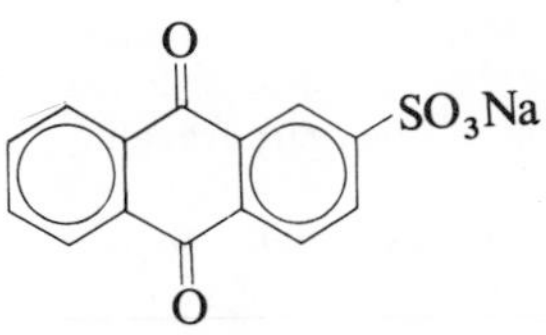

(8) 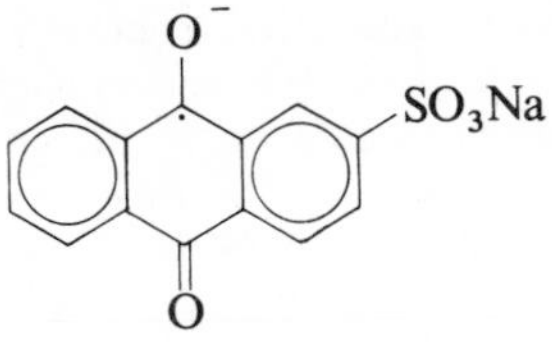 (9) 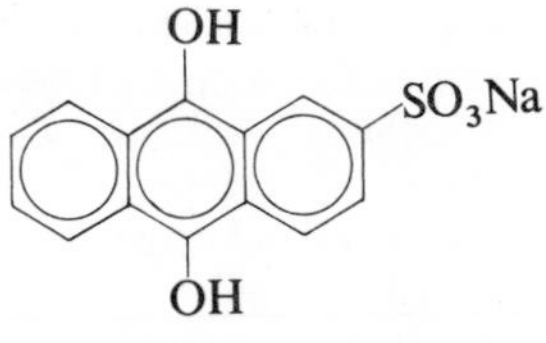 (10)

semiquinone (9) are the emitters in their triplet state, since the wavelengths of the emissions were higher than those of the corresponding singlet states[16]. Perhaps, although evidence is still lacking, the chemiluminescence in quinone/semiquinone/hydroquinone systems is a special case of CIEEL (chemically initiated electron exchange luminescence; for a general treatment of this general concept of chemiluminescence see Ref. 1).

Another case is the well-known Trautz-Schorigin reaction[17], i.e. the oxidation of pyrogallol with hydrogen peroxide and formaldehyde in aqueous alkaline solution is chemiluminescing with a red emission mainly caused by singlet oxygen[18, 19]. However, more detailed investigations demonstrated that an additional emission is observed in the range of 400–600 nm. This has been ascribed to oxidation products of pyrogallol (**11**) which are formed via purpurogalline (**12**) as the first intermediate[18], since purpurogalline itself, on oxidation with different oxidants such as O_2, ozone, hypochlorite, permanganate or electrolytic oxidation in aqueous alkaline solution (pH 9–11) was found to yield chemiluminescence at 400–600 nm with quantum yields of 10^{-7}–10^{-6} Einsteins mol^{-1}. No chemiluminescence occurred in acidic media (pH 3.5–7) with different oxidants, e.g. Fe^{3+}, $Fe(CN)_6^{3-}$, H_2O_2/peroxidase, or by heating. The rate-controlling step is suggested to be the nucleophilic attack of HOO^- ions on purpurogalloquinone (**13**):

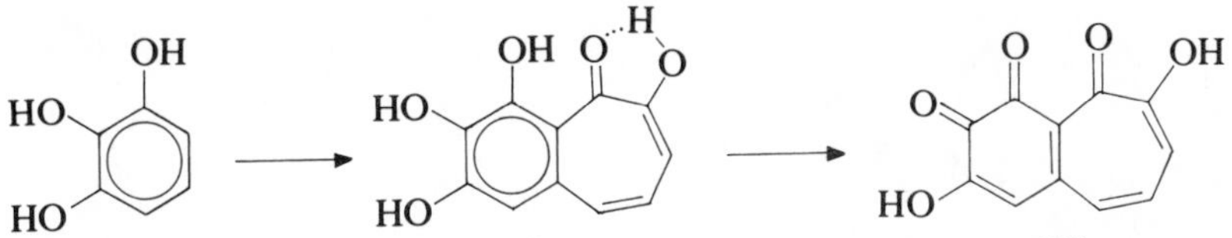

(11) (12) (13)

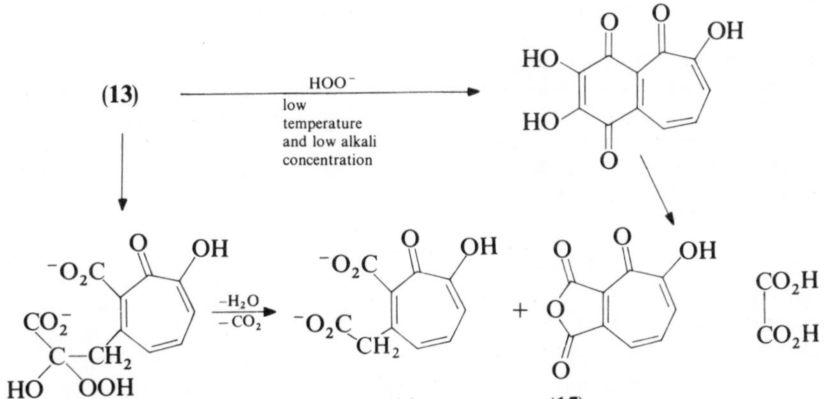

(13) $\xrightarrow{HOO^-}$ low temperature and low alkali concentration

$\xrightarrow[-CO_2]{-H_2O}$

(14) (15) tropolone anhydride

Light emission is produced by the oxidative degradation of the quinone **13** by which excited tropolone derivatives are formed. These compounds, **14**, **15** and perhaps similar tropolone derivatives fluoresce in the range of the spectrum of the purpurogalline chemiluminescence.

Formation of purpurogalloquinone has been established to be the key intermediate in pyrogallol oxidation[18, 20, 21]. This reaction has been studied as a model for the generally very weak chemiluminescence of polyhydroxyphenols regarded as models for humic acids.

C. Humic Acids, Melanins and other Quinone Polymers

Autoxidative polymerization of hydroquinone, purpurogalline or adrenochrome yields products which on photolysis yield weak chemiluminescence[22]: the irradiation is suggested to initiate electron transfer reactions which in the presence of oxygen produce excited molecular singlet oxygen (cf. Section II.B), whereas in the absence of oxygen some excited quinone is formed as emitting species.

Adrenochrome polymers[23] chemiluminesce after photo-oxidation in alkaline media with quantum yields of 10–510 photons per second, the maxima of this emission are observed at 480–500, 570 and 615–650 nm. Radical scavengers reduce the quantum yield by 20–97 %. The mechanism is suggested to be an oxidative ring-opening of *ortho*-semiquinone structures analogous to that of purpurogalline, a dioxetane being the key intermediate:

CP = starting polymer

The excited products formed in this way are postulated to be the emitters of the light in the spectral range lower than 600 nm whereas the longer wavelength emission (at 615–650 nm) stems from excited singlet oxygen.

Although quinone structures are very probably involved in this type of chemiluminescence, one has to be aware of the fact that a definite determination of the mechanism of these complex reactions will be extremely difficult since only a very small part of the molecules involved lead to electronically excited products. Above all, the exact structure of humic acids is only very approximately known at present.

III. ANTHRAQUINONE HYDRAZIDES

9,10-Anthraquinone-2,3-dicarboxylic acid hydrazide (**1**) chemiluminesces weakly on oxidation with O_2 in DMSO solution and potassium *t*-butylate as base[24, 25] but practically not in aqueous alkaline hydroperoxide[26]. The emission maximum of **1** in the aprotic system is at 580 nm, the quantum yield is ca. 4 % of that of a luminol standard[25].

The 1-, 5- and 6-amino derivatives of **1** exhibited an interesting behavior of the normally chemiluminescence-enhancing effect of the electron-donating amino group. It was found

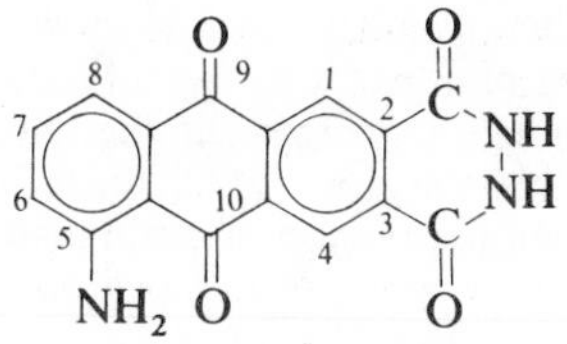

(16)

that only the 5-amino compound (**16**) was strongly chemiluminescent in the aqueous system (93 % light yield of luminol) as well as in the aprotic one (273 % that of luminol!), whereas the 1-isomer had only 0.37 % of the luminol light yield in the aprotic system and gave no light at all in the aqueous one.

The phenylimine derivative (**17**) gave less than 10 % of the quantum yield of its parent compound (**1**).

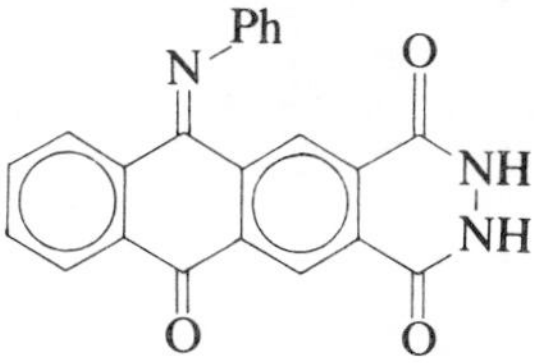

(17)

IV. DIAZAQUINONES

In the first paper on luminol chemiluminescence, Albrecht[27] had already suggested a diazaquinone (5-aminophthalazine-1,4-dione) (**18**) as an intermediate in the reaction mechanism, although such compounds were unknown at that time, and the role of the diazaquinone had been interpreted incorrectly, as shown later by further experimental evidence. The first diazaquinones were prepared by Clement[28] and by Kealy[29] but none of them was chemiluminescent, due to the fact that their oxidative hydrolysis produced non-fluorescent dicarboxylates.

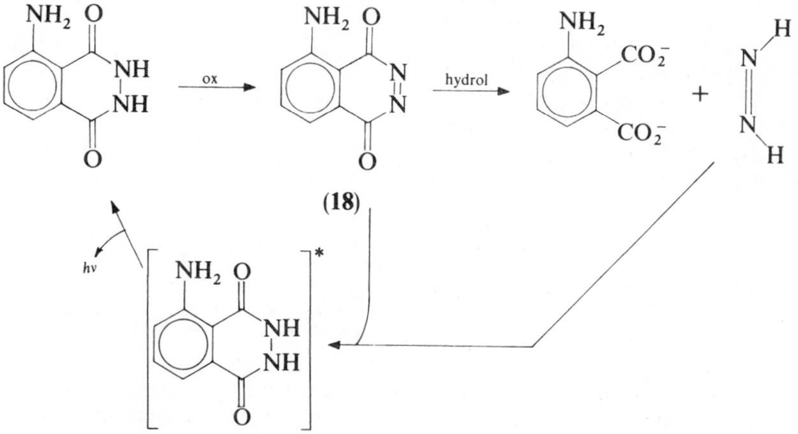

(18)

The diazaquinones **19** and **20** were synthesized by dehydrogenation of the corresponding hydrazides at temperatures of ca. $-50°C$[30, 31]. **19** and **20** are deeply violet-colored crystalline substances which undergo nitrogen elimination at 0°C. They are very prone to nucleophilic attack at their carbonyl groups. Oxidation by alkaline hydrogen peroxide results in chemiluminescence whose emission spectrum matches that of the corresponding hydrazide and the fluorescence spectrum of the dicarboxylates **21** and **22**, respectively.

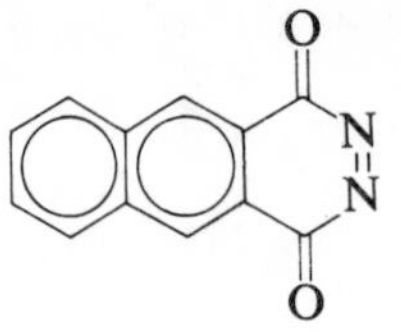

(19) 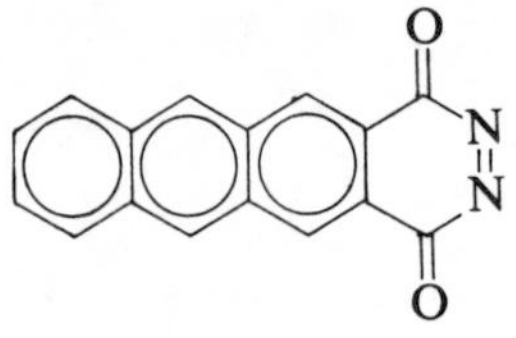 **(20)**

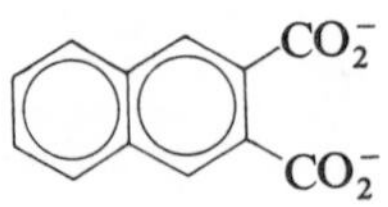

(21) 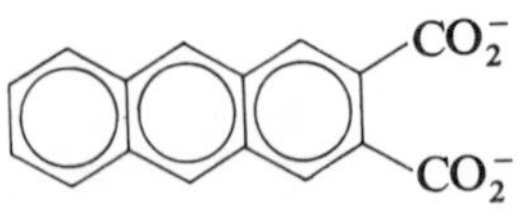 **(22)**

Whereas the chemiluminescence of benzophthalazine-1,4-dione (**19**) was investigated in an inhomogeneous aqueous system (where the diazaquinone is practically insoluble)[48] a very thorough investigation of the naphthophthalazine-1,4-dione (**20**) was performed in homogeneous dimethyl phthalate solution using anhydrous hydrogen peroxide and diethylamine as base[30]. Trapping the diazaquinone by cyclopentadiene with which a Diels–Alder adduct (**23**) is very rapidly formed it could be made probable that a diazaquinone is also formed as intermediate in the hydrazide chemiluminescence reaction[30].

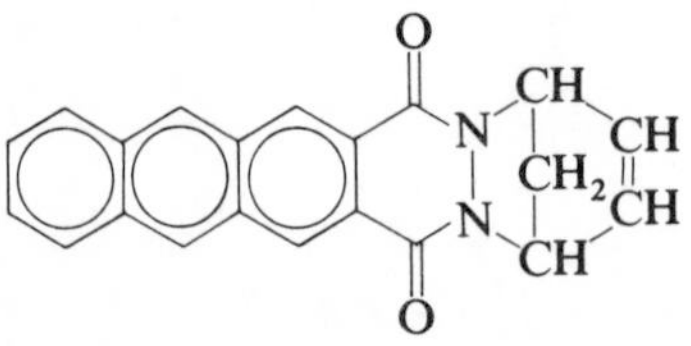

(23) 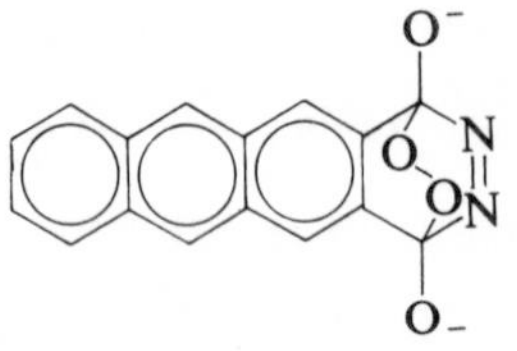 **(24)**

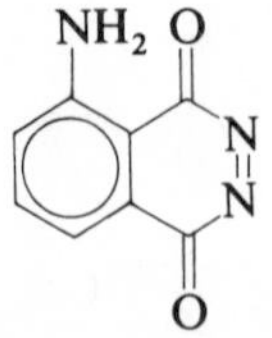

(18) 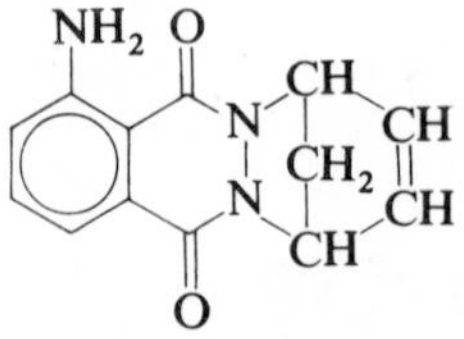 **(26)** 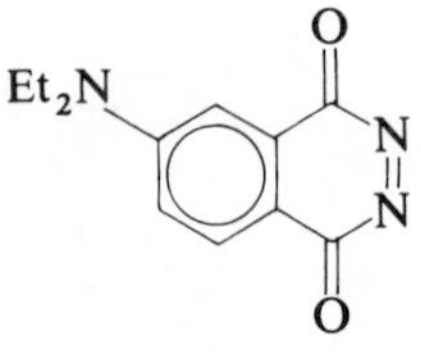 **(27)**

The requirement of an additional oxidant demonstrates that it is not the diazaquinone itself which on mere hydrolysis yields the chemiluminescence but an oxidation product of the diazaquinone, which is very probably the endoperoxide (**24**). The theories concerning its chemiluminescent decomposition are described in Section VI.

'Luminol diazaquinone' (5-aminophthalazine-1,4-dione, **18**) has not yet been obtained in analytically pure form probably due to the easy formation of by-products during the dehydrogenation of the hydrazide by oxidative degradation of the primary amino group. However, the corresponding Diels–Alder adduct (**26**) was isolated from luminol solutions undergoing ferricyanide oxidation in the presence of cyclopentadiene[31].

Recently the diazaquinone **27**, obtained by dehydrogenation of 7-diethylaminophthalic acid hydrazide, has been described[32].

V. QUINONE IMINES AND RELATED COMPOUNDS

A. Wurster's Red and Wurster's Blue

Compounds of the Wurster's Red (**28**) and Wurster's Blue type (**29**) are semiquinone analogs which chemiluminesce in reactions with aromatic or heterocyclic radical anions

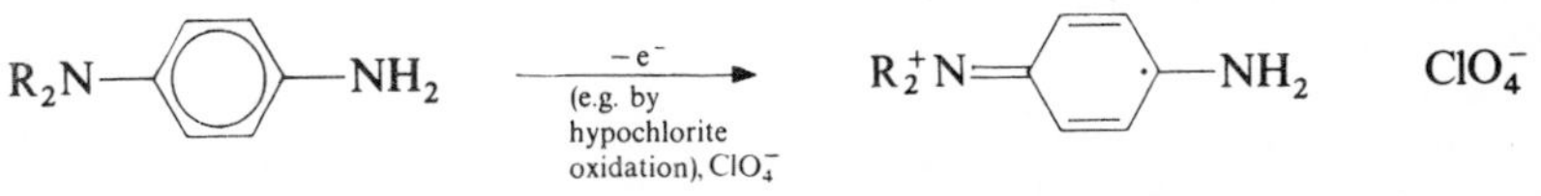

R_2N–C₆H₄–NH_2 $\xrightarrow[\text{(e.g. by hypochlorite oxidation), } ClO_4^-]{-e^-}$ $R_2\overset{+}{N}$=C₆H₄(·)–NH_2 ClO_4^-

R = Me (**28**)

$R_2\overset{+}{N}$=C₆H₄(·)–NR'_2

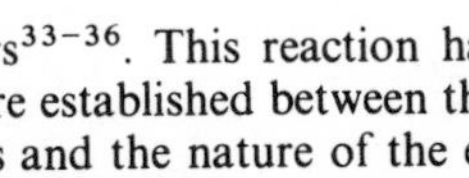

(**29**, R = R′ = Me)

acting as electron donors[33–36]. This reaction has been investigated very thoroughly: quantitative relations were established between the reduction potential of the respective electron donor molecules and the nature of the excited state formed from the electron donor. Irrespective of whether the latter was directly formed in its first excited (so-called 'energy-sufficient') singlet state, or via first excited triplet states which, e.g. by triplet–triplet annihilation reaction give rise to the first excited singlet state, the latter is in all cases studied the emitting species.

Some chemiluminescent systems of aromatic hydrocarbon radical anions[37] are shown below:

with Wurster's Red perchlorate:

Excited species.

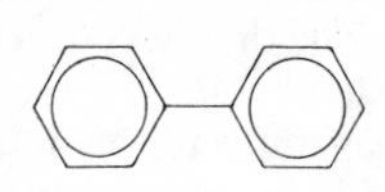

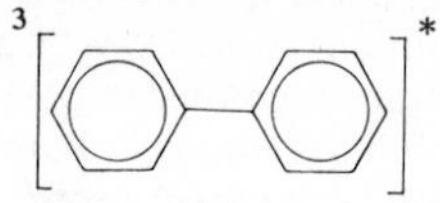

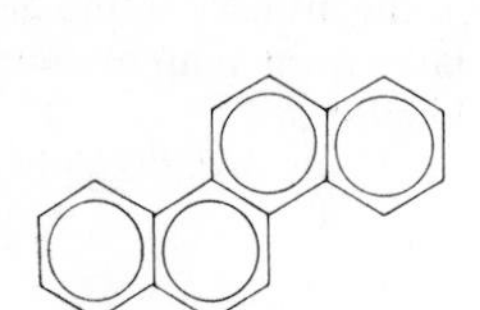

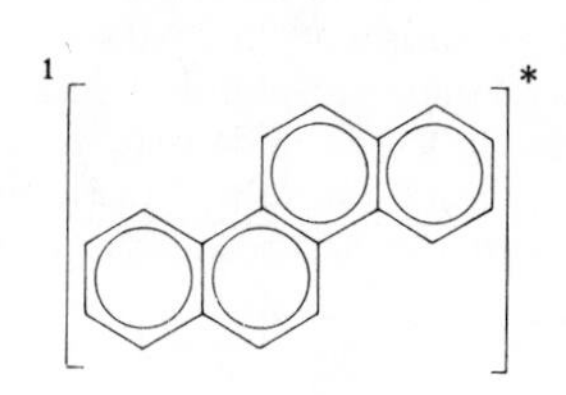

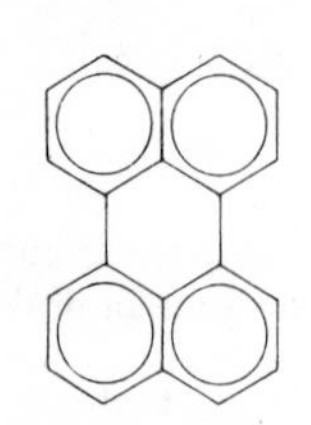

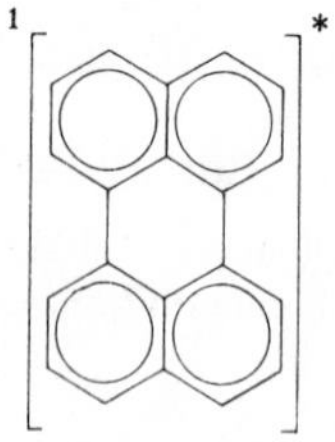

with Wurster's Blue perchlorate:

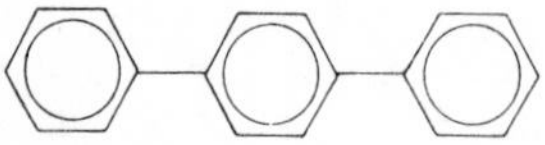

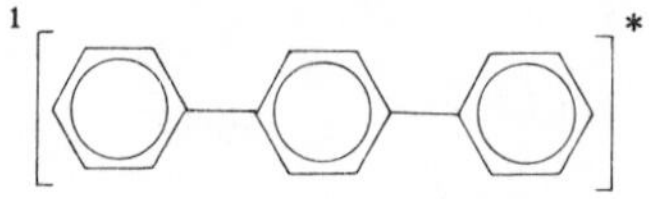

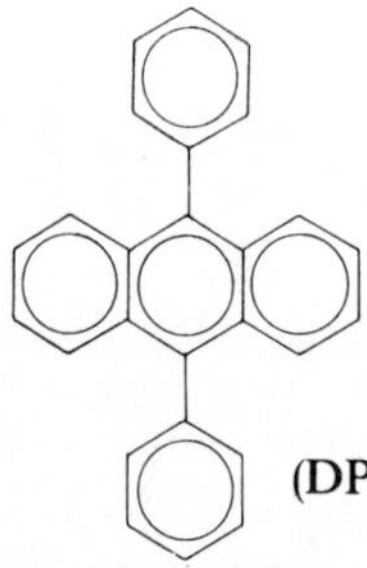

(DPA) ^{1}DPA*

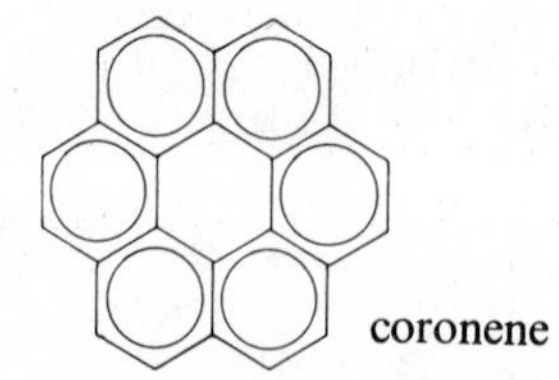

coronene ^{1}coronene*

The reaction between Wurster's Blue and chrysene radical anion may be quoted as an explicit example:

$$\mathrm{Me_2\overset{+}{N}{=}C_6H_4{-}\dot{N}Me_2\ ClO_4^- + [chrysene]^{-\cdot} \longrightarrow Me_2\overline{N}{-}C_6H_4{-}NMe_2 + [chrysene]^*}$$

B. *N*-Arylsulfonyl benzoquinone Imines

Condensation of *N*-arylsulfonyl benzoquinone imines of the type **30** with *N*-arylamino crotonates (**31**) yields chemiluminescence provided the crotonate is substituted with phenyl groups (R = Ph) whereas R = alkyl or cycloalkyl leads to non-chemiluminescent reactions[38]. This reaction leading to substituted indoles appears to

(**30**) ArSO₂N=C₆H₄=O + (**31**) HC(CO₂Et)=C(NHR)Me → [intermediate] → (−H₂O) ArSO₂NH–C₆H₃(OH)–C(CO₂Et)=C(NHR)Me, or (−H₂O) 1-R-2-Me-3-CO₂Et-indole

involve no radical intermediates—at least none detectable by ESR. The excitation mechanism is not yet clear. Interestingly the quinone imines (**30**) produce chemiluminescence also on reaction with aliphatic amines such as butyl amine, benzyl amine and diethyl amine[39]. This evidently demonstrates that the excitation mechanism is different from that of the indole formation for which as mentioned above phenyl groups must be present on the amino crotonate nitrogen atom. Moreover, in the chemiluminescent reaction of **30** with simple amines free radicals with rather high lifetimes were detected.

C. Flavins

The bioluminescence of bacteria involves the reaction

Dihydroflavin mononucleotide ($FMNH_2$ = **32**) + O_2 + RCHO $\xrightarrow{\text{enzyme}}$ FMN (**33**) + RCOOH + $h\nu$

(RCHO = long-chain aliphatic aldehyde)

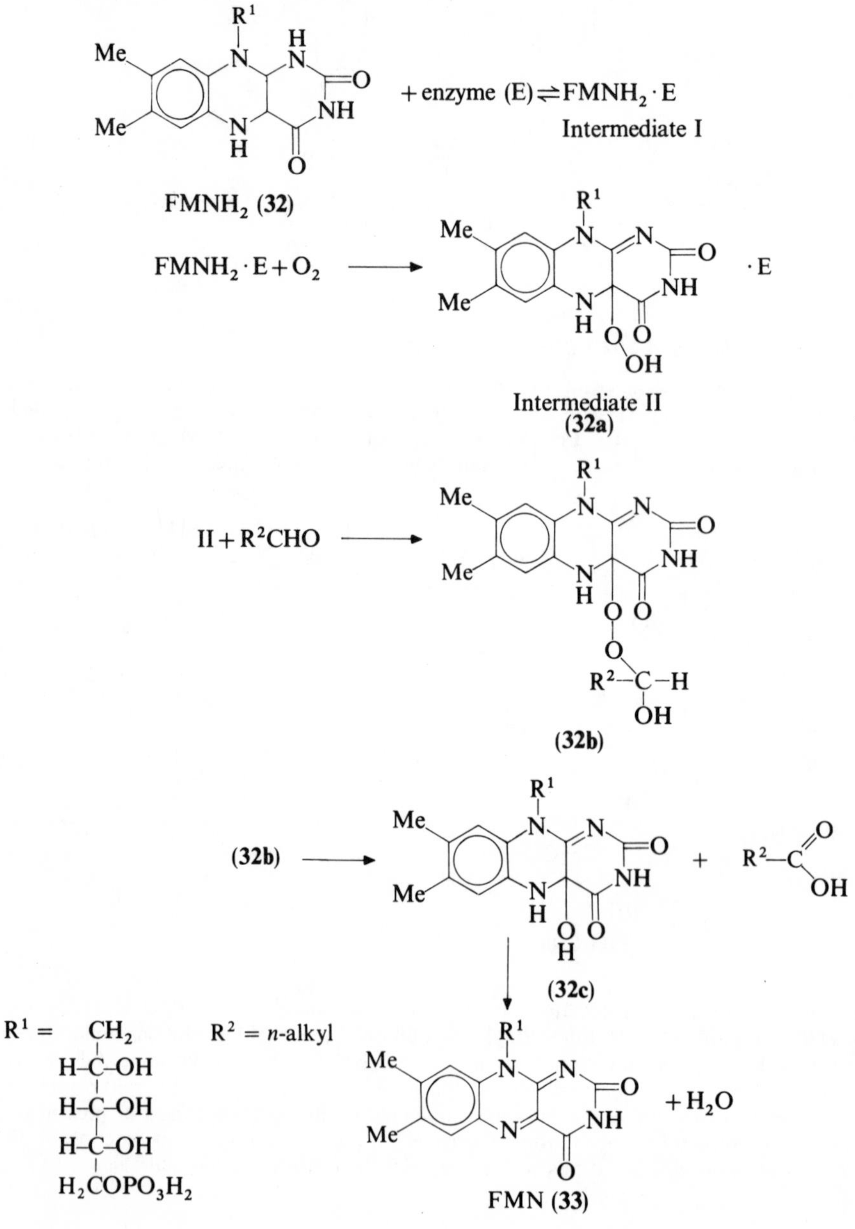

$FMNH_2$ (**32**) + enzyme (E) $\rightleftharpoons$ $FMNH_2 \cdot E$
Intermediate I

$FMNH_2 \cdot E + O_2$ ⟶ [structure] $\cdot$ E

Intermediate II
(**32a**)

II + R^2CHO ⟶ (**32b**)

(**32b**) ⟶ (**32c**) + R^2–C(=O)OH

(**32c**) ⟶ FMN (**33**) + H_2O

R^1 = CH_2–(H–C–OH)–(H–C–OH)–(H–C–OH)–$H_2COPO_3H_2$ R^2 = *n*-alkyl

The emitting species (maximum of the blue–green emission ca. 510 nm) is the 4a-hydroxyflavin intermediate **32c**[40]. Aside from all important biochemical and other implications concerning the mechanism of this reaction (for reviews see Refs 41, 42) the flavin system and its dihydro derivative as iso-alloxazine systems may be classified as 'extended quinone' imine derivatives. The first step in the oxidation of $FMNH_2$ involves the addition of molecular oxygen to the 4a position of the flavin ring[40], a peroxide (**32a**) being formed. The end products—in the luminescent reaction—are formed via a peroxy-semiacetal (**32b**) which results from the reaction of the peroxide with the long-chain aldehyde.

VI. QUINODIMETHANE DERIVATIVES

It appears to be solidly established that the oxidation of diazaquinones, under appropriate conditions, yields a chemiluminescence very similar to or even identical with that of the corresponding hydrazides showing among other features the same maximum of the emission spectrum and the same end product. However it is still uncertain whether the generally assumed peroxide (**34**) actually is the key intermediate whose decomposition into

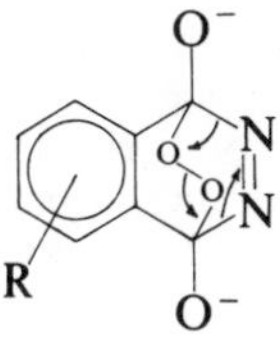

(**34**)

molecular nitrogen and *o*-dicarboxylate is the light-producing step. Since a synchronous electrocyclic decomposition of **34** (following the inserted arrows) is a thermally allowed process according to the Woodward–Hoffmann rules[43,44], it has been proposed by Michl[45] and by Schuster[2] that the decomposition of the peroxide **34a** (i.e., **34**, with R = 5-amino) proceeds in two steps:

(1) elimination of N_2 with formation of the quinodimethane compound (**35**);

(2) intramolecular electron transfer from the amino group of luminol (and from higher condensed aromatic systems such as anthracene) to the peroxide group, followed by cleavage of the latter and re-aromatization involving the return of the transferred electron to the amino group:

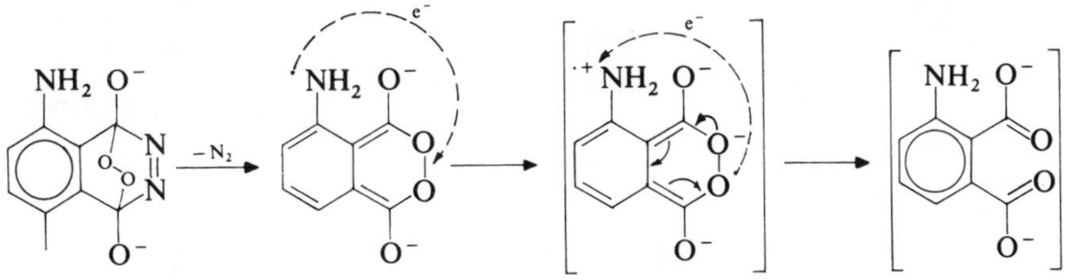

(**34a**) (**35**)

For the time being the existence of both postulated compounds **34a** and **35** is still hypothetical.

However, an example has been provided for an *o*-quinodimethane chemiluminescence[46,47]. The very unstable quinodimethane peroxide (**37**) could be trapped as its

O_2
methylene blue
$h\nu$

(36)

(37) **(38)**

Diels–Alder adduct with maleic anhydride (**38**). In the presence of 'activators' such as perylene, acridine orange or 9,10-diphenylanthracene, the photo-oxidation of the isocoumarine derivative (**36**) yields indirect chemiluminescence whose emission matches that of the fluorescence of the activator. An intramolecular electron transfer mechanism has been proposed as the cause of this chemiluminescence:

(**36**) → → (**37**) + activator (= Act) ⇌ [(**37**) · Act]

encounter complex

→ [$(\mathbf{37})^{-\cdot}$ $Act^{+\cdot}$] → [O^- $O\cdot$ $Act^{+\cdot}$]

radical ion annihilation

↓

+ Act*

The electron transfer steps take place in a solvent cage (symbolised by the brackets in the scheme above).

VII. REFERENCES

1. G. B. Schuster, B. Dixon, J.-Y. Koo, S. P. Schmidt and J. P. Smith, *Photochem. Photobiol.*, **30**, 17 (1979).
2. G. B. Schuster and S. P. Schmidt, *Adv. Phys. Org. Chem.*, **18**, 187 (1982).
3. L. Eberson, *Adv. Phys. Org. Chem.*, **18**, 79 (1982).
4. K.-D. Gundermann and F.Mc Capra, *Chemiluminescence in Organic Chemistry*, Springer Verlag, 1987.
5. B. Lachowitz, *Ber. dt. chem. Ges.*, **16**, 330 (1883).
6. R. B. Kurtz, *Trans. N. Y. Acad. Sci.*, **16**, 399 (1954).
7. (a) H. H. Wasserman and R. W. Murray (Eds), *Singlet Oxygen*, Academic Press, New York 1979; (b) C. Dufraisse and J. Le Bras, *Bull. Soc. Chim. France*, **4**, 349 (1937).
8. I. Saito and T. Matsuura, in *Singlet Oxygen* (Eds H. H. Wasserman and R. W. Murray), Academic Press, New York, 1979, p. 511.
9. R. J. Browne and E. A. Ogryzlo, *Can. J. Chem.*, **43**, 2915 (1965).
10. E. A. Ogryzlo and A. E. Pearson, *J. Phys. Chem.*, **72**, 2913 (1968).
11. S. R. Abbott, S. Ness and D. M. Hercules, *J. Am. Chem. Soc.*, **92**, 1128 (1970).
12. J. Stauff and H. Schmidkunz, *Z. Phys. Chem.* (*Frankfurt*), **35**, 295 (1962).
13. H. Salow and W. Steiner, *Z. Phys.*, **99**, 137 (1936).
14. S. Slawinskii and Y. M. Petrusevich, *Biol. Nauki*, **5**, 69 (1969).
15. J. Stauff, *Photochem. Photobiol.*, **4**, 1199 (1965).
16. J. Stauff and P. Bartolmes, *Angew. Chem. Int. Ed.*, **9**, 307 (1970).
17. (a) M. Trautz, *Z. Phys. Chem.*, **53**, 1 (1905); (b) M. Trautz and P. Schorigin, *Z. Wiss. Photogr. Photochem.*, **3**, 121 (1905).
18. J. Slawinski, *Photochem. Photobiol.*, **13**, 489 (1971).
19. K. Lichszeld and I. Kruk, *Z. Phys. Chem.* (*NF*), **108**, 167 (1977).
20. D. Slawinska, *Photochem. Photobiol.*, **28**, 453 (1978).
21. R. Nilsson, *Acta Chem. Scand.*, **18**, 389 (1964).
22. D. Slawinska and J. Slawinski, *Photochem. Photobiol.*, **21**, 393 (1975).
23. J. Slawinski, W. Puzyna and D. Slawinska, *Photochem. Photobiol.*, **28**, 459 (1978).
24. C. Röker, Dissertation, Techn. Univ. Clausthal, 1975.
25. K.-D. Gundermann, G. Klockenbring, C. Röker and H. Brinkmeyer, *Liebigs Ann. Chem.*, 1873 (1976).
26. R. Wegler, *J. prakt. Chem.*, **148**, 135 (1937).
27. H. O. Albrecht, *Z. Phys. Chem.*, **136**, 321 (1928).
28. R. A. Clement, *J. Org. Chem.*, **25**, 1724 (1960).
29. T. J. Kealy, *J. Am. Chem. Soc.*, **84**, 966 (1962).
30. K.-D. Gundermann, H. Unger and J. Stauff, *J. Chem. Res.* (*S*), 318 (1978).
31. Y. Omote, T. Miyake and N. Sugiyama, *Bull. Chem. Soc. Japan*, **40**, 2446 (1967).
32. A. N. Rusin, N. A. Leksin and A. L. Roshin, *Zh. Org. Khim.*, **16**, 209 1(980).
33. A. Weller and K. Zachariasse, *J. Chem. Phys.*, **46**, 4984 (1967).
34. A. Weller and K. Zachariasse, in *Molecular Luminescence* (Ed. E. C. Lim), Benjamin, New York, 1969, p. 895.
35. A. Weller and K. Zachariasse, *Chem. Phys. Lett.*, **10**, 590 (1971).
36. A. Weller and K. Zachariasse, *Chem. Phys. Lett.*, **10**, 197 (1971).
37. K. A. Zachariasse, Thesis, Vrije Universiteit Amsterdam, 1972.
38. G. N. Bogdanov, E. A. Titov, V. N. Shtol'ko and A. S. Grishchenko, *Izv. Akad. Nauk SSSR, Ser. Khim.*, **12**, 2814 (1972); *Chem. Abstr.*, **78**, 123 518^{t} (1973).
39. E. A. Titov, *Vop. Khim. Khim. Tekhnol.*, **30**, 43 (1973); *Chem. Abstr.*, **81**, 3509^{r} (1974).
40. J. W. Hastings, in *International Symposium on Analytical Applications of Bioluminescence and Chemiluminescence* (Eds. E. Schram and P. Stanley), State Printing & Publishing Inc., Westlake Village, Calif., 1979, pp. 1, 26.

41. M. De Luca, in *Clinical and Biochemical Luminescence* (Eds. L. J. Kricka and T. J. N. Carter), Marcel Dekker Inc., New York, 1982, p. 75.
42. J. W. Hastings, T. O. Baldwin and M. Z. Nicoli, in *Methods in Enzymology* (Ed. M. De Luca), Academic Press, New York, 1978, Vol. 57, p. 135.
43. H. Unger, Diplomarbeit, Techn. Univ. Clausthal, 1971.
44. E. H. White and R. B. Brundrett, in *Chemiluminescence and Bioluminescence* (Eds M. J. Cormier, D. M. Hercules and J. Lee), Plenum Press, New York–London, 1973, p. 231.
45. J. Michl, *J. Phys. Chem.*, **7**, 125 (1975); *Photochem. Photobiol.*, **25**, 141 (1977).
46. J. P. Smith, A. K. Schrock and G. B. Schuster, *J. Am. Chem. Soc.*, **104**, 1041 (1982); J. P. Smith and G. B. Schuster, *J. Am. Chem. Soc.*, **100**, 2564 (1978).
47. W. Adam and I. Erden, *J. Am. Chem. Soc.*, **101**, 5692 (1979).
48. E. H. White, F. G. Nash, D. R. Roberts and O. C. Zafiriou, *J. Am. Chem. Soc.*, **90**, 5932 (1968).

The Chemistry of Quinonoid Compounds, Vol. II
Edited by S. Patai and Z. Rappoport

CHAPTER 8

Recent advances in the synthesis of quinonoid compounds

YOSHINORI NARUTA and KAZUHIRO MARUYAMA
Department of Chemistry, Faculty of Science, Kyoto University, Kyoto 606, Japan

I. INTRODUCTION

After the publication of the early volumes on the quinonoid compounds in this series in 1974[1], several reviews and books treated this field[2–7]. Especially, references 5 and 6 comprehensively cover the literature published until the early 1970s. In the following decade, great progress took place in synthetic organic chemistry, and many sophisticated methodologies and new reagents have been developed and utilized for the synthesis of quinonoid compounds with complex structures. Particularly, the advances in organometallic chemistry extensively contributed to this field. Synthetic methods with high regio- and stereoselectivity have great importance in modern organic synthesis. Some of them have been applied to the synthesis of naturally occurring quinones, most of which possess versatile interesting physiological and biological activities. In Section VII, we discuss total and fragment synthesis of selected natural quinonoid products, as a good demonstration for these reactions in quinone synthesis.

II. OXIDATION METHODS

Quinones have an higher oxidation state than the corresponding synthetic precursors, i.e. the aromatic hydrocarbons, phenols, hydroquinones and catechols. Consequently, the oxidation method consists of one of the major routes to quinones which does not require a change in the carbon framework. The oxidation method offers the best procedure, especially for preparation of quinones with rather simple structure. In this section, we discuss the recent progress, which includes the finding of several new oxidation reagents and mainly the improvement and modification of reactivity and selectivity of well known methods. In order to obtain quinones in high efficiency from commercially available compounds, it is recommended to start with phenols as precursors rather than with hydrocarbons, due to the ease and high product selectivity of their oxidation. Several methods are compared with each other in order to enable the reader to choose an appropriate oxidation method.

In the synthesis of naturally occurring quinones, one can find many successful examples of oxidative quinone formation starting from substrates with rather complex functionality and compare the applicability of the various oxidation methods.

A. Aromatic Hydrocarbons

Only condensed polycyclic aromatic hydrocarbons have been reported to be successfully oxidized to the corresponding quinones. Because of severe oxidation conditions and the powerful reagents required in this conversion, most other functional groups as well as the produced quinones themselves frequently suffer further oxidation. However, for a large scale production, this route is attractive, because of the inexpensive supply of raw materials. On the other hand, under laboratory conditions other oxygenated precursors, such as phenols or aromatic ethers, have great advantage as starting materials due to their higher susceptibility toward oxidation, and they can be selectively converted to the corresponding quinones under mild conditions. Typical examples of reported procedures are listed in Table 1.

Among the many metallic oxidants, the high valent cerium, Ce(IV), gives satisfactory results, regardless of stoichiometric or catalytic conditions[8,9]. Ammonium cerium(IV) sulfate (CAS)[10] prepared from cerium sulfate and ammonium sulfate in diluted acid gives quinones in better yields than ammonium cerium(IV) nitrate (CAN)[11]. The major problem encountered in these types of oxidation is the low product selectivity. For example, 2-alkylnaphthalenes and phenanthrene give the corresponding two isomeric quinones, respectively, in comparable yields. A similar selectivity is also observed in oxidation with $Mn_2(SO_4)_3$, which usually gives better yield than CAN[12]. Anodic oxidation with cerium sulfate or CAN as a mediator is reported to give improved yields than those obtained under stoichiometric conditions[13]. A two-phase oxidation by ammonium persulfate in the presence of catalytic amounts of CAS, $AgNO_3$, and a surfactant also gives results similar to those mentioned above[14]. Chromic acid is a classically well known oxidant, but it requires large amounts of an acidic solution and prolonged reaction time. This defect is due to the low solubility of chromic acid in most organic solvents whereas the aromatic hydrocarbons are soluble in the reaction media. Both phase transfer method in the presence of a quaternary ammonium salt[15] and oxidation in organic solvent with pyridinium fluorochromate (Py-HF-CrO_3)[16] give much improved results, and oxidation of 2-methylnaphthalene and phenanthrene affords selectively 2-methylnaphthoquinone and 9,10-phenanthrenequinone, respectively. The latter oxidant, prepared from anhydrous chromic acid and hydrofluoric acid in pyridine, oxidizes these hydrocarbons in acetic acid media within a few hours. Oxidation of anthracene is one of the easiest examples and all of the aforementioned oxidants give excellent yields.

Electrophilic substitution by thallium trifluoroacetate (TTFA) occurs at a *para* position of a substituted benzene (**1**) and gives the arylthallium(III) derivative (**2**)[18–20], which can be reoxidized to the *p*-quinone (**3**) with pertrifluoroacetic acid in yields ranging from fair to good (equation 1 and Table 2)[21]. A 1,2-migration of the alkyl group from a phenolic *para*

(**1**) [R^1, R^3, R^4] $\xrightarrow[MeCO_2H]{Tl(OCOCF_3)_3}$ (**2**) [R^1, R^3, R^4, $Tl(OCOCF_3)_2$] $\xrightarrow{H_2O_2,\ CF_3CO_2H}$ (**3**) [R^2, R^3, R^4] (1)

position to the *meta* one is observed in this reaction. From the mechanistic standpoint, intermediary formation of **5** and then the phenol derivative (**6**) is proposed, and the latter would be reoxidized by regenerated TTFA to the corresponding quinol (**7**), which will afford the quinone (**8**) with or without migration of the alkyl group (Scheme 1). Scheme 2 is an example of a variant of the TTFA oxidation. The aromatic substitution of **9** by Tl(III) gives **10** which gives an intramolecular lactonization to **11**. Alchoholysis gives the phenol

TABLE 1. Oxidation of aromatic hydrocarbons

Aromatic hydrocarbon	Oxidant	Quinone (% yield)		Reference
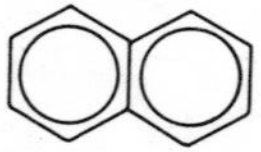 naphthalene	$Ce(SO_4)_2$, $2(NH_4)_2SO_4$ dil. H_2SO_4	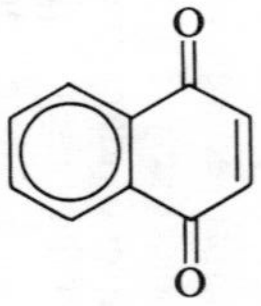 1,4-naphthoquinone (90–95)		10
	CAN[a]	(20)		11
	$Mn_2(SO_4)_3$	(75)		12
	H_2CrO_4, n-$Bu_4N^+HSO_4^-$ two-phase reaction	(39)		15
	e^-, $Ce(SO_4)_2$, HNO_3	(81)		13
2-R-naphthalene		2-R-1,4-naphthoquinone	6-R-1,4-naphthoquinone	
R = *t*-Bu	$Ce(SO_4)_2$, $2(NH_4)_2SO_4$ dil. H_2SO_4	R = *t*-Bu (45)	(26)	10
R = Me	$Mn_2(SO_4)_3$	R = Me (55)	(10)	12

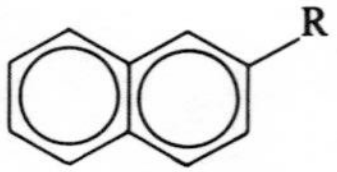

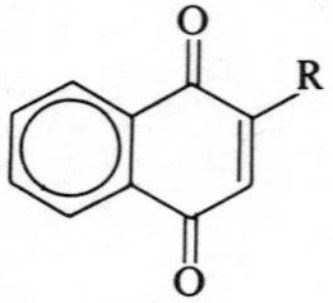

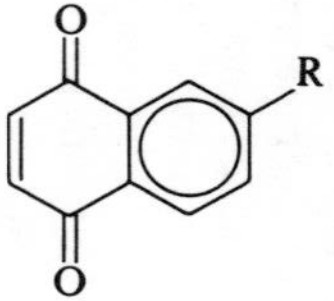

Substrate	Reagent	Product (yield, %)			Ref.
R = Me	H_2CrO_4, n-$Bu_4N^+HSO_4^-$ two-phase reaction	R = Me (56)	—		15
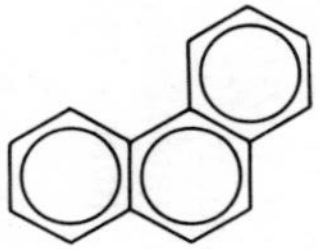		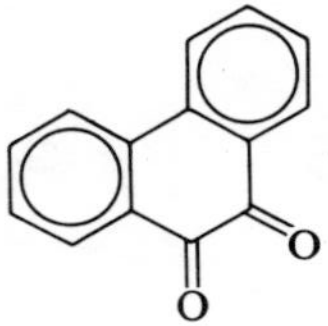	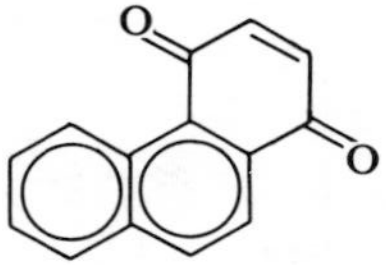		10
	$Ce(SO_4)_2$, $2(NH_4)_2SO_4$ dil. H_2SO_4	(60)	(15)		
	CAN[a]	(27.4)	(11.5)		11
	$Mn_2(SO_4)_3$	(32)	(19)		12
	pyridine·HF·CrO_3	(72)			16
	$PhIO_2$	(46)			17
	e^-, CAN[a]	(65)			13
	$PhIO_2$	(12)	(15)	(14)	17

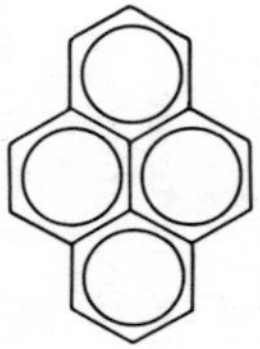

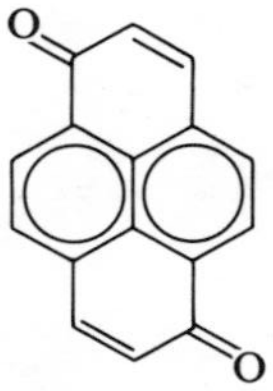

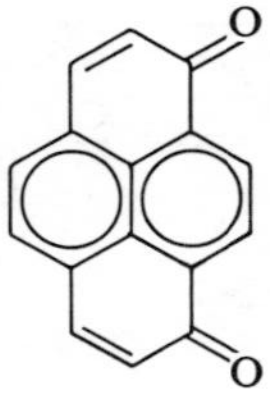

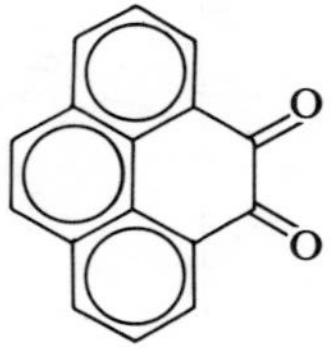

[a] CAN = ceric ammonium nitrate.

TABLE 2. Oxidation of aromatic hydrocarbons by TTFA[21]

Aromatic hydrocarbon (1)			Quinone (3)			
R^1	R^3	R^4	R^2	R^3	R^4	% yield
H	H	H	H	H	H	65
Me	H	H	Me	H	H	68
Et	H	H	Et	H	H	70
Me	Me	H	Me	Me	H	55
Me	Me	Me	H	Me	Me	50
t-Bu	H	H	H	H	H	50
			t-Bu	H	H	20
Cl	H	H	H	H	H	42
MeO	H	H	H	H	H	61
naphthalene			1,4-naphthoquinone			60

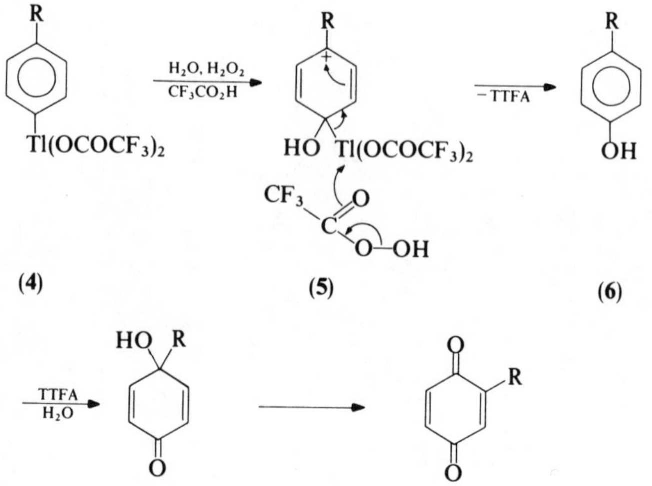

R
$Tl(OCOCF_3)_2$
(4)
H_2O, H_2O_2
CF_3CO_2H
R
HO $Tl(OCOCF_3)_2$
CF_3 O C O—OH
(5)
−TTFA
R
OH
(6)
TTFA
H_2O
HO R
O
(7)
O
R
O
(8)

SCHEME 1

derivative (**12**), which is further oxidized to **13**[22]. Aerobic oxidation of aromatic hydrocarbons is another attractive method relevant to industrial processes. Oxidation of anthracene with air by $RuCl(PPh_3)_3$ as a catalyst gives 9,10-anthraquinone in 55% yield[23]. In this process, the use of hydrogen peroxide, which is assumed to be generated *in situ* in the oxidation with air, instead of molecular oxygen as an oxidant gives the quinone in almost quantitative yield[23, 24].

Although anodic oxidation is expected to have great advantage for the preparation of large quantities of quinones, it has not been established as a reliable methodology yet. In most cases, overoxidation and/or oxidative decomposition is the obstacle to the general application of this method. Special cases are perfluorinated aromatic hydrocarbons. For example, hexafluorobenzene (**14**) is converted to the corresponding *p*-quinone (**15**) in 75% yield and octafluoronaphthalene gives the quinone in 60% yield[25] (equation 2).

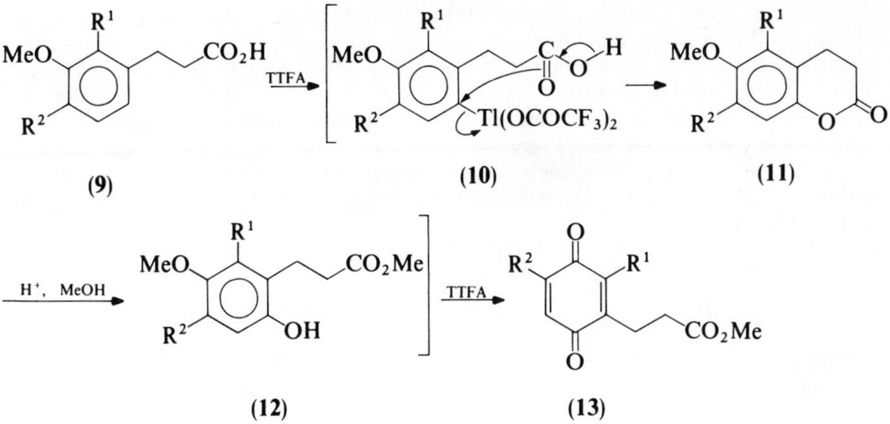

MeO, R^1, R^2, CO_2H —TTFA→ [MeO, R^1, R^2, C(=O)OH, $Tl(OCOCF_3)_2$] → MeO, R^1, R^2, O, O

(9) (10) (11)

—H^+, MeOH→ [MeO, R^1, R^2, CO_2Me, OH] —TTFA→ O, O, R^2, R^1, CO_2Me

(12) (13)

R^1 and/or R^2 = MeO

SCHEME 2

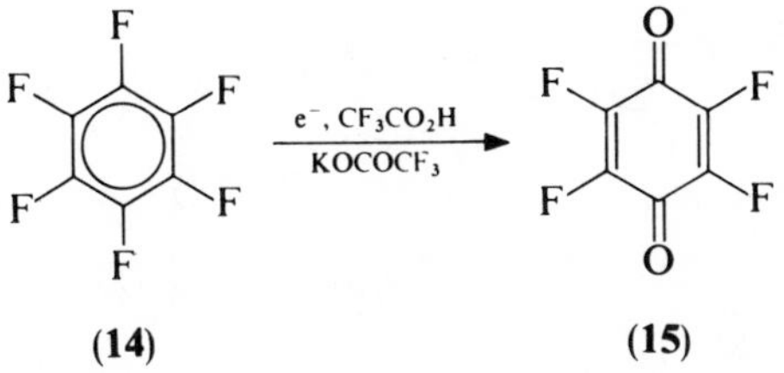

F, F, F, F, F, F —e^-, CF_3CO_2H / $KOCOCF_3$→ O, O, F, F, F, F (2)

(14) (15)

B. Monohydric Phenols

In the course of organic syntheses, it is frequently necessary to convert phenols to quinones. Fremy's salt ($\cdot ON(SO_3K)_2$) (**16**) is still used as a reliable reagent in this conversion[26]. It is suitable mainly for small scale experiments, since the preparation of Fremy's salt is rather tedious and it is difficult to prepare it in quantity due to its instability. Despite these defects, it is frequently a good choice from many other oxidants, because of its excellent *p*-quinone selectivity, wide applicability and the good yields obtained. It is worthwhile to mention the recent comparison of the oxidation by this reagent with other oxidation methods. Vaniline (**17**) cannot be oxidized by Fremy's salt, while the corresponding dimethyl acetal (**18**) affords a *p*-quinone (**19**) under the same conditions

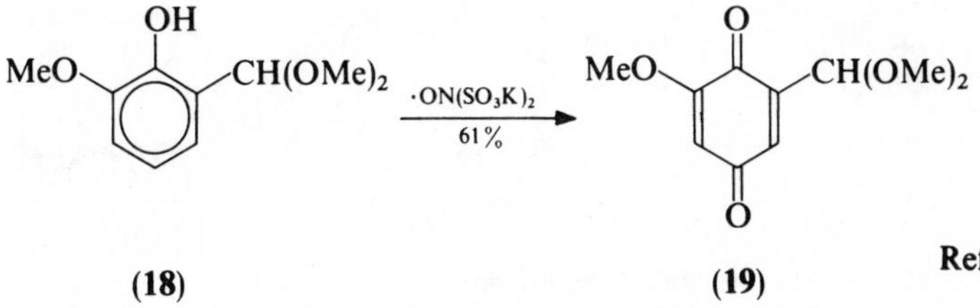

OH, MeO, $CH(OMe)_2$ —$\cdot ON(SO_3K)_2$ / 61%→ O, O, MeO, $CH(OMe)_2$ (3)

(18) (19) Ref. 28

(equation 3)[28]. This oxidation is much more efficiently performed by the O_2–salcomine system, while 5-hydroxy-6-methoxyquinoline (**20**) can be oxidized to the corresponding quinolinoquinone (**21**) by Fremy's salt in better yield than with O_2–salcomine (equation 4)[28]. Although Fremy's salt is known to be a specific oxidant to afford *p*-quinones (equation 5)[27], phenols **24**[28], **27**[29] and **30**[29] give the mixture of the corresponding *o*- and *p*-quinones in the oxidation with Fremy's salt, even when their *p*-position is unsubstituted (equations 6–8)[29]. This *o*-quinone selectivity can be elucidated either by change of the electron density of intermediary phenoxy radicals or by changing the steric environment around the *para* positions.

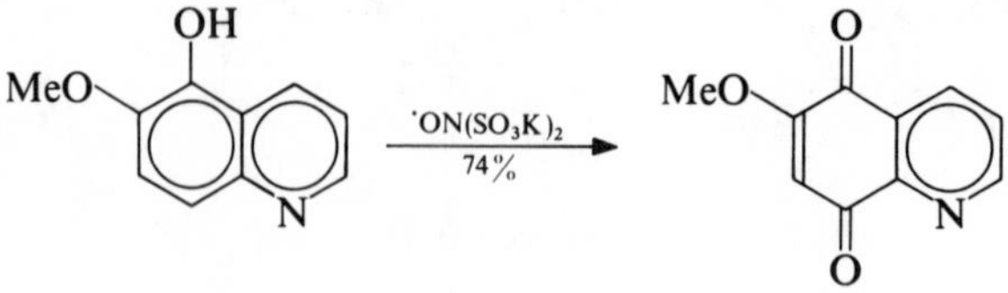

(20) —$\cdot ON(SO_3K)_2$, 74%→ (21) (4)

Ref. 28

(22) —$\cdot ON(SO_3K)_2$→ (23) (5)

X = O, S

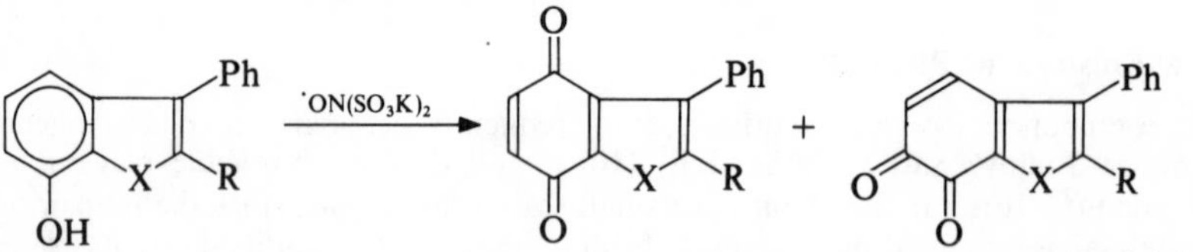

Ref. 29

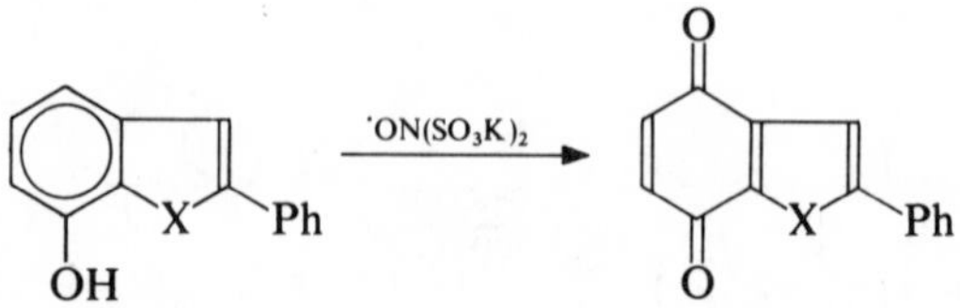

(24) —$\cdot ON(SO_3K)_2$→ (25) + (26) (6)

X = O, S, NH, NMe; R = H, Me

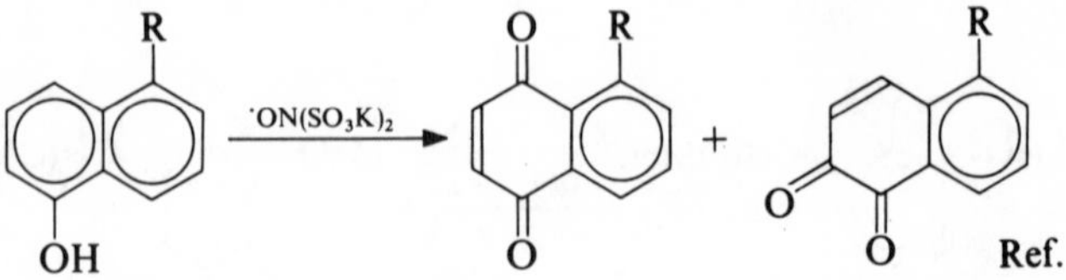

Ref. 29

(27) —$\cdot ON(SO_3K)_2$→ (28) + (29) (7)

R = alkyl, halogen, NO_2, NHAc

Ref. 29

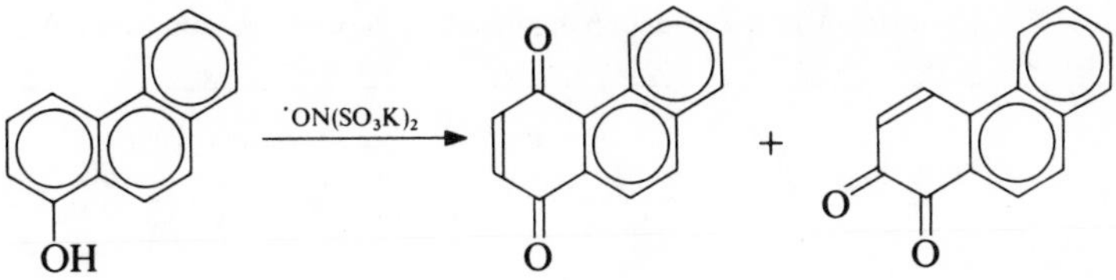

$$\xrightarrow{\cdot ON(SO_3K)_2}$$ (8)

(30) (31) 46.6% (32) 42.4% Ref. 29

Another defect encountered in the oxidation with Fremy's salt is its insolubility in organic solvents and since it requires an aqueous alcoholic medium, it cannot be applied to highly lipophylic substrates. In order to improve the solubility problem two modifications are reported. First, in the analogy to the reaction of Fremy's salt, organic nitroxides are used for the oxidation of phenols. *N*-Benzoyl- (**33**)[30] and *N*-(3,5-dinitrobenzoyl)-*t*-butylnitroxyls (**34**)[30, 31] are prepared and oxidation of phenols is effected in an organic

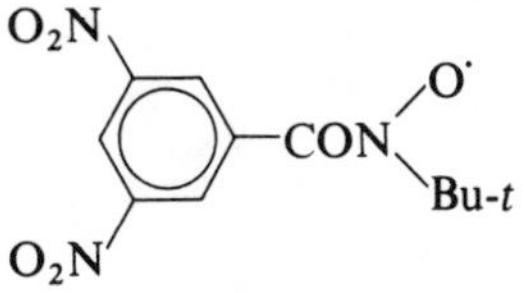

O_2N O_2N CON O· Bu-*t*

(**34**)

solvent. Simple phenols are oxidized in almost comparable yields to those obtained with Fremy's salt except in a few cases. As a typical exception, 1,4-benzoquinone itself cannot be obtained by the treatment of phenol with **33**. Comparison of the two oxidizing agents in phenol oxidation is shown in Table 3. **34** is considered as a more powerful oxidant than **33**,

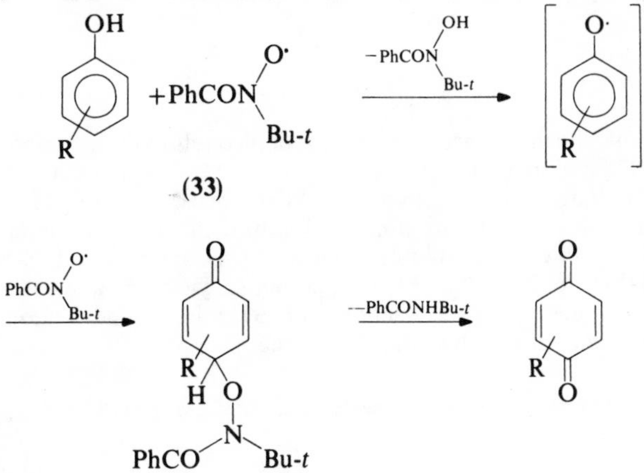

OH R + PhCON(O·)Bu-*t* (**33**) $\xrightarrow{-PhCON(OH)Bu\text{-}t}$ [O· R]

$\xrightarrow{PhCON(O\cdot)Bu\text{-}t}$ O R H O N PhCO Bu-*t* $\xrightarrow{-PhCONHBu\text{-}t}$ O R O

SCHEME 3

TABLE 3. Comparison of the oxidation of phenols to the corresponding quinones by nitroxides[30]

Phenol R^1	R^2	R^3	R^4	Quinone (% yield) with (Fremy's salt)	Quinone (% yield) with (PhCON(Bu-*t*)O·)
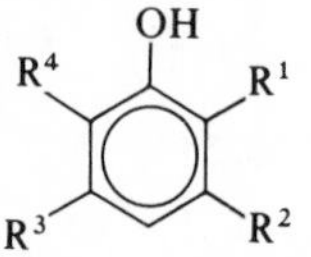 *p*-quinone series				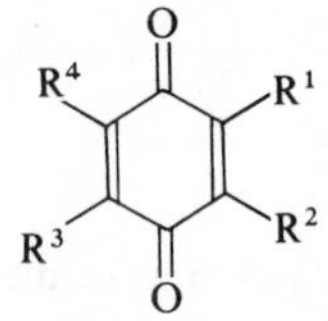	
H	H	H	H	81	0
Me	H	H	Me	75	86
H	Me	Me	H	99	0
Me	H	Me	H	87	42
H	*t*-Bu	H	H		60
MeO	H	H	MeO		60
H	MeO	MeO	H	76	60
Cl	H	H	Cl		46
Me	Me	Me	Me	87	67
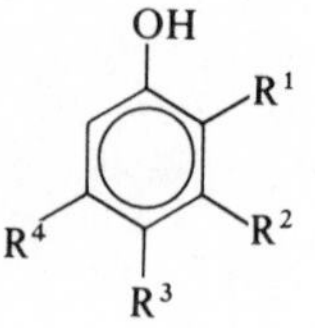 *o*-quinone series				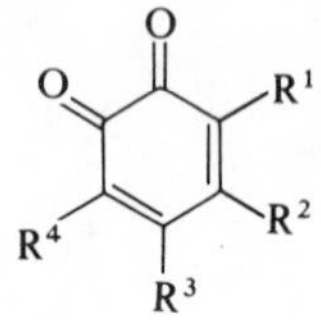	
H	*t*-Bu	H	H		70
H	H	*t*-Bu	H	80	51
t-Bu	H	*t*-Bu	H		87
1-Naphthol		1,4-Naphthoquinone		81	72
2-Naphthol		1,2-Naphthoquinone		91	84

and is capable of oxidizing even allylic or benzylic alcohols to the corresponding carbonyl compound. In the second modification of the oxidation by Fremy's salt, two equivalents of a phase-transfer quaternary ammonium salt catalyst are used to convert Fremy's salt to the lipophylic ammonium nitrosodisulfonate. Tricaproylmethylammonium chloride (**35**) gives **36** (equation 9), which is soluble in most organic solvents and **36** can oxidize the extremely lipophylic phenol **37** to **38** in a quantitative yield (equation 10), while Fremy's salt is completely ineffective on reaction with **37**[32]. This modification is potentially applicable to a wide range of hydrophobic phenols.

$$\underset{(\mathbf{16})}{\cdot ON(SO_3K)_2} + \underset{(\mathbf{35})}{2(n\text{-}C_8H_{17})_3N^+Me\ Cl^-} \longrightarrow \underset{(\mathbf{36})}{\cdot ON[SO_3^-N^+Me(C_8H_{17}\text{-}n)_3]_2} \tag{9}$$

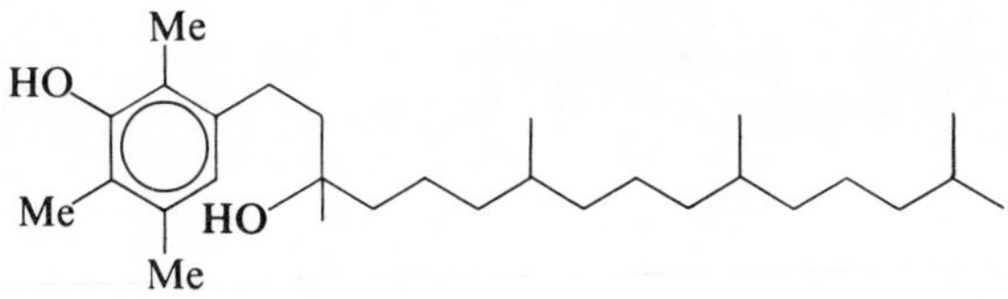

(37)

(10)

$+ \cdot ON[SO_3^- N^+Me(C_8H_{17}\text{-}n)_3]_2$

(36)

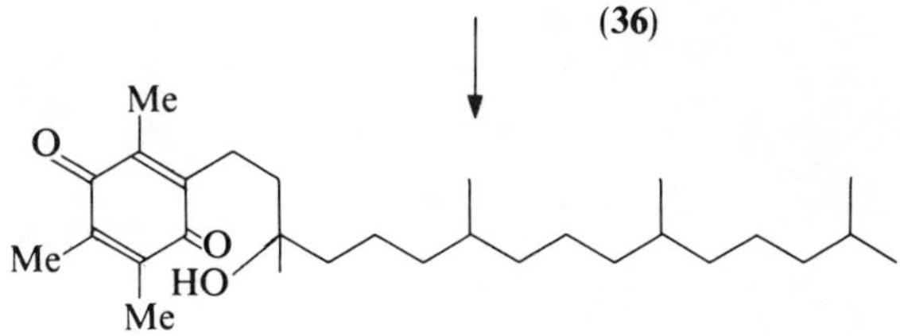

(38)

Benzeneselenic anhydride, $(PhSeO)_2O$ (**39**), oxidizes phenols to the corresponding *o*-quinones in good yields and high product selectivities, even when the *para* position is unsubstituted[33–36]. This method is especially useful for oxidation of hydroxypolycyclic aromatics (Table 4). In some cases, the *p*-quinone is obtained as in the oxidation of **40** to **41**. The *ortho* selectivity is ascribed to a rapid [2, 3] sigmatropic rearrangement of an intermediarily produced phenyl selenyl ether (**42**) and a successive loss of benzeneselenol (PhSeOH) (Scheme 4).

TABLE 4. Oxidation of phenols to *o*-quinones by phenylselenic anhydride

Phenol	*o*-Quinone	(% Yield)	Reference
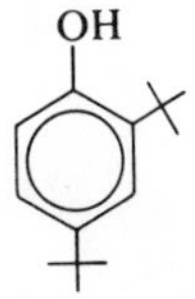	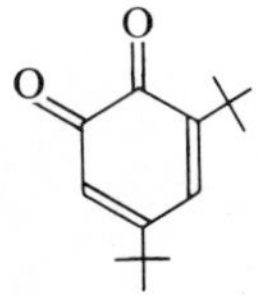	(68)	34
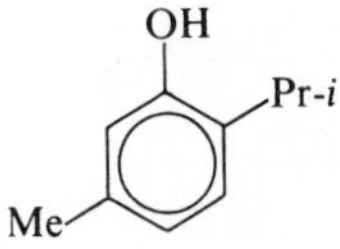	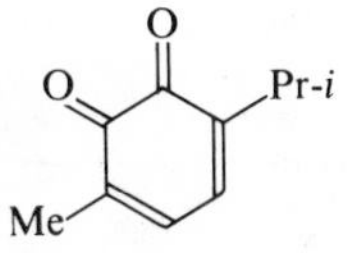	(59)[a]	34
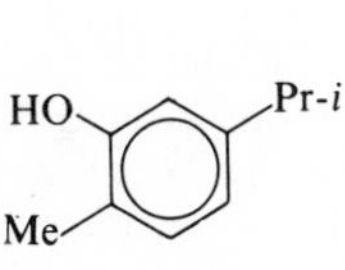	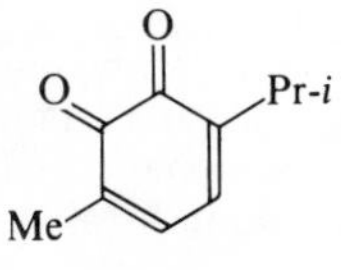	(60)	34

TABLE 4 (continued)

Phenol	*o*-Quinone	(% Yield)	Reference
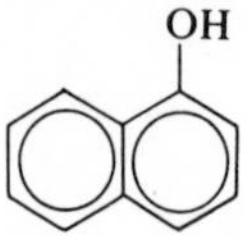	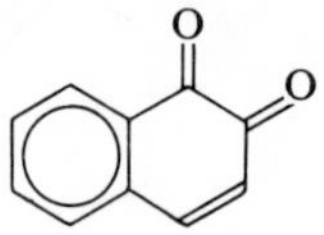	(62–65)[b]	34, 36
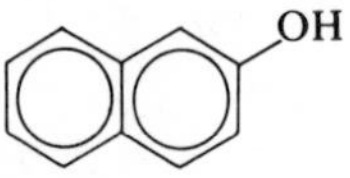	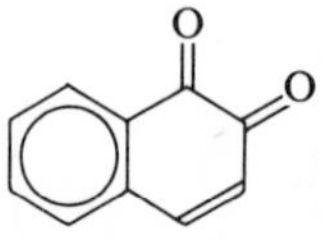	(73)	34, 35
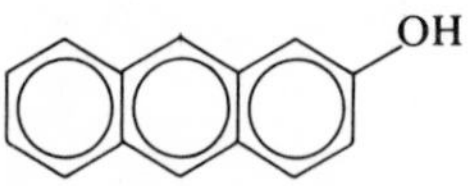	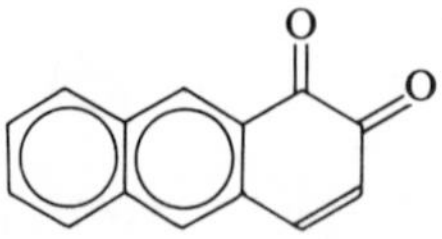	(54)	35
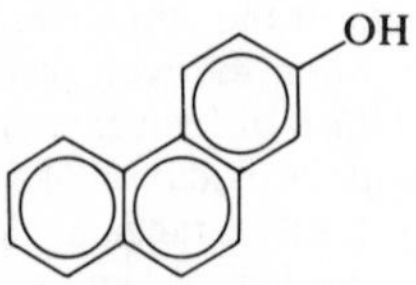	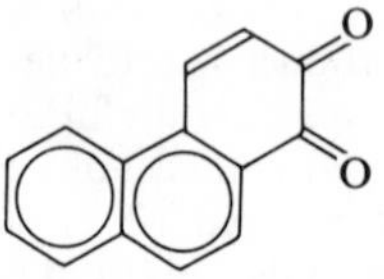	(56)	35
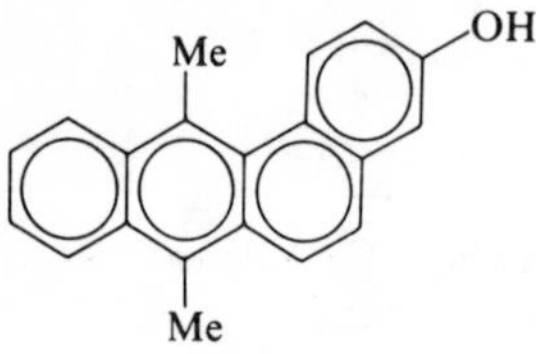	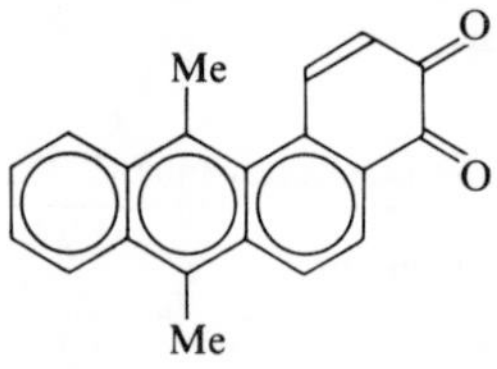	(80)	35
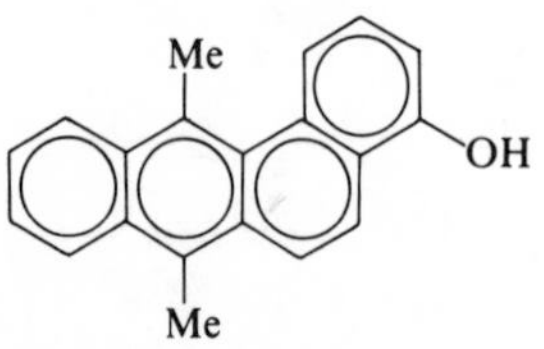 (**40**)	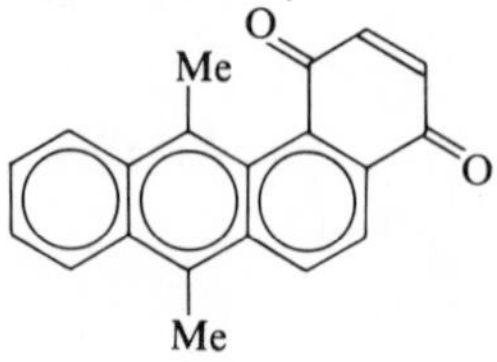 (**41**)	(60)	35

[a] The corresponding *p*-quinone is formed in 15% yield.
[b] 1,4-Naphthoquinone is formed in 5–10% yield.

Phenols **43**–**45** give complex reaction mixtures and phenol **46**, in which the *para* position is blocked, gives the corresponding *o*-quinone as a minor product accompanied by several side products. In the oxidation of paracyclophanes (**47**) the *o*-quinone **48** is sometimes accompanied by the corresponding *p*-quinones (**49**) (equation 11). Consequently, the

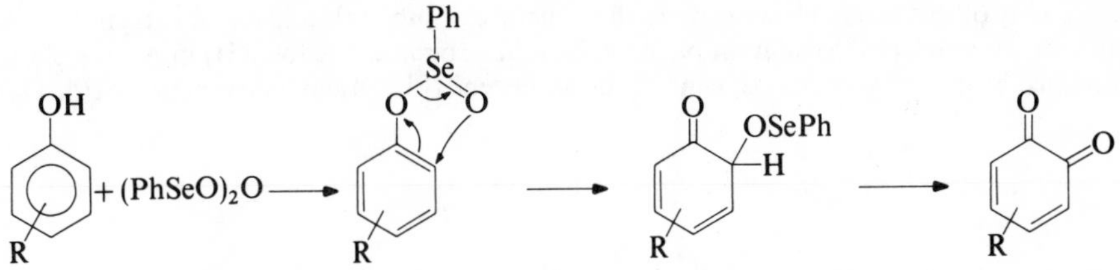

OH
+ $(PhSeO)_2O$ →
Ph
Se
O O
R
→
O OSePh
H
R
→
O
O
R

(39) (42)

SCHEME 4

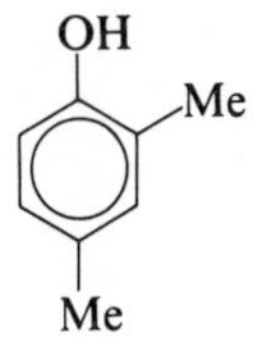

OH Me
Me
(43)

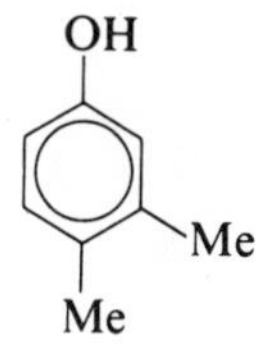

OH
Me
Me
(44)

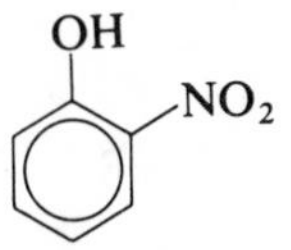

OH
NO_2
(45)

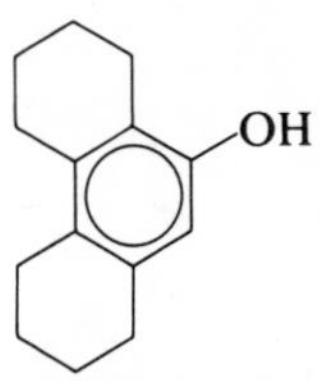

OH
(46)

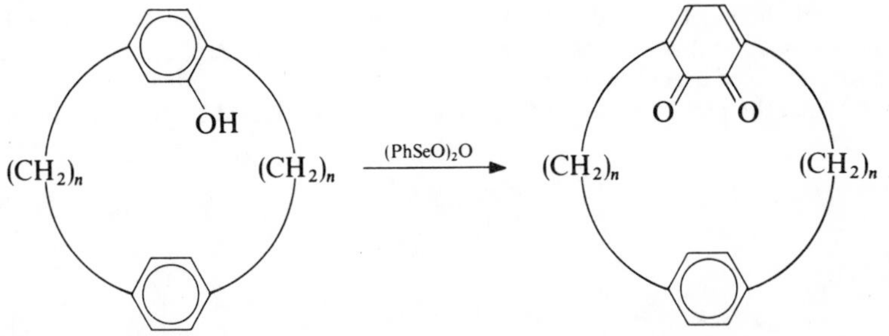

OH
$(CH_2)_n$ $(CH_2)_n$
$(PhSeO)_2O$ →
O O
$(CH_2)_n$ $(CH_2)_n$

(47) (48) (11)

$n = 2$ 37.3%

$n = 3$ 40.3%

O
O
+ $(CH_2)_n$ $(CH_2)_n$ + $(CH_2)_n$ $(CH_2)_n$
O SePh
O

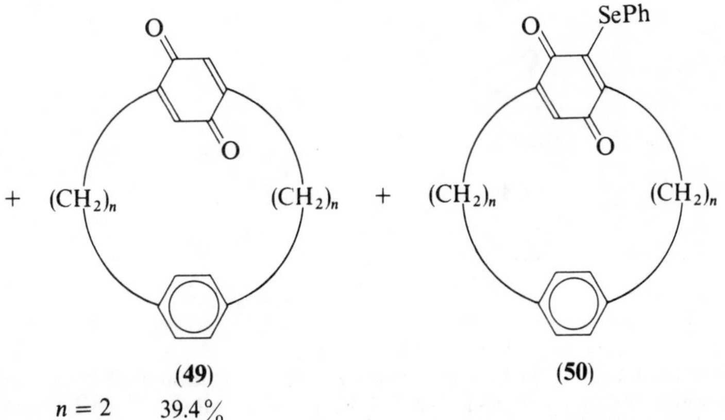

(49) (50)

$n = 2$ 39.4%

$n = 3$ 13.3%

selectivity of this reagent is sensitive to the structure of substrate and its substituent(s)[38]. A similar *ortho* selective oxidation of the polycyclic aromatic phenols (**51**) in good yield is attained by using molecular oxygen[37]. Iodosobenzene (PhIO) and iodoxybenzene ($PhIO_2$)

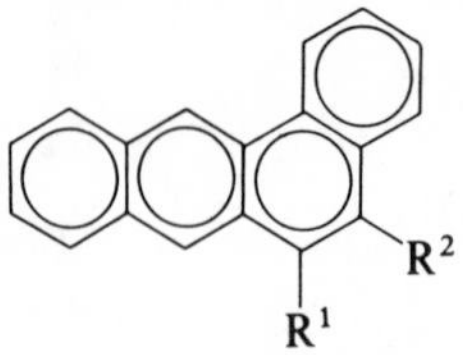
R²
R¹

(**51**)

$R^1 = H, R^2 = OH$

$R^1 = OH, R^2 = H$

also exhibit selective *o*-quinone formation in good yield in the presence of a protic acid or a Lewis acid catalyst[39]. In contrast, thallium(III) derivatives, especially TTFA, exhibit specific *para* oxidation of phenols[40–42]. Thallium trinitrate (TTN)[43] in methanol oxidizes 2,6-disubstituted phenols to the corresponding *p*-quinones in good yields (70–85%). The mechanism is similar to that supposed for TTFA oxidation of phenols, and involves the successive formation of quinol monomethyl ether (**52**) and the quinone monoacetal (**53**) (Scheme 5). When the *para* position of the phenol (**6**) is substituted by an alkyl group R,

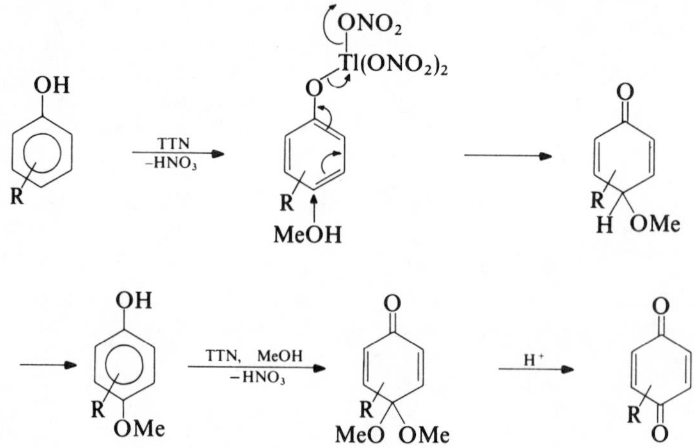
OH
R
TTN
-HNO₃
ONO₂
Tl(ONO₂)₂
O
R
MeOH
O
R
H OMe
OH
R
OMe
TTN, MeOH
−HNO₃
O
R
MeO OMe
H⁺
O
R
O

(**52**) (**53**)

SCHEME 5

TTN oxidation in MeOH gives the corresponding quinol (**54**) (R = alkyl, R′ = H)[43]. However, thallium triperchlorate, $Tl(ClO_4)_3$, gives 2-alkyl-1,4-benzoquinones (**8**) from the corresponding 4-alkylphenols (**6**) (R = Me, $(CH_2)_2OAc$, $(CH_2)_2OCOCH_2Cl$) in 60%

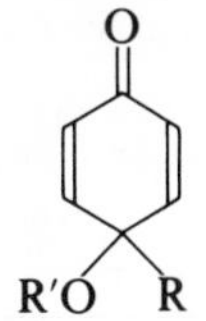

(54)

perchloric acid, a medium which accelerates the dienone–phenol rearrangement of the intermediary quinol **54** (R′ = H)[44].

One interesting problem is the selective oxidation of 1,5-dihydroxynaphthalene (**55**, R = H) or its monomethyl ether (**55**, R = Me), which affords juglone (**56**, R = H) or its methyl ether (**56**, R = Me) together with their *ortho* analogs (**57**) (equation 12). These

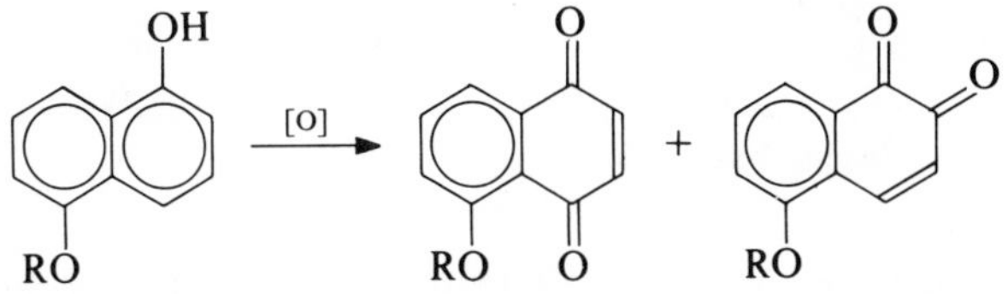

(12)

(55) (56) (57)

compounds are important precursors of many naturally occurring quinones, i.e. anthracyclinones and pyranonaphthoquinones (Section VII). For this reason, various reagents are examined in order to realize higher *para* selectivity. The comparison of oxidizing reagent and reaction conditions (Table 5) is useful for the choice of the appropriate oxidation reagent and conditions of related phenols. Interestingly, singlet oxygen[45] and TTN[48] specifically give the *p*-quinone (**56**) regardless of R group. When **55** (R = Me) is treated with TTN, ethylene glycol and methyl orthoformate, the corresponding quinone monoketal (**58**) becomes the major product. 2,3-Dichloro-5,6-dicyanobenzoquinone

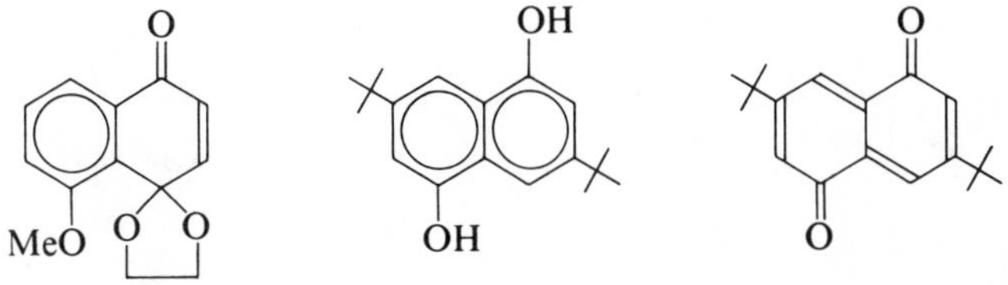

(58) (59) (60)

(DDQ) shows marked contrast to the other oxidants discussed above and affords **60** from **59** in quantitative yield[49].

Oxidation of phenols with molecular oxygen in the presence of *N,N′*-bis (salicylidene)ethylenediiminocobalt(II) (salcomine) (**61**, $n = 2$) gives juglone in good yield in spite of concomitant formation of the corresponding *o*-quinones (**57**)[50,52].

Transition metal catalyzed oxidation with molecular oxygen is extensively developed due to the simplicity of its manipulation and use of the following metal catalysts was

TABLE 5. Comparison of oxidation methods of 1,5-dihydroxynaphthalene and its derivatives

R in **55**	Oxidation method	% yield **56**	% yield **57**	Reference
H	CeO_2, H_2O_2	18		51
H	O_2-CuCl	80		61
H	$MeCO_3H$–$MeCO_2H$	45–55	28	46
H	Ⓟ-$C_6H_4SeO_2H$[a] t-BuOOH	70		47
H	$Tl(NO_3)_3$	64	26	48
Me	$Tl(NO_3)_3$-celite	72		48
H	O_2-salcomine	71	14	50, 52
Me	Fremy's salt	trace	91.6	29
H	$^1O_2$	70		45
Me	$^1O_2$	43		45
Me	$^1O_2$	68.8	17.2	50
Ac	$^1O_2$	30	35	50

[a] Polymer bound-phenylselenic acid.

reported: **61** ($n = 2$)[28,45,53–59], **61** ($n = 3$)[55], **62** (R = Me)[55], **62** (R = polystyrene)[55], **63**[55], **64a**[55], **64b**[55] and **64c**[54,60]. By using these metal catalysts, only 2,6-disubstituted phenols

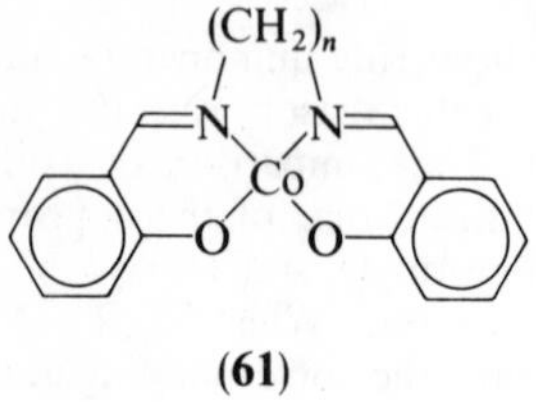

(61)

$n = 2$ (Co(salen))
$n = 3$ (Co(salpn))

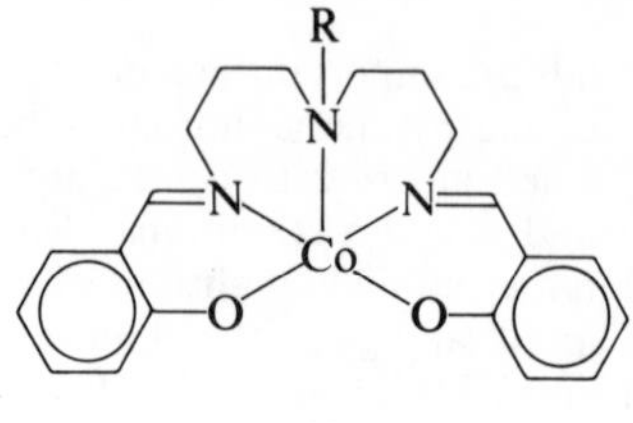

(62)

R = Me (Co(salen-Mdept))
R = polystyrene (Co-polysaldpt)

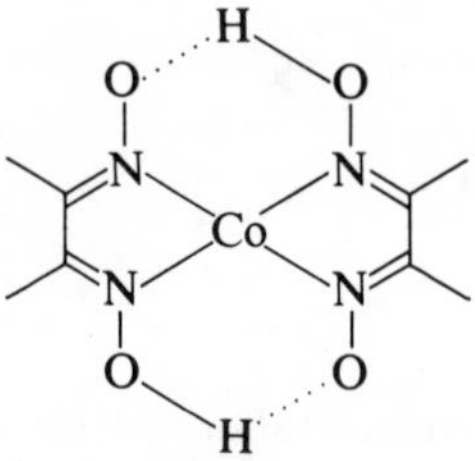

(63)

Co(dmgH)$_2$

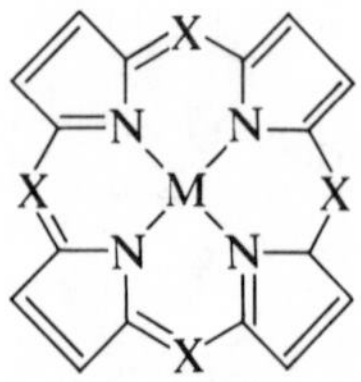

(64)

a, X = CPh, M = Co (CoTPP)
b, X = CPh, M = MnCl (MnTPPCl)
c, X = N, M = Co (CoPc)

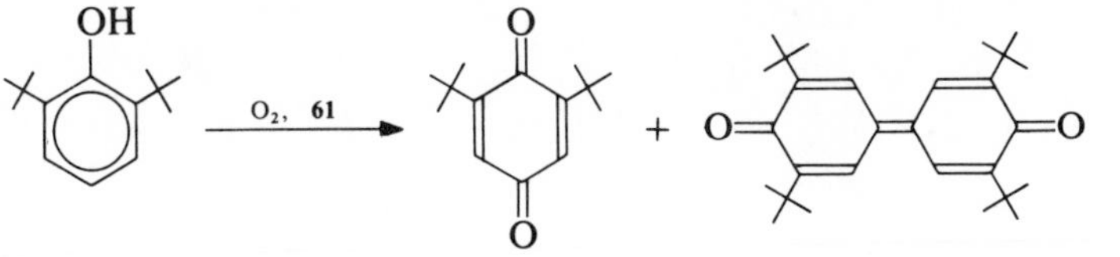

(65) —O_2, 61→ (66) + (67) (13)

	(66)	(67)
$MeOH-H_2O$[56]	44%	56%
DMF[53]	83%	0%

such as **65** give *p*-quinones in good to excellent yields (equation 13). A definite solvent effect is recognized on the product selectivity between diphenoquinone (**67**) and *p*-quinone, which can be exclusively obtained in DMF, whereas both compounds are formed in aqueous methanol. Similar results are obtained in the preparation of quinones in the oxidation of the phenols and naphthols **68**–**71**.

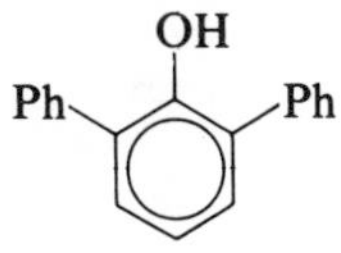

(**68**) 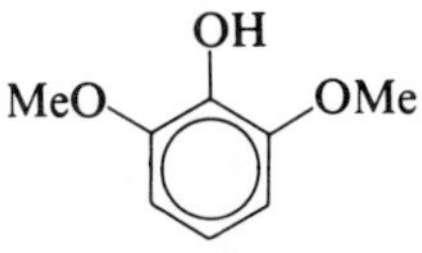 (**69**) 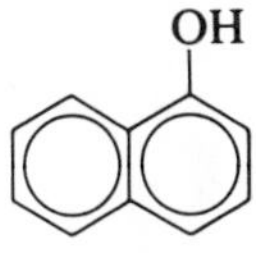 (**70**) 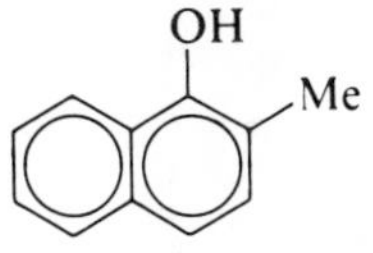 (**71**)

A μ-oxo cupric complex, $Cu_4Cl_4O_2(MeCN)$, which is prepared *in situ* from CuCl in acetonitrile, as well as the cobalt complexes shown above afford similar results[61].

The classical chromic acid oxidation of phenols, which is used under a modification of two-phase Jones oxidation conditions[62], gives *p*-quinones in reasonable yields. It becomes an alternative and practical method for Fremy's salt, TTN, or TTFA oxidation to obtain simple *p*-quinones in quantity.

p-Halogenated phenols are susceptible to oxidation to *p*-quinones by CAN[63,64] and H_5IO_6[65]. However, the latter reagent frequently affords the *o*-quinone instead of the *para* derivative.

Under electrolytic conditions, phenols substituted with electron donating group(s) at *o*- or *p*-positions generally give several products with a distribution which depends on the electrolytic conditions[67]. Electrolysis in aqueous acidic media is the favored method to avoid undesirable side reactions, such as phenol coupling and semiquinone formation, and to maximize the efficiency of direct quinone formation. The quinone formation is also very sensitive to the reaction conditions, involving the solvent, the supporting electrode, the anodic material, the applied potential, etc. For a large scale synthesis, this method is advantageous over Fremy's salt oxidation. Selected results are listed in Table 6.

The less common oxidation with lead tetraacetate (LTA) was applied to the preparation of the naphthazarin derivative (**75**) which is relevant to anthracycline synthesis from **72** (equation 14)[66]. The reaction is very sensitive to the purity of the reagent and there remain problems in reproducibility and applicability to other phenols. The mechanism may involve the intermediacy of either **73** or **74**.

Generally, the Fe^{3+} catalyzed H_2O_2 oxidation, i.e. by Fenton's reagent, is too drastic for oxidation of arenes and possesses little synthetic utility. However, the polymethoxybenzenes, **76** and **79**, are easily oxidized to afford the corresponding methoxy-*p*-quinones

TABLE 6. Anodic oxidation of phenols to the corresponding quinones

Phenol	Electrolytic conditions[a]	Product (% yield)		Reference
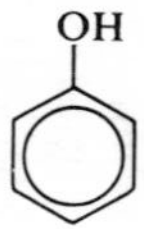 OH	1 M H_2SO_4/Pb	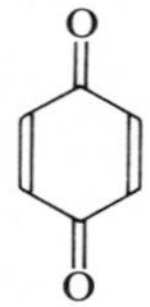 O, O	(68)	68
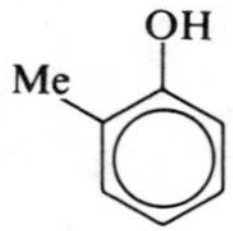 OH, Me	1 M H_2SO_4/Pb	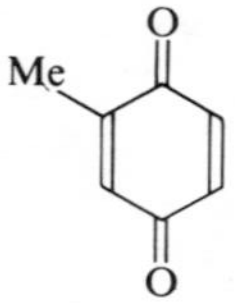 O, Me, O	(62)	68
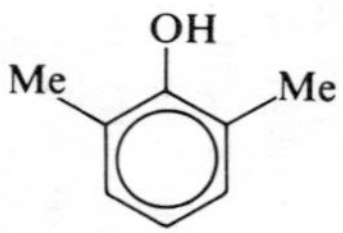 OH, Me, Me	1 M H_2SO_4/PbO_2	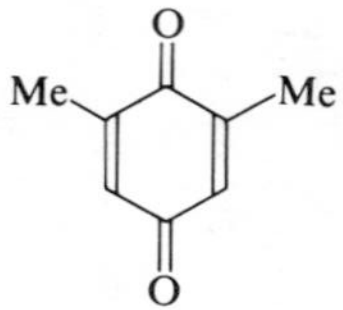 O, Me, Me, O	(100)	69
	MeOH/$LiClO_4$/C		(50)	74
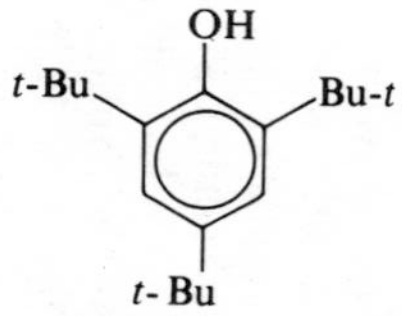 OH, t-Bu, Bu-t, t-Bu	aq. MeCN/$NaClO_4$/Pt	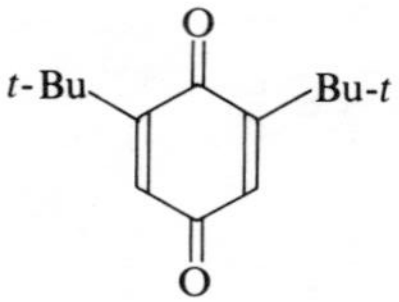 O, t-Bu, Bu-t, O	(85)	70

Substrate	Conditions[a]	Product	Yield (%)	Ref.
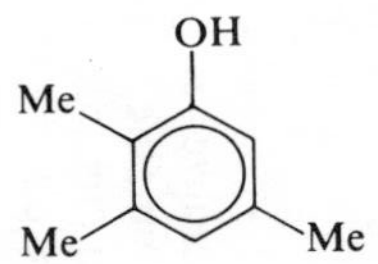	aq. MeOH/Na_2SO_4/Pt	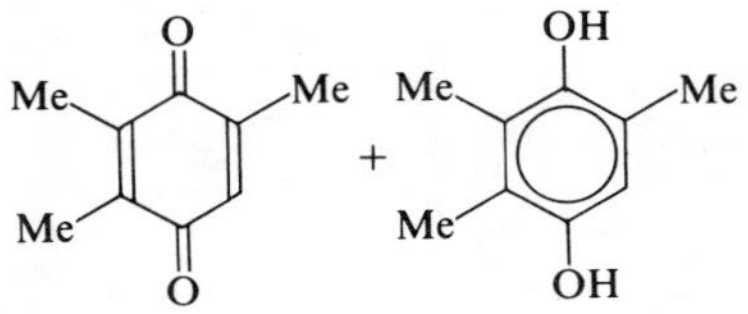 +	(total 51)	71
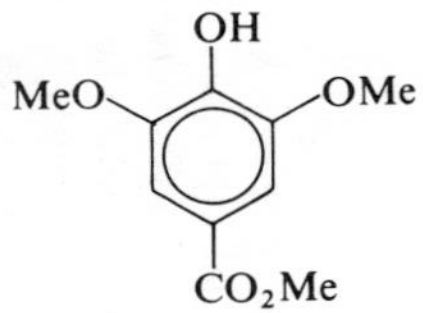	aq. MeCN/Et_4NClO_4/Pt	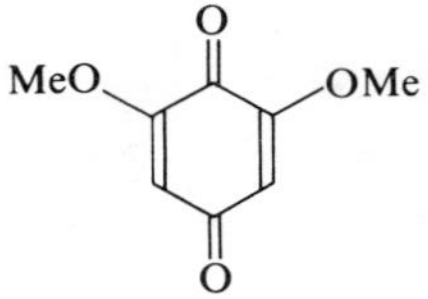	(48)	72
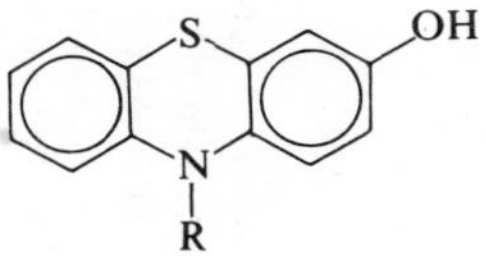	aq. citric acid/$NaHPO_4$/C	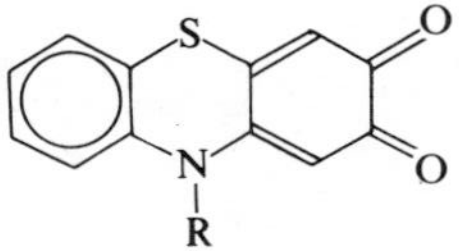	(33)	73

[a] Solvent/(supporting electrolyte)/anode.

(72) —LTA, AcOH→ (73) or (74) → (75) (14)

77, **78** and **80**, respectively, in moderate to good yields (equations 15 and 16)[75]. The quinones obtained are important starting materials for the synthesis of ubiquinone and related compounds.

(76) —H_2O_2, $K_3Fe(CN)_6$→ (77) + (78) (15)

(79) —H_2O_2, $K_3Fe(CN)_6$→ (80) (16)

C. Hydroquinones and Catechols

Interconversion of hydroquinones or catechols to quinones is known to be a famous redox reaction and the low oxidation/reduction potentials of these systems is utilized even in electron transport relay in biological systems. From the synthetic standpoint, the two components of the oxidation/reduction systems are considered to be equivalent regardless of their oxidation state. Classically well known and established oxidants, e.g. Ag_2O, Ag_2CO_3, $FeCl_3$, etc., still preserve their synthetic utility for these systems. In the last decade, many additional reagents have been examined and joined this group (Table 7). In this section, we discuss several oxidants, whose general applicability have been demonstrated. For other reagents listed in Table 7, further exploitation is necessary in order to establish their utility.

TABLE 7. Oxidizing agents applied for hydroquinones

Oxidizing agent	Reference	Oxidizing agent	Reference
KO_2, crown ether (two-phase system)	76	$BaMnO_4$	88
$NaOCl$, $R_4N^+X^-$ (two-phase system)	77	$[AgPy_2]MnO_4$	89
$PhIO$	78	$Fe_2(SO_4)_3$	90
$PhIO_2$	78	$Tl(NO_3)_3$	43
$Bu_4N^+IO_4^-$	78	$(p\text{-}MeOC_6H_4)_2TeO$	91, 92
$NaIO_4$; $NaIO_4$, $R_4N^+X^-$ (two-phase system)	78, 81	$(p\text{-}MeOC_6H_4)_2TeO_2$	93
$NaIO_4$ on silica gel	79	$HgBr_2$	94
Ⓟ-$C_6H_4CH_2N^+Me_3X^{-a}$ $X^- = IO_3^-, IO_4^-$	80	$Ce(NH_4)_2(NO_3)_6$	95
$PhI(OAc)_2$	82	$Ce(OH)_3OOH$	96
$PhI(OCOCF_3)_2$	83	*o*-Chloranil	97, 98
Ph_2SeO	84	DDQ	99
$(PhSeO)_2O$	35, 36	*t*-$BuN(COPh)O\cdot$	30
Ⓟ-$C_6H_4SeO_2H^a$, *t*-BuOOH	47	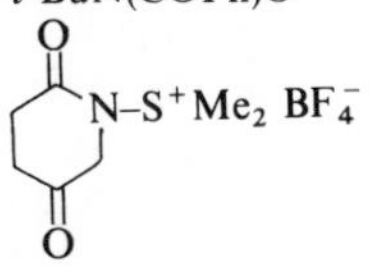 O, N–S^+Me_2 BF_4^-, O	100
$KMnO_4$, crown ether	85	NCS-Et_3N	101
MnO_2	86, 87	$Ph_2S[OC(CF_3)_2Ph]_2$	102
MnO_2-HNO_3	87	O_2-CuCl	61

[a] Polystylene bounded reagent.

Sodium hypochlorite in the presence of phase transfer catalyst oxidizes hydroquinones to the corresponding quinones in yields ranging from moderate to good and it gives acceptable results especially for polyalkylated hydroquinones[77]. This method, however, could not be applied to other quinones with inherent instability under basic conditions, since at the rather high pH value (pH = 8–10) of the applied aqueous solution, decomposition of the quinones will take place.

Hypervalent iodine oxides are the reagents of choice due to their general applicability and higher product selectivity[78]. Among them, tetrabutylammonium periodate (Bu_4NIO_4) and sodium periodate in the presence of a phase transfer catalyst give excellent yields for a wide range of hydroquinones[78]. Relevant to the homogeneous system, silica gel[79] and polymer-supported modifications[80] are equally useful for this purpose. The former reagent[79] can quantitatively oxidize the hydroquinone **81** to the corresponding quinone **82** while preserving the sulfide group intact. Commercially available[86] or freshly

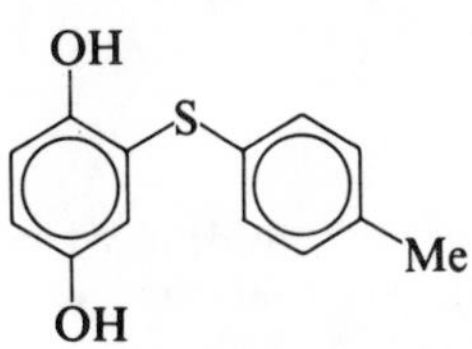

OH, S, Me, OH

(**81**)

O, S, Me, O

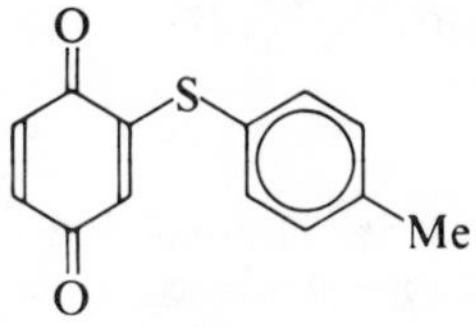

(**82**)

activated manganese dioxide[87], is an excellent and inexpensive reagent which is applicable for synthesis of wide range of quinones such as **8** (R = Me, CN, COMe, CO_2Me) and **83** (R^1, R^2 = H or Me; X = MeCO, PhCO, MeO_2C).

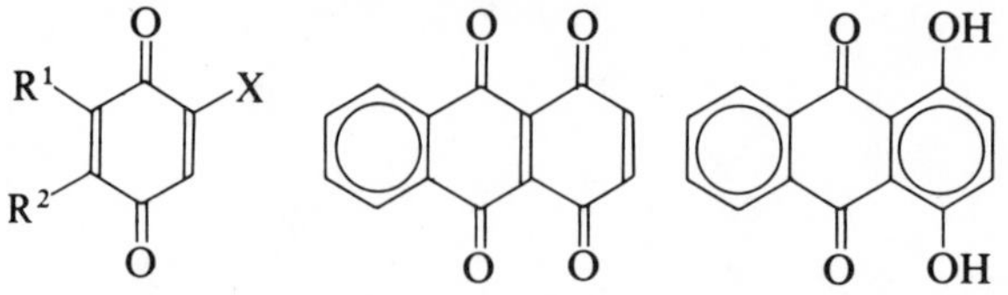

(**83**) (**84**) (**85**)

Nitric acid impregnated manganese dioxide, which is a more powerful oxidizing reagent than the forms mentioned above, is suitable for the preparation of highly reactive quinones, such as 1,2-benzoquinone and 9,10-anthraquinone from the corresponding hydroquinones, which can not be oxidized by activated MnO_2[87]. Quinone **84** is also obtained in 86% yield from **85**.

Ceric ammonium nitrate also possesses sufficient oxidizing ability toward hydroquinones and it demonstrated this conversion in high yields[95]. It is worth while to mention in this respect oxosulfonium cation (**86**)[103], known as Corey–Kim reagent, which can be prepared from *N*-chlorosuccinimide (NCS) and dimethyl sulfide in the presence of silver tetrafluoroborate. It can oxidize hydroquinones and catechols to the quinones in high yields at low temperature (-20 to -50 °C)[100]. An almost similar reaction with NCS and Et_3N can be achieved without significant loss of efficiency[101]. These reactions are thought to be an extension of the Swern oxidation[104] and in both cases a base, usually Et_3N, is essential for completion of the reaction. This is apparently connected with the formation of intermediate 1:1 complexes between the hydroquinone and the sulfonium cation which prevents both an overoxidation and a successive nucleophilic addition of the substrate to the quinone (Scheme 6). Some exceptional cases, which were reported in the oxidation of

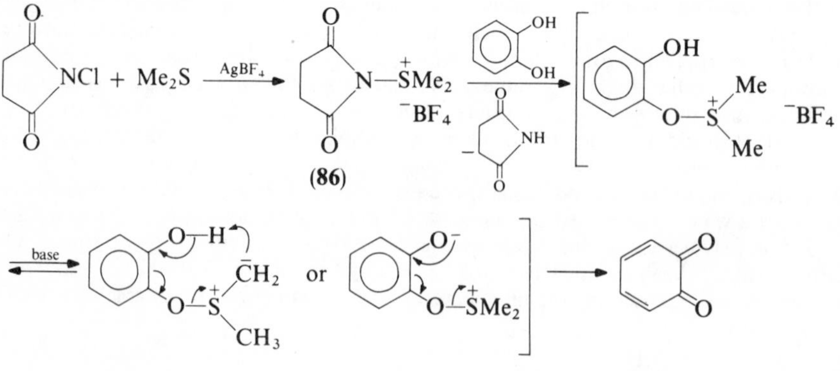

NCl + Me_2S $\xrightarrow{AgBF_4}$ N—$\overset{+}{S}Me_2$ $^-BF_4$ (**86**)

SCHEME 6

catechols, demonstrate the importance of the choice of the oxidant. When Fremy's salt[105] and Fe(III)[90] are applied to catechol derivatives, **87** and **89**, the corresponding *p*-quinones, **88** and **90**, respectively, are preferentially obtained, instead of the corresponding *o*-quinones (equations 17 and 18).

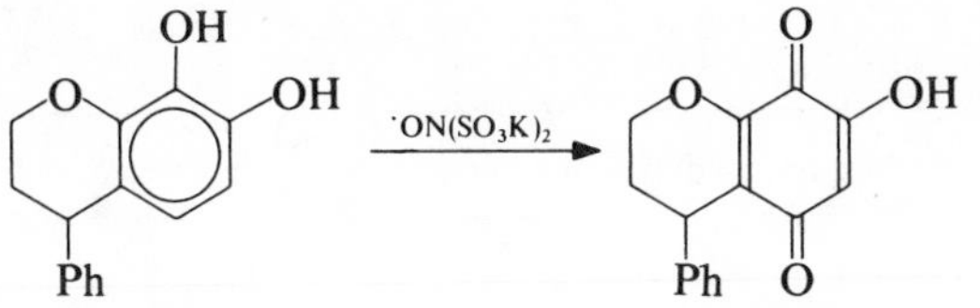

(87) —$\cdot ON(SO_3K)_2$→ (88) (17)

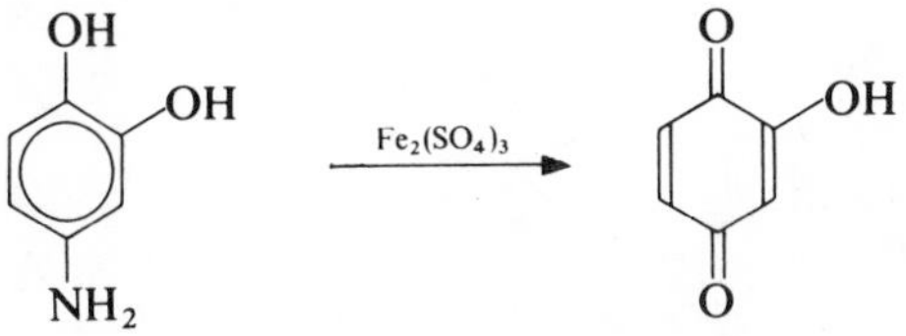

(89) —$Fe_2(SO_4)_3$→ (90) (18)

D. Aromatic Ethers and Hydroquinone Mono- and Diethers

For the synthesis of simple quinones, the title compounds are not the precursors of choice, since the corresponding phenols and hydroquinones are more easily oxidized under many established conditions. However, these materials are of great importance in quinonoid compound synthesis for the following reason. In the course of preparation of naturally occurring quinones, the reactive quinonoid moiety must frequently be protected against destruction under various reaction conditions. The hydroquinone dialkyl ether or quinone diacetal are very convenient and useful such synthons. Hydroquinone diether is oxidized to the corresponding quinone by silver (II) oxide (AgO)[106] or by cerium ammonium nitrate (CAN)[107] with high efficiency. These reagents are recognized now as the standard oxidizing reagents with general applicability, and they replace traditional oxidants such as nitric acid and chromic acid. The yields obtained in oxidations of various hydroquinone dimethyl ethers by CAN are given in Table 8. A similar mechanism is proposed for the oxidation by CAN or AgO. It involves two successive electron transfers and nucleophilic addition of water, as clarified by isotope labeling experiments with ^{18}O-enriched water (Scheme 7)[107]. There are two important differences in the oxidations by

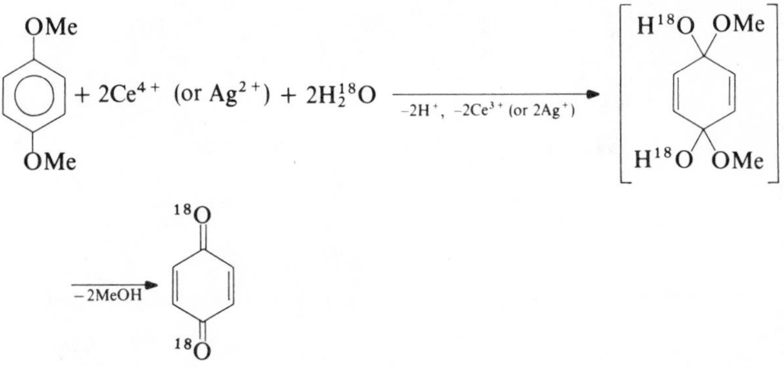

$$+ 2Ce^{4+} \text{ (or } Ag^{2+}) + 2H_2^{18}O \xrightarrow[-2H^+,\ -2Ce^{3+}\text{ (or } 2Ag^+)]{} \left[\ \right] \xrightarrow{-2MeOH}$$

SCHEME 7

TABLE 8. Oxidation of hydroquinone dimethyl ethers in aqueous acetonitrile with CAN[107]

Hydroquinone dimethyl ether		Quinone, % yield
R^1	R^2	
H	H	57
Me	Me	95
Me	CH_2OH	73
Me	$CH{=}C(Me)NO_2$	65
Me	$CH_2CH(Me)NHCO_2Bu$-*t*	61
MeO	$CH_2CH(Me)NHCOPh$	35
1,4-dimethoxynaphthalene		94

AgO and CAN: (1) CAN can be applied under mild conditions, while AgO requires strong acidic conditions; (2) AgO gives *o*-quinones in moderate yields from the corresponding catechol dimethyl ethers, while CAN usually affords a mixture of several oxidized products and is therefore less useful for oxidation of these substrates. For example, the acid-sensitive compound **91** affords inevitably an allylic rearranged quinone **92** accompanied by a normal oxidation product **93** when oxidized with AgO under acidic conditions (equation 19)[108]. Other acid sensitive groups such as epoxide can also not survive under acidic

$$\textbf{(91)} \xrightarrow{AgO,\ HNO_3} \textbf{(92)} + \textbf{(93)} \qquad (19)$$

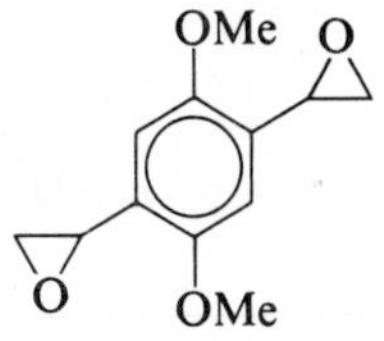

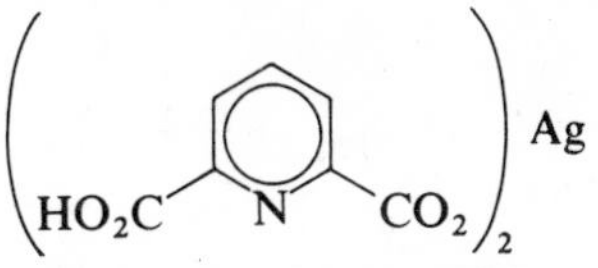

(94) (95)

conditions. To overcome these problems, silver dipicolinate (**95**) was examined as a substitute to AgO, and it oxidized **94** to the corresponding quinone in a moderate yield[109]. When monoalkyl- or monoalkoxy-hydroquinone dimethyl ethers, e.g. **96** and **98**, are treated with CAN, the corresponding dimeric products, i.e. **97** and **99**, respectively, are obtained in preference to the 'expected normal quinones' (equations 20 and 21)[157].

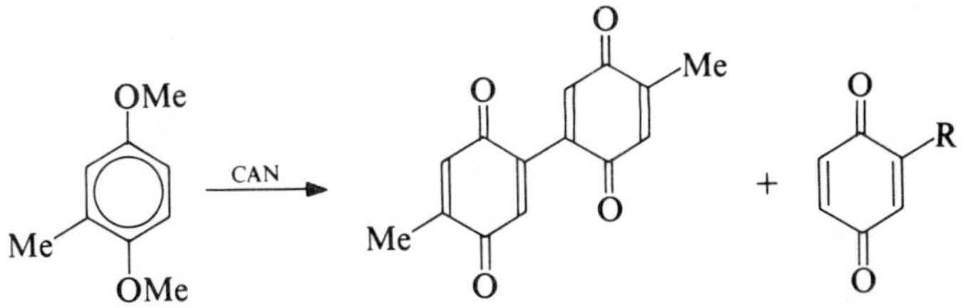

(96) —CAN→ (97) + (8, R = Me) (20)

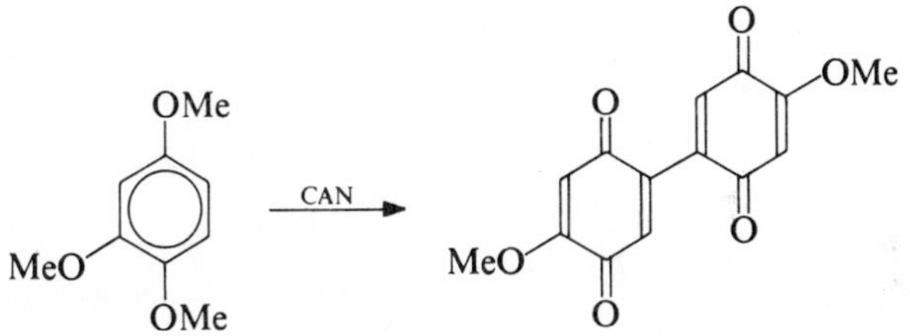

(98) —CAN→ (99) (21)

As a special example, vinylquinone **100** bearing a hydroxy group at the benzylic position gives the unusual cyclization product **101** instead of **102**[110]. This reaction would occur via cyclization to a pyran ring and concomitant hydroxylation of the generated benzylic cation, as shown in Scheme 8.

In the case of the quinoline or isoquinoline dimethyl ethers, e.g. **103** and **104**, when either R, R^1, or R^2 is an alkoxy group, the corresponding *o*-quinones are preferentially obtained with CAN[111]. Consequently, it is difficult to predict the product selectivity in the oxidation of polyalkoxybenzenes by CAN.

Nitric acid impregnated MnO_2 is also useful for the similar conversion and in many cases a good yield is achieved (equation 22)[112a]

Oxidation of 1,2,4-trimethoxybenzene derivatives usually gives a mixture of the corresponding *o*- and *p*-quinones. This problem is frequently encountered in the synthesis of naturally occurring quinones. For the selective *para* oxidation, nitric acid gives better results than other oxidation reagents[112b,c].

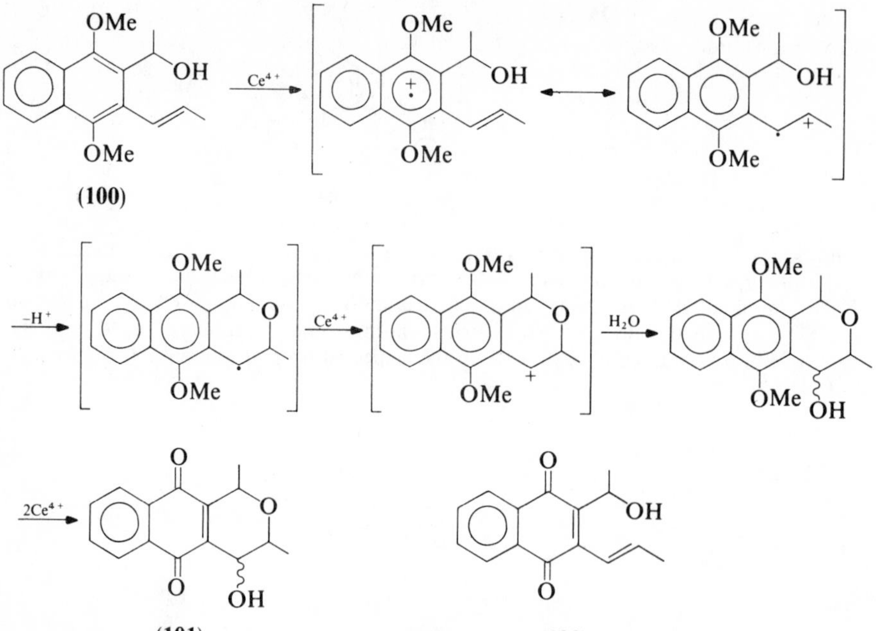

OMe OH OMe (100) Ce^{4+} OMe OH OMe OMe OH OMe

$-H^+$ OMe O OMe Ce^{4+} OMe O OMe H_2O OMe O OMe OH

$2Ce^{4+}$ O O O OH O OH O

(101) (102)

SCHEME 8

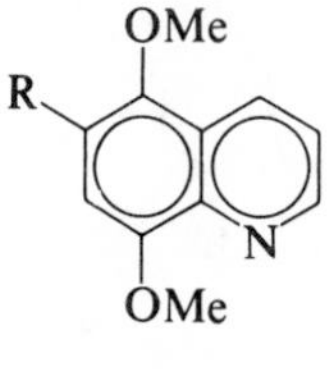

OMe R N OMe (103) OMe R^1 R^2 N OMe (104)

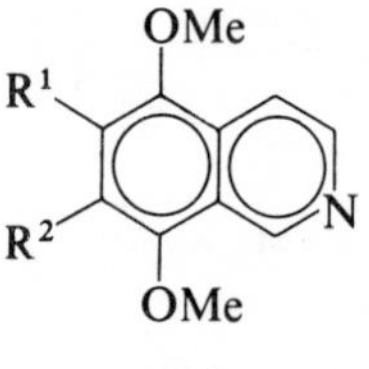

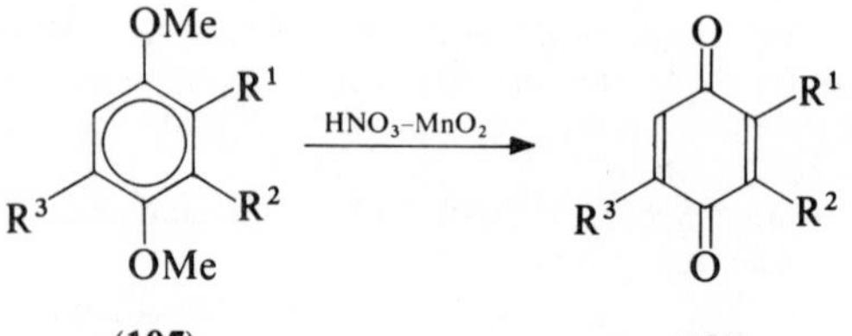

OMe R^1 R^3 R^2 OMe (105) HNO_3–MnO_2 O R^1 R^3 R^2 O (106) (22)

R^1, R^2, R^3 = H, Me, Et; R^1R^2 = –CH=CHCH=CH–

Indirect oxidation of phenol ethers and hydroquinone diethers via their quinone dialkyl acetals is realized by anodic oxidation in alkaline alcoholic media. Aromatic ethers and hydroquinone diethers are excellent precursors in anodic oxidation to quinone dialkyl acetals, which can be deprotected to the corresponding quinones in acidic media. Its applicability to a wide range of these precursors makes this method a standard one[113,114]. The following mechanistic pathway is established (Scheme 9). In basic alcoholic media,

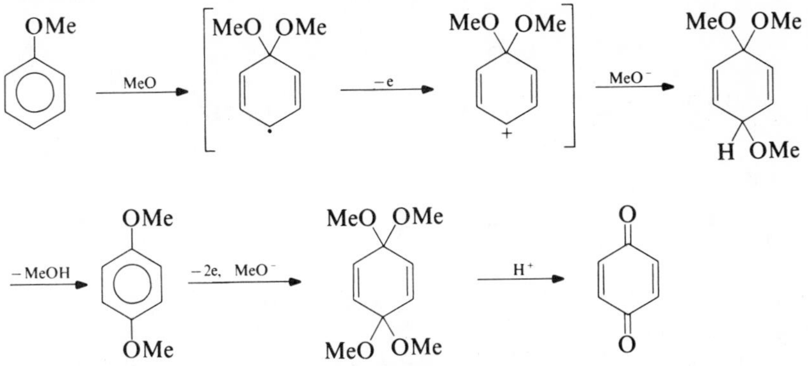

SCHEME 9

typically 1% KOH–MeOH, an alkoxy radical, which is produced from the alkoxy anion via single electron oxidation, will initially attack the *ipso* position to the alkoxy group of the aromatic ether. Successive oxidation and nucleophilic addition of alkoxide anion will afford the hydroquinone diether, which is ultimately oxidized to the corresponding quinone diacetal[115]. Consequently, both phenol ethers and hydroquinone diethers are considered to be equivalent to one another in this oxidation. Several relevant examples are shown in equations 23–25.

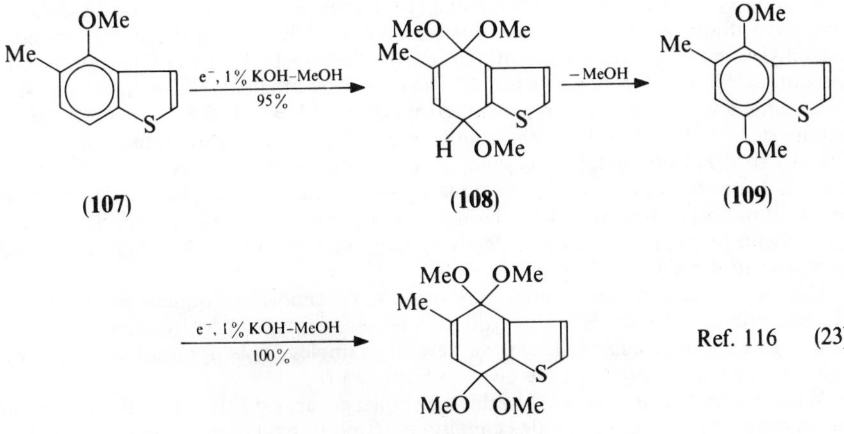

(107) —(e⁻, 1% KOH–MeOH, 95%)→ **(108)** —(−MeOH)→ **(109)** —(e⁻, 1% KOH–MeOH, 100%)→ **(110)**

Ref. 116 (23)

OMe, MeO (**111**) $\xrightarrow[100\%]{e^-,\ 1\%\ KOH\text{–}MeOH}$ MeO OMe, MeO H OMe (**112**) $\xrightarrow[84\%]{CF_3CO_2H}$ OMe, MeO OMe (**113**)

$\xrightarrow[100\%]{e^-,\ 1\%KOH\text{–}MeOH}$ MeO OMe, MeOMeO OMe (**114**) Ref. 117 (24)

OMe (**115**) $\xrightarrow[35\%]{(1)\ e^-,\ 1\%\ KOH\text{–}MeOH;\ (2)\ (COOH)_2}$ MeO OMe, O (**116**) $\xrightarrow[71\%]{(1)\ e^-,\ 1\%\ KOH\text{–}MeOH;\ (2)\ (COOH)_2}$ MeO, OMe (**117**) Ref. 118 (25)

4-Methoxy-5-methylbenzo[*b*]thiophene (**107**) undergoes a four-electron oxidation in one pot without isolation of intermediates **108** and **109**, to afford the corresponding dimethyl acetal (**110**), which is also obtained from the dimethyl ether (**109**) under the same conditions[116]. On the other hand, 1,5-dimethoxynaphthalene (**111**) under the same conditions gives the two-electron oxidized product **112**, which does not spontaneously aromatize to **113**[117]. In spite of these differences in the feasibility of aromatization of the initially oxidized product, phenol ethers and hydroquinone diethers are useful starting materials for this conversion. In most cases, both compounds afford the quinone dialkyl acetals in almost comparable yields. However, sometimes there is a considerable difference in the yields by two methods and the hydroquinone dialkyl ether shows a higher efficiency, as shown in the above example[118].

Dimethyl acetal formation from anodic oxidation of simple hydroquinone dialkyl ethers is summarized in Table 9[119]. This oxidation is also realized for hydroquinone dimethyl ethers with rather complex structures. Several examples of its application to the total synthesis of natural products are given in Section VII.

When a substrate carries an oxidizable functional group, e.g. OH, CHO, this group does not survive intact under the applied conditions of anodic oxidation as shown in equations 26–30. When a free hydroxy group is present at the proximal position, it will

TABLE 9. Anodic oxidation of hydroquinone dimethyl ethers[119]

$\xrightarrow{e^-,\ 1\%\ KOH\text{-}MeOH}$

R^1	R^2	R^3	% Yield
Br	H	H	78[a]
Br	H	Br	58[a]
Me	H	H	80[a]
Me_3Si	H	H	93[a]
Me	Me	Me	63[a]
CH(OMe)Me	H	H	92[a]
CH (1,3-dioxolan-2-yl)	H	H	88[a]
$CH_2CH{=}CH_2$	H	H	81[a]
NHAc	H	H	17[b]
$(CH_2)_2CO_2Me$	H	H	61[b]
$(CH_2)_2CONHMe$	H	H	68[b]
$(CH_2)_2CONH_2$	H	H	50[b]
$CH{=}CHCO_2Me$	H	H	46[b]

$\xrightarrow{e^-,\ 1\%\ KOH\text{-}MeOH}$

R^1	R^2	% Yield
H	H	74[a]
Me	H	75[a]
MeO	H	83[a]
Me	MeO	82[a]
Me	Me_3Si	80[a]
Br	H	84[b]
Br	Me	85[b]
Br	Br	50[b]
$-CH{=}CHCH{=}CH-$		32[a,c]

[a] Electrolysis in a single cell.
[b] Electrolysis in a divided cell.
[c] See Ref. 120.

(118) $\xrightarrow{e^-}$ (119) + (120) + (121) (26)

Ref. 121

(122) $\xrightarrow{e^-}$ (123) Ref. 121 (27)

(124) $\xrightarrow{e^-}$ (125) + (126) Ref. 121 (28)

(127) $\xrightarrow{e^-}$ (128) + (129) Ref. 122 (29)

R = H, Me, Br, MeO; (128) major; (129) minor

(130) $\xrightarrow[67\%]{e^-, \text{MeCN}}$ (131) Ref. 123 (30)

intramolecularly react to form a cyclic acetal[121, 122]. In neutral or aprotic media, anodic oxidation of phenol ethers requires a rather higher potential compared with the basic oxidation, and phenolic coupling to afford a biphenol derivative becomes a major pathway since a single-electron oxidation will preferentially occur at the aromatic ring to give the relatively stable cyclohexadienyl cation radical **132** (Scheme 10)[124]. This mechanistic difference between these two media is also reflected in the oxidized products shown in the example given in Scheme 11[115,123,125].

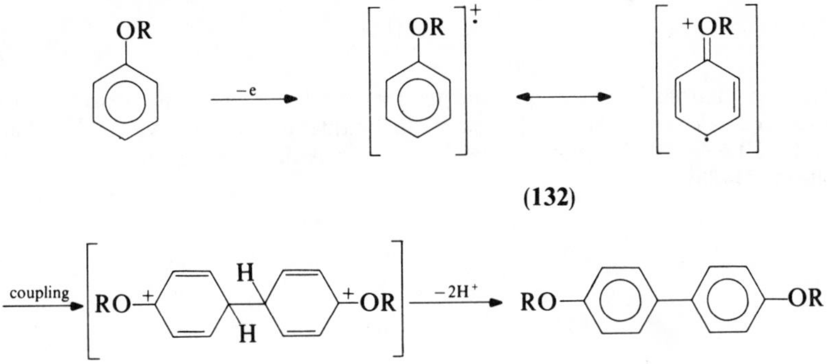

(132)

SCHEME 10

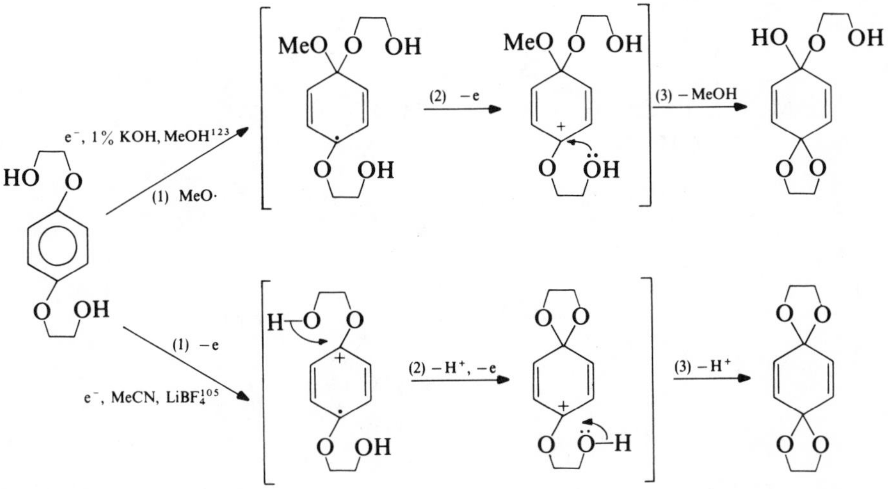

SCHEME 11

Another feature of the quinone bisacetals is their utility as synthetic equivalents of quinones with inverted charge affinity, i.e. 'umpolung'. The bisacetal of a haloquinone can be lithiated to the corresponding vinyl lithium derivative which reacts with electrophiles (Scheme 12)[126,127]. This conversion is capable of introducing various carbon functionalities into the quinone nucleus without oxidation after the nucleophilic reaction, and was applied to the synthesis of various naturally occurring quinones.

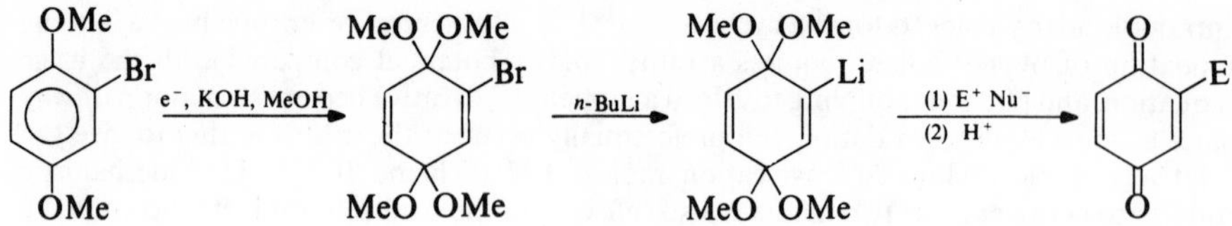

$E^+Nu^- = R^1R^2CO$, RCO_2Me, RCOCl, RBr

SCHEME 12

Similar oxidations of hydroquinone monoethers to quinone monoacetals can be achieved in an alcoholic solution by many other oxidants, such as CAN[128,129], thallium trinitrate (TTN)[43,48,130], DDQ[131] and $HgO–I_2$[132] as shown by the following examples (equations 31–38).

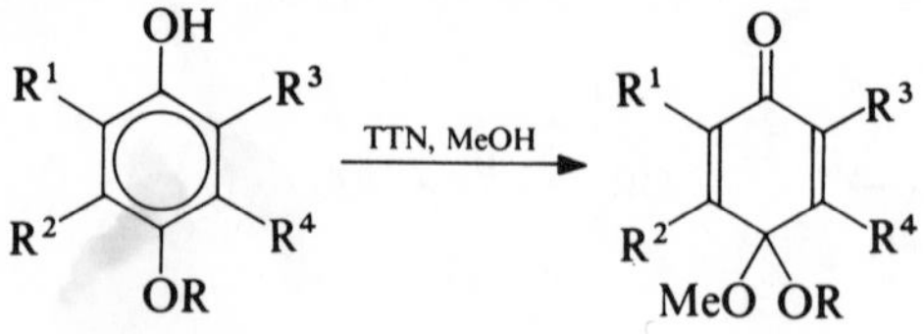

(133) TTN, MeOH (134) Ref. 43 (31)

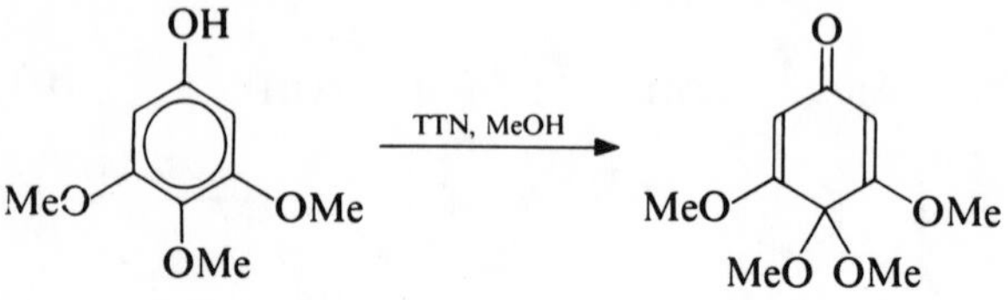

(135) TTN, MeOH (136) Ref. 43 (32)

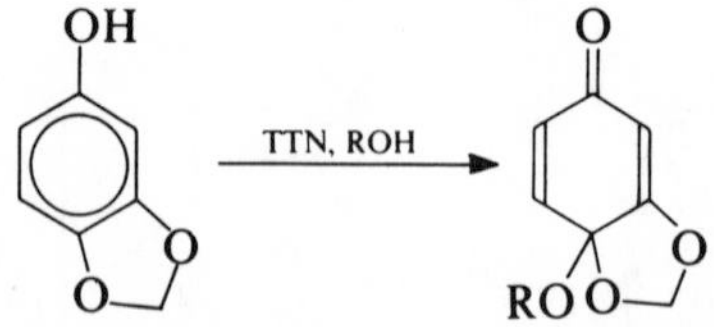

(137) TTN, ROH (138) Ref. 43 (33)

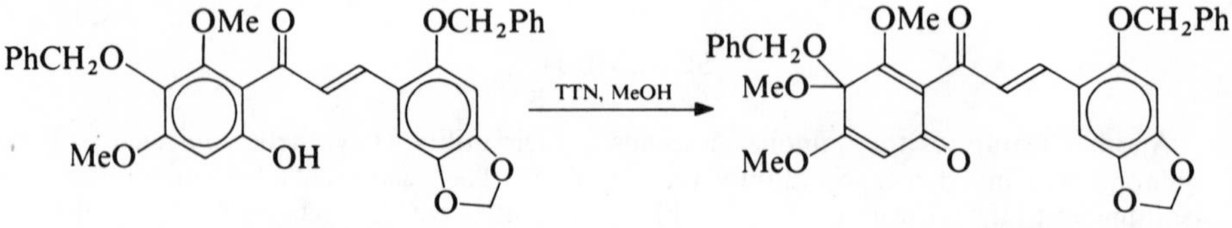

(139) TTN, MeOH (140) Ref. 43 (34)

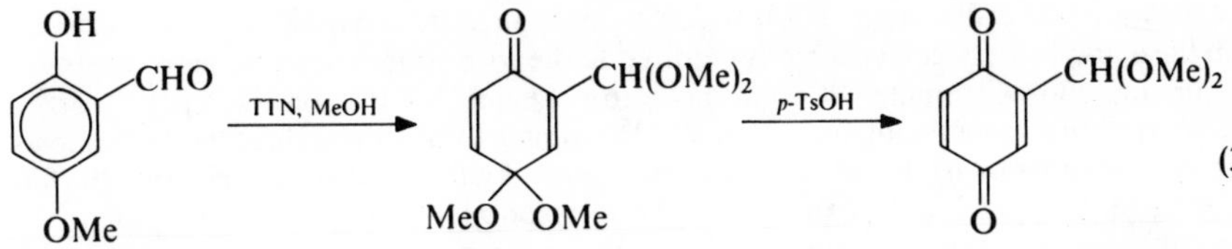

(141) —TTN, MeOH→ (142) —p-TsOH→ (143) (35)

Ref. 130

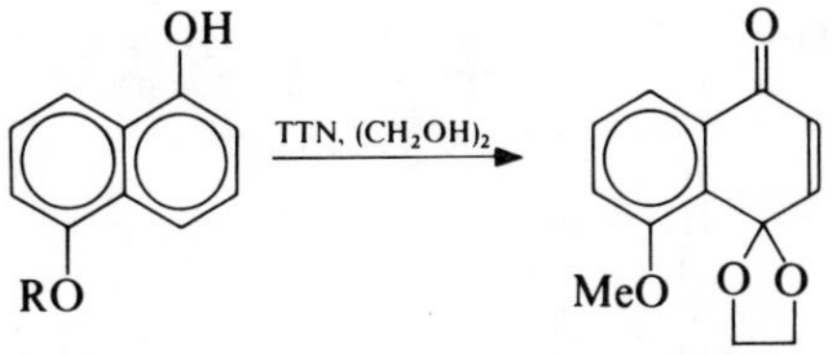

(55) R = Me —TTN, $(CH_2OH)_2$→ (58) Ref. 38 (36)

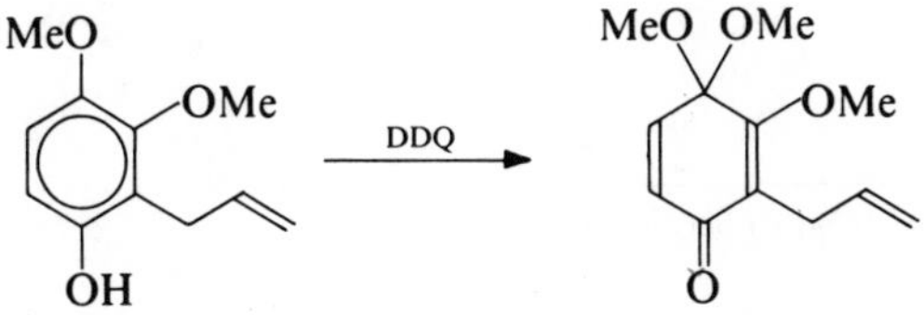

(144) —DDQ→ (145) Ref. 131 (37)

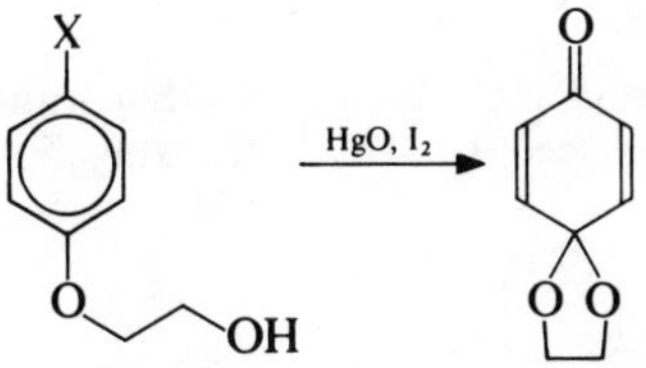

(146) —HgO, I_2→ (147) Ref. 132 (38)

X = OR, halogen

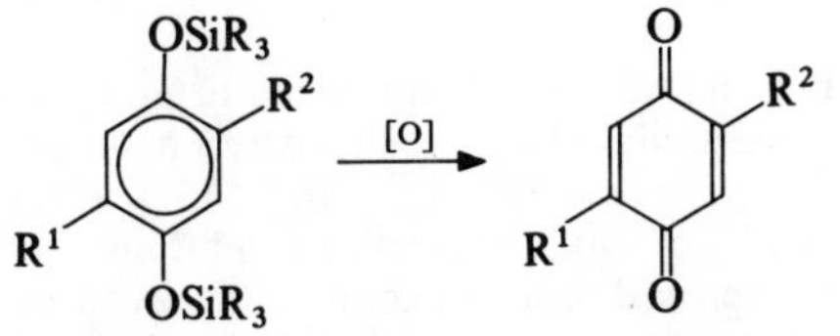

(148) —[O]→ (149) (39)

Hydroquinone disilyl ethers (**148**), which are used as protected quinones less frequently than the dialkyl ethers, can be directly oxidized to the corresponding quinones by means of pyridinium chlorochromate (PCC) in dichloromethane[133], $FeCl_3$ on silica gel[134], or by anodic oxidation in methanol (equation 39)[135]. Comparison of these three methods shows that PCC is superior to the other reagents as judged by the yields given in Table 10. The other reagents applied to the oxidation of the corresponding dialkyl ethers are potentially applicable for oxidation of the disilyl ethers which are more reactive than the corresponding dimethyl ethers.

TABLE 10. Direct oxidation of hydroquinone di-(trialkyl)silyl ethers

(**148**)			% Yield of quinone (**149**)		
R_3Si[a]	R^1	R^2	PCC[c,133]	$FeCl_3/SiO_2$[134]	e^-[d,135]
TMS	H	H	99	57	86
TBDMS	H	H	6	52	52
TMS	H	Me	62	76	
TBDMS	H	Me	90	59	
TMS	Me	Me	93	98	
TBDMS	Me	Me	80	99	
TMS	*t*-Bu	*t*-Bu	91		92
TBDMS	*t*-Bu	*t*-Bu	99	99	90
TMS	H	MeO	65		92
TBDMS	H	MeO	50		
TMS	H	Cl	[b]		92
TBDMS	H	Cl	[b]		

[a] TMS, trimethylsilyl; TBDMS, *t*-butyldimethylsilyl.
[b] No reaction.
[c] Pyridinium chlorochromate.
[d] Anodic oxidation.

E. Miscellaneous Reactions

o-Quinone (**151**) formation from the vicinal diol (**150**) without accompanying dehydration reaction can be accomplished by DDQ in refluxing dioxane[136]. More sensitive

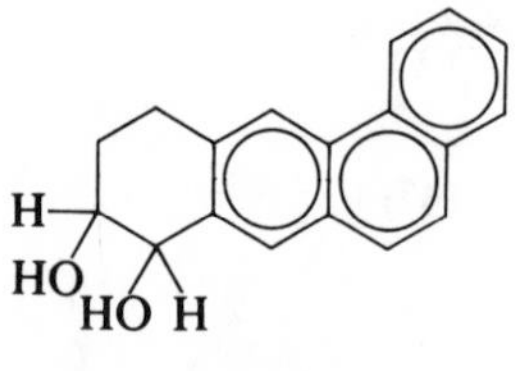

H
HO HO H

(**150**)

O
O

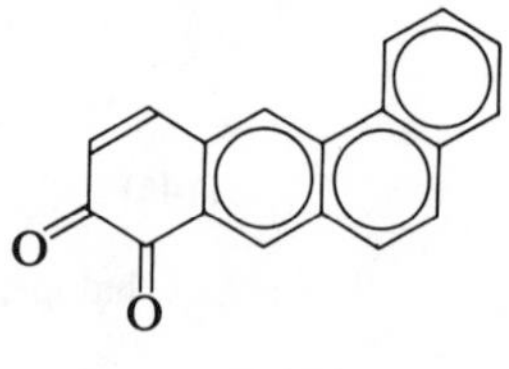

(**151**)

vicinal diols, such as **152** and **154**, which are prone to undergo both dehydration and oxidative cleavage, are successfully oxidized to the corresponding *o*-quinones, **153** and **155**, respectively, by means of Swern oxidation, i.e. with DMSO–SO_3 in pyridine–Et_3N[137] or with DMSO–Ac_2O[138] under carefully controlled conditions (equations 40 and 41). The former system gives an improved yield. In combination with the OsO_4 oxidation of a carbon–carbon double bond to a vicinal diol these methods allow the conversion of polycyclic aromatic hydrocarbons to the corresponding *o*-quinones.

(152) $\xrightarrow[\text{pyridine, Et}_3\text{N}]{\text{DMSO–SO}_3}$ (153) Ref. 137 (40)

(154) $\xrightarrow{\text{DMSO, Ac}_2\text{O}}$ (155) Ref. 138 (41)

The tetralone derivative **156** is oxidized to the corresponding *o*-quinone (**157**) by selenium dioxide (equation 42)[139]. However, the superoxide formed from potassium

(156) $\xrightarrow{\text{SeO}_2\text{, HOAc}}$ (157) (42)

superoxide and a crown ether oxidizes both α- (**158**) or β-tetralone (**159**) to 2-hydroxy-1,4-naphthoquinone (**160**) (equation 43)[140]. The same product is also obtained from 1,2- or 1,3-dihydroxynaphthalenes[76].

(158) or (159) $\xrightarrow{\text{KO}_2\text{, crown ether}}$ (160) (43)

Singlet oxygen is a reactive and unstable intermediate, which oxidizes 1-hydroxy- (**27**) and 1-alkoxynaphthalenes (**163**) to the quinone (**28**, R = H) and **165**, respectively[45]. The peroxides **161**, **162** and **164** are assumed to be the intermediates in these reactions (equations 44 and 45). The difference of the $^1O_2$ addition modes is attributed to the different electron densities of these substrates.

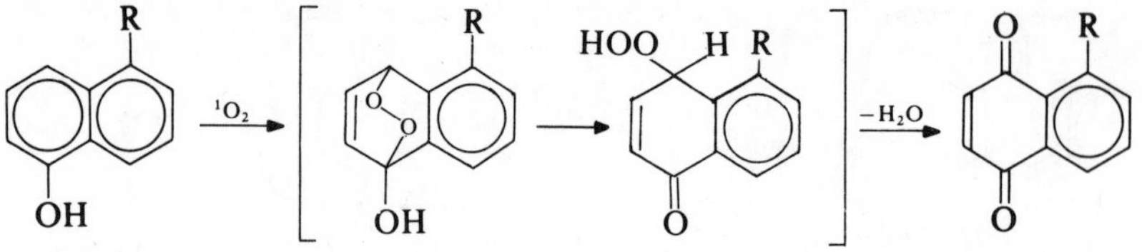

(27) $\xrightarrow{^1O_2}$ [(161) → (162)] $\xrightarrow{-H_2O}$ (28) R = H (44)

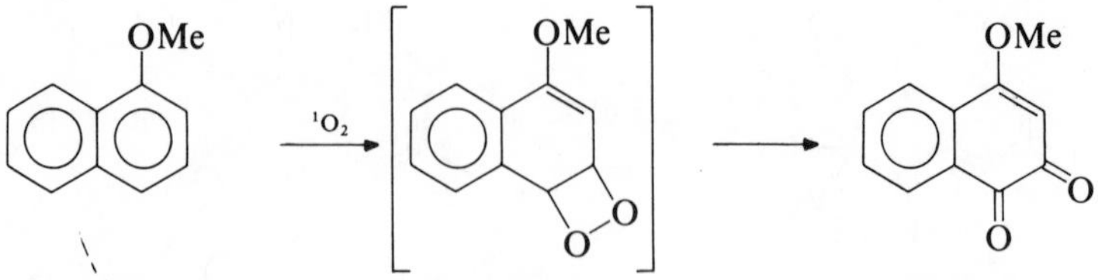

(163) $\xrightarrow{^1O_2}$ [(164)] → (165) (45)

Dehydration of the Diels–Alder adducts formed from conjugated dienes and quinones is one of the conventional and useful methods to prepare extended quinones with various functionalities. Activated manganese dioxide is found to affect the direct conversion of compounds **166**–**168** to the corresponding quinones with fully aromatized structures in moderate yields[141]. This reagent is claimed to be superior to other oxidants, such as

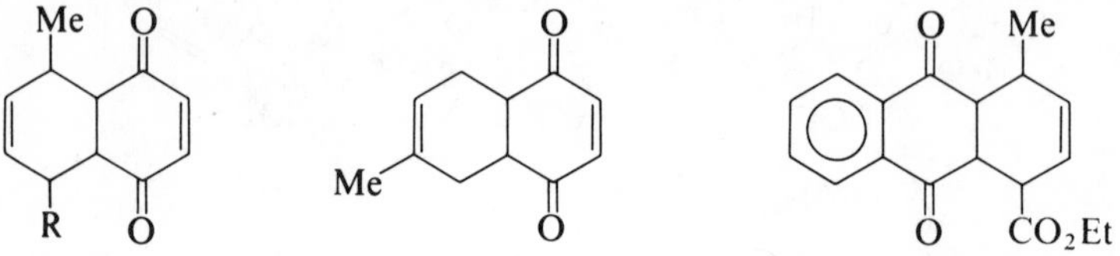

(166) (167) (168)

R = CO_2Et, CH_2OR

bromine, oxygen, DDQ, and chloranil, and is compatible with ether and carbonyl functionalities. In an alternative method for this transformation, it is also applicable for aromatization to the hydroquinone under basic conditions followed by a successive oxidation[142].

Although arylamines, as well as phenols, have been used as precursors to quinones, they are becoming less important, while satisfactory results are obtained by using traditional oxidizing agents (or methods) such as Fremy's salt[143], $H_2Cr_2O_7$[27,144,145], V_2O_5[145], $Fe_2(SO_4)_3$[90], $K_2Fe(CN)_6$[146], AgO[147], anodic oxidation[148] and photochemical oxidation[149].

III. ANNULATION METHODS

Annulation reactions are frequently used both for the synthesis of quinone itself and for extension of an aromatic ring to a present quinonoid structure. Many examples are available in the synthesis of naturally occurring quinones, especially anthracyclinones. Basic problems in annulation reactions are discussed in this section.

A. The Diels–Alder Reaction

A number of the combinations between quinones and conjugated dienes were reported for the extension of rings by means of the title reaction[150–152]. Recent progress is achieved in the use of heteroatom-substituted dienes and in the control of the regiochemistry between unsymmetrically substituted dienes and quinones, as shown in equation 46.

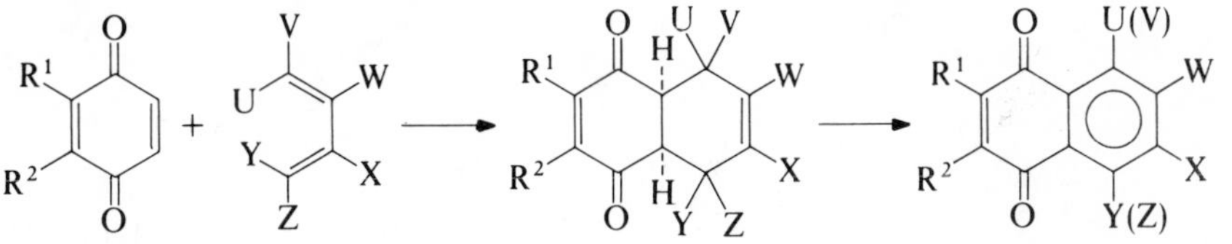

(169) (170) (171) (172)

U, V, W, X, Y, Z = H, alkyl, OR, OCOR, $OSiR_3$, SR, NR_2 (46)

Asymmetric induction in the annulation of quinone by diene with an optically active auxiliary is discussed in Section III.A.4. These results extend the applicability of the title reaction for the synthesis of quinonoid compounds.

1. *Heteroatom-substituted dienes*

We first treat here the use of heteroatom-substituted dienes including silyl enol ethers, ketene acetals, conjugated enamines, etc. A large number of heteroatom-substituted dienes possessing electron-donating groups show excellent reactivity in these annulation reactions and selected examples of them are listed in Chart 1. Other related dienes used in the same reaction are given in a comprehensive compilation in Refs 150–152. When a Diels–Alder adduct to quinone carries an alkoxy group or acetoxy group at the allylic

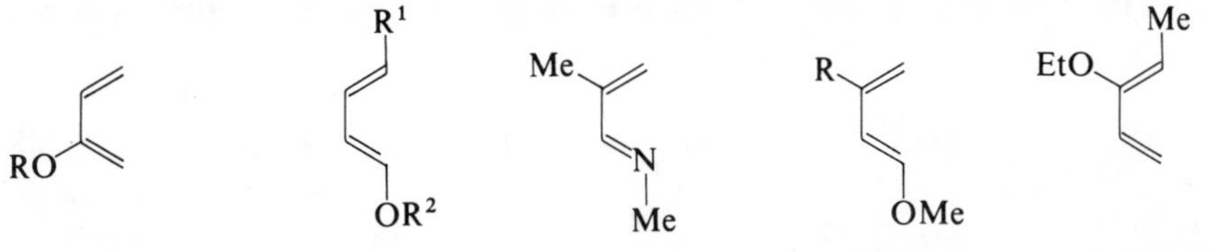

(173) (174)[153, 154] (175)[155] (176)[153, 156] (177)[157]

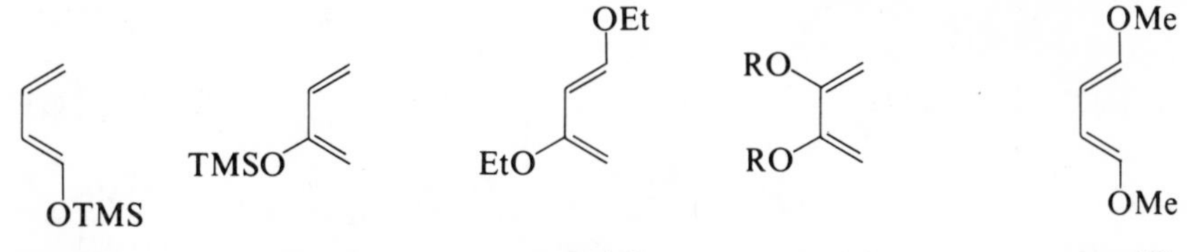

(178)[158] (179)[159] (180)[160] (181)[161] (182)[162]

CHART 1 (continued)

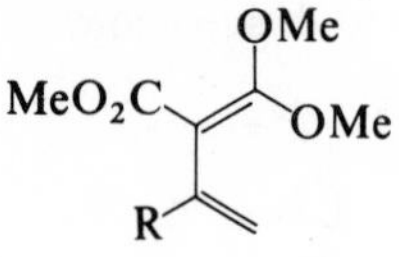

OMe, MeO_2C, OMe, R

(183)[162–165]

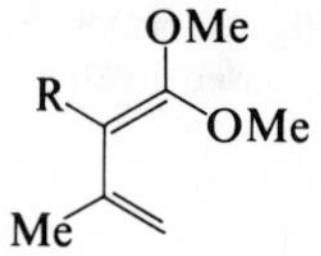

OMe, R, OMe, Me

(184)[162–164]

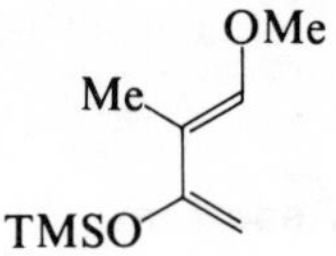

OMe, Me, TMSO

(185)[165]

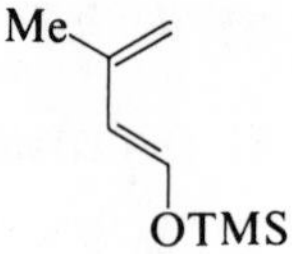

Me, OTMS

(186)[183]

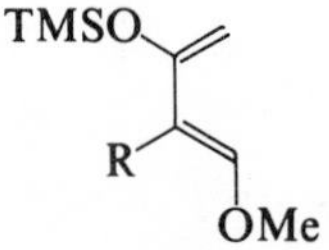

TMSO, R, OMe

(187)

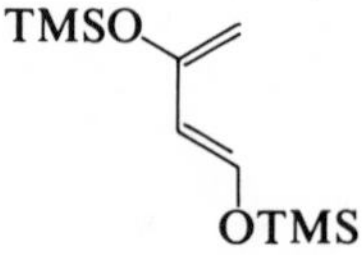

R = Me[160, 166, 167, 183]
R = Et[175]

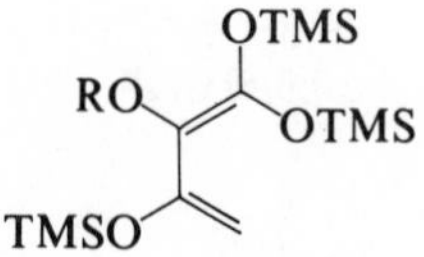

TMSO, OTMS

(188)[169]

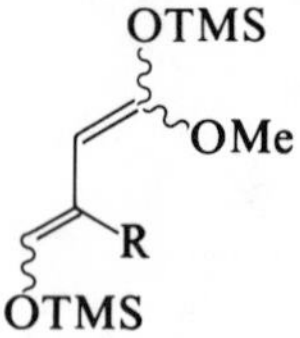

OTMS, OMe, R, OTMS

(189)[170]

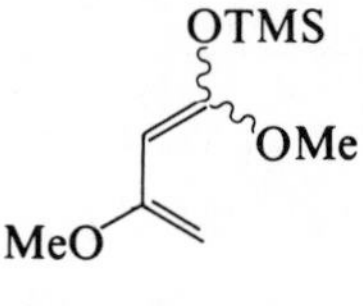

OTMS, OMe, MeO

(190)[171, 172]

OTMS, RO, OTMS, TMSO

(191)[173a]

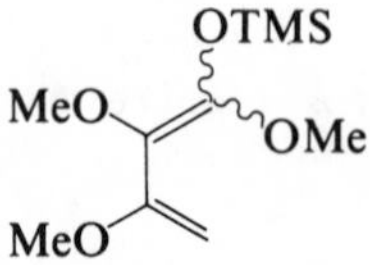

OTMS, MeO, OMe, MeO

(192)[173b]

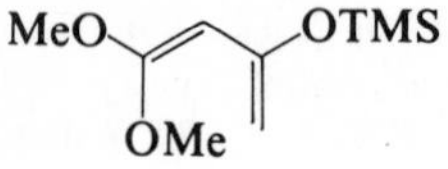

MeO, OTMS, OMe

(193)[171, 174, 175]

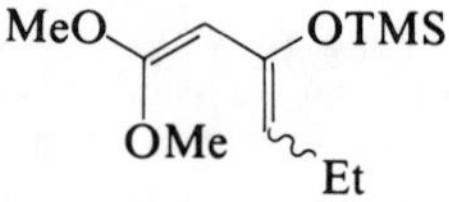

MeO, OTMS, OMe, Et

(194)[175]

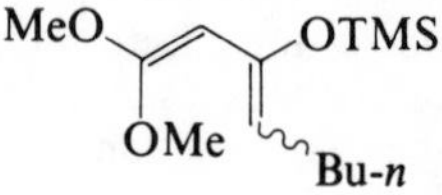

MeO, OTMS, OMe, Bu-*n*

(195)[175]

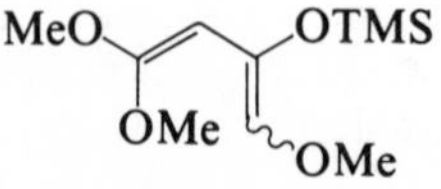

MeO, OTMS, OMe, OMe

(196)[164]

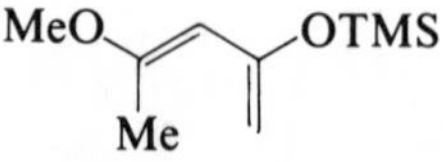

MeO, OTMS, Me

(197)[174, 176]

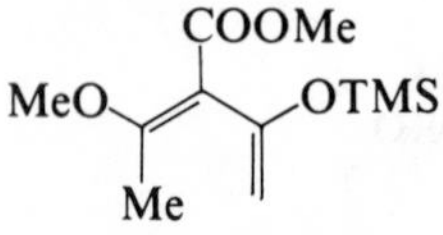

COOMe, MeO, OTMS, Me

(198)[174, 176]

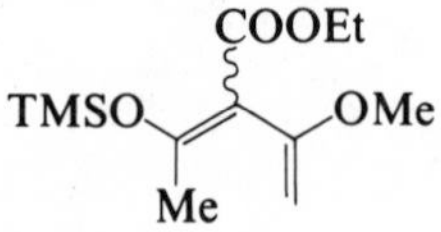

COOEt, TMSO, OMe, Me

(199)[177, 178]

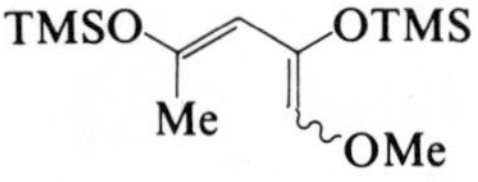

TMSO, OTMS, Me, OMe

(200)[171]

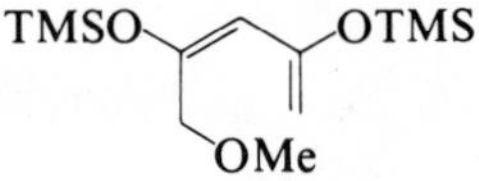

TMSO, OTMS, OMe

(201)[171]

CHART 1 (continued)

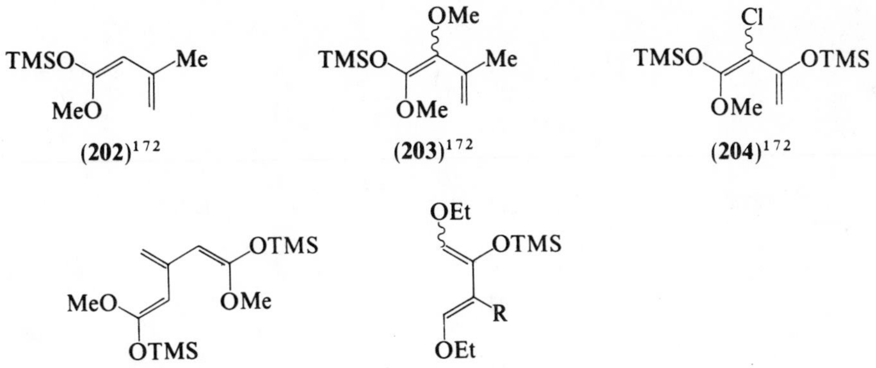

(202)[172] (203)[172] (204)[172]

(205)[179] (206)[168]

CHART 1

position, this group is easily eliminated during aromatization to the corresponding quinone under acidic conditions as shown in equations 47 and 48[180,181]. In some cases, this elimination spontaneously occurs during the Diels–Alder reaction. The ease of this

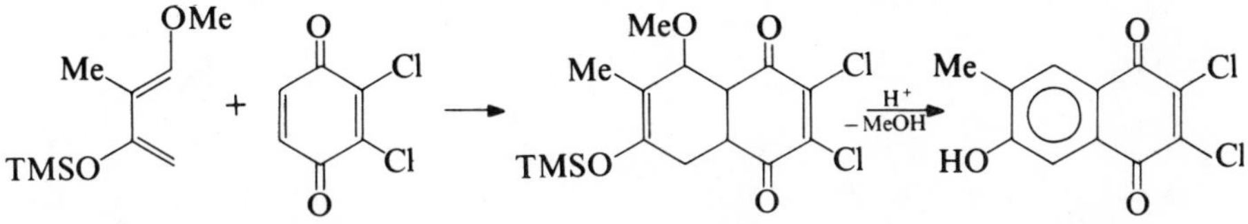

(185) (207) (208) (209)

Ref. 180 (47)

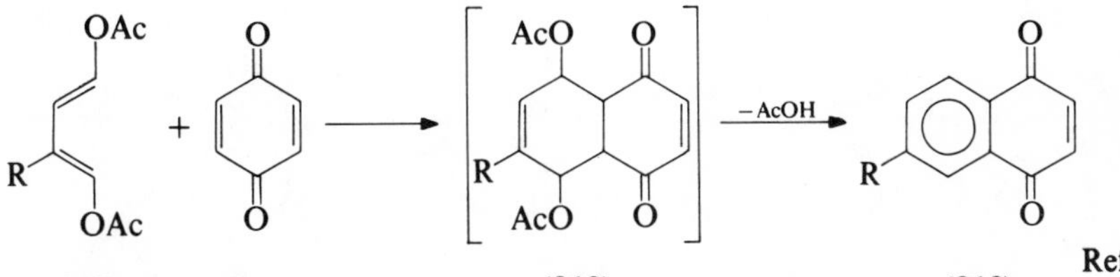

(210) R = Me (211) (212) R = Me (213) R = Me (48)

Ref. 181

elimination depends on the structure and the substitution pattern of the primary adduct. In the case of the ketene trimethylsilyl (TMS) acetal[182], **216** formed from **214** and **215** (equation 49) a hydroxyl group derived from acetal group remains in the product without elimination during the aromatization. However, after conducting the reaction of the 1-siloxydiene (**186**) under mild conditions, oxidation of the primary product with an appropriate oxidant, e.g. PCC, can preserve the oxygen functionality which originates

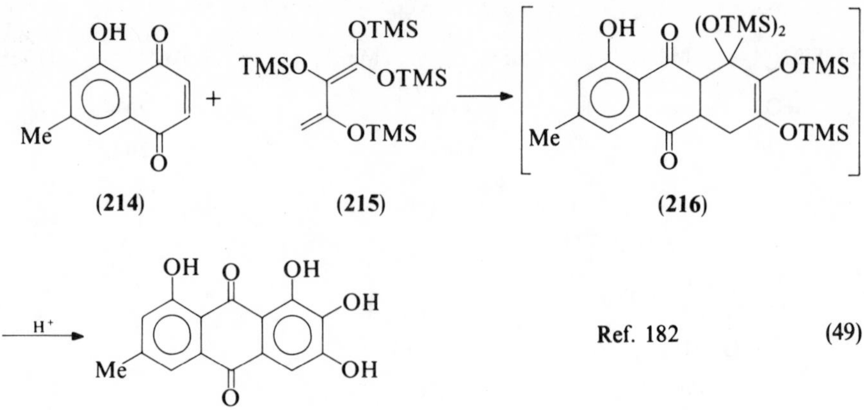

(214) (215) (216)

(217)

Ref. 182 (49)

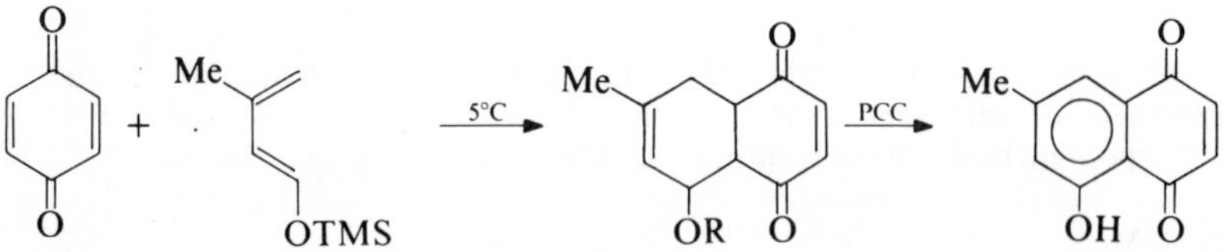

5°C PCC

R = TMS, H

(211) (186) (218) (214) (50)

Ref. 183

from the diene, without dehydrative aromatization (equation 50)[183]. When two groups which can be eliminated are present at both terminals of a conjugated diene, as in **220**[184], a competitive elimination is observed during aromatization. For example, the adduct **221** formed from **219** and **220** gives on aromatization both **222** and **223** and their ratio depends on the conditions of the work-up (equation 51). Under the mild conditions applied for the

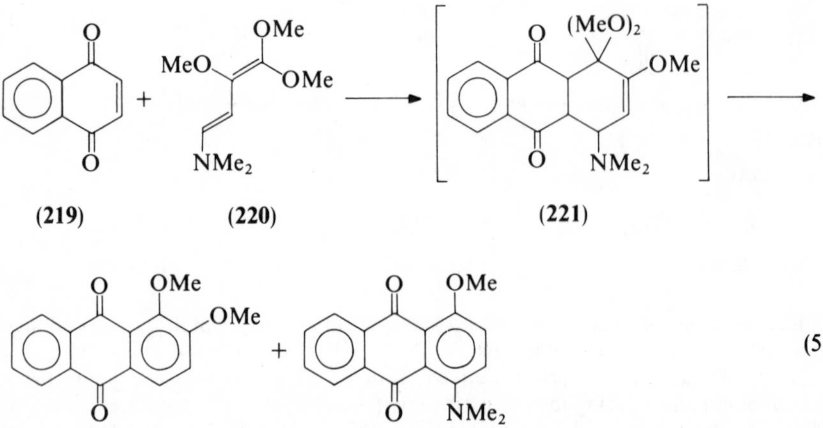

(219) (220) (221)

(222) (223) (51)

chromatographic separation, the two products **222** and **223** are obtained in almost equivalent yields, while under strongly acidic conditions the formation of **222** is preferred. Similarly, the reaction of **85** with **220** gives the two products **224** and **225** without any selectivity after chromatographic separation on silica gel[185] (equation 52). A few other examples of such an elimination from Diels–Alder adducts are given in equations 53, 54[164, 171].

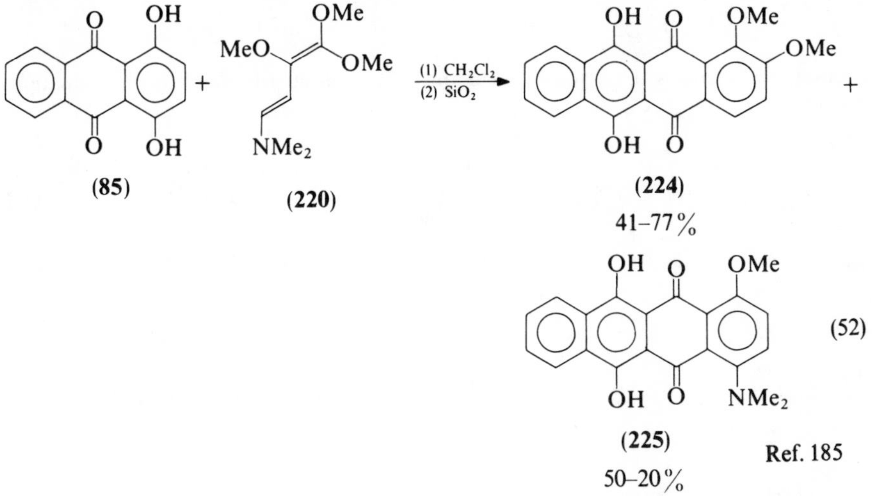

(1) CH_2Cl_2
(2) SiO_2

(85) **(220)** **(224)** 41–77%

(225) 50–20%

(52)

Ref. 185

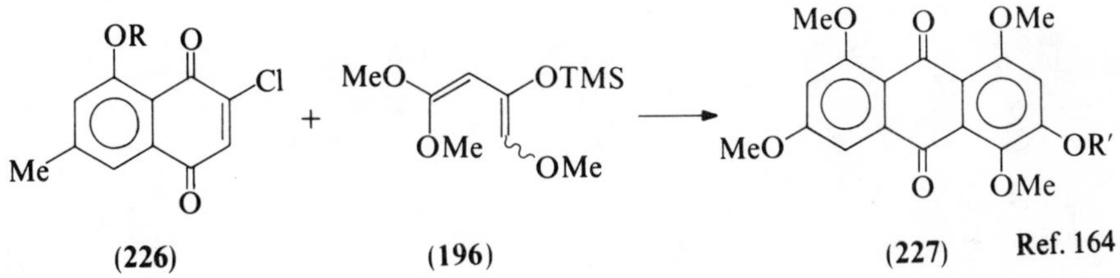

(226) R = Me **(196)** **(227)** Ref. 164

(53)

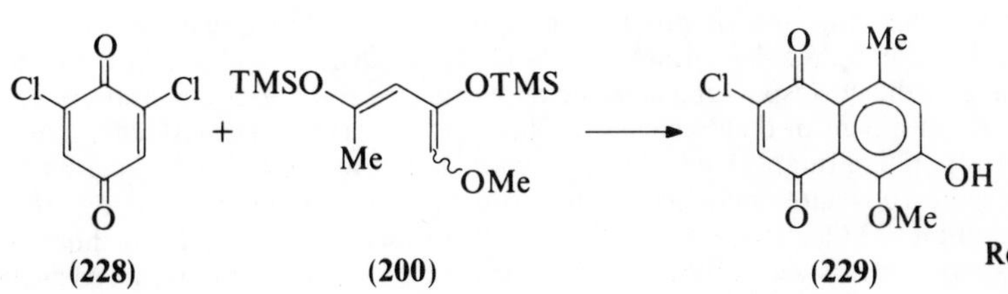

(228) **(200)** **(229)** Ref. 171

(54)

In the case of the α,β-unsaturated *N,N*-dimethylhydrazone (**230**), which possesses a conjugated imine structure, the cycloaddition with the quinone **219** spontaneously gives the corresponding aromatized product **231** by elimination of dimethylamine (equation 55)[186]. These functionalized dienes are frequently used for the synthesis of naturally

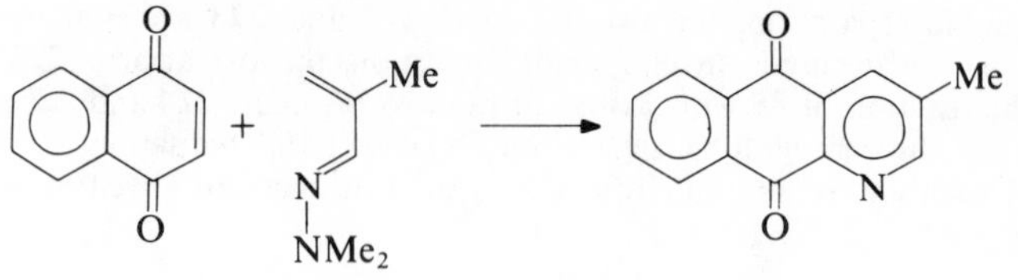

(219) + (230) → (231) (55)

Ref. 186

occurring pigments with polyoxygenated anthraquinone structure as shown in equations 56 and 57.

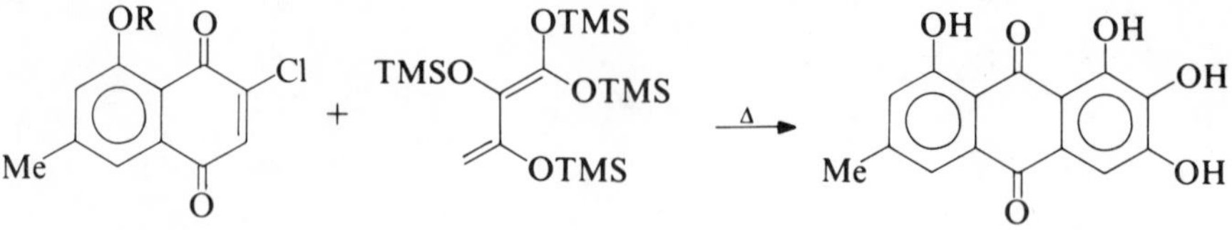

(226) R = H + (215) →Δ (217) 7-hydroxyemodine

Ref. 182 (56)

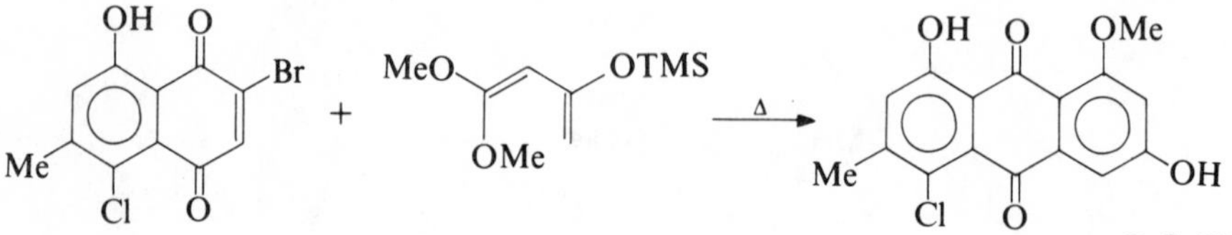

(232) + (193) →Δ (233)

Ref. 175 (57)

The quinone ring itself is prepared by annulation of functionalized dienes with an appropriate dienophile. 1,4-Disiloxyfurans (**234**), which are easily obtained by trimethylsilylation of the corresponding succinic anhydride derivatives, are considered to be the synthetic equivalent of the diketenes **235** and possess higher Diels–Alder reactivity than the corresponding furans. Diels–Alder reaction of **234** with dienophiles such as dimethyl acetylenedicarboxylate easily occurs to afford the corresponding quinone (**236**) and/or hydroquinone **237** (equation 58)[185–187]. The pyrolysis of the phthalide orthoesters (**238**) affords transiently isobenzofurans (**239**), which are trapped by dimethyl acetylenedicarboxylate to form **240** which spontaneously affords the corresponding hydroquinone monoether (**241**) (equation 59)[188]. Similarly, isobenzofuran (**244**) generated easily *in situ* from **242** via **243** cyclizes with a quinone to afford **245**. Acid treatment of **245** gives the extended quinone **246** after dehydration (equation 60)[189]. In an alternative synthesis involving 7-oxynorbornadiene intermediates such as **240**, *exo*-type tetraene (**248**), prepared in several steps from **247**[190], sequentially reacts with two moles of dienophiles (e.g. methyl vinyl ketone and benzyne (**250**)) to give first **249** and then **251**, which can be

OTMS, R^1, R^2, O, OTMS (234) ≡ [R^1, R^2, O, O (235)] —(1) $MeO_2CC\equiv CCO_2Me$ (2) F^-→ O, R^1, CO_2Me, R^2, CO_2Me, O (236) and/or

OH, R^1, CO_2Me, R^2, CO_2Me, OH (237) (58)

R, R, O, EtO OEt (238) —Δ, −EtOH→ [R, R, O, EtO (239)] —$MeO_2CC\equiv CCO_2Me$→

R, R, O, EtO, CO_2Me, CO_2Me (240) → OH, R, R, CO_2Me, CO_2Me, OEt (241) (59)

R = H, MeO

MeO, MeO, $CH(OMe)_2$, CH_2OH (242) —AcOH→ [MeO, MeO, OMe, O (243) —−MeOH→ MeO, MeO, O (244)]

—(1,4-naphthoquinone)→ MeO, MeO, O, O, O (245) —H^+→ MeO, MeO, O, O (246) (60)

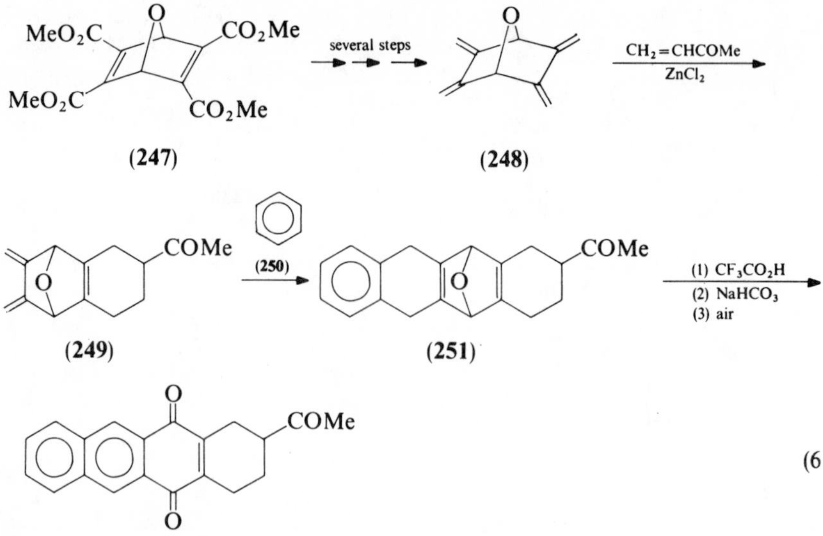

(247) → several steps → (248) → $CH_2{=}CHCOMe$, $ZnCl_2$ →

(249) → (250) → (251) → (1) CF_3CO_2H (2) $NaHCO_3$ (3) air →

(252) (61)

converted to anthraquinone (**252**) by treatment with acid (equation 61)[191]. Other unstable dienes, such as bisketene and *o*-quinodimethane, are also applied to the synthesis of quinones and for extension of the aromatic ring, respectively. Benzocyclobutene-1,2-dione (**253**) undergoes a Diels–Alder reaction with quinones or with electron-deficient olefins upon irradiation to afford Diels–Alder adducts such as **255**[192] but their yields are very low. This reaction is assumed to proceed via the intermediary formation of bisketene (**254**)

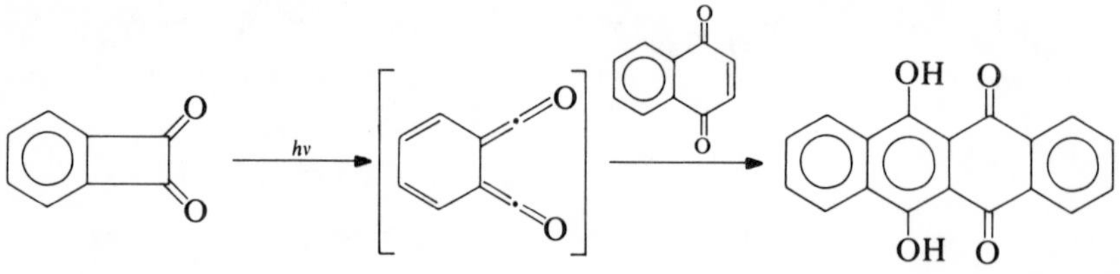

(253) → hν → [(254)] → (255) (62)

(equation 62). Extension of this reaction to the substituted benzocyclobutene-1,2-dione (**256**, R = Me) and quinones such as **257** gives mixture of the two possible regioisomers

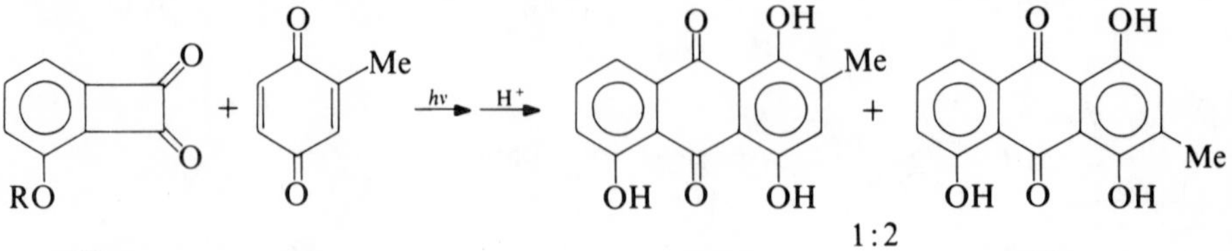

(256) + (257) → hν, H⁺ → (258) + (259), 1:2 (63)

258 and **259** (equation 63)[192]. The annulation of unsymmetrically substituted benzocyclobutene-1,2-dione and a quinone under photochemical conditions does not show high regioselectivity[192]. Contrary to **256** the benzocyclobutene-1,2-diol derivative (**260**) under pyrolytic conditions generates a difunctionalized *o*-quinodimethane (**261**), which can be trapped by 1,4-naphthoquinone to afford naphthacenequinone (**262**) in good

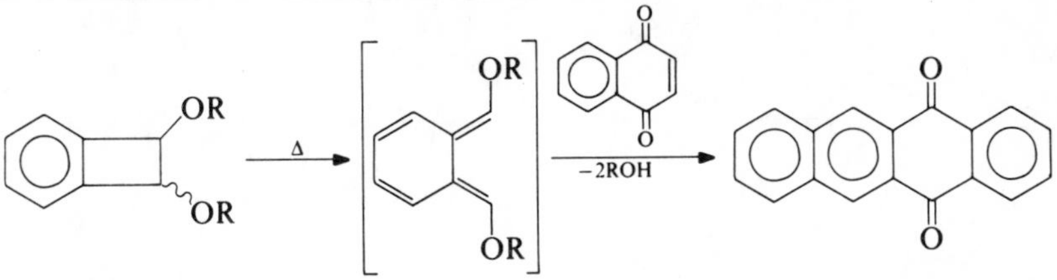

(**260**) (**261**) (**262**) (64)

R = Me, Ac

yield (equation 64)[193]. *o*-Quinodimethane (**264**) which is generated by a standard procedure of treatment of 1,2-bis(dibromomethyl)benzene with sodium iodide could be trapped by a quinone such as **169** or other dienophiles *in situ* to form extended quinones

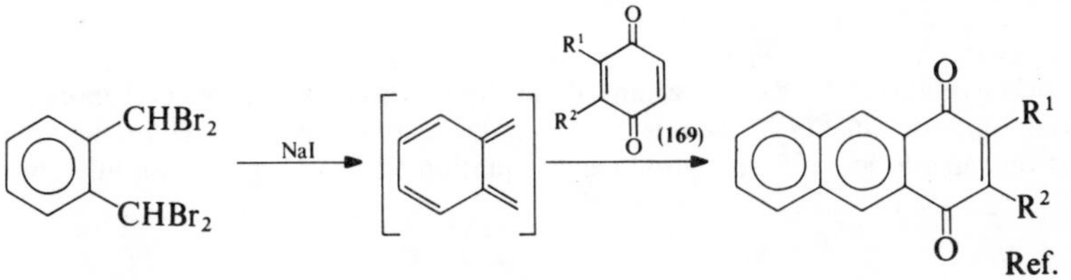

(**263**) (**264**) (**265**) (65)

Ref. 194

such as **265** (equation 65)[194, 195]. In a similar reaction, **267** which is generated from **266** is annulated with methyl vinyl ketone to **268** (equation 66). Under pyrolytic conditions, the

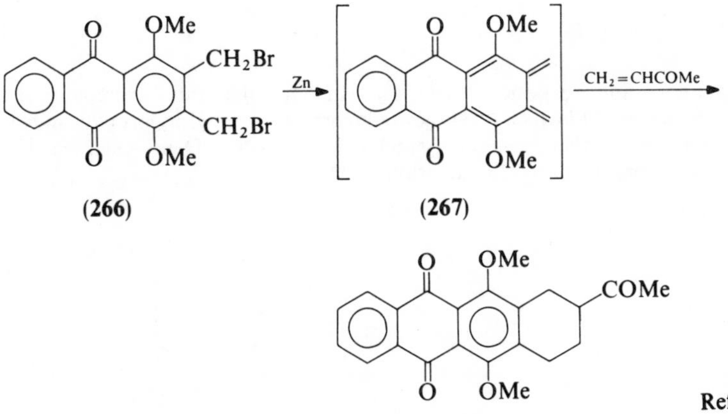

(**266**) (**267**)

(**268**) (66)

Ref. 195

cyclobutanone dimethyl acetal (**269**) gives an intermediary vinyl ketene acetal (**270**), which is very unstable compared with the aforementioned vinyl ketene acetals, and is trapped by a quinone such as **219** to afford **271** which gives the naphthacenequinone **272** after

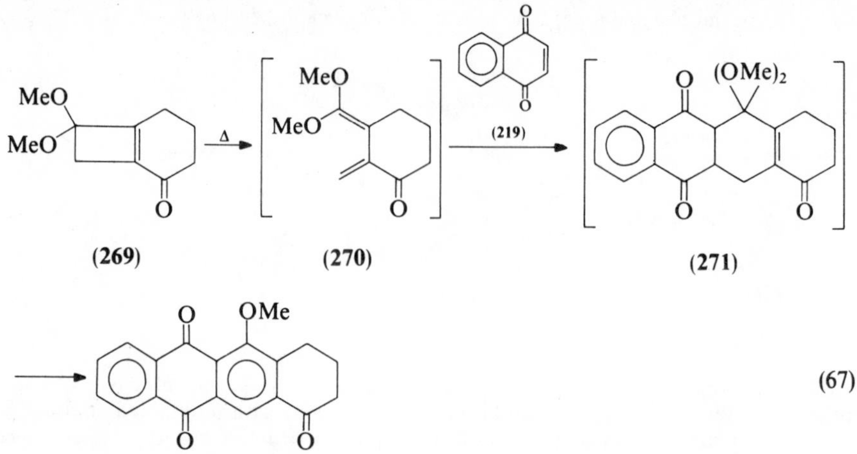

(269) (270) (271) (272) (67)

oxidation (equation 67)[196a]. As a variant of the above examples, hydroxycyclobutene with a masked quinone (cf. **274**) cyclizes in a similar manner with acrylonitrile to afford exclusively an adduct (**275**) in a good yield (equation 68)[196b]. Other dienophiles, however,

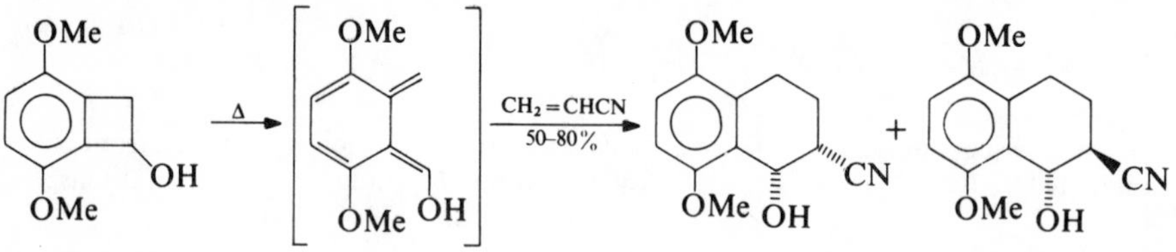

(273) (274) (275) (276) (68)

show low stereo- and product selectivity. In a relevant example, the (trimethylsilyl)vinylketene **277** which is synthesized from 1-(trimethylsily)propyne in two steps and is known to be a rather stable compound, also undergoes a Diels–Alder reaction with 1,4-naphthoquinone to form **278** (equation 69)[197].

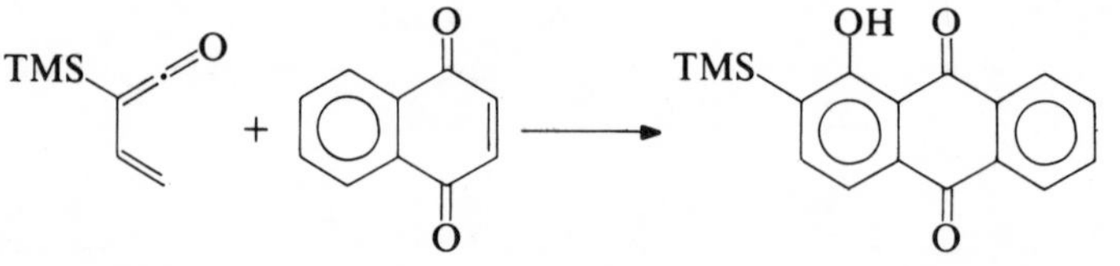

(277) (219) (278) (69)

2. *Halogen atom as a facile leaving group in control of regiochemistry*

Aromatization of the Diels–Alder adduct of quinones unsubstituted at the 2 and 3 positions with dienes usually requires rather strong basic conditions necessary for enolization of carbonyl groups. When a quinone carries a good leaving group, which preferably has electron-withdrawing character for the sake of acceleration of the annulation, at the 2 or 3 position, the regeneration of the quinone structure will proceed rather easily. For this purpose, chloro-, bromo-, or sulfinylquinones[198] are frequently used to construct naphthoquinone or anthraquinones in good yield, as shown by the following

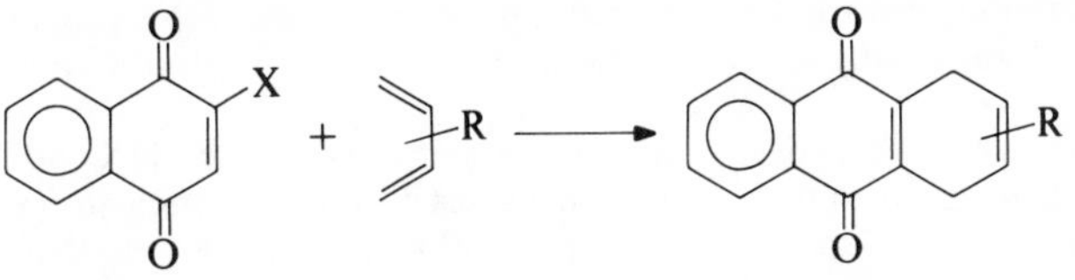

(279) + (280) → (281) (70)

X = Cl, Br, SO_2R

example (equation 70). Haloquinones, especially, were extensively investigated and utilized in this respect as shown in the previous section (Section II.A.1, see also Refs 151 and 152).

Another feature of the reaction of halogenated quinones in the Diels–Alder reaction is their influence on the regiochemical control of the addition by the electron-withdrawing character of the halogen atom. Typical examples are shown in equations 71–74. In every

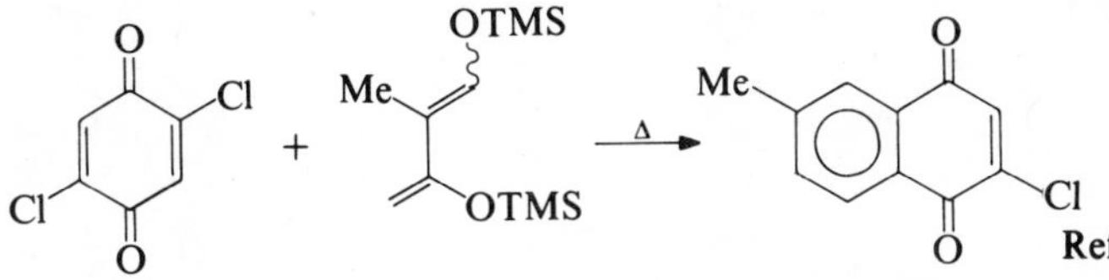

(282) + (283) →Δ (284) (71) Ref. 171

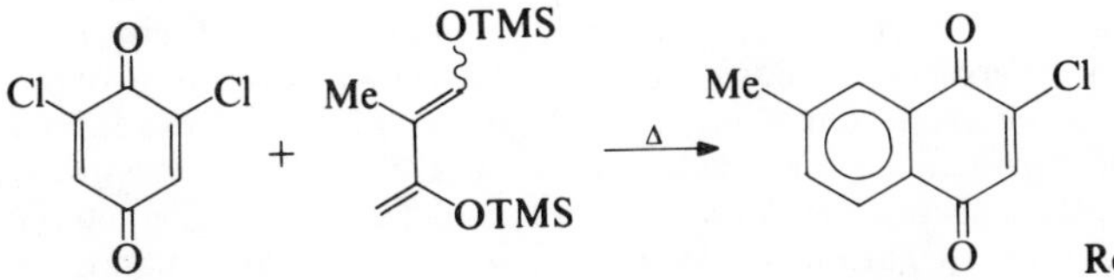

(228) + (283) →Δ (285) (72) Ref. 171

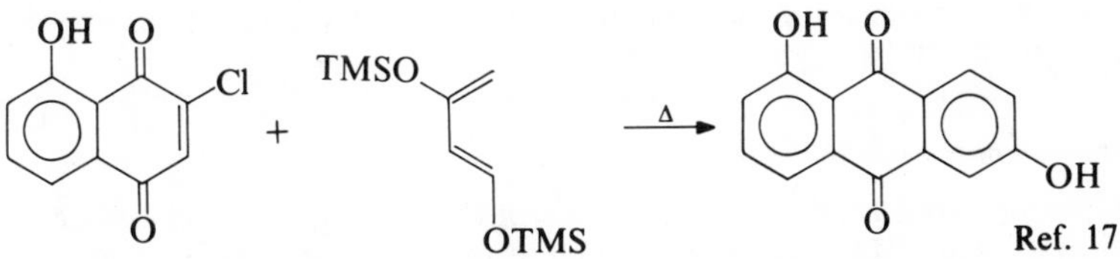

(286) + (188) →Δ (287) (73) Ref. 171

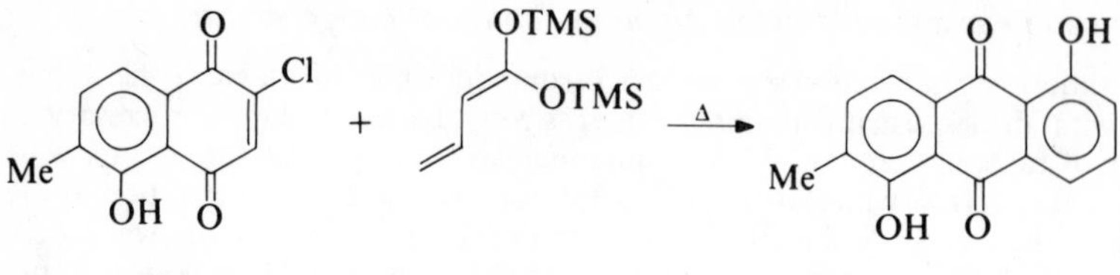

(288) + (289) —Δ→ (290) (74)

Ref. 181

case, the cyclization occurs in a very selective manner. This regiochemical outcome is well established in many examples and is rationalized by means of frontier molecular orbital theory[199].

During the course of the regeneration of the quinone structure, HX is eliminated from the primary Diels–Alder adduct, and in some cases it causes decomposition of other acid-sensitive groups in the molecule. For prevention of such an undesirable reaction, $SrCO_3$[179,200] or Et_3N[200] are used as acid scavengers. The latter base seems to be superior to the former[200] and aromatization would proceed by the route shown in Scheme 13.

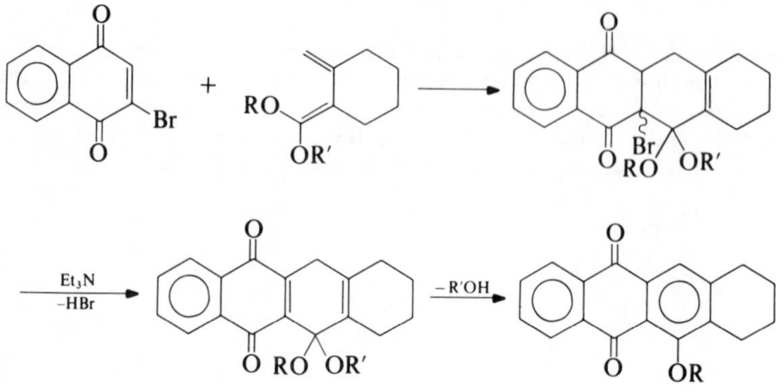

SCHEME 13

Sequential Diels–Alder reactions are developed in order to synthesize highly functionalized anthraquinones from dichlorobenzoquinones by taking advantage of the difference in reactivity between the starting quinone and the primary Diels–Alder adduct. For example, 2,6-dichlorobenzoquinone **228** reacts with diene **291** and the successive aromatization gives **292**. This is again treated with the second diene (**293**) to give tetramethylxantholaccaic acid B (**294**) after aromatization (equation 75)[201]. Similarly, successive treatment of diene **191** with quinone **228**, then with diene **295** gives ceroalbolinic acid (**296**) after deprotection (equation 76)[178].

Two dienes, **297** and **299**, with very similar structures to one another are sequentially annulated with 2,5-dichlorobenzoquinone to afford the precursor for vineomycinone B_2 (**302**) (equation 77)[202].

3. Other dienes

Elimination of small molecules, such as CO_2 or $CH_2{=}CH_2$, from Diels–Alder adducts is a useful variation of the annulation as well as the dehydrohalogenative aromatization processes shown in the previous section.

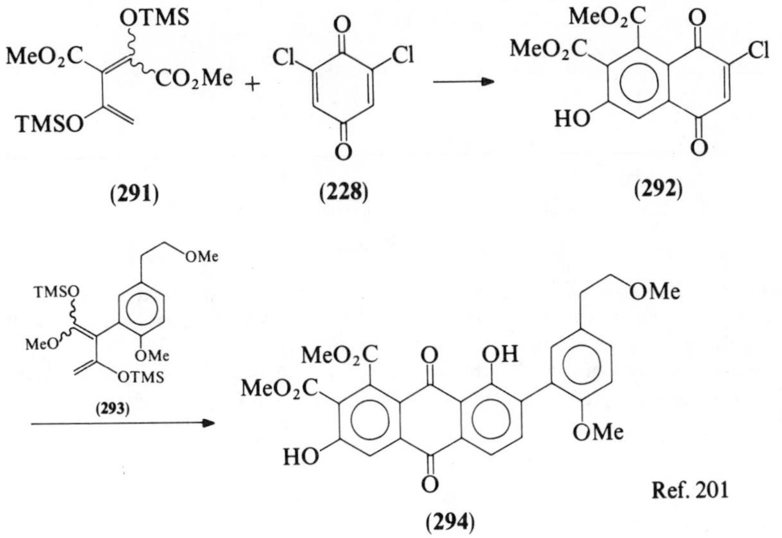

OTMS
MeO_2C
CO_2Me +
TMSO
O
Cl
Cl
O
→
MeO_2C
O
MeO_2C
Cl
HO
O

(291) (228) (292)

OMe
TMSO
MeO
OMe
OTMS
(293)
→
MeO_2C
O
OH
OMe
MeO_2C
HO
O
OMe

(294)

(75)

Ref. 201

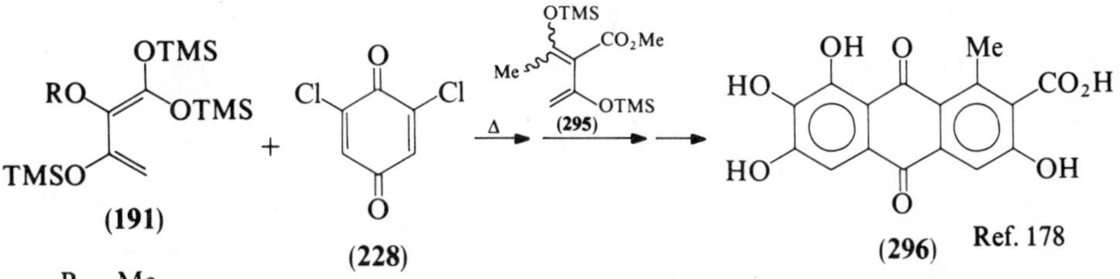

OTMS
RO
OTMS
TMSO
+
O
Cl
Cl
O
OTMS
Me
CO_2Me
OTMS
Δ
(295)
→ → →
OH
O
Me
HO
CO_2H
HO
O
OH

(191) (228) (296)

R = Me

Ref. 178

(76)

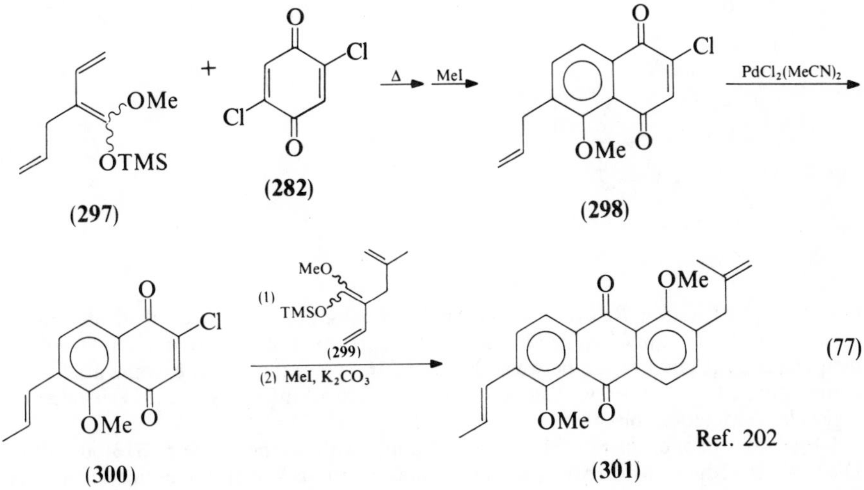

+
O
Cl
Cl
O
OMe
OTMS
Δ → MeI →
O
Cl
OMe
O
$PdCl_2(MeCN)_2$ →

(297) (282) (298)

O
Cl
OMe
O
MeO
(1) TMSO
(299)
(2) MeI, K_2CO_3
→
O
OMe
OMe
O

(300) (301)

(77)

Ref. 202

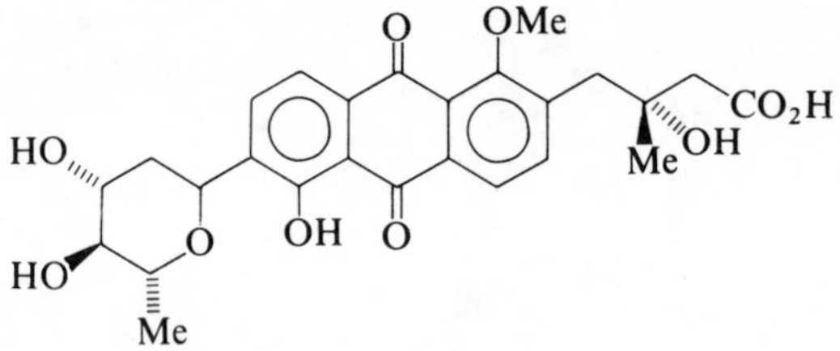

(302)

3-Hydroxy-2-pyrone (**303**) reacts with a 1,4-quinone (**211**) to afford a reduced form of juglone (**305**) probably via decarboxylation of the adduct (**304**) (equation 78)[203]. A similar result is obtained in the reaction of 6-methoxy-4-methyl-2-pyrone (**306**)[204] and extremely high regioselectivity is observed in the case of juglone (**56**, R = H) as a dienophile (equation 79)[205]. When juglone acetate instead of juglone is applied, no regioselectivity is observed in

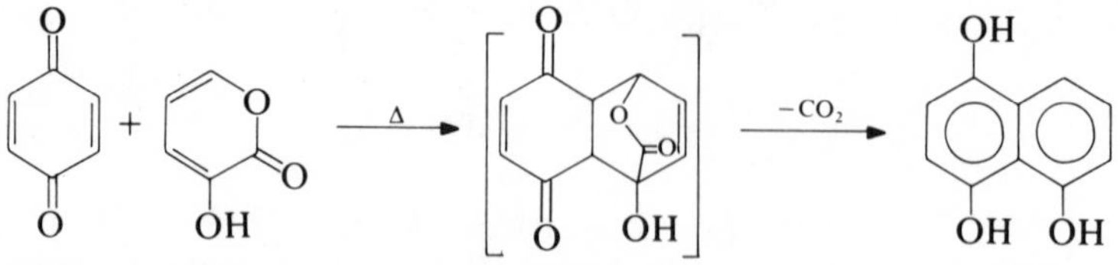

(211) + (303) —Δ→ [(304)] —$-CO_2$→ (305) (78)

(56) R = H + (306) —Δ→ [(307)] —$-CO_2$; (1) [O]; (2) HBr→ (308) (79)

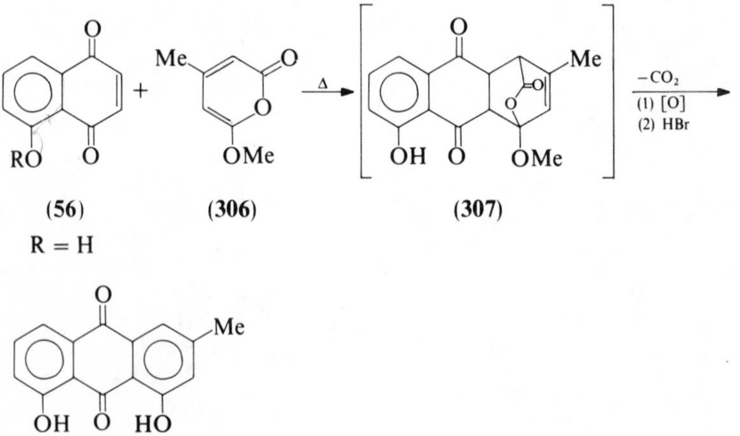

the analogous reaction. Relevant to pyrone derivatives, homophthalic anhydride (**309**) is considered as an equivalent synthone of the corresponding pyrone (**310**). At room temperature, the equilibrium between **309** and **310** favors the left side, while at elevated temperature **310** is trapped with juglone (**56**, R = H) to afford a quinone (**311**) in excellent regioselectivity (equation 80)[205].

1-Methoxycyclohexadiene (**312**) reacts readily with quinones (e.g. **313**) to afford Diels–Alder adducts, which are converted to the corresponding quinones in two steps, i.e.

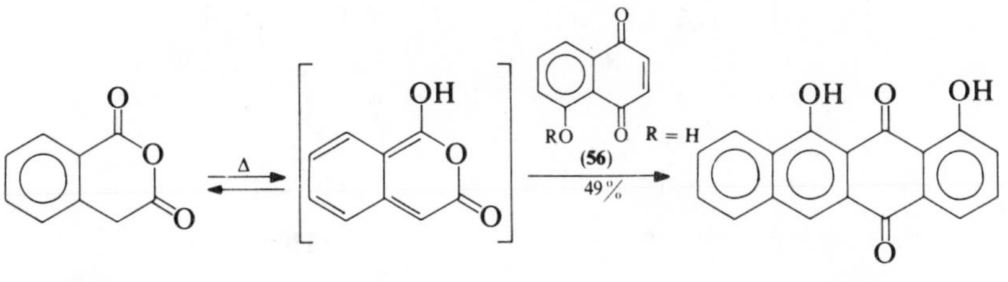

(309) (310) (311)

(80)

by enolization to the corresponding hydroquinone and then oxidation to the quinone. The successive pyrolysis gives 3-methyoxyjuglone methyl ether (**316**) (equation 81)[206,207]. When 2-methoxy-1,4-benzoquinone (**313**) is used as a dienophile, **316** is obtained stereoselectively, while 2-methoxy-3-methyl-1,4-benzoquinone (**169**, $R^1 = Me$, $R^2 = MeO$) gives two isomeric adducts (**317** and **318**) in $\simeq 4{:}1$ ratio (equation 82)[206].

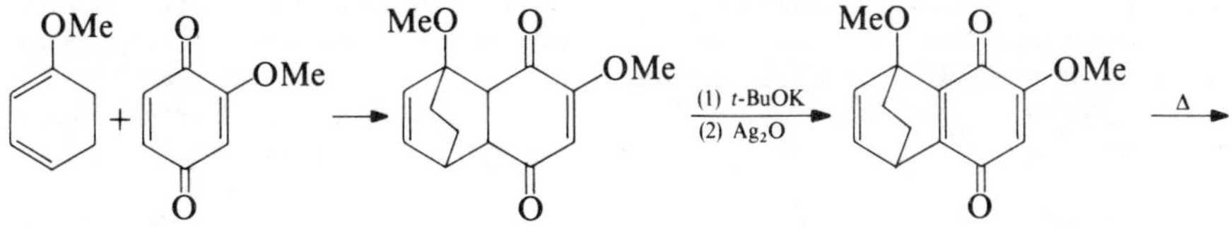

(312) (313) (314) (315)

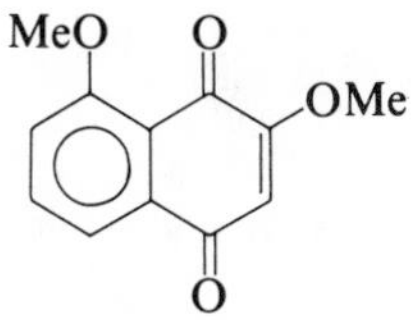

(316)

(81)

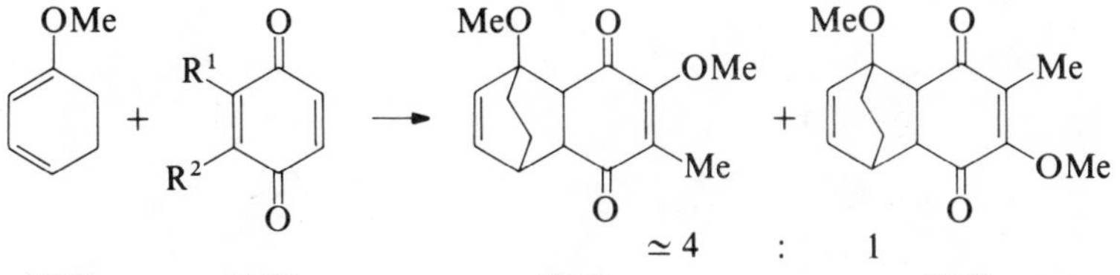

(312) (169) (317) (318)

$R^1 = Me, R^2 = MeO$

$\simeq 4 : 1$

(82)

Electron-deficient dienes are considered to have insufficient reactivity toward 'usual' dienophiles, which favor electron rich dienes. Diels–Alder reactions of such dienes (**319**) with *p*-benzoquinone (**211**) at high pressure give the adducts **320** in good yields (equation 83)[208]. This cycloaddition does not proceed at atmospheric pressure even at elevated temperatures.

(211) (319) $\xrightarrow[60 \sim 92\%]{1.5\ \mathrm{GPa,\ RT}}$ (320) (83)

R = Me, Et, R^1 = H, Et
R^2 = H, Me, CH_2OSiMe_2Bu-*t*

4. Regioselectivity and site-selectivity

When a diquinone, which possesses two reaction sites, undergoes a Diels–Alder reaction, the cyclization can occur at both sites as demonstrated by the formation of both **322** and **323** from **321** and **280** (equation 84). The site-selectivity is apparently dependent on the electronic nature of the applied diene. This problem is encountered at the Diels–Alder approach in anthracyclinone synthesis[5,209–215]. An electron-rich diene, such as isoprene, 1-alkoxy-1,3-butadiene (**174**), 2,3-dimethyl-1,3-butadiene, etc., attacks preferentially an internal double bond, while a less electron-rich diene, e.g. 1,3-butadiene, 1-acetoxy-1,3-butadiene (**174**, R^1 = H, R^2 = Ac) and 2,4-hexadiene, etc., cyclizes at the external double bond (equations 85 and 86). These phenomena are well explained by means of frontier

(321) + (280) R = Me → (322) + (323) (84) Ref. 209

(84) + (324) → (325) (85) Ref. 210

(326) + (174) → (327) (86) Ref. 211

R^1 = H, R^2 = Ac

molecular orbital (FMO) theory which considers the interactions between the lowest unoccupied molecular orbitals (LUMOs) of quinone and the highest occupied molecular orbital (HOMO) of diene[216]. The LUMO and the secondary LUMO (SLUMO) of the diquinone concentrate on the internal and external double bonds, respectively. The magnitude of LUMO–HOMO interaction between two reactants is proportional to $(IP^{diene} - EA^{quinone})^{-1}$ when IP and EA are the ionization potential and the electron affinity, respectively. Consequently, the electron-rich diene would preferentially interact with the LUMO of the quinone. There will not be an appreciable difference in the (IP – EA) value of a less electron-rich diene toward the LUMO and the SLUMO of a diquinone, and hence it should not contribute at all to the selectivity. The site-selectivity of the less electron-rich dienes will therefore be determined by the steric factor, since steric hindrance at the internal double bond of the diquinone is larger than at the external double bond.

To avoid the undesirable internal attack by an electron-rich diene, protection of an internal double bond by an oxirane ring, which can easily regenerate the aromatic skeleton, was developed as demonstrated in equation 87[212].

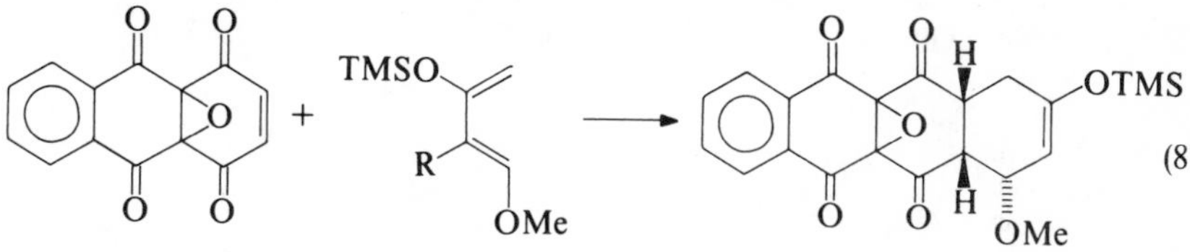

TMSO, R, OMe, OTMS, H, H, OMe (87)

(328) **(187)** R = H **(329)**

A substituent on a quinone has a strong directing effect on the Diels–Alder reaction. A large number of examples were reported and the results concerning the preferred site of attack are summarized in Chart 2[199, 217]. A large difference in the regioselectivity of the

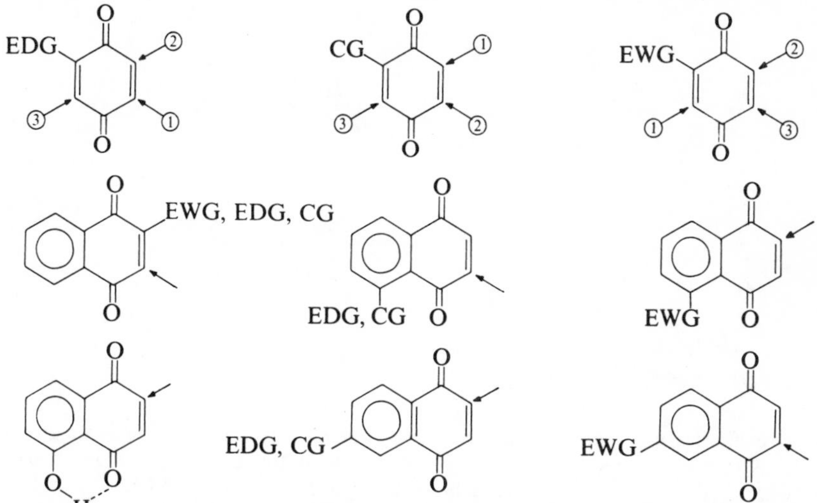

EDG ② ③ ① CG ① ③ ② EWG ② ① ③

EWG, EDG, CG

EDG, CG

EWG

O H

EDG, CG

EWG

CHART 2. Site of attack of the nucleophilic terminus of a diene on substituted benzoquinones and naphthoquinones: EDG, electron donating group; EWG, electron-withdrawing group; CG, conjugating group. The numbers show the relative order of reactivity

cyclization is observed between an electron-donating and an electron-withdrawing substituent. For example, the presence of an electron-withdrawing group on the benzoquinone generates a partial positive charge at the 3 position and an unsymmetrically substituted diene will cyclize with the quinone in such a way that its nucleophilic site will attack the positively charged center. Several examples are shown in equations 88–90. Other

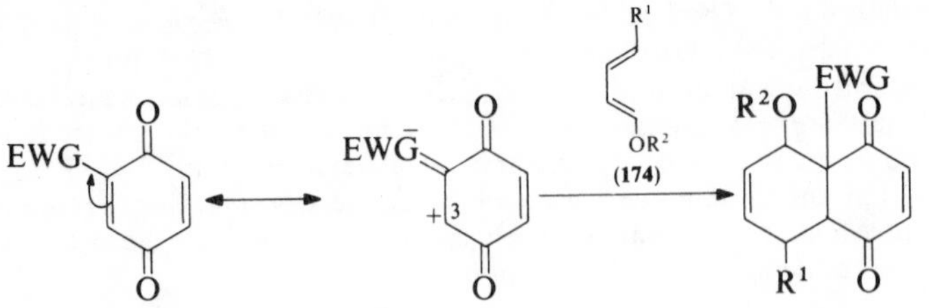

(330) ⟷ (331) —(174)→ (332) (88)

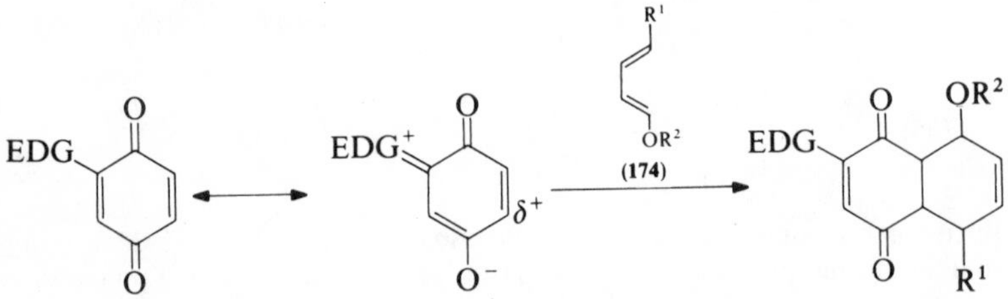

(333) ⟷ (334) —(174)→ (335) (89)

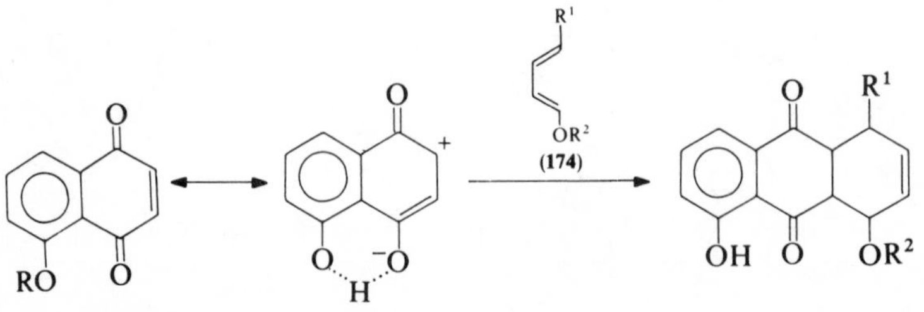

(56) R = H ⟷ (336) —(174)→ (337) (90)

selected examples, which are rationalized by the charge distribution argument are given in equations 91–97. The remote control of regioselectivity in the reaction of **344** is attributed to a weaker hydrogen bonding to the carbonyl group at the 4 position rather than to that at the 1 position. Since the acetylamino group has a weakly electron-donating character, quinone **355** shows a very high regioselectivity to afford adduct **356** as the exclusive product. Further examples related to this regioselectivity are reported in the literature[225,226]. All these effects are also elucidated by means of the FMO theory[199,216,217].

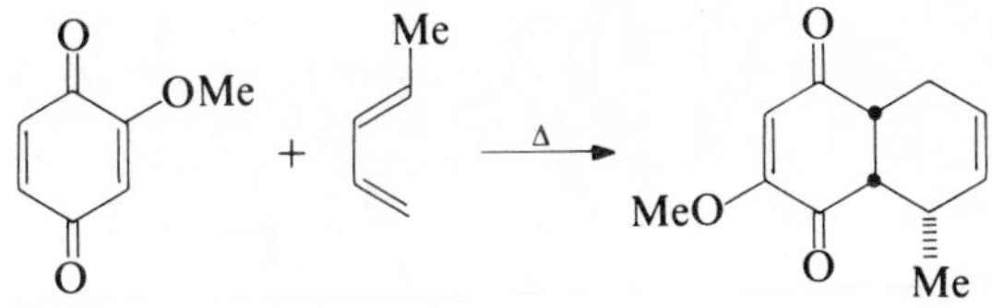

O OMe Me O
+ Δ Refs 218, 219 (91)
MeO
O Me
O
(313) (338) (339)

O O
Me TMSO R^1
+ or Δ (92)
R
RO O OTMS OMe OH O OR^2
(56) (186) (187) (340)
R = H R = H R^1 = Me, R^2 = OTMS[183]
R^1 = OTMS, R^2 = Me[220]

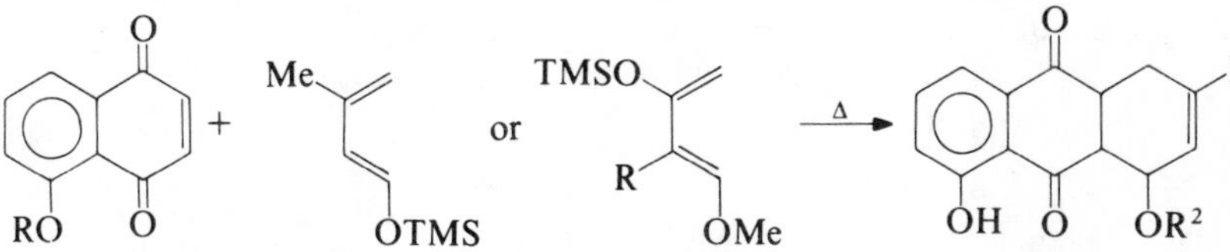

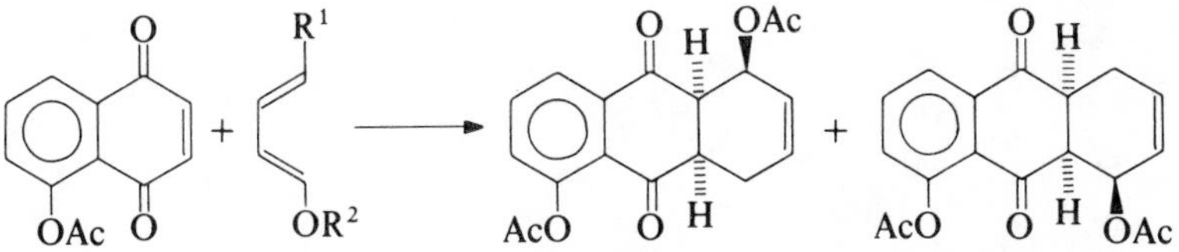

O R^1 O H OAc O H
+ + (93)
OAc O OR^2 AcO O H AcO O H OAc
3:1
(341) (174) (342) (343) Ref. 221
R^1 = H, R^2 = Ac

OH O OMe OH O OMe
1
+
4
MeO OH O MeO OH O
(344) (345) > 80%
(346)

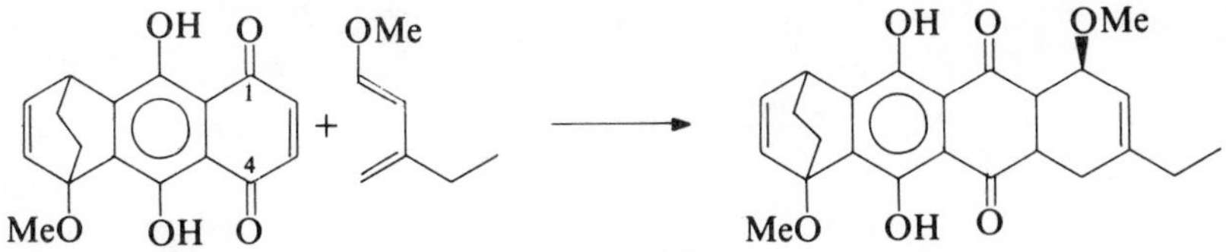

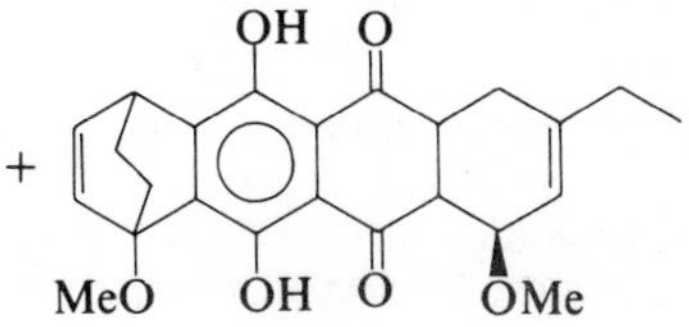

OH O
+ Ref. 222 (94)
MeO OH O OMe
< 20%
(347)

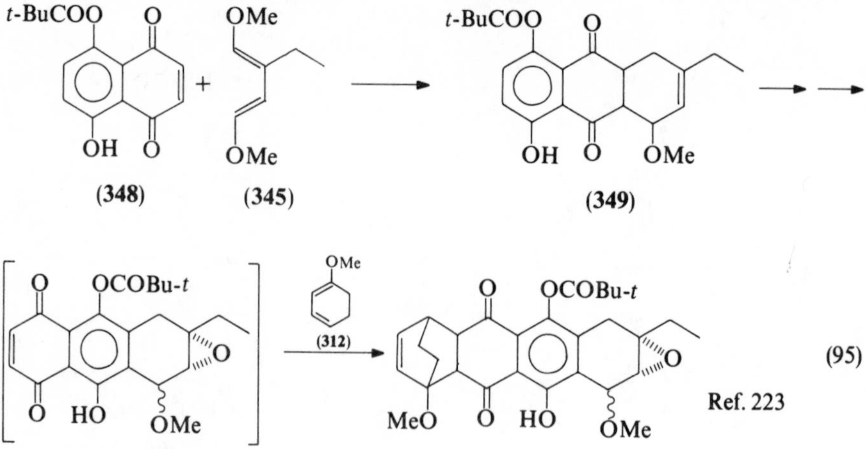

t-BuCOO O OMe t-BuCOO O

OH O OMe OH O OMe

(348) (345) (349)

O OCOBu-t OMe O OCOBu-t

(312)

O HO OMe MeO O HO OMe

(350) (351)

Ref. 223 (95)

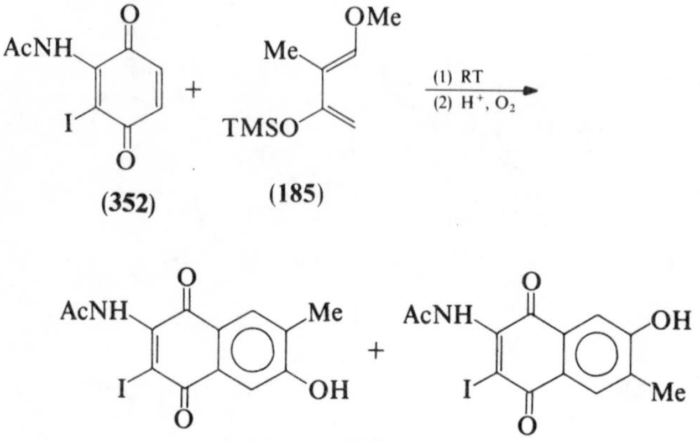

O OMe

AcNH Me

(1) RT
(2) H^+, O_2

I TMSO

O

(352) (185)

O O

AcNH Me AcNH OH

+ Ref. 224 (96)

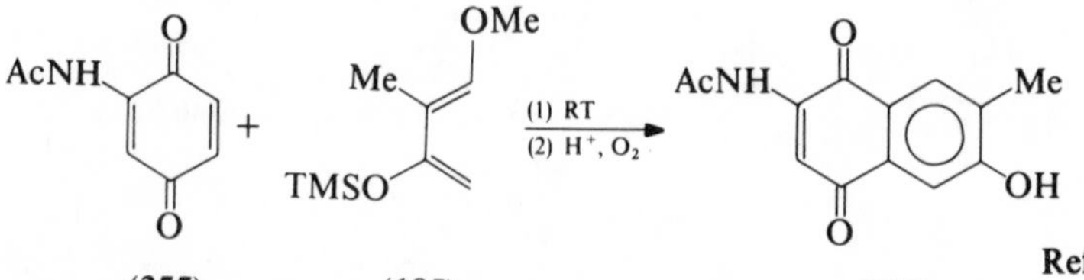

I OH I Me

O O

12:1

(353) (354)

O OMe

AcNH Me O

AcNH Me

(1) RT
(2) H^+, O_2

(97)

TMSO OH

O O

Ref. 224

(355) (185) (356)

Since its earlier days, addition of a Lewis acid to the Diels–Alder reaction mixture is known to be able to control regioselectivity[227] and in some cases it resulted in a reversed selectivity. Several remarkable effects of Lewis acid are especially observed in the annulation of juglone derivatives. The reported results in these syntheses are summarized in Table 11. The role of the Lewis acid which enhances or reverses the regioselectivity is

TABLE 11. Lewis acid mediated Diels–Alder annulation

Quinone	Diene	Product	Product ratio without Lewis acid	Product ratio with Lewis acid (equiv.)	Reference
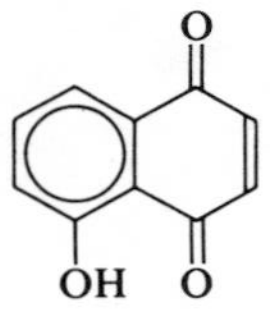		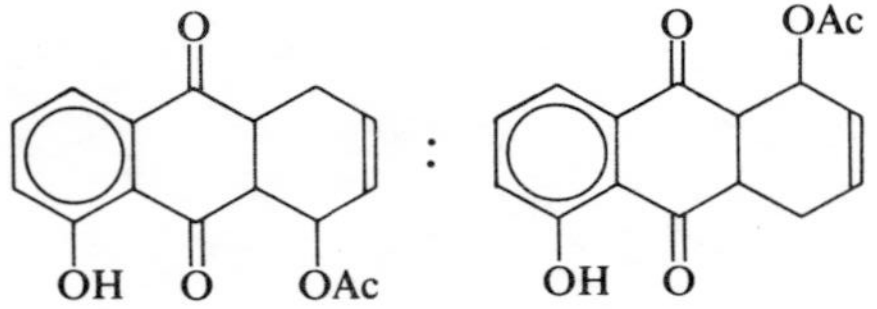	3:1	$BF_3 \cdot OEt_2$(0.04) $\geqslant 99.5 : \leqslant 0.5$ 100:0	228, 229 230 231, 232
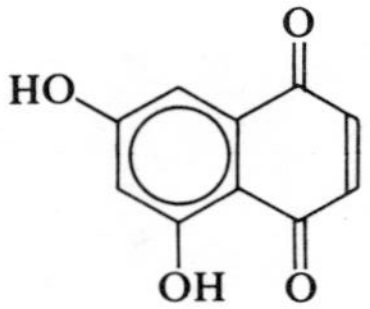	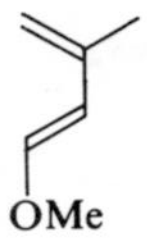	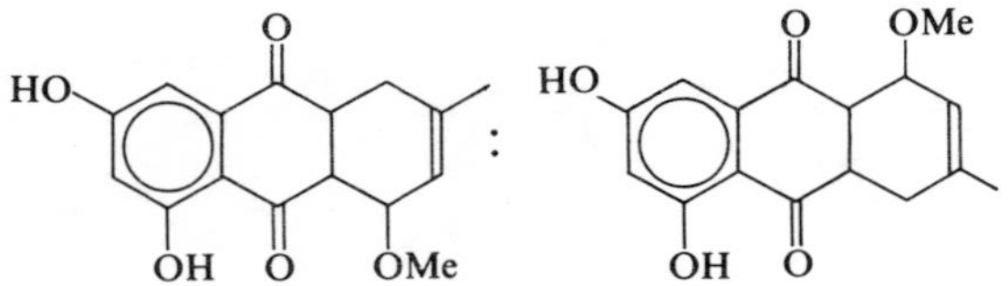	45:55	$BF_3 \cdot OEt_2$ (0.05) 55:45 $BF_3 \cdot OEt_2$(0.4) $\geqslant 95 : \leqslant 5$ $B(OAc)_3$ (2) $\geqslant 95 : \leqslant 5$	229 229 229
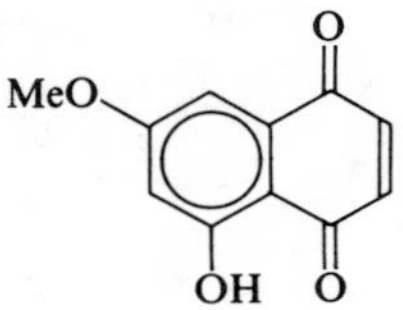	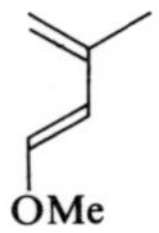	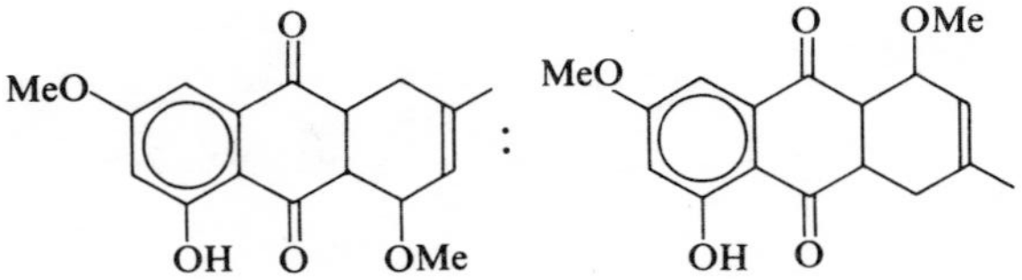	40:60	MgI_2(0.05) 35:65 MgI_2(0.5) $\leqslant 5 : \geqslant 95$	229 229
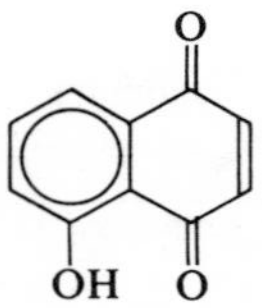		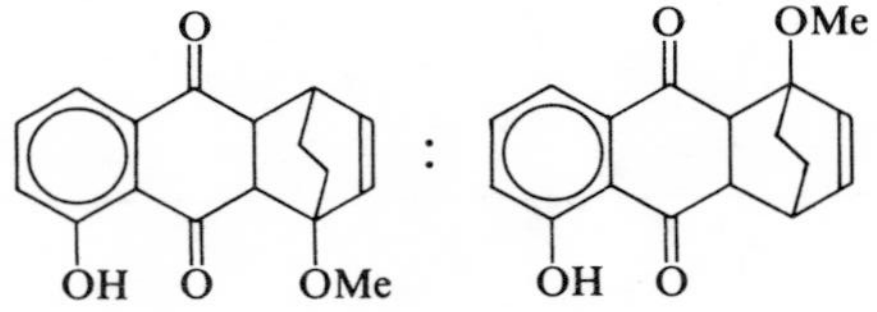	95:5	MgI_2(0.4) 15:85	229

TABLE 11 (continued)

Quinone	Diene	Product	Product ratio without Lewis acid	Product ratio with Lewis acid (equiv.)	Reference
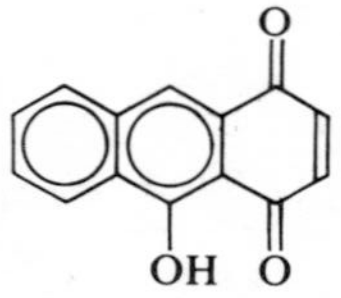 O OH O	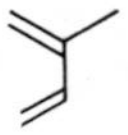	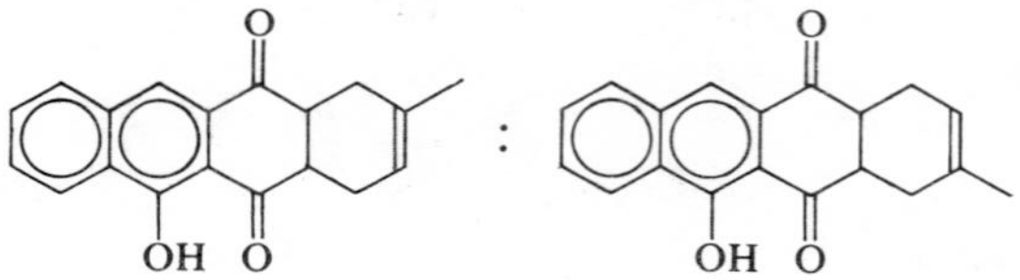 O OH O : O OH O	11:10	$B(OAc)_3$ (excess) 100:0	230
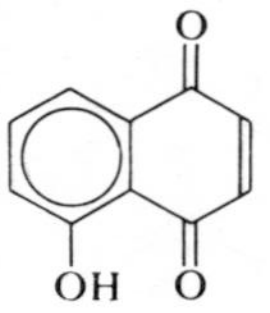 O OH O	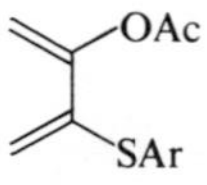 OAc SAr	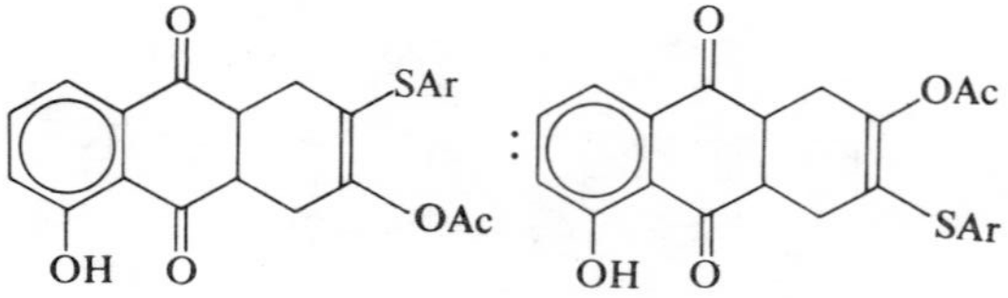 O SAr OH O OAc : O OAc OH O SAr	2:1	$BF_3 \cdot OEt_2$ (0.1–1.0) 0:100	231, 232
	 SPh OAc	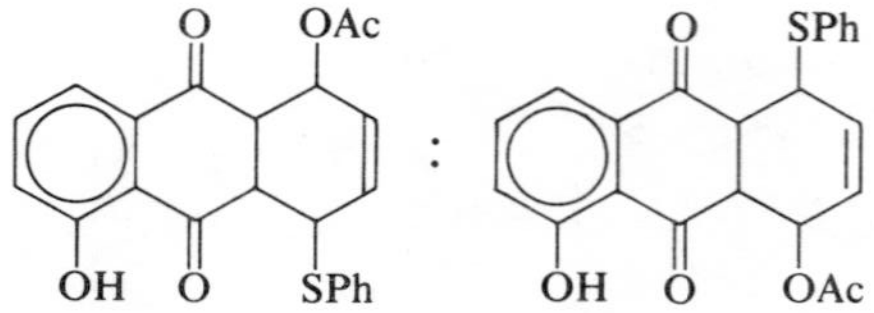 O OAc OH O SPh : O SPh OH O OAc	2.3:1	$BF_3 \cdot OEt_2$ (0.1–1.0) 0:100	231, 232
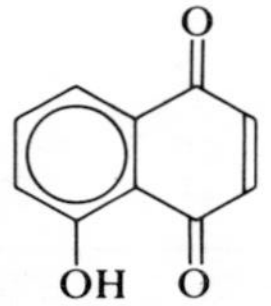 O OH O	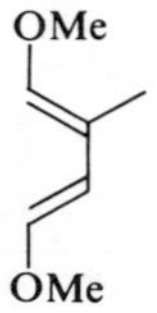 OMe OMe	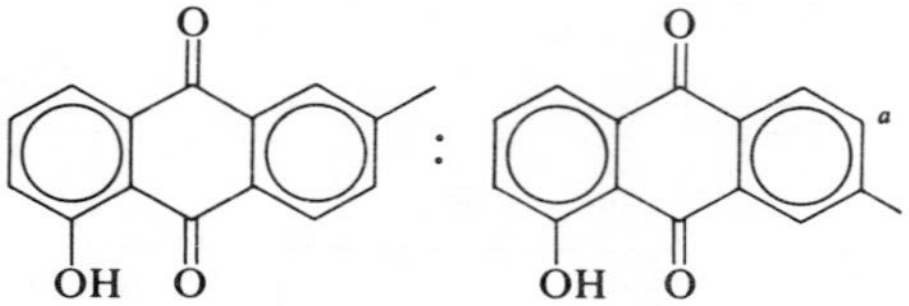 O OH O : O OH O [a]	1:1.8	$BF_3 \cdot OEt_2$(0.5) 1:4.5 $AlCl_3$(0.5) 4.7:1	160 160

Quinone	Diene	Products	Ratio	Catalyst / ratio	Ref.
	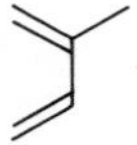	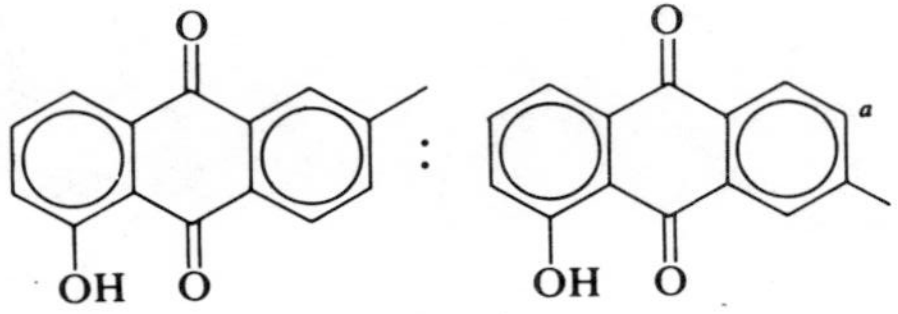 OH, O : OH, O [a]	1.2:1	$BF_3 \cdot OEt_2$(0.5) 1:7.3 $AlCl_3$(0.5) 1:4.5	160
	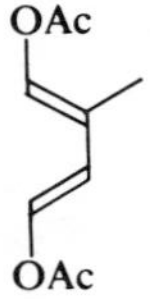 OAc, OAc	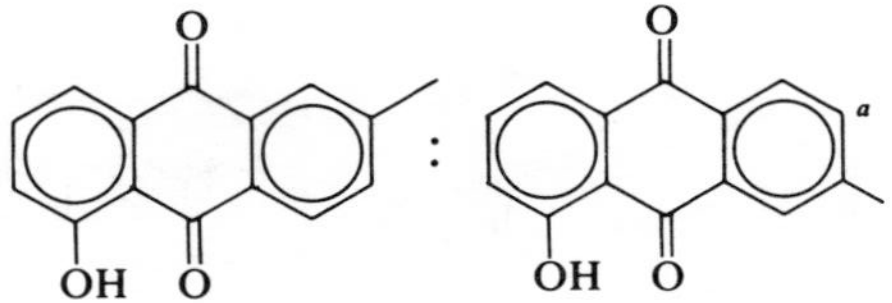 OH, O : OH, O [a]	1:1	$AlCl_3$(0.5) 1.2:1	160
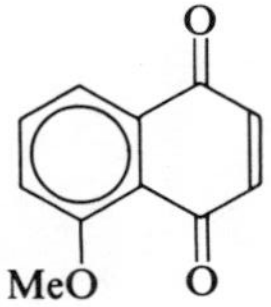 MeO, O	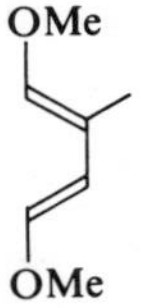 OMe, OMe	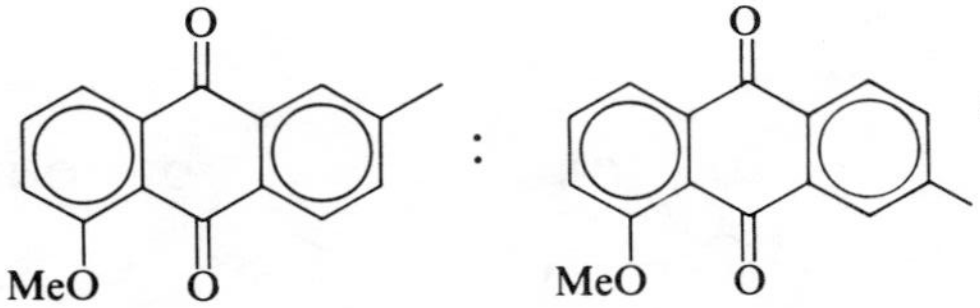 MeO, O : MeO, O [a]	1:1.9	$BF_3 \cdot OEt_2$(0.5) 1:2.4	160
		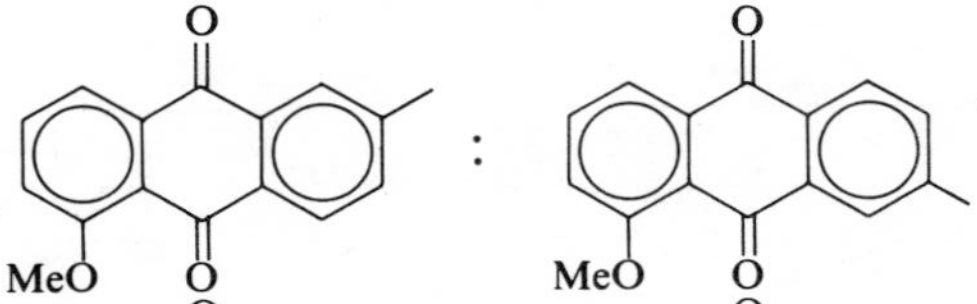 MeO, O : MeO, O [a]		$AlCl_3$(0.5) 1.3:1	160
	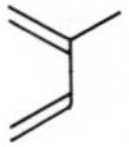	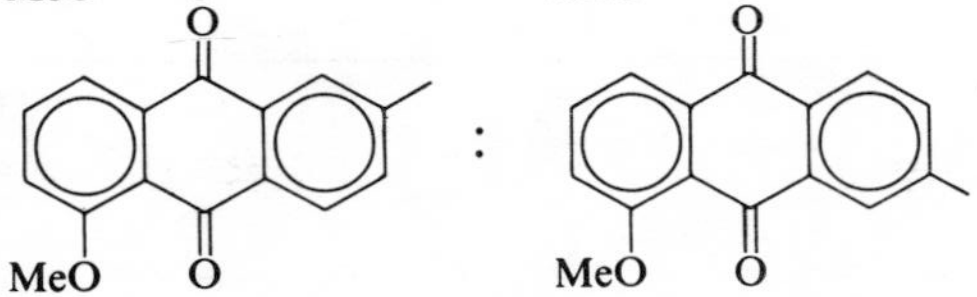 MeO, O : MeO, O [a]	1:1.1		160

TABLE 11 (continued)

Quinone	Diene	Product	Product ratio without Lewis acid	Product ratio with Lewis acid (equiv.)	Reference
	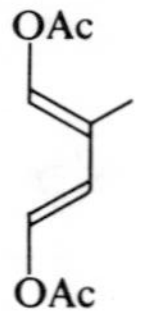 OAc, OAc	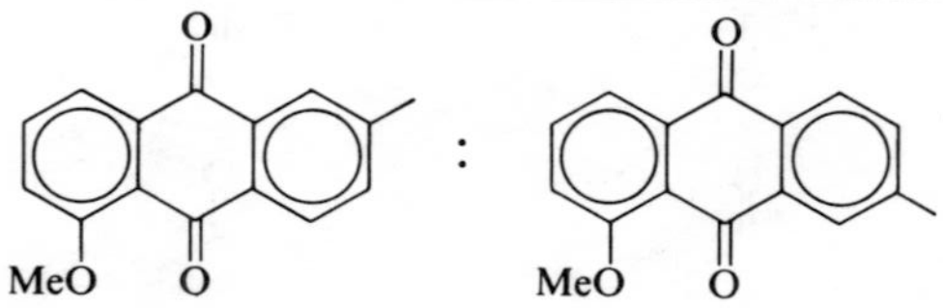 MeO, O, O : MeO, O, O [a]	1:1.1	$AlCl_3$(0.5) 1:1.1	160
AcO, O, O	OMe, OMe	AcO, O, O : AcO, O, O [a]	2:1	$BF_3 \cdot OEt_2$ (0.5) 4.7:1	160
				$AlCl_3$(0.5) 2.75:1	160
		AcO, O, O : AcO, O, O [a]	2.6:1	$BF_3 \cdot OEt_2$(0.5) 2.0:1	160
				$AlCl_3$(0.5) 1.4:1	160

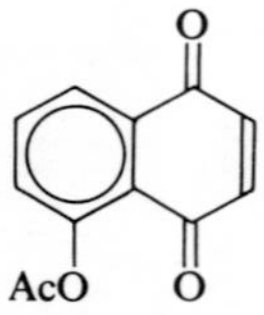

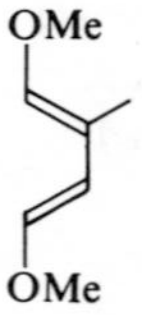

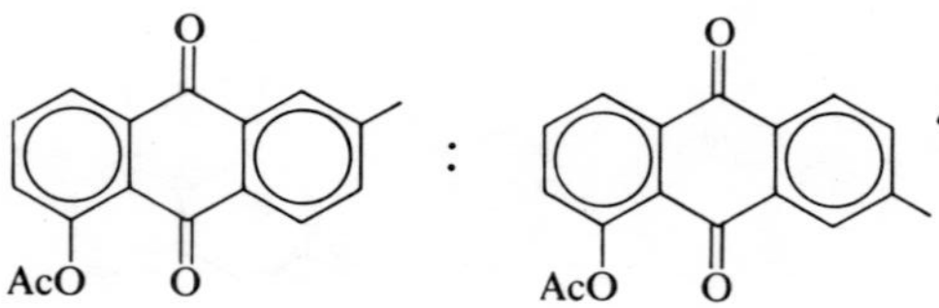

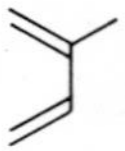

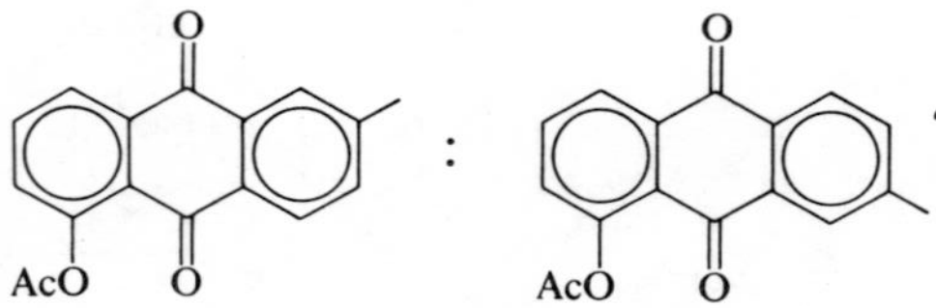

	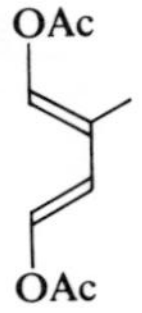 OAc, OAc	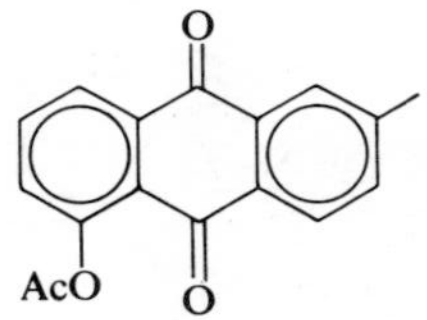 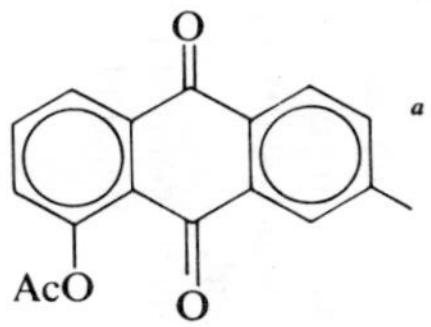 AcO, O, O : AcO, O, O [a]	2.1:1	$AlCl_3$(0.5) 3.1:1	160
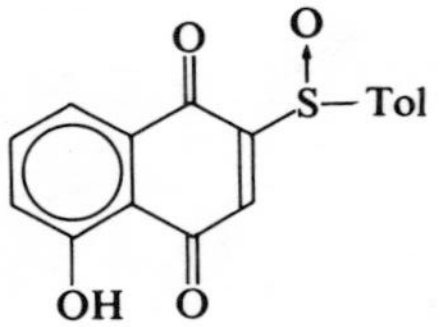 OH, O, O, O←S—Tol	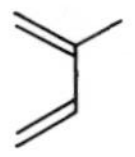	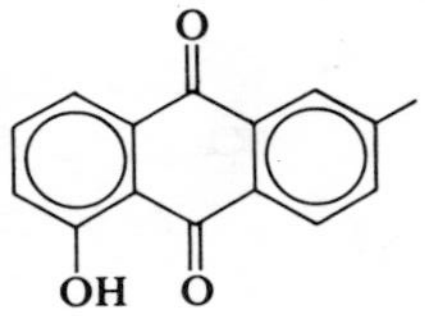 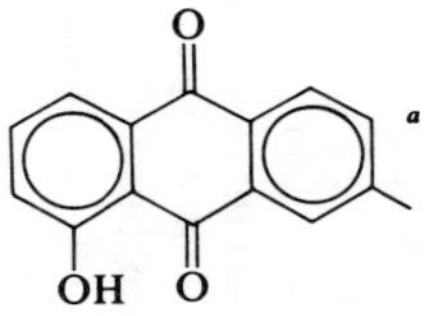 OH, O, O : OH, O, O [a]	2.9:1	$BF_3 \cdot OEt_2$ 5.4:1	160
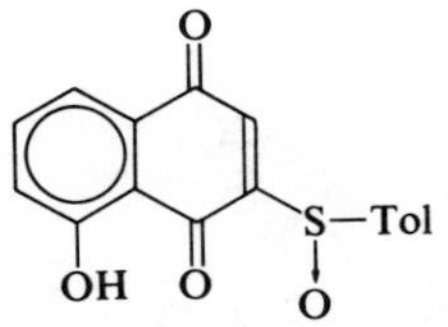 OH, O, O, S—Tol→O	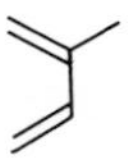	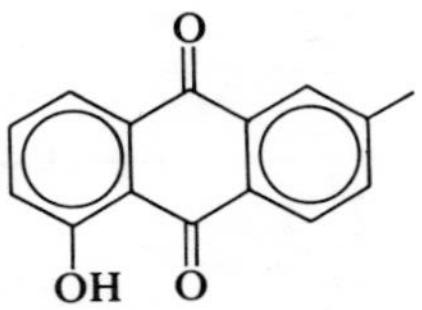 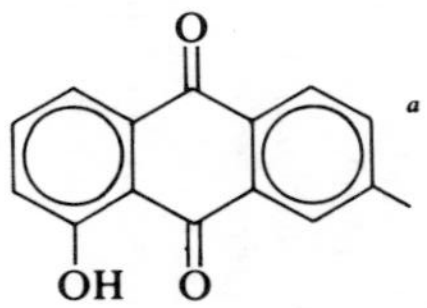 OH, O, O : OH, O, O [a]	1:2.0	$BF_3 \cdot OEt_2$ 1:4.3	160

[a] Products after aromatization

summarized as follows. (1) BF_3 and $B(OAc)_3$ easily afford complexes with juglone such as **357**, in which enhanced regioselectivity is observed due to the induced positive charge at the C(2). (2) In the case of MgI_2, a reversed regioselectivity is observed. It seems that

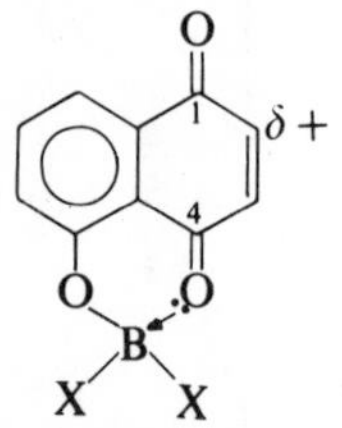

(357) 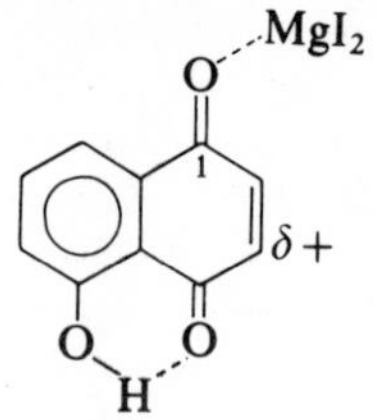 (358)

complexation at the C(1) carbonyl group (cf. **358**) dominates the selectivity rather than a bidentative complexation at the C(4) (cf. in **357**). This is apparently due to a higher steric hindrance and weaker acidity of MgI_2 than that of BX_3. (3) In the case of dienes substituted with sulfur and oxygen atom, the non-catalyzed reaction will be dominated by a sulfur substituent, while the Lewis acid catalyzed one will be controlled by the oxygen functionality, since the preferential complexation of Lewis acid to SR causes the decrease of electron-donating ability of the sulfide group. In some other examples, which show only a small change in the regioselectivity by the addition of Lewis acid, a systematic rationalization of the results is difficult, due to the presence of many competitive coordination sites for the Lewis acid among both the functionalized diene and the dienophile.

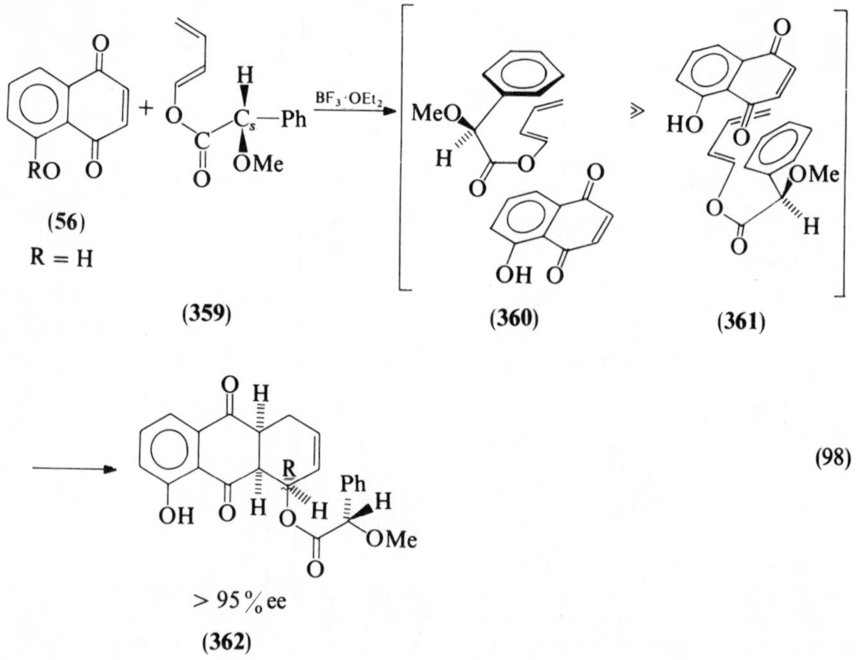

(56) R = H (359) (360) (361)

> 95% ee (362)

(98)

The high regiochemical control in the Diels–Alder reaction was extended to asymmetric induction and an almost complete transfer of chirality was observed by use of (*S*)-*o*-methylmalonyl derivative (**359**)[233]. The reaction of **56** (R = H) with **359** gives **362** with higher than 95% enantiomer excess (equation 98). This efficient asymmetric induction is interpreted in terms of π-stacking interactions, shown in **360** and **361**, which increase the charge transfer interaction between the diene and the dienophile. 2,4-Pentadienoates (**363**) with various chiral alcoholic functionalities were examined in order to evaluate the asymmetric induction in the Diels–Alder reaction with *p*-benzoquinone (**211**)[234]. The reaction (equation 99) affords adducts in a high yield, while the asymmetric induction gives a moderate enantiomer excess.

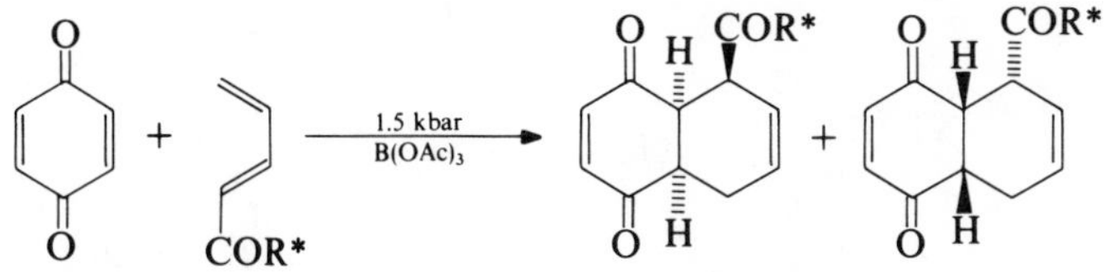

(211) + (363) —1.5 kbar, B(OAc)$_3$→ (364) + (365) (99)

2–50% ee

B. Metallacycles

Quinones or hydroquinones could be prepared from alkynes and metal carbonyls via metallacycles as shown in Scheme 14.

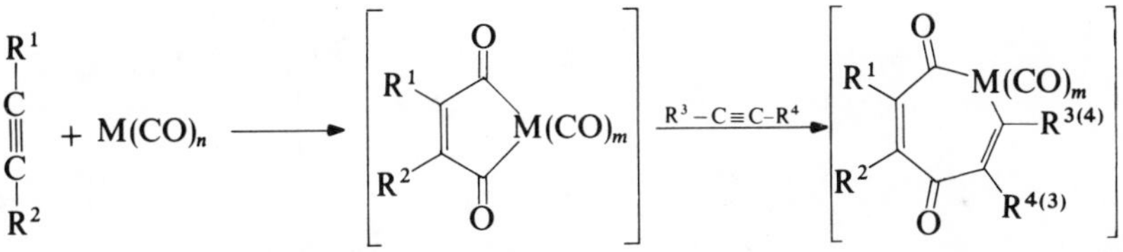

(366) + M(CO)$_n$ → (367) —R^3–C≡C–R^4→ (368)

M = Mn, Fe, Co, Ni, Mo, Rh, Pt

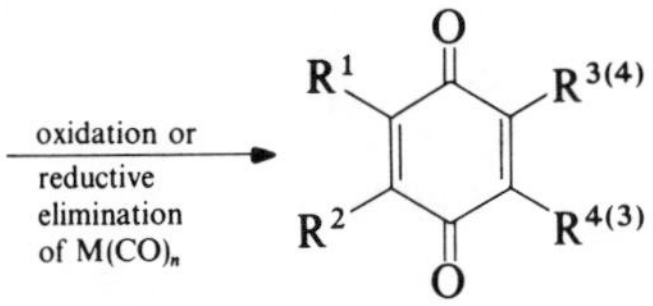

—oxidation or reductive elimination of M(CO)$_n$→ (369)

SCHEME 14

Numerous reports concerning this conversion have appeared since the 1940s, and the rather lower yields, the lack of general applicability, and the poor site-selectivity regardless of stoichiometric or catalytic use of metal carbonyls (**366**) did not draw much attention to the reaction, in spite of its potential for a facile construction of the quinone ring. The earlier reports in this field have appeared in Refs 235 and 236, and the successive reports are given in Refs 237–244.

Since the assumption of Maitlis and coworkers[245] that malonyl (or phthaloyl) metal complex (**367**) is formed as an intermediate, many attempts has been made to prepare them in pure form[246,247] because the establishment of a general synthetic route to these metallacycles will open a convenient route to highly substituted and functionalized 1,4-quinones (**369**). Most of the reported methods, however, lack synthetic efficiency. Recently, the reaction between benzocyclobutenedione and low valent metal complex gave the desired phthaloylmetal complex in excellent yield[248–250]. The relative stability of these complexes toward air and moisture makes the preparation of them easy (equations 100–102).

(**253**) + ML_n ⟶ (**370**) ML_{n-1} (100)

Ref. 249

ML_n = $Fe(CO)_5$, $ClCo(PPh_3)_3$, $CpCo(CO)_2$, $ClRh(PPh_3)_3$
ML_{n-1} = $Fe(CO)_4$, $ClCo(PPh_3)_2$, $CpCo(CO)$, $ClRh(PPh_3)_2$

(**371**) $Fe(CO)_4$ —[$R^1C{\equiv}CR^2$, MeCN, 100 °C]→ [(**372**) $Fe^{II}L_n$] —[H_3O^+, air]→ (**373**) (101)

(**374**) $Co(Cl)(PPh_3)_2$ —[$R^1C{\equiv}CR^2$, MeCN, 100 °C, $AgBF_4$]→ (**373**) (102)

Among the complexes, the iron and cobalt precursors are inexpensive sources of phthaloylmetal complexes, e.g. **371** and **374**, which afford, in turn, the corresponding naphthoquinones, e.g. **373**, in good to excellent yields on reaction with various kinds of alkynes at elevated temperature (Table 12)[249,250]. A complementarity in this annulation is observed between the iron and the cobalt complexes; for alkynes with electron donating groups R^1 and R^2 the cobalt complex gives superior results than the iron complexes. The generation of a six coordinate cationic cobalt complex is essential for the completion of this annulation. Silver tetrafluoroborate in acetonitrile gives with this cobalt complex the

TABLE 12. Annulation of phthaloylmetal complexes **371** and **374** with alkynes[249, 250]

R^1	R^2	% Yield of **373** From the iron complex **371**	% Yield of **373** From the cobalt complex **374**
Me	Me	99	73
Et	Et	95	90
Ph	Ph	88	68
Ph	Me	100	78
n-Bu	H	95	65
Ph	H	94	57
t-Bu	Me	37	72
Et	$CH_2CH{=}CH_2$	75	80
EtO	Et	—	80
n-Bu	$SiMe_3$	22	68
Ph	$(CH_2)_2OH$	81	27
Me	CO_2Et	74	0
Et	COMe	68	0

tris(acetonitrilo)cobalt derivative (**376**). The dimethylglyoxime complex (**378**) is also obtained in pyridine in high yield (equations 103 and 104)[251] and gives an excellent

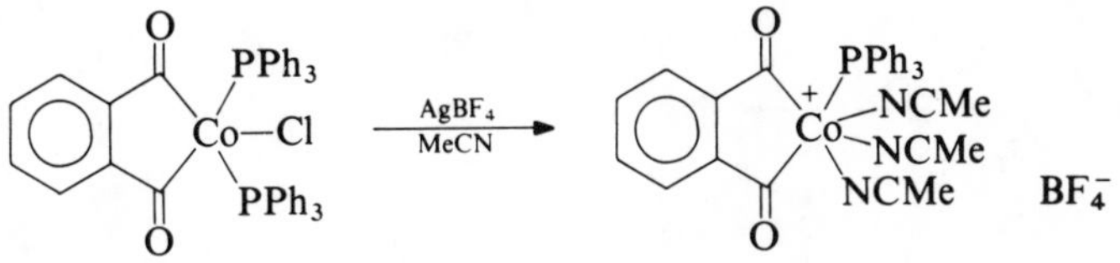

PPh3, Co—Cl, PPh3 —($AgBF_4$, MeCN)→ PPh3, Co+, NCMe, NCMe, NCMe BF_4^- (103)

(**375**) (**376**)

PPh3, Co—Cl, PPh3 —(HON NOH dimethylglyoxime (377), pyridine (Py))→ Py, Co, N, OH, N, OH, Cl (104)

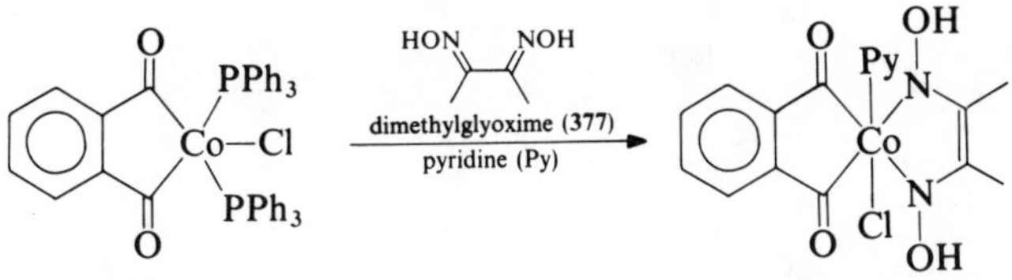

(**375**) (**378**)

TABLE 13. Annulation reaction with the phthaloylcobalt dimethylglyoxime complex (**378**)[251]

$R^1C{\equiv}CR^2$ R^1	R^2	Naphthoquinone (**373**) % Yield
EtO	Et	85
Et	Et	77
H	*n*-Bu	85
Me	CO_2Et	99
Me	$CH_2N(Ph)COMe$	70

annulation reaction toward alkynes ranging from electron-rich to electron-deficient, and the problems mentioned above concerning this annulation are overcome by this method[251]. Several examples are shown in Table 13.

Since the substituted benzocyclobutenediones **379** or **380** have been synthesized[252], this method is applicable to a wide variety of naphthoquinones.

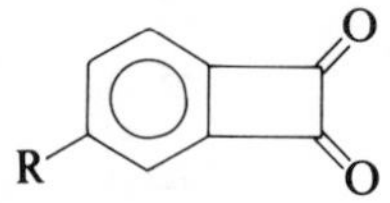

(379) R = H, Me

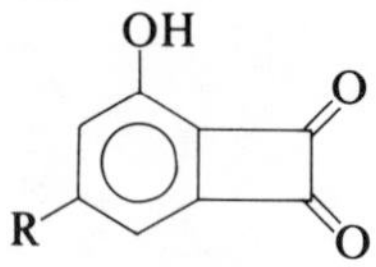

(380) R = H, Me

Benzoquinones as well as naphthoquinone are prepared by an extension of this methodology, in which the maleoylcobalt dimethylglyoxime complex (**383**) prepared from **381** via **382** gives the corresponding benzoquinone(s) in good yield(s) (Scheme 15)[253]. This

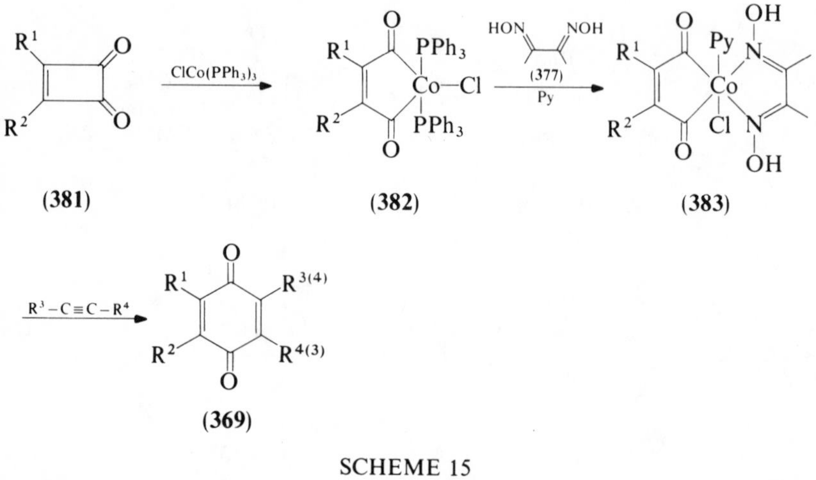

$ClCo(PPh_3)_3$; (377), Py ; $R^3-C\equiv C-R^4$

(381) **(382)** **(383)** **(369)**

SCHEME 15

method is extremely useful for the preparation of tri- and tetrasubstituted benzoquinones as shown in Table 14. Two limitations, however, are observed in this reaction. First, only a limited number of alkyl- or alkoxy-substituted cyclobutenediones are available. Second, the combination of an unsymmetrical cyclobutenedione with unsymmetrical alkynes results in the formation of the possible two regioisomers, usually in appreciable percentages. In order to overcome these defects, an intramolecular cyclization is examined in the following synthesis of a juglone derivative (equation 105)[254]. In this example, a covalently bound benzocyclobutenedione (**385**) is converted to the corresponding cobalt complex (**386**). The length of binding side chain of the alkyne controls the regiochemical course and only one isomer is obtained.

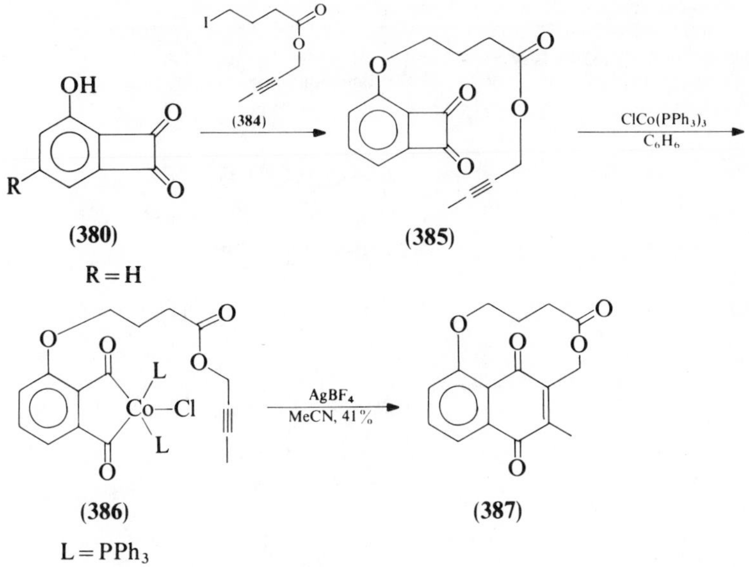

(384)

$ClCo(PPh_3)_3$, C_6H_6

(380) R = H

(385)

$AgBF_4$, MeCN, 41%

(386) $L = PPh_3$

(387)

(105)

TABLE 14. Preparation of substituted benzoquinones by the reaction of maleoylcobalt complexes (**383**) with alkynes[253]

Cyclobutenedione (**381**)	Alkyne, $R^1C{\equiv}CR^2$ R^1	R^2	Benzoquinone	% Yield
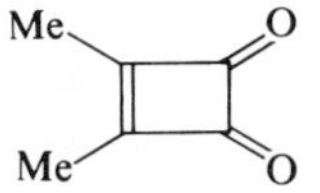 3,4-Me,Me-cyclobutenedione	Et	Et	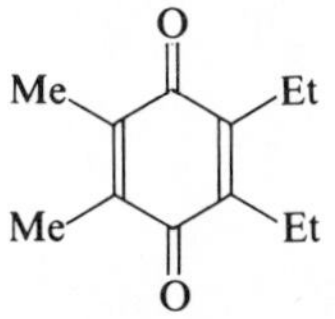 Me, Me, Et, Et benzoquinone	81
	H	*n*-Bu	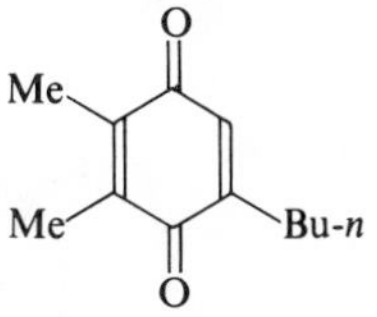 Me, Me, Bu-*n* benzoquinone	85
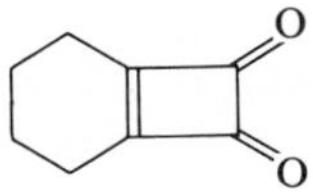 bicyclic (cyclohexene-fused) cyclobutenedione	Et	Et	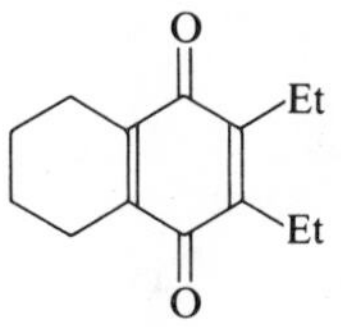 Et, Et tetrahydronaphthoquinone	85
	H	*n*-Bu	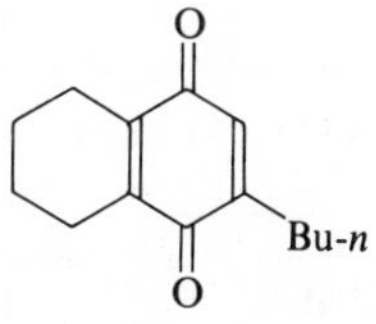 Bu-*n* tetrahydronaphthoquinone	79

TABLE 14 (continued)

Cyclobutenedione (**381**)	Alkyne, $R^1C{\equiv}CR^2$ R^1	R^2	Benzoquinone	% Yield
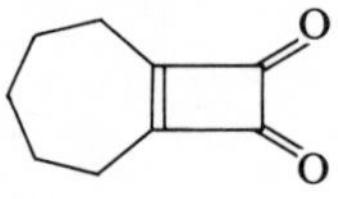	Et	Et	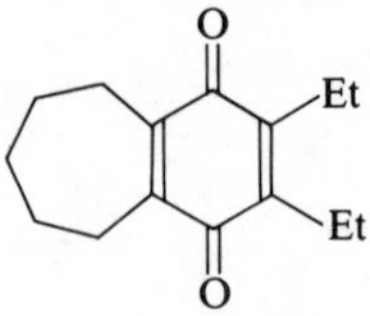	88
	H	*n*-Bu	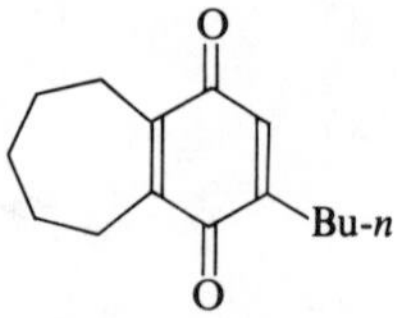	84
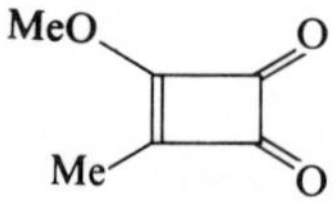	H	Me	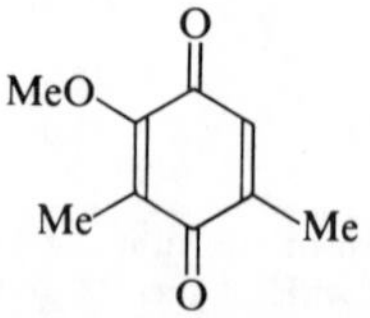 (5:1)[a]	81[b]
	H	*n*-Bu	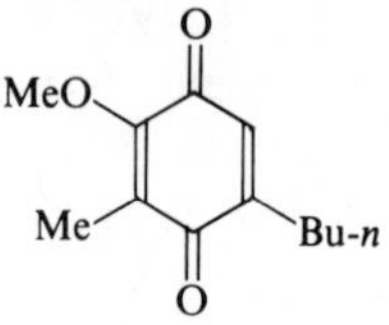 (3.7:1)[a]	89[b]
	Me	CO_2Et	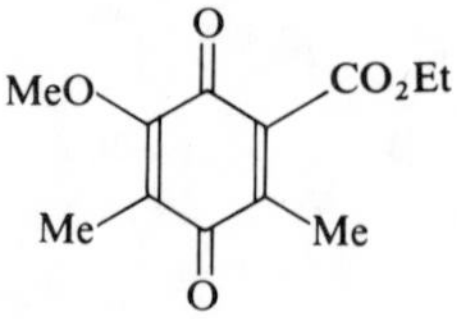 (3.7:1)[a]	64[b]
	Me	OEt	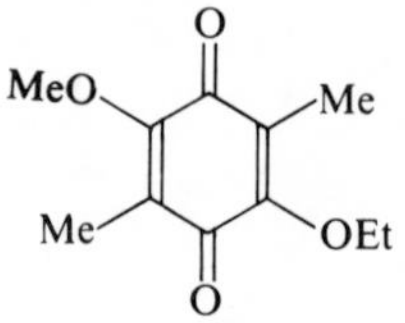 (13.5:1)[a]	81[b]

[a] The major isomer is shown in this table. Isomeric ratio is shown in parentheses.
[b] Total yield of two regioisomers.

C. Metal Carbene Complexes

In the presence of an alkyne, a carbonyl chromium complex (**388**) with a saturated carbene ligand easily affords the corresponding alkyne–carbene–carbonyl complex, which undergoes a cycloaddition reaction to give a chromium complex of a quinone monoalkyl ether (**390**) (Scheme 16)[255–257]. This π-bonded chromium complex (**390**) is easily

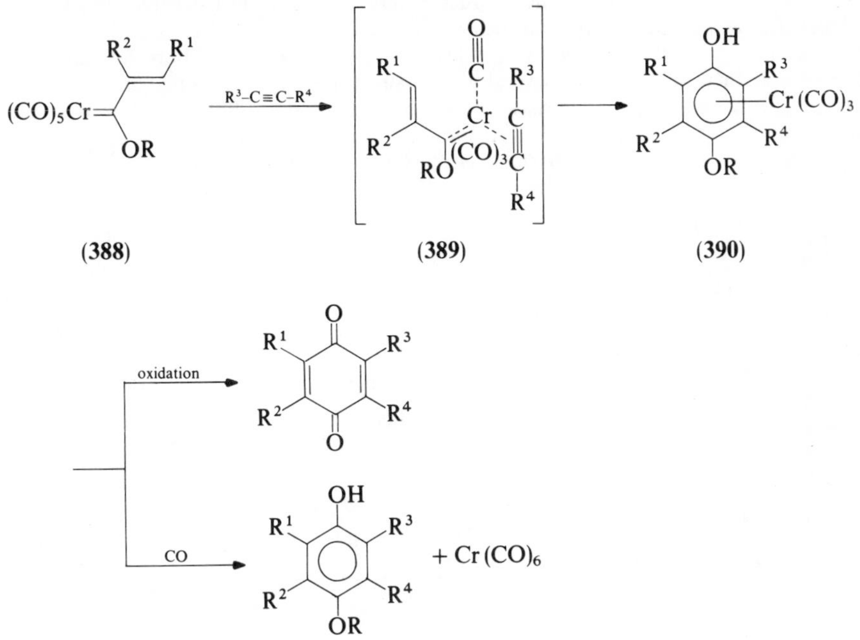

R^2 R^1 $(CO)_5Cr$ OR $R^3{-}C{\equiv}C{-}R^4$ R^1 O C R^3 C Cr C R^2 RO $(CO)_3$ C R^4 R^1 OH R^3 $Cr(CO)_3$ R^2 OR R^4

(**388**) (**389**) (**390**)

oxidation R^1 O R^3 R^2 O R^4

CO R^1 OH R^3 R^2 OR R^4 + $Cr(CO)_6$

SCHEME 16

demetallated either to a quinol monoether by treatment with CO or to the corresponding quinone by oxidative demetallation. The former method especially can regenerate the stable chromium hexacarbonyl which can be recycled in the preparation of the carbene complex as shown in equation 106. In this reaction, three components, i.e. α,β-unsaturated

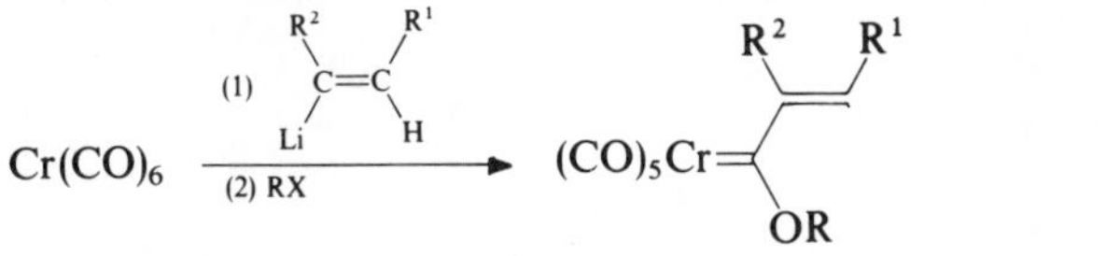

$$Cr(CO)_6 \xrightarrow[(2)\ RX]{(1)\ R^2(Li)C{=}C(H)R^1} (CO)_5Cr{=}C(OR)C(R^2){=}CHR^1 \quad (106)$$

(**388**)

alkoxycarbene, an alkyne and carbon monoxide, are cyclized together on the coordination sphere of chromium carbonyl to give **390**. This fundamental reaction has wide applicability for various alkynes and unsaturated metal carbene complexes, whose preparative method has already been established[258–262]. The yield is moderate to good and is affected by the structure, reactivity and stability of the applied chromium carbene complex as well as of the alkyne (Table 15).

Whereas the reaction of symmetrically substituted carbene complexes ($R^1 = R^2$ in Table 15) raises no regiochemical problem in the quinone product, the combination of an unsymmetrically substituted alkyne and an aryl carbene complex as well as of metallacycles gives intrinsically two isomeric quinones and the more bulky group preferentially ends up nearest to the free OH group in the product. Several examples are also shown in Table 16. The electronic effects of the two substituents on the acetylene have less importance on the regioselectivity, but the yield is severely decreased by an electron-withdrawing group on the alkyne. In the reaction of several alkoxy-substituted aryl carbene complexes with alkynes, the regioselectivity increases with increasing the steric difference between the two substituents on the acetylene[266]. A terminal acetylene gives one isomer in high regioselectivity (equation 107). *Meta* substitution on the carbene complex shows the two

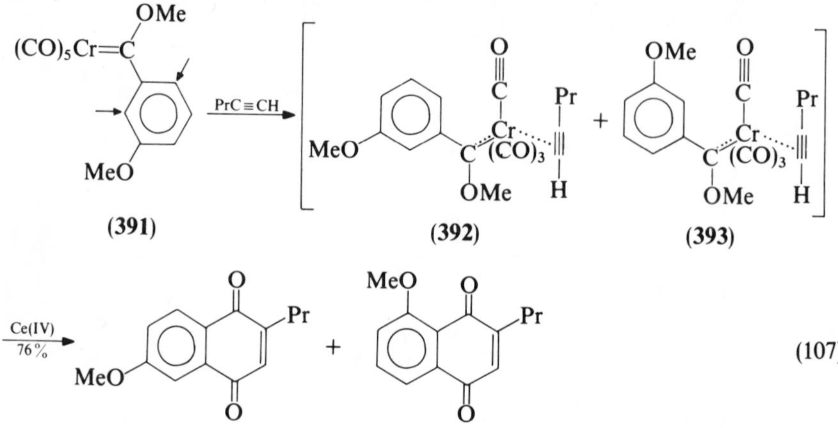

(391) —PrC≡CH→ [(392) + (393)] —Ce(IV), 76%→ (394) + (395) (107)

10.5 : 1.0

expected modes of cyclization at the sterically different *ortho* positions. From the regioselectivities observed in the cyclization of various systems with *meta* substituents of sterically and electronically different character (e.g. Me, F, CF_3), it can be concluded that both effects should be taken into account when explaining the product selectivity[266]. However, a systematic explanation has not yet been given.

In order to overcome these regiochemical problems in cyclization, an intramolecular reaction is designed (equation 108)[267, 268]. In **396**, the alkyne is covalently bound to the

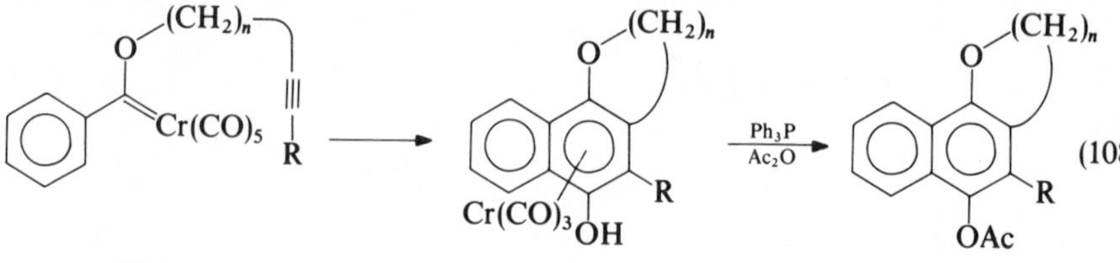

(396) → (397) —Ph_3P, Ac_2O→ (398) (108)

R = H, n = 2–4, 16–38%
R = Me, n = 2–4, 62–81%

TABLE 15. Coupling of unsaturated chromium carbene complexes with alkynes

(**388**) R = Me R^1	R^2	R^3–C≡C–R^4 R^3	R^4	Product (% Yield)	Reference
				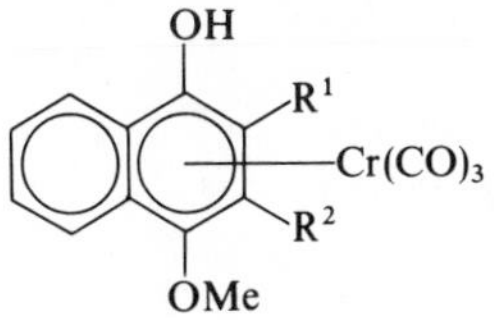 [structure: 1-OH-4-OMe-naphthalene with R^1, R^2, $Cr(CO)_3$]	
–CH=CH–CH=CH–		Et	H	(35)	263
		Me	Me	(68)	263
		n-Pr	H	(45)	263
		n-Bu	H	(45)	263
		n-Pr	Me	(58)	263
		Et	Et	(62)	256
				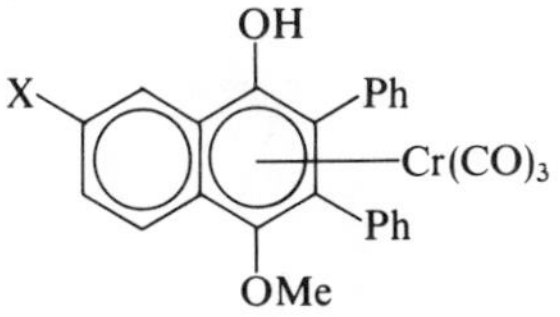 [structure: X-substituted 1-OH-4-OMe-2,3-diphenylnaphthalene $Cr(CO)_3$]	
–CH=CH–C(Me)=CH–		Ph	Ph	X = Me (40)	265
–CH=CH–C(CF_3)=CH–		Ph	Ph	X = CF_3(25)	265
–CH=C(OMe)–CH=CH–		Et	Et	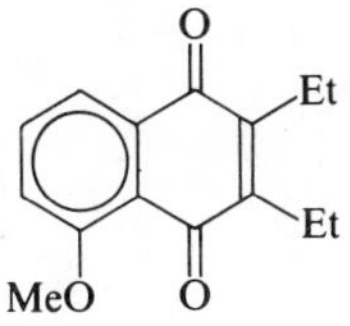 [structure: 2,3-diethyl-5-methoxy-1,4-naphthoquinone] (70)[b]	266
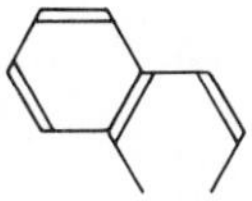 [structure]		Ph	Ph	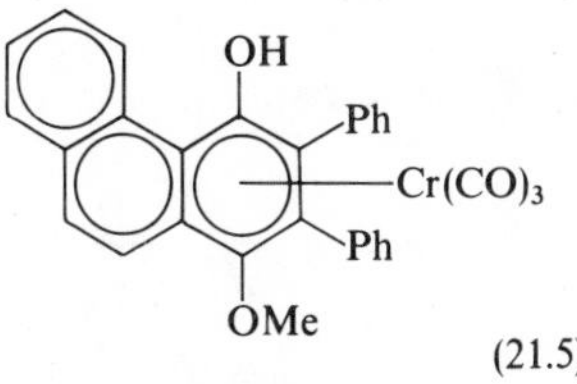 [structure: OH, OMe, 2 Ph, $Cr(CO)_3$ phenanthrene] (21.5)	265
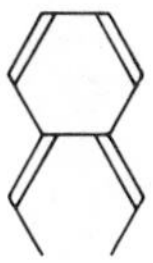 [structure]		Ph	Ph	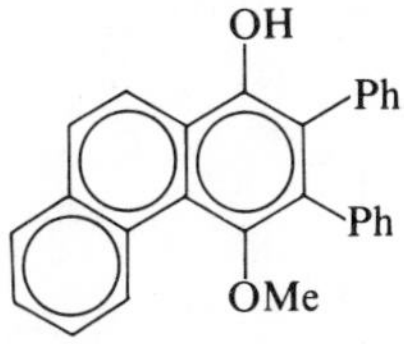 [structure: OH, OMe, 2 Ph phenanthrene] (19)	265

TABLE 15 (continued)

(**388**) R = Me R^1	R^2	R^3–C≡C–R^4 R^3	R^4	Product (% Yield)		Reference
				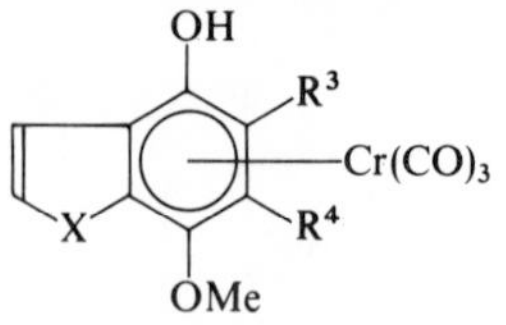		
–O–CH=CH–		Ph	Ph	X=O (19)		265
		n-Bu	H	X=O (23)		265
–S–CH=CH–		*n*-Bu	H	X=S (40)		265
		Ph	Ph	X=S (28)		265
				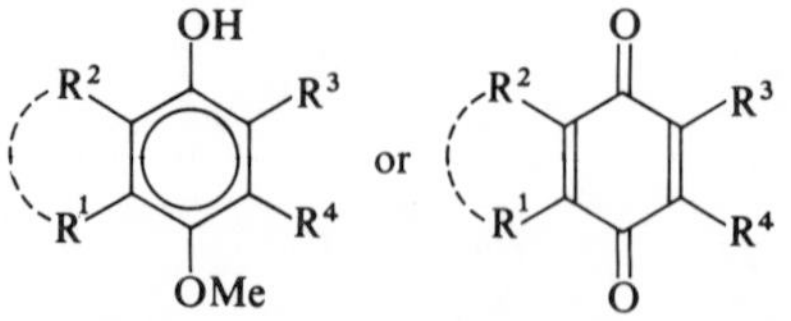		
$-(CH_2)_3-$		Ph	Ph	(36)[a]		264
		Ph	H	(76)[a]		264
		n-Pr	H	(54)[a]		264
		Et	Et		(37)[b]	264
$-(CH_2)_4-$		Et	Et	(64)[a]	(65)[b]	264
		n-Pr	H		(61)[b]	264
		$SiMe_3$	H	(71)[c]		264
		(*Z*)-MeOCH=CH–	H	(68)[c]		264
		$AcOCH_2$	H	(33)[a]		264
		MeO_2C	H	(22)[d]		264
		PhCO	H	(17)[d]		264
		MeO_2C	MeO_2C	(8)[d]		264
$-(CH_2)_5-$		*n*-Pr		(66)[a]		264
		$CH_2=CHC(Me)$	H	(57)[a]		264
$-O(CH_2)_3-$		Et	Et		(71)[b]	264
		$(EtO_2C)_2(NHAc)CCH_2$	H		(67)[b]	264
		n-Pr	H		(38)[b]	264
EtO	Me	Ph	Ph	(67)[a]		264
		n-Pr	H		(40)[b]	264
EtO	H	$Me_2C=CHCH_2$	H	(23)[e]		264
Me	H	*n*-Pr	H		(51)[b]	264
H	Me	*n*-Pr	H		(75)[b]	264

[a] After oxidation with $FeCl_3 \cdot DMF$ complex.
[b] After oxidation with CAN.
[c] After air oxidation.
[d] After oxidation with I_2.
[e] After treatment with tributylphosphine.

TABLE 16. Regioselectivity in the cyclization of a chromium carbene complex[266]

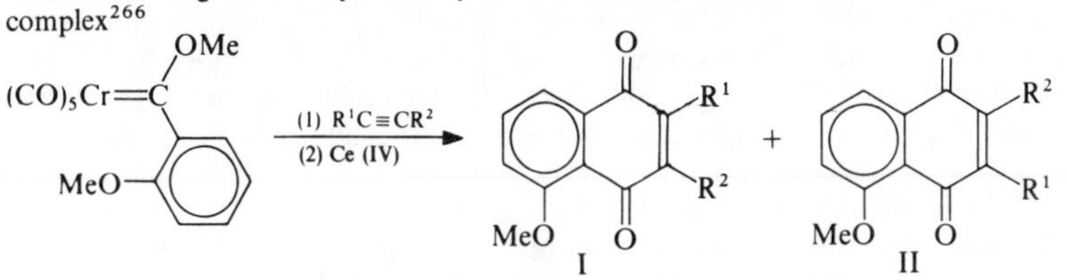

$(CO)_5Cr{=}C(OMe)$ (MeO) —(1) $R^1C{\equiv}CR^2$, (2) Ce (IV)→ I (O, O, R^1, R^2, MeO) + II (O, O, R^2, R^1, MeO)

R^1	R^2	% Yield (I + II)	Product ratio I/II
Et	Me	81	1.5
n-Pr	Me	64	2.9
n-Pr	H	74	≥ 111
i-Pr	Me	61	4.8
Ph	Me	78	*a*
$(CH_2)_3CO_2Bu$-*t*	H	66	*a*
$(CH_2)_3CONHBu$-*t*	H	70	*a*

[a] Only one isomer is obtained. The structure is not defined.

alkoxy group and the transition state geometry is restricted by this bonding of the two reactants. The single product **(397)** is expected from the metal complex and it is indeed obtained in moderate to good yield. This method is extraordinarily important for the synthesis of naturally occurring quinones, e.g. nanaomycin A and anthracyclinones.

As a variation and an extension of the above-mentioned reaction, a Diels–Alder reaction of an α,β-alkynyl carbene complex is developed to construct a new α,β-alkenyl complex[269].

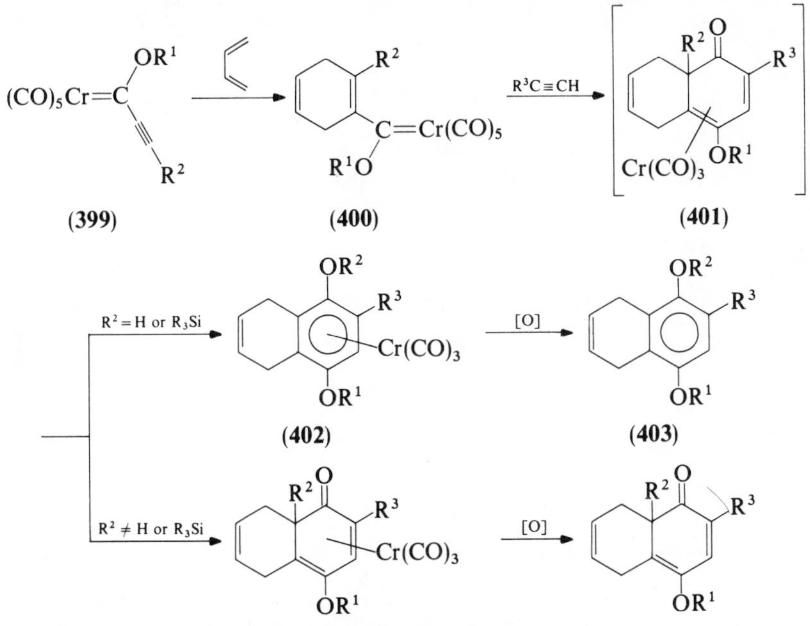

$(CO)_5Cr{=}C$ (OR^1, R^2) → (R^2, $C{=}Cr(CO)_5$, R^1O) —$R^3C{\equiv}CH$→ [R^2, O, R^3, OR^1, $Cr(CO)_3$]

(399) **(400)** **(401)**

$R^2 = H$ or R_3Si → (OR^2, R^3, $Cr(CO)_3$, OR^1) —[O]→ (OR^2, R^3, OR^1)

(402) **(403)**

$R^2 \neq H$ or R_3Si → (R^2, O, R^3, $Cr(CO)_3$, OR^1) —[O]→ (R^2, O, R^3, OR^1)

SCHEME 17

Due to the electron-withdrawing character of the metal carbene moiety, the alkynyl-carbene complex (**399**) is a good dienophile, and annulation with a conjugated diene proceeds easily to afford the corresponding carbene adduct (**400**), which is a good precursor for the cyclization to give **401**. When the terminal substituent R^2 on the alkynyl carbene (**399**) is a hydrogen or a trialkylsilyl group, **401** easily undergoes a dienone–phenol rearrangement to the corresponding hydroquinone derivative (**402**), which by oxidation gives **403** (Scheme 17)[269]. As the applied dienes and alkynes do not interfere in the successive reactions, this tandem reaction can be conducted in a one-pot without isolation of the intermediary alkenyl carbene complex (**400**). Several examples of this component reaction are shown in equations 109–112.

$(CO)_5Cr{=}C(OMe)C{\equiv}CSiMe_3$ (**404**) —[(**409**), $PrC{\equiv}CH$; 97%]→ (**405**) [OSiMe₃, Pr, OMe, $Cr(CO)_3$] (109)

(**404**) —[(1) (**410**), $PrC{\equiv}CH$; (2) SiO_2, air; 95%]→ (**406**) [OH, Pr, MeO] (110)

(**404**) —[(1) (**411**), $PrC{\equiv}CH$; (2) SiO_2, air; 56%]→ (**407**) [OH, Pr, OMe] (111)

(**404**) —[(1) TMSO-diene (**179**), $PrC{\equiv}CH$; (2) SiO_2, air; 58%]→ (**408**) [O, OH, Pr, OMe] (112)

D. Stabilized Carbanions

1. Tandem directed metalation

A tertiary benzamido group on an aromatic nucleus promotes *ortho* metalation. Condensation of the *ortho*-lithiated product with α,β-unsaturated, especially aromatic, aldehyde affords the benzyl alkoxide intermediate (**412**). In the presence of additional

TABLE 17. Synthesis of polycyclic quinones by tandem directed metalation[270]

Amide	Aldehyde	Product	% Yield
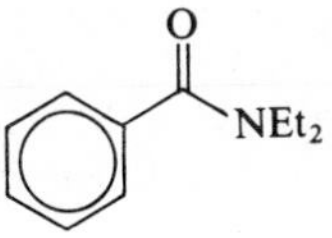 O, NEt_2	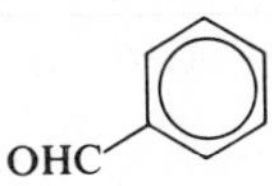 OHC	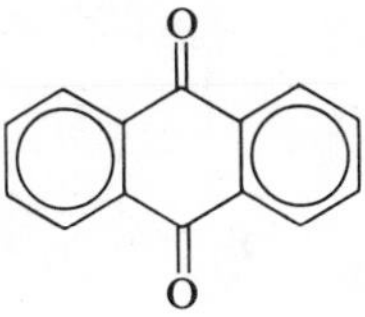 O, O	43
	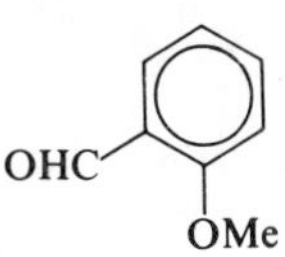 OHC, OMe	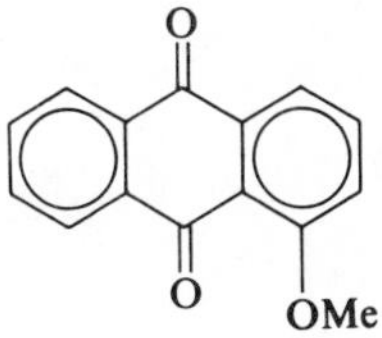 O, O, OMe	10
	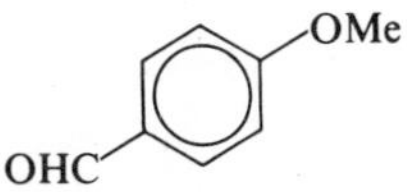 OMe, OHC	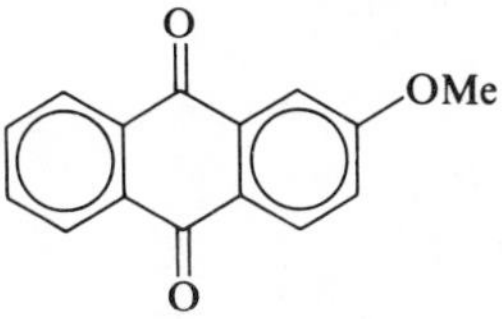 O, OMe, O	28
	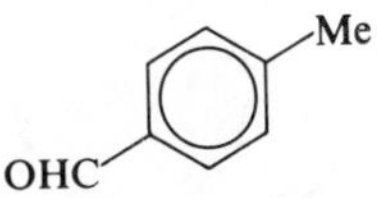 Me, OHC	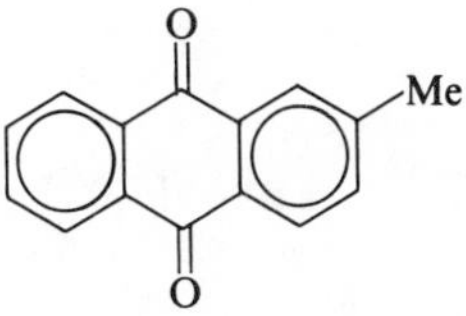 O, Me, O	15
	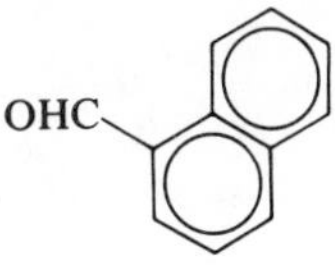 OHC	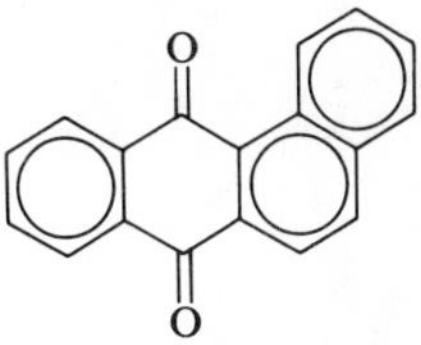 O, O	44[a]
	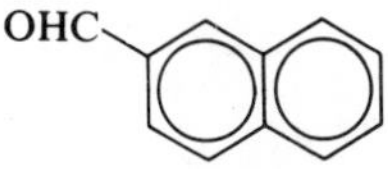 OHC	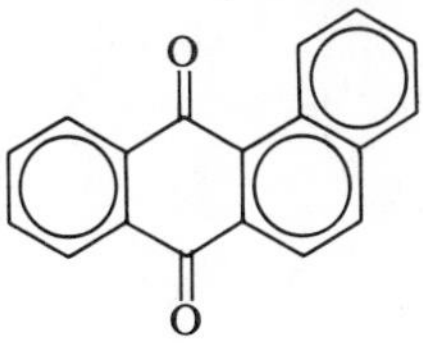 O, O	39[a]
		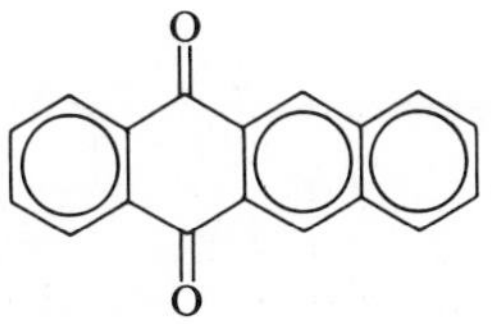 O, O	5[a]

TABLE 17 (continued)

Amide	Aldehyde	Product	% Yield
	OHC	O O	41[a]
O NEt_2	OHC	O O	10[a]
	OHC	O O	10[a]
O NEt_2	OHC	O O	2
		O O	2
O NEt_2	S OHC	O S O	35
O NEt_2	S OHC	O S O	37

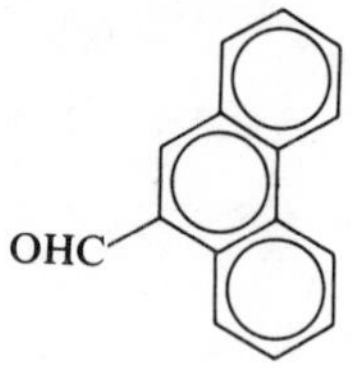

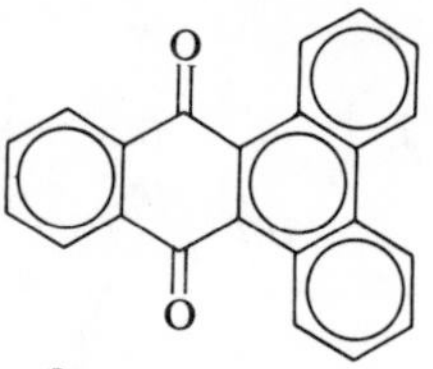

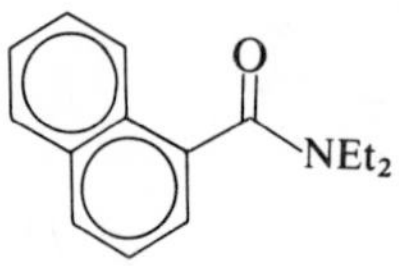

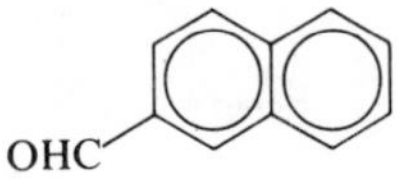

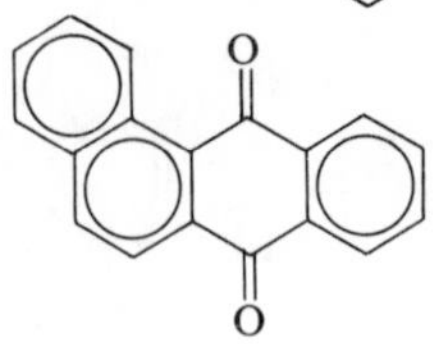

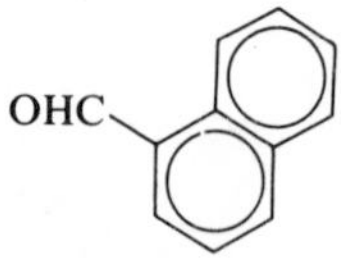

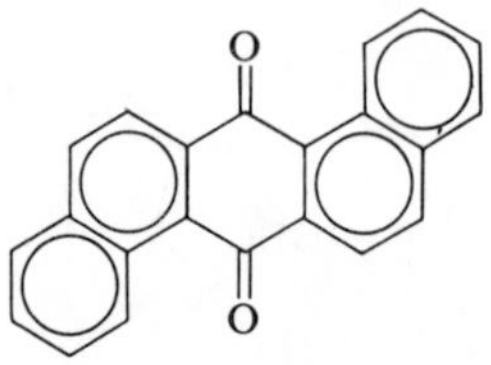

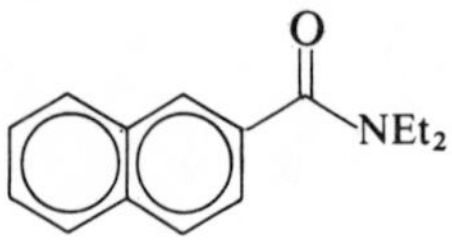

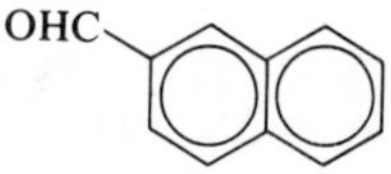

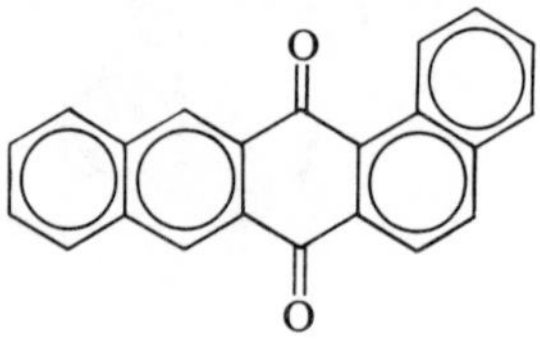

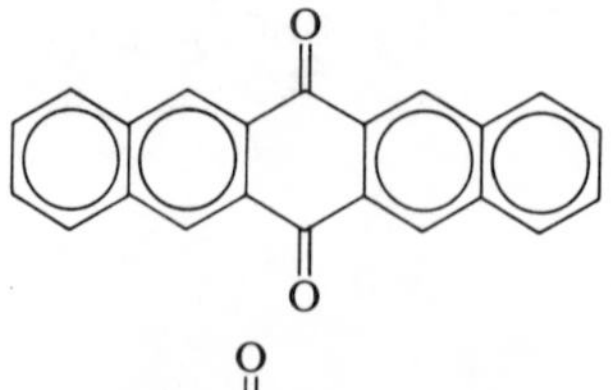

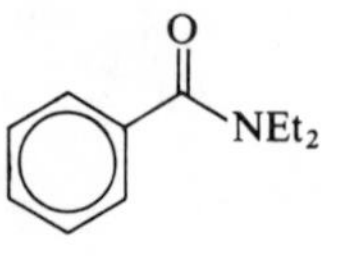

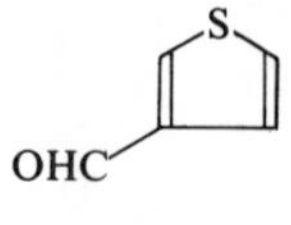

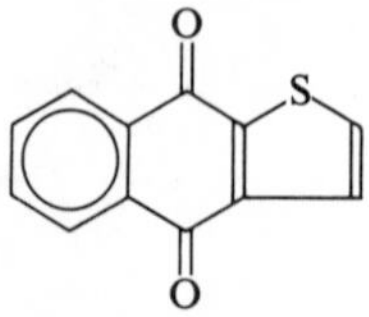

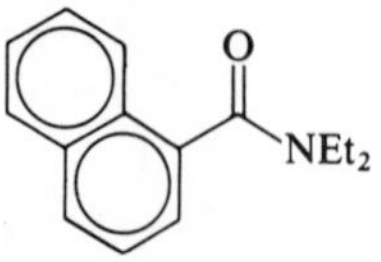

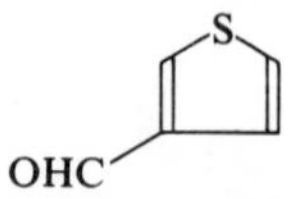

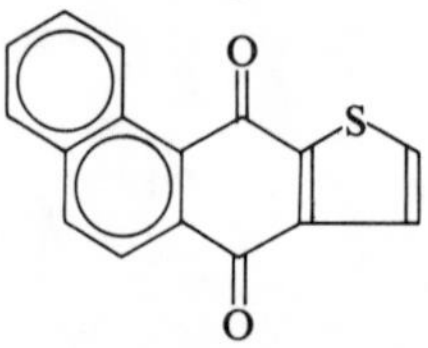

TABLE 17 (continued)

Amide	Aldehyde	Product	% Yield
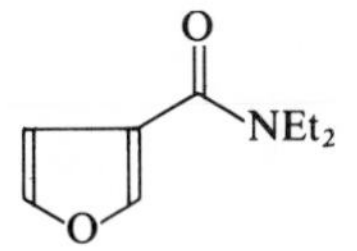 O, NEt_2, O	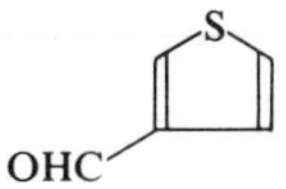 S, OHC	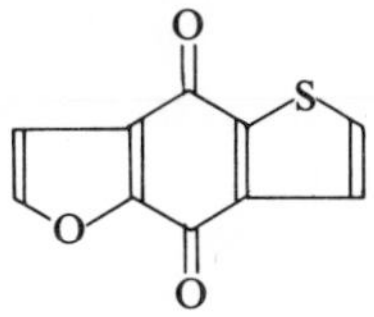 O, S, O, O	24
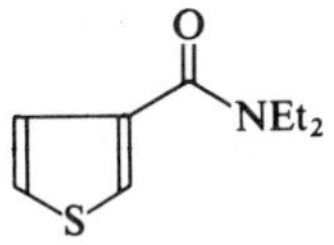 O, NEt_2, S	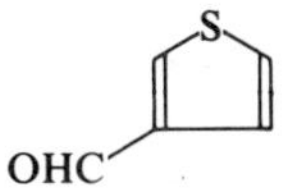 S, OHC	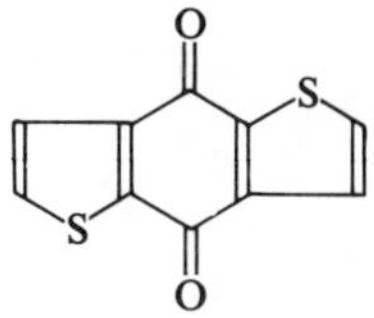 O, S, S, O	77
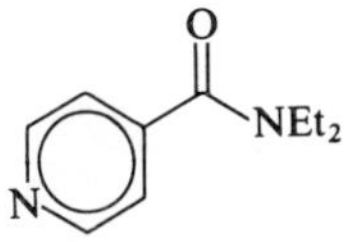 O, NEt_2, N	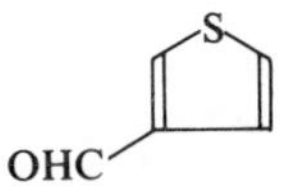 S, OHC	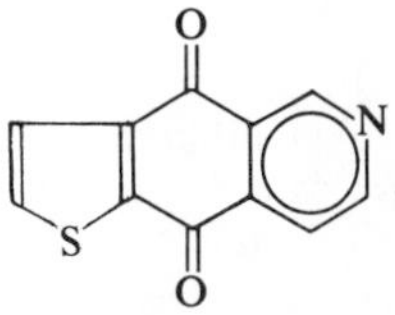 O, N, S, O	20
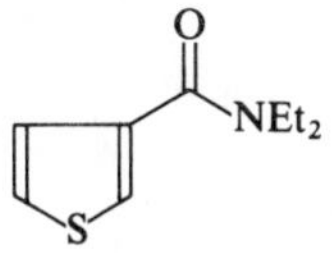 O, NEt_2, S	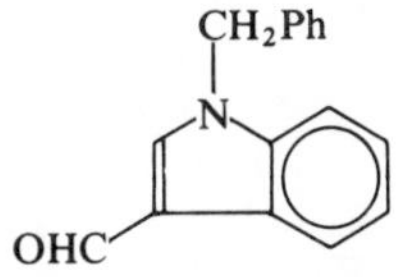 CH_2Ph, N, OHC	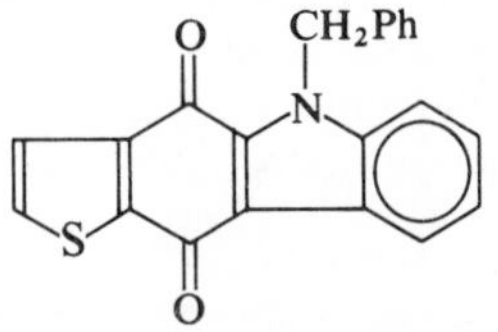 O, CH_2Ph, N, S, O	67
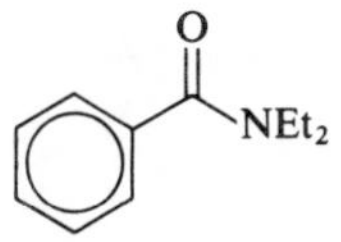 O, NEt_2	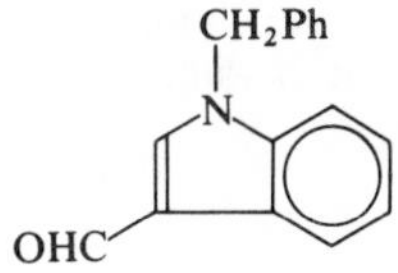 CH_2Ph, N, OHC	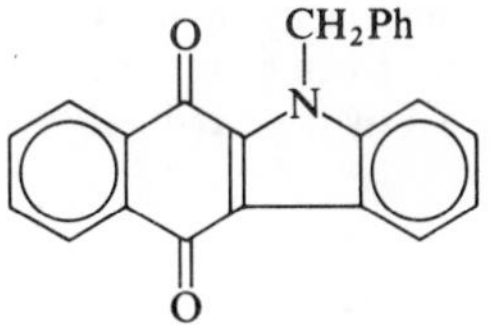 O, CH_2Ph, N, O	44
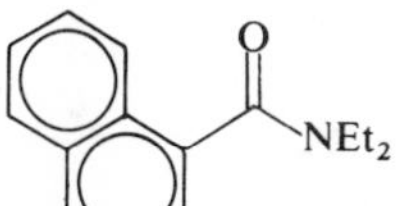 O, NEt_2	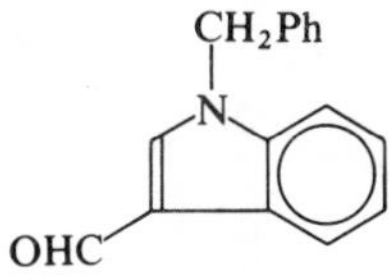 CH_2Ph, N, OHC	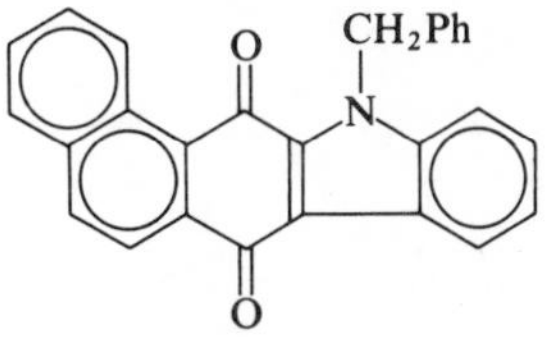 O, CH_2Ph, N, O	20
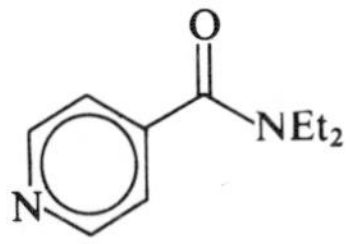 O, NEt_2, N	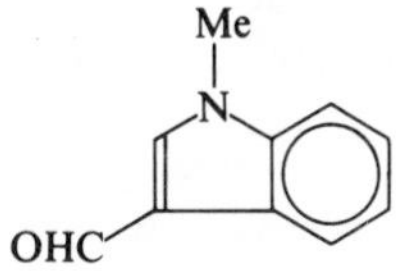 Me, N, OHC	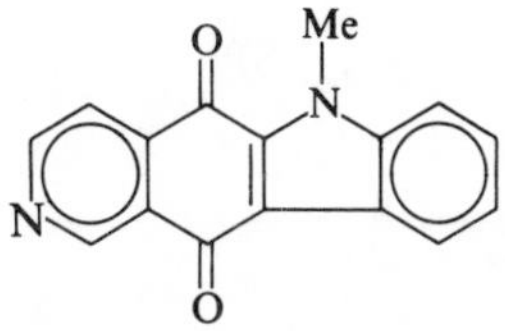 O, Me, N, N, O	76

[a] The corresponding phthalide is obtained as a major product owing to the incomplete lithiation.

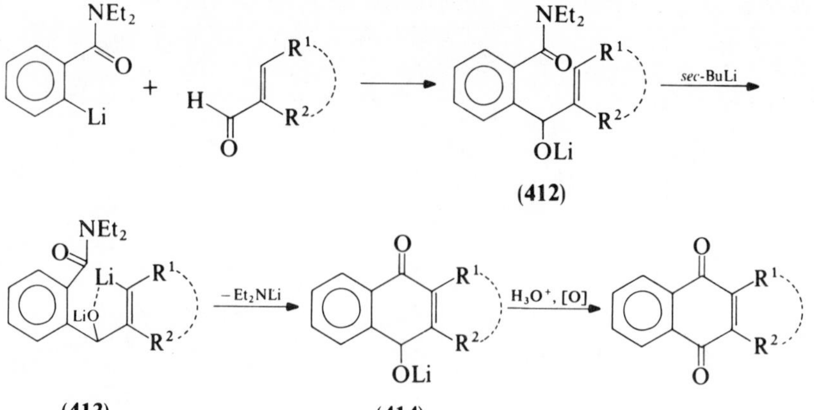

SCHEME 18

equivalent of alkyl lithium, **412** is further lithiated to **413**, which by intermolecular cyclization gives **414**, which in turn is converted to a quinone (Scheme 18)[270, 271]. This tandem reaction is useful for the preparation of polycyclic quinones, most of which are difficult to obtain selectively by other methods. The yields range from low to moderate (Table 17). The second lithiation is the key step in the reaction, as shown by the following facts. (1) Since the adduct from 2-naphthaldehyde possesses two positions capable of metalation, the corresponding two isomeric quinones are obtained. (2) Facile lithiation at the 2-position of furan, thiophene and pyrrole results in higher yields than in the other cases.

This methodology was extended to an α,β-unsaturated aldehyde which carries a β-thiophenyl group capable of stabilizing the formal vinyl anion (Scheme 19)[272]. Under these conditions, phenylthio-substituted quinones are obtained in fair yields. The auxiliary phenylthio group can be removed by oxidation of the sulfide by *m*-chloroperbenzoic acid followed by tin hydride reduction.

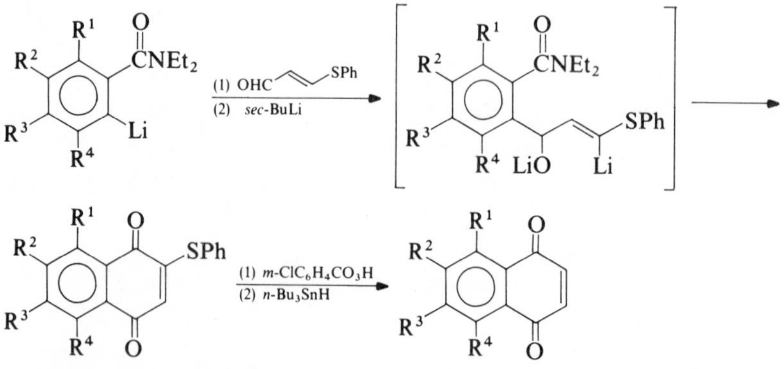

R^1, R^2, R^3, R^4 = H, Me, OMe

SCHEME 19

2. Annulation via isobenzofuran and related 1,4-dipole synthones

The carbanion derived from isophthalide (**415**) by treatment with lithium diisopropylamide (LDA) is considered to be a synthetic equivalent of **416** and reacts with α,β-unsaturated carbonyl compounds to give the corresponding acylnaphthoquinones

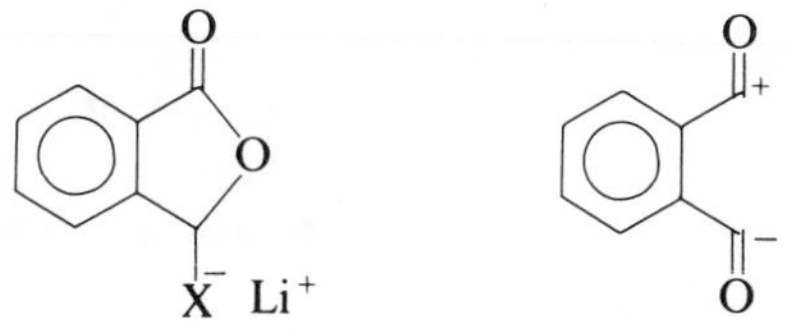

(**415**) (**416**)

X = α-carbanion-stabilizing group

$LiN(Pr\text{-}i)_2$ (**415**) $R^1CH{=}CHCOR^2$

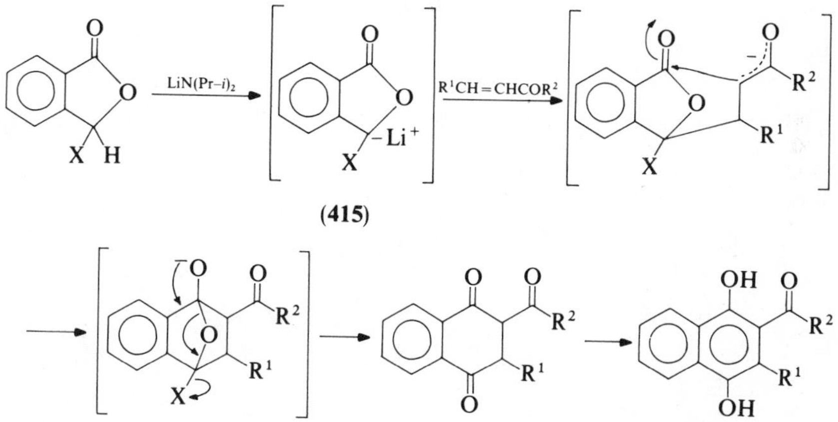

SCHEME 20

illustrated in Scheme 20. Although isophthalides (**417**) without any leaving group at the 3 position can be used in this cyclization[273], the obtained benzyl alcohol (**418**) is prone to dehydrate to the corresponding naphthol (**419**) under acidic conditions (equation 113).

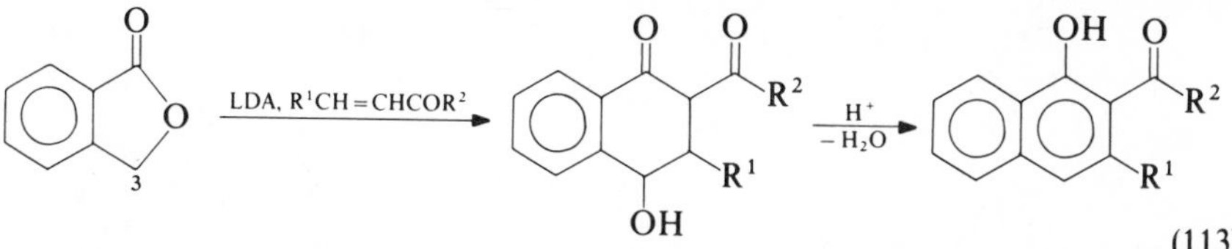

LDA, $R^1CH{=}CHCOR^2$ H^+, $-H_2O$

(**417**) (**418**) (**419**) (113)

Consequently, the introduction of a good leaving group at this position is necessary for the direct quinone synthesis. Since the use of cyanoisophthalide (**420**) was reported (equation 114)[274], many variations have been developed. Among related isophthalide derivatives, the sulfone (**422**) is extremely useful for this cyclization (equation 115), since the aryl sulfonyl group is both a good α-carbanion stabilizing group and a good leaving one[275, 276]. In most cases, acyl- or alkoxycarbonylhydroquinone is obtained in good yield.

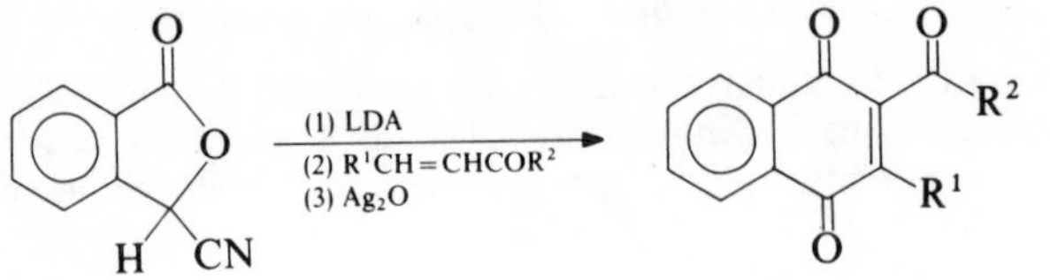

O
O
H CN
(1) LDA
(2) $R^1CH=CHCOR^2$
(3) Ag_2O
O O
R^2
R^1
O
(114)

(420) (421) Ref. 274

$R^1 = R^2 = Me$, 85%
$R^1 = H$, $R^2 = OEt$, 61%

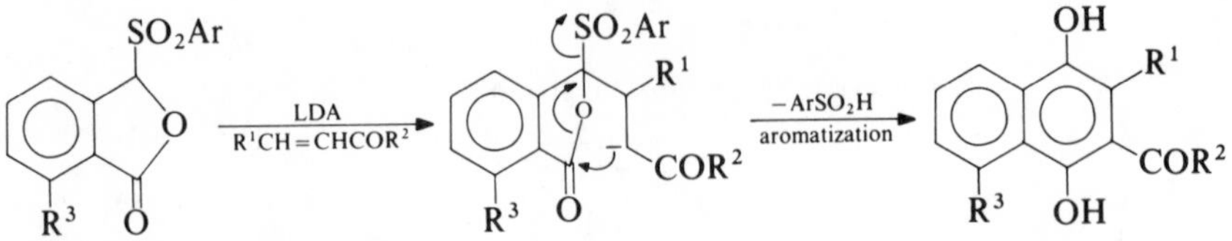

SO_2Ar
O
R^3 O
LDA
$R^1CH=CHCOR^2$
SO_2Ar
R^1
O
COR^2
R^3 O
$-ArSO_2H$
aromatization
OH
R^1
COR^2
R^3 OH

(422) (423) (424)

$R^1 = H$, Me, CH_2SMe, $R^2 = Me$, OEt, $R^3 = H$, OMe Refs 275, 276 (115)

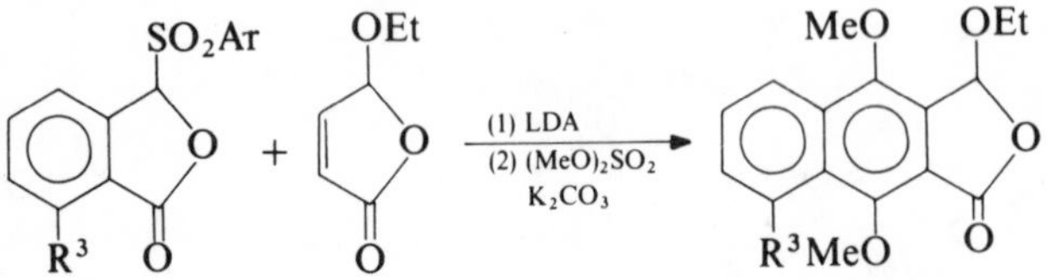

SO_2Ar
O
R^3 O
+
OEt
O
O
(1) LDA
(2) $(MeO)_2SO_2$
K_2CO_3
MeO OEt
O
R^3MeO O
(116)

(422) (425) (426) Ref. 277

SO_2Ar
O
R^3 O
+
R^1
R^2
O
t-BuOLi
OH
R^1
R^2
R^3 OH O
(117)

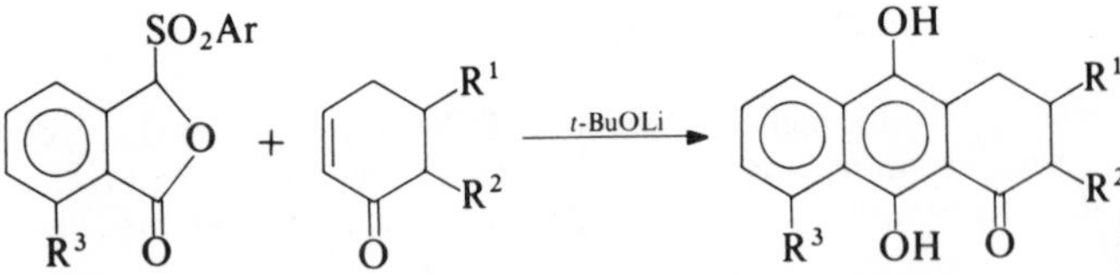

(422) (427) (428) Ref. 279

SO_2Ar
O
R^3 O
+
R^1
R^2 R^2
O
O
LDA
R^1
R^2 R^2
OH
O
R^3 OH O
(118)

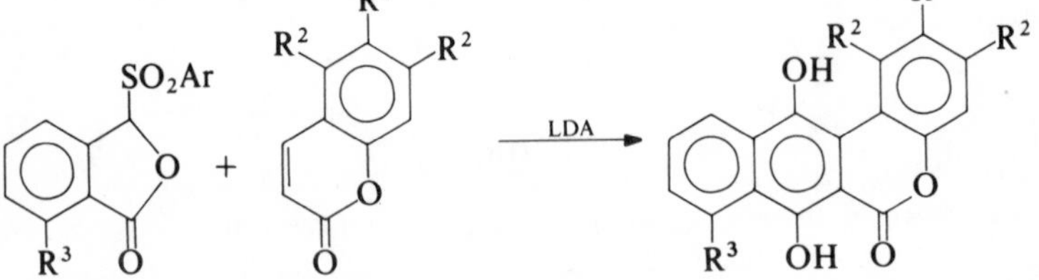

(422) (429) (430)

Ref. 281

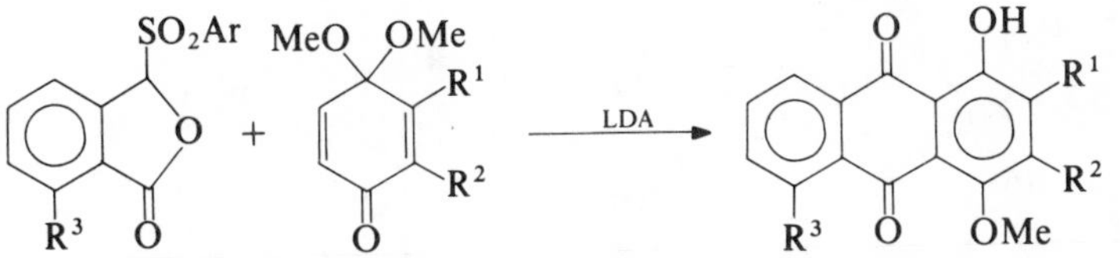

(422) + (431) $\xrightarrow{\text{LDA}}$ (432) (119)

Refs 282, 283

This reaction was applied to the synthesis of various naphthoquinone and anthraquinone derivatives. Furanone (**425**)[277, 278], cyclohexenone (**427**)[279, 280], benzo-2-pyrone (**429**)[281] and quinone monoacetal (**431**)[282, 283] were used as initial Michael acceptors, and gave **426**, **428**, **430** and **432**, respectively (equations 116–119). There are several impressive features in this reaction. (1) The regiochemistry of the annulation is predictable and there is no possibility of contamination with the regioisomers, which is frequently observed in the other annulations (i.e. a Diels–Alder reaction, metallacycle compounds, and metal carbene complexes). (2) A wide range of Michael acceptors can be applied in this reaction to afford the expected product in a satisfactory yield. (3) Versatile functional groups, which can survive under basic conditions, can be introduced in a regioselective manner to the produced aromatic compound.

As an extension of this method, the tetracyclic compound **436** was obtained by a repeated use of this reaction (equation 120)[275]. In addition to the α,β-unsaturated carbonyl

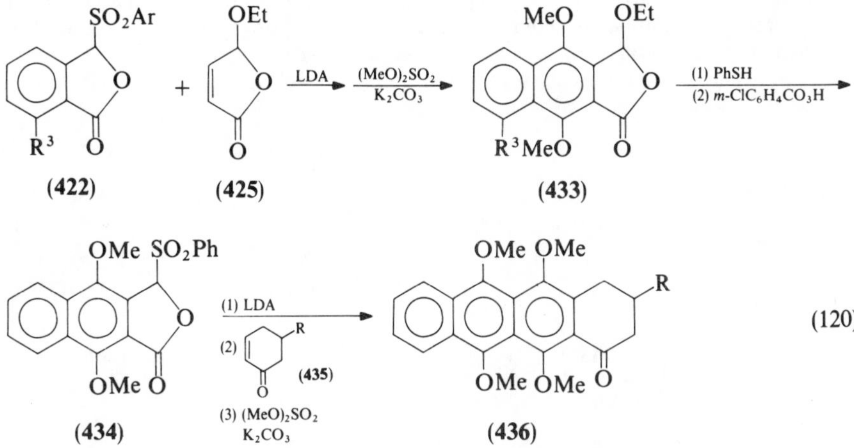

(422) + (425) $\xrightarrow{\text{LDA}}$ $\xrightarrow[K_2CO_3]{(MeO)_2SO_2}$ (433) $\xrightarrow[(2)\ m\text{-}ClC_6H_4CO_3H]{(1)\ PhSH}$

(434) $\xrightarrow[(3)\ (MeO)_2SO_2,\ K_2CO_3]{(1)\ LDA;\ (2)\ (435)}$ (436) (120)

R = H, 68%
R = COMe, 68%
R = CO_2Me, 87%

compounds, benzyne was also used as a trapping agent of the initially formed anion[282–284]. When a phthalide and a substituted bromobenzene are treated together with two equivalents of LDA, a nucleophilic addition of lithium phthalide to benzyne which is generated *in situ*, is followed by an intramolecular cyclization. When a phenylsulfonyl group is at the 3 position, the anthraquinone is obtained directly in good yield, whereas the hydroxy derivative is obtained with 3-*H*-isophthalide (Scheme 21).

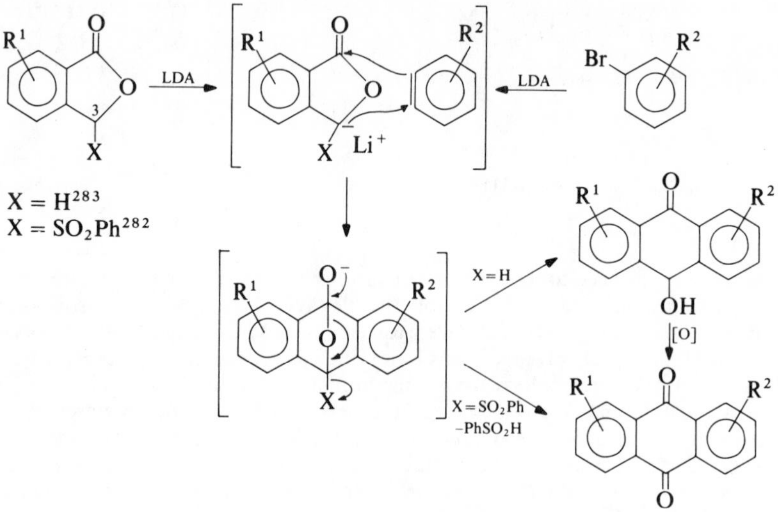

R1 O O 3 X LDA → [R1 O O R2 X Li+] ← LDA Br R2

$X = H$[283]
$X = SO_2Ph$[282]

[R1 O⁻ R2 O X] X=H → R1 O R2 OH [O] ; $X=SO_2Ph$, $-PhSO_2H$ → R1 O R2 O

SCHEME 21

Relevant to this annulation, the benzocyclobutenedione monoacetal (**437**) is also considered as a synthon of **416**[285, 286]. The reaction of **437** with lithiated quinone bisacetal (**438**) proceeds by an initial nucleophilic attack of the lithiated reagent on the ketone. The intermediate acetal anion (**440**), considered to be the charge inverted equivalent of a formyl anion, then cyclizes intramolecularly to **441** through Michael type addition (equation 121).

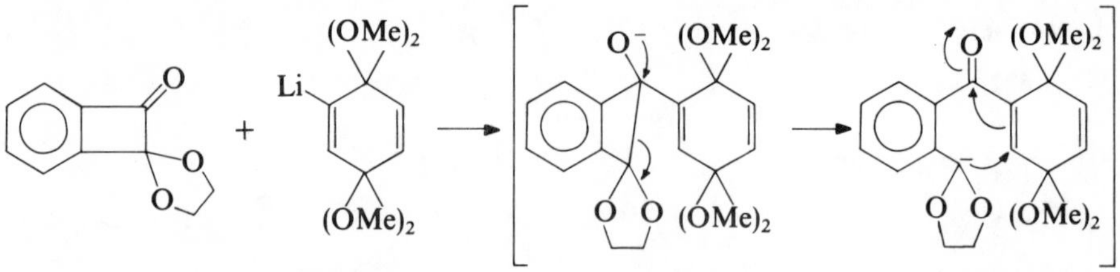

O O O + Li $(OMe)_2$ $(OMe)_2$ → [O⁻ $(OMe)_2$ O O $(OMe)_2$ → O $(OMe)_2$ O O $(OMe)_2$]

(**437**) (**438**) (**439**) (**440**)

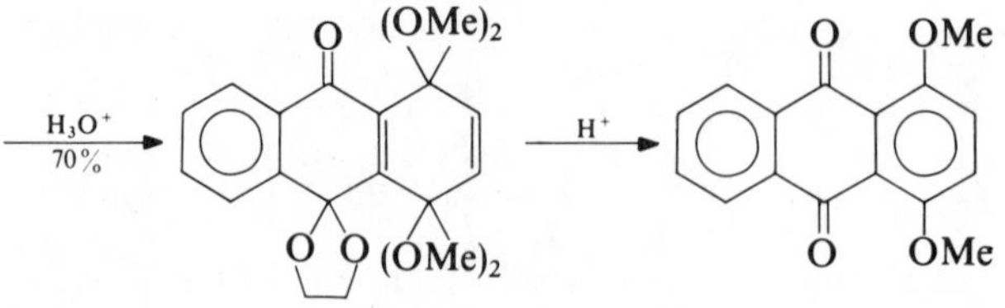

H_3O^+ 70% → O $(OMe)_2$ O O $(OMe)_2$ → H^+ → O OMe O OMe (121)

(**441**) (**442**)

E. Miscellaneous Reactions

1,3-Dipolar cycloaddition is utilized for the synthesis of heterocyclic quinones. Reactions of nitrile ylides (**443**) with *p*-quinones possessing free 2 and 3 positions give the

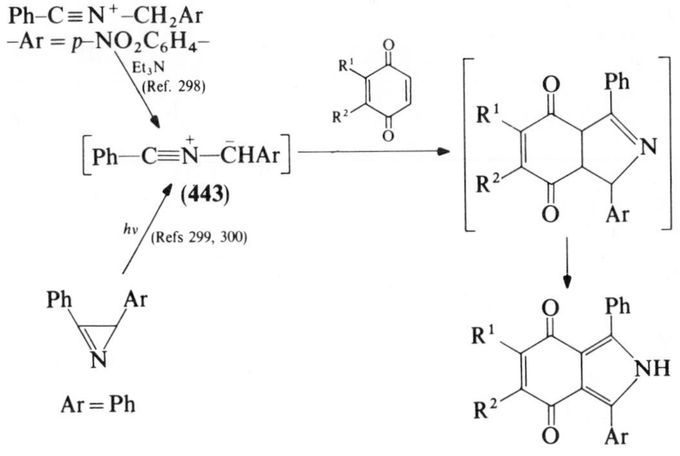

$Ph-C\equiv N^+-CH_2Ar$
$-Ar = p\text{-}NO_2C_6H_4-$

Et_3N (Ref. 298)

$[Ph-C\equiv \overset{+}{N}-\overset{-}{C}HAr]$ **(443)**

$h\nu$ (Refs 299, 300)

Ar = Ph

R^1, R^2 = H, Me, MeO, –CH=CHCH=CH–

SCHEME 22

corresponding isoindoloquinone (Scheme 22)[287–289]. Benzonitrile benzylide (**443**, Ar = Ph) reacts with wide range of *p*-quinones[288], while benzonitrile-2-propanide (**445**) gives the corresponding isoindoloquinone **446** with only 1,4-naphthoquinone (equation 122).

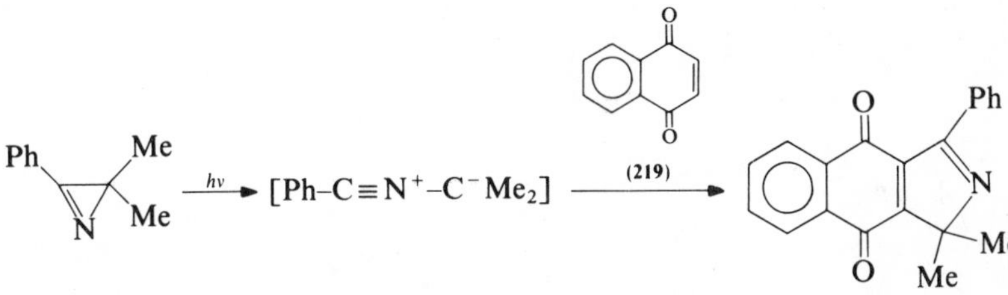

$$\xrightarrow{h\nu} [Ph-C\equiv N^+-C^-Me_2] \xrightarrow{(219)} \quad (122)$$

(444) **(445)** **(446)**

On the other hand, the reaction of **445** with benzoquinone arises at the quinone carbonyl group instead of the C–C double bond of the quinone. Since this reaction is very sensitive to the structure of the dipolarophiles, it does not have general applicability.

In the presence of sodium hydride, *N*-benzylidene glycine ester (**447**) forms *in situ* *N*-benzylidene imine (**448**), which cyclizes with a quinone monoacetal to give **451** after acid treatment (equation 123)[290]. **451** is also the promising precursor to the isoindoloquinone[290].

Recently, non-stabilized azomethine ylides, generated *in situ* by the treatment of cyanomethylaminosilanes with AgF, were found to cyclize with quinones to the corresponding isoindoloquinones (**453**) (equation 124)[291]. This reaction seems to have a rather wide applicability to various quinones and it gives the products in good yields. Another example is the 1,3-dipolar cycloaddition of quinones with oxazolium oxide (**455**), which is formed by dehydrative cyclization of the phenylglycine derivative (**454**) (Scheme 23)[292]. There is no regioselectivity in the reaction between an unsymmetrical quinone and

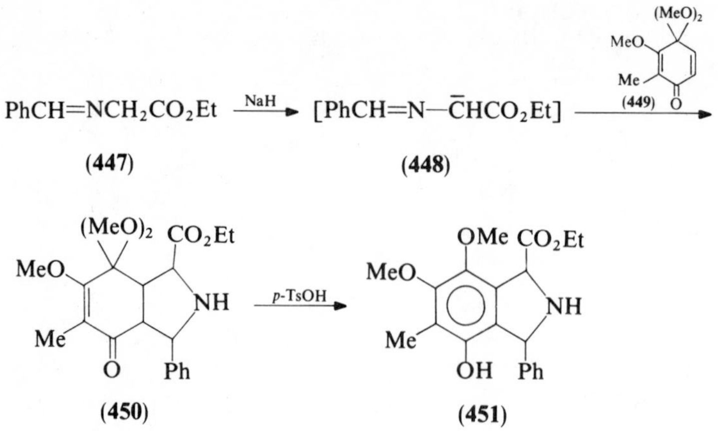

$PhCH{=}NCH_2CO_2Et \xrightarrow{NaH} [PhCH{=}N{-}\bar{C}HCO_2Et] \xrightarrow{(449)}$

(447) (448)

(450) $\xrightarrow{p\text{-TsOH}}$ (451) (123)

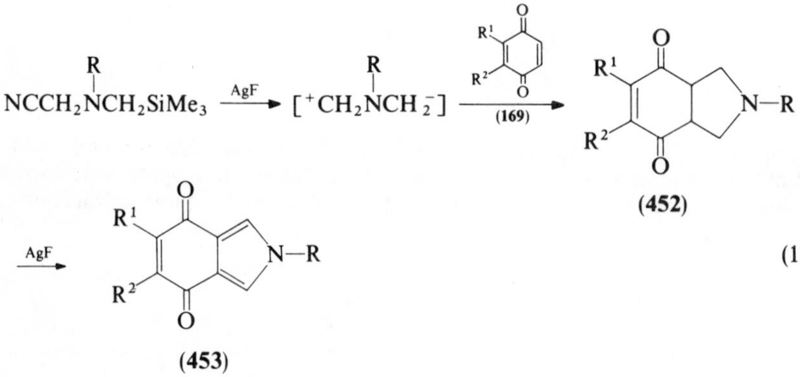

$NCCH_2N(R)CH_2SiMe_3 \xrightarrow{AgF} [^+CH_2N(R)CH_2^-] \xrightarrow{(169)}$ (452)

$\xrightarrow{AgF}$ (453) (124)

$R^1, R^2 = H, Me, MeO, 60–75\%$

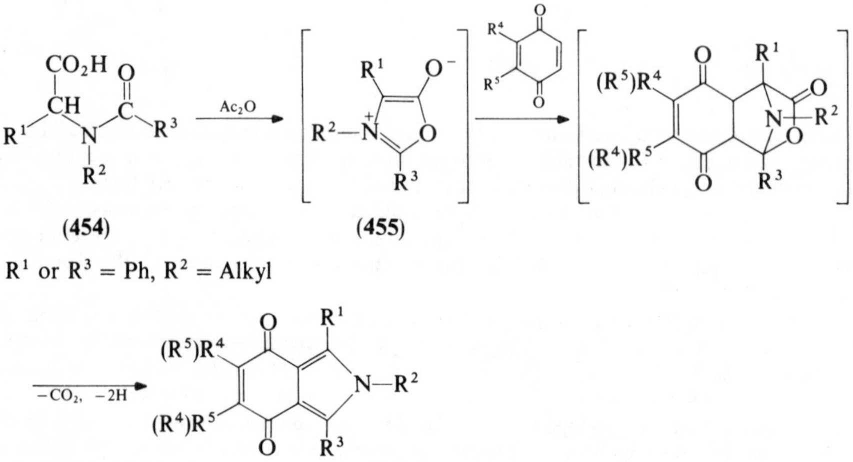

(454) $\xrightarrow{Ac_2O}$ (455)

R^1 or $R^3 = Ph$, $R^2 = Alkyl$

$\xrightarrow{-CO_2,\ -2H}$

SCHEME 23

an oxazolium oxide when $R^1 \neq Ph$. The reaction can be used to prepare the tricyclic systems **457** and **458** from the bicyclic oxazolium oxide **456** (equation 125).

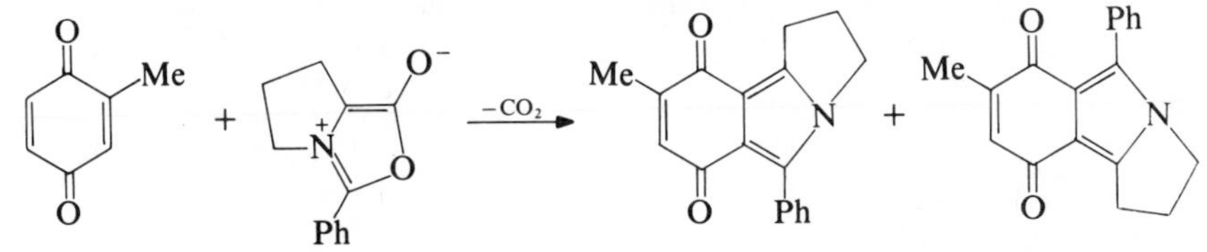

(257) (456) (457) (458)

$-CO_2$

(125)

As seen above, the number of 1,3-dipoles applied to the cycloaddition reaction with quinone as a dienophile is so far limited. Further exploitation is necessary in order to extend the scope of the reaction with many other 1,3-dipoles[293].

IV. CYCLIZATION AND CONDENSATION METHODS

A. The Friedel–Crafts Reaction

The Friedel–Crafts reaction is traditionally used for the synthesis of anthraquinone derivatives and has almost an equivalent value to that of the anionic annulation of a phthalide with α,β-unsaturated carbonyl compound (Section III.D.2). Several proton acids and Lewis acids, e.g. conc. H_2SO_4[294,295], $BF_3 \cdot OEt_2$[296], have been used for this reaction, but the yields are not always satisfactory. Trifluoroacetic anhydride (TFAA), trifluoromethanesulfonic acid and hydrofluoric acid are superior as acid catalysts and give cyclized products in good yields[121,297,298]. Several examples are compared in equations 126–130. As the carboxyl moiety, ester, acid chloride and acid anhydride as well as the free acid, are most commonly used, and are equally applied to this cyclization. Many other variations are known in the synthesis of anthracyclinones. As a special example, carbothioic acid derivative (**475**) is used as an alternative and the corresponding cyclized product (**476**) is obtained by using copper(I) trifrate (CuOTf) as the Lewis acid (equation 131)[299].

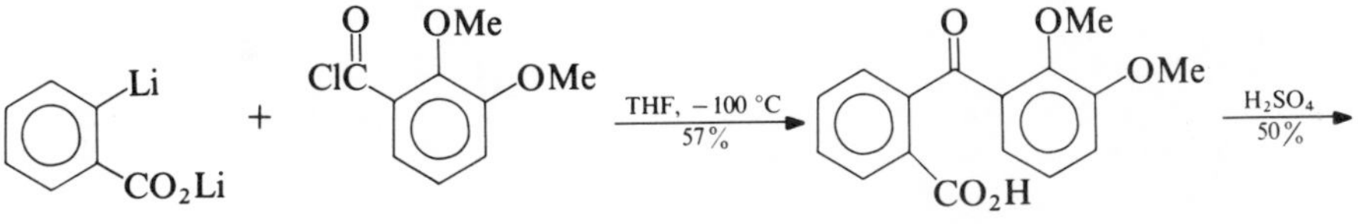

(459) (460) (461)

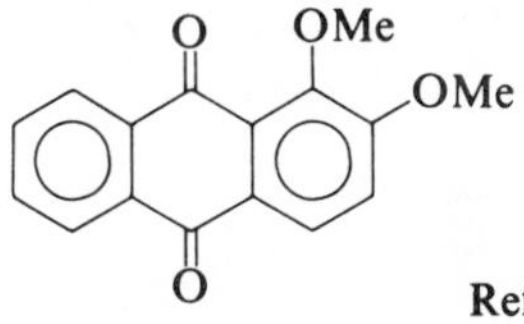

THF, −100 °C, 57%; H_2SO_4, 50%

(222)

(126)

Ref. 294

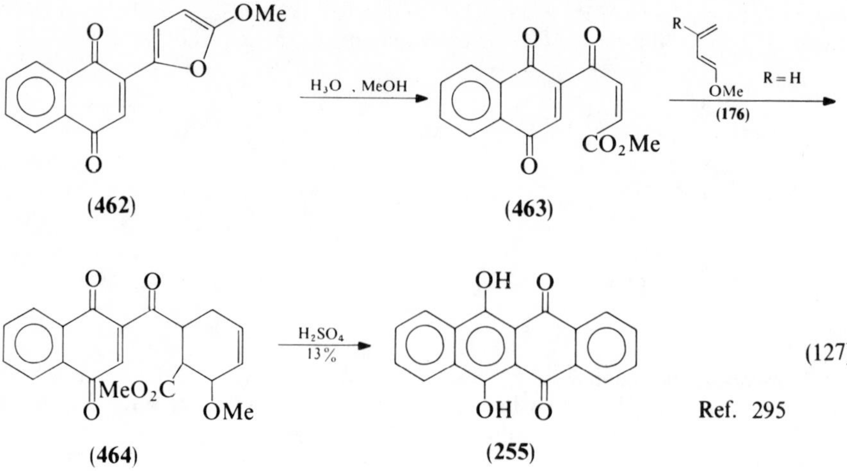

(462) $\xrightarrow{H_3O^+, MeOH}$ (463) $\xrightarrow[R = H]{(176)}$ (464) $\xrightarrow[13\%]{H_2SO_4}$ (255) (127)

Ref. 295

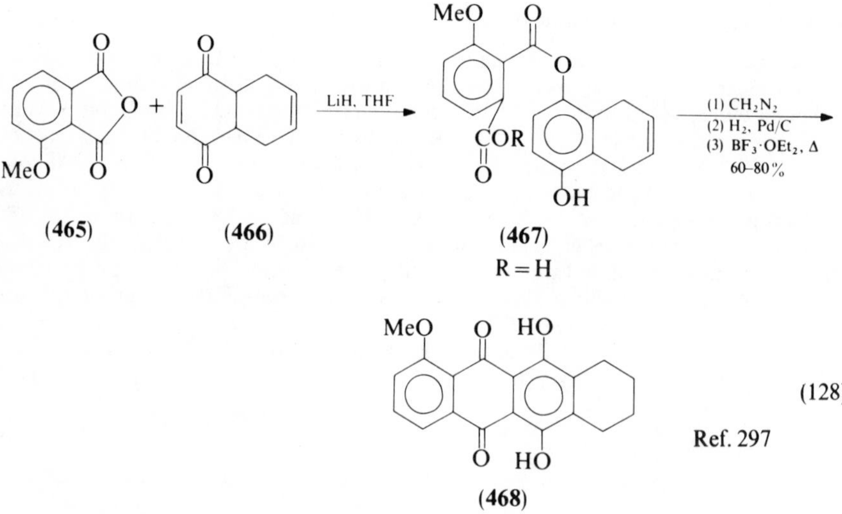

(465) + (466) $\xrightarrow{LiH, THF}$ (467) R = H $\xrightarrow[60\text{–}80\%]{(1)\ CH_2N_2;\ (2)\ H_2, Pd/C;\ (3)\ BF_3 \cdot OEt_2, \Delta}$ (468) (128)

Ref. 297

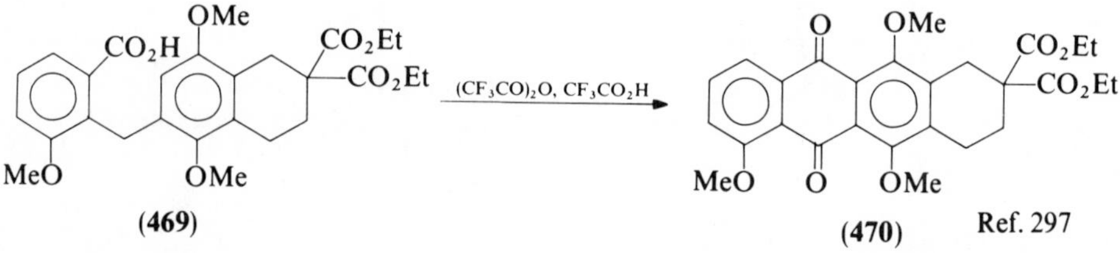

(469) $\xrightarrow{(CF_3CO)_2O, CF_3CO_2H}$ (470) Ref. 297

(129)

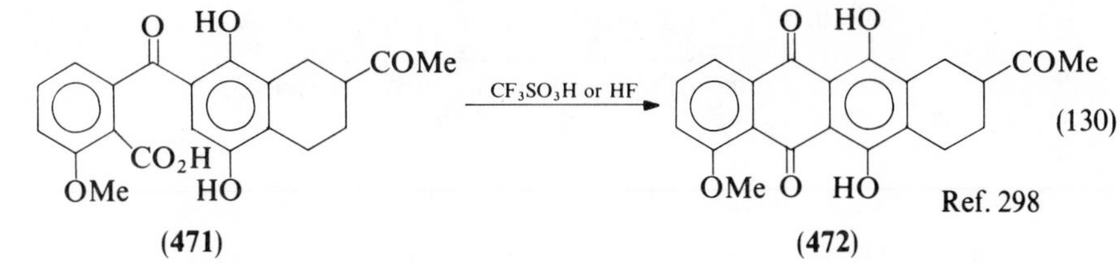

(471) $\xrightarrow{CF_3SO_3H \text{ or } HF}$ (472) (130)

Ref. 298

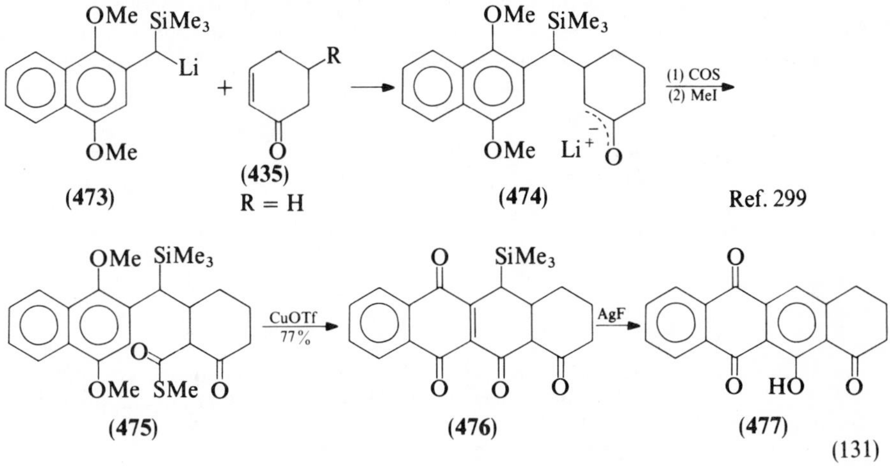

(473) + (435) R = H → (474) $\xrightarrow{(1)\ COS,\ (2)\ MeI}$

Ref. 299

(475) $\xrightarrow[77\%]{CuOTf}$ (476) $\xrightarrow{AgF}$ (477) (131)

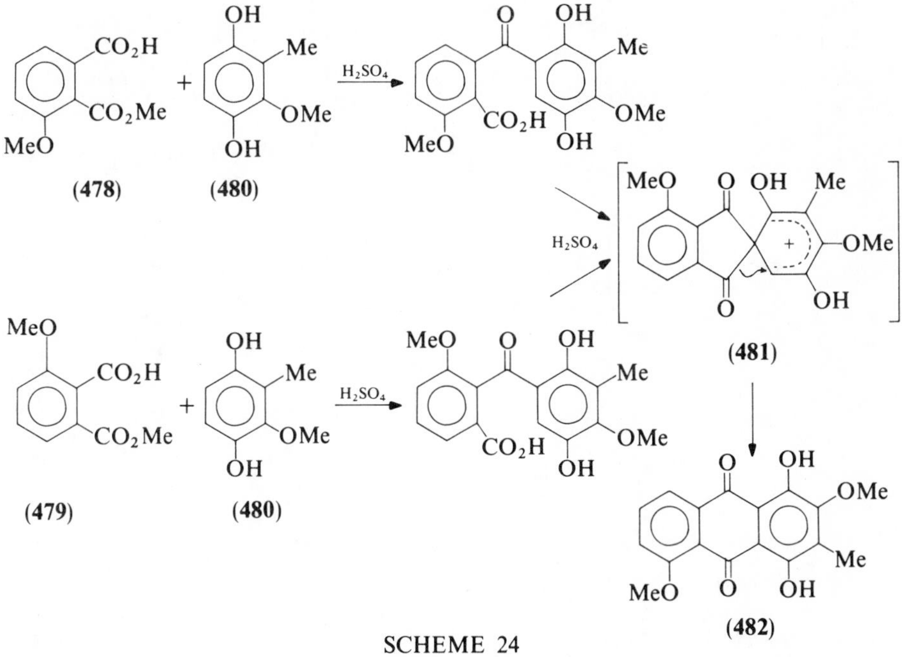

(478) + (480) $\xrightarrow{H_2SO_4}$ → $\xrightarrow{H_2SO_4}$ [(481)]

(479) + (480) $\xrightarrow{H_2SO_4}$ → $\xrightarrow{H_2SO_4}$ [(481)] → (482)

SCHEME 24

When an intermolecular Friedel–Crafts cyclization to anthraquinone is conducted with a combination of both unsymmetrical phthalic acid monoesters, **478** or **479**, and the hydroquinone **480**, a single product (**482**) is obtained. It was speculated that the reaction proceeds via the spiro-type intermediate (**481**) (Scheme 24)[300,301].

B. Miscellaneous Reactions

In the biosynthetic route to quinones, polyketide intermediates are postulated as the precursors of the aromatic nucleus. For a demonstration of this process, oxalyldiacetone (**483**) is treated with a concentrated base to afford the anthraquinone derivative **484** in 5.8% yield (equation 132)[302]. The low yield is attributed to the occurrence of random and unfavorable intermolecular condensation. Since diketene **486** is considered as a synthetic

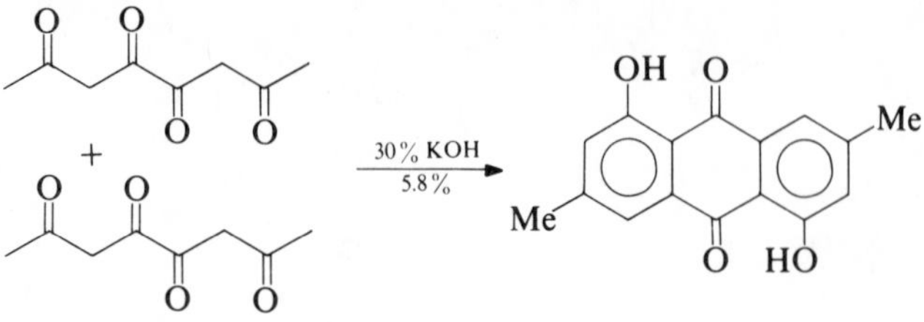

30% KOH, 5.8%   (132)

(**483**)   (**484**)

equivalent of the β-ketoacyl cation **487**, its condensation with the β-ketoacid ester **485** will give the polyketo intermediate **489**, which undergoes a Robinson type annulation to give **490** and then **491** (equation 133)[303].

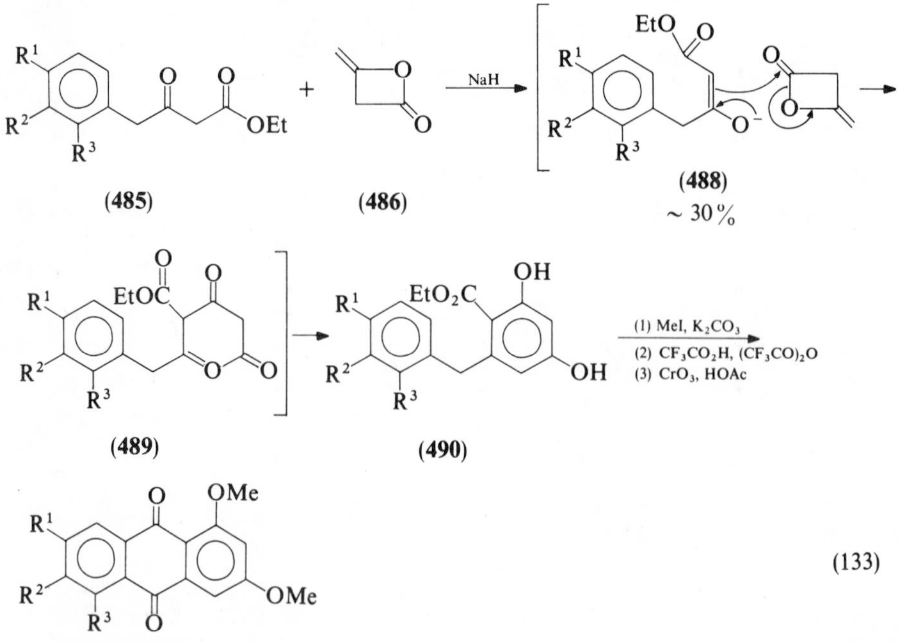

NaH

(**485**)   (**486**)   (**488**) ~ 30%

(1) MeI, K_2CO_3
(2) CF_3CO_2H, $(CF_3CO)_2O$
(3) CrO_3, HOAc

(**489**)   (**490**)

(133)

(**491**)

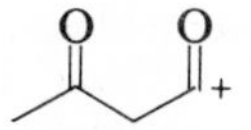

(487)

Nucleophilic attack of two moles of a ketene acetal on a haloquinone has been known to afford regiospecifically the corresponding 1,3-dialkoxyanthraquinone in varying yields[304]. This reaction was extensively examined with halonaphthazarines[163], 3-bromojuglones[163,305] and dihalo-*p*-benzoquinone (equations 134–136)[306]. Under optimized conditions, a moderate yield is obtained when five molar equivalents of the ketene acetal

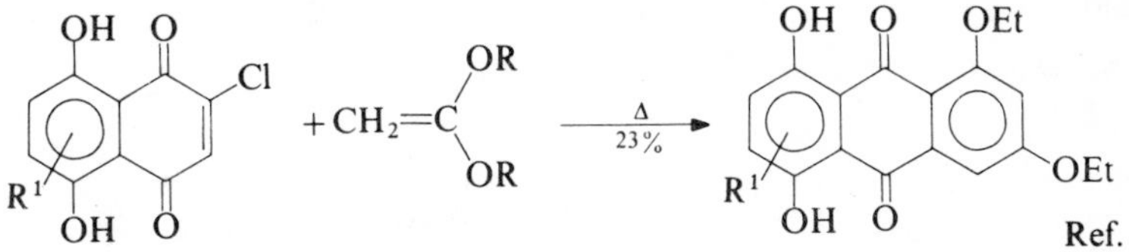

$$(493) + CH_2{=}C(OR)_2\ (492),\ R = Et \xrightarrow[23\%]{\Delta} (494) \quad (134)$$

Ref. 163

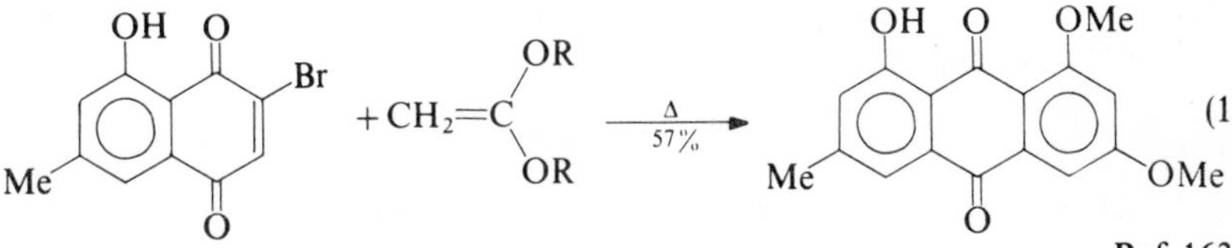

$$(495) + CH_2{=}C(OR)_2\ (492),\ R = Me \xrightarrow[57\%]{\Delta} (496) \quad (135)$$

Ref. 163

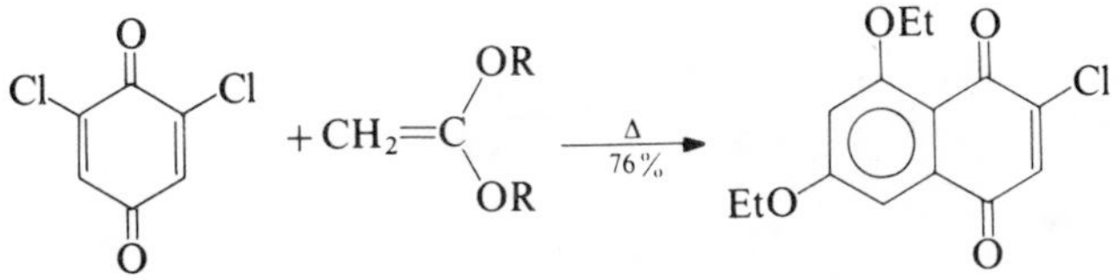

$$(228) + CH_2{=}C(OR)_2\ (492),\ R = Et \xrightarrow[76\%]{\Delta} (497) \quad (136)$$

Ref. 306

to haloquinones are used. The requirement for such a large excess of the ketene acetal is due to its decomposition by the acid and alcohol, formed during the reaction, although the detailed reaction mechanism has not been fully revealed. This method is useful for the synthesis of several specific quinones in a one-step procedure. A similar product is obtained in the reaction of a vinyl ketene acetal with a haloquinone (Section II.A).

Heterocycles, such as furan, thiophene and indole, substituted by 3,4-bis(bromoacetyl) groups (**498**) afford the corresponding cyclized product (**500**) by their treatment with a Zn–Cu alloy[307]. The biradical **499** is postulated as an intermediate in the cyclization reaction. The quinone **501** is obtained by further oxidation (equation 137).

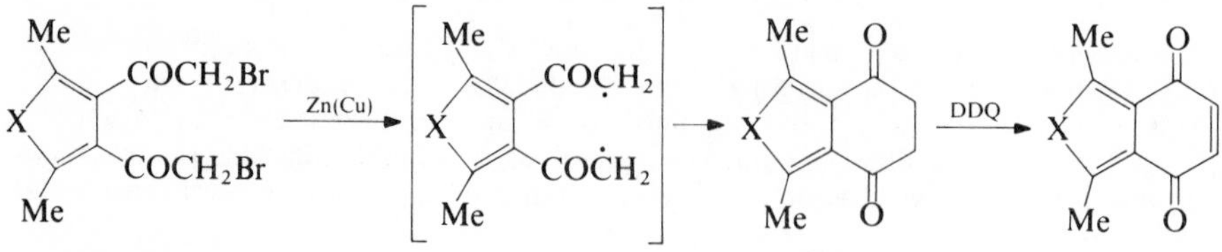

(**498**) (**499**) (**500**) (**501**)

X = O, S, NMe

(137)

A photochemical reaction of the halogenated naphthoquinone **502** with 1,1-diarylethylenes (**503**) gives the benzanthraquinones (**505**) in one step presumably via the intermediacy of **504** as shown in equation 138. This reaction is extensively explored for the

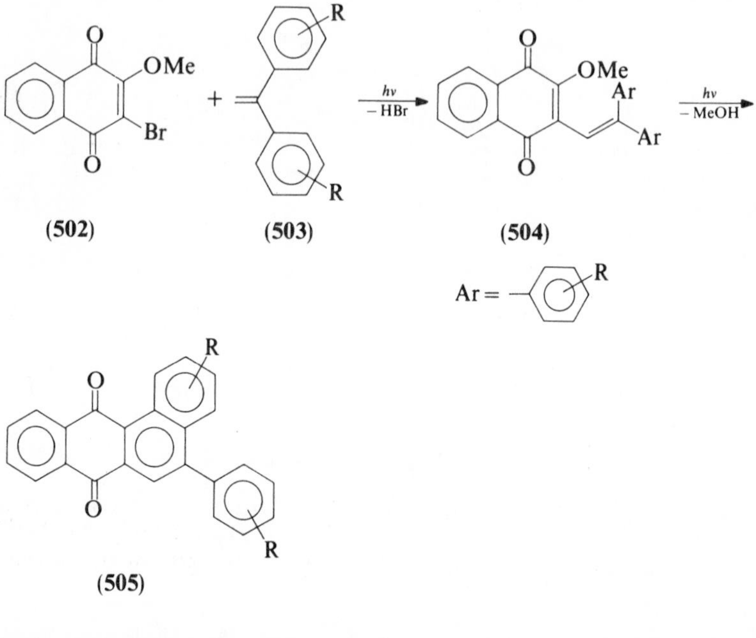

(**502**) (**503**) (**504**)

Ar =

(**505**)

(138)

combination between various halogenated quinones and diarylethylenes[308–316]. The best results are obtained in the reaction of 2-bromo-3-methoxy-1,4-quinone derivatives and diarylethylenes (**503**) substituted by electron-donating group R, since the reaction is initiated by a photo-excited single electron transfer from the olefin to the quinone. The applicability of the reaction is shown by the synthesis of the polycyclic quinones **506**–**508**[316]. When the two R substituents on the aryl group are different, the cyclization occurs preferentially at the electron-rich aryl group.

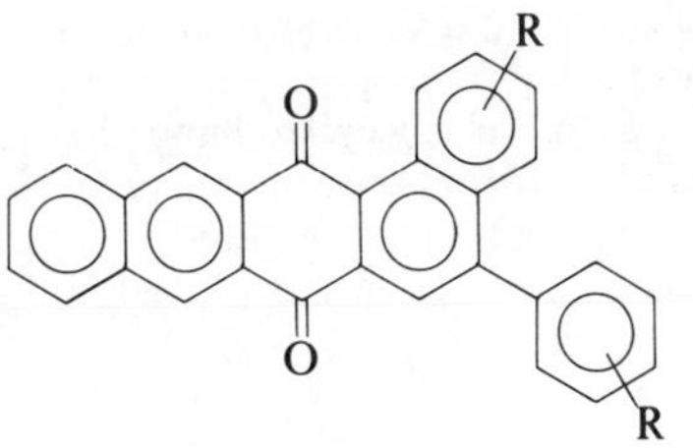

(506) 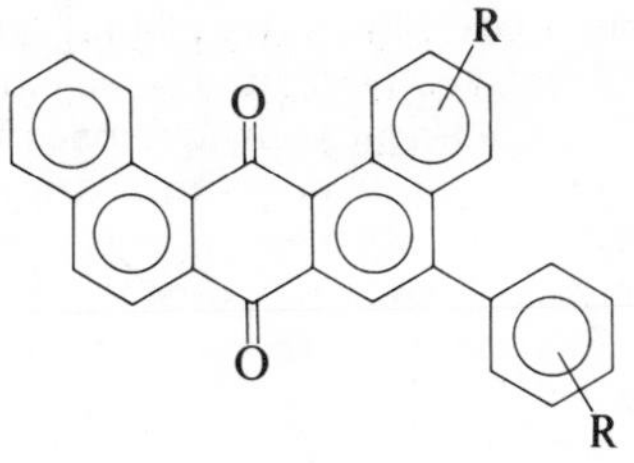 (507)

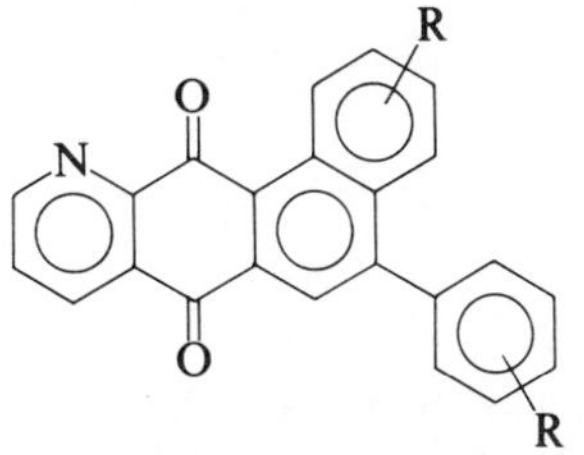

(508)

Azidoquinones show a versatile reactivity toward olefins and are considered to be good precursors for indoloquinones. Intramolecular cyclization of the 2-azido-3-alkenyl-1,4-naphthoquinone (**509**) upon heating gives the corresponding indoloquinone (**510**) in good yield (equation 139)[317]. A concerted mechanism rather than an intermediary formation of a nitrene was suggested in the reaction.

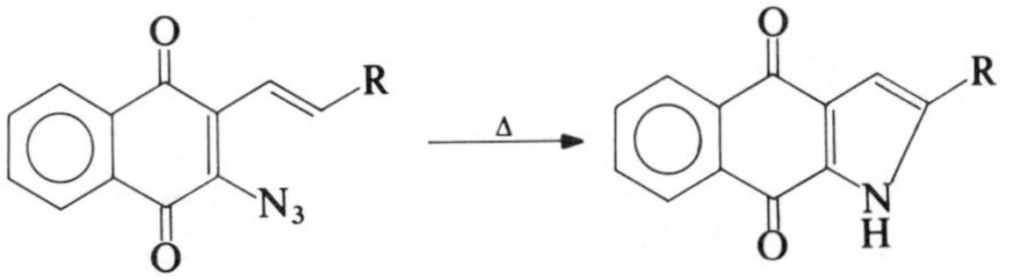

(509) → Δ → (510) (139)

R = Alkyl, Ph

Ref. 317

On the other hand, the 2-azidoquinones (**511**) afford cyclized products (**513**) with conjugated dienes upon irradiation with near UV light (equation 140)[318–320]. The reaction proceeds in a completely stereoselective manner. The relative stereochemistry at

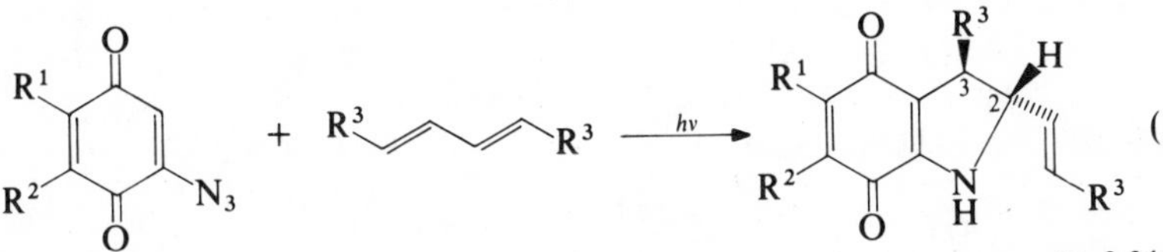

(511) + (512) → hν → (513) (140)

Ref. 319

the 2 and 3 positions is *trans* and it is established that a vinylic group at the 2 position preserves completely its original stereochemistry[319,320].

The 2-azido-3-pentadienylquinone (**514**) intramolecularly cyclizes upon heating with a copper catalyst directly to afford a tricyclic product (**516**) in reasonable yield[321,322]. This product is neither obtained under photochemical conditions nor without a copper catalyst. Mechanistically, it was proposed that an initial formation of unstable 1,2-addition product (**515**) is followed by a 1,3-rearrangement (equation 141).

(**514**) —Δ, 'Cu', 50–60%→ [(**515**)] → (**516**)

Refs 321, 322 (141)

Palladium-salt catalyzed cyclization of 2-allyl-3,6-diamino-5-methyl-1,4-benzoquinone (**517**) affords the quinolinoquinone **518** in a modest yield (equation 142), while 2-allyl-3,6-bis(allylamino)-5-methyl-1,4-benzoquinone (**519**) gives a mixture of the three indoloquinones **520**–**522** without selectivity (equation 143)[323].

(**517**) —$PdCl_2(MeCN)_2$, Et_3N, 46%→ (**518**) (142)

(**519**) —$PdCl_2(MeCN)_2$→ (**520**) 20% + (**521**) 15% + (**522**) 25% (143)

V. MISCELLANEOUS METHODS

Nucleophilic addition of organometallic reagents to quinones have been extensively studied[324, 325]. At a lower temperature, these reactive carbon nucleophiles react with quinone to afford the 1,2-addition product under conditions where simple reduction of the quinone to the hydroquinone does not take place. When the alkylation with alkyllithium is applied to 2,5-dialkoxy-1,4-benzoquinone (**523**), the product is 2,5-dialkyl-1,4-benzoquinone (**149**), whose carbonyl groups are formally rearranged as shown in equation 144[326]. Addition of the first equivalent of a lithium reagent R^1Li affords an adduct (**524**), to

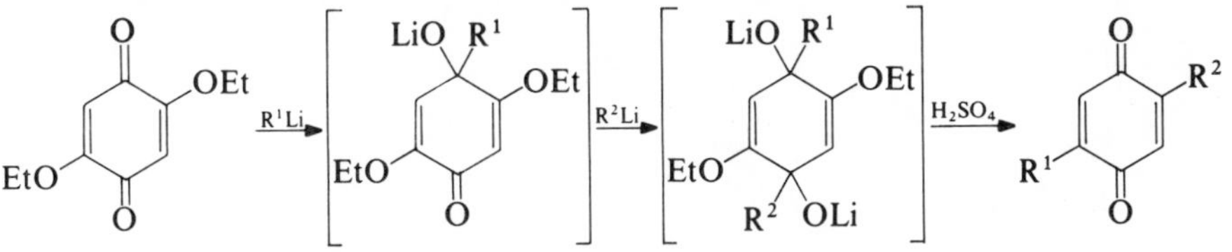

(**523**) —R^1Li→ [(**524**)] —R^2Li→ [(**525**)] —H_2SO_4→ (**149**) 52–62%

(144)

which the second lithium reagent R^2Li adds to give **525**. Hydrolysis and successive dehydration of **525** affords **149** in moderate yields. When R^1Li is a 'soft' nucleophile such as alkynyllithium, a selective monoalkylation is possible by control of the amount of the reagent, and successive treatment with the second R^2Li affords the quinone with two different carbon functionalities at the 2,5-positions. A limitation is that when R^1Li is an alkyl- or alkenyllithium they are too reactive to stop the reaction at the monoalkyl (or monoalkenyl) stage.

Similarly, 2,5-dichloro-3,6-dimethoxybenzoquinone (**526**) gives **527** by a similar reaction without loss of halogen atoms[326]. This methodology is also applicable to 4,5-dimethoxy-1,2-benzoquinone (**530**)[327], which gives 2-alkynyl-5-methoxy-1,4-benzoquinone (**530**) in good yield (equations 145 and 146).

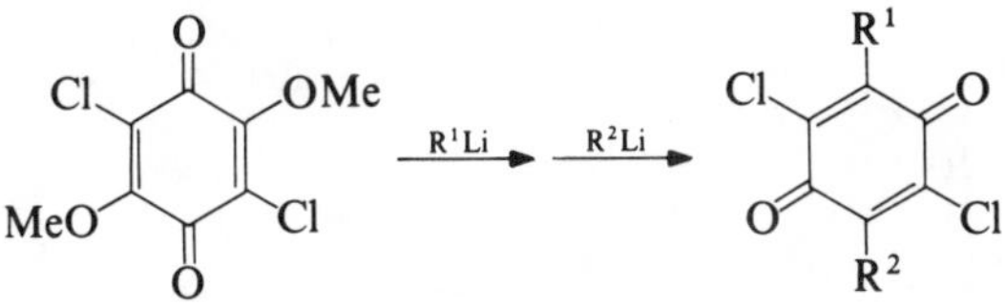

(**526**) —R^1Li→ —R^2Li→ (**527**) (145) Ref. 326

(**528**) —$LiC\equiv CR$→ (**529**) —H_3O^+→ (**530**) (146) Ref. 327

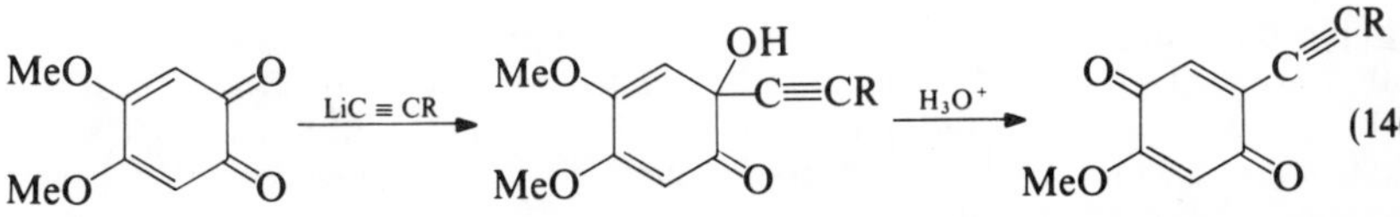

VI. PROTECTION OF QUINONES AND HYDROQUINONES

Since a quinone (or a hydroquinone) is a reactive species, it is frequently required to protect the carbonyl or the hydroxy groups against a reactive reagent or destructive reaction conditions. The criteria for the protecting method are (1) easy operation in the protection and the deprotection, (2) an excellent yield for both reactions, (3) high chemoselectivity, and (4) a wide applicability. Several protecting methods (or reagents) have been specifically developed for quinones.

Trimethylsilyl cyanide can give monocyanosilylation of quinones in the presence of a catalytic amount of a KCN–18-crown-6 complex (equation 147)[328, 329]. The site of the

$$\textbf{(106)} \underset{F^-}{\overset{Me_3SiCN,\ KCN,\ 18\text{-crown-6}}{\rightleftharpoons}} \textbf{(531)} + \textbf{(532)} \qquad (147)$$

cyanosilylation is dominated by the relative electrophilicity of the carbonyl groups. For electron-donating substituents it occurs mainly at the highly substituted site, except in the presence of severe steric hindrance by a *t*-butyl group (Table 18). Since this protecting

TABLE 18. Regioselective trimethylsilyl cyanation of quinones[328]

R^1	R^2	R^3	Product ratio **(531)**:**(532)**	% Yield[a]
H	H	H		80
H	H	Me	89: 11	92
H	H	MeO	100: 0	80
Me	H	MeO	100: 0	90
MeO	H	MeO	100: 0	65(100)
Me	H	Me	94: 6	(100)
H	*t*-Bu	*t*-Bu	0:100	(98)
–CH=CH–CH=CH–		H		75
–CH=CH–CH=CH–		Me	91: 9	96
–CH=CH–CH=CH–		MeO	100: 0	92
H	Br	Br	100: 0[b]	100[c]

[a] Yields in parentheses were determined by NMR; all others are yields of the purified adducts.
[b] The reaction was conducted in the presence of Ph_3P at 0°C in MeCN.
[c] Ref. 329a.

group is inherently unstable to both aqueous media and nucleophiles which easily react with the carbonyl group[329a], its use is rather limited. Deprotection of the cyanosilyl ether with AgF can regenerate the quinone in good yield. This reaction is applied before alkylation of a quinone with an organolithium[325] or a Grignard reagent[329b]. The related reaction with $Me_2Si(CN)_2$ is also effective for the regiospecific protection of juglone (**56**, R=H) (equation 148)[330].

$$(56)\ R = H \xrightarrow[\text{MeOH or AgF}]{Me_2Si(CN)_2,\ 62\%} (533) \qquad (148)$$

Quinone dialkyl acetals are extensively used as synthetic equivalents or as protected quinones. The details are given at Section II.D and Chapter 15. Since the quinone diacetals are obtained by electrolytic oxidation of the corresponding hydroquinone dialkyl ether, this method is not applicable to substrates carrying oxidizable functional group(s).

Protection of hydroquinones is much easier than that of the corresponding quinone, and in most cases, protecting groups which are applicable to phenols, are also suitable for this purpose[331, 332]. Typical groups are Me, $PhCH_2$, $MeOCH_2$ and trialkylsilyl groups, and the order of deprotectivity is $R_3Si > MeOCH_2 \simeq PhCH_2 > Me$. A trialkylsilyl group can be easily deblocked by aqueous protic acid when R = Me or by fluoride anion when R_3 = *t*-$BuMe_2$. Methoxymethyl and benzyl ethers are easily removed by a protic acid (typically *p*-TsOH) and by catalytic hydrogenation, respectively. Since hydroquinone dimethyl ether is very stable and can tolerate a wide range of reaction conditions, the only reliable method of its deprotection is oxidative demethylation by AgO or CAN (see Section II.D) and many applications are shown in Section VII. Both oxidizing reagents as well as the electrolytic acetal formation are reported to cause undesirable side reactions, when the substrate carries reactive functionalities.

Reductive silylation of quinones is also useful for the protection of hydroquinones (equation 149). Hexamethyldisilane is used as a silylating agent in the presence of a

$$(534) + Me_3SiSiMe_3 \xrightarrow{I_2\ (cat.)} (535) \qquad (149)$$

catalytic amount of iodine[333]. These conditions are applicable to a wide range of quinones and they provide hydroquinone disilyl ethers in better yields compared with the related reactions with chlorotrimethylsilane–alkali metal systems[334–337]. It is assumed that iodotrimethylsilane is formed in the course of this reaction. For the preparation of hydroquinone disilyl ethers from hydroquinones, conventional methods, such as the use of R_3SiCl-base or R_3SiOTf-base, are useful[331]. For example, after a hydrogenative reduction of a quinone in lutidine, the hydroquinone obtained is converted *in situ* to the corresponding silyl ether by addition of R_3SiOTf[338]. Even when a quinone is substituted by a double bond in the molecule, a reductive silylation can be conducted without hydrogenation of the double bond, presumably due to the much higher reduction rate of the quinone compared with that of the double bond in lutidine. This method was applied to the reductive silylation of **516** (equation 150), for which the reductive protection method is difficult due to its easy oxidizability.

H_2, Pd/C; R_3SiOTf / lutidine (150)

(516) (536)

In all these cases, the hydroquinone disilyl ether can be easily transferred either to the original hydroquinone by treatment with a protic acid or a fluoride anion, or to the corresponding quinone by an oxidative desilylation.

VII. SYNTHESIS OF NATURALLY OCCURRING QUINONES

A large number of quinones are found in pigments, bacterial metabolites, coenzymes and vitamins. A comprehensive coverage of the whole class of naturally occurring quinones is found in references 339 and 340, which treat their origin, distribution, structural determination and biological activities. Discussions in this review are concentrated on the synthesis of the title compounds which possess interesting biological activities. This field is considered especially as a *concours d'élégance* of recently developed new reactions and methodologies, where many 'state-of-art' syntheses were realized. Many applications of fundamental reactions which are discussed in Sections II–IV have also appeared.

Anthracyclines are interesting due to their broad range of antitumor activities, and many of them are used in the practice of cancer chemotherapy. A large number of synthetic works on anthracyclinones, the aglycones of anthracyclines, have appeared in recent years. Since most of reported methodologies applied to anthracyclinone synthesis are summarized in several reviews[341a–k], this topic is not included in this section in order to avoid overlap with these reviews.

A. Polyprenylated Quinones

Quinones belonging to the title compounds are mainly found in electron transport systems of animals, plants and microorganisms. The typical prenylated quinones are shown in Chart 3.

The most characteristic feature of these quinones, except for tocopherols, is that they have a polyprenyl side chain with an all-*trans* configuration. These isomers of phylloquinone (**540**) and menaquinone (**539**) show higher physiological activities than their *cis* isomers. The synthetic interest is mainly in two problems: first, the introduction of a polyprenyl group to a quinone (or a hydroquinone) with high *trans* stereoselectivity at the Δ^2 position of the side chain; second, in prevention of chroman ring formation, side chain cyclization and other unfavorable reactions. Classically, Friedel–Crafts type reaction by an acid-catalyzed alkylation of the corresponding polyprenyl alcohol (or halide) with a hydroquinone or its protected form has been performed (equation 151). A large number of variations has appeared, mostly in patents, and they can be found in the leading references[342–348]. Discussions in this review are concentrated on new methodologies[349]. Since the synthesis of tocopherols by means of Lewis acid-catalyzed prenylation of trimethylhydroquinone and concurrent formation of chroman ring has been well established, the asymmetric synthesis of **541** is discussed here.

The many efforts in this field are classified into the following four categories: (1) coupling reactions between quinones (or their protected analogues) and several

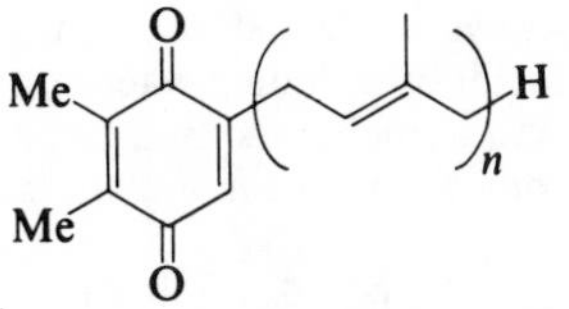

O Me Me O H *n*

(**537**) plastoquinone-*n*

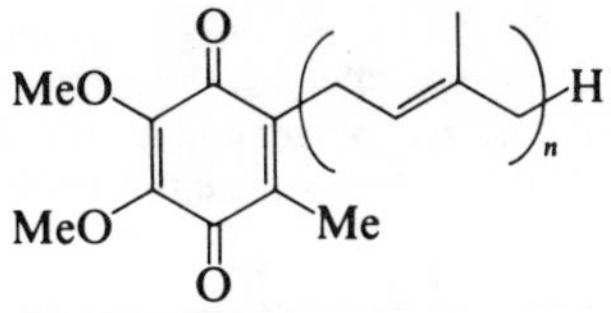

O MeO MeO O Me H *n*

(**538**) ubiquinone-*n* (coenzyme Q_n)

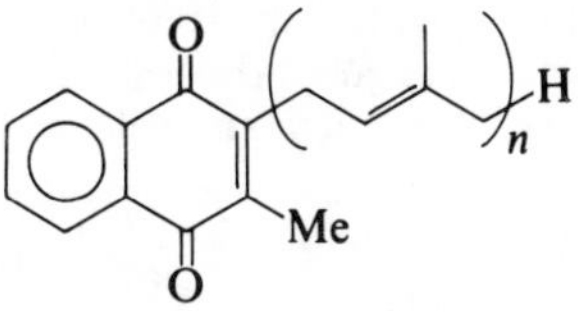

O O Me H *n*

(**539**) menaquinone-*n* (vitamin $K_{2(5n)}$)

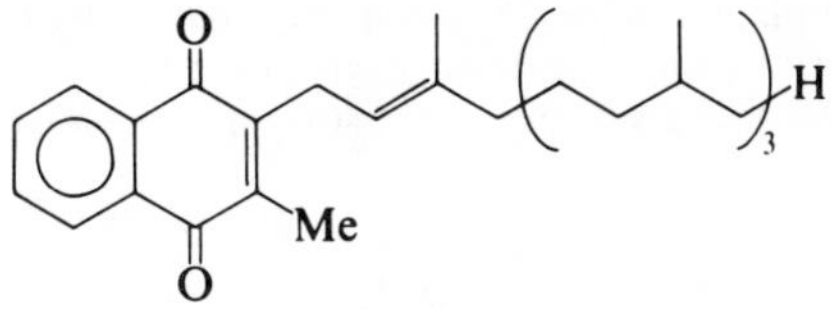

O O Me H 3

(**540**) phylloquinone (vitamin K_1)

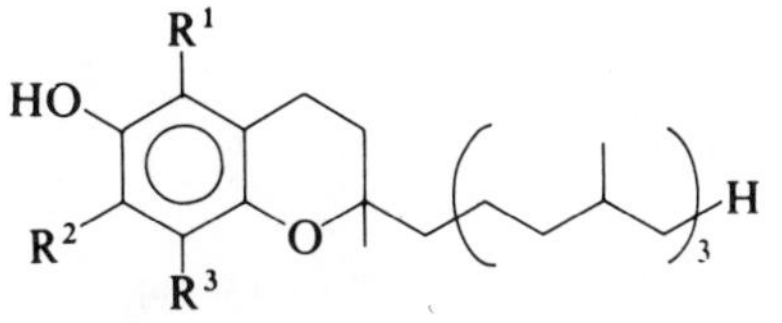

R^1 HO R^2 R^3 O H 3

(**541**) tocopherol (vitamin E)

α: $R^1 = R^2 = R^3 = Me$
β: $R^1 = R^3 = Me, R^2 = H$
γ: $R^1 = R^3 = H, R^2 = Me$
δ: $R^1 = R^2 = H, R^3 = Me$

CHART 3

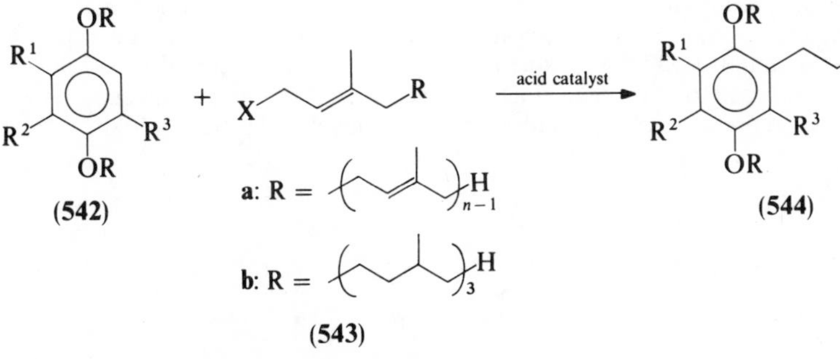

OR R^1 R^2 R^3 OR + X R $\xrightarrow{\text{acid catalyst}}$ OR R^1 R^2 R^3 OR R (151)

(**542**)

a: R = ( )$_{n-1}$H

b: R = ( )$_3$H

(**543**)

X = OH, halogen

(**544**)

organometallic reagents; (2) coupling reactions between prenyl halide and an arylmetal, which is the synthetic equivalent of the quinone; (3) other methods including extension of polyprenyl side chain, rearrangement of the prenyl group to quinone nucleus, etc.; and (4) construction of the quinone nucleus itself from a non-aromatic precursor with a prenyl side chain.

We will first discuss the coupling reaction of quinones of the corresponding hydroquinone diethyl ethers with allylic organometallics. A direct introduction of prenyl group to quinone seems to be the best method for maximizing the synthetic efficiency. The most successful method is a Lewis acid-catalyzed prenylation of a quinone with polyprenyltrialkylstannanes (**545a, b**)[350-354]. These substrates can be prepared by the coupling reaction of polyprenyl halide with R_3SnLi without loss of the stereoisomeric purity. Consequently, the reaction affords prenylated quinones in good to excellent yields with preservation of the original *trans* stereochemistry at Δ^2 (equation 152). The

R_3SnLi + X–(prenyl)–R (**543**) ⟶ R_3Sn–(prenyl)$_n$–H (**545a**) or R_3Sn–(prenyl)–(dihydroprenyl)$_3$–H (**545b**)

R = Me, *n*-Bu **a**: R = (prenyl)$_{n-1}$–H, n = 1–10 **b**: R = (dihydroprenyl)$_3$–H

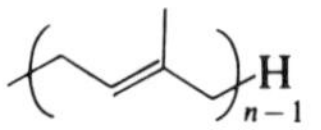

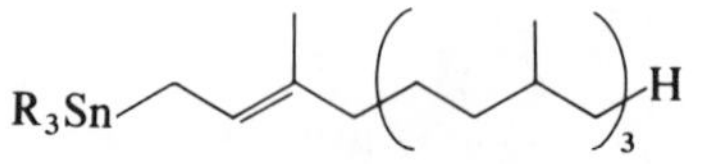

(**80**) MeO, MeO, Me-substituted quinone —(1) (**545a**), $BF_3 \cdot OEt_2$; (2) [O]→ (**538**) ubiquinone-*n* (coenzyme Q_n) (152)

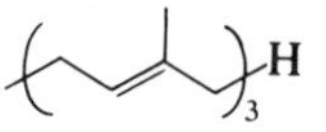

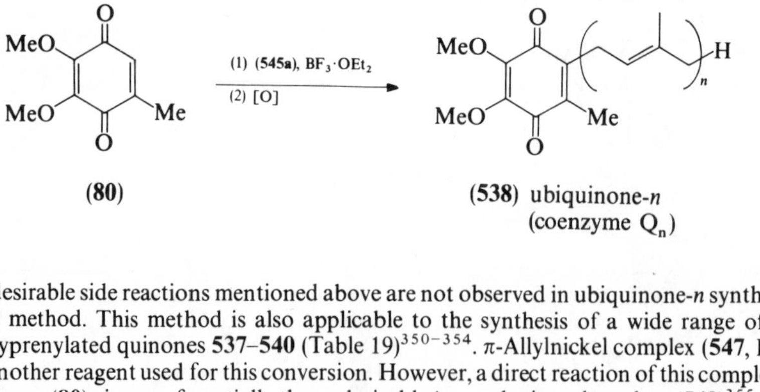

undesirable side reactions mentioned above are not observed in ubiquinone-*n* synthesis by this method. This method is also applicable to the synthesis of a wide range of other polyprenylated quinones **537**–**540** (Table 19)[350-354]. π-Allylnickel complex (**547**, R = H) is another reagent used for this conversion. However, a direct reaction of this complex with quinone (**80**) gives preferentially the undesirable *ipso*-substituted product (**548**)[355-358]. In

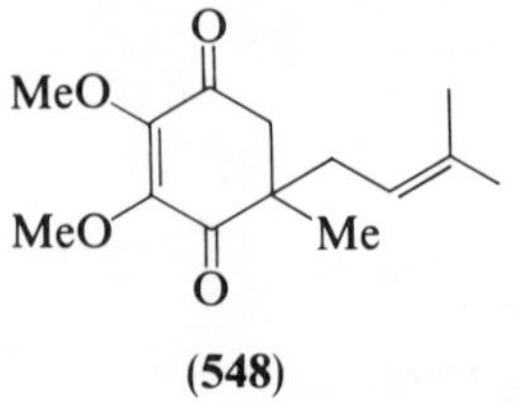

(**548**)

TABLE 19. Synthesis of polyprenylated quinones by the Lewis acid catalyzed allylation of quinones with polyprenylated stannanes[350, 351, 353]

Quinone	Polyprenylstannane (**545a**) R = Me, *n*	Product[a]	% yield	Stereochemistry at $\Delta^{2'}$, *trans*/*cis*
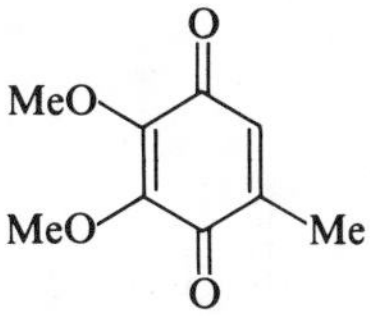	1	(**538**)	(75)	—
	2		90(65)	> 99/1
	2 ($\Delta^{2'}$ = *cis*)		79(70)	12/88[b]
	3		100(93)	99/1
	4		88(70)	99/1
	5		100(94)	99/1
	6		60(59)	98/2
	7		(40)	99/1
	8		(25)	98/2
	9		(51)	100/0
	10		(51)	86/14[c]
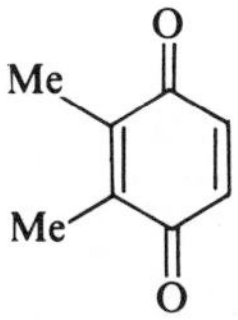	2	(**537**)	(23)	95/5
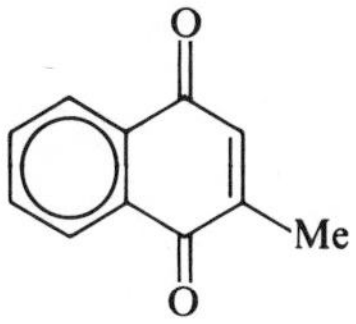	2	(**539**)	(48)	96/4
	2 (**545b**)	(**540**)	46(30)	95/5

[a] Yield in parentheses is of isolated material. All other yields were determined by ^{1}H-NMR.
[b] A 94:6 *cis*/*trans* mixture of nerylstannanes was used.
[c] The isomeric purity of the applied decaprenyl bromide is *trans*/*cis* = 82/18.

contrast, the coupling reaction of **547** with protected bromohydroquinone **546** gives the corresponding prenylated product **549** (equation 153), but the control of the stereochemistry at Δ^2 is difficult due to the inherent character of the complex **547**[359, 360].

The coupling reaction of an arylmetal derivative and polyprenyl halide was used successfully in several syntheses (equations 154 and 155). Both the hydroquinone monopotassium salt (**559**) and the arylmetal derivative (**552**) undergo a coupling reaction with polyprenyl halide (**543**, X = halogen) or polyprenyl phosphates to afford the desirable product with complete retention of the stereochemistry[361].

The arylstannyl derivative (**555**) causes coupling reaction with phytyl halide (**543b**, X = Br) in the presence of $ZnCl_2$ as a Lewis acid catalyst (equation 156)[362], but the

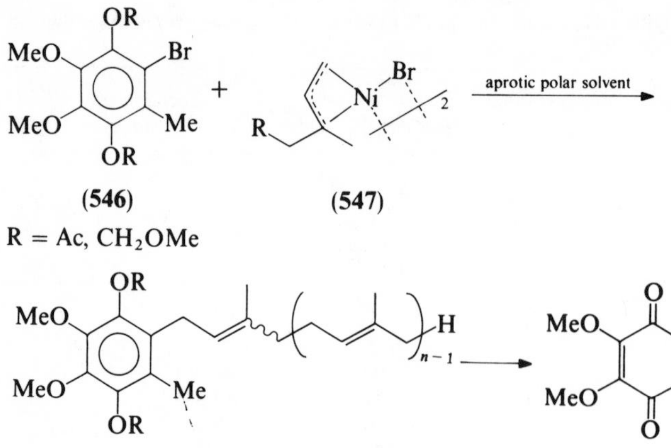

(546) + (547) —aprotic polar solvent→ (153)

$R = Ac, CH_2OMe$

(549) → (538)

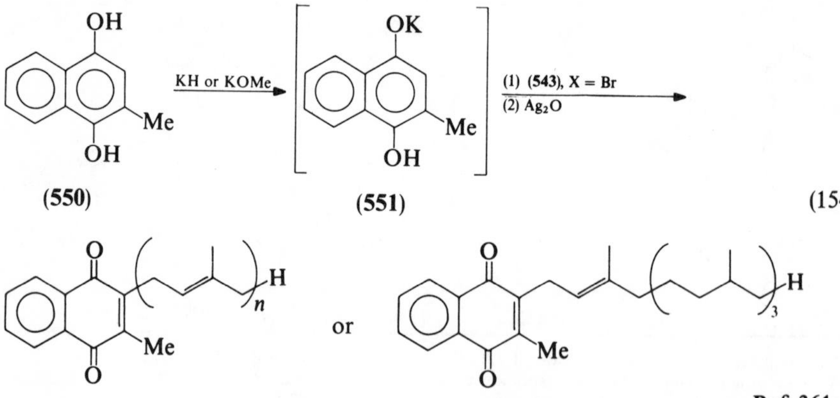

(550) —KH or KOMe→ [(551)] —(1) (543), X = Br; (2) Ag_2O→ (539) or (540) (154)

Ref. 361

stereoselectivity of the reaction is unclear. A quinone bisacetal, e.g. **566**, is considered to be a synthetic equivalent of a quinone with inverted charge affinity, and is easily converted to the corresponding cuprate (**557**) by successive treatment with BuLi/CuI. Nucleophilic substitution of **543b** (X = Br) by **557** gives phylloquinone (**540**) in good yield after deprotection of the bisacetal group (equation 157)[126]. This methodology was also applied to the synthesis of cymopol (**558**), a marine antibiotic with hydroquinone nucleus[363, 364].

A coupling reaction of a quinone or a hydroquinone with polyprenyl halides is conducted in the presence of Zn-amalgam[365] or metallic Zn[366]. The Zn-amalgam method gives ubiquinones and related polyprenylquinones in good *trans* stereoselectivity ($E/Z = 8/1–11/1$). Mechanistically, these reactions proceed by the formation of an

OMe, M, Me, OMe (552) M = Li, ½ CuLi, MgBr $\xrightarrow{(543),\ X=Br}$ OMe, R, Me, OMe (553) $\xrightarrow{AgO}$ (155)

O, H, n, Me, O (539) or O, H, 3, Me, O (540) Ref. 361

$OSiMe_2Bu$-*t*, Br, Me, $OSiMe_2Bu$-*t* (554) $\xrightarrow{(1)\ t\text{-BuLi};\ (2)\ Me_3SnLi}$ $OSiMe_2Bu$-*t*, $SnMe_3$, Me, $OSiMe_2Bu$-*t* (555) $\xrightarrow{10\%\ ZnCl_2,\ (543b),\ X = Br,\ 40\%}$ $\xrightarrow{PCC}$

O, H, 3, Me, O (540) Ref. 362 (156)

$(MeO)_2$, Br, Me, $(MeO)_2$ (556) $\xrightarrow{(1)\ BuLi;\ (2)\ CuI,\ Me_2S}$ [$(MeO)_2$, ½CuLi, Me, (MeO_2)] (557) $\xrightarrow{(1)\ (543b),\ X=Br;\ (2)\ H_3O^+}$

O, H, 3, Me, O (540) Ref. 124 (157)

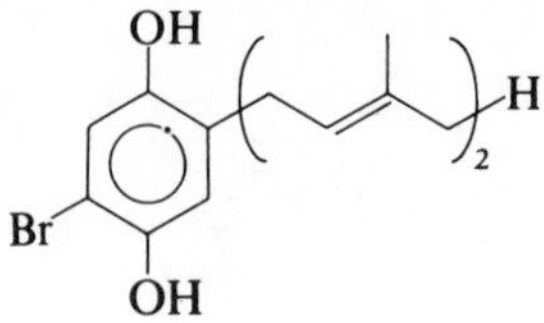

(558)

intermediate Zn-phenoxide derived from reduction of the quinone, followed by nucleophilic attack on the prenyl halide.

Several miscellaneous reactions belonging to the third category are shown below. A prenyl group can be selectively introduced to a trimethylsilyl cyanide protected quinone (**560**) by means of a Grignard reagent generated *in situ*. In the course of deprotection, the prenyl group rearranges in [3, 3] fashion to afford menaquinone-1 (**539**, $n = 1$) with concomitant elimination of MeOH (equation 158)[329b]. Since this reaction was only conducted with the simple prenyl group, the possibility of a stereoselective introduction of a polyprenyl group is unclear.

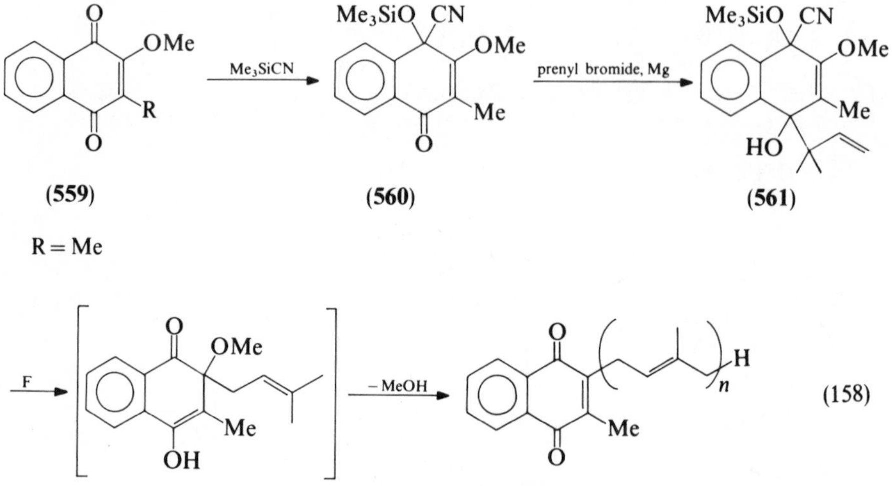

(559) R = Me → (Me_3SiCN) (560) → (prenyl bromide, Mg) (561)

→ (F^-) [(562)] → ($-$MeOH) (539) $n = 1$ (158)

A Lewis acid mediated Claisen-type [1, 3]-sigmatropic rearrangement of an aryl polyprenyl ether (**563**) is applied in the ubiquinone-*n* synthesis (equation 159)[367]. The rearrangement proceeds with complete retention of the stereochemistry for the *trans* isomer of **563**, but with the *cis* isomer, a slight contamination with the *trans* form is observed in this stage. This procedure was also applied for the synthesis of plastoquinone-*n*, and a similar high stereoselectivity was attained.

Many catalytic condensation reactions and their modification between hydroquinone and polyprenyl halide have been reported. Inclusion of 2-methylnaphthohydroquinone in a β-cyclodextrin cavity is applied for the selective coupling with prenyl bromide[368,369]. The reaction probably proceeds via nucleophilic attack of the monoanion of the

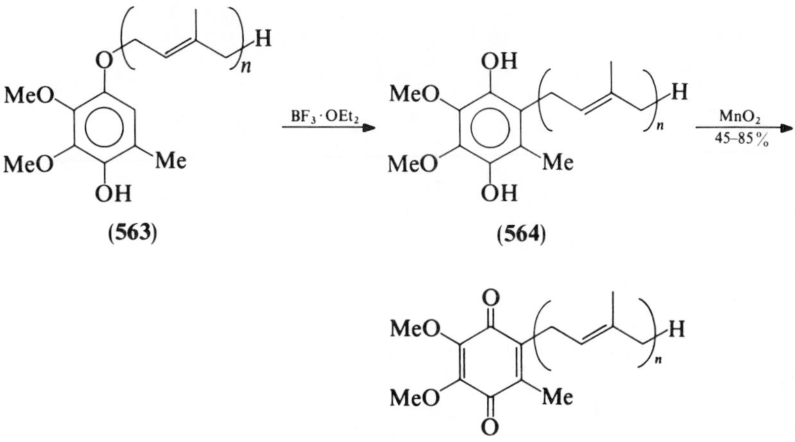

(563) $\xrightarrow{BF_3 \cdot OEt_2}$ (564) $\xrightarrow[45\text{–}85\%]{MnO_2}$ (538) (159)

hydroquinone on the bromide followed by aerial oxidation to prenylated quinone. β-Cyclodextrin plays a key role in controlling the regiochemistry of the nucleophilic attack and in prevention of successive side reactions of the prenylated product (equation 160).

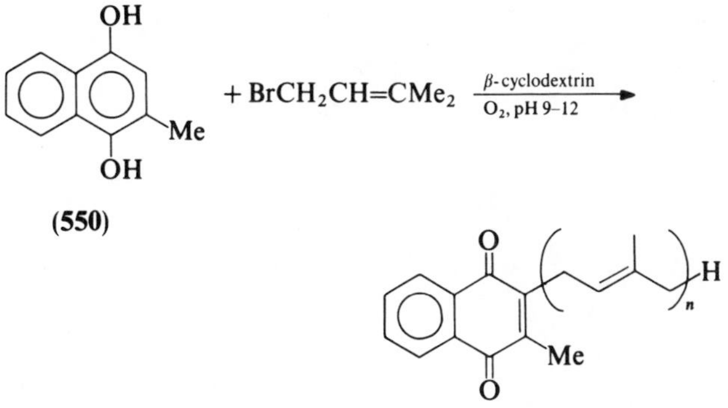

(550) + $BrCH_2CH{=}CMe_2$ $\xrightarrow[O_2,\ pH\ 9\text{–}12]{\beta\text{-cyclodextrin}}$ (539) $n = 1$ (160)

Quaternary ammonium and phosphonium salts are used as phase transfer catalysts in the synthesis of phylloquinone and menaquinones[370]. The phase transfer system as well as the cyclodextrin one plays a protection role of the lipophilic product from undesirable side reactions.

A chromium carbene complex is known to be an excellent reagent for construction of the prenylated quinone ring itself from these compounds (equation 161) (see Section II.C). The method is utilized for the synthesis of phylloquinone, menaquinones and tocopherol from alkynes substituted with the corresponding polyprenyl group[354,371,372]. Under these

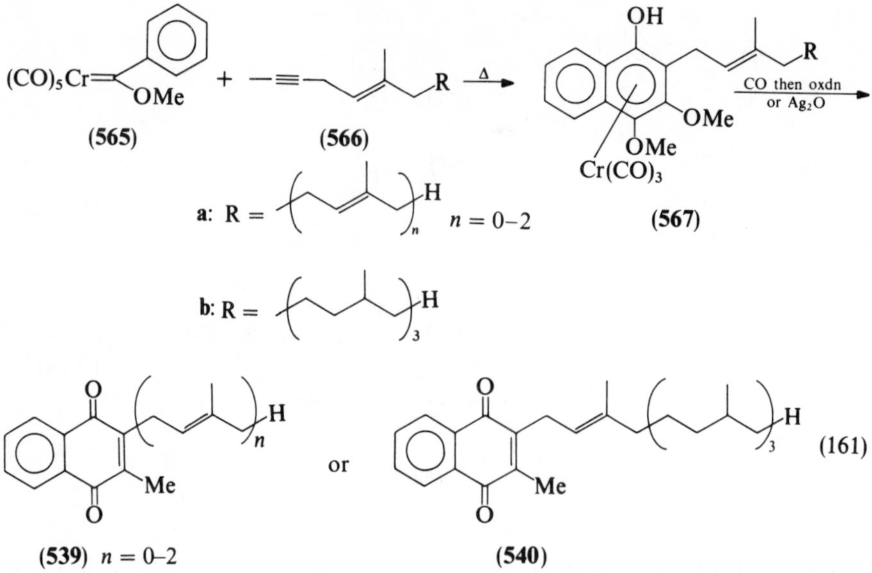

$(CO)_5Cr$ OMe (565) + (566) R $\xrightarrow{\Delta}$ OH, OMe, OMe, $Cr(CO)_3$ (567) $\xrightarrow[\text{or } Ag_2O]{\text{CO then oxdn}}$

a: R = ( )$_n$H $n = 0–2$

b: R = ( )$_3$H

O, O, Me, ( )$_n$H (539) $n = 0–2$ or O, O, Me, ( )$_3$H (540) (161)

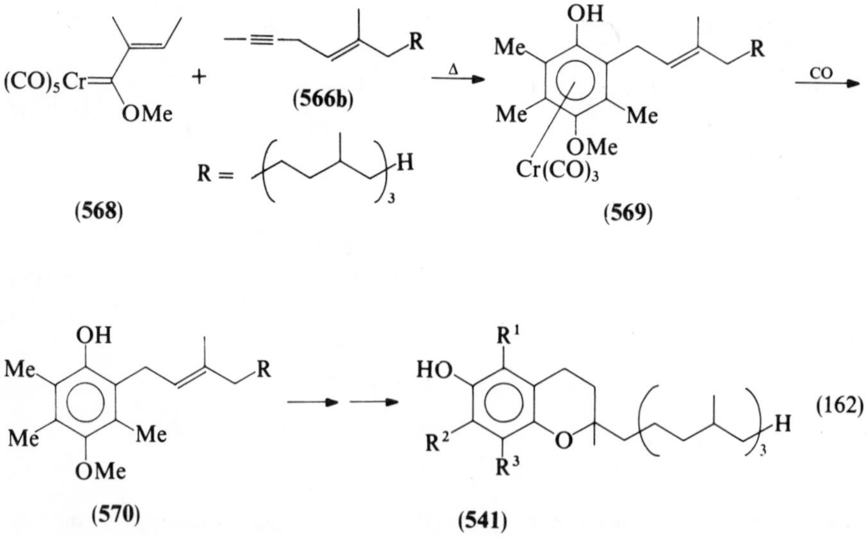

$(CO)_5Cr$ OMe (568) + (566b) R = ( )$_3$H $\xrightarrow{\Delta}$ OH, Me, Me, Me, OMe, $Cr(CO)_3$, R (569) $\xrightarrow{\text{CO}}$

OH, Me, Me, Me, OMe, R (570) → → HO, R^1, R^2, R^3, O, ( )$_3$H (541) (162)

conditions, the olefinic double bond is immune to isomerization and its *trans* stereochemistry at Δ^2 is completely preserved in the produced quinone.

Since some bacteria, e.g. *Candida utilis*, produce ubiquinone, e.g. **538** ($n = 7$), with a rather short polyprenyl chain, elongation of this unit is examined for the preparation of its higher homolog, which is more useful for therapeutic treatment (equations 163 and 164)[373–376].

OR, MeO, MeO, Me, OR, H, 7 — several steps →

$R = CH_2Ph$

(571) → (1) $LiN(Pr\text{-}i)_2$ (2) $SOCl_2$, Py → (572)

$l = 7$, $R = PhCH_2$, $X = Cl$

(1) $p\text{-}CH_3C_6H_4SO_2$–(isoprenyl)$_m$–H, n-BuLi, $m = 1\text{–}3$ (2) Li, $EtNH_2$ (3) O_2 → (538)

(573) (538) $n = l + m = 8\text{–}10$ Ref. 374 (163)

(572) + (574) → (1) Base (2) Li, $EtNH_2$[375] or Na, EtOH[376] (3) O_2 → (538) $n = 10$

{ $R = PhCH_2$, $l = 1$, $X = PhSO_2$[375] — $X = Br$

{ $R = MeOCH_2$, $l = 1$, $X = Br$[376] — $X = PhSO_2$ (164)

Synthesis of tocopherol itself is rather easy, since after the introduction of a phytyl group to trimethylhydroquinone in the presence of a strong (Lewis) acid it spontaneously cyclizes to the chromane ring. *d*-α-Tocopherol, with a 2*R*, 4′*R*, 8′*R* stereochemistry, has the highest physiological activity in the group. Consequently, many attempts for the enantioselective synthesis of *d*-α-tocopherol (**575**) have been made. The central problem in the synthesis is the control of the *R*-stereochemistry at the 2 position of **575**, since the absolute

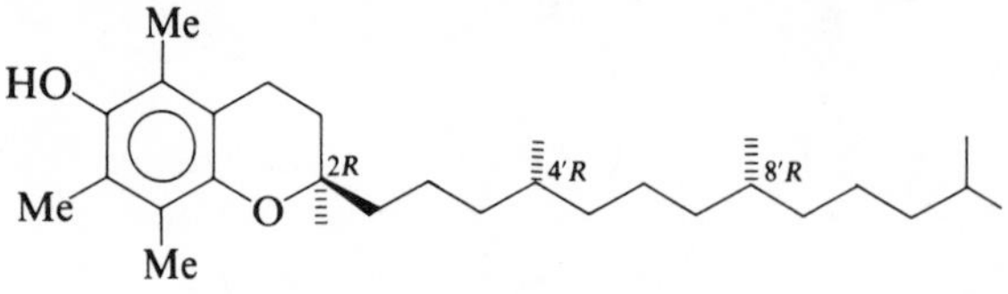

(**575**) 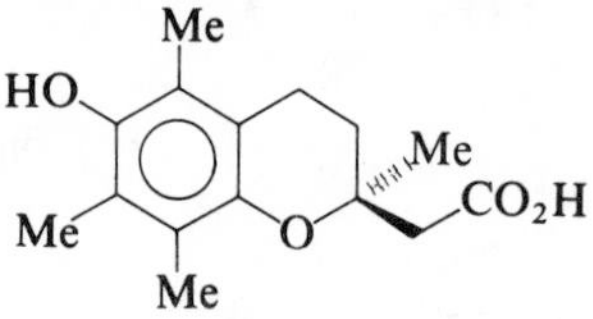 (**576**)

configuration at this position should affect biological activity. Most of the reported methods use (*S*)-2-chromanacetic acid (**576**) as a chiral chroman precursor which is obtained by optical resolution[377]. Two other methods toward synthesis of the chiral alcohol (**582**) related to the chiral acid (**576**) have been reported so far. In the first one, an optically active C_4 synthon (**578**) is coupled with Wittig reagent (**577**) and successive hydrogenation affords **580**. After several steps, (*S*)-chroman ethanol (**582**) is obtained exclusively (Scheme 25)[378]. An almost similar approach uses a different chiral synthon, i.e. (*S*)-(−)-2-methyl-5-oxotetrahydro-2-furoic acid (**583**), which can be derived as a racemic lactone acid from levulinic acid and successively resolved to its enantiomers[379]. The (*S*)-(−)-lactone acid (**583**) can be transferred to **586** in several steps, and annulation of **586** with dimethyl acetonedicarboxylate under basic conditions affords the phenol (**588**), which is

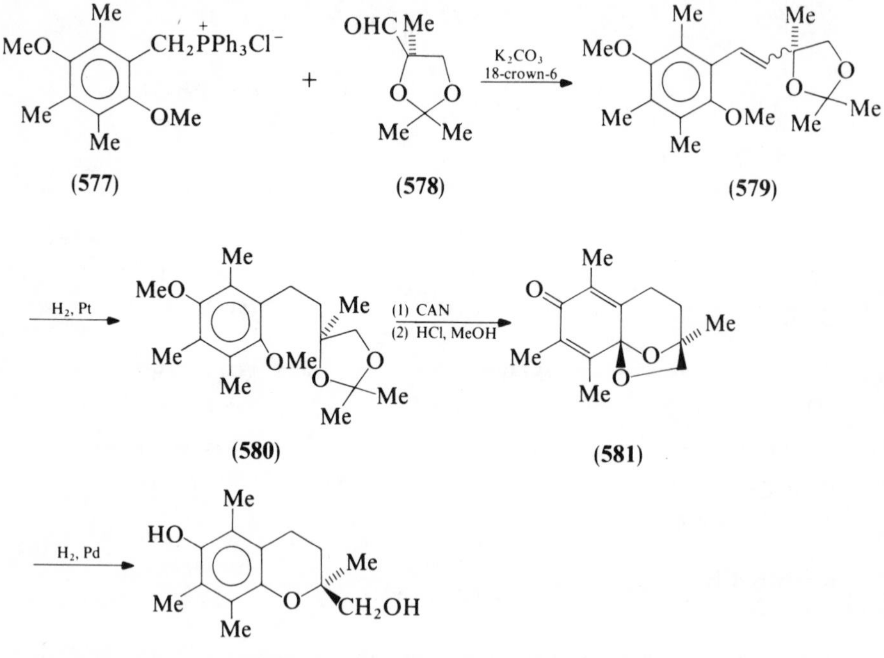

(**577**) (**578**) (**579**)

(**580**) (**581**)

(**582**)

SCHEME 25

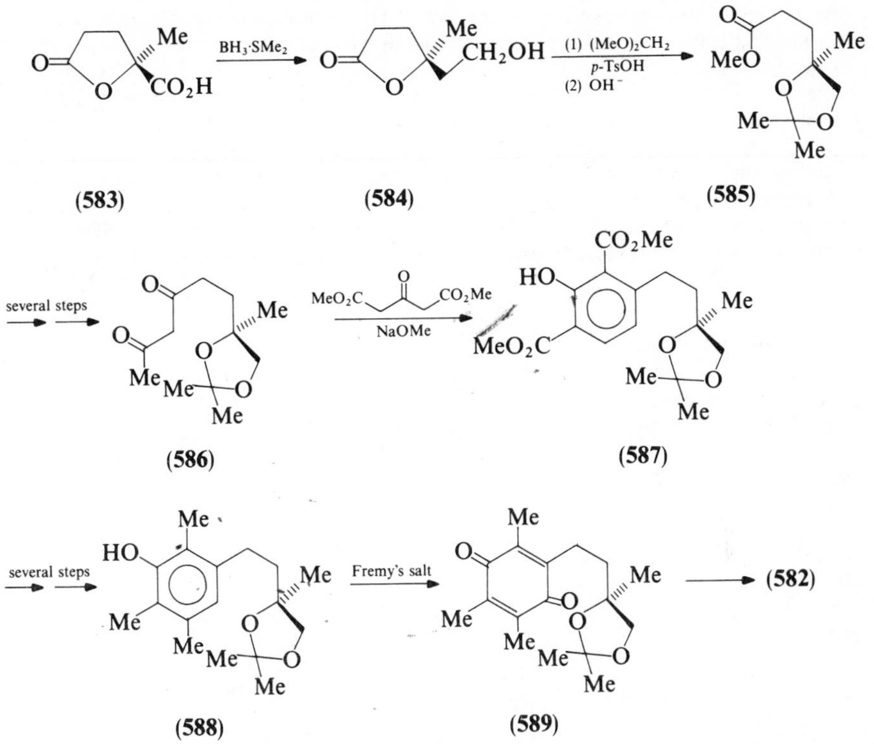

$BH_3 \cdot SMe_2$ (1) $(MeO)_2CH_2$, *p*-TsOH (2) OH^-

(583) (584) (585)

several steps

MeO_2C–CO–CO_2Me, NaOMe

(586) (587)

several steps

Fremy's salt → (582)

(588) (589)

SCHEME 26

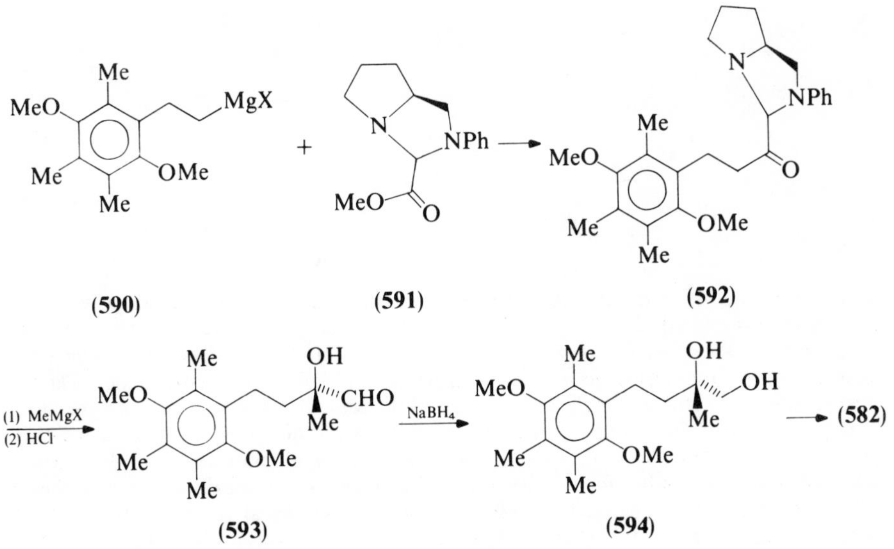

(590) + (591) → (592)

(1) MeMgX (2) HCl

NaBH$_4$ → (582)

(593) (594)

SCHEME 27

converted to **582** in a similar way (Scheme 26)[379]. This method is completely different from other methods, since it involves construction of the phenol ring from non-aromatic chiral starting material.

The second method involves an enantioselective methylation of the keto aminal (**592**) as the key step of asymmetric induction leading to the chiral diol (**594**) in > 95% optical purity (Scheme 27)[380]. The obtained chiral diol (**594**) is converted to (**582**) in a similar way.

The second major problem is to extend the side chain of the optically active chromanacetic acid **576**. The most straightforward approach is a coupling between the requisite optically active side chain and **595**, which is derived from either **576** or **582** (equation 165)[377]. The optically active C_{14} unit of **596** is derived from 7*R*,11*R*-phytol by a series of degradative reactions.

(**595**) + $Br^{-}Ph_3P^{+}$ (**596**) ⟶ (**597**) $\xrightarrow[(2)\ NaAlH_2(OR)_2]{(1)\ H_2,\ PtO_2}$ (**575**) (2*R*, 4′*R*, 8′*R*) (165)

The sequential synthesis of optically active side chain is examined in several ways. After the diastereoisomers of the propynyl alcohol (**598**), derived from the key intermediate (**576**) in a few steps, are separated as the corresponding 3,5-dinitrobenzoates **599** and **600**, each diastereomer is subjected to partial hydrogenation, followed by orthoester Claisen rearrangement to afford a single diastereomer of the acid ester **603** with 4′*R* configuration. Very interestingly, the chirality transfer from both **601** and **602** proceeds stereospecifically with almost 99% efficiency, allowing the synthesis of optically pure **603** from the mixture of diastereoisomers **601** and **602** (R = H) (Scheme 28). The optically active tocopherol (**575**) can be easily prepared from **603**[381]. The stereochemical course of this Claisen rearrangement is well established in the study of model systems and this method is also utilized for the synthesis of the optically active side chain fragment[382,383].

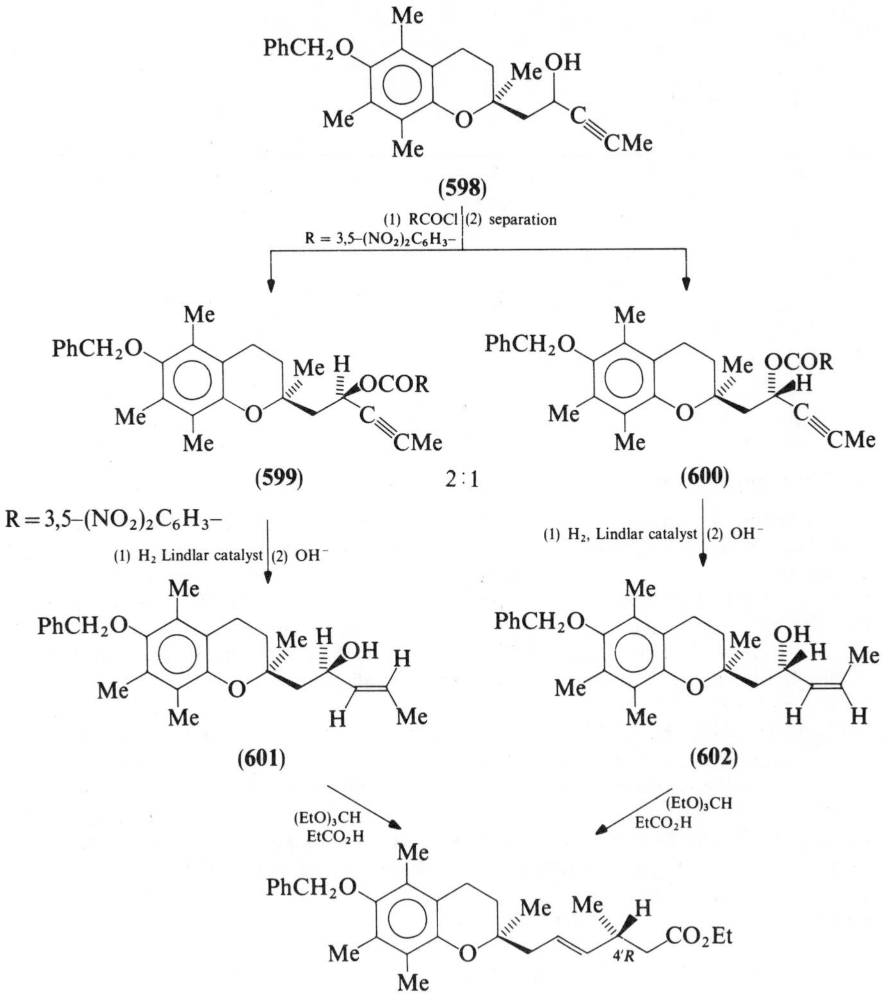

(598)

(1) RCOCl (2) separation
R = 3,5-$(NO_2)_2C_6H_3$–

(599) 2:1 (600)

R = 3,5-$(NO_2)_2C_6H_3$–

(1) H_2 Lindlar catalyst (2) OH^-

(1) H_2, Lindlar catalyst (2) OH^-

(601) (602)

$(EtO)_3CH$
$EtCO_2H$

(603)

SCHEME 28

As another chiral source, (−)-(*S*)-3-methyl-*γ*-butyrolactone (**604**) is used for the synthesis of the optically active side chain. The chiral aldehyde (**605**) derived from **604** is repeatedly used as a chiral fragment, and elongation of the chain by means of Wittig reaction furnishes C_{15} chloride (**607**). Another Wittig reaction with **595** and successive hydrogenation affords (−)-(*S*)-tocopherol acetate (Scheme 29)[384]. Variations of the synthesis of the optically active 2-chromanacetic acid (**576**)[385] and of the coupling of a side chain with **576**[386] were also reported.

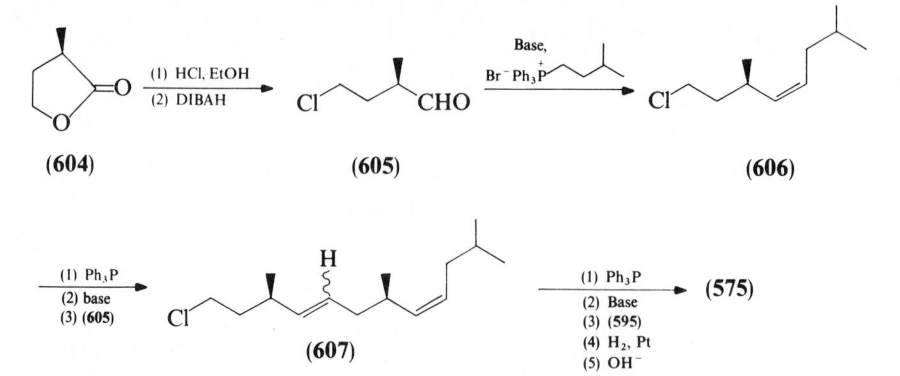
(1) HCl, EtOH
(2) DIBAH
Cl
CHO
Base,
$Br^- Ph_3\overset{+}{P}$
Cl
(604)
(605)
(606)
(1) Ph_3P
(2) base
(3) (605)
H
Cl
(607)
(1) Ph_3P
(2) Base
(3) (595)
(4) H_2, Pt
(5) OH^-
(575)

SCHEME 29

B. Pyranoquinones

Pyranoquinones are found in plants and in microbial metabolites. Since most of them have broad biological activities, many synthetic efforts have been made toward their preparation. The following compounds attracted special attention due to their strong biological activities; eleutherin (**608**)[387], isoeleutherin (**609**)[387], nanaomycin A (**610**)[388–390], nanaomycin D (**611**)[391], deoxyfrenolicin (**612**)[394], frenolicin (**613**)[395], kalafungin (**614**)[392], granaticin (**615**)[393,396], griceusin A (**616**)[397] and griceusin B (**617**)[397] (Chart 4). Many other quinones classified to this family have been found and they are listed in reference 340.

Eleutherin (**608**) and isoeleutherin (**609**) are the compounds with the simplest structure in this family. Since the first synthesis[398], allylation followed by benzopyran ring formation are the key steps in the synthesis. The acylquinone (**618**) is a good precursor for these quinones because the acyl group first activates the vicinal position toward nucleophilic Michael addition, and then it is utilized as an oxygen source for the pyran ring formation[399]. Allylation of the acylquinone **618** with allylstannane (**619**) in the presence of

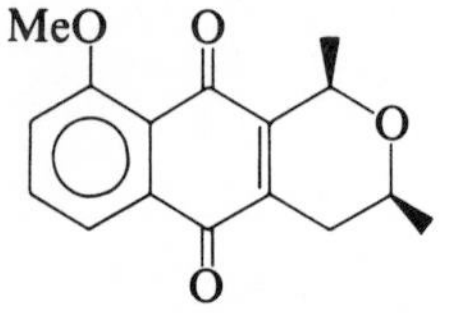
MeO
O
O
O

(608)

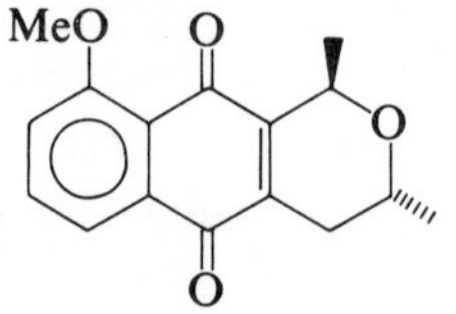
MeO
O
O
O

(609)

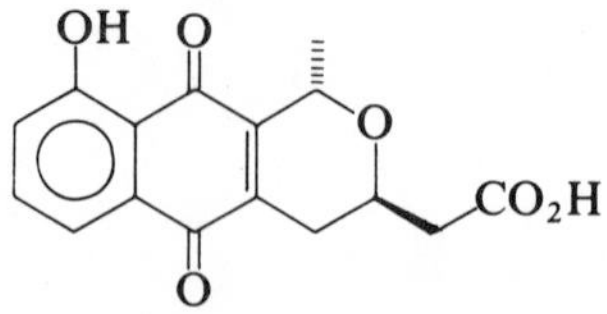
OH
O
O
CO_2H
O

(610)

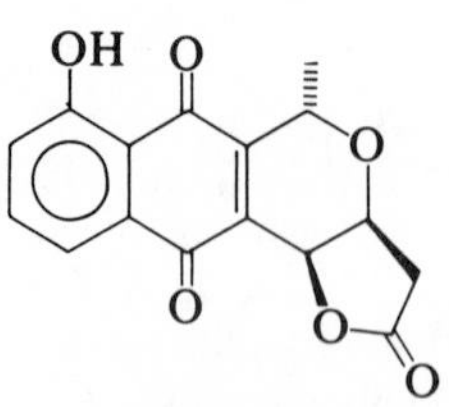
OH
O
O
O
O
O

(611)

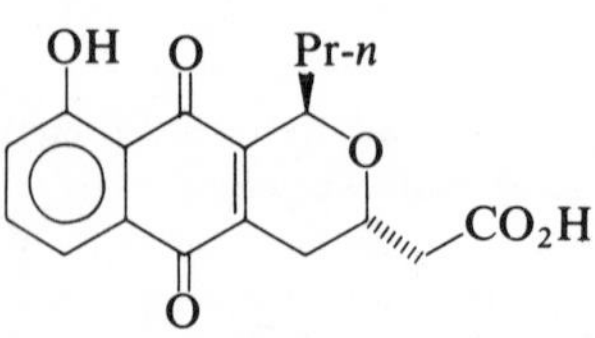
OH
O
Pr-*n*
O
CO_2H
O

(612)

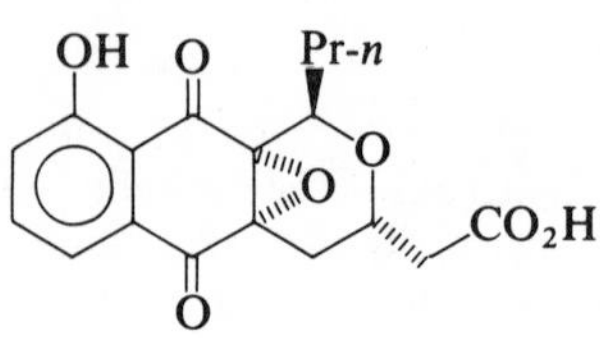
OH
O
Pr-*n*
O
O
CO_2H
O

(613)

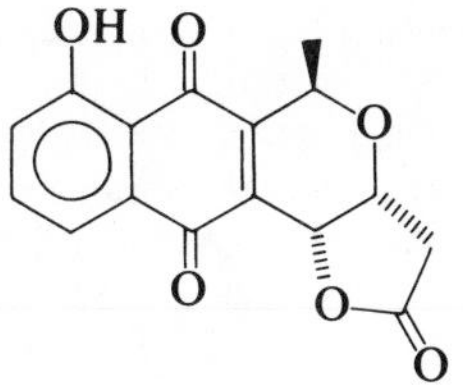

(614) 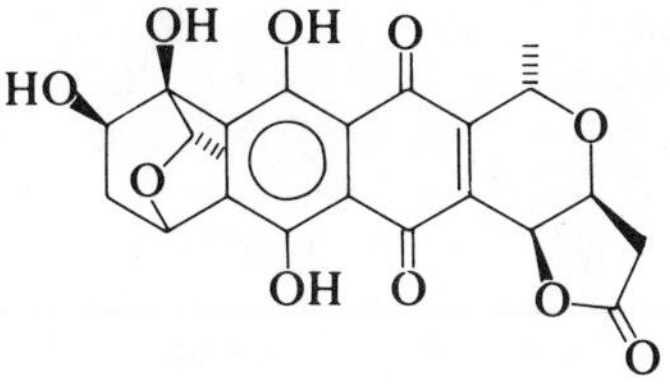 (615)

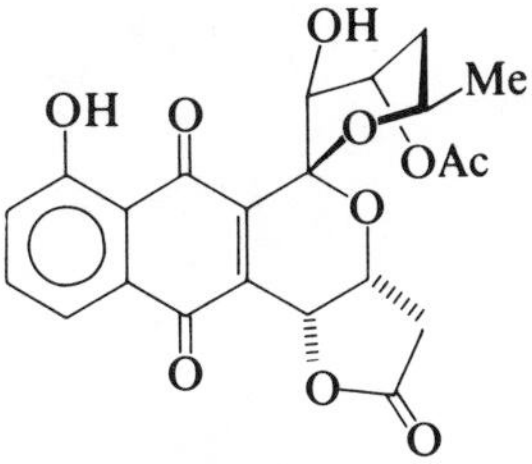

(616) 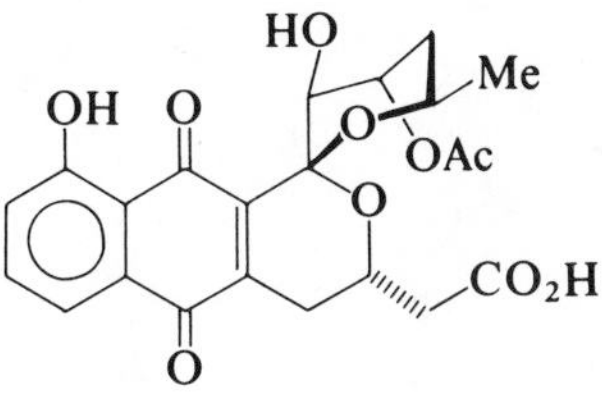 (617)

CHART 4

a Lewis acid affords allylated product (**620**) in 92% yield after etherification of the obtained hydroquinone. The successive reduction and pyran ring formation give racemic eleutherin (**608**) and isoeleutherin (**609**) (Scheme 30)[399]. According to an almost similar strategy to use **621** as the key compound, racemic **608** and **609** were prepared[110,400–402]. After treatment of **620** with potassium *t*-butoxide in DMF, the *trans* isomer of **622**, which can be oxidized to (±)-isoeleutherin (**609**), is obtained in high stereoselectivity[110].

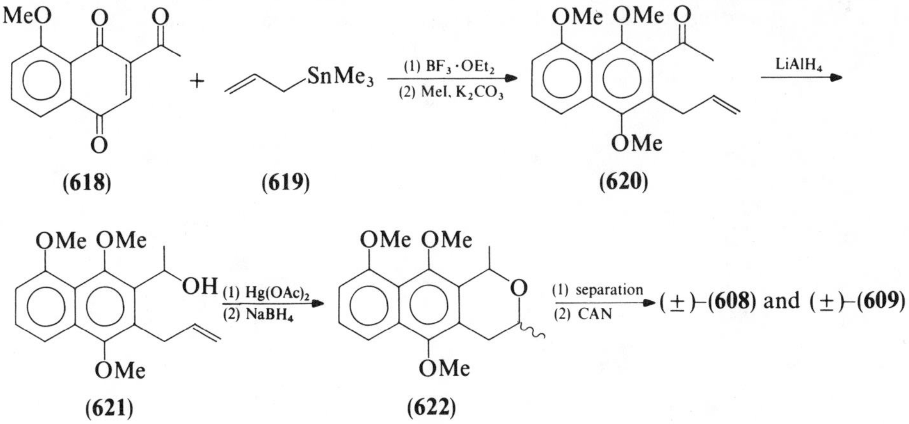

(618) + (619) —(1) $BF_3 \cdot OEt_2$ (2) MeI, K_2CO_3→ (620) —$LiAlH_4$→

(621) —(1) $Hg(OAc)_2$ (2) $NaBH_4$→ (622) —(1) separation (2) CAN→ (±)-(**608**) and (±)-(**609**)

isomeric ratio = ca. 1:1

SCHEME 30

Nanaomycin A (**610**) and deoxyfrenolicin (**612**) were prepared through versatile synthetic routes due to the interest in their strong and characteristic biological activity against mycoplasms, fungi and Gram-positive bacteria. The demethoxy analogs of (±)-nanaomycin A[403] and (±)kalafungin[404] were synthesized. In the former synthesis[403], the pathway similar to the previously described eleutherin synthesis[398] is utilized for the construction of the pyran ring, i.e. condensation of acetaldehyde with the hydroquinone (**623**, R = H) in the presence of HCl generates only the *cis* isomer (**624**, R = H) (equation 166). In spite of its thermodynamic instability compared with the *trans* isomer[404] this

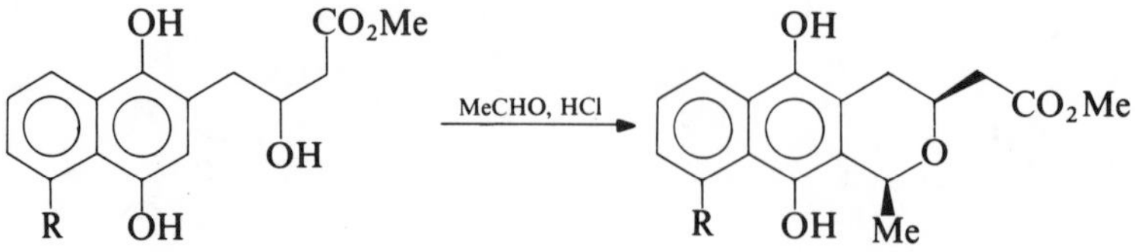

MeCHO, HCl (166)

(**623**) (**624**)

method is extended to the synthesis of (±)-nanaomycin A (**610**)[405,406] and (±)-kalafungin (**614**)[405], using **623** (R = OMe,[404] OH[406]) as the key intermediate. Both approaches to **623** or its oxidized form are illustrated in Schemes 31[405] and 32[406].

Another interesting approach to 9-deoxykalafungin includes Michael addition of the furan (**626**) (as the butenolide anion equivalent) to the acylquinone (**625**) (Scheme 33)[404]. The reduction of the acyl group in **627** followed by cyclization affords the pyranolactone (**628**) as the stereoisomeric mixture. A stepwise synthetic approach to **610** and **612** uses an intramolecular hemiacetal formation and a successive Wittig reaction as a key step (Scheme 34)[407]. The Wittig reaction generates two diastereoisomers of **630**, but at the final deprotection stage the acid treatment completely converts the isomeric mixture to the *trans* form. Attempted epoxidation of (±)-deoxyfrenolicin methyl ester (**612**) with *t*-BuOOH gives (±)-frenolicin (**613**) and (±)-epifrenolicin (**629**) in 1:1 ratio. This is the only example of the synthesis of (±)-frenolicin[407]. In an alternative approach to formation of the pyran ring, an intramolecular Michael addition of **631** is considered to be an efficient pathway, since the α,β-unsaturated acid ester side chain can play two roles as a Michael acceptor and

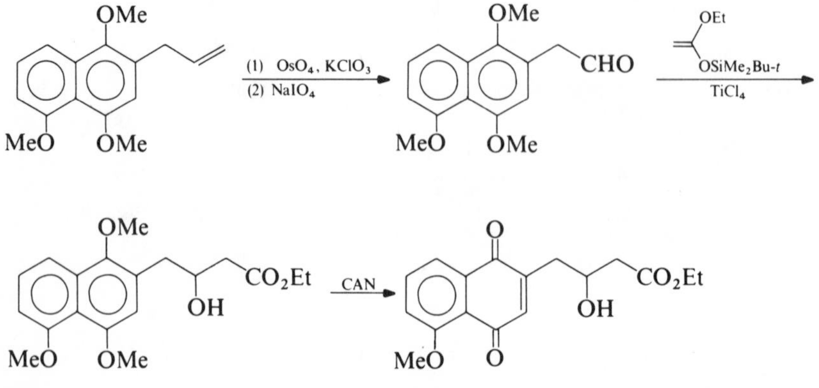

(1) OsO_4, $KClO_3$ (2) $NaIO_4$

$CH_2{=}C(OEt)OSiMe_2Bu\text{-}t$, $TiCl_4$

CAN

SCHEME 31

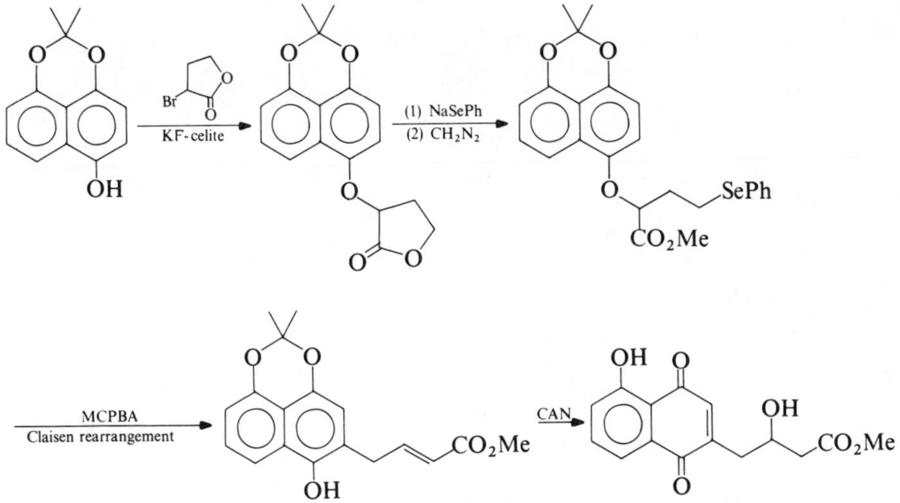

SCHEME 32

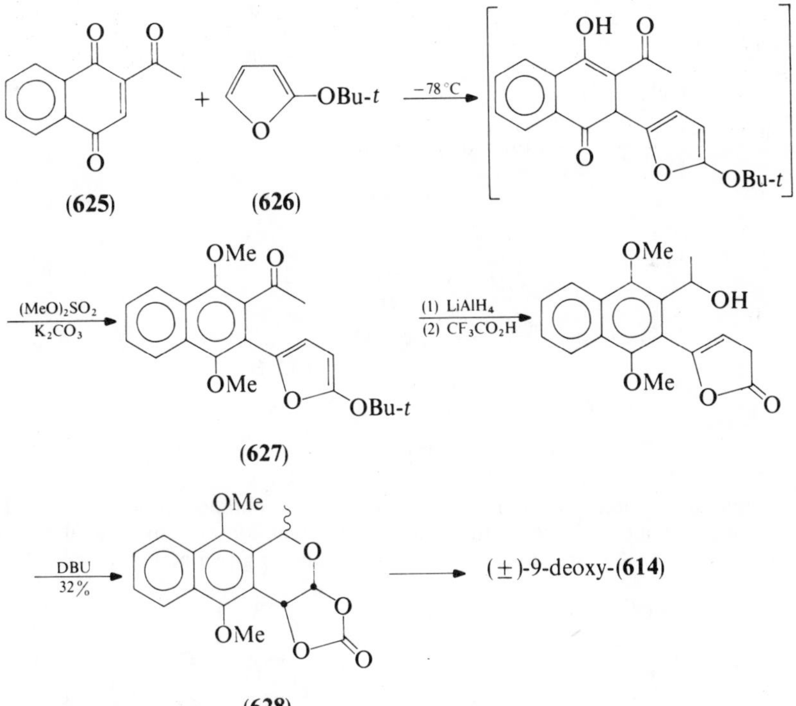

(625) (626)

(627)

(628)

isometric ratio = 2.7:1

(±)-9-deoxy-(614)

SCHEME 33

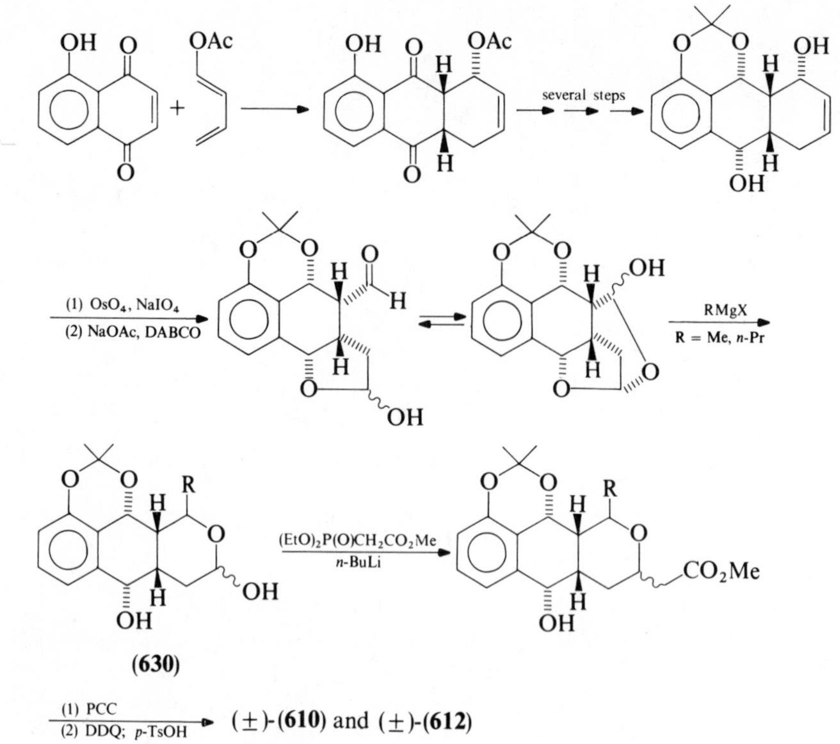

several steps

(1) OsO_4, $NaIO_4$
(2) NaOAc, DABCO

RMgX
R = Me, *n*-Pr

$(EtO)_2P(O)CH_2CO_2Me$
n-BuLi

(630)

(1) PCC
(2) DDQ; *p*-TsOH

(±)-(**610**) and (±)-(**612**)

SCHEME 34

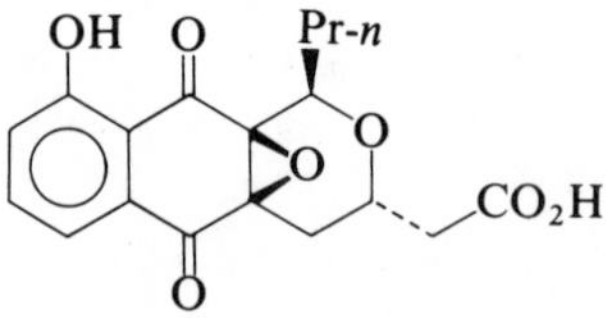

(629)

as an essential functionality for ring-closure to pyranonaphthoquinones (equation 167). The key intermediate (**631**) is synthesized by a combination of conventional reactions (Scheme 35)[400,401]. A more direct approach was found to give a higher overall yield. As

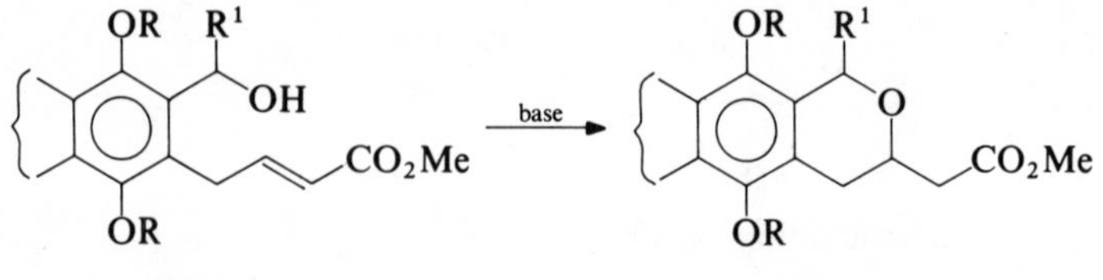

base (167)

(631) **(632)**

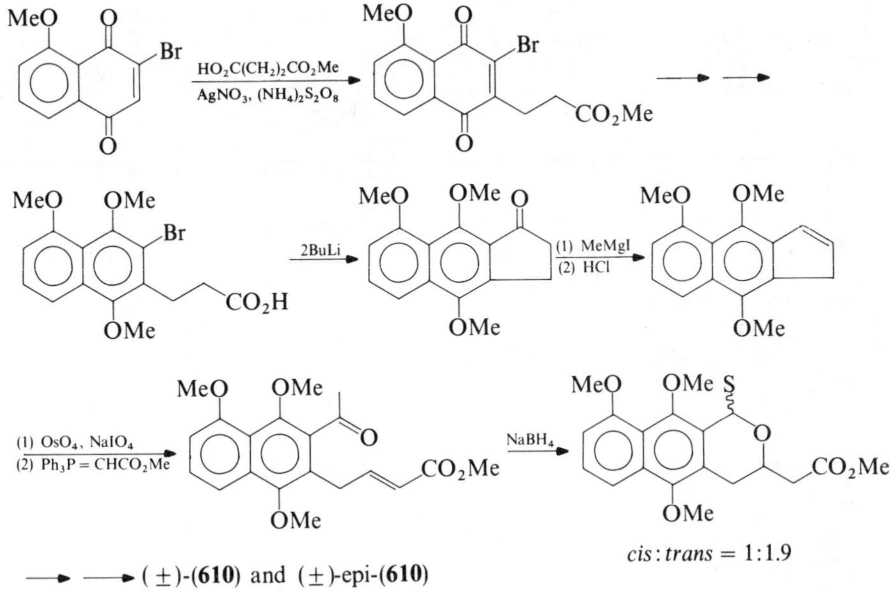

MeO O Br $\xrightarrow[AgNO_3,\ (NH_4)_2S_2O_8]{HO_2C(CH_2)_2CO_2Me}$ MeO O Br CO_2Me O → →

MeO OMe Br CO_2H OMe $\xrightarrow{2BuLi}$ MeO OMe O OMe $\xrightarrow[(2)\ HCl]{(1)\ MeMgI}$ MeO OMe OMe

$\xrightarrow[(2)\ Ph_3P=CHCO_2Me]{(1)\ OsO_4,\ NaIO_4}$ MeO OMe O CO_2Me OMe $\xrightarrow{NaBH_4}$ MeO OMe O CO_2Me OMe

cis:*trans* = 1:1.9

→ → (±)-(**610**) and (±)-epi-(**610**)

SCHEME 35

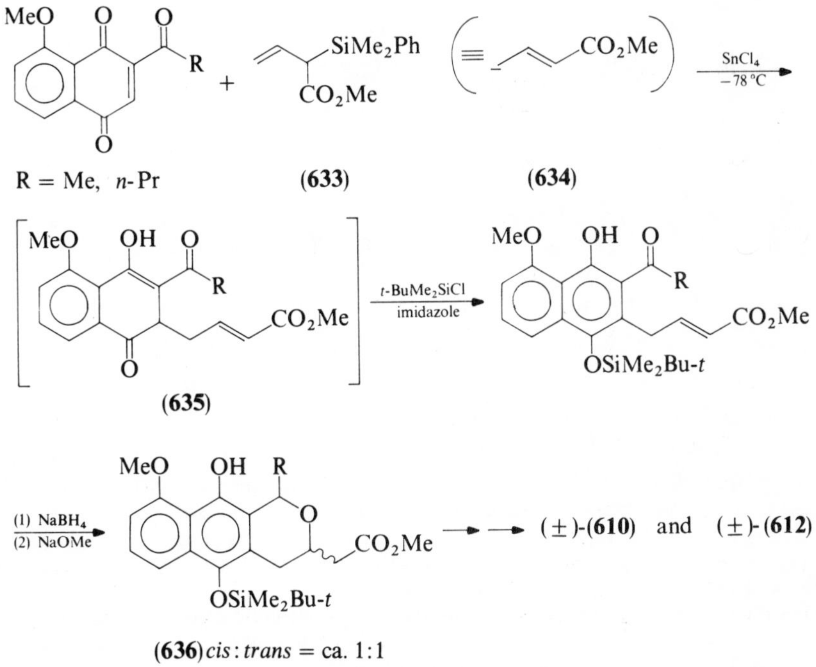

MeO O O R O + SiMe$_2$Ph CO_2Me (≡ CO_2Me) $\xrightarrow[-78\ ^\circ C]{SnCl_4}$

R = Me, *n*-Pr (**633**) (**634**)

[MeO OH O R CO_2Me O] (**635**) $\xrightarrow[imidazole]{t\text{-}BuMe_2SiCl}$ MeO OH O R CO_2Me OSiMe$_2$Bu-*t*

$\xrightarrow[(2)\ NaOMe]{(1)\ NaBH_4}$ MeO OH R O CO_2Me OSiMe$_2$Bu-*t* → → (±)-(**610**) and (±)-(**612**)

(**636**) *cis*:*trans* = ca. 1:1

SCHEME 36

shown in the eleutherin synthesis (Scheme 30), an acylquinone is an extremely good Michael acceptor, and its Lewis acid mediated reaction with methyl 1-dimethylphenylsilyl-2-butenoate (**633**) (a synthetic equivalent of methyl 2-butenoate anion, **634**) successfully affords a desired intermediate (**635**) in a quantitative yield. After protection of the hydroxy group in order to avoid reaction with the α,β-unsaturated ester moiety, successive reduction and cyclization furnish an equimolar mixture of two diastereoisomers (**636**). By this methodology, (±)-(**610**) and (±)-(**611**) are obtained in excellent overall yield (Scheme 36)[408–410].

The key intermediate (**631**) mentioned above is prepared in two different ways. One method for the required functionalization involves a Michael type addition of the lithium acylnickel carbonyl complex (**637**) to juglone monoacetal (**58**) and successive trapping of the generated enolate anion with allyl iodide. After reduction of the alkanoyl group and then oxidation, the key intermediate quinone (**638**), which is obtained more directly and conventionally by allylation of acylquinones with allylstannane[399,414] is prepared. Further reaction gives **639** (Scheme 37)[411].

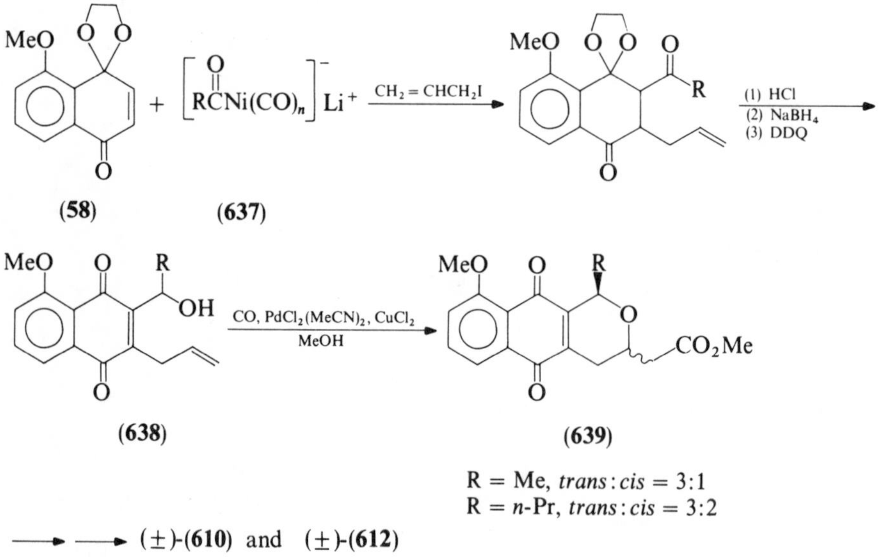

$$(58) + \left[\mathrm{RCNi(CO)}_n\right]^- \mathrm{Li}^+ \ (637) \xrightarrow{CH_2=CHCH_2I} \xrightarrow{(1)\ HCl;\ (2)\ NaBH_4;\ (3)\ DDQ} (638) \xrightarrow[MeOH]{CO,\ PdCl_2(MeCN)_2,\ CuCl_2} (639)$$

R = Me, *trans*:*cis* = 3:1
R = *n*-Pr, *trans*:*cis* = 3:2

⟶ ⟶ (±)-(**610**) and (±)-(**612**)

SCHEME 37

Annulation reaction of the chromium carbene complex (**640**) with an appropriate alkyl chain is applied to the synthesis of the similar key intermediate (Scheme 38)[412,413]. This synthetic strategy is based on a non-quinonoid compound as a precursor. From the standpoint of the total synthetic efficiency, there is no advantage in this procedure compared with other methods, but it demonstrates a regioselective intramolecular annulation of the chromium carbene complex.

The regiocontrolled intramolecular annulation of a phthaloylcobalt complex applies the same synthetic strategy (Scheme 39)[254]. The covalently bound precursor (**643**) is necessary for control of the regiochemistry in the annulation, as was the case with the chromium carbene complex.

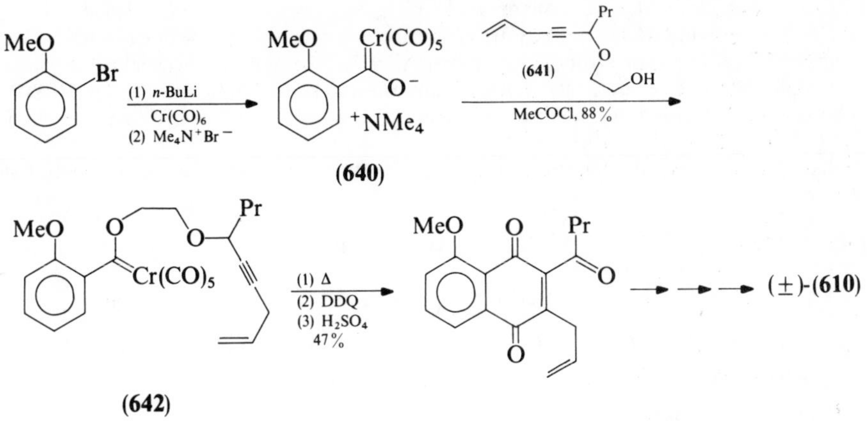

(1) n-BuLi, $Cr(CO)_6$; (2) $Me_4N^+Br^-$

(640)

(641)

MeCOCl, 88%

(1) Δ; (2) DDQ; (3) H_2SO_4; 47%

→ → → (±)-**(610)**

(642)

SCHEME 38

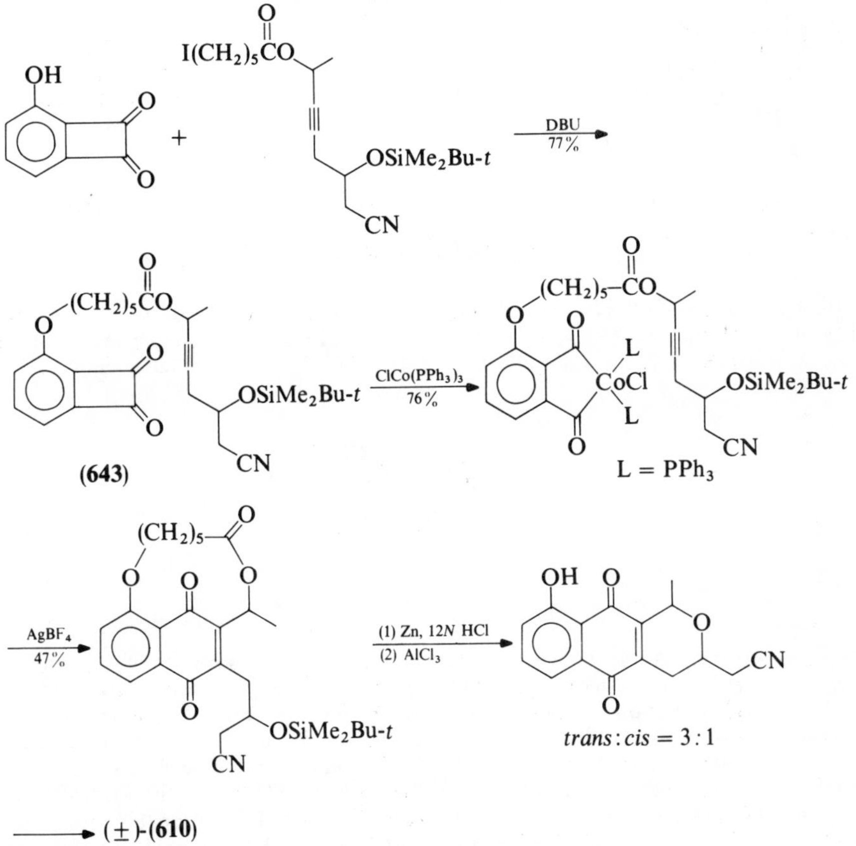

DBU, 77%

(643)

$ClCo(PPh_3)_3$, 76%

$L = PPh_3$

$AgBF_4$, 47%

(1) Zn, 12*N* HCl; (2) $AlCl_3$

trans:*cis* = 3:1

→ (±)-**(610)**

SCHEME 39

Enantiospecific synthesis of nanaomycins and related compounds is performed by using a chiral compound (**645**) derived from a carbohydrate source (Scheme 40)[415]. The requisite naphthopyran structure is constructed by the tandem cyclization between phthalide (**644**) and the optically active α,β-unsaturated ketone (**645**) (see Section III.D.2). The intermediates **646** and **647** are converted to enantiomerically pure nanaomycin D (**611**) and kalafungin (**614**), respectively. Consequently, one asymmetric center is transferred from **645** and the successive Wittig reaction gives two isomeric products, **646** and **647**.

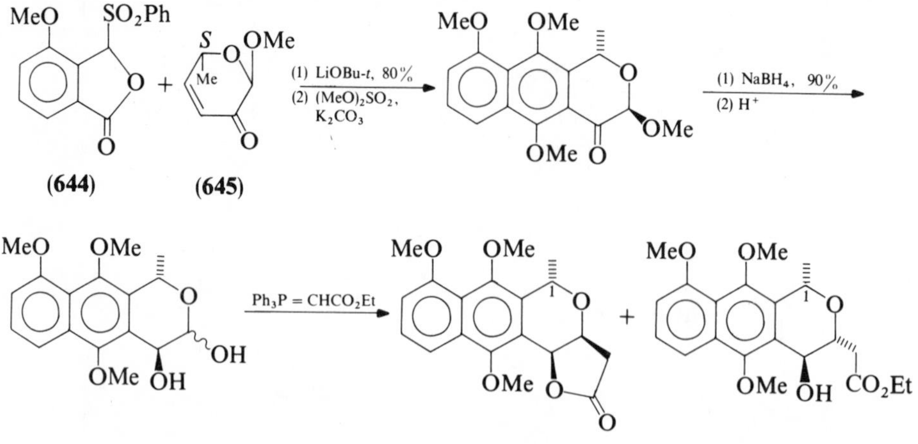

MeO SO2Ph (**644**) + S OMe Me O (**645**) (1) LiOBu-*t*, 80% (2) $(MeO)_2SO_2$, K_2CO_3 → (1) $NaBH_4$, 90% (2) H^+ → $Ph_3P=CHCO_2Et$ → (**646**) + (**647**)

SCHEME 40

Several partial syntheses toward granaticin have been reported so far. Its oxabicyclo[2.2.2]octene portion is a characteristic moiety in this family, and a similar one is found in sarubicin A (or antibiotic U-58,431) (**648**)[416]. As the initial attempt in the

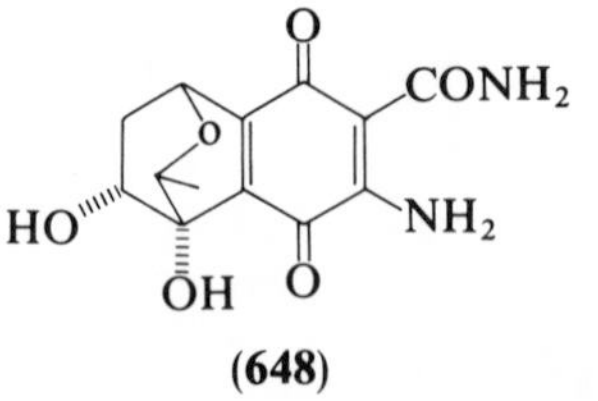

(**648**)

direction, sarubicin A was synthesized. The method used for construction of the common oxabicyclooctane ring system is potentially applicable to the synthesis of granaticin. It includes two key reactions: (1) preparation of a triol (**652**) by means of stereoselective diol formation by OsO_4, and (2) an oxabicyclooctane ring formation by an intramolecular cyclization of the triol to **652** (Scheme 41)[417,418]. The key intermediate (**655**) analogous to **651** is prepared in an alternative method[419]. The aldehyde (**653**) reacts with $TiMe(OPr\text{-}i)_3$ to give **655** in a very high stereoselectivity (equation 168). The Ti(IV) probably coordinates tridentatively to three oxygen functionalities in **653** and as in **654** methylation therefore occurs from the least hindered site.

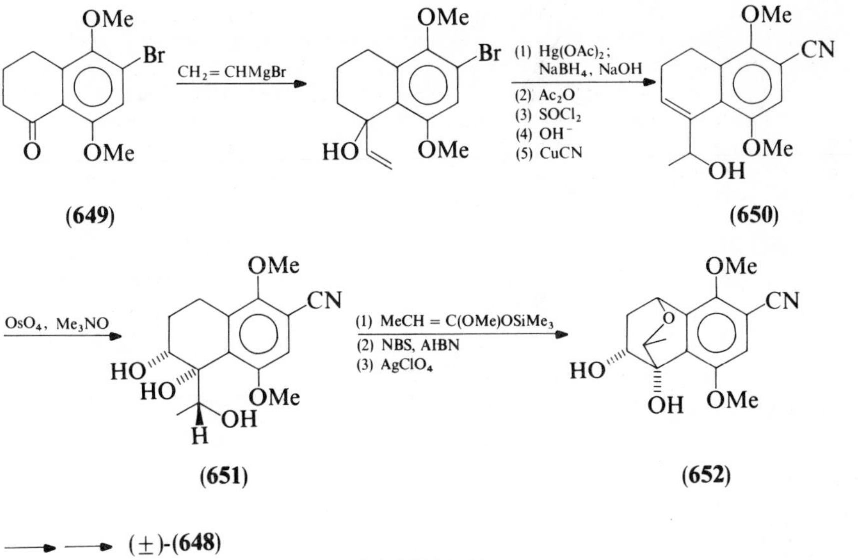

(649) $\xrightarrow{CH_2{=}CHMgBr}$ $\xrightarrow{(1)\ Hg(OAc)_2;\ NaBH_4,\ NaOH\ (2)\ Ac_2O\ (3)\ SOCl_2\ (4)\ OH^-\ (5)\ CuCN}$ (650)

$\xrightarrow{OsO_4,\ Me_3NO}$ (651) $\xrightarrow{(1)\ MeCH = C(OMe)OSiMe_3\ (2)\ NBS,\ AIBN\ (3)\ AgClO_4}$ (652)

$\longrightarrow\ \longrightarrow$ (±)-(**648**)

SCHEME 41

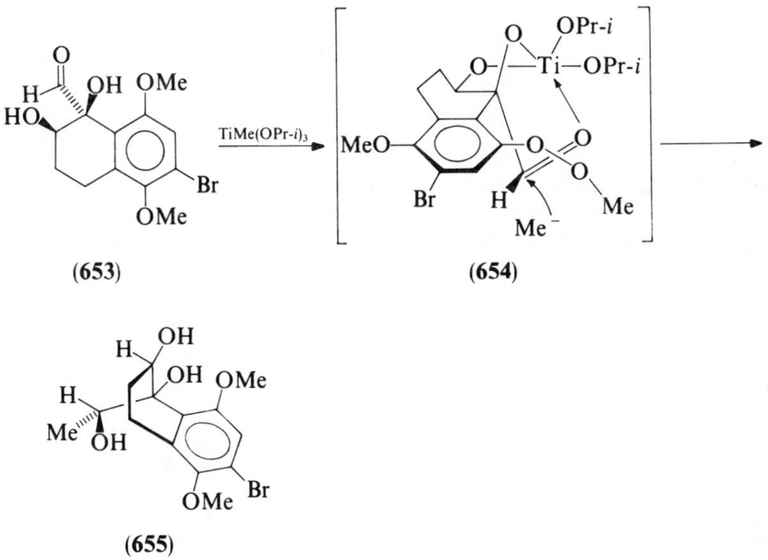

(**653**) $\xrightarrow{TiMe(OPr\text{-}i)_3}$ [(**654**)] $\longrightarrow$ (**655**) (168)

Griseusin A and B have a spiroketal moiety together with two pyran rings. Enantioselective synthesis of these compounds is achieved by using **658** as a chiral source, derived from sugar (Scheme 42)[420]. The nucleophilic coupling reaction of **656** with **657** and successive PCC oxidation afforded **658**. The requisite spiro ketal structure of **659** is accomplished by acid treatment of a triol (**658**). Since the final product in this synthetic

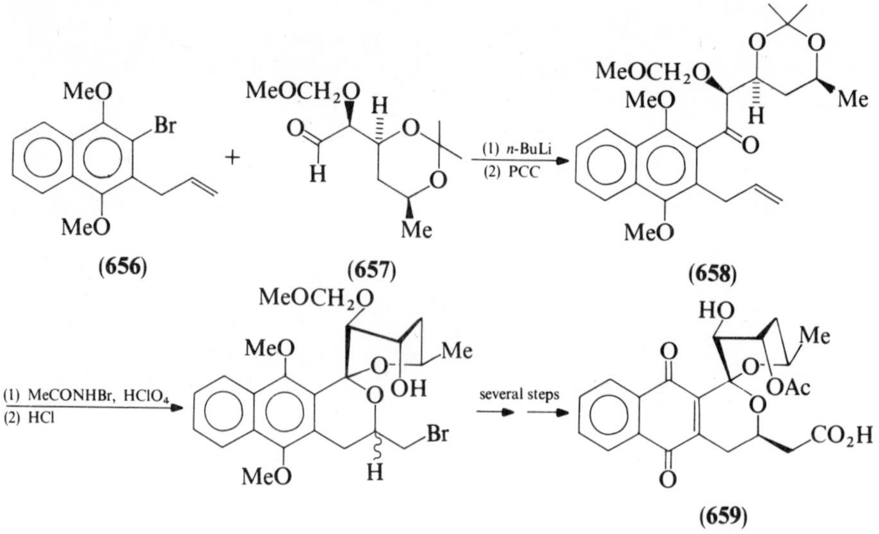

SCHEME 42

scheme is the antipode of the expected griseusin B, the correct absolute configuration of natural griseusin A and B was determined to be **616** and **617**, respectively, by comparison of the CD spectrum with those of the synthetic griseusins.

C. Ansamycins

A large number of ansamycin antibiotics have been found so far, and their structures are characterized by the presence of an *ansa* chain bridge attached to a naphthoquinone (or naphthohydroquinone) or to a benzoquinone nuclei. Examples are structures **660–663** given in Chart 5. The detailed chemistry of ansamycins is treated in leading reviews[421–424]. Due to the complexity of their structure, a limited number of ansamycin antibiotics have been synthesized. In this review, the synthesis of the quinonoid moiety is discussed, whereas for the preparation of *ansa* chains, several successful approaches have appeared in the literature[425–430].

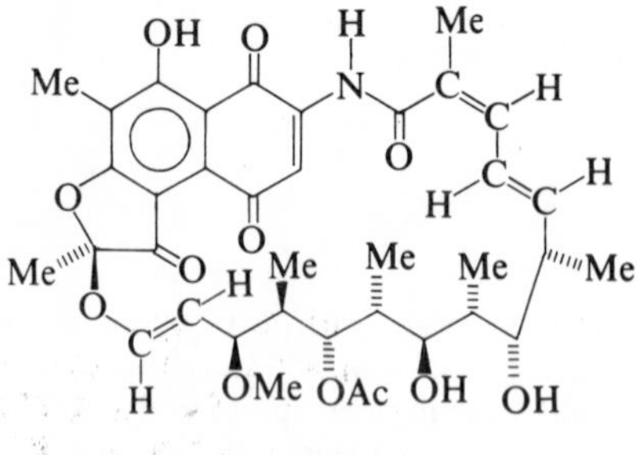

(660) 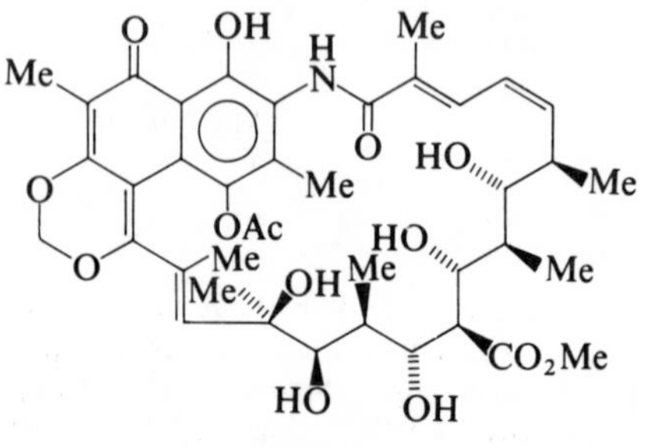 **(661)**

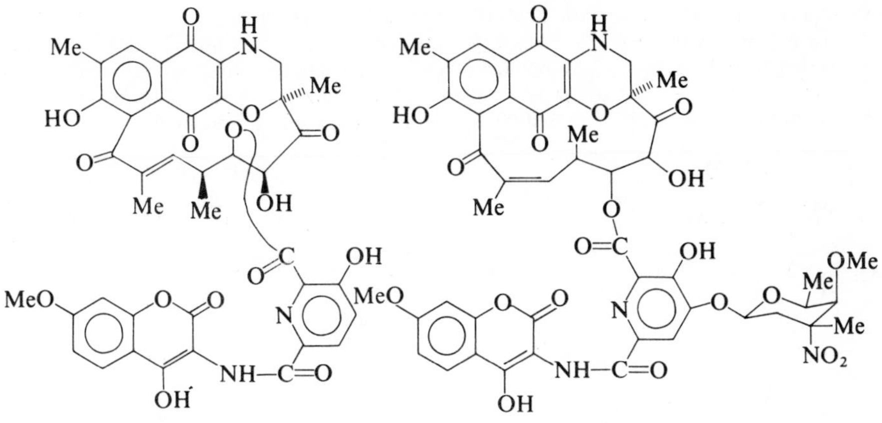

(662) (663)

CHART 5

For construction of required polyfunctionalized quinone (**668**), which is common to all the structures, the Diels–Alder approach seemed to be most efficient. Thus, the vinyl ketene acetal (**665**), derived from **664**, reacts with the bromoquinone (**666**) to afford **667** in 73%

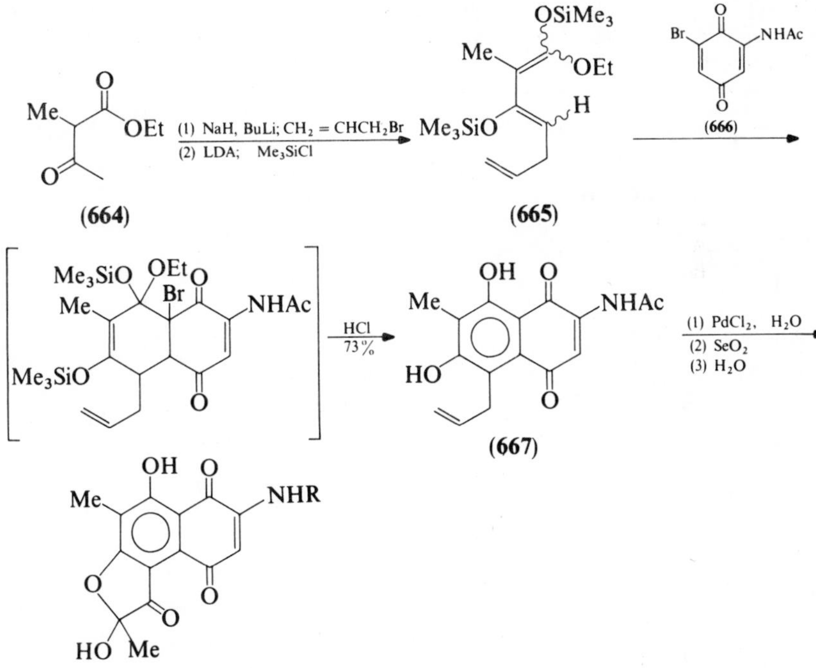

(664) (1) NaH, BuLi; $CH_2=CHCH_2Br$ (2) LDA; Me_3SiCl (665) (666) HCl 73% (667) (1) $PdCl_2$, H_2O (2) SeO_2 (3) H_2O (668)

R = Ac

SCHEME 43

yield after aromatization with acid (Scheme 43)[431]. This annulation proceeds in a regioselective manner as in the related reactions discussed in Section III.A.4. **667** can be converted to **668** in three steps including Wacker type and selenium dioxide oxidation.

A similar method is utilized for the synthesis of the aromatic segment of streptovaricin D as shown in equation 169[432]. The difference in the electron-donating ability between the

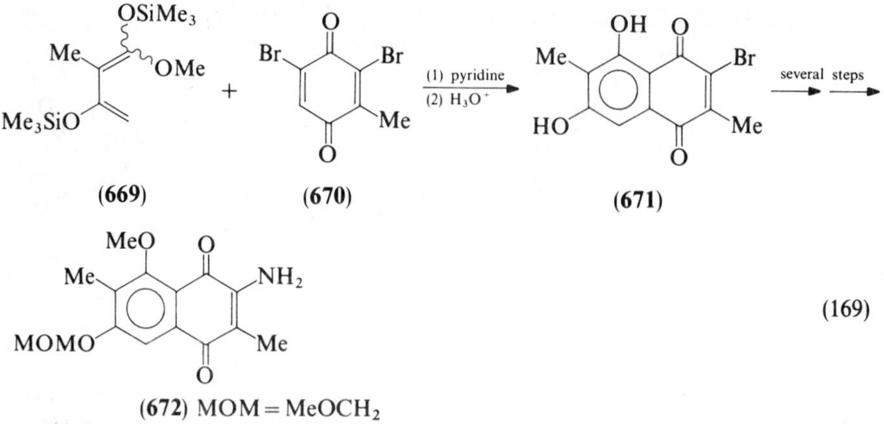

(1) pyridine
(2) H_3O^+

several steps

(669) **(670)** **(671)**

(672) MOM = $MeOCH_2$

(169)

ethereal oxygen and the amido nitrogen in **673**[224] is utilized for a regioselective Diels–Alder reaction with **674**, which is considered as a synthetic equivalent of **675** (Scheme 44)[433]. A Lewis acid mediated Claisen-type rearrangement from a leuco form of

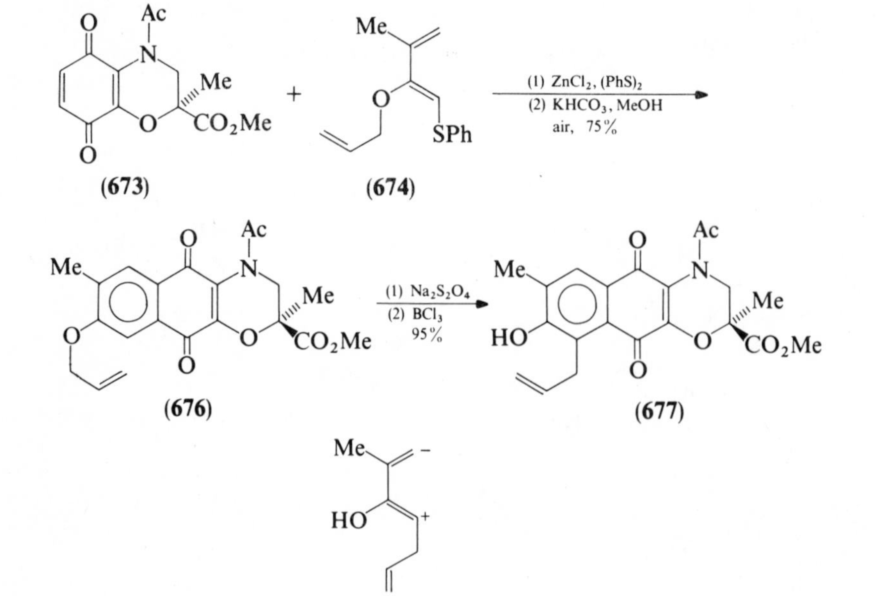

(1) $ZnCl_2$, $(PhS)_2$
(2) $KHCO_3$, MeOH air, 75%

(673) **(674)**

(1) $Na_2S_2O_4$
(2) BCl_3
95%

(676) **(677)**

(675)

SCHEME 44

676 affords **677**, which is a promising precursor to build the aromatic segment of rubradirins **662** and **663**.

The aromatic segment (**668**, R = H) is prepared in a rather indirect pathway. Diels–Alder reaction of furan with the benzyne, generated from **678**, affords **679** after successive acylation with EtCHO and oxidation. The oxabicyclo[2.2.1]heptene (**679**) is oxidized to **680** and then to **682** after following several steps including the intermediacy of **681**. Deprotection of the latter affords **668** in a good yield (Scheme 45)[434].

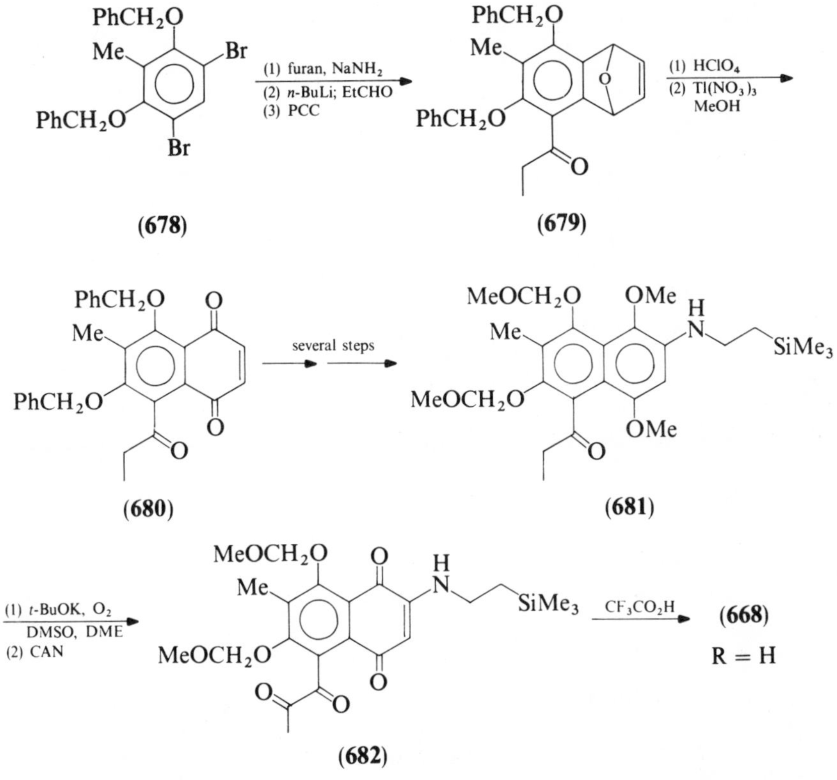

PhCH$_2$O, Me, Br, PhCH$_2$O, Br (**678**) —(1) furan, NaNH$_2$; (2) *n*-BuLi; EtCHO; (3) PCC→ PhCH$_2$O, Me, O, PhCH$_2$O, O (**679**) —(1) HClO$_4$; (2) Tl(NO$_3$)$_3$, MeOH→

PhCH$_2$O, O, Me, PhCH$_2$O, O, O (**680**) —several steps→ MeOCH$_2$O, OMe, H, N, SiMe$_3$, Me, MeOCH$_2$O, OMe, O (**681**)

—(1) *t*-BuOK, O$_2$, DMSO, DME; (2) CAN→ MeOCH$_2$O, O, H, N, SiMe$_3$, Me, MeOCH$_2$O, O, O, O (**682**) —CF$_3$CO$_2$H→ (**668**) R = H

SCHEME 45

An alternative route, not including a Diels–Alder reaction, is achieved by two groups. One approach is a route which mainly involves a modification of the benzocyclohexane ring of **683**. In this synthesis, an orthoquinone (**684**) is transferred to **688** via the corresponding quinone imine (**685**) and its tautomer (**686**) (Scheme 46)[435]. This aromatic segment (**688**) is successfully coupled with an *ansa* chain. For this coupling, the chlorothio ether (**689**) is used due to the low reactivity of the hydroxyl group of **688**. The coupling gives a mixture of four diastereoisomers with respect to the C(12), C(27) and C(29) positions. Two of them, which possess the same stereochemistry as the natural compound, were separated, and converted to the precursor quinone by a multi-step treatment: deblocking of the phenolic protection at C(1), olefination at C(28) and C(29) position, and

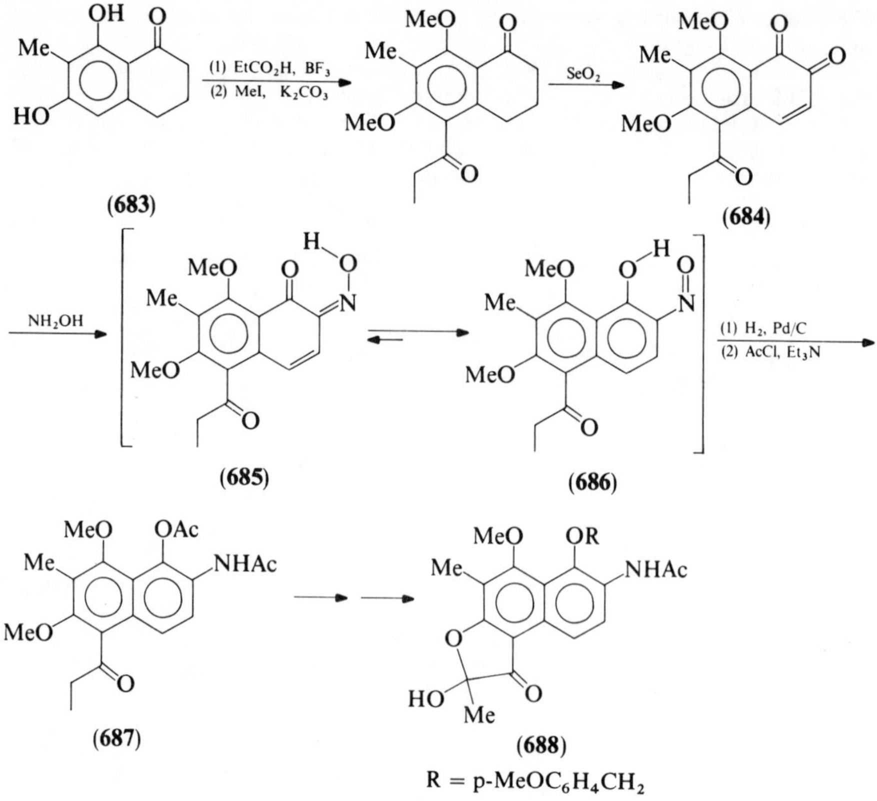

(683) (1) $EtCO_2H$, BF_3 (2) MeI, K_2CO_3 → SeO_2 → (684)

NH_2OH → [(685) ⇌ (686)] (1) H_2, Pd/C (2) AcCl, Et_3N →

(687) → → (688)

R = p-$MeOC_6H_4CH_2$

SCHEME 46

oxidation of the aromatic ring by Fremy's salt. Finally, the total synthesis of rifamycin S (**660**) is completed by forming the amide linkage in several steps (Scheme 47)[436].

The related quinone (**695**) used for the rifamycin synthesis is prepared by a sequence of nucleophilic Michael addition of a β-keto ester to quinone (**691**) followed by an intramolecular Claisen condensation of **693** to **694** and final oxidation by cerium ammonium nitrate (Scheme 48)[437].

D. Mitomycins

Since mitomycin A[438], B[438], C[439] and porfiromycin[440] are isolated from *Streptomyces* strains, they are recognized to be active against Gram-positive and Gram-negative bacteria. Mitomycin C, especially, has the broadest and strongest activity against a wide range of tumors. The whole range of chemistry and biological and clinical activities are treated in many reviews[441–448].

Recently, the absolute configurations were corrected and they are depicted in **696** and **697**[449,450]. They are reasonable and in agreement with the biosynthetic studies. Mitosenes (**698**, R^1, R^2 = ⋝NR), which can be generated by the elimination of either methanol or

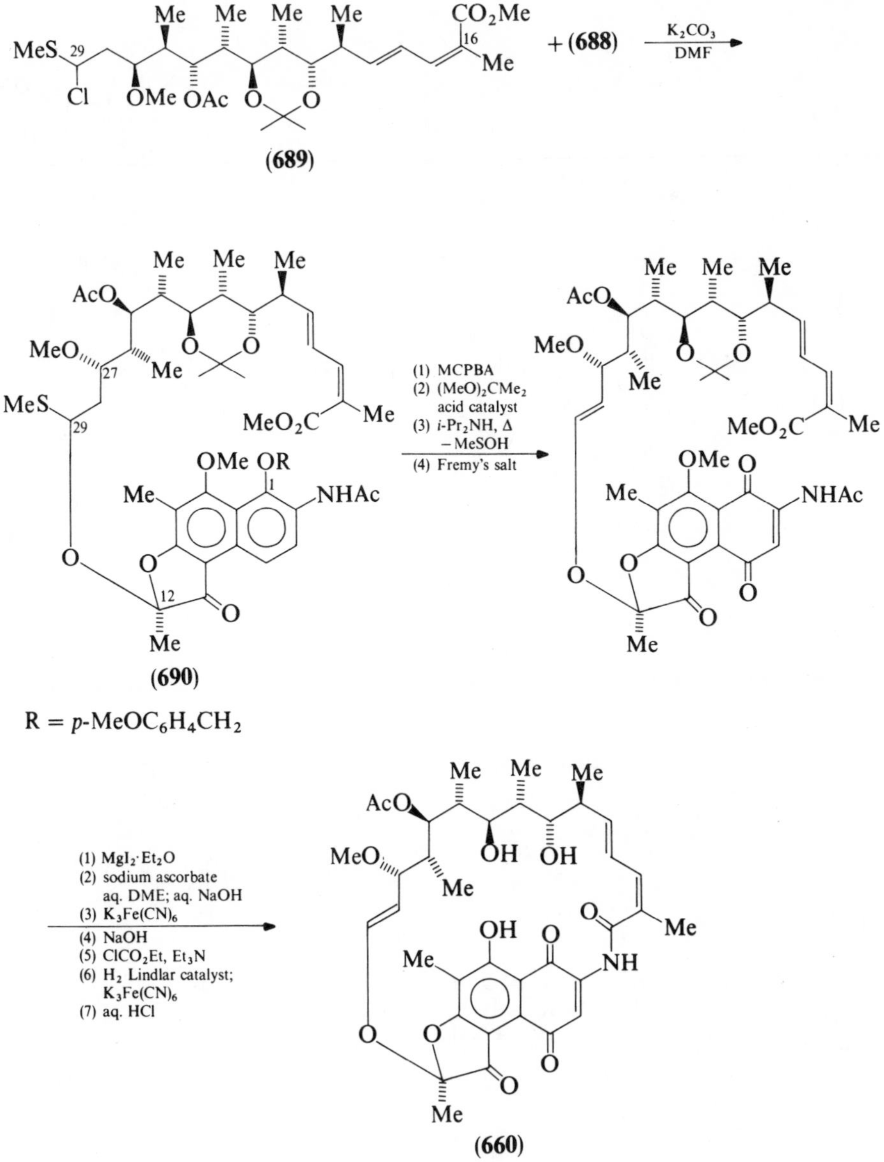

Me Me Me Me CO$_2$Me
MeS 29 16
Cl OMe OAc O O Me
+ (688) $\xrightarrow{K_2CO_3,\ DMF}$

(689)

Me Me Me
AcO
MeO 27 O O
Me
MeS 29
MeO$_2$C Me
OMe OR
Me 1 NHAc
O O
12 O
Me

(690)

$R = p\text{-}MeOC_6H_4CH_2$

(1) MCPBA
(2) $(MeO)_2CMe_2$ acid catalyst
(3) $i\text{-}Pr_2NH$, Δ $-MeSOH$
(4) Fremy's salt

Me Me Me
AcO
MeO O O
Me
MeO$_2$C Me
OMe O
Me NHAc
O O
O O
Me

(1) $MgI_2 \cdot Et_2O$
(2) sodium ascorbate aq. DME; aq. NaOH
(3) $K_3Fe(CN)_6$
(4) NaOH
(5) $ClCO_2Et$, Et_3N
(6) H_2 Lindlar catalyst; $K_3Fe(CN)_6$
(7) aq. HCl

Me Me Me
AcO
MeO OH OH
Me
O Me
OH O
Me NH
O O
O O
Me

(660)

SCHEME 47

water from mitomycins, possess an almost similar activity toward bacteria and tumors as the parent mitomycins, and also are the target compounds in the synthetic studies.

Since synthesis of these compounds involves many interesting points and challenges, many attempts have been reported so far. The earlier results are summarized in references

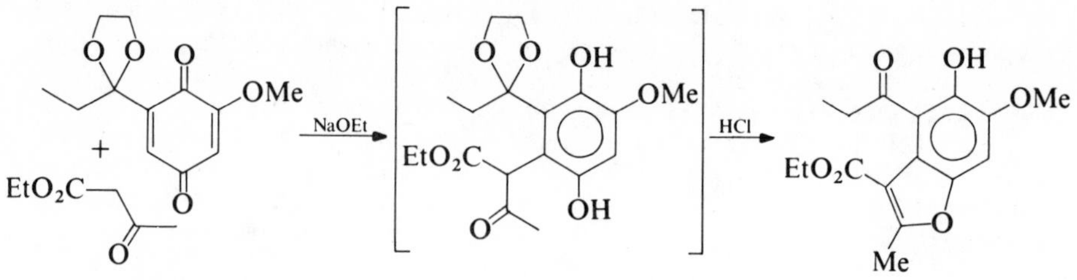

(691) (692) (693)

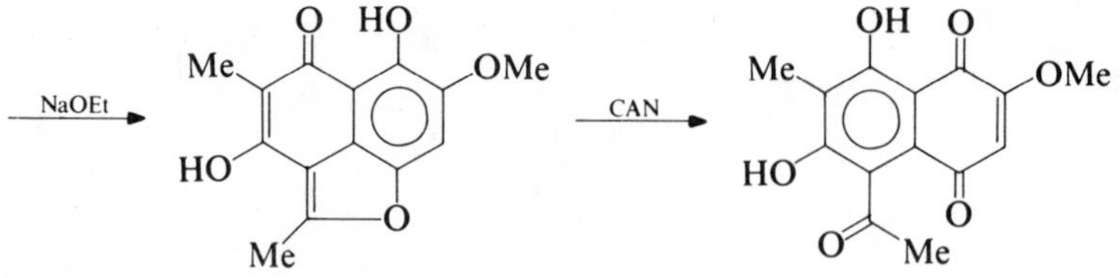

(694) (695)

SCHEME 48

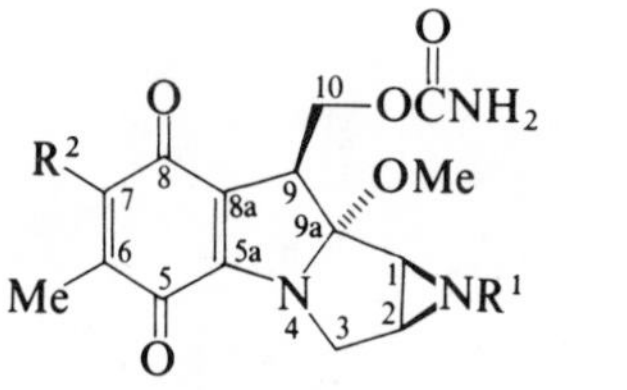

(696)

Mitomycin A, $R^1 = H$, $R^2 = OMe$
C, $R^1 = H$, $R^2 = NH_2$
Porfiromycin, $R^1 = Me$, $R^2 = NH_2$

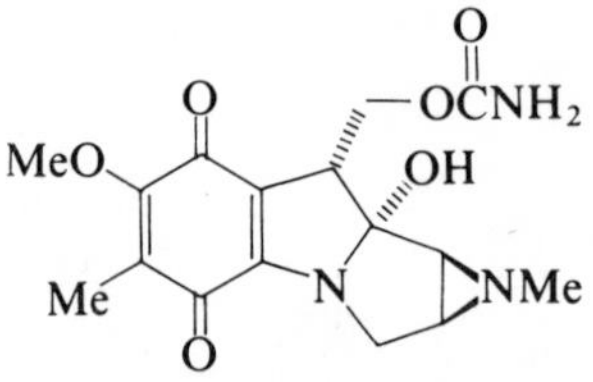

(697)

Mitomycin B

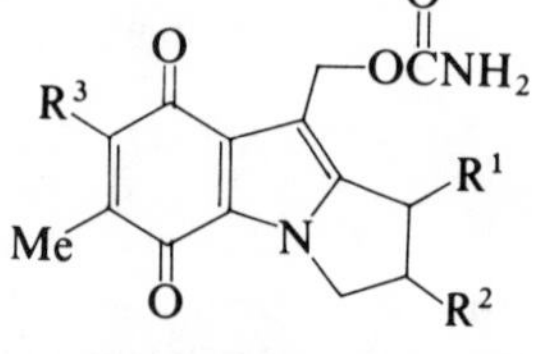

(698)

Mitosene

441–444, 446, 447 and here we treat the progress in this field after 1980. A total synthesis of the racemic forms of mitomycins A, C and porfiromycin was accomplished in 1977, by means of intramolecular amino cyclization. The central part of this methodology is briefly shown in Scheme 49 as a typical example for a series of related syntheses[451,452].

The most serious problem in the synthesis of mitomycins is how to preserve the fragile alkoxy group at the C(9a) position. This is solved by a transannular aminocyclization of **701** under acidic conditions to give **703** with a favorable stereochemistry (Scheme 49)[451,452].

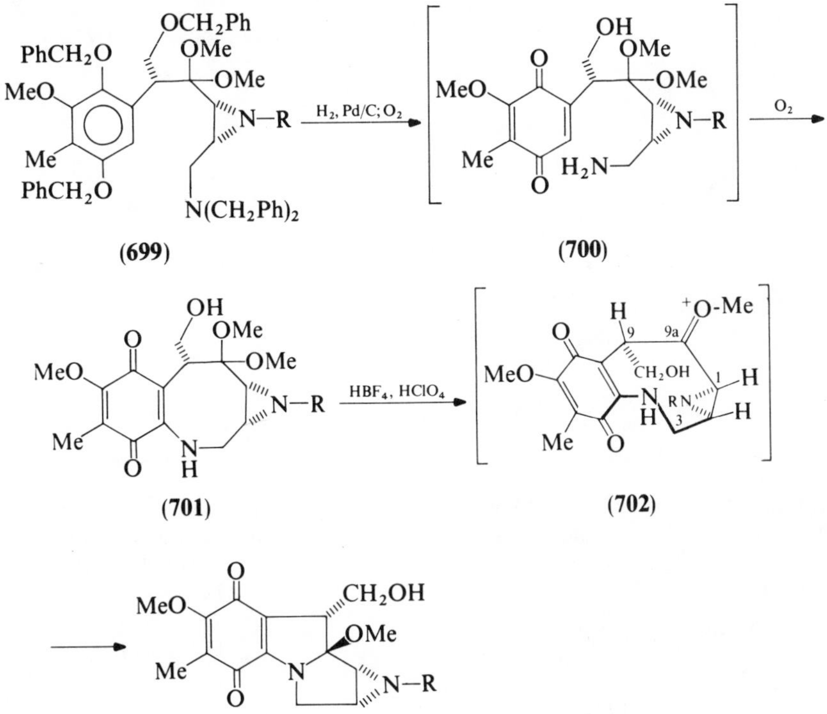

(**699**) $\xrightarrow{H_2, Pd/C; O_2}$ [(**700**)] $\xrightarrow{O_2}$ (**701**) $\xrightarrow{HBF_4, HClO_4}$ [(**702**)] $\longrightarrow$ (**703**)

SCHEME 49

Most of the reported methods are concerned with the synthesis of the pyrrolo[1,2-*a*] indole ring and of mitosene (**698**). The simplest synthesis of this ring system is accomplished by a copper catalyzed intramolecular double cyclization of 2-(2′,4′-

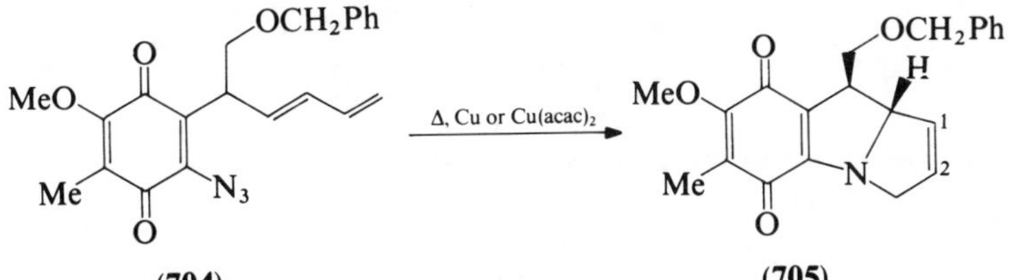

(**704**) $\xrightarrow{\Delta, \text{ Cu or Cu(acac)}_2}$ (**705**) (170)

pentadienyl)-3-azidoquinone (**704**) to **705** (equation 170)[321,322]. This method preserves the double bond at C(1)–C(2), which can be converted to an aziridine ring at a later stage.

Intermolecular photochemical reaction of the azidoquinone (**706**) with a symmetrically functionalized *Z*,*Z*-diene (**707**) gives stereospecifically the indole derivative (**708**), which can be easily cyclized to **709**[320]. The stereochemistry of the olefinic double bond of the diene (**707**) is completely preserved in the *cis* configuration in **708** and this is a great advantage for the following cyclization to **709** (Scheme 50)[320].

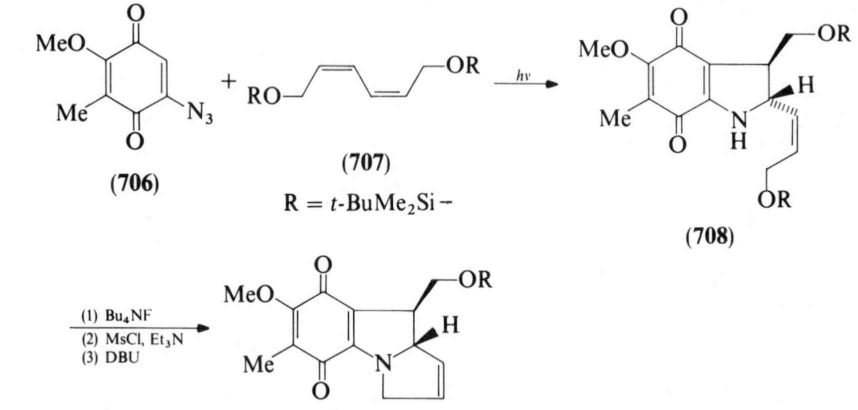

(**706**) + (**707**) $\xrightarrow{h\nu}$ (**708**)

R = *t*-BuMe$_2$Si–

(**708**) $\xrightarrow[(3)\ DBU]{(1)\ Bu_4NF;\ (2)\ MsCl,\ Et_3N}$ (**709**)

SCHEME 50

A similar target compound (**713**) is synthesized by several alternative ways. A stepwise preparation of the dihydroindole (**711**) is synthesized by means of intramolecular nucleophilic ring-opening of an activated cyclopropane ring. The conversion of the functionalities on the pyrrole ring from **712** to **713** is conducted in a five steps sequence (Scheme 51)[453].

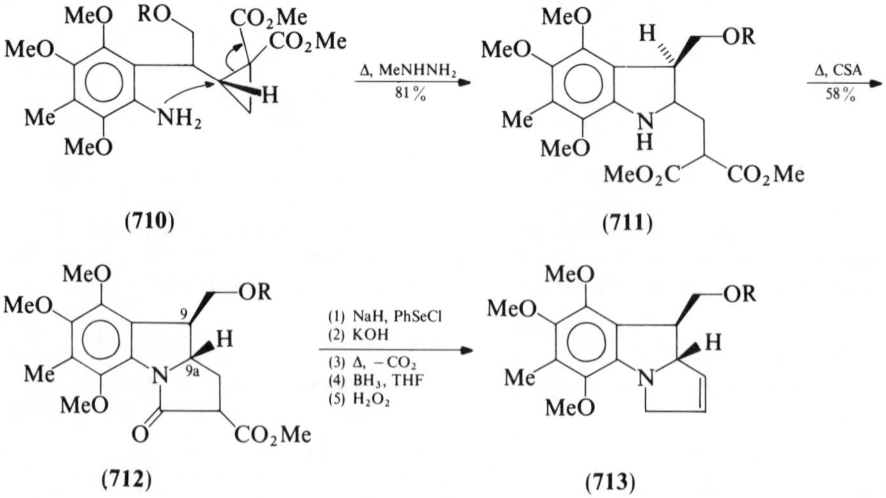

(**710**) $\xrightarrow[81\%]{\Delta,\ MeNHNH_2}$ (**711**) $\xrightarrow[58\%]{\Delta,\ CSA}$ (**712**) $\xrightarrow[(3)\ \Delta,\ -CO_2;\ (4)\ BH_3,\ THF;\ (5)\ H_2O_2]{(1)\ NaH,\ PhSeCl;\ (2)\ KOH}$ (**713**)

SCHEME 51

Intramolecular aminoselenation is one of the important methods to form dihydroindoloquinone and the related compound, **716**. In the cyclization of **714**, participation by the *N*-phenylselenophthalimide gives **716** in 31% yield after oxidative elimination. The latter can be converted to **717** (Scheme 52)[454]. Interestingly, all the doubly cyclized products,

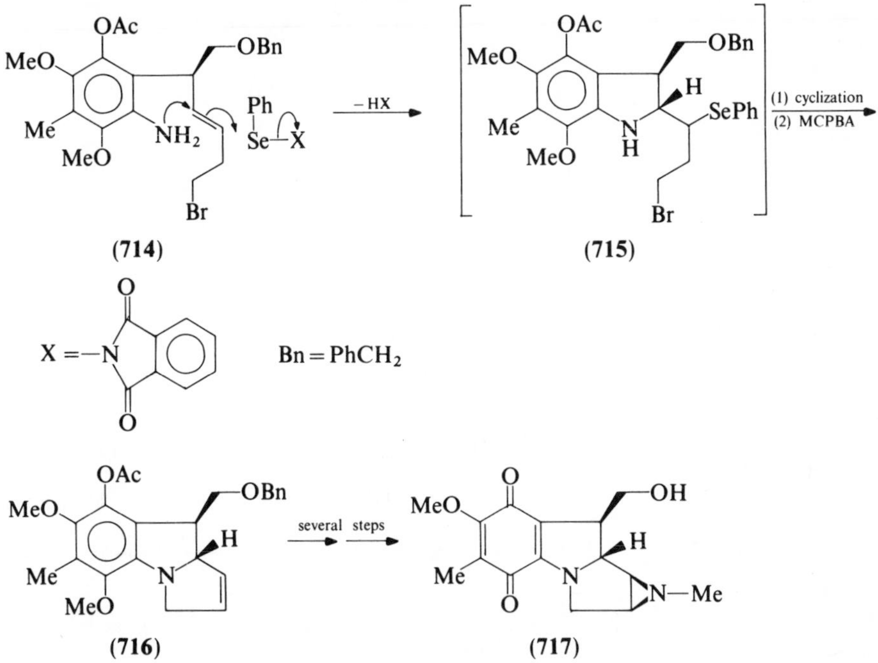

OAc MeO Me MeO OBn NH₂ Ph Se X Br (714) —HX→ [MeO Me MeO OAc OBn H SePh N H Br] (715) (1) cyclization (2) MCPBA →

X = —N(phthalimido) Bn = $PhCH_2$

OAc MeO Me MeO OBn H N (716) several steps → O MeO Me O OH H N N—Me (717)

SCHEME 52

705, **709**, **712** and **716**, obtained according to the schemes as shown above possess *trans* configuration at C(9)–C(9a), irrespective of the applied methodologies.

In a similar strategy, an intramolecular amino mercuration is used for the same purpose[455].

A double Michael addition of the exocyclic enamino ester (**719**) to the quinone monoacetal (**718**) gives the bicyclic product (**720**), which when subjected to acid catalysis rearranges to **721** (Scheme 53)[456]. The successive transformation gives indolequinone (**722**). A use of the quinone monoacetal instead of the quinone is the key step in the regiocontrolled cyclization.

Several photochemical cyclizations have been applied for the synthesis of the required carbon framework. Photochemical reaction of the pyrrolidinoquinone (**723**) gives the cyclized product (**727**) in a good yield (Scheme 54)[457,458]. By the photochemical reaction, the cyclized product (**724**) is first obtained via intramolecular hydrogen abstraction and cyclization. Upon silica gel chromatography, an intramolecular attack of a stabilized anion on the iminium salt in **725** is proposed to give **726**, which on acid hydrolysis and decarboxylation gives **727** (Scheme 54)[457,458]. A similar, but a non-photochemical intramolecular nucleophilic attack on an iminium salt has also been investigated (equation 171)[459]. The related intramolecular and electrophilic cyclization of an iminium salt (**729**) is

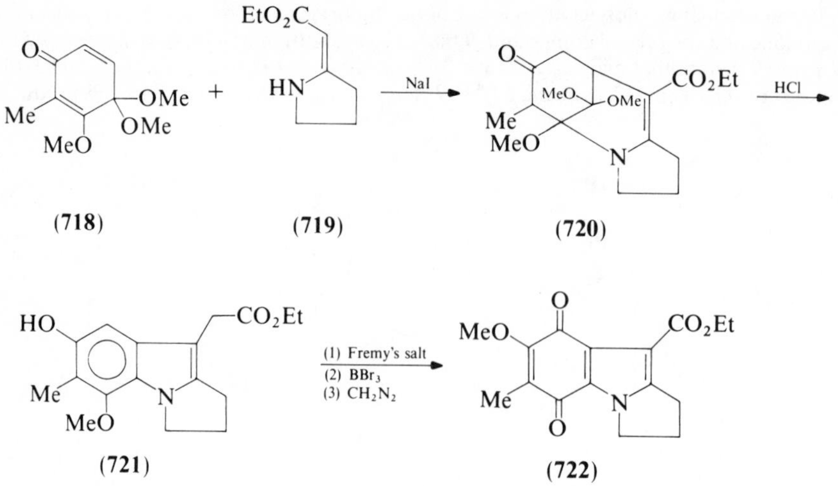

EtO_2C O OMe Me OMe MeO + HN NaI O CO_2Et MeO OMe Me MeO N HCl

(718) (719) (720)

HO CO_2Et Me MeO N (1) Fremy's salt (2) BBr_3 (3) CH_2N_2 O MeO CO_2Et Me O N

(721) (722)

SCHEME 53

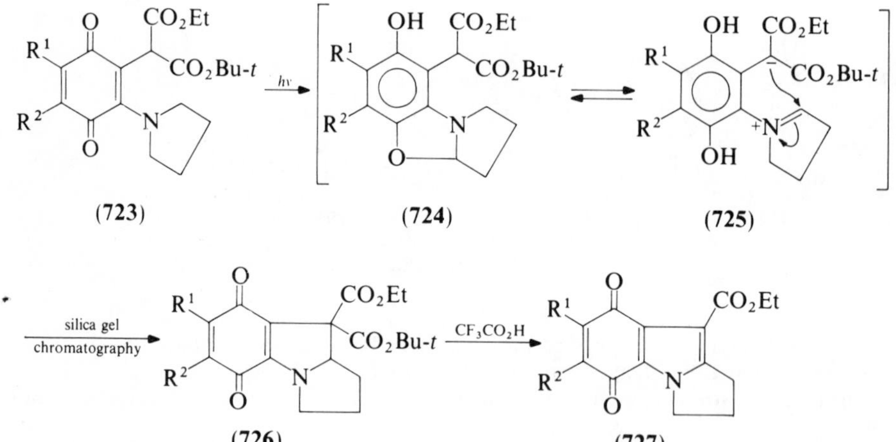

O CO_2Et R^1 CO_2Bu-*t* R^2 O N *hν* OH CO_2Et R^1 CO_2Bu-*t* R^2 N O ⇌ OH CO_2Et R^1 CO_2Bu-*t* R^2 ^+N OH

(723) (724) (725)

silica gel chromatography O R^1 CO_2Et CO_2Bu-*t* R^2 O N CF_3CO_2H O R^1 CO_2Et R^2 O N

(726) (727)

SCHEME 54

RO OMe OMe R^1 CO_2Me R^2 RO N Δ, $POCl_3$ OMe RO R^1 R^2 RO N^+ MeOH RO $CH(OMe)_2$ R^1 R^2 RO N

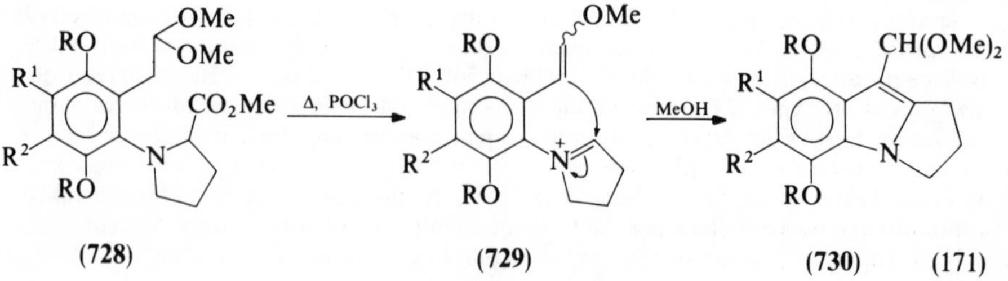

(728) (729) (730) (171)

also reported in the synthesis of the analogous compound (**730**) from **728**[460]. A similar photocyclization to that shown in Scheme 54 is repeatedly used, first for dehydrogenation and then for cyclization (Scheme 55)[461]. The first photochemical reaction of quinone (**731**)

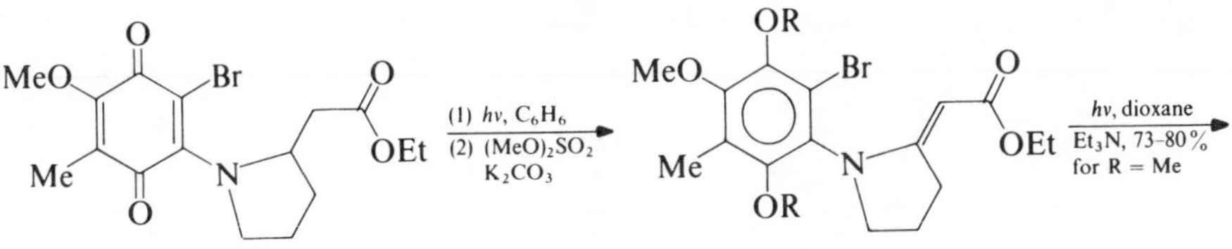

(1) $h\nu$, C_6H_6; (2) $(MeO)_2SO_2$, K_2CO_3

$h\nu$, dioxane; Et_3N, 73–80% for R = Me

(**731**) (**732**) R = H, Me

HNO_3

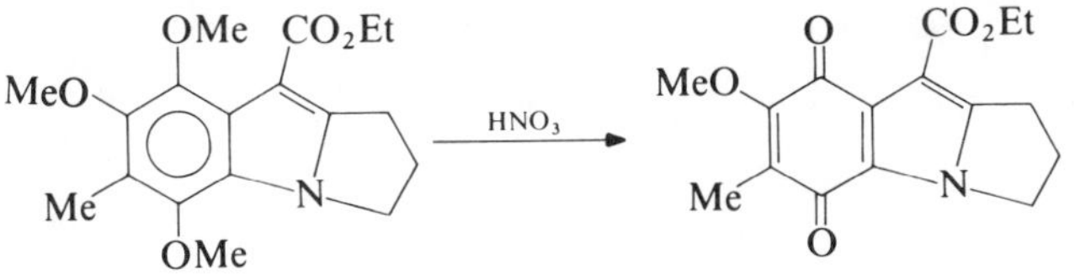

(**733**) (**734**)

SCHEME 55

gives the enamine (**732**, R = H) after intramolecular hydrogen abstraction. The second photochemical step of the corresponding dimethyl ether (**732**, R = Me) gives indole derivative **733** in a good yield, in a process which is analogous to the photocyclization of *N*-haloaryl substituted enamines. However, direct irradiation of hydroquinone (**732**, R = H) does not give **734**. As an alternative for the photochemical cyclization to the indole, palladium acetate[461] or cupric bromide[112b,462] gives **734** from **735** in an excellent yield (equation 172). This route was also applied to the synthesis of the chiral aziridinomitosene

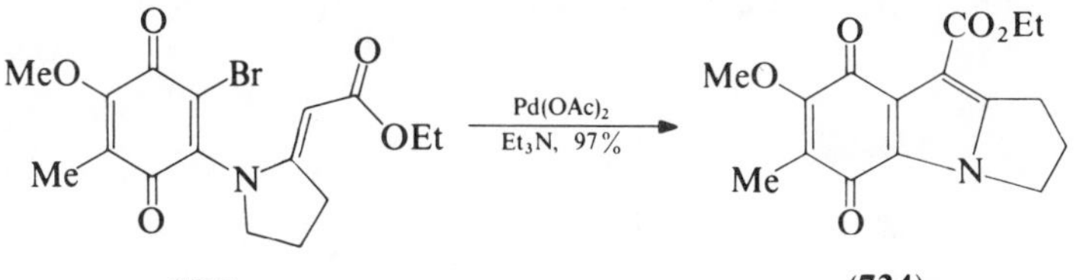

$Pd(OAc)_2$; Et_3N, 97% (172)

(**735**) (**734**)

(**740**), which has unnatural chirality at C(1) and C(2) (Scheme 56)[463]. The analogous intramolecular cyclization reaction of β-enamino ester (**743**) with its aromatic halide moiety is reported to give **744** in good yield. **744** is then utilized as a precursor to ($\pm$)-apomitomycin (**745**) (Scheme 57)[464].

The pyrrolizidine skeleton of mitosene including the quinonoid moiety itself is prepared by a 1,3-dipolar cycloaddition between dimethyl acetylene dicarboxylate and a dipole (**747**) generated from **746**[465]. The quinonoid part is then formed by a Dieckman cyclization of the product **748** to **749**. This tricyclic precursor is ultimately converted to mitosene (**750**) (Scheme 58).

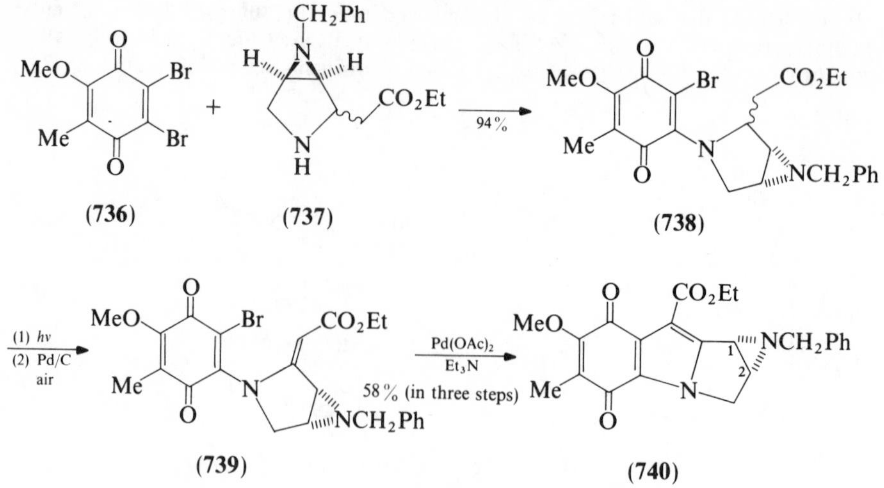

(736) (737) (738)

(739) (740)

SCHEME 56

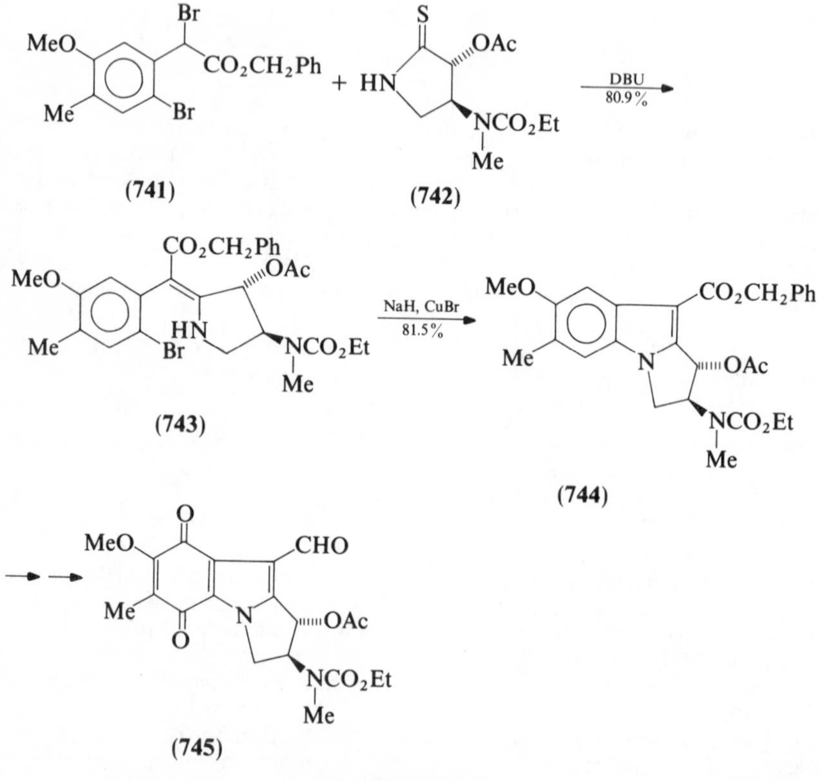

(741) (742)

(743) (744)

(745)

SCHEME 57

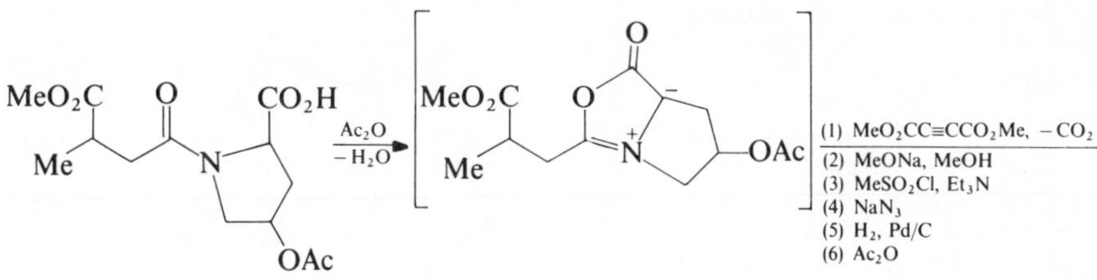

(746) (747) 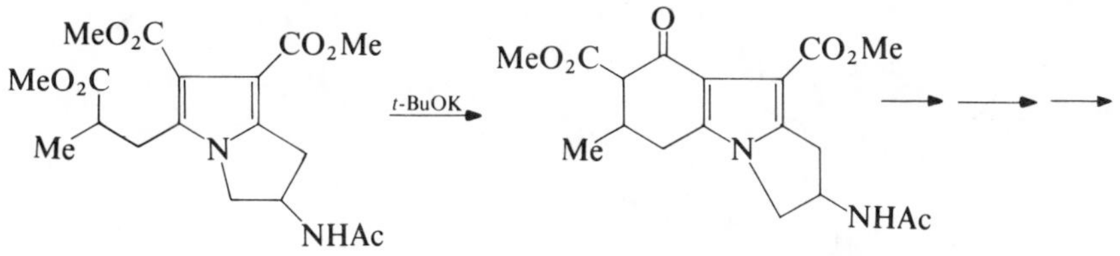 (748) (749) 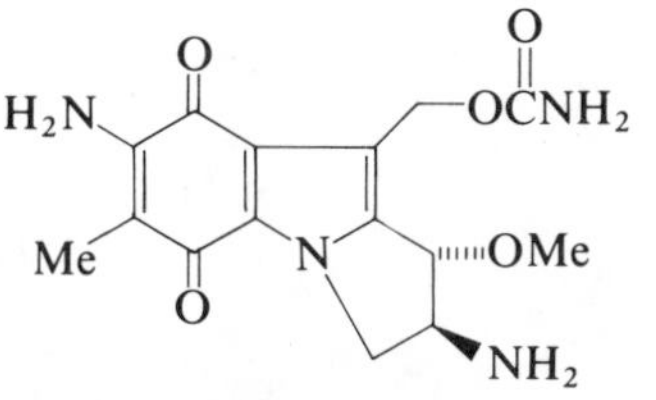 (750)

SCHEME 58

As another method of pyrroloindole ring synthesis, intramolecular Reformatsky reaction between the bromoester and the succinimido moieties in **751** stereoselectively gives **752** in a high yield under ultrasound irradiation (equation 173)[466]. Interestingly, the amido group of **752** seems to prevent the elimination of hydroxyl group at C(9a) to the corresponding iminium salt.

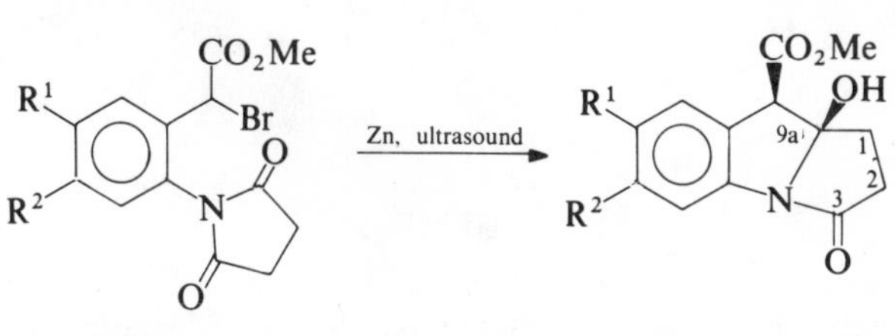

(173)

(751) (752)

In addition to the examples mentioned above, various attempts to prepare the required carbon skeleton by means of palladium catalyzed cyclization have been conducted[323,467,468].

E. Miscellaneous Quinones

1. Methoxatin

Methoxatin (**753**) is the cofactor for certain non-flavin or nicotinamide dependent bacterial dehydrogenases found in methylotrophic bacteria, which use one-carbon compounds as their sole source of energy[469-473]. It possesses an interesting *ortho*-quinone structure, which is rather uncommon among naturally occurring quinones. Three types of synthetic approach to the requisite rings were examined: (1) B → AB or BC → ABC, (2) AB or BC → ABC, and (3) A + C → ABC. The first approach is the extension of the A and C rings from ring B as the core compound. The synthesis along this pathway is shown in Scheme 59[474]. Formation of the indole ring is based on the

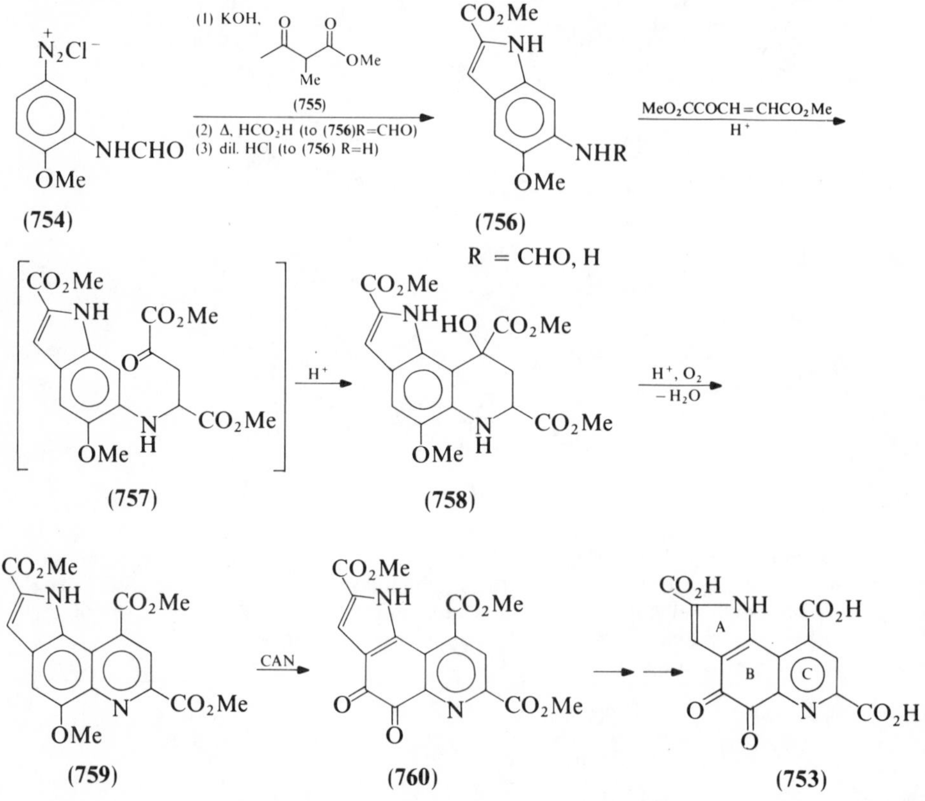

SCHEME 59

Japp–Klingemann reaction and a Fischer indolization. Condensation of the diazonium salt (**754**) with methyl α-methylacetoacetate (**755**) followed by acid treatment gives the corresponding hydrazone, which can be converted to **756** in 79% yield. After deprotection of the amino group of **756**, a Doebner–von Miller type annulation with dimethyl 2-oxaglutaconate affords **758** via the conjugate adduct (**757**). A following aromatization gives the pyrroloquinoline (**759**) in > 90% yield. Direct oxidation of **759** by CAN gives the *o*-

quinone (**769**) in 60% yield. Due to the instability of the *o*-quinonoid part in basic media, deprotection of the tricarboxylic acid of **760** could not be performed directly in this synthesis. However, it could be achieved via a monoacetal protection, saponification and deprotection sequence.

A different method for construction of the indole ring has been developed (Scheme 60)[475]. The methyl α-azidocinnamate (**762**) is obtained by aldol condensation of **761** with methyl azidoacetate and thermal decomposition of **762** gives **763** in a good yield. Additional quinoline ring is then prepared by just the same method shown in Scheme 59. However, oxidation of the phenolic group of **764** (R = H) with Fremy's salt is unsuccessful owing to the insolubility of **764** in aqueous media. An alternative method of oxidation with a lipophylic nitroxide (see Section II.B) gives the target *o*-quinone (**760**) in 93% yield.

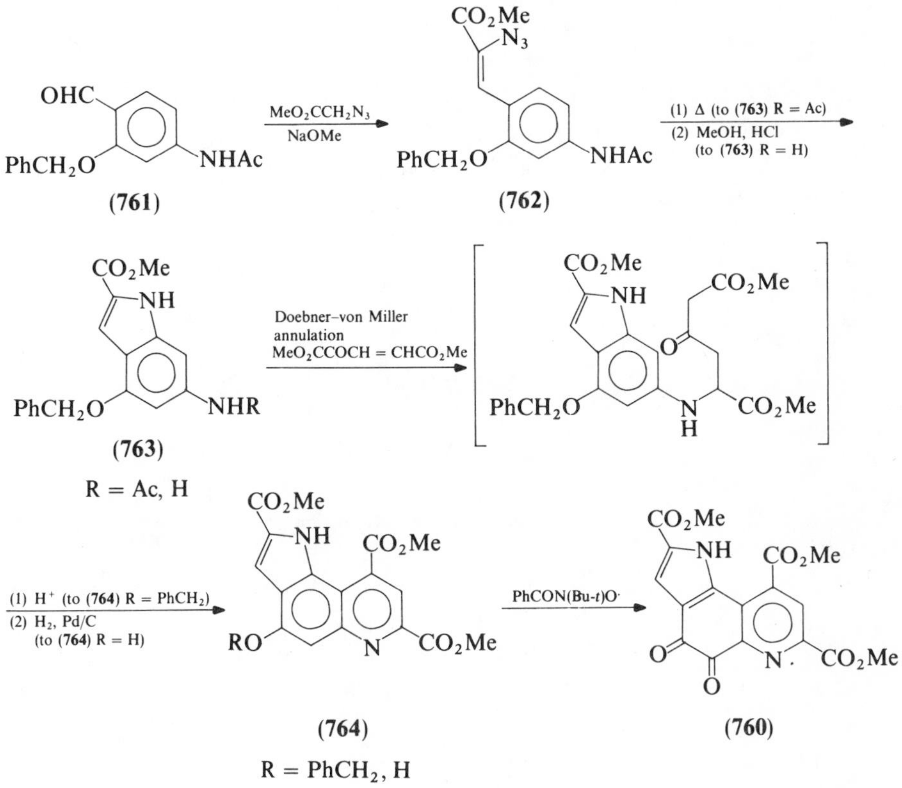

OHC, PhCH$_2$O, NHAc (**761**) → [$MeO_2CCH_2N_3$, NaOMe] → CO_2Me, N_3, PhCH$_2$O, NHAc (**762**) → [(1) Δ (to (**763**) R = Ac); (2) MeOH, HCl (to (**763**) R = H)]

CO_2Me, NH, PhCH$_2$O, NHR (**763**) R = Ac, H → [Doebner–von Miller annulation, $MeO_2CCOCH = CHCO_2Me$] → [CO_2Me, NH, CO_2Me, O, PhCH$_2$O, N H, CO_2Me]

→ [(1) H$^+$ (to (**764**) R = PhCH$_2$); (2) H$_2$, Pd/C (to (**764**) R = H)] → CO_2Me, NH, CO_2Me, RO, N, CO_2Me (**764**) R = PhCH$_2$, H → [PhCON(Bu-*t*)O·] → CO_2Me, NH, CO_2Me, O, O, N, CO_2Me (**760**)

SCHEME 60

An initial construction of the quinoline ring gives another approach to **753** (Scheme 61)[476]. The quinoline derivative (**765**) is synthesized by means of the Pfitzinger method from the corresponding aniline. Bromination at the benzylic position and nuclear nitration converts **765** to **766**, which is subjected to addition of methyl acetoacetate and then converted to the corresponding catechol (**767**) by Kozikowski's modification of the

Japp–Klingemann reaction in 64% yield in two steps. Hydrogenative reduction of the nitro group in **767** under acidic conditions spontaneously gives the indole (**768**) in 60% yield. Demethylative oxidation is successfully conducted by AgO, and finally the saponification of **768** is achieved by LiOH in aqueous THF. As mentioned above this hydrolysis was unsuccessful in the previous report (Scheme 61)[474].

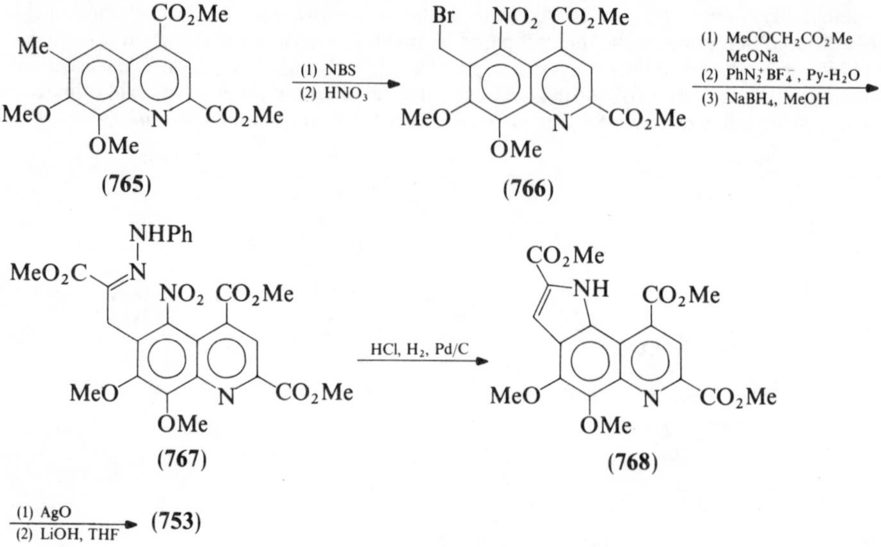
CO2Me
Me
MeO
N
CO2Me
OMe
(765)
(1) NBS
(2) HNO3
Br
NO2
CO2Me
MeO
N
CO2Me
OMe
(766)
(1) MeCOCH2CO2Me
MeONa
(2) PhN2+BF4-, Py-H2O
(3) NaBH4, MeOH
NHPh
MeO2C
N
NO2
CO2Me
MeO
N
CO2Me
OMe
(767)
HCl, H2, Pd/C
CO2Me
NH
CO2Me
MeO
N
CO2Me
OMe
(768)
(1) AgO
(2) LiOH, THF
(753)

SCHEME 61

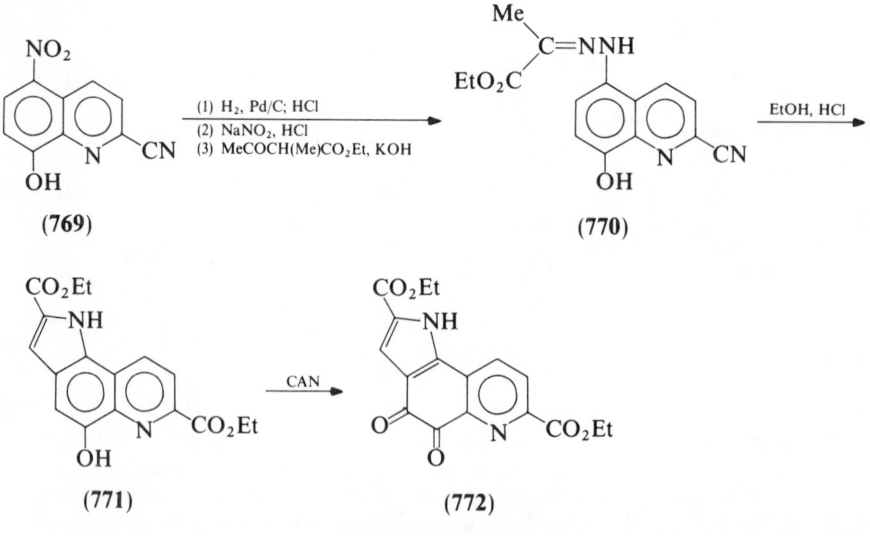
NO2
N
CN
OH
(769)
(1) H2, Pd/C; HCl
(2) NaNO2, HCl
(3) MeCOCH(Me)CO2Et, KOH
Me
C=NNH
EtO2C
N
CN
OH
(770)
EtOH, HCl
CO2Et
NH
N
CO2Et
OH
(771)
CAN
CO2Et
NH
O
N
CO2Et
O
(772)

SCHEME 62

9-Decarboxymethoxatin (**772**) is prepared by a more conventional route starting from 8-hydroxyquinoline, which is converted to **769** in several steps. **769** was converted to the corresponding diazonium salt via the aminoquinoline derivative and then was transferred to the corresponding hydrazone derivative (**770**) by using methyl α-methylacetoacetate in a basic alcohol in a low yield. Cyclization and the following transformation of the nitrile function to ester were conducted in an acidic alcohol to give **771**. The final oxidation of the phenol (**771**) to **772** can be achieved by using CAN (Scheme 62)[477]. A quinone imine → quinone methide → quinone interconversion sequence is utilized in the functionalization of the indole derivative (Scheme 63)[478]. In spite of its low overall yield, this is one

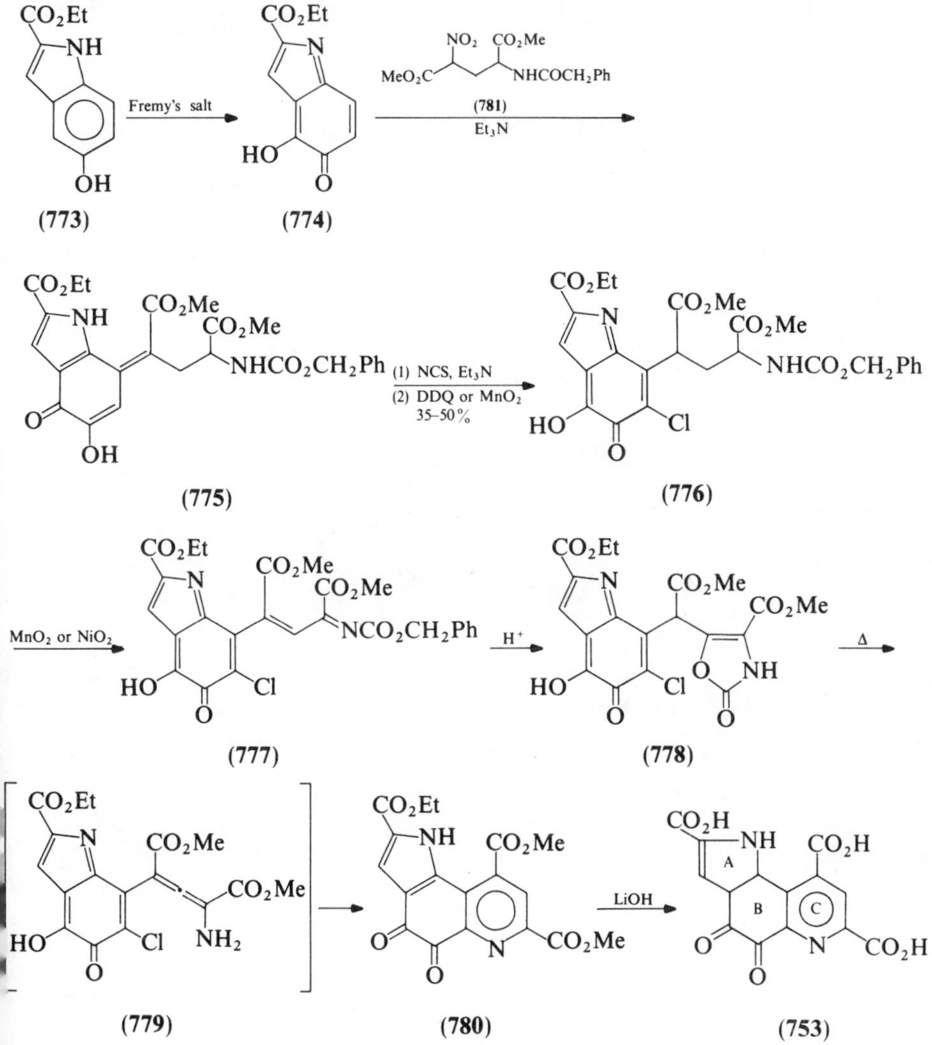

SCHEME 63

of the most interesting synthetic pathways, which fully utilizes the specific interconversion between the quinone imine and the indoloquinone. Fremy's salt oxidation of hydroxyindole (**773**) gives the corresponding *p*-quinone imine (**774**) in 94% yield. A subsequent Michael addition of **781** gives quinone methide **775** in 47% yield, and then stepwise chlorination and oxidation of the formed **775** affords **777** in a low yield (10–15%) from **774**. The final ring-closure to the quinolinoquinone (**780**) includes the intermediary formation of the allene (**779**), of which electrocyclic reaction furnishes **780** in a good yield.

The final example utilizes a photoinitiated oxidative cyclization reaction for construction of the B ring from the AC unit (Scheme 64)[479]. The photochemical cyclization of **782**

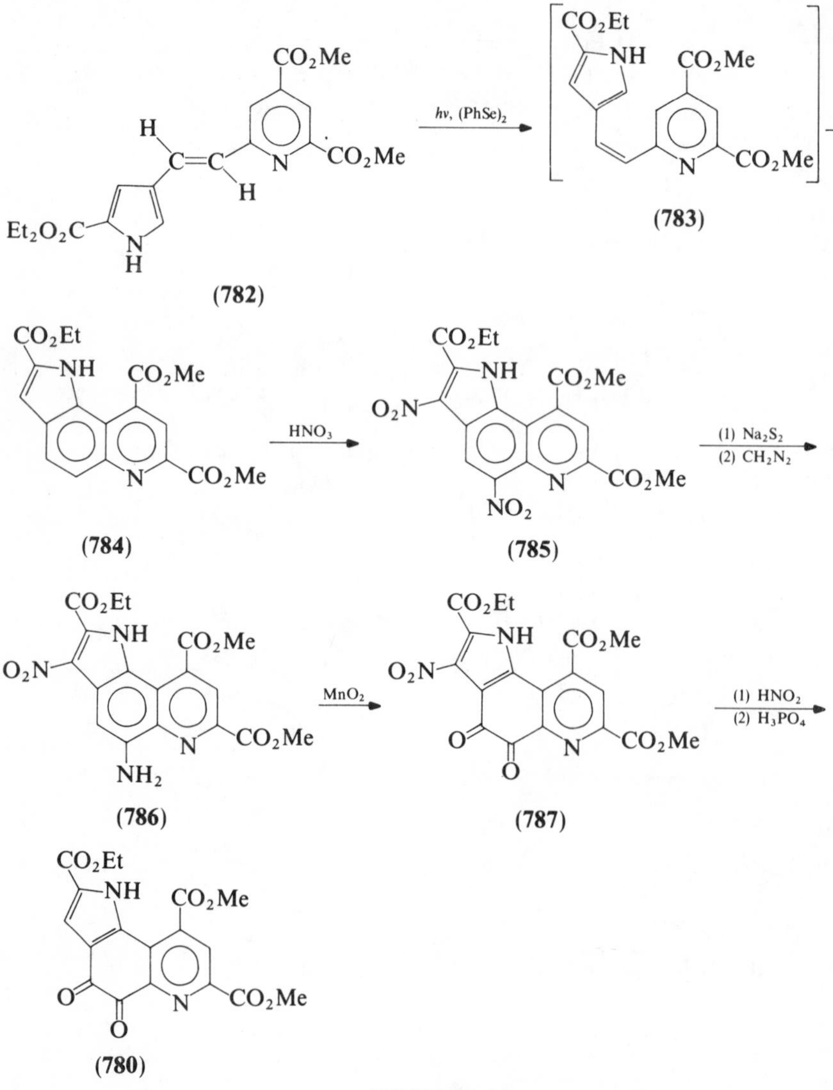

(**782**) $\xrightarrow{h\nu,\ (PhSe)_2}$ [(**783**)] $\longrightarrow$ (**784**) $\xrightarrow{HNO_3}$ (**785**) $\xrightarrow{(1)\ Na_2S_2,\ (2)\ CH_2N_2}$ (**786**) $\xrightarrow{MnO_2}$ (**787**) $\xrightarrow{(1)\ HNO_2,\ (2)\ H_3PO_4}$ (**780**)

SCHEME 64

to **784** presumably proceeds via an initial photochemical isomerization to **783** and then an oxidative electrocyclic reaction. It is worthwhile mentioning the other steps of the synthesis. The dinitro compound (**785**) obtained by nitration of **784** is reduced by means of a Zinin reduction to the amine (**786**). The latter is then oxidized by MnO_2 to the corresponding *o*-quinone in 94% yield, while Fremy's salt, as the first choice in such a transformation, is ineffective toward this conversion. Presumably, this can be attributed to the insufficient solubility of the substrate in the aqueous media.

2. Khellin

Khellin (**789**) is a member of the family of furochromanones and it possesses a lipid-altering and antiatherosclerotic activity[480,481]. Compounds related to **789** are known as khellinquinone (**788**) and khellinone (**790**), both of which can be derived from khellin. Due

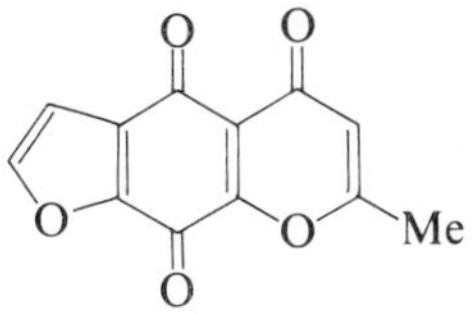

(**788**) 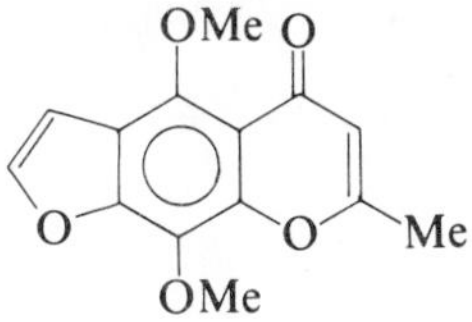 (**789**) 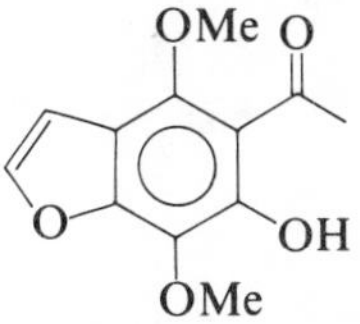 (**790**)

to its structural simplicity, many synthetic efforts to prepare it have appeared for the last 30 years. Several of the recent ones are interesting processes and will be discussed briefly here.

Khellinone is synthesized from the dihydrobenzofuran **791**[482]. For oxidation of the intermediate (**792**), TTN or $Pb(OAc)_4$ are used (equation 174). The Dieckman conden-

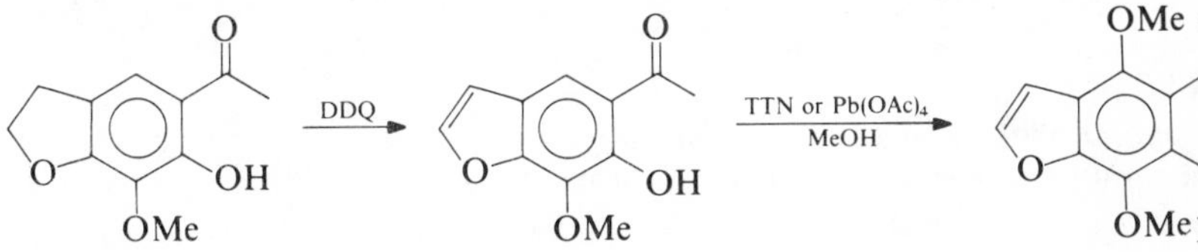

(**791**) →DDQ→ (**792**) →TTN or $Pb(OAc)_4$, MeOH→ (**790**) Ref. 482 (174)

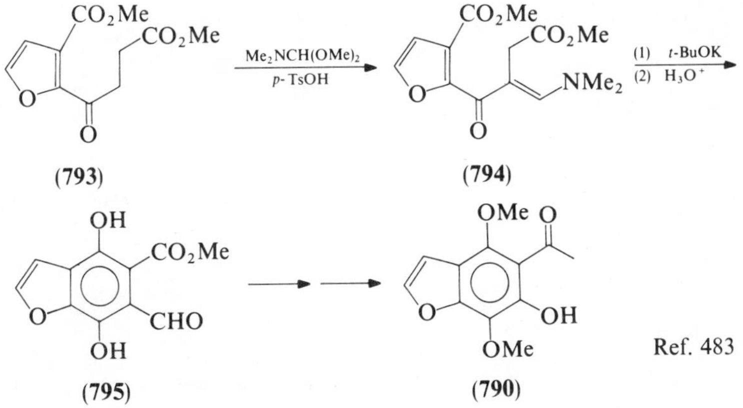

(**793**) →$Me_2NCH(OMe)_2$, *p*-TsOH→ (**794**) →(1) *t*-BuOK (2) H_3O^+→

(**795**) → → (**790**) Ref. 483 (175)

sation of **794** can also be used for the synthesis of the hydroquinone (**795**) with the formyl group, which can be converted to **790** (equation 175)[483]. Annulation of the chromium carbene complex (**796**) is another possibility to obtain the highly functionalized benzofuran[481]. The reaction of **796** with an unsymmetrical alkyne such as **797** or **798** proceeds in a completely regioselective manner (equations 176 and 177). The yields of **799** and **800**, however, are not satisfactory ones. Such a high selectivity is attributed to the steric effect in the annulation stage (see Section III.C). The annulation products **799** and **800** are converted to khellin in several steps.

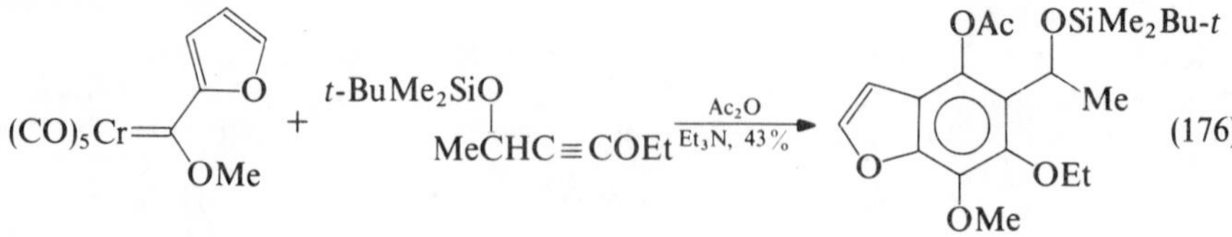

(**796**) + (**797**) $\xrightarrow[\text{Et}_3\text{N, 43\%}]{\text{Ac}_2\text{O}}$ (**799**) (176)

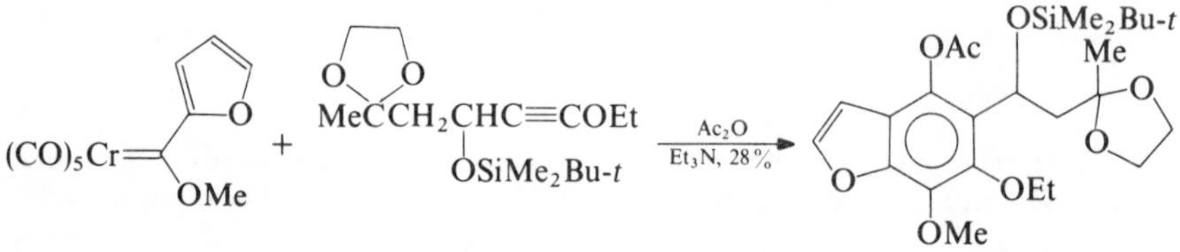

(**796**) + (**798**) $\xrightarrow[\text{Et}_3\text{N, 28\%}]{\text{Ac}_2\text{O}}$ (**800**) (177)

3. *Streptonigrin*

Streptonigrin (**801**) is isolated from *Streptomyces flocculus* and is recognized as an antitumor antibiotic. The whole range of its chemistry is covered by reference 484, which

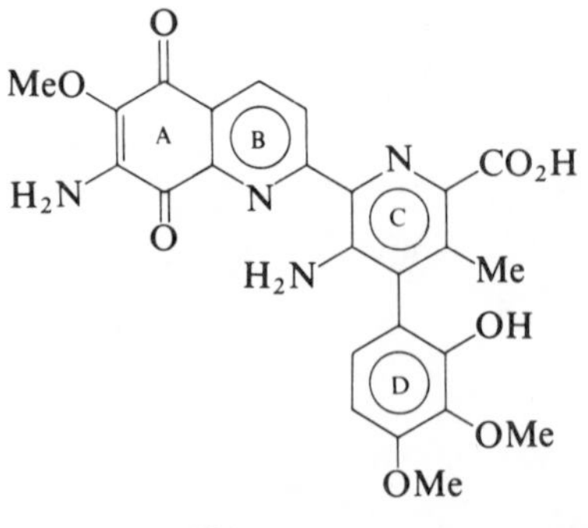

(**801**)

also includes most of the fragment or partial synthesis. Three approaches of the total synthesis are given here.

The major problem in the total synthesis is the binding of the fully functionalized four rings A–D. The first synthesis is due to Weinreb and coworkers[485]. Most of the reactions are concerned with the modification of the substituents on ring C. The order of construction of the four aromatic rings is D → CD → ACD → ABCD (Scheme 65). Ring C is synthesized by a Diels–Alder annulation of the diene (**802**) with a dienophile (**803**), which is generated *in situ* from **811**, as a major isomer and it is subsequently converted to the

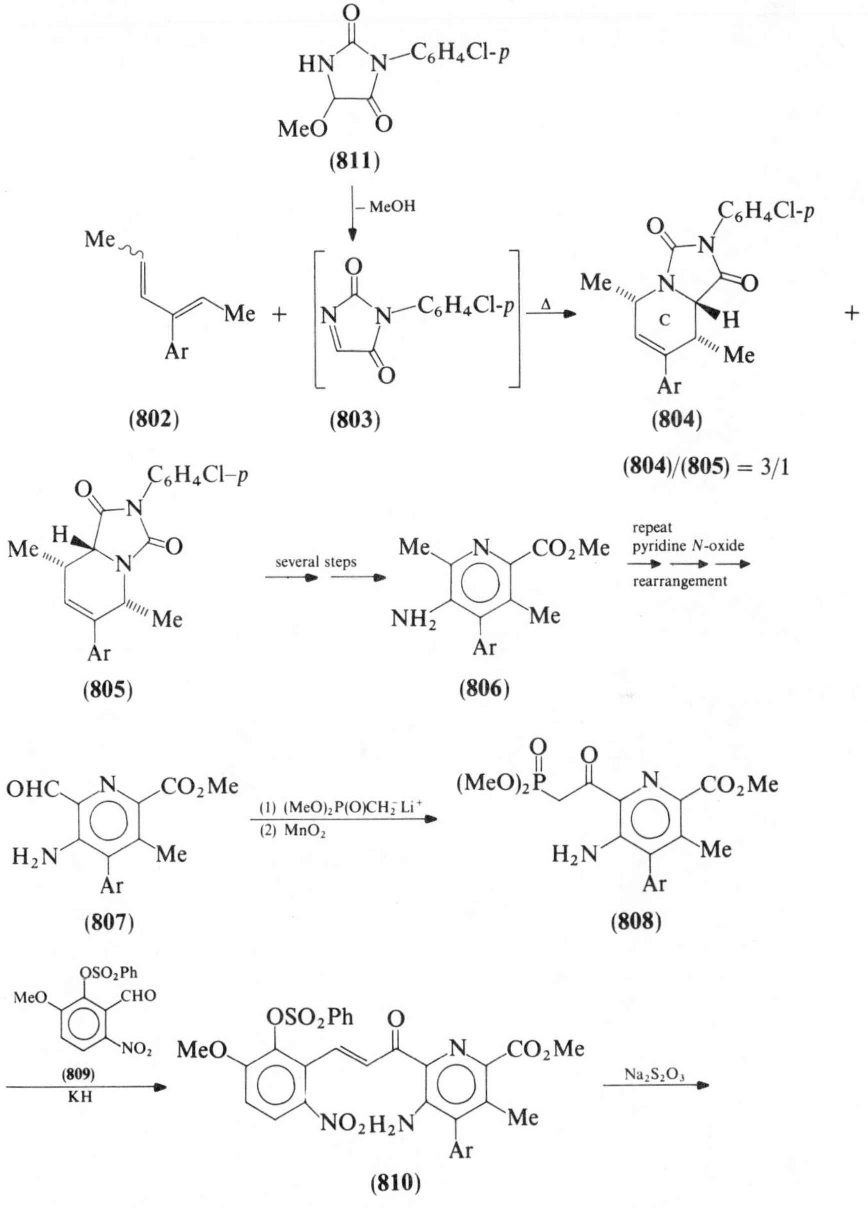

(**811**)

– MeOH

Δ

(**802**) (**803**) (**804**)

(**804**)/(**805**) = 3/1

several steps

repeat pyridine *N*-oxide rearrangement

(**805**) (**806**)

(1) $(MeO)_2P(O)CH_2^-Li^+$
(2) MnO_2

(**807**) (**808**)

(**809**)
KH

$Na_2S_2O_3$

(**810**)

SCHEME 65 (continued)

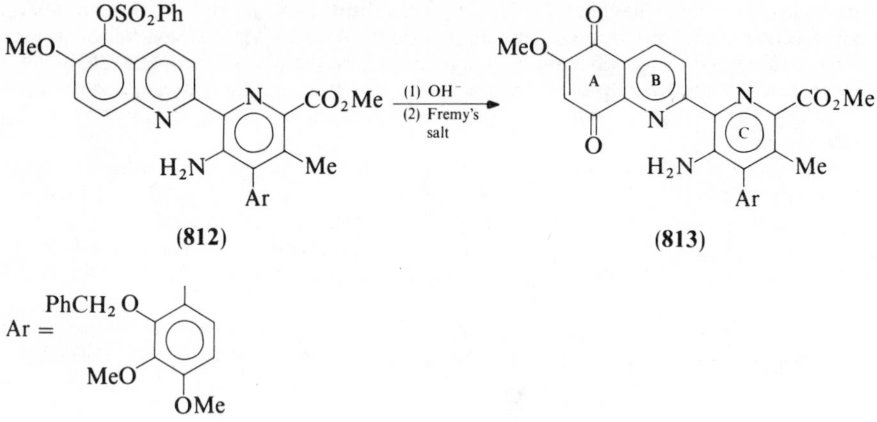

(812) → [(1) OH^-; (2) Fremy's salt] → (813)

OSO_2Ph, MeO, N, CO_2Me, H_2N, Me, Ar; A, B, C

Ar = $PhCH_2O$, MeO, OMe

SCHEME 65

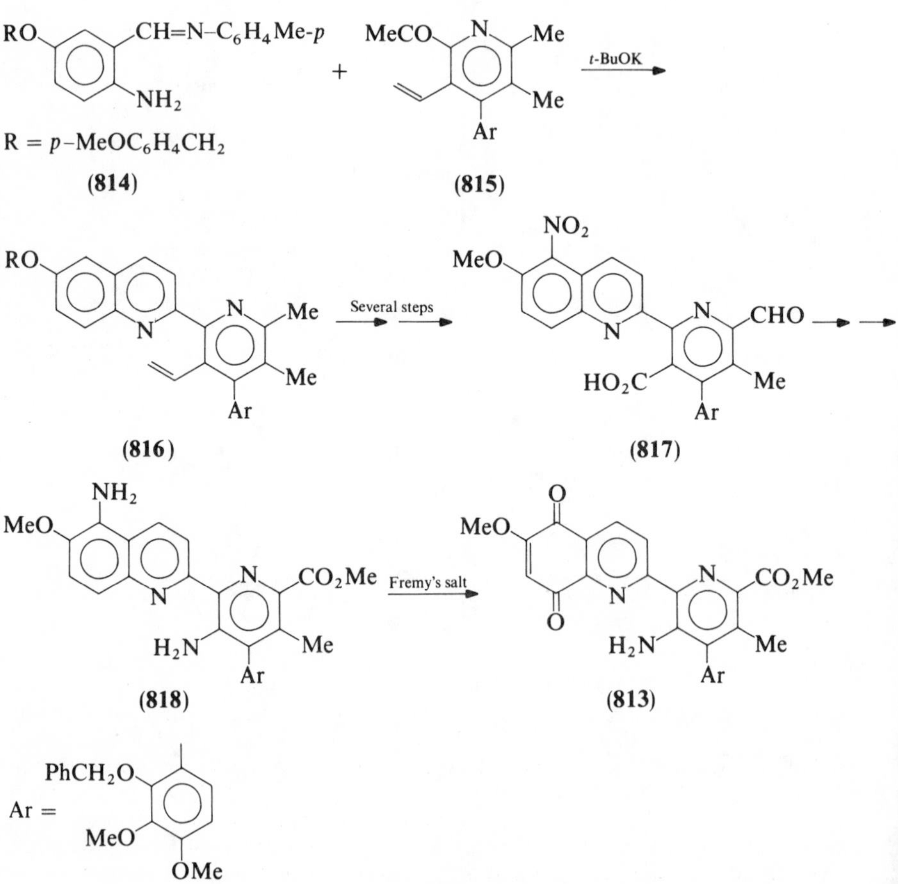

(814) + (815) → [t-BuOK]

RO, $CH{=}N{-}C_6H_4Me$-p, NH_2; $R = p{-}MeOC_6H_4CH_2$

MeCO, N, Me, Me, Ar

(816) → [Several steps] → (817) → →

RO, N, Me, Me, Ar

NO_2, MeO, N, CHO, HO_2C, Me, Ar

(818) → [Fremy's salt] → (813)

NH_2, MeO, N, CO_2Me, H_2N, Me, Ar

Ar = $PhCH_2O$, MeO, OMe

SCHEME 66

pyridine derivative (**806**). After the introduction of the remaining functionalities by means of a repeated use of pyridine *N*-oxide rearrangement, the nitrochalcone (**810**), derived via a Horner–Emmons reaction of **808**, is reductively cyclized to a quinoline (**812**) by a modification of Friedlander quinoline synthesis. The corresponding free phenol obtained from **812** is oxidized by Fremy's salt to give quinolinoquinone (**813**), which is converted to streptonigrine (Scheme 65)[485].

A modified Friedlander quinoline synthesis is also applied in the total synthesis given in Scheme 66[486,487]. Condensation of iminoaniline (**814**) with the acyl pyridine (**815**) in the presence of *t*-BuOK gives the expected quinoline (**816**) in excellent yield. The remaining operation is the conversion of the functionalities on the A and B rings of **817** to the appropriate ones in **818**. Finally, oxidation of the aminoquinoline (**818**) with Fremy's salt gives the same product (**813**) discussed above.

The third approach is based on the inverse electronic demands of azadienes in a Diels–Alder reaction for the construction of the C ring (Scheme 67)[488]. Diels–Alder

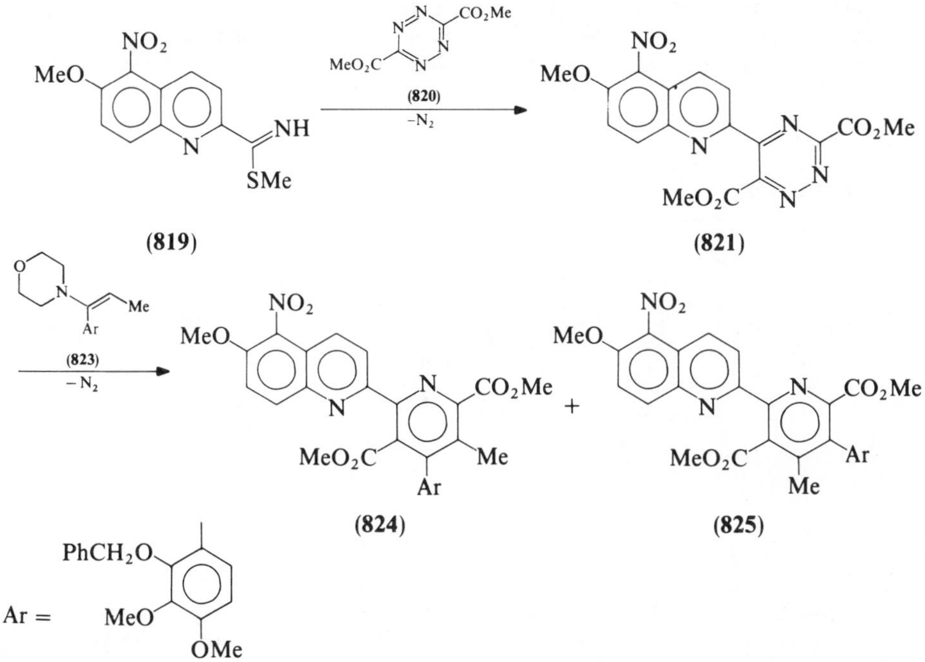

(**819**) (**820**) $-N_2$ (**821**)

(**823**) $-N_2$ (**824**) + (**825**)

SCHEME 67

reaction of the 1,2,4,5-tetrazene (**820**) as the electron-deficient component with the electron-rich thioimide (**819**) gives the corresponding 1,2,4-triazene (**821**). A following annulation reaction with the enamine (**823**) affords the two regioisomeric pyridines, **824** and **825**. The [4 + 2] annulation reaction gives under high pressure a higher proportion of the desired compound (**824**). This procedure is very elegant due to the simplicity of its strategy. A following reaction of **824** affords an important precursor to streptonigrin.

The synthesis of lavendamycin (**826**)[489,490], which is the hypothetical precursor in streptonigrin biosynthesis, has also been reported.

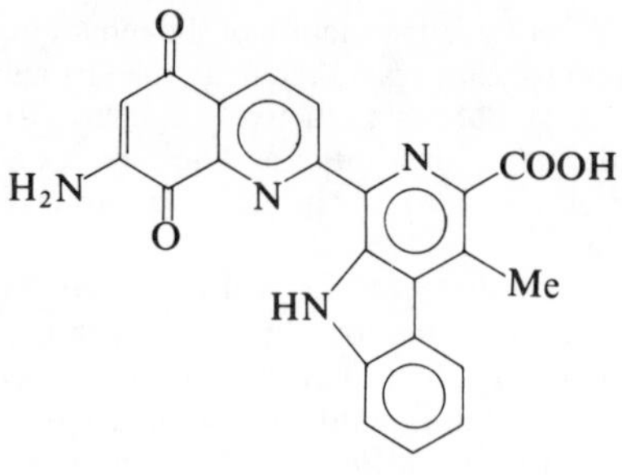

(826)

4. Saframycin

Saframycins are antitumor antibiotics isolated from *Streptomyces lavendulae*[491]. They have a characteristic isoquinoline unit and the structurally similar two components are bound at the central ring. The symmetrical structure is utilized in the total synthesis of

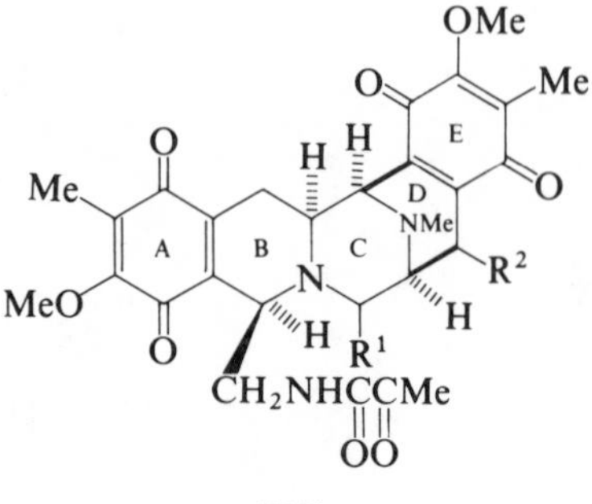

(827)

a: $R^1 = CN$, $R^2 = H$
b: $R^1 = R^2 = H$
c: $R^1 = H$, $R^2 = OMe$

827b (Scheme 68)[492]. The key fragments, **831** and **832**, both of which are prepared from the same aromatic aldehyde (**828**), are coupled to give amide **833a**. Aminoaldehyde **834**, which is obtained from **833a** in several steps, undergoes an intramolecular 'phenolic cyclization' between aromatic ring, then the condensation between the amino and the formyl groups gives **835** in a stereoselective manner. For the final construction of **B** ring from **836**, the same reaction with a protected glycyl aldehyde is efficiently utilized.

Partial synthesis of the right-hand half of the structure is obtained by means of a double cyclization method[112c]. The reaction proceeds by an electrophilic attack of the intermediate immonium salt on the aromatic nucleus (equation 178).

5. Naphtholidinomycin

Naphtholidinomycin (**842a**) is isolated from *Streptomyces lusitanus*, and its hexacyclic structure as well as those of other three homologues derivatives (**842b**)–(**842d**) is one of the

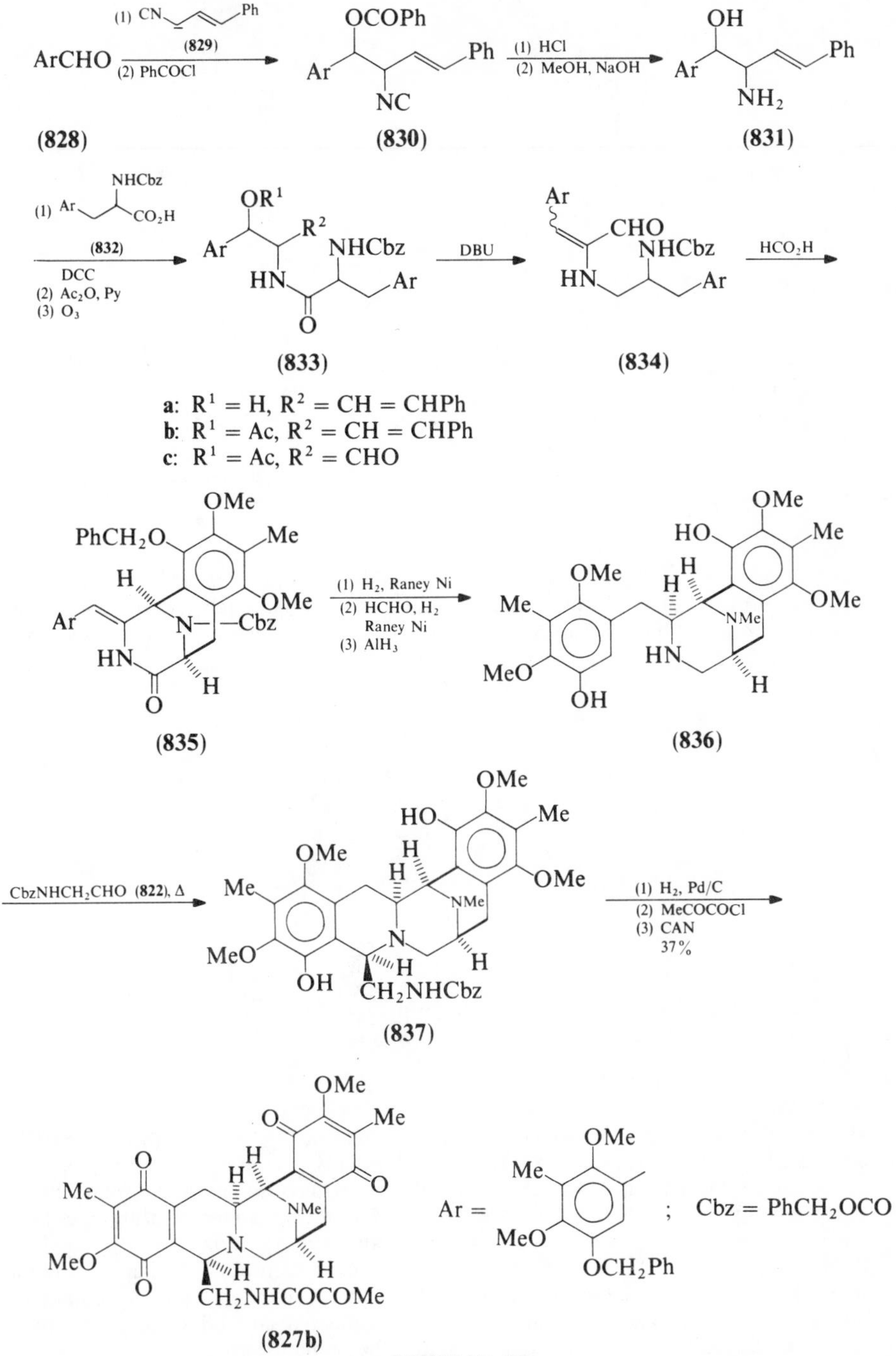

ArCHO (1) CN–CH=CH–Ph (829) (2) PhCOCl → (830) (1) HCl (2) MeOH, NaOH → (831)

(828) (830) (831)

(1) (832) DCC (2) Ac_2O, Py (3) O_3 → (833) DBU → (834) HCO_2H →

(833) (834)

a: $R^1 = H, R^2 = CH=CHPh$
b: $R^1 = Ac, R^2 = CH=CHPh$
c: $R^1 = Ac, R^2 = CHO$

(835) (1) H_2, Raney Ni (2) HCHO, H_2 Raney Ni (3) AlH_3 → (836)

(835) (836)

$CbzNHCH_2CHO$ (822), Δ → (837) (1) H_2, Pd/C (2) MeCOCOCl (3) CAN 37% →

(837)

(827b) Ar = [3-benzyloxy-5-methoxy-4-methyl-2-methoxyphenyl structure: OMe, Me, MeO, OCH_2Ph] ; Cbz = $PhCH_2OCO$

(827b)

SCHEME 68

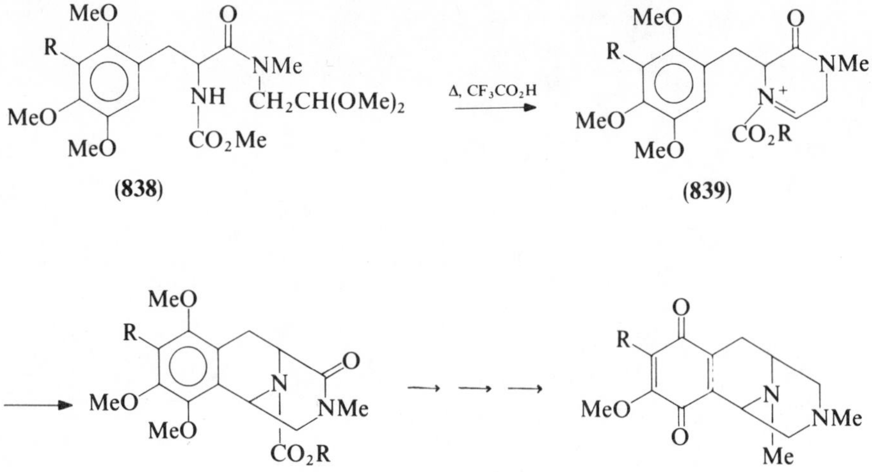

(838) $\xrightarrow{\Delta,\ CF_3CO_2H}$ (839) → (840) → → → (841)

(178)

most complex quinonoid metabolite structures isolated from bacteria[493-495]. Recently, they are of interest due to their marked antitumor and antibiotic activity. Three synthetic approaches to naphtholidinomycin are reported so far. The first attempt is based on a

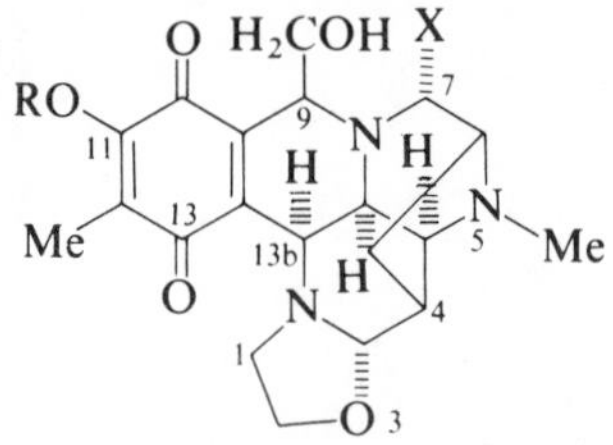

(842)

a: X = OH, R = Me naphthyridinomycin
b: X = CN, R = Me cyanocycline
c: X = OH, R = H
d: X = CN, R = H

double cyclization reaction of the tetrahydroisoquinoline derivative (**843**), which involves an amide formation between two corresponding fragments **843** and **844** (Scheme 69)[496]. This reaction was originally shown in the partial synthesis of saframycin[112c]. Oxidation of a benzylic hydroxyl group of **843** gives a mixture of diastereoisomers at C(6) and C(9) of the coupling product **845**. $BF_3 \cdot OEt_2$ treatment of **845** at elevated temperature gives **847**, whose stereochemistry at C(13c) is assigned to be an unnatural form, in 30% yield. This double cyclization reaction presumably involves deprotective immonium ion (**846**) formation and a subsequent nucleophilic cyclization between the enol and the immonium ion moieties. This cyclization occurs from the less hindered site of **846** due to the steric interaction between the vinyl group at C(9) and the *N*-methyl group.

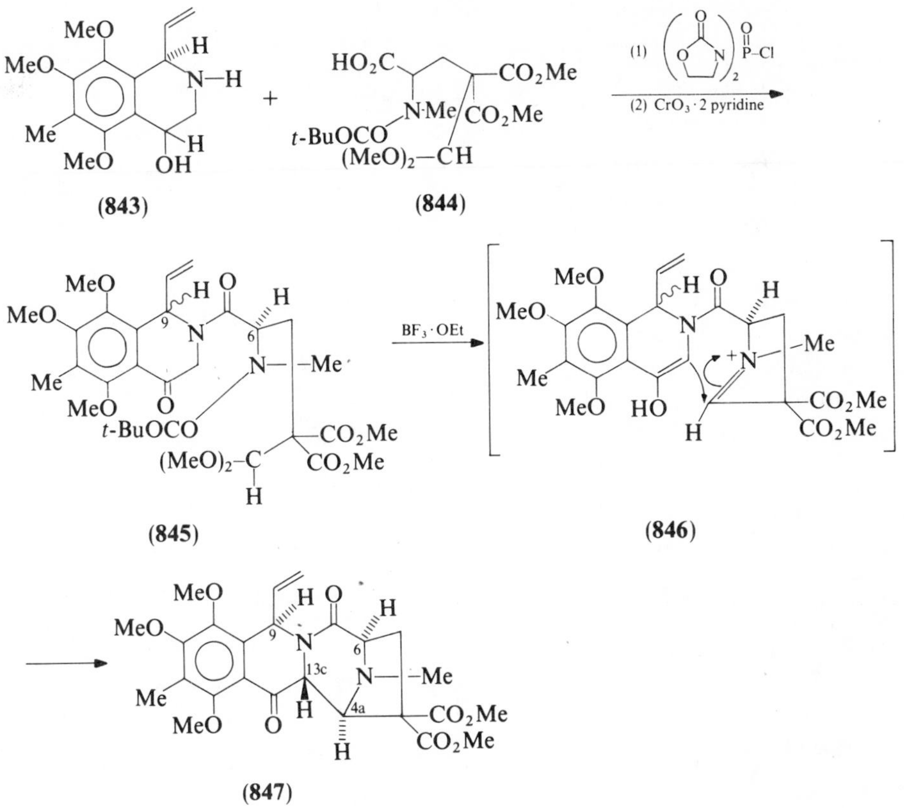

(843) (844)

(845) (846)

(847)

SCHEME 69

The second approach involves the formation of the isoquinoline ring from two components by means of an intramolecular amidoalkylation, which in the reaction of **848** affords stereoselectively **850a** in 56% yield (Scheme 70)[497]. This stereoselectivity is attributed to a diastereoface selective reaction of a Lewis acid complexed intermediary immonium salt. The benzyl ether of **850** is subjected to sequential ozonolysis and reductive workup to afford an isomeric mixture of the corresponding hemiacetals (**852**). A successive chlorination and Friedel–Crafts type cyclization gives **854**, which possesses most of the required functionalities and carbon skeleton, and is therefore considered as a promising synthetic precursor of (±)-naphtholidinomycin.

A total synthesis of the rather stable congener cyanocyclinone (**842b**) was accomplished by means of 'phenolic cyclization' of the immonium salt (**859**) (Scheme 71)[498]. **850** was converted to the dialdehyde (**857**), which was cyclized with the *O*-protected aminoethanol to give **858**. Acid treatment of **858** gave stereoselectively the cyclized product (**860**) in a fair yield, presumably via iminium salt **859**. Due to the difficulty of the direct and selective reduction of the amido group of **860**, the silatropic product (**861**) derived from (**860**) was stereoselectively reduced to **862** and then converted to the corresponding nitrile derivative (**863**), whose deprotection and an aereal oxidation gave racemic cyanocyclinone, (±)-**842b**.

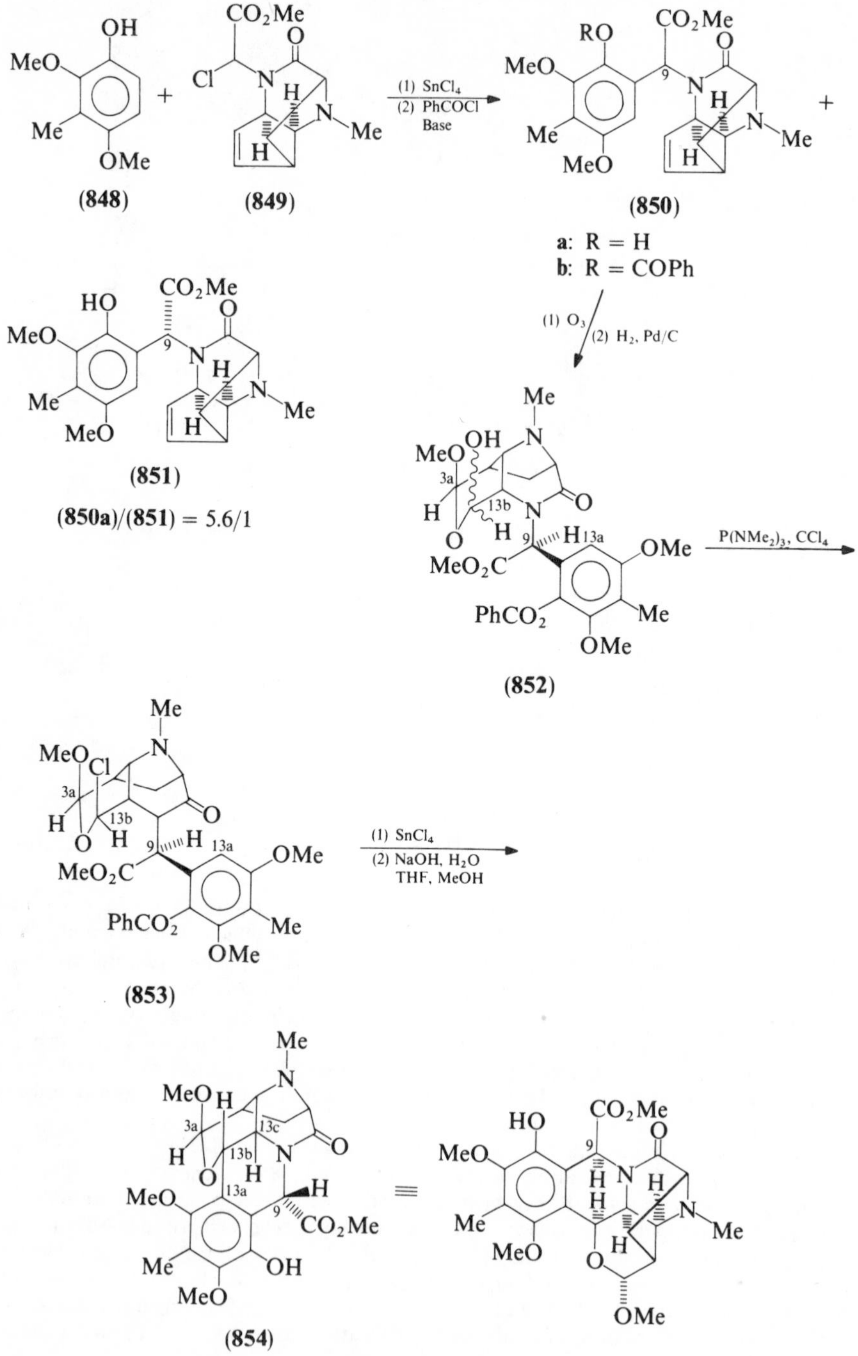

(848) + (849) → (1) $SnCl_4$ (2) PhCOCl Base → (850) + (851)

(850) a: R = H
b: R = COPh

(850a)/(851) = 5.6/1

(850) → (1) O_3 (2) H_2, Pd/C → (852)

(852) → $P(NMe_2)_3$, CCl_4 → (853)

(853) → (1) $SnCl_4$ (2) NaOH, H_2O THF, MeOH → (854)

SCHEME 70

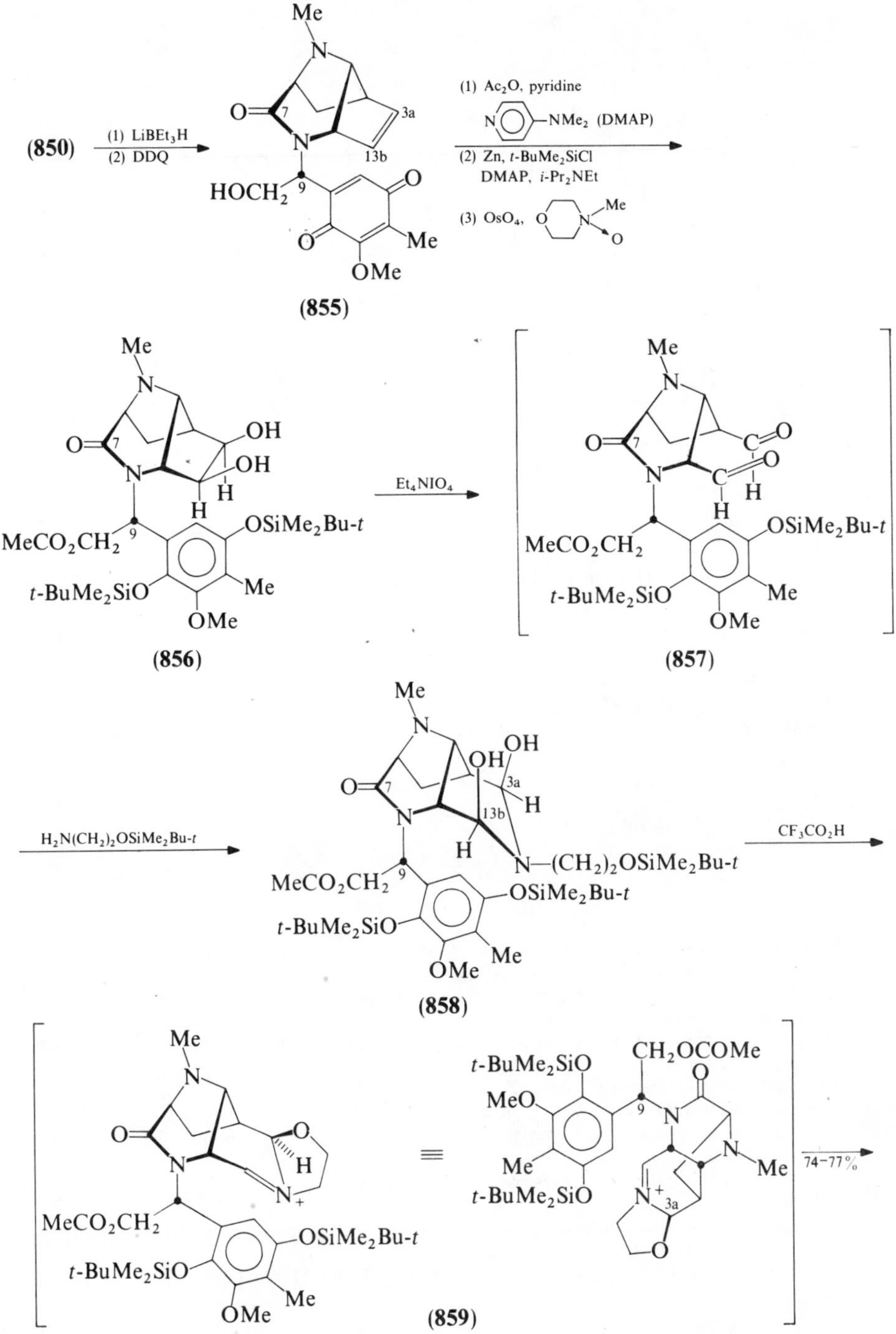

(850) —(1) $LiBEt_3H$ (2) DDQ→ (855) —(1) Ac_2O, pyridine, DMAP; (2) Zn, t-$BuMe_2SiCl$, DMAP, i-Pr_2NEt; (3) OsO_4, N-methylmorpholine N-oxide→ (856) —Et_4NIO_4→ [(857)] —$H_2N(CH_2)_2OSiMe_2Bu$-t→ (858) —CF_3CO_2H→ [(859)] —74–77%→

SCHEME 71 (continued)

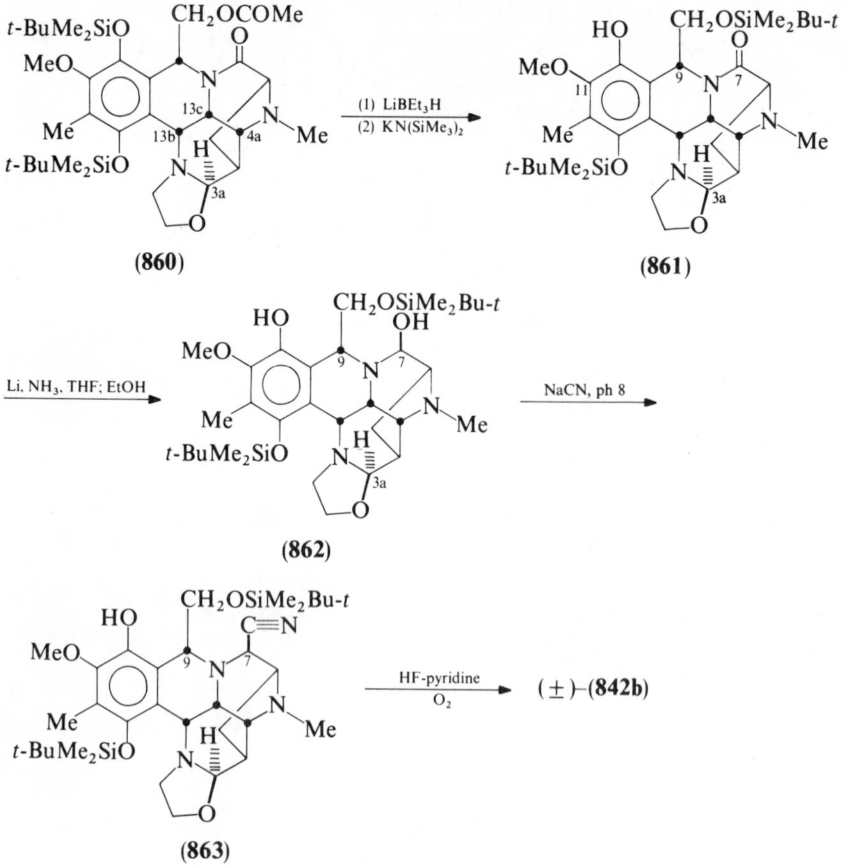

(860) —(1) $LiBEt_3H$ (2) $KN(SiMe_3)_2$→ (861)

—Li, NH_3, THF; EtOH→ (862) —NaCN, ph 8→

(863) —HF-pyridine, O_2→ (±)-(842b)

SCHEME 71

VIII. REFERENCES

1. R. H. Thomson, in *The Chemistry of the Quinonoid Compounds*, Part 1 (Ed. S. Patai), John Wiley and Sons, New York, 1974, pp. 111–161.
2. J. M. Bruice, in *Rodd's Chemistry of Carbon Compounds*, Vol. IIIB (Ed. S. Coffery), Elsevier, Amsterdam, 1974, pp. 1–176.
3. C. A. Buehler and D. E. Pearson, *Survey of Organic Synthesis*, John Wiley and Sons, New York, 1977, pp. 635–654.
4. T. Laid, in *Comprehensive Organic Chemistry*, Vol. 1 (Eds. D. R. H. Barton and W. D. Ollis), Pergamon Press, Oxford, 1979, pp. 1213–1227.
5. *Houben-Weyl Methoden der Organischen Chemie*, Vol. VII-3a, Chinon-I, George Thieme Verlag, Berlin, 1979.
6. *Houben-Weyl Methoden der Organischen Chemie*, Vol. IV-1b, George Thieme Verlag, Stuttgart, 1975, p. 153.
7. J. M. Bruice, in *Rodd's Chemistry of Carbon Compounds*, Suppl. Col. IIIB/IIIC (Ed. M. F. Ansell). Elsevier, Amsterdam, 1981, pp. 1–137.
8. I. P. Beletskaya and D. I. Makhonkov, *Usp. Khim.*, **50**, 1007 (1981); *Chem. Abstr.*, **95**, 114224 (1981).

9. L. Eberson, *Adv. Phys. Org. Chem.*, **18**, 79 (1982).
10. M. Periasamy and M. V. Bhatt, *Synthesis*, 330 (1977).
11. T.-L. Ho, T.-W. Hall and C. M. Wong, *Synthesis*, 206 (1973).
12. M. Periasamy and M. V. Bhatt, *Tetrahedron Lett.*, 4561 (1978).
13. Ger. Offen., 1,804,727 (1969); *Chem. Abstr.*, **71**, 81034e (1969).
14. J. Skarzewski, *Tetrahedron*, **40**, 4997 (1984).
15. D. Pletcher and S. J. D. Tait, *J. Chem. Soc. Perkin Trans. 2*, 788 (1979).
16. M. N. Bhattcharjee, M. K. Chaudhuri, H. S. Dasgupta, N. Roy and D. T. Khathing, *Synthesis*, 588 (1982).
17. S. Ranganathan, D. Ranganathan and P. V. Ramachandran, *Tetrahedron*, **40**, 3145 (1984).
18. E. C. Taylor and A. McKillop, *Acc. Chem. Res.*, **3**, 338 (1970).
19. A. McKillop, J. D. Hunt, M. J. Zelesko, J. S. Fowler, E. C. Taylor, G. McGillivray and F. Kienzle, *J. Am. Chem. Soc.*, **93**, 4841 (1971).
20. E. C. Taylor, F. Kienzle, R. L. Robey, A. McKillop and J. D. Hunt, *J. Am. Chem. Soc.*, **93**, 4845 (1971).
21. G. K. Chip and J. S. Grossert, *J. Chem. Soc. Perkin Trans. 1*, 1629 (1972).
22. E. C. Taylor, J. G. Andrade, G. J. H. Rall and A. McKillop, *J. Org. Chem.*, **43**, 3632 (1978).
23. P. Müller and C. Bobillier, *Tetrahedron Lett.*, **22**, 5157 (1981).
24. P. Müller and C. Bobillier, *Helv. Chim. Acta*, **68**, 450 (1985).
25. Z. Blum and K. Nijberg, *Acta Chem. Scand. B*, **33**, 73 (1979).
26. H. Zimmer, D. C. Lankin and S. W. Horgan, *Chem. Rev.*, **71**, 229 (1971).
27. L. Birladeanu, E. Chamot, W. E. Fristad, L. A. Paquett and S. Winstein, *J. Org. Chem.*, **42**, 3260 (1977).
28. S. Hibino and S. M. Weinreb, *J. Org. Chem.*, **42**, 232 (1977).
29. H. Ishii, T. Hanaoka, A. Asada, Y. Harada and N. Ikeda, *Tetrahedron*, **32**, 2693 (1976).
30. M. J. Perkins and P. Ward, *J. Chem. Soc. Chem. Commun.*, 883 (1973); S. A. Hussain, T. C. Jenkins and M. J. Perkins, *Tetrahedron Lett.*, 3199 (1977); S. A. Hussian, T. C. Jenkins and M. J. Perkins, *J. Chem. Soc. Perkin Trans. 2*, 2809 (1979).
31. P. F. Alewood, I. C. Calder and R. L. Richardson, *Synthesis*, 121 (1981).
32. G. L. Olson, H.-C. Cheung, K. Morgan and G. Saucy, *J. Org. Chem.*, **45**, 803 (1980).
33. D. H. R. Barton, A. G. Brewster, S. V. Ley and M. N. Rosenfeld, *J. Chem. Soc. Chem. Commun.*, 985 (1976).
34. D. H. R. Barton, S. V. Ley, P. D. Magnus and M. N. Rosenfeld, *J. Chem. Soc. Perkin Trans. 1*, 567 (1977).
35. D. H. R. Barton, A. G. Brewster, S. V. Ley, C. M. Read and M. N. Rosenfeld, *J. Chem. Soc. Perkin Trans. 1*, 1473 (1981).
36. K. B. Sukumaran and R. G. Harvey, *J. Am. Chem. Soc.*, **101**, 1357 (1979); K. B. Sukumaran and R. G. Harvey, *J. Org. Chem.*, **45**, 4407 (1980).
37. D. T. C. Yang and W. M. Trie, *Org. Prep. Proc. Int.*, **14**, 202 (1982).
38. Y. Miyahara, T. Inazu and T. Yashiro, *Tetrahedron Lett.* **23**, 2187 (1982).
39. D. H. R. Barton, C. R. A. Godfrey and J. E. Morzycki, *Tetrahedron Lett.*, **23**, 957 (1982).
40. S. Uemura, in *Synthetic Reagents*, Vol. 5 (Ed. J. S. Pizey), Ellis Horwood, Chichester, 1983, Chapter 5.
41. A. McKillop, B. P. Swann, M. J. Zelesko and E. C. Taylor, *Angew. Chem. Int. Ed. Engl.*, **9**, 74 (1970).
42. A. McKillop, B. P. Swann and E. C. Taylor, *Tetrahedron*, **26**, 4013 (1970).
43. A. McKillop, D. H. Perry, M. Edwards, S. Antus, L. Farkas, M. Nógrádi and E. C. Taylor, *J. Org. Chem.*, **41**, 282 (1976).
44. Y. Yamada and K. Hosokawa, *Synthesis*, 53 (1977).
45. J. Griffiths, K.-Y. Chu and C. Hawkins, *J. Chem. Soc. Chem. Commun.*, 676 (1976).
46. C. Grundmann, *Synthesis*, 644 (1977).
47. R. T. Taylor and L. A. Flood, *J. Org. Chem.*, **48**, 5160 (1983).
48. D. J. Crouse, M. M. Wheeler, M. Goemann, P. S. Tobin, S. K. Basu and D. M. S. Wheeler, *J. Org. Chem.*, **46**, 1814 (1981).
49. H. L. K. Schmand and P. Boldt, *J. Am. Chem. Soc.*, **97**, 447 (1975); H. L. K. Schmand, H. Kratzin and P. Boldt, *Justus Liebigs Ann. Chem.*, 1560 (1976).
50. T. Wakamatsu, T. Nishi, T. Ohnuma and Y. Ban, *Synth. Commun.*, **14**, 1167 (1984).

51. D. H. R. Barton, P. D. Magnus and J. C. Quinny, *J. Chem. Soc. Perkin Trans. 1*, 1610 (1975).
52. H. J. Duchstein, *Arch. Pharm.*, **318**, 177 (1985).
53. C. R. H. I. de Jonge, H. J. Hageman, G. Hoetjen and W. J. Mijs, *Org. Syn.*, **57**, 78 (1977).
54. T. J. Fullerton and S. P. Ahern, *Tetrahedron Lett.*, 139 (1976).
55. M. Frostin-Rio, D. Pujol, C. Bied-Charreton, M. Perréé-Fauvet and A. Gaudemer, *J. Chem. Soc. Perkin Trans. 1*, 1971 (1984).
56. Ger. Pat., 2,460,665 (1975); *Chem. Abstr.*, **83**, 192831e (1975).
57. Ger. Pat., 2,517,870 (1975); *Chem. Abstr.*, **84**, 58927h (1976).
58. A. Nishinaga, H. Tomita, K. Nishizawa and T. Matsuura, *J. Chem. Soc. Dalton Trans.*, 1504 (1981), and references cited therein.
59. A. Zombeck, R. S. Drago, B. B. Corden and J. H. Gaul, *J. Am. Chem. Soc.*, **103**, 7580 (1981), and references cited therein.
60. V. Kothari and J. J. Tazuma, *J. Catal.*, **41**, 180 (1976).
61. P. Capdevielle and M. Maumy, *Tetrahedron Lett.*, **24**, 5611 (1983).
62. D. Liotta, J. Arbiser, J. W. Short and M. Saindane, *J. Org. Chem.*, **48**, 2932 (1983).
63. M. B. Gopinathan and M. V. Bhatt, *Indian J. Chem., Sec. B*, **20B**, 71 (1981).
64. T. R. Kelly, A. Echavarren and M. Behforous, *J. Org. Chem.*, **48**, 3849 (1983).
65. P. T. Perumal and M. V. Bhatt, *Synthesis*, 205 (1979).
66. Y. Tamura, M. Sasho, S. Arai, A. Wada and Y. Kita, *Tetrahedron*, **40**, 4539 (1984).
67. J. S. Clarke, R. E. Ehigamusoe and A. T. Kuhn, *J. Electroanal. Chem. Interfacial Electrochem.*, **70**, 333 (1976).
68. A. Nilsson and A. R. Ronlán, *J. Chem. Soc. Perkin Trans. 1*, 2337 (1973).
69. A. Ronlán and V. D. Parker, *J. Chem. Soc. (C)*, 3214 (1971).
70. A. B. Suttie, *Tetrahedron Lett.*, 953 (1969).
71. N. Nojiri, Jpn. Pat., 58-40728 (1978).
72. C. Steelink and W. E. Britton, *Tetrahedron Lett.*, 2869 (1974).
73. M. Neptune and R. C. McCreary, *J. Med. Chem.*, **22**, 196 (1979).
74. A. Nilsson, U. Palmquist, T. Patterson and A. Ronlán, *J. Chem. Soc. Perkin Trans. 1*, 696 (1978).
75. M. Matsumoto, H. Kobayashi and Y. Hotta, *J. Org. Chem.*, **50**, 1776 (1985).
76. D. Vidril-Robert, M. T. Maurette, E. Oliveros, H. Hocquaux and B. Jacquet, *Tetrahedron Lett.*, **25**, 529 (1984).
77. F. Ishii and K. Kishi, *Synthesis*, 706 (1980).
78. T. Takata, R. Tajima and W. Ando, *J. Org. Chem.*, **48**, 4764 (1983).
79. D. N. Gupta, P. Hodge and J. E. Davis, *J. Chem. Soc. Perkin Trans. 1*, 2970 (1981).
80. C. R. Harrison and P. Hodge, *J. Chem. Soc. Perkin Trans. 1*, 509 (1982).
81. W. Schäfer and A. Aquado, *Angew. Chem. Int. Ed. Engl.*, **10**, 405 (1971).
82. A. T. Balaban, *Rev. Roumaine Chim.*, **14**, 1281 (1969).
83. S. Spyroudis and A. Varvoglis, *Synthesis*, 445 (1975).
84. J. P. Marino and A. Schwartz, *Tetrahedron Lett.*, 3253 (1979).
85. G. W. Gokel and H. D. Durst, *Aldrichimica Acta*, **9**, 3 (1976).
86. J. M. Bruce, S. F. Fitzjohn and R. T. Pardasiri, *J. Chem. Res. (S)*, 252 (1981).
87. R. Cassis and J. A. Valderrama, *Synth. Commun.*, **13**, 347 (1983).
88. H. Firouzabadi and Z. Mostafavipoor, *Bull. Chem. Soc. Jpn*, **56**, 914 (1983).
89. H. Firouzabadi, B. Vessal and M. Naderi, *Tetrahedron Lett.*, **23**, 1847 (1982).
90. G. Schill, C. Zürcher and W. Vetter, *Chem. Ber.*, **106**, 228 (1973).
91. D. H. R. Barton, S. V. Ley and C. A. Meerholz, *J. Chem. Soc. Chem. Commun.*, 755 (1979).
92. S. V. Ley, C. A. Meerholz and D. H. R. Barton, *Tetrahedron, Suppl. (9)*, 213 (1981).
93. L. Engman and M. P. Cava, *J. Chem. Soc. Chem. Commun.*, 164 (1982).
94. G. B. Deacon, M. F. O'Donoghue, A. McKillop and D. W. Young, *Synth. Commun.*, **10**, 615 (1980).
95. T.-L. Ho, T. W. Hall and C. M. Wong, *Chem. Ind. (Lond.)*, 729 (1972).
96. H. Firouzabadi and N. Iranpoor, *Synth. Commun.*, **14**, 875 (1984).
97. L. Horner and D. W. Baston, *Justus Liebigs Ann. Chem.*, 910 (1973); 1967 (1976).
98. R. Davis, *Tetrahedron Lett.*, 313 (1976).
99. M. Sato, N. Katsumata and S. Ebine, *Synthesis*, 685 (1984).
100. J. P. Marino and A. Schwartz, *J. Chem. Soc. Chem. Commun.*, 812 (1974).

101. H. D. Durst, M. P. Mack and F. Wudl, *J. Org. Chem.*, **40**, 268 (1975).
102. M. R. Ross, PhD Dissertation, Univ. Illinois, Urbrana-Champaign, Ill., 1981.
103. E. J. Corey and C. U. Kim, *J. Am. Chem. Soc.*, **94**, 7586 (1972).
104. A. J. Mancuso and D. Swern, *Synthesis*, 165 (1981).
105. S. V. Ali, J. W. A. Findlay and A. B. Turner, *J. Chem. Soc. Perkin Trans. 1*, 407 (1976).
106. C. D. Snyder and H. Rapoport, *J. Am. Chem. Soc.*, **94**, 227 (1972).
107. P. Jacob, III, P. S. Callery, A. T. Shulgin and N. Castagnoli, Jr, *J. Org. Chem.*, **41**, 3627 (1976).
108. C. D. Snyder, W. E. Bondinell and H. Rapoport, *J. Org. Chem.*, **36**, 3951 (1971).
109. K. Kloc, J. Młochowski and L. Syper, *Chem. Lett.*, 725 (1980).
110. R. G. F. Giles, I. R. Green, V. I. Hugo, P. R. K. Mitchell and C. S. Yorke, *J. Chem. Soc. Perkin Trans. 1*, 2383 (1984).
111. A. Kubo, Y. Kitahara, S. Nakahara and R. Numata, *Chem. Pharm. Bull.*, **31**, 341 (1983).
112. (a) B. Errazuriz, R. Tapia and J. A. Valderrama, *Tetrahedron Lett.*, **26**, 819 (1985); (b) J. R. Lury and H. Rapoport, *J. Am. Chem. Soc.*, **105**, 2859 (1983); (c) H. Kurihara and H. Mishima, *Tetrahedron Lett.*, **23**, 3639 (1982).
113. J. S. Swenton, *Acc. Chem. Res.*, **16**, 74 (1983).
114. S. Fujita, *Yuki Gosei Kagaku Kyokai Shi*, **40**, 307 (1982).
115. A. Nilsson, U. Palmquist, T. Pettersson and A. Ronlán, *J. Chem. Soc. Perkin Trans. 1*, 708 (1978).
116. B. L. Chenard and J. S. Swenton, *J. Chem. Soc. Chem. Commun.*, 1172 (1979).
117. D. K. Jackson and J. S. Swenton, *Synth. Commun.*, **7**, 333 (1977).
118. M. G. Dolson, D. K. Jackson and J. S. Swenton, *J. Chem. Soc. Chem. Commun.*, 327 (1979).
119. D. R. Henton, R. L. McCreery and J. S. Swenton, *J. Org. Chem.*, **45**, 369 (1980).
120. N. L. Weinberg and B. Belleau, *Tetrahedron*, **29**, 279 (1973).
121. J. S. Swenton and P. W. Raynolds, *J. Am. Chem. Soc.*, **100**, 6188 (1978).
122. M. G. Dolson and J. S. Swenton, *J. Org. Chem.*, **46**, 177 (1981).
123. P. Margaretha and P. Tissot, *Helv. Chim. Acta*, **58**, 933 (1975).
124. A. Ronlán, K. Bachgard and V. D. Parker, *Acta Chem. Scand.*, **27**, 2375 (1973).
125. M. G. Dolson and J. S. Swenton, *J. Am. Chem. Soc.*, **103**, 2361 (1981).
126. J. S. Swenton, D. K. Jackson, M. J. Manning and P. W. Raynolds, *J. Am. Chem. Soc.*, **100**, 6182 (1978).
127. M. J. Manning, P. W. Raynolds and J. S. Swenton, *J. Am. Chem. Soc.*, **98**, 5008 (1976).
128. W. Dückheimer and L. A. Cohen, *Biochemistry*, **3**, 1948 (1964).
129. W. Duckheimer and L. A. Cohen, *J. Am. Chem. Soc.*, **86**, 4388 (1964).
130. T. W. Hart and F. Scheinmann, *Tetrahedron Lett.*, **21**, 2295 (1980).
131. G. Büchi, P.-S. Chu, A. Hoppmann, C.-P. Mak and A. Pearce, *J. Org. Chem.*, **43**, 3983 (1978).
132. A. Goosen and C. W. McCleland, *J. Chem. Soc. Perkin Trans. 1*, 646 (1978).
133. J. P. Willis, K. A. Z. Gogins and L. L. Miller, *J. Org. Chem.*, **46**, 3215 (1981).
134. T. C. Jempty, K. A. Z. Gogins, Y. Mazur and L. L. Miller, *J. Org. Chem.*, **46**, 4545 (1981).
135. R. F. Stewart and L. L. Miller, *J. Am. Chem. Soc.*, **102**, 4999 (1980).
136. K. L. Platt and F. Oesch, *Tetrahedron Lett.*, **23**, 163 (1982).
137. R. G. Harvey, S. H. Goh and C. Cortez, *J. Am. Chem. Soc.*, **97**, 3468 (1975).
138. M. S. Newman and C. C. David, *J. Org. Chem.*, **32**, 66 (1967).
139. H. Nagaoka, C. Schmid, H. Iio and Y. Kishi, *Tetrahedron Lett.*, **22**, 899 (1981).
140. M. Hocquaux, B. Jacquet, D. Vidril-Robert, M.-T. Maurette, and E. Oliveros, *Tetrahedron Lett.*, **25**, 533 (1984).
141. S. Mashraqui and P. Keehn, *Synth. Commun.*, **12**, 637 (1982).
142. Reference 1, p. 150.
143. D. L. Frost, N. N. Ekwuribe and W. A. Remers, *Tetrahedron Lett.*, 131 (1973).
144 B. Eistert, J. Riedinger, F.Küffner and W. Lazik, *Chem. Ber.*, **106**, 727 (1973).
145. S. N. Holter, U. S. Pat., 3,708,509 (1972); *Chem. Abstr.*, **78**, 58055 (1973).
146. S. L. Goldstein and E. McNelis, *J. Org. Chem.*, **38**, 185 (1973).
147. K. A. Parker and S.-K. Kang, *J. Org. Chem.*, **44**, 1536 (1979).
148. R. L. Hand and R. F. Nelson, *J. Am. Chem. Soc.*, **96**, 850 (1974).
149. D. R. Crup, R. W. Franck, R. Gruska, A. A. Ozorio, M. Pagnotta, G. J. Suita and J. G. White, *J. Org. Chem.*, **42**, 105 (1977).
150. T. Wagner-Jauregg, *Synthesis*, 165 and 769 (1980).

151. M. Petrazilika and J. I. Grayson, *Synthesis*, 753 (1981).
152. P. Brownbridge, *Synthesis*, 85 (1983).
153. J. F. W. Keana, J. S. Bland, P. E. Eckler and V. Nelson, *J. Org. Chem.*, **41**, 2124 (1976).
154. T. R. Kelly, R. N. Goerner, T. W. Gillard and B. K. Prazak, *Tetrahedron Lett.*, 3869 (1976).
155. T. R. Kelly, *Tetrahedron Lett.*, 1387 (1978).
156. B. Serckx-Poncin, A.-M. Hesbain-Frisque and L. Ghosez, *Tetrahedron Lett.*, **23**, 3261 (1982).
157. C. Schmid, S. D. Sabnis, E. Schmidt and D. K. Taylor, *Can. J. Chem.*, **49**, 371 (1971).
158. Rhône-Poulenc, S. A., Belgian Pat., 670769 (1966); *Chem. Abstr.*, **65**, 5487 (1966).
159. M. E. Jung, *J. Chem. Soc. Chem. Commun.*, 956 (1974); J. F. W. Keana and P. E. Eckler, *J. Org. Chem.*, **41**, 2625 (1976); C. Girard, P. Amis, J. P. Barnier and J. M. Conia, *Tetrahedron Lett.*, 3329 (1974); M. E. Jung and A. McCombs, *Org. Synth.*, **58**, 163 (1978).
160. R. K. Boeckmann, T. M. Dolak and K. O. Culos, *J. Am. Chem. Soc.*, **100**, 7098 (1978).
161. E. McDonald, A. Suksamrarn and R. D. Wylie, *J. Chem. Soc. Perkin Trans. 1*, 1893 (1979).
162. J. Banville and P. Brassard, *J. Org. Chem.*, **41**, 3018 (1976).
163. J. Banville, J.-L. Grandmaison, G. Lang and P. Brassard, *Can. J. Chem.*, **52**, 80 (1974).
164. J.-L. Grandmaison and P. Brassard, *J. Org. Chem.*, **43**, 1435 (1978).
165. S. Danishefsky, C.-F. Yan, R. K. Singh, R. B. Gammill, P. M. McCurry, Jr, N. Fritsch and J. Clardy, *J. Am. Chem. Soc.*, **101**, 7001 (1979).
166. S. Danishefsky and T. Kitahara, *J. Am. Chem. Soc.*, **96**, 7808 (1974).
167. S. Danishefsky, T. Kitahara, C.-F. Yan and J. Morris, *J. Am. Chem. Soc.*, **101**, 6996 (1979).
168. R. W. Aben and H. W. Scheeren, *J. Chem. Soc. Perkin Trans. 1*, 3132 (1979).
169. K. Krohn and K. Tolkiehn, *Chem. Ber.*, **112**, 3453 (1979).
170. B. Simoneau, J. Savard and P. Brassard, *J. Org. Chem.*, **50**, 5434 (1985).
171. C. Brisson and P. Brassard, *J. Org. Chem.*, **46**, 1810 (1981).
172. J. Savard and P. Brassard, *Tetrahedron Lett.*, 4911 (1979).
173. (a) G. Roberge and P. Brassard, *J. Org. Chem.*, **46**, 4161 (1981); (b) V. Guay and P. Brassard, *Tetrahedron*, **40**, 5039 (1984).
174. G. Roberge and P. Brassard, *J. Chem. Soc. Perkin Trans. 1*, 1041 (1978).
175. J. Banville and P. Brassard, *J. Chem. Soc. Perkin Trans. 1*, 1852 (1976).
176. G. Roberge and P. Brassard, *Synthesis*, 148 (1979).
177. D. W. Cameron, G. I. Feutrill, P. G. Griffiths and D. J. Hodder, *J. Chem. Soc. Chem. Commun.*, 688 (1978).
178. D. W. Cameron, C. Conn and G. I. Feutrill, *Aust. J. Chem.*, **34**, 1945 (1981); D. W. Cameron, D. J. Deutscher, G. I. Feutrill and P. G. Griffiths, *Aust. J. Chem.*, **34**, 2401 (1981).
179. B. A. Pealmau, I. M. McNamara, I. Hasan, S. Hatakeyama, H. Sekizaki and Y. Kishi, *J. Am. Chem. Soc.*, **103**, 4248 (1981).
180. A. P. Kozikowski and K. Sugiyama, *Tetrahedron Lett.*, **21**, 3257 (1980).
181. J. F. W. Keana and P. E. Eckler, *J. Org. Chem.*, **41**, 2625 (1976).
182. R. Roberge and P. Brassard, *Synthesis*, 381 (1981).
183. K. Krohn, *Tetrahedron Lett.*, **21**, 3557 (1980).
184. R. Gompper and M. Sramek, *Synthesis*, 649 (1981).
185. P. Brownbridge and T.-H. Chan, *Tetrahedron Lett.*, **21**, 3423 (1980).
186. A. Murai, K. Takahashi, H. Taketsuru and T. Masamune, *J. Chem. Soc. Chem. Commun.*, 221 (1981).
187. P. Brownbridge and T.-H. Chan, *Tetrahedron Lett.*, **21**, 3427 and 3431 (1980).
188. L. Contreras, C. E. Slemon and D. B. MacLean, *Tetrahedron Lett.*, 4237 (1978).
189. B. A. Keay, D. K. W. Lee and R. Rodrigo, *Tetrahedron Lett.*, **21**, 3663 (1980).
190. P. Vogel and A. Florey, *Helv. Chim. Acta*, **57**, 200 (1974).
191. Y. Bessiere and P. Vogel, *Helv. Chim. Acta*, **63**, 232 (1980).
192. H. Staab and J. Ipaktschi, *Chem. Ber.*, **101**, 1457 (1968).
193. B. J. Arnold, P. G. Sammes and T. W. Wallace, *J. Chem. Soc. Perkin Trans. 1*, 415 (1974).
194. J. W. Wiseman, N. I. French, R. K. Hallmark and K. G. Chiong, *Tetrahedron Lett.*, 3765 (1978).
195. F. A. J. Kerdesky and M. P. Cava, *J. Am. Chem. Soc.*, **100**, 3635 (1978).
196. (a) R. K. Brockmann, M. H. Delton, T. Nagasaka and T. Watanabe, *J. Org. Chem.*, **42**, 2946 (1977); (b) M. Azadi-Ardakani and T. W. Walac, *Tetrahedron Lett.*, **24**, 1829 (1983).
197. R. L. Danheiser and H. Sard, *J. Org. Chem.*, **45**, 4810 (1980).

198. G. A. Kraus and S. H. Wo, *J. Org. Chem.*, **51**, 114 (1986).
199. K. N. Houk, L. N. Domelsmith, R. W. Strozier and R. T. Patterson, *J. Am. Chem. Soc.*, **100**, 6531 (1978).
200. J. G. Brauman, R. C. Hawley and H. Rapoport, *J. Org. Chem.*, **50**, 1569 (1985).
201. D. W. Cameron, G. I. Feutrill and P. Perlmutter, *Tetrahedron Lett.*, **22**, 3273 (1981).
202. S. Danishefsky, B. J. Uang and G. Quallich, *J. Am. Chem. Soc.*, **107**, 1285 (1985).
203. E. J. Corey and A. P. Kozikowski, *Tetrahedron Lett.*, 2389 (1975).
204. M. E. Jung and J. A. Lowe, III, *J. Chem. Soc. Chem. Commun.*, 95 (1978).
205. Y. Tamura, A. Wada, M. Sasho and Y. Kita, *Tetrahedron Lett.*, **22**, 4283 (1981).
206. R. G. F. Giles and G. H. P. Roos, *J. Chem. Soc. Perkin Trans. 1*, 1632 (1976).
207. R. G. F. Giles and G. H. P. Roos, *J. Chem. Soc. Perkin Trans. 1*, 2057 (1976).
208. W. G. Dauben and W. R. Baker, *Tetrahedron Lett.*, **23**, 2611 (1982).
209. T. R. Kelly and W.-G. Tsang, *Tetrahedron Lett.*, 4457 (1978).
210. W. W. Lee, A. P. Martinez, T. H. Smith and D. W. Henry, *J. Org. Chem.*, **41**, 2296 (1976).
211. A. S. Kende, T.-G. Tsay and J. E. Mills, *J. Am. Chem. Soc.*, **98**, 1967 (1976).
212. D. A. Jackson and R. J. Stoodly, *J. Chem. Soc. Chem. Commun.*, 478 (1981).
213. A. S. Kende, D. P. Curran, T.-G. Tsay and J. E. Mills, *Tetrahedron Lett.*, 3537 (1977).
214. J. Altman, E. Cohen, T. Maymon, J. B. Peterson, N. Reshef and D. Ginsburg, *Tetrahedron*, **25**, 5515 (1969).
215. R. N. Warrener, D. A. C. Evans and R. A. Russell, *Tetrahedron Lett.*, **25**, 4833 (1984).
216. M. D. Rozenboom, I.-M. Tegmo-Larsson and K. N. Houk, *J. Org. Chem.*, **46**, 2338 (1981).
217. I.-M. Tegmo-Larsson, M. D. Rozenboom, N. G. Rondan and K. N. Houk, *Tetrahedron Lett.*, **22**, 2047 (1981).
218. F. Bohlmann, W. Mather and H. Schwarz, *Chem. Ber.*, **110**, 2028 (1977).
219. J. S. Ton and W. Reusch, *J. Org. Chem.*, **45**, 5012 (1980).
220. W. B. Manning, T. R. Kelly and G. M. Muschik, *Tetrahedron Lett.*, **21**, 2629 (1980).
221. H. Muxfeldt, *Angew. Chem.*, **74**, 825 (1962).
222. T. R. Kelly, J. W. Gillard and R. N. Goerner, Jr, *Tetrahedron Lett.*, 3873 (1976).
223. T. R. Kelly, J. W. Gillard, R. N. Goerner, Jr and J. M. Lyding, *J. Am. Chem. Soc.*, **99**, 5513 (1977).
224. A. P. Kozikowski and K. Sugiyama, *Tetrahedron Lett.*, **21**, 4579 (1980).
225. R. K. Boeckman, Jr, M. H. Delton, T. M. Dolak, T. W. Watanabe and M. D. Glick, *J. Org. Chem.*, **44**, 4396 (1979).
226. T. R. Kelly and J. K. Saha, *J. Org. Chem.*, **50**, 3679 (1985).
227. Z. Stojanac, R. A. Dickinson, N. Stojanac, R. J. Woznow and Z. Valenta, *Can. J. Chem.*, **53**, 616 (1975) and references cited therein.
228. G. Stork and A. A. Hagedon, III, *J. Am. Chem. Soc.*, **100**, 3609 (1978).
229. T. R. Kelly and M. Montury, *Tetrahedron Lett.*, 4311 (1978).
230. R. A. Russell, G. J. Collin, M. Sterns and R. N. Warrener, *Tetrahedron Lett.*, 4229 (1979).
231. B. M. Trost, J. Ippen and W. C. Vladuchick, *J. Am. Chem. Soc.*, **99**, 8116 (1977).
232. B. M. Trost, V. C. Vladuchick and A. J. Bridges, *J. Am. Chem. Soc.*, **102**, 3554 (1980).
233. B. M. Trost, D. O'Krongly and J. L. Belletire, *J. Am. Chem. Soc.*, **102**, 7595 (1980).
234. M. Azadi-Ardakani and T. W. Walac, *Tetrahedron Lett.*, **24**, 1829 (1983).
235. W. Hubel, in *Organic Synthesis via Metal Carbonyls*, Vol. 1 (Eds. I. Wender and P. Pino), John Wiley and Sons, New York, 1968, p. 273.
236. P. Pino and G. Braca, in *Organic Synthesis via Metal Carbonyls*, Vol. 2 (Eds. I. Wender and P. Pino), John Wiley and Sons, New York, 1977, p. 419.
237. R. Victor, R. Ben-Shoshan and S. Sarel, *Tetrahedron Lett.*, 4211 (1973).
238. R. S. Dickson and H. P. Kirsch, *Aust. J. Chem.*, **27**, 61 (1974).
239. R. S. Dickson and S. H. Johnson, *Aust. J. Chem.*, **29**, 2186 (1976).
240. F. Wagner and H. Meier, *Tetrahedron*, **30**, 773 (1974).
241. R. L. Hillard, III and K. P. C. Vollhardt, *J. Am. Chem. Soc.*, **99**, 4058 (1977).
242. F. Canziani and M. C. Malatesta, *J. Organomet. Chem.*, **90**, 235 (1975).
243. J. L. Davidson, M. Green, F. G. A. Stone and A. J. Welch, *J. Chem. Soc. Dalton Trans.*, 738 (1976).
244. K. Maruyama, T. Shio and Y. Yamamoto, *Bull. Chem. Soc. Jpn*, **52**, 1877 (1979).
245. J. W. Kang, S. McVey and P. M. Maitlis, *Can. J. Chem.*, **46**, 3189 (1968).

246. N. A. Bailey, S. E. Kull, R. W. Jothan and S. F. Kettle, *J. Chem. Soc. Chem. Commun.*, 282 (1971).
247. F. W. Grevels, J. Buchkremer and E. A. Koerner von Gustorf, *J. Organomet. Chem.*, **111**, 235 (1976).
248. L. S. Liebeskind, S. L. Baysdon, M. S. South and J. F. Blount, *J. Organomet. Chem.*, **202**, C73 (1980).
249. L. S. Liebeskind, S. L. Baysdon, M. S. South, S. Iyer and J. P. Leeds, *Tetrahedron*, **41**, 5839 (1985).
250. L. S. Liebeskind, S. L. Baysdon and M. S. South, *J. Am. Chem. Soc.*, **102**, 7397 (1980).
251. L. S. Liebeskind, S. L. Baysdon, V. Geodken and R. Chidambaram, *Organometallics*, **5**, 1086 (1986).
252. M. S. South and L. S. Liebeskind, *J. Org. Chem.*, **47**, 3815 (1982).
253. L. S. Liebeskind, J. P. Leeds, S. L. Baysdon and S. Iyer, *J. Am. Chem. Soc.*, **106**, 6451 (1984).
254. M. S. South and L. S. Liebeskind, *J. Am. Chem. Soc.*, **106**, 4181 (1984).
255. K. H. Dötz, *Angew. Chem. Int. Ed. Engl.*, **14**, 644 (1975).
256. K. H. Dötz, *Angew. Chem. Int. Ed. Engl.*, **23**, 587 (1984).
257. W. D. Wulff, P.-C. Tang, K.-S. Chan, J. S. McCallum, D. C. Yang and S. R. Gilbertson, *Tetrahedron*, **41**, 5813 (1985).
258. H. Fischer, in *Transition Metal Carbene Complexes*, Georg Thieme Verlag, Weinheim, 1983, pp. 1–68 and 247–264; F. R. Kreissel, in *Transition Metal Carbene Complexes*, Georg Thieme Verlag, Weinheim, 1983, pp. 151–189; K. H. Dötz, in *Transition Metal Carbene Complexes*, Georg Thieme Verlag, Weinheim, 1983, pp. 191–226.
259. C. P. Casey, in *Reactive Intermediates*, Vol. 2 (Eds. M. Jones, Jr and R. A. Moss), John Wiley and Sons, New York, 1981, pp. 135–174.
260. F. J. Brown, *Prog. Inorg. Chem.*, **27**, 1 (1980).
261. H. Fischer, in *The Chemistry of the Metal–Carbon Bond* (Eds. F. R. Hartley and S. Patai), Vol. 1, John Wiley and Sons, Chichester, 1982, p. 233.
262. K. H. Dötz, in *Reactions of Co-ordinated Ligands* (Ed. P. Braterman), Plenum Publishing, New York, 1986, Ch. 4.
263. K. H. Dötz and R. Dietz, *Chem. Ber.*, **110**, 1555 (1977).
264. W. D. Wulff, K.-S. Chang and P.-C. Tang, *J. Org. Chem.*, **49**, 2293 (1984).
265. K. H. Dötz and R. Dietz, *Chem. Ber.*, **111**, 2517 (1978).
266. W. D. Wulff, P.-C. Tang and J. S. McCallum, *J. Am. Chem. Soc.*, **103**, 7677 (1981).
267. M. F. Semmelhack and J. J. Bozell, *Tetrahedron Lett.*, **23**, 2931 (1982).
268. M. F. Semmelhack, J. J. Bozell, L. Keller, T. Sato, E. J. Spiess, W. Wulff and A. Zask, *Tetrahedron*, **41**, 5803 (1985).
269. W. D. Wulff and D. C. Yang, *J. Am. Chem. Soc.*, **106**, 7565 (1984).
270. M. Watanabe and V. Snieckus, *J. Am. Chem. Soc.*, **102**, 1457 (1980).
271. P. Beak and V. Snieckus, *Acc. Chem. Res.*, **15**, 306 (1982).
272. M. Iwao and T. Kuraishi, *Tetrahedron Lett.*, **26**, 2613 (1985).
273. N. J. P. Broom and P. G. Sammes, *J. Chem. Soc. Chem. Commun.*, 162 (1978).
274. G. A. Kraus and H. Sugimoto, *Tetrahedron Lett.*, 2263 (1978).
275. F. M. Hauser and R. P. Rhee, *J. Org. Chem.*, **43**, 178 (1978).
276. F. M. Hauser and R. P. Rhee, *J. Am. Chem. Soc.*, **101**, 1628 (1979).
277. F. M. Hauser and S. Prasanna, *J. Org. Chem.*, **44**, 2596 (1979).
278. F. M. Hauser and S. Prasanna, *J. Am. Chem. Soc.*, **103**, 6378 (1981).
279. F. M. Hauser and S. Prasanna, *J. Org. Chem.*, **47**, 383 (1982).
280. F. M. Hauser and D. Mal, *J. Am. Chem. Soc.*, **106**, 1862 (1984).
281. F. M. Hauser and D. W. Combs, *J. Org. Chem.*, **45**, 4071 (1980).
282. R. A. Russell and R. N. Warrener, *J. Chem. Soc. Chem. Commun.*, 108 (1981).
283. P. G. Sammes and D. J. Dodsworth, *J. Chem. Soc. Chem. Commun.*, 33 (1979).
284. B. L. Chenard, M. G. Dolson, A. D. Stercel and J. S. Swenton, *J. Org. Chem.*, **49**, 318 (1984).
285. D. K. Jackson, L. Narasimhan and J. S. Swenton, *J. Am. Chem. Soc.*, **101**, 3989 (1979).
286. J. S. Swenton, D. K. Anderson, D. K. Jackson and L. Narasimhan, *J. Org. Chem.*, **46**, 4825 (1981).
287. R. Huisgen, H. Stangl, H. J. Sturm, R. Raab and K. Bunge, *Chem. Ber.*, **105**, 1258 (1972).
288. P. Gilgen, B. Jackson, H.-J. Hansen, H. Heimgantner and H. Schmid, *Helv. Chim. Acta*, **57**, 2634 (1974).

289. W. Stegmann, P. Uebelhart and H. Heimgartner, *Helv. Chim. Acta*, **66**, 2252 (1983).
290. K. A. Parker, I. D. Cohen and R. E. Babine, *Tetrahedron Lett.*, **25**, 3543 (1984).
291. K. A. Parker and I. D. Cohen, *Tetrahedron Lett.*, **25**, 4917 (1984).
292. J. A. Myers, L. D. Moore, Jr, W. L. Whitter, S. L. Council, R. M. Waldo, J. L. Lanier and B. U. Omoji, *J. Org. Chem.*, **45**, 1202 (1980).
293. A. Padwa (Ed.), *1,3-Dipolar Cycloaddition Chemistry*, John Wiley and Sons, New York, 1984.
294. W. E. Parham, C. K. Bradsher and K. J. Edger, *J. Org. Chem.*, **46**, 1057 (1981).
295. J. N. Bridson, S. M. Bennett and G. Butler, *J. Chem. Soc. Chem. Commun.*, 413 (1980).
296. D. G. Miller, S. Trenbeath and C. J. Sih, *Tetrahedron Lett.*, 1637 (1976).
297. K. S. Kim, E. Vanotti, A. Suarato and F. Johnson, *J. Am. Chem. Soc.*, **101**, 2483 (1979).
298. P. W. Raynolds, M. J. Manning and J. S. Swenton, *Tetrahedron Lett.*, 2383 (1977).
299. E. Vedejs and B. Nader, *J. Org. Chem.*, **47**, 3193 (1982).
300. A. S. Kende, J. L. Belletire, J. L. Herrmann, R. F. Romanet, E. L. Hume, R. H. Schlessinger, J. Fayos and J. C. Clardy, *Synth. Commun.*, **3**, 387 (1973)
301. R. D. Gleim, S. Trenbeath, F. Suzuki and C. J. Sih, *J. Chem. Soc. Chem. Commun.*, 242 (1978).
302. K. Balenović and M. Poje, *Tetrahedron Lett.*, 3427 (1975).
303. N. Katagiri, T. Kato and J. Nakano, *Chem. Pharm. Bull.*, **30**, 2440 (1982).
304. S. M. McElvain and H. Cohen, *J. Am. Chem. Soc.*, **64**, 260 (1942); S. M. McElvain and E. L. Engelhardt, *J. Am. Chem. Soc.*, **66**, 1077 (1944); S. M. McElvain, E. R. Degginger and J. D. Behun, *J. Am. Chem. Soc.*, **76**, 5736 (1954).
305. J. Banville and P. Brassard, *J. Chem. Soc. Perkin Trans. 1*, 613 (1976).
306. J.-L. Grandmaison and P. Brassard, *Tetrahedron*, **33**, 2047 (1977).
307. E. Ghera, Y. Gaoni and D. H. Perry, *J. Chem. Soc. Chem. Commun.*, 1034 (1974).
308. K. Maruyama and T. Otsuki, *Chem. Lett.*, 87 (1975).
309. K. Maruyama, T. Otsuki and K. Mitsui, *Bull. Chem. Soc. Jpn*, **49**, 3361 (1976).
310. K. Maruyama, T. Otsuki and K. Mitsui, *J. Org. Chem.*, **45**, 1424 (1980).
311. K. Maruyama, K. Mitsui and T. Tojo, *J. Heterocycl. Chem.*, **17**, 695 (1980).
312. K. Maruyama, T. Tojo, H. Iwamoto and T. Otsuki, *Chem. Lett.*, 827 (1980).
313. K. Maruyama, S. Tai and T. Otsuki, *Chem. Lett.*, 1565 (1981).
314. K. Maruyama, S. Tai and T. Otsuki, *Heterocycles*, **20**, 1031 (1983).
315. K. Maruyama, S. Tai and T. Otsuki, *Chem. Lett.*, 843 (1983).
316. K. Maruyama, T. Otsuki and S. Tai, *J. Org. Chem.*, **50**, 52 (1985).
317. P. Germeraad and H. W. Moore, *J. Org. Chem.*, **39**, 774 (1974).
318. P. Germeraad, W. Weyler, Jr and H. W. Moore, *J. Org. Chem.*, **39**, 781 (1974).
319. Y. Naruta, T. Yokota, N. Nagai and K. Maruyama, *J. Chem. Soc. Chem. Commun.*, 972 (1986).
320. Y. Naruta, T. Yokota, N. Nagai and K. Maruyama, *Chem. Lett.*, 1185 (1986).
321. Y. Naruta, Y. Arita, N. Nagai, H. Uno and K. Maruyama, *Chem. Lett.*, 1859 (1982).
322. Y. Naruta, N. Nagai and K. Maruyama, *Chem. Lett.*, 1385 (1983).
323. P. R. Weider, L. S. Hegedus, H. Asada and S. V. D'Andreq, *J. Org. Chem.*, **50**, 4276 (1985).
324. A. Fisher and G. N. Henderson, *Tetrahedron Lett.*, **21**, 701 (1980), and references cited therein.
325. D. Liotta, M. Saindane and C. Barnum, *J. Org. Chem.*, **46**, 3370 (1981).
326. H. W. Moore, Y.-L. L. Sing and R. S. Sidhu, *J. Org. Chem.*, **42**, 3321 (1977); H. W. Moore, Y.-L. L. Sing and R. S. Sidhu, *J. Org. Chem.*, **45**, 5057 (1980).
327. K. F. West and H. W. Moore, *J. Org. Chem.*, **47**, 3591 (1982).
328. D. A. Evans, J. M. Hoffman and J. K. Truesdale, *J. Am. Chem. Soc.*, **95**, 5822 (1973).
329. (a) D. A. Evans and R. Y. Wong, *J. Org. Chem.*, **42**, 350 (1977); (b) D. A. Evans and J. M. Hoffman, *J. Am. Chem. Soc.*, **98**, 1983 (1976).
330. I. Ryu, S. Murai, T. Horiike, A. Shinonaga and N. Sonoda, *J. Org. Chem.*, **43**, 780 (1978).
331. T. W. Greene, *Protective Groups in Organic Synthesis*, John Wiley and Sons, New York, 1981, pp. 87–113.
332. H. Haslam, in *Protective Groups in Organic Synthesis* (Ed. J. F. W. McOmie), Plenum Press, New York and London, 1973, pp. 145–182.
333. H. Matsumoto, S. Koike, I. Matsubara, T. Nakano and Y. Nagai, *Chem. Lett.*, 533 (1982).
334. H. Bouas-Laurent, R. Lopouyade, C. Brigand and J. P. Desvergne, *C. R. Acad. Sci. Paris, Ser. C*, **270**, 2167 (1970).
335. T. Murakawa, K. Fujii, S. Murai and S. Tatsumi, *Bull. Chem. Soc. Jpn*, **45**, 2520 (1972).
336. A. G. Beaumont, C. Eaborn and R. A. Jackson, *J. Chem. Soc. (D)*, 1624 (1970).

337. W. P. Newmann and G. Newmann, *J. Organomet. Chem.*, **25**, C59 (1970); W. P. Newmann and G. Newmann, *J. Organomet. Chem.*, **42**, 277 (1972).
338. Y. Naruta, N. Nagai and K. Maruyama, unpublished results.
339. R. A. Morton (Ed.), *Biochemistry of Quinones*, Academic Press, New York, 1965.
340. R. H. Thomson, *Naturally Occurring Quinones*, 2nd edn, Academic Press, New York, 1981; R. H. Thomson, *Naturally Occurring Quinones*; 3rd edn, Academic Press, New York (1987).
341. (a) F. Arcamone, *Lloydia*, **40**, 45 (1977); (b) T. R. Kelly, *Annu. Rep. Med. Chem.*, **14**, 288 (1979); (c) S. T. Crooke and S. D. Reich (Eds.), *Anthracyclines; Current Status and New Developments*, Academic Press, New York, 1980; (d) F. Arcamone, in *Anticancer Agents Based on Natural Product Models* (Eds. J. M. Cassady and J. D. Douros), Academic Press, New York, 1980, pp. 1–41; (e) F. Arcamone, *Doxorubicin, Anticancer Antibiotics*, Academic Press, New York, 1981; (f) S. Oki and T. Takeuchi, *Yuki Gosei Kagaku Kyokai Shi*, **40**, 2 (1982); (g) S. Terashima, *Yuki Gosei Kagaku Kyokai Shi*, **40**, 20 (1982); (h) E. El Khadem (Ed.), *Anthracycline Antibiotics*, Academic Press, New York, 1982; (i) F. M. Muggic, C. W. Young and S. K. Carter, *Developments in Oncology*, Vol. 10; *Anthracycline Antibiotics in Cancer Therapy*, Martinus Hijhoff Publishers, The Hague, 1982; (j) Reference 447, pp. 63–132; (k) T. R. Kelly (Ed.), *Recent Aspects of Anthracyclinone Chemistry*, Tetrahedron Symposia-in Print No. 17, *Tetrahedron*, **40**, 4537–4793 (1984).
342. A. F. Wagner and K. Folkers, *Vitamins and Coenzymes*, Interscience, New York, 1964.
343. R. S. Harris, I. G. Wool and J. A. Loraine (Eds.), *Vitamins and Hormones*, Vol. 24, Academic Press, New York, 1966.
344. W. H. Sebrell, Jr and R. S. Harris (Eds.), *The Vitamins*, Vols 3 and 5, Academic Press, New York, 1971 and 1972.
345. D. B. McCormick and L. D. Wright (Eds.), *Methods in Enzymology*, Vol. 18, Part C, Academic Press, New York, 1971, pp. 137–562.
346. H. Morimoto and I. Imada, in *Methodicium Chimicus*, Vol. 11, Part 2 (Ed. K. Korte), Georg Thieme Verlag, Stuttgart, 1977, pp. 69–145.
347. L. J. Machlin (Ed.), *Vitamin E*, Marcel Dekker, New York, 1980.
348. S. Yamada, T. Takeshita and J. Tanaka, *Yuki Gosei Kagaku Kyokai Shi*, **40**, 268 (1982).
349. Y. Naruta and K. Maruyama, *Yuki Gosei Kagaku Kyokai Shi*, **42**, 415 (1984).
350. Y. Naruta and K. Maruyama, *Chem. Lett.*, 881 (1979).
351. Y. Naruta and K. Maruyama, *Chem. Lett.*, 885 (1979).
352. Y. Naruta, *J. Am. Chem. Soc.*, **102**, 3774 (1980).
353. Y. Naruta, *J. Org. Chem.*, **45**, 4097 (1980).
354. K. Maruyama and Y. Naruta, *J. Org. Chem.*, **43**, 3796 (1978).
355. L. S. Hegedus and R. K. Stiverson, *J. Am. Chem. Soc.*, **96**, 7155 (1974).
356. L. S. Hegedus and E. L. Waterman, *J. Am. Chem. Soc.*, **96**, 6789 (1974).
357. L. S. Hegedus, B. R. Evans, D. E. Korte, E. L. Waterman and S. Sjoberg, *J. Am. Chem. Soc.*, **98**, 3901 (1976).
358. L. S. Hegedus and B. R. Evans, *J. Am. Chem. Soc.*, **100**, 3461 (1978).
359. K. Sato, S. Inoue and K. Saito, *J. Chem. Soc. Perkin Trans. 1*, 2289 (1973); K. Sato, S. Inoue and K. Saito, *J. Chem. Soc. Chem. Commun.*, 953 (1972).
360. K. Sato, S. Inoue and R. Yamaguchi, *J. Org. Chem.*, **37**, 1189 (1972); S. Inoue, R. Yamagami and K. Sato, *Bull. Chem. Soc. Jpn*, **47**, 3098 (1974).
361. C. D. Snyder and H. Rapoport, *J. Am. Chem. Soc.*, **96**, 8046 (1974).
362. J. P. Godschalx and J. K. Stille, *Tetrahedron Lett.*, **24**, 1905 (1983).
363. B. L. Chenard, M. J. Manning, P. W. Raynolds and J. S. Swenton, *J. Org. Chem.*, **45**, 378 (1980).
364. P. W. Raynolds, M. J. Manning and J. S. Swenton, *J. Chem. Soc. Chem. Commun.*, 499 (1977).
365. H. Sugihara, Y. Kawamatsu and H. Morimoto, *Justus Liebigs Ann. Chem.*, **763**, 128 (1972).
366. Y. Tachibana, *Chem. Lett.*, 901 (1977).
367. T. Yoshizawa, H. Toyofuku, K. Tachibana and T. Kuroda, *Chem. Lett.*, 1131 (1982).
368. I. Tabushi, K. Fujita and K. Kawakubo, *J. Am. Chem. Soc.*, **99**, 6456 (1977).
369. I. Tabushi, K. Yamamura, K. Fujita and H. Kawakubo, *J. Am. Chem. Soc.*, **101**, 1019 (1979).
370. I. Tabushi, H. Sugimoto and A. Yazaki, Ger, Offen., 2,907,864 (1979); *Chem. Abstr.*, **92**, 42164m.
371. K. H. Dötz and I. Pruskil, *J. Organomet. Chem.*, **209**, C4 (1981); K. H. Dotz, I. Pruskil and J. Muhlemeier, *Chem. Ber.*, **115**, 1278 (1982).

372. K. H. Dötz and W. Kuhn, *Angew. Chem. Int. Ed. Engl.*, **22**, 732 (1983).
373. Y. Fujita, T. Onishi and T. Nishida, Jpn. Pat., 54–160, 327 (1979); Y. Fujita, M. Ishiguro, T. Onishi and T. Nishida, *Synthesis*, 469 (1981).
374. S. Terao, K. Kato, M. Shiraishi and H. Morimoto, *J. Chem. Soc. Perkin Trans. 1*, 1101 (1978).
375. S. Terao, K. Kato, M. Shiraishi and H. Morimoto, *J. Org. Chem.*, **44**, 868 (1979).
376. K. Sato, O. Miyamoto, S. Inoue, T. Yamamoto and Y. Hirasawa, *J. Chem. Soc. Chem. Commun.*, 153 (1982).
377. J. W. Scott, F. T. Bizzarro, D. R. Parrish and G. Saucy, *Helv. Chim. Acta*, **59**, 290 (1976).
378. R. Barner and M. Schmid, *Helv. Chim. Acta*, **62**, 2384 (1979).
379. N. Cohen, R. J. Lopresti and G. Saucy, *J. Am. Chem. Soc.*, **101**, 6710 (1979).
380. Y. Sakito and G. Suzukamo, *Tetrahedron Lett.*, **23**, 4953 (1982).
381. K.-K. Chan, A. C. Specian, Jr and G. Saucy, *J. Org. Chem.*, **43**, 3435 (1978).
382. K.-K. Chan, N. Cohen, J. P. De Noble, A. C. Specian, Jr and G. Saucy, *J. Org. Chem.*, **41**, 3497 (1976).
383. N. Cohen, W. F. Eichel, R. J. Lopresti, C. Neukom and G. Saucy, *J. Org. Chem.*, **41**, 3505 (1976).
384. R. Zell, *Helv. Chim. Acta*, **62**, 474 (1979).
385. N. Cohen, J. W. Scott, F. T. Brizzaro, R. J. Lopresti, W. F. Eichel and G. Saucy, *Helv. Chim. Acta*, **61**, 873 (1978).
386. N. Cohen, C. G. Scott, C. Neukom, R. J. Lopresti, G. Weber and G. Saucy, *Helv. Chim. Acta*, **64**, 1158 (1981).
387. H. Schmid, A. Ebnöther and T. M. Meijer, *Helv. Chim. Acta*, **33**, 1751 (1950); H. Schmid and E. Ebnother, *Helv. Chim. Acta*, **34**, 561, 1041 (1951).
388. S. Omura, H. Tanaka, Y. Koyama, R. Oiwa and M. Katagiri, *J. Antibiot.*, **27**, 363 (1974).
389. H. Tanaka, H. Marumo, T. Nagai, M. Okada, K. Taniguchi and S. Omura, *J. Antibiot.*, **28**, 925 (1975).
390. H. Tanaka, Y. Koyama, J. Awaya, H. Marumo, R. Omura, M. Katagiri, T. Nagai and S. Oiwa, *J. Antibiot.*, **28**, 860 (1975).
391. S. Omura, H. Tanaka, Y. Okada and H. Marumo, *J. Chem. Soc. Chem. Commun.*, 320 (1976).
392. M. E. Bergy, *J. Antibiot.*, **21**, 454 (1968); H. Hoeksema and W. C. Kruger, *J. Antibiot.*, **29**, 704 (1976).
393. W. Keller-Schierlein, M. Brufani and S. Barcza, *Helv. Chim. Acta*, **51**, 1256 (1968); M. Brufani and M. Dobler, *Helv. Chim. Acta*, **51**, 1269 (1968), and references cited therein.
394. Y. Iwai, A. Kōra, Y. Takahashi, T. Hayashi, J. Awaya, R. Masuma, R. Oiwa and S. Omura, *J. Antibiot.*, **31**, 959 (1978).
395. J. C. Van Meter, M. Dann and N. Bohonos, in *Antimicrobial Agents Annual 1960*, Plenum Press, New York, 1961, p. 77; G. A. Ellestad, M. P. Kunstman, H. A. Whaley and E. L. Patterson, *J. Am. Chem. Soc.*, **90**, 1325 (1968).
396. C. E. Snipes, C. Chang and H. G. Floss, *J. Am. Chem. Soc.*, **101**, 701 (1979).
397. N. Tsuji, M. Kobayashi, Y. Yakisaka, Y. Kawamura, M. Mayama and K. Matsumoto, *J. Antibiot.*, **29**, 7 (1976); N. Tsuji, M. Kobayashi, Y. Terui and K. Tori, *Tetrahedron*, **32**, 2207 (1976).
398. W. Eisenhuth and H. Schmid, *Helv. Chim. Acta*, **41**, 2021 (1958).
399. Y. Naruta, H. Uno and K. Maruyama, *J. Chem. Soc. Chem. Commun.*, 127 (1981).
400. T. Kometani and E. Yoshii, *J. Chem. Soc. Perkin Trans. 1*, 1191 (1981).
401. T. Kometani, Y. Takeuchi and E. Yoshii, *J. Chem. Soc. Perkin Trans. 1*, 1197 (1981).
402. R. G. F. Giles, I. R. Green, V. I. Hugo and P. R. K. Mitchell, *J. Chem. Soc. Chem. Commun.*, 51 (1983).
403. J. St Pyrek, O. Achmatowicz, Jr and A. Zamojski, *Tetrahedron*, **33**, 673 (1977).
404. G. A. Kraus and B. Roth, *J. Org. Chem.*, **43**, 4923 (1978).
405. T.-T. Li and R. H. Ellison, *J. Am. Chem. Soc.*, **100**, 6263 (1978).
406. T. Kometani, Y. Takeuchi and E. Yoshii, *J. Org. Chem.*, **48**, 2631 (1983).
407. A. Ichihara, M. Ubukata, H. Oikawa, K. Murakami and S. Sakamura, *Tetrahedron Lett.*, **21**, 4469 (1980).
408. Y. Naruta, H. Uno and K. Maruyama, *Chem. Lett.*, 609 (1982).
409. H. Uno, *J. Org. Chem.*, **51**, 350 (1986).
410. Y. Naruta, H. Uno and K. Maruyama, *Chem. Lett.*, 961 (1982).

411. M. F. Semmelhack, L. Keller, T. Sato and E. Spiess, *J. Org. Chem.*, **47**, 4382 (1982).
412. M. F. Semmelhack, J. J. Bozell, J. Sato, W. Wulff, E. Spiess and A. Zask, *J. Am. Chem. Soc.*, **104**, 5850 (1982); M. F. Semmelhack and A. Zask, *J. Am. Chem. Soc.*, **105**, 2034 (1983).
413. M. F. Semmelhack, *Pure Appl. Chem.*, **53**, 2379 (1981).
414. Y. Naruta, H. Uno and K. Maruyama, *Tetrahedron Lett.*, **22**, 5221 (1981).
415. K. Tatsuta, K. Akimoto and M. Annaka, *Bull. Chem. Soc. Jpn*, **58**, 1699 (1985).
416. L. Slechta, C. G. Chidester and F. Reusser, *J. Antibiot.*, **33**, 919 (1980); G. Reinhardt, G. Bradler, K. Eckart, D. Tresselt and W. Ihn, *J. Antibiot.*, **33**, 787 (1980); D. Tresselt, K. Eckardt, W. Ihn, L. Radies and G. Reinhardt, *Tetrahedron*, **37**, 1961 (1981).
417. M. Sudani, Y. Takeuchi, E. Yoshii and T. Kometani, *Tetrahedron Lett.*, **22**, 4253 (1981).
418. Y. Takeuchi, M. Sudani and E. Yoshii, *J. Org. Chem.*, **48**, 4151 (1983).
419. M. F. Semmelhack, Y. Appapillai and T. Sato, *J. Am. Chem. Soc.*, **107**, 4577 (1985).
420. T. Kometani, Y. Takeuchi and E. Yoshii, *J. Org. Chem.*, **48**, 2311 (1983).
421. V. Prelog and W. Oppolzer, *Helv. Chim. Acta*, **56**, 2279 (1973).
422. W. Kump and H. Bickel, *Helv. Chim. Acta*, **56**, 2323 (1973).
423. W. Wehri, *Top. Curr. Chem.*, **72**, 21 (1977).
424. K. L. Rinehart, Jr and L. S. Shield, *Prog. Chem. Org. Nat. Prod.*, **33**, 231 (1976).
425. H. Nagaoka, W. Rutsch, G. Schmid, H. Iio, M. R. Johnson and Y. Kishi, *J. Am. Chem. Soc.*, **102**, 7962 (1980).
426. N. Nagaoka and Y. Kishi, *Tetrahedron*, **37**, 3873 (1981).
427. S. Masamune, B. Imperiali and G. S. Garvey, *J. Am. Chem. Soc.*, **104**, 5528 (1982).
428. S. Hanessian, J.-R. Pougny and I. K. Boessenkool, *J. Am. Chem. Soc.*, **104**, 6164 (1982).
429. M. Nakata, H. Takao, Y. Ikeyama, T. Sakai, K. Tatsuta and M. Kinoshita, *Bull. Chem. Soc. Jpn*, **54**, 1749 (1981), and references cited therein.
430. B. Fraser-Reid, L. Magdzinski and B. Molino, *J. Am. Chem. Soc.*, **106**, 931 (1984).
431. T. R. Kelly, M. Behforouz, A. Echavarren and J. Vaya, *Tetrahedron Lett.*, **24**, 2331 (1983).
432. B. M. Trost and W. H. Pearson, *Tetrahedron Lett.*, **24**, 269 (1983).
433. A. P. Kozikowski, K. Sugiyama and E. Huie, *Tetrahedron Lett.*, **22**, 3381 (1981); A. P. Kozikowski, K. Sugiyama and J. P. Springer, *J. Org. Chem.*, **46**, 2426 (1981).
434. M. Nakata, S. Wada, K. Tatsuta and M. Kinoshita, *Bull. Chem. Soc. Jpn*, **58**, 1801 (1985).
435. H. Nagaoka, G. Schmid, H. Iio and Y. Kishi, *Tetrahedron Lett.*, **22**, 3381 (1981); Y. Kishi, *Pure Appl. Chem.*, **53**, 1163 (1981).
436. H. Iio, H. Nagaoka and Y. Kishi, *J. Am. Chem. Soc.*, **102**, 7965 (1980).
437. K. A. Parker and J. J. Petrakis, *Tetrahedron Lett.*, **22**, 397 (1981).
438. T. Hata, Y. Sano, R. Sugawara, A. Matsumae, K. Tanamori, T. Shima and T. Hoshi, *J. Antibiot.*, *Tokyo*, *Ser. A*, **9**, 141 (1956).
439. S. Wakaki, H. Marumo, K. Tomioka, G. Shimizu, E. Kato, H. Kamada, S. Kudo and Y. Fujimoto, *Antibiotics and Chemotherapy*, **8**, 228 (1958).
440. R. R. Herr, M. E. Bergy, T. E. Ebler and H. K. Jahnke, *Antimicrobial Agent Ann.*, **17**, 23 (1960).
441. K. Kametani and K. Takahashi, *Heterocycles*, **9**, 293 (1978).
442. W. A. Remers, *The Chemistry of Antitumor Antibiotics*, Vol. 1, John Wiley and Sons, New York, 1979, pp. 221–276.
443. K. Takahashi and T. Kametani, *Heterocycles*, **13**, 411 (1979).
444. R. W. Franck, *Fortschr. Chem. Org. Naturst.*, **38**, 1 (1979).
445. S. K. Carter and S. T. Crooke (Eds.), *Mitomycin C; Current Status and New Developments*, Academic Press, New York, 1979.
446. T. Onuma and Y. Ban, *Yuki Gosei Kagaku Kyokai Shi*, **38**, 1054 (1980).
447. W. A. Remers, in *Anticancer Agents Based on Natural Product Models* (Eds. J. M. Cassady and J. D. Douros), Academic Press, New York, 1980, pp. 131–146.
448. W. N. Speckamp, *Heterocycles*, **21**, 211 (1984).
449. K. Shirahata and N. Hirayama, *J. Am. Chem. Soc.*, **105**, 7199 (1983).
450. U. Hornemann and M. J. Heins, *J. Org. Chem.*, **50**, 1301 (1985).
451. F. Nakatsubo, A. J. Cocuzza, D. E. Keeley and Y. Kishi, *J. Am. Chem. Soc.*, **99**, 4835 (1977); F. Nakatsubo, T. Fukuyama, A. J. Cocuzza and Y. Kishi, *J. Am. Chem. Soc.*, **99**, 8115 (1977); T. Fukuyama, F. Nakatsubo, A. J. Cocuzza and Y. Kishi, *Tetrahedron Lett.*, 4295 (1977).
452. Y. Kishi, *J. Nat. Prod.*, **42**, 549 (1979).
453. S. Danishefsky, J. Regan and R. Doehner, *J. Org. Chem.*, **46**, 5255 (1981).

454. S. Danishefsky, E. M. Berman, M. Ciufolini, S. J. Etheredge and B. E. Segmuller, *J. Am. Chem. Soc.*, **107**, 3891 (1985).
455. S. Danishefsky and J. Regan, *Tetrahedron Lett.*, **22**, 3919 (1981).
456. R. M. Coates and P. A. MacManus, *J. Org. Chem.*, **47**, 4822 (1982).
457. M. Akiba and T. Takada, *Heterocycles*, **6**, 1861 (1977); M. Akiba, K. Kosugi, M. Okuyama and T. Takada, *J. Org. Chem.*, **43**, 181 (1978); M. Akiba, S. Ikuta and T. Takada, *Heterocycles*, **16**, 1579 (1981).
458. M. Akiba, S. Ikuta and T. Takada, *J. Chem. Soc. Chem. Commun.*, 817 (1983).
459. S. N. Falling and H. Rapoport, *J. Org. Chem.*, **45**, 1260 (1980).
460. W. C. Dijksman, W. Verboom, R. J. M. Egberink and D. N. Reinhoudt, *J. Org. Chem.*, **50**, 3791 (1985).
461. J. R. Luly and H. Rapoport, *J. Org. Chem.*, **47**, 2404 (1982).
462. J. R. Lury and H. Rapoport, *J. Org. Chem.*, **49**, 1671 (1984).
463. K. J. Shaw, J. R. Lury and H. Rapoport, *J. Org. Chem.*, **50**, 4515 (1985).
464. T. Kametani, Y. Kigawa, H. Nemoto, M. Ihara and K. Fukumoto, *Heterocycles*, **14**, 799 (1980).
465. J. Rebek, Jr and S. H. Shaber, *Heterocycles*, **16**, 1173 (1981); J. Rebek, Jr, S. H. Shaber, Y.-Y. Shue, J.-C. Gehret and S. Zimmerman, *J. Org. Chem.*, **49**, 5164 (1984).
466. W. Flitsch and P. Russkamp, *Justus Liebigs Ann. Chem.*, 1398 and 1422 (1985); W. Flitsch, P. Russkamp and W. Langer, *Justus Liebigs Ann. Chem.*, 1413 (1985); M. Dartmann, W. Flitsch, B. Krebs and P. Russkamp, *Justus Liebigs Ann. Chem.*, 1437 (1985).
467. L. S. Hegedus, G. F., Allen, J. J. Bozell and E. L. Watermann, *J. Am. Chem. Soc.*, **100**, 5800 (1978); L. S. Hegedus, P. M. Winton and S. Varaprath, *J. Org. Chem.*, **46**, 2215 (1981); L. S. Hegedus, G. F. Allen and D. J. Olsen, *J. Am. Chem. Soc.*, **102**, 3583 (1980).
468. S. Danishefsky and E. Taniyama, *Tetrahedron Lett.*, **24**, 15 (1983).
469. C. Anthony and L. J. Zatman, *Biochem. J.*, **104**, 960 (1967).
470. S. Ohta, T. Fujita and J. Taobari, *J. Biochem.*, **90**, 205 (1981).
471. J. A. Duine and J. Frank, *Biochem. J.*, **187**, 213 (1980).
472. J. Amemiya, K. Matsushita, Y. Ohno, E. Shinagawa and O. Adachi, *FEBS Lett.*, **130**, 179 (1981).
473. J. Westerling, J. Frank and F. A. Duine, *Biochem. Biophys. Res. Commun.*, **87**, 719 (1979); R. de Beer, D. van Ormondt, M. A. van Ast, R. Banen, J. A. Duine and J. Frank, *J. Chem. Phys.*, **70**, 4491 (1979).
474. E. J. Corey and A. Trannontano, *J. Am. Chem. Soc.*, **103**, 5599 (1981).
475. A. R. Mackenzie, C. J. Moody and C. W. Rees, *J. Chem. Soc. Chem. Commun.*, 1372 (1983).
476. J. A. Gainor and S. M. Weinreb, *J. Org. Chem.*, **47**, 2833 (1982).
477. J. B. Naar, E. J. Rodriguez and T. C. Bruice, *J. Am. Chem. Soc.*, **107**, 7198 (1985).
478. G. Büchi, J. H. Botkin, G. C. M. Lee and K. Yakushijin, *J. Am. Chem. Soc.*, **107**, 5555 (1985).
479. J. B. Hendrickson and J. G. de Vries, *J. Org. Chem.*, **47**, 1150 (1982).
480. A. Mustafa, in *The Chemistry of Heterocyclic Compounds*, Vol. 23 (Ed. A. Weissberger), John Wiley and Sons, New York, 1967, p. 103.
481. A. Yamashita, *J. Am. Chem. Soc.*, **107**, 5823 (1985), and references cited therein.
482. R. B. Gammill, *Tetrahedron Lett.*, **26**, 1385 (1985).
483. R. B. Gammill and B. R. Hyde, *J. Org. Chem.*, **48**, 3865 (1983).
484. S. J. Gould and S. M. Weinreb, *Fortschr. Chem. Org. Naturst.*, **41**, 77 (1982).
485. F. Z. Basha, S. Hibino, D. Kim, W. E. Pye, T.-T. Wu and S. M. Weinreb, *J. Am. Chem. Soc.*, **102**, 3962 (1980); S. M. Weinreb, F. Z. Basha, S. Hibino, N. A. Khatri, D. Kim, W. E. Pye and T.-T. Wu, *J. Am. Chem. Soc.*, **104**, 536 (1982).
486. A. S. Kende, D. P. Lorah and R. J. Boatman, *J. Am. Chem. Soc.*, **103**, 1271 (1981).
487. A. S. Kende, F. H. Ebetino, R. Battista, R. J. Boatman, D. P. Lorah and E. Lodge, *Heterocycles*, **21**, 91 (1984).
488. D. L. Boger and J. S. Panek, *J. Am. Chem. Soc.*, **107**, 5745 (1985).
489. A. S. Kende and F. H. Ebetino, *Tetrahedron Lett.*, **25**, 923 (1984).
490. D. L. Boger, S. R. Duff, J. S. Panek and M. Yasuda, *J. Org. Chem.*, **50**, 579 (1985).
491. T. Arai, K. Takehashi and A. Kubo, *J. Antibiot.*, **30**, 1015 (1977); T. Arai, K. Takehashi, K. Ishiguro and T. Mikami, *Gann*, **71**, 790 (1980).
492. T. Fukuyama and R. A. Sachleben, *J. Am. Chem. Soc.*, **104**, 4957 (1982).

493. J. Sygusch, F. Brisse, S. Hanessian and D. Kluepfel, *Tetrahedron Lett.*, 4021 (1974) and 170 (1975); J. Sygusch, F. Brisse and S. Hanessian, *Acta Cryst.*, **B32**, 1139 (1976); D. Kluepfel, H. A. Baker, G. Piattoni, S. N. Sehgal, A. Sidowicz, S. Kartar and C. Vezina, *J. Antibiot.*, **28**, 497 (1975).
494. J. Itoh, S. Omote, S. Inouye, Y. Kodama, T. Hisamitsu, T. Niida and Y. Ogawa, *J. Antibiot.*, **35**, 642 (1982).
495. T. Hayashi, T. Noto, Y. Nawata, H. Okazaki, M. Sawada and K. Ando, *J. Antibiot.*, **35**, 771 (1982).
496. S. Danishefsky, B. T. O'Neil, E. Taniyama and K. Vaughan, *Tetrahedron Lett.*, **25**, 4199 (1984); S. Danishefsky, B. T. O'Neil, and J. Springer, *Tetrahedron Lett.*, **25**, 4203 (1984).
497. D. A. Evans and S. A. Biller, *Tetrahedron Lett.*, **26**, 1907 and 1911 (1985).
498. D. A. Evans, C. R. Illig and J. C. Saddler, *J. Am. Chem. Soc.*, **108**, 2478 (1986).

The Chemistry of Quinonoid Compounds, Vol. II
Edited by S. Patai and Z. Rappoport

CHAPTER **9**

ortho-Quinonoid compounds

COLIN W. G. FISHWICK and DAVID W. JONES
Organic Chemistry Department, The University, Leeds LS2 9JT, U.K.

I. INTRODUCTION

A. Reactivity Trends

o-Quinodimethane (*o*-xylylene) (**1**) is a reactive intermediate that dimerizes at −150°C to give the Diels–Alder (spiro-) dimer (**2**). This reactivity as a Diels–Alder diene is manifested to some extent by the variety of other molecules which contain a structural unit related to **1** and even extends to the polynuclear aromatic compound anthracene. Successive replacement of the exocyclic methylene groups of **1** by oxygen atoms leads to increasingly stable compounds. *o*-Quinone methide (**3**) appears to be stable[1] at −50°C but

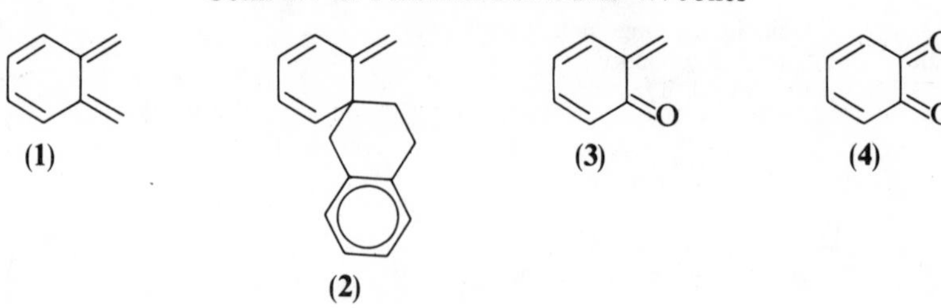

trimerizes at $-20\,^{\circ}$C, and *o*-benzoquinone (**4**) is an isolable crystalline solid which, however, dimerizes readily and can act as both a homo- and a heterodiene in cycloaddition reactions[2]. *o*-Quinodimethane (**1**) is less stable[3] than its bicyclic valence tautomer, benzocyclobutene (**5**), by 10.5 kcal mol^{-1}. The importance of the *o*-quinonoid tautomer in the equilibrium **6** $\rightleftharpoons$ **7** is expected to increase as the electronegativity of X increases and the contribution of the dipolar canonical form (**8**) to the structure of **6** becomes more likely[4].

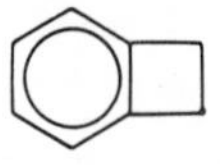

(**5**) 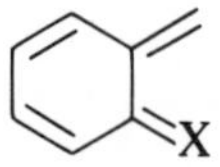 (**6**) 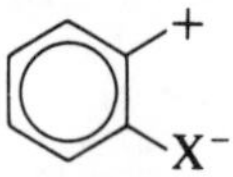 (**8**)

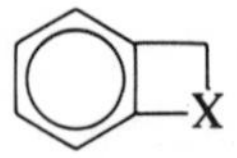

(**7**) 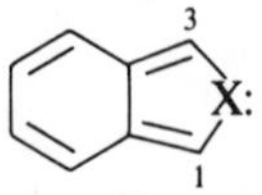 (**9**)

Indeed benzoxetes (**7**; X = O) are unknown whereas *o*-quinone methides (**6**; X = O) are well known both as reactive intermediates and in certain cases as stable isolable compounds[1]. Although there are no reports of the isolation of a benzoxete, benzothietes (**7**; X = S) and benzazetidines (**7**; X = NR) are stable compounds that can arise by ring-closure of *o*-thioquinone methides (**6**; X = S) and *o*-quinomethide imines (**6**; X = NR) respectively.

The 10π-electron aromatic compounds (**9**; X = O, NH, S) dimerize less readily than *o*-quinodimethane and can be obtained in crystalline form at low temperature. However, they are sensitive to oxygen and polymerize readily and none survives long at 20 °C. In contrast several stable compounds result by replacement of C(1) and C(3) in **9** by heteroatoms. Of the stable compounds **10** (Y = CH, X = O); **10** (Y = CH, X = S),

(**10**) ⟷ (**11**)

(**12**) ⟷ (**13**)

10 (Y = CH, X = NH) and **10** (Y = N, X = S) only the first is reported[5] to react as a Diels–Alder diene. No doubt the lack of pronounced *o*-quinonoid reactivity in these compounds is associated with important contributions from dipolar canonical forms like **11**. Likewise the stability of **12** (X = NH), an isolable crystalline solid[6] which prefers the quinolone tautomeric form (at least in polar solvents[6]) can be attributed to important contribution from the dipolar aromatic structure (**13**, X = NH). In contrast the *o*-quinonoid pyrone (**12**, X = O), for which the dipolar canonical form (**13**, X = O) would be expected to be less important, is stable only at low temperature[7].

B. Biradical Character

On many occasions triplet biradical structures like **14** have been considered for **1** and its derivatives or alternatively biradical character has been attributed to *o*-quinonoid compounds. The highest occupied molecular orbital (HOMO) of *o*-quinodimethane (**15**) would be expected to be of higher energy than the HOMO (**16**) of octatetraene as the result

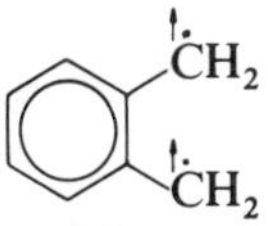

(**14**)

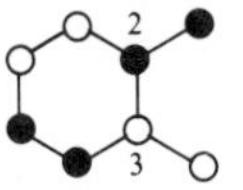

(**15**) $E = +0.295\beta$

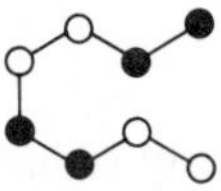

(**16**)

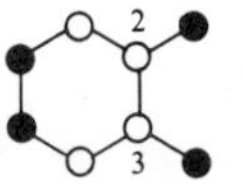

(**17**) $E = -0.295\beta$

of antibonding interaction between C(2) and C(3) in **15**. Similarly bonding C(2)–C(3) interaction in the lowest unoccupied MO (LUMO) of *o*-quinodimethane (**17**) would be expected to lead to a lower energy LUMO for **1** than for linear octatetraene. *o*-Quinodimethane would therefore be expected to have a lower HOMO–LUMO energy gap than its linearly conjugated counterpart. The energy gap calculated by simple Hückel calculations[8] (0.59 β) is 0.1 β smaller than that calculated for linear octatetraene. The high energy HOMO and low energy LUMO of **1** can account for its easy dimerization and its reactivity towards both electron-rich and electron-deficient olefins. However, the energy gap appears too large to allow serious consideration of a biradical structure for **1**. Even cyclobutadiene, for which the predicted energy gap is zero, escapes a biradical structure by Jahn–Teller distortion involving a rectangular structure with bond alternation. No experimental evidence for the biradical structure **14** has been obtained. Nor is such evidence forthcoming for 2,3-naphthoquinodimethane for which the predicted HOMO–LUMO energy gap is smaller (0.34 β).

Inden-2-one (**18**) is an even more extreme case. For **18** simple Hückel MO calculations make the alarming prediction that the orbital **19** is the HOMO ($E = +0.443\,\beta$) and **20** the LUMO ($E = +0.295\,\beta$). The orbital **19** can be seen to be the LUMO **17** of *o*-quinodimethane perturbed to lower energy by bonding combination with a carbonyl π^* orbital whilst the orbital **20** is essentially the unperturbed HOMO **15** of *o*-quinodimethane. The stabilization of the LUMO **17** of *o*-quinodimethane is unlikely to be so great that it becomes the HOMO of **18**; in related calculations localization of the double

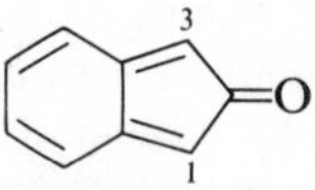

(18) 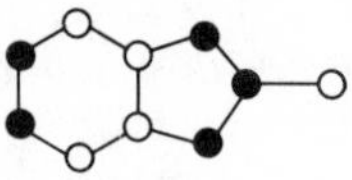 (19) 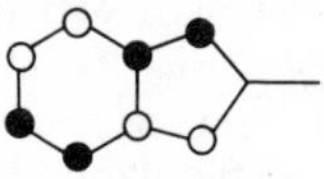 (20)

bonds prevents this situation actually arising[9]. Indeed derivatives of **18** add to one double bond of cyclopentadiene to give **21** rather than 1,4- to the diene system[10]. The observed mode is the mode of addition expected of a molecule with a HOMO of the symmetry **20** and a LUMO with the symmetry **19** rather than vice versa. Thus **20** is the HOMO and **19** the LUMO of inden-2-one (**18**). The LUMO of **18** will be of lower energy than that of *o*-quinodimethane but **1** and **18** will have HOMOs of similar energy. Accordingly **18**, with a

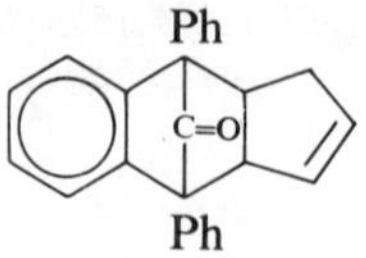

(21) 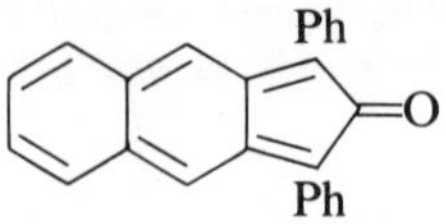 (22)

reduced HOMO–LUMO energy gap, would be expected to show more biradical character than **1**. Since both the 1,3-diphenyl derivative of **18** and the benz[*f*]inden-2-one (**22**) add stereospecifically to olefins[10] no evidence for biradical behaviour is forthcoming even in these auspicious cases.

C. Scope of the Review

In this review *o*-quinonoid compounds are discussed as a group of compounds which have many similar properties and methods of preparation. The observed differences in stability and properties form a fascinating study of reactivity. A disadvantage of this

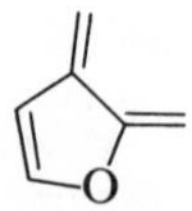

(23) 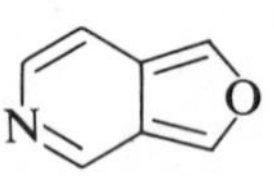 (24) 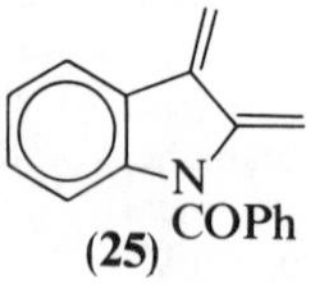 (25)

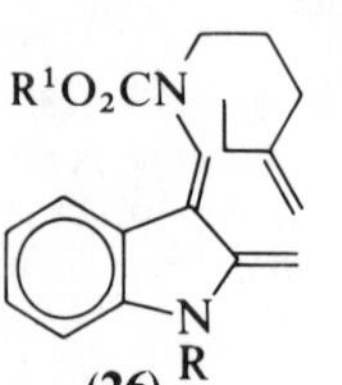

(26) 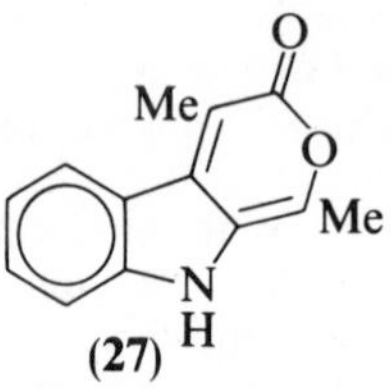 (27)

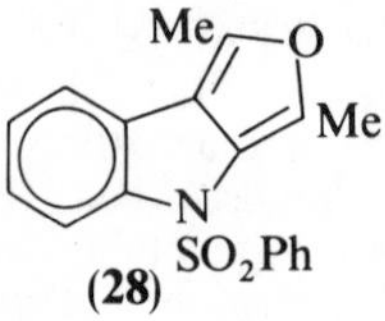

(28) 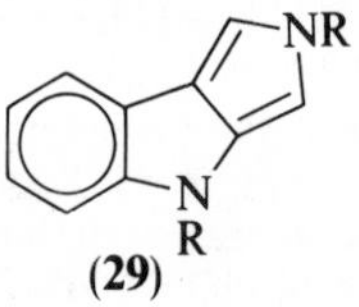 (29)

treatment is that not all routes to, or all reactions of a particular kind of *o*-quinonoid compound are described. For isobenzofurans (**9**, X=O)[11], isothianaphthenes (**9**, X=S)[12], isoindoles (**9**, X = NR)[13] and *o*-quinone methides (**3**)[1] this is not serious as the indicated reviews are available. Routes to benzocyclobutenes, which are important precursors to *o*-quinodimethanes (Section III.C), have also been reviewed[14]. Herein and in accord with the philosophy of this series attention is directed to *o*-quinodimethane, *o*-thioquinone methide and *o*-quinone methide imine derivatives which have not been the subject of review. Current interest in *o*-quinonoid compounds centers on their use in synthesis. Aspects of this subject which have not been extensively reviewed elsewhere are described in Section IV.B. Finally attention is drawn to the growing family of heterocyclic *o*-quinonoid compounds **23**[15], **24**[16], **25**[17], **26**[18], **27**[19, 20], **28**[21], and **29**[22, 23] whose chemistry is emerging rapidly. Space does not permit a detailed treatment of compounds of this type although they have considerable synthetic potential[18, 20].

II. *o*-QUINODIMETHANE, 2,3-NAPHTHOQUINODIMETHANE AND THEIR SIMPLE STABILIZED DERIVATIVES

o-Quinodimethane (**1**) has been characterized by the matrix-isolation technique[24]. The 1,4-dihydrophthalazine (**30**) rapidly loses nitrogen above $-40°C$ to give the spiro-dimer (**2**) of *o*-quinodimethane. Upon irradiation ($\lambda < 345$ nm) of a rigid glassy solution of **30** in a matrix of ether–isopentane–ethanol (EPA) at $-196°C$ the characteristic structured UV–visible band of **1** (λ_{max} 373 nm) appears. Upon melting the glassy solution ($-150°C$) this absorption is rapidly replaced by that due to **2**. Irradiation in the long wavelength band

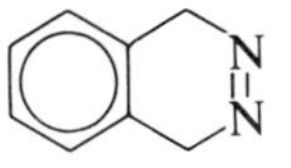

(**30**) 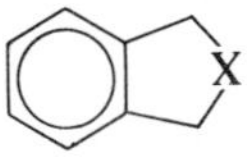 (**31**)

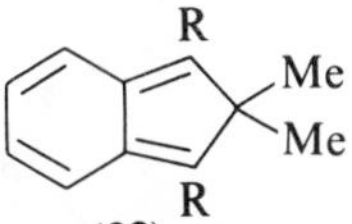

(**32**) 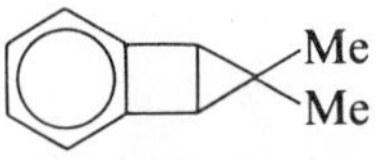 (**33**)

of **1** ($\lambda > 345$ nm) gave benzocyclobutene (**5**). Importantly, matrix-isolated **1** shows no half-field ESR signal and is therefore a ground state singlet; a triplet biradical structure (**14**) is ruled out for the ground state of the molecule. Irradiation of glassy solutions of several other compounds—benzocyclobutene (**5**), the sulphone (**31**, X = SO_2), indan-2-one (**31**, X = CO), the tosylhydrazone salt (**31**, X = $NNNaSO_2Tol$) and even *o*-xylene—all give **1**. However, **30** provides **1** most rapidly and cleanly. Incorporation of the reactive exocyclic diene moiety of **1** into a ring begins the design of sterically stabilized *o*-quinonoid compounds. This technique was probably first used by Alder and Fremery[25] who prepared the isoindene (2*H*-indene) (**32**, R = Ph). The stability of this compound is associated with its inability to undergo thermally allowed conrotatory electrocyclic ring-closure to a benzocyclobutene, the slow 1,5-sigmatropic shift of alkyl groups compared to hydrogen, as well as conjugative stabilization due to the phenyl groups and steric effects associated with both the phenyl and the methyl substituents. The derivative (**32**, R = H) lacking phenyl substituents was subsequently prepared[26–28]. In the absence of oxygen this compound is indefinitely stable at 20°C. Photolysis of **32** (R = H) at or below 0°C gives the pseudoindene **33**[26]. Generation of isoindene (**34**) gives dimeric species at $-60°C$[29] and indene by 1,5-hydrogen shift (**34**, arrows) at 0°C[28]. Steric shielding of the usually

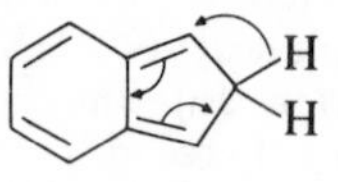

(34) 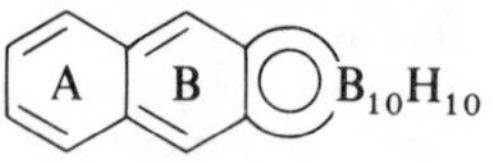 (35) 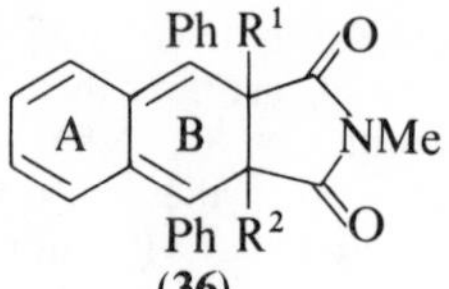 (36)

reactive ring-B diene system is held responsible for the remarkable stability of the isolable crystalline compounds **35**[30], **36** ($R^1 = R^2 =$ Me)[31] and **36** ($R^1, R^2 = (CH_2)_4$)[31]. Dienophiles react with all these compounds at the normally less reactive diene system in ring A. The similar long wavelength bands in the electronic spectra of **1**, **35**, and **36** ($R^1 = R^2$ = Me) shows that they are all *o*-quinodimethanes and that in **36** ($R^1 = R^2 =$ Me) the phenyl groups conjugate very little with the quinonoid system. Rather these groups are forced by steric factors to lie orthogonal to the quinonoid system in which position they provide substantial steric stabilization[31]. Steric shielding is probably also responsible for the greater stability of the *o*-quinodimethane **37** compared to **1**[32]. Irradiation of the *exo*-benzyne–norbornene adduct (**38**) in a degassed rigid matrix at 77 K gave **37** which persisted on warming the matrix to 20 °C. Since irradiation of **38** in solution failed to result in ring-opening, two photons of UV light may be involved in the conversion of **38** into **37**[32].

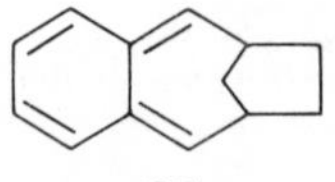

(37) 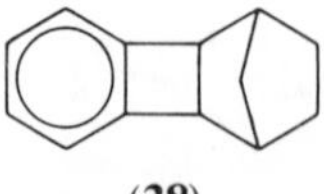 (38) 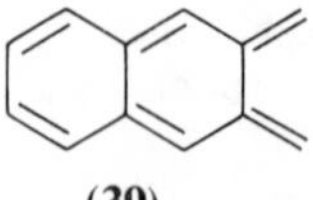 (39)

2,3-Naphthoquinodimethane (**39**) has also been observed in a matrix isolation study[33]. Irradiation of **40** in an EPA matrix at 77 K gave rise to a ruby colour (λ_{max} 541 nm) attributed to **39**. In the presence of oxygen the peroxide **41** was obtained. The latter had

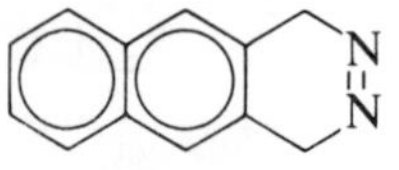

(40) 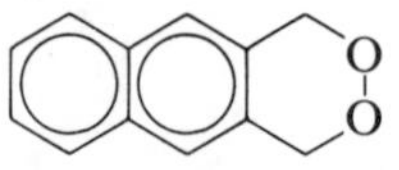 (41)

earlier been observed as a trapping product of **39**[34]. Photolysis of **42** (X = CH) in a rigid glass at -196 °C gave **43** (X = CH) which was stable in fluid solution to ≈ 200 K, and in a plastic medium (polyvinyl acetate matrix) briefly to 20°C. Destruction of **43** (X = CH) was purely intermolecular with no detectable reversion to **42** (X = CH). In contrast the diaza-analogue (**43**, X = N) underwent thermally forbidden disrotatory ring-closure to **42** (X = N) at 120 K[35]. The sterically stabilized 2,3-naphthoquinodimethanes analogous to **36** ($R^1 = R^2 =$ Me) and **36** ($R^1, R^2 = (CH_2)_4$) are long-lived in fluid solution at 20°C[10b].

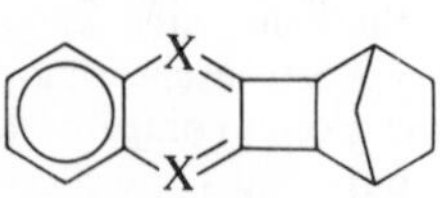

(42) 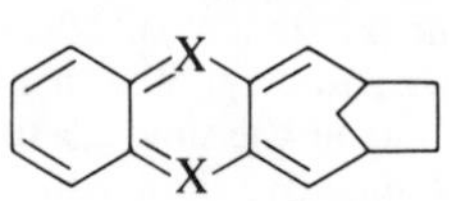 (43)

o-Quinodimethanes stabilized by push–pull resonance include **44**[36], the isobenzofulvene (**45**)[37], the highly coloured *o*-quinone methides **46**[38] and **47**[39], and the quinone methide imine **48**[40].

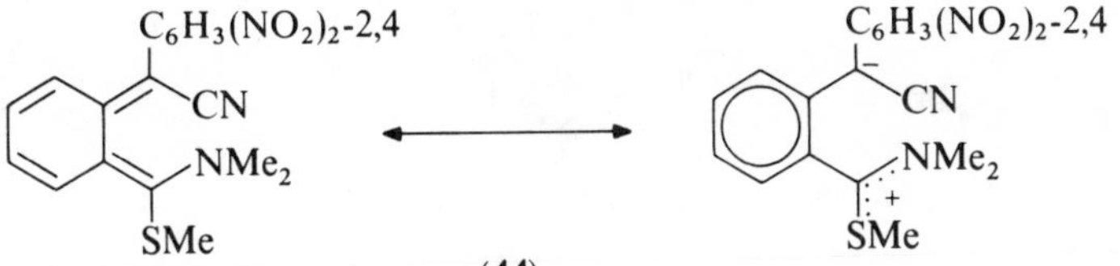

(44)

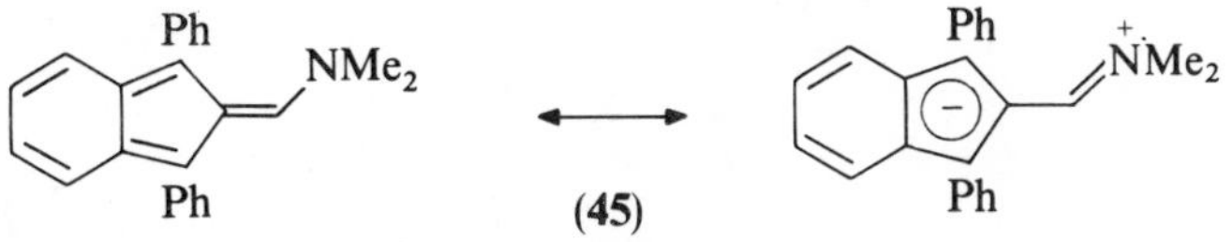

(45)

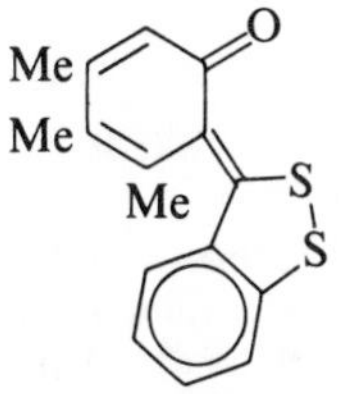

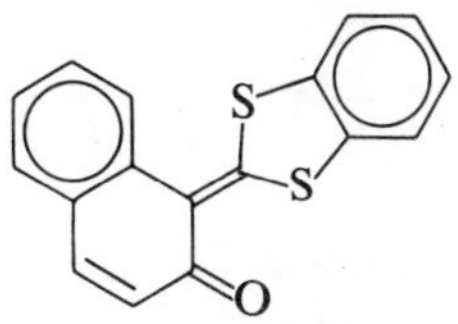

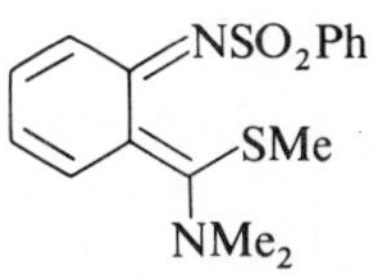

(46) **(47)** **(48)**

In common with other reactive intermediates *o*-quinonoid compounds have been isolated in the form of stable metal complexes. The first complexes of this type were **49** and **50**[41a].

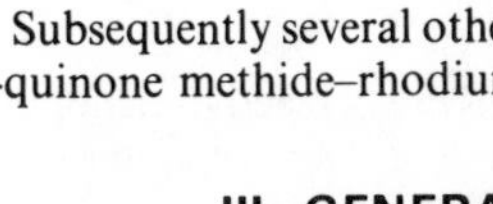

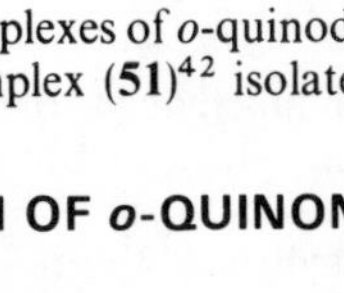

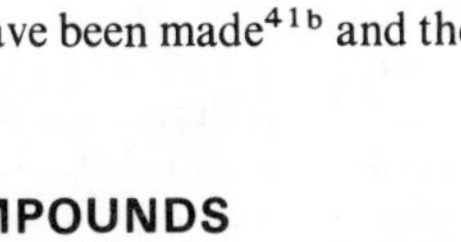

(49) **(50)** **(51)**

Subsequently several other complexes of *o*-quinodimethane have been made[41b] and the *o*-quinone methide–rhodium complex (**51**)[42] isolated.

III. GENERATION OF *o*-QUINONOID COMPOUNDS

A. 1,4-Elimination Reactions

The preparation of 1,2-dibromobenzocyclobutene (**52**) by reaction of *o*-xylylene tetrabromide (**53**) with sodium iodide is an early example of the 1,4-elimination route[43]. Elimination (**53**, arrows) leads to the *o*-quinodimethane (**54**) which gives **52** by conrotatory ring-closure; **54** can be trapped with *N*-phenylmaleimide. In contrast reaction of dimethyl α,α'-dibromo-*o*-phenylene diacetate with sodium iodide gives no monomeric benzocyclobutene analogous to **52**. Rather the spiro-dimer (**55**) of α,α'-bis(methoxycarbonyl)-*o*-quinodimethane is obtained in good yield. Trapping experiments with *N*-phenylmaleimide and dimethyl fumarate strongly indicate intermediacy of the *E,E*-*o*-

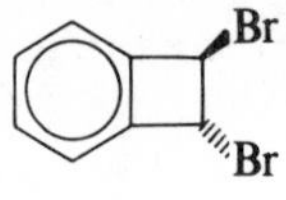

(52) 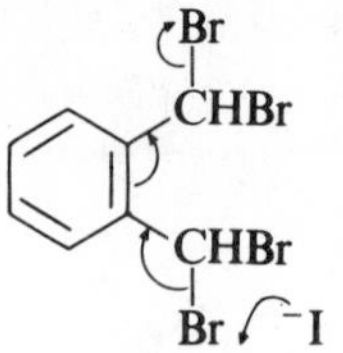 (53) 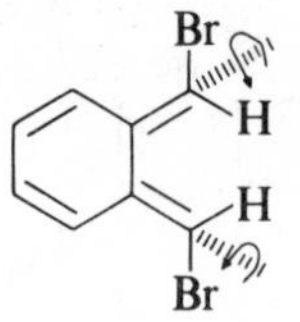 (54)

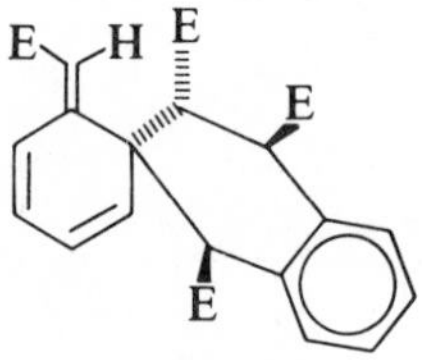

$E = CO_2Me$ (55) $E = CO_2Me$ (56) (57) 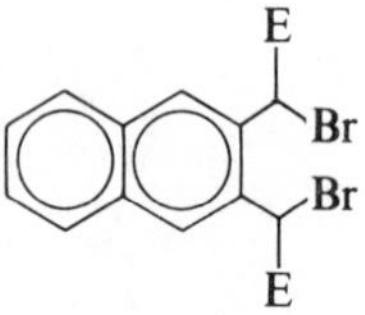 $E = CO_2Me$ (58)

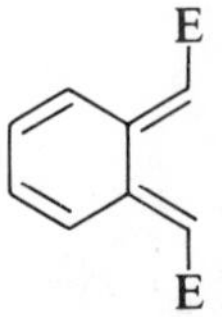

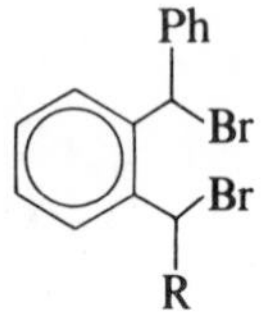

quinodimethane (**56**) in these eliminations[44]. Upon similar debromination **57** (R = H) and **58** give mixtures of products derived by ring-closure and dimerization of intermediate *o*-quinodimethanes, whereas **57** (R = Ph) gives mainly the benzocyclobutene[45]. A similar dehalogenation route has been used to generate indole-2,3-quinodimethanes, e.g. **25**[17] (Section I.C).

1,4-Dehalogenation using zinc was used to generate *o*-quinodimethane from *o*-xylylene dibromide; in the presence of maleic anhydride the adduct **59** was formed in good yield[25a]. The use of ultrasound has been recommended to improve the efficiency of these reactions; adduct **59** is then obtained in 89% yield[46].

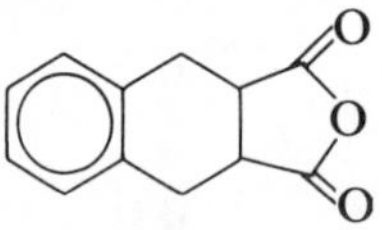

(59) 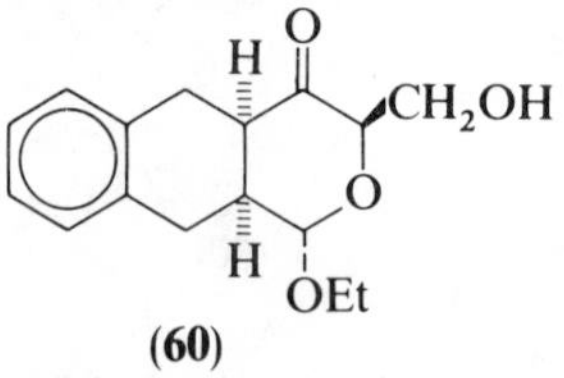 (60) 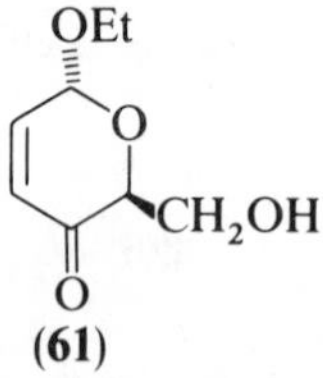 (61)

Similarly, the adduct (**60**) of *o*-quinodimethane with the moderately reactive dienophile (**61**) was obtained in 70% yield upon ultrasound-promoted zinc debromination of *o*-xylylene dibromide in the presence of **61**[47].

1,4-Dehydrohalogenation of *o*-methylbenzyl chlorides is conveniently accomplished by flash vacuum thermolysis; good yields of benzocyclobutenes are obtained under these

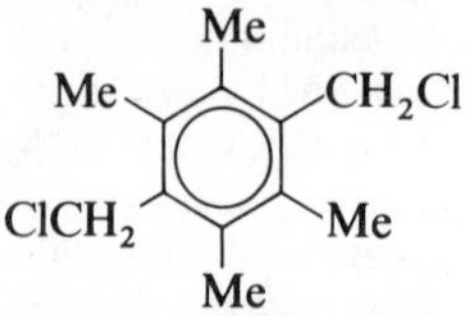

(62) 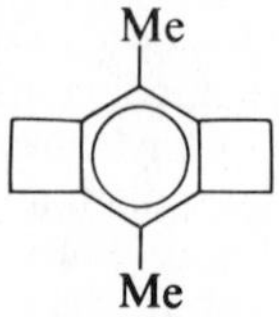 (63) 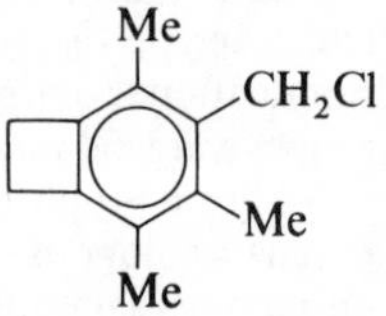 (64)

conditions and dimerization of the intermediate *o*-quinodimethane is minimized[48]. A particularly interesting example of this method[49] involves bis-dehydrohalogenation of **62** to **63** via **64**. Elimination of HCl from **65** (R = Me, X = Cl) and of HBr from **65** (R = Et, X = Br)[50] yields *o*-quinone methides **66** (R = Me) and **66** (R = Et) respectively; in the

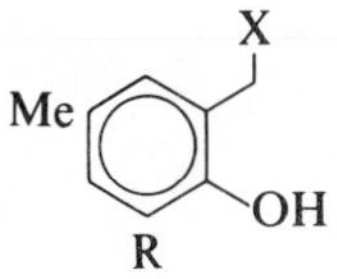

(65) 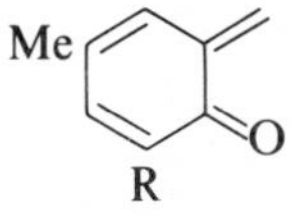 (66)

absence of traps these *o*-quinone methides form dimers. The novel *o*-quinone methide imine (**68**) has been proposed as an intermediate in the flash vacuum pyrolysis of imidoyl chloride (**67**) which affords 2-phenylindole (**69**) (76 %). This appears to be a general route to 2-phenylindoles as similar pyrolysis of **70** yields indole **72**, possibly via ring-closure of the *o*-quinone methide imine (**71**)[51].

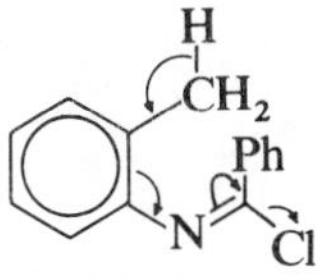

(67) 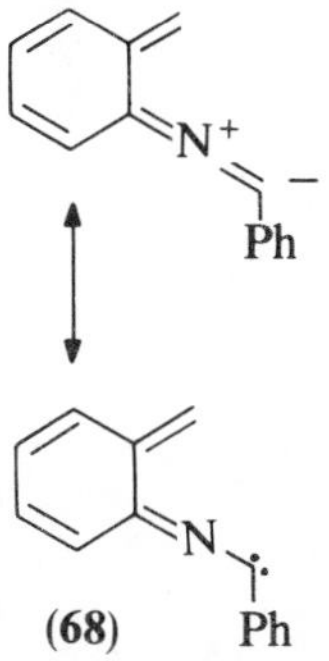 (68) 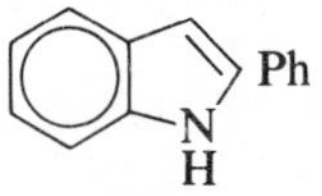 (69)

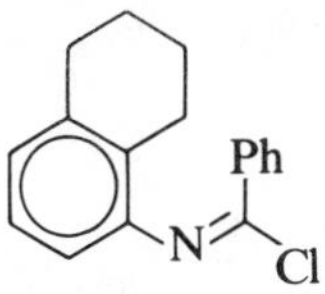

(70) 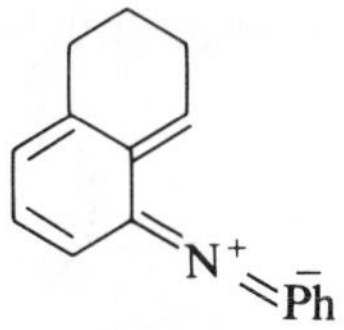 (71) 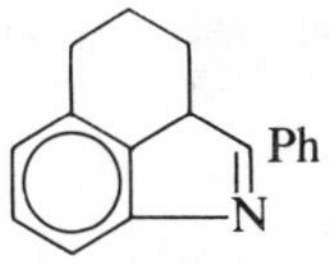 (72)

The synthetic use of *o*-quinodimethanes as reactive diene components in Diels–Alder additions has prompted the development of particularly mild 1,4-elimination methods. The fluoride ion induced elimination of the type (**73**, arrows) has been used in a neat estrone synthesis[52]. This elimination proceeds so rapidly at or below 20°C that the UV–visible band of *o*-quinodimethane can be observed (λ_{max} 367 nm) and its second order decay followed kinetically[53]. The same approach has been used to generate pyridine[52] and indole-2,3-quinodimethane[17b] systems. Similarly, *o*-quinone methide imine (**75**) can be produced at 20°C by this type of elimination of the salt (**74**). However, despite attempts to trap **75** only the dimer **76** could be isolated[54]. A related elimination has been utilized to generate *o*-quinone methides. Treatment of bis-trimethylsilyl ether (**77**) with fluoride ion produces cycloadduct (**79**) by intramolecular trapping of *o*-quinone methide (**78**)[55].

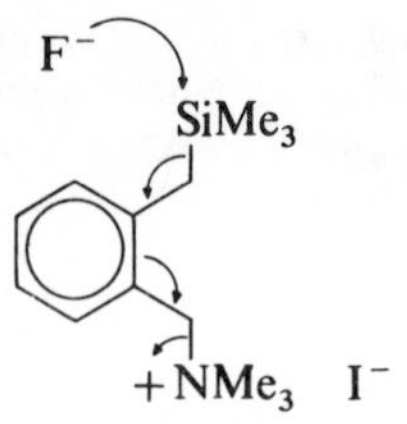

(73)

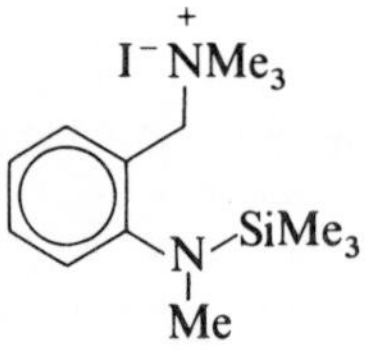

(74) 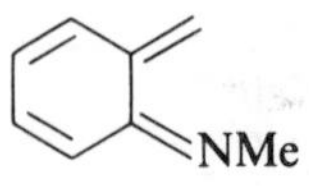 (75) 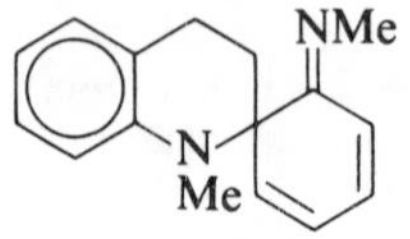 (76)

(77) 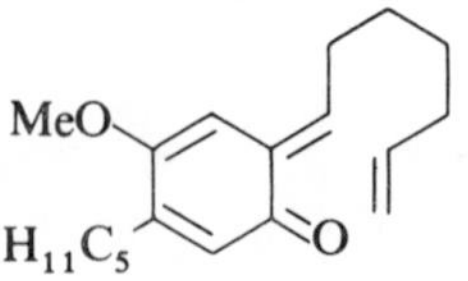 (78) 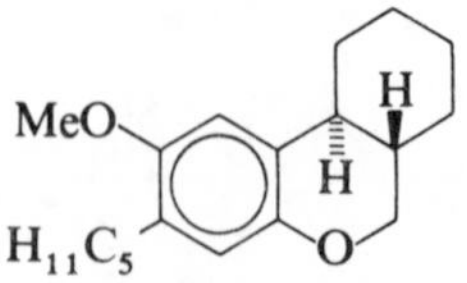 (79)

Rickborn and his collaborators have used 1,4-methanol elimination induced by a strong base to generate *o*-quinodimethanes[56]. Thus treatment of (**80**) with lithium tetramethylpiperidide (LTMP) gives both the spiro-dimer (**2**) and the linear dimer (**81**) of *o*-quinodimethane as well as a polymer. Trapping of the intermediate (**1**) can be carried out

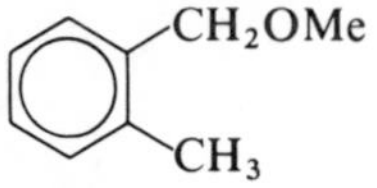

(80) 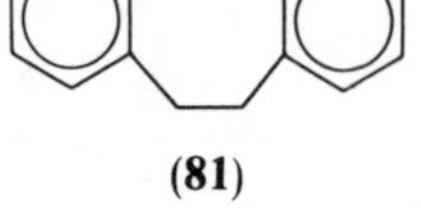 (81)

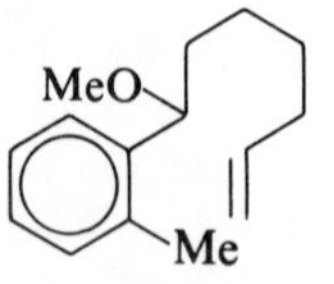

(82) 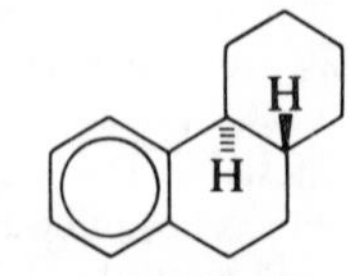 (83)

efficiently with norbornene (70%) and norbornadiene (56%). The process is also an efficient route to polycycles when trapping is intramolecular, e.g. **82** gives **83** in 71% yield on treatment with LTMP[56a].

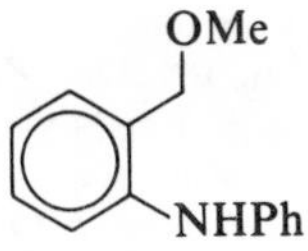

(**84**) 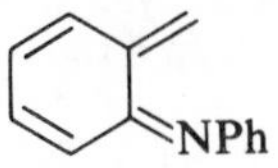 (**85**) 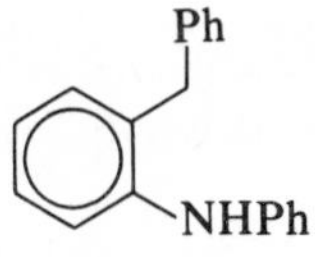 (**86**)

Elimination of methanol from amine ether (**84**) to yield *o*-quinone methide imine (**85**) was postulated to occur in the reaction of **84** with phenylmagnesium bromide. The ultimate formation of amine (**86**) was explained by nucleophilic attack of a second molecule of Grignard reagent on the intermediate (**85**)[57]. Methanol elimination from *o*-tolualdehyde dimethyl acetal generates 1-methoxy-*o*-quinodimethane[56b], and reaction of the cyclic acetal (**87**) with lithium di-isopropylamide (LDA) generates benzo[*c*]furan

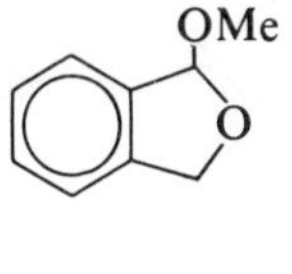

(**87**) 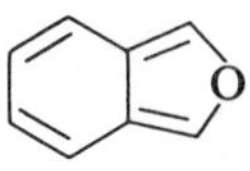 (**88**) 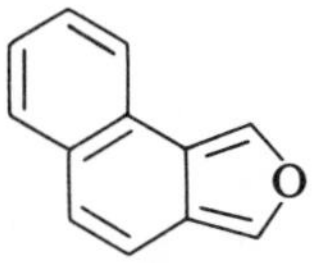 (**89**)

(**88**)[56c]. Acetic acid catalysed loss of two molecules of methanol from the dimethyl acetal of *o*-hydroxymethylbenzaldehyde also gives **88** which can be trapped with a variety of dienophiles[56d]. Both benz[*e*]isobenzofuran (**89**) and benz[*f*]isobenzofuran (**90**) can be

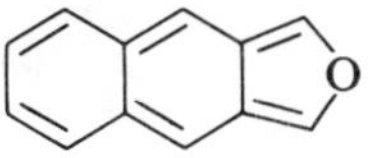

(**90**) 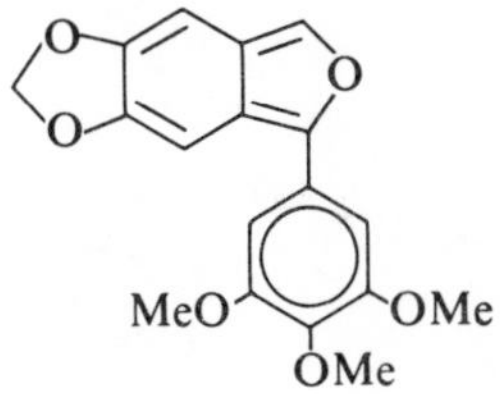 (**91**)

generated using the strong base elimination of acetals[58a,c] or the related acetic acid catalysed dehydration of cyclic hemiacetals[58b]; **89** is isolable[58a] whereas **90** is a reactive intermediate[58b,c]. Rodrigo and his collaborators used the acid catalysed route to generate the isobenzofuran (**91**) as part of their synthesis of podophyllotoxin[59]. A popular route to *N*-methyl-substituted isoindoles starts with addition of an alkyl-lithium to an *N*-methylphthalimidine (**92**); the adducts (**93**) undergo 1,4-elimination of water to give the isoindoles (**94**)[60]. Other *N*-alkylisoindoles have been prepared from dihydroisoindoles

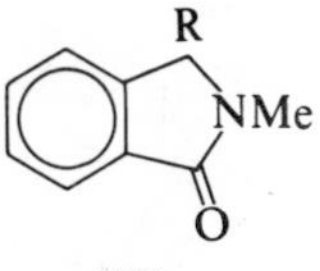

(**92**) 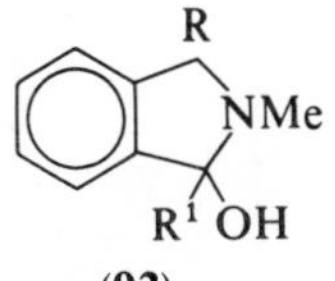 (**93**) 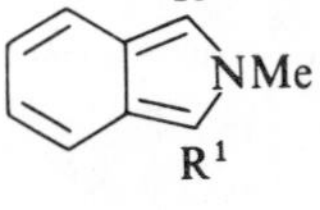 (**94**)

(95)[61] by reaction with hydrogen peroxide and dehydration of the *N*-oxides (**96**) either by heat or upon treatment with acetic anhydride. If this reaction involves Polonovski rearrangement, e.g. to the acetate (**97**) then it too is a 1,4-elimination. Related elimination

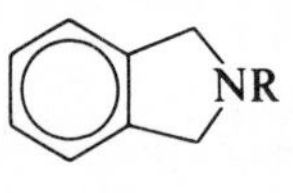

(**95**) 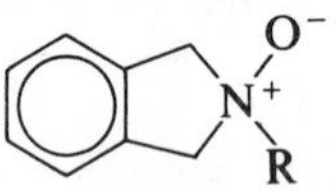 (**96**) 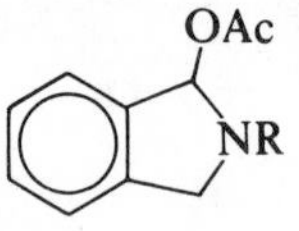 (**97**)

of water from sulphoxides is an excellent route to isothianaphthene (**98**)[62] and benz[*f*]isothianaphthene (**99**)[63]. The latter was deposited as a bright yellow solid at liquid nitrogen temperature following flash vacuum pyrolysis of the sulphoxide (**100**).

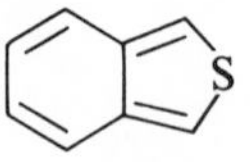

(**98**) 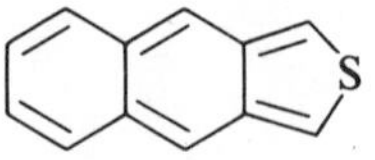 (**99**) 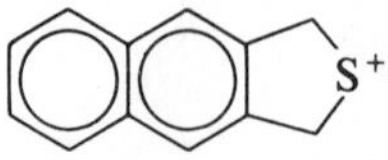 (**100**)

Benzo[*c*]selenophene (**101**) can be generated by treatment of the dibromide (**102**) with sodium hydroxide solution[64]. The benzopyrone (**103**) is generated by heating *o*-formylphenylacetic acid (**104**) in acetic anhydride[65]. If **104** is stirred with cold acetic

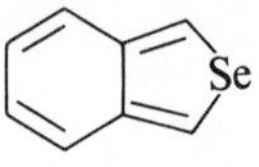

(**101**) 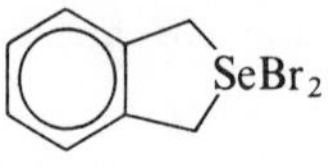 (**102**) 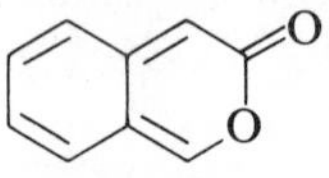 (**103**)

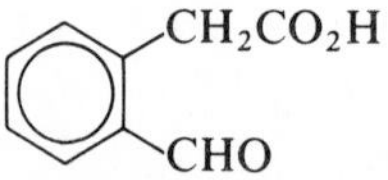

(**104**) 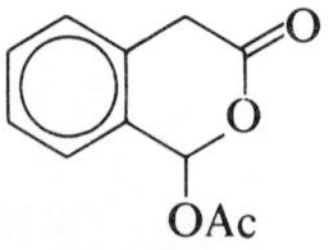 (**105**) 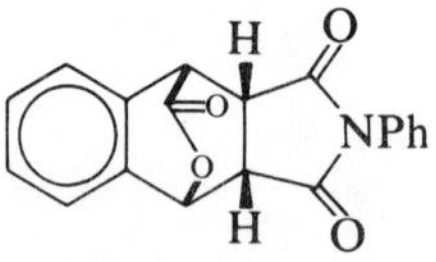 (**106**)

anhydride and the latter is removed under a high vacuum the residue has a PMR spectrum consistent with the presence of the pseudo-acid anhydride (**105**)[65b]. With *N*-phenylmaleimide in boiling xylene **105** undergoes 1,4-elimination of acetic acid to give the *o*-quinonoid pyrone **103** isolated as the adduct **106**[65b]. Likewise the 1,3-diphenyl derivative of inden-2-one (**18**) can be generated by acetic anhydride dehydration of **107** (R^1 = OH, R^2 = H). However, iodide reduction of **107** (R^1 = R^2 = Br) is a more efficient

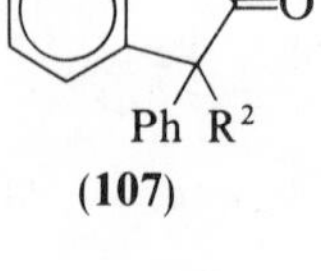

(**107**) 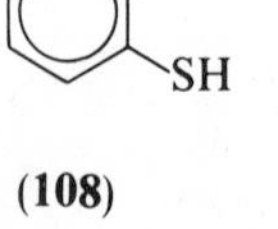 (**108**) 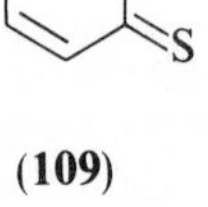 (**109**) 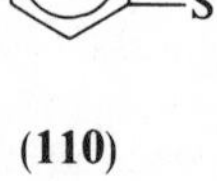 (**110**)

route to 1,3-diphenylinden-2-one[10]. 1,4-Dehydration has also been applied widely to the generation of other *o*-quinonoid species, although this usually involves the use of flash vacuum pyrolysis. For example such dehydration of thiol (**108**) produces a good yield of benzothiete (**110**) from ring-closure of *o*-thioquinone methide (**109**)[66]. Similarly, pyrolysis of carboxylic acid (**111**) at 840 °C produces naphthothioquinone methide (**112**) which

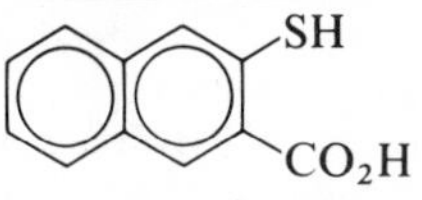

(**111**) 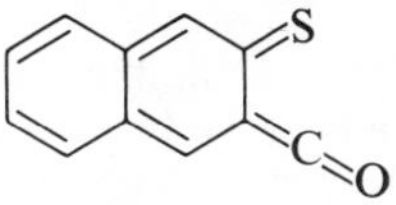 (**112**) 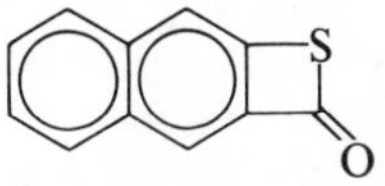 (**113**)

closes to the stable naphthothietone (**113**)[67]. *o*-Quinone methide imines, e.g. **85**[68,69] and *o*-quinone methide (**3**) itself[70], are produced by flash vacuum pyrolysis of amines, e.g. **114** (X = NPh) at 750 °C and the alcohol (**114**, X = O) at 600 °C respectively.

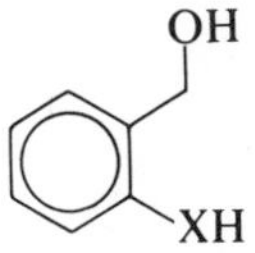

(**114**)

Gardener[1b] has reported that elimination of dimethylamine from amine (**115**) yields the highly unstable *o*-quinone methide (**116**) which promptly dimerizes to **117**. It should be

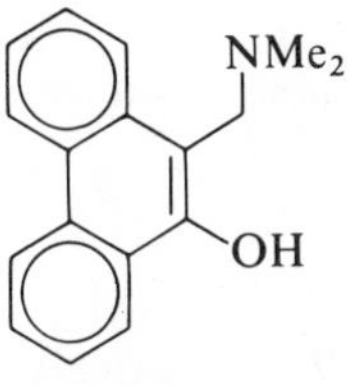

(**115**) 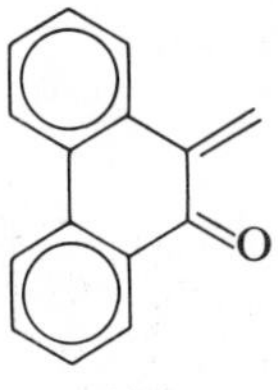 (**116**) 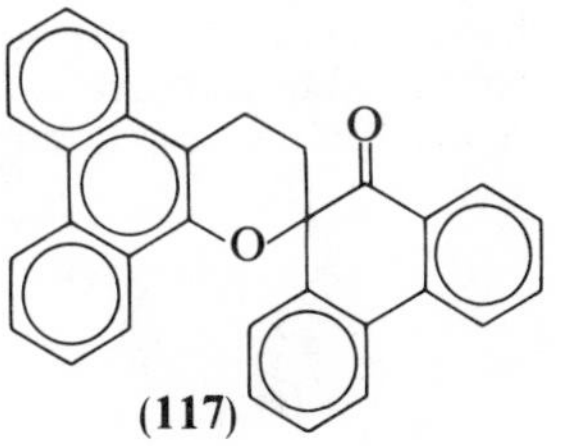 (**117**)

noted that Errede used vinylogous Hofmann elimination of the quaternary ammonium hydroxide (**118**) in the first detailed study of the generation of *o*-quinodimethane[71].

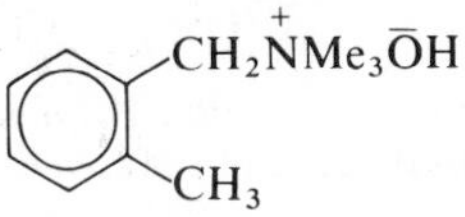

(**118**)

B. Extrusion Reactions and Reverse Diels–Alder Processes

Thermal or photochemical extrusion of a small molecule X from structures of the general type **119** is a popular route to *o*-quinodimethanes. Thus vapour phase pyrolysis of 1,3-dihydrobenzo[*c*]thiophene 2,2-dioxide (**119**, X = SO_2, R = H) gives benzocyclobutene (**120**, R = H), the linear dimer (**81**) of *o*-quinodimethane, and *o*-xylene in proportions that vary with concentration and temperature; in boiling diethyl phthalate

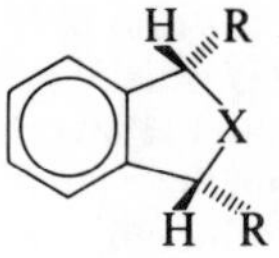

(119) 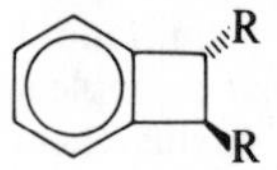 **(120)**

(300 °C) **119** ($X = SO_2$, R = H) gives **81** in 48 % yield but low pressure vapour phase pyrolysis (460–670 °C) gives mainly (59–63 %) benzocyclobutene[72]. Pyrolysis of **119** ($X = SO_2$, R = H) at 280 °C in the presence of *N*-phenylmaleimide gives the expected *o*-quinodimethane adduct **121**[72]. Thermolysis of the *cis*-dideuteriated sulphone (**119**, $X = SO_2$, R = D) gives **120** (R = D). This agrees with a suprafacial (*cis*) elimination of sulphur dioxide to give the *E,E*-(**122**) and *Z,Z*-dideuterio-*o*-quinodimethanes both of

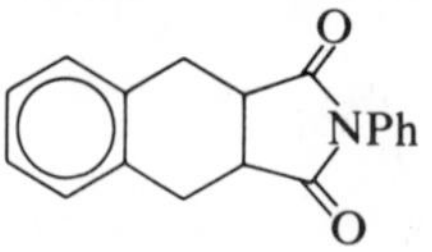

(121) 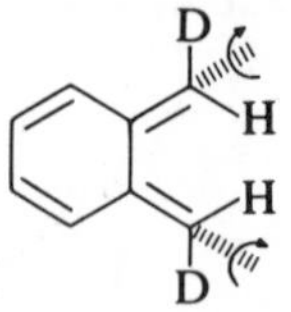 **(122)**

which undergo conrotatory ring-closure, e.g. **122** (arrows) to give **120**[73]. Although sulphur dioxide extrusion is a high temperature reaction it finds extensive use and can be very effective. Thus thermolysis of **119** ($X = SO_2$, R = H) in refluxing 1,3,5-trichlorobenzene in the presence of 1,2-dihydronaphthalene gave the adduct **123** in 92 % yield[74] and thermolysis of **124** (Ar = 3,4,5-trimethoxyphenyl) at 170–210 °C in the presence of dimethyl fumarate gave the adduct **125** (> 90 % yield)[75a]. The synthesis of aromatic steroids has also employed extrusion of sulphur dioxide from sulphones[76].

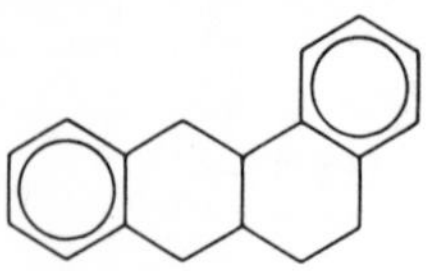

(123) 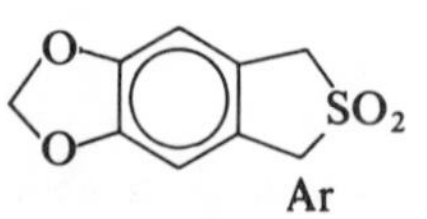 **(124)** 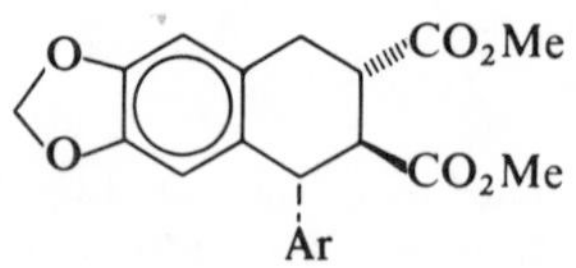 **(125)**

The benzo-3,6-dihydro-1,2-oxathi-in-2-oxide (**126**) extrudes sulphur dioxide at temperatures some 200 °C lower than those required for the related sulphone[77]. Heating **126** and maleic anhydride in benzene gave **59** (> 95 % yield). The availability of a convenient new route to sultines like **126** involving, e.g. reaction of the reagent **127** with phthalyl

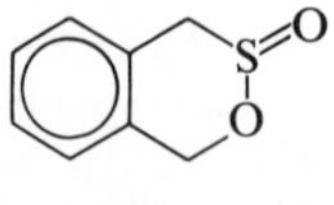

(126) 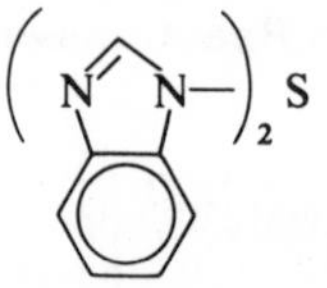 **(127)**

alcohol should make this route to *o*-quinodimethanes more popular[78]. It is noteworthy that in the addition of sulphur dioxide to *o*-quinodimethane the sultine **126** is kinetically favoured over sulphone (**119**, X = SO_2, R = H)[79] by a factor of 9:1.

In view of the high temperatures required for SO_2 extrusion from sulphones like **119** (X = SO_2) it is interesting that triethylamine catalyses SO_2 loss from strained cyclic sulphones like **128**. Although purely thermal extrusion of SO_2 from **128** requires heating

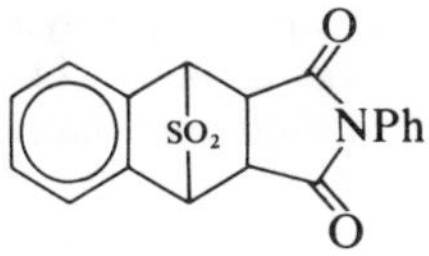

(**128**) 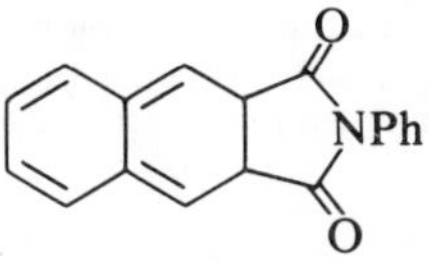 (**129**)

at 150–170 °C, addition of triethylamine produces **129** at 20 °C. The mechanism of this catalysis is not understood but does seem to require a strained sulphone; decomposition of **119** (X = SO_2, R = H) was not catalysed by triethylamine[80].

The extrusion of sulphur dioxide has also proven useful for the generation of other *o*-quinonoid intermediates. Photochemical elimination of SO_2 from **130** (X = S) yields *o*-thioquinone methide (**109**) which in the presence of *N*-phenylmaleimide gives the adduct **131** in 43 % yield. In the absence of a trap only polymeric material is obtained[81]. Photolysis

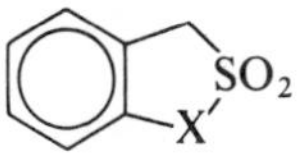

(**130**) 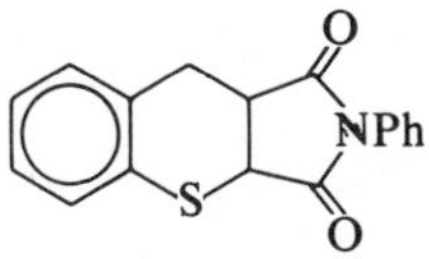 (**131**) 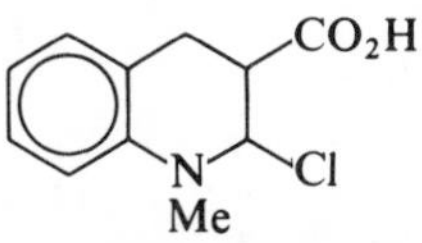 (**132**)

of the sultam (**130**, X = NMe) in the presence of *E*-chloroacrylic acid gives the Diels–Alder adduct (**132**) of *o*-quinone methide imine (**75**); in the absence of dienophiles, the benzazetidine (**133**) is formed in 62 % yield as a white crystalline solid. This is a rare example of the ring-closure of an *o*-quinone methide imine to the corresponding benzazetidine[82]. Photolysis of **134**, the oxygen analogue of **126**, yields *o*-quinone methide (**3**) which adds to 1,1-dimethoxyethylene to give adduct **135**[70]. However, thermal conversion of **134** to *o*-quinone methide was not reported.

(**133**) 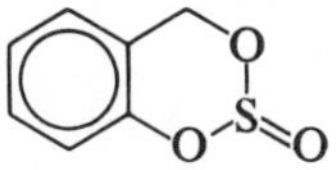 (**134**) 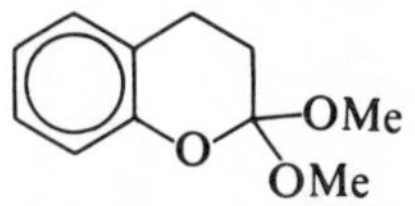 (**135**)

Thermal decarboxylation of isochromanones, e.g. **136** (R^1 = OMe, R^2 = H), by flash vacuum thermolysis at ca. 500 °C is a convenient route to certain benzocyclobutenes[83]. Photodecarboxylation of **137** (R = Ph) and **137** (R = Me) allowed observation of the UV–visible bands of the corresponding *o*-quinodimethanes[84]. 2-Benzopyran-3-one (**103**) can be generated either by thermal (ca. 150 °C) or photochemical decarboxylation of the bis-lactone (**138**). The photochemical method is useful in the low temperature characterization of **103**[7].

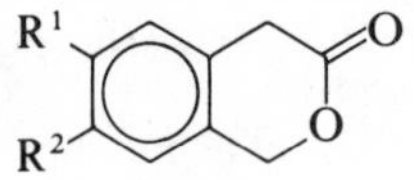

(136) 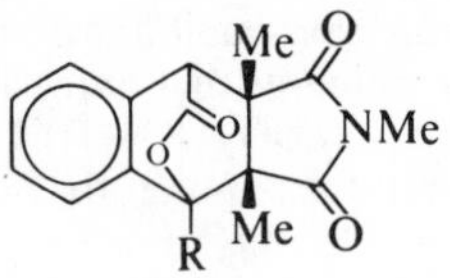 (137) 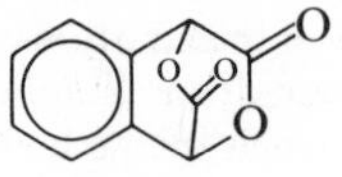 (138)

Generation of the *o*-quinone methide imine **139** by flash vacuum pyrolytic decarboxylation of the 1,4-dihydro-3,1-benzoxazin-2-one (**140**) gives acridine (**141**) by electrocyclization (**139**, arrows) and dehydrogenation. Similarly, the intramolecular Diels–Alder

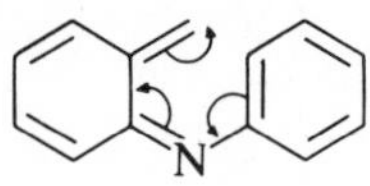

(139) 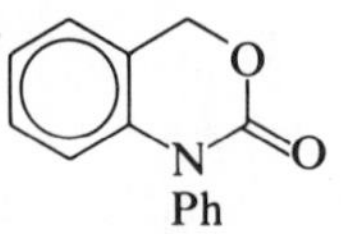 (140) 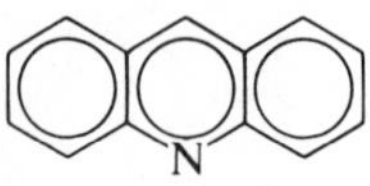 (141)

cycloadducts **143** ($n = 3$) and **143** ($n = 4$) may be obtained from pyrolysis of benzoxazinones **142** ($n = 3$) and **142** ($n = 4$) respectively, although in only moderate yield.

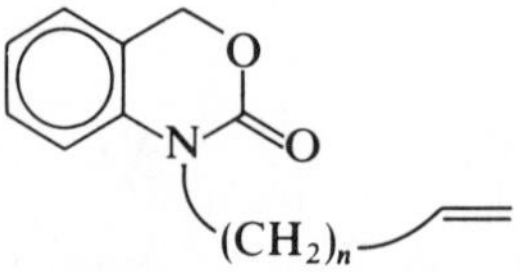

(142) 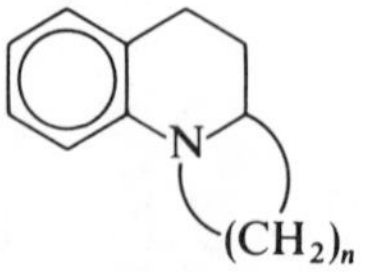 (143)

Interestingly, it was later reported that substantially higher yields of these cycloadducts can be obtained by conducting these pyrolyses in the presence of alumina or silica when the temperature required for decarboxylation is lowered by 400 °C[85].

Van Tilborg and Plomp[86] have reported that pyrolysis of thianaphthene 1,1-dioxide (**144**) at 1000 °C induces an unusual rearrangement to yield benzothiete (**110**) (42 % yield). They propose a mechanism involving rearrangement of **144** to **145**, followed by decarboxylation to **109** which ring closes to **110**.

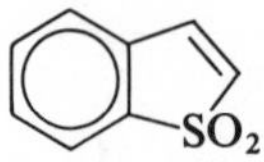

(144) 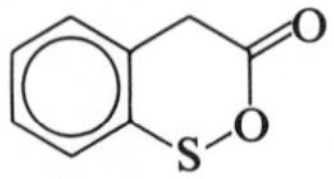 (145) 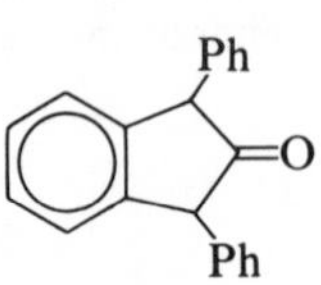 (146) 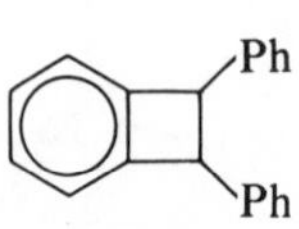 (147)

Photodecarbonylation of indan-2-ones (**146**) is the method of choice for the preparation of 1,2-diphenylbenzocyclobutenes (**147**)[87]. Irradiation of *cis*-**146** ($\lambda > 300$ mm) in ether solution gives *trans*-**147** (89 %) and only 11 % of the *cis*-isomer. Since photochemical *cis–trans* isomerization of *cis*-**146** and photodecarbonylation of *cis*-**146** involve the same singlet excited state a stepwise loss of carbon monoxide via a biradical intermediate is indicated[88a]. In agreement with loss of stereochemistry via such an intermediate, photodecarbonylation of *cis*-**146** in the crystalline state[88b] results in much greater retention of stereochemistry giving *cis*-**147** (95 %) and only 5 % of *trans*-**147**. The sterically stabilized *o*-quinodimethanes (**36**) were also prepared in high yield by photodecarbonylation[31]. Similarly, photolysis of **148** gives a mixture of products **149**, **150** and **151** clearly derived from the *o*-quinodimethane intermediate (**152**). The 1,5-hydrogen shift (**152**, arrows) leading to **151** is a thermal process that is suppressed below $-40\,^{\circ}C$[89].

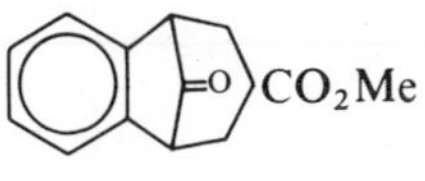

(**148**) 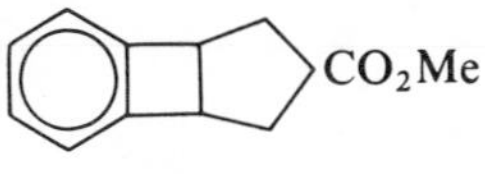 (**149**) 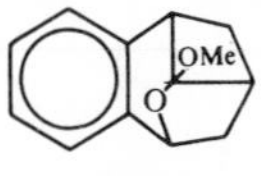 (**150**)

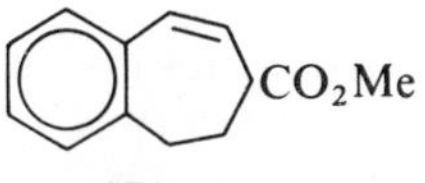

(**151**) 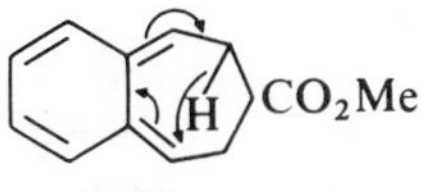 (**152**) 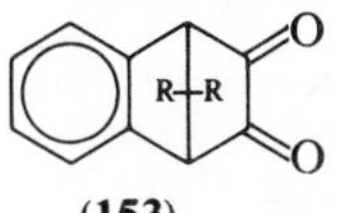 (**153**)

Photo-bisdecarbonylation of α-diketones is an effective route to isoindenes[28, 90] which, however, ring-close to pseudoindenes under the reaction conditions. Thus irradiation of **153** (R, R = $(CH_2)_4$) gives the pseudoindene **154**. Photodecarbonylation of **155** (X = O[70], S[91], NPh[92]) provides *o*-quinone methide (**3**), *o*-thioquinone methide (**109**) and *o*-quinone methide imine (**85**) respectively, and flash vacuum thermolysis of **155** (X = O) gives the trimer of **3** (68%) in addition to fulvene (5.5%) and benzene (22%)[93].

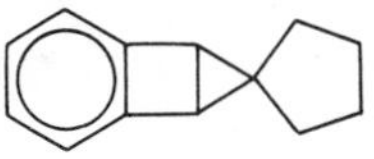

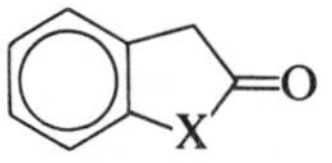

(**154**) (**155**)

Extrusion of nitrogen from dihydrophthalazines can be carried out photochemically, a process useful in matrix isolation studies (Section II), or under very mild thermal conditions (ca. $-40\,^{\circ}C$). The easy tautomerism and thermal decomposition of dihydrophthalazines limits their preparative use. The more stable benzotriazinones **156** (R = 2-pyridyl and 2-thiazolyl) also undergo thermal loss of nitrogen[94]. However, the expected benzazetidinones (**157**) ring-open to *o*-quinone methide imines, e.g. **158**, which ring-close, e.g. **158**, arrows to the observed fused products, e.g. **159**.

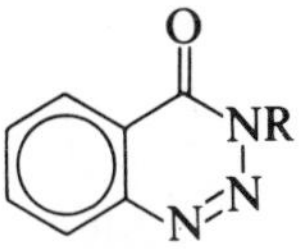

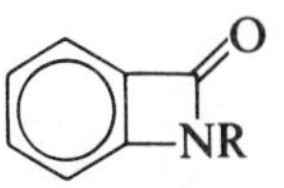

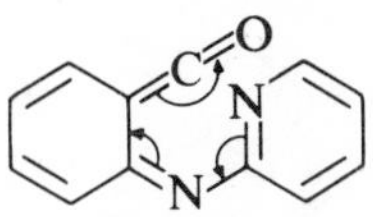

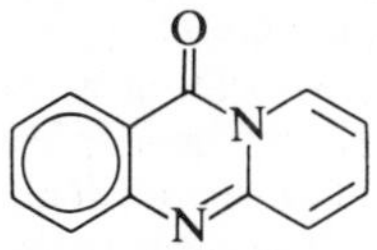

(**156**) (**157**) (**158**)

(**159**)

Oxidative loss of nitrogen from *N*-aminoisoindolenines (**160**) can be brought about by mercuric oxide or through reaction with *p*-toluenesulphonyl chloride in pyridine to give diphenylbenzocyclobutene (**147**)[95]. In either case it is believed that fragmentation of the

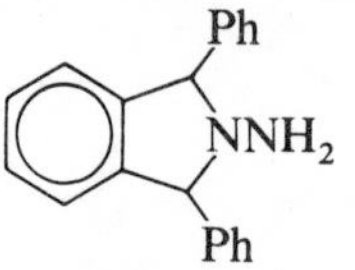

(**160**) 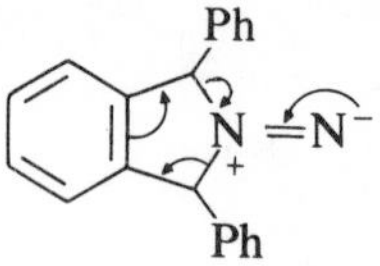 (**161**)

amino-nitrene (**161**) is involved. Although the preformed tosylate of *trans*-**160** gives *trans*-**147** (80% yield) the equally high yield conversion of the tosylate of *cis*-**160** gives *cis*-**147** and *trans*-**147** in a ratio of 3:1. Although the instability of *cis*-**147** under the reaction conditions (KOH/EtOH, 20 °C) is suggested to explain this lack of stereospecificity a non-cheletropic nitrogen loss from **161** is not ruled out.

Fragmentation involving a reverse Diels–Alder reaction has been extensively used to generate *o*-quinodimethanes of the type **162** from the adducts (**163**) readily available by addition of benzyne to the dienes **164** (X = O, X = CH_2, X = NR and X = C=CMe_2).

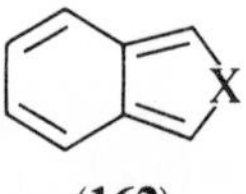

(**162**) 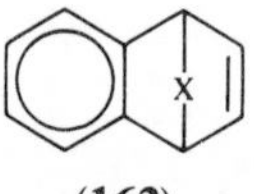 (**163**)  (**164**) 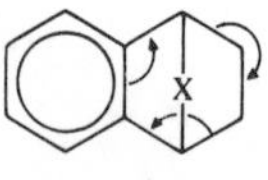 (**165**)

Formal removal of an acetylene unit from the benzyne adducts (**163**) to give the *o*-quinonoid compounds (**162**) can be accomplished in several ways. Most simply the reduced adducts (**165**) are subjected to flash vacuum pyrolysis which induces reverse Diels–Alder reaction (**165**, arrows). This method has allowed the preparation of both isobenzofuran (**162**; X = O)[96] and isoindole (**162**; X = NH)[97]. Fieser and Haddadin[98] reacted **163** (X = O) with α-pyrone to give the adduct **166**. In boiling toluene **166** loses carbon dioxide to

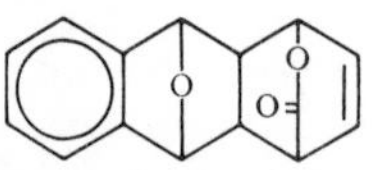

(**166**) 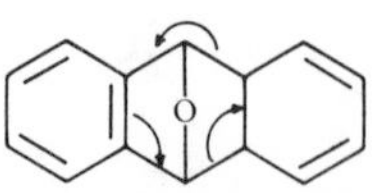 (**167**) 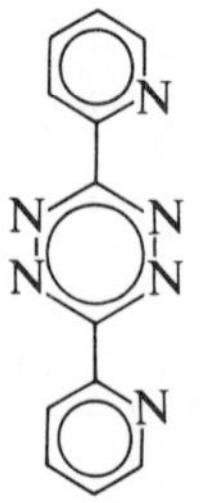 (**168**)  (**169**)

give **167** which then loses benzene (**167**, arrows) to give **162** (X = O). Vacuum thermolysis of **166** allowed isolation of both **162** (X = O) and the intermediate **167**[99]. Warrener[100] used the tetrazine (**168**), which behaves as an electron-deficient Diels–Alder diene, to convert **163** (X = NH) to isoindole. Reaction of **163** (X = NH) with **168** occurs at −25 °C to give **162** (X = NH) and the pyridazine (**169**). The intermediate (**170**, X = NH) which should be formed from **163** (X = NH) and **168** by Diels–Alder addition and nitrogen loss could not be detected, presumably because of its very rapid reverse Diels–Alder

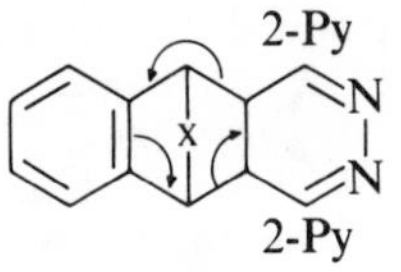

(170) 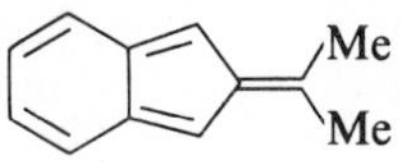 (171)

fragmentation (**170**, arrows). The oxygen analogue (**170**, X = O) although isolable dissociates to benzo[*c*]furan and **169** at 20 °C. This and related methods have been used to generate isoindene[101] and isobenzofulvene (**171**)[102].

A formal reverse Diels–Alder process is involved in the flash vacuum pyrolysis of benzoxazine (**172**) which gives benzaldehyde and a trimer of imine (**173**); the latter is

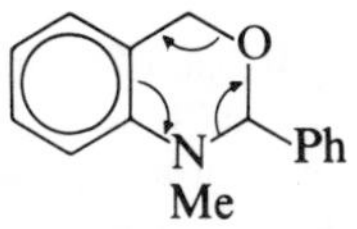

(172) 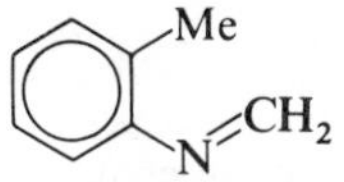 (173) 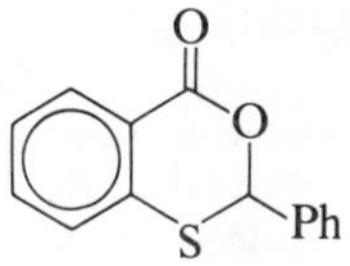 (174)

thought to arise by 1,5-hydrogen shift in the quinone methide imine (**75**) formed by the fragmentation (**172**, arrows)[69]. Similar photochemical fragmentation of **174** at 77 K gave rise to infrared absorptions consistent with formation of the thiolactone (**175**). Upon warming the product to − 40 °C a dimer of **176** was isolated[103].

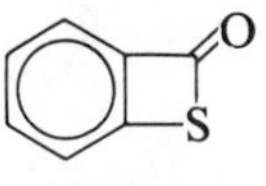

(175) 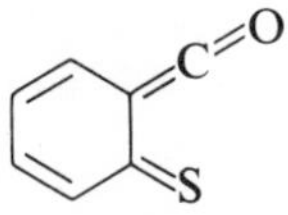 (176)

C. Electrocyclic Ring-opening Reactions of Benzocyclobutenes and Related Compounds

The ease of the electrocyclic ring-opening of benzocyclobutenes (**177**, arrows) is markedly dependent upon the substituents present on the four-membered ring[104]. The parent system (**177**, R = H) ring-opens on heating above 200 °C. When the substituent R

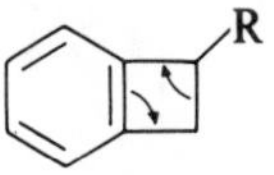

(177) 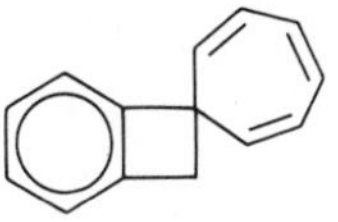 (178) 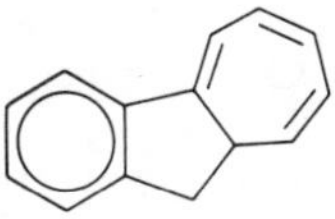 (179) 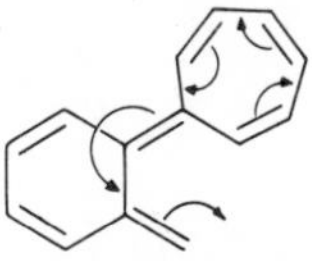 (180)

on the cyclobutene ring is capable of donating an electron pair the ease of ring-opening parallels the availability of the electron pair: **177** (R = NH_2) ring-opens at 25 °C, **177** (R = OH) at 80 °C, **177** (R = NHCOR′) at 110 °C and **177** (R = Alk) at 180 °C. The alkoxides related to **177** (R = O^-) produced by benzyne addition to enolate anions ring-open under very mild conditions[105]. Conjugating electron-withdrawing substituents also accelerate ring-opening; the ketones (**177**; R = COR′) ring-open at ca. 150 °C. Substituents which are only conjugating also accelerate ring-opening, e.g. **178** is converted to **179** at 40–60 °C. Ring-opening of **178** to **180** is followed by 10π-electron ring-closure (**180**, arrows) to **179**[106]. The *trans*-diphenylbenzocyclobutene (**181**) reacts with dienophiles at 20 °C to give adducts derived from the *E*, *E*-*o*-quinodimethane (**182**)[87b]. Huisgen's

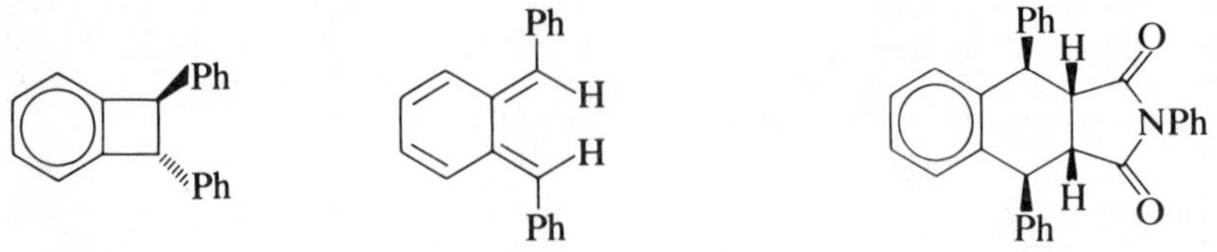

(**181**) (**182**) (**183**)

mechanistic investigation has shown[107] that an equilibrium between **181** and **182** precedes adduct formation. In solution at 20 °C the concentration of **182** is small but at 120 °C a yellow colour, probably due to **182** is observed. In agreement with orbital symmetry control the *trans* isomer (**181**) gives the *E*,*E*-diene (**182**), the product of least hindered conrotatory ring-opening. With *N*-phenylmaleimide this gives the *cis*-*endo*-adduct (**183**). The *cis* isomer of **181** undergoes slower (ca. 70 times at 80 °C) conrotatory ring-opening via a necessarily more hindered transition state (TS) involving inward rotation of one phenyl group towards the centre of the breaking cyclobutene ring. The resulting *Z*,*E*-*o*-quinodimethane (**184**) gives the adduct (**185**) with *N*-phenylmaleimide. The preference for

(**184**) (**185**)

outward rotation of a substituent on the benzocyclobutene ring extends to monosubstituted benzocyclobutenes and is particularly marked when the substituent is an electron donor. Thus benzocyclobutenol (**186**), available by photolysis of *o*-tolualdehyde[108a], gives only the adduct (**187**) derived by *endo*-addition of the *E*-dienol (**188**) with maleic

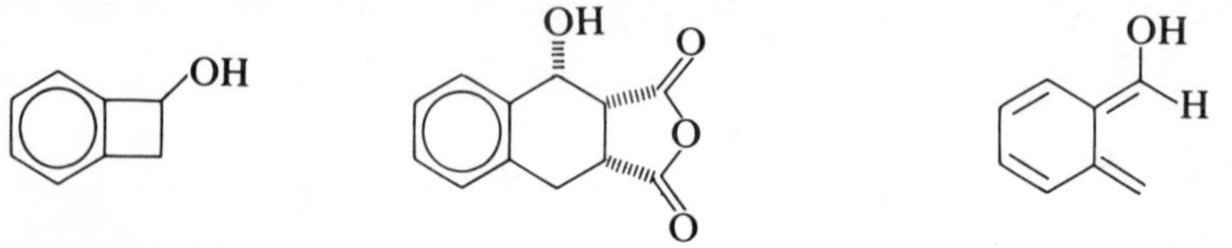

(**186**) (**187**) (**188**)

anhydride[108b]. The acetal (**189**) and the *cis*-1,2-bismethoxybenzocyclobutene (**190**) for which conrotatory ring-opening would necessarily involve inward rotation of one oxygen substituent fail to undergo ring-opening[108c]. The effect of substituents on the rate and stereochemistry of benzocyclobutene ring-opening noted above are paralleled in the ring-opening of simple cyclobutenes, and have been discussed and rationalized[109]. Steric effects

(189) **(190)**

are only partly responsible for the preferred outward rotation of alkoxyl groups in the ring-opening of simple cyclobutenes. Good quality calculations trace the electronic contribution to reduced destabilizing donor $\leftrightarrow \sigma$ interaction (**191a**) and improved stabilizing donor $\leftrightarrow \sigma^*$ interaction (**191b**) in the transition structure (**191**) for outward rotation in comparison with better destabilizing donor $\leftrightarrow \sigma$ overlap (**192a**) and poorer stabilizing

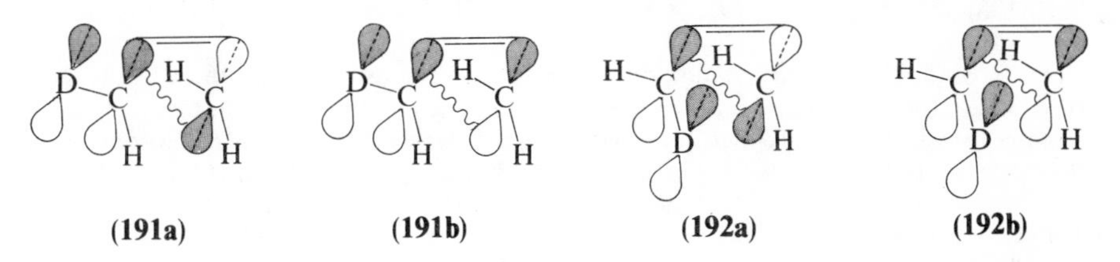

(191a) **(191b)** **(192a)** **(192b)**

D = donor atom; ∿∿∿ = breaking σ-system

donor $\leftrightarrow \sigma^*$ overlap (**192b**) in the transition structure (**192**) for inward rotation[109]. The HOMO and LUMO of the transition structures **191** and **192** are to a good approximation σ and σ^* of the breaking cyclobutene σ-bond.

Even when allowed conrotatory ring-opening of benzocyclobutenes is prevented by ring-fusion, *o*-quinodimethanes may be obtained under mild conditions when ring-strain is relieved, e.g. pseudo-indene (**33**) gives isoindene (**32**, R = H) above 0 °C[26]. Similarly **193** gives **194** at 180 °C. This conversion is dramatically accelerated by Ag(I) salts, occurring almost instantly at 20 °C. The presumed *o*-quinodimethane intermediate (**195**) is then

(193) **(194)** **(195)**

efficiently trapped by *N*-phenylmaleimide[110]. The novel *o*-quinodimethane precursor, Dewar *o*-quinodimethane (**196**), ring-opens[111] at 60 °C. The greater aromaticity loss accompanying ring-opening to give 2,3-naphthoquinonoid systems is reflected in the great

(196) **(197)**

difficulty of such reactions. Whereas *trans*-1,2-diphenylbenzocyclobutene reacts with *N*-phenylmaleimide at 20 °C, **197** requires fusion with the dienophile at 150 °C for adduct formation[112].

There are few simple examples of the electrocyclic ring-opening of benzothietes, benzazetidines and benzoxetes. In agreement with theoretical prediction[4] (Section I.A) benzoxetes appear to be much less stable than their ring-opened *o*-quinone methide tautomers and in several reactions where benzoxetes could have been produced the isolated products can be seen to be derived from the corresponding *o*-quinone methides. For instance, generation of benzyne in the presence of diphenylcyclopropenone (**198**) gives **199**

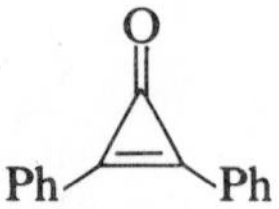

(**198**) 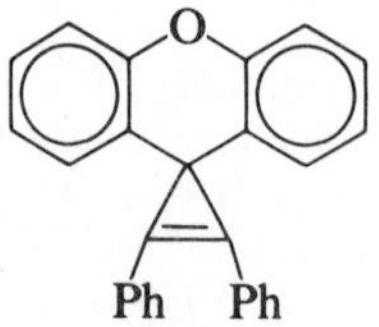 (**199**) 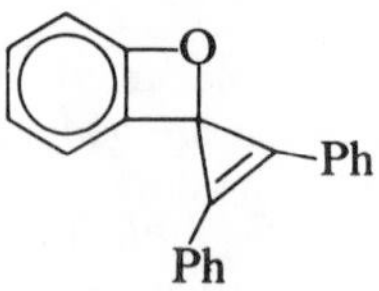 (**200**) 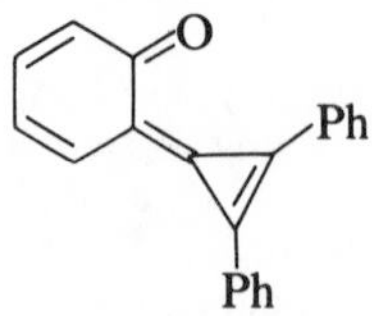 (**201**)

which is probably derived by addition of benzyne to the carbonyl group of **198** to give benzoxete (**200**) which undergoes rapid ring-opening to the *o*-quinone methide (**201**); addition of benzyne to **201** then affords **199**[113]. Similarly addition of tetrachlorobenzyne to cinnamaldehyde gives the chromene (**202**)[114]. In this case when the carbonyl carbon of

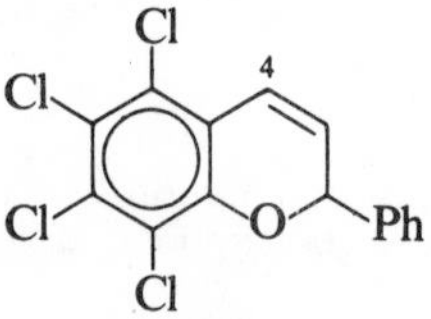

(**202**) 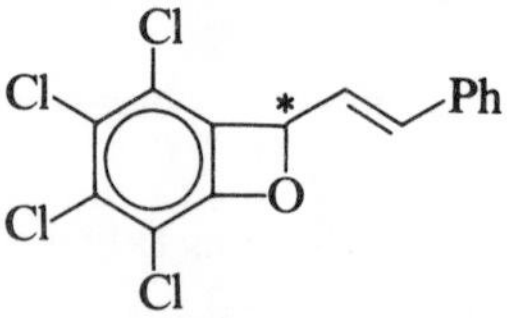 (**203**) 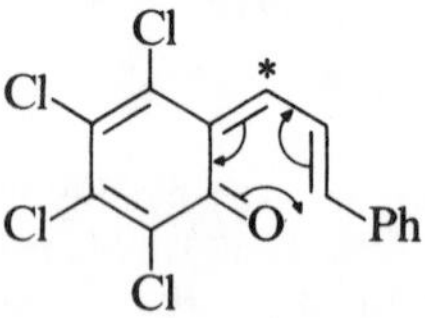 (**204**)

cinnamaldehyde is labelled the label appears at C(4) of the product (**202**). This agrees with intermediacy of the benzoxetene (**203**) and its rapid ring-opening to the quinone methide (**204**) which undergoes the indicated 6π-electron electrocyclic ring-closure to **202**.

In a similar attempt to prepare benzazetidines, benzyne was generated in the presence of amidine (**205**)[115]. This produced acridine (**206**) in place of the expected **207**. Initial

$PhN{=}CHNMe_2$

(**205**) (**206**) 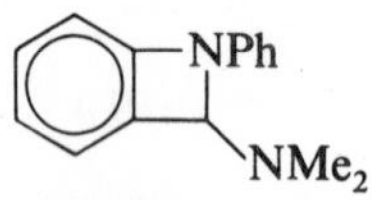 (**207**)

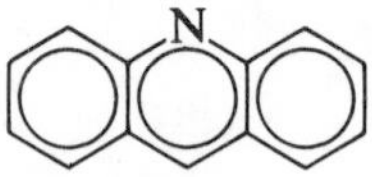

formation of **207** followed by ring-opening would give **208**, which upon 6π-electrocyclization (**208**, arrows) and loss of dimethylamine would give **206**. Rapid

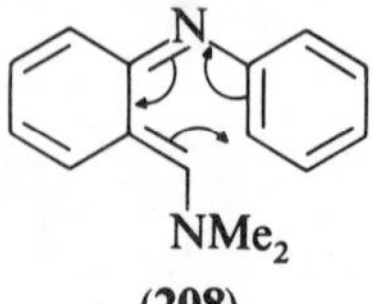

(**208**)

electrocyclic ring-opening of **207** may be associated with the presence of the strongly electron-donating NMe_2 group. Certainly other benzazetidines appear to be more stable;

thus **209** was heated with *N*-phenylmaleimide at 200 °C to form the adduct (**210**)[116]. Benzothiete **110** is also a stable isolable compound which undergoes ring-opening only if

(**209**) 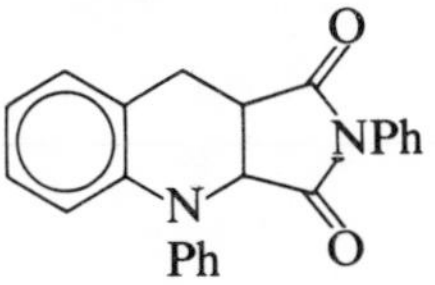 (**210**)

heated to 140 °C; the *o*-thioquinone methide produced can be trapped, and in the absence of traps forms a dimer. Although the benzothiete **211** is available via Wolff rearrangement of the diazo compound **212** by photolysis in methanol its ring-opening does not appear to

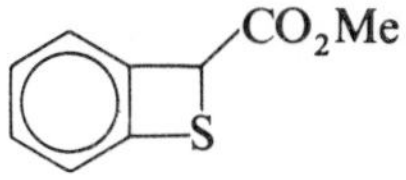

(**211**) 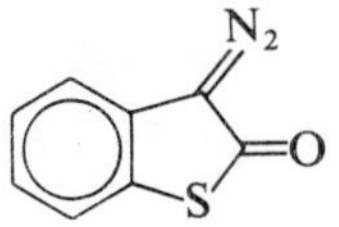 (**212**)

have been studied[117]. The thiete-1,1-dioxides **213**[118] and **214** (R = H and Ph)[119, 120] are also known but their ring-opening does not appear to have been achieved.

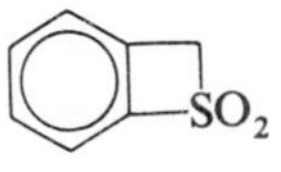

(**213**) 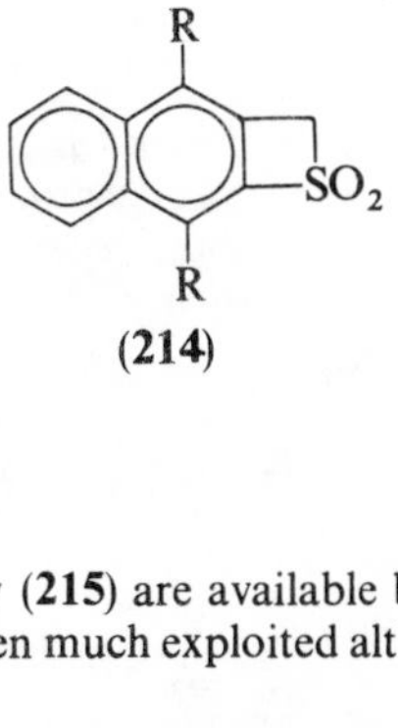 (**214**)

D. Sigmatropic Shifts

In principle *o*-quinodimethanes with *Z*-geometry (**215**) are available by 1,5-shift of a group M in the styrenes (**216**). This route has not been much exploited although at 170 °C

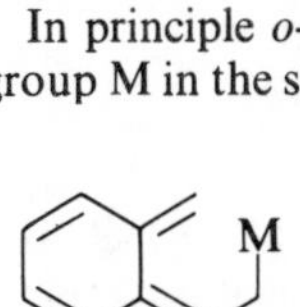

(**215**) 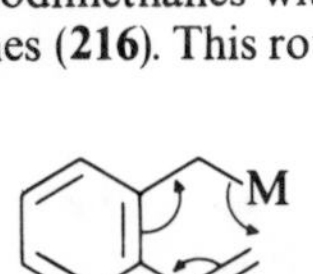 (**216**) 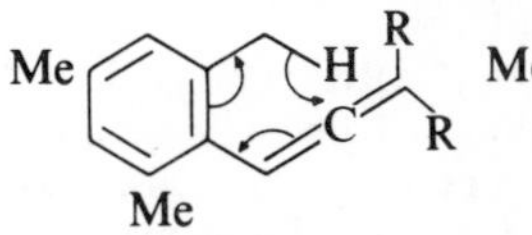 (**217**) 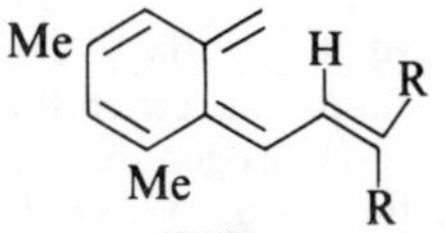 (**218**)

the allenes (**217**, R = H or Me) undergo a 1,5-hydrogen shift which involves as migration terminus the particularly reactive central carbon of the allene unit (**217**, arrows) to give the *o*-quinodimethanes (**218**, R = H or Me); **218** (R = H) cyclizes to dihydronaphthalene (**219**) and **218** (R = Me) undergoes antarafacial 1,7-hydrogen shift to the diene (**220**)[121]. An example of a photochemically driven process that is formally a 1,5-shift (**221**, arrows) converts **221** into **222** and **223**[122]. Here the expected *E*-*o*-quinodimethane (**224**) can account for the *trans* isomer (**222**) formed, by assuming an *exo*-selective internal

(219) (220)

(221) (222)

Diels–Alder reaction (**224**, arrows). However the 1,5-hydrogen shift product (**223**) indicates participation of the *Z*-isomer of **224**.

(223) (224)

The well established[123] photoenolization of *o*-tolualdehyde[108a] and related compounds is only a formal 1,5-sigmatropic process as it is believed to involve biradical intermediates which may have zwitterionic character[124]. A recent application[75b] involved trapping the photoenol (**225**) from *o*-benzylbenzaldehyde with sulphur dioxide to give the sulphone

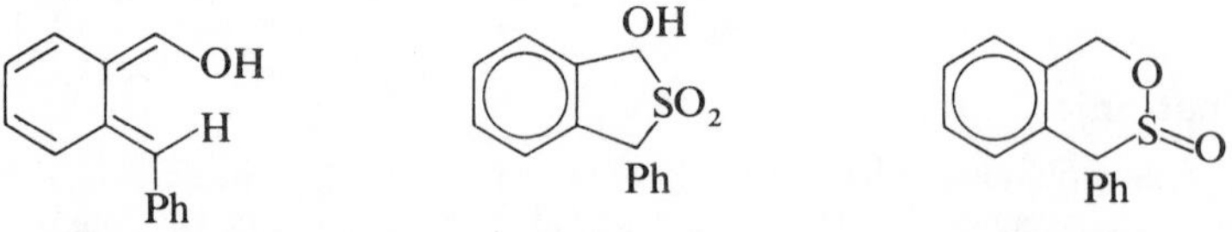

(225) (226) (227)

(**226**). Reduction of **226** ($NaBH_4$) and acid treatment led to the sultine (**227**) which was useful for the generation of 1-phenyl *o*-quinodimethane under mild thermal conditions.

1,5-Hydrogen shift in indene (**228**, arrows) was first[125] postulated to explain formation of the adduct (**229**) from indene and maleic anhydride at 180°C. Elegant D-labelling experiments support the intermediacy of isoindene (**230**) in this reaction. Thus after

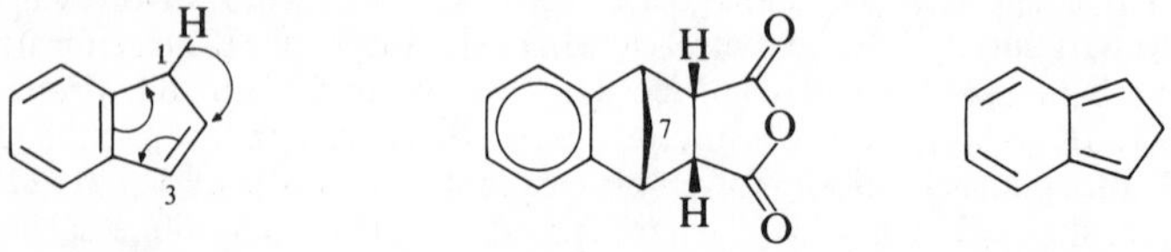

(228) (229) (230)

heating at 200°C 1-deuterioindene has the label statistically distributed over C(1), C(2) and C(3)[126] and 2-deuterioindene provides the adduct (**229**) in which 90% of the label is located in the bridge methylene[127]. On heating the optically active indene (**231**) the isomer (**232**) is formed in optically inactive form, showing that the isoindene (**233**) is involved[128]. A 1,3-hydrogen shift in **231** would have given optically active **232**.

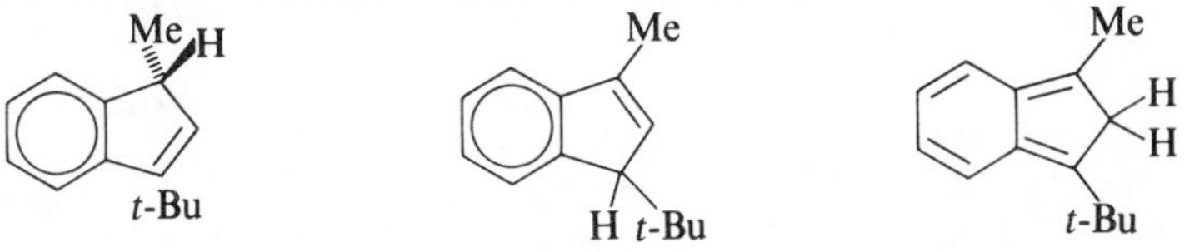

(**231**) (**232**) (**233**)

Despite their easy generation isoindene intermediates have not been extensively employed in synthesis. In one example the isoindene (**234**) generated by heating 5-methoxy-3-methoxycarbonylindene at 170°C was trapped efficiently with maleic anhydride to give **235**, a useful intermediate in the synthesis of diterpene alkaloids[129].

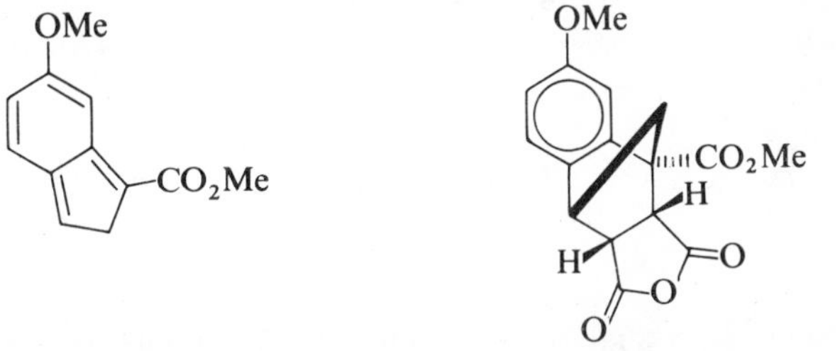

(**234**) (**235**)

A kinetic study[130] of the thermal rearrangement of the indenes (**236**) to their 2-substituted isomers (**237**) via the intermediate isoindenes (**238**) has shown the order of migratory aptitude is H > Ph > Me. It is interesting that isoindene intermediates are also

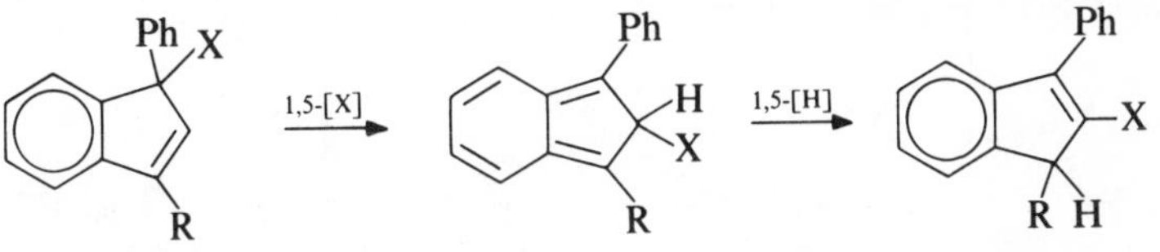

(**236**) (**238**) (**237**)

obtained by apparent photochemical 1,5-aryl migration. Using the flash photolysis technique the transients can be observed and their decay by 1,5-hydrogen shift to give simple indenes monitored[131]; they have half-lives of less than a minute at 20°C.

Migratory aptitudes of a variety of groups have been measured using the optically active 1,3-dimethylindenes (**239**, X = migrating group). Upon 1,5-shift of X the symmetric isoindenes are formed with consequent loss of optical activity. This method allowed the determination of accurate migratory tendencies for groups like the formyl group which migrate more rapidly than hydrogen. In such cases isoindene formation is reversible and the rate of formation of the 2-substituted indenes (**240**) underestimates the migratory

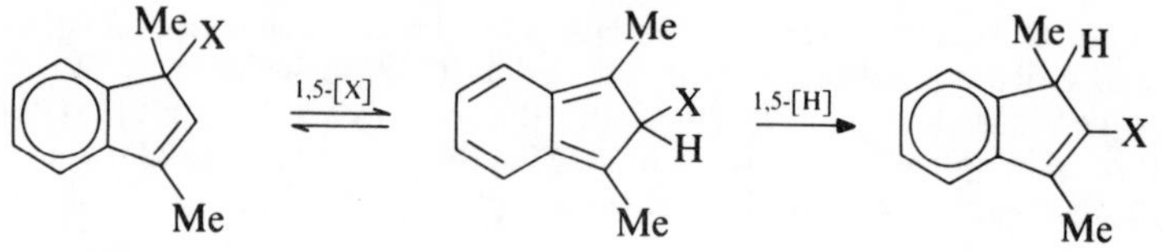

(239) (240)

aptitude of X. Migratory ability decreases in the order: HCO > PhCO and MeCO > H > vinyl > CONHMe and CO_2Ph > CO_2Me > CN and C ≡ CH > alkyl. The formyl group migrates ca. 10^5 times more rapidly than hydrogen whilst the CO_2Me group migrates ca. 10^2 times more slowly than hydrogen. The fast migration of formyl, benzoyl and acetyl groups has been attributed to secondary MO ↔ MO interaction involving the carbonyl π^* orbital of the migrating acyl group[132]. The migratory aptitude of the formyl group is comparable with that of trimethylsilyl and related groups. 1-Trimethylsilylindenes are in equilibrium with isoindenes produced by 1,5-silyl migration under mild thermal conditions[133]. Trimethylstannyl and trimethylgermyl groups are likewise very mobile in the indene system[133b, 134]. As yet the synthetic potential in the isoindene intermediates involved here has not been exploited.

1,5-Shifts of the type **216** (arrows) correspond to tautomerism in the mercaptobenzaldimines (**241**). In several cases the *o*-thioquinone methide tautomers (**242**) have been detected

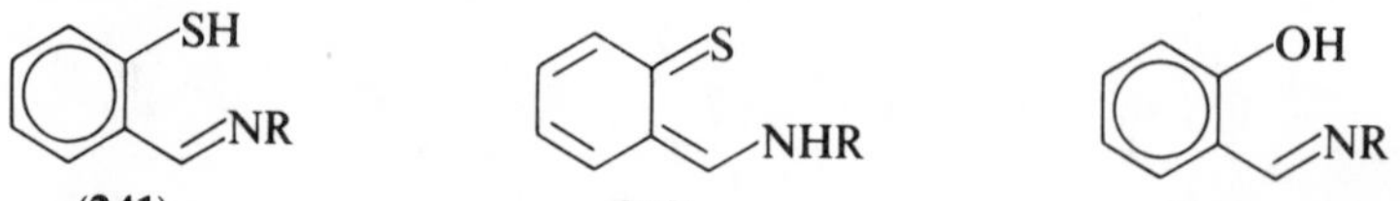

(241) (242) (243)

spectroscopically[135]. In contrast it appears that the oxygen analogues prefer to exist as the imine tautomers (**243**)[135].

E. Other Routes to *o*-Quinonoid Compounds

The allenes **244** and **245** undergo 6π-electron cyclization generating *o*-quinodimethane[136] and 2,3-naphthoquinodimethane[34] respectively. The allene **245** could be isolated following base treatment of the diacetylene (**246**)[34] or reaction of the

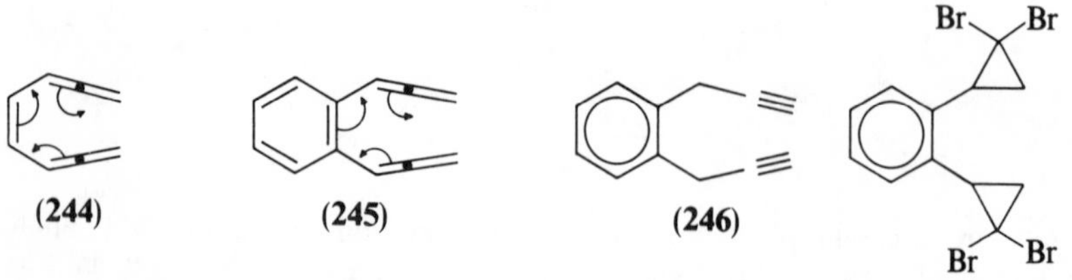

(244) (245) (246) (247)

tetrabromide (**247**) with methyl-lithium[137]. Reaction of **245** with dimethyl maleate and dimethyl fumarate is stereospecific in agreement with a singlet ground state for the intermediate 2,3-naphthoquinodimethane[34]. Oxidative generation is applicable to 2,3-naphthoquinones and *o*-quinone methides. Potassium iodate oxidation of naphthalene-2,3-diol (**248**, R = H) is believed to give transient 2,3-naphthoquinone (**249**, R = H) which

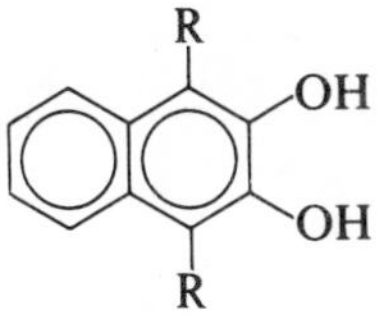

(248) 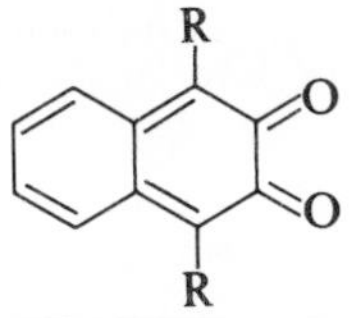 (249) 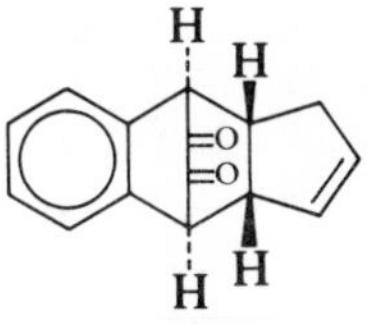 (250)

is trapped by a large excess of cyclopentadiene as the adduct (**250**)[138]. Low temperature oxidation of **248** (R = Ph) with lead tetra-acetate gives a green colour attributed to 1,4-diphenyl-2,3-naphthoquinone (**249**, R = Ph) which can be trapped with a variety of olefins; the green colour faded rapidly and disappeared after 45 min at -20°C[139].

In contrast the similarly generated green colour due to the sterically stabilized derivative (**249**, R = *o*-tolyl) remained unchanged after 2 h at -20°C; this compound was characterized by UV and IR spectroscopy, as the adduct (**251**) formed with norbornadiene,

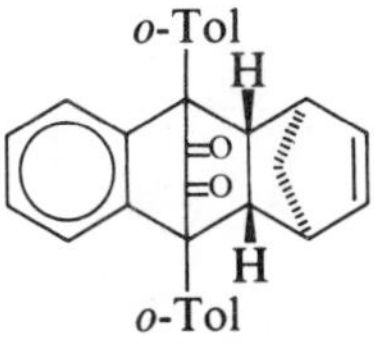

(251) 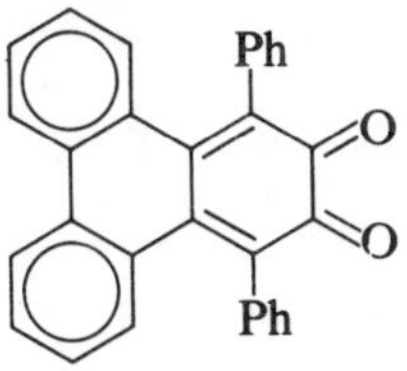 (252) 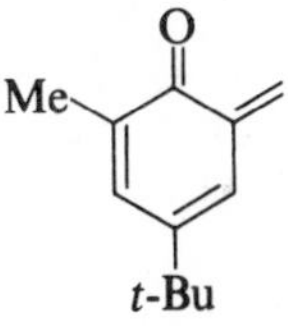 (253)

and by a comparison of its properties with those of the stable quinone (**252**); **249** (R = *o*-tolyl) could be isolated albeit in somewhat impure form[140].

Oxidation of a 2,6-dimethylphenol blocked in the *para* position with a group lacking α-hydrogen results in the formation of an *o*-quinone methide[141]. Thus oxidation of 4-*t*-butyl-2,6-dimethylphenol with silver oxide gives the *o*-quinone methide (**253**).

In principle the photochemical 6π-electron electrocyclic ring-opening of dihydronaphthalene (**254**) should provide the *o*-quinodimethane (**255**). However the observed product is **257** derived via ring-closure (**256**, arrows) of the transoid isomer (**256**) of **255**[142];

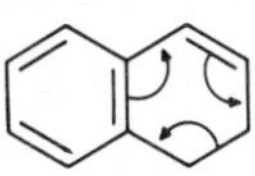

(254) 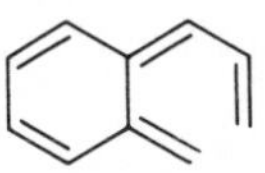 (255) 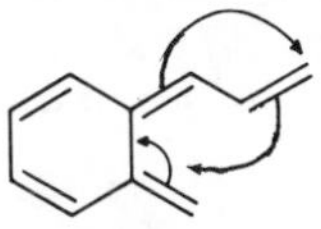 (256) 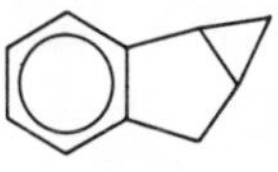 (257)

255 polymerized so readily even at low temperature that it could not be observed. In contrast irradiation of heterocyclic analogues of **254** produce observable *o*-quinonoid species. Thus **258** (X = S, R = H)[143] and **258** (X = O, R = H)[144] give the coloured *o*-quinonoid species **259** (X = S, R = H) and **259** (X = O, R = H) respectively. Similarly,

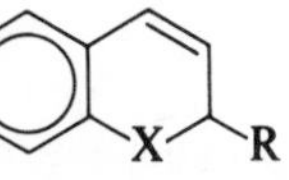

(258) 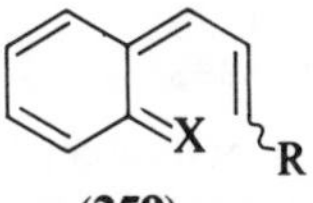 (259) 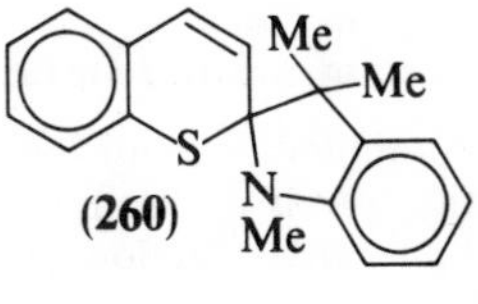 (260)

irradiation of **260** at 77 K gives the green *o*-thioquinone methide **261**[143] and **258** (X = NCO_2Et, R = CN) gives **259** (X = NCO_2Et, R = CN) upon irradiation at low temperature[145].

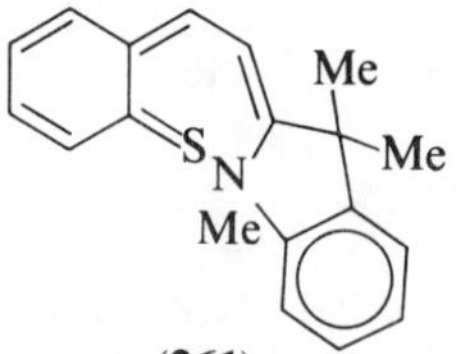

(**261**)

In an intriguing reaction, irradiation of **262** in the presence of olefins yields relatively stable, deep blue *o*-thioquinone methides, e.g. **263**[146]. Treatment of **263** with an ethylene

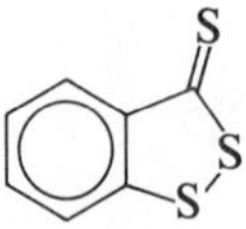

(**262**) 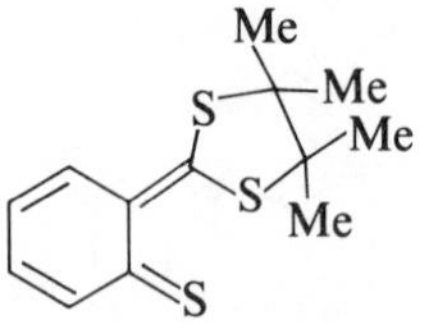 (**263**) 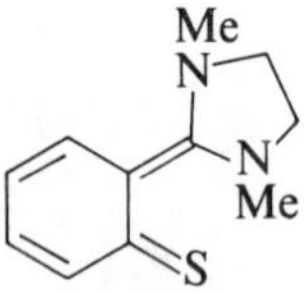 (**264**)

diamine gives the stable *o*-thioquinone methide (**264**). Stabilization of **264** by push–pull resonance is supported by its NMR spectrum[146]. The 2,3-naphthoquinone methide (**265**)

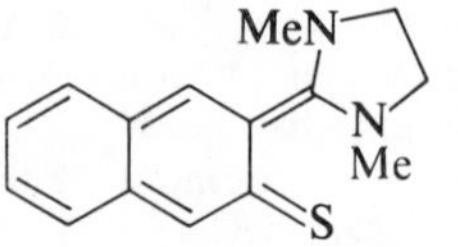

(**265**) 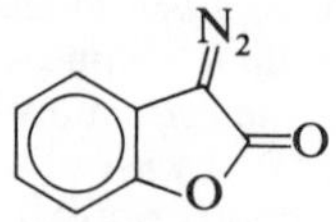 (**266**)

can also be prepared by this route[147]. Irradiation of **266** in an argon matrix at 8 K gave rise to two primary products, **267** and **268**, which could be interconverted photochemically. Prolonged irradiation of **267** led to decarbonylation to benzyne[148].

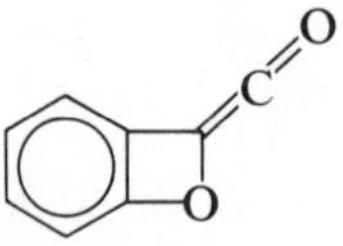

(**267**) 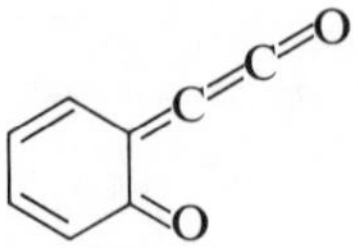 (**268**)

IV. REACTIONS OF *o*-QUINONOID COMPOUNDS

A. Dimerization and Related Processes

When *o*-quinodimethane is generated under mild conditions in solution the spiro-dimer (**2**) (Section I.A) is the main product[24]. However, in solution at higher temperatures the linear dimer (**81**) (Section III.A, B) is increasingly dominant[56a, 72]. In contrast 2,3-

naphthoquinodimethane appears to give no spiro-dimer analogous to **2**; instead in addition to the linear dimer (**269**) the $(6+4)\pi$-dimer (**270**) is formed[34b]. An analogous

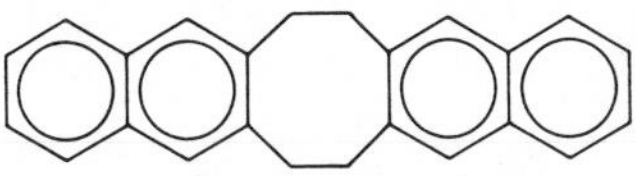

(**269**) 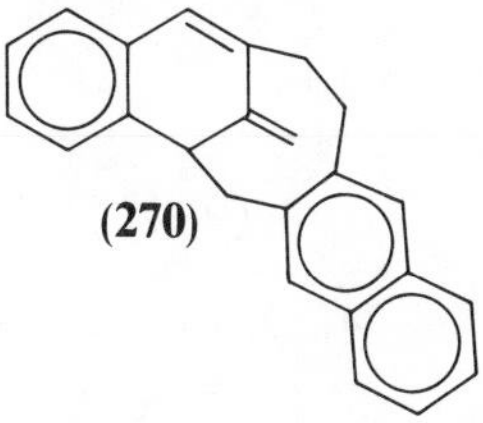 (**270**)

formal $(6+4)\pi$-dimer (**271**)[102] from 8,8-dimethylisobenzofulvene is known and the labile $(6+4)\pi$-dimer (**272**) of 1,3-diphenylinden-2-one (**273**, X = CO) is isolable together with

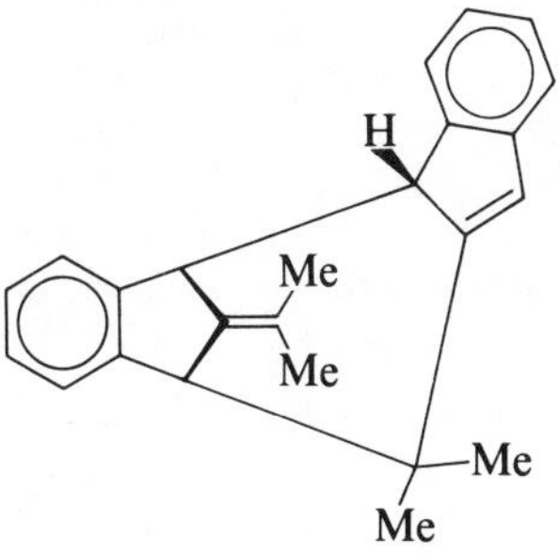

(**271**) 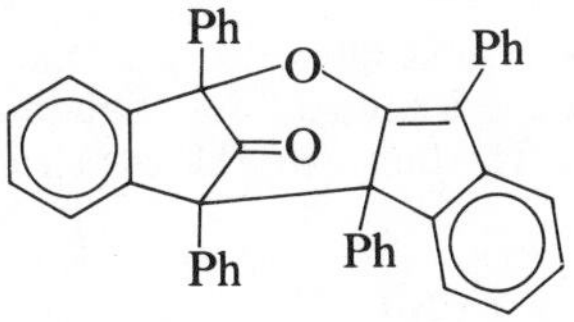 (**272**)

the major dimer (**274**) of linear type[10a] when **273**, X = CO is generated at low temperature; **272** rearranges to **274** on heating in benzene[65b]. The benz[*f*]derivative of **273** (X = CO) forms a linear dimer analogous to **274**[10b].

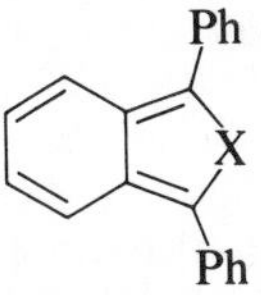

(**273**) 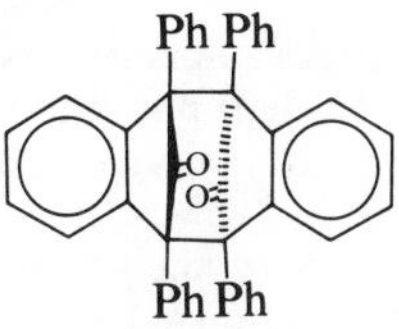 (**274**)

Anthracene *endo*-peroxide (**275**) rearranges upon irradiation to **276** which on heating undergoes ring-opening (**276**, arrows) to the novel *o*-quinodimethane (**277**). The latter decays by formal $(6+4)\pi$-dimerization[149]. Steric effects have a marked influence on the

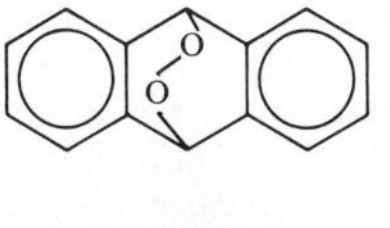

(**275**) 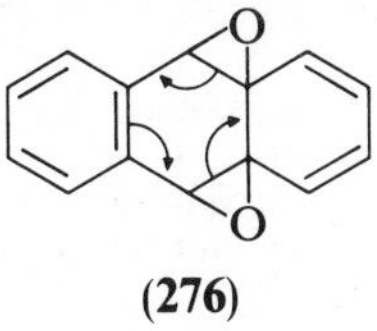 (**276**) 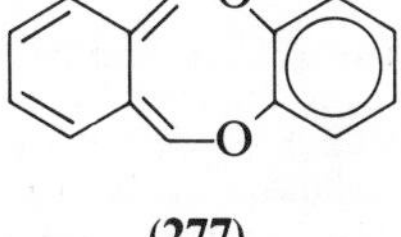 (**277**)

mode of dimerization of inden-2-ones for in contrast to the 1,3-diphenyl derivative (**273**, X = CO) the 1,3-di-*o*-tolyl derivative gives the formal $(6+6)\pi$-dimer (**278**)[140]. The tetraphenyl derivative of benz[*f*]inden-2-one, **279**, behaves analogously[10b].

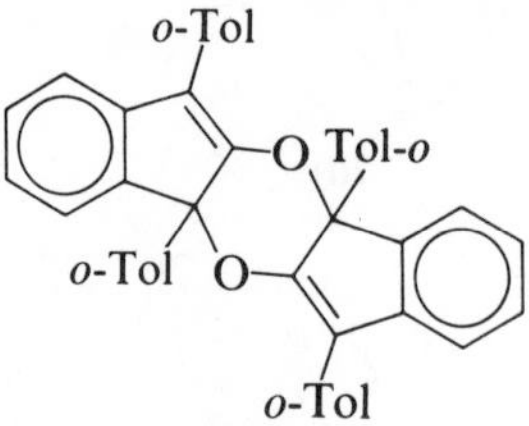

(**278**) 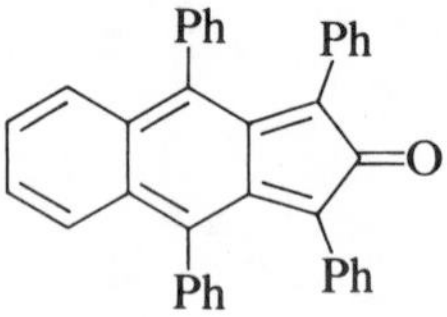 (**279**) 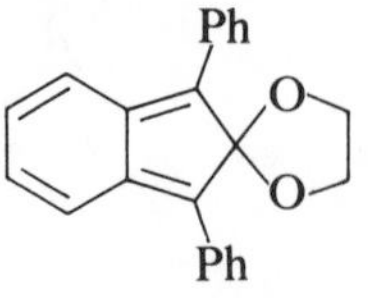 (**280**)

As the result of spiroconjugative interaction* between the oxygen lone pairs and the termini of the *o*-quinonoid system the ethylene acetal (**280**) dimerizes readily[150]. This is in contrast to the stability of the isolable isoindene (**273**, X = CMe_2)[25]. Spiroconjugative interaction is predicted to raise the HOMO of **280** whilst leaving the LUMO unperturbed. In agreement **280** shows λ_{max} 537 nm whilst **273** (X = CMe_2) has λ_{max} 444 nm. The higher energy HOMO of **280** will interact more strongly with the LUMO of a second molecule of **280** in a dimerization TS resulting in a more ready dimerization of **280** than of **273** (X = CMe_2). The dimer of **280** has structure **281**; steric effects associated with the acetal

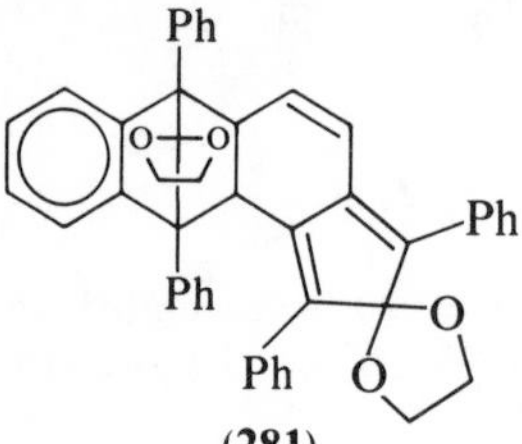

(**281**) 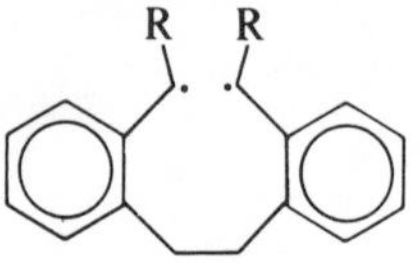 (**282**) 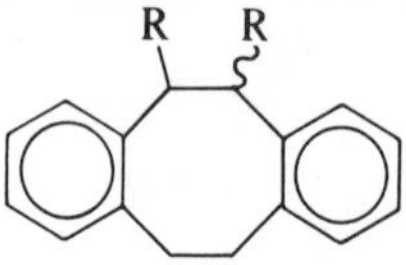 (**283**)

groups presumably prevent linkage of two molecules of **280** using the usually reactive diene systems of both. Related phosphole oxides, e.g. **273** (X = P(O)Ph) and **273** (X = P(O)Me)[150] and the sulphone (**273**, X = SO_2)[80], also absorb at much longer wavelength than **273** (X = CMe_2) and the phosphole oxides form dimers of the type **281**. These compounds are very likely also destabilized by spiroconjugative effects.

Of the several dimerization modes observed for *o*-quinonoid species the most intriguing are those which correspond to orbital symmetry forbidden processes. However, it is not clear even for *o*-quinodimethane itself whether the spiro-(**2**) and linear-(**81**) dimer are formed via a common biradical intermediate (**282**, R = H) or whether instead **2** is the product of an allowed Diels–Alder addition and **81** alone is formed via the biradical (**282**, R = H). Formation of the more stable biradical intermediate (**282**, R = substituent) will explain the head-to-head linear dimers (**283**) obtained from the *o*-quinodimethanes **284** (R = $PO(OR)_2$[151], R = SMe[152], R = CN[153] and R = CONHR′[154]). A biradical is also an

* When the termini of two conjugated systems are united to the same spiro-centre, the through-space interaction between the termini of the two systems is termed spiroconjugation. In acetals of cyclopentadienone a through-space interaction of two oxygen lone pairs constitutes one conjugated system and the diene system the other[9].

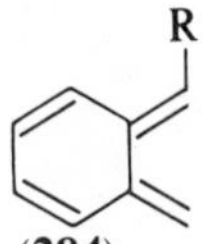

R

(**284**)

attractive intermediate to explain the reversible dimerization of **273** (X = CO), the easy conversion of **272** into **274** and formation of formal $(6+6)\pi$-dimers like **278**. Whether or not the formal $(6+4)\pi$-dimers **270**, **271** and **272** are formed directly by orbital symmetry allowed concerted cycloaddition is also unknown; **270** could form from the unobserved spiro-dimer **285** of 2,3-naphthoquinodimethane by a 1,5-benzyl shift (**285**, arrows) which

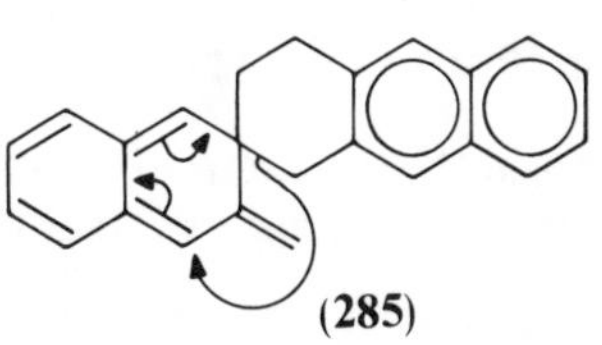

(**285**) 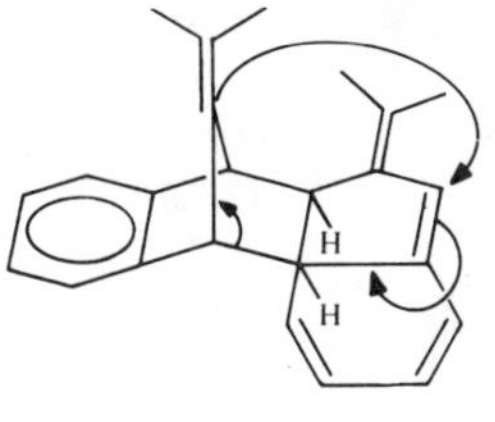 (**286**)

might be expected to occur readily. Similarly the formal $(6+4)\pi$-dimer (**271**) of 8,8-dimethylisobenzofulvene (**171**) could arise from the *exo*-$(4+2)\pi$-dimer **286** by Cope rearrangement (**286**, arrows). This would explain the unusual *endo* stereochemistry of the major apparent $(6+4)\pi$-dimer (*endo*:*exo* ratio, 3). There is good analogy for preferred *exo* adduction in the Diels–Alder additions of **171**. With dimethyl maleate **171** gives *exo* and *endo* adducts in a ratio of 2[102].

Formation of the spiro-dimer **55** of *E,E*-α,α′-bis(methoxycarbonyl)-*o*-quinodimethane is remarkably stereoselective[44]. If **55** is formed by concerted $(4+2)\pi$-addition the TS arrangement (**287**) must be preferred to the alternative (**288**). Since **287** allows a greater

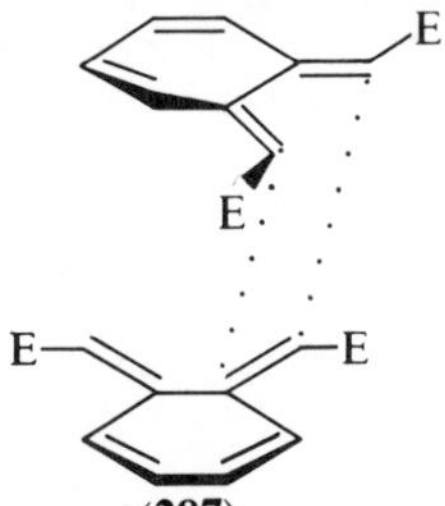

(**287**) 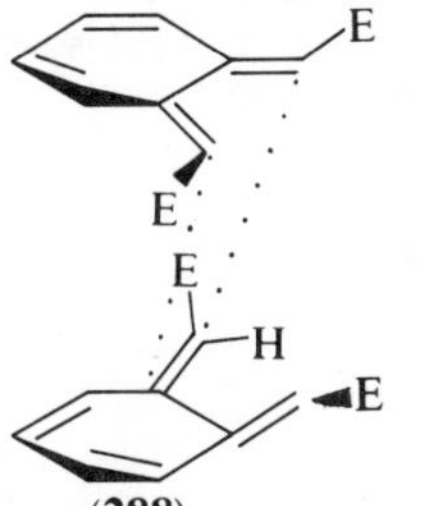 (**288**)

'accumulation of unsaturation' than **288** it is reasonable that **287** should be preferred[155] if dimerization is indeed a concerted $(4+2)\pi$-addition. This stereoselectivity is more difficult to explain if spiro-dimer formation proceeds via a biradical intermediate.

o-Quinone methide imines (**289**, X = NMe or NPh) resemble *o*-quinodimethane itself in providing spiro-dimers **290** (R = Me[54] or Ph[92,116]). Even the ketene (**291**) gives a simple $(4+2)\pi$-dimer (**292**) or one of its tautomers[156]. Similarly, *o*-quinone methides give no evidence for formation of linear dimers. The parent (**289**, X = O) forms the trimer (**293**)[1]

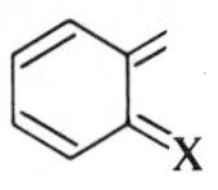

(289) 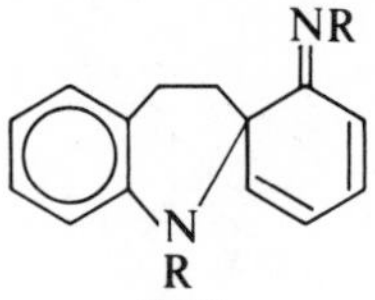 (290) 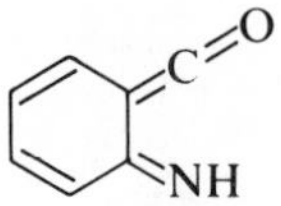 (291)

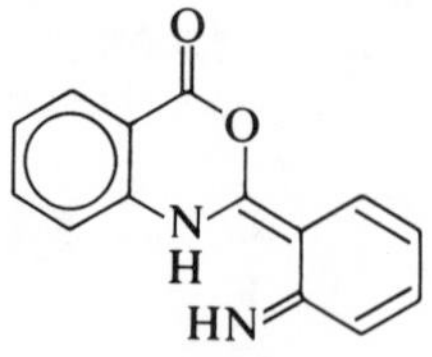

(292) 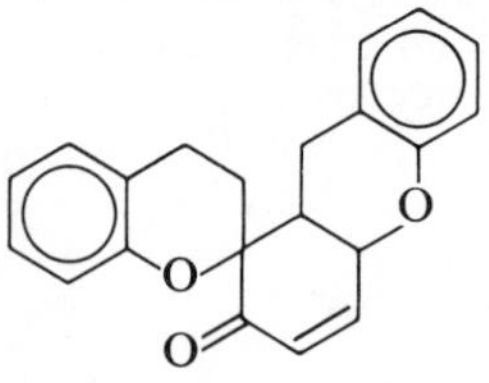 (293)

clearly derived by addition of a third molecule of **289** (X = O) to the spiro-dimer **294** (R = H). For more sterically hindered quinone methides, dimers rather than trimers, are

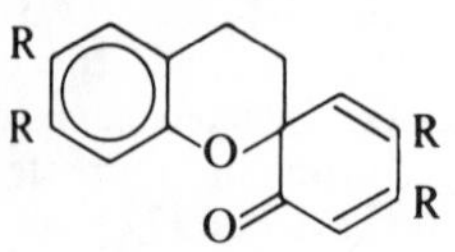

(294) 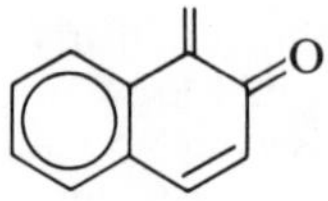 (295) 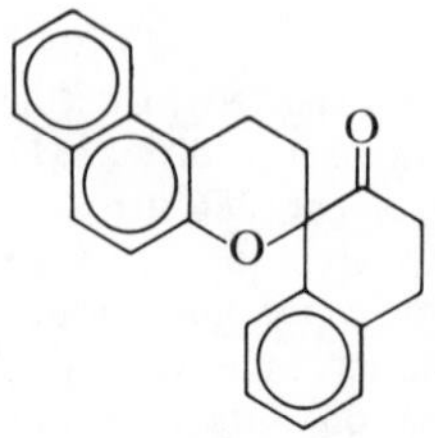 (296)

obtained, e.g. **294** (R = *t*-Bu)[1]. Naphthoquinone methide (**295**) also forms only the spiro-dimer (**296**), in this case reversibly[157]. However, generation of **295** by dissociation of **296** is complicated by formation of the isomeric dimer (**297**)[158]. *o*-Thioquinone methide (**289**, X = S) gives the head-to-tail linea dimer (**298**)[66, 86]. Indeed most other *o*-thioquinone

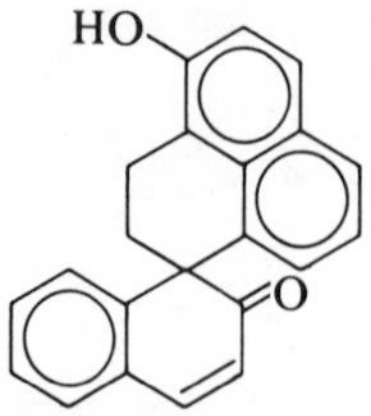

(297) 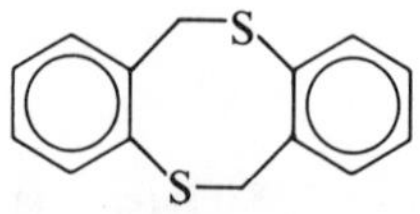 (298) 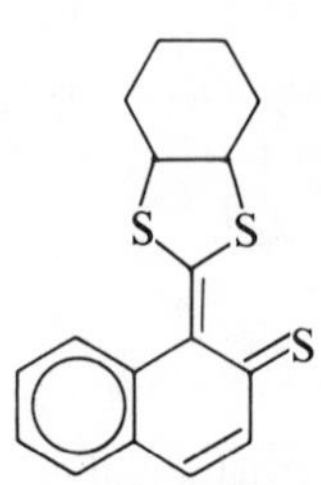 (299)

methides form linear dimers; steric factors may be responsible for the exceptional case of **299** which forms a spiro-dimer of uncertain structure[159]. Photolysis of **300** at 77 K provides a route to the ketene (**301**) which ring-closes to **302**. On warming to −40 °C the head-to-tail dimer (**303**) is obtained[103]. However, in an apparently contradictory report it

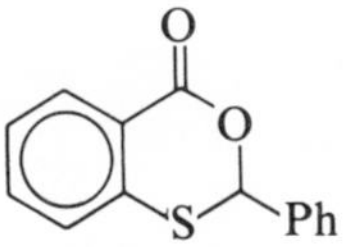

(300) 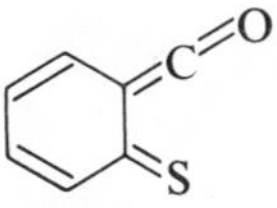 (301) 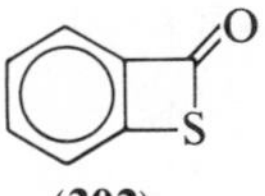 (302)

is noted that photolysis of **300** at 77 K yields the head-to-head dimer (**304**)[160]. It seems possible that **303** is the result of a thermal, and **304** the result of a photochemical

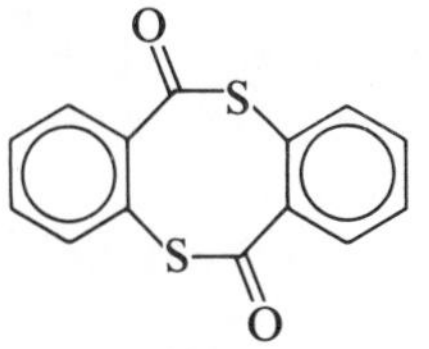

(303) 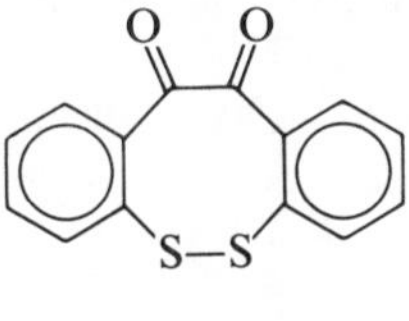 (304)

dimerization. Whether **263** gives a head-to-tail or a head-to-head linear dimer depends on the presence or absence of the hindered enamine (**305**)[161]!

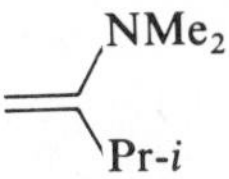

NMe_2

Pr-*i*

(305)

B. Diels–Alder Reactions

Apart from the numerous Diels–Alder additions carried out to intercept or characterize unstable *o*-quinonoid compounds, and the use of *o*-quinonoid compounds to intercept other transient species, the reactive *o*-quinonoid diene and heterodiene systems have found use in testing fundamental aspects of the Diels–Alder reaction. More recently the coupling of *o*-quinonoid reactivity with the advantages of the intramolecular Diels–Alder reaction has led to numerous elegant syntheses of complex structures of natural origin. In some cases the intermolecular Diels–Alder reactions of *o*-quinonoid compounds are also synthetically useful. The Diels–Alder reactivity of the isolable isobenzofuran (**306**) has long been exploited by organic chemists. Thus in addition to being a well-known aryne trap, **306** was used by Wittig and Wilson[162] who obtained evidence for cyclohexyne by isolation of the adduct **307** (43 %) from the reaction of 1,2-dibromocyclohexene (**308**) with magnesium

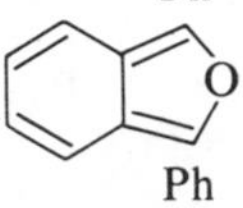

(306) 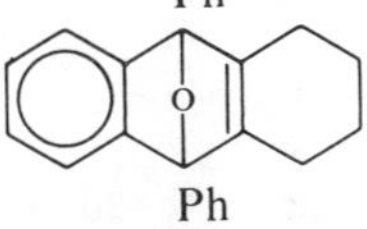 (307)  (308) 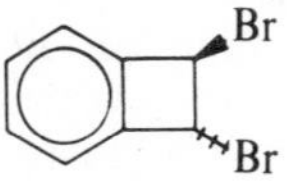 (309)

in the presence of **306**. Cava and his collaborators found **306** to be an efficient trap for transient benzocyclobutadienes. Thus reaction of *trans*-1,2-dibromobenzocyclobutene (**309**) with Li/Hg gave the adduct **310** (57 %) in the presence of **306**. Halogen derivatives of

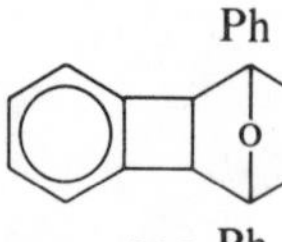

(310) (311) 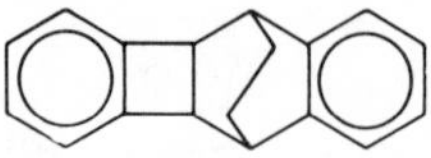 (312)

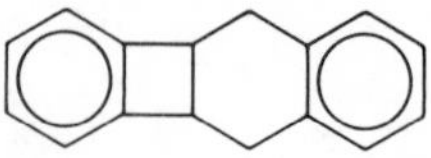

benzocyclobutadiene were even more efficiently trapped[163]. Interestingly cogeneration of benzocyclobutadiene and either *o*-quinodimethane or 2,3-dihydronaphthalene results in remarkably high yield mutual trapping giving **311** (44%) and **312** (65%) respectively[164].

More recently **306** has been used to probe the timing of formation of the two bonds in the Diels–Alder reaction[165]. When methyl *l*-bornyl fumarate adds to **306** the enantiomeric ratio is 1.53 in the *endo*-bornyl adduct (**313**) and 1.41 in the *exo*-bornyl adduct (**314**). If both

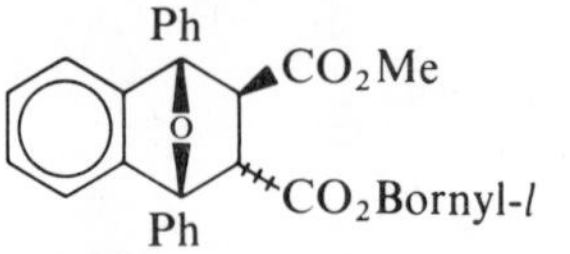

(313) 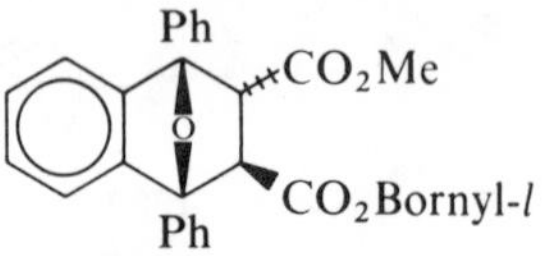 (314)

new bonds are formed to the same extent in the TS for addition of di-*l*-bornyl fumarate to **306** the enantiomeric ratio should be $1.41 \times 1.53 = 2.16$; the observed enantiomeric ratio was 2.08. The cooperativity of asymmetric induction observed here provides evidence for synchronous formation of the new σ bonds in this Diels–Alder reaction.

In contrast to **306**, 1,3-dimesitylisobenzofuran (**315**) is inert to several dienophiles even under forcing conditions. Newman[166] therefore proposed that the observed additions to 1-mesityl-3-phenylisobenzofuran (**316**) were two-step processes involving diradical or

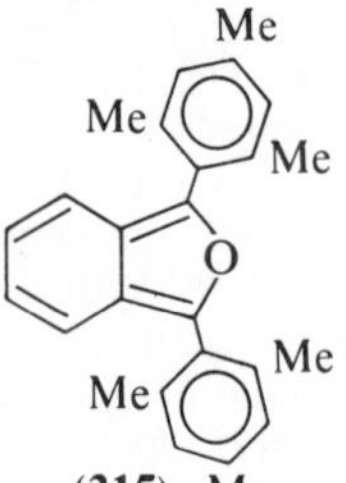

(315) (316) (317) (318)

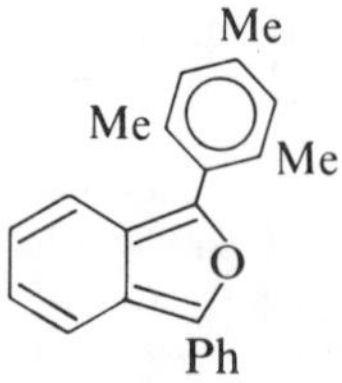

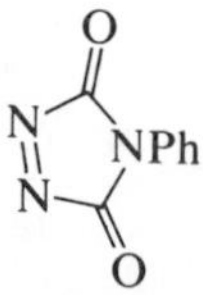

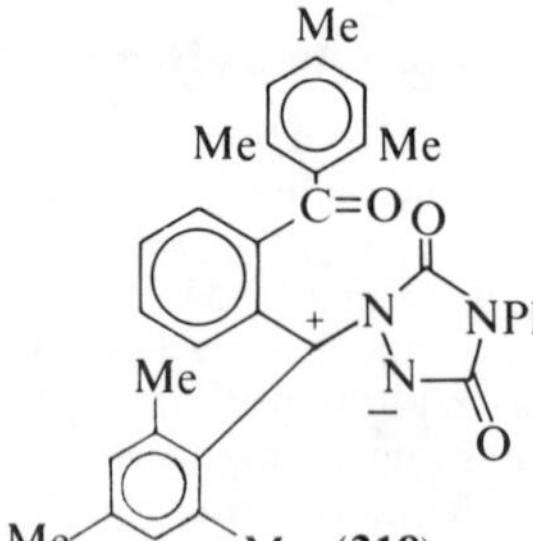

zwitterionic intermediates. Reaction of **315** with 4-phenyltriazoline-3,5-dione (**317**) gives the stable zwitterion (**318**)[167a] which is clearly derived (**319**, arrows) from the kind of dipolar intermediate (**319**) envisaged by Newman. Interestingly **306** is reported to react normally with **317**[167b].

The high reactivity of *o*-quinonoid dienes has been employed in a test of the FMO theory of regioselectivity in cycloadditions[168]. Whilst the theory concurs with the rule of '*ortho–para*' addition for most Diels–Alder additions, the addition of an electron-rich diene with an electron-rich dienophile provides an excellent test of the theory.

Matching the larger coefficients in both HOMO–LUMO pairs as in **320** and **321** predicts predominant formation of the '*meta*' isomer. Were a diradical an intermediate in such an addition the '*para*' isomer would be expected to predominate. Since Diels–Alder reactions between electron-rich partners were virtually unknown and would be expected to

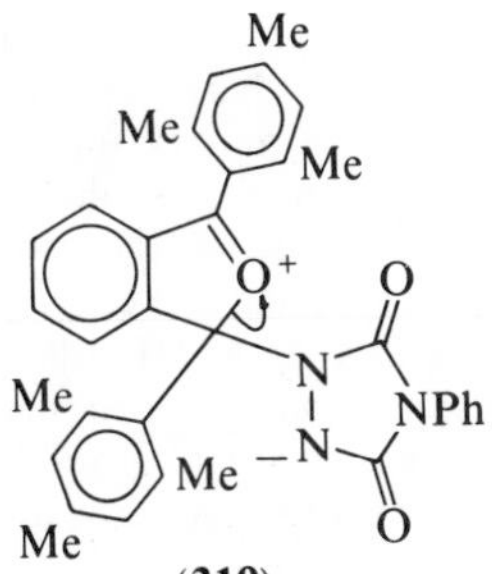

(319) 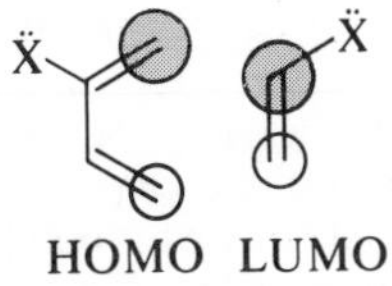 (320)  (321)

HOMO LUMO LUMO HOMO

require harsh conditions the reactive *o*-quinodimethanes **322** (R^1 = H, R^2 = Me) and **322** (R^1 = OMe, R^2 = H), generated by electrocyclic opening of benzocyclobutenes, were employed. Trapping with prop-1-yne and ethoxyacetylene gave in each case more of the unusual '*meta*' adduct, e.g. **323** and **324** were formed in a ratio of 1.56.

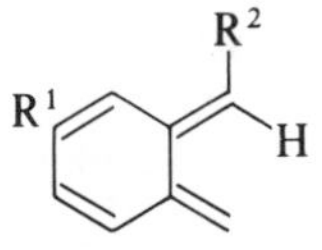

(322) 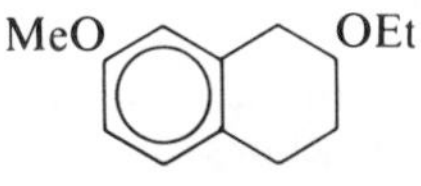 (323) 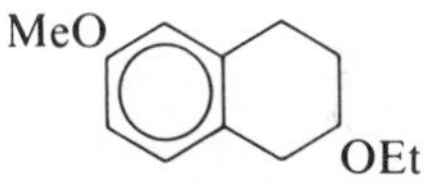 (324)

Diels–Alder addition of simple olefins like cyclopentene and *cis*-but-2-ene to the pyrones **325** (R^1 = R^2 = H) and **325** (R^1 = H, R^2 = Me) and the *o*-quinodimethanes **326** (R = CN) and **326** (R = CO_2Me) showed a preference for *endo* addition[153] varying from 6.5:1 to 2:1. This was taken as evidence for diene–alkyl group attraction in the TSs of the additions. Earlier it had been observed[169] that diene **327** gave almost equal quantities of

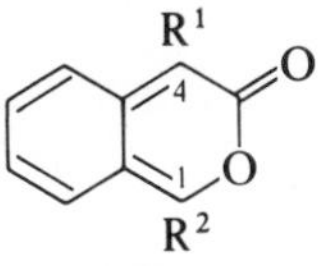

(325) 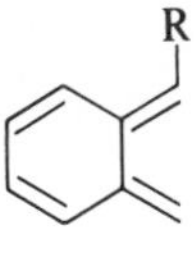 (326) 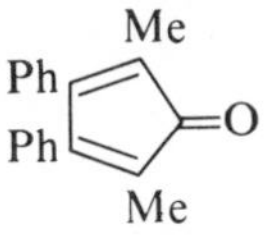 (327) 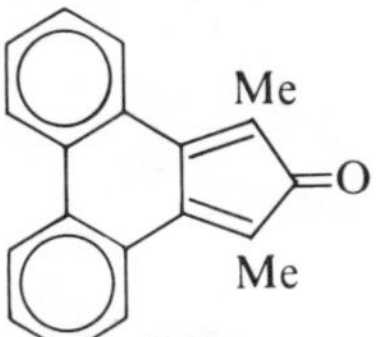 (328)

endo and *exo* adducts with cyclopentene, whereas with cyclopentadiene as dienophile, **327** gave mainly the *endo* adduct. This result had been taken to show the presence of secondary attractive forces in the cyclopentadiene addition and their absence in the *endo*-cyclopentene addition. The result was extrapolated to suggest the absence of attractive diene–alkyl group interactions in the Diels–Alder additions of simple olefins. Indeed a critical survey of mechanistic aspects of the Diels–Alder reaction[170] cites this work as demonstrating the importance of secondary interactions. Since it seemed likely that non-coplanarity of the phenyl groups and the diene system in **327** would impose a greater steric barrier to the *endo* addition of cyclopentene than cyclopentadiene the 9,10-phenanthroquinodimethane **328** in which the aryl groups are forced to lie in the same plane as the diene system was generated and added to several simple olefins[171]. In contrast to the addition of cyclopentene to **327**, corresponding addition to **328** in which the steric barrier is removed gives only the *endo* adduct. Thus diene **327** is probably a poor diene for use in testing for secondary interactions. The course of its additions to cyclopentene and cyclopentadiene probably reflect the greater number of out-of-plane hydrogens in cyclopentene as well as the somewhat greater importance of secondary interactions involving unsaturated than saturated groups. The existence of diene–alkyl group attraction is supported by other evidence[172]. Although the nature of the attractive interaction is unknown it could be of the orbital interaction type arising as a consequence of hyperconjugation between the allylic hydrogen atoms and the π system of the olefin. The allylic hydrogen atoms in the HOMOs of cyclopropene, cyclopentene and cyclobutene[173] carry small coefficients of appropriate phase for bonding interaction with C(2) and C(3) of a diene LUMO as shown in **329** for propene. This effect has been termed steric attraction,

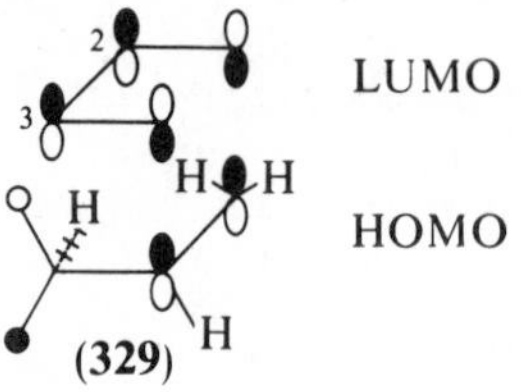

LUMO

HOMO

(**329**)

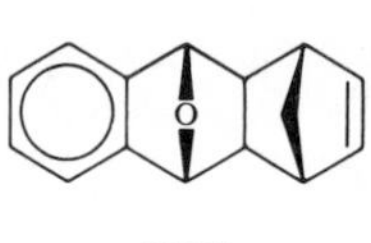

(**330**)

and used to explain certain aspects of carbene addition to olefins[174a]. Houk now seems to accept the existence of secondary interactions involving alkyl groups[174b]. Cyclopentene and cycloheptene also add to benzo[*c*]furan to give mostly *endo* adducts[153], but addition to norbornadiene gives mostly the *exo* adduct (**330**) (*exo*:*endo* ratio, 1.8:1). In contrast 1,3-diphenylbenzo[*c*]furan gives *only* the *exo* adduct with norbornadiene[175]. This difference reflects an effect due to phenyl substitution more strikingly observed in additions to the stable diphenyl-substituted pyrone (**325**, $R^1 = R^2 =$ Ph). In complete contrast to the parent pyrone (**325**, $R^1 = R^2 =$ H), the diphenyl derivative adds a range of dienophiles (butadiene, isoprene, cyclopentadiene, dimethyl maleate, norbornadiene and cyclopentene) with marked *exo* selectivity[176]. Although the origin of this effect is unclear it appears to be associated with the more congested environment of the olefinic substituents (R) in the

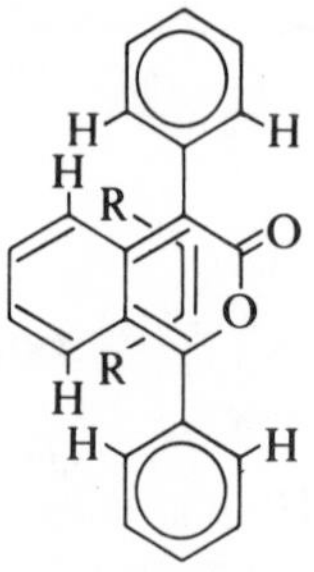

(**331**) 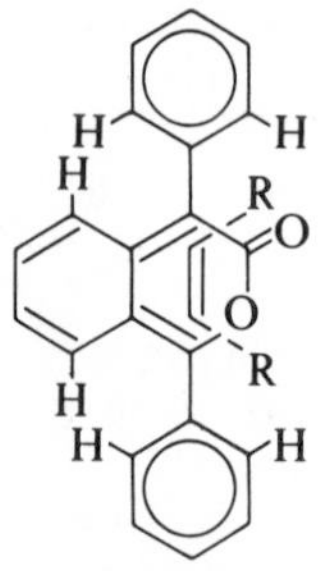 (**332**) 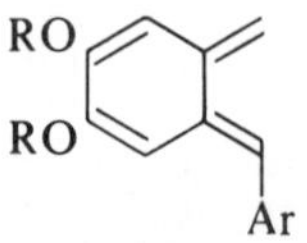 (**333**) 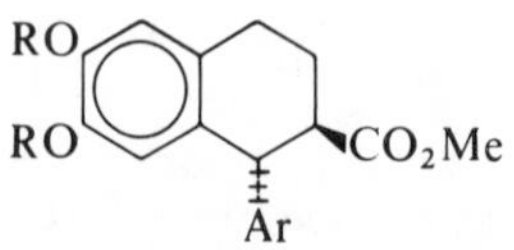 (**334**)

endo array (**331**) than in the *exo* array (**332**). Related effects are observed in additions to 1,4-diphenyl-2,3-naphthoquinone (**249**, R = Ph)[139] and *o*-quinodimethanes of type **333** which give mostly the *exo* adducts (**334**) with methyl acrylate[75]. The same trend is discernible in additions to *o*-quinone methides[177]. Thus **335** (R = Me) and *Z*-methyl propenyl ether (**336**) give the *endo* adduct **337** and its *exo* isomer in a ratio of 9:1 whereas the corresponding ratio for addition to **335** (R = Ph) is 2.33:1.

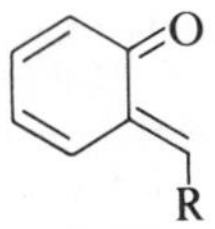

(335) 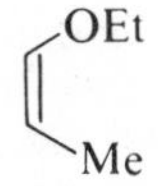 (336) 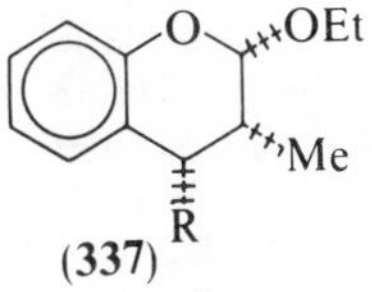 (337)

The benzo[*c*]heterocycles (**162**, X = O, N, S) all react with *N*-phenylmaleimide to give mixtures of *endo* and *exo* adducts[62, 96, 178]. Since this dienophile normally strongly prefers *endo* addition and reversibility of addition is unlikely to be a problem in at least two cases (**162**, X = O and NH)[96, 178] another explanation is required; secondary interaction involving the hetero atoms of diene and dienophile is one possibility. The *o*-quinodimethane **338** generated by benzocyclobutene ring-opening has been used to test diastereoface selection in additions to the olefins **339** (R = OEt), **339** (R = Me) and

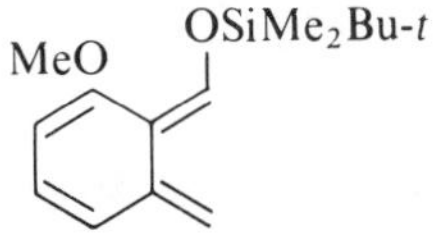

(338) 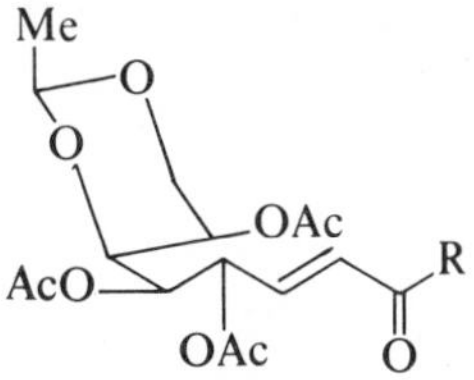 (339) 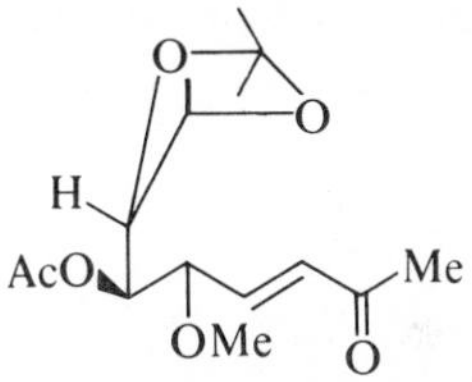 (340)

340[179]. Assuming a ground state conformation (**341**) for the olefins (R″ = sugar residue), with the small allylic C–H bond in the plane of the carbon–carbon double bond, and an *endo*-COR TS, steric factors associated with R″ might be expected to direct attack of the diene **338** to the *upper* face of the olefin leading to adducts (**342**) of *R*-configuration at C(3). In fact addition of **339** (R = OEt) gives mostly **343** derived by addition to the *lower* face of **341** (four parts) and only one part of **342**. Preferred attack of the diene on the face of **341** opposite the allylic oxygen substituent is also observed for the ketones **339** (R = Me) and **340**.

The ideas of Houk and his collaborators[180] which have in part appeared subsequent to these experimental observations are readily extended to accommodate them. The idea of staggering forming bonds with allylic bonds[180a, b] suggests that **344** rather than **341** is the

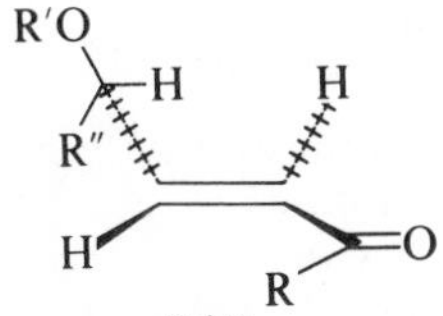

(341) 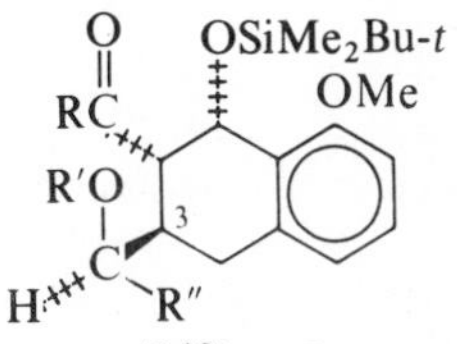 (342) 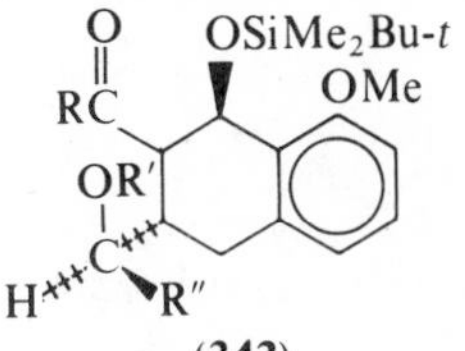 (343)

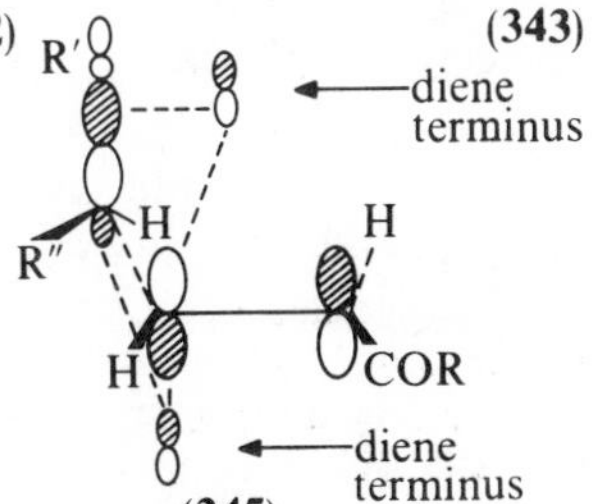

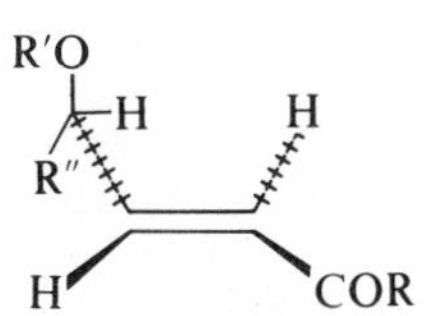

(344) (345)

more likely olefin conformation in the TS of the addition. In **344** better overlap of the C–O σ^* orbital with π and π^* of the double bond results in electron withdrawal from the double bond making it a more electron-deficient dienophile. Secondary interactions with the modified LUMO (**345**) ($= \sigma^* + \pi^*$) then determine which face of the olefin is preferentially attacked. As shown in **345** approach of a diene terminus from above gives rise to antibonding interaction involving the σ^* component whilst approach from below gives rise to bonding secondary interaction. These ideas help explain why (with one exception[181]) the most reactive conformation of an allylic ether changes[180a] when the reagent is electron-deficient. These effects are clearly of importance in planning chiral syntheses using Diels–Alder and other reactions.

The use of *o*-quinonoid compounds in intramolecular Diels–Alder reactions is detailed both in the extensive review literature of the intramolecular Diels–Alder reaction[104b, 182] and in reviews of the use of this process in steroid synthesis[183]. Nevertheless we must mention the stunning work of Vollhardt and his collaborators[184] who showed that co-oligomerization of **346** and **347** with $CpCo(CO)_2$ as catalyst led via the stereoisomeric

(**346**) 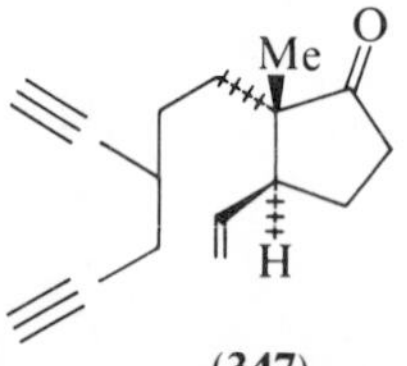 (**347**) 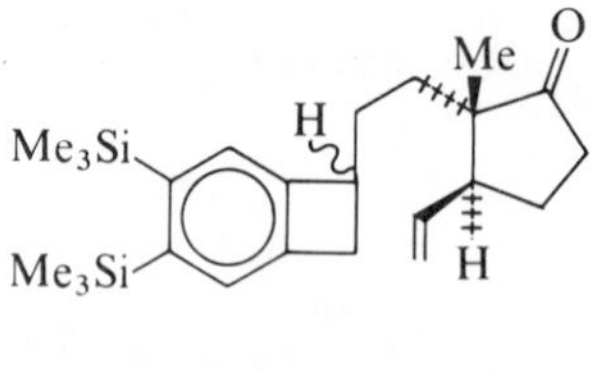 (**348**)

benzocyclobutenes (**348**) to the steroid (**349**); selective protodesilylation of **349** at C(2) (CF_3CO_2H) and oxidative [$Pb(OCOCF_3)_4$] removal of the remaining $SiMe_3$ group gave

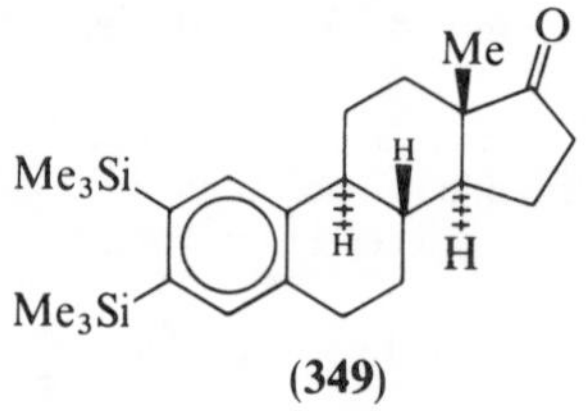

(**349**) 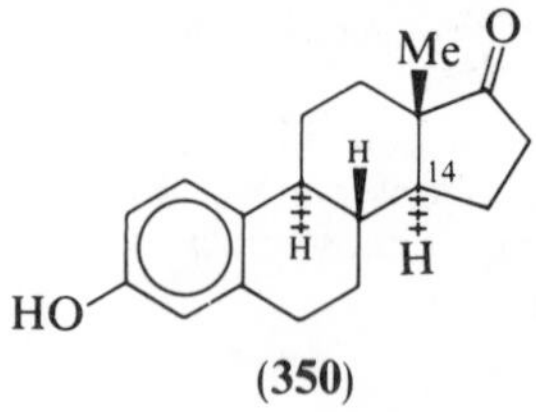 (**350**)

oestrone (**350**). The intramolecular cycloaddition leading to **349** involves the chair-like *exo*-TS arrangement **351**. The alternative *exo* array (**352**) which would have given a *cis* disposition of hydrogens at C(8) and C(14) is presumably destabilized by repulsion between bowsprit and flag-pole in the developing boat conformation of ring C. The synthesis of resistomycin by Rodrigo and Keay is also remarkable for the efficiency with

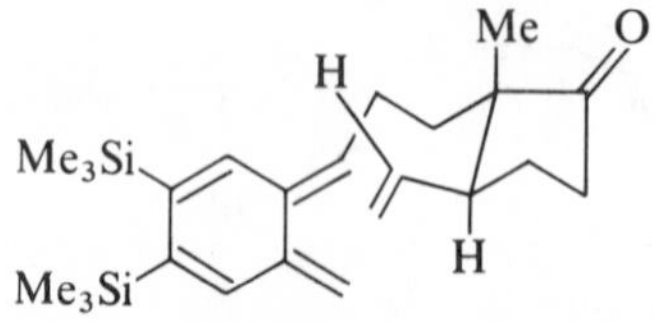

(**351**) 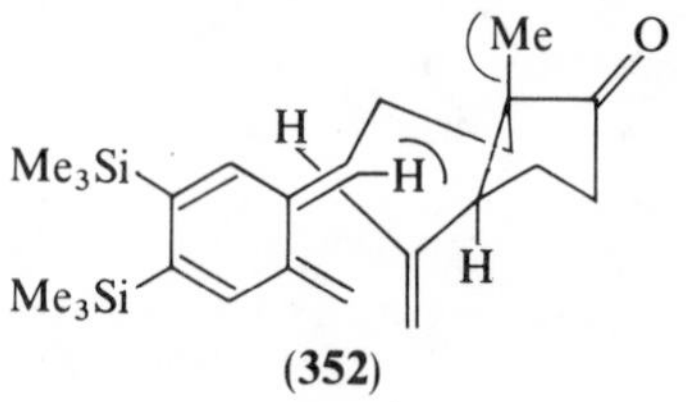 (**352**)

which a polycyclic molecule is assembled[185]. Reaction of **353** with iodoacetic acid in refluxing benzene gave the isobenzofuran (**354**) which underwent spontaneous intramolecular Diels–Alder reaction to **355**. Reaction of **355** with pyridine hydrochloride

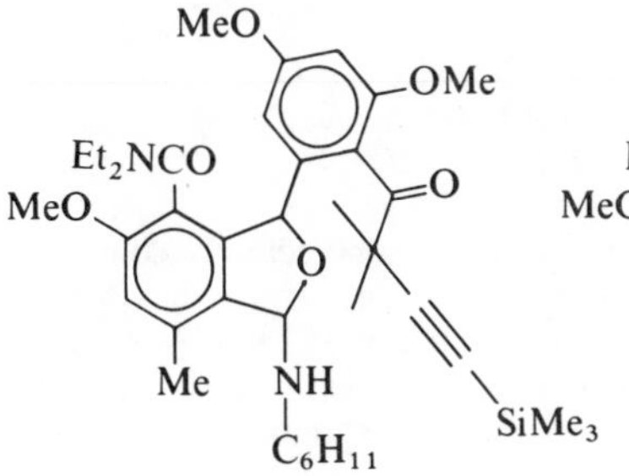

(**353**) 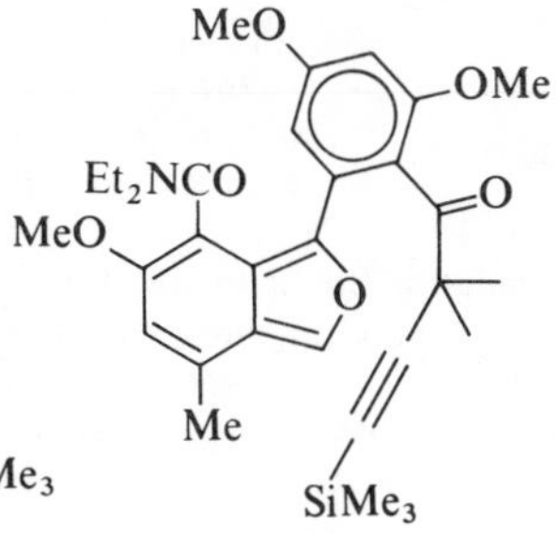 (**354**) 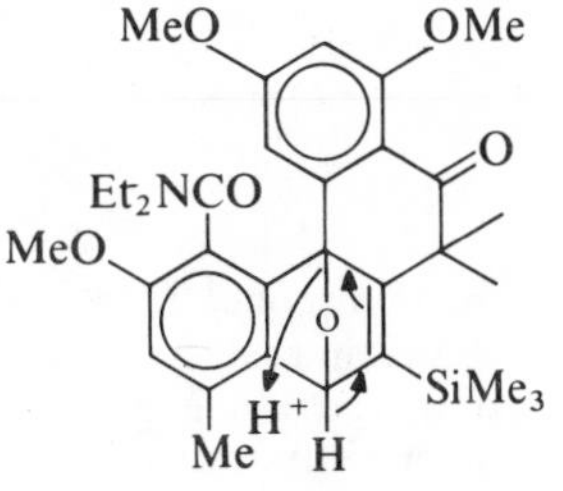 (**355**)

resulted in acid-catalysed aromatization (**355**, arrows), protodesilylation, demethylation and an intramolecular Friedel–Crafts acylation to give resistomycin (**356**) in high yield.

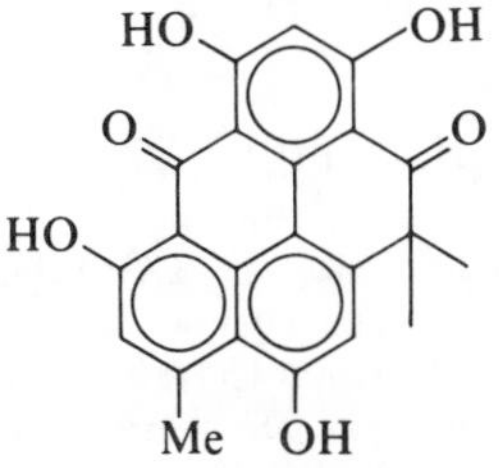

(**356**)

Intermolecular Diels–Alder reactions of *o*-quinodimethanes are also synthetically valuable. The *o*-quinodimethanes **357** (R = Me)[186] and **357** (R = H)[187] were generated by 1,4-elimination of bromine (NaI), and benzocyclobutene ring-opening respectively. In both cases trapping with the moderately reactive dienophile (**358**) gave the adducts (**359**) in

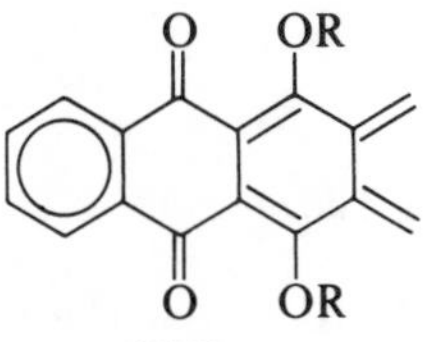

(**357**) 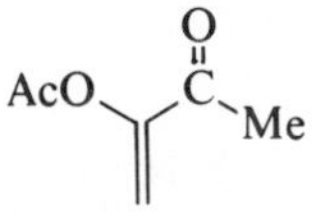 (**358**) 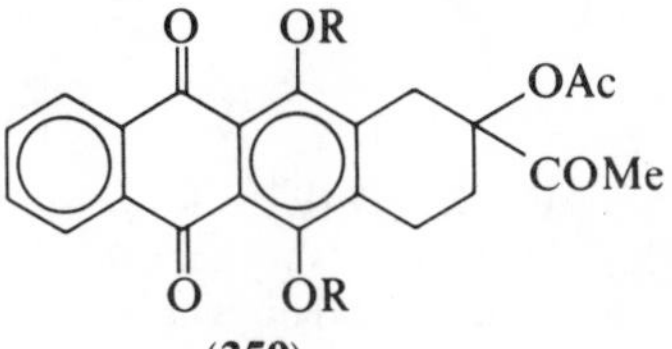 (**359**)

fair yield. These compounds are closely related to 4-deoxydaunomycinone, a compound with useful anti-cancer properties. The problem of regioselectivity which is generally greater in inter- than intramolecular additions is not acute for *o*-quinodimethanes that are differently substituted at the termini of the quinonoid system. Thus the *o*-quinodimethane **360** adds regio- and site-selectively to the quinone (**361**) to give after *in situ* dehydrobromination the quinone (**362a**) which could be converted to islandicin. On the other hand addition of **360** to **363** led to the regioisomeric quinone **362b** which could be converted to

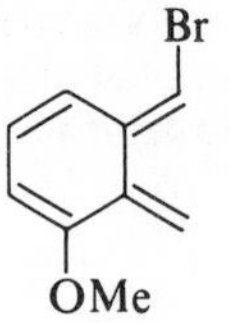

(360) 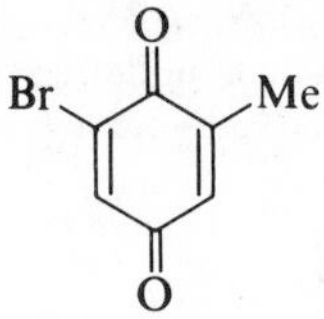 (361) 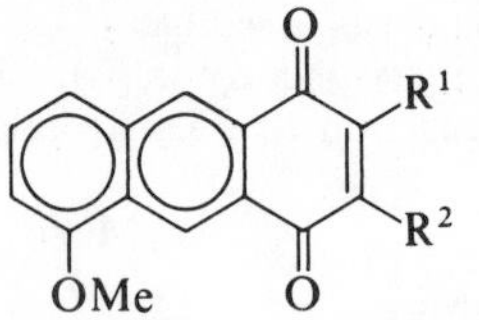 (362a, R^1 = Me, R^2 = H) (362b, R^1 = H, R^2 = Me) (363)

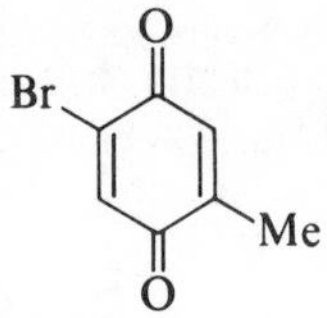

digitopurpone[188]. In these reactions the bromine atom in the quinone determined both regio- and site-selectivity and is subsequently removed. Such a substituent has been described as a 'ghost' substituent[189]. Addition of 1-methylisobenzofuran (**364**) as well as other 1-substituted isobenzofurans to quinone dimethyl acetals like **365** is also both site- and regioselective; **365** gives the tetracyclic *endo* adduct **366**, of interest in the synthesis of

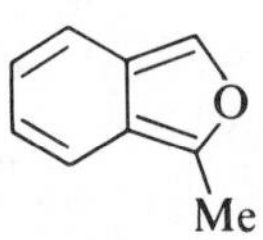

(364) 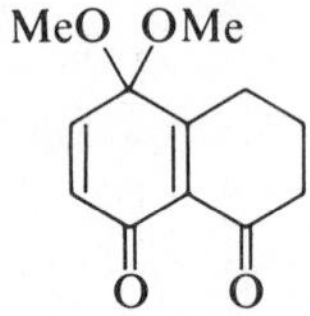 (365) 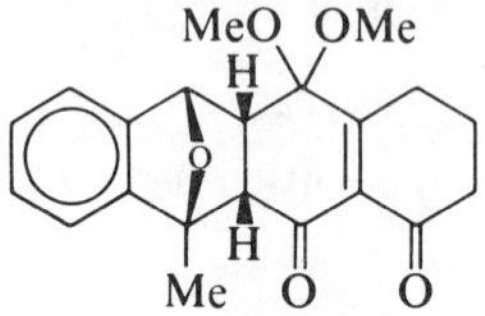 (366)

anthracyclines[190]. 2-Benzopyran-3-one (**325**, $R^1 = R^2$ = H) undergoes strongly regioselective additions to a range of electron-rich olefins[191, 192]. In such inverse electron demand Diels–Alder additions the HOMO-dienophile–LUMO-diene frontier orbital interaction should be more important. Simple Hückel calculations show that the LUMO of **325** ($R^1 = R^2$ = H) has a much larger coefficient at C(1) than at C(4) whereas the HOMO of **325** ($R^1 = R^2$ = H) is polarized in the opposite sense. Since the HOMOs of electron-rich olefins have the larger coefficient at the less substituted end, their reaction with **325** ($R^1 = R^2$ = H) should lead to adducts of the type **367**. This is indeed what is observed;

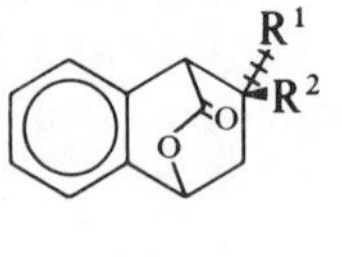

(367) 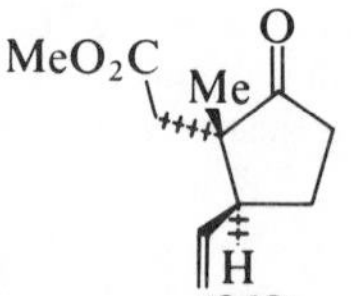 (368)

isobutene gave **367** ($R^1 = R^2$ = Me) as the only isolable product. The olefin (**368**) gave a 70% yield of adducts with those adducts (**369**, R = H) with the correct regiochemistry for steroid synthesis predominating (ratio 5.1:1). The adducts **369** (R = H) and **369** (R

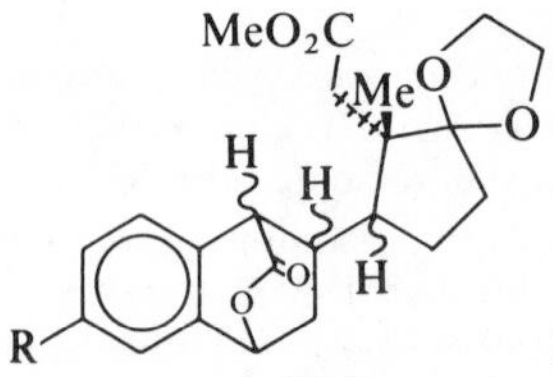

(369) 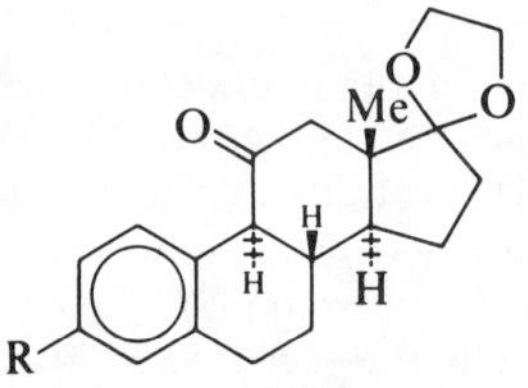 (370)

= OMe) were readily transformed into the 11-oxo aromatic steroids (**370**)[191]. The adduct **367** (R^1 = OMe, R^2 = Me) was prepared by Jung and his collaborators and shown to react with sodium methoxide to give **371**, an AB-ring analogue of the anthracyclines[192]. The

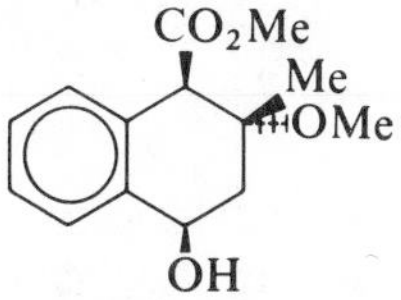

(**371**)

pyrone (**325**, $R^1 = R^2$ = H) appears less prone to dimerization and other self-destructive processes than related *o*-quinonoid compounds, allowing its efficient trapping even with simple olefins. In contrast α-cyano-*o*-quinodimethane (**326**, R = CN) shows a strong tendency to dimerize[193].

Oxa, aza and sulpha derivatives of *o*-quinodimethane have all been trapped as Diels–Alder adducts. Such reactions include the previously mentioned reactions of *o*-thioquinone methide with *N*-phenylmaleimide (**109** giving **131**), *N*-methyl-*o*-quinone methide imine with chloroacrylic acid (**75** giving **132**), and *o*-quinone methide with 1,1-dimethoxyethylene (**3** giving **135**). Certain stabilized *o*-thioquinone methides also readily undergo cycloaddition; **372** adds to enamines like **373** to give adducts like **374** in good yield, and probably by a stepwise, ionic reaction mechanism. With electron-deficient olefins like maleonitrile and fumaronitrile, the cycloadditions appear to be concerted as indicated by the retention of stereochemistry in the adducts[194, 195]. *o*-Quinone methides are

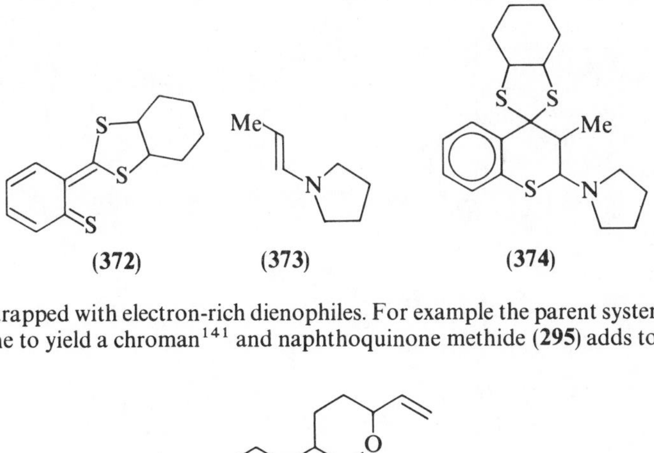

(**372**) (**373**) (**374**)

efficiently trapped with electron-rich dienophiles. For example the parent system (**3**) reacts with styrene to yield a chroman[141] and naphthoquinone methide (**295**) adds to butadiene

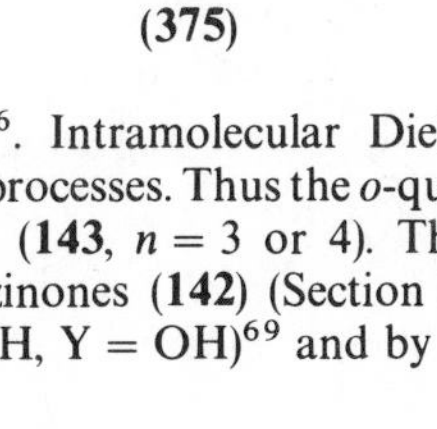

(**375**)

to give a good yield of **375**[196]. Intramolecular Diels–Alder additions to hetero-*o*-quinonoid species are also viable processes. Thus the *o*-quinone methide imines (**376**, *n* = 3 or 4) yield the tricyclic adducts (**143**, *n* = 3 or 4). The same adducts are formed by decarboxylation of the benzoxazinones (**142**) (Section III.B), by flash pyrolytic dehydration of the alcohols (**377**, X = H, Y = OH)[69] and by reaction of fluoride ion with the

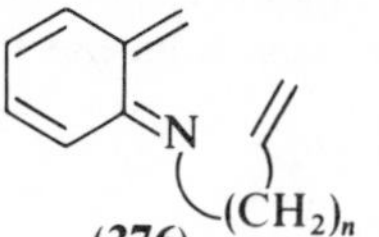

(376) 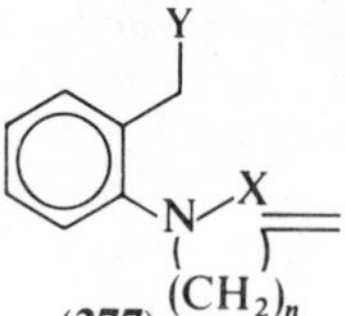 (377)

cations (377, X = $SiMe_3$, Y = $\overset{+}{N}Me_3$)[54]. The analogous carbon-substituted *o*-quinone methide imines (**378**, X = CH) generated by flash vacuum pyrolysis of either **379** (X = CH) or **380**, gave the cycloadducts (**381**, X = CH). Tricycle (**381**, X = CH, $n = 3$) was

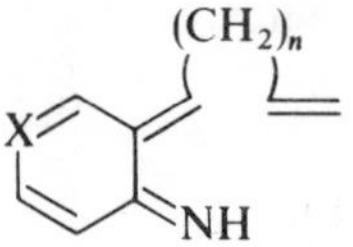

(378) 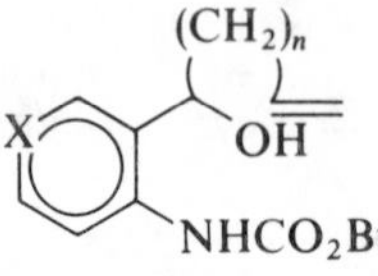 (379)

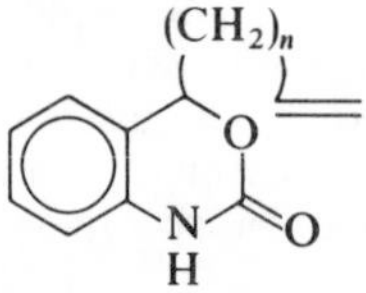

(380) 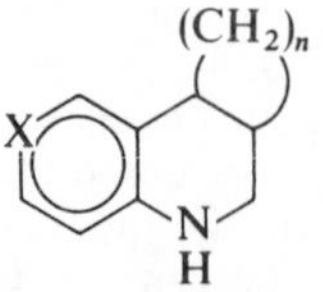 (381)

stereochemically pure and was tentatively assigned the *cis* ring junction stereochemistry, whereas **381** (X = CH, $n = 4$) was a 3:1 mixture of *cis* and *trans* isomers. The pyridine based *o*-quinone methide imine (**378**, X = N, $n = 3$) generated by flash pyrolysis of **379** (X = N, $n = 3$) similarly gave the adduct (**381**, X = N, $n = 3$) in 64% yield[197].

Mao and Boekelheide have reported[66] the intramolecular Diels–Alder trapping of *o*-quinone methide (**382**, R = H, $n = 1$) to give the *cis*-fused product (**383**, R = H, $n = 1$); **382** (R = H, $n = 1$) was generated by flash vacuum pyrolysis of diol (**384**, R = H, $n = 1$).

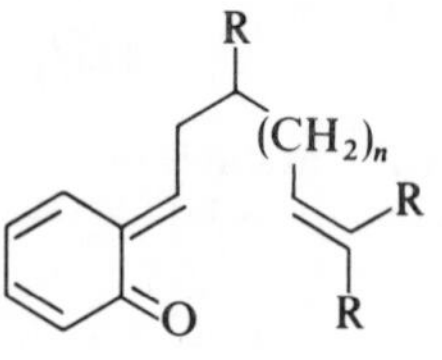

(382) 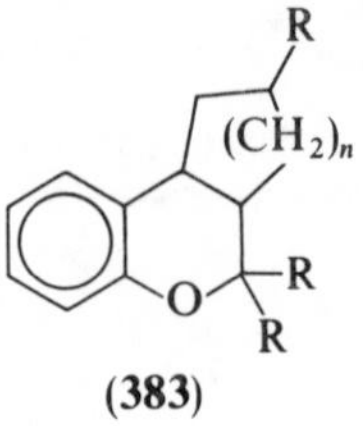 (383)

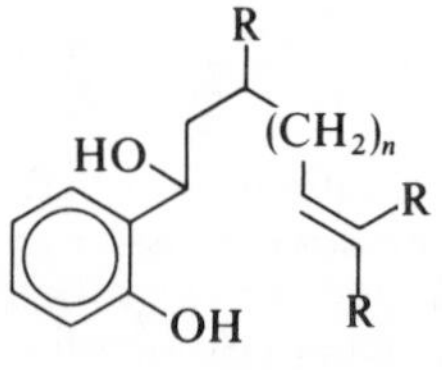

(384)

Thermolysis of diol (**384**, R = Me, $n = 2$) at 180°C is also reported[198] to yield the adduct (**383**, R = Me, $n = 2$). There have been a number of reports concerning the use of *o*-quinone methide imines and *o*-quinone methides for the synthesis of natural products. These include the synthesis of 9-azaoestrone (**386**) via cycloaddition of **385** followed by

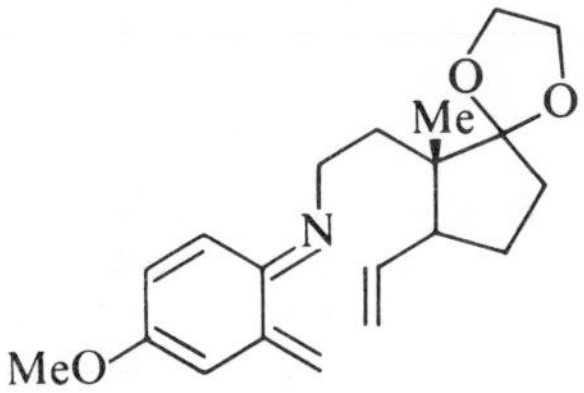

(**385**) 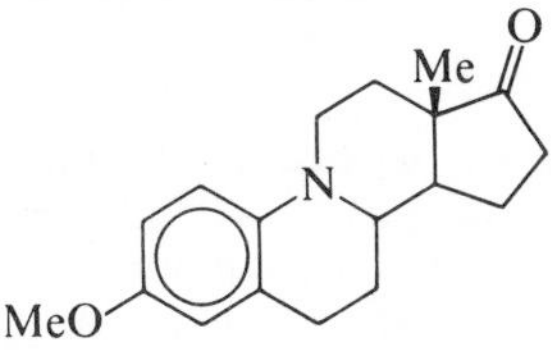 (**386**)

deprotection[54] and the synthesis of the lignan carpanone (**389**) via cycloaddition of *o*-quinone methide (**388**), generated by phenolic coupling of **387**[199].

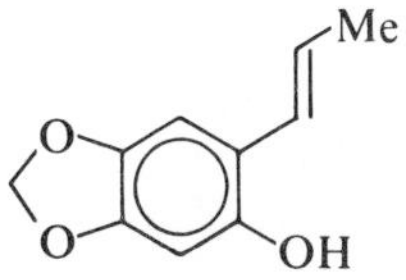

(**387**) 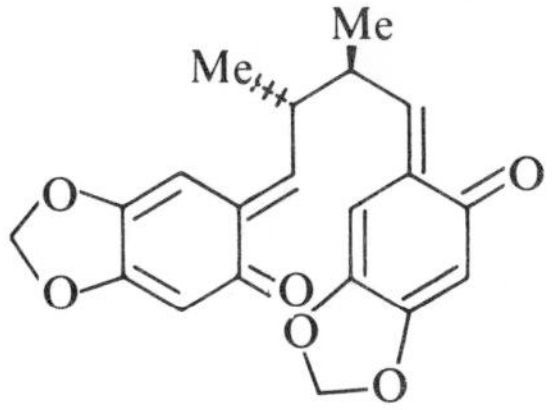 (**388**)

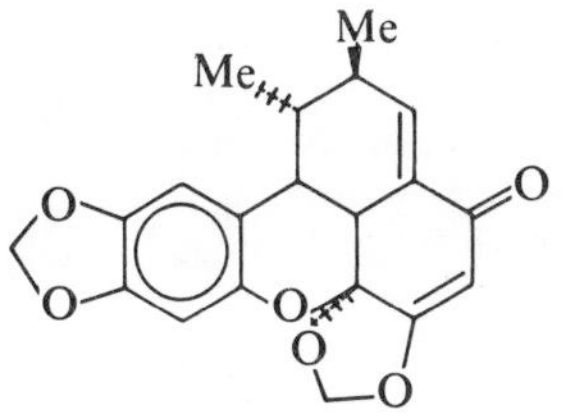

(**389**)

Certain *o*-quinodimethanes also form peroxides when exposed to oxygen, a process which formally involves a Diels–Alder reaction. However, although 2,3-naphthoquinodimethane forms a peroxide with atmospheric oxygen (Section II) a similar peroxide has not been reported for *o*-quinodimethane itself. It seems likely that similar peroxides contribute to the deterioration of many *o*-quinonoid compounds upon exposure to air and that such peroxides could be prepared if more rapid decay modes did not supervene. Thus the *o*-quinodimethane **390**, in which the usual decay modes of conrotatory ring closure, 1,5-hydrogen shifts and dimerization are suppressed, rapidly absorbs atmospheric oxygen to yield the isolable peroxide (**391**) which upon thermolysis eliminates norbornene (**391**, arrows) to give *o*-dibenzoylbenzene[200]. The explosive peroxide (**392**, X = O) obtained by photosensitized oxygenation of 1,3-diphenylisobenzofuran was the first isolated furan endoperoxide[201] and a similar peroxide (**392**, X = NPh) was obtained from 1,2,3-triphenylisoindole[202]. In solution both peroxides decompose to *o*-dibenzoylbenzene.

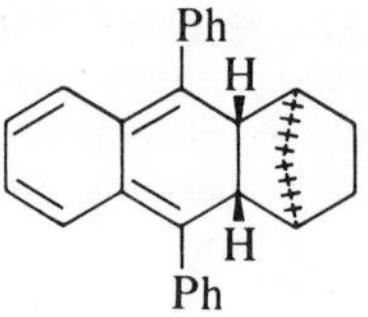

(390) 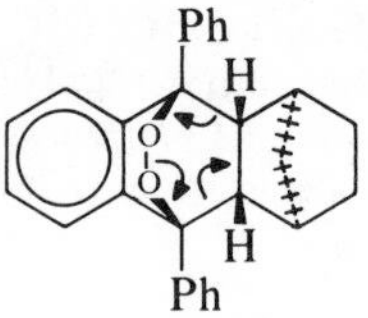 (391) 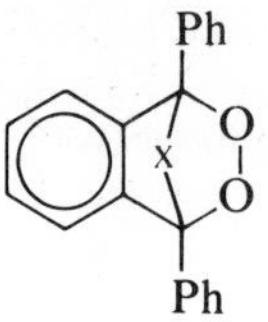 (392)

Photosensitized oxygenation[203] of the stable pyrone (**393**)[204] gives the peroxide (**394**). Thermolysis of **394** at ca. 110°C proceeds by two pathways, both of which give *o*-dibenzoylbenzene as the major product observed. Surprisingly, loss of CO_2 with cleavage of the O–O bond is the minor path, and CO_2 loss to give the remarkable *o*-quinodimethane **395** is the major process (ca. 70%). In the presence of maleic anhydride **395** can be efficiently trapped as the adduct (**396**).

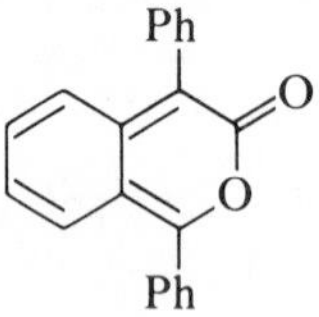

(393) 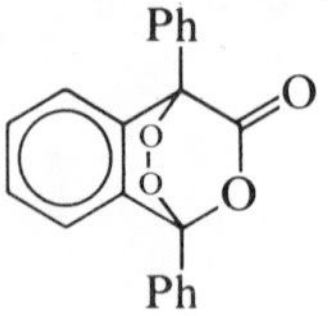 (394) 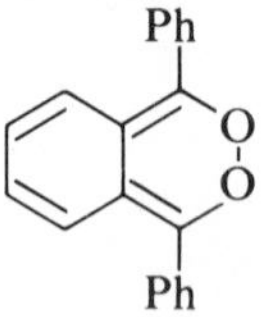 (395) 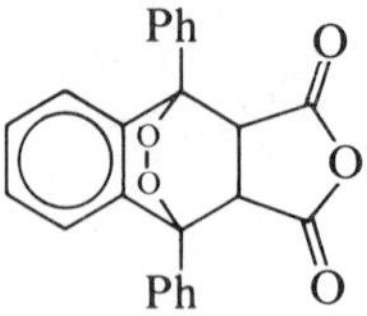 (396)

C. Other Reactions

Generation of *o*-quinodimethanes by benzocyclobutene ring-opening (Section III.C) and 1,5-sigmatropy (Section III.D) are equilibria which strongly favour the benzenoid isomers. Accordingly such benzenoid isomers are often encountered as products from reactions proceeding through transient *o*-quinodimethanes. Developing benzenoid aromaticity in the TS may also considerably ease the rearrangement of *o*-quinonoid compounds. Thus alkyl shifts in simple cyclopentadienes commonly require temperatures above 330°C but generation of the isoindene (**397**, R = Me) by thermolysis of **398** (R = Me) at 180°C affords products derived via the 1,5-methyl shift shown in **397**. In **397** (R = Et) the ethyl

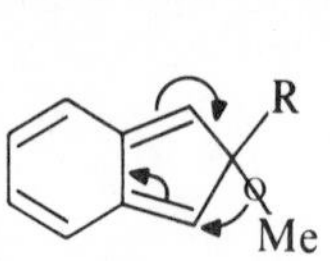

(397) 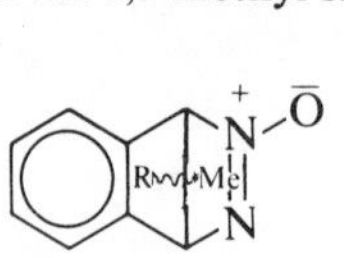 (398) 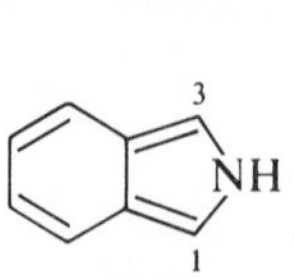 (399) 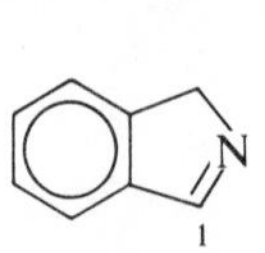 (400)

group migrates six times faster than the methyl group and benzyl migrates 56 times faster than methyl[26a].

When the *o*-quinonoid isomer is itself aromatic the position of equilibrium with a benzenoid isomer can be finely balanced. Thus isoindole (**399**) exists mostly as the 10π-electron aromatic *o*-quinonoid tautomer in both $(CD_3)_2CO$[178] and $CDCl_3$[205] but the isoindolenine form (**400**) can be detected even in the hydrogen-bonding solvent $(CD_3)_2CO$[205] and is quite important in $CDCl_3$[205]. For 1-aryl substituted isoindoles the importance of the isoindolenine tautomer increases as the donor ability of the aryl group increases and the electron deficiency at C(1) in **400** is diminished; for 1-ethoxyisoindolenine the isoindole form is not detectable[206]. The decomposition of isoindole has been associated with reaction between **399** which is electron rich at C(1) and C(3) and its tautomer (**400**) which is electron deficient at C(1)[178]. This agrees with the stability of both *N*-substituted isoindoles and those that exist exclusively in the isoindole form[207].

Less highly developed aromaticity in oxygen than nitrogen heterocycles can account for the complete conversion of **401** to **402** on standing in benzene or (rapidly) upon treatment

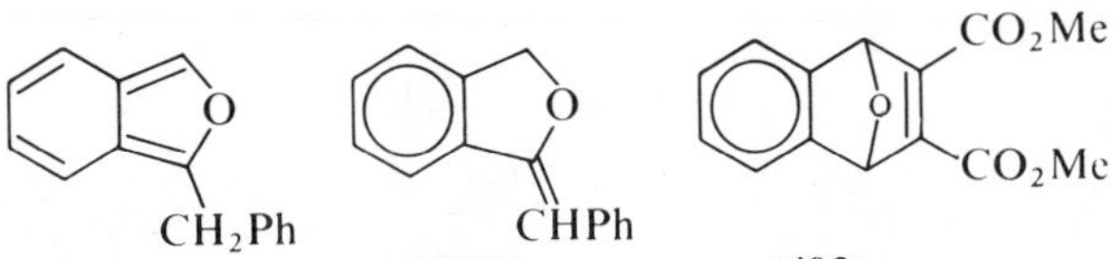

CO_2Me CO_2Me CH_2Ph CHPh

(**401**) (**402**) (**403**)

with trifluoroacetic acid[208]. Although the *o*-quinonoid tautomer (**401**) could not be detected by NMR spectroscopy it was trapped as the adduct (**403**) by heating **402** with dimethyl acetylenedicarboxylate and a trace of acid[209].

The effect of increasing *o*-quinonoid character upon tautomeric equilibrium is illustrated by the series **404**, **405** and **406**. For **404** the α-pyridone tautomer shown is overwhelmingly favoured. For **405** equilibrium with 3-hydroxyisoquinoline is more evenly balanced, the pyridone form (**405**) being favoured in ethanol and the pyridinol form in ether[6]. For **406** the pyridinol tautomer (**407**) is favoured (ca. 90%) even in ethanol[210].

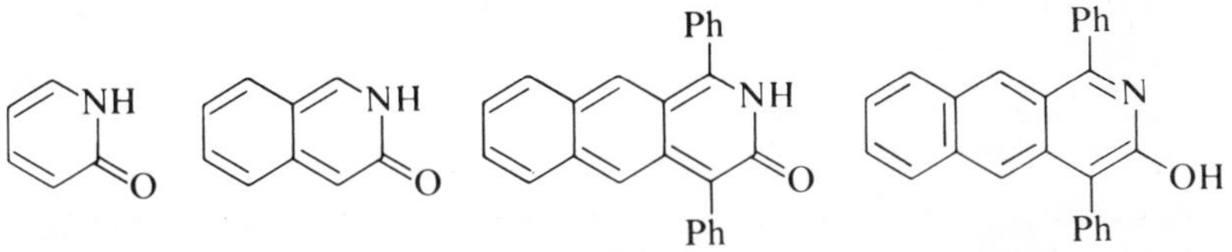

(**404**) (**405**) (**406**) (**407**)

As one might expect, the presence of the heteroatoms in the hetero-*o*-quinonoid species polarize the system and renders it prone to attack by nucleophiles. The numerous examples of this reaction include the formation of thiols (**408**, $R^1 = R^2 = i$-Pr) from attack of secondary amines[66] and **408** (R^1 = H, R^2 = alkyl) from primary amines. In the case of additions of primary amines, benzisothiazoles (**409**) are formed by autoxidation[211].

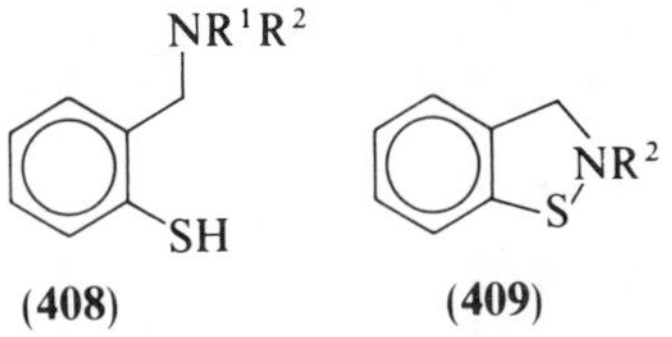

NR^1R^2 SH NR^2 S

(**408**) (**409**)

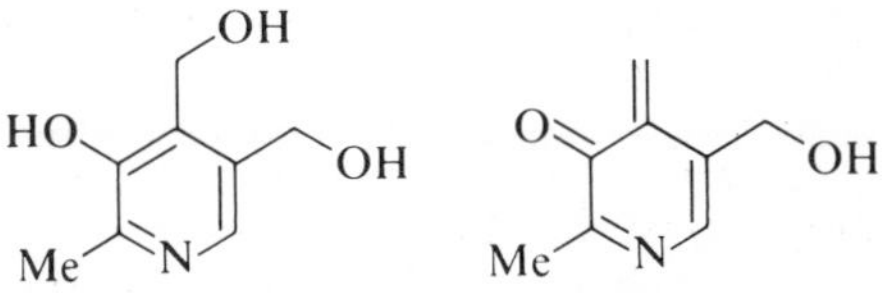

(**410**) (**411**)

This type of process has also been implicated in certain biological systems. For example, nucleophilic addition to the pyridine-quinone methide (**411**), formed from pyridoxine (**410**) has been reported, and it has been suggested that the potent alkylating activity of **411** may be involved in the enzymatic reactions of vitamin B_6 and in certain toxicological reactions induced by pyridoxine[212].

V. REFERENCES

1. (a) S. B. Cavitt, H. Sarrafizadeh and P. D. Gardner, *J. Org. Chem.*, **27**, 1211 (1962); *o*- and *p*-quinone methides have been reviewed: A. B. Turner, *Q. Rev.*, **18**, 347 (1964); H.-U. Wagner and R. Gompper, in *The Chemistry of Quinonoid Compounds*, Part 2 (Ed. S. Patai), Wiley, 1974, p. 1145; (b) P. D. Gardner and H. Sarrafizadeh, *J. Org. Chem.*, **25**, 641 (1960).
2. W. M. Horspool, *Q. Rev.*, **23**, 204 (1969).
3. W. R. Roth, M. Biermann, H. Dekker, R. Jochems, C. Mosselman and H. Hermann, *Chem. Ber.*, **111**, 3892 (1978).
4. H. Kolshorn and H. Meier, *Z. Naturforsch.*, **32A**, 780 (1977).
5. E. C. Taylor, D. R. Eckroth and J. Bartulin, *J. Org. Chem.*, **32**, 1899 (1967).
6. D. A. Evans, G. F. Smith and M. A. Wahid, *J. Chem. Soc.* (*B*), 590 (1967); D. W. Jones, *J. Chem. Soc.* (*C*), 1729 (1969).
7. D. A. Bleasdale, D. W. Jones, G. Maier and H.-P. Reissenauer, *J. Chem. Soc., Chem. Commun.*, 1095 (1983).
8. C. A. Coulson and A. Streitweiser, *Dictionary of π-Electron Calculations*, Pergamon Press, Oxford, 1965.
9. R. Hoffmann, in *The Chemistry of Non-benzenoid Aromatic Compounds* (Ed. M. Oki), Butterworths, London, 1971, p. 188.
10. (a) J. M. Holland and D. W. Jones, *J. Chem. Soc.* (*C*), 608 (1971); K. Blatt and R. W. Hoffmann, *Angew. Chem. Int. Ed. Engl.*, **8**, 606 (1969); (b) D. W. Jones, A. Pomfret and R. L. Wife, *J. Chem. Soc., Perkin Trans 1*, 459 (1983).
11. M. J. Haddadin, *Heterocycles*, **9**, 865 (1978); W. E. Friedrichsen, in *Advances in Heterocyclic Chemistry*, Vol. 26 (Ed. A. R. Katritzky and A. J. Boulton), Academic Press, New York, 1980, p. 135; U. E. Wiersum, *Aldrichimica Acta*, **14**, 53 (1981).
12. B. Iddon, in *Advances in Heterocyclic Chemistry*, Vol. 14 (Ed. A. R. Katritzky and A. J. Boulton), Academic Press, New York, 1972, p. 331.
13. J. D. White and M. E. Mann, in *Advances in Heterocyclic Chemistry*, Vol. 10 (Ed. A. R. Katritzky and A. J. Boulton), Academic Press, New York, 1969, p. 113.
14. R. P. Thummel, *Acc. Chem. Res.*, **13**, 70 (1980).
15. J. Jullien, J. M. Pechine, F. Perez and J. J. Piade, *Tetrahedron Lett.*, 3079 (1979).
16. U. E. Wiersum, C. D. Eldred, P. Vrijhof and H. C. van der Plas, *Tetrahedron Lett.*, 1741 (1977).
17. (a) B. Saroja and P. C. Srinivasan, *Tetrahedron Lett.*, **25**, 5429 (1984); (b) E. R. Marinelli, *Tetrahedron Lett.*, **23**, 2745 (1982).
18. P. Magnus, P. Brown and P. Pappalardo, *Acc. Chem. Res.*, **17**, 35 (1984); P. Magnus, P. M. Cairns and C. S. Kim, *Tetrahedron Lett.*, **26**, 1963 (1985); P. Magnus and N. L. Sear, *Tetrahedron*, **40**, 2795 (1984).
19. H. Pleininger, W. Müller and K. Weinerth, *Chem. Ber.*, **97**, 667 (1964).
20. C. J. Moody, *J. Chem. Soc., Chem. Commun.*, 925 (1984).
21. G. W. Gribble, M. G. Saulnier, M. P. Sibi and J. A. Obaza-Nutaitis, *J. Org. Chem.*, **49**, 4518 (1984).
22. C.-K. Sha, K.-S. Chuang and J.-J. Yaung, *J. Chem. Soc., Chem. Commun.*, 1552 (1984).
23. W. M. Welch, *J. Org. Chem.*, **41**, 2031 (1976).
24. C. R. Flynn and J. Michl, *J. Am. Chem. Soc.*, **96**, 3280 (1974); **95**, 5802 (1973).
25. (a) K. Alder and M. Fremery, *Tetrahedron*, **14**, 190 (1961); (b) E. Johansson and J. Skranstad, *J. Org. Chem.*, **46**, 3752 (1981).
26. (a) W. R. Dolbier, K. Matsui, J. Michl and D. V. Horak, *J. Am. Chem. Soc.*, **99**, 3876 (1977); W. R. Dolbier, K. E. Anapolle, L. McCullagh, K. Matsui, J. M. Riemann, and D. Rolison, *J. Org. Chem.*, **44**, 2845 (1979); (b) J. Dewey, W. R. Dolbier, K. Matsui, J. Michl and D. V. Horak, *J. Am. Chem. Soc.*, **101**, 2136 (1979).
27. F. J. Palenski and H. A. Morrison, *J. Am. Chem. Soc.*, **99**, 3507 (1977).
28. R. N. Warrener, P. A. Harrison and R. A. Russell, *J. Chem. Soc., Chem. Commun.*, 1134 (1982).

29. R. N. Warrener, R. A. Russell and T. S. Lee, *Tetrahedron Lett.*, 49 (1977); R. N. Warrener, R. A. Russell and P. A. Harrison, *Tetrahedron Lett.*, 2031 (1977).
30. D. S. Matteson and R. A. Davies, *J. Chem. Soc., Chem. Commun.*, 699 (1970); D. S. Matteson and N. K. Hota, *J. Am. Chem. Soc.*, **93**, 2893 (1971).
31. D. W. Jones and G. Kneen, *J. Chem. Soc., Perkin Trans 1*, 171 (1975).
32. R. D. Miller, J. Kolc and J. Michl, *J. Am. Chem. Soc.*, **98**, 8510 (1976).
33. M. Gisin and J. Wirz, *Helv. Chim. Acta*, **59**, 2273 (1976).
34. (a) C. M. Bowes, D. F. Montecalvo and F. Sondheimer, *Tetrahedron Lett.*, 3181 (1973); (b) T. W. Bell, C. M. Bower and F. Sondheimer, *Tetrahedron Lett.*, **21**, 3299 (1980).
35. R. P. Steiner, R. D. Miller, H. J. Dewey and J. Michl, *J. Am. Chem. Soc.*, **101**, 1820 (1979).
36. R. Gompper, E. Kutter and H. Kast, *Angew. Chem. Int. Ed. Engl.*, **6**, 171 (1967).
37. K. Hafner and W. Bauer, *Angew. Chem. Int. Ed. Engl.*, **7**, 297 (1968).
38. E. Klingsberg, *J. Org. Chem.*, **37**, 3226 (1972).
39. W. R. H. Hurtley and S. Smiles, *J. Chem. Soc.*, 534 (1927).
40. R. Gompper and H.-D. Lehmann, *Angew. Chem. Int. Ed. Engl.*, **7**, 74 (1968).
41. (a) W. R. Roth and J. D. Meier, *Tetrahedron Lett.*, 2053 (1967); (b) B. F. G. Johnson, J. Lewis and D. J. Thompson, *Tetrahedron Lett.*, 3789 (1974); W. H. Hersh and R. G. Bergmann, *J. Am. Chem. Soc.*, **103**, 6992 (1981); W. D. Jones and F. J. Feher, *J. Am. Chem. Soc.*, **106**, 1650 (1984), and cited references.
42. R. F. Heldewey and H. Hogeveen, *J. Am. Chem. Soc.*, **98**, 6040 (1976).
43. H. Finkelstein, Ph.D. dissertation, Strasbourg (1909); H. Finkelstein, *Chem. Ber.*, **43**, 1528 (1910); H. Finkelstein, *Chem. Ber.*, **92** (1959); M. P. Cava and D. R. Napier, *J. Am. Chem. Soc.*, **79**, 1701 (1957); M. P. Cava, A. A. Deana and K. Muth, *J. Am. Chem. Soc.*, **81**, 6458 (1959).
44. A. T. Blomquist, Y. C. Meinwald, C. G. Bottomley and P. W. Martin, *Tetrahedron Lett.*, 13 (1960); D. W. Jones and W. S. McDonald, *J. Chem. Soc., Perkin Trans 1*, 2257 (1982).
45. F. R. Jensen and W. E. Coleman, *J. Am. Chem. Soc.*, **80**, 6149 (1958).
46. B. H. Han and P. Boudjouk, *J. Org. Chem.*, **47**, 751 (1982).
47. S. Chew and R. J. Ferrier, *J. Chem. Soc., Chem. Commun.*, 911 (1984).
48. P. Schiess, M. Heitzmann, S. Rutschmann and R. Staheli, *Tetrahedron Lett.*, 4569 (1978); W. S. Trahanovsky and B. W. Turner, *J. Am. Chem. Soc.*, **107**, 4995 (1985).
49. R. Gray, L. G. Harruff, J. Krymowski, J. Peterson and V. Boekelheide, *J. Am. Chem. Soc.*, **100**, 2892 (1978).
50. Rubber-stitching, Dutch Patent No. 58, 664 (1946), *Chem. Abstr.*, **41**, 4672j (1947); K. Fries and E. Brandes, *Justus. Liebigs Ann. Chem.*, **542**, 48 (1939).
51. C. W. G. Fishwick, K. R. Randles and R. C. Storr, *Tetrahedron Lett.*, **26**, 3053 (1985).
52. Y. Ito, M. Nakatsuka and T. Saegusa, *J. Am. Chem. Soc.*, **104**, 7609 (1982); see too S. Djuric, T. Sarkar and P. Magnus, *J. Am. Chem. Soc.*, **102**, 6885 (1980).
53. J. R. Macias and W. S. Trahanovsky, *Abstracts of the American Chem. Soc. Meeting*, Chicago, Sept. 1985, Abstract 152.
54. Y. Ito, S. Miyata, M. Nakatsuka and T. Saegusa, *J. Am. Chem. Soc.*, **103**, 5250 (1981).
55. J. P. Marino and S. L. Dax, *J. Org. Chem.*, **49**, 3671 (1984).
56. (a) T. Tuschka, K. Naito and B. Rickborn, *J. Org. Chem.*, **48**, 70 (1983); (b) R. J. Moss and B. Rickborn, *J. Org. Chem.*, **49**, 3694 (1984); (c) K. Naito and B. Rickborn, *J. Org. Chem.*, **45**, 4061 (1980); (d) J. G. Smith and P. W. Dibble, *J. Org. Chem.*, **48**, 5361 (1983).
57. F. G. Mann and F. H. C. Stewart, *Chem. Ind. (Lond.)*, 373 (1954).
58. (a) J. J. Cornejo, S. Ghodsi, R. D. Johnson, R. Woodling and B. Rickborn, *J. Org. Chem.*, **48**, 3869 (1983); (b) J. G. Smith, P. W. Dibble and R. E. Sandborn, *J. Chem. Soc., Chem. Commun.*, 1197 (1983); (c) M.-M. Sadeghy and B. Rickborn, *J. Org. Chem.*, **48**, 2237 (1983).
59. R. Rodrigo, *J. Org. Chem.*, **45**, 4538 (1980); D. Rajapaksa and R. Rodrigo, *J. Am. Chem. Soc.*, **103**, 6208 (1981).
60. G. Wittig and H. Streib, *Annalen*, **584**, 1 (1953); W. Theilacker and H. Kalenda, *Annalen*, **584**, 87 (1953); W. Theilacker and W. Schmidt, *Annalen*, **597**, 95 (1955).
61. J. Thesing, W. Schäfer and D. Melchior, *Annalen*, **671**, 119 (1964); R. Keher and J. Seubert, *Angew. Chem. Int. Ed. Engl.*, **3**, 639 (1964); J. K. Kochi and E. A. Singleton, *Tetrahedron*, **24**, 4649 (1968).
62. M. P. Cava and N. M. Pollack, *J. Am. Chem. Soc.*, **88**, 4112 (1966).
63. J. Bornstein, R. P. Hardy and D. E. Remy, *J. Chem. Soc., Chem. Commun.*, 612 (1980).
64. M. P. Cava and L. E. Saris, *J. Am. Chem. Soc.*, **98**, 867 (1976).

65. (a) J. M. Holland and D. W. Jones, *J. Chem. Soc.*, (*C*), 536 (1970); (b) D. W. Jones, unpublished observations.
66. Y. L. Mao and V. Boekelheide, *Proc. Natl Acad. Sci. U.S.A.*, **77**, 1732 (1980).
67. C. Wentrup and G. Gross, *Angew. Chem.*, *Int. Ed. Engl.*, 543 (1983).
68. Y. L. Mao and V. Boekelheide, *J. Org. Chem.*, **45**, 1547 (1980).
69. R. D. Bowen, D. E. Davies, C. W. G. Fishwick, T. O. Glasby, S. J. Noyce and R. C. Storr, *Tetrahedron Lett.*, **23**, 4501 (1982).
70. O. L. Chapman and C. L. McIntosh, *J. Chem. Soc., Chem. Commun.*, 383 (1971).
71. L. A. Errede, *J. Am. Chem. Soc.*, **83**, 949 (1961).
72. M. P. Cava and A. A. Deana, *J. Am. Chem. Soc.*, **81**, 4266 (1959).
73. J. R. duManoir, J. F. King and R. R. Fraser, *J. Chem. Soc., Chem. Commun.*, 536 (1970).
74. L. A. Levy and L. Pruitt, *J. Chem. Soc., Chem. Commun.*, 227 (1980).
75. (a) J. Mann and S. E. Piper, *J. Chem. Soc., Chem. Commun.*, 430 (1982); (b) T. Durst, E. C. Kozma and J. L. Charlton, *J. Org. Chem.*, **50**, 4829 (1985); J. L. Charlton and T. Durst, *Tetrahedron Lett.*, **25**, 5287 (1984).
76. K. C. Nicolaou and W. E. Barnette, *J. Chem. Soc., Chem. Commun.*, 1119 (1979); W. Oppolzer and D. A. Roberts, *Helv. Chim. Acta*, **63**, 1703 (1980).
77. F. Jung, M. Molin, R. Van Den Elzen and T. Durst, *J. Am. Chem. Soc.*, **96**, 935 (1974).
78. D. N. Harpp, K. Steliou and T. H. Chan, *J. Am. Chem. Soc.*, **100**, 1222 (1978).
79. T. Durst and L. Tétreault-Ryan, *Tetrahedron Lett.*, 2353 (1978).
80. M. P. Cava and J. McGrady, *J. Org. Chem.*, **40**, 73 (1975).
81. A. G. Hortmann, A. J. Aron and A. K. Bhattacharyer, *J. Org. Chem.*, **43**, 3374 (1978).
82. M. Lancaster and D. J. H. Smith, *J. Chem. Soc., Chem. Commun.*, 471 (1980).
83. R. J. Spangler and B. G. Beckmann, *Tetrahedron Lett.*, 2517 (1976); R. J. Spangler and J. H. Kim, *Synthesis*, 107 (1973).
84. D. W. Jones and G. Kneen, *J. Chem. Soc., Perkin Trans. 1*, 175 (1975).
85. I. Hodgetts, S. J. Noyce and R. C. Storr, *Tetrahedron Lett.*, **25**, 5434 (1984).
86. W. J. M. Van Tilborg and R. Plomp, *J. Chem. Soc., Chem. Commun.*, 130 (1977).
87. (a) G. Quinkert, *Pure Appl. Chem.*, **9**, 607 (1964); (b) G. Quinkert, K. Opitz, W.-W. Wiersdorff and M. Finke, *Annalen*, **693**, 44 (1966).
88. (a) G. Quinkert, J. Palmowski, H.-P. Lorenz, W.-W. Wiersdorff and M. Finke, *Angew. Chem. Int. Ed. Engl.*, **10**, 198 (1971); (b) G. Quinkert, T. Tabata, E. A. J. Hickmann and W. Dobrat, *Angew. Chem. Int. Ed. Engl.*, **10**, 199 (1971).
89. D. S. Weiss, *J. Am. Chem. Soc.*, **97**, 2550 (1975).
90. R. N. Warrener, P. A. Harrison and R. A. Russell, *J. Chem. Soc., Chem. Commun.*, 107 (1973).
91. G. Jacquin, J. Nasielski, G. Billy and M. Reny, *Tetrahedron Lett.*, 3695 (1973).
92. M. Fischer and F. Wagner, *Chem. Ber.*, **102**, 3486 (1969).
93. C. Wentrup and P. Mueller, *Tetrahedron Lett.*, 2915 (1973).
94. T. C. McC. Paterson, R. K. Smalley, H. Suschitzky and A. J. Barker, *J. Chem. Soc., Perkin Trans. 1*, 633 (1980).
95. L. A. Carpino, *J. Chem. Soc., Chem. Commun.*, 494 (1966), and cited references.
96. U. E. Wiersum and W. J. Mijs, *J. Chem. Soc., Chem. Commun.*, 347 (1972).
97. J. Bornstein, D. E. Remy and J. E. Shields, *J. Chem. Soc., Chem. Commun.*, 1149 (1972).
98. L. F. Fieser and M. J. Haddadin, *Can. J. Chem.*, **43**, 1599 (1965).
99. D. Wege, *Tetrahedron Lett.*, 2337 (1971).
100. R. N. Warrener, *J. Am. Chem. Soc.*, **93**, 2346 (1971).
101. R. McCulloch, A. R. Rye and D. Wege, *Tetrahedron Lett.*, 5231 (1969).
102. M. N. Paddon-Row, P. L. Watson and R. N. Warrener, *Tetrahedron Lett.*, 1033 (1973); H. Tanida, T. Irie and K. Tori, *Bull. Chem. Soc. Japan*, **45**, 1999 (1972).
103. O. L. Chapman and C. L. McIntosh, *J. Am. Chem. Soc.*, **92**, 7001 (1970).
104. (a) T. Kametani, M. Tsubuki, Y. Shiratori, Y. Kato, H. Nemoto, M. Ihara, K. Fukumoto, F. Satoh and H. Inoue, *J. Org. Chem.*, **42**, 2672 (1977); (b) W. Oppolzer, *Synthesis*, 793 (1978).
105. P. Caubere, *Top. Curr. Chem.*, **73**, 72 (1978).
106. M. A. O'Leary, G. W. Richardson and D. Wege, *Tetrahedron*, **37**, 813 (1981).
107. R. Huisgen and H. Seidl, *Tetrahedron Lett.*, 3381 (1964).
108. (a) B. J. Arnold, S. M. Mellows, P. G. Sammes and T. W. Wallace, *J. Chem. Soc.*, 401 (1974); (b) B. J. Arnold, P. G. Sammes and T. W. Wallace, *J. Chem. Soc.*, 409 (1974); (c) B. J. Arnold, P. G. Sammes and T. W. Wallace, *J. Chem. Soc.*, 415 (1974).

109. N. G. Rondan and K. N. Houk, *J. Am. Chem. Soc.*, **107**, 2099 (1985).
110. W. Merk and R. Pettit, *J. Am. Chem. Soc.*, **89**, 4788 (1967).
111. F. R. Farr and N. L. Bauld, *J. Am. Chem. Soc.*, **92**, 6695 (1970); N. L. Bauld, F. R. Farr and C.-S. Chang, *Tetrahedron Lett.*, 2443 (1972).
112. M. P. Cava, B. Hwang and J. P. Van Meter, *J. Am. Chem. Soc.*, **85**, 4031 (1963).
113. R. W. Hoffmann, in *The Chemistry of Acetylenes* (Ed. H. G. Viehe), Dekker, New York, 1969.
114. H. Heaney, S. M. Jablonski and C. T. McCarty, *J. Chem. Soc., Perkin Trans. 1*, 2903 (1972).
115. C. W. G. Fishwick, R. C. Gupta and R. C. Storr, *J. Chem. Soc., Perkin Trans. 1*, 2827 (1984).
116. E. M. Burgess and L. McCullagh, *J. Am. Chem. Soc.*, **88**, 1580 (1966).
117. E. Noigt and H. Meier, *Angew. Chem. Int. Ed. Engl.*, **15**, 117 (1976).
118. D. C. Dittmer and F. A. Davis, *J. Org. Chem.*, **32**, 3872 (1967).
119. L. A. Paquette and T. R. Phillips, *J. Org. Chem.*, **30**, 3883 (1965).
120. D. C. Dittmer and N. Takashina, *Tetrahedron Lett.*, 3801 (1964).
121. H. Heimgartner, J. Zsindely, H.-J. Hansen and H. Schmid, *Helv. Chim. Acta*, **53**, 1212 (1970).
122. J. M. Hornback and R. D. Barrows, *J. Org. Chem.*, **48**, 90 (1983).
123. P. G. Sammes, *Tetrahedron*, **32**, 405 (1976).
124. Y. Ito, H. Nishimura, H. Shimizu and T. Matsuura, *J. Chem. Soc., Chem. Commun.*, 1110 (1983).
125. K. Alder, F. Pascher and H. Vagt, *Chem. Ber.*, **75**, 1501 (1942).
126. W. R. Roth, *Tetrahedron Lett.*, 1009 (1964).
127. N. S. Isaacs, *Can. J. Chem.*, **44**, 415 (1966).
128. J. Almy and D. J. Cram, *J. Am. Chem. Soc.*, **92**, 4316 (1970).
129. P.-T. Ho, S. Oida and K. Wiesner, *J. Chem. Soc., Chem. Commun.*, 883 (1972).
130. L. L. Miller and R. F. Boyer, *J. Am. Chem. Soc.*, **93**, 650 (1971); L. L. Miller, R. Griesinger and R. F. Boyer, *J. Am. Chem. Soc.*, **91**, 1578 (1969).
131. J. J. McCullough, *Acc. Chem. Res.*, 270 (1980); K. K. de Fonseka, C. Manning, J. J. McCullough and A. J. Yarwood, *J. Am. Chem. Soc.*, **99**, 8257 (1977).
132. D. J. Field, D. W. Jones and G. Kneen, *J. Chem. Soc., Perkin Trans. 1*, 1050 (1978); D. J. Field and D. W. Jones, *J. Chem. Soc., Perkin Trans.* 1, 714 (1980).
133. (a) A. J. Ashe III, *Tetrahedron Lett.*, 2105 (1970); (b) R. B. Larrabee and B. F. Dowden, *Tetrahedron Lett.*, 915 (1970); (c) M. N. Andrews, P. E. Rakita and G. A. Taylor, *Tetrahedron Lett.*, 1851 (1973).
134. P. E. Rakita and A. Davison, *Inorg. Chem.*, **8**, 1164 (1969).
135. V. I. Minkin, L. P. Olekhnovich and B. Y. Simkin, *Zh. Org. Khim.*, **7**, 2364 (1971); *Chem. Abstr.*, **76**, 71843g (1972).
136. D. A. Ben-Efraim and F. Sondheimer, *Tetrahedron Lett.*, 313 (1963).
137. L. Skatebøl, G. A. Abskharoun and T. Greibrakk, *Tetrahedron Lett.*, 1367 (1973); L. Skatebøl, *J. Org. Chem.*, **29**, 2951 (1964).
138. V. Horak, F. V. Foster and P. Svoronos, *Tetrahedron Lett.*, **22**, 3577 (1981).
139. D. W. Jones and R. L. Wife, *J. Chem. Soc., Perkin Trans. 1*, 1 (1974).
140. D. W. Jones and A. Pomfret, *J. Chem. Soc., Chem. Commun.*, 703 (1983).
141. D. A. Bolon, *J. Org. Chem.*, **35**, 715 (1970); J. G. Westra, W. G. B. Huysmans, W. J. Mijs, H. A. Gaur, J. Vriend and J. Smidt, *Recl. Trav. Chim. Pays-Bas*, **87**, 1121 (1968).
142. R. C. Cookson and S. M. de B. Costa, *J. Chem. Soc., Chem. Commun.*, 1272 (1969); K. Salisbury, *Tetrahedron Lett.*, 737 (1971).
143. R. S. Becker and S. Kok, *J. Phys. Chem.*, **72**, 197 (1968).
144. B. S. Lukyanov, M. I. Knazachanskii, I. V. Rovinskii, L. E. Nivorozhkim, and U. I. Minkin, *Tetrahedron Lett.*, 2007 (1973).
145. J. Kolc and R. S. Becker, *J. Am. Chem. Soc.*, **91**, 6513 (1969).
146. P. deMayo and H. Y. Ng, *Can. J. Chem.*, **55**, 3763 (1977); P. deMayo and H. Y. Ng, *J. Chem. Soc., Chem. Commun.*, 877 (1974).
147. K. T. Kang, R. Okazaki and N. Inamoto, *Bull. Chem. Soc. Japan*, **52**, 3640 (1979).
148. O. L. Chapman, C. C. Chang, J. Kolc, N. R. Rosenquist and H. Tomioka, *J. Am. Chem. Soc.*, **97**, 6586 (1975).
149. J. Rigaudy, J. Baranne-Lafont, A. Defoin and Nguyen Kim Cuong, *Tetrahedron*, **34**, 73 (1978); J. Guilham, J. Rigaudy, J. Baranne-Lafont and A. Defoin, *Tetrahedron*, **34**, 83 (1978); J. Rigaudy, A. Defoin and J. Baranne-Lafont, *Angew. Chem. Int. Ed. Engl.*, **18**, 413 (1979).
150. J. M. Holland and D. W. Jones, *J. Chem. Soc., Perkin Trans. 1*, 927 (1973).
151. T. H. Chan and K. T. Nue, *Tetrahedron Lett.*, 3601 (1973).

152. J. Bornstein, J. E. Shields and J. H. Sapple, *J. Org. Chem.*, **32**, 1499 (1967).
153. D. W. Jones and G. Kneen, *J. Chem. Soc., Perkin Trans. 1*, 1647 (1976).
154. W. Oppolzer, *J. Am. Chem. Soc.*, **93**, 3883 (1971).
155. cf. C. G. Cardenas, *J. Chem. Soc., Chem. Commun.*, 134 (1970).
156. R. K. Smalley and H. Suschitsky, *Tetrahedron Lett.*, 3465 (1966).
157. G. Catteral, *J. Chem. Soc., Chem. Commun.*, 41 (1974).
158. W. Waters, *J. Chem. Soc.*, 243 (1954).
159. M. S. Chauhan, F. M. Dean, D. Makkin and N. L. Robinson, *J. Chem. Soc., Perkin Trans. 1*, 120 (1973).
160. A. O. Pedersen, A. O. Lawessen, P. D. Kelmmensen and J. Kolc, *Tetrahedron*, **26**, 1157 (1970).
161. R. Okazaki and N. Inamoto, *Bull. Chem. Soc. Japan*, **52**, 3640 (1979).
162. G. Wittig and E. R. Wilson, *Chem. Ber.*, **98**, 451 (1965).
163. M. P. Cava and M. J. Mitchell, *Cyclobutadiene and Related Compounds*, Academic Press, New York, 1967, p. 206.
164. M. Avram, I. G. Dinulescu, D. Dinu and C. D. Nenitzescu, *Chem. Ind. (Lond.)*, 555 (1962).
165. L. M. Tolbert and M. B. Ali, *J. Am. Chem. Soc.*, **106**, 3806 (1984).
166. M. S. Newman, *J. Org. Chem.*, **26**, 2630 (1961).
167. (a) D. W. Jones, *J. Chem. Soc., Chem. Commun.*, 766 (1982); (b) J. A. Moore, R. Muth and R. Sorace, *J. Org. Chem.*, **39**, 3799 (1974).
168. I. Fleming, F. L. Gianni and T. Mah, *Tetrahedron Lett.*, 881 (1976).
169. K. N. Houk, *Tetrahedron. Lett.*, 2621 (1970).
170. J. Sauer and R. Sustmann, *Angew. Chem. Int. Ed. Engl.*, **19**, 779 (1980).
171. D. W. Jones, *J. Chem. Soc., Perkin Trans. 1*, 980 (1977).
172. Y. Kobuke, T. Fueno and J. Furukawa, *J. Am. Chem. Soc.*, **92**, 6548 (1970); Y. Kobuke, T. Sugimoto, J. Furukawa and T. Fueno, *J. Am. Chem. Soc.*, **94**, 3633 (1972).
173. W. L. Jorgensen and L. Salem, *The Organic Chemists Book of Orbitals*, Academic Press, New York, 1973.
174. (a) R. Hoffmann, C. C. Levin and R. A. Moss, *J. Am. Chem. Soc.*, **95**, 629 (1973); (b) K. N. Houk, in *Pericyclic Reactions*, Vol. II (Ed. A. P. Marchand and R. E. Lehr), Academic Press, New York, 1977, p. 229.
175. M. P. Cava and F. M. Scheel, *J. Org. Chem.*, **52**, 1304 (1967).
176. D. W. Jones and R. L. Wife, *J. Chem. Soc., Perkin Trans. 1*, 1654 (1976).
177. A. Arduini, A. Bosi, A. Pochini and R. Ungora, *Tetrahedron*, **41**, 3095 (1985).
178. R. Bonnet, R. F. C. Brown and R. G. Smith, *J. Chem. Soc., Perkin Trans. 1*, 1432 (1973).
179. R. W. Franck, T. V. John and K. Olejniczak, *J. Am. Chem. Soc.*, **104**, 1106 (1982).
180. (a) K. N. Houk, S. R. Moses, Y.-D. Wu, N. G. Rondan, V. Jäger, R. Schohe and F. R. Fronczek, *J. Am. Chem. Soc.*, **106**, 3880 (1984); (b) M. N. Paddon-Row, N. G. Rondan and K. N. Houk, *J. Am. Chem. Soc.*, **104**, 7162 (1982); (c) P. Caramella, N. G. Rondan, M. N. Paddon-Row and K. N. Houk, *J. Am. Chem. Soc.*, **103**, 2438 (1981).
181. A. P. Kozikowski and A. K. Ghosh, *J. Am. Chem. Soc.*, **104**, 5788 (1982).
182. E. Ciganek, *Org. React.*, **32**, 1 (1984); A. G. Fallis, *Can. J. Chem.*, **62**, 183 (1984); D. F. Taber, *Intramolecular Diels–Alder and Alder Ene Reactions*, Vol. 18 of *Reactivity and Structure Concepts in Organic Chemistry*, Springer-Verlag, Berlin, 1984; G. Desimoni, G. Tacconi, A. Barco and G. P. Pollini, *ACS Monograph* 180, Am. Chem. Soc., Washington DC, 1980.
183. R. L. Funk and K. P. C. Vollhardt, *Chem. Soc. Rev.*, **9**, 41 (1980); T. Kametani and H. Nemoto, *Tetrahedron*, **37**, 3 (1981).
184. R. L. Funk and K. P. C. Vollhardt, *J. Am. Chem. Soc.*, **102**, 5253 (1980).
185. B. A. Keay and R. Rodrigo, *J. Am. Chem. Soc.*, **104**, 4725 (1982).
186. R. J. Ardecky, D. Dominguez and M. P. Cava, *J. Org. Chem.*, **47**, 409 (1982).
187. T. Watanabe, Y. Takahashi and M. Oda, *Tetrahedron Lett.*, **24**, 5623 (1983).
188. J. R. Wiseman, J. J. Pendery, C. A. Otto and K. G. Chiong, *J. Org. Chem.*, **45**, 516 (1980).
189. R. A. Russell, E. G. Vikingur and R. N. Warrener, *Aust. J. Chem.*, **34**, 131 (1981).
190. R. N. Warrener, B. C. Hammer and R. A. Russell, *J. Chem. Soc., Chem. Commun.*, 942 (1981).
191. D. A. Bleasdale and D. W. Jones, *J. Chem. Soc., Chem. Commun.*, 1027 (1985).
192. M. E. Jung, R. W. Brown, J. A. Hagenah and C. E. Strouse, *Tetrahedron Lett.*, **25**, 3659 (1984).
193. R. W. Franck and T. V. John, *J. Org. Chem.*, **48**, 3269 (1983).
194. R. Okazaki, K. T. Kang, K. Sunagawa and N. Inamoto, *Chem. Lett.*, 55 (1978).
195. F. Ishii, R. Okazaki and N. Inamoto, *Tetrahedron Lett.*, 4283 (1976).

196. J. Brugidou and H. Christol, *Compt. rend.*, **256**, 3149 (1963).
197. C. W. G. Fishwick, R. C. Storr and P. W. Manley, *J. Chem. Soc., Chem. Commun.*, 1304 (1984).
198. J. S. Talley, *J. Org. Chem.*, **50**, 1695 (1985).
199. O. L. Chapman, M. R. Engel, J. P. Spinjer and J. C. Clardy, *J. Am. Chem. Soc.*, **93**, 6696 (1971).
200. A. Pomfret, Ph.D. Thesis, University of Leeds, 1982.
201. C. Dufraisse and S. Ecary, *C. R. Acad. Sci. Paris*, **233**, 735 (1946).
202. W. Theilacker and W. Schmidt, *Annalen*, **605**, 43 (1957).
203. J. P. Smith, A. K. Schrock and G. B. Schuster, *J. Am. Chem. Soc.*, **104**, 1041 (1982).
204. J. M. Holland and D. W. Jones, *J. Chem. Soc.* (*C*), 530 (1970).
205. J. Bornstein, D. E. Remy and J. E. Shields, *Tetrahedron Lett.*, 4247 (1974).
206. D. Veber and W. Lwowski, *J. Am. Chem. Soc.*, **86**, 4152 (1964).
207. J. Kopeky, J. E. Shields and J. Bornstein, *Tetrahedron Lett.*, 3669 (1967).
208. E. Chacko, D. J. Sardella and J. Bornstein, *Tetrahedron Lett.*, 2507 (1976).
209. J. G. Smith and R. T. Wickman, *J. Org. Chem.*, **39**, 3648 (1974).
210. D. W. Jones and R. L. Wife, *J. Chem. Soc., Perkin Trans. 1*, 2722 (1972).
211. K. Kanakoajan and H. Meier, *Angew. Chem. Int. Ed. Engl.*, **23**, 224 (1984).
212. M. Frater-Schroder and M. Mahrer-Basato, *Bioorg. Chem.*, **4**, 332 (1975).

The Chemistry of Quinonoid Compounds, Vol. II
Edited by S. Patai and Z. Rappoport

CHAPTER **10**

meta-Quinonoid compounds

JEROME A. BERSON
Department of Chemistry, Yale University, P.O. Box 6666, New Haven, Connecticut 06511-8118, USA

I. HISTORICAL BACKGROUND

The roots of the *meta*-quinone problem go deep into the intellectual history of chemistry. Kekulé proposed the cyclic formula for benzene in 1865, and in the following decades, legions of organic chemists explored the consequences of the theory. Only with effort can the modern scholar's imagination recapture the atmosphere of mingled excitement and conjecture which prevailed in that early morning of organic chemistry a century or more ago. Major elements of the physical basis of chemical theory, which today are commonplace, then were either bewilderingly new or had not been conceived. For example, the concepts of bond length and bond angle were vague, and the nature of the physical forces holding atoms together in molecules was unknown. A certain indulgence therefore is required if the reader of today is to appreciate the challenge of the structure of *m*-quinonoids as it existed then.

In the context of that time, the relevance of *m*-quinones to the problem of benzene was made clear by Zincke[1], whose own words (translated here from the German and augmented by structures drawn in the modern style) state the matter succinctly:

> By all means, the discovery of a *meta*-benzoquinone would be of significance for the theory of benzene; as the presence of a meta-bond in the former must be assumed, it can only be expressed by [one of] the formulas **1** and **2**.

Zincke implicitly assumed that Kekulé's standard valence rules were inviolable. From this it followed that a *m*-quinone could not be constructed wholly with bonds between adjacent (*ortho*) carbons on the six-membered ring.

Zincke continues:

> Since I have been able to convert catechol into derivatives of *ortho*-benzoquinone, I have not neglected attempts to prepare analogous derivatives of the *meta* series, but so far without success; it appears in fact as if derivatives of this kind are incapable of existence.

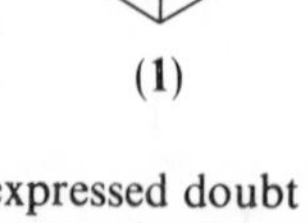

(**1**) 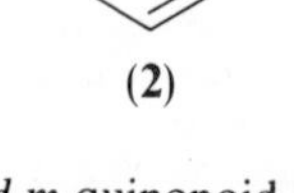 (**2**)

Hantzsch[2, 3] also expressed doubt that *uncharged m*-quinonoid substances could exist but believed that the *anions* of certain metallic salts obtained by the action of strong alkalis on *m*-nitrophenols might have such a structure (**3**). The methods of physical characterization

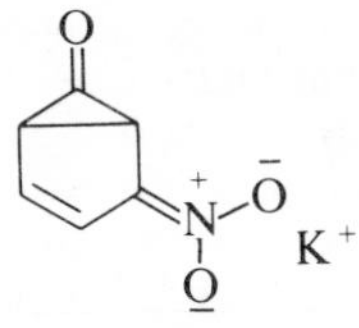

(3)

available at the time did not suffice to establish this assignment, and the elucidation by Meisenheimer[4] of the reactions of strong bases with aromatic nitro compounds suggests plausible alternative formulations.

In an extension of work of Liebermann and Dittler[5], Meyer and Desamari[6] assigned a *meta*-quinone structure (**4**) to 'tribromoresoquinone', the product of the pyrolysis of pentabromoresorcinol (**5**). However, Zincke and Schwabe[7] and later, in a second paper,

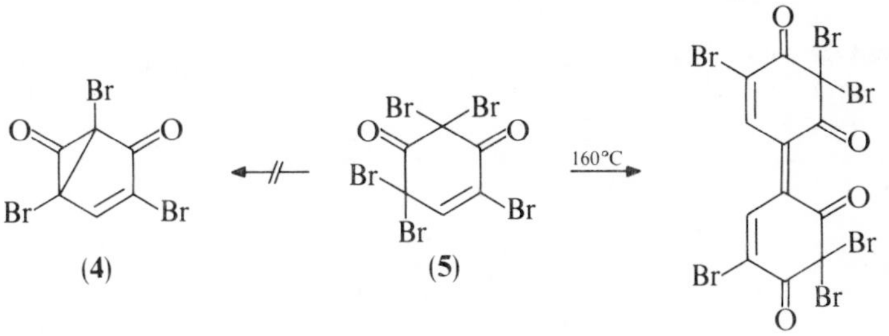

(4) (5) 160°C (6)

Meyer and Desamari[8] themselves showed that 'tribromoresoquinone' is dimeric and has the diphenoquinone structure **6**, not the *m*-quinone structure **4**. Zincke[7] took the occasion to reiterate his earlier opinion[1] by remarking that 'the existence of *meta*-quinones appears to us improbable; years ago, one of us made many attempts to obtain halogen derivatives of these quinones, but without success'.

Nevertheless, Stark and his coworkers[9–12] undertook further attempts to prepare a *meta*-quinonoid system. Modeling their experiments on those of Gomberg, whose preparation of triarylmethyl radicals was already widely known, they treated tetraphenyl-*m*-xylylene dichloride (**7**) with metals such as zinc or silver. Implicitly, they used analogical criteria to decide whether the reaction gave a bis-triphenylmethyl (**8**) or a *m*-quinonoid, e.g. **9**. Gomberg's work had established the extreme reactivity of triphenylmethyls to oxygen, whereas a true quinonoid substance, by analogy to Thiele's tetraphenyl-*p*-quinodimethane (**10**), Stark assumed, would be colored but relatively insensitive to oxygen.

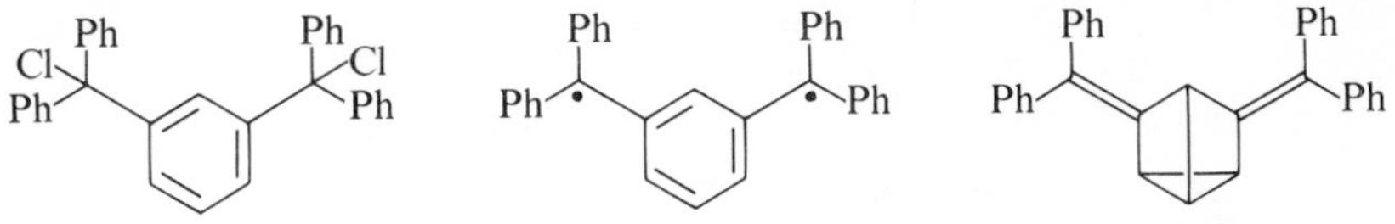

(7) (8) (9)

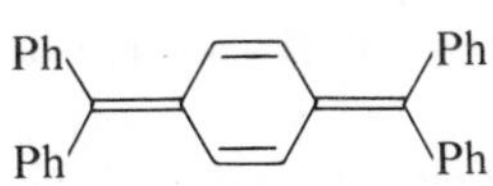

(10)

The materials obtained in Stark's experiments appeared to conform to the quinonoid criteria, but the results were brought under question by later studies of Schlenk and Brauns[13, 14]. The latter authors, by meticulous exclusion of air from their reaction mixtures and by the use of rigorously prepared and purified starting materials, demonstrated the sequential removal of the two chlorines of **7**, forming successively solutions of the yellow chlorotriphenylmethyl derivative **11** and the deep violet bis-triphenylmethyl **8**. Both species **11** and **8** were inordinately sensitive to oxygen, which

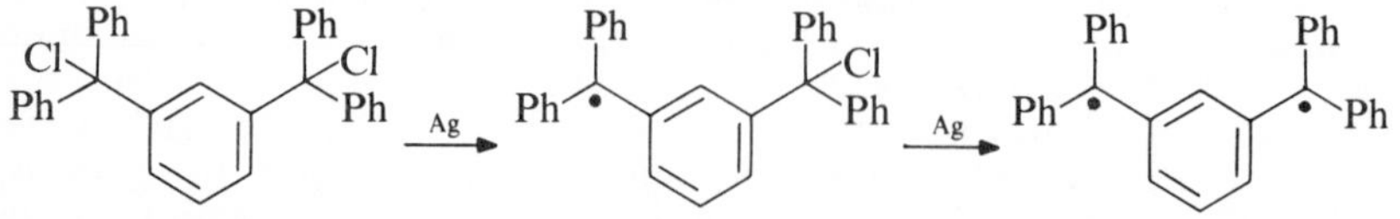

(7) (11) (8)

discharged the color instantaneously[13]. Schlenk and Brauns extended this work soon after to the preparation of the biradical **12** by similar methods[14]. This paper contains one of the first clear recognitions that the 'spatial relationship' (i.e., distance) of the atomic centers involved may preclude the formation of a bond (as in **13**).

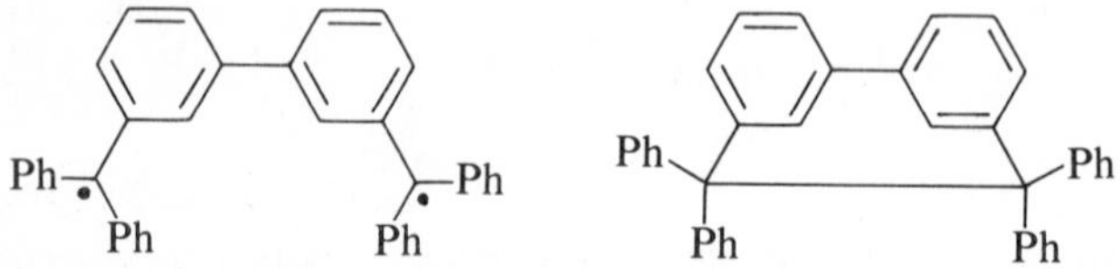

(12) (13)

With respect to the *m*-quinone problem, Schlenk and Brauns[13] concluded that 'the formation of a bis-triarylmethyl from tetraphenyl-*m*-xylylene dichloride is a new indication that *m*-quinonoid compounds do not exist'.

The dramatic series of refutations encountered to this point seemed to settle the matter in the minds of most chemists, for no further attempts to make *m*-quinonoids were reported for many years. They were considered to be somehow outside the structural theory. For example, in the 1940s, popular undergraduate textbooks[15, 16] discussed a very old structural problem also first formulated by Zincke[17] and much debated thereafter[18–23]: should *ortho*-quinones be formulated as ketones (**14**), in accord with their carbonyl group reactivity, or as peroxides (**15**), in accord with their oxidizing power?

(14) (15)

The decision rendered in 1941[15] relied on the following argument:

> It is a fact that only *para* and *ortho* quinones have been obtained. The peroxide theory allows *meta*-quinones to exist. But the keto structure is impossible for a *meta*-quinone, as the student will see if he tries to draw one with four bonds to each carbon.

Note that the reasoning again embodies the implicit assumptions that Kekulé's valence numbers are inviolable and that *m*-quinonoids cannot exist.

In the 1950s, Clar and coworkers invoked related arguments to explain the failure to observe the hypothetical hydrocarbon biradicals triangulene (**16**)[24] and dibenzopentacene (**17**)[25] when partially hydrogenated precursors were heated over palladium, conditions that had succeeded in generating fully aromatic polycycles in other cases. Thus,

> . . . the interpretation of this result is that Kekulé structures are of paramount importance when considering the stability of aromatic hydrocarbons. . . . This excludes the existence of *m*-quinonoids and similar systems which cannot be expressed in terms of Kekulé structures.[26]

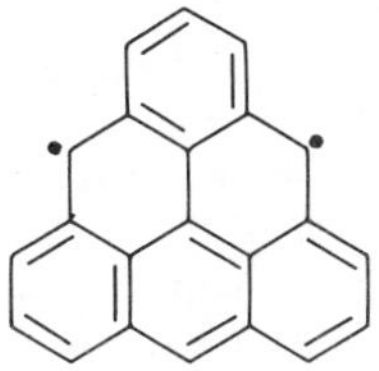

(**16**) 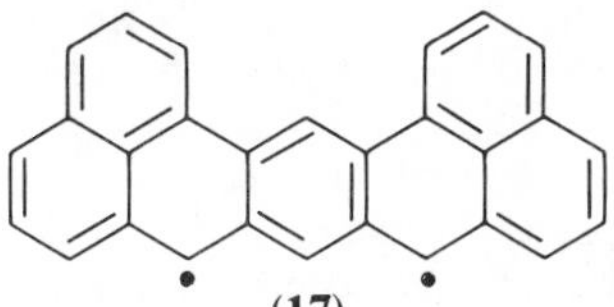 (**17**)

Clar did not take into account the two hydrocarbon biradicals **8** and **12** of Schlenk and Brauns, of which the former is closely related structurally to **16** and **17**. Neither **8** nor **12** can be expressed in terms of Kekulé structures.

It was not until 1979 that the first *m*-quinonoid pair of valency tautomers, the covalent (**18**) and biradical (**19**) forms of *m*-quinomethane, were prepared[27], 88 years after the problem had been presented by Zincke. The structural relationship of the bicyclic, fully covalent *m*-quinonoid **18** to **2**, one of Zincke's hypothetical *m*-quinones, will be immediately apparent. In modern terms, the valency tautomerism **18** → **19** involves cleavage of a carbon–carbon bond (C(1)–C(5)), planarization of the six-membered ring, and other adjustments of the atomic coordinates.

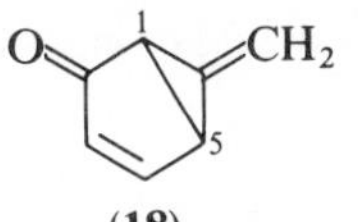

O 1 CH_2 5 Ȯ $\dot{C}H_2$

(**18**) 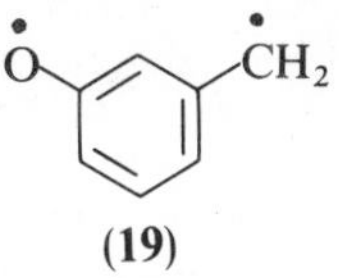 (**19**)

As has been noted here, the supposed non-existence of *meta*-quinones was a key logical element in deciding the diketone vs. peroxide controversy over the structure of *ortho*- and *para*-quinones. Although the choice of the diketone formula ultimately was proven by other means to be correct, the synthesis of **18** and other *m*-quinonoids invalidates the form of argument outlined above. The appeal to isomer number, so often invoked in structural organic chemistry, has only an experiential basis and therefore is vulnerable, should subsequent events establish the existence of an isomer previously thought to be impossible.

The struggles of earlier chemists with the *m*-quinone problem paralleled the growth and improvement of the structural theory of organic chemistry itself. In the realization that future generations probably will hold our theories in the same tolerant regard as we hold those of our predecessors, we do well to shun what E. P. Thompson (in another context) has called 'the enormous condescension of posterity'.

II. THEORETICAL ASPECTS OF HUND'S RULE

A. Quantum Mechanical Justification

In one way, the hydrocarbon biradicals **8** and **12** synthesized by Schlenk and Brauns in 1915[13, 14] became available far ahead of their time. One could justifiably argue that their true significance transcended the obvious fact that they were bifunctional examples of Gomberg's triphenylmethyl, which had been a discovery of the immediately preceding era around the turn of the century. Moreover, although the Schlenk–Brauns work must have influenced contemporary thinking in the context which the authors constructed for it, namely as the refutation of a supposed *m*-quinonoid structure, our vantage point in time allows us to see the more important questions to which these compounds were the first answers: is it possible to make molecules for which no full-valence Kekulé structure can be written, i.e. in Dewar's coinage[28], 'non-Kekulé' molecules? What physical and chemical properties should be expected of such species? The Schlenk–Brauns work thus looked forward to quantum mechanics, which was still a decade beyond the horizon.

Actually, more than two decades passed before the first important quantum mechanically inspired experiments on the Schlenk–Brauns systems were carried out. In 1936, Müller and Bunge[29] determined the magnetic susceptibility of these substances and found them both to be paramagnetic. This was a gratifying observation, for the application of Hund's rule[30] would predict exactly this property for biradicals. Since these pioneering observations, it has become clear that an appreciation of the sources of such paramagnetism is crucial to understanding not only the magnetic and spectroscopic behavior of non-Kekulé molecules but also their chemistry. A brief digression here is intended to provide a qualitative background to this subject.

Hund's rule, originally an empirical generalization based upon experimental atomic spectroscopy, states that of the terms of a given electron configuration, the one of lowest energy is the one of highest multiplicity. Although many of the key features of the multiplet structure of atomic spectra in applied magnetic fields (Zeeman effect) had been analyzed by the Russell–Saunders coupling model in the early 1920s, it was not until the concept of electron spin had been introduced by Uhlenbeck and Goudsmit in 1925 that the present interpretation of these phenomena became established[31]. The multiplicity is given by $2S+1$, where S is the total electron spin. In a simple example, Hund's rule would predict that the triplet state ($S = 1$) of a two-electron atom should be more stable than the singlet state ($S = 0$) of the same configuration.

To understand the origin of the singlet–triplet energy splitting, it is helpful to recall that the Pauli exclusion principle, as interpreted by Heisenberg and Dirac, requires that the total atomic (or molecular) wave function be antisymmetric (change into its own negative) upon an interchange of the space and spin coordinates[32–34]. For example, consider the lowest excited configuration of the helium atom, in which the 1s and 2s orbitals each are assigned one electron. If the electrons are identified as (1) and (2) and their spins are identified by the quantum symbols α or β, the total wave functions are products of a space wave function, which can be expressed as a product of a linear combination of the 'independent electron' wave functions (1s and 2s) and a spin wave function. The space wave functions are either symmetric or antisymmetric, according to whether the sign in the linear combination is positive or negative. The spin wave functions are symmetric if both electrons have the same spin or if the spin is represented as a positive linear combination of opposite spins; they are antisymmetric if the spin is represented as a negative linear combination of opposite spins. The only total wave functions for the He 1s2s electron configuration that satisfy the exclusion principle, therefore, are those shown in equations 1–4, where the subscripts s and a designate symmetric and antisymmetric, respectively, and the designations refer, in order, to the space and spin parts of the wave function.

$$\psi_{s,a}(1,2) = \overbrace{(1/\sqrt{2})[1s(1)2s(2)+2s(1)1s(2)]}^{\text{space part}}\overbrace{1/\sqrt{2}[\alpha(1)\beta(2)-\beta(1)\alpha(2)]}^{\text{spin part}} \quad (1)$$

$$\psi_{a,s}(1,2) = (1/\sqrt{2})[1s(1)2s(2)-2s(1)1s(2)] \begin{cases} \alpha(1)\alpha(2) & (2) \\ (1/\sqrt{2})[\alpha(1)\beta(2)+\beta(1)\alpha(2)] & (3) \\ \beta(1)\beta(2) & (4) \end{cases}$$

Thus, the excited He configuration 1s(1)2s(2) leads to *four* states, one of which has a symmetric space wave function (equation 1), and three of which have antisymmetric space wave functions (equations 2–4). In a spherically symmetrical system at zero applied magnetic field, the latter three states have identical energies. They are said to be the components of a triplet state. The unique state is said to be a singlet state.

The total wave functions can be derived more elegantly by the method of Slater determinants[31], but the presentation here may be more accessible to those organic chemists whose contact with the formal part of quantum mechanics is infrequent.

Because the spin part of the singlet wave function is antisymmetric (equation 1), the total spin of the singlet is zero. It is proper to imagine the physical basis for this as being the cancellation of equal and opposite z-components of the spin angular momentum vectors, although it may be easier to remember that the total electronic spin S is made up of the algebraic sum of the uncompensated electron spins, each with $S = |\frac{1}{2}|$. In the singlet, each spin of $+\frac{1}{2}$ is compensated for by one of $-\frac{1}{2}$.

In the case of the triplet, two of the spin wave functions (in equations 2 and 4) represent additions of 'up' or 'down' z-components, whereas the third (in equation 3) represents a spin vector that is perpendicular to the z-axis. When placed in a magnetic field strong enough to quantize the energies of the triplet sublevels, these states separate because of the Zeeman effect and are designated T_{+1}, T_{-1} and T_0, respectively, where the subscripts are quantum numbers. It is the summation of two uncompensated like electron spins that results in a value of $S = 1$ for the triplet state and in the physically observable properties associated with paramagnetism.

To avoid confusion over the term 'energy of the triplet state', it is necessary to keep in mind that the Zeeman splittings of the triplet sublevels in the earth's magnetic field or even in the much higher field of an electron paramagnetic resonance spectroscopic experiment are very small (of the order of a fraction of a small calorie per mol) compared to most singlet–triplet separations (often 10 000–20 000 cal mol^{-1})[35]. For many chemical purposes, therefore, it does no harm to think of the 'singlet–triplet separation' as having one value rather than three.

The energies, E_1 and E_3, associated with the singlet and triplet states can be found by application of the Schrödinger equation. The singlet–triplet splitting emerges in a very natural way when the two-electron Hamiltonian is used, that is, when the full set of physical interactions (neglecting spin-orbit terms) in the three-particle He system (nucleus, electron (1) and electron (2)) is taken into account (equations 5 and 6)[36]. The energies $\bar{E}_3$ and $\bar{E}_1$ are expectation values (average energies), since $\psi_{s,a}\psi_{a,s}$ are one-electron wave functions and are not eigenfunctions of $H(1,2)$[33].

$$H(1,2)\psi_{s,a} = \bar{E}_1\psi_{s,a} \quad (5)$$

$$H(1,2)\psi_{a,s} = \bar{E}_3\psi_{a,s} \quad (6)$$

The solutions of equations 5 and 6 are given[33] by equation 7, in which the energies of the singlet and triplet are quadrisected into the components E_{1s}, E_{2s}, J and K. The terms E_{1s} and E_{2s} are the atomic orbital energies.

$$\bar{E}_{\substack{1\\3}} = E_{1s} + E_{2s} + J \pm K \quad (7)$$

The Coulomb integral has the form of equation 8, where r_{12} is the interelectronic distance, and the differentials $dv(1)$ and $dv(2)$ refer to the spatial coordinates of the electrons. Since all the terms in the integrand of equation 8 are positive, J is positive. As the name Coulomb integral implies, this energy term is electrostatic in nature and arises from the mutual repulsion of two charge clouds, one associated with electron 1 and described by 1s*1s, and the other associated with electron 2 and described by 2s*2s.

$$J = \int \int 1s^*(1)2s^*(2)(1/r_{12})1s(1)2s(2)dv(1)dv(2) \tag{8}$$

The exchange integral has the form of equation 9. Note that the two product functions in the integrand differ by an exchange of electrons, hence the name. The integral K gives the energy of interaction of an electron 'distribution' described by 1s*2s with another electron distribution of the same kind[33]. Although these are mathematical functions, not physically realizable electron distributions, the energy K is a real energy. An important purpose of the particular quadrisection in equation 7 is to divide up the energy so as to focus attention on the component responsible for the singlet–triplet splitting, namely K.

$$K = \int \int 1s^*(1)2s^*(2)(1/r_{12})2s(1)1s(2)dv(1)dv(2) \tag{9}$$

K is also an electrostatic term which turns out to be positive[36–38]. In the context of equation 7, this leads to the conclusion that $\bar{E}_1 > \bar{E}_3$, that is Hund's rule is obeyed. The energy separation between the singlet and the triplet is $2K$.

B. Physical Interpretation of Hund's Rule[39]

It is frequently stated that the energetic preference for the triplet state in excited He (and by extension, in other open-shell chemical systems) is the greater average separation of the electrons (r_{12}) which minimizes the interelectronic repulsion energy in that state. This circumstance is traced to the form of the space part of the wave function of the triplet, which is required to be antisymmetric by the Pauli principle, as already has been outlined (see equations 2–4).

A simple demonstration of the effect of spatial symmetry vs. antisymmetry is provided by the two-electrons-in-a-box model[40]. Imagine a one-dimensional box of length L containing two electrons whose position coordinates, x_1 and x_2, can be represented as fractions of L. The charge density in quantum mechanics is given by the square of the wave function, which will be ψ_+^2 for the symmetric space function of the singlet and ψ_-^2 for the antisymmetric space function of the triplet.

A contour diagram of the charge density distribution for the singlet shows two maxima, each of which occurs in a region where the two electronic position coordinates are equal (Figure 1a). In the case of the triplet, however, the charge density distribution peaks at points where the two electronic position coordinates are unequal (Figure 1b). In physical terms, the quantum mechanically enforced propinquity of the electrons in the singlet will result in a greater interelectronic repulsion energy. Another way of stating the matter would be to say that, at this level of approximation, the antisymmetric nature of the triplet wave function ensures efficient correlation of the electronic motions by keeping the electrons in the triplet farther apart, on the average, than is possible with the singlet. The separation between the two charge density distributions in the triplet is sometimes called a 'Fermi hole'.

This simple picture surely serves as a useful starting point for understanding the nature of the singlet–triplet splitting, but one should keep in mind that it is based upon an approximation, which is that the same set of atomic orbitals is used to compare the energies of open-shell singlet and corresponding triplet configurations. Qualitatively, the consequences of this approximation can be apprehended by dissection of the electronic energy of our model 1s2s configuration of He into its most important components. These consist

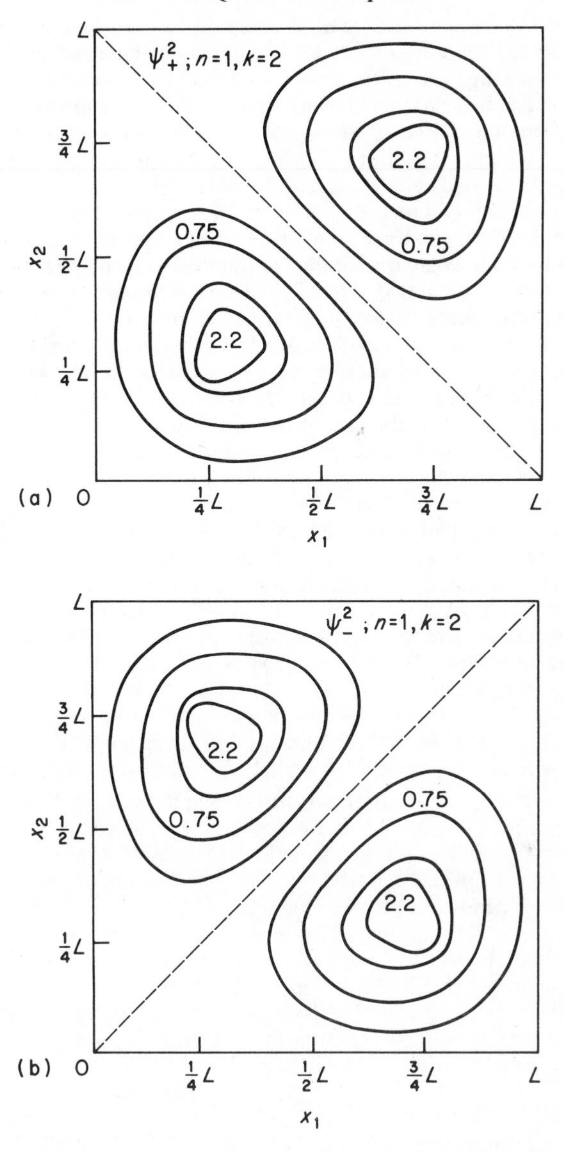

FIGURE 1. Contour diagrams of the charge density distribution for the singlet (a) and the triplet (b) of a hypothetical two-electrons-in-a-box system. *From Ref.* 35, *McGlynn, Azumi and Kinoshita,* Molecular Spectroscopy of the Triplet State, © 1969, *pp.* 69, 70. *Reprinted by permission of Prentice Hall, Englewood Cliffs, New Jersey*

of the two-electron portion (electron–electron repulsion) and the one-electron portions (kinetic energy and electron-nuclear attraction). Because of the approximation in which the atomic orbitals used for the singlet and triplet are the same, the one-electron components (kinetic energy and electron-nuclear attraction) remain the same for the two multiplets. As was outlined above, the only energy difference between the two is in the two-electron components. This difference is generated because of the Pauli principle and is expressed in the form of the exchange energy, K.

At this point, it can be seen[43] that the approximation has led to a violation of the Virial Theorem, which may be stated as $2T = -V$, where T stands for kinetic energy, and V stands for potential energy. The theorem enforces a zero-sum rule on the total energy of the system and states that one cannot change the potential energy without a concomitant change in the kinetic energy. To satisfy the Virial Theorem, the decrease in potential energy when the singlet becomes the triplet must be accompanied by an increase in the kinetic energy. The result of this would be that the optimized orbitals of the triplet would be more contracted than those of the singlet. This effect increases the electron–electron repulsion in the triplet but also increases electron-nuclear attraction.

In other words, a higher order of theory than the approximation first used would recognize the balance of forces affecting the spatial extent of the wave function as the system 'attempts' simultaneously to minimize kinetic energy and electron–electron repulsion, which is favored by dilation of the wave function, and to maximize electron-nuclear attraction, which is favored by its contraction. Several calculations of atomic multiplet states, including those of our model system, the 1s,2s configuration of excited He, show that the electron-nuclear attraction is the dominant factor in determining multiplet stability[41–48]. In the triplet state, the orbitals are contracted to the point where electron–electron repulsion actually is greater than in the singlet, but the destabilization resulting from this increased repulsion is more than counterbalanced by the increased electron-nuclear attraction. It should be noted that this relationship holds even in one of the few atomic systems where Hund's rule is violated[44]. Thus, in the lowest multiplet states of Mg, where ^{1}D falls below ^{3}D, a calculation shows[44] that electron repulsion is greater in the lower energy ^{1}D state.

Colpa[43] has suggested an instructive reformulation of the physical basis of Hund's rule in terms of several crucial inequalities. First, one performs an SCF calculation in a two-electron system with two singly occupied orbitals, A and B, using for the singlet state a wave function of the type $(1/\sqrt{2})[\mathrm{A}(1)\mathrm{B}(2) + \mathrm{A}(2)\mathrm{B}(1)] \times$ spin function. The results of this (and further) calculations are expressed with the notation S for singlet energy and T for triplet energy. A superscript stands for a calculation done using orbitals optimized for the indicated state; a subscript signifies the type of calculation. On this basis, the result of the above calculation is denoted $S^{\mathrm{S}}_{\mathrm{SCF}}$ (see Figure 2a).

Using these same orbitals (optimized for the singlet state), one then calculates $T^{\mathrm{S}}_{\mathrm{SCF}}$, the energy for the triplet state. Of course, had the calculation been done using the best triplet orbitals, one would have obtained a lower energy for the T state, $T^{\mathrm{T}}_{\mathrm{SCF}}$. $T^{\mathrm{S}}_{\mathrm{SCF}}$ is then an upper limit for the 'true' (i.e. best within the approximation) triplet energy. Therefore, if we take into account the singlet–triplet splitting for each set of optimized orbitals (see Figure 2 and equation 7), we may write equation 10:

$$T^{\mathrm{T}}_{\mathrm{SCF}} \leqslant S^{\mathrm{S}}_{\mathrm{SCF}} - 2K^{\mathrm{S}}_{\mathrm{AB}} \tag{10}$$

which may be rearranged to equation 11:

$$2K^{\mathrm{S}}_{\mathrm{AB}} \leqslant S^{\mathrm{S}}_{\mathrm{SCF}} - T^{\mathrm{T}}_{\mathrm{SCF}} \tag{11}$$

Similarly, if one were to perform the calculation using the best triplet orbitals in the triplet wave function, one would obtain the triplet energy, T^{T}_{SCF} (see Figure 2b). Using triplet-optimized orbitals to calculate the singlet energy, one would obtain the energy $S^{\mathrm{T}}_{\mathrm{SCF}}$. Again, this is an upper limit to the 'true' singlet energy, $S^{\mathrm{S}}_{\mathrm{SCF}}$. Therefore, we have the inequality of equation 12:

$$S^{\mathrm{S}}_{\mathrm{SCF}} \leqslant T^{\mathrm{T}}_{\mathrm{SCF}} + 2K^{\mathrm{T}}_{\mathrm{AB}} \tag{12}$$

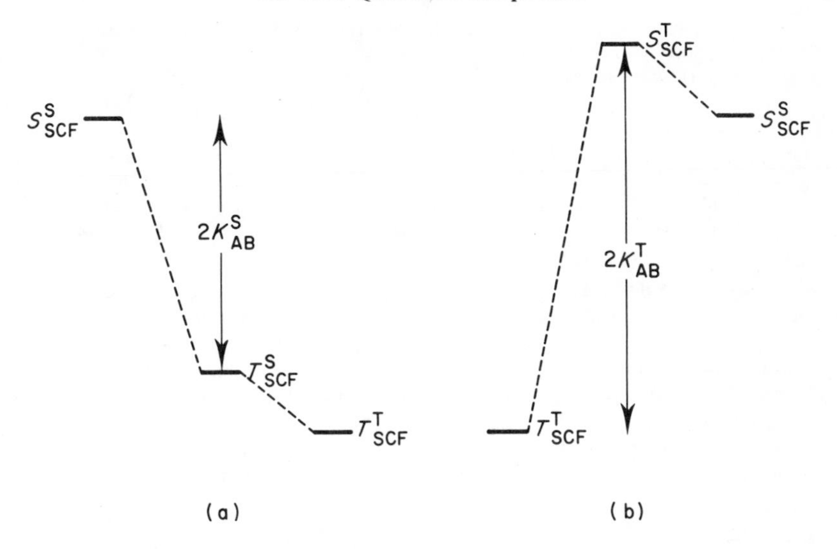

FIGURE 2. Relative energies of S and T states for different SCF calculations. K_{AB} is the exchange integral

Rearrangement of equation 12 gives equation 13:

$$2K^{T}_{AB} \geqslant S^{S}_{SCF} - T^{T}_{SCF} \tag{13}$$

Equations 11 and 13 now may be combined to give equation 14:

$$2K^{S}_{AB} \leqslant S^{S}_{SCF} - T^{T}_{SCF} \leqslant 2K^{T}_{AB} \tag{14}$$

In words, the singlet–triplet energy gap is larger than the exchange energy calculated with singlet-optimized orbitals and smaller than the exchange energy calculated with triplet-optimized orbitals. Since the exchange energy is necessarily positive[37], the 'true' singlet–triplet energy gap, $S^{S}_{SCF} - T^{T}_{SCF}$, must be positive, and Hund's rule is proved within this approximation. It is important that in this formulation, an assignment of the explicit origin of the difference in energy, that is, the relative contributions of electron–electron repulsion and electron-nuclear attraction, is not needed to derive the order of the states, which is attributable only to the fact that K is a positive quantity and that in calculating the energy of the spatially antisymmetric state, namely the triplet, we take the negative of K (see equation 7 and Figure 2).

For the present purpose, the effects of electron correlation have not been addressed. When this is done, the state of lower multiplicity, which is higher in energy, will be stabilized more than the state of higher multiplicity. This will have the effect of narrowing the S–T gap or even possibly inverting the order of states and leading to a violation of Hund's rule[49]. A further discussion, with emphasis on molecular rather than atomic systems, is given in Section III.F.

To strengthen the explication of Hund's rule, it may be useful to quote two succinct statements of the issues. According to Colpa and coworkers[45]

> . . . When we begin in an excited singlet state and go down in energy with frozen orbitals to an approximate triplet state, we use the argument that the triplet state has a lower energy solely because of its lower interelectron repulsion energy. The next step in our argument, the [orbital] relaxation, lowers the energy of the triplet state even further, but the electron repulsion may be raised so much

that there is no sign left of the original argument (lowering of repulsion) in the net result of the two steps taken together. In short, the second step strengthens the result of the first step but destroys the generality of the argument used in the first step.

To the same effect, McBride and Vary[50] point out

> First order stabilization [i.e., the decrease in energy of the higher multiplet due to the lesser extent of electron–electron repulsion assuming frozen orbitals] is very close to the true singlet–triplet difference, and its ultimate appearance in an unusual guise should not cloud the fact that the fundamental difference between the singlet and the triplet is the spatial antisymmetry of the latter, which keeps the electrons apart, other things being equal.

As long as the level of approximation is kept in mind, it would seem to be a matter of personal preference whether one chooses to ascribe the basis of Hund's rule to first-order electron–electron repulsion effects or to the more general formulation in terms of 'Pauli repulsion' as expressed by all of the components of the exchange integral K.

The profound and elegant systematization of atomic spectroscopy which culminated in Hund's rule was put forward 60 years ago[30]. The formulation of a simple qualitative understanding of its physical basis has developed slowly during the intervening time, and one must account the rule's author unduly modest when he remarks (in translation):

> The 'argument' for the validity of the rule (stronger antisymmetry of the space part of the eigenfunction is more favorable because of electron repulsion) first became apparent to me only later. Really, it should have been in *Z. physik*, **43**, 788 (1927). But I was just not that smart.[51]

III. THE *META*-QUINONE SERIES

A. Structure

General structural formulas for the biradical **19–21** and for the conceivable fully covalent isomers **22–27** of *m*-quinonoids are shown in Scheme I. In the cases of **22**, **26** and

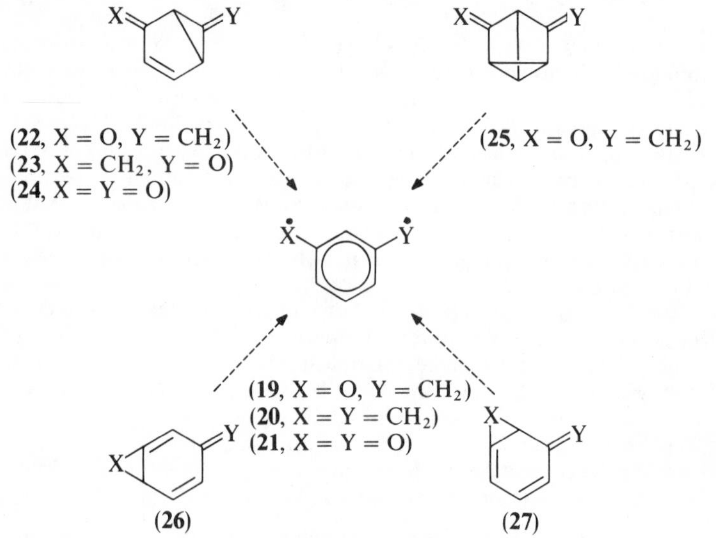

(**22**, X = O, Y = CH_2)
(**23**, X = CH_2, Y = O)
(**24**, X = Y = O)

(**25**, X = O, Y = CH_2)

(**19**, X = O, Y = CH_2)
(**20**, X = Y = CH_2)
(**21**, X = Y = O)

(**26**) (**27**)

X or Y = CR_2, NR, O, S, etc.

SCHEME 1

27, only one of a pair of isomers is shown: the other may be derived by interchange of X and Y. Structures **24** and **25** are equivalent to the original *m*-quinones hypothesized by Zincke (**1** and **2**). In principle, the biradical **19** could be derived by cleavage of one bond of **22**, **26** or **27**, or of both transannular bonds of **25**. Other fully covalent connectivities can be imagined, but they would violate Bredt's rule severely and are not shown. Structures **26** and **27** do not embody two exocyclic double bonds and thus have a dubious kinship to quinones. Although they are so far without example in the literature, Scheme I includes them as illustrations of potential alternative entry points onto the *m*-quinonoid energy surface.

Molecules of the group **22**–**27**, although they are expected to be strained and highly reactive, have conventional Kekulé structures and pose no special problems in valence theory. Structures **19**–**21**, however, are non-Kekulé species. They have enough atoms but not enough bonds to satisfy the standard rules of valence. These entities should be clearly differentiated from the more familiar reactive intermediates such as carbenes, carbanions, carbocations and radicals, which not only are bond-deficient but also are atom-deficient. Anticipating a later discussion, we note that the *m*-quinonoid non-Kekulé molecules are related in many ways to simpler biradicals such as trimethylenemethane (**28**) and tetramethyleneethane (**29**).

(**28**) 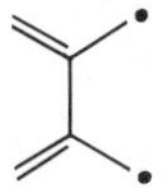 (**29**)

B. *meta*-Quinodimethane[32]

1. Theory

a. Hückel molecular orbital treatment. Theoretical description of *m*-quinonoids began in 1936 with Hückel's treatment[52] of the Schlenk–Brauns hydrocarbons and has continued to the present[28, 53–73]. As is the case with other biradicals[74], the *m*-quinonoid hydrocarbon **20** is characterized in the π-electron approximation by a pair of 'nearly degenerate', nominally non-bonding frontier molecular orbitals (NBMOs). Structures **20**, **28** and **29** are alternant, that is, the carbon atoms can be marked ('starred') in such a way that no two starred or unstarred positions are vicinal. For such systems (assumed for this purpose to be coplanar), Longuet-Higgins[56] showed by the application of Hückel π-electron theory that the number of NBMOs is always at least $N-2T$, where N is the number of π-electron centers, and T is the number of double bonds in the resonance structure with the maximum number of bonds. For most alternants, the number of NBMOs is actually predicted to be equal to $N-2T$. The only significant exceptions occur in the 4n annulene cases (for example, cyclobutadiene), where there are typically two NBMOs despite the fact that $N-2T=0$. Thus, all three molecules **20**, **28** and **29**, should have two NBMOs.

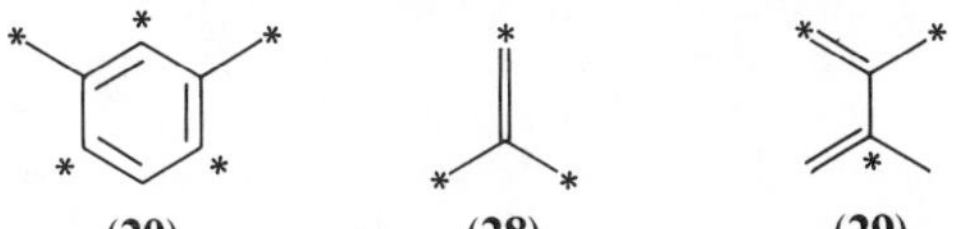

(**20**) (**28**) (**29**)

The number of π electrons in the neutral form of each of these systems is enough to provide double occupation for all the bonding MOs as the levels are filled by *aufbau*, but only two electrons are left to occupy the two NBMOs. There will be four zeroth-order configurations, two with single occupancy of each NBMO and either singlet or triplet spin configurations (Figure 3, a and b), and two with double occupancy of one or the other of the NBMOs (Figure 3, c and d). The latter two 'closed-shell' configurations will mix to produce two singlet configurations, whereas the 'open-shell' configurations will be distinguished by their spin as a singlet and a triplet.

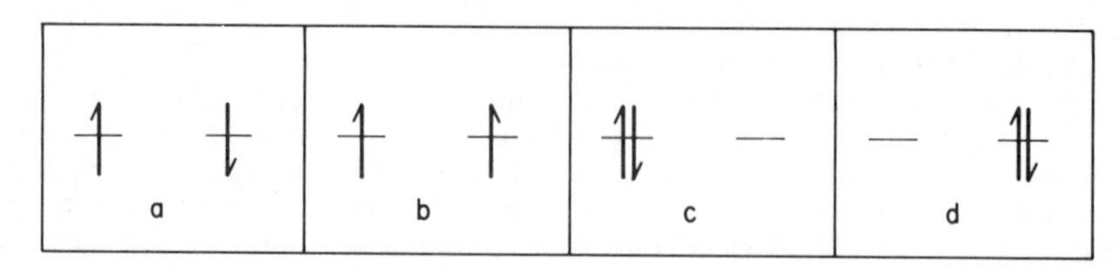

FIGURE 3. Zeroth-order states of a biradical. Configurations a and b correspond to 'open-shell' singlet and triplet, respectively, c and d are 'closed-shell' singlets

It now should be apparent why Longuet-Higgins's elegant theorem for predicting the number of NBMOs of alternant systems is significant. If Hund's rule applies to molecules, the diagram of Figure 3 would predict that the triplet would be the ground state of the neutral alternant system. In general, the total spin S of the ground state of such molecules will be half the number of NBMOs. The multiplicity, which is given by $2S + 1$, will be one more than the number of NBMOs.

Longuet-Higgins assumed that Hund's rule does apply to molecules, not only to atoms. Indeed, this assumption has a long history[51, 75–80], which can be traced back almost as far as the atomic Hund's rule itself. The satisfying explanation on this basis of the long-known paramagnetism of the ubiquitous molecule O_2 was one of the earliest triumphs of the molecular orbital theory[75–80]. In retrospect, one may say that Hund's rule sometimes was applied to molecules uncritically, but it is easy to understand that the O_2 case must have swept aside any lingering doubt. The combination of Longuet-Higgins's theorem with Hund's rule thus leads to the prediction that *m*-quinodimethane (**20**) should have a triplet ground state. A similar conclusion for the specific case of the Schlenk–Brauns tetraphenyl derivative (**8**) was reached by Hückel[52], although he did not generalize the result.

These predictions, of course, are based upon a very approximate level of theory. Except for the few cases in which the molecular symmetry is high enough to enforce it, the NBMO degeneracy of non-Kekulé alternants is 'accidental' and vanishes in any higher level calculation[81–84]. When even the crudest consideration is given explicitly to electron repulsion, the energies of the nominal NBMOs split. For example, even an INDO calculation on *m*-quinodimethane (**20**) shows the frontier orbitals to be separated by $\sim 15\,\text{kcal mol}^{-1}$. It is not obvious that Hund's rule should apply to such a case. Hoffmann[85] has suggested that unless the gap between the frontier orbitals is 1.5 eV ($34\,\text{kcal mol}^{-1}$) or more at the extended Hückel level of theory, the triplet should be the ground state. Otherwise, it would be energetically advantageous to put both electrons in the lower orbital, since the electron repulsion so generated would not suffice to offset the orbital separation. In that case, the ground state would be described as a closed-shell singlet. Corresponding rules of thumb for higher levels of calculation are not available, and the prediction of the relative energies of the singlet and triplet in non-Kekulé molecules is a major undertaking in computational quantum mechanics. Hückel theory, even in its extended form, does not explicitly distinguish the singlet and triplet energies, since it is a one-electron theory[85]. As already has been discussed in Section III.B.1, the exchange integral K, which determines the singlet–triplet splitting, emerges only when a two-electron (or multi-electron) Hamiltonian is used.

Since the analytical solution of the full Schrödinger equation for a molecule as large as *m*-quinodimethane (**20**) is not yet feasible, the task of theory becomes the choice of an approximate method. There are two older approaches, semiempirical[86] and *ab initio*[87]. More recently, a non-empirical Heisenberg Hamiltonian method has been applied to a few biradical systems[68].

b. Semiempirical treatments. The semiempirical molecular orbital methods all rely on a parameterization from experimental spectroscopic data to provide values for the needed core integrals and resonance integrals. Various levels of improvement can be achieved by configuration interaction (CI) calculations, in which singly, doubly, or even more highly excited electronic configurations are allowed to mix with the ground configuration. In most of the methods a reasonable molecular geometry is guessed by analogy to model structures.

The chief advantage of semiempirical MO methods for non-Kekulé problems are speed and economy. The disadvantages, however, are not to be overlooked. Perhaps the most fundamental of these is that the semiempirical nature of the wave functions means that they are not eigenfunctions of a correct molecular Hamiltonian but rather of an *effective* Hamiltonian. To derive energies, it is necessary to parameterize the integrals in order to include (implicitly) interelectronic effects. In such circumstances, the variational theorem does not apply, which has the serious consequence that one cannot be sure that a change of the parameterization scheme that leads to a lower calculated energy has necessarily produced an answer closer to the true energy.

Three semiempirical calculations of the energy separation between the lowest singlet and triplet states of *m*-quinodimethane are available. An SCF-CI calculation[60,61,88] with CI based on the singlet MOs using all singly and some doubly excited configurations leads to the conclusion that the triplet is the ground state by 7.8 kcal mol^{-1} [60]. Another approach[69] uses a full CI PPP (Pariser–Parr–Pople) calculation to predict a triplet ground state by 9 kcal mol^{-1}. Still another calculation[70] is based upon the INDO-S/CI program of Ridley and Zerner[89,90], which has been extensively checked[70] for reliability by comparison of its predictions to those of higher level theory or to experimental data for a number of non-Kekulé systems. Again, a triplet ground state for **33** is predicted. The lowest lying singlet state is $^1A_1$, which is calculated to be 15.0 kcal mol^{-1} above the $^3B_2$ ground state. Note that the $^1A_1$ state is largely of a closed-shell nature, despite the electron repulsion energy that one might imagine to be associated with a doubly occupied HOMO configuration. The reason for this can be found in the effect of CI. At the INDO/S level, before CI, the lowest singlet is found to be the open-shell type, $^1B_2$, but the CI treatment selectively stabilizes $^1A_1$ [70].

It should also be noted that CI sharply diminishes the S–T gap in the INDO-S/CI calculation by the selective stabilization of the singlet. This can be traced to the fact that singlet configurations can be of either open- or closed-shell types, but triplet configurations must be open-shell. Thus, there are many more excited singlet configurations in any practical CI active space. If other factors are equal, mixing of configurations of like multiplicity will be more effective in lowering the energy of that multiplet.

c. Non-empirical treatment. The so-called non-empirical Heisenberg valence bond Hamiltonian approach[66] extracts the effective Hamiltonian H^{eff} from a high quality *ab initio* calculation on ethylene, using extended basis sets of double zeta plus d quality and a CI scheme involving up to 10^5 configurations. This procedure provides a model for the σ potential and the effective exchange integral between adjacent atoms which can be applied to other conjugated hydrocarbons. Since the method does not incorporate experimental data, it would not be correct to refer to it as 'semiempirical', yet it is not *ab initio* either in a direct sense. The method actually uses parameters derived from an *ab initio* calculation, so that it occupies a territory between the two older schemes. It also makes provision for direct and efficient geometry optimization. This method predicts a triplet ground state for *m*-quinodimethane (**20**) 28 kcal mol^{-1} below the $^1A_1$ state. This separation is in the same direction as, but substantially larger than the values of 8, 9, 15 and 10 kcal mol^{-1} predicted by the semiempirical and *ab initio* methods.

d. Ab initio methods with large basis sets for a molecule as large as *m*-quinodimethane (**20**) require massive computational capacity and are far from commonplace. A major effort on **20**[67] employed the Dunning split valence basis functions and an extensive CI routine, after geometry optimization at the SCF level, to produce the predicted triplet ground state, 10 kcal mol^{-1} below the $^1A_1$ state. The (largely) open shell $^1B_2$ state is found[67] to be much higher in energy than the $^1A_1$ state, as is observed in the INDO-S/CI calculation[70].

e. Prospects. Although the semiempirical methods often give quite reliable results for non-Kekulé molecules, there does not seem to be general agreement[91, 92] on whether they are to be regarded as substitutes for the much more expensive high-level *ab initio* theory. At least, they seem destined to be used as a guide to experiment, an indicator of what magnetic and chemical properties might be reasonably expected of new molecules, and an aid in the choice of synthetic non-Kekulé targets which may serve as tests of theory. The non-empirical method has not yet been extensively tested but should ultimately afford many interesting comparisons with other methods. One looks forward also to the expansion of computational power that will permit *ab initio* calculations for many more non-Kekulé systems.

f. Non-planar geometries. All of the above calculations on *m*-quinodimethane have imposed a planar geometry on the entire π-electron system. Superficially, this might be expected to be the most stable geometry since it preserves the conjugative stabilization. However, it is known[82, 93–97] from *ab initio* calculations on another conjugated biradical, specifically trimethylenemethane (**28**), that the electron–electron repulsion relieved by twisting one methylene group of the singlet is sufficient to overcome the loss of conjugation energy. This results in a closely matched pair of singlet states, of which one is planar and the other, usually at a slightly lower energy, is non-planar. The triplet states of conjugated biradicals prefer to preserve their planarity[57, 93–98]. The difference in the degree of preference for the planar configuration by the two multiplets can be understood in terms of the better electron correlation in the planar triplet, which accomplishes with the Fermi hole what can only be achieved by a structural distortion in the singlet[82, 99].

Dewar and Holloway[100] have recently proposed on the basis of MNDO-CI semiempirical calculations that a biradical with the doubly twisted geometry (symmetry designation $^1B_2$) is the lowest energy singlet of the non-Kekulé molecule *m*-quinodimethane (**20**). The diagonalized MNDO force constant matrix shows this state to

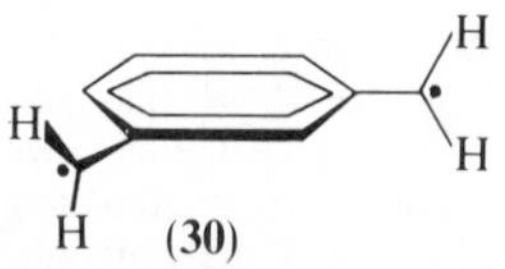

H H H H (**30**) Ȯ (**31**)

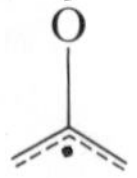

be in an energy minimum on the singlet surface. This is a most interesting result and opens the possibility that such species may contribute to the chemistry of *m*-quinodimethane. In the absence of experimental data, the reliability of this finding is difficult to evaluate, but it should be pointed out that a similar calculation[100] found planar singlet ($^1A_1$) oxyallyl (**31**) to be a saddle point on the energy surface, rather than a minimum, as was found in the most recent *ab initio* calculations of Osamura and coworkers[101].

2. *Spectroscopy*

a. Optical absorption and emission. Schlenk and Brauns[13] reported that the deep violet solution obtained by heating tetraphenyl-*m*-xylylene dichloride (**7**) in benzene with copper–bronze showed an intense absorption band between 576 and 636 nm. A likely candidate for the absorbing species is tetraphenyl-*m*-quinodimethane (**8**), although the spectrum does not provide much structural information. However, modern techniques have contributed significantly in this regard.

The optical spectroscopy of the unsubstituted *m*-quinodimethane (**20**) was first studied by Migirdicyan[102]. Porter and Strachan[103] had shown that ultraviolet photolysis of

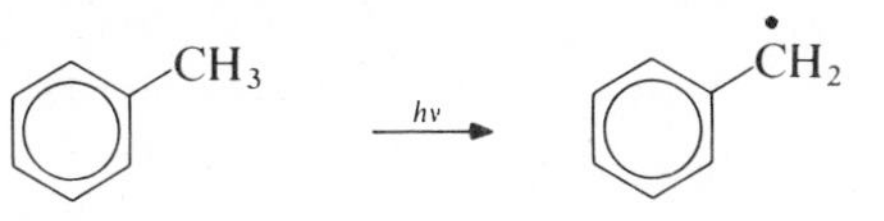

(32) → hν → (33)

(34) → hν → (20) ← hν ← (35)

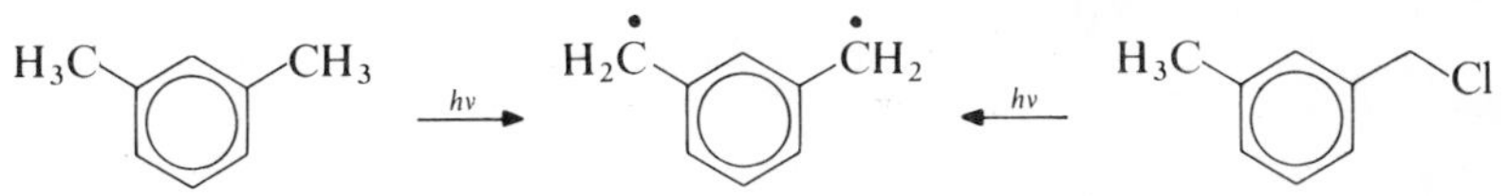

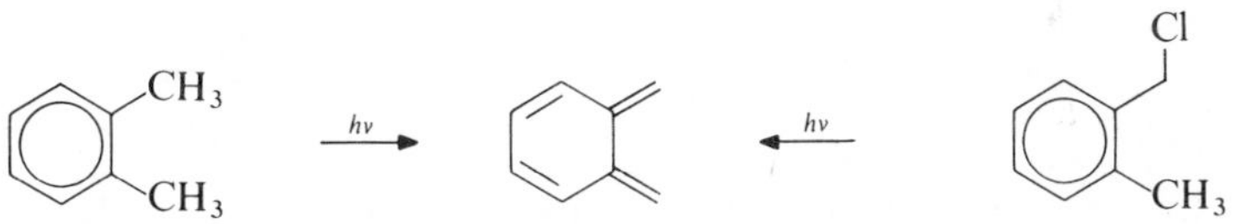

(36) → hν → (38) ← hν ← (37)

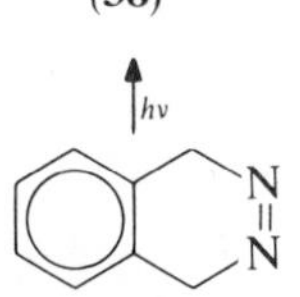

(39) → hν → (38)

toluene (**32**) in rigid media was capable of cleaving a CH_2–H bond to give benzyl radical (**33**). This species was stable when thus trapped in the solid and could be observed by absorption spectroscopy. Extending these observations, Migirdicyan found that photolysis of *m*-xylene (**34**) or α-chloro-*m*-xylene (**35**) in methylcyclohexane glass at 77 K gave a common trapped species whose emission spectrum was entirely different from those obtained from the *ortho* counterparts **36** and **37**. Photolysis of the latter two species gave the Kekulé polyene *o*-quinodimethane (**38**), which Flynn and Michl[104] later independently generated from the diazene **39** and other precursors and showed to have the same spectroscopic properties.

A subsequent report[61] described the photolysis of *m*-xylene (**34**) in polycrystalline *n*-pentane (Shpolskii matrix), a medium that favors the observation of sharp vibrationally resolved electronic spectra. The photolysis produces both *m*-quinodimethane (**20**) and the *m*-methylbenzyl (*m*-xylyl) radical. Although the latter species has a strong fluorescence spectrum lying to the red side of 4740 Å, the fluorescence of **20** fortunately can be observed in the narrow wavelength range 4400–4700 Å. Figure 4 displays the two spectra shifted with respect to each other so that their 0, 0 bands coincide. Notice that the vibrational progressions are very similar. Moreover, the two main vibronic bands of *m*-quinodimethane (at 530 and 988 cm^{-1}) and those of *m*-methylbenzyl radical (at 521 and

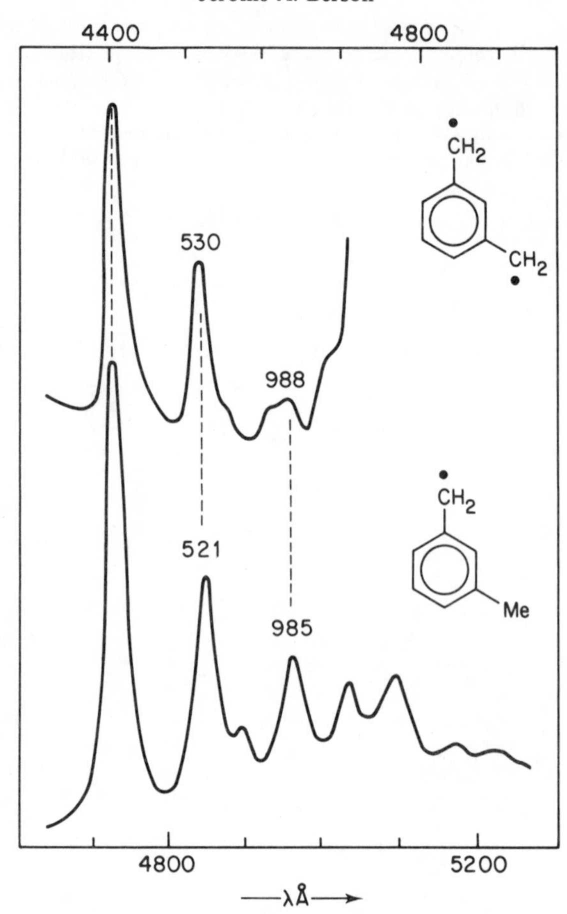

FIGURE 4. Fluorescence spectra of *m*-quinodimethane biradical (upper curve) and *m*-xylyl monoradical (lower curve, $\lambda_{exc} = 3230$ Å) produced by photolysis of *m*-xylene in pentane at 77 K. *Reproduced with permission from E. Migirdicyan and J. Baudet*, J. Am. Chem. Soc., **97**, 7400 (1975). *Copyright* (1975) *American Chemical Society*

985 cm^{-1}) have frequencies very similar to those of ground state vibrational modes of *m*-xylene. The optical spectroscopic data alone do not suffice to identify the spin states involved in the observed electronic transitions, and the quantum mechanical calculations[61] are not decisive in this regard. However, the subsequent experimental demonstration of a triplet ground state for **20** (see the next section) supports the assignment[105] of the electronic spectra to triplet–triplet transitions.

Further information on the excited triplet states of *m*-quinodimethane and its methylated derivatives is provided by site-selective laser fluorescence-excitation spectroscopy at 5–10 K. It is concluded that there are two close-lying excited states in these triplet molecules which can interact vibronically. Computational support (SCF MO/CI) for these assignments also is put forward[105–107].

b. Characterization of the spin state of non-Kekulé compounds. The experimental determination of the spin of the ground state of these species is not a routine matter. Techniques appropriate to the properties of specific molecules include chemical trapping, measurements of magnetic susceptibilities, chemically induced nuclear polarization and probably most often, electron paramagnetic resonance spectroscopy.

i. Magnetic susceptibility. The application of magnetic susceptibility measurements to the Schlenk–Brauns hydrocarbons has already been mentioned in Section II. A substance with one or more unpaired electrons will possess a permanent magnetic dipole moment because of the net electron spin. This causes it to have a positive magnetic susceptibility (paramagnetism) and therefore to be drawn into the more intense part of an inhomogeneous magnetic field[108, 109]. The paramagnetism will be proportional to the number of unpaired electron spins. However, the paired electrons in such a molecule (or any molecule) contribute a negative magnetic susceptibility (diamagnetism), which tends to push the sample out of the field. In most substances with net spin $S > 0$, the paramagnetism overbalances the diamagnetism, but a quantitative estimation of the paramagnetism requires that the observed net susceptibility be corrected for the diamagnetic contribution. The susceptibilities are usually measured with some form of Gouy balance, which is capable of determining the gain in apparent weight experienced by a paramagnetic substance in the magnetic field. For various reasons[108, 109], these measurements and corrections are difficult to carry out with a high degree of accuracy.

An even more serious problem with the magnetic susceptibility technique is that it is not sensitive to the structure of the paramagnetic species, since it measures only the net paramagnetism. Frequently (see below) the chemical or photochemical generation of a paramagnetic (e.g. triplet) species of interest is inadvertently accompanied by the formation of (often unidentified) radical impurities. The magnetic susceptibility measurement reports the total paramagnetism of the sample, a quantity from which it usually will not be possible to extract the information of interest, namely the paramagnetism of the triplet species.

ii. Electron paramagnetic resonance (EPR) spectroscopy[35, 110–115]. EPR spectroscopy, which became widely available in the 1950s, proved to be a powerful and decisive technique for characterization of the structure of paramagnetic species. Chemical substances with total electron spin $S > 0$ are paramagnetic and are required by quantum mechanics to exist in discrete spin substates in an applied magnetic field. The number of such substates (multiplicity) is given by $2S + 1$. Thus, a monoradical, with $S = \frac{1}{2}$, constitutes a doublet of states, usually referred to in abbreviated form as a 'doublet'. Although the outlines of EPR spectroscopy of doublet species probably are familiar to many chemists, a brief review will help to introduce the less familiar topic of EPR spectroscopy of high-spin species. The discussion that follows concerns entities that are adequately described by a time-independent spin Hamiltonian, a proviso that applies to the great majority (if not all) of the cases that will concern us here.

(a) The Zeeman effect. It should be obvious that substances with $S = 0$ cannot have EPR spectra, because a multiplicity of unity means that only one spin level exists at any field strength, and no other state can be populated by absorption of microwave radiation.

When $S = \frac{1}{2}$, two sublevels exist, whose energy separation increases with the strength (H) of the applied magnetic field (Zeeman effect, Figure 5). Because of the negative charge on the unpaired electron, the lower energy sublevel is the one with the electron's spin vector aligned against the field ($M_S = -\frac{1}{2}$), as can be seen from equation 15:

$$E = g\beta M_S H \tag{15}$$

The symbol g is the electronic g-factor, β is the Bohr magneton, H is the magnetic field strength and M_S is the spin quantum number of the doublet sublevel, which can take on the values $\pm\frac{1}{2}$. As H is increased, equation 15 predicts that the doublet sublevels should

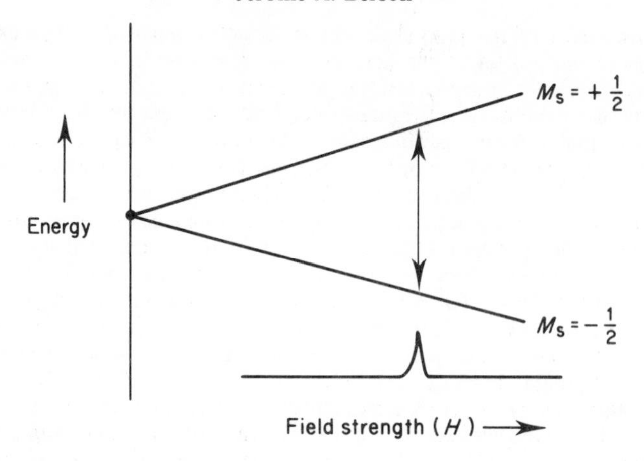

FIGURE 5. Energies of the sublevels of a doublet state as a function of field strength. The double-headed arrow shows the resonance condition for a particular microwave frequency. The transition represents a change in spin quantum number of $\Delta M_s = \pm 1$. *Adapted with permission from Ref.* 111, *p.* 13

separate in energy, the $M_S = -\frac{1}{2}$ state going down and the $M_S = +\frac{1}{2}$ state going up along lines whose slopes are $M_S g\beta$ (Figure 5).

The selection rule $\Delta M_S = \pm 1$, $\Delta M_I = 0$ (M_I is the nuclear spin quantum number) governs the transition. The ΔM_S rule is self-evident in the doublet case, but it applies also to higher spin systems. The probability of a transition depends upon the difference in populations of the two states, which usually (but not invariably) is controlled by a Boltzmann distribution.

For practical reasons, the resonance condition is met in a typical EPR experiment by holding the frequency of electromagnetic radiation constant and sweeping the field. Most of the experiments use X-band spectrometers, for which the typical operating parameters are a frequency in the microwave region (near 9 GHz) and a field strength up to about 10 000 gauss. At these field strengths, the sublevels are separated by a fraction of a small calorie per mol. A transition can occur, either by absorption of a microwave quantum by the lower state or by emission of one by the higher state, whenever the separation between the levels matches the energy of the quantum. Detailed information about the structure of the radical may be deduced from the hyperfine splitting of the electronic spin levels by the spin magnetic moments of the magnetically active nuclei (especially protons) in the molecule.

(*b*) *The spin–dipolar interaction and zero-field splitting.* In high-spin systems, in which two or more unpaired electrons are present, an additional complication is introduced, because each electron now is affected not only by the applied magnetic field but also by the field associated with the spin of the other unpaired electron(s). For reasons to be given, this is called the spin–dipolar interaction. The electron spin energies also are affected by another phenomenon, the spin–orbit interaction, but in most of the instances discussed here, this is a minor perturbation.

Although it adds conceptual complexity, the spin–dipolar interaction actually is a delightful boon, because it forms the physical basis for the so-called zero-field splitting (ZFS) of the sublevels of a high-spin molecular state. The ZFS is responsible for the characteristic features of multiplet EPR spectra which provide structural information about the molecular carrier of the signal. A physical picture of the spin–dipolar interaction may be developed by following the discussion of El-Sayed[116].

It is useful to think of a spinning electron as generating a spin dipole. Since the dipoles of the unpaired electrons of a high-spin system all have the same directionality, the dipole–dipole coupling is repulsive. Imagine an atom in a triplet state in which the distribution of the two unpaired electrons is spherical (state symmetry designation ^3S). Assume the absence of an applied magnetic field. We may select arbitrarily three mutually perpendicular planes, YZ, XZ and XY. The spin of the two electrons may be quantized so that the component of the spin angular momentum in the direction perpendicular to the chosen plane is zero. For example, if the two electrons are considered to have their spin axes confined to the XY plane, the component of spin angular momentum along the Z direction is zero. The quantum state so defined may be called the T_Z zero-field state. Similarly, the T_Y and T_X zero-field states correspond to magnetic sublevels in which the two unpaired electrons are spinning in (that is, have their spin axes confined to) the XZ and YZ planes, respectively, and have zero components of spin angular momentum in the Y and X directions. Because of the spherical symmetry of the electron distribution, the average distance of the two parallel spins in the different planes is the same. Therefore, the repulsive magnetic dipolar interaction between the two electrons is independent of the plane that happens to contain their spin angular momentum vectors. Consequently, the zero-field states T_Z, T_Y and T_X all have the same energy (Figure 6a) in the absence of an applied magnetic field.

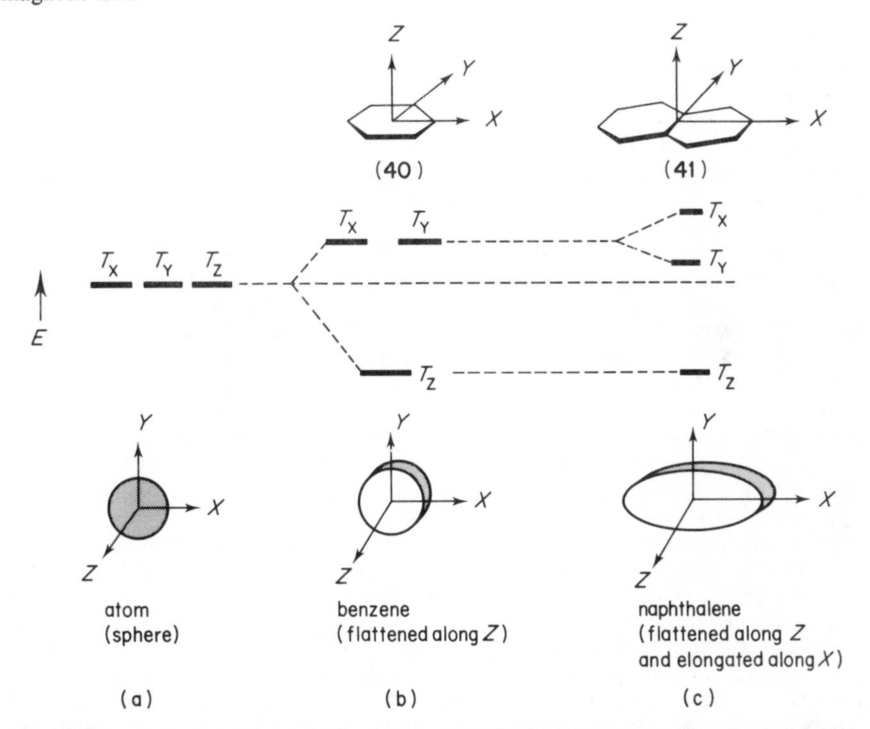

FIGURE 6. The origin of zero-field splitting in molecules. The average distance, and thus the magnetic dipolar couplings, will depend upon the plane that contains the spin angular momentum vectors of the two electrons. In the states T_X, T_Y and T_Z, these planes are respectively, YZ, XZ, and XY. The energy scale is schematic. *Taken from M. A. El-Sayed, Ref.* 116, *and copied with permission*

When two electrons with unpaired spins are present in a molecule rather than in a ^3S state of an atom, the spherical symmetry of the electron distribution is broken. In the triplet state of benzene **40** (Figure 6b), for example, the electron distribution can be modeled by the flattened tablet of Figure 6b. The average distance of the two parallel spins in plane YZ is the same as that in plane XZ, but the separation is greater in plane XY. Consequently, the state T_Z is stabilized by the resulting diminished dipolar repulsion, whereas states T_Y and T_X are destabilized. This splitting of the state energies is the zero-field splitting. It results directly from the anisotropic nature of the electron spin dipolar coupling in the molecule. In other words, the molecular geometry has quantized the spin levels. A moment's reflection makes obvious the potential utility of the converse form of this reasoning to deduce elements of molecular structure of high-spin molecules from ZFS data.

The flattening shown in Figure 6 is oblate and tends to spread out the electron distribution in the plane of the two equivalent axes (X and Y) and compress it along the unique direction. If the distortion were to be imposed in a prolate sense, by stretching the sphere along the unique axis Z, it would have the effect of concentrating the electron distribution in the XY plane, thereby raising the energy of T_Z and lowering that of T_X and T_Y. The state ordering shown in Figure 6 corresponds to a positive D-value (see below), whereas the prolate distortion would lead to a negative D-value. Examples of the latter kind in molecular triplets are not common, but several cases exist in radical pairs immobilized in rigid media[50, 117].

Figure 6b shows that a two-fold degeneracy of spin states at zero field persists in a hypothetically hexagonally symmetric benzene triplet as a consequence of the directional equivalence of the X and Y axes. This degeneracy is characteristic of any triplet species whose geometry embodies a magnetically isotropic plane (e.g. XY in Figure 6b) and an axis perpendicular to it of three-fold or higher symmetry.

Actually, the EPR and ENDOR spectra of the $^3B_{1u}$ photoexcited state of benzene (C_6H_6) in C_6D_6 host crystal cannot be interpreted in terms of species with the expected D_{6h} symmetry[118-120]. It appears that in this molecule and also in triplet mesitylene[118], under the stated conditions, slight geometric distortions lower the symmetry. The discussion based on Figure 6 therefore is idealized.

The remaining degeneracy vanishes when the molecule's shape is further distorted by elongation along one of the axes X or Y. Figure 6c shows the energies of the zero-field spin sublevels of naphthalene triplet, **41** (Figure 6c), whose π-electron distribution can be modeled by the lozenge shape obtained by elongation along X.

(c) *EPR spectroscopy of high-spin systems.* EPR is the most widely used technique for measuring the ZF energy separations. The details of procedures for extracting the ZFS parameters, D and E, from EPR systems of high-spin molecules have been reviewed[35, 110-115] and need not be repeated here, but for a qualitative understanding of these experiments, it will be useful to call attention to a few important elementary concepts.

We return to our simple atomic triplet system with a spherically symmetrical electron distribution. As we saw in Figure 6a, the three triplet levels are degenerate at zero field. What would happen if this species were placed in a magnetic field whose lines of force lie along the Z-axis? Then the electrons whose spin vectors are in the XY plane (perpendicular to Z) would remain with equal energy in all directions in this plane, but because of the negative charge on each electron, those in the YZ and XZ planes would be of higher energy when aligned with the field and of lower energy when aligned against it. Quantum mechanically, the energy is again given by equation 15, but now, M_S is the spin quantum number of the triplet sublevel, which can take on the values -1, 0 and $+1$ corresponding to spin alignments against, perpendicular to, and with the field direction. As H_Z is increased from zero, equation 15 predicts that the triplet sublevels should separate in energy. One of them, characterized by $M_S = 0$, will maintain its zero-field energy, but the $M_S = -1$ and

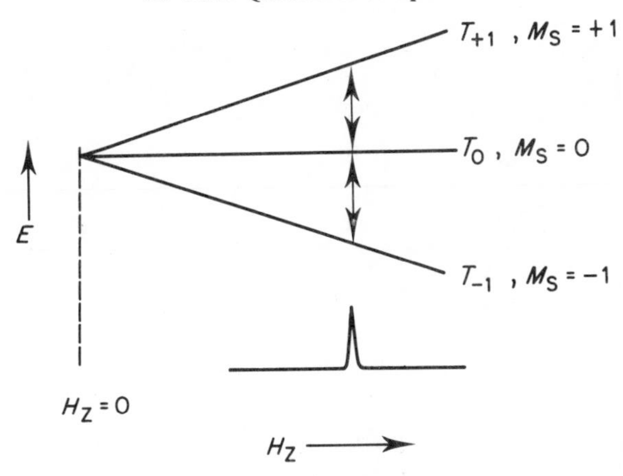

FIGURE 7. Zeeman splitting of the electronic spin sublevels of a spherically symmetrical triplet in a magnetic field of strength H_Z. The 'allowed' $\Delta M_S = \pm 1$ transitions coincide

$M_S = +1$ states will go down and up, respectively, along lines whose slopes are $M_S g\beta$ (see Figure 7). Because of the spherical symmetry, the splitting pattern will be identical to that in Figure 7 when the external magnetic field is applied along the X or Y axis. Thus, the EPR spectrum of an immobilized spherically symmetrical triplet will be isotropic.

As already has been mentioned in connection with doublet spectra, the selection rule for an EPR spectroscopic transition is $\Delta M_S = \pm 1$. Because the splitting of the spherically symmetrical triplet sublevels of Figure 7 is symmetrical about the zero-energy line, the $\Delta M_S = \pm 1$ transitions between either nearest-neighbor pair of levels will coincide, so that only one line will be observed.

This simple picture is changed in a triplet molecule where the electron distribution no longer is spherically symmetrical. As we have seen, this produces a separation of the triplet sublevels, even at zero field. Imagine that we have a rigidly oriented triplet benzene molecule and we apply a magnetic field, H_Z. What happens to the zero-field energies as the field is increased? To answer this question, we must be aware that (except for the case of a linear molecule in a field parallel to the axis), the zero-field states are not eigenfunctions of the high-field spin Hamiltonian. The high-field eigenfunctions are the basis functions of the triplet state, T_{-1}, T_{+1}, and T_0, but the zero-field eigenfunctions are mixtures of the basis functions (equations 16 and 17).

$$T_X = 1/\sqrt{2}\,(T_{-1} - T_{+1}) \tag{16}$$

$$T_Y = i/\sqrt{2}\,(T_{-1} + T_{+1}) \tag{17}$$

$$T_Z = T_0$$

Although the triplet basis functions are well described with the spin quantum numbers $M_S = \pm 1, 0$, this is not true of the zero-field states.

As the field increases, the major cause of the energy separation between the states gradually shifts from the molecular quantization associated with the spin–dipolar interaction to the field quantization associated with the Zeeman effect. Correspondingly, the zero-field eigenfunctions T_X, T_Y and T_Z evolve into the high-field eigenfunctions $T_{\pm 1}$ and T_0. This process is depicted in Figure 8. Note that the state T_Z which is destined to become T_0, does not change its energy. This property is similar to that of the T_0 state in the

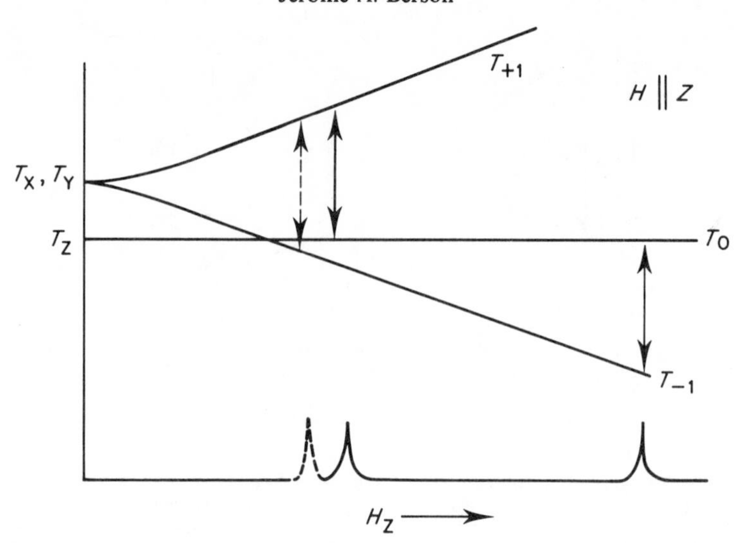

FIGURE 8. Zeeman splitting of the electronic spin sublevels of a triplet with $\geqslant$ 3-fold rotation axis (e.g. benzene) in a magnetic field of strength H_z. The $\Delta M_s = \pm 1$ transitions occur at different field strengths

spherically symmetrical system of Figure 7. The states T_X and T_Y which ultimately become $T_{\pm 1}$ are split symmetrically by the Zeeman effect, but because their zero-field energies are different from that of T_Z, the energy separation that corresponds to the microwave frequency occurs at a different field strength for T_{+1} and T_{-1}. Thus, two transitions will be observed at the positions marked by the full arrows, where the energy of the microwave just fits between sublevels whose spin quantum numbers differ by one unit (hence the name '$\Delta M_S = \pm 1$ transitions'). Note that there is one other position (marked by the dashed arrow) at which a fit can be made, but it connects two levels that differ by $\Delta M_S = 2$. This transition is formally forbidden if it falls at a field position high enough to make M_S a 'good' (i.e. well-defined) quantum number. Full field quantization may occur when the zero-field splitting in the triplet is very small, but frequently, it is not achieved throughout the experimental region, and the $\Delta M_S = \pm 2$ transitions are seen as a (usually weak) resonance at a field approximately half that of the center of gravity of the $\Delta M_S = \pm 1$ transitions. These 'half-field transitions' are especially diagnostic of a triplet state, because by definition only such a species has the three sublevels required for ΔM_S to change by two units.

If the magnetic field is imposed along one of the other two equivalent axes of the hypothetical benzene triplet, i.e. perpendicular to the Z axis, the Zeeman splitting shown in Figure 9 is observed. Note that the two $\Delta M_S = \pm 1$ transitions occur at field positions different from those in Figure 8. Therefore, for oriented samples of an axially symmetric triplet a total of four EPR lines may be expected, two in each of the canonical orientations. It is an extremely useful quirk of nature that these four lines also can be observed in randomly oriented samples in which the axes of the triplet molecules are disposed at all possible angles to the applied field (see below).

Using the principles outlined for the Zeeman effects in the axially symmetric case, it is quite straightforward to sketch the state energies as a function of field strength (Figure 10)

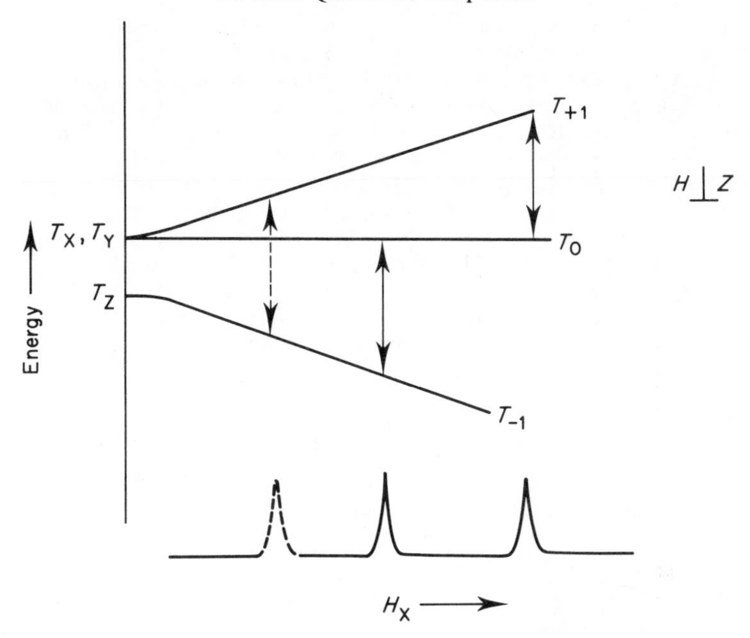

FIGURE 9. Similar to Figure 8, with the applied field perpendicular to the Z axis

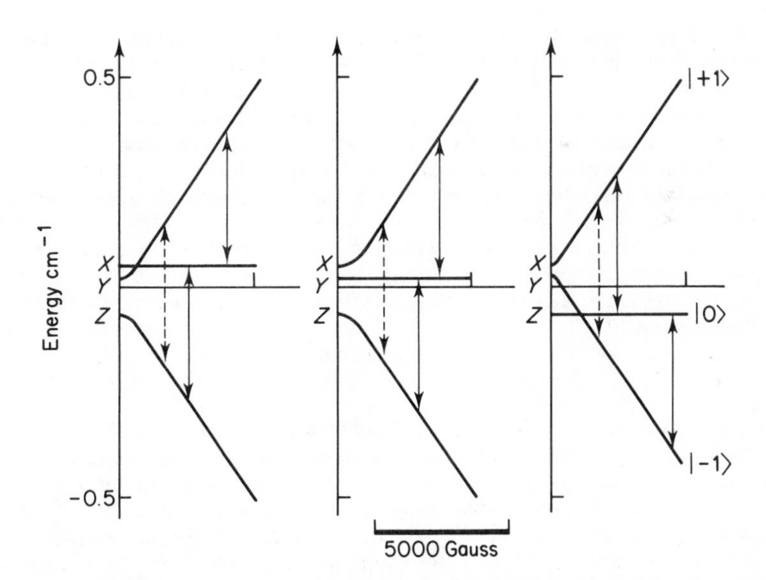

FIGURE 10. Zeeman splitting of the electronic spin sublevels of the triplet state of a molecule with $D > 0, E < 0$, and $|D| \neq 3|E|$. An example is naphthalene (see Figure 6c). *Reproduced with permission from A. Carrington and A. D. McLachlan,* Introduction to Magnetic Resonance, *2nd edn, Harper and Row, New York,* 1978, *p.* 121

expected for the three canonical orientations of the non-symmetrical case, e.g., naphthalene, whose zero-field energies are shown in Figure 6c.

It turns out that only two parameters, D and E, are needed to characterize the zero-field splittings. If the spin–dipolar approximation is appropriate for the spin system under observation (i.e. if spin-orbit coupling can be neglected) the parameter D measures the average propinquity of the spin dipoles. It is inversely proportional to the cube of the distance between them. The parameter E is related to the molecular symmetry. E is necessarily zero for an axially symmetric case (see Figures 4b and 8) but may be accidentally zero in other cases as well. The signs of D and E, which determine the state orderings, are not available directly from the EPR spectra, and, the parameters usually are reported in absolute magnitudes, $|D|$ and $|E|$. In the special circumstance that $|D| = 3|E|$, the spectrum collapses from six lines to three. These relationships are summarized in Table 1, where the number of lines listed is the sum of those observable in the three canonical orientations.

TABLE 1. Characteristics of EPR spectra triplets

Molecular symmetry	Number of lines from dipolar coupling
Axially symmetric	4
Unsymmetric $|D| \neq 3|E|$	6
Unsymmetric $|D| = 3|E|$	3

The first observations of the EPR transitions of a canonically oriented triplet molecule were reported in a landmark paper by Hutchison and Mangum[121]. By incorporating a small amount of naphthalene in a host single crystal of durene[122] and irradiating the sample in the EPR microwave cavity, they were able to obtain orientationally dependent spectra of excited triplet naphthalene from which the principal molecular axes could be deduced. The success of such experiments depends upon the similarity in shape between the guest and host molecules, which permits substitution of the guest in the crystal lattice of the host.

Superficially, it might be imagined that observation of EPR spectra of randomly oriented triplet molecules would be difficult. The effective magnetic field (H_{eff}) at one of the electrons (e_1) in such an experiment when the field is applied in a fixed direction (H_z), is given by equation 18,

$$H_{eff} = H_z + \frac{\mu_e (3\cos^2\theta - 1)}{r^3} \tag{18}$$

where θ is the angle between the applied field and the vector connecting the two electronic spin dipoles (e_1 and e_2) and r is the scalar magnitude of the vector (see Figure 11). The uncertainty principle acts to ensure that when the triplet molecules are tumbling rapidly on the time-scale of the EPR transition, the angularly dependent second term of equation 18 averages to zero, which has the effect of eliminating the spectroscopic consequences of the dipolar coupling. However, when the molecules are fixed, even if randomly oriented, the anisotropic contributions to H_{eff} still exist. Therefore, characteristic EPR spectra of high spin molecules in powders or frozen glassy or polycrystalline media can be observed, as was first noted by Burns[123] for the inorganic species Cr^{3+} ($S = 3/2$) and Mn^{2+} ($S = 5/2$), and for organic triplets by Yager and coworkers[124].

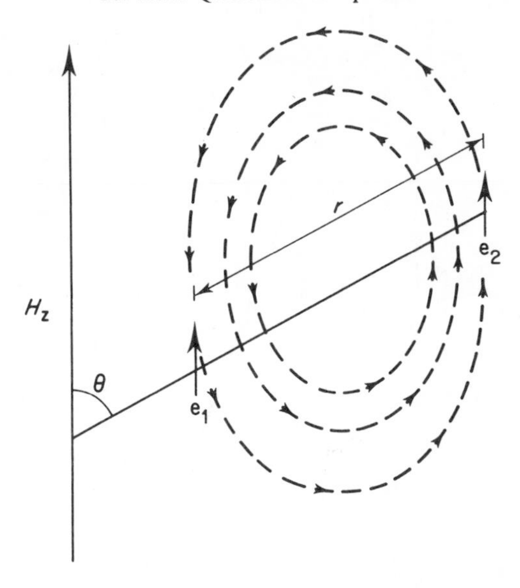

FIGURE 11. Spin–dipolar interaction of two electrons in an applied field H_z. The diagram shows magnetic lines of force emanating only from one side of electron 2, but a similar set is to be imagined on the other side

Although it is true that in a randomly oriented sample, all orientations (θ) of the interelectronic vector with respect to the applied field are equally probable, it does not follow that their contributions to H_{eff} are equal, because this contribution depends not directly on θ but rather on $3\cos^2\theta - 1$. It is a property of this function (see Figure 12) that its value changes very little near $\theta = 0°$ and near $\theta = 90°$. Near each of these angles there will be an unusually wide range of orientations all of which correspond to almost the same value of $3\cos^2\theta - 1$. These angles therefore tend to 'gather' resonances because the probability of a transition there is much higher than elsewhere. It is this 'lumpy' distribution of transition probabilities that preserves the spectral anisotropy, despite the 'smooth' distribution of orientations. These 'turning points' correspond to the transitions that would be observed in canonically oriented spectra, so that the aspects of the spectra predicted in Table 1 apply also to the randomly oriented case.

(*d*) *Ground state multiplicity from EPR spectroscopy*[125]. Although a substantial body of literature exists[126] on the experimental determination of the absolute and relative energies of the lowest singlet and triplet photochemically excited states of ordinary closed-shell molecules, very few measurements of these energies are available for biradical or multiradical low-lying or ground states. The reasons for this difference are not difficult to find, since the standard optical spectroscopic techniques and photosensitization experiments that provide the requisite data for closed-shell systems do not apply to the open-shell cases. Not only is the size of the singlet–triplet separation in biradicals usually difficult to determine, but even the ordering of the states is by no means straightforward.

In principle, the ordering and separation of the states should be accessible through the temperature dependence of the paramagnetism of the triplet. According to the Curie law, the paramagnetism (I) is inversely proportional to the absolute temperature (equation 19, where C is a constant). In some cases, the relationship has a non-zero intercept

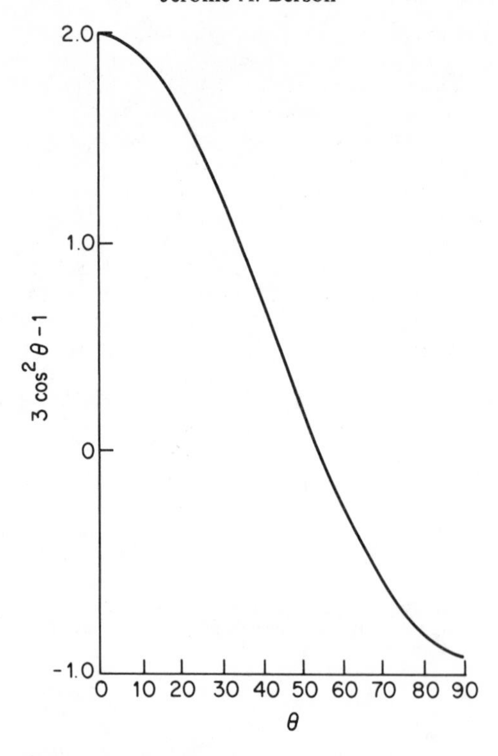

FIGURE 12. The value of the function $3\cos^2\theta - 1$ changes very little near $\theta = 0°$ and near $\theta = 90°$

(Curie–Weiss law), but this need not concern us here[109].

$$I = C/T \tag{19}$$

Although like any paramagnetic substance, the triplet follows the Curie law, the concentration of triplet will change with temperature (unless the singlet–triplet gap is accidentally zero). The behavior of a singlet–triplet pair in a Curie law study is predictable from thermodynamic considerations combined with the Curie law.

Consider a triplet and a singlet in thermal equilibrium and separated by an energy $\Delta E = E_T - E_S$. From the Boltzmann distribution of the population of states and the statistical factor of 3 in favor of the triplet, the relative concentration of the triplet in a given sample can be expressed as in equation 20, and the Curie law as in equation 21[127].

$$[\text{triplet}]_{\text{rel}} = \frac{3\,[\exp(-\Delta E/RT)]}{1 + 3\,[\exp(-\Delta E/RT)]} \tag{20}$$

$$IT = C\,[\text{triplet}]_{\text{rel}} \tag{21}$$

Because of the high chemical reactivity of non-Kekulé compounds, most of the EPR spectroscopic studies have been carried out on samples generated photochemically in isolation media at low temperature. Qualitatively, it is sometimes concluded that the ground state is triplet if the EPR signal persists in the dark at the lowest readily achievable temperature (usually 4.2 K, the boiling point of helium). This conclusion embodies the assumption that relaxation to the ground spin state will occur rapidly compared to the rate at which the investigator's patience evaporates. Moreover, the Boltzmann distribution ensures some concentration of the triplet species at equilibrium, even when the ground state is singlet. For example, if the triplet energy is 100 cal mol^{-1} greater than that of the singlet, the equilibrium mixture at 4.2 K will contain the triplet species to the extent of about 2 parts in 10^5. In a typical sample, this might correspond to an absolute concentration in the range of 10^{-8} M, which might be detectable by EPR spectroscopy[128].

In principle, the shape of the Curie plot is capable of giving a more quantitative measure of the sign and the magnitude of ΔE. The simplest case is one in which the triplet is favored as the ground state (ΔE is negative). With a large negative ΔE, the concentration of the triplet will change over the experimental temperature range by only an insignificant amount, and the Curie plot will be linear. If ΔE is small and negative, upward curvature of the Curie plot at low temperature might be expected, as the triplet concentration increases at the expense of the singlet. Actually, however, this curvature is almost imperceptible, and plots for the cases $\Delta E = -5$ to -50 cal mol^{-1} are essentially linear below 70 K. Of course, if $\Delta E = 0$, equations 20 and 21 require the plot to be linear.

If the singlet is the ground state and $\Delta E \geqslant +2000$ cal mol^{-1}, it may be difficult to observe any signal below 77 K because of insufficient triplet concentration. For a smaller gap, exemplified in Figure 13, where $\Delta E = +50$ cal mol^{-1} under appropriate experimental conditions, the curvature between 77 K and 12 K should be detectable (and in a very few cases[126], has been). However, the maximum near 16 K, which is the most dramatic region of such a curve (Figure 13) may not be easy to characterize because saturation of the EPR transitions at or just below this temperature begins to become pronounced. This imparts its own curvature to the Curie plot. Saturation usually can be recognized by a deviation

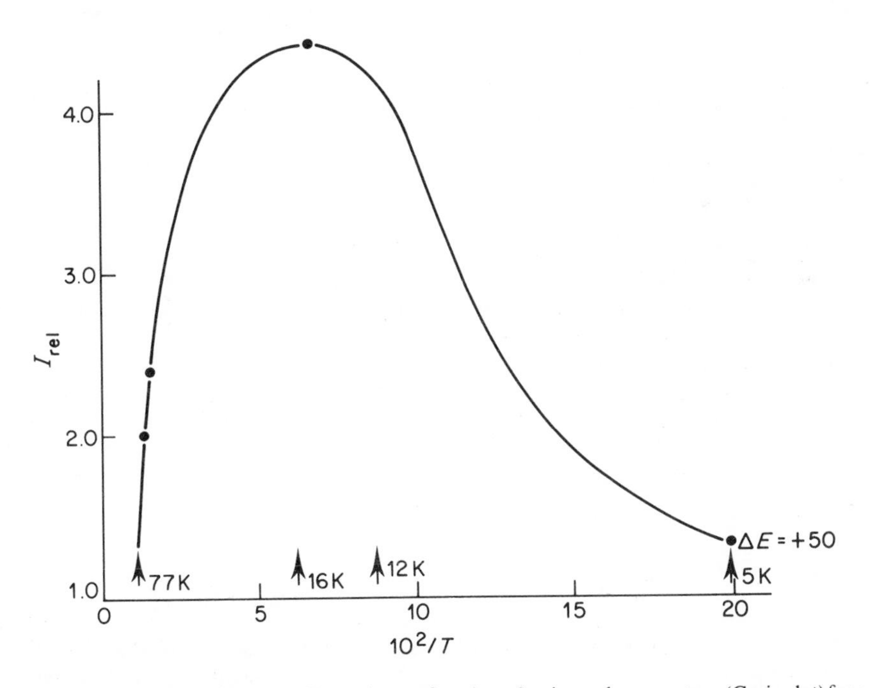

FIGURE 13 Relative EPR signal intensity as a function of reciprocal temperature (Curie plot) for a species whose triplet state lies above the singlet by 50 cal/mol

from linearity of the plot of I vs. square root of microwave power (at constant temperature)[111]. Smaller positive values of ΔE will lead to plots down to 16 K that appear linear within the experimental error.

As Figure 14 shows, the region above 77 K (boiling point of liquid nitrogen) can be essentially uninformative. Values of ΔE between $+100$ and -100 cal mol^{-1} give nearly linear plots. Larger positive values show curvature, but the relative intensity changes are small. For $\Delta E = +200$ cal mol^{-1}, for example, I_{rel} changes only 30% between 77 K and 130 K, so that the experimental error in each point becomes a significant fraction of the total range.

Measurements at temperatures higher than those of Figure 14 may be useful, if the chemical stability of the system permits (a big if!). In this region, one can populate thermally a higher triplet from a ground state singlet to a significant extent. For example, when $\Delta E = +2000$ cal mol^{-1}, the triplet will constitute about 2% of the total biradical concentration. Unfortunately, the Boltzmann effect of temperature, which raises the triplet population, is partially offset by the Curie effect of temperature, which lowers the paramagnetism. Figure 15 shows the lag between relative triplet concentration and the relative EPR signal intensity in this region.

The literature contains a number of qualitative assignments of triplet ground states based upon linear Curie plots in the region above 77 K. Although there is a strong likelihood that the assignments are correct in most cases, one should bear in mind that the range of circumstances in which the Curie plot can be used as the basis of a convincing assignment is quite limited and that measurements in the cryogenic region (< 77 K) are imperative.

The need for a more reliable, more broadly applicable method for the determination of

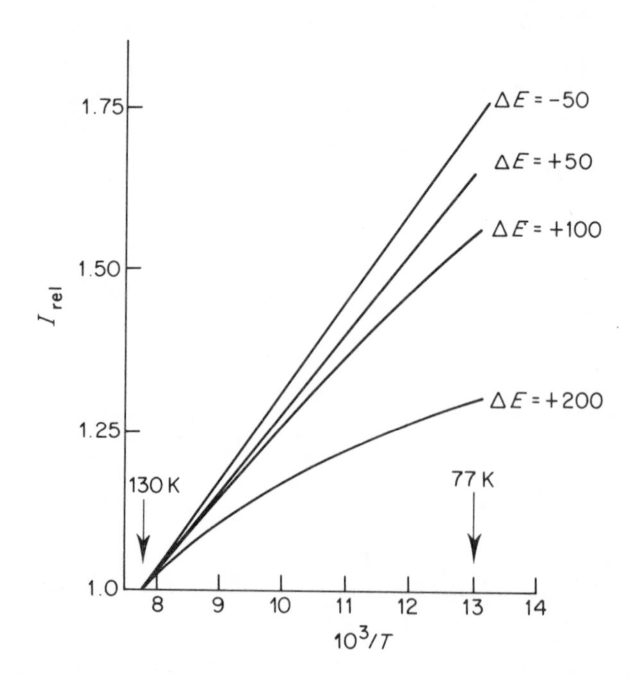

FIGURE 14. Relative EPR signal intensity (I_{rel}) as a function of reciprocal absolute temperature. The indicated values of $\Delta E = (E_T - E_S)$ are in cal mol^{-1}

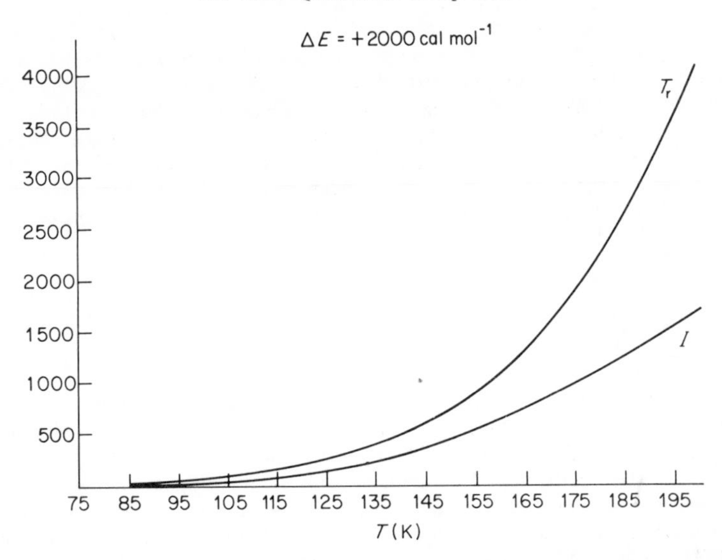

FIGURE 15. Relationship between absolute temperature and relative triplet concentration (T_r) or relative EPR signal intensity (I) for a singlet–triplet equilibrium system when $\Delta E = +2000$ cal mol^{-1} (singlet ground state)

ΔE should now be obvious. Perhaps no other single development would do more to advance the study of non-Kekulé molecules.

iii. Structural characterization of high spin m-quinonoids from the ZFS parameters D and E. The assignment of structure to high-spin states of non-Kekulé compounds would be facilitated by a reliable method for calculation of the ZFS parameters of the EPR spectrum. A quantitative comparison of predicted and observed ZFS requires an accurate knowledge of the electronic spin-density distribution, which in general is not readily available without wave functions derived from an *ab initio* molecular orbital calculation using a large basis set and extensive configuration interaction[129, 130].

In the spin–dipolar approximation, the Hamiltonian H_{SS} may be expressed in equation 22:

$$H_{SS} = \hat{\boldsymbol{S}} \cdot \hat{\boldsymbol{D}} \cdot \hat{\boldsymbol{S}} = \frac{g^2\beta^2}{2}[\hat{S}_x, \hat{S}_y, \hat{S}_z] \begin{bmatrix} \dfrac{r^2 - 3x^2}{r^5} & \dfrac{-3xy}{r^5} & \dfrac{-3xy}{r^5} \\ \dfrac{-3yx}{r^5} & \dfrac{r^2 - 3y^2}{r^5} & \dfrac{-3yz}{r^5} \\ \dfrac{-3zx}{r^5} & \dfrac{-3zy}{r^5} & \dfrac{r^2 - 3z^2}{r^5} \end{bmatrix} \begin{bmatrix} \hat{S}_x \\ \hat{S}_y \\ \hat{S}_z \end{bmatrix} \quad (22)$$

The terms of equation 22 are defined as follows: $\hat{\boldsymbol{S}}$ is the total spin operator of the two electrons ($= \hat{S}_1 + \hat{S}_2$), $\boldsymbol{D}$ is a traceless second rank tensor, $\hat{\boldsymbol{S}}_i$ is the component of the total spin operator in the i-direction ($i = x$, y or z), g is taken to be the g-factor of the free electron ($g_e = 2.0023$), β is the Bohr magneton, r is the interelectronic distance, and x, y, and z are the respective components of the interelectronic distance in a coordinate system that in principle may be arbitrarily oriented relative to the molecular framework.

The matrix elements of $\boldsymbol{D}$ are averaged over the electronic wave function. $\boldsymbol{D}$ is diagonalized to ${}^{d}\boldsymbol{D}$ by rotation of the arbitrarily chosen Cartesian coordinate system until it is parallel to the coordinate system defined by the spin–dipolar interaction, namely the principal axes of $\boldsymbol{D}$. This rotation of the coordinate axes causes the off-diagonal elements of $\boldsymbol{D}$ to become equal to zero after averaging over the electronic wave function. The diagonal elements D_{II} ($I = X$, Y, or Z) of ${}^{d}\boldsymbol{D}$ are the negatives of the relative energies of the three triplet sublevels in the absence of an applied magnetic field. For the coordinate system in which $\boldsymbol{D} = {}^{d}\boldsymbol{D}$, the ZFS parameters D and E take the simple forms of equations 23 and 24.

$$D = \frac{3g^2\beta^2}{4} \left\langle \frac{r^2 - 3z^2}{r^5} \right\rangle \tag{23}$$

$$E = \frac{3g^2\beta^2}{4} \left\langle \frac{y^2 - x^2}{r^5} \right\rangle \tag{24}$$

In a theoretically rigorous calculation[129–132], high-quality *ab initio* wave functions would be used to obtain the expectation values in these equations. Such calculations have been carried out for the excited triplet states of benzene and naphthalene and for the ground triplet state of trimethylenemethane[130, 131] at various levels of approximation. The expansion of the expectation value $\langle \psi H_{SS} \psi \rangle$ of the spin operator (ψ is the triplet wave function) sometimes can be simplified by neglect of the two-center and multicenter exchange integrals[131]. Calculation of the expectation value then requires the evaluation of a number of two-center Coulomb integrals over the entire wave function. However, even in its simplified form, the application of these methods to large non-Kekulé molecules of low symmetry is still a formidable task.

A semiempirical way of dealing with this difficulty has been examined[133–142]. It is based upon an approximation of the wave functions suggested by McWeeny[143], in which the positions of two point-half-charges, placed at the most probable distance of the electron from the nucleus in a 2p atomic orbital, replace ϕ_i, the atomic orbital function in ψ. The orbital interactions involved are shown in Figure 16. From idealized geometries and several different semiempirical methods for obtaining the wave functions, useful results have been obtained for a number of molecules of the *m*-quinone series as well as for several other cases[142]. The predictions for the *m*-quinonoids are presented in Table 2 along with the available experimental data.

The details of the extraction of accurate ZFS parameters from the observed spectra are described elsewhere[110–114], but an illustration of an approximate method based on the

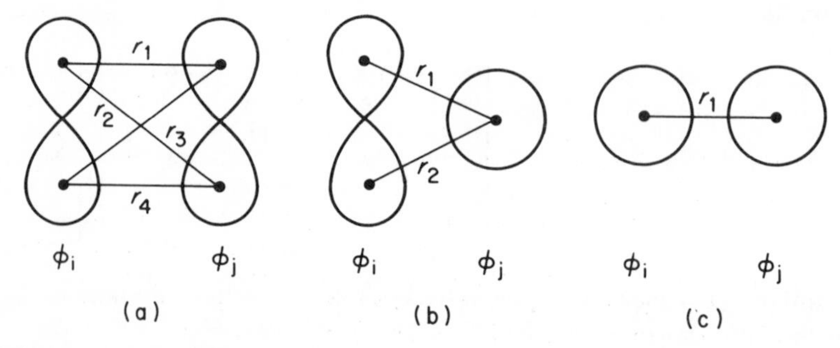

FIGURE 16. Orbital interactions used in approximate calculations of ZFS in triplet species. The diagrams show interactions in (a) p, p; (b) s, p; (c) s, s. The atomic orbital ϕ_i in the wave function ψ is approximated, according to McWeeny[143], by the position of a point-half-charge at the most probable distance of the electron from the nucleus in a 2p atomic orbital

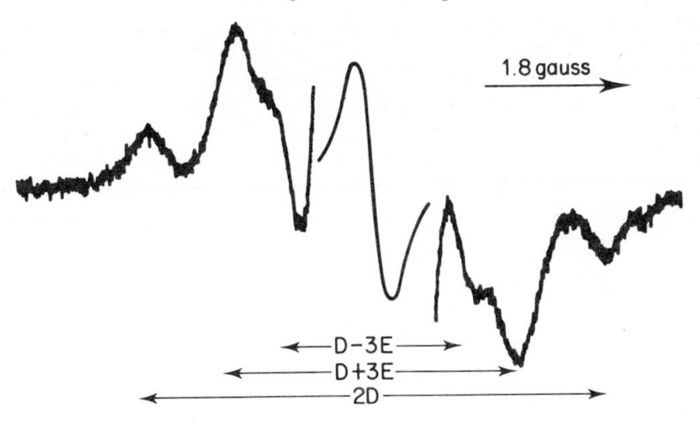

FIGURE 17. EPR spectrum of **8** in toluene glass at 173 K. The ZFS parameters are $|D|/hc = 0.0064\ \text{cm}^{-1}$, $|E|/hc = 0.0056\ \text{cm}^{-1}$. *Reproduced with permission from Ref.* 136

separations of the pairs of lines in a randomly oriented sample is given in Figure 17, which shows the spectrum[136] of the Schlenk hydrocarbon tetraphenyl-*m*-quinodimethane (**8**). A similar but not identical spectrum ($|D| = 0.0079\ \text{cm}^{-1}$, $|E| \leqslant 0.0005\ \text{cm}^{-1}$) is reported in an independent study[144].

The best calculational fit to the $|D|$ value for **8** is obtained on the assumption that the biradical is non-planar, with the benzhydryl moieties both twisted by 90° with respect to the central ring[136]. The other calculated values of Table 2 were based upon the assumption of planar molecular geometries.

Although the calculations of E fluctuate badly around the observed values (a deficiency found in higher level calculations also[131]), the calculated $|D|$ values in Table 2 are uniformly higher than the corresponding experimental values (where available) by about a factor of 2. The same factor applies to several comparisons (not shown here) in the trimethylenemethane and tetramethyleneethane series[140]. It seems likely that the empirical correction should be applied to $|D|$ values calculated for other members of these series, but it would be prudent to keep in mind that the physical basis for this apparently consistent discrepancy is not understood and that the range of reliability of the method is therefore not well demarcated.

The predictions nevertheless are useful for the anticipation of trends in $|D|$ as a function of structural change. For example, the calculations predict a trend in $|D|$ for the simple *m*-quinomethanoids **20** < **19** < **21** in agreement with experiment for the two available cases. This probably results from the increasingly concentrated electron density distribution caused by the electronegative oxygen atoms, which increases the average propinquity of the unpaired electrons. Similar trends are seen in the benzo series **43** < **45** or **44** < **47**. On the other hand, the comparable trend in the experimental data for the 1,8-naphthoquinonoids **48** < **49** is barely perceptible. Nevertheless, 1,8-naphthoquinone biradical itself (**50**) is predicted to have a large $|D|$ value, almost as large as that of *m*-quinone biradical (**21**). It should be emphasized that all of these calculations are based upon the spin–dipolar coupling approximation. Especially in the heteroatom-substituted systems, spin-orbit effects may contribute in the actual molecule, and the observed $|D|$ values may differ significantly from those predicted here.

The calculation produces the expected diminution in $|D|$ value when the electron density is spread over a broader area, as is observed experimentally in the available cases (compare **20** > **43**, **19** > **44** or **45**, **21** > **47**, and **42** > **46**).

TABLE 2. Calculated[a] and experimental ZFS parameters[g] for *m*-quinonoid non-Kekulé compounds

Compound	Experimental $\lvert D \rvert/hc$	Experimental $\lvert E \rvert/hc$	Calculated $\lvert D \rvert/hc$	Calculated $\lvert E \rvert/hc$
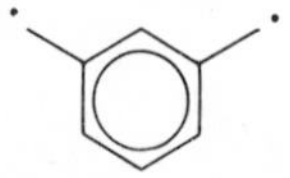 **(20)**	0.011	$\leqslant$ 0.001[h–j]	0.0323[b] 0.0371[d] 0.0338[e] 0.0378[f]	0.0047[b] 0.0028[d] 0.0035[e] 0.0025[f]
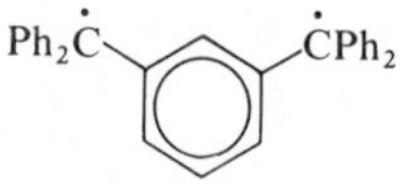 $Ph_2\dot{C}$–C₆H₄–$\dot{C}Ph_2$ **(8)**	0.0064[o] 0.0079[p]	0.00056[o] $\leqslant$ 0.0005[p]	0.0079[n]	0.00036[n]
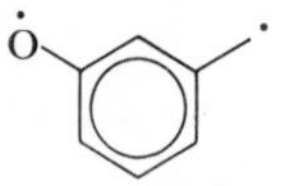 **(19)**	0.0266[j]	0.0074[j]	0.0543[c] 0.0572[f]	0.0087[c] 0.0099[f]
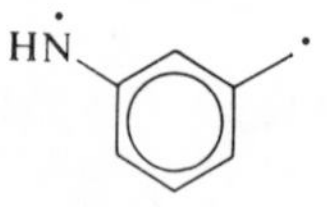 $H\dot{N}$ **(42)**			0.0459[f]	0.00427[f]
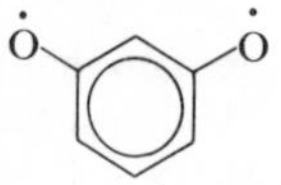 **(21)**			0.0873[d] 0.0871[f]	0.0068[d] 0.0037[f]
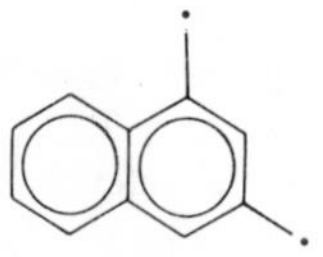 **(43)**			0.0262[b] 0.0320[d]	0.0023[b] 0.0012[d]
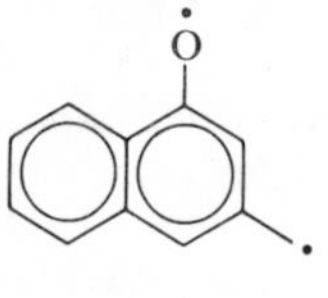 **(44)**	0.0204[k]	0.0052[k]	0.0438[d] 0.0477[f]	0.0076[d] 0.0081[f]
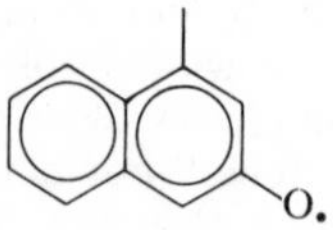 **(45)**			0.0390[d] 0.0388[f]	0.0089[d] 0.0099[f]
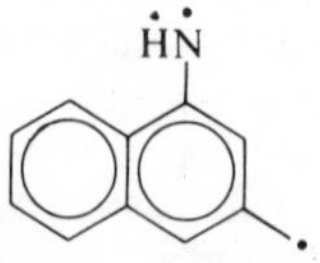 $H\dot{N}$ **(46)**			0.0393[f]	0.00387[f]
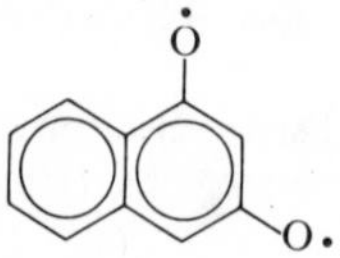 **(47)**			0.0708[d]	0.0052[d]

TABLE 2. (*Contd.*)

Compound		Experimental $|D|/hc$	Experimental $|E|/hc$	Calculated $|D|/hc$	Calculated $|E|/hc$
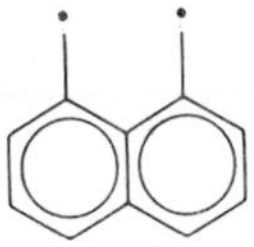	**(48)**	0.0218[l] 0.024[m]	0.0021[l] 0.001[m]	0.0401[b] 0.0377[d]	0.0076[b] 0.0111[d]
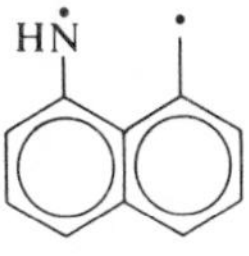 HN	**(49)**	0.0255[q]	0.0008[q]	0.0356[d]	0.0111[d]
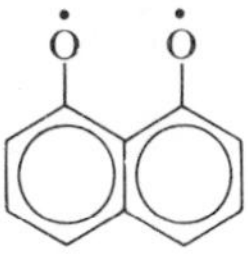 O O	**(50)**			0.0756[d]	0.0154[d]

[a] Except for **8**, calculated in Refs 139, 140, 142, assuming molecular planarity and using wave functions from the indicated method (footnotes *b*–*q*).
[b] Simple Hückel. [c] CNDO closed-shell singlet. [d] CNDO open-shell triplet. [e] INDO closed-shell singlet. [f] INDO open-shell triplet. [g] Absolute values in cm^{-1}. [h] Ref. 145. [i] Refs 146, 147. [j] Ref. 139, 140. [k] Refs 140, 148. [l] Ref. 149. [m] Refs 150, 151. [n] Calculated for a non-planar configuration in Ref. 136. [o] Ref. 136. [p] Ref. 144. [q] Ref. 151.

C. Chemistry of *m*-Quinonoids

1. *Synthesis and reactions of m-quinodimethane*

Although Migirdicyan and co-workers[61, 102, 105–107] identified the *m*-quinodimethane biradical **20** from the emission and excitation spectra of irradiated *m*-xylene preparations (see Section III.B.2.a), another attempt[145] to observe **20** by EPR spectroscopy under these conditions failed. Wright and Platz[145] made the first EPR observation of **20** using a different synthetic technique, which had been developed originally by Platz[152] in another context. The key reaction is a hydrogen transfer to a carbene from a nearby C–H bond as exemplified for an intramolecular case in the formation of 1,8-naphthoquinodimethane (**48**) from the photolysis at 77 K of 8-methyl-1-diazomethylnaphthalene (**51**), presumably via the carbene **52**. The EPR spectrum of **48** obtained in this way matched that of the triplet species obtained[149] from the diazene **53**. The carbene could not be isolated, even at 4 K. Senthilnathan and Platz[150] suggest that at low temperatures, the hydrogen transfer reaction occurs by quantum mechanical tunneling through a potential barrier.

An intermolecular variant of this hydrogen transfer leads to *m*-quinodimethane **20**[145]. The Bell Laboratories group[153] had generated the bis-carbene (**54**) by photolysis of the diazo compound (**55**). The bis-carbene (**54**) was stable at cryogenic temperatures and showed a beautifully detailed ESR spectrum which was identified as that of the quintet state of this species. Wright and Platz[145] reasoned that it might be possible to reduce **54** to the *m*-quinodimethane biradical **20** by hydrogen transfer from a more reactive medium. Indeed photolysis of the bis-diazo compound **55** in ethanol glass at 22 K gave the characteristic quintet spectrum of **54**, but when the temperature of the sample was raised to

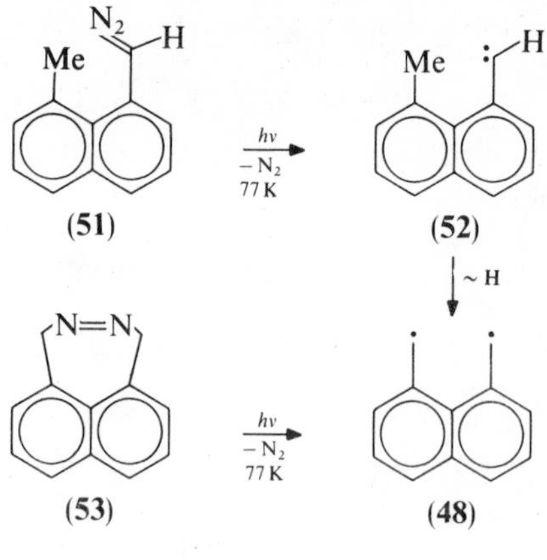

(51) (52)

(53) (48)

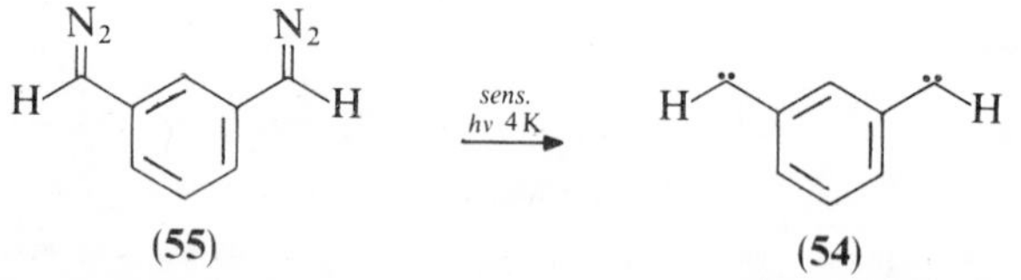

(55) (54)

77 K, the quintet spectrum faded and was replaced by a new spectrum of a randomly oriented triplet species, $|D| = 0.011\ \text{cm}^{-1}$, $|E| \leqslant 0.001\ \text{cm}^{-1}$, to which the structure **20** was assigned. The ZFS parameter $|D|$ is in reasonably good agreement with the predicted value[139] provided the empirical correction factor of 0.5 is applied to the latter (see Table 2). That a hydrogen transfer from the solvent is involved in the transformation of **54** was established by the observation that the quintet spectrum at 77 K is stable in ethanol-d_6 or perfluoroalkane glasses[145].

Significantly, the same spectrum of biradical **20** was obtained by Wright and Platz[145] by photolysis of *m*-xylylene dibromide (**56**) in the presence of diphenylamine (**57**). A completely different preparation of the same species was subsequently reported by Goodman and Berson[146, 147], who generated it from either of the ketones **58** or **59**, or the

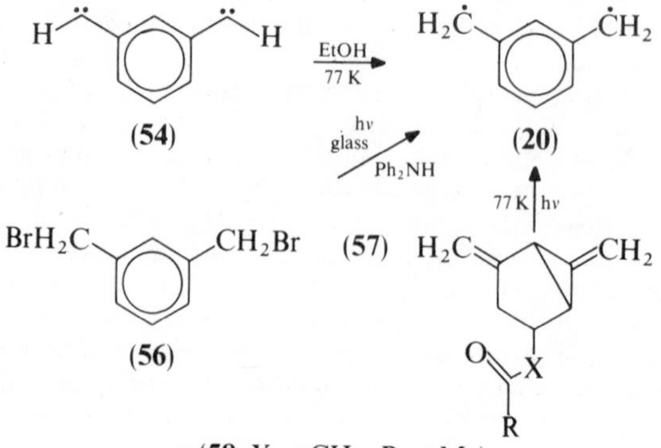

(54) (20) (57) (56)

(**58**, X = CH_2, R = Me)
(**59**, X = CH_2, R = Ph)
(**60**, X = O, R = Ph)

benzoate **60** (see below). Neither group has been able to observe the $\Delta M_S = 2$ transition, but this is not surprising in view of the small $|D|$ value (see Section III.B.2.b).

Although the EPR spectrum of **20** is not observed[145] under the conditions (photolysis of *m*-xylene in an alkane matrix, 77 K) which permitted detection of its optical emission spectroscopy[61] (see Section III.B.2.a), an EPR spectrum similar to that of **20** is generated[145] by photolysis of mesitylene (1,3,5-trimethylbenzene) in an alkane matrix. Presumably, the carrier of the EPR signal is 5-methyl-1,3-bis-methylenebenzene, a methyl-substituted **20**. Reasons for the differences in behavior between *m*-xylene and mesitylene remain to be elucidated.

The EPR signal intensity of *m*-quinodimethane is linear with $1/T$ over the range 30–77 K[145]. The triplet therefore is the ground state or within a few small calories of it. This result is compatible with the triplet ground state predicted by theory (see Section II.B.1).

Three methods of generating and detecting *m*-quinodimethanes in fluid solutions have been devised. All use the same overall strategy as the original covalent → biradical *m*-quinonoid valency tautomeric change (**22** → **19**, see below) designed for the preparation of *m*-quinomethane, the oxy analog of *m*-quinodimethane (**20**)[27].

However, the thermochemical–kinetic properties of the covalent–biradical pairs necessitate some significant changes in technique when the **22** → **19** type of reaction is adapted to the hydrocarbon case, **61** → **20**. Table 3 shows the relative energies of some of the covalent–biradical valency tautomeric parts in the *m*-quinone series[140]. Although these estimates are rough, they leave no doubt that the ring-opening, covalent → biradical, is thermochemically much more favorable in the hydrocarbon case **61** → **23**, (Table 3, entry 4) than in the monoketone case **22** → **19** (Table 3, entry 2). The reason for this is primarily the large dissociation energy of the carbonyl group's π bond ($\sim$ 88 kcal mol^{-1}), which stabilizes the covalent cyclic form in the ketone case.

TABLE 3. Relative energies (in kcal mol^{-1})[a] of the covalent and biradical forms of *m*-quinonoid valency tautomers[b]

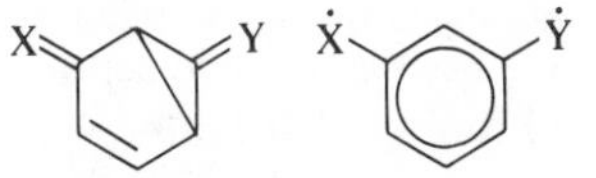

No.	X	Y	Covalent (ΔH_f°)	Biradical (ΔH_f°)	ΔH_f^0 (biradical) − ΔH_f^0 (covalent)
1	O	O	− 4.4	5.8	10.2
2	O	CH_2	44.1	39.9	− 4.2
3	CH_2	O	41.4	39.9	− 1.5
4	CH_2	CH_2	90.2	70.0	− 20.2

[a] The standard heat of formation values (ΔH_f^0) are calculated using the group equivalents of Benson[154] and the assumption that the strain energy in the covalent form is 45 kcal mol^{-1} estimated by analogy to other systems[155]. No correction for the spin state is included. The values shown are in kcal mol^{-1}.

[b] Adapted from Ref. 140 with permission.

Moreover, the covalent bicyclic hydrocarbon **61** should be less stable kinetically than the covalent oxygen analog **24**. The latter compound is a stable species and survives distillation or gas chromatography. Below 80°C it undergoes thermal reactions in which the rate-determining step is the ring-opening **22** → **19** only slowly, with the indicated Arrhenius parameters of Table 4[156]. In contrast, the hydrocarbon **61** is unknown. A rough estimate[157] based on analogies to model systems[158, 159] suggests an activation energy of

TABLE 4. Arrhenius activation parameters

		E_a (kcal mol^{-1})	log A (A in s^{-1})
 (22) → (19)	obsd.	30.6	14.0
 (61) → (20)	est.	10	14

only 10 kcal mol^{-1} for the ring-opening, which if correct would correspond to a lifetime in the microsecond range for **61** at room temperature (assuming $\Delta S^{\ddagger} \approx 0$ cal mol^{-1} deg^{-1}). Obviously, special methods are needed to study this system.

In one of these, dimethylvinylidene (**62**), generated by the potassium fluoride induced decomposition[160] of the silylated enol triflate **63**, reacts in CCl_4 or tetrahydrofuran solution with 6,6-dimethylfulvene (**64**) to give a 20% yield of the octamethyl-2,2-metacyclophane **66**[161]. A reasonable interpretation is that the carbenoid addition generates the tetramethyl derivative (**65**) of hydrocarbon **61**. Thermal ring-opening of **65** to tetramethyl-*m*-quinodimethane (**67**), followed by dimerization of the latter, would lead to the metacyclophane (**66**).

The same unstable hydrocarbon **65** probably is the intermediate in the thermal decomposition of the sodium tosylhydrazone **68**, which gives a 17% yield of the 2,2-metacyclophane (**66**)[162].

Neither of these synthetic methods has been applied to the generation of a *m*-quinodimethane biradical under conditions that would permit its chemical interception or its observation by EPR spectroscopy. Indeed, it might not be easy to find a trapping agent for biradical **67** that would be compatible with carbene **62**.

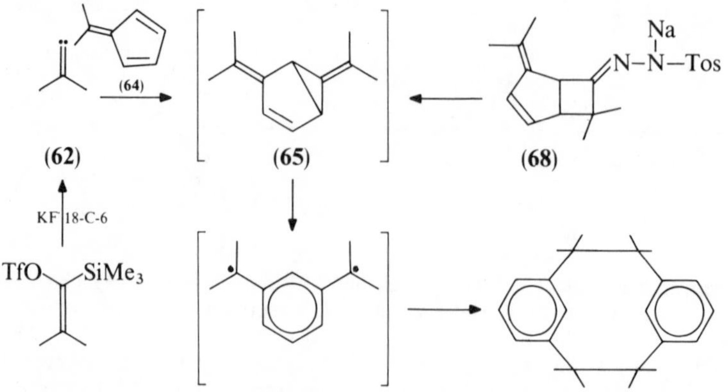

(64)

(62) (65) (68)

Na

N–N–Tos

KF 18-C-6

TfO SiMe$_3$

(63) (67) (66)

A third approach accomplishes both of these objectives by the generation of the covalent hydrocarbon **61** and from it the parent *m*-quinodimethane **20** in Norrish type II photofragmentations of the ester **60** or the ketones **58** or **59** (Scheme 2).

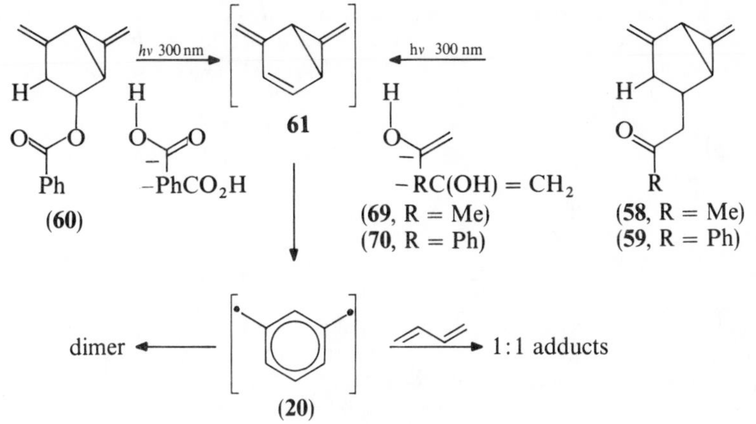

hν 300 nm
H
(60)
Ph
$-PhCO_2H$
61
$-RC(OH) = CH_2$
(69, R = Me)
(70, R = Ph)
(58, R = Me)
(59, R = Ph)
dimer
(20)
1:1 adducts

SCHEME 2

Photolyses of the ester **60** or the phenyl ketone **59** in 2-propanol glass at 77 K give rise to the characteristic six-line EPR signal of *m*-quinodimethane biradical **20**[146,147], essentially identical (except for a different doublet impurity pattern) with that generated by the bis-carbene reduction route of Wright and Platz[145] (**54** → **20**). These reactions constitute a second and a third independent synthesis of **20** and serve to strengthen its assignment as the carrier of the EPR triplet signal. Upon warming the matrix, 2,2-metacyclophane was identified in the reaction mixture [146,147].

Solution phase photolyses of the methyl ketone **58** or phenyl ketone **59** presumably again resulted in the elimination of the enol of the fragment ketone (**69** or **70**) and formation of the bicyclic *m*-quinomethanoid **61**. Although the enol of acetophenone (**70**, Scheme 2) was identified by low temperature nuclear magnetic resonance (NMR) spectroscopy of a solution that had been irradiated at −78°C, the spectrum did not reveal any trace of the hydrocarbon **61**, which thus appears to be thermally (or photochemically) unstable under the conditions of its birth[146,147].

In the absence of added trapping agents, the photolysis of methyl ketone **50** shown in Scheme 2 leads to 2,2-metacyclophane in low yields (5–15%), resulting from dimerization

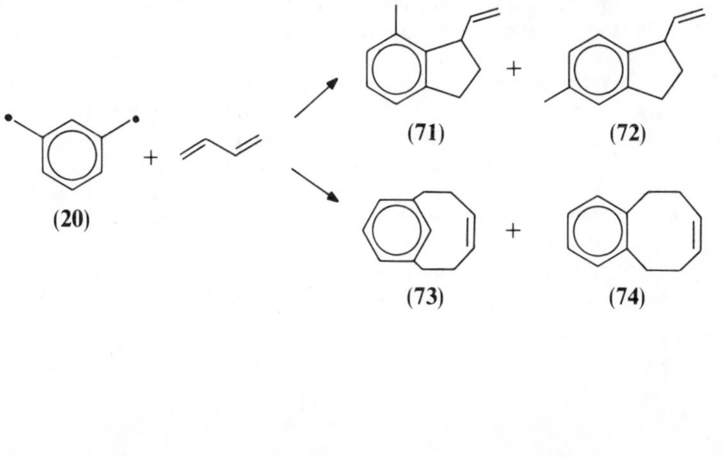

(20)
(71)
(72)
(73)
(74)

of *m*-quinodimethane biradical **20**. However, this species can be intercepted by incorporation of any of a variety of conjugated dienes or α,β-unsaturated carbonyl compounds in the reaction mixture. Photolysis of the methyl ketone **58** in the presence of a large excess of butadiene gives a 40–60% yield of four 1:1 adducts, whose structures are **71–74**.

Products **71** and **72** result from a 1,2-addition of **20** to a double bond of the diene and presumably arise from a primary prearomatic adduct, e.g. **75**, by a hydrogen shift. Product **73** results from a 1,4-addition to the diene, and product **74** results from a secondary photolysis of **73**[163].

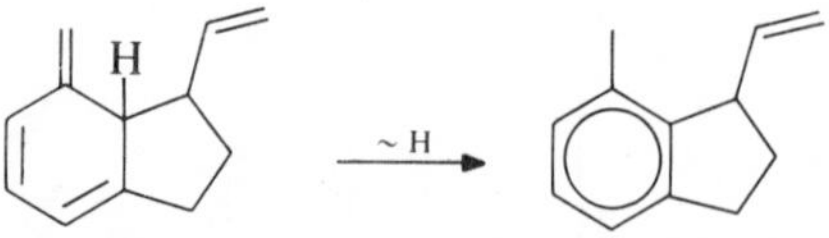

H ~H

(75) (71)

Several lines of evidence support the sequential mechanism of Scheme 2: Norrish type II photofragmentation of ketone **58** or **59** produces hydrocarbon **61** and then biradical **20**, which ultimately is captured by cycloaddition to 1,3-butadiene. The already described observation of acetophenone enol in the photolysis mixture strongly suggests the occurrence of some sort of Norrish type II reaction. However, this finding alone does not suffice to exclude an alternative mechanism (Scheme 3), in which cycloaddition of

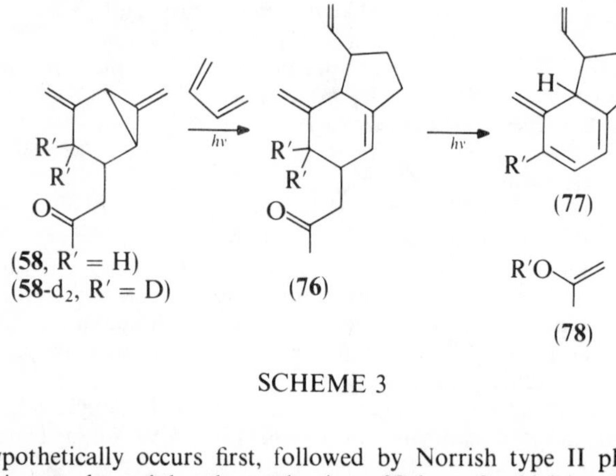

R′ R′ O hν R′ R′ O hν H R′ R′O

(**58**, R′ = H)
(**58**-d_2, R′ = D) (**76**) (**77**) (**78**)

SCHEME 3

butadiene hypothetically occurs first, followed by Norrish type II photoelimination. Further experiments showed that the mechanism of Scheme 2, which involves the biradical **20**, is to be preferred to that of Scheme 3, which bypasses **20** entirely. The choice is made by noting that the Scheme 3 mechanism consumes ketone by a direct cycloaddition pathway (**58** → **76**), whereas the Scheme 2 mechanism consumes ketone in a Norrish type II process involving a hydrogen transfer. The Scheme 2 mechanism therefore predicts a primary deuterium isotope effect on the quantum yield for consumption of the ketone (**58** vs. **58**-d_2), but the Scheme 3 mechanism predicts only a very small secondary isotope effect. The observed value, $\phi_H/\phi_D \sim 2.2$ for the singlet photoprocess, is compatible only with the first prediction and, moreover, is very similar in magnitude to isotope effects in singlet Norrish type II reactions in other systems[146, 147]

A second piece of evidence favoring Scheme 2 comes from an isotopic position labeling experiment (Scheme 4) starting with the ketone **79**, deuterium-substituted in only one of

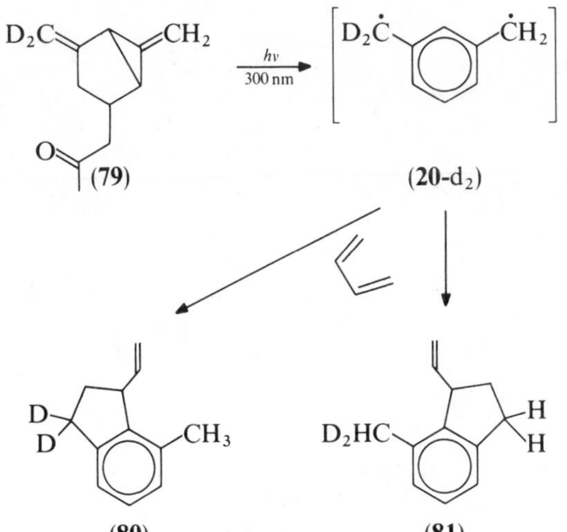

SCHEME 4

the two exocyclic methylene groups. The butadiene adduct from **79** contains deuterium essentially equally distributed between the benzylic methylene group and the methyl group, as would be required if the bilaterally symmetrical intermediate *m*-quinodimethane **20**-d_2 were on the reaction pathway[146, 147].

2. *Mechanism of the cycloadditions*

Both the 1,2 and the 1,4 cycloadditions of **20** to *cis,cis*- and *trans,trans*-2,4-hexadienes are most easily interpreted as stepwise rather than concerted processes. Either pure diene gives mixtures of stereoisomeric 1,2-adducts in which the original stereochemistry of the unreacted diene double bond is completely preserved, but that of the reacting double bond is not carried over into the product. These results are exemplified in Scheme 5, which shows

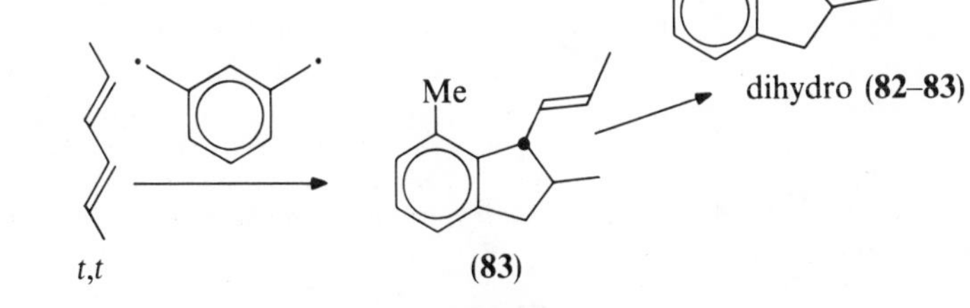

SCHEME 5

that in *both* of the dominant 1,2-*ortho* adducts, **82** and **83**, from *cis,cis* (*c,c*)- and *trans,trans* (*t,t*)-2,4-hexadiene, respectively, the stereochemistry of the ring substituents is *trans*. The adducts differ only in the side-chain stereochemistry, and each can be reduced to a common dihydro derivative, dihydro **82–83**[147].

Scheme 6 shows the suggested[147] stepwise mechanism applied to the reactions with the three 2,4-hexadienes. Addition of the biradical to the diene (say *trans, trans*) gives an adduct biradical (**86**), which can either cyclize to product **83**, or suffer internal rotation to another adduct biradical (**87**). Cyclization of **87** gives a stereoisomeric adduct **84**. The adduct biradical is assumed to preserve stereochemistry around the allylic unit, in analogy to model systems. Application of the same scheme to *cis,cis*-2,4-hexadiene predicts the formation of two new stereoisomeric adducts, **82** and **85**, whereas all four adducts should be formed from *cis,trans*-2,4-hexadiene, since that diene has two non-equivalent sites for 1,2-cycloaddition. The experimental data[147] are fully in accord with this scheme.

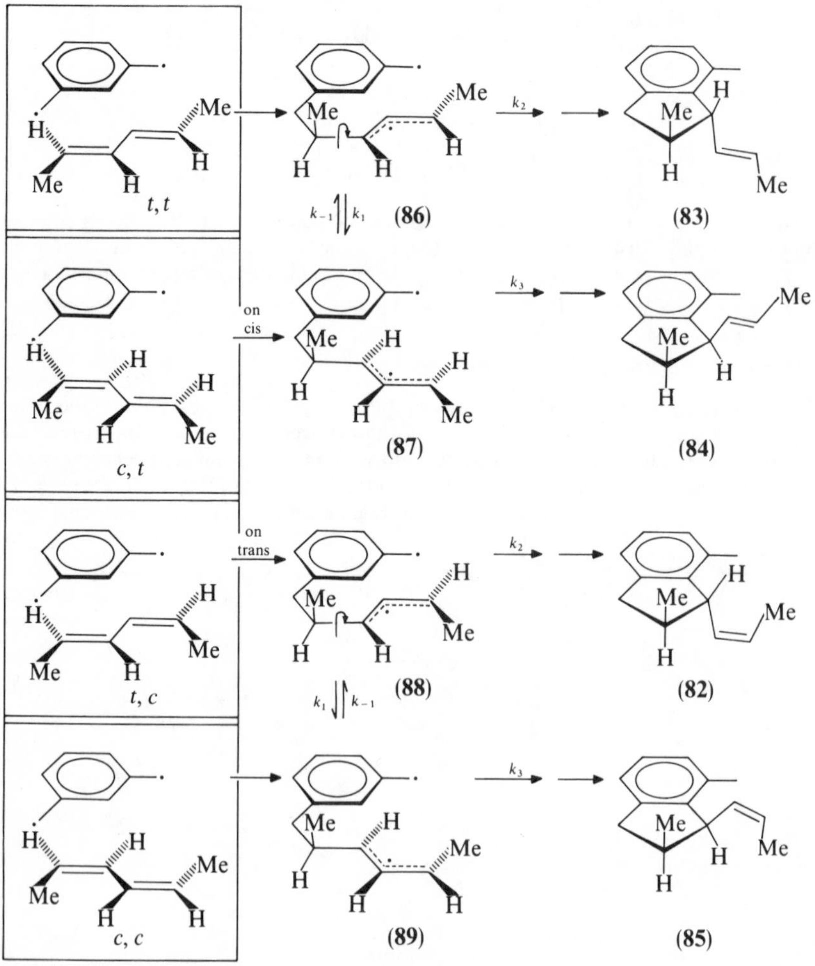

SCHEME 6

The competition between cyclization and internal rotation (Scheme 6) may be analyzed quantitatively[147] by treatment of the data with the steady-state approximation, a procedure copied directly from that used by Montgomery and coworkers[164] in a study of the additions of 1,1-dichloro-2,2-difluoroethene to the 2,4-hexadienes. Those reactions also involve a biradical adduct intermediate (e.g. **90**) in which a similar internal rotation vs. cyclization competition controls the relative stereochemistry of the ring carbons in the cyclobutane product (e.g. **91**).

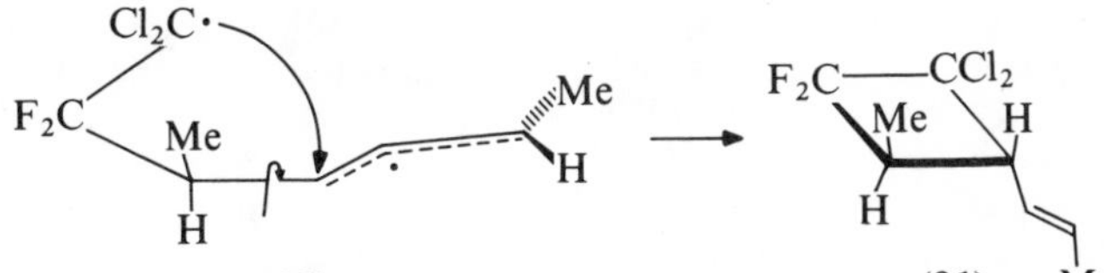

$Cl_2C\cdot$ F_2C Me H Me H **(90)** → F_2C—CCl_2 Me H H Me **(91)**

The analysis employs the assumption that common k_2 and k_3 values apply to the intermediates from stereoisomeric precursors. Applied to the additions of the *m*-quinodimethane biradical it permits the extraction of the competition ratios of Scheme 6, k_1/k_2 and k_{-1}/k_3, which turn out to be in remarkable agreement with the corresponding ratios in the 1,1-dichloro-2,2-difluoroethene additions. In both systems, the ratios may be expressed as $k_1/k_2 = 11.5 \pm 1.5$ and $k_{-1}/k_3 = 2.5 \pm 0.3$. The original paper[147] discusses the possibility that an 'entropy-controlled' cyclization may be the common feature that is responsible for the similarities in the competition ratios.

The Norrish type II photofragmentation which leads to biradical **20** (Scheme 2), when carried out in the presence of a large excess of diene, should occur in the n, π^* singlet state of the ketone precursor **58** or **59**. This could proceed adiabatically via singlet **61** to singlet **20** (**20-S**), which therefore is likely to be the first-formed *m*-quinodimethane biradical intermediate. However, as we have seen, singlet **20** is unstable with respect to triplet, **20-T**, and therefore should decay to it forthwith.

Scheme 7 shows this mechanism in which two sequentially formed reactive intermediates each can give rise to products by reaction with an olefin. If the intersystem crossing (isc) **20-S** → **20-T** is irreversible, the intermediates form a cascade of energy states, with the higher energy state being populated earlier in the mechanism. If the isc is fast and reversible, the intermediates may be said (preserving the aqueous analogy) to constitute a pool of energy states. Taken at face value, the theoretical calculations, which place **20-T** 10 kcal mol^{-1} below **20-S** (see Section III.B.1.d), suggest that the pool mechanism cannot be important, since the **20-T** → **20-S** reaction would be too slow to meet the requirement of

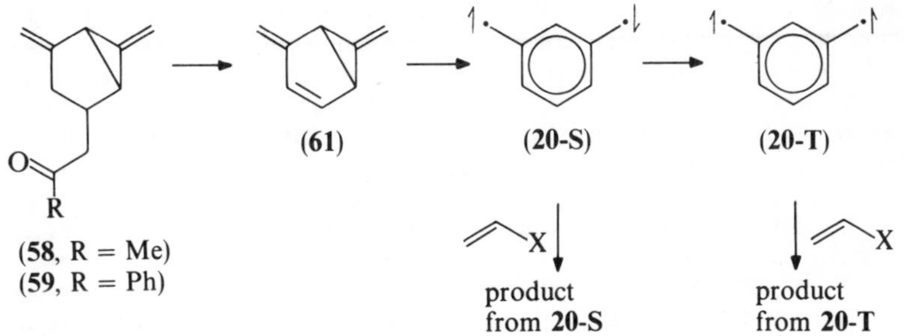

O R (**58**, R = Me) (**59**, R = Ph) → **(61)** → **(20-S)** → **(20-T)**

X product from **20-S**

X product from **20-T**

SCHEME 7

rapid reversible isc. Therefore, a cascade mechanism seems probable. Capture of **20-S** then would have to compete with isc. The competition would be dependent on the concentration of the trapping agent, since trapping is bimolecular, whereas isc is unimolecular. Provided that the ratios of rate constants were suitable, it might be possible to manipulate at will the fraction of the product derived from **20-S** or **20-T** by changing the concentration of the trapping agent. The use of dilution effects for this type of analysis is well known from the chemistry of carbenes, nitrenes and trimethylenemethanes[165, 166].

Experimentally, however, little or no dilution effect on the product ratios is observed in reactions of the *m*-quinodimethane system[146, 147]. This result does not necessarily exclude the cascade mechanism (Scheme 7), since it could be caused by a slow rate of capture of the first intermediate **20-S**, or an exceptionally fast rate of isc, or a product distribution from **20-T** that is the same as that from **20-S**.

The effect of added oxygen has provided another device for distinguishing singlet from triplet biradical chemistry in cycloadditions. The triplet state of 2-isopropylidene-cyclopentan-1,3-diyl, a trimethylenemethane derivative, is selectively scavenged by O_2, leaving the singlet to form its characteristic cycloadducts with olefinic trapping agents[165, 166]. In the case of the *m*-quinodimethane system, however, O_2 is without effect on the product composition[146, 147].

The negative results of the application of these two test criteria to biradical **20** leaves the investigators (for the present, at least) without a means of assigning the spin state of the reactive species[146, 147]. Currently, the mechanism of cycloadditions of **20**, rather than being characterized as a cascade or a pool, more accurately might be described as a swamp, blanketed by a miasma of ignorance, into whose murky depths our senses cannot yet penetrate.

D. *m*-Quinomethane and *m*-Naphthoquinomethane

1. Theory

At the simple Hückel π-electron level of calculation, the frontier orbitals of the hydrocarbon system *m*-quinodimethane are exactly degenerate. When combined with Hund's rule, this leads to the prediction of a triplet ground state (see Section III.B.1) in accord with the experimental observations (see Section III.C.2). However, rejoicing over this concordance should not be unrestrained, because the degeneracy vanishes at any level of theory that takes into account two-electron effects. As we have seen (Section III.B.1) it is not clear whether Hund's rule should be applied, although a *post hoc* justification for doing so comes from the results of the best *ab initio* calculations, which confirm the expectation of a triplet ground state.

What then should we predict for the spin of the ground state of the *m*-quinomethane non-Kekulé system **19**? At the INDO/1 level of calculation, the two frontier orbitals are separated by some 67 kcal mol^{-1}. Surely, one would think that so large a separation would place this molecule well outside the category of substances with degenerate or nearly degenerate orbitals to which one applies the recipe: 'If the separation of the molecular orbitals is sufficiently small, Hund's rule requires the triplet to be the ground state'[74]. Rather, one might expect the closed-shell zwitterionic singlet state, with both frontier electrons occupying the lower frontier orbital, to be preferred.

A higher level semiempirical calculation (INDO/S-CI)[70], however, predicts that although the most stable singlet is in fact one of largely closed-shell character (with small coefficients on the open-shell configurations in the CI wave function), the triplet state is the ground state and lies 12 kcal mol^{-1} lower than the singlet (Table 5). INDO/S-CI predicts[70] a similar result for the naphtho derivative **92**: the lowest singlet is of largely closed-shell character, but the triplet is more stable by about 17 kcal mol^{-1}.

TABLE 5. Relative energies of low-lying states of non-Kekulé molecules calculated by INDO/S-CI[70]

		E (kcal mol^{-1})
(20)	$^3B_2$	0.0
	$^1A'$	15.0
(19)	$^3A'$	0.0
	$^1A'$	12.0
(43)	$^3A'$	0.0
	$^1A'$	14.7
(92)	$^3A'$	0.0
	$^1A'$	17.0

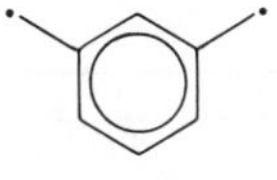

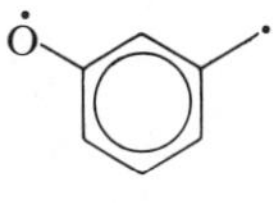

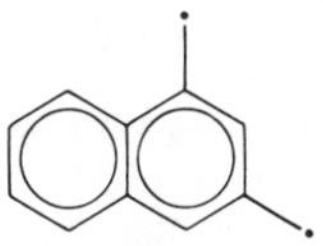

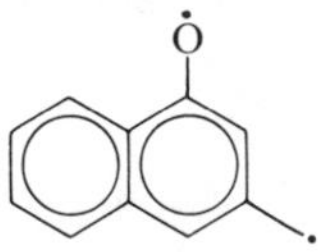

These non-intuitive predictions (retrodictions, actually) are confirmed by the experimental findings that both **19**[27, 140, 167] and **92**[140, 148], when generated photochemically from the covalent *m*-quinonoid precursors in glassy media at low temperatures, persist in the dark as paramagnetic species (see Figure 18) whose signal intensities follow the Curie law. These molecules therefore very probably have triplet ground states. Similarly 1-imino-8-methylenenaphthalene (**49**) has a triplet ground state, as judged by the same experimental criteria[151]. These findings stress the importance of electron repulsion effects in determining the state orderings in non-Kekulé molecules. One-electron theories, even when parameterized to include such effects, simply cannot be relied upon to deal with this problem, and it is not until CI is included that one approaches a proper account of the relative energies.

Moreover, the calculational and experimental results just described provide some quantitative justification for the assertion[63] that heteroatom-for-carbon substitution should not change the state ordering predicted for the hydrocarbon. In fact, INDO/S-CI predicts not only that the order of the states should survive a heteroatomic perturbation but also that their energy separation should not change much. This can be seen in the comparisons (Table 5) **20** vs. **19** and **45** vs. **92**.

What is the physical basis for these calculational results? Qualitatively, the splitting of the NBMOs at the one-electron level of theory (e.g., simple Hückel) is produced by the changes in the Coulomb and resonance integrals associated with the heteroatomic

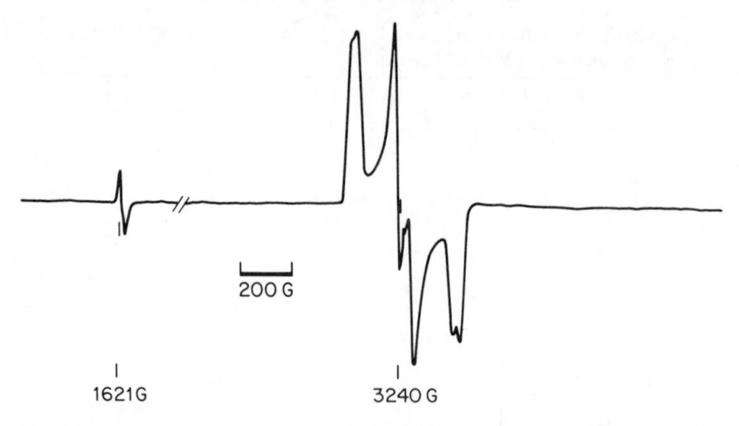

FIGURE 18. EPR spectrum of the triplet state of 3-methylenenaphthalene-1-oxy (**92**) obtained by irradiation of a 2-methyltetrahydrofuran matrix of ketone **93** at 77 K. The small absorption near 3245 G is due to cavity background. Note the $M_S = \pm 2$ transition at 1621 G. The triplet ZFS parameters are (in cm^{-1}) $|D|/hc = 0.0204 \pm 0.0002$ and $|E|/hc = 0.0052 \pm 0.0004$. The fit of the spectrum can be achieved only with the use of an anisotropic g- tensor, i.e., one in which the g-factor along one direction (coincident with the C–O bond) is significantly different from the free-spin value because of spin-orbit perturbation by the oxygen atom. *Reproduced with permission from D. E. Seeger, E. F. Hilinski and J. A. Berson,* J. Am. Chem. Soc., **103**, 720 (1981). *Copyright* (1981) *American Chemical Society*

perturbation. Because of the neglect of electron repulsion effects at this level, the aufbau principle forces double occupation of the lower-lying frontier level and hence a singlet ground state. If electron repulsion effects are large enough, however, the first-order NBMO splitting will play a minor role in determining the ordering and separation of the

(22) → **(19)**

(93) → **(92)**

N_3 CH_3 → $\dot{N}H$ $\dot{C}H_2$ **(49)**

states. Evidently, this is the case in the systems studied here, and by implication, in many others. The usual criterion for the application of Hund's rule, namely that the frontier orbitals be degenerate or nearly so, can be misleading.

One might hope to effect thermally a reversal of the photochemical ring-opening **22** → **19** detected by EPR spectroscopy, since entry 2 of Table 3 shows the biradical **19** to be only 4.2 kcal mol^{-1} lower in energy than the covalent structure **22**. Unless a substantial additional barrier is added on to this energy gap, thermal cyclization of **19** should occur well below room temperature. An attempt[140] to observe this reaction made use of the optical activity of **18**, which would be lost (**18a** → **18b**) during the photochemical–thermal cycle via achiral biradical **19**. Thus, a sample of optically active **18a** was cooled to 77 K in a glassy

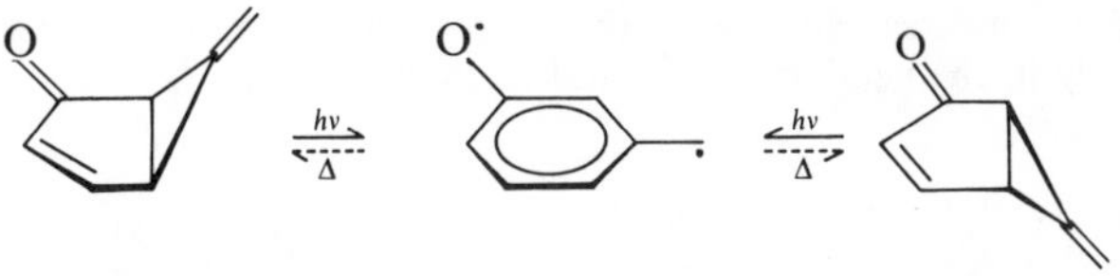

(**18a**) (**19**) (**18b**)

matrix, photolyzed (6% conversion to **19** by an EPR 'spin count') and the matrix was thawed by warming to room temperature. The sample was subjected to ten successive such cycles, but the optical activity of the ketone recovered from the reaction mixture was the same as that of the starting material. Although the products of the photoreaction are unknown, the result suggests that thermal return of the aromatic biradical **19** to the bicyclic ketone **18** is at least inefficient compared to side reactions that may occur during the warm-up part of the cycle and may be slow in the absolute sense because of a substantial energy barrier. Further experiments (see Section III.C.2) on the kinetics of the addition reactions of **18** support the hypothesis of a large barrier.

2. *Chemistry*

As would be expected from the thermochemical relationships of Table 3 (Section III.B.3), the ketone **18** seems to be much more stable than its unknown hydrocarbon counterpart **61**. Compound **18** can be prepared by the sequence shown and is

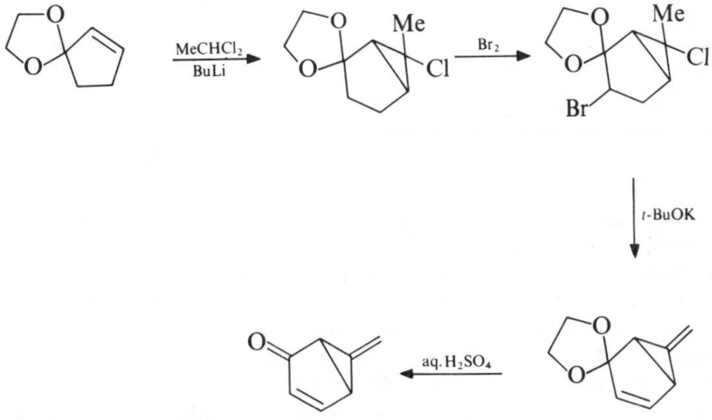

MeCHCl$_2$ / BuLi; Br$_2$; *t*-BuOK; aq. H$_2$SO$_4$

(**18**)

stable up to about 80°C in solution. At higher temperatures, it gives rise to a mixture of products which have not yet been identified. Flash vacuum pyrolysis of **18** gives fulvene, a product which may arise via a two-step sequence: 1,3-sigmatropic rearrangement to the

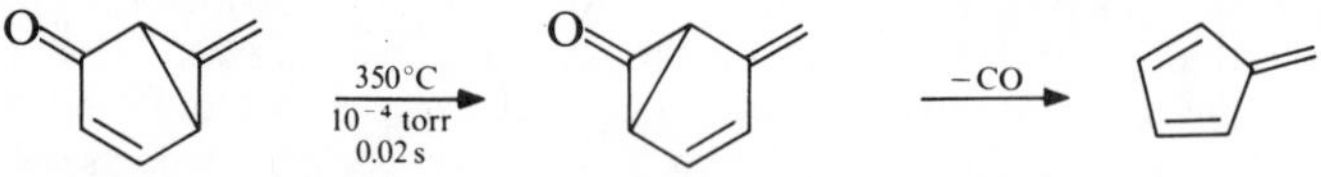

350°C, 10^{-4} torr, 0.02 s; −CO

(18) **(94)**

cyclopropanone **94** followed by cheletropic decarbonylation[168, 169]. Although **94** has not been detected in this rearrangement, the corresponding structural change has been observed directly in the ketal **95**[140, 170] and methyl ether **96**[170].

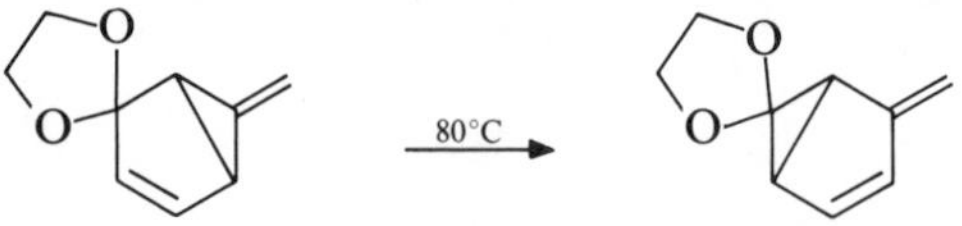

80°C

(95) **(97)**

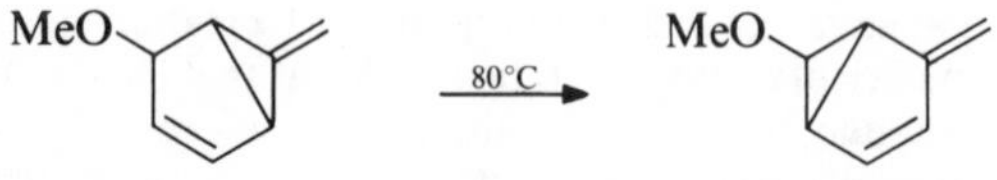

MeO 80°C MeO

(96) **(98)**

Stereochemical studies[170] implicate vinyltrimethylenemethane biradicals as true intermediates in the latter two rearrangements. Optically active ketal **95** gives completely racemic rearranged ketal **97**, hypothetically by way of the achiral intermediate **99**, and either epimer of the ether **96** gives the same mixture of epimers of **98** (predominantly the *syn*-methoxy isomer). The expected intermediate from the comparable cleavage of the bridge bond of **18** would be the non-Kekulé biradical **19**, or its zwitterionic counterpart.

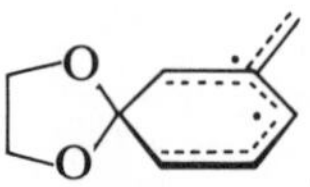

(99) 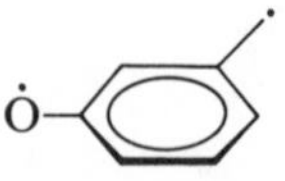 **(19)**

The first proposal of the occurrence of such a species in a solution phase reaction was made in 1964, when Leitich and Wessely[171] explained the formation of products **100** and **101** in the reactions of the acetoxydiene (**102**) as arising from the zwitterionic form **104**.

Subsequently, Seiler and Wirz[172] proposed the *m*-difluoromethylenephenoxyl biradical **105** as the key intermediate in the photohydrolysis of *m*-trifluoromethylphenol to *m*-hydroxybenzoic acid (**106**).

Capture of the zwitterionic species **19** and **107** of the *m*-quinomethane and *m*-naphthoquinomethane series is implied by the thermal and photochemical alcoholysis reactions of **18** and **93**, which give *m*-hydroxybenzyl ethers **108**[27, 140] and **109**[140, 148].

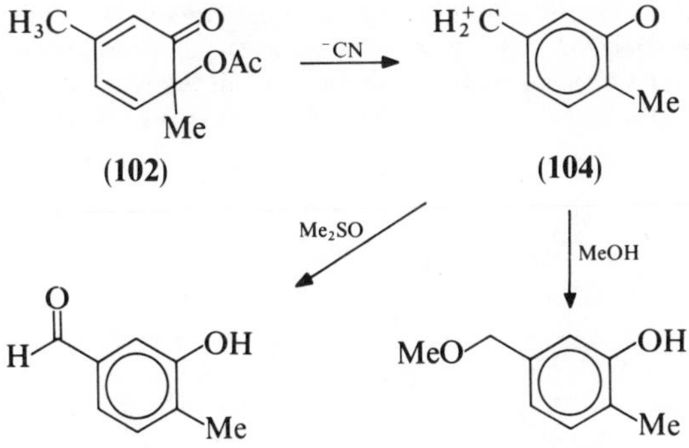

(102) (104)

(100) (101)

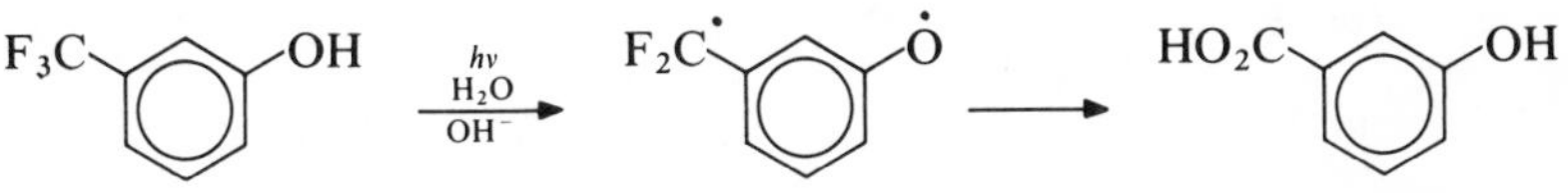

(105) (106)

These are not the types of products that would have been expected if the reactive intermediate behaved like an ordinary free radical, which should have displayed redox chemistry initiated by abstraction of hydrogen from the α-OH position of the alcohol.

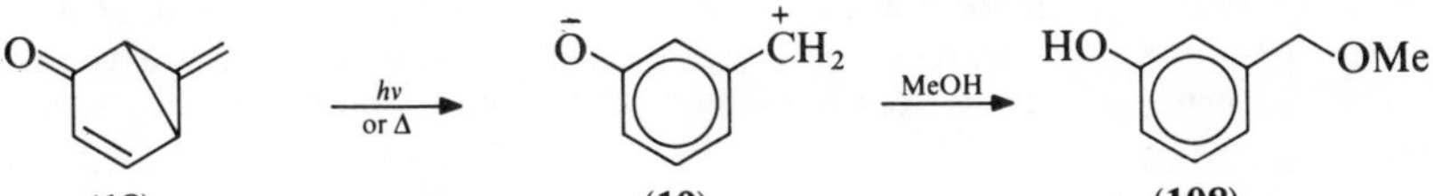

(18) (19) (108)

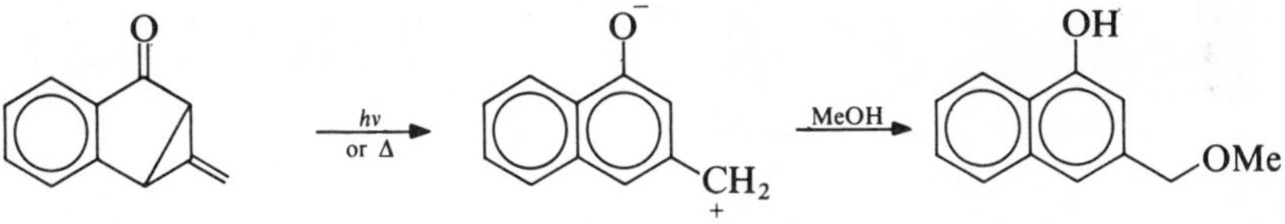

(93) (107) (109)

Similarly, the thermal or photochemical cycloaddition of **18** occurs smoothly with dienes or electron-rich olefins but not with electron-deficient ones[156, 167]. The products are 5- or 7-hydroxyindans, e.g. **110** and **111**. Also formed is the monocyclic olefin **112**.

Superficially, these reactions might be formulated as a Diels–Alder reaction and a vinylogous ene reaction, respectively, in which a σ bond plays a π-like role. However, these analogies imply a bimolecular mechanism, which is incompatible with the kinetics of the thermal reactions[156].

The thermal disappearance of the enone **18** is a first-order process, the rate being independent of the concentration of trapping agent (MeOH or isoprene). This suggests that *unimolecular* formation of a reactive intermediate is the rate-determining step. The Arrhenius parameters, $\log A$ (A in s^{-1}) = 14.0, E_a = 30.6 kcal mol^{-1}, are consistent with a spin-conservative transition state and a high barrier separating the reactant enone **18** from a singlet intermediate. Although practical circumstances make difficult an isothermal comparison of the behavior of the reactive species generated thermally (110°C) and photochemically (0°C), the product compositions from the two sets of conditions are the same and imply that a common reactive intermediate is involved.

(18) $\xrightarrow[\text{or } \Delta]{h\nu}$ (19)

(111) (110) (112)

The formation of a singlet as the initial intermediate which is less stable than the triplet ground state (see Section III.B) establishes the conditions for a cascade mechanism. Attempts to test for the cascade of intermediates using the usual[165, 166] effects of dilution and O_2 on the product composition fail. Both tests give essentially negative results, as is also the case with the hydrocarbon biradical *m*-quinodimethane (Section III.D). Again, one cannot exclude the existence of the cascade mechanism on this basis. This can be verified by an examination of the kinetic form of the ratio of products from the singlet and triplet intermediates, P_S and P_T, expressed in equation 25. The rate constants are defined as shown in the reaction scheme, and k_2 may contain a term in concentration of another reactant.

(19-S) $\xrightarrow{k_1}$ (19-T) $\xrightarrow{k_2}$ side products

(19-S) $\xrightarrow{k_S \text{ olefin}}$ P_S (19-T) $\xrightarrow{k_T \text{ olefin}}$ P_T

$$P_S/P_T = \frac{k_S[\text{olefin}]}{k_1} + \frac{k_S k_2}{k_1 k_T} \tag{25}$$

If the rate constant (k_T) for the capture of the triplet is small relative to that leading to side products (k_2), the product ratio always will be dominated by the concentration-independent term, and there will be no dilution effect.

It is possible, however, to construct a strong presumptive case for the singlet as the reactive intermediate based upon a correlation of optical and EPR spectroscopy of immobilized species with time-resolved spectroscopy of the transients in fluid media.

The photochemical generation of the triplet ground states of *m*-quinomethane (**19**) and *m*-naphthoquinomethane (**92**) in glassy media has been described in the preceding section.

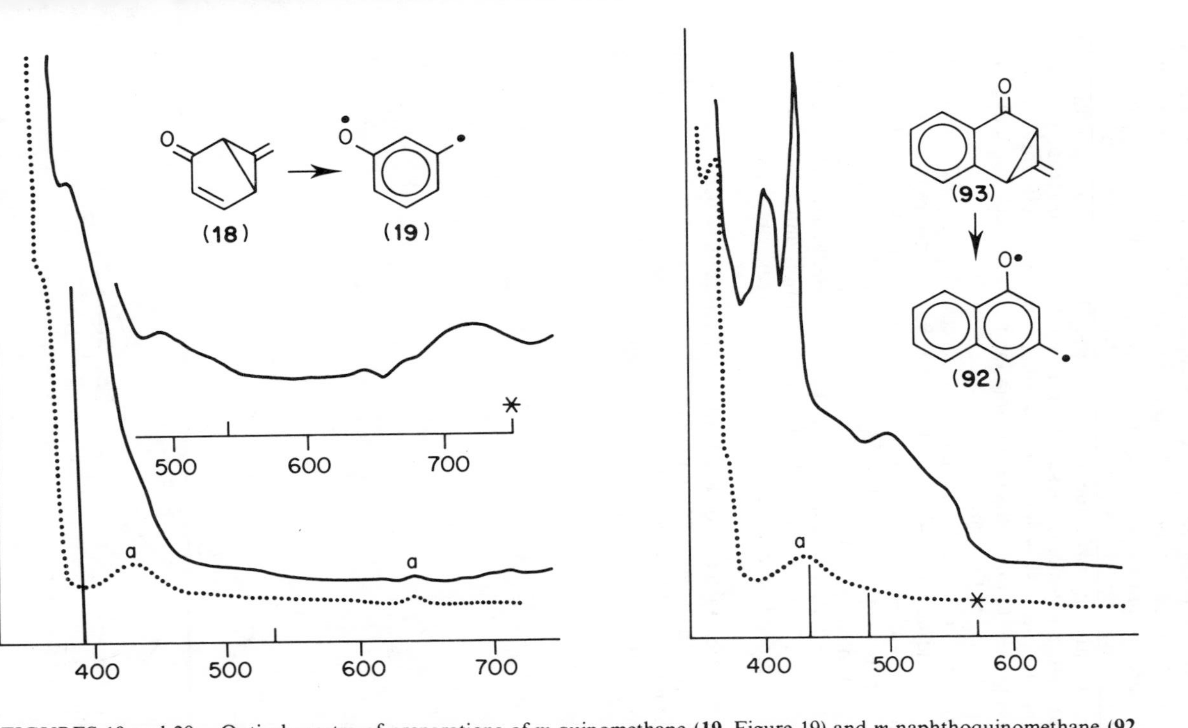

FIGURES 19 and 20. Optical spectra of preparations of *m*-quinomethane (**19**, Figure 19) and *m*-naphthoquinomethane (**92**, Figure 20) triplet states in EtOH glass at 15 K obtained by irradiation of ketones **18** and **93**, respectively. The vertical bars mark the locations (nm) and relative intensities calculated by INDO/S-CI theory. In Figure 19, the vertical scale is expanded for the uppermost curve. Asterisks (*) mark the predicted locations of the longest wavelength n, π^* band . . . . before irradiation; ——— after irradiation for one minute at 220–440 nm. The features marked 'a' are baseline artifacts. *Reproduced with permission from D. E. Seeger, P. M. Lahti, A. R. Rossi and J. A. Berson*, J. Am. Chem. Soc., **108**, 1251 (1986). *Copyright* (1986) *American Chemical Society*

Accompanying the appearance of the characteristic EPR signals of these biradicals are bands in the visible region of the optical absorption spectrum (Figures 19 and 20)[173]. If the large singlet–triplet splittings predicted by theory (Section III.E) are even approximately correct, the singlet will be present in too small a concentration to contribute to the optical absorption, which therefore must be ascribed essentially completely to the triplet.

Although the longest wavelength feature of the spectrum of *m*-quinomethane triplet (Figure 19) lies too far to the blue to escape being overlapped by the strong absorption tail of the starting ketone (**18**) and hence is not suitable for time-resolved spectroscopy, the *m*-naphthoquinomethane triplet has a well-defined maximum near 500 nm (Figure 20) which can be used for this purpose. In low-temperature isolation media the position of this band is insensitive to the polarity of the matrix[173].

The same species is deduced to be the second of two successive transients formed in the solution-phase picosecond flash photolysis of the ketone **93**[174]. This transient (called B) has an absorption maximum near 500 nm whose position is also insensitive to solvent polarity (Figure 22). It is long-lived on the picosecond time-scale. The primary photoproduct from the flash photolysis of **93** is a different species (A) which has a solvent-sensitive spectrum (Figure 21) and which decays to B by first-order kinetics with a solvent-sensitive lifetime of 0.25–2.74 nanoseconds. Transient A also can be intercepted by a large excess of methanol in a second-order process to give **109** which can compete with its decay to B.

The data are interpreted in terms of a cascade mechanism, in which A and B are identified, respectively, with the *m*-naphthoquinomethane zwitterionic singlet **107** and its biradical triplet counterpart[174]. Since the descriptive chemistry of *m*-quinomethane matches that of *m*-naphthoquinomethane, it seems highly likely that a similar cascade exists in both systems.

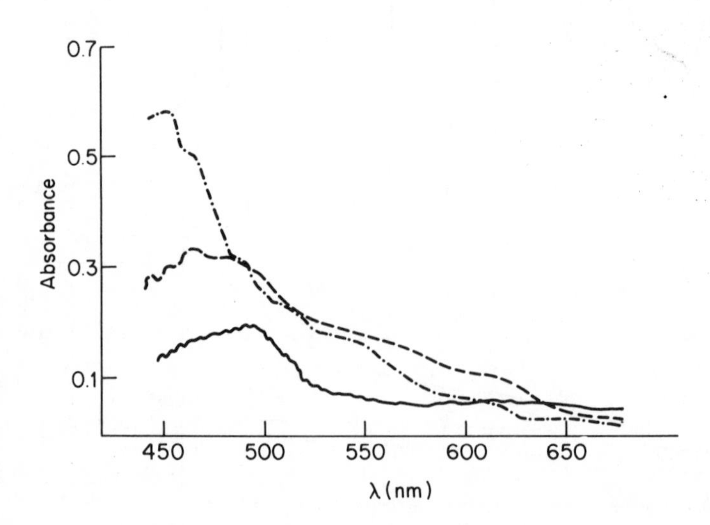

FIGURE 21. Spectra observed 25 ps after 355 nm excitation of ketone **93** in cyclohexane (——), benzene (– – –) and acetonitrile (–·–·–). *Reproduced with permission from J. L. Goodman, K. S. Peters, P. M. Lahti and J. A. Berson,* J. Am. Chem. Soc., **107**, 276 (1985). *Copyright* (1985) *American Chemical Society*

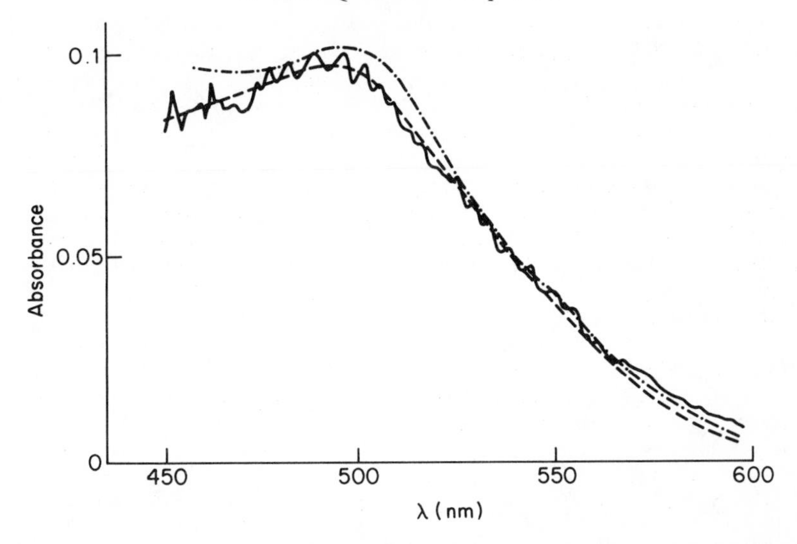

FIGURE 22. Spectra observed 9 ns after 355 nm excitation of ketone **93** in cyclohexane (——), benzene (– – –) and acetonitrile (· · ·). *Reproduced with permission from J. L. Goodman, K. S. Peters, P. M. Lahti and J. A. Berson,* J. Am. Chem. Soc., **107**, 276 (1985). *Copyright* (1985) *American Chemical Society*

(**93**) —$h\nu$, 25 ps→ singlet (**107-A**) —$8.5 \times 10^8\ s^{-1}$→ triplet (**107-B**)

(**107-A**) —MeOH, $3.6 \times 10^8\ M^{-1} s^{-1}$→ (**109**) (OH, CH_2OMe)

(**107-B**) —MeOH→ ?

3. *Stereochemistry and regiochemistry of m-quinomethane cycloadditions*

The cycloadditions of the trimethylenemethane (TMM) biradicals 2-methylenecyclopentane-1,3-diyl (**113**) and 2-isopropylidenecyclopentane-1,3-diyl (**114**) show all the signs expected of concerted reactions[165, 166]. They are stereospecifically *cis* and regiospecifically fused (reaction at a + b rather than at a + a′).

The regiospecificity has been ascribed[165, 166, 175, 176] to phase matching of the TMM HOMO and the olefin LUMO at the reactive sites (**115**). An extension of this reasoning to the case of *m*-quinomethane closed-shell singlet suggests that, if the reaction is to be concerted, only the orientation leading to the *ortho* product is allowed, whereas that leading to *para* product is forbidden. If these factors were determinative, the *ortho* adduct would be formed concertedly and stereospecifically, but the *para* adduct would be formed

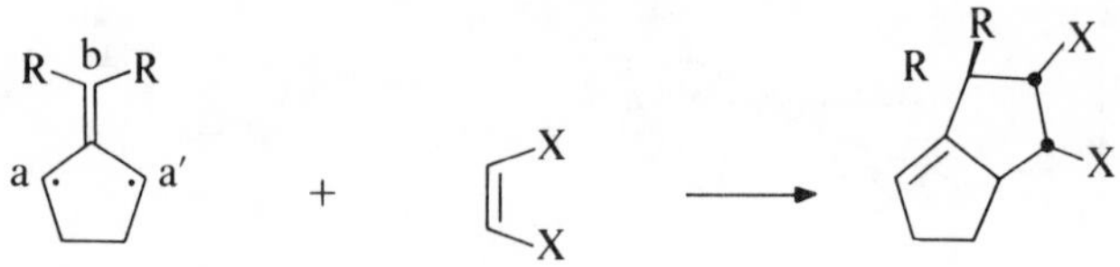

(113, R = H)
(114, R = Me)

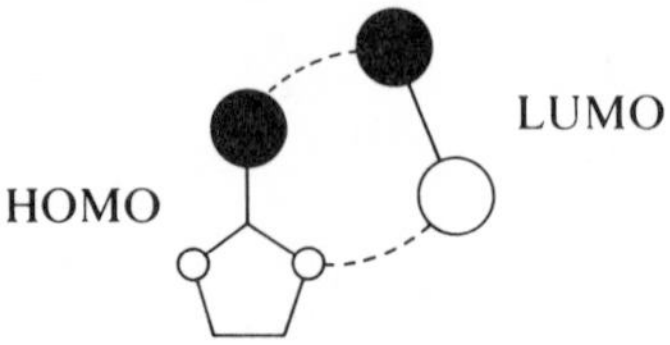

HOMO LUMO

(115)

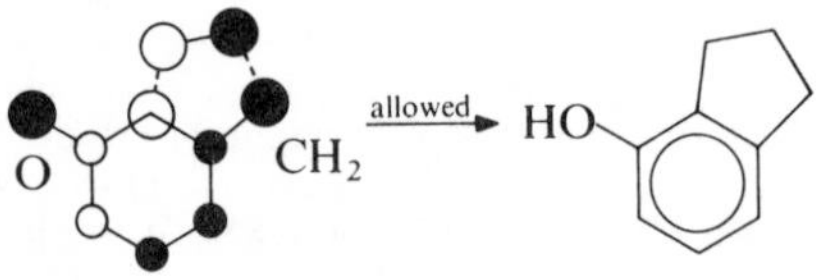

allowed HO

O CH$_2$

O CH$_2$ HO

forbidden

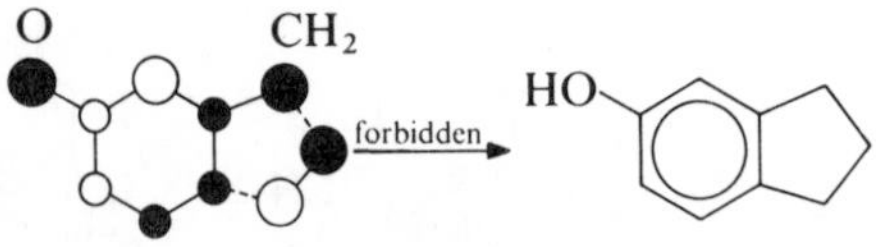

by a stepwise mechanism, which would lead to some loss of stereospecificity. The anticipated switch in stereospecificity would clearly signal a change in mechanism. However, as Table 6[167] shows, the cycloadditions of *m*-quinomethane to *cis*-1,2-dimethoxyethene show no significant differences in stereospecificity between the *ortho* and *para* adducts.

Although the overall stereospecificities of cycloaddition are somewhat lower with the 2,4-hexadienes than with *cis*-1,2-dimethoxyethene (e.g. 6:1 in the *ortho-cis* case as compared to 17:1), the stereospecificities in the *ortho* adduct series again do not differ significantly from those in the *para* series. With neither kind of trapping agent does the product stereospecificity depend markedly upon the orientation of the addition, a finding that argues against a dominant reaction under orbital symmetry control of a concerted *ortho* process[167].

TABLE 6. Product ratios in the reaction of **24** with *cis*-1,2-dimethoxyethene at 115°C

Product ratio	Solvent	
	Benzene	MeCN
o-cis/o-trans (**116**/**117**)	17.0 ± 0.6	17.1 ± 0.1
p-cis/p-trans (**118**/**119**)	13 ± 3	14.7 ± 1
ortho/para [(**116** + **117**)/(**118** + **119**)]	3.2 ± 0.2	1.5 ± 0.1

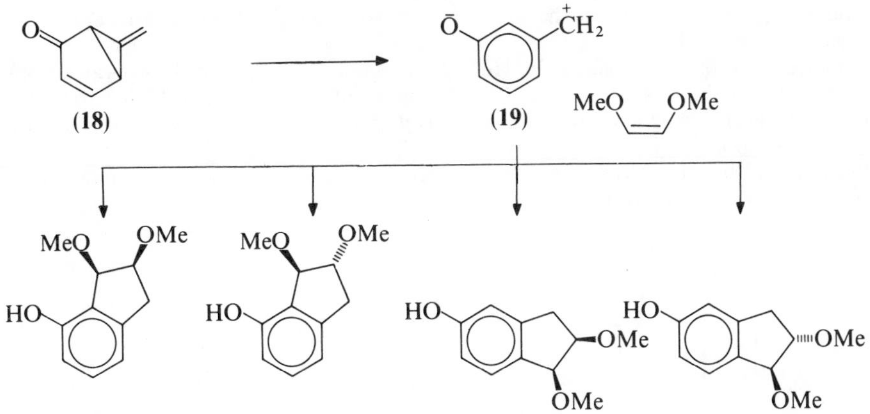

Solvent polarity has little effect on the *cis*/*trans* product ratios (Table 6) in the reaction with *cis*-1,2-dimethoxyethene. On the other hand, the *ortho*/*para* ratio responds noticeably, with the relative amount of *para* product increasing in the more polar solvent, MeCN. The effect may be rationalized as a selective stabilization of the more polar 'extended' transition state **121** relative to the less polar 'coiled' transition state **120** in a polar solvent[167]. This solvent effect is reminiscent of the increased preference for *endo* vs. *exo* stereochemistry of the Diels–Alder reaction with increasing solvent polarity, which was interpreted in a similar way[177].

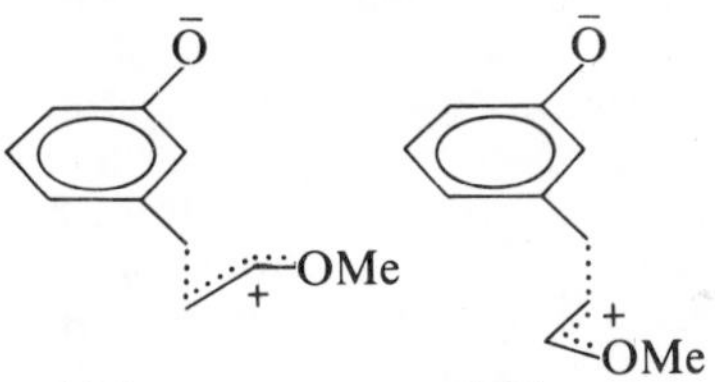

(120) (121)

E. Bis-*m*-Quinomethanes

1. Theory

The intrinsic interest in non-Kekulé molecules as examples of a highly unconventional form of matter would justify an unstructured examination of their synthesis and behavior. Such a program would be intellectually respectable in the same sense as was the exploration of aromatic chemistry of the late nineteenth century. One intuits that new chemical reactions and physical properties wait to be discovered in this field, and no more sophisticated motivation is needed. Nevertheless, because of their potential or actual open-shell nature, non-Kekulé molecules represent engaging theoretical problems (see Section III.B). The experimental organic chemist therefore finds motivation beyond the descriptive one in the design and construction of new non-Kekulé systems to stimulate the development of theory.

This interaction between theory and experiment actually began early in the study of non-Kekulé molecules. Section II has described briefly the observation of Müller and Bunge[29] that the two hydrocarbons **14** and **18** of Schlenk and Brauns[13, 14] are paramagnetic, as would be predicted from their half-filled degenerate NBMOs, if Hund's rule applied to molecules.

However, in the same year, 1936, Hückel[52] pointed out that good theoretical reasons existed to apply Hund's rule to **8** but not to the apparently closely related **12**. His argument derives from the recognition of fundamentally disparate connectivity patterns in the two molecules. Specifically, one can consider **12** to have been formed by a union at the *meta* positions of two triphenylmethyl fragments, whereas **8** is formed by a union at a triphenylmethyl *meta* position and a benzhydryl methine position.

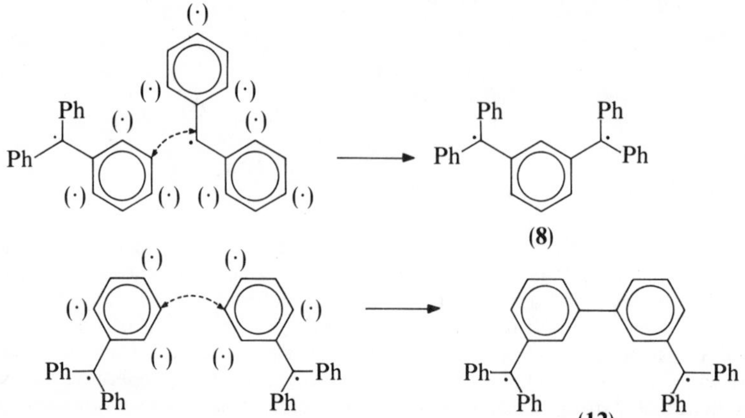

(8)

(12)

Because the carbon atoms at the **12**-forming union have zero HMO coefficients (the sites are 'inactive'[28]), the exchange interaction will be small. The quantum mechanical reasoning that leads to this conclusion is straightforward.

Assume that the interaction of the two triphenylmethyl fragments, A and B, can be modeled by the interaction of the two single electrons in the triphenylmethyl HMO NBMOs, orbital χ_{10} in each case. Denote NBMO χ_{10} on fragment A as orbital a and that on fragment B as b. Each HMO χ_j made up of $\pi -$ AOs ϕ_i may be expressed as in equation 26.

$$\phi_i = c_i p_\pi \tag{26a}$$

$$\chi_j = \sum_i \phi_i \tag{26b}$$

where c_i is the coefficient of the ith carbon p orbital in the π system. Expansion of the exchange integral (see equation 9) gives rise to products of the form in equation 27

$$\text{a}(1)\text{b}(1) = \left(\sum_i \phi_i\right)\cdot\left(\sum_k \phi_k\right) \tag{27}$$

where the summations over i and k represent the linear combinations of atomic orbitals that constitute the HMOs a and b. To a first approximation, products between terms of the two summations can make a contribution to the exchange energy only if the terms refer to carbon atoms that are adjacent and have non-zero coefficients. But the only linkage between the fragments A and B is the *meta–meta* junction in **12**. Since this junction links two sites with zero coefficients, the exchange energy vanishes at this level of approximation, and the normal basis for the application of Hund's rule vanishes with it. The argument is qualitative, since at a higher level of theory, the π coefficients at the junction will not be precisely zero. Moreover, the case of an inactive–active connection (as in trimethylenemethane, **28**) would be indistinguishable from that of an inactive–inactive one if the argument were to be taken literally. Nevertheless, the major point remains, namely, that an inactive–inactive junction in a π fragment analysis leads to the prediction of a small singlet–triplet gap.

Given the paucity of related experimental data and the primitive state of molecular quantum theory at the time, the immediate recognition[52] that **8** and **12**, although superficially similar in structure, were likely to be fundamentally disparate in magnetic properties, can only be regarded as remarkable[173].

Oddly, Hückel's prescient paper of 1936 seems to have made little impact. Longuet-Higgins does not cite it in his very influential work of 1950[56] and it is clear that he makes no distinction between non-Kekulé molecules with inactive–inactive connections and those without. For example, he applies Hund's rule to predict triplet ground states for both tetramethyleneethane (**29**) and trimethylenemethane (**28**), which are the parent substances of the former and the latter categories, respectively.

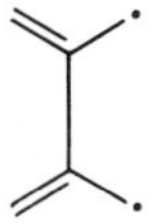

(**29**)  (**28**)

In fact, it took more than 40 years before attention was redirected to the problem Hückel had addressed in 1936. The seminal papers are those of Borden and Davidson[62], Misurkin and Ovchinnikov[64] and Ovchinnikov[63], which were followed by numerous other contributions[65–72, 178–180]. These theoretical studies have strengthened the conceptual basis for identifying which π-conjugated molecules might violate Hund's rule.

In the Borden–Davidson formalism[62], if the NBMOs are (or by linear combinations can be) confined to separate regions of the molecule, they are said to be disjoint. This property, to first order, erases the Coulombic repulsion that usually destabilizes the singlet in π-conjugated biradicals.

Disjoint character is exemplified by the NBMOs of cyclobutadiene (which is formally not a non-Kekulé molecule) and tetramethyleneethane (which is). The separation of the two NBMOs of cyclobutadiene to non-identical regions of the molecule is illustrated in **122a** and **122b**, and the corresponding NBMOs of tetramethyleneethane are shown as **29a** and **29b**. This property of separability of the NBMOs means that to first order, the peaks of

disjoint

(**122a**) 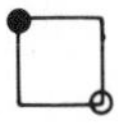 (**122b**) 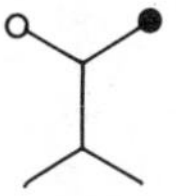 (**29a**) (**29b**)

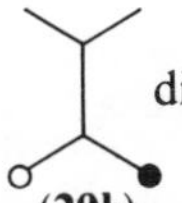

the two electron distributions in the singlet state of either of these disjoint systems are as well separated as those in the triplet. This is to be contrasted with the more usual non-disjoint type of biradical, exemplified by trimethylenemethane, whose NBMOs **28a** and **28b** cannot be confined to separate atoms by any linear combination.

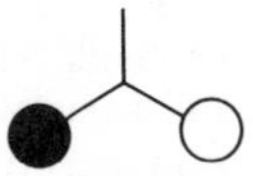

(**28a**) 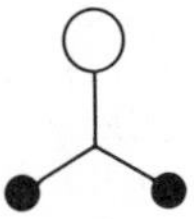 (**28b**)

A sufficient (see tetramethyleneethane) but not a necessary (see cyclobutadiene) criterion for disjoint character is met if the molecule can be constructed by inactive–inactive union of two radical fragments[62]. This is the intellectual bridge between Hückel's early insight and the more rigorous, computationally buttressed modern formulation.

At the SCF level of theory, the singlet and triplet states of a disjoint system are degenerate. A higher-order effect, dynamic spin polarization, then can selectively stabilize the singlet and produce a violation of Hund's rule. Two examples of this have been confirmed computationally at a reasonably high level: a square planar cyclobutadiene (**122**)[180–182] and planar tetramethyleneethane (**29**)[62].

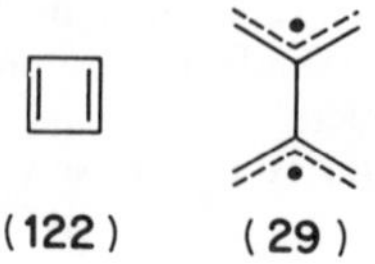

(122) (29)

The physical basis of dynamic spin polarization[180, 183–185] can be understood as an extension of the more familiar static spin polarization[81, 186–192]. In allyl radical, for example, simple HMO theory predicts zero spin density at C(2) because the NBMO ψ_2 has a node at that site. However, the EPR spectrum of allyl[193] shows a finite coupling constant (4.06 G) of the electron with the proton at C(2) (see Figure 23). This can be explained at a

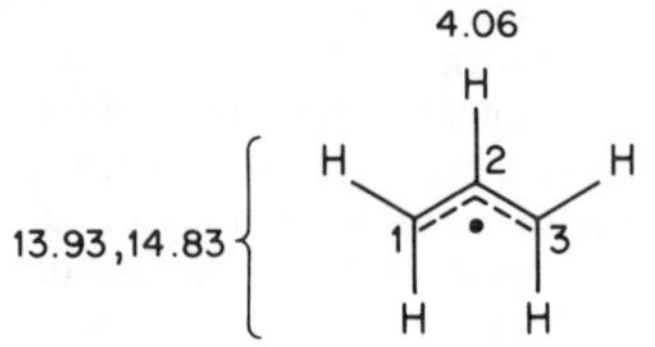

4.06

13.93,14.83

FIGURE 23. Proton hyperfine coupling constants in allyl radical (in gauss). *Reproduced with permission from Ref.* 110, *p.* 91

higher level of theory by taking electron correlation into account. One way of showing the interactions is to assign different orbitals, $\psi_1(\alpha)$ and $\psi_1(\beta)$, to the electrons of opposite spin in the lowest occupied π MO, ψ_1 (see Figure 24). Assume (arbitrarily) that the single electron in ψ_2 has α spin. This electron polarizes the electrons in ψ_1, that is, causes a perturbation in their spatial distribution, so as to minimize the energy. There are two conceivable polarizations: type A concentrates α spin at C(2) and β spin at C(1) and C(3), whereas type B concentrates the spins in the opposite sense. Both types of polarization will minimize the electron repulsion by localizing the electrons of ψ_1 to different sites. However, the type B polarization is preferred, because it concentrates electrons of like spin (α) at common sites (C(1) and C(3)), whereas type A polarization concentrates electrons of opposite spin at those sites. The physical basis of the preference for type B polarization is the same as that of Hund's rule (Section II.B): the exchange energy will favor the parallel spin configuration at a given site. The result is that, although most of the spin density does reside at C(1) and C(3) in the allyl radical, as would be predicted just from the electron distribution in ψ_2, a finite amount of spin density resides at C(2) because of the polarization of the electrons in ψ_1, and the spin at C(2) is opposite to that of C(1) and C(3).

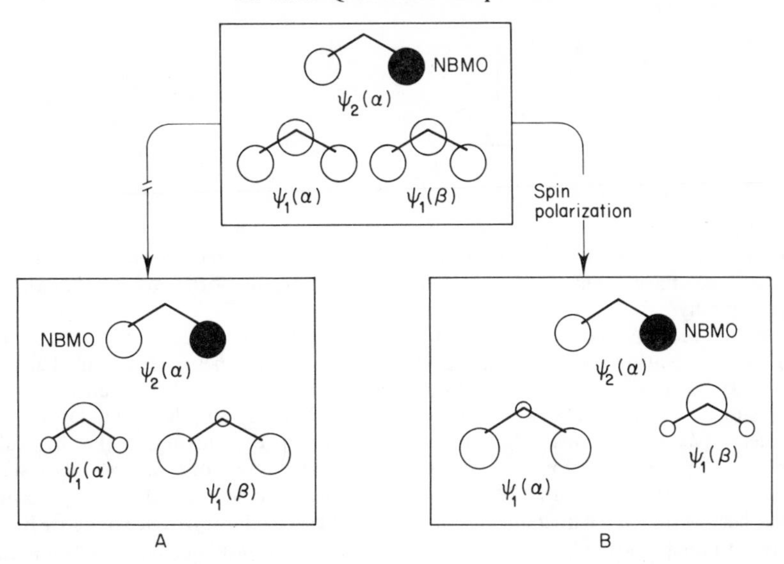

FIGURE 24. Two conceivable spin polarizations of the ψ_1 π electrons of allyl radical ψ_2 odd electrons. Type A is less favorable than type B

This static spin polarization arises from the existence of a net spin in the allyl radical, whereas in a singlet state of a biradical, e.g. **29**, or a 4n polyene, e.g. **122**, there is no net spin. However, a proper quantum mechanical description of such species employs configuration interaction, in which the ground electron configuration mixes with excited configurations, many of which have singly occupied orbitals. The electrons in such orbitals polarize the electrons in lower energy orbitals in the same way as described for allyl, so that the effect is similar. Hence the name[180] 'dynamic spin polarization'. Although a similar polarization might be imagined for the triplet, it turns out that the polarization is much less than in the singlet[180] and, in fact, to first order, there is no polarization in the triplet[185]. The net result is that dynamic spin polarization will conduce to a singlet ground state, in violation of Hund's rule, in disjoint π-electron molecules.

In the formalism of Ovchinnikov[63], a similar result can be obtained for the case of alternant π-conjugated molecules. These conform to a rule that each atom of the π system can be assigned a parity by starring in such a way that no two starred or unstarred positions are adjacent. Using a Heisenberg Hamiltonian within a valence-bond theoretical framework, Ovchinnikov[63] derived a formula which states that the total spin of the ground state will be given by half the difference between the numbers of starred and unstarred π centers (equation 28):

$$S = (n^* - n)/2 \tag{28}$$

Obviously, when $n^* = n$, the total spin is zero, and the ground state is singlet. Since it has been shown[62] that such equal parity systems are disjoint, but that not all disjoint alternant systems have equal parity, the $S = 0$ cases of Ovchinnikov are a subcategory of disjoint molecules.

The Ovchinnikov formulation[63] is especially attractive since it requires no calculations and reduces the task of finding candidate structures for low-spin ground states to a simple enumeration. How this works should be immediately obvious from the starring diagrams for trimethylenemethane (**31**) and tetramethyleneethane (**32**). Note that equation 28 is

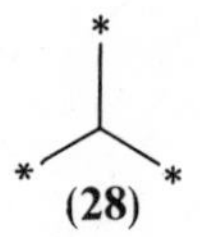

(28)
$S = (3 - 1)/2 = 1$
triplet

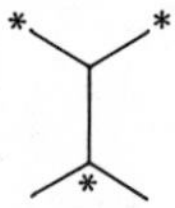

(29)
$S = (3 - 3)/2 = 0$
singlet

proposed[63] to hold even in heteroatom perturbed π-conjugated systems, an assumption that, as we have seen (Section III.B), is supported both computationally and experimentally.

Although reasons to expect Hund's rule violations thus have been recognized for more than 50 years, Nature has so far frustrated attempts to verify these predictions. Part of the blame may be placed upon the already described difficulties of identifying unstable species with singlet ground states. Moreover, the obvious simple candidates, cyclobutadiene (**122**) and tetramethyleneethane (**29**) test the theory only if they exist in idealized geometries, square planar for **122** and planar for **29**. The weight of evidence is against these geometries for **122**[194, 195] and **29**[95, 196, 197]. Similarly, the Schlenk–Brauns hydrocarbon **18**, has been assigned a triplet ground state on the basis of the adherence of its EPR signal intensity to the Curie law[198]. It seems probable that **12** is non-planar, so that its nominally equal-parity nature is not a reliable basis for predicting a low-spin ground state.

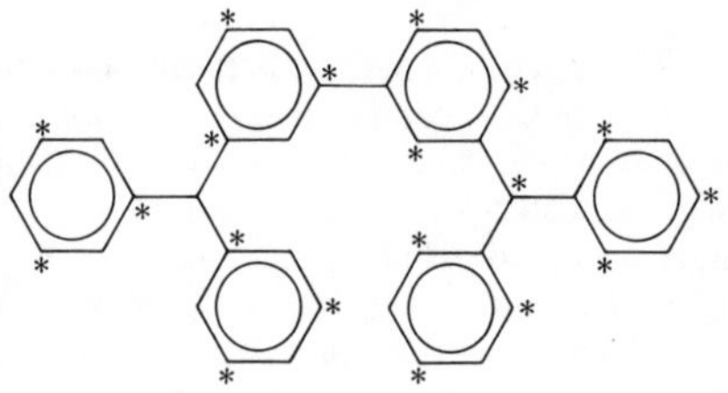

(12)
$S = (19 - 19)/2 = 0$

The tetramethyleneethane derivative **123** presumably is planar or nearly so, but the experimental evidence so far does not seem to permit an unequivocal assignment of the ground state[199–202].

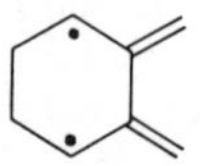

(123)

The currently available EPR techniques for measuring ground-state singlet-triplet energy separations (ΔE) do not work in most cases because curvature in the Curie plot usually is indiscernible (see Section III.B.2b). This forces the issue to be cast in binary form: is Hund's rule obeyed or not? It should be apparent, however, that two very different kinds of biradical, a non-disjoint triplet with a large ΔE and a disjoint one with a small ΔE, both might hide behind a 'linear' Curie plot. Hückel's conjecture then would be correct but experimentally unverified. At present, a historical accident makes us dependent upon an actual violation of Hund's rule for an experimental proof.

2. Experiment

Recent studies have used the bis-*m*-quinomethanoid **124** to test the equal parity criterion for low-spin ground states[203–205]. The biradical **124** has overall C_{2h} symmetry. As an equal parity system, it should have a singlet ground state. It is related to the simplest disjoint biradical tetramethyleneethane (**29**) in the sense that the Hückel NBMO electron density is confined to the two (ringed) pentadienyl moieties disjointly. The rigid conjugated framework prevents twisting about the connection between them. The isomeric molecule **125** has overall C_{2v} symmetry. It is non-disjoint and formally a tetraradical, for which the qualitative spin rule predicts a quintet ground state. Tetraradical **125** therefore serves as a control system to test the application of Hund's rule to a non-disjoint case closely related to the disjoint test **124**.

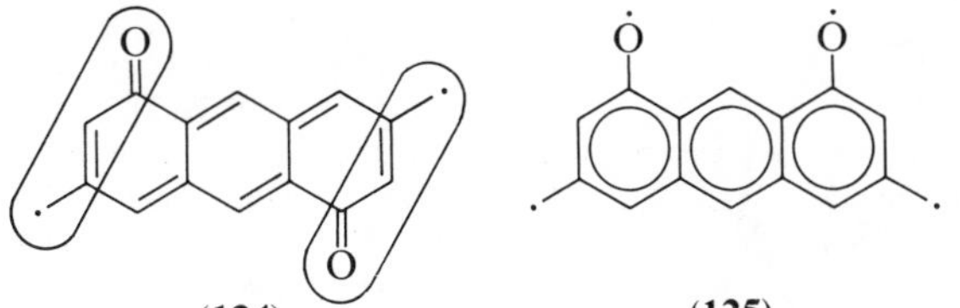

(**124**) $S = (9-9)/2 = 0$

(**125**) $S = (11-7)/2 = 2$

The synthesis of **124** and **125** are effected by successive photochemical ring-cleavages, **126** → **127** → **124** and **128** → **129** → **125**, of samples immobilized in rigid media. The

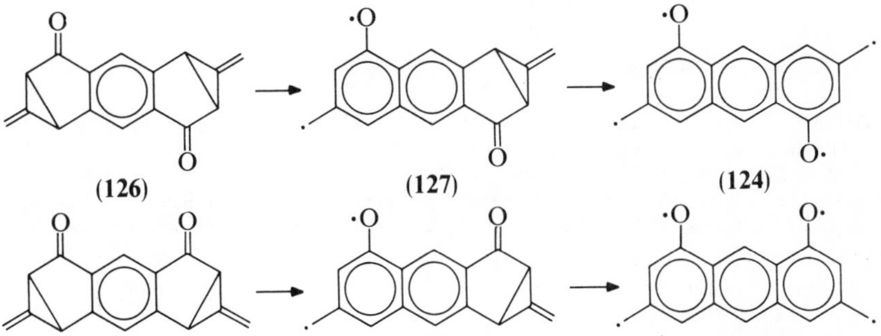

(**126**) → (**127**) → (**124**)

(**128**) → (**129**) → (**125**)

sequence of events can be monitored by both optical and EPR spectroscopy (Figures 25 and 26). Cleavage of one bridge bond in either **128** or **126** gives either of the orange species (Figures 25 and 26, respectively) formulated as substituted *m*-naphthoquinomethanes **129** and **127**. By analogy to *m*-naphthoquinomethane itself (Section III.E), these species are expected to be triplets and are in fact observable as such by EPR spectroscopy (Figures 27 and 28). The $|D|/hc$ values for **129** and **127**, $\sim 0.020\ \text{cm}^{-1}$, are in good agreement with that of *m*-naphthoquinomethane, $|D|/hc = 0.0204\ \text{cm}^{-1}$.

Irradiation at wavelengths beyond 420 nm of the immobilized sample of **129** causes the optical spectrum responsible for the orange color to fade and be replaced by a new absorption spectrum (Figure 25) associated with a visible red–purple color. A parallel change occurs in the EPR spectrum, as the primary photoproduct, the triplet **129**, is

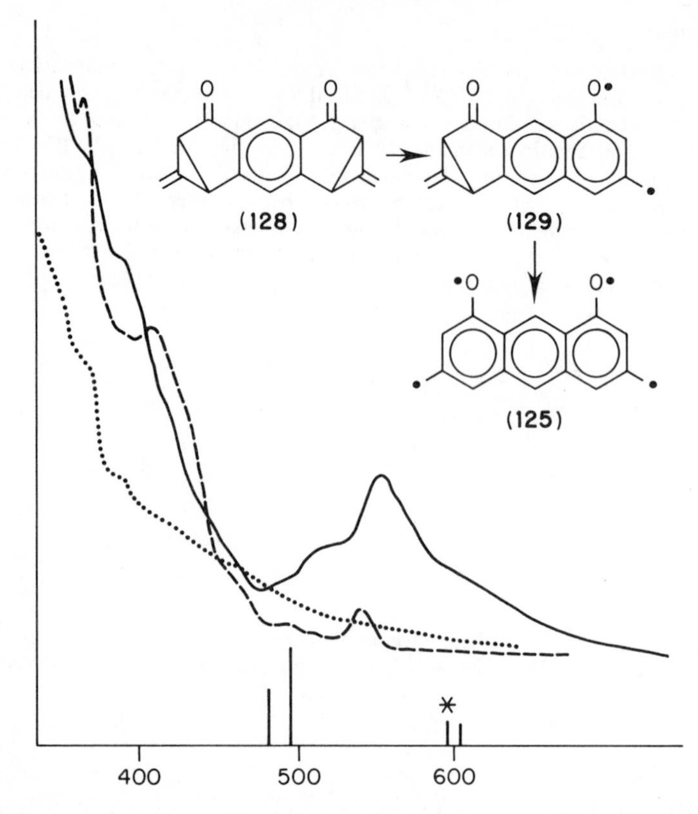

FIGURE 25. Optical spectra at 16 K in EtOH glass obtained by irradiation of the diketone **128** at 220–440 nm to give the triplet **129** and then at > 425 nm to give the quintet **125**. (· · ·) before irradiation; (– – –) primary photolysis product **129**; (——) secondary product **125**. Transitions shown by vertical bars are calculated for **125**. *Reproduced with permission from D. E. Seeger, P. M. Lahti, A. R. Rossi and J. A. Berson,* J. Am. Chem. Soc., **108**, 1251 (1986). *Copyright* (1986) *American Chemical Society*

converted to a new species, the quintet C_{2v} tetraradical **125** (Figure 27). The quintet is the ground spin state of the tricyclic structure **125**, as judged from a linear Curie plot between 15 and 70 K. Aside from the intensity variation, the EPR spectrum does not change in this temperature range, which is interpreted to mean that no nearby state of different multiplicity is populated or depleted. Similarly, the optical spectrum is temperature-insensitive. As expected, Hund's rule applies to this non-disjoint case.

Similar irradiation of the immobilized sample of the primary photoproduct **127** in the C_{2h} series causes the orange color to change to red–purple (Figure 26) and the triplet EPR spectrum to be replaced by that of another triplet species with a much smaller ZFS, $|D|/hc = 0.0025\ \text{cm}^{-1}$ (Figures 28 and 29). This is consistent with the much greater average separation of the unpaired electrons in **124** as compared to **127**. The Curie plot is linear between 15 and 70 K. Again, neither optical nor EPR spectroscopy gives evidence of a variable contribution from any other species in this temperature range. Thus, either the singlet and triplet are accidentally almost exactly degenerate ($\Delta E \leqslant 0.024\ \text{kcal mol}^{-1}$), or

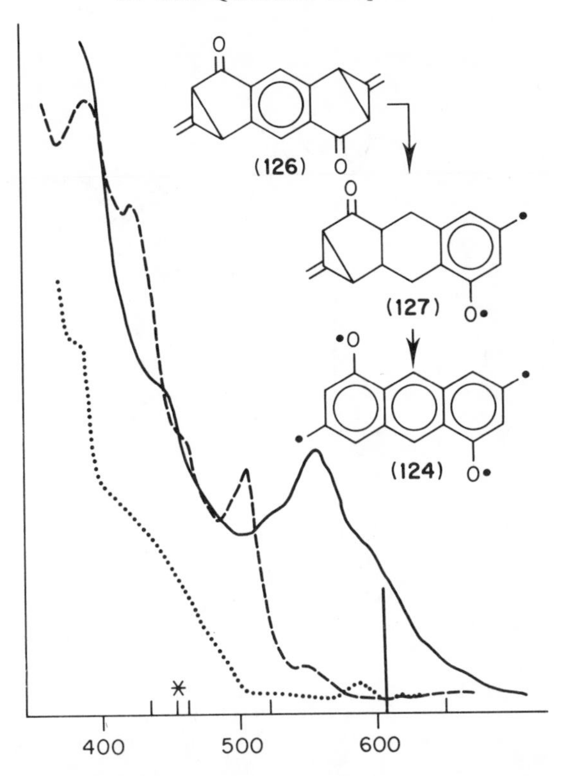

FIGURE 26. Similar to Figure **25** for the process **126** (· · ·) → **127** (–––) → **124** (——). Transitions shown by vertical bars are calculated for **124**. *Reproduced with permission from D. E. Seeger, P. M. Lahti, A. R. Rossi and J. A. Berson,* J. Am. Chem. Soc., **108**, 1251 (1986). *Copyright* (1986) *American Chemical Society*

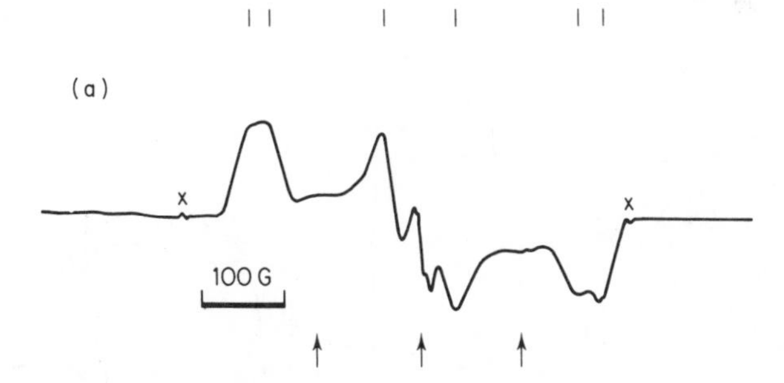

FIGURE 27. EPR spectra from photolysis of diketone **128** in 2-methyltetrahydrofuran (2-MTHF) glass at 36 K. The lines and arrows mark the positions of transitions associated with the triplet state of biradical **129** and the quintet state of tetraradical **125**, respectively. The symbol X marks absorption by H atoms. All the spectra were recorded with the same spectrometer control settings. (a) After 30 s irradiation at 305 nm $\leqslant \lambda \leqslant$ 525 nm; (b) after 10 m at $\lambda >$ 425 nm; (c) after 35 m at $\lambda >$ 425 nm; (d) after 180 m at $\lambda >$ 425 nm. *Reproduced with permission from D. E. Seeger, P. M. Lahti, A. R. Rossi and J. A. Berson,* J. Am. Chem. Soc., **108**, 1251 (1986). *Copyright* (1986) *American Chemical Society*

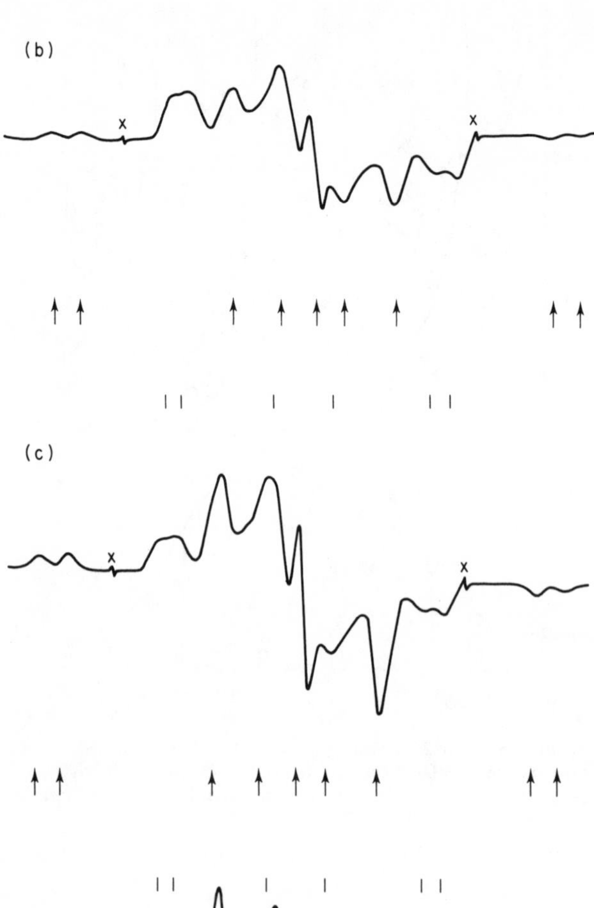

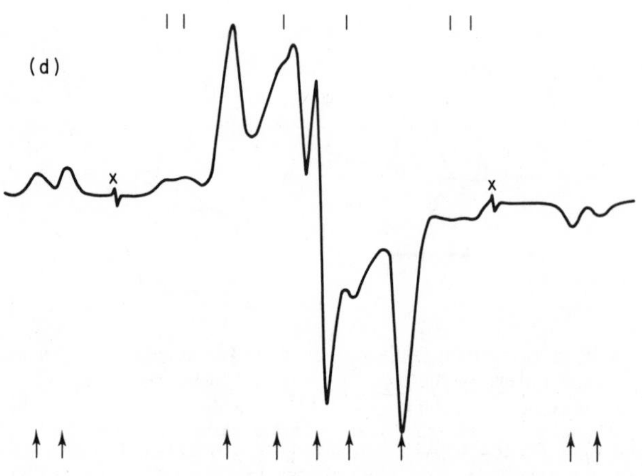

FIGURE 27. Continued

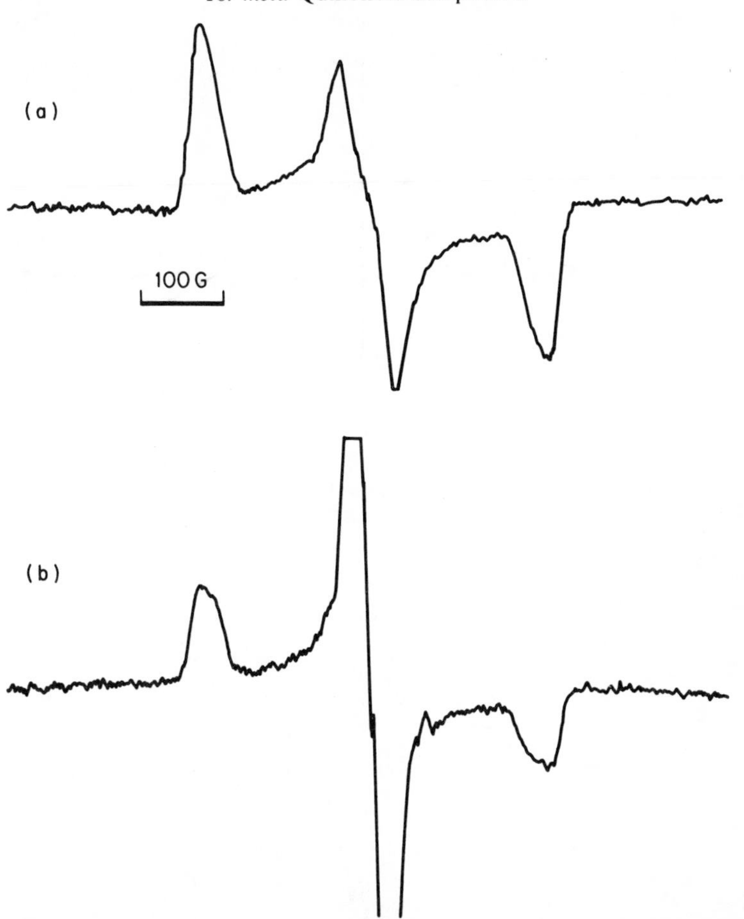

FIGURE 28. EPR spectra obtained from photolysis of diketone **126** in 2-MTHF glass at 77 K. Spectra (a)–(c) were recorded with the same spectrometer settings. Spectrum (d) is of the same preparation as (c), except that the x-axis has been expanded and the y-axis contracted. (a) After 30 s, $305 \leqslant \lambda \leqslant 525$ nm. The spectral contribution of a small amount of secondary photoproduct has been digitally subtracted. (b) After 4 m further photolysis, $\lambda \geqslant 425$ nm. (c) After 81 m, $\lambda \geqslant 425$ nm; (d) same preparation as (c) with altered scale. *Reproduced with permission from D. E. Seeger, P. M. Lahti, A. R. Rossi and J. A. Berson,* J. Am. Chem. Soc., **108**, 1251 (1986). *Copyright* (1986) *American Chemical Society*

more probably, the triplet is the ground state[173, 204, 205]. Thus, a high-spin ground state seems to be preferred, despite the disjoint character of the NBMOs associated with the equal-parity connectivity pattern of the C_{2h} biradical **124**.

The apparent breakdown of the qualitative theories in the case of **124** makes it clear that other factors can sometimes override disjoint character. There can hardly be any doubt that the main idea of the qualitative theories is correct, namely that the exchange interaction of the SOMO electrons in equal-parity (therefore disjoint) π-conjugated non-Kekulé molecules is small, and hence that a singlet should be the ground state or a very low-

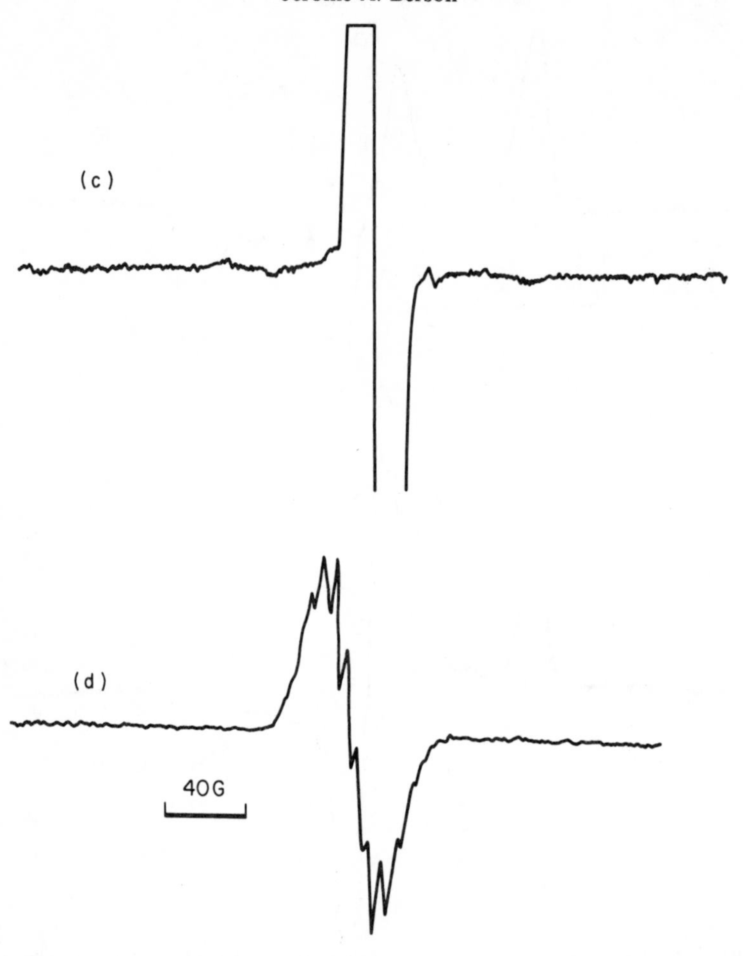

FIGURE 28. Continued

lying excited state. The problem at present is to perceive the structural features of formally disjoint test molecules that act to restore some of the exchange coupling and thereby contaminate the desired pure test situation.

In this connection, it is encouraging to note the outcome of an INDO/S-CI calculation[70], which places the triplet of **124** 4.3 kcal mol^{-1} lower in energy than the singlet, in qualitative agreement with the experimental observations. Although at this stage, these theoretical results cannot be said to have elucidated the underlying physical reasons for the unexpected adherence of this particular disjoint system to Hund's rule, they offer some basis for confidence that INDO/S-CI calculations may provide hints on which non-Kekulé molecules would constitute useful tests of spin state order and spacing. An example of such utility is given in the following section.

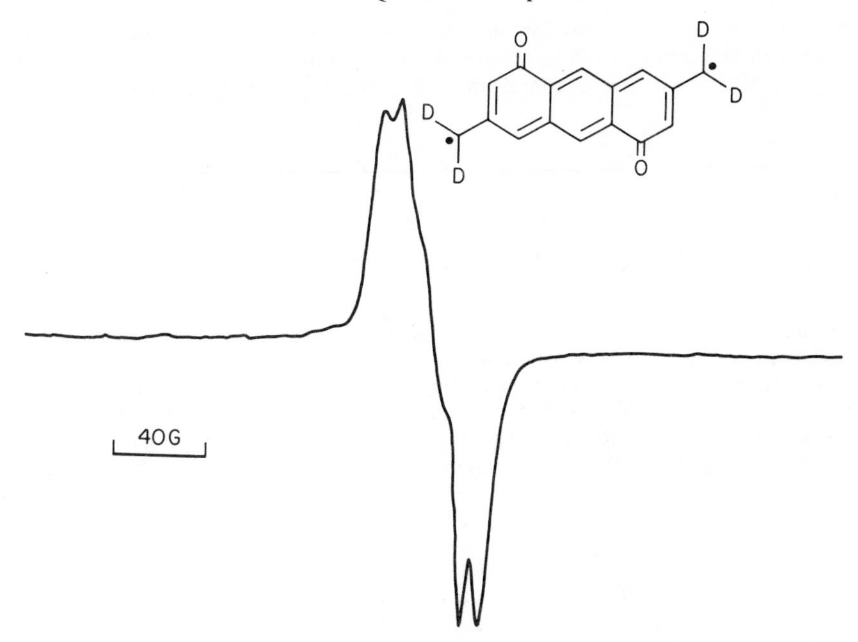

FIGURE 29. Secondary EPR spectrum obtained (similarly to that of Figure 27d) from the photolysis of **126**-d_4. The triplet ZFS parameters are $|D|/hc = 0.0025\ cm^{-1}$ and $|E|/hc = 0.0009\ cm^{-1}$. *Reproduced with permission from D. E. Seeger, P. M. Lahti, A. R. Rossi and J. A. Berson*, J. Am. Chem. Soc., **108**, 1251 (1986). *Copyright* (1986) *American Chemical Society*

F. 2,4- and 3,4-Dimethylenefuran and 3,4-Dimethylenethiophene

1. Theory

Analogous to the relationship of the *m*-quinonoids to benzene is that of the 2,4- and 3,4-dimethylene derivatives, **130**–**135**, to the five-membered heterocyclic aromatics furan, thiophene and pyrrole. These non-Kekulé substances are of special interest because the

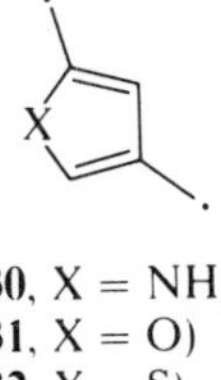

(**130**, X = NH)
(**131**, X = O)
(**132**, X = S)

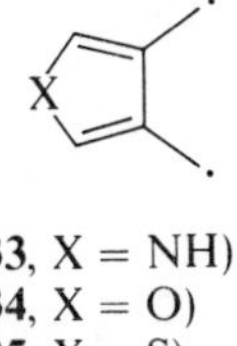

(**133**, X = NH)
(**134**, X = O)
(**135**, X = S)

odd number of ring members makes them non-alternant, so that Ovchinnikov's parity criterion for the spin of the ground state does not apply. INDO/S-CI calculations[70, 71] provide some insight on the properties that might be expected of these substances and are supported by *ab initio* calculations[206–208], which give similar predictions.

It should be clear from inspection of **130–135** that no non-ionic full-valence structure for these substances can be written without expansion of the octet of the heteroatom (**136** and **137**). Although this may be conceivable for the case of X = S, it is conventionally not

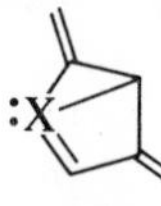

(**136**) 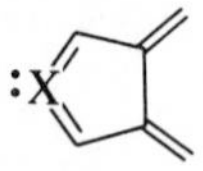 (**137**)

acceptable for atoms of the first row (N, O). In this sense, then the latter two systems can be considered to be of the non-Kekulé type.

In simple resonance terms, the contributing structures to the 2,4-dimethylene group, **130–132**, include, in addition to zwitterionic structures, a group of structures in which the odd electrons are assigned to different positions (e.g. **138–141**). The last two of these, **140** and **141**, emphasize the relationship of this system to trimethylenemethane (**31**). Similarly,

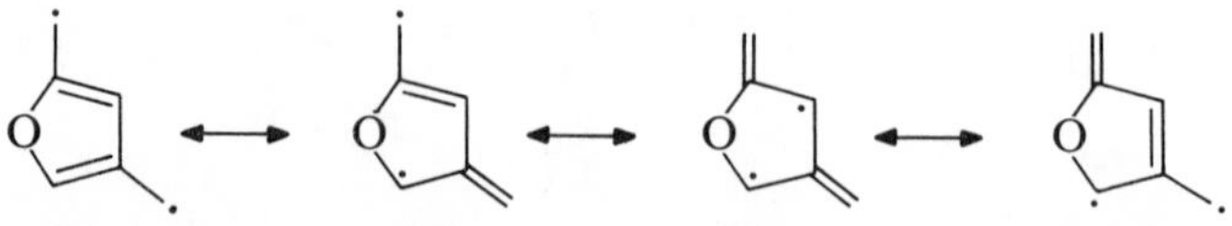

(**138**) (**139**) (**140**) (**141**)

(**31**)

the resonance forms of the 3,4-dimethylene system (e.g. **142–144**) suggest a relationship to tetramethyleneethane (**29**). To the extent that these analogies are valid, that is, if the oxygen

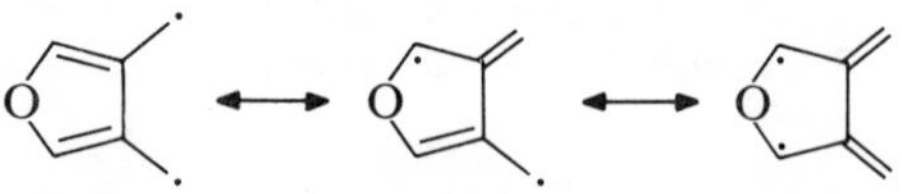

(**142**) (**143**) (**144**)

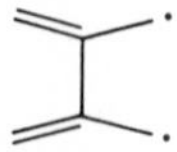

(**29**)

(or other heteroatom) introduces little perturbation, one might expect the ordering of the spin states in the 2,4-system to be the same as that in trimethylenemethane (triplet ground state), and that in the 3,4-system to be the same as in the tetramethyleneethane (singlet ground state). As Table 7 shows, the calculations predict exactly this result.

TABLE 7. Energies of some non-Kekulé molecular states

Structure	State	Calc. method	Relative energy (kcal mol^{-1})	Ref.
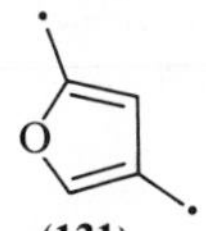 **(131)**	$^3A''$ $^1A''$	INDO/S-CI	0.0 13.3	70
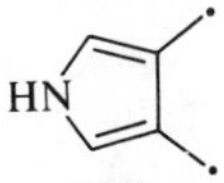 **(133)**	$^1A_1$ $^3B_2$	INDO/S-CI	0.0 4.3	206
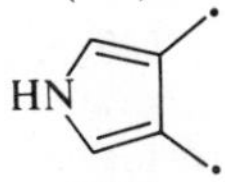 **(133)**	$^1A_1$ $^3B_2$	*a*	0.0 5.8–6.4	208
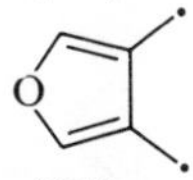 **(134)**	$^1A_1$ $^3B_2$	INDO/S-CI	0.0 0.5	70
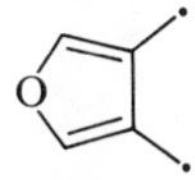 **(134)**	$^1A_1$ $^3B_2$	*b*	0.0 4.4	207
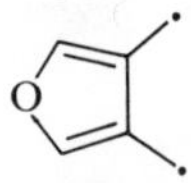 **(134)**	$^1A_1$ $^3B_2$	*c*	0.0 2.2–2.6	208
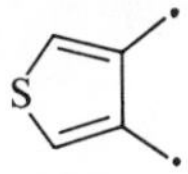 **(135)**	$^1A_1$ $^3B_2$	INDO/S-CI	0.0 0.4	206
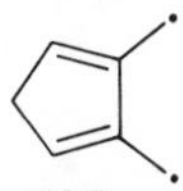 **(136)**	$^1A_1$ $^3B_2$	INDO/S-CI	0.4 0.0	206
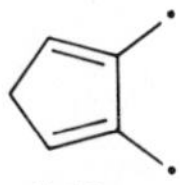 **(136)**	$^1A_1$ $^3B_2$	*d*	0.0 1.4–1.6	208

[a] Split valence (SV) SDTQ CI in the space of the conceptual minimal π basis set (5.8); SV multireference S CI in the full π space (6.4). [b] STO-3G/3-21G SDTQ CI, at 3-21G UHF triplet geometry. [c] SV SDTQ CI and SV MRSD CI. With polarization functions on the heavy atoms (SVP), the value is 1.7. [d] SV SDTQ CI (1.5), SV MRSD CI (1.6), SVP SDTQ CI (1.4).

In the 3,4-bismethylene series, the INDO-S/CI calculations consistently show a 2–3 kcal mol^{-1} less severe violation of Hund's rule than the *ab initio* results, but both calculations agree that the singlet should be the ground state or at worst a very low-lying excited state. On the other hand, the 2,4-bismethylenefuran (**131**) shows a decided preference for the triplet. For reference, Table 7 shows results of calculations on the cyclic hydrocarbon tetramethyleneethane (**136**), for which the INDO-S/CI and *ab initio* methods closely bracket a zero singlet–triplet gap.

2. *Chemistry of 3,4-dimethylenefuran and 3,4-dimethylenethiophene*

Vogel and Hardy showed that pyrolysis of 2,3-bismethylene-7-oxabicyclo[2.2.1]heptane (**137**) proceeded by two competitive pathways, one to give water and benzocyclobutene (**138**) and the other to give ethylene and 3-oxabicyclo[3.2.0]hepta-1,4-diene (furanocyclobutene; **139**). They pointed out the possibility that 3,4-bismethylenefuran biradical (**134**) might be an intermediate in the latter pathway[209].

The base-induced general reaction of bispropargyl ethers (**140a**), amines (**140b**) and sulfides (**140c**) to dimers **142** of putative bismethylene heterocycle (**141**) was discovered by Garratt and Neoh[210]. Further studies in the sulfur and oxygen series were provided soon after by Braverman and coworkers[211] and by the Garratt group[212,213]. When R = H or

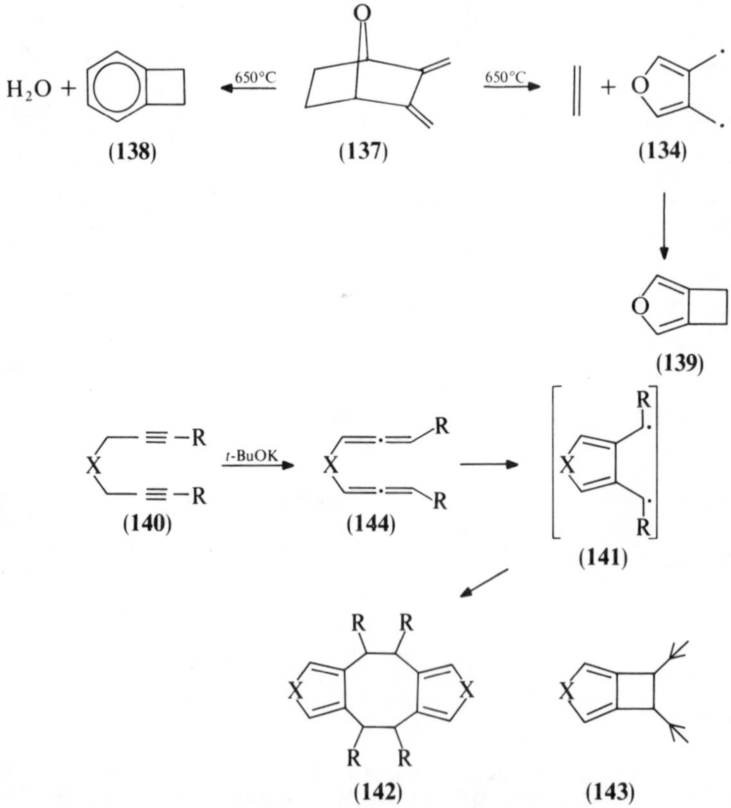

H_2O + (**138**) ← 650°C — (**137**) — 650°C → ‖ + (**134**)

(**134**) → (**139**)

(**140**) —*t*-BuOK→ (**144**) → [(**141**)] → (**142**)

(**143**)

a, X = O; **b**, X = NR′; **c**, X = S

Ph, the dimeric products **142** dominated, but in the cases where R = *t*-Bu, the cyclobutaheterocycle **143** was observed. Again, the bismethyleneheterocyclic biradical **141c** was invoked as an intermediate, supposedly being formed from the bis-allene **144c**, which was shown[211] to lead to dimer **142c** under the reaction conditions.

Attempts to trap the thiophene biradical **141c** with dimethyl maleate or fumarate were unsuccessful, but reaction in the presence of O_2 or maleic anhydride (M.A.) gave the peroxide **145** (in unspecified yield) or the adduct **145** (20% yield), respectively[212].

t-BuOK, 0°C

(**141c**)

O_2

(**146**) (**145**)

These reactions are plausibly formulated[212,213] as interceptions of the biradical **141c**. Kinetic evidence of a first-order rate-determining step in dimerization[213b] and in cycloadduct formation[213c] has been brought forward. The kinetic analysis[213c] uses known[214–220] criteria to determine that capture of biradical **141c** by olefins is much faster than return to bisallenyl sulfide.

A different approach to 3,4-bismethylenefuran and its thiophene analog **141a** and **141c** has been reported recently[213c,221a] in which the biradicals are generated from bicyclic azo compound precursors. The synthesis of the precursor is illustrated for the furanoid case.

The oxidation of the hydrazine **148** with most reagents gives the hydrazone **149**, but dimethyl azodicarboxylate smoothly affords the desired diazene **147**.

The diazene **147** is stable in solution below 0°C. It decomposes thermally or photochemically in degassed solvent to give two dimers, **150** and **151**, and a group of trimers of unknown structure. Oxygen, however, efficiently suppresses dimer formation and traps the biradical **141a** as the cyclic peroxide **158** in good yield. Thermal

(**148**)

X–N=N–X (3)

(4)

X—NH—NH—X +
(X = CO_2Me)

(**147**) (**149**)

Methods: (1) NaH, DMSO; (2) KOH, H_2O; (3) dimethyl azodicarboxylate, $CDCl_3$, −10°C; (4) other oxidants (HgO, MnO_2, *t*-BuOCl; O_2).

decomposition of **147** in the presence of olefins gives mixtures of fused and bridged products. The cycloadditions are highly stereospecific: maleonitrile gives *cis* adducts *c*-**152**, **153** and **154**, whereas fumaronitrile gives *t*-**152** and **155**.

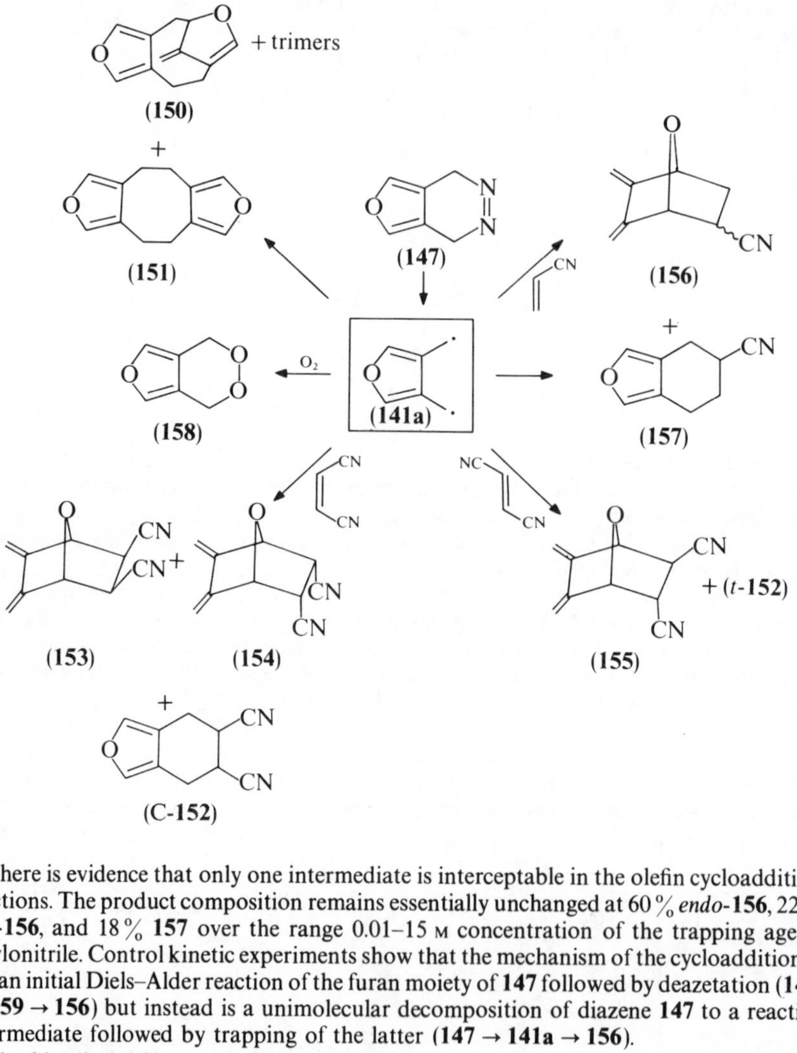

+ trimers

(**150**)

+

(**151**) (**147**) (**156**)

+

(**158**) (**141a**) (**157**)

(**153**) (**154**) (**155**) + (*t*-**152**)

+

(C-**152**)

There is evidence that only one intermediate is interceptable in the olefin cycloaddition reactions. The product composition remains essentially unchanged at 60 % *endo*-**156**, 22 % *exo*-**156**, and 18 % **157** over the range 0.01–15 M concentration of the trapping agent, acrylonitrile. Control kinetic experiments show that the mechanism of the cycloaddition is not an initial Diels–Alder reaction of the furan moiety of **147** followed by deazetation (**147** → **159** → **156**) but instead is a unimolecular decomposition of diazene **147** to a reactive intermediate followed by trapping of the latter (**147** → **141a** → **156**).

The biradical **141a** can be prepared in isolation media by photolysis of samples of diazene **147** in frozen solvents. Visually, these preparations are deep purple, and the optical absorption shows maxima at 338, 348 and 560 nm. The latter band is broad and intense, $E \geqslant 14000$. The same species is formed without delay during a 20 picosecond flash photolysis of solutions of **147**. The lifetime of the transient is about 3 microseconds in argon-purged media but is sharply curtained in the presence of O_2. Low temperature photolysis of **147** in a glassy medium of 2-methyltetrahydrofuran gives the intense purple color, but the sample shows no EPR spectrum in the range 12–80 K. In the absence of

(159) (147) (141a)

(156)

obvious reasons why a triplet state of **141a** should not give an EPR spectrum, the interpretation that the ground state of **141a** is a singlet must be entertained as at least a plausible working hypothesis. This would be in accord with the theoretical predictions of Table 7.

Further evidence in favor of a singlet ground state for **141a** is provided by the observation of the cross-polarization magic angle spinning proton-decoupled ^{13}C-NMR spectrum of isotopically enriched material at 77 K[221b]. The resonance of CH_2-labeled **141a** appears as a narrow single absorption near 100 ppm. Even a small amount of triplet in equilibrium with the singlet would have caused the NMR absorption to be so broadened or shifted by the Fermi contact interaction as to have precluded a detectable signal in the normal ^{13}C region. Thawing or photobleaching the purple sample causes both the color and the ^{13}C resonance to disappear.

The identity of the 3,4-dimethylenethiophene intermediates from the bis-allenyl sulfide (**144c**) and diazene (**147**, S instead of O) starting materials has been demonstrated by a 'fingerprinting' technique in which olefinic trapping agents are allowed to compete for the biradical. The quantitative values of the relative reactivities of the olefins derived in this way are the same from either precursor[213c].

Biradical **141a** does not readily cyclize to the cyclobutane **139**, since the purple color persists in toluene glass up to 160 K. Moreover, the measured lifetime in fluid medium, as reported above, is 3 microseconds. This corresponds to a maximum value for the unimolecular cyclization rate, since no **139** is observed among the products. The activation barrier for cyclization therefore is at least $\Delta G\ddagger \geqslant 10\ \text{kcal mol}^{-1}$. Among the factors contributing to the large barrier must be the strain energy of **139** and the necessity of twisting the two exocyclic methylene groups out of conjugation with the furan ring.

G. 1,8-Naphthoquinone Series

Of the ten possible bismethylenenaphthalenes, four (**160**–**162** and **50**) have aromatic non-Kekulé structures and hence may be considered to be relatives of the *m*-quinonoids. An oxy analog of **160**, *m*-naphthoquinomethane **92**, has been discussed in Section III.D. No derivatives of the types **161** or **162** have been reported, but the hydrocarbon 1,8-naphthoquinodimethane and a number of its tricyclic and heteroatom derivatives are well known, thanks largely to the important contributions from the laboratories of Michl, Pagni, Platz and Wirz. This work has been the subject of careful reviews[78, 222], so that a brief examination of most references will suffice for the present purposes.

Pagni and coworkers[223] have studied the thermal isomerization of the dideuteriated hydrocarbon **163** to phenalene-d_2, **164**. The observed distribution of deuterium in the

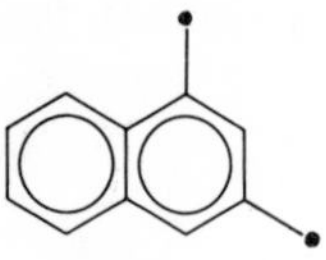

(160) 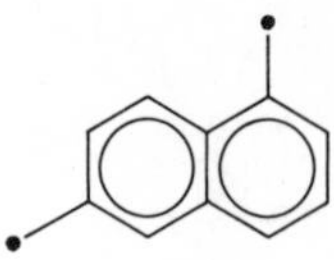 (161) 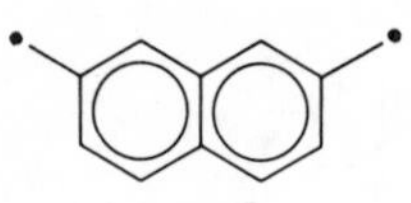 (162)  (48)

product is consistent with the intermediacy of the bridged 1,8-naphthoquinodimethane **165a**. In a faster but presumably related process studied by Roth and Enderer[224], the monodeuteriated analog **166** undergoes thermal epimerization. If the intermediate **165a–b** is invoked for the structural and stereochemical isomerizations, the ratio of rate constants for ring-closure vs. hydrogen shift is about 36 in the experimental temperature range

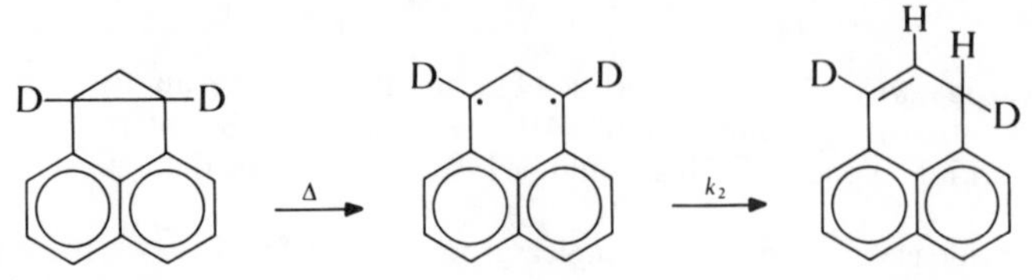

(163) (165a) (164)

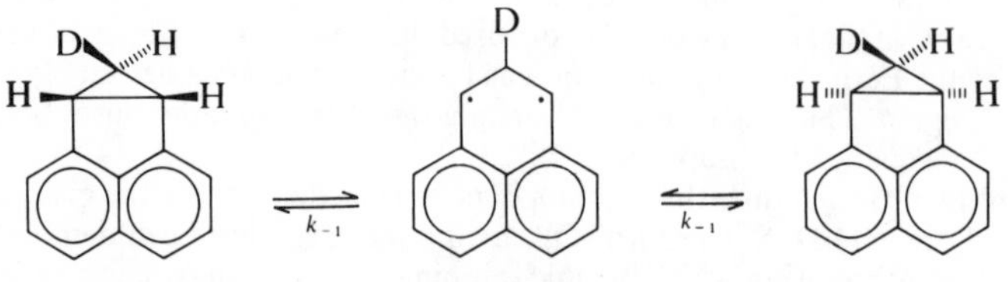

(166) (165b)

(140–170°C)[223]. Pagni points out that this is to be contrasted with the behavior of biradical **165** in frozen matrices (at −140 to −154°C), which had been observed by Michl and coworkers[225] to give rearrangement product **164** (H instead of D) exclusively. Although

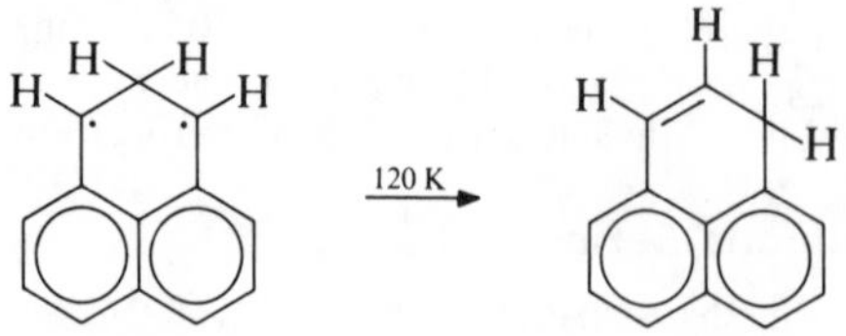

(165)

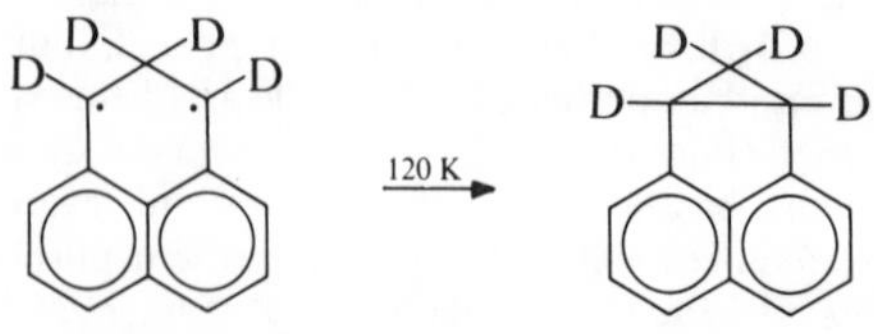

(165-d_4)

the difference in the results may be associated with a matrix effect[223], it is also possible that two different states of biradical **165** are reacting, the singlet at high temperature and the ground state triplet[226, 227] at low temperature. Michl and coworkers[227] recently have found a striking dependence of the products on the position of deuterium substitution:

The primary isotope effect on the hydrogen shift at the experimental temperatures apparently is large enough to retard the rate of formation of phenalene to the point where the ring-closure becomes the dominant reaction.

Michl and coworkers[227] also have partially photoreacted the two biradicals **165** and **165**-d_2 with linearly polarized light at two distinct visible wavelengths characterized by known mutually orthogonal electronic transition moment directions and have produced partially oriented samples. The infrared linear dichroism of these in stretched polyethylene is consistent only with a biradical of C_{2v} symmetry. The IR frequency of the α C–H stretching vibration is that expected for an sp^2 hybridized carbon, whereas that of the β C–H is at a strikingly low frequency, 111 cm^{-1}, shifted from its usual location.

Platz and coworkers[228] have shown by EPR spectroscopy that the members of the series of 4-substituted-1,8-naphthoquinodimethanes **167** all have triplet ground states.

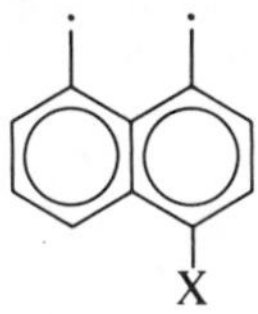

X

(**167**, X = H, Cl, Br, I, SPh, SO_2Ph)

Wirz has reviewed spectroscopic and kinetic investigations of conjugated biradicals, including 1,8-naphthoquinodimethanes[229], and he has published with Gisin[230] semi-empirical (PPP-SCF-CI) calculations and experimental observations of triplet–triplet optical absorptions.

Further work from the Wirz laboratory concerns the 2,2-dimethyl-1,3-perinaphthadiyl species **169**[231]. The biradicals can be formed by several pathways, as shown. The triplet

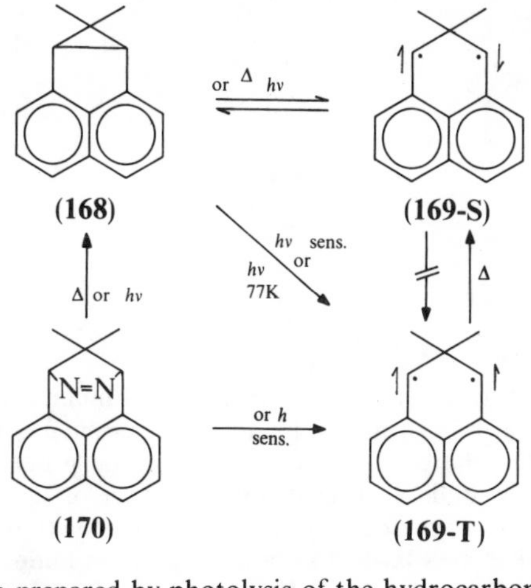

or Δ hν

(**168**) (**169-S**)

hν sens. or hν 77K

Δ or hν

Δ

N=N

or h sens.

(**170**) (**169-T**)

species **169-T** has been prepared by photolysis of the hydrocarbon **168** in EPA glass at 77 K and has been identified by optical and EPR spectroscopy. Curie law measurements suggest a triplet ground state.

The benzophenone triplet–triplet absorption ($\lambda_{max} = 530$ nm) is observed by nanosecond flash photolysis and is quenched by either hydrocarbon **168** or diazene **170** to form the spectrum characteristic of **169-T**. The unimolecular decay of **169-T** to the cyclization product can be followed in a viscous solvent, glycerol, in the temperature range -35 to 0°C. The activation parameters are log $A(s^{-1}) = 8.4$ and $E_a = 7$ kcal mol^{-1}. The low pre-exponential term is ascribed to the spin barrier. The triplet species can be trapped by O_2.

Using the known[224] rate of ring inversion for the parent compound **166** as a model, one may conclude that at 130°C during 40 hours, the dimethyl derivative would have passed through the singlet biradical **169-S** many times. Yet when the hydrocarbon **168** was heated in this way in the presence of 70 atm pressure of O_2, it was recovered unchanged[231]. Combined with the assumption that the exothermic spin-allowed reaction ^{1}(**169-S**) + $^3O_2 \rightarrow {}^3$(**169-T**) + $^1O_2$ would occur at a diffusion controlled rate, the absence of oxidation products led the authors[231] to conclude that an upper limit for the lifetime of **169-S** is 10^{-12} s.

Reasoning that a hypothetical concerted 12-electron process **171** → **172** or **173** → **172** would be orbital symmetry forbidden and hence that a biradical **174** is a likely intermediate, Wirz and coworkers[232] have shown that stereoquilibration of the isomers **171** and **173** is at least two orders of magnitude faster than formation of the polyene **172** despite the fact that the latter is estimated to be 12 kcal mol^{-1} more stable than **171** or **173**. Ring-closure of biradical **174** therefore is much faster than opening of the second cyclopropane ring bond.

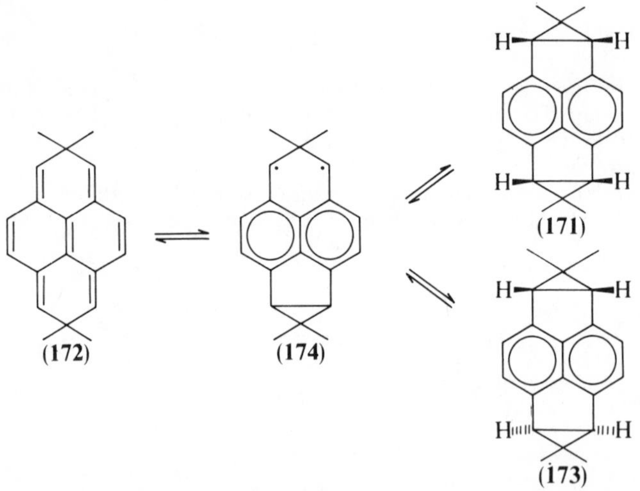

(172) (174) (171) (173)

H. Conclusions and Prospects

The non-Kekulé structure of the *m*-quinonoids has stimulated the development of new theoretical insights, which in turn have led to the experimental investigation of entirely novel forms of matter. Patterns of reactivity, magnetic properties and spectroscopic characterization can be studied in detail for a number of these systems and provide uniquely penetrating tests of present day theoretical understanding of molecules that exist outside the realm of Kekulé's laws of valence. The future undoubtedly will bring more sophisticated studies of reaction dynamics by fast kinetic techniques and more precise characterization by low temperature spectroscopy, methods which are already in widespread use. Gas phase characterization in molecular beam experiments using the newly developed techniques of mass-selected resonance enhanced multiphoton ionization spectroscopy combined with supersonic nozzle jet expansion may permit vibrational spectroscopy with unparalleled definition. It also seems likely that pulsed ion cyclotron resonance (ICR) spectroscopy can play an important role in gas phase studies of biradicals. One interesting application is the determination by Hehre and coworkers[233] of the heats of formation of the isomeric quinodimethanes by this technique: *para*, 50 kcal mol^{-1}; *ortho*, 53 kcal mol^{-1}; and *meta*, $\geqslant$ 76 kcal mol^{-1}. The basis of the experiment is outlined in the sequence of gas phase reactions **175** → **176** → **20**-d$_2$. The key value is the measurement of the proton affinity of the biradical (e.g. **20**) by ICR.

CH_2Br / CD_3 (**175**) $\xrightarrow{-Br^-}$ $\overset{+}{C}H_2$ / CD_3 (**176**) $\xrightarrow[-BD^+]{+Base\ (B)}$ $\dot{C}H_2$ / $\dot{C}D_2$ (**20**-d$_2$)

Beyond all this, the invention of new *m*-quinonoid and related non-Kekulé molecules and their application to synthesis is limited only by the imagination of the organic chemist.

IV. ACKNOWLEDGEMENTS

I am grateful to the following authors who have kindly provided me with reprints and preprints of their work and with helpful comments: Professors W. T. Borden, J. J. S. Dewar, A. Dreiding, J. J. Gajewski, J. Michl, E. Migirdicyan, R. M. Pagni, M. S. Platz, P. J. Stang and J. Wirz. I thank Professor F. Hund for a helpful letter of reminiscence, and for permission to cite from it.

Our own researches in the area covered by this essay received the generous support of the National Science Foundation, the National Institutes of Health and the Humphrey Chemical Company. Acknowledgement is made to the Petroleum Research Fund, administered by the American Chemical Society, for partial support of this research.

V. REFERENCES

1. T. Zincke, *Ann.*, **261**, 208 (1891).
2. A. Hantzsch, *Chem. Ber.*, **39**, 1095 (1906).
3. A. Hantzsch, *Chem. Ber.*, **40**, 330 (1907).
4. J. Meisenheimer, *Ann. Chem.*, **323**, 205 (1902).
5. C. Liebermann and A. Dittler, *Ann. Chem.*, **169**, 252 (1873).
6. R. Meyer and C. Desamari, *Chem. Ber.*, **41**, 2437 (1908).
7. T. Zincke and F. Schwabe, *Chem. Ber.*, **42**, 797 (1909).
8. R. Meyer and C. Desamari, *Chem. Ber.*, **42**, 2809 (1909).
9. O. Stark and O. Garben, *Chem. Ber.*, **46**, 659 (1913).
10. O. Stark and O. Garben, *Chem. Ber.*, **46**, 2252 (1913).
11. O. Stark, O. Garben and L. Klebahn, *Chem. Ber.*, **46**, 2542 (1913).
12. O. Stark and L. Klebahn, *Chem. Ber.*, **47**, 125 (1914).
13. W. Schlenk and M. Brauns, *Chem. Ber.*, **48**, 661 (1915).
14. W. Schlenk and M. Brauns, *Chem. Ber.*, **48**, 716 (1915).
15. F. E. Ray, *Organic Chemistry*, Lippincott, New York, 1941, p. 587 ff.
16. P. Karrer, *Organic Chemistry*, 2nd English edn, Elsevier, New York, 1946, p. 561 ff.
17. T. Zincke, *Chem. Ber.*, **20**, 1776 (1887).
18. R. Willstätter and F. Müller, *Chem. Ber.*, **37**, 4744 (1904).
19. R. Willstätter and F. Müller, *Chem. Ber.*, **41**, 2580 (1908).

20. R. Willstätter and F. Müller, *Chem. Ber.*, **44**, 2171 (1911).
21. F. Kehrmann, *Chem. Ber.*, **44**, 2632 (1911).
22. F. Kehrmann and M. Cordone, *Chem. Ber.*, **46**, 3009 (1913).
23. H. Haakh, *J. prakt. Chem.*, [2], **82**, 546 (1910).
24. E. Clar and D. G. Stewart, *J. Am. Chem. Soc.*, **75**, 2667 (1953).
25. E. Clar, W. Kemp and D. G. Stewart, *Tetrahedron*, **3**, 325 (1958).
26. See also E. Clar, *Aromatische Kohlenwasserstoffe*, Springer-Verlag, Berlin, 1941, p. 311; 2nd edn 1952, pp. 93, 461.
27. M. Rule, A. R. Matlin, D. A. Dougherty, E. F. Hilinski and J. A. Berson, *J. Am. Chem. Soc.*, **101**, 5098 (1979).
28. M. J. S. Dewar, *The Molecular Orbital Theory of Organic Chemistry*, McGraw-Hill, New York, 1969, p. 194.
29. E. Müller and W. Bunge, *Chem. Ber.*, **69**, 2168 (1936).
30. F. Hund, *Linienspektren und periodisches System der Elemente*, Springer-Verlag, Berlin, 1927, p. 124 ff.
31. For a lucid account of the history, see J. C. Slater, *Quantum Theory of Atomic Structure*, Vol. I, McGraw-Hill, New York, 1960, pp. 234–295.
32. Ref. 31, p. 282.
33. J. P. Lowe, *Quantum Chemistry*, Academic Press, 1978, p. 119 ff.
34. H. Eyring, J. Walter and G. E. Kimball, *Quantum Chemistry*, Wiley, New York, 1944, p. 129.
35. S. P. McGlynn, T. Azumi and M. Kinoshita, *Molecular Spectroscopy of the Triplet State*, Prentice-Hall, Englewood Cliffs, New Jersey, 1969.
36. Ref. 33, p. 123.
37. For a proof, see Ref. 31, p. 486.
38. Ref. 35, p. 69.
39. This treatment is based in part upon a discussion by D. E. Seeger, PhD thesis, Yale University, New Haven, Connecticut, 1983, p. 45 ff.
40. Ref. 35, p. 68.
41. J. Katriel and R. Pauncz, *Adv. Quantum Chem.*, **10**, 143 (1977).
42. J. Katriel, *Theor. Chim. Acta*, **23**, 309 (1972).
43. J. P. Colpa, *Mol. Physics.*, **28**, 581 (1974).
44. J. P. Colpa and R. E. Brown, *Mol. Phys.*, **26**, 1453 (1973).
45. J. P. Colpa, A. J. Thakkar, V. J. Smith Jr and P. Randle, *Mol. Phys.*, **29**, 1861 (1975).
46. E. R. Davidson, *J. Chem. Phys.*, **42**, 4199 (1965).
47. R. L. Snow and J. L. Bills, *J. Chem. Ed.*, **51**, 585 (1974).
48. R. P. Messmer and F. W. Birss, *J. Chem. Phys.*, **73**, 2085 (1969).
49. See *Diradicals* (Ed. W. T. Borden), Wiley-Interscience, New York, 1982, p. 9.
50. J. M. McBride and M. W. Vary, *Tetrahedron*, **38**, 765 (1982).
51. F. Hund, personal communication, September 27, 1985 (cited with permission).
52. E. Hückel, *Z. phys. Chem., Abt. B.*, **34**, 339 (1936).
53. C. A. Coulson, D. P. Craig, A. Maccoll and A. Pullman, *Disc. Faraday Soc.*, **2**, 36 (1947).
54. A. J. Namiot, M. E. Dyatkina and Y. K. Syrkin, *C. R. Acad. Sci., U.S.S.R.*, **48**, 267 (1945).
55. A. Pullman, G. Berthier and B. Pullman, *Bull. Soc. Chim. Fr.*, **15**, 450 (1948).
56. H. C. Longuet-Higgins, *J. Chem. Phys.*, **18**, 265 (1950).
57. N. C. Baird, *J. Am. Chem. Soc.*, **94**, 4941 (1972).
58. M. J. S. Dewar and G. J. Gleicher, *J. Chem. Phys.*, **44**, 759 (1966).
59. G. J. Gleicher, in *The Chemistry of the Quinonoid Compounds*, Part 1 (Ed. S. Patai), Wiley, New York, 1974, pp. 1–35.
60. J. Baudet, *J. Chim. Phys.-Chim. Biol.*, **68**, 191 (1971).
61. E. Migirdicyan and J. Baudet, *J. Am. Chem. Soc.*, **97**, 7400 (1975).
62. W. T. Borden and E. R. Davidson, *J. Am. Chem. Soc.*, **99**, 4587 (1977).
63. A. A. Ovchinnikov, *Theor. Chim. Acta*, **47**, 297 (1978).
64. I. A. Misurkin and A. A. Ovchinnikov, *Russ. Chem. Rev.* (English translation), **46**, 967 (1977).
65. D. J. Klein, C. J. Nelin, S. Alexander and F. A. Matsen, *J. Chem. Phys.*, **77**, 3101 (1982).
66. D. J. Klein, *Pure Appl. Chem.*, **55**, 299 (1983).
67. S. Kato, K. Morokuma, D. Feller, E. R. Davidson and W. T. Borden, *J. Am. Chem. Soc.*, **105**, 1791 (1983).
68. M. Said, J.-P. Malrieu and M.-A. Garcia Bach, *J. Am. Chem. Soc.*, **106**, 571 (1984).

69. D. Döhnert and J. Koutecky, *J. Am. Chem. Soc.*, **102**, 1789 (1980).
70. P. M. Lahti, A. R. Rossi and J. A. Berson, *J. Am. Chem. Soc.*, **107**, 2273 (1985).
71. P. M. Lahti, A. R. Rossi and J. A. Berson, *J. Am. Chem. Soc.*, **107**, 4362 (1985).
72. P. Du, D. Hrovat, W. T. Borden, P. M. Lahti, A. R. Rossi and J. A. Berson, *J. Am. Chem. Soc.*,
73. Some aspects are reviewed by M. S. Platz, in *Diradicals* (Ed. W. T. Borden), Wiley-Interscience, New York, 1982, Chapter 5.
74. L. Salem and C. Rowland, *Angew Chem. Intl. Ed. Engl.*, **11**, 92 (1972).
75. F. Hund, *Z. für Physik*, **51**, 759 (1928).
76. R. S. Mulliken, *Phys. Rev.*, **31**, 587 (1928).
77. F. Hund, personal communication as cited in Ref. 29.
78. M. J. S. Dewar and R. C. Dougherty, *The PMO Theory of Organic Chemistry*, Plenum Press, New York, 1975, pp. 12, 55.
79. J. E. Lennard-Jones, *Trans. Faraday Soc.*, **25**, 668 (1929).
80. C. Coulson, *Valence*, Oxford University Press, Oxford, 1952, p. 99.
81. L. Salem, *Molecular Orbital Theory of Conjugated Molecules*, W. A. Benjamin, Inc., New York, 1966.
82. E. R. Davidson and W. T. Borden, *J. Am. Chem. Soc.*, **99**, 2053 (1977).
83. W. T. Borden, *Modern Molecular Orbital Theory for Organic Chemists*, Prentice-Hall, Englewood Cliffs, New Jersey, 1975, p. 186.
84. J. K. Burdett, S. Lee and W. C. Sha, *Nouv. J. Chim.*, **9**, 757 (1985).
85. R. Hoffmann, *J. Am. Chem. Soc.*, **90**, 1475 (1968).
86. Cf. J. A. Pople and D. L. Beveridge, *Approximate Molecular Orbital Theory*, McGraw-Hill, New York, 1970, and references cited therein.
87. See for example G. Segal (Ed.), *Modern Theoretical Chemistry*, Plenum Press, New York, 1977, Vols. 8–9.
88. G. Berthier, J. Baudet and M. Suard, *Tetrahedron Suppl.*, 2, **19**, 1 (1963).
89. J. Ridley and M. Zerner, *Theor. Chim. Acta*, **32**, 11 (1973).
90. J. Ridley and M. Zerner, *Theor. Chim. Acta*, **42**, 233 (1976).
91. T. A. Halgren, D. A. Kleier and W. N. Lipscomb, *Science, N.Y.*, **190**, 591 (1975).
92. M. J. S. Dewar, *Science, N.Y.*, **190**, 591 (1975).
93. D. R. Yarkony and H. F. Schaefer III, *J. Am. Chem. Soc.*, **96**, 3754 (1974).
94. J. H. Davis and W. A. Goddard III, *J. Am. Chem. Soc.*, **99**, 4242 (1977).
95. D. A. Dixon, R. Foster, T. A. Halgren and W. N. Lipscomb, *J. Am. Chem. Soc.*, **100**, 1359 (1978).
96. D. A. Dixon, T. H. Dunning Jr, R. A. Eades and D. A. Kleier, *J. Am. Chem. Soc.*, **103**, 2878 (1981).
97. S. B. Auster, R. M. Pitzer and M. S. Platz, *J. Am. Chem. Soc.*, **104**, 3812 (1982).
98. P. M. Lahti, PhD Thesis, Yale University, New Haven, Connecticut, 1985, pp. 390–397.
99. Ref. 49, Chapter 1.
100. M. J. S. Dewar and M. K. Holloway, unpublished work cited with permission.
101. Y. Osamura, W. T. Borden and K. Morokuma, *J. Am. Chem. Soc.*, **106**, 5112 (1984).
102. E. Migirdicyan, *C. R. Hebd. Séances Adac. Sci.*, **266**, 758 (1968).
103. G. Porter and E. Strachan, *Trans. Faraday Soc.*, **54**, 1595 (1958).
104. C. R. Flynn and J. Michl, *J. Am. Chem. Soc.*, **95**, 5802 (1973); **96**, 3280 (1974).
105. V. Lejeune, A. Despres and E. Migirdicyan, *J. Phys. Chem.*, **88**, 2719 (1984).
106. V. Lejeune, A. Despres, E. Migirdicyan, J. Baudet and G. Berthier, *J. Am. Chem. Soc.*, **108**, 1853 (1986).
107. A. Despres, V. Lejeune and E. Migirdicyan, *J. Luminesc.*, **31**, 549 (1984).
108. For an introduction and references, see C. Walling, *Free Radicals in Solution*, John Wiley and Sons, New York, 1957, p. 8.
109. For an introduction and references, see F. A. Cotton and G. Wilkinson, *Advanced Inorganic Chemistry*, Wiley-Interscience, New York, 1972, p. 540 ff.
110. A. Carrington and A. D. McLachlan, *Introduction to Magnetic Resonances*, Harper and Row, New York, 1967.
111. J. E. Wertz and J. R. Bolton, *Electron Spin Resonance*, McGraw-Hill, New York, 1972.
112. C. P. Slichter, *Principles of Magnetic Resonance*, 2nd edn, Springer-Verlag, New York, 1978.
113. W. Weltner Jr, *Magnetic Atoms and Molecules*, Scientific and Academic Editions, New York, 1983.
114. E. Wasserman and R. S. Hutton, *Acc. Chem. Res.*, **10**, 27 (1977).

115. W. Gordy, *Theory and Applications of Electron Spin Resonance*, in *Techniques of Chemistry*, Vol. XV (Ed. W. West), Wiley, New York, 1980.
116. M. A. El-Sayed, *Pure Appl. Chem.*, **24**, 475 (1970).
117. J. M. McBride, personal communication.
118. M. S. de Groot, I. A. Hesselmann and J. H. van der Waals, *Mol. Phys.*, **16**, 45 (1969).
119. M. S. de Groot and J. H. van der Waals, *Mol. Phys.*, **6**, 545 (1963).
120. A. M. Ponte Goncalves and C. A. Hutchison Jr, *J. Chem. Phys.*, **49**, 4235 (1968).
121. C. A. Hutchison Jr and B. W. Mangum, *J. Chem. Phys.*, **34**, 908 (1961).
122. D. S. McClure, *J. Chem. Phys.*, **22**, 1668 (1954) had used this technique to study the polarized optical spectroscopy of oriented molecules.
123. G. Burns, *J. Appl. Phys.*, **32**, 2048 (1961).
124. W. A. Yager, E. Wasserman and R. M. R. Cramer, *J. Chem. Phys.*, **37**, 1148 (1962).
125. E. F. Hilinski, PhD Thesis, Yale University, New Haven, Connecticut, 1982.
126. N. J. Turro, *Modern Molecular Photochemistry*, Benjamin-Cummings, Menlo Park, California, 1978.
127. R. Breslow, H. W. Chang, R. Hill and E. Wasserman, *J. Am. Chem. Soc.*, **89**, 1112 (1967).
128. Ref. 111, p. 451.
129. S. R. Langhoff and C. W. Kern, in *Applications of Electronic Structure Theory*, Vol. 4 of *Modern Theoretical Chemistry* (Ed. H. F. Schaefer III), Plenum Press, New York, 1977, and references cited therein.
130. D. Feller, W. T. Borden and E. R. Davidson, *J. Chem. Phys.*, **74**, 2256 (1981) and references cited therein.
131. M. Godfrey, C. W. Kern and M. Karplus, *J. Chem. Phys.*, **44**, 4459 (1966).
132. J. H. van der Waals and G. ter Maten, *Mol. Phys.*, **8**, 301 (1964).
133. R. G. Shulman and R. O. Rahn, *J. Chem. Phys.*, **45**, 2940 (1966).
134. A. Pullman and E. Kochanski, *Int. J. Quantum Chem.*, **15**, 251 (1967).
135. A. Calder, A. R. Forester, P. G. James and G. R. Luckhurst, *J. Am. Chem. Soc.*, **91**, 3724 (1969).
136. G. R. Luckhurst, G. F. Pedulli and M. Tiecco, *J. Chem. Soc. (B)*, 329 (1971).
137. J. A. Berson, R. J. Bushby, J. M. McBride and M. J. Tremelling, *J. Am. Chem. Soc.*, **93**, 1544 (1971).
138. H. Vogler, *Mol. Phys.*, **43**, 83 (1981).
139. Ref. 125, p. 137.
140. M. Rule, A. R. Matlin, D. E. Seeger, E. F. Hilinski, D. A. Dougherty and J. A. Berson, *Tetrahedron*, **38**, 787 (1982).
141. R. S. Hutton and H. D. Roth, *J. Am. Chem. Soc.*, **104**, 7395 (1982).
142. E. F. Hilinski, D. A. Dougherty and J. A. Berson, unpublished results.
143. R. McWeeny, *J. Chem. Phys.*, **34**, 399 (1961).
144. G. Kothe, K.-H. Denkel and W. Sümmermann, *Angew. Chem. Intl. Ed. Engl.*, **9**, 906 (1970).
145. B. B. Wright and M. S. Platz, *J. Am. Chem. Soc.*, **105**, 628 (1983).
146. J. L. Goodman and J. A. Berson, *J. Am. Chem. Soc.*, **106**, 1867 (1984).
147. J. L. Goodman and J. A. Berson, *J. Am. Chem. Soc.*, **107**, 5409 (1985).
148. D. E. Seeger, E. F. Hilinski and J. A. Berson, *J. Am. Chem. Soc.*, **103**, 720 (1981).
149. R. Pagni, M. N. Burnett and J. R. Dodd, *J. Am. Chem. Soc.*, **99**, 1972 (1977).
150. V. P. Senthilnathan and M. S. Platz, *J. Am. Chem. Soc.*, **102**, 7637 (1980).
151. M. S. Platz and J. R. Burns, *J. Am. Chem. Soc.*, **101**, 4425 (1979).
152. M. S. Platz, *J. Am. Chem. Soc.*, **101**, 3398 (1979).
153. A. M. Trozzolo, R. W. Murray, G. Smolinsky, W. A. Yager and E. Wasserman, *J. Am. Chem. Soc.*, **85**, 2526 (1963).
154. S. W. Benson, *Thermochemical Kinetics*, 2nd edn, Wiley-Interscience, New York, 1976.
155. J. F. Liebman and A. Greenberg, *Strained Organic Molecules*, Academic Press, New York, 1978.
156. A. R. Matlin, T. A. Inglin and J. A. Berson, *J. Am. Chem. Soc.*, **104**, 4954 (1982).
157. J. L. Goodman, PhD Thesis, Yale University, New Haven, Connecticut, 1984, p. 149, as cited in Ref. 147.
158. J. J. Gajewski, *Hydrocarbon Thermal Isomerizations*, Academic Press, New York, 1981.
159. J. A. Berson, in *Rearrangements in Ground and Excited States* (Ed P. de Mayo), Academic Press, New York, 1980, p. 311.
160. Cf. P. J. Stang, *Acc. Chem. Res.*, **11**, 107 (1978), and references cited therein.
161. J. J. Gajewski, M. J. Chang, P. J. Stang and T. E. Fisk, *J. Am. Chem. Soc.*, **102**, 2096 (1980).
162. H. Stadler, M. Rey and A. S. Dreiding, *Helv. Chim. Acta*, **67**, 1379 (1984).

163. J. L. Goodman and J. A. Berson, *J. Am. Chem. Soc.*, **107**, 5424 (1985).
164. L. K. Montgomery, K. Schueller and P. D. Bartlett, *J. Am. Chem. Soc.*, **86**, 622 (1964).
165. J. A. Berson, *Acc. Chem. Res.*, **11**, 446 (1978).
166. J. A. Berson, in *Diradicals* (Ed. W. T. Borden), Wiley-Interscience, New York, 1982, p. 151, and references cited therein.
167. T. A. Inglin and J. A. Berson, *J. Am. Chem. Soc.*, **108**, (in press).
168. J. L. Goodman and J. A. Berson, as cited in Ref. 156.
169. J. L. Goodman, PhD Thesis, Yale University, New Haven, Connecticut, 1984.
170. S. Pikulin and J. A. Berson, *J. Am. Chem. Soc.*, **107**, 8274 (1985).
171. J. Leitich and F. Wessely, *Monatsh. Chem.*, **95**, 129 (1964).
172. P. Seiler and J. Wirz, *Helv. Chim. Acta*, **55**, 2693 (1972).
173. D. E. Seeger, P. M. Lahti, A. R. Rossi and J. A. Berson, *J. Am. Chem. Soc.*, **108**, 1251 (1986).
174. J. L. Goodman, K. S. Peters, P. M. Lahti and J. A. Berson, *J. Am. Chem. Soc.*, **107**, 276 (1985).
175. R. Siemionko, A. Shaw, G. O'Connell, R. D. Little, B. K. Carpenter, L. Shen and J. A. Berson, *Tetrahedron Lett.*, 3529 (1978).
176. R. L. Siemionko and J. A. Berson, *J. Am. Chem. Soc.*, **102**, 3870 (1980).
177. J. A. Berson, Z. Hamlet and W. A. Mueller, *J. Am. Chem. Soc.*, **84**, 297 (1962).
178. J. Koutecky, D. Döhnert, P. E. S. Wormer, J. Paldus and J. Cizek, *J. Chem. Phys.*, **80**, 2244 (1984).
179. P. Karafiloglou, *J. Chem. Phys.*, **82**, 3728 (1985).
180. H. Kollmar and V. Staemmler, *Theor. Chim. Acta*, **48**, 223 (1978).
181. R. J. Buenker and S. D. Peyerimhoff, *J. Chem. Phys.*, **48**, 354 (1968).
182. W. T. Borden, E. R. Davidson and P. Hart, *J. Am. Chem. Soc.*, **100**, 388 (1978).
183. W. T. Borden and E. R. Davidson, *Ann. Rev. Phys. Chem.*, **30**, 125 (1979).
184. Ref. 83, pp. 208–223.
185. W. T. Borden, *J. Am. Chem. Soc.*, **97**, 5968 (1975).
186. S. I. Weissman, *J. Chem. Phys.*, **25**, 890 (1956).
187. H. M. McConnell and D. B. Chesnut, *J. Chem. Phys.*, **28**, 107 (1958).
188. J. W. Linnett and O. Sovers, *Disc. Faraday Soc.*, **35**, 58 (1963).
189. R. Bersohn, *J. Chem. Phys.*, **24**, 1066 (1956).
190. Ref. 110, p. 91.
191. Ref. 81, Chapter 2.
192. Ref. 111, p. 112.
193. R. W. Fessenden and R. H. Schuler, *J. Chem. Phys.*, **39**, 2147 (1963).
194. S. Masamune, F. A. Souto-Bachiller, T. Machiguchi and J. E. Bertie, *J. Am. Chem. Soc.*, **100**, 4889 (1978).
195. D. W. Whitman and B. K. Carpenter, *J. Am. Chem. Soc.*, **102**, 4272 (1980).
196. (a) B. G. Odell, R. Hoffmann and A. Imamura, *J. Chem. Soc. B*, 1675 (1970); (b) W. W. Schoeller, *Tetrahedron Lett.*, 2043 (1973); (c) P. Du and W. T. Borden, *J. Am. Chem. Soc.*, **109**, 930 (1987); (d) The ground state of tetramethyleneethane **29** recently has been assigned as triplet by EPR spectroscopy: P. Dowd, W. Chang and Y. H. Paik, *J. Am. Chem. Soc.*, **108**, 7416 (1986). *Ab initio* theory (Ref. 196c) nevertheless still predicts a singlet ground for both the planar and optimally twisted geometries. A resolution of this discrepancy is awaited.
197. For a review, see Ref. 158, pp. 138–155.
198. R. Schmidt and H.-D. Brauer, *Angew. Chem. Intl. Ed. Engl.*, **10**, 506 (1971).
199. W. R. Roth and W. Erker, *Angew. Chem. Intl. Ed. Engl.*, **12**, 503 (1973).
200. W. Grimme and H. J. Rother, *Angew. Chem. Intl. Ed. Engl.*, **12**, 502 (1973).
201. N. L. Bauld and C.-S. Chang, *J. Am. Chem. Soc.*, **94**, 7594 (1972).
202. W. R. Roth and B. P. Scholz, *Chem. Ber.*, **115**, 1197 (1982).
203. D. E. Seeger and J. A. Berson, *J. Am. Chem. Soc.*, **105**, 5144 (1983).
204. D. E. Seeger and J. A. Berson, *J. Am. Chem. Soc.*, **105**, 5146 (1983).
205. Ref. 39, pp. 179–186.
206. P. M. Lahti, personal communication, February 17, 1986.
207. Ref. 98, p. 339.
208. P. Du, D. A. Hrovat and W. T. Borden, *J. Am. Chem. Soc.*, (in press).
209. P. Vogel and M. Hardy, *Helv. Chim. Acta*, **57**, 196 (1974).
210. P. J. Garratt and S. B. Neoh, *J. Am. Chem. Soc.*, **97**, 3255 (1975).
211. S. Braverman, Y. Duar and D. Segev, *Tetrahedron Lett.*, 3181 (1976).
212. P. J. Garratt, S. B. Neoh, Y. S. P. Chang and E. Dominguez, *Tetrahedron Lett.*, 691 (1978).

213. (a) P. J. Garratt and S. B. Neoh, *J. Org. Chem.*, **44**, 2667 (1979); (b) Y. S. P. Cheng, P. J. Garratt, S. B. Neoh and V. M. Rumjanek, *Israel J. Chem.*, **26**, 101 (1985); (c) M. M. Greenberg, S. C. Blackstock and J. A. Berson, submitted for publication.
214. R. Huisgen and F. Mietzsch, *Angew. Chem. Intl. Ed. Engl.*, **3**, 83 (1964).
215. W. J. Linn and R. E. Benson, *J. Am. Chem. Soc.*, **87**, 3657 (1965).
216. W. J. Linn, *J. Am. Chem. Soc.*, **87**, 3665 (1965).
217. P. Otto, L. A. Feiler and R. Huisgen, *Angew. Chem. Intl. Ed. Engl.*, **7**, 737 (1968).
218. M. R. Mazur and J. A. Berson, *J. Am. Chem. Soc.*, **103**, 684 (1981).
219. M. R. Mazur and J. A. Berson, *J. Am. Chem. Soc.*, **104**, 2217 (1982).
220. A. D. Sabatelli, R. F. Salinaro, J. A. Mondo and J. A. Berson, *Tetrahedron Lett.*, **26**, 5851 (1985).
221. (a) K. J. Stone, M. Greenberg, J. L. Goodman, K. S. Peters and J. A. Berson, *J. Am. Chem. Soc.*, **108**, 8088 (1986); (b) K. W. Zilm, R. A. Merrill, M. M. Greenberg and J. A. Berson, *J. Am. Chem. Soc.*, **109**, 1567 (1987).
222. M. S. Platz, G. Carrol, F. Pierrot, J. Zayas and S. Auster, *Tetrahedron*, **38**, 777 (1982).
223. R. M. Pagni, M. N. Burnett and H. M. Hassaneen, *Tetrahedron*, **38**, 843 (1982).
224. W. R. Roth and K. Enderer, *Liebig's Ann. Chem.*, **730**, 82 (1969).
225. J.-F. Muller, D. Muller, H. J. Dewey and J. Michl, *J. Am. Chem. Soc.*, **100**, 1629 (1978).
226. M. Gisin, E. Rommel, J. Wirz, M. N. Burnett and R. M. Pagni, *J. Am. Chem. Soc.*, **101**, 2216 (1979).
227. J. J. Fisher, J. H. Penn, D. Döhnert and J. Michl, *J. Am. Chem. Soc.*, submitted.
228. M. J. Fritz, E. L. Ramos and M. S. Platz, *J. Org. Chem.*, in press.
229. J. Wirz, *Pure Appl. Chem.*, **56**, 1289 (1984).
230. M. Gisin and J. Wirz, *Helv. Chim. Acta*, **66**, 1556 (1983).
231. E. Hasler, E. Gassmann and J. Wirz, *Helv. Chim. Acta*, **68**, 777 (1985).
232. J. Ackermann, H. Angliker, E. Hasler and J. Wirz, *Angew. Chem. Intl. Ed. Engl.*, **21**, 618 (1982).
233. S. K. Pollock, B. C. Raine and W. J. Hehre, *J. Am. Chem. Soc.*, **103**, 6308 (1981).

The Chemistry of Quinonoid Compounds, Vol. II
Edited by S. Patai and Z. Rappoport

CHAPTER 11

Quinones as synthones

K. THOMAS FINLEY
Department of Chemistry, State University College, Brockport, New York 14420, USA

I. INTRODUCTION

In 1971 I cited 686 references to describe what had been accomplished in quinone addition and substitution chemistry in the 124 years since Wöhler's effort in that area. The best I have been able to do for the past 15 years is 796; boiled down from over 1100 obvious candidates!

New and exciting reactants and reactions have sprung up alongside familiar schemes. Young investigators are proving to be worthy successors of Smith and Fieser. Research groups which contributed to mid-twentieth century quinone chemistry have continued and expanded their interests. In general this field has grown in more fundamental ways than simply numbers.

The quinone chemistry treated here has played various roles, since both theoretical and applied contributions are abundant and significant. While the majority of the studies conducted involve quinones as intermediate goals, many make direct and important contributions to our understanding of these colorful compounds. In several instances the centrality of the quinonoid nucleus has demanded careful studies of its involvement. Sadly, there remains a notable deficiency of detailed mechanistic studies.

II. HETEROATOM ADDITION REACTIONS OF QUINONES

A. Sulfur Addition

1. Sulfides from thiols

The nucleophilic addition of thiols to quinones continues to be of interest to chemists seeking more active drugs and searching for mechanistic detail in complex biochemical systems. There has also been a lively interest in the theoretical explanation of experimental observations. Perhaps the most detailed work in this area is happily understandable to the organic chemist[1]. Houk and his collaborators have applied their molecular orbital approach primarily to cycloaddition chemistry (see Section V.A.1) but they have provided a valuable overview of our severely limited quantitative data for nucleophilic additions. The past decade has also produced an expanded effort in the synthesis of sulfur heterocycles where quinone–sulfur interactions have proven to be of considerable utility.

The thioketal reported by Récsei[2] has been reinvestigated and shown to actually be a mixture of normal addition products (equation 1)[3].

$$\text{1,4-naphthoquinone} + EtSH \xrightarrow{HOAc/H^+} \text{2,3-bis(SEt)-1,4-naphthalenediol} + \text{2-SEt-1,4-naphthoquinone} \quad (1)$$

In a continuation of their search for antimetabolites of coenzyme Q, Folkers and coworkers prepared 7-alkylmercapto-6-hydroxy-5,8-quinolinequinones containing long-chain alkyl substituents (equation 2)[4]. The poor yields obtained (14–19 %) contrast sharply

$$\text{6-hydroxy-5,8-quinolinequinone} + RSH \xrightarrow{EtOH} \text{7-SR-6-hydroxy-5,8-quinolinequinone} \quad (2)$$

$R = n\text{-}C_{12}-C_{18}$

with their later work on analogous 2,3-dimethoxy-1,4-benzoquinone derivatives (equation 3)[5,6]. These studies demonstrate that optimum conditions for addition require collecting

$$\text{2,3-(MeO)}_2\text{-5-R}^1\text{-1,4-benzoquinone} + R^2SH \xrightarrow[Na_2Cr_2O_7 \text{ or } Ag_2O]{EtOH/C_6H_{14}} \text{2,3-(MeO)}_2\text{-5-R}^1\text{-6-SR}^2\text{-1,4-benzoquinone} \quad (3)$$

$R^1 = H, Me; R^2 = n\text{-}C_{12}-C_{18}$

the crude hydroquinone, reoxidizing and adding more thiol. Yields of 70–75 % were obtained after three such cycles. With soluble products chromatography on silica gel provided acceptable yields.

The addition of ω-mercaptoalkanoic acids to 1,4-naphthoquinones gives fair yields (36–67%) of the corresponding 2-substituted products (equation 4)[7]. No evidence is

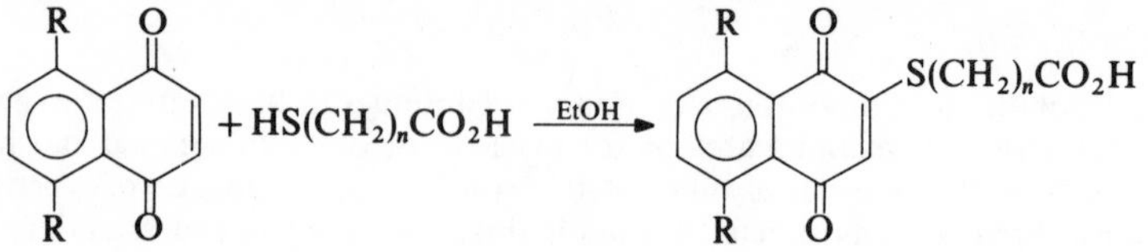

$$+ HS(CH_2)_nCO_2H \xrightarrow{EtOH} \quad S(CH_2)_nCO_2H \qquad (4)$$

R = H, Me, OH, Cl $n = 1$–3

presented for products derived from 8-chlorojuglone. Alkyl and alkenyl sulfides were prepared but in poor yield.

Unlike the carbon–carbon bond formation normally observed with 2-phenylindane-1,3-dione, its monothio analog (**1**) shows addition of the thioenol to quinones (equation 5)[8]. The product structure was assigned on the basis of an interesting nitrogen substitution

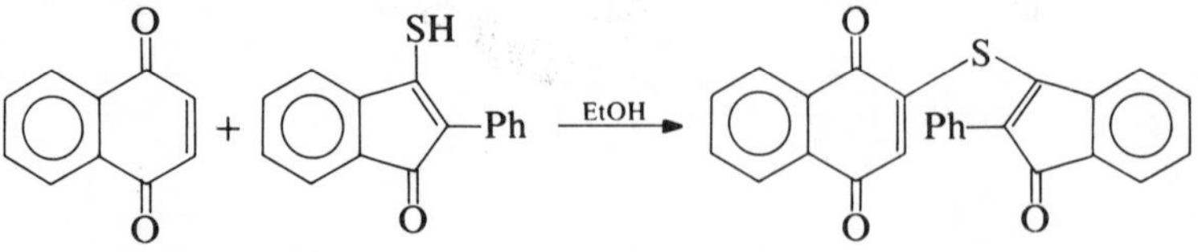

$$+ \xrightarrow{EtOH} \qquad (5)$$

(**1**)

reaction. Buggle and Power have studied the apparently analogous compounds **2** and **3** and report that reaction occurs at carbon to form a bridgehead nitrogen heterocycle rather than at sulfur (see Section VII)[9].

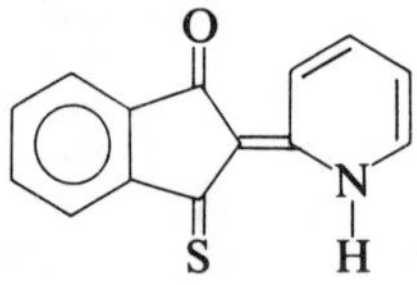

(**2**) 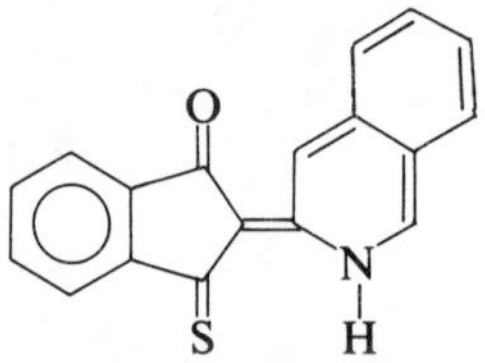 (**3**)

The addition of strongly acidic heterocycles like 1,4,5-trimethyl-1,2,4-triazolium-3-thiolate (**4**) to various quinones proceeds smoothly in aqueous acid (equation 6)[10]. Yields

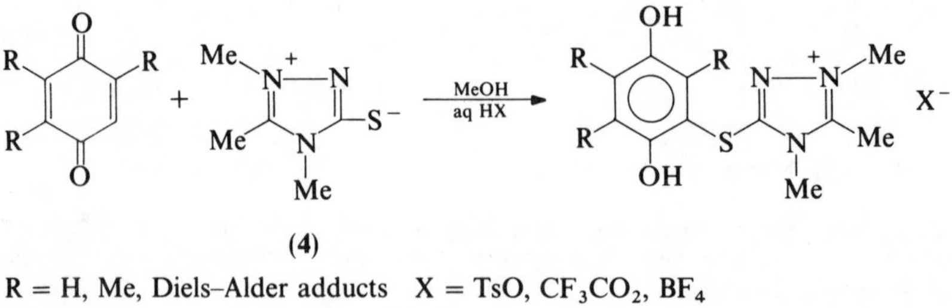

$$+ \xrightarrow[\text{aq HX}]{\text{MeOH}} \quad X^- \qquad (6)$$

(**4**)

R = H, Me, Diels–Alder adducts X = TsO, CF_3CO_2, BF_4

appear to depend rather strongly on the acid employed and divide into two distinct groups; 40–50% and 75–80%. This observation and some interesting structural effects should be followed up. The hydroquinone–sulfide products can be oxidized to the corresponding quinone in variable yields. The few unsymmetrically substituted quinones studied appear to react in a stereospecific fashion; for example, *o*-quinones give only the 4-substituted product, and 2-methoxy-1,4-benzoquinone gives a 78% yield of the 2,5-product.

The subject of possible addition reactions of protein thiol groups to nascent 1,2-benzoquinones has received a great deal of attention. Studies aimed at understanding the role of catecholamines in developing biochemical theories of mental illness have provided one of the major forces behind this work. Adams and his students have searched actively for information concerning the reactions which might be expected if 1,2-quinones are formed in brain tissue. They have measured the rates at which the model compounds 4-methyl- and 4-(2-aminoethyl)-1,2-benzoquinone (dopamine) react with external nucleophiles. Thiol groups like those present in cysteine and glutathione react three to four orders of magnitude faster than various amino acids lacking such a sulfur function. These addition reactions are also three orders of magnitude faster than the intramolecular cyclization of dopamine-1,2-benzoquinone (**5**, equation 7)[11].

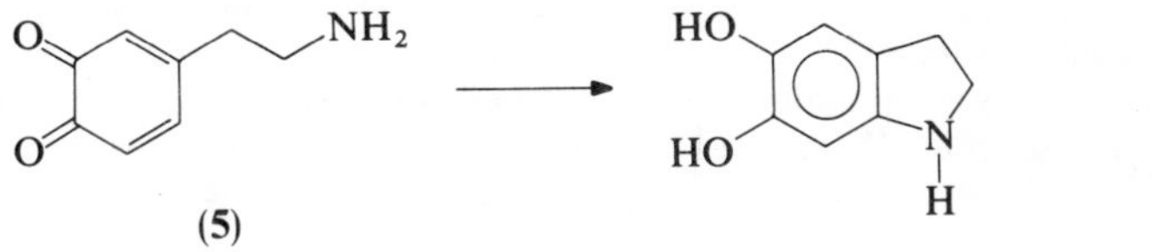

O O NH$_2$ → HO HO N H (7)

(**5**)

Evidence for such reactions with the quinone resulting from the oxidation of 6-hydroxydopamine was presented by Adams' group in 1975[12]. Later they reported the isolation and characterization of the sulfide product of this quinone with glutathione

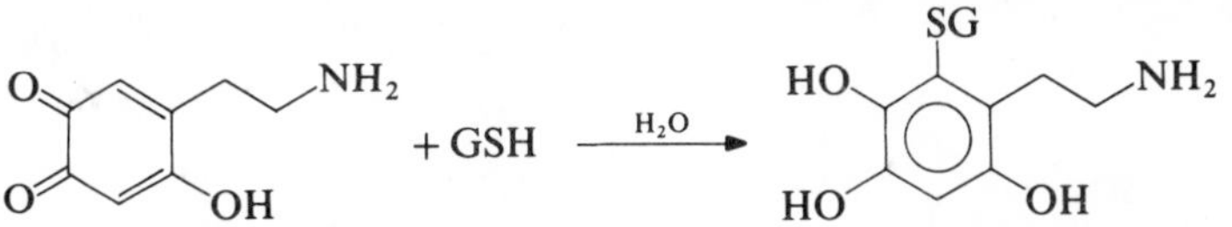

O O OH NH$_2$ + GSH $\xrightarrow{H_2O}$ SG HO HO OH NH$_2$ (8)

(**6**) ca. 70%

$$G = HO_2CCH_2NHC(=O)CH_2NH(C=O)(CH_2)_2CH(NH_2)CO_2H$$

(equation 8)[13]. It was also shown that further oxidation and cyclization can take place (equation 9). The same indole is formed when 6-hydroxydopamine is injected in rat brains.

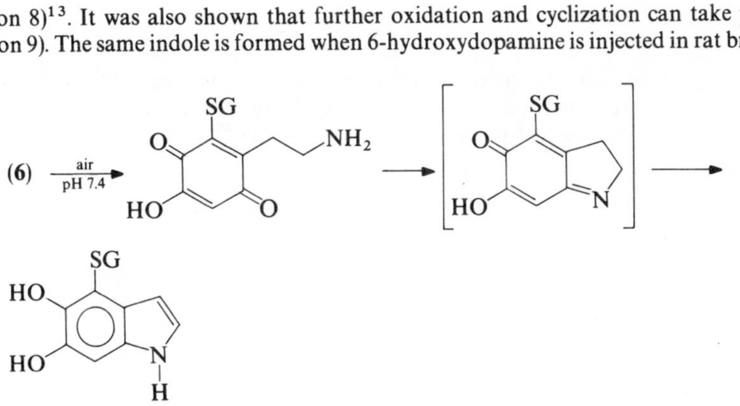

(**6**) $\xrightarrow[\text{pH 7.4}]{\text{air}}$ SG O HO O NH$_2$ → [SG O HO N] →

SG HO HO N H (9)

In a closely related study Vithayathil and Gupta report spectral evidence for the structure and chemistry of *o*-quinone–thiol adducts[14]. Unfortunately, much of the useful information is only found in Gupta's thesis from which this brief note is taken.

The addition chemistry of 3,5-di-*t*-butyl-1,2-benzoquinone and thiols has been suggested as a sensitive detection method in thin-layer chromatography[15]. Colored chelates are formed by spraying the catechols with iron(III) chloride solution. This procedure allows cysteine and reduced glutathione to be detected and differentiated at the 2 nanomole level and in the presence of a variety of other thiols.

Two sulfur-quinone studies deserve to be made more accessible. First, a polarographic investigation of the reactions of vitamin K with thiols confirmed the naphthohydroquinone intermediate in the absence of oxygen[16]. Admission of air changed the product to the quinone sulfide. Spectrophotometric and electrochemical measurements were used to confirm the mechanism with several different thiols. The reaction of cysteine gave more complicated polarograms suggesting that subsequent cyclization must be considered. These results contrast sharply with earlier reports on this system[17].

Second, some evidence concerning the relative reactivity of cyclic sulfides with quinones has been obtained[18]: 2-propyl or 2-methylthiophene > thiophene and 3-hexyl > 2-methyl > 2-methyl-4-pentyl > 4-pentylthiacyclohexane > thiacyclohexane. For the quinones studied the order was: *p*-benzoquinone > 2-(*N*-methylanilino)-*p*-benzoquinone > 5,8-quinolinedione.

In their study of the addition of thiols to 1,4-benzoquinones bearing a strong electron-withdrawing group Fariña and Valderrama obtained excellent yields showing a high degree of stereospecificity (equation 10)[19]. Two examples were found in which the oxidized

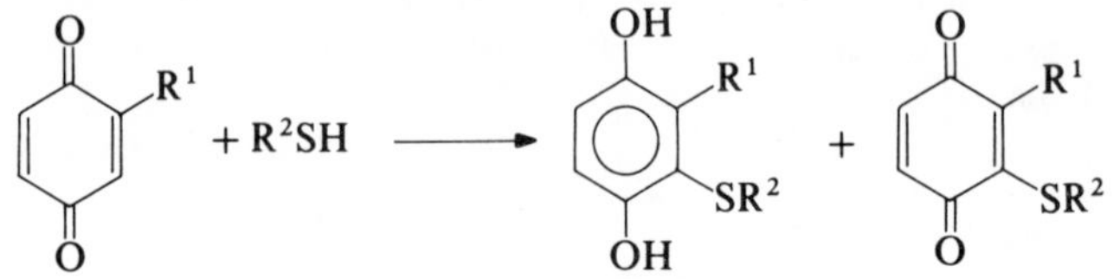

R^1-1,4-benzoquinone + R^2SH ⟶ 2-R^1-3-SR^2-hydroquinone + 2-R^1-3-SR^2-1,4-benzoquinone (10)

R^1 = OAc

R^2 =	% (hydroquinone)	% (quinone)
Et	81	100
Bu	67	
$PhCH_2$	74	
MeO_2CCH_2	98	

product could be cyclized (equation 11). When thioglycolic acid was added to 2-acetyl-1,4-benzoquinone, cyclization occurred (equation 12) involving the quinonoid carbonyl in a fashion consistent with that observed by Snell and Weissberger[20].

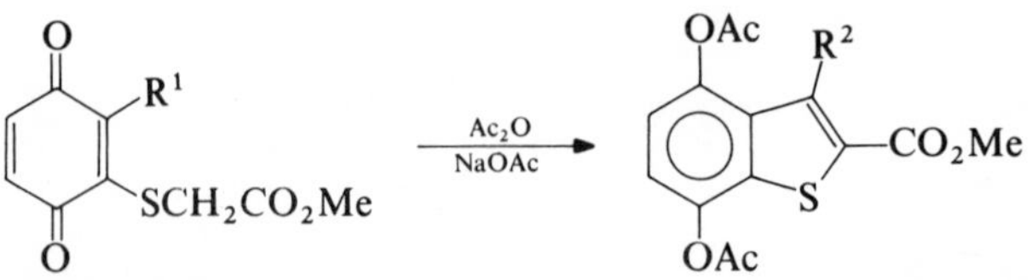

2-R^1-3-SCH_2CO_2Me-1,4-benzoquinone $\xrightarrow[NaOAc]{Ac_2O}$ 4,7-di-OAc-3-R^2-2-CO_2Me-benzothiophene (11)

R^1 = Ac, CO_2Me R^2 = Me (92%), OMe (97%)

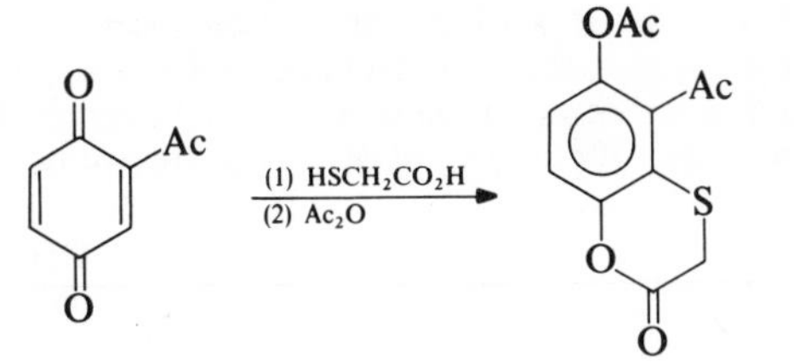

(12)

2. *Sulfones from sulfinic acids*

The extensive application of sulfinic acids and quinones in polymerization reactions makes their interactions important in detailed kinetic studies of such systems[21]. In other cases the sulfinic acid has proven to be an interesting reactant as a constituent of the polymer itself[22]. Included in this study is an example of the nascent quinone from alizarin (7; equation 13).

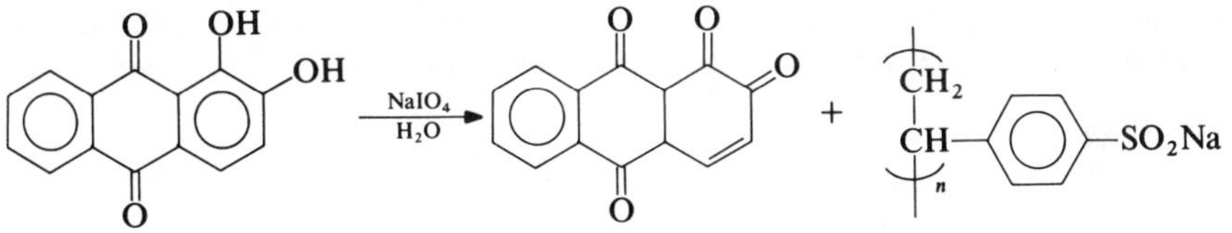

(7)

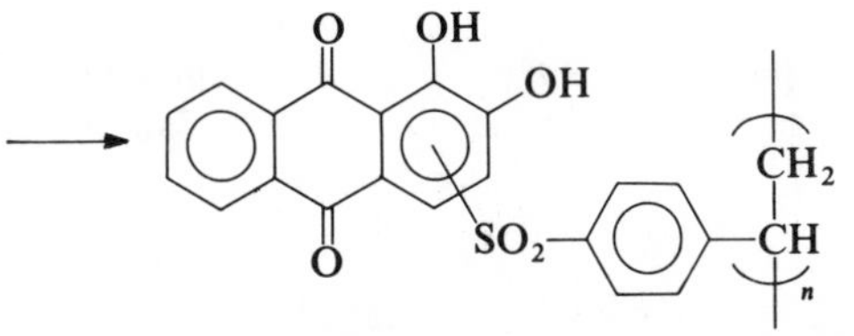

95 mole %

(13)

In their studies of the interactions of 3-pentadecylcatechol with proteins Castagnoli and coworkers found that benzenesulfinic acid is an inhibitor (equation 14)[23]. They obtained a

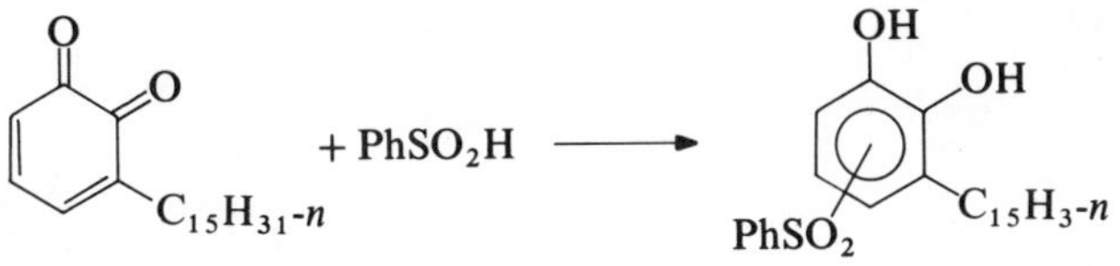

86%

(14)

single sulfone product from reaction with the corresponding quinone, but did not establish its structure.

The analytical application of benzenesulfinic acid as a trapping reagent for *o*-benzoquinones generated, or at least suspected, in enzymatic reactions has produced some conflicting observations. These problem areas have been reviewed by Davies and Pierpoint[24]. Their results illustrate the great importance of solvent and suggest that the enzyme itself may influence the structural outcome. For example, with even the simplest structures a complete change of product can be observed (equations 15,16). Unfortunately,

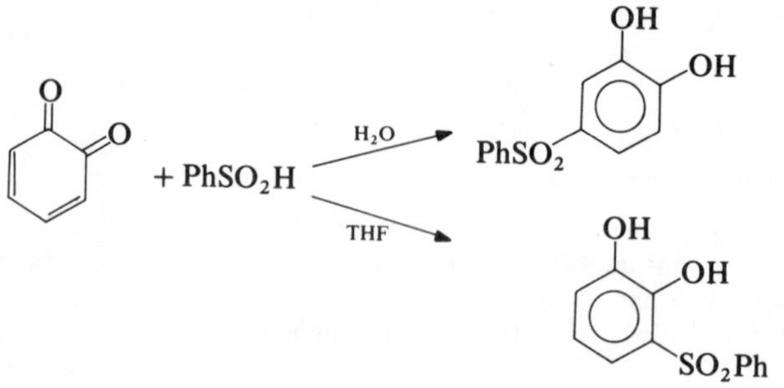

(15)

(16)

no yields are given. The more complicated caffeic acid (**8**) and its ester show the same regiospecificity in tetrahydrofuran (equation 17). Pierpoint has previously shown that the

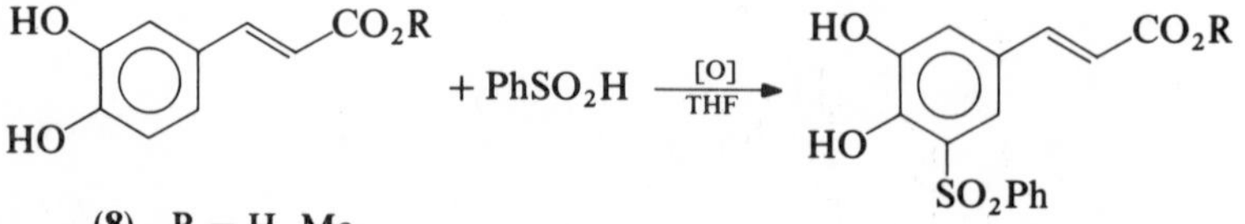

(17)

(**8**) R = H, Me

enzymatic oxidation of chlorogenic acid (**9**) leads to sulfinic acid addition in a different position (equation 18)[25]. The product structure (**10**) was demonstrated by Janes[26].

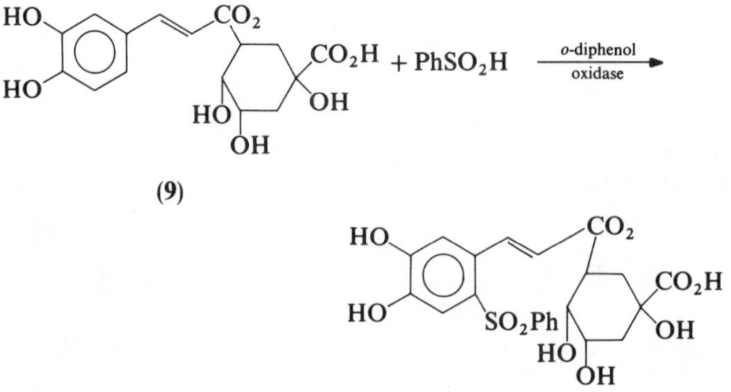

(**9**)

(18)

(**10**)

The oxidation of caffeic acid in the presence of benzenesulfinic acid in an aqueous buffer or with an enzyme produces **11**, the same sulfone as obtained from the hydrolysis of the chlorogenic acid product (**11**, equation 19).

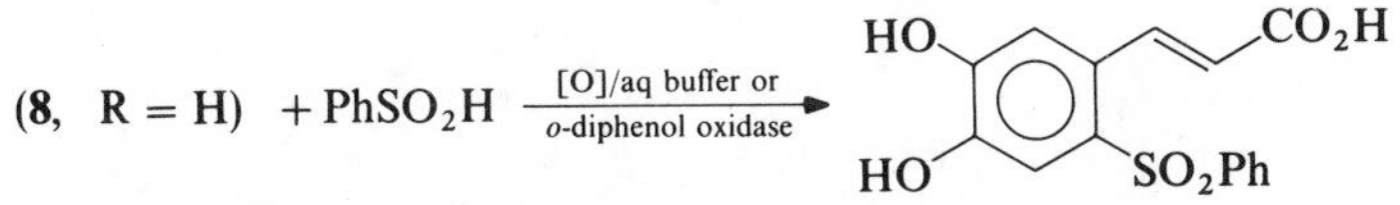

$$(\mathbf{8},\ R = H) + PhSO_2H \xrightarrow[o\text{-diphenol oxidase}]{[O]/\text{aq buffer or}} \mathbf{(11)} \quad (19)$$

In related oxidation experiments with a different enzyme Stom and his colleagues obtained products which appear to differ from those just described, but their structures were not assigned[27, 28].

Benzenesulfinic acid itself is a suitable reagent for the photometric determination of quinones[29]. When coupled with an extraction procedure aqueous solutions can be measured satisfactorily in the 1–5 $\mu g\, ml^{-1}$ range.

Finally, three unsual synthetic reports have been made. A method for the preparation of symmetrical sulfones involves the addition of hydroxymethanesulfinate to two molar equivalents of quinones (equation 20)[30]. The reaction also takes place with naphtho- and

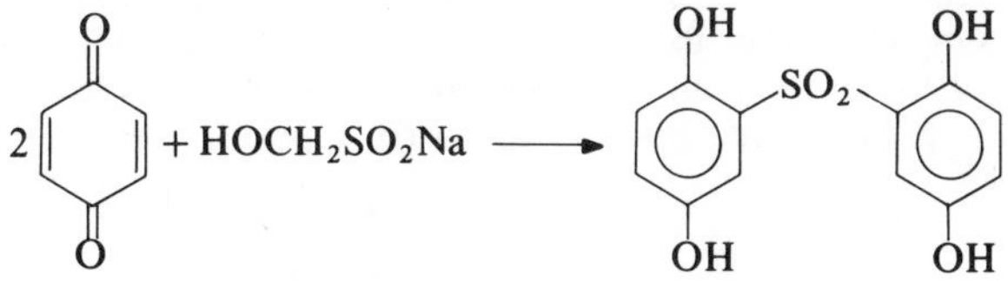

$$2\ \text{quinone} + HOCH_2SO_2Na \longrightarrow \mathbf{(12)} \quad (20)$$

1,2-quinones. While the yields are modest (45–66 %) they provide entry to a useful series of heterocyclic compounds by oxidation of the initial adduct (equation 21).

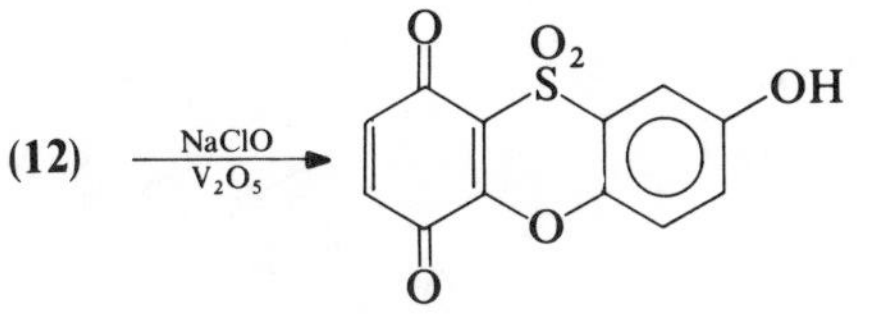

$$\mathbf{(12)} \xrightarrow[V_2O_5]{NaClO} \quad (21)$$

The addition of 2-chloroethanesulfinic acid to quinones provides a potential entry to vinyl quinones and redox polymers (equation 22)[31].

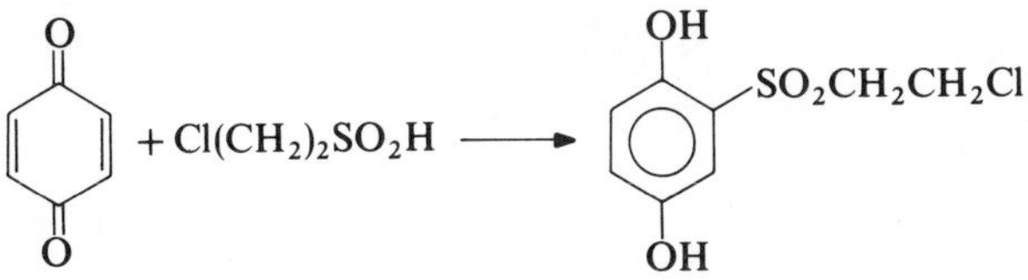

$$\text{quinone} + Cl(CH_2)_2SO_2H \longrightarrow \text{2-}(SO_2CH_2CH_2Cl)\text{hydroquinone} \quad (22)$$

In the course of their studies reported above Folkers and his collaborators found high yields of sulfones from the addition of the rarely encountered alkyl sulfinic acids to quinones (equation 23)[6].

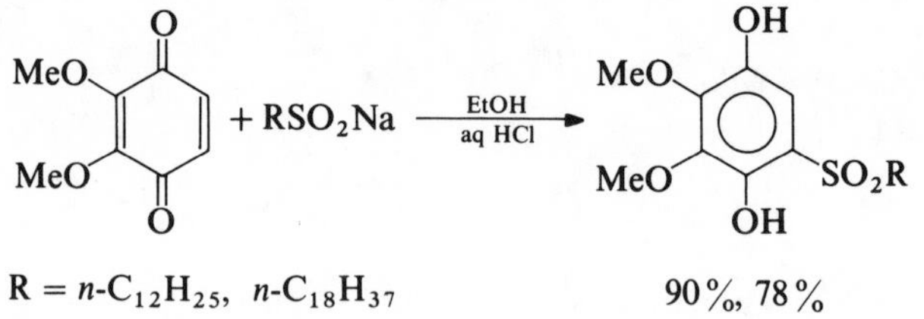

(23)

3. Cyclization reactions

An area of sulfur addition to quinones that has continued to show rapid growth during the past 15 years is that involving bidentate molecules and their subsequent cyclization. This vitality is entirely consistent with the great upsurge of interest in heterocyclic systems shown by organic chemists during this period.

Interest in L-cysteine has continued with a reinvestigation of Kuhn's earlier studies[32, 33]. Prota and Ponsiglione found the product previously reported but the complexity of the reaction mixture caused them to discontinue that phase of their work[34]. With L-cysteine ethyl ester two substances, probably diastereoisomers, are formed. Elemental analysis and NMR spectra of the more soluble isomer are in agreement with a bi-benzothiazine structure (**13**) and consistent with the chemistry reported for the *o*-quinone series (equation 24).

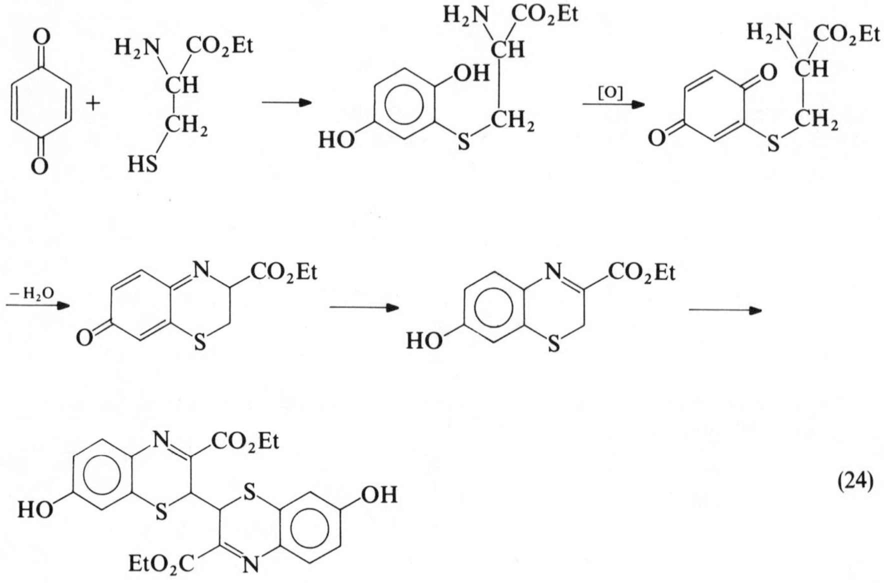

(24)

(**13**)

The same research group has demonstrated a useful synthetic application of a simpler model system (equation 25)[35]. Evidence for the predicted intermediate was obtained by carrying out the reaction under acidic conditions and treatment with acetic anhydride. The triacetate product after hydrolysis with dilute acid formed the dihydrobenzothiazine **14** in

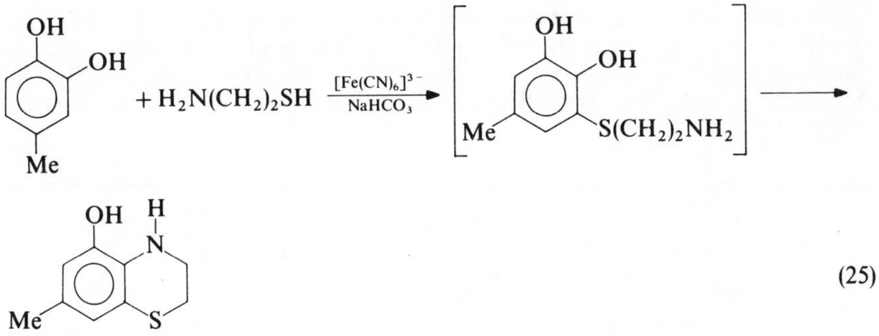

OH OH Me + $H_2N(CH_2)_2SH$ $\xrightarrow[NaHCO_3]{[Fe(CN)_6]^{3-}}$ [OH OH Me $S(CH_2)_2NH_2$] ⟶

OH H N Me S

(**14**) 50%

(25)

78% yield. If the initial reaction is carried out with an excess of potassium ferricyanide present, two significant colored products (**15** and **16**) are formed. Their yields can be

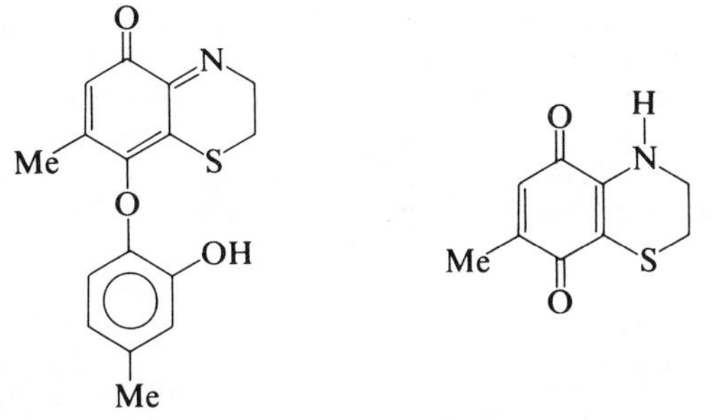

(**15**) (**16**)

greatly increased (24 and 33% respectively) by oxidizing the initial product (**14**) in the presence of 4-methylcatechol.

Phenothiazones because of their utility in a wide variety of medical applications have stimulated extensive study of their synthesis. In general it has proven difficult to prepare these compounds by conventional methods, and recently Ueno and coworkers have investigated photochemical routes employing 1,2-dimethyl-3*H*-phenothiazin-3-one (**17**) as a substrate (equation 26)[36]. This chemistry was used as an alternative route to

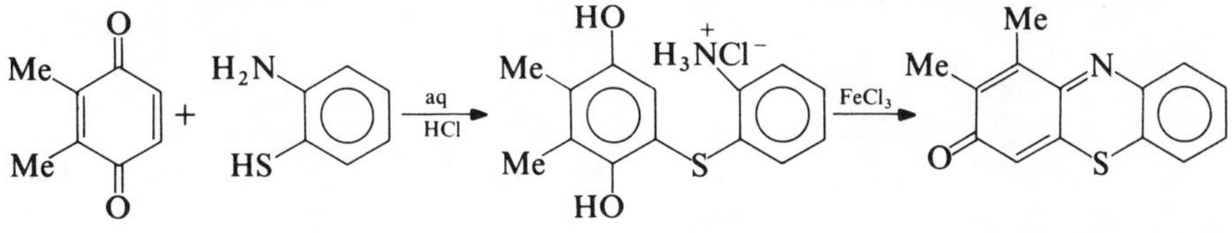

$\xrightarrow[HCl]{aq}$ $H_3\overset{+}{N}Cl^-$ $\xrightarrow{FeCl_3}$

(**17**) 50%

(26)

substantiate their structural assignment of analogous photochemical products. Ueno has also reported similar chemistry with di- and trimethyl-1,4-benzoquinones[37].

One of the most widely used reagents for the direct preparation of the phenothiazones is the zinc salt of 2-aminobenzenethiols. Most often these reagents are used in substitution reactions (see Section VIII.A.3), but one example of addition chemistry exists. Terdic and Smarandache prepared a series of seven dimethyl-3*H*-phenothiazin-3-ones in generally low yields (equation 27)[38].

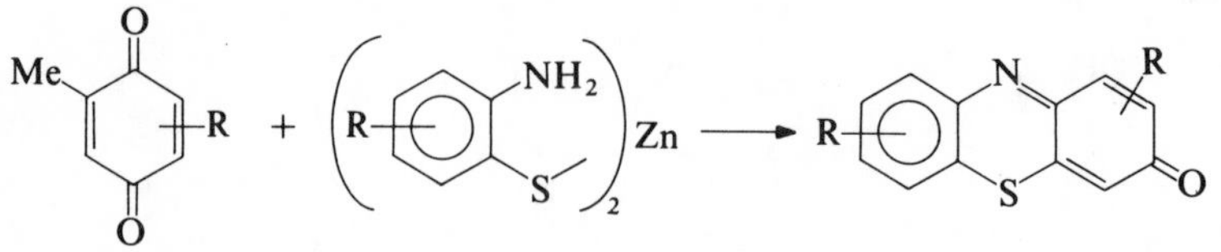

Me, O, O, R + R, NH_2, S, Zn → R, N, R, S, O (27)

R = H, Me

Me_2	%	Me_2	%
1,2	52	2,4	17
1,4	15	2,6	3.8
1,6	1.3	2,8	19
1,8	13		

An interesting sulfur heterocyclic synthesis involves the initial addition of sulfur followed by an unusual oxygen displacement (equation 28)[39]. Unlike an earlier report of

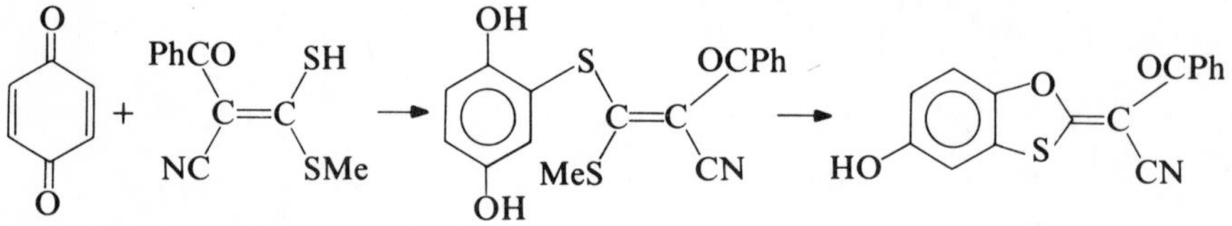

O, O + PhCO, SH, C=C, NC, SMe → OH, OH, S, C=C, MeS, OCPh, CN → HO, O, S, =C, OCPh, CN

82% 76% (28)

this reaction[40], involving the dithio enolate salt, the monothio ether prevents the formation of the dithio ring system. With 1,4-naphthoquinone the yield is 45% and the intermediate is not isolated.

The pioneering work on the addition of thioureas to quinones by Lau and his collaborators[41,42] has stimulated several studies which demonstrate important modifications, e.g., the addition of a wide variety of acylthioureas (equation 29)[43].

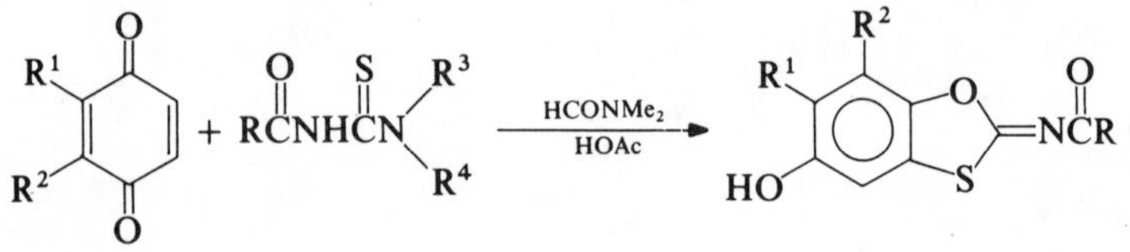

R^1, R^2, O, O + RC(O)NHC(S)NR^3R^4 $\xrightarrow[\text{HOAc}]{\text{HCONMe}_2}$ R^1, R^2, HO, O, S, =NC(O)R (29)

R^1, R^2 = H, carbocyclic
R = alkyl, aryl, heteroaryl, alkoxycarbonyl
R^3, R^4 = H, alkyl, aryl, N-ring

R = Ph, $R^1 = R^2$ = H 74%

Mann and his colleagues have expanded the earlier work to a series of monosubstituted 1,4-naphthoquinones (equation 30)[44]. The yields vary from modest to good (31–90%) over the range of electron-donating alkyl, alkoxy, and alkylthio substituents studied.

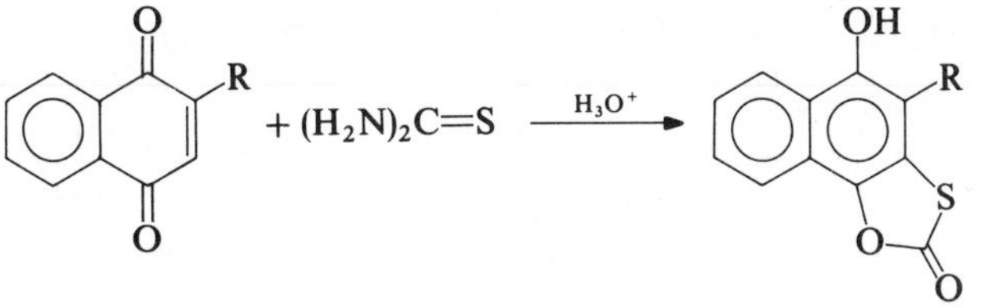

$$+ (H_2N)_2C{=}S \xrightarrow{H_3O^+} \qquad (30)$$

Horak and Manning have reported the addition of thioamides to 1,4-quinones and studied the simplest case of thioacetamide and 1,4-benzoquinone in greater detail (equation 31)[45]. The use of 2-methyl-1,4-benzoquinone produces an approximately

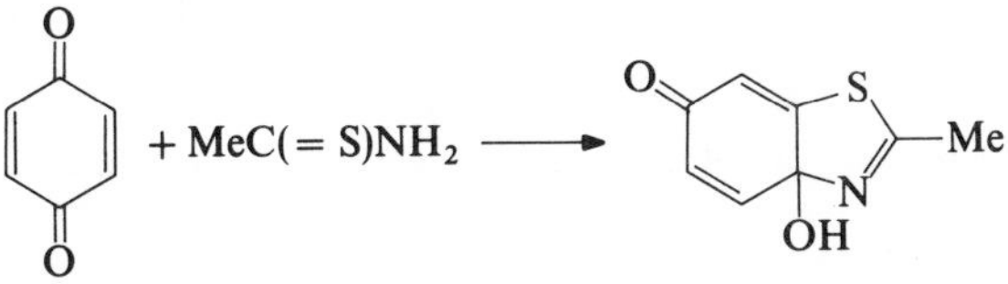

$$+ MeC({=}S)NH_2 \longrightarrow \qquad (31)$$

23% (35% by GC)

equimolar mixture of 2,4- and 2,5-dimethyl products (overall yield 26%). With thiobenzamide the 2-phenyl derivative is obtained in 51% yield. Possible explanations of the presence of several by-products is given. Significantly, more complicated quinones lead only to sulfur oxidation.

The use of *N,N'*-disubstituted thioureas with 2-methyl-1,4-naphthoquinone under the conditions developed by Lau gave fair yields of 1,N^2-disubstituted 2-amino-5-hydroxy-4-methylnaphtho[1,2-*d*]thiazolium salts (**18**, equation 32)[46]. Cyclic thioureas lead to the

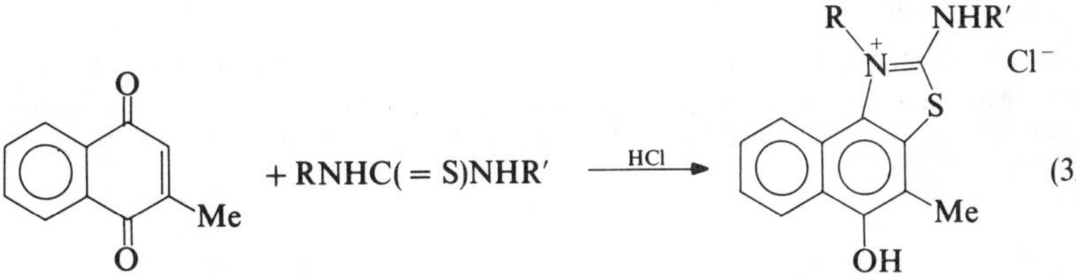

$$+ RNHC({=}S)NHR' \xrightarrow{HCl} \qquad (32)$$

R = alkyl, allyl, aryl

(**18**)

formation of tetracyclic naphthothiazolium compounds (equation 33). Interestingly, 1,4-naphthoquinone failed to give any identifiable product. The proposed mechanism which allows for a Michael attack on the quinonoid ring is consistent with this observation.

It is possible to add dithiocarbamic acids to 1,4-benzoquinone in good yield with rapid mixing of the aqueous/dimethylformamide and acetic acid solutions (equation 34)[47]. The initial addition products (**19**) can be converted to heterocyclic systems which vary depending on the solvent employed (equations 35, 36). The few examples given produced

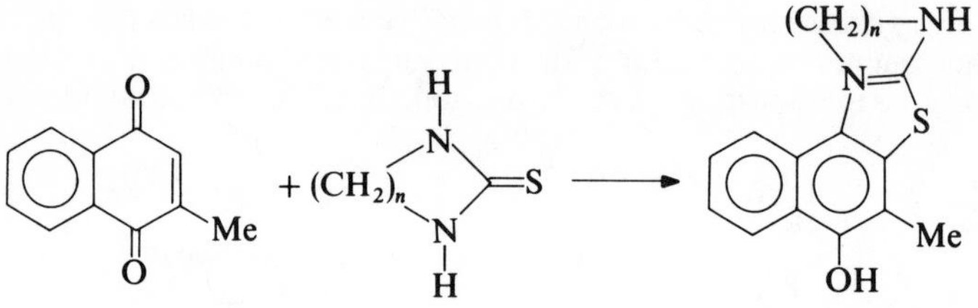

$n = 2, 3$ 56%, 27% (33)

$+ R_2NC(=S)SNa \xrightarrow{pH < 5}$ (34)

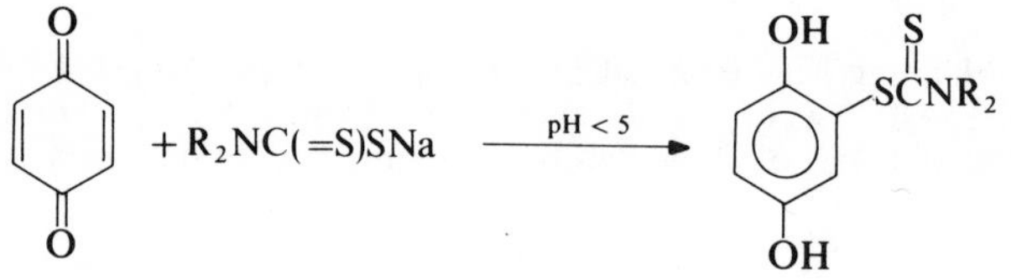

(19)

R = H, Me, Et, *i*-Pr, Bu, NH_2, cyclic 60–90%

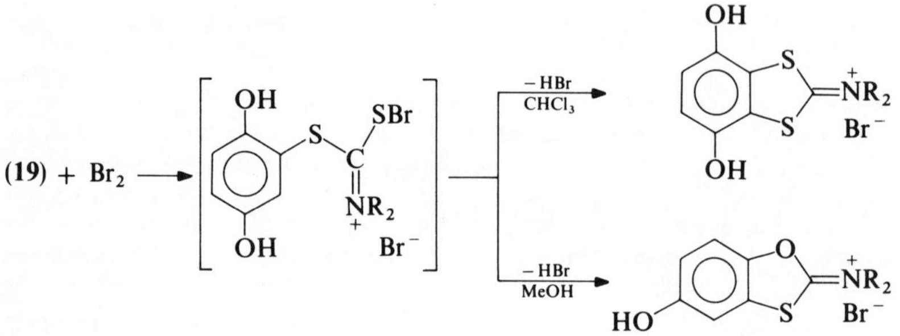

(19) + Br_2 → (35) (36)

yields of 50–80%, but the structural influence varies in the two reactions. Finally, thiobenzoic acid adds to 1,4-benzoquinone in 95% yield under these conditions.

More recently the addition of a dithiobenzoic acid salt has been reported (equation 37)[48].

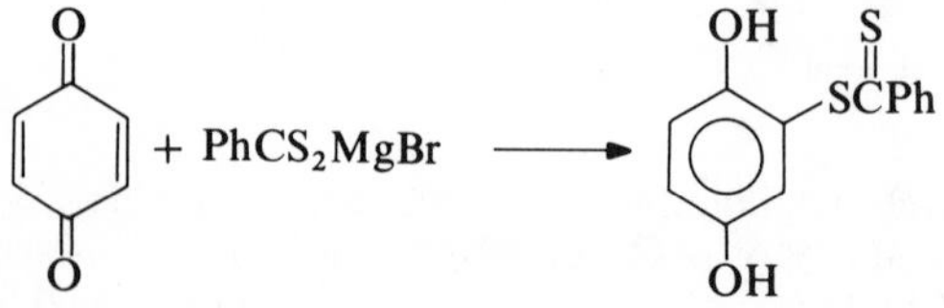

$+ PhCS_2MgBr \longrightarrow$ (37)

The condensation products from the reaction of 3-hydroxycoumarins and 1,4-benzoquinone (see Section VI) undergo thiourea addition reactions in approximately 75% yield (equation 38)[49].

$$\text{quinone-coumarin} + (H_2N)_2C{=}S \longrightarrow \text{product} \qquad (38)$$

X = H, 6-Cl and 6,8-Cl_2

In the addition of ethyl thioacetoacetate to 1,4-benzoquinone Campaigne and Abe found it possible to isolate and characterize all four combinations of geometric isomers and oxidation states (equation 39)[50]. The quinonoid form of each isomer was obtained as

$$\text{1,4-benzoquinone} + MeC(=S)CH_2CO_2Et \longrightarrow \text{(Z)-isomer } (58\%) + \text{(E)-isomer } (6\%) \qquad (39)$$

were the corresponding products of 1,4-naphthoquinone. In the latter case two separate addition experiments were carried out: one in refluxing alcohol and one completely below 40°C, including workup.

Two brief reports related to sulfur addition chemistry deserve mention. First, the formation of a polymeric quinonoid electrode from thiophenol and 1,4-benzoquinone offers strong evidence for the essential correctness of successive oxidation and addition as the preferred mechanistic pathway[51].

Second, adding phenylselenenyl chloride to alkenes produced excellent results except with the α,β-unsaturated carbonyl group where it failed completely (equations 40, 41)[52].

$$+ PhSeCl \longrightarrow \text{SePh/Cl adduct} + \text{SePh/Cl hydroquinone adduct} \quad \text{quantitative} \qquad (40)$$

$$+ PhSeCl \longrightarrow \text{no reaction after 6 h} \qquad (41)$$

B. Nitrogen Addition

1. Kinetics and mechanisms

In contrast to the paucity of data for sulfur reactions kinetic studies of the addition of nitrogen nucleophiles to quinones continue to appear at a modest pace. Amonkar and Ghosh presented a brief, but systematic rate study of the addition of glycine to several 1,4-benzoquinones (equation 42)[53]. The reaction is first order in each reactant. The product(s)

$$\text{R}_3\text{-1,4-benzoquinone} + H_2NCH_2CO_2H \longrightarrow \text{R}_3\text{-}(NHCH_2CO_2H)\text{-1,4-benzoquinone} \qquad (42)$$

R = H, Me

were not reported, and it is difficult to see how the same rate law can be followed by both 1,4-benzoquinone and its trimethyl derivative. Earlier[54] studies on less highly substituted quinones show disubstitution and a second-order dependence on primary amines. Possibly the rate of formation of some common intermediate was being followed in what is clearly a rather complex system. The order of reactivity observed for the quinones is certainly not consistent with a simple nucleophilic addition reaction. The proposal that the reaction involves semiquinones is made and potentiometric studies are promised.

Much more recently Talati and his coworkers have published a study of the oxidation behavior of products obtained from the addition of aniline to 1,4-benzoquinone[55]. They claim high yields (80–85%) of a disubstituted product (surely the 2,5-dianilino). This product oxidizes readily and its oxidation capacity, after reduction with hydrogen iodide or sodium dithionite solution, increases by 60%. Thus, there is a substantial amount of hydroquinone product formed in the second aniline addition. Diamines form redox polymers; some with and some without cross-linking. Evidence for tautomeric quinone–anils is presented for the later examples.

Much of the kinetic and mechanistic study of quinone–amine reactions has centered on the formation of a charge transfer complex and its contribution to the addition reaction sequence. Muralikrishna and Krishnamurthy have made a careful investigation of the kinetics shown by the reaction of 1,4-benzoquinone and piperidine in chloroform (equation 43)[56]. The rate of appearance of the charge transfer complex, which is

$$\text{1,4-benzoquinone} + \text{piperidine (excess)} \xrightarrow{CHCl_3} [CTC] \longrightarrow \text{2,5-dipiperidino-1,4-benzoquinone} \qquad (43)$$

reasonably stable, and the rate of formation of product were both measured. The latter is several times greater than the rate of decomposition of the complex. On this basis another intermediate is postulated and the complete mechanism written as:

$$\text{quinone} + \text{piperidine} \underset{}{\overset{\text{fast}}{\rightleftharpoons}} \text{charge transfer complex}$$

$$\text{charge transfer complex} + \text{piperidine} \xrightarrow{\text{fast}} \text{intermediate}$$

$$\text{intermediate} \xrightarrow{\text{slow}} \text{product}$$

These workers do not suggest a structure for the new intermediate, but on the basis of our general understanding of these reactions **20** seems reasonable. It should be noted that in

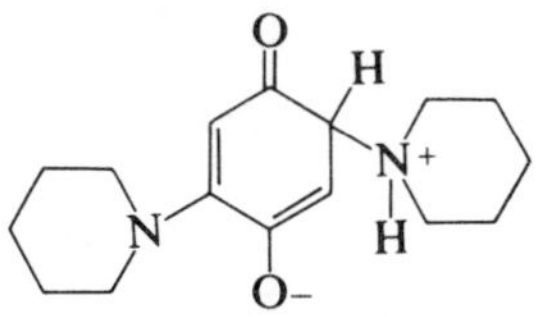

O
H
N+
N
H
O−

(20)

some instances there has been controversy concerning the evidence for charge transfer complex formation in the addition of anilines to quinones[57,58].

Sasaki and his collaborators have used high-pressure kinetic methods to investigate the reaction between 1,4-benzoquinone and dibutylamine[59,60]. In addition to chloroform they studied 1,2-dichloroethane and acetonitrile. There are differences among the solvents as reflected by all the activation parameters, but no detailed explanation could be offered on the basis of the present studies. Lowered activation energy coupled with increasing solvent dielectric constant suggests an electrostatic solvent–transition state interaction. The mystery lies in the failure of the activation volume to show a similar dependence.

Observations of greater interest in these studies concern the reaction's pressure dependence. Over several hours at one bar the rate is third order requiring two moles of amine for each quinone and only a monosubstituted product is obtained. With increased pressure a disubstitution product appears and it shows greater pressure sensitivity. The reaction sequence suggested involves a radical ion pair or a zwitterionic σ complex formed rapidly and reversibly. The following and rate-determining step is the abstraction of a proton by a second amine molecule.

A relevant observation was made by examining the proton and ^{13}C spectra of 1,4-benzoquinone and various chlorinated derivatives in liquid ammonia[61]. Excellent evidence was obtained for reversible mono- and dicarbonyl addition of ammonia. The 1,4-benzoquinone adducts are stable for about 12 hours after which irreversible reaction occurs. The chlorinated quinones react more rapidly. While the chemistry in such a large excess of ammonia is undoubtedly different in some fashion, it may be worthwhile to investigate such intermediates with amines as the nucleophile.

Pelizzetti and his collaborators in conjunction with studies of the oxidation of catecholamines (see Section II.A.1) have presented data relevant to the intramolecular cyclization reaction (equation 7)[62]. Their evidence demonstrates the complexity of the system at higher pH and provides support for earlier structure–reactivity correlations.

$$\text{adrenalin} \gg \text{L-dopa} > \text{dopamine}$$

Some product studies without kinetic data have advanced our knowledge of nitrogen addition to quinonoid systems. Baxter and Phillips have extended their earlier preliminary work concerning 2,5-di-*t*-butyl-1,4-benzoquinone (**21**)[63] to include *n*-butylamine and *t*-butylamine[64]. In experiments with the former they found product yields identical to those of propylamine. Predictably, no dark reaction occurred with *t*-butylamine. No comment was made about the *n*-butylamine reaction intermediate and there is every reason to expect that the quinol analogous to that reported previously is involved. When this amine reacts with **21** in the dark and with air present, an epoxide (**22**) similar to that reported earlier is obtained (equations 44, 45).

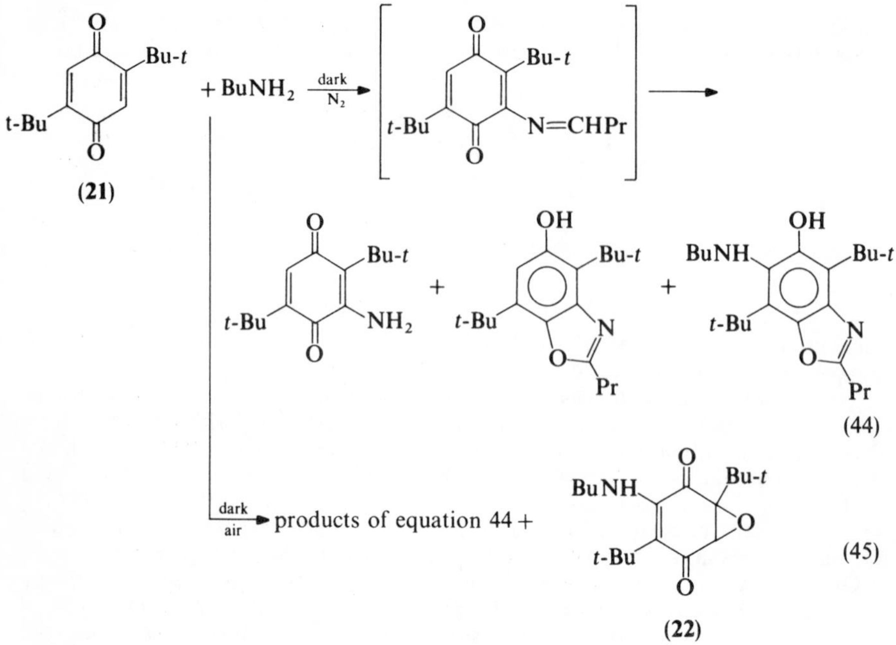

$$\text{(21)} + BuNH_2 \xrightarrow[N_2]{\text{dark}} \left[\text{intermediate: } t\text{-Bu, Bu-}t\text{, N=CHPr quinone}\right] \longrightarrow \text{products} \tag{44}$$

$$\text{(21)} + BuNH_2 \xrightarrow[\text{air}]{\text{dark}} \text{products of equation 44} + \text{(22)} \tag{45}$$

It is often useful to examine 'simple' systems as models for molecules of greater complexity and interest, for example, the cyclization of nascent *o*-quinonoid compounds related to dopamine (see Section II.A.2). Davies and Frahn reported that the four simplest primary amines react with 1,2-benzoquinone by such complex pathways that thin-layer chromatography shows the presence of at least 50 products plus polymers in each case[65]. Still, it is instructive to examine the structure of the two products which could be isolated and purified (equation 46). The authors realize that these structures are tautomers and are surprised that the proton transfers are slow enough to allow their isolation. The chemical behavior of each isomeric pair as well as extensive spectral evidence strongly supports the proposed structures. A test of the suggestion that peptide bonds might react with oxidized polyphenols[66] was carried out, but no evidence could be found to support a reaction between 1,2-benzoquinone and *N*-methylacetamide.

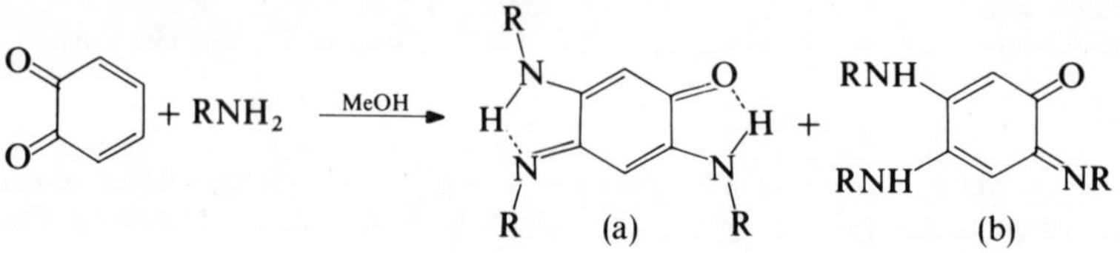

$$\text{1,2-benzoquinone} + RNH_2 \xrightarrow{\text{MeOH}} \text{(a)} + \text{(b)} \tag{46}$$

R	%a	%b
Me	8	11
Et	7	10
Pr	—	10
Bu	—	10

2. *Synthetic studies*

It is rather hard to distinguish a clear borderline between studies directed primarily toward mechanistic goals and those of a more preparatory nature. In fact the bulk of the research described as synthetic in this chapter is significant for its contribution to our understanding of chemical principles.

In a continuation of their studies of dihydroperimidione (**23**) Cameron and Samuel found that methylated 5-amino-1,4-naphthoquinones react with piperidine in a manner much like the heterocyclic quinone anil (equations 47, 48)[67]. For compound **23** lower rates

(**23**) R = H, Me + piperidine (excess) → 75%; 2,3-Me_2-**23** no reaction (47)

R = H, Me + piperidine → 60%, 70% (48)

are observed and there is no evidence for side-chain amination in the 2,3-dimethyl derivative. These models were also compared with the important 5-hydroxy-1,4-naphthoquinone (juglone **24**) where a mixture in which the 2-piperidino addition product predominated was obtained (equation 49). If the solvent is changed from excess piperidine to ethanol, the isomeric distribution is just reversed (equation 50). The explanation offered for the change in reaction position in substrates **23** and **24** referred to the combination of greater electron-donating capacity of the amino group with its poorer intramolecular hydrogen bonding ability. The solvent effect with juglone was not investigated beyond

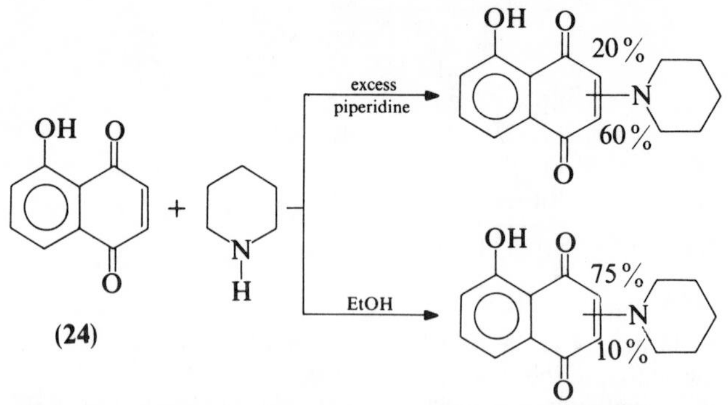

(49)

(50)

finding that it is not observed in the amino compounds. Surely this striking effect deserves attention other than merely stating that it might result from some combination of intramolecular hydrogen bonding, group polarity or ionization of the hydroxy group.

When 2,6- or 3,7-dimethyl-5-amino-1,4-naphthoquinone react with neat piperidine there is a strong preference for 3-substitution (equations 51, 52). There is no evidence of simple 2-piperidino addition for the 3,7-dimethyl substrate.

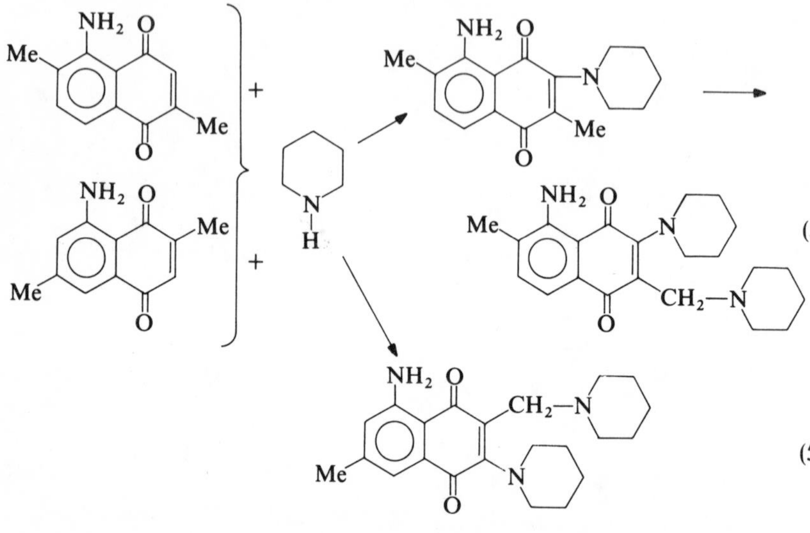

(51)

(52)

In a study of the addition of primary aliphatic amines to 1,4-benzoquinones Ott and his colleagues have discovered that the products are not as simple as was once thought and that solvent plays a significant role[68]. While in the addition of methylamine, 2-methylamino-1,4-benzoquinone is an intermediate and 2,5-bis(methylamino)-1,4-benzoquinone (**25**) is a product, the major product is 6-hydroxy-9-methyl-3-methylamino-1,4-carbazolquinone (**26**, equation 53). This result is consistent with the inability of Yamaoka and Nagakura to observe any spectra of 2-butylamino-1,4-benzoquinone in their detailed kinetic study of the butylamine system[69]. Chemical and spectral evidence combine to assure the correctness of the proposed structure. Furthermore, the carbazolquinone was prepared starting with 2,5,2′,5′-biphenyldiquinone (**27**, equation 54).

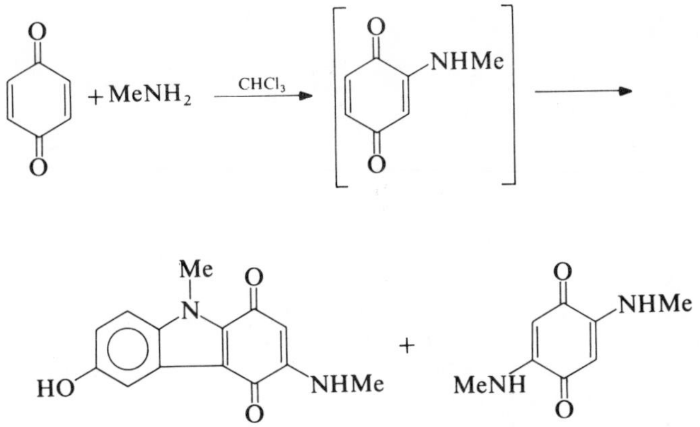

(26) 78% (25) 16% (53)

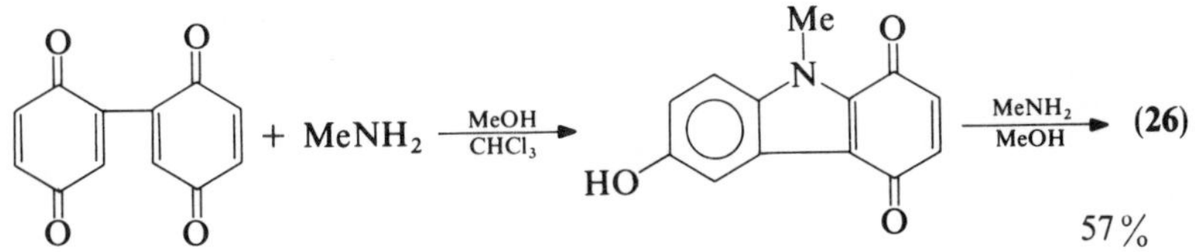

(27) 82% 57% (54)

Several other solvents were examined and with both piperidine and dimethylformamide the product distribution was exactly reversed (19% **26**; 62% **25**). Finally, the use of ethanol at low temperature gave a 77% yield of the monoaddition product.

From a synthetic point of view Ott and his group have developed a generally useful pathway to the carbazolquinones (equation 55).

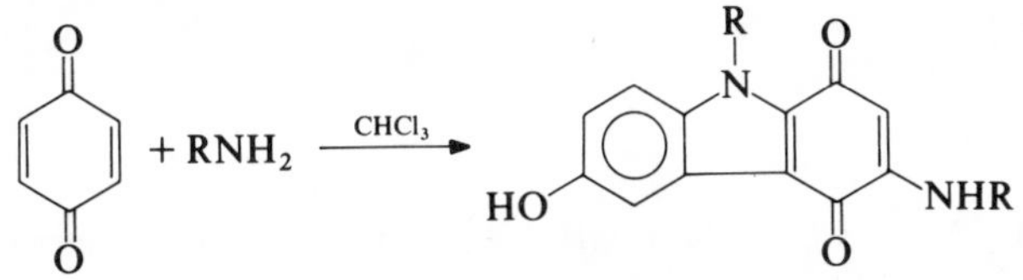

(55)

R	%	R	%	R	%
Et	47	Bu	19	$PhCH_2$	54
Pr	37	Hex	54	$HOCH_2CH_2$	43

In a subsequent publication Ott and his coworkers noted the presence of another unexpected product in these systems (equation 56)[70]. Lower yields of the two simple addition products were also found. The structures of the methylanilino phenols were supported by independent synthesis, and all product structures were consistent with various spectral studies. Changes in the acid component of the solvent did not effect the order of the products but did change their relative yield. These observations necessitate a

$$R^1 = H, Me;\ R^2 = Me, Et, Pr, Bu$$

ca. 28 %

(56)

careful re-examination of the role of preliminary carbonyl–amine interactions and their influence on the nature of the isolated quinone addition product. One is reminded of the work in liquid ammonia discussed earlier[61].

The addition of imidazole to 1,4-benzoquinone produces both the expected 2,5- and the unexpected 2,3-diamino products (equation 57)[71]. The authors examined several solvents (acetone, methylene chloride and acetonitrile) but were unable to obtain information concerning the steps leading to the charge transfer complex. No yields were reported.

< 0°C or ambient

(57)

From a synthetic point of view the direct addition of anilines to quinones has continued to generate interest. Both ethanol and ethanol–acetic acid proved satisfactory for the addition of *p*-substituted anilines to 1,4-benzoquinone and its 2,5-dichloro derivative (equation 58)[72–74]. Many of the compounds prepared contained a sulfonamido group

(58)

$$R^1 = H, Cl, Me, -CH{=}CHCH{=}CH-;\ R^2 = SO_2NHR^3\ (R^3 = \text{heterocyclic ring})$$

and a heterocyclic ring since these substituents often promote antitubercular or antimicrobial activity; 1,4-naphthoquinones were included for the same reason. The solvents selected produced uniformly high yields with the exception of reactions involving 2-methyl-1,4-naphthoquinone which were in the 20–50 % range. A massive study by Clark (129 substituted 1,4-naphthoquinones) is especially valuable for the extensive set of truly comparable melting points for this important class of compounds[75].

In a study of the tautomerism of 4-(*N*-arylamino)-1,2-naphthoquinones Biggs and Tedder prepared their compounds by direct addition and included two examples of *N*-substituted anilines and hexylamine (equation 59)[76]. The yields in methanol vary widely and are not given in some cases, but most lie in the 40–70 % range. The spectra of *N*-alkyl substituted cases are difficult to interpret, but the general pattern of 1,2-quinone structure in the solid or ethanol solution and 1,4-quinonoid structure in trifluoroacetic acid is clear.

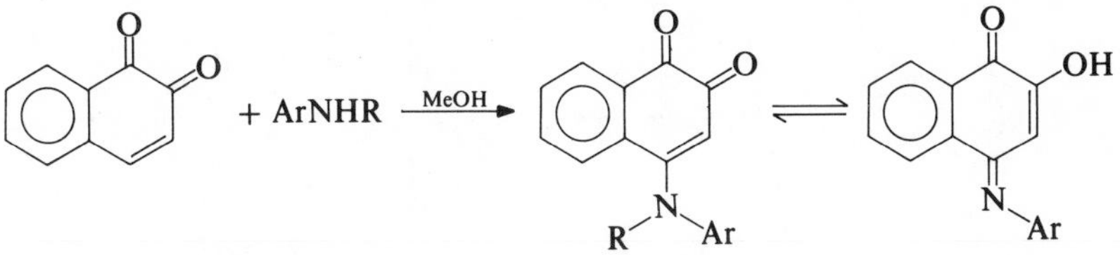

$$\text{Ar} = 2,5\text{-Cl}_2,\ 2,5\text{-Me}_2\text{C}_6\text{H}_3,\ 4\text{-An},\ 4\text{-Br},\ 4\text{-NH}_2,\ 4\text{-NHCOMeC}_6\text{H}_4$$

$$\text{R} = \text{H, Me, Et} \qquad (59)$$

(R = H)

An interesting rearrangement leading to carbon–carbon bond formation is found in a nascent 1,4-naphthoquinone addition chemistry (equation 60)[77]. The product obtained is

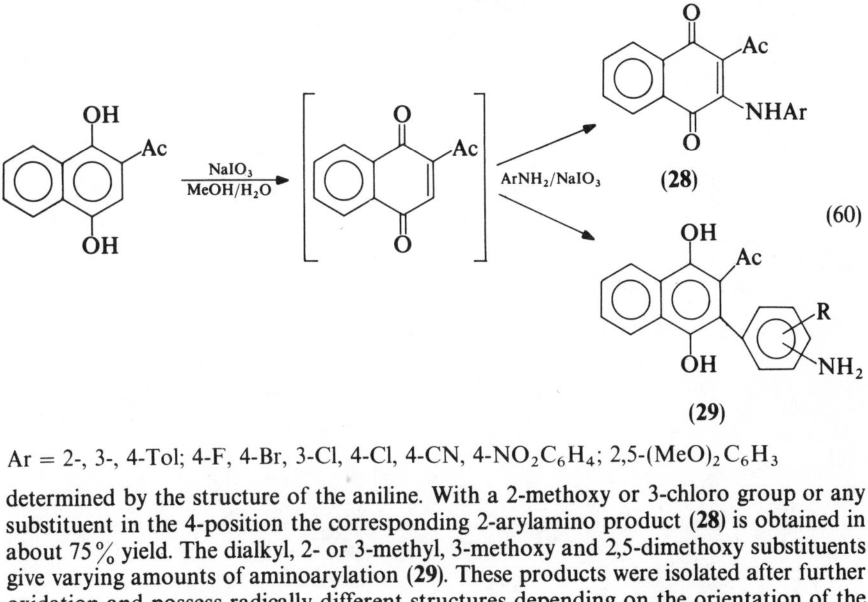

(28)

(60)

(29)

$$\text{Ar} = 2\text{-, }3\text{-, }4\text{-Tol; }4\text{-F, }4\text{-Br, }3\text{-Cl, }4\text{-Cl, }4\text{-CN, }4\text{-NO}_2\text{C}_6\text{H}_4\text{; }2,5\text{-(MeO)}_2\text{C}_6\text{H}_3$$

determined by the structure of the aniline. With a 2-methoxy or 3-chloro group or any substituent in the 4-position the corresponding 2-arylamino product (**28**) is obtained in about 75% yield. The dialkyl, 2- or 3-methyl, 3-methoxy and 2,5-dimethoxy substituents give varying amounts of aminoarylation (**29**). These products were isolated after further oxidation and possess radically different structures depending on the orientation of the amino group (equations 61, 62). With strong electron-donating groups these products

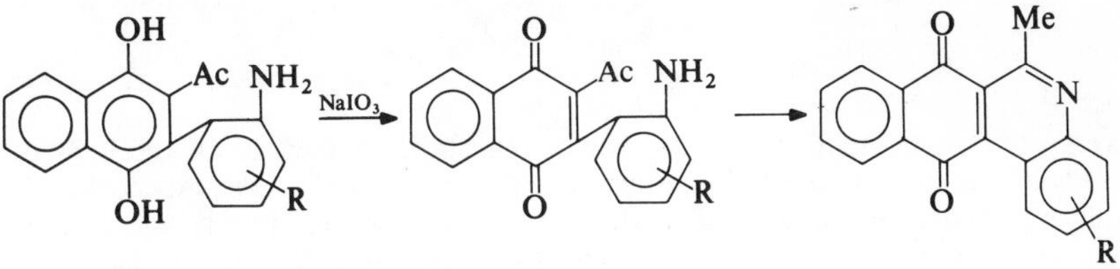

(61)

become important and in some cases are the only ones isolated. The reactions of the same anilines in chloroform and without added oxidant generally produce good yields of the direct amino addition. The 3-methoxyaniline and 2-aminonaphthalene are notable exceptions (equations 63, 64).

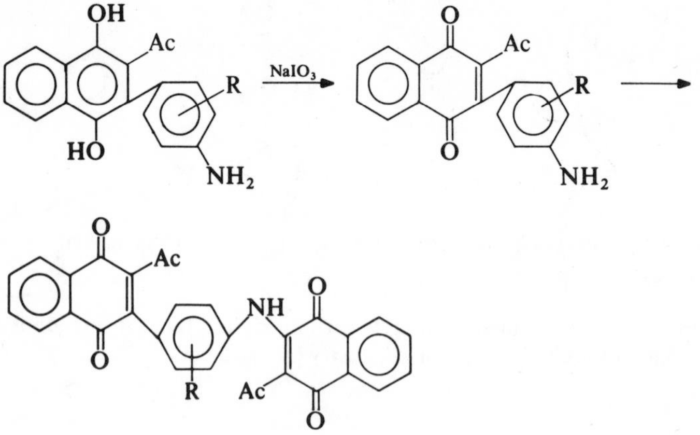

(62)

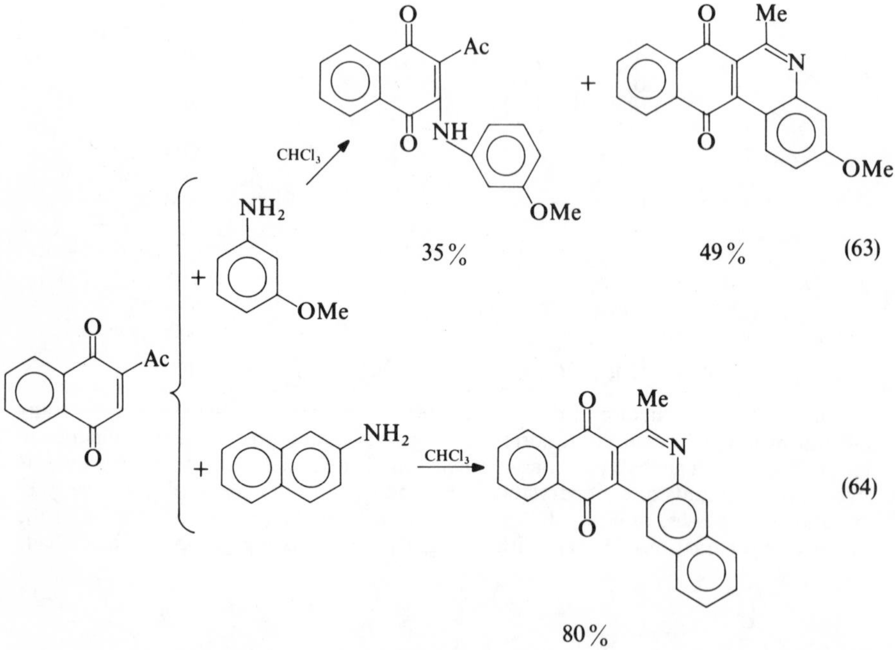

(63)

(64)

The addition of anilines to 1,2-quinones has been studied using the 4,4′-bi-(1,2-naphthoquinone) molecule[78]. For most substituents on aniline the expected addition product is obtained in good yield (equation 65). An exception is 2-pyridyl which gave only 27%. With phenyl and its methyl, methoxy and chloro derivatives oxidation to the quinone-anils is accomplished in generally good yields. The latter compounds were obtained directly with the halogenated anilines. No interpretation of these observations was presented.

Many studies involving the addition of nitrogen nucleophiles to quinones have been reported in connection with the search for dyestuffs, medicinals, etc. A few of these advance our understanding and raise questions of interest for further study. Folkers and his

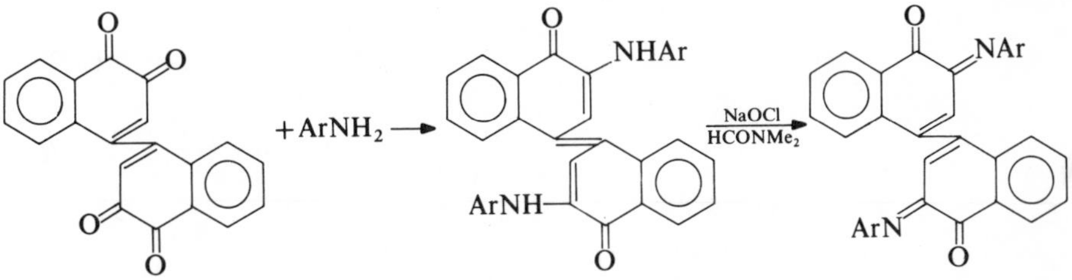

$$\text{4,4'-bi(1,2-naphthoquinone)} + ArNH_2 \longrightarrow \text{2,2'-bis(arylamino)-4,4'-binaphthoquinone} \xrightarrow[HCONMe_2]{NaOCl} \text{bis(arylimino) product}$$

Ar = Ph, 2,3,4-Tol, 2,4-An, 4-ClC_6H_4, 4-$NO_2C_6H_4$, 2,4-ClC_6H_3, 2- and 3-pyridyl (65)

collaborators prepared a 12 new 2,5-bis(alkylamino)-1,4-benzoquinones using a combination of addition and substitution chemistry (equation 66)[79]. The yields, which do not

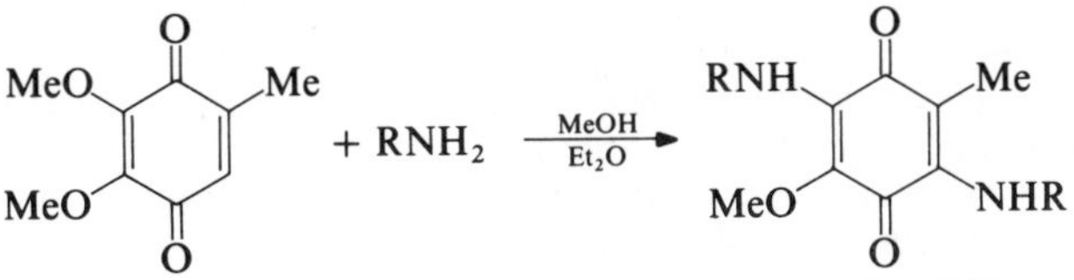

$$\text{2,3-(MeO)}_2\text{-5-Me-1,4-benzoquinone} + RNH_2 \xrightarrow[Et_2O]{MeOH} \text{2-RNH-3-MeO-5-NHR-6-Me-1,4-benzoquinone} \quad (66)$$

R = C_6 to C_{18} *n*-alkyl, C_5 to C_7 cycloalkyl, CH_2CH_2OH

appear to have been optimized, show a puzzling variation; six average 57% and four average 28%. Neither chain length nor reaction time appears to explain these observations. The reaction time was generally 16 hours, but was reduced to 3 without explanation in three instances. Ethanolamine produced only a 9% yield after 20 hours at room temperature.

Tindale applied standard methods of amine-quinone additions to certain biochemically interesting amines with significant results[80]. In ethanol solution at room temperature good, but unspecified, yields of 2,4-diamino addition products were obtained with the exception of histamine (**30**) which produced only the monoaddition product (equations 67, 68). Similar additions to 1,2-benzoquinones gave tar and many minor products. One

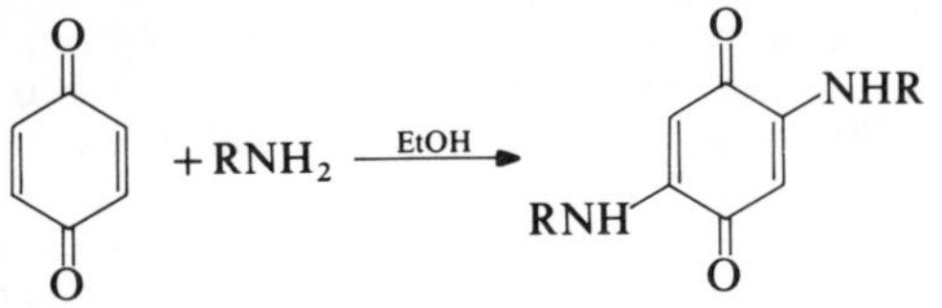

$$\text{1,4-benzoquinone} + RNH_2 \xrightarrow{EtOH} \text{2,5-bis(RNH)-1,4-benzoquinone} \quad (67)$$

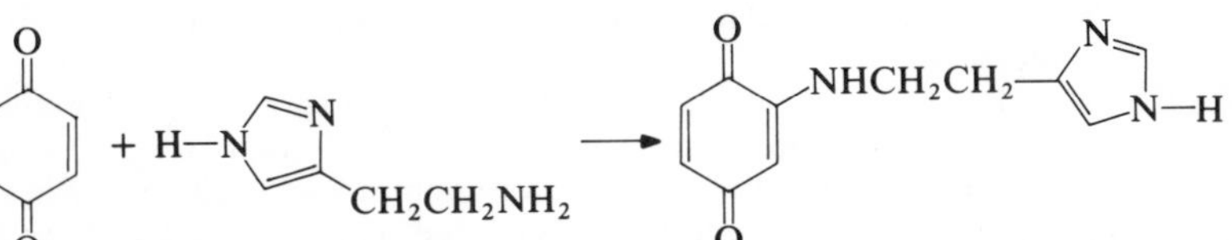

$$\text{1,4-benzoquinone} + \text{histamine } (\mathbf{30}) \longrightarrow \text{2-}(NHCH_2CH_2\text{-imidazolyl})\text{-1,4-benzoquinone} \quad (68)$$

R = Ph, $PhCH_2CH_2$, 4-$HOC_6H_4CH_2CH_2$

product was isolated from the reaction with 2-phenylethylamine, but it was not characterized. With 2,5-dihydroxy-1,4-benzoquinone all four amines gave 1,2-carbonyl addition in contrast to the earlier report of unusual displacement of the hydroxyl groups[81]. However, in boiling aniline the net result of substitution is obtained, probably by an addition–elimination mechanism (equation 69). On long standing in ethanol, or upon the

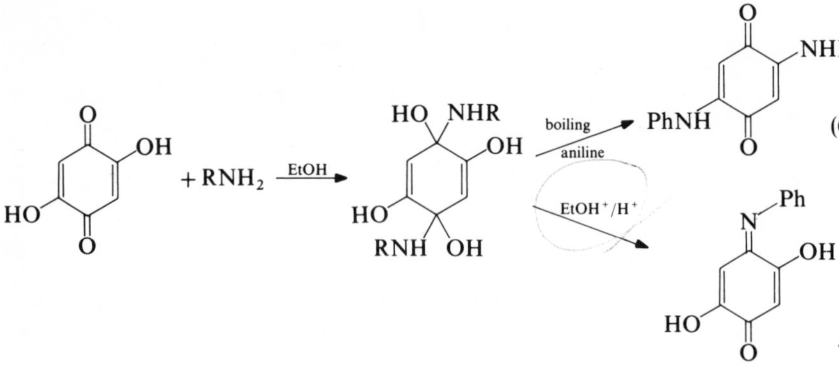

(69)

70)

addition of a trace of hydrochloric acid, the aniline monoimine is formed (equation 70). The other amines hydrolyzed, but did not form imines.

Naphthazarin has been of interest for some time both in the search for antibiotics and in the dye industry. A study of its 6,7-dichloro derivative (**31**) shows not only competition between addition and substitution, but also facile tautomerization (equation 71)[82].

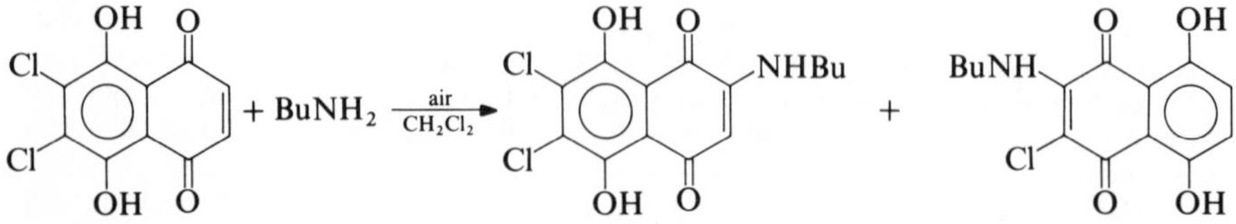

(**31**) 39% 14%

(71)

Ethanol gives the highest degree of selectivity with addition favored over substitution 76:1. This preference holds for the three other amines studied, but not for oxygen nucleophiles. Sodium methoxide gave a good yield of 2-methoxy-3-chloronaphthazarin while methanol or ethanol in the presence of triethylamine gave fair yields of both addition and substitution products analogous to those found with primary amines.

Kallmayer and Tappe have prepared a series of drugs related to desipramin through the addition of secondary amines containing polycyclic substituents to 1,4-benzoquinone (Table 1)[83]. The similar amounts of mono- and di-addition products in three cases is unlike most work in the field.

The results of Matsuoka and his colleagues[82] are especially surprising and promising in view of their earlier report on naphthazarin itself[84]. In ethanol solution excess butylamine gave a variety of butylamino addition and substitution products, all in very low yield. When the reaction was carried out in the presence of cupric acetate the product distribution, while still complex, changed markedly (equation 72). The structure of the product (**32**) was not determined, but it was discovered that the yield could be more than doubled (61%) using cupric chloride at ambient temperature. If the copper complex of naphthazarin is prepared and then allowed to react with butylamine, the yield of **32** also increases (41.8%) and potentially useful yields of a simpler array of by-products is obtained (equation 73).

Rapoport and Luly have shown that copper acetate promotes highly regioselective amination of unsymmetrical 1,4- and 1,2-benzoquinones (equation 74)[85]. Traces of the

TABLE 1. Distribution of 1,4-benzoquinone mono- and di-addition products from the synthesis of compounds related to the desipramin drugs[83]

Amine (R)	% Yield	
	Monoadduct	Diadduct
=CH(CH2)2NHMe	63	44
(CH2)3NHMe	55	41
(CH2)3NHMe	90	7
CH2NHMe	17	19

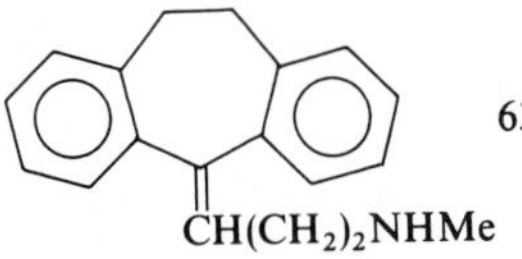

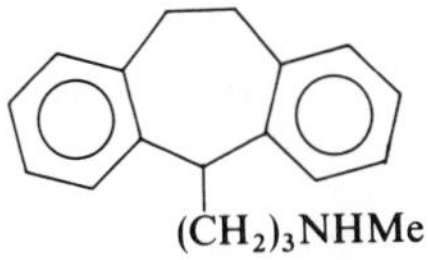

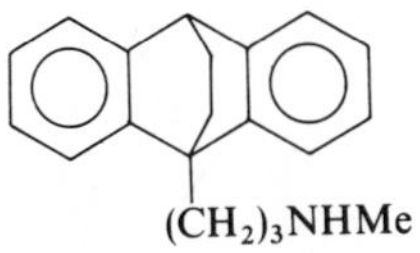

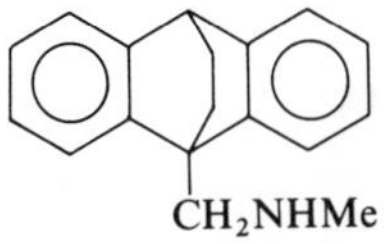

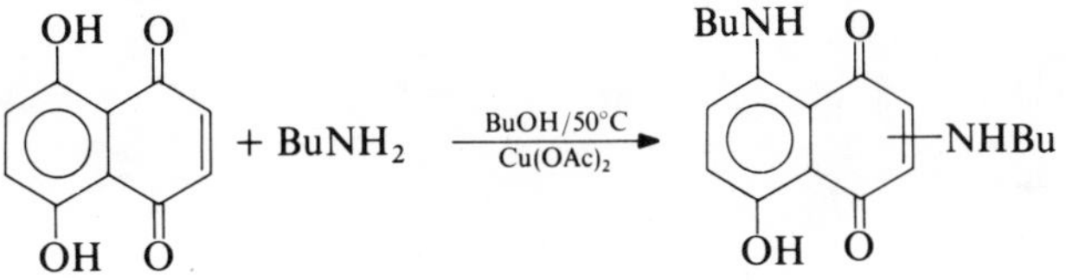

$$\text{1,4-dihydroxy-5,8-naphthoquinone} + BuNH_2 \xrightarrow[Cu(OAc)_2]{BuOH/50^{\circ}C} \textbf{(32)}\ 26.5\% \tag{72}$$

dipyrrolidine adduct and a demethylated analog resulting respectively from methoxy or methyl group displacement could be identified. In the absence of the copper salt the isomeric ratio became 2:1 showing the powerful effect of the complex formation.

The excellent regioselectivity found in the preceding model study was utilized by the same authors as the first step in a superb synthesis of 7-methoxymitosene (equation 75)[86]. As was expected from the more hindered and less nucleophilic proline methyl ester of **33** a product of higher isomeric purity was obtained; 96% with less than 1% of the demethylated product and no isomeric addition product. In the absence of the copper salt a major side product resulted from the displacement of the methoxy group.

3. *Cyclization reactions*

The study of heterocyclic ring-forming reactions has become one of the most active areas of synthetic interest in recent years. In addition to his several studies already described Ott

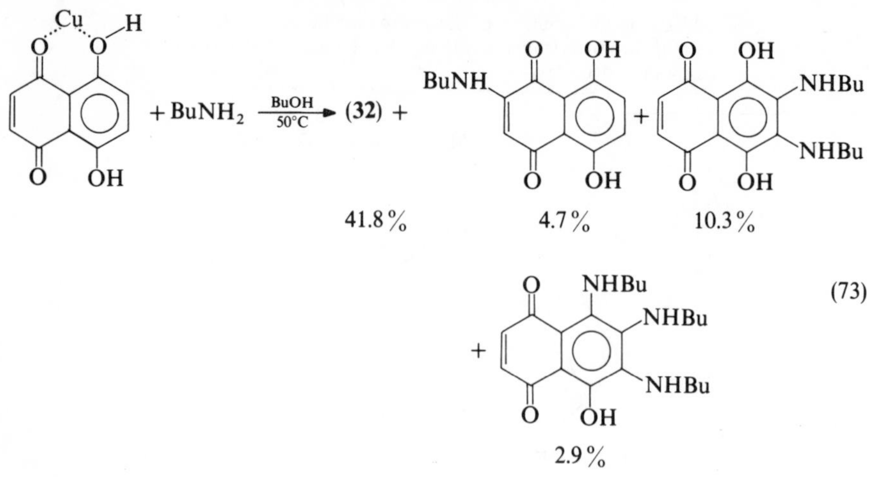

$$+ BuNH_2 \xrightarrow[50°C]{BuOH} (32) + \ldots$$

41.8% 4.7% 10.3%

2.9% (73)

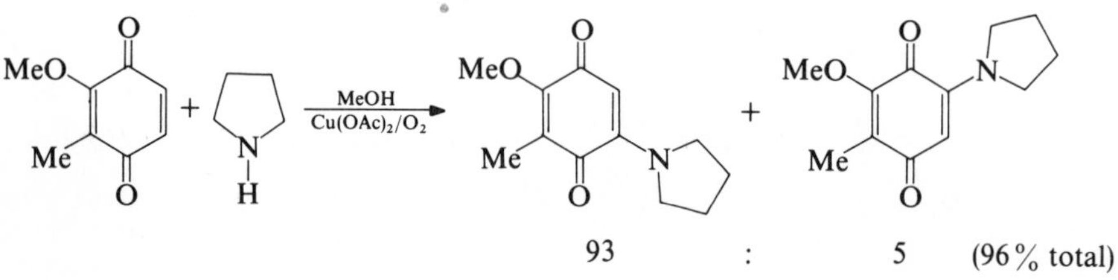

$$\xrightarrow[Cu(OAc)_2/O_2]{MeOH}$$

93 : 5 (96% total) (74)

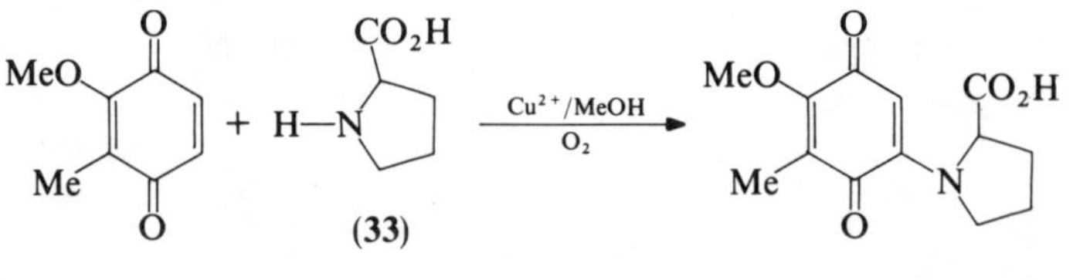

$$\xrightarrow[O_2]{Cu^{2+}/MeOH}$$

(33) (75)

has looked at the chemistry of *o*-phenylenediamine and 2-substituted 1,4-naphthoquinones (equations 76, 77)[87]. While a great amount of detail of information concerning the intermediate steps of the reaction is presented, the essence of this study is that the quinone substituent controls the course of the addition and the structure of the product. Oxygen and nitrogen substituents are easily displaced, and acetic acid is the solvent of choice with yields in excess of 90% being common. The ratio of quinone to diamine seems to be important and 1:2 is preferred, but no clear pattern was found. Alcohol fails to be a useful solvent in all cases. Some rather puzzling observations emerge from the data; for example, while both the amino and the dimethylamino groups produce excellent yields of **34**, methylamino gives only very small amounts. All anilino substituents, except the 2-hydroxy case, are poor leaving groups; this system gave excellent yields for both itself and the *N*-methyl derivative. Unlike these nitrogen substituents the acetamido group and a 2:1 excess of quinone gave a nearly quantitative yield of the product resulting from reaction at the 3-carbon and 4-carbonyl (**35**). This drastic reversal of reactant ratio is typical of the second pathway and easily explained by the proposed mechanism which

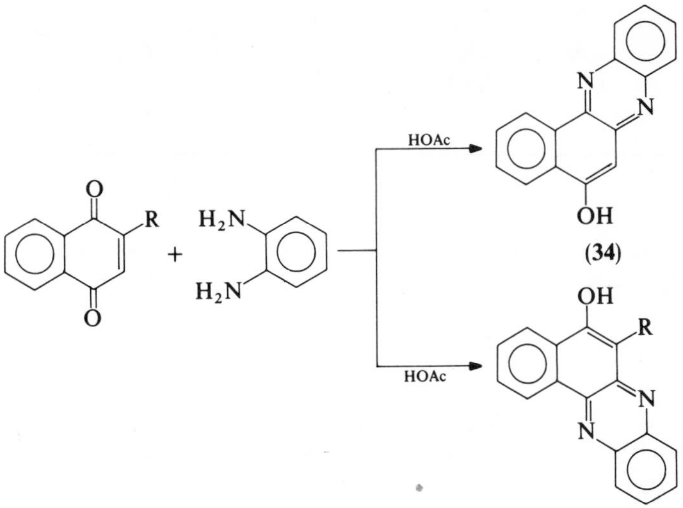

(76)

(34)

(77)

(35)

involves an additional oxidation step. Oxidized carbon or sulfur substituents are both successful in promoting this reaction direction; chloro groups somewhat less so. Alkyl and aryl, represented only by methyl and phenyl, also react in this fashion but give poor and fair yields respectively. For the acetyl group ethanol is an excellent solvent; water for the sulfonic acid.

The studies of Kallmayer and Seyfang complement those of Ott through an intensive examination of the addition of such aliphatic analogs as 1,2-ethylenediamine to naphthoquinones[88–91]. They have shown that 2-methyl-1,4-naphthoquinone reacts in the expected manner when either 2 or 3 methylene groups are present, but longer chains (4, 5, or 6) lead only to open chain products (equations 78, 79)[88]. In this particular study

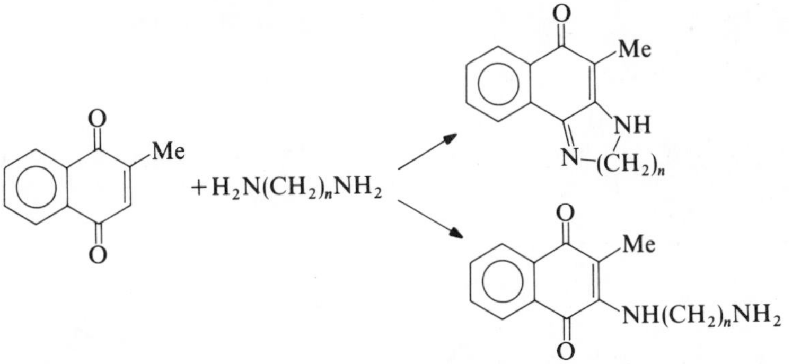

$+ H_2N(CH_2)_nNH_2$

(78) $n = 2, 3$

(79) $n = 4$–6

ethanolamine was examined and found to add in the usual fashion illustrating once again the great preference for nitrogen addition (equation 80).

An even closer analogy to Ott's work is found in the next paper in the series where the authors explore 1,4-naphthoquinones bearing a large range of 2- and/or 3-substituents[89]. No yields are given, but the cyclized products appear to be formed by substitution of one

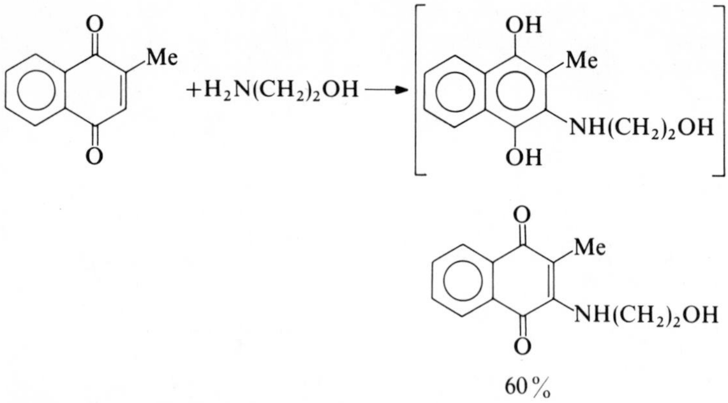

$$+H_2N(CH_2)_2OH \longrightarrow \longrightarrow \quad (80)$$

60%

amino group for the displaceable group. Using *N*-acetyl-1,2-ethylenediamine it was demonstrated (equations 81, 82) that the intermediate has the ring-addition structure (**36**) rather than that of an anil (**37**).

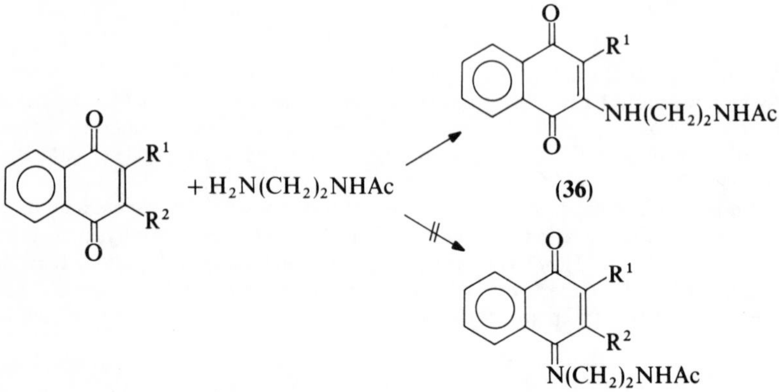

$$+H_2N(CH_2)_2NHAc \quad (81) \quad (82)$$

(**36**)

R^1 = H, alkyl, Ph, Br, MeO, AcO; R^2 = Me, Cl, Br (**37**)

These workers have tried to prepare the 5-hydroxy derivative of some heterocyclic products[90]. While the efforts failed some unusual chemistry of cyclization did emerge (equation 83).

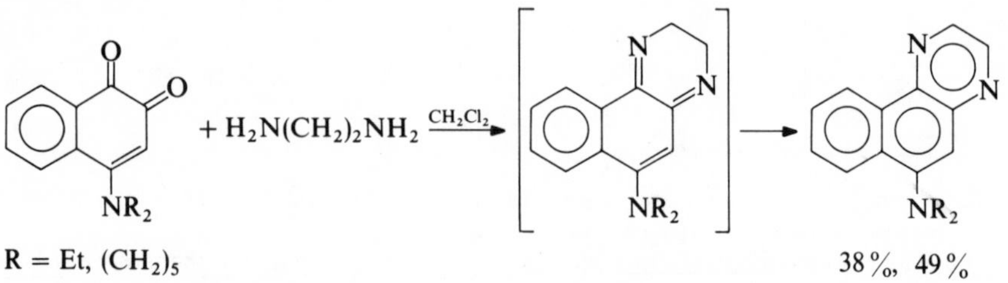

$$+H_2N(CH_2)_2NH_2 \xrightarrow{CH_2Cl_2} \longrightarrow$$

R = Et, $(CH_2)_5$

38%, 49%

(83)

When *N*-alkyl or *N*-arylamino ethylenediamines are added to 2-bromo or 2-methyl-1,4-naphthoquinones the yields are low, but highly selective (equations 84, 85)[91]. The cyclized products are not produced by alkylation of the unsubstituted benzoquinoxalinones. The reaction also takes place with trimethyl-1,4-benzoquinone, again in low yield (equation 86).

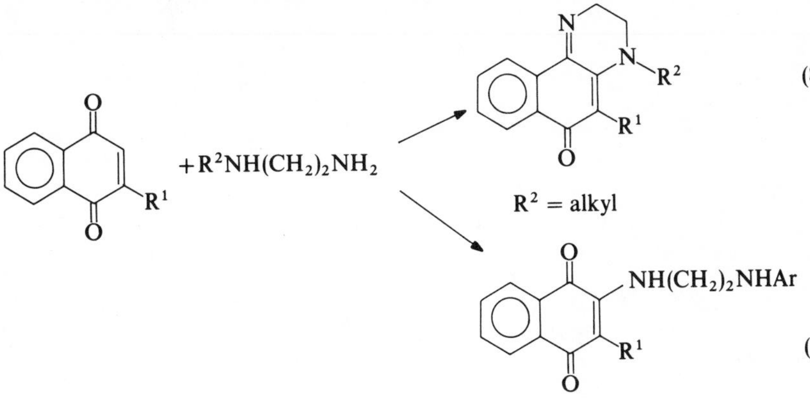

$+ R^2NH(CH_2)_2NH_2$

R^2 = alkyl (84)

$NH(CH_2)_2NHAr$

R^2 = aryl (85)

R^1 = Br, Me; R^2 = Me, Et, Ph, 1-Naphthyl

$+ MeNH(CH_2)_2NH_2 \longrightarrow$ (86)

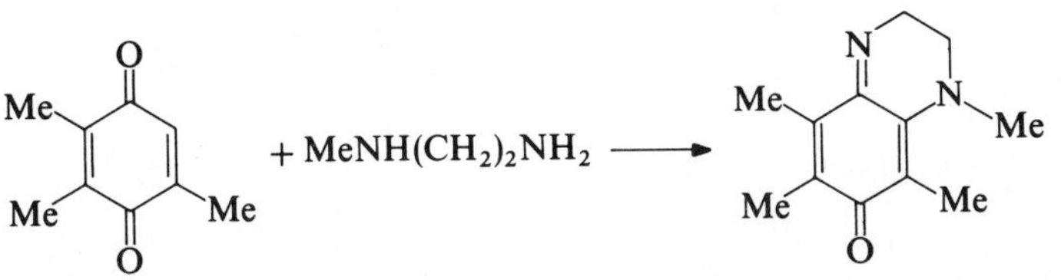

7%

Saxena and Soni have conducted an interesting series of investigations involving the initial addition of 2-amino heterocycles to 1,4-benzoquinone followed by ring-closure, creating three series of multi-heterocyclic products (equations 87–89)[92–94].

Chaaban and his collaborators have shown that the addition of 2-aminobenzoic acids followed by dehydration leads to acceptable yields of 2-halobenzo[*b*]acridine-6,11,12-triones (**38**, equation 90)[95].

The much studied competition between thiol and amino groups in their addition reactions with 1,2-quinones has produced an interesting example in the 1,4-benzoquinone series. In their model study of biosynthetic routes to a firefly luciferin McCapra and Razavi

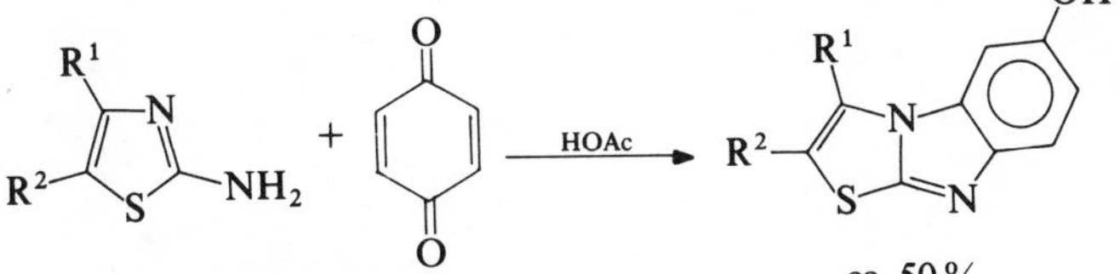

HOAc (87)

ca. 50%

R^1 = H, Me, Ph, 4-Tol, 4-BrC_6H_4 4-HOC_6H_4, 2-thienyl, 2-naphthyl

R^2 = H, Me, Ph, CO_2Et

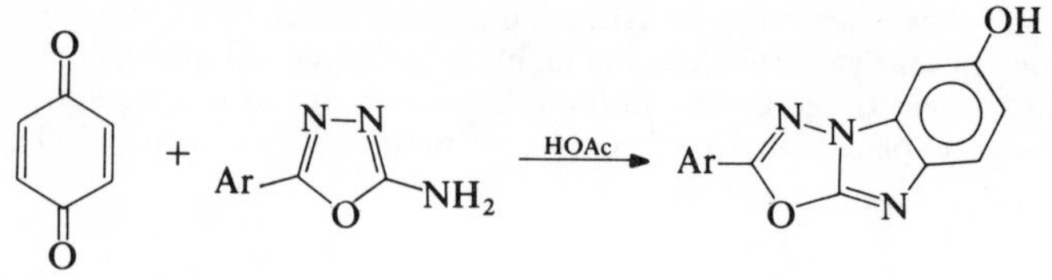

Ar = Ph, 4-Cl, 4-Me, or
$4\text{-}FC_6H_4$, $3,5\text{-}Br_2\text{-}2\text{-}HOC_6H_2$

ca. 25%

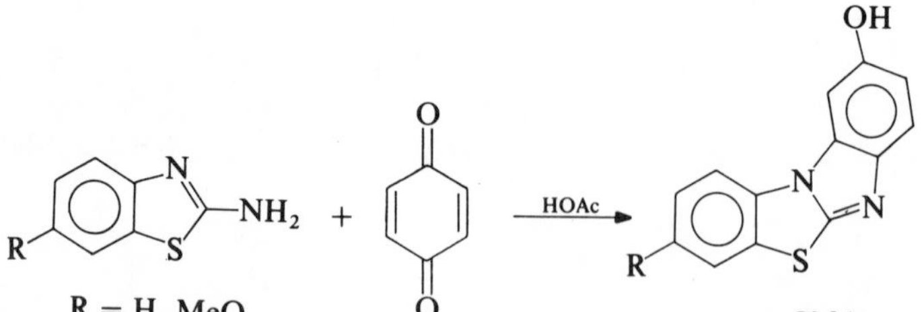

(88)

R = H, MeO

ca. 50%

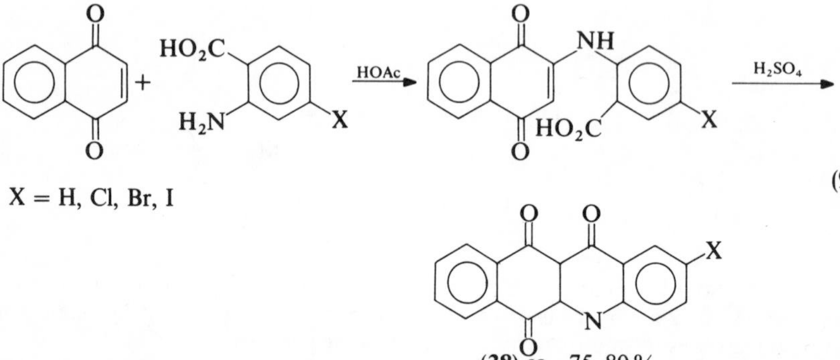

(89)

X = H, Cl, Br, I

(38) ca. 75–80%

(90)

showed that under acidic conditions both groups add (equation 91)[96]. Under basic, aerobic conditions the amino heterocycle undergoes ring contraction; this chemistry was utilized with a more elaborate model (equation 92),

In an attempt to use 3,5-di-*t*-butyl-1,2-benzoquinone in the Strecker degradation of a sensitive α-amino acid Vander Zwan and coworkers found an interesting synthesis of

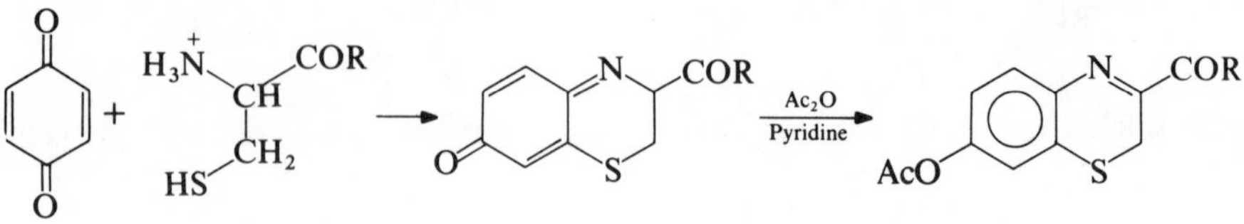

$R = NH_2$, OEt

(91)

$$\text{1,4-benzoquinone} + \text{H}_3\overset{+}{\text{N}}\text{CH(CH}_2\text{SH)C(=O)NHCH(CH}_2\text{SH)CO}_2\text{H} \xrightarrow{\text{several steps}} \text{2-(6-hydroxybenzothiazol-2-yl)-4,5-dihydrothiazole-4-carboxylic acid} \quad (92)$$

benzoxazoles (equation 93)[97]. While not many examples are given and the yield with phenylalanine is low, the method appears to be general for highly hindered *o*-quinones where ring-addition is restricted. The method is somewhat similar to Corey's preparation of benzoxazoles from this quinone and primary amines[98].

$$\text{3,5-di-}t\text{-Bu-}o\text{-benzoquinone} + \text{PhCH}_2\text{CH(NH}_2\text{)CO}_2\text{H} \xrightarrow[\text{MeONa}]{\text{MeOH}} \text{5,7-di-}t\text{-Bu-2-(CH}_2\text{Ph)benzoxazole} \;\; 32\% \quad (93)$$

Interest in tautomeric quinonoid systems is potentially useful in the construction of heterocyclic rings, for example, treatment of the natural product embelin (**39**) with 2-aminophenol produced 2-hydroxy-4-*n*-undecyl-3(*H*)-phenoxazone (**40**, equation 94)[99].

$$\text{(39)} \; [\text{Me(CH}_2)_{10}, \text{OH}, \text{HO}] + \text{2-aminophenol} \xrightarrow[\text{reflux/3 h}]{\text{glacial HOAc}} \text{(40)} \; [\text{OH}, (\text{CH}_2)_{10}\text{Me}] \quad (94)$$

Rao and Nageshwar have also studied the reaction of embelin with *o*-phenylenediamines to produce phenazines[100]. Of greater interest is the later study in which Rao and a group of collaborators examined the reactions of embelin with urea (equation 95)[101]. With an

$$\text{(39)} + (\text{H}_2\text{N})_2\text{C=O} \xrightarrow{\text{glacial HOAc}} \text{benzoxazolone quinone [HO, } (\text{CH}_2)_{10}\text{Me]} \;\; 38\% + \text{benzimidazolone quinone [HO, } (\text{CH}_2)_{10}\text{Me, N–H]} \;\; 60\% \quad (95)$$

overall yield of 98% and useful quantities of both products this chemistry surely ought to be examined more fully.

Schäfer and coworkers have used 2,5-bisanilino-1,4-benzoquinones as the first intermediates in a synthesis of the labile and important isoxazolequinones (**41**, equation 96)[102]. A naphthoquinone analog was also prepared and the yields for all reactions are in the 70–80% range.

$$\text{2-}CO_2R\text{-1,4-benzoquinone} + ArNH_2 \xrightarrow[NH_3/Pb(OAc)_4]{\text{three steps}} \mathbf{(41)} \tag{96}$$

R = Me, Et; Ar = Ph, 4-Tol

4. *Brief notes*

The reaction of α,β-unsaturated carbonyl compounds with *N*-ethoxycarbonyliminopyridinium ylide (**42**) in the presence of silicic acid gives good yields of ethoxycarbonylamino derivatives including those of quinones (equations 97, 98)[103]. The structure of the benzoquinone product was demonstrated by synthesis of an identical material from 2,5-diamino-1,4-benzoquinone and ethyl chloroformate.

$$\text{1,4-benzoquinone} + \mathbf{(42)} \xrightarrow{PhH} \text{2,5-bis}(NHCO_2Et)\text{-1,4-benzoquinone}\ (57\%) \tag{97}$$

$$\xrightarrow{\text{sealed tube}} \text{2-}NHCO_2Et\text{-1,4-naphthoquinone}\ (82\%) \tag{98}$$

An example of the addition of a lactam to quinones presented by Seth and Khanna deserves special note (equation 99)[104]. The importance of adjacent substituents is also clear in their attempted syntheses of certain quinazolones.

$$\xrightarrow{\text{Triton-B}} \tag{99}$$

R^1 = H, $-CH{=}CHCH{=}CH-$; R^2 = SEt, $N(C_2H_4)_2O$, Me   60%, 55%, 0%, 0%

The extensive synthetic efforts directed toward quinones with aziridinyl substituents has usually involved nucleophilic substitution of a halogen, but Yoshimoto and his collaborators have used direct addition of ethylenimine in the synthesis of complex 1,4-

benzoquinones for structure–activity studies (equation 100)[105]. While the yields are hardly exciting this route does offer an alternative that may warrant further study.

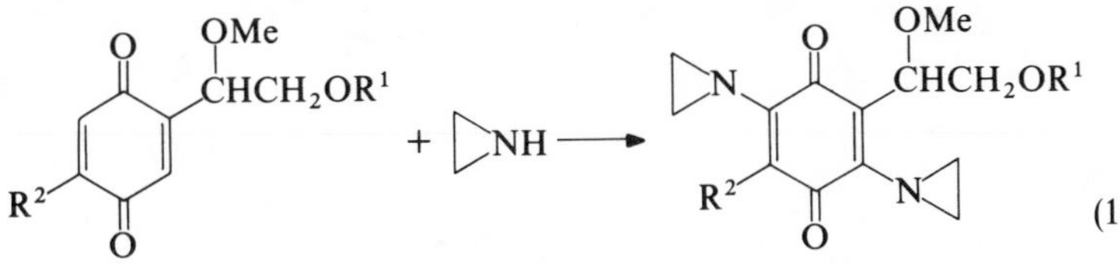

(100)

$R^1 = H, CONH_2; R^2 = CH(OMe)CH_2OR^1$ 15%, 29%

R^1 = 2,3,4,6-tetra-*o*-acetyl-β-D-glucosyl; R^2 = Me 7%

For a comparison of the antineoplastic activity of some aminoanthroquinones several naphthoquinone analogs have been prepared[106]. The point of interest lies in taking advantage of the much lower reactivity of a secondary amine or an alcohol to achieve specificity (equation 101). In a similar fashion 2-(dimethylamino)ethylamine added exclusively through the primary amino group (94%).

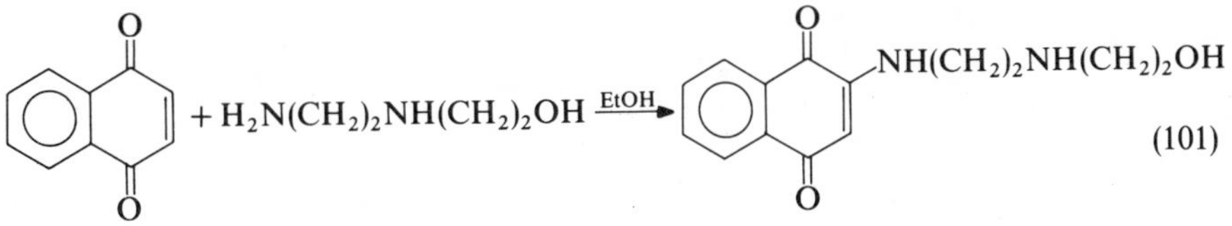

$+ H_2N(CH_2)_2NH(CH_2)_2OH \xrightarrow{EtOH}$ $NH(CH_2)_2NH(CH_2)_2OH$ (101)

54%

An exception to this observation concerning tertiary amines is found with pyridine and other aromatic nitrogen heterocycles whose hydrobromides add readily to quinones (equations 102, 103)[107]. This reaction was actually discovered when it was found that

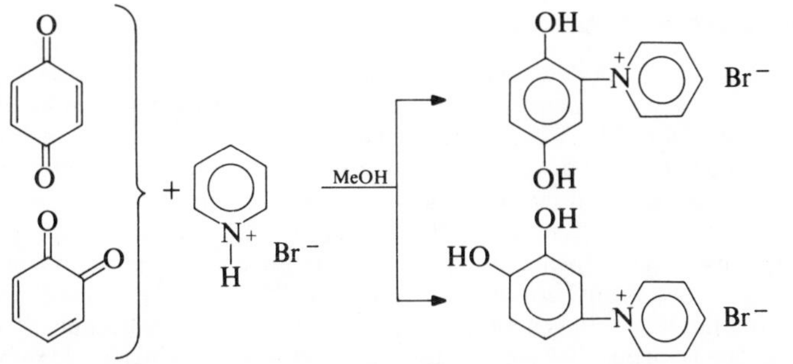

MeOH; Br^- (102); Br^- (103)

resorcinol fails to show evidence of nascent quinone formation under these reaction conditions. The yields are said to be about 60%.

A study of the nucleophilicity of arylsulfilimines included 1,4-naphthoquinone (equation 104)[108]. The proposed mechanism involves addition to the α, β-unsaturated ketone followed by a hydride shift and concomitant S–N bond cleavage (**43**).

In connection with a study of the interactions of catechols found in poison oak and ivy Castagnoli and coworkers examined the reaction of a model nucleophile, pentylamine, with their model substrate, 3-heptadecyl-1,2-benzoquinone (equation 105)[23], and obtained 32% of a red solid whose spectra were consistent with the quinonoid structure **44**. They showed an impressive similarity between the extinction coefficient of **44** and the reaction

$$Ph_2\overset{+}{S}\bar{N}H + \text{1,4-naphthoquinone} \xrightarrow[\text{ambient}]{CHCl_3} \text{2-amino-1,4-naphthoquinone} + Ph_2S \quad (104)$$

79%

(43)

$$\text{3-(CH}_2)_{16}\text{Me-1,2-benzoquinone} + C_5H_{11}NH_2 \xrightarrow[CH_2Cl_2]{\text{dry}} \text{3-(CH}_2)_{16}\text{Me-5-NH(CH}_2)_4\text{Me-1,2-benzoquinone} \quad (105)$$

(44)

product obtained by incubation of an equimolar mixture of the original quinone with human serum albumin.

C. Thiele–Winter Acetoxylation

McOmie and Blatchly, two of the most notable workers in this field, have presented a detailed review of the synthetic and mechanistic aspects of the reaction in 1972[109]. The historical introduction is typical of the care they lavished on this study. Their recognition of Winter's importance and of the actual use of the reaction led them and me to favor the revised name over the traditional Thiele acetylation. Their subsequent primary publications also bear this new designation. In the first of that series to appear following the review they present a complete survey of the reactions of 1,2- and 1,4-benzoquinones containing one or two *t*-butyl groups (equations 106–108)[110]. Nearly all of the earlier work is repeated and discussed. In general the findings previously reported were confirmed, the structures more rigorously demonstrated and minor by-products discovered. Several points are significant in these experiments: (1) de-*t*-butylation does in fact occur, (2) small amounts of product from reaction adjacent to the *t*-butyl substituent are found, (3) no simple substitution product was found in the case of 2,6-di-*t*-butyl-1,4-benzoquinone and (4) in the simplest cases one product is obtained in a synthetically useful yield. It should be noted concerning yields cited above that these are all minimum figures which often, as in the 2-*t*-butyl case, involve hydrolysis and methylation prior to fractionation. The authors state, for example, that they do not regard the greater yield of the more highly hindered product **45** as significant for these reasons.

Similar results were obtained for most examples of methyl, bromo and hydroxyl-*t*-butyl-1,4-benzoquinones. An interesting exception is 2-hydroxy-5-*t*-butyl-1,4-benzoquinone

$$\text{2-}t\text{-butyl-1,4-benzoquinone} + Ac_2O \xrightarrow[HOAc]{BF_3} \text{(52\%)} + \text{(45) 6\%} + \text{(3\%)} \qquad (106)$$

$$\text{2,6-di-}t\text{-butyl-1,4-benzoquinone} + Ac_2O \xrightarrow{H_2SO_4} \text{(61\%)} \qquad (107)$$

$$\text{2,5-di-}t\text{-butyl-1,4-benzoquinone} + Ac_2O \xrightarrow{H_2SO_4} \text{(50\%)} + \text{(8\%)} \qquad (108)$$

where Flaig and his coworkers had claimed, without reaction times or yields, both a de-*t*-butylation and a simple reductive esterification product (equation 109)[111]. In the present

$$\text{2-hydroxy-5-}t\text{-butyl-1,4-benzoquinone} + Ac_2O \xrightarrow{H_2SO_4} \text{1,2,4,5-tetraacetoxybenzene} + \text{1,2,4-triacetoxy-5-}t\text{-butylbenzene} \qquad (109)$$

study[110] only the former could be identified and that in 5% yield. As would be expected neither group found any evidence for a Thiele–Winter product The importance of catalyst selection is shown in several examples, but is best exemplified by 2-bromo-5-*t*-butyl-1,4-benzoquinone where sulfuric acid produces low yields of purifiable products while the more modern boron trifluoride and perchloric acid gives only tars. There is a minor inconsistency in that the narrative describes the product, which was not completely characterized, as 1,3,4-triacetoxy-2-bromo-5-*t*-butyl-1,4-benzoquinone while the table gives 1,2,4-triacetoxy-5-bromo-1,4-benzoquinone. The latter, a de-*t*-butylation product would certainly have been noted in the text.

The reactions of 2-methyl-5-*t*-butyl- and 2-methyl-6-*t*-butyl-1,4-benzoquinones illustrate the critical importance of substituent position on the outcome of the

Thiele–Winter acetoxylation (equations 110, 111). While both reactions proceed normally there is a ten-fold difference in the production of the isomeric products that deserves more attention.

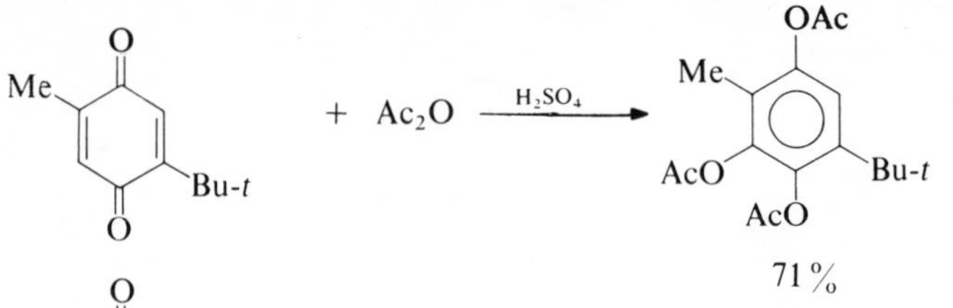

O, Me, Bu-*t*, O $+ \ Ac_2O \xrightarrow{H_2SO_4}$ OAc, Me, AcO, Bu-*t*, AcO (110)

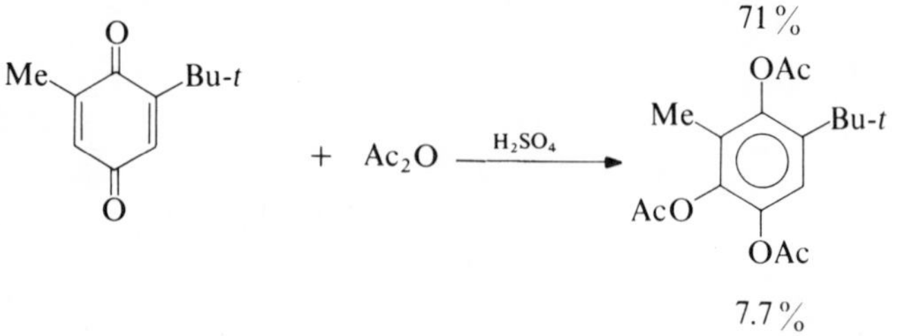

71%

O, Me, Bu-*t*, O $+ \ Ac_2O \xrightarrow{H_2SO_4}$ OAc, Me, Bu-*t*, AcO, OAc (111)

7.7%

Studies of the alkyl-1,2-benzoquinones provide, with the exception of the 3-methyl-5-*t*-butyl case, markedly higher product yields for reasons that are not at all clear (equations 112–114)[110]. This low yield surprised the authors who reinvestigated an earlier report[112]

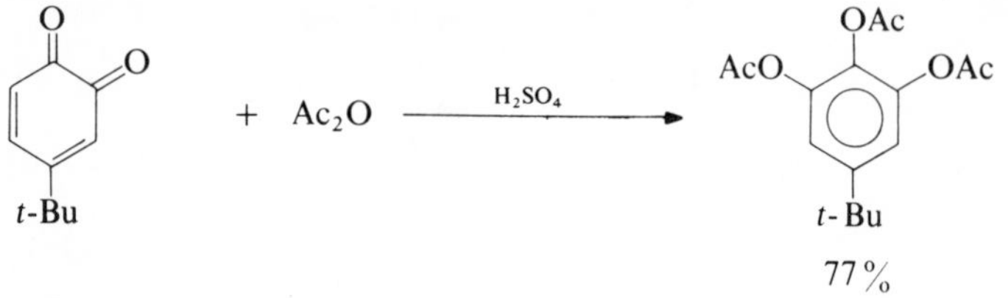

O, O, *t*-Bu $+ \ Ac_2O \xrightarrow{H_2SO_4}$ OAc, AcO, OAc, *t*-Bu (112)

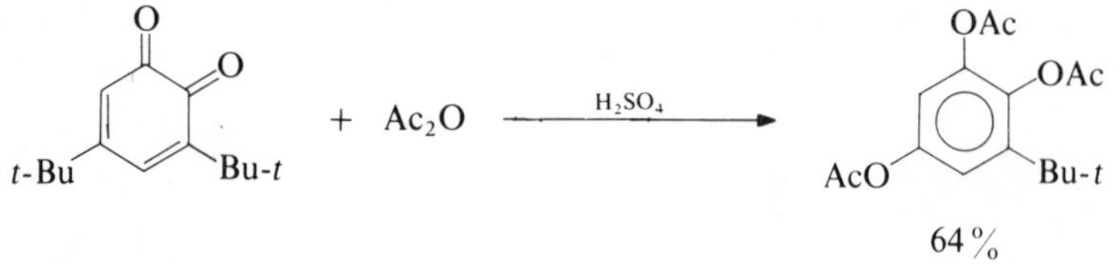

77%

O, O, *t*-Bu, Bu-*t* $+ \ Ac_2O \xrightarrow{H_2SO_4}$ OAc, OAc, AcO, Bu-*t* (113)

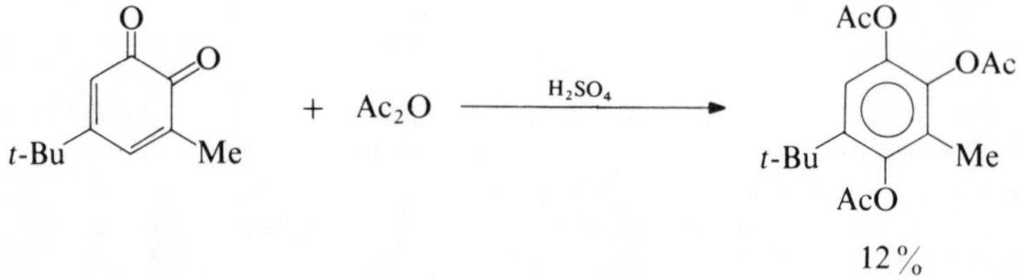

64%

O, O, *t*-Bu, Me $+ \ Ac_2O \xrightarrow{H_2SO_4}$ AcO, OAc, *t*-Bu, Me, AcO (114)

12%

concerning 5-methyl-3-*t*-butyl-1,2-benzoquinone and found a new product (**46**) resulting from reaction by a *p*-quinone methide (equation 115)[110].

When an oxygen or halogen substituent are present very low yields or decomposition are the rule; the only exception is the non-Thiele–Winter esterification of 3-hydroxy-4,6-di-*t*-butyl-1,2-benzoquinone which isn't even reduced under these conditions (equation 116).

$$\text{3-Me-5-}t\text{-Bu-1,2-benzoquinone} + Ac_2O \xrightarrow{H_2SO_4} \text{1,2,4-tri-OAc-5-Me-3-}t\text{-Bu-benzene}\ (20\%) + \text{AcOCH}_2\text{-C}_6\text{H}_2(\text{OAc})_2\text{-}t\text{-Bu}\ \mathbf{(46)}\ (14.5\%) \tag{115}$$

$$\text{3,5-di-}t\text{-Bu-6-OH-1,2-benzoquinone} + Ac_2O \xrightarrow{H_2SO_4} \text{3,5-di-}t\text{-Bu-6-OAc-1,2-benzoquinone}\ (45\%) \tag{116}$$

Several more complex quinonoid systems were examined and shown to exhibit interesting chemistry. For example, 3,3′, 5,5′-tetra-*t*-butyldiphenylquinone gave a nearly quantitative yield of the 1,8-addition product involving an alkyl group displacement (**47**, equation 117).

$$\text{3,3',5,5'-tetra-}t\text{-butyldiphenylquinone} + Ac_2O \xrightarrow[HOAc]{BF_3} \text{AcO-C}_6\text{H}_2(t\text{-Bu})_2\text{-C}_6\text{H}_2(\text{OAc})_2(\text{Bu-}t)\ \mathbf{(47)}\ \ 90\%\ \text{isolated} \tag{117}$$

Since *t*-butyl is the only alkyl group known to be displaced in these reactions a carbonium ion mechanism seems reasonable.

$$\text{2,5-di-}t\text{-Bu-1,4-benzoquinone} \xrightarrow{Ac^+} \text{cation (OAc, }^+\text{, Bu-}t) \xrightarrow{Ac_2O} \text{adduct (OAc, Bu-}t\text{, OAc; Ac}^+) \longrightarrow \text{1,2,4-tri-OAc-5-}t\text{-Bu-benzene} + Me_3C^+$$

In their continuing studies of the synthetic utility and mechanistic detail of Thiele–Winter acetoxylations Blatchly and McOmie have examined the products of phenyl and 4-substituted phenyl 1,4-benzoquinones in which there is also a hydroxy, halo, or methoxy group present (equations 118–120)[113]. The corresponding hydroxy quinones

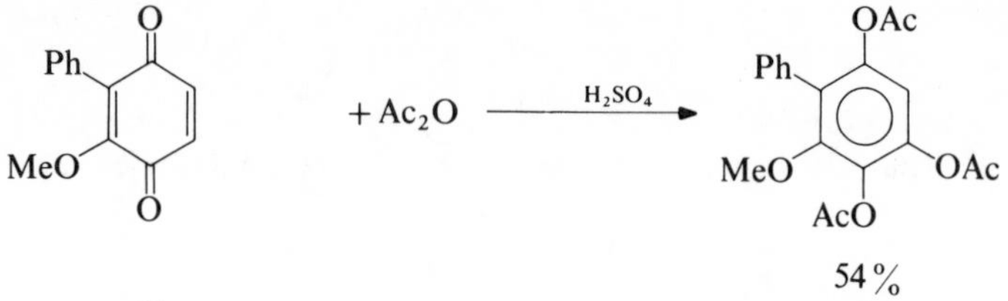

Ph, MeO, O, O $+ Ac_2O$ $\xrightarrow{H_2SO_4}$ OAc, Ph, MeO, OAc, AcO (118)

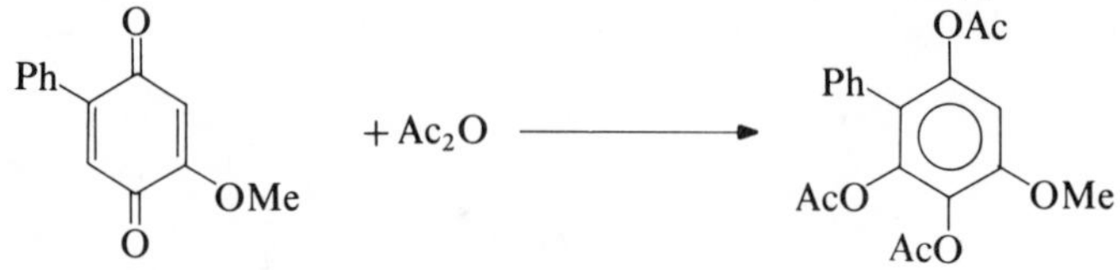

54%

Ph, OMe, O, O $+ Ac_2O$ $\longrightarrow$ OAc, Ph, AcO, OMe, AcO (119)

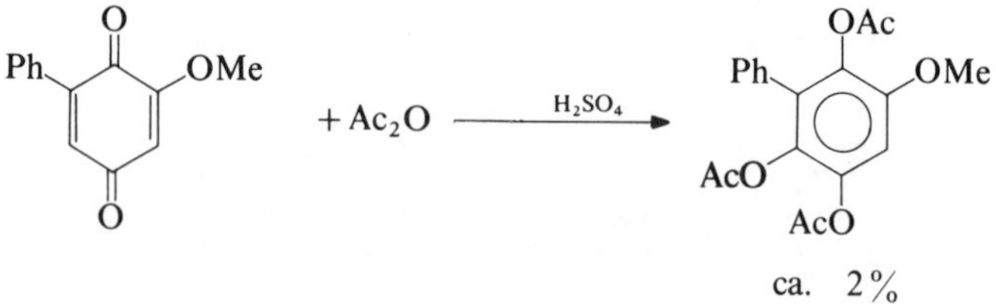

Ph, OMe, O, O $+ Ac_2O$ $\xrightarrow{H_2SO_4}$ OAc, Ph, OMe, AcO, AcO (120)

ca. 2%

are much less reactive and give comparable or better yields (equations 121, 122). In the first example the intermediate had been reported by Fieser[114].

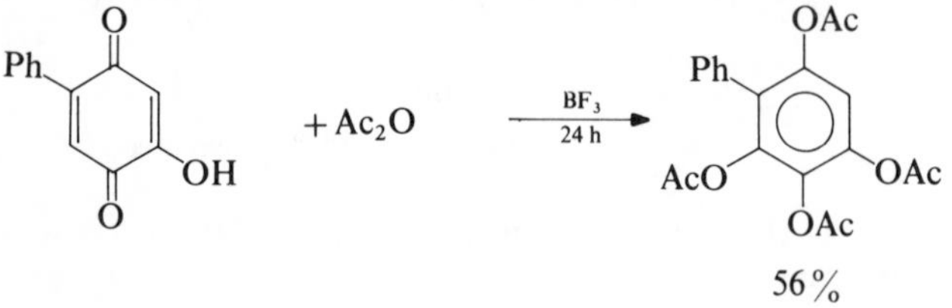

Ph, OH, O, O $+ Ac_2O$ $\xrightarrow[24\ h]{BF_3}$ OAc, Ph, AcO, OAc, OAc (121)

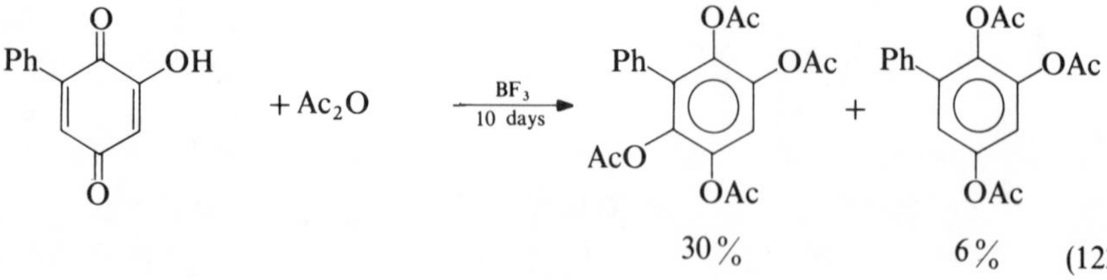

56%

Ph, OH, O, O $+ Ac_2O$ $\xrightarrow[10\ days]{BF_3}$ OAc, Ph, OAc, AcO, OAc + OAc, Ph, OAc, OAc

30% 6% (122)

The knowledge that strong electron-withdrawing groups, e.g. cyano, direct to the adjacent position (often incorrectly referred to as *ortho*) prompted an investigation of 2-(4-nitrophenyl)-1,4-benzoquinone. The results of this study extend the concurrent work of Wilgus and Gates[115] and confirm their conclusion that such *p*-substituents have a very small influence. While the question of the resonance effect of the 4′-nitro group does seem resolved there are other, perhaps minor, questions not answered. Thus, the yields of the 4′-nitro of bromo products are high (82% and 86% respectively) while the same can hardly be said for those bearing 4′-hydroxy or acetoxy groups (48% and 45% respectively). It is true that the authors point out[113] that the differences in yield are related to the relative solubility in ethanol since all values were reported after two recrystallizations. What is more bothersome is the implication that only one isomeric product is found even in the high yield cases. Wilgus and Gates reported significant amounts of more than one isomer in the three 2-aryl substituted examples studied including all three isomers for 4′-nitro-1,4-benzoquinone[115]. Surely, the stated purpose of this paper, to propose an improved mechanism for this important reaction, demands a careful consideration of such electronic influences.

In the final paper of this mini-series McOmie and his collaborators examine the influence of bromo and phenyl groups on the orientation and reactivity of 1,4-benzoquinones under Thiele–Winter conditions[116]. This group of compounds proved to be a most interesting selection since two of them yield a specific orientation (equations 123, 124), one, a mixture of isomers (equation 125), and two, reduction only (equations 126, 127). The product mixture from 2-bromo-3-phenyl-1,4-benzoquinone was found to be

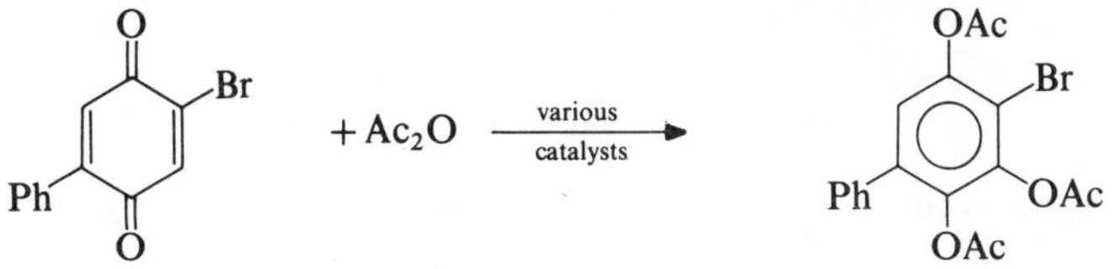
O
Br
Ph
O
$+ Ac_2O$
various
catalysts
OAc
Br
Ph
OAc
OAc

(123)

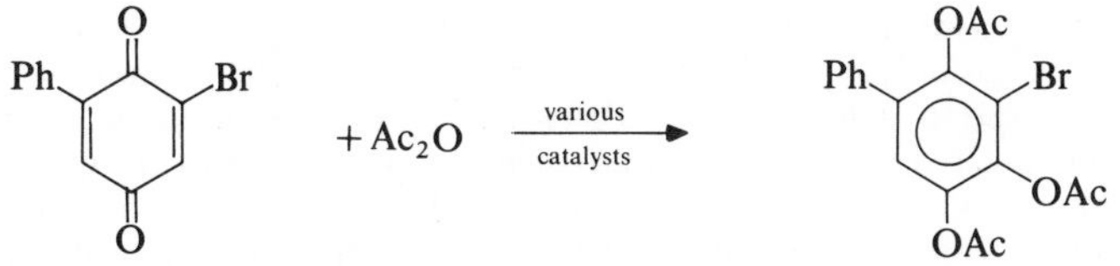
O
Ph
Br
O
$+ Ac_2O$
various
catalysts
OAc
Ph
Br
OAc
OAc

(124)

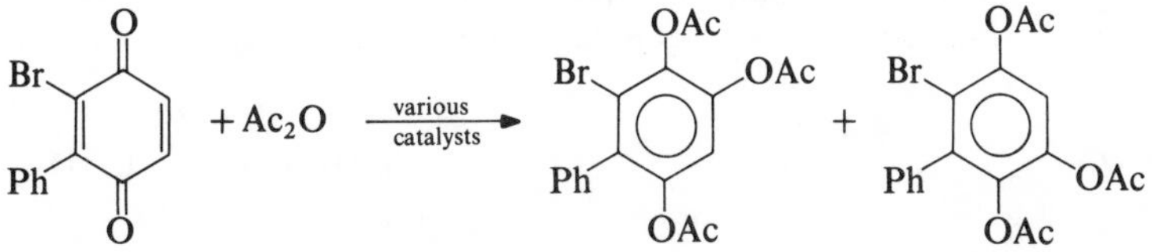
O
Br
Ph
O
$+ Ac_2O$
various
catalysts
OAc
Br
OAc
Ph
OAc
+
OAc
Br
Ph
OAc
OAc

125)

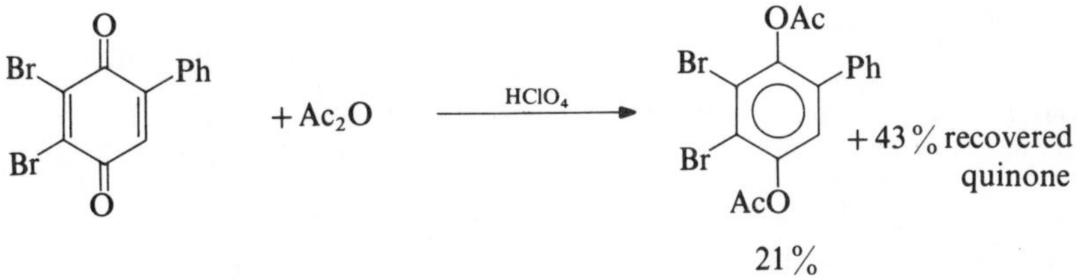
O
Br
Ph
Br
O
$+ Ac_2O$
$HClO_4$
OAc
Br
Ph
Br
AcO
+ 43 % recovered
quinone
21 %

(126)

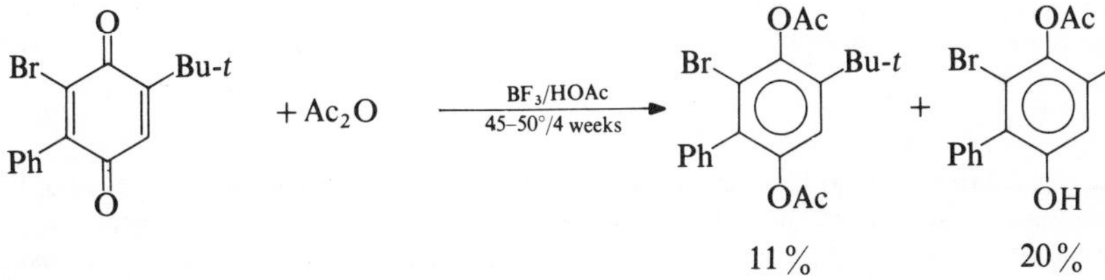
O
Br
Bu-t
Ph
O
$+ Ac_2O$
BF_3/HOAc
45–50°/4 weeks
OAc
Br
Bu-t
Ph
OAc
+
OAc
Br
Bu-t
Ph
OH
11 %
20 %

(127)

approximately 40:60, but the major component was not determined. The structure of the 3,4,6-triacetate was demonstrated by hydrolysis, methylation, debromination and comparison with an authentic sample. The comparison of catalysts presented in Table 2 is consistent with earlier work in showing no significant effect. It is hardly surprising that these two greatly hindered substrates do not undergo the reductive addition reactions.

TABLE 2. Effect of catalyst on the yield of Thiele–Winter acetoxylation product (equations 123, 124)[116]

	% Yield with		
Substituents	BF_3	H_2SO_4	$HClO_4$
2-Br-3-Ph	69	84	82
2-Br-5-Ph	77	80	76
2-Br-6-Ph	62	73	65

These studies taken together present a fairly complete picture of what can be expected from Thiele–Winter reactions in a wide variety of steric and electronic situations. They fail to reach the detailed mechanistic description one might desire for this important transformation. Kinetic studies are clearly needed to supplement the product and reactivity information available. There also remains completely untouched the rather fundamental question of the relationship between this clearly electrophilic reaction and the normally encountered nucleophilic reductive addition reactions of the quinones.

Several applications of Thiele–Winter chemistry have appeared in the recent literature. Interest in the role of resorcinols in certain groups of natural products prompted the study of the relationship of one such compound to quinonoid compounds[117]. During this study 2-methoxy-6-tridecyl-1,4-benzoquinone was treated with acetic anhydride and sulfuric acid (equation 128). The single product obtained illustrates once again just how reluctant

$$\text{2-MeO-6-}(C_{13}H_{27}\text{-}n)\text{-1,4-benzoquinone (0.36g)} + Ac_2O \xrightarrow[\text{ambient/overnight}]{H_2SO_4} \text{1,3,4-(OAc)}_3\text{-2-}(C_{13}H_{27}\text{-}n)\text{-6-MeO-benzene (0.36 g crude gum)} \qquad (128)$$

this reaction is to place the acetoxy group adjacent to a methoxy substituent. Nearly identical chemistry was reported by Marini-Bettolo and coworkers (equation 129)[118].

$$\text{2-MeO-6-}(C_5H_{11}\text{-}n)\text{-1,4-benzoquinone} + Ac_2O \longrightarrow \text{1,3,4-(OAc)}_3\text{-2-}(C_5H_{11}\text{-}n)\text{-6-MeO-benzene} \qquad (129)$$

The information gained from these experiments failed to provide the desired route for the proposed total synthesis because both an acetylenic and an alkenyl bond were attacked by acetic anhydride in preference to the quinonoid structures under all of the usual conditions (equation 130). Two completely reduced analogs were used successfully (equation 131)[117].

$$\text{MeO-quinone-(CH}_2)_7\text{C}\equiv\text{C(CH}_2)_7\text{-quinone-OMe} + \text{Ac}_2\text{O} \xrightarrow[\text{catalysts}]{\text{various}} \!\!\!\!/\!\!/ \quad \text{or} \quad \text{HC=CH} \quad \text{MeO-Ar(OAc)}_2\text{-(CH}_2)_7\text{C}\equiv\text{C(CH}_2)_7\text{-Ar(OAc)}_2\text{-OMe} \tag{130}$$

$$\text{MeO-quinone-(CH}_2)_{16}\text{-quinone-OMe} + \text{Ac}_2\text{O} \xrightarrow[\text{ambient/overnight}]{\text{H}_2\text{SO}_4} \text{MeO-Ar(OAc)}_3\text{-(CH}_2)_{16}\text{-Ar(OAc)}_3\text{-OMe} \tag{131}$$

An interesting case involving Thiele–Winter acetoxylation conditions produced reductive acetylation (equation 132)[119] reminiscent of Fieser's work with 2-hydroxy-5-phenyl-1,4-benzoquinone[114].

$$\text{HO,MeO-quinone-CH}_2\text{OH} + \text{Ac}_2\text{O} \xrightarrow{\text{H}^+} \text{AcO,MeO-Ar(OAc)}_2\text{-CH}_2\text{OAc} \tag{132}$$

Noteworthy are two reports demonstrating that an *o*-methoxy group in an attached or fused ring does not have the fatal effect observed by both Gates[115] and McOmie[113] for the 4′-methoxyphenyl group. For example, as one step in a study of arylated quinones Cameron and his collaborators carried out an acetoxylation in which the reaction site is *para* to an aryl substituent (equation 133)[120]. The crude acetylated mixture was reacted with dimethyl sulfate in methanol and the corresponding tetramethyl ether isolated.

The conversion of a 1,2-naphthoquinone to the corresponding 2-hydroxy-1,4-naphthoquinone was accomplished in overall 56% yield (equation 134)[121].

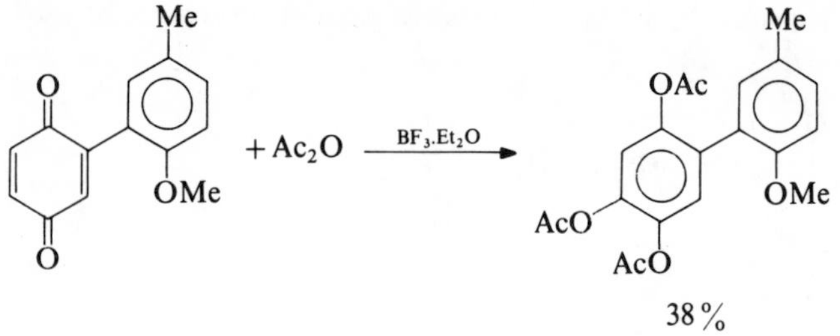

$$\text{quinone} + Ac_2O \xrightarrow{BF_3.Et_2O} \text{triacetate} \quad 38\% \qquad (133)$$

$$\text{naphthoquinone} + Ac_2O \xrightarrow{H_2SO_4} [\text{triacetate}] \xrightarrow[{[O]}]{OH^-} \text{hydroxynaphthoquinone} \qquad (134)$$

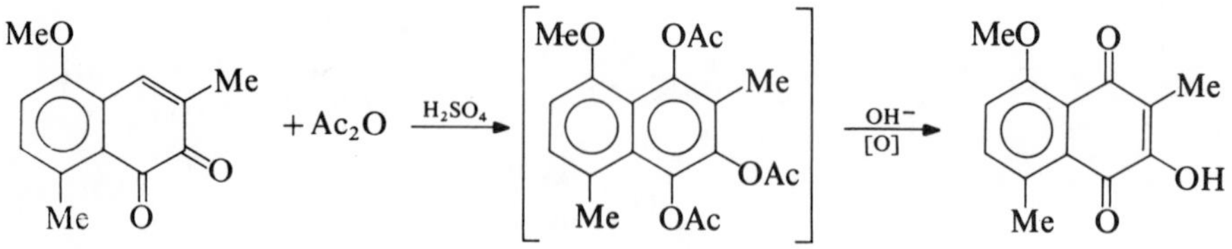

In their studies of quinones derived from *m*-diols in the naphtho [2,3-*c*]furan series Villessot and Lepage found that compound **48** undergoes reductive acetylation across the *m*-carbon atoms (equation 135)[122].

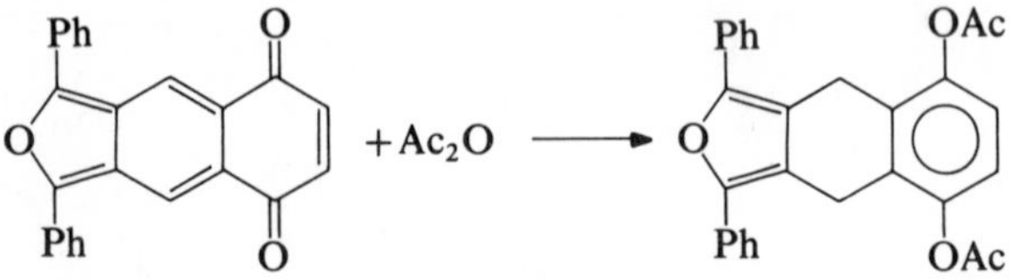

$$\mathbf{(48)} + Ac_2O \longrightarrow \text{diacetate} \qquad (135)$$

D. Addition of Inorganic Substances

1. Halogens and hydrogen halides

The haloquinones have played an increasingly important role in several areas (see especially Section V.A.2) and the reactions leading to such substrates have been examined in detail. Unlike most areas of Michael addition to quinones we have a rather detailed mechanistic picture for chlorine, bromine and bromochloride addition in acetic acid[123]. In elaborate product study de la Mare and his collaborators found the ratio of bromochloride/dibromide ratio to vary widely depending on the relative availability of bromine and chlorine. The only serious competing reaction is the addition of hydrogen halide, a very effective catalyst. No evidence of dichloro or acetoxy products could be obtained. Earlier reports of the kinetically controlled formation of the *trans*-dihalogeno-enedione were confirmed as well as extended to the mixed halogen. The proposed mechanism (equation 136) requires the first halogen to be attached nucleophilically while the second reacts electrophilically. The evidence does not require the involvement of a 1,2-adduct. The results obtained with 1,4-naphthoquinone show a similar pattern although competition from hydrogen halide addition appears to be less important. The authors point out that in synthetic applications there is no need to add the hydrohalic acid since its effect is very powerful and enough of it is supplied by other trace reactions.

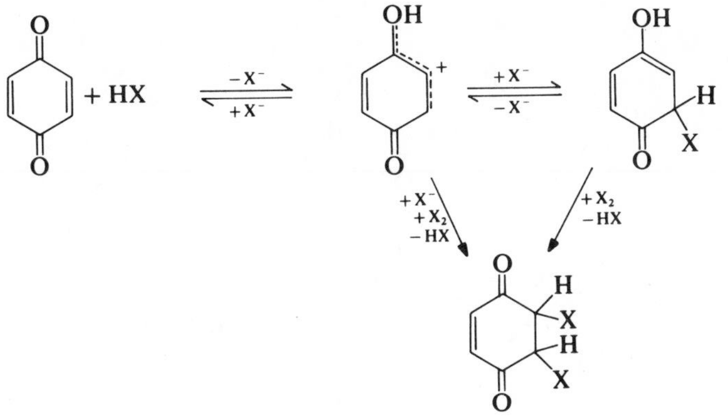

(136)

(49)

The ethanolysis kinetics of hydrogen halide elimination from the dihalide (**49**) show a reaction approaching an *E*1cB mechanism (equation 137). Thus, the loss of a proton to solvent becomes the dominant feature and accounts for the relative rates of loss of HCl and HBr.

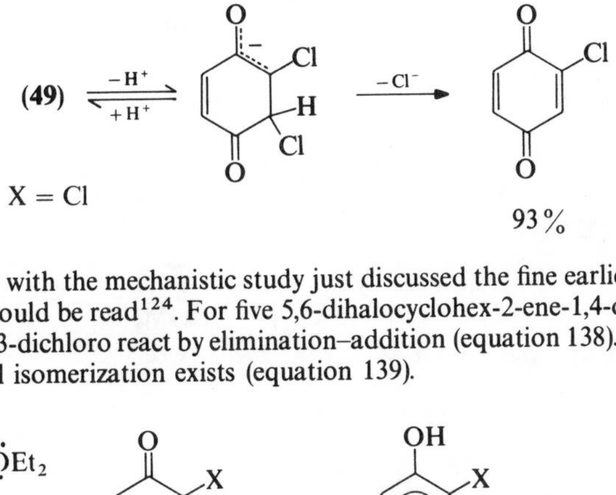

(137)

X = Cl

93%

In connection with the mechanistic study just discussed the fine earlier work of Norris and Sternhell should be read[124]. For five 5,6-dihalocyclohex-2-ene-1,4-diones they found that all except 2,3-dichloro react by elimination–addition (equation 138). For that example a true keto–enol isomerization exists (equation 139).

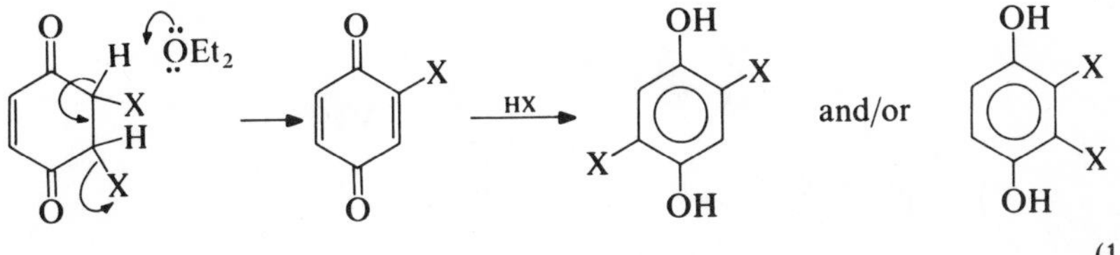

(138)

An earlier examination of the addition of hydrogen chloride to 1,4-benzoquinone and its 2-chloro derivative by Chaillet and his coworkers shows both an interesting involvement of solvent (methanol) and the utility of theoretical calculations[125]. In this solvent the normal addition product represents only a minor constituent (equation 140). Methanol also enters in a significant fashion the reaction with 2-chloro-1,4-benzoquinone (equation 141).

Consideration of the following array of potential intermediates (equation 142) and the application of Hückel, Hückel ω and Wheland polarization energy methods allowed good interpretations of the experimental results. Similar results were obtained for the 2-chloro

$$+BF_3 \longrightarrow \longrightarrow +Et_2OH^+ \quad (139)$$

$$\longrightarrow \xrightarrow{Et_2OH^+} +BF_3{\cdot}Et_2O$$

$$\xrightarrow{MeOH/HCl} \; 85\% \; + \; 10\% \; + \; 5\% \quad (140)$$

$$\xrightarrow{MeOH/HCl} \; 50\% \; + \; 35\% \; + \; 15\% \quad (141)$$

$$+H^+ \rightleftharpoons \; +MeOH \rightleftharpoons \; +H^+ \rightleftharpoons \quad (142)$$

$$+H_2O \xrightleftharpoons{MeOH}$$

derivative, and a later report describes an analogous study of 1,4-benzoquinone monooximes[126].

Some rather surprising evidence concerning the more highly chlorinated 1,4-benzoquinones came to light in a study of the hydrolysis of 2,4,4,6-tetrachloro-2,5-cyclohexadienone (**50**) which results from the further chlorination of 2,4,6-trichlorophenol (**51**; equation 143)[127]. In the absence of chloride **50** reacts with water or dilute sulfuric acid to

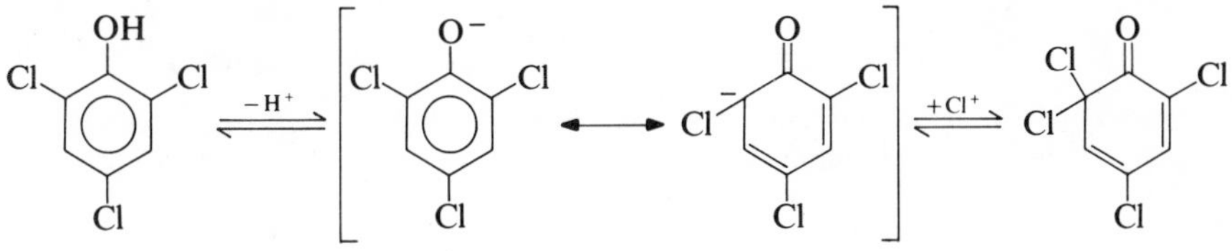

(**51**) $\underset{}{\overset{-H^+}{\rightleftharpoons}}$ [...] $\underset{}{\overset{+Cl^+}{\rightleftharpoons}}$ ... $\longrightarrow$ (**50**) (143)

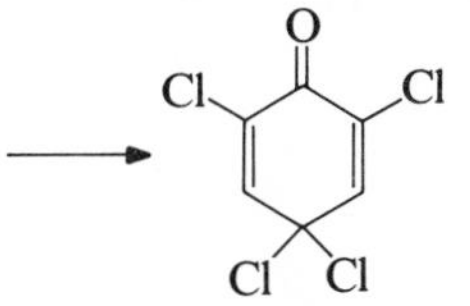

give a complicated mixture, but in hydrochloric acid significant amounts of **51** along with chlorinated quinones and hydroquinones can be identified (equation 144). The inter-

(**50**) $\xrightarrow{-Cl^-}$ (**51**) $\xrightarrow{+H_2O}$ ... $\xrightarrow{+HCl}$ ... (144)

mediates **51** and 2,6-dichloro-1,4-benzoquinone occur only at short reaction times and the concentration of 2,3,5-trichloro-1,4-benzoquinone remains fairly constant.

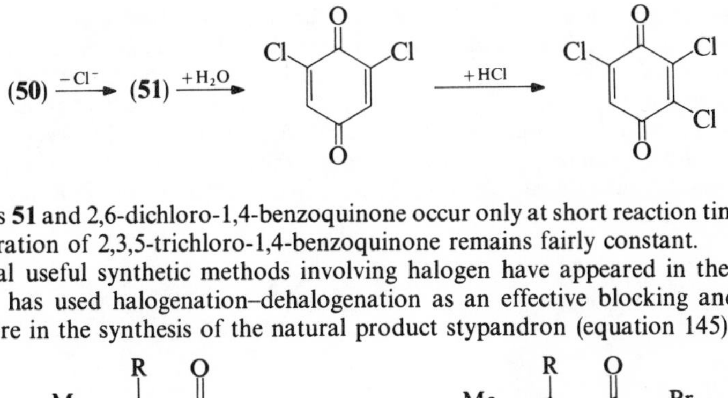

Several useful synthetic methods involving halogen have appeared in the literature. Laatsch has used halogenation–dehalogenation as an effective blocking and directing procedure in the synthesis of the natural product stypandron (equation 145)[128].

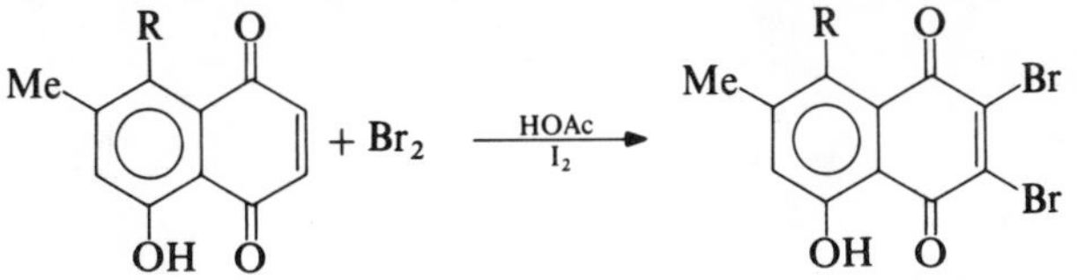

... $+ Br_2 \xrightarrow[I_2]{HOAc}$ ... (145)

R = H,Cl, Br

Fariña and Valderrama have provided additional examples of the strong directive influence of electron-withdrawing substituents on nucleophilic addition to quinones (equation 146)[129]. The excellent yields and high regioselectivity of both additions along with the possibility of reoxidation to quinones should be of synthetic use.

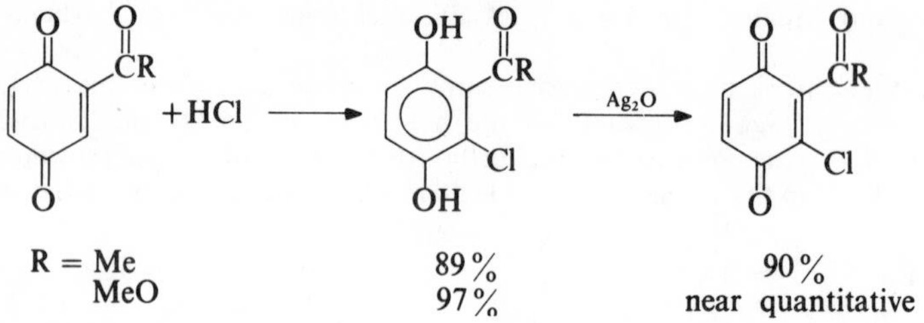

(146)

R = Me — 89% — 90%
MeO — 97% — near quantitative

Cameron and his colleagues have applied bromination imaginatively in the synthesis of naturally occurring anthraquinones (equations 147, 148)[130]. The 3-bromo isomer was

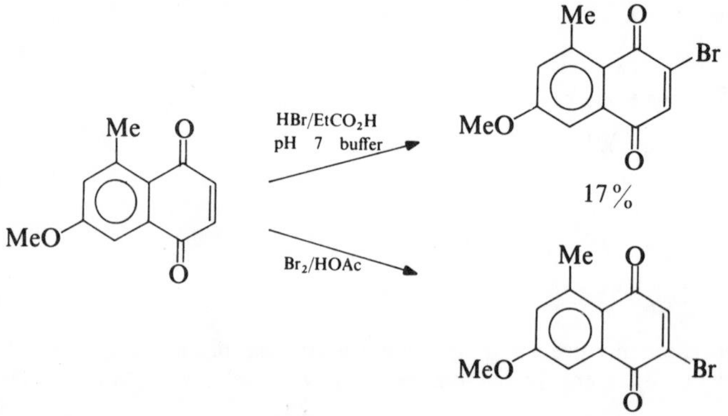

$HBr/EtCO_2H$ pH 7 buffer — 17% (147)

$Br_2/HOAc$ (148)

easily prepared, but in low yield, by hydrobromination and oxidation. Compounds with such groups present are of significance in directing subsequent cycloaddition chemistry (see Section V.A.2). This same group has studied these complimentary routes and confirmed that benzenoid substituents with strong electronic influence provide routes to the isomeric products[131]. Of greater importance is the observation that weaker groups, while not as discriminating, can still lead to synthetically useful methods (equations 149, 150). The 5-methyl substituent behaved differently and easily gave 60% 2-bromo product

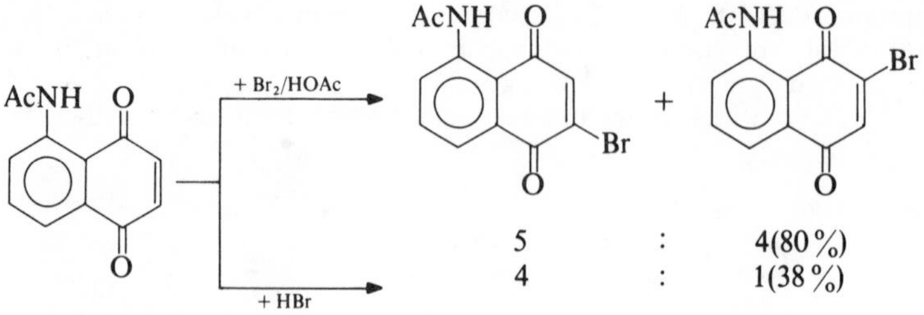

+ $Br_2/HOAc$ (149)

+ HBr (150)

5 : 4(80%)
4 : 1(38%)

from bromination while hydrobromination produced only 15% of an inseparable 1:1 mixture of 2- and 3-bromo products.

Other approaches to the synthesis of brominated naphthoquinones have been reported[132] and recently Jung and Hagenah have given evidence that the brominated naphthol **52** and the tribromonaphthalen-1-one **53** are both intermediates (equation 151)[133].

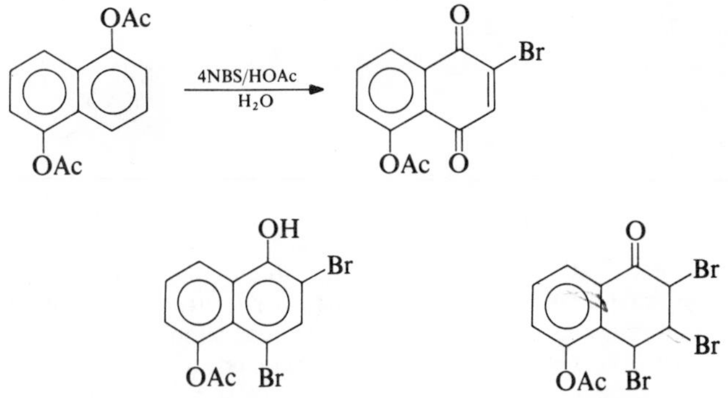

$$\text{1,5-diacetoxynaphthalene} \xrightarrow[\text{H}_2\text{O}]{\text{4NBS/HOAc}} \text{5-acetoxy-2-bromo-1,4-naphthoquinone} \tag{151}$$

(52) (53)

A promising synthetic method for chloro- and bromo-1,4-benzoquinones has been suggested (equation 152)[134]. The yields with the limited range of substituents studied are

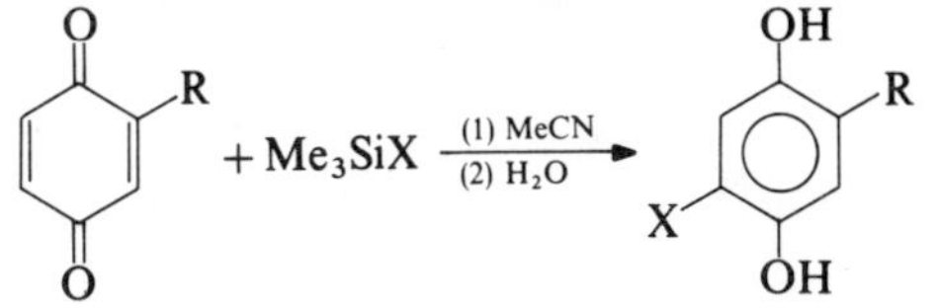

$$\text{quinone (R)} + Me_3SiX \xrightarrow[\text{(2) H}_2\text{O}]{\text{(1) MeCN}} \text{X,R-substituted hydroquinone} \tag{152}$$

R = H, Me, Cl, OMe

good to excellent and the reaction can be applied to nascent 1,2-benzoquinones. A further modification involves treatment of the unhydrolyzed product with hexamethyldisilazane to form the disilyl ether. The authors rightly point out that this work should be studied in greater detail.

The synthesis of chloranil in nearly quantitative yield has been achieved by treating 1,4-benzoquinone with antimony(V) chloride[135]. Both 2,5- and 2,6-dichloro-1,4-benzoquinones are intermediates in the reaction.

In the specific instance of 3,6-di-*t*-butyl-1,2-benzoquinone there is, probably because of the serious steric hindrance of the carbonyl groups, a tendency towards reduction rather than acid addition. However, it has been found that in the equally special case of the 4-methoxy derivative of this quinone (**54**), hydrolysis of the methoxy group and addition of hydrogen chloride are the principal reactions (equation 153)[136].

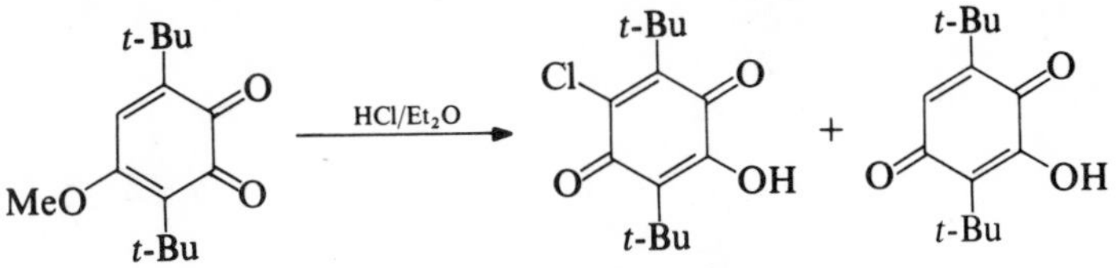

$$(\mathbf{54}) \xrightarrow{\text{HCl/Et}_2\text{O}} \text{Cl-product } (52\%) + \text{OH-product } (36\%) \tag{153}$$

2. *Azides*

The need of a 3-aminojuglone derivative led Parker and Sworin to repeat earlier work by Thomson and his colleagues[137] concerning the regioselective addition of hydrazoic acid to

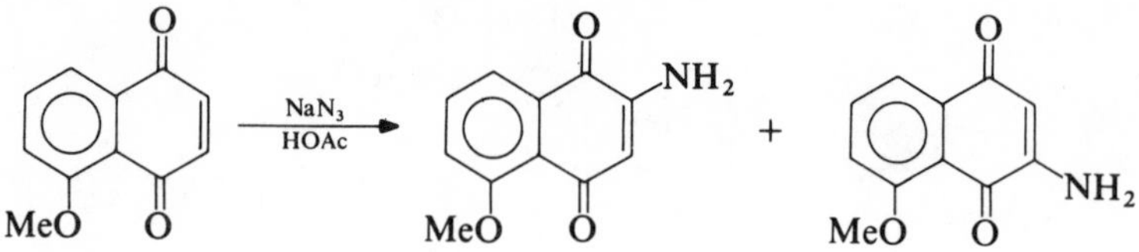

$$\xrightarrow[\text{HOAc}]{\text{NaN}_3}$$

40% 10% (154)

the 5-methyl ether (equation 154)[138]. They were able to show that both products are formed and that the 3-amino isomer predominates. The structures were established by unambiguous synthesis and detailed spectral evidence. Juglone and its acetate were also examined. The former compound gave only the 3-amino product (27%) and the latter a 4:1 preference for the same orientation. A very useful discussion of the confused history of these reactions is presented.

Rees and his collaborators found three of the four possible ring expansion products from the Schmidt reaction of 2-methoxy-5-methyl-1,4-benzoquinone and sodium azide (equation 155)[139]. All three compounds were isolated and studied, but only **55** showed appreciable aromatic character, i.e. can be considered an azatropolone.

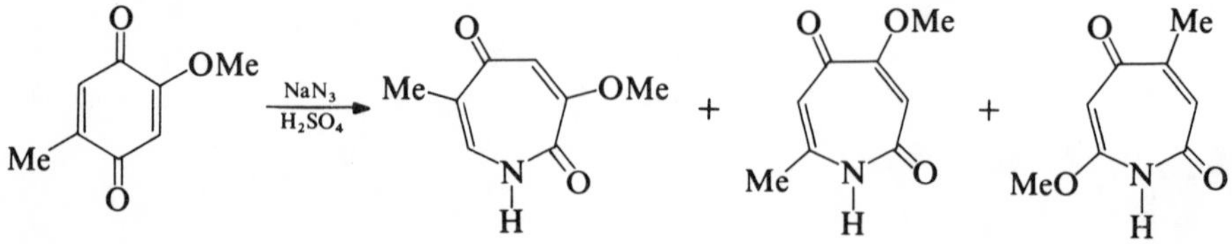

$$\xrightarrow[\text{H}_2\text{SO}_4]{\text{NaN}_3}$$

(**55**) 22% 16% 13%

(155)

Kende and Naegely have used iodine azide generated from iodine chloride and sodium azide to overcome a synthetic difficulty in the next to the last step of their model synthesis related to streptonigrin (equation 156)[140].

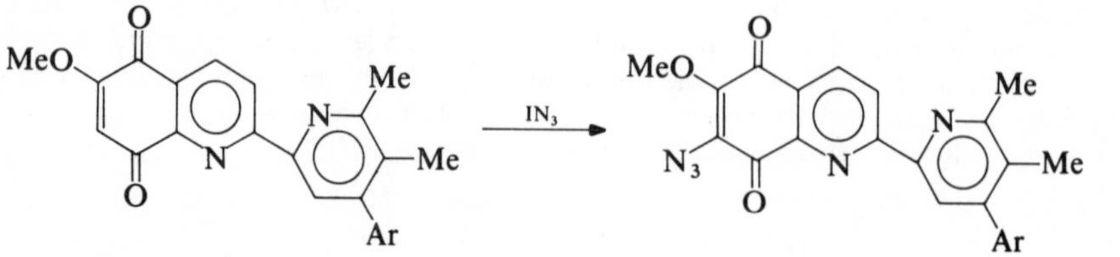

$$\xrightarrow{\text{IN}_3}$$

(156)

3. Inorganic sulfur

Nichols and Shulgin have used the method developed by Alcolay[141] as a starting point for the synthesis of some sulfur analogs of psychotomimetic amines (equation 157)[142].

An unusual heterocyclic system has been synthesized by the addition of S_2X_2 to the enamine bond in 4-amino-1,2-quinones (equation 158)[143].

Arai and his collaborators have carried out a series of mechanistic studies of the reactions between quinones and sodium sulfite[144–147] and sodium thiosulfate[148–150]. It is clear that, depending on the oxidation–reduction potentials of the quinone, either redox

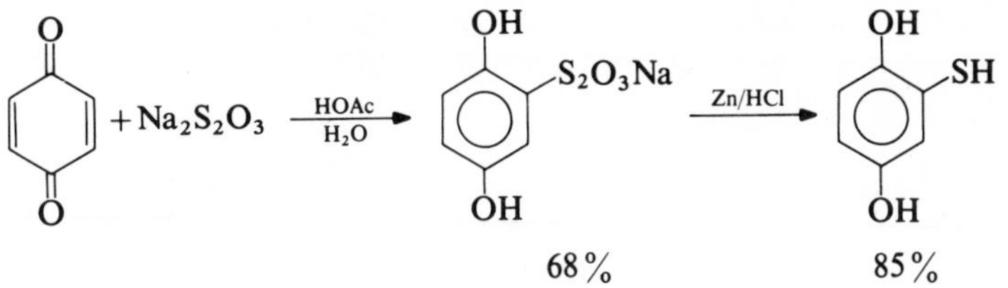

(157)

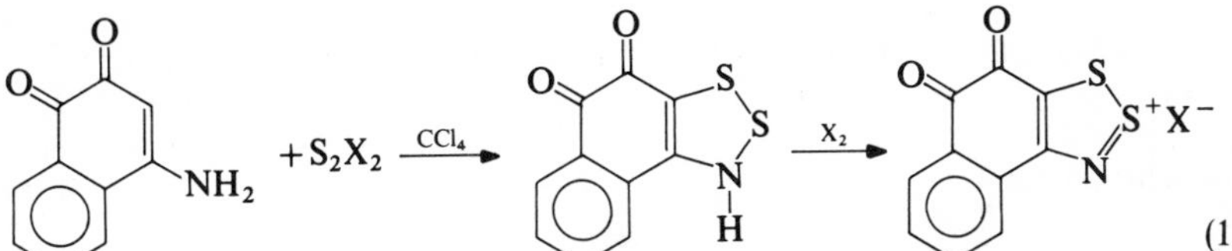

(158)

or addition chemistry is observed. Unfortunately these efforts are available only in Japanese at this time.

E. Oxygen Addition

1. The alkoxyquinones

The frequency with which quinones, or substances directly related to quinones, bearing oxygen functions occur in natural products has created surprisingly little interest in the study of alcohol addition. An exception to this sad picture is the useful investigation by Singh and Turner in which they examined methanol addition with and without Lewis acid catalysts (equation 159)[151]. Various 2-alkyl groups made very little difference in the

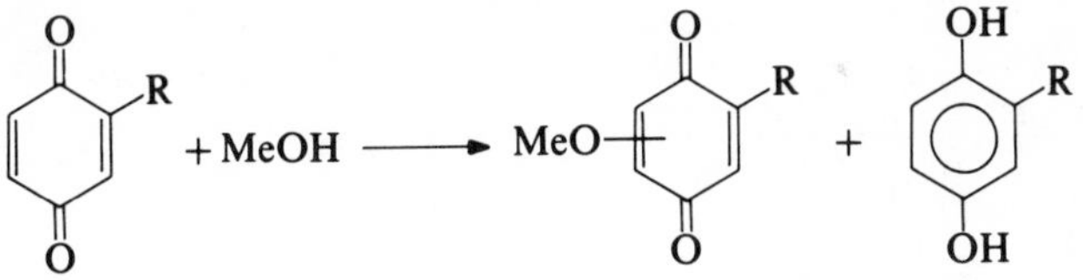

(159)

R = Me, Et, *s*-Bu, *t*-Bu, Bz

product distribution in the absence of catalyst. Similar results were obtained for the zinc chloride catalyzed reaction except for the two bulkiest substituents where only the 2,6-isomer was found. Under both conditions 1,4-benzoquinone gave the 2,5-dimethoxy product. In ethanol 2-methyl-1,4-benzoquinone showed essentially the same behavior as in methanol. Finally, 2-benzoyl-1,4-benzoquinone reacts with methanol under both experimental conditions to produce only the 3-methoxy product. Similar results have been reported by Fariña and Valderrama in their studies of the acetyl, methoxycarbonyl and cyano groups[129]. They also were able to add phenol and 4-methoxyphenol to 2-acetyl-1,4-benzoquinone in the presence of pyridine, but 4-nitrophenol failed to react. An earlier report by these investigators showed the best conditions for addition to be equimolar reactants in dry benzene[152].

A recent application of phenol addition to a nascent 1,2-benzoquinone provides an important intermediate in the alkylation of quinones (see Section IV.E)[153]. Displacement of the aryl ether by an allylic alcohol followed by a Claisen–Cope rearrangement allows the transformation of 1,2- to 1,4-quinones (equation 160).

2 MeO–C6H4–OH $\xrightarrow[\text{MeCN/25°/4h}]{Cu_2Cl_2/Cu/O_2}$ [MeO-1,2-benzoquinone] ⟶ MeO-1,2-benzoquinone–O–C6H4–OMe

$\xrightarrow{CH_2=CHCH(OH)(CH_2)_{11}Me}$ MeO-1,4-benzoquinone (OH, $CH_2CH=HC(CH_2)_{11}Me$) (160)

2. *Epoxidation of quinones*

The preparation of most simple epoxy-1,4-benzoquinones is complicated and produces low yields, but Ichihara and his coworkers have developed a system based on retro-Diels–Alder chemistry which has demonstrated great versatility (equation 161)[154]. The

1,4-benzoquinone + dimethylfulvene $\xrightarrow{EtOH}$ adduct $\xrightarrow[K_2CO_3]{H_2O_2}$ epoxide

$\xrightarrow[\text{(CH}_2\text{)}_4\text{O/sealed tube}]{\text{heat/Ph}_2\text{O or}}$ epoxy-1,4-benzoquinone (161)

diene of choice is dimethylfulvene which releases the epoxide under milder conditions than cyclopentadiene; the mixture of *endo* and *exo* adducts are epoxidized without separation. Only three substituents are reported and while the yields for 1,4-benzoquinone and the 2-methyl derivative are near quantitative, 2-hydroxymethyl-1,4-benzoquinone gave at best 43%. An earlier preliminary report of this chemistry indicated that the sealed tube method with lower boiling solvents gave improved results[155].

Pluim and Wynberg have made a detailed study of the synthesis of optically active epoxy-1,4-naphthoquinones (equation 162)[156]. With a variety of alkyl and a few aryl

$\xrightarrow[\text{benzylquininium chloride}]{30\% H_2O_2/NaOH/H_2O\text{–toluene}}$ (162)

R = H, alkyl, c-C_6H_{11}, Ph, Bz, $CH_2CH = CMe_2$
R^1 = H, 5-, 6-, 7-Me, 5-MeO

substituents they used a phase-transfer oxidation method to obtain good yields of epoxides with a substantial excess of one enantiomer. Both the yield and the enantiomeric excess are excellent in certain cases; for example, 2-phenyl-1,4-naphthoquinone was converted to its epoxides in 92% yield with a 45% excess of one enantiomer. Some discussion of the effect of substituent and the absolute configuration of the products is presented.

III. CHEMISTRY AT THE QUINONE CARBONYL

A. Nitrogen Addition

To a large extent the recent chemistry of quinonoid carbonyl groups has dealt with the preparation of non-quinonoid products, thus, a great many opportunities exist for profitable study. For example, Wright and Brown prepared several arylhydrazinouracils starting with 6-hydrazinouracil and 1,4-benzoquinones (equation 163)[157]. The yields from

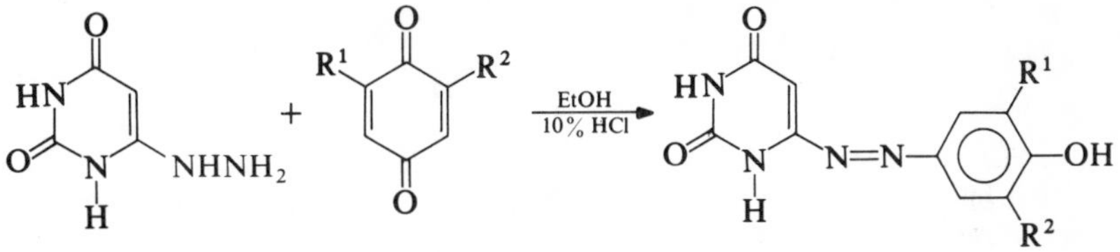

O HN O N H $NHNH_2$ + R^1 O R^2 O $\xrightarrow{\text{EtOH, 10\% HCl}}$ O HN O N H N=N R^1 OH R^2

R^1,R^2 = H, Me, Cl, Br (163)

unsymmetrically substituted quinones were approximately 60% and isomerically pure. By implication the more soluble isomer was not isolated.

Roushdi and Habib have shown that a wide variety of 2-hydroxy-1,4-naphthoquinones form hydrazines at the 1-carbonyl in good to excellent yields (equation 164)[158]. A similar reaction occurs with isoniazide (**56**, equation 165).

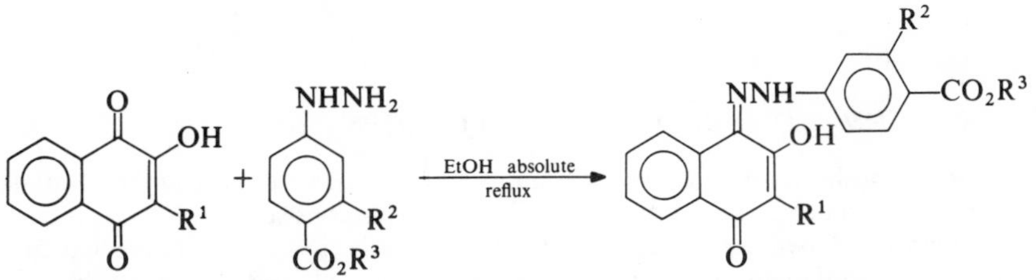

O OH R^1 O + $NHNH_2$ R^2 CO_2R^3 $\xrightarrow{\text{EtOH absolute, reflux}}$ NNH R^2 CO_2R^3 OH R^1 O

R^1 = alkyl, CO_2Et, R^2 = H, OH, R^3 = H, Et 59–94%

(164)

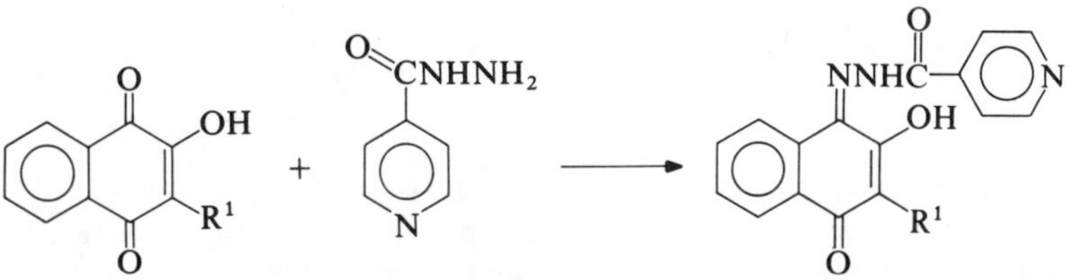

O OH R^1 O + O=$CNHNH_2$ N $\longrightarrow$ NNHC(O) N OH R^1 O (165)

(**56**)

Several phenyl-substituted 1,4-benzoquinones have been converted to the azo derivative through reaction with *p*-toluenesulfonhydrazide followed with aqueous potassium hydroxide (equation 166)[159]. The yields are good (ca. 85 %) except for the trisubstituted

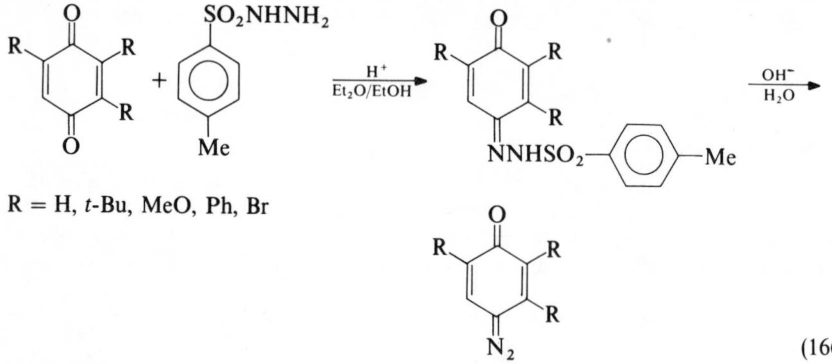

R = H, *t*-Bu, MeO, Ph, Br

(166)

cases in which two phenyl groups are present (30 %). The intermediate hydrazones were not isolated, but could have been if necessary. Boron trifluoride etherate is also satisfactory as a catalyst with dry tetrahydrofuran as solvent.

Synthetic dyes derived from 1,2-naphthoquinone have been prepared by reaction with substituted hydrazino-*s*-triazines (equation 167)[160]. The NMR spectra indicate the

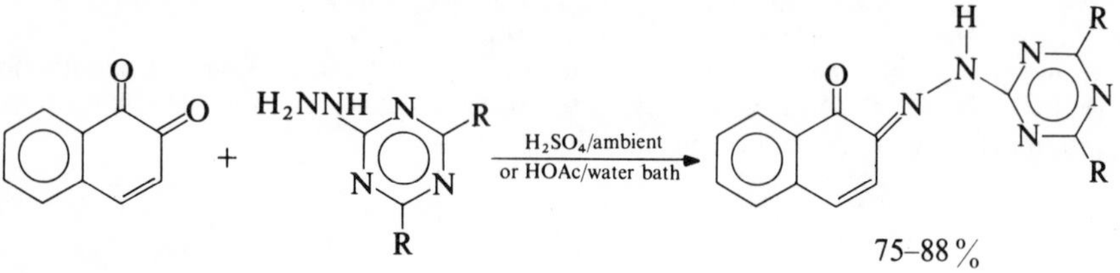

75–88 %

R = NHAc, NMe_2, OMe, $N(CH_2)_5$, $N(C_2H_4)_2O$, NHPh (167)

hydrazone form shown, but the mass spectra indicate conversion to azonaphthols under electron bombardment. The reaction at the 2-carbonyl is consistent with these spectra.

The method devised by Corey and Achiwa[98] has been applied to the oxidative deamination of aminoglucosides (equation 168)[161]. Yields of 30–35 % are typical.

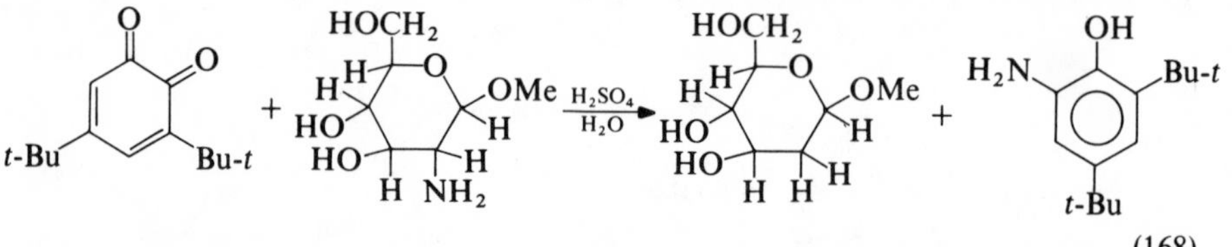

(168)

An analogous method involving a hindered 1,4-benzoquinone provides an attractive route for the conversion of a methylamino group into an amide (equation 169)[162]. The intermediate imine is found in yields of 40–80 %. The yield of amide varies greatly. The base obviously plays an important role, but from the small number of examples studied there is

$$\text{ArCH}_2\text{NH}_2 + \text{2,6-di-}t\text{-Bu-1,4-benzoquinone} \xrightarrow{\text{EtOH}} \text{ArCH=N-C}_6\text{H}_2(\text{Bu-}t)_2\text{-OH} \xrightarrow[\text{or KOH/EtOH}]{O_2,\ t\text{-BuOK}/t\text{-BuOH}}$$

$$\text{ArC(=O)NH}_2 + \text{O=C}_6\text{H}_2(\text{Bu-}t)_2\text{=O} \quad (169)$$

Ar = Ph, 4-Tol, 4-An, 4-ClC_6H_4, 3-pyridyl, 2-furyl

no clear pattern. With the proper selection of conditions good yields (60–85%) can be obtained.

Another application of the tosylhydrazones is found in a convenient synthesis of 1-(4-hydroxyphenyl)-2-methyldiazenes (equation 170)[163]. In addition to the synthetic work,

$$\text{R}_3\text{-1,4-benzoquinone} + p\text{-MeC}_6\text{H}_4\text{SO}_2\text{N(Me)NH}_2 \xrightarrow[\text{F}_3\text{CCO}_2\text{H}]{\text{MeOH}} \text{R}_3\text{C}_6\text{H(=O)(=NN(Me)SO}_2\text{C}_6\text{H}_4\text{Me)}\ (60\text{–}70\%) \xrightarrow{\text{H}_2\text{SO}_4} \text{HO-C}_6\text{HR}_3\text{-N=NMe}\ (40\text{–}60\%) \quad (170)$$

R = H, alkyl, halo

investigation of tautomerism in these compounds shows the presence of certain forms and the absence of others (equation 171). The same chemistry takes place with 2-methyl-1,4-naphthoquinone.

$$\text{RCH}_2\text{N=N-C}_6\text{H}_3(\text{R})\text{OH} \rightleftharpoons \text{RCH}_2\text{NHN=C}_6\text{H}_3(\text{R})\text{=O} \not\rightleftharpoons \text{RCH=NNH-C}_6\text{H}_3(\text{R})\text{OH} \quad (171)$$

B. Silyl and Organolithium Additions

Neumann and Neumann have reported in great detail their studies of the reduction of quinones with bis(trimethylsilyl)mercury and confirmed the radical mechanism of this efficient process[164]. Similar results were obtained by Adeleke and Wan for chemically generated triarylsilyl radicals[165]. The mechanism proposed is supported by other published observations (equation 172). With 2,6-di-*t*-butyl-1,4-benzoquinone the ESR

$$(t\text{-Bu})_2O_2 \xrightarrow[\text{PhH}]{h\nu} 2\ t\text{-BuO}^{\bullet}$$

$$t\text{-BuO}^{\bullet} + Ar_3SiH \longrightarrow t\text{-BuOH} + Ar_3Si^{\bullet}$$

$$Ar_3Si^{\bullet} + \text{R-substituted 1,4-benzoquinone} \longrightarrow \text{R-substituted } 4\text{-}(OSiAr_3)\text{phenoxyl radical } (O^{\bullet}) \qquad (172)$$

spectra of both isomeric radicals were observed unlike the 2,6-dimethyl case. It appeared to the authors that such substituted quinones are excellent spin traps for silyl radicals.

The application of reductive silylation in synthesis as a protective method is attractive for quinonoid systems (equation 173)[166]. The very high yields (85–100%) are coupled with

$$\text{2,3,5,6-R}_4\text{-1,4-benzoquinone} + Me_3SiSiMe_3 \xrightarrow[\text{PhH/N}_2/60\text{–}80^\circ]{I_2} \text{2,3,5,6-R}_4\text{-1,4-bis}(OSiMe_3)\text{benzene} \qquad (173)$$

R = H, Me, Cl, —CH=CHCH=CH—

mild conditions and reasonable reaction times. An alternative approach to these protected quinones has been suggested, but only two examples are given and the yields are somewhat lower[167].

A particularly exciting variant of the silylation theme has been advanced by Evans and his colleagues in the addition of trimethylsilyl cyanide (**57**, equation 174)[168]. Their work

$$\text{1,4-benzoquinone} + Me_3SiCN\ (\mathbf{57}) \xrightarrow[\text{18-crown-6}]{\text{KCN or}} \text{4-NC-4-}(OSiMe_3)\text{cyclohexa-2,5-dienone}\ (80\%) \qquad (174)$$

shows that only the quinones show selectivity towards catalysts, i.e. are inert towards Lewis acids. These studies have been reported in greater detail and applied to additional quinonoid substrates[169–171].

Some disadvantages of the use of cyanide might be overcome by its generation *in situ* and such a scheme has been developed (equation 175)[172]. Unfortunately, the one reaction with a quinone produced a mixture of products, but more work was promised.

$$\text{1,4-benzoquinone} + KCN + Me_3SiCl \xrightarrow[\text{pyridine}]{\text{MeCN/NaI}} \text{4-(NC)(Me}_3\text{SiO)-cyclohexa-2,5-dienone } (21\%) + \text{1,4-(NC)}_2\text{-1,4-(OSiMe}_3)_2\text{-cyclohexa-2,5-diene } (63\%\ (1{:}1;\ E{:}Z)) + C_6H_4(OSiMe_3)_2\ (16\%) \qquad (175)$$

Cyanosilylation followed by addition of organolithium reagents to the second carbonyl, and finally esterification is an efficient method of synthesis for *p*-quinol benzoates (equation 176)[173]. The reaction is carried out at low temperatures to avoid a facile

$$\text{4-(Me}_3\text{SiO)-4-(NC)-cyclohexa-2,5-dienone} \xrightarrow[\text{(2) PhCOCl/}-70^{\circ}\text{C}]{\text{(1) R}^1\text{R}^2\text{CHLi/THF/}-70^{\circ}\text{C}} \left[\text{Me}_3\text{SiO(NC)C}_6\text{H}_4\text{(OCOPh)(CHR}^1\text{R}^2)\right] \xrightarrow[\text{THF}]{\text{(3) NaF/H}_2\text{O}} \text{O=C}_6\text{H}_4\text{(OCOPh)(CHR}^1\text{R}^2) \ (50\text{–}70\%) \qquad (176)$$

R^1 = Ph, $PhCH_2$, 2-An, 3,4-$(MeO)_2C_6H_3$
R^2 = H, CO_2Me, CO_2Et, PhSO, $PhSO_2$

dienone–phenol rearrangement. With phenyl and sulfoxyphenyl substituents the two diastereomers are readily separated. Several related reagents were explored in a preliminary fashion and give evidence of a versatile synthetic method (equations 177–180).

In the past ten years great progress has been made on the 1,2-addition of carbanions, mostly organolithium compounds, to quinones. Such reactions are often highly selective with above average yields; an example is the direct synthesis of jacaranone (**58**, equation 181)[174].

Moore and his students made the earliest report of this method when they developed a clever synthesis of 2,5-disubstituted 1,4-benzoquinones (equation 182)[175]. The yields are very reasonable and the potential for the development of new synthetic pathways is considerable.

Fischer and Henderson showed that in 1,2-additions (equation 183)[176], the solvent plays an important role in directing the reaction to mono- or di-addition and that the major product in the latter case has *cis* stereochemistry. The yields are excellent although the selectivity of **59**/**60** is not very great. In a later paper the authors demonstrated that a variety of alkyl lithium compounds add to quinones in satisfactory yield[177].

Liotta and coworkers examined the problem of poor regioselectivity in unsymmetrical quinones[178]. They made the argument that the site of attack should be dependent on the size of the carbanion including its degree of solvation. The test they devised involved the methyl carbanion in two different systems; one sterically, the other electronically controlled (equations 184, 185). With a few limitations involving very bulky carbanions, similar results were demonstrated for several combinations of quinones and carbanions.

$$\text{4-cyano-4-(trimethylsilyloxy)cyclohexa-2,5-dienone (}Me_3SiO,\ CN\text{)} + PhCH_2(PhSO_2)CHLi \longrightarrow \text{4-(PhCOO)-4-[CH(CH_2Ph)SO_2Ph]cyclohexa-2,5-dienone}\quad 75\% \tag{177}$$

$$\text{4-cyano-4-(trimethylsilyloxy)cyclohexa-2,5-dienone} + PhCH_2MgBr \xrightarrow[\text{(2) PhCOCl; (3) NaF/}H_2O]{\text{(1) THF}/-70^{\circ}C} \text{4-hydroxy-4-benzylcyclohexa-2,5-dienone (}HO,\ CH_2Ph\text{)}\quad 25\% \tag{178}$$

$$\text{4-cyano-4-(trimethylsilyloxy)cyclohexa-2,5-dienone} + PhCH(Li)CO_2Et + Me_3SiCl \longrightarrow \left[\text{1-}Me_3SiO\text{-1-NC-4-}OSiMe_3\text{-4-}CH(Ph)CO_2Et\text{-cyclohexa-2,5-diene}\right]\ (\mathbf{57a})$$

$$(\mathbf{57a}) \xrightarrow[\text{THF/48 h}]{AgF/H_2O} \text{2-[}CH(Ph)CO_2Et\text{]-1,4-benzoquinone}\quad 39\% \tag{179}$$

$$(\mathbf{57a}) \xrightarrow[\text{THF}]{NaF/H_2O} \text{4-}OSiMe_3\text{-4-[}CH(Ph)CO_2Et\text{]cyclohexa-2,5-dienone}\quad 65\% \tag{180}$$

$$\text{1,4-benzoquinone} + CH_2{=}C(OLi)OMe \xrightarrow[\text{(2) }H_2O]{\text{(1) }Et_2O/-110^{\circ}C/N_2} \text{4-hydroxy-4-(}CH_2CO_2Me\text{)cyclohexa-2,5-dienone}\ (\mathbf{58})\quad 55\% \tag{181}$$

Moore and his coworkers continued their original study cited above and determined that for the synthesis of unsymmetrical quinones the first substituent introduced must be alkynyl[179]. This requirement is not much of a limitation when one considers the possibilities inherent in the carbon–carbon triple bond. Additional groups studied include 4-methoxyphenyl, phenylalkynyl and 3-hydroxypropylalkynyl. Yields in the 45–65% range were realized for either symmetrical or unsymmetrical quinones.

The utility of such an approach to 2,5-disubstituted-1,4-benzoquinones was illustrated in the synthesis of a ring system containing important elements of mitomycin (**61**, equation

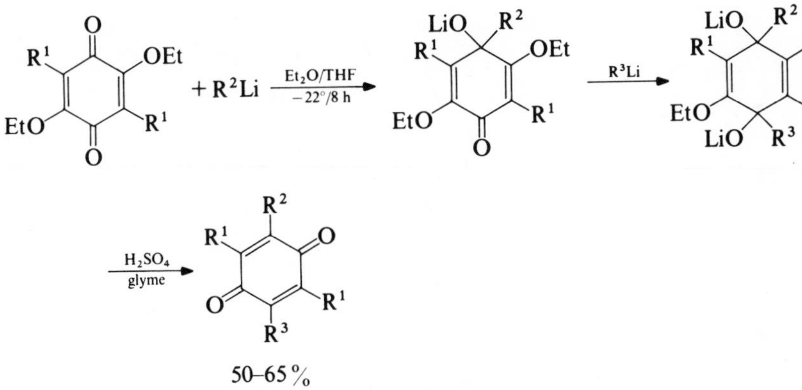

$R^1 = H, Cl; R^2 = R^3 = Me, Ph, PhCH_2OCH_2C{\equiv}C$

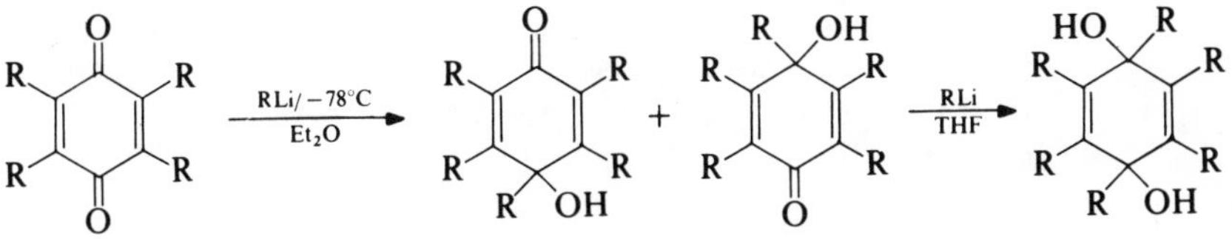

(182)

50–65%

(59) (60)

R = Me, Bu, *t*-Bu, Ph

(183)

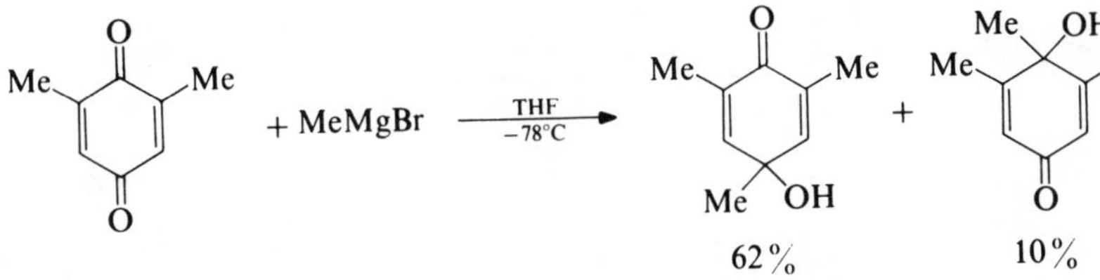

62% 10% (184)

87% 9%

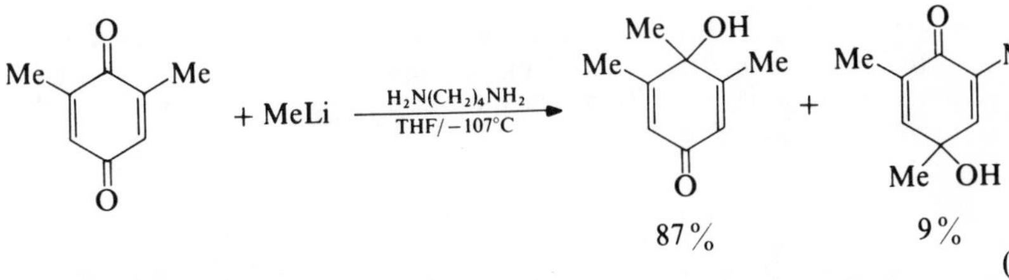

(185)

186). Some aspects of the stereoselectivity shown are certainly worthy of further investigation along with more extensive application of the overall strategy.

A useful variant of these methods involves 1,2-benzoquinone (equation 187)[180]. The importance of the 2-oxygen-5-hydrocarbon pattern in naturally occurring quinones and its potential in drug synthesis is widely recognized.

One especially promising method for the introduction of an α,β-unsaturated nitrile substituent failed completely with 1,4-benzoquinone (equation 188)[181]. It certainly deserves additional attention.

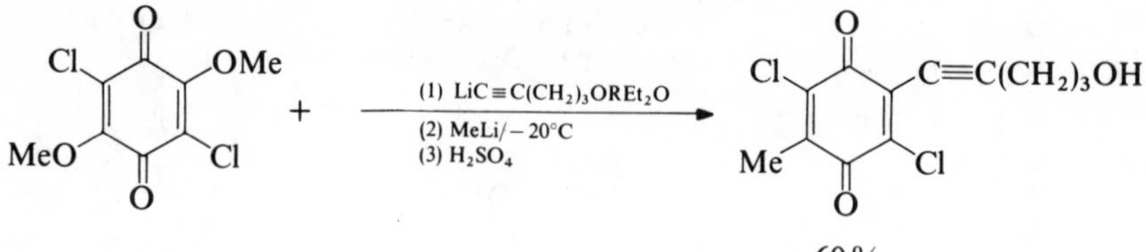

(1) $LiC{\equiv}C(CH_2)_3OREt_2O$
(2) $MeLi/-20°C$
(3) H_2SO_4

$C{\equiv}C(CH_2)_3OH$

60%

(186)

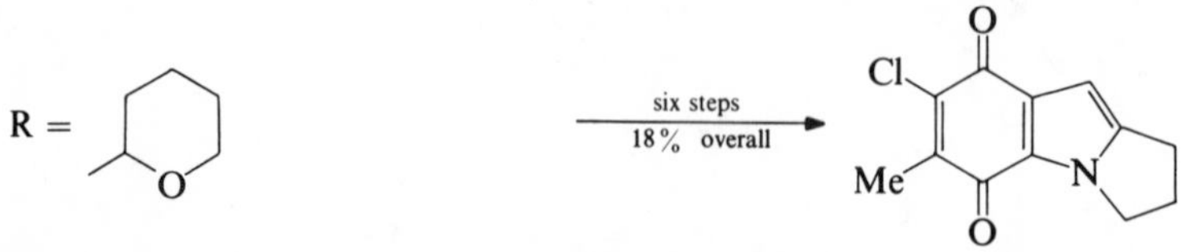

R =

six steps
18% overall

(61)

(1) $LiC{\equiv}CR$
(2) NH_4Cl

H_3O^+

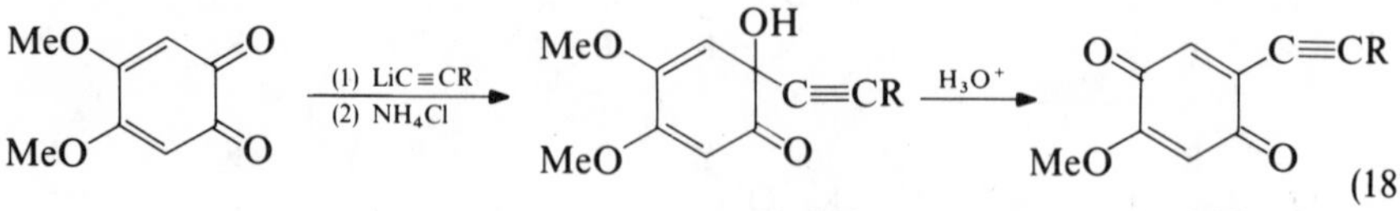

(62) 70–95%

85–95%

(187)

R = Bu, Ph, CO_2Et, $MeC{=}CH_2$, $(CH_2)_4C{\equiv}CH$, OMe, CH_2NEt_2

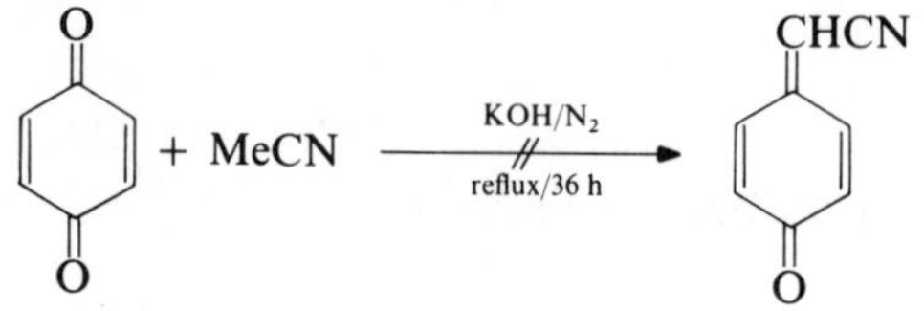

+ MeCN

KOH/N_2
reflux/36 h

CHCN

(188)

C. Ramirez and Wittig Phosphorous Chemistry

An important modification of the pioneering work by Ramirez and his students involves the reductive addition of tris(trimethylsilyl) phosphite (**63**) to 1,4-benzo- and naphthoquinones (equation 189)[182]. The combination of high yields and ease of hydrolysis of the silylated ether suggests further study.

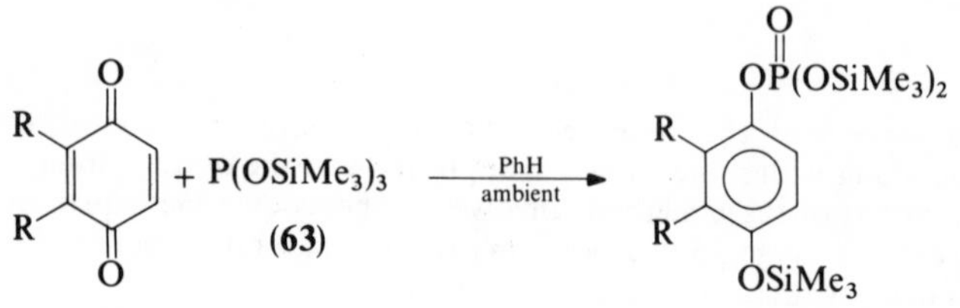

+ $P(OSiMe_3)_3$

(**63**)

PhH
ambient

$OP(O)(OSiMe_3)_2$

$OSiMe_3$

86%, 92%

(189)

R = H, —CH=CHCH=CH—

Duthaler and his coworkers examined the reaction of ten monosubstituted 1,4-benzoquinones with trimethyl phosphites (equation 190)[183]. Control of regioselectivity is

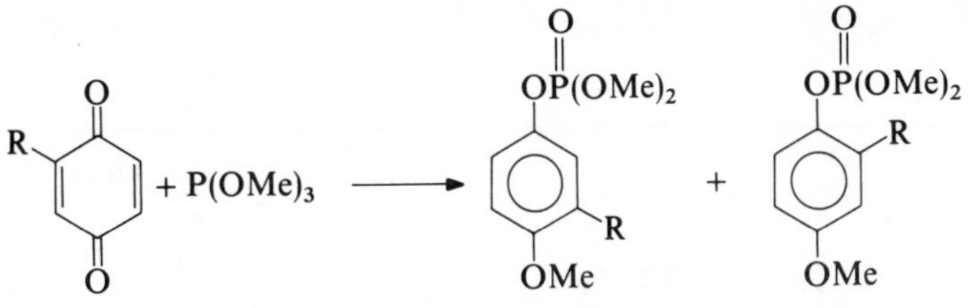

$$\text{R-1,4-benzoquinone} + P(OMe)_3 \longrightarrow \mathbf{(64)} + \mathbf{(65)} \qquad (190)$$

determined by the expected variables; for example, bulky groups such as *t*-butyl and trimethylsilyl give good yields of the 3-substituted product **64**, while the electron-donating methoxy group directed largely to the 2-substituted product (**65**). Solvent also plays an important role in controlling the product distribution with the non-polar benzene giving the best selectivity. With juglone methyl ether and in cases involving strong electron-withdrawing groups modest yields of ring phosphorylation are observed (equations 191, 192). This new understanding was put to use in a critical intermediate step directed toward

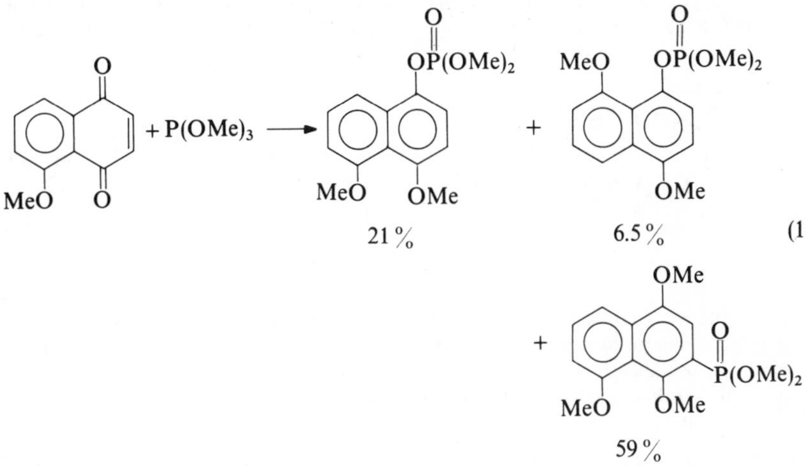

$$+ P(OMe)_3 \longrightarrow 21\% + 6.5\% + 59\% \qquad (191)$$

$$+ P(OMe)_3 \longrightarrow \qquad (192)$$

R = COPh, CO_2Me 43%, 48%

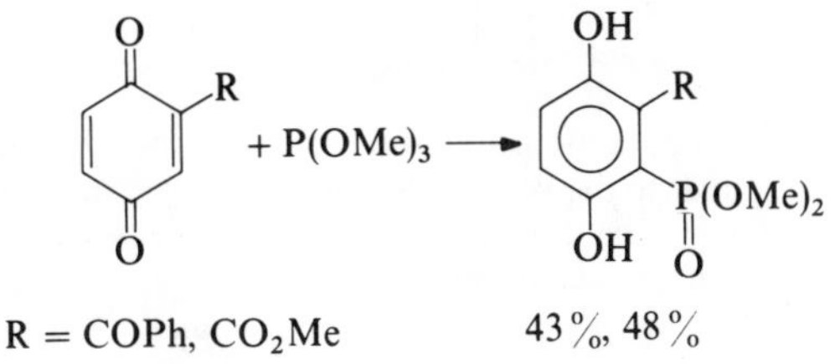

the synthesis of the antibiotic Lysolipin I. The appropriate quinone formed the desired product (**66**) without any detectable amount of its regioisomer or the tetrahydrodibenzofuran (**67**; equations 193, 194).

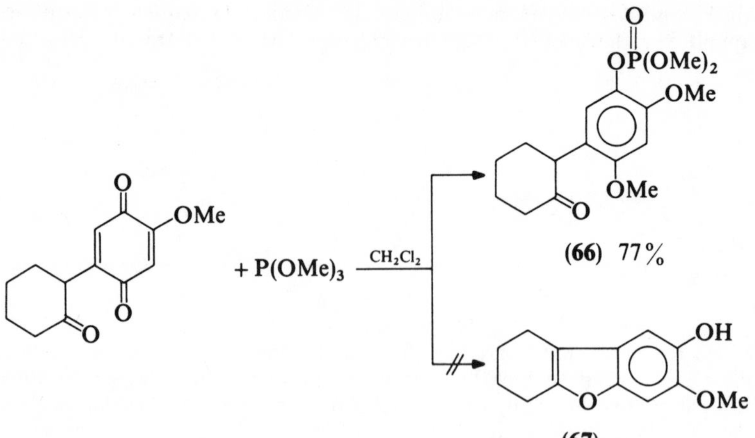

(193)

(66) 77%

(194)

(67)

The extensive interest in synthesis of polycyclic aromatic compounds has made the bis-Wittig reaction attractive in spite of low yields and difficult reaction conditions. Minsky and Rabinovitz have shown that under phase-transfer catalysis conditions the phosphonium salt acts both as reagent and catalyst (equation 195)[184]. The yields are not

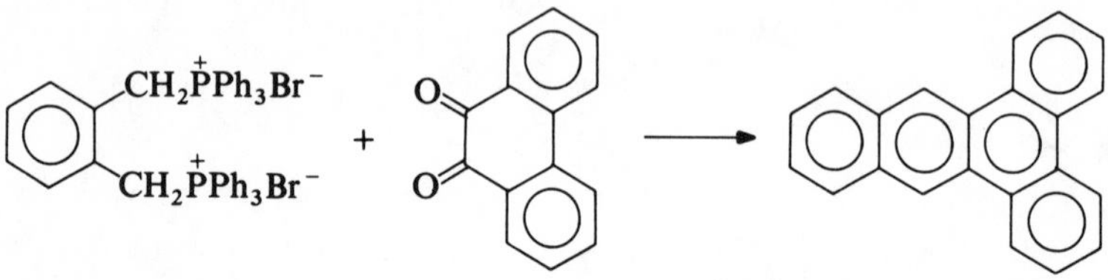

(195)

24%

unreasonable when compared with methods far more demanding in terms of maintaining an inert atmosphere and anhydrous conditions. The chief limitation is with quinones sensitive to aqueous base. A complementary procedure under non-basic conditions extends the synthetic application of this chemistry to these more delicate substrates (equation 196)[185].

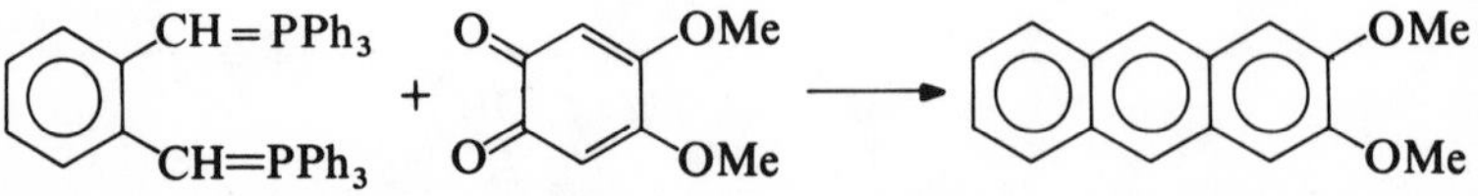

(196)

Frøyen has examined the reactions of triphenylarsine phenylimine (**68**) with quinones and found chemistry similar to the Wittig reaction resulting in the production of mono- or diimines (equation 197)[186]. The differences in reactivity between these two reagents are discussed; unfortunately no yields are given.

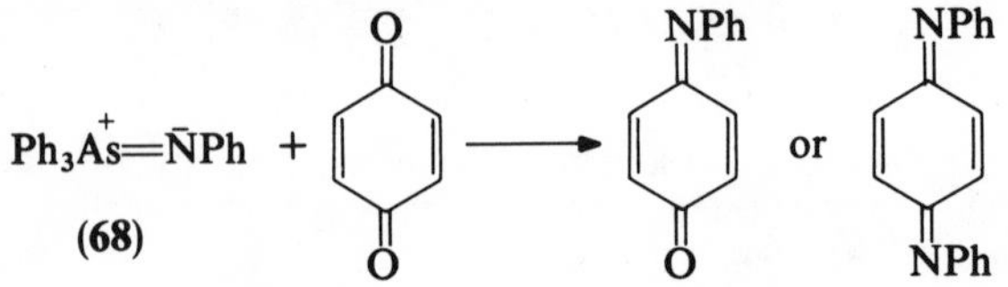

$$Ph_3\overset{+}{As}{=}\bar{N}Ph \quad (68)$$

(197)

A very active Russian research group led by Pudovik has made important contributions in the study of phosphinous amides with quinones (equation 198)[187]. They obtained

$$(RO)_2PNHAr + \text{1,4-benzoquinone} \longrightarrow 4\text{-}[(RO)_2P{=}NAr]C_6H_4OH \quad (198)$$

R = Et, Pr, Bu 74–90%
Ar = Ph, 4-An, 4-Tol, Naphthyl

excellent yields of esters as opposed to carbon–phosphorous bond formation. These intermediates undergo thermal rearrangement to the analogous 4-ethers. In a later publication the occurrence of one-electron transfer and the formation of radical ions was demonstrated by ESR studies[188].

The addition of diethyl trimethylsilyl phosphite (**69**) to 1,4-benzoquinone produces an 81% yield of the phosphate ester containing the 4-trimethylsilyl ether group (equation 199)[189].

$$\underset{(\mathbf{69})}{(EtO)_2POSiMe_3} + \text{1,4-benzoquinone} \longrightarrow 4\text{-}[(EtO)_2P(O)O]C_6H_4OSiMe_3 \quad (199)$$

Finally, the reaction of these versatile reagents with the halo-*o*-quinones generally results in the formation of the phosphorous heterocyclic system 2,2-dihydro-1,3,2-benzodioxaphosphole (**70**; equation 200)[190]. If the starting material contains an acetamido

$$(RO)_2PNHAr + \text{tetrahalo-}o\text{-quinone (X)} \longrightarrow \underset{(\mathbf{70})\ 70\text{–}75\%}{(RO)_2(ArNH)P(O_2C_6X_4)} \quad (200)$$

group the initial adduct loses acetonitrile to form an intermediate which is slowly transformed to a hydroxyphosphate (**71**; equation 201).

$$(EtO)_2PNHAc + \text{tetrabromo-}o\text{-quinone} \xrightarrow{-MeCN} \underset{(\mathbf{71})}{(EtO)_2(HO)P(O_2C_6Br_4)} \quad (201)$$

Arbuzov and his colleagues studied the reductive addition of trimethyl phosphite to 1,4-benzoquinones bearing strong electron-withdrawing groups (equation 202) and found the more highly hindered arylphosphate ester formed; no yields are given[191].

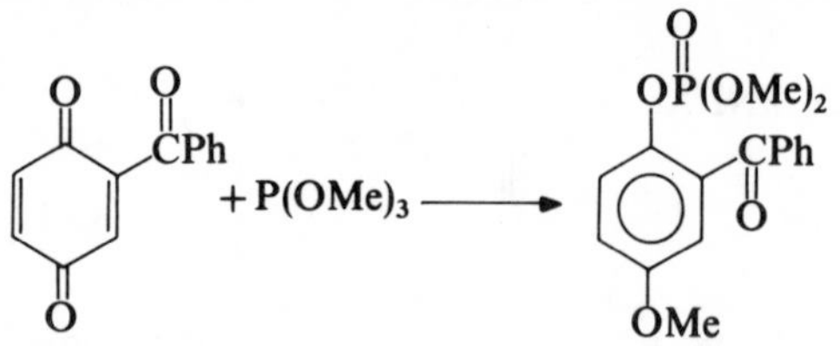

$$\text{2-benzoyl-1,4-benzoquinone} + P(OMe)_3 \longrightarrow \text{2-}(MeO)_2P(O)O\text{-5-MeO-}C_6H_3COPh \quad (202)$$

Mixed phosphite esters generally give good yields of the ether corresponding to the more stable carbonium ion (equation 203)[192]. For example, with the ethyl and propyl cations the

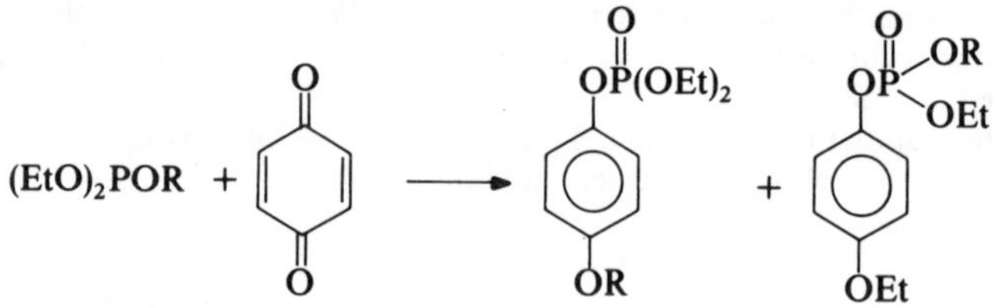

$$(EtO)_2POR + \text{1,4-benzoquinone} \longrightarrow p\text{-}(EtO)_2P(O)OC_6H_4OR + p\text{-}(RO)(EtO)P(O)OC_6H_4OEt \quad (203)$$

R = Pr, Allyl, Bz, *t*-Bu

product ratio is 3:1; not too unlike the 2:1 ratio of the alkyl groups. When the alkyl group other than ethyl would be expected to form a much more stable cation the *para*-ether formed from it is generally formed in good yield. An exception is the *t*-butyl case where a significant quantity of 2-methylpropene is also formed.

D. Brief Notes

During the course of a study of the bis-Wittig reaction described earlier[185] Nicolaides and Litinas were surprised to find that reaction between the ylide and *o*-chloranil produced the bisbenzodioxole (**72**, equation 204). This has also been reported for both *o*-chloranil

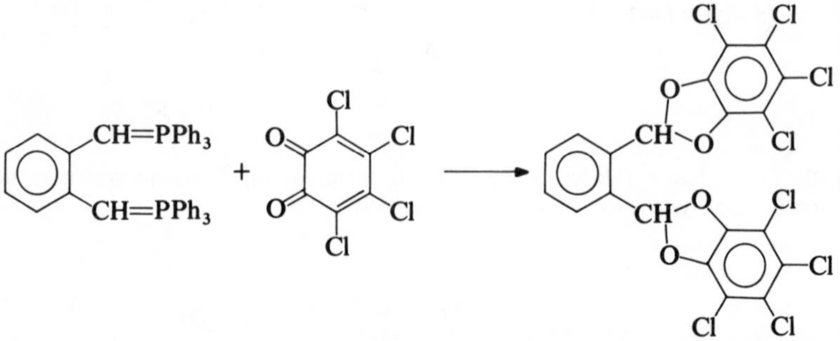

$$o\text{-}C_6H_4(CH{=}PPh_3)_2 + o\text{-chloranil} \longrightarrow \textbf{(72)} \quad (204)$$

(**72**)

and *o*-bromanil in reactions with several mono-triphenylphosphoranes (equation 205)[193]. The authors note the similarity between these reactions and that of diazomethane with 1,2-quinones.

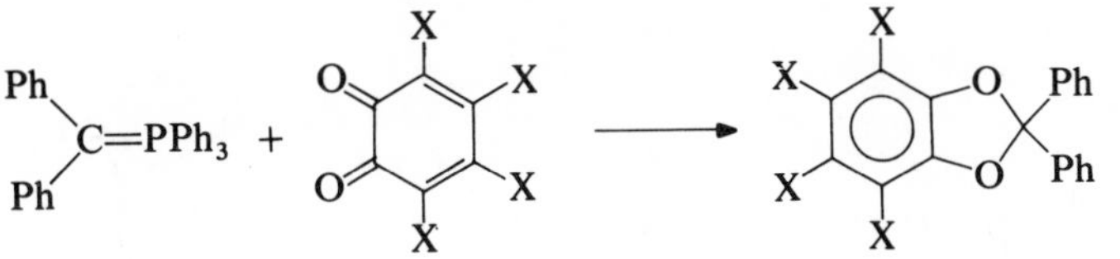

Ph₂C=PPh₃ + tetrahalo-1,2-benzoquinone → 2,2-diphenyl-4,5,6,7-tetrahalo-1,3-benzodioxole (205)

X = Cl, Br

92%, 95%

A new class of spiranes has been prepared by Latif and his colleagues using heterocyclic thiones and halogenated 1,2-benzoquinones (equations 206, 207)[194, 195]. The yield is not as

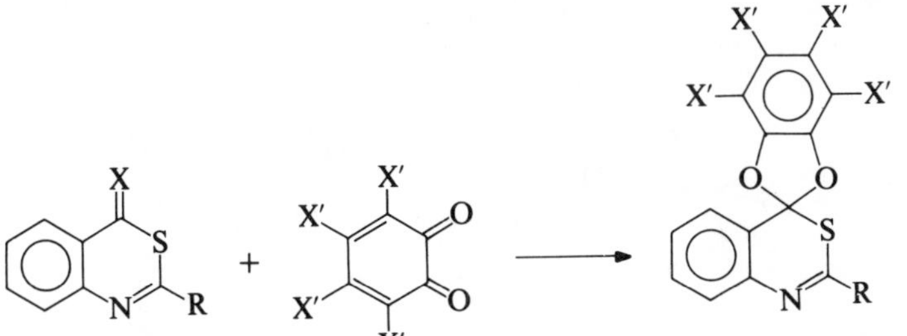

(206)

X = O,S

X′ = Cl,Br

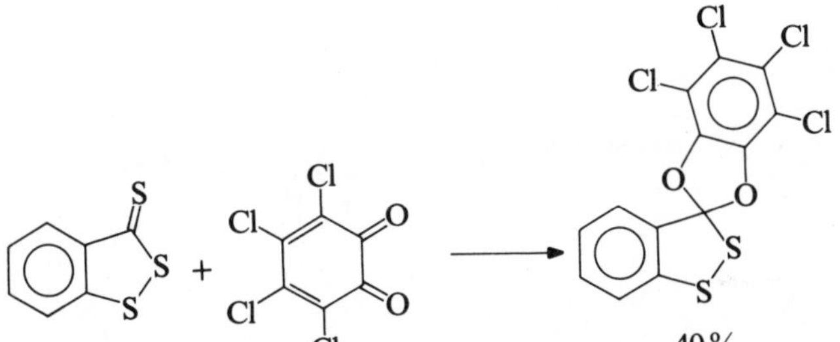

40–50%

(207)

(73)

40%

high in the case of 3*H*-1,2-benzodithiole-3-thione (**73**), but all products are only obtainable by this route and show promising chemical attributes beyond the range of this review.

More recently Zeid and colleagues have provided experimental evidence of the similarity between the reactions of *o*-chloranil with a thione and diazoalkanes (equation 208)[196]. The range of chemistry displayed by the spiro products was also usefully expanded.

Awad and his coworkers have studied the Darzens reaction of quinones and shown that while both phenanthrenequinone and chrysenequinone react with 4-nitrobenzyl chloride they give rather different chemical outcomes (equations 209, 210)[197, 198].

In an effort to prepare fluorine-substituted quinodimethanes bis(trifluoromethyl) ketene was added to 1,4-benzoquinone (equation 211)[199]. The initial product undergoes the dienone–phenol rearrangement and adds a second mole of the ketene to produce the observed γ-lactone in 60% yield. Attempted reactions with chloranil, fluoranil and anthraquinone which could be expected not to rearrange failed even at high temperatures.

Cl Cl O O Cl Cl + X Cl O → Cl Cl Cl Cl O O Cl O (208)

$X = S, N_2$

O O $\xrightarrow{4\text{-}O_2NC_6H_4CH_2Cl}$ $4\text{-}NO_2C_6H_4$ CH O O (209)

O O $\xrightarrow{4\text{-}O_2NC_6H_4CH_2Cl}$ $4\text{-}NO_2C_6H_4$ CH O O + $C_6H_4NO_2\text{-}4$ H O O

(210)

O O $+ \ (CF_3)_2C{=}C{=}O$ → O O $(CF_3)_2$

$\xrightarrow{(CF_3)_2C=C=O}$ O O CF_3 CF_3 $OCCH(CF_3)_2$ O

(211)

A more successful effort to prepare mono-carbon analogs of quinones is found in the work of Verboom and Bos who added *t*-butyl-1,2-benzoquinones to acetylenes in good yield (equation 212)[200].

$$t\text{-Bu-1,2-benzoquinone} + XC{\equiv}CNR_2 \longrightarrow t\text{-Bu-quinone methide }(=C(X)-C(=O)-NR_2) \qquad (212)$$

X = Ph, SMe; R = Et, *i*-Pr, pyrrolyl

The addition of a cyclic diphosphirane (**74**) to quinones gives low yields of the intriguing cyclic phosphirane derivatives (**75**; equation 213)[201].

$$\underset{(\mathbf{74})}{Ph_3P\!-\!PPh_3\!>\!CCl_2} + \text{1,4-benzoquinone} \longrightarrow \underset{(\mathbf{75})}{\text{spiro}[Ph_3P\!-\!CCl_2]\text{cyclohexadienone}} \qquad (213)$$

IV. THE ALKYLATION OF QUINONES

A. Mechanism of Radical Additions

The need to attach hydrocarbon chains to quinones for the synthesis of natural products continues to stimulate activity in the study of radical alkylations. In a detailed review of radical substitution reactions Dou and his collaborators included a useful table of relative reactivities for quinones with methyl radicals (Table 3)[202].

TABLE 3. Reactivity of various quinones toward methyl radicals at 65 °C[202, 203]

Quinone	Relative rates
1,4-Benzoquinone	15 200
2-Methyl-1,4-benzoquinone	10 400
2,5-Dimethyl-1,4-benzoquinone	6500
Duroquinone	790
2-Methoxy-1,4-benzoquinone	8000
2-Chloro-1,4-benzoquinone	27 000
2,5-Dichloro-1,4-benzoquinone	39 200
2,6-Dichloro-1,4-benzoquinone	38 400
Chloranil	300
1,4-Naphthoquinone	3400
2-Methyl-1,4-naphthoquinone	4100
2,7-Dimethyl-1,4-naphthoquinone	4100
2,3-Dimethyl-1,4-naphthoquinone	550
2,3-Dichloro-1,4-naphthoquinone	90
2-*t*-Butylanthraquinone	90

Citterio has been especially active, along with his colleagues in the determination of rates of addition of alkyl radicals to 1,4-benzoquinone[204]. They have demonstrated the importance of nucleophilic polar effects on these reactions using the cyclopentylmethyl radical formed from the 5-hexenyl radical (equation 214). The very high rate observed

$$(CH_2=CH(CH_2)_4CO_2)_2 \xrightarrow{Cu^+} CH_2=CH(CH_2)_3CH_2^{\bullet} \longrightarrow \text{cyclo-}C_5H_9CH_2^{\bullet}$$

$$\text{cyclo-}C_5H_9CH_2^{\bullet} + \text{1,4-benzoquinone} \longrightarrow \text{2-(cyclopentylmethyl)-1,4-benzoquinone} \quad (214)$$

($9 \times 10^6\ \text{M}^{-1}\ \text{s}^{-1}$ at 40°C) is not found with electrophilic radicals of the amino or oxygen types. These observations are further supported by the increased reactivity of quinones bearing electron-withdrawing substituents. Several examples are given which extend the synthetic utility of the method.

More evidence for the importance of polar effects in quinone–radical reactions is found in a detailed comparison with several conjugated alkenes[205]. Discussion of the probable contribution by a charge transfer interaction is phrased in terms of recent orbital theory calculations.

The synthetic utility of radical benzylation of quinones has been significantly increased by Citterio who showed that either homogeneous or two-phase reaction conditions can be employed (equations 215, 216)[206]. With 2,5-dimethyl-1,4-benzoquinone and 1,4-naph-

$$\text{1,4-benzoquinone} + ArCH_2Y \xrightarrow[\text{2-phase}]{Ag^+} \text{2-(CHArY)-1,4-benzoquinone} \quad 70\text{–}90\% \quad (215)$$

$$\text{1,4-benzoquinone} + ArCH_2Y \xrightarrow[\text{homogeneous}]{Na_2S_2O_8} \text{2-(CHArY)-1,4-benzoquinone} \quad 50\text{–}80\% \quad (216)$$

Ar = Ph, 2-Tol, 4-Tol, 4-An, 4-*i*-PrC_6H_4
Y = H, Me, OH, Ac

thoquinone the results are not as promising because of lower yields and less discrimination between mono- and dibenzylation; no unsymmetrical quinones were studied.

The competition between alkyl and aryl radicals has been studied by Citterio[626] and Asmus[627]; these data will be discussed in connection with the arylation of quinones (see Section VI).

A detailed kinetic study of the *n*-decane–dicyclohexylperoxydicarbonate system, which produces *s*-decyl radicals, has been made[207]. The inclusion of di-*t*-butyl and diphenyl quinones makes this treatment an important complement to earlier studies. The rate constants and inhibition coefficients determined are in agreement with analogous studies of simpler systems and account for the widespread use of these quinones as inhibitors in radical polymerizations.

Finally, the extensive use of azobisisobutyronitrile and 1,4-benzoquinones for initiation and inhibition of polymerization reactions has continued to create interest in their interactions[208]. Both non-polar (toluene or chlorobenzene) and polar (acetonitrile) solvents were employed and the nature of the products changed dramatically. The former systems produced simple crystalline products after work-up (equation 217) while the latter

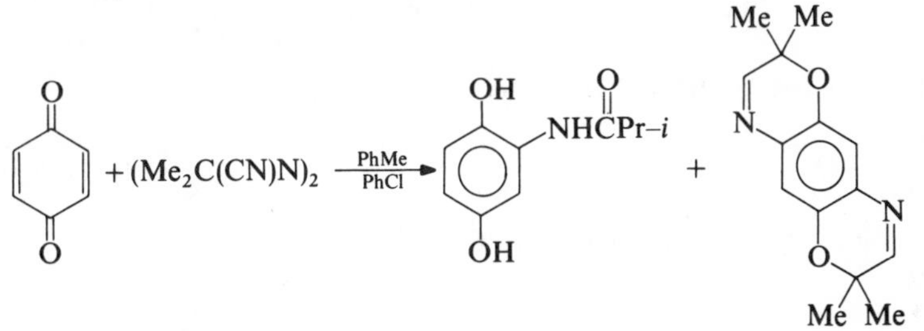

$$+ (Me_2C(CN)N)_2 \xrightarrow[PhCl]{PhMe} \quad \text{OH, OH, NHCOPr-}i \quad + \quad \text{(Me Me, O, N; N, O, Me Me)} \qquad (217)$$

gave only quinonoid resins. In both cases there was no evidence of a nitrile or ether in the products and only hydroquinone–nitrogen bonds could be detected. This evidence requires that the 2-cyano-2-isopropyl radicals react in the dimethylketenimine form (**76**,

$$Me_2\dot{C}CN \longleftrightarrow Me_2C{=}C{=}N^{\cdot} \qquad (218)$$

(76)

equation 218). It also casts serious doubt on the earlier reports of hydroquinone mono- and diethers from this reaction[209].

B. Alkylation with Organotin, Nickel, or Silicon Reagents

In their efforts to synthesize members of the vitamin K and coenzyme Q families, the group of Maruyama, Naruta and Uno, have made a major contribution. The basic chemistry involves alkylation with trialkylallyl or polyprenylstannanes in the presence of a Lewis acid (equations 219, 220)[210, 211]. Good to excellent yields of product are isolated; however, no unsymmetrical quinones were reported. Boron trifluoride etherate is clearly superior to the other Lewis acids examined and starting the reaction at low temperature is also important.

Naruta[210] has provided compelling evidence that the initial product involves 1,2-carbonyl addition (equation 221) followed by a dienone–phenol rearrangement. He also examined the regioselectivity of unsymmetrical allylic groups and found that steric effects in either the quinone or the allylic reagent lead to a marked preference for α-addition (equation 222).

This chemistry has been extended to the synthesis of 4-allyl-1,2-naphthoquinones where the regioselectivity depends on the nature of the quinone 3-substituent (equations 223, 224)[212]. Unlike the 1,4-quinones these substrates show either preliminary 1,2-addition or direct 1,4-addition to the enone (**77**, equation 225). Both electronic and steric factors are significant in determining the preferred course of the reaction.

$$\text{R}_3\text{-1,4-benzoquinone} + Bu_3SnCH_2CH{=}CH_2 \xrightarrow[CH_2Cl_2,\ -78^\circ C/\text{ambient}]{BF_3\cdot Et_2O} \text{R}_3\text{-hydroquinone-}CH_2CH{=}CH_2 \quad (219)$$

$$\text{R}_3\text{-1,4-benzoquinone} + Me_3Sn(CH_2CH{=}C(Me)CH_2)_nH \longrightarrow \text{R}_3\text{-hydroquinone-}(CH_2CH{=}C(Me)CH_2)_nH \quad n = 2\text{–}5$$

R = H, Me, —CH=CHCH=CH—, MeO (220)

$$\text{Me}_3\text{-1,4-benzoquinone} + R_3SnCH_2CH{=}CH_2 \xrightarrow{BF_3\cdot Et_2O} \text{4-HO-4-}(CH_2CH{=}CH_2)\text{-2,3,5-Me}_3\text{-cyclohexa-2,5-dienone } (23\%) + \text{4-HO-4-}(CH_2CH{=}CH_2)\text{-2,3,6-Me}_3\text{-cyclohexa-2,5-dienone } (5\%) \quad (221)$$

$$\text{1,4-benzoquinone} + R_3SnCH_2CH{=}CHMe \xrightarrow{BF_3\cdot Et_2O} \text{2-}(CH_2CH{=}CHMe)\text{-hydroquinone } (\alpha) + \text{2-}(CH(Me)CH{=}CH_2)\text{-hydroquinone } (\gamma) \quad (222)$$

These workers have successfully applied these reactions, with many useful extensions, to the synthesis of a variety of complex, naturally occurring quinones[213–218]. Especially noteworthy contributions include: the inclusion of 1,2-benzoquinones[216], the addition of conjugated dienes without accompanying Diels–Alder chemistry[217] and addition to quinones containing the highly reactive formyl, acetyl, or methoxycarbonyl groups[218].

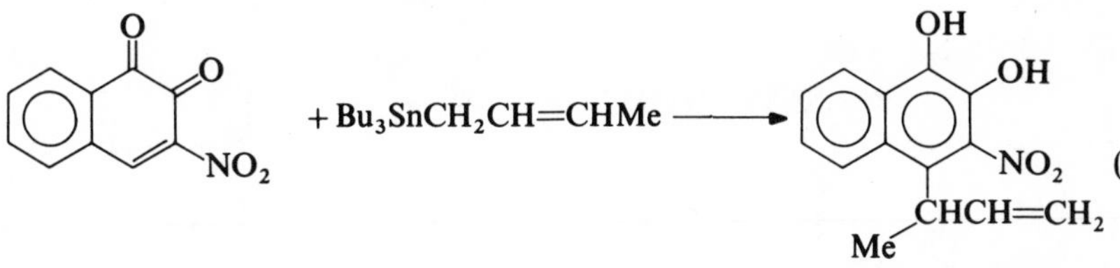

$$\text{1,2-naphthoquinone, 3-NO}_2 + Bu_3SnCH_2CH{=}CHMe \longrightarrow \text{1,2-dihydroxy-3-NO}_2\text{-4-(CH(Me)CH{=}CH}_2\text{)naphthalene} \quad (223)$$

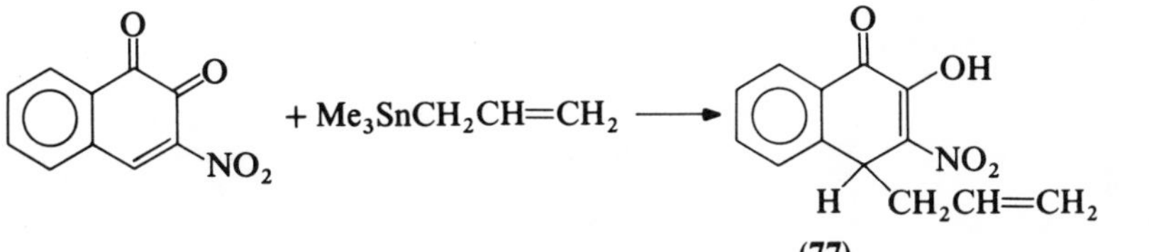

91% (89% total)

$$\text{1,2-naphthoquinone, 3-Me} + Bu_3SnCH_2CH{=}CHMe \longrightarrow \text{1,2-dihydroxy-3-Me-4-(CH}_2\text{CH{=}CHMe)naphthalene} \quad (224)$$

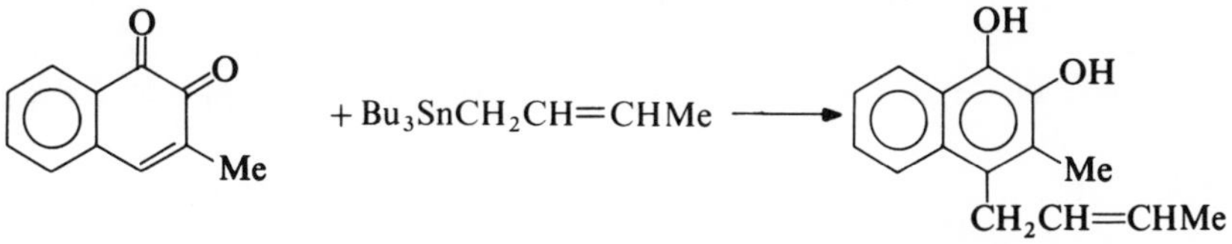

93% (75%total)

$$\text{1,2-naphthoquinone, 3-NO}_2 + Me_3SnCH_2CH{=}CH_2 \longrightarrow \textbf{(77)} \; (\text{OH, NO}_2\text{, H, CH}_2\text{CH{=}CH}_2) + (\text{O, OH, CH}_2\text{CH{=}CH}_2\text{, NO}_2) \quad (225)$$

An interesting modification of these tin reagents is found in the allylation of 2-acetyl-1,4-naphthoquinone with allyltriphenylsilane (equation 226)[219]. The spontaneous cyclization

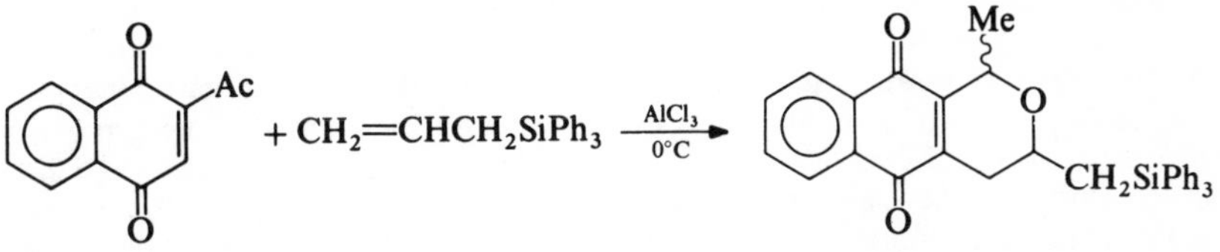

$$\text{2-Ac-1,4-naphthoquinone} + CH_2{=}CHCH_2SiPh_3 \xrightarrow[0^\circ C]{AlCl_3} \text{pyranonaphthoquinone (Me, O, CH}_2\text{SiPh}_3\text{)} \quad (226)$$

61%

of the presumed intermediate makes this reaction of special interest in the synthesis of pyranonaphthoquinones. Hosomi and Sakurai have also investigated the use of silicon compounds in the allylation of quinones (equation 227)[220]. They found average yields of hydroquinone products except for 2,6-dimethoxy-1,4-benzoquinone which gave excellent yields of the more hindered *p*-allylquinol.

An especially intriguing reaction was carried out by Mori and coworkers involving an unconjugated allylic tin diene which also contained a siloxy group (**78**, equation 228)[221].

$$\text{2,3,5,6-}R_4\text{-1,4-benzoquinone} + Me_3SiCH_2CH{=}CH_2 \xrightarrow[\text{0 to } -78^\circ C]{TiCl_4,\ CH_2Cl_2} \text{2,3,5,6-}R_4\text{-... hydroquinone with } CH_2CH{=}CH_2 \quad (227)$$

R = H, Me, MeO, —CH=CHCH=CH—

$$\text{1,4-benzoquinone} + t\text{-}BuMe_2SiOCH_2C(Me){=}CH(CH_2)_2C(Me){=}CHCH_2SnBu_3 \ \textbf{(78)} \xrightarrow{BF_3\cdot Et_2O} \quad (228)$$

$$\text{hydroquinone-}CH_2CH{=}C(Me)(CH_2)_2CH{=}C(Me)CH_2OSiMe_2Bu\text{-}t$$

91%

Tin(II) compounds containing a SnOC system have been prepared and react with 1,4-benzoquinone to give essentially quantitative yields of a reductive addition polymer (equation 229)[222].

$$\text{1,4-benzoquinone} + Sn(OR)_2 \xrightarrow{PhH} \left(Sn(OR)_2O\text{—}C_6H_4\text{—}O \right)_n \quad (229)$$

R = $(CH_2)_2NMe_2$, C(Me)=CHCOMe

Hegedus has examined a complimentary route to allylhydroquinones using π-nickel bromide complexes (**79**, equation 230)[223–225]. In all cases equal amounts of allylation and reduction are observed and quinones with a reduction potential more negative than − 70 V do not react. The correlation of the allylation site with the carbon of highest spin density makes it likely that electron transfer is followed immediately by coordination and alkyl transfer. Convincing evidence for allylquinol intermediates is presented in a subsequent and more detailed report[226]. The properties of these isolated, but relatively unstable, compounds was determined and the range of quinones extended. In conjunction with other alkylation methods the procedure proved to be especially versatile[227]. This chemistry has been applied to the synthesis of some naphthoquinone antibiotics and methods for trapping enolate intermediates with carbon electrophiles developed[228].

C. Alkylation by Radicals from Decarboxylation Reactions

Torssell and Jacobsen have developed a useful method related to the much earlier work of Fieser. Using silver ion and peroxydisulfate they were able to generate alkyl radicals from carboxylic acids (equation 231)[229]. These reactive species are effective alkylation

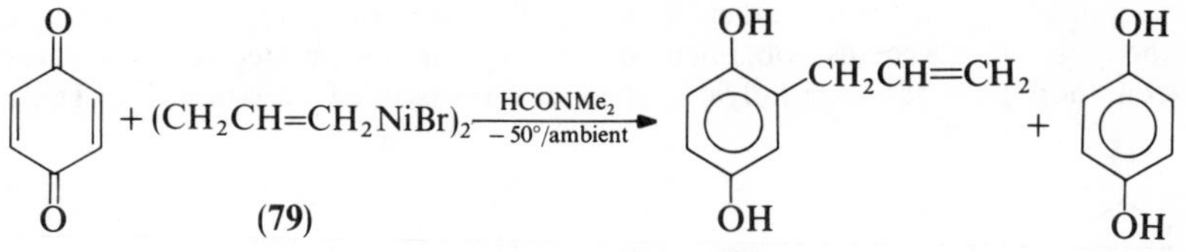

$$\text{1,4-benzoquinone} + (CH_2CH{=}CH_2NiBr)_2 \xrightarrow[-50°/\text{ambient}]{HCONMe_2} \text{2-allylhydroquinone (48\%)} + \text{hydroquinone (47\%)}$$

(79)

(230)

$$S_2O_8^{2-} + Ag^+ \rightarrow Ag^{2+} + SO_4^{2-} + SO_4^{-}$$
$$Ag^{2+} + RCO_2^- \rightarrow Ag^+ + CO_2 + R^{\cdot}$$

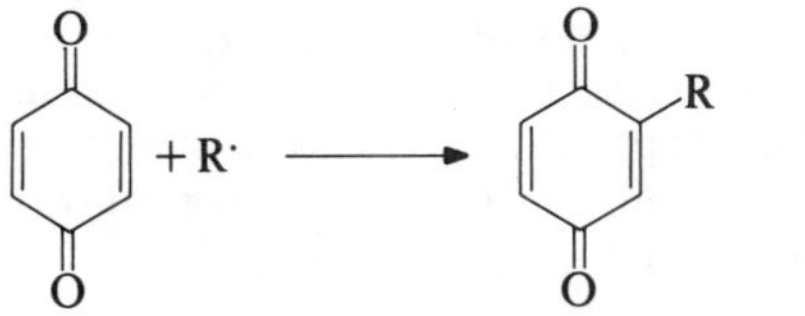

$$\text{1,4-benzoquinone} + R^{\cdot} \longrightarrow \text{2-R-1,4-benzoquinone}$$

40–80% (231)

R = *i*-Pr, *t*-Bu, $EtOCH_2$, Bz, $PhOCH_2$

agents for quinones. Only one unsymmetrically substituted quinone, 2-methyl-1,4-benzoquinone, was reported and it gave roughly random amounts of the three isomeric products.

TABLE 4. Isomer distribution in the radical pentylation of monosubstituted 1,4-benzoquinones[229, 230]

Substituent	2-Pentyl, 3-R	2-Pentyl, 5-R	2-Pentyl, 6-R
Me	25	40	35
MeO	5	68	27

In a later paper they reported the preparation of the natural product primin (**80**, equation 232)[230]. Comparison of the isomeric distribution for monosubstituted 1,4-benzoquinones shows the stronger influence of the methoxy group and indicated the

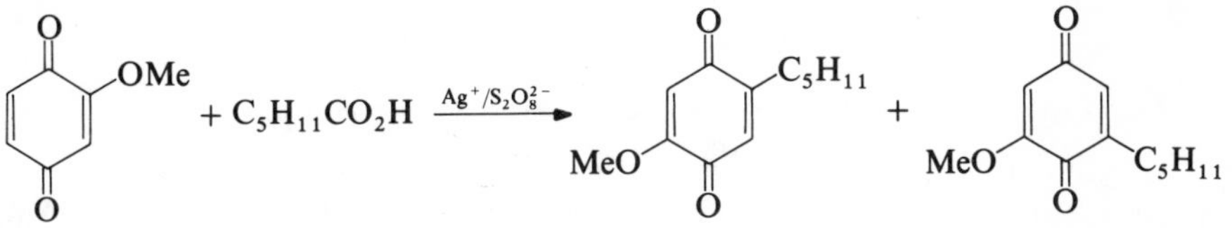

$$\text{2-methoxy-1,4-benzoquinone} + C_5H_{11}CO_2H \xrightarrow{Ag^+/S_2O_8^{2-}} \text{2-}C_5H_{11}\text{-5-MeO-1,4-benzoquinone (34\%)} + \text{2-}C_5H_{11}\text{-6-MeO-1,4-benzoquinone } \mathbf{(80)}\ (13\%)$$

(232)

nucleophilic character of the radical involved (Table 4). Low yields of all three dipentyl 2-methoxyquinones were also obtained. Where the radical generated could be expected to react in more than one form only one product was observed (equation 233). These yields

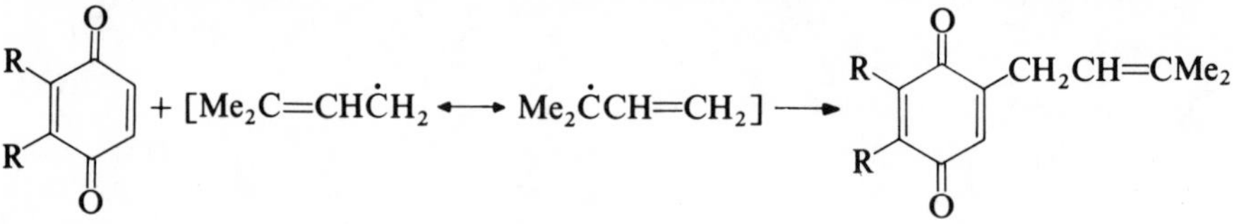

$$\text{R}_2\text{-benzoquinone} + [\text{Me}_2\text{C}{=}\text{CH}\dot{\text{C}}\text{H}_2 \longleftrightarrow \text{Me}_2\dot{\text{C}}\text{CH}{=}\text{CH}_2] \longrightarrow \text{R}_2\text{-2-(CH}_2\text{CH}{=}\text{CMe}_2\text{)-benzoquinone} \qquad (233)$$

R = H, —CH=CHCH=CH— 34%, 58%

leave some questions to be answered as does the observation that the decarboxylation of α, β-unsaturated acids in the presence of quinones gave only recovered starting material.

These studies have been extended to include cycloalkyl radicals where it was found that steric hindrance is extremely important. Fair yields of simple ring substitution products are obtained (equation 234), but only one product of reasonably complicated structure was produced in quite a few attempts (equation 235)[231].

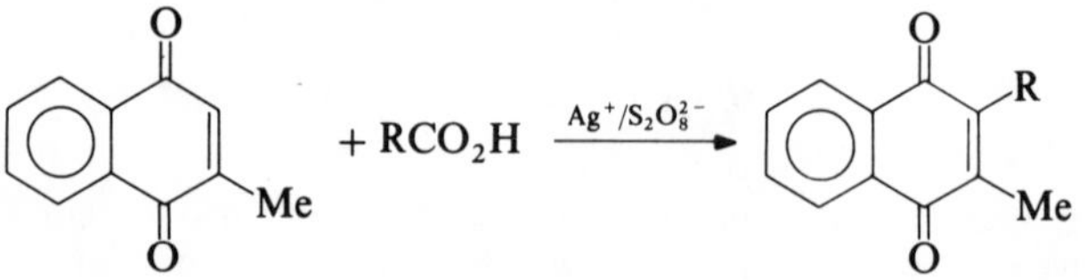

$$\text{2-Me-1,4-naphthoquinone} + \text{RCO}_2\text{H} \xrightarrow{\text{Ag}^+/\text{S}_2\text{O}_8^{2-}} \text{2-R-3-Me-1,4-naphthoquinone} \qquad (234)$$

R = c-Pr, c-Bu, c-Pen, c-Hex 37–56%

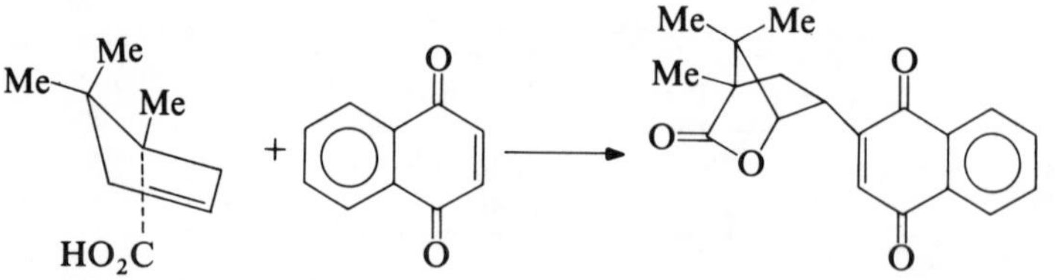

$$\text{(trimethylcyclopentenyl)-HO}_2\text{C} + \text{1,4-naphthoquinone} \longrightarrow \text{(lactone-substituted)-1,4-naphthoquinone} \qquad (235)$$

41% (apparently *endo*)

Efforts to use this radical chemistry for the introduction of chains containing an ester group using the monoesters of oxalic acid met with limited success with naphthoquinone (equation 236)[232]. Such intermediates provide an attractive alternative entry to the naphthacene quinones.

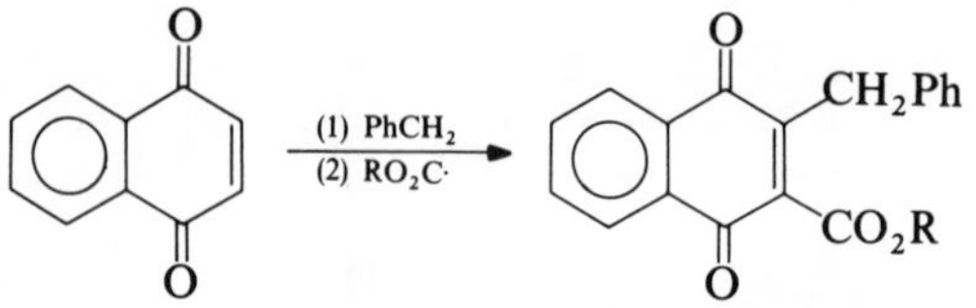

$$\text{1,4-naphthoquinone} \xrightarrow[\text{(2) RO}_2\text{C}\cdot]{\text{(1) PhCH}_2} \text{2-CH}_2\text{Ph-3-CO}_2\text{R-1,4-naphthoquinone} \qquad (236)$$

R = Me, Et 64%, 80%

A variety of natural product syntheses using this decarboxylation chemistry have been reported[233-238]. In general yields have been satisfactory and certain developmental efforts are noteworthy: dicarboxylic acid esters[235], cyclopropyl radicals[236] and additional quinone substituents[238].

D. Quinones and Organoaluminums

Studies with aluminum compounds represent a recent interest. The chief reaction product with most alkylaluminum reactants is the corresponding hydroquinone[239]. Depending on the number of the alkyl groups present in the aluminum compound, ring substitution or monoether formation can occur (equation 237, 238). The authors propose

$$\text{1,4-benzoquinone} + i\text{-Bu}_3\text{Al} \xrightarrow[-78^\circ\text{C}]{\text{Et}_2\text{O}} \text{hydroquinone (25\%)} + \text{2-}i\text{-butylhydroquinone (28\%)} + \text{4-}i\text{-butoxyphenol (0\%)} \quad (237)$$

$$\text{1,4-benzoquinone} + i\text{-BuAlCl}_2 \longrightarrow \text{(50\%)} \quad \text{(0\%)} \quad \text{(15\%)} \quad (238)$$

and support a carbanionic 1,4-addition mechanism for the carbon–carbon bond product and a radical 1,6-addition for ether formation. With chloranil higher yields of ether are obtained and the product ratio is dependent on both the structure of the aluminum reagent and the temperature (Table 5)[240]. Traces of ring substitution products are obtained. This aspect of the chemistry has been expanded to include a variety of alkyl groups and quinones (Table 5)[241]. The clear trends show yields dependent on the electron affinities of the quinone and that methylaluminum dichloride leads only to reduction.

TABLE 5. Yields of reduction and 4-alkoxyhydroquinones from the reactions of alkylaluminums with 1,4-benzoquinones[240, 241]

Quinone	Aluminum	T (°C)	Hydroquinone (%)	Ether (%)
Unsubst.	Et_3	−78	64	6
Unsubst.		0	56	41
Unsubst.		25	65	18
Unsubst.		100	86	0
Unsubst.	Et_2Cl	−78	51	27
Unsubst.	$EtCl_2$	−78	23	53
Unsubst.	$BuCl_2$	−78	35	43
Unsubst.	i-$BuCl_2$	−78	52	12
2,5-Cl_2	$EtCl_2$	−78	9	59
2,6-Cl_2			12	60
2,3,5-Cl_3			13	67
2,3,5,6-Cl_4			23	58
1,4-Naphtho-			26	44

Alberola and his coworkers have made extensive investigations of the interaction of organoaluminum compounds and quinones[242]. They have added a great deal of detail to our understanding of this chemistry and shown that in certain instances useful yields of ring alkylation can be obtained[243]. In addition these workers have examined triphenylaluminum[244] and tribenzylaluminum[245], finding similar chemistry and raising questions about its possible future development along synthetic lines.

E. Miscellaneous Metal Catalyzed Alkylations

Tachibana studied the synthesis of vitamins K_1 and K_2 using phytyl and geranyl chlorides and various metallic dusts (equation 239)[246]. He found that good yields (60–

$$\text{2-methyl-1,4-naphthoquinone} + ClCH_2CH{=}C(Me)R \xrightarrow{\text{metal dust (Zn)}} \text{2-Me-3-}(CH_2CH{=}C(Me)R)\text{-1,4-naphthoquinone} \qquad (239)$$

$R = C_6H_{11}, C_{16}H_{33}$

65%) of the desired product can be obtained with zinc in tetrahydrofuran and that no reduction of the quinone occurs. The by-product from reaction at the 2-methyl position is well limited in this process. Finally, the order of reactivity of various metal dusts is: Zn > Pd > Fe > Ce > Sn > Cu.

Further application of this chemistry has been reported and a variety of naturally occurring quinones prepared[247]. The range of quinone substituents was increased and while the yields are not great the starting materials and process make it an attractive approach.

Good to excellent yields of tocopherols, including vitamin E, are obtained using copper–zinc powder in the presence of formic acid (equation 240)[248]. Various isoprenyl alcohol-related compounds were used.

$$\text{2,6-R}_2\text{-3-Me-1,4-benzoquinone} + CH_2{=}CHC(Me)(OH)((CH_2)_3CH(Me))_nMe \longrightarrow \text{chromanol (HO, R, R, Me; 2-Me, 2-}((CH_2)_3CH(Me))_n\text{Me)} \qquad (240)$$

$n = 0–3$
$R = H, Me$

The use of zinc-amalgam or palladium have both been proposed for the prenylation of 1,4-benzoquinones or their corresponding hydroquinones[249]. Yields appear to be comparable to the other related studies already cited, but the possibility of working with either the oxidized or reduced starting material suggests this approach should be examined further.

Palladium has been shown to be an effective catalyst for the methylation of quinones by methylcobalt complexes[250]. The authors claim yields as high as 70%, but 30% or less is the rule.

A particularly exciting application of organolithium chemistry has been proposed and deserves more extensive study and application (equation 241)[251]. The oxidation of 2-bromo-1,4-dimethoxybenzene provides a good route to the quinone ketal (**81**) which reacts smoothly with lithium compounds.

(81) + BuLi $\xrightarrow{HCONMe_2}$ (Li derivative) $\xrightarrow{RCOY}$ (CR(OH)Y derivative) $\xrightarrow{HCl/H_2O}$ (quinone with CR(OH)Y) or hydrolysis product (241)

$Y = H, R', OR, Cl, NR_2$

F. Hydroboration of Quinones

Kabalka has extended his earlier study to include chloro, cyano, methoxy, and ester groups in the alkylating agent (equation 242)[252]. The yields of substituted hydroquinone

$$\text{benzoquinone} + B((CH_2)_nY)_3 \xrightarrow[\text{trace } O_2]{H_2O} \text{2-}(CH_2)_nY\text{-hydroquinone} \qquad (242)$$

$n = 3, 4, 10$; $Y = Cl, CO_2Et, CO_2Ph, CN, MeO$

are all nearly quantitative, but the syntheses of the required trialkylboranes are subject to electronic effects which produce isomers. The products are generally easy to separate by gas chromatography and the difficulty can be minimized using one of several alternative hydroborating agents available.

An attempt to prepare 3-substituted 2,5-dihydroxy-1,4-benzoquinones by hydroboration showed that the quinone rather than hydroquinone is produced (equation 243)[253].

$$\text{2,5-dihydroxy-1,4-benzoquinone} + R_2B\text{-cyclohexyl} \xrightarrow[-60°C]{HCONMe_2} \text{3-cyclohexyl-2,5-dihydroxy-1,4-benzoquinone} \qquad (243)$$

The yield of the desired product is low as a result of polyalkylation and hydroxyl group displacement. However, the presence of a large excess of trialkylborane in dimethylformamide or protection of the hydroxy groups gave good results. The latter approach resulted in the reduced product.

The hydroboration procedure is also applicable to the preparation of mole scale amounts of alkylhydroquinones[254]. Yields of 70 to < 90% were obtained for alkylations involving ethyl through dodecyl chains. Several branched chain groups were also prepared; however, it should be noted that both 2,4,4-trimethylpentene-1 and 2,4,4-trimethylpentene-2 gave the 1-pentyl product. Much longer reaction and workup times are required for the higher homologs.

V. CYCLOADDITION TO QUINONES

A. The Diels–Alder Reaction

1. Theoretical studies

In recent years there has been a great deal of interest in the synthesis of quinonoid natural products and the Diels–Alder reaction has been employed to great advantage. This utility has had the happy consequence of supporting a continued high level of interest in the mechanistic details of this family of reactions. As a result of these two concurrent efforts, in several cases by the same research group, one finds quinones playing a role in the theoretical study of the process in general. The number of literature references to quinones in Diels–Alder reactions is so large and diverse that many difficult choices had to be made. In spite of efforts to select the most detailed and significant studies the reader must refer to the usually extensive citations in the original papers to be sure of seeing all relevant material.

Houk and Paquette presented thoughtful and readable accounts of their efforts in cycloaddition chemistry[255,256]. Houk's interest is in general much more directly concerned with quinone chemistry while Paquette has simply used quinones where appropriate in his emphasis on gaining full appreciation of Diels–Alder chemistry. While Houk's review was published much earlier it is especially valuable as an introduction to the concepts of frontier molecular orbitals as an approach to understanding chemical processes. In their more recent publications both authors have essentially left quinone chemistry in an effort to resolve fine, perhaps important, details concerning stereoselectivity in cycloadditions[257,258].

Of greatest significance in quinone–cycloaddition chemistry are the studies by Houk and his collaborators which deal directly with theoretical applications. The unifying potential of frontier molecular orbital theory has already been mentioned in discussing the reductive Michael addition so typical of the quinones[1]. The basic assumption is that reactivity depends on the interaction of the lowest unoccupied molecular orbital (LUMO) of the quinone and the highest occupied molecular orbital (HOMO) of the nucleophile[255]. The closer these orbitals are in energy, the more strongly they interact. The computational results are presented in detail and comparisons with published results, both experimental and theoretical, are made along with predictions for unexamined cases. The prediction that substituent effects can be transmitted through many bonds is coupled with an implied plea for more carefully organized studies.

Shortly after the appearance of this general study Houk and his collaborators published the results of their efforts to account for cycloaddition reactions in which the products represent regioselectivity opposite to that predicted by consideration of LUMO coefficients[259]. Three different dienes reacted with 2,5-dimethyl-1,4-benzoquinone (equations 244–246) and showed similar product compositions also with 2,6-dimethyl-, 2,5-, or 2,6-dimethoxy-1,4-benzoquinone and 2-methyl- or 2-methoxy-1,4-naphthoquinone. Houk argues that since all of these examples lead to the more nucleophilic end of the diene reacting preferentially at the less highly substituted carbon of the quinone it is likely that

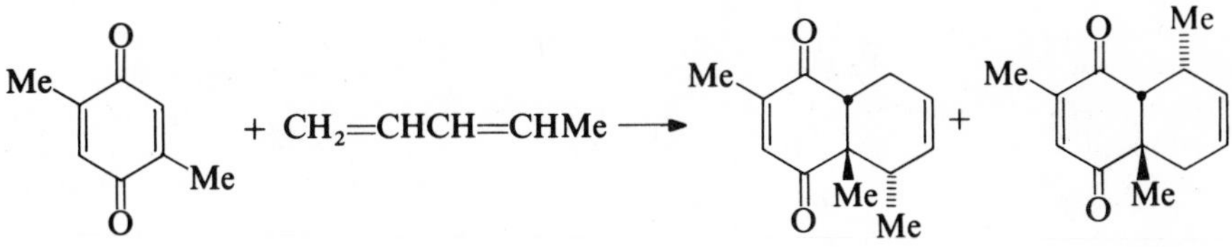

O Me O Me O
Me Me Me
$$+ CH_2{=}CHCH{=}CHMe \longrightarrow$$ +
Me
O Me O Me O
Me

5 : 1 (75%) (total yield)

(244)

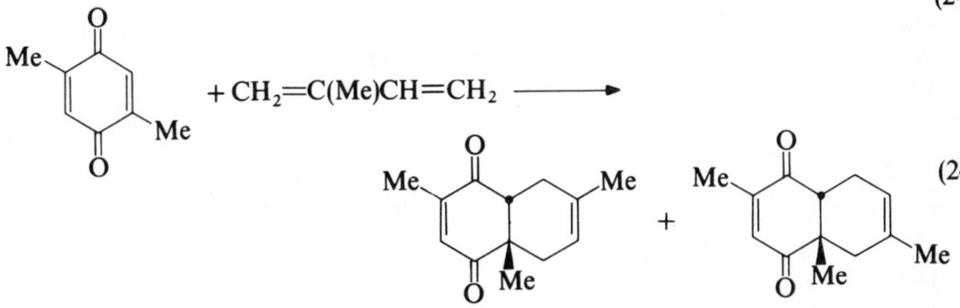

O
Me
$$+ CH_2{=}C(Me)CH{=}CH_2 \longrightarrow$$
Me
O
O O
Me Me Me
+
Me
O Me O Me

1 : 1 (70%)

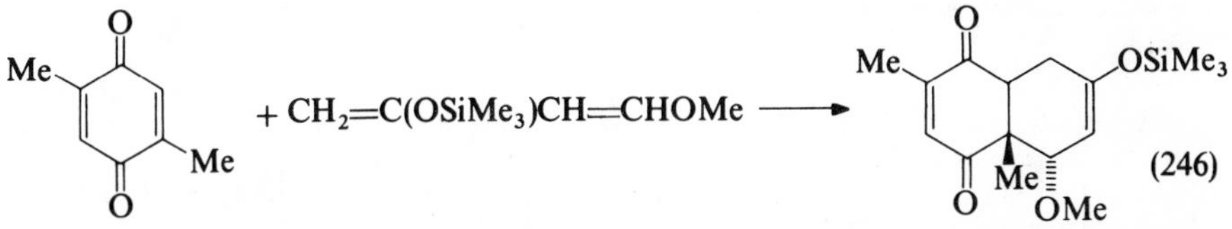

(245)

O
Me O
Me OSiMe₃
$$+ CH_2{=}C(OSiMe_3)CH{=}CHOMe \longrightarrow$$
Me
Me
O O OMe

90%

(246)

secondary orbital repulsive effects offer the most attractive rationale for the experimental observations. These ideas are generalized and the applicability of the frontier molecular orbital approach expanded in a subsequent publication[260].

Others have examined the importance of secondary orbital overlap in Diels–Alder reactions; for example, Ginsburg and his coworkers have added specific evidence from the reaction of propellanes (**82**, **83**) with 1,4-benzoquinone (equation 247)[261]. In contrast to

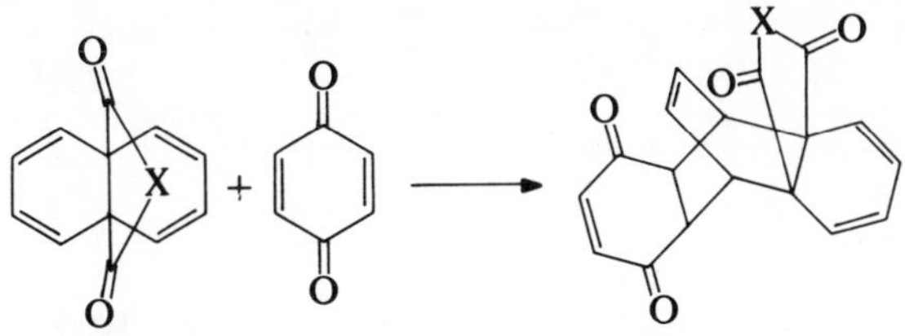

X O
O O
O O
X + ⟶
O O O

(247)

(**82**, X = O) 68%,
(**83**, X = NMe) 88%

N=N dienophiles which add to the *syn*-face of compounds like **82** and **83**, the C=C system of quinones add exclusively to the *anti*-face.

Kanematsu and his collaborators made an extensive examination of the kinetics of phencyclone (**84**) with various dienophiles including 1,4-benzoquinone (equation 248)[262].

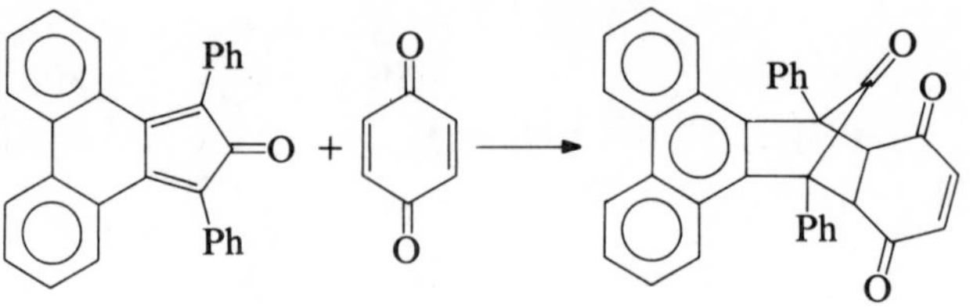

(248)

(84)

They classify those reactions in which the HOMO–LUMO separations are similar and either electron attraction or release should increase reactivity as neutral or type B. They too draw the conclusion that secondary orbital interactions are important, especially in cases like this where the planarity of the diene is nearly perfect.

In his early work Houk stated that Lewis acid-catalyzed Diels–Alder reactions promote stronger secondary orbital interactions[263]. However, it has been found that with the rather unusual diene spiro[bicyclo[2.2.2]octane-2,1′-[2,4]cyclopentadiene] (**85**) no change in the *syn*/*anti* ratio occurs in the two reactions (equation 249)[264]. The rates of the two processes

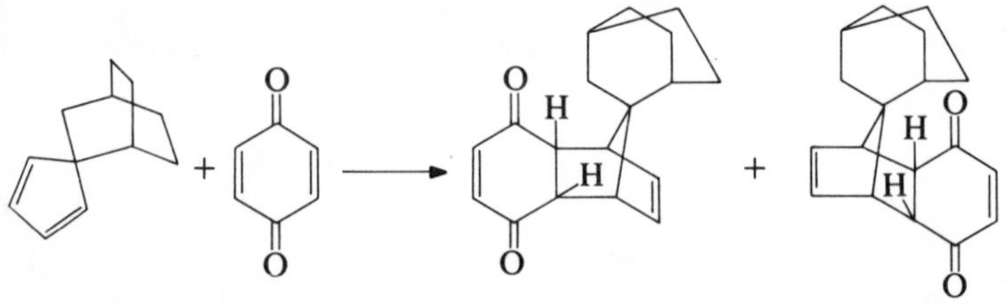

(249)

(85)	thermal	86	:	14 (100%)
	catalysed	90	:	10 (95%)

do change dramatically; several hours at 0 °C versus 30 minutes at −78 °C in the presence of boron trifluoride etherate. The authors point out that their results do not necessarily demand more than a modification of Houk's hypothesis. They suggest that the *endo* reaction transition state may simply lie further along the reaction coordinate than the corresponding *exo* reaction.

The effect of solvent on the rates of the Diels–Alder reaction of 2,3-dimethyl-1,3-butadiene and 1,4-naphthoquinone (equation 250) has been studied[265]. Seventeen solvents

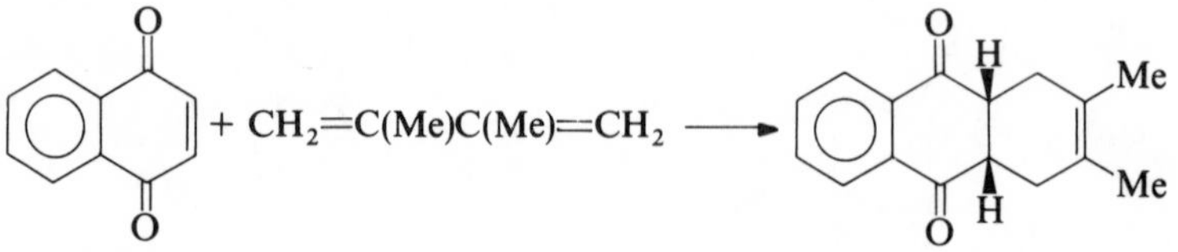

$+ CH_2{=}C(Me)C(Me){=}CH_2 \longrightarrow$

100% (250)

ranging from cyclohexane to acetic acid were used and all of the traditional correlation techniques failed to rationalize the small (31-fold) change observed. The much less widely applied donor/acceptor interactions of Mulliken not only served to explain the data, but the associated acceptor numbers correlate with the LUMOs of the solvent. The frontier molecular orbitals interactions of these LUMOs and the HOMO of the quinone leads to a qualitative explanation of the rate increases. A detailed treatment of the hyperbolic shape of the plot of log second-order rate constants versus acceptor numbers is also given. This study appears entirely consistent with a reaction in which weakly polar reactants, transition states and products are dominant.

In a related study a comparison of 1,4-benzoquinone with several other dienophiles also ruled out a zwitterionic intermediate; the evidence involving substituent and solvent effects favored a concerted mechanism[266]. By contrast a less detailed examination of the reaction of 1,3-butadiene with 1,4-naphthoquinone was interpreted as favoring a charge separated transition state[267].

In an effort to apply the logic of isoselective relationships to the Diels–Alder reaction, competitions between 1,4-benzoquinone and maleic anhydride were carrried out with dienes representing normal and reverse electron demands[268]. The evidence shows clearly that these reactions do possess a common isoselective relationship and therefore a common reaction mechanism.

The instability of simple quinonoid systems toward water is well documented and may be part of the reason so few representatives are known beyond those contained exclusively in a single six-membered ring. Boldt and his coworkers used perturbation molecular orbital theory to examine both of these questions[269]. On the basis of their theoretical studies a 1,7-naphthoquinone was selected for preparation and a 13-step synthesis was carried out! This quinone, the 3,7-di-*t*-butyl-8-methyl derivative, proved to be unstable, thus supporting the thesis that the alkyl substituents which provide shielding from water also cause great steric strain.

2. Stereochemistry and catalysis

In addition to the considerable theoretical interest in the Diels–Alder reaction there has continued a steady stream of more experimentally oriented studies relating to the details of this important process. For example, various questions concerning stereochemical requirements are of central concern for the application of cyclization reactions to natural product synthesis. It has been observed that Lewis acids can change the product orientation dramatically and that reaction under high pressure provides a route to otherwise inaccessible compounds. The chemistry of quinones makes one aware of the modern trend to all but erase any clear distinction between mechanistic and synthetic studies.

Jurczak has shown that quinones greatly restrict freedom in the transition state and enhance the utility of asymmetric induction as a synthetic tool[270]. In the reaction of *d,l*-1′-acetoxyethyl-1,4-benzoquinone with 2,3-dimethyl-1,3-butadiene, cyclization takes place in a completely regiospecific manner and the two diastereomers are produced in a ratio of 70:30, i.e. one is produced in 40% excess (equation 251). Reaction with 1,3-cyclohexadiene

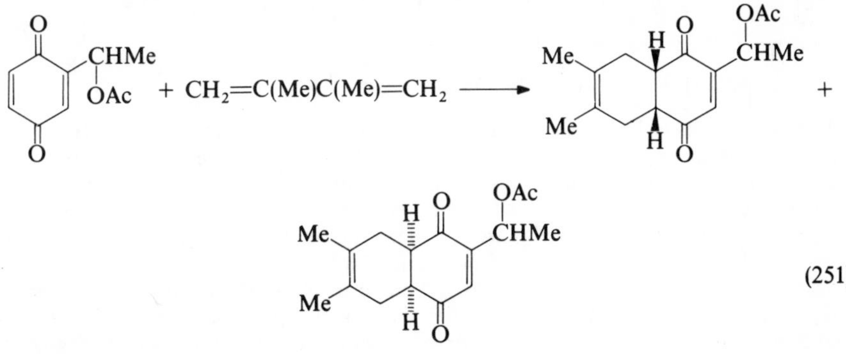

$$\text{quinone-CH(Me)OAc} + CH_2{=}C(Me)C(Me){=}CH_2 \longrightarrow \text{adduct (H, H)} + \text{adduct (H, H)} \qquad (251)$$

gave identical results. The application of these observations to a model proposed earlier is discussed[271].

In the study just described the author points out that such differentiation is sensitive to the reaction pressure. Along with their coworkers Jurczak and Eugster have shown that 1,4-benzoquinones react smoothly with 3,4-dimethoxyfuran to give a highly stereospecific product (equation 252)[272]. With increasing pressure the *endo/exo* ratio decreased

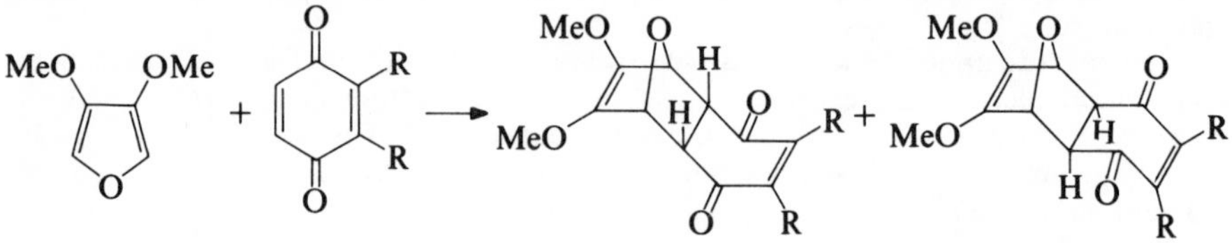

$t = 25°C$.

	R = H, H *endo/exo*	R = H, Me *endo/exo*
7 kbar	89/11	89/11
11 kbar	83/17	76/24
19 kbar	54/46	68/32

(252)

dramatically. Raising the temperature produced an even larger change in the same direction, but the relationship between the two was not investigated. The situation for 2,3-dimethyl- and 2,3-dimethoxy-1,4-benzoquinones appears still more complicated and needs further work. The former gave only a 46% yield of the *exo* isomer at the lowest pressure and was not investigated further. The latter showed only a very small change in product composition with changing pressure. These reactions were first reported by Eugster and coworkers and apparently only the isomer which crystallized was considered[273]. A kinetic study shows that substituents in the furan ring play an important role in determining the thermodynamic stability of the adducts formed[274]. Activation energies for retro-Diels–Alder reactions of such adducts show the rather small effect of substituents in the quinone ring and make clear the necessity of the high pressure modification for success in their synthesis.

Application of the Diels–Alder reactions at high pressure can provide a synthesis of unstable adducts such as those of simple 1,4-benzoquinones with furan (equation 253)[275]. The two stereoisomers have different stabilities, but neither survive 12 hours at 5°C.

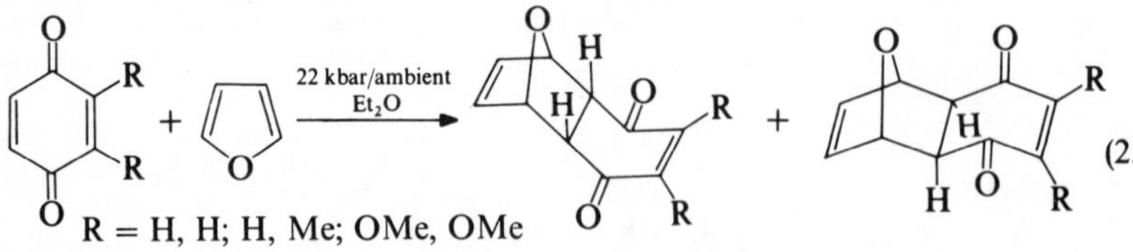

22 kbar/ambient Et_2O

R = H, H; H, Me; OMe, OMe

(253)

Dauben and Baker have also examined Diels–Alder chemistry at high pressures for the synthesis of heat-sensitive quinones[276]. Of particular interest is the use of electron-poor dienes and the avoidance of aromatization (equation 254). Unless the diene has an unfavorable stereochemistry or an electron-withdrawing substituent at its terminus, adequate to excellent isolated yields are obtained. The *t*-butyldimethylsilyloxy group is an exception (84%), but it required a 65-hour reaction time.

Dauben and Bunce have applied this technique to the synthesis of quinonoid chiral esters (equation 255)[277]. The bulk of the ester is the principal determinant of the enantiomeric excess obtained. A variety of examples representing intermediate degrees of

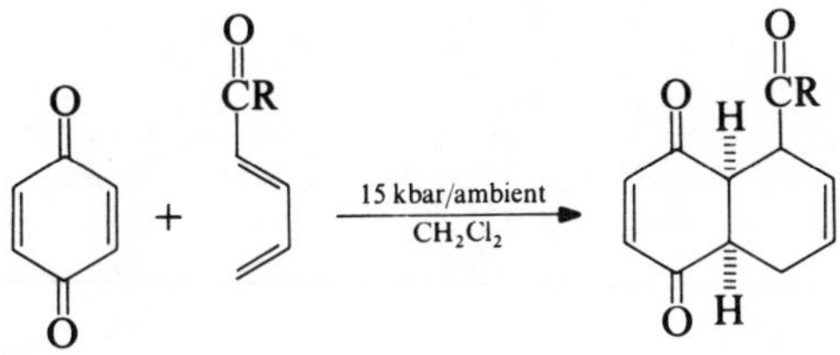

$$\text{p-benzoquinone} + \text{CH}_2\!=\!\text{CH-CH}\!=\!\text{CH-COR} \xrightarrow[\text{CH}_2\text{Cl}_2]{\text{15 kbar/ambient}} \text{adduct} \qquad (254)$$

R = OMe, OEt, (−)-menthyl

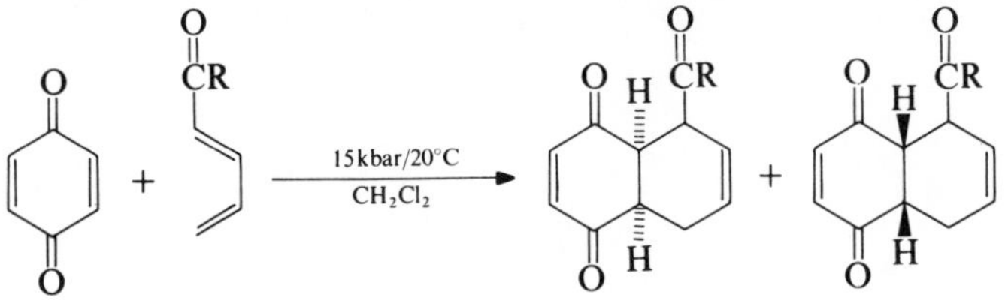

$$\text{p-benzoquinone} + \text{CH}_2\!=\!\text{CH-CH}\!=\!\text{CH-COR} \xrightarrow[\text{CH}_2\text{Cl}_2]{\text{15kbar/20°C}} \text{adduct} + \text{adduct} \qquad (255)$$

R = (+) O—C(H)(Me)CH$_2$Ph 98% yield 2% enantiomeric excess

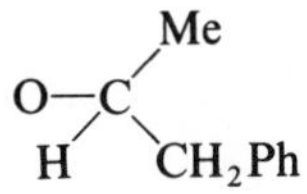

(−) 8-phenylmenthyl (O—C$_6$H$_9$(Me)—C(Me)$_2$Ph) 62% yield 50% enantiomeric excess

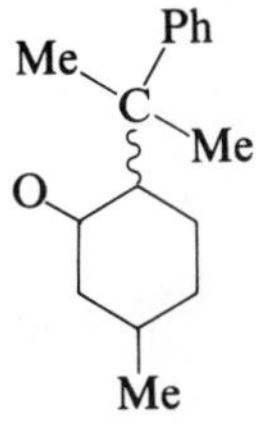

asymmetric induction are given. In all cases the adducts could not have been prepared by classical methods.

Trost and his coworkers have given an especially clear example of the control possible in molecules of great significance to the synthesis of natural products (equation 256)[278]. This

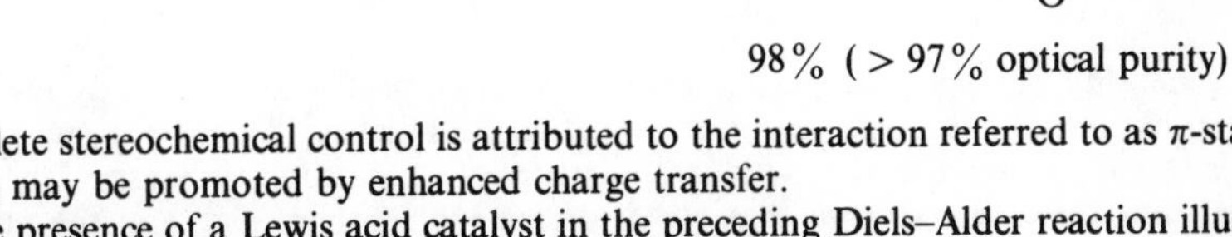

$$\text{5-hydroxy-1,4-naphthoquinone} + \text{CH}_2\!=\!\text{CH-CH}\!=\!\text{CH-OC(O)-C(H)(OMe)Ph} \xrightarrow[\text{CHCl}_3/0°\text{C}]{\text{BF}_3\cdot\text{Et}_2\text{O}} \text{adduct} \qquad (256)$$

98% (> 97% optical purity)

complete stereochemical control is attributed to the interaction referred to as π-stacking which may be promoted by enhanced charge transfer.

The presence of a Lewis acid catalyst in the preceding Diels–Alder reaction illustrates what may well be the most important single synthetic discovery in the quinonoid field during the past decade and a half. Valenta and coworkers reported that in the presence of boron trifluoride etherate a key step in the synthesis of a steroid showed almost complete

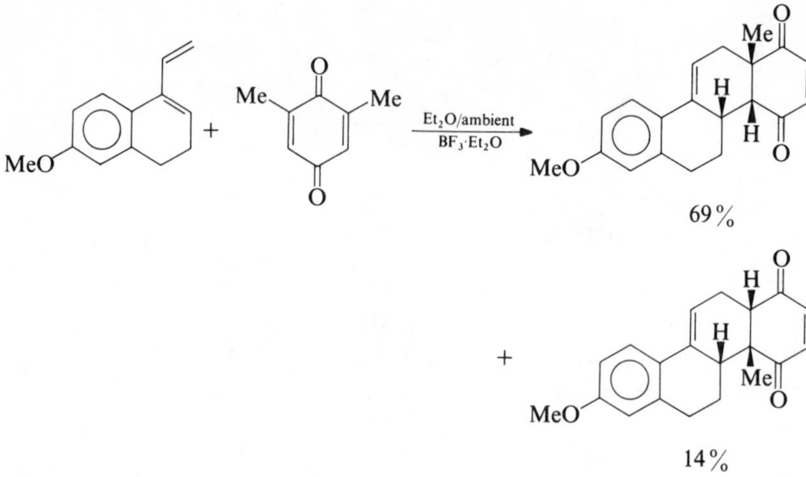

MeO Me Me O O Et$_2$O/ambient BF$_3$·Et$_2$O Me O Me H H O MeO

69% (257)

\+ H O H Me Me O MeO

14%

reversal of product orientation (equation 257)[279]. The minor product is the only isomer obtained from reaction in refluxing benzene. In this first publication the authors indicated that similar exciting results had been found for an example in the saturated steroids (equation 258). No yields were given and the structures were only proposed, but neither of the products is found in the uncatalyzed reaction.

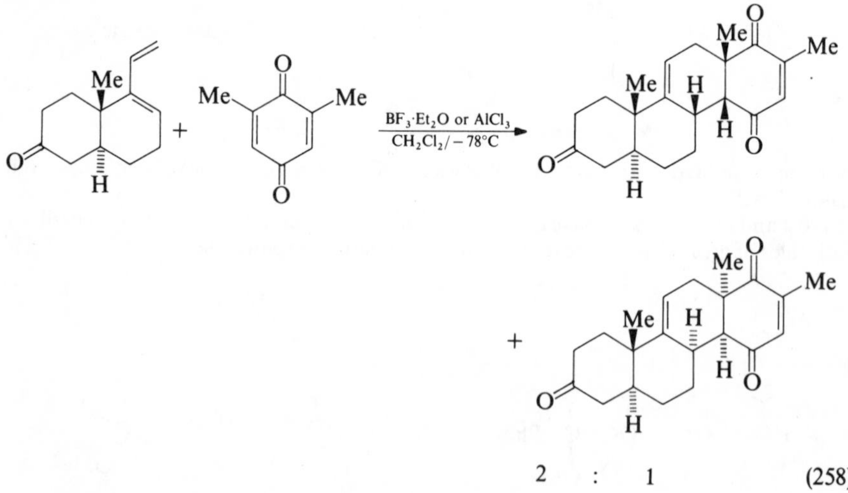

Me O H + Me Me O O BF$_3$·Et$_2$O or AlCl$_3$ CH$_2$Cl$_2$/−78°C Me O Me Me H H O O H

\+ Me O Me Me H H O O H

2 : 1 (258)

The generality of such great regioselectivity was demonstrated by a study involving simple (equation 259) as well as complex (equation 260) dienes[280]. In the latter example both the starting material and product are mixtures of stereoisomers. No specific figures are cited, but the authors claim that all yields are higher than 80% and the product represents complete reversal of orientation. They have rigorously demonstrated the stereochemistry and advance tentative arguments based on the frontier molecular orbital model.

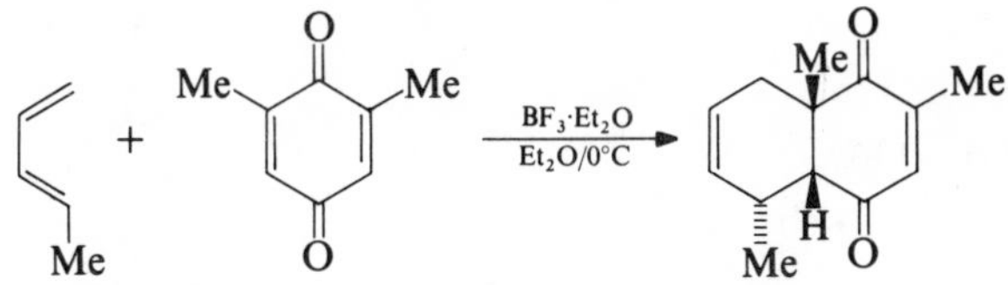

(259)

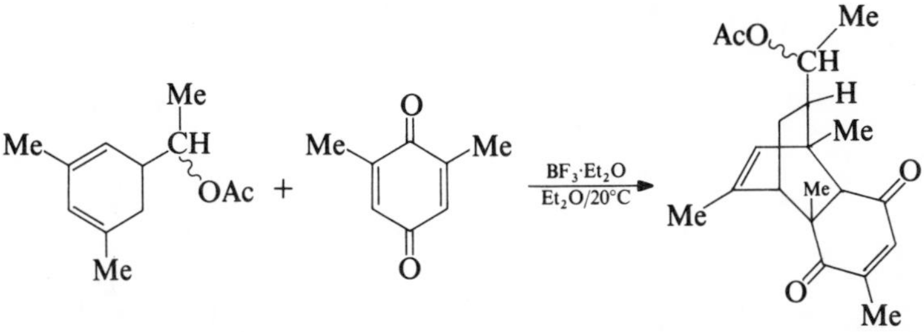

(260)

The chemistry just described is not as completely stereoselective with monosubstituted 1,4-benzoquinones, but even there the shift in product orientation is impressive (equation 261). The choice of solvent, with or without catalyst, provides 60–70% yields of either isomer.

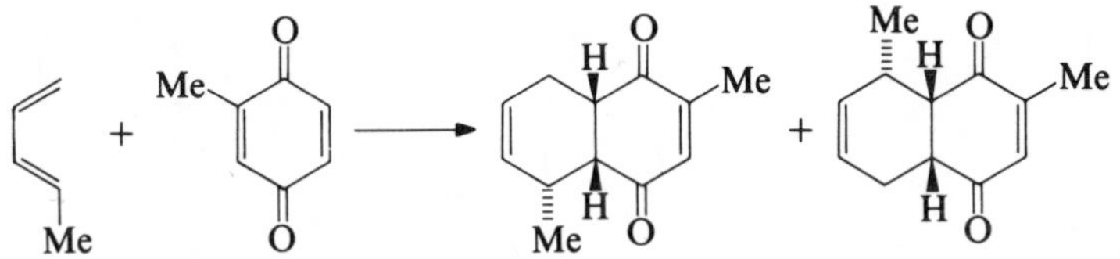

(261)

toluene or acetone	2	:	1
dioxane	1	:	1
toluene/$BF_3 \cdot Et_2O$	1	:	2

Valenta and his collaborators have used their method to advantage in several natural product syntheses. For example, the molecule quassin (**86**) presents a serious stereo-

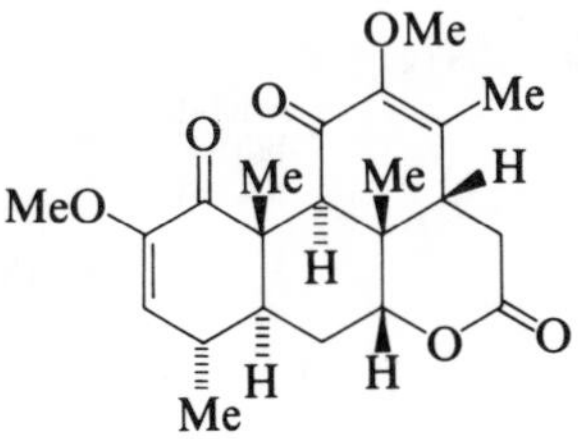

(**86**)

chemical challenge with seven chiral centers[281]. The application of a Lewis acid catalyzed Diels–Alder reaction gave an excellent yield of **87** which shows exactly the opposite stereochemistry to that predicted and obtained from the thermal reaction (equation 262).

A later publication gives a brief overview of the utility of Diels–Alder based steroid syntheses followed by an impressive set of examples of the complementary possibilities of

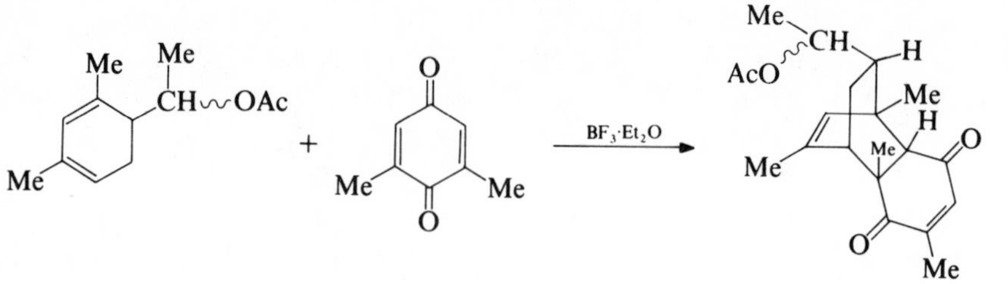

(87) 80% (262)

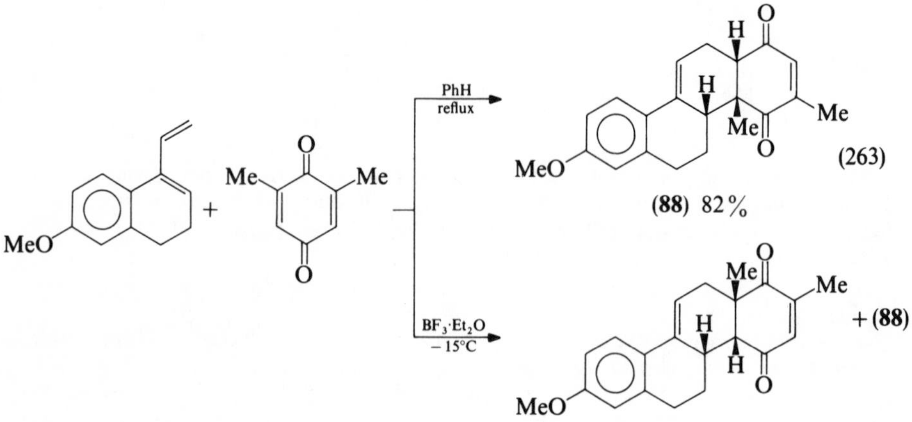

(88) 82% (263)

69% 11% (264)

the thermal and catalyzed reactions (equations 263, 264)[282]. One compound, 2-methoxy-6-methyl-1,4-benzoquinone proved to be an exception in that under both reaction conditions a much lower yield of the thermal product is obtained (equation 265).

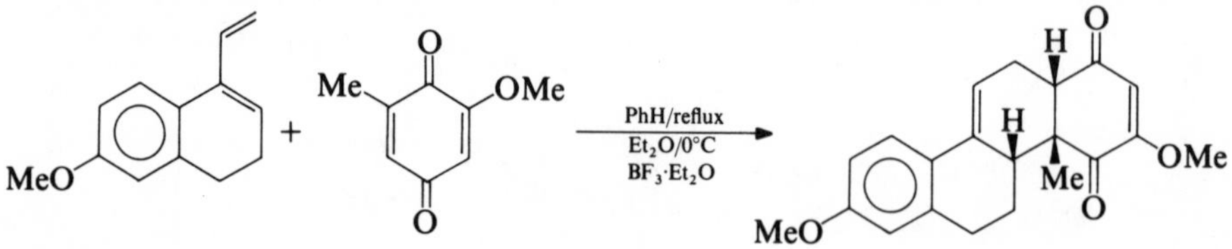

48%, 31% (265)

A variety of fragmentary experimental evidence exists and the authors discuss the problems involved in interpretation of these extraordinary results. The overall conclusion is that a great deal of additional detailed research is required before any useful mechanistic conclusions can be reached. Unfortunately, the expected later studies have not yet appeared, but the synthetic applicability of the reaction is clear and available.

With methyl-methoxy quinones the choice of Lewis acid catalyst can have an impressive effect on the product structure (equation 266)[283]. These variables were also applied to the

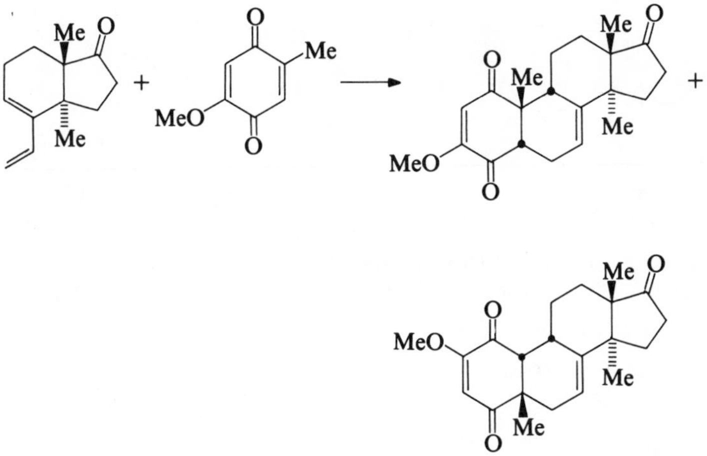

(266)

PhH/reflux	24%	16%
$SnCl_4/CH_2Cl_2/0\,°C$	0%	75%
$BF_3.Et_2O/-16\,°C$	55%	< 10%

reactions of 2-methoxy-6-methyl-1,4-benzoquinone with piperylene or isoprene with very similar results.

Several research groups have been active in relating these catalyzed Diels–Alder reactions to a variety of synthetic problems. A great deal of important chemistry has been produced and several mechanistic proposals have been put forward, but the debate has hardly reached a point of substantial agreement. For the moment the best approach seems to be to collect the experimental data and await the design of critical experiments.

Trost and his collaborators have been among the most active investigators of the potential of the two paths of Diels–Alder chemistry combining them with their own elegant studies of the utility of sulfur and oxygen for regiospecific control[284]. Their work, like much of recent chemistry in this field, has involved the naturally occurring quinone juglone (**89**) as a substrate (equation 267). Thus, with oxygen alone in the diene the Lewis acid

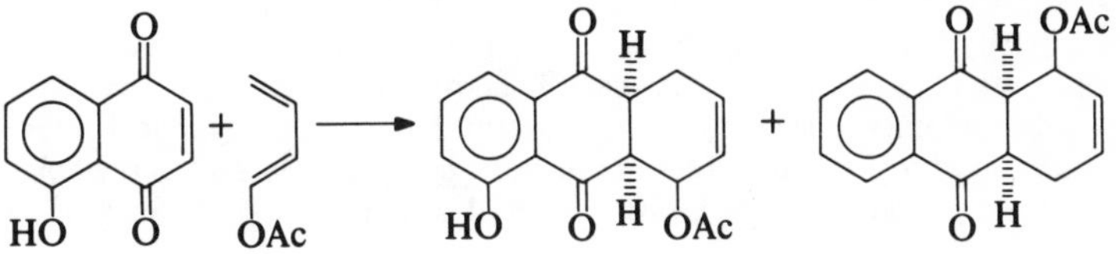

(**89**)

(267)

thermal addition	3	:	1
$BF_3.Et_2O$	> 20	:	1

catalyst greatly enhances the formation of the major isomer. When a diene substituted with both oxygen and sulfur is employed thermal addition shows a marked preference for sulfur control of the regiochemistry, but a Lewis acid can completely reverse the product distribution (equation 268). When these observations are coupled with the wealth of sulfur chemistry available the procedure becomes extremely versatile.

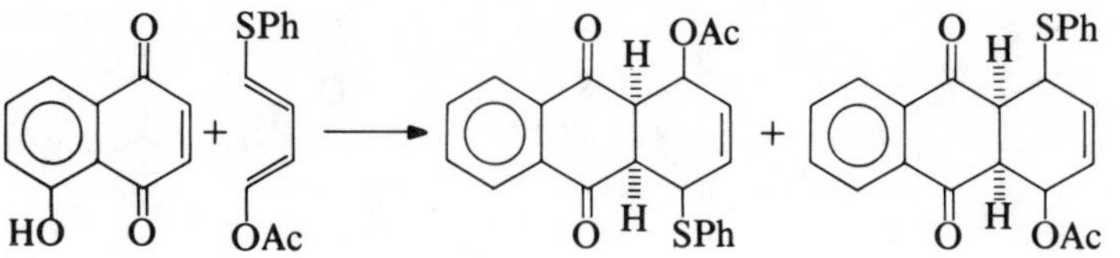

(268)

thermal addition	2.3	:	1
$BF_3.Et_2O$			sole product

The use of a 2,3-substituted 1,3-butadiene allows the same extraordinary control in both thermal and catalyzed reactions (equation 269). In a subsequent full paper these

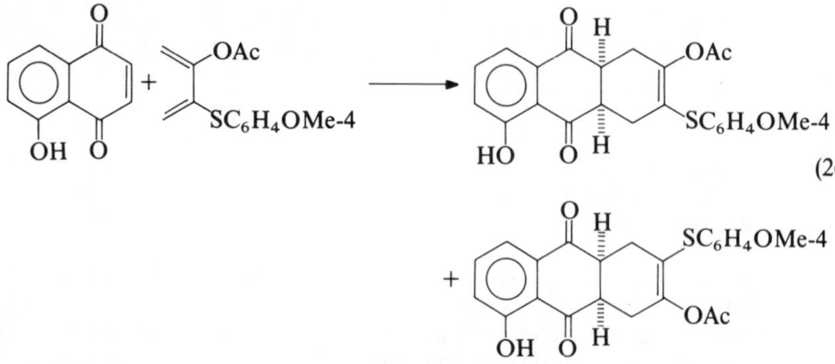

(269)

thermal addition	1	:	2
$BF_3.Et_2O$	sole product		

observations are described in greater detail and the product structures assigned unambiguously[285]. The synthetic result is clear; the minor isomer of the thermal reaction (oxygen control) becomes the exclusive product of the catalyzed reaction.

The original ideas advanced by Valenta[280] dealt with the effect of catalyst on the quinone dienophile while the evidence advanced by Trost places emphasis on the diene structure. Boeckman and his collaborators have examined the importance of the diene and catalyst in Diels–Alder reactions of juglone and its derivatives[286]. Like Trost, they find that the polarity of the diene is of the greatest importance in regiochemical control. Like all workers in this area, they conclude that the complete explanation is very complex and much additional experimentation is required. Of immediate practical interest is their observation that for juglone and certain dienes a complete reversal of regiochemistry can be obtained by a change in the Lewis acid catalyst (equation 270). Such an observation

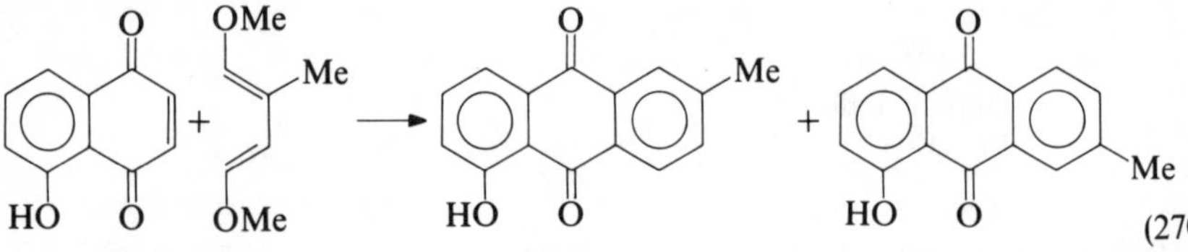

(270)

thermal addition	1	:	1.8
$BF_3 \cdot Et_2O$	1	:	4.5
$AlCl_3$	4.7	:	1

recalls the work of Tou and Reusch[283]. Both Trost and Boeckman have expanded their investigations to much more complex dienes and applied the results to the synthesis of tetracyclic antibiotics[287, 288].

The synthesis of these important pharmaceuticals with their complex functionality and stereochemistry has occupied the time and talents of several research groups over the past decade. One of the pioneers is Kelly who, with his collaborators, has significantly enhanced our practical and theoretical knowledge of quinone–Diels–Alder chemistry.

In early work[289] Kelly suggested that the regiochemical outcome of the reaction of 5-acetoxy-1,4-naphthoquinone (acetyl juglone) with 1-acetoxy-1,3-butadiene (equation 271)

1 : 3 (271)

can be rationalized on the basis of resonance electron donation. While the idea led to useful predictions[290,291] it is not in accord with the Hammett treatment showing the acetoxy group as electron-withdrawing. This situation resulted in a series of experiments designed to remove the substituents in question to a more remote site and avoid alternative explanations based on direct interactions (equation 272)[292]. The results shown are clearly

$Me_2C=O\text{-}d_6$, 25°C (272)

R		
H	75	25
Me	75	25
Ac	60	40

consistent with the acetoxy group behaving as an electron donor, a result which is supported by a more refined treatment of its electronic properties. The authors also document the fact that 5-acetoxy-1,4-naphthoquinone shows identical regiospecificity in this reaction. Extensive additional experimental support and applications published by Kelly's group allow one to follow the development of these ideas and to imagine useful new extensions of them[293–299].

Another impressive example of the change in regiochemistry made possible by selecting the appropriate catalyst is found in the work of Stoodley and coworkers (equation 273)[300]. They also repeated the work of Boeckman[286] using boron triacetate and observed a complete reversal of orientation (equation 274).

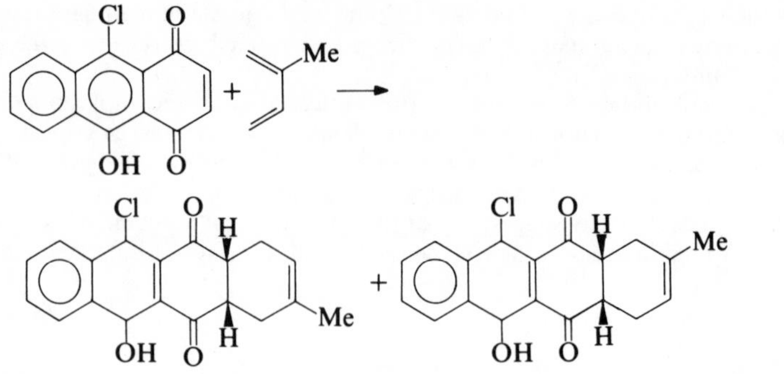

(273)

PhH/reflux	1	:	1	(88%)
$BF_3 \cdot Et_2O$	sole product			(70%)
$AlCl_3$	sole product			(68%)
$B(OAc)_3$	1	:	10	(72%)

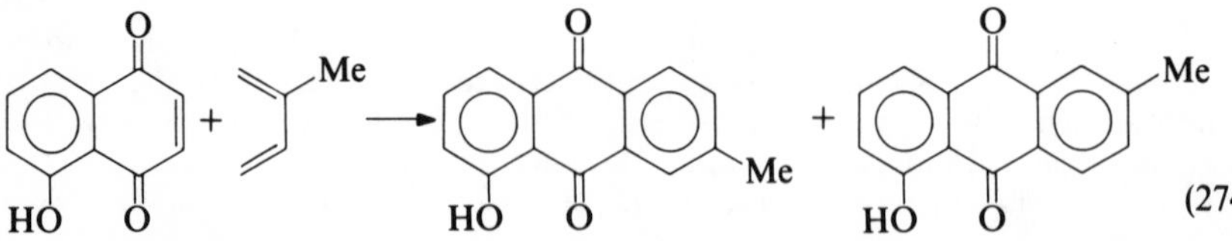

(274)

$BF_3 \cdot Et_2O$	8	:	1 (66%)
$B(OAc)_3$	1	:	7 (63%)

Olah has employed the resinsulfonic acid Nafion-H as a catalyst for Diels–Alder cyclizations (equation 275)[301]. The excellent yields and easy workup make this an

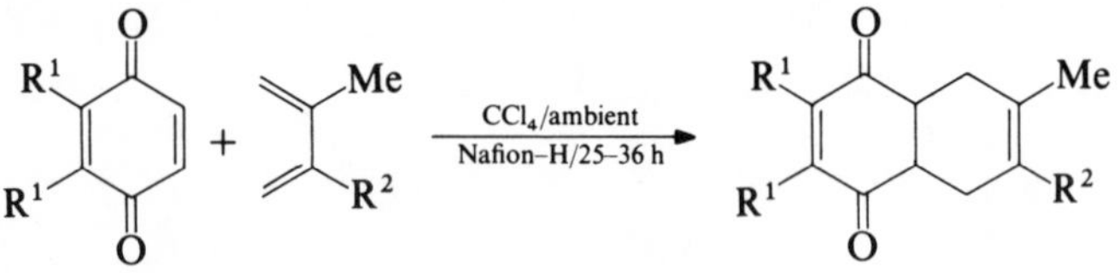

CCl_4/ambient
Nafion-H/25–36 h

(275)

R^1 = H, ; R^2 = H, 80%
–CH=CHCH=CH– ; Me 93%

especially attractive reaction for further study in spite of the fairly long reaction times required. Several dienophiles, including 1,4-benzoquinone, react with anthracene in high (92%) yield. Such products have importance in the study of stereochemical questions by NMR spectroscopy[302].

The need for reactive dienes which can easily be converted to oxygen functions is of major synthetic importance in preparing many classes of natural products. Danishefsky's group has made investigations of silyl enol ethers in Diels–Alder reactions. His account[303] is an interesting survey of these important reagents but fails to mention their relationships to quinone chemistry. A limited number of quinonoid reactions is presented by Brownbridge in his extended synthetic review[304].

Danishefsky and Kitahara described the requirements of an improved diene, prepared *trans*-1-methoxy-3-trimethylsilyloxy-1,3-butadiene (**90**) and demonstrated its reaction with 1,4-benzoquinone (equation 276)[305].

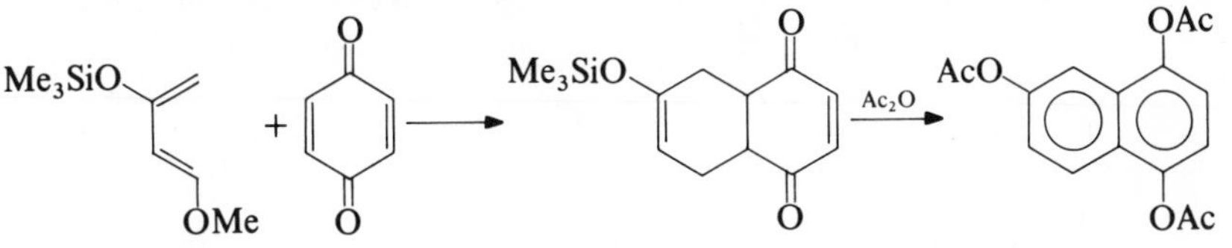

(**90**) 87% (276)

Later, Brassard and his group prepared 1,1-dimethoxy-3-trimethylsilyloxy-1,3-butadiene (**91**) and some longer chain derivatives[306]. They found that reaction with 2-bromo-5-chloro-8-hydroxy-6-methyl-1,4-naphthoquinone took place rapidly and cleanly (equation 277). After pyrolysis and hydrolysis, 87% yield of the corresponding

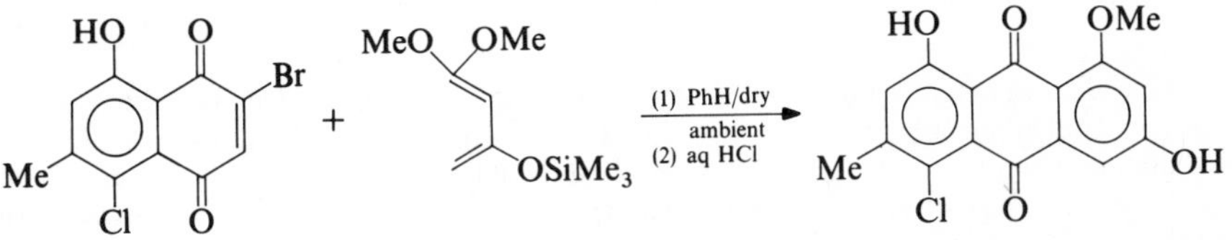

(1) PhH/dry ambient (2) aq HCl

(**91**) 87% (277)

anthraquinone was obtained. The presence of the hydroxy group is not necessary and the corresponding methyl ether formed its anthraquinone in 90% yield. Failure to heat strongly prior to hydrolysis leads to a mixture of alcohol addition products. Three naturally occurring anthraquinones were prepared in excellent yields using this diene and its ethyl and butyl homologs. Subsequently, Danishefsky and coworkers were making a more detailed study of the range of Diels–Alder chemistry possible with this new reactive diene[307]. Reaction with 1,4-benzoquinone continued to support the general observation that the initial adducts are unstable and best isolated after conversion to an aromatic derivative (equation 278). Two competitive experiments showed that this diene is more reactive than the monomethoxy compound reported earlier[305].

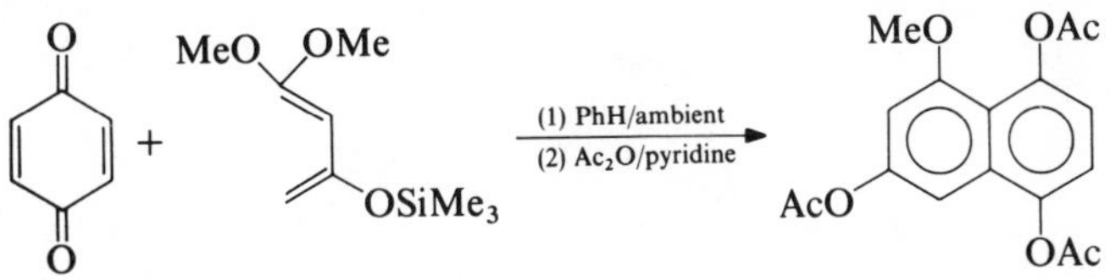

(1) PhH/ambient (2) Ac_2O/pyridine (278)

78%

If the intermediate adduct first formed is alkylated with a bulkier group such as isopropyl or benzyl, the major product retains one free hydroxy substituent (equation 279)[308].

Further work by Danishefsky and his colleagues greatly expanded the practical application of this synthetic method without introducing novel quinonoid chemistry[309–313]. They have also been able to demonstrate that in certain instances enol ether

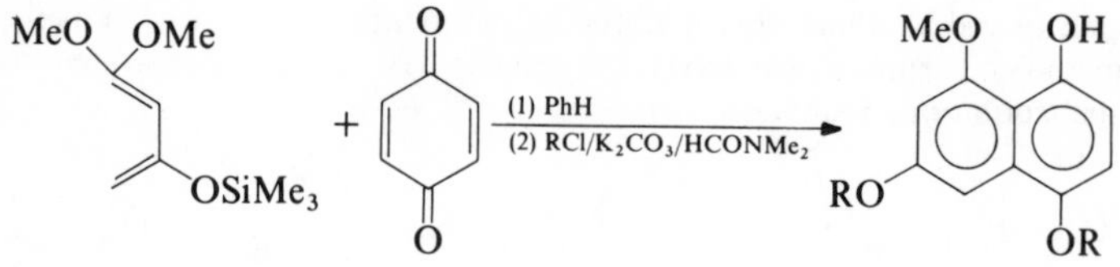

MeO OMe, OSiMe$_3$ + benzoquinone $\xrightarrow[\text{(2) RCl/K}_2\text{CO}_3\text{/HCONMe}_2]{\text{(1) PhH}}$ MeO, OH, RO, OR (279)

R = *i*-Pr, Bz 69%, 42%

dienes can display extraordinary regioselectivity (equation 280)[314]. Many other workers have recognized the importance of polymethoxy or polysiloxy dienes in Diels–Alder chemistry[315–330].

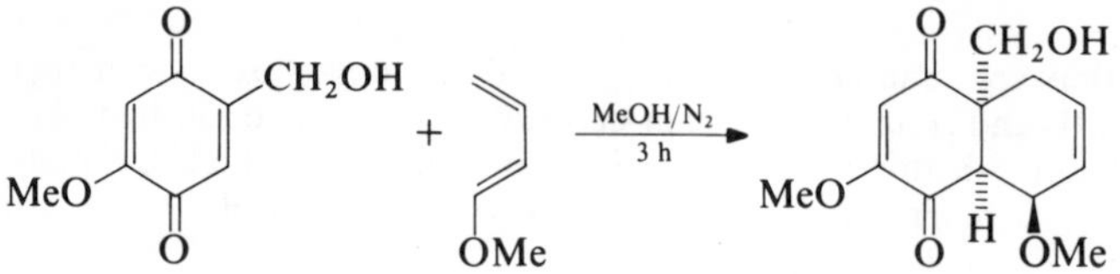

MeO, CH$_2$OH quinone + OMe diene $\xrightarrow[\text{3 h}]{\text{MeOH/N}_2}$ CH$_2$OH, MeO, H, OMe (280)

94%

Brassard and his collaborators developed different routes and explored modifications of the silyl enol ethers. One major thrust of their research has been the exploration of ketene acetals (see Section V.C.1) and this has often led them to examine silylated derivatives. One such reagent has proven especially useful in synthetic efforts directed toward the anthragallols (equation 281)[331]. These compounds are somewhat less sensitive than the

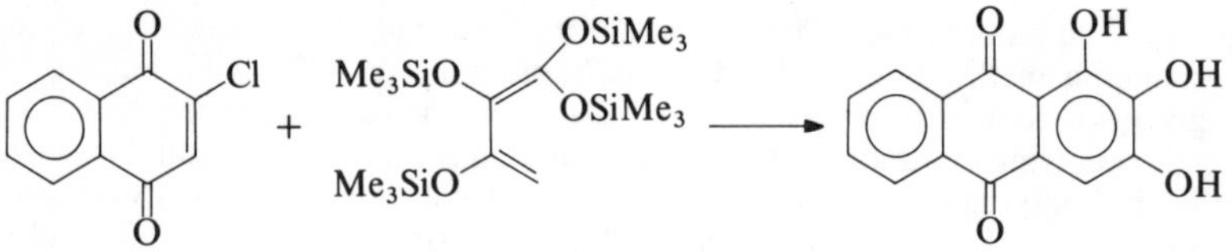

Cl naphthoquinone + Me$_3$SiO, OSiMe$_3$, OSiMe$_3$, Me$_3$SiO ⟶ O, OH, OH, OH, O

82% (281)

corresponding naphthoquinones which could not be obtained by this chemistry. The reaction takes place with great regiospecificity as indicated by the synthesis of the tetramethyl ether of the natural product copareolatin (**92**, equation 282). This reagent and

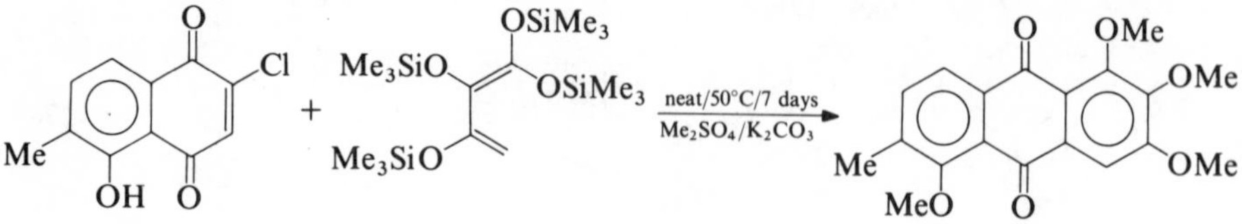

Cl, Me, OH naphthoquinone + Me$_3$SiO, OSiMe$_3$, OSiMe$_3$, Me$_3$SiO $\xrightarrow[\text{Me}_2\text{SO}_4\text{/K}_2\text{CO}_3]{\text{neat/50°C/7 days}}$ O, OMe, OMe, OMe, Me, MeO, O

(**92**) (282)

the analogous 2-methoxy derivative were employed in the synthesis of a variety of naturally occurring anthraquinones[332]. This work also allows some useful comments on the factors affecting the reaction mechanism. The yields and ease of reaction for various halojuglones studied present a consistent picture of the electronic demands of these reactions. While no analysis is given it seems clear that the trends from these resonance arguments correspond closely to those of frontier molecular orbital calculations, a comparison outlined fully by Houk (see Section V.A.1).

A major shortcoming of the siloxylated 1,3-butadienes is their failure to provide an entry to the naphthoquinones. While they apparently react easily with 1,4-benzoquinones, the hydrolysis to highly hydroxylated naphthoquinones leads to extensive decomposition. In a series of explorations Brassard and his collaborators have steadily improved the efficiency of these synthons by modifying the balance between silyloxy and methoxy substituents in the 1,3-butadienes[333-338]. There are still limitations to be overcome[339], but the great progress made is documented by the vastly improved syntheses of a broad range of naturally occurring quinonoid compounds. A number of other research efforts have recognized the importance of Brassard's contributions[340-342].

Fleming and his collaborators[343], while their major interest has not been in quinone reactions, also provided a great deal of insight to these reactions; see especially the detailed survey of allyl compounds[344].

3. *Synthetic survey*

The amount of synthetic studies in the past 15 years is such that a systematic treatment is not possible within the limitations of this chapter. Also, these reports are so diverse that their proper contribution to understanding the mechanistic implications demand many carefully constructed and controlled future studies. In order to promote such work the available chemistry has been grouped in what seem to be logical subdivisions coupled with illustrative equations and with leading references.

a. *Pyrones as dienes*

Several groups, most notably that of Jung and his collaborators[345-347], have used various alkylated and alkoxylated derivatives of these lactones in effective approaches to the anthracycline antitumor antibiotics (equation 283)[348].

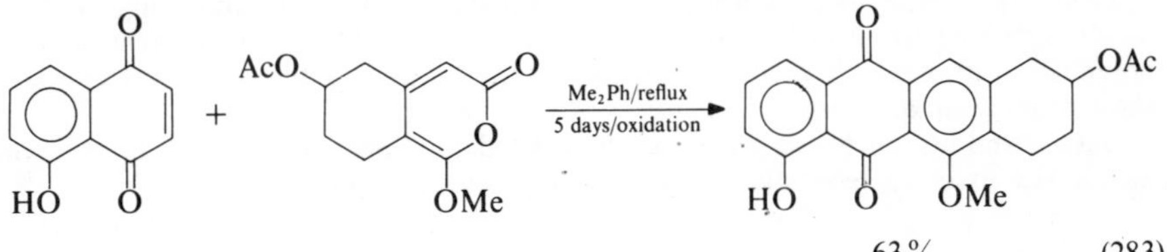

Me_2Ph/reflux, 5 days/oxidation

63% (283)

b. *Heteroatom bridges from furans and thiophenes*

As a complement to their extensive investigations of the cycloaddition chemistry of phenylcyclone[349,350] Sasaki and his coworkers have studied the related furotropone (**93**, equation 284)[351]. Warrener[352,353], Rickborn[354] and Smith[355,356], with various collabor-

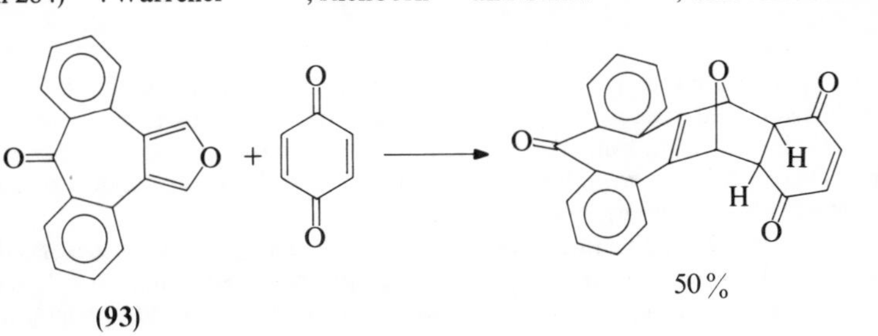

50% (284)

(**93**)

ators, have greatly expanded the range of isobenzofuran cycloaddition chemistry. Still other studies have involved the analogous sulfur heterocycles[357,358] and the *in situ* generation of the furan[359]. One interesting report concerns the formation of a cyclopropane ring in an apparently similar reaction (equation 285)[360].

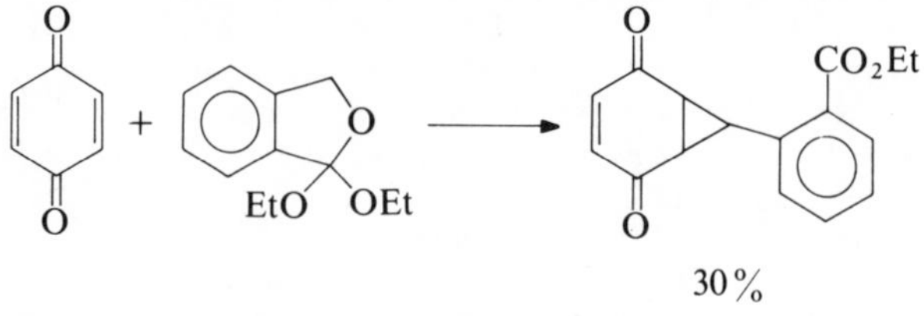

(285)

30%

c. Exocyclic dienes

The polycyclic structure of so many natural products makes the ability to create a new ring with predetermined structure particularly attractive. Thus, the application of exocyclic dienes in Diels–Alder reactions has attracted a great deal of interest. Of special importance are recent investigations of *o*-quinodimethanes generated from benzocyclobutenes and related compounds (equation 286)[361–365].

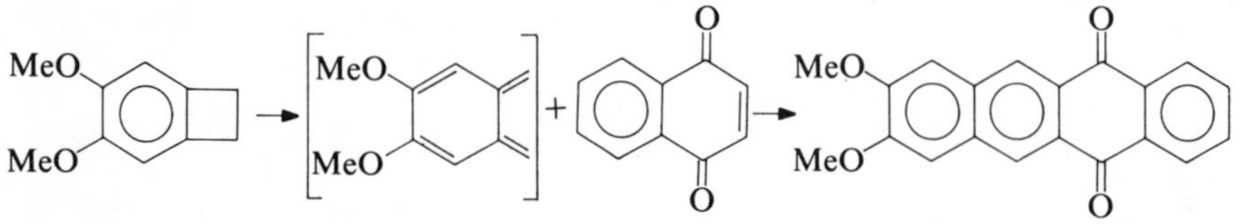

75% (286)

Yields are generally much lower than this example, and some odd competing reactions do occur, but the variety of functionality that can be introduced in a single step make this a potentially important route. Other methods for the generation of the reactive intermediate show promise of good regioselectivity[366].

Gesson and his coworkers have used siloxylated enol ethers in the form of exocyclic dienes with excellent results in the synthesis of anthracyclines (equation 287)[367–369].

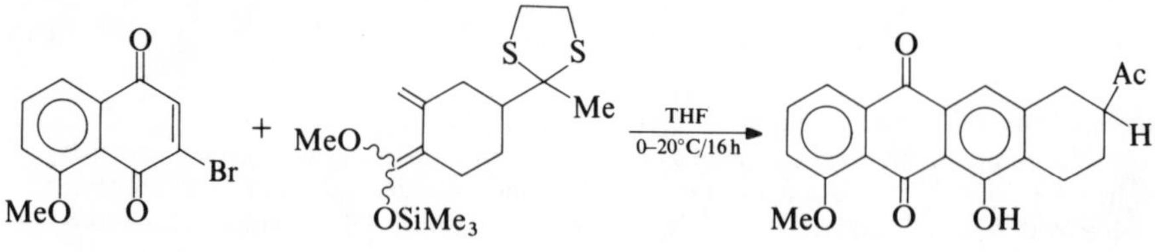

71% (287)

A variety of related dienes have been prepared and used in imaginative ways[370–377]. The cyclization of diynes with dicyclopentadienyltitanium leads to five-, six-, or seven-membered rings bearing both the exocyclic diene system and various other functional groups. These products with 1,4-naphthoquinone allow the production of some very interesting products (equation 288)[378].

A final point in the application of exocyclic dienes to Diels–Alder syntheses is the report by Scharf and his collaborators which shows the importance of 1,4-distances in the reactivity of such compounds[379]. The reaction rates, activation parameters, ionization

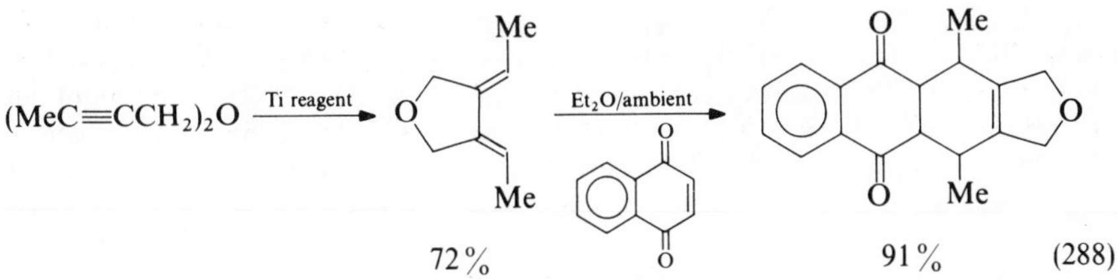

$$(MeC{\equiv}CCH_2)_2O \xrightarrow{\text{Ti reagent}} \quad \xrightarrow{Et_2O/\text{ambient}}$$

72% 91% (288)

potentials, and molecular geometries combined to reduce the importance of the dihedral angle in such systems. While quinones played only a secondary role in these studies, the implications for them are clear.

d. Cyclic dienes

For most of the reasons cited in the preceding section the Diels–Alder chemistry of conjugated alicyclic dienes occupies an extremely important position in modern natural product synthesis. These substrates have also been frequently involved in more theoretical investigations. The simplest representative of the series, cyclobutadiene, has been generated from its iron tricarbonyl complex and reacts with highly hindered quinones (equation 289)[380]. The tetramethyl analog shows similar reactivity toward quinones[381].

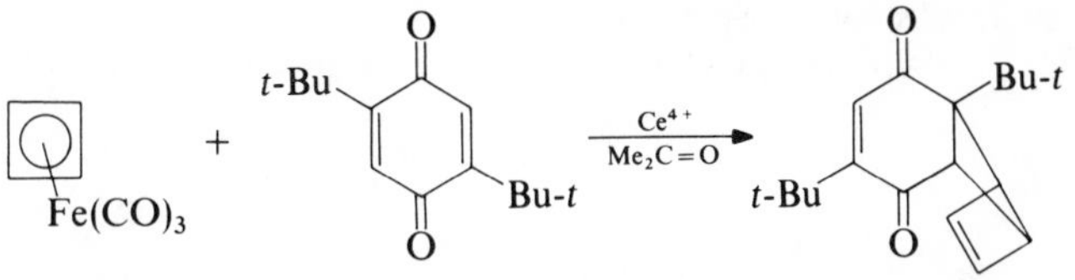

$$Fe(CO)_3 + \xrightarrow[Me_2C=O]{Ce^{4+}} \tag{289}$$

O'Connor and Rosen have applied Diels–Alder reactions of cyclopentadienes bearing alkoxy and chloro substituents to the synthesis of interesting aromatic quinones (equation 290)[382,383]. Yates and Gupta demonstrated that quinones are capable of reacting

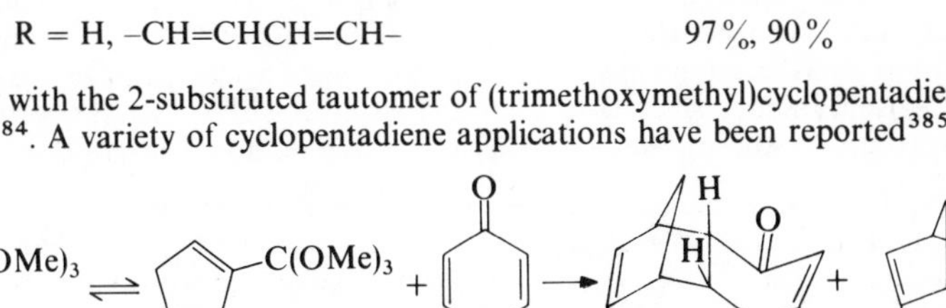

$$\xrightarrow[18\text{ h}]{PhH/\text{reflux}} \tag{290}$$

R = H, –CH=CHCH=CH– 97%, 90%

selectively with the 2-substituted tautomer of (trimethoxymethyl)cyclopentadienes (equation 291)[384]. A variety of cyclopentadiene applications have been reported[385–391].

$$C(OMe)_3 \rightleftharpoons C(OMe)_3 + \longrightarrow C(OMe)_3 + C(OMe)_3$$

3 : 2 (291)

In the cyclohexadiene series 1-methoxy or silyloxy derivatives have occupied a major position[392–398] along with the related cyclohexadienones[399,400]. Applications have proven fruitful in the study of steric strain[401], cage compounds[402,403] and intramolecular hydride transfer[404]. In the latter two citations some useful comparisons of the influence of ring size are made. One of the more exotic examples of these compounds involves two spirocyclopropyl substituents (**94**, equation 292)[405]. The delicate balance observed

(**94**) + benzoquinone —(dry THF, 70°C/16 h)→ adduct 70% (292)

between electronic and steric effects in this cycloaddition suggests a great deal of future interest.

e. Vinyl pyrroles, indoles and benzothiophenes

Hiremath and his colleagues have contributed to the synthesis of complex polyaromatic heterocyclic molecules with their discovery of the utility of an added nitro substituent (equation 293)[406–408]. Porter and his coworkers have extended the range of vinylindoles

N-Me pyrrole–CH=CH–NO_2 + naphthoquinone —(Me_2Ph/reflux, 48 h)→ product 64% (293)

and investigated the analogous chemistry of benzothiophenes with considerable success[409–411]. This latter work has been applied by others to more complex[412] and to simpler systems[413].

f. Sorbates and aqueous reactions

Grieco and his collaborators have had an interest in aqueous Diels–Alder reactions and employed the salts of sorbic and related acids in these studies[414–417]. Very high yields, mild conditions, short reaction times, and great regioselectivity characterize these reactions (equation 294) and stand in sharp contrast to the results obtained with the corresponding esters in hydrocarbon solvent. Methods of avoiding competition from complex sequential

2,6-dimethylbenzoquinone + diene–CH_2CO_2Na —((1) H_2O/25°C/1 h; (2) CH_2N_2)→ adduct (CH_2CO_2Me, Me, H) 77% (294)

Michael additions have been found[414] and a promising allenic acid has been developed (equation 295)[417]. To obtain the same yield using the methyl ester in benzene, 21 hours at

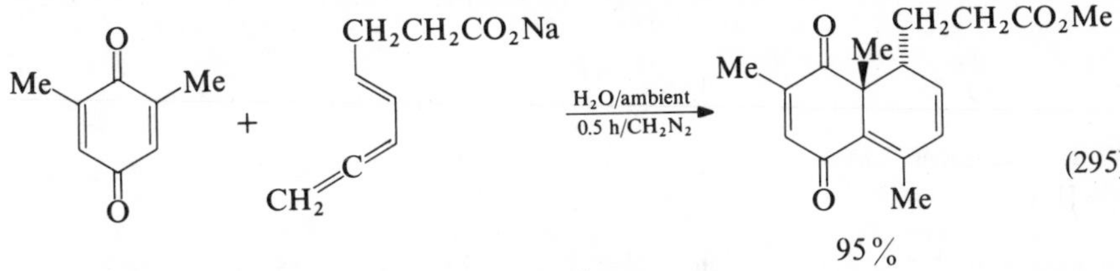

(295)

50°C is required. On the other hand Kanematsu and his colleagues have used exactly this non-aqueous reaction to investigate the synthesis of stable enols (equation 296)[418,419]. It

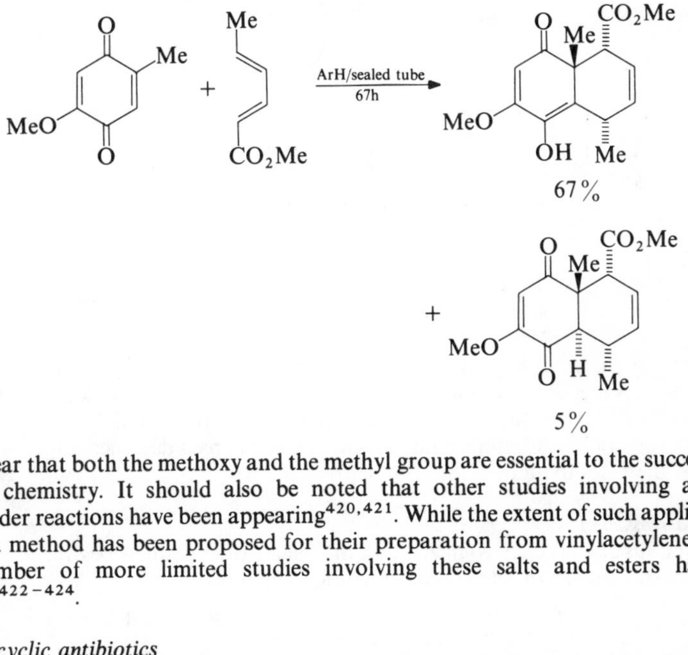

(296)

seems clear that both the methoxy and the methyl group are essential to the success of this unusual chemistry. It should also be noted that other studies involving allenes in Diels–Alder reactions have been appearing[420,421]. While the extent of such applications is limited a method has been proposed for their preparation from vinylacetylene[420].

A number of more limited studies involving these salts and esters have been reported[422–424].

g. Tetracyclic antibiotics

Beyond all doubt the synthesis of derivatives of the anthracycline antibiotic compounds, which also show great promise for the treatment of various tumors, has exerted a huge influence on the application and development of Diels–Alder chemistry. Gesson and coworkers have studied especially ketene acetals (equation 297)[425–428]. Much of this work has been summarized[427] and applied to a total synthesis[428].

Other research groups have made approaches to the synthesis of these stereochemically demanding molecules[429–442]. Some of the chemistry described will show the way to broad improvements in synthetic methods, but for the moment they remain isolated examples awaiting further systematic development.

h. Polycyclic aromatic syntheses

The great interest in polycyclic aromatic compounds as carcinogens has provided an ideal area for the application of Diels–Alder chemistry. One research group has dominated

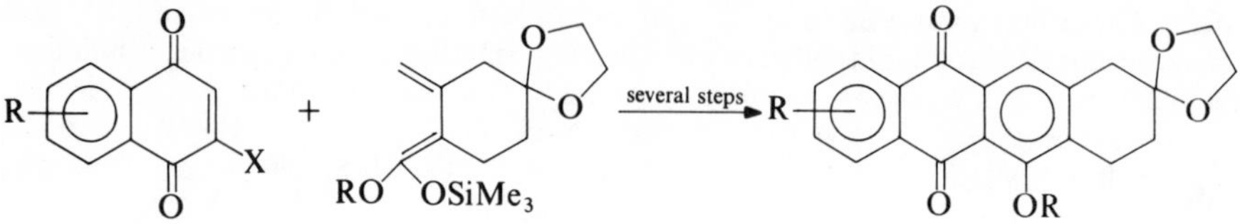

several steps

(297)

R = H, OH, OMe, OAc
X = H, Cl, Br

the work and obtained results of the highest quality. In five short years Manning, Muschik and Tomaszewski produced a formidable number of syntheses which represent a major contribution to our understanding of the mechanism of these reactions[290,443–452]. The two most important aspects of their work may be illustrated in the general synthesis of benz[*a*]anthracene-7,12-diones (equation 298)[444].

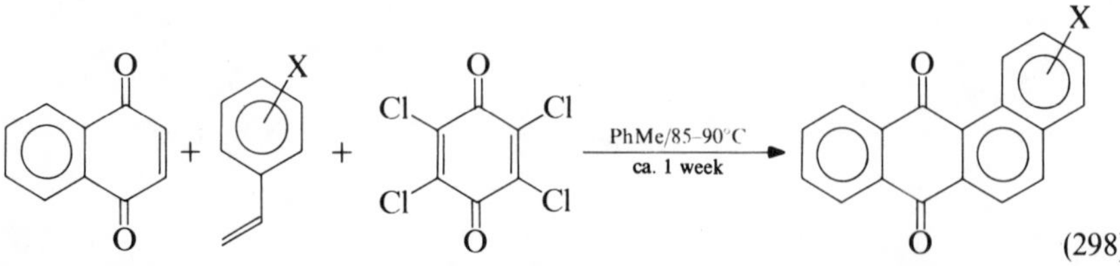

PhMe/85–90°C
ca. 1 week

(298)

30–40%

X = Me, 2,5-Me_2, F, Cl, Br, OMe

The use of substituted styrenes allows the preparation of specific product isomers, and the presence of chloranil improved the yields significantly.

Rosen and Weber have applied the same methods to the reactions of 1,4-phenanthraquinones[453]. Here too Manning's group improved the original work and showed that steric interaction, while important, does not by itself preclude a successful Diels–Alder reaction in these compounds[451].

i. Acyl rearrangements

Bruce and his coworkers recognized that the Diels–Alder adducts bearing angular acyl substituents can undergo a very specific migration (equation 299)[454–457]. The importance

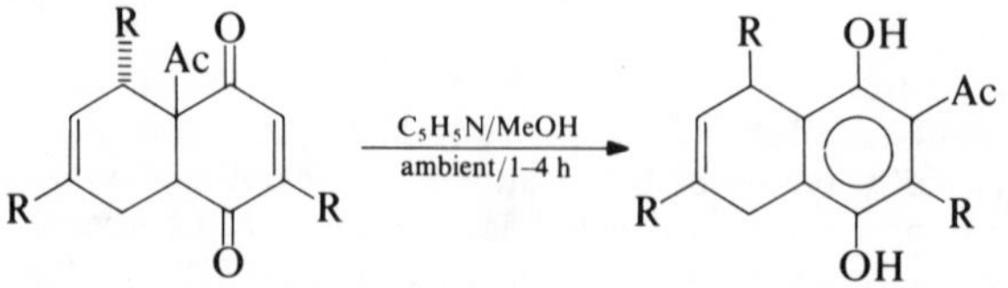

C_5H_5N/MeOH
ambient/1–4 h

(299)

R = H, Me

50–80%

of substituents in both the quinone and the diene are fully developed and the intramolecular nature of the shift demonstrated. The reasonable yields and mild conditions coupled with high regiospecificity make this an attractive synthesis. This rearrangement was first observed by Cooper and Sammes[458] who later published a detailed study of the effects of substituents and conditions on the yield and ease of the reaction[459]. The general mechanistic conclusion seems to demand at least a partial retro-diene process[457].

j. Unusual isolated observations

Some more reports represent work that deserves to be developed in greater detail. For example, there is a report that palladium metal in the presence of triphenylphosphine dimer acts as a powerful catalyst for the conversion of terminal alkenes into conjugated dienes which undergo addition and aromatization with 1,4-benzoquinone (equation 300)[460].

$$RCH_2CH_2CH{=}CH_2 + \text{1,4-benzoquinone} \xrightarrow{Pd(Ph_3P)_2} \text{R-substituted anthraquinone (R on each outer ring)} \quad (300)$$

R = Me, Et, Bu 69–75%

Keana and Eckler[461] continued study of an interesting report by Jung[462] concerning the detailed effect of enol–ester diastereomers on reactivity in Diels–Alder chemistry. Three of the four isomers are reported and in view of the importance of these reactive dienes this investigation should be made more quantitative. There also seems to be a lack of clear communication among the authors and some irritating typographical problems occur in the first article.

While the *in situ* generation of 1,2-quinones has been a common practice, this method is rare with the corresponding 1,4-quinones. Kraus and Taschner have applied the procedure to quinones bearing destabilizing electron-withdrawing substituents (equation 301) with excellent results[463].

$$\text{2-}R^1\text{-hydroquinone} + \text{diene (Me, }R^2) \xrightarrow[\text{dark}/10^\circ C]{Ag_2O/PhH} \text{adduct (}R^1, R^2, Me) \quad (301)$$

94–100%

$R^1 = CO_2Me$, CHO, Ac; $R^2 = OSiMe_3$, CH_2CO_2Et, $(CH_2)_2OBz$

An extraordinary formation of a polycyclic molecule deserves attention (equation 302)[464].

k. Non-Diels–Alder products

In some instances the formation of an unexpected product can be the most interesting outcome. The addition of 8-methoxyheptafulvene to 1,4-naphthoquinone produces a good yield of the azulene derivative (**95**, equation 303)[465]. This chemistry has been developed more fully showing the range of quinones which can be used[466].

There are several potential diene reaction sites in the indene **96**. A low yield of cycloadduct is obtained when 1,4-benzoquinone reacts at what might be considered the least likely of these sites (equation 304)[467].

1. Miscellaneous Diels–Alder reactions

The reactions reported here represent a great many suggestions of worthwhile studies, but at the moment must be considered only as interesting examples of the still unrealized potential of this reaction. The vastness of the literature was made clearer to me when I

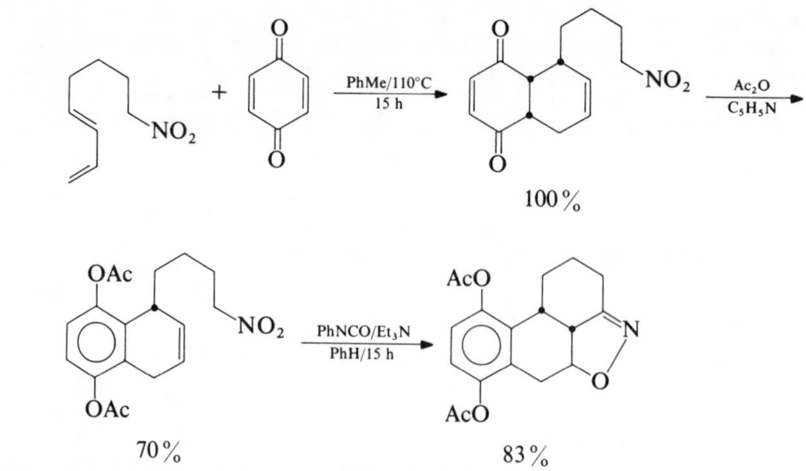

PhMe/110°C, 15 h; 100%; Ac_2O / C_5H_5N; 70%; PhNCO/Et_3N / PhH/15 h; 83% (302)

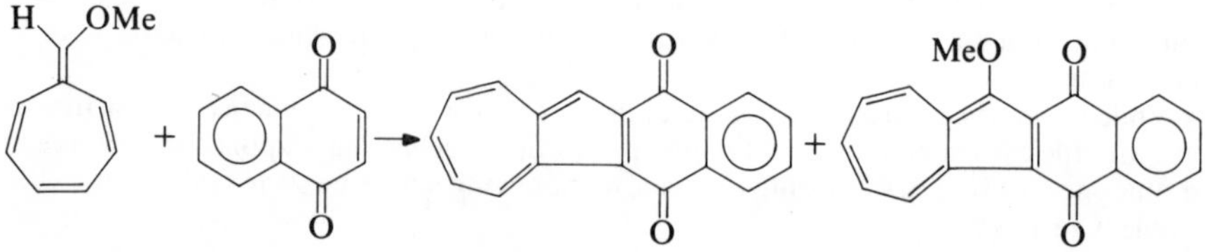

(95) 65 : 35 (67%) (303)

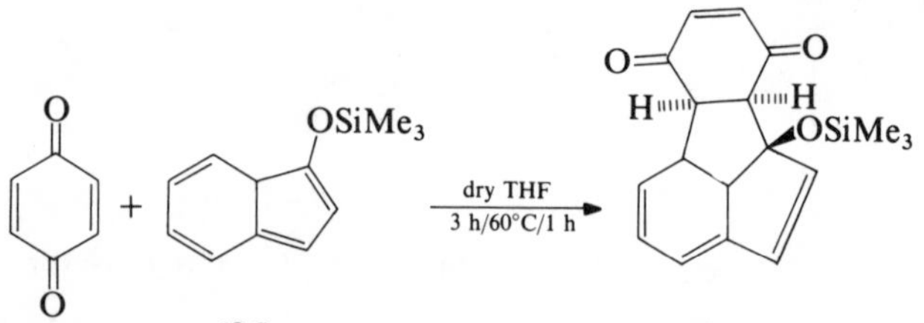

(96); dry THF, 3 h/60°C/1 h; 10% (304)

realized that my earlier intensive search of Fieser's work had failed to uncover a long and rather important paper from 1935[468]! The reaction product failed to form a semicarbazone and was assigned a structure shown to be incorrect by modern methods (equation 305)[469].

Additional references in this section note the wide variety of modifications that have resulted from the stimulus of Diels–Alder–quinonoid chemistry[470–503].

4. *Diels–Alder chemistry of* ortho-*quinones*

This area of cyclization chemistry has received much attention during the past decade. For example, Diels–Alder products involving 1,2-quinones are often of interest as unusual substrates for subsequent photochemical transformations. Paquette and his collaborators have introduced a novel synthetic approach to 1,3-disubstituted cyclooctatetraenes involving the adduct of 3,5-di-*t*-butyl-1,2-benzoquinone and cyclobutadiene[504, 505]. Both

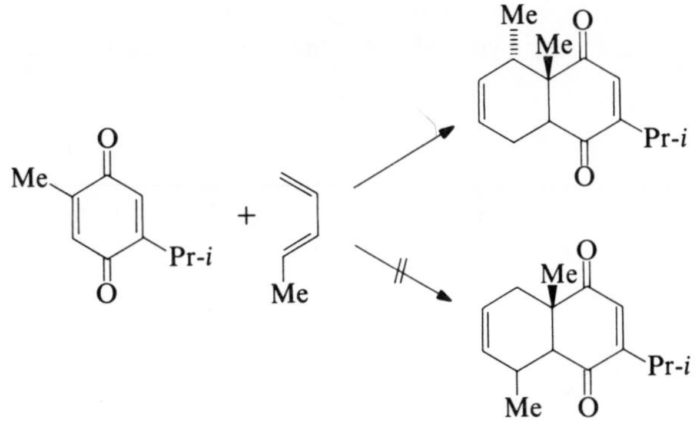

(305)

starting materials were generated *in situ* and the *endo*-adduct was obtained in 51% yield (equation 306).

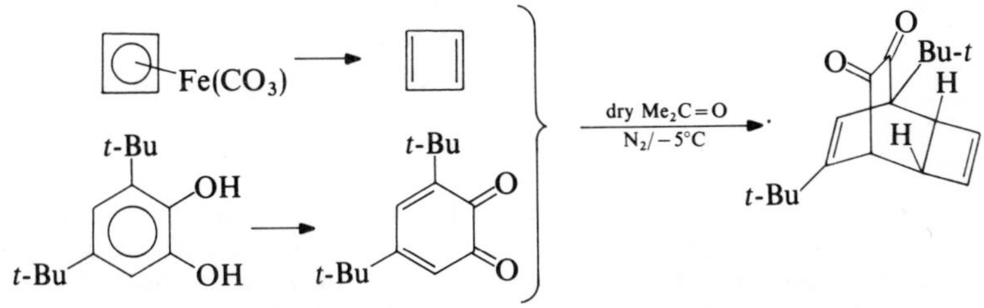

(306)

Unsymmetrical chlorobiphenyls for toxicological studies have been prepared from the adducts of *o*-chloranil and appropriate arylacetylenes (equation 307)[506]. In an earlier study

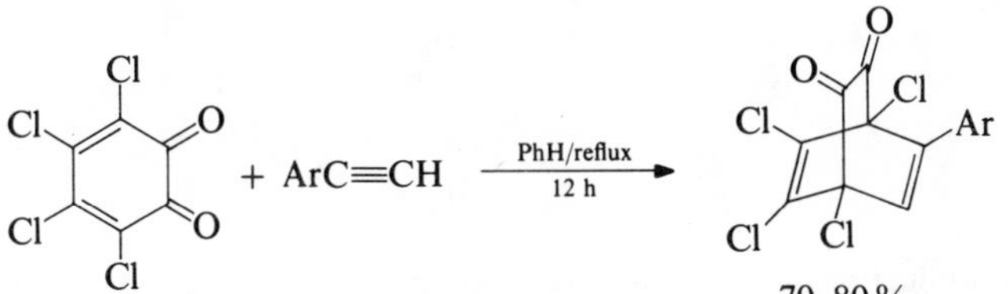

$$\text{ArC}\equiv\text{CH} \xrightarrow[\text{12 h}]{\text{PhH/reflux}} \quad 70\text{–}80\% \tag{307}$$

Ar = 3- or 4-ClC_6H_4, 2,4- or 2,5-$Cl_2C_6H_3$

Pyle and his colleagues had repeated a still earlier Diels–Alder reaction of *o*-chloranil with benzyne followed by basic hydrolysis to produce a glyoxylic acid (equation 308)[507].

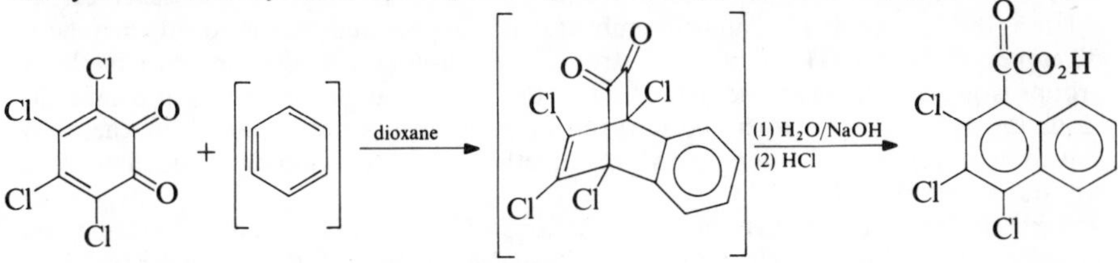

53%

(308)

Realizing the synthetic potential of this sequence they employed phenylacetylene and obtained a nearly quantitative yield of two isomeric biphenylglyoxylic acids (**97** and **98**, equation 309).

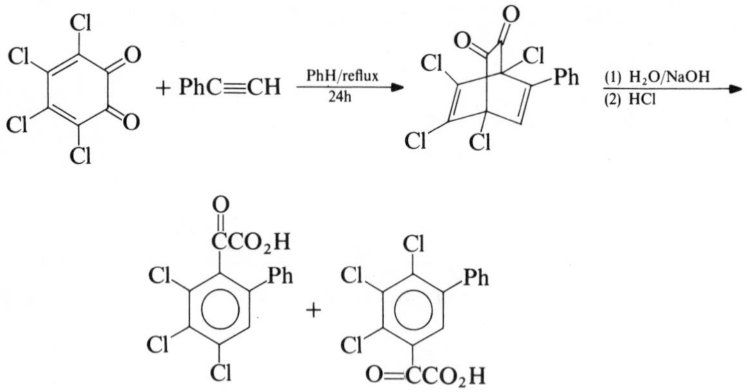

$$\text{tetrachloro-1,2-benzoquinone} + PhC{\equiv}CH \xrightarrow[24h]{PhH/reflux} \text{bicyclic adduct} \xrightarrow[(2)\ HCl]{(1)\ H_2O/NaOH} (97) + (98) \quad (309)$$

(**97**) 3 : 4 (**98**) 99%

The Diels–Alder chemistry of 1,2-quinones with alkynes has been the subject of several reports. For example, given a choice between reaction at alkyne or alkene linkages the tetrahalo-1,2-benzoquinones react as dienes and the alkene as a dienophile (equation 310)[508]. The yields are moderate, and no evidence is presented for the absence of the

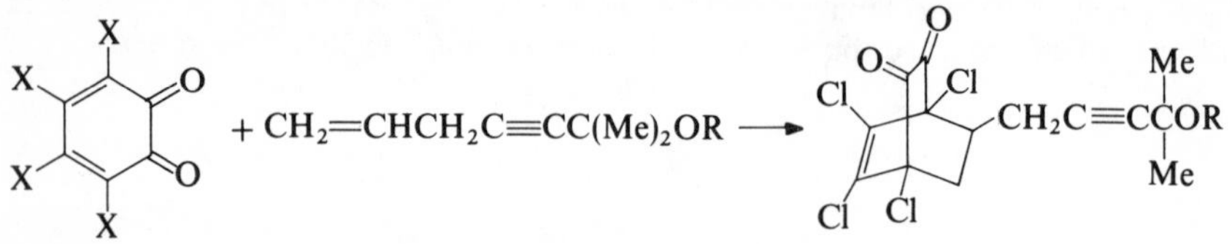

$$\text{tetrahalo-1,2-benzoquinone} + CH_2{=}CHCH_2C{\equiv}CC(Me)_2OR \longrightarrow \text{bicyclic adduct-}CH_2C{\equiv}CC(Me)_2OR \quad (310)$$

X = Cl, Br R = H, Me 60–90%

alternative product or bis-addition. A later, greatly expanded report shows that terminal conjugated dienes behave analogously, giving the adducts in high yield. The quinones proved to be the most reactive dienes studied, and a detailed account of their molecular orbitals is presented[509].

An interesting observation illustrates the relationship between the structure of the quinonoid starting material and that of the product[510]. Using dimethyl acetylenedicarboxylate as the dienophile alkyl and allyl substituted 1,2-benzoquinones gave only bicyclic α-diketones (equation 311). When the strong electron-donating effect of two methoxy groups is present the probable initial adduct decomposes under the reaction conditions (equation 312). Finally, when both types of substituents are involved in the fully substituted quinone, 3,6-di-*n*-propyl-4,5-dimethoxy-1,2-benzoquinone, decomposition without addition is observed.

Verboom and Bos have studied the reactions of 1,2-quinones with cyclooctyne and found several different product structures depending upon both the reaction conditions and the exact nature of the quinone[511]. The presence of the Lewis acid catalyst is essential for the 9,10-phenanthrenequinone reaction (equation 313) and leads to complete conver-

$$MeO_2CC{\equiv}CCO_2Me + \text{[3,4,5,6-tetra-R-1,2-benzoquinone]} \longrightarrow \text{[bicyclic adduct]} \quad (311)$$

R = H, Pr, allyl, CH=CHMe — variable 20–70%

$$\text{[4,5-dimethoxy-1,2-benzoquinone]} + MeO_2CC{\equiv}CCO_2Me \longrightarrow [\text{adduct}] \longrightarrow \text{[dimethyl 4,5-dimethoxyphthalate]} + 2CO_2 \quad (312)$$

42%

sion of the expected Diels–Alder product to the rearranged isomers shown with 3,5-di-*t*-butyl-1,2-benzoquinone (equation 314). The amount of catalyst, temperature and reaction time are important to the product distribution in equations 314 and 315. The second carbonyl bridged product (**100**) is not formed by rearrangement of either the first (**99**) or the normal Diels–Alder adduct.

A ring-expansion reaction has been used in attempts to synthesize an intermediate related to the highly oxygenated natural product purpurogallin (**101**) (equation 316)[512]. While the product is not one expected from the Diels–Alder reaction, its relationship to those obtained by Bos and Verboom might be investigated.

The addition of cyclobutadiene to 1,2-naphthoquinone occurred in good yield (equation 317)[513].

Danishefsky and his colleagues have added to their extensive investigations of regiospecificity in Diels–Alder reactions of oxygenated dienes and dienophiles with a few examples of 1,2-benzoquinones[514–516]. The use of 4-methoxy derivatives is generally successful and extremely structure dependent (equations 318, 319). Of even greater interest is their demonstration that the first reaction is very solvent dependent while the second is not. With either solvent, or their mixtures, a small amount of the other product was obtained, but the influence of the change in environment is far more dramatic than might be expected.

A kinetic study of the addition of substituted styrenes to the halo-1,2-benzoquinones (equation 320) shows normal Diels–Alder adducts in which the quinone acts as a diene and the characteristics of a neutral mechanism[517]. The results are discussed in terms of molecular orbitals.

A much more detailed orbital treatment accompanies the reported addition of cyclopropenes to 1,2-benzoquinone and its tetrachloro derivative (equations 321, 322)[518, 519]. When *o*-chloranil reacts with 3-methyl-3-phenyl- or 3,3-diphenylcyclo-

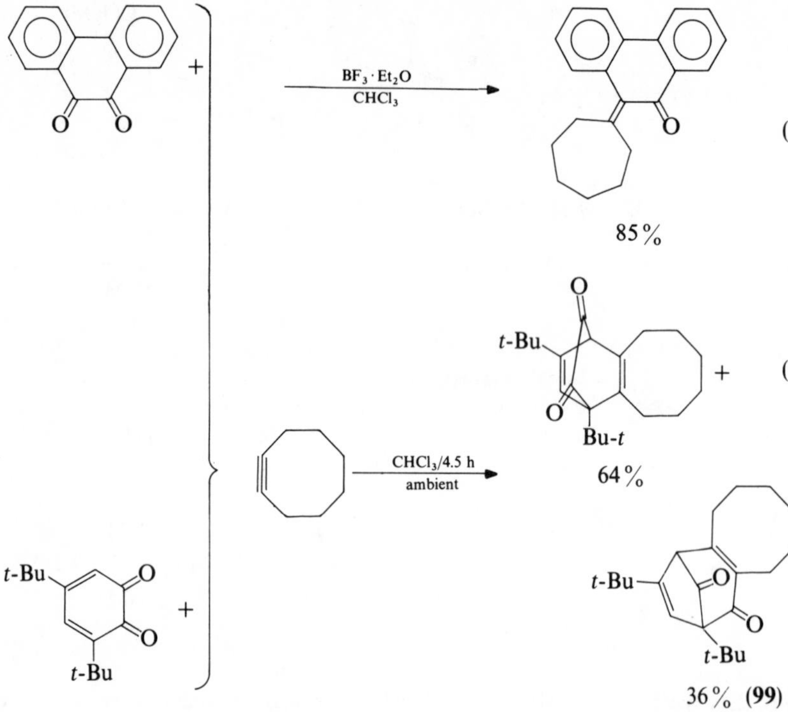

+ $\xrightarrow[CHCl_3]{BF_3 \cdot Et_2O}$ O O (313)

85%

O

t-Bu O + (314)

O Bu-t

$\xrightarrow[\text{ambient}]{CHCl_3/4.5\ h}$ 64%

t-Bu O + t-Bu O O

t-Bu t-Bu

36% (99)

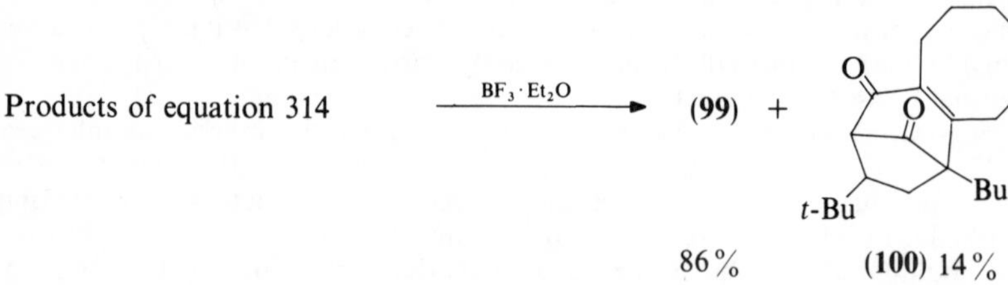

Products of equation 314 $\xrightarrow{BF_3 \cdot Et_2O}$ (99) + O O O (315)

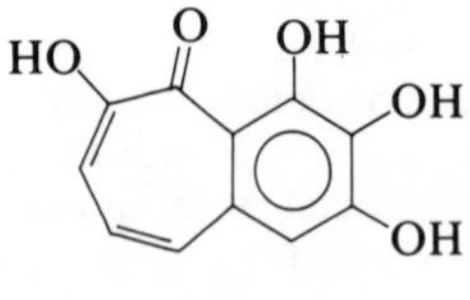

t-Bu Bu-t

86% (100) 14%

HO O OH OH OH

(101)

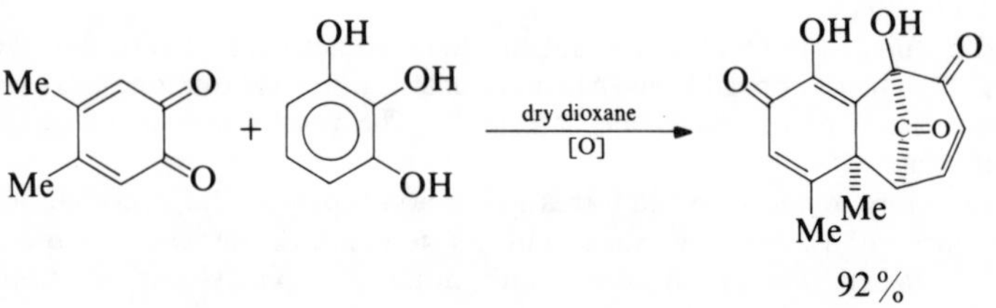

Me O + OH OH $\xrightarrow[\text{[O]}]{\text{dry dioxane}}$ O OH OH O C=O (316)

Me O OH Me Me

92%

$Fe(CO)_3$ Ce^{+4} $Me_2C=O$ 75% (317)

PhH/reflux 6 h MeO OMe OH 37% MeO Me OMe (318)

MeOH/reflux 6 h MeO Me H OMe 67%

MeO CH_2OAc PhH or MeOH MeO CH_2OAc HO 81% (319)

X Y CH_2ClCH_2Cl Ar (320)

Me Ph CH_2Cl_2 Me Ph (321)

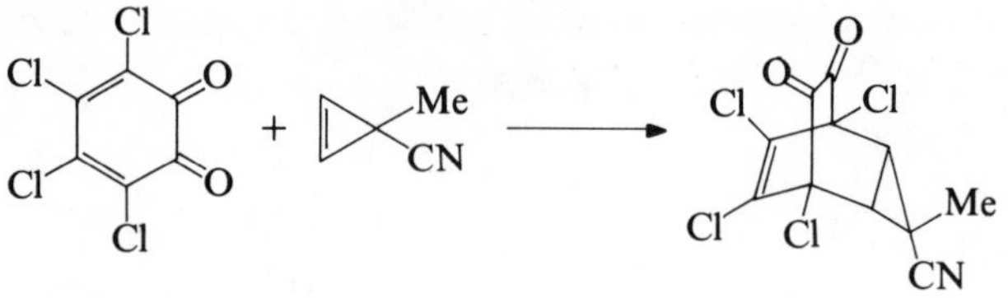

(322)

propene a very different product structure results (equation 323). A mechanism involving a charge transfer complex is presented and supported by ionization potential measurements.

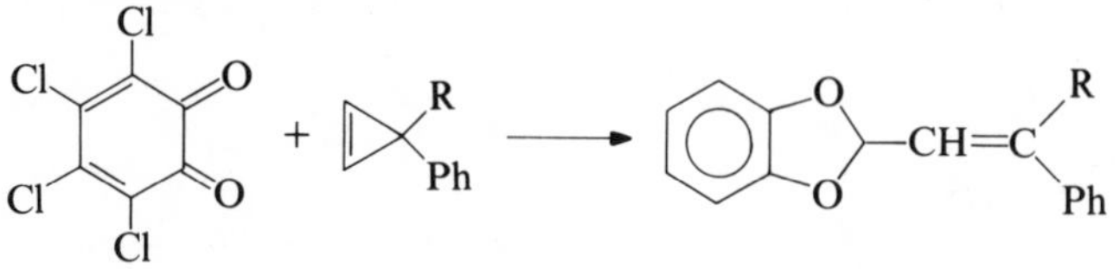

(323)

R = Me, Ph

Simple 1,2-benzoquinones produce normal Diels–Alder adducts in which the quinone reacts as a carbon diene[520–524]. For example, both 1,2-benzoquinone and its tetrachloro analog add readily to the electron-rich double bond of benzvalene (**102**) to form the expected pentacyclic adduct (equation 324)[521]. The presence of four chlorine atoms led to

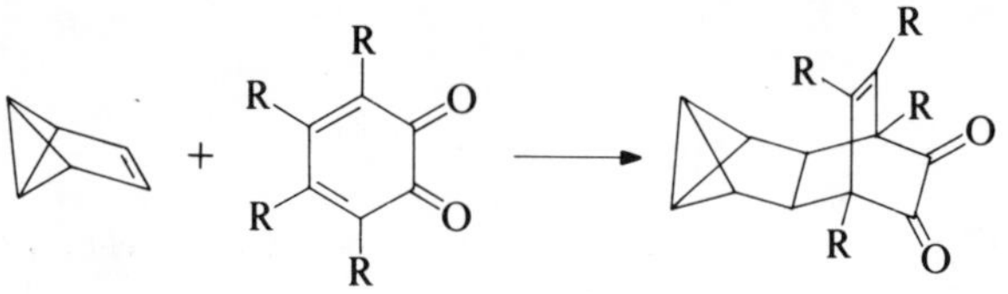

(324)

(**102**) R = H, Cl

a much higher isolated yield (90% compared to 58%) of product. The structures were determined using ^{13}C-NMR spectra.

When these fragile, but versatile, compounds contain strong electron-withdrawing substituents they can serve as carbon dienophiles. Al-Hamdany and coworkers have examined this chemistry using nascent 1,2-benzoquinones bearing such substituents in the 3- or 4-position. Their first report suggested a mechanistic pathway for the observed diadducts (**103**) summarized in Table 6 (equation 325)[525].

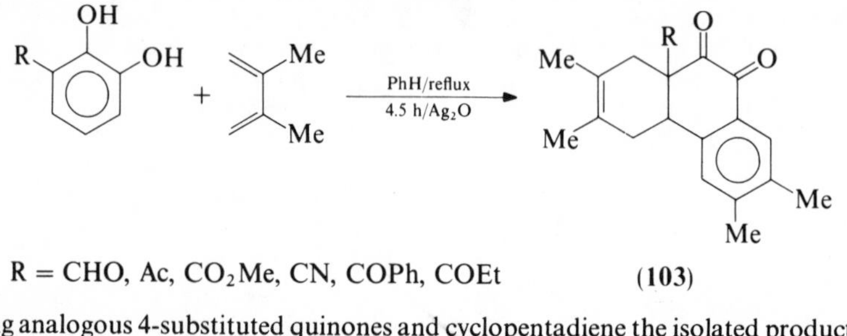

PhH/reflux 4.5 h/Ag_2O (325)

R = CHO, Ac, CO_2Me, CN, COPh, COEt (**103**)

Using analogous 4-substituted quinones and cyclopentadiene the isolated product (**104**) would be the result of the quinone acting as the dienophile (equation 326)[526]. An

Table 6. Product yields from Diels–Alder chemistry of nascent 1,2-benzoquinones with electron-withdrawing 3- or 4-substituents[525–527]

R	Yield (%) 103	104
3-CHO	25	60
4-CHO		45
3-COMe	51	40
4-COMe		55
4-COEt		—
3-COPh	65	46
4-COPh		51
3-CO_2Me	78	63
4-CO_2Me		69
3-CN	43	62
4-CN		58

dry Et_2O/dark, Ag_2O/ambient

(105) (104) (326)

intermediate, the adduct **105** is postulated, but no evidence is presented. Yields similar to those just cited are also noted in Table 6. These results contrast markedly with those reported for 1,2-benzoquinones bearing electron-withdrawing groups in the 3-position (equation 327)[527]. While the authors are wrong in reporting that Ansell and his

dry Et_2O/dark, MnO_2/ambient (327)

collaborators[528] did not study electron-withdrawing substituents, their studies do expand our understanding of those important compounds.

The α-dicarbonyl structure of 1,2-quinones provides a useful variant from the diene system, and it too has been the subject of several Diels–Alder studies, for example, the reactions of 3-substituted indoles with 1,2-benzoquinone occur in low yield (equation 328)[529]. However, when both the 3- and 4-position are part of a fused ring system, good yields are obtained.

$CHCl_3$/ambient, trace HOAc/1 h (328)

Latif and his collaborators, in a study with *o*-chloranil showed that with furfurylidenemalononitrile a mono-adduct involving the heterocyclic ring was obtained (equation 329); while with furfurylidenecyanoacetic ester a di-adduct was obtained (equation 330)[530, 531].

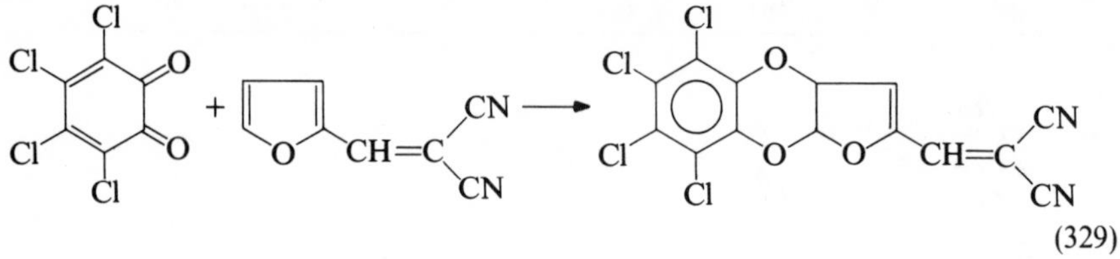

Cl Cl O O Cl Cl + O CH=C(CN)CN → Cl Cl Cl Cl O O O CH=C(CN)CN

(329)

Cl Cl O O Cl Cl + O CH=C(CO$_2$R)CN → Cl Cl Cl Cl O O NC CO$_2$R O O O Cl Cl Cl Cl

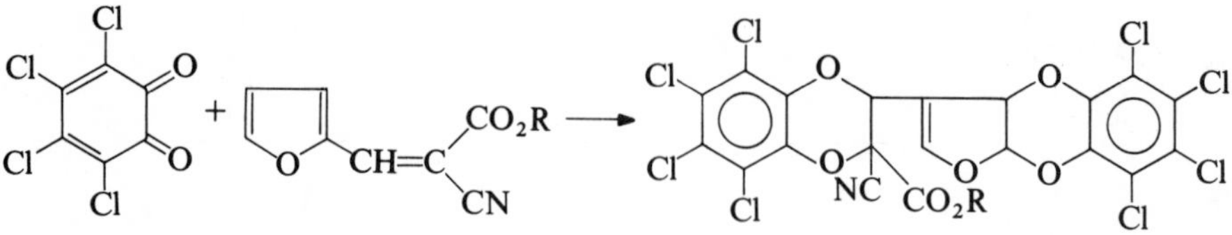

(330)

Furans substituted by a 2-nitrovinyl group cyclized in fair to good yield and only at the furan ring with either chloranil or bromanil (equation 331)[531].

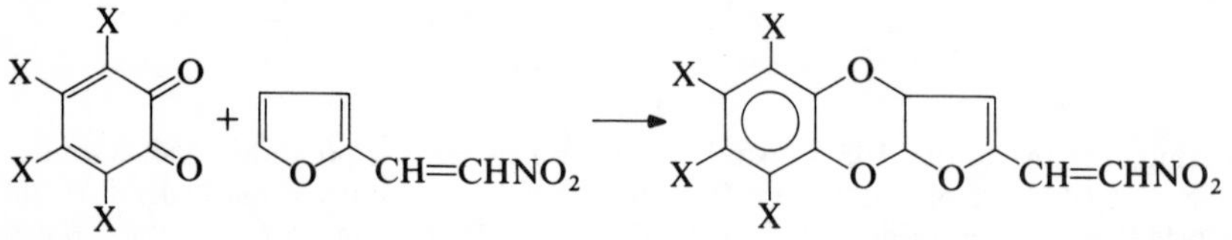

X X O O X X + O CH=CHNO$_2$ → X X X X O O O CH=CHNO$_2$

X = Cl, Br (331)

These studies have been extended in several directions. The thiophene ring (equation 332) and the furan system bearing acrylophenone substituents (equation 333) have both

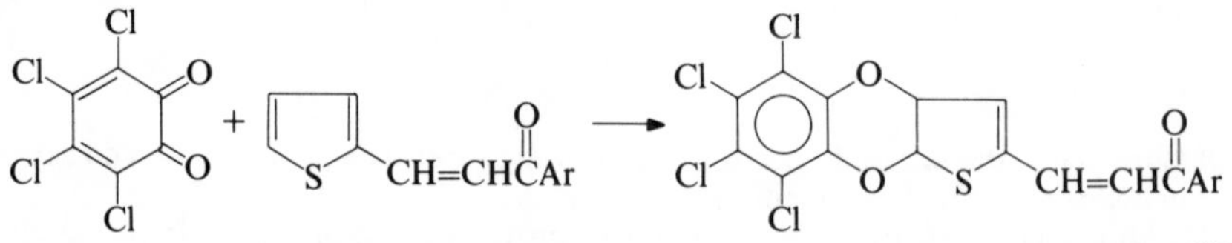

Cl Cl O O Cl Cl + S CH=CHC(O)Ar → Cl Cl Cl Cl O O S CH=CHC(O)Ar

Ar = Ph, 4-ClC_6H_4, 4-An (332)

Cl Cl O O Cl Cl + O CH=CHC(O)Ar → Cl Cl Cl Cl O O O CH=CHC(O)Ar

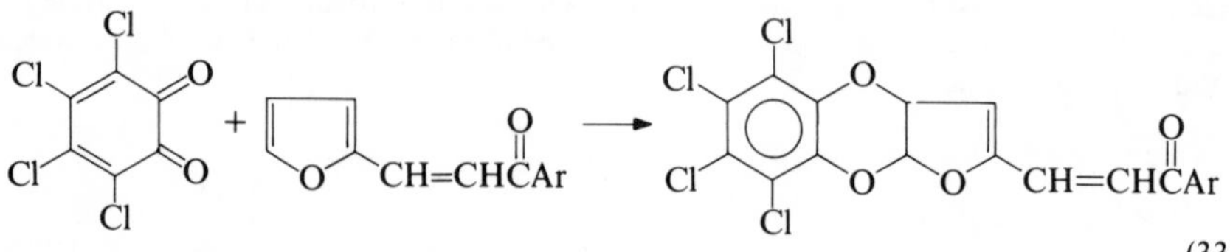

(333)

proven fruitful[532]. The yields reported are generally above average and the difference in reactivity is marked. The apparently analogous reaction of bromanil follows a different path; only cyclization by dehydrohalogenation between two moles of quinone is observed (equation 334)[533].

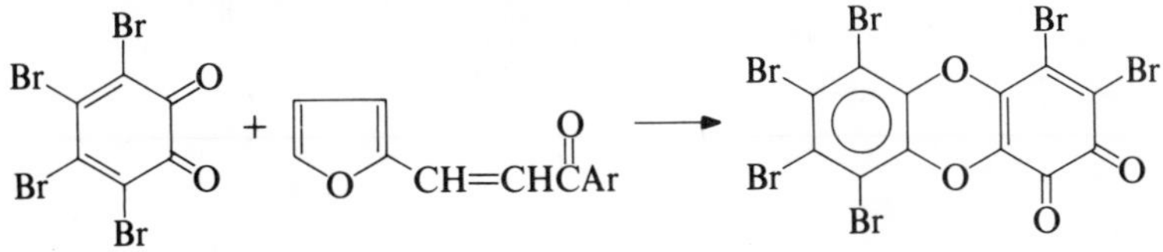

(334)

Members of this same team have examined the situation when both the furan and the thiophene ring are present in the substrate (equation 335)[534]. In both cases the mono-

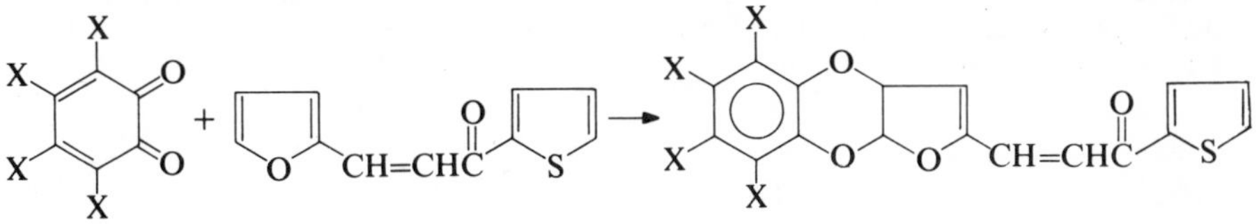

X = Cl, Br

(335)

adduct corresponded to furan ring addition. Even the more drastic conditions required to force addition of *o*-chloranil to the alkene failed to produce any reaction at thiophene. The di-adduct from *o*-bromanil was too unstable to isolate.

Latif and his colleagues have applied this useful dioxane preparation to other synthetic efforts[535] as have a number of other workers in studies ranging from rather theoretical[536–539] to the preparation of complex heterocyclic molecules[540, 541]. Friedrichsen and colleagues have continued to experiment with reactions of 1,2-benzoquinone and isobenzofurans[542]. In addition to the eight-membered heterocycle (**106**) previously reported, they obtained a product which is possibly related to that found in certain reactions of cyclopropenes[520] (equation 336). The products distribution is discussed in

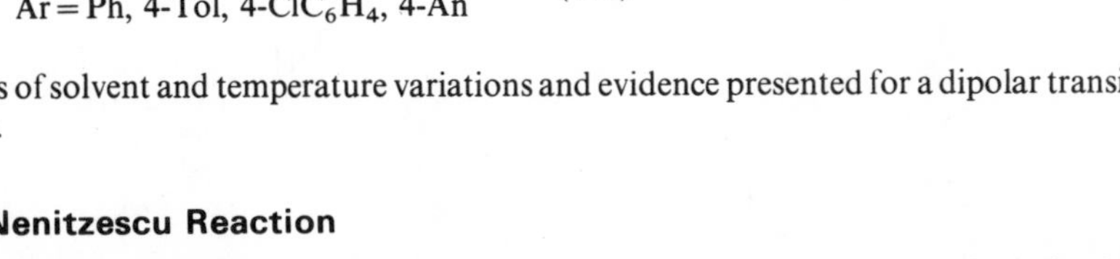

Ar = Ph, 4-Tol, 4-ClC_6H_4, 4-An

(**106**)

(336)

terms of solvent and temperature variations and evidence presented for a dipolar transition state.

B. Nenitzescu Reaction

Allen has not only made extensive contributions to the understanding and application of this singular synthesis of 5-hydroxyindoles, but has written a detailed review of the field[543]. Several additional studies have appeared by Kuckländer and his collaborators[544–550].

Studies in acetic and propionic acids show that a variety of *n*-arylindoles can be prepared, but in rather modest yields (equation 337)[544–547]. A more detailed investigation

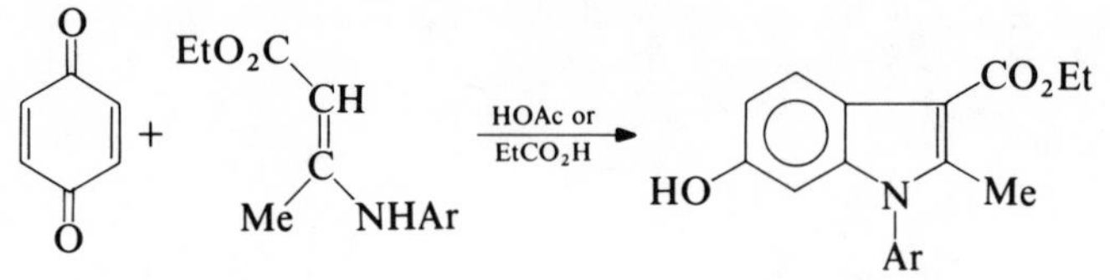

(337)

20–35%

Ar = Ph, 3- or 4-Tol, 4-XC_6H_4, 4-$NO_2C_6H_4$, 4-NCC_6H_4, 4-An

of the by-products from addition of two equivalents of enamine was made and a direct cyclization mechanism suggested (equation 338)[548]. These same workers have shown that acyl migration can take place under Nenitzescu conditions (equation 339)[549] and one is reminded of the studies of Bruce[457]

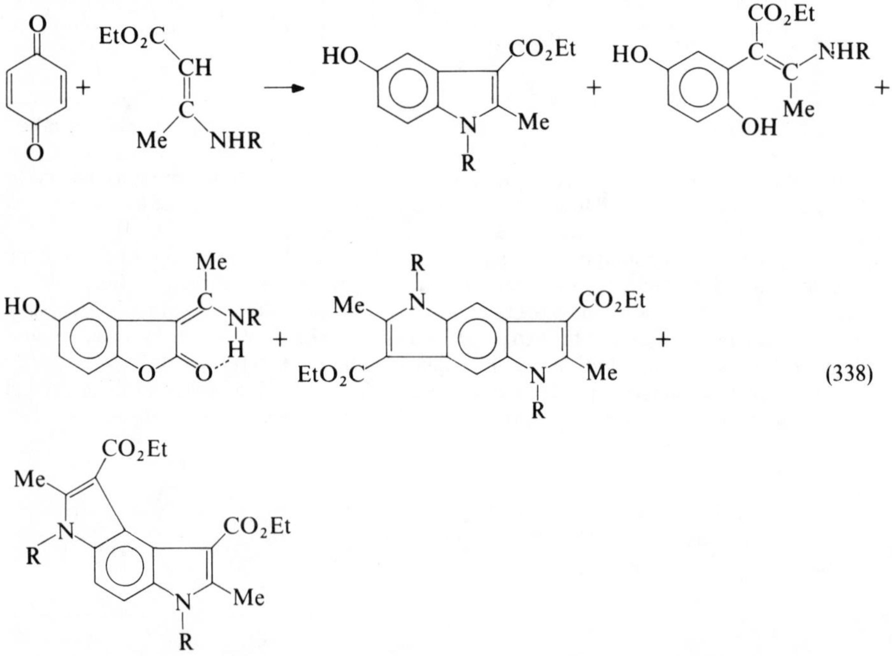

(338)

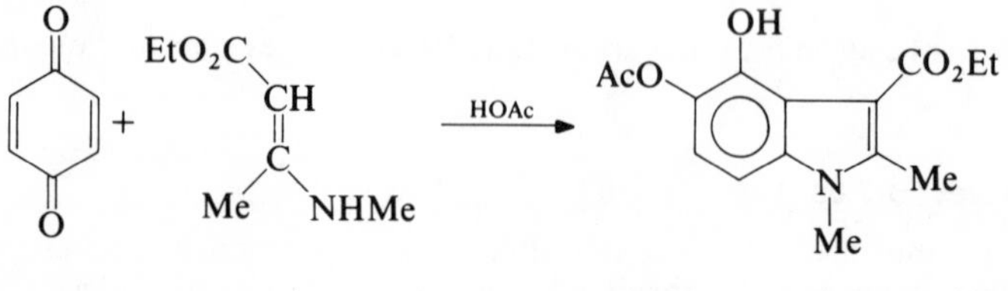

(339)

Kuckländer also studied the formation of furans derived from 2-acetoxy-1,4-naphthoquinone (equation 340)[547]. An acyl migration in this reaction leads to the formation of the minor by-product (**107**).

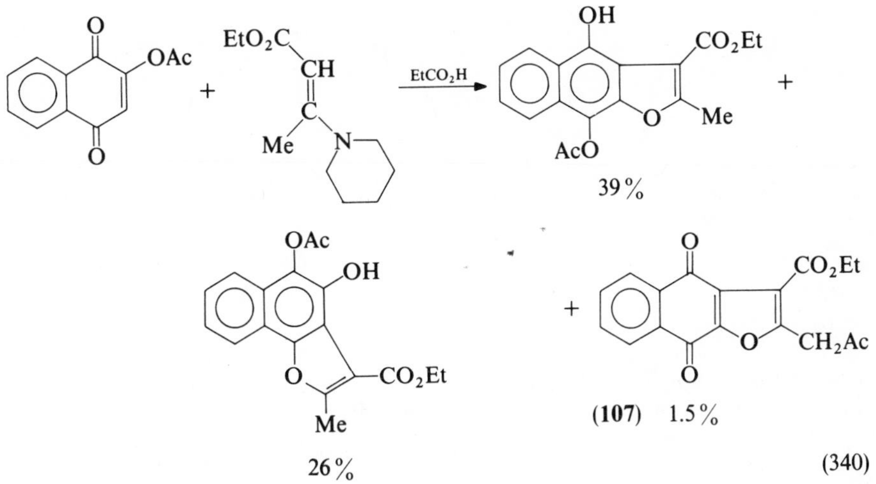

39%

26%

(107) 1.5%

(340)

The reaction of cyclohexenones with quinones to produce carbazoles is known but when the enamine is external to the ketone ring, an interesting ring-expansion reaction takes place (equation 341)[549]. While the yields are low (ca. 20%) the method provides a simple

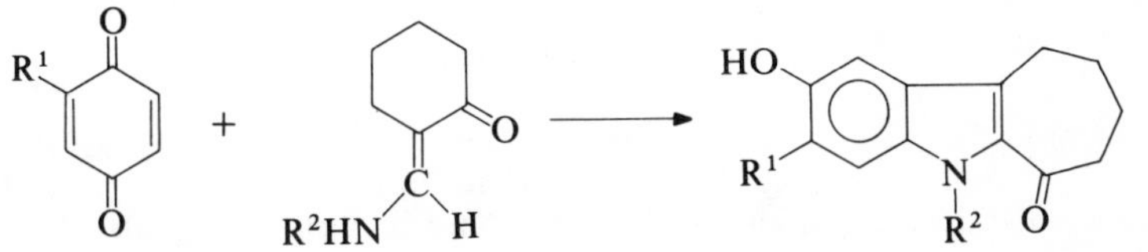

(341)

R^1 = Me, Bz, Ph, 4-Tol, 4-FC_6H_4, 4-An; R^2 = H, Me

entry to rather complex ring systems. The apparent regiospecificity remains to be deomonstrated.

More recently Kuckländer and Töberich have examined the frequently observed formation of benzofurans under Nenitzescu conditions (equation 342)[550]. With 2,3- and 2,6-dichloro-1,4-benzoquinones the yields are exceptionally high.

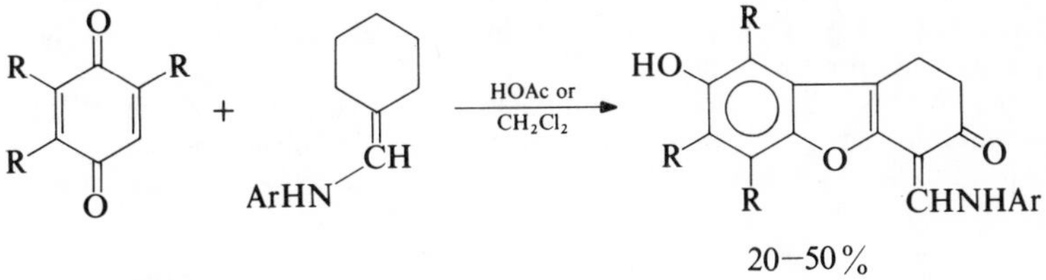

(342)

20–50%

R = H, Cl; Ar = 4-FC_6H_4, 4-An

Grinev and his colleagues have continued their studies of the influence of nitrogen substituents on the course of the condensation[551]. Hydrogen or a methyl group leads to 5-hydroxyindoles (equation 343) while phenyl or *p*-tolyl groups produced the corresponding 6-hydroxyindoles (equation 344). They have also examined the addition of enamines to

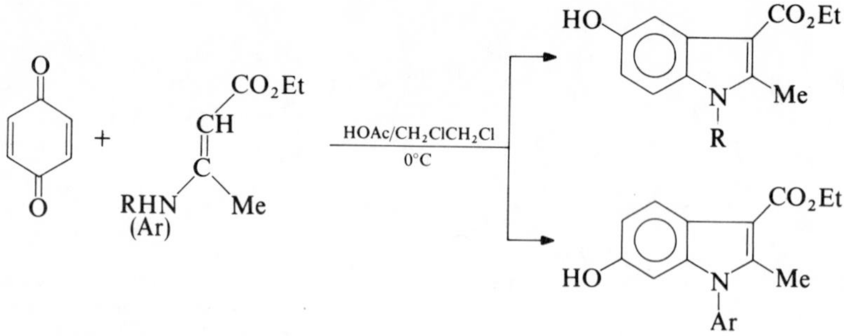

$$\text{benzoquinone} + \text{RHN(Ar)C(Me)=CHCO}_2\text{Et} \xrightarrow[0^\circ C]{HOAc/CH_2ClCH_2Cl} \text{5-HO-indole (N-R, 2-Me, 3-CO}_2\text{Et)} \quad (343) \;+\; \text{6-HO-indole (N-Ar, 2-Me, 3-CO}_2\text{Et)} \quad (344)$$

R = H, Me; Ar = Ph, 4-Tol

quinones bearing strong electron-withdrawing groups and found that the initial adduct leads to benzofuran but not to indole (equation 345)[552].

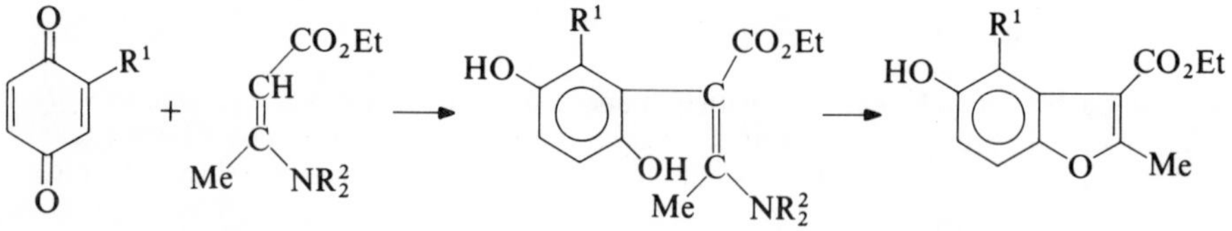

$$\text{2-R}^1\text{-benzoquinone} + \text{MeC(NR}^2_2\text{)=CHCO}_2\text{Et} \longrightarrow \text{HO-C}_6\text{H}_2(\text{R}^1)(\text{OH})\text{-C(CO}_2\text{Et)=C(Me)NR}^2_2 \longrightarrow \text{5-HO-4-R}^1\text{-benzofuran (2-Me, 3-CO}_2\text{Et)}$$

$R^1 = SO_2Ph,\ SO_2C_6H_4Me\text{-}4;\ R^2 = H, Me, Ph$ (345)

The chemistry of enaminones has attracted some interest. Kozerski noted that the unusual 2*H*-1,5-benzodioxepine ring system (**108**) can be formed in 50% yield under Nenitzescu conditions (equation 346)[553]. In a later publication he offers interesting

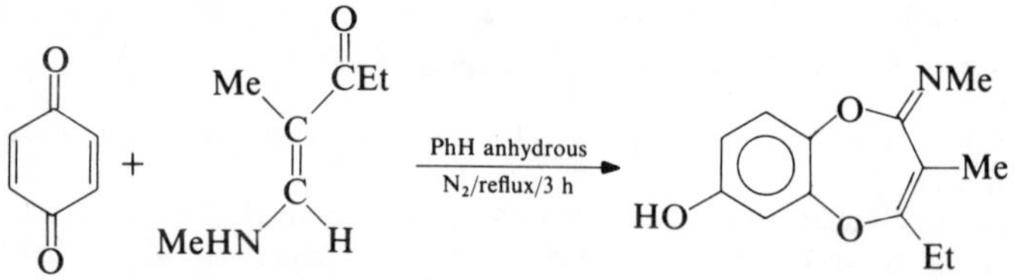

$$\text{benzoquinone} + \text{MeHN-CH=C(Me)C(O)Et} \xrightarrow[N_2/\text{reflux}/3\ h]{\text{PhH anhydrous}} \text{HO-benzodioxepine (=NMe, Me, Et)} \quad (346)$$

(**108**)

information concerning acyl rearrangements in intermediates but unfortunately fails to develop the earlier promising chemistry[554].

Siddappa and his colleagues have examined an example of the traditional cinnamate enamine (equation 347)[555, 556] and some newer unsaturated phenones (equation 348)[557]. In both instances the yields with few exceptions lie in the 30–40% range.

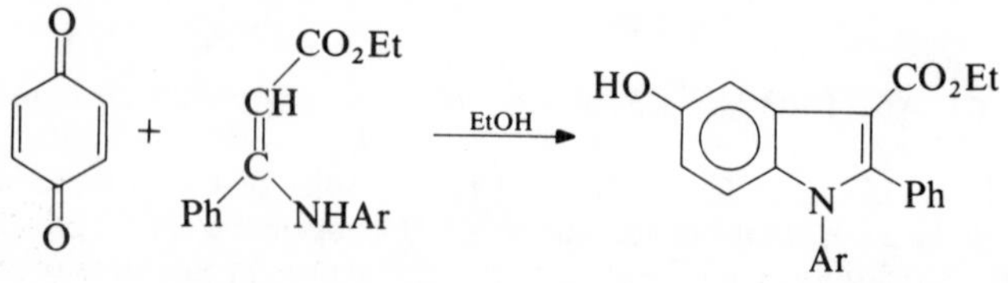

$$\text{benzoquinone} + \text{PhC(NHAr)=CHCO}_2\text{Et} \xrightarrow{\text{EtOH}} \text{5-HO-indole (N-Ar, 2-Ph, 3-CO}_2\text{Et)} \quad (347)$$

$Ar = Ph,\ 4\text{-}BrC_6H_4$

$$\text{1,4-naphthoquinone} + \text{PhC(NHR)=CH-C(O)Ph} \xrightarrow[\text{ambient-1 h/reflux 4 h}]{\text{absolute EtOH}} \text{HO-benz[g]indole (N-R, 2-Ph, 3-C(O)Ph)} \quad (348)$$

R = Pr, Bu, *i*-Bu, Ph, Bz

A number of research groups have reported isolated examples of the Nenitzescu reaction[558–562]. Of broader interest is the development of a procedure for the synthesis of enamines from *S,S*-acetals of ketene which expands the range of available starting materials showing average reactivity toward quinones[563]. Another extention is the use of anilines as the enamine reactant (equation 349)[564]. The yields are low, but important by-

$$\text{1,4-benzoquinone} + \text{H}_2\text{N-C}_6\text{H}(R^1)_2R^2 \xrightarrow[\text{F}_3\text{CCO}_2\text{H/0.5 h}]{\text{HOAc/4 h or}} \text{HO-carbazole (N-H; } R^1, R^2, R^1) \quad (349)$$

R^1 = H, Me; R^2 = NO_2, CN

products are obtained. This chemistry should be examined in more detail. A novel modification has been reported by Patrick and Saunders[565]. Using nitromethane as the solvent and methyl crotonates as enamines, they were able to obtain 2–3-fold improvement in yields with few exceptions, e.g. *N*-phenyl. The reasons for this will require additional study.

Allen suggested that studies of *o*-quinones with enamines should be undertaken[543] and one such report is available (equation 350)[566]. In spite of the very low yield this is a truly

$$\text{4-Me-catechol} + \text{NH}_2\text{CH=CHCO}_2\text{Me} \xrightarrow[\text{MeOH/3 h}]{\text{Ag}_2\text{CO}_3} \text{dioxo-indoline (Me, CO}_2\text{Me, N-CH}_2\text{CH}_2\text{CO}_2\text{Me)} \quad (350)$$

novel approach, and the authors indicate that it has an obvious relationship to the cross-linking of proteins by such quinones.

C. Other Examples of Cycloaddition Chemistry

1. Nucleophilic alkenes

In 1944 Gates studied reactions between quinones and unsymmetrical diaryl ethenes and reported the formation of the expected Michael addition products (equation 351)[567]. Nearly 40 years later modern instrumentation and synthetic methods enabled him to show that a cyclization reaction to produce compound **109** had taken place[568]. The corresponding reaction of 1,2-naphthoquinone does lead to the Michael products originally proposed (equation 352). Gates suggests that differences in charge density at the reactive carbons

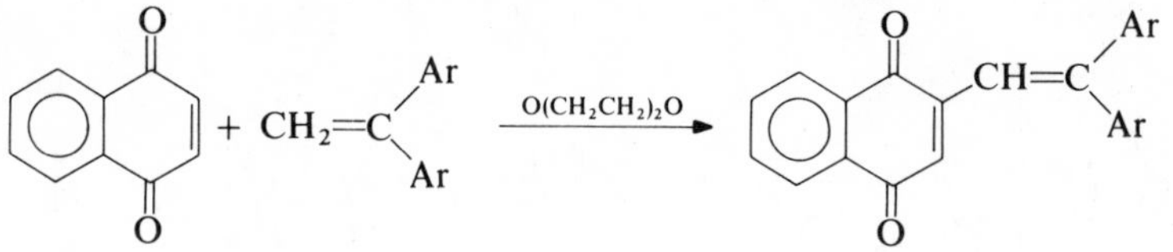

$+ CH_2{=}CAr_2 \xrightarrow{O(CH_2CH_2)_2O}$ (351)

$Ar = 4\text{-}Me_2NC_6H_4$, 4-An

59%, 40%

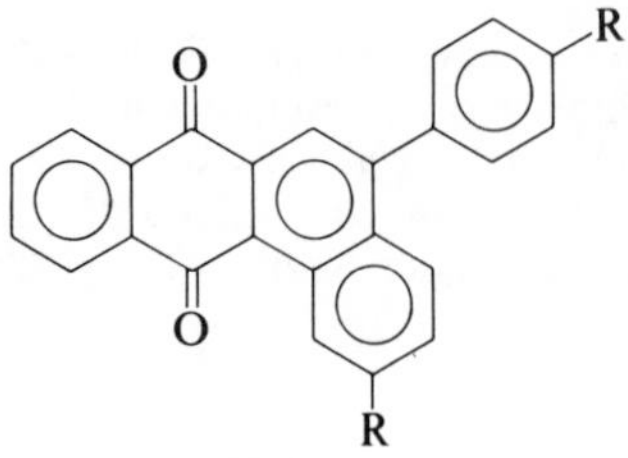

(109)

$R = NMe_2$, OMe

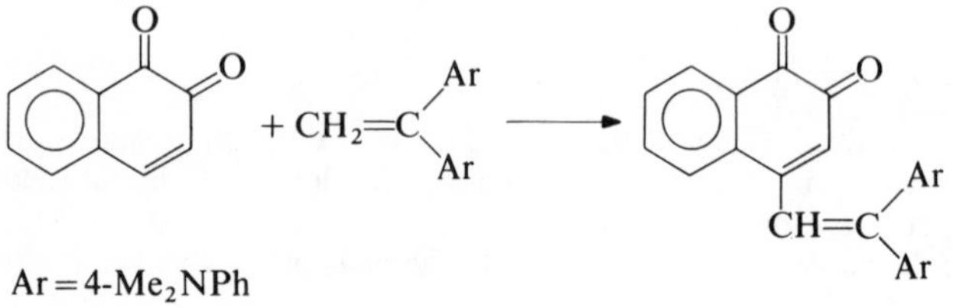

$+ CH_2{=}CAr_2 \longrightarrow$ (352)

$Ar = 4\text{-}Me_2NPh$

84%

may account for this potentially useful distinction. At least one other group has noted and utilized this chemistry[569].

Other highly polarized alkenes have been explored extensively by Brassard and Cameron and their research groups. In 1973 Brassard's group published the first of a series of studies on the cycloaddition of ketene acetals (**110**) chiefly to halogenated juglones (equation 353)[570]. Their greatest contributions lie in determining the regiospecificity of

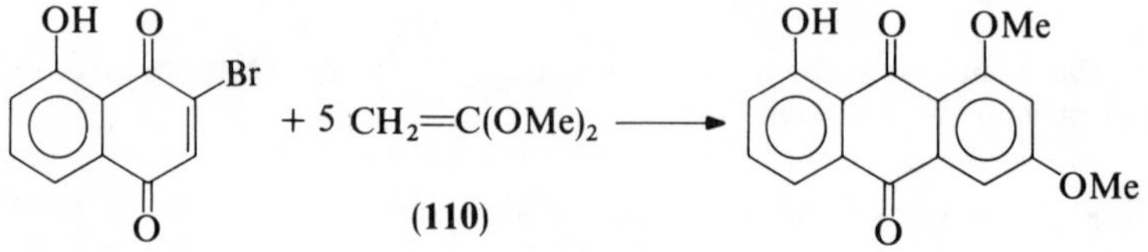

$+ 5\ CH_2{=}C(OMe)_2 \longrightarrow$ (353)

(**110**)

60%

these reactions. When 2-bromojuglone is used the isomeric anthraquinone is obtained (equation 354). In both experiments no evidence of the isomeric product was found. The reaction was also successfully applied to halonaphthazarins and isopropenylketene acetal can also be used (equation 355). An attempt to produce chloroanthraquinones showed that chloroketene acetals are too unreactive to be useful.

$$\text{5-hydroxy-2-bromo-1,4-naphthoquinone} + 5\ CH_2{=}C(OMe)_2 \longrightarrow \text{anthraquinone (HO, OMe, OMe)}\quad 53\% \tag{354}$$

$$\text{5,8-dihydroxy-2-chloro-1,4-naphthoquinone} + \text{diene (OEt, OEt, Me)} \longrightarrow \text{anthraquinone (HO, HO, OEt, Me)}\quad 72\% \tag{355}$$

The extension of this chemistry to the benzoquinone series has been accomplished in low yield but does avoid the long reaction sequences previously required (equation 356)[571].

$$\text{dichloro-dihydrobenzoquinone} + CH_2{=}C(OMe)_2 \xrightarrow{\text{anhydrous PhH}} \text{naphthoquinone (MeO, MeO, Cl)}\ 11\% + \text{hydroquinone (OH, OH, Cl, } CH_2CO_2Me)\ 42.5\% \tag{356}$$

The major product also presents interesting potential applications. The use of acetic acid leads to good yields of the desired naphthoquinones (equation 357)[572].

$$\text{2,6-dichlorobenzoquinone} + CH_2{=}C(OEt)_2 \xrightarrow{\text{HOAc}} \text{naphthoquinone (EtO, EtO, Cl)}\quad 76\% \tag{357}$$

In later work Brassard and Banville found that juglone methyl ether, which gives low yields of the expected anthraquinone, reacts slowly with the ketene acetal to produce a cyclobutane derivative (equation 358)[573]. Neither acid nor base provide any catalysis and

$$\text{naphthoquinone (MeO, Me, Cl, Br)} + CH_2{=}C(OMe)_2 \longrightarrow \text{cyclobutane adduct (MeO, Me, Cl, Br, OMe, OMe)}\quad 78\% \tag{358}$$

the cyclobutyl compound does not seem to be an intermediate in the anthraquinone reaction.

The same two scholars have found a method of converting the ketene acetals to vinylketene acetals showing promise as efficient dienes for a Diels–Alder synthesis of the anthraquinone system of important natural products (see Section V.A.2) (equation 359)[574]

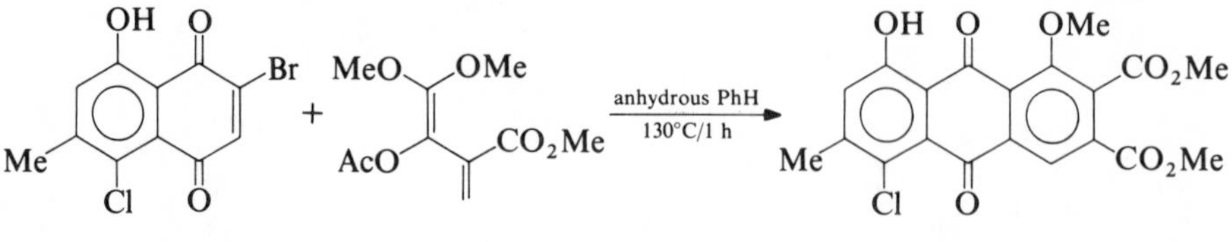

anhydrous PhH, 130°C/1 h

89% (359)

This chemistry has been applied to the preparation of a broad range of natural products through the introduction of the trimethylsilyloxy substituent (equation 360)[575].

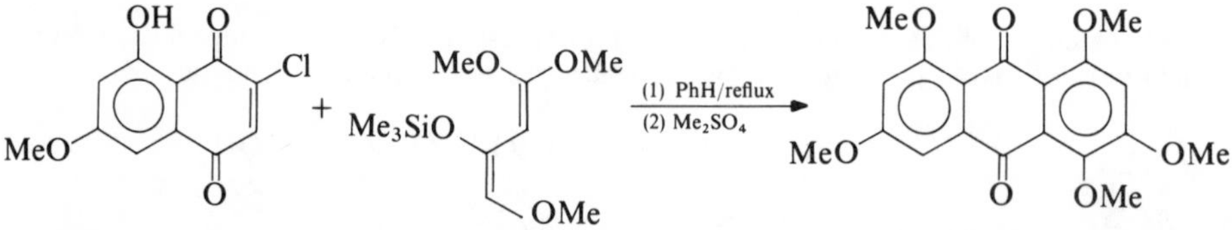

(1) PhH/reflux (2) Me_2SO_4

78% (360)

Cameron and his coworkers have not only made excellent use of the chemistry of nucleophilic alkenes in the synthesis of natural products but have added their own original touches. Some of this chemistry is closely associated with Diels–Alder[120,130] and bromination reactions[131] already cited. They have found dimethyl sulfoxide to be a useful solvent for these reactions (equation 361)[576,577].

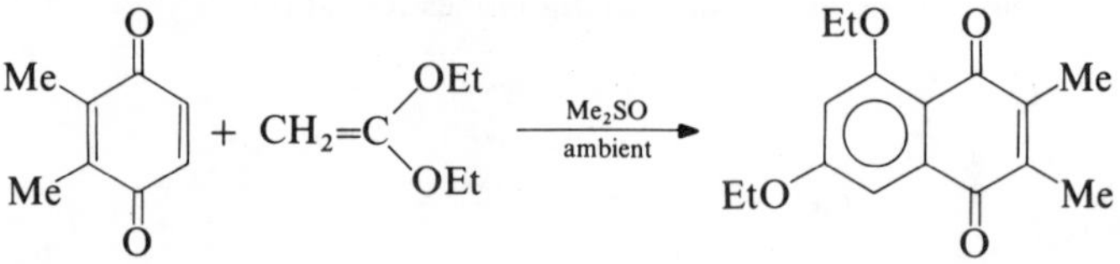

Me_2SO, ambient (361)

56%

In addition to its synthetic work[578] Cameron's group has contributed to our understanding of the mechanism of these cyclization reactions[579]. It has long been recognized that a competing 1:1 addition leads to benzofuran products and the presence of a zwitterionic intermediate (**111**) is strongly supported by the isolation of the initial adduct from both paths (equations 362, 363).

Cameron and Crossley used dimethyl sulfoxide to good advantage and complemented Brassard's studies by exploring the addition of nucleophilic alkenes to benzoquinones (equation 364)[580]. In benzene they were able to duplicate the earlier work which leads to benzofuran (equation 365). They were even able to isolate the initial orthoester adduct (**112**).

When 2-acetyl-1,4-benzoquinone was employed as the substrate rather different chemistry was found (equation 366)[581]. The overall yield was only 48% but the promise of this new route deserves to be explored.

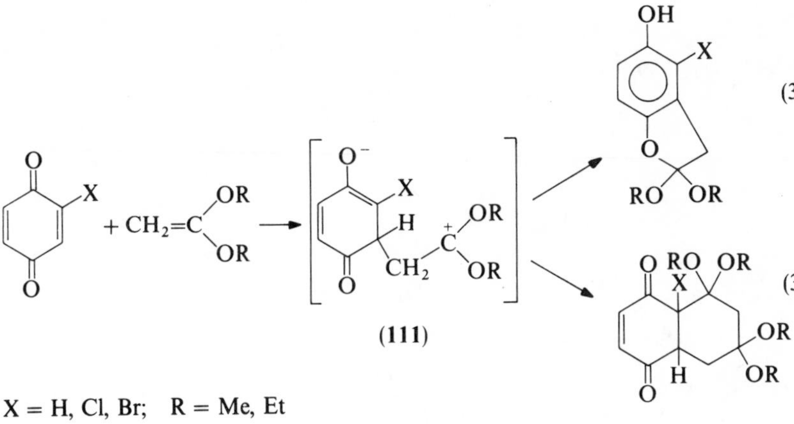

$$\text{(362), (363)}$$

$X = H, Cl, Br;\quad R = Me, Et$

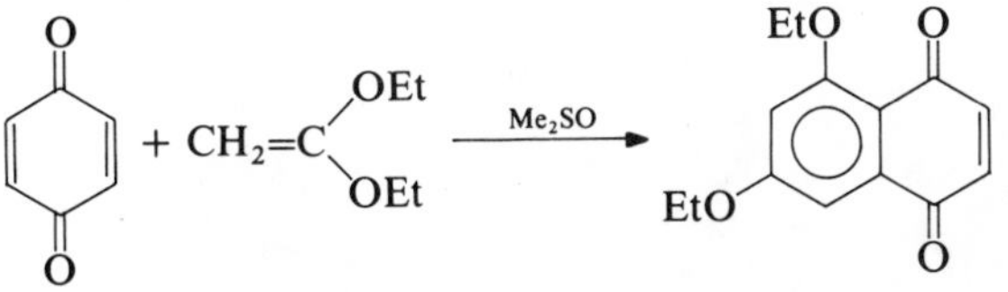

(111)

$$\text{(364)}$$

63%

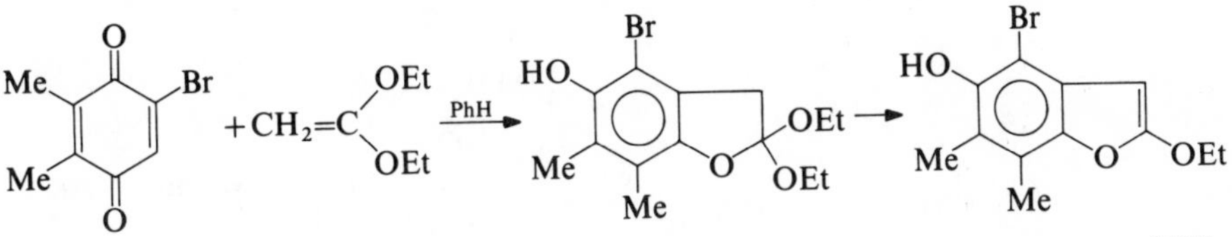

(112)

$$\text{(365)}$$

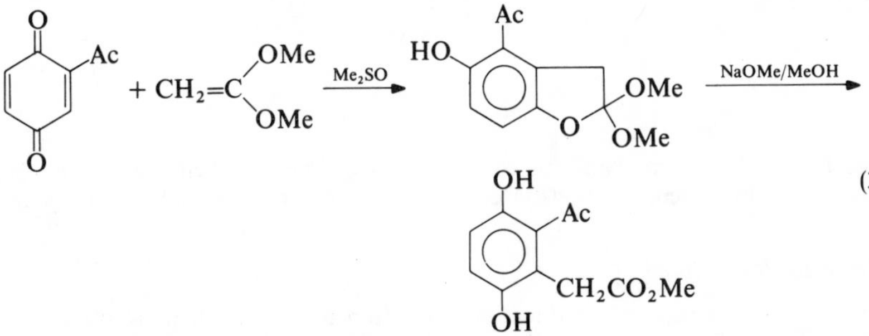

$$\text{(366)}$$

The addition of 1,1-dimethoxyethene to 1,2-naphthoquinones has also been carried out and as one might expect the product mixture is more complex and only the two major products were examined in detail (equation 367)[582].

In an attempt to expand this chemistry a 1,1-dianilinoethene was employed and good yields of a substitution product (**113**) and a dimeric quinone (**114**) were obtained (equation 368)[583a].

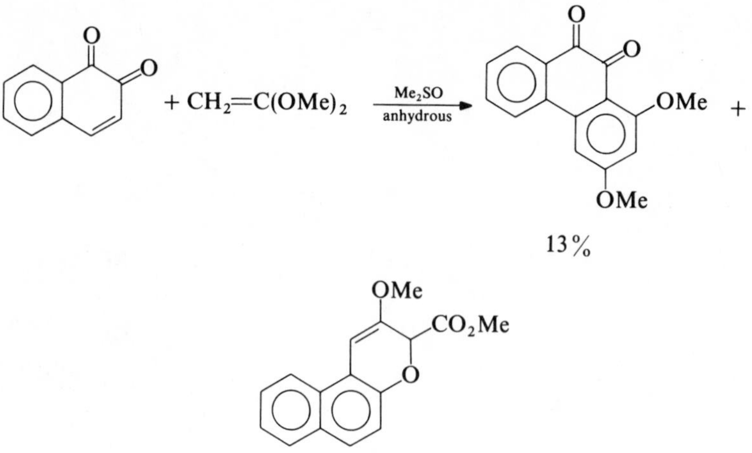

$$+ CH_2{=}C(OMe)_2 \xrightarrow[\text{anhydrous}]{Me_2SO}$$

OMe +

OMe

13%

(367)

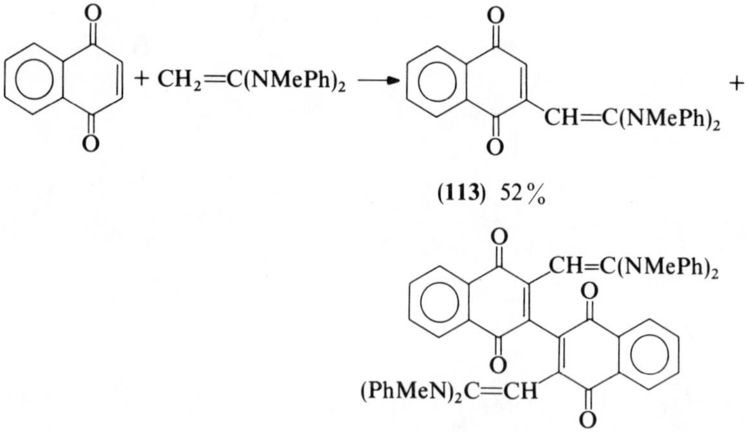

OMe

CO_2Me

O

21%

$$+ CH_2{=}C(NMePh)_2 \longrightarrow$$

$CH{=}C(NMePh)_2$ +

(**113**) 52%

$CH{=}C(NMePh)_2$ (368)

$(PhMeN)_2C{=}CH$

(**114**) 16%

Like Brassard, Cameron and his colleagues have extended their work to analogous alkoxy and silyloxy dienes in Diels–Alder approaches to their natural product goals[583b].

2. *Diazo cycloadditions*

Interest has continued in the direct cycloaddition of diazomethane to quinones with Eistert and his colleagues making substantial contributions. Further studies on the von Pechmann reaction show that the unsymmetrical isomer is formed in either the addition or substitution route to the initial unstable intermediate (equation 369)[584]. Analogous chemistry is found with diazoethane and ethyl diazoacetate. Yields are much improved (35–50%) by the addition of alkali with the greatest change (ca. 85%) found in substitution chemistry. A later study showed that when three moles of diazoalkanes are added to 2,6-dichloro-1,4-benzoquinone, one of the carbonyl groups is epoxidized and the unsym-

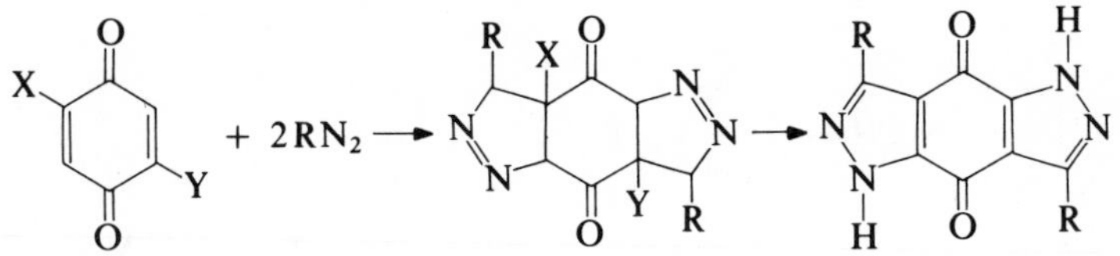

(369)

X, Y = H, Cl, Br; R = H, Me, Et, CO_2Et

metrical adduct is formed[585]. The chemistry has been extended to 2,3-dichloro- and 2-anilino-1,4-benzoquinones (equation 370)[586, 587]. In all cases initial addition takes place at

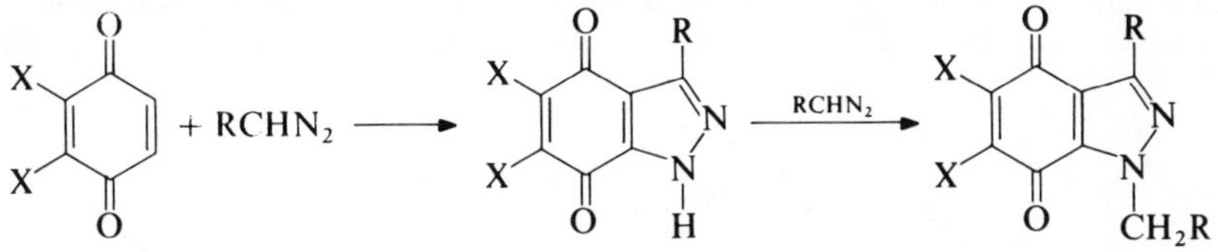

R = H, Me; Ar = Ph, 4-Tol; X = H, Cl, ArNR (370)

the unsubstituted carbon–carbon double bond and more diazoalkane attacks the pyrazole nitrogens.

An interesting and unexpected example of the ease with which diazomethane addition can overcome serious steric hindrance has been noted (equation 371)[588].

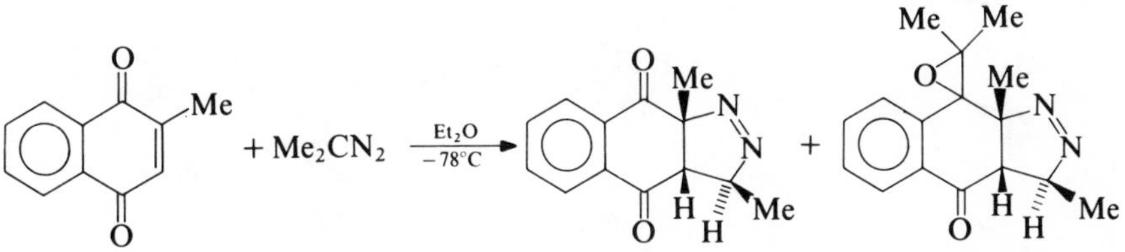

60% 12% (371)

A detailed study of the regiospecificity of diazomethane addition to 2-methyl-1,4-benzoquinone shows that one isomer is formed in significantly greater yield (equation 372)[589].

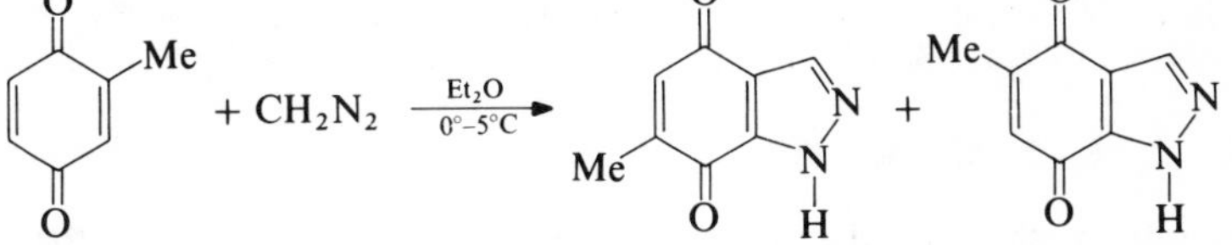

(372)

85 : 15 (30%)

Laatsch, using the dimeric naphthoquinones to synthesize diazo addition products of known structure, has been able to investigate the isomer distribution when diazomethane is added to naphthoquinones substituted in the benzenoid ring (equation 373)[590]. These experiments provided data for a sensitive test of the predictive value of HMO calculations.

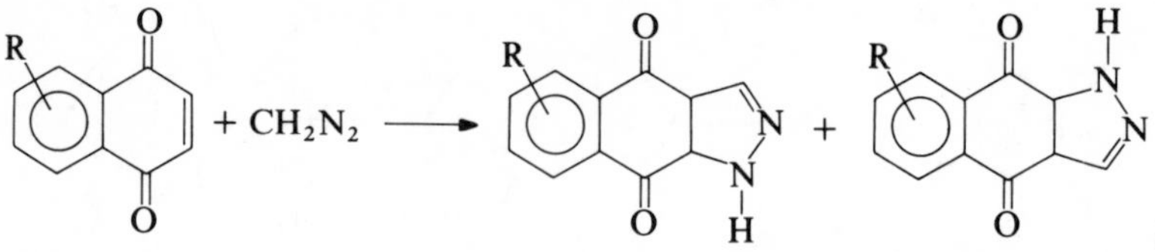

$+ CH_2N_2 \longrightarrow$ (373)

The following sequence of group effects was observed: 5-OAc > 5-OMe > 5-Cl > 5-Me > 6-Me > 6-Ac > 5-OH.

From a synthetic point of view the addition of diazoalkanes to quinones has been employed in a number of instances. For example, the adduct of 2-methyl-1,4-naphthoquinone served as a useful intermediate in the synthesis of 2-(anilinomethyl) derivatives (equation 374)[591]. Under acidic conditions the diazoalkane adducts of 2,6-di-*t*-

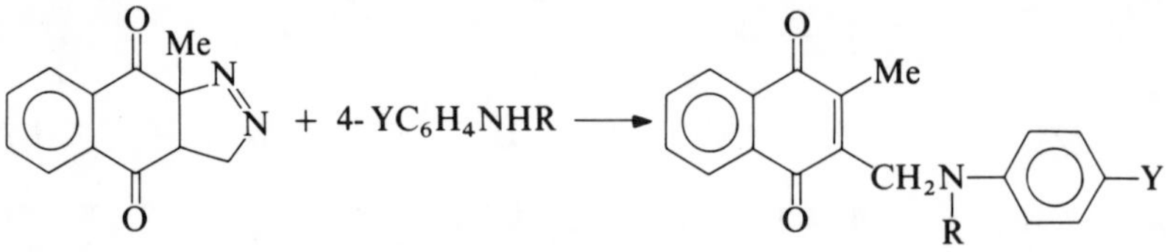

$+ 4\text{-}YC_6H_4NHR \longrightarrow$

R = H, Me, Cl; Y = Me, MeO, NMe_2 (374)

butyl-1,4-benzoquinone suffer dealkylation (equation 375)[592]. The addition of vinyl-diazomethane to various 1,4-benzoquinones provided monomers for the investigation of redox polymers (equation 376)[593].

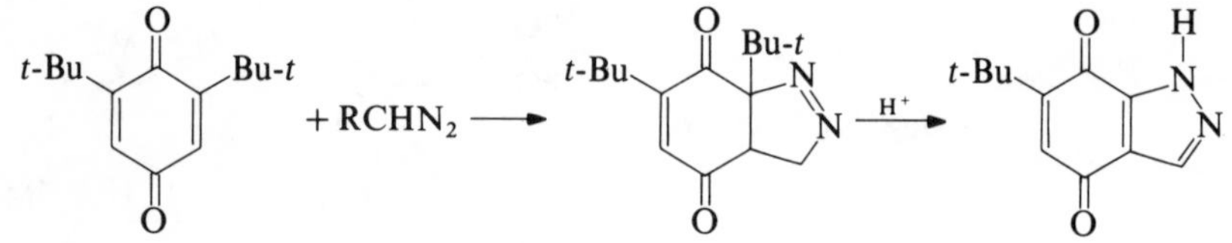

$+ RCHN_2 \longrightarrow \xrightarrow{H^+}$

R = H, Me (375)

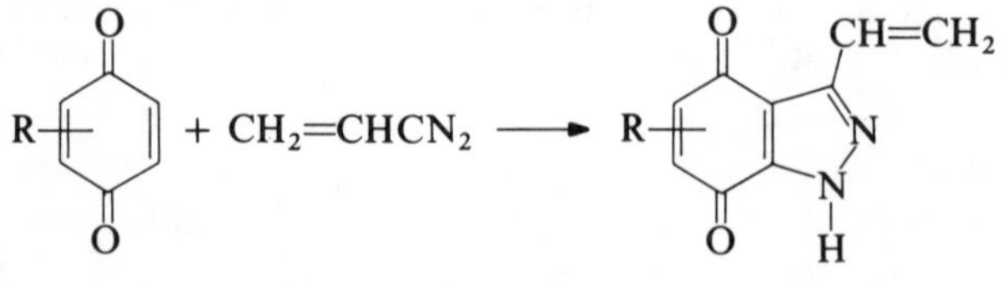

$+ CH_2{=}CHCN_2 \longrightarrow$ (376)

R = H, Cl, Me, OMe

In general the reaction of tetrahalo-1,2-quinones with diazo groups leads to attack at a carbonyl group and the formation of a dioxole. Eistert and his coworkers trapped the 'diazo aldol' from 3,3-diphenylindane-1,2-dione (**115**) by reaction with *o*-bromanil (equation 377)[594].

Latif and Meguid have used this method to prepare a series of furyl and thienyl derivatives (equation 378)[595] The ease of the reaction and the good yields as compared with their earlier experience with analogous acetophenones suggest that the heteroatoms enhance the nucleophilic character of the carbanion intermediate.

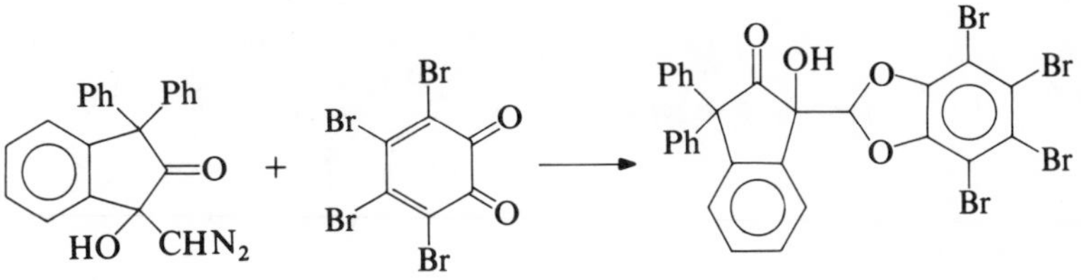

(115) 50% (377)

X=O, S; X′=Cl, Br 50–70% (378)

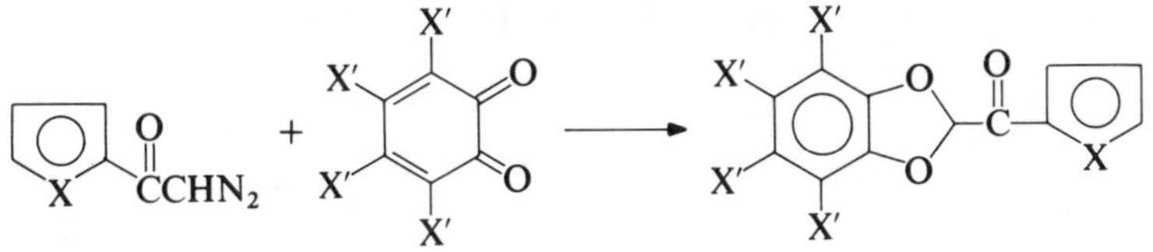

The highly strained diazoketone of a bridged bicyclobutane forms a dioxole in good yield under mild conditions (equation 379)[596]. The authors point out that this observation

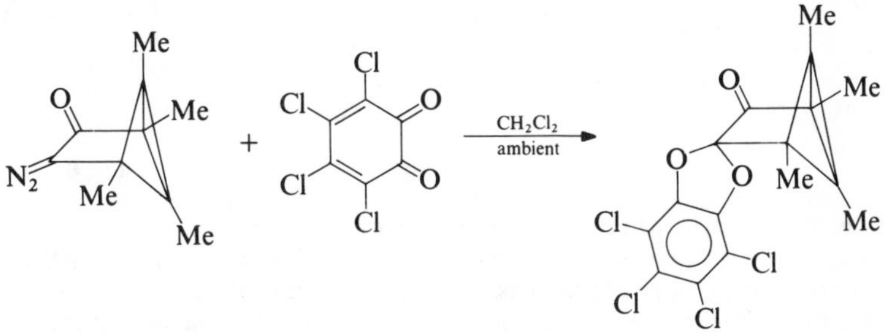

73% (379)

is more consistent with a zwitterionic intermediate than a carbene since the latter would be likely to undergo intramolecular rearrangements.

Ershov and his colleagues have shown that the actual addition product obtained with 1,2-quinones depends heavily on the ratio of starting materials[597]. When 3,6-di-*t*-butyl-1,2-benzoquinone reacts with an equimolar quantity of diazomethane only the indazole **116** is obtained (equation 380). With additional diazomethane one of the carbonyl groups

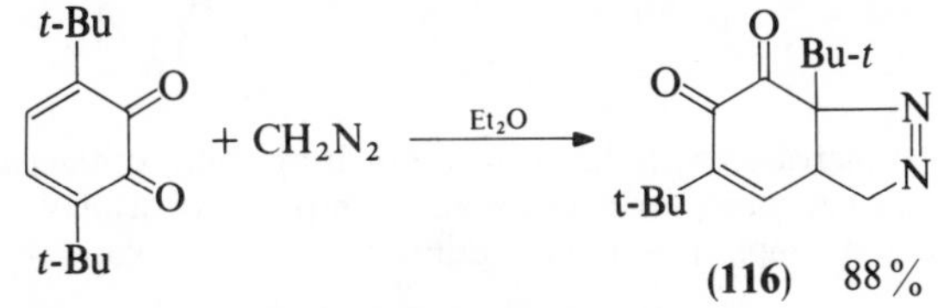

$$+ CH_2N_2 \xrightarrow{Et_2O}$$

(**116**) 88% (380)

is converted to a spirooxirane (**117**, equation 381). Finally, with a 5–10-fold excess of diazomethane both **117** and the dioxole **118** are found (equation 382).

Unfortunately, while there are some excellent mechanistic studies in this field, most concern charge transfer complexes with chloranil and are of limited direct interest. Nagai

$$(\mathbf{116}) + CH_2N_2 \xrightarrow{Et_2O} (\mathbf{117})\ 100\% \quad (381)$$

$$(\mathbf{116}) + 5\,CH_2N_2 \xrightarrow{Et_2O} (\mathbf{117})\ 38\% + (\mathbf{118})\ 55\% \quad (382)$$

and Oshima have shown that aryldiazomethanes react with a carbonyl of chloranil to give modest amounts of spirooxetanes (**119**, equation 383)[598]. The required stereochemistry of

$$\text{chloranil} + ArCHN_2 \xrightarrow{CH_2ClCH_2Cl} ArCH{=}CHAr + (\mathbf{119}) \quad (383)$$

the products is important in the discussion of probable transition states consistent with that observed. When these workers expanded their studies to diaryldiazomethanes they observed a dramatic change of products. The total absence of alkenes and spirooxetanes is attributed to the greater steric hindrance of the two aryl substituents and the decreased nucleophilicity of the diazo carbon atom[599]. Only polyethers were found as products from these reactions.

Oshima and Nagai examined the products and kinetics of the reactions of various diaryldiazomethanes with 2,5-dichloro-1,4-benzoquinone (equation 384)[600]. The second-

$$\text{2,5-dichloro-1,4-benzoquinone} + Ar_2CN_2 \xrightarrow[30^\circ C]{CH_2ClCH_2Cl} \text{cyclopropane adduct} + \text{polyethers} \quad (384)$$

order rate constants increased with the electron-donating ability of the *para* substituent. A Hammett plot of the constants for the cycloaddition pathway display an excellent linear dependence indicating simple 1,3-dipolar addition.

3. 1,3-Dipolar cycloadditions

The greatest interest in these additions has centered on the mesoionic oxazolium-5-oxide system. Myers and his colleagues showed that the 3-methyl-2,4-diphenyl derivative (**120**) adds in good yield to the 1,4-benzoquinone–cyclopentadiene Diels–Alder adduct (equation 385)[601]. On refluxing in benzene this hexacyclic system loses carbon dioxide and opens

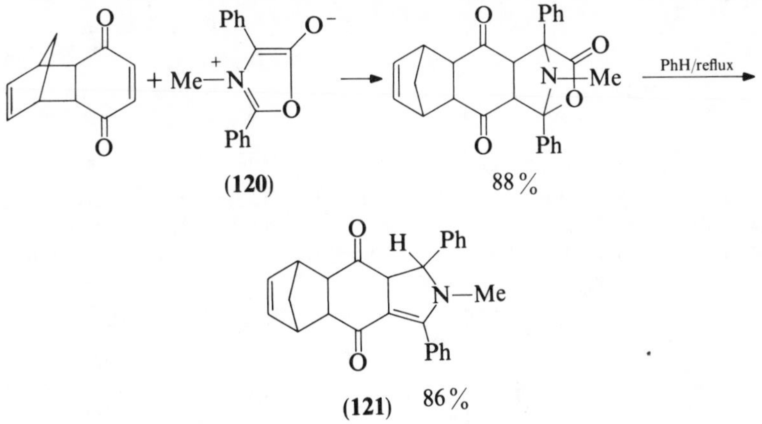

(120)

88%

PhH/reflux

(385)

(121) 86%

a new pathway to the important 2*H*-isoindole-4,7-diones (**121**). Friedrichsen and his collaborators have been able to isolate the acid intermediate of the decarboxylation reaction and they obtained both the mono- and bis-adducts with 1,4-benzoquinone itself[602].

With 1,2-benzoquinones the reaction of either **120** or its sulfur analog takes place through an open-chain ketene form and a lactone results (equation 386)[603].

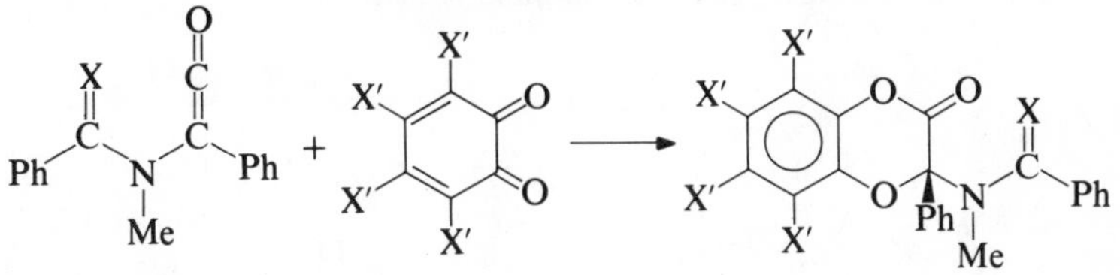

(386)

X = O, S X′ = H, Cl

Matsukubo and Kato have questioned the structure of a pyrolysis product of the initial adduct and at the same time confirmed the structures of several important quinonoid products[604].

Matsumoto and coworkers have examined the more elaborate system, anhydro-5-hydroxyoxazolium hydroxide (**122**) and found useful yields of heterocyclic quinones (equation 387)[605].

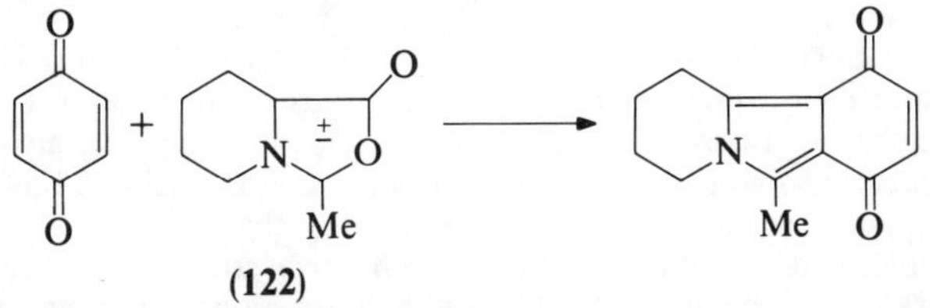

(387)

(122)

Myers has expanded the work of his group to include several more sensitive reactants by generating them *in situ* (equation 388)[606]. The yields are lower but the relative ease of making these heterocyclic systems makes the route attractive.

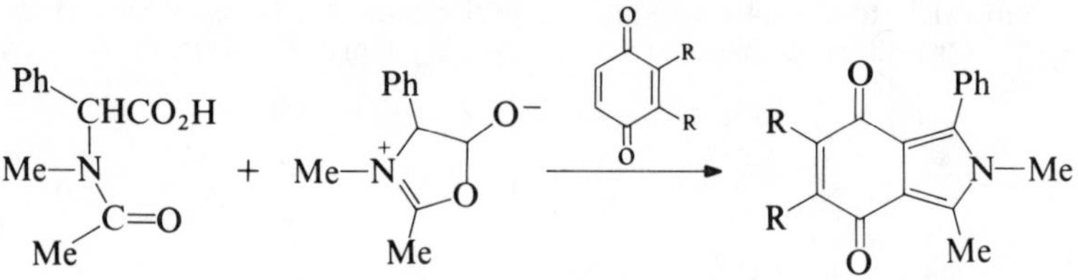

(388)

R = H, Me, OMe

The reaction of 6-oxo-6*H*-1,3-oxazin-3-ium-4-olate (**123**) with 1,2-benzoquinone or *o*-chloranil gives a product which must arise from a much more complex sequence of steps (equation 389)[607]. The high yields and the polycyclic structure make this too a promising

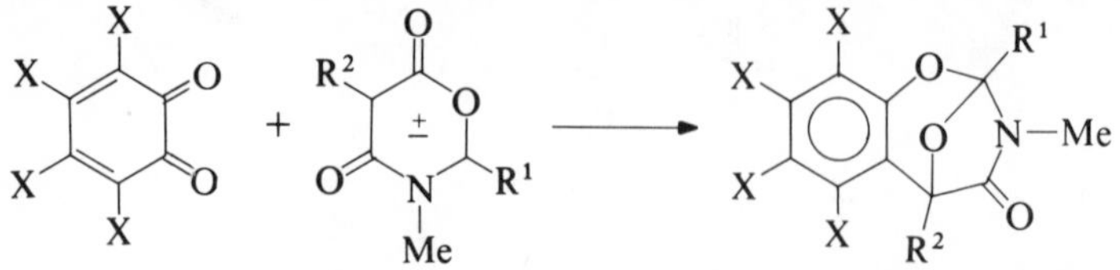

(**123**) 60–90%

X = H, Cl; R^1 = Me, Ph, Bz, 4-Tol 4-An; R^2 = Ph, Bz (389)

route worthy of further investigation. A perhaps distantly related addition reaction involves 2,6-piperazinediones (**124**) and chloranil (equation 390)[608].

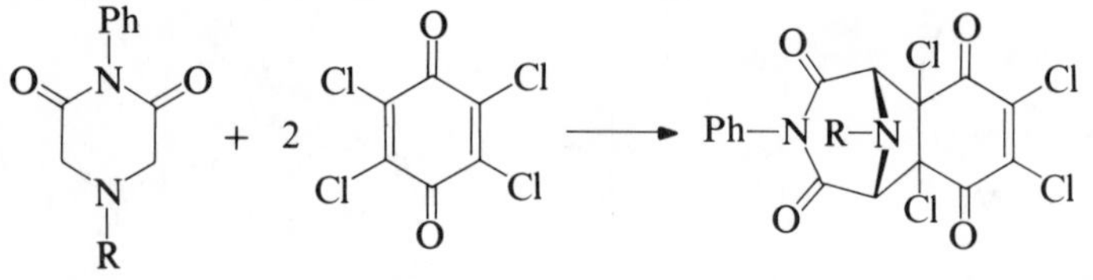

(**124**)

R = Me, Ph 59%, 31% (390)

A recently published route to the isoindole quinonoid structure involves the addition of an azomethine ylide generated in the presence of a quinone (equation 391)[609]. Once again the ease of the reaction and the good yields make this an attractive route to an important class of compounds.

Still another route to the 4,7-isoindolediones involves the addition of aryl isocyanides to 1,4-benzoquinones (equations 392, 393)[610]. The yields of mono-adduct are quite good (30–70%) and the two isomeric di-adducts are formed in approximately equal yields. A zwitterionic intermediate involving carbon–carbon bond formation followed by reaction with a second equivalent of isocyanide is proposed as the mechanistic pathway.

The 1,3-dipolar addition of nitrile *N*-oxides have been studied extensively by Shiraishi and his coworkers. In reactions with chloranil only reactions at the carbonyl groups are observed (equation 394)[611]. Both mono- and di-adducts are obtained and the yields are generally excellent. With less highly substituted quinones both carbonyl and ring-carbon addition take place (equation 395)[612]. The dimethyl quinones give mainly isoxazolines.

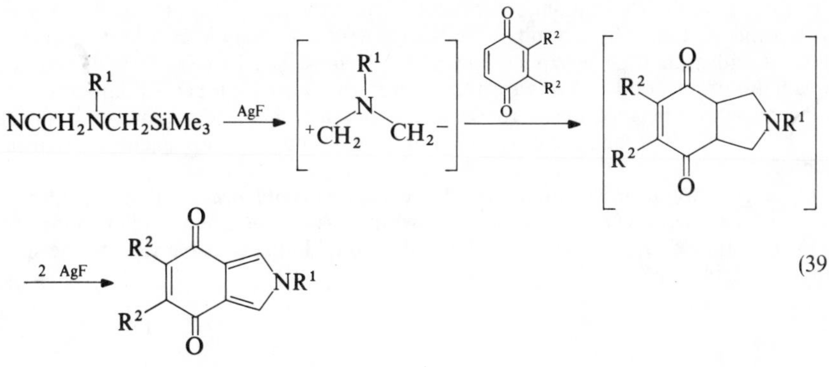

$$NCCH_2N(R^1)CH_2SiMe_3 \xrightarrow{AgF} [\,^{+}CH_2N(R^1)CH_2^{-}\,] \longrightarrow [\,R^2, R^2, O, O, NR^1\,] \xrightarrow{2\ AgF}$$

R^2 R^2 O O NR^1

60–75%

R^1 = Me, Bz; R^2 = H, Me, MeO

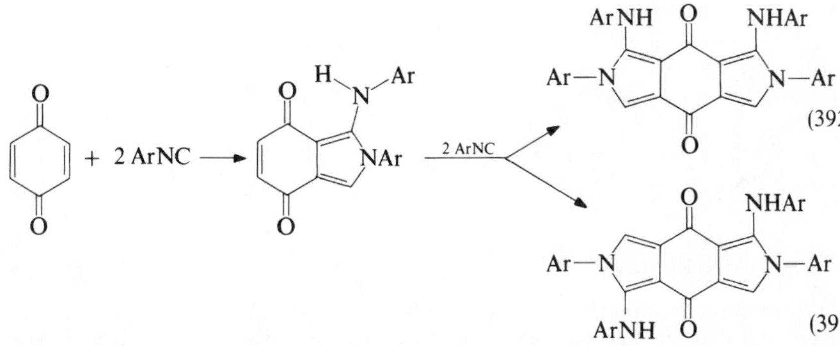

(391)

O O + 2 ArNC ⟶ H N Ar, O O NAr —2 ArNC→

ArNH O NHAr

Ar—N N—Ar

O

(392)

O NHAr

Ar—N N—Ar

ArNH O

(393)

X X X X O O $+ ArC{\equiv}N^{+}{-}O^{-} \longrightarrow$ X X N—O O Ar X X =O + X X N—O O O—N O Ar X X Ar

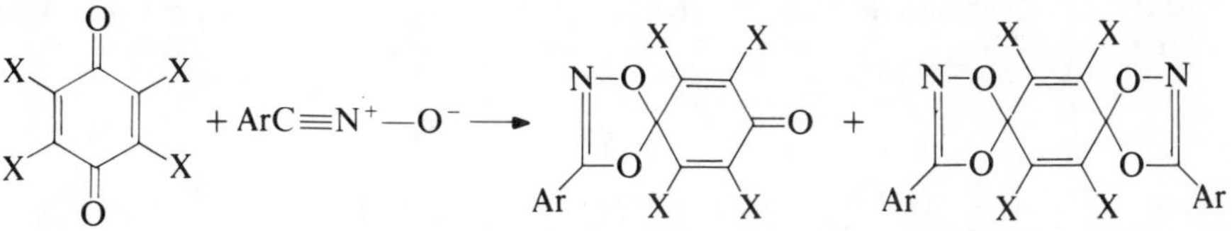

X = F, Cl, Br, I; Ar = Ph, 2,4,6-$Me_3C_6H_2$, 2,3,5,6-Me_4C_6H isomers 1:1 (394)

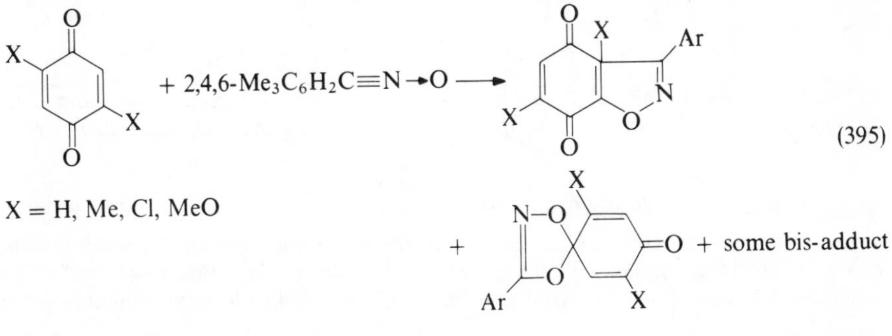

X X O O $+ 2,4,6\text{-}Me_3C_6H_2C{\equiv}N{\rightarrow}O \longrightarrow$ O X Ar N O X O

(395)

X = H, Me, Cl, MeO

\+ X N—O O Ar X =O + some bis-adduct

Electronic effects of the quinone substituents explain the reactivity of the carbonyl groups toward this reagent. These results have been extended to include a variety of alkyl and alkylhalo substituted 1,4-benzoquinones and the regiospecificity described[613]. The yields remain uniformly good, and it is apparent that steric effects are much less important than electronic ones. A detailed analysis of these reactions in terms of frontier molecular orbitals has been presented and some predictions confirmed by subsequent experimental studies[614].

Trifluoroacetonitrile *N*-sulfide, which bears a resemblance to the *N*-oxides just discussed, has also been prepared and reacts with 1,4-naphthoquinone and juglone in poor yields (equation 396)[615]. Such compounds also do add in reasonable yield to the simple quinones (equation 397)[616].

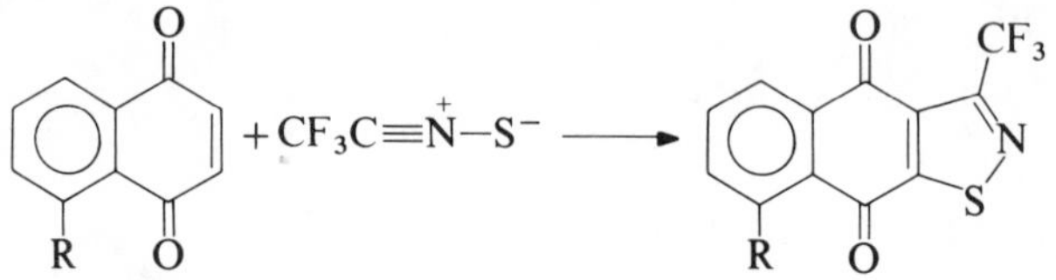

$$+ CF_3C{\equiv}\overset{+}{N}{-}S^- \longrightarrow \quad (396)$$

R = H, OH

$$\xrightarrow{\text{heat}} ArC{\equiv}\overset{+}{N}{-}S^- \longrightarrow \quad (397)$$

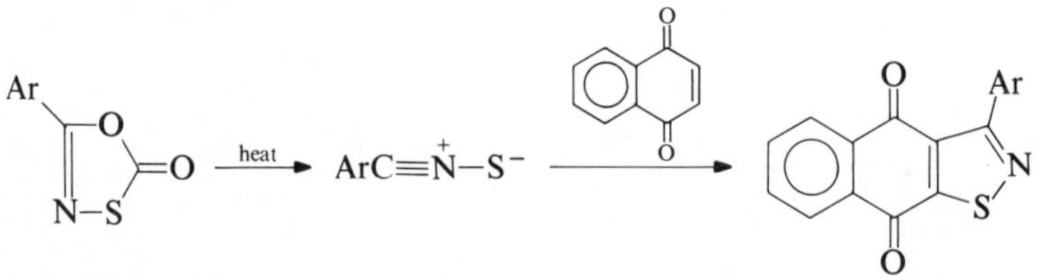

Ar = Ph, 4-An

36%, 42%

Two interesting sulfur dipolar reagents have been reported. Sasaki and his colleagues have found that the highly strained bisspiroadamantane thiadiazine **125** loses nitrogen smoothly in refluxing xylene and the resulting ylide adds to 1,4-benzoquinone (equation 398)[617].

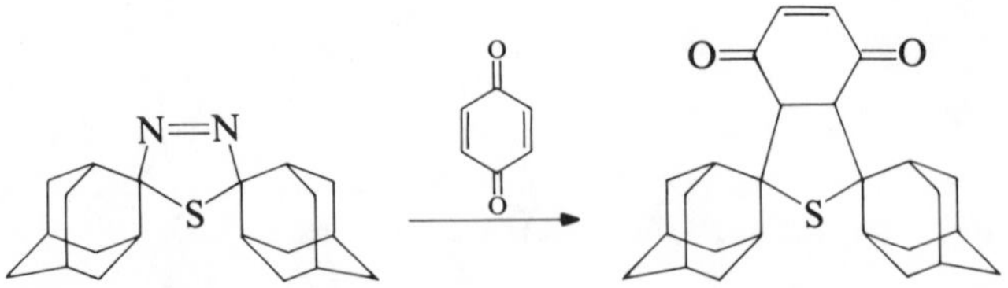

$$\longrightarrow \quad (398)$$

(125) 59%

The mesoionic 1,3-dithiolylium-4-olate (**126**) adds to 1,4-benzoquinone and loses carbon oxysulfide to form the corresponding thiophene quinone (equation 399)[618].

4. *Homophthalic anhydride cyclization*

A new route to anthracyclinones was developed by Tamura and his collaborators by the condensation of quinones with homophthalic anhydrides (equation 400)[619,620]. This scheme has been applied successfully to the generation of key tetracyclic intermediates

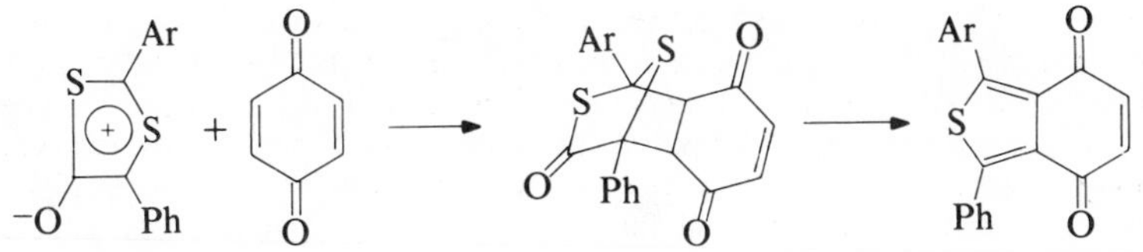

(126)
Ar = Ph, 4-Tol

60%, 55%   21%, 42%   (399)

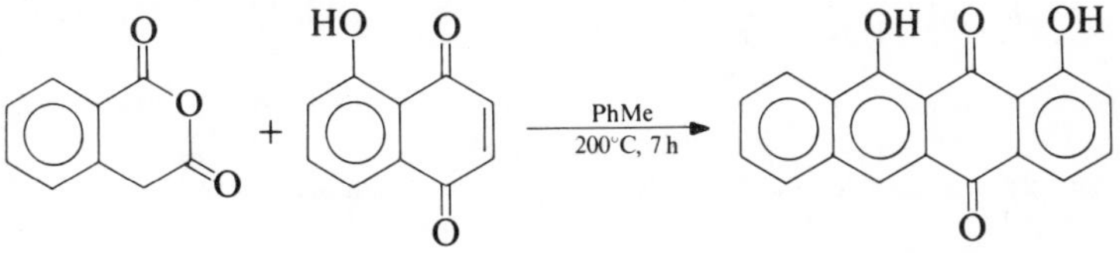

PhMe, 200°C, 7 h

49%   (400)

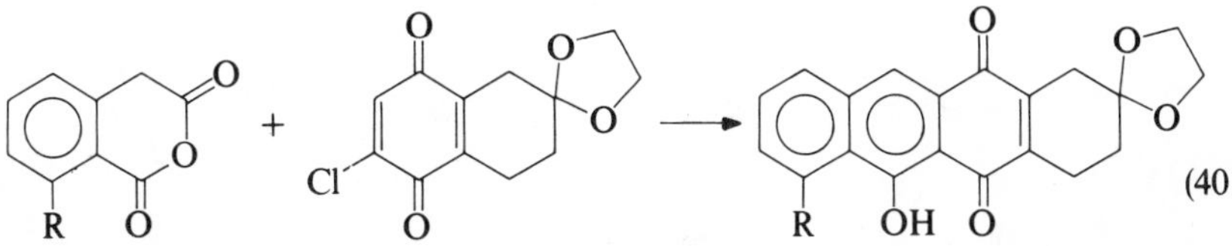

R = H, OMe   43%   (401)

(equation 401). In the case of the 8-methoxy compound cycloaddition took place in poor yield but was greatly improved (65%) using the lithium salt of the anhydride. This technique has been explored and found to give excellent results[621, 622]. The optimum method appears to be treatment of the anhydride with sodium hydride in dry tetrahydrofuran (equation 402). The application of this powerful new method to the synthesis of complex natural products continues at a high level[623, 624].

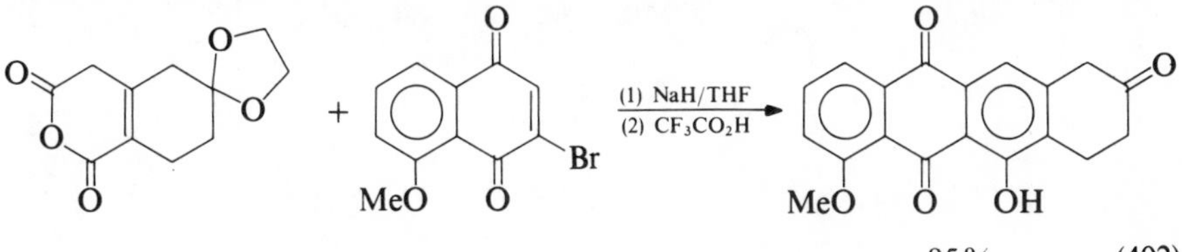

(1) NaH/THF
(2) CF_3CO_2H

95%   (402)

VI. ARYLATION OF QUINONES

A lively interest has continued in joining aromatic rings to quinones. The use of diazonium salts, the Meerwein reaction, represents the method of choice in most instances, and a detailed review of it appeared[625], as well as a report of the absolute rate constants for the addition of aryl radicals to 1,4-benzoquinone (equation 403)[626]. The radicals show near diffusion-controlled rates and are not affected by significant polar considerations. The rate constant obtained for the phenyl radical ($8.8 \times 10^8\,dm^3 mol^{-1} s^{-1}$) agrees well with that reported for radicals generated by radiolysis of diphenyl sulfoxide

$$ArN_2^+ \xrightarrow{Cu/Me_2SO} ArN_2^\cdot \longrightarrow Ar^\cdot$$

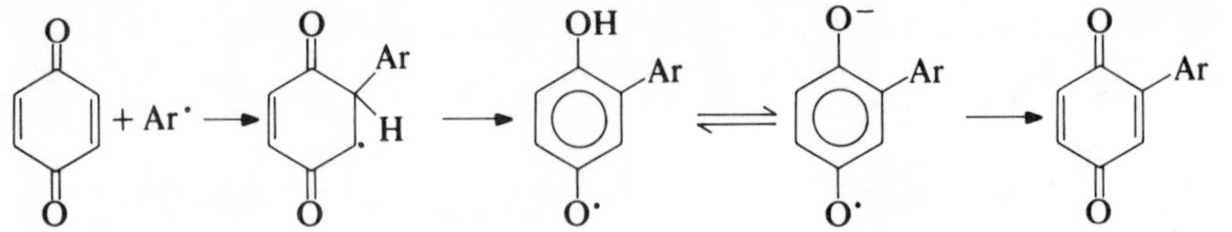

O, O + Ar· → O, Ar, H, O → OH, Ar, O· ⇌ O⁻, Ar, O· → O, Ar, O

$Ar = 4\text{-}XC_6H_4$ (X = H, Cl, Me, MeO, NO_2) (403)

($1.2 \times 10^9\ dm^3 mol^{-1} s^{-1}$)[627]. An electrochemical investigation supports the importance of the semiquinone radical ion as an intermediate[628].

A few purely synthetic applications have appeared in connection with studies of natural products. Mondon and Krohn have used some highly oxygenated diazonium salts to good advantage (equation 404)[629]. The low yield with a free hydroxyl group is not surprising

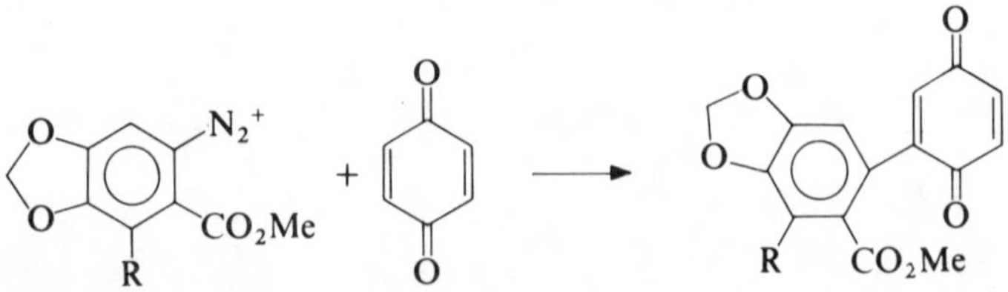

O, O, N_2^+, CO_2Me, R + O, O → O, O, O, O, R, CO_2Me (404)

R = H, OH, OBz 87%, 24%, 68%

and it is easily overcome as indicated. Cameron and his coworkers have observed that a trace of hydroquinone is a significant catalyst for the Meerwein reaction[120]. They found a large difference in yield for two closely related reactions but did not offer any explanation (equation 405).

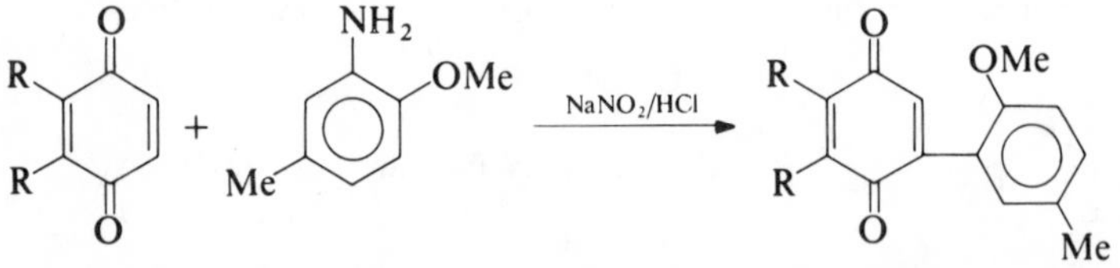

O, R, R, O + NH_2, OMe, Me $\xrightarrow{NaNO_2/HCl}$ O, R, R, O, OMe, Me (405)

R = H, Me 53%, 20%

Tertiary aromatic amines can be effective arylating reagents (equation 406)[630] The yields are variable (10–66%) and the corresponding substitution chemistry with 2,3-dichloro-1,4-naphthoquinone was especially disappointing.

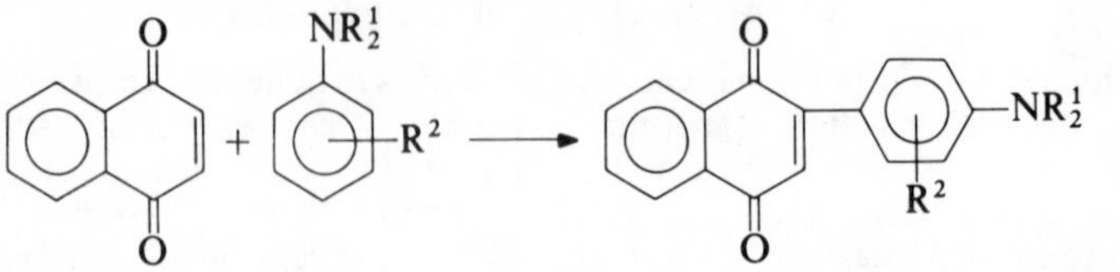

O, O + NR^1_2, R^2 → O, O, NR^1_2, R^2 (406)

R^1 = Me, Et; R^2 = 2-Me, 2-MeO, 2-Cl, 2,6-Me_2

Reactions between aromatic ethers and quinones constitute a means of arylation and have received a good deal of study, particularly by Musgrave and his collaborators. They have examined in great detail the reactions of veratrole (**127**) with 2,5- and 2,6-dichloro-1,4-benzoquinone (equation 407)[631]. The quinone products were obtained after oxidation

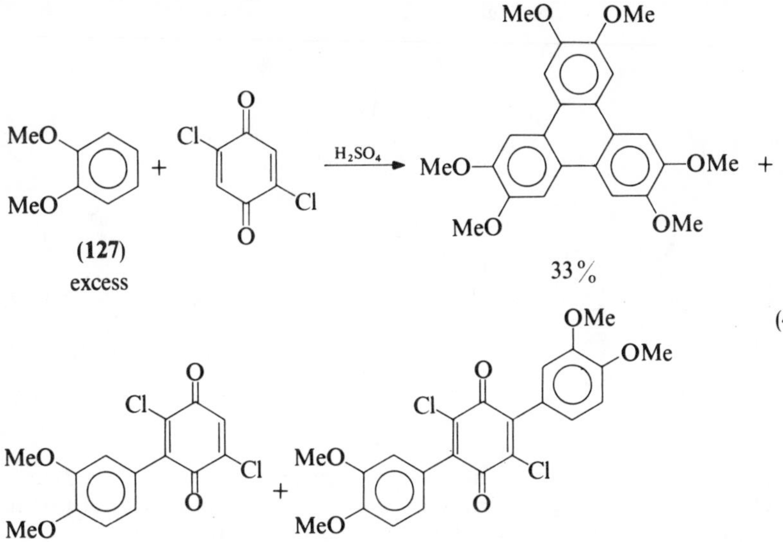

MeO OMe

MeO Cl + O Cl O Cl $\xrightarrow{H_2SO_4}$ MeO— —OMe +

MeO MeO OMe

(**127**) excess 33%

OMe (407)

OMe

Cl O Cl O

MeO Cl MeO Cl

+

MeO O MeO O

50% (**128**) 13%

with ferric chloride. If the ratio of reactants is reversed the principal products are the diarylquinone (**128**) and a dibenzofuran (**129**) both formed in 38% yield. The chemistry of

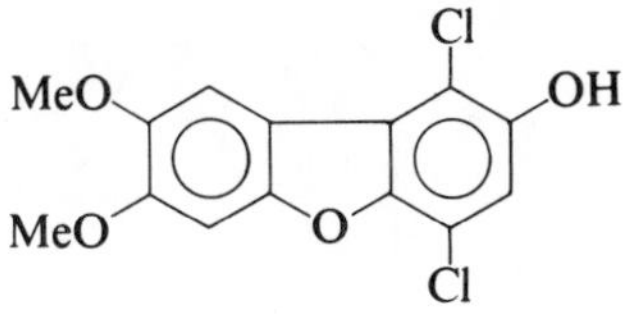

Cl

MeO OH

MeO O

Cl

(**129**)

the 2,6-dichloro analog produces a closely related product distribution, and show similar results with 1,4-benzoquinone and its 2-chloro derivative[632, 633]. In the latter case aluminum chloride in carbon disulfide is an effective catalyst while several other traditional Lewis acids fail completely. Evidence is presented that the arylbenzoquinones isolated from the reaction mixtures are true intermediates in the formation of the major product which is a triphenylenediquinone (**130**). In demonstrating the structure of these quinonoid products several applications of the Meerwein reaction have been employed[634, 635].

By extending these studies to naphthoquinones and aryl ethers Buchan and Musgrave found also two other product types, a dimeric quinone (**131**) and a dinaphthofuran with the 1,2-quinonoid structure (**132**, equation 408)[636]. The latter structure was demonstrated by the unambiguous synthesis of the isomeric 1,4-quinone.

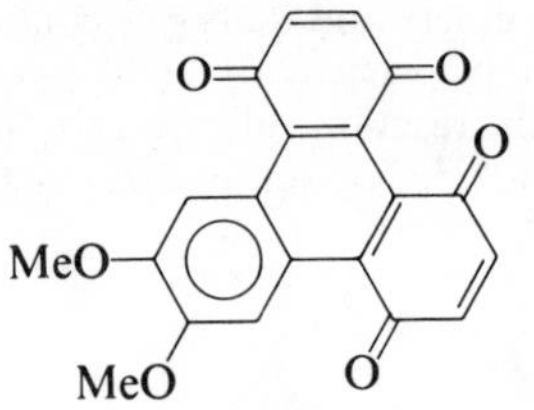

(130)

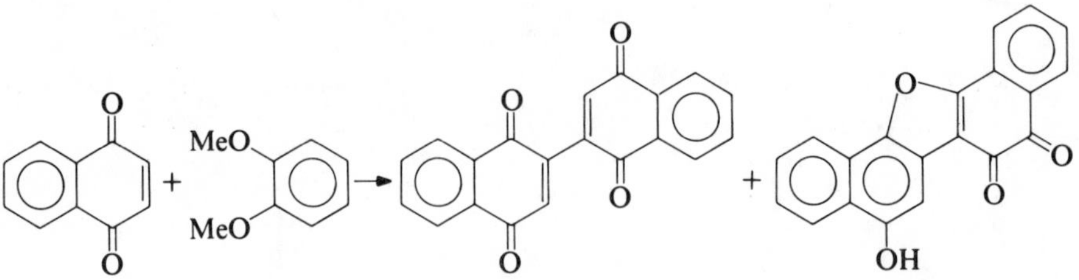

(131) 18%

(132) 5%

(408)

The synthesis and study of dimeric naphthoquinones has been a principal interest of Laatsch. Working with juglone and various methoxynaphthoquinones in pyridine or acetic acid, he has been able to obtain synthetically useful yields of dimers and tricyclic quinones analogous to those found by Musgrave (equations 409, 410)[637] These results are

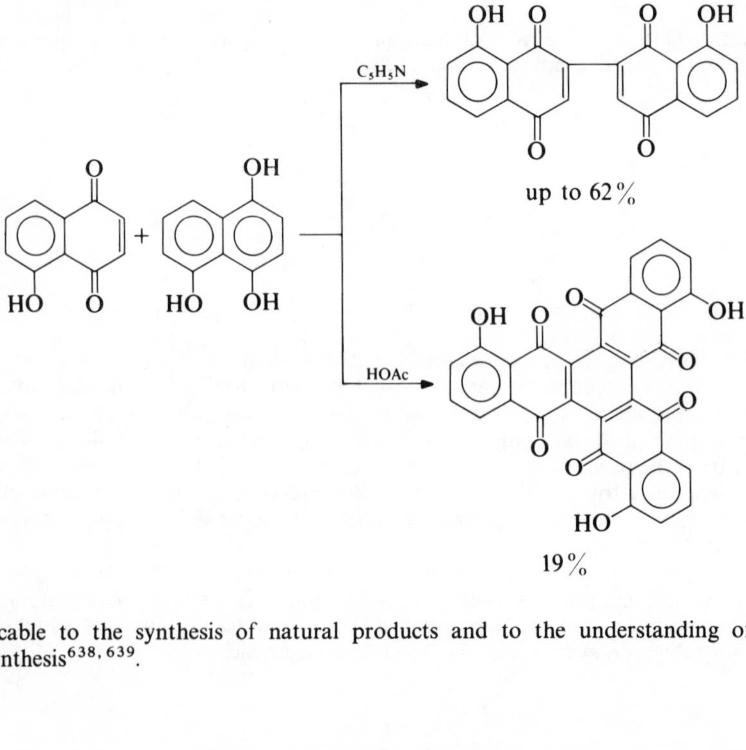

C_5H_5N (409)

up to 62%

HOAc (410)

19%

applicable to the synthesis of natural products and to the understanding of their biosynthesis[638, 639].

The frequent presence of benzofurans in natural products makes the preparation of this heterocyclic system especially desirable and the quinone–phenol route attractive. Both benzo-[640] and naphthoquinones[641] react with phenols to give key intermediates in reasonable yields (equation 411). The analogous reactions with resorcinol have been

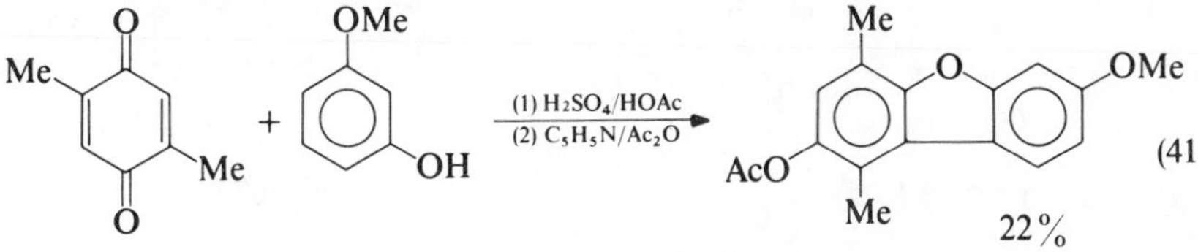

(1) H_2SO_4/HOAc
(2) C_5H_5N/Ac_2O

22% (411)

studied in some detail; an ionic intermediate and the alkyl substitution pattern of the quinone appear to explain satisfactorily the widely differing yields observed[642].

The arylation of quinones by substituted furans also plays a potentially important role in developing new synthetic methods. Kraus and his coworkers have examined the use of butenolide anions for this purpose (equation 412)[643, 644]. The presence of the 2-acetyl

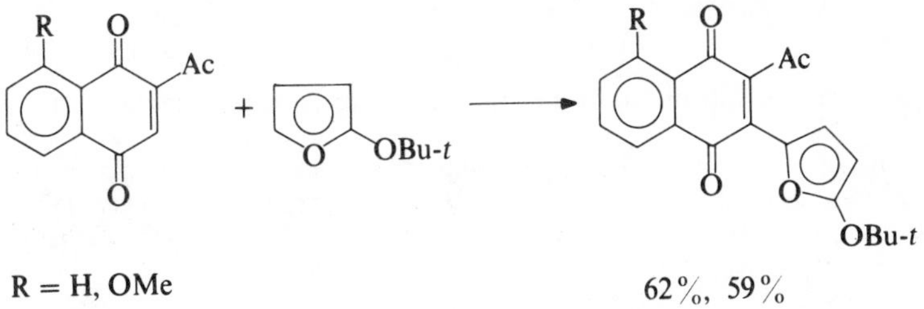

R = H, OMe 62%, 59% (412)

group is essential since in its absence no reaction took place even with added Lewis acid.

4-Hydroxycoumarin (**133**) has been studied by several groups in efforts to prepare compounds with potentially useful medicinal activity. This lactone adds to 1,4-naphthoquinone and the product has a furan ring (**134**, equation 413)[645]. Slower reactions of 3-

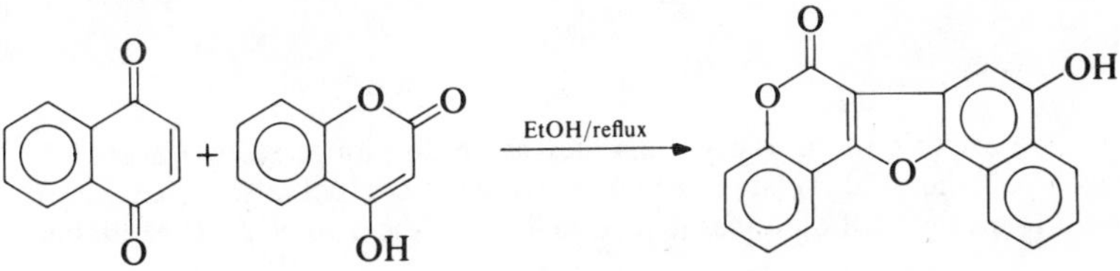

EtOH/reflux (413)

(**133**) (**134**)

phenylcoumarin with either 1,4-benzo- or 1,4-naphthoquinone produce the initial adduct (**135**) in lower (41 and 25%) yields. Similar chemistry with 1,4-benzoquinones (equa-

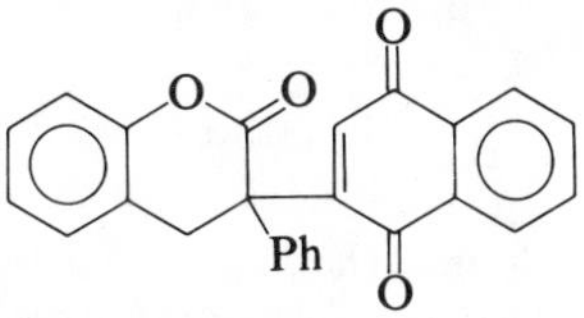

(**135**)

tion 414)[646] gives very reasonable yields. The addition of 3-hydroxycoumarin to 1,4-benzoquinone has also been demonstrated (equation 415)[647]. The statement is made that

$$\text{4-hydroxycoumarin} + \text{R-1,4-benzoquinone} \xrightarrow{Me_2C{=}O} \text{product } (40\text{–}50\%) \qquad R = H, Me, Cl \tag{414}$$

$$\text{3-hydroxycoumarin} + \text{1,4-benzoquinone} \xrightarrow[H_2O/0.5\,h]{Me_2C{=}O} \text{product } (70\%) \tag{415}$$

the reaction can be used with substituted quinones and 1,2-benzoquinone but no data are presented.

The addition of 4-hydroxy-5-methylcoumarin to 2- and 7-methyljuglones led to the efficient new synthesis of two natural products[648]. The former reaction is notable for the ease with which two bulky groups are introduced at adjacent positions (equation 416).

$$\text{4-hydroxy-5-methylcoumarin} + \text{juglone (R)} \longrightarrow \text{product} \qquad R = H, Me \tag{416}$$

A new and exciting route to aryl quinones has been pioneered by Itahara[649,650]. In acetic acid, palladium(II) acetate reacts with aromatic compounds and couples them in good yield to a variety of quinones (equation 417). In the case of 1,4-benzoquinone, 2,5-

$$\text{1,4-benzoquinone} + Pd(OAc)_2 + ArH \xrightarrow[\text{reflux}/14\,h]{HOAc/N_2} \text{2-Ar-1,4-benzoquinone} \tag{417}$$

$Ar = Ph, 2,5\text{-}Me_2C_6H_3, 2,5\text{-}Cl_2C_6H_3$

and 2,6-diaryl products are also found. The single instance in which a monosubstituted quinone was studied produced a mixture of two isomeric products (equation 418). Still the result shows that unsymmetrical diarylquinones can be made using this chemistry. Itahara

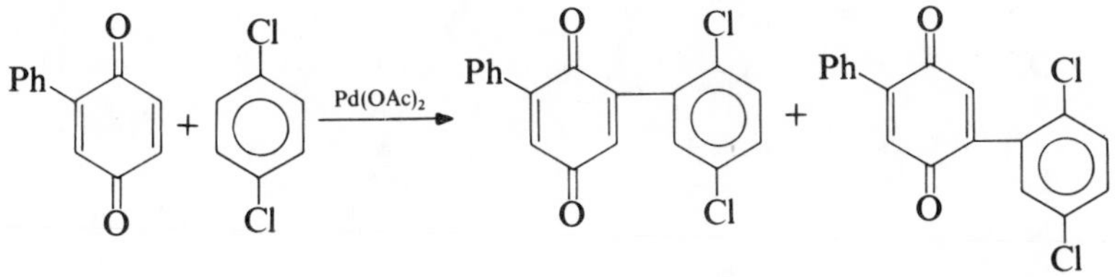

$$\text{(418)}$$

36% 19% (418)

has also made some progress in the use of reoxidants to improve the efficiency of the palladium[651]. Naphthoquinones and 1,2-quinones are useful in this reaction and its further study seems important, especially in the synthesis of heterocycles bearing quinonoid substituents where it gives excellent results over a broad range of structures (equation 419)[652]. Itahara overlooked an earlier report describing similar chemistry with

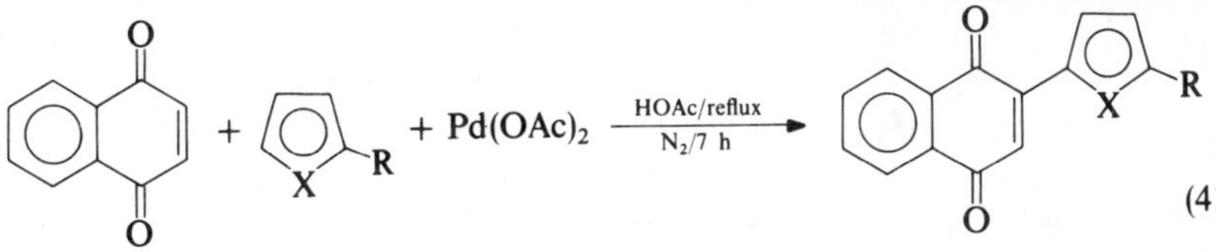

$$+ \; Pd(OAc)_2 \xrightarrow[N_2/7\ h]{HOAc/reflux} \qquad (419)$$

$X = O, S, NSO_2Ph$; $R = CHO, Ac, H$ 50–70%

palladium dichloride and sodium acetate[653], but he did make a large improvement in yields and has investigated the range of the reaction in detail.

Brief reports of several other organometallic arylation reactions of quinones have appeared. The treatment of 1,4-benzoquinone with diphenylcadmium (equation 420)[654] and with triphenylstibine (equation 421)[655] are examples. The change in degree of

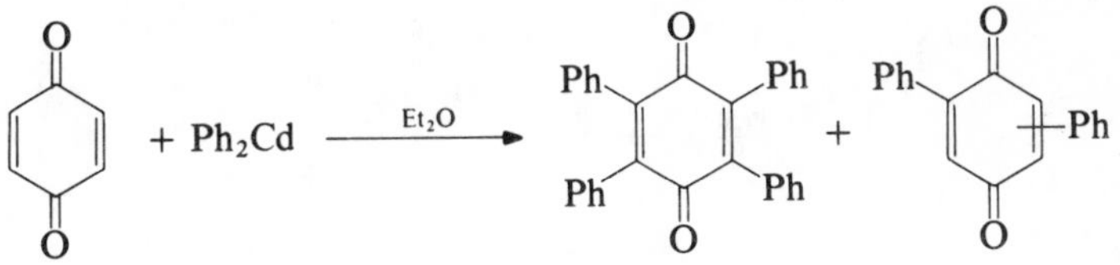

$$+ \; Ph_2Cd \xrightarrow{Et_2O} \qquad (420)$$

10% 3%

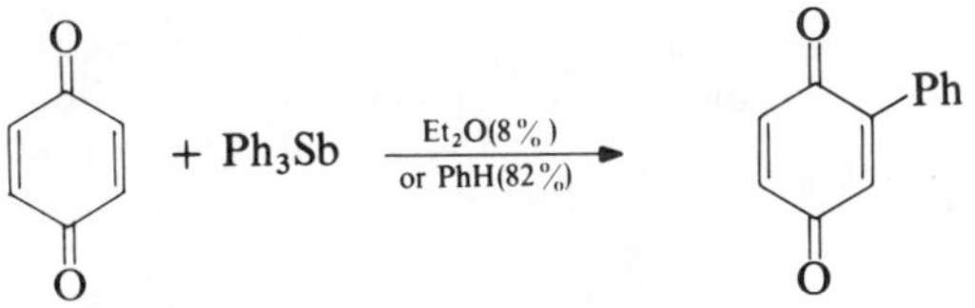

$$+ \; Ph_3Sb \xrightarrow[\text{or } PhH(82\%)]{Et_2O(8\%)} \qquad (421)$$

substitution and the influence of solvent in the latter reaction are both remarkable and deserve further study.

The use of a carbonyl protection reaction and boron trifluoride etherate as a catalyst produces good yields of biaryls with several potentially useful functional groups (equation 422)[656].

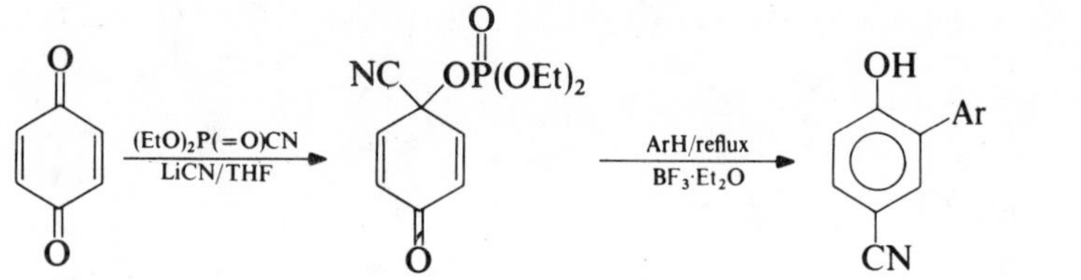

$$\xrightarrow[\text{LiCN/THF}]{(EtO)_2P(=O)CN} \quad \xrightarrow[BF_3 \cdot Et_2O]{\text{ArH/reflux}} \tag{422}$$

78% (Ar = Ph)

Ar = Ph, 4-Tol, 2,3- or 2,4-Xylyl, 4-HOC_6H_4, 4-An, 4-ClC_6H_4

While studying the oxidative demethylation synthesis of quinones Valderrama and his colleagues obtained significant yields of arylated products under certain conditions (equation 423)[657]. These results which were obtained when the reaction mixture was shaken slowly have not yet been further developed.

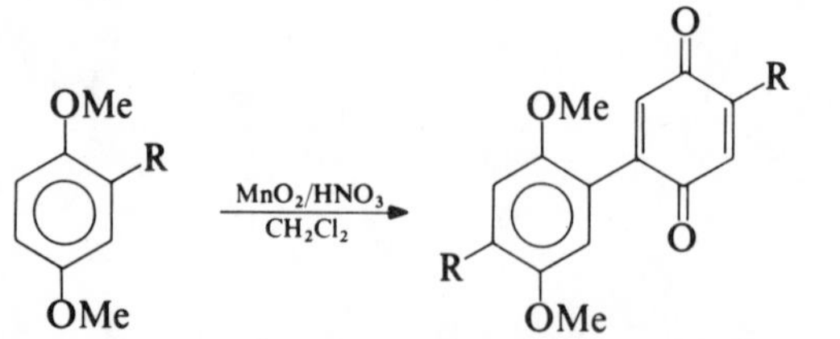

$$\xrightarrow[CH_2Cl_2]{MnO_2/HNO_3} \tag{423}$$

R = Me, MeO 39%, 24%

In a study of the thermal addition of cycloheptatriene to 1,4-benzoquinones a surprising arylation product is obtained along with the normal Diels–Alder product (equation 424)[658]. The formation of the crowded 2,3-product was studied and a stepwise

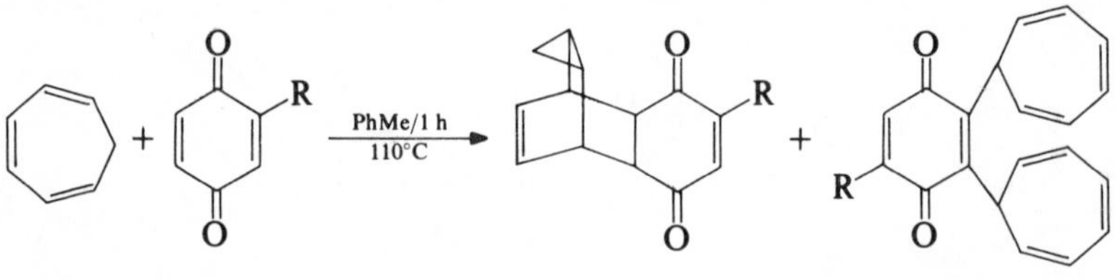

$$\xrightarrow[110°C]{PhMe/1\ h}$$

R = H, Me, Ph Me 16% 18% (424)

mechanistic pathway proposed. In a subsequent publication the chemistry has been extended to 1,4-naphthoquinone and the expected monotropyl product isolated[659].

Buggle and her collaborators added 2-phenylindane-1,3-dione (**136**) to a variety of 1,4-naphthoquinones (equation 425)[660]. The ease of addition as opposed to dimerization of

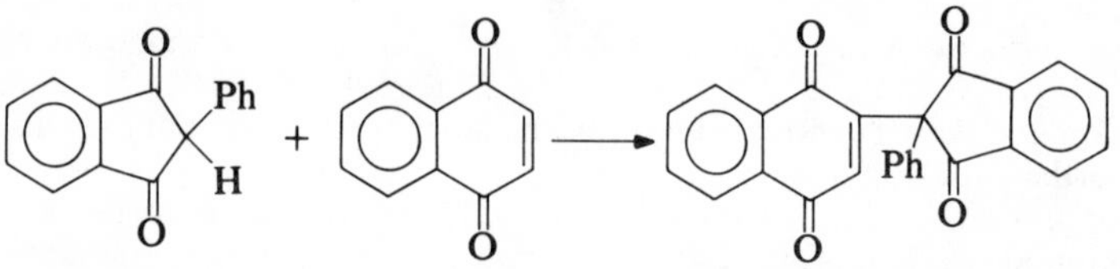

$$(136) + \longrightarrow \tag{425}$$

136 appears to depend on the redox potential of the quinone, its steric requirements and the solvent.

A potentially useful route to nitrogen bridged heterocyclic compounds is found in the reaction of indolizines (**137**) with 1,4-benzoquinone (equation 426)[661].

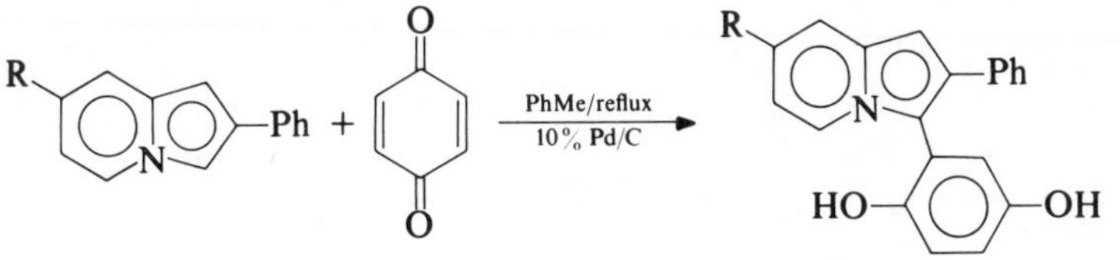

(426)

(**137**) R = H, Me — PhMe/reflux, 10% Pd/C — 42%, 59%

VII. ACTIVE METHYLENE QUINONE CHEMISTRY

In contrast to the situation in the first half of this century no one person dominates this field. It is a tribute to Lee Irvin Smith and his numerous students to notice the number of recent publications in which their work is given credit for its pioneering spirit.

With the profound effect of modern instrumentation on structure determination it is hardly surprising that changes of the earlier work have been proposed. Under basic conditions 2,3,5-trimethyl-1,4-benzoquinone reacts with acetylacetone to give the hydroquinone half ester **138**, not the simple addition product previously reported (equation 427)[662]. In a similar fashion cyanoacetamide forms an indole rather than a furan (equation 428)[663]. In this case the earlier workers had pointed out that they could not rigorously

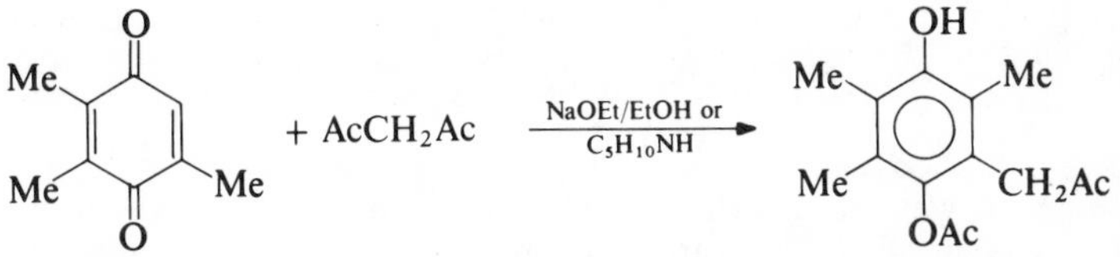

(427)

+ $AcCH_2Ac$ — NaOEt/EtOH or $C_5H_{10}NH$ — (**138**) 67%

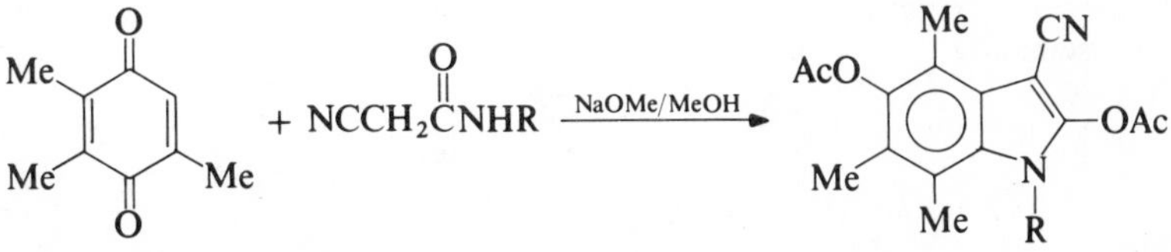

(428)

+ $NCCH_2C(O)NHR$ — NaOMe/MeOH — 78%, 71%

R = H, Me

exclude this structure although they favored the furan. The closely related active methylene addition–cyclization sequence of ethyl cyanoacetate does lead to 2-aminofurans (equation 429)[664]. This chemistry is also observed, in slightly lower yield, with 2-methyl-1,4-naphthoquinone and 2-methyl-5-isopropyl-1,4-benzoquinone. The use of 1,3-cyclohexanedione leads to partially reduced dibenzofurans bearing the useful carbonyl group (equation 430)[665].

Wikholm has expanded the study of the addition of the acetylacetone anion to 1,4-benzoquinones in two directions. First, after confirming the report cited above and its

$$\text{2,3,5-trimethyl-1,4-benzoquinone} + NCCH_2CO_2Et \xrightarrow[C_5H_{10}NH]{NaOEt/EtOH\ or} \text{ethyl 2-amino-5-hydroxy-4,6,7-trimethylbenzofuran-3-carboxylate}\ (70\%) \tag{429}$$

$$\text{2,3,5-trimethyl-1,4-benzoquinone} + \text{3-hydroxycyclohex-2-enone} \xrightarrow[MeOH]{C_5H_{10}NH} \text{hydroxy-trimethyl-tetrahydrodibenzofuranone}\ (52\%) \tag{430}$$

proposed new structure for the initial adduct, he shows that similar chemistry takes place with less highly substituted quinones (equation 431)[666]. The bis-adduct (**139**) also

$$\text{2,5-R}_2\text{-1,4-benzoquinone} + AcCH_2Ac \xrightarrow[MeOH]{BzN^+Me_3OH^-} \text{OH, OAc, Me, Me, }CH_2Ac\text{ arene}\ (60\text{–}75\%) + \text{OAc, OAc, }AcCH_2\text{, Me, Me, }CH_2Ac\text{ arene}\ (\mathbf{139})<10\% \tag{431}$$

R = Me, Ph

undergoes cyclization in refluxing acidic methanol. The second route involves the dichloroquinones which do give the simple Michael product analogous to those described by Smith (equation 432). The rearranged product can be obtained quantitatively by treating the initial adduct with Triton B in tetrahydrofuran. Apparently the combination of high oxidation potential and limited solubility prevent further reaction in this case and for the 2,6-dichloro isomer as well. A simple reoxidation–recycling procedure improves the yields substantially.

$$\text{2,5-dichloro-1,4-benzoquinone} + AcCH_2Ac \xrightarrow[(2)\ Ag_2O]{(1)\ \text{Triton B}} \text{hydroquinone with Cl, Cl, }(Ac)_2CH\text{, }CH(Ac)_2\ (78\%) + (\mathbf{139})\ (\text{trace}) \tag{432}$$

A very new product type results when the anion of ethyl cyanoacetate reacts with 2-chloro-3-methyl-1,4-naphthoquinone. In the presence of ammonia the simple products (**140**, equation 433) is formed but under slightly modified conditions two epimeric fused cyclopropyl products are obtained (equation 434)[667]. These compounds are especially interesting in that the initial attack of the carbanion must be at the alkyl-quinone carbon

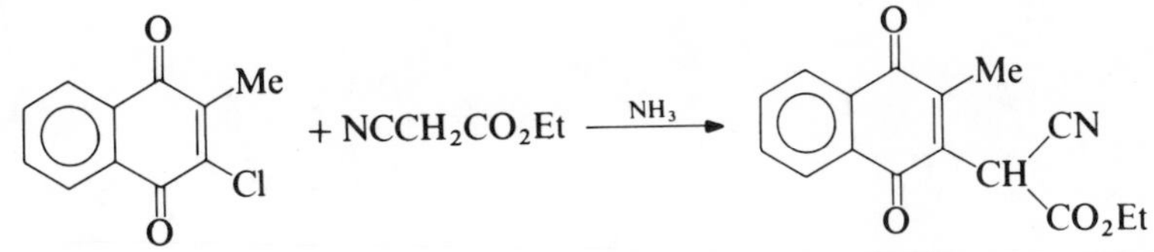

$$\text{2-chloro-3-methyl-1,4-naphthoquinone} + NCCH_2CO_2Et \xrightarrow{NH_3} \textbf{(140)} \quad (433)$$

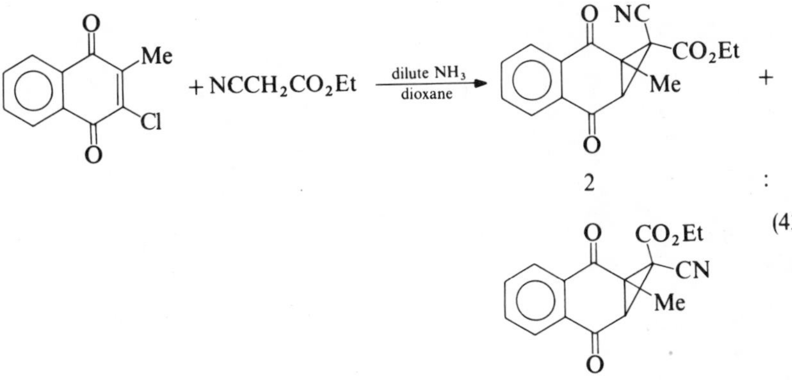

$$\text{2-chloro-3-methyl-1,4-naphthoquinone} + NCCH_2CO_2Et \xrightarrow[\text{dioxane}]{\text{dilute } NH_3} 2 : 1\ (75\%) \quad (434)$$

site followed by chloride displacement–cyclization. These compounds are not intermediates in the synthesis of the simple chlorine displacement product **140**. Another carbanion synthesis of a cyclopropyl ring-fused dihydro-1,4-naphthoquinone system has been reported (equation 435)[668].

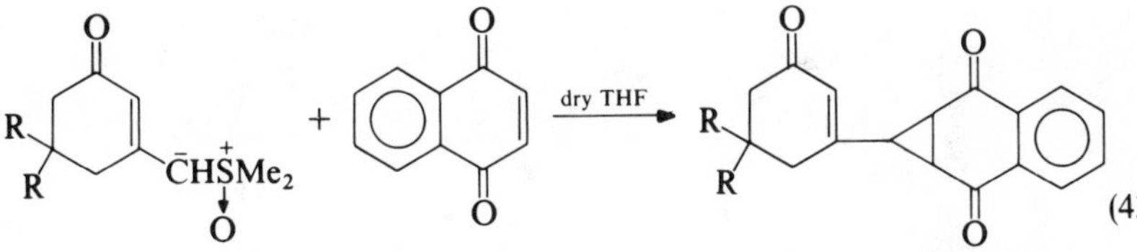

$$\text{R} = \text{H, Me} \xrightarrow{\text{dry THF}} 26\%,\ 36\% \quad (435)$$

Dean and his colleagues studied the chemistry of the 3-methyl-1,4-naphthoquinone-2-ylmethyl carbanion which they have generated in the presence of 2,6-dimethyl-1,4-benzoquinone[669]. The chemistry and the product structures are complex but surprisingly good yields of some can be obtained (equation 436). The authors attempted to obtain a xanthen derivative (**141**) or to trap its unstable carbanion with an excess of the quinone. Such an experiment (equations 437, 438) produced only the bridged trione **143** resulting from the carbanion of the diquinone product (**142**).

With 2,5-dimethyl-1,4-benzoquinone they were able to produce a cage compound (equation 439)[670]. Similar results are obtained but with greater difficulty in the cases of 2,6-dimethyl-1,4-benzoquinone and the combination of 2,5-dimethyl- and 2,3,5-trimethyl-1,4-benzoquinone. The last compound does not form a cage with itself indicating that steric effects are very important in these systems.

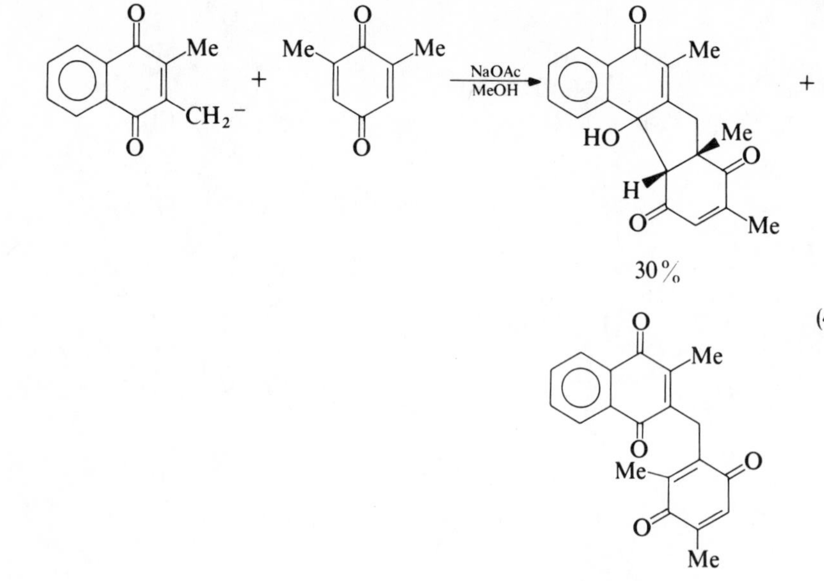

30%

(436)

(142) 43%

(437)

excess

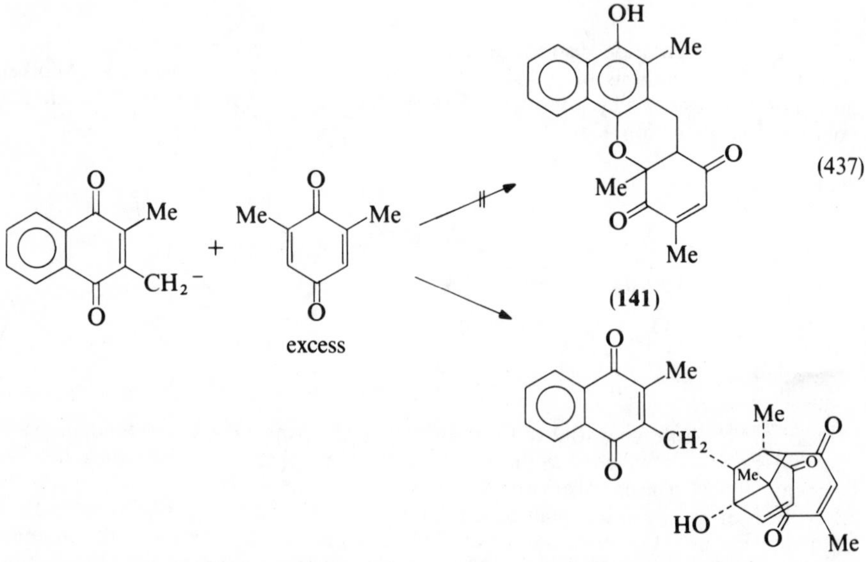

(141)

(143) 70% (438)

An extension of these studies to the analogous carbanion of 2-ethyl-3-methyl-1,4-naphthoquinone showed that while the chemistry does differ somewhat from that previously observed, these differences are not as great as might be inferred from earlier studies[671]. The chief difference lies in the regiospecificity of addition to 2-methyl-1,4-

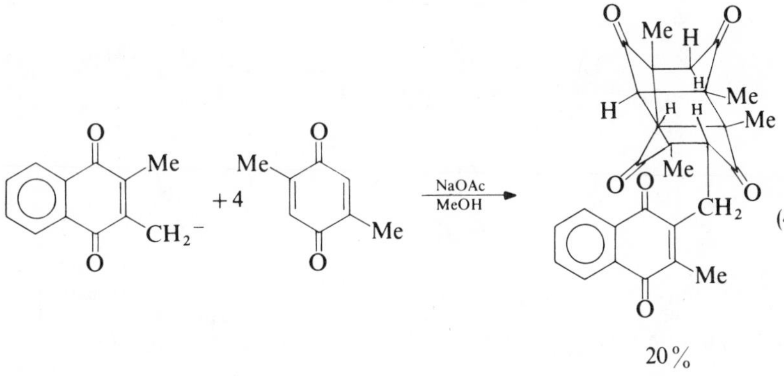

(439)

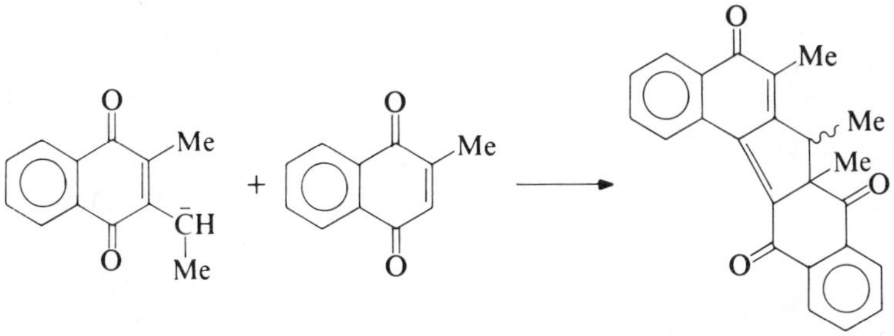

20%

(440)

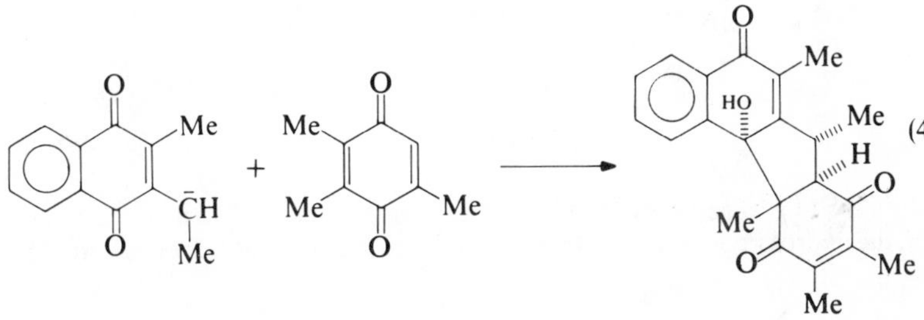

(441)

naphthoquinone (equation 440) and 2,3,5-trimethyl-1,4-benzoquinone (equation 441). In the former case the orientation corresponds to that often observed before, while in the latter a unique arrangement is found. In a later study these workers sought to assess the generality of the cage-forming reactions of quinones and quinonoid carbanions[672]. They were able to verify the negative results with a variety of carbanions reported by several different laboratories. Even with an excess of quinone designed to optimize cage compound formation, these reactions are not observed. The conclusion is that such reactions are indeed rare.

A particularly surprising reaction is reported by Thomson and his colleagues in their search for naphthoquinone methide lactones[673]. The aluminum chloride catalyzed reaction of juglone with chloroformylacetate (**144**) produced neither of the expected products (equation 442). A detailed study of related compounds shows that the carbanion reaction is at the C(8) position and suggests that the reduced quinone may first form an ester with the acid chloride. The results thus far are not conclusive, but further studies are promised.

(144) 45% (442)

A different type of lactone results when carbanions are produced bearing a single activating group (equation 443)[674]. It is possible to hydrolyze the ester and the chlorine in near quantitative yields (equation 444). When the acid (**146**) is heated under nitrogen,

$[Me_2CH]_2NLi$ (443)

(**145**) 50%

(**145**) aq HCl / $MeOCH_2OMe$ → (**146**) 94% KOH / MeOH → (**148**) 92% (444)

evolution of hydrogen chloride takes place and the lactone **147** is formed in 35% yield; from the hydroxy compound (**148**) the yield is more than doubled (equation 445). A minor

(**146**) or (**148**) heat / N_2 → (**147**) 35%, 74% (445)

product in these reactions is the analogous spirolactone of 1,2-naphthoquinone (**149**) but it can be produced in good yield from the initial ester (**145**, equation 446).

An example of an allyl carbanion has been explored by Maruyama and his coworkers (equation 447)[675, 676]. The high yields, mild conditions and excellent regiospecificity make this a potentially important route. Once again we note the requirement for a Lewis acid catalyst as the synthetic alternative to the traditional base-induced carbanion–Michael chemistry. This chemistry is also applicable to 2-acetyl-1,4-benzoquinone.

In the preparation of a substituted quinone for a photosynthetic study Akiba and his colleagues made an observation concerning the metal cation involved in these active

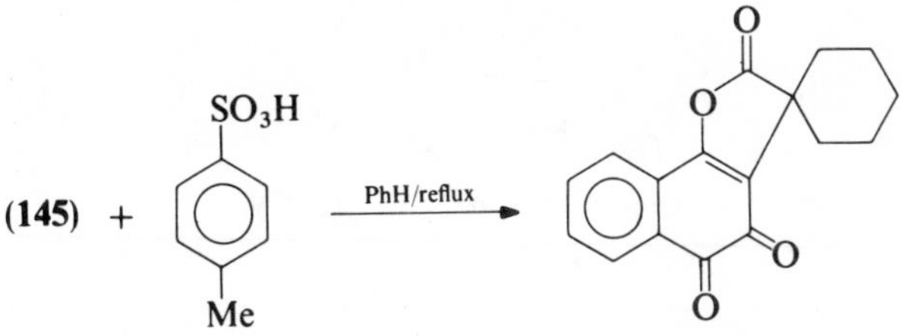
(145) +
SO_3H
Me
PhH/reflux
O

(149) 64%

(446)

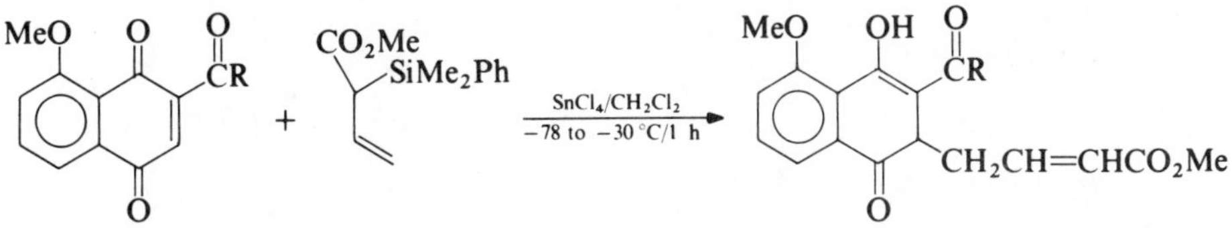
MeO
O
CR
CO_2Me
$SiMe_2Ph$
+
$SnCl_4/CH_2Cl_2$
−78 to −30 °C/1 h
OH
$CH_2CH{=}CHCO_2Me$

R = Me, Pr

74%, 82%

(447)

methylene reactions. Diethyl sodium malonate reacts somoothly with 2,3-dibromo-5-6-dimethyl-1,4-benzoquinone (equation 448) while the 2-bromo-5-methyl- and 2-bromo-6-

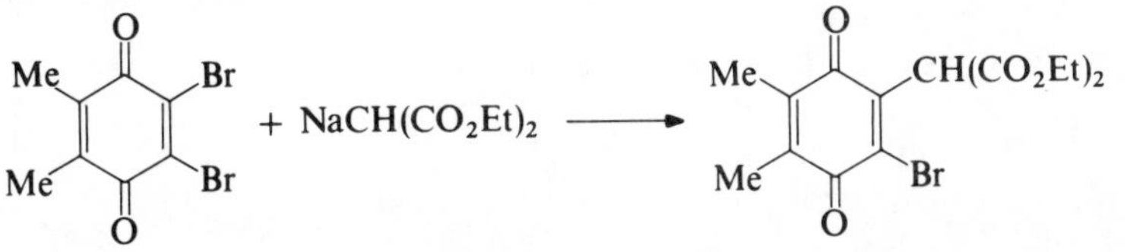
Me
Br
O
+ $NaCH(CO_2Et)_2$
$CH(CO_2Et)_2$

(448)

methyl- analogs gave only intractable mixtures[677] Using the thallium malonate salt moderate yields of the desired quinones are obtained (equation 449). Notice that some

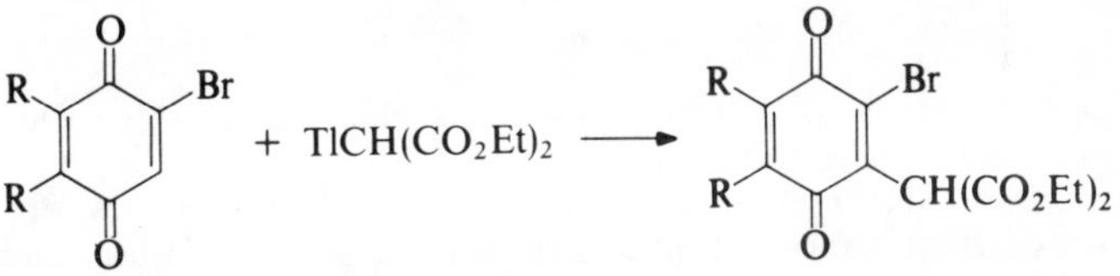
R
Br
O
+ $TlCH(CO_2Et)_2$
$CH(CO_2Et)_2$

(449)

R = H, Me

intriguing questions arise concerning the competition between addition and substitution; what would happen with 2-bromo-5,6-dimethyl-1,4-benzoquinone?

A reinvestigation of the reaction between 2,3,5,6-tetrahydroxy-1,4-benzoquinone and malononitrile in the absence of either acid or base shows that the principal product results from formal displacement of two hydroxyl groups (equation 450)[678]. While the yield is not great the functionality of this representative of the furan–quinone system promises useful chemistry.

A more traditional method for the synthesis of quinones with a fused furan ring involves the use of potassium fluoride (equation 451)[679]. Unfortunately, chloranil failed to react.

A third route involves the condensation of benzoylacetonitrile with 2,3-dichloro-1,4-naphthoquinones in the presence of pyridine (equation 452)[680]. An analogous product

$$\text{2,3,5,6-tetrahydroxy-1,4-benzoquinone} + CH_2(CN)_2 \xrightarrow[80\,^\circ C/2\,h]{HCONMe_2} [\text{2,5-bis(dicyanomethyl)-3,6-dihydroxy-1,4-benzoquinone}] \longrightarrow \text{2,6-diamino-3,7-dicyanobenzo[1,2-b:4,5-b']difuran-4,8-dione} \quad 48\% \tag{450}$$

$$\text{1,4-naphthoquinone} + R^1COCH_2R^2 \xrightarrow[60\text{–}65\,^\circ C]{HCONMe_2/KF} \text{2-}R^1\text{-3-}R^2\text{-naphtho[2,3-b]furan-4,9-dione} \quad 32\text{–}97\% \tag{451}$$

R^1 = Me, Ph; R^2 = CN, Ac, Bz, CO_2Et

$$\text{6,7-dimethyl-2,3-dichloro-1,4-naphthoquinone} + PhCOCH_2CN \xrightarrow{C_5H_5N} \text{3-cyano-6,7-dimethyl-2-phenylnaphtho[2,3-b]furan-4,9-dione} \quad 20\% \tag{452}$$

from 6-methyl-5-phenyl-1,4-naphthoquinone was obtained in 48% yield but its regiochemistry was not determined.

Benzofurans are another related structural type derived from quinones. McPherson and Ponder have shown that various benzoylacetates condense with 1,4-benzoquinone in the presence of a Lewis acid catalyst (equation 453)[681]. They have also examined the addition

$$\text{1,4-benzoquinone} + ArCOCH_2CO_2Et \xrightarrow[ZnCl_2]{EtOH/reflux} \text{ethyl 5-hydroxy-2-Ar-benzofuran-3-carboxylate} \tag{453}$$

Ar = 2-An, 2,4-, 2,5-$(MeO)_2C_6H_3$

of ethyl acetoacetate to 1,4-benzoquinone and its 2-chloro and 2-methyl derivatives (equation 454)[682]. The use of an excess of quinone or its rapid addition tends to increase the yield of the difuran. The reinvestigation of several earlier preparations of these

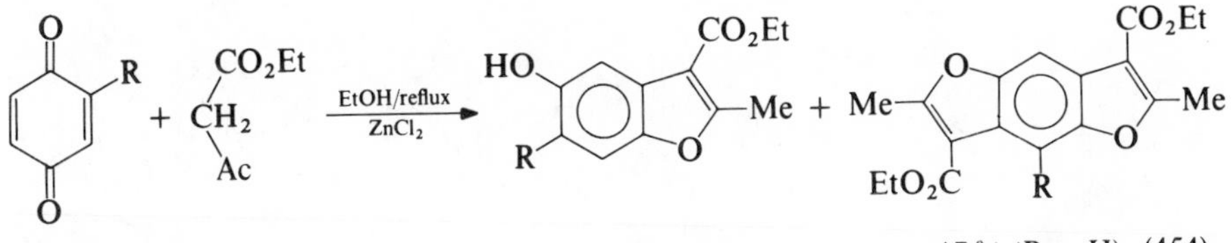

$$\text{quinone (R)} + CH_2(CO_2Et)(Ac) \xrightarrow[ZnCl_2]{EtOH/reflux} \text{benzofuran} + \text{benzodifuran} \quad (454)$$

R = H, Me, Cl

25% (R = H) 17% (R = H) (454)

compounds showed that the structures proposed without modern instrumental methods are correct in most instances.

A search for dyestuffs has produced a novel preparation of benzodifuranones (equation 455)[683]. The yields are modest (up to 35%) but the range is quite broad. The authors attribute the poor yields to the instability of the mandelic acid and promise further studies.

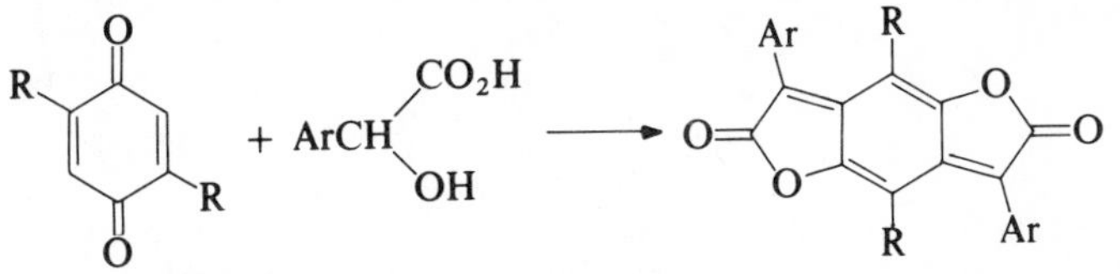

$$\text{quinone} + ArCH(CO_2H)(OH) \longrightarrow \text{benzodifuranone} \quad (455)$$

R = H, Me, Cl; Ar = Ph, 4-HOC_6H_4, 4-$AcOC_6H_4$, 4-An

Certain α,β-unsaturated ketones add well to 2-hydroxy-1,4-naphthoquinone (equation 456)[684]. When the carbonyl group is reduced it is possible to cyclize by two paths depending on the choice of catalyst (equations 457, 458)[685].

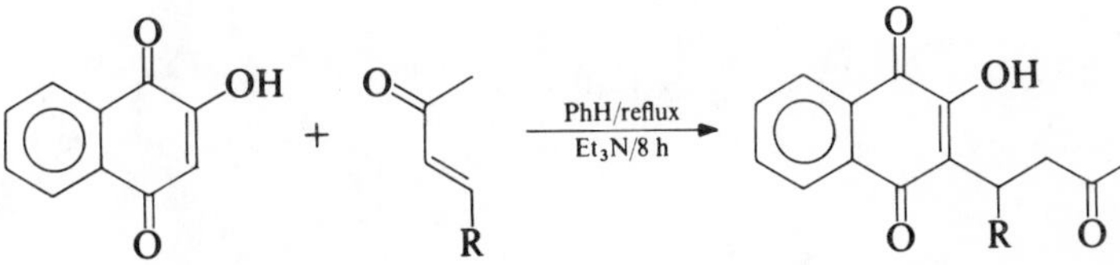

$$\xrightarrow[Et_3N/8\ h]{PhH/reflux} \quad (456)$$

R = H, Me, Ph

68%, 28%, 35%

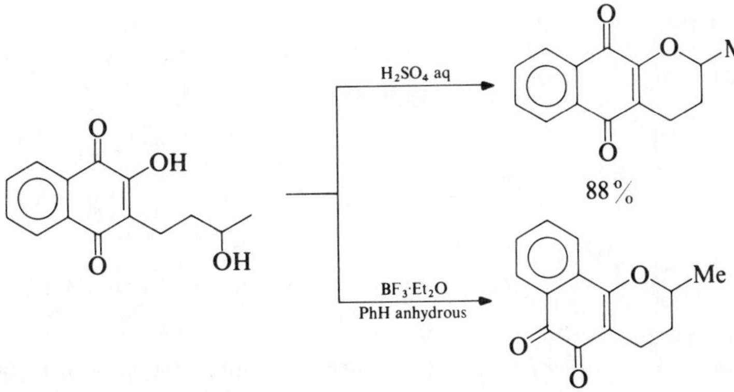

$$\xrightarrow{H_2SO_4\ aq} \quad 88\% \quad (457)$$

$$\xrightarrow[PhH\ anhydrous]{BF_3 \cdot Et_2O} \quad 98\% \quad (458)$$

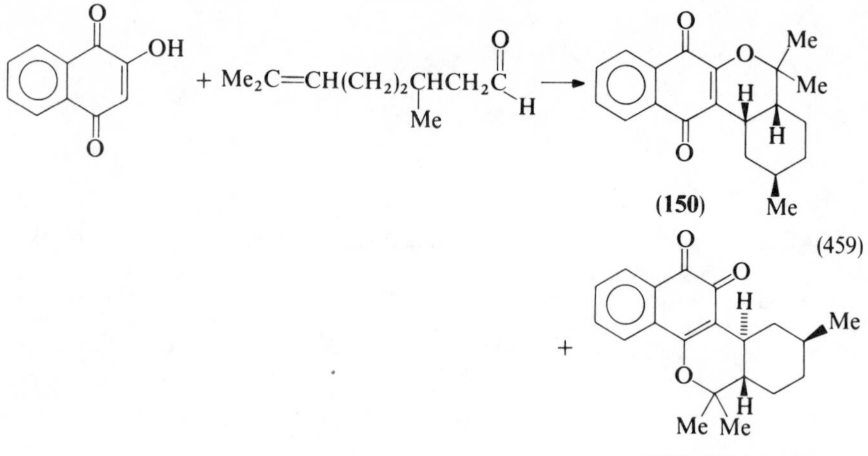

$$+ Me_2C{=}CH(CH_2)_2CH(Me)CH_2CHO \longrightarrow$$

(150) (459)

(151) 70% total

Quite similar compounds have been prepared in a one-step reaction (equation 459)[686]. The observation that **150** can be converted into **151** by treatment with sulfuric acid at 0°C deserves attention in the light of the study discussed above[684, 685].

In some instances the presence of a nitrogen base can produce rather complex polyheterocyclic quinones of importance in the dye industry (equation 460)[687]. This chemistry has been extended to the use of quinoline (equation 461), but again no yields are

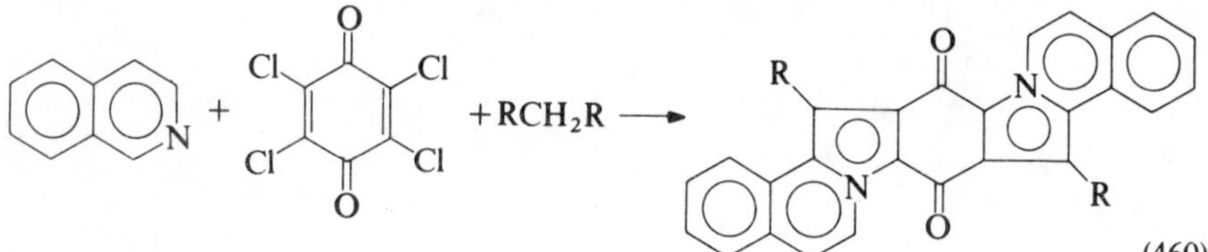

$$+ RCH_2R \longrightarrow \qquad (460)$$

R = Ac, Bz, CO_2Et

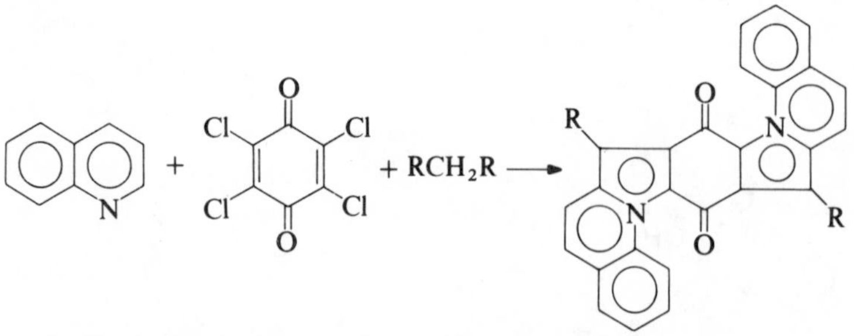

$$+ RCH_2R \longrightarrow \qquad (461)$$

given[688]. A related reaction involving a homophthalimide and pyridine has been presented (equation 462)[689]. The chemistry is discussed in detail and a reasonable account of the product mixture presented.

A report of an analogous heterocyclic quinone system has been made by Buggle and Power (equation 463)[9]. In view of her earlier work this failure to react at the thiol group is exciting. Another interesting note is that the quinoline derivative of **152** fails to add to

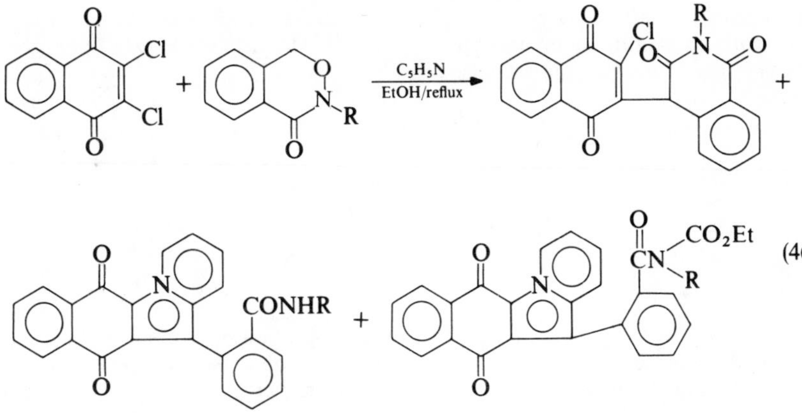

C₅H₅N EtOH/reflux

(462)

EtOH/CHCl₃ reflux/4 h

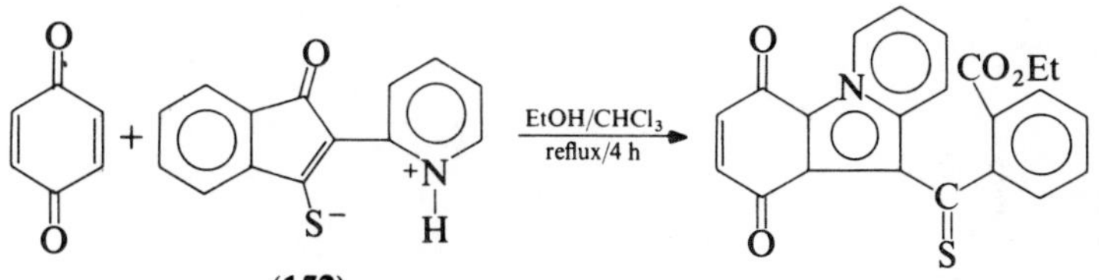

(463)

(152)

quinones while the isoquinoline does. Presumably for similar steric reasons **152** fails to react with 2,3-dichloro-1,4-naphthoquinone.

Three short notes are of interest for their promising hints of future development. Wheeler and his coworkers obtained a small amount of the unusual spirobutenolide **153** when coumalic acid was added to 1,4-benzoquinone (equation 464)[690]. All stated components of the reaction mixture must be present for this reaction to take place.

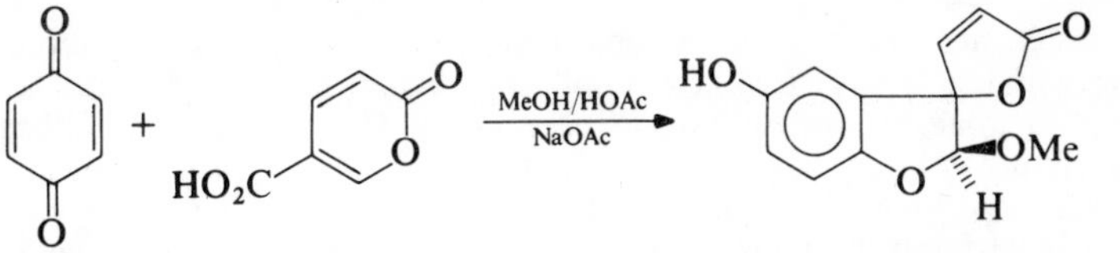

MeOH/HOAc NaOAc

(464)

(153) 22%

Bloomer and Damodaran have published a synthesis of homogentisic acid (**154**) with an intermediate lactone of additional interest (equation 465)[691].

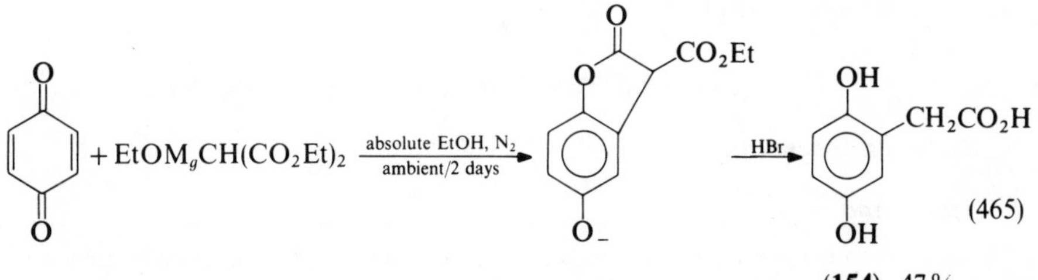

$+ EtOMgCH(CO_2Et)_2$ absolute EtOH, N_2 ambient/2 days; HBr

(465)

(154) 47%

In the process of trying to prepare a dihydropyran system Rieck and Grunwell found a novel synthesis of anthraquinones (equation 466)[692]. The yield needs to be optimized and the generality of the reaction established.

(466) [Reaction: 2-butyryl-1,4-naphthoquinone + diethyl glutaconate (CO_2Et … CO_2Et), MeOH / aq NH_3 → 1-Pr-anthraquinone with 2-CO_2Et and 4-CO_2Et substituents, 33%]

Parker has given a good introduction to the application of quinone monoacetals to quinonoid active methylene chemistry[693]. She and Kang obtained very good yields of both the initial adduct and the aromatic monoether with diethyl malonate (equation 467).

(467) [Reaction: 4,4-dimethoxycyclohexa-2,5-dienone (MeO, OMe) + $CH_2(CO_2Et)_2$, NaOEt/EtOH, ambient → adduct bearing $CH(CO_2Et)_2$, 91%; then 4-$MeC_6H_4SO_3H$ → 4-methoxy-3-[$CH(CO_2Et)_2$]phenol (OMe, OH), 88%]

Similar results are obtained with a naphthoquinone acetal and ethyl cyanopropionate. However, with keto esters most of the products are bicyclic bridged compounds which can be converted by acid to benzofurans (equation 468). The yields are generally adequate.

Chan and Brownbridge have added the 2,5-bis(trimethylsiloxy)furan (**155**) to the acetals of 1,4-benzo- and 1,4-naphthoquinone (equations 469, 470)[694]. The nature of the isolated adducts can certainly be understood on the basis of oxidation potentials. The final products appear to be mixtures of cyclic and open tautomers (equation 471). Semmelhack and his coworkers[695], using butyl lithium and nickel tetracarbonyl, were able to acylate naphthoquinone acetal (equation 472) and to trap the presumed enolate as well (**156**, equation 473). This chemistry has been used in the synthesis of two natural products with significant antibiotic activity.

VIII. THE SUBSTITUTION CHEMISTRY OF QUINONES

A. Nitrogen Substitution

1. Kinetics and mechanisms

In contrast to the general trend, the chemistry of nucleophilic substitution in quinones has produced little in the way of truly new knowledge or technique during the past decade

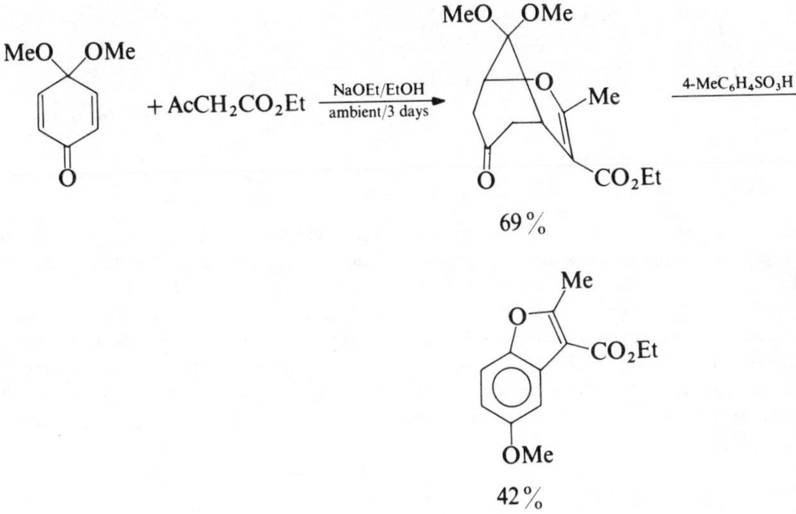

MeO OMe

$+ AcCH_2CO_2Et$ $\xrightarrow[\text{ambient/3 days}]{\text{NaOEt/EtOH}}$

MeO OMe O Me O CO_2Et

69%

$\xrightarrow{4\text{-}MeC_6H_4SO_3H}$

Me O CO_2Et OMe

42%

(468)

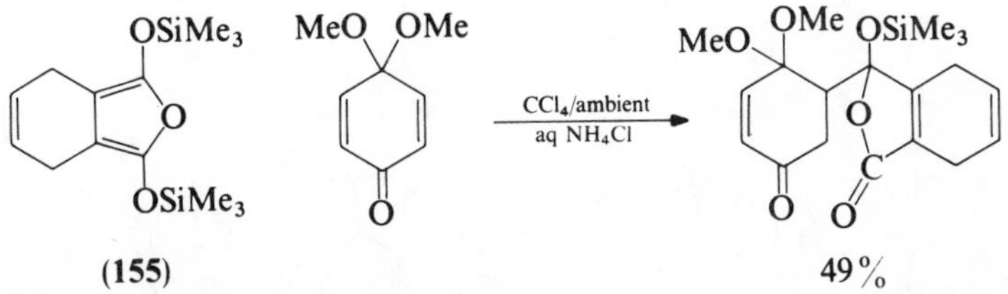

$OSiMe_3$ O $OSiMe_3$ (155) MeO OMe O $\xrightarrow[\text{aq } NH_4Cl]{CCl_4\text{/ambient}}$ MeO OMe $OSiMe_3$ O O C O

49%

(469)

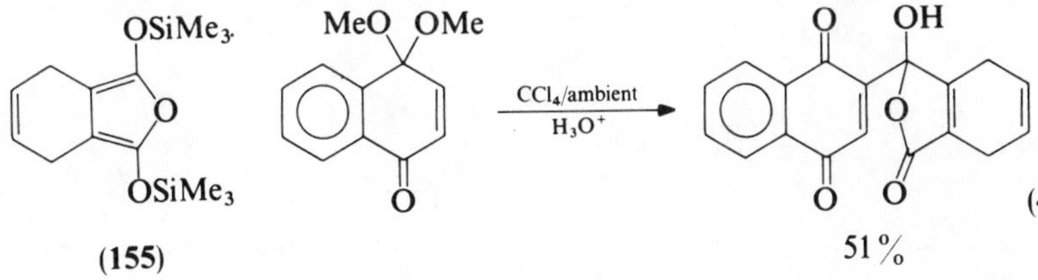

$OSiMe_3$ O $OSiMe_3$ (155) MeO OMe O $\xrightarrow[H_3O^+]{CCl_4\text{/ambient}}$ O OH O O O

51%

(470)

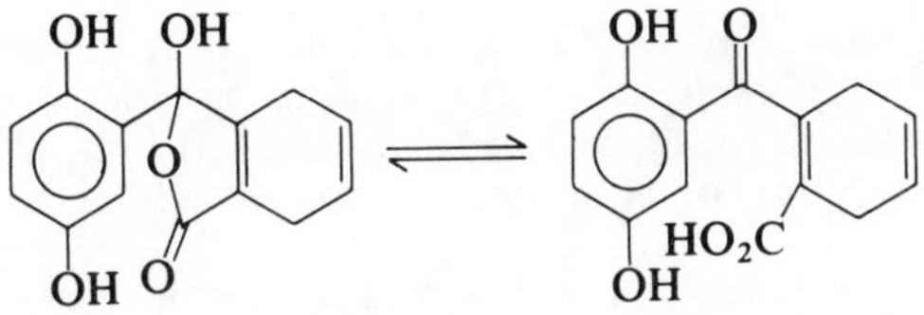

OH OH O OH O $\rightleftharpoons$ OH O HO_2C OH

(471)

MeO OMe O $+ BuLi + Ni(CO)_4$ $\longrightarrow$ MeO OMe O C Bu O

91%

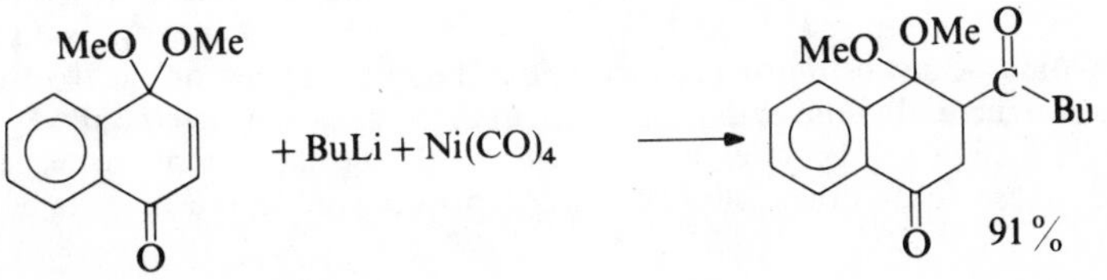

(472)

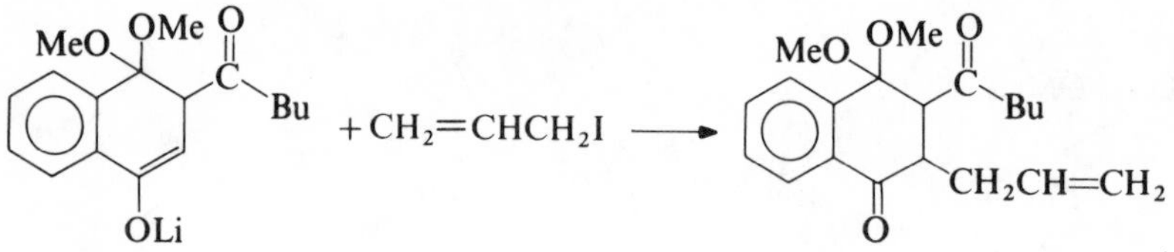

$$\text{(156)} + CH_2{=}CHCH_2I \longrightarrow \text{product } (85\%) \quad (473)$$

and a half. An important exception is apparent with carbanions (see Section VII). Kinetic studies have concentrated exclusively on the charge transfer intermediate with chloranil although anilines[696], tertiary amines[697] and primary amines[698] have been included.

A detailed study of the behavior of substituted anilines allowed the recognition of several distinct classes of mechanistic pathways[696]. Diphenylamine and *N,N,N′,N′*-tetramethyl-*p*-phenylenediamine form such stable charge transfer complexes with chloranil that no further reaction takes place. Aniline and *m*-phenylenediamine form substitution products after passing through both an outer and an inner complex (equation 474). Finally, *o*-

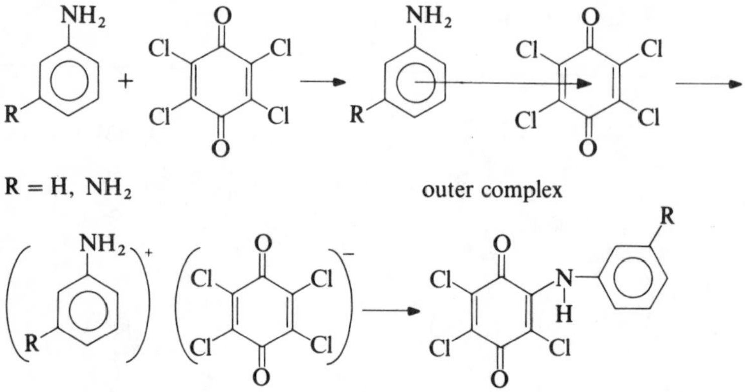

R = H, NH_2 — outer complex — inner complex (474)

phenylenediamine proceeds to product through the outer complex and an ionic species (equation 475). The differences are discussed on the basis of electron density of the amino substituent.

Shah and Murthy have reinvestigated the interaction of triethylamine with chloranil (equation 476) in a variety of solvents[697]. They confirm the earlier suggestion of a radical anion intermediate and find this species to be unreactive with more highly branched alkyl groups. The chemistry of this system changes in hydroxylic solvents but the products were not identified.

Similar evidence for a charge transfer intermediate was presented for the reaction of primary amines with chloranil (equation 477)[698]. This is found in the fast reaction forming monosubstitution product **157**, but there is no evidence for an analogous complex in the slow formation of disubstituted product (**158**). The rates depend on the electron donor strengths of structurally similar amines except in those cases in which the alkyl groups are significantly different. The rates also increase with solvent polarity while activation energies decrease. Clearly there must be charge separation in the transition state leading to product.

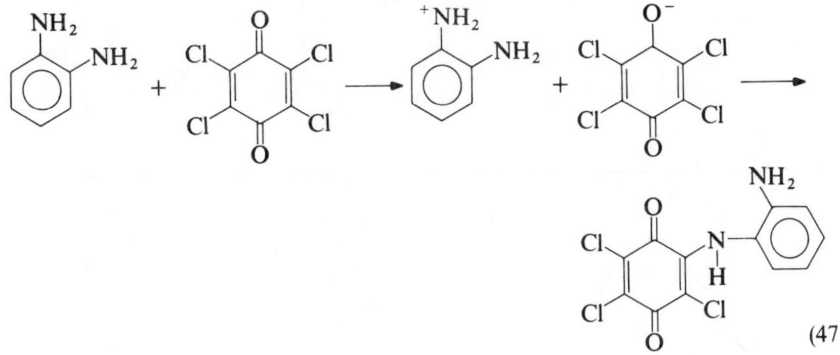

(475)

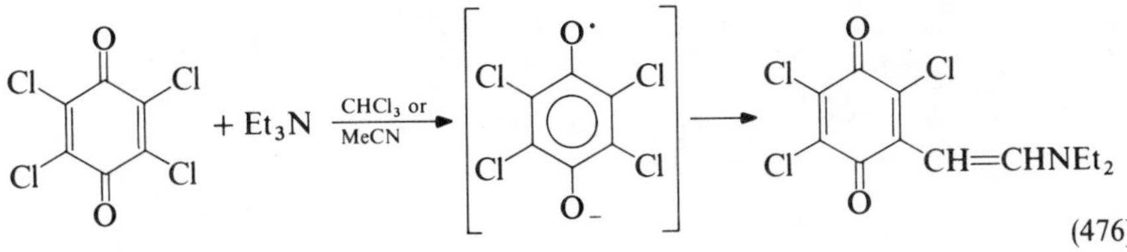

(476)

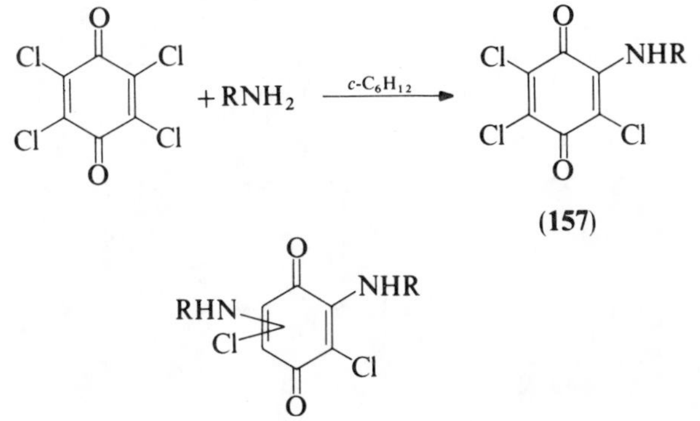

(**157**)

(477)

(**158**)

R = *s*-Bu, *i*-Bu, *i*-Pr, allyl, *c*-C_6H_{12}

Cameron and his colleagues have established the regiospecificity in the reaction 2-^{13}C-2-chloro-1,4-naphthoquinone with a variety of nucleophiles[699]. In principle two different paths might be followed by reaction at the halogenated or *ipso* carbon (equation 478) or at the vicinal carbon (equation 479). They found that the amines, being neither hard nor soft nucleophiles, show a striking solvent dependence (equation 480).

A product and kinetic study of reactions between various heterocyclic secondary amines and sodium 1,2-naphthoquinone-4-sulfonate has been carried out (equation 481)[700]. The products obtained from reaction with pyrrole are unique in this series since they involve carbon–carbon bond formation (equation 482). The authors attribute this outcome to the low electron density of nitrogen in the quasi-aromatic ring. The product with 1,2,4-triazole is also unexpected (equation 483).

The second-order rates were determined polarographically and the pH-rate profile showed a maximum at pH 10. This suggests a typical nucleophilic substitution

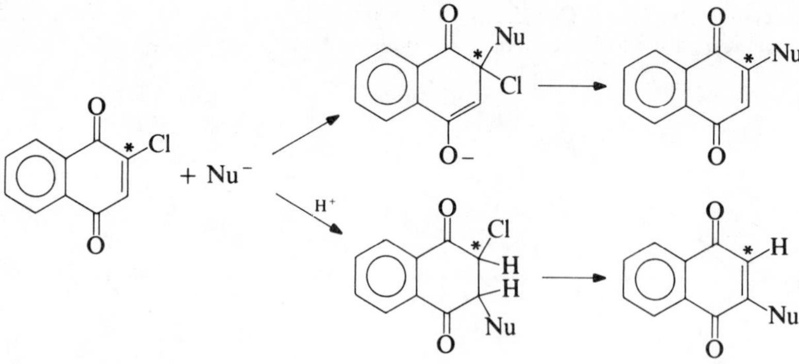

(478)

(479)

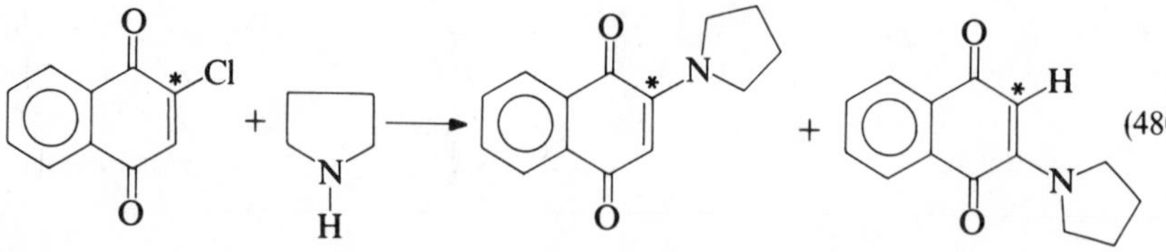

(480)

Solvent	%	%
MeCN	25	70
PhH	8	80
MeOH	84	10

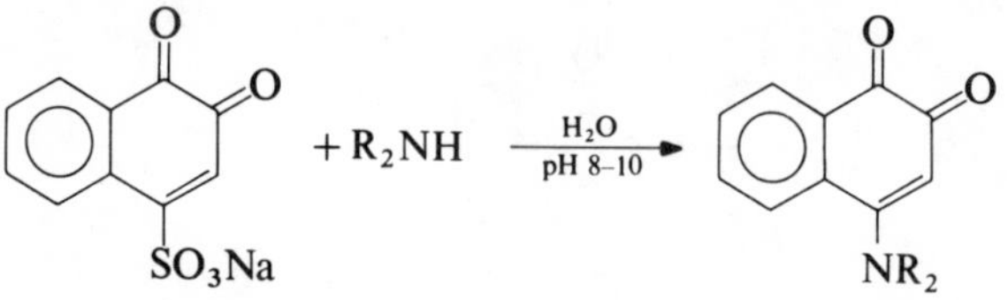

(481)

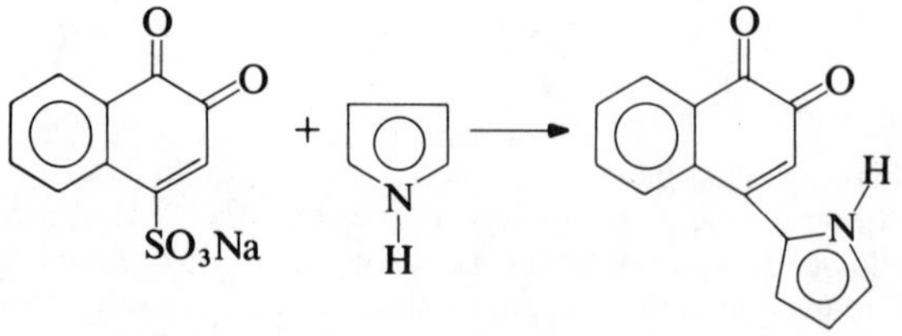

(482)

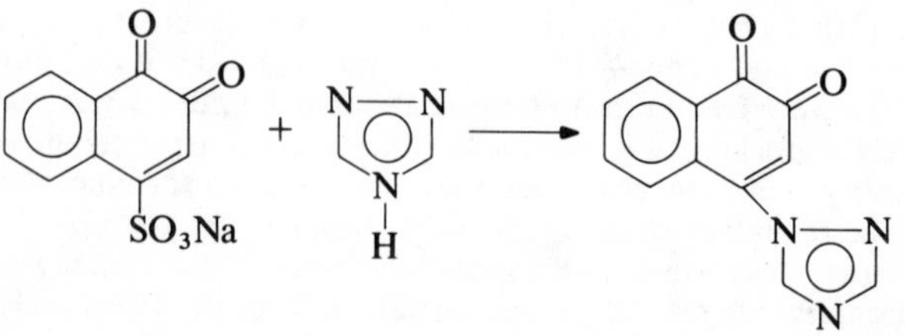

(483)

intermediate (**159**). The displaced sodium hydrogen sulfite reacts in a successive second-order reaction with another equivalent of quinone to give the 4,4-disulfonate **160**.

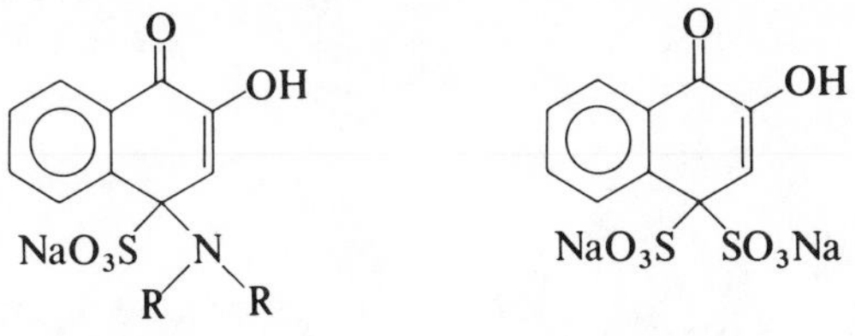

(159) **(160)**

2. Synthetic studies with mechanistic implications

A reinvestigation of the reactions of amino acids with 1,4-benzoquinones reveals that earlier claims for charge transfer intermediates are unfounded[701]. It seems clear that all of the observed absorptions are explained by the formation of mono- and disubstitution products (equation 484).

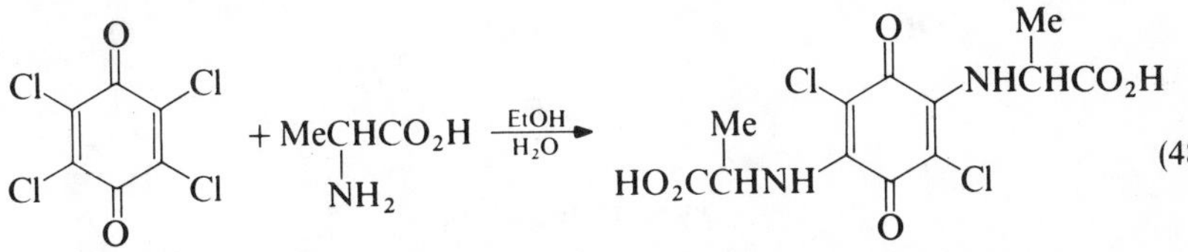

(484)

Hewgill and Mullings followed their important study of quinone–oxygen substitution reactions[702] with a parallel examination using dimethylamine[703]. Strong evidence is presented for the addition–elimination pathway (equations 485, 486). Unlike the charged

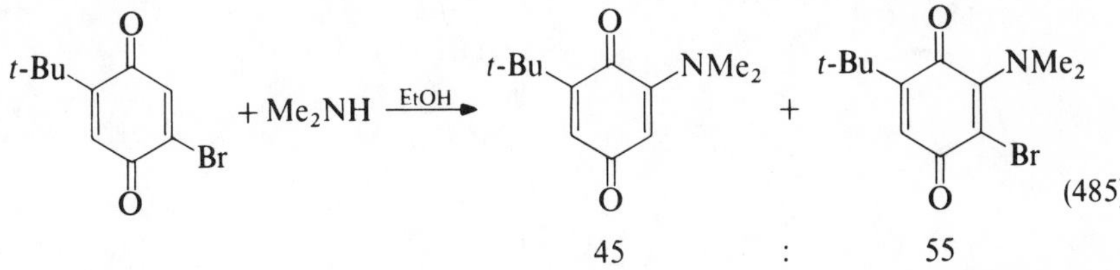

(485)

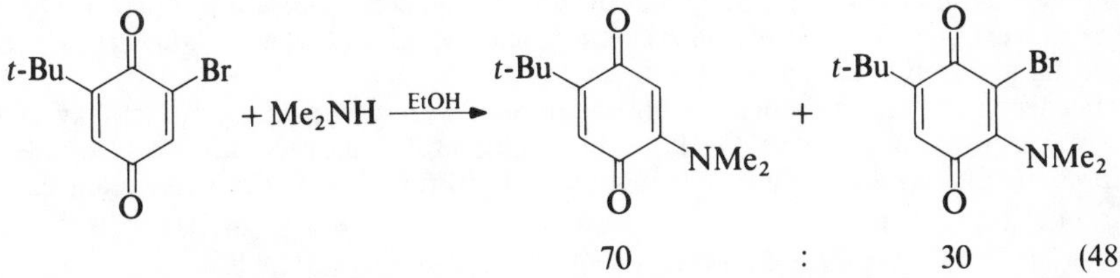

45 : 55

70 : 30 (486)

methoxide ion case these reactions show very little influence of the solvent on the product composition. The absence of a direct substitution product makes the following addition intermediates, common to both products, attractive (equation 487).

The displacement of hydroxy substituents from quinones by primary amines is much less usual. Joshi and Kamat studied the reactions of embelin (**161**) with a variety of primary

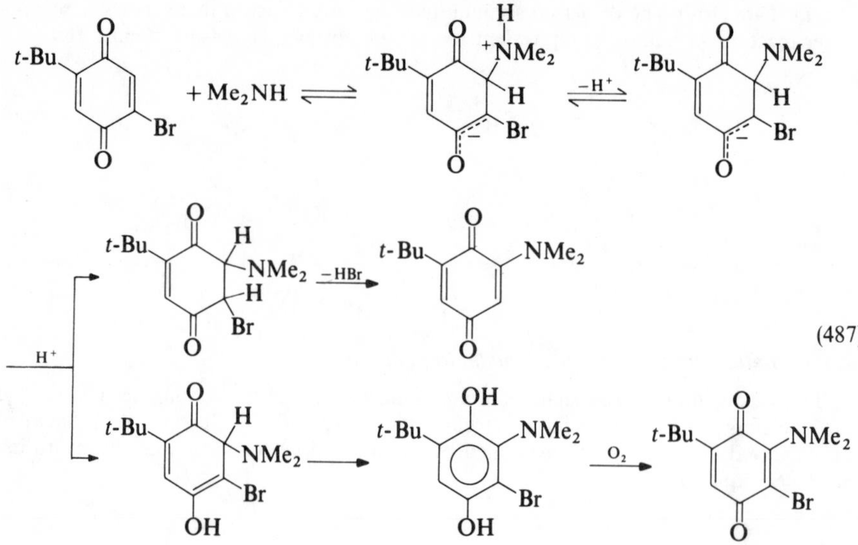

(487)

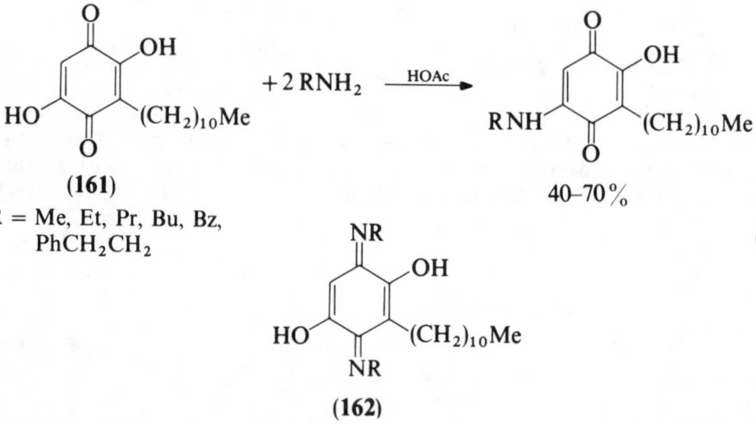

(161)

R = Me, Et, Pr, Bu, Bz, $PhCH_2CH_2$

40–70%

(488)

(162)

amines and found generally good yields of substitution at the 5-position (equation 488)[704]. A greater excess of methylamine leads to the bismethylimine (**162**) which appears to involve the ring nitrogen adduct as an intermediate.

Higher yields of 2-amino-1,4-naphthoquinones were obtained using methoxy as the leaving group (equation 489)[705]. The cyanide anion proved unreactive as a nucleophile but represents a good leaving group (equation 490). In this report, and subsequent papers,

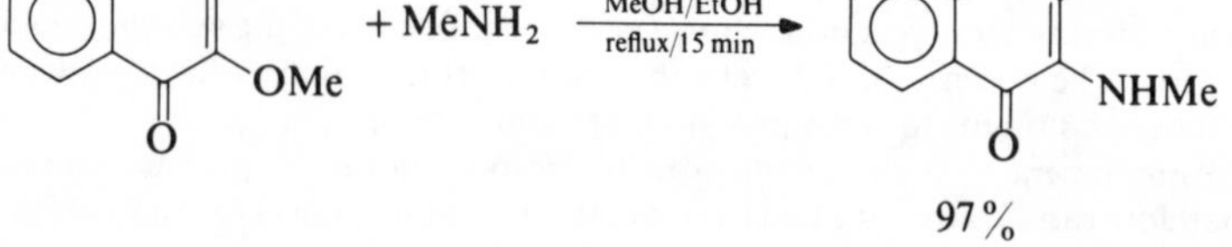

$$+ MeNH_2 \xrightarrow[\text{reflux/15 min}]{\text{MeOH/EtOH}} \quad (489)$$

97%

$$\text{2,3-dicyano-1,4-naphthoquinone} + MeNH_2 \xrightarrow[\text{1 h}]{\text{EtOH/reflux}} \text{2-cyano-3-(NHMe)-1,4-naphthoquinone} \quad (85\%) \qquad (490)$$

similar chemistry was extended to substituents in the aromatic ring[706, 707]. The one truly unexpected outcome of this chemistry is the reaction of aniline with 5-amino-2,3-dicyano-1,4-naphthoquinone (equation 491).

$$\text{5-amino-2,3-dicyano-1,4-naphthoquinone} + PhNH_2 \longrightarrow \text{2,3-dicyano-1,4-naphthoquinone} \qquad (491)$$

The halo atoms of 2,3-dichloro- or 2,3-dibromo-1,4-naphthoquinone continue to represent the major point of departure for many synthetic routes involving nucleophilic substitution. Parr and Reiss applied this method to the synthesis of enaminones for use in the Nenitzescu reaction[708]. Belitskaya and Kolesnikov studied the influence of oxygen substituents in the aromatic ring on such chemistry[709a]. Their results suggest that the selective replacement of each halogen is possible depending on the exact nature of the oxygen substituent at position five, but no truly comparable data are presented (equation 492). An earlier report by this same group had shown that aniline substitutes only at the 2-position with 5-hydroxy **163** and at the 3-position with the 5-acetoxy **163**[709b].

$$\text{amine} + \text{5-RO-2,3-dichloro-1,4-naphthoquinone } (\mathbf{163}) \longrightarrow \text{5-RO-2-amino-3-chloro-1,4-naphthoquinone} + \text{5-RO-3-amino-2-chloro-1,4-naphthoquinone} \qquad (492)$$

morpholine, R = H	77%	17%
aniline, R = Me	5%	89%

This chemistry has been applied and extended by workers interested in the synthesis of the mitomycin antibiotics. Rapoport and Falling tried a variety of routes involving the sequential displacement of the 2,3-bromines in 1,4-naphthoquinone, for example (equation 493)[710]. They found that the best route involved the proper balance of quinone and hydroquinone chemistry, but the specific quinonoid reactions they found add substantially to our appreciation of their possibilities.

Okamoto and Ohta made a similar approach to the same class of compounds (equation 494)[711]. The formation of the lactam and the successful active methylene reaction on it are both noteworthy. Unfortunately, no yields are given.

There has been little recent activity in the synthesis of ethylenimine substituted quinones. However, Driscoll and his coworkers have expanded the range of possibilities

$$\text{2,3-dibromo-1,4-naphthoquinone} + \text{pyrrolidine} \longrightarrow \text{2-bromo-3-pyrrolidinyl-1,4-naphthoquinone (91\%)} + \text{2-pyrrolidinyl-1,4-naphthoquinone} \tag{493}$$

91% product $\not\rightarrow$ 2-(CH_2CO_2Me)-3-pyrrolidinyl-1,4-naphthoquinone (24%)

$$\text{2,3-dibromo-1,4-naphthoquinone} + CH_2{=}C(OMe)_2 \longrightarrow \text{2-}(CH_2CO_2Me)\text{-3-bromo-1,4-naphthoquinone (90\%)} \longrightarrow \text{24\% product}$$

$$\text{2,3-dichloro-1,4-naphthoquinone} + H_2N(CH_2)_nCO_2H \longrightarrow \text{2-chloro-3-}NH(CH_2)_nCO_2H\text{-1,4-naphthoquinone} \tag{494}$$

$n = 1,2,3,5$ 93–100%

$$\xrightarrow{n=3} \text{2-chloro-3-(2-oxopyrrolidin-1-yl)-1,4-naphthoquinone} \xrightarrow[\text{K/phthalimide}]{CH_2(CO_2Et)_2} \text{2-}CH(CO_2Et)_2\text{-3-(2-oxopyrrolidin-1-yl)-1,4-naphthoquinone}$$

available[712,713]. Using either chloranil or fluoranil they were able to prepare a large variety of 2,5-diaziridinyl-3,6-diamino-1,4-benzoquinones (equation 495). The two substituents may be introduced in sequence permitting the preparation of unsymmetrical products. The yields are generally good regardless of the order of introduction. Methyl- and 2,2-dimethylaziridines undergo analogous reactions.

A limited number of studies involving 1,2-quinones have appeared. The displacement of the 4-sulfonate group from 1,2-naphthoquinone by primary amines leads to a complex dye mixture in which the principal component is **164** (equation 496)[714]. This structure was confirmed by independent synthesis.

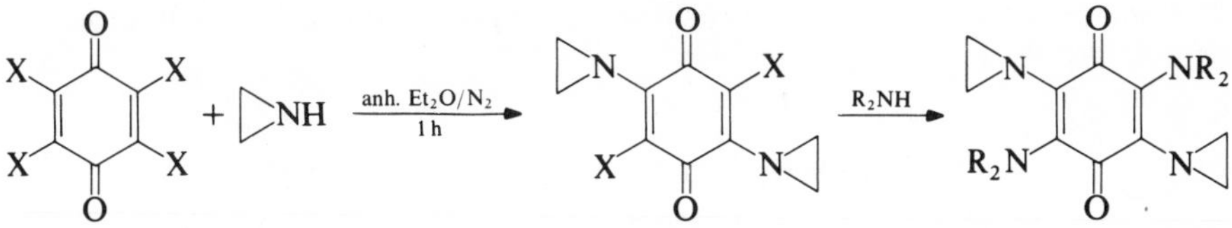

X = F, Cl
R = H, Me, Et, Pr, Bu, $HO(CH_2)_2$, morpholinyl, piperidyl, pyrrolidinyl (495)

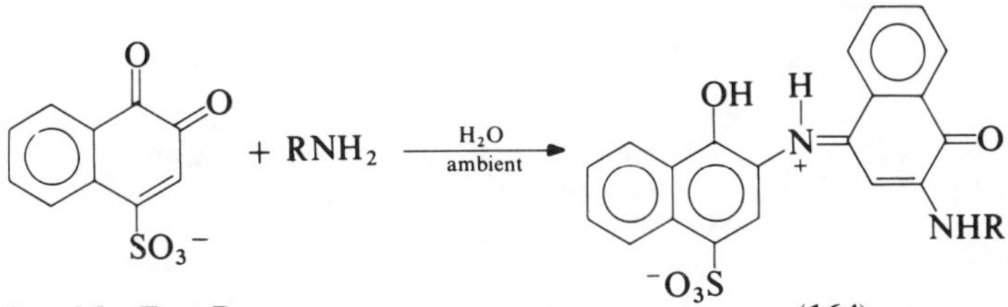

R = Me, Et, *i*-Pr (164) (496)

In the transamination of 1,2-naphthoquinone[715], in the absence of a catalyst, such as a calcium ion, a 1,4-quinonoid monoimine (**165**) is formed. The catalyst, however, leads to the simple primary amine substitution product (**166**, equations 497, 498). The same

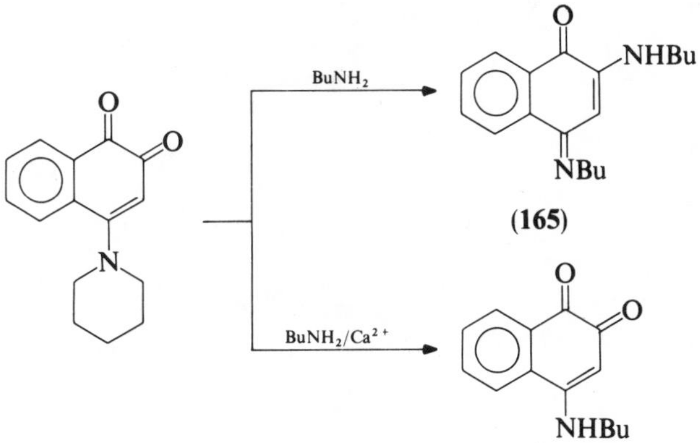

(**165**) (497)

(**166**) (498)

chemistry obtains with anilines if triethylamine is added. This method represents a valuable synthesis of compounds that are usually rather difficult to prepare, and it has been applied to several heterocyclic systems. Gauss and his colleagues have reported the complete substitution of chloranil and 2,3-dichloro-1,4-naphthoquinone by nitrogen aromatics (equation 499)[716]. The yields for these hindered compounds are surprisingly high.

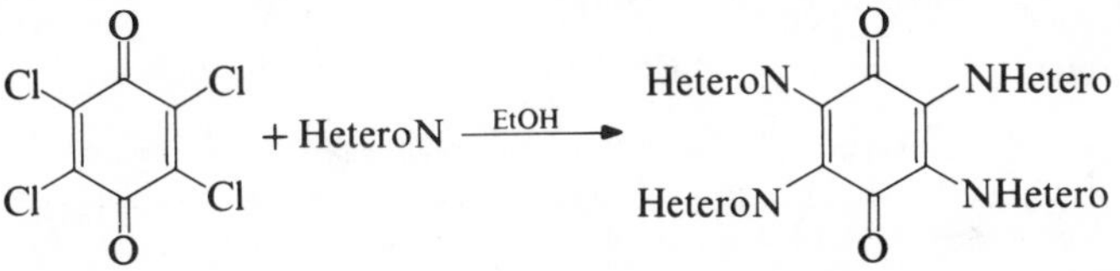

(499)

HeteroN = pyrazolyl, imidazolyl, 1,2,4-triazolyl 66%, 83%, 72%

A brief report of an entirely new carbon skeleton is noteworthy and should be developed (equation 500)[717].

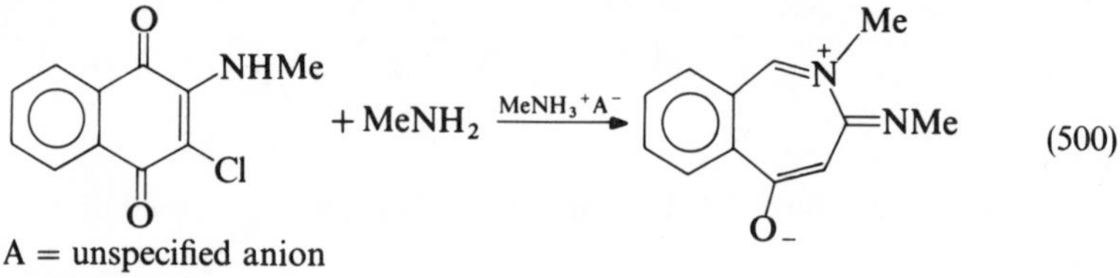

$$\text{2-(methylamino)-3-chloro-1,4-naphthoquinone} + MeNH_2 \xrightarrow{MeNH_3^+A^-} \text{benzazepine product (=NMe, } \overset{+}{N}\text{-Me, O}^-) \qquad (500)$$

A = unspecified anion

3. Heterocyclic syntheses

As was observed in the section on nitrogen addition chemistry the fastest growing area of quinone chemistry in the past decade and a half is the synthesis of heterocyclic compounds. This has been especially true with respect to substitution reactions. While much of the work has dealt with routine synthesis there are instances of exciting new chemical thought.

The two most heavily studied systems are the triphenodithiazines (**167**) and the triphenodioxazines (**168**). In both of these structural classes Mital and his collaborators have been exceedingly active[718]. The best chemical routes to these dyestuffs are quite similar (equations 501, 502). Considering the complexity of the products the yields of **168**

$$(\text{2-aminophenylthio})_2Zn + \text{tetrahalo-1,4-benzoquinone} \xrightarrow{HOAc} \textbf{(167)} \qquad (501)$$

$$2\ \text{2-aminophenol} + \text{tetrahalo-1,4-benzoquinone} \xrightarrow{PhCO_2H} \textbf{(168)} \qquad (502)$$

X = Cl, Br

are high (40–80%) but very low (10–13%) for **167**. Potentially useful intermediates can be isolated in good yield from both reactions with anhydrous ethanol as the solvent. This observation was developed further, and pyridine was found to be the preferred solvent (equation 503)[719].

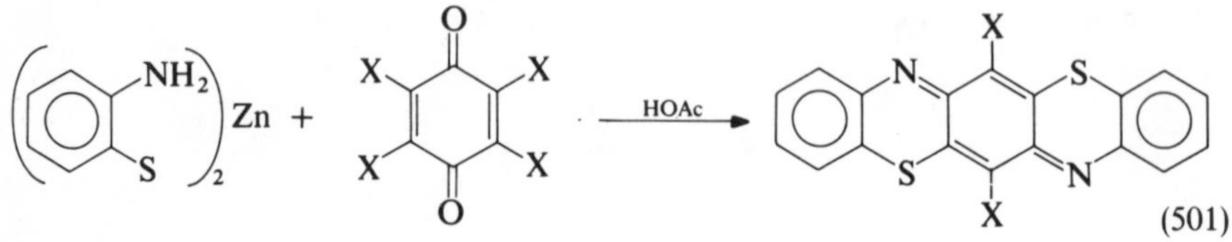

$$(\text{R-2-aminophenylthio})_2Zn + \text{2-methoxy-1,4-benzoquinone} \xrightarrow{C_5H_5N} \text{R-substituted OMe-phenothiazin-3-one} \qquad (503)$$

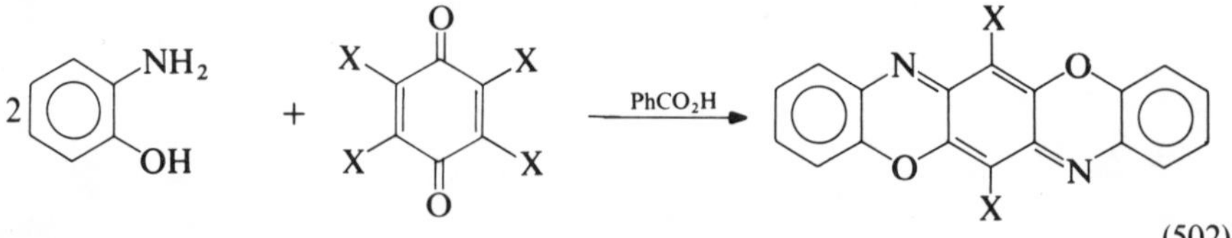

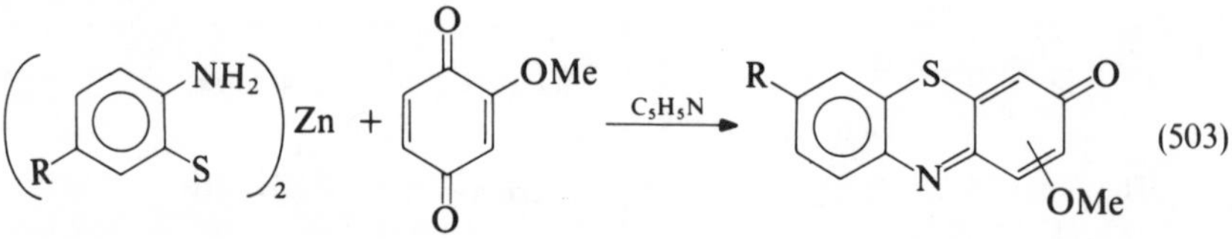

R = Cl, Br, Me, OMe, OEt

89–97%

With 2,3-dichloro-1,4-naphthoquinone Mital and Agrawal found that the monocyclization product (**169**, equation 504) alone is obtained using the zinc salt but that the free thiol leads to bis-product (**170**, equation 504)[720]. Yields are good to excellent in both instances.

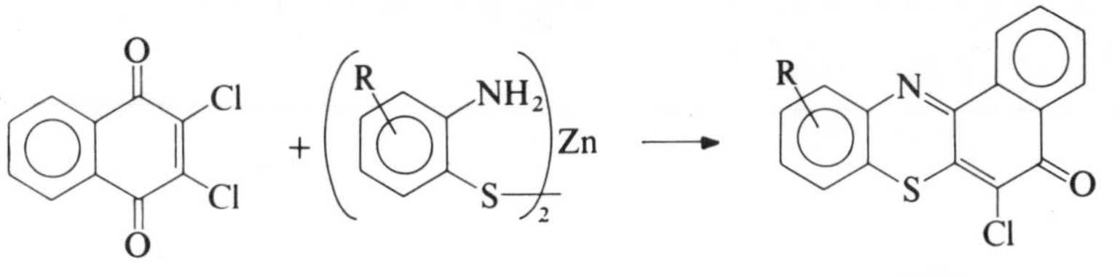

(504)

(**169**)

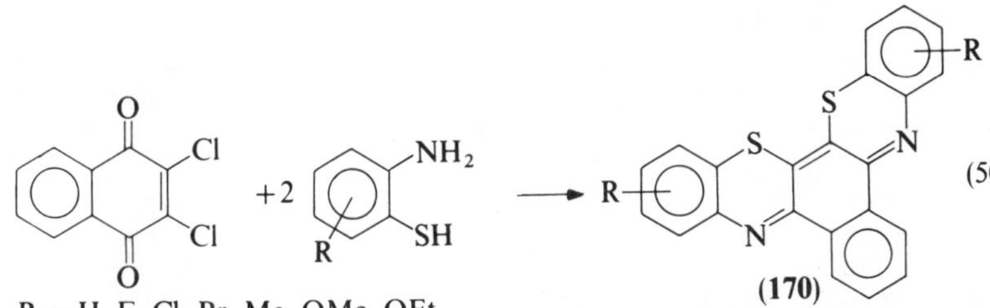

(505)

(**170**)

R = H, F, Cl, Br, Me, OMe, OEt

This chemistry has been applied to a large number of additional examples, and the resulting compounds studied extensively[721–734]. Other laboratories have followed Mital's lead and expanded the range of substituents and ring systems[735–743].

Similar advantage of the synthesis of the triphenodioxazines (**168**) is apparent in the work of Gupta and his colleagues (equation 506)[744–747]. Their approach has concentrated

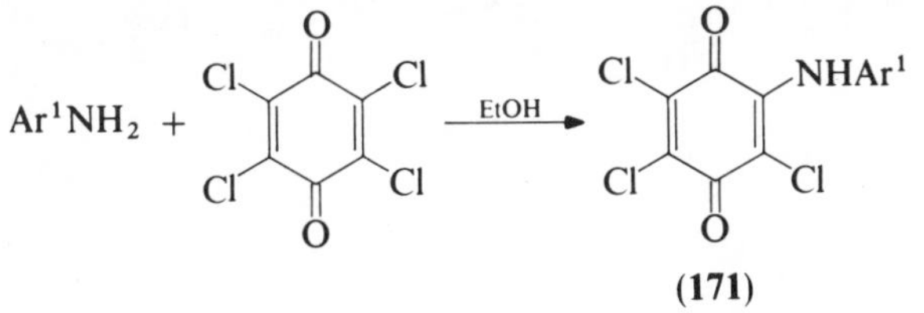

(**171**)

(506)

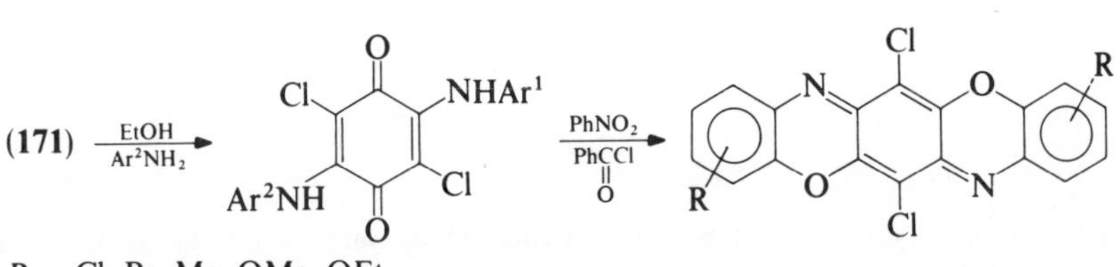

R = Cl, Br, Me, OMe, OEt

on the use of anilines and chloranil; also suggested by Mital[719]. By isolation of the monosubstituted product (**171**) they have been able to prepare unsymmetrical products in quite good yields.

Ueno and his coworkers have extended this chemistry to the 1,4-naphthoquinone series[748] again following the pioneering efforts of Mital (equation 507)[749].

The phenazines represent another series in which the quinones offer an attractive entry. Ried and Schaefer showed that the quinone–halogen addition products with which they had been working react with *o*-diamino aromatics in excellent yield (equation 508)[750]

R = H, Cl, NO_2 — EtOH, anh. KOAc — 57–80% (507)

R = H, Me, –CH=CHCH=CH– — MeOH, 1 h — 79%, 67% (508)

Here, too, Mital has made a contribution by showing that with pyridine as the solvent a facile conversion to the corresponding 1,2-quinone takes place (equation 509)[751]. Rao

C_5H_5N — 80%; HNO_3 — 80% (509)

has shown that hydroxyl groups can be displaced by various 1,2-phenylenediamines (equation 510)[752].

glacial HOAc, reflux/2 h

R = H, Me, Cl, Br, OMe, NH_2 (510)

Nakazumi and his coworkers have developed some novel approaches to these compounds. For example, they reversed the usual method using 2,3,5-triamino-1,4-naphthoquinone and glyoxal (equation 511)[753]. They have also employed a novel ring-closure reaction using an azide substitution (equation 512)[754]. The low yield may be

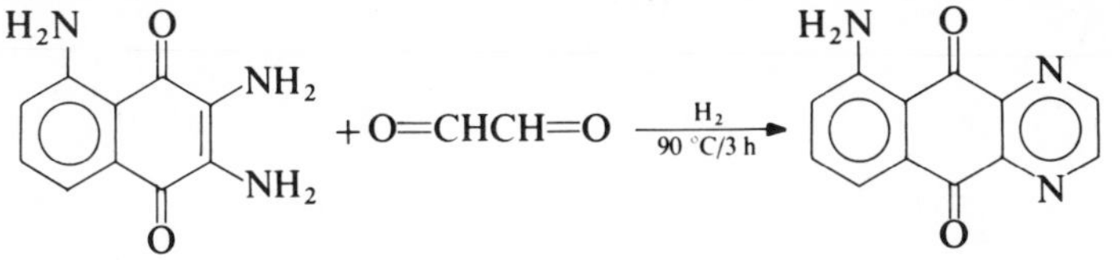

$$+\ O{=}CHCH{=}O \xrightarrow[90\,^{\circ}C/3\,h]{H_2} \quad 92\% \qquad (511)$$

$$\xrightarrow[HCONMe_2]{NaN_3} \quad 29\% \quad + \quad 42\% \qquad (512)$$

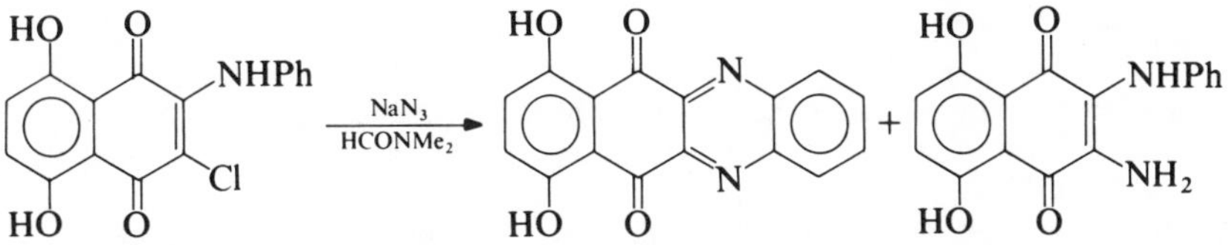

improved but the difficulty of other preparations for this compound or its nitrogen substitution products makes this an attractive route.

Still another slightly explored and potentially valuable synthesis has been presented by Schelz and Rotzler (equation 513)[755].

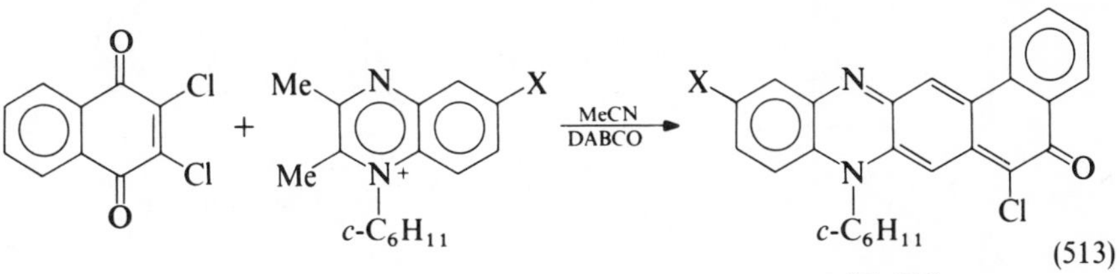

$$\xrightarrow[DABCO]{MeCN} \quad 25\text{–}55\% \qquad (513)$$

X = H, Me, F, Cl, Br, OMe, NHAc, CO_2Et, CF_3

Syntheses involving bridgehead nitrogen atoms have attracted quinone chemists. Two complementary approaches make use of substitution by thiols (equation 514)[756] or amino (equation 515)[757, 758] groups attached to heterocyclic rings which can then undergo ring

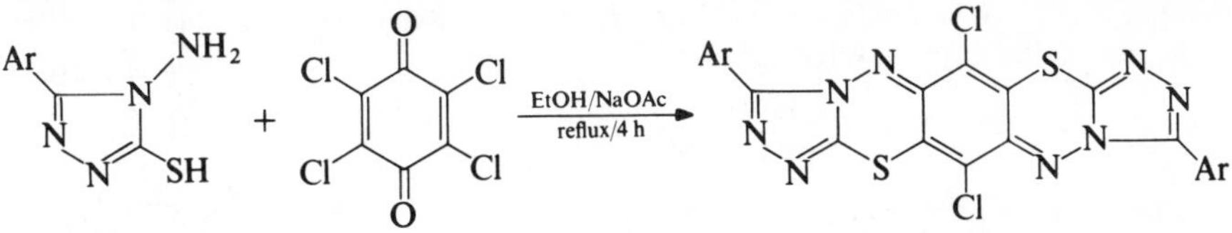

$$\xrightarrow[reflux/4\,h]{EtOH/NaOAc} \quad 50\text{–}80\%$$

Ar = Ph, 4-BrC_6H_4, 4-ClC_6H_4, 3-$O_2NC_6H_4$; 3,4-$(HO)_2C_6H_3$ (514)

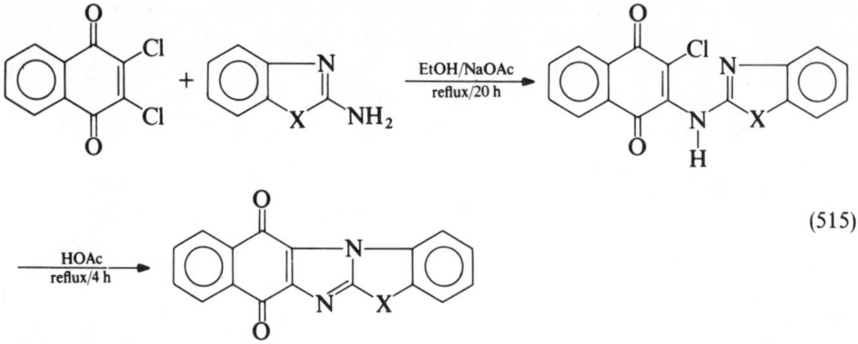

O Cl Cl O + N X NH_2 $\xrightarrow[\text{reflux/20 h}]{\text{EtOH/NaOAc}}$ O Cl O N H N X

$\xrightarrow[\text{reflux/4 h}]{\text{HOAc}}$ O O N N X (515)

X = N, S

fusion. Vernin and his colleagues have applied the phase transfer technique to a wide range of these reactions with extraordinary results; yields are not lower than 60% and most are 75% or greater[759].

A number of other laboratories have extended the range of substitution patterns and ring systems available by this substitution–ring closure chemistry. Bhakta and Chattopadhyay corrected a structure (**172**) presented earlier and prepared a series of related compounds (equation 516)[760]. Tilak and his colleagues have made more extensive

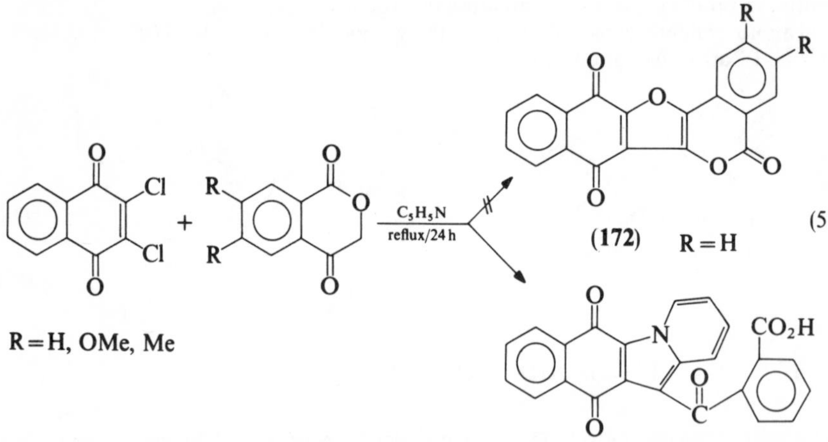

O Cl Cl O + R R O O O $\xrightarrow[\text{reflux/24 h}]{C_5H_5N}$ O O O O O R R (**172**) R = H (516)

R = H, OMe, Me

O O N O C CO_2H

efforts along these lines using α-picoline, isoquinoline, and ethyl isonicotinate as well as pyridine[761, 762]. When thiols or sulfides are employed the major product has often undergone sulfur expulsion (equation 517)[763].

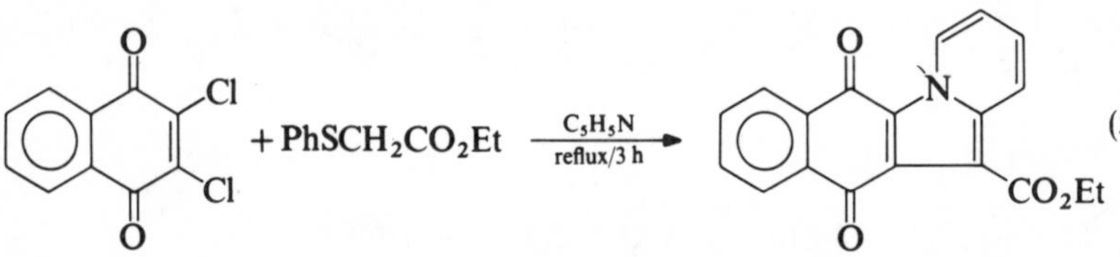

O Cl Cl O $+ PhSCH_2CO_2Et \xrightarrow[\text{reflux/3 h}]{C_5H_5N}$ O N O CO_2Et (517)

Soni and Saxena have published extensions and applications of several of these synthetic techniques[764–767], for example, the reaction of chloranil with substituted dithiocarbamates (equation 518)[765].

$$\text{chloranil} + 2ArNHCS^- \xrightarrow[\text{reflux/6 h}]{\text{EtOH/NaOAc}} \text{bis(thiazolethione)quinone (Ar)} \quad 30\text{–}60\% \tag{518}$$

Under the influence of pyridine alone or as a cosolvent anilines can react with bromanil or 2,5-dichloro-1,4-benzoquinone to produce fair to good yields of diindoloquinones (equation 519)[768].

$$\text{tetra-X-benzoquinone} + \text{4-R-aniline} \xrightarrow[O(CH_2CH_2)_2O]{C_5H_5N} \text{diindoloquinone (R)} \quad 50\text{–}85\% \tag{519}$$

X = H, Cl, Br; R = H, Me, MeO, OH, NH_2, Cl, CO_2H, NO_2

Hammam and his coworkers have added amides[769] and thioamides[770] to 2,3-dichloro-1,4-naphthoquinone to produce two related heterocyclic dione series in excellent yields (equation 520).

$$\text{2,3-dichloro-1,4-naphthoquinone} + RC(=X)NH_2 \xrightarrow[NaHCO_3]{\text{EtOH or } HOCH_2CH_2OH} \text{naphtho-fused azole-dione (R, X)} \quad 70\text{–}90\% \tag{520}$$

X = O, S; R = Me, NH_2

In an intense reinvestigation Schäfer and Agarwal have corrected earlier structural assignments in the reaction of 2,3-dichloro-1,4-naphthoquinone with 2-aminophenols (equation 521)[771–773]. In the process of these studies Schäfer and coworkers have developed useful methods for the synthesis of phenothiazinones[774] and acridine-

$$\text{2,3-dichloro-1,4-naphthoquinone} + \text{R-2-aminophenol} \xrightarrow[\text{anh. KOAc}]{\text{MeOH or EtOH}} \text{(Et)MeO-substituted benzo-phenoxazinone (R, N–H, Cl)} \tag{521}$$

R = H, Me, Cl

quinones[775]. They have studied the exchange of anilino substituents by amino acids and developed a fine method for the synthesis of 5-hydroxybenzoxazoles (equation 522)[776].

$$+ RCH(NH_2)CO_2H \xrightarrow[Et_3N]{Me_2C=O/H_2O} \quad 60\text{–}95\% \tag{522}$$

R = alkyl, Ph, Bz, PhCO, $CH_2=CH$

Kumar and Bhaduri also used 2-acyl-3,6-dianilino-1,4-benzoquinones in syntheses of indazolequinones and benzisoxazoles[777]. Schäfer and his colleagues have made similar synthetic efforts in connection with their studies of tautomerism of phenylazo-hydroquinones and hydrazinoquinones (equation 523)[778, 779].

$$+ H_2NNH_2 \xrightarrow{MeOH} \quad + H_3NNH_2 \quad 96\% \tag{523}$$

B. Other Substitution Reactions

Interest in the synthetic applications of thiol substitution of quinones has continued to produce useful modifications of the well studied standard reactions. For example, in an effort to prepare 2,3-dialkylthio-1,4-benzoquinones Wladislaw and his collaborators employed the reversible Diels–Alder addition of cyclopentadiene (equation 524)[780].

$$+ 2\ RSNa \xrightarrow[ambient]{MeOH} \xrightarrow[0.2\ torr]{180°C} \tag{524}$$

R = Me, Et, Bu   87%, 41%, 35%   88%, 77%, 80%

A dithio heterocyclic quinone is obtained in 90% yield from the reaction of potassium cyanodithioimidocarbonate with 2,3-dichloro-1,4-naphthoquinone (equation 525)[781].

$$+ (KS)_2C{=}NCN \xrightarrow[25\text{–}30\ °C/1\ day]{Me_2C=O/H_2O} \tag{525}$$

Somewhat similar chemistry has been used to introduce ^{14}C- and/or ^{35}S-labelled atoms in 2,3-dicyano-1,4-dithia-9,10-anthraquinone (equation 526)[782].

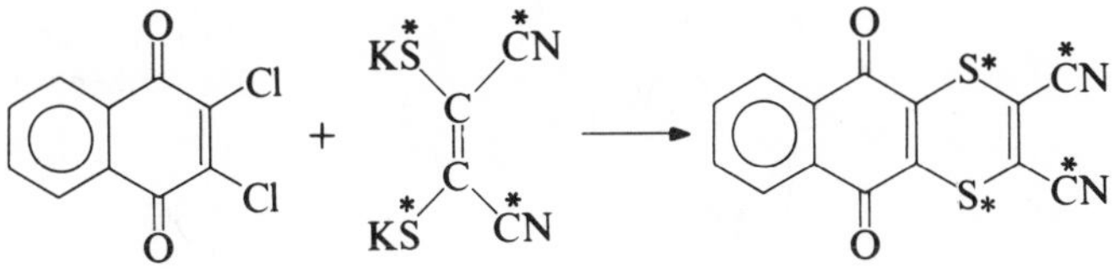

(526)

Oediger and Joop have made a series of polyfunctional quinonoid compounds beginning with the substitution of 2,3-dihaloquinones by mercaptoacetic acid (equation 527)[783]. The readily formed bislactones react with secondary or other strongly basic amines to provide good yields of these interesting molecules.

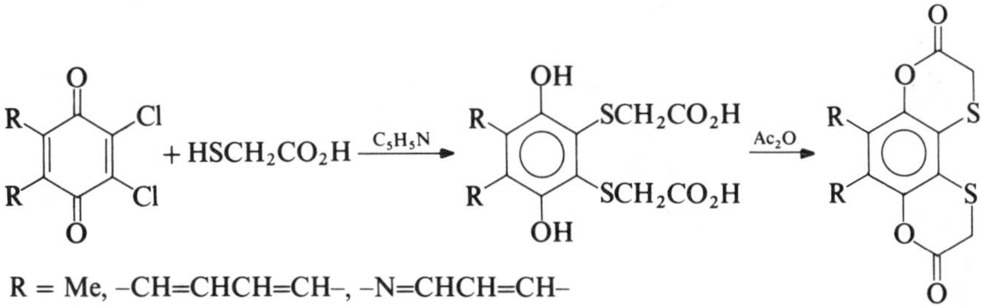

R = Me, –CH=CHCH=CH–, –N=CHCH=CH–

(527)

So many natural products are oxygenated quinones that it is surprising how little attention has been paid to their synthesis by nucleophilic substitution. Brassard has examined several aspects of this question[784]. He and Huot verified that the substitution of chloranil by methanol gives mixtures which are easily separable and provides useful amounts of several compounds. A significant improvement in the yield of 2,3-dimethoxy-1,4-naphthoquinone was obtained by an indirect approach (equation 528). In agreement

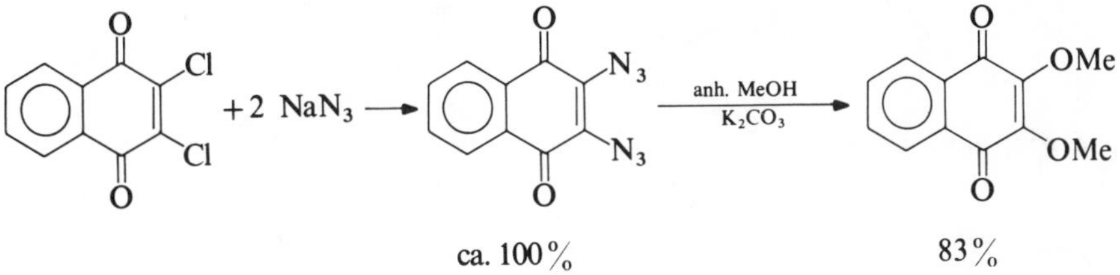

ca. 100%   83%

(528)

with other studies they found that the methanolysis of dichloronaphthazarin is not a useful preparative method; the corresponding diether works well as a substrate.

The substitution of chlorine by phenoxide has been shown to be a highly regiospecific method for the preparation of the 1,4-dihydroxyxanthones (**173**, equation 529)[785]. 2-Chloro-6-methyl-1,4-benzoquinone gave a large number of products, but it is the only compound showing erratic behavior. In an attempt to obtain samples of benzobisbenzofuranquinones for comparison with some natural products chloranil was allowed to react with a sodium aryloxide–ethoxide mixture (equation 530)[786]. Incredibly some members of this extensive series actually produced useful amounts of the desired product!

The great ease with which certain quinonoid halogen atoms can be replaced is illustrated by the attempt to recrystallize the Diels–Alder adduct of fluoranil and cyclopentadiene (equation 531)[787].

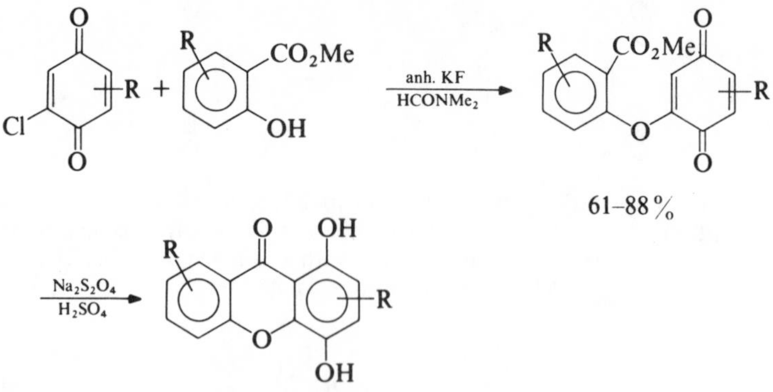

$\xrightarrow[HCONMe_2]{anh.\ KF}$ 61–88%

$\xrightarrow[H_2SO_4]{Na_2S_2O_4}$ (173) 45–77%

R = H, Me, MeO, OTs

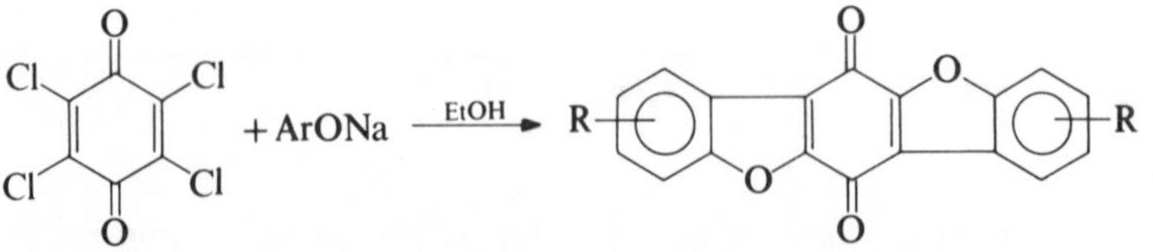

(529)

+ ArONa $\xrightarrow{EtOH}$

Ar = 3-An, 3,4- and 3,5-$(MeO)_2C_6H_3$ 30%, 16%, 10%

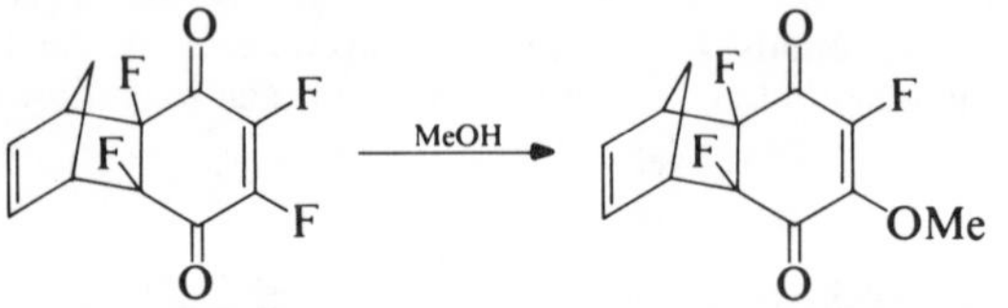

(530)

$\xrightarrow{MeOH}$

(531)

The introduction of a fluorine on a quinone ring has been a problem, but Cameron and his colleagues have offered some help in the naphthoquinone series (equation 532)[788]. This

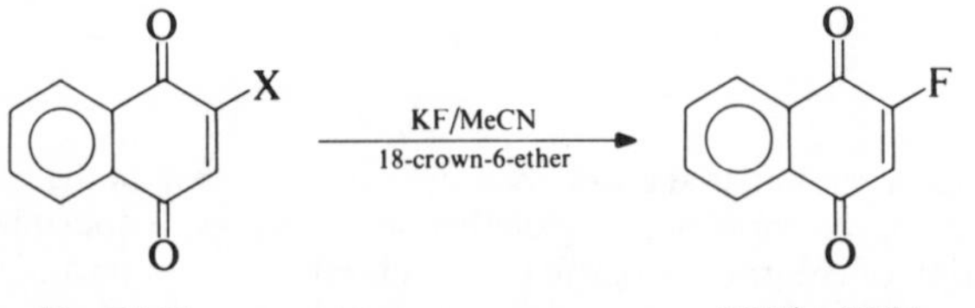

$\xrightarrow[18\text{-crown-6-ether}]{KF/MeCN}$

X=Cl, Br 96%, 90%

(532)

chemistry is not successful with benzoquinones, as was noted earlier[789]. However, Feiring and Sheppard were able to obtain some substitution of 2-chloro-1,4-benzoquinone using silver fluoride.

Buckle and his coworkers used thionyl chloride to chlorinate a 2-hydroxy-1,4-naphthoquinone which was then converted to the corresponding azido derivative (equation 533)[790]. Moore and his collaborators have often introduced the azido group by nucleophilic substitution in their extensive studies of such quinones. Most of these reports

$$\text{2-hydroxy-1,4-naphthoquinone} + SOCl_2 \longrightarrow \text{2-chloro-1,4-naphthoquinone (76\%)} \xrightarrow{\text{aq } NaN_3} \text{2-azido-1,4-naphthoquinone (75\%)} \quad (533)$$

are beyond the scope of the present review but provide excellent background for determining optimum substitution conditions[791–793]. Moore and Lee do report chemistry of immediate interest in showing that with an adjacent alkenyl group cyclization takes place giving an indolequinone (equation 534)[794]. The reaction involving two quinonoid rings leads to an entirely new class of fused ring systems.

$$\xrightarrow{\text{heat}} \quad 88\% \quad (534)$$

Another closely related anion is cyanide. Friedrich and Bucsis have examined the nucleophilic substitution of various halogenated 1,4-benzoquinones by cyanide and found modest yields of the corresponding 2,3-dicyanohydroquinone and much smaller amounts of tetracyanohydroquinone (equation 535)[795].

$$+ KCN \longrightarrow \text{(33\%, 32\%)} + \text{(7\%, 5\%)} \quad (535)$$

X = Cl, Br

A final example of nucleophilic substitution at quinonoid carbon is represented by the introduction of acetylenic groups with a palladium complex catalyst (equation 536)[796].

$$+ HC{\equiv}CPh \xrightarrow[CuI/C_5H_5N]{Pd(PPh_3)_2Cl_2} \text{2-R-3-(C}{\equiv}\text{CPh)-1,4-naphthoquinone}$$

R = Me_2N, morpholinyl, pyrrolidinyl, piperidinyl 40–80% (536)

IX. REFERENCES

1. M. D. Rozeboom, I.-M. Tegmo-Larsson and K. N. Houk, *J. Org. Chem.*, **46**, 2338 (1981).
2. A. Recsei, *Chem. Ber.*, **60**, 1836 (1927).
3. E. J. McInnis, B. Grant and E. Arcebo, *Tetrahedron Lett.*, **22**, 3807 (1981).
4. T. H. Porter, C. M. Bowman and K. Folkers, *J. Med. Chem.*, **16**, 115 (1973).
5. R. J. Wikholm, Y. Iwamoto, C. B. Bogentoft, T. H. Porter and K. Folkers, *J. Med. Chem.*, **17**, 893 (1974).
6. J. Vorkapić-Furač, T. Kishi, H. Kishi, T. H. Porter and K. Folkers, *Acta Pharm. Suec.*, **14**, 171 (1977).
7. R. Singh, V. K. Tandon, J. M. Khanna and N. Anand, *Indian J. Chem.*, **15B**, 970 (1977).
8. K. Buggle, D. O'Sullivan and N. D. Ryan, *Chem. Ind. (London)*, 164 (1974).
9. K. Buggle and J. Power, *J. Chem. Soc., Perkin Trans. 1*, 1070 (1980).
10. H. W. Atland and B. F. Briffa Jr, *J. Org. Chem.*, **50**, 433 (1985).
11. D. C. S. Tse, R. L. McCreery and R. N. Adams, *J. Med. Chem.*, **19**, 37 (1976).
12. Y.-O. Liang, R. M. Wightman, P. Plotsky and R. N. Adams, in *Chemical Tools in Catecholamine Research*, Vol. 1 (Eds. G. Jonsson, T. Malmfors and C. Sachs), Elsevier, New York, 1975, pp. 15–22.
13. Y.-O. Liang, P. M. Plotsky and R. N. Adams, *J. Med. Chem.*, **20**, 581 (1977).
14. P. J. Vithayathil and M. N. Gupta, *Indian J. Biochem. Biophys.*, **18**, 82 (1981).
15. S. Ito and K. Fujita, *J. Chromatogr.*, **187**, 418 (1980).
16. Y. Hayakawa and K. Takamura, *Yakugaku Zasshi*, **95**, 1173 (1975); *Chem. Abstr.*, **84**, 164514f (1976).
17. H. Burton and S. B. David, *J. Chem. Soc.*, 2193 (1952).
18. S. Khushvakhtova and I. U. Numanov, *Dokl. Akad. Nauk Tadzh. SSR*, **23**, 717 (1980); *Chem. Abstr.*, **95**, 6134e (1981).
19. F. Fariña and J. Valderrama, *Anal. Quim.*, **72**, 902 (1976).
20. J. M. Snell and A. Weissberger, *J. Am. Chem. Soc.*, **61**, 450 (1939).
21. D. Ades and M. Fontanille, *J. Appl. Polym. Sci.*, **23**, 11 (1979).
22. H. Kamogawa, H. Inoue, H. Fukuyama and M. Nanasawa, *Bull. Chem. Soc. Japan*, **52**, 3010 (1979).
23. D. J. Liberato, V. S. Byers, R. G. Dennick and N. Castagnoli, Jr, *J. Med. Chem.*, **24**, 28 (1981).
24. R. Davies and W. S. Pierpoint, *Biochem. Soc. Trans.*, **3**, 671 (1975).
25. W. S. Pierpoint, *Biochem. J.*, **98**, 567 (1966).
26. N. F. Janes, *Biochem. J.*, **112**, 617 (1969).
27. D. I. Stom, A. V. Kalabina, E. F. Kolmakova and S. S. Timofeeva, *Dokl. Akad. Nauk SSSR*, **205**, 989 (1972).
28. D. I. Stom, E. F. Kolmakova and S. S. Timofeeva, *Prikl. Biokhim. Mikrobiol.*, **8**, 112 (1972).
29. A. S. Stadnik, Y. Y. Lur'e and Y. M. Dedkov, *Zh. Anal. Khim. (Eng. Transl.)*, **32**, 1426 (1977).
30. R. Kerber and W. Gestrich, *Chem. Ber.*, **106**, 798 (1973).
31. A. Etienne and G. Lonchambon, *C. R. Acad. Sci. Paris*, **275**, 375 (1972).
32. R. Khun and H. Beinert, *Chem. Ber.*, **77**, 606 (1944).
33. R. Khun and I. Hammer, *Chem. Ber.*, **84**, 91 (1951).
34. G. Prota and E. Ponsiglione, *Tetrahedron Lett.*, 1327 (1972).
35. G. Prota, O. Petrillo, C. Santacroce and D. Sica, *J. Heterocycl. Chem.*, **7**, 555 (1970).
36. Y. Ueno, Y. Takeuchi, J. Koshitani and T. Yoshida, *J. Heterocycl. Chem.*, **18**, 645 (1981).
37. Y. Ueno, *Pharmazie*, **39**, 355 (1984).
38. M. H. Terdic and V. A. Smarandache, *Rev. Roumaine Chim.*, **30**, 133 (1985).
39. W.-D. Rudorf, E. Günther and M. Augustin, *Tetrahedron*, **40**, 381 (1984).
40. K. Klemm and B. Geiger, *Justus Liebigs Ann. Chem.*, **726**, 103 (1969).
41. P. T. S. Lau and M. Kestner, *J. Org. Chem.*, **33**, 4426 (1968).
42. P. T. S. Lau and T. E. Gompf, *J. Org. Chem.*, **35**, 4103 (1970).
43. J. Arnold and H. Hartmann, *Ger. (East) Pat.*, 133,670 (1979); *Chem. Abstr.*, **91**, 74588y (1979).
44. G. Mann, H. Wilde, S. Hauptmann, M. Naumann and P. Lepom, *J. Prakt. Chem.*, **323**, 776 (1981).
45. V. Horak and W. B. Manning, *J. Org. Chem.*, **44**, 120 (1979).
46. P. Ulrich and A. Cerami, *J. Med. Chem.*, **25**, 654 (1982).
47. R. N. L. Harris and L. T. Oswald, *Aust. J. Chem.*, **27**, 1309 (1974).

48. J. Ramachandran, S. R. Ramadas and C. N. Pillai, *Org. Prep. Proced. Int.*, **13**, 71 (1981); *Chem. Abstr.*, **95**, 61684j (1981).
49. K. Srihari and V. Sundarmurthy, *Indian J. Chem.*, **18B**, 80 (1979).
50. E. Campaigne and Y. Abe, *Bull. Chem. Soc. Japan*, **49**, 559 (1976).
51. G. Arai and M. Furui, *Nippon Kagaku Kaishi*, **5**, 673 (1984); *Chem. Abstr.*, **101**, 109968b (1984).
52. D. Liotta and G. Zima, *Tetrahedron Lett.*, 4977 (1978).
53. K. S. Amonkar and B. N. Ghosh, *Curr. Sci.*, **42**, 718 (1973).
54. A. Hikosaka, *Bull. Chem. Soc. Japan*, **43**, 3928 (1970).
55. A. M. Talati, N. D. Godhwani. A. D. Sheth, K. B. Shah and Y. K. Joshi, *Indian J. Tech.*, **22**, 468 (1984).
56. U. Muralikrishna and M. Krishnamurthy, *Indian J. Chem.*, **22A**, 858 (1983).
57. J. B. Chattopadhyay, M. N. Deshmukh and C. I. Jose, *J. Chem. Soc., Faraday Trans. 2*, **71**, 1127 (1975).
58. C. E. Alciaturi, *J. Chem. Soc., Faraday Trans. 2*, **73**, 433 (1977).
59. M. Sasaki, M. Bando, Y.-I. Inagaki, F. Amita and J. Osugi, *J. Chem. Soc. Chem. Commun.*, 725 (1981).
60. Y.-I. Inagaki, J. Osugi and M. Sasaki, *J. Chem. Soc., Perkin Trans. 2*, 115 (1985).
61. J. A. Chudek, R. Foster and F. J. Reid, *J. Chem. Soc. Chem. Commun.*, 726 (1983).
62. E. Pelizzetti, E. Mentasti and E. Pramauro, *J. Chem. Soc., Perkin Trans. 2*, 1651 (1976).
63. I. Baxter and W. R. Phillips, *J. Chem. Soc. Chem. Commun.*, 78 (1972).
64. I. Baxter and W. R. Phillips, *J. Chem. Soc., Perkin Trans. 1*, 268 (1973).
65. R. Davies and J. L. Frahn, *J. Chem. Soc., Perkin Trans. 1*, 2295 (1977).
66. T. Horigome, *Eiyo To Shokurgo*, **26**, 259 (1973); *Chem. Abstr.*, **80**, 35952r (1974).
67. D. W. Cameron and E. L. Samuel, *Austral. J. Chem.*, **30**, 2081 (1977).
68. R. Ott, E. Pinter and P. Kajtna, *Monatsh. Chem.*, **110**, 51 (1979).
69. T. Yamaoka and S. Nagakura, *Bull. Chem. Soc. Japan*, **44**, 2971 (1971).
70. R. Ott, E. Pinter and P. Kajtna, *Monatsh. Chem.*, **111**, 813 (1980).
71. K. Kouno, C. Ogawa, Y. Shimomura, H. Yano and Y. Veda, *Chem. Pharm. Bull.*, **29**, 301 (1981).
72. I. Shaaban and R. Soliman, *Pharmazie*, **33**, 642 (1978).
73. S. A. A. Osman, A. A. Abdalla and M. O. Alaib, *J. Pharm. Sci.*, **72**, 68 (1983).
74. I. Chaaban, A. Mohsen, M. E. Omar, F. A. Ashour and M. A. Mahran, *Sci. Pharm.*, **52**, 59 (1984).
75. N. G. Clark, *Pest. Sci.*, **16**, 23 (1985).
76. I. D. Biggs and J. M. Tedder, *Tetrahedron*, **34**, 1377 (1978).
77. M. Pardo, K. Joos and W. Schäfer, *Justus Liebigs Ann. Chem.*, 503 (1979).
78. H. Wittmann and H. Jeller, *Monatsch. Chem.*, **111**, 199 (1980).
79. T. H. Porter, T. Kishi and K. Folkers, *Acta Pharm. Suec.*, **16**, 74 (1979).
80. C. R. Tindale, *Aust. J. Chem.*, **37**, 611 (1984).
81. R. Nietzki and F. Schmidt, *Chem. Ber.*, **22**, 1653 (1889).
82. M. Matsuoka, K. Hamano, T. Kitao and K. Takagi, *Synthesis*, 953 (1984).
83. H.-J. Kallmayer and C. Tappe, *Arch. Pharm. (Weinheim)*, **318**, 569 (1985).
84. M. Matsuoka, T. Takei and T. Kitao, *Chem. Lett.*, 627 (1979).
85. J. R. Luly and H. Rapoport, *J. Org. Chem.*, **46**, 2745 (1981).
86. J. R. Luly and H. Rapoport, *J. Org. Chem.*, **47**, 2404 (1982).
87. R. Ott and R. Lachnit, *Monatsch. Chem.*, **104**, 15 (1973).
88. H.-J. Kallmayer, *Arch. Pharm.*, **307**, 806 (1974).
89. H.-J. Kallmayer and K. Seyfang, *Arch. Pharm.*, **313**, 603 (1980).
90. H.-J. Kallmayer and K. Seyfang, *Arch. Pharm.*, **316**, 283 (1983).
91. H.-J. Kallmayer and K. Seyfang. *Arch. Pharm.*, **317**, 743 (1984).
92. R. P. Soni and J. P. Saxena, *Bull. Chem. Soc. Japan*, **52**, 3096 (1979).
93. R. P. Soni, *J. Prakt. Chem.*, **323**, 853 (1981).
94. R. P. Soni and J. P. Saxena, *Bull. Chem. Soc. Japan*, **55**, 1681 (1982).
95. I. M. Roushdi, A. A. Mikhail and I. Chaaban, *Pharmazie*, **31**, 406 (1976); **32**, 269 (1977).
96. F. McCapra and Z. Razavi, *J. Chem. Soc. Chem. Commun.*, 42 (1975).
97. M. C. Vander Zwan, F. W., Hartner, R. A. Reamer and R. Tull, *J. Org. Chem.*, **43**, 509 (1978).
98. E. J. Corey and K. Achiwa, *J. Am. Chem. Soc.*, **91**, 1429 (1969).
99. Y. V. D. Nageswar and T. V. Padmanabha Rao, *Indian J. Chem.*, **18B**, 559 (1979).

100. Y. V. D. Nageshwar and T. V. Padmanabha Rao, *Indian J. Chem.*, **19B**, 624 (1980).
101. M. S. Rao, R. S. Kumar, V. R. Rao, K. R. Raju, S. M. Reddy and T. V.Padmanabha Rao, *Indian J. Chem.*, **23B**, 483 (1984).
102. T. Torres, S. V. Eswaran and W. Schäfer, *J. Heterocycl. Chem.*, **22**, 697 (1985).
103. T. Sasaki, K. Kanematsu and A. Kakehi, *Tetrahedron*, **28**, 1469 (1972).
104. M. Seth and N. M. Khanna, *Indian J. Chem.*, **14B**, 536 (1976).
105. M. Yoshimoto, H. Miyazawa, H. Nakao, K. Shinkai and M. Arakawa, *J. Med. Chem.*, **22**, 491 (1979).
106. R. K.-Y. Zee-Cheng, E. G. Podrebarac, C. S. Menon and C. C. Cheng, *J. Med. Chem.*, **22**, 501 (1979).
107. M. L. Jain, R. P. Soni and J. P. Saxena, *Indian J. Chem.*, **19B**, 718 (1980).
108. Y. Tamura, K. Sumoto, H. Matsushima, H. Taniguchi and M. Ikeda, *J. Org. Chem.*, **38**, 4324 (1973).
109. J. F. W. McOmie and J. M. Blatchly, in *Organic Reaction*, Vol. 19, Wiley, New York, 1972, pp. 199–277.
110. J. M. Blatchly, R. J. S. Green and J. F. W. McOmie, *J. Chem. Soc., Perkin Trans. 1*, 2286 (1972).
111. W. Flaig, T. Ploetz and H. Biergans, *Justus Liebigs Ann. Chem.*, **597**, 196 (1985).
112. F. Takacs, *Monatsch Chem.*, **95**, 961 (1964).
113. J. M. Blatchly, R. J. S. Green, J. F. W. McOmie and S. A. Saleh, *J. Chem. Soc. Perkin Trans. 1*, 309 (1975).
114. L. F. Fieser, *J. Am. Chem. Soc.*, **70**, 3165 (1948).
115. H. S. Wilgus, III and J. W. Gates Jr, *Can. J. Chem.*, **45**, 1976 (1967).
116. J. F. W. McOmie, J. B. Searle and S. A. Saleh, *J. Chem. Soc. Perkin Trans. 1*, 314 (1975).
117. J. A. Croft, E. Ritchie and W. C. Taylor, *Aust. J. Chem.*, **29**, 1979 (1976).
118. G. B. Marini-Bettolo, F. Delle Monache, O. Goncalves da Lima and S. De Bartas Coelho, *Gazz. Chim. Ital.*, **101**, 41 (1971).
119. P. T. Grant, P. A. Plack and R. H. Thomson, *Tetrahedron Lett.*, **21**, 4043 (1980).
120. D. W. Cameron, G. L. Feutrill, A. F. Patti, P. Perlmutter and M. A. Sefton, *Aust. J. Chem.*, **35**, 1501 (1982).
121. B. Achari, S. Bandyopadhyay, K. Basu and S. C. Pakrashi, *Tetrahedron*, **41**, 107 (1985).
122. D. Villessot and Y. Lepage, *J. Chem. Res. (S)*, 464; *(M)*, 5538 (1978).
123. R. C. Atkinson, P. B. D. de la Mare and D. S. Larsen, *J. Chem. Soc. Perkin Trans. 2*, 271 (1983).
124. R. K. Norris and S. Sternhell, *Austral. J. Chem.*, **26**, 333 (1973).
125. A. Dargelos, J. Migliaccio and M. Chaillet, *Tetrahedron*, **27**, 5673 (1971).
126. A. Dargelos and M. Chaillet, *Tetrahedron*, **28**, 3333 (1972).
127. P. Švec and M. Zbirovský, *Coll. Czech. Chem. Commun.*, **40**, 3029 (1975).
128. H. Laatsch, *Tetrahedron Lett.*, 3345 (1978).
129. F. Fariña and J. Valderrama, *Anal. Quim.*, **70**, 258 (1974).
130. D. W. Cameron, M. J. Crossley, G. I. Feutrill and P. G. Griffiths, *Austral. J. Chem.*, **31**, 1363 (1978).
131. D. W. Cameron, G. I. Feutrill and P. G. Griffiths, *Austral. J. Chem.*, **34**, 1513 (1981).
132. S. N. Heinzman and J. R. Grunwell, *Tetrahedron Lett.*, **21**, 4305 (1980).
133. M. E. Jung and J. A. Hagenah, *J. Org. Chem.*, **48**, 5359 (1983).
134. L. L. Miller and R. F. Stewart, *J. Org. Chem.*, **43**, 3078 (1978).
135. G. Rettig and H. P. Latscha, *Z. Naturforsch.*, **35b**, 399 (1980).
136. I. A. Novikova, V. B. Vol'eva, N. L. Komissarova, I. S. Belostotskaya and V. V. Ershov, *Izv. Akad. Nauk SSSR, Ser. Khim. (Eng. Transl.)*, 1732 (1982).
137. A. Forrester, A. S. Ingram, I. L. John and R. H. Thomson, *J. Chem. Soc. Perkin Trans. 1*, 1115 (1975).
138. K. A. Parker and M. E. Sworin, *J. Org. Chem.*, **46**, 3218 (1981).
139. C. G. Hughes, E. G. Lewars and A. H. Rees, *Can. J. Chem.*, **52**, 3327 (1974).
140. A. S. Kende and P. C. Naegely, *Tetrahedron Lett.*, 4775 (1978).
141. W. Alcolay, *Helv. Chim. Acta*, **30**, 578 (1947).
142. D. E. Nichols and A. T. Shulgin, *J. Pharm. Sci.*, **65**, 1554 (1976).
143. M. G. Voronkov, *Z. Chem.*, **21**, 136 (1981).
144. G. Arai, *Nippon Kagaku Kaishi*, **4**, 508 (1977); *Chem. Abstr.*, **87**, 38564w (1978).
145. G. Arai and M. Onozuka, *Nippon Kagaku Kaishi*, **11**, 1665 (1977); *Chem. Abstr.*, **88**, 36877x (1978).

146. G. Arai and M. Onozuka, *Nippon Kagaku Kaishi*, **2**, 243 (1979); *Chem. Abstr.*, **90**, 167692x (1979).
147. G. Arai and A. Suzuki, *Nippon Kagaku Kaishi*, **4**, 465 (1983); *Chem. Abstr.*, **99**, 52701t (1984).
148. G. Arai, Y. Suyama and H. Horie, *Kanagawa Daigaku Kogakubu Kenkyu Hokoku*, **16**, 39 (1978); *Chem. Abstr.*, **89**, 89881w (1979).
149. G. Arai, *Nippon Kagaku Kaishi*, **4**, 579 (1982); *Chem. Abstr.*, **96**, 225259y (1982).
150. G. Arai, *Nippon Kagaku Kaishi*, **4**, 551 (1983); *Chem. Abstr.*, **99**, 52662f (1984).
151. J. M. Singh and A. B. Turner, *J. Chem. Soc. Perkin Trans. 1*, 2556 (1974).
152. F. Fariña and J. Valderrama, *Synthesis*, 315 (1971).
153. O. Reinaud, P. Capdevielle and M. Maumy, *Tetrahedron Lett.*, **26**, 3993 (1985).
154. A. Ichihara, M. Kobayashi, K. Oda, S. Sakamura and R. Sakai, *Tetrahedron*, **35**, 2861 (1979).
155. A. Ichihara, M. Kobayashi, K. Oda and S. Sakamura, *Tetrahedron Lett.*, 4231 (1974).
156. H. Pluim and H. Wynberg, *J. Org. Chem.*, **45**, 2498 (1980).
157. G. E. Wright and N. C. Brown, *J. Med. Chem.*, **17**, 1277 (1974).
158. I. M. Roushdi and N. S. Habib, *Pharmazie*, **32**, 562 (1977).
159. G. A. Nikiforov, L. G. Plekhanova, K. De Jonge and V. V. Ershov, *Izv. Akad. Nauk SSSR. Ser. Khim. (Eng. Transl.)*, 2455 (1978).
160. N. R. Ayyangar, R. J. Lahoti and D. R. Wagle, *Indian J. Chem.*, **18B**, 196 (1979).
161. B. Lengstad and J. Lönngren, *Carbohydrate Res.*, **72**, 312 (1979).
162. A. Nishinaga, T. Shimizu and T. Matsuura, *J. Chem. Soc. Chem. Commun.*, 970 (1979).
163. E. Hofer, *Chem. Ber.*, **112**, 2913 (1979).
164. G. Neumann and W. P. Neumann, *J. Organomet. Chem.*, **42**, 277 (1972).
165. B. B. Adeleke and J. K. S. Wan, *J. Chem. Soc. Perkin Trans. 2*, 225 (1980).
166. H. Matsumoto, S. Koike, I. Matsubara, T. Nakano and Y. Nagai, *Chem. Lett.*, 533 (1982).
167. J. K. Rasmussen, L. R. Krepski, S. M. Heilmann, H. K. Smith, II and M. L. Tumey, *Synthesis*, 457 (1983).
168. D. A. Evans and L. K. Truesdale, *Tetrahedon Lett.*, 4929 (1973).
169. D. A. Evans, L. K. Truesdale and G. L. Carroll, *J. Chem. Soc. Chem. Commun.*, 55 (1973).
170. D. A. Evans, J. M. Hoffman and L. K. Truesdale, *J. Am. Chem. Soc.*, **95**, 5822 (1973).
171. D. A. Evans and R. Y. Wong, *J. Org. Chem.*, **42**, 350 (1977).
172. F. Duboudin, Ph. Cazeau, F. Moulines and O. Laporte, *Synthesis*, 212 (1982).
173. A. J. Guildford and R. W. Turner, *Synthesis*, 46 (1982).
174. M.-C. Lasne, J.-L. Ripoll and A. Thuillier, *Chem. Ind. (London)*, 830 (1980).
175. H. W. Moore, Y. L. Sing and R. S. Sidhu, *J. Org. Chem.*, **42**, 3320 (1977).
176. A. Fischer and G. N. Henderson, *Tetrahedron Lett.*, **21**, 701 (1980).
177. A. Fischer and G. N. Henderson, *Tetrahedron Lett.*, **24**, 131 (1983).
178. D. Liotta, M. Saindane and C. Barnum, *J. Org. Chem.*, **46**, 3369 (1981).
179. H. W. Moore, Y.-L. L. Sing and R. S. Sidhu, *J. Org. Chem.*, **45**, 5057 (1980).
180. K. F. West and H. W. Moore, *J. Org. Chem.*, **47**, 3591 (1982).
181. S. A. DiBiase, B. A. Lipisko, A. Haag, R. A. Wolak and G. W. Gokel, *J. Org. Chem.*, **44**, 4640 (1979).
182. T. Hata, M. Sekine and N. Ishikawa, *Chem. Lett.*, 645 (1975).
183. R. O. Duthaler, P. A. Lyle and C. Heuberger, *Helv. Chim. Acta*, **67**, 1406 (1984).
184. A. Minsky and M. Rabinovitz, *Synthesis*, 497 (1983).
185. D. N. Nicolaides and K. E. Litinas, *J. Chem. Res. (S)*, 57; *(M)*, 658 (1983).
186. P. Frøyen, *Acta Chem. Scand.*, **27**, 141 (1973).
187. A. N. Pudovik, É. S. Batyeva, V. D. Nesterenko and N. P. Anoshina, *Zh. Obshch. Khim. (Eng. Transl.)*, **43**, 29 (1973).
188. A. N. Pudovik, É. S. Batyeva, A. V. Il'yasov, V. D. Nesterenko, A. Sh. Mukhtarov and N. P. Anoshina, *Zh. Obshch. Khim. (Eng. Transl.)*, **43**, 1442 (1973).
189. A. N. Pudovik, A. M. Kibardin, A. P. Pashinkin, Yu. I. Sudarev and T. Kh. Gazizov, *Zh. Obshch. Khim. (Eng. Transl.)*, **44**, 500 (1974).
190. A. N. Pudovik, É. S. Batyeva, V. D. Nesterenko and E. I. God'dfarb, *Zh. Obshch. Khim. (Eng. Transl.)*, **44**, 976 (1974).
191. B. A. Arbuzov, N. A. Polezhaeva and V. S. Vinogradova, *Dokl. Akad. Nauk SSSR (Eng. Transl.)*, **201**, 881 (1971).
192. Yu. K. Gusev, V. N. Chistokletov and A. A. Petrov, *Zh. Org. Khim. (Eng. Transl.)*, **11**, 2369 (1975).

193. M. M. Sidky and L. S. Boulos, *Phos. Sulf.*, **19**, 27 (1984).
194. N. Latif, I. F. Zeid, N. Mishriky and F. M. Assad, *Tetrahedron Lett.*, 1355 (1974).
195. N. Latif, A. Nada, H. M. El-Namaky and B. Haggag, *Indian J. Chem.*, **18B**, 131 (1979).
196. I. Zeid, I. El-Sakka, S. Yassin and A. Abass, *Justus Liebigs Ann. Chem.*, 196 (1984).
197. W. I. Awad and R. R. Al-Sabti, *J. Prakt. Chem.*, **320**, 986 (1978).
198. W. I. Awad, Z. S. Salih and Z. H. Aiube, *J. Chem. Soc. Perkin Trans. 1*, 1280 (1977).
199. E. G. Ter-Gabriélyan, N. P. Gambaryan and I. K. Knunyants, *Izv. Akad. Nauk SSSR. Ser. Khim. (Eng. Transl.)*, 2049 (1972).
200. W. Verboom and H. J. T. Bos, *Recl. Trav. Chim. Pays-Bas*, **98**, 559 (1979).
201. S. Verma, N. M. Kansal, R. S. Mishra and M. M. Bokadia, *Heterocycles*, **16**, 1537 (1981).
202. H. J.-M. Dou, G. Vernin and J. Metzger, *Bull. Soc. Chim. France*, 4602 (1971).
203. A. Rembaum and M. Szwarc, *J. Am. Chem. Soc.*, **77**, 4468 (1955).
204. A. Citterio, *Tetrahedron Lett.*, 2701 (1978).
205. A. Citterio, A. Arnoldi and F. Minisci, *J. Org. Chem.*, **44**, 2674 (1979).
206. A. Citterio, *Gazz. Chim. Ital.*, **110**, 253 (1980).
207. I. A. Shlyapnikova, V. A. Roginskii and V. B. Miller, *Izv. Akad. Nauk SSSR. Ser. Khim. (Eng. Transl.)*, 2215 (1978).
208. A. A. Yassin and N. A. Rizk, *Eur. Polym. J.*, **12**, 393 (1976).
209. A. F. Bickel and W. A. Waters, *J. Chem. Soc.*, 1764 (1950).
210. Y. Naruta, *J. Am. Chem. Soc.*, **102**, 3774 (1980).
211. Y. Naruta, *J. Org. Chem.*, **45**, 4097 (1980).
212. A. Takuwa, Y. Naruta, O. Soga and K. Maruyama, *J. Org. Chem.*, **49**, 1857 (1984).
213. K. Maruyama and Y. Naruta, *J. Org. Chem.*, **43**, 3796 (1978).
214. Y. Naruta and K. Maruyama, *Chem. Lett.*, 885 (1979).
215. Y. Naruta, H. Uno and K. Maruyama, *J. Chem. Soc. Chem. Commun.*, 1277 (1981).
216. K. Maruyama, A. Takuwa, Y. Naruta, K. Satao and O. Soga, *Chem. Lett.*, 47 (1981).
217. Y. Naruta, N. Nagai, Y. Arita and K. Maruyama, *Chem. Lett.*, 1683 (1983).
218. H. Uno, *J. Org. Chem.*, **51**, 350 (1986).
219. Y. Naruta, H. Uno and K. Maruyama, *Tetrahedron Lett.*, **22**, 5221 (1981).
220. A. Hosomi and H. Sakurai, *Tetrahedron Lett.*, 4041 (1977).
221. K. Mori, M. Sakakibara and M. Waku, *Tetrahedron Lett.*, **25**, 1085 (1984).
222. I. Wakeshima and I. Kijima, *Bull. Chem. Soc. Japan*, **54**, 2345 (1981).
223. L. S. Hegedus, E. L. Waterman and J. Catlin, *J. Am. Chem. Soc.*, **94**, 7155 (1972).
224. L. S. Hegedus and E. L. Waterman, *J. Am. Chem. Soc.*, **96**, 6789 (1974).
225. L. S. Hegedus, B. R. Evans, D. E. Korte, E. L. Waterman and K. Sjöberg, *J. Am. Chem. Soc.*, **98**, 3901 (1976).
226. L. S. Hegedus and B. R. Evans, *J. Am. Chem. Soc.*, **100**, 3461 (1978).
227. L. S. Hegedus, R. R. Odle, P. M. Winton and P. R. Weider, *J. Org. Chem.*, **47**, 2607 (1982).
228. M. F. Semmelhack, L. Keller, T. Sato, E. J. Spiess and W. Wulff, *J. Org. Chem.*, **50**, 5566 (1985).
229. N. Jacobsen and K. Torssell, *Justus Liebigs Ann. Chem.*, **763**, 135 (1972).
230. N. Jacobsen and K. Torssell, *Acta Chem. Scand.*, **27**, 3211 (1973).
231. J. Goldman, N. Jacobsen and K. Torssell, *Acta Chem. Scand. B*, **28**, 492 (1974).
232. S. C. Sharma and K. Torssell, *Acta Chem. Scand. B*, **32**, 347 (1978).
233. P. M. Brown and R. H. Thomson, *J. Chem. Soc., Perkin Trans. 1*, 997 (1976).
234. R. G. F. Giles and G. H. P. Roos, *J. Chem. Soc. Perkin Trans. 1*, 1632 (1976).
235. R. G. F. Giles and G. H. P. Roos, *Tetrahedron Lett.*, 3093 (1977).
236. R. G. F. Giles, P. R. K. Mitchell and G. H. P. Roos, *S.-Afr. Tydskr. Chem.*, **32**, 131 (1979); *Chem. Abstr.*, **93**, 71351w (1980).
237. C. S. James, R. S. T. Loeffler and D. Woodcock, *Pest. Sci.*, **12**, 1 (1981).
238. P. Boehm, K. Cooper, A. T. Hudson, J. P. Elphick and N. McHardy, *J. Med. Chem.*, **24**, 295 (1981).
239. Z. Florjanczyk, W. Kuran, S. Pasynkiewicz and G. Kwas, *J. Organomet. Chem.*, **112**, 21 (1976).
240. Z. Florjanczyk, W. Kuran, S. Pasynkiewicz and A. Krasnicka, *J. Organomet. Chem.*, **145**, 21 (1978).
241. Z. Florjanczyk and E. Szymanska-Zachara, *J. Organomet. Chem.*, **259**, 127 (1983).
242. A. Alberola, A. M. G. Nogal, J. A. M. De Ilarduya and F. J. Pulido, *Anal. Quim.*, **78**, 175 (1982).
243. A. Alberola, A. M. G. Nogal and J. A. M. De Ilarduya, *Anal. Quim.*, **76**, 123 (1980).
244. A. Alberola, A. M. G. Nogal and F. J. Pulido, *Anal. Quim.*, **76**, 199 (1980).

245. A. Alberola, A. M. G. Nogal, J. A. M. De Ilarduya and F. J. Pulido, *Anal. Quim.*, **78**, 166 (1982).
246. Y. Tachibana, *Chem. Lett.*, 901 (1977).
247. N. K. Kapoor, R. B. Gupta and R. N. Khanna, *Indian J. Chem.*, **21B**, 189 (1982).
248. M. Kajiwara, O. Sakamoto and S. Ohta, *Heterocycles*, **14**, 1995 (1980).
249. H. Sugihara, Y. Kawamatsu and H. Morimoto, *Justus Liebigs Ann. Chem.*, **763**, 128 (1972).
250. J. Y. Kim, H. Yamamoto and T. Kwan, *Chem. Pharm. Bull.*, **23**, 1091 (1975).
251. M. J. Manning, P. W. Raynolds and J. S. Swenton, *J. Am. Chem. Soc.*, **98**, 5008 (1976).
252. G. W. Kabalka, *Tetrahedron*, **29**, 1159 (1973).
253. K. Maruyama, K. Saimoto and Y. Yamamoto, *J. Org. Chem.*, **43**, 4895 (1978).
254. J. Majnusz and R. W. Lenz, *Eur. Polym. J.*, **21**, 565 (1985).
255. K. N. Houk, *Acc. Chem. Res.*, **11**, 361 (1975).
256. R. Gleiter and L. A. Paquette, *Acc. Chem. Res.*, **16**, 328 (1983).
257. F. K. Brown and K. N. Houk, *J. Am. Chem. Soc.*, **107**, 1971 (1985).
258. L. A. Paquette, T. M. Kravetz and L-Y. Hsu, *J. Am. Chem. Soc.*, **107**, 6598 (1985).
259. I.-M. Tegmo-Larsson, M. D. Rozeboom and K. N. Houk, *Tetrahedron Lett.*, **22**, 2043 (1981).
260. I.-M. Tegmo-Larsson, M. D. Rozeboom, N. G. Rondan and K. N. Houk, *Tetrahedron Lett.*, **22**, 2047 (1981).
261. M. Kaftory, M. Peled and D. Ginsburg, *Helv. Chim. Acta*, **62**, 1326 (1979).
262. M. Yasuda, K. Harano and K. Kanematsu, *J. Org. Chem.*, **45**, 659 (1980).
263. K. N. Houk and R. W. Strozier, *J. Am. Chem. Soc.*, **95**, 4094 (1973).
264. D. J. Burnell, H. B. Goodbrand, S. M. Kaiser and Z. Valenta, *Can. J. Chem.*, **62**, 2398 (1984).
265. A. C. Coda, G. Desimoni, E. Ferrari, P. P. Righetti and G. Tacconi, *Tetrahedron*, **40**, 1611 (1984).
266. F. P. Ballistreri, E. Maccarone, G. Perrini, G. A. Tomaselli and M. Torre, *J. Chem. Soc. Perkin Trans. 2,* 273 (1982).
267. K. Neelakantan, S. Bhatia and J. K. Gehlawat, *Indian J. Tech.*, **19**, 522 (1981).
268. B. Giese, J. Stellmach and O. Exner, *Tetrahedron Lett.*, 2343 (1979).
269. K.-H. Menting, W. Eichel, K. Riemenschneider, H. L. K. Schmand and P. Boldt, *J. Org. Chem.*, **48**, 2814 (1983).
270. J. Jurczak, *Polish J. Chem.*, **53**, 209 (1979).
271. J. Jurczak and A. Zamojski, *Tetrahedron*, **28**, 1505 (1972).
272. J. Jurczak, T. Kożluk, M. Tkacz and C. H. Eugster, *Helv. Chim. Acta*, **66**, 218 (1983).
273. A. A. Hofmann, I. Wyrsch-Walraf, P. X. Iten and C. H. Eugster, *Helv. Chim. Acta*, **62**, 2211 (1979).
274. J. Jurczak, A. L. Kawczyński and T. Kożluk, *J. Org. Chem.*, **50**, 1106 (1985).
275. J. Jurczak, T. Kozluk and S. Filipek, *Helv. Chim. Acta*, **66**, 222 (1983).
276. W. G. Dauben and W. R. Baker, *Tetrahedron Lett.*, **23**, 2611 (1982).
277. W. G. Dauben and R. A. Bunce, *Tetrahedron Lett.*, **23**, 4875 (1982).
278. B. M. Trost, D. O'Krongly and J. L. Belletire, *J. Am. Chem. Soc.*, **102**, 7595 (1980).
279. R. A. Dickinson, R. Kubela, G. A. MacAlpine, Z. Stojanac and Z. Valenta, *Can. J. Chem.*, **50**, 2377 (1972).
280. Z. Stojanac, R. A. Dickinson, N. Stojanac, R. J. Woznow and Z. Valenta, *Can. J. Chem.*, **53**, 616 (1975).
281. N. Stojanac, A. Sood, Z. Stojanac and Z. Valenta, *Can. J. Chem.*, **53**, 619 (1975).
282. J. Das, R. Kubela, G. A. MacAlpine, Z. Stojanac and Z. Valenta, *Can. J. Chem.*, **57**, 3308 (1979).
283. J. S. Tou and W. Reusch, *J. Org. Chem.*, **45**, 5012 (1980).
284. B. M. Trost, J. Ippen and W. C. Vladuchick, *J. Am. Chem. Soc.*, **99**, 8116 (1977).
285. B. M. Trost, W. C. Vladuchick and A. J. Bridges, *J. Am. Chem. Soc.*, **102**, 3554 (1980).
286. R. K. Boeckman, Jr, T. M. Dolak and K. O. Culos, *J. Am. Chem. Soc.*, **100**, 7098 (1978).
287. B. M. Trost, C. G. Caldwell, E. Murayama and D. Heissler, *J. Org. Chem.*, **48**, 3252 (1983).
288. R. K. Boeckman, M. H. Delton, T. M. Dolak, T. Watanabe and M. D. Glick, *J. Org. Chem.*, **44**, 4396 (1979).
289. T. R. Kelly, J. Vaya and L. Ananthasubramanian, *J. Am. Chem. Soc.*, **102**, 5983 (1980).
290. W. B. Manning, *Tetrahedron Lett.*, 1661 (1979).
291. T. R. Kelly, J. Vaya and L. Ananthasubramanian, *J. Am. Chem. Soc.*, **102**, 5983 (1980).
292. T. R. Kelly and N. D. Parekh, *J. Org. Chem.*, **47**, 5009 (1982).
293. T. R. Kelly, R. N. Goerner, Jr, J. W. Gillard and B. K. Prazak, *Tetrahedron Lett.*, 3869 (1976).

294. T. R. Kelly, J. W. Gillard and R. N. Goerner, Jr, *Tetrahedron Lett.*, 3873 (1976).
295. T. R. Kelly, *Tetrahedron Lett.*, 1387 (1978).
296. T. R. Kelly and M. Montury, *Tetrahedron Lett.*, 4309 (1978).
297. T. R. Kelly and M. Montury, *Tetrahedron Lett.*, 4311 (1978).
298. T. R. Kelly, J. A. Magee and F. R. Weibel, *J. Am. Chem. Soc.*, **102**, 798 (1980).
299. T. R. Kelly, L. Ananthasubramanian, K. Borah, J. W. Gillard, R. N. Goerner, Jr, P. F. King, J. M. Lyding, W.-G. Tsang and J. Vaya, *Tetrahedron*, **40**, 4569 (1984).
300. R. C. Gupta, D. A. Jackson and R. J. Stoodley, *J. Chem. Soc. Chem. Commun.*, 929 (1982).
301. G. A. Olah, D. Meidar and A. P. Fung, *Synthesis*, 270 (1979).
302. S. Mahanti, H. Maurya and S. M. Verma, *J. Indian Chem. Soc.*, **61**, 1034 (1984).
303. S. Danishefsky, *Acc. Chem. Res.*, **14**, 400 (1981).
304. P. Brownbridge, *Synthesis*, 85 (1983).
305. S. Danishefsky and T. Kitahara, *J. Am. Chem. Soc.*, **96**, 7807 (1974).
306. J. Banville and P. Brassard, *J. Chem. Soc. Perkin Trans. 1*, 1852 (1976).
307. S. Danishefsky, R. K. Singh and R. B. Gammill, *J. Org. Chem.*, **43**, 379 (1978).
308. R. G. F. Giles, S. C. Yorke, I. R. Green and V. I. Hugo, *J. Chem. Soc. Chem. Commun.*, 554 (1984).
309. S. Danishefsky, T. Kitahara, C. F. Yan and J. Morris, *J. Am. Chem. Soc.*, **101**, 6996 (1979).
310. S. Danishefsky, C.-F. Yan, R. J. Singh, R. B. Gammill, P. M. McCurry, Jr, N. Fritsch and J. Clardy, *J. Am. Chem. Soc.*, **101**, 7001 (1979).
311. S. Danishefsky and T. A. Craig, *Tetrahedron*, **37**, 4081 (1981).
312. S. Danishefsky, B. J. Uang and G. Quallich, *J. Am. Chem. Soc.*, **106**, 2453 (1984).
313. S. Danishefsky, B. J. Uang and G. Quallich, *J. Am. Chem. Soc.*, **107**, 1285 (1985).
314. S. Danishefsky, P. Schuda and K. Kato, *J. Org. Chem.*, **41**, 1081 (1976).
315. D. G. Batt and B. Ganem, *Tetrahedron Lett.*, 3323 (1978).
316. D. R. Anderson and T. H. Koch, *J. Org. Chem.*, **43**, 2726 (1978).
317. R. W. Aben and H. W. Scheeren, *J. Chem. Soc. Perkin Trans. 1*, 3132 (1979).
318. A. P. Kozikowski, K. Sugiyama and J. P. Springer, *Tetrahedron Lett.*, **21**, 3257 (1980).
319. S. V. Ley, W. L. Mitchell, T. V. Radhakrishnan and D. H. R. Barton, *J. Chem. Soc. Perkin Trans. 1*, 1582 (1981).
320. R. Gompper and M. Sramek, *Synthesis*, 649 (1981).
321. D. W. Cameron, C. Conn and G. I. Feutrill, *Aust. J. Chem.*, **34**, 1945 (1981).
322. E. M. Acton and G. L. Tong, *J. Heterocycl. Chem.*, **18**, 1141 (1981).
323. W. A. Ayer, D. E. Ward, L. M. Browne, L. T. Delbaere and Y. Hoyano, *Can. J. Chem.*, **59**, 2665 (1981).
324. D. W. Cameron, G. I. Feutrill and P. Perlmutter, *Tetrahedron Lett.*, **22**, 3273 (1981).
325. P. R. Brook, B. Devadas and P. G. Sammes, *J. Chem. Res. (S)*, 134 (1982).
326. D. W. Cameron, G. I. Feutrill and P. G. McKay, *Aust. J. Chem.*, **35**, 2095 (1982).
327. B. M. Trost and W. H. Pearson, *Tetrahedron Lett.*, **24**, 269 (1983).
328. R. P. Potman, N. J. M. L. Janssen, J. W. Scheeren and R. J. F. Nivard, *J. Org. Chem.*, **49**, 3628 (1984).
329. Y. Kita, H. Yasuda, O. Tamura and Y. Tamura, *Tetrahedron Lett.*, **25**, 1813 (1984).
330. J. Kallmerten, *Tetrahedron Lett.*, **25**, 2843 (1984).
331. G. Roberge and P. Brassard, *Synthesis*, 381 (1981).
332. G. Roberge and P. Brassard, *J. Org. Chem.*, **46**, 4161 (1981).
333. G. Roberge and P. Brassard, *J. Chem. Soc. Perkin Trans. 1*, 1041 (1978).
334. J. Savard and P. Brassard, *Tetrahedron Lett.*, 4911 (1979).
335. C. Brisson and P. Brassard, *J. Org. Chem.*, **46**, 1810 (1981).
336. V. Guay and P. Brassard, *J. Org. Chem.*, **49**, 1853 (1984).
337. J. Savard and P. Brassard, *Tetrahedron*, **40**, 3455 (1984).
338. V. Guay and P. Brassard, *Tetrahedron*, **40**, 5039 (1984).
339. B. Simoneau, J. Savard and P. Brassard, *J. Org. Chem.*, **50**, 5433 (1985).
340. N. Benfaremo and M. P. Cava, *J. Org. Chem.*, **50**, 139 (1985).
341. G. J. O'Malley, R. A. Murphy, Jr and M. P. Cava, *J. Org. Chem.*, **50**, 5533 (1985).
342. G. A. Kraus and S. H. Woo, *J. Org. Chem.*, **51**, 114 (1986).
343. I. Fleming and A. Percival, *J. Chem. Soc. Chem. Commun.*, 681 (1976).
344. M. J. Carter, I. Fleming and A. Percival, *J. Chem. Soc. Perkin Trans. 1*, 2415 (1981).
345. M. E. Jung and J. A. Lowe, *J. Chem. Soc. Chem. Commun.*, 95 (1978).

346. M. E. Jung and R. W. Brown, *Tetrahedron Lett.*, **22**, 3355 (1981).
347. M. E. Jung, M. Node, R. W. Pfluger, M. A. Lyster and J. A. Lowe, III, *J. Org. Chem.*, **47**, 1152 (1982).
348. P. Cano, A. Echavarren, P. Prados and F. Fariña, *J. Org. Chem.*, **48**, 5373 (1983).
349. T. Sasaki, K. Kanematsu and K. Iizuka, *J. Org. Chem.*, **41**, 1105 (1976).
350. T. Sasaki, K. Kanematsu, K. Iizuka and N. Izumichi, *Tetrahedron*, **32**, 2879 (1976).
351. T. Sasaki, K. Kanematsu, K. Iizuka and I. Ando, *J. Org. Chem.*, **41**, 1425 (1976).
352. R. N. Warrener and B. C. Hammer, *J. Chem. Soc. Chem. Commun.*, 942 (1981).
353. R. N. Warrener, D. A. C. Evans and R. A. Russell, *Tetrahedron Lett.*, **25**, 4833 (1984).
354. M. A. Makhlouf and B. Rickborn, *J. Org. Chem.*, **46**, 2734 (1981).
355. J. G. Smith, S. S. Welankiwar, B. S. Shantz, E. H. Lai and N. G. Chu, *J. Org. Chem.*, **45**, 1817 (1980).
356. J. G. Smith and P. W. Dibble, *J. Org. Chem.*, **48**, 5361 (1983).
357. K. Torssell, *Acta Chem. Scand. B*, **30**, 353 (1976).
358. L. Lepage and Y. Lepage, *J. Heterocycl. Chem.*, **15**, 793 (1978).
359. B. A. Keay, H. P. Plaumann, D. Rajapaksa and R. Rodrigo, *Can. J. Chem.*, **61**, 1987 (1983).
360. L. Contreras, D. B. MacLean, R. Faggiani and C. J. L. Lock, *Can. J. Chem.*, **59**, 1247 (1981).
361. T. Kametani, T. Takahashi, M. Kajiwara, Y. Hiral, C. Ohtsuka, F. Satoh and K. Fukumoto, *Chem. Pharm. Bull.*, **22**, 2159 (1974).
362. T. Kametani, M. Chihiro, M. Takeshita, K. Takahashi, K. Fukumoto and S. Takano, *Chem. Pharm. Bull.*, **26**, 3820 (1978).
363. A. Amaro, M. C. Carreño and F. Fariña, *Tetrahedron Lett.*, 3983 (1979).
364. M. C. Carreño, F. Fariña, J. L. G. Ruano and L. Puebla, *J. Chem. Res. (S)*, 288 (1984); *(M)*, 2623 (1984).
365. Y. Kanao and M. Oda, *Bull. Chem. Soc. Japan*, **57**, 615 (1984).
366. J. R. Wiseman, J. J. Pendery, C. A. Otto and K. G. Chiong, *J. Org. Chem.*, **45**, 516 (1980).
367. J.-P. Gesson and M. Mondon, *J. Chem. Soc. Chem. Commun.*, 421 (1982).
368. J.-P. Gesson, J. C. Jacquesy and B. Renoux, *Tetrahedron Lett.*, **24**, 2757 (1983).
369. J.-P. Gesson, J. C. Jacquesy and B. Renoux, *Tetrahedron Lett.*, **24**, 2761 (1983).
370. J. F. W. McOmie and D. H. Perry, *Synthesis*, 416 (1973).
371. N. Oda, S. Nagaiand and I. Ito, *Chem. Pharm. Bull.*, **27**, 2229 (1979).
372. F. Fariña, J. Primo and T. Torres, *Chem. Lett.*, 77 (1980).
373. J. G. Bauman, R. B. Barber, R. D. Gless and H. Rapoport, *Tetrahedron Lett.*, **21**, 4777 (1980).
374. J. Tamariz, L. Schwager, J. H. A. Stibbard and P. Vogel, *Tetrahedron Lett.*, **24**, 1497 (1983).
375. T. Sasakai, K. Shimizu and M. Ohno, *Chem. Pharm. Bull.*, **32**, 1433 (1984).
376. C. L. Kirkemo and J. D. White, *J. Org. Chem.*, **50**, 1316 (1985).
377. R. A. Russell, E. G. Vikingur and R. A. Warrener, *Aust. J. Chem.*, **34**, 131 (1981).
378. W. A. Nugent and J. C. Calabrese, *J. Am. Chem. Soc.*, **106**, 6422 (1984).
379. H.-D. Scharf, H. Plum, J. Fleischhauer and W. Schleker, *Chem. Ber.*, **112**, 862 (1979).
380. E. G. Georgescu and M. D. Gheorghiu, *Rev. Roumaine Chim.*, **22**, 907 (1977).
381. V. U. Griebsch and H. Hoberg, *Angew. Chem.*, **90**, 1014 (1978).
382. U. O'Connor and W. Rosen, *Tetrahedron Lett.*, 601 (1979).
383. U. O'Connor and W. Rosen, *J. Org. Chem.*, **45**, 1824 (1980).
384. P. Yates and I. Gupta, *J. Chem. Soc. Chem. Commun.*, 449 (1981).
385. R. G. Pews, C. W. Roberts, C. R. Hand and T. E. Evans, *Tetrahedron*, **29**, 1259 (1973).
386. A. P. Marchand and T.-C. Chou, *J. Chem. Soc., Perkin Trans. 1*, 1948 (1973).
387. A. P. Marchand and R. W. Allen, *J. Org. Chem.*, **39**, 1596 (1974).
388. A. Ichihara, M. Ubukata and S. Sakamura, *Agric. Biol. Chem.*, **44**, 211 (1980).
389. M. C. Lasne and J. L. Ripoll, *Tetrahedron*, **37**, 503 (1981).
390. G. Mehta, D. S. Reddy and A. V. Reddy, *Tetrahedron Lett.*, **25**, 2275 (1984).
391. P. E. Eaton, C. Giordano and U. Vogel, *J. Org. Chem.*, **41**, 2236 (1976).
392. R. G. F. Giles and G. H. P. Roos, *Tetrahedron Lett.*, 4159 (1975).
393. L. R. Nassimbeni, G. E. Jackson, R. G. F. Giles and G. H. P. Roos, *Acta Cryst.*, **B34**, 298 (1978).
394. K. Krohn, H-H. Ostermeyer and K. Tolkiehn, *Chem. Ber.*, **112**, 2640 (1979).
395. K. Krohn, *Tetrahedron Lett.*, **21**, 3557 (1980).
396. W. C. Still and M.-Y. Tsai, *J. Am. Chem. Soc.*, **102**, 3654 (1980).
397. K. Tolkiehn and K. Krohn, *Chem. Ber.*, **113**, 1575 (1980).

398. K. T. Potts, D. Bhattacharjee and E. B. Walsh, *J. Chem. Soc. Chem. Commun.*, 114 (1984).
399. K. Holmberg, H. Kirudd and G. Westin, *Acta Chem. Scand. B*, **28**, 913 (1974).
400. G. Andersson, *Acta Chem. Scand. B*, **30**, 403 (1976).
401. L. A. Paquette and G. Kretschmer, *J. Am. Chem. Soc.*, **101**, 4655 (1979).
402. P. G. Gassman and R. Yamaguchi, *J. Org. Chem.*, **43**, 4654 (1978).
403. J. C. Barborak, D. Khoury, W. F. Maier, P. v. R. Schleyer, E. C. Smith, W. F. Smith, Jr and C. Wyrick, *J. Org. Chem.*, **44**, 4761 (1979).
404. G.-A. Craze and I. Watt, *J. Chem. Soc., Perkin Trans. 2*, 175 (1981).
405. D. Kaufmann and A. de Meijere, *Chem. Ber.*, **116**, 1897 (1983).
406. R. S. Hosmane, S. P. Hiremath and S. W. Schneller, *J. Chem. Soc., Perkin Trans. 1*, 2450 (1973).
407. S. P. Hiremath and M. G. Purohit, *Indian J. Chem.*, **12**, 493 (1974).
408. S. P. Hiremath, S. S. Kaddargi and M. G. Purohit, *J. Indian Chem. Soc.*, **55**, 156 (1977).
409. R. Bergamasco, Q. N. Porter and C. Yap, *Aust. J. Chem.*, **31**, 1841 (1978).
410. W. H. Cherry and Q. N. Porter, *Aust. J. Chem.*, **32**, 145 (1979).
411. J. D. Lambert and Q. N. Porter, *Aust. J. Chem.*, **34**, 1483 (1981).
412. I. R. Trehan, R. Inder and D. V. L. Rewal, *Indian J. Chem.*, **14B**, 210 (1976).
413. Y. Tominaga, R. N. Castle and M. L. Lee, *J. Heterocycl. Chem.*, **19**, 1125 (1982).
414. P. A. Grieco, P. Garner, K. Yoshida and J. C. Huffman, *Tetrahedron Lett.*, **24**, 3807 (1983).
415. P. A. Grieco, K. Yoshida and P. Garner, *J. Org. Chem.*, **48**, 3137 (1983).
416. P. A. Grieco, K. Yoshida and Z.-M. He, *Tetrahedron Lett.*, **25**, 5715 (1984).
417. K. Yoshida and P. A. Grieco, *Chem. Lett.*, 155 (1985).
418. K. Hayakawa, K. Ueyama and K. Kanematsu, *J. Chem. Soc. Chem. Commun.*, 71 (1984).
419. K. Hayakawa, K. Ueyama and K. Kanematsu, *J. Org. Chem.*, **50**, 1963 (1985).
420. F. Scott, G. Cahiez, J. F. Normant and J. Villieras, *J. Organomet. Chem.*, **144**, 13 (1978).
421. G. Schön and H. Hopf, *Justus Liebigs Ann. Chem.*, 165 (1981).
422. G. Boffa, G. Pieri and N. Mazzaferro, *Gazz. Chim. Ital.*, **102**, 697 (1972).
423. H. Irikawa, T. Koyama and Y. Okumura, *Bull. Chem. Soc. Japan*, **52**, 637 (1979).
424. A. Rougny, H. Fillion, C. Laharotte and M. Daudon, *Tetrahedron Lett.*, **25**, 829 (1984).
425. J. P. Gesson, J. C. Jacquesy and M. Mondon, *Tetrahedron Lett.*, **21**, 3351 (1980).
426. J. P. Gesson, J. C. Jacquesy and M. Mondon, *Tetrahedron Lett.*, **22**, 1337 (1981).
427. J. P. Gesson, J. C. Jacquesy and M. Mondon, *Nouv. J. Chim.*, **4**, 205 (1983).
428. J. P. Gesson, J. C. Jacquesy and B. Renoux, *Tetrahedron*, **40**, 4743 (1984).
429. K. Krohn and K. Tolkiehn, *Tetrahedron Lett.*, 4023 (1978).
430. A. Echavarren, P. Prados and F. Fariña, *Tetrahedron*, **40**, 4561 (1984).
431. H. Muxfeldt, G. Haas, G. Hardtmann, F. Kathawala, J. B. Mooberry and E. Vedejs, *J. Am. Chem. Soc.*, **101**, 689 (1979).
432. A. V. R. Rao, V. H. Deshpande and N. L. Reddy, *Tetrahedron Lett.*, **21**, 2661 (1980).
433. L. K. Bee and P. J. Garratt, *J. Chem. Res. (S)*, 368; *(M)*, 4301 (1981).
434. B.-M. G. Gáveby, J. C. Huffmann and P. Magnus, *J. Org. Chem:*, **47**, 3779 (1982).
435. S. D. Nero and P. Lombardi, *Gazz. Chim. Ital.*, **113**, 125 (1983).
436. G. A. Flynn, M. J. Vaal, K. T. Stewart, D. L. Wenstrup, D. W. Beight and E. H. Bohme, *J. Org. Chem.*, **49**, 2252 (1984).
437. W. M. Acton, K. J. Ryan and M. Tracy, *Tetrahedron Lett.*, **25**, 5743 (1984).
438. J. M. McNamara and Y. Kishi, *Tetrahedron*, **40**, 4685 (1984).
439. T.-T. Li, Y. L. Wu and T. C. Walsgrove, *Tetrahedron*, **40**, 4701 (1984).
440. M. E. Jung, J. A. Lowe, III, M. A. Lyster, M. Node, R. W. Pfluger and R. W. Brown, *Tetrahedron*, **40**, 4751 (1984).
441. J. G. Bauman, R. C. Hawley and H. Rapoport, *J. Org. Chem.*, **50**, 1569 (1985).
442. A. V. R. Rao, V. H. Deshpande, K. M. Sathaye and S. M. Jaweed, *Indian J. Chem.*, **24B**, 697 (1985).
443. J. E. Tomaszewski, W. B. Manning and G. M. Muschik, *Tetrahedron Lett.*, 971 (1977).
444. W. B. Manning, J. E. Tomaszewski, G. M. Muschik and R. I. Sato, *J. Org. Chem.*, **42**, 3465 (1977).
445. W. B. Manning, G. M. Muschik and J. E. Tomaszewski, *J. Org. Chem.*, **44**, 699 (1979).
446. G. M. Muschik, J. E. Tomaszewski, R. I. Sato and W. B. Manning, *J. Org. Chem.*, **44**, 2150 (1979).
447. R. K. Hallmark, W. B. Manning and G. M. Muschik, *J. Label. Comp. and Radiopharm.*, **18**, 331 (1980).

448. W. B. Manning and G. M. Muschik, *J. Chem. Eng. Data*, **25**, 289 (1980).
449. W. B. Manning, T. P. Kelly and G. M. Muschik, *Tetrahedron Lett.*, **21**, 2629 (1980).
450. W. B. Manning and D. J. Wilbur, *J. Org. Chem.*, **45**, 733 (1980).
451. W. B. Manning, T. P. Kelly and G. M. Muschik, *J. Org. Chem.*, **45**, 2536 (1980).
452. W. B. Manning, *Tetrahedron Lett.*, **22**, 1571 (1981).
453. B. I. Rosen and W. P. Weber, *J. Org. Chem.*, **42**, 3463 (1977).
454. F. B. H. Ahmad, J. M. Bruce, J. Khalafy, V. Pejanović, K. Sabetian and I. Watt, *J. Chem. Soc. Chem. Commun.*, 166 (1981).
455. F. B. H. Ahmad, J. M. Bruce, J. Khalafy and K. Sabetian, *J. Chem. Soc. Chem. Commun.*, 169 (1981).
456. R. Al-Hamdany, J. M. Bruce, R. T. Pardasani and I. Watt, *J. Chem. Soc. Chem. Commun.*, 171 (1981).
457. R. L. Beddoes, J. M. Bruce, H. Finch, L. M. J. Heelam, I. D. Hunt and O. S. Mills, *J. Chem. Soc., Perkin Trans. 1*, 2670 (1981).
458. S. C. Cooper and P. G. Sammes, *J. Chem. Soc. Chem. Commun.*, 633 (1980).
459. S. C. Cooper and P. G. Sammes, *J. Chem. Soc., Perkin Trans. 1*, 2407 (1984).
460. P. Roffia, F. Conti, G. Gregorio, G. F. Pregaglia and R. Ugo, *J. Organomet. Chem.*, **56**, 391 (1973).
461. J. F. W. Keana and P. E. Eckler, *J. Org. Chem.*, **41**, 2625 (1976).
462. M. E. Jung, *J. Chem. Soc. Chem. Commun.*, 956 (1974).
463. G. A. Kraus and M. J. Taschner, *J. Org. Chem.*, **45**, 1174 (1980).
464. A. P. Kozikowski, K. Hiraga, J. P. Springer, B. C. Wang and Z.-B. Xu, *J. Am. Chem. Soc.*, **106**, 1845 (1984).
465. M. Baier, J. Daub, A. Hasenhündl, A. Merz and K. M. Rapp, *Angew. Chem.*, **93**, 196 (1981).
466. J. Bindl, J. Daub, A. Hasenhündl, M. Meinert and K. M. Rapp, *Chem. Ber.*, **116**, 2408 (1983).
467. R. McCague, C. J. Moody, C. W. Rees and D. J. Williams, *J. Chem. Soc. Perkin Trans. 1*, 909 (1984).
468. L. F. Fieser and A. M. Seligman, *Chem. Ber.*, **68B**, 1747 (1935).
469. J. J. Sims and V. K. Honwad, *Tetrahedron Lett.*, 2155 (1973).
470. H. Christol, F. Pietrasanta, Y. Pietrasanta and J.-C. Rousselou, *Bull. Soc. Chim. France*, 2770 (1972).
471. G. W. K. Cavill and R. J. Quinn, *Aust. J. Chem.*, **26**, 595 (1973).
472. C. Schmidt, *Can. J. Chem.*, **51**, 3989 (1973).
473. J. L. Charlton and R. Agagnier, *Can. J. Chem.*, **51**, 1852 (1973).
474. M. Hudlicky, *J. Org. Chem.*, **39**, 3460 (1974).
475. M. Zander, *Chem. Ber.*, **108**, 367 (1975).
476. W. Lange, *Holz als Roh-und Werkstoff (Berlin)*, **34**, 101 (1976).
477. K. D. Paull, R. K. Y. Zee-Cheng and C. C. Cheng, *J. Med. Chem.*, **19**, 337 (1976).
478. R. Bergamasco and Q. N. Porter, *Aust. J. Chem.*, **30**, 1523 (1977).
479. J. R. Scheffer, R. E. Gayler, T. Zakouras and A. A. Dzakpasu, *J. Am. Chem. Soc.*, **99**, 7726 (1977).
480. V. K. Tandon, R. Singh, J. M. Khanna and N. Anand, *Indian J. Chem.*, **15B**, 839 (1977).
481. J.-L. Gras, *Tetrahedron Lett.*, 4117 (1977).
482. F. Bohlmann, W. Mathar and H. Schwarz, *Chem. Ber.*, **110**, 2028 (1977).
483. S. H. Mashraqui and G. K. Trivedi, *Indian J. Chem.*, **16B**, 1062 (1978).
484. H. Iwamura and K. Makino, *J. Chem. Soc. Chem. Commun.*, 720 (1978).
485. K. Krohn and A. Rösner, *Tetrahedron Lett.*, 353 (1978).
486. R. N. Ferguson, J. F. Whidby, E. B. Sanders, R. J. Levins, T. Katz, J. F. DeBardeleben and W. N. Einolf, *Tetrahedron Lett.*, 2645 (1978).
487. N. Oda, K. Kobayashi, T. Ueda and I. Ito, *Chem. Pharm. Bull.*, **26**, 2578 (1978).
488. M. Maienthal, W. R. Benson, E. B. Sheinin and T. D. Doyle, *J. Org. Chem.*, **43**, 972 (1978).
489. E. McDonald, A. Suksamrarn and R. D. Wylie, *J. Chem. Soc., Perkin Trans. 1*, 1893 (1979).
490. F. Bohlmann and E. Eickeler, *Chem. Ber.*, **113**, 1189 (1980).
491. A. Hosomi, M. Saito and H. Sakurai, *Tetrahedron Lett.*, **21**, 355 (1980).
492. A. Ichihara, M. Ubukata, H. Oikawa, K. Murakami and S. Sakamura, *Tetrahedron Lett.*, **21**, 4469 (1980).
493. T. O. Criodain, M. O'Sullivan, M. J. Meegan and D. M. X. Donnelly, *Phytochem.*, **20**, 1089 (1981).

494. D. P. G. Hamon and P. R. Spurr, *Synthesis*, 873 (1981).
495. R. R. Schmidt and A. Wagner, *Synthesis*, 273 (1981).
496. M. J. Tanga and E. J. Reist, *J. Org. Chem.*, **47**, 1365 (1982).
497. H. Hiranuma and S. I. Miller, *J. Org. Chem.*, **47**, 5083 (1982).
498. R. R. Schmidt and A. Wagner, *Synthesis*, 958 (1982).
499. K. T. Potts and D. Bhattacharjee, *Synthesis*, 31 (1983).
500. F. Zutterman and A. Krief, *J. Org. Chem.*, **48**, 1135 (1983).
501. Z. Q. Jiang, J. R. Scheffer, A. S. Secco, J. Trotter and Y.-F. Wong, *J. Chem. Soc., Chem. Commun.*, 773 (1983).
502. F. Bohlmann and T. Trantow, *Justus Liebigs Ann. Chem.*, 1689 (1983).
503. H. Hiranuma and S. I. Miller, *J. Org. Chem.*, **48**, 3096 (1983).
504. G. Wells, Y. Hanzawa and L. A. Paquette, *Angew. Chem.*, **91**, 578 (1979).
505. L. A. Paquette, Y. Hanzawa, K. J. McCullough, B. Tagle, W. Swenson and J. Clardy, *J. Am. Chem. Soc.*, **103**, 2262 (1981).
506. J. L. Pyle, A. A. Shaffer and J. S. Cantrell, *J. Org. Chem.*, **46**, 115 (1981).
507. J. L. Pyle, R. A. Lunsford and J. S. Cantrell, *J. Org. Chem.*, **44**, 2391 (1979).
508. M. G. Veliev, M. M. Guseinov, E. S. Mamedov and R. F. Gakhramanov, *Synthesis*, 337 (1984).
509. M. G. Veliev, M. M. Guseinov, L. A. Yanovskaya and K. Ya. Burstein, *Tetrahedron*, **41**, 749 (1985).
510. C. C. Liao, H. S. Lin and J. T. Lin, *J. Chinese Chem. Soc.*, **27**, 87 (1980).
511. W. Verboom and H. J. T. Bos, *Recl. Trav. Chim. Pays-Bas*, **100**, 207 (1981).
512. W. Dürckheimer and E. F. Paulus, *Angew. Chem.*, **97**, 219 (1985).
513. E. Vedejs and E. S. C. Wu, *J. Am. Chem. Soc.*, **97**, 4706 (1975).
514. S. Mazza, S. Danishefsky and P. McCurry, *J. Org. Chem.*, **39**, 3610 (1974).
515. S. Danishefsky, P. F. Schuda, S. Mazza and K. Kato, *J. Org. Chem.*, **41**, 3468 (1976).
516. S. Danishefsky, P. F. Schuda and W. Caruthers, *J. Org. Chem.*, **42**, 2179 (1977).
517. A. I. Konovalov, B. N. Solomonov and O. Yu. Chertov, *Zh. Org. Khim. (Eng. Transl.)*, **11**, 107 (1975).
518. V. V. Plemenkov, Kh. Z. Giniyatov, Ya. Ya. Villem, N. V. Villem, L. S. Surmina and I. G. Bolesov, *Dokl. Akad. Nauk SSSR (Eng. Transl.)*, **254**, 456 (1980).
519. V. V. Plemenkov, M. M. Latypova, I. G. Bolesov, Ya. Ya. Villem and N. V. Villem, *Dokl. Akad. Nauk SSSR (Eng. Transl.)*, **272**, 343 (1983).
520. W. Friedrichsen, E. Büldt, M. Betz and R. Schmidt, *Tetrahedron Lett.*, 2469 (1974).
521. M. Christl, H.-J. Lüddeke, A. Nagyrevi-Neppel and G. Freitag, *Chem. Ber.*, **110**, 3745 (1977).
522. W. Pritschins and W. Grimme, *Tetrahedron Lett.*, 4545 (1979).
523. D. D. Weller and E. P. Stirchak, *J. Org. Chem.*, **48**, 4873 (1983).
524. S. Knapp and S. Sharma, *J. Org. Chem.*, **50**, 4996 (1985).
525. R. Al-Hamdany and B. Ali, *J. Chem. Soc., Chem. Commun.*, 397 (1978).
526. R. Al-Hamdany and S. Salih, *J. Iraqi Chem. Soc.*, **6**, 53 (1981).
527. R. Al-Hamdany and S. Salih, *J. Iraqi Chem. Soc.*, **6**, 61 (1981).
528. M. F. Ansell, A. J. Bignold, A. F. Gosden, V. J. Leslie and R. A. Murray, *J. Chem. Soc. (C)*, 1414 (1971).
529. T. Komatsu, T. Nishio and Y. Omote, *Chem. and Ind. (London)*, 95 (1978).
530. N. Latif, N. S. Girgis and F. Michael, *Tetrahedron*, **26**, 2765 (1970).
531. N. Latif, N. Mishriky, N. S. Guirguis and A. Hussein, *J. Prakt Chem.*, **315**, 419 (1973).
532. N. Latif, N. Mishriky and N. S. Girgis, *J. Chem. Soc. Perkin Trans. 1*, 1052 (1975).
533. N. Latif, N. Mishriky and N. S. Girgis, *Indian J. Chem.*, **15B**, 118 (1977).
534. N. Mishriky, N. S. Girgis and G. A. M. Nawwar, *Egyptian J. Chem.*, **24**, 289 (1981).
535. N. Latif, N. Mishriky, N. S. Girgis and S. Arnos, *Indian J. Chem.*, **19B**, 301 (1980).
536. M. A. Battiste and M. Visnick, *Tetrahedron Lett.*, 4771 (1978).
537. T. Sasaki, K. Hayakawa, T. Manabe and S. Nishida, *J. Am. Chem. Soc.*, **103**, 565 (1981).
538. T. Sasaki, K. Hayakawa, T. Manabe, S. Nishida and E. Wakabayashi, *J. Org. Chem.*, **46**, 2021 (1981).
539. K. Saito, *Heterocycles*, **19**, 1197 (1982).
540. G. C. Rovnyak and V. Shu, *J. Org. Chem.*, **44**, 2518 (1979).
541. H. Gotthardt and J. Blum, *Chem. Ber.*, **118**, 2079 (1985).
542. W. Friedrichsen, I. Kallweit and R. Schmidt, *Justus Liebigs Ann. Chem.*, 116 (1977).
543. G. R. Allen, Jr, in *Organic Reactions*, Vol. 20, Wiley, New York, 1973, pp. 337–454.

544. U. Kuckländer, *Tetrahedron*, **29**, 921 (1973).
545. F. Eiden and U. Kuckländer, *Archiv. Pharm.*, **306**, 446 (1973).
546. U. Kuckländer and W. Hühnermann, *Archiv. Pharm.*, **312**, 515 (1979).
547. U. Kuckländer, *Justus Liebigs Ann. Chem.*, 140 (1978).
548. U. Kuckländer and H. Töberich, *Chem. Ber.*, **114**, 2238 (1981).
549. U. Kuckländer, *Tetrahedron*, **31**, 1631 (1975).
550. U. Kuckländer and H. Töberich, *Chem. Ber.*, **116**, 152 (1983).
551. V. I. Shvedov, E. K. Panisheva, T. F. Vlasova and A. N. Grinev, *Khim. Geterotsikl. Soedin. (Eng. Transl.)*, 1354 (1973).
552. F. A. Trofimov, N. G. Tsyshkova, T. F. Vlasova, and A. N. Grinev, *Khim. Geterotsikl. Soedin. (Eng. Transl.)*, 46 (1976).
553. L. Kozerski, *Polish J. Chem.*, **53**, 2393 (1979).
554. L. Kozerski, E. Czerwinska and T. Pobiedzinska, *Tetrahderon*, **38**, 621 (1982).
555. G. S. Gadaginamath and S. Siddappa, *J. Indian Chem. Soc.*, **53**, 17 (1976).
556. G. S. Gadaginamath, L. D. Basanagoudar and S. Siddappa, *J. Indian Chem. Soc.*, **54**, 709 (1977).
557. G. S. Gadaginamath and S. Siddappa, *J. Indian Chem. Soc.*, **52**, 330 (1975).
558. H. Mazarguil and A. Lattes, *Bull. Soc. Chim. France*, 3874 (1972).
559. M. D. Menachery, J. M. Saá and M. P. Cava, *J. Org. Chem.*, **46**, 2584 (1981).
560. R. O. Duthaler, P. Mathies, W. Petter, C. Heuberger and V. Scherrer, *Helv. Chim. Acta*, **67**, 1217 (1984).
561. R. W. Parr and J. A. Reiss, *Aust. J. Chem.*, **37**, 1263 (1984).
562. M, L. Casner, W. A. Remers and W. T. Bradner, *J. Med. Chem.*, **28**, 921 (1985).
563. V. Aggarwal, A. Kumar, H. Ila and H. Junjappa, *Synthesis*, 157 (1981).
564. J.-L. Bernier, J.-P. Hénichart, C. Vaccher and R. Houssin, *J. Org. Chem.*, **45**, 1493 (1980).
565. J. B. Patrick and E. K. Saunders, *Tetrahedron Lett.*, 4009 (1979).
566. M. G. Peter and F. Speckenbach, *Helv. Chim. Acta*, **65**, 1279 (1982).
567. M. Gates, *J. Am. Chem. Soc.*, **66**, 124 (1944).
568. M. Gates, *J. Org. Chem.*, **47**, 578 (1982).
569. S. B. Awad, A. B. Sakla, N. F. Abdul-Malik and N. Ishak, *Indian J. Chem.*, **17B**, 219 (1979).
570. J. Banville, J.-L. Grandmaison, G. Lang and P. Brassard, *Can. J. Chem.*, **52**, 80 (1974).
571. A. Castonguay and P. Brassard, *Synth. Commun.*, **5**, 377 (1975).
572. J.-L. Grandmaison and P. Brassard, *Tetrahedron*, **33**, 2047 (1977).
573. J. Banville and P. Brassard, *J. Chem. Soc., Perkin Trans. 1*, 613 (1976).
574. J. Banville and P. Brassard, *J. Org. Chem.*, **41**, 3018 (1976).
575. J.-L. Grandmaison and P. Brassard, *J. Org. Chem.*, **43**, 1435 (1978).
576. D. W. Cameron, M. J. Crossley and G. I. Feutrill, *J. Chem. Soc., Chem. Commun.*, 275 (1976).
577. H. J. Banks, D. W. Cameron, M. J. Crossley and E. L. Samuel, *Aust. J. Chem.*, **29**, 2247 (1976).
578. D. W. Cameron, G. I. Feutrill, P. G. Griffiths and D. J. Hodder, *J. Chem. Soc., Chem. Commun.*, 688 (1978).
579. D. W. Cameron, M. J. Crossley, G. I. Feutrill and P. G. Griffiths, *Aust. J. Chem.*, **31**, 1335 (1978).
580. D. W. Cameron and M. J. Crossley, *Aust. J. Chem.*, **31**, 1353 (1978).
581. D. W. Cameron, G. I. Feutrill and M. A. Sefton, *Aust. J. Chem.*, **31**, 2099 (1978).
582. D. W. Cameron, K. R. Deutscher and G. I. Feutrill, *Aust. J. Chem.*, **31**, 2259 (1978).
583. (a) D. W. Cameron, G. I. Feutrill and J. M. Thiel, *Aust. J. Chem.*, **34**, 453 (1981); (b) D. W. Cameron, D. J. Deutscher, G. I. Feutrill, and P. G. Griffiths, *Aust. J. Chem.*, **34**, 2401 (1981).
584. B. Eistert, K. Pfleger and P. Donath, *Chem. Ber.*, **105**, 3915 (1972).
585. B. Eistert, J. Riedinger, G. Küffner and W. Lazik, *Chem. Ber.*, **106**, 727 (1973).
586. B. Eistert, K. Pfleger, T. J. Arackal and G. Holzer, *Chem. Ber.*, **108**, 693 (1975).
587. B. Eistert, L. S. B. Goubran, C. Vamvakaris and T. J. Arackal, *Chem. Ber.*, **108**, 2941 (1975).
588. M. F. Aldersley, F. M. Dean and B. E. Mann, *J. Chem. Soc., Chem. Commun.*, 107 (1983).
589. G. A. Conway and L. J. Loeffler, *J. Heterocycl. Chem.*, **20**, 1315 (1983).
590. H. Laatsch, *Justus Liebigs Ann. Chem.*, 251 (1985).
591. Ya. Ya. Dregeris, D. Ya. Murnietse and Ya. F. Freimanis, *Zh. Obshch. Khim. (Eng. Transl.)*, **42**, 600 (1972).
592. G. F. Bannikov, G. A. Nikiforov and V. V. Ershov, *Izv. Akad. Nauk SSSR, Ser. Khim. (Eng. Transl.)*, 1807 (1979).

593. G. Manecke, W. Hübner and H.-J. Kretzschmar, *Angew. Chem., Int. Ed. Engl.*, **11**, 338 (1972).
594. B. Eistert, I. Mussler, H.-K. Witzmann and O. Ganster, *Chem. Ber.*, **105**, 234 (1972).
595. N. Latif and S. A. Meguid, *Indian J. Chem.*, **19B**, 975 (1980).
596. J. Elzinga, H. Hogeveen and E. P. Schudde, *J. Org. Chem.*, **45**, 4337 (1980).
597. N. L. Komissarova, I. S. Belostotskaya, V. B. Vol'eva, E. V. Dzhuaryan, I. A. Novikova and V. V. Ershov, *Izv. Akad. Nauk SSSR, Ser. Khim. (Eng. Transl.)*, 2360 (1981).
598. T. Oshima and T. Nagai, *Bull. Chem. Soc. Japan*, **53**, 726 (1980).
599. T. Oshima and T. Nagai, *Bull. Chem. Soc. Japan*, **53**, 3284 (1980).
600. T. Oshima and T. Nagai, *Bull. Chem. Soc. Japan*, **55**, 551 (1982).
601. J. A. Myers, W. W. Wilkerson and S. L. Council, *J. Org. Chem.*, **40**, 2875 (1975).
602. W. Friedrichsen and W.-D. Schröer, *Tetrahedron Lett.*, 1603 (1977).
603. W. Friedrichsen and I. Schwarz, *Tetrahedron Lett.*, 3581 (1977).
604. H. Matsukubo and H. Kato, *Bull. Chem. Soc. Japan*, **49**, 3333 (1976).
605. T. Uchida, S. Tsubokawa, K. Harihara and K. Matsumoto, *J. Heterocycl. Chem.*, **15**, 1303 (1978).
606. J. A. Myers, L. D. Moore, Jr, W. L. Whitter, S. L. Council, R. M. Waldo, J. L. Lanier and B. U. Omoji, *J. Org. Chem.*, **45**, 1202 (1980).
607. W. Friedrichsen, C. Krüger, E. Kujath, G. Liebezeit and S. Mohr, *Tetrahedron Lett.*, 237 (1979).
608. T. Tanaka, *Chem. Lett.*, 161 (1976).
609. K. A. Parker, I. D. Cohen, A. Padwa and W. Dent, *Tetrahedron Lett.*, **25**, 4917 (1984).
610. W. Ott, V. Formaček and H.-M. Seidenspinner, *Justus Liebigs Ann. Chem.*, 1003 (1984).
611. S. Shiraishi, S. Ikeuchi, M. Senō and T. Asahara, *Bull. Chem. Soc. Japan*, **50**, 910 (1977).
612. S. Shiraishi, S. Ikeuchi, M. Senō and T. Asahara, *Bull. Chem. Soc. Japan*, **51**, 921 (1978).
613. S. Shiraishi, B. S. Holla and K. Imamura, *Bull. Chem. Soc. Japan*, **56**, 3457 (1983).
614. T. Hayakawa, K. Araki and S. Shiraishi, *Bull. Chem. Soc. Japan*, **57**, 1643 (1984).
615. M. J. Sanders and J. R. Grunwell, *J. Org. Chem.*, **45**, 3753 (1980).
616. R. M. Paton, J. F. Ross and J. Crosby, *J. Chem. Soc., Chem. Commun.*, 1194 (1980).
617. T. Sasaki, S. Eguchi and Y. Hirako, *Heterocycles*, **4**, 1901 (1976).
618. H. Gotthardt, C. M. Weisshuhn and B. Christl, *Chem. Ber.*, **111**, 3037 (1978).
619. Y. Tamura, A. Wada, M. Sasho and Y. Kita, *Tetrahedron Lett.*, **22**, 4283 (1981).
620. Y. Tamura, A. Wada, M. Sasho, K. Fukunaga, H. Maeda and Y. Kita, *J. Org. Chem.*, **47**, 4376 (1982).
621. Y. Tamura, S. Akai, M. Sasho and Y. Kita, *Tetrahedron Lett.*, **25**, 1167 (1984).
622. Y. Tamura, M. Sasho, K. Nakagawa, T. Tsugoshi and Y. Kita, *J. Org. Chem.*, **49**, 473 (1984).
623. Y. Tamura, M. Sasho, H. Ohe, S. Akai and Y. Kita, *Tetrahedron Lett.*, **26**, 1549 (1985).
624. Y. Tamura, F. Fukata, M. Sasho, T. Tsugoshi and Y. Kita, *J. Org. Chem.*, **50**, 2273 (1985).
625. C. S. Rondestvedt, Jr, in *Organic Reactions*, Vol. 24, Wiley, New York, 1976, pp. 225–259.
626. A. Citterio, E. Vismara and R. Bernardi, *J. Chem. Res. (S)*, 88; *(M)*, 876 (1983).
627. D. Veltwisch and K.-D. Asmus, *J. Chem. Soc., Perkin Trans. 2*, 1147 (1982).
628. Y. A. Ilan, G. Czapski and D. Meisel, *Biochim. Biophys. Acta*, **430**, 209 (1976).
629. K. Krohn and A. Mondon, *Chem. Ber.*, **109**, 855 (1976).
630. C. Blackburn and J. Griffiths, *J. Chem. Res. (S)*, 320; *(M)*, 3457 (1982).
631. O. C. Musgrave and C. J. Webster, *J. Chem. Soc., Perkin Trans. 1*, 2260 (1974).
632. R. Buchan and O. C. Musgrave, *J. Chem. Soc., Perkin Trans. 1*, 811 (1975).
633. R. Buchan and O. C. Musgrave, *J. Chem. Soc., Perkin Trans. 1*, 2185 (1975).
634. O. C. Musgrave and C. J. Webster, *J. Chem. Soc., Perkin Trans. 1*, 2263 (1974).
635. R. Buchan and O. C. Musgrave, *J. Chem. Soc., Perkin Trans. 1*, 568 (1975).
636. R. Buchan and O. C. Musgrave, *J. Chem. Soc., Perkin Trans. 1*, 90 (1980).
637. H. Brockmann and H. Laatsch, *Justus Liebigs Ann. Chem.*, 433 (1983).
638. H. Laatsch, *Z. Naturforsch. B*, **39B**, 244 (1984).
639. H. Laatsch, *Justus Liebigs Ann. Chem.*, 1847 (1985).
640. H.-E. Högberg and M. Hjalmarsson, *Tetrahedron Lett.*, 5215 (1978).
641. H.-E. Högberg, *Acta Chem. Scand.*, **27**, 2559 (1973).
642. H.-E. Högberg and P. Komlos, *Acta Chem. Scand. B*, **33**, 271 (1979).
643. G. A. Kraus and B. Roth, *J. Org. Chem.*, **43**, 4923 (1978).
644. G. A. Kraus, H. Cho, S. Crowley, B. Roth, H. Sugimoto and S. Prugh, *J. Org. Chem.*, **48**, 3439 (1983).

645. K. Buggle, J. A. Donnelly and L. J. Maher, *Chem. Ind. (London)*, 88 (1972).
646. U. M. Wagh and R. N. Usgaonkar, *Indian J. Chem.*, **14B**, 861 (1976).
647. N. Someswari, K. Srihari and V. Sundaramurthy, *Synthesis*, 609 (1977).
648. J. A. D. Jeffreys, M. B. Zakaria, P. G. Waterman and S. Zhong, *Tetrahedron Lett.*, **24**, 1085 (1983).
649. T. Itahara, *J. Chem. Soc., Chem. Commun.*, 859 (1981).
650. T. Itahara, K. Kawasaki and F. Ouseto, *Bull. Chem. Soc. Japan*, **57**, 3488 (1984).
651. T. Itahara, *Chem. Ind. (London)*, 599 (1982).
652. T. Itahara, *J. Org. Chem.*, **50**, 5546 (1985).
653. M. Pardhasaradhi and B. M. Choudary, *Indian J. Chem.*, **17B**, 79 (1979).
654. M. Gocmen, G. Soussan and P. Fréon, *Bull. Soc. Chim. France*, 562 (1973).
655. A. Alberola, A. M. Gonzalez and F. J. Pulido, *Rev. Roumaine Chim.*, **29**, 441 (1984).
656. S. Harusawa, M. Miki, J. Hirai and T. Kurihara, *Chem. Pharm. Bull.*, **33**, 899 (1985).
657. B. Errazuriz, R. Tapia and J. A. Valderrama, *Tetrahedron Lett.*, **26**, 819 (1985).
658. H. Takeshita, A. Mori and H. Mametsukam, *Chem. Lett.*, 445 (1976).
659. H. Takeshita, H. Mametsuka and A. Mori, *Chem. Lett.*, 881 (1976).
660. K. Buggle, J. A. Donnelly and L. J. Maher, *J. Chem. Soc., Perkin Trans. 1*, 1006 (1973).
661. M. Masumura and Y. Yamashita, *Heterocycles*, **12**, 787 (1979).
662. V. P. Makovetskii, I. B. Dzvinchuk and A. A. Svishchuk, *Ukr. Khim. Zh. (Eng. Transl.)*, **44**, 1311 (1978).
663. V. P. Makovetskii, I. B.Dzvinchuk, Yu. M. Volovenko and A. A. Svishchuk, *Khim. Geterotsikl. Soedin. (Eng. Transl.)*, 129 (1979).
664. V. P. Makovetskii, I. B. Dzvinchik, V. V. Yaremenko and A. A. Svishchuk, *Ukr. Khim. Zh. (Eng. Transl.)*, **45**, 652 (1979).
665. V. P. Makovetskii, I. B. Dzvinchuk and A. A. Svishchik, *Ukr. Khim. Zh. (Eng. Transl.)*, **46**, 839 (1980).
666. R. J. Wikholm, *J. Org. Chem.*, **50**, 382 (1985).
667. P. H. Boyle, M. J. O'Mahony and C. J. Cardin, *J. Chem. Soc., Perkin Trans. 1*, 593 (1984).
668. Y. Tamuru, T. Miyamoto, H. Kiyokawa and Y. Kita, *J. Chem. Soc., Perkin Trans. 1*, 1125 (1974).
669. F. M. Dean, G. H. Mitchell, B. Parvizi and C. Thebtaranonth, *J. Chem. Soc., Perkin Trans. 1*, 2067 (1976).
670. F. M. Dean, L. E. Houghton, R. Nayyir-Mazhir and C. Thebtaranonth, *J. Chem. Soc., Perkin Trans. 1*, 159 (1979).
671. F. M. Dean, L. E. Houghton, R. Nayyir-Mazhir and C. Thebtaranonth, *J. Chem. Soc., Perkin Trans. 1*, 1994 (1980).
672. M. F. Aldersley, F. M. Dean and R. Nayyir-Mazhir, *J. Chem. Soc., Perkin Trans. 1*, 1763 (1983).
673. J. L. Carey, C. A. Shand, R. H. Thomson and C. W. Greenhalgh, *J. Chem. Soc., Perkin Trans. 1*, 1957 (1984).
674. A. T. Hudson and M. J. Pether, *J. Chem. Soc., Perkin Trans. 1*, 35 (1983).
675. Y. Naruta, H. Uno and K. Maruyama, *Chem. Lett.*, 609 (1982).
676. Y. Naruta, H. Uno and K. Maruyama, *Chem Lett.*, 961 (1982).
677. M. Akiba, S. Ikuta and T. Takada, *Heterocycles*, **16**, 1579 (1981).
678. R. Peltzmann, B. Unterweger and H. Junek, *Monatsch. Chem.*, **110**, 739 (1979).
679. H.-S. Kuo, K. Hotta, M. Yogo and S. Yoshina, *Synthesis*, 188 (1979).
680. G. Kumar and A. P. Bhaduri, *Indian J. Chem.*, **14B**, 496 (1976).
681. H. L. McPherson and B. W. Ponder, *J. Heterocycl. Chem.*, **13**, 909 (1976).
682. H. L. McPherson and B. W. Ponder, *J. Heterocycl. Chem.*, **15**, 43 (1978).
683. C. W. Greenhalgh, J. L. Carey and D. F. Newton, *Dyes Pigments*, **1**, 103 (1980).
684. H. Rafart, J. Valderrama and J. C. Vega, *An. Quim.*, **72**, 804 (1976).
685. R. Cassis, R. Tapia and J. Valderrama, *J. Heterocycl. Chem.*, **19**, 381 (1982).
686. V. F. Ferreira, L. C. Coutada, M. C. F. R. Pinto and A. V. Pinto, *Syn. Commun.*, **12**, 195 (1982).
687. R. P. Soni and J. P. Saxena, *J. Indian Chem. Soc.*, **55**, 470 (1978).
688. R. P. Soni and J. P. Saxena, *J. Indian Chem. Soc.*, **55**, 232 (1978).
689. N. R. Ayyangar, P. Y. Kolhe and B. D. Tilak, *Indian J. Chem.*, **19**, 836 (1980).
690. C.-T. Lfu, D. M. S. Wheeler and C. S. Day, *Syn. Commun.*, **11**, 983 (1981).
691. J. L. Bloomer and K. M. Damodaran, *Synthesis*, 111 (1980).

692. J. A. Rieck and J. R. Grunwell, *J. Org. Chem.*, **45**, 3512 (1980).
693. K. A. Parker and S.-K. Kang, *J. Org. Chem.*, **45**, 1218 (1980).
694. P. Brownbridge and T.-H. Chan, *Tetrahedron Lett.*, **21**, 3431 (1980).
695. M. F. Semmelhack, L. Keller, T. Sato and E. Spiess, *J. Org. Chem.*, **47**, 4382 (1982).
696. T. Nogami, K. Yoshihara and S. Nagakura, *Bull. Chem. Soc. Japan*, **45**, 122 (1972).
697. S. B. Shah and A. S. N. Murthy, *J. Phys. Chem.*, **79**, 322 (1975).
698. P. C. Dwivedi and R. Agarwal, *Indian J. Chem.*, **24A**, 100 (1985).
699. D. W. Cameron, P. J. Chalmers and G. I. Feutrill, *Tetrahedron Lett.*, **25**, 6031 (1984).
700. Y. Asahi, M. Tanaka and K. Shinozaki, *Chem. Pharm. Bull.*, **32**, 3093 (1984).
701. R. Foster, N. Kulevsky and D. S. Wanigasekera, *J. Chem. Soc., Perkin Trans. 1*, 1318 (1974).
702. F. R. Hewgill and L. R. Mullings, *J. Chem. Soc. (B)*, 1155 (1969).
703. F. R. Hewgill and L. R. Mullings, *J. Aust. Chem.*, **28**, 2561 (1975).
704. B. S. Joshi and V. N. Kamat, *J. Chem. Soc., Perkin Trans. 1*, 327 (1975).
705. K. Y. Chu and J. Griffiths, *J. Chem. Soc., Perkin Trans. 1*, 1083 (1978).
706. K. Y. Chu and J. Griffiths, *J. Chem. Res. (S)* 180; *(M)* 2319 (1978).
707. K. Y. Chu and J. Griffiths, *J. Chem. Soc., Perkin Trans. 1*, 696 (1979).
708. R. W. Parr and J. A. Reiss, *J. Aust. Chem.*, **37**, 389 (1984).
709. (a) L. D. Belitskaya and V. T. Kolesnikov, *Zh. Org. Khim. (Eng. Transl.)*, **20**, 1920 (1984); (b) V. T. Kolesnikov, L. D. Belitskaya and B. L. Litvin, *Zh. Org. Khim. (Eng. Transl.)*, **18**, 901 (1982).
710. S. N. Falling and H. Rapoport, *J. Org. Chem.*, **45**, 1260 (1980).
711. M. Okamoto and S. Ohta, *Chem. Pharm. Bull.*, **28**, 1071 (1980).
712. F. Chou, A. H. Khan and J. S. Driscoll, *J. Med. Chem.*, **19**, 1302 (1976).
713. A. H. Khan and J. S. Driscoll, *J. Med. Chem.*, **19**, 313 (1976).
714. K. Hartke and U. Lohmann, *Chem. Lett.*, 693 (1983).
715. Y. S. Tsizin, S. A. Chernyak and N. I. Kornienko, *Zh. Org. Khim. (Eng. Transl.)*, **20**, 2381 (1983).
716. W. Gauss, H. Heitzer and S. Petersen, *Justus Liebigs Ann. Chem.*, **764**, 131 (1972).
717. V. S. Kuznetsov, E. A. Korkhova, Y. A. Ignat'ev and A. V. El'tsov, *Zh. Org. Khim. (Eng. Transl.)*, **10**, 1557 (1974).
718. P. C. Taunk and R. L. Mital, *Ann. Soc. Sci. Bruxelles*, **88**, 259 (1974).
719. S. K. Jain and R. L. Mital, *Indian J. Chem.*, **12**, 780 (1974).
720. N. A. Agrawa and R. L. Mital, *J. Chem. Eng. Data*, **20**, 199 (1975).
721. S. K. Saxena, A. P. Shukla, M. D. Sadhanani and R. L. Mital, *Indian J. Chem.*, **14B**, 808 (1976).
722. N. L. Agarwal and R. L. Mital, *Indian J. Chem.*, **14B**, 381 (1976).
723. S. K. Saxena, A. P. Shukla, M. D. Sadhnani and R. L. Mital, *Z. Naturforsch. B*, **31B**, 1673 (1976).
724. A. R. Goyal and R. L. Mital, *Natural and Applied Sci. Bull. (Philippines)*, **29**, 37 (1977).
725. M. Mittal and R. L. Mital, *Z. Naturforsch. B*, **33B**, 336 (1978).
726. N. Bhargava and R. L. Mital, *Gazz. Chim. Ital.*, **109**, 201 (1979).
727. A. R. Goyal and R. L. Mital, *Gazz. Chim. Ital.*, **109**, 205 (1979).
728. J. P. Tiwari and R. L. Mital, *Indian J. Chem.*, **17B**, 408 (1979).
729. N. Bhargava and R. L. Mital, *Acta Ciencia Indica*, **VIc**, 95 (1980).
730. G. K. Oberoi, R. Agrawal, and R. L. Mital, *Indian J. Chem.*, **21B**, 584 (1982).
731. J. P. Tiwari and R. L. Mital, *Acta Ciencia Indica*, **VIc**, 27 (1980).
732. R. K. Jain and R. L. Mital, *Rev. Roumaine Chim.*, **25**, 697 (1980).
733. S. B. Tambi, R. K. Jain and R. L. Mital, *Rev. Roumaine Chim.*, **26**, 477 (1981).
734. V. Taneja, C. L. Gupta and R. L. Mital, *Chim. Ind. (London)*, 187 (1984).
735. J. K. Suzuki, A. Zirnis and A. A. Manian, *J. Heterocycl, Chem.*, **13**, 1067 (1976).
736. N. L. Agarwal and S. K. Jain, *Synthesis*, 437 (1978).
737. N. K. Goswami, R. R. Gupta and S. K. Jain, *Indian J. Chem.*, **19B**, 307 (1980).
738. M. H. Terdic, *Rev. Roumaine Chim.*, **29**, 489 (1981).
739. R. R. Gupta, K. G. Ojha, G. S. Kalwania and M. Kumar, *Ann. Soc. Sci. Bruxelles*, **97**, 101 (1983).
740. H. Nishi, Y. Hatada and K. Kitahara, *Bull. Chem. Soc. Japan*, **56**, 1482 (1983).
741. N. L. Agarwal and C. K. Atal, *J. Heterocycl. Chem.*, **20**, 1741 (1983).
742. N. L. Agarwal, S. Ghosh, A. K. Tripathi and C. K. Atal, *J. Heterocycl. Chem.*, **21**, 509 (1984).
743. K. Takagi and M. Kawabe, *Dyes and Pigments*, **6**, 177 (1985).

744. R. R. Gupta, S. K. Jain and N. K. Goswami, *Indian J. Chem.*, **17B**, 272 (1979).
745. S. K. Jain, K. G. Ojha and R. R. Gupta, *Indian J. Chem.*, **17B**, 278 (1979).
746. S. K. Jain, N. K. Goswami and R. R. Gupta, *Ann. Soc. Sci. Bruxelles*, **94**, 63 (1980).
747. R. R. Gupta, K. G. Ojha, M. Kumar and G. S. Kalwania, *Ann. Soc. Sci. Bruxelles*, **95**, 127 (1981).
748. Y. Ueno, Y. Takeuchi, J. Koshitani and T. Yoshida, *J. Heterocycl. Chem.*, **18**, 259 (1981).
749. N. L. Agarwal and R. L. Mital, *Indian J. Chem.*, **14B**, 382 (1976).
750. W. Ried and H.-J. Schaefer, *Justus Liebigs Ann. Chem.*, **756**, 139 (1972).
751. N. L. Agarwal and R. I. Mital, *Philippine J. Sci.*, 125 (1976).
752. Y. V. D. Nageswar and T. V. P. Rao, *Indian J. Chem.*, **19B**, 624 (1980).
753. H. Nakazumi, T. Agrawa and T. Kitao, *Bull. Chem. Soc. Japan*, **52**, 2445 (1979).
754. H. Nakazumi, K. Kondo and T. Kitao, *Synthesis*, 878 (1982).
755. D. Schelz and N. Rotzler, *Dyes and Pigments*, **5**, 37 (1984).
756. V. K. Chadha and G. R. Sharma, *J. Indian Chem. Soc.*, **57**, 1112 (1980).
757. A. M. Simonov and V. N. Komissarov, *Khim. Geterotsikl. Soedin. (Eng. Transl.)*, 654 (1976).
758. I. M. Issa, A. A. El-Samahy, R. M. Issa, G. El-Naggar and H. S. El-Kashef, *Indian J. Chem.*, **15B**, 356 (1977).
759. A. K. El-Shafei, A. Sultan and G. Vernin, *Heterocycles*, **19**, 333 (1982).
760. C. Bhakta and A. K. Chattopadhyay, *J. Indian Chem. Soc.*, **52**, 86 (1975).
761. N. R. Ayyangar, A. G. Lugade and B. D. Tilak, *Indian J. Chem.*, **18B**, 236 (1979).
762. N. R. Ayyangar, P. Y. Kolhe, A. G. Lugade and B. D. Tilak, *Indian J. Chem.*, **19B**, 24 (1980).
763. J. N. Shah and B. D. Tilak, *Indian J. Chem.*, **19B**, 29 (1980).
764. R. P. Soni and J. P. Saxena, *Curr. Sci.*, **46**, 774 (1977).
765. R. P. Soni and J. P. Saxena, *J. Indian Chem. Soc.*, **56**, 733 (1979).
766. R. P. Soni and J. P. Saxena, *J. Indian Chem. Soc.*, **59**, 668 (1982).
767. R. P. Soni, *Proc. Indian Natn. Sci. Acad.*, **48**, 108 (1982).
768. A. M. Osman, A. S. Hammam and A. Th. Salah, *Indian J. Chem.*, **15B**, 1118 (1977).
769. A. S. Hammam and A. M. Osman, *J. Prakt. Chem.*, **319**, 254 (1977).
770. A. S. Hammam and B. E. Bayoumy, *Collect. Czech. Chem. Commun.*, **50**, 71 (1985).
771. N. A. Agarwal and W. Schäfer, *J. Org. Chem.*, **45**, 2155 (1980).
772. N. A. Agarwal and W. Schäfer, *J. Org. Chem.*, **45**, 5139 (1980).
773. N. L. Agarwal and W. Schäfer, *J. Org. Chem.*, **45**, 5144 (1980).
774. I. Oprean and W. Schäfer, *Justus Liebigs Ann. Chem.*, **765**, 1 (1972).
775. K. Joos, M. Pardo and W. Schäfer, *J. Chem. Res. (S)*, 406; *(M)*, 4901 (1978).
776. W. Schäfer, M. Terdic and P. Bartl, *Synthesis*, 438 (1975).
777. G. Kumar and A. P. Bhaduri, *Indian J. Chem.*, **14B**, 575 (1976).
778. M. Pardo, K. Joos and W. Schäfer, *Justus Liebigs Ann. Chem.*, 99 (1982).
779. N. L. Agarwal, H. Bohnstengel and W. Schäfer, *J. Heterocycl. Chem.*, **21**, 825 (1984).
780. B. Wladislaw, L. Marzorati and C. Di Vitta, *Synthesis*, 464 (1983).
781. P. G. Ruminski, L. A. Suba and J. J. D'Amico, *Phos. Sulfur*, **19**, 335 (1984).
782. J. Seda, M. Fuad and R. Tykva, *J. Labelled Comp. Radiopharm.*, **14**, 673 (1978).
783. H. Oediger and N. Joop, *Justus Liebigs Ann. Chem.*, **758**, 1 (1971).
784. R. Huot and P. Brassard, *Can. J. Chem.*, **52**, 838 (1974).
785. B. Simoneau and P. Brassard, *J. Chem. Soc., Perkin Trans. 1*, 1507 (1984).
786. L. H.Briggs, R. C.Cambie, I. C. Dean, R. Hodges, W. B Ingram and P. S. Rutledge, *Aust. J. Chem.*, **29**, 179 (1976).
787. R. M. Wilson, *J. Org. Chem.*, **48**, 707 (1983).
788. D. W. Cameron, G. I. Feutrill, P. G. Griffiths and K. R. Richards, *Aust. J. Chem.*, **35**, 1509 (1982).
789. A. E. Feiring and W. A. Sheppard, *J. Org. Chem.*, **40**, 2543 (1975).
790. D. R. Buckle, B. C. C. Cantello, H. Smith and B. A. Spicer, *J. Med. Chem.*, **20**, 265 (1977).
791. W. Weyler, Jr, P. Germeraad and H. W. Moore, *J. Org. Chem.*, **38**, 3865 (1973).
792. H. W. Moore and D. S. Wilbur, *J. Org. Chem.*, **45**, 4483 (1980).
793. H. W. Moore, *Acc. Chem. Res.*, **12**, 125 (1979).
794. S. P. Lee and H. W. Moore, *Heterocycles*, **19**, 2019 (1982).
795. L. Bucsis and K. Friedrich, *Chem. Ber.*, **109**, 2462 (1976).
796. V. S. Romanov, A. A. Moroz and M. S. Shvartsberg, *Izv. Akad. Nauk SSSR, Ser. Khim. (Eng. Transl.)*, 772 (1984).

The Chemistry of Quinonoid Compounds, Vol. II
Edited by S. Patai and Z. Rappoport

CHAPTER **12**

Electrochemistry of quinones

JAMES Q. CHAMBERS
Department of Chemistry, University of Tennessee, Knoxville, TN 37996, USA

I. INTRODUCTION

Quinone/hydroquinone (Q/QH_2) couples have been widely used in electrochemical studies because they are readily available, easily handled under ordinary experimental conditions, and often exhibit 'well-behaved' electrochemistry. In addition they provide good models for biological redox processes, many of which involve quinone couples of varying degrees of complexity. The simplest quinone couple, *p*-benzoquinone/*p*-benzohydroquinone or BQ/BQH_2, is perhaps the most thoroughly studied organic or inorganic redox couple. As a result, the subject of 'quinone electrochemistry' covers many quite diverse electrochemical investigations.

Electrochemistry itself is an eclectic science that draws on and has applications in many disciplines ranging into solid state physics, molecular biology, electrical engineering, synthetic organic chemistry and others. A review of quinone electrochemistry, consequently, touches on recent developments in many areas, as well as current thinking in fundamental electrochemistry. The thread of quinonoid redox chemistry in the recent literature outlines and underscores the chemistry in electrochemical science.

A major theme that emerges from this review of the last 15 years is that surface chemistry is of prime importance to the understanding of electrode reactions. While this is evident in many of the contributions surveyed below, the important work of Hubbard and coworkers stands out (Section IV.A.1). Furthermore, understanding of the surface of solid electrodes, in particular the quinonoid nature of carbon electrodes, has played a critical role in the design of electrodes at the molecular level. The applications to electrocatalysis from the laboratories of Kuwana and Wrighton can be cited in this regard (Section IV.B.1). Quinones have also played a prominent role in the fabrication of more complex electrode interfaces such as the polymer modified electrodes described in Section IV.B.2.

Much progress in the understanding of the electrode kinetics of simple quinone couples has been made in the last decade. The contributions of Laviron (Section IV.A.2) for aqueous and Rüssel and Jaenicke (Section III.A) for non-aqueous solutions have rationalized several confusing and conflicting aspects of the older literature. These contributions, and those of others, have substantially lessened the distinctions between aqueous and non-aqueous quinone electrochemistry.

Nonetheless, this review, which covers the period from ca. 1973 to the end of 1985, divides the literature (more or less) into non-aqueous and aqueous studies, much in the same format as that in the previous volume of this series[1]. The reader interested in a more complete survey of quinone electrochemistry, with more extensive older references and explanatory material, should consult both review chapters. This is especially true for the sections on electron transfer kinetics (III.A), proton donor effects (III.B), metal ion effects (III.C), electrode kinetics in aqueous solutions (IV.A.2) and coupled chemical reactions in aqueous solutions (IV.C).

The following abbreviations have been employed: BQ, *p*-benzoquinone; NQ, 1,4-naphthoquinone; AQ, 9,10-anthraquinone; DQ, 2,3,5,6-tetramethyl-1,4-benzoquinone (duroquinone). The corresponding hydroquinones are indicated by BQH_2, NQH_2, AQH_2 and DQH_2. The term Q/QH_2 will refer to a generic quinone couple with a formal standard potential $E^{o\prime}$. The half-wave potential, $E_{1/2}$, of quinone couples is usually a good approximation of $E^{o\prime}$ since the diffusion coefficients of quinone–hydroquinone pairs are approximately equal. Also the following common abbreviations for non-aqueous solvents are used: DMF, *N,N*-dimethylformamide; DMSO, dimethyl sulfoxide; HMPA, hexamethylphosphoramide; PC, propylene carbonate; THF, tetrahydrofuran.

II. HALF-WAVE POTENTIALS

A. Aqueous Solutions

Since quinone–hydroquinone couples often are electrochemically reversible, the half-wave potentials carry thermodynamic or quasi-thermodynamic information. Of particular importance in aqueous solutions is the pH dependence of $E_{1/2}$ which reflects the pK_a values for the acid–base dissociation equilibria of the species that interconnect Q and QH_2 in the classical 3 × 3 nine-membered scheme[2]. In this scheme the quinone (Q) and the hydroquinone (QH_2) species are related by an array of electron and proton transfer steps. Simple $E_{1/2}$ measurements by polarographic or voltammetric techniques permit the construction of the Pourbaix diagram which shows the regions in which various members of the square scheme are the predominant solution species at different potentials. These $E^{o\prime}$–pH diagrams appear frequently in the older potentiometric literature[3,4], but not in the more modern polarographic or voltammetric literature where the data are more easily obtained. The review of Evans[5] contains several examples and a listing of $E^{o\prime}$ values for Q/QH_2 couples from Clark's monograph[3]. A useful compilation of mediator $E^{o\prime}$ values, including those of many quinone species, has also been published[6].

Bailey and colleagues[7] have recently constructed the pH–$E_{1/2}$ diagram for the *p*-benzoquinone–hydroquinone couple using data from a variety of sources, a slightly simplified and modified version of which is shown in Figure 1. This diagram has been drawn with the ideas of Laviron[8–10] in mind, as well as the measurements of Parker[11] in strongly acidic media (Section III.B). In Figure 1 the solid lines represent the equilibrium values for the borderlines between the areas in which the various members of the nine-membered scheme (seven out of the nine are included in this Figure) are the principal species in aqueous solution. As is well known, the BQ/BQH_2 couple is a two-electron, two-proton couple over a wide pH range, ca. −5 to 10. The dashed lines in the Pourbaix diagram represent the reversible formal potentials for the first and second steps of the BQ reduction, E_{r1} and E_{r2} in Laviron's nomenclature[9]. The relative position of these borderlines shows that the semiquinone intermediates are unstable with respect to disproportionation for pH < ca. 10.

There are several open questions concerning both the acidic and basic regions of Figure 1. The pK_a value for the BQH^+ species used by Bailey and coworkers[7] is probably too

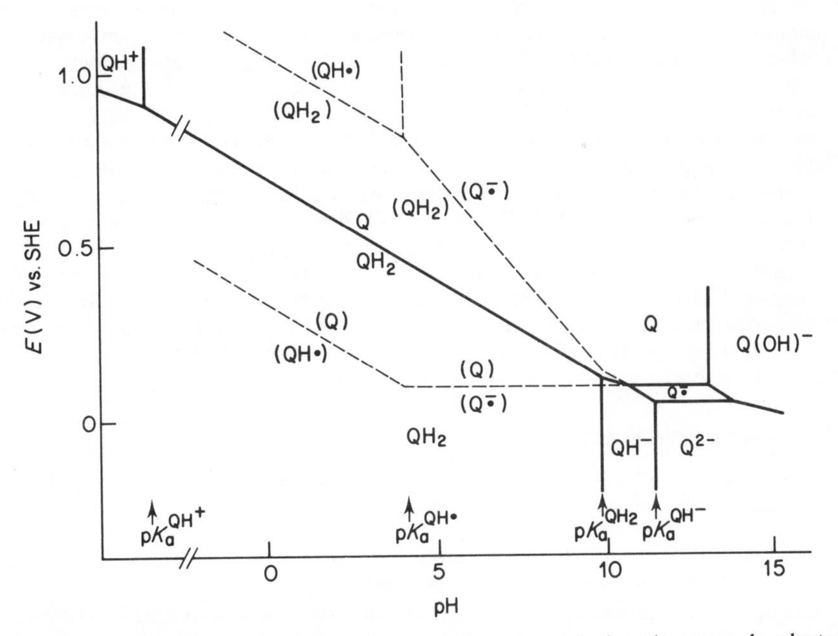

FIGURE 1. Predominance area diagram for the *p*-benzoquinone–hydroquinone couple, adopted from Bailey and coworkers, Ref. 7; SHE, standard hydrogen electrode

large. The value passed down in the literature is for protonated duroquinone, a much weaker acid than BQH^+. Laviron[10] uses the more reasonable value of -7 for the pK_a of BQH^+. It is possible to extend the diagram to the left using the $E^{o'}$ values recently reported by Parker[11] for the $BQH_2^{2+}/BQH_2^{+\cdot}/BQH_2$ couples in highly acidic non-aqueous media, although this hasn't been done here owing to uncertainties in solvent effects and liquid junction potential differences. In the basic region the borderlines are not completely consistent with the $E^{o'}$ values for the $BQ/BQ^{-\cdot}/BQ^{2-}$ couples in non-aqueous solvents, which are ca. 0.5 V more negative than indicated in Figure 1 for aqueous solutions. This region is difficult to study experimentally for several reasons, including the air sensitivity of the anionic species and the reversible addition of hydroxide to BQ that occurs at pH 13[7]. This latter reaction forms the $BQ(OH)^-$ species which complicates the Pourbaix diagram in the basic region. (It is worth noting that an OH^- adduct of AQ has recently been reported in non-aqueous solvents[12].) In the pH range, $4.2 < pH < 8.7$, the $BQO^-/BQ(OH)H_2$ couple exhibits a 90 mV pH dependence corresponding to a two-electron, three proton overall process[13]. A further example of the complex chemistry of quinone species in basic media is an ESR study showing that alkyl substituted quinone radical anions decay to secondary radicals including the anion radical of 2,3,5-trihydroxytoluene and diquinone radicals in aqueous KOH[14]. On the other hand, for quinones which are stable above pK_{a2} for the respective QH_2 species such as 9,10-anthraquinone-2-sulfonate or 9,10-phenanthraquinone-3-sulfonate[13], the Pourbaix diagrams feature a simple $Q^{0/1-/2-}$ 'EE' region at high pH values.

Pourbaix diagrams have been constructed for 9,10-anthraquinone and several substituted anthraquinones (1-amino-, 1-chloro-, 1,5-dihydroxy-, 1,4-dimethyl-, 2-sulfonate, and 2,6-disulfonate) using $E_{1/2}$ values obtained in a thin-layer electrochemical cell at pyrolytic graphite working electrodes[15]. While the $E_{1/2}$ values obtained in this fashion may contain contributions from adsorption terms, the accuracy possible is sufficient to yield much useful information. Furthermore, it is likely that the $E_{1/2}$ values are measured for quinone couples in solution at electrodes covered with adsorbed electroinactive species, in which case adsorption contributions may be minimal. This simple methodology was also applied to the Q/QH_2 couples derived from *p*-chloranil, *p*-chloranilinic acid, *o*-chloranil and *p*-fluoranil[16]. The latter species hydrolyzes in aqueous solutions limiting the pH range over which meaningful values can be obtained. Acid dissociation pK_a values for 1,4-dichlorotetrahydroxybenzene in solution and adsorbed on graphite electrodes have been reported for the pH range 0–10[17]. $E_{1/2}$ vs. pH diagrams have been constructed for several catechols in the pH range, 0–8[18]. The full $E_{1/2}$–pH diagram for the biologically important ubiquinone (UQ) molecule has been determined using the graphite electrode thin-layer cell technique[19]. Above pH 10 the leuco form of UQ must be considered since the pK_a for its formation, $UQ + H_2O = UQOH^- + H^+$, is 10.7. These measurements have been expanded to include other ubiquinones: CoQ_6, CoQ_9 and CoQ_{10} with isoprenoid side chains of varying length[20]. The effect of adsorbed phospholipid layers on the $E_{1/2}$ vs. pH diagrams has been determined in this study as well.

B. Non-aqueous Solutions

In non-aqueous solvents quinones are reduced in successive one-electron steps to form the radical anion and dianion in the absence of added acids.

$$Q \underset{E_1^{o'}}{\overset{e^-}{\rightleftharpoons}} Q^{-\cdot} \underset{E_2^{o'}}{\overset{e^-}{\rightleftharpoons}} Q^{2-} \qquad (1)$$

Under ordinary voltammetric conditions, the half-wave potentials for the two waves that result are excellent estimates of the formal potentials ($E^{o'}$s) for the two couples of equation 1. Several studies have been directed at the question of solvent effects on these $E^{o'}$ values.

Jaworski and coworkers[21] in an extensive study of BQ, NQ, AQ and 9,10-phenanthrenequinone in eight solvents (pyridine, acetone, HMPA, DMF, *N,N*-dimethylacetamide, acetonitrile, DMSO and PC) found a linear correlation with the solvent acceptor numbers of Gutmann[22]. (The $E_{1/2}$ values were also found to correlate with the spin density at oxygen of the semiquinone radical anions.) Solvent effects on $\Delta E_{1/2}$ ($= E_1^{o\prime} - E_2^{o\prime}$) have also been correlated with the Gutmann numbers[23] and with the solvent acidity as measured by the solvent autoionization constant or the potential difference when strong acid ($HClO_4$) or strong base (NEt_4OH) is added to the solvent[24].

In an even more extensive study than that of Jaworski and coworkers[21], Wilford and Archer[25] measured the $E_{1/2}$ values of BQ in 15 solvents and looked for correlations with 18 different solvent parameters. They found the best linear fit with Swain's A + B function[26] for solvents ranging from THF to water. Others have also reported poor correlations with Gutmann's donor numbers[27]. The A + B function is a measure of solvent polarity where A is the anion-solvating tendency and B is the cation-solvating tendency.

The temperature dependence of the $E_{1/2}$ values for BQ, NQ, AQ, and 5,8-dihydroxy-1,4-naphthoquinone has been determined in DMF and propionitrile by Nagaoka and Okazaki[28]. The temperature shift is most pronounced for BQ, which is ascribed to increased solvation of the radical anion being most important for the smallest quinone. The smallest temperature dependence was found for 5,8-dihydroxy-1,4-naphthoquinone where internal hydrogen bonding is possible. Finally, careful electromotive force measurements of the quinhydrone electrode in acetonitrile–water mixtures have been reported[29].

C. Substituent Effects

In the last 12 years substituent effects on $E_{1/2}$ values have been correlated with a variety of parameters, including the Hammett sigma substituent constants, as developed by Zuman in his classic monograph[30]. Studies of this type have included a series of more than 30 1,4-naphthoquinone derivatives substituted with arylamino groups where intermolecular charge transfer complexes are possible[31–33]. For these compounds a Hammett–Zuman substitutent effect treatment adequately described the results without invoking donor–acceptor complexes. Other correlations of this type have been noted for several ureido- and guanidino-substituted naphthoquinones[34], as well as anthraquinones[35], 6,11-dihydroxy-5,12-naphthacenequinones[36], 2*H*-benzo[*f*]isoindole-4,9-quinones[37] and several anthraquinones[38]. In the latter series solvent-dependent positive deviations from the Hammett–Zuman equation were seen when intramolecular hydrogen bonding was operative. Solvent effects have been noted on the Hammett–Streitwieser reaction constant, which increases with the Lewis acidity of the solvent[39].

Half-wave potentials have been reported for 20 quinones in 75% aqueous dioxane[40], several furanquinones in acetonitrile[41], a variety of anthraquinones in DMF[42,43] and fluoro- and trifluoromethylbenzohydroquinone[44]. In the latter case the strong electron-withdrawing nature of the $-CF_3$ group leads to a +73 mV increase in the $E_{1/2}$ value.

SCF molecular orbital calculations carried out on 36 quinones showed good correlation with $E_{1/2}$ and spectral data[45]. Correlations with electronic absorption energies have been observed[46,47] and $E_{1/2}$ values and charge transfer transition energies used to estimate electron affinities for NQ and derivatives[48] in the usual fashion[49]. Substituent effects on infrared frequencies have been noted[50,51]. In the work of Clark and Evans, a flow cell was used to obtain the IR spectra of the radical anions and dianions of several quinones at 0.001 M concentration levels[51].

Ring strain phenomena have been observed to influence the $E_{1/2}$ values of certain naphthoquinones[52]. The naphthoquinone with a tetramethylene bridge fused to the 2,3 bond is more difficult to reduce than the derivative with a fused four-membered ring by ca.

0.16 V. The former has an $E_{1/2}$ close to that of 2,3-dimethylnaphthoquinone, while the latter is reduced at almost the same potential as the parent unsubstituted compound. The results were interpreted using a hybridization model within the Huckel framework. In a related study, Breslow and coworkers used naphthoquinone as a probe to assess the energies of ring systems fused to the 2,3 bond in structures like **1** where oxidation of the dianion leads to an unstable, possibly antiaromatic, ring system[53]. With the aid of fast-

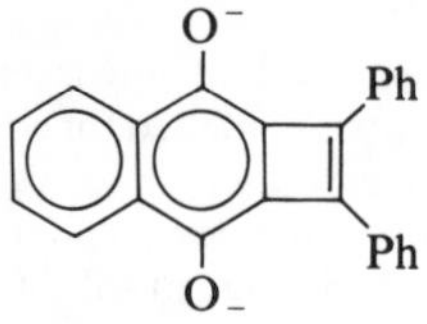

(1) 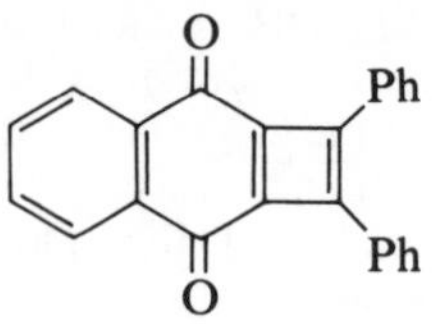 (2)

sweep cyclic voltammetry, evidence was obtained for successive one-electron oxidation (EE behavior) of dianion species like **1** to highly unstable quinones with 'antiaromatic' structures like that of **2**. The $E_{1/2}$ for the $Q^{1-/2-}$ couple is 0.60 V more positive than that of the unsubstituted naphthoquinone. In order to correct for inductive and strain effects of the above type, the dianion **3**, which can be oxidized (see equation 2) to a relatively stable quinone, **4**, was synthesized and found to exhibit classical EE behavior with $E_{1/2}$ values

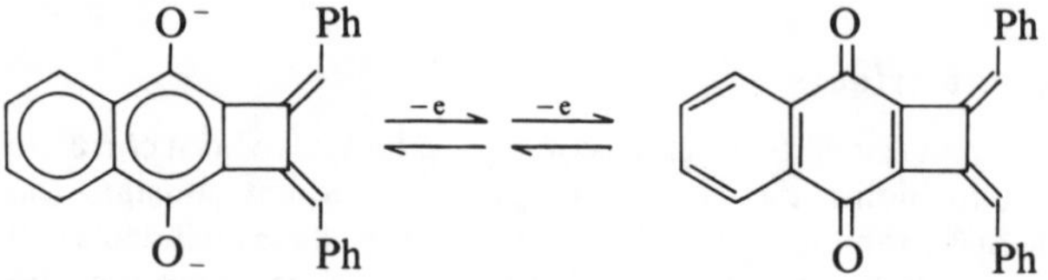

(3) $\underset{}{\overset{-e}{\rightleftharpoons}}$ $\underset{}{\overset{-e}{\rightleftharpoons}}$ (4) (2)

intermediate between **1** and the dianion of naphthoquinone. After careful analysis, the authors conclude that cyclobutadiene is destabilized by at least 12–16 kcal mol^{-1}, and is clearly antiaromatic.

Breslow and coworkers have also reported the electrochemical reduction of several 'cyclobutadienoquinones'[54]. For these molecules it is found that fusion of two anti-aromatic ring systems stabilizes the reduced dianion state. In another study the EE behavior of several tetradehydrol[18]annulenediones was established indicating that they can be considered as quinones of an aromatic system[55]. The application of electrochemistry to questions of this type has been reviewed[56].

Ring strain effects of the type noted by Rieke and coworkers[52] (see above) have also been reported by Iyoda and Oda[57, 58] for BQ and 5,8-dihydroxy-1,4-naphthoquinone with fused four-membered ring substituents. For the series, **5**–**8**, potential shifts of 0.00, −0.03, −0.16 and −0.30 V vs. $E_{1/2}$ for *p*-benzoquinone were observed.

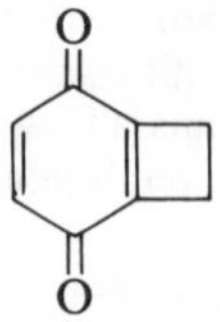

(5) 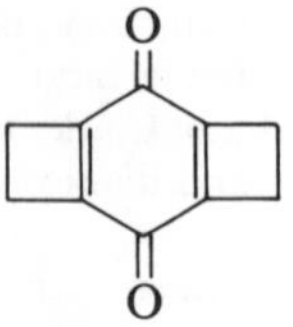 (6) 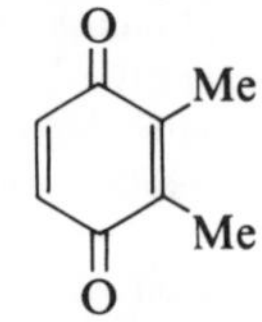 (7) 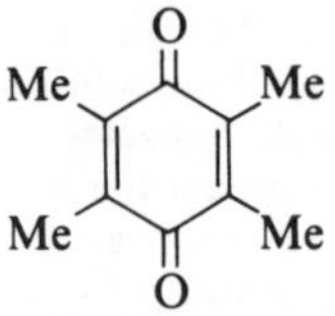 (8)

D. Quinone Analogs

Half-wave potentials have been measured and used to establish quinonoid character for a variety of interesting molecules. These include two acepleiadylenediones[59], *p*-tropoquinone[60], 1,2,5-benzotropoquinone[61,62], and the croconate dianion and related dicyanomethylene derivatives[63]. $E_{1/2}$ values and ESR spectra have been reported for several di- and tri-quinones[64–67]. The ESR spectra of the electrochemically generated radical anions of triptycene bis- and tris(quinones) indicated diradical formation and the absence of intramolecular interactions[66].

The most fascinating quinone analogs of recent years are the polyquinocycloalkanes synthesized by West and his students[68–75]. These molecules have structures of the following type:

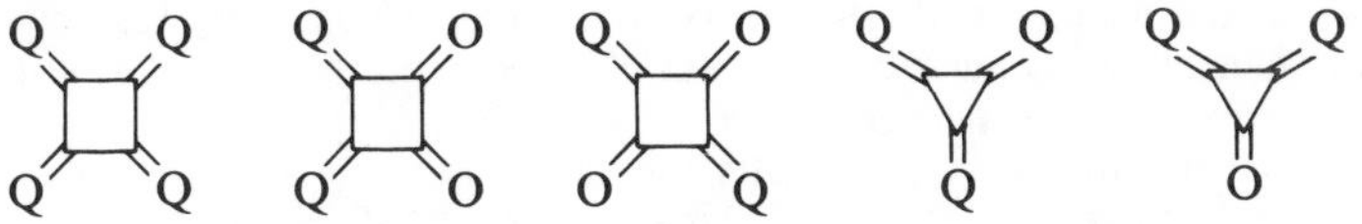

Q =

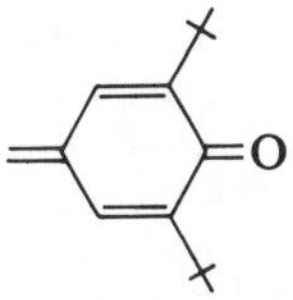

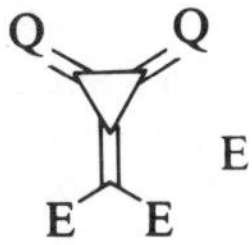

E = CN, CO_2Et, COMe or EE =

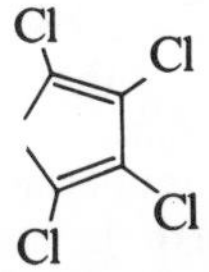

They exhibit EE quinone-like electrochemistry with $E_{1/2}$ values mostly well positive of *p*-benzoquinone. As is to be expected for the strong electron-withdrawing groups in West's molecules, the $E_{1/2}$ values approach those of DDQ (dichlorodicyano-*p*-benzoquinone) in several cases. In one example, where EE is a 10-(1,1-dicyanomethylidene)anthrylidene-9-ylidene electron-withdrawing group, the first and second reduction waves overlap, resulting in a two-electron voltammetric wave[74]. Reversible conformational changes (see Section III.D below) occurring during the reduction are suggested to account for this behavior.

III. ELECTROCHEMISTRY IN NON-AQUEOUS SOLVENTS

A. Electron Transfer Kinetics

Heterogeneous electrode kinetics of simple quinone couples in non-aqueous solvents has been a lively topic in recent years. There have been some real and some apparent discrepancies in the measured values for the rate constant k_1° of the $BQ^{0/-}$ couple in non-aqueous solvents such as acetonitrile or *N*,*N*-dimethylformamide. Electrode material, solvent, electrolyte, nature of the quinone couple, adsorption of impurities or reactants, and temperature can all influence an heterogeneous rate constant. In retrospect, the solvent and electrolyte have much more significant influences on the k° values than was initially suspected.

Rosanske and Evans[76] measured k_1° values for $BQ^{0/-}$, $AQ^{0/-}$, $NQ^{0/-}$ and the 3,5-di-*t*-butyl-*o*-quinone couple in 0.3 M NEt_4ClO_4/acetonitrile or DMF at Hg, Au and Pt electrodes. They measured apparent rate constants in the range, 0.1–1.3 cm s^{-1}, considerably larger values than reported previously by Capon and Parsons[77]. The latter workers reported k_1^o values for $BQ^{0/-}$ in DMF/NEt_4ClO_4 of ca. 0.005 cm s^{-1} at six metal electrodes (Pt, Pd, Rh, Ir, Au, Hg). The values of Capon and Parsons were approximately independent of electrode material and of the same order of magnitude as those measured for the BQ/BQH_2 couple in aqueous media[78,79]. Others have measured kinetics for fast quinone couples in non-aqueous solvents[80–83]. The paper of Howell and Wightman is especially noteworthy. Employing microelectrodes, they obtained exceptionally good cyclic voltammetric data at sweep rates up to 10^4 V s^{-1} for the $BQ^{0/-}$, $NQ^{0/-}$ and $AQ^{0/-}$ couples in acetonitrile/NEt_4ClO_4 at Pt and Au surfaces. They found a significant dependence on molecular size: $k_1^o = 0.39 \pm 0.10$, 0.73 ± 0.12, 1.78 ± 0.35 cm s^{-1}, respectively, for the above couples. (The 1.8 cm s^{-1} apparent rate constant for the AQ couple, measured using the widely employed theory of Nicholson[84], is very fast by current experimental standards.) Samuelsson and Sharp[81] also reported a similar, but not so dramatic, molecular size dependence.

It appears that in polar aprotic solvents the electrode material does not affect the quinone electron transfer rates markedly. This statement should be taken with a grain of salt since double layer corrections are seldom taken into account. However, in some cases[85], but not all[82], surface layer modification can decrease the rates considerably. Activation enthalpy values have been measured in DMF/NBu_4ClO_4 as a function of electrode material[28]. At glassy carbon, Pt and Au electrodes, values of 3.1 ± 0.5, 3.3 ± 1.0, and 5.0 ± 0.7 kcal mol^{-1} were obtained for the $BQ^{0/-}$ couple.

Much of the confusion in this area has been removed by the excellent work of Rüssel and Jaenicke[23,27,86–88]. Upon careful measurement of the k_1^o values for $BQ^{0/-}$ in nine non-aqueous solvents containing three different tetraalkylammonium perchlorate electrolytes, they found that the values varied by 3–4 orders of magnitude. The activation energies, ΔG^*, were a non-linear function of the solvent polarity and increased with the size of the NR_4^+ cation. The results have been interpreted in terms of the Marcus theory, equation 3, where λ_i and λ_o are the inner and outer sphere reorganization energies that accompany an electron transfer reaction, and the ideas of Fawcett[89,90] on the variation of the location of the electron exchange site at an electrode surface with the size of the counter ion. (In the simplified form of equation 3, the potential drop across the diffuse double layer has been neglected.) For polar solvents Marcus theory describes the results with ΔG^* increasing with solvent polarity, and k^o decreasing, since the outer sphere reorganizational energy, λ_o, is directly proportional to solvent polarity. In opposition to theory, however, ΔG^* increases in non-polar solvents like chloroform and slow electron transfer rates are observed. Ion association in the non-polar solvents is a possible cause of this behavior. As a result the largest rate constants are observed in solvents of intermediate polarity, e.g. cyclohexanone, and for supporting electrolyte cations with the smallest ionic radii. This behavior is shown in Figure 2[87], in which the solid lines were calculated using a three-centered model for the quinone and the ionic radii of the tetraalkylammonium ions to estimate distance parameters in the calculation of ΔG^*. While the rate constant data are only partially explained by this treatment, the trends evident in Figure 2 rationalize much discordant data in previous literature. Earlier, Sharp also had noted marked solvent effects on the k_1^o for the tetracyanoquinodimethane couple[91].

$$k^o = A \exp\{-\Delta G^*/RT\} = A \exp\{-(\lambda_i + \lambda_o)/4RT\} \tag{3}$$

Rüssel and Jaenicke have extended their study of the quinone electron transfer kinetics to the $Q^{-/2-}$ couple[23]. Exchange rates for the second step are much slower than the first, and a value of ca. 1.2 was determined for the ratio of the activation energies for the two processes in acetonitrile. Part of the decrease could be due to ion association since they find that the quinone dianions are complexed with NEt_4^+ and NBu_4^+, but not $NOct_4^+$ in

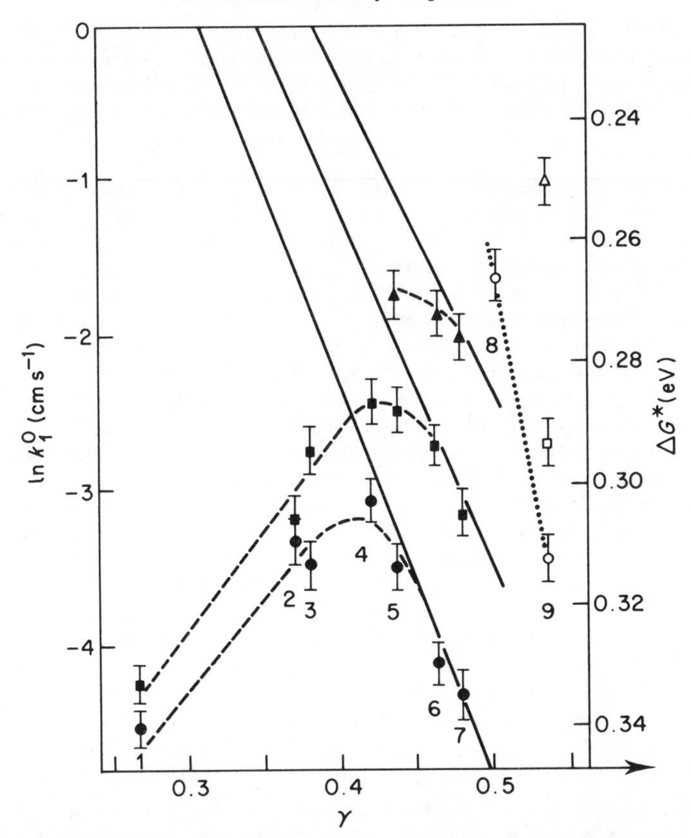

FIGURE 2. Apparent heterogeneous rate constant, ln k', (cm s^{-1}) and activation energy, ΔG^* (eV), at 298 K as a function of solvent polarity, γ. Parameter, supporting electrolyte (0.1 M): (▲, Δ) tetraethylammonium perchlorate; (■, □) tetra-*n*-butylammonium perchlorate; (●, o) tetra-*n*-octylammonium perchlorate. Solvents: filled symbols: (1) $CHCl_3$, (2) THF, (3) CH_2Cl_2, (4) cyclohexanone, (5) DMSO, (6) DMF, (7) PC; open symbols: (8) propionitrile, (9) acetonitrile, solid lines: theoretical curves. ***Reproduced by permission of Elsevier Science Publishers from Rüssel and Jaenicke,*** J. Electroanal. Chem., **180**, 205 (1984)

acetonitrile. In a very recent study of 13 quinone couples in DMF/NEt_4ClO_4, they found that $\Delta G_2^* - \Delta G_1^* = 0.78$ eV[88]. For the $BQ^{0/-\cdot/2-}$ couples in this system equation 3 describes the results with $k_1^\circ = 0.065$ cm s^{-1} ($\Delta G_1^* = 0.292$ eV), $k_2^\circ = 0.00083$ cm s^{-1} ($\Delta G_2^* = 0.404$ eV), and $A = 6040$ cm s^{-1}.

B. Proton Donor and Lewis Acid Effects

The effect of acids on the electrochemistry of quinones in non-aqueous solvents continues to be documented in the literature. Protonation of Q to form the QH^+ species, or of $Q^{-\cdot}$ to form $QH^\cdot$, opens mechanistic pathways for the quinone electrode reactions and gives rise to new voltammetric waves. Prewaves have been observed in the camphorquinone–benzoic acid system[92]; polarography of AQ-phenol has been reported[93]; protonation and disproportionation of quinone radical anions in formic acid[94] and in $MeOH/NEt_4ClO_4$[95], have been proposed; the decay of haloquinone radical anions in ethanol and acetone has been studied[96]; protonation of $AQ^{-\cdot}$ or $BQ^{-\cdot}$ radical anions in 1,2-dimethoxyethane by AQH_2 or BQH_2 gives rise to the following overall electrode process[97], while formation of a Q–QH_2 donor–acceptor complex has been reported for

$$Q + QH_2 + 2e^- = 2QH^- \tag{4}$$

acenazulenedione, **9**, in acetonitrile/NEt_4ClO_4[98]. Proton donors disrupt charge transfer complexes between NQ and 2,4,6-trinitrobenzene with the result that H^+EE electrochemical behavior is observed[99]. Polarograms of alizarin (1,2-dihydroxyanthraquinone) and alizarin-S display two waves in MeOH/LiCl, which coalesce, either into one wave at more positive potential in the presence of benzoic acid, or into one wave at more negative potential in the presence of NEt_4OH[100]. An interesting observation is that surface protonation by carboxylate functionalities of oxidized carbon fiber electrodes gives rise to quinone prewaves in $MeOH/LiClO_4$ solutions[101]. Electrochemical oxidation of NQH_2 in acetonitrile or acetic acid/acetic anhydride mixtures has been suggested as a means of generating non-aqueous acid titrant solutions[102].

The absorption spectra of electrochemically generated $BQ^{-\cdot}$, $BQH^\cdot$ and BQH^- species in DMSO have recently been published[103]. A broad band at 670 nm and one at 410 nm are assigned to the $BQH^\cdot$ species. The $BQ^{-\cdot}$ radical anion was found to be stable in neutral water solutions as previously reported[104].

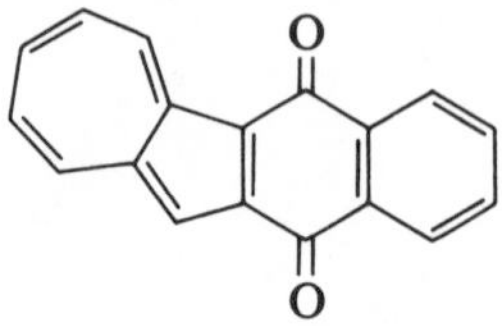

(9)

In an interesting approach, Keita and coworkers[105] studied protonation of AQ radical anions at *n*- and *p*-type silicon semiconductor electrodes in acetonitrile/acetic acid solutions. True ECE pathways are followed, since the radical anions are formed via photoinjection of electrons into the conduction band, and the protonated radicals are reduced by hole injection into the conduction band.

Increasing the basic character of the quinone by means of appropriate substituents clarifies the mechanistic pathways in the presence of proton donors considerably. Bessard and coworkers[106] examined the effect of proton donors on the voltammetry of 2,6-dimethoxy-*p*-benzoquinone, 3,3′,5,5′-tetramethoxy-*p*-biphenoquinone, and related species in acetonitrile and nitromethane in the presence of $HClO_4$. The protonated forms of these quinones are stable permitting measurement of the absorption spectra and careful characterization of the QH^+ reduction processes. The QH^+/QH_2 reduction waves are generally quasi-reversible and appear at potentials more than 1 V positive of the $Q/Q^{-\cdot}$ waves in neutral medium. Similar large potential shifts were found for the reduction of ubiquinone in acetonitrile with and without added $HClO_4$[107]. In the presence of 0.01 M $HClO_4$, the tetramethoxy-*p*-biphenoquinone couple displays a *reversible* two-electron voltammetric wave[106]. This study nicely demonstrates that the reversibility of quinone couples is strongly dependent on the nature of the substituents and the acidity of the medium.

The long-standing observation that radical cations form under acidic conditions can be rationalized by the following coproportionation reaction[106], equation 5:

$$QH^+ + QH_2 + H^+ = 2QH_2^{+\cdot} \tag{5}$$

Hammerich and Parker[11] have observed this reaction in $CH_2Cl_2/HFSO_3$ solution at -50°C and suggest that the reaction proceeds via the QH_2^{2+} species. They estimate that the $E^{o\prime}$ value for the $QH_2^{2+}/QH_2^{+\cdot}$ couple is 1.8 V vs. SCE in this medium. The $QH_2^{+\cdot}/QH_2$ couple is reversible in CH_2Cl_2 in the presence of strong acids such as methylsulfonic or fluorosulfonic acid with an $E^{o\prime}$ of 1.3 V vs. SCE. These measurements fill in the upper part of the square scheme of Jacq[2].

The square scheme is also operative in molten salt media where the Lewis acidity can be varied permitting the observation of novel acid–base species. Cheek and Osteryoung have examined the electrochemical and spectroscopic behavior of AQ and chloranil in a room temperature molten salt mixture of $AlCl_3$ and *n*-butylpyridinium chloride[108,109]. At high mole fractions of $AlCl_3$, Lewis adducts of the type $AQ(AlCl_3)_2$ are observed. Voltammetric reduction waves are assigned to different complexes suggesting that, in contrast to protonation steps in more common solvents, the acid–base equilibria are sluggish in these media. For chloranil, which has previously been studied in an $AlCl_3$/NaCl melt at 175°C[110], HF solution[111] and HF/SbF_5[112], complexation by up to six (!) $AlCl_3$ molecules is observed[109]. Presumably the complexation sites are the two carbonyl groups and the four chloride substituents on the chloranil ring.

The electrochemistry of catechol, 3,5-di-*t*-butylcatechol, tetrachlorocatechol and tetrafluorocatechol has been examined in several non-aqueous solvents where proton donor effects of water have been noted[113]. The respective quinones all present typical EE voltammetric behavior except for 3,5-di-*t*-butyl-*o*-quinone where the second reduction wave is broadened and decreased in peak height. Water-catalyzed disproportionation of the radical anion (equations 6 and 7),

$$Q^{-\cdot} + H_2O = QH^{\cdot} + OH^- \tag{6}$$

$$QH^{\cdot} + Q^{-\cdot} = Q + QH^- \tag{7}$$

and reduction of residual water by the substituted catechol dianion (equation 8),

$$Q^{2-} + H_2O = 1/2\,H_2 + OH^- + Q^{-\cdot} \tag{8}$$

are invoked to explain the results. Extensive voltammetric data in the form of half-wave potentials and $E^{o\prime}$ data for the *reduction* of the catechols (QH_2 species) are given.

Protonation of the radical anion of 1,8-dihydroxy-AQ (DAQ) by benzoic acid in *N,N*-dimethylformamide has been examined using stopped-flow methods by Wightman and colleagues[114] in a careful study marked by attention to detail. For a 1:1 to 50:1 molar excess of benzoic acid to $DAQ^{-\cdot}$, only the following reaction (equation 9), is observed.

$$2\,DAQ^{-\cdot} + PhCOOH = HDAQ^- + DAQ + PhCOO^- \tag{9}$$

The $DAQH_2$ molecule, which is partially ionized even in the presence of $HClO_4$, is a very strong acid and is not formed. For the above protonation, the kinetic analysis indicates that neither a classical ECE nor a simple disproportionation mechanism can explain the results. The authors make a strong case that a heterogeneous acid–base dimer intermediate, $\{DAQ{-}{-}{-}PhCOOH\}^{-\cdot}$, plays the key role in the reduction process. Protonation of the closely related quinizarin radical anion (1,4-dihydroxy-9,10-anthraquinone) by benzoic acid in *N,N*-dimethylformamide has also been studied[115]. At a 5:1 ratio of benzoic acid to quinone the $AQH_2(OH)_2$ species was reported to be formed.

Hydroxy-substituted anthraquinones are strong enough acids to protonate the superoxide species[116]. As a consequence, the radical anions of these species can react with trace amounts of O_2 in non-aqueous solvents even though the $E_{1/2}$ values are positive of the $E_{1/2}$ for the $O_2/O_2^{-\cdot}$ couple since the process can be driven by the proton transfer step.

C. Metal Ion Effects

It is convenient to separate effects of metal ions that can form ion associates with quinone anions from proton donor effects, although the fundamental interactions are closely related. Furthermore, the observed electrochemical responses are similar: voltammetric or polarographic waves shift to more positive potentials as the reduced species are complexed by the metal ions added to the solution in accord with the Nernst equation[117]. The magnitude of the effect is dependent on the nature of the quinone species, the charge and size of the metal or counter ion, and the solvent. In recent years several researchers have placed these phenomena on sound experimental footing.

Fujinaga and coworkers[118] looked at 21 compounds, 15 of them quinones, in DMF in the presence of K^+, Na^+ and four alkaline earth dications. They found the following order of increasing strength for the $Q^{-\cdot}$---M^{n+} ion pair: *para*-quinones = α-diketones < aromatic ketones < *ortho*-quinones. This group has also examined the dependence of the ion association formation constant on the size of the quinone anion for various Li^+, $Q^{-\cdot}$ pairs[119]. For *para*-quinones in DMF, formation constants for both ion pairs and ion triples ($QLi_2^{+\cdot}$) were found to correlate with the Fuoss equation[120], which is a simple electrostatic model that treats the ions as spherical charges in a dielectric medium. The dependence of the quinone $E_{1/2}$ values on the ionic radius of the cation of the supporting electrolyte was also noted by Kalinowski[121]. The variation of the $E_{1/2}$ values with the ionic potential of the cation in a linear fashion is an indication that the ion pairs are of the contact type and not solvent separated[121]. This is the situation for the alkali metal ion pairs formed with the radical anion of 1,2-naphthoquinone in DMSO[122]. Triple ions are also seen in this system. Similar conclusions were reached in an ESR study[123]. Ion association of the 1,2-naphthoquinone radical anion has also been studied by ESR, absorption spectrometry and electrochemistry in acetonitrile[124]. *Ortho*-quinones were also reported to be most strongly ion paired by Kalinowski and Tenderende-Guminska[125], who studied BQ, NQ, AQ, 1,2-naphthoquinone, 9,10-phenanthrenequinone and acenaphthenequinone. These authors also varied the solvent and suggested a correlation with Gutmann's donor numbers.

Solvent effects on the formation constant of the $BQ^{-\cdot}Li^+$ ion pair in acetonitrile, PC, DMF, *N,N*-dimethylacetamide, DMSO and *N,N*-diethylformamide, relative to HMPA where the $BQ^{-\cdot}$ anion is free, have been reported[126]. The term, $\log K_f$, where K_f is the formation constant of the ion pair, was found to be linear in Gutmann's donor numbers: $\log K_f = -0.155(\mathrm{DN}) + 5.994$. The reasonable implication is that as the Lewis base character of the solvent increases, the ion pairing becomes weaker. The dependence of the $E_{1/2}$ values on the solvent basicity had been noted earlier[127].

The ion association of tetraalkylammonium ions with the dianion of benzoquinone has been observed[23] where it plays a role in governing the rate of electron transffer (see Section II.A). In the presence of $NOct_4^+$ counter ions, the quinone dianion is free, but it is associated for smaller cations.

Ion association has also been reported in recent years for $AQ^{-\cdot}$ with dications[128] and alkali metal ions in *N,N*-dimethylformamide[129], and for the $BQ^{-\cdot}$ anion with Li^+ in acetonitrile[130]. Reduction of *o*-chloranil in acetonitrile/$NaClO_4$ solution, which results in a deep blue adsorbed $Na^+Q^{-\cdot}$ film, has been suggested as an electrochromic system[131]. The polyquinone **10**, nonylbenzohexaquinone, takes up six Li^+ cations per molecule, one

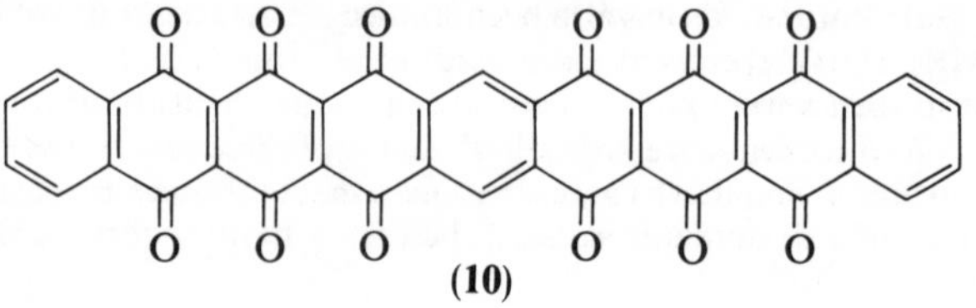

(10)

for each quinone function, upon reduction in one-to-one PC-dimethoxyethane/$LiClO_4$[132].

The voltammetric behavior of catechol complexes of several divalent and trivalent metal ions has been surveyed[113, 133]. For these systems, where both metal center and $Q/Q^{-\cdot}$ ligand center electron transfer reactions occur, the cyclic voltammetric behavior is complex. The paper by Sawyer and coworkers[133] reviews the inorganic literature in this area. In this context, it can be mentioned that zero valent transition metal complexes of quinones exist and their reduction potentials are occasionally reported[134, 135]. These species are more pertinent to the electrochemical behavior of quinones strongly adsorbed on metal electrodes (see Section IV.A) than quinone ion pairs in solution.

The electrochemistry of an interesting quinonoid compound containing the bipyridyl structure **11** has been examined and compared to that of **12**[136, 137]. Compound **11** binds Ni^{2+}, Co^{2+} and Zn^{2+}, but not Ca^{2+}, Mg^{2+}, Mn^{2+} and Pb^{2+}.

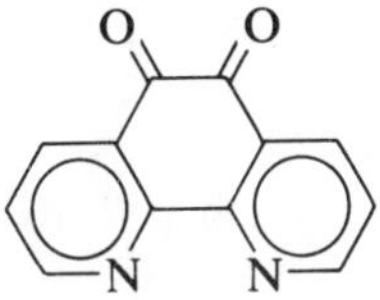

(11) 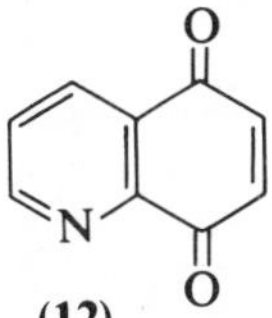 **(12)**

Bifunctional molecular species containing redox and specific complexing groups of a crown ether type have significant implications in biological electron transport and other energy conversion mechanisms. Several quinones have been incorporated into molecules of this type. Bock and colleagues[138] prepared compound **13** and found that upon reduction by metal in THF, the metal ion is bound via a contact ion pair in the solvate cage of azacoronandnaphthoquinone radical anion. Addition of excess free crown generates the ESR spectrum of the electrochemically produced radical anion.

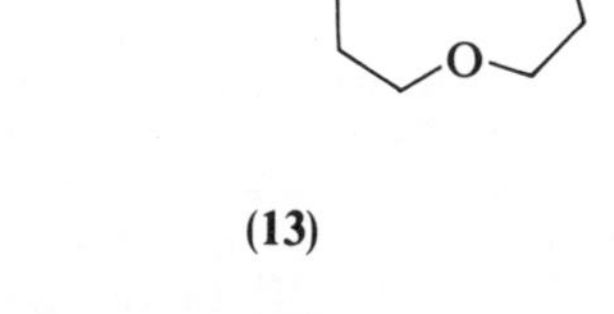

(13) 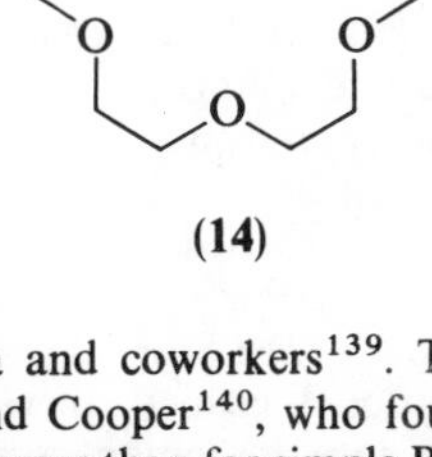 **(14)**

A related molecule, **14**, was synthesized by Sugihara and coworkers[139]. The electrochemistry of this molecule was examined by Wolf and Cooper[140], who found that the $E_{1/2}$ shifts in the presence of Li^+, Na^+ and K^+ were larger than for simple $BQ/BQ^{-\cdot}$ couples *and in the opposite order of that expected for ion association.* Thus in this molecule the selectivity of the crown ether moiety for the potassium ion is coupled to the electron transfer reaction of the quinone. Related quinone crowns that exhibit enhanced Li^+ binding have been reported recently[141].

D. Conformational Effects

The elegant work of Evans and his students[142, 143] on the electrochemistry of bianthrone, **15**, can be viewed in the context of quinone electrochemistry. Bianthrone,

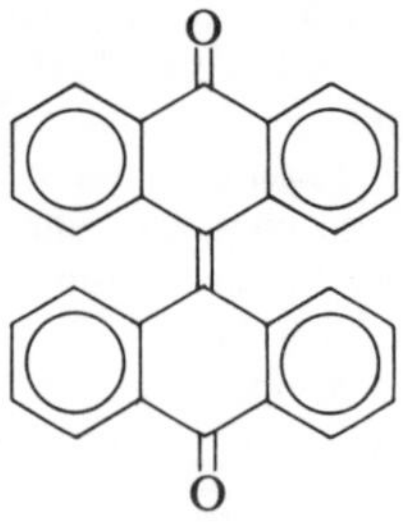

(15)

which exists in a doubly folded butterfly conformation, conformation A, at room temperature, undergoes an irreversible two-electron voltammetric reduction to a dianion, which exists in a twisted conformation (B). This conformational change can be viewed as a result of the decrease in the double bond character in the *exo*-bond upon reduction to the dianion which allows the anthracene rings to adopt the twisted form. Oxidation of the twisted dianion (B^{2-}) proceeds by an EE pathway at fast sweep rates (equation 10) to form

$$\begin{array}{ccccc} A & \overset{e^-}{\rightleftharpoons} & A^{-\cdot} & & \\ \upharpoonleft\downharpoonright & & \upharpoonleft\downharpoonright & & \\ B & \overset{e^-}{\rightleftharpoons} & B^{-\cdot} & \overset{e^-}{\rightleftharpoons} & B^{2-} \end{array} \qquad (10)$$

the unstable B conformation of bianthrone. The system was nicely studied using double potential step spectroelectrochemical and cyclic voltammetric techniques and rate constants determined for the transformation of $A^{-\cdot}$ to $B^{-\cdot}$ and B to A. Reduction of bianthrone is suggested to occur by the sequence, $A \rightleftharpoons A^{-\cdot} \rightarrow B^{-\cdot} \rightleftharpoons B^{2-}$, although other pathways cannot be ruled out. Remarkably, the same conformational square scheme is operative and on the anodic side when bianthrone and related molecules are oxidized[144]. The $[\text{bianthrone}]^{2+/+/0/-/2-}$ system, spanning five oxidation states, thus affords one of the most dramatic examples of conformational effects on electrochemistry in the literature.

Parker and coworkers have also made significant contributions in this area[145, 146]. They measured the rate constant for the twist to butterfly conformational change using Parker's derivative cyclic voltammetry technique and report an activation energy of 15.3 kcal mol^{-1} for the B to A process[146].

Evans and coworkers have extended their examination of the bianthrone reduction process using homogeneous mediator catalysts[147, 148]. By selection of mediator couples (they actually used quinones such as duroquinone) with $E^{o\prime}$ values between the equilibrium $E^{o\prime}$ for the A/B^{2-} couple and the irreversible two-electron reduction of A, the reduction process can be carried out by the reduced mediator (i.e. $DQ^{-\cdot}$) in a thin reaction layer next to the electrode surface. Armed with the theory of Saveant[149, 150], rate constants can be determined for the coupled chemical reactions. For fast reactions such as the $A^{-\cdot}$ to $B^{-\cdot}$ conformational change, there are experimental advantages to this 'electrocatalytic' approach since the concentration of electrogenerated mediator can be controlled both by the $E^{o\prime}$ of the mediator couple and the applied potential. Evans and Naixian[147] employed six $Q/Q^{-\cdot}$ couples as mediators for the process and found the expected qualitative dependence on their $E_{1/2}$ values.

Finally, thin-layer coulometry (at 100°C in DMF) has been carried out to obtain the equilibrium potential for the A/B^{2-} couple of 1,1-dimethylbianthrone[151]. With the readily determined $E_{1/2}$ values for the $B^{0/-/2-}$ steps, this permits determination of the equilibrium constant for the $A \rightleftharpoons B$ conformational change: a value of 8×10^{-6} was found.

E. Coupled Chemical Reactions

Miller and colleagues[152] have documented the reductive cleavage of several 2-hydroxymethyl ester derivatives of AQ in DMF:

(11)

(16)

The reduction process was postulated to proceed via the quinone methide, **16**, shown in equation 11. A similar process has been found to occur for the reduction of the anthracycline drug, daunomycin, when it is reduced to the quinone dianion in DMF[115, 153]. Elimination of a sugar residue gives a quinone methide intermediate which was characterized by spectroelectrochemistry. Quite similar pathways were suggested by Plambeck and coworkers for the reduction of daunomycin in pH 7.1 phosphate buffers[154, 155]; this is discussed in Section IV.C.

The mechanism and stereochemistry of the reductive alkylation of AQ and acenaphthenequinone in DMF/NBu_4I has been examined[156]. Misra and Yadav also reported the reductive alkylation of AQ[157]. Reductive silylation of quinones takes place in the presence of *t*-butyldimethylsilyl chloride[158]. The synthesis of a diquinone bridged by a $-CH_2CH_2-$ group and its reductive cyclization via an internal radical coupling reaction has been reported[67]. In an interesting study, Rieke and coworkers[159] studied the electrocatalytic dimerization of duroquinone in DMF/NBu_4ClO_4 leading to the cyclyzed dimeric product, compound **17**. Since this process (equation 12), can be initiated by the radical anion of duroquinone itself acting as an electrogenerated base (EGB), overall *n*-values less than 0.1 electron per duroquinone were observed.

The anthraquinone dianion will initiate styrene polymerization[160].

On the anodic side, oxidation of 1,4-dimethoxybenzene in wet acetonitrile leads to diquinones via radical cation coupling reactions and elimination of MeOH[65]. Good yields have been obtained for the electrosyntheses of vitamin K_3 by four-electron oxidation of a partially saturated precursor in 9:1 acetonitrile/*t*-butyl alcohol/$LiClO_4$[161] and for tetrafluorobenzoquinone and perfluorohydrocarbons in trifluoroacetic acid/trifluoroacetic anhydride media[162]. Anodic pyridination of several hydroquinones in pyridine–water has been reported[163], in agreement with previous work[164].

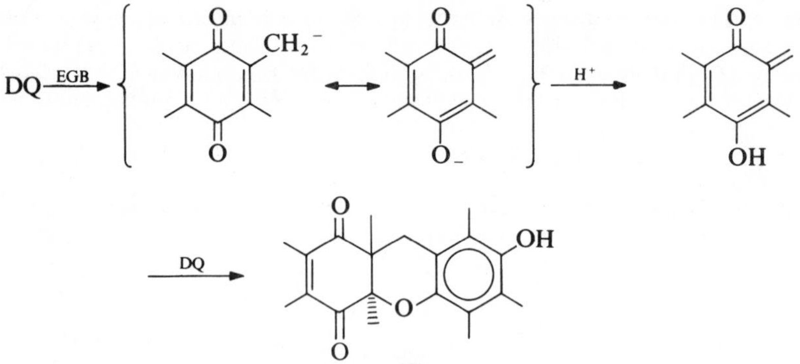

(12)

(17)

IV. ELECTROCHEMISTRY IN AQUEOUS SOLUTIONS

A. Electrochemistry at Solid Electrodes

1. Adsorbed Q–QH$_2$ layers

While there have been several significant contributions to our understanding of electrochemical kinetics of quinone couples at solid electrodes in recent years, the monumental contributions of Hubbard and coworkers dominate the literature of the 1975–1986 period. Using the sophisticated high-vacuum techniques of surface science, the simple methodology of thin-layer electrochemistry and penetrating insight into the nature of electrode reactions, Hubbard's group has revealed the atomic and molecular nature of electrode processes at new levels of detail. Quinone/hydroquinone couples have figured predominantly in this work, to an extent that it is difficult to review Hubbard's contributions in this area in reasonable space. Fortunately, Hubbard has reviewed his work in part[165, 166], but his research is on-going, making an up-to-date survey difficult at this time.

In an important 1973 paper[167], Lane and Hubbard first pointed out that adsorbed hydroquinones and catechols oxidize irreversibly at platinum in 1 M $HClO_4$, while the quinone couple in solution remains reversible at the surface covered with the adsorbed layer. Compounds **18** and **19** were among the molecules studied by Lane and Hubbard

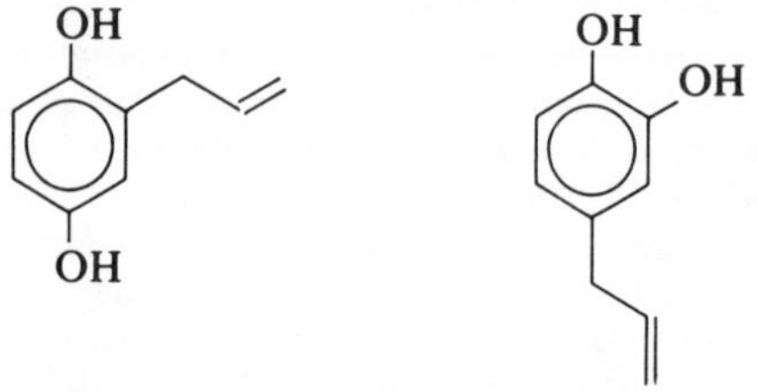

(18) **(19)**

with the expectation that adsorption would take place via the unsaturated sidearm and that surface orientation effects could be seen in the electrochemical response. They were on the right track as remarkable surface orientation effects are found for edge-on vs. flat configurations of the quinonoid and aromatic rings (see below). The ideas of Lane and

Hubbard also presage the great activity of the last decade on chemically modified electrodes.

The simple thin-layer electrochemical technique for measurement of adsorption of Q/QH_2 and related species from aqueous solutions (typically 1 M strong acid or acidic buffers) has been described by Soriaga and Hubbard[168]. A polished polycrystalline Pt(111)–Pt(100) electrode, cleaned between runs by electrochemical cycling in $HClO_4$, was employed in a thin-layer cell of ca. 4×10^{-6} litre volume. Thin-layer voltammograms of the initial filling of the cell show a symmetrical cyclic voltammogram for the Q/QH_2 couple in solution at potentials negative of a massive irreversible oxidation of the adsorbed layer (Figure 3). When the potential range is restricted to leave the adsorbed layer intact, subsequent fillings of the thin-layer cell give rise to markedly increased peak heights and areas for the reversible Q/QH_2 wave. This occurs because the bulk concentration of the hydroquinone is not decreased in the second or subsequent fillings of the cell by adsorption on the already saturated surface. For the initial filling of the thin-layer cell, the amount of substrate left in solution is given by $VC = VC^{\circ} - A\Gamma$, where V is the volume, C is the solution concentration, C° is the initial concentration, A is the electrode area, and Γ is the surface coverage (mol cm^{-2}). The charge under the reversible wave is given by equation 13,

$$Q_1 - Q_b = nF(VC^{\circ} - A\Gamma) \tag{13}$$

and upon repeating the scan after refilling the cell, by equation 14:

$$Q - Q_b = nFVC^{\circ} \tag{14}$$

In these equations Q_1 is the measured charge after the initial filling of the cell, Q is the charge after subsequent fillings, and Q_b is the background charge. Only slightly more refined applications of Faraday's law to this thin-layer coulometry experiment allow variation of Q_b and Q upon successive fillings of the cell to be taken into account[169, 170]. Solving equations 13 and 14 for Γ gives $\Gamma = (Q - Q_1)/nFA$. In terms of area per molecule (σ in $Å^2$) this translates into $10^{16}/N_A\Gamma$, where N_A is Avogadro's number. Surface cleanliness is favoured in thin-layer cells by the large surface to volume ratios that can be achieved, and in the hands of Hubbard and coworkers, a high degree of absolute accuracy in surface electroanalysis has been realized.

Initial application of the technique to 40 quinone or hydroquinone species present in solution at the 10^{-4} M concentration level revealed that most were adsorbed in a flat orientation on the platinum surface[168]. For example, hydroquinone was found to have a surface coverage of 51.6 ± 1.4 $Å^2$, in good agreement with calculations for a flat orientation based on molecular models using covalent and van der Waals radii of Pauling. (Benzoquinone gave the same value, 52.7 ± 1.4 $Å^2$; it is likely that BQH_2 adsorbs on the platinum surface in the quinone form via an oxidative process[171].) Accordingly, the flat (η^6), edgewise (η^2 bonding to the 2,3 bond), and endwise (η^1 bonding to the oxygen) orientations have calculated coverages, respectively, of 53.8, 28.6 and 21.8 $Å^2$. Optical monitoring of the QH_2 absorbance in a long optical path length thin-layer spectroelectrochemical cell gives the same result[172].

As stated by Soriaga and Hubbard, the orientation of an adsorbed molecule is potentially a function of the molecular structure of the adsorbate, electrolyte, solute concentration, solvent, electrode potential, temperature, pH, electrode material, surface structure, co-absorbates and perhaps other factors. These thin-layer measurements confirm, for example, the above expectations for compounds **18** and **19** which have surface coverages significantly less than the values calculated for η^6 orientations. Other deviations from flat orientations were observed for alkyl-substituted ring systems where tilted η^2 orientations were suggested as possibilities[168, 173].

The existence of different orientations of the adsorbed molecules gives rise to 'packing density plateaus' in the adsorption isotherms that correspond to specific orien-

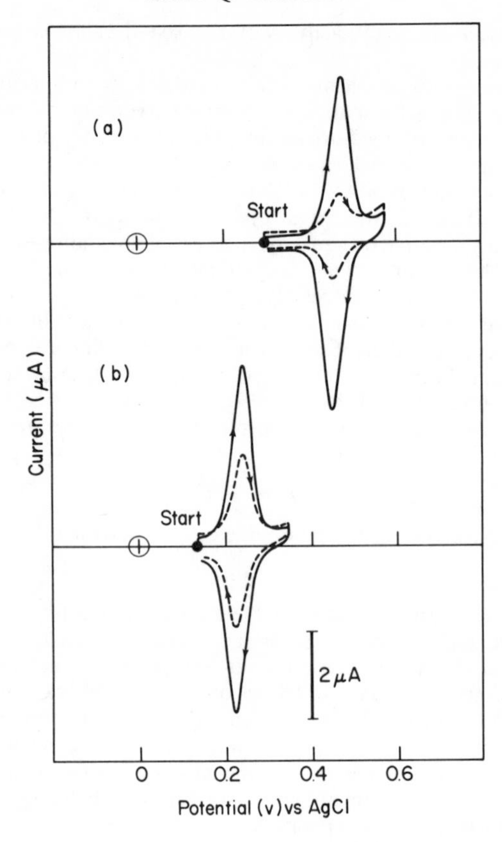
(a)
Start
(b)
Start
2 μA
Current (μA)
0
0.2
0.4
0.6
Potential (v) vs AgCl

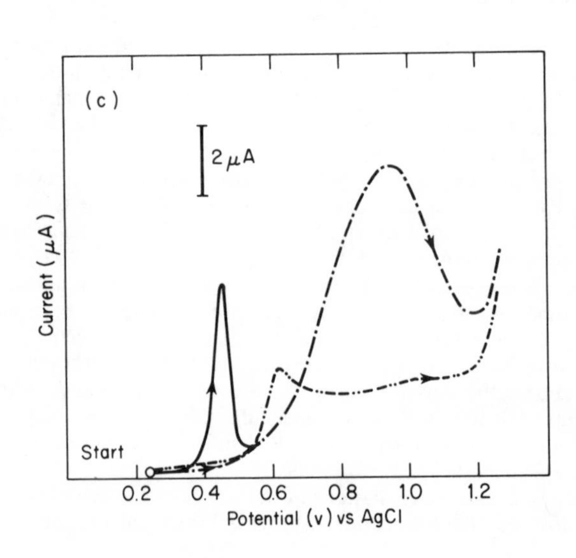
(c)
2 μA
Current (μA)
Start
0.2
0.4
0.6
0.8
1.0
1.2
Potential (v) vs AgCl

tations[169, 171]. For example, BQH_2 undergoes a transition from the flat to the edgewise orientation between 1×10^{-4} and 3×10^{-4} M. Chirality was found to influence the absorbate orientation in the case of L-DOPA (L-3-(3,4-dihydroxyphenyl)alanine) and a racemic mixture of D,L-DOPA. At concentrations where the edgewise orientation predominates, the pure enantiomer packs more densely than the racemic mixture[174]. Surface coverage isotherms depicting this behavior for the adsorption of 26 Q or QH_2 molecules on platinum are shown in Figure 4.

The temperature dependence of the surface coverages reveals additional features of the molecular surface[175]. For Q/QH_2 species with unsubstituted rings that tend to absorb in the flat configuration, the transitions become sharper at low temperature (e.g. 5 °C), and at higher temperatures additional surface rearrangements are suggested by the isotherms. For molecules that are attached to the platinum surface by a functional group other than a benzene ring (e.g. 2,5-dihydroxythiophenol or 2,3-dihydroxypyridine), the packing densities tend to be independent of temperature.

The ability to determine the molecular orientation of the adsorbed quinone couples allowed Hubbard to make a fascinating observation. The massive irreversible oxidation of the adsorbed layers is dependent on the surface orientation[176]. For flat configurations, essentially complete oxidation of the aromatic rings takes place in strong acid media yielding *n*-values up to more than 40 (!) electrons per molecule in some cases. Oxidation of the edgewise orientation requires fewer electrons, but the oxidation is still a multi-electron process. As a result, there is a distinct and striking correlation between the *n*-values and the surface coverages in the isotherms. The *n*-values for the oxidation of the edgewise orientation (e.g. for oxidation of adsorbed NQH_2) are strongly temperature dependent[177]. Kinetic barriers to reorientation of adsorbed layers from the flat to the edgewise orientation have been noted[165].

The latter observation is one of the reasons advanced for the fact that these effects have not been noted in previous studies of the electro-oxidation of aromatic compounds. Common practice is to immerse the electrode under study in a blank solution, and increase the concentration of adsorbate by successive addition of aliquots of a standard solution. This procedure can trap the surface in the $Pt-\eta^6$ adsorption configuration. The thin-layer cell area/volume ratio advantage mentioned above is another significant feature of the Hubbard methodology. Adsorbed impurities can influence the packing densities in subtle ways and obscure the transitions between flat, edgewise and other orientations. As emphasized by Hubbard and coworkers[170], this is a fundamental error in the technique of hydrogen codeposition that many have used to measure adsorption isotherms on solid electrodes. Another feature of the Hubbard approach is that relatively smooth electrode surfaces are used. For surface roughness factors above ca. 5, the concentration-dependent packing factor transition for adsorbed BQH_2 disappears, presumably due to suppression of the vertical adsorption mode[178]. This explains why these effects are not observed by other techniques, e.g. radiotracer methods, where platinized platinum electrodes are used.

The effect of the platinum electrode potential on surface coverage has also been studied[179]. The surface coverage shows little variation with potential in the double-layer region, but decreases considerably at positive potentials where coadsorption of oxygen can take place and at more negative potentials where hydrogen adsorbs. Once an adsorbed

FIGURE 3. Thin-layer current–potential curves for (A) hydroquinone, (B) durohydroquinone and (C) hydroquinone at a polycrystalline platinum electrode: (– – –) first filling; (——) presaturated surface; (·—·—) presaturated surface rinsed to remove dissolved reactant; (- -··- -) clean surface. The solutions contained initially 0.15 mM reactant and 1 M $HClO_4$ (thin-layer volume, V, 4.08 μl; platinum electrode area, A, 1.18 cm^2; rate of potential sweep, r, 2.00 $mV\,s^{-1}$; solution temperature, T, 296 $\pm$ 1 K). *Reprinted with permission from Soriaga and Hubbard,* J. Am. Chem. Soc., **104**, *2735(1982)*

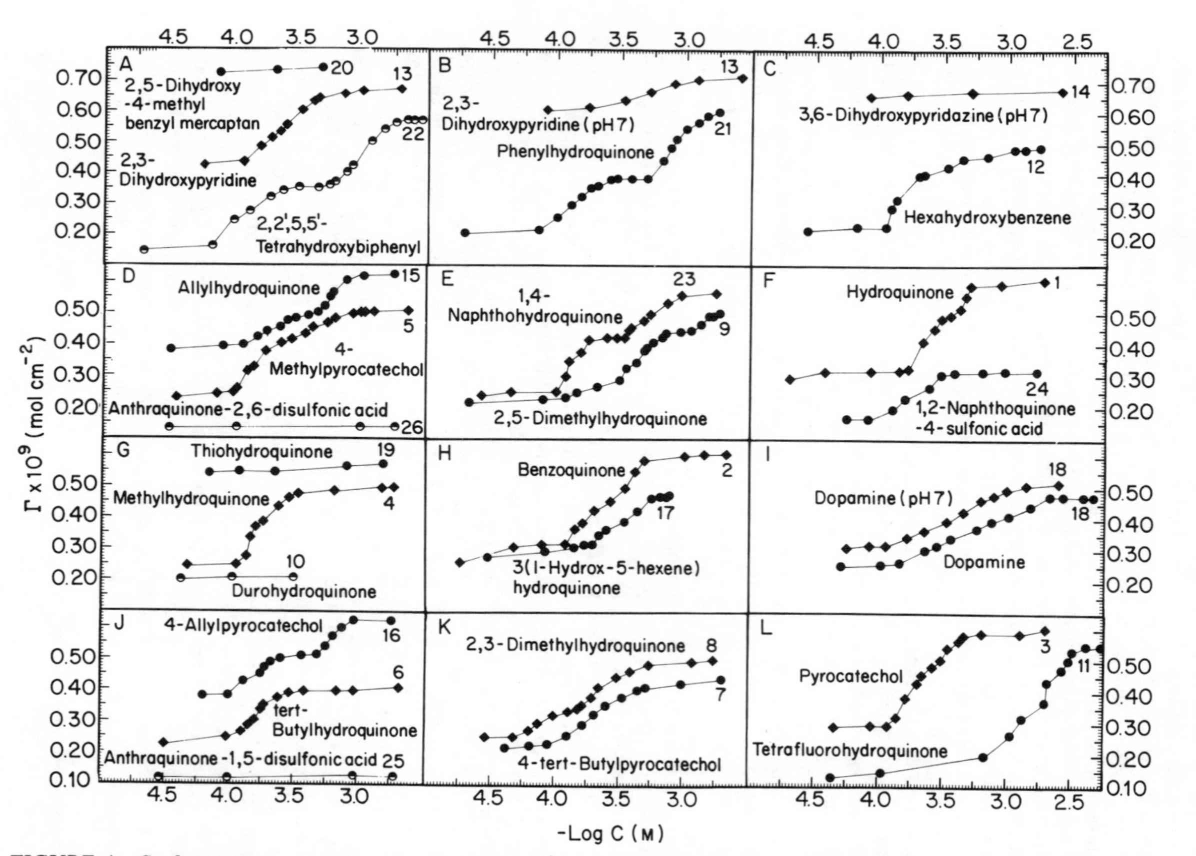

FIGURE 4. Surface coverages, expressed as Γ (mol cm^{-2}), as a function of solute concentration for the adsorption of 26 quinonoid compounds on atomically smooth platinum electrodes from 1 M $HClO_4$[171]. *Reproduced by permission of Elsevier Science Publishers from Soriaga* et al., J. Electroanal. Chem., **142**, *317 (1982)*

layer is formed, the coverage is relatively insensitive to the electrode potential within the range where faradaic processes do not occur.

The Q/QH_2 adsorption is markedly dependent on the presence of halide ions[180-182]. Treatment of the platinum electrodes with 5×10^{-4} M NaI, for example, totally prevented the adsorption of BQH_2. Treatment with 2×10^{-3} M NaBr severely attenuated the η^6 adsorption at low BQH_2 concentrations, but not the η^2 edgewise adsorption at higher concentration. Chloride ions had only a moderate influence on the adsorption isotherms. Other electrolytes such as $HClO_4$, $NaClO_4$, $CsClO_4$, H_2SO_4, H_3PO_4, $NaPF_6$, or NaF had little effect on packing density, oxidation n-values, or orientation. Other examples of these phenomena have occasionally been reported; for example, adsorbed BQH_2 is displaced from gold electrodes at pH 10.5 by sulfite ions[183].

Exchange and insertion reactions in coadsorbate layers have been examined recently by Hubbard and coworkers[184,185] in a very pretty fashion. For NQH_2 and 2,2′,5,5′-tetrahydroxybiphenyl, for example, which oxidize (in the non-adsorbed form) at well separated potentials, it is possible to measure the competitive coverage of both adsorbates by the thin-layer technique[185]. Ordering effects are seen in the adsorbed layers which are attributed to hydrogen-bonding interactions between adsorbates. Small mole fractions of one component, e.g. Ar′ in equations 15 and 16 below, were shown to introduce a significant degree of disorder in the adsorbed layer. For electrodes pretreated with BQH_2 (or with NQH_2), the surface reactions described in equations 15 and 16 were documented[184].

$$\eta^6\text{–Ar} + \text{Ar}'_{aq} \xrightarrow{C^o > 1\ \text{mM}} \eta^2\text{–Ar} + \eta^2\text{–Ar}' \tag{15}$$

$$\eta^2\text{–Ar} + \text{Ar}'_{aq} \xrightarrow{C^o > 1\ \text{mM}} \eta^2\text{–Ar} + \eta^2\text{–Ar}' + \text{Ar}_{aq} \tag{16}$$

The extent of molecular detail obtained with these simple experiments is indeed impressive.

Adsorption of quinones on other electrodes such as graphite from aqueous solutions has been known for some time[186-188]. Symmetrical, well behaved surface waves were obtained for the NQ, 9,10-phenanthrenequinone and 9,10-anthraquinone-2-sulfonate/hydroquinone couples adsorbed on the basal plane of pyrolytic graphite[186]. All three adsorbed quinones yielded voltammetric surface waves indicative of non-ideal, destabilizing interactions. Of the three, NQ, which was adsorbed at the highest coverage corresponding to the edgewise orientation based on the data of Soriaga and Hubbard[168], exhibited the most ideal behavior. Finally, the adsorption of 9,10-phenanthrenequinone and oxoapomorphine on carbon paste electrodes has been examined with a view toward analytical applications[189]. Baldwin has also reported the cyclic and differential pulse voltammetry of surface adsorbed adriamycin on carbon paste at pH 4.5[190, 191]; sensitivity for determination of this drug in urine samples was 10^{-8} M, with a linear concentration range of 10^{-5} to 10^{-7} M for the latter technique.

2. *Electrode kinetic studies*

The kinetics of simple Q/QH_2 couples at solid electrodes is a difficult problem and many questions remain open in this area. Clearly, adsorption of the quinone species, electrolyte components and possible impurities in solution must be taken into account. In addition, the protonation and disproportionation reactions embodied in the square scheme cannot be ignored. Significant observations on this problem have been made by several groups in recent years. These should be put in context of previous work, which is briefly reviewed in Volume I of this series[1].

As is well known, experimentalists in electrode kinetics must deal with the difficulty of adsorption of impurities that might alter the rate of the reactions[192]. A particularly dramatic example of the effect of surface adsorbed layers on the rate of Q/QH_2 couples at Pt comes again from the laboratory of Hubbard and coworkers[193]. The BQ, NQ, DQ and *t*-butyl-*p*-benzoquinone couples were examined at pH 4 where a minimum exists in the exchange rate of the BQ/BQH_2 couple. This pH range is just below the transition from the HeHe to the eHeH mechanism in the classical study of Vetter[78]. (This notation specifies the sequence in which protons and electrons are transferred in the reduction of Q to QH_2.) For this system addition of CN^- ion retards the rate of the quinone couples to total irreversibility as measured by the anodic to cathodic peak potential separation in the thin-layer cyclic voltammograms. For sulfonated BQ/BQH_2 couple, the indication is that not only does the rate decrease, but that the rate-determining step changes and the mechanism becomes eeHH[194]. Furthermore, the BQ/BQH_2 rate at pH 4 can be enhanced to near reversibility by the presence of adsorbed iodine. As emphasized by Hubbard and colleagues, these results show that the kinetics can be manipulated at will by the addition or removal of surface adsorbed components.

More interestingly, with the above thin-layer electrochemical methodology of Hubbard to control the surface layer structure available, new questions can be addressed that should give further insight into the Q/QH_2 kinetics. For example, significantly different rates are observed for the BQ/BQH_2 couple at pyridine-coated platinum electrodes when pyridine is adsorbed via the nitrogen (Pt–η^1 bonding) or via a C(4)–C(5) ring bond (Pt–η^2 bonding). The peak potential separations are 151 mV and 68 mV, respectively, under the same experimental conditions[193].

Others have observed the effect of adsorbed species on the kinetics of the BQ/BQH_2 couple in recent years using more classical techniques. Schmid and Holmes in a careful study found that the apparent rate constant was decreased exponentially with surface coverage of benzo[*f*]quinoline at low coverages[195]. The kinetics at oxygen covered rhodium electrodes has been reported[196], where the rate was found to be faster at the unblocked electrodes. Sulfite has been found to inhibit BQH_2 oxidation on gold at pH 8.1[197]. In some cases rate enhancements have been reported: Lacaze and coworkers[198] made the interesting observation that a specific polymer film, poly(2-hydroxymethyl-1,4-phenylene) oxide, coated on a platinum electrode increased the rate of BQ/BQH_2 to reversibility in $HClO_4$ solutions. Underpotential deposition of Bi, Tl and Pb on platinum was also found to improve the reversibility of this system[199]; however, deposition of Tl on silver single crystals had no effect on the BQ/BQH_2 electrochemistry[200].

Laviron has developed theoretical treatments of electrode kinetics that take into account fast protonation steps coupled to one- and two-electron processes in the context of the square scheme (Scheme 1). In an important paper[10], he has rationalized several puzzling aspects of the literature by means of theory which assumes that the coupled proton transfer reactions are at equilibrium. In this view, breaks in the Tafel slopes in Vetter's experimental results at low pH on the cathodic segment and at high pH on the anodic segment are due, respectively, to rate control by the second and the first electron transfer steps. In the square scheme reproduced, the apparent electron transfer rate constants that Laviron cites, and the pK_a values of the protonated species, are indicated. (The value of -28 for the pK_a of QH_2^{2+} is estimated from the data of Hammerich and Parker[146].) It is significant that this treatment predicts the minimum in the rate of the BQ/BQH_2 couple that is observed in the region of pH 4. The view that the protonation steps are fast for Q/QH_2 couples is supported by other studies, e.g. that of Albery and coworkers[201] discussed below. In passing it can be mentioned that the Tafel plots for the methylbenzoquinone couple[202] are nearly identical to those for BQ/BQH_2 in the pH 0–3.14 range; accordingly, they are also consistent with Laviron's analysis. Also, very recently Laviron's theoretical treatment has been nicely used to sort out the mechanistic details of several catechol/quinone couples at carbon electrodes[203].

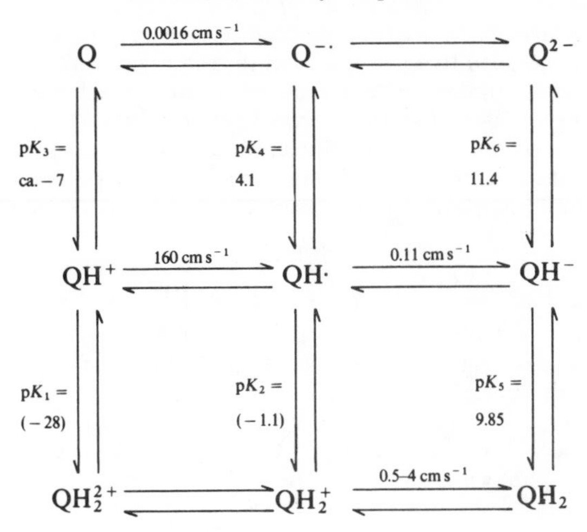

SCHEME 1

The extremely large rate constant given for the $QH^+/QH^{\cdot}$ couple, 160 cm s^{-1}, deserves comment. It is the correspondingly low value for the charge transfer resistance in this arm of the square scheme that allows the mechanism to follow the HeHe pathway at low pH. Whether or not rate constants of this magnitude are physically meaningful is uncertain. On the other hand, the value of 0.0016 cm s^{-1} for the apparent rate constant of the $Q/Q^{-\cdot}$ couple is not too far out of line with the values reported in non-aqueous solvents (see Section III.A). This is particularly true if one considers the results of Rüssel and Jaenicke[23, 27, 87] on the solvent and electrolyte dependence of k_1° *and* the picture of the Pt/solution interface that emerges from Hubbard's work. As realized by Laviron[10, 204], in aqueous buffers, electron transfer must occur through a compact adsorbed layer of quinone, which should decrease the observed rate constant. This unification of the literature is a satisfying aspect of Laviron's contribution.

As is often the case, however, the situation may be even more complex. In a more recent paper[205], theory has been developed that considers the three direct disproportionation reactions in the square scheme and the eight 'cross' disproportionation reactions, e.g. equation 17.

$$QH^{\cdot} + QH_2^{+\cdot} \rightleftharpoons QH^+ + QH_2 \qquad (17)$$

As has been realized for some time, the second electron transfer step in ECE mechanisms often occurs via a solution redox reaction in the diffusion layer. Laviron's analysis[205], for example, indicates that at low pH (< -1), the HeHe pathway actually takes place via equation 17 above. Around pH 5, equation 18 takes place in the diffusion layer; and at higher pH,

$$QH^{\cdot} + Q^{-\cdot} \rightleftharpoons QH^- + Q \qquad (18)$$

disproportionation of $Q^{-\cdot}$ is indicated. The important point here is that the overall rate is determined by a complex function of the pK_as, the individual apparent rate constants and the concentration ratios. In principle since quinone redox chemistry in solution may involve direct hydride transfer[206, 207], additional complexities may be operative.

Others have postulated rate-determining electron transfer steps between adsorbed states of quinone species coupled to equilibrium protonation reactions[188, 208, 209].

In an ingenious experiment, the electrochemistry of photochemically generated QH˙ radicals was studied at an optically transparent Pt electrode[201]. Owing to their favourable solubility, disulfonated anthraquinone couples were employed in the pH range 0–6. In the presence of both the quinone and hydroquinone forms, neutral QH˙ radicals could be generated in the diffusion layer by the sequence of equations 19 and 20,

$$AQ + h\nu \rightarrow AQ^* \tag{19}$$

$$AQ^* + AQH_2 \rightarrow 2\,AQH^{\cdot} \tag{20}$$

and their current–potential curves measured. Plots of E^* vs. pH, where E^* is the zero photocurrent potential, exhibited breaks corresponding to the transition between eHHe and eHeH mechanisms at pH = 3.2 and 3.9 for 9,10-anthraquinone-2,6-disulfonate and -1,5-disulfonate, respectively. At pH 1.12, where the mechanism is eHHe, a reasonable value of 0.0015 cm s^{-1} was obtained for k_1°. In an analogous experiment, the pH dependence of the photocurrent at a *p*-type GaP semiconductor indicated a pK_a-like transitions for BQ solutions at pH 5.5[210]. It was suggested that the initial step involves electron transfer from the conduction band of the photocathode and protonation to form BQH, followed by protonation and charge transfer by hole injection into the valence band, i.e. a HeHe or eHHe mechanism.

Coulostatic techniques have been applied to the 9,10-phenanthrenequinone couple adsorbed on graphite electrodes in 1 M $HClO_4$[188]. The analysis indicated that the rate-determining step is the electron transfer for the $QH^+/QH^{\cdot}$ couple, presumably in the adsorbed state, and that the proton transfers are fast. (It can be noted that the slow process may involve adsorption–desorption steps bracketing electron transfer to a solvated species.) On the other hand, for noradrenaline in 1 M H_2SO_4 at carbon paste and glassy carbon electrodes the rate-determining step in the two-electron oxidation was reported to be the deprotonation of the radical cation[211].

B. Electrochemsitry of Modified Electrodes

1. *Covalently bonded surface quinones*

In the last decade, modified electrodes have been the focus of attention in many electrochemical laboratories. The authoritative reviews of Royce W. Murray[212, 213] should be consulted for details. Quinone couples have occupied an important position in the efforts to modify electrode surfaces with electroactive groups owing to their well behaved and characterized electrochemical response, and to their possible chemical specificity. The promises of the proponents of modified electrodes, which have often involved combination of these properties, have been realized in part in several examples involving surface quinone functionalities.

The distinction between this section and the above (IV.A.1) on adsorbed Q/QH_2 layers is obviously diffuse. Qualitatively, one can imagine increasing levels of interaction with the electrode surface as follows: physical adsorption < chemical adsorption < covalent bonding. The latter includes the situation where the quinone functionality is part of the surface structure of the electrode itself, which can be the case for ordinary carbon materials. The electrochemical responses for the above modes of confinement at the electrode/solution interface can be quite similar.

Tse and Kuwana[214] demonstrated electrocatalysis of dihydronicotinamide adenosine diphosphate (NADH) oxidation at pyrolytic graphite electrodes modified by covalently attached *ortho*-quinones (3,4-dihydroxybenzylamine or dopamine), via equations 21 and 22.

$$\text{-}QH_2 \xrightarrow{-2e^-} \text{-}Q + 2H^+ \tag{21}$$

$$\text{-}Q + NADH + H^+ \rightarrow \text{-}QH_2 + NAD^+ \tag{22}$$

In this manner, oxidation of the NADH substrate takes place near the reversible potential of the Q/QH_2 couple, ca. 0.2 V more negative than the irreversible oxidation of NADH under the same conditions (pH 7, phosphate buffer, 0.05 V s^{-1} sweep rate) at a bare electrode. This paper is an important conceptual and experimental advance that drew on previous work of Kuwana's group, and that of others. Previously, they had reported that surface quinone functions could electrocatalyze ascorbate oxidation[215] and the above catechol derivatives were found to enter into a homogeneous electrocatalytic cycle with NADH. The next step was therefore to attach the catechol directly to the electrode surface.

Tse and Kuwana[214] achieved electroactive surface coverages of the attached quinone of 0.13 nmol cm^{-2} (130 A^2 per site) by the following procedure. Radiofrequency plasma oxidation of the surface of a pyrolytic graphite electrode[216] produced surface carboxylic groups; this surface was then treated with the amine using dicyclohexylcarbodiimide as a coupling agent[217] to form $\text{-}C(O)NHCH_2QH_2$ surface groups. In later work[218], coupling to surface acid chloride groups has been exploited to attach 2-*N*-methylanthraquinone groups to glassy carbon surfaces at a slightly higher coverage.

Other attempts to attach quinones to surfaces have been less successful[219, 220]. In one report[220] a possible orientation effect of the Lane and Hubbard type[167] was indicated. It must be mentioned that repeated cycling of the Tse and Kuwana electrodes leads to decreased electrocatalytic activity, presumably due to fouling by the products of the NADH oxidation, which limits their utility[221].

Other approaches to the construction of an electrocatalytic interface have been based on the above scheme. Graphite electrodes modified with a high surface coverage (ca. 20 A^2 per molecule) adsorbed layer of 4-[2-(2-naphthyl)vinyl]catechol exhibit catalytic activity for NADH oxidation[222], as do carbon paste electrodes containing dissolved *ortho*-quinones in the pasting material[223]. Electrodes modified with polymers containing electroactive pendant hydroquinone groups also show electrocatalysis of NADH oxidation in the manner of Tse and Kuwana[214]. It is difficult to evaluate the relative merits of these approaches, and the use of homogeneous couples, to electrocatalysis on the basis of the existing literature.

There is some indication that the behavior of certain redox couples at carbon electrodes is related to Q/QH_2 groups on the graphite surface. As mentioned above[215], the oxidation of ascorbic acid was illustrated using surface Q/QH_2 groups; hydrogen reduction of the carbon used to prepare carbon paste electrodes increases the irreversibility of the BQ/BQH_2 couple[225]; and glassy carbon modified with carbon black shows electrocatalysis for NADH oxidation, presumably via surface Q/QH_2 groups[226]. Very fast surface exchange rates have been recently reported for quinone groups on oxidized carbon electrodes[227]. In this study it was shown that the electrode kinetics of quinone and phenazine couples in solution were markedly influenced by the oxidation state of the carbon electrodes. It can be speculated that interactions of this type may be responsible for the differences between the electrochemistry of adsorbed quinone layers on graphite and platinum electrodes.

The presence of electroactive quinone functions on carbon electrodes has been known for some time[228–231]. The paper of Hallum and Drushel, for example, accounts for the polarography of carbon black suspensions in DMF by means of reduction of surface quinone functions to $Q^{-\cdot}$ and QH^{-}[229]. Some of the initial experiments on the chemical modification of glassy carbon and graphite electrodes were conceived with hydroquinone-like surface groups in mind[232], and 1,2-naphthoquinone groups were invoked to explain the ESCA spectra and electrochemistry of pyrolytic graphite[216]. The paper of Schreurs and coworkers[227] is especially interesting. They detected surface quinone functions using phase-sensitive AC voltammetry and were able to identify 1,2-naphthoquinone and 9,10-phenanthenequinone surface groups. Poorly defined waves for $Q/QH^-/QH_2$ couples were seen at −0.05 V, −0.16 V and −0.5 V vs SCE in pH 7 phosphate buffer. In a convincing experiment the quinone waves were replaced by phenazine surface waves after the electrodes were refluxed in alcohol solutions of *o*-phenylenediamine. This result has been confirmed by Kuwana and his colleagues who also presented ESCA evidence that the quinone groups represent only a minor fraction of the surface oxygen functionality[233, 234].

Another example of electrocatalysis with quinone modified electrodes is that of Wrighton and coworkers on the production of hydrogen peroxide in aqueous solutions[235, 236]. These workers attached a naphthoquinone group to tungsten, platinum, and *p*-WS_2 electrodes by means of the derivative **20** possessing a surface active $Si(OMe)_3$ group. The resulting electrodes showed dramatic enhancement of the two-electron

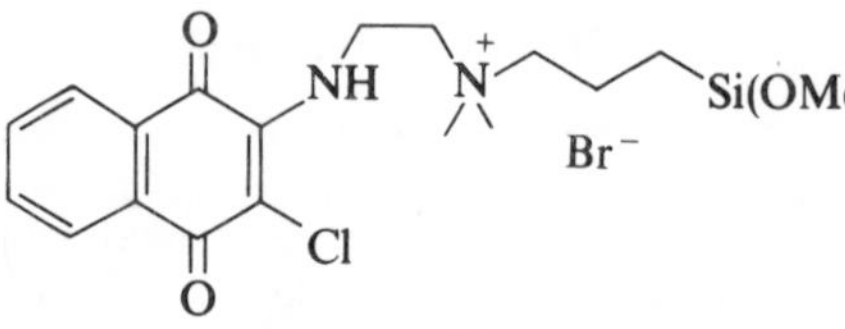

(20)

reduction wave of O_2 at the potential of the attached quinone couple, indicating that a scheme analogous to that of Tse and Kuwana[214] was operative. The solution reaction of oxygen with electrogenerated hydroquinone had been studied earlier[237]. Greater than 90% coulomb efficiency was reported for the generation of H_2O_2 at the quinone-modified electrode. Using the *p*-WS_2 photocathode modified with the surface quinone, O_2 reduction was seen at potentials 0.6 V more positive than the formal potential of the O_2/H_2O_2 couple upon illumination with visible light. In a novel extension of this approach, silica and alumina particles were derivatized using **20**, and oxygen reduction carried out via electrogeneration of mediators at more conventional working electrodes[236]. This approach retains the advantages of electrocatalysis using homogeneous mediators and the ability to separate the particles containing the catalytic sites from the electrolysis solution. Up to 0.1 M H_2O_2 in 0.1 M KCl was produced in this system.

The catalytic reduction of oxygen by AQ/AQH_2 couples has also been reported at *p*-type semiconductor electrodes, *p*-Si and *p*-WSe by Keita and Nadjo[238], who found high efficiencies for both the photochemical and electrochemical reduction, the latter at glassy carbon electrodes. On the other hand, low efficiency was reported for the formation of H_2O_2 at *p*-Si photocathode using 2-*t*-butyl-9,10-anthraquinone in acetonitrile[237].

Others have seen electrocatalysis for the two-electron reduction of oxygen to hydrogen peroxide at quinone modified electrodes. These include the crystalline chloranil or 2,6-diamino-9,10-anthraquinone electrodes studied by Roullier and colleagues[239] and several of the quinone polymer modified electrodes described in the next section.

2. Polymeric quinone films

One strategy for the fabrication of chemically specific interfaces is to coat electrode surfaces with thin films of redox polymers containing electroactive groups, either pendant to, or as part of the polymer backbone. The electrochemistry of several polymeric quinones has been studied in this manner in the last few years[240–244]. Electrocatalytic activity has been exhibited in some cases, e.g. for the reduction of O_2 at electrodes modified by a polymer containing pendant AQ groups attached to an ethenimine backbone[242], reduction of O_2 in H_2SO_4 at a poly(NQ) surface[245] and the NADH oxidation at surfaces containing dopamine groups attached to poly(methyl methacryloylate) mentioned above[224].

In solution, electroactive polymers can exhibit electrochemistry characteristic of the redox groups incorporated into the polymer structure. In the limiting case of negligible interaction between the groups, reversible voltammetric responses are observed for polymers containing repeating units derived from reversible monomer couples[246], although diffusion currents will be attenuated on account of the smaller diffusion coefficients of macromolecular species. This is the situation for anthraquinone-substituted polymers of known molecular weight in DMF solution[247]. For this system, log(peak current) was a linear function of the molecular weight in accord with the Mark–Houwink equation.

When electroactive polymers are coated on electrodes, however, charge transport through the polymer film may be controlled by electron hopping steps between redox sites, ion migration, polymer motions, or combinations of these phenomena. Accordingly, polymers containing reversible quinone functionalities may exhibit only limited electroactivity, or be entirely inactive when coated as a film on a substrate electrode. This is demonstrated by the catechol polymer of Degrand and Miller[224], a copolymer of methacrylic acid and methacrylamide with pendant catechol groups, **21** ($x = 0.41$), for which only a few equivalent monolayers are electroactive unless small mediator molecules are present that can penetrate the polymer network and shuttle electrons between redox sites[248, 249]. The catalytic efficiency of these electrodes is dependent on electrode loading, increasing for thin films and then decreasing for thick films[224].

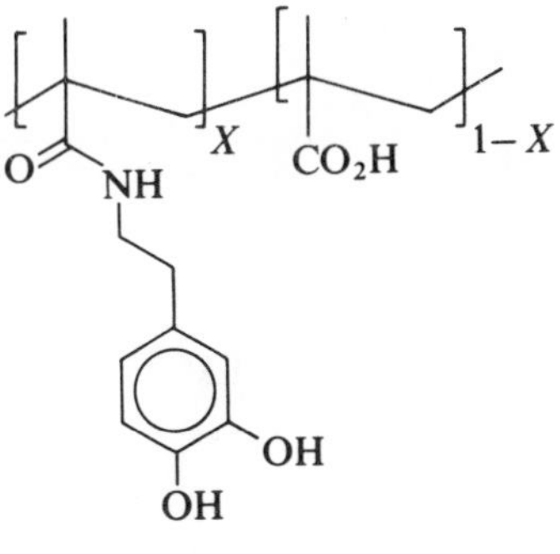

(21)

The need for a good solvent to impart polymer chain flexibility, and thereby facilitate charge transport, was emphasized by Degrand and Miller[241]. This has repeatedly been noted by researchers studying electroactive polymer quinone films[248, 250, 251]. Also, solution pH will dramatically alter the extent of electroactivity of quinone polymer films[241, 250, 252, 253]. It is likely that these pH effects are due to the complex nature of the Q/QH_2 redox reactions that are required to transport charge through the films (see the above discussion of Laviron's analysis[10, 205] of the BQ/BQH_2 kinetics in Section IV.A.2). In the related example of a tetracyanoquinodimethane polymer film[253], the pH dependence of the charge transport process was correlated with pK values of the reduced redox sites.

Substrate surface effects have been noted[245, 252, 254]. Roughening the surface or anodization can increase the charge capacity of adsorbed polymeric quinones[252] and greater electroactivity is sometimes seen on carbon substrates than on platinum[245].

DuBois and coworkers[245] have reported that oxidation of 5-hydroxy-1,4-naphthoquinone in MeOH/NaOH leads to insoluble electroactive poly(quinone) films that display good electrochemistry when transferred to other solutions. This is significant since *in situ* electropolymerization is a very attractive method for preparation of electroactive polymer films in their conducting forms, which has been widely exploited since Diaz and coworkers first reported the electrosynthesis of metallic-like polypyrrole films[255]. Importantly, the film obtained from 5-hydroxy-2-methyl-1,4-naphthoquinone is stable in H_2SO_4 and alkaline solutions[245].

Examples of chemical and electrochemical modification of polymer electrodes have appeared in recent years. Arai and coworkers[256] have described an elegant synthetic procedure, Scheme 2, for the conversion of a polymeric quinone surface film to a mercaptohydroquinone polymer. The resulting surface shows selectivity for metal ions and serves as a potentiometric sensor for Ag^+, Hg^{2+}, Cu^{2+}, Cd^{2+} and Pb^{2+} at the 10^{-3}–10^{-5} M level. Interestingly, the polymer film quinone couple remains electroactive at ca. 0.3 V vs. SCE after formation of the metal ion complexes. Polymer films of this type should find applications in electroanalysis following the ideas of Abruna[257].

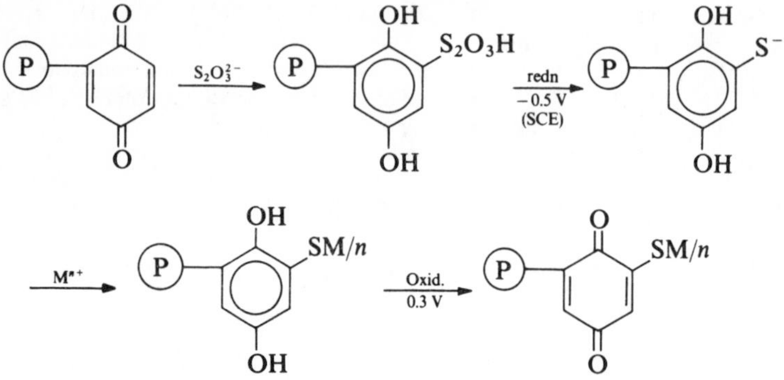

SCHEME 2

3. *Electrochemistry at other modified electrodes*

BQH_2 oxidation in neutral solution has been carried out at Nafion polymer film electrodes loaded with $Ru(bipyridyl)_3^{2+}$ ions[258]. Decreased overpotentials for the oxidation of catechol derivatives were observed at glassy carbon electrodes containing micron size alumina particles on the surface[259]. The alumina particles probably function in part as getter materials for removal of surface contaminants. Catechol phosphate is oxidized under diffusion control at enzyme electrodes prepared by attaching alkaline phosphatase covalently to a carbon surface[260]. The cyanuric acid method of Bourdillon[261] was employed for the surface modification.

The reaction of BQH_2 at a AgX 'electrode'

$$2AgX + BQH_2 \rightarrow 2Ag + BQ + 2HX \qquad (23)$$

consisting of a photographic film disk (equation 23), was monitored by the reduction of BQ at a ring in a ring–disk electrode assembly[262].

4. Composite electrodes

Several reports have appeared describing electrodes containing Q/QH_2 couples, pasted or bonded with conducting materials, and used as working electrodes in voltammetric or electrolysis experiments. Typical of these, which are termed 'composite electrodes' in this review, are the Alt electrodes[263–266]. These are constructed from quinone–graphite mixtures (ca. 1:1 wt ratio) pressed between graphite felt disks. In aqueous electrolyte solutions very large, and often quite sharp, voltammetric peaks are observed. Sizeable coulombic charges are measured indicating that a significant number of equivalent monolayers of quinone molecules are electrolyzed in a voltammetric sweep. Since insoluble quinone couples are generally employed, the electrolyzed layers remain intact upon electrochemical cycling. In some cases, double peaks are observed which have been attributed to the formation of quinhydrone, or possibly semiquinone, intermediates. These and related electrodes have usually been touted as positive electrodes for battery applications[266, 267].

Other quinone composite electrodes have been fabricated that exhibit similar behavior. These include electrodes containing various quinones and carbon black[267–270], 2,3-dichloro-5,6-dicyanobenzoquinone bonded with poly(styrene)[271], poly(aminoquinone) and graphite[272] and hydroquinone–phenol–formaldehyde resins mixed with acetylene black and Teflon and compacted on a Ni mesh[273]. The latter two electrodes could also be classified as polymer electrodes.

The above composite electrodes containing insoluble monomeric quinones are closely related to the crystal deposit electrodes studied by Laviron and coworkers[239, 274]. Similar voltammetric responses are observed, which are attributed to strong interactions between adsorbate molecules for the better defined crystalline electrodes. The voltammograms are characterized by marked hysteresis between the anodic and cathodic waves, which is independent of the sweep rate. It is suggested that the electrode process involves dissolution, electron transfer to the quinone presumably in some partially solvated state and then recrystallization[274].

C. Coupled Chemical Reactions

An interesting literature has developed concerning the electrochemistry of antitumor antibiotics (**22**) and related anthracycline derivatives. Usually mercury working electrodes are employed for these studies, and often pronounced adsorption effects are evident in the published voltammograms. At the dropping mercury electrode in pH 7.1 phosphate buffer,

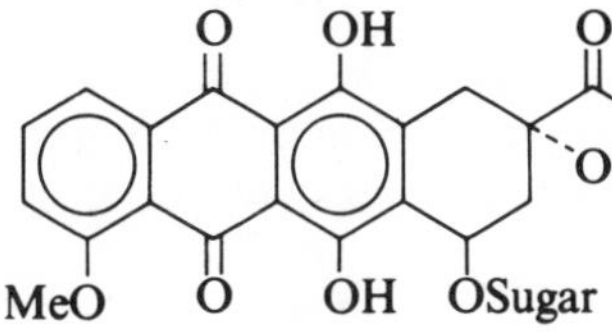

O OH O
OH
MeO O OH OSugar

(22)

loss of the sugar residue occurs upon reduction to the hydroquinone form[154,155]. The mechanism of the glycoside elimination has been studied further under similar conditions by Malatesta and coworkers[275], who found that the reaction was under stereochemical control. Glycoside bonds that have small dihedral angles with respect to the π-orbitals of the AQ ring system undergo rapid cleavage. The mechanism of this follow-up reaction has also been studied in DMF[115,153], as mentioned in Section III.E.

The surface nature of adriamycin electrochemistry at mercury in acetate buffer has been examined by Kano and colleagues[276,277], who analyzed the cyclic voltammograms in terms of a surface disproportionation reaction of the semiquinone form. Extensive adsorption on the mercury surface was indicated, corresponding to a maximum coverage of ca. 150 A^2 per molecule ascribed to a flat configuration. The pH dependence of the surface wave permitted evaluation of the pK_as for the dissociation of a hydroxyl group for the adsorbed quinone, semiquinone and hydroquinone forms; the values obtained, respectively, were 8.53 ± 0.09, 6.93 ± 0.08 and 6.83 ± 0.09. Wave shape changes were noted when the electrode was set at potentials negative of the $E_{1/2}$ value, which may be related to the loss of the sugar residue.

Anthracycline drugs such as adriamycin bond strongly to DNA molecules[278]. This property and the surface electrochemical behavior of the drugs permit electrochemical DNA binding studies to be carried out by measuring the decrease of the quinone reduction wave in the presence of DNA[279,280].

Mitomycin B and C, heterocyclic quinone antibiotics containing an aziridine ring, have been studied in aqueous solution at 37.5°C. The opening of the aziridine ring in the reduced form to generate a new Q/QH_2 couple was followed by cyclic voltammetry[281,282]. The air oxidation of vitamin K_5, 4-amino-2-methyl-1-naphthol hydrochloride, to H_2O_2 and 2-methylnaphthoquinone has been studied in aqueous buffers by polarography[283]. Ubiquinone-10 solubilized in sodium dodecyl sulfate micelles gives diffusion controlled polarographic waves which permits an estimate of the radii of the micelles to be made[284].

Young and coworkers have published several papers on the intramolecular cyclization of electrogenerated *ortho*-quinones derived from dopamine and related molecules[285–287]. These catechol amines oxidize in aqueous acidic buffers at carbon paste electrodes via an ECC mechanism, where the second two-electron transfer takes place in the diffusion layer. Scheme 3, based on that proposed by Hawley and colleagues[288] in a pioneering paper,

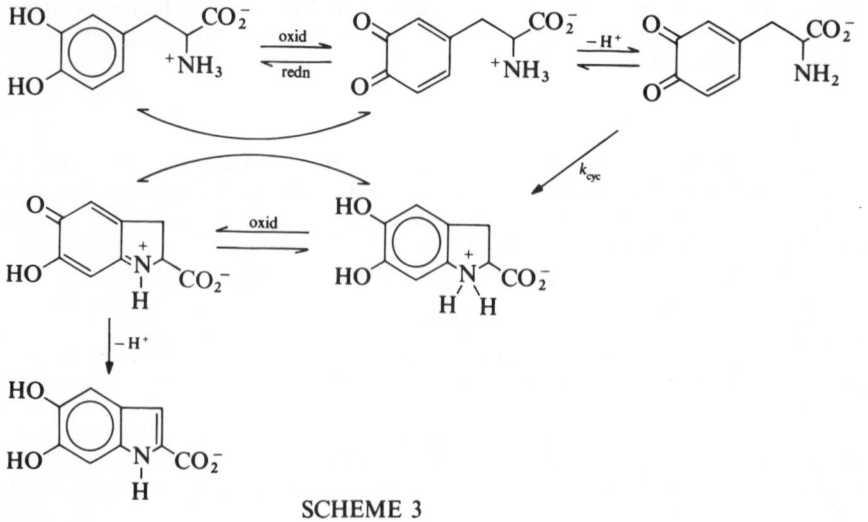

SCHEME 3

shows the pathway from DOPA to 5,6-dihydroxyindole[285]. Kinetic parameters of the cyclizations, which occur on the cyclic voltammetric time-scale, have been reported for several derivatives of DOPA[286, 287]. This intramolecular cyclization of dopaquinone has also been studied by Brun and Rosset[289] above pH 4 by voltammetric techniques. In $HClO_4$ media hydration of the *ortho*-quinones generated from DOPA, dopamine, adrenalin and pyrocatechol was observed. The reaction of the chloropromazine radical cation with several catechols, including dopamine, was studied using glancing angle incidence reflection techniques by Mayansky and McCreery[290]. The rate-determining step was found to be the initial electron transfer between the catechol and the radical cation.

Electrochemically generated *ortho*-quinones will react with nucleophiles other than water purposely added to the electrolysis solution. Adams and his students[291] measured the rates of addition of various nucleophiles with the quinones derived from dopamine and 4-methylcatechol and discussed their relationship (or lack thereof) to metabolic pathways. Adams has previously reviewed the relevance of these electrochemical results to problems in pharmaceutical chemistry[292].

In some cases, e.g. the use of nucleophiles such as dimedone in equation 24[293], the electrode reactions can have synthetic utility. In the absence of nucleophiles, dimeric products can be formed in related electrode processes[294].

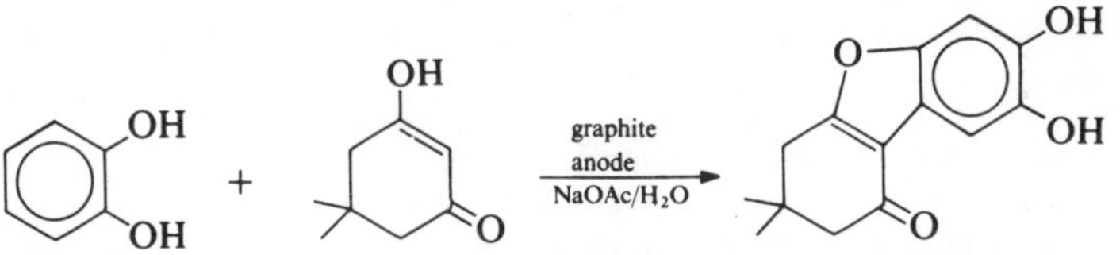

graphite anode, $NaOAc/H_2O$ (24)

At pH 2 a two-electron, two-proton oxidation of tetrahydroxy-1,4-benzoquinone (equation 25), occurs on mercury electrodes to give rhodizonic acid, **23**, which exists as the dihydrate **24**[295]. At pH < 1, a four-electron reduction of rhodizonic acid is seen, which

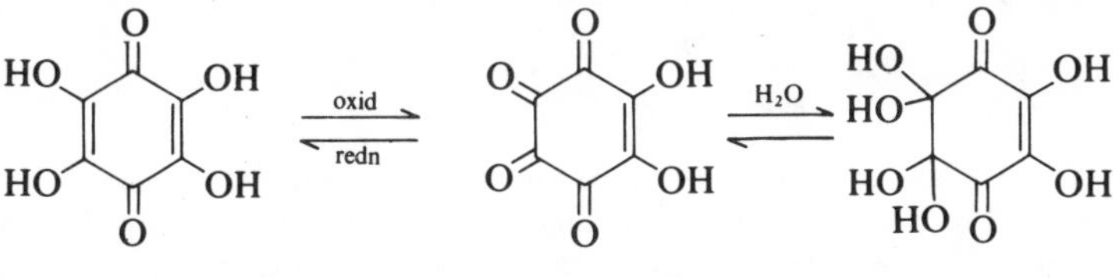

oxid / redn; H_2O (25)

(23) **(24)**

proceeds to hexahydroxybenzene[295]. Similar electrode reactions have been reported for 1,2,3,4-tetraoxonaphthalene[296].

Oxidation of BQH_2 in the presence of sulfite leads to the monosulfonated quinone in an ECE process[297]. For the 1,4-addition of thiols to the 3 position of 2-methyl-1,4-naphthoquinone in aqueous ethanolic buffers, good agreement was obtained between kinetic measurements using rotating disk and stopped-flow methods[298]. Berg and coworkers have continued their studies of β-hydroxyalkylaminobenzoquinone/quinol cyclization equilibria using classical Koutecky polarographic analysis and temperature-jump methods[299].

On platinum or silver cathodes, AQ is reduced to anthrone in concentrated H_2SO_4 with ca. 70% current efficiency[300].

V. PHOTOELECTROCHEMISTRY OF QUINONES

Quinone couples have often been employed in experimental schemes in which electromagnetic radiation is coupled to the flow of electrical current in electrochemical cells. For the sake of completeness several recent studies of this type will be mentioned here. However, photoelectrochemistry, which is a diverse area of increasing activity, will not be treated in any depth in this section. The reader will note that several examples of photoelectrochemical effects were discussed above when there was direct relevance to the conventional electrochemistry of quinones.

Quinones can function as electron acceptors, and hydroquinones as electron donors (or positive hole acceptors), in photochemical reactions where separation of charge occurs. One or both of these steps is generally involved when photocurrents are observed at semiconductor or dye modified electrodes in the classical manner of Tributsch and Gerischer[301]. The coupling of BQH_2 oxidation to the photon flux for ZnO–dye and ZnO–chlorophyll electrodes was demonstrated several years ago[302, 303], and photochemical oxidation of BQH_2 has been exploited at chlorophyll monolayer SnO_2 electrodes[304]. Honda and coworkers have also shown the photoelectrochemical oxidation of BQH_2 at *n*-CdS(single crystal)/rhodamine B dye sensitized disk electrodes[305, 306]. In these studies, the disk was illuminated from the backside and the quinone product was detected at a gold ring electrode. Polycrystalline TiO_2 photoanodes have been employed for the oxidation of BQH_2 and several other substrates with relatively high current efficiency[307].

The use of quinones as electron acceptors in conjunction with chlorophyll (Chl) modified electrodes has been a popular subject in part due to the relevance to photosynthesis where naturally occurring quinones function as acceptor molecules. Takahashi and colleagues have employed this strategy in several studies in which chlorophyll–quinone modified electrodes were prepared with water insoluble quinones that are miscible with chlorophyll[308–310]. Upon illumination of these electrodes, the following photoelectrochemical process can take place (equation 26–28)[310],

$$\mathrm{Chl} \xrightarrow{h\nu} \mathrm{Chl}^* \xrightarrow{\mathrm{Q}} \mathrm{Chl}^+ + \mathrm{Q}^{-\cdot} \tag{26}$$

$$n\mathrm{Chl}^+ + \mathrm{R} \rightarrow n\,\mathrm{Chl} + \mathrm{Ox} \tag{27}$$

$$\mathrm{Q}^{-\cdot} \rightleftharpoons \mathrm{Q} + e^- \tag{28}$$

where Ox and R are components of a redox couple present in solution. For different quinones, or different redox couples, these chlorophyll electrodes can also function as photocathodes[308], as was found earlier for an aqueous Pt(Chl)//quinhydrone(Pt) divided cell[311]. Related to these studies, are the pyranthrone electrodes of Ahuja and coworkers[312], for which both anodic and cathodic photocurrents were observed. An interesting application of these concepts is that of Janzen and Bolton[313] who constructed monolayer assemblies of chlorophyll-A and quinone acceptors sandwiched between Hg and Al electrodes. Upon illumination, electron ejection to the Al electrode took place.

Quinone reduction has been coupled to solvent oxidation, namely methanol conversion to formaldehyde, at illuminated rutile catalysts[314]. Quinone couples in non-aqueous solvents have been used extensively by Bard and coworkers to map the limits of the band gap in a variety of semiconductor materials[315–319].

A description has appeared of a rechargeable solar cell based on the uphill generation of 9,10-anthrahydroquinone-2,6-disulfonate driven by the oxidation of iodide at a semiconductor photoanode (*n*-WSe_2 or *n*-MoS_2) in an electrochemical cell[320, 321]. For light of 632 nm wavelength, conversion efficiencies of ca. 10% were achieved. Coupling to the O_2 reduction to hydrogen peroxide was also shown in this study.

Several quinones have been used to scavenge photoemitted electrons from mercury electrodes in aqueous and non-aqueous solvents[322].

The known photochemistry of the excited state of AQ has been coupled to electrode reactions, e.g. in the work of Bobbitt and Willis, where increased selectivity for photosynthetic reactions is possible by the controlled-potential generation of the ground state precursor[323]. The photochemical oxidation of 2-propanol to acetone has long been known. Fujihira and colleagues[218, 324] achieved photoassisted electrolysis of this process using glassy carbon electrodes modified with covalently attached anthraquinone groups. The quantum efficiency was low, however, presumably due to quenching of the excited state of AQ by the electrode. Studies in this area should be aided by the recently reported solution of the diffusion equation boundary value problem for the photogeneration of AQH_2 in a chronoamperometry mode[325].

VI. REFERENCES

1. J. Q. Chambers, in *The Chemistry of the Quinonoid Compounds* (Ed. S. Patai), John Wiley and Sons, New York, 1974, Chapter 14, pp. 737–791.
2. J. Jacq, *Electrochim. Acta*, **12**, 1345 (1967).
3. W. M. Clark, *Oxidation–Reduction Potentials of Organic Systems*, Williams & Wilkins, Baltimore, 1960.
4. L. Michaelis, *Oxidations–Reductions Potentials*, 2nd edn, Springer Verlag, Berlin, 1933.
5. D. H. Evans, in *The Encyclopedia of the Electrochemistry of the Elements* (Ed. A. J. Bard), Marcel Dekker, New York, 1978, pp. 3–259.
6. M. L. Fultz and R. A. Durst, *J. Electroanal. Chem.*, **140**, 1 (1982).
7. S. I. Bailey, I. M. Ritchie and F. R. Hewgill, *J. Chem. Soc., Perkin Trans. 2*, 645 (1983).
8. E. Laviron, *J. Electroanal. Chem.*, **124**, 9 (1981).
9. E. Laviron, *J. Electroanal. Chem.*, **146**, 15 (1983).
10. E. Laviron, *J. Electroanal. Chem.*, **164**, 213 (1984).
11. O. Hammerich and V. D. Parker, *Acta Chem. Scand.*, **B36**, 63 (1982).
12. J. L. Roberts Jr, H. Sugimoto, W. C. Barrette Jr and D. T. Sawyer, *J. Am. Chem. Soc.*, **107**, 4556 (1985).
13. S. I. Bailey and I. M. Ritchie, *Electrochim. Acta*, **30**, 3 (1985).
14. J. Pilar, I. Buben and J. Pospisil, *Coll. Czech. Chem. Commun.*, **37**, 3599 (1972).
15. O. S. Ksenzhek, S. A. Petrova, S. V. Oleinik, M. V. Kolodyazhnyi and V. Z. Moskovskii, *Electrokhim.*, **13**, 182 (1977).
16. O. S. Ksenzhek, S. A. Petrova, S. V. Oleinik, M. V. Kolodyazhnyi and V. Z. Moskovskii, *Electrokhim.*, **13**, 48 (1977).
17. H. Huck, *Ber. Bunsenges. Phys. Chem.*, **87**, 945 (1983).
18. G. M. Proudfoot and I. M. Ritchie, *Aust. J. Chem.*, **36**, 885 (1983).
19. O. S. Ksenzhek, S. A. Petrova and S. V. Oleinik, *Elektrokhim.*, **14**, 1816 (1978).
20. O. S. Ksenzhek, S. A. Petrova and M. V. Kolodyazhny, *J. Electroanal. Chem.*, **141**, 167 (1982).
21. J. S. Jaworski, E. Lesniewska and M. K. Kalinowski, *J. Electroanal. Chem.*, **105**, 329 (1979).
22. V. Gutmann, *The Donor–Acceptor Approach to Molecular Interactions*, Plenum Press, New York, 1978.
23. C. Rüssel and W. Jaenicke, *J. Electroanal. Chem.*, **199**, 139 (1986).
24. L. Ya. Kheifets and N. A. Sobina, *Elektrokhim*, **11**, 208 (1975).
25. J. H. Wilford and M. D. Archer, *J. Electroanal. Chem.*, **190**, 271 (1985).
26. C. G. Swain, M. S. Swain, A. L. Powell and S. Alumni, *J. Am. Chem. Soc.*, **105**, 502 (1983).
27. C. Rüssel and W. Jaenicke, *Z. Phys. Chem.*, **139**, 97 (1984).
28. T. Nagaoka and S. Okazaki, *J. Phys. Chem.*, **89**, 2340 (1985).
29. S. Rondinini, C. Confalonieri, P. Longhi and T. Mussini, *Electrochim. Acta*, **30**, 981 (1985).
30. P. Zuman, *Substituent Effects on Organic Polarography*, Plenum Press, New York, 1967, Chapter VIII.
31. V. T. Glezer, Ya. F. Freimanis, Ya. P. Stradyn and Ya. Ya. Dregeris, *Zh. Obshch. Khim.*, **49**, 1379 (1979).
32. V. T. Glezer, J. Stradins, J. Freimanis and L. Baider, *Electrochim. Acta*, **28**, 87 (1983).

33. L. Pospisil and V. T. Glezer, *J. Electroanal. Chem.*, **153**, 263 (1983).
34. A. Heesing, U. Kappler and W. Rauk, *Liebigs Ann. Chem.*, 2222 (1976).
35. E. Muller and W. Dilger, *Chem. Ber.*, **106**, 1643 (1973).
36. C.-Y. Li, M. L. Caspar and D. W. Dixon Jr, *Electrochim. Acta*, **25**, 1135 (1980).
37. E. Muller and W. Dilger, *Chem. Ber.*, **107**, 3946 (1974).
38. V. D. Bezuglyi, V. A. Shapovalov and V. Ya. Fain, *Zh. Obshch. Khim.*, **46**, 696 (1976).
39. J. S. Jaworski, E. Lesniewska and M. K. Kalinowski, *Pol. J. Chem.*, **54**, 1313 (1980).
40. G. Klopman and N. Doddapaneni, *J. Phys. Chem.*, **78**, 1820 (1974).
41. J. E. Kuder, D. Wychik, R. L. Miller and M. S. Walker, *J. Phys. Chem.*, **78**, 1714 (1974).
42. V. D. Bezuglyi, L. V. Shkodina, V. A. Shapovalov, A. F. Korunova and V. Ya. Fain, *Zh. Obshch. Khim.*, **48**, 2086 (1978).
43. V. A. Shapovalov, V. Ya. Fain and V. D. Bezuglyi, *Zh. Obshch. Khim.*, **46**, 2296 (1976).
44. A. E. Feiring and W. A. Sheppard, *J. Org. Chem.*, **40**, 2543 (1975).
45. G. J. Gleicher, D. F. Church and J. C. Arnold, *J. Am. Chem. Soc.*, **96**, 2403 (1974).
46. S. D. Popescu and E. Barbacaru, *Anal. Lett.*, **18A**, 947 (1985).
47. P. Carsky, P. Hobza and R. Zahradnik, *Coll. Czech. Chem. Commun.*, **36**, 1291 (1971).
48. S. Chatterjee, *J. Chem. Soc. (B)*, 2194 (1971).
49. K. M. C. Davis, P. R. Hammond and M. E. Peover, *Trans. Faraday Soc.*, **61**, 1516 (1965).
50. E. Muller and W. Dilger, *Chem. Ber.*, **107**, 3957 (1974).
51. B. R. Clark and D. H. Evans, *J. Electroanal. Chem.*, **69**, 181 (1976).
52. R. D. Rieke, W. E. Rich and J. H. Ridgeway, *J. Am. Chem. Soc.*, **93**, 1962 (1971).
53. R. Breslow, D. R. Murayama, S. Murahashi and R. Grubbs, *J. Am. Chem. Soc.*, **95**, 6688 (1973).
54. H. N. C. Wong, F. Sondheimer, R. Goodin and R. Breslow, *Tetrahedron Lett.*, 2715 (1976).
55. R. Breslow, D. R. Murayama, R. Drury and F. Sondheimer, *J. Am. Chem. Soc.*, **96**, 249 (1974).
56. R. Breslow, *Pure Appl. Chem.*, **40**, 493 (1974).
57. M. Iyoda and M. Oda, *Tetrahedron Lett.*, **24**, 1727 (1983).
58. T. Watabe and M. Oda, *Chem. Lett.*, 1791 (1984).
59. J. Tsunetsugu, T. Ikeda, N. Suzuki, M. Yaguchi, M. Sata, S. Ebine and K. Morinaga, *J. Chem. Soc., Chem. Commun.*, 28 (1983).
60. M. Hirama, Y. Koyama, Y. Shoji and S. Ito, *Tetrahedron Lett.*, 2289 (1978).
61. M. Hirama, Y. Fukazawa and S. Ito, *Tetrahedron Lett.*, 1299 (1978).
62. S. Ito, Y. Shoji, H. Takeshita, M. Hirama and H. Takahashi, *Tetrahedron Lett.*, 1075 (1975).
63. L. M. Doane and A. J. Fatiadi, *Angew. Chem. Int. Ed.*, **21**, 635 (1982).
64. H. Heberer, H. Schubert, H. Matschiner and B. Lukowczyk, *J. Prakt. Chem.*, **318**, 635 (1976).
65. R. P. Buck and D. E. Waggoner, *J. Electroanal. Chem.*, **115**, 89 (1980).
66. G. A. Russell, N. K. Suleman, H. Iwamura and O. W. Webster, *J. Am. Chem. Soc.*, **103**, 1560 (1981).
67. L. Mandell, S. M. Cooper, B. Rubin, C. F. Campana and R. A. Day Jr, *J. Org. Chem.*, **48**, 3132 (1983).
68. K. Komaster and R. West, *J. Chem. Soc., Chem. Commun.*, 570 (1976).
69. L. A. Wendling and R. West, *J. Org. Chem.*, **43**, 1577 (1978).
70. L. A. Wendling and R. West, *J. Org. Chem.*, **43**, 1573 (1978).
71. R. West and D. C. Zecher, *J. Am. Chem. Soc.*, **92**, 155 (1970).
72. S. K. Koster and R. West, *J. Org. Chem.*, **40**, 2300 (1975).
73. J. L. Benham and R. West, *J. Am. Chem. Soc.*, **102**, 5054 (1980).
74. D. E. Wellman and R. West, *J. Am. Chem. Soc.*, **106**, 355 (1984).
75. D. E. Wellman, K. R. Lassila and R. West, *J. Org. Chem.*, **49**, 965 (1984).
76. T. W. Rosanske and D. H. Evans, *J. Electroanal. Chem.*, **72**, 277 (1976).
77. A. Capon and R. Parsons, *J. Electroanal. Chem.*, **46**, 215 (1973).
78. K. J. Vetter, *Z. Elektrochem.*, **56**, 797 (1952).
79. K. J. Vetter, *Electrochemical Kinetics*, Academic Press, New York, 1967, p. 483.
80. H. Kojima and A. J. Bard, *J. Am. Chem. Soc.*, **97**, 6317 (1975).
81. R. Samuelsson and M. Sharp, *Electrochim. Acta*, **23**, 315 (1978).
82. M. Sharp, M. Petersson and K. Edström, *J. Electroanal. Chem.*, **95**, 123 (1979).
83. J. O. Howell and R. M. Wightman, *Anal. Chem.*, **56**, 524 (1984).
84. R. S. Nicholson, *Anal. Chem.*, **37**, 1351 (1965).

85. A. F. Diaz, F. A. O. Rosales, J. P. Rosales and K. K. Kanazawa, *J. Electroanal. Chem.*, **103**, 233 (1979).
86. G. Grampp and W. Jaenicke, *Ber. Bunsenges. Phys. Chem.*, **88**, 325 (1984).
87. C. Rüssel and W. Jaenicke, *J. Electroanal. Chem.*, **180**, 205 (1984).
88. C. Rüssel and W. Jaenicke, *J. Electroanal. Chem.*, **200**, 249 (1986).
89. W. R. Fawcett and Yu. I. Kherkats, *J. Electroanal. Chem.*, **47**, 413 (1973).
90. A. Baranski and W. R. Fawcett, *J. Electroanal. Chem.*, **100**, 185 (1979).
91. M. Sharp, *J. Electroanal. Chem.*, **88**, 193 (1978).
92. E. Ouziel and Ch. Yarnitzky, *J. Electroanal. Chem.*, **78**, 257 (1977).
93. V. D. Bezuglyi, L. V. Shkodina, T. A. Alekseeva and V. A. Shapovalov, *Zh. Obshch. Khim.*, **53**, 1889 (1983).
94. M. Arnac and G. Verboom, *Anal. Chem.*, **49**, 806 (1977).
95. A. B. Sullivan and G. F. Reynolds, *J. Phys. Chem.*, **80**, 2671 (1976).
96. A. I. Krynkov, E. P. Platonova and V. A. Krasnova, *Zh. Obshch. Khim.*, **48**, 2583 (1978).
97. V. J. Koshy, V. Swayambunathan and N. Periasamy, *J. Electrochem. Soc.*, **127**, 2761 (1980).
98. M. Baier, J. Duab, H. Hasenhundl, A. Merz and K. M. Rapp, *Angew. Chem. Int. Ed.*, **20**, 198 (1981).
99. I. Turovskis, J. Stradins, J. Volke, J. Freimanis and V. Glezer, *Coll. Czech. Chem. Commun.*, **43**, 909 (1978).
100. J. M. Abd El Kader, A. M. Shams El Din, B. Kastening and L. Holleck, *J. Electroanal. Chem.*, **26**, 41 (1970).
101. E. Theodoridou and A. D. Jannakoudakis, *Z. Phys. Chem. (Wiesbaden)*, **132**, 175 (1982).
102. V. J. Vajgand and R. P. Mihajlovic, *Anal. Chim. Acta*, **152**, 275 (1983).
103. C.-H. Pyun and S.-M. Park, *J. Electrochem. Soc.*, **132**, 2426 (1985).
104. S.-I. Fukuzumi, Y. Ono and T. Keii, *Bull. Chem. Soc. Japan*, **46**, 3353 (1973).
105. B. Keita, I. Kawenoki, J. Kossanyi and L. Nadjo, *J. Electroanal. Chem.*, **145**, 311 (1983).
106. J. Bessard, G. Cauquis and D. Serve, *Electrochim. Acta*, **25**, 1187 (1980).
107. G. Cauquis and G. Marback, *Biochim. Biophys. Acta*, **283**, 239 (1972).
108. G. T. Cheek and R. A. Osteryoung, *J. Electrochem. Soc.*, **129**, 2488 (1982).
109. G. T. Cheek and R. A. Osteryoung, *J. Electrochem. Soc.*, **129**, 2739 (1982).
110. D. E. Bartak and R. A. Osteryoung, *J. Electroanal. Chem.*, **74**, 69 (1976).
111. J.-P. Masson, J. Devynck and B. Tremillon, *J. Electroanal. Chem.*, **64**, 175 (1975).
112. J. Devynck, A. Hadid, P. L. Fabre and B. Tremillon, *Anal. Chim. Acta*, **100**, 343 (1978).
113. M. D. Stallings, M. M. Morrison and D. T. Sawyer, *Inorg. Chem.*, **20**, 2655 (1981).
114. R. M. Wightman, J. R. Cockrell, R. W. Murray, J. N. Burnett and S. B. Jones, *J. Am. Chem. Soc.*, **98**, 2562 (1976).
115. A. Anne and J. Moiroux, *Nouv. J. Chim.*, **9**, 83 (1985).
116. A. Anne and J. Moiroux, *Nouv. J. Chim.*, **8**, 259 (1984).
117. M. E. Peover and J. D. Daves, *J. Electroanal. Chem.*, **6**, 46 (1963).
118. T. Nagaoka, S. Okazaki and T. Fujinaga, *J. Electroanal. Chem.*, **133**, 89 (1982).
119. T. Nagaoka, S. Okazaki and T. Fujinaga, *Bull. Chem. Soc. Japan*, **55**, 1967 (1982).
120. J. O'M. Bockris and A. K. N. Reddy, *Modern Electrochemistry*, Plenum, New York, 1973, pp. 261–265.
121. M. K. Kalinowski, *Chem. Phys. Lett.*, **7**, 55 (1970).
122. J. S. Jaworski and M. K. Kalinowski, *J. Electroanal. Chem.*, **76**, 301 (1977).
123. E. Warhurst and A. M. Wilde, *Trans. Faraday Soc.*, **66**, 2124 (1970).
124. T. Fujinaga, S. Okazaki and T. Nagaoka, *Bull. Chem. Soc. Japan*, **53**, 2241 (1980).
125. M. K. Kalinowski and B. Tenderende-Guminska, *J. Electroanal. Chem.*, **55**, 277 (1974).
126. E. Lesniewska-Lada and M. K. Kalinowski, *Electrochim. Acta*, **28**, 1415 (1983).
127. T. M. Krygowski, *J. Electroanal. Chem.*, **35**, 436 (1972).
128. T. Nagaoka, S. Okazaki and T. Fujinaga, *J. Electroanal. Chem.*, **131**, 387 (1982).
129. V. D. Bezuglyi, L. V. Shkodina and T. A. Alekseeva, *Elektrokhim.*, **19**, 1430 (1983).
130. P. D. Jennakoudakis, P. Karabinas and E. Theodoridou, *Z. Phys. Chem. (N.F.)*, **131**, 89 (1982).
131. A. Desbene-Monvernay, A. Cherigui, P. C. Lacaze and J. E. Dubois, *J. Electroanal. Chem.*, **164**, 157 (1984).
132. T. Boschi, R. Pappa, G. Pistoia and M. Tocci, *J. Electroanal. Chem.*, **176**, 235 (1984).
133. S. E. Jones, L. E. Leon and D. T. Sawyer, *Inorg. Chem.*, **21**, 3692 (1982).

134. A. N. Nesmeyanov, T. A. Pegenova, L. I. Denisovich, L. S. Isaeva and A. G. Makarovskaya, *Dokl. Akad. Nauk, S.S.S.R.*, **230**, 1114 (1976).
135. M. Hiramatsu, K. Shiozaki, T. Fuginami and S. Sakai, *J. Organomet. Chem.*, **246**, 203 (1983).
136. D. H. Evans and D. A. Griffith, *J. Electroanal. Chem.*, **134**, 301 (1982).
137. D. H. Evans and D. A. Griffith, *J. Electroanal. Chem.*, **136**, 149 (1982).
138. H. Bock, B. Hierholzer, F. Vögtle and G. Hollmann, *Angew. Chem. Int. Ed.*, **23**, 57 (1984).
139. K. Sugihara, H. Kamiya, M. Yamaguchi, T. Kaneda and S. Misumi, *Tetrahedron Lett.*, **22**, 1619 (1981).
140. R. W. Wolf Jr and S. R. Cooper, *J. Am. Chem. Soc.*, **106**, 4646 (1984).
141. K. Maruyama, H. Sohmiya and H. Tsukube, *Tetrahedron Lett.*, **26**, 3583 (1985).
142. B. A. Olsen and D. H. Evans, *J. Am. Chem. Soc.*, **103**, 839 (1981).
143. B. A. Olsen, D. H. Evans and I. Agranat, *J. Electroanal. Chem.*, **136**, 139 (1982).
144. D. H. Evans and R. W. Basch, *J. Am. Chem. Soc.*, **104**, 5057 (1982).
145. E. Ahlberg, O. Hammerich and V. D. Parker, *J. Am. Chem. Soc.*, **103**, 844 (1981).
146. O. Hammerich and V. D. Parker, *Acta Chem. Scand.*, **B35**, 395 (1981).
147. D. H. Evans and X. Naixian, *J. Electroanal. Chem.*, **133**, 367 (1982).
148. D. H. Evans and N. Xie, *J. Am. Chem. Soc.*, **105**, 315 (1983).
149. J. M. Saveant, *Acc. Chem. Res.*, **13**, 323 (1980).
150. C. P. Andrieux, C. Blocman, J. M. Dumas-Bouchiat, F. M'Halla and J. M. Saveant, *J. Electroanal. Chem.*, **113**, 19 (1980).
151. D. H. Evans and A. Fitch, *J. Am. Chem. Soc.*, **106**, 3039 (1984).
152. R. L. Blankespoor, A. N. K. Lau and L. L. Miller, *J. Org. Chem.*, **49**, 4441 (1984).
153. A. Anne, F. Bennani, J.-C. Florent, J. Moiroux and C. Monneret, *Tetrahedron Lett.*, **26**, 2641 (1985).
154. G. M. Rao, J. W. Lown and J. A. Plambeck, *J. Electrochem. Soc.*, **125**, 534 (1978).
155. G. M. Rao, J. W. Lown and J. A. Plambeck, *J. Electrochem. Soc.*, **125**, 540 (1978).
156. J. Simonet and H. Lund, *Bull. Soc. Chim., France*, 2547 (1975).
157. R. A. Misra and A. K. Yadav, *Electrochim. Acta*, **25**, 1221 (1980).
158. R. F. Stewart and L. L. Miller, *J. Am. Chem. Soc.*, **102**, 4999 (1980).
159. R. D. Rieke, T. Saji and N. Kujundzic, *J. Electroanal. Chem.*, **102**, 397 (1979).
160. V. D. Bezuglyi and V. A. Shapovalov, *Elektrokhim.*, **13**, 984 (1977).
161. S. Torii, H. Tanaka and S. Nakane, *Bull. Chem. Soc., Japan*, **55**, 1673 (1982).
162. Z. Blum and K. Nyberg, *Acta Chem. Scand.*, **B33**, 73 (1979).
163. J.-L. Brisset, *J. Electroanal. Chem.*, **60**, 217 (1975).
164. W. R. Turner and P. J. Elving, *J. Electrochem. Soc.*, **112**, 1215 (1965).
165. A. T. Hubbard, J. L. Stickney, M. P. Soriaga, V. K. F. Chia, S. D. Rosasco, B. C. Schardt, T. Solomun, D. Song, J. H. White and A. Wieckowski, *J. Electroanal. Chem.*, **168**, 43 (1984).
166. A. T. Hubbard, *Acc. Chem. Res.*, **13**, 177 (1980).
167. R. F. Lane and A. T. Hubbard, *J. Phys. Chem.*, **77**, 1401 (1973).
168. M. P. Soriaga and A. T. Hubbard, *J. Am. Chem. Soc.*, **104**, 2735 (1982).
169. M. P. Soriaga and A. T. Hubbard, *J. Am. Chem. Soc.*, **104**, 3937 (1982).
170. M. P. Soriaga and A. T. Hubbard, *J. Electroanal. Chem.*, **167**, 79 (1984).
171. M. P. Soriaga, P. H. Wilson, A. T. Hubbard and C. S. Benton, *J. Electroanal. Chem.*, **142**, 317 (1982).
172. Y.-P. Gui, M. D. Porter and T. Kuwana, *Anal. Chem.*, **57**, 1474 (1985).
173. M. P. Soriaga and A. T. Hubbard, *J. Phys. Chem.*, **88**, 1089 (1984).
174. V. K. F. Chia, M. P. Soriaga, A. T. Hubbard and S. E. Anderson, *J. Phys. Chem.*, **87**, 232 (1983).
175. M. P. Soriaga, J. H. White and A. T. Hubbard, *J. Phys. Chem.*, **87**, 3048 (1983).
176. M. P. Soriaga, J. L. Stickney and A. T. Hubbard, *J. Electroanal. Chem.*, **144**, 207 (1983).
177. M. P. Soriaga and A. T. Hubbard, *J. Phys. Chem.*, **88**, 1758 (1984).
178. J. H. White, M. P. Soriaga and A. T. Hubbard, *J. Electroanal. Chem.*, **177**, 89 (1984).
179. V. K. F. Chia, M. P. Soriaga and A. T. Hubbard, *J. Electroanal. Chem.*, **167**, 97 (1984).
180. M. P. Soriaga and A. T. Hubbard, *J. Am. Chem. Soc.*, **104**, 2742 (1982).
181. M. P. Soriaga, J. H. White, D. Song and A. T. Hubbard, *J. Phys. Chem.*, **88**, 2284 (1984).
182. M. P. Soriaga, J. H. White, D. Song and A. T. Hubbard, *J. Electroanal. Chem.*, **171**, 359 (1984).
183. W. Jaenicke and H. Kobayashi, *Electrochim. Acta*, **28**, 245 (1983).
184. M. P. Soriaga, D. Song and A. T. Hubbard, *J. Phys. Chem.*, **89**, 285 (1985).

185. D. Song, M. P. Soriaga and A. T. Hubbard, *J. Electroanal. Chem.*, **193**, 255 (1985); *Langmuir*, **2**, 20 (1986).
186. A. P. Brown and F. C. Anson, *Anal. Chem.*, **49**, 1589 (1977).
187. A. P. Brown, C. Koval and F. C. Anson, *J. Electroanal. Chem.*, **72**, 379 (1976).
188. A. P. Brown and F. C. Anson, *J. Electroanal. Chem.*, **92**, 133 (1978).
189. H.-Y. Cheng, L. Falat and R.-L. Li, *Anal. Chem.*, **54**, 1384 (1982).
190. R. P. Baldwin, D. Packett and T. M. Woodcock, *Anal. Chem.*, **53**, 540 (1981).
191. E. N. Chaney Jr and R. P. Baldwin, *Anal. Chem.*, **54**, 2556 (1982).
192. P. Delahay, *Double Layer and Electrode Kinetics*, Interscience, New York, 1965.
193. J. H. White, M. P. Soriaga and A. T. Hubbard, *J. Phys. Chem.*, **89**, 3227 (1985).
194. J. H. White, M. P. Soriaga and A. T. Hubbard, *J. Electroanal. Chem.*, **185**, 331 (1985).
195. G. M. Schmid and T. A. Holmes, *J. Electrochem. Soc.*, **128**, 2582 (1981).
196. K. G. Everett, L. A. Drew, S. J. Ericson and G. M. Schmid, *J. Electrochem. Soc.*, **125**, 389 (1978).
197. W. Jaenicke and H. Kobayashi, *Electrochim. Acta*, **28**, 249 (1983).
198. P. C. Lacaze, M. C. Pham, M. Delamar and J. E. Dubois, *J. Electroanal. Chem.*, **108**, 9 (1980).
199. G. Kokkinidis, *J. Electroanal. Chem.*, **172**, 265 (1984).
200. A. M. Abd. El-Halim, K. Jüttner and W. J. Lorenz, *J. Electroanal. Chem.*, **106**, 193 (1980).
201. W. J. Albery, M. D. Archer, N. J. Field and A. D. Turner, *Faraday Disc. Chem. Soc.*, **56**, 28 (1973).
202. F. Kornfeil, *J. Electrochem. Soc.*, **119**, 1674 (1972).
203. M. R. Deakin, P. M. Kovach, K. J. Stutts and R. M. Wightman, *Anal. Chem.*, **58**, 1474 (1986).
204. E. Laviron, *J. Electroanal. Chem.*, **124**, 19 (1981).
205. E. Laviron, *J. Electroanal. Chem.*, **169**, 29 (1984).
206. B. W. Carlson and L. L. Miller, *J. Am. Chem. Soc.*, **107**, 479 (1985).
207. M. P. Youngblood, *J. Am. Chem. Soc.*, **107**, 6987 (1985).
208. W. J. Plieth, I. Stellmacher and B. Quast, *Electrochim. Acta*, **20**, 335 (1975).
209. T. K. Zolotova, I. V. Shelepin and Y. B. Vasil'ev, *Elektrokhim.*, **9**, 1211 (1973).
210. R. Memming and F. Mollers, *Ber. Bunsenges. Phys. Chem.*, **76**, 609 (1972).
211. W. J. Albery, T. W. Beck, W. N. Brooks and M. Fillenz, *J. Electroanal. Chem.*, **125**, 205 (1981).
212. R. W. Murray, *Acc. Chem. Res.*, **13**, 135 (1980).
213. R. W. Murray, in *Electroanalytical Chemistry*, Vol. 13 (Ed. A. J. Bard), M. Dekker, New York, 1984.
214. D. C.-S. Tse and T. Kuwana, *Anal. Chem.*, **50**, 1315 (1978).
215. J. F. Evans, T. Kuwana, M. T. Henne and G. P. Royer, *J. Electroanal. Chem.*, **80**, 409 (1977).
216. J. F. Evans and T. Kuwana, *Anal. Chem.*, **49**, 1632 (1977).
217. M. Fujihara, A. Tamura and T. Osa, *Chem. Lett.*, 361 (1977).
218. M. Fujihira, S. Tasaki, T. Osa and T. Kuwana, *J. Electroanal. Chem.*, **137**, 163 (1982).
219. S. Mazur, T. Matusinovic and K. Cammann, *J. Am. Chem. Soc.*, **99**, 3888 (1977).
220. M. Sharp, *Electrochim. Acta*, **23**, 287 (1978).
221. C. Ueda, D. C. Tse and T. Kuwana, *Anal. Chem.*, **54**, 850 (1982).
222. H. Jaegfeldt, A. B. C. Torstensson, L. G. O. Gorton and G. Johansson, *Anal. Chem.*, **53**, 1979 (1981).
223. K. Ravichandran and R. P. Baldwin, *J. Electroanal. Chem.*, **126**, 293 (1981).
224. C. Degrand and L. L. Miller, *J. Am. Chem. Soc.*, **102**, 5728 (1980).
225. C. Urbaniczky and K. Lundstrom, *J. Electroanal. Chem.*, **176**, 169 (1984).
226. N. K. Cenas, J. J. Kanapieniene and J. J. Kulys, *J. Electroanal. Chem.*, **189**, 163 (1985).
227. J. Schreurs, J. van den Berg, A. Wonders and E. Barendrecht, *Recl. Trav. Chim., Pays-Bas*, **103**, 251 (1984).
228. V. A. Garten and D. E. Weiss, *Aust. J. Chem.*, **8**, 68 (1955).
229. J. V. Hallum and H. V. Drushel, *J. Phys. Chem.*, **62**, 110 (1958).
230. B. D. Epstein, E. Dalle-Molle and J. S. Mattson, *Carbon*, **9**, 609 (1971).
231. K. F. Blurton, *Electrochim. Acta*, **18**, 869 (1973).
232. C. M. Elliot and R. W. Murray, *Anal. Chem.*, **48**, 1247 (1976).
233. D. H. Karweik, I.-F. Hu, S. Weng and T. Kuwana, in *Catalyst Characterization Science* (Eds M. L. Deviney and J. L. Gland), ACS Symposium Series No. 288, American Chemical Society, 1985, p. 582.

234. I.-F. Hu, D. H. Karweik and T. Kuwana, *J. Electroanal. Chem.*, **188**, 59 (1985).
235. G. S. Calabrese, R. M. Buchanan and M. S. Wrighton, *J. Am. Chem. Soc.*, **104**, 5786 (1982).
236. G. S. Calabrese, R. M. Buchanan and M. S. Wrighton, *J. Am. Chem. Soc.*, **105**, 5594 (1983).
237. C. S. Calabrese and M. S. Wrighton, *J. Electrochem. Soc.*, **128**, 1014 (1981).
238. B. Keita and L. Nadjo, *J. Electroanal. Chem.*, **145**, 431 (1983).
239. L. Roullier, E. Waldner and E. Laviron, *J. Electroanal. Chem.*, **191**, 59 (1985).
240. C. Degrand and L. L. Miller, *J. Electroanal. Chem.*, **117**, 267 (1981).
241. C. Degrand and L. L. Miller, *J. Electroanal. Chem.*, **132**, 163 (1982).
242. C. Degrand, *J. Electroanal. Chem.*, **169**, 259 (1984).
243. A. N. K. Lau and L. L. Miller, *J. Am. Chem. Soc.*, **105**, 5271 (1983).
244. R. W. Day, G. Inzelt, J. F. Kinstle and J. Q. Chambers, *J. Am. Chem. Soc.*, **104**, 6804 (1982).
245. M.-C. Pham and J.-E. Dubois, *J. Electroanal. Chem.*, **199**, 153 (1986).
246. J. B. Flanagan, S. Margel, A. J. Bard and F. C. Anson, *J. Am. Chem. Soc.*, **100**, 4284 (1978).
247. B. L. Funt and P. M. Hoang, *J. Electrochem. Soc.*, **131**, 2295 (1984).
248. M. Fukui, A. Kitani, C. Degrand and L. L. Miller, *J. Am. Chem. Soc.*, **104**, 28 (1982).
249. C. Degrand, L. Roullier, L. L. Miller and B. Zinger, *J. Electroanal. Chem.*, **178**, 101 (1984).
250. B. L. Funt and P. M. Hoang, *J. Electroanal. Chem.*, **154**, 229 (1983).
251. P. Joo and J. Q. Chambers, *J. Electrochem. Soc.*, **132**, 1345 (1985).
252. L. L. Miller, B. Zinger and C. Degrand, *J. Electroanal. Chem.*, **178**, 87 (1984).
253. G. Inzelt, J. Q. Chambers, J. F. Kinstle and R. W. Day, *J. Am. Chem. Soc.*, **106**, 3396 (1984).
254. G. Inzelt, J. Q. Chambers, J. F. Kinstle, R. W. Day and M. A. Lange, *Anal. Chem.*, **56**, 301 (1984).
255. K. K. Kanazawa, A. F. Diaz, R. H. Geiss, W. D. Gill, J. F. Kwak, J. A. Logan, J. F. Rabolt and G. B. Street, *J. Chem. Soc., Chem. Commun.*, 854 (1979).
256. G. Arai, A. Fujii and I. Yasumori, *Chem. Lett.*, 1091 (1985).
257. A. R. Guadalupe and H. D. Abruna, *Anal. Chem.*, **57**, 142 (1985).
258. M. Krishnan, X. Zhang and A. J. Bard, *J. Am. Chem. Soc.*, **106**, 7371 (1984).
259. J. Zak and T. Kuwana, *J. Electroanal. Chem.*, **150**, 645 (1983).
260. V. Razumas, J. Jasitis and J. Kulys, *J. Electroanal. Chem.*, **155**, 427 (1983).
261. C. Bourdillon, J. P. Bourgeois and D. Thomas, *J. Am. Chem. Soc.*, **102**, 4231 (1980).
262. A. Fujishima, F. Karasawa and K. Honda, *J. Electroanal. Chem.*, **134**, 187 (1982).
263. A. Alt, H. Binder, A. Köhling and G. Sandstede, *Angew. Chem. Int. Ed.*, **10**, 514 (1971).
264. H. Binder, A. Köhling and G. Sandstede, *Ber. Bunsenges. Phys. Chem.*, **80**, 66 (1976).
265. G. Sanstede, *Z. Phys. Chem. (Frankfurt)*, **98**, 389 (1975).
266. H. Alt, H. Binder, A. Köhling and G. Sandstede, *Electrochim. Acta*, **17**, 873 (1972).
267. G. Matricali, M. M. Dieng, J. F. Dufeu and M. Guillou, *Electrochim. Acta*, **21**, 943 (1976).
268. G. Matricali, A. Kergreis, B. Auclair and M. Guillou, *C.R. Acad. Sci., Ser. C*, **278**, 829 (1974).
269. G. Matricali, M. Dieng, J. Dufeu and M. Guillou, *C.R. Acad. Sci., Ser. C*, **284**, 301 (1977).
270. M. Dieng, G. Matricali, J. F. Dufeu and M. Guillou, *C.R. Acad. Sci., Paris*, **284**, 607 (1977).
271. M. Hiroi and N. Ichikawa, *J. Appl. Electrochem.*, **12**, 147 (1982).
272. V. N. Storozhenko, A. S. Strizhko, F. E. Dinkevich and V. Z. Moskovskii, *Electrokhim.*, **18**, 517 (1982).
273. T. Boschi, R. Pappa, G. Pistoia and M. Tocci, *J. Electroanal. Chem.*, **189**, 339 (1985).
274. L. Roullier and E. Laviron, *J. Electroanal. Chem.*, **134**, 181 (1982).
275. V. Malatesta, S. Penco, N. Sacchi, L. Valentini, A. Vigevani and F. Arcamone, *Can. J. Chem.*, **62**, 2845 (1984).
276. K. Kano, T. Konse, N. Nishimura and T. Kubota, *Bull. Chem. Soc. Japan*, **57**, 2383 (1984).
277. K. Kano, T. Konse and T. Kubota, *Bull. Chem. Soc. Japan*, **58**, 424 (1985).
278. W. D. Wilson and R. L. Jones, *Adv. Pharm. Chemother.*, **18**, 177 (1981).
279. G. Molinier-Jumel, B. Malfoy, J. A. Reynaud and G. Aubel-Sadron, *Biochem. Biophys. Res. Commun.*, **84**, 441 (1978).
280. J. A. Plambeck and J. W. Lown, *J. Electrochem. Soc.*, **131**, 2556 (1984).
281. G. M. Rao, J. W. Lown and J. A. Plambeck, *J. Electrochem. Soc.*, **124**, 195 (1977).
282. G. M. Rao, A. Begleiter, J. W. Lown and J. A. Plambeck, *J. Electrochem. Soc.*, **124**, 199 (1977).
283. K. Takamura and F. Watanabe, *Anal. Biochem.*, **74**, 512 (1976).
284. T. Erabi, H. Hiura and M. Tanaka, *Bull. Chem. Soc., Japan*, **48**, 1354 (1975).
285. T. E. Young, J. R. Griswald and M. H. Hulbert, *J. Org. Chem.*, **39**, 1980 (1974).

286. T. E. Young, B. W. Babbitt and L. A. Wolfe, *J. Org. Chem.*, **45**, 2899 (1980).
287. T. E. Young and B. W. Babbitt, *J. Org. Chem.*, **48**, 562 (1983).
288. M. D. Hawley, S. V. Tatawadi, S. Piekarski and R. N. Adams, *J. Am. Chem. Soc.*, **89**, 447 (1967); *J. Am. Chem. Soc.*, **90**, 1093 (1968).
289. A. Brun and R. Rosset, *J. Electroanal. Chem.*, **49**, 287 (1974).
290. J. S. Mayansky and R. L. McCreery, *Anal. Chem.*, **55**, 308 (1983).
291. A. W. Sternson, R. McCreery, B. Feinberg and R. N. Adams, *J. Electroanal. Chem.*, **46**, 313 (1973).
292. R. N. Adams, *J. Pharm. Sci.*, **58**, 1171 (1969).
293. Z. Grujic, I. Tabakovic and M. Trkovnik, *Tetrahedron Lett.*, 4823 (1976).
294. M. D. Ryan, A. Yueh and W.-Y. Chen, *J. Electrochem. Soc.*, **127**, 1489 (1980).
295. J. Moiroux, D. Escourrou and M. B. Fleury, *J. Electroanal. Chem.*, **116**, 333 (1980).
296. A. B. Grigor'ev, M. K. Polievktov and A. N. Grinev, *Zh. Obshch. Khim.*, **43**, 1372 (1973).
297. G. Platz, H. Raithel, U. Nickel and W. Jaenicke, *Z. Phys. Chem. (N.F.)*, **98**, 407 (1975).
298. K. Takamura, M. Sakamoto and Y. Hayakawa, *Anal. Chim. Acta*, **106**, 261 (1979).
299. K. Weber, W. Förster and H. Berg, *J. Electroanal. Chem.*, **100**, 135 (1979).
300. C. Comninellis and E. Plattner, *J. Appl. Electrochem.*, **15**, 771 (1985).
301. H. Tributsch and H. Gerischer, *Ber. Bunsenges. Phys. Chem.*, **72**, 437 (1968); **73**, 251 (1969); **73**, 850 (1969).
302. H. Tributsch, *Ber. Bunsenges. Phys. Chem.*, **73**, 582 (1969).
303. H. Tributsch and M. Calvin, *Photochem. Photobiol.*, **14**, 95 (1971).
304. T. Miyasaka, T. Watanabe, A. Fujishima and K. Honda, *J. Am. Chem. Soc.*, **100**, 6657 (1978).
305. A. Fujishima, T. Iwase, T. Watanabe and K. Honda, *J. Am. Chem. Soc.*, **97**, 4134 (1975).
306. T. Watanabe, A. Fujishima and K. Honda, *Ber. Bunsenges. Phys. Chem.*, **79**, 1213 (1975).
307. S. N. Frank and A. J. Bard, *J. Am. Chem. Soc.*, **99**, 4667 (1977).
308. F. Takahashi and R. Kikuchi, *Biochim. Biophys. Acta*, **430**, 490 (1976).
309. F. Takahashi and R. Kikuchi, *Bull. Chem. Soc., Japan*, **49**, 3394 (1976).
310. F. Takahashi, K. Seki, R. Kaneko, K. Sato and Y. Kusumoto, *Bull. Chem. Soc. Japan*, **54**, 387 (1981).
311. F. K. Fong and N. Winograd, *J. Am. Chem. Soc.*, **98**, 2287 (1976).
312. R. C. Ahuja, K. Hauffe, R. Schumacher and R. N. Schindler, *Ber. Bunsenges. Phys. Chem.*, **86**, 167 (1982).
313. A. F. Janzen and J. R. Bolton, *J. Am. Chem. Soc.*, **101**, 6342 (1979).
314. M. Miyake, H. Yoneyama and H. Tamura, *Electrochim. Acta*, **22**, 319 (1977).
315. P. A. Kohl and A. J. Bard, *J. Am. Chem. Soc.*, **99**, 7531 (1977).
316. D. Laser and A. J. Bard, *J. Phys. Chem.*, **80**, 459 (1976).
317. S. N. Frank and A. J. Bard, *J. Am. Chem. Soc.*, **97**, 7427 (1975).
318. P. A. Kohl and A. J. Bard, *J. Electrochem. Soc.*, **126**, 59 (1979).
319. F. DiQuarto and A. J. Bard, *J. Electroanal. Chem.*, **127**, 43 (1981).
320. B. Keita and L. Nadjo, *J. Electroanal. Chem.*, **151**, 283 (1983).
321. B. Keita and L. Nadjo, *J. Electroanal. Chem.*, **163**, 171 (1984).
322. J. H. Richardson, L. J. Kovalenko, S. B. Deutscher, J. E. Harrar and S. P. Perone, *J. Electroanal. Chem.*, **106**, 263 (1980).
323. J. M. Bobbitt and J. P. Willis, *J. Org. Chem.*, **42**, 2347 (1977).
324. M. Fujihira, S. Tasaki, T. Osa and T. Kuwana, *J. Electroanal. Chem.*, **150**, 665 (1983).
325. M. Fujihira, *J. Electroanal. Chem.*, **130**, 351 (1981).

The Chemistry of Quinonoid Compounds, Vol. II
Edited by S. Patai and Z. Rappoport

CHAPTER **13**

Recent advances in the photochemistry of quinones

KAZUHIRO MARUYAMA and ATSUHIRO OSUKA
Department of Chemistry, Faculty of Science, Kyoto University, Kitashirakawa Oiwakecho, Kyoto 606, Japan

---

I. INTRODUCTION

In recent years the photochemistry of quinones has been a subject of extensive investigations. As a result, a large number of papers have appeared describing novel aspects of quinone photochemistry. Our knowledge of the mechanism of photochemical reactions of quinones has expanded rapidly. Particularly, charge transfer or electron transfer interactions of photoexcited quinones with other substrates have been widely recognized as the important primary process in numerous photoreactions such as reduction, cycloaddition, substitution and decomposition. The advances in nanosecond and pico-second time-resolved spectroscopy and magnetic resonance techniques, particularly those of CIDEP and CIDNP, have enabled a much more systematic study to be undertaken and have led to a much more precise description of the reaction mechanism.

Because of the availability of excellent reviews on the photochemistry of quinones covering the literature prior to 1973[1–4], this chapter will focus mainly on the recent developments in the photochemistry of quinones. Recent developments are now summarized annually in *The Chemical Society Specialist Periodical Reports on Photochemistry*, volumes 1–16 (1968–1984)[5]. Additionally, two reviews appeared in 1974[6] and 1977[7], mainly dealing with voluminous contributions reported in the USSR.

Representative quinones are expressed by the abbreviations summarized below, and the abbreviations Q, $Q^{-\cdot}$, $QH^{\cdot}$ and QH_2 are used to indicate, respectively, the quinone, the semiquinone anion radical, the neutral semiquinone radical, and the hydroquinone.

Quinone	Abbreviation
1,4-benzoquinone	BQ
tetramethyl-1,4-benzoquinone	DQ
tetrachloro-1,4-benzoquinone	CQ
1,4-naphthoquinone	NQ
2-methyl-1,4-naphthoquinone	MQ
9,10-anthraquinone	AQ
9,10-phenanthraquinone	PQ
plastoquinone-*n*	PLQ-*n*
9,10-anthraquinone-2-sulfonate	ASQ
9,10-anthraquinone-2,6-disulfonate	ADSQ

II. SPECTRA AND EXCITED STATES

Recent advances in the main features of the absorption and emission spectra of quinones in relation to their photochemistry are discussed in this section.

Experimental and theoretical studies concerning the electronic structure and absorption and emission spectra of quinones continue to appear[8–23]. These spectroscopic studies have shown that at least two excited (n, π^*) states of B_{1g} and A_u symmetry lie within 250–300 cm^{-1} for both the singlet and triplet excited state of BQ. The pure electronic $S_0 \rightarrow T_1$ transition of NQ single crystals has been investigated in electric and magnetic fields[24]. Highly resolved phosphorescence spectra of AQ and AQ-d_8 in crystals and solution were reported[25–28]. Photoelectron spectra of BQ and NQ and their derivatives are extensively studied[29–32]. Highly structured fluorescence and fluorescence excitation spectra of **1** ($R = NH_2$) and **4** ($R^1 = R^2 = OH$) are measured in low-temperature Shpol'skii matrices[33, 34].

In contrast to the lowest (n, π^*) triplet state of AQ and 2-chloroanthraquinone (**2**, R = Cl), 1-halogenoanthraquinones (**1**, R = F, Cl, Br) have a lowest triplet state of (π, π^*) character with short lifetimes[35]. The short lifetimes of the lowest triplet state and the small phosphorescence quantum yield of **1** (R = halogen) indicate the presence of a rapid radiationless decay process through the interaction of the carbonyl group with the 1-halogen atom[35]. The relative contributions of (n, π^*) character in the lowest triplet states of polyhalogenoanthraquinones are assessed on the basis of the magnitudes of their quenching rate constants by benzene molecule[36]. A new absorption with a rise time of 35 ps and decay time of 700 ps in the flash photolysis of **5** ($R^1 = R^2 = Cl$) is observed in toluene and is ascribed to the $T_2 \rightarrow T_n$ transition[37].

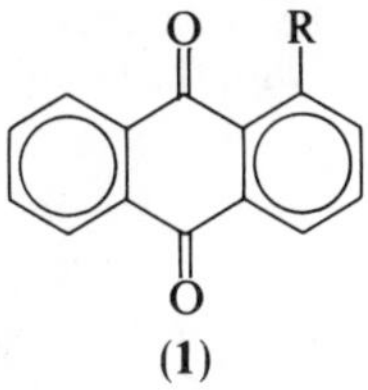

(**1**) 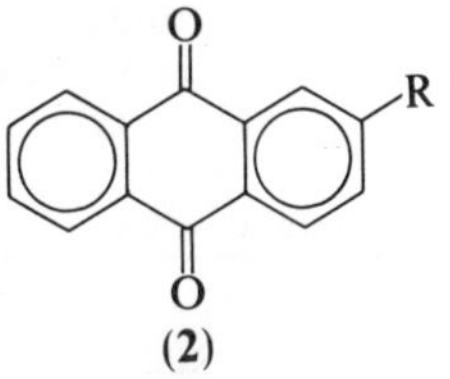 (**2**) 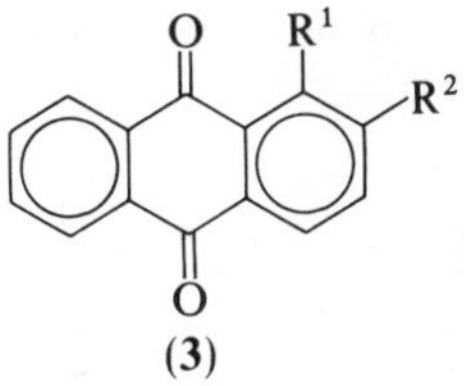 (**3**)

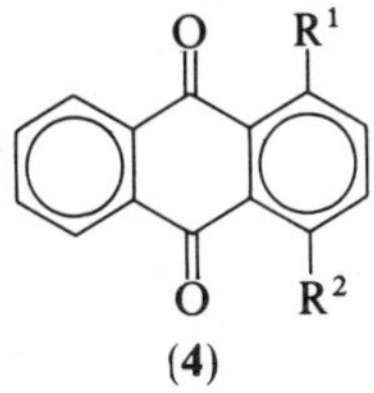

(**4**) 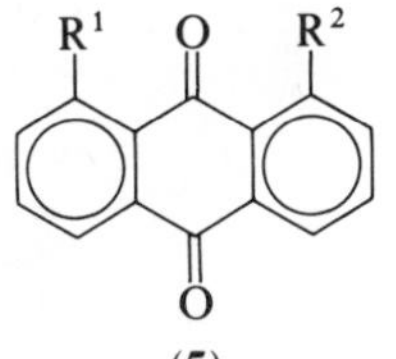 (**5**)

In general, introduction of electron-donating substituents such as amino and hydroxy groups to AQ changes the electronic nature of the lowest excited state to the intramolecular CT state[38, 39]. The photochemistry of amino- and hydroxy-substituted anthraquinone derivatives has been extensively studied because of their structural relationship to vat dyes, which can promote phototendering of textiles. Quantum yields for intersystem crossing (ϕ_{isc}) of amino-substituted anthraquinones are very low (Table 1). The decrease in ϕ_{isc} on substitution of amino and hydroxy groups must result from the change in the nature of the excited state involved and also from the increased singlet–triplet energy gap, which for amino-substituted anthraquinones ranges from 5000 cm^{-1} to 7000 cm^{-1}, compared with a value of 1600 cm^{-1} for AQ itself[45]. The quantum yields for

TABLE 1. Quantum yields for intersystem crossing of quinones

Quinone	Solvent	ϕ_{isc}	Ref.
AQ	Benzene	0.90	40
MQ	Water	0.66	41
DQ	Cyclohexane	1.0	42
	Ethanol	1.0	42
	Water	1.0	42
CQ[a]	DCE[b]	1.0	43
ASQ	Water	>0.9	44
ADSQ	Water	1.0	44
4 ($R^1 = R^2 = NH_2$)	Benzene	0.015	45
4 ($R^1 = NH_2$, $R^2 = NHMe$)	Benzene	0.02	45
4 ($R^1 = R^2 = NHMe$)	Benzene	0.015	45
4 ($R^1 = NH_2$, $R^2 = NHPh$)	Benzene	0.025	45
4 ($R^1 = NH_2$, $R^2 = OH$)	Benzene	0.01	45

[a] The rate of intersystem crossing from $^1(CQ)^*$ to $^3(CQ)^*$ is $3.0 \times 10^{10}\ s^{-1}$ (Ref. 46).
[b] 1,2-Dichloroethane.

photodecomposition of these amino-substituted anthraquinones, particularly for oxidation of the amino group and photodealkylation of the alkylamino group, have been determined to be 10^{-4} or less[47,48], which suggests that the photodecomposition may be initiated through the triplet state. Thus one may attribute the light-stability of amino and hydroxy-anthraquinones largely to rapid deactivation within the singlet manifold[49b].

Fluorescence lifetimes, quantum yields, and solvent shifts of amino and hydroxy-anthraquinones were studied by several groups (Table 2). The wavelength of the fluorescence maximum is red-shifted in polar solvents, consistent with the notion of the

TABLE 2. Fluorescence parameters of anthraquinones

Anthraquinone	Solvent	$\phi_f{}^a$	λ_{max} (nm)[b]	τ_f (ns)[c]	Ref.
1-Amino	DMC[d]	0.24			50
	n-Hexane	0.07	530		49a
	Benzene	0.022	548		51
	Benzene	0.058		1.75	52
	Ethyl acetate	0.008	567		51
	Furan	0.07	558		49a
	2-Propanol	0.003	590		49a
	Ethanol	0.0082		0.46	52
	Ethanol-d_1	0.0498		3.03	52
	Acetonitrile	0.034	600	0.7	53
	Acetonitrile	0.01		0.66	52
2-Amino	Benzene	0.24			54
	Benzene	0.08	548		51
	Benzene	0.21		6.5	52
	n-Hexane	None	None		49a
	Ethyl acetate	0.024	563		51
	Furan	0.05	532		49a
	2-Propanol	$< 10^{-5}$	None		49a
	Ethanol	0.00059		0.054	52
	Ethanol-d_1	0.0014		0.109	52
	Acetonitrile	0.008		0.85	52
	Acetonitrile	0.019	620	0.7	53

TABLE 2. (*Continued.*)

Anthraquinone	Solvent	ϕ_f^a	λ_{max} (nm)^b	τ_f (ns)^c	Ref.
1-Piperidino	*n*-Hexane	$< 10^{-5}$	None		49a
2-Piperidino	Cyclohexane	0.19			39b
	n-Hexane	0.27	510, 545		49a
	Benzene	0.085		7.7	52
	Furan	0.04	615		49a
	2-Propanol	10^{-5}	630		49a
	Ethanol	0.00012		0.040	52
	Ethanol-d_1	0.0006		0.068	52
	Acetonitrile	0.0012		0.3	52
1-Methylamino	*n*-Hexane	0.003	560		49a
	Benzene	0.0076		0.33	52
	Furan	0.04	615		49a
	2-Propanol	10^{-4}	612		49a
	Ethanol	0.0014		0.094	52
	Acetonitrile	0.0016		0.19	52
N-1-Chloroacetylamino	*n*-Hexane	$< 10^{-5}$	None		55
	Ethanol	0.0013	509		55
	2-Propanol	0.0011	515		55
1-Acetylamino	*n*-Hexane	10^{-5}	500		49a
	Benzene	0.017		0.41	52
	Ethyl acetate	0.0068	515		55
	Furan	0.02	505		49a
	2-Propanol	0.004	505		49a
	Ethanol	0.0041		0.066	52
	Acetonitrile	0.0082		0.32	52
1-Benzoylamino	*n*-Hexane	10^{-5}	505		49a
	Furan	0.005	510		49a
	2-Propanol	10^{-5}	528		49a
1,4-Diamino	Benzene	10^{-5}	612, 635		51
	n-Hexane	10^{-5}	610, 630		49a
	Furan	10^{-5}	612, 645		51
	Ethyl acetate	10^{-5}	614, 640		51
	2-Propanol	$< 10^{-5}$	None		51
	Acetonitrile	0.03	660	0.6	53
1,5-Diamino	Benzene	0.013	550		51
	Ethyl acetate	0.007	555		51
	2-Propanol	0.0047	575		51
1,2,3,4-Tetraamino	Benzene	0.004	672		51
	Ethyl acetate	0.0024	672		51
	2-Propanol	0.002	684		51
1-Amino-2-chloro	*n*-Hexane	0.021	540		55
	Toluene	0.018	550		55
	Ethyl acetate	0.016	563		55
	2-Propanol	0.0084	583		55
1-Amino-3-chloro	*n*-Hexane	0.017	528		55
	Toluene	0.015	540		55
	Ethanol	0.014	563		55
	2-Propanol	0.0077	585		55
2-Amino-3-chloro	Toluene	0.085	512		55
	Ethanol	0.018	545		55
	2-Propanol	0.0022	558		55

TABLE 2. (*Continued*)

Anthraquinone	Solvent	ϕ_f[a]	λ_{max} (nm)[b]	τ_f (ns)[c]	Ref.
1-Amino-2-carboxy	Ether	0.0024	562		55
	Ethyl acetate	0.015	572		55
	2-Propanol	0.004	590		55
1-Amino-3-carboxy	Ether	0.0078	558		55
	Ethanol	0.01	565		55
	2-Propanol	0.0035	585		55
2-Amino-3-carboxy	Ether	0.06	520		55
	Ethanol	0.05	535		55
	2-Propanol	0.0014	573		55
1-Anilino		None			49a
1-*N*-Methylanilino		None			49a
1,5-Dipiperidino		None			49a
1,4-Bis(methylamino)	*n*-Hexane	10^{-5}	665, 670		49a
	Furan	10^{-5}	670, 710		49a
	2-Propanol	$< 10^{-5}$	None		49a
2-Dimethylamino	*n*-Hexane	0.2	495, 535		49a
	Furan	0.04	590		49a
	2-Propanol	10^{-5}	610		49a
1-Amino-4-hydroxy	Acetonitrile	0.041	630	1.0	53
1-Amino-4-hydroxy-2-Methoxy	Acetonitrile	0.12	616	1.9	53
1,8-Diamino	Acetonitrile	0.032	596	0.8	53
1-Amino-2,4-dibromo	Acetonitrile	0.003	594		53
1,5-Bis-*N*-ethylanilino	Acetonitrile	0.013	630		53
1-Hydroxy	Acetonitrile	0.013	608	0.6	53
1,2-Dihydroxy	Acetonitrile	0.004	660		53
1,2-Dihydroxy-3-$CH_2N(COOH)_2$	Acetonitrile	0.003	660		53
1,4-Dihydroxy	Acetonitrile	0.126	575	2.8	53
1,8-Dihydroxy	Acetonitrile	0.017	585	0.8	53
1,2,5,8-Tetrahydroxy	Acetonitrile	0.025	590		53
1,2-Diethoxy	Acetonitrile	0.009	646		53
1-Methylamino-4-bromo	Benzene	0.0032		0.19	52
	Ethanol	0.00087		0.053	52
	Acetonitrile	0.0011		0.19	52
2-Bis(dibutylamino)	Benzene	0.15		9.8	52
	Ethanol	0.00021		0.076	52
	Acetonitrile	0.0053		0.68	52

[a] Quantum yields for fluorescence.
[b] Fluorescence maxima.
[c] Fluorscence lifetimes.
[d] 1,4-Dimethylcyclohexane.

first singlet CT state. Of these, **2** (R = NH_2) is considerably much more sensitive to solvent polarity than the others. Amines quench the fluorescence by an electron transfer mechanism. Alcohols also quench the fluorescence but by a different mechanism involving hydrogen bonding[39]. The almost no difference between the fluorescence lifetimes of **3** ($R^1 = R^2 = OH$) (0.25 ns) and **3** ($R^1 = R^2 = OEt$) (0.30 ns) is interpreted in terms of a radiationless decay by intramolecular electron transfer from the substituent to the carbonyl group rather than as being due to hydrogen bonding[56]. However, solvent isotope effects on the radiationless decay rate constant, k_d(EtOH)/k_d(EtOD), 9.0 for **1** (R = NH_2), 2.1 for **2** (R = NH_2), and 1.7 for **2** (R = piperidino) are regarded as evidence for a radiationless deactivation of S_1 (CT) through intra- and intermolecular hydrogen bonding[52]. These fluorescence data as well as the intersystem crossing yields and the phosphorescence data are often used in the interpretation of light fastness properties of anthraquinone dyes in both polyester and nylon textiles.

Extensive pulse radiolysis studies have been used to characterize the spectra and the formation and decay kinetics of the radicals and their triplet states of quinones[57–61].

Charge transfer (CT) bands due to π complexes between quinones and electron-donating molecules are often observed[4, 43, 62, 63]. The comparison of the intramolecular CT absorptions of **6** and **7** shows very large differences; whereas a strong CT absorption is observed for **7** in the 350–600 nm region (λ_{max} = 495 nm, ε = 1600), **6** shows only a weak absorption (λ_{max} = 515 nm, ε = 170)[64, 65]. The differences in absorption due to the

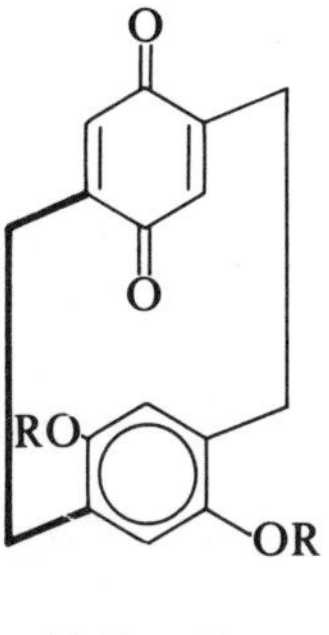

(**6**, R = H)
(**10**, R = Me)

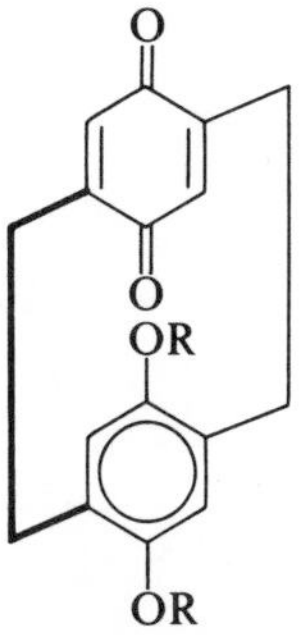

(**7**, R = H)
(**11**, R = Me)

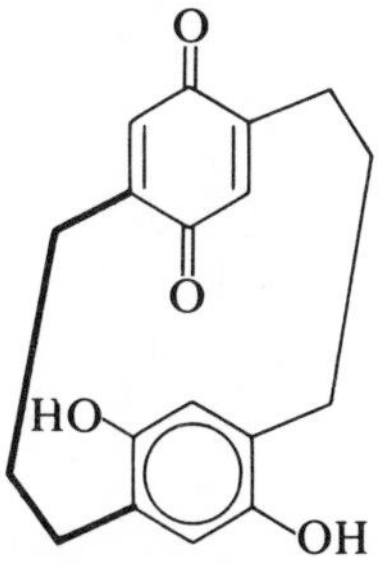

(**8**)

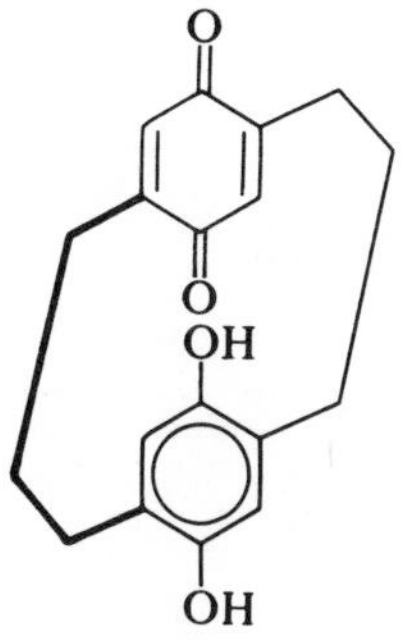

(**9**)

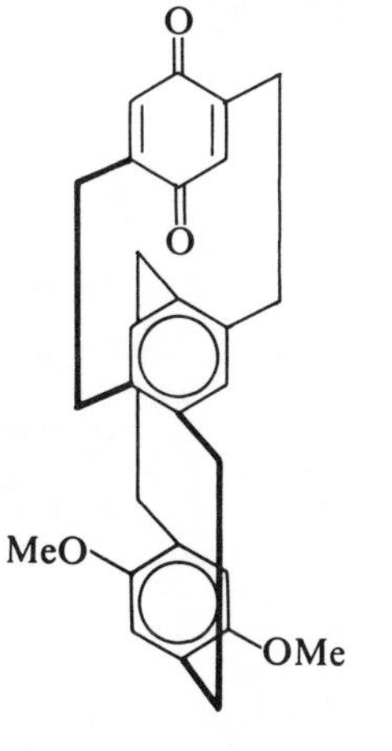

(**12**)

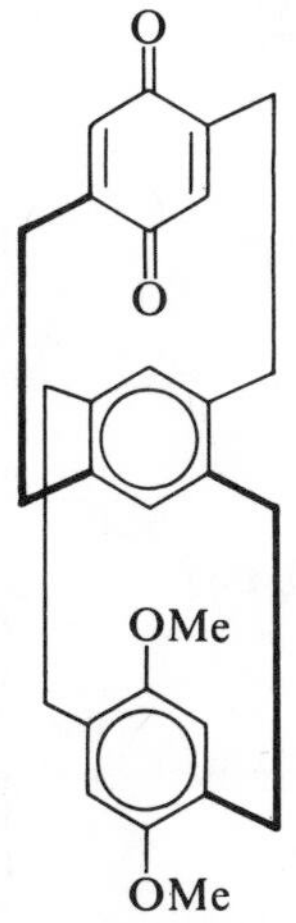

(**13**)

different donor–acceptor orientations are much more pronounced for the different connectivity pair **8**/**9**; whereas a strong CT absorption is observed for **9** (λ_{max} = 462 nm, ε = 3210), **8** shows only a weak absorption shoulder around 500 nm with ε = 105[66]. Similar orientation-dependent CT absorptions are also observed for **10** and **11**[65,66]. Intramolecular CT absorptions, which seem to be indirectly transmitted through the intercalated aromatic π system, are observed in triple-layered [2,2]paracyclophanes **12** and **13**; however, the orientation dependence of the CT absorptions is almost lost in these compounds: **12** (λ_{max} = 415 nm, ε = 2010), **13** (λ_{max} = 437 nm, ε = 2270)[66]. Quantum mechanical calculations indicate that the intramolecular CT interaction arises only from through-space interaction[67,68].

In relation to photochemical hole burning (PHB) phenomena, the photochemistry of **14** was studied in *n*-alkane (non-hydrogen bonding), polymethylmethacrylate (PMMA) (proton accepting), and ethanol and polyvinyl alcohol (PVA) (strong hydrogen bonding) matrices at low temperature. The PHB observation is moderate in PMMA and prominent in PVA matrices but only scarce in the *n*-alkane matrix[69]. Formation of a rotamer **15** in which the intramolecular hydrogen bond is broken and an intermolecular hydrogen bond is newly formed with a hydrogen bond accepting matrix molecule (A) is suggested for this PHB process[69] (equation 1).

(**14**) A $\xrightleftharpoons{h\nu}$ (**15**) (1)

TABLE 3. Triplet lifetimes (τ_t) of quinones

Quinone	Solvent	τ(s)	Ref.
BQ	Ethanol	< 0.01	70
	Ethanol/water (1:1)	< 0.01	70
	Water	< 0.01	70
	Water	0.5	71
TQ[a]	Ethanol	< 0.01	70
	Ethanol/water (1:1)	< 0.01	70
	Water	0.3	70
	Water	0.37	71
2,5-DMBQ[b]	Ethanol	< 0.01	70
	Ethanol/water (1:1)	0.03	70
	Water	0.32	70
	Water	0.87	71
2,3-DMBQ[c]	Ethanol	0.045	70
	Ethanol/water (1:1)	0.045	70
	Ethanol/benzene (1:1)	0.1	70
	Water	1.2	70
	Water	5.9	71

TABLE 3. (*Contd.*)

Quinone	Solvent	τ(s)	Ref.
TMBQ[d]	Ethanol	0.45	70
	Ethanol/water (1:1)	2.4	70
	Water	1.9	70
DQ	Benzene	25	72
	Benzene	10	70
	Benzene[k]	0.5	70
	Cyclohexane	21	42b
	Cyclohexane	8.3	70
	Cyclohexane[k]	0.25	70
	Liquid paraffin	29	70
	Ethanol	15	42b
	Ethanol	9	70
	Ethanol[k]	0.3	70
	Ethanol/water (1:1)	7.7	70
	Water	2.9	70
	Water	5.9	71
	Water	5.9	42b
	2-Propanol[k]	0.3	73
	2-Propanol	45	74
CQ	Cyclohexane	2.0	77
	DCE[l]	5.6	75
	Ethanol	1.2	77
	Butyronitrile	6.25	76
AQ	Benzene	0.11	44
	Benzene	0.125	61
	Benzene	0.13	36
TFAQ[e]	Benzene	0.20	36
PFAQ[f]	Benzene	0.17	36
TCAQ[g]	Benzene	3.0	36
ADSQ	Water	< 1	44
ASQ	Water	< 0.1	61
ASQ	Water	0.1	78
ASQ	Acetonitrile	12.5	78
MQ	Water	1.5	41
2-PAQ[h]	Benzene	20	61
1-AAQ[i]	Benzene	5	61
2-AAQ[j]	Benzene	5	61

[a] TQ, 2-methyl-1,4-benzoquinone; [b] 2,5-DMBQ, 2,5-dimethyl-1,4-benzoquinone; [c] 2,3-DMBQ, 2,3-dimethyl-1,4-benzoquinone; [d] TMBQ, trimethyl-1,4-benzoquinone; [e] TFAQ, 1,2,3,4-tetrafluoro-9,10-anthraquinone; [f] PFAQ, perfluoro-9,10-anthraquinone; [g] TCAQ, 1,2,3,4-tetrachloro-9,10-anthraquinone; [h] 2-PAQ, 2-piperidino-9,10-anthraquinone; [i] 1-AAQ, 1-amino-9,10-anthraquinone; [j] 2-AAQ, 2-amino-9,10-anthraquinone; [k] air saturated; [l] DCE, 1,2-dichloroethane.

III. PHOTOREDUCTION

A. Electron Transfer

The excited states of quinones decay by numerous reactive and non-reactive pathways such as emission (fluorescence and phosphorescence), energy transfer, charge or electron transfer, hydrogen (H) abstraction, bimolecular dimerization and cycloaddition, unimolecular rearrangement, fragmentation and many others. Of these, photoreduction, which can proceed via either electron transfer or H abstraction, is the major mode of photochemical reaction of quinones. When the triplet state of the quinone lies below the triplet state of the molecule being quenched, electronic energy transfer has been shown to occur in many cases. However, if charge transfer or electron transfer reaction is energetically feasible, such reaction is often favored over energy transfer. For instance, quenching of triplet *N*-methylindole by AQ, for which the quenching rate is $8 \times 10^9\ M^{-1}\ s^{-1}$ in benzene solution, does not give $^3(AQ)^*$ but produces triplet exciplex with greater charge transfer nature even though its energy is about 7.2 kcal mol^{-1} below that of triplet *N*-methylindole[79]. Photoreduction of quinones can occur by two mechanisms. One is the two-step mechanism of electron transfer followed by proton transfer via the ion-pair state. Another is the direct H-atom transfer in a non-relaxed encounter complex. Since most of the triplet quinones are strong electron and hydrogen acceptors, it is not easy to distinguish which mechanism is predominant in their H-abstraction reactions. The coexistence of the ion-pair formation and the rapid H-atom transfer in the non-relaxed encounter complexes has recently been claimed by nanosecond and picosecond absorption spectroscopy for the systems of 2,6-diphenylbenzoquinone (**16**)/diphenylamine[80] and CQ/durene[43], respectively. Since photoexcited quinones have

(**16**)

strong oxidizing power, their bimolecular quenching processes are, in general, more or less charge transfer in nature. Many evidences indicate that the quenching efficiency is determined by a charge transfer process from an electron donor to photoexcited quinone which will induce, or not, a chemical reaction, depending on the medium. The charge transfer interaction of excited quinone with electron donor produces an exciplex whose reactivity is sometimes quite different from that of the parent quinone in its excited state. For example, AQ/ammonia triplet exciplex abstracts a primary hydrogen from *t*-butanol or acetonitrile to give a cross adduct with much more efficiency and selectivity compared with AQ itself (see Section III.B)[81,82].

The energy of the charge transfer or electron transfer state can be estimated by the excited energy of the quinone and electrochemical redox potentials[83]. Here, extensive data on the family are available[84,85].

Reported results of the quenching rate constants of photoexcited quinones by electron donors and oxygen are summarized in Table 4.

TABLE 4. Quenching rate constants of photoexcited quinones by electron donors and oxygen

Quinone	Quencher	Solvent	$k_q(M^{-1}s^{-1}) \times 10^{-9}$	Ref.
$^3(DQ)^*$	$Fe(CN)_6^{4-}$	Water[a]	2.6×10^9	86
	Fe^{2+}	Water[a]	1.3	86
	Aniline	Water[a]	3	86

Table 4. (*Continued*)

Quinone	Quencher	Solvent	$k_q(M^{-1}s^{-1}) \times 10^{-9}$	Ref.
	TMB[b]	Water[a]	3.3	86
	CO_3^{2-}	Water[a]	7.3×10^{-2}	86
	Ethanol	Water[a]	2.8×10^{-5}	86
	Acetone	Acetone	5×10^{-5}	86
	Cl^-	Methanol[c]	1×10^{-2}	87
	Br^-	Methanol[c]	4	87
	I^-	Methanol[c]	9	87
	BA[d]	Methanol[c]	8×10^{-4}	87
	OH^-	Methanol[e]	1.5	87
	TPA[f]	Benzene	0.18	42a
	TPA	MeCN	10	42a
	DEA[g]	MeCN	9.4	42a
	TEA[h]	MeCN	6.0	42a
	TEA	Cyclohexane	2.0	42a
	DEA	Cyclohexane	8	42a
	c-ST[i]	Water[k]	1.5	88
	t-ST[j]	Water[k]	1.6	88
$^3(DQH^+)^*$	Cl^-	Methanol[l]	1	87
	BA	Methanol[l]	2.7×10^{-2}	87
$^3(CQ)^*$	Naphthalene	BN[m]	7	76
	AN[n]	DCE[o]	3.1×10^{-5}	89
	MMA[p]	DCE	5.1×10^{-3}	89
	MMA	MeCN	6.7×10^{-3}	89
^{3}(**16**)*	TMPD[q]	Toluene	18	80a
	TMPD	DBP[r]	1.1	80a
	DPA	Toluene	7.9	80a
	DPA	DBP	5.1	80a
	TPA	Toluene	5.0	80a
	TPA	DBP	0.33	80a
$^3(AQ)^*$	NMI[s]	Benzene	10.7	79
$^3(MQ)^*$	MQ	Water	4.5	41
	Thymine	Water	2.7	41
	Uracil	Water	3.0	41
	MU[t]	Water	3.2	41
$^3(AQS)$	Cl^-	Water	0.5	78, 90
	N_3^-	Water	3.1	78, 90
	NO_2^-	Water	3.2	78, 90
	Br^-	Water	3.8	78, 90
	NSC^-	Water	3.9	78, 90
	I^-	Water	4.2	78, 90
$^3(BQ)^*$	O_2	Benzene	2.4	71
$^3(MQ)^*$	O_2	Water	1.2	41
$^3(DQ)^*$	O_2	Toluene	2.1	412
$^3(AQ)^*$	O_2	Benzene	1.5	61
^{3}(2-PAQ[u])*	O_2	Benzene	1.9	61

[a] Ethanol/water = 1:2; [b] TMB, 1,3,5-trimethoxybenzene; [c] pH = 7; [d] BA, benzyl alcohol; [e] pH = 10–12; [f] TPA, triphenylamine; [g] DEA, diethylamine; [h] TEA, triethylamine; [i] *c*-ST, *cis*-stilbene; [j] *t*-ST, *trans*-stilbene; [k] in the presence of sodium dodecylsulfate; [l] pH = −2; [m] BN, butyronitrile; [n] AN, acrylonitrile; [o] DCE, 1,2-dichloroethane; [p] MMA, methyl methacrylate; [q] TMPD, tetramethylphenylenediamine; [r] DBP, dibutyl phthalate; [s] NMI, *N*-methylindole; [t] MU, 6-methyluracil; [u] 2-PAQ, 2-piperidino-9,10-anthraquinone.

1. Aromatic hydrocarbons

The laser photolysis at the charge transfer band of CQ and durene in 1,2-dichloroethane does not give any transient absorption over the time range from picosecond to several microseconds. These findings suggest that the singlet excited state of the charge transfer complex decays so rapidly through internal conversion that its detection is not possible even by means of the picosecond laser photolysis. Excitation of uncomplexed CQ at 347 nm results in the formation of $^3(CQ)^*$, which is rapidly quenched by durene. Interestingly, the H-atom transfer leading to production of $CQH^{\cdot}$ is shown to proceed through two distinct mechanisms; a sequential electron and proton transfer and a more rapid direct H-atom transfer competing with the electron transfer process (Scheme 1)[43].

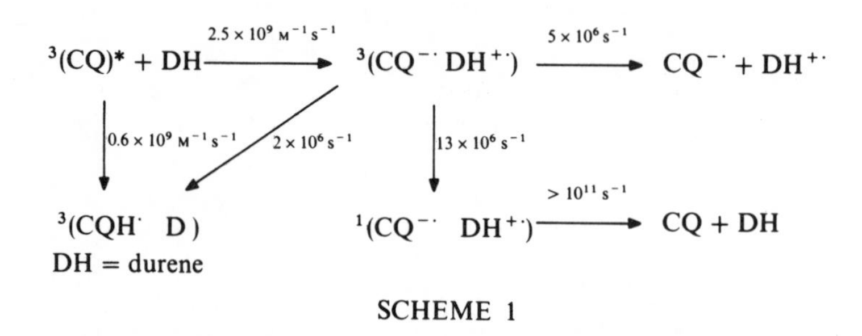

SCHEME 1

Excitation of CQ in butyronitrile at room temperature leads to a rapid production of $^3(CQ)^*$ which decays predominantly to $CQH^{\cdot}$ with $k_d = 1.6 \times 10^5$ $M^{-1} s^{-1}$. Observation of a photo-induced current suggests simultaneous production of $CQ^{-\cdot}$, formed by electron transfer quenching of $^3(CQ)^*$ by the medium. Added naphthalene (NA) quenches $^3(CQ)^*$ with $k_q = 7.0 \times 10^9$ $M^{-1} s^{-1}$; the cation radical of naphthalene is unambiguously identified as product of the electron transfer process[76]. This electron transfer process follows a Weller-type[83] quenching mechanism; the yields and lifetime of exciplex composed of CQ and NA are reduced in polar solvents as a result of ionic dissociation of the CQ/NA pair. The same CQ/NA pair has been investigated by picosecond spectroscopy in order to elucidate a more detailed mechanism of the electron transfer[46]. The intermediate non-relaxed exciplex, $^3(CQ^-NA^+)^{**}$, may be produced by interaction between the molecular components of the encounter complex, $^3(CQ \cdots NQ)$. This non-relaxed complex then may minimize its total energy by changing the relative positions of CQ and NA and also by solvent reorientations. Depending upon such factors as solvent polarity and relative oxidation and reduction potentials of CQ and NA, this non-relaxed intermediate may either form an exciplex, $^3(CQ^-, NA^+)^*$, or solvent-shared exciplex, $^3(CQ^- \cdots NA^+)^*$. However, the existence of a discrete encounter complex, unrelaxed exciplex, and/or exciplex is not revealed even in the picosecond time region[46].

Undoubtedly, most photochemical reactions of quinones proceed via the triplet state, since the intersystem crossing of the excited singlet of quinone is a rapid process. On the other hand, anthraquinones bearing amino and hydroxy groups have the lowest singlet (CT) state whose lifetimes are sometimes long enough for chemical reactions as well as fluorescence (Table 2). Electron transfer to these states occurs from electron-donating molecules whose oxidation potential lies between 0.35 V (*N, N*-diphenylethylenediamine) and 1.76 V (anisole)[56, 91]. Usually the quenching of the excited singlet quinone forms no detectable separate ions, contrary to the ion pair arising from the triplet excited state[56, 91].

2. Amines

Electron transfer is the generally observed mode of reaction between a photoexcited quinone and an amine in polar solvents, while exciplex formation is frequently detected and precedes the photoreduction of quinones by amines in non-polar solvents. In either case, the rate constant for the quenching is close to the diffusion-controlled limit and subsequent proton transfer from the amine cation radical to $Q^{-\cdot}$ results in the formation of $QH^{\cdot}$. Numerous reactions of this type have been reported, and studied by a variety of methods including laser flash photolysis[42, 72, 80, 86, 87, 92, 93], CIDNP[80b, 94, 95], CIDEP[73,96], steady-state irradiation[97] and photoconductivity measurements[86, 98]. Electron transfer-induced fragmentation of the aminoalcohol **17** to benzaldehyde and *N,N*-dimethylaniline is effectively sensitized by AQ[99] (equation 2).

AQ + PhN(Me)CH$_2$CH(OH)Ph (**17**) → (**18**) [Ph–N(Me)–CH$_2$–CH(Ph)–O–H ··· $AQ^{\bar{\cdot}}$]

$$\rightarrow \text{PhN(Me)CH}_2\cdot + \text{PhCHO} + \text{AQH}\cdot \qquad (2)$$

In the presence of triphenylamine (TPA), no products from $^3(DQ)^*$ photoreduction are detected by nanosecond laser flash photolysis in benzene. However, $^3(DQ)^*$ decays with a rate constant having a much higher value than in the absence of TPA. This is ascribed to the exciplex deactivation[42a]. Quenching of the triplet of **16** by TPA in non-polar solvents also occurs with the formation of triplet exciplex. In this case, however, characteristic absorption due to triplet exciplex, which possesses similar maxima to that of TPA^+ but substantially differs from the spectrum of $TPA^{+\cdot}$ has been successfully detected[80]. Quenching of the **16** triplet with the secondary aromatic amine, diphenylamine (DPA), in low-polar solvents yields, besides the short-lived triplet exciplex, a long-lived neutral radical $Ph_2N\cdot$. Interestingly, the triplet exciplex decay is not accompanied by $Ph_2N\cdot$ radical formation, indicating that proton transfer does not occur in the triplet ion pair (Scheme 2),

$$^3(Q)^* + Ph_2NH \longrightarrow {}^3(Q\text{- - - - -}Ph_2NH)$$

$$^3(Q\text{- - - - -}Ph_2NH) \xrightarrow{\text{H-atom transfer}} {}^3(QH^{\cdot}\ Ph_2N^{\cdot})$$

$$^3(Q\text{- - - - -}Ph_2NH) \xrightarrow{\text{electron transfer}} {}^3(Q^{-\cdot}\ Ph_2NH^{+\cdot}) \longrightarrow Q^{-\cdot} + Ph_2NH^{+\cdot}$$

$$^3(Q^{-\cdot}\ Ph_2NH^{+\cdot}) \rightleftarrows {}^1(Q^{-\cdot}\ Ph_2NH^{+\cdot}) \longrightarrow Q + Ph_2NH$$

SCHEME 2

contrary to the normally accepted mechanism of photoreduction of quinones by amines[80a]. This seemingly important finding, analogous to the mechanism of CQ/durene system, may provide a direct experimental proof that H-atom transfer competes with electron transfer and both occur from the non-relaxed triplet encounter complex[80]. In a series of works by the same authors, activation energies higher than 7 kcal mol^{-1} for H shift in the non-relaxed encounter complex have been estimated[80b,100]. However, prototropic equilibrium prior to H-atom shift was recently considered by the same authors[93].

3. *Olefins*

Photo-induced addition of electron-rich diarylethylenes to 2-bromo-3-methoxy-1,4-naphthoquinone apparently proceeds via a charge transfer or an electron transfer interaction (see Section IV.B). *E–Z* isomerizations of 1-phenylpropene and stilbene are brought about by the electron transfer interaction with photoexcited quinones[88,101]. In connection with the photosensitized polymerization by CQ, the electron transfer reaction from a variety of vinyl monomers to excited triplet state of CQ has been studied by laser flash photolysis[75,102,103]. Cation radical of photogenerated *N*-vinylcarbazole undergoes cyclodimerization in polar solvents while cationic polymerization takes place in non-polar solvents without any induction period. ^{3}(CQ)* abstracts an electron even from electron-deficient vinyl monomers such as acrylonitrile and methyl methacrylate[44].

4. *Strained hydrocarbons*

p-Chloranil (CQ) has a lowest triplet (π, π^*) excited state as well as an exceedingly low reduction potential, rendering an electron transfer reaction strongly exothermic toward a wide range of substrates. By using CQ as an efficient electron acceptor, the structures and reactivities of short-lived radical cations are extensively studied by the photo-CIDNP method[94]. The method is very simple but provides valuable information about the electronic structures of the cation radicals derived from cyclopropanes[104a–e], norbornadiene[104a], quadricyclane[104a], methanobridged cyclophane[105a], bicyclo[1.1.0]butane[105b,c], methylenecyclopropane[106], 3,3′-dimethylbicyclopropenyl[104i], hexamethyl Dewar benzene[104g], pentacyclic caged cyclobutane[104f], methylenebicyclo[2.2.0]hexene[104h], dicyclopentadiene[107] and fulvene derivatives[108]. These radical ion-pair induced CIDNP effects are derived from competitive hyperfine coupling dependent in-cage electron return and separation by diffusion in triplet radical ion pair (scheme 3)[94]. According to the generally accepted radical pair theory of

$$^3(\mathrm{CQ})^* + \mathrm{D} \longrightarrow {}^3\overline{\mathrm{CQ}^{-\cdot}\,\mathrm{D}^{+\cdot}} \quad (3)$$

$$^3\overline{\mathrm{CQ}^{-\cdot}\,\mathrm{D}^{+\cdot}} \rightleftharpoons {}^1\overline{\mathrm{CQ}^{-\cdot}\,\mathrm{D}^{+\cdot}} \quad (4)$$

$$^1\overline{\mathrm{CQ}^{-\cdot}\,\mathrm{D}^{+\cdot}} \longrightarrow \mathrm{CQ} + \mathrm{D}\ddagger \quad (5)$$

$$^3\overline{\mathrm{CQ}^{-\cdot}\,\mathrm{D}^{+\cdot}} \longrightarrow \mathrm{CQ}^{-\cdot} + \mathrm{D}^{+\cdot} \quad (6)$$

D = strained hydrocarbon

SCHEME 3

CIDNP[109], the directions and the intensities of these effects are determined by several parameters, including the spin multiplicity of the precursor (μ), the magnetic properties (electron g-factor(Δg), hyperfine coupling constants (a_i)) of the radical intermediates, and the type of reaction (ε) by which the polarized product is formed. Since three ($\Delta g, \mu, \varepsilon$) of the four parameters are always the same, the directions and magnitudes of the polarization may reflect the electronic structure of the radical intermediates (a_i). Irradiation of an acetone solution of CQ and 2,2-dianisylmethylenecyclopropane (**19**) leads to the formation of cycloadducts **20** and **21** through the intermediacy of a radical ion pair (equation 7)[106]. Photo-induced monomerization of dimethylthymine dimers (**22**) by ASQ to **23** proceeds also by an initial one-electron transfer from the strained cyclobutane ring to the photoexcited quinone (equation 8)[110].

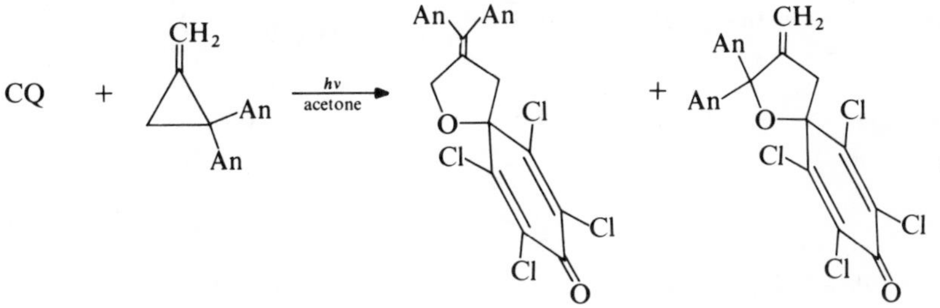

(**19**) (**20**) (**21**) (7)

An = 4-methoxyphenyl

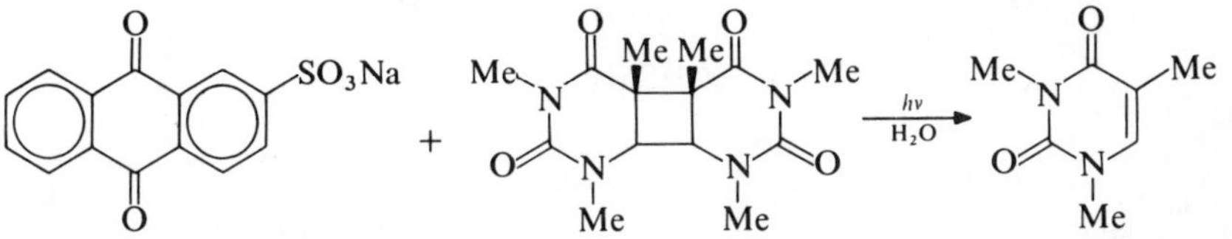

ASQ (**22**) (**23**) (8)

5. *Alcohols and hydroquinones*

A laser photolysis technique combined with fast conductance measurements and kinetic spectroscopy are used to monitor the electron transfer reactions of ^{3}(DQ)* with ethanol in water/ethanol (2:1, v/v) and with acetone[86]. Similarly, electron transfer from benzyl alcohol to ^{3}(DQ)* proceeds with $k_q = 8 \times 10^5\ \mathrm{M}^{-1}\,\mathrm{s}^{-1}$ in water/methanol (4:1, v/v)[87]. 2-Hydroxy-4-phenyldibenzofuran (**27**) is the principal product in the photolysis of **16** in polar solvents (Scheme 4)[111, 112]. CIDNP effects due to the reactants and the cyclized product **27** are interpreted as arising from the radical ion pair **24** or neutral radical pair **25** composed of semiquinone radical and ethoxy radical. These results are consistent with the mechanism in Scheme 4, where the primary process for the cyclization is a single electron transfer from ethanol to **16**[113]. Many observations are given as evidence for the primary electron transfer in the photoreduction of BQ in pure alcohol, in spite of the purely (n, π*) nature of the lowest triplet state of BQ (see Section III.B.1).

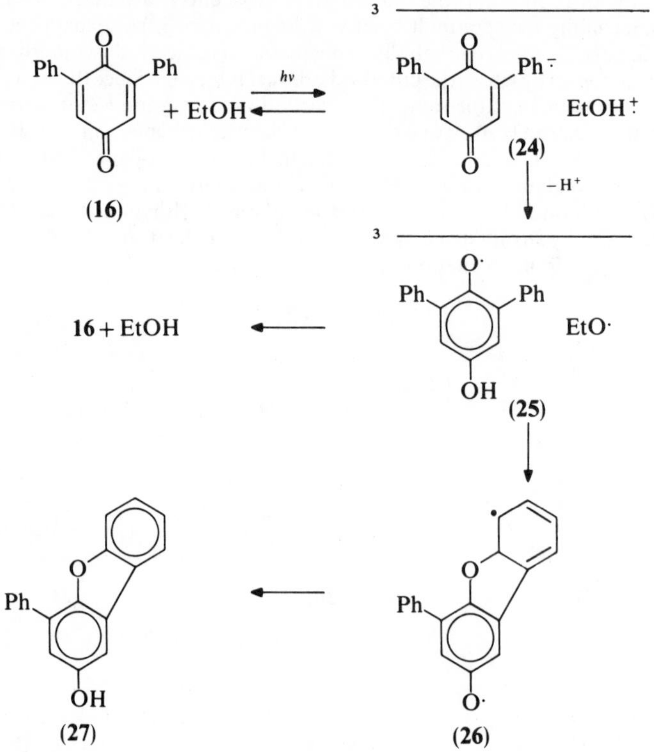

SCHEME 4

6. *Ions*

Organic anions[114] (alkoxides[115–118], $MeCO_2^-$, $CCl_3CO_2^-$), inorganic anions[6,78,90,114,119] (OH^-, SO_4^{2-}, SO_3^{2-}, SeO_3^{2-}, SCN^-, CO_3^{2-}, PO_4^{2-}, NO_3^-, X^-) can reduce photoexcited quinones via a single electron transfer. Photo-induced sulfonation of 1-aminoanthraquinone in aqueous solution containing Na_2SO_3 has been shown to involve electron transfer from SO_3^{2-} ion to the photoexcited quinone (see Section V). AQ, ASQ and **2** (R = piperidino) are readily photoreduced by hydroxide ion to produce the corresponding anions as the stable products[39,114,120,121], but the excited state involved is $T_1(n, \pi^*)$ for AQ and ASQ and T_2 (n, π^*) for **2** (R = piperidino)[120]. Recently, interest in this area has been heightened by suggestions for utilizing anthraquinone sulfonates, ASQ and ADSQ, as photocatalysts in solar energy storage, as in photooxidation of chloride ion to chlorine[122] and splitting of water into H_2 and H_2O_2[123]. Photogenerated anion radical from ASQ can reduce methyl viologen to its cation radical, which, in turn, splits water in the presence of colloidal platinum (Scheme 5).

In connection with the possible increase of the sensitivity of diazo-type light-sensitive materials, the photosensitized radical chain decomposition of arenediazonium salts has been extensively studied[124,125]. ASQ or ADSQ photosensitization leads to a considerable increase of ca. 10 times the rate of decomposition of the diazo compounds in the presence of appropriate electron and hydrogen donors. The following reaction mechanism is

$$\begin{array}{l} H_2O \rightleftarrows H^+ + HO^- \\ H_2\dot{O}_2 \leftarrow HO^{\cdot} \end{array} \overset{h\nu}{\rightleftarrows} \begin{array}{c} ASQ \\ ASQ^{-\cdot} \end{array} \rightleftarrows \begin{array}{c} MV^+ \\ MV^{2+} \end{array} \underset{\text{Pd-colloid}}{\rightleftarrows} \begin{array}{c} H^+ \\ H_2 \end{array}$$

SCHEME 5

suggested in the case of formate as electron and hydrogen donor:

$$^3(ASQ)^* + HCOO^- \rightarrow ASQ^{-\cdot} + HCOO^{\cdot} \quad (9)$$

$$ASQ^{-\cdot} + ArN_2^+ \rightarrow ASQ + Ar^{\cdot} + N_2 \quad (10)$$

$$Ar^{\cdot} + HCOO^- \rightarrow ArH + COO^{-\cdot} \quad (11)$$

$$COO^{-\cdot} + ArN_2^+ \rightarrow CO_2 + Ar^{\cdot} + N_2 \quad (12)$$

Photogenerated $ASQ^{-\cdot}$ (equation 9) reduces the diazonium salt, giving rise to an aryl radical (equation 10), which propagates the chain decomposition of the diazonium salt (equations 11 and 12). The chain length depends on the rate of initiation and on the nature of the donor molecules and is stated to reach 500 in solution and 20 in plasticized poly(vinyl alcohol) films[126].

7. *Miscellaneous*

The triplet state of DQ has a pK_a of -0.1. The protonated triplet state shows a different absorption spectrum (max 390 and 470 nm) from that of the neutral form (max 440 nm) and reacts with benzyl alcohol and Cl^- ion with rate constants which are about two orders of magnitude higher than those of the neutral form[87].

It is well known that thymine (**28**) and the other constituents of DNA are the principal targets for inactivation of cells by short-wavelength UV light. 2-Methyl-1,4-naphthoquinone (MQ) is also a target chromophore for the growth delay of *Bacillus subtilis* induced by near-UV light[127]. MQ and its bisulfate derivatives are among the most efficient radiation potentiating drugs as photosensitizers of mammalian cell killing by near-UV (365 nm) irradiation. The single-strand breaks in the supercoiled DNA of bacteriophage ϕX-174 are sensitized by MQ.

Photoexcitation of the quinone at 365 nm in oxygenated solutions leads to a number of products of **28**, including the *cis* and *trans* isomers of 5-hydroperoxy-6-hydroxy-5,6-dihydrothymine (**31**), the *cis*- and *trans*-thymine glycols **32**, 5-hydroxy-5-methylhydantoin (**33**) and *N*-formyl-*N'*-pyruvylurea (**34**), which are also formed by hydroxy radical attack upon **28** by ionizing irradiation (scheme 6)[41, 128]. Interaction of $^3(MQ)^*$ in water with pyrimidine bases such as thymine, uracil, 6-methyluracil and orotic acid involves electron transfer to give the pyrimidine cation radical and the $MQ^{-\cdot}$ with a nearly diffusion-controlled rate (Table 4). Therefore the thymine photoproducts are considered to arise from the reaction of **29** with water and oxygen. Oxygen also readily quenches $^3(MQ)^*$; the usual reaction involves energy transfer to give ground state MQ and $^1O_2$. In dilute oxygenated solutions of MQ, this is the main process and MQ is subsequently altered by the reaction with $^1O_2$. When **28** is also present in the solution, however, it competes efficiently with oxygen towards $^3(MQ)^*$, the yield of $^1O_2$ is considerably decreased, and MQ is thus protected against its self-sensitized photodynamic oxidation[41]. Photochemical reductions of dibromothymoquinone in ethanol are studied in relation to its inhibitory function in photosynthesis[129].

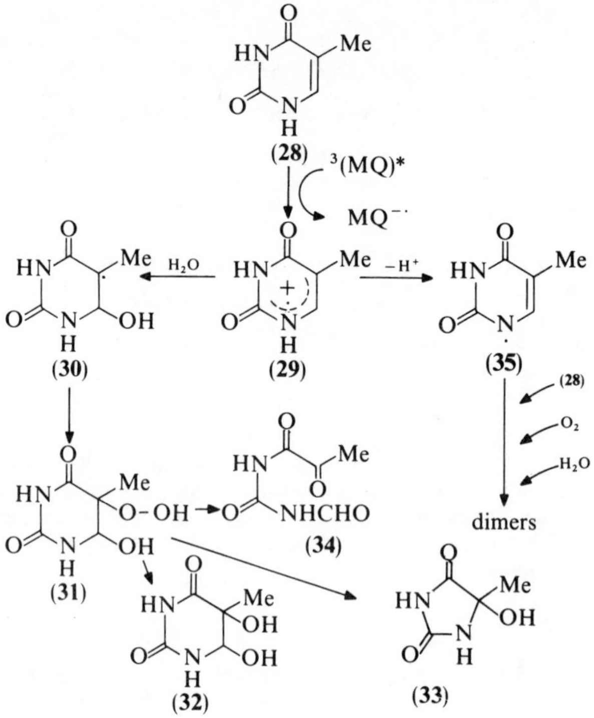

SCHEME 6

Triplet states of polycyclic aromatic hydrocarbons are efficiently quenched by quinones via charge transfer interactions[130, 131]. The magnitudes of the quenching rate constants k_q are found to be correlated with the energy difference between the triplet state of the aromatic hydrocarbon donor and the triplet charge transfer state assuming electron transfer from the donor to the quinone. The deactivation route is shown to involve intersystem crossing from this intermediate ^{3}(CT)* state to the almost isoenergetic corresponding ^{1}(CT)* state, followed by a subsequent internal conversion to the weakly bound ground state of the charge transfer complex.

Quinones have often been used merely as electron acceptors in photosensitized electron transfer reactions. Both singlet and triplet excited chlorophyll molecules are efficiently quenched by quinones[410]. There is clear evidence that the triplet exciplex between chlorophyll a and quinone produce the chlorophyll a cation radical and semiquinone anion radical. However, no separate ions are detected from the singlet quenching[132].

B. Hydrogen Abstraction

The mechanism of H abstraction by photoexcited quinones has received considerable experimental and theoretical attention. Quantum yields for the reduction of 0.5–1.0 and even higher values have been reported for quinones with low-lying (n, π*) electronic states[133]. Substitution of the quinones by electron-donating groups gives rise to low-lying intramolecular charge transfer (CT) or (π, π*) states in these molecules with a concomitant decrease in photoreduction reactivities.

The hydrogen abstraction reactions of triplet states of AQ have been studied theoretically as radiationless transitions in terms of the tunnel-effect theory, in which the rates of reactions are controlled by the Frank–Condon factors of the vibrational stretching motions of the carbonyl group in the excited AQ and the C–H oscillators of the substrates and the C–O and O–H vibrations in the ground state of the final products. The rates are dependent on the electronic energy, the nature of the excited state, the vibrational frequencies, the reduced mass of the oscillators, the C–H bond strength and the distances of the reactive bonds of substrates[134]. The effects of substituents on the reaction rates can be accounted for by change in the reduction potentials of the quinones, and/or by the nature and energy of the excited states. It is predicted theoretically that (π, π^*) states have an intrinsic reactivity for H abstraction which is 10^{-2} to 10^{-4} times lower than that of (n, π^*) states, but when both levels are energetically close, the observed reactivity is due to the thermally equilibrated population of the two states[135].

As noted in Section III.A, photochemical H abstraction can occur by a two-step mechanism involving electron and proton transfer. As another mechanism, H abstraction by a triplet exciplex has been known. AQ/ammonia triplet exciplex abstracts a hydrogen from *t*-butanol or acetonitrile to give adduct **36** (equation 13) or **37** (equation 14), respectively[81, 82]. The photoreaction of AQ in *t*-butanol-benzene (4:1, v/v) containing no ammonia or water proceeds much less rapidly, giving **36** in 29% yield[81]. In acetonitrile, the formation of the substituted anthrone (**37**) is not observed in the absence of ammonia, and is strongly suppressed by the presence of oxygen, which favors the amination of anthraquinone nucleus (see Section V.D.1)[82].

AQ —[$h\nu$; t-BuOH–C_6H_6–29% aqueous NH_3 (20:4:1)]→ HO $CH_2C(CH_3)_2OH$ (**36**, 74%) (13)

—[$h\nu$/Ar; CH_3CN–NH_3]→ HO CH_2CN (**37**, 54%) + NH_2 (1%) + NH_2 (6%) (14)

Rate constants for H abstraction reaction of photoexcited quinones with hydrogen donors (i.e. quenching rate constants) are listed in Table 5. Generally, the resulting semiquinone radicals QH˙ rapidly decay via disproportionation to QH_2 and Q with diffusion-controlled rates (Table 6).

1. *In alcohol*

Although the photoreduction of quinones has long been studied, there is still controversy about whether the primary reaction of quinone triplets is an electron transfer or a direct H transfer. Photoreduction of quinones with the lowest lying (n, π^*) triplet state in basic alcoholic media, in general, is considered to involve H abstraction from the solvent by the quinone (equation 15), followed by dissociation of the semiquinone radical (QH˙)

TABLE 5. Quenching rate constants of triplet quinone by hydrogen donors

Quinone	Hydrogen donor	k_q ($M^{-1}s^{-1}$)	Ref.
BQ	Ethanol	$> 6 \times 10^6$	70
TQ[a]	Ethanol	$> 6 \times 10^6$	70
DMBQ[b]	Ethanol	$> 6 \times 10^6$	70
TMBQ[c]	Ethanol	1.3×10^6	70
DQ	Ethanol	$0.64–1.6 \times 10^4$	70
	Ethanol	3.6×10^3	42b
	2-Propanol	1.8×10^4	70
	2-Propanol	6×10^3	74
	Phenol	2×10^8,[h]	136a
	2-MP[d]	4×10^8,[h]	136a
	PCP[e]	2×10^8,[h]	136a
	TEA[f]	3.6×10^8	95
	DQH_2	3.7×10^9,[h]	70
	DQH_2	3.9×10^9,[h]	42c
	DQH_2	2×10^9,[h]	74
DBQ[g]	2-Propanol	2×10^6	136c
CQ	Dioxane	2.6×10^5 (3.8×10^4)[j]	75
	CQH_2	1.7×10^9,[k] (0.9×10^9)[j]	137
	CQH_2	3.6×10^9,[l] (1.9×10^9)[j]	137
	CQH_2	4.7×10^9,[m] (4.3×10^9)[j]	137
	$MeCH_2CN$	1.6×10^5	138
AQ	Ethanol	$< 2 \times 10^7$	61
	Ethanol	3.3×10^6	138
	2-Propanol	4.0×10^6	61
	Hexane	8.6×10^5	138
	Benzene	4.8×10^5	138
1 (R = Cl)	Ethanol	2.3×10^6	89
6 ($R^1 = R^2$ = Cl)	Ethanol	1.6×10^4	89
ASQ	Ethanol	1×10^8,[n]	139

[a] TQ, toluquinone; [b] DMBQ, 2,5-dimethyl-1,4-benzoquinone; [c] TMBQ, trimethyl-1,4-benzoquinone; [d] 2-MP, 2-methylphenol; [e] PCP, pentachlorophenol; [f] TEA, triethylamine; [g] DBQ, 2,6-di-*t*-butyl-1,4-benzoquinone; [h] in ethanol; [i] in 2-propanol; [j] net hydrogen atom transfer rate; [k] in dioxane; [l] in 1,2-dichloroethane; [m] in acetonitrile; [n] in water/ethanol mixture.

TABLE 6. Second-order rate constant for decay of semiquinone radicals

Quinone	k_2 ($M^{-1} s^{-1} \times 10^{-9}$)	Solvent	Ref.
BQ	1.5	2-Propanol	137
	5.4	Dioxane	137
DQ	0.7	2-Propanol	137
	2.9	Dioxane	137
	2.1	—	74
NQ	0.23	2-Propanol	137
	0.90	Dioxane	137
CQ	0.17	2-Propanol	137
	0.76	Dioxane	137
FQ[a]	3.2	2-Propanol	140
DBQ[b]	1.7×10^{-3}	2-Propanol	136c
DMBQ[c]	0.3	2-Propanol	136c
AQ	1.3	—	138b
	1.2	Dioxane	137
AQ^-	0.049	2-Propanol	137
1 (R = Cl)	1.2	Ethanol	89
6 ($R^1 = R^2$ = Cl)	0.58	Ethanol	89
ASQ	3.0	Water	61
ADSQ	3.2	Water	61
1 (R = piperidino)	0.7	Methanol	61
2 (R = piperidino)	0.62	Methanol	61

[a] FQ, tetrafluoro-1,4-benzoquinone; [b] DBQ, 2,6-di-*t*-butyl-1,4-benzoquinone; [c] DMBQ, 2,6-dimethyl-1,4-benzoquinone.

to the semiquinone anion radical (equation 16). In the absence of base the subsequent disproportionation reaction results in the formation of hydroquinone (QH_2) (equation 17).

$$^3Q^* + RCH_2OH \rightarrow QH^{\cdot} + R\dot{C}HOH \tag{15}$$

$$QH^{\cdot} \rightleftharpoons Q^{-\cdot} + H^+ \tag{16}$$

$$2QH^{\cdot} \rightarrow QH_2 + Q \tag{17}$$

The flash photolysis technique has provided evidence favoring the direct H-abstraction mechanism by detection of $QH^{\cdot}$ as the primary intermediate[4,6,7,42b]. Formation of $QH^{\cdot}$ has also been indicated by ESR studies under steady-state[4,6,7] or flash photolysis[141,142] conditions. Even DQ, whose lowest triplet excited state is entirely of (π, π^*) character, shows a high H-abstraction reactivity toward ethanol and cyclohexane[42b]. However, a nanosecond flash photolysis combined with a fast flash conductance measurement reveals that $^3(DQ)^*$ in a 2:1 v/v mixture of water/ethanol does not abstract hydrogen but an electron from ethanol[86]. Both $BQ^{-\cdot}$ and $BQH^{\cdot}$ are observed by ESR during the continuous photolysis of a flowing solution of BQ in ethanol at room temperature[143]. Based on the concentration effects of BQ and flowing rate upon their relative yields, it is proposed that $BQ^{-\cdot}$ is formed directly by a photo-induced electron transfer from the solvent to BQ (equation 18, Q = BQ) and that $BQH^{\cdot}$ is formed by hydrogen abstraction of BQ from BQH_2[143,144]. The latter is produced by disproportionation of $BQH^{\cdot}$ followed by protonation (equation 20, Q = BQ). Halogenated benzoquinones behave similarly[143b]. Further, the formation of free ethoxy radical which appears to arise from the deprotonation of cation radical of ethanol (equation 19) gives support for the anionic reduction pathways[113,143c,145,146]. But one must take the possible complication from

CIDEP effects of the system into consideration in interpreting the steady-state ESR data, since the steady-state ESR signal is not a simple measure of the concentration of radicals in a system where CIDEP effect operates[96].

$$Q^* + MeCH_2OH \rightarrow Q^{-\cdot} + MeCH_2OH^{+\cdot} \qquad (18)$$

$$MeCH_2OH^{+\cdot} + MeCH_2OH \rightarrow MeCH_2O^{\cdot} + MeCH_2OH^{+} \qquad (19)$$

$$2Q^{-\cdot} \rightarrow Q^{2-} + Q \qquad (20)$$

However, the results obtained by CIDEP[96,141,147–151,408] and CIDNP[96,138,147] support a direct H abstraction as the main pathway in the photoreduction of BQ and NQ in pure alcohols. Initial polarizations observed in the CIDEP experiments provide a great deal of information on the primary radical intermediates; this technique appears particularly suitable for the determination which of the two, electron transfer or H abstraction, occurs in the reaction[148–150]. Polarized emissions for $BQH^{\cdot}$ and $NQH^{\cdot}$ are observed in the photoreduction of BQ and NQ in pure 2-propanol and 2-butanol, as well as in the presence of pyridine[96,151], but no emission from the anion $BQ^{-\cdot}$ nor $NQ^{-\cdot}$ can be detected. These are attributed to the formation of optically spin polarized triplets of the parent quinone and their subsequent H abstraction with retention of the polarization in the resultant semiquinone radicals. The spin polarization of the triplet quinone is believed to be a consequence of the spin selective intersystem crossing from the excited singlet. Further study using a benzene solution of BQ or NQ in the presence of 2,6-di-*t*-butyl-4-methylphenol as the hydrogen donor has established that both the semiquinone radical and the corresponding phenoxy radical exhibit total emission, consistent with the photochemical triplet polarization theory[148,149,152]. Polyhalogenated anthraquinones such as perfluoroanthraquinone (**38**), 1,2,3,4-tetrafluoroanthraquinone (**39**) and 1,2,3,4-tetrachloroanthraquinone (**40**) have the lowest triplet excited states having mixed $(n,\pi^*)-(\pi,\pi^*)$ character or pure (π,π^*) character and are likely to be photoreduced in ethanol via electron transfer from ethanol[36]. On the other hand, **1** (R = Cl) and **5** ($R^1 = R^2$ = Cl), which have a lowest triplet excited state quite similar to **38–40**, are reported to be photoreduced via H abstraction. No evidence for electron transfer from ethanol to **1**(R = Cl) and **5**($R^1 = R^2$ = Cl) is obtained[37]. Prolonged irradiation of **5**($R^1 = R^2$ = Cl) gives AQH_2 as a final product via a consecutive H abstraction and dechlorination sequence[153,154] (equation 21).

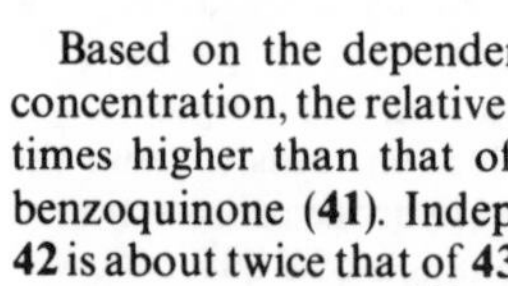

(**38**) 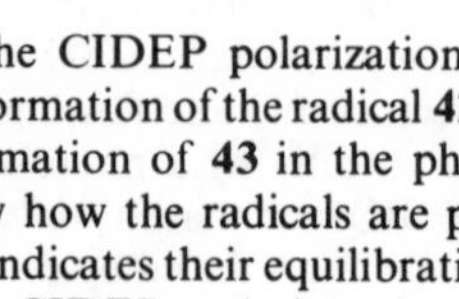 (**39**) 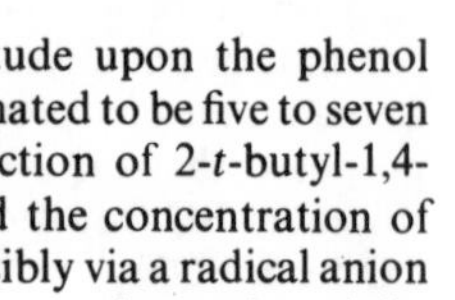 (**40**)

Based on the dependency of the CIDEP polarization magnitude upon the phenol concentration, the relative rate of formation of the radical **42** is estimated to be five to seven times higher than that of the formation of **43** in the photoreduction of 2-*t*-butyl-1,4-benzoquinone (**41**). Independently how the radicals are produced the concentration of **42** is about twice that of **43**, which indicates their equilibration, possibly via a radical anion of **41** (equation 22)[152d]. Further, CIDEP technique is utilized to evaluate the triplet quenching rate by H donors[136a,140].

Visible-light irradiation of the yellow polycyclic quinone **44** in acidic alcoholic solvents leads to a blue doubly reduced salt (**45**), which upon exposure to an oxidant (air or H_2SO_4) is converted back to **44** (equation 23)[155]. This represents a new, recyclable photochromic

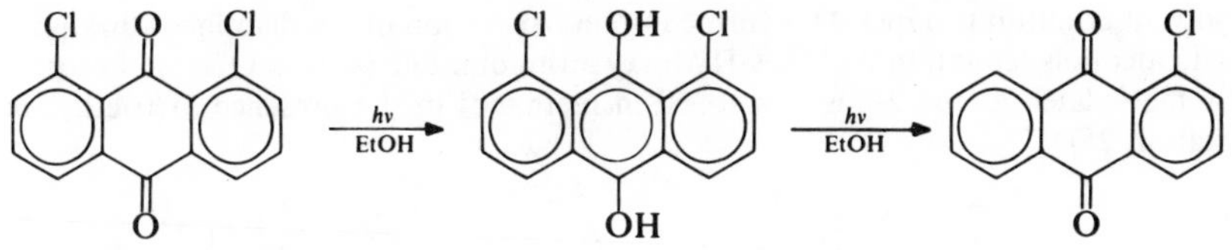

(5, $R^1 = R^2 = Cl$)

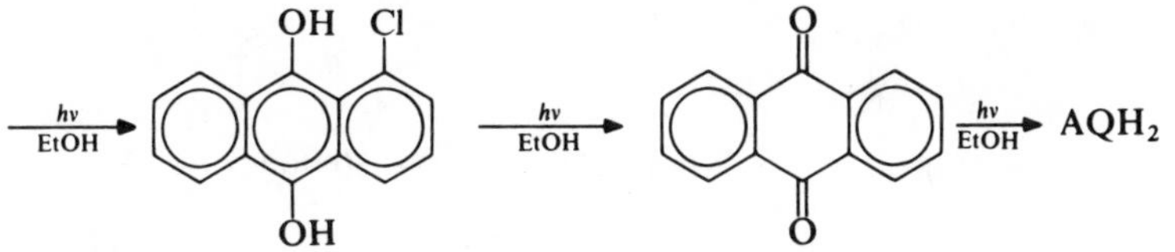

$\xrightarrow[EtOH]{h\nu}$ AQH_2 (21)

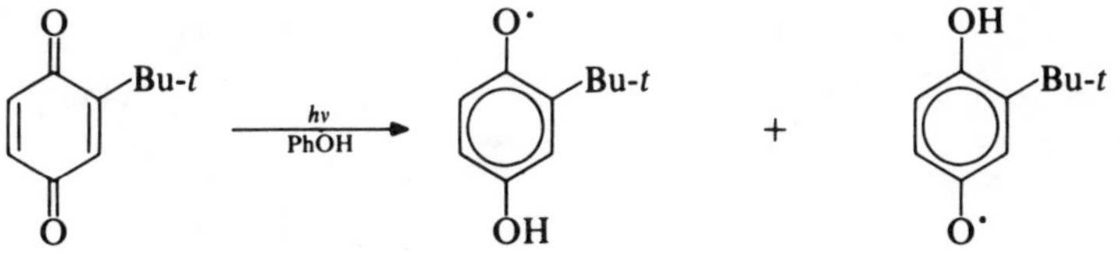

(41) $\xrightarrow[PhOH]{h\nu}$ (42) + (43) (22)

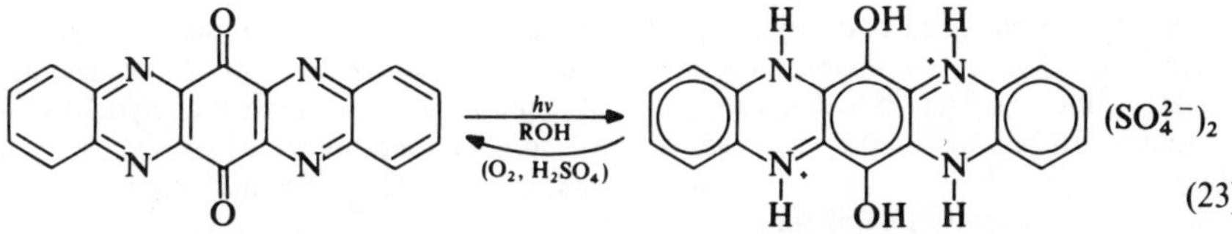

yellow (λ_{max} 342, 360, 380 nm) (44) $\underset{(O_2, H_2SO_4)}{\overset{h\nu, ROH}{\rightleftharpoons}}$ blue (λ_{max} 630 nm). (45) $(SO_4^{2-})_2$ (23)

system on an hour time-scale. Polycyclic furanonaphthoquinones **46** and **47** are photoreduced in basic alcohol solution with quantum yields of 10^{-3} to 10^{-2}, consistent with the (π, π^*) nature of the low-lying electronic states of these furanoquinones[156].

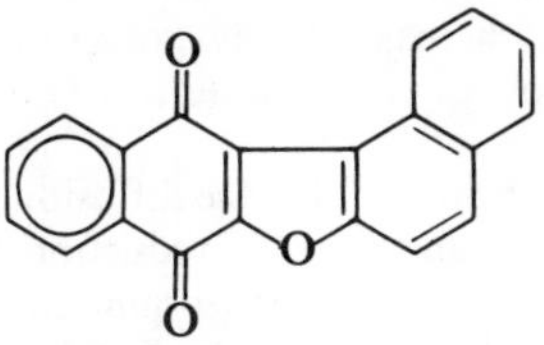

(46) 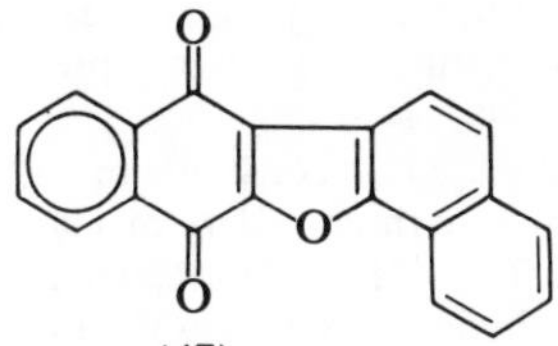 (47)

These structures can be represented by

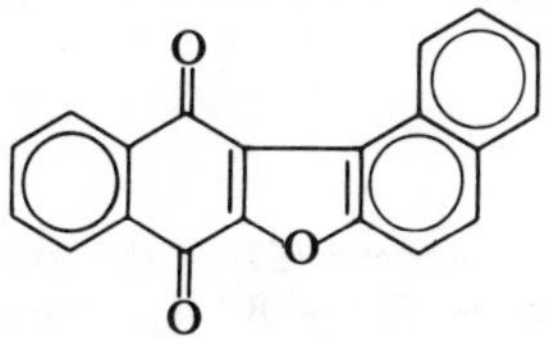

or

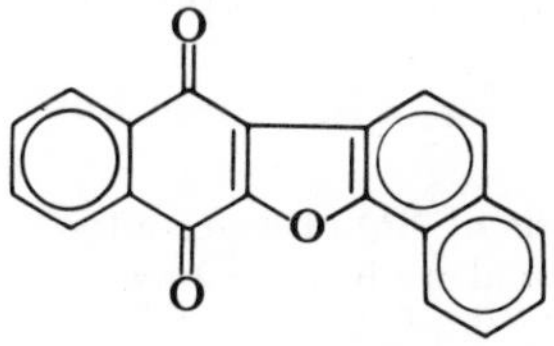

Solvent addition product **49** is formed upon irradiation of 1,4-diaminoanthraquinone (**48**) in alcohols (equation 24)[157]. 1-Hydroxyanthraquinone (**1**, R = OH) is photochemically methylated at the 2-position to **50** in methanol in the presence of triethylamine (equation 25)[409].

$$\mathbf{(48)} \xrightarrow[\text{RCH}_2\text{OH}]{h\nu} \mathbf{(49,\ 30\text{–}40\%)} \qquad \text{R = H, Me} \tag{24}$$

$$\text{1-hydroxyanthraquinone} \xrightarrow[\text{MeOH}]{h\nu/\text{Et}_3\text{N}} \mathbf{(50,\ 54\%)} \tag{25}$$

ASQ-photosensitized hydroxylation of benzene to phenol is accomplished in water/ethanol (1:1) solution in the presence of Cu^{2+} under aerobic conditions. Photogenerated ASQH· reduces oxygen to hydroperoxy radical (equations 26 and 27), which hydroxylates aromatic compounds in the presence of Cu^{2+} ion. In the absence of oxygen, metallic copper is formed by the reactions shown in equations 28 and 29. The maximum quantum yield for the formation of phenol is 0.04[139].

$$^3(\text{ASQ})^* + \text{MeCH}_2\text{OH} \rightarrow \text{ASQH}^{\cdot} + \text{Me}\dot{\text{C}}\text{HOH} \tag{26}$$

$$\text{ASQH}^{\cdot} + \text{O}_2 \rightarrow \text{ASQ} + \text{HOO}^{\cdot} \tag{27}$$

$$\text{ASQH} + \text{Cu}^{2+} \rightarrow \text{ASQ} + \text{Cu}^{+} + \text{H}^{+} \tag{28}$$

$$2\text{Cu}^{+} \rightarrow \text{Cu}^{0} + \text{Cu}^{2+} \tag{29}$$

Nitro-substituted anthraquinones are reduced to the corresponding aminoanthraquinones upon irradiation in alcohols, whereas carbonyl group is not reduced at all. The reduction probably proceeds via the lowest triplet state (n, π^*) of the nitro group[158–160].

Normally the rate constants of $^3(Q)^*$ photoreduction by QH_2 are diffusion controlled and are two or three orders of magnitudes higher than those for reduction by alcohol (Table 5). The ^{19}F CIDNP signals of both tetrafluoro-1,4-benzoquinone (FQ) and tetrafluorohydroquinone (FQH_2) observed in the photolysis of the FQ/FQH_2 pair in benzene are interpreted in terms of the triplet mechanism[136b, 161]. At relatively low magnetic field, two other mechanisms, a radical pair route and a biradical-induced polarization, are suggested[161]. Clear evidence for the primary electron transfer from QH_2 to Q is provided by CIDNP[162].

2. *In water*

Photolysis of BQ in water affords 1,2,4-trihydroxybenzene (**51**) as the sole primary photochemical product and 2-hydroxy-1,4-benzoquinone (**52**) and BQH_2 as the secondary products via **53** (equation 30)[163]. Formation of **51** is not affected by addition of a radical scavenger such as *p*-nitro-*N*,*N*-dimethylaniline and the isotope effect in the reaction ($k_H/k_D = 1.15$) is small[164]. These results are not compatible with the radical mechanism involving H abstraction from water and suggest an ionic mechanism for the ring hydroxylation of BQ in water[165, 166]. Quantum yields for the photoreduction in water are 0.55 at 313 nm and 0.31 at 436 nm, indicating the participation of the upper (π, π^*) triplet state of BQ. Accordingly, a direct nucleophilic addition of water to the (π, π^*) excited state of BQ is postulated (equation 30)[165]. Photolysis of BQ in water in the presence of micelles[166] or cyclodextrin[167] results in the increase of the quantum yields for the formation of BQH_2 along with the decrease of the quantum yield for the formation of **51** through the inclusion of BQ into the micelle or the cyclodextrin. This is consistent with the polar mechanism[167].

$$BQ \xrightarrow[H_2O]{h\nu} (BQ, \text{upper } \pi, \pi^*) \longrightarrow \textbf{(53)} \longrightarrow \textbf{(51)} \xrightarrow{BQ} \textbf{(52)} + BQH_2 \qquad (30)$$

On the other hand, laser flash photolysis experiments have shown the formation of both short-lived and long-lived transients upon irradiation of BQ in water[71]; the short-lived one (λ_{max} 410 nm) is assigned to $^3(BQ)^*$ which gives rise to the long-lived $BQH^{\cdot}$ and $BQ^{-\cdot}$ transients, as well as to a species considered to be the benzoquinone–hydroxy radical adduct ($BQOH^{\cdot}$). Based on these results, H-atom abstraction by $^3(BQ)^*$ from water is suggested by the authors[71]. Photooxidation of water by BQ is also shown by Raman technique[168].

Increases in the triplet lifetimes of trimethyl-1,4-benzoquinone, 2,6-dimethoxy-1,4-benzoquinone and DQ in water are certainly related to an increase in the (π, π^*) character of the lowest triplet of these compounds.

3. *From alkyl aromatics*

Photochemical reactions of BQ, NQ and PQ with alkyl aromatics have been extensively studied by CIDNP technique, which can provide definitive evidence for the caged radical pair and unstable photoproducts[169–172]. Preparative irradiation of a benzene solution of BQ and xanthene (**54**) leads to a clean reduction of BQ to BQH_2 with the concurrent formation of 9,9′-bixanthenyl (**60**) (Scheme 7). The CIDNP effects observed during the irradiation indicate the intermediate formation of unstable adducts such as **55**, **56** and **57** as the in-cage recombination products of BQH and the 9-xanthenyl radical (**59**) formed via (**58**) (Figure 1)[170]. Polarized signal 1 in Figure 1a was due to the ring hydrogen of BH_2.

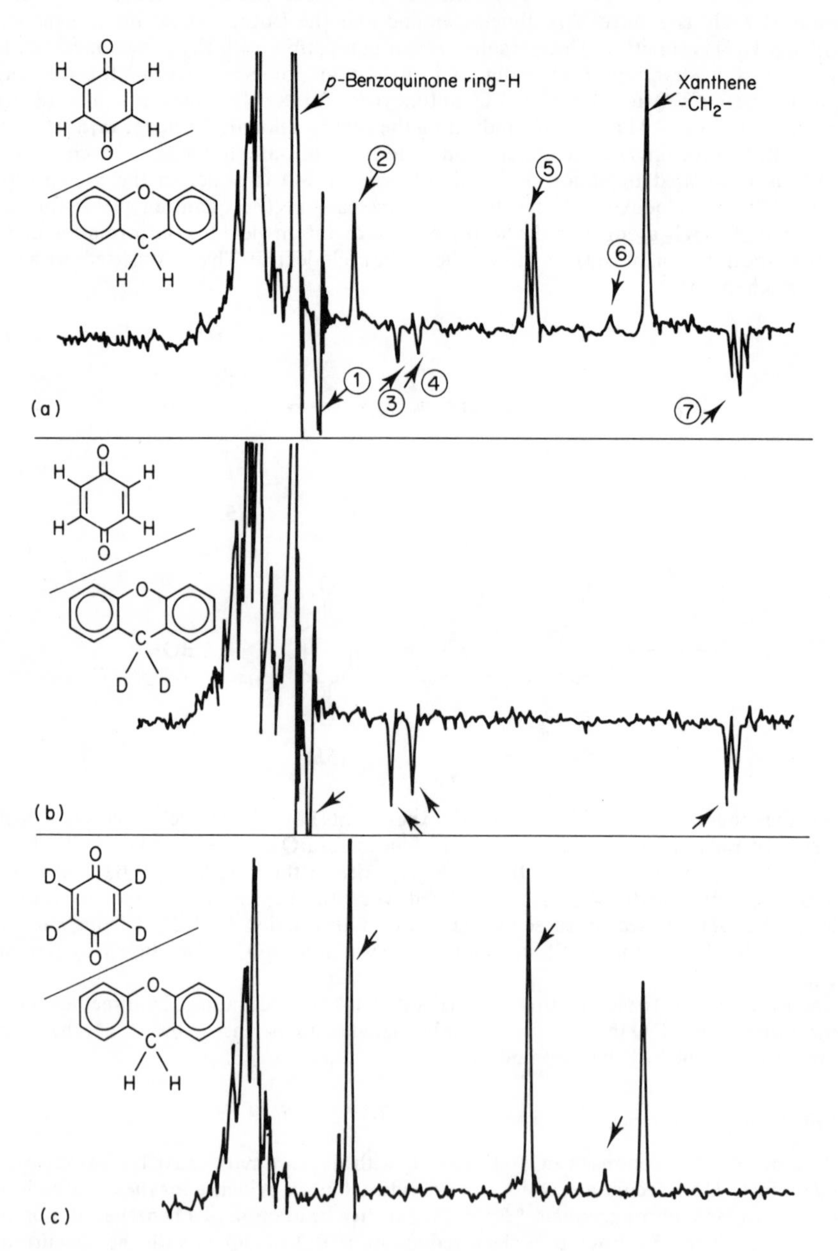

FIGURE 1. ^{1}H-NMR spectra observed in the photochemical reaction of BQ with xanthene in CCl_4 at 22°C: (a) BQ/xanthene, (b) BQ/xanthene-9,9-d_2, (c) BQ-d_4/xanthene

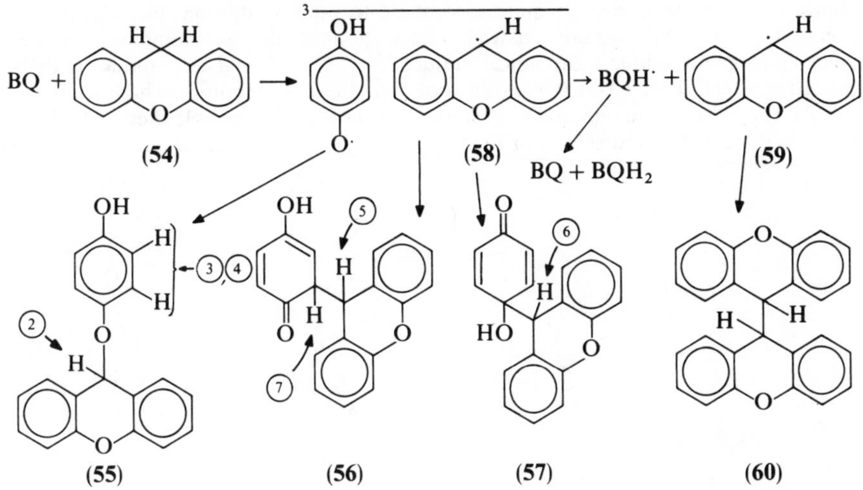

SCHEME 7

Based on the CIDNP effects obtained with xanthene-9,9-d_2 and BQ-d_4, polarized signals 2, 3, 4, 5, 6 and 7 indicated in Figure 1a were assigned as shown in Scheme 7. Alkyl-substituted naphthoquinones are photochemically reduced to the corresponding hydronaphthoquinone by **54**, while 2,3-dichloro-1,4-naphthoquinone (**61**) gives cross-coupling product **62** and a green-colored stable radical **63** in addition to the corresponding hydronaphthoquinones (equation 31)[170]. Photochemical reaction of AQ with toluene in the presence of *t*-butyl peroxide gives an adduct **64** in 43% yield (equation 32)[173].

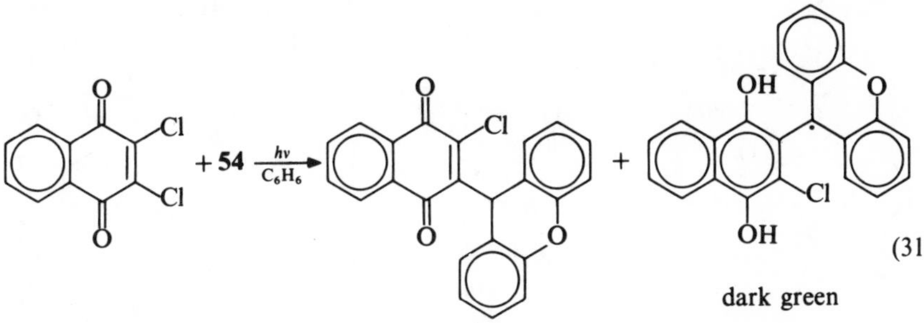

$$\textbf{61} + \textbf{54} \xrightarrow[C_6H_6]{h\nu} \textbf{62}\ (24\%) + \textbf{63}\ (17\%,\ \text{dark green}) \qquad (31)$$

$$\text{AQ} + \text{PhMe} \xrightarrow[C_6H_6/t\text{-BuOH}]{h\nu} \textbf{64}\ (43\%) \qquad (32)$$

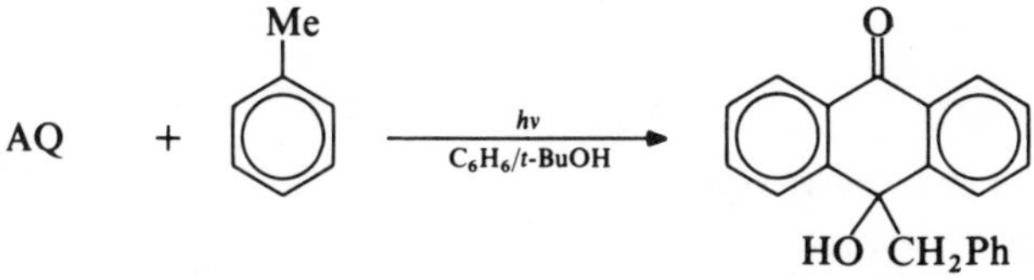

Since the quinone nucleus is quite susceptible to radical addition, photochemically generated organic and inorganic radicals attack quinone and semiquinone molecules, giving diamagnetic adducts or radical adducts[142, 169–170, 174]. These adducts can be detected by CIDNP or CIDEP, regardless of their thermal stabilities. The 9-xanthenyl radical is shown to add to semiquinone radicals at almost all the possible sites, as disclosed by CIDNP technique and shown in Figure 2.

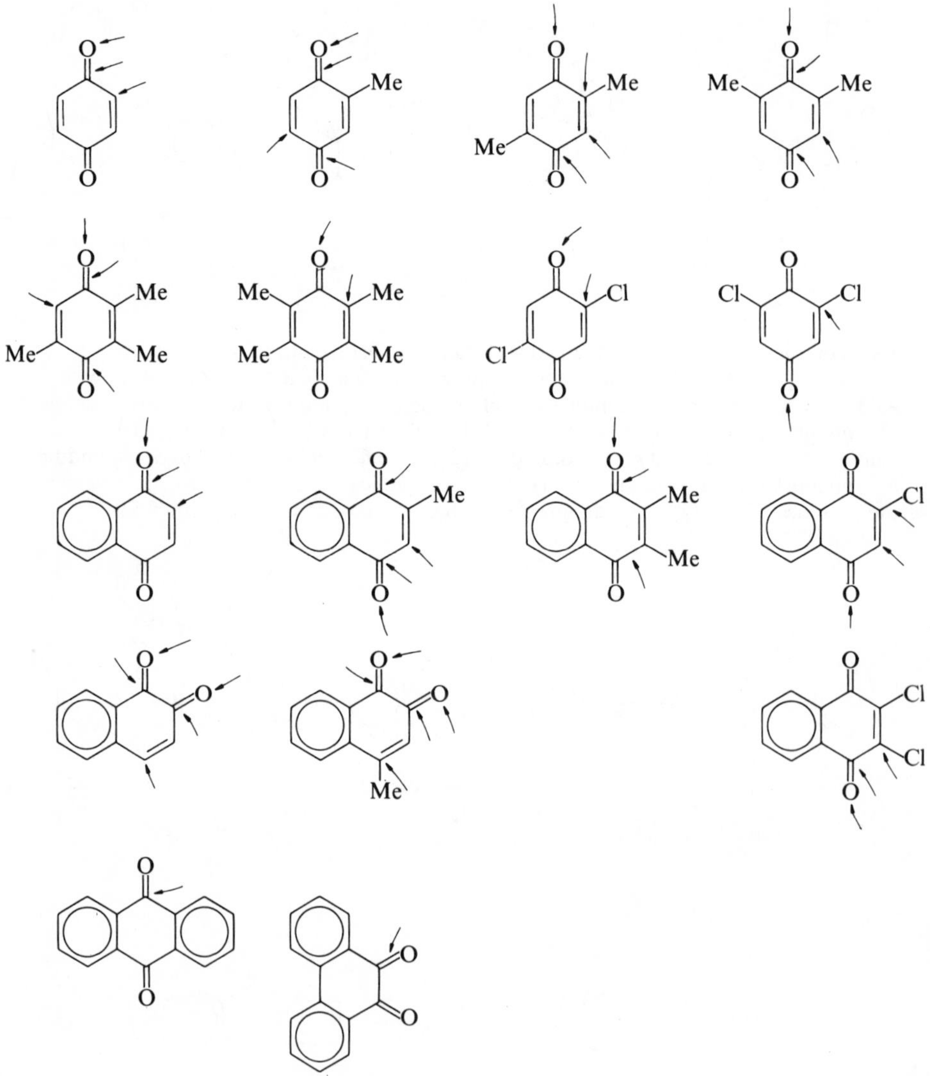

FIGURE 2. 9-Xanthenyl radical addition sites to semiquinone radicals

In the photochemical reaction with alkyl aromatics, PQ affords the 1,2-adduct **65** and the 1,4-adduct **66** with a ratio which depends on the structure of the hydrogen donor. During the course of this reaction, the methine hydrogens of **65** and **66** are strongly polarized but in opposite directions to each other. The formation of **66** has been explained in terms of a mechanism involving the homolysis of vibrationally excited **67** and subsequent in-cage recombination in singlet radical pair **68** (Scheme 8)[169, 172].

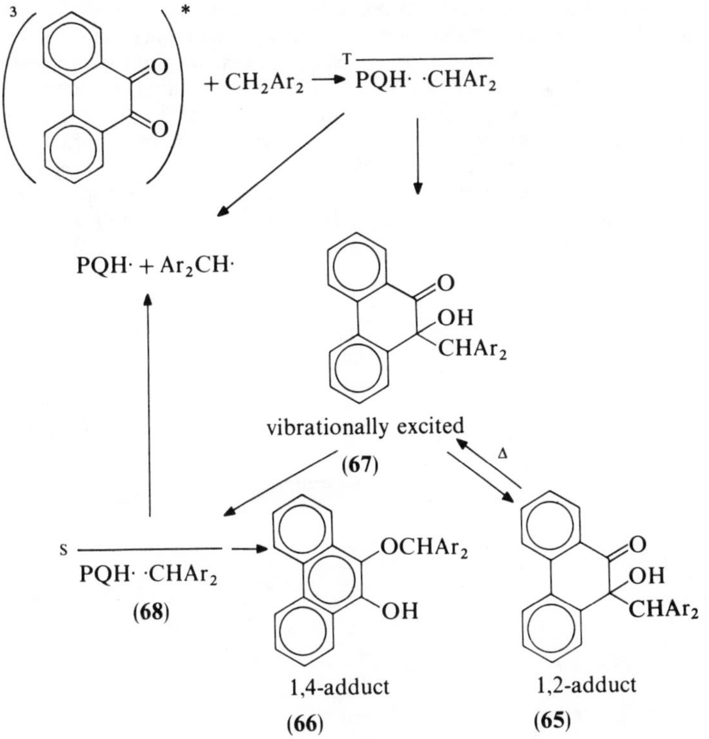

$$^{3}(PQ)^{*} + CH_2Ar_2 \longrightarrow \overline{PQH\cdot \ \cdot CHAr_2}^{T}$$

PQH· + Ar_2CH·

vibrationally excited **(67)**

$\overline{PQH\cdot \ \cdot CHAr_2}^{S}$ **(68)**

1,4-adduct **(66)**

1,2-adduct **(65)**

SCHEME 8

In the photochemical reaction of *o*-NQ with xanthene, a 1:1 adduct (**69**) is exclusively formed (equation 33), while 4-substituted *o*-NQ gives *o*-NQH_2 as a major product. Investigation by means of the CIDNP technique indicates that **69** is formed via in-cage

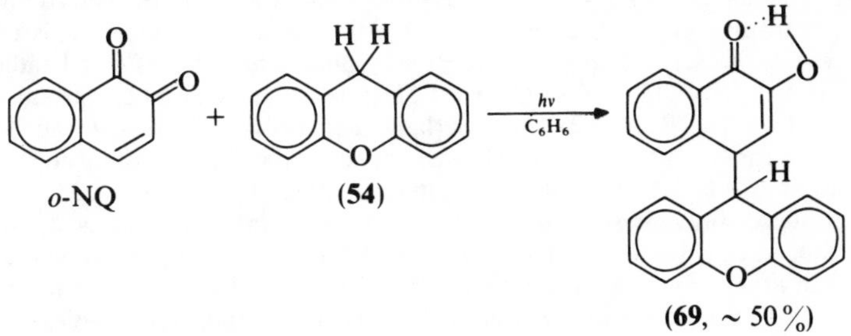

$$o\text{-NQ} + (\mathbf{54}) \xrightarrow[C_6H_6]{h\nu} (\mathbf{69}, \sim 50\%) \qquad (33)$$

recombination of a triplet radical pair and that an unstable adduct is also formed but does not accumulate in the reaction mixture[175,176].

4. From aldehydes

Products of the photochemical H-abstraction reaction of quinones from aldehydes are generally the acyl hydroquinone by C attack and the hydroquinone monoester by O attack. Although it was previously proposed that these products arose via scavenging of acyl radicals by a ground-state quinone[2,4], an in-cage recombination of acyl radical and semiquinone radical is now well recognized as a main pathway for their formation. This is due to two reasons: (*a*) CIDNP effects due to in-cage coupling products are generally observed[172,177] and (*b*) thermal generation of an acyl radical in the presence of quinone leads to reaction products which are quite different from those obtained from the photochemical reaction. For example, the thermal reaction of the acetyl radical with ground state PQ gives isomeric dimers **70** (equation 34), in contrast to a clean production of the acetate **71** from the photochemical reaction of PQ with acetaldehyde (equation 35)[1–4,178].

$$PQ + MeCHO + (t\text{-BuOO–}\overset{O}{\overset{\|}{C}})_2 \xrightarrow[C_6H_6]{\Delta} \quad (34)$$

(**70**, 90%)

$$PQ + MeCHO \xrightarrow[C_6H_6]{h\nu} \quad (35)$$

(quant.)

(**71**)

Irradiation of benzene solution of *o*-NQ and aliphatic aldehydes gives 3-acyl-1,2-naphthalenediol **72** and 1,2-naphthalenediol monoacyl esters **73a** (R = alkyl) and **73b** (R = alkyl) (equation 36)[179–182]. On the other hand, only **73a** (R = aryl) and **73b** (R = aryl) are formed in the reaction of *o*-NQ with aromatic aldehydes. This remarkable regioselectivity is ascribed to decreasing nucleophilic character of aroyl radical in the order: $CH_3\dot{C}{=}O > CH_2 = CH{-}\dot{C}{-}O > C_6H_5\dot{C}{=}O$[182]. With phenylacetaldehyde, *o*-NQ gives **74**, **75**, **76**, **77** and 1,2-diphenylethane (equation 37)[180]. Benzyl and 9-xanthenyl radical attack the 4-position of *o*-NQH˙, whereas acyl radicals attack the 3-position and the oxygen atom of *o*-NQH˙. The different pattern of attack by alkyl and acyl radicals on *o*-NQH˙ are successfully interpreted by taking into consideration the charge densities of the radicals concerned (Figure 3)[183]. The charge on C(4) of *o*-NQH is positive, while both C(3) and oxygen atom have negative charge. Therefore, benzyl radical (which has negative charge at the benzylic position) attacks the 4-position, and acetyl radical (which has

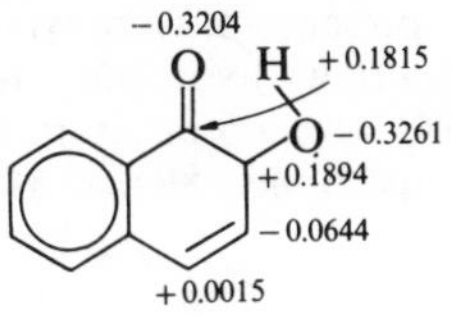

(a)

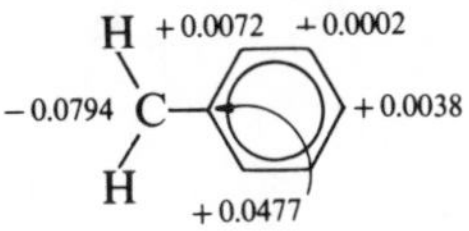

(b) 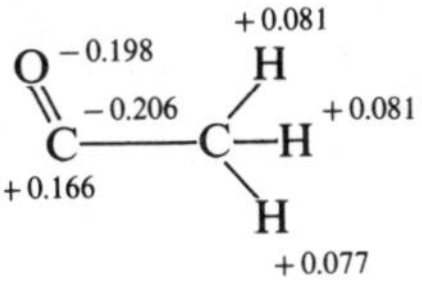 (c)

FIGURE 3. Charge densities of (a) 1,2-naphthosemiquinone radical, (b) benzyl radical and (c) acetyl radical

positive charge) attacks both the 3-position and the 3-position and the oxygen atom in *o*-NQH˙. The photochemical reaction of *o*-NQ with acetaldehyde was investigated in order to assess the relative contribution of 'in-cage' and 'out-of-cage' mechanisms. At a low temperature, almost all parts of the reaction proceed via an 'in-cage' mechanism but at 20°C at least 6.7% of the reaction proceeds via an 'out-of-cage' mechanism[184].

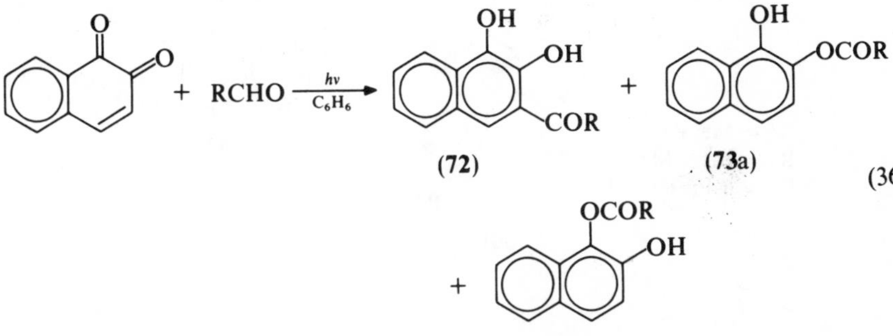

$$+ \ RCHO \xrightarrow[C_6H_6]{h\nu} \textbf{(72)} + \textbf{(73a)} + \textbf{(73b)} \qquad (36)$$

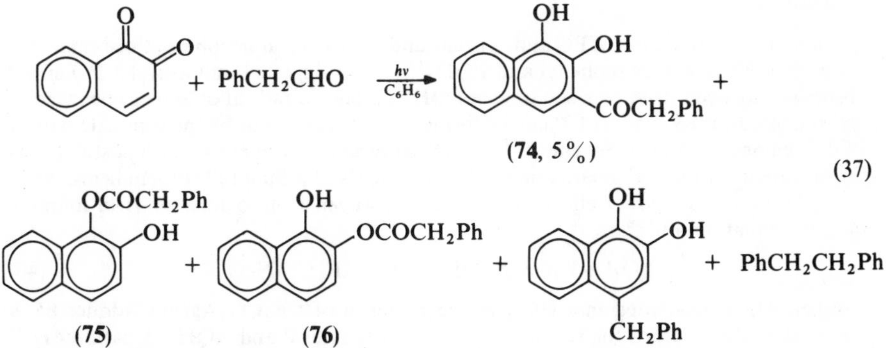

$$+ \ PhCH_2CHO \xrightarrow[C_6H_6]{h\nu} \textbf{(74, 5\%)} + \underbrace{\textbf{(75)} + \textbf{(76)}}_{(12\%)} + \textbf{(77, 27\%)} + PhCH_2CH_2Ph \qquad (37)$$

Irradiation of *o*-CQ in the presence of isobutyraldehyde affords **78** which is likely to arise from abstraction of the tertiary hydrogen α to the formyl group (equation 38)[185]. Photochemical reaction of NQ with α,β-unsaturated aldehydes gives adducts **79**, which are converted into α- and β-lapachones, **80** and **81** (equation 39)[186].

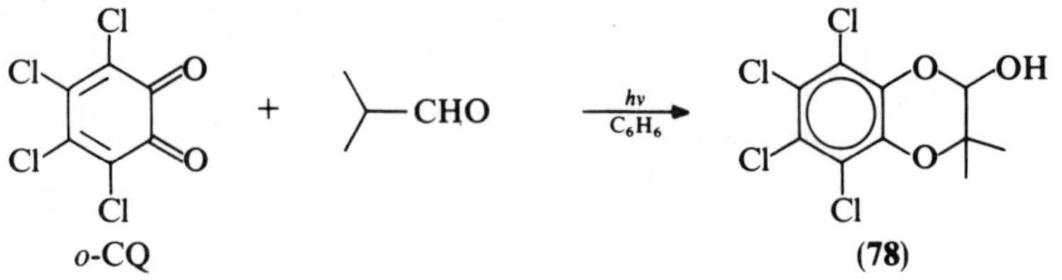

o-CQ + CHO —($h\nu$, C_6H_6)→ (**78**) (38)

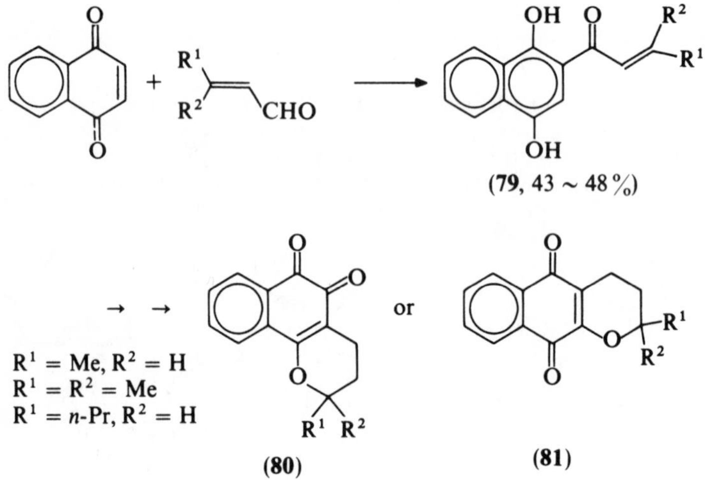

(**79**, 43 ~ 48%)

$R^1 = Me, R^2 = H$
$R^1 = R^2 = Me$
$R^1 = n\text{-}Pr, R^2 = H$

(**80**) or (**81**) (39)

5. *Miscellaneous*

Photochemical reaction of PQ and benzene under nitrogen atmosphere affords mainly a 1:1 adduct **82** as well as biphenyl and PQH_2[187]. It is suggested that a triplet PQ either abstracts hydrogen from benzene to give PQH and phenyl radical or adds to benzene to give the biradical **83**. Adduct **82** may be formed either from **83** or by the geminate pair of $PQH^{\cdot}$ and phenyl radical (Scheme 9)[187]. A similar biradical adduct **84** is also postulated as the precursor for the ^{19}F polarization in the photolysis of *p*-fluoranil (FQ) in benzene[161].

Triplet excited quinones efficiently abstract hydrogen from oximes to yield iminoxy radicals (equation 40)[188].

$$^3(Q)^* + R_2C{=}N{-}OH \rightarrow QH^{\cdot} - R_2C{=}N{-}O^{\cdot} \qquad (40)$$

When AQ is irradiated in $CHCl_3$ in the presence of CF_3CO_2Ag, the adduct **85** is obtained as the result of the recombination of CCl_3 radical and $AQH^{\cdot}$ (equation 41)[73]. This reaction is suppressed by the presence of oxygen or the radical scavenger 2,6-di-*t*-butylphenol.

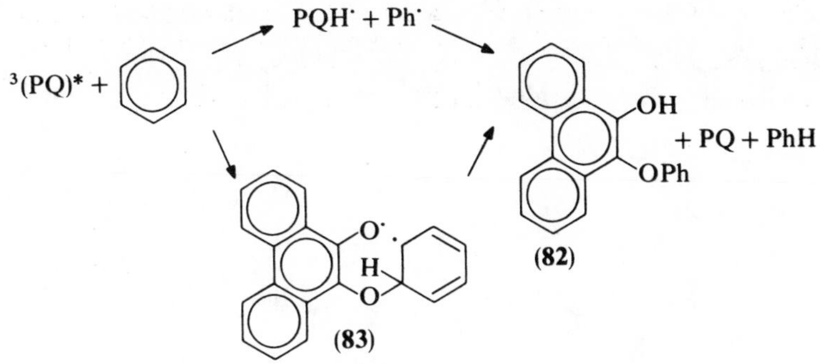

$^3(PQ)^* +$ benzene → $PQH^{\bullet} + Ph^{\bullet}$ → (82) + PQ + PhH

(83)

(82)

SCHEME 9

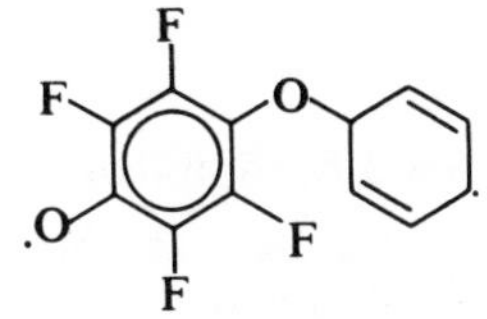

(84)

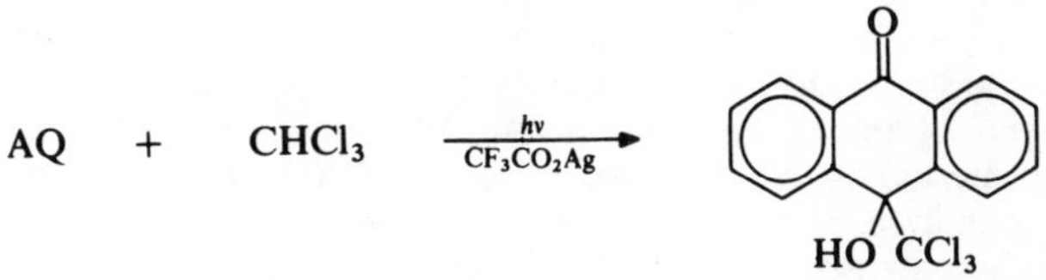

$$AQ + CHCl_3 \xrightarrow[CF_3CO_2Ag]{h\nu} \text{(85, 30–40\%)} \quad (41)$$

Photolysis of various quinones in the presence of organotin and organolead compounds yields paramagnetic organometal adducts of the quinones. Time-resolved CIDEP observations indicate that organometals having relatively low ionization potentials quench the carbonyl triplet via a charge transfer mechanism[174]. Photochemically generated trimethylstannyl radical adds to the carbonyl oxygen of 2,6-di-*t*-butyl-1,4-benzoquinone to give **86** (R = $SnMe_3$), while $Mn(CO)_5$ radical attacks only at the carbon–carbon double bond of the quinone to give **87** (R = $Mn(CO)_5$)[174]. Trimethylsilyl radical adds both to the carbonyl oxygen and to the ring carbon to give **88** (R = $SiMe_3$) and **89** (R = $SiMe_3$) (equations 42 and 43)[142].

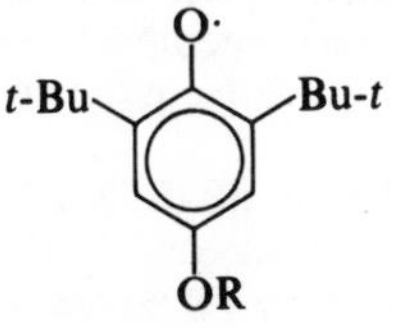

(86) 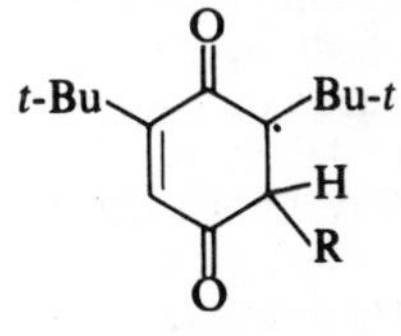 (87)

t-Bu, Bu-t (2,6-di-t-butyl-1,4-benzoquinone) + Me_3SiH $\xrightarrow{h\nu}$ t-Bu, Bu-t, O·, OH (phenoxyl radical) + $Me_3Si\cdot$ (42)

$Me_3Si\cdot$ + t-Bu, Bu-t (quinone) ⟶ t-Bu, Bu-t, O·, $OSiMe_3$ (**88**, R = $SiMe_3$) + t-Bu, Bu-t, ·, H, $SiMe_3$ (**89**, R = $SiMe_3$) (43)

C. Photoreduction in Organized Assemblies

Photoreductions of quinones in organized assemblies have attracted considerable interest in view of their potential use for an efficient charge separation as well as because of the structural similarity of these assemblies to natural membranes. Photochemical solar energy conversion, in which quinones are often used as photosensitizers as well as electron acceptors, has been extensively treated in recent books and reviews[132,189]. The most important role of the microscopic heterogeneous environment imposed by micelles, microemulsions, vesicles and polyelectrolytes may be the significant retardation of the back electron transfer in the photochemically generated ion pairs[83].

In the presence of anionic and non-ionic micelles, irradiation of ASQ and ADSQ in aerobic phosphate buffer (pH 8.0) gives $ASQ^{-\cdot}$ and $ADSQ^{-\cdot}$, respectively. These are considerably long-lived as compared with irradiation without surfactants[166,190,191]. Interestingly, photohydroxylation of ASQ does not occur, since the hydroxy radicals are efficiently scavenged by the surfactants (see Section V.D.2). The remarkable stability of $ASQ^{-\cdot}$ and $ADSQ^{-\cdot}$ under micellar conditions has been ascribed to the formation of association complex between the quinone and the surfactant[190]. Photoreductions of 9,10-phenanthraquinone-3-sulfonate and ASQ are greatly enhanced by the presence of cationic surfactants, whereas photoreduction of 1,2-naphthoquinone-4-sulfonate is highly retarded[12,192]. Apparently, the anionic quinone forms a complex with the cationic head of the micelles, whose surface is also expected to be covered with excess hydroxide ion, and therefore supplies effective high concentration of hydroxide ion for the photoreduction of the quinone. An irreversible electron transport system is developed in a cationic reversed micelle by using ASQ as the photosensitizer (Scheme 10)[121]. The key step is a photoinduced electron transfer from $ASQ^{-\cdot}$ to benzonitrile which can be induced with red light ($\lambda > 450$ nm).

In the presence of anionic sodium lauryl sulfate micelles, electron transfer from $^3(DQ)^*$ to Fe^{3+} ion is markedly accelerated[86]. Laser photolysis experiments have shown that the intramolecular electron transfer proceeds via electron flux from the donor on the surface to the acceptor inside the micelle with a rate constant of $2.5 \times 10^7\ s^{-1}$ (Scheme 11)[86]. Photoreduction of ASQ and ADSQ was also studied in plasticized and non-plasticized polymer films[193,194].

Photolysis of AQ in aqueous sodium dodecyl sulphate (SDS) led to the formation of AQH_2 and the surfactant–AQ adduct **90**, according to Scheme 12[195–197,201]. The triplet

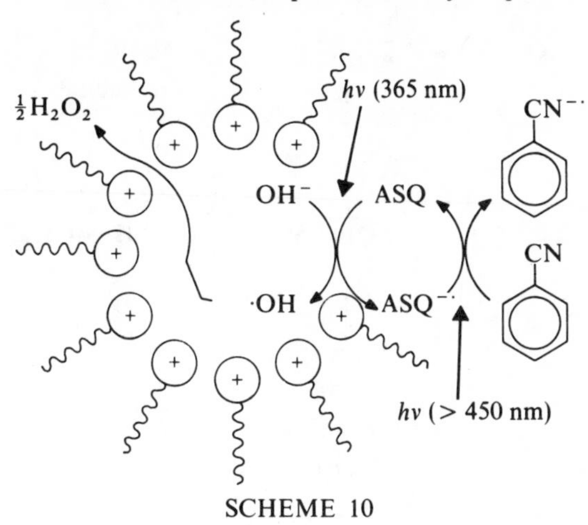

SCHEME 10

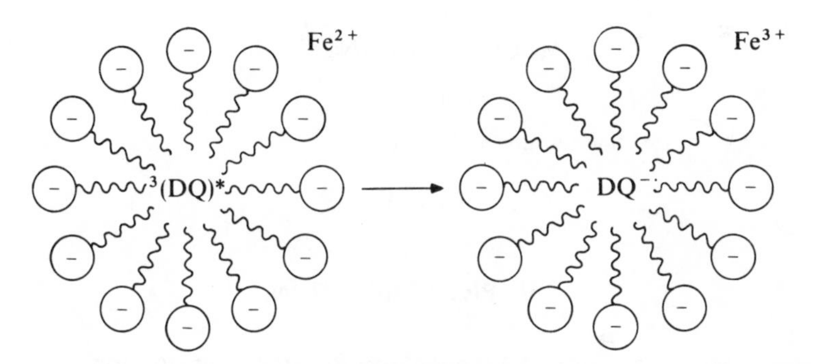

SCHEME 11

radical pair of AQH˙ and the surfactant radical (S˙) undergoes competitive intersystem crossing (ISC) to the singlet pair and separation by diffusion. The triplet–singlet ISC of the radical pair occurs via electron-nuclear hyperfine interaction and should be reduced in a magnetic field because of the Zeeman splitting of triplet sublevels. Consistent with this theory, an external magnetic field (2600 G) increases the yield of free radicals at the expense of the competing ISC[195, 197, 201]. Similar external magnetic field effects are observed in BQ[198], DQ[198], 2,5-dimethyl-1,4-benzoquinone[198, 199], 1,2-benzanthraquinone[198], anthraquinone-2-carboxylic acid[198] and ASQ[198] in the SDS micelle as well as in the ion radical pair photochemically generated from DQ and diphenylamine[200]. However, external magnetic field effects which are still observed at much higher field (1.34 T) apparently require another mechanism (e.g. relaxation mechanism)[202b] instead of the hyperfine coupling mechanism. CIDEP studies on the NQ/SDS system reveal the total emission pattern of the spectra due to NQH˙ and surfactant radical at an early stage after pulsed excitation, and this should be correlated with the triplet mechanism[92, 202a].

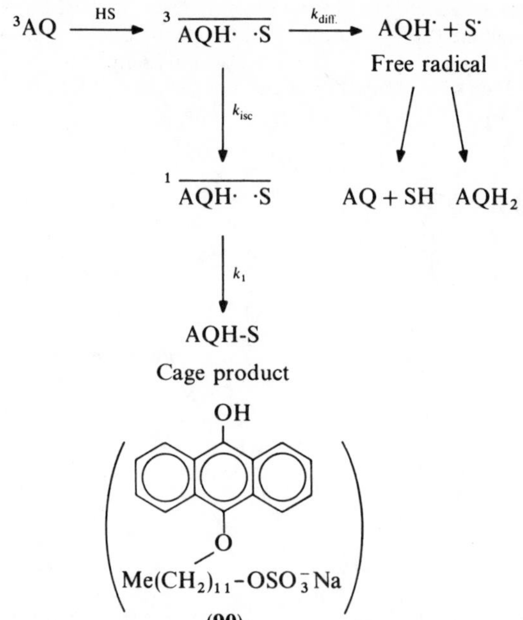

$$^{3}AQ \xrightarrow{HS} {}^{3}\overline{AQH\cdot \ \cdot S} \xrightarrow{k_{diff.}} AQH^{\cdot} + S^{\cdot}$$

Free radical

$\downarrow k_{isc}$

$${}^{1}\overline{AQH\cdot \ \cdot S} \qquad AQ + SH \quad AQH_2$$

$\downarrow k_1$

AQH-S

Cage product

OH

$Me(CH_2)_{11}-OSO_3^-Na$

(90)

HS, Sodium dodecyl sulfate

SCHEME 12

IV. PHOTOADDITION

A. Cycloaddition to Simple Alkenes and Related Compounds

In common with other saturated and unsaturated carbonyl compounds, *p*-quinones afford cycloaddition products upon irradiation with alkenes. With BQ itself, which has a lowest triplet state of (n, π^*) configuration, addition to the carbonyl group is the only mode of cyclization that has so far been observed, although reaction at the olefinic bond to give a cyclobutane might have been possible. Indeed, cyclobutane formation was found with other *p*-quinones such as CQ and DQ having lowest triplet (π, π^*) states[4].

Therefore it may be attractive to propose that *p*-quinones whose lowest triplet state is (n, π^*) should react with alkenes to give oxetanes, while cyclobutanes would result from the corresponding reactions of quinones whose lowest triplet state is (π, π^*)[4, 203].

A theoretical calculation predicts that progressive introduction of methyl groups into the nucleus of 1,4-benzoquinone will lower the energy level of $^{3}B_{3q}(\pi, \pi^*)$ and raise that of $^{3}B_{1g}(n, \pi^*)$[204].

Of particular interest in this regard are quinones such as 1,4-naphthoquinones that are capable of forming both kinds of adducts simultaneously[203, 205]. Calculations on NQ suggest that the lowest triplet is again of (n, π^*) character, in agreement with experiment. An examination of these calculations and the photochemistry of NQ, however, suggests that several excited triplets lie very close in energy, and that all of them may be populated within the lifetime of the lowest triplet state[204].

Support for the intermediacy of a preoxetane biradical is provided by two observations: (1) *cis*- and *trans*-2-butene add to photochemically excited BQ to afford the same mixture of spiro oxetanes **91a** and **91b**[206]; (2) 1,2,4-trioxane (**92**) and sulfone **93** are formed by trapping of the preoxetane biradicals by molecular oxygen[207, 208] and sulfur dioxide[209], respectively.

(91a)  **(91b)** 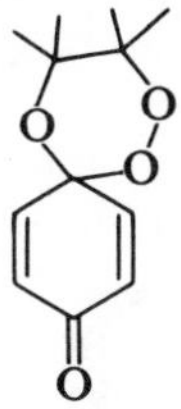 **(92)** 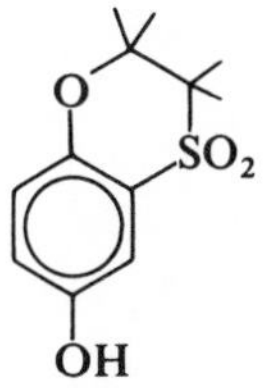 **(93)**

Elegant and extensive studies on the biradical trapping by molecular oxygen have led to three conclusions.

(1) The yields of 1,2,4-trioxane adducts are markedly raised by the use of monochromatic light from an argon ion laser in non-polar solvents (CCl_4 and $CFCl_3$) at high oxygen pressure (ca. 11 atm).

(2) Under these conditions the biradical trapping is quite general in both inter- and intramolecular photocycloaddition reactions of *p*-quinones.

(3) The intermediates which are trapped by oxygen are frequently not the preoxetane biradicals but rather the charge transfer exciplex species[210].

The last conclusion, which is of most mechanistic importance, is obtained from the study of photocycloaddition of BQ to unsymmetrical olefins such as *t*-butylethylene (Scheme 13)[210, 211]. When this reaction is conducted under anaerobic conditions, oxetanes **95** and **96** are formed with a **95**/**96** ratio of 0.25. Reactions conducted under increasing pressure of oxygen lead to the formation of a single trioxane (**94**), which appears to result from oxygen trapping of the more stable biradical that gives rise to **96**. The independence of the oxetane isomer ratio (**95**/**96**) of the oxygen pressure implies that oxygen is not trapping the more

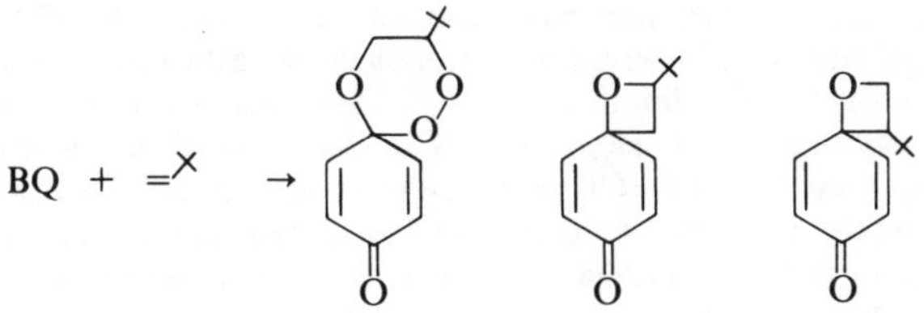

BQ + =X →

	(94)	**(95)**	**(96)**
No O_2	0	25	75(%)
11 atom O_2	43	13	44(%)

SCHEME 13

stable preoxetane biradical, but trapping instead a species **97** that precedes the partitioning to the two preoxetane biradicals **98** and **99** (Scheme 14)[210]. Triplet exciplex may be trapped by oxygen or proton, leading to a biradical (**100**) or to a cation (**101**). Presumed triplet

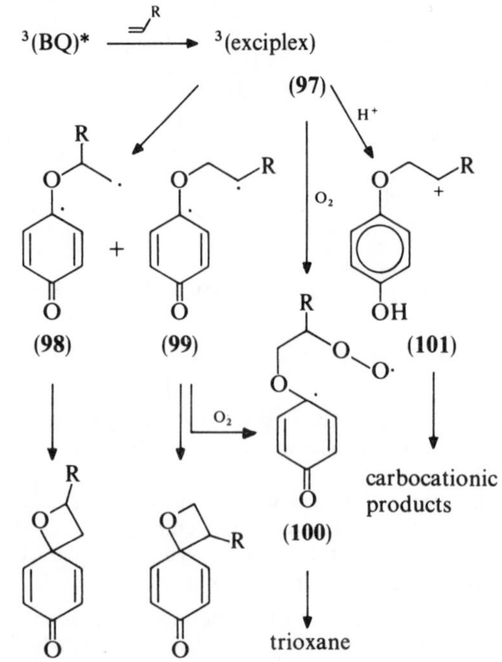

SCHEME 14

BQ–olefin charge transfer exciplex apparently has a polar character as it can be trapped by a carboxyl group both intermolecularly and intramolecularly. Thus, argon ion laser irradiation of BQ in the presence of 4-pentenoic acid (**102**, R = H) or *trans*-5-phenyl-4-pentenoic acid (**102**, R = Ph) in non-polar media affords δ-lactones **103**[212]. Formation of these products is explained in terms of intramolecular carboxylic acid trapping of the charge transfer complex (cf. **104**) (Scheme 15)[212]. Analogous products **108** and **110** are produced in the photochemical reactions of BQ with *trans*-5-phenyl-4-pentenamide (**106**) (equation 44) and 4-penten-1-ol (**109**) (equation 45), respectively. This type of collapse in a triplet charge transfer exciplex would require a tight radical ion pair, and indeed these reactions occur only in very non-polar solvents such as CCl_4 and CCl_3F[212]. On the other hand, the photoreaction of BQ with 5-phenyl-4-pentynoic acid (**111**) leads to the formation of ketolactone **114** via **113**, indicating that the polarity of the quinone–acetylene exciplex is insufficient to induce trapping by carboxylic acid to give **112** (Scheme 16)[212]. Irradiation of BQ along with cyclooctatetraene **115** in acetic acid under anaerobic conditions leads to the formation of **116** and **117** which appear to arise from a carbocationic intermediate (**118**)[209] (Scheme 17). CNDO calculations for the triplet BQ–ethylene system predict the existence of a quite stable charge transfer exciplex, in line with a mechanism involving a triplet exciplex between BQ and olefin[211].

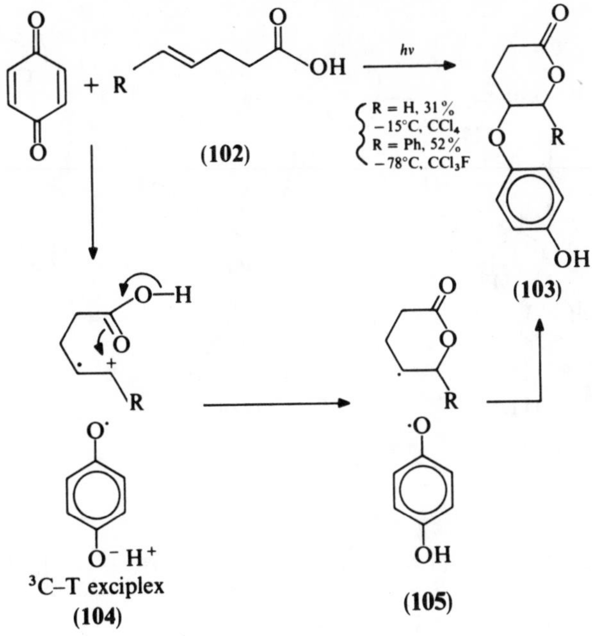
hν
OH
R
(102)
R = H, 31%
−15°C, CCl_4
R = Ph, 52%
−78°C, CCl_3F
OH
(103)
O—H
R
$O^- H^+$
^{3}C–T exciplex
(104)
R
OH
(105)

SCHEME 15

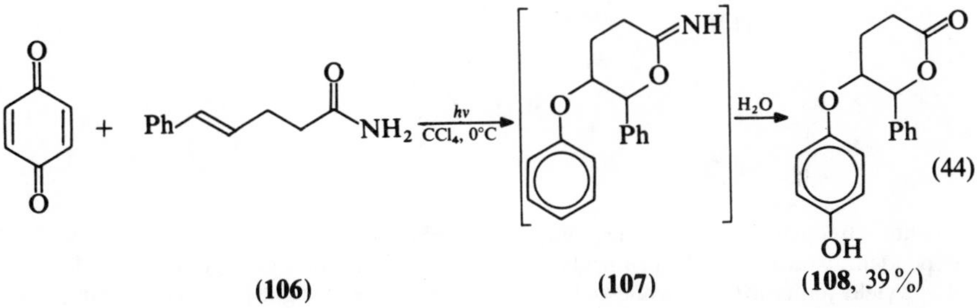
Ph
NH_2
hν
CCl_4, 0°C
NH
Ph
H_2O
Ph
OH
(106)
(107)
(108, 39%)

(44)

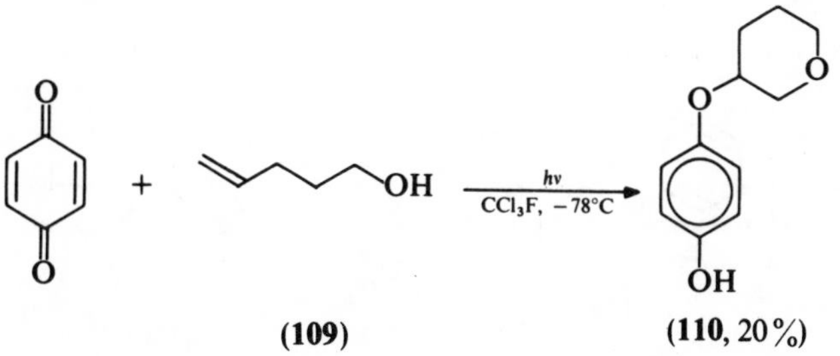
OH
hν
CCl_3F, −78°C
OH
(109)
(110, 20%)

(45)

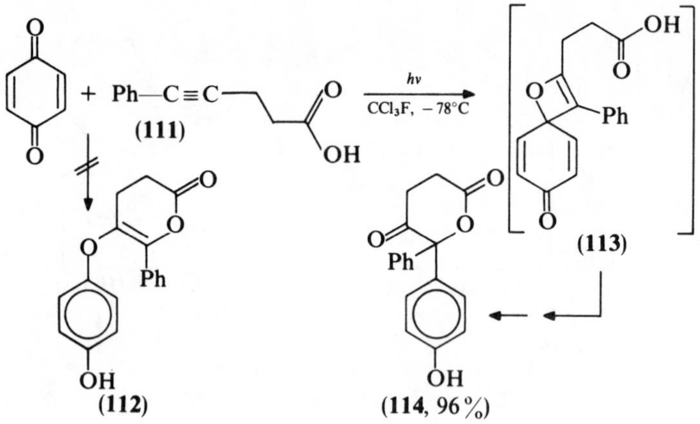

+ Ph—C≡C— ... OH (111) $\xrightarrow[CCl_3F,\ -78°C]{h\nu}$ (113) → (114, 96%) ; ↛ (112)

SCHEME 16

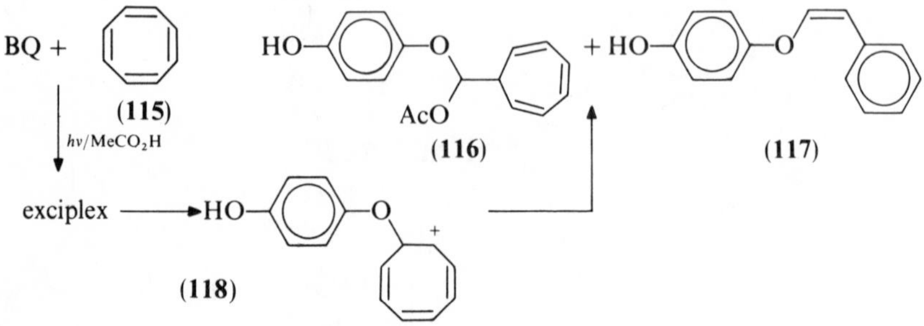

BQ + (115) $\xrightarrow{h\nu/MeCO_2H}$ exciplex → (118) → (116) + (117)

SCHEME 17

Photochemistry of cyclophane compounds **119a** and **119b** is apparently configuration-dependent; irradiation of **119a** leads to an efficient formation of cage compounds **120** in 90 % yield (equation 46), while only a higher molecular weight product is formed from **119b**[66, 213]. Photoaddition of BQ to the nitro allyl compound **121** proceeds with

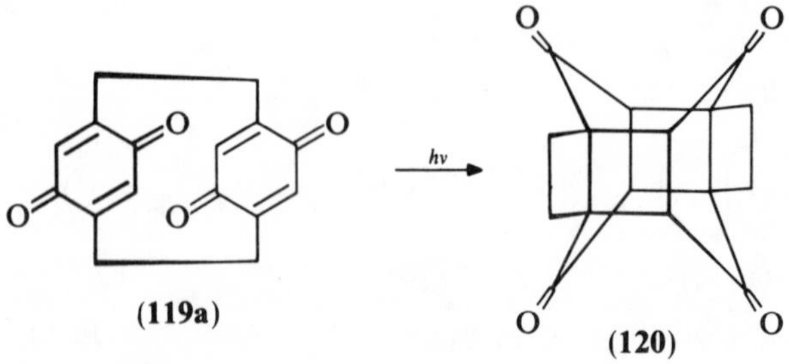

(119a) $\xrightarrow{h\nu}$ (120) (46)

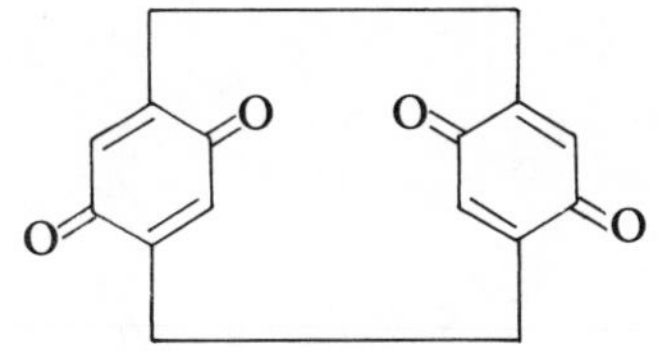

(119b)

migration of nitro group to furnish **122** in a low yield (equation 47)[214]. Photocycloaddition of BQ to norbornadiene (**123**) gives the isomeric 1:1 adducts **124**, **125**, **126** and **127** in a ratio of 48:16:21:15 (equation 48), whereas irradiation of a solution of BQ and quadricyclane under the same conditions yields a mixture of the two *exo* adducts **124** and **125** in a ratio of 56:44[215].

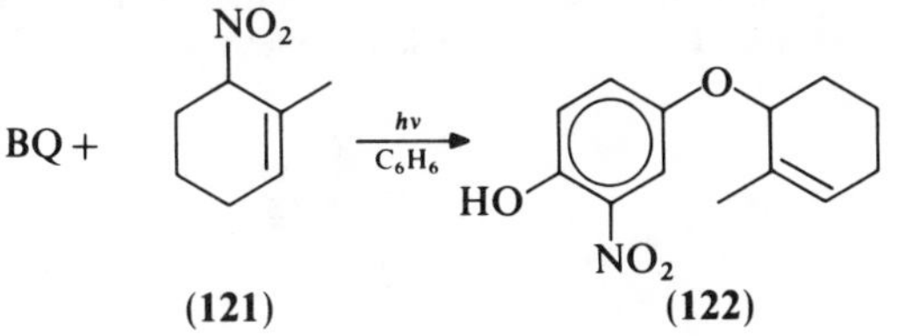

BQ + (121) $\xrightarrow[C_6H_6]{h\nu}$ (122) (47)

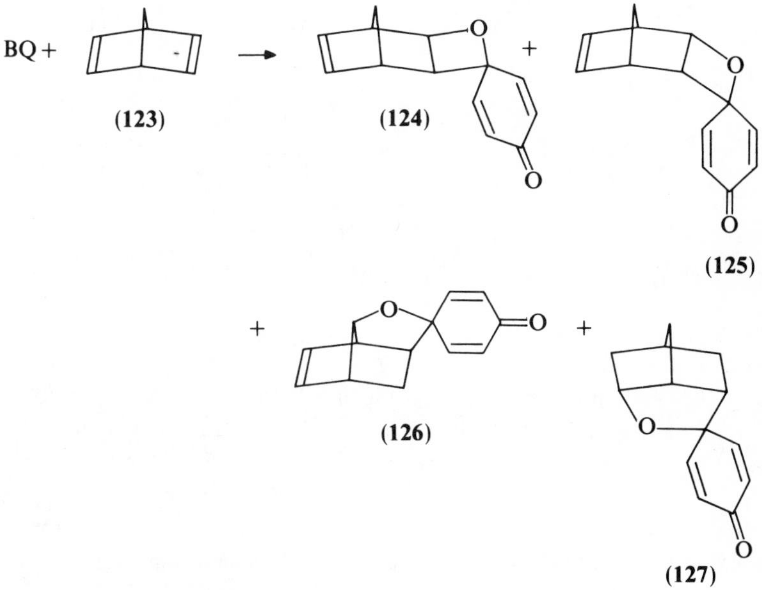

BQ + (123) → (124) + (125) + (126) + (127) (48)

An interesting example of the formation of the 2:1 adduct **128** together with the monoadduct **129** has been reported in the photoreaction of NQ with norbornene (equation 49)[216]. Although the reaction mechanism for the formation of this unusual adduct has not been revealed yet, it may be formed via the reaction of triplet exciplex between NQ and norbornene with another NQ molecule.

NQ + [norbornene] $\xrightarrow[C_6H_6]{h\nu}$ (128) + (129) (49)

Substituted naphthoquinones such as MQ and 2-chloro-1,4-naphthoquinone (**132**) undergo facile photocycloadditions with electron-rich olefins to give the corresponding cyclobutanes **130**, **133** and **136** in good yields (equations 50–52). These cyclobutane

MQ + =/OAc $\xrightarrow[C_6H_6]{h\nu}$ (**130**) (98%) $\xrightarrow[THF]{H_2SO_4}$ (**131**) (92%) (50)

(**132**) + = $\xrightarrow[C_6H_6]{h\nu}$ (**133**) $\xrightarrow[CHCl_3]{Et_3N}$ (**134**) (51)

MQ + (**135**) $\xrightarrow[K_2CO_3, C_6H_6]{h\nu > 370\ nm}$ (**136**) $\xrightarrow[AcOH-H_2O]{CAN}$ (**137**)

R′ = Me or $SiMe_3$ CAN, cerium (IV) ammonium nitrate (52)

photoproducts are photo-, acid-, or base-labile, and can be transformed, depending upon their substitution patterns, into a variety of useful compounds such as naphthofurans (**131**), 1,2-dihydrocyclobuta[*b*]naphthalene-3,8-dione (**134**) and *β*-oxoalkyl-substituted 1,4-naphthoquinones (**137**)[217–220]. Introduction of a *β*-oxoalkyl chain into the *p*-benzoquinone nucleus is also achieved by this photocycloaddition method (**138** → **139**) in

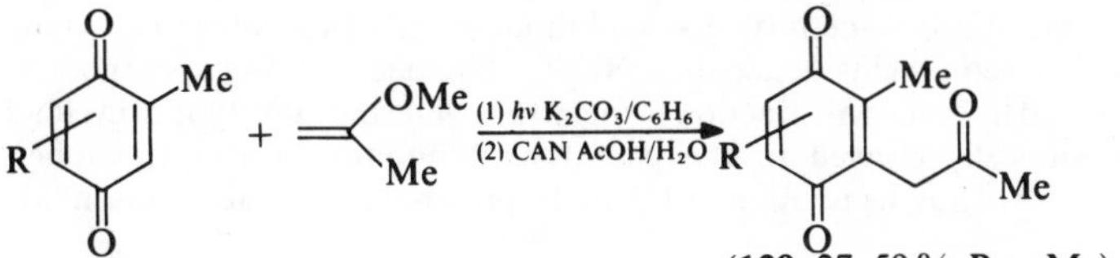

$$\mathbf{138} + CH_2{=}C(OMe)Me \xrightarrow[\text{(2) CAN AcOH/H}_2\text{O}]{\text{(1) }h\nu\ K_2CO_3/C_6H_6} \mathbf{139} \qquad (53)$$

(138)

(139, 27–59 %; R = Me)

CAN, cerium (IV) ammonium nitrate

moderated yields (equation 53)[221]. Analogous β-oxoalkyl-substituted naphthoquinones (**141**) are directly formed in the photoreaction of **140** and 2-methoxy-1-alkenes (equation 54)[221]. 2-Allyl-1,4-naphthoquinone (**144**) is synthesized by a two-step sequence including

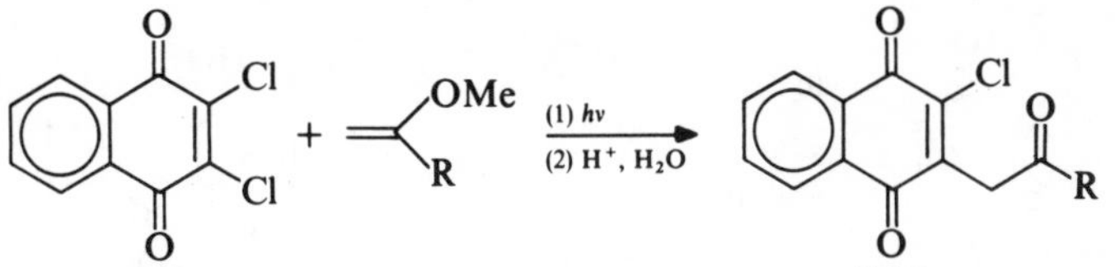

$$\mathbf{140} + CH_2{=}C(OMe)R \xrightarrow[\text{(2) H}^+\text{, H}_2\text{O}]{\text{(1) }h\nu} \mathbf{141} \qquad (54)$$

(140) (141)

photocycloaddition of NQ to allyltrimethylsilane (**142**) and subsequent treatment of the cyclobutane adduct **143** with a Lewis acid (equation 55)[222]. The same reaction sequence with olefin **145** leads to the formation of the tricyclic compound **147** via the cyclobutane **146** (equation 56)[222].

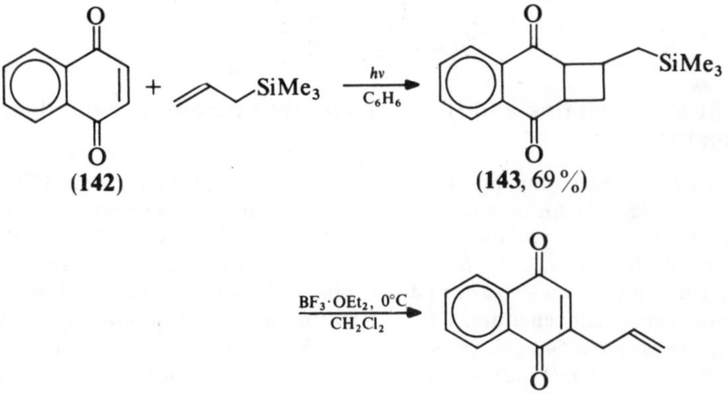

$$\mathbf{142} + CH_2{=}CHCH_2SiMe_3 \xrightarrow[C_6H_6]{h\nu} \mathbf{143} \xrightarrow[CH_2Cl_2]{BF_3\cdot OEt_2,\ 0°C} \mathbf{144} \qquad (55)$$

(142) (143, 69 %) (144, 92 %)

$$\text{NQ} + \mathbf{145} \longrightarrow \mathbf{146} \longrightarrow \mathbf{147} \qquad (56)$$

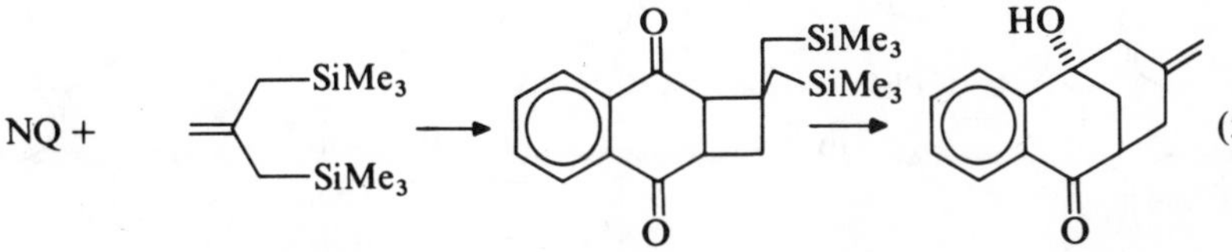

(145) (146) (147)

Upon irradiation, 2-methoxy-1,4-naphthoquinone (**148**) adds to styrene and 1,1-diphenylethylene in yields better than NQ[223]. Cycloadduct **149** undergoes a secondary photochemical hydrogen abstraction to yield **151**, while the same type of cycloadduct (**150**) is photochemically cleaved to the parent naphthoquinone and 1,1-diphenylethylene (Scheme 18)[223]. Thus, irradiation of **150** in the presence of styrene results in the formation of **151**.

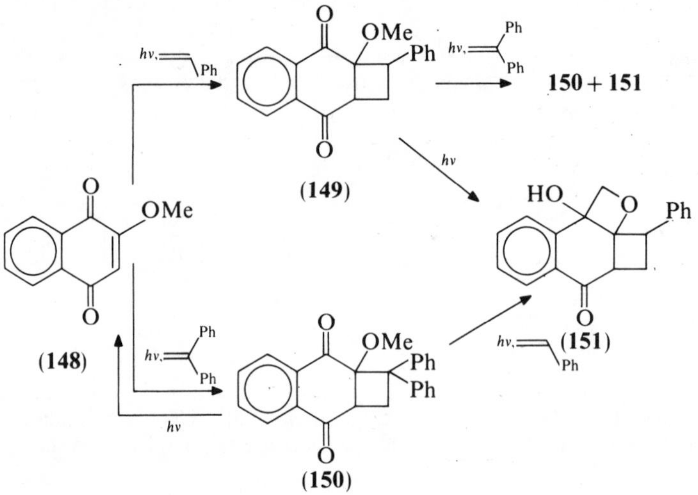

SCHEME 18

B. Addition to Olefins and Heteroaromatic Compounds via Electron Transfer

Photochemical reactions of electron-deficient 1,4-naphthoquinone **152** with 1,1-diarylethylene **153** constitute a novel synthetic route to polycyclic aromatic compounds **155** (equation 57)[224]. Typically, dinaphtho[1,2-*c*; 2,3-*e*]pyrene (**159**) is synthesized by three steps starting from **156** and **157** via this photocyclization (equation 58)[225]. The photocyclization involves two key photochemical reactions, (1) the initial substitution of an electronegative halogen atom, Y in **152**, by the ethylenic group to give an adduct **154** and (2) the subsequent $[4_\pi + 2_\pi]$ cyclization of **154** with the liberation of HX to give **155**. The first step is apparently induced by charge transfer interaction between the quinone and diarylethylene and is highly dependent upon the reduction potential of quinone and the oxidation potential of the diarylethylene. Therefore the first step competes with usual

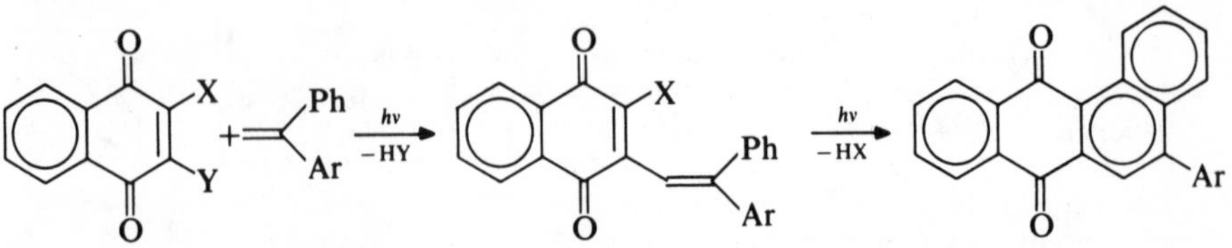

(152) (153) (154) (155) (57)

(156) (157) (158)

(58)

(1) $h\nu/I_2$ (2) $LiAlH_4$

(159)

cyclobutane formation when the charge transfer interactions are weak (Scheme 19)[226]. Cyclobutane **162** is not a precursor of **161**, since it does not give **161** under the same photolytic conditions. A definitive evidence for an electron transfer process is provided by CIDNP in the reaction of **140** with **160** (X = OMe)[227]. Besides 1,1-diarylethylenes, other electron-rich heterocyclic aromatic compounds[62, 228, 229], silyl enol ether[230, 231] and dicyclopropyl ether[232] can be employed in this type of electron transfer reaction. In the photochemical reaction of **156** with dicyclopropyl ethylene (**210**), compounds **211** and **212** are obtained as the coupling products (equation 59), indicating that the coupling takes place via the ionic intermediate **213** rather than via the biradical **214**[232]. Irradiation of naphthoquinones **140**, **156** and **195** with electron-rich heterocyclic aromatic compounds such as furans, thiophenes and pyrroles give rise to the coupling of heterocycles to the naphthoquinone nucleus[62, 228, 229]. In non-polar solvents such as benzene, **140** and furan or pyrrole form charge transfer complexes, which upon photoexcitation give coupling products. Interestingly, selective excitation at the CT bands gives a better yield of the coupling products[62], in contrast to the CQ/durene and **16**/DPA systems. Preferential coupling at the α-position of these heterocycles can be explained in terms of formation of their cation radical intermediates[62, 228, 229].

Since the original report[233a] on the photochemical synthesis of 5-phenyl-7,12-benz[*a*] anthracenedione from **156** and 1,1-diphenylethylene which appeared in 1975, more than 150 new polycyclic aromatic and heteroaromatic compounds have been synthesized by this reaction. These results are summarized in Table 7.

C. Cycloaddition to Conjugated Dienes and Acetylenes

Photocycloadditions of quinones to cyclic dienes and trienes afford a variety of products, depending upon the structure of substrate[209, 240–242]. Cage compound **216** is isolated in the reaction of DQ with cycloheptatriene together with **215** (equation 60)[243]. Irradiation of furanobenzoquinone **217** along with 1,3-pentadiene produces cyclobutane **218** in good yield (equation 61)[413]. Irradiation of BQ in the presence of a ketenimine gave an imino-oxetane **219** as the primary cycloadduct and β-lactam **221** presumably via ring-opening to the diradical **220** as the secondary photoproduct. The latter photoproducts are

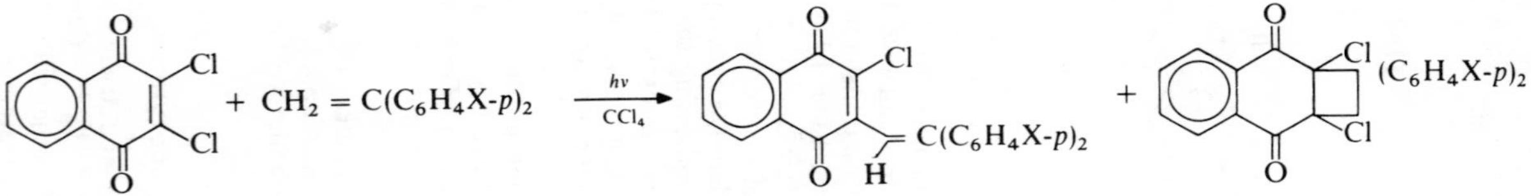

(140) + $CH_2 = C(C_6H_4X\text{-}p)_2$ (160) $\xrightarrow[CCl_4]{h\nu}$ (161) $C(C_6H_4X\text{-}p)_2$ + (162) $(C_6H_4X\text{-}p)_2$

X	Product distribution (%)[a] (161)	(162)
F	17	71
Cl	11	15
H	41	0
Me	9	0
OMe	100	0

[a] Isolated yields

SCHEME 19

TABLE 7. Synthesis of polycyclic quinonoid compounds by photochemical reaction of electron-deficient quinones with double bond containing substrates

Quinone	Double bond substrate	Products (% yield)	Ref.
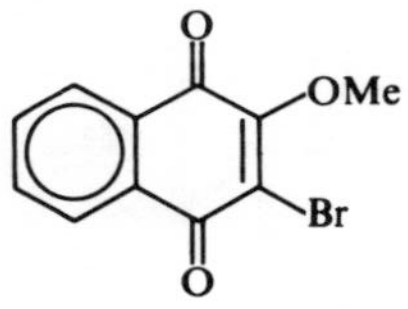 OMe, Br **(156)**	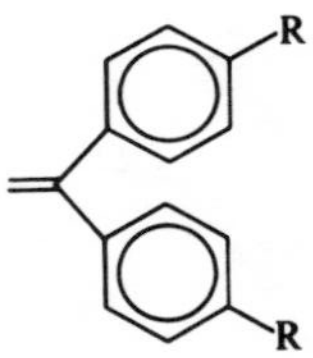 R, R R, see text **(160)**	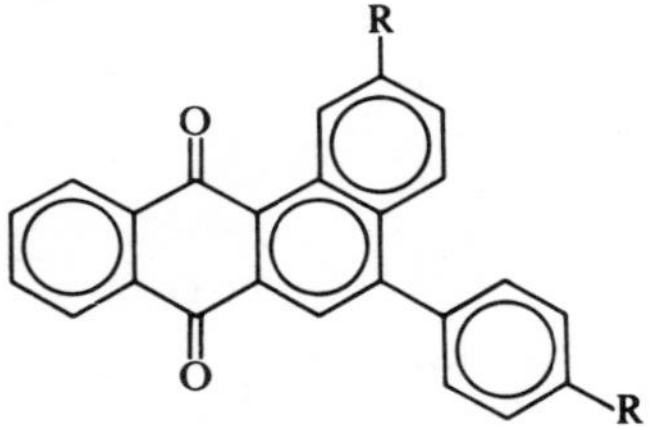 R, R **(161)** (22–68)	233
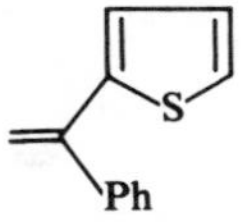	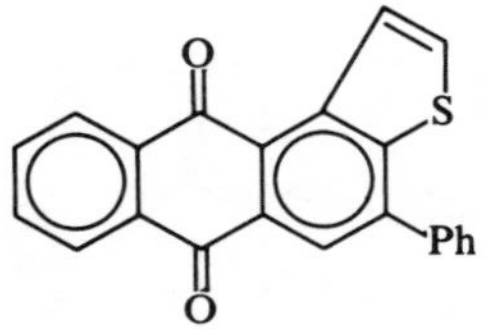 S, Ph **(162)**	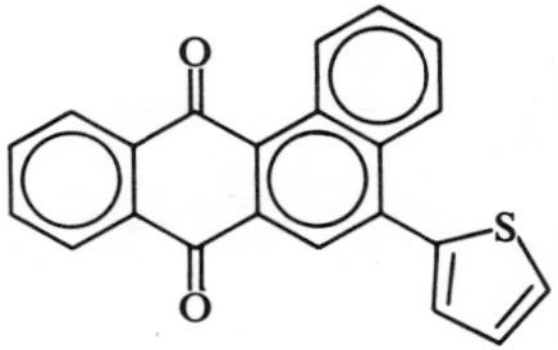 S, Ph **(163)** (62); S **(164)** (8)	234, 235
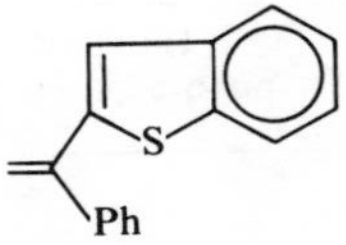	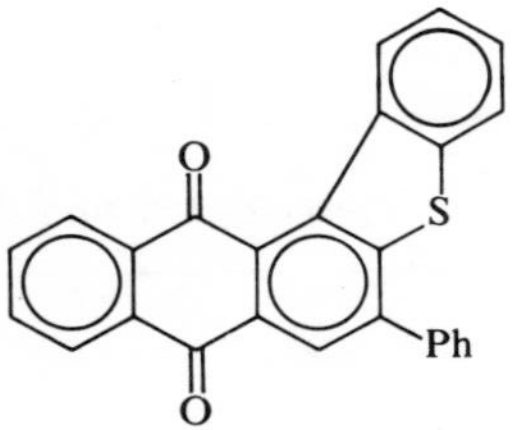 S, Ph **(165)**	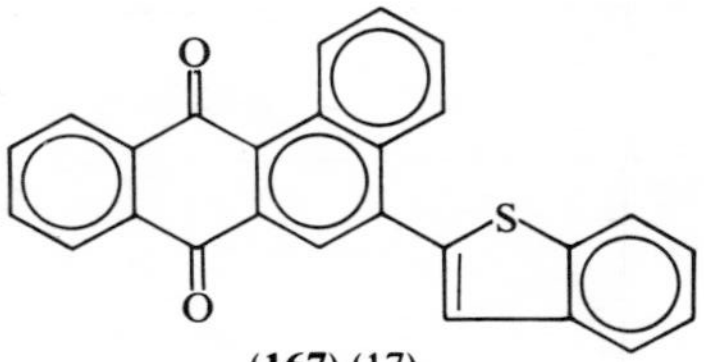 S, Ph **(166)** (32); S **(167)** (17)	235

TABLE 7. (*Contd.*)

Quinone	Double bond substrate	Products (% yield)	Ref.
	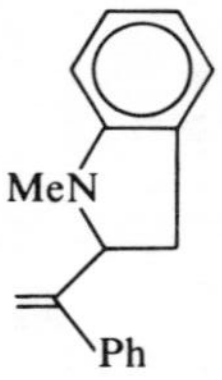 MeN, Ph (**168**)	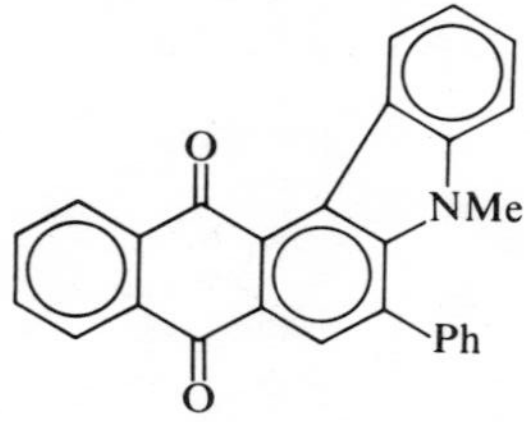 O, O, NMe, Ph (**169**) (61)	235, 236
	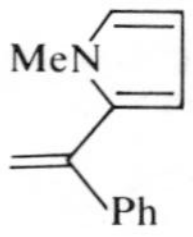 MeN, Ph (**170**)	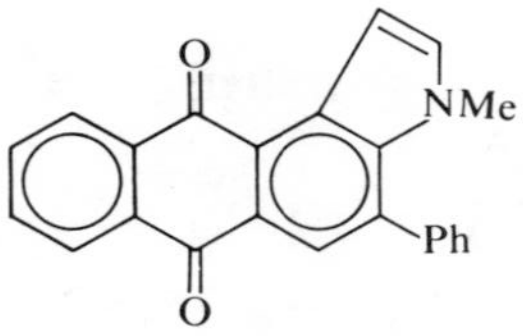 O, O, NMe, Ph (**171**) (50)	235
	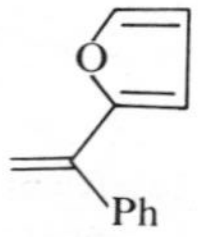 O, Ph (**172**)	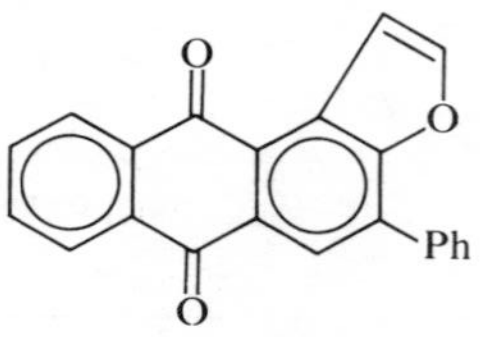 O, O, O, Ph (**173**) (30)	235

Quinone	Substrate	Product (yield %)	Ref.
	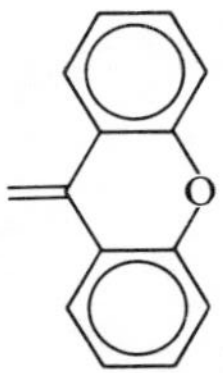 (174)	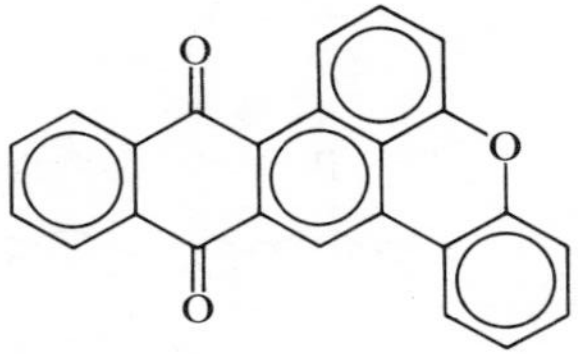 (175) (4)	235
	Ph—≡—H (176)	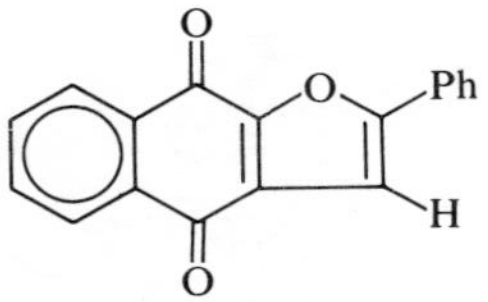 Ph, H (177) (63)	237
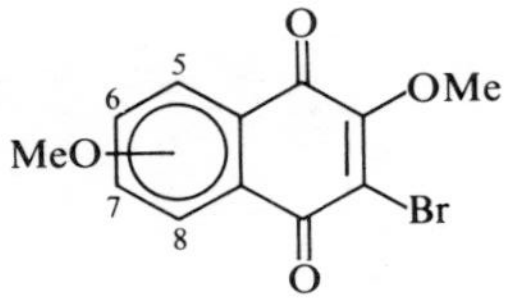 MeO, 5, 6, 7, 8, OMe, Br (178)	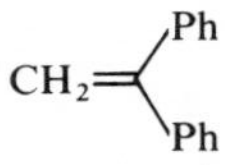 CH_2=C(Ph)Ph	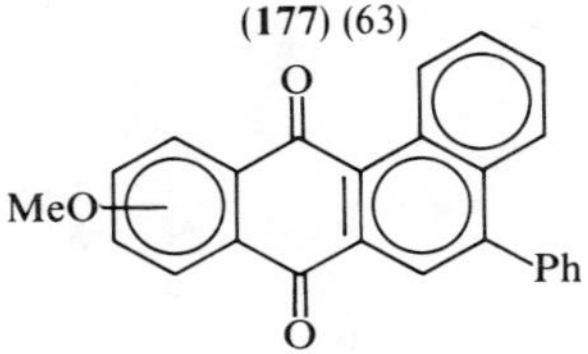 MeO, Ph (179) (36–51)	238
(156)	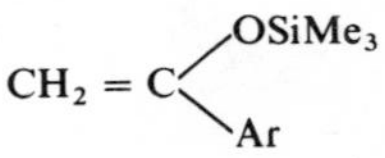 $CH_2 = C(OSiMe_3)Ar$ (180)	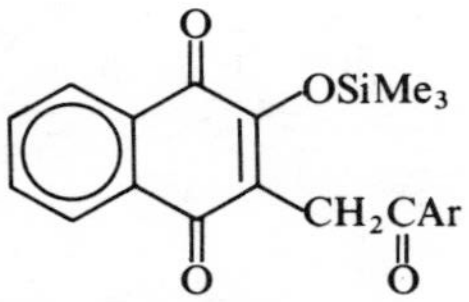 $OSiMe_3$, $CH_2C(O)Ar$ (181) (21–28); 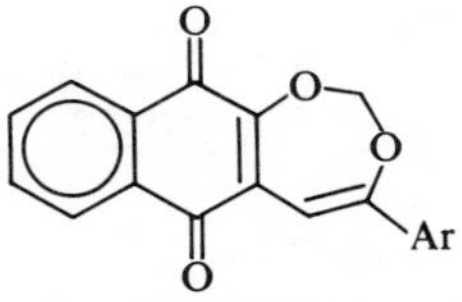 Ar (182) (12–21)	230

TABLE 7. (*Contd.*)

Quinone	Double bond substrate	Products (% yield)	Ref.
	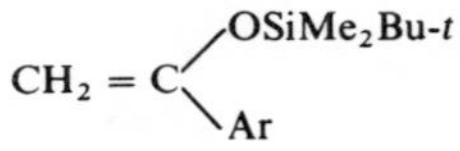 $CH_2=C(OSiMe_2Bu\text{-}t)Ar$ (**183**)	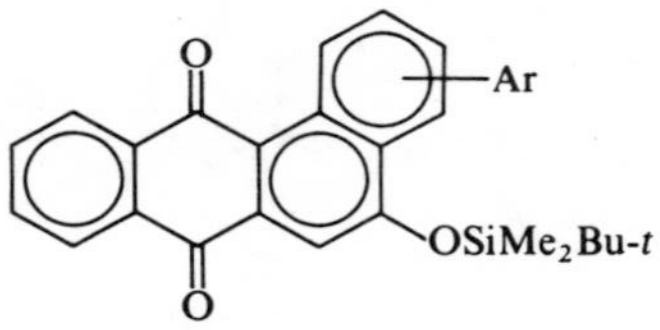 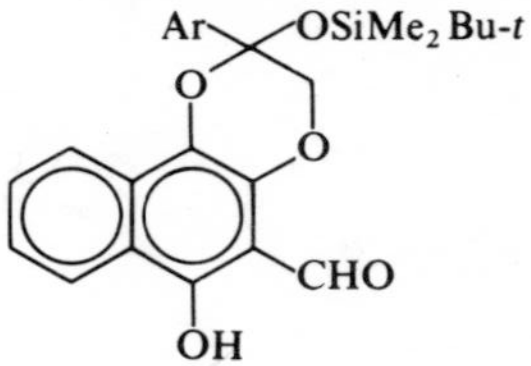 (**184**) (12–24); (**185**) (4–10)	231
	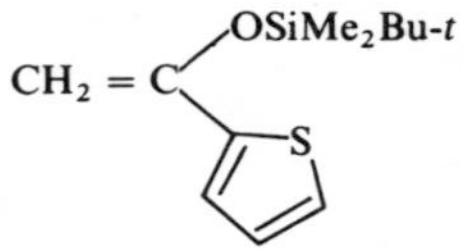 $CH_2=C(OSiMe_2Bu\text{-}t)(2\text{-thienyl})$ (**186**)	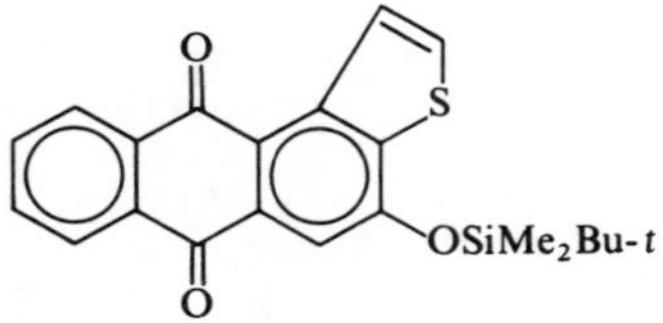 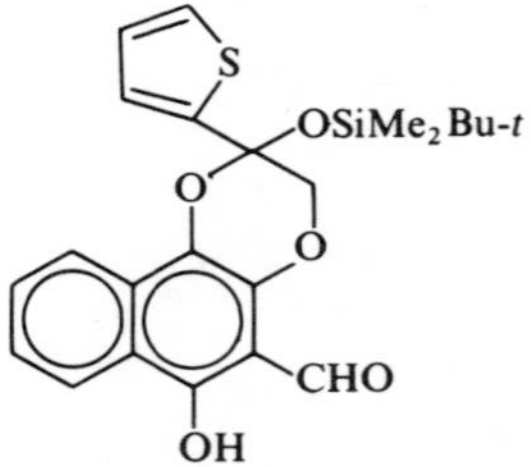 (**187**) (24); (**188**) (5)	231
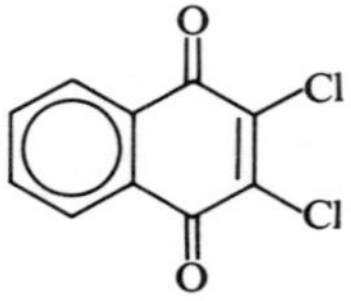 (**140**)	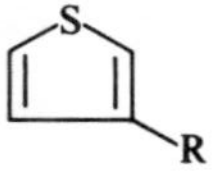 (**189**)	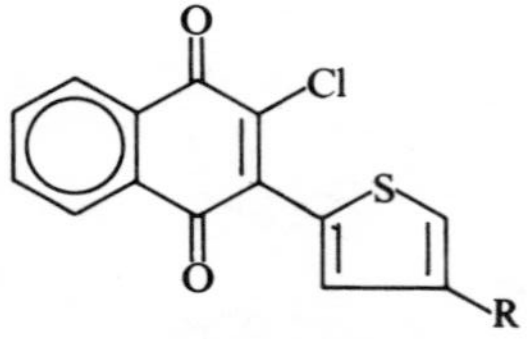 (**190**) (R = H, 57) (R = Me, 53)	228

	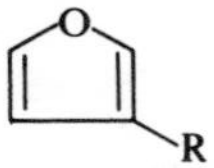 (191)	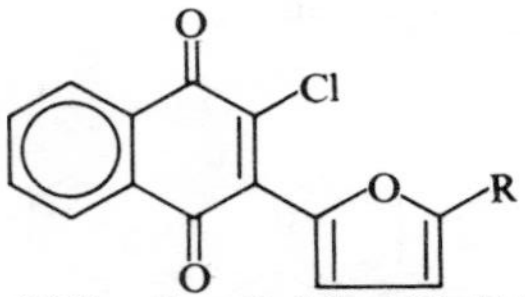 (192) R = H, Me, CH_2OMe, CH_2OEt (22–92)	62
(156)	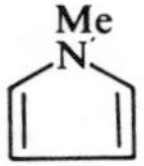 (193)	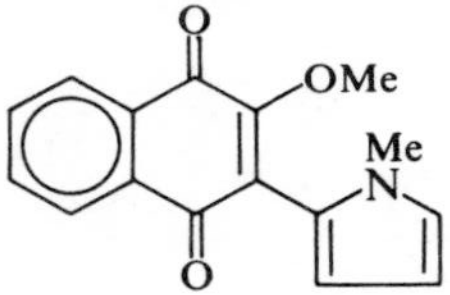 (194) (64)	229
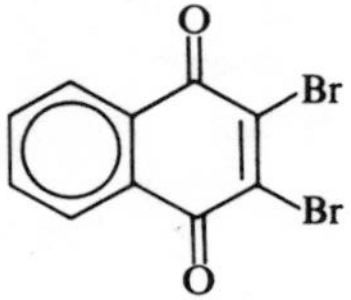 (195)	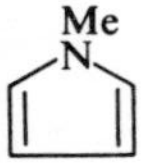 (193)	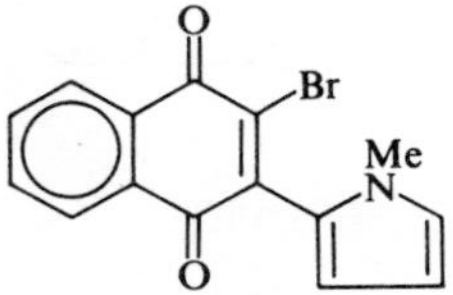 (196) (75)	229
(156)	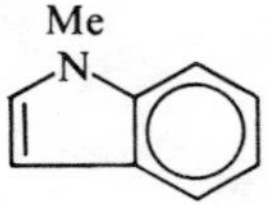 (197)	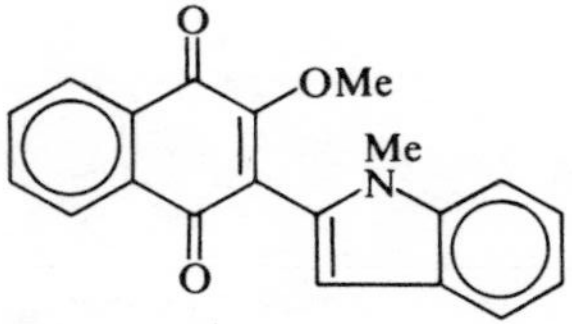 (198) (50) 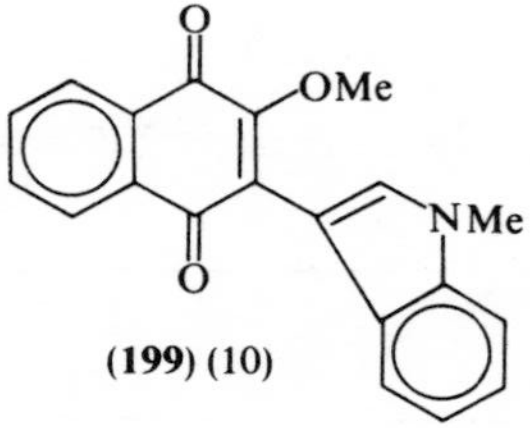 (199) (10)	229

TABLE 7. (*Contd.*)

Quinone	Double bond substrate	Products (% yield)	Ref.
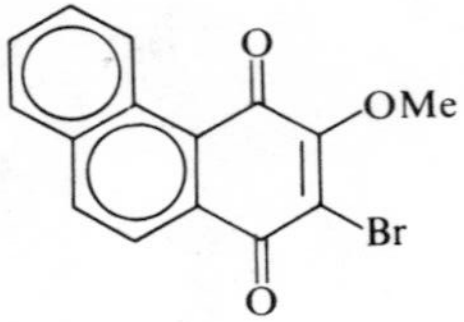 O, O, OMe, Br (**200**)	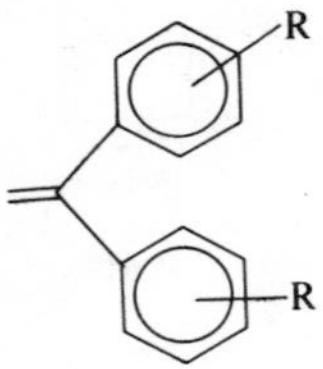 R, R (**160**) R = H, *o*-Me, *m*-Me, *p*-Me, *m*-OMe, *p*-OMe	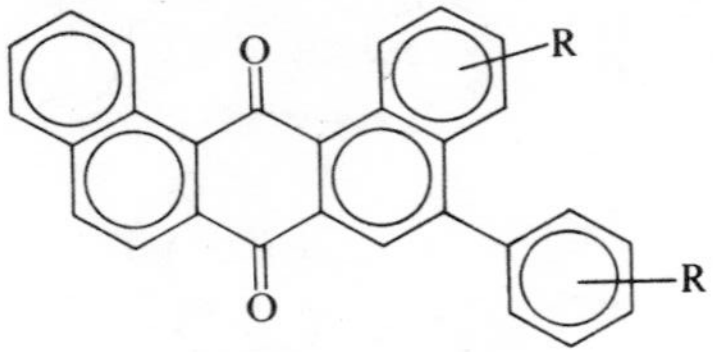 O, O, R, R (**201**) (14–23)	226
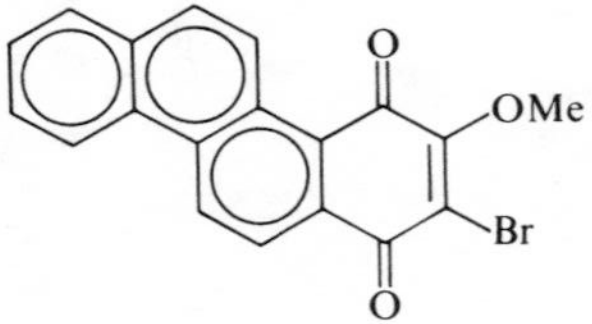 O, O, OMe, Br (**202**)	(**160**)	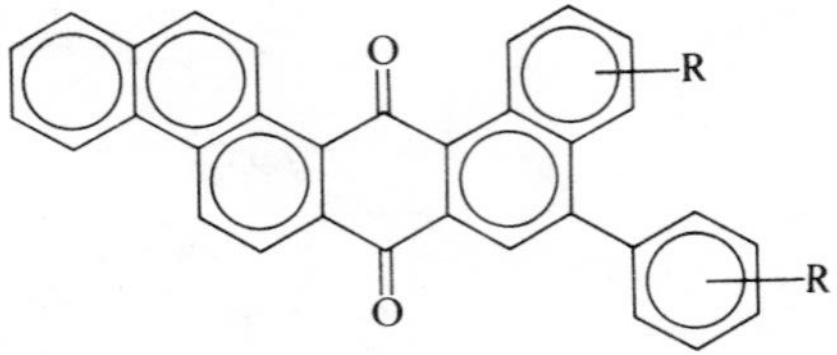 O, O, R, R (**203**) (19–42)	226
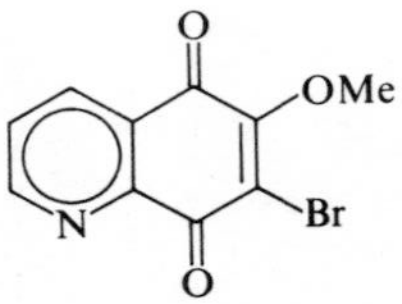 O, O, N, OMe, Br (**204**)	(**160**)	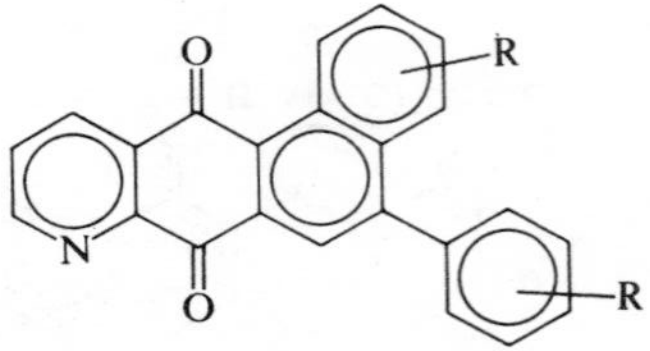 O, O, N, R, R (**205**) (17–44)	226 239b

(206)	(160)	(207) (7–13)	226 239b
(208)	(160)	(209) (4–23)	226

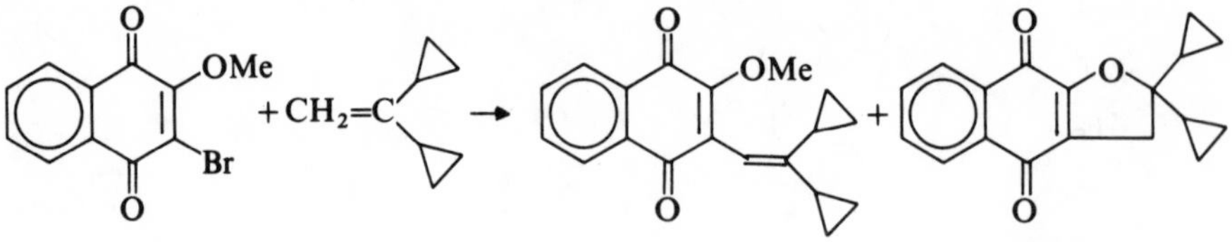

(156) (210) (211) (212) (59)

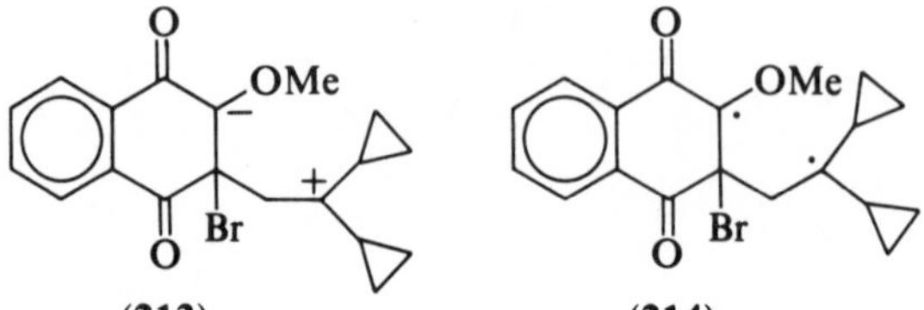

(213) (214)

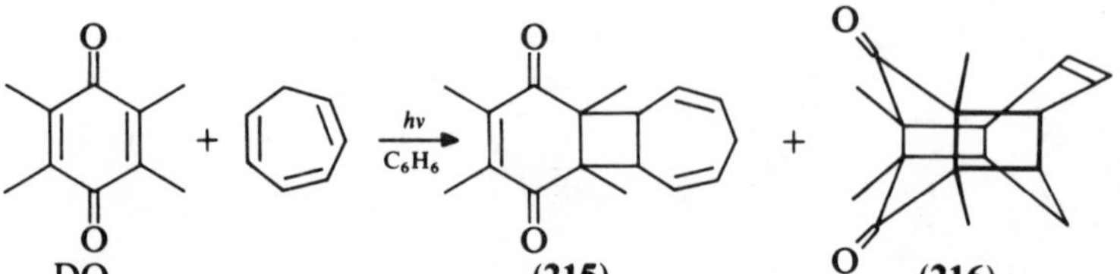

DQ (215) (216) (60)

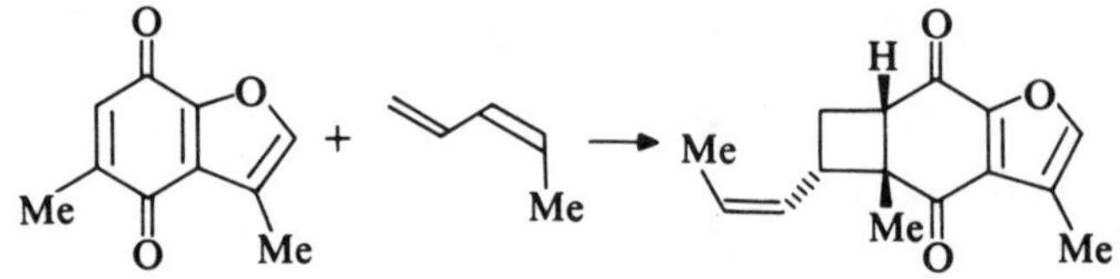

(217) (218, 65 %) (61)

extremely acid-labile, being transformed into **222** upon treatment with a Lewis acid (equation 62), and are isolated by direct crystallization from the reaction mixture[244, 245]. On irradiation in the presence of 1,1-dimethylallene, DQ affords cyclobutane **223** exclusively (equation 63), whereas photocycloaddition to triphenylketenimine occurs at the carbonyl function of DQ to give **224** in spite of its (π, π^*) triplet excited state[246]. Prolonged irradiation of **224** leads to the formation of duroquinone diphenylmethide (**225**) (equation 64). Analogous iminooxetane adduct **226** is formed in the reaction of DQ with dimethyl-*N*-phenylketenimine. However, the photochemistry of **226** is quite different from either that of **218** or **223**, in that it gives lactoneimine (**228**) upon irradiation presumably via **227**[246]. Hydrolysis gives **229** (equation 65). Oxetanone adduct **230** is produced (together with the cyclobutanones **231** in the case of DQ) in the photochemical reaction of tetra-substituted quinones such as DQ, CQ and AQ, with diphenylketene, although their lowest triplet excited states are (π, π^*) (Scheme 20). These results are explained in terms of dipole–dipole interaction of the reactants[247].

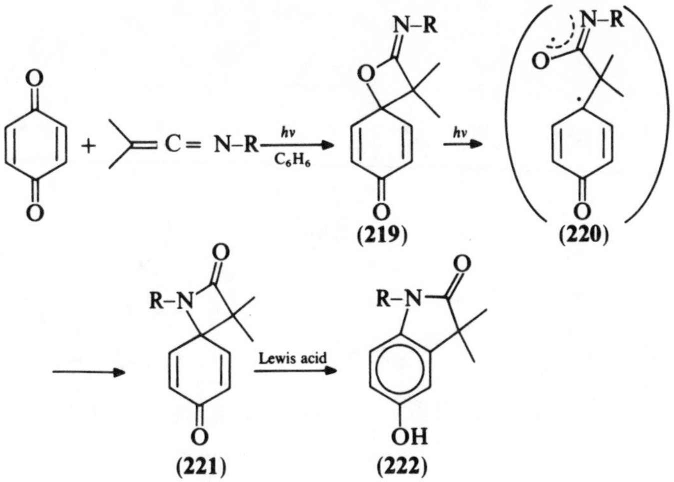

(62)

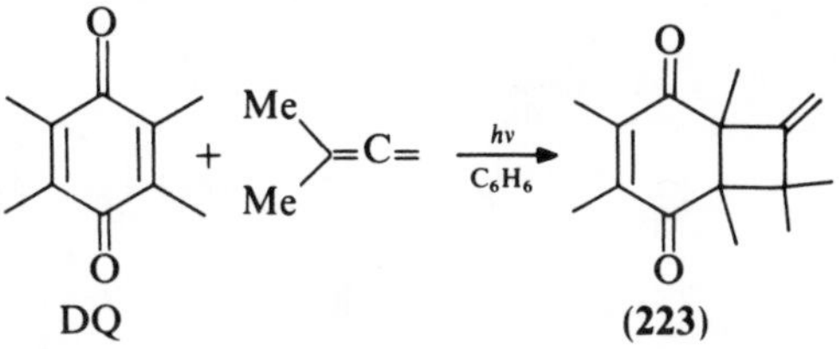

(63)

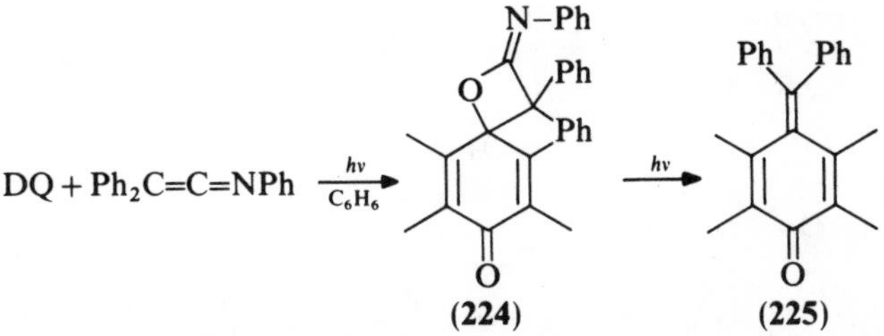

(64)

Anthraquinone **1** (R = NH_2), whose lowest excited state is an intramolecular charge transfer (π, π^*) state, undergoes a cycloaddition reaction with dienes in ethanol to produce the corresponding oxetane (**232**) when subjected to visible-light irradiation (equation 66)[248–250]. Apparently this reaction is not a conventional photocycloaddition via the (n, π^*) triplet state of the quinone, since it is not affected by the presence of triplet quenchers such as dissolved oxygen and anthracene[249]. Singlet exciplex between the aminoanthraquinone and the diene is proposed as the key intermediate, based on the fluorescence quenching of **1** (R = NH_2) by the diene[250].

Naphthoquinones (**233**) undergo photocycloaddition with acetylenes to give cyclobutenes **234** (equation 67)[251,252].

$$\text{DQ} + \text{Me}_2\text{C}{=}\text{C}{=}\text{NPh} \xrightarrow[\text{C}_6\text{H}_6]{h\nu} (\textbf{226}) \xrightarrow{h\nu} (\textbf{227}) \rightarrow (\textbf{228}) \xrightarrow{\text{H}_2\text{O}} (\textbf{229}) \tag{65}$$

$$\text{Quinone} + \text{Ph}_2\text{C}{=}\text{C}{=}\text{O} \xrightarrow[\text{C}_6\text{H}_6]{h\nu} (\textbf{230}) + (\textbf{231})$$

Quinone	Yield (%) 230	231
DQ R = Me	49	25
CQ R = Cl	38	—
AQ R − R = −(CH = CH)$_2$	53	—

SCHEME 20

$$(\textbf{1}, \text{R} = \text{NH}_2) + \text{CH}_2{=}\text{CH}{-}\text{CH}{=}\text{CR}^1\text{R}^2 \xrightarrow[\text{EtOH}]{h\nu} (\textbf{232}) \tag{66}$$

D. Cycloaddition of *o*-Quinones to Alkenes and Alkynes

Photocycloaddition of PQ to olefins gives mainly two types of products, the keto oxetane **235** and the 1,4-dioxine **236** along with H-abstraction products (equation 68)[253–255]. Keto oxetane **235** predominates in the reaction with highly strained olefins[256]. With 1-methoxypropyne, PQ affords a 1,3-dioxole adduct **238** via a 1,2-H shift

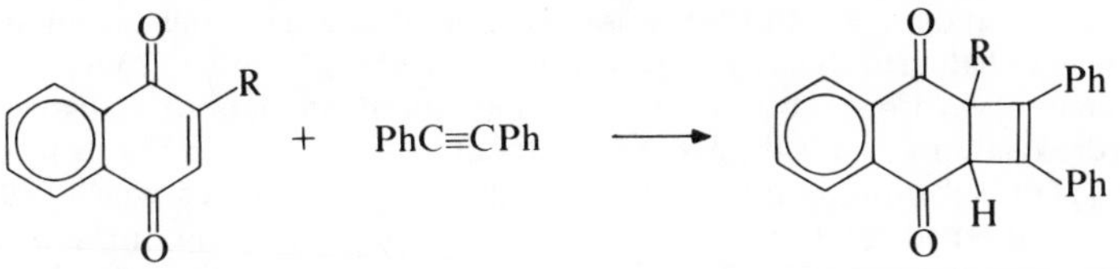

R = OMe, Cl, Br

(233) + PhC≡CPh → (234) (67)

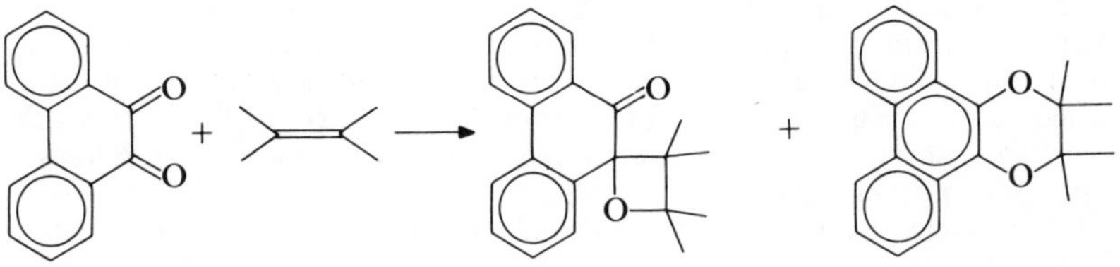

PQ + tetramethylethylene → (235) + (236) (68)

in the intermediate **237** (equation 69)[257], while α,β-unsaturated keto ester **240** is formed in the reaction with methoxyphenylacetylene, presumably by way of keto oxetane **239** (equation 70)[258]. In the reaction with conjugated alkenynes, PQ and *o*-CQ add to the double bond predominantly rather than to the triple bond[259]. Thermal and photochemical additions of PQ to alkylthio- and dialkylamino-substituted acetylenes were reported[260].

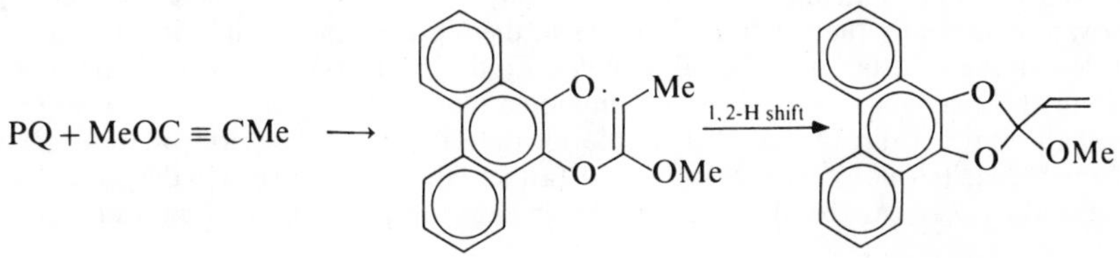

PQ + MeOC ≡ CMe → (237) —1,2-H shift→ (238) (69)

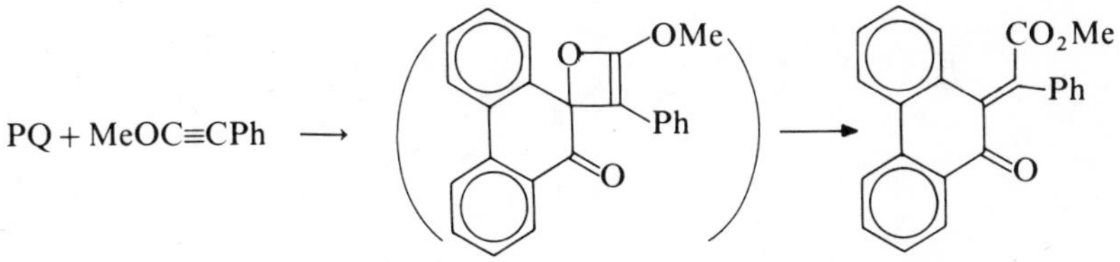

PQ + MeOC≡CPh → (239) → (240) (70)

V. PHOTOSUBSTITUTION

Many types of photochemical substitution reactions have been discovered mainly in anthraquinone systems, which have been extensively studied because of their structural relationship to vat dyes. Photosubstitution reactions discussed in this section include the displacement of halogen, alkoxy group, or a sulfonate group by other groups as well as the displacement of hydrogen by an amino group (photoamination), a hydroxy group (photohydroxylation), or a sulfonate group (photosulfonation). In the displacement of hydrogen, photosubstitution often gives predominantly one isomer, which is difficult to synthesize by thermal reactions. In order to clarify the mechanism of the photosubstitution, one must examine many parameters such as the nature of the excited states of the quinones (singlet or triplet), the dependence of the reaction rates or the product distributions on the presence of additives such as oxygen or amines, on the solvent polarity, and on the HOMO or LUMO coefficients of the reactive intermediates involved. The photosubstitution mechanisms are very versatile, and involve a direct addition of a nucleophile to the excited quinone, an electron transfer or a charge transfer induced substitution, or aromatic radical substitution by photogenerated free radicals. Photosubstitutions of quinones are classified to two main groups according to the nature of the excited state of the quinone. Group 1 is a photosubstitution of quinones having the lowest (n, π^*) excited state (strong sensitizer according to the classification of Phillips[39]) and group 2 is a photosubstitution of quinones having the lowest (π, π^*) or CT state (weak sensitizer). In the former case, a direct nucleophilic substitution in the excited state is difficult, since the excitation energy is localized at the carbonyl group. Normally, an electron transfer or a charge transfer reaction precedes the photosubstitution of group 1. In contrast, many examples of direct nucleophilic substitution have been reported for group 2. The results are summarized in Table 8.

A. Substitution of a Sulfonate Group

The irradiation of anthraquinone-1-sulfonate (**1**, R = SO_3^-) in aqueous solutions containing chloride ions or ammonia gives rise to substitution of the sulfonate group by a chlorine atom or by an amino group, respectively[261–264]. Two hypotheses have been put forward concerning the reaction pathways in these processes. The first involves a direct nucleophilic substitution of the sulfonate group by Cl^- or NH_3 by way of intermediate **241** or the analogous zwitterion with ammonia (equation 71a). The second involves an aromatic radical substitution, where a chloride radical generated by electron transfer from Cl^- to the excited molecule of **1** (R = SO_3^-) attacks the 1-position of **1** in the ground state. The adduct **241** can be then formed by in-cage recombination of radical ion-pairs

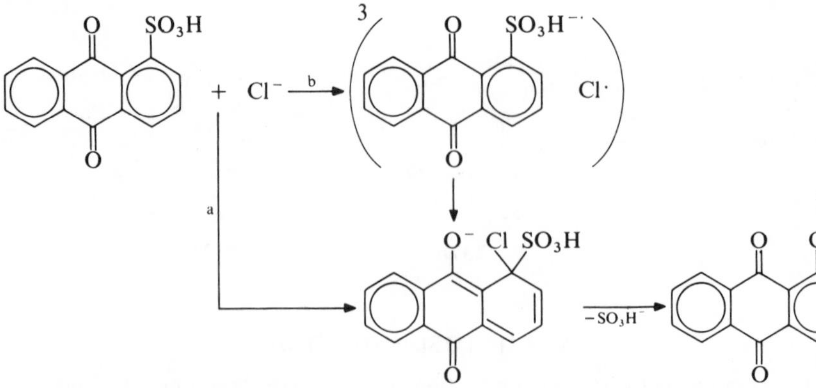

(71)

(**241**)

TABLE 8. Photosubstitution reactions of quinones

Reaction mode	Starting quinone	Conditions	Product quinone (isolated yield, %)		Ref.
$SO_3H \rightarrow Cl$	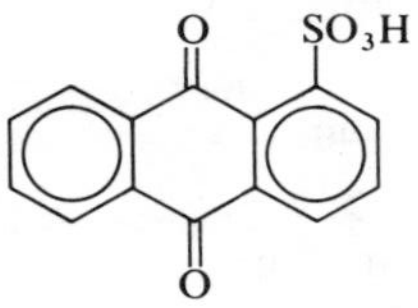 O, SO_3H, O	Cl^-/H_2O	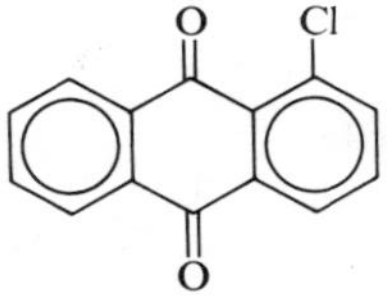 O, Cl, O		6, 7
$SO_3H \rightarrow NH_2$	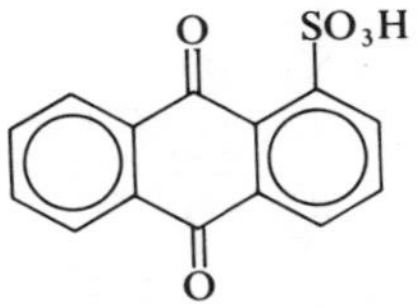 O, SO_3H, O	NH_3/H_2O	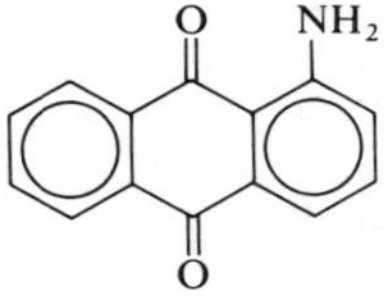 O, NH_2, O	(45)	6, 7, 264
	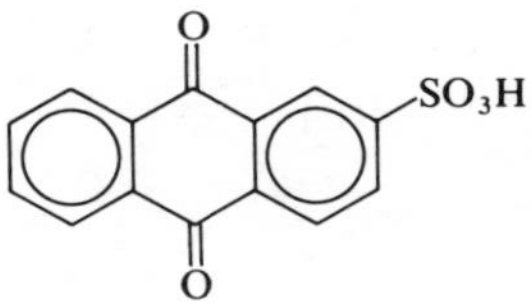 O, SO_3H, O	NH_3/H_2O	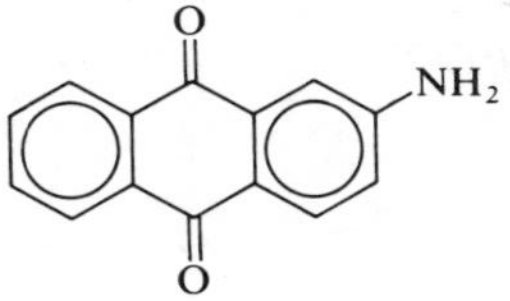 O, NH_2, O	(60)	264
$Cl \rightarrow OR$	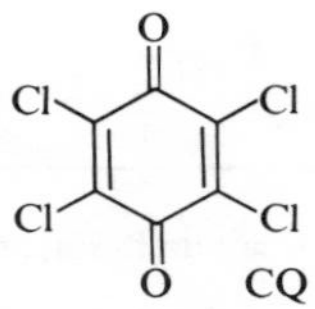 O, Cl, Cl, Cl, Cl, O CQ	EtOH	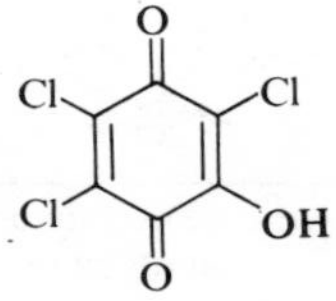 O, Cl, Cl, Cl, OH, O	(28)	300

TABLE 8. (*Contd.*)

Reaction mode	Starting quinone	Conditions	Product quinone (isolated yield, %)	Ref.
	CQ	PhOH	2,3,5-trichloro-6-phenoxy-1,4-benzoquinone (Cl, Cl, Cl, OPh) (30)	300
Br → NHR	(242) 1-NH_2-2-SO_3H-4-Br-anthraquinone	RNH_2/O_2 $MeCN/H_2O$	(243) 1-NH_2-2-SO_3H-4-NHR-anthraquinone	265–267
	(244) 1-NH_2-2-Br-4-Br-anthraquinone	$BuNH_2/O_2$ $MeCN/C_6H_6$	(245) 1-NH_2-2-Br-4-NHBu-anthraquinone (41.7)	269
	(246) 1-NHCOMe-4-Br-anthraquinone	$BuNH_2$ $EtOH/C_6H_6$	(247) 1-NHCOMe-4-NHBu-anthraquinone (92)	268, 270

Reaction	Substrate	Conditions	Product (yield, %)		Ref.
	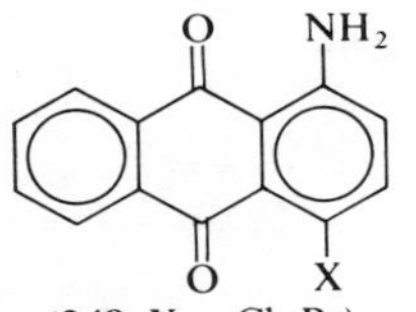 O NH_2 O X (**248**, X = Cl, Br)	Aqueous NH_3 EtOH	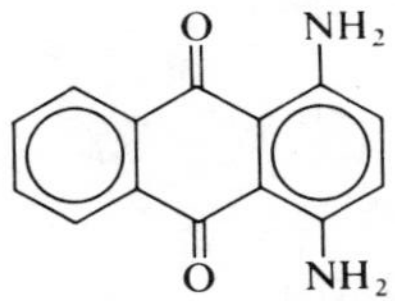 O NH_2 O NH_2		301
Cl → OH	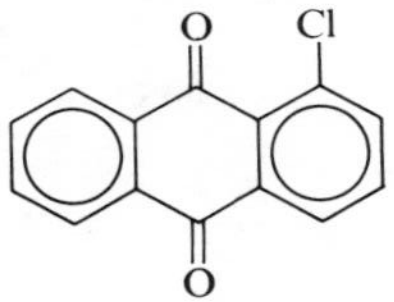 O Cl O	97% H_2SO_4	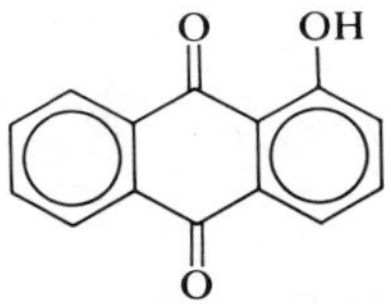 O OH O + other hydroxylated product (16)	(62)	271
OMe → NHR	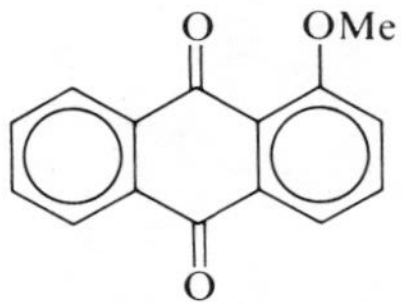 O OMe O (**249**)	NH_3 H_2O/MeCN (1:1)	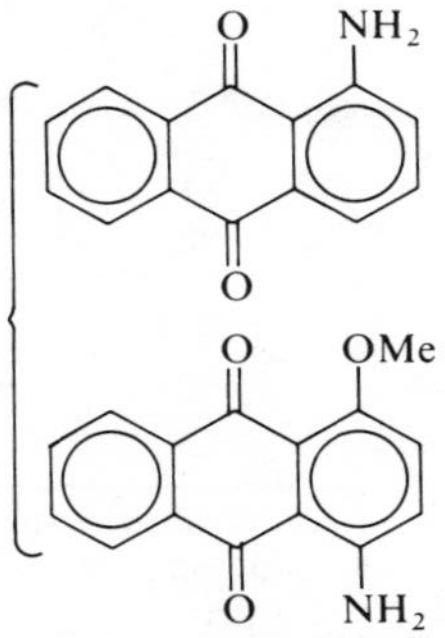 O NH_2 O	(96)	272, 273
			O OMe O NH_2	(0.5)	

TABLE 8. (*Contd.*)

Reaction mode	Starting quinone	Conditions	Product quinone (isolated yield, %)	Ref.
	(249)	$MeNH_2$ $H_2O/MeCN$ (1:1)	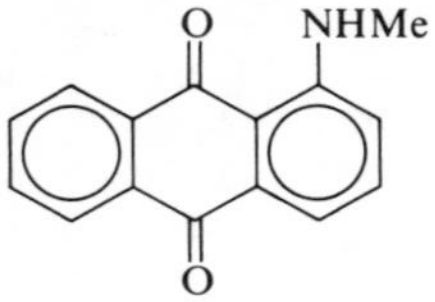 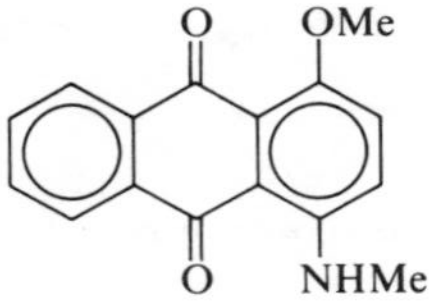 1-NHMe-anthraquinone (81) 1-OMe-4-NHMe-anthraquinone (7)	272, 273
	(249)	^{-}NCO $H_2O/MeCN$ (1:1)	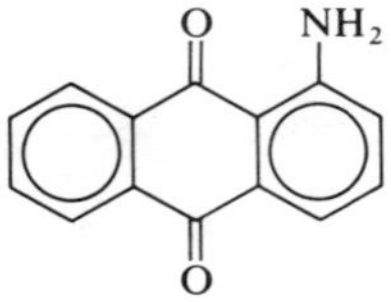 1-NH_2-anthraquinone (87)	273
	(249)	$HOCH_2CH_2NH_2$ $H_2O/MeCN$ (1:1)	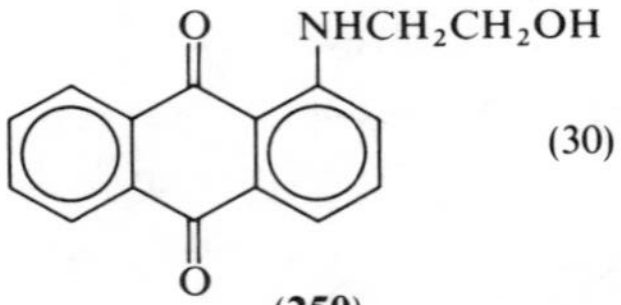 1-$NHCH_2CH_2OH$-anthraquinone **(250)** (30)	273

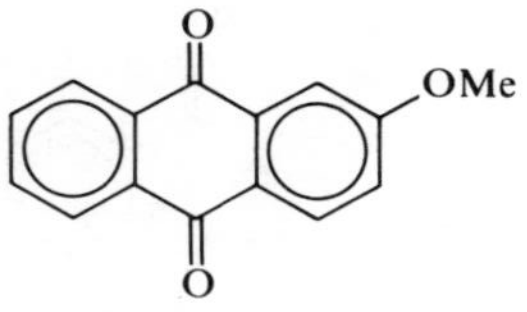

Quinone	Conditions	Products (yield, %)	Ref.
(251) (OMe)	NH_3/air H_2O/MeCN	(NH_2) (22)	273
	NH_3/N_2	(NH_2, OMe) (18) (252)	
(251)	$MeNH_2$/air H_2O/MeCN	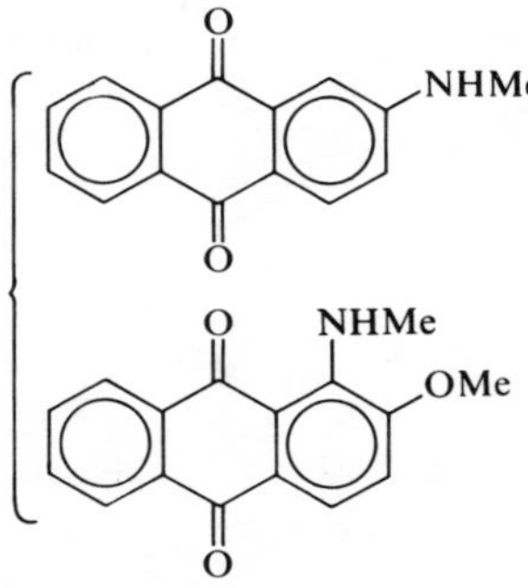 (NHMe) (4)	273
		(NHMe, OMe) (34) (255)	
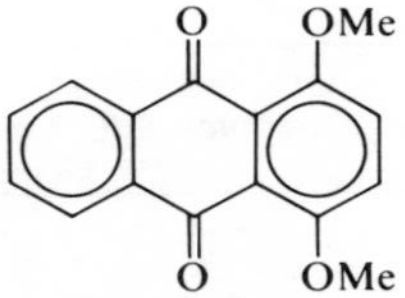 (256) (OMe, OMe)	RNH_2/CH_2Cl_2	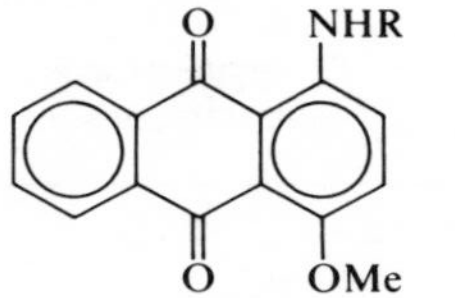 (NHR, OMe) (2–51) (257)	302

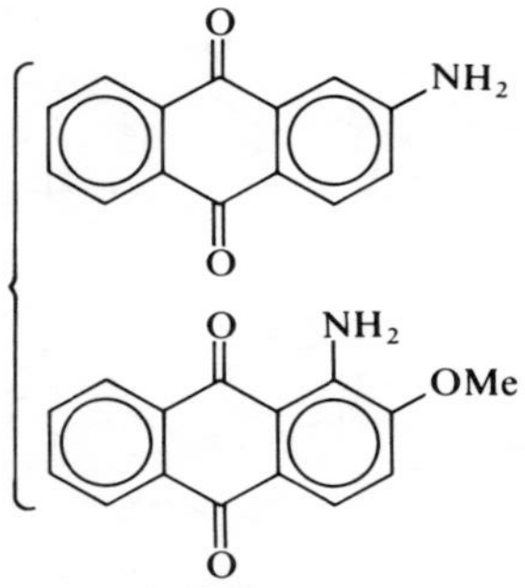

TABLE 8. (*Contd.*)

Reaction mode	Starting quinone	Conditions	Product quinone (isolated yield, %)	Ref.
	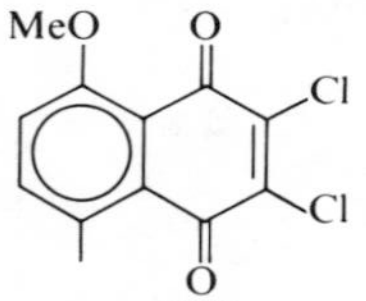 (258)	RNH_2/CH_2Cl_2	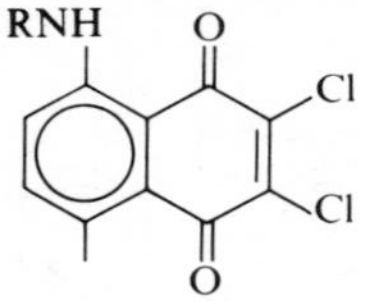 (259) (38)	274
	(260)	$MeNH_2/CH_2Cl_2$	(261) (53); (262) (30)	303, 304
OMe → OH	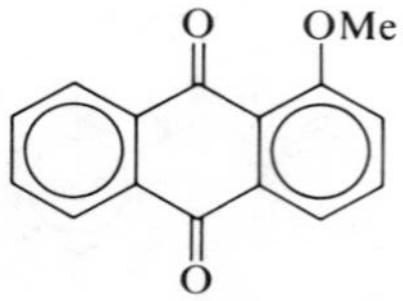 (249)	^{-}OH, air $H_2O/MeCN$ (1:1)	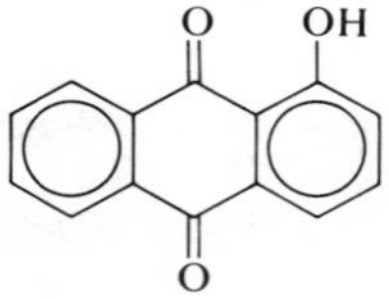 (84)	273

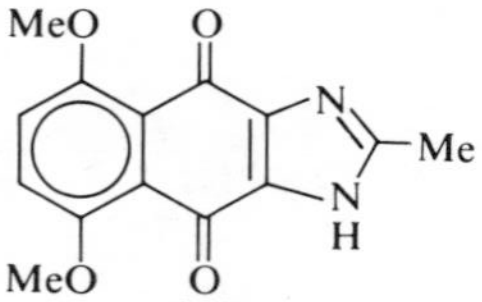

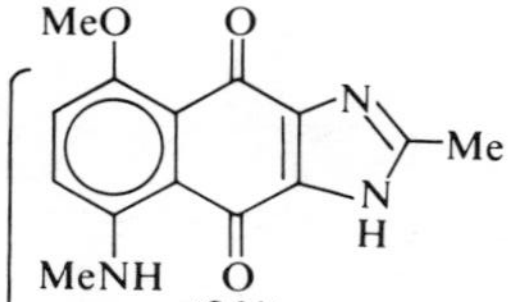

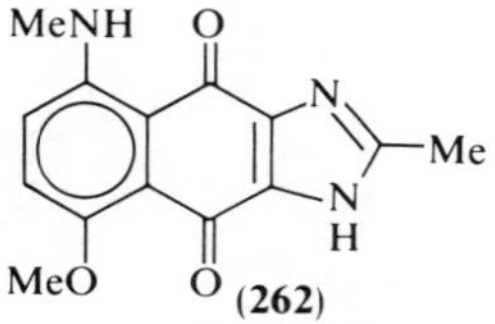

	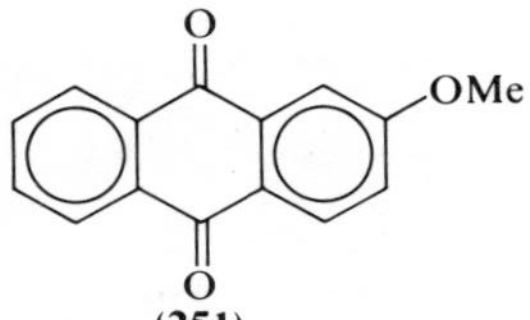 (**251**) (OMe)	^{-}OH, air H_2O/MeCN (1:1)	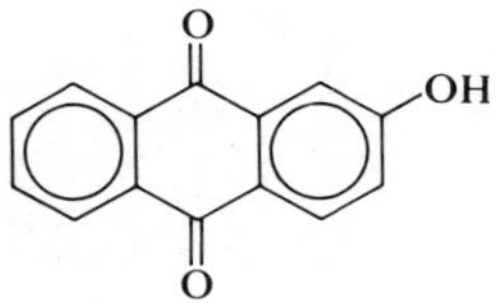 OH	(7)	273
H → Cl	(**249**)	Cl^{-} H_2O/MeCN (1:1)	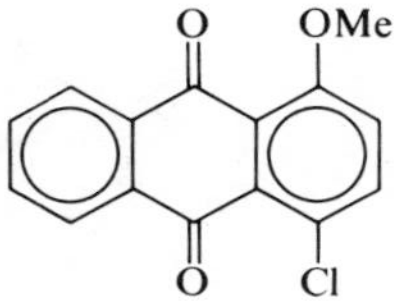 OMe, Cl	(7)	273
H → NHR	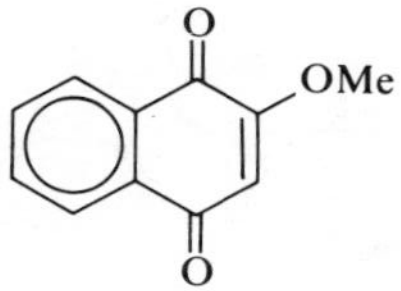 OMe	$MeNH_2$ −40°C CH_2Cl_2	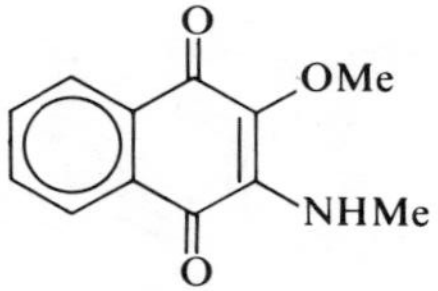 OMe, NHMe	(66)	275
	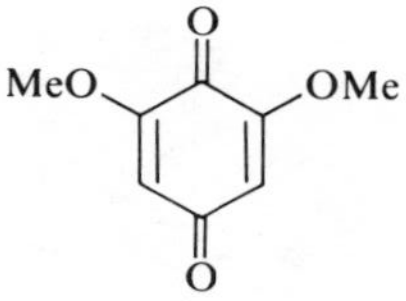 MeO, OMe	$MeNH_2$ −40°C CH_2Cl_2	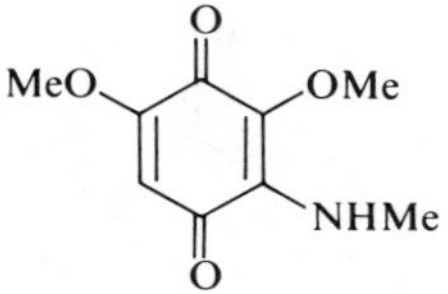 MeO, OMe, NHMe	(3)	275
	AQ	Pyridine	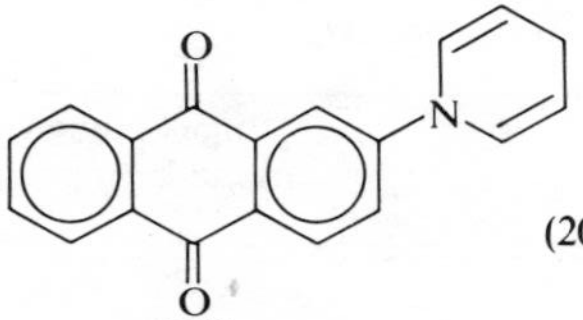 (**268**) (N)	(20)	126

TABLE 8. (*Contd.*)

Reaction mode	Starting quinone	Conditions	Product quinone (isolated yield, %)	Ref.
	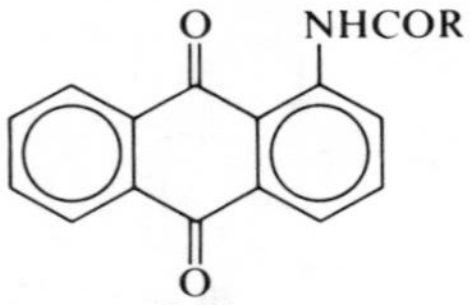 O, NHCOR, O (263)	$BuNH_2/O_2$ C_6H_6	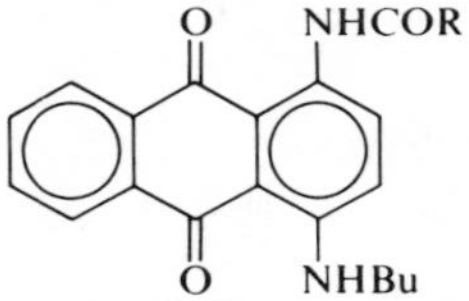 O, NHCOR, O, NHBu (264) (R = Ph, 64) (R = p-$NO_2C_6H_4$, 58) (R = Me, 71) (R = Et, 77) (R = n-Pr, 78) (R = PhCH = CH, 60) (R = EtO, 60)	276
	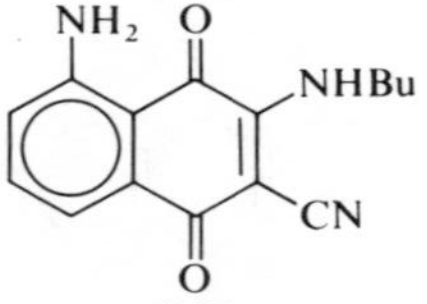 NH_2, O, NHBu, CN, O (265)	$BuNH_2/O_2$ C_6H_6	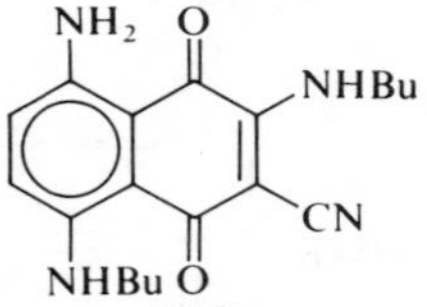 NH_2, O, NHBu, CN, NHBu O (266) (91)	277
	AQ	NH_3/O_2 H_2O-2-PrOH	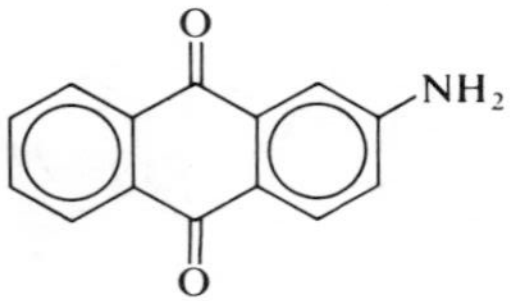 O, NH_2, O (33)	278 279
H → OH	AQ	conc. H_2SO_4 100%	1 (R = OH) (30–32) 2 (R = OH) (65–66)	282
		96% H_2SO_4[a]	1 (R = OH) (7)[c], (11)[d] 2 (R = OH) (74)[c], (85)[d]	282
		90% H_2SO_4[b]	1 (R = OH) (14)[c], (16)[d] 2 (R = OH) (71)[c], (81)[d]	282
		98% H_2SO_4	1 (R = OH) (68)	281

	Substrate	Conditions	Product (yield %)	Ref.
	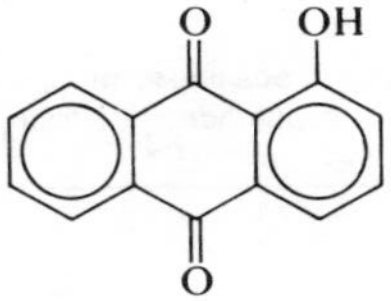	77% H_2SO_4	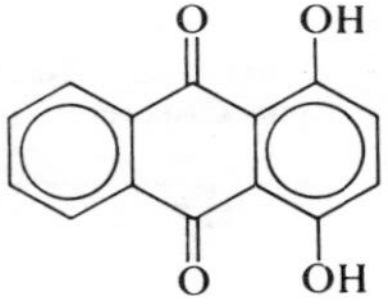 (81)	305
	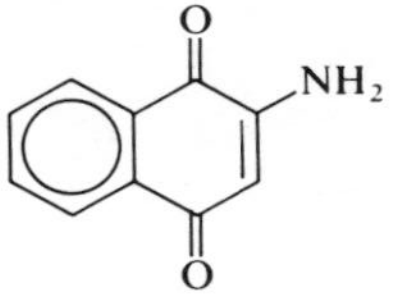	H_2O	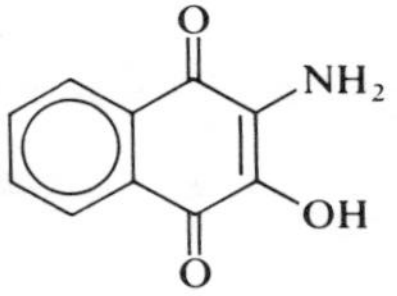	306
	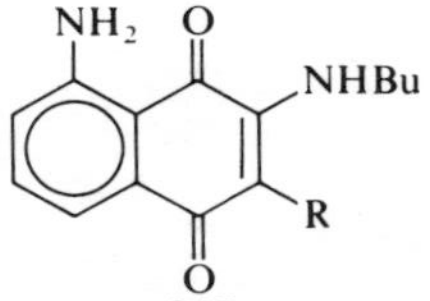 (265)	t-$BuNH_2/O_2$ C_6H_6	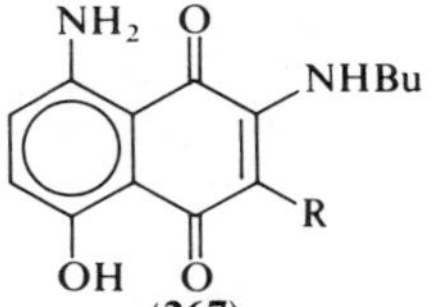 (267) (81)	277
$H \rightarrow SO_3H$	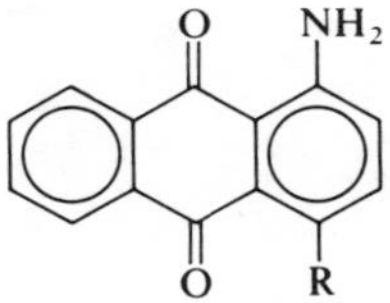	Na_2SO_3 Pyridine/H_2O	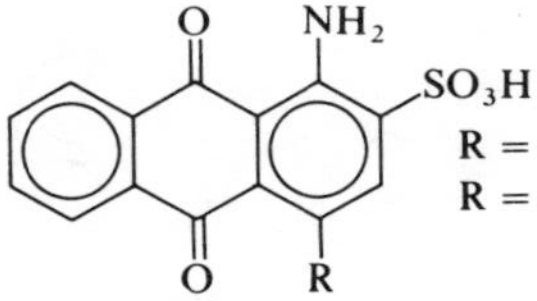 (269) R = H, (92.6) R = Cl, (96)	294, 295
	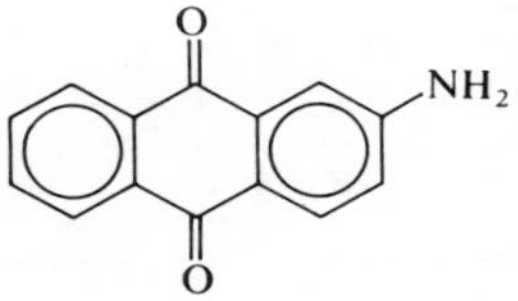	Na_2SO_3 Pyridine/H_2O	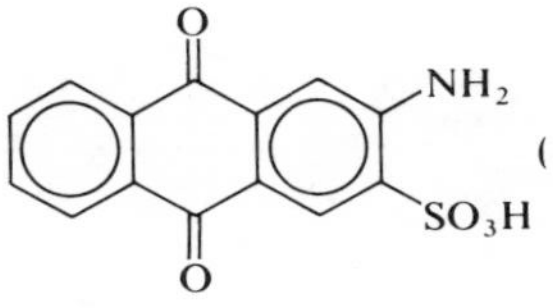 (18)	294, 295

TABLE 8. (*Contd.*)

Reaction mode	Starting quinone	Conditions	Product quinone (isolated yield, %)	Ref.
	O, OH	Na_2SO_3 Pyridine/H_2O	O, OH, SO_3H (**271**) (31)	295
			O, OH, O, SO_3H (**272**) (18)	
			O, OH, SO_3H, O, SO_3H (**273**) (24)	
H → SNa	O, NH_2, O	Na_2S Pyridine/H_2O	O, NH_2, SNa, O (**274**) (70)	295

[a] Besides quinone products, anthrone and bianthronyl were formed in 6 and 13% yields, respectively under N_2 atmosphere.
[b] Besides quinone products, anthrone and bianthronyl were formed in 10 and 5% yields, respectively under. N_2 atmosphere, and bianthronyl was formed in 2% yield even under O_2 atmosphere.
[c] Under N_2 atmosphere.
[d] Under O_2 atmosphere.

generated by electron transfer (equation 71b). However, the radical substitution mechanism does not seem to explain fully the high regioselectivity of the substitution reactions.

B. Substitution of Halogen Atom

Photosubstitutions proceed very well in the photolysis of anthraquinones whose lowest singlet or triplet excited states are (π, π^*). The displacement of the bromine from 1-amino-4-bromoanthraquinone-2-sulfonic acid (**242**) by ammonia or alkylamines has been reported [265–267]. Under similar conditions, 1-bromoanthraquinone whose lowest singlet or triplet excited state is (n, π^*) is inert. The fact that the rate of the photosubstitution increases as the polarity of the solvent is increased is evidence for a polar intermediate. Interestingly, the photosubstitution does not occur without oxygen. Oxygen probably does not act as a simple oxidizing agent but forms exciplex with the singlet (π, π^*) of **242**[267]. 1-Amino-2,4-dibromoanthraquinone (**244**) is similarly photoaminated to give **245** but the multiplicity of the (CT) states responsible for the reaction is mainly singlet in acetonitrile and in ethanol but triplet in benzene[268–270]. In the former solvents, oxygen increases the reaction rates through the formation of exciplex but it acts as a triplet quencher in the latter solvent. This change in nature of the reactive excited state is attributed to a change in the disposition of the energy levels of the substrate when the solvent polarity is altered (Figure 4).

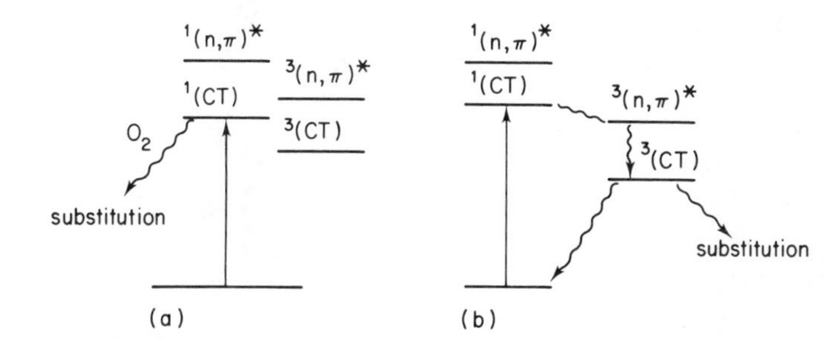

FIGURE 4. Energy diagram of **244**; (a) in MeCN or in EtOH; (b) in C_6H_6

Photolysis of 1-chloroanthraquinone (**1**, R = Cl) in 97 % sulfuric acid affords mainly 1-hydroxyanthraquinone (**1**, R = OH) along with minor hydroxylated products, while 2-chloro and 2-bromoanthraquinone give no dehalogenated products[271]. Reaction of **1** (R = Cl) with Fenton's reagent gives mainly hydroxydeprotonated products, but does not give the dehalogenated hydroxylation product, precluding a simple aromatic substitution by a free hydroxy radical.

C. Substitution of Alkoxy Group

Photosubstitution of 1-methoxyanthraquinone (**249**) with ammonia gives 1-amino-anthraquinone in 96 % yield (equation 72), while under identical conditions 2-methoxyanthraquinone (**251**) reacts less rapidly to give 2-aminoanthraquinone and 1-amino-2-methoxyanthraquinone (**252**) whose relative proportions depend on the availability of oxygen during the reaction (equation 73)[272, 273]. These substitution reactions are not affected by the presence of a radical scavenger. Formation of **252** involves formal replacement of hydrogen by an amino group and a possible mechanism is shown in Scheme 21. The leuco-compound **253** formed by the direct nucleophilic addition of

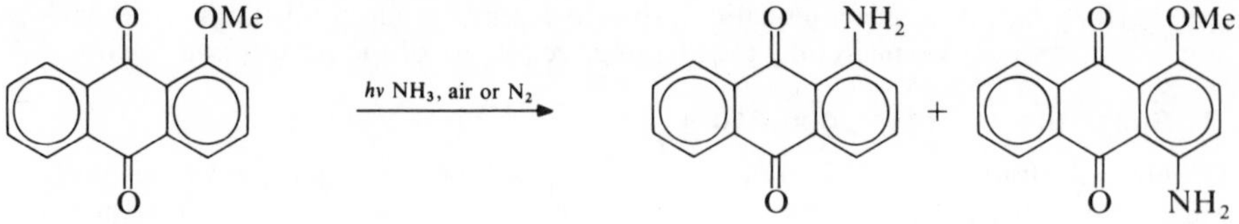

(249) $\xrightarrow{h\nu\ NH_3,\ \text{air or}\ N_2}$ + (72)

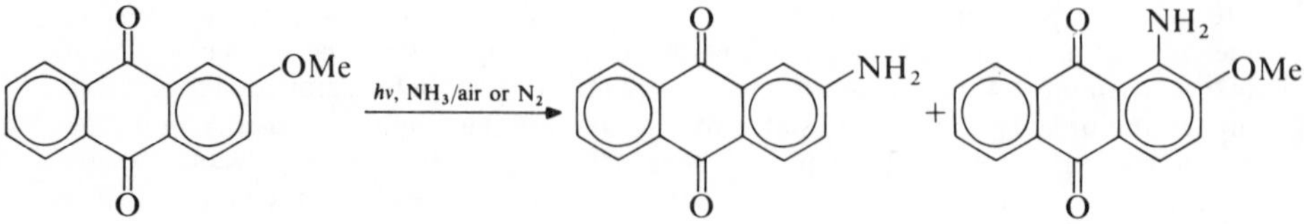

(251) $\xrightarrow{h\nu,\ NH_3/\text{air or}\ N_2}$ + (252) (73)

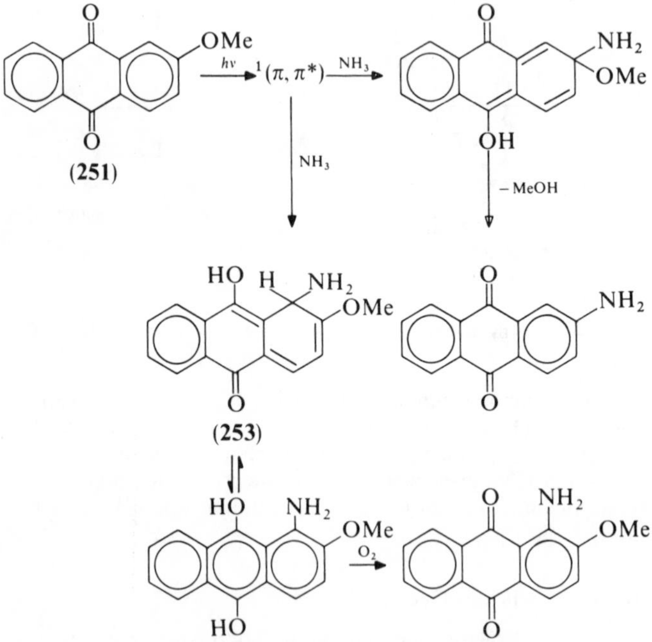

(251) $\xrightarrow{h\nu}$ $^1(\pi, \pi^*)$ $\xrightarrow{NH_3}$; $-$MeOH ; NH₃ → (253) ⇌ (254) $\xrightarrow{O_2}$ (252)

SCHEME 21

ammonia to the singlet (π, π^*) excited state of methoxyanthraquinone would be oxidized by dissolved oxygen, in agreement with the observation that oxygen favors the formation of **252**. Whereas thermal reaction of 2,3-dichloro-5,8-dimethoxy-1,4-naphthoquinone (**258**) with methylamine leads to halogen replacement, the photochemical reactions result in methoxy replacement to give **259** only[274]. Irradiation of 2-methoxy-1,4-naphthoquinone in the presence of methylamine results in the formation of 2-methoxy-3-methylamino-1,4-naphthoquinone in contrast to the exclusive replacement of the 2-methoxy group by the thermal reaction (equation 74)[275]. 2,6-Dimethoxy-1,4-

$$\text{2-methoxy-1,4-naphthoquinone} \xrightarrow[\text{rt}]{MeNH_2/CH_2Cl_2} \text{2-methylamino-1,4-naphthoquinone (quant)}$$

$$\text{2-methoxy-1,4-naphthoquinone} \xrightarrow[MeNH_2/CH_2Cl_2]{h\nu(-40^{\circ}C)} \text{2-methoxy-3-methylamino-1,4-naphthoquinone (66\%)} \qquad (74)$$

benzoquinone shows a similar reactivity[275]. These results are consistent with a direct nucleophilic addition of amines to the (π, π^*) excited state of the quinone, since there is a significantly larger decrease in a calculated electron density at the 3-position than the 2-position of 2-methoxy-1,4-naphthoquinone in its lowest excited (π, π^*) state.

D. Substitution of Hydrogen

1. Photoamination

Photosubstitutions involving replacement of hydrogen by an amino group are of apparent synthetic importance. Photochemical reaction of 1-acylaminoanthraquinones (**263**, R = Ph, *p*-$O_2NC_6H_4$, Me, Et, *n*-Pr, PhCH = CH, EtO) with primary aliphatic amines gives the corresponding 1-acylamino-4-alkylaminoanthraquinones **264** in 56–78 % yields[276]. Based on the sensitivity of the reaction to the steric hindrance of the amines used and the need for oxygen, a mechanism involving a HOMO-controlled addition of alkylamine to the triplet excited (CT) of **263**, followed by oxidation to give **264** was proposed. The 5-amino-naphthoquinone **265** is selectively alkylaminated or hydroxylated at the 8-position by photochemical reaction[277]. Alkylamination occurs on irradiation of a benzene solution of **265** and a primary alkylamine under a nitrogen atmosphere, while hydroxylation has been achieved upon irradiation in the presence of a tertiary alkylamine in the presence of oxygen. Tertiary alkylamines do not add to **265** due to steric hindrance, but are essential for the hydroxylation[277] (Table 8). Anthraquinone itself[278, 279] as well as chlorinated anthraquinones[308] can be photochemically aminated in aqueous ammonia solutions containing organic solvents such as 2-propanol or dioxane under aerated conditions. Electron transfer from ammonia to triplet AQ followed by in-cage combination would give the intermediate **268** which would then be oxidized by oxygen to the aminoanthraquinone (equation 75). When a similar photolysis is conducted in acetonitrile–aqueous ammonia solution, adduct **37** is formed along with the direct amination product (equation 14)[82].

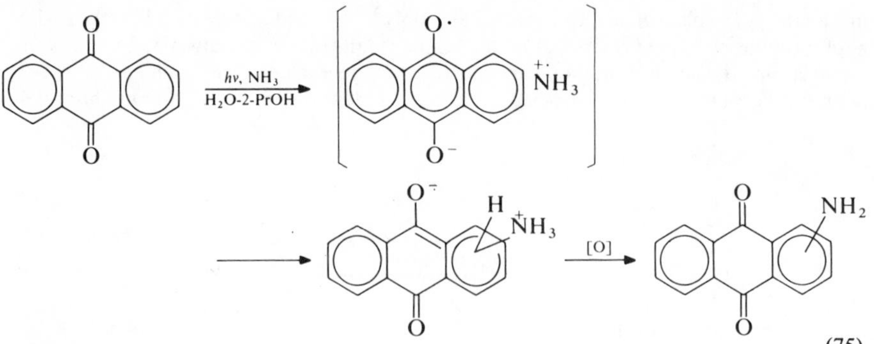

(75)

(268)

2. *Photohydroxylation*

AQ is readily photohydroxylated in concentrated sulfuric acid to give 2-hydroxyanthraquinone **2** (R = OH) as a major product[280–283]. Nucleophilic addition of an hydrogen sulfate ion to the photoexcited conjugated acid of AQ is suggested (Scheme 22)[281]. However, detailed examination of this reaction has shown that besides **2**

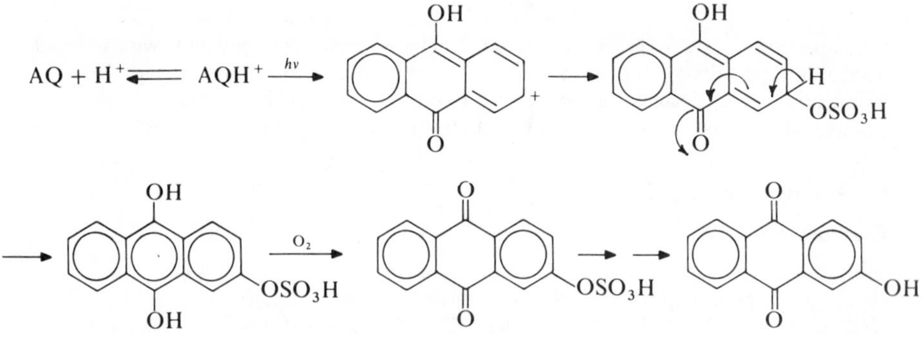

$AQ + H^+ \rightleftharpoons AQH^+ \xrightarrow{h\nu}$

SCHEME 22

(R = OH) other minor products such as 1-hydroxyanthraquinone, polyhydroxyanthraquinone and reductive products such as anthrone and bianthronyl are also formed and that the yields of these minor products are markedly influenced by the presence of water and oxygen in the sulfuric acid (equation 76)[282, 284]. It seems therefore that the mechanism of photohydroxylation of AQ in sulfuric acid is more complex than that shown in Scheme 22. An alternative mechanism initiated by an electron transfer from a bisulfate ion to the photoexcited AQ is also proposed. In this mechanism, hydroxylated products would arise by oxidation of the sulfate radical–AQ adduct, and polyhydroxy derivatives would result from interception of the sulfate radical adduct by oxygen. However, a definite conclusion regarding the mechanism of the overall reaction has not yet been established.

The photohydroxylation of ASQ and ADSQ is a topic that has attracted the interest of several groups over the years[44, 78, 90, 283, 285–290]. A considerable amount of work has been devoted to this reaction under aerobic and anaerobic conditions and three alternative

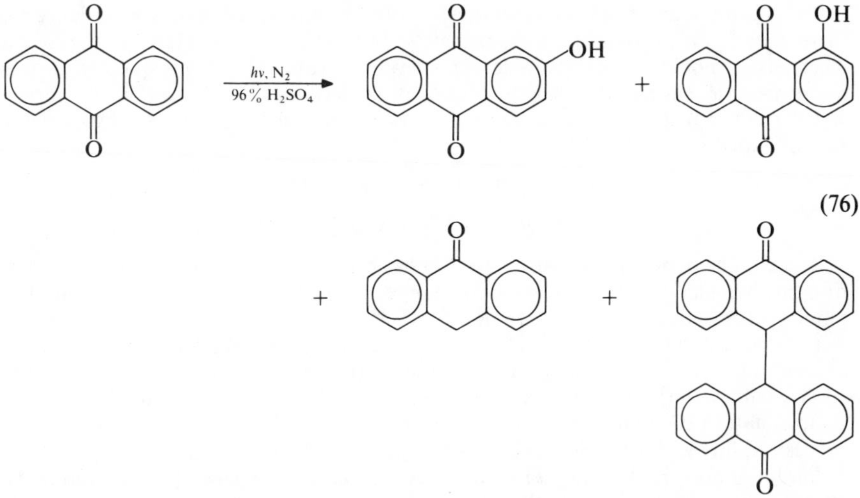

$h\nu$, N_2, 96% H_2SO_4

(76)

mechanisms have been proposed. Since the intersystem crossing is highly efficient[44], all three mechanisms start with the triplet ASQ. The first mechanism, which is favored by the Russian school[6, 7], involves a free hydroxy radical produced by the reactions of equations 77 and 78, which attacks ground state ASQ to give an hydroxy–ASQ radical adduct. The involvement of a hydroxylated radical adduct ASQOH$^{\cdot}$ and its decay to permanent hydroxyanthraquinone (equations 80 and 81) is a common feature of all the three mechanisms.

$$^3(ASQ)^* + H_2O \rightarrow ASQ^{-\cdot} + HO^{\cdot} + H^+ \quad (77)$$

$$^3(ASQ)^* + HO^- \rightarrow ASQ^{-\cdot} + HO^{\cdot} \quad (78)$$

$$ASQ + HO^{\cdot} \rightarrow ASQOH^{\cdot} \quad (79)$$

$$ASQOH^{\cdot} + O_2 \rightarrow ASQOH + {}^{\cdot}OOH \quad (80)$$

$$2ASQOH^{\cdot} \rightarrow ASQOH + ASQ + H_2O \quad (81)$$

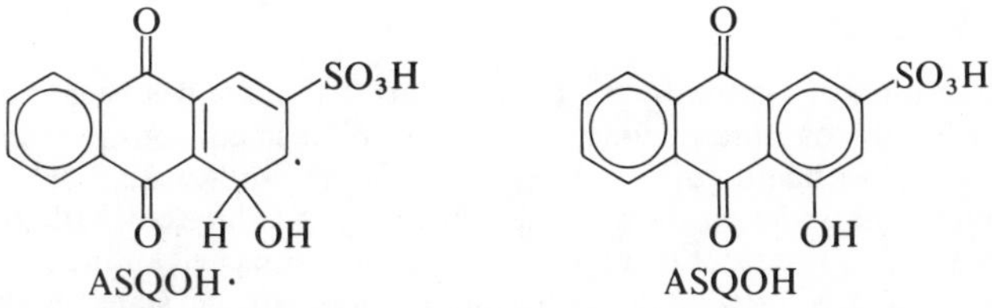

ASQOH·　　　ASQOH

In the second mechanism, it has been assumed, on the basis of the dependence of the quantum yield on the ASQ concentration, that electron transfer between triplet excited ASQ and ground state ASQ occurs as the primary step and subsequent oxidation of water by the ASQ$^{+\cdot}$ generates the hydroxy radical (equations 82 and 83).

$$^3(ASQ)^* + ASQ \rightarrow ASQ^{+\cdot} + ASQ^{-\cdot} \quad (82)$$

$$ASQ^{+\cdot} + H_2O \rightarrow ASQ + HO^{\cdot} + H^+ \quad (83)$$

However, a comparative study of the effects of pH and various hydroxy radical scavengers on the photolysis and γ-radiolysis of aqueous solution of ASQ seems to be at variance with these free hydroxy radical mechanisms[124, 283, 289, 291, 292]. In addition, the second mechanism should be unfavorable on energetic grounds[6,7]. Consequently, a third mechanism is suggested, whereby $^3(AQS)^*$ undergoes photosolvation to form a water–ASQ adduct which, upon reaction with ground state ASQ, gives the hydroxylated radical (equations 84 and 85).

$$^3(ASQ)^* + H_2O \rightarrow (ASQ \cdots H_2O) \quad (84)$$

$$(ASQ \cdots H_2O) + ASQ \rightarrow ASQOH + ASQ^{-\cdot} + H^+ \quad (85)$$

Recently the photohydroxylation of ASQ has been subjected to an intensive investigation involving laser photolysis, steady-state reactions and study of the inhibition of the photohydroxylation by inorganic anions[44, 78, 90]. $^3(ASQ)^*$ and two other intermediates were detected and the decay of the former was found to be independent of the initial ASQ concentration whereas the latter products were formed in parallel reactions of the triplet species with water[78]. One of the intermediates ($\lambda_{max} = 600$ nm) appears to react with ground-state ASQ, and this reaction is considered to lead to the photohydroxylation process, in line with the third mechanism (equations 84 and 85)[78].

Anthraquinone itself is photohydroxylated in water–2-propanol (1:2) solutions. The efficiency of the reaction is greatly reduced by deaeration of the reaction solutions and is enhanced by addition of hydroxide ion[124, 291].

Another type of photohydroxylation has recently been described, whereby 2-acylaminoanthraquinone is hydroxylated in either benzene or methanol with high selectivity at the 1-position under aerated conditions (equation 86)[293]. Apparently the hydroxylation takes place on account of oxygen dissolved in the organic solvent, although the detailed mechanism has not been studied.

O NHCOR O → ($h\nu$, O_2 / C_6H_6 or MeOH) O OH NHCOR O (86)

R = Ph, Me or CF_3

3. *Photosulfonation*

Photosulfonation of 1-amino- or 1-hydroxyanthraquinone has been reported[294, 295]. Irradiation of 1-aminoanthraquinone and sodium sulfite in aqueous pyridine with visible light results in the formation of 1-aminoanthraquinone-2-sulfonate (**269**) in 92.6% yield (Table 8). When the *ortho* position to the amino group is blocked, little or no reaction occurs. Initial electron transfer from the sulfite ion to the excited anthraquinone will give an SO_3 radical which will attack the anthraquinone in the ground state. On the other hand, 1-hydroxyanthraquinone gives a mixture of the 2-sulfonate **271** (31%), the 4-sulfonate **272** (18%), and the 2,4-disulfonate **273** (24%). The different substitution pattern observed for 1-aminoanthraquinone and 1-hydroxyanthraquinone has been rationalized by application of semiempirical molecular orbital theory[295].

E. Miscellaneous

Photochemical transformation of 1-amino- and 1-methylamino-4-nitroanthraquinone **275** to 1-amino- and 1-methylamino-4-hydroxyanthraquinone **277** has been reported[296].

A nitro–nitrite photorearrangement **275** → **276** was suggested to explain these reductive substitutions[296]. However, in aqueous 2-propanol, a reduction of the 4-nitro group to give **278** occurs concurrently with the nitro–nitrite rearrangement (equation 87). It seems likely that the former process occurs from the (n, π^*; NO_2) electronic configuration. Support for this interpretation have been obtained by several sensitized experiments[158].

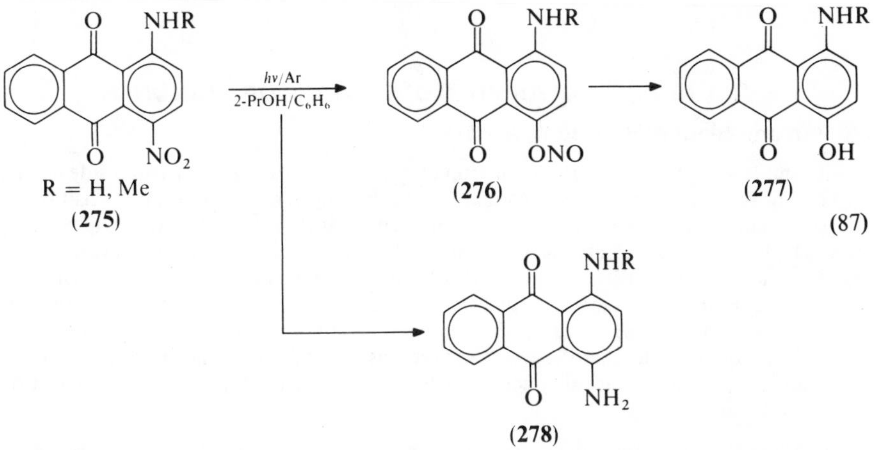

O NHR
O NO_2
R = H, Me
(**275**)

$h\nu$/Ar
2-PrOH/C_6H_6

O NHR
O ONO
(**276**)

O NHR
O OH
(**277**)

(87)

O NHR
O NH_2
(**278**)

An interesting example of the photochemical formation of **277** from **279** has been realized in concentrated sulfuric acid[280,297]. It is suggested that **279** is photochemically reduced to 1-hydroxylaminoanthraquinone (**280**), which rearranges into **277** in an acid-catalyzed dark reaction (equation 88).

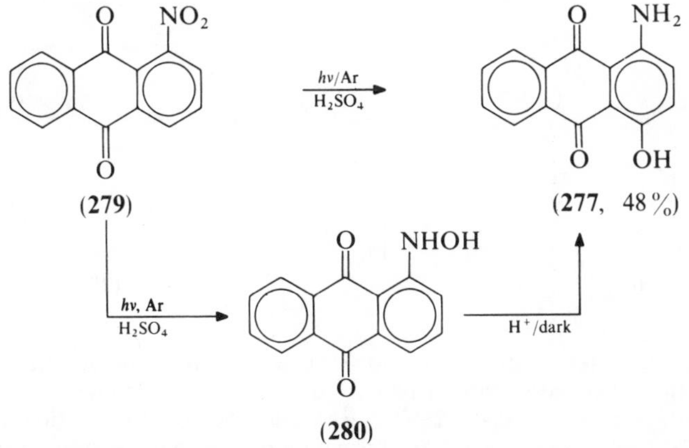

O NO_2
O
(**279**)

$h\nu$/Ar
H_2SO_4

O NH_2
O OH
(**277**, 48%)

(88)

$h\nu$, Ar
H_2SO_4

O NHOH
O
(**280**)

H^+/dark

Anthraquinones **242** and **244** and 1-hydroxy and 1-aminoanthraquinones are photochemically oxidized under irradiation with visible light in an aerated alkaline solution to give phthalic acid as the main oxidation product (equation 89)[298,299]. A singlet exciplex between the anthraquinone and oxygen is suggested to be involved in this photooxidation process.

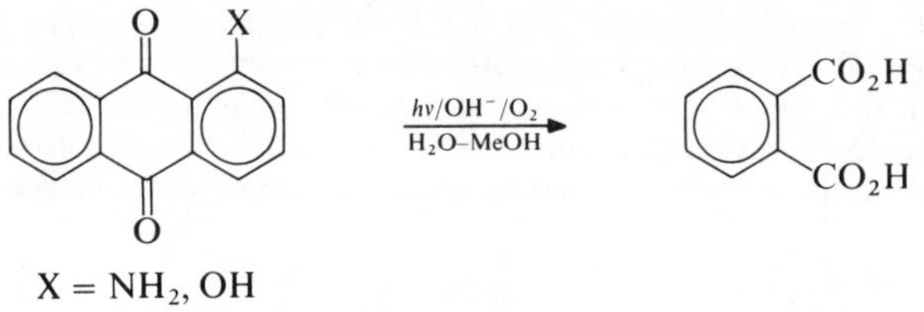

$X = NH_2, OH$ (89)

VI. UNIMOLECULAR PHOTOREACTION OF QUINONES

A. Intramolecular Hydrogen Abstraction

Intramolecular photochemical reduction of quinones bearing hydrocarbon side chains (**281**) almost always involves the abstraction of a hydrogen atom from the 2′-position of the side chain via a readily achieved six-membered transition state, similar to the Norrish type II photoreaction of aromatic ketones. Therefore, quinones whose lowest triplet excited states are (n, π^*) are reactive in this intramolecular H-atom abstraction reaction. However, quinones with a lowest triplet (π, π^*) excited state can also undergo this reaction via a charge transfer or an electron transfer when heteroatoms such as nitrogen or sulfur are present in the side chain. Several characteristic features of the intramolecular H abstraction of quinones are distinct from the usual Norrish type II photoreaction of aromatic ketones.

(1) Since the ketyl radical site in the intermediate biradical **282** is stabilized as a semiquinone radical, the main decay pathway is the cyclization to the dihydrobenzofuran (**284**). Formation of the cyclobutanol **287** is very rare with the exception of the photolysis of **288**, which gives **289** and **290** in 45% and 13% yields, respectively (equation 90)[307]. Fragmentation, which is usually the main decay process of a Norrish type II biradical, is apparently unfavorable and has never been reported.

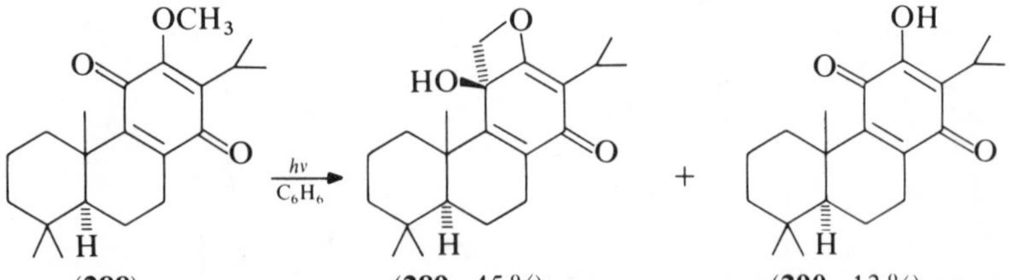

(**288**) (**289**, 45%) (**290**, 13%) (90)

(2) Since the double bonds of the quinone ring are susceptible to radical attack, a radical center in the alkyl side chain may add to the quinone double bond to give a spirocyclopropane intermediate **285**, which gives the solvent addition product **286** (Scheme 23). This is exemplified in the formation of side chain rearranged products **292** and **293** together with **291** in the photolysis of 2,6-di-*t*-butyl-1,4-benzoquinone (Scheme 24)[112]. In nucleophilic solvents, the solvent adduct **293** is favored over **292**.

(3) When the alkyl radical center is stabilized by an adjacent heteroatom such as nitrogen or sulfur, a zwitterionic character becomes significant in the intermediate.

(4) Abstraction of an allylic hydrogen from the side chain often leads to a unique structure of *o*-quinone methide intermediates.

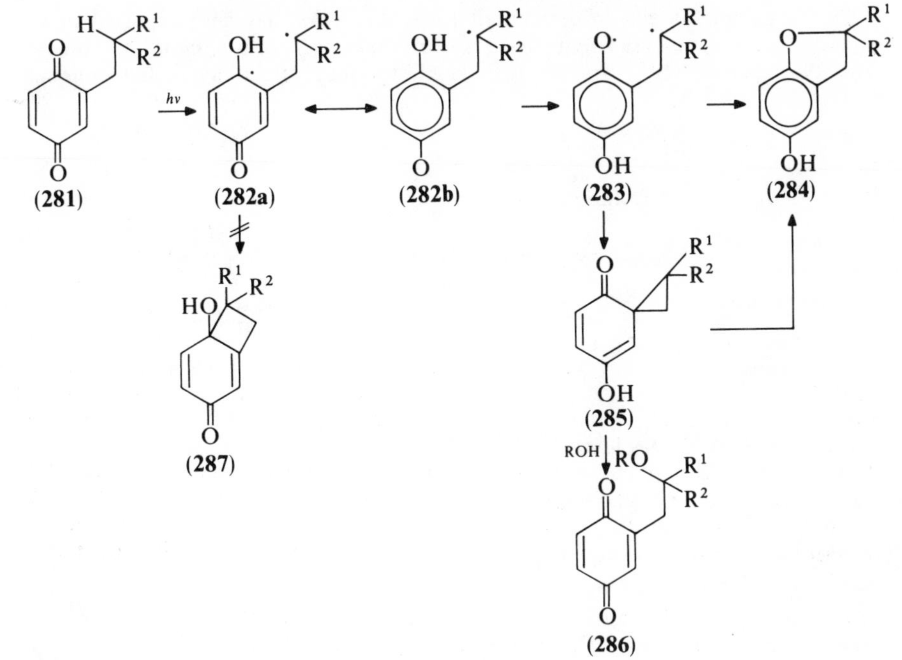

(281) (282a) (282b) (283) (284) (287) (285) (286)

SCHEME 23

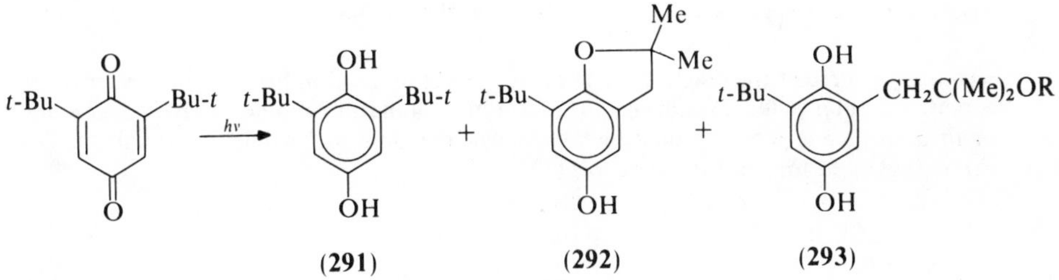

(291) (292) (293)

Solvent	Product distribution (%) (291)	(292)	(293)
MeOH	0	0	92
EtOH	40	—	58
2-PrOH	25	—	62
t-BuOH	10	42	48
AcOH	—	37	52* (as acetate)
Toluene	75	25	—
Benzene	14	36	—
C_6F_6	33	66	—

SCHEME 24

The formation of dihydrobenzofuran derivatives (**295**) has been reported in the photolysis of quinones (**294**) (equation 91), while the rearranged product **297** is obtained from **296** (equation 92)[309]. Apparently the stability of the resultant radical determines the

(294) $\xrightarrow[C_6H_6]{h\nu}$ (295, 65–75%) (91)

R = Me, Et, CO_2Me, $(CH_2)_3N$(phthalimido)

(296) $\xrightarrow[C_6H_6]{h\nu}$ (297, 41%) (92)

secondary course of the reaction. In contrast, **298** is unusual in being stable under these conditions, but it yields ethanol and 45–50% of the coumarin **299** when a catalytic amount of air or oxygen is present (equation 93). An 'ene reaction' with singlet oxygen formed *in situ* is suggested for the formation of **299**[310].

(298) $\xrightarrow[C_6H_6]{h\nu/N_2}$ no reaction

(298) $\xrightarrow[C_6H_6]{h\nu/O_2}$ → → → (299) (93)

Irradiation of the pale yellow 5-methyl-1,4-naphthoquinone **300** yields the blue photoenol **301** which is stable at 77 K. At room temperature the enol retautomerizes back to the starting material, the reaction rate being strongly dependent on the hydrogen-bond-accepting ability of the solvent. The enol is trapped by acetic acid to give **303** (equation 94)[311]. Analogous reversible photoenolization is observed for 1,4-dimethylanthraquinone

(300) $\xrightarrow[\Delta]{h\nu}$ (301) $\xrightarrow[\text{slow}]{H^+}$ (302) $\xrightarrow{\text{AcOH}}$ (303) (94)

at 2-methyltetrahydrofuran matrix at 2 K[69a, 312]. *o*-Quinone methide **304** is apparently an intermediate in the photodimerization of DQ in aqueous solution, but its formation is not a major pathway for decay of the DQ triplet state[70, 313]. Flash photolysis of lactone **305** gives rise to a transient with the broad visible absorption spectrum (λ_{max} (EtOH) 445 nm), which is assigned to **304** (equation 95)[313a]. Bicyclic diketone **306** was isolated in the photolysis of 2,5-di-*t*-butyl-1,4-benzoquinone in benzene with UV light (equation 96)[314].

DQ $\xrightarrow{h\nu}$ (304) $\xleftarrow[-CO]{h\nu}$ (305) (95)

$\xrightarrow[C_6H_6]{h\nu}$ (306) + others (96)

The photochemistry of the naturally occurring diterpenoid quinone royleanone **290** and its derivatives has been examined with the hypothesis that barbatusin (**307**) and cyclobutatusin (**308**) in which the common isopropyl side chain appears as a spirocyclopropyl unit, could be formed from photochemical transformation of royleanone-like precursors in the plants[307]. However, intramolecular hydrogen abstraction takes place at the other carbonyl group to give products having a cyclic ether linkage such as **309** (equation 97).

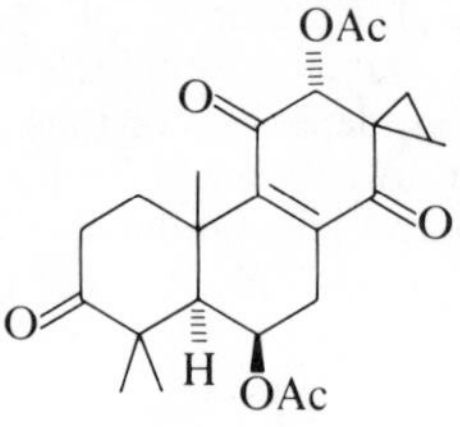

(307)

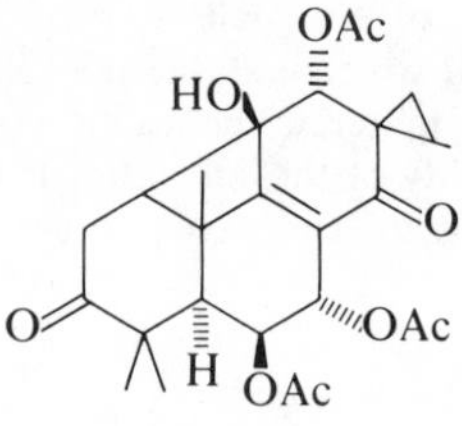

(308)

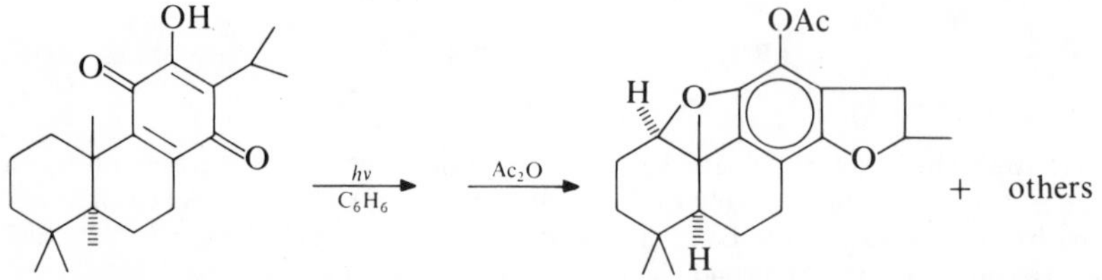

(290) → hν, C_6H_6 → Ac_2O → (309) + others (97)

The photolysis of 2-methoxy- or 2-ethoxy-1,4-naphthoquinone leads to the formation of cyclobutane dimers **310** and **311** (equation 98)[315], while under similar conditions 3-substituted 2-alkoxy-1,4-naphthoquinones (**312**, R = Me, Br) give, after treatment with acetic anhydride, the ring-closed compounds **313** (equation 99)[315, 316]. The differences are

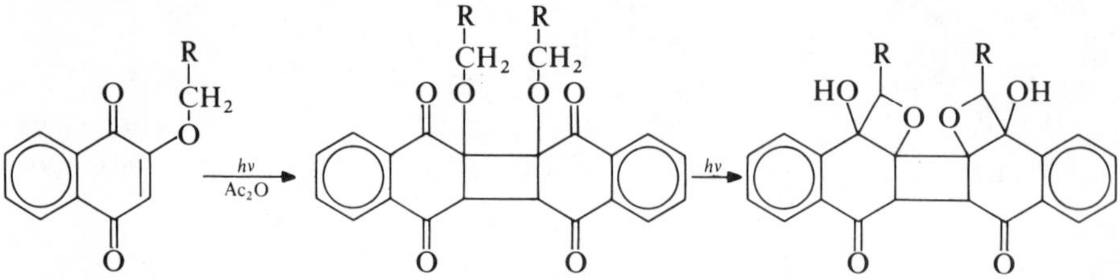

R = H, Me → hν, Ac_2O → (310) → hν → (311) (98)

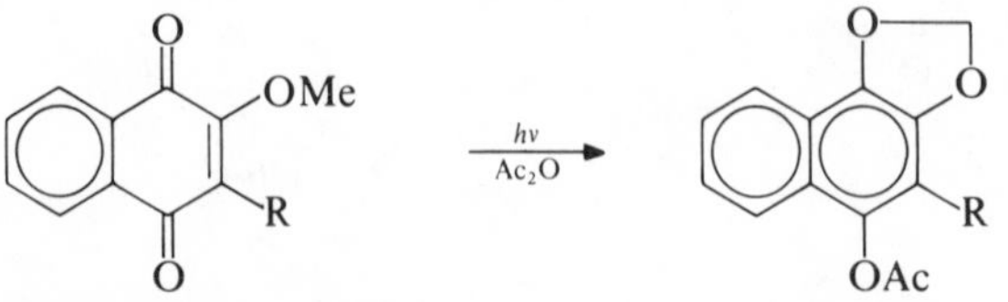

(312, R = Me, Br) → hν, Ac_2O → (313) (99)

obviously due to the buttressing effects rather than to the electronic effects of 3-substituents. Similar photolysis of 2-isopropoxy-1,4-naphthoquinone produces none of the dimer but a cyclized product **314** via intramolecular hydrogen abstraction (equation 100)[315]. Photochemical intramolecular redox reaction of **315** to **316** (equation 101) occurs smoothly[316].

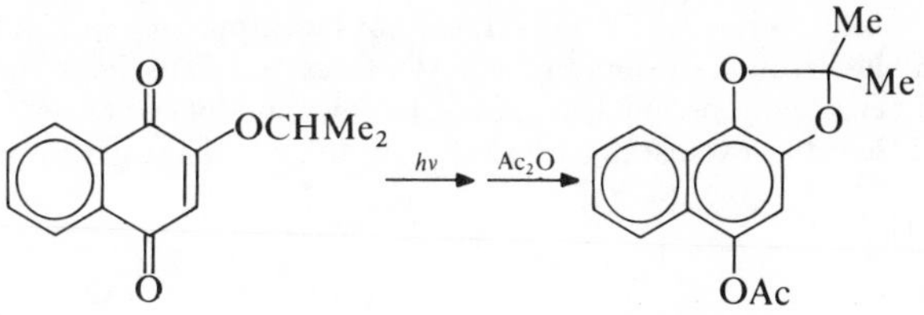

O, $OCHMe_2$, O $\xrightarrow{h\nu}$ $\xrightarrow{Ac_2O}$ Me, Me, O, O, OAc (100)

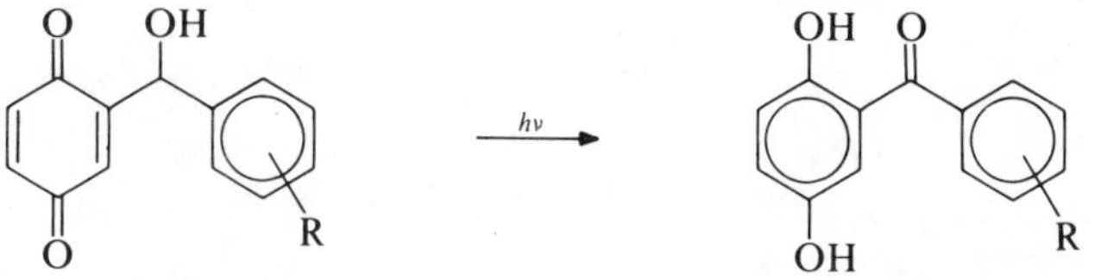

(314)

O, OH, O, R $\xrightarrow{h\nu}$ OH, O, OH, R (101)

(315) (316)

Irradiation of 2-acyl-1,4-benzoquinone **317** in non-polar solvents gives dimer **318** with high regio- and stereoselectivity (equation 102)[317–319]. The structure of **318** was

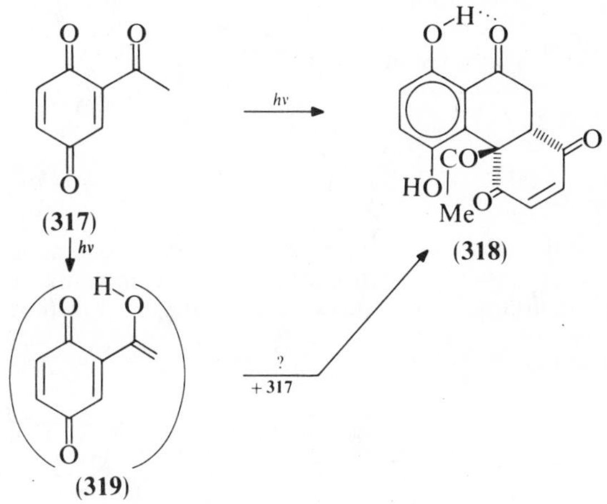

O, O, O $\xrightarrow{h\nu}$ O, H, O, O, CO, HO, MeO

(317) (318) (102)

$h\nu$

H, O, O, O $\xrightarrow[+317]{?}$

(319)

determined by X-ray analysis[318, 319]. However, irradiation of isobutyryl-1,4-benzoquinone **320** gives the dihydrobenzofuranone product **321** (equation 103)[319]. It seems reasonable to assume that the dimerization proceeds through the *o*-quinone

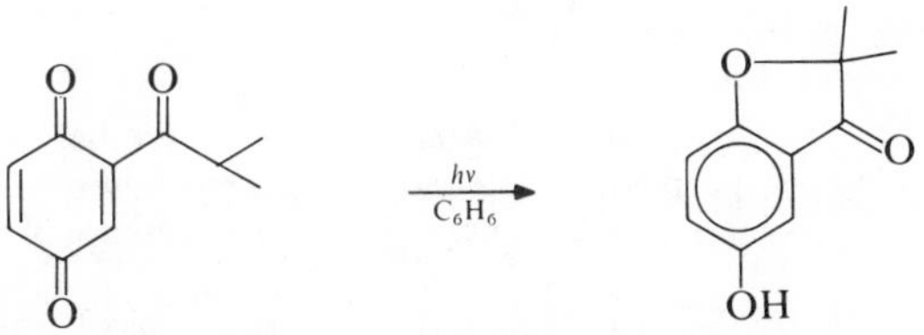

O, O, O $\xrightarrow[C_6H_6]{h\nu}$ O, O, OH (103)

(320) (321)

methide intermediate **319**, which, however, cannot be trapped by any other dienophiles such as *N*-phenylmaleimide or dimethyl acetylenedicarboxylate. Even in a system (**322**) where two acyl benzoquinone units are connected by a methylene bridge, intramolecular regio- and stereoselective cycloaddition takes place to give **323** (equation 104)[319]. On the

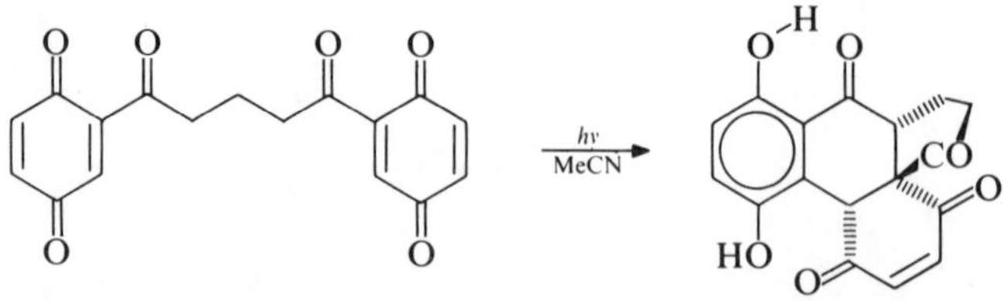

$h\nu$, MeCN

(**322**) (**323**) (104)

other hand, a different dimer (**324**) is formed in the Rose Bengal sensitized photoreaction of **317** (equation 105)[320, 321]. In the latter dimerization reaction, electron transfer from

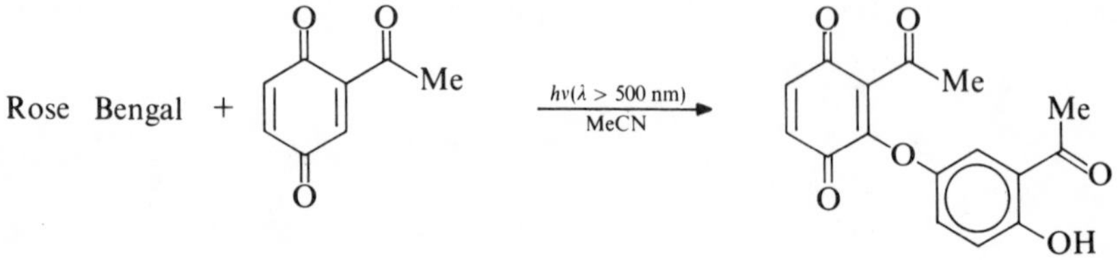

Rose Bengal + (**317**) $\xrightarrow[\text{MeCN}]{h\nu(\lambda > 500\text{ nm})}$ (**324**, 60%) (105)

singlet excited Rose Bengal to the electron-deficient acylbenzoquinone seems to be the initial process[322]. Porphyrin-sensitized electron transfer reactions with quinones as electron acceptors are utilized in the oxidative dimerization of 4-methoxyphenol (**325**) to

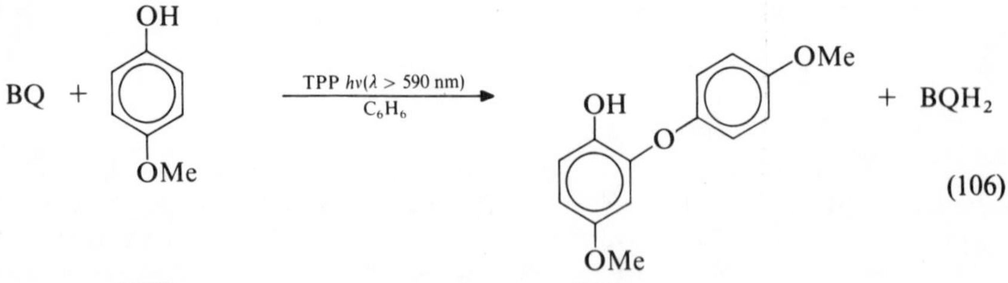

BQ + (**325**) $\xrightarrow[C_6H_6]{\text{TPP } h\nu(\lambda > 590\text{ nm})}$ (**326**) + BQH_2 (106)

(TPP: meso-tetrakisphenylporphyrin)

326 (equation 106)[323]. In an extension of this process, a novel synthesis of exotic quinone-linked and quinone-capped porphyrins **329** and **330** has recently been achieved by self-photosensitization of tyrosine-linked prophyrin **327** in the presence of halogenoquinone (equation 107)[324].

Photo-induced intramolecular hydrogen abstraction reactions of aminoquinones have attracted considerable interest due to their potential use in synthetic routes toward

(327) + (328) $\xrightarrow[CHCl_3]{h\nu > 590\ nm}$ (329) $\xrightarrow{h\nu > 590\ nm}$ (330) (107)

mitomycins[325–332]. The photolysis of **331** with a sunlamp to produce **332** proceeds cleanly. The oxazole **332** is characterized by NMR and is isolated as **333** after oxidation with silver oxide (equation 108)[328]. After allowing an irradiated solution of **334** to stand

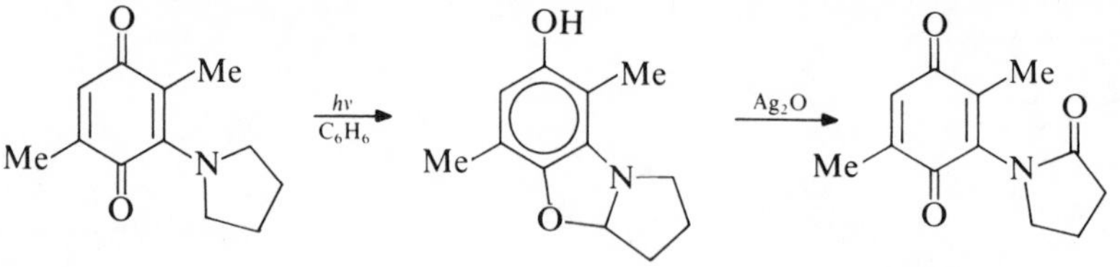

O
Me
Me
N
O
$h\nu$
C_6H_6
OH
Me
Me
N
O
Ag_2O
O
Me
O
Me
N
O

(**331**) (**332**) (**333**) (108)

for more than 24 h at room temperature, ring-closed quinone **337** has been obtained in high yield (equation 109). An aziridine-containing pyrroloindoloquinone **339** is synthesized by analogous intramolecular photoreaction of **338** (equation 110)[329].

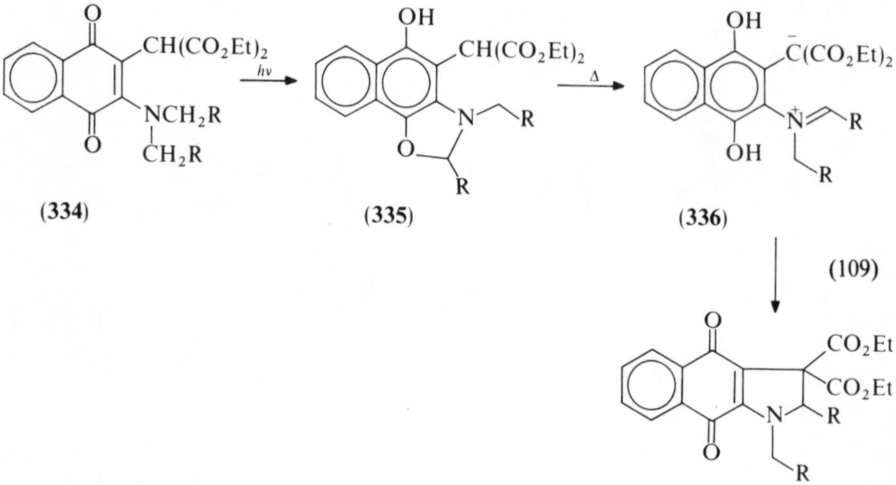

O
$CH(CO_2Et)_2$
$h\nu$
NCH_2R
O
CH_2R
OH
$CH(CO_2Et)_2$
Δ
N
R
O
R
OH
$\bar{C}(CO_2Et)_2$
$\overset{+}{N}$
R
OH
R

(**334**) (**335**) (**336**)

(109)

O
CO_2Et
CO_2Et
N
R
O
R

(**337**, 55–78%)

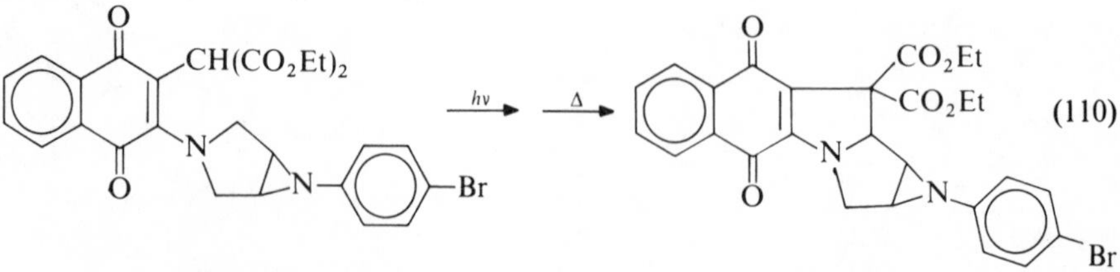

O
$CH(CO_2Et)_2$
N
O
N
Br
$h\nu$
Δ
O
CO_2Et
CO_2Et
(110)
N
O
N
Br

(**338**) (**339**, 63%)

Isomerization of **335** to zwitterionic intermediate **336**, followed by intramolecular nucleophilic attack, is proposed to explain a formation of **337**[329]. Analogous irradiation of aminobenzoquinone **340** in ethanol followed by contact with SiO_2 for a few days gives rise

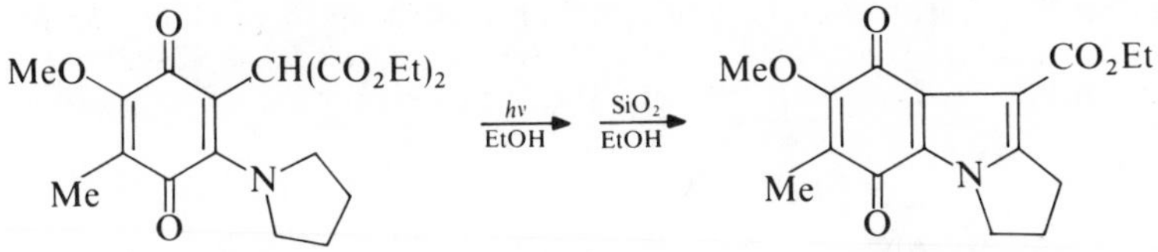

(340) (341, 68%) (111)

to pyrro[1,2-*b*] indoloquinone **341** (equation 111)[325]. 2-Diethylamino-3-phenylethynyl-1,4-naphthoquinone (**342**) is transformed into tricyclic quinone **343** by a two-step photoreaction (equation 112)[333].

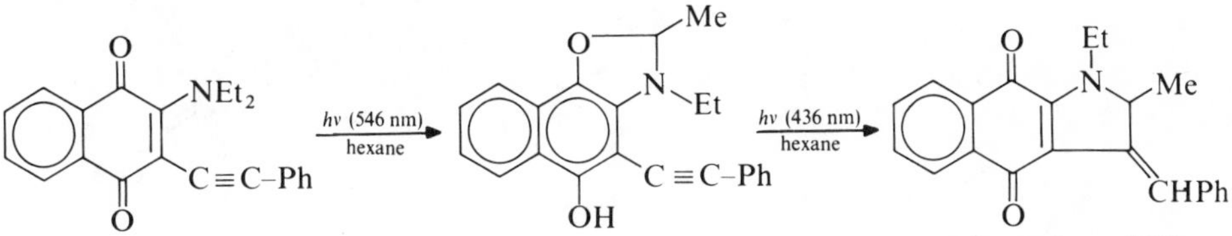

(342) (343, 45%) (112)

The photolysis of naphthoquinone **344** gives 2-pyrryl-1,4-naphthoquinone (**345**) after oxidation with air (equation 113)[330], while the main product obtained in the irradiation of **346** is the ring-opened aminonaphthoquinone **348** (equation 114). The latter product is

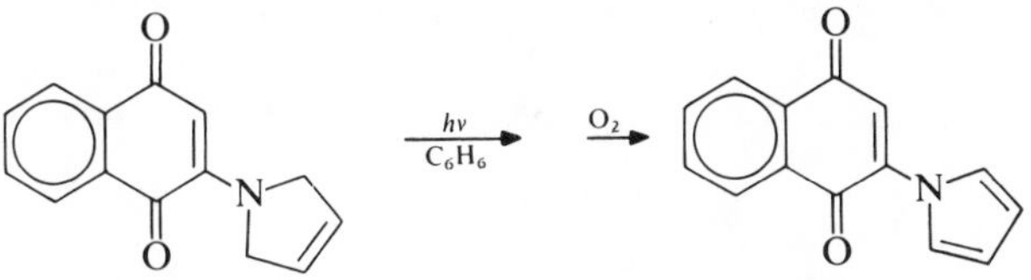

(344) (345) (113)

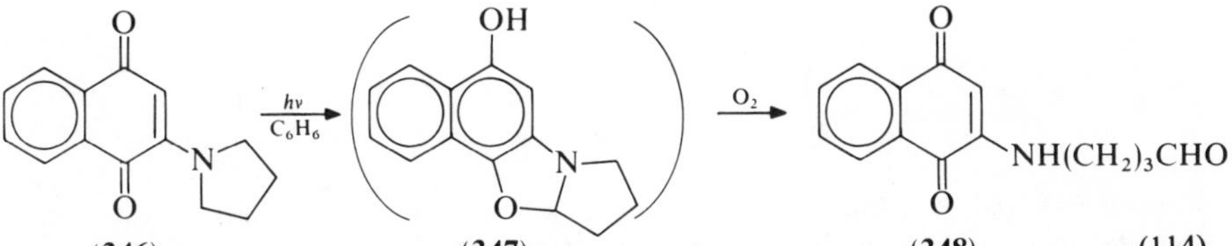

(346) (347) (348) (114)

shown to arise from a secondary dark reaction of the cyclized product **347** with oxygen[330]. Photolysis of **349** affords quantitatively isoindolylnaphthohydroquinone **350**, which can be trapped by suitable dienophiles to produce compound **351** (equation 115)[331].

Zwitterionic intermediate **353** (with a contributing diradical hybrid **354**) is reported to be detected in the photolysis of an ethanol solution of **352** at 77 K. Upon warming to room temperature, **353** cyclizes to give **355** (equation 116)[334, 335]. Irradiation of aminomethyl-substituted naphthoquinone **356** in heptane gives the photosensitive 1,4-dihydroxy-3-methyl-2-naphthaldehyde **357** as the main photolysis product along with the secondary

(349) $\xrightarrow[C_6H_6]{h\nu}$ (350) (115)

$\xrightarrow{MeO_2CC\equiv CCO_2Me}$ (351)

(352) $\xrightarrow[EtOH]{h\nu/77\ K}$ (353) $\xrightarrow{\Delta}$ (355) (116)

(353) ↔ (354)

products **358** and **359**. Initial electron transfer and subsequent hydrogen transfer are suggested to give the immonium intermediate **360** which will be hydrolyzed by water to **357** (equation 117)[336].

B. Photoreaction of Quinones Bearing Side-chain Double Bonds

The plastoquinones (PLQ) are a class of trisubstituted isoprenyl-1,4-benzoquinones found in algae and in the chloroplasts of green plants, and PLQ-9 is believed to function in photosynthetic electron transport and coupled photophosphorylation[337, 338]. The photochemistry of PLQ-9 and related electron-transport quinones is of considerable interest for three main reasons.

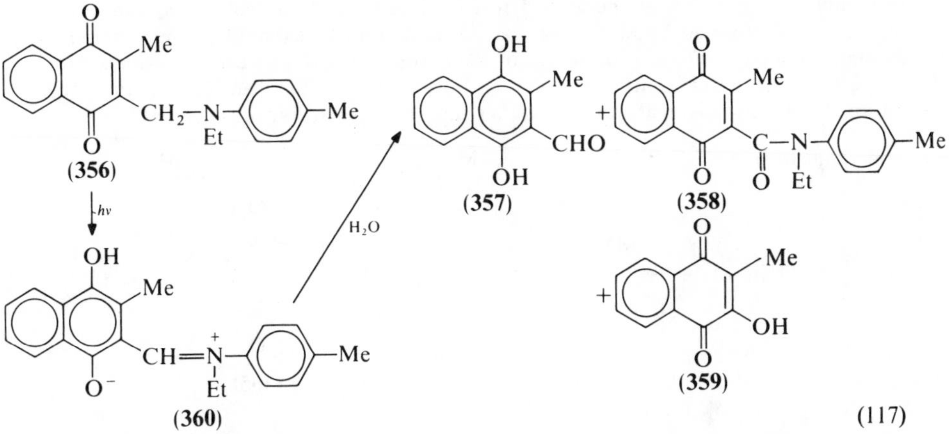

(356) (357) (358) (359) (360) (117)

(1) It may provide some insight into the function of these quinones *in vivo*.

(2) The *in vivo* destruction of electron-transport quinones may be responsible for some of the biological effects of UV irradiation on living cells.

(3) A comparison of the photoreactions of quinones and other cellular constituents *in vitro* and *in vivo* may give some indications of the molecular environment of the photolyzed molecules *in vivo*.

So far only two photoproducts of PLQ-9 have been reported in cells; plastohydroquinone and a partially characterized dimer[339]. The dimer may arise from the addition of the quinone ring of one molecule to one of the nine double bonds of the second molecule. However, it is noted that it may be an artifact of the isolation procedure.

In vitro photolysis of isoprenoid quinones usually leads to a mixture of many products. Detailed study on the photolysis of PLQ-1 and PLQ-9 in benzene in the presence of methanol[340], revealed the formation of the isomeric dihydrobenzofuran **361** as the major product along with the minor formation of chromenol (**362**) and benzoxepin (**363**) (equation 118). Two mechanisms were proposed for the formation of **361**; (1) addition of

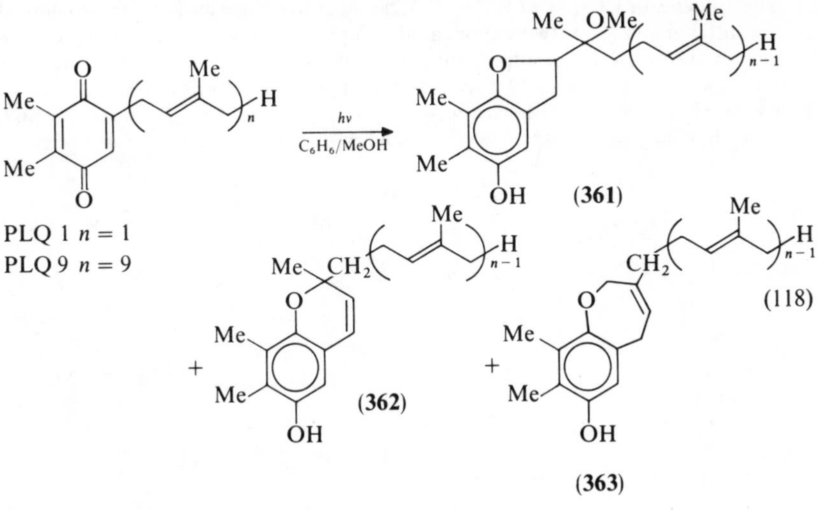

PLQ 1 $n = 1$
PLQ 9 $n = 9$

hv, $C_6H_6/MeOH$

(361) (362) (363) (118)

methanol to the zwitterionic species **365**, which can be formed via electron reorganization of the initially formed biradical **364**, and (2) addition of methanol to the *o*-quinone methides **366** and **367**. The formation of the chromenol **362** was used as evidence for supporting the latter route (Scheme 25). Formation of **363** can be explained by intramolecular hydrogen abstraction via an eight-membered transition state.

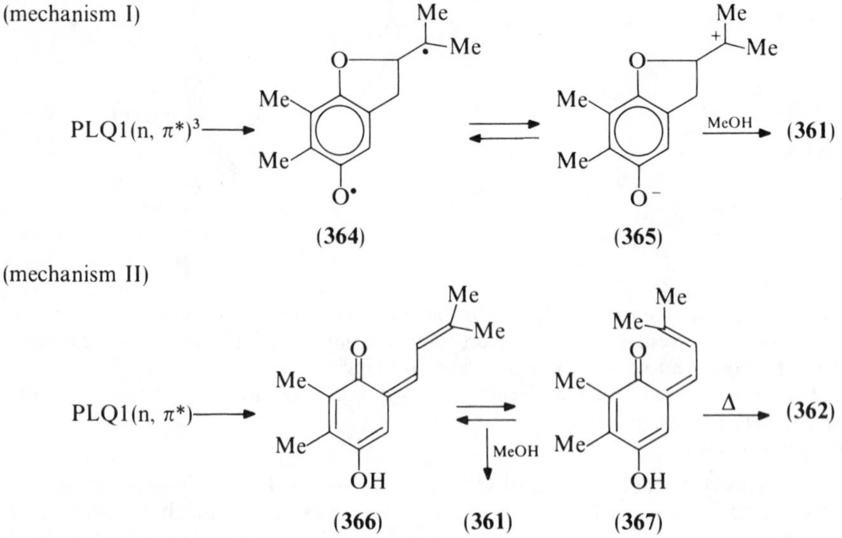

(mechanism I)

PLQ1(n, π^*)^3 → (**364**) ⇌ (**365**) $\xrightarrow{MeOH}$ (**361**)

(mechanism II)

PLQ1(n, π^*) → (**366**) ⇌ (**367**) $\xrightarrow{\Delta}$ (**362**)

(**366**) $\xrightarrow{MeOH}$ (**361**)

SCHEME 25

Irradiation of PLQ-1 in ethanol leads to the formation of a short-lived species (λ_{max} = 420 nm) which decays to a long-lived species (λ_{max} = 500 nm); the former is tentatively assigned to the biradical **368** and the latter is assigned to a mixture of cisoid and transoid *o*-quinone methide[341]. Similar transients are observed in the photolysis of vitamin K_1 in heptane, toluene, or CCl_4 at 77 K[341, 342]. The short-lived species (λ_{max} = 440 nm) whose decay can be analyzed by two exponential components has also been studied by ESR. From ESR spectra, a dipole–dipole interaction constant, $D = 200 \pm 3$ *G*, is determined; this corresponds to the mean distance between unpaired electrons of r = 5.2 Å. Based on these observations, the short-lived transients are suggested to be the biradicals **369** and **370**, and they may be transformed into cisoid and transoid *o*-quinone methides.

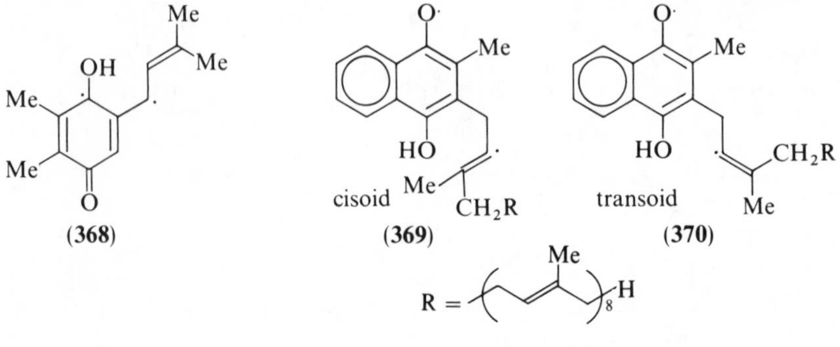

(**368**)   cisoid (**369**)   transoid (**370**)

$R = (CH_2CH{=}C(Me)CH_2)_8H$

Argon ion laser excitation of 3-methyl-2-butenyl-1,4-benzoquinone **371** in $CFCl_3$ under high pressure of oxygen (11 atom) at $-80\,°C$ affords the trioxane adduct **372** quantitatively (equation 119)[210]. Under similar conditions, 2-methyl-3-(3-methyl-2-butenyl)-1,4-naphthoquinone (**373**) gives trioxane (**375**) and hydroperoxide **376**. The

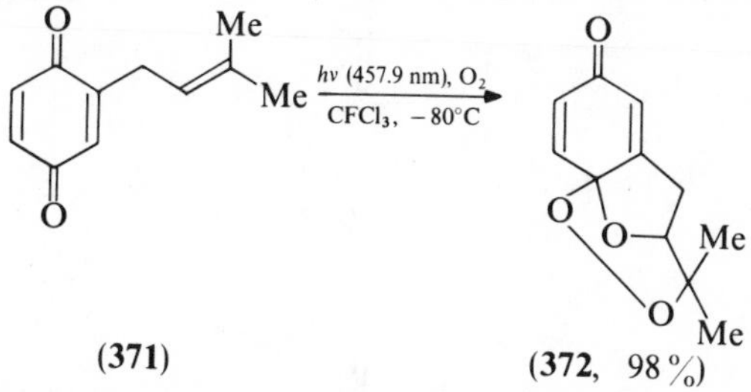

$$\xrightarrow[\text{CFCl}_3,\ -80°\text{C}]{h\nu\ (457.9\ \text{nm}),\ \text{O}_2} \qquad (119)$$

(**371**) (**372**, 98%)

trioxanes **372** and **375** are thermally very stable but extremely photolabile; UV irradiation of **375** induces a cleavage to acetone (89 %) and a very photolabile quinone aldehyde **377** (24 %)[343]. Formation of **372** and **375** can be most easily rationalized as a trapping of a 1,4-preoxetane biradical **374** (or an intramolecular charge transfer complex between quinone and olefin) by oxygen (Scheme 26). However, a different report shows that in

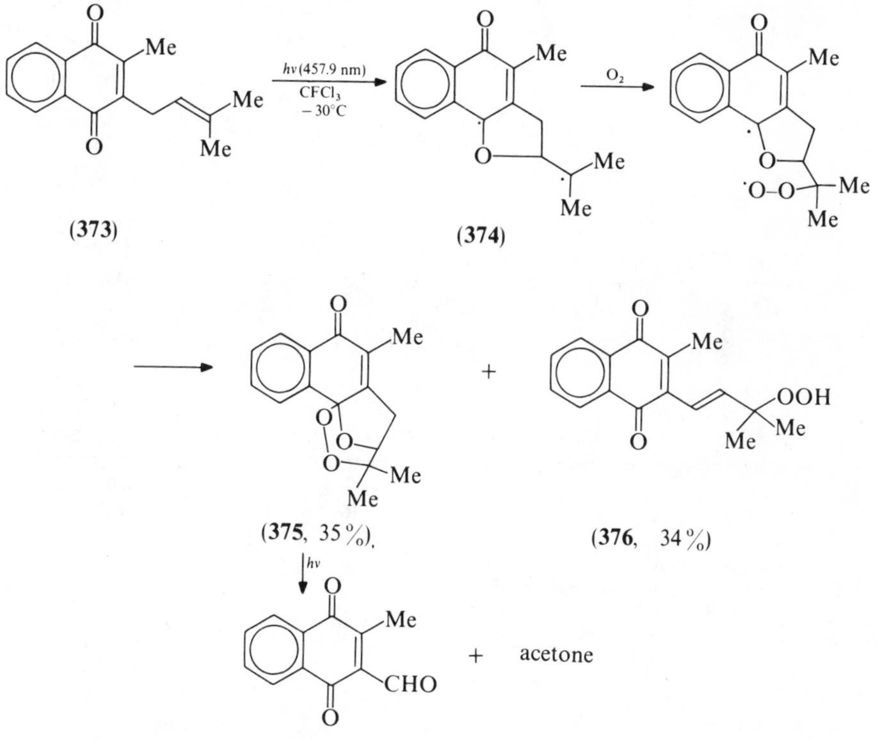

$$\xrightarrow[\text{CFCl}_3,\ -30°\text{C}]{h\nu\ (457.9\ \text{nm})} \qquad \xrightarrow{\text{O}_2}$$

(**373**) (**374**)

(**375**, 35%), + (**376**, 34%)

$\downarrow h\nu$

(**377**) + acetone

SCHEME 26

ethanol **376** is formed in the dark reaction of metastable ground-state intermediate (λ_{max} 550 nm), probably *o*-quinone methide **378a** or **378b** with oxygen (Scheme 27)[344, 345]. Even at atmospheric oxygen pressure, quinone–oxygen adducts **382** are

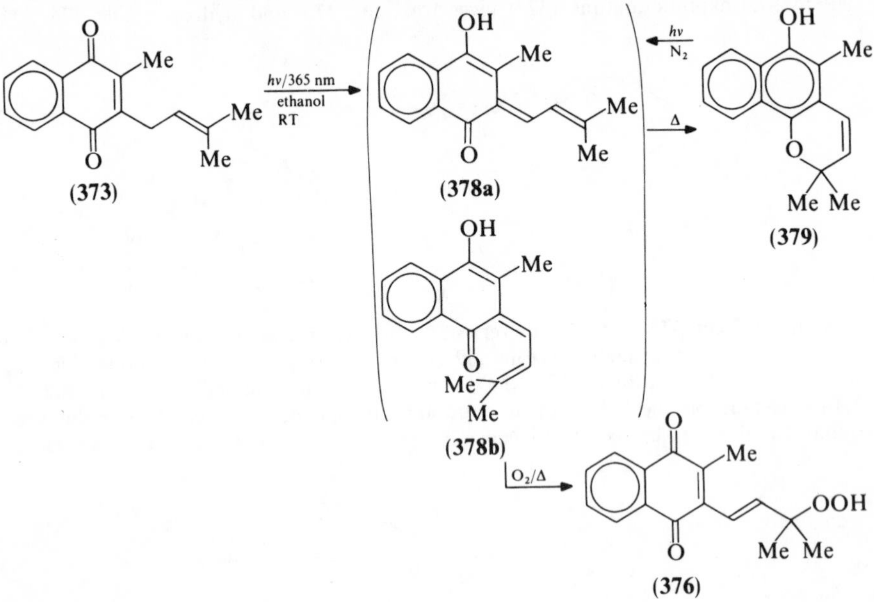

SCHEME 27

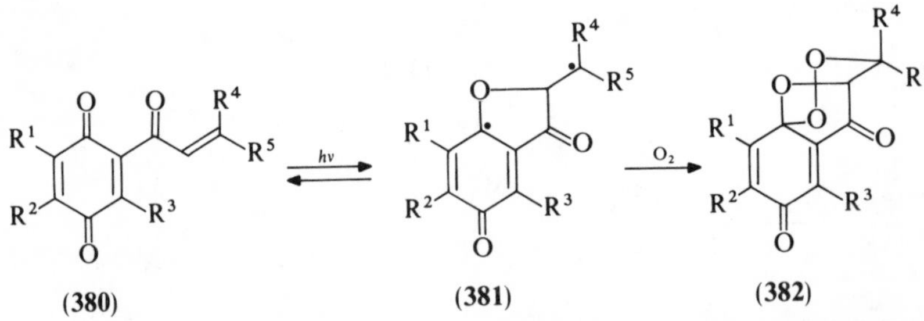

	Recovery (%)	Isolated yields (%) of **382**
a $R^1 = R^2 = R^3 = R^5 = Me, R^4 = H$	30	25
b $R^1 = R^2 = R^3 = R^4 = R^5 = Me$	14	31
c $R^1 = R^3 = R^4 = H, R^2 = R^5 = Me,$	0	36
d $R^1 = R^2 = -CH=CH-CH=CH-$, $R^3 = R^4 = H, R^5 = Me$	0	83
e $R^1 = R^2 = -CH=CH-CH=CH-$, $R^3 = H, R^4 = R^5 = Me$	0	81
f $R^1 = R^2 = -CH=CH-CH=CH-$, $R^3 = R^4 = H, R^5 = Ph$	0	88

SCHEME 28

formed in the photochemical reaction of 2-alkenoyl-1,4-quinones **380** in good yields[346]. Reversible interconversion of the precursor quinone **380** and the biradical **381** is evident by the *E–Z* isomerization of the olefinic side chain when the photolysis is conducted under anaerobic conditions (Scheme 28)[346].

In more polar solvents, the zwitterionic structure **385** appears to be much favored over **384**, since dihydrobenzofuranone **386** and alkenyl ether **387** are produced in the photochemical reaction of **383** in alcoholic solvents (Scheme 29)[347]. Nucleophilic addition

(383) (384) (385) (386)

(385) (387)

383	R^1	R^2	Isolated yields (%) in ethanol 386	in ethanol 387	in methanol 386	in methanol 387
a	H	Me	9	87	11	86
b	Me	Me	73	21	87	9
c	H	*p*-$MeOC_6H_4$	93	0	97	0
d	H	*p*-MeC_6H_4	97	1	99	1
e	H	Ph	75	23	82	17
f	H	*p*-ClC_6H_4	66	32	77	21
g	H	*p*-$NO_2C_6H_4$	9	84	18	74

SCHEME 29

of alcohol to the carbocationic center or to ketone carbonyl group may lead to **386** or **387**, respectively. The ratios of **386** to **387** are dependent upon the substituents of R^1 and R^2 as well as on the solvent alcohols. In line with a mechanism involving the intermediacy of **385**, the yields of **386** are raised when the carbocationic center is better stabilized such as in **383c** and **383d**[347]. Another support for the zwitterionic intermediates is provided in the reaction of **388**, where regio-isomeric mixtures of chromene derivatives **393** and **395** are formed in addition to **390** and **391** (Scheme 30)[348]. On the other hand, photolysis of **396** results in hydrogen abstraction from methoxy group, producing a tricyclic compound **397** (equation 120)[349].

When the olefinic moiety is directly conjugated with the quinone chromophore, the photochemical reactivity is greatly altered. Irradiation of isopropenyl-1,4-benzoquinone

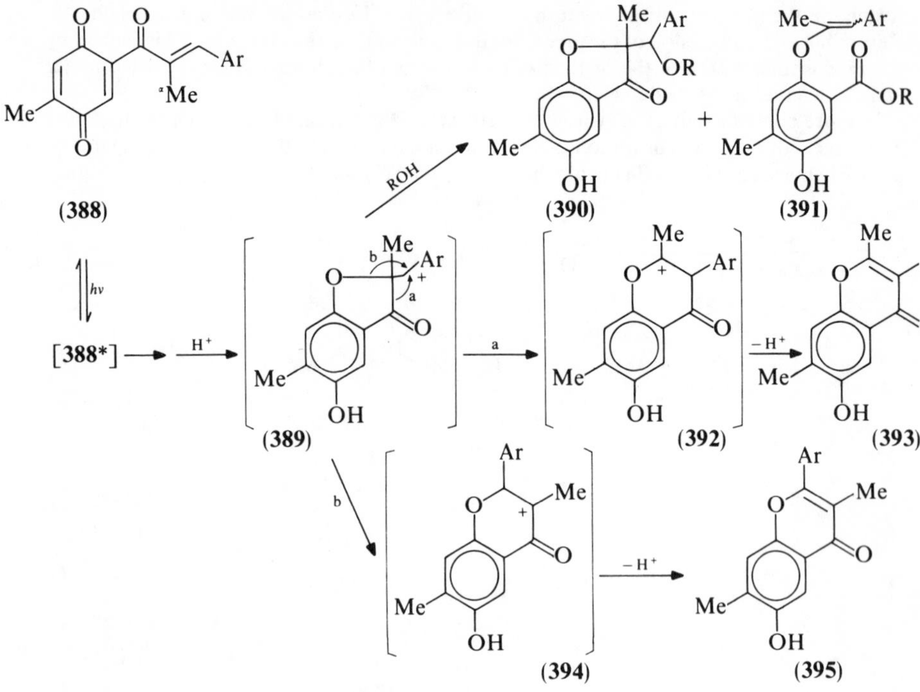

SCHEME 30

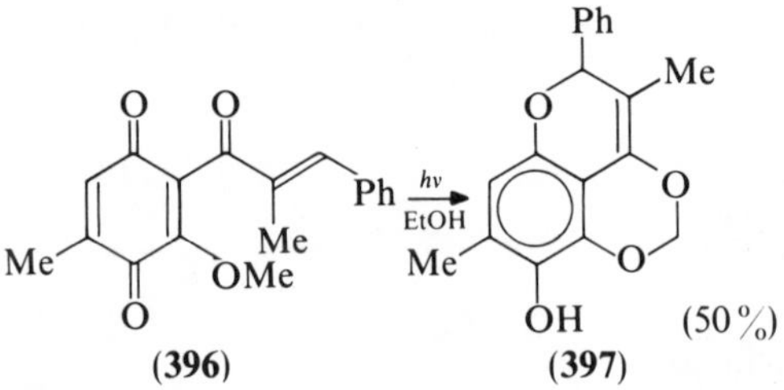

(120)

398 (R = Me) gives 5-hydroxy-3-methylbenzofuran 399 (R = Me) (equation 121)[310]. The labeled quinone 398 (R = CD_3) behaves analogously, and the mass and ^{1}H-NMR spectra of the product indicate that there has been neither loss nor scrambling of the label. This rules out a mechanism requiring hydrogen abstraction from the allylic position of the side chain, and therefore indirectly supports a mechanism involving intramolecular addition to the olefinic double bond, possibly to give **400**. Phenylvinylquinone **398** (R = Ph) gives the corresponding benzofuran **399** (R = Ph) in 33 % yield, whereas vinyl-1,4-benzoquinone itself affords only amorphous materials[310].

As discussed in Section IV.B, compounds **401** are photochemically cyclized to benz[*a*]anthracene-7,12-dione derivatives. When one of the aryl groups is an electron-rich heterocyclic aromatic ring, the heteroaromatic ring predominantly cyclizes to give a

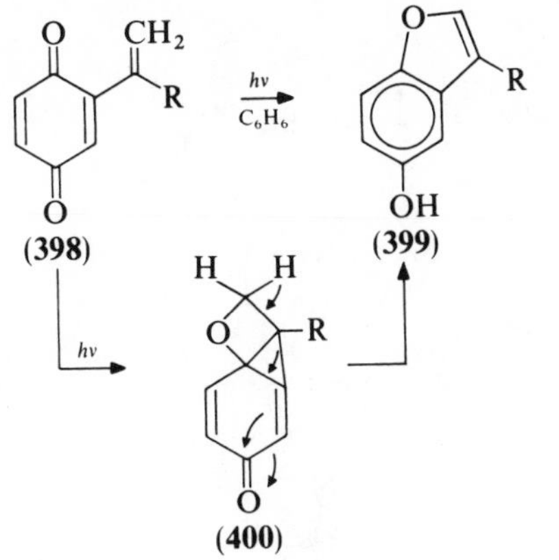

(398) $\xrightarrow[C_6H_6]{h\nu}$ (399)

(398) $\xrightarrow{h\nu}$ (400) → (399)

(121)

product **402** over **403** (Scheme 31)[226b]. Intramolecular charge transfer interaction between the electron-rich heterocyclic ring and the electron-deficient quinone moiety in **401** may induce this selectivity.

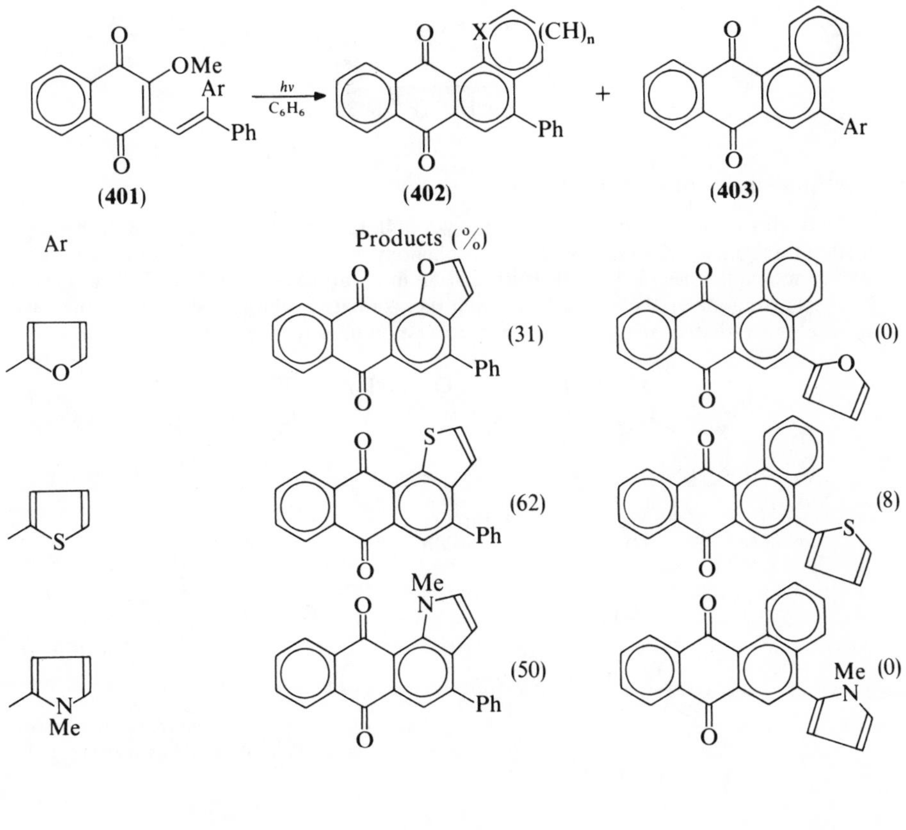

(401) $\xrightarrow[C_6H_6]{h\nu}$ (402) + (403)

Ar	Products (%)	
2-furyl	(31)	(0)
2-thienyl	(62)	(8)
N-methyl-2-pyrrolyl	(50)	(0)

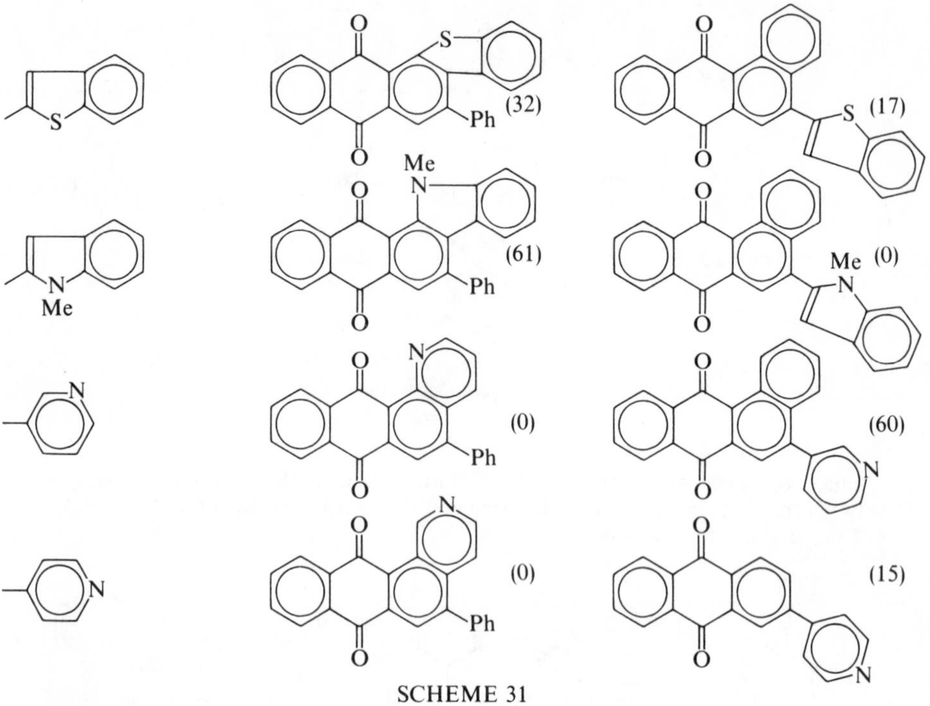

(32) (17)

(61) Me (0)

(0) (60)

(0) (15)

SCHEME 31

C. 'Ana'-quinone Rearrangement

Photochromism is observed in 1-phenoxy- (**404**), 1,5-diphenoxy- (**405**), 1,8-diphenoxy-9,10-anthraquinones (**406**) and in 6-phenoxy- (**407**) and 6,11-diphenoxy-5,12-naphthacenequinones (**408**) on their irradiation in non-polar solvents[350–357]. The spectra of the starting materials are restored when the irradiated solutions are kept in the dark or are re-irradiated with visible light corresponding to the new absorption band

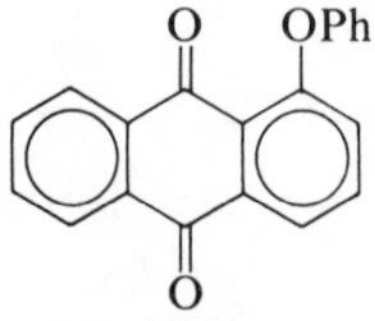

(**404**) 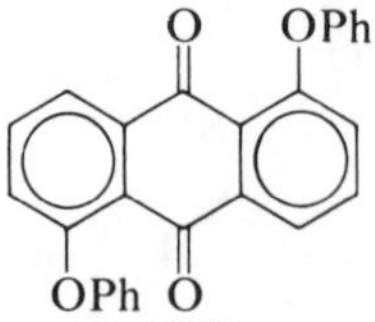 (**405**) 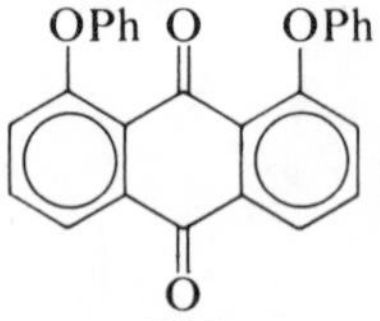 (**406**)

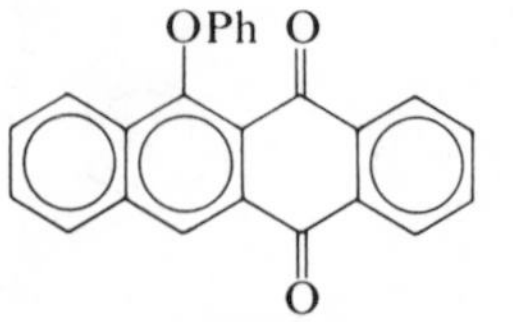

(**407**) 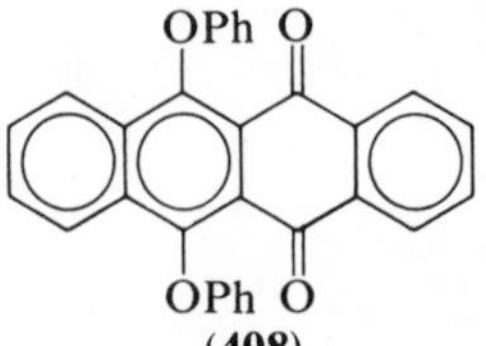 (**408**)

($\lambda_{max} > 440$ nm). This photochromism is due to the photoisomerization to an 'ana-quinone' structure **409**, formed through 1,5-phenyl migration in e.g. **404** (equation 122). In many cases, these 'ana-quinones' are isolated as stable crystals and characterized[350, 356].

$$\mathbf{(404)} \underset{h\nu_2,\ \Delta}{\overset{h\nu_1}{\rightleftharpoons}} \mathbf{(409)} \tag{122}$$

The ability of such 1-phenoxy-9,10-anthraquinones to rearrange to 9-phenoxy-1,10-anthraquinones depends on the position and electronic nature of the substituents[350]. For example, a crystalline 4-amino-9-phenoxy-1,10-anthraquinone (**411**) is isolated on irradiation of 1-amino-4-phenoxy-9,10-anthraquinone (**410**). The unusual absorption spectrum of this compound is likely to be due to the existence of two tautomeric forms (**411** and **412**) (equation 123)[350]. Under UV irradiation, 5-phenoxy-6,13-pentacenequinone (**413**) isomerizes in benzene to 6-phenoxy-5,13-pentacenedione (**414**) (λ_{max} 461, 490 nm), which irreversibly reacts with aniline to give **415** (equation 124). Recently, an analogous photochemical 1,5-acetyl migration was detected upon irradiation of **416** at low temperature matrix (equation 125)[356, 358, 359].

$$\mathbf{(410)} \xrightarrow[C_6H_6]{h\nu} \mathbf{(411,\ 85.5\%)} \longleftrightarrow \mathbf{(412)} \quad (\lambda_{max}\ 543\ \text{nm},\ 586) \tag{123}$$

$$\mathbf{(413)} \longrightarrow \mathbf{(414)} \xrightarrow{PhNH_2} \mathbf{(415)} \tag{124}$$

(416) → (417) (125)

D. Photoreaction of Azidoquinones

2-Azido-1,4-quinones **418**, **421**, **423**, **425** and **427** react with acyclic and cyclic dienes (e.g. **419**) upon irradiation with 360 nm light to give 2-alkenyl-2,3-dihydroindole-4,7-diones **420**, **422**, **424**, **426** and **428** (Scheme 32 and equations 126–129)[360, 361]. This reaction has

(418) + (419) → (420)

R^1	R^2	R^3	R^4	R^5	R^6	(%)
H	*t*-Bu	Me	H	H	H	96
H	*t*-Bu	H	Me	H	H	73
H	*t*-Bu	Me	H	H	Me	66
H	*t*-Bu	H	Me	Me	H	53
H	Me	Me	H	H	H	40
H	Me	H	Me	Me	H	72
–CH=CH–CH=CH–		Me	H	H	H	72
–CH=CH–CH=CH–		H	Me	Me	H	82

SCHEME 32

(421) + cyclohexadiene → (422, 53%) (126)

(423) + hexadiene → (424a, 60%) $E > 99\%$ + (424b, 18%) $E > 99\%$ (127)

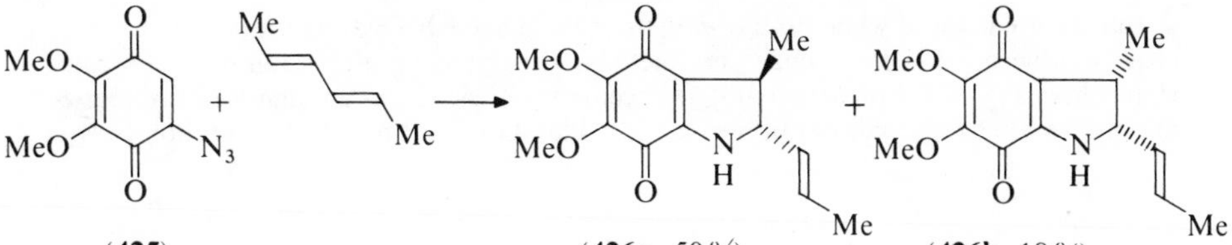

(425) (426a, 50%) (426b, 18%)

(128)

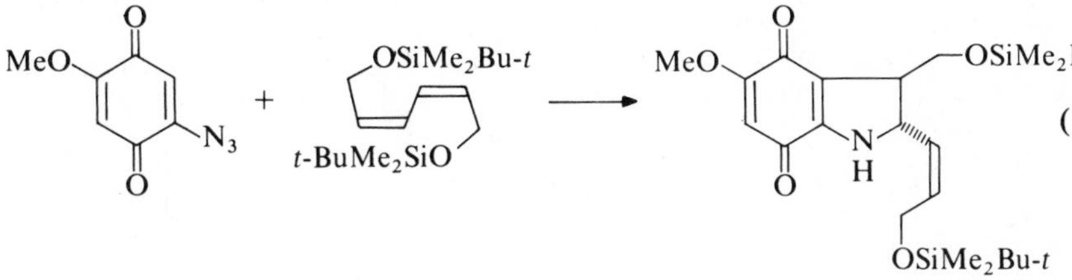

(129)

(427) (428, 43%)

wide applicability and a synthetic utility in relevance to mitomycin synthesis. Contrary to the original report[360], the stereochemical outcomes of this cyclization are as follows: (1) regardless of the stereochemistry of the used diene, the stereochemistry at the 2,3-positions in the cyclized product is predominantly *trans* (equations 127–129), and this

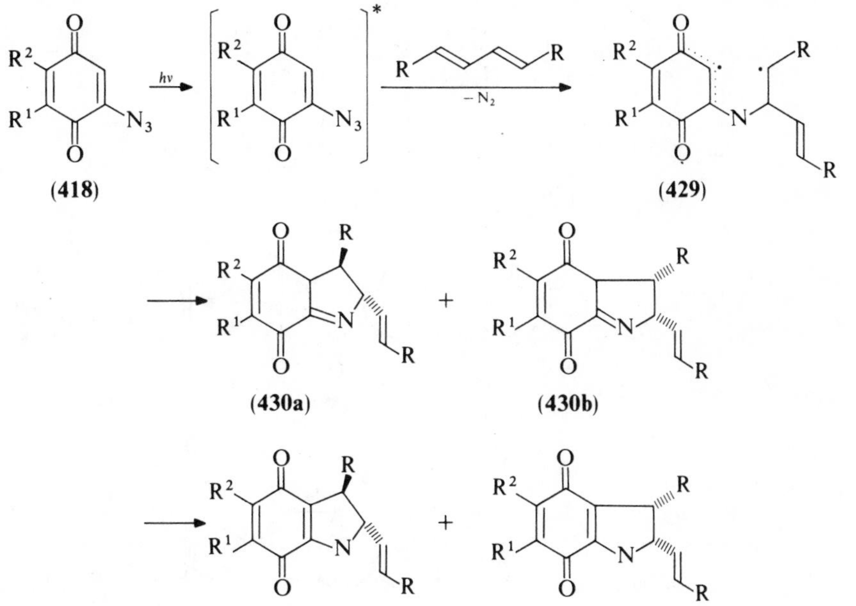

(418) (429)

(430a) (430b)

(431a) (431b)

SCHEME 33

selectivity is enhanced when both terminals of the diene have bulky groups; (2) the side-chain double bond in the cyclized product shows complete preservation of the original stereochemistry of the diene; and (3) the recovered diene shows no significant isomerization, even after repeated uses in the reaction. Based on these results, the reaction course

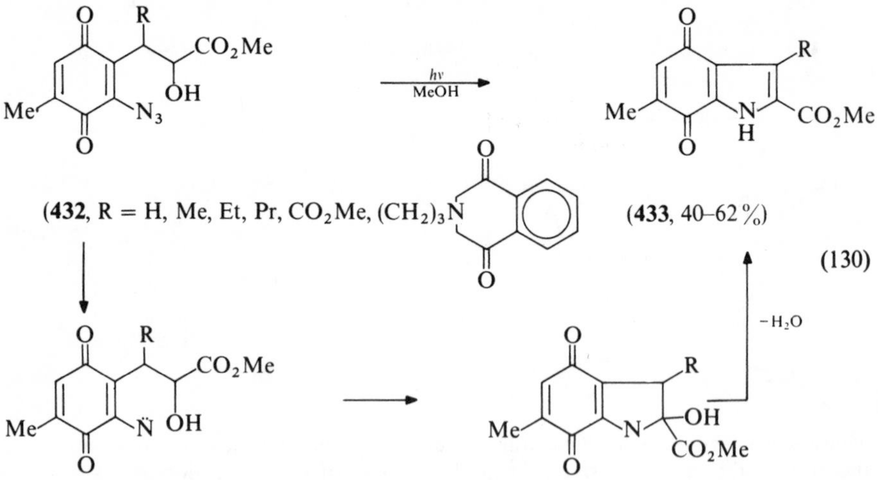

(**432**, R = H, Me, Et, Pr, CO_2Me, $(CH_2)_3N$ (phthalimido)) → hν / MeOH → (**433**, 40–62%)

(130)

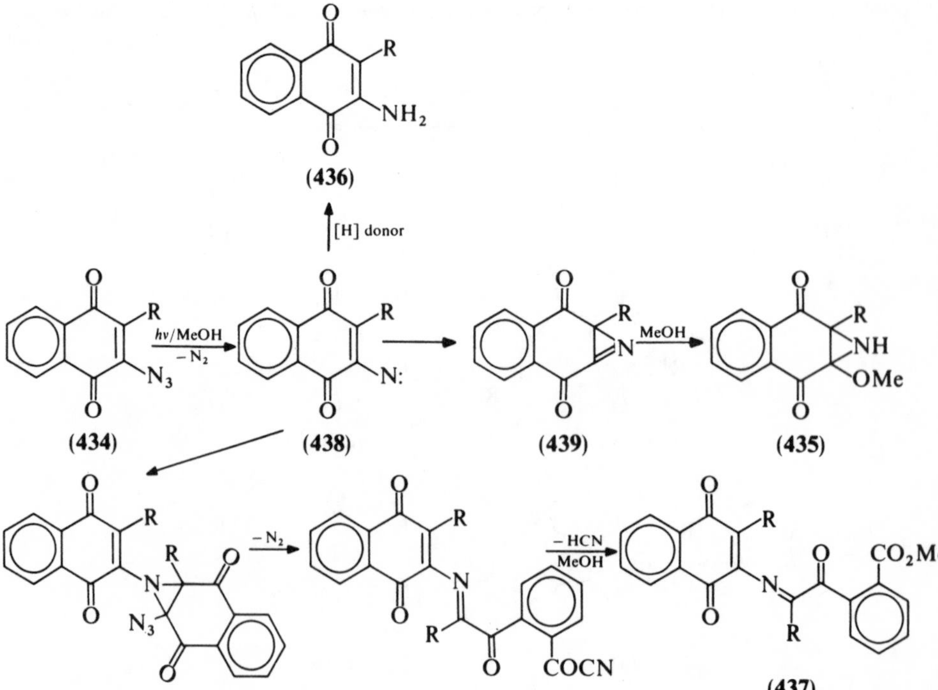

$-H_2O$

(**436**)

[H] donor

(**434**) hν/MeOH, $-N_2$ (**438**) (**439**) MeOH (**435**)

$-N_2$

$-HCN$ MeOH

(**437**)

SCHEME 34

shown in Scheme 33 is proposed. Attack of a photoexcited azidoquinone on a conjugated diene with loss of nitrogen affords a biradical intermediate **429**, which will cyclize to **430a** and **430b**. At this stage, the bulky group R will favor the *trans* isomer **430a**[361a].

Irradiation of a methanol solution of azidoquinones (**432**) gives indoloquinones (**433**) in moderate yields. Apparently the reaction proceeds via the nitrene intermediate (equation 130)[362]. On the other hand, irradiation of 2-alkyl-3-azido-1,4-naphthoquinones (**434**) in methanol yields aziridinonaphthoquinones (**435**), aminonaphthoquinones (**436**) and compounds **437**, all of which seem to arise via a nitrene intermediate (**438**) (Scheme 34)[363]. Of these, the formation of aziridinonaphthoquinone **435** is most interesting from both synthetic and mechanistic points of view, since it has a unique aziridine-condensed naphthoquinone structure and is undoubtedly a product of methanol addition to the azirine **439**.

Brief irradiation of 2-azidoanthraquinones (**440**) in dioxane give 3-arylthio[1,9-*cd*]isoxazol-6-ones (**441**) in good yields. These, in turn, are transformed into phenothiazine derivatives (**442**) upon prolonged irradiation. The nitrenes **443** were considered as intermediates in these transformations (equation 131)[364, 365]. Irradiation of 1-alkylamino-2-azidocarbonyl anthraquinone **444** gives **445** in good yield via a photo-induced Curtius rearrangement (equation 132)[366].

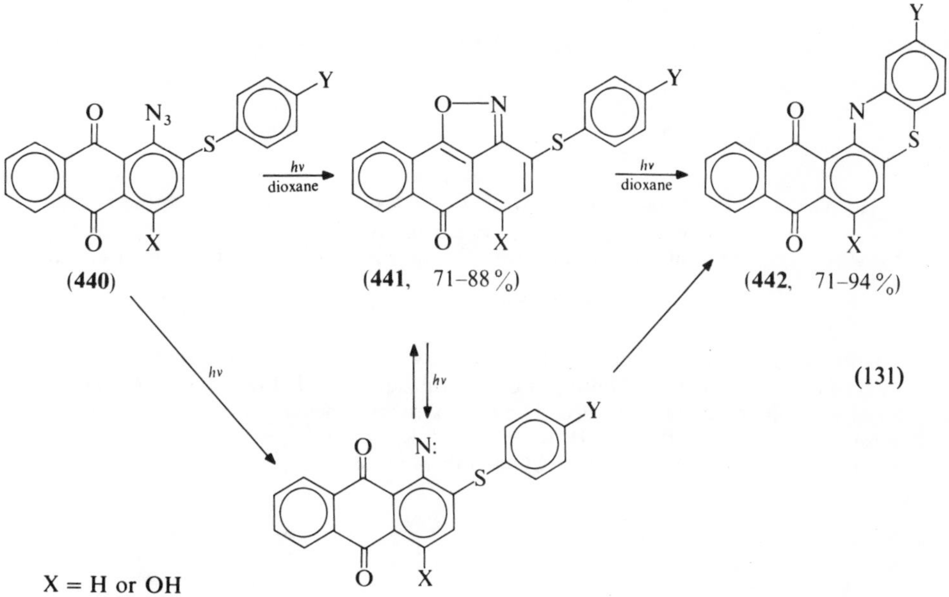

(**440**) —$h\nu$, dioxane→ (**441**, 71–88%) —$h\nu$, dioxane→ (**442**, 71–94%)

(**440**) —$h\nu$→ (**443**) ⇌ (**441**); (**443**) → (**442**)

X = H or OH
Y = alkyl, halogen

(131)

(**444**) —$h\nu$, $-N_2$→ (**445**, 83%) (132)

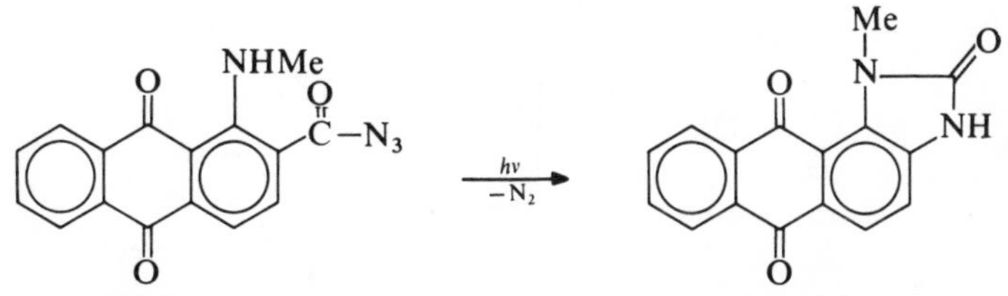

E. Miscellaneous

The heterocyclic naphthoquinone **446** (R^1 = Ph) is photochemically converted to **447** (R^1 = Ph, R^2 = H) in methanol, whereas *N*-methylated product **447** (R^1 = Ph, R^2 = Me) is formed from the same compound upon irradiation in dioxane in the presence of dimethylamine (equation 133)[367]. Photolysis of 1,2-naphthoquinone **448** also results in the cleavage of isoxazole ring to produce **449** (equation 134)[367].

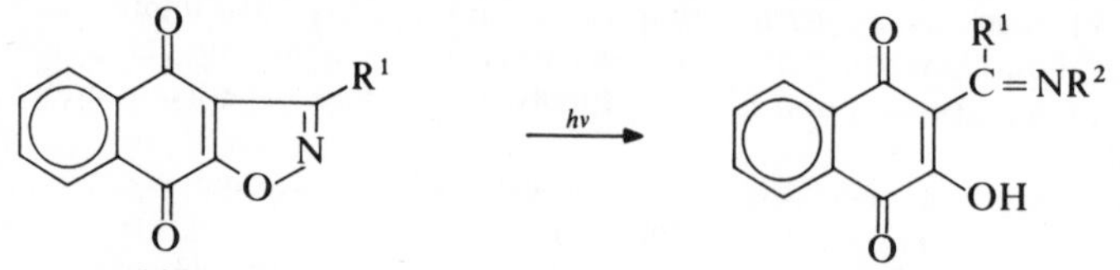

(**446**) $\xrightarrow{h\nu}$ (**447**) (133)

(**448**) $\xrightarrow[\text{Me}_2\text{NH/dioxane}]{h\nu}$ (**449**, 30%) (134)

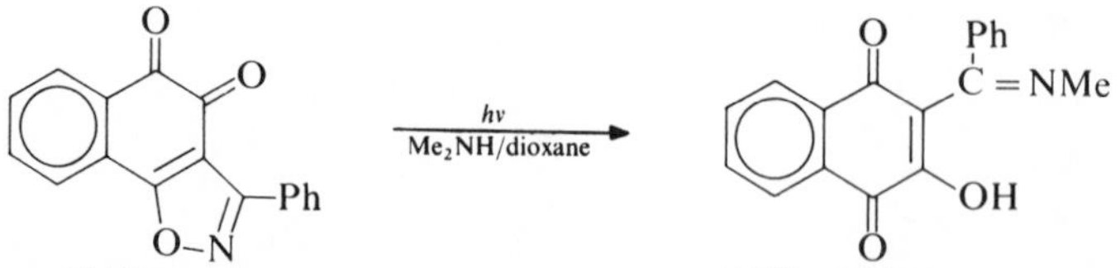

The photolysis of BQ in nitrogen matrix at 12 K leads to a slow decomposition of the quinone, which is accompanied by the appearance of the IR bands of acetylene (3268 and 750 cm^{-1}) and CO (2140 cm^{-1}) (equation 135), while NQ and AQ are inert under similar

$$\text{BQ} \xrightarrow{h\nu} 2\text{HC}\equiv\text{CH} + 2\text{CO} \qquad (135)$$

conditions. The authors suggest the possibility of a two-photon process for this fragmentation reaction[368]. Photolysis of 1,2-benzoquinone **450** (R^1 = R^2 = *t*-butyl) in nujol at 77 K results in the decarbonylation to cyclopentadienone **451**. With less substituted quinone **450** (R^1 = H, R^2 = *t*-butyl; R^1 = Me, R^2 = H), irradiation brings about the bis-ketene **452** as well as CO and dienone **451** (equation 136)[369–371].

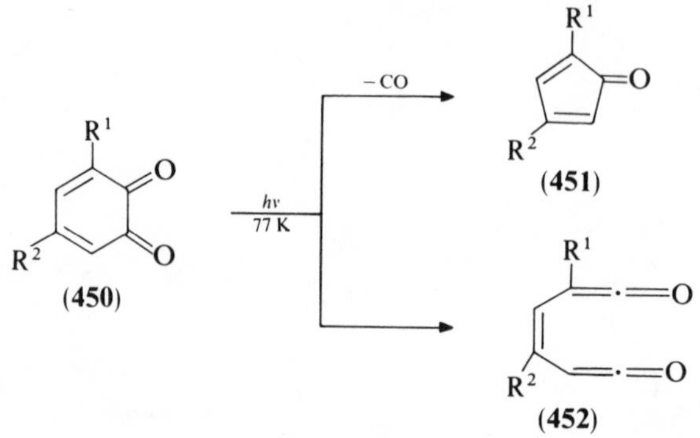

(**450**) $\xrightarrow[77\ \text{K}]{h\nu}$ (**451**) (−CO) + (**452**) (136)

VII. PHOTOCHEMISTRY OF QUINONE RELATED COMPOUNDS

A. Diphenoquinone

Diphenoquinones are generally strong electron acceptors capable of being reduced to semiquinone radicals. The ion-radical salts of diphenoquinones have been found to be organic conductors[372]. Investigations of the photochemical reactions of diphenoquinones have been stimulated by their expected usefulness in silver-free photography as well as by the molecular level understanding of the degrading effect of light on certain biological substrates. Diphenoquinones are intensively colored compounds that are reduced and decolorized on photoexcitation in the presence of appropriate hydrogen donors[373–377].

Irradiation of parent diphenoquinone **453** itself in acetaldehyde gives 4,4′-dihydroxybiphenyl (**454**), 4-acetoxy-4′-hydroxybiphenyl (**455**) and 3-acetyl-4,4′-dihydroxybiphenyl (**456**) (equation 137)[378]. Under identical conditions, 3,3′,5,5′-tetramethyldiphenoquinone

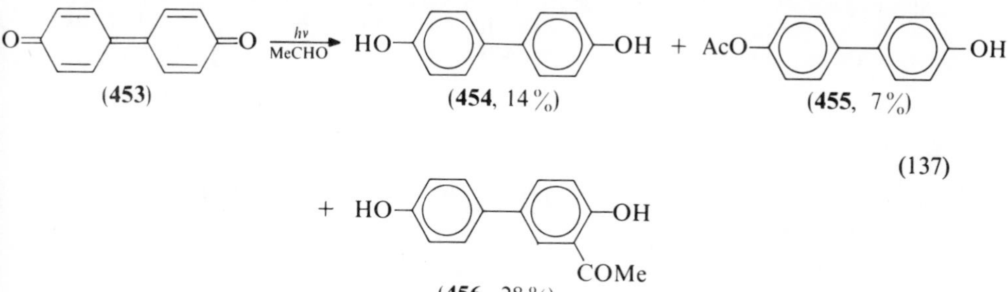

(**453**) $\xrightarrow[\text{MeCHO}]{h\nu}$ (**454**, 14%) + (**455**, 7%) + (**456**, 28%) (137)

is reduced to the corresponding diphenol but sterically hindered 3,3′,5,5′-tetra-*t*-butyldiphenoquinone is inert[378]. Photoexcited 2,2′,6,6′-tetramethyldiphenoquinone (**457**) abstracts a hydrogen even from benzene to give **458** and biphenyl (equation 138)[376].

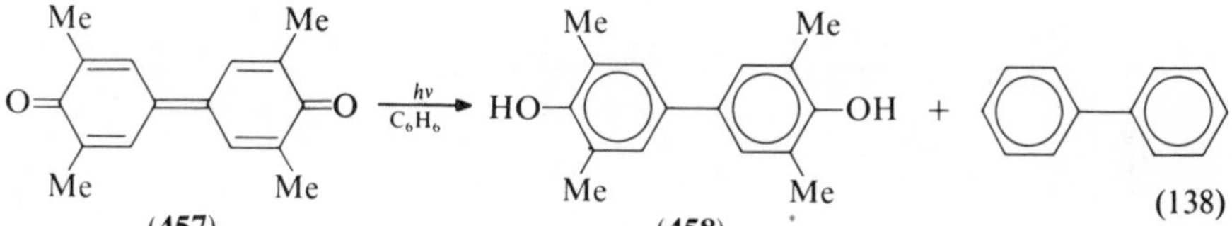

(**457**) $\xrightarrow[C_6H_6]{h\nu}$ (**458**) + biphenyl (138)

B. Quinone Methide and Quinone Imine

The ethylene acetal of 4-(2-formylbenzyl)-2-phenylnaphtho[1,8-*bc*]furan-5-one (**459**) has been cyclized under acid-catalyzed photolytic conditions to **460** (equation 139)[379].

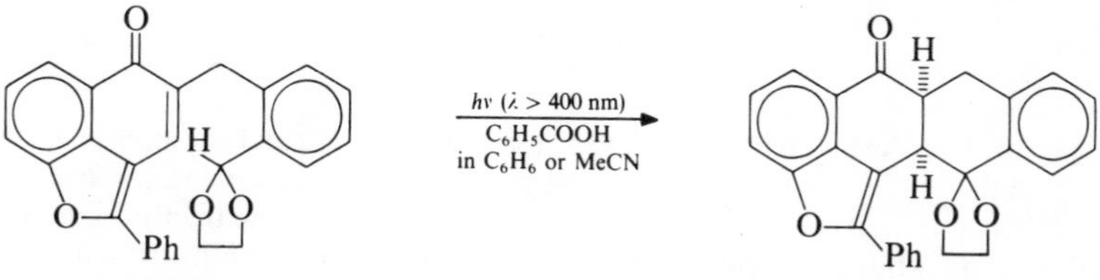

(**459**) $\xrightarrow[\text{C}_6\text{H}_5\text{COOH in C}_6\text{H}_6 \text{ or MeCN}]{h\nu\ (\lambda > 400\text{ nm})}$ (**460**, 78%) (139)

Improved procedure has recently been used for the photocyclization of **461** to **462** (equation 140)[380]. In relation to these reactions, the kinetics of the photophysical decay

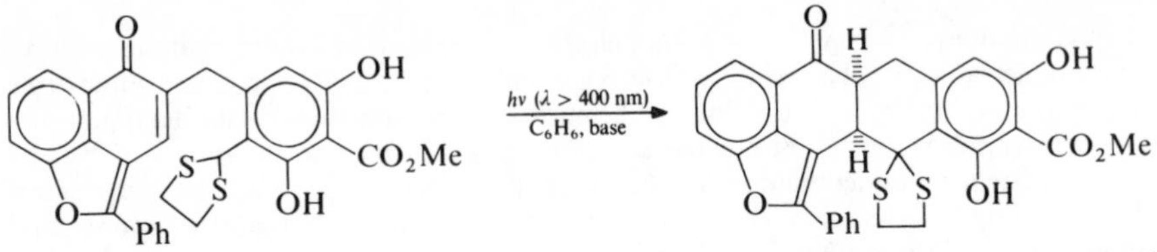

(**461**) $\xrightarrow[C_6H_6,\ base]{h\nu\ (\lambda > 400\ nm)}$ (**462**, 68 %) (140)

and the photochemical reactivity of the parent **463** and its conjugated acid **464** have been investigated (Scheme 35)[381]. The fluorescence of **463** in benzene is very weak, whereas the conjugated acid **464** exhibits an intense green fluorescence ($\phi_f = 0.97 \pm 0.1$)[381]. The lowest (π, π^*) triplet is an intermediate in the photoreduction of **463** in 2-propanol, which leads to the formation of dimer **466** via a semiquinone methide radical **465**[381].

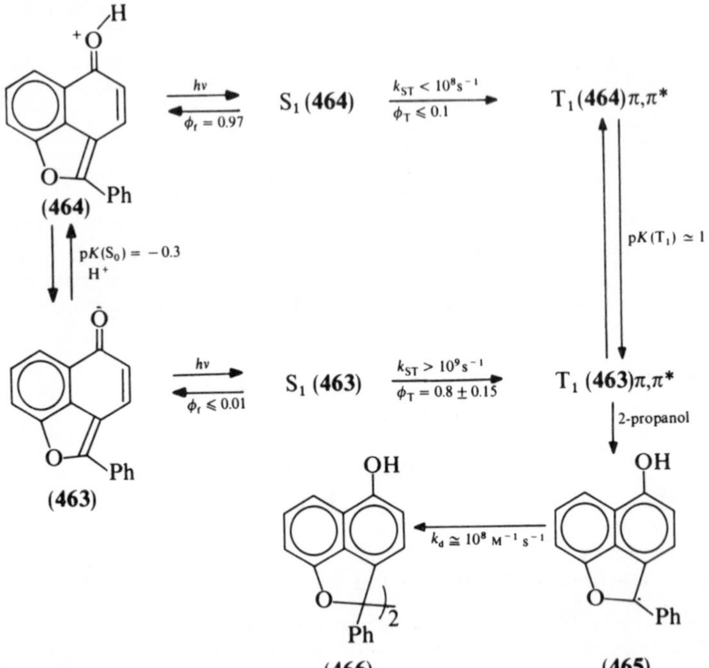

(**464**) $\underset{\phi_f = 0.97}{\overset{h\nu}{\rightleftarrows}}$ S_1(**464**) $\xrightarrow[\phi_T \leqslant 0.1]{k_{ST} < 10^8 s^{-1}}$ T_1(**464**)π,π^*

(**464**) ⇅ $pK(S_0) = -0.3$, H^+ (**463**); T_1(**464**) ⇅ $pK(T_1) \simeq 1$ T_1(**463**)

(**463**) $\underset{\phi_f \leqslant 0.01}{\overset{h\nu}{\rightleftarrows}}$ S_1(**463**) $\xrightarrow[\phi_T = 0.8 \pm 0.15]{k_{ST} > 10^9 s^{-1}}$ T_1(**463**)π,π^* $\xrightarrow{2\text{-propanol}}$ (**465**) $\xrightarrow{k_d \cong 10^8\ M^{-1} s^{-1}}$ (**466**)

SCHEME 35

The photochromism of bianthrones (e.g. the parent **467**) has long been a subject of extensive study[382, 383]. The structure of the photochromic isomer ($\lambda_{max} = 650$ nm) has been established as that of the twisted conformer **468** in which the bianthronyl system is twisted by ca. 57° about the central double bond and each of its anthrone half-moiety is planar[384]. Photochromism takes place via the triplet state of the starting isomer and gives

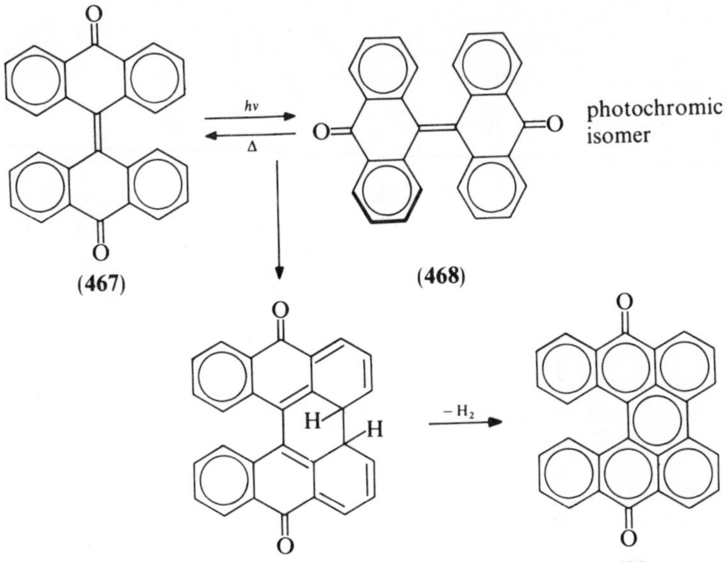

photochromic isomer

(467) (468)

(469) (470)

SCHEME 36

470 via 469 (Scheme 36)[385, 386]. A photochromic system which is based on the photoreversible addition of singlet oxygen into an arene derivative has been reported (equation 141)[387, 388]. Photocleavage of 472, which is reported to occur via the S_2 (π, π^*) state, produces 471 in the ground state and O_2 in an electronically excited singlet state.

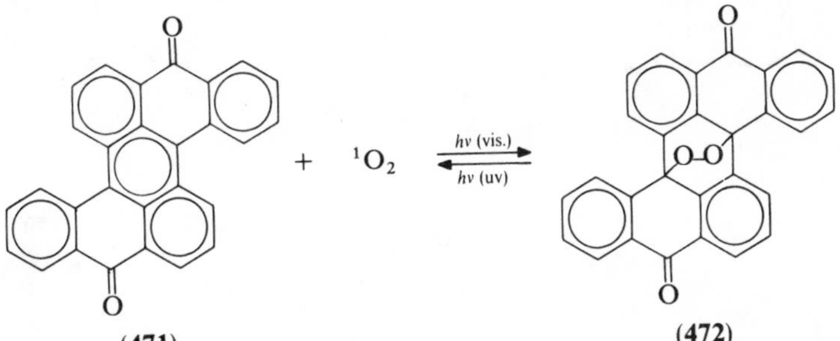

(471) + $^1O_2$ $\underset{h\nu\ (\text{uv})}{\overset{h\nu\ (\text{vis.})}{\rightleftharpoons}}$ (472) (141)

Illumination of a solution of **473** in CCl_4–methanol using a tungsten lamp produces a red quinone (**474**) and a blue quinone (**475**), the latter of which undergoes photocyclization and aerial oxidation to give the former (equation 142)[389]. Primary photochemical processes of phototendering vat dyes **476** and **477** are shown to be H abstraction and electron transfer, respectively[118].

Irradiation ($\lambda_{max} = 365$ nm) of benzene solutions of quinone methide **478** produces the benzofuran derivative **479**[390]. The route envisaged is shown in equation 143, whereby the photoexcited carbonyl group adds to the benzene ring, yielding the intermediate spiro-biradical **480**. Subsequent bonding to give **481** and ring-opening of the cyclopropane moiety yields **479**[390].

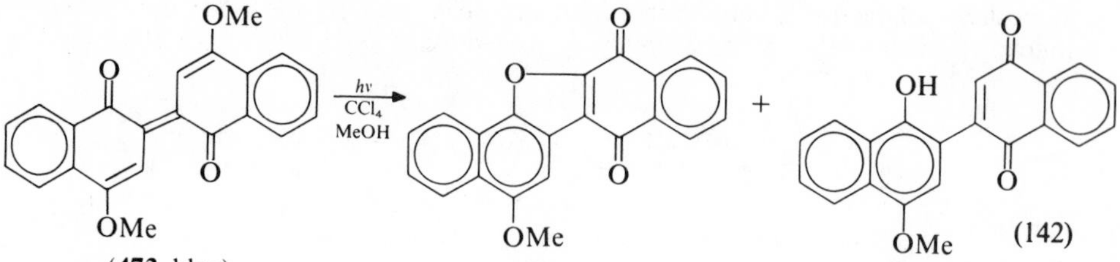

(**473**, blue) $\xrightarrow[\text{MeOH}]{h\nu,\ CCl_4}$ (**474**, red) + (**475**, blue) (142)

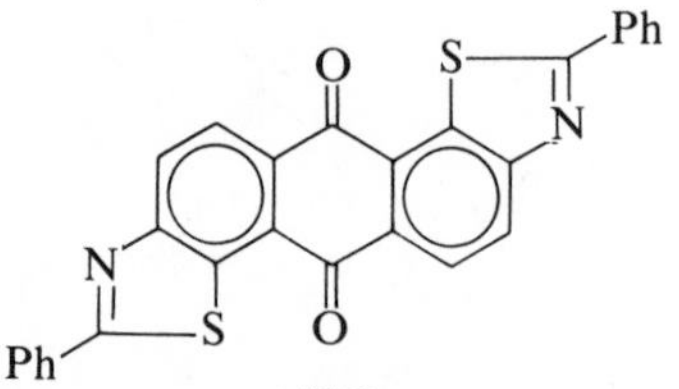

(**476**)

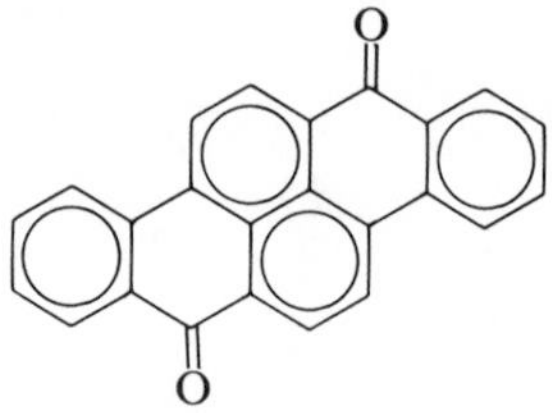

(**477**)

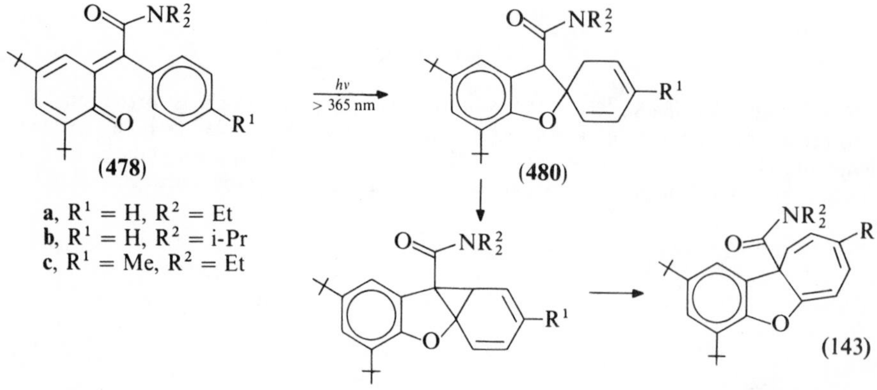

(**478**) $\xrightarrow{h\nu,\ >365\ \text{nm}}$ (**480**) → (**481**) → (**479**) (143)

a, $R^1 = H$, $R^2 = Et$
b, $R^1 = H$, $R^2 = i\text{-}Pr$
c, $R^1 = Me$, $R^2 = Et$

(**479**)
a, 85%
b, 55%
c, 75%

Azamerocyanine (**482**) undergoes a reversible ring-closure on photoexcitation, forming colorless 2*H*-1,4-oxazine (**483**) (equation 144)[391]. Phenothiazine quinone imine (**484**) is

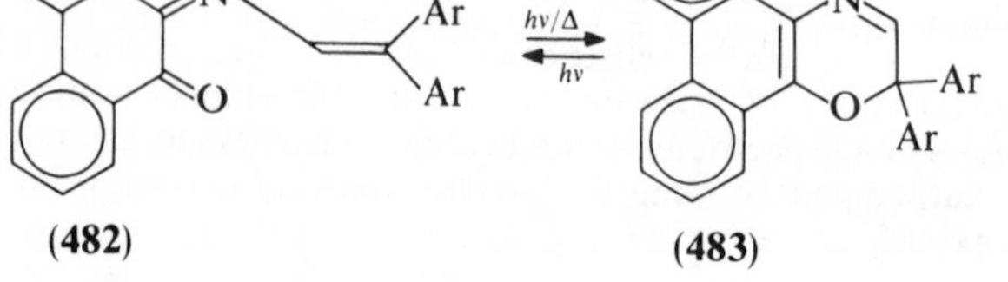

(**482**) $\underset{h\nu}{\overset{h\nu/\Delta}{\rightleftharpoons}}$ (**483**) (144)

photochemically active toward hydrogen donors such as thiols and aldehydes, giving coupling products **485** and **486**, respectively (equations 145 and 146)[392]. An analogous

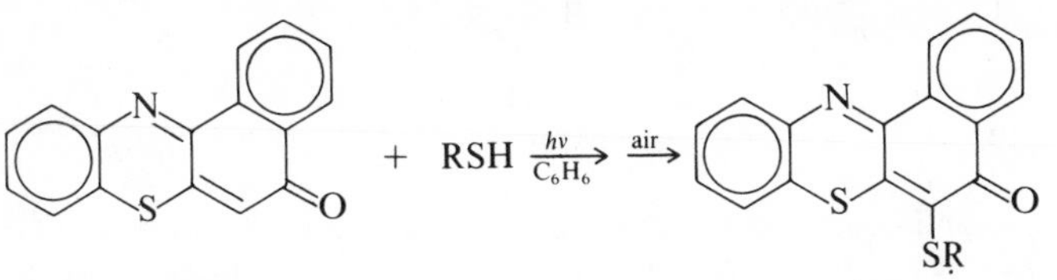

(484) + RSH $\xrightarrow[C_6H_6]{h\nu}$ $\xrightarrow{air}$ (485) (145)

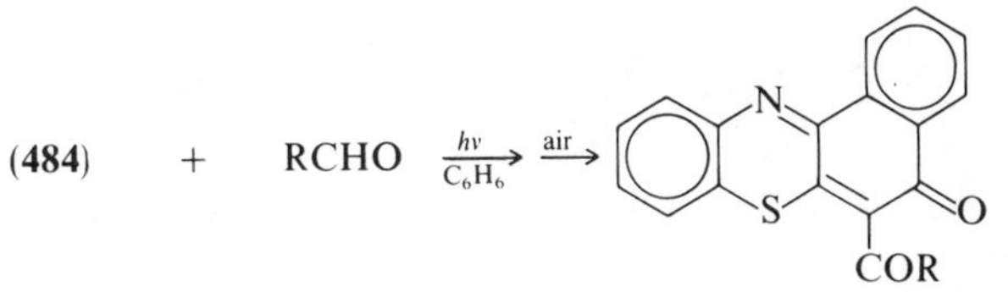

(484) + RCHO $\xrightarrow[C_6H_6]{h\nu}$ $\xrightarrow{air}$ (486) (146)

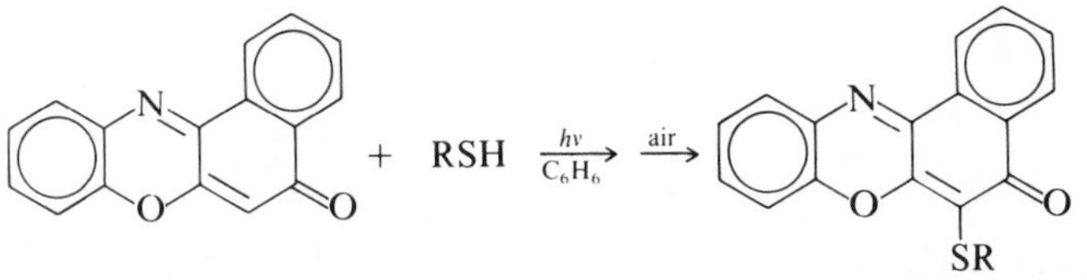

(487) + RSH $\xrightarrow[C_6H_6]{h\nu}$ $\xrightarrow{air}$ (488) (147)

reaction of phenoxazine quinone imine (**487**) with thiol gives **488** (equation 147). 2-Aza-3-aryl-1,4-naphthoquinone (**489**) has the lowest excited state of a (n, π^*) character and is photochemically reduced to **490**, **491** and **492** in the presence of hydrogen donors (equation 148)[393].

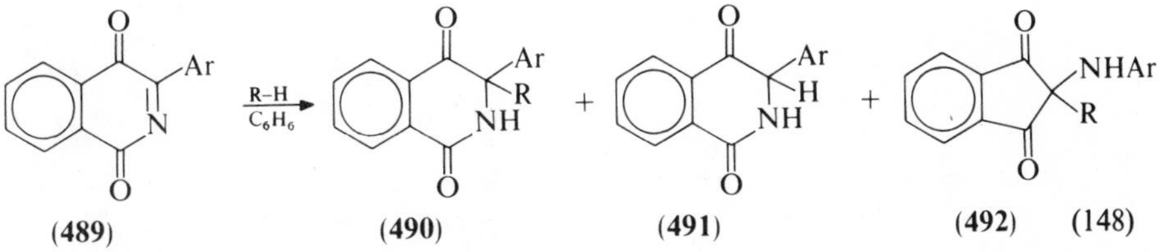

(489) $\xrightarrow[C_6H_6]{R-H}$ (490) + (491) + (492) (148)

R–H = Xanthene, toluene, and tetrahydrofuran

C. One-atom Homologs of 1,4-Naphthoquinone

The photochemistry of one-atom homologs of 1,4-naphthoquinone such as methanonaphthoquinone (**493**), epoxynaphthoquinone (**494**) and iminonaphthoquinone (**495**), in which the carbon–carbon double bond of the quinone ring is replaced with cyclopropane, oxirane and aziridine ring, respectively, has been extensively studied.

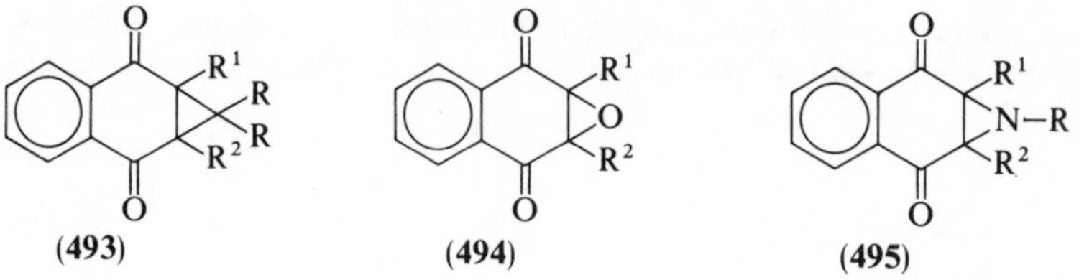

(493) (494) (495)

Methanonaphthoquinone **496** has a usual (n, π^*) reactivity quite similar to that of aromatic ketones. Thus, inter- and intramolecular H-abstraction reactions proceed efficiently (equations 149 and 150)[394–396]. Irradiation of methanonaphthoquinone **499** in

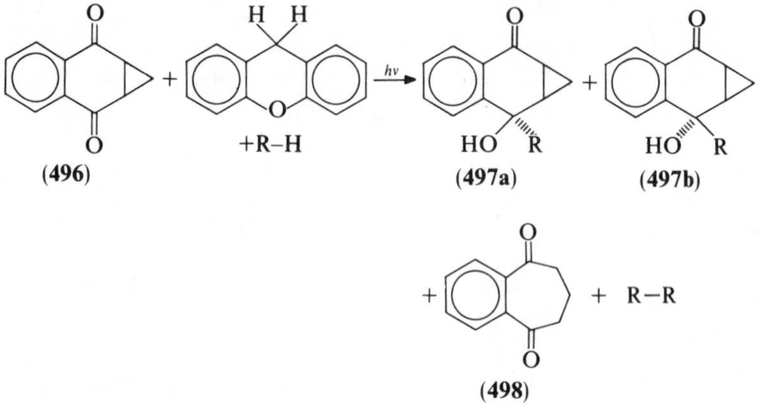

(496) +R–H $\xrightarrow{h\nu}$ (497a) + (497b) + (498) + R–R (149)

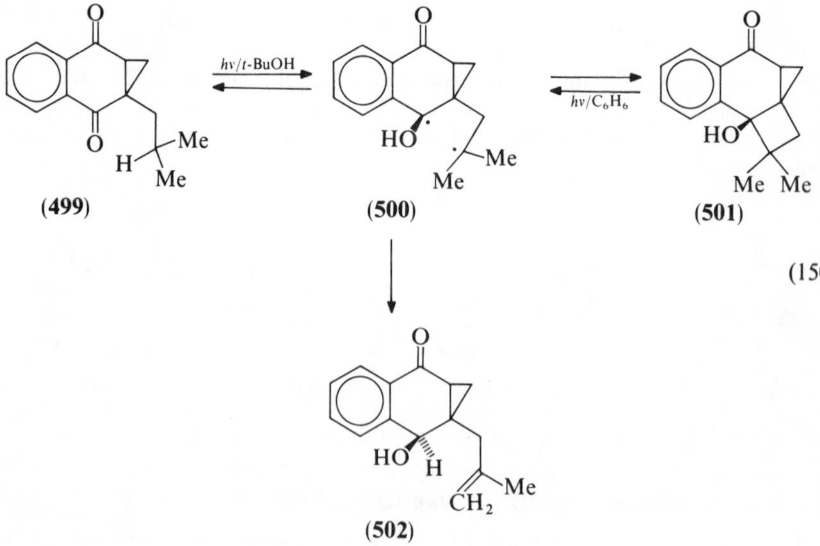

(499) $\xrightarrow{h\nu/t\text{-BuOH}}$ (500) $\xleftarrow{h\nu/C_6H_6}$ (501); (500) → (502) (150)

t-butanol gives cyclobutanol **501** efficiently. Upon irradiation in benzene, **501** is converted back to **499** via a 1,4-biradical **500**[395, 396]. The internal carbon–carbon bond of the cyclopropane ring of **503** is photochemically cleaved to produce **504** and **505** (equation 151)[394b]. Diphenyl-substituted methanonaphthoquinone **506** is quite stable even in the presence of good hydrogen donors under completely deaerated conditions but it cleaves readily to MQ and benzophenone in the presence of both oxygen and hydrogen donors[394b]. Reversible formation of biradical **507** and its interception by oxygen to form **508** which abstracts hydrogen to form **509** are suggested for this oxidative fragmentation reaction (equation 152).

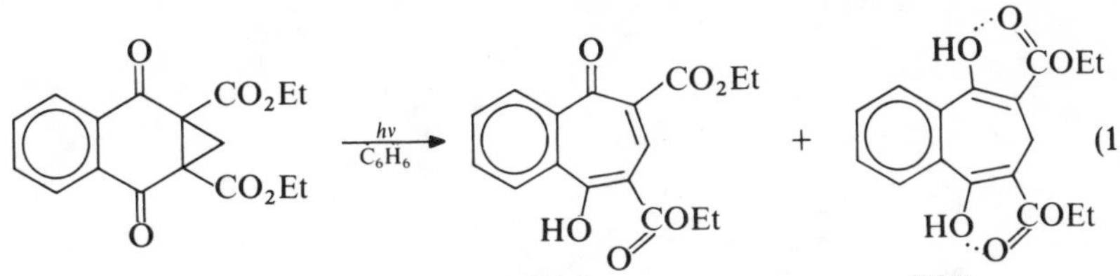

(151)

(503) (504) (505)

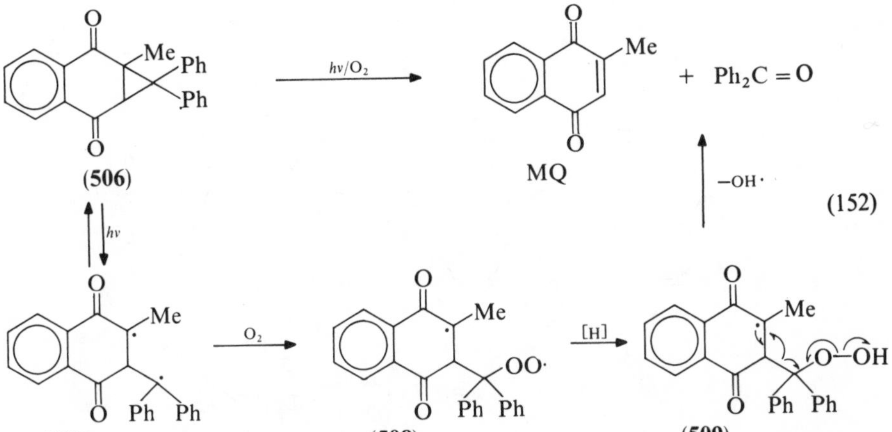

(152)

(506) MQ (507) (508) (509)

Epoxyquinones are metabolites of biologically important quinones, and thus their photochemical reaction may be related to some photodamaging effect of quinonoid compounds. Since intersystem crossing of epoxynaphthoquinones is very efficient and their lowest triplet state has a characteristic (n, π^*) nature, all photochemical reactions of epoxynaphthoquinone may be considered to start from the lowest triplet (n, π^*) state[397]. Irradiation of dimethyl-substituted epoxynaphthoquinone (**510**) results in a reversible formation of carbonyl ylide **511** or the 1,3-biradical **512**, which are trapped by olefins[398], aldehydes and ketones[399], singlet oxygen[400] and alcohols to form products **513**–**520** (Scheme 37)[401]. Excited triplet state of **510** is quenched by electron-rich olefins via exciplex deactivation; this quenching competes with the decaying pathway leading to **511** or **512**[398]. Photochemical generation of a carbonyl ylide or a 1,3-biradical is also reported in the photolysis of 2-aryl and 2,3-diaryl substituted epoxynaphthoquinones[402, 403]. Photolysis of epoxynaphthoquinones **521** and **524** gives **522** and **525**, respectively[404–406]. These photoisomerizations are explained in terms of a mechanism starting with α-cleavage (equations 153 and 154).

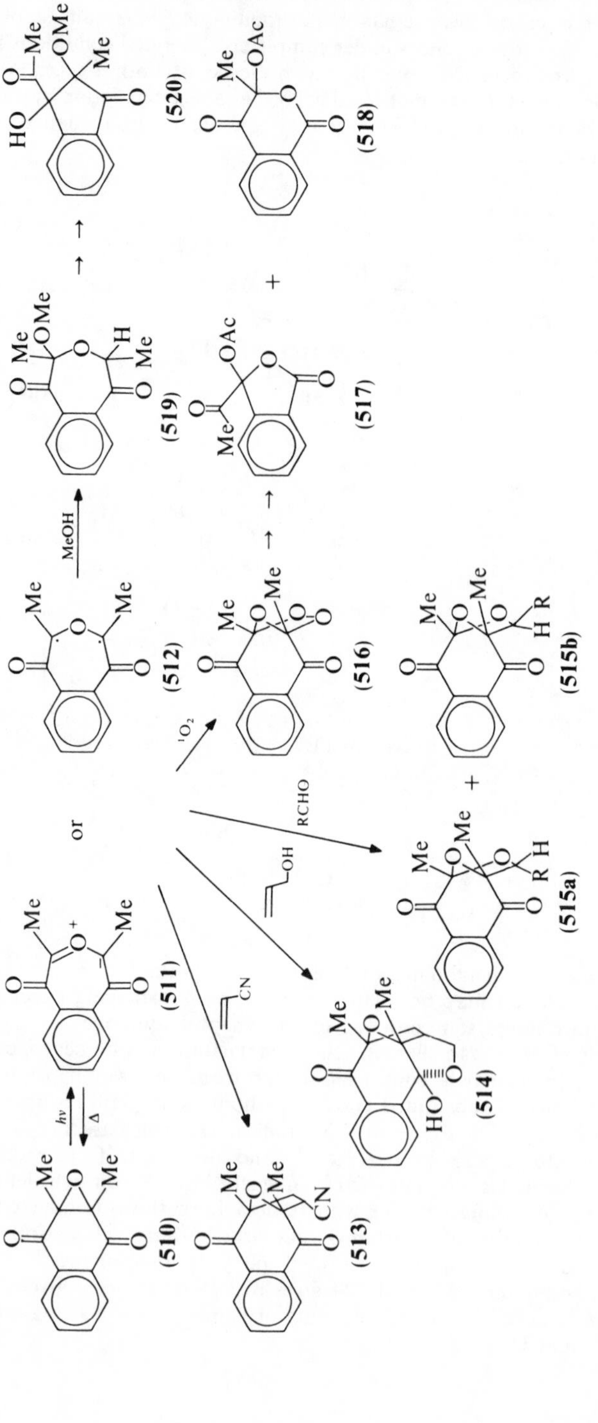

(510) (511) (512) (513) (514) (515a) (515b) (516) (517) (518) (519) (520)

SCHEME 37

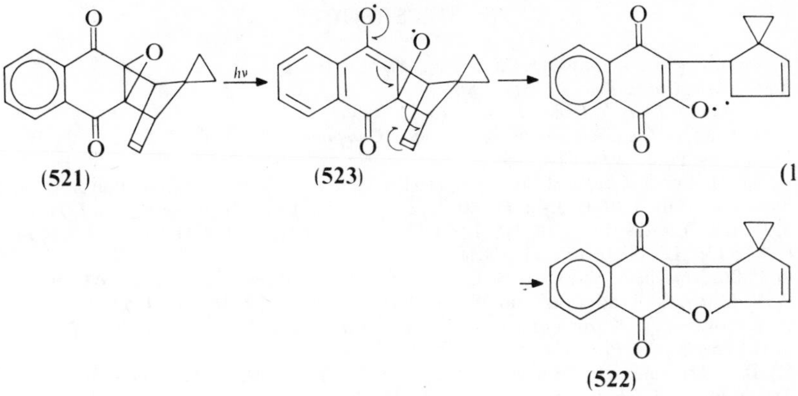

(521) (523) (153)

(522)

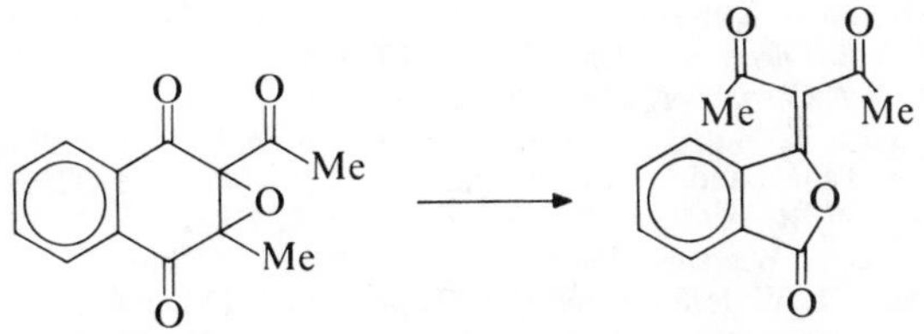

(524) (525) (154)

The lowest triplet state of iminonaphthoquinone has a mixed (n, π^*) and CT character. Irradiation of **526** in the presence of an olefin gives cycloadduct **529** via intermediate **527** or **528** (equation 155), similarly but much less efficiently compared with epoxynaphthoquinone[407].

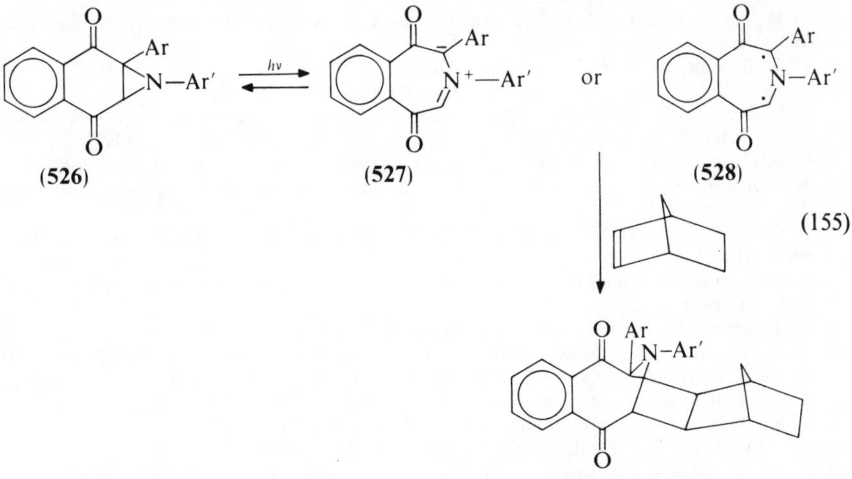

(526) (527) or (528) (155)

(529)

VIII. REFERENCES

1. A. Schönberg and A. Mustafa, *Chem. Rev.*, **40**, 181 (1947).
2. J. M. Bruce, *Quart. Rev.*, **21**, 405 (1967).
3. M. B. Rubin, *Fortsch. Chem. Forsch.*, **13**, 251 (1969).
4. J. M. Bruce, in *The Chemistry of the Quinonoid Compounds*, Part I (Ed. S. Patai), John Wiley & Sons, New York, 1974, Chapt. 9.
5. W. H. Horspool, Chemical Society Specialist Periodical Reports (Senior Editor D. Bryce-Smith), **1** (1970); **2** (1971); **3**, 423 (1972); **4**, 513 (1973); **5**, 303 (1974); **6**, 348 (1975); **7**, 246 (1976); **8**, 262 (1977); **9**, 279 (1978); **10**, 298 (1979); **11**, 301 (1980); **12**, 214 (1981); **13**, 241 (1982); **14**, 207 (1983); **15**, 212 (1984); **16**, 233 (1985).
6. O. P. Studzinskii, A. V. El'tov, N. I. Rtischev and G. V. Fomin, *Usp. Khim.*, **43**, 401 (1974).
7. A. V. El'tov, O. P. Studzinskii and V. M. Grebenkina, *Usp. Khim.*, **46**, 185 (1977).
8. A. Kuboyama, Y. Kozima and J. Maeda, *Bull. Chem. Soc. Japan*, **55**, 3635 (1982).
9. P. E. Stevenson, *J. Phys. Chem.*, **76**, 2424 (1972).
10. (a) H. P. Trommsdorff, *J. Chem. Phys.*, **56**, 5358 (1972); (b) M. F. Merienne-Lafore and H. P. Trommsdorff, *J. Chem. Phys.*, **64**, 3791 (1976).
11. G. Olbrich, O. E. Polansky and M. Zander, *Ber. Bunsenges. Phys. Chem.*, **81**, 692 (1977).
12. M. H. Wood, *Theor. Chim. Acta*, **36**, 345 (1975).
13. S. Y. Matsuzaki and A. Kuboyama, *Bull. Chem. Soc. Japan*, **51**, 2264 (1978).
14. A. Kuboyama, *Bull. Chem. Soc. Japan*, **51**, 2771 (1978).
15. A. Kuboyama, *Bull. Chem. Soc. Japan*, **52**, 329 (1979).
16. A. Kuboyama and S. Y. Matsuzaki, *Bull. Chem. Soc. Japan*, **54**, 3635 (1981)
17. Y. Miyagi, M. Koyanagi and Y. Kanda, *Bull. Chem. Soc. Japan*, **53**, 2502 (1980).
18. R. L. Martin and W. R. Wadt, *J. Phys. Chem.*, **86**, 2382 (1982).
19. H. Veenvliet and D. A. Wiersma, *Chem. Phys. Lett.*, **22**, 87 (1973).
20. R. M. Hochstrasser, L. W. Johnson and H. P. Trommsdorff, *Chem. Phys. Lett.*, **21**, 251 (1973).
21. D. O. Cowan, R. Gleiter, J. A. Hashmall, E. Heilbronner and V. Hornung, *Angew. Chem. Int. Ed. Engl.*, **10**, 401 (1971).
22. D. Dougherty and S. P. McGlynn, *J. Am. Chem. Soc.*, **99**, 3234 (1977).
23. S. G. Semenov, V. V. Redchenko, Ya. F. Freimanis, V. S. Bannikov and V. A. Gailite, *Zh. Obshch. Khim.*, **52**, 2626 (1982).
24. J. P. Galaup, J. Megel and H. P. Trommsdorff, *J. Chem. Phys.*, **69**, 1030 (1978).
25. O. S. Khalil and L. Goodman, *J. Phys. Chem.*, **80**, 2170 (1976).
26. T. Narisawa, M. Sano and Y. J. I'haya, *Chem. Lett.*, 1289 (1975).
27. J. P. Galaup, J. Megel and H. P. Trommsdorff, *Chem. Phys. Lett.*, **41**, 397 (1976).
28. K. E. Drabe, H. Veenvliet and D. A. Wiersma, *Chem. Phys. Lett.*, **35**, 469 (1975).
29. D. W. Turner, C. Baker, A. D. Baker and C. R. Brundle, *Molecular Photoelectron Spectroscopy*, Wiley-Interscience, London, 1970.
30. C. R. Brungle, M. B. Robin and N. A. Kuebler, *J. Am. Chem. Soc.*, **94**, 1466 (1972).
31. T. Kobayashi, *J. Electron. Spectrosc. Relat. Phenom.*, **7**, 349 (1975).
32. V. V. Redchenko, Ya. F. Freimanis and Ya. Ya. Dreigeris, *Zh. Obshch. Khim.*, **50**, 1847 (1980).
33. T. P. Carter, G. D. Gillispie and M. A. Connolly, *J. Phys. Chem.*, **86**, 192 (1982).
34. T. P. Carter, M. H. Van Benthem and G. D. Gillispie, *J. Phys. Chem.*, **87**, 1891 (1983).
35. K. Hamanoue, Y. Kajiwara, T. Miyake, T. Nakayama, S. Hirase and H. Teranishi, *Chem. Phys. Lett.*, **94**, 36 (1983).
36. H. Inoue, K. Ikeda, H. Mihara, M. Hida, N. Nakashima and K. Yoshihara, *Chem. Phys. Lett.*, **95**, 60 (1983).
37. K. Hamanoue, K. Nakajima, Y. Kajiwara, T. Nakayama and T. Teranishi, *Chem. Phys. Lett.*, **110**, 178 (1984).
38. H. Inoue and M. Hida, *J. Synth. Org. Japan*, **32**, 348 (1978).
39. (a) A. K. Davies, J. F. McKellar and G. O. Phillips, *Proc. R. Soc. Lond. A*, **323**, 69 (1971); (b) A. K. Davies, G. A. Gee, J. F. McKellar and G. O. Philips, *J. Chem. Soc., Perkin Trans. 2*, 1742 (1973).
40. A. A. Lamola and G. S. Hammond, *J. Chem. Phys.*, **43**, 2129 (1965).
41. G. J. Fisher and E. J. Land, *Photochem. Photobiol.*, **37**, 27 (1983).
42. (a) E. Amouyal and R. Bensasson, *J. Chem. Soc., Faraday Trans. 1*, **73**, 1561 (1977). (b) *J. Chem. Soc., Faraday Trans. 1*, **72**, 1274 (1976).

43. H. Kobashi, M. Funabashi, T. Kondo, T. Morita, T. Okada and N. Mataga, *Bull. Chem. Soc. Japan*, **57**, 3557 (1984).
44. A. Harriman and A. Mills, *Photochem. Photobiol.*, **33**, 619 (1981).
45. J. McVie, R. S. Sinclair and T. G. Truscott, *Photochem. Photobiol.*, **29**, 395 (1979).
46. E. F. Hilinski, S. V. Milton and P. M. Rentzepis, *J. Am. Chem. Soc.*, **105**, 5193 (1983).
47. (a) C. H. Giles and R. S. Sinclair, *J. Soc. Dyers Colour*, **88**, 109 (1972); (b) C. H. Giles, B. J. Hojiwala and C. D. Shah, *J. Soc. Dyers Colour*, **88**, 403 (1972).
48. E. McAlpine and R. S. Sinclair, *Textile Res. J.*, **47**, 283 (1977).
49. (a) N. S. Allen, P. Bentley and J. F. McKellar, *J. Photochem.*, **5**, 225 (1976); (b) N. S. Allen and J. F. McKellar, *J. Photochem.*, **5**, 317 (1975).
50. N. A. Sheglova, R. A. Nezhl'skaya, D. N. Shigorin, L. V. Galitsina and N. S. Dokunikhin, *Russ. J. Phys. Chem.*, **48**, 166 (1974).
51. N. S. Allen and J. F. McKellar, *J. Photochem.*, **7**, 107 (1977).
52. H. Inoue, M. Hida, N. Nakashima and K. Yoshihara, *J. Phys. Chem.*, **86**, 3184 (1982).
53. J. C. Andre, I. Kawenoki, J. Kossanyi and P. Valat, *J. Photochem.*, **19**, 139 (1982).
54. B. Ya Dain, S. M. Ermenko and V. A. Kalibabchuk, *Theor. Exp. Chem.*, **8**, 40 (1972).
55. N. S. Allen, B. Harwood and J. F. McKellar, *J. Photochem.*, **9**, 559, 565 (1978); **10**, 187, 193 (1979).
56. I. Kawenoki, B. Keita, J. Kossanyi and L. Nadjo, *Bull. Soc. Chim. Fr.*, II, 104 (1982).
57. K. B. Patel and R. L. Wilson, *J. Chem. Soc., Faraday Trans. 1*, 814 (1973).
58. E. J. Land, E. McAlpine, R. S. Sinclair and T. G. Truscott, *J. Chem. Soc., Faraday Trans. 1*, **72**, 2091 (1976).
59. E. McAlpine, R. S. Sinclair, T. G. Truscott and E. J. Land, *J. Chem. Soc., Faraday Trans. 1*, **74**, 597 (1978).
60. J. H. Fendler and E. J. Fendler, in *The Chemistry of Quinonoid Compounds* (Ed. S. Patai), John Wiley and Sons, New York, 1974, Chapt. 10.
61. B. E. Hulme, E. J. Land and G. O. Phillips, *J. Chem. Soc., Faraday Trans. 1*, **68**, 1992 and 2003 (1972).
62. K. Maruyama and T. Otsuki, *Bull. Chem. Soc. Japan*, **50**, 3429 (1977).
63. R. Foster and M. I. Foreman, in *The Chemistry of the Quinonoid Compounds*, Pt. I (Ed. S. Patai), John Wiley and Sons, New York, 1974, Chapt. 6.
64. (a) W. Rebafka and H. A. Staab, *Angew. Chem. Int. Ed. Engl.*, **12**, 776 (1973); **13**, 203 (1974); (b) A. Staab, C. P. Herz and H. E. Henke, *Tetrahedron Lett.*, 4393 (1974).
65. H. A. Staab and W. Rebafka, *Chem. Ber.*, **110**, 3333 (1977).
66. (a) H. A. Staab and C. P. Herz, *Angew. Chem. Int. Ed. Engl.*, **16**, 799 (1977); (b) H. A. Staab, U. Zapf and A. Gurke, *Angew. Chem. Int. Ed. Engl.*, **16**, 801 (1977).
67. H. Vogler, C. Ege and H. A. Staab, *Tetrahedron*, **31**, 2441 (1975).
68. H. Vogler, *Tetrahedron Lett.*, **24**, 2159 (1983).
69. (a) F. Graf, H.-K. Hong, A. Nazzal and D. H. Haarer, *Chem. Phys. Lett.*, **59**, 217 (1978); (b) J. Friedrich and D. Haarer, *Chem. Phys. Lett.*, **74**, 503 (1980); (c) J. Friedrich and D. Haarer, *J. Chem. Phys.*, **76**, 61 (1982); (d) W. Breinl, J. Friedrich and D. Haarer, *J. Chem. Phys.*, **81**, 3915 (1984).
70. D. R. Kemp and G. Porter, *Proc. R. Soc. Lond. A*, **326**, 117 (1971).
71. J.-C. Ronfard-Haret, R. V. Besasson and E. Amouyal, *J. Chem. Soc., Faraday Trans. 1*, **76**, 2432 (1980).
72. Y. Yamaguchi, T. Miyashita and M. Matsuda, *J. Phys. Chem.*, **85**, 1369 (1981).
73. P. W. Atkins, A. J. Dobbs, G. T. Evans, K. A. McLauchlan and P. W. Parcival, *Mol. Phys.*, **27**, 769 (1974).
74. J. Nafisi-Movaghar and F. Wilkinson, *Trans. Faraday Soc.*, **66**, 2268 (1970).
75. (a) H. Kobashi, H. Gyoda and T. Morita, *Bull. Chem. Soc. Japan*, **50**, 1731 (1977); (b) H. Kobashi, Y. Tomioka and T. Morita, *Bull. Chem. Soc. Japan*, **52**, 1568 (1979); (c) H. Kobashi, T. Nagumo and T. Morita, *Chem. Phys. Lett.*, **57**, 369 (1978).
76. R. Gschwind and E. Haselbach, *Helv. Chim. Acta*, **62**, 941 (1979).
77. G. Porter and M. R. Topp, *Proc. R. Soc.*, **A315**, 163 (1970).
78. I. Loeff, A. Treinin and H. Linschitz, *J. Phys. Chem.*, **87**, 2536 (1983).
79. F. Wilkinson and A. Garner, *J. Chem. Soc., Faraday Trans. 2*, **73**, 222 (1973).
80. (a) V. A. Kuz'min, A. P. Darmanyan and P. P. Levin, *Chem. Phys. Lett.*, **63**, 509 (1979); (b) P. P. Levin, A. P. Darmanyan, V. A. Kuz'min, A. Z. Yankelevich and V. M. Kuznets, *Izv. Akad.*

Nauk SSSR, Ser. Khim., 2744 (1980).
81. G. G. Wubbels, W. J. Monaco, D. E. Johnson and R. S. Meredith, *J. Am. Chem. Soc.*, **98**, 1036 (1976).
82. S. M. Lukonina, S. V. Mikhailova, I. K. Korobeinicheva, V. A. Loskutov and E. P. Fokin, *Izv. Akad. Nauk SSSR, Ser. Khim.*, 2648 (1983).
83. D. Rehm and A. Weller, *Isr. J. Chem.*, **8**, 259 (1970).
84. C. K. Monn and K. K. Barnes, *Electrochemical Reactions in Nonaqueous Solvents*, Marcel Dekker, New York, 1970, Chapter 6.
85. J. Q. Chambers, in *The Chemistry of Quinonoid Compounds*, Part I (Ed. S. Patai), John Wiley and Sons, New York, 1974, p. 737.
86. R. Scheere and M. Gratzel, *J. Am. Chem. Soc.*, **99**, 865 (1977).
87. J. C. Scaiano and P. Neta, *J. Am. Chem. Soc.*, **102**, 1608 (1980).
88. Y. Tanimoto, M. Takayama, S. Shima and M. Itoh, *Bull. Chem. Soc. Japan*, **58**, 3641 (1985).
89. K. Hamanoue, K. Yokoyama, Y. Kajiwara, T. Nakayama and H. Teranishi, *Chem. Phys. Lett.*, **110**, 25 (1984).
90. I. Loeff, A. Treinin and H. Linschitz, *J. Phys. Chem.*, **88**, 4931 (1984).
91. I. Kawenoki, B. Keita and J. Kossanyi, *Nouv. J. Chim.*, **6**, 387 (1982).
92. Y. Sakaguchi, H. Hayashi, H. Murai and Y. Ihaya, *Chem. Phys. Lett.*, **110**, 275 (1984).
93. P. P. Levin, T. A. Kokrashvili and V. A. Kuz'min, *Izv. Akad. Nauk SSSR, Ser. Khim.*, 284 (1983).
94. H. D. Roth, in *Chemically Induced Magnetic Polarization* (Ed. L. T. Muus), Reidel, Dordrecht, The Netherlands, 1977, p. 39.
95. P. W. Atkins, A. J. Dobbs and K. A. McLauchlan, *Chem. Phys. Lett.*, **29**, 616 (1974).
96. S. K. Wong, *J. Am. Chem. Soc.*, **100**, 5488 (1978).
97. G. V. Fomin and M. M. Shabarchina, *Zh. Fiz. Khim.*, **55**, 2166 (1981).
98. J. Nakata, T. Imura and K. Kawabe, *Bull. Chem. Soc. Japan*, **48**, 701 (1975).
99. R. S. Davidson and S. P. Orton, *J. Chem. Soc., Chem. Commun.*, 209 (1974).
100. P. P. Levin and T. A. Kokrashvili, *Izv. Akad. Nauk SSSR, Ser. Khim.*, 1234 (1981).
101. H. D. Roth and M. L. M. Schilling, *J. Am. Chem. Soc.*, **101**, 1898 (1979); **102**, 4303 (1980).
102. Y. Shirota, K. Kawai, N. Yamamoto, K. Tada, T. Shida, H. Mikawa and H. Tsubomura, *Bull. Chem. Soc. Japan*, **45**, 2683 (1972).
103. (a) K. Tada, Y. Shirota, S. Kusabayashi and H. Mikawa, *Chem. Commun.*, 1169 (1971); (b) K. Tada, Y. Shirota and H. Mikawa, *Macromolecules*, **6**, 9 (1973).
104. (a) H. D. Roth and M. L. M. Schilling, *J. Am. Chem. Soc.*, **103**, 1246 and 7210 (1981); (b) *J. Am. Chem. Soc.*, **102**, 7956 (1980); (c) H. D. Roth and M. L. M. Schilling, *Can. J. Chem.*, **61**, 1027 (1983); (d) H. D. Roth and M. L. M. Schilling, *J. Am. Chem. Soc.*, **105**, 6805 (1983); (e) H. D. Roth, M. L. M. Schilling and F. C. Schilling, *J. Am. Chem. Soc.*, **107**, 4152 (1985); (f) H. D. Roth, M. L. M. Schilling, T. Mukai and T. Miyashi, *Tetrahedron Lett.*, **24**, 5815 (1983); (g) H. D. Roth, M. L. M. Schilling and K. Raghavachari, *J. Am. Chem. Soc.*, **106**, 253 (1984); (h) H. D. Roth, M. L. M. Schilling and C. C. Wamser, *J. Am. Chem. Soc.*, **106**, 5023 (1984); (i) C. J. Abert and H. D. Roth, *J. Am. Chem. Soc.*, **107**, 3840 (1985).
105. (a) H. D. Roth, M. L. M. Schilling, R. S. Hutton and E. A. Truesdale, *J. Am. Chem. Soc.*, **105**, 153 (1983); (b) H. D. Roth, M. L. M. Schilling, P. G. Gassman and J. L. Smith, *J. Am. Chem. Soc.*, **106**, 2711 (1984); (c) C. J. Abert, H. D. Roth and M. L. M. Schilling, *J. Am. Chem. Soc.*, **107**, 4148 (1985).
106. T. Miyashi, Y. Takahashi, T. Mukai, H. D. Roth and M. L. M. Schilling, *J. Am. Chem. Soc.*, **107**, 1079 (1985).
107. H. D. Roth and M. L. M. Schilling, *J. Am. Chem. Soc.*, **107**, 716 (1985).
108. C. J. Abert and H. D. Roth, *J. Am. Chem. Soc.*, **107**, 6814 (1985).
109. (a) R. Kaptein, *J. Chem. Soc., Chem. Commun.*, 732 (1971); (b) R. Kaptein, *J. Am. Chem. Soc.*, **94**, 6251 (1974).
110. (a) H. D. Roth and A. A. Lamola, *J. Am. Chem. Soc. Japan*, **94**, 1013 (1972); (b) E. Ben-Hur and I. Rosenthal, *Photochem. Photobiol.*, **11**, 163 (1970).
111. H. G. Hageman and W. B. G. Huysmans, *Chem. Commun.*, 837 (1969).
112. B. D. Sviridov, L. P. Gryzunova, V. M. Kuznets, G. A. Nikiforov, K. De Jonge, H. J. Hageman and V. V. Ershov, *Izv. Akad. Nauk SSSR, Ser. Khim.*, 2160 (1978).
113. V. M. Kuznetz, P. P. Levin, I. V. Khudyakov and V. A. Kuz'min, *Izv. Akad. Nauk SSSR, Ser. Khim.*, 1284 (1978).

114. V. A. Kuzmin and A. K. Chivisov, *Chem. Commun.*, 1559 (1971).
115. O. P. Studzinskii, N. I. Rtishchev, A. V. Devekki, G. V. Fomin, L. M. Gurdzhiyan and A. V. El'tov, *Zh. Org. Khim.*, **8**, 2130 (1972).
116. O. P. Studzinskii, A. V. El'tov, N. I. Rtishchev, G. V. Fomin, A. V. Devekki and L. M. Gurdzhiyan, *Zh. Org. Khim.*, **9**, 1932 (1973).
117. G. O. Phillips, A. K. Davies and J. F. McKellar, *Chem. Commun.*, 519 (1970).
118. A. K. Davies, R. Ford, G. A. Gee, J. F. McKellar and G. O. Phillips, *J. Chem. Soc., Chem. Commun.*, 873 (1972).
119. K. Kano and T. Matsuo, *Tetrahedron Lett.*, 4323 (1974).
120. (a) H. Inoue, K. Kawabe, N. Kitamura and M. Hida, *Chem. Lett.*, 987 (1977); (b) H. Inoue, K. Kawabe, N. Kitamura and M. Hida, *Bull. Chem. Soc. Japan*, **55**, 1874 (1982).
121. (a) H. Inoue and M. Hida, *Chem. Lett.*, 107 (1979); (b) H. Inoue and M. Hida, *Bull. Chem. Soc. Japan*, **55**, 1880 (1982).
122. H.-D. Scharf and R. Weitz, *Tetrahedron*, **35**, 2255 (1979).
123. I. Okura and N. Kim-Thuan, *Chem. Lett.*, 1569 (1980).
124. O. P. Studzinskii and A. V. El'tsov, *Zh. Org. Khim.*, **16**, 2117 (1980).
125. (a) G. V. Fomin, P. I. Mordvintsev, R. A. Mkhitarov and T. A. Gordina, *Zh. Fiz. Khim.*, **54**, 238, 240 and 242 (1980); (b) G. V. Fomin and P. I. Mordvintsev, *Zh. Fiz. Khim.*, **57**, 1295 (1983).
126. V. A. Loskutov, S. M. Lukonina, A. V. Konstantinov and E. P. Fokin, *Zh. Org. Khim.*, **17**, 587 (1981).
127. H. Taber, J. Pomerantz and G. N. Halfenger, *Photochem. Photobiol.*, **28**, 191 (1978).
128. J. R. Wagner, J. Cadet and G. J. Fisher, *Photochem. Photobiol.*, **40**, 589 (1984).
129. M. A. Stidnam and J. N. Siedow, *Photochem. Photobiol.*, **38**, 537 (1983).
130. (a) R. Livington and D. W. Tanner, *Trans. Faraday Soc.*, **54**, 765 (1958); (b) F. Wilkinson and J. Schroeder, *J. Chem. Soc., Faraday Trans. 2*, **75**, 441 (1979).
131. J. Schroeder and F. Wilkinson, *J. Chem. Soc., Faraday Trans. 2*, **75**, 896 (1979).
132. (a) G. Porter, *Proc. R. Soc. Lond. A*, **362**, 281 (1978); (b) J. H. Fendler and E. J. Fendler, *Catalysis in Micellar and Macromolecular Systems*, Academic Press, New York, 1975; (c) J. H. Fendler, *Membrane Mimetic Chemistry*, John Wiley & Sons, New York, 1982; (d) M. Calvin, *Acc. Chem. Res.*, **11**, 369 (1978); (e) D. G. Whitten, *Acc. Chem. Res.*, **13**, 83 (1980); (f) M. Gratzel, *Angew. Chem. Int. Ed. Engl.*, **19**, 981 (1980).
133. R. Mitzner, D. Frosch and H. Dorst, *Z. Phys. Chem., Leipzig*, **258**, S. 845 (1977).
134. S. J. Formosinho, *J. Chem. Soc., Faraday Trans. 2*, **72**, 1313 and 1332 (1976).
135. S. J. Formosinho, *J. Chem. Soc., Faraday Trans. 2*, **74**, 1978 (1978).
136. (a) A. J. Elliot and J. K. S. Wan, *J. Phys. Chem.*, **82**, 444 (1978); (b) H. M. Vyas and J. K. S. Wan, *Chem. Phys. Lett.*, **34**, 470 (1975); (c) H. M. Vyas and J. K. S. Wan, *Can. J. Chem.*, **54**, 979 (1976). (1976).
137. S. K. Wong, W. Sytnk and J. K. S. Wan, *Can. J. Chem.*, **50**, 3052 (1972).
138. (a) S. A. Carlson and D. M. Hercules, *J. Am. Chem. Soc.*, **93**, 5611 (1971); (b) S. A. Carlson and D. M. Hercules, *Photochem. Photobiol.*, **17**, 123 (1973).
139. S. I. Skuratova, P. I. Mordvintsev and G. V. Fomin, *Zh. Fiz. Khim.*, **56**, 2093 (1982).
140. H. M. Vyas and J. K. S. Wan, *Int. J. Chem. Kinet.*, **6**, 125 (1974).
141. J. B. Pedersen, C. E. M. Hansen, H. Parbo and L. T. Muus, *J. Chem. Phys.*, **63**, 2398 (1975).
142. M. T. Craw, A. Alberti, M. C. Depew and J. K. S. Wan, *Bull. Chem. Soc. Japan*, **58**, 3657 (1985).
143. (a) Y. Kambara and H. Yoshida, *Bull. Chem. Soc. Japan*, **50**, 1367 (1977); (b) Y. Kambara, H. Yoshida and B. Ranby, *Bull. Chem. Soc. Japan*, **50**, 2554 (1977); (c) S. Noda, T. Doba, T. Mizuta, M. Miura and H. Yoshida, *J. Chem. Soc., Perkin Trans. 2*, 61 (1980).
144. V. E. Kholmogorov, *Theor. Eksp. Khim.*, **5**, 826 (1969).
145. K. A. McLauchlan and R. C. Sealy, *J. Chem. Soc., Chem. Commun.*, 115 (1976).
146. J. R. Harbour and G. Tollin, *Photochem. Photobiol.*, **20**, 387 (1974).
147. D. A. Hutchinson, H. M. Vyas, S. K. Wong and J. K. S. Wan, *Mol. Phys.*, **29**, 1767 (1975).
148. J. K. S. Wan, S. K. Wong and D. A. Hutchinson, *Acc. Chem. Res.*, **7**, 58 (1974).
149. J. K. S. Wan and A. J. Elliot, *Acc. Chem. Res.*, **10**, 161 (1979).
150. P. J. Hore, C. G. Joslin and K. A. McLauchlan, *Chem. Soc. Rev.*, **8**, 29 (1979).
151. (a) H. M. Vyas, S. K. Wong, B. B. Adeleke and J. K. S. Wan, *J. Am. Chem. Soc.*, **97**, 1385 (1975); (b) V. M. Kuznetz, D. N. Shigorin, A. L. Buchachenko, G. A. Val'kova, A. Z. Yankelevich and N. N. Shapet'ko, *Izv. Akad. Nauk SSSR, Ser. Khim.*, 62 (1978).
152. (a) S. K. Wong and J. K. S. Wan, *J. Am. Chem. Soc.*, **94**, 7197 (1972); (b) S. K. Wong, D. A.

Hutchinson and J. K. S. Wan, *J. Am. Chem. Soc.*, **95**, 622 (1973); (c) S. K. Wong, D. A. Hutchinson and J. K. S. Wan, *Can. J. Chem.*, **52**, 251 (1974); (d) T. Foster, A. J. Elliot, B. B. Adeleke and J. K. S. Wan, *Can. J. Chem.*, **56**, 869 (1978).
153. K. Hamanoue, K. Yokoyama, T. Miyake, T. Kasuya, T. Nakayama and H. Teranishi, *Chem. Lett.*, 1967 (1982).
154. H. Inoue, K. Ikeda and M. Hida, *Nippon Kagaku Kaishi*, 381 (1984).
155. M. A. Fox and T. A. Voynick, *J. Org. Chem.*, **46**, 1235 (1981).
156. M. S. Walker, M. A. Abkowitz, R. W. Bigelow and J. H. Sharp, *J. Phys. Chem.*, **77**, 987 (1973).
157. K. Yamada, H. Shosenji, S. Fukunaga and K. Hirahara, *Bull. Chem. Soc. Japan*, **49**, 3701 (1976).
158. A. V. El'tsov, O. P. Studzinskii, Y. K. Levental' and M. V. Florinskaya, *Zh. Org. Khim.*, **13**, 1061 (1977).
159. O. P. Studzinskii, Yu. K. Levental' and A. V. El'tsov, *Zh. Org. Khim.*, **14**, 2150 (1978).
160. A. V. El'tsov, O. P. Studzinskii, M. V. Sendyurev and Y. K. Levental', *Zh. Obshch. Khim.*, **48**, 2799 (1978).
161. R. S. Hutton, H. D. Roth, M. L. M. Schilling, A. M. Trozzolo and T. M. Leslie, *J. Am. Chem. Soc.*, **104**, 5878 (1982).
162. K. Maruyama and H. Kato, *Mem. fac. Sci., Kyoto Univ., Ser. A*, **36**, 463 (1985).
163. K. C. Kurie and R. A. Robins, *J. Chem. Soc. (B)*, 855 (1970).
164. M. Shirai, T. Awatsuji and M. Tanaka, *Bull. Chem. Soc. Japan*, **48**, 1329 (1975).
165. S. Hashimoto, H. Takashima and M. Onohara, *Nippon Kagaku Kaishi*, 1019 (1975).
166. K. Kano and T. Matsuo, *Chem. Lett.*, 1127 (1973).
167. K. Yamada, S. Kohmoto and H. Iida, *Bull. Chem. Soc. Japan*, **49**, 1171 (1976).
168. S. M. Beck and L. E. Brus, *J. Am. Chem. Soc.*, **104**, 1103 (1982).
169. (a) K. Maruyama, H. Shindo and T. Maruyama, *Bull. Chem. Soc. Japan*, **44**, 585 (1971); (b) K. Maruyama, T. Otsuki, H. Shindo and T. Maruyama, *Bull. Chem. Soc. Japan*, **44**, 2000 (1971); (c) K. Maruyama, H. Shindo, T. Otsuki and T. Maruyama, *Bull. Chem. Soc. Japan*, **44**, 2756 (1971); (d) H. Shindo, K. Maruyama, T. Otsuki and T. Maruyama, *Bull. Chem. Soc. Japan*, **44**, 2789 (1971); (e) K. Maruyama and T. Otsuki, *Bull. Chem. Soc. Japan*, **44**, 2885 (1971).
170. (a) K. Maruyama, T. Otsuki and A. Takuwa, *Chem. Lett.*, 131 (1973); (b) K. Maruyama, T. Otsuki, A. Takuwa and S. Arakawa, *Bull. Chem. Soc. Japan*, **46**, 2470 (1973); (c) K. Maruyama and S. Arakawa, *Bull. Chem. Soc. Japan*, **47**, 1960 (1974).
171. K. Maruyama and G. Takahashi, *Chem. Lett.*, 295 (1973).
172. K. Maruyama, T. Otsuki and Y. Naruta, *Bull. Chem. Soc. Japan*, **49**, 791 (1976).
173. V. A. Loskutov, A. V. Konstantinova and E. P. Fokin, *Izv. Akad. Nauk SSSR, Ser. Khim.*, 1142 (1981).
174. (a) T. Foster, K. S. Chen and J. K. S. Wan, *J. Organomet. Chem.*, **184**, 113 (1980); (b) J. K. Kochi, K. S. Chen and J. K. S. Wan, *Chem. Phys. Lett.*, **73**, 557 (1980); (c) S. Emori, D. Weri and J. K. S. Wan, *Chem. Phys. Lett.*, **84**, 512 (1981).
175. (a) K. Maruyama and A. Takuwa, *Chem. Lett.*, 135 (1972); (b) K. Maruyama and A. Takuwa, *Bull. Chem. Soc. Japan*, **46**, 1529 (1973).
176. K. Maruyama, A. Takua and O. Soga, *J. Chem. Soc., Perkin Trans. 2*, 255 (1979).
177. K. Maruyama and Y. Miyagi, *Bull. Chem. Soc. Japan*, **47**, 1303 (1974).
178. K. Maruyama, H. Sakurai and T. Otsuki, *Bull. Chem. Soc. Japan*, **50**, 2777 (1977).
179. (a) K. Maruyama and A. Takuwa, *Chem. Lett.*, 471 (1974); (b) A. Takuwa, *Bull. Chem. Soc. Japan*, **49**, 2790 (1976).
180. K. Maruyama, A. Takuwa and O. Soga, *Chem. Lett.*, 1097 (1979).
181. A. Takuwa, H. Iwamoto, O. Soga and K. Maruyama, *Bull. Chem. Soc. Japan*, **55**, 3657 (1982).
182. A. Takuwa, *Bull. Chem. Soc. Japan*, **50**, 2973 (1977).
183. A. Takuwa, O. Soga and K. Maruyama, *J. Chem. Soc., Perkin Trans. 2*, 409 (1985).
184. K. Maruyama, A. Takuwa, S. Matsukiyo and O. Soga, *J. Chem. Soc., Perkin Trans. 1*, 1414 (1980).
185. K. Maruyama, T. Miyazawa and Y. Kishi, *Chem. Lett.*, 721 (1974).
186. K. Maruyama and Y. Naruta, *Chem. Lett.*, 847 (1977).
187. M. B. Rubin and Z. Neuwirth-Weiss, *J. Am. Chem. Soc.*, **94**, 6048 (1972).
188. T.-S. Lin, S. H. Mastin and N. Ohkaku, *J. Am. Chem. Soc.*, **95**, 6845 (1973).
189. (a) M.-P. Pileni, *Chem. Phys. Lett.*, **75**, 540 (1980); (b) M.-P. Pileni, *Chem. Phys. Lett.*, **71**, 317 (1980); (c) I. Willner and Y. Degani, *J. Chem. Soc., Chem. Commun.*, 761 (1982); (d) M.-P. Pileni

and M. Gratzel, *J. Phys. Chem.*, **84**, 1822 (1980); (e) C. Wolff and M. Gratzel, *Chem. Phys. Lett.*, **52**, 542 (1977).
190. (a) K. Kano and T. Matsuo, *Chem. Lett.*, 11 (1974); (b) K. Kano and T. Matsuo, *Bull. Chem. Soc. Japan*, **47**, 2836 (1974).
191. G. V. Fomin, M. M. Shabarchina and Yu. Sh. Moshkovskii, *Zh. Fiz. Khim.*, **53**, 2101 and 2102 (1979).
192. K. Kano, Y. Takada and T. Matsuo, *Bull. Chem. Soc. Japan*, **48**, 3215 (1975).
193. G. V. Fomin, P. I. Mordvintsev, M. I. Cherkashin and V. V. Shibanov, *Zh. Fiz. Khim.*, **52**, 2901 (1978).
194. G. V. Fomin and P. I. Mordvintsev, *Zh. Fiz. Khim.*, **57**, 762 and 764 (1983).
195. Y. Taminoto, K. Shimizu and M. Itoh, *Photochem. Photobiol.*, **39**, 511 (1984).
196. V. Swyambunathan and N. Priasamy, *J. Photochem.*, **13**, 325 (1980).
197. Y. Tanimoto, H. Udagawa and M. Itoh, *J. Phys. Chem.*, **87**, 724 (1983).
198. Y. Tanimoto, M. Takashima and M. Itoh, *Chem. Lett.*, 1981 (1984).
199. Y. Tanimoto and M. Itoh, *Chem. Phys. Lett.*, **83**, 626 (1981).
200. Y. Tanimoto, K. Shimizu, H. Udagawa and M. Itoh, *Chem. Lett.*, 353 (1983).
201. Y. Tanimoto, K. Shimizu and M. Itoh, *Chem. Phys. Lett.*, **112**, 217 (1984).
202. (a) Y. Sakaguchi and H. Hayashi, *J. Phys. Chem.*, **88**, 1437 (1984); (b) H. Hayashi and S. Nagakura, *Bull. Chem. Soc. Japan*, **57**, 332 (1984).
203. J. A. Barltrop and B. Hesp, *J. Chem. Soc. (C)*, 1625 (1967).
204. N. J. Bunce, J. E. Ridley and M. C. Zerner, *Theor. Chim. Acta*, **45**, 283 (1977).
205. K. Maruyama, T. Otsuki, A. Takuwa and S. Kato, *Bull. Ins. Chem. Res., Kyoto Univ.*, **50**, 344 (1972).
206. N. J. Bunce and M. Hadley, *Can. J. Chem.*, **53**, 3240 (1975).
207. E. J. Gardner, R. H. Squire, R. C. Elder and R. M. Wilson, *J. Am. Chem. Soc.*, **95**, 1694 (1973).
208. R. M. Wilson and S. W. Wunderly, *J. Chem. Soc., Chem. Commun.*, 461 (1974).
209. R. M. Wilson and S. W. Wunderly, *J. Am. Chem. Soc.*, **96**, 7350 (1974); (b) R. M. Wilson, E. J. Gardner, R. C. Elder, R. H. Squire and L. R. Florian, *J. Am. Chem. Soc.*, **96**, 2955 (1974).
210. R. M. Wilson, S. W. Wunderly, T. F. Walsh, A. K. Musser, R. Outcaet, F. Geiser, S. K. Gee, W. Brabender, L. Yerino Jr, T. T. Conrad and G. A. Tharp, *J. Am. Chem. Soc.*, **104**, 4429 (1982).
211. R. M. Wilson, R. Outcaet and H. H. Jaffe, *J. Am. Chem. Soc.*, **100**, 301 (1978).
212. R. M. Wilson and A. K. Musser, *J. Am. Chem. Soc.*, **102**, 1720 (1980).
213. H. Irngartinger, R.-D. Acker, W. Rebafka and H. A. Staab, *Angew. Chem. Int. Ed. Engl.*, **13**, 674 (1974).
214. A. Hassner and D. J. Blythin, *J. Org. Chem.*, **37**, 4209 (1972).
215. E. A. Fehnel and F. C. Brokaw, *J. Org. Chem.*, **45**, 578 (1980).
216. (a) K. Maruyama, T. Otsuki and Y. Naruta, *Chem. Lett.*, 641 (1973); (b) K. Maruyama, Y. Naruta and T. Otsuki, *Bull. Chem. Soc. Japan*, **48**, 1153 (1975).
217. H. J. Lin and W. H. Chan, *Can. J. Chem.*, **58**, 2196 (1980).
218. T. Naito, Y. Makita and C. Kaneko, *Chem. Lett.*, 921 (1984).
219. K. Maruyama and N. Narita, *Bull. Chem. Soc. Japan*, **53**, 757 (1980).
220. K. Maruyama, T. Otsuki and S. Tai, *Chem. Lett.*, 371 (1984).
221. K. Maruyama and S. Tai, *Chem. Lett.*, 681 (1985).
222. M. Ochiai, M. Arimoto and E. Fujita, *J. Chem. Soc., Chem. Commun.*, 460 (1981).
223. T. Otsuki, *Bull. Chem. Soc. Japan*, **49**, 2596 (1976).
224. T. Otsuki and K. Maruyama, *J. Synth. Org. Chem. Japan*, **36**, 206 (1978).
225. K. Maruyama, T. Otsuki and K. Mitsui, *J. Org. Chem.*, **45**, 1424 (1980).
226. (a) K. Maruyama, S. Tai and T. Otsuki, *Nippon Kagaku Kaishi*, 90 (1984); (b) K. Maruyama, T. Otsuki and S. Tai, *J. Org. Chem.*, **50**, 52 (1985).
227. K. Maruyama, S. Tai and T. Otsuki, *Chem. Lett.*, 843 (1983).
228. K. Maruyama and T. Otsuki, *Chem. Lett.*, 851 (1977).
229. K. Maruyama, T. Otsuki and H. Tamiaki, *Bull. Chem. Soc. Japan*, **58**, 3049 (1985).
230. K. Maruyama, M. Tojo, K. Mastumoto and T. Otsuki, *Chem. Lett.*, 859 (1980).
231. K. Maruyama, S. Tai, M. Tojo and T. Otsuki, *Heterocycles*, **16**, 1963 (1981).
232. K. Maruyama, M. Tojo and T. Otsuki, *Bull. Chem. Soc. Japan*, **53**, 567 (1980).
233. (a) K. Maruyama and T. Otsuki, *Chem. Lett.*, 87 (1975); (b) K. Maruyama, T. Otsuki and K. Mitsui, *Bull. Chem. Soc. Japan*, **49**, 3361 (1976).
234. K. Maruyama, K. Mitsui and T. Otsuki, *Chem. Lett.*, 853 (1977).

235. K. Maruyama, T. Otsuki, K. Mitsui and M. Tojo, *J. Heterocycl. Chem.*, **17**, 695 (1980).
236. K. Maruyama, K. Mitsui and T. Otsuki, *Chem. Lett.*, 323 (1978).
237. T. Otsuki, *Bull. Chem. Soc. Japan*, **49**, 3713 (1976).
238. K. Maruyama, M. Tojo, H. Iwamoto and T. Otsuki, *Chem. Lett.*, 827 (1980).
239. (a) K. Maruyama, S. Tai and T. Otsuki, *Chem. Lett.*, 1565 (1981); (b) K. Maruyama, S. Tai and T. Otsuki, *Heterocycles*, **20**, 1031 (1983).
240. A. Mori and H. Takeshita, *Chem. Lett.*, 599 (1975).
241. H. Takeshita, A. Mori, M. Funakura and H. Mametsuka, *Bull. Chem. Soc. Japan*, **50**, 315 (1977).
242. K. Ogino, T. Matsumoto, T. Kawai and S. Kozuka, *J. Chem. Soc., Chem. Commun.*, 644 (1979).
243. (a) K. Ogino, T. Minami and S. Kozuka, *J. Chem. Soc., Chem. Commum.*, 480 (1980); (b) K. Ogino, T. Minami, S. Kozuka and T. Kinshita, *J. Org. Chem.*, **45**, 4694 (1980).
244. N. Ishibe and Y. Yamaguchi, *J. Chem. Soc., Perkin Trans. 1*, 2618 (1973).
245. K. Ogino, S. Yamashina, T. Matsumnto and S. Kozuka, *J. Chem. Soc., Perkin Trans. 1*, 1552 (1979).
246. K. Ogino, T. Matsumoto, T. Kawai and S. Kozuka, *J. Org. Chem.*, **44**, 3352 (1979).
247. K. Ogino, T. Matsumoto and S. Kozuka, *J. Chem. Soc., Chem. Commun.*, 643 (1979).
248. H. Inoue, A. Ezaki, H. Tomono and M. Hida, *J. Chem. Soc., Chem. Commun.*, 860 (1979).
249. H. Inoue, A. Ezaki, D. Nakajima, H. Tomono and M. Hida, *J. Chem. Soc., Perkin Trans. 1*, 1771 (1982).
250. H. Inoue, A. Ezaki and M. Hida, *J. Chem. Soc., Perkin Trans. 2*, 833 (1982).
251. M. E. Kuehne and H. Linde, *J. Org. Chem.*, **37**, 4031 (1972).
252. R. Breslow, D. R. Murayama, S. Murahashi and R. Grubbs, *J. Am. Chem. Soc.*, **95**, 6683 (1973).
253. (a) Y. L. Chow, T. C. Joseph, H. H. Quon and J. N. S. Tam, *Can. J. Chem.*, **48**, 3045 (1970); (b) W. Scott, T. C. Joseph and Y. L. Chow, *J. Org. Chem.*, **41**, 2223 (1976).
254. (a) K. Maruyama, T. Iwai and Y. Naruta, *Chem. Lett.*, 1219 (1975); (b) K. Maruyama, T. Iwai, Y. Naruta, T. Otsuki and Y. Miyagi, *Bull. Chem. Soc. Japan*, **51**, 2052 (1978).
255. T. Sasaki, K. Kanematsu, I. Ando and O. Yamashita, *J. Am. Chem. Soc.*, **99**, 871 (1977).
256. K. Maruyama, M. Muraoka and Y. Naruta, *J. Org. Chem.*, **46**, 983 (1981).
257. H. J. T. Bos, H. Palman and P. F. E. Montfort, *J. Chem. Soc., Chem. Commun.*, 188 (1973).
258. A. Mosterd, L. J. de Noten and H. T. J. Bos, *Recl. Trav. Chim. Pays-Bas*, **96**, 16 (1977).
259. R. J. C. Koster, D. G. Streefkerk, J. O. Veen and H. T. J. Bos, *Recl. Trav. Chim. Pays-Bas*, **93**, 157 (1974).
260. A. Mosterd, R. E. L. J. Lecluize and H. J. T. Bos, *Recl. Trav. Chim. Pays-Bas*, **94**, 72 and 220 (1975).
261. A. Eckert, *Chem. Ber.*, **60**, 1691 (1927).
262. (a) N. I. Rtishchev, O. P. Studzinkii and A. V. El'tsov, *Zh. Org. Khim.*, **8**, 1054 (1972); (b) A. V. El'tsov, O. P. Stuzinskii, O. V. Kul'bitskaya, N. V. Ogal'tosova and L. S. Efros, *Zh. Org. Khim.*, **6**, 638 (1970).
263. O. P. Studzinskii, N. I. Rtishchev, A. V. El'tsov and A. V. Devekki, *Zh. Org. Khim.*, **8**, 74 (1972).
264. G. G. Wubbels, D. M. Tollefsen, R. S. Meredith and L. A. Hewraldt, *J. Am. Chem. Soc.*, **95**, 3820 (1973).
265. H. Inoue, T. D. Tuong, M. Hida and T. Murata, *J. Chem. Soc., Chem. Commun.*, 1347 (1971).
266. H. Inoue, M. Hida, T. D. Tuong and T. Murata, *Bull. Chem. Soc. Japan*, **46**, 1759 (1973).
267. H. Inoue, K. Nakamura, S. Kato and M. Hida, *Bull. Chem. Soc. Japan*, **48**, 2872 (1975).
268. A. V. El'tov and O. P. Studinskii, *Zh. Org. Khim.*, **9**, 847 (1973).
269. (a) H. Inoue and M. Hida, *Chem. Lett.*, 255 (1974); (b) H. Inoue and M. Hida, *Bull. Chem. Soc. Japan*, **51**, 1793 (1978); (c) H. Inoue, T. Shinoda and M. Hida, *Bull. Chem. Soc. Japan*, **53**, 154 (1980).
270. M. Tajima, H. Inoue and M. Hida, *Nippon Kagaku Kaishi*, 1728 (1979).
271. K. Seguchi and H. Ikeyama, *Chem. Lett.*, 1493 (1980).
272. J. Griffiths and C. Hawkins, *J. Chem. Soc., Chem. Commun.*, 111 (1973).
273. J. Griffiths and C. Hawkins, *J. Chem. Soc., Perkin Trans. 1*, 2283 (1974).
274. K.-Y. Chu and J. Griffiths, *J. Chem. Res.*, (S) 180, (M) 2319 (1978).
275. S. M. Drew, J. Griffiths and A. J. King, *J. Chem. Soc., Chem. Commun.*, 1037 (1979).
276. (a) K. Yoshida, T. Okugawa and Y. Yamashita, *Chem. Lett.*, 335 (1981); (b) K. Yoshida, T. Okugawa, E. Nagamatsu, Y. Yamashita and M. Matsuoka, *J. Chem. Soc., Perkin Trans. 1*, 529 (1984).

277. (a) M. Matsuoka, K. Takagi, K. Ueda, H. Tajima and T. Kitao, *J. Chem. Soc., Chem. Commun.*, 521 (1983); (b) M. Matsuoka, K. Takagi, H. Tajima, K. Ueda and T. Kitao, *J. Chem. Soc., Perkin Trans. 1*, 1297 (1984).
278. V. A. Loskutov, S. M. Lukonina and E. P. Fokin, *Izv. Sib. Otd. Akad. Nauk SSSR, Ser. khim. Nauk*, 135 (1978).
279. O. P. Studzinskii, A. V. El'tsov and Y. K. Levental', *Zh. Obshch. Khim.*, **50**, 435 (1980).
280. A. V. El'tsov, Yu. K. Levental' and O. P. Studzinskii, *Zh. Org. Khim.*, **12**, 2483 (1976).
281. G. G. Mihai, P. G. Tarassoff and N. Filipescu, *J. Chem. Soc., Perkin Trans. 1*, 1374 (1975).
282. (a) A. D. Broadbent and J. M. Stewart, *J. Chem. Soc., Chem. Commun.*, 676 (1980); (b) A. D. Broadbent and J. M. Stewart, *Can. J. Chem.*, **61**, 1965 (1983).
283. K. P. Clark and H. I. Stonehill, *J. Chem. Soc., Faraday Trans. 1*, **68**, 1676 (1972).
284. O. P. Stuzinskii and A. V. El'tsov, *Zh. Org. Khim.*, **18**, 1904 (1982).
285. G. O. Phillips, N. W. Worthington, J. F. McKeller and R. R. Sharpe, *J. Chem. Soc. (A)*, 767 (1969).
286. K. P. Clark and H. I. Stonehill, *J. Chem. Soc., Faraday Trans. 1*, **68**, 578 (1972).
287. A. D. Broadbent and R. P. Newton, *Can. J. Chem.*, **50**, 381 (1972).
288. A. D. Broadbent, H. B. Matheson and R. P. Newton, *Can. J. Chem.*, **53**, 826 (1975).
289. J. L. Charlton, R. G. Smerchanski and C. E. Burchill, *Can. J. Chem.*, **54**, 512 (1976).
290. H. I. Stonehill and K. P. Clark, *Can. J. Chem.*, **54**, 516 (1976).
291. A. V. El'tsov and O. P. Studzinskii, *Zh. Org. Khim.*, **15**, 2219 (1979).
292. A. Roy, D. Battacharya and S. Aditya, *J. Indian Chem. Soc.*, **59**, 585 (1982).
293. V. Ya. Denisov and N. A. Pirogova, *Zh. Org. Khim.*, **18**, 2397 (1982).
294. J. Morley, *J. Chem. Soc., Chem. Commun.*, 88 (1976).
295. K. Hamilton, J. A. Hunter, P. N. Preston and J. O. Morley, *J. Chem. Soc. Perkin Trans. 2*, 1544 (1980).
296. O. P. Studzinskii and A. V. El'tsov, *Zh. Org. Khim.*, **15**, 2597 (1979).
297. A. V. El'tsov and O. P. Studzinskii, *Zh. Org. Khim.*, **15**, 1104 (1979).
298. S. Kato, H. Inoue and M. Hida, *Nippon Kagaku Kaishi*, 1411 (1978).
299. S. Kato, H. Inoue and M. Hida, *Nippon Kagaku Kaishi*, 96 (1979).
300. M. H. Fisch and W. M. Hemmerlin, *Tetrahedron Lett.*, 3125 (1972).
301. O. P. Studzinskii, N. I. Rtishchev and A. V. El'tsov, *Zh. Org. Khim.*, **11**, 1133 (1975).
302. A. P. Krapcho, K. J. Shaw, J. J. Landi Jr and D. G. Phinney, *J. Org. Chem.*, **49**, 5253 (1984).
303. G. Green-Buckley and J. Griffiths, *J. Chem. Soc., Chem. Commun.*, 396 (1977).
304. G. Green-Buckley and J. Griffiths, *J. Chem. Soc., Perkin Trans. 1*, 702 (1977).
305. O. P. Studzinskii and A. V. El'tsov, *Zh. Org. Khim.*, **16**, 1101 (1980).
306. S. Hashimoto and H. Hashimoto, *Nippon Kagaku Kaishi*, 1645 (1976).
307. O. E. Edwards and P-T. Ho, *Can. J. Chem.*, **56**, 733 (1978).
308. (a) S. M. Lukonina, V. A. Loskutov and E. P. Fokin, *Izv. Sib. Otd. Akad. Nauk SSSR, Ser. Khim. Nauk*, 106 (1981); (b) S. M. Lukonina, V. A. Loskutov and E. P. Fokin, *Izv. Sib. Otd. Akad. Nauk SSSR, Ser. Khim. Nauk*, 112 (1982).
309. K. Maruyama and T. Kozuka, *Chem. Lett.*, 341 (1980).
310. (a) J. M. Bruce, D. Creed and K. Dawes, *J. Chem. Soc., Chem. Commun.*, 3749 (1971); (b) J. M. Bruce, A. Chaudhry and K. Dawes, *J. Chem. Soc., Perkin Trans. 1*, 288 (1974).
311. E. Rommel and J. Wirz, *Helv. Chim. Acta*, **60**, 38 (1977).
312. N. P. Gritsan, V. A. Rogov, N. M. Bazhin, V. V. Russkikh and E. P. Fokin, *Izv. Akad. Nauk SSSR, Ser. Khim.*, 89 (1980).
313. (a) D. Creed, *J. Chem. Soc., Chem. Commun.*, 121 (1976); (b) F. Wilkinson, G. M. Seddon and K. Tickle, *Ber. Bunsengs. Phys. Chem.*, **73**, 315 (1968); (c) E. S. Land, *Trans. Faraday Soc.*, **65**, 2815 (1969).
314. T. J. King, A. R. Forrester, M. M. Ogilvy and R. Thomson, *J. Chem. Soc., Chem. Commun.*, 844 (1973).
315. (a) J. V. Ellis and J. E. Jones, *J. Org. Chem.*, **40**, 485 (1975); (b) J. E. Baldwin and J. E. Brown, *J. Chem. Soc., Chem. Commun.*, 167 (1969).
316. K. A. Abdulla, R. Al-Hamdany and Z. Y. Al-Saigh, *J. Prakt. Chem.*, **324**, 498 (1982).
317. Y. Miyagi, K. Kitamura, K. Maruyama and Y. L. Chow, *Chem. Lett.*, 33 (1978).
318. Y. Miyagi, K. Maruyama, H. Ishii, S. Mizuno, M. Kakudo, N. Tanaka, Y. Matsuura and S. Harada, *Bull. Chem. Soc. Japan*, **53**, 3019 (1979).
319. Y. Miyagi, K. Maruyama and S. Yoshinmoto, *Bull. Chem. Soc. Japan*, **53**, 2962 (1980).

320. K. Maruyama, N. Narita and Y. Miyagi, *Chem. Lett.*, 1033 (1978).
321. K. Maruyama and N. Narita, *Chem. Lett.*, 1211 (1979).
322. K. Maruyama and N. Narita, *J. Org. Chem.*, **45**, 1421 (1980).
323. K. Maruyama and H. Furuta, *Chem. Lett.*, 243 (1986).
324. (a) K. Maruyama, H. Furuta and A. Osuka, *Chem. Lett.*, 475 (1986); (b) A. Osuka, H. Furuta and K. Maruyama, *Chem. Lett.*, 479 (1986).
325. M. Akiba, S. Ikuta and T. Takada, *J. Chem. Soc., Chem. Commun.*, 817 (1983).
326. M. Akiba, S. Ikuta and T. Takada, *Heterocycles*, **16**, 1579 (1981).
327. M. Akiba, Y. Kosugi, M. Okuyama and T. Takada, *Heterocycles*, **6**, 1113 (1977).
328. K. J. Falci, R. W. Franck and G. P. Smith, *J. Org. Chem.*, **42**, 3317 (1977).
329. M. Akiba, Y. Kosugai, M. Okuyama and T. Takada, *J. Org. Chem.*, **43**, 181 (1978).
330. K. Maruyama, T. Kozuka and T. Otsuki, *Bull. Chem. Soc. Japan*, **50**, 2170 (1977).
331. K. Maruyama, T. Kozuka, T. Otsuki and Y. Naruta, *Chem. Lett.*, 1125 (1977).
332. M. Akiba, Y. Kosugi and T. Takada, *J. Org. Chem.*, **43**, 4472 (1978).
333. N. P. Gritsan, V. M. Shvartsberg, V. S. Romanov and M. S. Shvartsberg, *Izv. Akad. Nauk SSSR, Ser. Khim.*, 433 (1984).
334. N. P. Gritsan and N. M. Bazhin, *Izv. Akad. Nauk SSSR, Ser. Khim.*, 1275 (1980).
335. N. P. Gritsan and N. M. Bazhin, *Izv. Akad. Nauk SSSR, Ser. Khim.*, 118 and 280 (1981).
336. V. A. Gailite, Y. F. Freimanis, Y. Y. Dregeris, E. E. Liepin'sh and I. B. Mazheika, *Zh. Org. Khim.*, **14**, 2118 (1978).
337. R. A. Morton (Ed.), *Biochemistry of Quinones*, Academic Press, New York, 1965.
338. B. L. Trumpower (Ed.), *Function Quinones in Energy Conserving Systems*, Academic Press, New York, 1982.
339. (a) P. Schmidt-Mende and B. Rumberg, *Z. Naturforsch.*, 23B, 225 (1968); (b) V. H. Eck and A. Trebst, *Z. Naturforsch.*, **18B**, 46 (1963).
340. D. Creed, H. Werbin and T. Strom, *Tetrahedron*, **30**, 2037 (1974).
341. D. Creed, B. J. Hales and G. Porter, *Proc. R. Soc. Lond. A*, **334**, 505 (1973).
342. G. G. Lazarev, M. V. Serdobov and N. G. Khrapova, *Izv. Akad. Nauk SSSR, Ser. Khim.*, 1409 (1979).
343. R. M. Wilson, T. F. Walsh and S. K. Gee, *Tetrahedron Lett.*, **21**, 3459 (1980).
344. G. Leary and G. Porter, *J. Chem. Soc. (A)*, 2273 (1970).
345. D. Creed, *Tetrahedron Lett.*, **22**, 2039 (1981).
346. K. Maruyama, M. Muraoka and Y. Naruta, *J. Chem. Soc., Chem. Commun.*, 1282 (1980).
347. K. Maruyama, H. Iwamoto, O. Soga and A. Takuwa, *Chem. Lett.*, 1343 (1984).
348. K. Maruyama, H. Iwamoto, O. Soga, A. Takuwa and A. Osuka, *Chem. Lett.*, 595 (1985).
349. K. Maruyama, H. Iwamoto, O. Soga, A. Takuwa and A. Osuka, *Chem. Lett.*, 1675 (1985).
350. Yu. E. Gerashimenko, N. T. Poteleshchenko and V. V. Romanov, *Zh. Org. Khim.*, **14**, 2387 (1978).
351. Yu. E. Gerashimenko and N. T. Poteleshchenko, *Zh. Vses. Khim. Obshch. im. D. I. Mendeleeva*, **16**, 105 (1971).
352. Yu. E. Gerashimenko and N. T. Poteleshchenko, *Zh. Org. Khim.*, **7**, 2413 (1971).
353. Yu. E. Gerashimenko and N. T. Poteleshchenko, *Zh. Org. Khim.*, **9**, 2392 (1973); **15**, 393 (1979).
354. Yu. E. Gerashimenko, N. T. Poteleshchenko and V. V. Romanov, *Zh. Org. Khim.*, **16**, 1938 (1980).
355. E. P. Fokin, S. A. Russkikh and L. S. Klimenko, *Zh. Org. Khim.*, **13**, 2010 (1977).
356. S. A. Russkikh, L. S. Klimenko, N. P. Grsitsan and E. P. Fokin, *Zh. Org. Khim.*, **18**, 2224 (1982); **20**, 1949 (1984).
357. N. P. Gritsan, S. A. Russkikh, L. S. Klimenko and V. F. Plyusnin, *Teor. Eksp. Khim.*, **19**, 455 and 577 (1983).
358. E. P. Fokin, S. A. Russkikh and L. S. Klimenko, *Izv. Sib. Otd. Akad. Nauk SSSR, Ser. Khim. Nauk*, 117 (1979).
359. E. P. Fokin, S. L. Klimenko and V. V. Russkikh, *Izv. Sib. Otd. Akad. Nauk SSSR, Ser. Khim. Nauk*, 110 (1978).
360. P. Germeraad, W. Weyler Jr and H. W. Moore, *J. Org. Chem.*, **39**, 781 (1974).
361. (a) Y. Naruta, T. Yokota, N. Nagai and K. Maruyama, *J. Chem. Soc., Chem. Commun.*, in press (1986); (b) Y. Naruta, N. Nagai, T. Yokota and K. Maruyama, *Chem. Lett.*, in press (1986).
362. T. Kozuka, *Bull. Chem. Soc. Japan*, **55**, 2922 (1982).
363. T. Ogawa, Ph.D. Thesis, Kyoto University, 1984.

364. L. M. Gornostaev, V. A. Levdanskii and E. P. Fokin, *Zh. Org. Khim.*, **15**, 1692 (1979).
365. L. M. Gornostaev and V. A. Levdanskii, *Zh. Org. Khim.*, **16**, 2209 (1980).
366. L. M. Gornostaev and T. I. Lavrikov, *Zh. Org. Khim.*, **18**, 339 (1982).
367. M. Ogata, H. Matsumoto, H. Kano and H. Yukinaga, *J. Chem. Soc., Chem. Commun.*, 218 (1973).
368. I. R. Dunkin and J. G. MacDonald, *Tetrahedron Lett.*, **23**, 5201 (1982).
369. H. Tomioka, H. Fukao and Y. Izawa, *Bull. Chem. Soc. Japan*, **51**, 540 (1978).
370. R. C. Deselms and W. R. Schleigh, *Synthesis*, 614 (1973).
371. V. B. Vol'eva, V. V. Ershov, I. S. Belostotskaya and N. L. Komissaraova, *Izv. Akad. Nauk SSSR, Ser. Khim.*, 739 (1974).
372. Y. Matsunaga and Y. Narita, *Bull. Chem. Soc. Japan*, **45**, 408 (1972).
373. G. V. Fomin, O. B. Lantratova and I. E. Porkrovskaya, *Izv. Akad. Nauk SSSR, Ser. Khim.*, 946 (1980).
374. I. V. Khudyakov, S. F. Burlatsky, B. L. Tumansky and V. A. Kuzmin, *Izv. Akad. Nauk SSSR, Ser. Khim.*, 2153 (1976).
375. O. B. Lantratov, A. I. Prokof'ev, I. V. Khudyakov, V. A. Kuzmin and I. E. Pokrovskaya, *Nouv. J. Chim.*, **6**, 365 (1982).
376. S. Tsuruya and T. Yonezawa, *J. Org. Chem.*, **39**, 2438 (1974).
377. G. A. Val'kova, D. A. Shigorin, V. M. Gebel, N. S. Dkunikhim, N. N. Artmonova and L. M. Gaeva, *Zh. Fiz. Khim.*, **48**, 1259 (1974).
378. J. M. Bruce and A. Chaudry, *J. Chem. Soc., Perkin Trans. 1*, 295 (1974).
379. D. H. R. Barton, D. L. J. Clive, P. D. Magnus and G. Smith, *J. Chem. Soc. (C)*, 2193 (1971).
380. D. H. R. Barton, M. T. Bielska, J. M. Cardoso, N. J. Cussans and S. V. Ley, *J. Chem. Soc., Perkin Trans. 1*, 1840 (1981).
381. J. Wirz, *Helv. Chim. Acta*, **57**, 1283 (1974).
382. K. H. Gschwind and U. P. Wild, *Helv. Chim. Acta*, **56**, 809 (1973).
383. T. Bercovici, R. Korenstein, K. A. Muszkat and E. Fischer, *Pure Appl. Chem.*, **24**, 531 (1970).
384. R. Korenstein, K. A. Muszkat and Sh. Sharafi-Ozeri, *J. Am. Chem. Soc.*, **95**, 6177 (1973).
385. T. Bercovici and E. Fischer, *Helv. Chim. Acta*, **56**, 1114 (1973).
386. T. Bercovici, R. Korenstein, G. Fischer and E. Fischer, *J. Phys. Chem.*, **80**, 108 (1976).
387. H.-D. Brauer, W. Drews and R. Schmidt, *J. Photochem.*, **12**, 293 (1980).
388. R. Schmidt, W. Drews and H.-D. Brauer, *J. Am. Chem. Soc.*, **102**, 2791 (1980).
389. O. C. Musgrave and D. Skoyle, *J. Chem. Soc., Perkin Trans. 1*, 2679 (1979).
390. W. Verboom and H. J. T. Bos, *Tetrahedron Lett.*, 1229 (1978).
391. U.-W. Grummt, M. Reichenbacher and R. Paetzold, *Tetrahedron Lett.*, **22**, 3945 (1981).
392. Y. Ueno, Y. Takeuchi, J. Koshitani and T. Yoshida, *J. Heterocycl. Chem.*, **18**, 295 and 645 (1981); **19**, 167 (1982).
393. K. Maruyama, T. Iwai, T. Otsuki, Y. Naruta and Y. Miyagi, *Chem. Lett.*, 1127 (1977).
394. (a) K. Maruyama and S. Tanioka, *Bull. Chem. Soc. Japan*, **49**, 2647 (1976); (b) K. Maruyama and S. Tanioka, *J. Org. Chem.*, **43**, 310 (1978).
395. A. Osuka, M. H. Chiba, H. Shimizu, H. Suzuki and K. Maruyama, *J. Chem. Soc., Chem. Commun.*, 919 (1980).
396. (a) A. Osuka, H. Shimizu, H. Suzuki and K. Maruyama, *Chem. Lett.*, 329 (1982); (b) A. Osuka, H. Shimizu, M. H. Chiba, H. Suzuki and K. Maruyama, *J. Chem. Soc., Perkin Trans 1*, 2037 (1983).
397. (a) K. Maruyama, A. Osuka and H. Suzuki, *J. Chem. Soc., Chem. Commun.*, 323 (1980); (b) M. Jimentz, L. Rodrriguez-Hahn and J. Romo, *Lationam. Quim.*, **5**, 184 (1974); (c) A. Osuka, *J. Org. Chem.*, **47**, 3131 (1982).
398. (a) K. Maruyama, A. Osuka and H. Suzuki, *Chem. Lett.*, 919 (1979); (b) A. Osuka, H. Suzuki and K. Maruyama, *J. Chem. Soc., Perkin Trans 1*, 2671 (1982).
399. (a) K. Maruyama and A. Osuka, *Chem. Lett.*, 77 (1979); (b) K. Maruyama, A. Osuka and H. Suzuki, *Chem. Lett.*, 1477 (1979).
400. K. Maruyama, A. Osuka and H. Suzuki, *J. Chem. Soc., Chem. Commun.*, 723 (1980).
401. A. Osuka, H. Suzuki and K. Maruyama, *Chem. Lett.*, 201 (1981).
402. (a) K. Maruyama, S. Arakawa and T. Otsuki, *Tetrahedron Lett.*, 2433 (1975); (b) S. Arakawa, *J. Org. Chem.* **42**, 3800 (1977).
403. (a) H. Kato, K. Yamaguchi and H. Tezuka, *Chem. Lett.*, 1089 (1974); (b) H. Kato, H. Tezuka, K. Yamaguchi, K. Nowada and Y. Nakamura, *J. Chem. Soc., Perkin Trans 1*, 1029 (1978).

404. (a) R. G. F. Giles and I. R. Green, *J. Chem. Soc., Chem. Commun.*, 1334 (1972); (b) R. G. Giles, I. R. Green, P. R. K. Mitchell, C. L. Raston and A. H. White, *J. Chem. Soc., Perkin Trans. 1*, 719 (1979).
405. K. Maruyama and S. Arakawa, *J. Org. Chem.*, **42**, 3793 (1977).
406. (a) K. Maruyama and S. Arakawa, *Chem. Lett.*, 719 (1974); (b) S. Arakawa, *Mem. Fac. Sci., Kyoto Univ., Ser. A*, **35**, 327 (1980).
407. (a) K. Maruyama and T. Ogawa, *Chem. Lett.*, 1027 (1981); (b) K. Maruyama and T. Ogawa, *J. Org. Chem.*, **48**, 4968 (1983).
408. L. T. Muus, S. Frydkjaer and K. B. Nielsen, *Chem. Phys.*, **30**, 163 (1978).
409. M. Tajima, H. Inoue and M. Hida, *Bull. Chem. Soc. Japan.*, **57**, 305 (1984).
410. (a) G. Tollin, *J. Phys. Chem.*, **80**, 2274 (1976); (b) A. A. Lamola, M. L. Manion, H. D. Roth and G. Tollin, *Proc. Nat. Acad. Sci. USA*, **72**, 3265 (1975); (c) W. E. Ford and G. Tollin, *Photochem. Photobiol.*, **38**, 441 (1983) and references cited therein.
411. P. Kertesz and J. Reisch, *Arch. Pharm.*, **313**, 476 (1980).
412. A. P. Darmanyan, *Chem. Phys. Lett.*, **96**, 383 (1983).
413. F. Bohlmann and H. J. Föster, *Chem. Ber.*, **110**, 2016 (1977).